SOLUTIONS MANUAL

LUCIO GELMINI   ROBERT W. HILTS
Grant MacEwan College

PETRUCCI   HARWOOD   HERRING   MADURA

GENERAL CHEMISTRY

PRINCIPLES & MODERN APPLICATIONS

NINTH EDITION

Upper Saddle River, NJ 07458

Project Manager: Kristen Kaiser
Senior Editor: Kent Porter-Hamann
Editor-in-Chief, Science: Dan Kaveney
Executive Managing Editor: Kathleen Schiaparelli
Assistant Managing Editor: Karen Bosch
Production Editor: Robert Merenoff
Supplement Cover Manager: Paul Gourhan
Supplement Cover Designer: Christopher Kossa
Manufacturing Buyer: Ilene Kahn
Manufacturing Manager: Alexis Heydt-Long

© 2007 Pearson Education, Inc.
Pearson Prentice Hall
Pearson Education, Inc.
Upper Saddle River, NJ 07458

Printed in the United States of America

10 9 8 7 6 5 4

ISBN 0-13-199734-3

Pearson Education Ltd., *London*
Pearson Education Australia Pty. Ltd., *Sydney*
Pearson Education Singapore, Pte. Ltd.
Pearson Education North Asia Ltd., *Hong Kong*
Pearson Education Canada, Inc., *Toronto*
Pearson Educación de Mexico, S.A. de C.V.
Pearson Education—Japan, *Tokyo*
Pearson Education Malaysia, Pte. Ltd.

ACKNOWLEDGEMENTS

First and foremost, we thank our families for the unflagging love and support they gave us during the time we were working on this latest incarnation of our solutions manual. Special thanks are extended to the Science Department at Grant MacEwan College for playing an instrumental role in the completion of the manual by giving us access to departmental resources. For sharing their mathematical expertise with us, and for providing the solutions to some tricky integrals, we thank Dr. David McLaughlin and Cornelia Bica. Dr. Roy Jensen warrants a very special thank you for critically evaluating several of our answers. Kent Porter-Mann, Carole Snyder and Kristen Kaiser, all of Pearson Education, are commended for their timely editorial advice and steady support over the past year. Finally, we are especially indebted to Dr. Ralph Petrucci, who has helped us in innumerable ways, including dozens of clarifications, many highly constructive suggestions and prompt delivery of essential information. Thank you so much, Ralph! We could not have completed this colossal project without your tireless efforts!

Dr. Lucio Gelmini
Dr. Robert Hilts

Grant MacEwan College
Edmonton, Alberta, Canada
February 2006

Table of Contents

Chapter 1   Matter—Its Properties and Measurement   1

Chapter 2   Atoms and the Atomic Theory   20

Chapter 3   Chemical Compounds   44

Chapter 4   Chemical Reactions   79

Chapter 5   Introduction to Reactions in Aqueous Solutions   107

Chapter 6   Gases   137

Chapter 7   Thermochemistry   173

Chapter 8   Electrons in Atoms   204

Chapter 9   The Periodic Table and Some Atomic Properties   239

Chapter 10   Chemical Bonding I: Basic Concepts   265

Chapter 11   Chemical Bonding II: Additional Aspects   316

Chapter 12   Liquids, Solids, and Intermolecular Forces   362

Chapter 13   Solutions and Their Physical Properties   402

Chapter 14   Chemical Kinetics   444

Chapter 15   Principles of Chemical Equilibrium   489

Chapter 16   Acids and Bases   534

Chapter 17   Additional Aspects of Acid–Base Equilibria   592

Chapter 18   Solubility and Complex Ion Equilibria   677

Chapter 19   Spontaneous Change: Entropy and Free Energy   725

Chapter 20   Electrochemistry   765

Chapter 21   Chemistry of the Main Group Elements I: Groups 1, 2, 13, an 14   815

Chapter 22   Main-Group Elements II: Groups 18, 17, 16, 15 and Hydrogen   841

Chapter 23   The Transition Elements   876

Chapter 24   Complex Ions and Coordination Compounds   899

Chapter 25   Nuclear Chemistry   927

Chapter 26   Organic Chemistry   949

Chapter 27   Chemistry of the Living State   990

CHAPTER 1
MATTER—ITS PROPERTIES AND MEASUREMENT
PRACTICE EXAMPLES

1A   Convert the Fahrenheit temperature to Celsius and compare.

$$^{\circ}C = \left(^{\circ}F - 32^{\circ}F\right)\frac{5^{\circ}C}{9^{\circ}F} = \left(350\,^{\circ}F - 32^{\circ}F\right)\frac{5^{\circ}C}{9^{\circ}F} = 177\,^{\circ}C.$$

1B   We convert the Fahrenheit temperature to
Celsius. $^{\circ}C = \left(^{\circ}F - 32^{\circ}F\right)\frac{5^{\circ}C}{9^{\circ}F} = \left(-15^{\circ}F - 32^{\circ}F\right)\frac{5^{\circ}C}{9^{\circ}F} = -26\,^{\circ}C$. The antifreeze only protects to

$-22^{\circ}C$ and thus it will not offer protection to temperatures as low as $-15^{\circ}F = -26.1^{\circ}C$.

2A   The mass is the difference between the mass of the full and empty flask.

$$\text{density} = \frac{291.4\ g - 108.6\ g}{125\ mL} = 1.46\ g\,/\,mL$$

2B   First determine the volume required. $V = (1.000 \times 10^{3}\ g) \div (8.96\ g\ cm^{-3}) = 111.\underline{6}\ cm^{3}$.
Next determine the radius using the relationship between volume of a sphere and radius.

$$V = 4/3\pi r^{3} = 111.\underline{6}\ cm^{3} = 4/3(3.1416)r^{3} \qquad r = \sqrt[3]{\frac{111.6 \times 3}{4(3.1416)}} = 2.987 cm$$

3A   The volume of the stone is the difference between the level in the graduated cylinder with
the stone present and with it absent.

$$\text{density} = \frac{\text{mass}}{\text{volume}} = \frac{28.4\ g\ rock}{44.1\ mL\ rock\ \&\ water - 33.8\ mL\ water} = 2.76\ g\,/\,mL = 2.76\ g/cm^{3}$$

3B   The water level will remain unchanged. The mass of the ice cube displaces the same mass
of liquid water. A 10.0 g ice cube will displace 10.0 g of water. When the ice cube melts,
it simply replaces the displaced water, leaving the liquid level unchanged.

4A   The mass of ethanol can be found using dimensional analysis.

$$\text{ethanol mass} = 25\ L\ gasohol \times \frac{1000\ mL}{1\ L} \times \frac{0.71\ g\ gasohol}{1\ mL\ gasohol} \times \frac{10\ g\ ethanol}{100\ g\ gasohol} \times \frac{1\ kg\ ethanol}{1000\ g\ ethanol}$$

$$= 1.8\ kg\ ethanol$$

4B   We use the mass percent to determine the mass of the 25.0-mL sample.

$$\text{rubbing alcohol mass} = 15.0\ g\ (\text{2-propanol}) \times \frac{100.0\ g\ rubbing\ alcohol}{70.0\ g\ (\text{2-propanol})} = 21.4\underline{3}\ g\ rubbing\ alcohol$$

$$\text{rubbing alcohol density} = \frac{21.4\ g}{25.0\ mL} = 0.857\ g/mL$$

1

5A   For this calculation, the value 0.00456 has the least precision (three significant figures), thus the final answer must also be quoted to three significant figures.

$$\frac{62.356}{0.000456 \times 6.422 \times 10^3} = 21.3$$

5B   For this calculation, the value 1.3×10^{-3} has the least precision (two significant figures), thus the final answer must also be quoted to two significant figures.

$$\frac{8.21 \times 10^4 \times 1.3 \times 10^{-3}}{0.00236 \times 4.071 \times 10^{-2}} = 1.1 \times 10^6$$

6A   The number in the calculation that has the least precision is 102.1 (± 0.1), thus the final answer must be quoted to just one decimal place. $0.236 + 128.55 - 102.1 = 26.7$

6B   This is easier to visualize if the numbers are not in scientific notation.

$$\frac{\left(1.302 \times 10^3\right) + 952.7}{\left(1.57 \times 10^2\right) - 12.22} = \frac{1302 + 952.7}{157 - 12.22} = \frac{2255}{145} = 15.6$$

Exercises

The Scientific Method

1.   One theory is preferred over another if it can correctly predict a wider range of phenomena and if it has fewer assumptions.

2.   No. The greater the number of experiments that conform to the predictions of the law, the more confidence we have in the law. There is no point at which the law is ever verified with absolute certainty.

3.   For a given set of conditions, a cause, is expected to produce a certain result or effect. Although these cause-and-effect relationships may be difficult to unravel at times ("God is subtle"), they nevertheless do exist ("he is not malicious").

4.   As opposed to scientific laws, legislative laws are voted on by people and thus are subject to the whims and desires of the electorate. Legislative laws can be revoked by a grass roots majority, whereas scientific laws can only be modified if they do not account for experimental observations. As well, legislative laws are imposed on people, who are expected to modify their behaviors, whereas, scientific laws cannot be imposed on nature, nor will nature change to suit a particular scientific law that is proposed.

5.   The experiments should be carefully set up so as to create a controlled situation in which one can make careful observations after altering the experimental parameters, preferably one at a time. The results must be reproducible (to within experimental error) and, as more and more experiments are conducted, a pattern should begin to emerge, from which a comparison to the current theory can be made.

6. For a theory to be considered as plausible, it must, first and foremost, agree with and/or predict the results from controlled experiments. It should also involve the fewest number of assumptions (i.e. follow Occam's Razor). The best theories predict new phenomena that are subsequently observed after the appropriate experiments have been performed.

Properties and Classification of Matter

7. When an object displays a physical property it retains its basic chemical identity. By contrast, the display of a chemical property is accompanied by a change in composition.

   (a) Physical: The iron nail is not changed in any significant way when it is attracted to a magnet. Its basic chemical identity is unchanged.

   (b) Chemical: The paper is converted to ash, $CO_2(g)$ and $H_2O(g)$ along with the evolution of considerable energy.

   (c) Chemical: The green patina is the result of the combination of water, oxygen, and carbon dioxide with the copper in the bronze to produce basic copper carbonate.

   (d) Physical: Neither the block of wood nor the water has changed its identity.

8. When an object displays a physical property it retains its basic chemical identity. By contrast, the display of a chemical property is accompanied by a change in composition.

   (a) Chemical: The change in the color of the apple indicates that a new substance (oxidized apple) has formed by reaction with air.

   (b) Physical: The marble slab is not changed into another substance by feeling it.

   (c) Physical: The sapphire retains its identity as it displays its color.

   (d) Chemical: After firing, the properties of the clay have changed from soft and pliable to rigid and brittle. New substances have formed. (Many of the changes involve driving off water and slightly melting the silicates that remain. These molten substances cool and harden when removed from the kiln.)

9. (a) Homogeneous mixture: Air is a mixture of nitrogen, oxygen, argon, and traces of other gases. By "fresh," we mean no particles of smoke, pollen, etc. are present. Such species would produce a heterogeneous mixture.

   (b) Heterogeneous mixture: A silver plated spoon has a surface coating of the element silver and an underlying baser metal (typically iron). This would make the coated spoon a heterogeneous mixture.

   (c) Heterogeneous mixture: Garlic salt is simply garlic power mixed with table salt. Pieces of garlic can be distinguished from those of salt by careful examination.

   (d) Substance: Ice is simply solid water (assuming no air bubbles).

10. (a) Heterogeneous mixture: We can clearly see air pockets within the solid matrix. On close examination, we can distinguish different kinds of solids by their colors.

    (b) Homogeneous mixture: Modern inks are solutions of dyes in water. Older inks often were heterogeneous mixtures: suspensions of particles of carbon black (soot) in water.

(c)   Substance: This is assuming that no gases or organic chemicals are dissolved in the water.

(d)   Heterogeneous mixture: The pieces of orange pulp can be seen through a microscope. Most "cloudy" liquids are heterogeneous mixtures; the small particles impede the transmission of light.

11.   (a)   If a magnet is drawn through the mixture, the iron filings will be attracted to the magnet and the wood will be left behind.

(b)   When the glass-sucrose mixture is mixed with water, the sucrose will dissolve, whereas the glass will not. The water can then be boiled off to produce pure sucrose.

(c)   Olive oil will float to the top of a container and can be separated from water, which is more dense. It would be best to use something with a narrow opening that has the ability to drain off the water layer at the bottom (i.e. buret).

(d)   The gold flakes will settle to the bottom if the mixture is left undisturbed. The water then can be decanted (i.e. carefully poured off).

12.   (a)   Physical: This is simply a mixture of sand and sugar (i.e. not chemically bonded).

(b)   Chemical: Oxygen needs to be removed from the iron oxide.

(c)   Physical: Seawater is a solution of various substances dissolved in water.

(d)   Physical: The water-sand slurry is simply a heterogeneous mixture.

Exponential Arithmetic

13.   (a) $8950 = 8.950 \times 10^3 \, (8.95 \times 10^3 \text{ if only 3 sig. fig.})$

(b) $10,700 = 1.0700 \times 10^4 \, (1.07 \times 10^4 \text{ if 3 sig. fig.})$ (c) $0.0240 = 2.40 \times 10^{-2}$

(d) $0.0047 = 4.7 \times 10^{-3}$ (e) $938.3 = 9.383 \times 10^2$ (f) $275,482 = 2.75482 \times 10^5$

14.   (a) $3.21 \times 10^{-2} = 0.0321$ (b) $5.08 \times 10^{-4} = 0.000508$

(c) $121.9 \times 10^{-5} = 0.001219$ (d) $16.2 \times 10^{-2} = 0.162$

15.   (a) $34,000$ centimeters / second $= 3.4 \times 10^4$ cm/s

(b)   six thousand three hundred seventy eight kilometers $= 6378$ km $= 6.378 \times 10^3$ km

(c)   (trillionth $= 1 \times 10^{-12}$) hence, 74×10^{-12} m or 7.4×10^{-11} m

(d) $\dfrac{(2.2 \times 10^3) + (4.7 \times 10^2)}{5.8 \times 10^{-3}} = \dfrac{2.7 \times 10^3}{5.8 \times 10^{-3}} = 4.6 \times 10^5$

16.   (a)   173 thousand trillion watts $= 173,000,000,000,000,000$ W $= 1.73 \times 10^{17}$ W

(b)   one ten millionth of a meter $= 1 \div 10,000,000$ m $= 1 \times 10^{-7}$ m

(c)   (trillionth $= 1 \times 10^{-12}$) hence, 142×10^{-12} m or 1.42×10^{-10} m

(d) $\dfrac{5.07 \times 10^4 \times \left(1.8 \times 10^{-3}\right)^2}{0.065 + \left(3.3 \times 10^{-2}\right)} = \dfrac{0.16}{0.098} = 1.6$

Significant figures

17. **(a)** An exact number—500 sheets in a ream of paper.

 (b) Pouring the milk into the jug is a process that is subject to error; there can be slightly more or slightly less than one liter of milk in the jug. This is a measured quantity.

 (c) The distance between any pair of planetary bodies can only be determined through certain astronomical measurements, which are subject to error.

 (d) Measured quantity: the internuclear separation quoted for O_2 is an estimated value derived from experimental data, which contains some inherent error.

18. **(a)** The number of pages in the text is determined by counting; the result is an exact number.

 (b) An exact number. Although the number of days can vary from one month to another (say from January to February), the month of January always has 31 days.

 (c) The area is determined by calculations based on measurements. These measurements are subject to error.

 (d) Measured quantity: Average internuclear distance for adjacent atoms in gold medal is an estimated value derived from X-ray diffraction data, which contain some inherent error.

19. Each of the following is expressed to four significant figures.

 (a) $3984.6 \approx 3985$ **(b)** $422.04 \approx 422.0$ **(c)** $186{,}000 = 1.860 \times 10^5$

 (d) $33{,}900 \approx 3.390 \times 10^4$ **(e)** 6.321×10^4 is correct   **(f)** $5.0472 \times 10^{-4} \approx 5.047 \times 10^{-4}$

20. **(a)** 450 has two or three significant figures; trailing zeros left of the decimal are indeterminate, if no decimal point is present.

 (b) 98.6 has three significant figures; non-zero digits are significant.

 (c) 0.0033 has two significant digits; leading zeros are not significant.

 (d) 902.10 has five significant digits; trailing zeros to the right of the decimal point are significant, as are zeros flanked by non-zero digits.

 (e) 0.02173 has four significant digits; leading zeros are not significant.

 (f) 7000 can have anywhere from one to four significant figures; trailing zeros left of the decimal are indeterminate, if no decimal point is shown.

 (g) 7.02 has three significant figures; zeros flanked by non-zero digits are significant.

 (h) 67,000,000 can have anywhere from two to eight significant figures; there is no way to determine which, if any, of the zeros are significant, without the presence of a decimal point.

21. **(a)** $0.406 \times 0.0023 = 9.3 \times 10^{-4}$ **(b)** $0.1357 \times 16.80 \times 0.096 = 2.2 \times 10^{-1}$

 (c) $0.458 + 0.12 - 0.037 = 5.4 \times 10^{-1}$ **(d)** $32.18 + 0.055 - 1.652 = 3.058 \times 10^{1}$

22.   **(a)** $\dfrac{320 \times 24.9}{0.080} = \dfrac{3.2 \times 10^2 \times 2.49 \times 10^1}{8.0 \times 10^{-2}} = 1.0 \times 10^5$

      (b) $\dfrac{432.7 \times 6.5 \times 0.002300}{62 \times 0.103} = \dfrac{4.327 \times 10^2 \times 6.5 \times 2.300 \times 10^{-3}}{6.2 \times 10^1 \times 1.03 \times 10^{-1}} = 1.0$

      (c) $\dfrac{32.44 + 4.9 - 0.304}{82.94} = \dfrac{3.244 \times 10^1 + 4.9 - 3.04 \times 10^{-1}}{8.294 \times 10^1} = 4.47 \times 10^{-1}$

      (d) $\dfrac{8.002 + 0.3040}{13.4 - 0.066 + 1.02} = \dfrac{8.002 + 3.040 \times 10^{-1}}{1.34 \times 10^1 - 6.6 \times 10^{-2} + 1.02} = 5.79 \times 10^{-1}$

23.   **(a)** 2.44×10^4 **(b)** 1.5×10^3 **(c)** 40.0

      (d) 2.131×10^3 **(e)** 4.8×10^{-3}

24.   **(a)** 7.5×10^1 **(b)** 6.3×10^{12} **(c)** 4.6×10^3

      (d) 1.058×10^{-1} **(e)** 4.18×10^{-3} (quadratic equation solution)

25.   **(a)**   The average speed is obtained by dividing the distance traveled (in miles) by the elapsed time (in hours). First, we need to obtain the elapsed time, in hours.

$$9 \text{ days} \times \frac{24 \text{ h}}{1 \text{d}} = 216.000 \text{ h} \quad 3 \text{ min} \times \frac{1 \text{ h}}{60 \text{ min}} = 0.050 \text{ h} \quad 44 \text{ s} \times \frac{1 \text{ h}}{3600 \text{ s}} = 0.012 \text{ h}$$

Total time $= 216.000 \text{ h} + 0.050 \text{ h} + 0.012 \text{ h} = 216.062 \text{ h}$

$$\text{average speed} = \frac{25{,}012 \text{ mi}}{216.062 \text{ h}} \times \frac{1.609344 \text{ km}}{1 \text{ mi}} = 186.30 \text{ km/h}$$

      (b)   First compute the mass of fuel remaining

$$\text{mass} = 14 \text{ gal} \times \frac{4 \text{ qt}}{1 \text{ gal}} \times \frac{0.9464 \text{ L}}{1 \text{ qt}} \times \frac{1000 \text{ mL}}{1 \text{ L}} \times \frac{0.70 \text{ g}}{1 \text{ mL}} \times \frac{1 \text{ lb}}{453.6 \text{ g}} = 82 \text{ lb}$$

Next determine the mass of fuel used, and then finally, the fuel consumption. Notice that the initial quantity of fuel is not known precisely, perhaps at best to the nearest 10 lb, certainly ("nearly 9000 lb") not to the nearest pound.

$$\text{mass of fuel used} = (9000 \text{ lb} - 82 \text{ lb}) \times \frac{0.4536 \text{ kg}}{1 \text{ lb}} \cong 4045 \text{ kg}$$

$$\text{fuel consumption} = \frac{25{,}012 \text{ mi}}{4045 \text{ kg}} \times \frac{1.609344 \text{ km}}{1 \text{ mi}} = 9.95 \text{ km/kg or} \sim 10 \text{ km/kg}$$

26.   If the proved reserve truly was an estimate, rather than an actual measurement, it would have been difficult to estimate it to the nearest trillion cubic feet. A statement such as 2,911,000 trillion cubic feet (or even 3×10^{18} ft^3) would have more realistically reflected the precision with which the proved reserve was known.

Units of Measurement

27. **(a)** $0.127 \text{ L} \times \dfrac{1000 \text{ mL}}{1 \text{ L}} = 127 \text{ mL}$ **(b)** $15.8 \text{ mL} \times \dfrac{1 \text{ L}}{1000 \text{ mL}} = 0.0158 \text{ L}$

      (c) $981 \text{ cm}^3 \times \dfrac{1 \text{ L}}{1000 \text{ cm}^3} = 0.981 \text{ L}$ **(d)** $2.65 \text{ m}^3 \times \left(\dfrac{100 \text{ cm}}{1 \text{ m}}\right)^3 = 2.65 \times 10^6 \text{ cm}^3$

28. **(a)** $1.55 \text{ kg} \times \dfrac{1000 \text{ g}}{1 \text{ kg}} = 1.55 \times 10^3 \text{ g}$ **(b)** $642 \text{ g} \times \dfrac{1 \text{ kg}}{1000 \text{ g}} = 0.642 \text{ kg}$

      (c) $2896 \text{ mm} \times \dfrac{1 \text{ cm}}{10 \text{ mm}} = 289.6 \text{ cm}$ **(d)** $0.086 \text{ cm} \times \dfrac{10 \text{ mm}}{1 \text{ cm}} = 0.86 \text{ mm}$

29. **(a)** $68.4 \text{ in.} \times \dfrac{2.54 \text{ cm}}{1 \text{ in.}} = 174 \text{ cm}$ **(b)** $94 \text{ ft} \times \dfrac{12 \text{ in.}}{1 \text{ ft}} \times \dfrac{2.54 \text{ cm}}{1 \text{ in.}} \times \dfrac{1 \text{ m}}{100 \text{ cm}} = 29 \text{ m}$

      (c) $1.42 \text{ lb} \times \dfrac{453.6 \text{ g}}{1 \text{ lb}} = 644 \text{ g}$ **(d)** $248 \text{ lb} \times \dfrac{0.4536 \text{ kg}}{1 \text{ lb}} = 112 \text{ kg}$

      (e) $1.85 \text{ gal} \times \dfrac{4 \text{ qt}}{1 \text{ gal}} \times \dfrac{0.9464 \text{ dm}^3}{1 \text{ qt}} = 7.00 \text{ dm}^3$ **(f)** $3.72 \text{ qt} \times \dfrac{0.9464 \text{ L}}{1 \text{ qt}} \times \dfrac{1000 \text{ mL}}{1 \text{ L}} = 3.52 \times 10^3 \text{ mL}$

30. **(a)** $1.00 \text{ km}^2 \times \left(\dfrac{1000 \text{ m}}{1 \text{ km}}\right)^2 = 1.00 \times 10^6 \text{ m}^2$

      (b) $1.00 \text{ m}^3 \times \left(\dfrac{100 \text{ cm}}{1 \text{ m}}\right)^3 = 1.00 \times 10^6 \text{ cm}^3$

      (c) $1.00 \text{ mi}^2 \times \left(\dfrac{5280 \text{ ft}}{1 \text{ mi}} \times \dfrac{12 \text{ in.}}{1 \text{ ft}} \times \dfrac{2.54 \text{ cm}}{1 \text{ in.}} \times \dfrac{1 \text{ m}}{100 \text{ cm}}\right)^2 = 2.59 \times 10^6 \text{ m}^2$

31. Express both masses in the same units for comparison. $3245 \mu\text{g} \times \left(\dfrac{1 \text{g}}{10^6 \mu\text{g}}\right) \times \left(\dfrac{10^3 \text{ mg}}{1 \text{g}}\right) = 3.245 \text{ mg}$, which is larger than 0.00515 mg.

32. Express both masses in the same units for comparison. $0.00475 \text{ kg} \times \dfrac{1000 \text{ g}}{1 \text{ kg}} = 4.75 \text{ g}$, which is larger than $3257 \text{ mg} \times \dfrac{1 \text{ g}}{10^3 \text{ mg}} = 3.257 \text{ g}$.

33. $\text{height} = 15 \text{ hands} \times \dfrac{4 \text{ in.}}{1 \text{ hand}} \times \dfrac{2.54 \text{ cm}}{1 \text{ in.}} \times \dfrac{1 \text{ m}}{100 \text{ cm}} = 1.5 \text{ m}$

34. A mile is defined as being 5,280 ft in length. We must use this conversion factor to find the length of a link in inches.

$$1.00 \text{ link} \times \frac{1 \text{ chain}}{100 \text{ links}} \times \frac{1 \text{ furlong}}{10 \text{ chains}} \times \frac{1 \text{ mile}}{8 \text{ furlongs}} \times \frac{5280 \text{ ft}}{1 \text{ mi}} \times \frac{12 \text{ in.}}{1 \text{ ft}} \times \frac{2.54 \text{ cm}}{1 \text{ in}} = 20.1 \text{ cm.}$$

35. **(a)** We use the speed as a conversion factor, but need to convert yards into meters.

$$\text{time} = 100.0 \text{ m} \times \frac{9.3 \text{ s}}{100 \text{ yd}} \times \frac{1 \text{ yd}}{36 \text{ in.}} \times \frac{39.37 \text{ in.}}{1 \text{ m}} = 10. \text{ s}$$

The final answer can only be quoted to a maximum of two significant figures.

(b) We need to convert yards to meters.

$$\text{speed} = \frac{100 \text{ yd}}{9.3 \text{ s}} \times \frac{36 \text{ in.}}{1 \text{ yd}} \times \frac{2.54 \text{ cm}}{1 \text{ in.}} \times \frac{1 \text{ m}}{100 \text{ cm}} = 9.8\underline{3} \text{ m/s}$$

(c) The speed is used as a conversion factor.

$$\text{time} = 1.45 \text{ km} \times \frac{1000 \text{ m}}{1 \text{ km}} \times \frac{1 \text{ s}}{9.8\underline{3} \text{ m}} \times \frac{1 \text{ min}}{60 \text{ s}} = 2.5 \text{ min}$$

36. **(a)** $\text{mass (mg)} = 2 \text{ tablets} \times \dfrac{5.0 \text{ gr}}{1 \text{ tablet}} \times \dfrac{1.0 \text{ g}}{15 \text{ gr}} \times \dfrac{1000 \text{ mg}}{1 \text{ g}} = 6.7 \times 10^2 \text{ mg}$

(b) $\text{dosage rate} = \dfrac{6.7 \times 10^2 \text{ mg}}{155 \text{ lb}} \times \dfrac{1 \text{ lb}}{453.6 \text{ g}} \times \dfrac{1000 \text{ g}}{1 \text{ kg}} = 9.5 \text{ mg aspirin/kg body weight}$

(c) $\text{time} = 1.0 \text{ lb} \times \dfrac{453.6 \text{ g}}{1 \text{ lb}} \times \dfrac{2 \text{ tablets}}{0.67 \text{ g}} \times \dfrac{1 \text{ day}}{2 \text{ tablets}} = 6.8 \times 10^2 \text{ days}$

37. $1 \text{ hectare} = 1 \text{ hm}^2 \times \left(\dfrac{100 \text{ m}}{1 \text{ hm}} \times \dfrac{100 \text{ cm}}{1 \text{ m}} \times \dfrac{1 \text{ in.}}{2.54 \text{ cm}} \times \dfrac{1 \text{ ft}}{12 \text{ in.}} \times \dfrac{1 \text{ mi}}{5280 \text{ ft}} \right)^2 \times \dfrac{640 \text{ acres}}{1 \text{ mi}^2}$

$1 \text{ hectare} = 2.47 \text{ acres}$

38. Here we must convert pounds per cubic inch into grams per cubic centimeter:

$$\text{density for metallic iron} = \frac{0.284 \text{ lb}}{\text{in}^3} \times \frac{454 \text{ g}}{1 \text{ lb}} \times \frac{(1 \text{ in.})^3}{(2.54 \text{ cm})^3} = 7.87 \frac{\text{g}}{\text{cm}^3}$$

39. $\text{pressure} = \dfrac{32 \text{ lb}}{1 \text{ in.}^2} \times \dfrac{453.6 \text{ g}}{1 \text{ lb}} \times \left(\dfrac{1 \text{ in.}}{2.54 \text{ cm}} \right)^2 = 2.2 \times 10^3 \text{ g/cm}^2$

$\text{pressure} = \dfrac{2.2 \times 10^3 \text{ g}}{1 \text{ cm}^2} \times \dfrac{1 \text{ kg}}{1000 \text{ g}} \times \left(\dfrac{100 \text{ cm}}{1 \text{ m}} \right)^2 = 2.2 \times 10^4 \text{ kg/m}^2$

40. First we will calculate the radius for a typical red blood cell using the equation for the volume of a sphere. $V = 4/3\pi r^3 = 9.00 \times 10^{-11}$ cm^3

$r^3 = 2.15 \times 10^{-11}$ cm^3 and $r = 2.78 \times 10^{-4}$ cm

Thus, the diameter is $2 \times r = 2 \times 2.78 \times 10^{-4}$ cm $\times \dfrac{10 \text{ mm}}{1 \text{ cm}} = 5.56 \times 10^{-3}$ mm

Expressed in inches: 5.56×10^{-4} cm $\times \dfrac{1 \text{ in}}{2.54 \text{ cm}} = 2.19 \times 10^{-4}$ inches in diameter

Temperature Scales

41. low: $^\circ F = \frac{9^\circ F}{5^\circ C}(^\circ C) + 32^\circ F = \frac{9^\circ F}{5^\circ C}\left(-10^\circ C\right) + 32^\circ F = 14^\circ F$

high: $^\circ F = \frac{9^\circ F}{5^\circ C}(^\circ C) + 32^\circ F = \frac{9^\circ F}{5^\circ C}\left(50^\circ C\right) + 32^\circ F = 122^\circ F$

42. high: $^\circ C = \left(^\circ F - 32^\circ F\right)\frac{5^\circ C}{9^\circ F} = \left(118^\circ F - 32^\circ F\right)\frac{5^\circ C}{9^\circ F} = 47.8\,^\circ C \approx 48\,^\circ C$

low: $^\circ C = \left(^\circ F - 32^\circ F\right)\frac{5^\circ C}{9^\circ F} = \left(17^\circ F - 32^\circ F\right)\frac{5^\circ C}{9^\circ F} = -8.3\,^\circ C$

43. Let us determine the Fahrenheit equivalent of absolute zero.

$^\circ F = \frac{9^\circ F}{5^\circ C}(^\circ C) + 32^\circ F = \frac{9^\circ F}{5^\circ C}\left(-273.15^\circ C\right) + 32^\circ F = -459.7^\circ F$

A temperature of $-465^\circ F$ cannot be achieved because it is below absolute zero.

44. Determine the Celsius temperature that corresponds to the highest Fahrenheit temperature, $240^\circ F$. $^\circ C = \left(^\circ F - 32^\circ F\right)\frac{5^\circ C}{9^\circ F} = \left(240^\circ F - 32^\circ F\right)\frac{5^\circ C}{9^\circ F} = 116\,^\circ C$

Because $116^\circ C$ is above the range of the thermometer, this thermometer cannot be used in this candy making assignment.

45. **(a)** From the data provided we can write down the following relationship:$-38.9^\circ C = 0\,^\circ M$ and $356.9\,^\circ C = 100\,^\circ M$. To find the mathematical relationship between these two scales, we can treat each relationship as a point on a two-dimensional Cartesian graph

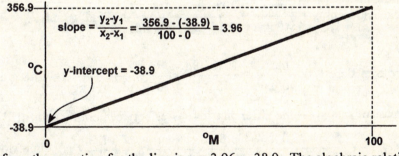

Therefore, the equation for the line is $y = 3.96x - 38.9$ The algebraic relationship between the two temperature scales is

$t(^\circ C) = 3.96(^\circ M) - 38.9$ or rearranging, $\quad t(^\circ M) = \dfrac{t(^\circ C) + 38.9}{3.96}$

9

Alternatively, note that the change in temperature in °C corresponding to a change of 100 °M is [356.9 – (-38.9)] = 395.8 °C, hence, (100 °M/395.8 °C) = 1 °M/3.96 °C. This factor must be multiplied by the number of degrees Celsius above zero on the M scale. This number of degrees is t(°C) + 38.9, which leads to the general equation t(°M) = [t(°C) + 38.9]/3.96.

The boiling point of water is 100 °C, corresponding to $t(°M) = \dfrac{100 + 38.9}{3.96} = 35.1°M$

(b) $t(°M) = \dfrac{-273.15 + 38.9}{3.96} = -59.2 °M$ would be the absolute zero on this scale.

46. **(a)** From the data provided we can write down the following relationship: -77.75 = 0 °A and -33.35 °C = 100 °A. To find the mathematical relationship between these two scales, we can treat each relationship as a point on a two-dimensional Cartesian graph

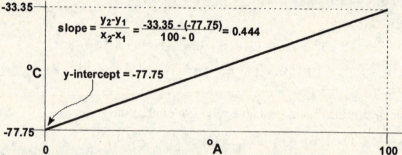

Therefore, the equation for the line is y = 0.444x - 77.75

The algebraic relationship between the two temperature scales is

$t(°C) = 0.444(°A) - 77.75$ or rearranging $t(°A) = \dfrac{t(°C) + 77.75}{0.444}$

The boiling point of water(100 °C) corresponds to $t(°A) = \dfrac{100 + 77.75}{0.444} = 400. °A$

(b) $t(°A) = \dfrac{-273.15 + 77.75}{0.444} = -440. °A$

Density

47. Butyric acid $\text{density} = \dfrac{\text{mass}}{\text{volume}} = \dfrac{2088 \text{ g}}{2.18 \text{ L}} \times \dfrac{1 \text{ L}}{1000 \text{ mL}} = 0.958 \text{ g} / \text{mL}$

48. Chloroform $\text{density} = \dfrac{\text{mass}}{\text{volume}} = \dfrac{22.54 \text{ kg}}{15.2 \text{ L}} \times \dfrac{1 \text{ L}}{1000 \text{ mL}} \times \dfrac{1000 \text{ g}}{1 \text{ kg}} = 1.48 \text{ g/mL}$

49. The mass of acetone is the difference in masses between empty and filled masses.

$\text{Density} = \dfrac{437.5 \text{ lb} - 75.0 \text{ lb}}{55.0 \text{ gal}} \times \dfrac{453.6 \text{ g}}{1 \text{ lb}} \times \dfrac{1 \text{ gal}}{3.785 \text{ L}} \times \dfrac{1 \text{ L}}{1000 \text{ mL}} = 0.790 \text{ g} / \text{mL}$

50. Density is a conversion factor. $\text{volume} = (283.2 \text{ g filled} - 121.3 \text{ g empty}) \times \dfrac{1 \text{ mL}}{1.59 \text{ g}} = 102 \text{ mL}$

51. Acetone mass $= 7.50$ L antifreeze $\times \dfrac{1000 \text{ mL}}{1 \text{ L}} \times \dfrac{0.9867 \text{ g antifreeze}}{1 \text{ mL antifreeze}} \times \dfrac{8.50 \text{ g acetone}}{100.0 \text{ g antifreeze}}$

$$\times \dfrac{1 \text{ kg}}{1000 \text{ g}} = 0.629 \text{ kg acetone}$$

52. Solution mass $= 1.00$ kg sucrose $\times \dfrac{1000 \text{ g sucrose}}{1 \text{ kg sucrose}} \times \dfrac{100.00 \text{ g solution}}{10.05 \text{ g sucrose}} = 9.95 \times 10^3$ g solution

53. Fertilizer mass $= 225$ g nitrogen $\times \dfrac{1 \text{ kg N}}{1000 \text{ g N}} \times \dfrac{100 \text{ kg fertilizer}}{21 \text{ kg N}} = 1.0\underline{7}$ kg fertilizer

54. $m_{\text{acetic acid}} = 1.00$ L vinegar $\times \dfrac{1000 \text{ mL vinegar}}{1 \text{ L}} \times \dfrac{1.006 \text{ g vinegar}}{1 \text{ mL vinegar}} \times \dfrac{5.4 \text{ g acetic acid}}{100 \text{ g vinegar}}$

$m_{\text{acetic acid}} = 54.\underline{3}$ g acetic acid

55. The calculated volume of the iron block is converted to its mass by using the provided density.

$$\text{Mass} = 52.8 \text{ cm} \times 6.74 \text{ cm} \times 3.73 \text{ cm} \times 7.86 \dfrac{\text{g}}{\text{cm}^3} = 1.04 \times 10^4 \text{ g iron}$$

56. The calculated volume of the steel cylinder is converted to its mass by using the provided density.

$$\text{Mass} = V(\text{density}) = \pi r^2 h(d) = 3.14159 \left(1.88 \text{ cm}\right)^2 18.35 \text{ cm} \times 7.75 \dfrac{\text{g}}{\text{cm}^3} = 1.58 \times 10^3 \text{ g steel}$$

57. We start by determining the mass of each item.

(1) mass of iron bar $= \left(81.5 \text{ cm} \times 2.1 \text{ cm} \times 1.6 \text{ cm}\right) \times 7.86 \text{ g/cm}^3 = 2.2 \times 10^3$ g iron

(2) mass of Al foil $= \left(12.12 \text{ m} \times 3.62 \text{ m} \times 0.003 \text{ cm}\right) \times \left(\dfrac{100 \text{ cm}}{1 \text{ m}}\right)^2 \times 2.70 \text{ g Al/cm}^3 = 4 \times 10^3$ g Al

(3) mass of water $= 4.051$ L $\times \dfrac{1000 \text{ cm}^3}{1 \text{ L}} \times 0.998 \text{ g}/\text{cm}^3 = 4.04 \times 10^3$ g water

In order of increasing mass, the items are: iron bar < aluminum foil < water. Please bear in mind however, that, strictly speaking, the rules for significant figures do not allow us to distinguish between the masses of aluminum and water.

58. Total volume of 125 pieces of shot

$$V = 8.9 \text{ mL} - 8.4 \text{ mL} = 0.5 \text{ mL}; \qquad \dfrac{\text{mass}}{\text{shot}} = \dfrac{0.5 \text{ mL}}{125 \text{ shot}} \times \dfrac{1 \text{ cm}^3}{1 \text{ mL}} \times \dfrac{8.92 \text{ g}}{1 \text{ cm}^3} = 0.04 \text{ g/shot}$$

<u>59.</u> First determine the volume of the aluminum foil, then its area, and finally its thickness.

$$\text{volume} = 2.568 \text{ g} \times \frac{1 \text{ cm}^3}{2.70 \text{ g}} = 0.951 \text{ cm}^3; \qquad \text{area} = (22.86 \text{ cm})^2 = 522.6 \text{ cm}^2$$

$$\text{thickness} = \frac{\text{volume}}{\text{area}} = \frac{0.951 \text{ cm}^3}{522.6 \text{ cm}^2} \times \frac{10 \text{ mm}}{1 \text{ cm}} = 1.82 \times 10^{-2} \text{ mm}$$

60. The vertical piece of steel has a volume $= 12.78 \text{ cm} \times 1.35 \text{ cm} \times 2.75 \text{ cm} = 47.4 \text{ cm}^3$
The horizontal piece of steel has a volume $= 10.26 \text{ cm} \times 1.35 \text{ cm} \times 2.75 \text{ cm} = 38.1 \text{ cm}^3$
$V_{\text{total}} = 47.4 \text{ cm}^3 + 38.1 \text{ cm}^3 = 85.5 \text{ cm}^3$. Mass $= 85.5 \text{ cm}^3 \times 7.78 \text{ g/cm}^3 = 665 \text{ g of steel}$

<u>61.</u> Here we are asked to calculate the number of liters of whole blood that must be collected in order to end up with 0.5 kg of red blood cells. Each red blood cell has a mass of
$9.00 \times 10^{-11} \text{ cm}^3 \times 1.096 \text{ g cm}^{-3} = 9.864 \times 10^{-11} \text{ g}$

$$\text{red blood cells (mass per mL)} = \frac{9.864 \times 10^{-11} \text{ g}}{1 \text{ cell}} \times \frac{5.4 \times 10^9 \text{ cells}}{1 \text{ mL}} = \frac{0.533 \text{ g red blood cells}}{1 \text{ mL of blood}}$$

For 0.5 kg or 5×10^2 g of red blood cells, we require

$$= 5 \times 10^2 \text{ g red blood cells} \times \frac{1 \text{ mL of blood}}{0.533 \text{ g red blood cells}} = 9 \times 10^2 \text{ mL of blood or 0.9 L blood}$$

62. The mass of the liquid mixture can be found by subtracting the mass of the full bottle from the mass of the empty bottle $= 15.4448 \text{ g} - 12.4631 \text{ g} = 2.9817 \text{ g liquid}$. Similarly, the total mass of the water that can be accommodated in the bottle is $13.5441 \text{ g} - 12.4631 \text{ g} = 1.081 \text{ g H}_2\text{O}$. The volume of the water and hence the internal volume for the bottle is equal to

$$1.081 \text{ g H}_2\text{O} \times \frac{1 \text{ mL H}_2\text{O}}{0.9970 \text{ g H}_2\text{O}} = 1.084 \text{ mL H}_2\text{O} \ (25 \text{ °C})$$

Thus, the density of the liquid mixture $= \dfrac{2.9817 \text{ g liquid}}{1.084 \text{ mL}} = 2.751 \text{ g mL}^{-1}$

Since the calcite just floats in this mixture of liquids, it must have the same density as the mixture. Consequently, the solid calcite sample must have a density of 2.751 g mL^{-1} as well.

Percent Composition

<u>63.</u> The percent of students with each grade is obtained by dividing the number of students with that grade by the total number of students.　$\%A = \dfrac{7 \text{ A's}}{76 \text{ students}} \times 100\% = 9.2\% \text{ A}$

$$\%B = \frac{22 \text{ B's}}{76 \text{ students}} \times 100\% = 28.9\% \text{ B} \qquad\qquad \%C = \frac{37 \text{ C's}}{76 \text{ students}} \times 100\% = 48.7\% \text{ C}$$

$$\%D = \frac{8 \text{ D's}}{76 \text{ students}} \times 100\% = 11\% \text{ D} \qquad\qquad \%F = \frac{2 \text{ F's}}{76 \text{ students}} \times 100\% = 3\% \text{ F}$$

Note that the percentages add to 101% due to rounding effects.

64. The number of students with a certain grade is determined by multiplying the total number of students by the fraction of students who earned that grade.

$$\text{no. of A's} = 84 \text{ students} \times \frac{18 \text{ A's}}{100 \text{ students}} = 15 \text{ A's}$$

$$\text{no. of B's} = 84 \text{ students} \times \frac{25 \text{ B's}}{100 \text{ students}} = 21 \text{ B's} \quad \text{no. of C's} = 84 \text{ students} \times \frac{32 \text{ C's}}{100 \text{ students}} = 27 \text{ C's}$$

$$\text{no. of D's} = 84 \text{ students} \times \frac{13 \text{ D's}}{100 \text{ students}} = 11 \text{ D's} \quad \text{no. of F's} = 84 \text{ students} \times \frac{12 \text{ F's}}{100 \text{ students}} = 10 \text{ F's}$$

65. Use the percent composition as a conversion factor.

$$\text{mass of sucrose} = 3.50 \text{ L} \times \frac{1000 \text{ mL}}{1 \text{ L}} \times \frac{1.118 \text{ g soln}}{1 \text{ mL}} \times \frac{28.0 \text{ g sucrose}}{100 \text{ g soln}} = 1.10 \times 10^3 \text{ g sucrose}$$

66. Again, percent composition is used as a conversion factor. We are careful to label both the numerator and denominator for each factor.

$$V_{\text{solution}} = 2.25 \text{ kg sodium hydroxide} \times \frac{1000 \text{ g}}{1 \text{ kg}} \times \frac{100.0 \text{ g soln}}{12.0 \text{ g sodium hydroxide}} \times \frac{1 \text{ mL}}{1.131 \text{ g soln}}$$

$$V_{\text{solution}} = 1.66 \times 10^4 \text{ mL soln} \times \frac{1 \text{ L}}{1000 \text{ mL}} = 16.6 \text{ L soln}$$

INTEGRATIVE AND ADVANCED EXERCISES

67. 99.9 is known to 0.1 part in 99.9, or 0.1%. 1.008 is known to 0.001 part in 1.008 or 0.1%. The product 100.7 also is known to 0.1 part in 100.7 or 0.1%, which is the same precision as the two factors. On the other hand, the three-significant-figure product, 101, is known to 1 part in 101 or 1%, which is ten times less precise than either of the two factors. Thus, the result is properly expressed to four significant figures.

68. $1.543 = 1.5794 - 1.836 \times 10^{-3}(t-15) \quad 1.543 - 1.5794 = -1.836 \times 10^{-3}(t-15) = 0.036\underline{4}$

$$(t-15) = \frac{-0.036\underline{4}}{-1.836 \times 10^{-3}} = 19.\underline{8} \text{ °C} \qquad t = 19.\underline{8} + 15 = 34.\underline{8} \text{ °C}$$

69. $\text{volume needed} = 18{,}000 \text{ gal} \times \frac{4 \text{ qt}}{1 \text{ gal}} \times \frac{0.9464 \text{ L}}{1 \text{ qt}} \times \frac{1000 \text{ mL}}{1 \text{ L}} \times \frac{1.00 \text{ g}}{1 \text{ mL}} \times \frac{1 \text{ g Cl}}{10^6 \text{ g water}}$

$$\times \frac{100 \text{ g soln}}{7 \text{ g Cl}} \times \frac{1 \text{ mL soln}}{1.10 \text{ g soln}} \times \frac{1 \text{ L soln}}{1000 \text{ mL soln}} = 0.9 \text{ L soln}$$

70. We first determine the volume of steel needed. This volume, divided by the cross-sectional area of the cylinder, gives the length of the steel bar needed.

$$V = 1.000 \text{ kg steel} \times \frac{1000 \text{ g steel}}{1 \text{ kg steel}} \times \frac{1 \text{ cm}^3 \text{ steel}}{7.70 \text{ g steel}} = 129.\underline{87} \text{ cm}^3 \text{ of steel}$$

For an equilateral triangle of length s, $\text{Area} = \dfrac{s^2\sqrt{3}}{4} = \dfrac{(2.50 \text{ in})^2 \sqrt{3}}{4} = 2.70\underline{6} \text{ in}^2$

$\text{length} = \dfrac{\text{volume}}{\text{area}} = \dfrac{129.\underline{87} \text{ cm}^3}{2.70\underline{6} \text{ in}^2} \times \left(\dfrac{1 \text{ in.}}{2.54 \text{ cm}}\right)^2 \times \dfrac{1 \text{ in.}}{2.54 \text{ cm}} = 2.93 \text{ in.}$

71. $\text{NaCl mass} = 330{,}000{,}000 \text{ mi}^3 \times \left(\dfrac{5280 \text{ ft}}{1 \text{ mi}} \times \dfrac{12 \text{ in.}}{1 \text{ ft}} \times \dfrac{2.54 \text{ cm}}{1 \text{ in}}\right)^3 \times \dfrac{1 \text{ mL}}{1 \text{ cm}^3} \times \dfrac{1.03 \text{ g}}{1 \text{ mL}}$

$\times \dfrac{3.5 \text{ g sodium chloride}}{100.0 \text{ g sea water}} \times \dfrac{1 \text{ lb}}{453.6 \text{ g}} \times \dfrac{1 \text{ ton}}{2000 \text{ lb}} = 5.5 \times 10^{16} \text{ tons}$

72. First, we find the volume of the wire, then its cross-sectional area, and finally its length. We carry an additional significant figure through the early stages of the calculation to help avoid rounding errors.

$V = 1 \text{ lb} \times \dfrac{453.6 \text{ g}}{1 \text{ lb}} \times \dfrac{1 \text{ cm}^3}{8.92 \text{ g}} = 50.8\underline{5} \text{ cm}^3$

Note: $\text{Area} = \pi r^2$

$\text{area} = 3.1416 \times \left(\dfrac{0.05082 \text{ in.}}{2} \times \dfrac{2.54 \text{ cm}}{1 \text{ in.}}\right)^2 = 0.01309 \text{ cm}^2$

$\text{length} = \dfrac{\text{volume}}{\text{area}} = \dfrac{50.8\underline{5} \text{ cm}^3}{0.01309 \text{ cm}^2} \times \dfrac{1 \text{ m}}{100 \text{ cm}} = 38.8 \text{ m}$

73. $V_{seawater} = 1.00 \times 10^5 \text{ ton Mg} \times \dfrac{2000 \text{ lb Mg}}{1 \text{ ton Mg}} \times \dfrac{453.6 \text{ g Mg}}{1 \text{ lb Mg}} \times \dfrac{1000 \text{ g seawater}}{1.4 \text{ g Mg}} \times \dfrac{0.001 \text{ L}}{1.025 \text{ g seawater}}$

$\times \dfrac{1 \text{ m}^3}{1000 \text{ L}} = 6 \times 10^7 \text{ m}^3 \text{ seawater}$

74. (a) $\text{dustfall} = \dfrac{10 \text{ ton}}{\text{mi}^2 \cdot \text{mo}} \times \left(\dfrac{1 \text{ mi}}{5280 \text{ ft}} \times \dfrac{1 \text{ ft}}{12 \text{ in.}} \times \dfrac{39.37 \text{ in.}}{1 \text{ m}}\right)^2 \times \dfrac{2000 \text{ lb}}{1 \text{ ton}} \times \dfrac{454 \text{ g}}{1 \text{ lb}} \times \dfrac{1000 \text{ mg}}{1 \text{ g}}$

$= \dfrac{3.5 \times 10^3 \text{ mg}}{\text{m}^2 \cdot \text{mo}} \times \dfrac{1 \text{ month}}{30 \text{ d}} \times \dfrac{1 \text{ d}}{24 \text{ h}} = \dfrac{5 \text{ mg}}{\text{m}^2 \cdot \text{h}}$

(b) This problem is solved by the conversion factor method, starting with the volume that deposits on each square meter, 1 mm deep.

$\dfrac{(1.0 \text{ mm} \times 1 \text{ m}^2)}{1 \text{ m}^2} \times \dfrac{1 \text{ cm}}{10 \text{ mm}} \times \left(\dfrac{100 \text{ cm}}{1 \text{ m}}\right)^2 \times \dfrac{2 \text{ g}}{1 \text{ cm}^3} \times \dfrac{1000 \text{ mg}}{1 \text{ g}} \times \dfrac{1 \text{ m}^2 \cdot \text{h}}{4.9 \text{ mg}}$

$= 4.1 \times 10^5 \text{ h} = 5 \times 10^1 \text{ y}$ About half a century to accumulate a depth of 1 mm.

75. **(a)** $\text{volume} = 3.54 \times 10^6 \text{ acre - feet} \times \dfrac{1\,\text{mi}^2}{640\,\text{acre}} \times \left(\dfrac{5280\,\text{ft}}{1\,\text{mi}}\right)^2 = 1.54 \times 10^{11} \text{ ft}^3$

(b) $\text{volume} = 1.54 \times 10^{11} \text{ ft}^3 \times \left(\dfrac{12\,\text{in.}}{1\,\text{ft}} \times \dfrac{2.54\,\text{cm}}{1\,\text{in.}} \times \dfrac{1\,\text{m}}{100\,\text{cm}}\right)^3 = 4.36 \times 10^9 \text{ m}^3$

(c) $\text{volume} = 4.36 \times 10^9 \text{ m}^3 \times \dfrac{1000\,\text{L}}{1\,\text{m}^3} \times \dfrac{1\,\text{gal}}{3.785\,\text{L}} = 1.15 \times 10^{12} \text{ gal}$

76. Let F be the Fahrenheit temperature and C be the Celsius temperature. $C = (F - 32)\frac{5}{9}$

(a) $F = C - 49 \qquad C = (C - 49 - 32)\frac{5}{9} = \frac{5}{9}(C - 81) \qquad C = \frac{5}{9}C - \frac{5}{9}(81) \qquad C = \frac{5}{9}C - 45$

$\frac{4}{9}C = -45 \qquad \text{Hence: } C = -101.\underline{25}$

When it is $\sim -101\,°C$, the temperature in Fahrenheit is $-150.\,°F$ (49° lower).

(b) $F = 2C \qquad\qquad C = (2C - 32)\frac{5}{9} = \frac{10}{9}C - 17.8 \qquad\qquad 17.8 = \frac{10}{9}C - C = \frac{1}{9}C$

$C = 9 \times 17.8 = 160.\,°C \qquad\qquad F = \frac{9}{5}C + 32 = \frac{9}{5}(160.) + 32 = 320.\,°F$

(c) $F = C/8 \qquad C = (C/8 - 32)\frac{5}{9} = \frac{5}{72}C - 17.8 \qquad 17.8 = \frac{5}{72}C - C = -\frac{67}{72}C$

$C = -\dfrac{72 \times 17.8}{67} = -19.1\,°C \qquad F = \frac{9}{5}C + 32 = \frac{9}{5}(-19.1) + 32 = -2.4\,°F$

(d) $F = C + 300 \qquad C = (C + 300 - 32)\frac{5}{9} = \frac{5}{9}C + 148.9 \qquad 148.9 = \frac{4}{9}C$

$C = \dfrac{9 \times 148.9}{4} = 335\,°C \qquad F = \frac{9}{5}C + 32 = \frac{9}{5}(335) + 32 = 635\,°F$

77. We will use the density of diatomaceous earth, and its mass in the cylinder, to find the volume occupied by the diatomaceous earth.

$\text{diatomaceous earth volume} = 8.0\,\text{g} \times \dfrac{1\,\text{cm}^3}{2.2\,\text{g}} = 3.6\,\text{cm}^3$

The added water volume will occupy the remaining volume in the graduated cylinder.

$\text{water volume} = 100.0\,\text{mL} - 3.6\,\text{mL} = 96.4\,\text{mL}$

78. We will use the density of water, and its mass in the pycnometer, to find the volume of liquid held by the pycnometer.

$\text{pycnometer volume} = (35.552\,\text{g} - 25.601\,\text{g}) \times \dfrac{1\,\text{mL}}{0.99821\,\text{g}} = 9.969\,\text{mL}$

the mass of the methanol and the pyncnometer's volume determine liquid density

$\text{density of methanol} = \dfrac{33.490\,\text{g} - 25.601\,\text{g}}{9.969\,\text{mL}} = 0.7914\,\text{g/mL}$

79. We use the density of water, and its mass in the pycnometer, to find the volume of liquid held by the pycnometer.

$$\text{pycnometer volume} = (35.552\,\text{g} - 25.601\,\text{g}) \times \frac{1\,\text{mL}}{0.99821\,\text{g}} = 9.969\,\text{mL}$$

the mass of the ethanol and the pyncnometer's volume determine liquid density

$$\text{density of ethanol} = \frac{33.470\,\text{g} - 25.601\,\text{g}}{9.969\,\text{mL}} = 0.7893\,\text{g/mL}$$

The difference in the density of pure methanol and pure ethanol is 0.0020 g/mL. If the density of the solution is a linear function of the volume-percent composition, we would see that the maximum change in density (0.0020 g/mL) corresponds to a change of 100% in the volume percent. This means that the absolute best accuracy that one can obtain is a differentiation of 0.0001 g/mL between the two solutions. This represents a change in the volume % of ~ 5%. Given this apparatus, if the volume percent does not change by at least 5%, we would not be able to differentiate based on density (probably more like a 10 % difference would be required, given that out error when measuring two solutions is more likely ± 0.0002 g/mL).

80. We first determine the pycnometer's volume.

$$\text{pycnometer volume} = (35.55\,\text{g} - 25.60\,\text{g}) \times \frac{1\,\text{mL}}{0.9982\,\text{g}} = 9.97\,\text{mL}$$

Then we determine the volume of water present with the lead.

$$\text{volume of water} = (44.83\,\text{g} - 10.20\,\text{g} - 25.60\,\text{g}) \times \frac{1\,\text{mL}}{0.9982\,\text{g}} = 9.05\,\text{mL}$$

Difference between the two volumes is the volume of lead, which leads to the density of lead.

$$\text{density} = \frac{10.20\,\text{g}}{(9.97\,\text{mL} - 9.05\,\text{mL})} = 11\,\text{g/mL}$$

Note that the difference in the denominator has just two significant digits.

81. $$\text{Water used(in kg/week)} = 1.8 \times 10^{6}\,\text{people} \times \left(\frac{750\text{L}}{\text{day}}\right) \times \left(\frac{7\,day}{1\text{week}}\right) \times \frac{1\,\text{kg}}{1\,\text{L}} = 9.45 \times 10^{9}\,\text{kg water/week}$$

Given: Sodium hypochlorite is NaClO

$$\text{mass of NaClO} = 9.45 \times 10^{9}\,\text{kg water} \left(\frac{1\,\text{kg chlorine}}{1 \times 10^{6}\,\text{kg water}}\right) \times \left(\frac{100\,\text{kg NaClO}}{47.62\,\text{kg chlorine}}\right)$$

$$= 1.98 \times 10^{4}\,\text{kg sodium hypochlorite}$$

82. $$\frac{1.77\,\text{lbs}}{1\,\text{L}} \times \frac{1\,\text{kg}}{2.2046\,\text{lbs}} = 0.803\,\text{kg L}^{-1}; \quad 22,300\,\text{kg of fuel are required, hence:}$$

$$22,300\;\text{kg fuel} \times \frac{1\,\text{L}}{0.803\;\text{kg}} = 2.78 \times 10^{4}\,\text{L of fuel}$$

(Note, the plane had 7682 L of fuel left in the tank)

Hence, the volume of fuel that should have been added = $2.78 \times 10^{4}\,\text{L} - 0.7682\,\text{L} = 2.01 \times 10^{4}\,\text{L}$

83. **a)** Density of water at 10. °C:

$$\text{density} = \frac{0.99984 + (1.6945 \times 10^{-2}(10.) - (7.987 \times 10^{-6}(10.)^2)}{1 + (1.6880 \times 10^{-2}(10.))} = 0.9997 \text{ g cm}^{-3} \text{ (4 sig fig)}$$

b) set a = 0.99984, b = 1.6945 × 10^2, c = 7.987 × 10^{-6}, d = 1.6880 × 10^{-2} (for simplicity)

$$0.99860 = \frac{0.99984 + (1.6945 \times 10^{-2}(t) - (7.987 \times 10^{-6}(t)^2)}{1 + (1.6880 \times 10^{-2}t)} = \frac{a + bt - ct^2}{1 + dt}$$

multiply both sides by (1 + dt): 0.99860(1 + dt) = 0.99860 + 0.99860dt = a + bt – ct^2
bring all terms to the left hand side: 0 = a + bt – ct^2 – 0.99860 – 0.99860dt
collect terms 0 = a – 0.99860 + bt – 0.99860dt – ct^2
substitute in for a, b, c and d:
0 = 0.99984 – 0.99860 + 1.6945 × 10^{-2}t – 0.99860(1.6880 × 10^{-2})t – 7.987 × 10^{-6}t^2
simplify: 0 = 0.00124 + 0.000088623t – 7.987 × 10^{-6}t^2
solve the quadratic equation: t = 19.188 °C

c) **i)** Maximum density by estimation:
Determine density every 5 °C then narrow down the range to the degree.
First set of data suggests ~5 °C, the second set of data suggests ~4 °C
and the final set of data suggests about 4.1 °C (+/-0.1 °C)

1ˢᵗ data set		2ⁿᵈ data set		3ʳᵈ data set	
0 °C	0.999840	3 °C	0.999965	3.6 °C	0.999972
5 °C	0.999968	4 °C	0.999974	3.8 °C	0.999973
10 °C	0.999736	5 °C	0.999968	4.0 °C	0.999974
15 °C	0.999216	6 °C	0.999948	4.2 °C	0.999974
20 °C	0.998464	7 °C	0.999914	4.4 °C	0.999973

ii) Graphical method shown below:

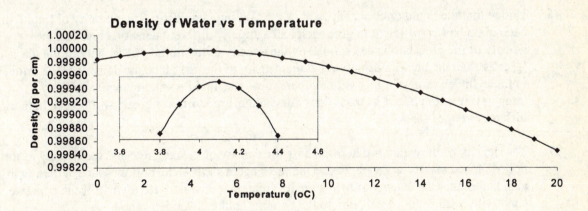

17

iii)   Method based on differential calculus: set the first derivative equal to zero

set a = 0.99984, b = 1.6945×10^2, c = 7.987×10^{-6}, d = 1.6880×10^{-2} (simplicity)

$$f(t) = \frac{a + bt - ct^2}{1 + dt} \qquad\qquad f'(t) = \frac{(b - 2ct)(1 + dt)\ (a + bt\ ct^2)d}{(1 + dt)^2} \quad \text{(quotient rule)}$$

$$f'(t) = \frac{b + bdt - 2ct - 2cdt^2 - ad - bdt + cdt^2}{(1 + dt)^2} = \frac{b - 2ct - cdt^2 - ad}{(1 + dt)^2} = 0 \quad \text{(max)}$$

We need to set the first derivative = 0, hence consider the numerator = 0
Basically we need to solve a quadratic: $0 = -cdt^2 - 2ct + b\text{-}ad$

$$t = \frac{2c \pm \sqrt{(2c)2 - 4(-cd)(b - ad)}}{-2cd} \quad \text{only the positive solution is acceptable.}$$

Plug in a = 0.99984, b = 1.6945×10^2, c = 7.987×10^{-6}, d = 1.6880×10^{-2}
By solving the quadratic equation, one finds that a temperature of 4.09655 °C has the highest density (0.999974 g cm^{-3}). Hence, ~ 4.1 °C is the temperature where water has its maximum density.

FEATURE PROBLEMS

84. All of the pennies minted before 1982 weigh more than 3.00 g, while all of those minted after 1982 weigh less than 2.60 g. It would not be unreasonable to infer that the composition of a penny changed in 1982. In fact, pennies minted prior to 1982 are composed of almost pure copper (about 96% pure). Those minted after 1982 are composed of zinc with a thin copper cladding. Both types of pennies were minted in 1982.

85. After sitting in a bathtub that was nearly full and observing the water splashing over the side, Archimedes realized that the crown—when submerged in water—would displace a volume of water equal to its volume. Once Archimedes determined the volume in this way and determined the mass of the crown with a balance, he was able to calculate the crown's density. Since the gold-silver alloy has a different density (it is lower) than pure gold, Archimedes could tell that the crown was not pure gold.

86. Notice that the liquid does not fill each of the floating glass balls. The quantity of liquid in each glass ball is sufficient to give each ball a slightly different density. Note that the density of the glass ball is determined by the density of the liquid, the density of the glass (greater than the liquid's density), and the density of the air. Since the density of the liquid in the cylinder varies slightly with temperature—the liquid's volume increases as temperature goes up, but its mass does not change, ergo, different balls will be buoyant at different temperatures.

87. The density of the canoe is determined by the density of the concrete and the density of the hollow space inside the canoe, where the passengers sit. It is the hollow space, (filled with air), that makes the density of the canoe less than that of water (1.0 g/cm^3). If the concrete canoe fills with water, it will sink to the bottom, unlike a wooden canoe.

88. In sketch (a), the mass of the plastic block appears to be 50.0 g.
In sketch (b), the fact that the plastic block is clearly visible on the bottom of a beaker filled with ethanol, shows that it is both insoluble in and more dense than ethanol (i.e. > 0.789 g/cm³).
In sketch (c), because the plastic block floats on bromoform, the density of the plastic must be less than that for bromoform (i.e. < 2.890 g/cm³). More over, because the block is ~ 40% submerged, the volume of bromoform having the same 50.0 g mass as the block is only about 40% of the volume of the block. Thus using the expression $V = m/d$, we can write
Volume of displaced bromoform ~ $0.40 \times V_{block}$

$$\frac{\text{Mass of bromoform}}{\text{density of bromoform}} = \frac{0.40 \times \text{mass of block}}{\text{density of plastic}} = \frac{50.0 \text{ g of bromoform}}{2.890 \frac{\text{g bromoform}}{\text{cm}^3}} = 0.40 \times \frac{50.0 \text{ g of plastic}}{\text{density of plastic}}$$

$$\text{density of plastic} \approx \frac{2.890 \frac{\text{g bromoform}}{\text{cm}^3}}{50.0 \text{ g of bromoform}} \times 0.40 \times 50.0 \text{ g of plastic} \approx 1.16 \frac{\text{g}}{\text{cm}^3}$$

The information provided in sketch (d) provides us with an alternate method for estimating the density of the plastic (use the fact that the density of water is 0.99821 g/cm³ at 20 °C).
Mass of water displaced = 50.0 g – 5.6 g = 44.4 g

$$\text{Volume of water displaced} = 44.4 \text{ g} \times \frac{1 \text{ cm}^3}{0.99821 \text{ g}} = 44.5 \text{ cm}^3$$

$$\text{Therefore the density of the plastic} = \frac{mass}{volume} = \frac{50.0 \text{ g}}{44.5 \text{ cm}^3} = 1.12 \frac{\text{g}}{\text{cm}^3}$$

This is reasonably close to the estimate based on the information in sketch (c).

89. One needs to convert (lb of force) into (Newtons) 1 lb of force = 1 slug × 1 ft s⁻²

$$\text{(1 slug = 14.59 kg). Therefore, 1 lb of force} = \frac{14.59 \text{ kg} \cdot 1 \text{ ft}}{s^2}$$

$$= \left(\frac{14.59 \text{ kg} \times 1 \text{ ft}}{s^2}\right)\left(\frac{12 \text{ in}}{1 \text{ ft}}\right)\left(\frac{2.54 \text{ cm}}{1 \text{ in}}\right)\left(\frac{1 \text{ m}}{100 \text{ cm}}\right) = \frac{4.45 \text{ kg} \cdot \text{m}}{s^2} = 4.45 \text{ Newtons}$$

From this result it is clear that 1 lb of force = 4.45 Newtons.

CHAPTER 2
ATOMS AND THE ATOMIC THEORY
PRACTICE EXAMPLES

1A The total mass must be the same before and after reaction.
mass before reaction = 0.382 g magnesium + 2.652 g nitrogen = 3.034 g
mass after reaction = magnesium nitride mass + 2.505 g nitrogen=3.034 g
magnesium nitride mass = 3.034 g − 2.505 g = 0.529 g magnesium nitride

1B Again, the total mass is the same before and after the reaction.
mass before reaction = 7.12g magnesium +1.80g bromine = 8.92g
mass after reaction = 2.07 g magnesium bromide + magnesium mass = 8.92 g
magnesium mass = 8.92 g − 2.07 g = 6.85 g magnesium

2A In Example 2-2 we are told that 0.500 g MgO contains 0.301 g of Mg. With this information, we can determine the mass of magnesium needed to form 2.000 g magnesium oxide.

$$\text{mass of Mg} = 2.000 \text{ g MgO} \times \frac{0.301 \text{ g Mg}}{0.500 \text{ g MgO}} = 1.20 \text{ g Mg}$$

The remainder of the 2.00 g of magnesium oxide is the mass of oxygen
mass of oxygen = 2.00g magnesium oxide −1.20g magnesium = 0.80 g oxygen

2B In Example 2-2, we see that a 0.500 g sample of MgO has 0.301 g Mg, hence, it must have 0.199 g O_2. From this we see that if we have equal masses of Mg and O_2, the oxygen is in excess. First we find out how many grams of oxygen reacts with 10.00 g of Mg.

$$\text{mass}_{oxygen} = 10.00 \text{ g Mg} \times \frac{0.199 \text{ g O}_2}{0.301 \text{ g Mg}} = 6.61 \text{ g O}_2 \text{ (used up)}$$

Hence, 10.00 g - 6.61 g = 3.39 g O_2 unreated. Mg is the limiting reactant.
MgO(s) mass = mass Mg + Mass O_2 = 10.00 g + 6.61 g = 16.61 g MgO.
There are only two substances present, 16.61 g of MgO (product) and 3.39 g of unreacted O_2

3A Silver has 47 protons. If the isotope in question has 62 neutrons, then it has a mass number of 109. This can be represented as $^{109}_{47}Ag$

3B Tin has 50 electrons and 50 protons when neutral, while a neutral Cadmium atom has 48 electrons. This means that we are dealing with Sn^{2+}. We do not know how many neutrons tin has so there can be more than one answer. For instance, $^{116}_{50}Sn^{2+}$, $^{117}_{50}Sn^{2+}$, $^{118}_{50}Sn^{2+}$, $^{119}_{50}Sn^{2+}$, $^{120}_{50}Sn^{2+}$ are all possible answers.

4A The ratio of the masses of ^{202}Hg and ^{12}C is: $\frac{^{202}Hg}{^{12}C} = \frac{201.97062 \text{ u}}{12 \text{ u}} = 16.8308848$

4B Atomic mass is 12 u × 13.16034 = 157.9241 u. The isotope is $^{158}_{64}Gd$. Using an atomic mass of 15.9949 u for ^{16}O, the mass of $^{158}_{64}Gd$ relative to ^{16}O is

$$\text{relative mass to oxygen-16} = \frac{157.9241 \text{ u}}{15.9949 \text{ u}} = 9.87340$$

5A The average atomic mass of boron is 10.811, which is closer to 11.009305 than to 10.012937. Thus, boron-11 is the isotope that is present in greater abundance.

5B The average atomic mass of indium is 114.82, one isotope is known to be ^{113}In. Since the weight average atomic mass is almost 115, the second isotope must be larger than both In-113 and In-114. Clearly then, the second isotope must be In-115(^{115}In).

6A Weighted average atomic mass of Si = $(27.97693\ u \times 0.9223) \rightarrow 25.80\underline{3}\ u$

$\qquad\qquad\qquad\qquad\qquad\qquad\ (28.97649\ u \times 0.0467) \rightarrow \quad 1.35\underline{3}\ u$

$\qquad\qquad\qquad\qquad\qquad\underline{(29.97376\ u \times 0.0310) \rightarrow \quad 0.929\ u}$

$\qquad\qquad\qquad\qquad\qquad\qquad\qquad\qquad\qquad\qquad\qquad 28.08\underline{5}\ u$

We should report the weighted atomic mass of Si as 28.08 u

6B We let x be the fractional abundance of lithium-6.

$$6.941\ u = \left[x \times 6.01513\ u \right] + \left[(1-x) \times 7.01601\ u \right] = x \times 6.01513\ u + 7.01601\ u - x \times 7.01601\ u$$

$$6.941\ u - 7.01601\ u = x \times 6.01513\ u - x \times 7.01601\ u = -x \times 1.00088\ u$$

$$x = \frac{6.941\ u - 7.01601\ u}{-1.00088\ u} = 0.075 \quad \text{Percent abundances}: 7.5\%\ \text{lithium-6},\ 92.5\%\ \text{lithium-7}$$

7A We assume that atoms lose or gain relatively few electrons to become ions. Thus, elements that will form cations will be on the left-hand side of the periodic table, while elements that will form anions will be on the right-hand side. The number of electrons "lost" when a cation forms is usually equal to the last digit of the periodic group number; the number of electrons added when an anion forms is typically eight minus the last digit of the group number.

Li is in group 1(1A); it should form a cation by losing one electron: Li^+.
S is in group 6(6A); it should form an anion by adding two electrons: S^{2-}.
Ra is in group 2(2A); it should form a cation by losing two electrons: Ra^{2+}.
F and I are both group 17(7A); they should form anions by gaining an electron: F^- and I^-.
Al is in group 13(3A); it should form a cation by losing three electrons: Al^{3+}.

7B Main group elements are in the "A" families, while transition elements are in the "B" families. Metals, nonmetals, metalloids, and noble gases are color coded in the periodic table inside the front cover of the textbook.

Na is a main-group metal in group 1(1A).

S is a main-group nonmetal in group 16(6A).

Kr is a noble gas in group 18(8A).

U is an inner transition metal, an actinide.

B is a main-group nonmetal in group 13(3A).

As is a main-group metalloid in group 15(5A).

Re is a transition metal in group 7

I is a main-group nonmetal in group 17.

Mg is a main-group metal in group 2.

Si is a main-group metalloid in group 14.

Al is a main-group metal in group 13.

H is a main-group nonmetal in group 1.
(H is a metal at extremely high pressures)

8A This is similar to Practice Examples 2-8A and 2-8B.

$$\text{Cu mass} = 2.35 \times 10^{24} \text{ Cu atoms} \times \frac{1 \text{mol Cu}}{6.022 \times 10^{23} \text{ atoms}} \times \frac{63.546 \text{g Cu}}{1 \text{mol Cu}} = 248 \text{ g Cu}$$

8B Of all lead atoms, 24.1% are lead-206, or 241 ^{206}Pb atoms in every 1000 lead atoms. First we need to convert a 22.6 gram sample of lead into moles of lead (below) and then by using Avogadro's constant, and the percent natural abundance, we can determine the number of ^{206}Pb atoms.

$$n_{Pb} = 22.6 \text{ g Pb} \times \frac{1 \text{ mole Pb}}{207.2 \text{ g Pb}} = 0.109 \text{ mol Pb}$$

$$^{206}\text{Pb atoms} = 0.109 \text{ mol Pb} \times \frac{6.022 \times 10^{23} \text{Pb atoms}}{1 \text{mol Pb}} \times \frac{241 \text{ }^{206}\text{Pb atoms}}{1000 \text{Pb atoms}} = 1.58 \times 10^{22} \text{ }^{206}\text{Pb atoms}$$

9A Both the density and the molar mass of Pb serve as conversion factors.

$$\text{atoms of Pb} = 0.105 \text{cm}^3 \text{ Pb} \times \frac{11.34 \text{g}}{1 \text{cm}^3} \times \frac{1 \text{mol Pb}}{207.2 \text{g}} \times \frac{6.022 \times 10^{23} \text{ Pb atoms}}{1 \text{mol Pb}} = 3.46 \times 10^{21} \text{ Pb atoms}$$

9B First we find the number of rhenium atoms in 0.100 mg of the element.

$$0.100 \text{mg} \times \frac{1 \text{g}}{1000 \text{mg}} \times \frac{1 \text{mol Re}}{186.207 \text{ g Re}} \times \frac{6.022 \times 10^{23} \text{ Re atoms}}{1 \text{ mol Re}} = 3.23 \times 10^{17} \text{ Re atoms}$$

$$\% \text{ abundance } ^{187}\text{Re} = \frac{2.02 \times 10^{17} \text{ atoms } ^{187}\text{Re}}{3.23 \times 10^{17} \text{ Re atoms}} \times 100\% = 62.5\%$$

REVIEW QUESTIONS

Law of Conservation of Mass

1. The observations cited do not necessarily violate the law of conservation of mass. The oxide formed when iron rusts is a solid and remains with the solid iron, increasing the mass of the solid by an amount equal to the mass of the oxygen that has combined. The oxide formed when a match burns is a gas and will not remain with the solid product (the ash); the mass of the ash thus is less than that of the match. We would have to collect all reactants and all products and weigh them to determine if the law of conservation of mass is obeyed or violated.

2. The magnesium that is burned in air combines with some of the oxygen in the air and this oxygen (which, of course, was not weighed when the magnesium metal was weighed) adds its mass to the mass of the magnesium, making the magnesium oxide product weigh more than did the original magnesium. When this same reaction is carried out in a photoflash bulb, the oxygen (in fact, some excess oxygen) that will combine with the magnesium is already present in the bulb before the reaction. Consequently, the product contains no unweighed oxygen.

3. By the law of conservation of mass, all of the magnesium initially present and all of the oxygen that reacted are present in the product. Thus, the mass of oxygen that has reacted is obtained by difference. mass of oxygen = 0.674g MgO − 0.406g Mg = 0.268g oxygen

4. Reaction: $2 K(s) + Cl_2(g) \rightarrow 2KCl(s)$ Mass of Cl_2 reacted $= 8.178\ g - 6.867\ g = 1.311\ g\ Cl_2(g)$

$$m_{KCl} = 1.311\ g\ Cl_2 \times \frac{1\ mol\ Cl_2}{70.9054\ g\ Cl_2} \times \frac{2\ mol\ KCl}{1\ mol\ Cl_2} \times \frac{74.551\ g\ KCl}{1\ mol\ KCl} = 2.757\ g\ KCl$$

5. We need to compare the mass before reaction (initial) with that after reaction (final) to answer this question.

initial mass $= 10.500$ g calcium hydroxide $+ 11.125$ g ammonium chloride $= 21.625$ g

final mass $= 14.336$ g solid residue $+ (69.605 - 62.316)$ g of gases $= 21.625$ g

These data support the law of conservation of mass. Note that the gain in the mass of water is equal to the mass of gas absorbed by the water.

6. We compute the mass of the reactants and compare that with the mass of the products to answer this question.

reactant mass $=$ mass of calcium carbonate $+$ mass of hydrochloric acid solution

$$= 10.00g\ calcium\ carbonate + 100.0\ mL\ soln \times \frac{1.148g}{1mL\ soln}$$

$$= 10.00g\ calcium\ carbonate + 114.8g\ solution\ = 124.8g\ reactants$$

product mass $=$ mass of solution $+$ mass of carbon dioxide

$$= 120.40\ g\ soln + 2.22L\ gas \times \frac{1.9769\ g}{1L\ gas} = 120.40\ g\ soln + 4.39\ g\ carbon\ dioxide$$

$$= 124.79\ g\ products \left(\begin{array}{l} \text{The same mass within experimental error,} \\ \text{Thus, the law of conservation of mass obeyed} \end{array} \right)$$

Law of Constant Composition

7. **(a)** Ratio of O:MgO by mass $= \dfrac{(0.755 - 0.455)g}{0.755\ g} = 0.397$

(b) Ratio of O:Mg in MgO by mass $= \dfrac{0.300\ g}{0.455\ g} = 0.659$

(c) Percent magnesium by mass $= \dfrac{0.455\ g\ Mg}{0.755\ g\ MgO} \times 100\% = 60.3\%$

8. **(a)** We can determine that carbon dioxide has a fixed composition by finding the % C in each sample. (In the calculations below, the abbreviation "cmpd" is short for compound.)

$$\%C = \frac{3.62g\ C}{13.26g\ cmpd} \times 100\% = 27.3\%\ C \qquad \%C = \frac{5.91g\ C}{21.66g\ cmpd} \times 100\% = 27.3\%\ C$$

$$\%C = \frac{7.07g\ C}{25.91g\ cmpd} \times 100\% = 27.3\%\ C$$

Since all three samples have the same percent of carbon, these data do establish that carbon dioxide has a fixed composition.

(b) Carbon dioxide contains only carbon and oxygen. As determined above, carbon dioxide is 27.3 % C by mass. The percent of oxygen in carbon dioxide is obtained by difference. %O $= 100.0\ \% - (27.3\ \%C) = 72.7\ \%O$

9. In the first experiment, 2.18 g of sodium produces 5.54 g of sodium chloride. In the second experiment, 2.10 g of chlorine produces 3.46 g of sodium chloride. The amount of sodium contained in this second sample of sodium chloride is given by

mass of sodium = 3.46 g sodium chloride − 2.10 g chlorine = 1.36 g sodium.

We now have sufficient information to determine the % Na in each of the samples of sodium chloride.

$$\%Na = \frac{2.18\,g\,Na}{5.54\,g\,cmpd} \times 100\% = 39.4\%\,Na \qquad \%Na = \frac{1.36\,g\,Na}{3.46\,g\,cmpd} \times 100\% = 39.3\%\,Na$$

Thus, the two samples of sodium chloride have the same composition. Recognize that, based on significant figures, each percent has an uncertainty of ±0.1%.

10. If the two samples of water have the same % H, the law of constant composition is demonstrated. Notice that, in the second experiment, the mass of the compound is equal to the sum of the masses of the elements produced from it.

$$\%H = \frac{3.06\,g}{27.35\,g\,H_2O} \times 100\% = 11.2\%\,H \qquad \%H = \frac{1.45\,g\,H}{(1.45+11.51)\,g\,H_2O} \times 100\% = 11.2\%\,H$$

Thus, the results are consistent with the law of constant composition.

11. The mass of sulfur (0.312 g) needed to produce 0.623 g sulfur dioxide provides the information required for the conversion factor.

$$sulfur\ mass = 0.842\,g\,sulfur\,dioxide \times \frac{0.312\,g\,sulfur}{0.623\,g\,sulfur\,dioxide} = 0.422\,g\,sulfur$$

12. (a) From the first experiment we see that 1.16 g of compound is produced per gram of Hg. These masses enable us to determine the mass of compound produced from 1.50 g Hg.

$$mass\ of\ cmpd = 1.50\,g\,Hg \times \frac{1.16\,g\,cmpd}{1.00\,g\,Hg} = 1.74\,g\,cmpd$$

(b) Since the compound weighs 0.24 g more than the mass of mercury (1.50 g) that was used, 0.24 g of sulfur must have reacted. Thus, the unreacted sulfur has a mass of 0.76 g (= 1.00 g initially present − 0.24 g reacted).

Law of Multiple Proportions

13. By dividing the mass of the oxygen per gram of sulfur in the second sulfur-oxygen compound (compound 2) by the mass of oxygen per gram of sulfur in the first sulfur-oxygen compound (compound 1), we obtain the ratio (shown to the right):

$$\frac{\dfrac{1.497\,g\,of\,O}{1.000\,g\,of\,S}\,(cpd\,2)}{\dfrac{0.998\,g\,of\,O}{1.000\,g\,of\,S}\,(cpd\,1)} = \frac{1.500}{1}$$

To get the simplest whole number ratio we need to multiply both the numerator and the denominator by 2. This gives the simple whole number ratio 3/2. In other words, for a given mass of sulfur, the mass of oxygen in the second compound (SO_3) relative to the mass of oxygen in the first compound (SO_2) is in a ratio of 3:2. These results are entirely consistent with the Law of Multiple Proportions because the same two elements, sulfur and oxygen in this case, have reacted together to give <u>two</u> different compounds that have masses of oxygen that are in the ratio of small positive integers for a fixed amount of sulfur.

14. This question is similar to question 13 in that two elements, phosphorus and chlorine in this case, have combined to give two different compounds. This time, however, different masses have been used for both of the elements in the second reaction. To see if the Law of Multiple Proportions is being followed, the mass of one of the two elements must be set to the same value in both reactions. This can be achieved by dividing the masses of both phosphorus and chlorine in reaction 2 by 2.500:

"normalized" mass of phosphorus = $\dfrac{2.500 \text{ g phosphorus}}{2.500}$ = 1.000 g of phosphorus

"normalized" mass of chlorine = $\dfrac{14.308 \text{ g chlorine}}{2.500}$ = 5.723 g of chlorine

Now the mass of phosphorus for both reactions is fixed at 1.000 g. Next, we will divide each amount of chlorine by the fixed mass of phosphorus with which they are combined. This gives

$$\dfrac{\dfrac{3.433 \text{ g of Cl}}{1.000 \text{ g P}} \text{(reaction 1)}}{\dfrac{5.723 \text{ g of Cl}}{1.000 \text{ g P}} \text{(reaction 2)}} = 0.600 = 6{:}10 \text{ or } 3{:}5$$

15. (a) First of all we need to fix the mass of nitrogen in all three compounds at 1.000 g. This can be accomplished by multiplying the masses of hydrogen and nitrogen in compound A by 2 and the amount of hydrogen and nitrogen in compound C by 4/3 (1.333):

Cmpd. A: "normalized" mass of nitrogen = 0.500 g N × 2 = 1.000 g N
"normalized" mass of hydrogen = 0.108 g H × 2 = 0.216 g H

Cmpd. C: "normalized" mass of nitrogen = 0.750 g N × 1.333 = 1.000 g N
"normalized" mass of hydrogen = 0.108 g H × 1.333 = 0.144 g H

Next, we divide the mass of hydrogen in each compound by the smallest mass of hydrogen, namely, 0.0720 g. This gives 3.000 for compound A, 1.000 for compound B and 2.00 for compound C. The ratio of the amounts of hydrogen in the three compounds is 3 (cmpd A) : 1 (cmpd B) : 2 (cmpd C)

These results are consistent with the Law of Multiple Proportions because the masses of hydrogen in the three compounds end up in a ratio of small whole numbers when the mass of nitrogen in all three compounds is normalized to a simple value (1.000 g here).

(b) The text states that compound B is N_2H_2. This means that, based on the relative amounts of hydrogen calculated in part (a), compound A might be N_2H_6 and compound C, N_2H_4. Actually, compound A is NH_3, but we have no way of knowing this from the data. Note that the H:N ratio in NH_3 and N_2H_6 are the same, 3H:1N.

16. (a) As with the previous problem, one of the two elements must have the same mass in all of the compounds. This can be most readily achieved by setting the mass of iodine in all four compounds to 1.000 g. With this approach we only need to manipulate the data for compounds B and C. To normalize the amount of iodine in compound B to 1.000 g, we need to multiply the masses of both iodine and fluorine by 2. To accomplish the analogous normalization of compound C, we must multiply by 4/3 (1.333).

Cmpd. B: "normalized" mass of iodine   = 0. 500 g I × 2   = 1.000 g I
           "normalized" mass of fluorine = 0.2246 g F × 2   = 0.4492 g F

Cmpd. C: "normalized" mass of iodine   = 0.750 g I   × 1.333 = 1.000 g I
           "normalized" mass of fluorine = 0.5614 g F × 1.333 = 0.7485 g F

Next we divide the mass of fluorine in each compound by the smallest mass of fluorine, namely, 0.1497 g. This gives 1.000 for compound A, 3.001 for compound B, 5.000 for compound C and 7.001 for compound D. The ratios of the amounts of fluorine in the four compounds A : B : C : D is 1 : 3 : 5 : 7. These results are consistent with the law of multiple proportions because for a fixed amount of iodine (1.000 g), the masses of fluorine in the four compounds are in the ratio of small whole numbers.

(b) As with the preceding problem, we can figure out the empirical formulas for the four iodine-fluorine containing compounds from the ratios of the amounts of fluorine that were determined in 38(a): Comp A: IF   Comp B: IF_3 Comp C: IF_5 Comp C: IF_7

17. One oxide of copper has about 20% oxygen by mass. If we assume a 100 gram sample, then ~ 20 grams of the sample is oxygen (~1.25 moles) and 80 grams is copper (~1.26 moles). This would give an empirical formula of CuO (copper (II) oxide). The second oxide has less oxygen by mass, hence the empirical formula must have less oxygen or more copper (Cu:O ratio greater than 1). If we keep whole number ratios of atoms, a plausible formula would be Cu_2O (copper (I) oxide), where the mass percent oxygen is ≈11%.

18. Assuming the intermediate is "half-way" between CO (oxygen: carbon mass ratio = 16:12 or 1.333) and CO_2 (oxygen: carbon mass ratio = 32:12 or 2.6667), then the oxygen: carbon ratio would be 2:1, or O:C = 24:12. This mass ratio gives a mole ratio of O:C = 1.5:1. Empirical formulas are simple whole number ratios of elements; hence, a formula of C_3O_2 must be the correct empirical formula for this carbon oxide. (Note: C_3O_2 is called tricarbon dioxide or carbon suboxide)

Fundamental Charges and Mass-to-Charge Ratios

19. We can calculate the charge on each drop, express each in terms of 10^{-19} C, and finally express each in terms of $e = 1.6 \times 10^{-19}$ C.

drop 1: 1.28×10^{-18} $= 12.8 \times 10^{-19}$ C $= 8e$

drops 2 & 3: $1.28 \times 10^{-18} \div 2 = 0.640 \times 10^{-18}$ C $= 6.40 \times 10^{-19}$ C $= 4e$

drop 4: $1.28 \times 10^{-18} \div 8 = 0.160 \times 10^{-18}$ C $= 1.60 \times 10^{-19}$ C $= 1e$

drop 5: $1.28 \times 10^{-18} \times 4 = 5.12 \times 10^{-18}$ C $= 51.2 \times 10^{-19}$ C $= 32e$

We see that these values are consistent with the charge that Millikan found for that of the electron, and he could have inferred the correct charge from these data, since they are all multiples of e.

20. We calculate each drop's charge; express each in terms of 10^{-19} C, and then in terms of $e = 1.6 \times 10^{-19}$ C.

drop 1:	6.41×10^{-19} C	$= 6.41 \times 10^{-19}$ C	$= 4e$
drop 2:	$6.41 \times 10^{-19} \div 2 = 3.21 \times 10^{-19}$ C	$= 3.21 \times 10^{-19}$ C	$= 2e$
drop 3:	$6.41 \times 10^{-19} \times 2 = 1.28 \times 10^{-18}$ C	$= 12.8 \times 10^{-19}$ C	$= 8e$
drop 4:	1.44×10^{-18}	$= 14.4 \times 10^{-19}$ C	$= 9e$
drop 3:	$1.44 \times 10^{-18} \div 3 = 4.8 \times 10^{-19}$ C	$= 4.8 \times 10^{-19}$ C	$= 3e$

We see that these values are consistent with the charge that Millikan found for that of the electron. He could have inferred the correct charge from these values, since they are all multiples of e, and have no other common factor.

21. **(a)** Determine the ratio of the mass of a hydrogen atom to that of an electron. We use the mass of a proton plus that of an electron for the mass of a hydrogen atom.

$$\frac{\text{mass of proton} + \text{mass of electron}}{\text{mass of electron}} = \frac{1.0073\,\text{u} + 0.00055\,\text{u}}{0.00055\,\text{u}} = 1.8 \times 10^3$$

or $\qquad \dfrac{\text{mass of electron}}{\text{mass of proton} + \text{mass of electron}} = \dfrac{1}{1.8 \times 10^3} = 5.6 \times 10^{-4}$

(b) The only two mass-to-charge ratios that we can determine from the data in Table 2-1 are those for the proton, a hydrogen ion, H^+; and that for the electron.

For the proton : $\qquad \dfrac{\text{mass}}{\text{charge}} = \dfrac{1.673 \times 10^{-24}\,\text{g}}{1.602 \times 10^{-19}\,\text{C}} = 1.044 \times 10^{-5}\,\text{g/C}$

For the electron : $\qquad \dfrac{\text{mass}}{\text{charge}} = \dfrac{9.109 \times 10^{-28}\,\text{g}}{1.602 \times 10^{-19}\,\text{C}} = 5.686 \times 10^{-9}\,\text{g/C}$

The hydrogen ion is the lightest positive ion available. We see that the mass-to-charge ratio for a positive particle is considerably larger than that for an electron.

22. We do not have the precise isotopic masses for the two ions. The values of the mass-to-charge ratios are only approximate. This is because some of the mass is converted to energy (binding energy), that holds all of the positively charged protons in the nucleus together. Consequently, we have used a three-significant figure mass for a nucleon, rather than the more precisely known proton and neutron masses. (Recall that the term "nucleon" refers to a nuclear particle— either a proton or a neutron.)

$^{127}I^-$ $\quad \dfrac{m}{e} = \dfrac{127\ \text{nucleons}}{1\ \text{electron}} \times \dfrac{1\,e}{1.602 \times 10^{-19}\,\text{C}} \times \dfrac{1.67 \times 10^{-24}\,\text{g}}{1\ \text{nucleon}} = 1.32 \times 10^{-3}\,\text{g/C}\ (7.55 \times 10^2\,\text{C/g})$

$^{32}S^{2-}$ $\quad \dfrac{m}{e} = \dfrac{32\ \text{nucleons}}{2\ \text{electrons}} \times \dfrac{1\,e}{1.602 \times 10^{-19}\,\text{C}} \times \dfrac{1.67 \times 10^{-24}\,\text{g}}{1\ \text{nucleon}} = 1.67 \times 10^{-4}\,\text{g/C}\ (6.00 \times 10^3\,\text{C/g})$

23. **(a)** cobalt-60 $^{60}_{27}\text{Co}$ **(b)** phosphorus-32 $^{32}_{15}\text{P}$ **(c)** iron-59 $^{59}_{26}\text{Fe}$ **(d)** radium-226 $^{226}_{88}\text{Ra}$

24. The nucleus of $^{202}_{80}\text{Hg}$ contains 80 protons and $(202 - 80) = 122$ neutrons.

Thus, the percent of nucleons that are neutrons is given by

$$\% \text{ neutrons} = \frac{122 \text{ neutrons}}{202 \text{ nucleons}} \times 100 = 60.4\% \text{ neutrons}$$

25.

Name	Symbol	# of protons	# of electrons	# of neutrons	Mass number
sodium	$^{23}_{11}\text{Na}$	11	11	12	23
silicon	$^{28}_{14}\text{Si}$	14	14[a]	14	28
rubidium	$^{85}_{37}\text{Rb}$	37	37[a]	48	85
potassium	$^{40}_{19}\text{K}$	19	19	21	40
arsenic[a]	$^{75}_{33}\text{As}$	33[a]	33	42	75
neon	$^{20}_{10}\text{Ne}^{2+}$	10	8	10	20
bromine[b]	$^{80}_{35}\text{Br}$	35	35	45	80
lead[b]	$^{208}_{82}\text{Pb}$	82	82	126	208

[a] This result assumes that a neutral atom is involved.

[b] Insufficient data. Does not characterize a specific nuclide; several possibilities exist. The minimum information needed is the atomic number (or some way to obtain it: the name or the symbol of the element involved), the number of electrons (or some way to obtain it, such as the charge on the species), and the mass number (or the number of neutrons).

26. (a) Since all of these species are neutral atoms, the number of electrons are the atomic numbers, the subscript numbers. The symbols must be arranged in order of increasing value of these subscripts. $^{40}_{18}\text{Ar} < ^{39}_{19}\text{K} < ^{58}_{27}\text{Co} < ^{59}_{29}\text{Cu} < ^{120}_{48}\text{Cd} < ^{112}_{50}\text{Sn} < ^{122}_{52}\text{Te}$

(b) The number of neutrons is given by the difference between the mass number and the atomic number, $A - Z$. This is the difference between superscripted and subscripted values and are provided (in parentheses) after each element in the following list.

$^{39}_{19}\text{K}(20) < ^{40}_{18}\text{Ar}(22) < ^{59}_{29}\text{Cu}(30) < ^{58}_{27}\text{Co}(31) < ^{112}_{50}\text{Sn}(62) < ^{122}_{52}\text{Te}(70) < ^{120}_{48}\text{Cd}(72)$

(c) Here the nuclides are arranged by increasing mass number, given by the superscripts.

$^{39}_{19}\text{K} < ^{40}_{18}\text{Ar} < ^{58}_{27}\text{Co} < ^{59}_{29}\text{Cu} < ^{112}_{50}\text{Sn} < ^{120}_{48}\text{Cd} < ^{122}_{52}\text{Te}$

27. (a) A ^{108}Pd atom has 46 protons, and 46 electrons. The atom described is neutral, hence, the number of electrons must equal the number of protons. Since there are 108 nucleons in the nucleus, the number of neutrons is 62 $(= 108 \text{ nucleons} - 46 \text{ protons})$.

(b) The ratio of the two masses is determined as follows: $\dfrac{^{108}\text{Pd}}{^{12}\text{C}} = \dfrac{107.90389 \text{ u}}{12 \text{ u}} = 8.9919908$

28. **(a)** The atomic number of Ra is 88 and equals the number of protons in the nucleus. The ion's charge is 2+ and, thus, there are two more protons than electrons: no. protons = no. electrons + 2 = 88; no. electrons = 88 − 2 = 86. The mass number (228) is the sum of the atomic number and the number of neutrons: 228 = 88+ no. neutrons; Hence, the number of neutrons = 228 − 88 = 140 neutrons.

(b) The mass of ^{16}O is 15.9994 u. $$\text{ratio} = \frac{\text{mass of isotope}}{\text{mass of}\,^{16}O} = \frac{228.030u}{15.9994u} = 14.2524$$

29. The mass of ^{16}O is 15.9994 u. Isotopic mass = 15.9949 u × 6.68374 = 106.936 u

30. The mass of ^{16}O is 15.9994 u.

mass of heavier isotope = 15.9994 u × 7.1838 = 114.93$\underline{6}$ u = mass of ^{115}In

$$\text{mass of lighter isotope} = \frac{114.93\underline{6}\,u}{1.0177} = 112.94\,u = \text{mass of}\,^{113}In$$

31. Each isotopic mass must be divided by the isotopic mass of ^{12}C, 12 u, an exact number.
   (a) $^{35}Cl \div {}^{12}C = 34.96885u \div 12u = 2.914071$
   (b) $^{26}Mg \div {}^{12}C = 25.98259u \div 12u = 2.165216$
   (c) $^{222}Rn \div {}^{12}C = 222.0175u \div 12u = 18.50146$

32. We need to work through the mass ratios in sequence to determine the mass of ^{81}Br.

mass of ^{19}F = mass of $^{12}C \times 1.5832 = 12\,u \times 1.5832 = 18.998\,u$

mass of ^{35}Cl = mass of $^{19}F \times 1.8406 = 18.998\,u \times 1.8406 = 34.968\,u$

mass of ^{81}Br = mass of $^{35}Cl \times 2.3140 = 34.968\,u \times 2.3140 = 80.916\,u$

33. First, we determine the number of protons, neutrons, and electrons in each species.

species:	$^{24}_{12}Mg^{2+}$	$^{47}_{24}Cr$	$^{60}_{27}Co^{3+}$	$^{35}_{17}Cl^-$	$^{120}_{50}Sn^{2+}$	$^{226}_{90}Th$	$^{90}_{38}Sr$
no. protons	12	24	27	17	50	90	38
no. neutrons	12	23	33	18	70	136	52
no. electrons	10	24	24	18	48	90	38

   (a) The number of neutrons and electrons is equal for $^{35}_{17}Cl^-$.
   (b) $^{60}_{27}Co^{3+}$ has protons (27), neutrons (33), and electrons (24) in the ratio 9:11:8.
   (c) The species $^{124}_{50}Sn^{2+}$ has a number of neutrons (74) equal to its number of protons (50) plus one-half its number of electrons $(48 \div 2 = 24)$.

34. **(a)** Atoms with equal numbers of protons and neutrons will have mass numbers that are approximately twice the size of their atomic numbers. The following species are approximately suitable (with numbers of protons and neutrons in parentheses).

$^{24}_{12}Mg^{2+}$ (12 p$^+$, 12 n), $^{47}_{24}Cr$ (24 p$^+$, 23 n), $^{60}_{27}Co^{3+}$ (27 p$^+$, 33 n), and $^{35}_{17}Cl^-$ (17 p$^+$, 18 n). Of these four nuclides, only $^{24}_{12}Mg^{2+}$ has equal numbers of protons and neutrons.

(b) A species in which protons have more than 50% of the mass must have a mass number smaller than twice the atomic number. Of these species, only in $^{47}_{24}Cr$ is more than 50% of the mass contributed by the protons.

(c) A species with about 50% more neutrons than protons will have a mass number that is at least 2.5 times greater than the atomic number. $^{226}_{90}Th$ has just slightly greater than 50% more neutrons than protons.

35. If we let n represent the number of neutrons and p represent the number of protons, then $p + 4 = n$. The mass number is the sum of the number of protons and the number of neutrons: $p+n = 44$. Substitution of $n = p + 4$ yields $p + p + 4 = 44$. From this relation, we see p = 20. Reference to the periodic table indicates that 20 is the atomic number of the element calcium.

36. We will use the same type of strategy and the same notation as we used previously in question 35 to come up with the answer.
$n = p + 1$ There is one more neutron than the number of protons.
$n + p = 9 \times 3 = 27$ The mass number equals nine times the ion's charge of 3+.
Substitute the first relationship into the second, and solve for p. $27 = (p+1) + p = 2p + 1$

$p = \dfrac{27-1}{2} = 13$ Thus this is the +3 cation of the isotope Al-27 $\rightarrow$ $^{27}_{13}Al^{3+}$

Atomic Mass Units, Atomic Masses

37. There are no chlorine atoms that have a mass of 35.4527 u. The masses of individual atoms are close to integers and this mass is about midway between two integers. It is an average atomic mass, the result of averaging two (or more) isotopic masses, each weighted by its natural abundance.

38. It is exceedingly unlikely that another nuclide would have an exact integral mass. The mass of carbon-12 is *defined* as precisely 12 u. Each nuclidic mass is close to integral, but none that we have encountered in this chapter are precisely integral. The reason is that each nuclide is composed of protons, neutrons, and electrons, none of which have integral masses, and there is a small quantity of the mass of each nucleon (nuclear particle) lost in the binding energy holding the nuclides together. It would be highly unlikely that all of these contributions would add up to a precisely integral mass.

39. To determine the average atomic mass, we use the following expression:
average atomic mass $= \sum (\text{isotopic mass} \times \text{fractional natural abundance})$
Each of the three percents given is converted to a fractional abundance by dividing it by 100.
Mg atomic mass $= (23.985042\,u \times 0.7899) + (24.985837\,u \times 0.1000) + (25.982593\,u \times 0.1101)$
$= 18.95\,u + 2.499\,u + 2.861\,u = 24.31\,u$

40. To determine the average atomic mass, we use the following expression
average atomic mass $= \sum (\text{isotopic mass} \times \text{fractional natural abundance})$
Each of the three percents given is converted to a fractional abundance by dividing it by 100.
Cr atomic mass $= (49.9461 \times 0.0435) + (51.9405 \times 0.8379) + (52.9407 \times 0.0950) + (53.9389 \times 0.0236)$
$= 2.17\,u + 43.52\,u + 5.03\,u + 1.27\,u = 51.99\,u$

41. We will use the expression to determine the weighted-average atomic mass.

$$107.868\,u = (106.905092\,u \times 0.5184) + \left(^{109}Ag \times 0.4816\right) = 55.42\,u + 0.4816\,^{109}Ag$$

$$107.868\,u - 55.42\,u = 0.4816\,^{109}Ag = 52.45\,u \qquad ^{109}Ag = \frac{52.45\,u}{0.4816} = 108.9\,u$$

42. The percent abundances of the two isotopes must add to 100.00%, since there are only two naturally occurring isotopes of bromine. Thus, we can determine the percent natural abundance of the second isotope by difference. % second isotope = 100.00% − 50.69% = 49.31%

From the periodic table, we see that the weighted-average atomic mass of bromine is 79.904 u. We use this value in the expression for determining the weighted-average atomic mass, along with the isotopic mass of ^{79}Br and the fractional abundances of the two isotopes (the percent abundances divided by 100).

$$79.904\,u = (0.5069 \times 78.918336\,u) + (0.4931 \times \text{other isotope}) = 40.00\,u + (0.4931 \times \text{other isotope})$$

$$\text{other isotope} = \frac{79.904\,u - 40.00\,u}{0.4931} = 80.92\,u = \text{mass of } ^{81}Br, \text{ the other isotope}$$

43. Since the three percent abundances total 100%, the percent abundance of ^{40}K is found by difference. % ^{40}K = 100.0000% − 93.2581% − 6.7302% = 0.0117%

Then the expression for the weighted-average atomic mass is used, with the percent abundances converted to fractional abundances by dividing by 100. Note that the average atomic mass of potassium is 39.0983 u.

$$39.0983\,u = (0.932581 \times 38.963707\,u) + (0.000117 \times 39.963999\,u) + \left(0.067302 \times ^{41}K\right)$$

$$= 36.3368\,u + 0.00468\,u + \left(0.067302 \times ^{41}K\right)$$

$$\text{mass of } ^{41}K = \frac{39.0983\,u - (36.3368\,u + 0.00468\,u)}{0.067302} = 40.962\,u$$

44. We use the expression for determining the weighted-average atomic mass, where x represents the fractional abundance of ^{10}B and $(1-x)$ the fractional abundance of ^{11}B

$$10.811\,u = (10.012937\,u \times x) + \left[11.009305 \times (1-x)\right] = 10.012937x + 11.009305 - 11.009305x$$

$$10.811 - 11.009305 = -0.198 = 10.012937x - 11.009305x = -0.996368x$$

$$x = \frac{0.198}{0.996368} = 0.199 \qquad \therefore 19.9\% \,^{10}B \quad \text{and} \quad (100.0 - 19.9) = 80.1\% \,^{11}B$$

Mass spectrometry

45. **(a)**

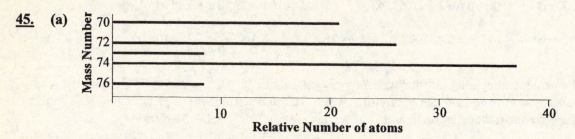

(b) As before, we multiply each isotopic mass by its fractional abundance; after which, we sum these products to obtain the (average) atomic mass for the element.

$$(0.205 \times 70) + (0.274 \times 72) + (0.078 \times 73) + (0.365 \times 74) + (0.078 \times 76)$$

$$14 + 20. + 5.7 + 27 + 5.9 = 72._6 = \text{average atomic mass of germanium}$$

The result is only approximately correct because the isotopic masses are given to only two significant figures. Thus, only a two-significant-figure result can be quoted.

46. **(a)** Six unique HCl molecules are possible (called isotopomers):

$^1H^{35}Cl$, $^2H^{35}Cl$, $^3H^{35}Cl$, $^1H^{37}Cl$, $^2H^{37}Cl$, and $^3H^{37}Cl$

The mass numbers of the six different possible types of molecules are obtained by summing the mass numbers of the two atoms in each molecule:

$^1H^{35}Cl$ has $A = 36$	$^2H^{35}Cl$ has $A = 37$	$^3H^{35}Cl$ has $A = 38$
$^1H^{37}Cl$ has $A = 38$	$^2H^{37}Cl$ has $A = 39$	$^3H^{37}Cl$ has $A = 40$

(b) The most abundant molecule contains the isotope for each element that is most abundant. It is $^1H^{35}Cl$. The second most abundant molecule is $^1H^{37}Cl$. The relative abundance of each type of molecule is determined by multiplying together the fractional abundances of the two isotopes present. Relative abundances of the molecules are as follows.

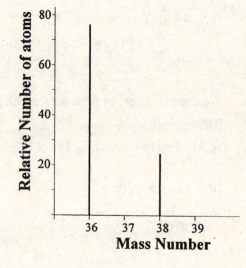

$^1H^{35}Cl : 75.76\%$ $^1H^{37}Cl : 24.23\%$

$^2H^{35}Cl : < 0.011\%$ $^2H^{37}Cl : < 0.0036\%$

$^3H^{35}Cl : < 0.0008\%$ $^3H^{37}Cl : < 0.0002\%$

47. **(a)** Ge is in group 14 and in the fourth period.

(b) Other elements in group 16(6A) are similar to S: O, Se, and Te. Most of the elements in the periodic table are unlike S, but particularly metals such as Na, K, and Rb.

(c) The alkali metal (group 1), in the fifth period is Rb.

(d) The halogen (group 17) in the sixth period is At.

48. **(a)** Au is in group 11 and in the sixth period.

(b) Ar, $Z = 18$, a noble gas. Xe is a noble gas with atomic number (54) greater than 50.

(c) If an element forms a stable anion with charge 2-, it is in group 16.

(d) If an element forms a stable cation with charge 3+, it is in group 13.

49. If the seventh period of the periodic table is 32 members long, it will be the same length as the sixth period. Elements in the same family (vertical group), will have atomic numbers 32 units higher. The noble gas following radon will have atomic number $= 86 + 32 = 118$. The alkali metal following francium will have atomic number $= 87 + 32 = 119$.

50. There are a number of interchanges: Ar/K, Co/Ni, Te/I, Th/Pa, U/Np, Pu/Am, Sg/Bh
The reverse order is necessary because the periodic table lists elements in order of increasing atomic number (protons in the nucleus) and not in order of increasing atomic masses.

The Avogadro Constant and the Mole

51. **(a)** atoms of Fe $= 15.8$ mol Fe $\times \dfrac{6.022 \times 10^{23} \text{ atoms Fe}}{1 \text{ mol Fe}} = 9.51 \times 10^{24}$ atoms Fe

(b) atoms of Ag $= 0.000467$ mol Ag $\times \dfrac{6.022 \times 10^{23} \text{ atoms Ag}}{1 \text{ mol Ag}} = 2.81 \times 10^{20}$ atoms Ag

(c) atoms of Na $= 8.5 \times 10^{-11}$ mol Na $\times \dfrac{6.022 \times 10^{23} \text{ atoms Na}}{1 \text{ mol Na}} = 5.2 \times 10^{13}$ atoms Na

52. Since the molar mass of nitrogen is 14.0 g/mol, 25.0 g N is almost two moles (1.79 mol N), while 6.02×10^{23} Ni atoms is about one mole, and 52.0 g Cr (52.00 g/mol Cr) is also almost one mole. Finally, 10.0 cm^3 Fe (55.85 g/mol Fe) has a mass of about 79 g, and contains about 1.4 moles of atoms. Thus, 25.0 g N contains the greatest number of atoms. Note: Even if you take nitrogen as N$_2$, the answer is the same.

53. **(a)** moles of Zn $= 415.0$ g Zn $\times \dfrac{1 \text{ mol Zn}}{65.39 \text{ g Zn}} = 6.347$ mol Zn

(b) # of Cr atoms $= 147,400$ g Cr $\times \dfrac{1 \text{ mol Cr}}{51.9961 \text{ g Cr}} \times \dfrac{6.022 \times 10^{23} \text{ atoms Cr}}{1 \text{ mol Cr}} = 1.707 \times 10^{27}$ atoms Cr

(c)   mass Au $= 1.0 \times 10^{12}$ atoms Au $\times \dfrac{1 \text{ mol Au}}{6.022 \times 10^{23} \text{ atoms Au}} \times \dfrac{196.967 \text{ g Au}}{1 \text{ mol Au}} = 3.3 \times 10^{-10}$ g Au

(d)   mass of F atom $= \dfrac{18.9984 \text{ g F}}{1 \text{ mol F}} \times \dfrac{1 \text{ mol F}}{6.022 \times 10^{23} \text{ atoms F}} = \dfrac{3.155 \times 10^{-23} \text{ g F}}{1 \text{ atom F}}$

54.   **(a)**   number Kr atoms $= 5.25$ mg Kr $\times \dfrac{1 \text{ g Kr}}{1000 \text{ mg Kr}} \times \dfrac{1 \text{ mol Kr}}{83.80 \text{ g Kr}} \times \dfrac{6.022 \times 10^{23} \text{ atoms Kr}}{1 \text{ mol Kr}}$

$= 3.77 \times 10^{19}$ atoms Kr

(b)   Molar mass is defined as the mass per mole of substance. Mass $= 2.09$ g.
This calculation requires that the number of moles be determined.

moles $= 2.80 \times 10^{22}$ atoms $\times \dfrac{1 \text{ mol}}{6.022 \times 10^{23} \text{ atoms}} = 0.0465$ mol

Molar mass $= \dfrac{\text{mass}}{\text{moles}} = \dfrac{2.09 \text{ g}}{0.0465 \text{ mol}} = 44.9$ g/mol   The element is Sc, Scandium.

(c)   mass P $= 44.75$ g Mg $\times \dfrac{1 \text{ mol Mg}}{24.3050 \text{ g Mg}} \times \dfrac{1 \text{ mol P}}{1 \text{ mol Mg}} \times \dfrac{30.9738 \text{ g P}}{1 \text{ mol P}} = 57.03$ g P

Note: same answer is obtained if you assume phosphorus is P_4 instead of P.

55.   Determine the mass of Cu in the jewelry, then convert to moles and finally to the number
of atoms. If sterling silver is 92.5% by mass Ag, it is 100-92.5 = 7.5% by mass Cu.

number of Cu atoms $= 33.24$ g sterling $\times \dfrac{7.5 \text{ g Cu}}{100.0 \text{ g sterling}} \times \dfrac{1 \text{ mol u}}{63.546 \text{ g Cu}} \times \dfrac{6.022 \times 10^{23} \text{ atoms Cu}}{1 \text{ mol Cu}}$

number of Cu atoms $= 2.4 \times 10^{22}$ Cu atoms

56.   We first need to determine the amount in moles of each metal.

amount of Pb $= 75.0 \text{ cm}^3$ solder $\times \dfrac{9.4 \text{ g solder}}{1 \text{ cm}^3} \times \dfrac{67 \text{ g Pb}}{100 \text{ g solder}} \times \dfrac{1 \text{ mol Pb}}{207.2 \text{ g Pb}} = 2.3$ mol Pb

amount of Sn $= 75.0 \text{ cm}^3$ solder $\times \dfrac{9.4 \text{ g solder}}{1 \text{ cm}^3} \times \dfrac{33 \text{ g Sn}}{100 \text{ g solder}} \times \dfrac{1 \text{ mol Sn}}{118.7 \text{ g Sn}} = 2.0$ mol Sn

total atoms $= (2.3 \text{ mol Pb} + 2.0 \text{ mol Sn}) \times \dfrac{6.022 \times 10^{23} \text{ atoms}}{1 \text{ mol}} = 2.6 \times 10^{24}$ atoms

57.   We first need to determine the number of Pb atoms of all types in 1.57 g of Pb, and then
use the percent abundance to determine the number of ^{204}Pb atoms present.

^{204}Pb atoms $= 215$ mg Pb $\times \dfrac{1 \text{ g}}{1000 \text{ mg}} \times \dfrac{1 \text{ mol Pb}}{207.2 \text{ g Pb}} \times \dfrac{6.022 \times 10^{23} \text{ atoms}}{1 \text{ mol Pb}} \times \dfrac{14 \; ^{204}\text{Pb atoms}}{1000 \text{ Pb atoms}}$

$= 8.7 \times 10^{18}$ atoms ^{204}Pb

58. $\text{mass of alloy} = 6.50 \times 10^{23} \text{ Cd atoms} \times \dfrac{1 \text{mol Cd}}{6.022 \times 10^{23} \text{ Cd atoms}} \times \dfrac{112.4 \text{ g Cd}}{1 \text{mol Cd}} \times \dfrac{100.0 \text{ g alloy}}{8.0 \text{ g Cd}}$

$= 1.5 \times 10^{3} \text{ g alloy}$

59. We will use the average atomic mass of lead, 207.2 g/mol to answer this question.

(a) $\dfrac{30 \ \mu\text{g Pb}}{1 \text{ dL}} \times \dfrac{1 \text{ dL}}{0.1 \text{ L}} \times \dfrac{1 \text{ g Pb}}{10^{6} \ \mu\text{g Pb}} \times \dfrac{1 \text{ mol Pb}}{207.2 \text{ g}} = 1.4\underline{5} \times 10^{-6} \text{ mol Pb/L}$

(b) $\dfrac{1.4\underline{5} \times 10^{-6} \text{ mol Pb}}{\text{L}} \times \dfrac{1 \text{ L}}{1000 \text{ ml}} \times \dfrac{6.022 \times 10^{23} \text{ atoms}}{1 \text{ mol}} = 8.7 \times 10^{14} \text{ Pb atoms/mL}$

60. The concentration of Pb in air provides the principal conversion factor. Other conversion factors are needed to convert to and from its units, beginning with the 0.500-L volume, and ending with the number of atoms.

$\text{no. Pb atoms} = 0.500 \text{ L} \times \dfrac{1 \text{ m}^3}{1000 \text{ L}} \times \dfrac{3.01 \ \mu\text{g Pb}}{1 \text{ m}^3} \times \dfrac{1 \text{ g Pb}}{10^{6} \ \mu\text{g Pb}} \times \dfrac{1 \text{ mol Pb}}{207.2 \text{ g Pb}} \times \dfrac{6.022 \times 10^{23} \text{ Pb atoms}}{1 \text{ mol Pb}}$

$= 4.37 \times 10^{12} \text{ Pb atoms}$

61. To answer this question we simply need to calculate the ratio of the mass (in grams) of each sample to its molar mass. Whichever elemental sample gives the largest ratio will be the one that has the greatest number of atoms.

Iron sample: $10 \text{ cm} \times 10 \text{ cm} \times 10 \text{ cm} \times 7.86 \text{ g cm}^{-3} = 7860 \text{ g Fe}$

$7860 \text{ g Fe} \times \dfrac{1 \text{ mol Fe}}{55.847 \text{ g Fe}} = 141 \text{ moles of Fe atoms}$

Hydrogen sample: $\dfrac{1.00 \times 10^{3} \text{ g H}_2}{2 \times (1.00794 \text{ g H})} \times 1 \text{ mol H} = 496 \text{ mol of H}_2 \text{ molecules} = 992 \text{ mol of H atoms}$

Mercury sample: $76 \text{ lb Hg} \times \dfrac{454 \text{ g Hg}}{1 \text{ lb Hg}} \times \dfrac{1 \text{ mol Hg}}{200.6 \text{ g Hg}} = 172 \text{ mol of Hg atoms}$

Sulfur sample: $\dfrac{2.00 \times 10^{4} \text{ g S}}{32.066 \text{ g S}} \times 1 \text{ mol S} = 624 \text{ moles of S atoms}$

Clearly then, it is the 1.00 kg sample of hydrogen that contains the greatest number of atoms.

62. **(a)** 23 g Na = 1 mol with a density ~ 1 g/cm^3. 1 mole = 23 g, so volume of 25.5 mol ~ 600 cm^3
(b) Liquid bromine occupies 725 mL or 725 cm^3 (given).
(c) 1.25×10^{25} atoms Cr is ~20 moles. At ~50 g/mol, this represents approximately 1000 g. Given the density of 9.4 g/cm^3, this represents about 100 cm^3 of volume
(d) 2150 g solder at 9.4 g/cm^3 represents approximately 200 cm^3

From this we can see that the liquid bromine would occupy the largest volume.

INTEGRATIVE AND ADVANCED EXERCISES

63. (a) Total mass(40 °C) = 2.50 g + 100.0 mL $\times \dfrac{0.9922\,g}{1\,mL}$ = 2.50 g + 99.22 g = 101.72 g

Mass of solution (20 °C) = 100 mL $\times \dfrac{1.0085\,g}{1\,mL}$ = 100.85 g

Solid crystallized = total mass(40 °C) − solution mass(20°C) = 101.72 g − 100.85 g = 0.87 g

(b) The answer cannot be more precise because both the initial mass and the subtraction only allows the reporting of masses to ± 0.01 g. For a more precise answer, more significant figures would be required for the initial mass (2.50 g) and the densities of the water and solution.

64. The atomic masses cited is 24.3 for magnesium and 35.5 for chlorine (remember these are actually weight averaged of isotopic masses). These isotopic masses individually are quite close to whole numbers, that is, multiples of the mass of a hydrogen atom. In this sense (that of integral masses), each nuclide can be considered to be built up of hydrogen atoms, in agreement with Proust's hypothesis.

65. Each atom of ^{19}F contains 9 protons (1.0073 u each), 10 neutrons (1.0087 u each) and 9 electrons (0.0005486 u each). The mass of each atom should be the sum of the masses of these particles.

Total mass = $\left(9\ protons \times \dfrac{1.0073\ u}{1\ proton} \right) + \left(10\ neutrons \times \dfrac{1.0087\ u}{1\ proton} \right) + \left(9\ electrons \times \dfrac{0.0005486\ u}{1\ electron} \right)$

= 9.0657 u + 10.087 u + 0.004937 u = 19.158 u

This compares with a mass of 18.998 u given in the periodic table. The difference, 0.160 u per atom, is called the mass defect and represents the energy that holds the nucleus together, the nuclear binding energy. This binding energy is released when 9 protons and 9 neutrons fuse to give a fluorine-19 nucleus.

66. volume of nucleus(single proton) $= \dfrac{4}{3}\pi\ r^3 = 1.3333 \times 3.14159 \times (0.5 \times 10^{-13}\ cm\ )^3 = 5 \times 10^{-40}\ cm^3$

density $= \dfrac{1.673 \times 10^{-24}\ g}{5 \times 10^{-40}\ cm^3} = 3 \times 10^{15}\ g/cm^3$

67. This method of establishing Avogadro's number uses fundamental constants.

$1.000\,g\ ^{12}C \times \dfrac{1\,mol\ ^{12}C}{12\ g\ ^{12}C} \times \dfrac{6.022 \times 10^{23}\ ^{12}C\,atoms}{1\,mol\ ^{12}C} \times \dfrac{12\ u}{1\ ^{12}C\,atom} = 6.022 \times 10^{23}\ u$

68. Let Z = # of protons, N = # of neutrons, E = # of electrons, and A = # of nucleons = $Z + N$.

(a) $Z + N = 234$ The mass number is 234 and the species is an atom.
$N = 1.600\,Z$ The atom has 60.0% more neutrons than protons.
Next we will substitute the second expression into the first and solve for Z.

$Z + N = 234 = Z + 1.600\,Z = 2.600\,Z \quad Z = \dfrac{234}{2.600} = 90$ protons.

Thus this is an atom of the isotope ^{234}Th.

(b) $Z = E + 2$ The ion has a charge of +2. $Z = 1.100\,E$
There are 10.0% more protons than electrons. By equating these two expressions and solve for E, we can find the number of electrons. $E + 2 = 1.100\,E$

$$2 = 1.100\,E - E = 0.100\,E \quad E = \frac{2}{0.100} = 20 \text{ electrons} \quad Z = 20 + 2 = 22, \text{ (titanium).}$$

The ion is Ti^{2+}. There is not enough information given to determine the mass number.

(c) $Z + N = 110$ The mass number is 110. $Z = E + 2$ The species is a cation with a charge of +2.
$N = 1.25\,E$ Thus, there are 25.0% more neutrons than electrons. By substituting the 2nd and 3rd expressions into the first, and we can solve for E, the number of electrons.

$$(E + 2) + 1.25\,E = 110 = 2.25\,E + 2 \quad 108 = 2.25\,E \quad E = \frac{108}{2.25} = 48$$

Then $Z = 48 + 2 = 50$, (the element is Sn) $N = 1.25 \times 48 = 60$ Thus $^{110}Sn^{2+}$

69. Because the net ionic charge (2+) is one-tenth of its the nuclear charge, the nuclear charge is 20+. This is also the atomic number of the nuclide, which means the element is calcium. The number of electrons is 20 for a neutral calcium atom, but only 18 for a calcium ion with a net 2+ charge. Four more than 18 is 22 neutrons. The ion described is $^{42}_{20}Ca^{2+}$.

70. $A = Z + N = 2.50\,Z$ The mass number is 2.50 times the atomic number
The neutron number of selenium-82 equals $82 - 34 = 48$, since $Z = 34$ for Se. The neutron number of isotope Y also equals 48, which equals 1.33 times the atomic number of isotope Y.

Thus $48 = 1.33 \times Z_Y$ $Z_Y = \frac{48}{1.33} = 36$

The mass number of isotope Y = $48 + 36 = 84$ = the atomic number of E, and thus, the element is Po. Thus, from the relationship in the first line, the mass number of $E = 2.50\,Z = 2.50 \times 84 = 210$ The isotope E is ^{210}Po.

71. As a result of the redefinition, all masses will decrease by a factor of $35.00000/35.4527 = 0.987231$.

(a) atomic mass of He $4.00260 \times 0987231 = 3.95149$
atomic mass of Na $22.9898 \times 0.987231 = 22.6962$
atomic mass of I $126.905 \times 0.987231 = 125.285$

(b) These three elements have ~integral atomic masses based on C-12 because these three elements and C-12 all consist mainly of one stable isotope, rather than a mixture of two or more stable isotopes, with each being present in significant amounts (10% or more), as is the case with chlorine.

72. To solve this question represent the fractional abundance of ^{14}N by x and that of ^{14}N by $(1 - x)$.
Then use the expression for determining average atomic mass.
$14.0067 = 14.0031\,x + 15.0001(1 - x)$
$14.0067 - 15.0001 = 14.0031\,x - 15.0001\,x$ OR $-0.9934 = -0.9970\,x$

$x = \frac{0.9934}{0.9970} \times 100\% = 99.64\% = $ percent abundance of ^{14}N. Thus, $0.36\% = $ percent abundance of ^{15}N.

73. In this case we will use the expression for determining average atomic mass; the sum of products of nuclidic mass times fractional abundances (from Figure 2-15) to answer the question.

^{196}Hg :   195.9658 u × 0.00146 = 0.286 u

^{198}Hg :   197.9668 u × 0.1002 = 19.84 u ^{199}Hg :   198.9683 u × 0.1684 = 33.51 u

^{200}Hg :   199.9683 u × 0.2313 = 46.25 u ^{201}Hg :   200.9703 u × 0.1322 = 26.57 u

^{202}Hg :   201.9706 u × 0.2980 = 60.19 u ^{204}Hg :   203.9735 u × 0.0685 = 14.0 u

Atomic weight = 0.286 u + 19.84 u + 33.51 u + 46.25 u + 26.57 u + 60.19 u + 14.0 u = 200.6 u

74. The sum of the percent abundances of the two minor isotopes equals 100.00% − 84.68% = 15.32%. Thus, we denote the fractional abundance of ^{73}Ge as x, and the other as $(0.1532 - x)$. These fractions are then used in the expression for average atomic mass.

Atomic mass = 72.61 u = (69.92425 u × 0.2085) + (71.92208 u × 0.2754) + (73.92118 u × 0.3629)

$$+(72.92346 \text{ u} \times x) + [75.92140 \text{ u} \times (0.1532 - x)]$$

72.61 u = (14.58 u) + (19.81 u) + (26.83 u) + (72.92346 ux) + [11.63 u - 75.92140 ux)]

0.24 = 2.99794x Hence: $x = 0.080$.

^{73}Ge has 8.0 % natural abundance and ^{76}Ge has 7.3 % natural abundance

From the calculations, we can see that the number of significant figures drops from four in the percent natural abundances supplied to only two significant figures owing to the imprecision in the supplied values of the percent natural abundance.

75. From the question we may initially infer the following:
(a) Assume percent abundance of ^{84}Kr ~ 55% as a start (somewhat more than 50)
(b) Let percent abundance of ^{82}Kr = x %; percent abundance ^{83}Kr ~ ^{82}Kr = x %
(c) ^{86}Kr = 1.50(percent abundance of ^{82}Kr) = 1.50(x %)
(d) ^{80}Kr = 0.195(percent abundance of ^{82}Kr) = 0.195(x %)
(e) ^{78}Kr = 0.030(percent abundance of ^{82}Kr) = 0.030(x %)

100 % = %^{78}Kr + %^{80}Kr + %^{82}Kr + %^{83}Kr + %^{84}Kr + %^{86}Kr
100 % = 0.030(x %) + 0.195(x %) + x % + x % + %^{84}Kr + 1.50(x %)
100 % = 3.725(x %) + %^{84}Kr

The problem here is the inaccuracy in of the percent abundance for ^{84}Kr, which is crudely estimated to be ~ 55%.

For the purpose of solving this problem, we will assume the mass percent of ^{82}Kr to an exact percent, say 12% and calculate all other percent abundances for the other isotopes exactly (including ^{84}Kr).

For instance, if we set percent abundance of ^{82}Kr = 12 %(exactly), we can determine, using 100 % = 3.725(%^{82}Kr) + %^{84}Kr, that the percent abundance ^{84}Kr = 55.3%

weighted average isotopic mass = 0.030(12%)(77.9204 u) + 0.195(12%)(79.9164 u) + 12%(81.9135 u) + 12%(82.9141 u) + 55.3%(83.9115 u) + 1.50(12%)(85.9106 u) = 83.79864 u

Our answer is close, however, the answer is relatively insensitive to the percent abundance of ^{82}Kr. Tabulated below are the calculated weighted average isotopic mass for various percent abundances of ^{82}Kr.

Percent abundance ^{82}Kr.	Weighted average isotopic mass
15 %	83.768 u
14 %	83.777 u
13 %	83.787 u
12 %	83.797 u
11 %	83.806 u
10 %	83.816 u
9 %	83.826 u
8 %	83.835 u

From this table we can see the answer is somewhere between 11% to 12% (Remember that the weighted average atomic mass of Kr is given as 83.80 u)

A value of 11.67 % yields an answer of 83.80 u. This is very close to the actual value of 11.58 %.

76. Four molecules are possible, given below with their calculated molecular masses.
^{35}Cl-^{79}Br   mass = 34.9689 u + 78.9183 u = 113.8872 u
^{35}Cl-^{81}Br   mass = 34.9689 u + 80.9163 u = 115.8852 u
^{37}Cl-^{79}Br   mass = 36.9658 u + 78.9183 u = 115.8841 u
^{37}Cl-^{81}Br   mass = 36.9658 u + 80.9163 u = 117.8821 u

Each molecule has a different intensity pattern (relative number of molecules), based on the natural abundance of the isotopes making up each molecule. If we divide all of the values by the lowest ratio, we can get a better idea of the relative ratio of each molecule.
^{35}Cl-^{79}Br   Intensity = (0.7577) x (0.5069) = 0.3841÷0.1195 = 3.214
^{35}Cl-^{81}Br   Intensity = (0.7577) x (0.4931) = 0.3736÷0.1195 = 3.129
^{37}Cl-^{79}Br   Intensity = (0.2423) x (0.5069) = 0.1228÷0.1195 = 1.028
^{37}Cl-^{81}Br   Intensity = (0.2423) x (0.4931) = 0.1195÷0.1195 = 1.000

A plot of intensity versus atomic mass reveals the following pattern under ideal circumstances(high resolution mass spectrometry). The second mass spectrum shows the pattern that would be revealed using a lower resolution mass spectrometer.

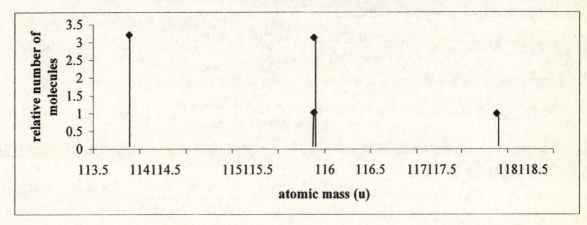

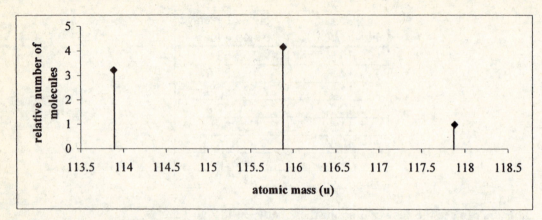

77. Let's begin by finding the volume of copper metal.

Wire diameter (cm) = 0.03196 in $\times \dfrac{2.54 \text{ cm}}{1 \text{ in}} = 0.08118$ cm

The radius in cm is 0.08118 cm $\times$ 1/2 = 0.04059 cm

The volume of Cu(cm^3) = (0.04059 cm)$^2 \times (\pi) \times$ (100 cm) = 0.5176 cm^3

So, the mass of Cu = 0.5176 cm$^3 \times \dfrac{8.92 \text{ g Cu}}{1 \text{ cm}^3} = 4.62$ g Cu

The number of moles of Cu = 4.62 g Cu $\times \dfrac{1 \text{ mol Cu}}{63.546 \text{ g Cu}} = 0.0727$ mol Cu

Cu atoms in the wire = 0.0727 mol Cu $\times \dfrac{6.022 \times 10^{23} \text{ atoms Cu}}{1 \text{ mol Cu}} = 4.38 \times 10^{22}$ atoms

78. Volume = (15.0 cm $\times$ 12.5 cm $\times$ 0.300 cm) $-$ (3.1416) $\times \left(\dfrac{2.50 \text{ cm}}{2}\right)^2 \times$ (0.300 cm)

Volume = (56.2$\underline{5}$ cm $-$ 1.47 cm) = 54.8 cm^3

Mass of object = 54.8 cm$^3 \times \dfrac{8.80 \text{ g}}{1 \text{ cm}^3} = 482$ g Monel metal

Then determine the number of silicon atoms in this quantity of alloy.

482 g Monel metal $\times \dfrac{2.2 \times 10^{-4} \text{g Si}}{1.000 \text{g metal}} \times \dfrac{1 \text{mol Si}}{28.0855 \text{ g Si}} \times \dfrac{6.022 \times 10^{23} \text{ Si atoms}}{1 \text{mol Si}} = 2.3 \times 10^{21}$Si atoms

Finally determine the number of ^{30}Si atoms in this quantity of silicon.

number of ^{30}Si atoms = (2.3$\times 10^{21}$ Si atoms) $\times \left(\dfrac{310 \ ^{30}\text{Si atoms}}{10000 \text{ Si atoms}}\right) = 7.1 \times 10^{19} \ ^{30}$Si

79. The percent natural abundance of deuterium of 0.015% means that in a sample of 100,000 H atoms, only 15 ^{2}H are present.

mass H$_2$ = 2.50$\times 10^{21}$ atoms ^{2}H $\times \dfrac{100,000 \text{ H atoms}}{15 \text{ atoms } ^2\text{H}} \times \dfrac{1 \text{ mol H}}{6.022 \times 10^{23} \text{H atoms}} \times \dfrac{1 \text{ mol H}_2}{2 \text{ mol H}} \times \dfrac{2.0158 \text{ g H}_2}{1 \text{ mol H}_2}$

mass H$_2$ = 27.9 g H$_2$ are required such that we have 2.50$\times 10^{21}$ atoms ^{2}H

80. The numbers total to 21 (= 10 + 6 + 5). Thus, in one mole of the alloy there is $^{10}/_{21}$ mol Bi, $^6/_{21}$ mol Pb, and $^5/_{21}$ mol Sn. The mass of this mole of material is figured in a similar fashion to computing an average atomic mass from isotopic masses.

$$\text{mass of alloy} = \left(\frac{10}{21}\text{ mol Bi} \times \frac{209.0\text{ g}}{1\text{ mol Bi}}\right) + \left(\frac{6}{21}\text{ mol Pb} \times \frac{207.2\text{ g}}{1\text{ mol Pb}}\right) + \left(\frac{5}{21}\text{ mol Sn} \times \frac{118.7}{1\text{ mol Sn}}\right)$$

$$= 99.52\text{ g Bi} + 59.20\text{ g Pb} + 28.26\text{ g Sn} = 186.98\text{ g alloy}$$

81. The atom ratios, of course, are the same as the mole ratios. We first determine the mass of alloy that contains 5.00 mol Ag, 4.00 mol Cu, and 1.00 mol Zn.

$$\text{mass} = 5.00\text{ mol Ag} \times \frac{107.868\text{ g Ag}}{1\text{ mol Ag}} + 4.00\text{ mol Cu} \times \frac{63.546\text{ g Cu}}{1\text{ mol Cu}} + 1.00\text{ mol Zn} \times \frac{65.39\text{ g Zn}}{1\text{ mol Zn}}$$

$$= 539.\underline{3}\text{ g Ag} + 254.\underline{2}\text{ g Cu} + 65.4\text{ g Zn} = 858.\underline{9}\text{ g alloy} = 859\text{ g alloy}$$

Then, for 1.00 kg of the alloy, (1000 g), we need 1000/859 moles. (859 g is the alloy's "molar mass.")

$$\text{mass of Ag} = 1000\text{ g alloy} \times \frac{539\text{ g Ag}}{859\text{ g alloy}} = 627\text{ g Ag}$$

$$\text{mass of Cu} = 1000\text{ g alloy} \times \frac{254\text{ g Cu}}{859\text{ g alloy}} = 296\text{ g Cu}$$

$$\text{mass of Zn} = 1000\text{ g alloy} \times \frac{65.4\text{ g Zn}}{859\text{ g alloy}} = 76.1\text{ g Zn}$$

82. The relative masses of Sn and Pb are 207.2 g Pb (assume one mole of Pb) to (2.73 × 118.710 g/mol Sn =) 324 g Sn. Then the mass of cadmium, on the same, scale, is 207.2/1.78 = 116 g Cd.

$$\%\text{Sn} = \frac{324\text{ g Sn}}{207.2 + 324 + 116\text{ g alloy}} \times 100\% = \frac{324\text{ g Sn}}{647\text{ g alloy}} \times 100\% = 50.1\%\text{ Sn}$$

$$\%\text{Pb} = \frac{207.2\text{ g Pb}}{647\text{ g alloy}} \times 100\% = 32.0\%\text{ Pb} \qquad \%\text{Cd} = \frac{116\text{ g Cd}}{647\text{ g alloy}} \times 100\% = 17.9\%\text{ Cd}$$

FEATURE PROBLEMS

83. The product mass differs from that of the reactants by $(5.62 - 2.50 =)$ 3.12 grains. In order to determine the percent gain in mass, we need to convert the reactant mass to grains.

$$13\text{ onces} \times \frac{8\text{ gros}}{1\text{ once}} = 104\text{ gros} \times (104 + 2)\text{ gros} \times \frac{72\text{ grains}}{1\text{ gros}} = 7632\text{ grains}$$

$$\%\text{ mass increase} = \frac{3.12\text{ grains increase}}{(7632 + 2.50)\text{ grains original}} \times 100\% = 0.0409\%\text{ mass increase}$$

The sensitivity of Lavoisier's balance can be as little as 0.01 grain, which seems to be the limit of the readability of the balance; or it can be as large as 3.12 grains, which assumes that all of the error in the experiment is due to the (in)sensitivity of the balance. Let us convert 0.01 grains to a mass in grams.

$$\text{minimum error} = 0.01 \text{ gr} \times \frac{1 \text{ gros}}{72 \text{ gr}} \times \frac{1 \text{ once}}{8 \text{ gros}} \times \frac{1 \text{ livre}}{16 \text{ once}} \times \frac{30.59 \text{ g}}{1 \text{ livre}} = 3 \times 10^{-5} \text{g} = 0.03 \text{ mg}$$

$$\text{maximum error} = 3.12 \text{ gr} \times \frac{3 \times 10^{-5} \text{ g}}{0.01 \text{ gr}} = 9 \times 10^{-3} \text{ g} = 9 \text{ mg}$$

The maximum error is close to that of a common modern laboratory balance, which has a sensitivity of 1 mg. The minimum error is approximated by a good quality analytical balance. Thus we conclude that Lavoisier's results conform closely to the law of conservation of mass.

84. One way to determine the common factor of which all 13 numbers are multiples is to first divide all of them by the smallest number in the set. The ratios thus obtained may either be integers or they may be rational numbers whose decimal equivalents are easy to recognize.

Obs.	1	2	3	4	5	6	7	8	9	10	11	12	13
Quan.	19.66	24.60	29.62	34.47	39.38	44.42	49.41	53.91	59.12	63.68	68.65	78.34	83.22
Ratio	1.000	1.251	1.507	1.753	2.003	2.259	2.513	2.742	3.007	3.239	3.492	3.984	4.233
Mult.	4.000	5.005	6.026	7.013	8.012	9.038	10.05	10.97	12.03	12.96	13.97	15.94	16.93
Int.	4	5	6	7	8	9	10	11	12	13	14	16	17

The row labeled "Mult." is obtained by multiplying the row "ratio" by 4.000. In the row labeled "Int." we give the integer closest to each of these multipliers. It is obvious that each of the 13 measurements is exceedingly close to a common quantity multiplied by an integer.

85. In a 50-year-old chemistry textbook the atomic mass for oxygen would be exactly 16 u because chemists assigned precisely 16 u as the atomic mass of the naturally occurring mixture of oxygen isotopes. This value is slightly higher than the value of 15.9994 in modern chemistry textbooks. Thus, we would expect all other atomic masses to be slightly higher as well in the older textbooks.

86. We begin with the amount of reparations and obtain the volume in cubic kilometers with a series of conversion factors.

$$V = \$28.8 \times 10^9 \times \frac{1 \text{ troy oz Au}}{\$21.25} \times \frac{31.103 \text{ g Au}}{1 \text{ troy oz Au}} \times \frac{1 \text{ mol Au}}{196.97 \text{ g Au}} \times \frac{6.022 \times 10^{23} \text{ atoms Au}}{1 \text{ mol Au}}$$

$$\times \frac{1 \text{ ton sea water}}{4.67 \times 10^{17} \text{ Au atoms}} \times \frac{2000 \text{ lb sea water}}{1 \text{ ton sea water}} \times \frac{453.6 \text{ g sea water}}{1 \text{ lb sea water}} \times \frac{1 \text{ cm}^3 \text{ sea water}}{1.03 \text{ g sea water}}$$

$$\times \left(\frac{1 \text{ m}}{100 \text{ cm}} \times \frac{1 \text{ km}}{1000 \text{ m}} \right)^3 = 2.43 \times 10^5 \text{ km}^3$$

87. We start by using the percent natural abundances for ^{87}Rb and ^{85}Rb along with the data in the "spiked" mass spectrum to find the total mass of Rb in the sample. Then, we calculate the Rb content in the rock sample in ppm by mass by dividing the mass of Rb by the total mass of the rock sample, and then multiplying the result by 10^6 to convert to ppm.

^{87}Rb = 27.83 % natural abundance ^{85}Rb = 72.17 % natural abundance

Therefore, $\dfrac{^{87}\text{Rb(natural)}}{^{85}\text{Rb(natural)}} = \dfrac{27.83\,\%}{72.17\,\%} = 0.3856$

For the ^{87}Rb(spiked) sample, the ^{87}Rb peak in the mass spectrum is 1.12 times as tall as the

^{85}Rb peak. Thus, for this sample $\dfrac{^{87}\text{Rb(natural)} + {}^{87}\text{Rb(spiked)}}{^{85}\text{Rb(natural)}} = 1.12$

Using this relationship, we can now find the masses of both ^{85}Rb and ^{87}Rb in the sample.

So, $\dfrac{^{87}\text{Rb(natural)}}{^{85}\text{Rb(natural)}} = 0.3856;$ $^{85}\text{Rb(natural)} = \dfrac{^{87}\text{Rb(natural)}}{0.3856}$

$^{87}\text{Rb(natural)} + {}^{87}\text{Rb(spiked)} = \dfrac{1.12 \times {}^{87}\text{Rb (natural)}}{0.3856} = 2.905\ {}^{87}\text{Rb(natural)}$

$^{87}\text{Rb(spiked)} = 1.905\ {}^{87}\text{Rb(natural)}$

and $\dfrac{^{87}\text{Rb(natural)} + {}^{87}\text{Rb(spiked)}}{^{85}\text{Rb(natural)}} = \dfrac{^{87}\text{Rb(natural)} + {}^{87}\text{Rb(spiked)}}{\dfrac{^{87}\text{Rb(natural)}}{0.3856}} = 1.12$

Since the mass of ^{87}Rb(spike) is equal to 29.45 μg, the mass of ^{87}Rb(natural) must be

$\dfrac{29.45\ \mu g}{1.905} = 15.46\ \mu g$ of ^{87}Rb(natural)

So, the mass of $^{85}\text{Rb(natural)} = \dfrac{15.46\ \mu g\ \text{of}\ {}^{87}\text{Rb(natural)}}{0.3856} = 40.09\ \mu g$ of ^{85}Rb(natural)

Therefore, the total mass of Rb in the sample = 15.46 μg of ^{87}Rb(natural) + 40.09 μg of ^{85}Rb(natural) = 55.55 μg of Rb convert to grams:

$= 55.55\ \mu g\ \text{of Rb} \times \dfrac{1\ g\ \text{Rb}}{1 \times 10^6\ \mu g\ \text{Rb}} = 5.555 \times 10^{-5}\ g\ \text{Rb}$

Rb content (ppm) $= \dfrac{5.555 \times 10^{-5}\ g\ \text{Rb}}{0.350\ g\ \text{of rock}} \times 10^6 = 159\ \text{ppm Rb}$

CHAPTER 3
CHEMICAL COMPOUNDS
PRACTICE EXAMPLES

1A First we convert the number of chloride ions to the mass of $MgCl_2$.

$$\text{mass}_{MgCl_2} = 5.0 \times 10^{23} \, Cl^- \times \frac{1 \, MgCl_2}{2 \, Cl^-} \times \frac{1 \, mol \, MgCl_2}{6.022 \times 10^{23} \, MgCl_2} \times \frac{95.211 \, g \, MgCl_2}{1 \, mol \, MgCl_2} = 4.0 \times 10^1 \, g \, MgCl_2$$

1B First we convert mass $Mg(NO_3)_2$ to moles $Mg(NO_3)_2$ and formula units $Mg(NO_3)_2$ then finally to NO_3^- ions.

$$1.00 \, \mu g \, Mg(NO_3)_2 \times \frac{1 \, g \, Mg(NO_3)_2}{1,000,000 \, \mu g \, Mg(NO_3)_2} \times \frac{1 \, mol \, Mg(NO_3)_2}{148.3148 \, g \, Mg(NO_3)_2} \times \frac{6.022 \times 10^{22} \, \text{formula units} \, Mg(NO_3)_2}{1 \, mol \, Mg(NO_3)_2}$$

$$= 4.06 \times 10^{15} \, \text{formula units} \, Mg(NO_3)_2 \times \frac{2 \, NO_3^- \, \text{ions}}{1 \, \text{formula unit} \, Mg(NO_3)_2} = 8.12 \times 10^{15} \, NO_3^- \, \text{ions}$$

Next determine the number of oxygen atoms by multiplying by the appropriate ratio.

$$\text{\# atoms O} = 4.06 \times 10^{15} \, \text{formula units} \, Mg(NO_3)_2 \times \frac{6 \, \text{atoms O}}{1 \, \text{formula unit} \, Mg(NO_3)_2} = 2.44 \times 10^{16} \, O$$

2A The volume of gold is converted to its mass and then to the amount in moles.

$$\text{\# Au atoms} = (2.50 \, cm)^2 \times \left(0.100 \, mm \times \frac{1 \, cm}{10 \, mm}\right) \times \frac{19.32 \, g}{1 \, cm^3} \times \frac{1 \, mol \, Au}{196.97 \, g \, Au} \times \frac{6.022 \times 10^{23} \, \text{atoms}}{1 \, mol \, Au}$$

$$= 3.69 \times 10^{21} \, \text{Au atoms}$$

2B We need the molar mass of ethyl mercaptan for one conversion factor.
Volume of room: $62 \, ft \times 35 \, ft \times 14 \, ft = 3.0\underline{4} \times 10^4 \, ft^3$. We also need to convert ft^3 to m^3

$$3.0\underline{4} \times 10^4 \, ft^3 \times \left(\frac{12 \, in}{1 \, ft}\right)^3 \times \left(\frac{2.54 \, cm}{1 \, in}\right)^3 \times \left(\frac{1 \, m}{100 \, cm}\right)^3 = 8.6 \times 10^2 \, m^3$$

$$M = (2 \times 12.011 \, g \, C) + (6 \times 1.008 \, g \, H) + (1 \times 32.066 \, g \, S) = 62.136 \, g/mol \, C_2H_6S$$

$$[C_2H_6S] = \frac{1.0 \, \mu L \, C_2H_6S}{8.6 \times 10^2 \, m^3} \times \frac{1L}{1 \times 10^6 \, \mu L} \times \frac{1000 \, mL}{1L} \times \frac{0.84 \, g}{1 \, mL} \times \frac{1 \, mol \, C_2H_6S}{62.136 \, g} \times \frac{10^6 \, \mu mol}{1 \, mol}$$

$$= 0.016 \, \mu mol/m^3 > 0.9 \times 10^{-3} \, \mu mol/m^3 = \text{the detectable limit}$$

Thus, the vapor will be detectable.

3A The molar mass of halothane is given in Example 3-3 *in the textbook* as 197.4 g/mol. The rest of the solution uses conversion factors to change units.

$$\text{mass Br} = 25.0\,\text{mL C}_2\text{HBrClF}_3 \times \frac{1.871\,\text{g C}_2\text{HBrClF}}{1\,\text{mL C}_2\text{HBrClF}} \times \frac{1\,\text{mol C}_2\text{HBrClF}_3}{197.4\,\text{g C}_2\text{HBrClF}} \times \frac{1\ \text{mol Br}}{1\,\text{mol C}_2\text{HBrClF}_3}$$

$$\times \frac{79.904\ \text{g Br}}{1\ \text{mol Br}} = 18.9\ \text{g Br}$$

3B Again, the molar mass of halothane is given in Example 3-3 *in the textbook* as 197.4 g/mol.

$$V_{\text{halothane}} = 1.00 \times 10^{24}\,\text{Br} \times \frac{1\,\text{mol Br}}{6.022 \times 10^{23}\,\text{Br}} \times \frac{1\,\text{mol C}_2\text{HBrClF}_3}{1\,\text{mol Br}} \times \frac{197.4\,\text{g C}_2\text{HBrClF}_3}{1\,\text{mol C}_2\text{HBrClF}_3} \times \frac{1\,\text{mL}}{1.871\,\text{g}}$$

$$= 175\ \text{mL C}_2\text{HBrClF}_3$$

4A We use the same technique as before: determine the mass of each element in a mole of the compound. Their sum is the molar mass of the compound. The percent composition is determined by comparing the mass of each element with the molar mass of the compound.

$$M = (10 \times 12.011\,\text{g C}) + (16 \times 1.008\,\text{g H}) + (5 \times 14.01\,\text{g N}) + (3 \times 30.97\,\text{g P}) + (13 \times 15.999\,\text{g O})$$

$$= 120.11\,\text{g C} + 16.13\,\text{g H} + 70.05\,\text{g N} + 92.91\,\text{g P} + 207.99\,\text{g O} = 507.19\,\text{g ATP/mol}$$

$$\%C = \frac{120.11\,\text{g C}}{507.19\,\text{g ATP}} \times 100\% = 23.681\%\,\text{C} \quad \%H = \frac{16.13\,\text{g H}}{507.19\,\text{g ATP}} \times 100\% = 3.180\%\ \text{H}$$

$$\%N = \frac{70.05\,\text{g N}}{507.19\,\text{g ATP}} \times 100\% = 13.81\%\,\text{N} \quad \%P = \frac{92.91\,\text{g P}}{507.19\,\text{g ATP}} \times 100\% = 18.32\%\ \text{P}$$

$$\%O = \frac{207.99\,\text{g O}}{507.19\,\text{g ATP}} \times 100\% = 41.008\%\,\text{O} \text{ (NOTE: the mass percents sum to 99.999\%)}$$

4B Both (b) and (e) have the same empirical formula, that is CH_2O. These two molecules have the same percent oxygen by mass.

5A Once again, we begin with a 100.00 g sample of the compound. In this way, each elemental mass in grams is numerically equal to its percent. We convert each mass to an amount in moles, and then determine the simplest integer set of molar amounts. This determination begins by dividing all three molar amounts by the smallest.

$$39.56\,\text{g C} \times \frac{1\,\text{mol C}}{12.011\,\text{g C}} \ = 3.294\,\text{mol C} \div 3.294 \ \rightarrow 1.000\,\text{mol C} \ \times 3.000 \ = 3.000\,\text{mol C}$$

$$7.74\,\text{g H} \times \frac{1\,\text{mol H}}{1.008\,\text{g H}} \ = 7.68\,\text{mol H} \div 3.294 \ \rightarrow 2.33\,\text{mol H} \ \times 3.000 \ = 6.99\,\text{mol H}$$

$$52.70\,\text{g O} \times \frac{1\,\text{mol O}}{15.9994\,\text{g O}} = 3.294\ \text{mol O} \div 3.294 \ \rightarrow 1.000\ \text{mol O} \ \times 3.000 \ = 3.000\ \text{mol O}$$

Thus, the empirical formula of the compound is $C_3H_7O_3$. The empirical molar mass of this compound is:

$$(3 \times 12.01\,\text{g C}) + (7 \times 1.008\,\text{g H}) + (3 \times 16.00\,\text{g O}) = 36.03\ \text{g} + 7.056\ \text{g} + 48.00\ \text{g} = 91.09\,\text{g/mol}$$

The empirical mass is almost precisely one half the reported molar mass, leading to the conclusion that the molecular formula must be twice the empirical formula in order to double the molar mass. Thus, the molecular formula is $C_6H_{14}O_6$.

5B To answer this question, we start with a 100.00 g sample of the compound. In this way, each elemental mass in grams is numerically equal to its percent. We convert each mass to an amount in moles, and then determine the simplest integer set of molar amounts. This determination begins by dividing all molar amounts by the smallest number of moles in the group of four, viz. 1.1025 moles. Multiplication of the resulting quotients by eight produces the smallest possible set of whole numbers.

$$21.51g\,C \times \frac{1\,mol\,C}{12.011g\,C} = 1.791 \ mol\,C \qquad \div 1.102\underline{5} \rightarrow 1.624 \ mol\,C \times 8 \ = 12.99\,mol\,C$$

$$2.22 \ g\,H \times \frac{1\,mol\,H}{1.00794g\,H} = 2.20 \ mol\,H \qquad \div 1.102\underline{5} \rightarrow 2.00 \ mol\,H \times 8 \ = 16.0 \ mol\,H$$

$$17.64 \ g\,O \times \frac{1\,mol\,O}{15.9994g\,O} = 1.102\underline{5}\,mol\,O \qquad \div 1.1025 \rightarrow 1.000\,mol\,O \times 8 \ = 8.000\,mol\,O$$

$$58.63g\,Cl \times \frac{1\,mol\,Cl}{35.453g\,Cl} = 1.654 \ mol\,Cl \qquad \div 1.102\underline{5} \rightarrow 1.500 \ mol\,C \times 8 \ = 12.00\,mol\,Cl$$

Thus, the empirical formula of the compound is $C_{13}H_{16}O_8Cl_{12}$. The empirical molar mass of this compound is 725.7 g/mol.

The empirical mass is almost precisely the same as the reported molar mass, leading to the conclusion that the molecular formula must be the same as the empirical formula. Thus, the molecular formula is $C_{13}H_{16}O_8Cl_{12}$.

6A We calculate the amount in moles of each element in the sample (determining the mass of oxygen by difference) and transform these molar amounts to the simplest integral amounts, by first dividing all three by the smallest.

$$2.726g\,CO_2 \times \frac{1\,mol\,CO_2}{44.010g\,CO_2} \times \frac{1\,mol\,C}{1\,mol\,CO_2} = 0.06194\,mol\,C \times \frac{12.011g\,C}{1\,mol\,C} = 0.7440g\,C$$

$$1.116g\,H_2O \times \frac{1\,mol\,H_2O}{18.015g\,H_2O} \times \frac{2\,mol\,H}{1\,mol\,H_2O} = 0.1239\,mol\,H \times \frac{1.008g\,H}{1\,mol\,H} = 0.1249g\,H$$

$$(1.152g\,cmpd - 0.7440g\,C - 0.1249g\,H) = 0.283g\,O \times \frac{1\,mol\,O}{16.00g\,O} = 0.0177\,mol\,O$$

$$\left. \begin{array}{l} 0.06194\,mol\,C \div 0.0177 \quad \rightarrow 3.50 \\ 0.1239\,mol\,H \ \div 0.0177 \quad \rightarrow 7.00 \\ 0.0177\,mol\,O \ \div 0.0177 \quad \rightarrow 1.00 \end{array} \right\}$$ All of these amounts in moles are multiplied by 2 to make them integral. Thus, the empirical formula of isobutyl propionate is $C_7H_{14}O_2$.

6B   Notice that we do not have to obtain the mass of any element in this compound by difference; there is no oxygen present in the compound. We calculate the amount in mole of each element in the sample and transform these molar amounts to the simplest integral amounts, by first dividing all three by the smallest.

$$3.149\,g\,CO_2 \times \frac{1\,mol\,CO_2}{44.010\,g\,CO_2} \times \frac{1\,mol\,C}{1\,mol\,CO_2} = 0.07155\,mol\,C \quad \div\,0.01789 = 3.999\,mol\,C$$

$$0.645\,g\,H_2O \times \frac{1\,mol\,H_2O}{18.015\,g\,H_2O} \times \frac{2\,mol\,H}{1\,mol\,H_2O} = 0.0716\,mol\,H \quad \div\,0.01789 = 4.00\,mol\,H$$

$$1.146\,g\,SO_2 \times \frac{1\,mol\,SO_2}{64.065\,g\,SO_2} \times \frac{1\,mol\,S}{1\,mol\,SO_2} = 0.01789\,mol\,S \quad \div\,0.01789 = 1.000\,mol\,S$$

Thus, the empirical formula of thiophene is C_4H_4S.

7A $\underline{S}_8$ For an atom of a free element, the oxidation state is 0 (rule 1).

     $\underline{Cr}_2O_7^{2-}$ The sum of all the oxidation numbers in the ion is -2 (rule 2). The O.S. of each oxygen is -2 (rule 6). Thus, the total for all seven oxygens is -14. The total for both chromiums must be $+12$. Thus, each Cr has an O.S. $= +6$.

     $\underline{Cl}_2O$ The sum of all oxidation numbers in the compound is 0 (rule 2). The O.S. of oxygen is -2 (rule 6). The total for the two chlorines must be $+2$. Thus, each chlorine must have O.S. $= +1$.

     $K\underline{O}_2$ The sum for all the oxidation numbers in the compound is 0 (rule 2). The O.S. of potassium is $+1$ (rule 3). The sum of the oxidation numbers of the two oxygens must be -1. Thus, each oxygen must have O.S. $= -1/2$.

7B $\underline{S}_2O_3^{2-}$ The sum of all the oxidation numbers in the ion is -2 (rule 2). The O.S. of oxygen is -2 (rule 6). Thus, the total for three oxygens must be -6. The total for both sulfurs must be $+4$. Thus, each S has an O.S. $= +2$.

     $\underline{Hg}_2Cl_2$ The O.S. of each Cl is -1 (rule 7). The sum of all O.S. is 0 (rule 2). Thus, the total for two Hg is $+2$ and each Hg has O.S. $= +1$.

     $K\underline{Mn}O_4$ The O.S. of each O is -2 (rule 6). Thus, the total for 4 oxygens must be -8. The K has O.S. $= +1$ (rule 3). The total of all O.S. is 0 (rule 2). Thus, the O.S. of Mn is $+7$.

     $H_2\underline{C}O$ The O.S. of each H is $+1$ (rule 5), producing a total for both hydrogens of $+2$. The O.S. of O is -2 (rule 6). Thus, the O.S. of C is 0, because total of all O.Ss. is 0 (rule 2).

8A   In each case, we determine the formula *with its accompanying charge* of each ion in the compound. We then produce a formula for the compound in which the total positive charge equals the total negative charge.

lithium oxide	Li^+ and O^{2-}	*two* Li^+ and *one* O^{2-}	Li_2O
tin(II) fluoride	Sn^{2+} and F^-	*one* Sn^{2+} and *two* F^-	SnF_2
lithium nitride	Li^+ and N^{3-}	*three* Li^+ and *one* N^{3-}	Li_3N

8B Using a similar procedure as that provided in **8A**

aluminum sulfide	Al^{3+} and S^{2-}	*two* Al^{3+} and *three* S^{2-}	Al_2S_3
magnesium nitride	Mg^{2+} and N^{3-}	*three* Mg^{2+} and *two* N^{3-}	Mg_3N_2
vanadium(III) oxide	V^{3+} and O^{2-}	*two* V^{3+} and *three* O^{2-}	V_2O_3

9A The name of each of these ionic compounds is the name of the cation followed by that of the anion. Each anion name is a modified (with the ending "ide") version of the name of the element. Each cation name is the name of the metal, with the oxidation state appended in Roman numerals in parentheses if there is more than one type of cation for that metal.

CsI cesium iodide

CaF_2 calcium fluoride

FeO The O.S. of $O = -2$ (rule 6). Thus, the O.S. of $Fe = +2$ (rule 2).
   The cation is iron(II). The name of the compound is iron(II) oxide.

$CrCl_3$ The O.S. of $Cl = -1$ (rule 7). Thus, the O.S. of $Cr = +3$ (rule 2).
   The cation is chromium (III). The compound is chromium (III) chloride.

9B The name of each of these ionic compounds is the name of the cation followed by that of the anion. Each anion name is a modified (with the ending "ide") version of the name of the element. Each cation name is the name of the metal, with the oxidation state appended in Roman numerals in parentheses if there is more than one type of cation for that metal.

CaH_2 calcium hydride Ag_2S silver(I) sulfide

In the next two compounds, the oxidation state of chlorine is −1 (rule 7) and thus the oxidation state of the metal in each cation must be +1 (rule 2).

CuCl copper(I) chloride Hg_2Cl_2 mercury(I) chloride

10A SF_6 Both S and F are nonmetals. This is a binary molecular compound: sulfur hexafluoride.

HNO_2 The NO_2^- ion is the nitrite ion. Its acid is nitrous acid.

$Ca(HCO_3)_2$ HCO_3^- is the bicarbonate ion or the hydrogen carbonate ion.
      This compound is calcium bicarbonate or calcium hydrogen carbonate.

$FeSO_4$ The SO_4^{2-} ion is the sulfate ion. The cation is Fe^{2+}, iron(II).
      This compound is iron(II) sulfate.

10B NH_4NO_3 The cation is NH_4^+, ammonium ion. The anion is NO_3^-, nitrate ion.
      This compound is ammonium nitrate.

PCl_3 Both P and Cl are nonmetals. This is a binary molecular compound: phosphorus trichloride.

HBrO BrO^- is hypobromite, this is hypobromous acid.

$AgClO_4$ The anion is perchlorate ion, ClO_4^-. The compound is silver(I) perchlorate.

$Fe_2(SO_4)_3$ The SO_4^{2-} ion is the sulfate ion. The cation is Fe^{3+}, iron(III).
      This compound is iron(III) sulfate.

11A boron trifluoride — Both elements are nonmetals. This is a binary molecular compound: BF_3

potassium dichromate — Potassium ion is K^+, and dichromate ion is $Cr_2O_7^{2-}$. This is $K_2Cr_2O_7$.

sulfuric acid — The anion is sulphate, SO_4^{2-}. There must be two H^+s. This is H_2SO_4.

calcium chloride — The ions are Ca^{2+} and Cl^-. There must be one Ca^{2+} and two Cl^-s: $CaCl_2$.

11B aluminum nitrate — Aluminum is Al^{3+}; the nitrate ion is NO_3^-. This is $Al(NO_3)_3$.

tetraphosphorus decoxide — Both elements are nonmetals. This is a binary molecular compound; P_4O_{10}

chromium(III) hydroxide — Chromium(III) ion is Cr^{3+}; the hydroxide ion is OH^-. This is $Cr(OH)_3$.

iodic acid — The halogen "ic" acid has the halogen in a +5 oxidation state. This is HIO_3.

12A **(a)** Not isomers-molecular formulas are different (C_8H_{18} vs C_9H_{20}).

(b) Molecules are isomers (same formula C_7H_{16})

12B **(a)** Molecules are isomers (same formula C_7H_{14})

(b) Not isomers-molecular formulas are different (C_4H_8 vs C_5H_{10}).

13A **(a)** The carbon to carbon bonds are all single bonds in this hydrocarbon. This compound is an alkane.

(b) In this compound, there are only single bonds, and a Cl atom has replaced one H atom. This compound is a chloroalkane.

(c) The presence of the carboxyl group (—CO_2H) in this molecule means that the compound is a carboxylic acid.

(d) There is a carbon to carbon double bond in this hydrocarbon. This is an alkene.

13B **(a)** The presence of the hydroxyl group (—OH) in this molecule means that this compound is an alcohol.

(b) The presence of the carboxyl group (—CO_2H) in this molecule means that the compound is a carboxylic acid. This molecule also contains the hydroxyl group(—OH).

(c) The presence of the carboxyl group (—CO_2H) in this molecule means that the compound is a carboxylic acid. As well, a Cl atom has replaced one H atom. This compound is a chloroalkane. The compound is a chloro carboxylic acid.

(d) There is a carbon to carbon double bond in this compound; hence, it is an alkene. There is also one H atom that has been replaced by a Br atom. This compound is also a bromoalkene.

14A **(a)** The structure is that of an alcohol with the hydroxyl group on the second carbon atom of a three carbon chain. The compound is 2-propanol (commonly isopropyl alcohol).

(b) The structure is that of an iodoalkane molecule with the I atom on the first carbon of a three-carbon chain. The compound is called 1-iodopropane.

(c) The carbon chain in this structure is four carbon atoms long with the end C atom in a carboxyl group. There is also a methyl group on the third carbon in the chain. The compound is 3-methylbutanoic acid.

(d) The structure is that of a three carbon chain that contains a carbon to carbon double bond. This compound is propene.

14B **(a)** 2-chloropropane   **(b)** 1,4-dichlorobutane   **(c)** 2-methyl propanoic acid

15A **(a)** pentane: $CH_3(CH_2)_3CH_3$ **(b)** ethanoic acid: CH_3CO_2H
   (c) 1-iodooctane: $ICH_2(CH_2)_6CH_3$ **(d)** 1-pentanol: $CH_2(OH)(CH_2)_3CH_3$

15B **(a)** propene: CH_3CHCH_2 **(b)** 1-heptanol: $CH_2(OH)(CH_2)_5CH_3$
   (c) chloroacetic acid: CH_2ClCO_2H **(d)** hexanoic acid: $CH_3(CH_2)_4CO_2H$

EXERCISES

Representing Molecules

1. **(a)** H_2O_2 **(b)** CH_3CH_2Cl **(c)** P_4O_{10}
   (d) $CH_3CH(OH)CH_3$ **(e)** HCO_2H

2. **(a)** N_2H_4 **(b)** CH_3CH_2OH **(c)** P_4O_6
   (d) $CH_3(CH_2)_3CO_2H$ **(e)** $CH_2(OH)CH_2Cl$

3. **(b)** CH_3CH_2Cl **(d)** $CH_3CH(OH)CH_3$ **(e)** HCO_2H

```
      H  H                    H OH H                     OH
      |  |                    |  |  |                    |
   H—C—C—Cl                H—C—C—C—H                  O=C
      |  |                    |  |  |                    |
      H  H                    H  H  H                    H
```

4. **(b)** CH_3CH_2OH **(d)** $CH_3(CH_2)_3CO_2H$ **(e)** $CH_2(OH)CH_2Cl$

```
     H  H              H  H  H  H  O                H  H
     |  |              |  |  |  |  ||               |  |
  H—C—C—OH          H—C—C—C—C—C—OH             HO—C—C—Cl
     |  |              |  |  |  |                    |  |
     H  H              H  H  H  H                    H  H
```

The Avogadro Constant and the Mole

5. **(a)** A trinitrotoluene molecule, $CH_3C_6H_2(NO_2)_3$, contains 7 C atoms, 5 H atoms, 3 N atoms, and 3×2 O atoms = 6 O atoms, for a total of $7 + 5 + 3 + 6 = 21$ atoms.

(b) CH$_3$(CH$_2$)$_4$CH$_2$OH contains 6 C atoms, 14 H atoms, and 1 O atom, for a total of 21 atoms.

$$\text{# of atoms} = 0.00102\,\text{mol C}_6\text{H}_{14}\text{O} \times \frac{6.022 \times 10^{23}\,\text{C}_6\text{H}_{14}\text{O molecules}}{1\,\text{mol C}_6\text{H}_{14}\text{O}} \times \frac{21\,\text{atoms}}{1\,\text{C}_6\text{H}_{14}\text{O molecule}}$$

$$= 1.29 \times 10^{22}\,\text{atoms}$$

(c) $$\text{number of F atoms} = 12.15\,\text{mol C}_2\text{HBrClF}_3 \times \frac{3\,\text{mol F}}{1\,\text{mol C}_2\text{HBrClF}_3} \times \frac{6.022 \times 10^{23}\,\text{F atoms}}{1\,\text{mol F atoms}}$$

$$= 2.195 \times 10^{25}\,\text{F atoms}$$

6. (a) To convert the amount in moles to mass, we need the molar mass of N$_2$O$_4$ (92.02 g/mol).

$$\text{mass N}_2\text{O}_4 = 7.34\,\text{mol N}_2\text{O}_4 \times \frac{92.02\,\text{g N}_2\text{O}_4}{1\,\text{mol N}_2\text{O}_4} = 675\,\text{g N}_2\text{O}_4$$

(b) $$\text{mass of O}_2 = 3.16 \times 10^{24}\,\text{O}_2\,\text{molecules} \times \frac{1\,\text{mol O}_2}{6.022 \times 10^{23}\,\text{molecules O}_2} \times \frac{32.00\,\text{g O}_2}{1\,\text{mol O}_2} = 168\,\text{g O}_2$$

(c) $$\text{mass of CuSO}_4 \cdot 5\text{H}_2\text{O} = 18.6\,\text{mol} \times \frac{249.7\,\text{g CuSO}_4 \cdot 5\text{H}_2\text{O}}{1\,\text{mol CuSO}_4 \cdot 5\text{H}_2\text{O}} = 4.64 \times 10^3\,\text{g CuSO}_4 \cdot 5\text{H}_2\text{O}$$

(d) $$\text{mass of C}_2\text{H}_4(\text{OH})_2 = 4.18 \times 10^{24}\,\text{molecules of C}_2\text{H}_4(\text{OH})_2 \times \frac{1\,\text{mole C}_2\text{H}_4(\text{OH})_2}{6.022 \times 10^{23}\,\text{molecules of C}_2\text{H}_4(\text{OH})_2}$$

$$\times \frac{62.07\,\text{g C}_2\text{H}_4(\text{OH})_2}{1\,\text{mole C}_2\text{H}_4(\text{OH})_2} = 431\,\text{grams C}_2\text{H}_4(\text{OH})_2$$

7. (a) molecular mass (mass of one molecule) of C$_5$H$_{11}$NO$_2$S is:

$$(5 \times 12.011\,\text{u C}) + (11 \times 1.0079\,\text{u H}) + 14.0067\,\text{u N} + (2 \times 15.9994\,\text{u O}) + 32.066\,\text{u S}$$
$$= 149.213\,\text{u/C}_5\text{H}_{11}\text{NO}_2\text{S molecule}$$

(b) Since there are 11 H atoms in each C$_5$H$_{11}$NO$_2$S molecule, there are 11 moles of H atoms in each mole of C$_5$H$_{11}$NO$_2$S molecules

(c) $$\text{mass C} = 1\,\text{mol C}_5\text{H}_{11}\text{NO}_2\text{S} \times \frac{5\,\text{mol C}}{1\,\text{mol C}_5\text{H}_{11}\text{NO}_2\text{S}} \times \frac{12.011\,\text{g C}}{1\,\text{mol C}} = 60.055\,\text{g C}$$

(d) $$\text{no. C atoms} = 9.07\,\text{mol C}_5\text{H}_{11}\text{NO}_2\text{S} \times \frac{5\,\text{mol C}}{1\,\text{mol C}_5\text{H}_{11}\text{NO}_2\text{S}} \times \frac{6.022 \times 10^{23}\,\text{atoms}}{1\,\text{mol C}}$$

$$= 2.73 \times 10^{25}\,\text{C atoms}$$

8. (a) $$\text{amount of Br}_2 = 8.08 \times 10^{22}\,\text{Br}_2\,\text{molecules} \times \frac{1\,\text{mole Br}_2}{6.022 \times 10^{23}\,\text{Br}_2\,\text{molecules}}$$

$$= 0.134\,\text{mol Br}_2$$

(b) $$\text{amount of Br}_2 = 2.17 \times 10^{24}\,\text{Br atoms} \times \frac{1\,\text{Br}_2\,\text{molecule}}{2\,\text{Br atoms}} \times \frac{1\,\text{mole Br}_2}{6.022 \times 10^{23}\,\text{Br}_2\,\text{molecules}}$$

$$= 1.80\,\text{mol Br}_2$$

(c) $\text{amount of Br}_2 = 11.3 \text{ kg Br}_2 \times \dfrac{1000 \text{ g}}{1 \text{ kg}} \times \dfrac{1 \text{ mol Br}_2}{159.8 \text{ g Br}_2} = 70.7 \text{ mol Br}_2$

(d) $\text{amount of Br}_2 = 2.65 \text{ L Br}_2 \times \dfrac{1000 \text{ mL}}{1 \text{ L}} \times \dfrac{3.10 \text{ g Br}_2}{1 \text{ mL Br}_2} \times \dfrac{1 \text{ mol Br}_2}{159.8 \text{ g Br}_2} = 51.4 \text{ mol Br}_2$

9. The greatest number of N atoms is found in the compound with the greatest number of moles of N. The molar mass of $\text{N}_2\text{O} = (2 \text{ mol N} \times 14.0 \text{ g N}) + (1 \text{ mol O} \times 16.0 \text{ g O}) = 44.0 \text{ g/mol N}_2\text{O}$. Thus, 50.0 g N_2O is slightly more than 1 mole of N_2O, and contains slightly more than 2 moles of N. Each mole of N_2 contains 2 moles of N. The molar mass of NH_3 is 17.0 g. Thus, there is 1 mole of NH_3 present, which contains 1 mole of N. The molar mass of pyridine is $(5 \text{ mol C} \times 12.0 \text{ g C}) + (5 \text{ mol H} \times 1.01 \text{ g H}) + 14.0 \text{ g N} = 79.1 \text{ g/mol}$. Because each mole of pyridine contains 1 mole of N, we need slightly more than 2 moles of pyridine to have more N than is present in the N_2O. But that would be a mass of about 158 g pyridine, and 150 mL has a mass of less than 150 g. Thus, the greatest number of N atoms is present in 50.0 g N_2O.

10. The greatest number of S atoms is contained in the compound with the greatest number of moles of S. The solid sulfur contains $8 \times 0.12 \text{ mol} = 0.96$ mol S atoms. There are 0.50×2 mol S atoms in 0.50 mol S_2O. There is slightly greater than 1 mole (64.1 g) of SO_2 in 65 g, and thus a bit more than 1 mole of S atoms. The molar mass of thiophene is 84.1 g and thus contains less than 1 mole of S. This means that 65 g SO_2 has the greatest number of S atoms.

11. **(a)** $\text{moles N}_2\text{O}_4 = 115 \text{ g N}_2\text{O}_4 \times \dfrac{1 \text{ mol N}_2\text{O}_4}{92.02 \text{ g N}_2\text{O}_4} = 1.25 \text{ mol N}_2\text{O}_4$

(b) $\text{moles N} = 43.5 \text{ g Mg(NO}_3)_2 \times \dfrac{1 \text{ mol Mg(NO}_3)_2}{148.33 \text{ g}} \times \dfrac{2 \text{ mol N}}{1 \text{ mol Mg(NO}_3)_2} = 1.43 \text{ mol N atoms}$

(c) $\text{moles N} = 12.4 \text{ g C}_6\text{H}_{12}\text{O}_6 \times \dfrac{1 \text{ mol C}_6\text{H}_{12}\text{O}_6}{180.16 \text{ g}} \times \dfrac{6 \text{ mol O}}{1 \text{ mol C}_6\text{H}_{12}\text{O}_6} \times \dfrac{1 \text{ mol C}_7\text{H}_5(\text{NO}_2)_3}{6 \text{ mol O}}$

$\times \dfrac{3 \text{ mol N}}{1 \text{ mol C}_7\text{H}_5(\text{NO}_2)_3} = 0.206 \text{ mol N}$

12. **(a)** $\text{P mass} = 6.25 \times 10^{-2} \text{ mol P}_4 \times \dfrac{4 \text{ mol P}}{1 \text{ mol P}_4} \times \dfrac{30.974 \text{ g P}}{1 \text{ mol P}} = 7.74 \text{ g P}$

(b) $\text{Stearic acid mass} = 4.03 \times 10^{24} \text{ molecules} \times \dfrac{1 \text{ mole}}{6.022 \times 10^{23} \text{ molecules}} \times \dfrac{284.483 \text{ g C}_{18}\text{H}_{36}\text{O}_2}{1 \text{ mole C}_{18}\text{H}_{36}\text{O}_2}$

$= 1.90 \times 10^3 \text{ g stearic acid}$.

(c) $\text{mass} = 3.03 \text{ mol N} \times \dfrac{1 \text{ mol C}_6\text{H}_{14}\text{N}_2\text{O}_2}{2 \text{ mol N}} \times \dfrac{146.19 \text{ g C}_6\text{H}_{14}\text{N}_2\text{O}_2}{1 \text{ mol C}_6\text{H}_{14}\text{N}_2\text{O}_2} = 221.\underline{5} \text{ g C}_6\text{H}_{14}\text{N}_2\text{O}_2$

13. Number of Fe atoms in 6 L of blood can be found using dimensional analysis.

$$= 6\,\text{L blood} \times \frac{1000\,\text{mL}}{1\,\text{L}} \times \frac{15.5\,\text{g Hb}}{100\,\text{mL blood}} \times \frac{1\,\text{mol Hb}}{64,500\,\text{g Hb}} \times \frac{4\,\text{mol Fe}}{1\,\text{mol Hb}} \times \frac{6.022 \times 10^{23}\,\text{atoms Fe}}{1\,\text{mol Fe}}$$

$$= 3 \times 10^{22}\,\text{Fe atoms}$$

14. **(a)** Volume $= \pi r^2 \times h = 3.1416 \times \left(\dfrac{1.22\ \text{cm}}{2}\right)^2 \times 6.50\ \text{cm} = 7.60\ \text{cm}^3$

$$\text{mol P}_4 \doteq 7.60\ \text{cm}^3 \times \frac{1.823\ \text{g P}_4}{1\ \text{cm}^3} \times \frac{1\ \text{mol P}_4}{123.895\ \text{g P}_4} = 0.112\ \text{mol P}_4$$

(b) # P atoms $= 0.112\ \text{mol P}_4 \times \dfrac{4\ \text{mol P}}{1\ \text{mol P}_4} \times \dfrac{6.022 \times 10^{23}\ \text{P}}{1\ \text{mol P}} = 2.70 \times 10^{23}$ P atoms

Chemical Formulas

15. For glucose (blood sugar), $C_6H_{12}O_6$,

(a) FALSE The percentages by mass of C and O are *different* than in CO. For one thing, CO contains no hydrogen.

(b) TRUE In dihydroxyacetone, $(CH_2OH)_2 CO$ or $C_3H_6O_3$, the ratio of C:H:O $= 3:6:3$ or $1:2:1$. In glucose, this ratio is C:H:O $= 6:12:6 = 1:2:1$. Thus, ratios are the same.

(c) FALSE The proportions, by number of atoms, of C and O are the same in glucose. Since, however, C and O have different molar masses, their proportions by mass must be *different*.

(d) FALSE Each mole of glucose contains $(12 \times 1.01 =) 12.1$ g H. But each mole also contains 72.0 g C and 96.0 g O. Thus, the highest percentage, by mass, is that of O. The highest percentage, by number of atoms, is that of H.

16. For sorbic acid, $C_6H_8O_2$,

(a) FALSE The C:H:O mole ratio is 3:4:1, but the mass ratio differs because moles of different elements have different molar masses.

(b) TRUE Since the two compounds have the same empirical formula, they have the same mass percent composition.

(c) TRUE Aspidinol, $C_{12}H_{16}O_4$, and sorbic acid have the same empirical formula, C_3H_4O.

(d) TRUE The ratio of H atoms to O atoms is $8:2 = 4:1$. Thus, the mass ratio is $(4\,\text{mol H} \times 1\,\text{g H}) : (1\,\text{mol O} \times 16.0\,\text{g O}) = 4\,\text{g H} : 16\,\text{g O} = 1\,\text{g H} : 4\,\text{g O}$.

17. **(a)** $Cu(UO_2)_2(PO_4)_2 \cdot 8H_2O$ has 1 Cu, 2 U, 2 P, 20 O, 16 H or a total of 41 atoms.

(b) By number, $Cu(UO_2)_2(PO_4)_2 \cdot 8H_2O$ has a H to O ratio of 16:20 or 4:5 or 0.800 H atoms/O atom.

(c) By number $Cu(UO_2)_2(PO_4)_2 \cdot 8H_2O$ has a Cu to P ratio of 1:2 The mass ratio of Cu:P is

$$\frac{1\,\text{mol Cu} \times \dfrac{63.546\ \text{g Cu}}{1\,\text{mol Cu}}}{2\,\text{mol P} \times \dfrac{30.9738\ \text{g P}}{1\,\text{mol P}}} = 1.026$$

(d) With a mass percent slightly greater than 50%, U has the largest mass percent with oxygen coming in at ~34%.

(e) $1.00 \text{ g P} \times \dfrac{1 \text{ mol P}}{30.9738 \text{ g P}} \times \dfrac{1 \text{ mol Cu(UO}_2)_2(\text{PO}_4)_2 \cdot 8\text{H}_2\text{O}}{2 \text{ mol P}} \times \dfrac{937.666 \text{ g Cu(UO}_2)_2(\text{PO}_4)_2 \cdot 8\text{H}_2\text{O}}{1 \text{ mol Cu(UO}_2)_2(\text{PO}_4)_2 \cdot 8\text{H}_2\text{O}}$

$= 15.1 \text{ g of Cu(UO}_2)_2(\text{PO}_4)_2 \cdot 8\text{H}_2\text{O}$

18. **(a)** A formula unit of $\text{Ge}\left[\text{S}(\text{CH}_2)_4\,\text{CH}_3\right]_4$ contains:

1 Ge atom   4 S atoms $4(4+1) = 20$ C atoms $4\left[4(2)+3\right] = 44$ H atoms

For a total of $1+4+20+44 = 69$ atoms per formula unit

(b) $\dfrac{\text{no. C atoms}}{\text{no. H atoms}} = \dfrac{20 \text{ C atoms}}{44 \text{ H atoms}} = \dfrac{5 \text{ C atoms}}{11 \text{ H atoms}} = 0.455 \text{ C atom/H atom}$

(c) $\dfrac{\text{mass Ge}}{\text{mass S}} = \dfrac{1\,\text{mol Ge} \times \dfrac{72.61 \text{ g Ge}}{1 \text{ mol Ge}}}{4\,\text{mol S} \times \dfrac{32.066 \text{ g S}}{1 \text{ mol S}}} = \dfrac{72.61 \text{ g Ge}}{128.264 \text{ g S}} = 0.566 \text{ g Ge/g S}$

(d) $\text{mass of S} = 1 \text{ mol Ge}\left[\text{S}(\text{CH}_2)_4\,\text{CH}_3\right]_4 \times \dfrac{4 \text{ mol S}}{1 \text{ mol Ge}\left[\text{S}(\text{CH}_2)_4\,\text{CH}_3\right]_4} \times \dfrac{32.066 \text{ g S}}{1 \text{ mol S}} = 128.264 \text{ g S}$

(e) $\text{no. C atoms} = 33.10 \text{ g cmpd} \times \dfrac{1 \text{ mol cmpd}}{485.6 \text{ g cmpd}} \times \dfrac{20 \text{ mol C}}{1 \text{ mol cmpd}} \times \dfrac{6.022 \times 10^{23} \text{ C atoms}}{1 \text{ mol C}}$

$= 8.210 \times 10^{23} \text{ C atoms}$

Percent Composition of Compounds

19. The information obtained in the course of calculating the molar mass is used to determine the mass percent of H in decane.

$\text{molar mass C}_{10}\text{H}_{22} = \left(\dfrac{10 \text{ mol C}}{1 \text{ mol C}_{10}\text{H}_{22}} \times \dfrac{12.011 \text{ g C}}{1 \text{ mol C}}\right) + \left(\dfrac{22 \text{ mol H}}{1 \text{ mol C}_{10}\text{H}_{22}} \times \dfrac{1.00794 \text{ g H}}{1 \text{ mol H}}\right)$

$= \dfrac{120.11 \text{ g C}}{1 \text{ mol C}_{10}\text{H}_{22}} + \dfrac{22.1747 \text{ g H}}{1 \text{ mol C}_{10}\text{H}_{22}} = \dfrac{142.28 \text{ g}}{1 \text{ mol C}_{10}\text{H}_{22}}$

$\%\text{H} = \dfrac{22.1747 \text{ g H/mol decane}}{142.28 \text{ g C}_{10}\text{H}_{22}/\text{mol decane}} \times 100\% = 15.585\% \text{ H}$

20. Determine the mass of O in a mol of $\text{Cu}_2(\text{OH})_2\text{CO}_3$ and the molar mass of $\text{Cu}_2(\text{OH})_2\text{CO}_3$.

$\text{mass O/mol Cu}_2(\text{OH})_2\text{CO}_3 = \dfrac{5 \text{ mol O}}{1 \text{ mol Cu}_2(\text{OH})_2\text{CO}_3} \times \dfrac{16.00 \text{ g O}}{1 \text{ mol O}} = 80.00 \text{ g O/mol Cu}_2(\text{OH})_2\text{CO}_3$

$\text{molar mass Cu}_2(\text{OH})_2\text{CO}_3 = (2 \times 63.55 \text{ g Cu}) + (5 \times 16.00 \text{ g O}) + (2 \times 1.01 \text{ g H}) + 12.01 \text{ g C}$

$= 221.13 \text{ g/mol Cu}_2(\text{OH})_2\text{CO}_3$

$\text{Percent oxygen in sample} = \dfrac{80.00 \text{ g}}{221.13 \text{ g}} \times 100\% = 36.18 \% \text{ O}$

21. C(CH$_3$)$_3$CH$_2$CH(CH$_3$)$_2$ has a molar mass of 114.231 g/mol and one mole contains 18.143 g of H.

Percent hydrogen in sample $= \dfrac{18.143\,\text{g}}{114.231\,\text{g}} \times 100\% = 15.88\;\%\text{H}$

22. Determine the mass of a mole of $Cr(NO_3)_3 \cdot 9H_2O$, and then the mass of water in a mole.

molar mass $Cr(NO_3)_3 \cdot 9H_2O = 51.9961\text{g Cr} + (3\times14.0067\text{g N}) + (18\times15.9994\text{g O})$

$+ (18\times1.00794\text{g H}) = 400.148\,\text{g/mol } Cr(NO_3)_3 \times 9H_2O$

$\text{mass H}_2\text{O} = \dfrac{9\,\text{mol H}_2\text{O}}{1\,\text{mol }Cr(NO_3)_3 \times 9H_2O} \times \dfrac{18.0153\,\text{g H}_2\text{O}}{1\,\text{mol H}_2\text{O}} = 162.14\,\text{g H}_2\text{O/mol }Cr(NO_3)_3 \cdot 9H_2O$

$\dfrac{162.14\,\text{g H}_2\text{O/mol }Cr(NO_3)_3 \cdot 9H_2O}{400.148\,\text{g/mol }Cr(NO_3)_3 \cdot 9H_2O} \times 100\% = 40.52\;\%\,H_2O$

23. molar mass $= (20\,\text{mol C} \times 12.011\text{g C}) + (24\,\text{mol H} \times 1.00794\text{g H}) + (2\,\text{mol N} \times 14.0067\text{g N})$

$+ (2\,\text{mol O} \times 15.9994\text{g O}) = 324.42\,\text{g/mol}$

$\%\text{C} = \dfrac{240.22}{324.42} \times 100\% = 74.046\%\text{C}$ $\%\text{H} = \dfrac{24.1906}{324.42} \times 100\% = 7.4566\%\text{H}$

$\%\text{N} = \dfrac{28.0134}{324.42} \times 100\% = 8.6349\%\text{N}$ $\%\text{O} = \dfrac{31.9988}{324.42} \times 100\% = 9.8634\%\text{O}$

24. The molar mass of Cu(C$_{18}$H$_{33}$O$_2$)$_2$ is 626.464 g/mol. One mole contains 66 H (66.524 g H), 36 C (432.396 g C), 4 O (63.998 g O) and 1 Cu (63.546 g Cu).

$\%\text{C} = \dfrac{432.396\,\text{g}}{626.464\,\text{g}} \times 100\% = 69.0217\%\text{C}$ $\%\text{H} = \dfrac{66.524\,\text{g}}{626.464\,\text{g}} \times 100\% = 10.619\%\text{H}$

$\%\text{Cu} = \dfrac{63.546\,\text{g}}{626.464\,\text{g}} \times 100\% = 10.144\%\text{Cu}$ $\%\text{O} = \dfrac{63.998\,\text{g}}{626.464\,\text{g}} \times 100\% = 10.216\%\text{O}$

25. In each case, we first determine the molar mass of the compound, and then the mass of the indicated element in one mole of the compound. Finally, we determine the percent by mass of the indicated element to four significant figures.

(a) molar mass $Pb(C_2H_5)_4 = 207.2\text{g Pb} + (8\times12.011\text{g C}) + (20\times1.00794\text{g H})$

$= 323.447\,\text{g/mol } Pb(C_2H_5)_4$

$\text{mass Pb/mol }Pb(C_2H_5)_4 = \dfrac{1\,\text{mol Pb}}{1\,\text{mol }Pb(C_2H_5)_4} \times \dfrac{207.2\,\text{g Pb}}{1\,\text{mol Pb}} = 207.2\,\text{g Pb/mol }Pb(C_2H_5)_4$

$\%\text{Pb} = \dfrac{207.2\,\text{g Pb}}{323.447\,\text{g }Pb(C_2H_5)_4} \times 100\% = 64.06\;\%\,\text{Pb}$

(b) molar mass $Fe_4\left[Fe(CN)_6\right]_3 = (7 \times 55.847\,g\,Fe) + (18 \times 12.011\,g\,C) + (18 \times 14.0067\,g\,N)$

$$= 859.248\,g/mol\,Fe_4\left[Fe(CN)_6\right]_3$$

$$\frac{mass\,Fe}{mol\,Fe_4\left[Fe(CN)_6\right]_3} = \frac{7\,mol\,Fe}{1\,mol\,Fe_4\left[Fe(CN)_6\right]_3} \times \frac{55.847\,g\,Fe}{1\,mol\,Fe} = 390.929\,g\,Fe/mol\,Fe_4\left[Fe(CN)_6\right]_3$$

$$\%Fe = \frac{390.929\,g\,Fe}{859.248\,g\,Fe_4[Fe(CN)_6]_3} \times 100\% = 45.497\,\%\,Fe$$

(c) molar mass $C_{55}H_{72}MgN_4O_5$

$$= (55 \times 12.011\,g\,C) + (72 \times 1.00794\,g\,H) + (1 \times 24.305\,g\,Mg) + (4 \times 14.0067\,g\,N) + (5 \times 15.9994\,g\,O)$$

$$= 893.505\,g/mol\,C_{55}H_{72}MgN_4O_5$$

$$\frac{mass\,Mg}{mol\,C_{55}H_{72}MgN_4O_5} = \frac{1\,mol\,Mg}{1\,mol\,C_{55}H_{72}MgN_4O_5} \times \frac{24.305\,g\,Mg}{1\,mol\,Mg} = \frac{24.305\,g\,Mg}{mol\,C_{55}H_{72}MgN_4O_5}$$

$$\%Mg = \frac{24.305\,g\,Mg}{893.505\,g\,C_{55}H_{72}MgN_4O_5} \times 100\% = 2.7202\,\%\,Mg$$

26. **(a)** $\%Zr = \dfrac{1\,mol\,Zr}{1\,mol\,ZrSiO_4} \times \dfrac{1\,mol\,ZrSiO_4}{183.31\,g\,ZrSiO_4} \times \dfrac{91.224\,g\,Zr}{1\,mol\,Zr} \times 100\% = 49.765\%\,Zr$

(b) $\%Be = \dfrac{3\,mol\,Fe}{1\,mol\,Be_3Al_2Si_6O_{18}} \times \dfrac{1\,mol\,Be_3Al_2Si_6O_{18}}{537.502\,g\,Be_3Al_2Si_6O_{18}} \times \dfrac{9.01218\,g\,Be}{1\,mol\,Be} \times 100\%$

$\%\,Be = 5.03004\,\%\,Be$

(c) $\%Fe = \dfrac{3\,mol\,Fe}{1\,mol\,Fe_3Al_2Si_3O_{12}} \times \dfrac{1\,mol\,Fe_3Al_2Si_3O_{12}}{497.753\,g\,Fe_3Al_2Si_3O_{12}} \times \dfrac{55.847\,g\,Fe}{1\,mol\,Fe} \times 100\%$

$\%Fe = 33.659\%\,Fe$

(d) $\%S = \dfrac{1\,mol\,S}{1\,mol\,Na_4SSi_3Al_3O_{12}} \times \dfrac{1\,mol\,Na_4SSi_3Al_3O_{12}}{481.219\,g\,Na_4SSi_3Al_3O_{12}} \times \dfrac{32.066\,g\,S}{1\,mol\,S} \times 100\%$

$\%S = 6.6635\%\,S$

27. Oxide with the largest %Cr will have the largest number of moles of Cr per mole of oxygen.

$CrO: \dfrac{1\,mol\,Cr}{1\,mol\,O} = 1\,mol\,Cr/mol\,O$ $Cr_2O_3 : \dfrac{2\,mol\,Cr}{3\,mol\,O} = 0.667\,mol\,Cr/mol\,O$

$CrO_2 : \dfrac{1\,mol\,Cr}{2\,mol\,O} = 0.500\,mol\,Cr/mol\,O$ $CrO_3 : \dfrac{1\,mol\,Cr}{3\,mol\,O} = 0.333\,mol\,Cr/mol\,O$

Arranged in order of increasing %Cr: $CrO_3 < CrO_2 < Cr_2O_3 < CrO$

28. For SO_2 and Na_2S, a mole of each contains a mole of S and two moles of another element; in the case of SO_2 the other element (oxygen) has a smaller atomic mass than the other element in Na_2S (Na), causing SO_2 to have a higher mass percent sulfur. For S_2Cl_2 and $Na_2S_2O_3$, a mole of each contains two moles of S; for S_2Cl_2 the rest of the mole has a mass of 71.0 g, while for $Na_2S_2O_3$ it would be $(2 \times 23) + (3 \times 16) = 94$ g. Sulfur makes up the greater proportion of the mass in S_2Cl_2, giving it the larger percent of S. Now we compare SO_2 and S_2Cl_2: S_2O_4 has the same mass proportions as does SO_2 but also has the two moles of S, as does S_2Cl_2. In S_2Cl_2 the remainder of a mole has a mass of 71.0 g, while in S_2O_4 the remainder of a mole would be $4 \times 16.0 = 64.0$ g. Thus, SO_2 has the highest percent of S so far. For CH_3CH_2SH compared to SO_2, we see that both compounds have one S-atom, SO_2 has two O-atoms (each with a molar mass of ~16 g mol^{-1}) and CH_3CH_2SH effectively has two CH_3 groups (each CH_3 group with a mass of ~15 g mol^{-1}). Thus, CH_3CH_2SH has the highest percentage sulfur by mass of the compounds listed.

Chemical Formulas from Percent Composition

<u>29.</u> SO_3 (40.05% S) and S_2O (80.0 % S) (2 O atoms ≈ 1 S atom in terms of atomic masses)
Note the molar masses are quite close (within 0.05 g/mol).

30. The element chromium has an atomic mass of 52.0 u. Thus, there can only be one chromium atom per formula unit of the compound. (Two atoms of chromium have a mass of 104 u, more than the formula mass of the compound.) Three of the four remaining atoms in the formula unit must be oxygen. Thus, the oxide is CrO_3, chromium(VI) oxide.

<u>31.</u> Determine the % oxygen by difference. $\%O = 100.00\% - 45.27\% C - 9.50\% H = 45.23\% O$

$$\text{mol O} = 45.23\,g \times \frac{1\,\text{mol O}}{16.00\,g\ O} = 2.827\,\text{mol O} \quad \div 2.827 \rightarrow 1.000\,\text{mol O}$$

$$\text{mol C} = 45.27\,g\ C \times \frac{1\,\text{mol C}}{12.01\,g\ C} = 3.769\,\text{mol C} \quad \div 2.827 \rightarrow 1.333\,\text{mol C}$$

$$\text{mol H} = 9.50\,g\ H \times \frac{1\,\text{mol H}}{1.008\,g\ H} = 9.42\,\text{mol H} \quad \div 2.827 \rightarrow 3.33\,\text{mol H}$$

Multiply all amounts by 3 to obtain integers. Empirical formula is $C_4H_{10}O_3$.

32. We base our calculation on 100.0 g of monosodium glutamate.

$$13.6\,g\ Na \times \frac{1\,\text{mol Na}}{22.99\,g\ Na} = 0.592\,\text{mol Na} \quad \div 0.592 \rightarrow 1.00\,\text{mol Na}$$

$$35.5\,g\ C \times \frac{1\,\text{mol C}}{12.01\,g\ C} = 2.96\,\text{mol C} \quad \div 0.592 \rightarrow 5.00\,\text{mol C}$$

$$4.8\,g\ H \times \frac{1\,\text{mol H}}{1.01\,g\ H} = 4.8\,\text{mol H} \quad \div 0.592 \rightarrow 8.1\,\text{mol H}$$

$$8.3\,g\ N \times \frac{1\,\text{mol N}}{14.01\,g\ N} = 0.59\,\text{mol N} \quad \div 0.592 \rightarrow 1.0\,\text{mol N}$$

$$37.8\,g\ O \times \frac{1\,\text{mol O}}{16.00\,g\ O} = 2.36\,\text{mol O} \quad \div 0.592 \rightarrow 3.99\,\text{mol O} \qquad \text{Empirical formula}: NaC_5H_8NO_4$$

33. **(a)**

$$74.01\,\text{g C} \times \frac{1\,\text{mol C}}{12.01\,\text{g C}} = 6.162\,\text{mol C} \div 1.298 \rightarrow 4.747\,\text{mol C}$$

$$5.23\,\text{g H} \times \frac{1\,\text{mol H}}{1.01\,\text{g H}} = 5.18\,\text{mol H} \div 1.298 \rightarrow 3.99\,\text{mol H}$$

$$20.76\,\text{g O} \times \frac{1\,\text{mol O}}{16.00\,\text{g O}} = 1.298\,\text{mol O} \div 1.298 \rightarrow 1.00\,\text{mol O}$$

Multiply each of the mole numbers by 4 to obtain an empirical formula of $C_{19}H_{16}O_4$.

(b)

$$39.98\,\text{g C} \times \frac{1\,\text{mol C}}{12.011\,\text{g C}} = 3.328\underline{6}\,\text{mol C} \div 0.7397 \rightarrow 4.500\,\text{mol C}$$

$$3.73\,\text{g H} \times \frac{1\,\text{mol H}}{1.00794\,\text{g H}} = 3.70\,\text{mol H} \div 0.7397 \rightarrow 5.00\,\text{mol H}$$

$$20.73\,\text{g N} \times \frac{1\,\text{mol N}}{14.0067\,\text{g N}} = 1.480\,\text{mol N} \div 0.7397 \rightarrow 2.001\,\text{mol N}$$

$$11.84\,\text{g O} \times \frac{1\,\text{mol O}}{15.9994\,\text{g O}} = 0.7400\,\text{mol O} \div 0.7397 \rightarrow 1.000\,\text{mol O}$$

$$23.72\,\text{g S} \times \frac{1\,\text{mol S}}{32.066\,\text{g S}} = 0.7397\,\text{mol S} \div 0.7397 \rightarrow 1.000\,\text{mol S}$$

Multiply by 2 to obtain the empirical formula

$C_9H_{10}N_4O_2S_2$

34. **(a)**

$$95.21\,\text{g C} \times \frac{1\,\text{mol C}}{12.01\,\text{g C}} = 7.928\,\text{mol C} \div 4.74 \rightarrow 1.67\,\text{mol C}$$

$$4.79\,\text{g H} \times \frac{1\,\text{mol H}}{1.01\,\text{g H}} = 4.74\,\text{mol H} \div 4.74 \rightarrow 1.00\,\text{mol H}$$

Multiply each of the mole numbers by 3 to obtain an empirical formula of C_5H_3.

(b) Each percent is numerically equal to the mass of that element present in 100.00 g of the compound. These masses then are converted to amounts of the elements, in moles.

$$\text{amount C} = 38.37\,\text{g C} \times \frac{1\,\text{mol C}}{12.01\,\text{g C}} = 3.195\,\text{mol C} \div 0.491 \rightarrow 6.51\,\text{mol C}$$

$$\text{amount H} = 1.49\,\text{g H} \times \frac{1\,\text{mol H}}{1.01\,\text{g H}} = 1.48\,\text{mol H} \div 0.491 \rightarrow 3.01\,\text{mol H}$$

$$\text{amount Cl} = 52.28\,\text{g Cl} \times \frac{1\,\text{mol Cl}}{35.45\,\text{g Cl}} = 1.475\,\text{mol Cl} \div 0.491 \rightarrow 3.004\,\text{mol Cl}$$

$$\text{amount O} = 7.86\,\text{g O} \times \frac{1\,\text{mol O}}{16.0\,\text{g O}} = 0.491\,\text{mol O} \div 0.491 \rightarrow 1.00\,\text{mol O}$$

Multiply each number of moles by 2 to obtain the empirical formula: $C_{13}H_6Cl_6O_2$.

35. Convert each percentage into the mass in 100.00 g, and then to the moles of that element.

$$94.34 \text{ g C} \times \frac{1 \text{ mol C}}{12.011 \text{ g C}} = 7.854 \text{ mol C} \qquad \div 5.61\underline{5} = 1.40 \text{ mol C} \times 5 = 7.00$$

$$5.66 \text{ g H} \times \frac{1 \text{ mol H}}{1.00794 \text{ g H}} = 5.61\underline{5} \text{ mol H} \qquad \div 5.61\underline{5} = 1.00 \text{ mol H} \times 5 = 5.00$$

Multiply by 5 to achieve whole number ratios. The empirical formula is C_5H_7, and formula mass $[(7 \times 12.011 \text{ g C}) + (5 \times 1.00794 \text{ g H})] = 89.117$ u. Since this empirical molar mass is one-half of the 178 u, the correct molecular mass, the molecular formula must be twice the empirical formula. Molecular formula: $C_{14}H_{10}$

36. The percent of selenium in each oxide is found by difference.

First oxide: $\%\text{Se} = 100.0\% - 28.8\% \qquad O = 71.2\% \text{ Se}$
A 100 gram sample would contain 28.8 g O and 71.2 g Se

$$28.8 \text{ g O} \times \frac{1 \text{ mol O}}{16.0 \text{ g O}} = 1.80 \text{ mol O} \qquad \div 0.901 \qquad \rightarrow 2.00 \text{ mol O}$$

$$71.2 \text{ g Se} \times \frac{1 \text{ mol Se}}{79.0 \text{ g Se}} = 0.901 \text{ mol Se} \qquad \div 0.901 \qquad \rightarrow 1.00 \text{ mol Se}$$

The empirical formula is SeO_2. An appropriate name is selenium dioxide.

Second oxide: $\%\text{Se} = 100.0\% - 37.8\% O = 62.2\% \text{ Se} \%$
A 100 gram sample would contain 37.8 g O and 62.2 g Se

$$37.8 \text{ g O} \times \frac{1 \text{ mol O}}{16.0 \text{ g O}} = 2.36 \text{ mol O} \qquad \div 0.787 \rightarrow 3.00 \text{ mol O}$$

$$62.2 \text{ g Se} \times \frac{1 \text{ mol Se}}{79.0 \text{ g Se}} = 0.787 \text{ mol Se} \div 0.787 \rightarrow 1.00 \text{ mol Se}$$

The empirical formula is SeO_3. An appropriate name is selenium trioxide.

37. Determine the mass of oxygen by difference. Then convert all masses to amounts in moles.
oxygen mass $= 100.00 \text{ g} - 73.27 \text{ g C} - 3.84 \text{ g H} - 10.68 \text{ g N} = 12.21 \text{ g O}$

$$\text{amount C} = 73.27 \text{ g C} \times \frac{1 \text{ mol C}}{12.011 \text{ g C}} = 6.100 \text{ mol C} \qquad \div 0.7625 \qquad \rightarrow 8.000 \text{ mol C}$$

$$\text{amount H} = 3.84 \text{ g H} \times \frac{1 \text{ mol H}}{1.008 \text{ g H}} = 3.81 \text{ mol H} \qquad \div 0.7625 \qquad \rightarrow 5.00 \text{ mol H}$$

$$\text{amount N} = 10.68 \text{ g N} \times \frac{1 \text{ mol N}}{14.007 \text{ g N}} = 0.7625 \text{ mol N} \qquad \div 0.7625 \qquad \rightarrow 1.000 \text{ mol N}$$

$$\text{amount O} = 12.21 \text{ g O} \times \frac{1 \text{ mol O}}{15.999 \text{ g O}} = 0.7632 \text{ mol O} \qquad \div 0.7625 \qquad \rightarrow 1.001 \text{ mol O}$$

The empirical formula is C_8H_5NO, which has an empirical mass of 131 u. This is almost exactly half the molecular mass of 262.3 u. Thus, the molecular formula is twice the empirical formula and is $C_{16}H_{10}N_2O_2$.

38. Convert each percentage into the mass in 100.00 g, and then to the moles of that element.

$$\text{amount C} = 44.45 \text{ g C} \times \frac{1 \text{mol C}}{12.011 \text{g C}} = 3.701 \text{mol C} \quad \div\, 3.70 \quad \rightarrow 1.00 \text{ mol C}$$

$$\text{amount H} = 3.73 \text{g H} \times \frac{1 \text{mol H}}{1.00794 \text{g H}} = 3.70 \text{mol H} \quad \div\, 3.70 \quad \rightarrow 1.00 \text{ mol H}$$

$$\text{amount N} = 51.82 \text{ g N} \times \frac{1 \text{mol N}}{14.0067 \text{g N}} = 3.700 \text{mol N} \quad \div\, 3.70 \quad \rightarrow 1.00 \text{ mol N}$$

The empirical formula is CHN, which has an empirical mass of 27.025$\underline{6}$ u. This is exactly one fifth the molecular mass of 135.14 u. Thus, the molecular formula is five times greater than the empirical formula and is $C_5H_5N_5$.

39. The molar mass of element X has the units of grams per mole. We can determine the amount, in moles of Cl, and convert that to the amount of X, equivalent to 25.0 g of X.

$$\text{molar mass} = \frac{25.0 \text{g X}}{75.0 \text{g Cl}} \times \frac{35.453 \text{g Cl}}{1 \text{mol Cl}} \times \frac{4 \text{mol Cl}}{1 \text{mol X}} = \frac{47.3 \text{g X}}{1 \text{mol X}}$$

The atomic mass is 47.3 u. This atomic mass is close to that of the element titanium, which therefore is identified as element X.

40. The molar mass of element X has the units of grams per mole. We can determine the amount, relative to the mass percent Cl. Assume 1 mole of compound. This contains 15.9994 g O and 70.905 g Cl. The following relation must hold true.

$$\text{Mass percent Cl} = 59.6 \% = 0.596 = \frac{\text{mass Cl in one mole}}{\text{mass of one mole of XOCl}_2} = \frac{70.905 \text{ g}}{(X + 15.9994 + 70.905) \text{ g}}$$

$$0.596 = \frac{70.905}{(X + 86.905)} \quad \text{or} \quad 0.596X + 51.795 = 70.905 \qquad 0.596X = 19.110$$

$$\text{Hence, } X = \frac{19.110}{0.596} = 32.0\underline{6} \quad \text{The atomic mass of } X = 32.0\underline{6} \text{ g mol}^{-1} \qquad X \text{ is the element sulfur}$$

41. Consider 100 g of chlorophyll, which contains 2.72 g is Mg. To answer this problem we must take note of the fact that 1 mole of Mg contains 1 mole of chlorophyll.

$$\frac{100 \text{g chlorophyl}}{2.72 \text{g Mg}} \times \frac{24.305 \text{g Mg}}{1 \text{mol Mg}} \times \frac{1 \text{mol Mg}}{1 \text{mol chlorophyll}} = 894 \text{ g mol}^{-1}$$

Therefore, the molecular mass of chlorophyll is 894 u

42. Compound I has a molecular mass of 137 u. We are told that chlorine constitutes 77.5 % of the mass, so the mass of chlorine in each molecule is $137 \text{ u} \times \dfrac{77.5}{100} = 106$ u.

This corresponds to three chlorine atoms (106 u ÷ 35.453 u/Cl atom = 2.99 or 3 Cl atoms). The remaining 31 u, (137 u - 106 u), is the mass for element X in one molecule of Compound I. Compound II has 85.1 % chlorine by mass, so the mass of chlorine in each

molecule of Compound II is $208 \text{ u} \times \dfrac{85.1}{100} = 177$ u.

This corresponds to five Cl atoms (177 u ÷ 35.453 u/Cl atom = 4.99 ~5 chlorine atoms). The remaining mass is 31 u (208 u - 177 u), which is very close to the mass of X found in each molecule of compound I. Thus, we have two compounds: X_nCl_3, which has a molecular mass of 137 u, and X_nCl_5, which has a molecular mass of 208 u

(We also know that the mass of X in both molecular species is ~31 u). If we assume that n = 1 in the formulas above, then element X must be phosphorus (30.974 u) and the formulas for the compounds are PCl_3 (Compound I) and PCl_5 (Compound II).

Combustion Analysis

43. **(a)** First we determine the mass of carbon and of hydrogen present in the sample. Remember that a hydrocarbon contains only hydrogen and carbon.

$$0.6260\,g\ CO_2 \times \frac{1\,mol\ CO_2}{44.01g\ CO_2} \times \frac{1\,mol\ C}{1\,mol\ CO_2} = 0.01422\,mol\ C \times \frac{12.011g\ C}{1\,mol\ C} = 0.17085\,g\ C$$

$$0.1602\,g\ H_2O \times \frac{1\,mol\ H_2O}{18.0153g\ H_2O} \times \frac{2\,mol\ H}{1\,mol\ H_2O} = 0.017785\,mol\ H \times \frac{1.00794\ g\ H}{1\,mol\ H} = 0.017926g\ H$$

Then the % C and % H are found.

$$\%C = \frac{0.17085}{0.1888\ g\ cmpd} \times 100\% = 90.49\ \%C \quad \%H = \frac{0.017926\,g\ H}{0.1888\ g\ cmpd} \times 100\% = 9.495\ \%H$$

(b) Use the moles of C and H from part (a), and divide both by the smallest value, namely 0.01422 mol. Thus   0.01422 mol C ÷ 0.01422 mol = 1 mol H;
0.017785 mol H ÷ 0.01422 mol = 1.251 mol H.
The empirical formula is obtained by multiplying these mole numbers by 4. It is C_4H_5.

(c) The molar mass of the empirical formula C_4H_5 $\left[(4\times12.0g\ C)+(5\times1.0g\ H)\right] = 53.0$ g/mol. This value is 1/2 of the actual molar mass. The molecular formula is twice the empirical formula. ∴ Molecular formula: C_8H_{10}.

44. Determine the mass of carbon and of hydrogen present in the sample.

$$1.1518g\ CO_2 \times \frac{1\,mol\ CO_2}{44.010g\ CO_2} \times \frac{1\,mol\ C}{1\,mol\ CO_2} = 0.026171\,mol\ C \times \frac{12.011g\ C}{1\,mol\ C} = 0.3143\ g\ C$$

$$0.2694\ g\ H_2O \times \frac{1\,mol\ H_2O}{18.0153g\ H_2O} \times \frac{2\,mol\ H}{1\,mol\ H_2O} = 0.02991\,mol\ H \times \frac{1.00794g\ H}{1\,mol\ H} = 0.030145g\ H$$

(a) The percent composition can be determined using the masses of C and H.

$$\%C = \frac{0.3144g\ C}{0.4039g\ cmpd} \times 100\% = 77.83\ \%C \quad \%H = \frac{0.030145g\ H}{0.4039g\ cmpd} \times 100\% = 7.4636\ \%H$$

%O = 100 – 77.83 % - 7.4636 % = 14.706 %
These percents can be used in determining the empirical formula if one wishes.

(b) To find the empirical formula determine the mass of oxygen by difference, and its amount in moles. Mass O $= 0.4039 - 0.3144 - 0.030145 = 0.05936$ g

$$0.05936 \text{ g O} \times \frac{1 \text{ mol O}}{15.9994 \text{ g O}} = \quad 0.00371 \text{ mol O} \div 0.00371 \rightarrow 1.00 \text{ mol O}$$

$$0.026177 \text{ mol C} \div 0.00371 \rightarrow 7.06 \text{ mol C}$$

$$0.02991 \text{ mol H} \div 0.00371 \rightarrow 8.06 \text{ mol H}$$

Empirical formula is C_7H_8O
Note: error limits in mass of O determines the error limits for the calculation.

(c) The molecular formula is found by realizing that a mole of empirical units has a mass of $(7 \times 12.0 \text{ g C} + 8 \times 1.0 \text{ g H} + 16.0 \text{ g O}) = 108.0$ g. Since this agrees with the molecular mass, the molecular formula is the same as the empirical formula: C_7H_8O.

45. First determine the mass of carbon and hydrogen present in the sample.

$$0.458 \text{ g CO}_2 \times \frac{1 \text{ mol CO}_2}{44.01 \text{ g CO}_2} \times \frac{1 \text{ mol C}}{1 \text{ mol CO}_2} = 0.0104 \text{ mol C} \times \frac{12.011 \text{ g C}}{1 \text{ mol C}} = 0.125 \text{ g C}$$

$$0.374 \text{ g H}_2\text{O} \times \frac{1 \text{ mol H}_2\text{O}}{18.0153 \text{ g H}_2\text{O}} \times \frac{2 \text{ mol H}}{1 \text{ mol H}_2\text{O}} = 0.0415 \text{ mol H} \times \frac{1.00794 \text{ g H}}{1 \text{ mol H}} = 0.04185 \text{ g H}$$

Then, the mass of N that this sample would have produced is determined.
(Note that this is also the mass of N_2 produced in the reaction.)

$$0.226 \text{ g N}_2 \times \frac{0.312 \text{ g 1st sample}}{0.486 \text{ g 2nd sample}} = 0.145 \text{ g N}_2$$

From which we can calculate the mass of N in the sample.

$$0.145 \text{ g N}_2 \times \frac{1 \text{ mol N}_2}{28.0134 \text{ g N}_2} \times \frac{2 \text{ mol N}}{1 \text{ mol N}_2} \times \frac{14.0067 \text{ g N}}{1 \text{ mol N}} = 0.145 \text{ g N}$$

We may alternatively determine the mass of N by difference:
$0.312 \text{ g} - 0.125 \text{ g C} - 0.04185 \text{ g H} = 0.145 \text{ g N}$

Then, we can calculate the relative number of moles of each element.

$$0.145 \text{ g N} \times \frac{1 \text{ mol N}}{14.0067 \text{ g N}} = \quad 0.01035 \text{ mol N} \div 0.01035 \rightarrow 1.00 \text{ mol N}$$

$$0.0104 \text{ mol C} \div 0.01035 \rightarrow 1.00 \text{ mol C}$$

$$0.0415 \text{ mol H} \div 0.01035 \rightarrow 4.01 \text{ mol H}$$

Thus, the empirical formula is CH_4N

46. Thiophene contains only carbon, hydrogen, and sulfur, so there is no need to determine the mass of oxygen by difference. We simply determine the amount of each element from the mass of its combustion product.

$$2.7224 \text{ g CO}_2 \times \frac{1 \text{ mol CO}_2}{44.010 \text{ g CO}_2} \times \frac{1 \text{ mol C}}{1 \text{ mol CO}_2} = 0.061859 \text{ mol C} \div 0.01548 \rightarrow 3.996 \text{ mol C}$$

$$0.5575 \text{ g H}_2\text{O} \times \frac{1 \text{ mol H}_2\text{O}}{18.0153 \text{ g H}_2\text{O}} \times \frac{2 \text{ mol H}}{1 \text{ mol H}_2\text{O}} = 0.06189 \text{ mol H} \div 0.01548 \rightarrow 3.999 \text{ mol H}$$

$$0.9915 \text{ g SO}_2 \times \frac{1 \text{ mol SO}_2}{64.0648 \text{ g SO}_2} \times \frac{1 \text{ mol S}}{1 \text{ mol SO}_2} = 0.01548 \text{ mol S} \div 0.01548 \rightarrow 1.00 \text{ mol S}$$

The empirical formula of thiophene is C_4H_4S.

47. Each mole of CO_2 is produced from a mole of C. Therefore, the compound with the largest number of moles of C per mole of the compound will produce the largest amount of CO_2 and, thus, also the largest mass of CO_2. Of the compounds listed, namely $CH_4, C_2H_5OH, C_{10}H_8$ and C_6H_5OH, $C_{10}H_8$ has the largest number of moles of C per mole of the compound and will produce the greatest mass of CO_2 per mole on complete combustion.

48. The compound that produces the largest mass of water per gram of the compound will have the largest amount of hydrogen per gram of the compound. Thus, we need to compare the ratios of amount of hydrogen per mole to the molar mass for each compound.
Note that C_2H_5OH has as much H per mole as does C_6H_5OH, but C_6H_5OH has a higher molar mass. Thus, C_2H_5OH produces more H_2O per gram than does C_6H_5OH.
Notice also that CH_4 has 4 H's per C, while $C_{10}H_8$ has 8 H's per 10 C's or 0.8 H per C. Thus CH_4 will produce more H_2O than will $C_{10}H_8$. Thus, we are left with comparing CH_4 to C_2H_5OH. The O in the second compound has about the same mass (16 u) as does C (12 u). Thus, in CH_4 there are 4 H's per C, while in C_2H_5OH there are about 2 H's per C.
Thus CH_4 will produce the most water per gram on combustion, of all four compounds.

49. Here we will use the fact that $C_4H_{10}O$ has a molar mass of 74.1228 g/mol to calculate the masses of CO_2 and H_2O:

Mass of CO_2:

$$1.562 \text{ g C}_4\text{H}_{10}\text{O} \times \frac{1 \text{ mol C}_4\text{H}_{10}\text{O}}{74.1228 \text{ g C}_4\text{H}_{10}\text{O}} \times \frac{4 \text{ mol C}}{1 \text{ mol C}_4\text{H}_{10}\text{O}} \times \frac{1 \text{ mol CO}_2}{1 \text{ mol C}} \times \frac{44.010 \text{ g CO}_2}{1 \text{ mol CO}_2} = 3.710 \text{ g CO}_2$$

Mass H_2O:

$$1.562 \text{ g C}_4\text{H}_{10}\text{O} \times \frac{1 \text{ mol C}_4\text{H}_{10}\text{O}}{74.1228 \text{ g C}_4\text{H}_{10}\text{O}} \times \frac{10 \text{ mol H}}{1 \text{ mol C}_4\text{H}_{10}\text{O}} \times \frac{1 \text{ mol H}_2\text{O}}{2 \text{ mol H}} \times \frac{18.0153 \text{ g H}_2\text{O}}{1 \text{ mol H}_2\text{O}} = 1.898 \text{ g H}_2\text{O}$$

50. moles of $C_2H_6S = 3.15$ mL $\times \dfrac{0.84 \text{ g } C_2H_6S}{1 \text{ mL } C_2H_6S} \times \dfrac{1 \text{ mol } C_2H_6S}{62.1356 \text{ g } C_2H_6S} = 0.042\underline{6}$ mol C_2H_6S

Thus, the mass of CO_2 expected is

$= 0.042\underline{6}$ mol $C_2H_6S \times \dfrac{2 \text{ mol } CO_2}{1 \text{ mol } C_2H_6S} \times \dfrac{44.010 \text{ g } CO_2}{1 \text{ mol } CO_2} = 3.7\underline{5}$ g of $CO_2(g)$

The mass of $SO_2(g)$ expected from the complete combustion is

$= 0.042\underline{6}$ mol $C_2H_6S \times \dfrac{1 \text{ mol } SO_2}{1 \text{ mol } C_2H_6S} \times \dfrac{64.0648 \text{ g } SO_2}{1 \text{ mol } H_2O} = 2.7\underline{3}$ g of $SO_2(g)$

The mass of $H_2O(l)$ expected from the complete combustion is

$= 0.042\underline{6}$ mol $C_2H_6S \times \dfrac{6 \text{ mol } H}{1 \text{ mol } C_2H_6S} \times \dfrac{1 \text{ mol } H_2O}{2 \text{ mol } H} \times \dfrac{18.0153 \text{ g } H_2O}{1 \text{ mol } H_2O} = 2.3$ g of $H_2O(l)$

Oxidation States

51. The oxidation state (O.S.) is given first, followed by the explanation for its assignment.

(a) $C = -4$ in CH_4 H has an oxidation state of +1 in its non-metal compounds (Remember that the sum of the oxidation states in a neutral compound equals 0.)

(b) $S = +4$ in SF_4 F has O.S. $= -1$ in its compounds.

(c) $O = -1$ in Na_2O_2 Na has O.S. $= +1$ in its compounds.

(d) $C = 0$ in $C_2H_3O_2^-$ H has O.S. $= +1$ in its non-metal compounds; that of $O = -2$ (usually). (Remember that the sum of the oxidation states in a polyatomic ion equals the charge on that ion.)

(e) $Fe = +6$ in FeO_4^{2-} O has O.S. $= -2$ in most of its compounds (especially metal containing compounds).

52. The oxidation state of sulfur in each species is determined below. Remember that the oxidation state of O is -2 in its compounds. And the sum of the oxidation states in an ion equals the charge on that ion.

(a) $S = +4$ in SO_3^{2-} **(b)** $S = +2$ in $S_2O_3^{2-}$ **(c)** $S = +7$ in $S_2O_8^{2-}$

(d) $S = +6$ in HSO_4^- **(e)** $S = -2.5$ in $S_4O_6^{2-}$

53. Remember that the oxidation state of oxygen is usually -2 in its compounds. Cr^{3+} and O^{2-} form Cr_2O_3, chromium(III) oxide. Cr^{4+} and O^{2-} form CrO_2, chromium (IV) oxide. Cr^{6+} and O^{2-} form CrO_3, chromium(VI) oxide.

54. Remember that oxygen usually has an oxidation state of -2 in its compounds.

$N = +1$ in N_2O, dinitrogen monoxide

$N = +2$ in NO, nitric oxide or nitrogen monoxide

$N = +3$ in N_2O_3, dinitrogen trioxide

$N = +4$ in NO_2, nitrogen dioxide

$N = +5$ in N_2O_5, dinitrogen pentoxide

Nomenclature

55. (a) SrO — strontium oxide   (b) ZnS — zinc sulfide

(c) K_2CrO_4 — potassium chromate   (d) Cs_2SO_4 — cesium sulfate

(e) Cr_2O_3 — chromium(III) oxide   (f) $Fe_2(SO_4)_3$ — iron(III) sulfate

(g) $Mg(HCO_3)_2$ — magnesium hydrogen carbonate   (h) $(NH_4)_2HPO_4$ — ammonium hydrogen phosphate

(i) $Ca(HSO_3)_2$ — calcium hydrogen sulfite   (j) $Cu(OH)_2$ — copper(II) hydroxide

(k) HNO_3 — nitric acid   (l) $KClO_4$ — potassium perchlorate

(m) $HBrO_3$ — bromic acid   (n) H_3PO_3 — phosphorous acid

56. (a) $Ba(NO_3)_2$ — barium nitrate   (b) HNO_2 — nitrous acid

(c) CrO_2 — chromium(IV) oxide   (d) KIO_3 — potassium iodate

(e) $LiCN$ — lithium cyanide   (f) KIO — potassium hypoiodite

(g) $Fe(OH)_2$ — iron(II) hydroxide   (h) $Ca(H_2PO_4)_2$ — calcium dihydrogen phosphate

(i) H_3PO_4 — phosphoric acid   (j) $NaHSO_4$ — sodium hydrogen sulfate

(k) $Na_2Cr_2O_7$ — sodium dichromate   (l) $NH_4C_2H_3O_2$ — ammonium acetate

(m) MgC_2O_4 — magnesium oxalate   (n) $Na_2C_2O_4$ — sodium oxalate

57. (a) CS_2 — carbon disulfide   (b) SiF_4 — silicon tetrafluoride

(c) ClF_5 — chlorine pentafluoride   (d) N_2O_5 — dinitrogen pentoxide

(e) SF_6 — sulfur hexafluoride   (f) I_2Cl_6 — diiodine hexachloride

58. (a) ICl — iodine monochloride   (b) ClF_3 — chlorine trifluoride

(c) SF_4 — sulfur tetrafluoride   (d) BrF_5 — bromine pentafluoride

(e) N_2O_4 — dinitrogen tetroxide   (g) S_4N_4 — tetrasulfur tetranitride

59. (a) $Al_2(SO_4)_3$ aluminum sulfate   (b) $(NH_4)_2Cr_2O_7$ ammonium dichromate

    (c) SiF_4 silicon tetrafluoride   (d) Fe_2O_3 iron(III) oxide

    (e) C_3S_2 tricarbon disulfide   (f) $Co(NO_3)_2$ cobalt(II) nitrate

    (g) $Sr(NO_2)_2$ strontium nitrite   (h) HBr(aq)   hydrobromic acid

    (i) HIO_3 iodic acid   (j) PCl_2F_3 phosphorus dichloride trifluoride

60. (a) $Mg(ClO_4)_2$ magnesium perchlorate   (b) $Pb(C_2H_3O_2)_2$ lead(II) acetate

    (c) SnO_2 tin(IV) oxide   (d) HI(aq)   hydroiodic acid

    (e) $HClO_2$ chlorous acid   (f) $NaHSO_3$ sodium hydrogen sulfite

    (g) $Ca(H_2PO_4)_2$ calcium dihydrogen phosphate   (h) $AlPO_4$ aluminum phosphate

    (i) N_2O_4 dinitrogen tetroxide   (j) S_2Cl_2 disulfur dichloride

61. (a) Ti^{4+} and Cl^- produce $TiCl_4$ (b) Fe^{3+} and SO_4^{2-} produce $Fe_2(SO_4)_3$

    (c) Cl^{7+} and O^{2-} produce Cl_2O_7 (d) S^{7+} and O^{2-} produce $S_2O_8^{2-}$

62. (a) N^{5+} and O^{2-} produce N_2O_5 (b) N^{3+}, O^{2-} and H^+ produce HNO_2

    (c) $C^{+4/3}$ and O^{2-} produce C_3O_2 (d) $S^{+2.5}$ and O^{2-} produce $S_4O_6^{2-}$

63. (a) $HClO_2$ chlorous acid   (b) H_2SO_3 sulfurous acid

    (c) H_2Se hydroselenic acid   (d) HNO_2 nitrous acid

64. (a) HF (aq)   hydrofluoric acid   (b) HNO_3 nitric acid

    (c) H_3PO_3 phosphorous acid   (d) H_2SO_4 sulfuric acid

Hydrates

65. The hydrate with the greatest mass percent H_2O is the one that gives the largest result for the number of moles of water in the hydrate's empirical formula, divided by the mass of one mole of the anhydrous salt for the hydrate.

$$\frac{5H_2O}{CuSO_4} = \frac{5\,mol\,H_2O}{159.6\,g} = 0.03133 \qquad \frac{6H_2O}{MgCl_2} = \frac{6\,mol\,H_2O}{95.2\,g} = 0.0630$$

$$\frac{18H_2O}{Cr_2(SO_4)_3} = \frac{18\,mol\,H_2O}{392.3\,g} = 0.04588 \qquad \frac{2H_2O}{LiC_2H_3O_2} = \frac{2\,mol\,H_2O}{66.0\,g} = 0.0303$$

The hydrate with the greatest % H_2O therefore is $MgCl_2 \cdot 6H_2O$

66. A mole of this hydrate will contain about the same mass of H_2O and of Na_2SO_3.

Molar mass $Na_2SO_3 = (2 \times 23.0 \text{g Na}) + 32.1 \text{g S} + (3 \times 16.0 \text{g O}) = 126.1 \text{g/mol}$

$$\text{no. mol } H_2O = 126.1 \text{g} \times \frac{1 \text{mol } H_2O}{18.0 \text{g } H_2O} = 7.01 \text{mol } H_2O$$

Thus, the formula of the hydrate is $Na_2SO_3 \cdot 7H_2O$.

67. Molar mass $CuSO_4 = 63.546 \text{ g Cu} + 32.066 \text{ g S} + (4 \times 15.9994 \text{ g O}) = 159.61 \text{ g } CuSO_4/\text{mol}$.

Note that each $CuSO_4$ will pick up 5 equivalents of H_2O to give $CuSO_4 \cdot 5H_2O$.

$$\text{mass of required } CuSO_4 = 12.6 \text{g } H_2O \times \frac{1 \text{mol } H_2O}{18.0153 \text{ g } H_2O} \times \frac{1 \text{mol } CuSO_4}{5 \text{mol } H_2O} \times \frac{159.61 \text{ g } CuSO_4}{1 \text{mol } CuSO_4}$$

$= 22.3 \text{ g } CuSO_4$ is the minimum amount required to remove all the water

68. The increase in mass of the solid is the result of each mole of the solid absorbing ten moles of water.

$$\text{increase in mass} = 24.05 \text{ g } Na_2SO_4 \times \frac{1 \text{mol } Na_2SO_4}{142.043 \text{ g } Na_2SO_4} \times \frac{10 \text{mol } H_2O \text{ added}}{1 \text{mol } Na_2SO_4} \times \frac{18.0153 \text{ g } H_2O}{1 \text{mol } H_2O}$$

$= 30.50 \text{ g } H_2O$ added

69. We start by converting to molar amounts for each element based on 100.0g:

$$20.3 \text{ g Cu} \times \frac{1 \text{ mol Cu}}{63.546 \text{ g Cu}} = 0.319 \text{ mol Cu} \quad \div 0.319 \quad \rightarrow 1.00 \text{ mol Cu}$$

$$8.95 \text{ g Si} \times \frac{1 \text{ mol Si}}{28.0855 \text{ g Si}} = 0.319 \text{ mol Si} \quad \div 0.319 \rightarrow 1.00 \text{ mol Si}$$

$$36.3 \text{ g F} \times \frac{1 \text{ mol F}}{18.9984 \text{ g F}} = 1.91 \text{ mol F} \quad \div 0.319 \rightarrow 5.99 \text{ mol F}$$

$$34.5 \text{ g } H_2O \times \frac{1 \text{ mol } H_2O}{18.0153 \text{g}} = 1.91\underline{5} \text{ mol } H_2O \div 0.319 \rightarrow 6.00 \text{ mol } H_2O$$

Thus the empirical formula for the hydrate is $CuSiF_6 \cdot 6H_2O$.

70. Let's start by looking at the data provided.

mass of anhydrous compound $= 3.967$ g

mass of water $= 8.129 \text{ g} - 3.967 \text{ g} = 4.162 \text{ g}$

$$\text{moles of anhydrous compound} = 3.967 \text{ g } MgSO_4 \times \frac{1 \text{ mol } MgSO_4}{120.37 \text{ g}} = 0.03296 \text{ mol}$$

$$\text{moles of } H_2O = 4.162 \text{ g} \times \frac{1 \text{ mol } H_2O}{18.0153 \text{ g } H_2O} = 0.2310 \text{ mol } H_2O$$

$$\text{setting up proportions} \quad \frac{0.2310 \text{ mol } H_2O}{0.03296 \text{ mol anhydrous compound}} = \frac{x \text{ mol } H_2O}{1.00 \text{ mol anhydrous compound}}$$

$x = 7.009$ Thus, the formula of the hydrate is $MgSO_4 \cdot 7H_2O$.

Organic Compounds and Organic Nomenclature

71. Answer is (b), 2-butanol is the most appropriate name for this molecule. It has a four carbon atom chain with a hydroxyl group on the carbon second from the end.

72. Answer (c), butanoic acid is the most appropriate name for this molecule. It has a four carbon atom chain with an acid group on the 1st carbon (terminal carbon atom)

73. Molecules (a), (b), (c) and (d) are structural isomers. They share a common formula, namely $C_5H_{12}O$, but have different molecular structures. Molecule (e) has a different chemical formula ($C_6H_{14}O$) and hence cannot be classified as an isomer. It should be pointed out that molecules (a) and (c) are identical as well as being isomers of (b).

74. Molecules (a), (b) and (c) are structural isomers. They share a common formula, namely $C_5H_{11}Cl$, but have different molecular structures. Molecule (d) has a different chemical formula ($C_6H_{13}Cl$) and hence cannot be classified as an isomer.

75. (a) $CH_3(CH_2)_5CH_3$ (b) $CH_3CH_2CO_2H$
 (c) $CH_3CH_2CH_2CH(CH_3)CH_2OH$ (d) CH_3CH_2F

76. (a) $CH_3(CH_2)_6CH_3$ (b) $CH_3(CH_2)_5CO_2H$
 (c) $CH_3(CH_2)_2CH(OH)CH_2CH_3$ (d) $CH_3CHClCH_2CH_3$

77. (a) Methanol; CH_3OH; Molecular mass = 32.04 u
 (b) 2-chlorohexane; $CH_3(CH_2)_3CHClCH_3$ Molecular mass = 120.6 u
 (c) pentanoic acid; $CH_3(CH_2)_3CO_2H$ Molecular mass = 102.1 u
 (d) 2-methyl-1-propanol; $CH_3CH(CH_3)CH_2OH$ Molecular mass = 74.12 u

78. (a) 2-pentanol; $CH_3CH_2CH_2CH(OH)CH_3$ Molecular mass = 88.15 u
 (b) Propanoic acid; $CH_3CH_2CO_2H$ Molecular mass = 74.08 u
 (c) 1-bromobutane; $CH_3(CH_2)_2CH_2Br$ Molecular mass = 137.0 u
 (d) 3-chlorobutanoic acid; $CH_3CHClCH_2CO_2H$ Molecular mass = 122.6 u

Integrative and Advanced Exercises

79. molar mass = (1×6.941 g Li) + (1×26.9815 g Al) + (2×28.0855 g Si) + (6×15.9994 g O) = 186.09 g/mol

$$\text{no. Li} - 6 \text{ atoms } = 518 \text{ g spodumene} \times \frac{1 \text{ mol spodumene}}{186.09 \text{ g spodumene}} \times \frac{1 \text{ mol Li}}{1 \text{ mol spodumene}} \times \frac{7.40 \text{ mol Li} - 6}{100.00 \text{ mol total Li}}$$

$$\times \frac{6.022 \times 10^{23} \text{ Li} - 6 \text{ atoms}}{1 \text{ mol Li} - 6} = 1.24 \times 10^{23} \text{ Li} - 6 \text{ atoms}$$

80. Determine the mass of each element in the sample.

$$\text{mass Sn} = 0.245 \text{ g SnO}_2 \times \frac{1 \text{ mol SnO}_2}{150.71 \text{ g SnO}_2} \times \frac{1 \text{ mol Sn}}{1 \text{ mol SnO}_2} \times \frac{118.71 \text{ g Sn}}{1 \text{ mol Sn}} = 0.193 \text{ g Sn}$$

$$\text{mass Pb} = 0.115 \text{ g PbSO}_4 \times \frac{1 \text{ mol PbSO}_4}{303.26 \text{ g PbSO}_4} \times \frac{1 \text{ mol Pb}}{1 \text{ mol PbSO}_4} \times \frac{207.2 \text{ g Pb}}{1 \text{ mol Pb}} = 0.0786 \text{ g Pb}$$

$$\text{mass Zn} = 0.246 \text{ g Zn}_2\text{P}_2\text{O}_7 \times \frac{1 \text{ mol Zn}_2\text{P}_2\text{O}_7}{304.72 \text{ g Zn}_2\text{P}_2\text{O}_7} \times \frac{2 \text{ mol Zn}}{1 \text{ mol Zn}_2\text{P}_2\text{O}_7} \times \frac{65.39 \text{ g Zn}}{1 \text{ mol Zn}} = 0.106 \text{ g Zn}$$

Then determine the % of each element in the sample.

$$\% \text{ Sn} = \frac{0.193 \text{ g Sn}}{1.1713 \text{ g brass}} \times 100\% = 16.5\% \text{ Sn} \quad \% \text{ Pb} = \frac{0.0786 \text{ g Pb}}{1.1713 \text{ g brass}} \times 100\% = 6.71\% \text{ Pb}$$

$$\% \text{ Zn} = \frac{0.106 \text{ g Zn}}{1.1713 \text{ g brass}} \times 100\% = 9.05\% \text{ Zn}$$

The % Cu is found by difference. $\%\text{Cu} = 100\% - 16.5\% \text{ Sn} - 6.71\% \text{ Pb} - 9.05\% \text{ Zn} = 67.7\% \text{ Cu}$

81. 1 lb = 16 oz = 453.59237 g or 1 oz = 28.35 g

$$2.52 \text{ oz meat} \times \frac{28.35 \text{ g meat}}{1 \text{ oz meat}} \times \frac{0.10 \text{ g C}_7\text{H}_5\text{O}_2\text{Na}}{100 \text{ g meat}} \times \frac{22.9898 \text{ g Na}}{144.105 \text{ g C}_7\text{H}_5\text{O}_2\text{Na}} \times \frac{1000 \text{ mg Na}}{1 \text{ g Na}} = 11.\underline{4} \text{ mg Na}$$

82. First we determine the amount of each mineral necessary to obtain 1 kg or 1000 g of boron.

$$1000 \text{ g B} \times \frac{1 \text{ mol B}}{10.811 \text{ g B}} \times \frac{1 \text{ mol Na}_2\text{B}_4\text{O}_7 \cdot 4\text{H}_2\text{O}}{4 \text{ mol B}} \times \frac{273.28 \text{ g Na}_2\text{B}_4\text{O}_7 \cdot 4\text{H}_2\text{O}}{1 \text{ mol Na}_2\text{B}_4\text{O}_7 \cdot 4\text{H}_2\text{O}} = 6,319.\underline{5} \text{ g Na}_2\text{B}_4\text{O}_7 \cdot 4\text{H}_2\text{O}$$

$$1000 \text{ g B} \times \frac{1 \text{ mol B}}{10.811 \text{ g B}} \times \frac{1 \text{ mol Na}_2\text{B}_4\text{O}_7 \cdot 4\text{H}_2\text{O}}{4 \text{ mol B}} \times \frac{381.372 \text{ g Na}_2\text{B}_4\text{O}_7 \cdot 10\text{H}_2\text{O}}{1 \text{ mol Na}_2\text{B}_4\text{O}_7 \cdot 4\text{H}_2\text{O}} = 8,819.\underline{1} \text{ g Na}_2\text{B}_4\text{O}_7 \cdot 10\text{H}_2\text{O}$$

The difference between these two masses is the required additional mass. Hence,

$8819.\underline{1}$ g - $6319.\underline{5}$ g = $2499.\underline{6}$ g. Thus, an additional 2.500 kg mass is required.

83. $N_A = \dfrac{9.64853415 \times 10^4 \text{ C}}{1 \text{ mol Ag}} \times \dfrac{1 \text{ mol Ag}}{1 \text{ mol e}^-} \times \dfrac{1 \text{ e}^-}{1.602176462 \times 10^{-19} \text{C}} = \dfrac{6.0221422 \times 10^{23} \text{ e}^-}{\text{mole e}^-}$

84. The hydrocarbon with the greatest ratio of the amount of hydrogen to the amount of carbon will produce the greatest mass of H_2O per gram of CO_2. This hydrocarbon is CH_4, and its produces 2 mol H_2O for every mol CO_2. From this information we can determine the maximum ratio of mass H_2O/mass CO_2.

$$\frac{\text{mass H}_2\text{O}}{\text{mass CO}_2} = \frac{2 \text{ mol H}_2\text{O}}{1 \text{ mol CO}_2} \times \frac{18.02 \text{ g H}_2\text{O}}{1 \text{ mol H}_2\text{O}} \times \frac{1 \text{ mol CO}_2}{44.01 \text{ g CO}_2} = 0.8189 \text{ g H}_2\text{O} / \text{g CO}_2$$

Thus no hydrocarbon exists that yields a greater mass of H_2O than of CO_2.

85. We determine the masses of CO_2 and H_2O produced by burning the C_3H_8.

$$mass_{H_2O} = 6.00 \text{ g } C_3H_8 \times \frac{1 \text{ mol } C_3H_8}{44.0965 \text{ g } C_3H_8} \times \frac{8 \text{ mol H}}{1 \text{ mol } C_3H_8} \times \frac{1 \text{ mol } H_2O}{2 \text{ mol H}} \times \frac{18.0153 \text{ g } H_2O}{1 \text{ mol } H_2O} = 9.80\underline{5} \text{ g } H_2O$$

$$mass_{CO_2} = 6.00 \text{ g } C_3H_8 \times \frac{1 \text{ mol } C_3H_8}{44.0965 \text{ g } C_3H_8} \times \frac{3 \text{ mol C}}{1 \text{ mol } C_3H_8} \times \frac{1 \text{ mol } CO_2}{1 \text{ mol C}} \times \frac{44.01 \text{ g } CO_2}{1 \text{ mol } CO_2} = 17.9\underline{6} \text{ g } CO_2$$

Then, from the masses of CO_2 and H_2O in the unknown compound, we determine the amounts of C and H in that compound and finally its empirical formula.

$$amount \ C = (29.0 - 17.9\underline{6}) \text{g } CO_2 \times \frac{1 \text{ mol } CO_2}{44.01 \text{ g } CO_2} \times \frac{1 \text{ mol C}}{1 \text{ mol } CO_2} = 0.251 \text{ mol C}$$

$$amount \ H = (18.8 - 9.80\underline{5}) \text{ g } H_2O \times \frac{1 \text{ mol } H_2O}{18.0153 \text{ g } H_2O} \times \frac{2 \text{ mol H}}{1 \text{ mol } H_2O} = 0.998\underline{6} \text{ mol H}$$

The empirical formula of the unknown compound is CH_4. The C:H ratio is $0.9986/0.251 = 3.98$

The molecular formula can be calculated by knowing that we have 0.251 moles which accounts for the 4.00 g of hydrocarbon (40 % of 10.0 g). This gives a molar mass of $4.00 \div 0.251 = 15.9$ g/mol. This is nearly the same as the molar mass of the empirical formula CH_4 (16.04 g/mol)

86. (a) We determine the mass of CO_2 produced from the mixture, with x representing the mass of CH_4, and then solve for x.

$$n_{carbon} = x \text{ g } CH_4 \times \frac{1 \text{ mol } CH_4}{16.043 \text{ g } CH_4} \times \frac{1 \text{ mol C}}{1 \text{ mol } CH_4} + (0.732-x) \text{g } C_2H_6 \times \frac{1 \text{ mol } C_2H_6}{30.070 \text{ g } C_2H_6} \times \frac{2 \text{ mol C}}{1 \text{ mol } C_2H_6}$$

$$2.064 \ CO_2 = \left(\frac{x}{16.043} + \frac{2(0.732-x)}{30.070} \right) \text{mol C} \times \frac{1 \text{mol } CO_2}{1 \text{ mol C}} \times \frac{44.010 \text{g } CO_2}{1 \text{ mol } CO_2}$$

$$2.064 \ CO_2 = 2.7433x + 2.142 - 2.9272x = -0.1839x + 2.142$$

$$x = \frac{2.142 - 2.064}{0.1839} = 0.4\underline{2} \text{ g } CH_4$$

$$\%CH_4 = \frac{0.4\underline{1} \text{ g } CH_4}{0.732 \text{ g mixture}} \times 100\% = 57\% \ CH_4 \approx 60\% \ CH_4 \text{ and } 4\underline{3} \% \ C_2H_6 \sim 40\% \ C_2H_6$$

(b) In 100 g of mixture there are the following amounts.

$$amount \ CH_4 = 5\underline{7} \text{ g } CH_4 \times \frac{1 \text{ mol } CH_4}{16.043 \text{ g } CH_4} = 3.\underline{6} \text{ mol } CH_4$$

$$amount \ C_2H_6 = 4\underline{3} \text{ g } C_2H_6 \times \frac{1 \text{ mol } C_2H_6}{30.070 \text{ g } C_2H_6} = 1.\underline{4} \text{ mol } C_2H_6$$

$$mol \ \% \ CH_4 = \frac{3\underline{6} \text{ mol } CH_4}{5\underline{0} \text{ mol total}} \times 100\% = 7\underline{2} \text{ mol\% } CH_4 \text{ and } 2\underline{8} \text{ mol \% } C_2H_6$$

87. (a) Make the assumption that 1 mL H_2O = 1 g H_2O

$$225 \text{ g } H_2O \times \frac{1 \text{ g CHCl}_3}{1,000,000,000 \text{ g } H_2O} \times \frac{1 \text{ mol CHCl}_3}{119.377 \text{ g CHCl}_3} \times \frac{6.022 \times 10^{23} \text{ CHCl}_3}{1 \text{ mol CHCl}_3} = 1.14 \times 10^{15} \text{ CHCl}_3$$

(b) $1.14 \times 10^{15} \text{ CHCl}_3 \times \dfrac{1 \text{ mol CHCl}_3}{6.022 \times 10^{23} \text{ CHCl}_3} \times \dfrac{119.377 \text{ g CHCl}_3}{1 \text{ mol CHCl}_3} = 2.25 \times 10^{-7} \text{ g CHCl}_3$

Alternatively, $225 \text{ g } H_2O \times \dfrac{1 \text{ g CHCl}_3}{1,000,000,000 \text{ g } H_2O} = 2.25 \times 10^{-7} \text{ CHCl}_3$

This amount would not be detected with an ordinary analytical balance. You would require something that was at least 450 times more sensitive.

88. We can determine both the number of moles of M and the mass of M in 0.1131 g MSO_4. Their quotient is the atomic mass of M.

$$\text{mol } M^{2+} = 0.2193 \text{ g BaSO}_4 \times \frac{1 \text{ mol BaSO}_4}{233.39 \text{ g BaSO}_4} \times \frac{1 \text{ mol SO}_4^{2-}}{1 \text{ mol BaSO}_4} \times \frac{1 \text{ mol } M^{2+}}{1 \text{ mol SO}_4^{2-}} = 0.0009396 \text{ mol } M^{2+}$$

$$\text{mass SO}_4^{2-} = 0.0009396 \text{ mol } M^{2+} \times \frac{1 \text{ mol SO}_4^{2-}}{1 \text{ mol } M^{2+}} \times \frac{96.064 \text{ g SO}_4^{2-}}{1 \text{ mol SO}_4^{2-}} = 0.09026 \text{ g SO}_4^{2-}$$

$$\text{mass M} = \text{mass MSO}_4 - \text{mass SO}_4^{2-} = 0.1131 \text{ g MSO}_4 - 0.09026 \text{ g SO}_4^{2-} = 0.0228 \text{ g M}$$

$$\text{atomic mass M} = \frac{\text{mass M}}{\text{moles M}} = \frac{0.0228 \text{ g M}}{0.0009396 \text{ mol M}} = 24.3 \text{ g M/mol}$$

M is the element magnesium.

89. $\text{mass SO}_4^{2-} = 1.511 \text{ g BaSO}_4 \times \dfrac{1 \text{ mol BaSO}_4}{233.39 \text{ g BaSO}_4} \times \dfrac{1 \text{ mol SO}_4^{2-}}{1 \text{ mol BaSO}_4} \times \dfrac{96.064 \text{ g SO}_4^{2-}}{1 \text{ mol SO}_4^{2-}} = 0.6219 \text{ g SO}_4^{2-}$

$$\text{amount M} = 0.006474 \text{ mol SO}_4^{2-} \times \frac{2 \text{ mol } M^{3+}}{3 \text{ mol SO}_4^{2-}} = 0.004316 \text{ mol } M^{3+}$$

$$\text{mass M} = 0.738 \text{ g } M_2(SO_4)_2 - 0.6219 \text{ g SO}_4^{2-} = 0.116 \text{ g M}$$

$$\text{atomic mass of M} = \frac{0.116 \text{ g M}}{0.004316 \text{ mol M}} = 26.9 \text{ g M/mol} \qquad \text{M is the element aluminum.}$$

90. Set up an equation in the usual conversion-factor format, to determine what mass of MS can be obtained from the given mass of M_2O_3. Of course, the mass of MS is not unknown; it is 0.685 g. What is unknown is the atomic mass of the element M; Let's call this x and solve for x.

$$0.685 \text{ g MS} = 0.622 \text{ g } M_2O_3 \times \frac{1 \text{ mol } M_2O_3}{[2x + (3 \times 16.0)] \text{ g } M_2O_3} \times \frac{2 \text{ mol M}}{1 \text{ mol } M_2O_3} \times \frac{1 \text{ mol MS}}{1 \text{ mol M}} \times \frac{(x + 32.1) \text{ g MS}}{1 \text{ mol MS}}$$

$$0.685 \text{ g MS} = \frac{0.622 \times 2 \times (x + 32.1)}{2x + 48.0} \qquad 1.244x + 39.9 = 1.37x + 32.9 \quad 0.126x = 7.0$$

$x = 56$ is the atomic mass of M $\qquad\qquad$ M is element 56, iron (Fe)

91. Let $x = $ mol $MgCl_2$ in the sample, and $y = $ mol NaCl. Set up two equations.

$$0.6110 = \frac{2x \times 35.4527 \text{ g Cl} + 35.4527y \text{ g Cl}}{0.5200 \text{ g sample}}$$

$0.5200 \text{ g} = 95.2104x \text{ g MgCl}_2 + 58.4425y \text{ g NaCl}$

We then solve these two equations for x.

$0.6110 \times 0.5200 = 0.3177 = 70.9054x + 35.4527y$ $\qquad$ $\dfrac{0.3177}{35.4527} = 0.008961 = 2x + y$

$y = 0.008961 - 2x \, 0.5200 = 95.210x + 58.4425(0.008961 - 2x)$

$\quad = 95.210x + 0.5237 - 116.885x$

$0.5200 - 0.5237 = -0.0037 = (95.2104 - 116.885)x = -21.675x \qquad x = \dfrac{0.0037}{21.675}$

$\quad = 1.7 \times 10^{-4} \text{ mol MgCl}_2$

Then determine the value of y.

$y = 0.008961 - 2x = 0.008961 - 2 \times 0.00017 = 0.00862 \text{ mol NaCl}$

$$\text{mass MgCl}_2 = 1.7 \times 10^{-4} \text{ mol MgCl}_2 \times \frac{95.210 \text{ g MgCl}_2}{1 \text{ mol MgCl}_2} = 0.016 \text{ g MgCl}_2$$

$$\text{mass NaCl} = 0.00862 \text{ mol NaCl} \times \frac{58.4425 \text{ g NaCl}}{1 \text{ mol NaCl}} = 0.504 \text{ g NaCl}$$

$$\%\text{MgCl}_2 = \frac{0.016 \text{ g MgCl}_2}{0.5200 \text{ g sample}} \times 100\% = 3.1\% \text{ MgCl}_2 \qquad \% \text{ NaCl} = \frac{0.504 \text{ g NaCl}}{0.5200 \text{ g sample}} \times 100\% = 96.9\% \text{ NaCl}$$

The precision of the calculation is poor because there is only a small % $MgCl_2$ in the mixture, which is a result of the subtraction of one large number from another large number.

92. First we determine the mass of Pb in 2.750 g Pb_3O_4.

$$\text{mass Pb} = 2.750 \text{ g Pb}_3\text{O}_4 \times \frac{1 \text{ mol Pb}_3\text{O}_4}{685.598 \text{ g Pb}_3\text{O}_4} \times \frac{3 \text{ mol Pb}}{1 \text{ mol Pb}_3\text{O}_4} \times \frac{207.2 \text{ g Pb}}{1 \text{ mol Pb}}$$

$$= 2.493 \text{ g Pb}$$

Then we determine the amounts of O and Pb in the second oxide. From these, we determine the empirical formula of the second oxide.

$$\text{amount O} = (2.686 \text{ g} - 2.493 \text{ g}) \, \text{O} \times \frac{1 \text{ mol O}}{16.00 \text{ g O}} = 0.0121 \text{ mol O}$$

$$\text{amount Pb} = 2.493 \text{ g Pb} \times \frac{1 \text{ mol Pb}}{207.2 \text{ g Pb}} = 0.0120 \text{ mol Pb}$$

Thus, the empirical formula of the second oxide is PbO.

93. If we determine the mass of anhydrous $ZnSO_4$ in the hydrate, we then can determine the mass of water, and the formula of the hydrate.

$$\text{mass } ZnSO_4 = 0.8223 \text{ g } BaSO_4 \times \frac{1 \text{ mol } BaSO_4}{233.391 \text{ g}} \times \frac{1 \text{ mol } ZnSO_4}{1 \text{ mol } BaSO_4} \times \frac{161.454 \text{ g } ZnSO_4}{1 \text{ mol } ZnSO_4}$$

$$= 0.5688 \text{ g } ZnSO_4$$

The water present in the hydrate is obtained by difference.

mass H_2O = 1.013 g hydrate -0.5688 g $ZnSO_4$ = 0.444 g H_2O

The hydrate's formula is determined by a method similar to that for obtaining an empirical formula.

$$\text{amt. } ZnSO_4 = 0.5688 \text{ g} \times \frac{1 \text{ mol } ZnSO_4}{161.454 \text{ mol } ZnSO_4} = 0.003523 \text{ mol } ZnSO_4 \div 0.003523 \longrightarrow 1.00 \text{ mol } ZnSO_4$$

$$\text{amt. } H_2O = 0.444 \text{ g} \times \frac{1 \text{ mol } H_2O}{18.0153 \text{ g } H_2O} = 0.02465 \text{ mol } H_2O \div 0.003523 \longrightarrow 7.00 \text{ mol } H_2O$$

Thus, the formula of the hydrate is $ZnSO_4 \cdot 7H_2O$

94. $\dfrac{1.552 \text{ g MI}}{1.186 \text{ g I}} \times \dfrac{126.904 \text{ g I}}{1 \text{ mol I}} \times \dfrac{1 \text{ mol I}}{1 \text{ mol MI}} = \dfrac{166.1 \text{ g MI}}{1 \text{ mol MI}}$ (This is the molar mass of MI)

Subtract the mass of 1 mol of I to obtain the molar mass of M

molar mass M = $(166.1 - 126.904)$ g mol^{-1} = 39.2 g mol^{-1}

The cation is probably K (39.0983 g mol^{-1})

Alternatively

Find mass of M in sample: 1.552 g MI $-$ 1.186 g I = 0.366 g M

$$\dfrac{0.366 \text{ g M}}{1.186 \text{ g I}} \times \dfrac{126.904 \text{ g I}}{1 \text{ mol I}} \times \dfrac{1 \text{ mol I}}{1 \text{ mol M}} = 39.2 \text{ g mol}^{-1}$$

95. 13 atoms $\times \dfrac{15.38 \text{ atoms E}}{100 \text{ atoms in formula unit}} = \dfrac{1.999 \text{ atom E}}{\text{formula unit}}$ $\therefore H_xE_2O_z$ (Note: x + z = 11)

34.80 % E by mass, hence, 65.20% H and O by mass

178 u $\times$ 0.3480 E = 61.944 u for 2 atoms of E, $\therefore$ E = 30.972 u Probably P (30.9738 u)

H and O in formula unit = 178 u $-$ 30.972 u = 116 u

x + z = 11 or x = 11 $-$ z and x(1.00794 u) + z(15.9994 u) = 116 u

Substitute and solve for z: (11 $-$ z)(1.00794 u) + z(15.9994 u) = 116 u

11.08734 u $-$ 1.00794 u(z) + 15.9994 u(z) = 116 u Divide through by u and collect terms

105 = 14.9915(z) or z = 7 and x = 11- z = 11- 7 = 4.

Therefore the formula is $H_4P_2O_7$ (As a check, 13 atoms and 177.975 u ~ 178 u)

96. First find the mass of carbon, hydrogen, chlorine and oxygen. From the molar ratios, we determine the molecular formula.

$$2.094 \text{ g CO}_2 \times \frac{1 \text{ mol CO}_2}{44.001 \text{ g CO}_2} \times \frac{1 \text{ mol C}}{1 \text{ mol CO}_2} = 0.04759 \text{ mol C} \times \frac{12.011 \text{ g C}}{1 \text{ mol C}} = 0.5716 \text{ g C}$$

$$0.286 \text{ g H}_2\text{O} \times \frac{1 \text{ mol H}_2\text{O}}{18.0153 \text{ g H}_2\text{O}} \times \frac{2 \text{ mol H}}{1 \text{ mol H}_2\text{O}} = 0.03175 \text{ mol H} \times \frac{1.00794 \text{ g H}}{1 \text{ mol H}} = 0.0320 \text{ g H}$$

$$\text{Moles of chlorine} = \frac{mol \text{ C}}{2} = \frac{0.04759}{2} = 0.02380 \text{ mol Cl}$$

$$\text{mass of Cl} = 0.02380 \text{ mol Cl} \times \frac{35.4527 \text{g Cl}}{1 \text{ mol Cl}} = 0.8436 \text{ g Cl}$$

Mass of oxygen obtained by difference. $1.510 \text{ g} - 0.8438 \text{ g} - 0.5716 \text{ g} - 0.0320 \text{ g} = 0.063 \text{ g O}$

$$\text{Moles of oxygen} = 0.063 \text{ g O} \times \frac{1 \text{ mol O}}{15.9994 \text{ g O}} = 0.00394 \text{ mol O}$$

Divide the number of moles of each element by 0.00394 to give an empirical formula of $C_{12.1}H_{8.06}Cl_{6.04}O_{1.00}$ owing to the fact that the oxygen mass is obtained by difference, and it has only two significant digits and thus a higher degree of uncertainty

The empirical formula is $C_{12}H_8Cl_6O$, with a molecular mass of 381 u has the same molecular mass of the molecular formula. Hence, this empirical formula is also the molecular formula.

97. 1.271 g Na_2SO_4 absorbs 0.387 g H_2O

$$\text{mass}_{Na_2SO_4 \cdot 10 H_2O} = 0.387 \text{ g H}_2\text{O} \times \frac{1 \text{ mol H}_2\text{O}}{18.0153 \text{ g H}_2\text{O}} \times \frac{1 \text{ mol Na}_2SO_4 \cdot 10H_2O}{10 \text{ mol H}_2\text{O}} \times \frac{322.196 \text{ g Na}_2SO_4 \cdot 10H_2O}{1 \text{ mol Na}_2SO_4 \cdot 10H_2O}$$

$$= 0.692 \text{ g Na}_2SO_4 \cdot 10H_2O$$

$$\text{mass percent Na}_2SO_4 \cdot 10 H_2O = \frac{0.692 \text{ g}}{(1.271 \text{ g} + 0.387 \text{ g})} \times 100\% = 41.7\%$$

98. Let X = molar mass of Bi and Y = moles of Bi_2O_3

Molar mass of $Bi(C_6H_5)_3 = X + (18 \times 12.011 \text{ g mol}^{-1} + 15 \times 1.008 \text{ g mol}^{-1}) = X + 231.318 \text{ g mol}^{-1}$
Molar mass of $Bi_2O_3 = 2X + (3 \times 15.9994 \text{ g mol}^{-1} = 2X + 47.998 \text{ g mol}^{-1}$
Consider the reaction: $2 \text{ Bi}(C_6H_5)_3 \rightarrow Bi_2O_3$
$5.610 \text{ g} = 2XY + (2Y)231.318 \text{ g mol}^{-1}$ and $2.969 \text{ g} = 2XY + (Y)47.998 \text{ g mol}^{-1}$
Rearrange $5.610 \text{ g} = 2XY + (Y)231.318 \text{ g mol}^{-1}$ to $2XY = 5.610 \text{ g} - (2Y)231.318 \text{ g mol}^{-1}$
Substitute for 2XY in $2.969 = 2XY + (Y)47.998 \text{ g mol}^{-1}$
$2.969 \text{ g} = 5.610 \text{ g} - (2Y)231.318 \text{ g mol}^{-1} + (Y)47.998 \text{ g mol}^{-1} = 5.610 \text{ g} - (Y)414.638 \text{ g mol}^{-1}$
Collect terms and solve for Y.
$5.610 \text{ g} - 2.969 \text{ g} = (Y)414.638 \text{ g mol}^{-1} = 2.641 \text{ g}$
$Y = 2.641 \text{ g} \div 414.638 \text{ g mol}^{-1} = 0.0063694 \text{ mol}$
Substitute Y in $2.969 \text{ g} = 2XY + (Y)47.998 \text{ g mol}^{-1}$ and solve for X, the molar mass of Bi
$2.969 \text{ g} = 2X(0.0063694 \text{ mol}) + (0.0063694 \text{ mol}) 47.998 \text{ g mol}^{-1}$

$$X = \frac{2.969 \text{ g} - (0.0063694 \text{ mol}) 47.998 \text{ g mol}^{-1}}{(2)0.0063694 \text{ mol}} = 209.1 \text{ g mol}^{-1} \quad (\text{Actually it is } 208.98 \text{ g mol}^{-1})$$

FEATURE PROBLEMS

99. **(a)** "5-10-5" fertilizer contains 5.00 g N (that is, 5.00% N), 10.00 g P_2O_5, and 5.00 g K_2O in 100.00 g fertilizer. We convert the last two numbers into masses of the two elements.

(1) $\%P = 10.00\% P_2O_5 \times \dfrac{1\,mol\,P_2O_5}{141.9\,g\,P_2O_5} \times \dfrac{2\,mol\,P}{1\,mol\,P_2O_5} \times \dfrac{30.97\,g\,P}{1\,mol\,P} = 4.37\%\,P$

(2) $\%K = 5.00\% K_2O \times \dfrac{1\,mol\,K_2O}{94.20\,g\,K_2O} \times \dfrac{2\,mol\,K}{1\,mol\,K_2O} \times \dfrac{39.10\,g\,K}{1\,mol\,K} = 4.15\%\,K$

(b) First, we determine %P and then convert it to $\% P_2O_5$, given that 10.0% P_2O_5 is equivalent to 4.37% P.

(1) $\% \, P_2O_5 = \dfrac{2\,mol\,P}{1\,mol\,Ca(H_2PO_4)_2} \times \dfrac{30.97\,g\,P}{1\,mol\,P} \times \dfrac{1\,mol\,Ca(H_2PO_4)_2}{234.05\,g\,Ca(H_2PO_4)_2} \times 100\%$

$\times \dfrac{10.0\% P_2O_5}{4.37\% P} = 60.6\%\,P_2O_5$

(2) $\% \, P_2O_5 = \dfrac{1\,mol\,P}{1\,mol\,(NH_4)_2 HPO_4} \times \dfrac{30.97\,g\,P}{1\,mol\,P} \times \dfrac{1\,mol\,(NH_4)_2 HPO_4}{132.06\,g\,(NH_4)_2 HPO_4} \times 100\%$

$\times \dfrac{10.0\% P_2O_5}{4.37\% P} = 53.7\%\,P_2O_5$

(c) If the mass ratio of $(NH_4)_2HPO_4$ to KCl is set at 5.00:1.00, then for every 5.00 g of $(NH_4)_2HPO_4$ in the mixture there must be 1.00 g of KCl. Let's start by finding the %N, %P and %K for the fertilizer mixture.

$\%N(by\,mass) = \dfrac{2\,mol\,N}{1\,mol\,(NH_4)_2 HPO_4} \times \dfrac{1\,mol\,(NH_4)_2 HPO_4}{132.06\,g\,(NH_4)_2 PO_4} \times \dfrac{14.0067\,g\,N}{1\,mol\,N} \times \dfrac{5.00\,g\,(NH_4)_2 HPO_4}{6.00\,g\,mixture} \times 100\%$

$= 17.7\%\,N$

$\%P(by\,mass) = \dfrac{1\,mol\,P}{1\,mol\,(NH_4)_2 HPO_4} \times \dfrac{1\,mol\,(NH_4)_2 HPO_4}{132.06\,g\,(NH_4)_2 HPO_4} \times \dfrac{30.9738\,g\,P}{1\,mol\,P} \times \dfrac{5.00\,g\,(NH_4)_2 HPO_4}{6.00\,g\,of\,mixture} \times 100\%$

$= 19.5\%\,P$

$\%K(by\,mass) = \dfrac{1\,mol\,K}{1\,mol\,KCl} \times \dfrac{1\,mol\,KCl}{74.55\,g\,KCl} \times \dfrac{39.0983\,g\,K}{1\,mol\,K} \times \dfrac{1.00\,g\,KCl}{6.00\,g\,mixture} \times 100\%$

$= 8.74\%\,K$

Next we convert %P to P_2O_5 and %K to $\%K_2O$

$\%P_2O_5 = 19.5\%\,P \times \dfrac{10.0\,\% P_2O_5}{4.37\,\% P} = 44.\underline{6}\%\,P_2O_5$

$\%K_2O = 8.\underline{74}\%\,K \times \dfrac{5.00\% K_2O}{4.15\%\,K} = 10.\underline{5}\%\,K_2O$

Thus, the combination of 5.00 g $(NH_4)_2HPO_4$ with 1.00 g KCl affords a "17.7-44.6-10.5" fertilizer, that is, 17.7% N, a percentage of phosphorus expressed as 44.6% P_2O_5, and a percentage of potassium expressed as 10.5% K_2O.

(d)   A "5-10-5" fertilizer must possess the mass ratio 5.00 g N: 4.37 g P: 4.15 g K per 100 g of fertilizer. Thus a "5-10-5" fertilizer requires an N:P relative mass ratio of 5.00 gN:4.37 g P = 1.00 gN:0.874 g P. Note specifically that the fertilizer has a somewhat *greater* mass of N than of P.

If all of the N and P in the fertilizer comes solely from $(NH_4)_2HPO_4$, then the atom ratio of N relative to P will remain fixed at 2 N:1 P. Whether or not an inert non-fertilizing filler is present in the mix is immaterial. The relative N:P mass ratio is (2×14.01) g N:30.97 g P, that is, 0.905 g N:1.00 g P. Note specifically that $(NH_4)_2HPO_4$ has a somewhat *lesser* mass of N than of P. Clearly, it is impossible to make a "5-10-5" fertilizer if the only fertilizing components are $(NH_4)_2HPO_4$ and KCl.

100.   **(a)**   First calculate the mass of water that was present in the hydrate prior to heating.
Mass of H_2O = 2.574 g $CuSO_4 \cdot x\ H_2O$ - 1.647 g $CuSO_4$ = 0.927 g H_2O
Next we need to find the number of moles of anhydrous copper(II) sulfate and water that were initially present together in the original hydrate sample.

$$\text{Moles of } H_2O = 0.927 \text{ g } H_2O \times \frac{1 \text{mol } H_2O}{18.015 \text{ g } H_2O} = 0.05146 \text{ moles of water}$$

The empirical formula is obtained by dividing the number of moles of water by the number of moles of $CuSO_4$ (x = ratio of moles of water to moles of $CuSO_4$)

$$x = \frac{0.05146 \text{ moles } H_2O}{0.01032 \text{ moles } CuSO_4} = 4.99 \sim 5 \text{ The empirical formula is } CuSO_4 \cdot 5\ H_2O.$$

(b)   Mass of water present in hydrate = 2.574 g - 1.833 g = 0.741 g H_2O

$$\text{moles of water} = 0.741 \text{ g } H_2O \times \frac{1 \text{mol } H_2O}{18.015 \text{ g } H_2O} = 0.0411 \text{ moles of water}$$

Mass of $CuSO_4$ present in hydrate = 1.833 g $CuSO_4$

$$\text{moles of } CuSO_4 = 1.833 \text{ g } CuSO_4 \times \frac{1 \text{mol } CuSO_4}{159.61 \text{ g } CuSO_4} = 0.0115 \text{ mol } CuSO_4$$

The empirical formula is obtained by dividing the number of moles of water by the number of moles of $CuSO_4$ (x = ratio of moles of water to moles of $CuSO_4$)

$$x = \frac{0.0411 \text{ moles } H_2O}{0.0115 \text{ moles } CuSO_4} = 3.58 \sim 4.$$

Since the hydrate has not been completely dehydrated, there is no problem with obtaining non-integer "garbage" values.

So, the empirical formula is $CuSO_4 \bullet 4H_2O$.

(c)   When copper(II) sulfate is strongly heated, it decomposes to give $SO_3(g)$ and $CuO(s)$. The black residue formed at 1000 °C in this experiment is probably CuO. The empirical formula for copper(II) oxide is CuO. Let's calculate the percentages of Cu and O by mass for CuO:

$$\text{Mass percent copper} = \frac{63.546 \text{ g Cu}}{79.545 \text{ g CuO}} \times 100 \% = 79.89 \% \text{ by mass Cu}$$

$$\text{Mass percent oxygen} = \frac{15.9994 \text{ g O}}{79.545 \text{ g CuO}} \times 100 \% = 20.11 \% \text{ by mass O}$$

The number of moles of CuO formed (by reheating to 1000 °C)

$$= 0.812 \text{ g CuO} \times \frac{1 \text{ mol CuO}}{79.545 \text{ g CuO}} = 0.0102 \text{ moles of CuO}$$

This is very close to the number of moles of anhydrous $CuSO_4$ formed at 400. °C. Thus, it would appear that upon heating to 1000 °C, the sample of $CuSO_4$ was essentially completely converted to CuO.

101.   **(a)**   The formula for stearic acid, obtained from the molecular model, is $CH_3(CH_2)_{16}CO_2H$. The number of moles of stearic acid in 10.0 grams is

$$= 10.0 \text{ g stearic acid} \times \frac{1 \text{ mol stearic acid}}{284.48 \text{ g stearic acid}} = 3.51\underline{5} \times 10^{-2} \text{ mol of stearic acid.}$$

The layer of stearic acid is one molecule thick. According to the figure provided with the question, each stearic acid molecule has a cross-sectional area of ~0.22 nm². In order to find the stearic acid coverage in square meters, we must multiply the total number of stearic acid molecules by the cross-sectional area for an individual stearic acid molecule. The number of stearic acid molecules is:

$$= 3.51\underline{5} \times 10^{-2} \text{ mol of stearic acid} \times \frac{6.022 \times 10^{23} \text{ molecules}}{1 \text{ mol of stearic acid}} = 2.11\underline{7} \times 10^{22} \text{ molecules}$$

$$\text{Area in m}^2 = 2.11\underline{7} \times 10^{22} \text{ molecules of stearic acid} \times \frac{0.22 \text{ nm}^2}{\text{molecule}} \times \frac{(1 \text{ m})^2}{(1 \times 10^9 \text{ nm})^2}$$

The area in m² = 4657 m² or 4.7×10^3 m² (with correct number of sig. fig.)

(b)   The density for stearic acid is 0.85 g cm⁻³. Thus, 0.85 grams of stearic acid occupies 1 cm³. Find the number of moles of stearic acid in 0.85 g of stearic acid

$$= 0.85 \text{ grams of stearic acid} \times \frac{1 \text{ mol stearic acid}}{284.48 \text{ g stearic acid}} = 3.0 \times 10^{-3} \text{ mol of stearic}$$

acid. This number of moles of acid occupies 1 cm³ of space. So, the number of stearic acid molecules in 1 cm³

$$= 3.0 \times 10^{-3} \text{ mol of stearic acid} \times \frac{6.022 \times 10^{23} \text{ molecules}}{1 \text{ mol of stearic acid}}$$

$$= 1.8 \times 10^{21} \text{ stearic acid molecules.}$$

Thus, the volume for a single stearic acid molecule in nm³

$$= 1 \text{ cm}^3 \times \frac{1}{1.8 \times 10^{21} \text{ molecules stearic acid}} \times \frac{(1.0 \times 10^7 \text{ nm})^3}{(1 \text{ cm})^3} = 0.55\underline{6} \text{ nm}^3$$

The volume of a rectangular column is simply its area of the base multiplied by its height (i.e. V = area of base (in nm^2) × height (in nm)).

So, the average height of a stearic acid molecule $= \dfrac{0.556 \text{ nm}^3}{0.22 \text{ nm}^2} = 2.5 \text{ nm}$

(c)   The density for oleic acid = 0.895 g mL^{-1}. So, the concentration for oleic acid is

$$= \frac{0.895 \text{ g acid}}{10.00 \text{ mL}} = 0.0895 \text{ g mL}^{-1} \text{ (solution 1)}$$

This solution is then divided by ten, three more times to give a final concentration of $8.9\underline{5} \times 10^{-5}$ g mL^{-1}. A 0.10 mL sample of this solution contains:

$$= \frac{8.95 \times 10^{-5} \text{ g acid}}{1.00 \text{ mL}} \times 0.10 \text{ mL} = 8.9\underline{5} \times 10^{-6} \text{ g of acid.}$$

The number of acid molecules $= 85 \text{ cm}^2 \times \dfrac{1}{4.6 \times 10^{-15} \text{cm}^2 \text{ per molecule}}$

$$= 1.8\underline{5} \times 10^{16} \text{ oleic acid molecules.}$$

So, 8.95×10^{-6} g of oleic acid corresponds to 1.85×10^{16} oleic acid molecules.

The molar mass for oleic acid, $C_{18}H_{34}O_2$, is 282.47 g mol^{-1}.

The number of moles of oleic acid is

$$= 8.9\underline{5} \times 10^{-6} \text{ g} \times \frac{1 \text{ mol oleic acid}}{282.47 \text{ g}} = 3.1\underline{7} \times 10^{-8} \text{ mol}$$

So, Avogadro's number here would be equal to:

$$= \frac{1.8\underline{5} \times 10^{16} \text{ oleic acid molecules}}{3.1\underline{7} \times 10^{-8} \text{ oleic acid moles}} = 5.8 \times 10^{23} \text{ molecules per mole of oleic acid.}$$

CHAPTER 4
CHEMICAL REACTIONS

PRACTICE EXAMPLES

1A **(a)** Unbalanced reaction: $H_3PO_4(aq) + CaO(s) \rightarrow Ca_3(PO_4)_2(aq) + H_2O(l)$
Balance Ca & PO_4^{3-}: $2\,H_3PO_4(aq) + 3\,CaO(s) \rightarrow Ca_3(PO_4)_2(aq) + H_2O(l)$
Balance H atoms: $2\,H_3PO_4(aq) + 3\,CaO(s) \rightarrow Ca_3(PO_4)_2(aq) + 3\,H_2O(l)$
Self Check: $6\,H + 2\,P + 11\,O + 3\,Ca \rightarrow 6\,H + 2\,P + 11\,O + 3\,Ca$

    (b) Unbalanced reaction: $C_3H_8(g) + O_2(g) \rightarrow CO_2(g) + H_2O(g)$
Balance C & H: $C_3H_8(g) + O_2(g) \rightarrow 3\,CO_2(g) + 4\,H_2O(g)$
Balance O atoms: $C_3H_8(g) + 5\,O_2(g) \rightarrow 3\,CO_2(g) + 4\,H_2O(g)$
Self Check: $3\,C + 8\,H + 10\,O \rightarrow 3\,C + 8\,H + 10\,O$

1B **(a)** Unbalanced reaction: $NH_3(g) + O_2(g) \rightarrow NO_2(g) + H_2O(g)$
Balance N and H: $NH_3(g) + O_2(g) \rightarrow NO_2(g) + 3/2\,H_2O(g)$
Balance O atoms: $NH_3(g) + 7/4\,O_2(g) \rightarrow NO_2(g) + 3/2\,H_2O(g)$
Multiply by 4 (whole #): $4\,NH_3(g) + 7\,O_2(g) \rightarrow 4\,NO_2(g) + 6\,H_2O(g)$
Self Check: $4\,N + 12\,H + 14\,O \rightarrow 4\,N + 12\,H + 14\,O$

    (b) Unbalanced reaction: $NO_2(g) + NH_3(g) \rightarrow N_2(g) + H_2O(g)$
Balance H atoms: $NO_2(g) + 2\,NH_3(g) \rightarrow N_2(g) + 3\,H_2O(g)$
Balance O atoms: $3/2\,NO_2(g) + 2\,NH_3(g) \rightarrow N_2(g) + 3\,H_2O(g)$
Balance N atoms: $3/2\,NO_2(g) + 2\,NH_3(g) \rightarrow 7/4\,N_2(g) + 3\,H_2O(g)$
Multiply by 4 (whole #) $6\,NO_2(g) + 8\,NH_3(g) \rightarrow 7\,N_2(g) + 12\,H_2O(g)$
Self Check: $14\,N + 24\,H + 12\,O \rightarrow 14\,N + 24\,H + 12\,O$

2A Unbalanced reaction: $HgS(s) + CaO(s) \rightarrow CaS(s) + CaSO_4(s) + Hg(l)$
Balance 0 atoms: $HgS(s) + 4\,CaO(s) \rightarrow CaS(s) + CaSO_4(s) + Hg(l)$
Balance Ca atoms: $HgS(s) + 4\,CaO(s) \rightarrow 3\,CaS(s) + CaSO_4(s) + Hg(l)$
Balance S atoms: $4\,HgS(s) + 4\,CaO(s) \rightarrow 3\,CaS(s) + CaSO_4(s) + Hg(l)$
Balance Hg atoms: $4\,HgS(s) + 4\,CaO(s) \rightarrow 3\,CaS(s) + CaSO_4(s) + 4\,Hg(l)$
Self Check: $4\,Hg + 4\,S + 4\,O + 4\,Ca \rightarrow 4\,Hg + 4\,S + 4\,O + 4\,Ca$

2B Unbalanced reaction: $C_7H_6O_2S(l) + O_2(g) \rightarrow CO_2(g) + H_2O(l) + SO_2(g)$
Balance C atoms: $C_7H_6O_2S(l) + O_2(g) \rightarrow 7\,CO_2(g) + H_2O(l) + SO_2(g)$
Balance S atoms: $C_7H_6O_2S(l) + O_2(g) \rightarrow 7\,CO_2(g) + H_2O(l) + SO_2(g)$
Balance H atoms: $C_7H_6O_2S(l) + O_2(g) \rightarrow 7\,CO_2(g) + 3\,H_2O(l) + SO_2(g)$
Balance O atoms: $C_7H_6O_2S(l) + 8.5\,O_2(g) \rightarrow 7\,CO_2(g) + 3\,H_2O(l) + SO_2(g)$
Multiply by 2 (whole #): $2\,C_7H_6O_2S(l) + 17\,O_2(g) \rightarrow 14\,CO_2(g) + 6\,H_2O(l) + 2\,SO_2(g)$
Self Check: $14\,C + 12\,H + 2\,S + 38\,O \rightarrow 14\,C + 12\,H + 2\,S + 38\,O$

3A   The balanced chemical equation provides the factor needed to convert from moles $KClO_3$

to moles O_2. Amount $O_2 = 1.76 \, \text{mol} \, KClO_3 \times \dfrac{3 \, \text{mol} \, O_2}{2 \, \text{mol} \, KClO_3} = 2.64 \, \text{mol} \, O_2$

3B   First, find the molar mass of Ag_2O.

$\left(2 \, \text{mol} \, Ag \times 107.87 \, \text{g} \, Ag \right) + 16.00 \, \text{g} \, O = 231.74 \, \text{g} \, Ag_2O / \text{mol}$

amount $Ag = 1.00 \, \text{kg} \, Ag_2O \times \dfrac{1000 \, \text{g}}{1.00 \, \text{kg}} \times \dfrac{1 \, \text{mol} \, Ag_2O}{231.74 \, \text{g} \, Ag_2O} \times \dfrac{2 \, \text{mol} \, Ag}{1 \, \text{mol} \, Ag_2O} = 8.63 \, \text{mol} \, Ag$

4A   The balanced chemical equation provides the factor to convert from amount of Mg to amount of Mg_3N_2. First we must determine the molar mass of Mg_3N_2.

molar mass $= \left(3 \, \text{mol} \, Mg \times 24.305 \, \text{g} \, Mg \right) + \left(2 \, \text{mol} \, N \times 14.007 \, \text{g} \, N \right) = 100.93 \, \text{g} \, Mg_3N_2$

mass $Mg_3N_2 = 3.82 \, \text{g} \, Mg \times \dfrac{1 \, \text{mol} \, Mg}{24.31 \, \text{g} \, Mg} \times \dfrac{1 \, \text{mol} \, Mg_3N_2}{3 \, \text{mol} \, Mg} \times \dfrac{100.93 \, \text{g} \, Mg_3N_2}{1 \, \text{mol} \, Mg_3N_2} = 5.29 \, \text{g} \, Mg_3N_2$

4B   The pivotal conversion is from $H_2 \, (g)$ to $CH_3OH \, (l)$. For this we use the balanced equation, which requires that we use the amounts in moles of both substances. The solution involves converting to and from amounts, using molar masses.

mass $H_2 \, (g) = 1.00 \, \text{kg} \, CH_3OH(l) \times \dfrac{1000 \, \text{g}}{1 \, \text{kg}} \times \dfrac{1 \, \text{mol} \, CH_3OH}{32.04 \, \text{g} \, CH_3OH} \times \dfrac{2 \, \text{mol} \, H_2}{1 \, \text{mol} \, CH_3OH} \times \dfrac{2.016 \, \text{g} \, H_2}{1 \, \text{mol} \, H_2}$

mass $H_2 \, (g) = 126 \, \text{g} \, H_2$

5A   The equation for the cited reaction is: $2 \, NH_3 \, (g) + 1.5 \, O_2 \, (g) \longrightarrow N_2 \, (g) + 3 \, H_2O \, (l)$

The pivotal conversion is from one substance to another, in moles with the balanced chemical equation providing the conversion factor.

mass $NH_3 \, (g) = 1.00 \, \text{g} \, O_2 \, (g) \times \dfrac{1 \, \text{mol} \, O_2}{32.00 \, \text{g} \, O_2} \times \dfrac{2 \, \text{mol} \, NH_3}{1.5 \, \text{mol} \, O_2} \times \dfrac{17.0305 \, \text{g} \, NH_3}{1 \, \text{mol} \, H_2} = 0.710 \, \text{g} \, NH_3$

5B   The equation for the combustion reaction is: $C_8H_{18} \, (l) + \dfrac{25}{2} O_2 \, (g) \rightarrow 8 \, CO_2 \, (g) + 9 \, H_2O \, (l)$

mass $O_2 = 1.00 \, \text{g} \, C_8H_{18} \times \dfrac{1 \, \text{mol} \, C_8H_{18}}{114.23 \, \text{g} \, C_8H_{18}} \times \dfrac{12.5 \, \text{mol} \, O_2}{1 \, \text{mol} \, C_8H_{18}} \times \dfrac{32.00 \, \text{g} \, O_2}{1 \, \text{mol} \, O_2} = 3.50 \, \text{g} \, O_2 \, (g)$

6A   We must convert mass $H_2 \rightarrow$ amount of $H_2 \rightarrow$ amount of $Al \rightarrow$ mass of $Al \rightarrow$ mass of alloy $\rightarrow$ volume of alloy. The calculation is performed as follows: each arrow in the preceding sentence requires a conversion factor.

$V_{\text{alloy}} = 1.000 \, \text{g} \, H_2 \times \dfrac{1 \, \text{mol} \, H_2}{2.016 \, \text{g} \, H_2} \times \dfrac{2 \, \text{mol} \, Al}{3 \, \text{mol} \, H_2} \times \dfrac{26.98 \, \text{g} \, Al}{1 \, \text{mol} \, Al} \times \dfrac{100.0 \, \text{g} \, \text{alloy}}{93.7 \, \text{g} \, Al} \times \dfrac{1 \, \text{cm}^3 \text{alloy}}{2.85 \, \text{g} \, \text{alloy}}$

Volume of alloy $= 3.34 \, \text{cm}^3 \, \text{alloy}$

6B In the example, $0.207\,\text{g}\ H_2$ is collected from 1.97 g alloy; the alloy is 6.3% Cu by mass. This information provides the conversion factors we need.

$$\text{mass Cu} = 1.31\,\text{g}\ H_2 \times \frac{1.97\,\text{g alloy}}{0.207\,\text{g}\ H_2} \times \frac{6.3\,\text{g Cu}}{100.0\,\text{g alloy}} = 0.79\,\text{g Cu}$$

Notice that we do not have to consider each step separately. We can simply use values produced in the course of the calculation as conversion factors.

7A The cited reaction is $2\,Al(s) + 6\,HCl(aq) \rightarrow 2\,AlCl_3\,(aq) + 3\,H_2\,(g)$. The HCl(aq) solution has a density of 1.14 g/mL and contains 28.0% HCl. We need to convert between the substances HCl and H_2; the important conversion factor comes from the balanced chemical equation. The sequence of conversions is: volume of HCl(aq) $\rightarrow$ mass of HCl(aq) $\rightarrow$ mass of pure HCl $\rightarrow$ amount of HCl $\rightarrow$ amount of H_2 $\rightarrow$ mass of H_2. In the calculation below, each arrow in the sequence is replaced by a conversion factor.

$$\text{mass } H_2 = 0.05\,\text{mL HCl(aq)} \times \frac{1.14\,\text{g sol}}{1\,\text{mL soln}} \times \frac{28.0\,\text{g HCl}}{100.0\,\text{g soln}} \times \frac{1\,\text{mol HCl}}{36.46\,\text{g HCl}} \times \frac{3\,\text{mol }H_2}{6\,\text{mol HCl}} \times \frac{2.016\,\text{g }H_2}{1\,\text{mol }H_2}$$

$$\text{mass } H_2 = 4 \times 10^{-4}\,\text{g }H_2\,(g) = 0.4\,\text{mg }H_2\,(g)$$

7B Density is necessary to determine the mass of the vinegar, and then the mass of acetic acid.

$$\text{mass } CO_{2\,(g)} = 5.00\,\text{mL vinegar} \times \frac{1.01\,\text{g}}{1\,\text{mL}} \times \frac{0.040\,\text{g acid}}{1\,\text{g vinegar}} \times \frac{1\,\text{mol } HC_2H_3O_2}{60.05\,\text{g } HC_2H_3O_2} \times \frac{1\,\text{mol } CO_2}{1\,\text{mol } HC_2H_3O_2} \times \frac{44.01\,\text{g } CO_2}{1\,\text{mol } CO_2}$$

$$= 0.15\,\text{g } CO_2$$

8A Determine the amount in moles of acetone and the volume in liters of the solution.

$$\text{molarity of acetone} = \frac{22.3\,\text{g}\,(CH_3)_2\,CO \times \dfrac{1\,\text{mol}\,(CH_3)_2\,CO}{58.08\,\text{g}\,(CH_3)_2\,CO}}{1.25\,\text{L soln}} = 0.307\,\text{M}$$

8B The molar mass of acetic acid, $HC_2H_3O_2$, is 60.05 g/mol. We begin with the quantity of acetic acid in the numerator and that of the solution in the denominator, and transform to the appropriate units for each.

$$\text{molarity} = \frac{15.0\,\text{mL } HC_2H_3O_2}{500.0\,\text{mL soln}} \times \frac{1000\,\text{mL}}{1\,\text{L soln}} \times \frac{1.048\,\text{g } HC_2H_3O_2}{1\,\text{mL } HC_2H_3O_2} \times \frac{1\,\text{mol } HC_2H_3O_2}{60.05\,\text{g } HC_2H_3O_2} = 0.524\,\text{M}$$

9A The molar mass of $NaNO_3$ is 84.99 g/mol. We recall that "M" stands for "mol /L soln."

$$\text{mass } NaNO_3 = 125\,\text{mL soln} \times \frac{1\,\text{L}}{1000\,\text{mL}} \times \frac{10.8\,\text{mol } NaNO_3}{1\,\text{L soln}} \times \frac{84.99\,\text{g } NaNO_3}{1\,\text{mol } NaNO_3} = 115\,\text{g } NaNO_3$$

9B We begin by determining the molar mass of $Na_2SO_4 \cdot 10H_2O$. The amount of solute needed is computed from the concentration and volume of the solution.

$$\text{mass } Na_2SO_4 \cdot 10H_2O = 355\,\text{mL soln} \times \frac{1\,\text{L}}{1000\,\text{mL}} \times \frac{0.445\,\text{mol } Na_2SO_4}{1\,\text{L soln}} \times \frac{1\,\text{mol } Na_2SO_4 \cdot 10H_2O}{1\,\text{mol } Na_2SO_4}$$

$$\times \frac{322.21\,\text{g } Na_2SO_4 \cdot 10H_2O}{1\,\text{mol } Na_2SO_4 \cdot 10H_2O} = 50.9\,\text{g } Na_2SO_4 \cdot 10H_2O$$

10A The amount of solute in the concentrated solution doesn't change when the solution is diluted. We take advantage of an alternate definition of molarity to answer the question: millimoles of solute/milliliter of solution.

$$\text{amount } K_2CrO_4 = 15.00\,mL \times \frac{0.450\,mmol\ K_2CrO_4}{1\,mL\ soln} = 6.75\,mmol\ K_2CrO_4$$

$$K_2CrO_4 \text{ molarity, dilute solution} = \frac{6.75\,mmol\ K_2CrO_4}{100.00\,mL\ soln} = 0.0675\,M$$

10B We know the initial concentration (0.105 M) and volume (275 mL) of the solution, along with its final volume (237 mL). The final concentration equals the initial concentration times a ratio of the two volumes.

$$c_f = c_i \times \frac{V_i}{V_f} = 0.105\,M \times \frac{275\,mL}{237\,mL} = 0.122\,M$$

11A The balanced equation is $K_2CrO_4\,(aq) + 2\,AgNO_3\,(aq) \rightarrow Ag_2CrO_4\,(s) + 2\,KNO_3\,(aq)$. The molar mass of Ag_2CrO_4 is 331.73 g/mol. The conversions needed are mass $Ag_2CrO_4 \rightarrow$ amount Ag_2CrO_4 (moles) $\rightarrow$ amount K_2CrO_4 (moles) $\rightarrow$ volume K_2CrO_4 (aq).

$$V_{K_2CrO_4} = 1.50\,g\ Ag_2CrO_4 \times \frac{1\,mol\ Ag_2CrO_4}{331.73\,g\ Ag_2CrO_4} \times \frac{1\,mol\ K_2CrO_4}{1\ mol\ Ag_2CrO_4} \times \frac{1\,L\ soln}{0.250\,mol\ K_2CrO_4}$$

$$\times \frac{1000\ mL\ solution}{1\,L\ solution} = 18.1\,mL$$

11B Balanced reaction: $2\,AgNO_3(aq) + K_2CrO_4(aq) \rightarrow Ag_2CrO_4(s) + 2\,KNO_3(aq)$

moles of $K_2CrO_4 = C \times V = 0.0855\,M \times 0.175\,L\ sol = 0.01496$ moles K_2CrO_4

$$\text{moles of } AgNO_3 = 0.01496\ mol\ K_2CrO_4 \times \frac{2\ mol\ AgNO_3}{1\ mol\ K_2CrO_4} = 0.0299\ mol\ AgNO_3$$

$$V_{AgNO_3} = \frac{n}{C} = \frac{0.0299\ mol\ AgNO_3}{0.150\ \frac{mol}{L}\ AgNO_3} = 0.1995\,L \text{ or } 2.00 \times 10^2\ mL\ (0.200\ L) \text{ of } AgNO_3$$

$$\text{Mass of } Ag_2CrO_4 \text{ formed} = 0.01496\ \text{moles } K_2CrO_4 \times \frac{1\,mol\ Ag_2CrO_4}{1\ mol\ K_2CrO_4} \times \frac{331.73\ g\ Ag_2CrO_4}{1\,mol\ Ag_2CrO_4}$$

Mass of Ag_2CrO_4 formed = 4.96 g Ag_2CrO_4

12A Reaction: $P_4\,(s) + 6\,Cl_2\,(g) \rightarrow 4\,PCl_3\,(l)$. We must determine the mass of PCl_3 formed by each reactant.

$$\text{mass } PCl_3 = 215\ g\,P_4 \times \frac{1\,mol\ P_4}{123.90\,g\,P_4} \times \frac{4\ mol\ PCl_3}{1\,mol\ P_4} \times \frac{137.33\,g\ PCl_3}{1\ mol\ PCl_3} = 953\,g\ PCl_3$$

$$\text{mass } PCl_3 = 725\ g\,Cl_2 \times \frac{1\,mol\ Cl_2}{70.91\,g\,Cl_2} \times \frac{4\ mol\ PCl_3}{6\,mol\ Cl_2} \times \frac{137.33\,g\ PCl_3}{1\ mol\ PCl_3} = 936\,g\ PCl_3$$

Thus, a maximum of $936\,g\ PCl_3$ can be produced; there is not enough Cl_2 to produce any more.

12B Since data are supplied and the answer is requested in kilograms (thousands of grams), we can use kilomoles (thousands of moles) to solve the problem. We calculate the amount in kilomoles of $POCl_3$ that would be produced if each of the reactants were completely converted to product. The smallest of these amounts is the one that is actually produced. (This is a limiting reactant question).

$$\text{amount } POCl_3 = 1.00 \text{ kg } PCl_3 \times \frac{1 \text{ kmol } PCl_3}{137.33 \text{ kg } PCl_3} \times \frac{10 \text{ kmol } POCl_3}{6 \text{ kmol } PCl_3} = 0.0121 \text{ kmol } POCl_3$$

$$\text{amount } POCl_3 = 1.00 \text{ kg } Cl_2 \times \frac{1 \text{ kmol } Cl_2}{70.905 \text{ kg } Cl_2} \times \frac{10 \text{ kmol } POCl_3}{6 \text{ kmol } Cl_2} = 0.0235 \text{ kmol } POCl_3$$

$$\text{amount } POCl_3 = 1.00 \text{ kg } P_4O_{10} \times \frac{1 \text{ kmol } P_4O_{10}}{283.89 \text{ kg } P_4O_{10}} \times \frac{10 \text{ kmol } POCl_3}{1 \text{ kmol } P_4O_{10}} = 0.0352 \text{ kmol } POCl_3$$

Thus, a maximum of 0.0121 kmol $POCl_3$ can be produced.

We next determine the mass of the product.

$$\text{mass } POCl_3 = 0.0121 \text{ kmol } POCl_3 \times \frac{153.33 \text{ kg } POCl_3}{1 \text{ kmol } POCl_3} = 1.86 \text{ kg } POCl_3$$

13A The 725 g Cl_2 limits the mass of product formed. The $P_4(s)$ therefore is the reactant in excess. From the quantity of excess reactant we can find the amount of product formed: 953 g PCl_3 − 936 g PCl_3 = 17 g PCl_3. We calculate how much P_4 this is, both in the traditional way and by using the initial $(215 \text{ g } P_4)$ and final $(953 \text{ g } PCl_3)$ values of the previous calculation.

$$\text{mass } P_4 = 17 \text{ g } PCl_3 \times \frac{1 \text{ mol } PCl_3}{137.33 \text{ g } PCl_3} \times \frac{1 \text{ mol } P_4}{4 \text{ mol } PCl_3} \times \frac{123.90 \text{ g } P_4}{1 \text{ mol } P_4} = 3.8 \text{ g } P_4$$

13B Find the amount of $H_2O(l)$ formed by each reactant, to determine the limiting reactant.

$$\text{amount } H_2O = 12.2 \text{ g } H_2 \times \frac{1 \text{ mol } H_2}{2.016 \text{ g } H_2} \times \frac{2 \text{ mol } H_2O}{2 \text{ mol } H_2} = 6.05 \text{ mol } H_2O$$

$$\text{amount } H_2O = 154 \text{ g } O_2 \times \frac{1 \text{ mol } O_2}{32.00 \text{ g } O_2} \times \frac{2 \text{ mol } H_2O}{1 \text{ mol } O_2} = 9.63 \text{ mol } H_2O$$

Since H_2 is limiting, we must compute the mass of O_2 needed to react with all of the H_2

$$\text{mass } O_2 \text{ reacting} = 6.05 \text{ mol } H_2O \text{ produced} \times \frac{1 \text{ mol } O_2}{2 \text{ mol } H_2O} \times \frac{32.00 \text{ g } O_2}{1 \text{ mol } O_2} = 96.8 \text{ g } O_2 \text{ reacting}$$

mass O_2 remaining = 154 g originally present − 96.8 g O_2 reacting = 57 g O_2 remaining

14A **(a)** The theoretical yield is the calculated maximum mass of product expected if we were to assume that the reaction has no losses (100% reaction).

$$\text{mass } CH_2O(g) = 1.00 \text{ mol } CH_3OH \times \frac{1 \text{ mol } CH_2O}{1 \text{ mol } CH_3OH} \times \frac{30.03 \text{ g } CH_2O}{1 \text{ mol } CH_2O} = 30.0 \text{ g } CH_2O$$

(b) The actual yield is what is obtained experimentally: 25.7 g CH_2O (g).

(c) The percent yield is the ratio of actual yield to theoretical yield, multiplied by 100%:

$$\% \text{ yield} = \frac{25.7 \text{ g } CH_2O \text{ produced}}{30.0 \text{ g } CH_2O \text{ calculated}} \times 100 \% = 85.6 \% \text{ yield}$$

14B First determine the mass of product formed by each reactant.

$$\text{mass PCl}_3 = 25.0\,\text{g P}_4 \times \frac{1\,\text{mol P}_4}{123.90\,\text{g P}_4} \times \frac{4\,\text{mol PCl}_3}{1\,\text{mol P}_4} \times \frac{137.33\,\text{g PCl}_3}{1\,\text{mol PCl}_3} = 111\,\text{g PCl}_3$$

$$\text{mass PCl}_3 = 91.5\,\text{g Cl}_2 \times \frac{1\,\text{mol Cl}_2}{70.91\,\text{g Cl}_2} \times \frac{4\,\text{mol PCl}_3}{6\,\text{mol Cl}_2} \times \frac{137.33\,\text{g PCl}_3}{1\,\text{mol PCl}_3} = 118\,\text{g PCl}_3$$

Thus, the limiting reactant is P_4, and $111\,\text{g PCl}_3$ should be produced. This is the theoretical maximum yield. The actual yield is $104\,\text{g PCl}_3$. Thus, the percent yield of the reaction is

$$\frac{104\,\text{g PCl}_3\ \text{produced}}{111\,\text{g PCl}_3\ \text{calculated}} \times 100\% = 93.7\%\ \text{yield.}$$

15A The reaction is $2\,NH_3(g) + CO_2(g) \rightarrow CO(NH_2)_2(s) + H_2O(l)$. We need to distinguish between mass of urea produced (actual yield) and mass of urea predicted (theoretical yield).

$$\text{mass CO}_2 = 50.0\,\text{g CO}(NH_2)_2\ \text{produced} \times \frac{100.0\,\text{g predicted}}{87.5\,\text{g produced}} \times \frac{1\,\text{mol CO}(NH_2)_2}{60.1\,\text{g CO}(NH_2)_2} \times \frac{1\,\text{mol CO}_2}{1\,\text{mol CO}(NH_2)_2}$$

$$\times \frac{44.01\,\text{g CO}_2}{1\,\text{mol CO}_2} = 41.8\,\text{g CO}_2\ \text{needed}$$

15B Care must be taken to use the proper units/labels in each conversion factor. Note, you cannot calculate the molar mass of an impure material or mixture.

$$\text{mass C}_6H_{11}OH = 45.0\,\text{g C}_6H_{10}\ \text{produced} \times \frac{100.0\,\text{g C}_6H_{10}\ \text{cal'd}}{86.2\,\text{g C}_6H_{10}\ \text{produc'd}} \times \frac{1\,\text{mol C}_6H_{10}}{82.1\,\text{g C}_6H_{10}} \times \frac{1\,\text{mol C}_6H_{11}OH}{1\,\text{mol C}_6H_{10}}$$

$$\times \frac{100.2\,\text{g pure C}_6H_{11}OH}{1\,\text{mol C}_6H_{11}OH} \times \frac{100.0\,\text{g impure C}_6H_{11}OH}{92.3\,\text{g pure C}_6H_{11}OH} = 69.0\ \text{g impure C}_6H_{11}OH$$

16A We can trace the nitrogen through the sequence of reactions. We notice that 4 moles of N (as 4 mol NH_3) are consumed in the first reaction, and 4 moles of N (as 4 mole NO) are produced. In the second reaction, 2 moles of N (as 2 mol NO) are consumed and 2 moles of N (as 2 mol NO_2) are produced. In the last reaction, 3 moles of N (as 3 mol NO_2) are consumed and just 2 moles of N (as 2 mol HNO_3) are produced.

$$\text{mass HNO}_3 = 1.00\,\text{kg NH}_3 \times \frac{1000\,\text{g NH}_3}{1\,\text{kg NH}_3} \times \frac{1\,\text{mol NH}_3}{17.03\,\text{g NH}_3} \times \frac{4\,\text{mol NO}}{4\,\text{mol NH}_3} \times \frac{2\,\text{mol NO}_2}{2\,\text{mol NO}}$$

$$\times \frac{2\,\text{mol HNO}_3}{3\,\text{mol NO}_2} \times \frac{63.01\,\text{g HNO}_3}{1\,\text{mol HNO}_3} = 2.47 \times 10^3\ \text{g HNO}_3$$

16B $\text{mass H}_2\,(\text{from Al}) = 0.710\,\text{g alloy} \times \dfrac{0.700\,\text{g Al}}{1.000\,\text{g alloy}} \times \dfrac{1\,\text{mol Al}}{26.98\,\text{g Al}} \times \dfrac{3\,\text{mol H}_2}{2\,\text{mol Al}} \times \dfrac{2.016\,\text{g H}_2}{1\,\text{mol H}_2} = 0.0557\ \text{g}$

$\text{mass H}_2\,(\text{from Mg}) = 0.710\,\text{g alloy} \times \dfrac{0.300\,\text{g Mg}}{1.000\,\text{g alloy}} \times \dfrac{1\,\text{mol Mg}}{24.31\,\text{g Mg}} \times \dfrac{1\,\text{mol H}_2}{1\,\text{mol Mg}} \times \dfrac{2.016\,\text{g H}_2}{1\,\text{mol H}_2} = 0.0177\,\text{g}$

total mass of $H_2 = 0.0557\,\text{g H}_2$ from Al $+\ 0.0177\,\text{g H}_2$ from Mg $= 0.0734\,\text{g H}_2$

EXERCISES

Writing and Balancing Chemical Equations

1. (a) $2 SO_3 \longrightarrow 2 SO_2 + O_2$

   (b) $Cl_2O_7 + H_2O \longrightarrow 2 HClO_4$

   (c) $3 NO_2 + H_2O \longrightarrow 2 HNO_3 + NO$

   (d) $PCl_3 + 3 H_2O \longrightarrow H_3PO_3 + 3 \ HCl$

2. (a) $3 P_2H_4 \longrightarrow 4 PH_3 + \frac{1}{2} P_4$ or $6 P_2H_4 \longrightarrow 8 PH_3 + P_4$

   (b) $P_4 + 6 Cl_2 \longrightarrow 4 PCl_3$

   (c) $2 FeCl_3 + 3 H_2S \longrightarrow Fe_2S_3 + 6 HCl$

   (d) $Mg_3N_2 + 6 H_2O \longrightarrow 3 Mg(OH)_2 + 2 \ NH_3$

3. (a) $3 PbO + 2 NH_3 \longrightarrow 3 Pb + N_2 + 3 H_2O$

   (b) $2 FeSO_4 \longrightarrow Fe_2O_3 + 2 SO_2 + \frac{1}{2} O_2$ or $4 FeSO_4 \longrightarrow 2 Fe_2O_3 + 4 SO_2 + O_2$

   (c) $6 S_2Cl_2 + 16 NH_3 \longrightarrow N_4S_4 + 12 NH_4Cl + S_8$

   (d) $C_3H_7CHOHCH(C_2H_5)CH_2OH + \frac{23}{2} O_2 \longrightarrow 8 CO_2 + 9 H_2O$

     or $\quad 2 C_3H_7CHOHCH(C_2H_5)CH_2OH + 23 O_2 \longrightarrow 16 CO_2 + 18 H_2O$

4. (a) $SO_2Cl_2 + 8 HI \rightarrow H_2S + 2 H_2O + 2 HCl + 4 I_2$

   (b) $FeTiO_3 + 2 H_2SO_4 + 5 H_2O \longrightarrow FeSO_4 \cdot 7H_2O + TiOSO_4$

   (c) $2 Fe_3O_4 + 12 HCl + 3 \ Cl_2 \longrightarrow 6 FeCl_3 + 6 H_2O + O_2$

   (d) $C_6H_5CH_2SSCH_2C_6H_5 + \frac{39}{2} O_2 \longrightarrow 14 CO_2 + 2 SO_2 + 7 H_2O$

     or $\quad 2 C_6H_5CH_2SSCH_2C_6H_5 + 39 \ O_2 \longrightarrow 28 CO_2 + 4 SO_2 + 14 H_2O$

5. (a) $2 Mg(s) + O_2(g) \rightarrow 2 MgO(s)$

   (b) $2 NO(g) + O_2(g) \rightarrow 2 NO_2(g)$

   (c) $2 C_2H_6(g) + 7 O_2(g) \rightarrow 4 CO_2(g) + 6 H_2O(l)$

   (d) $Ag_2SO_4(aq) + BaI_2(aq) \rightarrow BaSO_4(s) + 2 AgI(s)$

6. (a) $3 Mg(s) + N_2(g) \rightarrow Mg_3N_2(s)$

   (b) $KClO_3(s) \longrightarrow KCl(s) + \frac{3}{2} O_2(g)$ *or* $\quad 2 KClO_3(s) \longrightarrow 2 KCl(s) + 3 O_2(g)$

   (c) $NaOH(s) + NH_4Cl(s) \longrightarrow NaCl(s) + NH_3(g) + H_2O(g)$

   (d) $2 Na(s) + 2 H_2O(l) \longrightarrow 2 NaOH(aq) + H_2(g)$

7. **(a)** $2C_4H_{10}(l) + 13O_2(g) \rightarrow 8CO_2(g) + 10H_2O(l)$

   (b) $2\ CH_3CH(OH)CH_3(l) + 9\ O_2(g) \rightarrow 6\ CO_2(g) + 8\ H_2O(l)$

   (c) $CH_3CH(OH)COOH(s) + 3O_2(g) \rightarrow 3CO_2(g) + 3H_2O(l)$

8. **(a)** $2C_3H_6(g) + 9O_2(g) \rightarrow 6CO_2(g) + 6H_2O(l)$

   (b) $2CH_2OHCH(OH)CH_2OH(l) + 7O_2(g) \rightarrow 6CO_2(g) + 8H_2O(l)$

   (c) $C_6H_5COSH(s) + 9O_2(g) \rightarrow 7CO_2(g) + 3H_2O(l) + SO_2(g)$

9. **(a)** $NH_4NO_3(s) \xrightarrow{\Delta} N_2O(g) + 2H_2O(g)$

   (b) $Na_2CO_3(aq) + 2HCl(aq) \rightarrow 2NaCl(aq) + H_2O(l) + CO_2(g)$

   (c) $2CH_4(g) + 2NH_3(g) + 3O_2(g) \rightarrow 2HCN(g) + 6H_2O(g)$

10. **(a)** $2SO_2(g) + O_2(g) \rightarrow 2SO_3(g)$

   (b) $CaCO_3(s) + H_2O(l) + CO_2(aq) \rightarrow Ca(HCO_3)_2(aq)$

   (c) $4NH_3(g) + 6NO(g) \rightarrow 5N_2(g) + 6H_2O(g)$

11.

Unbalanced reaction:	$N_2H_4(g) + N_2O_4(g)$	$\rightarrow$	$H_2O(g) + N_2(g)$
Balance H atoms:	$N_2H_4(g) + N_2O_4(g)$	$\rightarrow$	$2\ H_2O(g) + N_2(g)$
Balance O atoms:	$N_2H_4(g) + 1/2\ N_2O_4(g)$	$\rightarrow$	$2\ H_2O(g) + N_2(g)$
Balance N atoms:	$N_2H_4(g) + 1/2\ N_2O_4(g)$	$\rightarrow$	$2\ H_2O(g) + 3/2\ N_2(g)$
Multiply by 2 (whole #)	$2\ N_2H_4(g) + N_2O_4(g)$	$\rightarrow$	$4\ H_2O(g) + 3\ N_2(g)$
Self Check:	$6\ N + 8\ H + 4\ O$	$\rightarrow$	$6\ N + 8\ H + 4\ O$

12.

Unbalanced reaction:	$NH_3(g) + O_2(g)$	$\rightarrow$	$H_2O(g) + NO(g)$
Balance H atoms:	$2\ NH_3(g) + O_2(g)$	$\rightarrow$	$3\ H_2O(g) + NO(g)$
Balance N atoms:	$2\ NH_3(g) + O_2(g)$	$\rightarrow$	$3\ H_2O(g) + 2\ NO(g)$
Balance O atoms:	$2\ NH_3(g) + 5/2\ O_2(g)$	$\rightarrow$	$3\ H_2O(g) + 2\ NO(g)$
Multiply by 2 (whole #)	$4\ NH_3(g) + 5\ O_2(g)$	$\rightarrow$	$6\ H_2O(g) + 4\ NO(g)$
Self Check:	$4\ N + 12\ H + 10\ O$	$\rightarrow$	$4\ N + 12\ H + 10\ O$

Stoichiometry of Chemical Reactions

13. The conversion factor is obtained from the balanced chemical equation.

$$\text{moles } FeCl_3 = 7.26\ \text{mol } Cl_2 \times \frac{2\ \text{mol } FeCl_3}{3\ \text{mol } Cl_2} = 4.84\ \text{mol } FeCl_3$$

14. Each calculation uses the stoichiometric coefficients from the balanced chemical equation and the molar mass of the reactant.

$$\text{mass } Cl_2 = 0.337\ \text{mol } PCl_3 \times \frac{6\ \text{mol } Cl_2}{4\ \text{mol } PCl_3} \times \frac{70.91\ \text{g } Cl_2}{1\ \text{mol } Cl_2} = 35.8\ \text{g } Cl_2$$

$$\text{mass } P_4 = 0.337\ \text{mol } PCl_3 \times \frac{1\ \text{mol } P_4}{4\ \text{mol } PCl_3} \times \frac{123.9\ \text{g } P_4}{1\ \text{mol } P_4} = 10.4\ \text{g } P_4$$

15. **(a)** $\text{mol } O_2 = 32.8\,\text{g KClO}_3 \times \dfrac{1\,\text{mol KClO}_3}{122.6\,\text{g KClO}_3} \times \dfrac{3\,\text{mol } O_2}{2\,\text{mol KClO}_3} = 0.401\,\text{mol } O_2$

(b) $\text{mass KClO}_3 = 50.0\,\text{g } O_2 \times \dfrac{1\,\text{mol } O_2}{32.00\,\text{g } O_2} \times \dfrac{2\,\text{mol KClO}_3}{3\,\text{mol } O_2} \times \dfrac{122.6\,\text{g KClO}_3}{1\,\text{mol KClO}_3} = 128\,\text{g KClO}_3$

(c) $\text{mass KCl} = 28.3\,\text{g } O_2 \times \dfrac{1\,\text{mol } O_2}{32.00\,\text{g } O_2} \times \dfrac{2\,\text{mol KCl}}{3\,\text{mol } O_2} \times \dfrac{74.55\,\text{g KCl}}{1\,\text{mol KCl}} = 44.0\,\text{g KCl}$

16. **(a)** $\text{amount } H_2 = 42.7\,\text{g Fe} \times \dfrac{1\,\text{mol Fe}}{55.85\,\text{g Fe}} \times \dfrac{4\,\text{mol } H_2}{3\,\text{mol Fe}} = 1.02\,\text{mol } H_2$

(b) $\text{mass } H_2O = 63.5\,\text{g Fe} \times \dfrac{1\,\text{mol Fe}}{55.85\,\text{g Fe}} \times \dfrac{4\,\text{mol } H_2O}{3\,\text{mol Fe}} \times \dfrac{18.02\,\text{g } H_2O}{1\,\text{mol } H_2O} = 27.3\,\text{g } H_2O$

(c) $\text{mass Fe}_3O_4 = 7.36\,\text{mol } H_2 \times \dfrac{1\,\text{mol Fe}_3O_4}{4\,\text{mol } H_2} \times \dfrac{231.54\,\text{g Fe}_3O_4}{1\,\text{mol Fe}_3O_4} = 426\,\text{g Fe}_3O_4$

17. Balance the given equation, and then solve the problem.

$$2\,\text{Ag}_2\text{CO}_3\,(s) \xrightarrow{\ \Delta\ } 4\text{Ag}\,(s) + 2\,\text{CO}_2\,(g) + O_2\,(g)$$

$\text{mass Ag}_2\text{CO}_3 = 75.1\,\text{g Ag} \times \dfrac{1\,\text{mol Ag}}{107.87\,\text{g Ag}} \times \dfrac{2\,\text{mol Ag}_2\text{CO}_3}{4\,\text{mol Ag}} \times \dfrac{275.75\,\text{g Ag}_2\text{CO}_3}{1\,\text{mol Ag}_2\text{CO}_3} = 96.0\,\text{g Ag}_2\text{CO}_3$

18. The balanced equation is $\text{Ca}_3(\text{PO}_4)_2(s) + 4\,\text{HNO}_3(aq) \rightarrow \text{Ca}(\text{H}_2\text{PO}_4)_2(s) + 2\,\text{Ca}(\text{NO}_3)_2(aq)$

$\text{mass HNO}_3 = 125\,\text{kg Ca(H}_2\text{PO}_4)_2 \times \dfrac{1\,\text{kmol Ca(H}_2\text{PO}_4)_2}{234.05\,\text{kg Ca(H}_2\text{PO}_4)_2} \times \dfrac{4\,\text{kmol HNO}_3}{1\,\text{kmol Ca(H}_2\text{PO}_4)_2} \times \dfrac{63.01\,\text{kg HNO}_3}{1\,\text{kmol HNO}_3}$

$\text{mass HNO}_3 = 135\,\text{kg HNO}_3$

19. The balanced equation is $\text{CaH}_2(s) + 2\,\text{H}_2\text{O}(l) \rightarrow \text{Ca(OH)}_2(s) + 2\,\text{H}_2(g)$

(a) $\text{amount } H_2 = 127\,\text{g CaH}_2 \times \dfrac{1\,\text{mol CaH}_2}{42.094\,\text{g CaH}_2} \times \dfrac{2\,\text{mol } H_2}{1\,\text{mol CaH}_2} = 6.03\ \text{mol } H_2$

(b) $\text{mass } H_2O = 56.2\ \text{g CaH}_2 \times \dfrac{1\,\text{mol CaH}_2}{42.094\ \text{g CaH}_2} \times \dfrac{2\ \text{mol } H_2O}{1\,\text{mol CaH}_2} \times \dfrac{18.0153\ \text{g } H_2O}{1\ \text{mol } H_2O} = 48.1\ \text{g } H_2O$

(c) $\text{mass CaH}_2 = 8.12\times10^{24}\ \text{molecules } H_2 \times \dfrac{1\,\text{mol } H_2}{6.022\times10^{23}\ \text{molecules } H_2} \times \dfrac{1\,\text{mol CaH}_2}{2\,\text{mol } H_2} \times \dfrac{42.094\ \text{g CaH}_2}{1\,\text{mol CaH}_2}$

$\text{mass CaH}_2 = 284\ \text{g CaH}_2$

20. **(a)** $\text{amount } O_2 = 156\,\text{g CO}_2 \times \dfrac{1\,\text{mol CO}_2}{44.01\,\text{g CO}_2} \times \dfrac{3\,\text{mol } O_2}{2\,\text{mol CO}_2} = 5.32\,\text{mol } O_2$

(b) $\text{mass KO}_2 = 100.0\,\text{g CO}_2 \times \dfrac{1\,\text{mol CO}_2}{44.01\,\text{g CO}_2} \times \dfrac{4\,\text{mol KO}_2}{2\,\text{mol CO}_2} \times \dfrac{71.10\,\text{g KO}_2}{1\,\text{mol KO}_2} = 323.1\,\text{g KO}_2$

(c) $\text{no. } O_2 \text{ molecules} = 1.00\,\text{mg KO}_2 \times \dfrac{1\,\text{g KO}_2}{1000\,\text{mg}} \times \dfrac{1\,\text{mol KO}_2}{71.10\,\text{g KO}_2} \times \dfrac{3\,\text{mol } O_2}{4\,\text{mol KO}_2}$

$\times \dfrac{6.022\times10^{23}\ \text{molecules}}{1\,\text{mol } O_2} = 6.35\times10^{18}\,O_2 \text{ molecules}$

21. The balanced equation is $Fe_2O_3(s) + 3C(s) \xrightarrow{\Delta} 2Fe(l) + 3CO(g)$

$$\text{mass } Fe_2O_3 = 523 \text{ kg Fe} \times \frac{1 \text{ kmol Fe}}{55.85 \text{ kg Fe}} \times \frac{1 \text{ kmol } Fe_2O_3}{2 \text{ kmol Fe}} \times \frac{159.7 \text{ kg } Fe_2O_3}{1 \text{ kmol } Fe_2O_3} = 748 \text{ kg } Fe_2O_3$$

$$\% \ Fe_2O_3 \text{ in ore} = \frac{748 \text{ kg } Fe_2O_3}{938 \text{ kg ore}} \times 100\% = 79.7\% \ Fe_2O_3$$

22. The following reaction occurs: $2Ag_2O(s) \xrightarrow{heat} 4Ag(s) + O_2(g)$

$$\text{mass } Ag_2O = 0.187 \text{g } O_2 \times \frac{1 \text{mol } O_2}{32.0 \text{g } O_2} \times \frac{2 \text{mol } Ag_2O}{1 \text{mol } O_2} \times \frac{231.7 \text{g } Ag_2O}{1 \text{mol } Ag_2O} = 2.71 \text{g } Ag_2O$$

$$\% \ Ag_2O = \frac{2.71 \text{g } Ag_2O}{3.13 \text{g sample}} \times 100\% = 86.6\% \ Ag_2O$$

23. $2Al(s) + 6HCl(aq) \rightarrow 2AlCl_3(aq) + 3H_2(g)$. First determine the mass of Al in the foil.

$$\text{mass Al} = (10.25 \text{cm} \times 5.50 \text{cm} \times 0.601 \text{mm}) \times \frac{1 \text{cm}}{10 \text{mm}} \times \frac{2.70 \text{g}}{1 \text{cm}^3} = 9.15 \text{g Al}$$

$$\text{mass } H_2 = 9.15 \text{g Al} \times \frac{1 \text{mol Al}}{26.98 \text{g Al}} \times \frac{3 \text{mol } H_2}{2 \text{mol Al}} \times \frac{2.016 \text{g } H_2}{1 \text{mol } H_2} = 1.03 \text{g } H_2$$

24. $2Al(s) + 6HCl(aq) \rightarrow 2AlCl_3(aq) + 3H_2(g)$

$$\text{mass } H_2 = 225 \text{mL soln} \times \frac{1.088 \text{g}}{1 \text{mL}} \times \frac{18.0 \text{g HCl}}{100.0 \text{g soln}} \times \frac{1 \text{mol HCl}}{36.46 \text{g HCl}} \times \frac{3 \text{mol } H_2}{6 \text{mol HCl}} \times \frac{2.016 \text{g } H_2}{1 \text{mol } H_2}$$

$$= 1.22 \text{g } H_2$$

25. First write the balanced chemical equation for each reaction.

$2Na(s) + 2HCl(aq) \rightarrow 2NaCl(aq) + H_2(g)$ $Mg(s) + 2HCl(aq) \rightarrow MgCl_2(aq) + H_2(g)$

$2Al(s) + 6HCl(aq) \rightarrow 2AlCl_3(aq) + 3H_2(g)$ $Zn(s) + 2HCl(aq) \rightarrow ZnCl_2(aq) + H_2(g)$

Three of the reactions—those of Na, Mg, and Zn—produce 1 mole of $H_2(g)$. The one of these three that produces the most hydrogen per gram of metal is the one for which the metal's atomic mass is the smallest, remembering to compare twice the atomic mass for Na. The atomic masses are: 2×23 u for Na, 24.3 u for Mg, and 65.4 u for Zn. Thus, among these three, Mg produces the most H_2 per gram of metal, specifically 1 mol H_2 per 24.3 g Mg. In the case of Al, 3 moles of H_2 are produced by 2 moles of the metal, or 54 g Al. This reduces as follows: 3 mol $H_2 / 54$ g Al = 1 mol $H_2 / 18$ g Al. Thus, Al produces the largest amount of H_2 per gram of metal.

26. In order for a substance to yield the same mass of $CO_2(g)$ per gram of compound as does ethanol when combusted in excess oxygen, the substance must have the same empirical formula. Compound (d), CH_3OCH_3, is the only compound that fits the description. In fact, compound (d) has the same formula as ethanol, CH_3CH_2OH, as they are structural isomers, and they should give the same amount of $CO_2(g)$ when combusted in excess O_2.

27. (a) CH_3OH molarity (M) $= \dfrac{2.92 \text{ mol } CH_3OH}{7.16 \text{ L}} = 0.408 \text{ M}$

(b) C_2H_5OH molarity (M) $= \dfrac{7.69 \text{ mmol } C_2H_5OH}{50.00 \text{ mL}} = 0.154 \text{ M}$

(c) $CO(NH_2)_2$ molarity (M) $= \dfrac{25.2 \text{ g } CO(NH_2)_2}{275 \text{ mL}} \times \dfrac{1 \text{ mol } CO(NH_2)_2}{60.06 \text{ g } CO(NH_2)_2} \times \dfrac{1000 \text{ mL}}{1 \text{ L}} = 1.53 \text{ M}$

28. (a) C_2H_5OH molarity (M) $= \dfrac{2.25 \times 10^{-4} \text{ mol } C_2H_5OH}{125 \text{ mL}} \times \dfrac{1000 \text{ mL}}{1 \text{ L}} = 0.00180 \text{ M}$

(b)

$(CH_3)_2 CO$ molarity (M) $= \dfrac{57.5 \text{ g } (CH_3)_2 CO}{525 \text{ mL}} \times \dfrac{1 \text{ mol } (CH_3)_2 CO}{58.08 \text{ g } (CH_3)_2 CO} \times \dfrac{1000 \text{ mL}}{1 \text{ L}} = 1.88\underline{6} \text{ M}$

(c) $C_3H_5(OH)_3$ molarity (M) $= \dfrac{18.5 \text{ mL } C_3H_5(OH)_3}{375 \text{ mL soln}} \times \dfrac{1.26 \text{ g}}{1 \text{ mL}} \times \dfrac{1 \text{ mol } C_3H_5(OH)_3}{92.09 \text{ g } C_3H_5(OH)_3} \times \dfrac{1000 \text{ mL}}{1 \text{ L}} = 0.675 \text{ M}$

Molarity

29. (a) $[C_{12}H_{22}O_{11}] = \dfrac{150.0 \text{ g } C_{12}H_{22}O_{11}}{250.0 \text{ mL soln}} \times \dfrac{1000 \text{ mL}}{1 \text{ L}} \times \dfrac{1 \text{ mol } C_{12}H_{22}O_{11}}{342.3 \text{ g } C_{12}H_{22}O_{11}} = 1.753 \text{ M}$

(b) $[CO(NH_2)_2] = \dfrac{98.3 \text{ mg solid}}{5.00 \text{ mL soln}} \times \dfrac{97.9 \text{ mg } CO(NH_2)_2}{100 \text{ mg solid}} \times \dfrac{1 \text{ mmol } CO(NH_2)_2}{60.06 \text{ mg } CO(NH_2)_2}$

$= 0.320 \text{ M } CO(NH_2)_2$

(c) $[CH_3OH] = \dfrac{125.0 \text{ mL } CH_3OH}{15.0 \text{ L soln}} \times \dfrac{0.792 \text{ g}}{1 \text{ mL}} \times \dfrac{1 \text{ mol } CH_3OH}{32.04 \text{ g } CH_3OH} = 0.206 \text{ M}$

30. (a) $[H_2C_4H_5NO_4] = \dfrac{0.405 \text{ g } H_2C_4H_5NO_4}{100.0 \text{ mL}} \times \dfrac{1000 \text{ mL}}{1 \text{ L}} \times \dfrac{1 \text{ mol } H_2C_4H_5NO_4}{133.10 \text{ g } H_2C_4H_5NO_4} = 0.0304 \text{ M}$

(b) $[C_3H_6O] = \dfrac{35.0 \text{ mL } C_3H_6O}{425 \text{ mL soln}} \times \dfrac{1000 \text{ mL}}{1 \text{ L}} \times \dfrac{0.790 \text{ g } C_3H_6O}{1 \text{ mL}} \times \dfrac{1 \text{ mol}}{58.08 \text{ g } C_3H_6O} = 1.12 \text{ M}$

(c) $[(C_2H_5)_2 O] = \dfrac{8.8 \text{ mg } (C_2H_5)_2 O}{3.00 \text{ L soln}} \times \dfrac{1 \text{ g}}{1000 \text{ mg}} \times \dfrac{1 \text{ mol } (C_2H_5)_2 O}{74.12 \text{ g } (C_2H_5)_2 O} = 4.0 \times 10^{-5} \text{ M}$

31. (a) mass $C_6H_{12}O_6 = 75.0 \text{ mL soln} \times \dfrac{1 \text{ L}}{1000 \text{ mL}} \times \dfrac{0.350 \text{ mol } C_6H_{12}O_6}{1 \text{ L soln}} \times \dfrac{180.16 \text{ g } C_6H_{12}O_6}{1 \text{ mol } C_6H_{12}O_6} = 4.73 \text{ g}$

(b) $V_{CH_3OH} = 2.25 \text{ L soln} \times \dfrac{0.485 \text{ mol}}{1 \text{ L}} \times \dfrac{32.04 \text{ g } CH_3OH}{1 \text{ mol } CH_3OH} \times \dfrac{1 \text{ mL}}{0.792 \text{ g}} = 44.1 \text{ mL } CH_3OH$

32. (a) $V_{C_2H_5OH} = 200.0 \text{ L soln} \times \dfrac{1.65 \text{ mol } C_2H_5OH}{1 \text{ L}} \times \dfrac{46.07 \text{ g } C_2H_5OH}{1 \text{ mol } C_2H_5OH} \times \dfrac{1 \text{ mL}}{0.789 \text{ g}} \times \dfrac{1 \text{ L}}{1000 \text{ mL}} = 19.3 \text{ L}$

(b) $V_{HCl} = 12.0 \text{ L} \times \dfrac{0.234 \text{ mol } HCl}{1 \text{ L}} \times \dfrac{36.46 \text{ g } HCl}{1 \text{ mol } HCl} \times \dfrac{100 \text{ g soln}}{36.0 \text{ g } HCl} \times \dfrac{1 \text{ mL soln}}{1.18 \text{ g}} = 241 \text{ mL}$

33. First we determine each concentration in moles per liter and find the 0.500 M solution.

(a) $[KCl] = \dfrac{0.500 \text{ g KCl}}{1 \text{ mL}} \times \dfrac{1 \text{ mol KCl}}{74.551 \text{ g KCl}} \times \dfrac{1000 \text{ mL}}{1 \text{ L}} = 6.71 \text{ M KCl}$

(b) $[KCl] = \dfrac{36.0 \text{ g KCl}}{1 \text{ L}} \times \dfrac{1 \text{ mol KCl}}{74.551 \text{ g KCl}} = 0.483 \text{ M KCl}$

(c) $[KCl] = \dfrac{7.46 \text{ mg KCl}}{1 \text{ mL}} \times \dfrac{1 \text{ g KCl}}{1000 \text{ mg KCl}} \times \dfrac{1 \text{ mol KCl}}{74.551 \text{ g KCl}} \times \dfrac{1000 \text{ mL}}{1 \text{ L}} = 0.100 \text{ M KCl}$

(d) $[KCl] = \dfrac{373 \text{ g KCl}}{10.00 \text{ L}} \times \dfrac{1 \text{ mol KCl}}{74.551 \text{ g KCl}} = 0.500 \text{ M}$

Solution (d) is a 0.500 M KCl solution.

34. By inspection, we see that (b) and (c) are the only two that are not per volume of solution. These two solutions need not be considered. A close inspection of the remaining choices reveals that the units for (a) are equivalent to those for (d), that is g NaCl per liter of solution is equivalent to mg NaCl per mL of solution (the mass:volume ratio is the same).

35. We determine the molar concentration for the 46% by mass sucrose solution.

$$[C_{12}H_{22}O_{11}] = \dfrac{46 \text{ g } C_{12}H_{22}O_{11} \times \dfrac{1 \text{ mol } C_{12}H_{22}O_{11}}{342.3 \text{ g } C_{12}H_{22}O_{11}}}{100 \text{ g soln} \times \dfrac{1 \text{ mL}}{1.21 \text{ g soln}} \times \dfrac{1 \text{ L}}{1000 \text{ mL}}} = 1.6 \text{ M}$$

The 46% by mass sucrose solution is the more concentrated.

36. Here we must calculate the $[C_2H_5OH]$ in the white wine and compare it with 1.71 $M \, C_2H_5OH$, the concentration of the solution described in Example 4-8.

$$[C_2H_5OH] = \dfrac{11 \text{ g } C_2H_5OH}{100.0 \text{ g soln}} \times \dfrac{0.95 \text{ g soln}}{1 \text{ mL}} \times \dfrac{1000 \text{ mL}}{1 \text{ L}} \times \dfrac{1 \text{ mol } C_2H_5OH}{46.1 \text{ g } C_2H_5OH} = 2.3 \text{ M } C_2H_5OH$$

Thus, the white wine has a greater ethyl alcohol content.

37. $[KNO_3] = \dfrac{0.01000 \text{ L conc'd soln} \times \dfrac{2.05 \text{ mol } KNO_3}{1 \text{ L}}}{0.250 \text{ .L diluted solution}} = 0.0820 \text{ M}$

38. Volume of concentrated $AgNO_3$ solution

$V_{AgNO_3} = 250.0 \text{ mL dilute soln} \times \dfrac{0.425 \text{ mmol } AgNO_3}{1 \text{ mL dilute soln}} \times \dfrac{1 \text{ mL conc. soln.}}{0.750 \text{ mmol } AgNO_3} = 142 \text{ mL conc. soln}$

39. Both the diluted and concentrated solutions contain the same number of moles of K_2SO_4 is present in both solutions. This number is given in the numerator of the following expression.

$$K_2SO_4 \text{ molarity} = \dfrac{0.125 \text{ L} \times \dfrac{0.198 \text{ mol } K_2SO_4}{1 \text{ L}}}{0.105 \text{ L}} = 0.236 \text{ M } K_2SO_4$$

40. $[HCl] = \dfrac{0.500 \text{ L dilute sol'n} \times \dfrac{0.085 \text{ mol HCl}}{1 \text{ L soln}}}{0.0250 \text{ L}} = 1.7 \text{ M}$

41. Let us compute how many mL of dilute ($_d$) solution we obtain from each mL of concentrated ($_c$) solution. $V_c \times C_c = V_d \times C_d$ becomes 1.00 mL $\times 0.250$M $= x$ mL $\times 0.0125$ M and Thus the ratio of the volume of the volumetric flask to that of the pipet would be 20:1. We could use a 100.0-mL flask and a 5.00-mL pipet, a 1000.0-mL flask and a 50.00-mL pipet, or a 500.0-mL flask and a 25.00-mL pipet. There are many combinations that could be used.

42. First we must determine the amount of solute in the final solution and then the volume of the initial, more concentrated, solution that must be used.

volume conc'd soln $= 250.0 \text{ mL} \times \dfrac{0.175 \text{ mmol KCl}}{1 \text{ mL dil soln}} \times \dfrac{1 \text{ mL conc'd soln}}{0.496 \text{ mmol KCl}} = 88.2$ mL

Thus the instructions are as follows: Place 88.2 mL of 0.496 M KCl in a 250-mL volumetric flask. Dilute to the mark with distilled water, stopping to mix thoroughly several times during the addition of water.

Chemical Reactions in Solutions

43. **(a)** mass $Na_2S = 27.8 \text{ mL} \times \dfrac{1 \text{ L}}{1000 \text{ mL}} \times \dfrac{0.163 \text{ mol AgNO}_3}{1 \text{ L soln}} \times \dfrac{1 \text{ mol Na}_2S}{2 \text{ mol AgNO}_3}$

$\times \dfrac{78.05 \text{ g Na}_2S}{1 \text{ mol Na}_2S} = 0.177 \text{ g Na}_2S$

(b) mass $Ag_2S = 0.177 \text{ g Na}_2S \times \dfrac{1 \text{ mol Na}_2S}{78.05 \text{ g Na}_2S} \times \dfrac{1 \text{ mol Ag}_2S}{1 \text{ mol Na}_2S} \times \dfrac{247.80 \text{ g Ag}_2S}{1 \text{ mol Ag}_2S} = 0.562 \text{ g Ag}_2S$

44. The balanced chemical equation provides us with the conversion factor between the two compounds.

(a) mass $NaHCO_3 = 525 \text{ mL solution} \times \dfrac{1 \text{ L solution}}{1000 \text{ mL solution}} \times \dfrac{0.220 \text{ mol Cu(NO}_3)_2}{1 \text{ L solution}} \times \dfrac{2 \text{ mol NaHCO}_3}{1 \text{ mol Cu(NO}_3)_2}$

$\times \dfrac{84.00694 \text{ g NaHCO}_3}{1 \text{ mol NaHCO}_3} = 19.4 \text{ g NaHCO}_3$

(b) mass $CuCO_3 = 525 \text{ mL solution} \times \dfrac{1 \text{ L solution}}{1000 \text{ mL solution}} \times \dfrac{0.220 \text{ mol Cu(NO}_3)_2}{1 \text{ L}} \times \dfrac{1 \text{ mol CuCO}_3}{1 \text{ mol Cu(NO}_3)_2}$

$\times \dfrac{123.6 \text{ g CuCO}_3}{1 \text{ mol CuCO}_3} = 14.3 \text{ g CuCO}_3$

45. The molarity can be expressed as millimoles of solute per milliliter of solution.

$V_{K_2CrO_4} = 415 \text{ mL} \times \dfrac{0.186 \text{ mmol AgNO}_3}{1 \text{ mL soln}} \times \dfrac{1 \text{ mmol K}_2CrO_4}{2 \text{ mmol AgNO}_3} \times \dfrac{1 \text{ mL K}_2CrO_4 \text{ (aq)}}{0.650 \text{ mmol K}_2CrO_4}$

$V_{K_2CrO_4} = 59.4 \text{ mL K}_2CrO_4$

46. (a) $\text{mass Ca(OH)}_2 = 415 \text{ mL} \times \dfrac{1 \text{ L}}{1000 \text{ mL}} \times \dfrac{0.477 \text{ mol HCl}}{1 \text{ L soln}} \times \dfrac{1 \text{ mol Ca(OH)}_2}{2 \text{ mol HCl}}$

$\times \dfrac{74.1 \text{ g Ca(OH)}_2}{1 \text{ mol Ca(OH)}_2} = 7.33 \text{ g Ca(OH)}_2$

(b) $\text{mass Ca(OH)}_2 = 324 \text{ L} \times \dfrac{1.12 \text{ kg}}{1 \text{ L}} \times \dfrac{24.28 \text{ kg HCl}}{100.00 \text{ kg soln}} \times \dfrac{1 \text{ kmol HCl}}{36.46 \text{ kg HCl}}$

$\times \dfrac{1 \text{ kmol Ca(OH)}_2}{2 \text{ kmol HCl}} \times \dfrac{74.10 \text{ kg Ca(OH)}_2}{1 \text{ kmol Ca(OH)}_2} = 89.5 \text{ kg Ca(OH)}_2$

47. (a) We know that the Al forms the $AlCl_3$.

$\text{mol AlCl}_3 = 1.87 \text{ g Al} \times \dfrac{1 \text{ mol Al}}{26.98 \text{ g Al}} \times \dfrac{1 \text{ mol AlCl}_3}{1 \text{ mol Al}} = 0.0693 \text{ mol AlCl}_3$

(b) $[\text{AlCl}_3] = \dfrac{0.0693 \text{ mol AlCl}_3}{23.8 \text{ mL}} \times \dfrac{1000 \text{ mL}}{1 \text{ L}} = 2.91 \text{ M AlCl}_3$

48. The balanced chemical reaction indicates that 4 mol $NaNO_2$ are formed from 2 mol Na_2CO_3.

$[\text{NaNO}_2] = \dfrac{138 \text{ g Na}_2\text{CO}_3}{1.42 \text{ L soln}} \times \dfrac{1 \text{ mol Na}_2\text{CO}_3}{106.0 \text{ g Na}_2\text{CO}_3} \times \dfrac{4 \text{ mol NaNO}_2}{2 \text{ mol Na}_2\text{CO}_3} = 1.83 \text{ M NaNO}_2$

49. The volume of solution determines the amount of product.

$\text{mass Ag}_2\text{CrO}_4 = 415 \text{ mL} \times \dfrac{1 \text{ L}}{1000 \text{ mL}} \times \dfrac{0.186 \text{ mol AgNO}_3}{1 \text{ L soln}} \times \dfrac{1 \text{ mol Ag}_2\text{CrO}_4}{2 \text{ mol AgNO}_3} \times \dfrac{331.73 \text{ g Ag}_2\text{CrO}_4}{1 \text{ mol Ag}_2\text{CrO}_4}$

$\text{mass Ag}_2\text{CrO}_4 = 12.8 \text{ g Ag}_2\text{CrO}_4$

50. $V_{\text{KMnO}_4} = 9.13 \text{ g KI} \times \dfrac{1 \text{ mol KI}}{166.0023 \text{ g KI}} \times \dfrac{2 \text{ mol KMnO}_4}{10 \text{ mol KI}} \times \dfrac{1 \text{ L KMnO}_4}{0.0797 \text{ mol KMnO}_4} = 0.138 \text{ L KMnO}_4$

51. $\text{mass Na} = 155 \text{ mL soln} \times \dfrac{1 \text{ L}}{1000 \text{ mL}} \times \dfrac{0.175 \text{ mol NaOH}}{1 \text{ L soln}} \times \dfrac{2 \text{ mol Na}}{2 \text{ mol NaOH}} \times \dfrac{22.99 \text{ g Na}}{1 \text{ mol Na}}$

$= 0.624 \text{ g Na}$

52. We determine the amount of HCl present initially, and the amount desired.

$\text{amount HCl present} = 250.0 \text{ mL} \times \dfrac{1.023 \text{ mmol HCl}}{1 \text{ mL soln}} = 255.8 \text{ mmol HCl}$

$\text{amount HCl desired} = 250.0 \text{ mL} \times \dfrac{1.000 \text{ mmol HCl}}{1 \text{ mL soln}} = 250.0 \text{ mmol HCl}$

$\text{mass Mg} = (255.8 - 250.0) \text{ mmol HCl} \times \dfrac{1 \text{ mmol Mg}}{2 \text{ mmol HCl}} \times \dfrac{24.3 \text{ mg Mg}}{1 \text{ mmol Mg}} = 70. \text{ mg Mg}$

53. The mass of oxalic acid enables us to determine the amount of NaOH in the solution.

$$[NaOH] = \frac{0.3126\,g\,H_2C_2O_4}{26.21\,mL\,soln} \times \frac{1000\,mL}{1\,L\,soln} \times \frac{1\,mol\,H_2C_2O_4}{90.04\,g\,H_2C_2O_4} \times \frac{2\,mol\,NaOH}{1\,mol\,H_2C_2O_4} = 0.2649\,M$$

54. The total amount of HCl present is the amount that reacted with the $CaCO_3$ plus the amount that reacted with the $Ba(OH)_2$ (aq).

$$\frac{\text{moles HCl from}}{CaCO_3\,\text{reaction}} = 0.1000\,g\,CaCO_3 \times \frac{1\,mol\,CaCO_3}{100.09\,g\,CaCO_3} \times \frac{2\,mol\,HCl}{1\,mol\,CaCO_3} \times \frac{1000\,mmol}{1\,mol}$$

$$= 1.998\,mmol\,HCl$$

$$\frac{\text{moles HCl from}}{Ba(OH)_2\,\text{reaction}} = 43.82\,mL \times \frac{0.01185\,mmol\,Ba(OH)_2}{1\,mL\,soln} \times \frac{2\,mmol\,HCl}{1\,mmol\,Ba(OH)_2}$$

$$= 1.039\,mmol\,HCl$$

The HCl molarity is this total mmol of HCl divided by the total volume of 25.00 mL.

$$[HCl] = \frac{(1.998 + 1.039)\,mmol\,HCl}{25.00\,mL} = 0.1215\,M$$

Determining the Limiting Reactant

55. The limiting reactant is NH_3. For every mole of NH_3(g) that reacts, a mole of NO(g) forms. Since 3.00 moles of NH_3(g) reacts, 3.00 moles of NO(g) forms (1:1 mole ratio).

56. The reaction of interest is: CaH_2(s) + 2 H_2O(l) $\rightarrow$ $Ca(OH)_2$(s) + 2 H_2(g)
The limiting reactant is H_2O(l). The mole ratio between water and hydrogen gas is 1:1. Hence, if 1.54 moles of H_2O(l) reacts, 1.54 moles of H_2(g) forms (1:1 mole ratio).

57. First we must determine the number of moles of NO produced by each reactant. The one producing the smaller amount of NO is the limiting reactant.

$$mol\,NO = 0.696\,mol\,Cu \times \frac{2\,mol\,NO}{3\,mol\,Cu} = 0.464\,mol\,NO$$

$$mol\,NO = 136\,mL\,HNO_3\,(aq) \times \frac{1\,L}{1000\,mL} \times \frac{6.0\,mol\,HNO_3}{1\,L} \times \frac{2\,mol\,NO}{8\,mol\,HNO_3} = 0.204\,mol\,NO$$

Since HNO_3(aq) is the limiting reactant, it will be completely consumed, leaving some Cu unreacted.

58. First determine the mass of H_2 produced from each of the reactants. The smaller mass is that produced by the limiting reactant, which is the mass that should be produced.

$$mass\,H_2\,(Al) = 1.84\,g\,Al \times \frac{1\,mol\,Al}{26.98\,g\,Al} \times \frac{3\,mol\,H_2}{2\,mol\,Al} \times \frac{2.016\,g\,H_2}{1\,mol\,H_2} = 0.206\,g\,H_2$$

$$mass\,H_2\,(HCl) = 75.0\,mL \times \frac{1\,L}{1000\,mL} \times \frac{2.95\,mol\,HCl}{1\,L} \times \frac{3\,mol\,H_2}{6\,mol\,HCl} \times \frac{2.016\,g\,H_2}{1\,mol\,H_2} = 0.223\,g\,H_2$$

Thus, 0.206 g H_2 should be produced.

59. First we need to determine the amount of Na_2CS_3 produced from each of the reactants.

$$n_{Na_2CS_3 \text{ (from } CS_2)} = 92.5 \,\text{mL } CS_2 \times \frac{1.26 \,\text{g}}{1 \,\text{mL}} \times \frac{1 \,\text{mol } CS_2}{76.14 \,\text{g } CS_2} \times \frac{2 \,\text{mol } Na_2CS_3}{3 \,\text{mol } CS_2} = 1.02 \,\text{mol } Na_2CS_3$$

$$n_{Na_2CS_3 \text{ (from NaOH)}} = 2.78 \,\text{mol NaOH} \times \frac{2 \,\text{mol } Na_2CS_3}{6 \,\text{mol NaOH}} = 0.927 \,\text{mol } Na_2CS_3$$

Thus, the mass produced is 0.927 mol $Na_2CS_3 \times \dfrac{154.2 \,\text{g } Na_2CS_3}{1 \,\text{mol } Na_2CS_3} = 143 \,\text{g } Na_2CS_3$

60. Since the two reactants combine in an equimolar basis, the one present with the fewer number of moles is the limiting reactant and determines the mass of the products.

$$\text{mol } ZnSO_4 = 315 \,\text{mL} \times \frac{1 \,\text{L}}{1000 \,\text{mL}} \times \frac{0.275 \,\text{mol } ZnSO_4}{1 \,\text{L soln}} = 0.0866 \,\text{mol } ZnSO_4$$

$$\text{mol BaS} = 285 \,\text{mL} \times \frac{1 \,\text{L}}{1000 \,\text{mL}} \times \frac{0.315 \,\text{mol BaS}}{1 \,\text{L soln}} = 0.0898 \,\text{mol BaS}$$

Thus, $ZnSO_4$ is the limiting reactant and 0.0866 mol of each of the products will be produced.

$$\text{mass products} = \left(0.0866 \,\text{mol } BaSO_4 \times \frac{233.4 \,\text{g } BaSO_4}{1 \,\text{mol } BaSO_4} \right) + \left(0.0866 \,\text{mol } ZnS \times \frac{97.46 \,\text{g } ZnS}{1 \,\text{mol } ZnS} \right)$$

$$= 28.7 \,\text{g product mixture (lithopone)}$$

61. $$Ca(OH)_2(s) + 2 NH_4Cl(s) \rightarrow CaCl_2(aq) + 2 H_2O(l) + 2 NH_3(g)$$

First compute the amount of NH_3 formed from each reactant in this limiting reactant problem.

$$n_{NH_3 \text{ (from NH}_4\text{Cl)}} = 33.0 \,\text{g } NH_4Cl \times \frac{1 \,\text{mol } NH_4Cl}{53.49 \,\text{g } NH_4Cl} \times \frac{2 \,\text{mol } NH_3}{2 \,\text{mol } NH_4Cl} = 0.617 \,\text{mol } NH_3$$

$$n_{NH_3 \text{ (from Ca(OH)}_2)} = 33.0 \,\text{g } Ca(OH)_2 \times \frac{1 \,\text{mol } Ca(OH)_2}{74.09 \,\text{g } Ca(OH)_2} \times \frac{2 \,\text{mol } NH_3}{1 \,\text{mol } Ca(OH)_2} = 0.891 \,\text{mol } NH_3$$

Thus, 0.617 mol NH_3 should be produced as NH_4Cl is the limiting reagent.

$$\text{mass } NH_3 = 0.617 \,\text{mol } NH_3 \times \frac{17.03 \,\text{g } NH_3}{1 \,\text{mol } NH_3} = 10.5 \,\text{g } NH_3$$

Now we will determine the mass of reactant in excess, $Ca(OH)_2$.

$$Ca(OH)_2 \text{ used} = 0.617 \,\text{mol } NH_3 \times \frac{1 \,\text{mol } Ca(OH)_2}{2 \,\text{mol } NH_3} \times \frac{74.09 \,\text{g } Ca(OH)_2}{1 \,\text{mol } Ca(OH)_2} = 22.9 \,\text{g } Ca(OH)_2$$

excess mass $Ca(OH)_2 = 33.0$ g $Ca(OH)_2 - 22.9$ g $Ca(OH)_2 = 10.1$ g excess $Ca(OH)_2$

62. The balanced chemical equation is: $Ca(OH)_2(s) + 4 HCl(aq) \rightarrow CaCl_2(aq) + 2 H_2O(l)$

$$n_{Cl_2 \text{ (from Ca(OCl)}_2)} = 50.0 \,\text{g } Ca(OCl)_2 \times \frac{1 \,\text{mol } Ca(OCl)_2}{142.98 \,\text{g } Ca(OCl)_2} \times \frac{2 \,\text{mol } Cl_2}{1 \,\text{mol } Ca(OCl)_2} = 0.699 \,\text{mol } Cl_2$$

$$n_{Cl_2 \text{ (from HCl)}} = 275 \,\text{mL} \times \frac{1 \,\text{L}}{1000 \,\text{mL}} \times \frac{6.00 \,\text{mol HCl}}{1 \,\text{L soln}} \times \frac{2 \,\text{mol } Cl_2}{4 \,\text{mol HCl}} = 0.825 \,\text{mol } Cl_2$$

Thus, mass Cl_2 expected $= 0.699\,\text{mol } Cl_2 \times \dfrac{70.91\,\text{g } Cl_2}{1\,\text{mol } Cl_2} = 49.6\,\text{g } Cl_2$

The excess reactant is the one that produces the most Cl_2, namely HCl(aq). The quantity of excess HCl(aq) is determined from the amount of excess Cl_2(g) it theoretically could produce (if it were the limiting reagent).

$V_{\text{excess HCl}} = (0.825\text{-}0.699)\,\text{mol } Cl_2 \times \dfrac{4\,\text{mol HCl}}{2\,\text{mol } Cl_2} \times \dfrac{1000\,\text{mL}}{6.00\,\text{mol HCl}}$

$V_{\text{excess HCl}} = 42.0\,\text{mL excess 6.00 M HCl(aq)}$

mass excess HCl $= (0.825 - 0.699)\,\text{mol } Cl_2 \times \dfrac{4\,\text{mol HCl}}{2\,\text{mol } Cl_2} \times \dfrac{36.46\,\text{g HCl}}{1\,\text{mol HCl}} = 9.19\,\text{g excess HCl}$

Theoretical, Actual and Percent Yields

63. **(a)** Since the stoichiometry indicates that 1 mole CCl_2F_2 is produced per mole CCl_4, the use of 1.80 mole CCl_4 should produce 1.80 mole CCl_2F_2. This is the theoretical yield of the reaction.

(b) The actual yield of the reaction is the amount actually produced, 1.55 mol CCl_2F_2.

(c) % yield $= \dfrac{1.55\,\text{mol } CCl_2F_2 \text{ obtained}}{1.80\,\text{mol } CCl_2F_2 \text{ calculated}} \times 100\% = 86.1\%$ yield

64. **(a)** mass $C_6H_{10} = 100.0\,\text{g } C_6H_{11}OH \times \dfrac{1\,\text{mol } C_6H_{11}OH}{100.16\,\text{g } C_6H_{11}OH} \times \dfrac{1\,\text{mol } C_6H_{10}}{1\,\text{mol } C_6H_{11}OH}$

$\times \dfrac{82.146\,\text{g } C_6H_{10}}{1\,\text{mol } C_6H_{10}} = 82.01\,\text{g } C_6H_{10} = $ theoretical yield

(b) percent yield $= \dfrac{64.0\,\text{g } C_6H_{10} \text{ produced}}{82.01\,\text{g } C_6H_{10} \text{ calculated}} \times 100\% = 78.0\%$ yield

(c) mass $C_6H_{11}OH = 100.0\,\text{g } C_6H_{10} \text{ produced} \times \dfrac{1.000\,\text{g calculated}}{0.780\,\text{g produced}} \times \dfrac{1\,\text{mol } C_6H_{10}}{82.15\,\text{g } C_6H_{10}}$

$\times \dfrac{1\,\text{mol } C_6H_{11}OH}{1\,\text{mol } C_6H_{10}} \times \dfrac{100.2\,\text{g } C_6H_{11}OH}{1\,\text{mol } C_6H_{11}OH} = 156\,\text{g } C_6H_{11}OH$ are needed

65. **(a)** We first need to solve the limiting reactant problem involved here.

$n_{C_4H_9Br \,(\text{from } C_4H_9OH)} = 15.0\,\text{g } C_4H_9OH \times \dfrac{1\,\text{mol } C_4H_9OH}{74.12\,\text{g } C_4H_9OH} \times \dfrac{1\,\text{mol } C_4H_9Br}{1\,\text{mol } C_4H_9OH} = 0.202\,\text{mol } C_4H_9Br$

$n_{C_4H_9Br \,(\text{from NaBr})} = 22.4\,\text{g NaBr} \times \dfrac{1\,\text{mol NaBr}}{102.9\,\text{g NaBr}} \times \dfrac{1\,\text{mol } C_4H_9Br}{1\,\text{mol NaBr}} = 0.218\,\text{mol } C_4H_9Br$

$n_{C_4H_9Br \,(\text{from } H_2SO_4)} = 32.7\,\text{g } H_2SO_4 \times \dfrac{1\,\text{mol } H_2SO_4}{98.1\,\text{g } H_2SO_4} \times \dfrac{1\,\text{mol } C_4H_9Br}{1\,\text{mol } H_2SO_4} = 0.333\,\text{mol } C_4H_9Br$

Thus the theoretical yield of $C_4H_9Br = 0.202\,\text{mol } C_4H_9Br \times \dfrac{137.0\,\text{g } C_4H_9Br}{1\,\text{mol } C_4H_9Br} = 27.7\,\text{g } C_4H_9Br$

(b) The actual yield is the mass obtained, 17.1 g C_4H_9Br.

(c) Then, % yield = $\dfrac{17.1\,\text{g C}_4\text{H}_9\text{Br produced}}{27.7\,\text{g C}_4\text{H}_9\text{Br expected}} \times 100\% = 61.7\%$ yield

66. (a) Again, we solve the limiting reactant problem first.

$$n_{(C_6H_5N)_2}\,^{(\text{from }C_6H_5NO_2)} = 0.10\,\text{L C}_6\text{H}_5\text{NO}_2 \times \dfrac{1000\,\text{mL}}{1\,\text{L}} \times \dfrac{1.20\,\text{g}}{1\,\text{mL}} \times \dfrac{1\,\text{mol C}_6\text{H}_5\text{NO}_2}{123.1\,\text{g C}_6\text{H}_5\text{NO}_2}$$

$$\times \dfrac{1\,\text{mol}(C_6H_5N)_2}{2\,\text{mol C}_6\text{H}_5\text{NO}_2} = 0.49\,\text{mol}(C_6H_5N)_2$$

$$n_{(C_6H_5N)_2}\,^{(\text{from }C_6H_{14}O_4)} = 0.30\,\text{L C}_6\text{H}_{14}\text{O}_4 \times \dfrac{1000\,\text{mL}}{1\,\text{L}} \times \dfrac{1.12\,\text{g}}{1\,\text{mL}} \times \dfrac{1\,\text{mol C}_6\text{H}_{14}\text{O}_4}{150.2\,\text{g C}_6\text{H}_{14}\text{O}_4}$$

$$\times \dfrac{1\,\text{mol}(C_6H_5N)_2}{4\,\text{mol C}_6\text{H}_{14}\text{O}_4} = 0.56\,\text{mol}(C_6H_5N)_2$$

Thus the theoretical yield for $(C_6H_5N)_2 = 0.49\,\text{mol}(C_6H_5N)_2 \times \dfrac{182.2\,\text{g}(C_6H_5N)_2}{1\,\text{mol}(C_6H_5N)_2} = 89\,\text{g}(C_6H_5N)_2$

(b) actual yield $= 55\,\text{g}(C_6H_5N)_2$ produced

(c) percent yield $= \dfrac{55\,\text{g}(C_6H_5N)_2 \text{ produced}}{89\,\text{g}(C_6H_5N)_2 \text{ expected}} \times 100\% = 62\%$ yield

67. Balanced equation: $3\,C_2H_4O_2 + PCl_3 \rightarrow 3\,C_2H_3OCl + H_3PO_3$

mass acid $= 75\,\text{g C}_2\text{H}_3\text{OCl} \times \dfrac{100.0\,\text{g calculated}}{78.2\,\text{g produced}} \times \dfrac{1\,\text{mol C}_2\text{H}_3\text{OCl}}{78.5\,\text{g C}_2\text{H}_3\text{OCl}} \times \dfrac{3\,\text{mol C}_2\text{H}_4\text{O}_2}{3\,\text{mol C}_2\text{H}_3\text{OCl}}$

$$\times \dfrac{60.1\,\text{g pure C}_2\text{H}_4\text{O}_2}{1\,\text{mol C}_2\text{H}_4\text{O}_2} \times \dfrac{100\,\text{g commercial}}{97\,\text{g pure C}_2\text{H}_4\text{O}_2} = 76 \text{ g commercial C}_2\text{H}_4\text{O}_2$$

68. mass $CH_2Cl_2 = 112\,\text{g CH}_4 \times \dfrac{1\,\text{mol CH}_4}{16.04\,\text{g CH}_4} \times \dfrac{1\,\text{mol CH}_3\text{Cl}}{1\,\text{mol CH}_4} \times \dfrac{0.92\,\text{mol CH}_3\text{Cl produced}}{1.00\,\text{mol CH}_3\text{Cl expected}}$

$$\times \dfrac{1\,\text{mol CH}_2\text{Cl}_2}{1\,\text{mol CH}_3\text{Cl}} \times \dfrac{84.93\,\text{g CH}_2\text{Cl}_2}{1\,\text{mol CH}_2\text{Cl}_2} \times \dfrac{0.92\,\text{g CH}_2\text{Cl}_2 \text{ produced}}{1.00\,\text{g CH}_2\text{Cl}_2 \text{ calculated}} = 5.0 \times 10^2\,\text{g CH}_2\text{Cl}_2$$

69. A less-than-100% yield of desired product in synthesis reactions is always the case. This is because of side reactions that yield products other than those desired and because of the loss of material on the glassware, on filter paper, etc. during the various steps of the procedure. A main criterion for choosing a synthesis reaction is how economically it can be run. In the analysis of a compound, on the other hand, it is essential that all of the material present be detected. Therefore, a 100% yield is required; none of the material present in the sample can be lost during the analysis. Therefore analysis reactions are carefully chosen to meet this 100 % yield criterion; they need not be economical to run.

70. The theoretical yield is 2.07 g Ag_2CrO_4. If the mass actually obtained is less than this, it is likely that some of the pure material was not recovered, perhaps stuck to the walls of the flask in which the reaction occurred, or left suspended in the solution. Thus while it is almost a certainty that less than 2.07 g will be obtained, the absolute maximum mass of Ag_2CrO_4 expected is 2.07 g. If the precipitate weighs more than 2.07 g, the extra mass must be impurities (e.g., the precipitate was not thoroughly dried).

Consecutive Reactions, Simultaneous Reactions

71. Hence we must determine the amount of HCl needed to react with each component of the mixture.

$$Mg(OH)_2 (s) + 2\,HCl(aq) \longrightarrow MgCl_2 (aq) + 2\,H_2O(l)$$

$$MgCO_3 (s) + 2\,HCl(aq) \longrightarrow MgCl_2 (aq) + H_2O(l) + CO_2 (g)$$

$$n_{HCl\,(consumed\,by\,MgCO_3)} = 425\,g\,mixt. \times \frac{35.2\,g\,MgCO_3}{100.0\,g\,mixt.} \times \frac{1\,mol\,MgCO_3}{84.3\,g\,MgCO_3} \times \frac{2\,mol\,HCl}{1\,mol\,MgCO_3} = 3.55\,mol\,HCl$$

$$n_{HCl\,(consumed\,by\,Mg(OH)_2)} = 425\,g\,mixt. \times \frac{64.8\,g\,Mg(OH)_2}{100.0\,g\,mixt.} \times \frac{1\,mol\,Mg(OH)_2}{58.3\,g\,Mg(OH)_2} \times \frac{2\,mol\,HCl}{1\,mol\,MgCO_3} = 9.45\,mol\,HCl$$

$$mass\,HCl = (3.55 + 9.45)\,mol\,HCl \times \frac{36.46\,g\,HCl}{1\,mol\,HCl} = 474\,g\,HCl$$

72. Here we need to determine the amount of CO_2 produced from each reactant.

$$C_3H_8 (g) + 5\,O_2 (g) \longrightarrow 3\,CO_2 (g) + 4\,H_2O(l)$$

$$2\,C_4H_{10} (g) + 13\,O_2 (g) \longrightarrow 8\,CO_2 (g) + 10\,H_2O(l)$$

$$n_{CO_2\,(from\,C_3H_8)} = 406\,g\,mixt. \times \frac{72.7g\,C_3H_8}{100.0\,g\,mixt.} \times \frac{1\,mol\,C_3H_8}{44.10g\,C_3H_8} \times \frac{3\,mol\,CO_2}{1\,mol\,C_3H_8} = 20.1\,mol\,CO_2$$

$$n_{CO_2\,(from\,C_4H_{10})} = 406\,g\,mixt. \times \frac{27.3g\,C_4H_{10}}{100.0\,g\,mixt} \times \frac{1\,mol\,C_4H_{10}}{58.12g\,C_4H_{10}} \times \frac{8\,mol\,CO_2}{2\,mol\,C_4H_{10}} = 7.63\,mol\,CO_2$$

$$mass\,CO_2 = (20.1 + 7.63)\,mol\,CO_2 \times \frac{44.01g\,CO_2}{1\,mol\,CO_2} = 1.22 \times 10^3\,g\,CO_2$$

73. The molar ratios given by the stoichiometric coefficients in the balanced chemical equations are used in the solution, namely $\quad CH_4(g) + 4\,Cl_2(g) \rightarrow CCl_4(g) + 4\,HCl(g)$

$$CCl_4(g) + 2\,HF\,(g) \rightarrow CCl_2F_2(g) + 2\,HCl(g)$$

$$amount\,Cl_2 = 2.25 \times 10^3\,g\,CCl_2F_2 \times \frac{1\,mol\,CCl_2F_2}{120.91g\,CCl_2F_2} \times \frac{1\,mol\,CCl_4}{1\,mol\,CCl_2F_2} \times \frac{4\,mol\,Cl_2}{1\,mol\,CCl_4} = 74.4\,mol\,Cl_2$$

74. Balanced Equations: $\quad C_2H_6(g) + \frac{7}{2}\,O_2(g) \rightarrow 2\,CO_2(g) + 3\,H_2O(l)$

$$CO_2(g) + Ba(OH)_2(aq) \rightarrow BaCO_3(s) + H_2O(l)$$

$$mass_{C_2H_6} = 0.506\,g\,BaCO_3 \times \frac{1\,mol\,BaCO_3}{197.3\,g\,BaCO_3} \times \frac{1\,mol\,CO_2}{1\,mol\,BaCO_3} \times \frac{2\,mol\,C_2H_6}{4\,mol\,CO_2} \times \frac{30.07\,g\,C_2H_6}{1\,mol\,C_2H_6} = 0.0386\,g\,C_2H_6$$

75.

$NaI(aq) + AgNO_3(aq)$	$\rightarrow$	$AgI(s) + NaNO_3(aq)$	(multiply by 4)
$2\,AgI(s) + Fe(s)$	$\rightarrow$	$FeI_2(aq) + 2\,Ag(s)$	(multiply by 2)
$2\,FeI_2(aq) + 3\,Cl_2(g)$	$\rightarrow$	$2\,FeCl_3(aq) + 2\,I_2(s)$	(unchanged)

$$4NaI(aq) + 4AgNO_3(aq) + 2Fe(s) + 3Cl_2(g) \rightarrow 4NaNO_3(aq) + 4Ag(s) + 2FeCl_3(aq) + 2I_2(s)$$

For every 4 moles of $AgNO_3$, 2 moles of $I_2(s)$ are produced. The mass of $AgNO_3$ required

$$= 1.00\,kg\,I_2(s) \times \frac{1000\,g\,I_2(s)}{1\,kg\,I_2(s)} \times \frac{1\,mol\,I_2(s)}{253.809\,g\,I_2(s)} \times \frac{4\,mol\,AgNO_3(s)}{2\,mol\,I_2(s)} \times \frac{169.873\,g\,AgNO_3(s)}{1\,mol\,AgNO_3(s)}$$

$$= 1338.\underline{59}\,g\,AgNO_3\,per\,kg\,of\,I_2\,produced\,or\,1.34 \times 10^3\,g\,AgNO_3\,per\,kg\,of\,I_2\,produced$$

76.

$Fe + Br_2 \qquad\qquad \rightarrow FeBr_2 \qquad\qquad$ (multiply by 3)

$3\,FeBr_2 + Br_2 \qquad\quad \rightarrow Fe_3Br_8$

$\underline{Fe_3Br_8 + 4\,Na_2CO_3 \rightarrow 8\,NaBr + 4\,CO_2 + Fe_3O_4}$

$3\,Fe + 4\,Br_2 + 4\,Na_2CO_3 \;\rightarrow 8\,NaBr + 4\,CO_2 + Fe_3O_4$

Hence, 3 moles Fe(s) forms 8 mol NaBr

$$Mass_{Fe\ consumed} = 2.50{\times}10^3 \text{ kg NaBr} \times \frac{1000 \text{ g NaBr}}{1 \text{ kg NaBr}} \times \frac{1 \text{ mol NaBr}}{102.894 \text{ g NaBr}} \times \frac{3 \text{ mol Fe}}{8 \text{ mol NaBr}} \times \frac{55.847 \text{ g Fe}}{1 \text{ mol Fe}}$$

$$= 509{\times}10^3 \text{ g Fe} \times \frac{1 \text{ kg Fe}}{1000 \text{ g Fe}} = 509 \text{ kg Fe required to produce } 2.5 \times 10^3 \text{ kg KBr}$$

Integrative and Advanced Exercises

77. **(a)** $CaCO_3(s) \xrightarrow{\;\Delta\;} CaO(s) + CO_2(g)$

(b) $2\,ZnS(s) + 3\,O_2(g) \xrightarrow{\;\Delta\;} 2\,ZnO(s) + 2\,SO_2(g)$

(c) $C_3H_8(g) + 3\,H_2O(g) \longrightarrow 3\,CO(g) + 7\,H_2(g)$

(d) $4\,SO_2(g) + 2\,Na_2S(aq) + Na_2CO_3(aq) \longrightarrow CO_2(g) + 3\,Na_2S_2O_3(aq)$

78. **(a)** $4\,Ca_3(PO_4)_2 + 12\ SiO_2(s) + 20\ C(s) \xrightarrow{\;\Delta\;} 12\ CaSiO_3(s) + 2\,P_4(g) + 20\ CO(g)$

$P_4(s) + 6\ Cl_2(g) \longrightarrow 4\,PCl_3(g)$

$PCl_3(g) + 3\,H_2O(l) \longrightarrow H_3PO_3(aq) + 3\,HCl(aq)$

(b) $2\,Cu(s) + O_2(g) + CO_2(g) + H_2O(g) \longrightarrow Cu_2(OH)_2CO_3(s)$

(c) $P_4(s) + 5\ O_2(g) \longrightarrow P_4O_{10}(s) \xrightarrow{\;6\,H_2O\;} 4\ H_3PO_4(aq)$

(d) $3\,Ca(H_2PO_4)_2(aq) + 8\,NaHCO_3(aq) \longrightarrow Ca_3(PO_4)_2(aq) + 4\,Na_2HPO_4(aq) + 8\,CO_2(g) + 8\,H_2O(l)$

79. $\text{mass } CaCO_3 = 0.981 \text{ g } CO_2 \times \dfrac{1\,\text{mol } CO_2}{44.01\,\text{g } CO_2} \times \dfrac{1\,\text{mol } CaCO_3}{1\,\text{mol } CO_2} \times \dfrac{100.1\,\text{g } CaCO_3}{1\,\text{mol } CaCO_3} = 2.23 \text{ g } CaCO_3$

$\% \ CaCO_3 = \dfrac{2.23 \text{ g } CaCO_3}{3.28 \text{ g sample}} \times 100\% = 68.0\% \ \ CaCO_3 \text{ (by mass)}$

80. We determine the empirical formula, basing our calculation on 100.0 g of the compound.

$\text{amount Fe} = 72.3 \text{ g Fe} \times \dfrac{1 \text{ mol Fe}}{55.85 \text{ g Fe}} = 1.29 \text{ mol Fe} \qquad \div 1.29 \longrightarrow 1.00 \text{ mol Fe}$

$\text{amount O} = 27.7 \text{ g O} \times \dfrac{1 \text{ mol O}}{16.00 \text{ g O}} = 1.73 \text{ mol O} \qquad \div 1.29 \longrightarrow 1.34 \text{ mol O}$

The empirical formula is Fe_3O_4 and the balanced equation is as follows.

$3\,Fe_2O_3(s) + H_2(g) \longrightarrow 2\,Fe_3O_4(s) + H_2O(g)$

81. Assume 100g of the compound Fe_xS_y, then:

Number of moles of S atoms = 36.5g/32.066 g S/mol = 1.13$\underline{8}$ moles

Number of moles of Fe atoms = 63.5g/ 55.847g Fe/mol = 1.13$\underline{7}$ moles

So the empirical formula for the iron –containing reactant is FeS

Assume 100g of the compound Fe_xO_y, then:

Number of moles of O atoms = 27.6g/16.0 g O/mol = 1.72$\underline{5}$ moles
Number of moles of Fe atoms = 72.4g/ 55.847g Fe/mol = 1.29$\underline{6}$ moles
So the empirical formula for the iron-containing product is Fe_3O_4
Balanced equation: $3\,FeS + 5\,O_2 \rightarrow Fe_3O_4 + 3\,SO_2$

82. $\dfrac{1.52\,g\,Na}{1\times10^6\,g\,sol} \times \dfrac{1\,g\,sol}{1\,mL\,sol} \times \dfrac{1000\,mL}{1\,L} \times \dfrac{1\,mol\,Na}{22.9898\,g\,Na} \times \dfrac{1\,mol\,NaCl}{1\,mol\,Na} \times \dfrac{58.4425\,g\,NaCl}{1\,mol\,NaCl}$

$= 0.003864\,M$ NaCl

83. $mass\,Ca(NO_3)_2 = 50.0\,L\,soln \times \dfrac{1000\,mL}{1\,L} \times \dfrac{1.00\,g\,soln}{1\,mL\,soln} \times \dfrac{2.35\,g\,Ca}{10^6\,g\,soln} \times \dfrac{1\,mol\,Ca}{40.08\,g\,Ca}$

$\times \dfrac{1\,mol\,Ca(NO_3)_2}{1\,mol\,Ca} \times \dfrac{164.09\,g\,Ca(NO_3)_2}{1\,mol\,Ca(NO_3)_2} \times \dfrac{1000\,mg}{1\,g} = 481\,mg\,Ca(NO_3)_2$

84. We can compute the volume of Al that reacts with the given quantity of HCl.

$V_{Al} = 0.05\,mL \times \dfrac{1\,L}{1000\,mL} \times \dfrac{12.0\,mol\,HCl}{1\,L} \times \dfrac{2\,mol\,Al}{6\,mol\,HCl} \times \dfrac{27.0\,g\,Al}{1\,mol\,Al} \times \dfrac{1\,cm^3}{2.70\,g\,Al} = 0.002\,cm^3$

$area = \dfrac{volume}{thickness} = \dfrac{0.002\,cm^3}{0.10\,mm} \times \dfrac{10\,mm}{1\,cm} = 0.2\,cm^2$

85. Here we need to determine the amount of HCl before and after reaction; the difference is the amount of HCl that reacted.

initial amount HCl $= 0.05000\,L \times \dfrac{1.035\,mol\,HCl}{1\,L} = 0.05175\,mol\,HCl$

final amount HCl $= 0.05000\,L \times \dfrac{0.812\,mol\,HCl}{1\,L} = 0.0406\,mol\,HCl$

mass Zn $= (0.05175 - 0.0406)\,mol\,HCl \times \dfrac{1\,mol\,Zn}{2\,mol\,HCl} \times \dfrac{65.39\,g\,Zn}{1\,mol\,Zn} = 0.365\,g\,Zn$

86. Let us first determine the moles of NH_4NO_3 in the dilute solution.

$mass_{NH_4NO_3} = 1000\,mL \times \dfrac{2.37\times10^{-3}\,g\,N}{1\,mL} \times \dfrac{1\,mol\,N}{14.007\,g\,N} \times \dfrac{1\,mol\,NH_4NO_3}{2\,mol\,N} = 0.0846\,mol\,NH_4NO_3$

volume of solution $= 0.0846\,mol\,NH_4NO_3 \times \dfrac{1\,L\,soln}{0.715\,mol\,NH_4NO_3} \times \dfrac{1000\,mL}{1\,L} = 118\,mL\,soln$

87. First we determine the molarity of seawater.

concentration $= \dfrac{2.8\,g\,NaCl}{100.0\,g\,soln} \times \dfrac{1.03\,g}{1\,mL\,soln} \times \dfrac{1000\,mL}{1\,L\,soln} \times \dfrac{1\,mol\,NaCl}{58.44\,g\,NaCl} = 0.49\,M\,NaCl$

Then the volume of the final solution: $c_1V_1 = c_2V_2$ $\quad V_2 = \dfrac{0.49\,M \times 1.00\times10^6\,L}{5.45\,M} = 9.0\times10^4\,L$

Volume to be evaporated $= 1.00\times10^6\,L - 9.0\times10^4\,L = 9.1\times10^5\,L$ of water to be evaporated

88. Here we must determine the amount of PbI_2 produced from each solute in this limiting reactant problem.

$$n_{PbI_2 \text{ (from KI)}} = 99.8 \text{ mL} \times \frac{1.093 \text{ g}}{1 \text{ mL soln}} \times \frac{0.120 \text{ g KI}}{1 \text{ g soln}} \times \frac{1 \text{ mol KI}}{166.00 \text{ g KI}} \times \frac{1 \text{ mol PbI}_2}{2 \text{ mol KI}} = 0.0394 \text{ mol}$$

$$n_{PbI_2 \text{ (from Pb(NO}_3)_2)} = 96.7 \text{ mL} \times \frac{1.134 \text{ g}}{1 \text{ mL soln}} \times \frac{0.140 \text{ g Pb(NO}_3)_2}{1 \text{ g soln}} \times \frac{1 \text{ mol Pb(NO}_3)_2}{331.2 \text{ g}} \times \frac{1 \text{ mol PbI}_2}{1 \text{ mol Pb(NO}_3)_2}$$

$$= 0.0464 \text{ mol PbI}_2$$

Then the mass of PbI_2 is computed from the smaller amount produced.

$$\text{mass PbI}_2 = 0.0394 \text{ mol PbI}_2 \times \frac{461.0 \text{ g PbI}_2}{1 \text{ mol PbI}_2} = 18.2 \text{ g PbI}_2$$

89. $CaCO_3(s) + 2 HCl(aq) \longrightarrow CaCl_2(aq) + H_2O + CO_2(g)$

$$\text{Amount of HCl reacted} = 45.0 \text{ g CaCO}_3 \times \frac{1 \text{ mol CaCO}_3}{100.09 \text{ g CaCO}_3} \times \frac{2 \text{ mol HCl}}{1 \text{ mol CaCO}_3} = 0.899 \text{ mol HCl}$$

$$\text{Init. amt. of HCl} = 1.25 \text{ L soln} \times \frac{1000 \text{ mL}}{1 \text{ L}} \times \frac{1.13 \text{ g}}{1 \text{ mL}} \times \frac{0.257 \text{ g HCl}}{1 \text{ g soln}} \times \frac{1 \text{ mol HCl}}{36.46 \text{ g HCl}} = 9.96 \text{ mol HCl}$$

$$\text{Final HCl concentration} = \frac{9.96 \text{ mol HCl} - 0.899 \text{ mol HCl}}{1.25 \text{ L soln}} = 7.25 \text{ M HCl}$$

90. We allow the mass of Al in the alloy to be represented by x. We then set up an expression for determining the mass of H_2 and solve this expression for x.

$$0.105 \text{ g H}_2 = \left((2.05 - x) \text{ g Fe} \times \frac{1 \text{ mol Fe}}{55.8 \text{ g Fe}} \times \frac{1 \text{ mol H}_2}{1 \text{ mol Fe}} \times \frac{2.02 \text{ g H}_2}{1 \text{ mol H}_2} \right)$$

$$+ \left(x \text{ g Al} \times \frac{1 \text{ mol Al}}{27.0 \text{ g Al}} \times \frac{3 \text{ mol H}_2}{2 \text{ mol Al}} \times \frac{2.02 \text{ g H}_2}{1 \text{ mol H}_2} \right)$$

$0.105 \text{ g H}_2 = (2.05 - x)0.0362 + 0.112x = 0.0742 + (0.112 - 0.0362)x = 0.0742 + 0.076x$

$0.076 \, x = 0.105 - 0.0742 = 0.031 \qquad x = \dfrac{0.031}{0.076} = 0.41 \text{ g Al}$

$\%\text{Al} = \dfrac{0.41 \text{ g Al}}{2.05 \text{ g alloy}} = 20.\% \text{ Al (by mass)}$ and the alloy is also 80. % Fe (by mass).

91. $Mg(s) + 2 HCl(aq) \longrightarrow MgCl_2(aq) + H_2(g)$

$2 Al(s) + 6 HCl(aq) \longrightarrow 2 AlCl_3(aq) + 3 H_2(g)$

Let x represent the mass of Mg.

$$0.0163 \text{ g H}_2 \times \frac{1 \text{ mol H}_2}{2.016 \text{ g H}_2} = x \text{ g Mg} \times \frac{1 \text{ mol Mg}}{24.305 \text{ g Mg}} \times \frac{1 \text{ mol H}_2}{1 \text{ mol Mg}} + (0.155 - x) \text{ g Al} \times \frac{1 \text{ mol Al}}{26.982 \text{ g Al}} \times \frac{3 \text{ mol H}_2}{2 \text{ mol Al}}$$

$0.00809 = 0.041144x + 0.00862 - 0.055593x \qquad x = \dfrac{0.00862 - 0.00809}{0.055593 - 0.041144} = 0.0367 \text{ g Mg}$

$\% \text{ Mg} = \dfrac{0.0367 \text{ g Mg}}{0.155 \text{ g alloy}} \times 100\% = 23.7\% \text{ Mg or} \sim 24\% \text{ Mg (by mass)}.$

92. One way to solve this problem would be to calculate the mass of CO_2 produced from a 0.220 g sample of each alcohol. The results are 0.303 g CO_2 from 0.220 g CH_3OH and 0.421 g CO_2 from 0.220 g C_2H_5OH. Obviously a mixture has been burned. But we have sufficient information to determine the composition of the mixture. First we need the balanced equations for the combustion reactions. Then, we represent the mass of CH_3OH by x.

$$2\,CH_3OH(l) + 3\,O_2(g) \rightarrow 2\,CO_2(g) + 4\,H_2O(l) \qquad C_2H_5OH(l) + 3\,O_2(g) \rightarrow 2\,CO_2(g) + 3\,H_2O(l)$$

$$\text{mass } CO_2 = 0.352 \text{ g } CO_2 = \left(x \text{ g } CH_3OH \times \frac{1 \text{ mol } CH_3OH}{32.04 \text{ g } CH_3OH} \times \frac{2 \text{ mol } CO_2}{2 \text{ mol } CH_3OH} \times \frac{44.01 \text{ g } CO_2}{1 \text{ mol } CO_2} \right)$$

$$+ \left((0.220 - x) \text{ g } C_2H_5OH \times \frac{1 \text{ mol } C_2H_5OH}{46.07 \text{ g } C_2H_5OH} \times \frac{2 \text{ mol } CO_2}{1 \text{ mol } C_2H_5OH} \times \frac{44.01 \text{ g } CO_2}{1 \text{ mol } CO_2} \right)$$

$$= 1.374x + (0.220 - x)1.911 = 0.421 - 0.537x$$

$$-0.537x = 0.352 - 0.420 = -0.068 \text{ or } x = \frac{-0.068}{-0.537} = 0.127 \text{ g } CH_3OH$$

by difference, the mass of C_2H_5OH is 0.220 g $-$ 0.127 g $=$ 0.093 g CH_3OH

93. $C_2H_5OH(l) + 3O_2(g) \longrightarrow 2\,CO_2(g) + 3\,H_2O(l)$

$(C_2H_5)_2O\,(l) + 6\,O_2(g) \longrightarrow 4\,CO_2(g) + 5\,H_2O(l)$

Since this is classic mixture problem, we can use the systems of equations method to find the mass percents. First we let x be the mass of $(C_2H_5)_2O$ and y be the mass of C_2H_5OH. Thus,

$x + y = 1.005$ g or $y = 1.005$ g $- x$

We then construct a second equation involving x that relates the mass of carbon dioxide formed to the masses of ethanol and diethyl ether., viz.

$$1.963 \text{ g } CO_2 \times \frac{1 \text{ mol } CO_2}{44.010 \text{ g } CO_2} = x \text{ g } (C_2H_5)_2O \times \frac{1 \text{ mol } (C_2H_5)_2O}{74.123 \text{ g } (C_2H_5)_2O} \times \frac{4 \text{ mol } CO_2}{1 \text{ mol } (C_2H_5)_2O}$$

$$+ (1.005 - x) \text{ g } C_2H_5OH \times \frac{1 \text{ mol } C_2H_5OH}{46.07 \text{ g } C_2H_5OH} \times \frac{2 \text{ mol } CO_2}{1 \text{ mol } C_2H_5OH}$$

$$0.04460 = 0.05396x + 0.04363 - 0.04341x \qquad x = \frac{0.04460 - 0.04363}{0.05396 - 0.04341} = 0.092 \text{ g } (C_2H_5)_2O$$

$$\%\,(C_2H_5)_2O \text{ (by mass)} = \frac{0.092 \text{ g } (C_2H_5)_2O}{1.005 \text{ g mixture}} \times 100\% = 9.2\%\,(C_2H_5)_2O$$

$\%\,C_2H_5OH$ (by mass) $= 100.0\% - 9.2\%\,(C_2H_5)_2O = 90.8\%\,C_2H_5OH$

94. (a) $\text{mol } Cu^{2+} = 48.7 \text{ g } Cu^{2+} \times \dfrac{1 \text{ mol } Cu^{2+}}{63.55 \text{ g } Cu^{2+}} = 0.766 \text{ mol } Cu^{2+} \quad \div 0.307 \longrightarrow 2.50 \text{ mol } Cu^{2+}$

$\text{mol } CrO_4^{2-} = 35.6 \text{ g } CrO_4^{2-} \times \dfrac{1 \text{ mol } CrO_4^{2-}}{115.99 \text{ g}} = 0.307 \text{ mol } CrO_4^{2-} \quad \div 0.307 \longrightarrow 1.00 \text{ mol } CrO_4^{2-}$

$\text{mol } OH^- = 15.7 \text{ g } OH^- \times \dfrac{1 \text{ mol } OH^-}{17.01 \text{ g } OH^-} = 0.923 \text{ mol } OH^- \quad \div 0.307 \longrightarrow 3.01 \text{ mol } OH^-$

Empirical formula $\qquad Cu_5(CrO_4)_2(OH)_6$

(b) $5\,CuSO_4(aq)\ +\ 2\,K_2CrO_4(aq)\ +\ 6\,H_2O\,(l)$

$$\downarrow$$

$$Cu_5(CrO_4)_2(OH)_6(s)\ +\ 2\,K_2SO_4(aq) + 3\,H_2SO_4(aq)$$

95. We first need to compute the empirical formula of malonic acid.

$$34.62\,g\,C \times \frac{1\,mol\,C}{12.01\,g\,C} = 2.883\,mol\,C \qquad \div 2.883 \longrightarrow 1.000\,mol\,C$$

$$3.88\,g\,H \times \frac{1\,mol\,O}{1.01\,g\,H} = 3.84\,mol\,O \qquad \div 2.883 \longrightarrow 1.33\,mol\ H$$

$$61.50\,g\,O \times \frac{1\,mol\,O}{16.00\,g\,O} = 3.844\,mol\ O \qquad \div 2.883 \longrightarrow 1.333\,mol\ O$$

Multiply each of these mole numbers by 3 to obtain the empirical formula $C_3H_4O_4$.

Combustion reaction: $C_3H_4O_4(l) + 2\ O_2(g) \longrightarrow 3\,CO_2(g)\ +\ 2\,H_2O(l)$

96. (a) $2\,H_2(g) + O_2(g) \longrightarrow 2\,H_2O(l)$

(b) Determination of the Limiting Reagent

$$\text{mol }H_2O(\text{if }H_2 \text{ is lim. reagent}) = 4.800\ g\,H_2 \times \frac{1\,mol\ H_2}{2.016\quad g\,H_2} \times \frac{1\,mol\ H_2O}{1\quad mol\,H_2} = 2.381\,mol\ H_2O$$

$$\text{mol }H_2O(\text{ if }O_2 \text{ is lim. reagent}) = 36.40\,g\,O_2 \times \frac{1\,mol\ O_2}{32.00\,g\ O_2} \times \frac{2\,mol\ H_2O}{1\,mol\ O_2} = 2.275\,mol\,H_2O$$

Thus: $\quad \text{mass }H_2O = 2.275\,mol\,H_2O \times \dfrac{18.02\,g\,H_2O}{1\,mol\,H_2O} = 41.00\,g\,H_2O$

All of the 36.40 g of O_2 is used up; only H_2 remains. Since the mass of the products must equal the mass of the reactants, the unreacted mass of H_2 is computed as follows.
mass unreacted H_2 = mass of reactants – mass H_2O produced

$$= (4.800\,g\,H_2 + 36.40\,g\,O_2) - 41.00\,g\,H_2O = 0.20\,g\,H_2$$

(c) We used the principle of equality of masses of products and reactants to solve part (b). We can demonstrate the validity of this principle by computing the mass of H_2 unreacted in another way.

$$\text{mass }H_2 \text{ unreacted} = (2.381 - 2.275)\,mol\,H_2O \times \frac{1\ mol\ H_2}{1\ mol\ H_2O} \times \frac{2.016\ g\ H_2}{1\ mol\ H_2} = 0.214\ g\,H_2$$

The agreement of this result with the one from part (b) demonstrates the validity of the principle of equality of the masses of products and reactants.

97. Compute the amount of $AgNO_3$ in the solution on hand and the amount of $AgNO_3$ in the desired solution. the difference is the amount of $AgNO_3$ that must be added; simply convert this amount to a mass.

$$\text{amount AgNO}_3 \text{ present} = 50.00 \text{ mL} \times \frac{0.0500 \text{ mmol AgNO}_3}{1 \text{ mL soln}} = 2.50 \text{ mmol AgNO}_3$$

$$\text{amount AgNO}_3 \text{ desired} = 100.0 \text{ mL} \times \frac{0.0750 \text{ mmol AgNO}_3}{1 \text{ mL soln}} = 7.50 \text{ mmol AgNO}_3$$

$$\text{mass AgNO}_3 = (7.50 - 2.50) \text{ mmol AgNO}_3 \times \frac{1 \text{ mol AgNO}_3}{1000 \text{ mmol AgNO}_3} \times \frac{169.9 \text{ g Ag NO}_3}{1 \text{ mol AgNO}_3}$$

$$= 0.850 \text{ g AgNO}_3$$

98. The balanced equation for the reaction is: $S_8(s) + 4 Cl_2(g) \rightarrow 4 S_2Cl_2(l)$
Both the top left and top right boxes are consistent with the stoichiometry of this equation. Neither bottom row box is valid. Box (c) does not account for all the S_8, since we started out with 3 molecules, but end up with 1 S_8 molecule and 4 S_2Cl_2 molecules. Box (d) shows a yield of 2 S_8 molecules and 8 S_2Cl_2 molecules so we ended up with more sulfur atoms than we started with. This, of course, violates the Law of Conservation of Mass.

99. The ammonium dichromate reaction is an example of an internal redox reaction. Both the oxidizing agent ($Cr_2O_7^{2-}$) and the reducing agent (NH_4^+) are found in the compound in the correct stoichiometry.

$$2 NH_4^+ \rightarrow N_2 + 8 H^+ + 6 e^-$$
$$8 H^+ + 6 e^- + Cr_2O_7^{2-} \rightarrow Cr_2O_3 + 4 H_2O$$
$$\overline{\qquad\qquad\qquad\qquad\qquad\qquad\qquad\qquad}$$
$$(NH_4)_2Cr_2O_7(s) \rightarrow Cr_2O_3(s) + 4 H_2O(l) + N_2(g)$$

$$1000 \text{ g } (NH_4)_2Cr_2O_7 \times \frac{1 \text{ mol } (NH_4)_2Cr_2O_7}{252.065 \text{ g } (NH_4)_2Cr_2O_7} \times \frac{1 \text{ mol } N_2}{1 \text{ mol } (NH_4)_2Cr_2O_7} \times \frac{28.0134 \text{ g } N_2}{1 \text{ mol } N_2} = 111.1 \text{ g } N_2$$

100. There are many ways one can go about answering this question. We must use all of the most concentrated solution and dilute this solution down using the next most concentrated solution. Hence start with 345 mL of 01.29 M then add x mL of the 0.775 M solution. The value of x is obtained by solving the following equation.

$$1.25 \text{ M} = \frac{(1.29 \text{ M} \times 0.345 \text{ L}) + (0.775 \text{ M} \times x)}{(0.345 + x) \text{ L}}$$

$$1.25 \text{ M} \times (0.345 + x) \text{ L} = (1.29 \text{ M} \times 0.345 \text{ L}) + (0.775 \text{ M} \times x)$$

$$043125 + 1.25x = 0.44505 + 0.775x \quad \text{Thus,} \quad 0.0138 = 0.475x$$

$$x = 0.029 \text{ L or } 29 \text{ mL}$$
A total of (29 mL + 345 mL) = 374 mL may be prepared this way.

101. $1.00 \times 10^3 \text{ kg FeTiO}_3 \times \dfrac{1 \text{ kmol FeTiO}_3}{151.725 \text{ kg FeTiO}_3} \times \dfrac{1 \text{ kmol FeSO}_4 \cdot 7 H_2O}{1 \text{ kmol FeTiO}_3} \times \dfrac{278.018 \text{ kg FeSO}_4 \cdot 7 H_2O}{1 \text{ kmol FeSO}_4 \cdot 7 H_2O}$

$$= 1.83 \times 10^3 \text{ kg FeSO}_4 \cdot 7 H_2O$$

102. $\text{mass of } Fe_2O_3 = 1.00 \times 10^3 \text{ kg } FeSO_4 \cdot 7\,H_2O \times \dfrac{1 \text{ kmol } FeSO_4 \cdot 7\,H_2O}{278.018 \text{ kg } FeSO_4 \cdot 7\,H_2O} \times \dfrac{1 \text{ kmol } Fe_2O_3}{2 \text{ kmol } FeSO_4 \cdot 7\,H_2O}$

$\times \dfrac{159.692 \text{ kg } Fe_2O_3}{1 \text{ kmol } Fe_2O_3} = 287 \text{ kg kg } Fe_2O_3$

103. $1.00 \times 10^3 \text{ kg } FeTiO_3 \times \dfrac{1 \text{ kmol } FeTiO_3}{151.725 \text{ kg } FeTiO_3} \times \dfrac{1 \text{ kmol } FeSO_4 \bullet 7\,H_2O}{1 \text{ kmol } FeTiO_3} \times \dfrac{1 \text{ kmol } H_2SO_4}{1 \text{ kmol } FeSO_4 \bullet 7\,H_2O}$

$\times \dfrac{98.0795 \text{ kg } H_2SO_4}{1 \text{ kmol } H_2SO_4} = 646 \text{ kg } H_2SO_4$

104. (a) $6\,CO(NH_2)_2(l) \rightarrow 6\,HNCO(l) + 6\,NH_3(g) \rightarrow C_3N_3(NH_2)_3(l) + 3\,CO_2(g)$

(b) $\text{mass } C_3N_3(NH_2) = 100.0 \text{ kg } CO(NH_2)_2 \times \dfrac{1 \text{ kmol } CO(NH_2)_2}{60.063 \text{ kg } CO(NH_2)_2} \times \dfrac{1 \text{ kmol } C_3N_3(NH_2)_3}{6 \text{ kmol } CO(NH_2)_2}$

$\times \dfrac{126.121 \text{ kg } C_3N_3(NH_2)_3}{1 \text{ kmol } C_3N_3(NH_2)_3} \times \dfrac{84 \text{ g actual yield}}{100 \text{ g theoretical yield}} = 29.\underline{4} \text{ kg } C_3N_3(NH_2)_3$

105. (a) $2\,C_3H_6(g) + 2\,NH_3(g) + 3\,O_2(g) \rightarrow 2\,C_3H_3N(l) + 6\,H_2O(l)$

(b) For every pound of propylene we get 0.73 pound of acrylonitrile; we can also say that for every gram of propylene we get 0.73 gram of acrylonitrile. One gram of propylene is $0.023\underline{8}$ mol of propylene. The corresponding quantity of NH_3 is $0.023\underline{8}$ mol or $0.40\underline{5}$ g; then because NH_3 and C_3H_6 are required in the same molar amount (2:2) for the reaction, $0.40\underline{5}$ of a pound of NH_3 will be required for every 0.73 of a pound of acrylonitrile. To get 2000 lb of acrylonitrile we need, by simple proportion, $2000 \times (0.40\underline{5})/0.73 = 1.1 \times 10^3$ pounds NH_3.

FEATURE PROBLEMS

106. I f the sample that was caught is representative of all fish in the lake, there are five marked fish for every 18 fish. Thus, the total number of fish in the lake is determined.

$\text{total fish} = 100 \text{ marked fish} \times \dfrac{18 \text{ fish}}{5 \text{ marked fish}} = 360 \text{ fish} \cong 4 \times 10^2 \text{ fish}$

107. (a) The graph obtained is one of two straight lines, meeting at a peak of about 2.50 g $Pb(NO_3)_2$, corresponding to about 3.5 g PbI_2. Maximum mass of PbI_2 (calculated)

$= 2.503 \text{ g KI} \times \dfrac{1 \text{ mol KI}}{166.0 \text{ g KI}} \times \dfrac{1 \text{ mol } PbI_2}{2 \text{ mol KI}} \times \dfrac{461.01 \text{ g } PbI_2}{1 \text{ mol } PbI_2} = 3.476 \text{ g } PbI_2$

(b) The total quantity of reactant is limited to 5.000 g. If either reactant is in excess, the amount in excess will be "wasted," because it cannot be used to form product. Thus, we obtain the maximum amount of product when neither reactant is in excess (i.e., when there is a stoichiometric amount of each present). The balanced chemical equation for this reaction, $2\,KI + Pb(NO_3)_2 \rightarrow 2\,KNO_3 + PbI_2$, shows that stoichiometric quantities are two moles of KI (166.00 g/mol) for each mole of $Pb(NO_3)_2$ (331.21 g/mol). If we have 5.000 g total, we can let the mass of KI equal x g, so that the mass of

$$Pb(NO_3)_2 = (5.000 - x) \text{ g. and the amount } KI = x \text{ g KI} \times \frac{1 \text{ mol KI}}{166.00 \text{ g}} = \frac{x}{166.00}$$

$$\text{amount } Pb(NO_3)_2 = (5.000 - x) \text{ g } Pb(NO_3)_2 \times \frac{1 \text{ mol } Pb(NO_3)_2}{331.21 \text{ g}} = \frac{5.000 - x}{331.21}$$

At the point of stoichiometric balance, amount $KI = 2 \times$ amount $Pb(NO_3)_2$

$$\frac{x}{166.00} = 2 \times \frac{5.000 - x}{331.21} \quad \text{OR} \quad 331.21x = 10.00 \times 166.00 - 332.00x$$

$$x = \frac{1660.0}{331.21 + 332.00} = 2.503 \text{ g KI} \times \frac{1 \text{ mol KI}}{166.00 \text{ g KI}} = 0.01508 \text{ mol KI}$$

$$5.000 - x = 2.497 \text{ g } Pb(NO_3)_2 \times \frac{1 \text{ mol } Pb(NO_3)_2}{331.21 \text{ g } Pb(NO_3)_2} = 0.007539 \text{ mol } Pb(NO_3)_2$$

As a mass ratio we have: $\dfrac{2.503 \text{ g KI}}{2.497 \text{ g } Pb(NO_3)_2} = \dfrac{1.002 \text{ g KI}}{1 \text{ g } Pb(NO_3)_2}$

As a molar ratio we have: $\dfrac{0.01508 \text{ mol KI}}{0.007539 \text{ mol } Pb(NO_3)_2} = \dfrac{2 \text{ mol KI}}{1 \text{ mol } Pb(NO_3)_2}$

(c) The molar ratio just determined in part (b) is the same as the ratio of the coefficients for KI and $Pb(NO_3)_2$ in the balanced chemical equation. To determine the proportions precisely, we simply use the balanced chemical equation.

108. (a) For the balanced equation, the order is immaterial; the relative amount of each is important. $20 \text{ rd} + 20 \text{ bl} + 30 \text{ gr} \rightarrow 1 \text{ necklace}$

(b) This is similar to a limiting reactant problem. We determine how many necklaces can be made from each quantity of beads.

$$\text{number of necklaces} = 10.0 \text{ kg beads} \times \frac{1000 \text{ g}}{1 \text{ kg}} \times \frac{1 \text{ rd bead}}{1.98 \text{ g}} \times \frac{1 \text{ necklace}}{20 \text{ rd beads}} = 252._5 \text{ necklaces}$$

$$\text{number of necklaces} = 10.0 \text{ kg beads} \times \frac{1000 \text{ g}}{1 \text{ kg}} \times \frac{1 \text{ bl bead}}{3.05 \text{ g}} \times \frac{1 \text{ necklace}}{20 \text{ bl beads}} = 163._9 \text{ necklaces}$$

$$\text{number of necklaces} = 10.0 \text{ kg beads} \times \frac{1000 \text{ g}}{1 \text{ kg}} \times \frac{1 \text{ gr bead}}{1.82 \text{ g}} \times \frac{1 \text{ necklace}}{30 \text{ gr beads}} = 183._1 \text{ necklaces}$$

We have expressed each result with an additional significant figure, written as a subscript, so that we can see the effect of rounding. With the beads available, we can produce 163 necklaces, since we are unable to produce a fraction of a necklace.

(c) Because the mass of a bead, and the total mass available of each type of bead, both are known to just three significant figures, our results are only known that well. The best we can state is that we can make at least 163 necklaces, because 164 is uncertain by one unit. We should not be surprised if we actually made just 161 necklaces, or if we produced 165 of them. More precise masses would help.

109. The re action is: $2 NaOH(aq) + Cl_2(g) + 2 NH_3(aq) \rightarrow N_2H_4(aq) + 2 NaCl(aq) + 2 H_2O(l)$

   (a) The theoretical maximum $= \dfrac{32.045 \text{ g per mole reaction}}{184.9607 \text{ g per mole reaction}} \times 100\% = 17.32\ \%$

   (b) The actual AE is less owing to side reactions that lower the yield of the product (N_2H_4).

   (c) The addition of acetone changes the mechanism, resulting in the elimination of the side reactions between N_2H_4 and NH_2Cl. This result in an increase in yield from ~70% to nearly 100%

   (d) $2 H_2 + N_2 \rightarrow N_2H_4$ has an AE of 100%, neglecting any side reactions.

110. The more HCl used, the more impure the sample (compared to $NaHCO_3$, twice as much HCl is needed to neutralize Na_2CO_3).
Sample from trona: 6.93 g sample forms 11.89 g AgCl or 1.72 g AgCl per gram sample.
Sample derived from manufactured sodium bicarbonate: 6.78 g sample forms 11.77 g AgCl or 1.74 g AgCl per gram sample.
Thus the trona sample is purer (i.e., it has the greater mass percent $NaHCO_3$).

CHAPTER 5
INTRODUCTION TO REACTIONS
IN AQUEOUS SOLUTIONS
PRACTICE EXAMPLES

1A In determining total $\left[Cl^-\right]$, we recall the definition of molarity: moles of solute per liter of solution.

$$\text{from } NaCl, \left[Cl^-\right] = \frac{0.438 \text{ mol } NaCl}{1 \text{ L soln}} \times \frac{1 \text{ mol } Cl^-}{1 \text{ mol } NaCl} = 0.438 \text{ M } Cl^-$$

$$\text{from } MgCl_2, \left[Cl^-\right] = \frac{0.0512 \text{ mol } MgCl_2}{1 \text{ L soln}} \times \frac{2 \text{ mol } Cl^-}{1 \text{ mol } MgCl_2} = 0.102 \text{ M } Cl^-$$

$$\left[Cl^-\right] \text{ total} = \left[Cl^-\right] \text{ from } NaCl + \left[Cl^-\right] \text{ from } MgCl_2 = 0.438 \text{ M} + 0.102 \text{ M} = 0.540 \text{ M } Cl^-$$

1B **(a)** $\dfrac{1.5 \text{ mg } F^-}{L} \times \dfrac{1 \text{ g } F^-}{1000 \text{ mg } F^-} \times \dfrac{1 \text{ mol } F^-}{18.998 \text{ g } F^-} = 7.9 \times 10^{-5} \text{ M } F^-$

(b) $1.00 \times 10^6 \text{ L} \times \dfrac{7.9 \times 10^{-5} \text{ mol } F^-}{1 L} \times \dfrac{1 \text{ mol } CaF_2}{2 \text{ mol } F^-} \times \dfrac{78.075 \text{ g } CaF_2}{1 \text{ mol } CaF_2} \times \dfrac{1 \text{ kg}}{1000 \text{ g}} = 3.1 \text{ kg } CaF_2$

2A In each case we use the solubility rules to determine whether either product is insoluble. The ions in each product compound are determined by simply "switching the partners" of the reactant compounds. The designation "(aq)" on each reactant indicates that it is soluble.

(a) Possible products are potassium chloride, KCl, which is soluble, and aluminum hydroxide, $Al(OH)_3$, which is not. Net ionic equation: $Al^{3+}(aq) + 3 \text{ } OH^-(aq) \rightarrow Al(OH)_3(s)$

(b) Possible products are iron(III) sulfate, $Fe_2(SO_4)_3$, and potassium bromide, KBr, both of which are soluble. No reaction occurs.

(c) Possible products are calcium nitrate, $Ca(NO_3)_2$, which is soluble, and lead(II) iodide, PbI_2, which is insoluble. The net ionic equation is: $Pb^{2+}(aq) + 2 \text{ } I^-(aq) \rightarrow PbI_2(s)$

2B **(a)** Possible products are sodium chloride, NaCl, which is soluble, and aluminum phosphate, $AlPO_4$, which is insoluble. Net ionic equation: $Al^{3+}(aq) + PO_4^{3-}(aq) \rightarrow AlPO_4(s)$

(b) Possible products are aluminum chloride, $AlCl_3$, which is soluble, and barium sulfate, $BaSO_4$, which is insoluble. Net ionic equation: $Ba^{2+}(aq) + SO_4^{2-}(aq) \rightarrow BaSO_4(s)$

(c) Possible products are ammonium nitrate, NH_4NO_3, which is soluble, and lead (II) carbonate, $PbCO_3$, which is insoluble. Net ionic equation: $Pb^{2+}(aq) + CO_3^{2-}(aq) \rightarrow PbCO_3(s)$

3A Propionic acid is a weak acid, not dissociated completely in aqueous solution. Ammonia similarly is a weak base. The acid and base react to form a salt solution of ammonium propionate.

$$NH_3(aq) + HC_3H_5O_2(aq) \rightarrow NH_4^+(aq) + C_3H_5O_2^-(aq)$$

3B Since acetic acid is a weak acid, it is not dissociated completely in aqueous solution (except at infinite dilution); it is misleading to write it in ionic form. The products of this reaction are the gas carbon dioxide, the covalent compound water, and the ionic solute calcium acetate. Only the latter exists as ions in aqueous solution.

$$CaCO_3(s) + 2\,HC_2H_3O_2(aq) \rightarrow CO_2(g) + H_2O(l) + Ca^{2+}(aq) + 2\,C_2H_3O_2^-(aq)$$

4A **(a)** This is a metathesis or double displacement reaction. Elements do not change oxidation states during this reaction. It is not an oxidation–reduction reaction.

(b) The presence of $O_2(g)$ as a product indicates that this is an oxidation–reduction reaction. Oxygen is oxidized from O.S. = -2 in NO_3^- to O.S. = 0 in $O_2(g)$. Nitrogen is reduced from O.S. = +5 in NO_3^- to O.S. = +4 in NO_2.

4B Vanadium is oxidized from O.S. = +4 in VO^{2+} to an O.S. = +5 in VO_2^+ while manganese is reduced from O.S. = +7 in MnO_4^- to O.S. = +2 in Mn^{2+}.

5A Aluminum is oxidized (from an O.S. of 0 to an O.S. of +3), while hydrogen is reduced (from an O.S. of +1 to an O.S. of 0).

Oxidation: $\{Al(s) \rightarrow Al^{3+}(aq) + 3\,e^-\}\ \times 2$

Reduction: $\{2\,H^+(aq) + 2\,e^- \rightarrow H_2(g)\} \times 3$

Net equation: $2\,Al(s) + 6\,H^+(aq) \rightarrow 2\,Al^{3+}(aq) + 3\,H_2(g)$

5B Bromide is oxidized (from -1 to 0) while chlorine is reduced (from 0 to -1).

Oxidation: $2\,Br^-(aq) \rightarrow Br_2(l) + 2\,e^-$

Reduction: $Cl_2(g) + 2\,e^- \rightarrow 2\,Cl^-(aq)$

Net equation: $2\,Br^-(aq) + Cl_2(g) \rightarrow Br_2(l) + 2\,Cl^-(aq)$

6A Step 1: Write the two skeleton half-equations.

$MnO_4^-(aq) \rightarrow Mn^{2+}(aq)$ *and* $Fe^{2+}(aq) \rightarrow Fe^{3+}(aq)$

Step 2: Balance each skeleton half-equation for O (with H_2O) and for H atoms (with H^+).

$MnO_4^-(aq) + 8\,H^+(aq) \rightarrow Mn^{2+}(aq) + 4\,H_2O(l)$ *and* $Fe^{2+}(aq) \rightarrow Fe^{3+}(aq)$

Step 3: Balance electric charge by adding electrons.

$MnO_4^-(aq) + 8\,H^+(aq) + 5\,e^- \rightarrow Mn^{2+}(aq) + 4\,H_2O(l)$ *and* $Fe^{2+}(aq) \rightarrow Fe^{3+}(aq) + e^-$

Step 4: Combine the two ½-reactions

$\{Fe^{2+}(aq) \rightarrow Fe^{3+}(aq) + e^-\} \times 5$

$MnO_4^-(aq) + 8\,H^+(aq) + 5\,e^- \rightarrow Mn^{2+}(aq) + 4\,H_2O(l)$

$MnO_4^-(aq) + 8\,H^+(aq) + 5\,Fe^{2+}(aq) \rightarrow Mn^{2+}(aq) + 4\,H_2O(l) + 5\,Fe^{3+}(aq)$

6B   Step 1:   Uranium is oxidized and chromium is reduced in this reaction. The "skeleton"
half-equations are: $UO^{2+}(aq) \rightarrow UO_2^{2+}(aq)$ *and* $Cr_2O_7^{2-}(aq) \rightarrow Cr^{3+}(aq)$

Step 2:   First, balance the chromium skeleton half-equation for chromium atoms:

$Cr_2O_7^{2-}(aq) \rightarrow 2\,Cr^{3+}(aq)$

Next, balance oxygen atoms with water molecules in each half-equation:

$UO^{2+}(aq) + H_2O(l) \rightarrow UO_2^{2+}(aq)$ *and* $Cr_2O_7^{2-}(aq) \rightarrow 2Cr^{3+}(aq) + 7H_2O(l)$

Then, balance hydrogen atoms with hydrogen ions in each half-equation:

$UO^{2+}(aq) + H_2O(l) \rightarrow UO_2^{2+}(aq) + 2\,H^+(aq)$

$Cr_2O_7^{2-}(aq) + 14H^+(aq) \rightarrow 2Cr^{3+}(aq) + 7H_2O(l)$

Step 3:   Balance the charge of each half-equation with electrons.

$UO^{2+}(aq) + H_2O(l) \rightarrow UO_2^{2+}(aq) + 2\,H^+(aq) + 2\,e^-$

$Cr_2O_7^{2-}(aq) + 14\,H^+(aq) + 6\,e^- \rightarrow 2\,Cr^{3+}(aq) + 7\,H_2O(l)$

Step 4:   Multiply the uranium half-equation by 3 and add the chromium half-equation to it.

$\left\{ UO^{2+}(aq) + H_2O(l) \rightarrow UO_2^{2+}(aq) + 2\,H^+(aq) + 2\,e^- \right\} \times 3$

$Cr_2O_7^{2-}(aq) + 14\,H^+(aq) + 6\,e^- \rightarrow 2\,Cr^{3+}(aq) + 7\,H_2O(l)$

———————————————————————————

$3\,UO^{2+}(aq) + Cr_2O_7^{2-}(aq) + 14\,H^+(aq) + 3\,H_2O(l) \longrightarrow 3\,UO_2^{2+}(aq) + 2\,Cr^{3+}(aq) + 7\,H_2O(l) + 6\,H^+(aq)$

Step 5:   SIMPLIFY. Subtract 3 H_2O (l) and 6 H^+ (aq) from each side of the equation.

$3\,UO^{2+}(aq) + Cr_2O_7^{2-}(aq) + 8\,H^+(aq) \rightarrow 3\,UO_2^{2+}(aq) + 2\,Cr^{3+}(aq) + 4\,H_2O(l)$

7A   Step 1:   Write the two skeleton half-equations.

$S(s) \rightarrow SO_3^{2-}(aq)$ *and* $OCl^-(aq) \rightarrow Cl^-(aq)$

Step 2:   Balance each skeleton half-equation for O (with H_2O) and for H atoms (with H^+).

$3\,H_2O(l) + S(s) \rightarrow SO_3^{2-}(aq) + 6\,H^+$

$OCl^-(aq) + 2H^+ \rightarrow Cl^-(aq) + H_2O(l)$

Step 3:   Balance electric charge by adding electrons.

$3\,H_2O(l) + S(s) \rightarrow SO_3^{2-}(aq) + 6\,H^+(aq) + 4\,e^-$

$OCl^-(aq) + 2H^+(aq) + 2e^- \rightarrow Cl^-(aq) + H_2O(l)$

Step 4:   Change from an acidic medium to a basic one by adding OH^- to eliminate H^+.

$3H_2O(l) + S(s) + 6\,OH^-(aq) \rightarrow SO_3^{2-}(aq) + 6\,H^+(aq) + 6\,OH^-(aq) + 4\,e^-$

$OCl^-(aq) + 2\,H^+(aq) + 2\,OH^-(aq) + 2\,e^- \rightarrow Cl^-(aq) + H_2O(l) + 2\,OH^-(aq)$

Step 5:   Simplify by removing the items present on both sides of each half-equation, and
combine the half-equations to obtain the net redox equation.

$\{ S(s) + 6\,OH^-(aq) \rightarrow SO_3^{2-}(aq) + 3\,H_2O(l) + 4\,e^- \} \times 1$

$\{ OCl^-(aq) + H_2O(l) + 2\,e^- \rightarrow Cl^-(aq) + 2\,OH^-(aq) \} \times 2$

———————————————————————————

$S(s) + 6\,OH^-(aq) + 2\,OCl^-(aq) + 2H_2O(l) \rightarrow SO_3^{2-}(aq) + 3\,H_2O(l) + 2\,Cl^-(aq) + 4OH^-$

Simplify by removing the species present on both sides.

Net Ionic equation: $S(s) + 2\ OH^-(aq) + 2\ OCl^-(aq) \rightarrow SO_3^{2-}(aq) + H_2O(l) + 2\ Cl^-(aq)$

7B   Step 1:   Write the two skeleton half-equations.

$MnO_4^-(aq) \rightarrow MnO_2(s)$ *and* $SO_3^2(aq) \rightarrow SO_4^2(aq)$

Step 2:   Balance each skeleton half-equation for O (with H_2O) and for H atoms (with H^+).

$MnO_4^-(aq) + 4\ H^+(aq) \rightarrow MnO_2(s) + 2\ H_2O(l)$

$SO_3^{2-}(aq) + H_2O(l) \rightarrow SO_4^{2-}(aq) + 2H^+(aq)$

Step 3:   Balance electric charge by adding electrons.

$MnO_4^-(aq) + 4\ H^+(aq) + 3\ e^- \rightarrow MnO_2(s) + 2\ H_2O(l)$

$SO_3^{2-}(aq) + H_2O(l) \rightarrow SO_4^{2-}(aq) + 2\ H^+(aq) + 2\ e^-$

Step 4:   Change from an acidic medium to a basic one by adding OH^- to eliminate H^+.

$MnO_4^-(aq) + 4\ H^+(aq) + 4\ OH^-(aq) + 3\ e^- \rightarrow MnO_2(s) + H_2O(l) + 4\ OH^-(aq)$

$SO_3^{2-}(aq) + H_2O(l) + 2\ OH^-(aq) \rightarrow SO_4^{2-}(aq) + 2\ H^+(aq) + 2OH^-(aq) + 2\ e^-$

Step 5:   Simplify by removing species present on both sides of each half-equation, and combine the half-equations to obtain the net redox equation.

$\{MnO_4^-(aq) + 2\ H_2O(l) + 3\ e^- \rightarrow MnO_2(s) + 4\ OH^-(aq)\} \times 2$

$\{SO_3^{2-}(aq) + 2\ OH^-(aq) \rightarrow SO_4^{2-}(aq) + H_2O(l) + 2\ e^-\} \times 3$

$2\,MnO_4^-(aq) + 3SO_3^{2-}(aq) + 6\ OH^-(aq) + 4\ H_2O(l) \rightarrow$

$2\ MnO_2(s) + 3SO_4^{2-}(aq) + 3\ H_2O(l) + 8\ OH^-(aq)$

Simplify by removing species present on both sides.

Net ionic equation: $2\,MnO_4^-(aq) + 3SO_3^{2-}(aq) + H_2O(l) \rightarrow 2\,MnO_2(s) + 3SO_4^{2-}(aq) + 2\ OH^-(aq)$

8A   Since the oxidation state of H is 0 in H_2 (g) and is +1 in both NH_3(g) and H_2O(g), hydrogen is oxidized. A substance that is oxidized is called a reducing agent. In addition, the oxidation state of N in NO_2 (g) is +4, while it is −3 in NH_3; the oxidation state of the element N decreases during this reaction, meaning that NO_2 (g) is reduced. The substance that is reduced is called the oxidizing agent.

8B   In $\left[Au(CN)_2\right]^-$ (aq), gold has an oxidation state of +1; Au has been oxidized and, thus, Au(s) (oxidization state = 0), is the reducing agent. In OH^- (aq), oxygen has an oxidation state of -2; O has been reduced and thus, O_2(g) (oxidation state = 0) is the oxidizing agent.

9A   We first determine the amount of NaOH that reacts with 0.500 g KHP.

$n_{NaOH} = 0.5000\ \text{g KHP} \times \dfrac{1\ \text{mol KHP}}{204.22\ \text{g KHP}} \times \dfrac{1\ \text{mol OH}^-}{1\ \text{mol KHP}} \times \dfrac{1\ \text{mol NaOH}}{1\ \text{mol OH}^-} = 0.002448\ \text{mol NaOH}$

$[NaOH] = \dfrac{0.002448\ \text{mol NaOH}}{24.03\ \text{mL soln}} \times \dfrac{1000\ \text{mL}}{1\ \text{L}} = 0.1019\ \text{M}$

9B The net ionic equation when solid hydroxides react with a strong acid is $OH^- + H^+ \rightarrow H_2O$. There are two sources of OH^-: NaOH and $Ca(OH)_2$. We compute the amount of OH^- from each source and add the results.

moles of OH^- from NaOH:

$$= 0.235 \text{ g sample} \times \frac{92.5 \text{ g NaOH}}{100.0 \text{ g sample}} \times \frac{1 \text{ mol NaOH}}{39.997 \text{ g NaOH}} \times \frac{1 \text{ mol OH}^-}{1 \text{ mol NaOH}} = 0.00543 \text{ mol OH}^-$$

moles of OH^- from $Ca(OH)_2$:

$$= 0.235 \text{ g sample} \times \frac{7.5 \text{ g Ca(OH)}_2}{100.0 \text{ g sample}} \times \frac{1 \text{ mol Ca(OH)}_2}{74.093 \text{ g Ba(OH)}_2} \times \frac{2 \text{ mol OH}^-}{1 \text{ mol Ca(OH)}_2} = 0.00048 \text{ mol OH}^-$$

total amount $OH^- = 0.00543$ mol from NaOH $+ 0.00048$ mol from $Ca(OH)_2 = 0.00591$ mol OH^-

$$[HCl] = \frac{0.00591 \text{ mol OH}^-}{45.6 \text{ mL HCl soln}} \times \frac{1 \text{ mol H}^+}{1 \text{ mol OH}^-} \times \frac{1 \text{ mol HCl}}{1 \text{ mol H}^+} \times \frac{1000 \text{ mL soln}}{1 \text{ L soln}} = 0.130 \text{ M}$$

10A First, determine the mass of iron that has reacted as Fe^{2+} with the titrant. The balanced chemical equation provides the essential conversion factor to answer this question.
Namely: $5 Fe^{2+}(aq) + MnO_4^-(aq) + 8 H^+(aq) \longrightarrow 5 Fe^{3+}(aq) + Mn^{2+}(aq) + 4 H_2O(l)$

$$\text{mass Fe} = 0.04125 \text{ L titrant} \times \frac{0.02140 \text{ mol MnO}_4^-}{1 \text{ L titrant}} \times \frac{5 \text{ mol Fe}^{2+}}{1 \text{ mol MnO}_4^-} \times \frac{55.847 \text{ g Fe}}{1 \text{ mol Fe}^{2+}} = 0.246 \text{ g Fe}$$

Then determine the % Fe in the ore. $\quad \text{\% Fe} = \dfrac{0.246 \text{ g Fe}}{0.376 \text{ g ore}} \times 100\% = 65.4\% \text{ Fe}$

10B The balanced equation provides us with the stoichiometric coefficients needed for the solution.
Namely: $5 C_2O_4^{2-}(aq) + 2 MnO_4^-(aq) + 16 H^+(aq) \longrightarrow 10 CO_2(g) + 2 Mn^{2+}(aq) + 8 H_2O(l)$

$$\text{amount MnO}_4^- = 0.2482 \text{ g Na}_2C_2O_4 \times \frac{1 \text{ mol Na}_2C_2O_4}{134.00 \text{ g Na}_2C_2O_4} \times \frac{1 \text{ mol C}_2O_4^{2-}}{1 \text{ mol Na}_2C_2O_4} \times \frac{2 \text{ mol MnO}_4^-}{5 \text{ mol C}_2O_4^{2-}}$$

$$= 0.0007409 \text{ mol MnO}_4^-$$

$$[KMnO_4] = \frac{0.0007409 \text{ mol MnO}_4^-}{23.68 \text{ mL soln}} \times \frac{1000 \text{ mL}}{1 \text{ L}} \times \frac{1 \text{ mol KMnO}_4}{1 \text{ mol MnO}_4^-} = 0.03129 \text{ M KMnO}_4$$

EXERCISES

Strong Electrolytes, Weak Electrolytes, and Nonelectrolytes

1. **(a)** Because its formula begins with hydrogen, HC_6H_5O is an acid. It is not listed in Table 5-1, so it is a weak acid. A weak acid is a *weak electrolyte*.

(b) Li_2SO_4 is an ionic compound, that is, a salt. A salt is a *strong electrolyte*.

(c) MgI_2 also is a salt, a *strong electrolyte*.

(d) $(CH_3CH_2)_2O$ is a covalent compound whose formula does not begin with H. Thus, it is neither an acid nor a salt. It also is not built around nitrogen, and thus it does not behave as a weak base. This is a *nonelectrolyte*.

(e) $Sr(OH)_2$ is a *strong electrolyte*, one of the strong bases listed in Table 5-1.

2. **(a)** The best electrical conductor is the solution of the strong electrolyte: 0.10 M NaCl. In each liter of this solution, there are 0.10 mol Na^+ ions and 0.10 mol Cl^- ions.

(b) The poorest electrical conductor is the solution of the nonelectrolyte: 0.10 M C_2H_5OH. In this solution, the concentration of ions is negligible.

3. HCl is practically 100% dissociated into ions. The apparatus should light up brightly. A solution of both HCl and $HC_2H_3O_2$ will yield similar results. In strongly acidic solutions, the weak acid $HC_2H_3O_2$ is molecular and does not contribute to the conductivity of the solution. However, the strong acid HCl is practically dissociated into ions and is unaffected by the presence of the weak acid $HC_2H_3O_2$. The apparatus should light up brightly.

4. NH_3 (aq) is a weak base; $HC_2H_3O_2$ (aq) is a weak acid. The reaction produces a solution of ammonium acetate, $NH_4C_2H_3O_2$ (aq), a salt and a strong electrolyte.

$$NH_3(aq) + HC_2H_3O_2(aq) \rightarrow NH_4^+(aq) + C_2H_3O_2^-(aq)$$

5. **(a)** Barium bromide-strong electrolyte **(b)** Propionic acid-weak electrolyte
(c) Ammonia-weak electrolyte

6. Sodium chloride(strong electrolyte)   hypochlorous acid (weak electrolyte)

Na^+	Cl^-	Na^+	Cl^-
Cl^-	Na^+	Cl^-	Na^+

$H-O-Cl$ $H-O-Cl$ $[\overline{O}-\overline{Cl}]^-$
$H-O-Cl$ H_3O^+ $H-O-Cl$

Ammonium chloride   (strong electrolyte)   Methanol   (non electrolyte)

$$\left[\begin{matrix}H\\|\\H-N-H\\|\\H\end{matrix}\right]^+ Cl^- \left[\begin{matrix}H\\|\\H-N-H\\|\\H\end{matrix}\right]^+ Cl^- \left[\begin{matrix}H\\|\\H-N-H\\|\\H\end{matrix}\right]^+ Cl^-$$

$H-O-CH_3$ $H-O-CH_3$
        $H-O-CH_3$
$H-O-CH_3$ $H-O-CH_3$

Ion Concentrations

7. **(a)** $\left[K^+\right] = \dfrac{0.238 \text{ mol } KNO_3}{1 \text{ L soln}} \times \dfrac{1 \text{ mol } K^+}{1 \text{ mol } KNO_3} = 0.238 \text{ M } K^+$

(b) $\left[NO_3^-\right] = \dfrac{0.167 \text{ mol } Ca(NO_3)_2}{1 \text{ L soln}} \times \dfrac{2 \text{ mol } NO_3^-}{1 \text{ mol } Ca(NO_3)_2} = 0.334 \text{ M } NO_3^-$

(c) $\left[Al^{3+}\right] = \dfrac{0.083 \text{ mol } Al_2(SO_4)_3}{1 \text{ L soln}} \times \dfrac{2 \text{ mol } Al^{3+}}{1 \text{ mol } Al_2(SO_4)_3} = 0.166 \text{ M } Al^{3+}$

(d) $\left[Na^+\right] = \dfrac{0.209 \text{ mol } Na_3PO_4}{1 \text{ L soln}} \times \dfrac{3 \text{ mol } Na^+}{1 \text{ mol } Na_3PO_4} = 0.627 \text{ M } Na^+$

8. For all these solutes but one— $Al_2(SO_4)_3$ —there is one sulfate ion per formula unit. Consequently, the concentration of the compound and the sulfate ion concentration in that compound's aqueous solution will be the same. This makes 0.22 M $Mg(SO_4)$ the solution with the highest $[SO_4^{2-}]$ among these four. But there are three sulfate ions per formula unit of $Al_2(SO_4)_3$. Thus, $[SO_4^{2-}]$ in the $Al_2(SO_4)_3$ solution is three times the concentration of the solute, or $[SO_4^{2-}] = 3 \times 0.080 \text{ M} = 0.24 \text{ M}$; therefore this solution has the highest $[SO_4^{2-}]$.

9.

$$\left[OH^-\right] = \frac{0.132 \text{ g } Ba(OH)_2 \cdot 8H_2O}{275 \text{ mL soln}} \times \frac{1000 \text{ mL}}{1 \text{ L}} \times \frac{1 \text{ mol } Ba(OH)_2 \cdot 8H_2O}{315.5 \text{g } Ba(OH)_2 \cdot 8H_2O} \times \frac{2 \text{ mol } OH^-}{1 \text{ mol } Ba(OH)_2 \cdot 8 \text{ H}_2O}$$

$$= 3.04 \times 10^{-3} \text{ M } OH^-$$

10. $\left[K^+\right] = \dfrac{0.126 \text{ mol } KCl}{1 \text{ L soln}} \times \dfrac{1 \text{ mol } K^+}{1 \text{ mol } KCl} = 0.126 \text{ M } K^+$

$\left[Mg^{2+}\right] = \dfrac{0.148 \text{ mol } MgCl_2}{1 \text{ L soln}} \times \dfrac{1 \text{ mol } Mg^{2+}}{1 \text{ mol } MgCl_2} = 0.148 \text{ M } Mg^{2+}$

Now determine the amount of Cl^- in 1.00 L of the solution.

$\text{mol } Cl^- = \left(\dfrac{0.126 \text{ mol } KCl}{1 \text{ L soln}} \times \dfrac{1 \text{ mol } Cl^-}{1 \text{ mol } KCl}\right) + \left(\dfrac{0.148 \text{ mol } MgCl_2}{1 \text{ L soln}} \times \dfrac{2 \text{ mol } Cl^-}{1 \text{ mol } MgCl_2}\right)$

$= 0.126 \text{ mol } Cl^- + 0.296 \text{ mol } Cl^- = 0.422 \text{ mol } Cl^-$

$\left[Cl^-\right] = \dfrac{0.422 \text{ mol } Cl^-}{1 \text{ L soln}} = 0.422 \text{ M } Cl^-$

11. **(a)** $[Ca^{2+}] = \dfrac{14.2 \text{ mg } Ca^{2+}}{1 \text{ L solution}} \times \dfrac{1 \text{ g } Ca^{2+}}{1000 \text{ mg } Ca^{2+}} \times \dfrac{1 \text{ mol } Ca^{2+}}{40.078 \text{ g } Ca^{2+}} = 3.54 \times 10^{-4} \text{ M } Ca^{2+}$

(b) $[K^+] = \dfrac{32.8 \text{ mg } K^+}{100 \text{ mL solution}} \times \dfrac{1 \text{ g } K^+}{1000 \text{ mg } K^+} \times \dfrac{1000 \text{ mL solution}}{1 \text{ L solution}} \times \dfrac{1 \text{ mol } K^+}{39.0983 \text{ g } K^+} = 8.39 \times 10^{-3} \text{ M } K^+$

(c) $[Zn^{2+}] = \dfrac{225\ \mu g\ Zn^{2+}}{1\ mL\ solution} \times \dfrac{1\ g\ Zn^{2+}}{1\times10^{6}\ \mu g\ Zn^{2+}} \times \dfrac{1000\ mL\ solution}{1\ L\ solution} \times \dfrac{1\ mol\ Zn^{2+}}{65.39\ g\ Zn^{2+}} = 3.44\times10^{-3}\ M\ Zn^{2+}$

12. $[NaF] = \dfrac{0.9\ mg\ F^{-}}{1\ L} \times \dfrac{1\ g}{1000\ mg} \times \dfrac{1\ mol\ F^{-}}{19.00\ g\ F^{-}} \times \dfrac{1\ mol\ NaF}{1\ mol\ F^{-}} = 5\times10^{-5}\ M\ NaF$

<u>13</u> In order to determine the solution with the largest concentration of K^{+}, we begin by converting each concentration to a common concentration unit, namely, molarity of K^{+}.

$0.0850\ M\ K_2SO_4 \dfrac{0.0850\ M\ K_2SO_4}{1\ L\ solution} \times \dfrac{2\ mol\ K^{+}}{1\ mol\ K_2SO_4} = 0.17\ M\ K^{+}$

$\dfrac{1.25\ g\ KBr}{100\ mL\ solution} \times \dfrac{1000\ mL\ solution}{1\ L\ solution} \times \dfrac{1\ mol\ KBr}{119.0023\ g\ KBr} \times \dfrac{1\ mol\ K^{+}}{1\ mol\ KBr} = 0.105\ M\ K^{+}$

$\dfrac{8.1\ mg\ K^{+}}{1\ mL\ solution} \times \dfrac{1000\ mL\ solution}{1\ L\ solution} \times \dfrac{1\ g\ K^{+}}{1000\ mg\ K^{+}} \times \dfrac{1\ mol\ K^{+}}{39.0983\ g\ K^{+}} = 0.207\ M\ K^{+}$

Clearly, the solution containing $8.1\ mg\ K^{+}$ per mL gives the largest K^{+} of the three solutions.

14. NH_3 is a weak base and would have an exceedingly low $\left[H^{+}\right]$; the answer is not $1.00\ M\ NH_3$.

$HC_2H_3O_2$ is a very weak acid; $0.011\ M\ HC_2H_3O_2$ would have a low $\left[H^{+}\right]$.

H_2SO_4 has two ionizable protons per mole while HCl has but one. Thus, H_2SO_4 would have the highest $\left[H^{+}\right]$ in a 0.010 M aqueous solution.

<u>15.</u> Determine the amount of I^{-} in the solution as it now exists, and the amount of I^{-} in the solution of the desired concentration. The different in these two amounts is the amount of I^{-} that must be added. Convert this amount to a mass of MgI_2 in grams.

moles of I^{-} in final solution $= 250.0\ mL \times \dfrac{1\ L}{1000\ mL} \times \dfrac{0.1000\ mol\ I^{-}}{1\ L\ soln} = 0.02500\ mol\ I^{-}$

moles of I^{-} in KI solution $= 250.0\ mL \times \dfrac{1\ L}{1000\ mL} \times \dfrac{0.0876\ mol\ KI}{1\ L\ soln} \times \dfrac{1\ mol\ I^{-}}{1\ mol\ KI} = 0.0219\ mol\ I^{-}$

mass MgI_2 required $= (0.02500 - 0.0219)\ mol\ I^{-} \times \dfrac{1\ mol\ MgI_2}{2\ mol\ I^{-}} \times \dfrac{278.11\ g\ MgI_2}{1\ mol\ MgI_2} \times \dfrac{1000\ mg}{1\ g}$

$\qquad = 4.3\times10^{2}\ mg\ MgI_2$

16. The final volume is $97\underline{5}$ mL. We can use dimensional analysis to obtain the $[K^{+}]$

$[K^{+}] = \dfrac{\dfrac{12.0\ mg\ K_2SO_4}{1\ mL} \times 1000\ mL}{0.97\underline{5}\ L\ solution} \times \dfrac{1\ g\ K_2SO_4}{1000\ mg\ K_2SO_4} \times \dfrac{1\ mol\ K_2SO_4}{174.26\ g\ K_2SO_4} \times \dfrac{2\ mol\ K^{+}}{1\ mol\ K_2SO_4} = 0.14\underline{1}\ M\ K^{+}$

17. Moles of Chloride ion

$$= \left(0.225 \text{ L} \times \frac{0.625 \text{ mol KCl}}{1 \text{ L soln}} \times \frac{1 \text{ mol Cl}^-}{1 \text{ mol KCl}} \right) + \left(0.615 \text{ L} \times \frac{0.385 \text{ mol MgCl}_2}{1 \text{ L soln}} \times \frac{2 \text{ mol Cl}^-}{1 \text{ mol MgCl}_2} \right)$$

$$= 0.141 \text{ mol Cl}^- + 0.474 \text{ mol Cl}^- = 0.615 \text{ mol Cl}^- \quad \left[\text{Cl}^- \right] = \frac{0.615 \text{ mol Cl}^-}{0.225 \text{ L} + 0.615 \text{ L}} = 0.732 \text{ M}$$

18. amount of NO_3^- ion =

$$\left(0.275 \text{ L} \times \frac{0.283 \text{ mol KNO}_3}{1 \text{ L soln}} \times \frac{1 \text{ mol NO}_3^-}{1 \text{ mol KNO}_3} \right) + \left(0.328 \text{ L} \times \frac{0.421 \text{ mol Mg}(NO_3)_2}{1 \text{ L soln}} \times \frac{2 \text{ mol NO}_3^-}{1 \text{ mol Mg}(NO_3)_2} \right)$$

$$= 0.0778 \text{ mol NO}_3^- + 0.276 \text{ mol NO}_3^- = 0.354 \text{ mol NO}_3^-$$

$$\left[NO_3^- \right] = \frac{0.354 \text{ mol NO}_3^-}{0.275 \text{ L} + 0.328 \text{ L} + 0.784 \text{ L}} = 0.255 \text{ M}$$

Predicting Precipitation Reactions

19. In each case, each available cation is paired with the available anions, one at a time, to determine if a compound is produced that is insoluble, based on the solubility rules of Chapter 5. Then a net ionic equation is written to summarize this information.

(a) $Pb^{2+}(aq) + 2 \text{ Br}^-(aq) \rightarrow PbBr_2(s)$

(b) No reaction occurs(all are spectator ions).

(c) $Fe^{3+}(aq) + 3 \text{ OH}^-(aq) \rightarrow Fe(OH)_3(s)$

20. **(a)** $Ca^{2+}(aq) + CO_3^{2-}(aq) \rightarrow CaCO_3(s)$

(b) $Ba^{2+}(aq) + SO_4^{2-}(aq) \rightarrow BaSO_4(s)$

(c) No precipitate forms. All ions stay in solution.

21.

	Mixture	Result (net ionic equation)
(a)	$HI(a) + Zn(NO_3)_2 (aq)$:	No reaction occurs.
(b)	$CuSO_4(aq) + Na_2CO_3(aq)$:	$Cu^{2+}(aq) + CO_3^{2-}(aq) \rightarrow CuCO_3(s)$
(c)	$Cu(NO_3)_2(aq) + Na_3PO_4(aq)$:	$3Cu^{2+}(aq) + 2PO_4^{3-}(aq) \rightarrow Cu_3(PO_4)_2(s)$

22.

	Mixture	Result (net ionic equation)
(a)	$AgNO_3(aq) + CuCl_2(aq)$:	$Ag^+(aq) + Cl^-(aq) \rightarrow AgCl(s)$
(b)	$Na_2S(aq) + FeCl_2(aq)$:	$S^{2-}(aq) + Fe^{2+}(aq) \rightarrow FeS(s)$
(c)	$Na_2CO_3(aq) + AgNO_3(aq)$:	$CO_3^{2-}(aq) + 2 Ag^+(aq) \rightarrow Ag_2CO_3(s)$

23. **(a)** Add K_2SO_4 (aq); $BaSO_4$ (s) will form and $MgSO_4$ will not precipitate.

$$BaCl_2(s) + K_2SO_4(aq) \rightarrow BaSO_4(s) + 2\ KCl(aq)$$

(b) Add $H_2O(l)$; Na_2CO_3 (s) dissolves, $MgCO_3$ (s) will not dissolve (appreciably).

$$Na_2CO_3(s) \xrightarrow{\text{water}} 2\ Na^+(aq) + CO_3^{2-}(aq)$$

(c) Add KCl(aq); AgCl(s) will form, while $Cu(NO_3)_2$ (s) will dissolve.

$$AgNO_3(s) + KCl(aq) \rightarrow AgCl(s) + KNO_3(aq)$$

24. **(a)** Add H_2O. $Cu(NO_3)_2$ (s) will dissolve, while $PbSO_4$(s) will not dissolve (appreciably).

$$Cu(NO_3)_2(s) \xrightarrow{\text{water}} Cu^{2+}(aq) + 2NO_3^-(aq)$$

(b) Add HCl(aq). $Mg(OH)_2$ (s) will dissolve, but $BaSO_4$ (s) will not dissolve (appreciably).

$$Mg(OH)_2(s) + 2\ HCl(aq) \rightarrow MgCl_2(aq) + 2\ H_2O(l)$$

(c) Add HCl(aq). Both carbonates dissolve, but $PbCl_2$(s) will precipitate while $CaCl_2(aq)$ remains dissolved.

$$PbCO_3(s) + 2\ HCl(aq) \rightarrow PbCl_2(s) + H_2O(l) + CO_2(g)$$

$$CaCO_3(s) + 2\ HCl(aq) \rightarrow CaCl_2(aq) + H_2O(l) + CO_2(g)$$

25.

Mixture	Net ionic equation
(a) $Sr(NO_3)_2(aq) + K_2SO_4(aq)$:	$Sr^{2+}(aq) + SO_4^{2-}(aq) \rightarrow SrSO_4(s)$
(b) $Mg(NO_3)_2(aq) + NaOH(aq)$:	$Mg^{2+}(aq) + 2\ OH^-(aq) \rightarrow Mg(OH)_2(s)$
(c) $BaCl_2(aq) + K_2SO_4(aq)$:	$Ba^{2+}(aq) + SO_4^{2-}(aq) \rightarrow BaSO_4(s)$ (upon filtering, KCl (aq) is obtained)

26.

Mixture	Net ionic equation
(a) $BaCl_2(aq) + K_2SO_4(aq)$:	$Ba^{2+}(aq) + SO_4^{2-}(aq) \rightarrow BaSO_4(s)$
(b) $NaCl(aq) + Ag_2SO_4(s)$:	$2\ AgCl(s) + 2Na^+(aq) + SO_4^{2-}(aq)$
alternatively $BaCl_2(aq) + Ag_2SO_4(s)$	(a mixed precipitate): $AgCl(s) + BaSO_4(s)$
(c) $Sr(NO_3)_2(aq) + K_2SO_4(aq)$:	$Sr^{2+}(aq) + SO_4^{2-}(aq) \rightarrow SrSO_4(s)$ (upon filtering, KNO_3 (aq) is obtained)

Acid–Base Reactions

27. The type of reaction is given first, followed by the net ionic equation.

    (a) Neutralization: $OH^- (aq) + HC_2H_3O_2 (aq) \rightarrow H_2O(l) + C_2H_3O_2^- (aq)$

    (b) No reaction occurs. This is the physical mixing of two acids.

    (c) Gas evolution: $FeS(s) + 2\ H^+ (aq) \rightarrow H_2S(g) + Fe^{2+} (aq)$

    (d) Gas evolution: $HCO_3^- (aq) + H^+ (aq) \rightarrow "H_2CO_3 (aq)" \rightarrow H_2O(l) + CO_2 (g)$

    (e) Redox: $Mg(s) + 2\ H^+ (aq) \rightarrow Mg^{2+} (aq) + H_2 (g)$

    (f) No reaction occurs, based on the information in Table 5-3 (Cu cannot reduce H^+ to H_2).

28. (a) $NaHCO_3 (s) + H^+ (aq) \qquad \rightarrow Na^+ (aq) + H_2O(l) + CO_2 (g)$

    (b) $CaCO_3 (s) + 2\ H^+ (aq) \qquad \rightarrow Ca^{2+} (aq) + H_2O(l) + CO_2 (g)$

    (c) $Mg(OH)_2 (s) + 2\ H^+ (aq) \qquad \rightarrow Mg^{2+} (aq) + 2\ H_2O(l)$

    (d) $Mg(OH)_2 (s) + 2\ H^+ (aq) \qquad \rightarrow Mg^{2+} (aq) + 2\ H_2O(l)$

    $Al(OH)_3 (s) + 3\ H^+ (aq) \qquad \rightarrow Al^{3+} (aq) + 3\ H_2O(l)$

    (e) $NaAl(OH)_2 CO_3 (s) + 4\ H^+ (aq) \rightarrow Al^{3+} (aq) + Na^+ (aq) + 3\ H_2O(l) + CO_2 (g)$

29. As a salt: $\qquad NaHSO_4 (aq) \rightarrow Na^+ (aq) + HSO_4^- (aq)$

    As an acid: $\qquad HSO_4^- (aq) + OH^- (aq) \rightarrow H_2O(l) + SO_4^{2-} (aq)$

30. Because all three compounds contain an ammonium cation, all are formed by the reaction of an acid with aqueous ammonia. The identity of the anion determines which acid present.

    $2\ NH_3 (aq) + H_3PO_4 (aq) \rightarrow (NH_4)_2 HPO_4 (aq)$

    $NH_3 (aq) + HNO_3 (aq) \qquad \rightarrow NH_4NO_3 (aq)$

    $2\ NH_3 (aq) + H_2SO_4 (aq) \rightarrow (NH_4)_2 SO_4 (aq)$

31. Use (b) $NH_3(aq)$: NH_3 affords the OH^- ions necessary to form $Mg(OH)_2(s)$

    Applicable reactions: $\quad \{NH_3(aq) + H_2O(l) \rightarrow NH_4^+(aq) + OH^-(aq)\} \times 2$

    $\qquad\qquad\qquad\qquad MgCl_2(aq) \rightarrow Mg^{2+}(aq) + 2\ Cl^-(aq)$

    $\qquad\qquad\qquad\qquad Mg^{2+}(aq) + 2\ OH^-(aq) \rightarrow Mg(OH)_2(s)$

32. $HCl(aq)$ reacts with $KHSO_3(aq)$ to give $SO_2(g)$ via the thermodynamically unstable intermediate sulfurous acid (H_2SO_3).

    Net ionic equation: $H^+(aq) + HSO_3^-(aq) \rightarrow "H_2SO_3(aq)" \rightarrow H_2O(l) + SO_2(g)$

117

Oxidation-Reduction (Redox) Equations

33. (a) The O.S. of H is $+1$, that of O is -2, that of C is $+4$, and that of Mg is $+2$ on each side of this equation. This is not a redox equation.

(b) The O.S. of Cl is 0 on the left and -1 on the right side of this equation. The O.S. of Br is -1 on the left and 0 on the right side of this equation. This is a redox reaction.

(c) The O.S. of Ag is 0 on the left and $+1$ on the right side of this equation. The O.S. of N is $+5$ on the left and $+4$ on the right side of this equation. This is a redox reaction.

(d) On both sides of the equation the O.S. of O is -2, that of Ag is $+1$, and that of Cr is $+6$. Thus, this is not a redox equation.

34. (a) In this reaction, iron is reduced from Fe^{3+} (aq) to Fe^{2+} (aq) *and* manganese is reduced from a $+7$ O.S. in MnO_4^- (aq) to a $+2$ O.S. in Mn^{2+} (aq). Thus, there are two reductions and no oxidation, which is an impossibility.

(b) In this reaction, chlorine is oxidized from an O.S. of 0 in Cl_2 (aq) to an O.S. of $+1$ in ClO^- (aq) *and* oxygen is oxidized from an O.S. of -1 in H_2O_2 (aq) to an O.S. of 0 in O_2 (g). Consequently there are two oxidation reactions and no reduction reactions, also an impossibility.

35. (a) Reduction: $2SO_3^{2-}(aq) + 6\ H^+(aq) + 4\ e^- \rightarrow\ S_2O_3^{2-}(aq) + 3\ H_2O(l)$

(b) Reduction: $2NO_3^-(aq) + 10\ H^+(aq) + 8\ e^- \rightarrow\ N_2O(g) + 5\ H_2O(l)$

(c) Oxidation: $Al(s) + 4\ OH^-(aq) \rightarrow Al(OH)_4^-(aq) + 3\ e^-$

36. (a) $H_2C_2O_4 \xrightarrow{\text{acidic}} 2\ CO_2 + 2\ H^+ + 2e^- \qquad\qquad$ Oxidation

(b) $6\ e^- + 14\ H^+ + Cr_2O_7^{2-} \xrightarrow{\text{acidic}} 2\ Cr^{3+} + 7\ H_2O \quad$ Reduction

(c) $2\ H_2O + 3\ e^- + MnO_4^- \xrightarrow{\text{basic}} MnO_2 + 4\ OH^- \quad$ Reduction

37. (a) Oxidation: $\{2\ I^-(aq) \rightarrow\ I_2(s) + 2\ e^- \qquad\qquad\qquad\qquad\} \times 5$

Reduction: $\{MnO_4^-(aq) + 8\ H^+(aq) + 5\ e^- \rightarrow\ Mn^{2+}(aq) + 4\ H_2O(l) \quad\} \times 2$

Net: $10\ I^-(aq) + 2\ MnO_4^-(aq) + 16\ H^+(aq) \rightarrow 5\ I_2(s) + 2\ Mn^{2+}(aq) + 8\ H_2O(l)$

(b) Oxidation: $\{N_2H_4(l) \rightarrow\ N_2(g) + 4\ H^+(aq) + 4\ e^- \qquad\qquad\} \times 3$

Reduction: $\{BrO_3^-(aq) + 6\ H^+(aq) + 6\ e^- \rightarrow\ Br^-(aq) + 3\ H_2O(l) \qquad\} \times 2$

Net: $3\ N_2H_4(l) + 2\ BrO_3^-(aq) \rightarrow 3\ N_2(g) + 2\ Br^-(aq) + 6\ H_2O(l)$

(c) Oxidation: $Fe^{2+}(aq) \rightarrow\ Fe^{3+}(aq) +\ e^-$

Reduction: $VO_4^{3-}(aq) + 6\ H^+(aq) +\ e^- \rightarrow\ VO^{2+}(aq) + 3\ H_2O(l)$

Net: $Fe^{2+}(aq) +\ VO_4^{3-}(aq) + 6\ H^+(aq) \rightarrow\ Fe^{3+}(aq) +\ VO^{2+}(aq) + 3\ H_2O(l)$

(d) Oxidation: $\{ UO^{2+}(aq) + H_2O(l) \rightarrow UO_2^{2+}(aq) + 2\ H^+(aq) + 2\ e^- \qquad \} \times 3$

Reduction: $\{ NO_3^-(aq) + 4\ H^+(aq) + 3\ e^- \rightarrow NO(g) + 2\ H_2O(l) \qquad \} \times 2$

Net: $\quad 3\ UO^{2+}(aq) + 2\ NO_3^-(aq) + 2\ H^+(aq) \rightarrow 3\ UO_2^{2+}(aq) + 2\ NO(g) + H_2O(l)$

38. **(a)** Oxidation: $\{ P_4(s) + 16\ H_2O(l) \rightarrow 4\ H_2PO_4^-(aq) + 24\ H^+(aq) + 20\ e^- \} \times 3$

Reduction: $\{ NO_3^-(aq) + 4\ H^+(aq) + 3\ e^- \rightarrow NO(g) + 2\ H_2O(l) \qquad \} \times 20$

Net: $\quad 3\ P_4(s) + 20\ NO_3^-(aq) + 8\ H_2O + 8\ H^+(aq) \rightarrow 12\ H_2PO_4^-(aq) + 20\ NO(g)$

(b) Oxidation: $\{ S_2O_3^{2-}(aq) + 5\ H_2O(l) \rightarrow 2\ SO_4^{2-}(aq) + 10\ H^+(aq) + 8\ e^- \qquad \} \times 5$

Reduction: $\{ MnO_4^-(aq) + 8\ H^+(aq) + 5\ e^- \rightarrow Mn^{2+}(aq) + 4\ H_2O(l) \qquad \} \times 8$

Net: $\quad 5\ S_2O_3^{2-}(aq) + 8\ MnO_4^-(aq) + 14\ H^+(aq) \rightarrow 10\ SO_4^{2-}(aq) + 8\ Mn^{2+}(aq) + 7\ H_2O(l)$

(c) Oxidation: $2\ HS^-(aq) + 3\ H_2O(l) \rightarrow S_2O_3^{2-}(aq) + 8\ H^+(aq) + 8\ e^-$

Reduction: $\{ 2\ HSO_3^-(aq) + 4\ H^+(aq) + 4\ e^- \rightarrow S_2O_3^{2-}(aq) + 3\ H_2O(l) \} \times 2$

Net: $\quad 2\ HS^-(aq) + 4\ HSO_3^-(aq) \rightarrow 3\ S_2O_3^{2-}(aq) + 3\ H_2O(l)$

(d) Oxidation: $\quad 2\ NH_3OH^+(aq) \rightarrow N_2O(g) + H_2O(l) + 6\ H^+(aq) + 4\ e^-$

Reduction: $\{ Fe^{3+}(aq) + e^- \rightarrow Fe^{2+}(aq) \qquad \} \times 4$

Net: $\quad 4\ Fe^{3+}(aq) + 2\ NH_3OH^+(aq) \rightarrow 4\ Fe^{2+}(aq) + N_2O(g) + H_2O(l) + 6\ H^+(aq)$

39. **(a)** Oxidation: $\{ MnO_2(s) + 4\ OH^-(aq) \rightarrow MnO_4^-(aq) + 2\ H_2O(l) + 3\ e^- \} \times 2$

Reduction: $ClO_3^-(aq) + 3\ H_2O(l) + 6\ e^- \rightarrow Cl^-(aq) + 6\ OH^-(aq)$

Net: $\quad 2\ MnO_2(s) + ClO_3^-(aq) + 2\ OH^-(aq) \rightarrow 2MnO_4^-(aq) + Cl^-(aq) + H_2O(l)$

(b) Oxidation: $\{ Fe(OH)_3(s) + 5\ OH^-(aq) \rightarrow FeO_4^{2-}(aq) + 4\ H_2O(l) + 3\ e^- \qquad \} \times 2$

Reduction: $\{ OCl^-(aq) + H_2O(l) + 2\ e^- \rightarrow Cl^-(aq) + 2OH^-(aq) \qquad \} \times 3$

Net: $2\ Fe(OH)_3(s) + 3\ OCl^-(aq) + 4\ OH^-(aq) \rightarrow 2FeO_4^{2-}(aq) + 3\ Cl^-(aq) + 5\ H_2O(l)$

(c) Oxidation: $\{ ClO_2(aq) + 2\ OH^-(aq) \rightarrow ClO_3^-(aq) + H_2O(l) + e^- \} \times 5$

Reduction: $ClO_2(aq) + 2\ H_2O(l) + 5\ e^- \rightarrow Cl^-(aq) + 4\ OH^-(aq)$

Net: $\quad 6\ ClO_2(aq) + 6\ OH^-(aq) \rightarrow 5ClO_3^-(aq) + Cl^-(aq) + 3\ H_2O(l)$

40. **(a)** Oxidation: $\{ CN^-(aq) + 2\ OH^-(aq) \rightarrow CNO^-(aq) + H_2O + 2\ e^- \qquad \} \times 3$

Reduction: $\{ MnO_4^-(aq) + 2\ H_2O(l) + 3\ e^- \rightarrow MnO_2(s) + 4\ OH^-(aq) \qquad \} \times 2$

Net: $3\ CN^-(aq) + 2\ MnO_4^-(aq) + H_2O(l) \rightarrow 3\ CNO^-(aq) + 2\ MnO_2(s) + 2\ OH^-(aq)$

(b) Oxidation: $N_2H_4(l) + 4\ OH^-(aq) \rightarrow N_2(g) + 4\ H_2O(l) + 4\ e^-$

Reduction: $\{[Fe(CN)_6]^{3-}(aq) + e^- \rightarrow [Fe(CN)_6]^{4-}(aq)\ \ \ \ \ \ \ \} \times 4$

Net: $4[Fe(CN)_6]^{3-}(aq) + N_2H_4(l) + 4OH^-(aq) \rightarrow 4[Fe(CN)_6]^{4-}(aq) + N_2(g) + 4H_2O(l)$

(c) Oxidation: $\{Fe(OH)_2(s) + OH^-(aq) \rightarrow Fe(OH)_3(s) + e^-\ \ \ \ \ \ \ \} \times 4$

Reduction: $O_2(g) + 2\ H_2O(l) + 4\ e^- \rightarrow 4\ OH^-(aq)$

Net: $\ \ \ \ \ \ \ \ \ 4\ Fe(OH)_2(s) + O_2(g) + 2\ H_2O(l) \rightarrow 4\ Fe(OH)_3(s)$

(d) Oxidation: $\{C_2H_5OH(aq) + 5\ OH^-(aq) \rightarrow C_2H_3O_2^-(aq) + 4\ H_2O(l) + 4\ e^-\ \ \ \ \} \times 3$

Reduction: $\{MnO_4^-(aq) + 2\ H_2O(l) + 3\ e^- \rightarrow MnO_2(s) + 4\ OH^-(aq)\ \ \ \ \ \ \ \} \times 4$

Net: $3\ C_2H_5OH(aq) + 4\ MnO_4^-(aq) \rightarrow 3\ C_2H_3O_2^-(aq) + 4\ MnO_2(s) + OH^-(aq) + 4\ H_2O(l)$

41. (a) Oxidation: $Cl_2(g) + 12\ OH^-(aq) \rightarrow 2\ ClO_3^-(aq) + 6\ H_2O(l) + 10\ e^-$

Reduction: $\{Cl_2(g) + 2\ e^- \rightarrow 2\ Cl^-(aq)\ \ \ \ \ \ \ \ \ \ \ \ \ \ \ \ \ \ \ \} \times 5$

Net: $\ \ \ \ 6\ Cl_2(g) + 12\ OH^-(aq) \rightarrow 10\ Cl^-(aq) + 2\ ClO_3^-(aq) + 6\ H_2O(l)$

Or: $\ \ \ \ 3\ Cl_2(g) + 6\ OH^-(aq) \rightarrow 5\ Cl^-(aq) + ClO_3^-(aq) + 3\ H_2O(l)$

(b) Oxidation: $S_2O_4^{2-}(aq) + 2\ H_2O(l) \rightarrow 2\ HSO_3^-(aq) + 2\ H^+(aq) + 2\ e^-$

Reduction: $S_2O_4^{2-}(aq) + 2\ H^+(aq) + 2\ e^- \rightarrow S_2O_3^{2-}(aq) + H_2O(l)$

Net: $\ \ \ \ 2\ S_2O_4^{2-}(aq) + H_2O(l) \rightarrow 2\ HSO_3^-(aq) + S_2O_3^{2-}(aq)$

42. (a) Oxidation: $\{MnO_4^{2-}(aq) \rightarrow MnO_4^-(aq) + e^-\ \ \ \ \ \ \ \ \ \ \ \ \ \ \ \ \ \ \} \times 2$

Reduction: $MnO_4^{2-}(aq) + 2\ H_2O(l) + 2\ e^- \rightarrow MnO_2(s) + 4\ OH^-(aq)$

Net: $\ \ \ 3\ MnO_4^{2-}(aq) + 2\ H_2O(l) \rightarrow 2\ MnO_4^-(aq) + MnO_2(s) + 4\ OH^-(aq)$

(b) Oxidation: $\{P_4(s) + 8\ OH^-(aq) \rightarrow 4\ H_2PO_2^-(aq) + 4\ e^-\ \ \ \ \ \ \ \ \ \ \} \times 3$

Reduction: $P_4(s) + 12\ H_2O(l) + 12\ e^- \rightarrow 4\ PH_3(g) + 12\ OH^-(aq)$

Net: $\ \ \ \ 4\ P_4(s) + 12\ OH^-(aq) + 12\ H_2O(l) \rightarrow 12\ H_2PO_2^-(aq) + 4\ PH_3(g)$

(c) Oxidation: $S_8(s) + 24\ OH^-(aq) \rightarrow 4\ S_2O_3^{2-}(aq) + 12\ H_2O(l) + 16\ e^-$

$\ S_8(s) + 16\ e^- \rightarrow 8\ S^{2-}(aq)$

Reduction:

Net: $\ \ \ \ 2\ S_8(s) + 24\ OH^-(aq) \rightarrow 8\ S^{2-}(aq) + 4\ S_2O_3^{2-}(aq) + 12\ H_2O(l)$

(d) Oxidation: $As_2S_3(s) + 40\ OH^-(aq) \rightarrow 2\ AsO_4^{3-}(aq) + 3\ SO_4^{2-}(aq) + 20\ H_2O + 28\ e^-$

Reduction: $\{H_2O_2(aq) + 2\ e^- \rightarrow 2\ OH^-(aq)\ \ \ \ \ \ \ \ \ \ \ \ \ \ \ \ \} \times 14$

Net: $As_2S_3(s) + 12\ OH^-(aq) + 14\ H_2O_2(aq) \rightarrow 2\ AsO_4^{3-}(aq) + 3\ SO_4^{2-}(aq) + 20\ H_2O(l)$

43. **(a)** Oxidation: $\{ NO_2^- (aq) + H_2O(l) \rightarrow NO_3^- (aq) + 2\ H^+ (aq) + 2\ e^- \qquad \} \times 5$

Reduction: $\{ MnO_4^- (aq) + 8\ H^+ (aq) + 5\ e^- \rightarrow Mn^{2+} (aq) + 4\ H_2O(l) \qquad \} \times 2$

Net: $5\ NO_2^- (aq) + 2\ MnO_4^- (aq) + 6\ H^+ (aq) \rightarrow 5\ NO_3^- (aq) + 2\ Mn^{2+} (aq) + 3\ H_2O(l)$

(b) Oxidation: $\{ Mn^{2+} (aq) + 4\ OH^- (aq) \rightarrow MnO_2 (s) + 2\ H_2O (l) + 2\ e^- \ \} \times 3$

Reduction: $\{ MnO_4^- (aq) + 2\ H_2O (l) + 3\ e^- \rightarrow MnO_2 (s) + 4\ OH^- (aq) \ \} \times 2$

Net: $3\ Mn^{2+} (aq) + 2\ MnO_4^- (aq) + 4\ OH^- (aq) \rightarrow 5\ MnO_2 (s) + 2\ H_2O (l)$

(c) Oxidation: $\{ C_2H_5OH \rightarrow CH_3CHO + 2\ H^+ (aq) + 2\ e^- \qquad \} \times 3$

Reduction: $Cr_2O_7^{2-} (aq) + 14\ H^+ (aq) + 6\ e^- \rightarrow 2\ Cr^{3+} (aq) + 7\ H_2O(l)$

Net: $Cr_2O_7^{2-} (aq) + 8\ H^+ (aq) + 3\ C_2H_5OH \rightarrow 2\ Cr^{3+} (aq) + 7\ H_2O(l) + 3\ CH_3CHO$

44. **(a)** Oxidation: $\{ Na(s) \rightarrow Na^+ (aq) + e^- \qquad \} \times 2$

Reduction: $\{ 2\ HI(aq) + 2\ e^- \rightarrow 2\ I^- (aq) + H_2 (g)$

$2\ Na(s) + 2\ HI(aq) \rightarrow 2\ NaI(aq) + H_2(g)$

(b) Oxidation: $Zn(s) \rightarrow Zn^{2+} (aq) + 2\ e^-$

Reduction: $\{ VO^{2+} (aq) + 2\ H^+ (aq) + e^- \rightarrow V^{3+} (aq) + H_2O(l) \ \} \times 2$

Net: $Zn(s) + 2\ VO^{2+} (aq) + 4\ H^+ (aq) \rightarrow Zn^{2+} (aq) + 2\ V^{3+} (aq) + 2\ H_2O(l)$

(c) Oxidation: $H_2O + CH_3OH \rightarrow CO_2 + 6\ H^+ + 6\ e^-$

Reduction: $\{ ClO_3^- (aq) + 2\ H^+ (aq) + e^- \rightarrow ClO_2 (aq) + H_2O(l) \} \times 6$

Net: $CH_3OH + 6\ ClO_3^- (aq) + 6\ H^+ \rightarrow 6\ ClO_2 (aq) + 5\ H_2O(l) + CO_2$

45. For the purpose of balancing its redox equation, each of the reactions is treated as if it takes place in acidic aqueous solution.

(a) $2\ H_2O(g) + CH_4(g) \rightarrow CO_2(g) + 8\ H^+(g) + 8\ e^-$

$\{ 2\ e^- + 2\ H^+(g) + NO(g) \rightarrow \frac{1}{2}\ N_2(g) + H_2O(g) \quad \} \times 4$

$CH_4(g) + 4\ NO(g) \rightarrow 2\ N_2(g) + CO_2(g) + 2\ H_2O(g)$

(b) $\{ H_2S(g) \rightarrow 1/8\ S_8(s) + 2\ H^+(g) + 2\ e^- \qquad \} \times 2$

$4\ e^- + 4\ H^+(g) + SO_2(g) \rightarrow 1/8\ S_8(s) + 2\ H_2O(g)$

$2\ H_2S(g) + SO_2(g) \rightarrow 3/8\ S_8(s) + 2\ H_2O(g)$ or

$16\ H_2S(g) + 8\ SO_2(g) \rightarrow 3\ S_8(s) + 16\ H_2O(g)$

(c) $\{ Cl_2O(g) + 2\ NH_4^+(aq) + 2\ H^+(aq) + 4\ e^- \rightarrow 2\ NH_4Cl(s) + H_2O(l) \quad \} \times 3$

$\{ 2\ NH_3(g) \rightarrow N_2(g) + 6\ e^- + 6\ H^+(aq) \qquad\qquad \} \times 2$

$6\ NH_3(g) + 6\ H^+(aq) \rightarrow 6\ NH_4^+(aq)$

$10\ NH_3(g) + 3\ Cl_2O(g) \rightarrow 6\ NH_4Cl(s) + 2\ N_2(g) + 3\ H_2O(l)$

46. For the purpose of balancing its redox equation, each of the reactions is treated as if it takes place in acidic aqueous solution.

(a) $CH_4(g) + NH_3(g) \rightarrow HCN(g) + 6\ e^- + 6\ H^+$

$\{2\ e^- + 2\ H^+(g) + \frac{1}{2}\ O_2(g) \rightarrow H_2O(g) \qquad\qquad\} \times 3$

$\overline{CH_4(g) + NH_3(g) + 3/2\ O_2(g) \rightarrow HCN(g) + 3\ H_2O(g)}$

(b) $\{H_2(g) \rightarrow 2\ H^+(aq) + 2\ e^- \qquad\qquad\} \times 5$

$2\ NO(g) + 10\ H^+(aq) + 10\ e^- \rightarrow 2\ NH_3(g) + 2\ H_2O(l)$

$\overline{5\ H_2(g) + 2\ NO(g) \rightarrow 2\ NH_3(g) + 2\ H_2O(g)}$

(c) $\{Fe(s) \rightarrow Fe^{3+}(aq) + 3\ e^- \qquad\qquad\} \times 4$

$\{4\ e^- + 2\ H_2O(l) + O_2(g) \rightarrow 4\ OH^-(aq) \qquad\} \times 3$

$\overline{4\ Fe(s) + 6\ H_2O(l) + 3\ O_2(g) \rightarrow 4\ Fe(OH)_3(s)}$

Oxidizing and Reducing Agents

47. The oxidizing agents experience a decrease in the oxidation state of one of their elements, while the reducing agents experience an increase in the oxidation state of one of their elements.

(a) $SO_3^{2-}(aq)$ is the reducing agent; the O.S. of $S = +4$ in SO_3^{2-} and $= +6$ in SO_4^{2-}.

$MnO_4^-(aq)$ is the oxidizing agent; the O.S. of $Mn = +7$ in MnO_4^- and $+2$ in Mn^{2+}.

(b) $H_2(g)$ is the reducing agent; the O.S. of $H = 0$ in $H_2(g)$ and $= +1$ in $H_2O(g)$.

$NO_2(g)$ is the oxidizing agent; the O.S. of $N = +4$ in $NO_2(g)$ and -3 in $NH_3(g)$.

(c) $\left[Fe(CN)_6\right]^{4-}(aq)$ is the reducing agent; the O.S. of $Fe = +2$ in $\left[Fe(CN)_6\right]^{4-}$

and $= +3$ in $\left[Fe(CN)_6\right]^{3-}$. $H_2O_2(aq)$ is the oxidizing agent; the O.S. of $O = -1$ in H_2O_2 and $= -2$ in H_2O.

48. **(a)** $2\ S_2O_3^{2-}(aq) + I_2(s) \rightarrow S_4O_6^{2-}(aq) + 2\ I^-(aq)$

(b) $S_2O_3^{2-}(aq) + 4\ Cl_2(g) + 5\ H_2O(l) \rightarrow 2\ HSO_4^-(aq) + 8\ Cl^-(aq) + 8\ H^+(aq)$

(c) $S_2O_3^{2-}(aq) + 4\ OCl^-(aq) + 2\ OH^-(aq) \rightarrow 2\ SO_4^{2-}(aq) + 4\ Cl^-(aq) + H_2O(l)$

Neutralization and Acid–Base Titrations

49. The problem is most easily solved with amounts in millimoles.

$$V_{NaOH} = 10.00\ mL\ HCl(aq) \times \frac{0.128\ mmol\ HCl}{1\ mL\ HCl(aq)} \times \frac{1\ mmol\ H^+}{1\ mmol\ HCl} \times \frac{1\ mmol\ OH^-}{1\ mmol\ H^+}$$

$$\times \frac{1\ mmol\ NaOH}{1\ mmol\ OH^-} \times \frac{1\ mL\ NaOH(aq)}{0.0962\ mmol\ NaOH} = 13.3\ mL\ NaOH(aq)\ soln$$

50.

$$[\text{NaOH}] = \frac{10.00 \text{ mL acid} \times \dfrac{0.1012 \text{ mmol H}_2\text{SO}_4}{1 \text{ mL acid}} \times \dfrac{2 \text{ mmol NaOH}}{1 \text{ mmol H}_2\text{SO}_4}}{23.31 \text{ mL base}} = 0.08683 \text{ M}$$

51. The net reaction is $\text{OH}^- (\text{aq}) + \text{HC}_3\text{H}_5\text{O}_2 (\text{aq}) \rightarrow \text{H}_2\text{O(l)} + \text{C}_3\text{H}_5\text{O}_2{}^- (\text{aq})$

$$V_{\text{base}} = 25.00 \text{ mL acid} \times \frac{0.3057 \text{ mmol HC}_3\text{H}_5\text{O}_2}{1 \text{ mL acid}} \times \frac{1 \text{ mmol KOH}}{1 \text{ mmol HC}_3\text{H}_5\text{O}_2} \times \frac{1 \text{ mL base}}{2.155 \text{ mmol KOH}}$$

$$= 3.546 \text{ mL KOH solution}$$

52. Titration reaction: $\text{Ba(OH)}_2 (\text{aq}) + 2 \text{ HNO}_3 (\text{aq}) \rightarrow \text{Ba(NO}_3)_2 (\text{aq}) + 2 \text{ H}_2\text{O(l)}$

$$V_{\text{base}} = 50.00 \text{ mL acid} \times \frac{0.0526 \text{ mol HNO}_3}{1 \text{ mL acid}} \times \frac{1 \text{ mmol Ba(OH)}_2}{2 \text{ mmol HNO}_3} \times \frac{1 \text{ mL base}}{0.0844 \text{ mmol Ba(OH)}_2}$$

$$= 15.6 \text{ mL Ba(OH)}_2 \text{ solution}$$

53. $\text{NaOH(aq)} + \text{HCl(aq)} \rightarrow \text{NaCl(aq)} + \text{H}_2\text{O(l)}$ is the titration reaction.

$$[\text{NaOH}] = \frac{0.02834 \text{ L} \times \dfrac{0.1085 \text{ mol HCl}}{1 \text{ L soln}} \times \dfrac{1 \text{ mol NaOH}}{1 \text{ mol HCl}}}{0.02500 \text{ L sample}} = 0.1230 \text{ M NaOH}$$

54. $$[\text{NH}_3] = \frac{28.72 \text{ mL acid} \times \dfrac{1.021 \text{ mmol HCl}}{1 \text{ mL acid}} \times \dfrac{1 \text{ mmol H}^+}{1 \text{ mmol HCl}} \times \dfrac{1 \text{ mmol NH}_3}{1 \text{ mmol H}^+}}{5.00 \text{ mL sample}} = 5.86 \text{ M NH}_3$$

55. The mass of acetylsalicylic acid is converted to the amount of NaOH, in millimoles, that will react with it.

$$[\text{NaOH}] = \frac{0.32 \text{ g HC}_9\text{H}_7\text{O}_4}{23 \text{ mL NaOH(aq)}} \times \frac{1 \text{ mol HC}_9\text{H}_7\text{O}_4}{180.2 \text{ g HC}_9\text{H}_7\text{O}_4} \times \frac{1 \text{ mol NaOH}}{1 \text{ mol HC}_9\text{H}_7\text{O}_4} \times \frac{1000 \text{ mmol NaOH}}{1 \text{ mol NaOH}}$$

$$= 0.077 \text{ M NaOH}$$

56. (a) $\text{vol conc. acid} = 20.0 \text{ L} \times \dfrac{0.10 \text{ mol HCl}}{1 \text{ L soln}} \times \dfrac{36.5 \text{ g HCl}}{1 \text{ mol HCl}} \times \dfrac{100 \text{ g conc soln'}}{38 \text{ g HCl}} \times \dfrac{1 \text{ mL}}{1.19 \text{ g conc soln'}}$

$$= 1.6 \times 10^2 \text{ mL conc. acid}$$

(b) The titration reaction is $\text{HCl(aq)} + \text{NaOH(aq)} \rightarrow \text{NaCl(aq)} + \text{H}_2\text{O(l)}$

$$[\text{HCl}] = \frac{20.93 \text{ mL base} \times \dfrac{0.1186 \text{ mmol NaOH}}{1 \text{ mL base}} \times \dfrac{1 \text{ mmol HCl}}{1 \text{ mmol NaOH}}}{25.00 \text{ mL acid}} = 0.09929 \text{ M HCl}$$

(c) First of all, the volume of the dilute solution (20 L) is known at best to a precision of two significant figures. Secondly, HCl is somewhat volatile (we can smell its odor above the solution) and some will likely be lost during the process of preparing the solution.

57. The equation for the reaction is $HNO_3(aq) + KOH(aq) \rightarrow KNO_3(aq) + H_2O(l)$.

This equation shows that equal numbers of moles are needed for a complete reaction. We compute the amount of each reactant.

$$mmol\ HNO_3 = 25.00\ mL\ acid \times \frac{0.132\ mmol\ HNO_3}{1\ mL\ acid} = 3.30\ mmol\ HNO_3$$

$$mmol\ KOH = 10.00\ mL\ acid \times \frac{0.318\ mmol\ KOH}{1\ mL\ base} = 3.18\ mmol\ KOH$$

There is more acid present than base. Thus, the resulting solution is acidic.

58. Here we compute the amount of acetic acid in the vinegar and the amount of acetic acid needed to react with the sodium carbonate. If there is more than enough acid to react with the solid, the solution will remain acidic.

$$Acetic\ Acid\ in\ vinegar = 125\ mL \times \frac{0.762\ mmol\ HC_2H_3O_2}{1\ mL\ vinegar} = 95.3\ mmol\ HC_2H_3O_2$$

$$Na_2CO_3(s) + 2\ HC_2H_3O_2(aq) \rightarrow 2\ NaC_2H_3O_2(aq) + H_2O(l) + CO_2(g)$$

Acetic acid required for solid:

$$= 7.55\ g\ Na_2CO_3 \times \frac{1000\ mmol\ Na_2CO_3}{106.0\ g\ Na_2CO_3} \times \frac{2\ mmol\ HC_2H_3O_2}{1\ mmol\ Na_2CO_3} = 143\ mmol\ HC_2H_3O_2$$

Clearly there is not enough acetic acid present to react with all of the sodium carbonate. The resulting solution will not be acidic. In fact, the solution will contain only a trace amount of acetic acid ($HC_2H_3O_2$).

59. $V_{base} = 5.00\ mL\ vinegar \times \frac{1.01\ g\ vinegar}{1\ mL} \times \frac{4.0\ g\ HC_2H_3O_2}{100.0\ g\ vinegar} \times \frac{1\ mol\ HC_2H_3O_2}{60.0\ g\ HC_2H_3O_2}$

$$\times \frac{1\ mol\ NaOH}{1\ mol\ HC_2H_3O_2} \times \frac{1\ L\ base}{0.1000\ mol\ NaOH} \times \frac{1000\ mL}{1\ L} = 34\ mL\ base$$

60. The titration reaction is $2\ NaOH(aq) + H_2SO_4(aq) \rightarrow Na_2SO_4(aq) + 2\ H_2O(l)$

It is most convenient to consider molarity as millimoles per milliliter when solving this problem.

$$[H_2SO_4] = \frac{49.74\ mL\ base \times \dfrac{0.935\ mmol\ NaOH}{1\ mL\ base} \times \dfrac{1\ mmol\ H_2SO_4}{2\ mmol\ NaOH}}{5.00\ mL\ battery\ acid} = 4.65\ M\ H_2SO_4$$

Thus, the battery acid is *not* sufficiently concentrated.

61. Answer is (d): 120 % of necessary titrant added in titration of NH_3

$$\left.\begin{array}{l} 5\ NH_3 \\ + \\ 5\ HCl \\ + \end{array}\right\} \begin{array}{l} \text{required for} \\ \text{equivalence} \\ \text{point} \end{array} \longrightarrow \begin{array}{l} 5\ NH_4^+ + 6\ Cl^- + H_3O^+ \\ \text{(depicted in question's drawing)} \end{array}$$

$$1\ HCl \left.\right\} \ 20\ \%\ \text{excess}$$

62. **(a)** $H_2O(l) + K^+(aq) + Cl^-(aq)$

(b) $CH_3COOH(aq) + CH_3COO^-(aq) + H_2O(l) + Na^+(aq)$

Stoichiometry of Oxidation–Reduction Reactions

63.

$$[MnO_4^-] = \frac{0.1078\,g\,As_2O_3 \times \dfrac{1\,mol\,As_2O_3}{197.84\,g\,As_2O_3} \times \dfrac{4\,mol\,MnO_4^-}{5\,mol\,As_2O_3} \times \dfrac{1\,mol\,KMnO_4}{1\,mol\,MnO_4^-}}{22.15\,mL \times \dfrac{1\,L}{1000\,mL}} = 0.01968\ \text{M KMnO}_4$$

64. The balanced equation for the titration is:

$$5\ SO_3^{2-}(aq) + 2\ MnO_4^-(aq) + 6\ H^+(aq) \rightarrow 5\ SO_4^{2-}(aq) + 2\ Mn^{2+}(aq) + 3\ H_2O(l)$$

$$\left[SO_3^{2-}\right] = \frac{31.46\,mL \times \dfrac{0.02237\,mmol\,KMnO_4}{1\,mL\,soln} \times \dfrac{1\,mmol\,MnO_4^-}{1\,mmol\,KMnO_4} \times \dfrac{5\,mmol\,SO_3^{2-}}{2\,mmol\,MnO_4^-}}{25.00\ mL\ SO_3^{2-}\ \text{soln}} = 0.07038\ \text{M SO}_3^{2-}$$

65. First, we will determine the mass of Fe, then the percentage of iron in the ore.

$$\text{mass Fe} = 28.72\ \text{mL} \times \frac{1\ L}{1000\ mL} \times \frac{0.05051\ mol\ Cr_2O_7^{2-}}{1\ L\ soln} \times \frac{6\ mol\ Fe^{2+}}{1\ mol\ Cr_2O_7^{2-}} \times \frac{55.85\ g\ Fe}{1\ mol\ Fe^{2+}}$$

$$\text{mass Fe} = 0.4861\ \text{g Fe} \qquad \%\ Fe = \frac{0.4861\,g\,Fe}{0.9132\,g\,ore} \times 100\% = 53.23\%\ \text{Fe}$$

66. First balance the titration equation.

Oxidation: $\{\ Mn^{2+}(aq) + 4\,OH^-(aq) \rightarrow MnO_2(s) + 2\,H_2O(l) + 2e^-\quad\} \times 3$

Reduction: $\{\ \underline{MnO_4^-(aq) + 2\,H_2O^-(l) + 3e^- \rightarrow MnO_2(s) + 4\,OH^-(aq)}\ \} \times 2$

Net: $\quad 3\,Mn^{2+}(aq) + 2\,MnO_4^-(aq) + 4\,OH^-(aq) \rightarrow 5\,MnO_2(s) + 2\,H_2O(l)$

$$\left[Mn^{2+}\right] = \frac{37.21\,mL\,\text{titrant} \times \dfrac{0.04162\,mmol\,MnO_4^-}{1\,mL\,\text{titrant}} \times \dfrac{3\,mmol\,Mn^{2+}}{2\,mmol\,MnO_4^-}}{25.00\ mL\ \text{soln}} = 0.09292\ \text{M Mn}^{2+}$$

67. First balance the titration equation:

Oxidation: $\{ C_2O_4^{2-}(aq) \rightarrow 2\ CO_2(g) + 2\ e^- \qquad\qquad\qquad\} \times 5$

Reduction: $\{ MnO_4^-(aq) + 8\ H^+(aq) + 5\ e^- \rightarrow\ Mn^{2+}(aq) + 4\ H_2O(l)\quad \} \times 2$

Net: $5\ C_2O_4^{2-}(aq) + 2\ MnO_4^-(aq) + 16\ H^+(aq) \rightarrow 10\ CO_2(g) + 2\ Mn^{2+}(aq) + 8\ H_2O(l)$

$$mass_{Na_2C_2O_4} = 1.00\ \text{L satd soln Na}_2\text{C}_2\text{O}_4 \times \frac{1000\ \text{mL}}{1\ \text{L}} \times \frac{25.8\ \text{mL satd soln KMnO}_4}{5.00\ \text{mL satd soln Na}_2\text{C}_2\text{O}_4} \times \frac{0.02140\ \text{mol KMnO}_4}{1000\ \text{mL KMnO}_4}$$

$$\times \frac{1\ \text{mol MnO}_4^-}{1\ \text{mol KMnO}_4} \times \frac{5\ \text{mol C}_2\text{O}_4^{2-}}{2\ \text{mol MnO}_4^-} \times \frac{1\ \text{mol Na}_2\text{C}_2\text{O}_4}{1\ \text{mol C}_2\text{O}_4^{2-}} \times \frac{134.0\ \text{g Na}_2\text{C}_2\text{O}_4}{1\ \text{mol Na}_2\text{C}_2\text{O}_4}$$

$mass_{Na_2C_2O_4} = 37.0\ \text{g Na}_2\text{C}_2\text{O}_4$

68. Balanced equation:

$$3\ S_2O_4^{2-}(aq) + 2\ CrO_4^{2-}(aq) + 4\ H_2O(l) \rightarrow 6\ SO_3^{2-}(aq) + 2\ Cr(OH)_3(s) + 2\ H^+(aq)$$

(a)

$$mass\ Cr(OH)_3 = 100.\ \text{L soln} \times \frac{0.0126\ \text{mol CrO}_4^{2-}}{1\ \text{L soln}} \times \frac{2\ \text{mol Cr(OH)}_3}{2\ \text{mol CrO}_4^{2-}} \times \frac{103.0\ \text{g Cr(OH)}_3}{1\ \text{mol Cr(OH)}_3}$$

$$= 130\ \text{g Cr(OH)}_3$$

(b)

$$mass\ Na_2S_2O_4 = 100.\ \text{L soln} \times \frac{0.0126\ \text{mol CrO}_4^{2-}}{1\ \text{L soln}} \times \frac{3\ \text{mol S}_2\text{O}_4^{2-}}{2\ \text{mol CrO}_4^{2-}} \times \frac{1\ \text{mol Na}_2\text{S}_2\text{O}_4}{1\ \text{mol S}_2\text{O}_4^{2-}}$$

$$\times \frac{174.1\ \text{g Na}_2\text{S}_2\text{O}_4}{1\ \text{mol Na}_2\text{S}_2\text{O}_4} = 329\ \text{g Na}_2\text{S}_2\text{O}_4$$

Integrative and Advanced Exercises

69. (a) $2\ Na(s) + 2\ H_2O(l) \longrightarrow 2NaOH(aq) + H_2(g)$

(b) $Fe^{3+}(aq) + 3OH^-(aq) \longrightarrow Fe(OH)_3(s)$

(c) $Fe(OH)_3(s) + 3\ H_3O^+(aq) \longrightarrow Fe^{3+}(aq) + 6\ H_2O(l)$

or $Fe(OH)_3(s) + 3\ H^+(aq) \longrightarrow Fe^{3+}(aq) + 3\ H_2O(l)$

70. (a) $2\ HCl(aq) + FeS(s) \longrightarrow FeCl_2(aq) + H_2S(g)$

(b) Oxidation: $2\ Cl^-(aq) \longrightarrow Cl_2(aq) + 2\ e^-$

Reduction: $MnO_2(s) + 4\ H^+(aq) + 2\ e^- \longrightarrow Mn^{2+}(aq) + 2\ H_2O(l)$

Net: $2\ Cl^-(aq) + MnO_2(s) + 4\ H^+(aq) \longrightarrow Cl_2(g) + Mn^{2+}(aq) + 2\ H_2O(l)$

+ Spectator Ions: $4\ HCl(aq) + MnO_2(s) \longrightarrow Cl_2(g) + MnCl_2(aq) + 2\ H_2O(l)$

(c) Because $NH_3(aq)$ is a weak base, the reaction takes place in alkaline solution.

Oxidation : $2 NH_3(aq) \longrightarrow N_2(g) + 6 H^+ + 6 e^-$

Reduction : $\{Br_2 + 2 e^- \longrightarrow 2 Br^-(aq)\}$ $\qquad\qquad\qquad\qquad \times 3$

Net : $2 NH_3(aq) + 3 Br_2 \longrightarrow N_2(g) + 6 H^+ + 6 Br^-(aq)$

The spectator ion is $NH_4^+(aq)$; first add 6 $NH_3(aq)$ on each side, to "neutralize" H^+.

$2 NH_3(aq) + 3 Br_2 + 6 NH_3(aq) \longrightarrow N_2(g) + 6 H^+ + 6 Br^-(aq) + 6 NH_3(aq)$

Then recognize that $NH_3(aq) + H^+(aq) \longrightarrow NH_4^+(aq)$, and $NH_4Br(aq)$ is really

$NH_4^+(aq) + Br^-(aq)$

$8 NH_3(aq) + 3 Br_2 \longrightarrow N_2(g) + 6 NH_4Br(aq)$

(d) $Ba(ClO_2)_2(s) + H_2SO_4(aq) \longrightarrow 2 HClO_2(aq) + BaSO_4(s)$

71. A possible product, based on solubility rules, is $Ca_3(PO_4)_2$. We determine the % Ca in this compound.

molar mass $= 3 \times 40.078 \text{ g Ca} + 2 \times 30.974 \text{ g P} + 8 \times 15.999 \text{ g O}$

$\qquad\qquad\quad = 120.23 \text{ g Ca} + 61.948 \text{ g P} + 127.99 \text{ g O} = 310.17 \text{ g}$

$\% \text{ Ca} = \dfrac{120.23 \text{ g Ca}}{310.17 \text{ g } Ca_3(PO_4)_2} \times 100\% = 38.763\%$

Thus, $Ca_3(PO_4)_2$ is the predicted product. The net ionic equation follows.

$3 Ca^{2+}(aq) + 2 HPO_4^{2-}(aq) \longrightarrow Ca_3(PO_4)_2(s) + 2 H^+(aq)$

72. We can calculate the initial concentration of OH^-

$$[OH^-] = \frac{0.0250 \text{ mol Ba(OH)}_2}{1 \text{ L soln}} \times \frac{2 \text{ mol OH}^-}{1 \text{ L soln}} = 0.0500 \text{ M}$$

We can determine the ratio of the dilute (volumetric flask) to the concentrated (pipet) solutions.

$V_c \times C_c = V_d \times C_d = V_c \times 0.0500 \text{ M} = V_d \times 0.0100 \text{ M}$ $\qquad \dfrac{V_d}{V_c} = \dfrac{0.0500 \text{ M}}{0.0100 \text{ M}} = 5.00$

If we pipet 0.0250 M $Ba(OH)_2$ with a 50.00-mL pipet into a 250.0-mL flask, and fill this flask, with mixing, to the mark with distilled water, the resulting solution will be 0.0100 M OH^-.

73. (a) The first reason why a small percentage of $Na_2CO_3(s)$ mixed in with $NaOH(s)$ would not affect the usefulness of the $NaOH(aq)$ prepared, is that the $NaOH(aq)$ solution will be standardized. But that explanation is suitable no matter what the percentage of $Na_2CO_3(s)$. Let us assume that the amount of NaOH is determined by weighing rather than by standardization. Determine the mass of solid that reacts with 1.000 mol HCl. The reactions are

$$NaOH(s) + HCl(aq) \longrightarrow NaCl(aq) + H_2O\ (l)\ \text{AND}$$

$$Na_2CO_3(s) + 2\,HCl(aq) \longrightarrow 2\,NaCl(aq) + CO_2(g) + H_2O(l)$$

$$\text{mass NaOH} = 1.00\ \text{mol HCl} \times \frac{1\ \text{mol NaOH}}{1\ \text{mol HCl}} \times \frac{40.0\ \text{g NaOH}}{1\ \text{mol NaOH}} = 40.0\ \text{g NaOH}$$

$$\text{mass Na}_2\text{CO}_3 = 1.00\ \text{mol HCl} \times \frac{1\ \text{mol Na}_2\text{CO}_3}{2\ \text{mol HCl}} \times \frac{106\ \text{g Na}_2\text{CO}_3}{1\ \text{mol Na}_2\text{CO}_3} = 53.0\ \text{g Na}_2\text{CO}_3$$

Thus, a small percentage of Na_2CO_3 with the NaOH has a small effect on the final result.

(b) As the proportion of $Na_2CO_3(s)$ grows, the error it introduces becomes more significant and makes an unstandardized solution unusable for precise work.

74. Let us first determine the mass of Mg in the sample analyzed.

$$\text{mass Mg} = 0.0549\ \text{g Mg}_2\text{P}_2\text{O}_7 \times \frac{1\ \text{mol Mg}_2\text{P}_2\text{O}_7}{222.55\ \text{g Mg}_2\text{P}_2\text{O}_7} \times \frac{2\ \text{mol Mg}}{1\ \text{mol Mg}_2\text{P}_2\text{O}_7} \times \frac{24.305\ \text{g}}{1\ \text{mol Mg}} = 0.0120\ \text{g Mg}$$

$$\text{ppm Mg} = 10^6\ \text{g sample} \times \frac{0.0120\ \text{g Mg}}{110.520\ \text{g sample}} = 108\ \text{ppm Mg}$$

75. Let V represent the volume of added 0.248 M $CaCl_2$ that must be added.

We know that $[Cl^-] = 0.250$ M, but also,

$$[Cl^-] = \frac{0.335\ \text{L}\ \dfrac{0.186\ \text{mol KCl}}{1\ \text{L soln}} \times \dfrac{1\ \text{mol Cl}^-}{1\ \text{mol KCl}} + V \times \dfrac{0.248\ \text{mol CaCl}_2}{1\ \text{L soln}} \times \dfrac{2\ \text{mol Cl}^-}{1\ \text{mol CaCl}_2}}{0.335\ \text{L} + V}$$

$$0.250\,(0.335 + V) = 0.0838 + 0.250\,V = 0.0623 + 0.496\,V \qquad V = \frac{0.0838 - 0.0623}{0.496 - 0.250} = 0.0874\ \text{L}$$

76. Cu^{2+} would produce a colored solid, while for NH_4^+/Na^+, no solids are expected(these cations form very soluble salts), $\therefore\ Cu^{2+}, NH_4^+$ and Na^+ are not present. Thus the possible cations are: Ba^{2+} and Mg^{2+} (both give colorless solutions)

Gas evolution when the solid reacts with $HCl(aq)$ suggests the presence of carbonate (CO_3^{2-})

$$2\,H^+(aq) + CO_3^{2-}(aq) \rightarrow \text{"}H_2CO_3(aq)\text{"} \rightarrow H_2O(l) + CO_2(g)$$

If the solid is indeed a mixture of carbonates, the net ionic equation for the reaction of the solid with HCl(aq) would be:

$$BaCO_3(s) + MgCO_3(s) + 4\ H^+(aq) \rightarrow Ba^{2+}(aq) + Mg^{2+}(aq) + 2\ H_2O(l) + 2\ CO_2(g)$$

The solution above contains $Ba^{2+} + Mg^{2+}$ ions, then the addition of $(NH_4)_2SO_4(aq)$ should result in the formation of the insoluble sulfates of these two metal ions. This is consistent with the observations.

Net ionic: $Ba^{2+}(aq) + Mg^{2+}(aq) + 2\ SO_4^{2-} \rightarrow BaSO_4(s) + MgSO_4(s)$.

If the solution above the solid should contain small quantities of all of the ions present in the solid, that is, it should contain $Ba^{2+}(aq)$, $Mg^{2+}(aq)$ and $CO_3^{2-}(aq)$. Addition of KOH should result in the formation of the hydroxide of the two metal ions and potassium carbonate. Note: K_2CO_3 and $Ba(OH)_2$ are both relatively soluble species, hence, neither of these species should precipitate out of solution. However, $Mg(OH)_2$ is relatively insoluble. A precipitate ($Mg(OH)_2(s)$ which is white) is expected to form. This is consistent with the observations provided. We can conclude that the solid is likely a mixture of $BaCO_3(s)$ and $MgCO_3(s)$. Without additional data, this conclusion would indeed explain all of the observations provided

77. (a) Oxidation: $\{\ IBr(aq) + 3\ H_2O(l) \longrightarrow IO_3^-(aq) + Br^-(aq) + 6\ H^+(aq) + 4\ e^-\}$ $\times 3$

Reduction: $\{\ BrO_3^- + 6\ H^+(aq) + 6\ e^- \longrightarrow Br^-(aq) + 3\ H_2O\ \}$ $\times 2$

Net: $3\ IBr(aq) + 3\ H_2O(l) + 2\ BrO_3^-(aq) \longrightarrow 3\ IO_3^-(aq) + 5\ Br^-(aq) + 6\ H^+(aq)$

(b) Oxidation : $\{Sn(s) \longrightarrow Sn^{2+}(aq) + 2\ e^-\}$ $\times 3$

Reduction: $C_2H_5NO_3(aq) + 6\ H^+(aq) + 6\ e^- \longrightarrow C_2H_5OH(aq) + NH_2OH(aq) + H_2O(l)$

Net: $3\ Sn(s) + C_2H_5NO_3(aq) + 6\ H^+(aq) \longrightarrow 3\ Sn^{2+}(aq) + C_2H_5OH(aq) + NH_2OH(aq) + H_2O(l)$

(c) Oxidation: $\{As_2S_3(s) + 8\ H_2O(l) \longrightarrow 2\ H_3AsO_4(aq) + 3\ S(s) + 10\ H^+(aq) + 10\ e^-\ \}$ $\times 3$

Reduction : $\{NO_3^-(aq) + 4\ H^+(aq) + 3\ e^- \longrightarrow NO(g) + 2\ H_2O\}$ $\times 10$

Net: $3\ As_2S_3(s) + 4\ H_2O(l) + 10\ NO_3^-(aq) + 10\ H^+(aq) \longrightarrow 6\ H_3AsO_4(aq) + 9\ S(s) + 10\ NO(g)$

(d) Oxidation: $I_2(aq) + 6\ H_2O(l) \longrightarrow 2\ IO_3^-(aq) + 12\ H^+(aq) + 10\ e^-$ $\times 1$

Reduction : $\{H_5IO_6(aq) + H^+(aq) + 2\ e^- \longrightarrow IO_3^-(aq) + 3\ H_2O\}$ $\times 5$

Net : $I_2(aq) + 5\ H_5IO_6(aq) \longrightarrow 7\ IO_3^-(aq) + 9\ H_2O(l) + 7\ H^+(aq)$

(e) Oxidation: $\{2\ S_2F_2(g) + 6\ H_2O(l) \longrightarrow H_2S_4O_6(aq) + 4\ HF(aq) + 6\ H^+(aq) + 6\ e^-\}$ $\times 4$

Reduction: $\{4\ S_2F_2(g) + 8\ H^+(aq) + 8\ e^- \longrightarrow S_8(s) + 8\ HF(aq)\ \}$ $\times 3$

Net: $20\ S_2F_2(g) + 24\ H_2O(l) \longrightarrow 4\ H_2S_4O_6(aq) + 3\ S_8(s) + 40\ HF(aq)$

78. (a) Oxidation : $\{Fe_2S_3(s) + 6\,OH^-(aq) \longrightarrow 2\,Fe(OH)_3(s) + 3\,S(s) + 6\,e^-\} \times 2$

Reduction : $\{O_2(g) + 2\,H_2O(l) + 4\,e^- \longrightarrow 4\,OH^-(aq) \qquad\} \qquad \times 3$

Net : $2\,Fe_2S_3(s) + 3\,O_2(g) + 6\,H_2O(l) \longrightarrow 4\,Fe(OH)_3(s) + 6\,S(s)$

(b) Oxidation : $\{4\,OH^-(aq) \longrightarrow O_2(g) + 2\,H_2O(l) + 4\,e^-\} \qquad \times 3$

Reduction : $\{O_2^-(aq) + 2\,H_2O(l) + 3\,e^- \longrightarrow 4\,OH^-(aq)\} \qquad \times 4$

Net : $4\,O_2^-(aq) + 2\,H_2O(l) \longrightarrow 3\,O_2(g) + 4\,OH^-(aq)$

(c) Oxidation: $\{CrI_3(s) + 32\,OH^-(aq) \longrightarrow CrO_4^{2-}(aq) + 3\,IO_4^-(aq) + 16\,H_2O + 27\,e^-\} \quad \times 2$

Reduction: $\{H_2O_2(aq) + 2\,e^- \longrightarrow 2\,OH^-(aq)\} \qquad\qquad \times 27$

Net : $2\,CrI_3(s) + 10\,OH^-(aq) + 27\,H_2O_2(aq) \longrightarrow 2\,CrO_4^{2-}(aq) + 6\,IO_4^-(aq) + 32\,H_2O$

(d) Oxidation : $\{Ag(s) + 2\,CN^-(aq) \longrightarrow [Ag(CN)_2]^-(aq) + e^-\} \times 4$

Reduction: $2H_2O(l) + O_2(g) + 4\,e^- \longrightarrow 4\,OH^-(aq)$

Net : $4\,Ag(s) + 8\,CN^-(aq) + 2\,H_2O + O_2(g) \longrightarrow 4[Ag(CN)_2]^-(aq) + 4\,OH^-(aq)$

(e) Oxidation: $B_2Cl_4(aq) + 8\,OH^-(aq) \longrightarrow 2\,BO_2^-(aq) + 4\,Cl^-(aq) + 4\,H_2O + 2\,e^-$

Reduction: $2\,H_2O(l) + 2\,e^- \longrightarrow H_2(g) + 2\,OH^-(aq)$

Net: $B_2Cl_4(aq) + 6\,OH^-(aq) \longrightarrow 2\,BO_2^-(aq) + 4\,Cl^-(aq) + 2\,H_2O(l) + H_2(g)$

(f) Oxidation : $\{C_2H_5OH(aq) + 5\,OH^-(aq) \longrightarrow C_2H_3O_2^-(aq) + 4\,H_2O(l) + 4\,e^-\} \qquad \times 3$

Reduction : $\{MnO_4^-(aq) + 2\,H_2O(l) + 3\,e^- \longrightarrow MnO_2(s) + 4\,OH^-(aq)\} \qquad \times 4$

Net : $3\,C_2H_5OH(aq) + 4\,MnO_4^-(aq) \longrightarrow 3\,C_2H_3O_2^-(aq) + 4\,MnO_2(s) + OH^-(aq) + 4\,H_2O(l)$

79. Oxidation: $\{P_4(s) + 16\,H_2O(l) \longrightarrow 4\,H_3PO_4(aq) + 20\,H^+(aq) + 20\,e^-\} \qquad \times 3$

Reduction : $\{P_4(s) + 12\,H^+(aq) + 12\,e^- \longrightarrow 4\,PH_3(g) \qquad\} \qquad \times 5$

Net: $8\,P_4(s) + 48\,H_2O(l) \longrightarrow 12\,H_3PO_4(aq) + 20\,PH_3(g)$

Or: $2\,P_4(s) + 12\,H_2O(l) \longrightarrow 3\,H_3PO_4(aq) + 5\,PH_3(g)$

80. (a) Oxidation: $\{FeS_2(s) + 8\,H_2O(l) \longrightarrow Fe^{2+}(aq) + 2\,SO_4^{2-}(aq) + 16\,H^+(aq) + 14\,e^-\} \quad \times 2$

Reduction : $\{O_2(g) + 4\,H^+(aq) + 4\,e^- \longrightarrow 2\,H_2O \qquad\} \qquad\qquad \times 7$

Net: $2\,FeS_2(s) + 2\,H_2O(l) + 7\,O_2(g) \longrightarrow 2\,Fe^{2+}(aq) + 4\,SO_4^{2-}(aq) + 4\,H^+(aq)$

(b) Oxidation: $\{FeS_2(s) + 8\ H_2O(l) \longrightarrow Fe^{2+}(aq) + 2\ SO_4^{2-}(aq) + 16\ H^+(aq) + 14\ e^-\}$

Reduction : $\{Fe^{3+}(aq) + e^- \longrightarrow Fe^{2+}(aq)\quad\}$ $\qquad\qquad\qquad\qquad\times 14$

Net: $\quad 14\ Fe^{3+}(aq) + FeS_2(s) + 8\ H_2O(l) \longrightarrow 15\ Fe^{2+}(aq) + 2\ SO_4^{2-}(aq) + 16\ H^+(aq)$

81. The neutralization reaction is $H_2SO_4(aq) + Ba(OH)_2(aq) \longrightarrow BaSO_4(s) + 2\ H_2O(l)$

First we determine the molarity of H_2SO_4 in the 10.00 mL of diluted acid.

$$\text{molarity of } H_2SO_4 = \frac{32.44\ \text{mL base} \times \dfrac{0.00498\ \text{mmol } Ba(OH)_2}{1\ \text{mL base}} \times \dfrac{1\ \text{mmol } H_2SO_4}{1\ \text{mmol } Ba(OH)_2}}{10.00\ \text{mL acid}} = 0.0162\ M\ H_2SO_4$$

Now we determine the molarity of H_2SO_4 in the concentrated solution, using $V_c \times C_c = V_d \times C_d$.

$$1.00\ \text{mL} \times C_c = 250.0\ \text{mL} \times 0.0162\ M \quad C_c = \frac{250.0\ \text{mL} \times 0.0162\ M}{1.00\ \text{mL}} = 4.05\ M\ H_2SO_4$$

$$\%H_2SO_4 = \frac{4.05\ \text{mol } H_2SO_4}{1\ \text{L soln}} \times \frac{98.08\ \text{g } H_2SO_4}{1\ \text{mol } H_2SO_4} \times \frac{1\ L}{1000\ \text{mL}} \times \frac{1\ \text{mL}}{1.239\ \text{g}} \times 100\% = 32.1\%\ H_2SO_4$$

82. The titration reaction is: $CaCO_3(s) + 2\ H^+(aq) \longrightarrow Ca^{2+}(aq) + H_2O(l) + CO_2(g)$

Now we calculate the mass of the marble, through several steps, as follows.

$$\text{initial moles HCl} = 2.00\ L \times \frac{2.52\ \text{mol HCl}}{1\ \text{L soln}} = 5.04\ \text{mol HCl}$$

$NaOH(aq) + HCl(aq) \longrightarrow NaCl(aq) + H_2O$ is the titration reaction.

$$\text{final molarity of HCl} = \frac{0.02487\ L \times \dfrac{0.9987\ \text{mol NaOH}}{1\ L} \times \dfrac{1\ \text{mol HCl}}{1\ \text{mol NaOH}}}{0.01000\ L} = 2.484\ M$$

$$\text{final amount HCl} = 2.00\ L \times \frac{2.484\ \text{mol HCl}}{1\ \text{L soln}} = 4.968\ \text{mol HCl}$$

$$\text{mass } CaCO_3 = (5.04 - 4.968)\ \text{mol HCl} \times \frac{1\ \text{mol } CaCO_3}{2\ \text{mol HCl}} \times \frac{100.1\ \text{g } CaCO_3}{1\ \text{mol } CaCO_3} = 3._6\ \text{g } CaCO_3 \sim 4\text{g } CaCO_3$$

There are two reasons why the final result can be determined to but one significant figure. The first is the limited precision of the data given, as in the three-significant-figure limitation of the initial concentration and the total volume of the solution. The second is that the initial solution is quite concentrated for the job it must do. If it were one-tenth as concentrated, the final result could be determined more precisely because the concentration of the solution would have decreased to a greater extent.

83. Oxidation : $\{2\,Cl^-\,(aq) \longrightarrow Cl_2\,(g) + 2\,e^-\}$ ×3

Reduction : $Cr_2O_7^{\,2-}\,(aq) + 14\,H^+\,(aq) + 6\,e^- \longrightarrow 2\,Cr^{3+}\,(aq) + 7\,H_2O$

Net : $6\,Cl^-\,(aq) + Cr_2O_7^{\,2-}\,(aq) + 14\,H^+\,(aq) \longrightarrow 2\,Cr^{3+}\,(aq) + 7\,H_2O + 3\,Cl_2\,(g)$

We need to determine the amount of $Cl_2(g)$ produced from each of the reactants. The limiting reactant is the one that produces the lesser amount of Cl_2..

$$\text{amount } Cl_2 = 325\ \text{mL} \times \frac{1.15\ \text{g}}{1\ \text{mL}} \times \frac{30.1\ \text{g HCl}}{100.\ \text{g soln}} \times \frac{1\ \text{mol HCl}}{36.46\ \text{g HCl}} \times \frac{1\ \text{mol Cl}^-}{1\ \text{mol HCl}} \times \frac{3\ \text{mol Cl}_2}{6\ \text{mol Cl}^-}$$

$$= 1.54\ \text{mol } Cl_2$$

$$\text{amount } Cl_2 = 62.6\ \text{g} \times \frac{98.5\ \text{g K}_2\text{Cr}_2\text{O}_7}{100.\ \text{g sample}} \times \frac{1\ \text{mol K}_2\text{Cr}_2\text{O}_7}{294.2\ \text{g K}_2\text{Cr}_2\text{O}_7} \times \frac{1\ \text{mol Cr}_2\text{O}_7^{\,2-}}{1\ \text{mol K}_2\text{Cr}_2\text{O}_7} \times \frac{3\ \text{mol Cl}_2}{1\ \text{mol Cr}_2\text{O}_7^{\,2-}}$$

$$= 0.629\ \text{mol } Cl_2, \text{ the amount produced from the limiting reactant}$$

Then we determine the mass of $Cl_2(g)$ produced. $= 0.629\ \text{mol } Cl_2 \times \dfrac{70.91\ \text{g Cl}_2}{1\ \text{mol Cl}_2} = 44.6\ \text{g Cl}_2$

84. Oxidation: $\{As_2O_3\,(s) + 5\,H_2O(l) \longrightarrow 2H_3AsO_4\,(aq) + 4\,H^+\,(aq) + 4\,e^-\}$ ×5

Reduction: $\{MnO_4^-\,(aq) + 8\,H^+\,(aq) + 5\,e^- \longrightarrow Mn^{2+}\,(aq) + 4\,H_2O(l)\}$ ×4

Net: $As_2O_3(s) + 9\,H_2O(l) + 4\,MnO_4^-(aq) + 12\,H^+(aq) \longrightarrow 10\,H_3AsO_4(aq) + 4\,Mn^{2+}(aq)$

$KMnO_4$ molarity = 0.02140 M, as in Example 5-10.

$$\text{Soln. volume} = 0.1304\ \text{g} \times \frac{99.96\ \text{g As}_2\text{O}_3}{100.00\ \text{g sample}} \times \frac{1\ \text{mol As}_2\text{O}_3}{197.84\ \text{g As}_2\text{O}_3} \times \frac{4\ \text{mol MnO}_4^-}{5\ \text{mol As}_2\text{O}_3} \times \frac{1\ \text{mol KMnO}_4}{1\ \text{mol MnO}_4^-}$$

$$\times \frac{1\ \text{L soln}}{0.02140\ \text{mol KMnO}_4} \times \frac{1000\ \text{mL}}{1\ \text{L}} = 24.63\ \text{mL soln}$$

85. $Cl_2\,(g) + NaClO_2\,(aq) \longrightarrow NaCl(aq) + ClO_2\,(g)$ (not balanced)

$Cl_2\,(g) + 2\,NaClO_2\,(aq) \longrightarrow 2\,NaCl(aq) + 2\,ClO_2\,(g)$

$$\text{amount } ClO_2 = 1\ \text{gal} \times \frac{3.785\ \text{L}}{1\ \text{gal}} \times \frac{2.0\ \text{mol NaClO}_2}{1\ \text{L soln}} \times \frac{2\ \text{mol ClO}_2}{2\ \text{mol NaClO}_2} \times \frac{67.45\ \text{g ClO}_2}{1\ \text{mol ClO}_2}$$

$$\times \frac{97\ \text{g ClO}_2\ \text{produced}}{100\ \text{g ClO}_2\ \text{calculated}} = 5.0 \times 10^2\ \text{g ClO}_2\,(g)$$

86. $CaCO_3(s) + 2\,HCl(aq) \rightarrow CaCl_2(aq) + H_2O(l) + CO_2(g)\backslash$

$HCl(aq) + NaOH(aq) \rightarrow H_2O(l) + NaCl(aq)$

$n_{HCl(initial)} = 0.05000 \text{ L} \times 0.5000 \text{ M} = 0.02500 \text{ mol HCl}$

$n_{OH^-}\ 0.04020 \text{ L} \times 0.2184 \text{ M} = 0.00870 \text{ mol OH}^-$

$n_{HCl(excess)} = 0.00870 \text{ mol OH}^- \times \dfrac{1 \text{ mol HCl}}{1 \text{ mol OH}^-} = 0.00870 \text{ mol HCl}$

$n_{HCl(reacted)} = n_{HCl(initial)} - n_{HCl(excess)} = 0.02500 \text{ mol HCl} - 0.00870 \text{ mol HCl} = 0.01622 \text{ mol HCl}$

$\text{mass Ca}^{2+} = 0.01622 \text{ mol HCl} \times \dfrac{1 \text{ mol CaCO}_3}{2 \text{ mol HCl}} \times \dfrac{1 \text{ mol Ca}^{2+}}{1 \text{ mol CaCO}_3} \times \dfrac{40.078 \text{ g Ca}^{2+}}{1 \text{ mol Ca}^{2+}} \times \dfrac{1000 \text{ mg Ca}^{2+}}{1 \text{ g Ca}^{2+}} = 325 \text{ mg Ca}^{2+}$

87. Let X_{KOH} = mass of KOH in grams and X_{LiOH} be the mass of LiOH in grams.

(Note: Molar masses: KOH = 56.1056 g mol^{-1} and LiOH = 23.9483 g mol^{-1})

Moles of HCl = C×V = 0.3520 M × 0.02828 L = 0.009956 mol HCl

We can set up two equations for the two unknowns:

$X_{KOH} + X_{LiOH} = 0.4324$ g and since moles of HCl = moles of OH$^-$ (Stoichiometry is 1:1)

$0.009956 \text{ mol OH}^- = \dfrac{X_{KOH}}{56.1056} + \dfrac{X_{LiOH}}{23.9483}$

Make the substitution that $X_{KOH} = 0.4324$ g - Y_{LiOH}

$0.009956 \text{ mol OH}^- = \dfrac{(0.4324 - X_{LiOH})}{56.1056} + \dfrac{X_{LiOH}}{23.9483} = \dfrac{0.4324}{56.1056} - \dfrac{X_{LiOH}}{56.1056} + \dfrac{X_{LiOH}}{23.9483}$

Collect terms: $0.009956 \text{ mol OH}^- = 0.007707 \text{ mol OH}^- + 0.02393 X_{LiOH} \text{ mol OH}^-$

$0.009956 \text{ mol OH}^- - 0.007707 \text{ mol OH}^- = 0.02393 X_{LiOH} \text{ mol OH}^- = 0.002249 \text{ mol OH}^-$

$X_{LiOH} = 0.09397$ g LiOH hence, $X_{KOH} = 0.4324$ g - 0.09397 g = 0.3384 g

Mass % LiOH = $\dfrac{0.09397\,g}{0.4324\,g} \times 100\% = 21.73\%$ Mass % KOH = $\dfrac{0.3384\,g}{0.4324\,g} \times 100\% = 78.26\%$

88. First balance the redox equations needed for the calculation.

Oxidation: $\{ \text{HSO}_3^-\text{ (aq)} + \text{H}_2\text{O(l)} \rightarrow \text{SO}_4^{2-}\text{ (aq)} + 3 \text{ H}^+\text{ (aq)} + 2 \text{ e}^- \}$ ×3

Reduction: $\underline{\text{IO}_3^-\text{ (aq)} + 6 \text{ H}^+\text{ (aq)} + 6 \text{ e}^- \rightarrow \text{ I}^-\text{(aq)} + 3 \text{ H}_2\text{O(l)}}$ ×1

Net: $3 \text{ HSO}_3^-\text{ (aq)} + \text{IO}_3^-\text{ (aq)} \rightarrow 3 \text{ SO}_4^{2-}\text{ (aq)} + 3 \text{ H}^+\text{ (aq)} + \text{I}^-\text{ (aq)}$

The solution volume of 5.00 L contains 29.0 g NaIO$_3$. This represents

29.0 g/197.9g/mol NaIO$_3$ = 0.147 mol NaIO$_3$

From the above equation we need 3 times that molar amount of NaHSO$_3$, which is

  3(0.147 mol) = 0.441 mol NaHSO$_3$; molar mass of NaHSO$_3$ is 104.06 g/mol

The required mass then is 0.441(104.06) = 45.9 g.

For the second process:

Oxidation: $\{2\ I^-(aq) \rightarrow\ I_2(aq)\ +2\ e^-\}$ x5

Reduction: $2\ IO_3^-\ (aq) + 12\ H^+\ (aq) + 10\ e^- \rightarrow\ I_2(aq) + 6\ H_2O(l)$ x1

Net: $5\ I^-(aq)\ +\ IO_3^-\ (aq) + 6\ H^+\ (aq)\ \rightarrow\ 3\ I_2(aq) + 3\ H_2O(l)$

In step 1, we produced 1 mol $2\ I^-$ for every mole of IO_3^- reactant; therefore we had 0.147 mol I^-.

In step 2, we require 1/5 mol IO_3^- for every mol of I^-.

We require only 1.00 L of the solution in the question instead of the 5.00 L in the first step.

FEATURE PROBLEMS

89. From the volume of titrant we can calculate both the amount in moles of NaC_5H_5 and (through its molar mass of 88.08 g/mol) the mass of NaC_5H_5 in a sample. The remaining mass in a sample is that of C_4H_8O (72.11 g/mol) whose amount in moles we calculate. The ratio of the molar amount of C_4H_8O in the sample to the molar amount of NaC_5H_5 is the value of x.

$$\text{moles of } NaC_5H_5 = 0.01492\ L \times \frac{0.1001\ \text{mol HCl}}{1\ \text{L soln}} \times \frac{1\ \text{mol NaOH}}{1\ \text{mol HCl}} \times \frac{1\ \text{mol } NaC_5H_5}{1\ \text{mol NaOH}}$$

$$= 0.001493\ \text{mol } NaC_5H_5$$

$$\text{mass of } C_4H_8O = 0.242\ \text{g sample} - \left(0.001493\ \text{mol } NaC_5H_5 \times \frac{88.08\ \text{g } NaC_5H_5}{1\ \text{mol } NaC_5H_5}\right)$$

$$= 0.110\ \text{g } C_4H_8O$$

$$x = \frac{0.110\,\text{g } C_4H_8O \times \dfrac{1\,\text{mol } C_4H_8O}{72.11\,\text{g } C_4H_8O}}{0.001493\ \text{mol } NaC_5H_5} = 1.02$$

For the second sample, parallel calculations give 0.001200 mol NaC_5H_5, 0.093 g C_4H_8, $x = 1.1$. There is rounding error in this second calculation because it is limited to two significant figures. The best answer is from the first run $x \sim 1.02$ or 1. The formula is $NaC_5H_5(THF)_1$.

90. First, we balance the two equations.

Oxidation: $H_2C_2O_4\,(aq) \rightarrow 2\ CO_2\,(g) + 2\ H^+\,(aq) + 2\ e^-$

Reduction: $MnO_2\,(s) + 4\ H^+\,(aq) + 2\ e^- \rightarrow\ Mn^{2+}\,(aq) + 2\ H_2O(l)$

Net: $H_2C_2O_4\,(aq) + MnO_2\,(s) + 2\ H^+\,(aq) \rightarrow 2\ CO_2\,(g) + Mn^{2+}\,(aq) + 2\ H_2O(l)$

Oxidation: $\{\ H_2C_2O_4\,(aq) \rightarrow 2\ CO_2\,(g) + 2\ H^+\,(aq) + 2\ e^-$ $\} \times 5$

Reduction: $\{MnO_4^-(aq) + 8\ H^+(aq) + 5\ e^- \rightarrow\ Mn^{2+}(aq) + 4\ H_2O(l)\quad\} \times 2$

Net: $5\ H_2C_2O_4(aq) + 2\ MnO_4^-(aq) + 6\ H^+(aq) \rightarrow 10\ CO_2(g) + 2\ Mn^{2+}(aq) + 8\ H_2O(l)$

Now we determine the mass of the excess oxalic acid.

$$\text{mass } H_2C_2O_4 \cdot 2H_2O = 0.03006\,L \times \frac{0.1000\,mol\,KMnO_4}{1\,L} \times \frac{1\,mol\,MnO_4^-}{1\,mol\,KMnO_4^-} \times \frac{5\,mol\,H_2C_2O_4}{2\,mol\,MnO_4^-}$$

$$\times \frac{1\ mol\ H_2C_2O_4 \cdot 2H_2O}{1\ mol\ H_2C_2O_4} \times \frac{126.07\ g\ H_2C_2O_4 \cdot 2H_2O}{1\ mol\ H_2C_2O_4 \cdot 2H_2O} = 0.9474\ g\ H_2C_2O_4 \cdot 2H_2O$$

The mass of $H_2C_2O_4 \cdot 2H_2O$ that reacted with $MnO_2 = 1.651$ g $- 0.9474$ g $= 0.704$ g $H_2C_2O_4 \cdot 2H_2O$

$$\text{mass } MnO_2 = 0.704\ g\ H_2C_2O_4 \cdot 2H_2O \times \frac{1\ mol\ H_2C_2O_4}{126.07\ g\ H_2C_2O_4 \cdot 2H_2O} \times \frac{1\ mol\ MnO_2}{1\ mol\ H_2C_2O_4} \times \frac{86.9\ g\ MnO_2}{1\ mol\ MnO_2}$$

$$= 0.485\ g\ MnO_2$$

$$\%\,MnO_2 = \frac{0.485\,g\,MnO_2}{0.533\,g\,sample} \times 100\% = 91.0\%\ MnO_2$$

91. Reactions: $2\ NH_3 + H_2SO_4 \rightarrow (NH_4)_2SO_4$ and $HCl + NH_3 \rightarrow NH_4Cl$

n_{OH^-} (used to find moles H_2SO_4 in excess) $= 0.03224\ L \times 0.4498\ M = 0.01450\ mol\ OH^-$

$$n_{H_2SO_4}\ (\text{in excess}) = 0.01450\ mol\ OH^- \times \frac{1\ mol\ H_2SO_4}{2\ mol\ OH^-} = 0.00725\underline{108}\ mol\ H_2SO_4$$

n_{OH^-} (used to find moles H_2SO_4 in separate unreacted sample) $= 0.02224\ L \times 0.4498\ M$

$$= 0.0100\underline{035}\ mol\ OH^-$$

$$n_{H_2SO_4}\ (\text{initial}) = 0.0100\underline{035}\ mol\ OH^- \times \frac{1\ mol\ H_2SO_4}{2\ mol\ OH^-} = 0.005002\ mol\ H_2SO_4$$

NOTE: this was in a 25.00 mL sample. Need to scale up to 50.00 mL

Hence, $n_{H_2SO_4}$ (initial) $= 2 \times\ 0.005002\ mol\ H_2SO_4 = 0.0100\underline{035}\ mol\ H_2SO_4$

$n_{H_2SO_4}$ (reacted) $= n_{H_2SO_4}$ (initial) $-\ n_{H_2SO_4}$ (excess)

$$= 0.0100\underline{035}\ mol\ H_2SO_4 - 0.00725\underline{108}\ mol\ H_2SO_4 = 0.00275\ mol\ H_2SO_4$$

$$n_{NH_3} = 0.00275\ mol\ H_2SO_4 \times \frac{2\ mol\ NH_3}{1\ mol\ H_2SO_4} = 0.00550\underline{5}\ mol\ NH_3$$

$$\text{mass}_{NH_3} = 0.00550\underline{5}\ mol\ NH_3 \times \frac{1\ mol\ N}{1\ mol\ NH_3} \times \frac{14.0067\ g\ N}{1\ mol\ N} = 0.0771\ g\ N\ \text{in sample}$$

$$\text{mass protein in sample} = 0.0771\ g\ N\ \text{in sample} \times \frac{100\ g\ protein}{16\ g\ N} = 0.48\underline{2}\ g\ \text{protein in sample}$$

$$\text{Percent protein in sample} = \frac{0.48\underline{2}\ g\ \text{protein in sample}}{1.250\ g\ sample} \times 100\% = 38.\underline{6}\ \%\ \text{protein}$$

92.

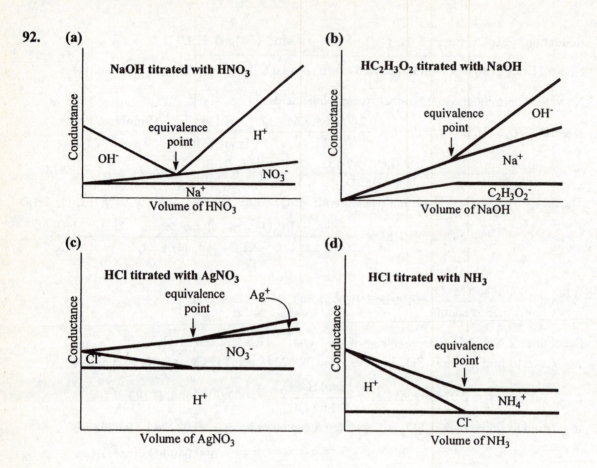

CHAPTER 6
GASES

PRACTICE EXAMPLES

1A   The pressure measured by each liquid must be the same. They are related through $P = g\,h\,d$ Thus, we have the following $g\,h_{DEG}\,d_{DEG} = g\,h_{Hg}\,d_{Hg}$. The g's cancel; we substitute known values: 9.25 m DEG $\times$ 1.118 g/cm^3 DEG = $h_{Hg} \times$ 13.6 g/cm^{3}Hg

$$h_{Hg} = 9.25\,\text{m} \times \frac{1.118\,\text{g/cm}^3}{13.6\,\text{g/cm}^3} = 0.760\,\text{m Hg}, \ P = 0.760\,\text{m Hg} = 760.\,\text{mmHg}$$

1B   The solution is found through the expression relating density and height: $h_{TEG}d_{TEG} = h_{Hg}d_{Hg}$

We substitute known values and solve for triethylene glycol's density:
9.14 m TEG $\times d_{TEG}$ = 757 mmHg $\times$ 13.6 g/cm^3 Hg. Using unit conversions, we get

$$d_{TEG} = \frac{0.757\,\text{m}}{9.14\,\text{m}} \times 13.6\,\text{g/cm}^3 = 1.13\,\text{g/cm}^3$$

2A   We know that $P_{gas} = P_{bar} + \Delta P$ with P_{bar} = 748.2 mmHg. We are told that ΔP = 7.8 mmHg. Thus, P_{gas} = 748.2 mmHg + 7.8 mmHg = 756.0 mmHg

2B   The difference in pressure between the two levels must be the same just expressed in different units. Hence, this problem is almost a repetition of Practice Example 6-1. h_{Hg}=748.2 mmHg – 739.6 mmHg=8.6 mmHg. Again we have $g\,h_g\,d_g$=$g\,h_{Hg}\,d_{Hg}$. This becomes $h_g \times$ 1.26 g/cm^3 glycerol = 8.6 mmHg $\times$13.6 g/cm^3 Hg

$$h_g = 8.6\,\text{mmHg} \times \frac{13.6\,\text{g} / \text{cm}^3\text{Hg}}{1.26\,\text{g} / \text{cm}^3\text{glycerol}} = 93\,\text{mm glycerol}$$

3A $A = \pi r^2$ (here r = ½(2.60 cm $\times \dfrac{1\,\text{m}}{100\,\text{cm}}$) = 0.0130 m)

$A = \pi(0.0130\,\text{m})^2 = 5.31 \times 10^{-4}\,\text{m}^2$
$F = m \times g = (1.000\,\text{kg})(9.81\,\text{m s}^{-2}) = 9.81\,\text{kg m s}^{-2} = 9.81\,\text{N}$

$$P = \frac{F}{A} = \frac{9.81\,\text{N}}{5.31 \times 10^{-4}\,\text{m}^2} = 18475\,\text{N m}^{-2} \text{ or } 1.85 \times 10^4\,\text{Pa}$$

P (torr) = 1.85×10^4 Pa $\times$ = 139 torr

3B   Final pressure = 100 mb. 100 mb $\times \dfrac{101,325\,\text{Pa}}{1013.25\,\text{mb}} = 1.000 \times 10^4$ Pa
The area of the cylinder is unchanged from that in Example 6-3, (1.32×10^{-3} m^2).

$$P = \frac{F}{A} = 1.000 \times 10^4\,\text{Pa} = \frac{F}{1.32 \times 10^{-3}\,\text{m}^2}$$

Solving for F, we find F = 13.2 (Pa)m^2 = 13.2 (N m^{-2})m^2 = 13.2 N
$F = m \times g = 13.2$ kg m s^{-2} = m $\times$ 9.81 m s^{-2}

Total mass = mass of cylinder + mass added weight = m = $\dfrac{F}{g} = \dfrac{13.2 \text{ kg m s}^{-2}}{9.81 \text{ m s}^{-2}} = 1.35$ kg

An additional 350 grams must be added to the top of the 1.000 kg (1000 g) red cylinder to increase the pressure to 100 mb. It is not necessary to add a mass with the same cross sectional area. The pressure will only be exerted over the area that is the base of the cylinder on the surface beneath it.

4A Boyle's Law relates the pressure-volume product. $P_1V_1 = P_2V_2$

$5.25\,\text{atm} \times V_1 = 1.85\,\text{atm} \times 12.5\,\text{L} \quad V_1 = \dfrac{1.85\,\text{atm} \times 12.5\,\text{L}}{5.25\,\text{atm}} = 4.40\,\text{L}$

4B Use Boyle's law, solved for the final pressure. After that, the pressure is converted to mmHg.

$P_2 = P_1 \times \dfrac{V_1}{V_2} = 2.25\,\text{atm} \times \dfrac{1.50\,\text{L}}{8.10\,\text{L}} = 0.417\,\text{atm} \times \dfrac{760 \text{ mmHg}}{1 \text{ atm}} = 317\,\text{mmHg}$

5A Charles's law states that the volume/temperature ratio is constant (with the temperature in kelvins).

$\dfrac{V_1}{T_1} = \dfrac{V_2}{T_2} = \dfrac{0.250\,\text{L}}{(25+273.15)\,\text{K}} = \dfrac{1.65\,\text{L}}{T_2}; \qquad T_2 = \dfrac{298\,\text{K} \times 1.65\,\text{L}}{0.250\,\text{L}} = 1.97 \times 10^3\,\text{K} = 1.69 \times 10^3\,^\circ\text{C}$

5B For oxygen: The increase in the volume will be directly proportional to the temperature increase (Kelvin scale). Assuming that the pressure remains constant,

$\dfrac{V_i}{T_i} = \dfrac{nR}{P} = \dfrac{V_f}{T_f}$ or $\dfrac{V_f}{V_i} = \dfrac{T_f}{T_i} = \dfrac{(25.0 + 273.15)}{(-13.5 + 273.15)} = 1.15$ (or a 15 % increase in volume)

For nitrogen: We expect the volume to increase 15% as well.

$V_f = V_i \times 1.15 = 50.5 \text{ mL} \times 1.15 = 58.1 \text{ mL}$

The temperature in Kelvin should also increase by 15%, i.e.

$T_f = T_i \times 1.15 = (33.4 + 273.15) \times 1.15 = 353 \text{ K or } 79 \,^\circ\text{C}$

6A The STP molar volume of 22.414 L enables us to determine the amount in moles of propane, from which we find the mass with the use of the molar mass.

mass propane = $30.0 \text{ L} \times \dfrac{1 \text{ mol}}{22.414 \text{ L}} \times \dfrac{44.10 \text{ g C}_3\text{H}_8}{1 \text{ mol C}_3\text{H}_8} = 59.0 \text{ g C}_3\text{H}_8$

6B A gas's STP molar volume is 22.414 L. In addition, we need the molar mass of $CO_2(g)$, 44.01 g/mol.

$CO_2(g)$ volume = $128 \text{ g} \times \dfrac{1 \text{ mol CO}_2}{44.01 \text{ g CO}_2} \times \dfrac{22.414 \text{ L at STP}}{1 \text{ mol CO}_2} = 65.2 \text{ L CO}_2(g)$ at STP

7A The ideal gas equation is solved for volume. Conversions are made within the equation.

$V = \dfrac{nRT}{P} = \dfrac{\left(20.2 \text{ g NH}_3 \times \dfrac{1 \text{ mol NH}_3}{17.03 \text{ g NH}_3}\right) \times \dfrac{0.08206\,\text{L} \cdot \text{atm}}{\text{mol} \cdot \text{K}} \times (-25 + 273)\text{K}}{752 \text{ mmHg} \times \dfrac{1\,\text{atm}}{760 \text{ mmHg}}} = 24.4 \text{ L NH}_3$

7B The amount of $Cl_2(g)$ is 0.193 mol Cl_2 and the pressure is 0.980 atm, as they are in Example 6–7. This information is substituted into the ideal gas equation after it has been solved for temperature.

$$T = \frac{PV}{nR} = \frac{0.980\,\text{atm} \times 7.50\,\text{L}}{0.193\,\text{mol} \times 0.08206\,\text{L atm mol}^{-1}\,\text{K}^{-1}} = 464\,\text{K}$$

8A The ideal gas equation is solved for amount and the quantities are substituted.

$$n = \frac{PV}{RT} = \frac{10.5\,\text{atm} \times 5.00\,\text{L}}{\dfrac{0.08206\,\text{L}\cdot\text{atm}}{\text{mol}\cdot\text{K}} \times (30.0 + 273.15)\,\text{K}} = 2.11\,\text{mol He}$$

8B

$$n = \frac{PV}{RT} = \frac{\left(6.67 \times 10^{-7}\ \text{Pa} \times \dfrac{1\,\text{atm}}{101325\,\text{Pa}}\right)\left(3.45\,\text{m}^3 \times \dfrac{1000\text{L}}{1\text{m}^3}\right)}{(0.08206\,\text{L atm K}^{-1}\,\text{mol}^{-1})\,(25 + 273.15)\text{K}} = 9.28 \times 10^{-10}\ \text{moles of N}_2$$

$$\text{molecules of N}_2 = 9.28 \times 10^{-10}\ \text{mol N}_2 \times \frac{6.022 \times 10^{23}\ \text{molecules of N}_2}{1\ \text{mole N}_2}$$

$$\text{molecules of N}_2 = 5.59 \times 10^{14}\ \text{molecules N}_2$$

9A The general gas equation is solved for volume, after the constant amount in moles is cancelled. Temperatures are converted to kelvin.

$$V_2 = \frac{V_1 P_1 T_2}{P_2 T_1} = \frac{1.00\,\text{mL} \times 2.14\,\text{atm} \times (37.8 + 273.2)\,\text{K}}{1.02\,\text{atm} \times (36.2 + 273.2)\,\text{K}} = 2.11\,\text{mL}$$

9B The flask has a volume of 1.00 L and initially contains $O_2(g)$ at STP. The mass of $O_2(g)$ that must be released is obtained from the difference in the amount of $O_2(g)$ at the two temperatures, 273 K and 373 K. We also could compute the masses separately and subtract them.

$$\text{mass released} = \left(n_{\text{STP}} - n_{100°C}\right) \times M_{O_2} = \left(\frac{PV}{R273\,\text{K}} - \frac{PV}{R373\,\text{K}}\right) \times M_{O_2} = \frac{PV}{R}\left(\frac{1}{273\,\text{K}} - \frac{1}{373\,\text{K}}\right) \times M_{O_2}$$

$$= \frac{1.00\,\text{atm} \times 1.00\,\text{L}}{0.08206\,\text{L atm mol}^{-1}\,\text{K}^{-1}}\left(\frac{1}{273\,\text{K}} - \frac{1}{373\,\text{K}}\right) \times \frac{32.00\,\text{g}}{1\,\text{mol O}_2} = 0.383\,\text{g O}_2$$

10A The volume of the vessel is 0.09841 L. We substitute other values into the expression for molar mass.

$$M = \frac{mRT}{PV} = \frac{(40.4868\,\text{g} - 40.1305\,\text{g}) \times \dfrac{0.08206\,\text{L}\cdot\text{atm}}{\text{mol}\cdot\text{K}} \times (22.4 \times 273.2)\,\text{K}}{\left(772\,\text{mmHg} \times \dfrac{1\,\text{atm}}{760\,\text{mmHg}}\right) \times 0.09841\text{L}} = 86.5\,\text{g/mol}$$

10B The gas's molar mass is its mass (1.27 g) divided by the amount of the gas in moles. The amount can be determined from the ideal gas equation.

$$n = \frac{PV}{RT} = \frac{\left(737\,\text{mm Hg} \times \dfrac{1\,\text{atm}}{760\,\text{mm Hg}}\right) \times 1.07\,\text{L}}{\dfrac{0.08206\,\text{L atm}}{\text{mol} \cdot \text{K}} \times (25+273)\,\text{K}} = 0.0424\,\text{mol gas}$$

$$M = \frac{1.27\,\text{g}}{0.0424\,\text{mol}} = 30.0\,\text{g/mol}$$

This answer is in good aggrement with the molar mass of NO, 30.006 g/mol.

11A The molar mass of He is 4.003 g/mol. This is substituted into the expression for density.

$$d = \frac{MP}{RT} = \frac{4.003\,\text{g mol}^{-1} \times 0.987\,\text{atm}}{0.08206\,\text{L} \cdot \text{atm mol}^{-1}\text{K}^{-1} \times 298\,\text{K}} = 0.162\,\text{g/L}$$

When compared to the density of air under the same conditions (1.16 g/L, based on the "average molar mass of air"=28.8g/mol) the density of He is only about one seventh as much. Thus, helium is less dense ("lighter") than air.

11B The suggested solution is a simple one; we merely need to solve for the temperature.

$$T = \frac{MP}{Rd} = \frac{\dfrac{32.00\,\text{g O}_2}{1\,\text{mol O}_2}\left(745\,\text{mmHg} \times \dfrac{1\,\text{atm}}{760\,\text{mmHg}}\right)}{\dfrac{0.08206\,\text{L atm}}{\text{mol K}} \times \dfrac{1.00\,\text{g}}{1\,\text{L}}} = 382\,\text{K}$$

However, suppose that you have forgotten the convenient formula for the density of an ideal gas? You can still solve a problem such as this one. The density of 1.00 g/L indicates a mass of 1.00 g of gas in a 1.00-L volume. Of course, the mass of a gas, given its identity (oxygen in this case) enables us to determine the amount in moles of the gas (*n* in the ideal gas equation). Then, we can solve the ideal gas equation for the desired property, namely temperature, as follows:

$$T = \frac{PV}{nR} = \frac{\left(745\,\text{mmHg} \times \dfrac{1\,\text{atm}}{760\,\text{mmHg}}\right) \times 1.00\,\text{L}}{\left(1.00\,\text{g O}_2 \times \dfrac{1\,\text{mol O}_2}{32.00\,\text{g O}_2}\right) \times \dfrac{0.08206\,\text{L atm}}{\text{mol K}}} = 382\,\text{K}$$

12A The balanced equation is $2\,\text{NaN}_3(s) \xrightarrow{\Delta} 2\,\text{Na}(l) + 3\,\text{N}_2(g)$

$$\text{moles N}_2 = \frac{PV}{RT} = \frac{\left(776\,\text{mmHg} \times \dfrac{1\,\text{atm}}{760\,\text{mmHg}}\right) \times 20.0\,\text{L}}{\dfrac{0.08206\,\text{L atm}}{\text{mol K}} \times (30.0+273.2)\,\text{K}} = 0.821\,\text{mol N}_2$$

Now, solve the stoichiometry problem.

$$\text{mass NaN}_3 = 0.821\,\text{mol N}_2 \times \frac{2\,\text{mol NaN}_3}{3\,\text{mol N}_2} \times \frac{65.01\,\text{g NaN}_3}{1\,\text{mol NaN}_3} = 35.6\,\text{g NaN}_3$$

12B Here we are not dealing with gaseous reactants; the law of combining volumes cannot be used. From the ideal gas equation we determine the amount of $N_2(g)$ per liter under the specified conditions. Then we determine the amount of Na(l) produced simultaneously, and finally the mass of that Na(l).

$$\text{mass of Na(l)} = \frac{\left(751\,\text{mmHg} \times \dfrac{1\,\text{atm}}{760\,\text{mmHg}}\right) \times 1.000\,\text{L}}{\dfrac{0.08206\,\text{L atm}}{\text{mol K}} \times (25+273)\,\text{K}} \times \frac{2\,\text{mol Na}}{3\,\text{mol N}_2} \times \frac{22.99\,\text{g Na}}{1\,\text{mol Na}} = 0.619\,\text{g Na(l)}$$

13A The law of combining volumes permits us to use stoichiometric coefficients for volume ratios.

$$O_2 \text{ volume} = 1.00\,\text{L NO(g)} \times \frac{5\,\text{L O}_2}{4\,\text{L NO}} = 1.25\,\text{L O}_2(g)$$

13B The first task is to balance the chemical equation. There must be three moles of hydrogen for every mole of nitrogen in both products (because of the formula of NH_3) and reactants: $N_2(g) + 3\,H_2(g) \rightarrow 2\,NH_3(g)$. The volumes of gaseous reactants and products are related by their stoichiometric coefficients, as long as all gases are at the same temperature and pressure.

$$\text{volume NH}_3(g) = 225\,\text{L H}_2(g) \times \frac{2\,\text{L NH}_3(g)}{3\,\text{L H}_2(g)} = 150.\ \text{L NH}_3$$

14A We can work easily with the ideal gas equation, with a the new temperature of $T = (55+273)\,\text{K} = 328\,\text{K}$. The amount of Ne added is readily computed.

$$n_{Ne} = 12.5\,\text{g Ne} \times \frac{1\,\text{mol Ne}}{20.18\,\text{g Ne}} = 0.619\,\text{mol Ne}$$

$$P = \frac{n_{total}RT}{V} = \frac{(1.75+0.619)\,\text{mol} \times \dfrac{0.08206\,\text{L atm}}{\text{mol K}} \times 328\,\text{K}}{5.0\,\text{L}} = 13\,\text{atm}$$

14B The total volume initially is $2.0\,\text{L} + 8.0\,\text{L} = 10.0\,\text{L}$. These two mixed ideal gases then obey the general gas equation as if they were one gas.

$$P_2 = \frac{P_1 V_1 T_2}{V_2 T_1} = \frac{1.00\,\text{atm} \times 10.0\,\text{L} \times 298\,\text{K}}{2.0\,\text{L} \times 273\,\text{K}} = 5.5\,\text{atm}$$

15A The partial pressures are proportional to the mole fractions.

$$P_{H_2O} = \frac{n_{H_2O}}{n_{tot}} \times P_{tot} = \frac{0.00278\,\text{mol H}_2\text{O}}{0.197\,\text{mol CO}_2 + 0.00278\,\text{mol H}_2\text{O}} \times 2.50\,\text{atm} = 0.0348\,\text{atm H}_2\text{O(g)}$$

$$P_{CO_2} = P_{tot} - P_{H_2O} = 2.50\,\text{atm} - 0.0348\,\text{atm} = 2.47\,\text{atm CO}_2(g)$$

15B Expression (6.17) indicates that, in a mixture of gases, the mole percent equals the volume percent, which in turn equals the pressure percent. Thus, we can apply these volume percents—converted to fractions by dividing by 100—directly to the total pressure.

N_2 pressure = 0.7808×748 mmHg $= 584$ mmHg,

O_2 pressure = 0.2095×748 mmHg $= 157$ mmHg,

CO_2 pressure = 0.00036×748 mmHg = 0.27 mmHg,

Ar pressure = 0.0093×748 mmHg $= 7.0$ mmHg

16A First compute the moles of $H_2(g)$, then use stoichiometry to convert to moles of HCl.

$$\text{amount HCl} = \frac{\left((755-25.2)\,\text{torr} \times \dfrac{1\,\text{atm}}{760\,\text{mmHg}}\right) \times 0.0355\,\text{L}}{\dfrac{0.08206\,\text{L atm}}{\text{mol K}} \times (26+273)\,\text{K}} \times \frac{6\,\text{mol HCl}}{3\,\text{mol H}_2} = 0.00278\,\text{mol HCl}$$

16B The volume occupied by the $O_2(g)$ at its partial pressure is the same as the volume occupied by the mixed gases: water vapor and $O_2(g)$. The partial pressure of $O_2(g)$ is found by difference.

O_2 pressure = 749.2 total pressure -23.8 mmHg$(H_2O$ pressure$) = 725.4$ mmHg

The mass of Ag_2O is related to the amount of $O_2(g)$ produced.

$$\text{moles O}_2 = 8.07\,\text{g sample} \times \frac{88.3\,\text{g Ag}_2\text{O}}{100.0\,\text{g sample}} \times \frac{1\,\text{mol Ag}_2\text{O}}{231.74\,\text{g Ag}_2\text{O}} \times \frac{1\,\text{mol O}_2}{2\,\text{mol Ag}_2\text{O}} = 0.0154\,\text{mol O}_2$$

The volume of $O_2(g)$ is found with the ideal gas law.

$$V = \frac{nRT}{P} = \frac{0.0154\,\text{mol O}_2 \times 0.08206 \dfrac{\text{L}\cdot\text{atm}}{\text{mol}\cdot\text{K}} \times (273+25)\,\text{K}}{725.4\,\text{mmHg} \times \dfrac{1\,\text{atm}}{760\,\text{mmHg}}} = 0.395\,\text{L O}_2$$

17A The gas with the smaller molar mass, NH_3 at 17.0 g/mol, has the greater root-mean-square speed

$$u_{\text{rms}} = \sqrt{\frac{3RT}{M}} = \sqrt{\frac{3 \times 8.3145\,\text{kg m}^2\,\text{s}^{-2}\,\text{mol}^{-1}\text{K}^{-1} \times 298\,\text{K}}{0.0170\,\text{kg mol}^{-1}}} = 661\,\text{m/s}$$

17B

$$\text{bullet speed} = \frac{2180\,\text{mi}}{1\,\text{h}} \times \frac{1\,\text{h}}{3600\,\text{s}} \times \frac{5280\,\text{ft}}{1\,\text{mi}} \times \frac{12\,\text{in.}}{1\,\text{ft}} \times \frac{2.54\,\text{cm}}{1\,\text{in.}} \times \frac{1\,\text{m}}{100\,\text{cm}} = 974.5\,\text{m/s}$$

Solve the rms-speed equation (6.20) for temperature by first squaring both sides.

$$\left(u_{\text{rms}}\right)^2 = \frac{3RT}{M} \qquad T = \frac{\left(u_{\text{rms}}\right)^2 M}{3R} = \frac{\left(\dfrac{974.5\,\text{m}}{1\,\text{s}}\right)^2 \times \dfrac{2.016 \times 10^{-3}\,\text{kg}}{1\,\text{mol H}_2}}{3 \times \dfrac{8.3145\,\text{kg m}^2}{\text{s}^2\,\text{mol K}}} = 76.75\,\text{K}$$

We expected the temperature to be lower than 298 K. Note that the speed of the bullet is about half the speed of a H_2 molecule at 298 K. To halve the speed of a molecule, its temperature must be divided by four.

18A The only difference are the gases molar mass. 2.2×10^{-4} mol N_2 effuses through the orifice in 105 s.

$$\frac{? \text{ mol } O_2}{2.2 \times 10^{-4} \text{ mol } N_2} = \sqrt{\frac{M_{N_2}}{M_{O_2}}} = \sqrt{\frac{28.014 \text{ g/mol}}{31.999 \text{ g/mol}}} = 0.9357$$

moles $O_2 = 0.9357 \times 2.2 \times 10^{-4} = 2.1 \times 10^{-4}$ mol O_2

18B Rates of effusion are related by the square root of the ratio of the molar masses of the two gases. H_2, effuses faster (by virtue of being lighter), and thus requires a shorter time for the same amount of gas to effuse.

$$\text{time}_{H_2} = \text{time}_{N_2} \times \sqrt{\frac{M_{H_2}}{M_{N_2}}} = 105 \text{ s} \times \sqrt{\frac{2.016 \text{ g } H_2/\text{mol } H_2}{28.014 \text{ g } N_2/\text{mol } N_2}} = 28.2 \text{ s}$$

19A Effusion times are related as the square root of the molar mass. It requires 87.3 s for Kr to effuse.

$$\frac{\text{unknown time}}{\text{Kr time}} = \sqrt{\frac{M_{unk}}{M_{Kr}}} \qquad \text{substitute in values} \qquad \frac{131.3 \text{ s}}{87.3 \text{ s}} = \sqrt{\frac{M_{unk}}{83.80 \text{ g/mol}}} = 1.50$$

$$M_{unk} = (1.504)^2 \times 83.80 \text{ g/mol} = 1.90 \times 10^2 \text{ g/mol}$$

19B This problem is solved in virtually the same manner as Practice Example 18B. The lighter gas is ethane, with a molar mass of 30.07 g/mol.

$$C_2H_6 \text{ time} = \text{Kr time} \times \sqrt{\frac{M(C_2H_6)}{M(Kr)}} = 87.3 \text{ s} \times \sqrt{\frac{30.07 \text{ g } C_2H_6/\text{mol } C_2H_6}{83.80 \text{ g Kr/mol Kr}}} = 52.3 \text{ s}$$

20A Because one mole of gas is being considered, the value of $n^2 a$ is numerically the same as the value of a, and the value of nb is numerically the same as the value of b.

$$P = \frac{nRT}{V-nb} - \frac{n^2 a}{V^2} = \frac{1.00 \text{ mol} \times \dfrac{0.08206 \text{ L atm}}{\text{mol K}} \times 273 \text{ K}}{(2.00 - 0.0427) \text{ L}} - \frac{3.59 \text{ L}^2 \text{ atm}}{(2.00 \text{ L})^2} = 11.4_3 \text{ atm} - 0.898 \text{ atm}$$

$$= 10.5 \text{ atm } CO_2(g) \quad \text{compared with 9.9 atm for } Cl_2(g)$$

$$P_{ideal} = \frac{nRT}{V} = \frac{1.00 \text{ mol} \times \dfrac{0.08206 \text{ L atm}}{\text{mol K}} \times 273 \text{ K}}{2.00 \text{ L}} = 11.2 \text{ atm}$$

$Cl_2(g)$ shows a greater deviation from ideal gas behavior than does $CO_2(g)$.

20B Because one mole of gas is being considered, the value of $n^2 a$ is numerically the same as the value of a, and the value of nb is numerically the same as the value of b.

$$P = \frac{nRT}{V-nb} - \frac{n^2 a}{V^2} = \frac{1.00 \text{ mol} \times \dfrac{0.08206 \text{ L atm}}{\text{mol K}} \times 273 \text{ K}}{(2.00 - 0.0399) \text{ L}} - \frac{1.49 \text{ L}^2 \text{ atm}}{(2.00 \text{ L})^2} = 11.4_3 \text{ atm} - 0.373 \text{ atm}$$

$$= 11.1 \text{ atm } CO(g)$$

compared to 9.9 atm for $Cl_2(g)$, 11.2 atm for an ideal gas, and 10.5 atm for $CO_2(g)$.

Thus, $Cl_2(g)$ displays the greatest deviation from ideality.

EXERCISES

Pressure and Its Measurement

<u>**1.**</u> **(a)** $P = 736 \text{ mmHg} \times \dfrac{1 \text{ atm}}{760 \text{ mmHg}} = 0.968 \text{ atm}$

  (b) $P = 58.2 \text{ cm Hg} \times \dfrac{1 \text{ atm}}{76 \text{ cmHg}} = 0.766 \text{ atm}$

  (c) $P = 892 \text{ torr} \times \dfrac{1 \text{ atm}}{760 \text{ torr}} = 1.17 \text{ atm}$

  (d) $P = 225 \text{ kPa} \times \dfrac{1000 \text{ Pa}}{1 \text{ kPa}} \times \dfrac{1 \text{ atm}}{101,325 \text{ Pa}} = 2.22 \text{ atm}$

2. **(a)** $h = 0.984 \text{ atm} \times \dfrac{760 \text{ mmHg}}{1 \text{ atm}} = 748 \text{ mmHg}$ **(b)** $h = 928 \text{ torr} = 928 \text{ mmHg}$

  (c) $h = 142 \text{ ft H}_2\text{O} \times \dfrac{12 \text{ in}}{1 \text{ ft}} \times \dfrac{2.54 \text{ cm}}{1 \text{ in}} \times \dfrac{10 \text{ mmH}_2\text{O}}{1 \text{ cm H}_2\text{O}} \times \dfrac{1 \text{ mmHg}}{13.6 \text{ mmH}_2\text{O}} \times \dfrac{1 \text{ mHg}}{1000 \text{ mmHg}} = 3.18 \text{ mHg}$

<u>**3.**</u> We use: $h_{\text{bnz}} d_{\text{bnz}} = h_{\text{Hg}} d_{\text{Hg}}$

  $h_{\text{bnz}} = 0.970 \text{ atm} \times \dfrac{0.760 \text{ m Hg}}{1 \text{ atm}} \times \dfrac{13.6 \text{ g/cm}^3 \text{ Hg}}{0.879 \text{ g/cm}^3 \text{ benzene}} = 11.4 \text{ m benzene}$

4. We use: $h_{\text{gly}} d_{\text{gly}} = h_{\text{CCl}_4} d_{\text{CCl}_4}$

  $h_{\text{gly}} = 3.02 \text{ m CCl}_4 \times \dfrac{1.59 \text{ g/cm}^3 \text{ CCl}_4}{1.26 \text{ g/cm}^3 \text{ glycerol}} = 3.81 \text{ m glycerol}$

<u>**5.**</u> The atmospheric pressure is less than the pressure of the gas.
  The difference in pressures is $\Delta P = 16.5 \text{ mmHg} - 7.9 \text{ mmHg} = 8.6 \text{ mmHg}$

  $P_{\text{gas}} = P_{\text{atm}} + \Delta P = 744 \text{ mmHg} + 8.6 \text{ mmHg} = 753 \text{ mmHg}$

6. The mercury level difference equals the difference in pressure between the container and the atmosphere. This mercury level difference is
  $\Delta P = 276 \text{ mmHg} - 49 \text{ mmHg} = 227 \text{ mmHg}$. Since the mercury level in the arm open to the atmosphere is higher than in the arm connected to the container, the pressure in the container is higher than atmospheric pressure.
  $P = \Delta P + P_{\text{atm}} = 227 \text{ mmHg} + 749 \text{ mmHg} = 976 \text{ mmHg}$

7. $F = m \times g$ and $1 \text{ atm} = 101325 \text{ Pa} = 101325 \text{ kg m}^{-1} \text{s}^{-2} = P = \dfrac{F}{A} = \dfrac{m \times 9.81 \text{ m s}^{-2}}{1 \text{ m}^2}$

mass (per m²) $= \dfrac{101325 \text{ kg m}^{-1}\text{s}^{-2} \times 1 \text{ m}^2}{9.81 \text{ m s}^{-2}} = 10329 \text{ kg}$

(Note:$1 \text{ m}^2 = (100 \text{ cm})^2 = 10{,}000 \text{ cm}^2$)

$P \text{ (kg cm}^{-2}) = \dfrac{m}{A} = \dfrac{10329 \text{ kg}}{10{,}000 \text{ cm}^2} = 1.03 \text{ kg cm}^{-2}$

8. $\dfrac{1.03 \text{ kg}}{1 \text{ cm}^2} \times \dfrac{(2.54 \text{ cm})^2}{(1 \text{ in})^2} \times \dfrac{2.2046 \text{ lb}}{1 \text{ kg}} = 14.6 \dfrac{\text{lb}}{\text{in}^2}$

The Simple Gas Laws

9. **(a)** $V = 26.7 \text{ L} \times \dfrac{762 \text{ mmHg}}{385 \text{ mmHg}} = 52.8 \text{ L}$

(b) $V = 26.7 \text{ L} \times \dfrac{762 \text{ mmHg}}{3.68 \text{ atm} \times \dfrac{760 \text{ mmHg}}{1 \text{ atm}}} = 7.27 \text{ L}$

10. Apply Charles's law: $V = kT$. $\quad T_i = 26 + 273 = 299 \text{ K}$

(a) $T = 273 + 98 = 371 \text{ K}$ $\qquad V = 886 \text{ mL} \times \dfrac{371 \text{ K}}{299 \text{ K}} = 1.10 \times 10^3 \text{ mL}$

(b) $T = 273 - 20 = 253 \text{ K}$ $\qquad V = 886 \text{ mL} \times \dfrac{253 \text{ K}}{299 \text{ K}} = 7.50 \times 10^2 \text{ mL}$

11. Apply Charles's law. $T_f = (22 + 273) \text{ K} \times \dfrac{165 \text{ mL}}{57.3 \text{ mL}} = 849 \text{ K} = 576°\text{C}$

12. $P_f = P_i \times \dfrac{V_i}{V_f} = \left(105 \text{ kPa} \times \dfrac{0.725 \text{ L}}{2.25 \text{ L}} \right) = 33.8 \text{ kPa } (0.334 \text{ atm})$

13. $P_i = P_f \times \dfrac{V_f}{V_i} = \left(721 \text{ mmHg} \times \dfrac{35.8 \text{ L} + 1875 \text{ L}}{35.8 \text{L}} \right) \times \dfrac{1 \text{ atm}}{760 \text{ mm H}_2\text{O}} = 50.6 \text{ atm}$

14. We let P represent barometric pressure, and solve the Boyle's law expression below for P.

$P \times 42.0 \text{ mL} = (P + 85 \text{ mmHg}) \times 37.7 \text{ mL}$ $\qquad 42.0P = 37.7P + 3.2 \times 10^3$

$P = \dfrac{3.2 \times 10^3}{42.0 - 37.7} = 7.4 \times 10^2 \text{ mmHg}$

15. Assuming pressure and moles of gas is kept constant, volume is directly proportional to the temperature in kelvins (Charles' Law).

$1\ ^\circ C \rightarrow 2\ ^\circ C$ Volume increase $= \dfrac{V_f}{V_i} = \dfrac{T_f}{T_i} = \dfrac{(2 + 273.15)}{(1 + 273.15)} = 1.0036\underline{5}$ (increase of 0.37 %)

$10\ ^\circ C \rightarrow 20\ ^\circ C$ Volume increase $= \dfrac{(20 + 273.15)}{(10 + 273.15)} = 1.035$ (increase of 3.5 %)

The volume doubles when the temperature in Kelvin doubles (i.e. 273 K $\rightarrow$ 546 K)

16. Convert 71 °F to Celsius then to a temperature on the Kelvin scale.

$(71 - 32) \times \dfrac{5}{9} = 22\ ^\circ C + 273.15 = 295\ K$ $-120\ ^\circ C = (-120 + 273.15) = 153\ K$

Volume decrease is proportional to the ratio of the temperatures in Kelvin

Volume contraction $= \dfrac{153\ K}{295\ K} = 0.52$ (volume will be about ½ of its original volume).

17. STP: P = 1 atm and T = 273.15

$\text{mass}_{Ar} = \dfrac{1\ atm \times 0.0750\ L}{0.08206\ \dfrac{L\ atm}{K\ mol} \times 273.15\ K} \times \dfrac{39.948\ g\ Ar}{1\ mol\ Ar} = 0.134\ g\ Ar$

18. STP: P = 1 atm and T = 273.15 K (note: one mole of gas at STP occupies 22.414 L).

$V = \dfrac{nRT}{P} = \dfrac{\left(250.0\ g\ Cl_2 \times \dfrac{1\ mol\ Cl_2}{70.906\ g}\right)\left(0.08206\ \dfrac{L\ atm}{K\ mol}\right)273.15\ K}{1\ atm} = 79.03\ L\ Cl_2$

Alternatively $V = n \times V_m = \left(250.0\ g\ Cl_2 \times \dfrac{1\ mol\ Cl_2}{70.906\ g}\right) \times \dfrac{22.414\ L\ at\ STP}{1\ mol} = 79.03\ L\ Cl_2$

19. **(a)**

$\text{mass} = 27.6\ mL \times \dfrac{1\ L}{1000\ mL} \times \dfrac{1\ mol}{22.414\ L\ STP} = 0.00123\ mol\ PH_3 \times \dfrac{34.0\ g\ PH_3}{1\ mol\ PH_3} \times \dfrac{1000\ mg}{1\ g}$

$= 41.8\ mg\ PH_3$

(b)   number of molecules of $PH_3 = 0.00123\ mol\ PH_3 \times \dfrac{6.022 \times 10^{23}\ molecules}{1\ mol\ PH_3}$

number of molecules of $PH_3 = 7.41 \times 10^{20}\ molecules$

20. **(a)** $\text{mass} = 5.0 \times 10^{17}\ atoms \times \dfrac{1\ mol\ Rn}{6.022 \times 10^{23}\ atoms} = 8.30 \times 10^{-7}\ mol \times \dfrac{222\ g\ Rn}{1\ mol} \times \dfrac{10^6\ \mu g}{1\ g}$

$\text{mass} = 1.8 \times 10^2\ \mu g\ Rn\,(g)$

(b) $\text{volume} = 8.30 \times 10^{-7}\ mol \times \dfrac{22.414\ L}{1\ mol} \times \dfrac{10^6\ \mu L}{1\ L} = 19\ \mu L\ Rn\,(g)$

21. At the higher elevation of the mountains, the atmospheric pressure is lower than at the beach. However, the bag is virtually leak proof; no gas escapes. Thus, the gas inside the bag expands in the lower pressure until the bag is filled to near bursting. (It would have been difficult to predict this result. The temperature in the mountains is usually lower than at the beach. The lower temperature would *decrease* the pressure of the gas.)

22. Based on densities, $1 \text{ mHg} = 13.6 \text{ mH}_2\text{O}$. For 30 m of water,

$$30 \text{ mH}_2\text{O} \times \frac{1 \text{ mHg}}{13.6 \text{ mH}_2\text{O}} = 2.2 \text{ mHg} \qquad P_{water} = 2.2 \text{ mHg} \times \frac{1000 \text{ mm}}{1 \text{ m}} \times \frac{1 \text{ atm}}{760 \text{ mmHg}} = 2.9 \text{ atm}$$

To this we add the pressure of the atmosphere above the water:

$P_{total} = 2.9 \text{ atm} + 1.0 \text{ atm} = 3.9 \text{ atm}$.

When the diver rises to the surface, she rises to a pressure of 1.0 atm. Since the pressure is about one fourth of the pressure below the surface, the gas in her lungs attempts to expand to four times the volume of her lungs. It is quite likely that her lungs would burst.

General Gas Equation

23. Because the number of moles of gas does not change, $\dfrac{P_i \times V_i}{T_i} = nR = \dfrac{P_f \times V_f}{T_f}$ is obtained

from the ideal gas equation. This expression can be rearranged as follows.

$$V_f = \frac{V_i \times P_i \times T_f}{P_f \times T_i} = \frac{4.25 \text{ L} \times 748 \text{ mmHg} \times (273.2 + 26.8)\text{K}}{742 \text{ mmHg} \times (273.2 + 25.6)\text{K}} = 4.30 \text{ L}$$

24. We first compute the pressure at $25°\text{C}$ as a result of the additional gas. Of course, the gas pressure increases proportionally to the increase in the mass of gas, because the mass is

proportional to the number of moles of gas. $P = \dfrac{12.5 \text{ g}}{10.0 \text{ g}} \times 762 \text{ mmHg} = 953 \text{ mmHg}$

Now we compute the pressure resulting from increasing the temperature.

$$P = \frac{(62 + 273) \text{ K}}{(25 + 273) \text{ K}} \times 953 \text{ mmHg} = 1.07 \times 10^3 \text{ mmHg}$$

25. Volume and Pressure are constant hence $n_i T_i = \dfrac{PV}{R} = n_f T_f$

$$\frac{n_f}{n_i} = \frac{T_i}{T_f} = \frac{(21 + 273.15) \text{ K}}{(210 + 273.15) \text{ K}} = 0.609 \quad (60.9 \text{ \% of the gas remains})$$

Hence, 39.1% of the gas must be released   Mass of gas released $= 12.5 \text{ g} \times \dfrac{39.1}{100} = 4.89 \text{ g}$

26. First determine the mass of O_2 in the cylinder under the final conditions.

$$\text{mass } O_2 = n \times M = \frac{PV}{RT} M = \frac{1.15 \text{ atm} \times 34.0 \text{ L} \times 32.0 \text{ g/mol}}{0.08206 \dfrac{\text{L atm}}{\text{mol K}}(22 + 273)\text{K}} = 51.7 \text{ g } O_2$$

mass of O_2 to be released $= 305 \text{ g} - 51.7 \text{ g} = 253 \text{ g } O_2$

Ideal Gas Equation

<u>27.</u> Assume that the $CO_2(g)$ behaves ideally and use the ideal gas law: $PV = nRT$

$$V = \frac{nRT}{P} = \frac{\left(89.2 \text{ g} \times \dfrac{1 \text{ mol } CO_2}{44.01 \text{ g}}\right) 0.08206 \dfrac{\text{L atm}}{\text{mol K}} (37 + 273.2) \text{ K}}{737 \text{ mmHg} \times \dfrac{1 \text{ atm}}{760 \text{ mmHg}}} \times \frac{1000 \text{ mL}}{1 \text{ L}} = 5.32 \times 10^4 \text{ mL}$$

<u>28.</u> $$P = \frac{nRT}{V} = \frac{\left(35.8 \text{ g } O_2 \times \dfrac{1 \text{ mol } O_2}{32.00 \text{ g } O_2}\right) \times \dfrac{0.08206 \text{ L atm}}{\text{mol K}} \times (46 + 273.2) \text{ K}}{12.8 \text{ L}} = 2.29 \text{ atm}$$

<u>29.</u> $\text{mass} = n \times M = \dfrac{PV}{RT}$ $\qquad M = \dfrac{11.2 \text{ atm} \times 18.5 \text{ L} \times 83.80 \text{ g/mol}}{0.08206 \dfrac{\text{L atm}}{\text{mol K}} (28.2 + 273.2) \text{ K}} = 702 \text{ g Kr}$

<u>30.</u> $$T = \frac{PV}{nR} = \frac{3.50 \text{ atm} \times 72.8 \text{ L}}{1.85 \text{ mol} \times 0.08206 \dfrac{\text{L atm}}{\text{mol K}}} = 1.68 \times 10^3 \text{ K}$$

$t(°C) = (1.68 \times 10^3 - 273) = 1.41 \times 10^3 \text{ °C}$

<u>31.</u> $n_{gas} = 5.0 \times 10^9 \text{ molecules gas} \times \dfrac{1 \text{ mol gas}}{6.022 \times 10^{23} \text{ molecules gas}} = 8.3 \times 10^{-15} \text{ mol gas}$

We next determine the pressure that the gas exerts at 25 °C in a cubic meter

$$P = \frac{8.3 \times 10^{-15} \text{ mol gas} \times 0.08206 \dfrac{\text{L atm}}{\text{K mol}} \times 298.15 \text{ K}}{1 \text{ m}^3 \times \left(\dfrac{10 \text{ dm}}{1 \text{ m}}\right)^3 \times \left(\dfrac{1 \text{ L}}{1 \text{ dm}^3}\right)} \times \frac{101,325 \text{ Pa}}{1 \text{ atm}} = 2.1 \times 10^{-11} \text{ Pa}$$

<u>32.</u> $n_{CO_2} = 1242 \text{ g } CO_2 \times \dfrac{1 \text{ mol } CO_2}{44.010 \text{ g } CO_2} = 28.22 \text{ mol } CO_2$ $\qquad T = 273.15 + (-25) = 248.15 \text{ K}$

$V_{CO_2} = \pi r^2 h = 3.1416 \times \left(\dfrac{0.250 \text{ m}}{2}\right)^2 1.75 \text{ m} \times \dfrac{1000 \text{ L}}{1 \text{ m}^3} = 85.9 \text{ L}$

$$P = \frac{28.22 \text{ mol } CO_2 \times 0.08206 \dfrac{\text{L atm}}{\text{K mol}} \times 248.15 \text{ K} \times \dfrac{101,325 \text{ Pa}}{1 \text{ atm}}}{85.9 \text{ L}} = 6.78 \times 10^5 \text{ Pa}$$

Determining Molar Mass

33. Use the ideal gas law to determine the amount in moles of the given quantity of gas.

$$M = \frac{mRT}{PV} = \frac{0.418 \text{ g} \times 0.08206 \dfrac{\text{L atm}}{\text{mol K}} \times 339.5 \text{ K}}{\left(743 \text{ mmHg} \times \dfrac{1 \text{ atm}}{760 \text{ mmHg}}\right) \times 0.115 \text{ L}} = 104 \text{ g mol}^{-1}$$

Alternatively

$$n = \frac{PV}{RT} = \frac{\left(743 \text{ mmHg} \times \dfrac{1 \text{ atm}}{760 \text{ mmHg}}\right)\left(115 \text{ mL} \times \dfrac{1 \text{ L}}{1000 \text{ mL}}\right)}{0.08206 \dfrac{\text{L atm}}{\text{mol K}}(273.2 + 66.3) \text{ K}} = 0.00404 \text{ mol gas} \qquad M = \frac{0.418 \text{ g}}{0.00404 \text{ mol}} = 103 \text{ g / mol}$$

34. Use the ideal gas law to determine the amount in moles in 1 L of gas.

$$n = \frac{PV}{RT} = \frac{\left(728 \text{ mmHg} \times \dfrac{1 \text{ atm}}{760 \text{ mmHg}}\right) \times 1 \text{ L}}{0.08206 \dfrac{\text{L atm}}{\text{mol K}} \times 415 \text{ K}} = 0.02813 \text{ mol gas} \qquad M = \frac{\text{mass}}{\text{moles}} = \frac{0.841 \text{ g}}{0.02813 \text{ mol}} = 29.9 \text{ g mol}^{-1}$$

35. First we determine the empirical formula for the sulfur fluoride. Assume 100 g sample of S_xF_y.

$$\text{moles S} = 29.6 \text{ g S} \times \frac{1 \text{ mol S}}{32.064 \text{ g S}} = 0.923 \text{ mol S} \qquad \text{moles F} = 70.4 \text{ g F} \times \frac{1 \text{ mol F}}{18.9984 \text{ g F}} = 3.706 \text{ mol F}$$

Dividing the number of moles of each element by 0.923 moles gives the empirical formula SF_4. To find the molar mass we use the relationship:

$$\text{Molar mass} = \frac{dRT}{P} = \frac{4.5 \text{ g L}^{-1} \times 0.08206 \text{ L atm K}^{-1} \text{ mol}^{-1}}{1.0 \text{ atm}} = 108 \text{ g mol}^{-1}$$

Thus, molecular formula = (empirical Formula) $\times \left(\dfrac{\text{Molecular formula mass}}{\text{Empirical formula mass}}\right) = \dfrac{108 \text{ g mol}^{-1}}{108.06 \text{ g mol}^{-1}} \times SF_4 = SF_4$

36. We first determine the molar mass of the gas.

$$T = 24.3 + 273.2 = 297.5 \text{ K} \qquad\qquad P = 742 \text{ mmHg} \times \frac{1 \text{ atm}}{760 \text{ mmHg}} = 0.976 \text{ atm}$$

$$M = \frac{mRT}{PV} = \frac{2.650 \text{ g} \times 0.08206 \text{ L atm mol}^{-1}\text{K}^{-1} \times 297.5 \text{ K}}{0.976 \text{ atm} \times \left(428 \text{ mL} \times \dfrac{1 \text{ L}}{1000 \text{ mL}}\right)} = 155 \text{ g/mol}$$

Then we determine the empirical formula of the gas, based on a 100.0-g sample.

$$\text{mol C} = 15.5 \text{ g C} \times \frac{1 \text{ mol C}}{12.01 \text{ g C}} = 1.29 \text{ mol C} \div 0.649 \rightarrow 1.99 \text{ mol C}$$

$$\text{mol Cl} = 23.0 \text{ g Cl} \times \frac{1 \text{ mol Cl}}{35.45 \text{ g Cl}} = 0.649 \text{ mol Cl} \div 0.649 \rightarrow 1.00 \text{ mol Cl}$$

$$\text{mol F} = 61.5 \text{ g F} \times \frac{1 \text{ mol F}}{19.00 \text{ g F}} = 3.24 \text{ mol F} \div 0.649 \rightarrow 4.99 \text{ mol F}$$

Thus, the empirical formula is C_2ClF_5, which has a molar mass of 154.5 g/mol. This is the same as the experimentally determined molar mass. Hence, the molecular formula is C_2ClF_5.

37. **(a)** $M = \dfrac{mRT}{PV} = \dfrac{0.231\ \text{g} \times 0.08206\ \dfrac{\text{L atm}}{\text{mol K}} \times (23+273)\ \text{K}}{\left(749\ \text{mmHg} \times \dfrac{1\ \text{atm}}{760\ \text{mmHg}}\right) \times \left(102\ \text{mL} \times \dfrac{1\ \text{L}}{1000\ \text{mL}}\right)} = 55.8\ \text{g/mol}$

(b) The formula contains 4 atoms of carbon. (5 atoms of carbon gives a molar mass of at least 60-too high-and 3 C atoms gives a molar mass of 36-too low to be made up by adding H's.) To produce a molar mass of 56 with 4 carbons requires the inclusion of 8 atoms of H in the formula of the compound. Thus the formula is C_4H_8.

38. First, we obtain the mass of acetylene, and then acetylene's molar mass.

mass of acetylene = 56.2445 g – 56.1035 g = 0.1410 g acetylene

$M = \dfrac{mRT}{PV} = \dfrac{0.1410\ \text{g} \times 0.08206\ \dfrac{\text{L atm}}{\text{mol K}} \times (20.02+273.15)\ \text{K}}{\left(749.3\ \text{mmHg} \times \dfrac{1\ \text{atm}}{760\ \text{mmHg}}\right) \times \left(132.10\ \text{mL} \times \dfrac{1\ \text{L}}{1000\ \text{mL}}\right)} = 26.04\ \text{g/mol}$

The formula contains 2 atoms of carbon (3 atoms of carbon gives a molar mass of at least 36-too high and 1C atom gives a molar mass of 12-too low to be made up by adding H's.) To produce a molar mass of 26 with 2 carbons requires the inclusion of 2 atoms of H in the formula. Thus the formula is C_2H_2.

Gas Densities

39. $d = \dfrac{MP}{RT} \quad \rightarrow \quad P = \dfrac{dRT}{M} = \dfrac{1.80\ \text{g/L} \times 0.08206\ \dfrac{\text{L atm}}{\text{mol K}} \times (32+273)\ \text{K}}{28.0\ \text{g/mol}} \times \dfrac{760\ \text{mmHg}}{1\ \text{atm}}$

$P = 1.21 \times 10^3\ \text{mmHg}$

40. $M = \dfrac{dRT}{M} = \dfrac{2.56\ \text{g/L} \times 0.08206\ \dfrac{\text{L atm}}{\text{mol K}} \times (22.8+273.2)\ \text{K}}{756\ \text{mmHg} \times \dfrac{1\ \text{atm}}{760\ \text{mmHg}}} = 62.5\ \text{g/mol}$

41. **(a)** $d = \dfrac{MP}{RT} = \dfrac{28.96\ \text{g/mol} \times 1.00\ \text{atm}}{0.08206\ \dfrac{\text{L atm}}{\text{mol K}} \times (273+25)\text{K}} = 1.18\ \text{g/L air}$

(b) $d = \dfrac{MP}{RT} = \dfrac{44.0\ \text{g/mol}\ CO_2 \times 1.00\ \text{atm}}{0.08206\ \dfrac{\text{L atm}}{\text{mol K}} \times (273+25)\text{K}} = 1.80\ \text{g/L}\ CO_2$

Since this density is greater than that of air, the balloon will not rise in air when filled with CO_2 at 25°C, instead it will sink!

42. $d = \dfrac{MP}{RT}$ becomes $T = \dfrac{MP}{RT} = \dfrac{44.0\ \text{g/mol} \times 1.00\ \text{atm}}{0.08206\ \dfrac{\text{L atm}}{\text{mol K}} \times 1.18\ \text{g/L}} = 454\text{K} = 181°C$

43. $d = \dfrac{MP}{RT}$ becomes $M = \dfrac{dRT}{P} = \dfrac{2.64 \text{ g/L} \times 0.08206 \dfrac{\text{L atm}}{\text{mol K}} \times (310+273)\text{K}}{775 \text{ mmHg} \times \dfrac{1 \text{ atm}}{760 \text{ mmHg}}} = 124 \text{ g/mol}$

Since the atomic mass of phosphorus is 31.0, the formula of phosphorus molecules in the vapor must be P_4. (4 atoms / molecule $\times$ 31.0 = 124)

44. We first determine the molar mass of the gas, then its empirical formula. These two pieces of information are combined to obtain the molecular formula of the gas.

$$M = \dfrac{dRT}{P} = \dfrac{2.33 \text{ g/L} \times 0.08206 \dfrac{\text{L atm}}{\text{mol K}} \times 296 \text{ K}}{746 \text{ mmHg} \times \dfrac{1 \text{ atm}}{760 \text{ mmHg}}} = 57.7 \text{ g/mol}$$

$$\text{mol C} = 82.7 \text{ g C} \times \dfrac{1 \text{ mol C}}{12.0 \text{ g C}} = 6.89 \text{ mol C} \div 6.89 \rightarrow 1.00 \text{ mol C}$$

$$\text{mol H} = 17.3 \text{ g H} \times \dfrac{1 \text{ mol H}}{1.01 \text{ g H}} = 17.1 \text{ mol H} \div 6.89 \rightarrow 2.48 \text{ mol H}$$

Multiply both of these mole numbers by 2 to obtain the empirical formula, C_2H_5, which has an empirical molar mass of 29.0 g/mol. Since the molar mass (calculated as 57.7 g/mol above) is twice the empirical molar mass, twice the empirical formula is the molecular formula, namely, C_4H_{10}.

Gases in Chemical Reactions

45. Balanced equation: $C_3H_8(g) + 5 O_2(g) \rightarrow 3 CO_2(g) + 4 H_2O(l)$

Use the law of combining volumes. O_2 volume = 75.6 L $C_3H_8 \times \dfrac{5 \text{ L } O_2}{1 \text{ L } C_3H_8} = 378 \text{ L } O_2$

46. Each mole of gas occupies 22.4 L at STP.

H_2 STP volume = 1.000 g Al $\times \dfrac{1 \text{ mol Al}}{26.98 \text{ g Al}} \times \dfrac{3 \text{ mol } H_2(g)}{2 \text{ mol Al}(s)} \times \dfrac{22.4 \text{ L } H_2(g) \text{ at STP}}{1 \text{ mol } H_2}$

$= 1.246 \text{ L } H_2(g)$

47. Determine the moles of $SO_2(g)$ produced and then use the ideal gas equation.

2.7×10^6 lb coal $\times \dfrac{3.28 \text{ lb S}}{100.00 \text{ lb coal}} \times \dfrac{454 \text{ g S}}{1 \text{ lb S}} \times \dfrac{1 \text{ mol S}}{32.1 \text{ g S}} \times \dfrac{1 \text{ mol } SO_2}{1 \text{ mol S}} = 1.3 \times 10^6 \text{ mol } SO_2$

$$V = \dfrac{nRT}{P} = \dfrac{1.3 \times 10^6 \text{ mol } SO_2 \times 0.08206 \dfrac{\text{L atm}}{\text{mol K}} \times 296 \text{ K}}{738 \text{ mmHg} \times \dfrac{1 \text{ atm}}{760 \text{ mmHg}}} = 3.3 \times 10^7 \text{ L } SO_2$$

48. Determine first the amount of $CO_2(g)$ that can be removed. Then use the ideal gas law.

$$\text{mol } CO_2 = 1.00 \text{ kg LiOH} \times \frac{1000 \text{ g}}{1 \text{ kg}} \times \frac{1 \text{ mol LiOH}}{23.95 \text{ g LiOH}} \times \frac{1 \text{ mol } CO_2}{2 \text{ mol LiOH}} = 20.9 \text{ mol } CO_2$$

$$V = \frac{nRT}{P} = \frac{20.9 \text{ mol} \times 0.08206 \dfrac{\text{L atm}}{\text{mol K}} \times (25.9 + 273.2) \text{ K}}{751 \text{ mmHg} \times \dfrac{1 \text{ atm}}{760 \text{ mmHg}}} = 519 \text{ L } CO_2(g)$$

49. Determine the moles of O_2, and then the mass of $KClO_3$ that produced this amount of O_2.

$$\text{mol } O_2 = \frac{\left(738 \text{ mmHg} \times \dfrac{1 \text{ atm}}{760 \text{ mmHg}}\right) \times \left(119 \text{ mL} \times \dfrac{1 \text{ L}}{1000 \text{ mL}}\right)}{0.08206 \dfrac{\text{L atm}}{\text{mol K}} \times (22.4 + 273.2)\text{K}} = 0.00476 \text{ mol } O_2$$

$$\text{mass } KClO_3 = 0.00476 \text{ mol } O_2 \times \frac{2 \text{ mol } KClO_3}{3 \text{ mol } O_2} \times \frac{122.6 \text{ g } KClO_3}{1 \text{ mol } KClO_3} = 0.389 \text{ g } KClO_3$$

$$\% \, KClO_3 = \frac{0.389 \text{ g } KClO_3}{3.57 \text{ g sample}} \times 100\% = 10.9\% \, KClO_3$$

50. Determine the moles and volume of O_2 liberated. $2 \, H_2O_2(aq) \rightarrow 2 \, H_2O(l) + O_2(g)$

$$\text{mol } O_2 = 10.0 \text{ mL soln} \times \frac{1.01 \text{ g}}{1 \text{ mL}} \times \frac{0.0300 \text{ g } H_2O_2}{1 \text{ g soln}} \times \frac{1 \text{ mol } H_2O_2}{34.0 \text{ g } H_2O_2} \times \frac{1 \text{ mol } O_2}{2 \text{ mol } H_2O_2}$$

$$= 0.00446 \text{ mol } O_2$$

$$V = \frac{0.00446 \text{ mol } O_2 \times 0.08206 \dfrac{\text{L atm}}{\text{mol K}} \times (22 + 273) \text{ K}}{752 \text{ mmHg} \times \dfrac{1 \text{ atm}}{760 \text{ mmHg}}} \times \frac{1000 \text{ mL}}{1 \text{ L}} = 109 \text{ mL } O_2$$

51. First we need to find the number of moles of $CO(g)$

Reaction is $3 \, CO(g) + 7 \, H_2(g) \longrightarrow C_3H_8(g) + 3 \, H_2O(l)$

$$n_{CO} = \frac{PV}{RT} = \frac{28.5 \text{ L} \times 760 \text{ torr} \times \dfrac{1 \text{ atm}}{760 \text{ tor}}}{0.08206 \dfrac{\text{L atm}}{\text{K mol}} \times 273.15 \text{ K}} = 1.27 \text{ moles CO}$$

$$V_{H_2 \text{ (required)}} = \frac{n_{H_2} RT}{P} = \frac{1.27 \text{ mol CO} \times \dfrac{7 \text{ mol } H_2}{3 \text{ mol CO}} \times 0.08206 \dfrac{\text{L atm}}{\text{K mol}} \times 299 \text{ K}}{751 \text{ mmHg} \times \dfrac{1 \text{ atm}}{760 \text{ mmHg}}} = 73.6 \text{ L} H_2$$

52. (a) Volume $NH_3 = 313$ L $H_2 \times \dfrac{2 \text{ L } NH_3}{3 \text{ L } H_2} = 209$ L NH_3

(b) Moles of NH_3 (@ 315 °C and 5.25 atm)

$$= \frac{5.25 \text{ atm} \times 313 \text{ L}}{0.08206 \dfrac{\text{L atm}}{\text{mol K}} \times (315 + 273)\text{K}} = 3.41 \times 10^1 \text{ mol } H_2 \times \frac{2 \text{ mol } NH_3}{3 \text{ mol } H_2} = 2.27 \times 10^1 \text{ mol } NH_3$$

$$V(\text{@25°C, 727mmHg}) = \frac{2.27 \times 10^1 \text{ mol } NH_3 \times 0.08206 \dfrac{\text{L atm}}{\text{mol K}} \times 298 \text{ K}}{727 \text{ mmHg} \times \dfrac{1 \text{ atm}}{760 \text{ mmHg}}} = 5.80 \times 10^2 \text{ L } NH_3$$

Mixture of Gases

53. Determine the total amount of gas; then use the ideal gas law, assuming that the gases behave ideally.

$$\text{moles gas} = \left(15.2 \text{ g Ne} \times \frac{1 \text{ mol Ne}}{20.18 \text{ g Ne}} \right) + \left(34.8 \text{ g Ar} \times \frac{1 \text{ mol Ar}}{39.95 \text{ g Ar}} \right)$$

$$= 0.753 \text{ mol Ne} + 0.871 \text{ mol Ar} = 1.624 \text{ mol gas}$$

$$V = \frac{nRT}{P} = \frac{1.624 \text{ mol} \times 0.08206 \dfrac{\text{L atm}}{\text{mol K}} \times (26.7 + 273.2) \text{ K}}{7.15 \text{ atm}} = 5.59 \text{ L gas}$$

54. 2.24 L H_2(g) at STP is 0.100 mol H_2(g). After 0.10 mol He is added, the container holds 0.20 mol gas.

$$V = \frac{nRT}{P} = \frac{0.20 \text{ mol} \times 0.08206 \dfrac{\text{L atm}}{\text{mol K}} (273 + 100)\text{K}}{1.00 \text{ atm}} = 6.1 \text{ L gas}$$

55. The two pressures are related as are the number of moles of N_2(g) to the total number of moles of gas.

$$\text{moles } N_2 = \frac{PV}{RT} = \frac{28.2 \text{ atm} \times 53.7 \text{ L}}{0.08206 \dfrac{\text{L atm}}{\text{mol K}} \times (26 + 273) \text{ K}} = 61.7 \text{ mol } N_2$$

$$\text{total moles of gas} = 61.7 \text{ mol } N_2 \times \frac{75.0 \text{ atm}}{28.2 \text{ atm}} = 164 \text{ mol gas}$$

$$\text{mass Ne} = \left(164 \text{ mol total} - 61.7 \text{ mol } N_2 \right) \times \frac{20.18 \text{ g Ne}}{1 \text{ mol Ne}} = 2.06 \times 10^3 \text{ g Ne}$$

56. Solve a Boyle's law problem for each gas and add the resulting partial pressures.

$$P_{H_2} = 762 \text{ mmHg} \times \frac{2.35 \text{ L}}{5.52 \text{ L}} = 324 \text{ mmHg} \qquad P_{He} = 728 \text{ mmHg} \times \frac{3.17 \text{ L}}{5.52 \text{ L}} = 418 \text{ mmHg}$$

$$P_{\text{total}} = P_{H_2} + P_{He} = 324 \text{ mmHg} + 418 \text{ mmHg} = 742 \text{ mmHg}$$

57. Initial Pressure of the cylinder

$$P = \frac{nRT}{V} = \frac{(1.60 \text{ g O}_2 \times \dfrac{1 \text{ mol O}_2}{31.998 \text{ g O}_2})(0.08206 \text{ L atm K}^{-1} \text{ mol}^{-1})(273.15 \text{ K})}{2.24 \text{ L}} = 0.500 \text{ atm}$$

We need to quadruple the pressure from 0.500 atm to 2.00 atm.

The mass of O_2 needs to quadruple as well from $1.60 \text{ g} \rightarrow 6.40 \text{ g}$ or add 4.80 g O_2

(this answer eliminates answer (a) and (b) as being correct)

One could also increase the pressure by adding the same number of another gas (e.g. He)
Mass of He = $n_{He} \times MM_{He}$

(note: moles of O_2 needed = $4.80 \text{ g} \times \dfrac{1 \text{ mol O}_2}{31.998 \text{ g O}_2} = 0.150 \text{ moles} = 0.150 \text{ moles of He}$)

Mass of He = $0.150 \text{ moles} \times \dfrac{4.0026 \text{ g He}}{1 \text{ mol He}} = 0.600 \text{ g He}$ ((d) is correct, add 0.600 g of He)

58. **(a)** First determine the moles of each gas, then the total moles, and finally the total pressure.

$$\text{moles H}_2 = 4.0 \text{ g H}_2 \times \frac{1 \text{ mol H}_2}{2.02 \text{ g H}_2} \qquad\qquad \text{moles He} = 10.0 \text{ g He} \times \frac{1 \text{ mol He}}{4.00 \text{ g He}}$$

$$= 2.0 \text{ mol H}_2 \qquad\qquad\qquad\qquad = 2.50 \text{ mol He}$$

$$P = \frac{nRT}{V} = \frac{(2.0 + 2.50) \text{ mol} \times 0.08206 \dfrac{\text{L atm}}{\text{mol K}} \times 273 \text{ K}}{4.3 \text{ L}} = 23 \text{ atm}$$

(b) $P_{H_2} = 23 \text{ atm} \times \dfrac{2.0 \text{ mol H}_2}{4.5 \text{ mol total}} = 10 \text{ atm}$ $\qquad\qquad P_{He} = 23 \text{ atm} - 10 \text{ atm} = 13 \text{ atm}$

59. **(a)** $P_{ben} = \dfrac{nRT}{V} = \dfrac{\left(0.728 \text{ g} \times \dfrac{1 \text{ mol C}_6\text{H}_6}{78.11 \text{ g C}_6\text{H}_6}\right) \times 0.08206 \dfrac{\text{L atm}}{\text{mol K}} \times (35 + 273) \text{ K}}{2.00 \text{ L}} \times \dfrac{760 \text{ mmHg}}{1 \text{ atm}}$

$= 89.5 \text{ mmHg}$

$P_{total} = 89.5 \text{ mmHg C}_6\text{H}_6(g) + 752 \text{ mmHg Ar}(g) = 842 \text{ mmHg}$

(b) $P_{benzene} = 89.5 \text{ mmHg}$ $\qquad\qquad P_{Ar} = 752 \text{ mmHg}$

60. **(a)** The $\%CO_2$ in ordinary air is 0.036%, while from the data of this problem, the $\%CO_2$ in expired air is 3.8%.

$$\frac{P\{CO_2 \text{ expired air}\}}{P\{CO_2 \text{ ordinary air}\}} = \frac{3.8\% \text{ CO}_2}{0.036\% \text{ CO}_2} = 1.1 \times 10^2 \text{ CO}_2 \text{ (expired air to ordinary air)}$$

(b/c) Density should be related to average molar mass. We expect the average molar mass of air to be between the molar masses of its two principal constituents, N_2 (28.0 g/mol) and O_2 (32.0 g/mol). The average molar mass of normal air is approximately 28.9 g/mol. Expired air would be made more dense by the presence of more CO_2 (44.0 g/mol) and less dense by the presence of more H_2O (18.0 g/mol). The change might be minimal. In fact, it is, as the following calculation shows.

$$M_{exp.air} = \left(0.742 \text{ mol N}_2 \times \frac{28.013 \text{ g N}_2}{1 \text{ mol N}_2}\right) + \left(0.152 \text{ mol O}_2 \times \frac{31.999 \text{ g O}_2}{1 \text{ mol O}_2}\right)$$

$$+ \left(0.038 \text{ mol CO}_2 \times \frac{44.01 \text{ g CO}_2}{1 \text{ mol CO}_2}\right) + \left(0.059 \text{ mol H}_2O \times \frac{18.02 \text{ g H}_2O}{1 \text{ mol H}_2O}\right)$$

$$+ \left(0.009 \text{ mol Ar} \times \frac{39.9 \text{ g Ar}}{1 \text{ mol Ar}}\right) = 28.7 \text{ g/mol of expired air}$$

Since the average molar mass of expired air is less than the average molar mass of ordinary air, expired air is less dense than ordinary air. Calculating the densities:

$$d(\text{expired air}) = \frac{(28.7 \text{ g/mol})(1.00 \text{ atm})}{(0.0821 \text{ L} \cdot \text{atm/K} \cdot \text{mol})(310 \text{ K})} = 1.13 \text{ g/L}$$

$$d(\text{ordinary air}) = 1.14 \text{ g/L}$$

61. $1.00 \text{ g H}_2 \approx 0.50 \text{ mol H}_2$ $\qquad 1.00 \text{ g He} \approx 0.25 \text{ mol He}$
Adding 1.00 g of He to a vessel, which only contains 1.00 g of H_2 results in the number of moles of gas being increased by 50%. Situation (b) best represents the resulting mixture as the volume has increased by 50%

62.

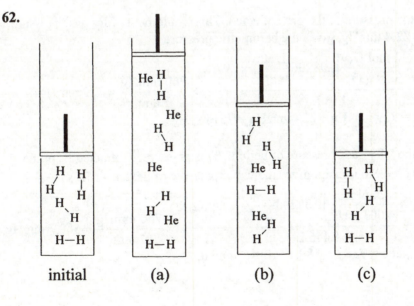

Collecting Gases over Liquids

63. The pressure of the liberated $H_2(g)$ is 744 mmHg $- 23.8$ mmHg $= 720.$ mmHg

$$V = \frac{nRT}{P} = \frac{\left(1.65 \text{ g Al} \times \dfrac{1 \text{ mol Al}}{26.98 \text{ g}} \times \dfrac{3 \text{ mol } H_2}{2 \text{ mol Al}}\right) 0.08206 \dfrac{\text{L atm}}{\text{mol K}} (273 + 25)\text{K}}{720. \text{ mmHg} \times \dfrac{1 \text{ atm}}{760 \text{ mmHg}}} = 2.37 \text{ L } H_2(g)$$

This is the total volume of both gases, each with a different partial pressure.

64. (a) The total pressure is the sum of the partial pressures of $O_2(g)$ and the vapor pressure of water.

$$P_{total} = P_{O_2} + P_{H_2O} = 756 \text{ mmHg} = P_{O_2} + 19 \text{ mmHg}$$

$$P_{O_2} = (756 - 19)\text{mmHg} = 737 \text{ mmHg}$$

(b) The volume percent is equal to the pressure percent.

$$\%O_2(g) \text{ by volume} = \frac{V_{O_2}}{V_{total}} \times 100\% = \frac{P_{O_2}}{P_{total}} \times 100\% = \frac{737 \text{ mmHg of } O_2}{756 \text{ mm Hg total}} \times 100\% = 97.5\%$$

(c) Determine the mass of $O_2(g)$ collected by multiplying the amount of O_2 collected, in moles, by the molar mass of $O_2(g)$.

$$\text{mass}_{O_2} = \frac{PV}{RT} = \frac{\left(737. \text{ mmHg} \times \dfrac{1 \text{ atm}}{760 \text{ mmHg}}\right)\left(89.3 \text{ mL} \times \dfrac{1 \text{ L}}{1000 \text{ mL}}\right)}{0.08206 \dfrac{\text{L atm}}{\text{mol K}} (21.3 + 273.2)\text{K}} \times \frac{32.00 \text{ g } O_2}{1 \text{ mol } O_2} = 0.115 \text{ g } O_2$$

65. We first determine the pressure of the gas collected. This would be its "dry gas" pressure and, when added to 22.4 mmHg, gives the barometric pressure.

$$P = \frac{nRT}{V} = \frac{\left(1.46 \text{ g} \times \dfrac{1 \text{ mol } O_2}{32.0 \text{ g } O_2}\right) 0.08206 \dfrac{\text{L atm}}{\text{mol K}} \times 297 \text{ K}}{1.16 \text{ L}} \times \frac{760 \text{ mmHg}}{1 \text{ atm}} = 729 \text{ mmHg}$$

barometric pressure $= 729$ mm Hg $+ 22.4$ mmHg $= 751$ mmHg

66. We first determine the "dry gas" pressure of helium. This pressure, subtracted from the barometric pressure of 738.6 mmHg, gives the vapor pressure of hexane at $25°C$.

$$P = \frac{nRT}{V} = \frac{\left(1.072 \text{ g} \times \dfrac{1 \text{ mol He}}{4.003 \text{ g He}}\right) 0.08206 \dfrac{\text{L atm}}{\text{mol K}} \times 298.2 \text{ K}}{8.446 \text{ L}} \times \frac{760 \text{ mmHg}}{1 \text{ atm}} = 589.7 \text{ mmHg}$$

vapor pressure of hexane $= 738.6 - 589.7 = 148.9$ mmHg

Kinetic Molecular Theory

67. Recall that $1 J = kg\ m^2\ s^{-2}$

$$u_{rms} = \sqrt{\frac{3RT}{M}} = \sqrt{\frac{3 \times 8.3145 \dfrac{J}{mol\ K} \times \dfrac{1\ kg\ m^2\ s^{-2}}{1\ J} \times 303\ K}{\dfrac{70.91 \times 10^{-3}\ kg\ Cl_2}{1\ mol\ Cl_2}}} = 326\ m/s$$

68.

$$2 = \frac{\sqrt{\dfrac{3RT_1}{M}}}{\sqrt{\dfrac{3RT_2}{M}}} = \sqrt{\frac{T_1}{T_2}} = \sqrt{\frac{T_1}{273\ K}}$$ Square both sides and solve for T_1. $T_1 = 4 \times 273\ K = 1092\ K$

Alternatively, recall that $1 J = kg\ m^2\ s^{-2}$

$$u_{rms} = \sqrt{\frac{3RT}{M}} = 1.84 \times 10^3\ m/s$$ Solve this equation for temperature with u_{rms} doubled.

$$T = \frac{Mu_{rms}^2}{3R} = \frac{2.016 \times 10^{-3}\ kg/mol\ (2 \times 1.84 \times 10^3\ m/s)^2}{3 \times 8.3145 \dfrac{J}{mol\ K} \times \dfrac{1\ kg\ m^2\ s^{-2}}{1\ J}} = 1.09 \times 10^3\ K$$

69. $M = \dfrac{3RT}{(u_{rms})^2} = \dfrac{3 \times 8.3145 \dfrac{J}{mol\ K} \times 298\ K}{\left(2180 \dfrac{mi}{hr} \times \dfrac{1\ hr}{3600\ sec} \times \dfrac{5280\ ft}{1\ mi} \times \dfrac{12\ in.}{1\ ft} \times \dfrac{1\ m}{39.37\ in.}\right)^2} = 0.00783\ kg/mol = 7.83\ g/mol.$ or $7.83\ u.$

70. A noble gas with molecules having u_{rms} at 25°C greater than that of a rifle bullet will have a molar mass less than 7.8 g/mol. Helium is the only possibility. A noble gas with a slower u_{rms} will have a molar mass greater than 7.8 g/mol; any one of the other noble gases will have a slower u_{rms}.

71. We equate the two expressions for root mean square speed, cancel the common factors, and solve for the temperature of Ne. Note that the units of molar masses do not have to be in kg/mol in this calculation; they simply must be expressed in the same units.

$$\sqrt{\frac{3RT}{M}} = \sqrt{\frac{3R \times 300\ K}{4.003}} = \sqrt{\frac{3R \times T_{Ne}}{20.18}}$$ square both sides: $\dfrac{300\ K}{4.003} = \dfrac{T_{Ne}}{20.18}$

Solve for T_{Ne}: $T_{Ne} = 300\ K \times \dfrac{20.18}{4.003} = 1.51 \times 10^3\ K$

72. u_m, the modal speed, is the speed that occurs most often, 55 mi/h

$$\text{Average speed} = \frac{38 + 44 + 45 + 48 + 50 + 55 + 55 + 57 + 58 + 60}{10} = 51.0\ mi/h = \bar{u}$$

$$u_{rms} = \sqrt{\frac{38^2 + 44^2 + 45^2 + 48^2 + 50^2 + 55^2 + 55^2 + 57^2 + 58^2 + 60^2}{10}} = \sqrt{\frac{26472}{10}} = 51.5\ \frac{miles}{h}$$

Diffusion and Effusion of Gases

73. $\dfrac{\text{rate (NO}_2)}{\text{rate (N}_2\text{O)}} = \sqrt{\dfrac{M(\text{N}_2\text{O})}{M(\text{NO}_2)}} = \sqrt{\dfrac{44.02}{46.01}} = 0.9781 = \dfrac{x \text{ mol NO}_2/t}{0.00484 \text{ mol N}_2\text{O}/t}$

mol $\text{NO}_2 = 0.00484 \text{ mol} \times 0.9781 = 0.00473 \text{ mol NO}_2$

74. $\dfrac{\text{rate (N}_2)}{\text{rate (unknown)}} = \dfrac{\text{mol (N}_2)/38\,\text{s}}{\text{mol (unknown)}/64\,\text{s}} = \dfrac{64\,\text{s}}{38\,\text{s}} = 1.68 = \sqrt{\dfrac{M(\text{unknown})}{M(\text{N}_2)}}$

$M(\text{unknown}) = (1.68)^2 M(\text{N}_2) = (1.68)^2 (28.01\,\text{g/mol}) = 79\,\text{g/mol}$

75. **(a)** $\dfrac{\text{rate (N}_2)}{\text{rate (O}_2)} = \sqrt{\dfrac{M(\text{O}_2)}{M(\text{N}_2)}} = \sqrt{\dfrac{32.00}{28.01}} = 1.07$ **(b)** $\dfrac{\text{rate (H}_2\text{O)}}{\text{rate (D}_2\text{O)}} = \sqrt{\dfrac{M(\text{D}_2\text{O})}{M(\text{H}_2\text{O})}} = \sqrt{\dfrac{20.0}{18.02}} = 1.05$

   (c) $\dfrac{\text{rate (}^{14}\text{CO}_2)}{\text{rate (}^{12}\text{CO}_2)} = \sqrt{\dfrac{M(^{12}\text{CO}_2)}{M(^{14}\text{CO}_2)}} = \sqrt{\dfrac{44.0}{46.0}} = 0.978$ **(d)** $\dfrac{\text{rate (}^{235}\text{UF}_6)}{\text{rate (}^{238}\text{UF}_6)} = \sqrt{\dfrac{M(^{238}\text{UF}_6)}{M(^{235}\text{UF}_6)}} = \sqrt{\dfrac{352}{349}} = 1.004$

76. $\dfrac{\text{Rate of effusion O}_2}{\text{Rate of effusion SO}_2} = \dfrac{\sqrt{M(\text{SO}_2)}}{\sqrt{M(\text{O}_2)}} = \dfrac{\sqrt{64}}{\sqrt{32}} = \sqrt{2} = 1.4$

O_2 will effuse at 1.4 times the rate of effusion of SO_2 The situation depicted in (c) best represents the distribution of the molecules. If 5 molecules of SO_2 effuse, we would expect that 1.4 times as many O_2 molecules to effuse over the same period of time

$(1.4 \times 5 = 7 \text{ molecules of O}_2)$. This is depicted in situation (c).

Nonideal Gases

77. For $\text{Cl}_2(g), n^2 a = 6.49\,\text{L}^2$ atm and $nb = 0.0562\,\text{L}$. $P_{\text{vdw}} = \dfrac{nRT}{V - nb} - \dfrac{n^2 a}{V^2}$

At 0°C, $P_{\text{vdw}} = 9.90$ atm and $P_{\text{ideal}} = 11.2$ atm, off by 1.3 atm or + 13%

(a) At 100°C $P_{\text{ideal}} = \dfrac{nRT}{V} = \dfrac{1.00\,\text{mol} \times \dfrac{0.08206\,\text{L atm}}{\text{mol K}} \times 373\,\text{K}}{2.00\,\text{L}} = 15.3\,\text{atm}$

$P_{\text{vdw}} = \dfrac{1.00\,\text{mol} \times \dfrac{0.08206\,\text{L atm}}{\text{mol K}} \times T}{(2.00 - 0.0562)\,\text{L}} - \dfrac{6.49\,\text{L}^2\,\text{atm}}{(2.00\,\text{L})^2} = 0.0422\,T\,\text{atm} - 1.62\,\text{atm}$

$= 0.0422 \times 373\,\text{K} - 1.62 = 14.1\,\text{atm}$ P_{ideal} is off by 1.2 atm or +8.5%

(b) At 200°C $P_{\text{ideal}} = \dfrac{nRT}{V} = \dfrac{1.00\,\text{mol} \times \dfrac{0.08206\,\text{L atm}}{\text{mol K}} \times 473\,\text{K}}{2.00\,\text{L}} = 19.4\,\text{atm}$

$P_{\text{vdw}} = 0.0422_2 \times 473\,\text{K} - 1.62_3 = 18.3_5\,\text{atm}$ P_{ideal} is off by 1.0 atm or +5.7%

(c) At 400°C $P_{ideal} = \dfrac{nRT}{V} = \dfrac{1.00\,mol \times \dfrac{0.08206\,L\,atm}{mol\,K} \times 673\,K}{2.00\,L} = 27.6\,atm$

$P_{vdw} = 0.0422 \times 673\,K - 1.62 = 26.8\,atm$ P_{ideal} is off by 0.8 atm or +3.0%

78. (a) V = 100.0 L,

$P_{vdw} = \dfrac{nRT}{V-nb} - \dfrac{n^2a}{V^2} = \dfrac{1.50\,mol \times \dfrac{0.08206\,L\,atm}{mol\,K} \times 298\,K}{(100.0 - 1.50 \times 0.0564)L} - \dfrac{1.50^2 \times 6.71\,L^2\,atm}{(100.0\,L)^2}$

$= 0.365\underline{5}\,atm\,SO_2\,(g)$

$P_{ideal} = \dfrac{1.50\,mol \times \dfrac{0.08206\,L\,atm}{mol\,K} \times 298\,K}{100.0\,L} = 0.367\,atm$

The two pressures are almost equal.

(b) V = 50.0 L, $P_{vdw} \doteq 0.729\,atm$; $P_{ideal} = 0.734\,atm$.
Here the two pressures agree within a few percent.

(c) V = 20.0 L, $P_{vdw} = 1.80\,atm$; $P_{ideal} = 1.84\,atm$

(d) V = 10.0 L, $P_{vdw} = 3.55\,atm$; $P_{ideal} = 3.67\,atm$

Integrative and Advanced Exercises

79. The standard atmosphere is defined as the pressure exerted by a certain height (760 mm) of mercury. The density of mercury (13.5951 g/cm^3) needs to be included in this definition and, because the density of mercury, like that of other liquids, varies somewhat with temperature, the pressure will vary if the density varies. The acceleration due to gravity (9.80665 m/s^2) needs to be included because pressure is a *force* per unit area, not just a mass per unit area, and this force depends on the acceleration applied to the given mass, i.e. force = mass × acceleration.

80. Of course, we express the temperatures in Kelvins:

$0\,°C = 273\,K, -100\,°C = 173\,K, -200\,°C = 73\,K, -250\,°C = 23\,K,$ and $-270\,°C = 3\,K$.

(A) $\dfrac{V}{T} = k = \dfrac{10.0\,cm^3}{400\,K} = 0.00250\,cm^3\,K^{-1}$ $V = 0.0250\,cm^3\,K^{-1} \times 273\,K = 6.83\,cm^3$

$V = 0.0250\,cm^3\,K^{-1} \times 173\,K = 4.33\,cm^3$ $V = 0.0250\,cm^3\,K^{-1} \times 73\,K = 1.8\,cm^3$

$V = 0.0250\,cm^3\,K^{-1} \times 23\,K = 0.58\,cm^3$ $V = 0.0250\,cm^3\,K^{-1} \times 3\,K = 0.08\,cm^3$

(B) $\dfrac{V}{T} = k = \dfrac{20.0\,\text{cm}^3}{400\,\text{K}} = 0.0500\,\text{cm}^3\,\text{K}^{-1}$ $V = 0.0500\,\text{cm}^3\,\text{K}^{-1} \times 273\,\text{K} = 13.7\,\text{cm}^3$

$V = 0.0500\,\text{cm}^3\,\text{K}^{-1} \times 173\text{K} = 8.65\,\text{cm}^3$ $\quad$ $V = 0.0500\,\text{cm}^3\,\text{K}^{-1} \times 73\text{K} = 3.7\,\text{cm}^3$

$V = 0.0500\,\text{cm}^3\,\text{K}^{-1} \times 23\text{K} = 1.2\,\text{cm}^3$ $\quad$ $V = 0.0500\,\text{cm}^3\,\text{K}^{-1} \times 3\text{K} = 0.2\,\text{cm}^3$

(C) $\dfrac{V}{T} = k = \dfrac{40.0\,\text{cm}^3}{400\,\text{K}} = 0.100\,\text{cm}^3\,\text{K}^{-1}$ $\qquad$ $V = 0.100\,\text{cm}^3\,\text{K}^{-1} \times 273\,\text{K} = 27.3\,\text{cm}^3$

$V = 0.100\,\text{cm}^3\,\text{K}^{-1} \times 173\text{K} = 17.3\,\text{cm}^3$ $\qquad$ $V = 0.100\,\text{cm}^3\,\text{K}^{-1} \times 73\text{K} = 7.3\,\text{cm}^3$

$V = 0.100\,\text{cm}^3\,\text{K}^{-1} \times 23\text{K} = 2.3\,\text{cm}^3$ $\qquad$ $V = 0.100\,\text{cm}^3\,\text{K}^{-1} \times 3\text{K} = 0.3\,\text{cm}^3$

As expected, in all three cases the volume of each gas goes to zero at 0 K.

81. Initial sketch shows 1 mole of gas at STP(760 torr, 273 K) with a volume of ~22.4 L. Velocity of particles = v_i. Note velocity increases as $\sqrt{T}$ (i.e. quadrupling temperature, results in a doubling of the velocity of the particles).

a) Pressure drops to 1/3 of its original value (760 torr → 253 torr). The number of moles and temperature are constant. Volume of the gas should increase from 22.4 L → ≈ 67 L while the velocity remains unchanged at v_i.

b) Pressure stays constant at its original value (760 torr). The number of moles is unchanged, while the temperature drops to half of its value (273 K → 137 K). Volume of the gas should decrease from 22.4 L → ≈ 11.2 L while the velocity drops to about 71% to $0.71v_i(\sqrt{T} \to \sqrt{0.5 \times T})$.

c) Pressure drops to 1/2 of its original value (760 torr → 380 torr). The number of moles remains constant, while the temperature is increased to twice it original value (273 K → 546 K). Volume of the gas should quadruple, increasing the volume from 22.4 L → ≈ 90 L. Velocity of the molecules increases by 41% to $1.41v_i(\sqrt{T} \to \sqrt{2 \times T})$.

d) Pressure increases to 2.25 times its original value (1710 torr) owing to a temperature increase of 50% (273 K → ≈ 410 K) and a 50% increase in the number of particles (1 mole → 1.5 moles). The volume of the gas should remain unchanged at 22.4 L, while the velocity increases by 22% to $1.22v_i(\sqrt{T} \to \sqrt{1.5 \times T})$.

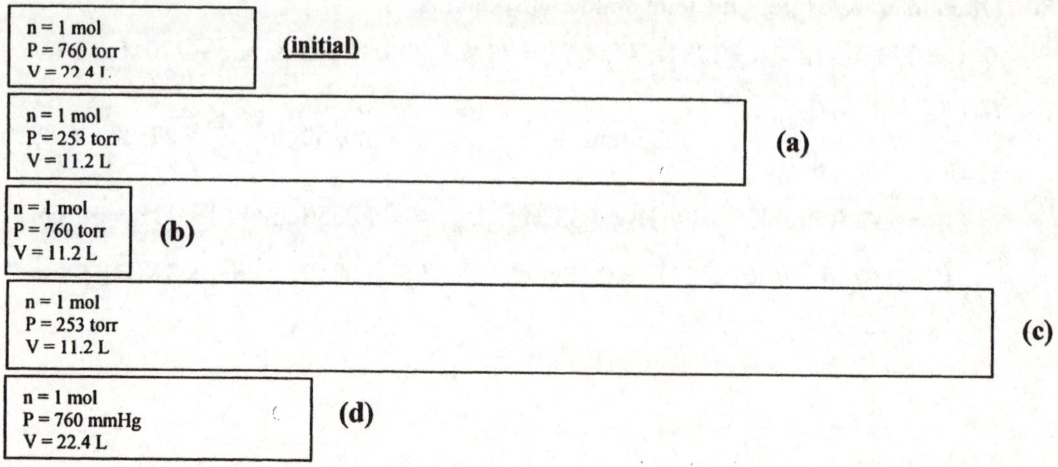

```
n = 1 mol
P = 760 torr        (initial)
V = 22.4 L
```

```
n = 1 mol
P = 253 torr                                    (a)
V = 11.2 L
```

```
n = 1 mol
P = 760 torr    (b)
V = 11.2 L
```

```
n = 1 mol
P = 253 torr                                    (c)
V = 11.2 L
```

```
n = 1 mol
P = 760 mmHg    (d)
V = 22.4 L
```

82. First we determine the molar mass of the hydrocarbon.

$$M = \frac{mRT}{PV} = \frac{0.288\,g \times 0.08206\,\dfrac{L \cdot atm}{mol \cdot K}(24.8 + 273.2)\,K}{\left(753\ mmHg \times \dfrac{1\ atm}{760\ mmHg}\right) \times 0.131\ L} = 54.3\ g/mol$$

Now determine the empirical formula. A hydrocarbon contains just hydrogen and carbon.

$$\text{amount } C = 3.613\,g\ CO_2 \times \frac{1\ mol\ CO_2}{44.01\ g\ CO_2} \times \frac{1\ mol\ C}{1\ mol\ CO_2} = 0.08210\ mol\ C \div 0.08209 \rightarrow 1.000\ mol\ C$$

$$\text{amount } H = 1.109\,g\ H_2O \times \frac{1\ mol\ H_2O}{18.02\ g\ H_2O} \times \frac{2\ mol\ H}{1\ mol\ H_2O} = 0.1231\ mol\ H \div 0.08209 \rightarrow 1.500\ mol\ H$$

The empirical formula is $CH_{1.5}$ or C_2H_3. This gives an empirical molar mass of 27.0 g/mol, almost precisely one-half the experimental molar mass. The molecular formula is therefore C_4H_6.

83. First we compute the STP volume of the gas. We use this volume along with the STP molar volume of 22.414 L/mol to determine the number of moles of He and then the reduced pressure.

$$V = 2.15\ ft^3 \times \left(\frac{12\ in.}{1\ ft} \times \frac{2.54\ cm}{1\ in.}\right)^3 \times \frac{1\ L}{1000\ cm^3} = 60.9\ L \qquad n = 60.9\ L \times \frac{1\ mol}{22.414\ L} = 2.72\ mol$$

$$P = \frac{nRT}{V} = \frac{2.72\ mol \times 0.08206\ L\ atm\ mol^{-1}\ K^{-1} \times 253\ K}{155\,L} = 0.364\ atm$$

Yet another way to solve this problem is to realize that the pressure decreases as the volume increases, and the pressure also decreases as the temperature decreases.

$$P = 1.00\ atm \times \frac{60.9\ L}{155\ L} \times \frac{253\ K}{273\ K} = 0.364\ atm$$

84. Trapped within the diver toy is a small bubble of air. When we squeeze the bottle, we increase the pressure everywhere inside the bottle and that small bubble of air is compressed. Air is much less dense than water but the diver under pressure contains a smaller volume of air. Thus, the pressurized diver is more dense (because of the decrease in air volume) than the unpressurized diver. The denser diver sinks in the bottle.

85. This is an impossible situation if the balloons are identical, for then the blue balloon should inflate when the stopcock is opened. But the observation is understandable if we assume that the blue balloon is made of stiffer rubber than the red one. Then it takes greater gas pressure to inflate the darker balloon. We then understand that with equal gas pressures, the blue balloon will not inflate as much as the red balloon.

86. Note that three moles of gas are produced for each mole of NH_4NO_3 that decomposes.

$$\text{amount of gas} = 3.05 \text{ g } NH_4NO_3 \times \frac{1 \text{ mol } NH_4NO_3}{80.04 \text{ g } NH_4NO_3} \times \frac{3 \text{ moles of gas}}{1 \text{ mol } NH_4NO_3} = 0.114 \text{ mol gas}$$

$$T = 250 + 273 = 523 \text{ K}$$

$$P = \frac{nRT}{V} = \frac{0.114 \text{ mol} \times 0.08206 \text{ L atm mol}^{-1} \text{ K}^{-1} \times 523 \text{ K}}{2.18 \text{ L}} = 2.24 \text{ atm}$$

87. **(a)** $n_{H_2} = 1.00 \text{ g}/2.02 \text{ g/mol} = 0.495 \text{ mol } H_2$; $n_{O_2} = 8.60 \text{ g}/32.0 \text{ g/mol} = 0.269 \text{ mol } O_2$

$$P_{total} = P_{O_2} + P_{H_2} = \frac{n_{total}RT}{V} = \frac{0.769 \text{ mol} \times 0.08206 \text{ L atm mol}^{-1} \text{ K}^{-1} \times 298 \text{ K}}{1.500 \text{ L}} = 12.5 \text{ atm}$$

The limiting reagent in the production of water is H_2 (0.495 mol) with O_2 (0.269 mol) in excess.

	$2 H_2 (g)$	$+$	$O_2 (g)$	$\rightarrow$	$2 H_2O(l)$
(initial)	0.495 mol		0.269 mol		0
after reaction	~0		0.021 mol		0.495 mol

so $P_{total} = P_{O_2} + P_{H_2O}$

$$P_{O_2} = \frac{n_{total}RT}{V} = \frac{0.021 \text{ mol} \times 0.08206 \text{ L atm mol}^{-1} \text{ K}^{-1} \times 298 \text{ K}}{1.50 \text{ L}} = 0.342 \text{ atm}$$

$$P_{total} = 0.342 \text{ atm} + \frac{23.8 \text{ mm Hg}}{760 \text{ mm Hg/atm}} = 0.37 \text{ atm}$$

88. **(a)** $\text{mL } O_2/\text{min} = \frac{4.0 \text{ L air}}{1 \text{ min}} \times \frac{1000 \text{ mL}}{1 \text{ L}} \times \frac{3.8 \text{ mL } CO_2}{100 \text{ mL air}} \times \frac{1 \text{ mL } O_2}{2 \text{ mL } CO_2} = 76 \text{ mL } O_2/\text{min}$

(b) We first determine the amount of O_2 produced per minute, then the rate of consumption of Na_2O_2.

$$P = 735 \text{ mmHg} \times \frac{1 \text{ atm}}{760 \text{ mmHg}} = 0.967 \text{ atm} \qquad T = 25 + 273 = 298 \text{ K}$$

$$n = \frac{PV}{RT} = \frac{0.967 \text{ atm} \times 0.076 \text{ L}}{0.08206 \text{ L atm mol}^{-1} \text{ K}^{-1} \times 298 \text{ K}} = 0.0030 \text{ mol } O_2$$

$$\text{rate} = \frac{0.0030 \text{ mol } O_2}{1 \text{ min}} \times \frac{2 \text{ mol } Na_2O_2}{1 \text{ mol } O_2} \times \frac{77.98 \text{ g } Na_2O_2}{1 \text{ mol } Na_2O_2} \times \frac{60 \text{ min}}{1 \text{ h}} = 28 \text{ g } Na_2O_2/\text{h}$$

89. Determine relative numbers of moles, and the mole fractions of the 3 gases in the 100.0 g gaseous mixture.

amount $N_2 = 46.5$ g $N_2 \times \dfrac{1\ \text{mol}\ N_2}{28.01\ \text{g}\ N_2} = 1.66$ mol N_2 $\div 2.86$ mol $\longrightarrow 0.580$

amount $Ne = 12.7$ g $Ne \times \dfrac{1\ \text{mol}\ Ne}{20.18\ \text{g}\ Ne} = 0.629$ mol Ne $\div 2.86$ mol $\longrightarrow 0.220$

amount $Cl_2 = 40.8$ g $Cl_2 \times \dfrac{1\ \text{mol}\ Cl_2}{70.91\ \text{g}\ Cl_2} = 0.575$ mol Cl_2 $\div 2.86$ mol $\longrightarrow 0.201$

total amount $= 1.66$ mol $N_2 + 0.629$ mol $Ne + 0.575$ mol $Cl_2 = 2.86$ mol

Since the total pressure of the mixture is 1.000 atm, the partial pressure of each gas is numerically equal to its mole fraction. Thus, the partial pressure of Cl_2 is 0.201 atm or 153 mmHg

90. Let us first determine the molar mass of this mixture.

$$M = \frac{dRT}{P} = \frac{0.518\ \text{g}\ L^{-1} \times 0.08206\ \ L\ \text{atm}\ \text{mol}^{-1}\ K^{-1} \times 298\ K}{721\ \text{mmHg} \times \dfrac{1\ \text{atm}}{760\ \text{mmHg}}} = 13.4\ \text{g/mol}$$

Then we let x be the mole fraction of He in the mixture.

13.4 g/mol $= x \times 4.003$ g/mol $+ (1.000 - x) \times 32.00$ g/mol $= 4.003x + 32.00 - 32.00x$

$$x = \frac{32.00 - 13.4}{32.00 - 4.003} = 0.664$$

Thus, one mole of the mixture contains 0.664 mol He. We determine the mass of that He and then the % He by mass in the mixture.

mass He $= 0.664$ mol He $\times \dfrac{4.003\ \text{g}\ He}{1\ \text{mol}\ He} = 2.66$ g He $\qquad$ %He $= \dfrac{2.66\ \text{g}\ He}{13.4\ \text{g}\ \text{mixture}} \times 100\% = 19.9\%$ He

91. The volume percents in a mixture of gases also equal mole percents, which can be converted to mole fractions by dividing by 100.

$$M_{\text{air}} = \left(0.7808\ \text{mol}\ N_2 \times \frac{28.013\ \text{g}\ N_2}{1\ \text{mol}\ N_2} \right) + \left(0.2095\ \text{mol}\ O_2 \times \frac{31.999\ \ \text{g}\ \ O_2}{1\ \text{mol}\ O_2} \right)$$

$$+ \left(0.00036\ \text{mol}\ CO_2 \times \frac{44.010\ \text{g}\ CO_2}{1\ \text{mol}\ CO_2} \right) + \left(0.0093\ \text{mol}\ Ar \times \frac{39.948\ \text{g}\ Ar}{1\ \text{mol}\ Ar} \right) = 28.96\ \text{g/mol}$$

92. The amount of $N_2(g)$ plus the amount of $He(g)$ equals the total amount of gas. We use this equality, and substitute with the ideal gas law, letting V symbolize the volume of cylinder B.

$n_{N_2} + n_{He} = n_{total}$ becomes $P_{N_2} \times V_{N_2} + P_{He} \times V_{He} = P_{total} \times V_{total}$ since all temperatures are equal. Substitution gives $(8.35 \text{ atm} \times 48.2 \text{ L}) + (9.50 \text{ atm} \times V) = 8.71 \text{ atm } (48.2 \text{ L} + V)$

$$402 \text{ L atm} + 9.50\,V = 420 \text{ L atm} + 8.71\,V \qquad V = \frac{420 - 402}{9.50 - 8.71} = 23 \text{ L}$$

93. To construct a closed-end monometer. Fill the tube with mercury, making sure to remove any air bubbles. Invert the tube results in a column of mercury with a vacuum in the space directly above the column.

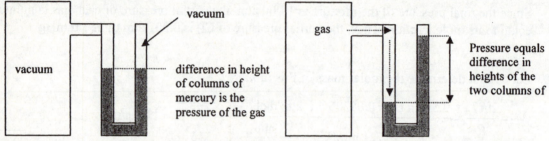

When the container is evacuated, the heights of both mercury columns are equal. As the gas pressure in the container increases, the height of the mercury columns will change. Because the system is closed, P_{bar} is not needed. One drawback of this system is that gas pressure much larger than one atmosphere cannot easily be handled unless a very long column is used. (> 1 meter). Otherwise, the column of mercury would be pushed up to the closed end of the monometer at a critical pressure, and remain unchanged with increasing pressure.

An open-ended monometer, would be better suited to measuring higher pressure than lower pressure. A gas pressure of one atmosphere results in the columns of mercury being equal. A sidearm of 760 mm pressure would allow one to measure pressures up to 2 atmospheres.

94. (a) $M_{av} = \dfrac{8.0}{100}\left(44.01\dfrac{g}{mol}\right) + \dfrac{23.2}{100}\left(28.01\dfrac{g}{mol}\right) + \dfrac{17.7}{100}\left(2.016\dfrac{g}{mol}\right) + \dfrac{1.1}{100}\left(16.043\dfrac{g}{mol}\right) + \dfrac{50.5}{100}\left(28.01\dfrac{g}{mol}\right)$

$M_{av} = 24.5\underline{6} \text{ g/mol}$

$\text{Density} = \dfrac{P\,M}{R\,T} = \dfrac{\left(763 \text{ mmHg} \times \dfrac{1 \text{ atm}}{760 \text{ mmHg}}\right) 24.5\underline{6}\dfrac{g}{mol}}{\left(0.08206\dfrac{\text{L atm}}{\text{K mol}}\right) 296 \text{ K}} = 1.01\underline{5}\dfrac{g}{L}$

(b) $P_{co} = P_{total} \times \dfrac{V\%}{100\%} = 763 \text{ mmHg} \times \dfrac{23.2\%}{100\%} = 177 \text{ mmHg or } 0.233 \text{ atm}$

(c) $CO(g) + \frac{1}{2} O_2(g) \rightarrow CO_2(g)$
$H_2(g) + \frac{1}{2} O_2(g) \rightarrow H_2O(g)$
$CH_4(g) + 2 O_2(g) \rightarrow CO_2(g) + 2 H_2O(g)$

Using the fact that volume is directly proportional to moles when the pressure and temperature are constant. 1000 L of producer gas contains:

232 L CO	which requires	116 L O_2
177 L H_2	which requires	88.5 L O_2
11 L CH_4	which requires	22 L O_2
		226.5 L O_2 (Note: air is 20.95 % O_2 by volume)

Thus, the reaction requires $\dfrac{226.5 \, L}{0.2095} = 1.08 \times 10^3 \, L$

95. First recognize that 3 moles of gas are produced from every 2 moles of water, and compute the number of moles of gas produced. Then determine the partial pressure these gases would exert.

$$\text{amount of gas} = 1.32 \text{ g } H_2O \times \frac{1 \text{ mol } H_2O}{18.02 \text{ g } H_2O} \times \frac{3 \text{ mol gas}}{2 \text{ mol } H_2O} = 0.110 \text{ mol gas}$$

$$P = \frac{nRT}{V} = \frac{0.110 \text{ mol} \times 0.08206 \text{ L atm mol}^{-1} \text{ K}^{-1} \times 303 \text{ K}}{2.90 \text{ L}} \times \frac{760 \text{ mmHg}}{1 \text{ atm}} = 717 \text{ mmHg}$$

Then the vapor pressure (partial pressure) of water is determined by difference.

$P_{water} = 748 \text{ mmHg} - 717 \text{ mmHg} = 31 \text{ mmHg}$

96. The total pressure of the mixture of O_2 and H_2O is 737 mmHg, and the partial pressure of H_2O is 25.2 mmHg.

(a) The percent of water vapor by volume equals its percent pressure.

$$\% \text{ } H_2O = \frac{25.2 \text{ mmHg}}{737 \text{ mmHg}} \times 100\% = 3.42\% \text{ } H_2O \text{ by volume}$$

(b) The % water vapor by number of molecules equals its percent pressure, 3.42% by number

(c) One mole of the combined gases contains 0.0342 mol H_2O and 0.9658 mol O_2.

$$\text{molar mass} = 0.0342 \text{ mol } H_2O \times \frac{18.02 \text{ g } H_2O}{1 \text{ mol } H_2O} + 0.9658 \text{ mol } O_2 \times \frac{31.999 \text{ g } O_2}{1 \text{ mol } O_2}$$

$$= 0.616 \text{ g } H_2O + 30.90 \text{ g } O_2 = 31.52 \text{ g}$$

$$\% \text{ } H_2O = \frac{0.616 \text{ g } H_2O}{31.52 \text{ g}} \times 100\% = 1.95 \% \text{ } H_2O \text{ by mass}$$

97. The reaction of interest is: $CH_4(g) + 2 O_2(g) \rightarrow CO_2(g) + 2 H_2O(g)$
Note: CH_4 is at the same temperature but not the same pressure and 100 L air contains 20.95 L O_2.

$$V_{O_2} \text{ (22 °C and 3.55 atm)} = 1.00 \text{ L } CH_4 \times \frac{2 \text{ L } O_2}{1 \text{ L } CH_4} = 2.00 \text{ L } O_2$$

$$V_{O_2} \text{ (22°C and 745 mmHg)} = 2.00 \text{ L } O_2 \times \frac{3.55 \text{ atm}}{745 \text{ mmHg} \times \dfrac{1 \text{ atm}}{760 \text{ mmHg}}} = 7.24 \text{ L } O_2$$

$$V_{air} \text{ (22°C and 745 mmHg)} = 7.24 \text{ L } O_2 \times \frac{100 \text{ L air}}{20.95 \text{ L } O_2} = 34.6 \text{ L}$$

98. 1 mol of the mixture at STP occupies a volume of 22.414 L. It contains 0.79 mol He and 0.21 mol O_2.

$$\text{STP density} = \frac{\text{mass}}{\text{volume}} = \frac{0.79 \text{ mol He} \times \dfrac{4.003 \text{ g He}}{1 \text{ mol He}} + 0.21 \text{ mol } O_2 \times \dfrac{32.0 \text{ g } O_2}{1 \text{ mol } O_2}}{22.414 \text{ L}} = 0.44 \text{ g/L}$$

25°C is a temperature higher than STP. This condition increases the 1.00-L volume containing 0.44 g of the mixture at STP. We calculate the expanded volume with the combined gas law.

$$V_{final} = 1.00 \text{ L} \times \frac{(25 + 273.2) \text{ K}}{273.2 \text{ K}} = 1.09 \text{ L} \qquad \text{final density} = \frac{0.44 \text{ g}}{1.09 \text{ L}} = 0.40 \text{ g/L}$$

We determine the apparent molar masses of each mixture by multiplying the mole fraction (numerically equal to the volume fraction) of each gas by its molar mass, and then summing these products for all gases in the mixture.

$$M_{air} = \left(0.78084 \times \frac{28.01 \text{ g } N_2}{1 \text{ mol } N_2}\right) + \left(0.20946 \times \frac{32.00 \text{ g } O_2}{1 \text{ mol } O_2}\right) + \left(0.00934 \times \frac{39.95 \text{ g Ar}}{1 \text{ mol Ar}}\right)$$

$$= 21.87 \text{ g } N_2 + 6.703 \text{ g } O_2 + 0.373 \text{ g Ar} = 28.95 \text{ g/mol air}$$

$$M_{mix} = \left(0.79 \times \frac{4.003 \text{ g He}}{1 \text{ mol He}}\right) + \left(0.21 \times \frac{32.00 \text{ g } O_2}{1 \text{ mol } O_2}\right) = 3.2 \text{ g He} + 6.7 \text{ g } O_2 = \frac{9.9 \text{ g mixture}}{\text{mol}}$$

In order to prepare two gases with the same density, the volume of the gas of smaller molar mass must be smaller by a factor equal to the ratio of the molar masses. According to Boyle's law, this means that the pressure on the less dense gas must be larger by a factor equal to a ratio of molar masses.

$$P_{mix} = \frac{28.95}{9.9} \times 1.00 \text{ atm} = 2.9 \text{ atm}$$

99. To convert from volume percent (numerically the same as mole percent) to mass percent, each volume fraction is multiplied by the molar mass of that component and then divided by the apparent molar mass of the gas. This means that substances with higher molar masses (such as O_2, 32.00 g/mol) will have higher mass percents than volume percents, compared to substances with lower molar masses (such as N_2, 28.01 g/mol).

100. First compute the pressure of the water vapor at 30.1°C

$$P_{H_2O} = \frac{\left(0.1052 \text{ g} \times \dfrac{1 \text{ mol H}_2\text{O}}{18.015 \text{ g H}_2\text{O}}\right) \times \dfrac{0.08206 \text{ L atm}}{\text{mol K}} \times (30.1 + 273.2) \text{ K}}{8.050 \text{ L}} \times \frac{760 \text{ mmHg}}{1 \text{ atm}} = 13.72 \text{ mmHg}$$

The water vapor is kept in the 8.050-L container, which means that its pressure is proportional to the absolute temperature in the container. Thus, for each of the six temperatures, we need to calculate two numbers: (1) the pressure due to this water (because gas pressure varies with temperature), and (2) 80% of the vapor pressure. The temperature we are seeking is where the two numbers agree.

$$P_{water}(T) = 13.72 \text{ mmHg} \times \frac{(T + 273.2) \text{ K}}{(30.1 + 273.2) \text{ K}}$$

For example, $P(20°C) = 13.72 \text{ mmHg} \times \dfrac{(20. + 273.2) \text{ K}}{(30.1 + 273.2) \text{ K}} = 13.3 \text{ mmHg}$

T	20. °C	19. °C	18. °C	17. °C	16. °C	15. °C
P_{water}, mmHg	13.3	13.2	13.2	13.1	13.1	13.0
80.0% v.p., mmHg	14.0	13.2	12.4	11.6	10.9	10.2

At approximately 19°C, the relative humidity of the air will be 80.0%.

101.

P_{gas} (mmHg)	$1/A \propto 1/V$ (cm^{-1})
810.8	0.0358
739.8	0.0328
896.8	0.0394
996.8	0.0437
1123	0.0493
1278	0.0562
1494	0.0658
1796	0.0787
2216	0.0980
2986	0.1316

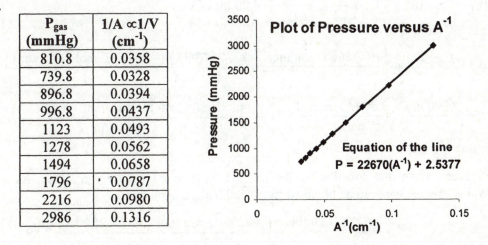

Factors that would affect the slope of this straight line are related to deviations real gases exhibit from ideality. At higher pressures, real gases tend to interact more, exerting forces of attraction and repulsion that Boyle's Law does not take into account.

102. (a) $T = 10\,°C = 283\,K$ $\qquad$ $M = 28.96\,g\,mol^{-1}$ (from question 99) or $0.02896\,kg\,mol^{-1}$

$R = 8.314472\,\dfrac{kg\,m}{s^2\,K\,mol}$ $\quad$ $g = 9.80665\,\dfrac{m}{s^2}$ $\quad$ $h = 14494\,ft \times \dfrac{12\,in}{1\,ft} \times \dfrac{2.54\,cm}{1\,in} \times \dfrac{1\,m}{100\,cm} = 4417.8\,m$

$$P = P_o \times 10^{\frac{-Mgh}{2.303\,RT}} = 760\,mmHg \times 10^{\dfrac{-\left(0.02896\frac{kg}{mol}\right)\left(9.80665\frac{m}{s^2}\right)4417.8\,m}{2.303\left(8.314472\frac{kg\,m}{s^2\,K\,mol}\right)283\,K}} = 760 \times 10^{-0.2314} = 446\,mmHg$$

(b) $h = 900\,ft \times \dfrac{12\,in}{1\,ft} \times \dfrac{2.54\,cm}{1\,in} \times \dfrac{1\,m}{100\,cm} = 274.3\,m$

$$P = P_o \times 10^{\dfrac{-\left(0.02896\frac{kg}{mol}\right)\left(9.80665\frac{m}{s^2}\right)274.3\,m}{2.303\left(8.314472\frac{kg\,m}{s^2\,K\,mol}\right)283\,K}} = 0.967455\,P_o = \frac{29}{30}P_o \quad or \quad \frac{1}{30}\,smaller$$

103. $\quad$ **(a)** $\quad$ First multiply out the left-hand side of the equation.

$$\left(P + \frac{an^2}{V^2}\right)(V - nb) = nRT = PV - Pnb + \frac{an^2}{V} - \frac{abn^3}{V^2}$$

Now multiply the entire equation through by V^2, and collect all terms on the right-hand side.

$$0 = -nRTV^2 + PV^3 - PnbV^2 + an^2V - abn^3$$

Finally, divide the entire equation by P and collect terms with the same power of V, to obtain:

$$0 = V^3 - n\left(\frac{RT + bP}{P}\right)V^2 + \left(\frac{n^2a}{P}\right)V - \frac{n^3ab}{P} = 0$$

(b) $\quad n = 185\,g\,CO_2 \times \dfrac{1\,mol\,CO_2}{44.0\,g\,CO_2} = 4.20\,mol\,CO_2$

$$0 = V^3 - 4.20\,mol\left(\frac{(0.08206\,L\,atm\,mol^{-1}\,K^{-1} \times 286\,K) + (0.0427\,L/mol \times 12.5\,atm)}{12.5\,atm}\right)V^2$$

$$+ \left(\frac{(4.20\,mol)^2\,3.59\,L^2\,atm\,mol^{-2}}{12.5\,atm}\right)V - \frac{(4.20\,mol)^3\,3.59\,L^2\,atm\,\,mol^{-2} \times 0.0427\,L/mol}{12.5\,atm}$$

$$= V^3 - 8.06\,V^2 + 5.07\,V - 0.909$$

We can solve this equation by the method of successive approximations. As a first value, we use the volume obtained from the ideal gas equation:

$$V = \frac{nRT}{P} = \frac{4.20\,mol \times 0.08206\,L\,atm\,mol^{-1}\,K^{-1} \times 286\,K}{12.5\,atm} = 7.89\,L$$

	$V = 7.89\,L$	$(7.89)3 - 8.06\,(7.89)2 + (5.07 \times 7.89) - 0.909 = 28.5$	> 0
Try	$V = 7.00\,L$	$(7.00)3 - 8.06\,(7.00)2 + (5.07 \times 7.00) - 0.909 = -17.4$	< 0
Try	$V = 7.40\,L$	$(7.40)3 - 8.06\,(7.40)2 + (5.07 \times 7.40) - 0.909 = 0.47$	> 0

Try $V = 7.38$ L $(7.38)3 - 8.06 (7.38)2 + (5.07 \times 7.38) - 0.909 = -0.53$ <0
Try $V = 7.39$ L $(7.39)3 - 8.06 (7.39)2 + (5.07 \times 7.39) - 0.909 = -0.03$ < 0

The volume of CO_2 is very close to 7.39 L.

A second way is to simply disregard the last term and to solve the equation

$0 = V^3 - 8.06\,V^2 + 5.07\,V$

This equation simplifies to the following quadratic equation:

$0 = V^2 - 8.06\,V + 5.07$, which is solved with the quadratic formula.

$$V = \frac{8.06 \pm \sqrt{(8.06)^2 - 4 \times 5.07}}{2} = \frac{+8.06 \pm 6.68}{2} = \frac{14.74}{2} = 7.37 \text{ L}$$

The other root, 0.69 L does not appear to be reasonable, due to its small size.

104. (a) $P\overline{V} = RT\left\{1 + \dfrac{B}{\overline{V}} + \dfrac{C}{\overline{V}^2}\right\}$ $\overline{V} = 500 \text{ cm}^3\text{mol}^{-1}$ $B = -21.89 \text{ cm}^3 \text{ mol}^{-1}$
   $T = 273$ K $C = 1230 \text{ cm}^6 \text{ mol}^{-2}$

$$P\left(0.500\ \frac{L}{mol}\right) = \left(0.08206\ \frac{L\ atm}{K\ mol}\right)273\ K\left\{1 + \frac{-21.89\ cm^3\ mol^{-1}}{500\ cm^3 mol^{-1}} + \frac{1230\ cm^6\ mol^{-2}}{\left(500\ cm^3 mol^{-1}\right)^2}\right\}$$

$$P\left(0.500\ \frac{L}{mol}\right) = 22.40\,\frac{L\ atm}{mol}\{0.961\}$$ $$P = \frac{22.4\,\dfrac{L\ atm}{mol}\{0.961\}}{\left(0.500\ \dfrac{L}{mol}\right)} = 43.1 \text{ atm}$$

(b) The result is consistent with Figure 6-22. In Figure 6-22, the compressibility factor is slightly below 1 at pressures between 0 and 400 atm. At a pressure of ~50 atm, the compressibility factor is just slightly below 1. Using the data provided, a pressure of 43.1 atm has a compressibility factor of 0.961, which is slightly below 1. The data is consistent with the information provided in Fig 6-22.

105. $n_{H_2} = n_{H_2}$ (from Al) + n_{H_2} (from Mg) Let x = mass of Al in g, therefore $0.156 - x$ = mass Mg in grams

$$n_{H_2} = 7.99 \times 10^{-3} \text{ mol } H_2 = \left(x \times \frac{1 \text{ mol Al}}{26.98 \text{ g Al}} \times \frac{3 \text{ mol } H_2}{2 \text{ mol Al}}\right) + \left((0.156 - x) \times \frac{1 \text{ mol Mg}}{24.305 \text{ g Mg}} \times \frac{1 \text{ mol } H_2}{1 \text{ mol Mg}}\right)$$

$7.99 \times 10^{-3} \text{ mol } H_2 = 0.0556x + 0.006418 - 0.04114x$

$0.001572 = 0.01446x$ $x = 0.108\underline{7}$ g Al

$\therefore$ Mass % Al = $\dfrac{0.108\underline{7} \text{ g Al}}{0.156 \text{ g mixture}} \times 100\% = 69.7 \%$ Al

$\therefore$ Mass % Mg = $100\% - 69.7\% = 30.3\%$ Mg

FEATURE PROBLEMS

106. Boyle's Law relates P and V, i.e. $P \times V = $ constant. If V is proportional to the value of A, and $P_{gas} = P_{bar} + P_{Hg}$ (i.e. the pressure of the gas equals the sum of the barometric pressure and the pressure exerted by the mercury column), then a comparison of individual $A \times P$ products should show a consistent result or a constant.

A (cm)	P_{bar} (atm)	P_{Hg} (mmHg)	P_{gas} (mmHg)	$A \times P_{gas}$ (cm × mmHg)
27.9	739.8	71	810.8	22621
30.5	739.8	0	739.8	22564
25.4	739.8	157	896.8	22779
22.9	739.8	257	996.8	22827
20.3	739.8	383	1123	22793
17.8	739.8	538	1278	22745
15.2	739.8	754	1494	22706
12.7	739.8	1056	1796	22807
10.2	739.8	1476	2216	22601
7.6	739.8	2246	2986	22692

Since consistent $A \times P_{gas}$ results are observed, that these data conform reasonably well (within experimental uncertainty) to Boyle's Law.

107. Nitryl Fluoride $\quad 65.01 \text{ u} \left(\dfrac{49.4}{100}\right) = 32.1 \text{ u of X}$

Nitrosyl Fluoride $\quad 49.01 \text{ u} \left(\dfrac{32.7}{100}\right) = 16.0 \text{ u of X}$

Thionyl Fluoride $\quad 86.07 \text{ u} \left(\dfrac{18.6}{100}\right) = 16.0 \text{ u of X}$

Sulfuryl Fluoride $\quad 102.07 \text{ u} \left(\dfrac{31.4}{100}\right) = 32.0 \text{ u of X}$

The atomic mass of X is 16 u which corresponds to the element oxygen.
The number of atoms of X (oxygen) in each compound is given below:
Nitryl Fluoride $\quad$ = 2 atoms of O $\qquad$ Nitrosyl Fluoride $\quad$ = 1 atom of O
Thionyl Fluoride $\quad$ = 1 atom of O $\qquad$ Sulfuryl Fluoride $\quad$ = 2 atoms of O

108. (a) The $N_2(g)$ extracted from liquid air has some $Ar(g)$ mixed in. Only $O_2(g)$ was removed from liquid air in the oxygen-related experiments.

(b) Because of the presence of $Ar(g)$ [39.95 g/mol], the $N_2(g)$ [28.01 g/mol] from liquid air will have a greater density than $N_2(g)$ from nitrogen compounds.

(c) Magnesium will react with molecular nitrogen $\left[3\,Mg(s) + N_2(g) \rightarrow Mg_3N_2(s) \right]$ but not with Ar. Thus, magnesium reacts with all the nitrogen in the mixture, but leaves the relatively inert $Ar(g)$ unreacted.

(d) The "nitrogen" remaining after oxygen is extracted from each mole of air (Rayleigh's mixture) contains $0.78084 + 0.00934 = 0.79018$ mol and has the mass calculated below.

mass of gaseous mixture $= (0.78084 \times 28.013 \text{ g/mol } N_2) + (0.00934 \times 39.948 \text{ g/mol Ar})$

mass of gaseous mixture $= 21.874 \text{ g } N_2 + 0.373 \text{ g Ar} = 22.247 \text{ g mixture}$.

Then, the molar mass of the mixture can be computed: 22.247 g mixture / 0.79018 mol = 28.154 g/mol. Since the STP molar volume of an ideal gas is 22.414 L, we can compute the two densities.

$$d(N_2) = \frac{28.013 \text{ g/mol}}{22.414 \text{ L/mol}} = 1.2498 \text{ g/mol} \quad d(\text{mixture}) = \frac{28.154 \text{ g/mol}}{22.414 \text{ L/mol}} = 1.2561 \text{ g/mol}$$

These densities differ by 0.50%.

109. (a) First convert pressures from mmHg to atm:

density (g/L)	pressure (atm)	density/pressure (g/L·atm)
1.428962	1.0000	$1.428962 \cong 1.4290$
1.071485	0.75000	$1.428647 \cong 1.4286$
0.714154	0.50000	$1.428308 \cong 1.4283$
0.356985	0.25000	$1.42794 \cong 1.4279$

average = 1.4285 g/L·atm

(b) $M_{O_2} = \dfrac{d}{P} RT$ $M_{O_2} = 1.4285 \text{ g/L} \cdot \text{atm} \times 0.082057 \text{ L} \cdot \text{atm/mol} \cdot \text{K} \times 273.15 \text{ K}$

$M_{O_2} = 32.0182 \text{ g/mol}$ Thus, the atomic mass of $O_2 = M_{O_2}/2 = 16.0009$.

This compares favorably with the value of 15.9994 given in the front of the textbook.

110. Total mass = mass of payload + mass of balloon + mass of H_2

Use ideal gas law to calculate mass of H_2: $PV = nRT = \dfrac{\text{mass}(m)}{\text{Molar mass}(M)} RT$

$$m = \frac{PVM}{RT} = \frac{1\,\text{atm}\left(120\text{ft}^3 \times \dfrac{(12\text{in})^3}{(1\text{ft})^3} \times \dfrac{(2.54\text{cm})^3}{(1\text{in})^3} \times \dfrac{1\times10^{-3}\text{L}}{1\text{cm}^3}\right)2.016\dfrac{\text{g}}{\text{mol}}}{\left(0.08206\dfrac{\text{L atm}}{\text{K mol}}\right)273\text{K}} = 306\text{ g}$$

Total mass = 1200 g + 1700 g + 306 g ≈ 3200 g

We know at the maximum height, the balloon will be 25 ft in diameter. Need to find out what mass of air is displaced. We need to make one assumption – the volume percent of air is unchanged with altitude. Hence we use an apparent molar mass for air of 29 g mol^{-1} (question 99). Using the data provided, we find the altitude at which the balloon displaces 3200 g of air.

Note: balloon radius $= \dfrac{25}{2} = 12.5$ ft. Volume $= \dfrac{4}{3}(\pi)r^3 = \dfrac{4}{3}(3.1416)(12.5)^3 = 8181$ ft^3

Convert to liters: $8181 \text{ft}^3 \times \dfrac{(12 \text{in})^3}{(1 \text{ft})^3} \times \dfrac{(2.54 \text{cm})^3}{(1 \text{in})^3} \times \dfrac{1 \times 10^{-3} \text{L}}{1 \text{cm}^3} = 231,660 \text{L}$

At 10 km: $m = \dfrac{PVM}{RT} = \dfrac{\left(2.7 \times 10^2 \text{mb} \times \dfrac{1 \text{atm}}{1013.25 \text{mb}}\right)(231,660 \text{L})\left(29 \dfrac{\text{g}}{\text{mol}}\right)}{\left(0.08206 \dfrac{\text{L atm}}{\text{K mol}}\right)223\text{K}} = 97,\underline{827} \text{ g}$

At 20 km: $m = \dfrac{PVM}{RT} = \dfrac{\left(5.5 \times 10^1 \text{mb} \times \dfrac{1 \text{atm}}{1013.25 \text{mb}}\right)(231,660 \text{L})\left(29 \dfrac{\text{g}}{\text{mol}}\right)}{\left(0.08206 \dfrac{\text{L atm}}{\text{K mol}}\right)217\text{K}} = 20,\underline{478} \text{ g}$

At 30 km: $m = \dfrac{PVM}{RT} = \dfrac{\left(1.2 \times 10^1 \text{mb} \times \dfrac{1 \text{atm}}{1013.25 \text{mb}}\right)(231,660 \text{L})\left(29 \dfrac{\text{g}}{\text{mol}}\right)}{\left(0.08206 \dfrac{\text{L atm}}{\text{K mol}}\right)230\text{K}} = 4,2\underline{12} \text{ g}$

At 40 km: $m = \dfrac{PVM}{RT} = \dfrac{\left(2.9 \times 10^0 \text{mb} \times \dfrac{1 \text{atm}}{1013.25 \text{mb}}\right)(231,660 \text{L})\left(29 \dfrac{\text{g}}{\text{mol}}\right)}{\left(0.08206 \dfrac{\text{L atm}}{\text{K mol}}\right)250\text{K}} = 93\underline{7} \text{ g}$

The lifting power of the balloon will allow it to rise to an altitude of just over 30 km.

CHAPTER 7
THERMOCHEMISTRY
PRACTICE EXAMPLES

1A The heat absorbed is the product of the mass of water, its specific heat $(4.18 \text{ J g}^{-1} \text{°C}^{-1})$, and the temperature change that occurs.

$$\text{heat energy} = 237 \text{ g} \times \frac{4.18 \text{ J}}{\text{g °C}} \times (37.0\text{°C} - 4.0\text{°C}) \times \frac{1 \text{ kJ}}{1000 \text{ J}} = 32.7 \text{ kJ of heat energy}$$

1B The heat absorbed is the product of the amount of mercury, its molar heat capacity, and the temperature change that occurs.

$$\text{heat energy} = \left(2.50 \text{ kg} \times \frac{1000 \text{ g}}{1 \text{ kg}} \times \frac{1 \text{ mol Hg}}{200.59 \text{ g Hg}}\right) \times \frac{28.0 \text{ J}}{\text{mol °C}} \times [-6.0 - (-20.0)]\text{°C} \times \frac{1 \text{ kJ}}{1000 \text{ J}}$$

$$= 4.89 \text{ kJ of heat energy}$$

2A First calculate the quantity of heat lost by the lead. This heat energy must be absorbed by the surroundings (water). We assume 100% efficiency in the energy transfer.

$$q_{\text{lead}} = 1.00 \text{ kg} \times \frac{1000 \text{ g}}{1 \text{ kg}} \times \frac{0.13 \text{ J}}{\text{g °C}} \times (35.2\text{°C} - 100.0\text{°C}) = -8.4 \times 10^3 \text{ J} = -q_{\text{water}}$$

$$8.4 \times 10^3 \text{ J} = m_{\text{water}} \times \frac{4.18 \text{ J}}{\text{g °C}} \times (35.2\text{°C} - 28.5\text{°C}) = 28 m_{\text{water}} \qquad m_{\text{water}} = \frac{8.4 \times 10^3 \text{ J}}{28 \text{ J g}^{-1}} = 3.0 \times 10^2 \text{ g}$$

2B We use the same equation, equating the heat lost by the copper to the heat absorbed by the water, except now we solve for final temperature.

$$q_{\text{Cu}} = 100.0 \text{ g} \times \frac{0.385 \text{ J}}{\text{g °C}} \times (x\text{°C} - 100.0\text{°C}) = -50.0 \text{ g} \times \frac{4.18 \text{ J}}{\text{g °C}} \times (x\text{°C} - 26.5\text{°C}) = -q_{\text{water}}$$

$$38.5x - 3850 = -209x + 5539 \text{ J} \qquad 38.5x + 209x = 5539 + 3850 \rightarrow 247.5x = 9389$$

$$x = \frac{9389 \text{ J}}{247.5 \text{ J °C}^{-1}} = 37.9\text{°C}$$

3A The molar mass of $C_8H_8O_3$ is 152.15 g/mol. The calorimeter has a heat capacity of 4.90 kJ / °C

$$q_{\text{calor}} = \frac{4.90 \text{ kJ °C}^{-1} \times (30.09\text{°C} - 24.89\text{°C})}{1.013 \text{ g}} \times \frac{152.15 \text{ g}}{1 \text{ mol}} = 3.83 \times 10^3 \text{ kJ / mol}$$

$$\Delta H_{\text{comb}} = -q_{\text{calor}} = -3.83 \times 10^3 \text{ kJ / mol}$$

3B The heat that is liberated by the benzoic acid's combustion serves to raise the temperature of the assembly. We designate the calorimeter's heat capacity by C.

$$q_{rxn} = 1.176 \text{ g} \times \frac{-26.42 \text{ kJ}}{1 \text{ g}} = -31.07 \text{ kJ} = -q_{calorim}$$

$$q_{calorim} = C\Delta t = 31.07 \text{ kJ} = C \times 4.96°C \qquad C = \frac{31.07 \text{ kJ}}{4.96°C} = 6.26 \text{ kJ} /°C$$

4A The heat that is liberated by the reaction raises the temperature of the reaction mixture. We assume that this reaction mixture has the same density and specific heat as pure water.

$$q_{calorim} = \left(200.0 \text{ mL} \times \frac{1.00 \text{ g}}{1 \text{ mL}} \right) \times \frac{4.18 \text{ J}}{\text{g}°C} \times (30.2 - 22.4)°C = 6.5 \times 10^3 \text{ J} = -q_{rxn}$$

Owing to the 1:1 stoichiometry of the reaction, the number of moles of AgCl(s) formed is equal to the number of moles of AgNO$_3$(aq) in the original sample.

$$\text{moles AgCl} = 100.0 \text{ mL} \times \frac{1 \text{ L}}{1000 \text{ mL}} \times \frac{1.00 \text{ M AgNO}_3}{1 \text{ L}} \times \frac{1 \text{ mol AgCl}}{1 \text{ mol AgNO}_3} = 0.100 \text{ mol AgCl}$$

$$q_{rxn} = \frac{-6.5 \times 10^3 \text{ J}}{0.100 \text{ mol}} \times \frac{1 \text{ kJ}}{1000 \text{ J}} = -65. \text{ kJ/mol}$$

Because q_{rxn} is a negative quantity, the precipitation reaction is exothermic.

4B The assumptions include no heat loss to the surroundings or to the calorimeter, a solution density of 1.00 g/mL, a specific heat of $4.18 \text{ J g}^{-1}°C^{-1}$, and that the initial and final solution volumes are the same. The equation for the reaction that occurs is $NaOH(aq) + HCl(aq) \rightarrow NaCl(aq) + H_2O(l)$. Since the two reactants combine in a one to one mole ratio, the limiting reactant is the one present in smaller amount (I.e. the one with a smaller molar quantity).

$$\text{amount HCl} = 100.0 \text{ mL} \times \frac{1.020 \text{ mmol HCl}}{1 \text{ mL soln}} = 102.0 \text{ mmol HCl}$$

$$\text{amount NaOH} = 50.0 \text{ mL} \times \frac{1.988 \text{ mmol NaOH}}{1 \text{ mL soln}} = 99.4 \text{ mmol NaOH}$$

Thus, NaOH is the limiting reactant.

$$q_{neutr} = 99.4 \text{ mmol NaOH} \times \frac{1 \text{ mmol H}_2\text{O}}{1 \text{ mmol NaOH}} \times \frac{1 \text{ mol H}_2\text{O}}{1000 \text{ mmol H}_2\text{O}} \times \frac{-56 \text{ kJ}}{1 \text{ mol H}_2\text{O}} = -5.5_7 \text{ kJ}$$

$$q_{calorim} = -q_{neutr} = 5.5_7 \text{ kJ} = (100.0 + 50.0) \text{ mL} \times \frac{1.00 \text{ g}}{1 \text{ mL}} \times \frac{4.18 \text{ J}}{\text{g}°C} \times \frac{1 \text{ kJ}}{1000 \text{ J}} \times (t - 24.52°C)$$

$$= 0.627t - 15.3_7 \qquad t = \frac{5.5_7 + 15.3_7}{0.627} = 33.4°C$$

5A $\quad w = -P\Delta V = -0.750 \text{ atm}(+1.50 \text{ L}) = -1.12\underline{5} \text{ L atm} \times \frac{101.33 \text{ J}}{1 \text{ L atm}} = -114 \text{ J}$

114 J of work is done by system

5B Determine the initial number of moles:

$$n = 50.0 \text{ g N}_2 \times \frac{1 \text{ mol N}_2}{28.014 \text{ g N}_2} = 1.78\underline{5} \text{ moles of N}_2$$

$$V = \frac{nRT}{P} = \frac{(1.785 \text{ mol N}_2)(0.08206 \text{ Latm K}^{-1}\text{mol}^{-1})(293.15 \text{ K})}{2.50 \text{ atm}} = 17.2 \text{ L}$$

$$\Delta V = 17.2 - 75.0 \text{ L} = -57.8 \text{ L}$$

$$w = -P\Delta V = -2.50 \text{ atm}(-57.8 \text{ L}) \times \frac{101.33 \text{ J}}{1 \text{ L atm}} \times \frac{1 \text{ kJ}}{1000 \text{ J}} = +14.6 \text{ kJ} \text{ work done on system.}$$

6A The work is $w = +355$ J. The heat flow is $q = -185$ J. These two are related to the energy change of the system by the first law equation: $\Delta U = q + w$, which becomes

$$\Delta U = +355 \text{ J} - 185 \text{ J} = +1.70 \times 10^2 \text{ J}$$

6B The internal energy change is $\Delta U = -125$ J. The heat flow is $q = +54$ J. These two are related to the work done on the system by the first law equation: $\Delta U = q + w$, which becomes $-125 \text{ J} = +54 \text{ J} + w$. The solution to this equation is $w = -125 \text{ J} - 54 \text{ J} = -179 \text{ J}$, which means that 179 J of work is done by the system to the surroundings.

7A Heat that is given off has a negative sign. In addition, we use the molar mass of sucrose, 342.30 g/mol.

$$\text{sucrose mass} = -1.00 \times 10^3 \text{ kJ} \times \frac{1 \text{ mol C}_{12}\text{H}_{22}\text{O}_{11}}{-5.65 \times 10^3 \text{ kJ}} \times \frac{342.30 \text{ g C}_{12}\text{H}_{22}\text{O}_{11}}{1 \text{ mol C}_{12}\text{H}_{22}\text{O}_{11}} = 60.6 \text{ g C}_{12}\text{H}_{22}\text{O}_{11}$$

7B Although the equation does not say so explicitly, the reaction of $\text{H}^+(aq) + \text{OH}^-(aq) \rightarrow \text{H}_2\text{O}(l)$ gives off 56 kJ of heat per mole of water formed. The equation then is the source of a conversion factor.

$$\text{heat flow} = 25.0 \text{ mL} \times \frac{1 \text{ L}}{1000 \text{ mL}} \times \frac{0.1045 \text{ mol HCl}}{1 \text{ L soln}} \times \frac{1 \text{ mol H}_2\text{O}}{1 \text{ mol HCl}} \times \frac{56 \text{ kJ evolved}}{1 \text{ mol H}_2\text{O}}$$

$$\text{heat flow} = 0.15 \text{ kJ heat evolved}$$

8A $V_{ice} = (2.00 \text{ cm})^3 = 8.00 \text{ cm}^3$

$m_{ice} = m_{water} = 8.00 \text{ cm}^3 \times 0.917 \text{ g cm}^{-3} = 7.34 \text{ g ice} = 7.34 \text{ g H}_2\text{O}$

$$\text{moles of ice} = 7.34 \text{ g ice} \times \frac{1 \text{ mol H}_2\text{O}}{18.015 \text{ g H}_2\text{O}} = 0.407 \text{ moles of ice}$$

$q_{overall} = q_{ice}(-10 \text{ to } 0 \text{ °C}) + q_{fus} + q_{water}(0 \text{ to } 23.2 \text{ °C})$

$q_{overall} = m_{ice}(\text{sp. ht.})_{ice}\Delta T + n_{ice}\Delta H_{fus} + m_{water}(\text{sp. ht.})_{water}\Delta T$

$q_{overall} = 7.34 \text{ g}(10.0 \text{ °C})(2.01 \frac{\text{J}}{\text{g °C}}) + 0.407 \text{ mol ice}(6.01 \frac{\text{kJ}}{\text{mol}}) + 7.34 \text{ g}(23.2 \text{ °C})(4.184 \frac{\text{J}}{\text{g °C}})$

$q_{overall} = 0.148 \text{ kJ} + 2.45 \text{ kJ} + 0.712 \text{ kJ}$

$q_{overall} = +3.31 \text{ kJ (the system absorbs this much heat)}$

8B $5.00 \times 10^3 \text{ kJ} = q_{ice}(-15 \text{ to } 0 \text{ °C}) + q_{fus} + q_{water} (0 \text{ to } 25 \text{ °C}) + q_{vap}$

$5.00 \times 10^3 \text{ kJ} = m_{ice}(\text{sp. ht.})_{ice}\Delta T + n_{ice}\Delta H_{fus} + m_{water}(\text{sp. ht.})_{water}\Delta T + n_{water}\Delta H_{vap}$

$5.00 \times 10^6 \text{ J} = m(15.0 \text{ °C})(2.01 \dfrac{\text{J}}{\text{g °C}}) + (\dfrac{m}{18.015 \text{ g H}_2\text{O/mol H}_2\text{O}} \times 6.01 \times 10^3 \dfrac{\text{J}}{\text{mol}})$

$\qquad + m(25.0 \text{ °C})(4.184 \dfrac{\text{J}}{\text{g °C}}) + \dfrac{m}{18.015 \text{ g H}_2\text{O/mol}}(44.0 \times 10^3 \dfrac{\text{J}}{\text{mol}})$

$5.00 \times 10^6 \text{ J} = m(30.1\underline{5} \text{ J/g}) + m(333.\underline{6} \text{ J/g}) + m(104.\underline{5} \text{ J/g}) + m(2.4\underline{4} \times 10^3 \text{ J/g})$

$5.00 \times 10^6 \text{ J} = m(2.91 \times 10^3 \text{ J/g}) \qquad m = \dfrac{5.00 \times 10^6 \text{ J}}{2.91 \times 10^3 \text{ J/g}} = 1718 \text{ g or } 1.72 \text{ kg H}_2\text{O}$

9A We combine the three combustion reactions to produce the hydrogenation reaction.

$C_3H_6(g) + \tfrac{9}{2}O_2(g) \rightarrow 3CO_2(g) + 3H_2O(l) \qquad \Delta H_{comb} = \Delta H_1 = -2058 \text{ kJ}$

$H_2(g) + \tfrac{1}{2}O_2(g) \rightarrow H_2O(l) \qquad\qquad\qquad \Delta H_{comb} = \Delta H_2 = -285.8 \text{ kJ}$

$3CO_2(g) + 4H_2O(l) \rightarrow C_3H_8(g) + 5O_2(g) \qquad -\Delta H_{comb} = \Delta H_3 = +2219.9 \text{ kJ}$

$\overline{C_3H_6(g) + H_2(g) \rightarrow C_3H_8(g) \qquad\qquad\qquad \Delta H_{rxn} = \Delta H_1 + \Delta H_2 + \Delta H_3 = -124 \text{ kJ}}$

9B The combustion reaction has propanol and $O_{2(g)}$ as reactants; the products are $CO_2(g)$ and $H_2O(l)$. Reverse the reaction given and combine it with the combustion reaction of $C_3H_6(g)$.

$C_3H_7OH(l) \rightarrow C_3H_6(g) + H_2O(l) \qquad\qquad \Delta H_1 = +52.3 \text{ kJ}$

$C_3H_6(g) + \tfrac{9}{2}O_2(g) \rightarrow 3CO_2(g) + 3H_2O(l) \qquad \Delta H_2 = -2058 \text{ kJ}$

$\overline{C_3H_7OH(l) + \tfrac{9}{2}O_2(g) \rightarrow 3CO_2(g) + 4H_2O(l) \quad \Delta H_{rxn} = \Delta H_1 + \Delta H_2 = -2006 \text{ kJ}}$

10A The enthalpy of formation is the enthalpy change for the reaction in which one mole of the product, $C_6H_{13}O_2N(s)$, is produced from appropriate amounts of the reference forms of the elements(in most cases, the most stable form of the elements).

$6 \text{ C}(\text{graphite}) + \tfrac{13}{2}H_2(g) + O_2(g) + \tfrac{1}{2}N_2(g) \rightarrow C_6H_{13}O_2N(s)$

10B The enthalpy of formation is the enthalpy change for the reaction in which one mole of the product, $NH_3(g)$, is produced from appropriate amounts of the reference forms of the elements, in this case from 0.5 mol $N_2(g)$ and 1.5 mol $H_2(g)$, that is, for the reaction:

$\tfrac{1}{2}N_2(g) + \tfrac{3}{2}H_2(g) \rightarrow NH_3(g)$

The specified reaction is twice the reverse of the formation reaction, and its enthalpy change is minus two times the enthalpy of formation of $NH_3(g)$:

$-2 \times (-46.11 \text{ kJ}) = +92.22 \text{ kJ}$

11A $\Delta H_{rxn}^{o}=2\times\Delta H_{f}^{o}\left[CO_{2}\left(g\right)\right]+3\times\Delta H_{f}^{o}\left[H_{2}O\left(l\right)\right]-\Delta H_{f}^{o}\left[CH_{3}CH_{2}OH\left(l\right)\right]-3\times\Delta H_{f}^{o}\left[O_{2}\left(g\right)\right]$

$=\left[2\times\left(-393.5\text{ kJ}\right)\right]+\left[3\times\left(-285.8\text{ kJ}\right)\right]-\left[-277.7\text{ kJ}\right]-\left[3\times0.00\text{ kJ}\right]=-1367\text{ kJ}$

11B We write the combustion reaction for each compound, and use that reaction to determine the compound's heat of combustion.

$C_{3}H_{8}(g)+5O_{2}(g)\rightarrow3CO_{2}(g)+4H_{2}O(l)$

$\Delta H_{combustion}^{o}=3\times\Delta H_{f}^{o}\left[CO_{2}\left(g\right)\right]+4\times\Delta H_{f}^{o}\left[H_{2}O\left(l\right)\right]-\Delta H_{f}^{o}\left[C_{3}H_{8}\left(g\right)\right]-5\times\Delta H_{f}^{o}\left[O_{2}\left(g\right)\right]$

$=\left[3\times\left(-393.5\text{ kJ}\right)\right]+\left[4\times\left(-285.8\text{ kJ}\right)\right]-\left[-103.8\text{ kJ}\right]-\left[5\times0.00\text{ kJ}\right]$

$=-1181\text{ kJ}-1143\text{ kJ}+103.8-0.00\text{ kJ}=-2220.\text{ kJ/mol }C_{3}H_{8}$

$C_{4}H_{10}(g)+\frac{13}{2}O_{2}(g)\rightarrow4CO_{2}(g)+5H_{2}O(l)$

$\Delta H_{combustion}^{o}=4\times\Delta H_{f}^{o}\left[CO_{2}\left(g\right)\right]+5\times\Delta H_{f}^{o}\left[H_{2}O\left(l\right)\right]-\Delta H_{f}^{o}\left[C_{4}H_{10}\left(g\right)\right]-6.5\times\Delta H_{f}^{o}\left[O_{2}\left(g\right)\right]$

$=\left[4\times\left(-393.5\text{ kJ}\right)\right]+\left[5\times\left(-285.8\text{ kJ}\right)\right]-\left[-125.6\right]-\left[6.5\times0.00\text{ kJ}\right]$

$=-1574\text{ kJ}-1429\text{ kJ}+125.6\text{ kJ}-0.00K\text{ kJ}=-2877\text{ kJ/mol }C_{4}H_{10}$

In 1.00 mole of the mixture there are 0.62 mol $C_{3}H_{8}(g)$ and 0.38 mol $C_{4}H_{10}(g)$.

$\text{heat of combustion}=\left(0.62\text{ mol }C_{3}H_{8}\times\frac{-2220.\text{ kJ}}{1\text{ mol }C_{3}H_{8}}\right)+\left(0.38\text{ mol }C_{4}H_{10}\times\frac{-2877\text{ kJ}}{1\text{ mol }C_{4}H_{10}}\right)$

$=-1.4\times10^{3}\text{ kJ}-1.1\times10^{3}\text{ kJ}=-2.5\times10^{3}\text{ kJ/mole of mixture}$

12A $6\text{ CO}_2(g)+6\text{ H}_2O(l)\rightarrow C_6H_{12}O_6(s)+6\text{ O}_2(g)$

$\Delta H^{o}{}_{rxn}=2803\text{ kJ}=\Sigma\Delta H_{f}^{o}\text{ products}-\Sigma\Delta H_{f}^{o}\text{ reactants}$

$2803\text{ kJ}=[1\text{ mol}(\Delta H_{f}^{o}[C_6H_{12}O_6(s)])+6\text{ mol}(0\frac{\text{kJ}}{\text{mol}})]-[6\text{ mol}(-393.5\frac{\text{kJ}}{\text{mol}})+6\text{ mol}(-285.8\frac{\text{kJ}}{\text{mol}})]$

$2803\text{ kJ}=\Delta H_{f}^{o}[C_6H_{12}O_6(s)]-[-4075.8\text{ kJ}]$. Thus, $\Delta H_{f}^{o}[C_6H_{12}O_6(s)]=-1273\text{ kJ/mol }C_6H_{12}O_6(s)$

12B $\Delta H^{o}{}_{comb}[CH_3OCH_3(g)]=-31.70\frac{\text{kJ}}{\text{g}}$ Molar Mass of $CH_3OCH_3=46.069$ g mol^{-1}

$\Delta H^{o}{}_{comb}[CH_3OCH_3(g)]=-31.70\frac{\text{kJ}}{\text{g}}\times46.069\frac{\text{g}}{\text{mol}}=-1460\frac{\text{kJ}}{\text{mol}}\text{kJ}=\Delta H^{o}{}_{rxn}$

$\Delta H^{o}{}_{rxn}=\Sigma\Delta H_{f}^{o}\text{ products}-\Sigma\Delta H_{f}^{o}\text{ reactants}$ Reaction: $CH_3OCH_3(g)+3\text{ O}_2(g)\rightarrow2\text{ CO}_2(g)+3\text{ H}_2O(l)$

$-1460\text{ kJ}=[2\text{ mol}(-393.5\frac{\text{kJ}}{\text{mol}})+3\text{ mol}(-285.8\frac{\text{kJ}}{\text{mol}})]-[1\text{ mol}(\Delta H_{f}^{o}[CH_3OCH_3(g)])+3\text{ mol}(0\frac{\text{kJ}}{\text{mol}})]$

$-1460\text{ kJ}=-1644.4\text{ kJ}-\Delta H_{f}^{o}[CH_3OCH_3(g)]$

Hence, $\Delta H_{f}^{o}[CH_3OCH_3(g)]=-184\text{ kJ/mol }CH_3OCH_3(g)$

13A The net ionic equation is: $Ag^+(aq) + I^-(aq) \rightarrow AgI(s)$ and we have the following:

$$\Delta H_{rxn}^\circ = \Delta H_f^\circ\left[AgI(s)\right] - \left[\Delta H_f^\circ\left[Ag^+(aq)\right] + \Delta H_f^\circ\left[I^-(aq)\right]\right]$$

$$= -61.84 \text{ kJ/mol} - \left[(+105.6 \text{ kJ/mol}) + (-55.19 \text{ kJ/mol})\right] = -112.3 \text{ kJ/mol AgI(s) formed}$$

13B $2\,Ag^+(aq) + CO_3^{2-}(aq) \rightarrow Ag_2CO_3(s)$

$\Delta H_{rxn}^\circ = -39.9 \text{ kJ} = \Sigma \Delta H_f^\circ \text{ products} - \Sigma \Delta H_f^\circ \text{ reactants} =$

$-39.9 \text{ kJ} = \Delta H_f^\circ[Ag_2CO_3(s)] - [2\text{ mol}(105.6\ \dfrac{\text{kJ}}{\text{mol}}) + 1\text{ mol}(-677.1\ \dfrac{\text{kJ}}{\text{mol}})]$

$-39.9 \text{ kJ} = \Delta H_f^\circ[Ag_2CO_3(s)] + 465.9 \text{ kJ}$

Hence, $\Delta H_f^\circ[Ag_2CO_3(s)] = -505.8 \text{ kJ/mol } Ag_2CO_3(s)$ formed.

EXERCISES

Heat Capacity (Specific Heat)

1. **(a)** $q = 9.25 \text{ L} \times \dfrac{1000 \text{ cm}^3}{1 \text{ L}} \times \dfrac{1.00 \text{ g}}{1 \text{ cm}^3} \times \dfrac{4.18 \text{ J}}{1 \text{ g}^\circ\text{C}} \times \dfrac{1 \text{ kJ}}{1000 \text{ J}}\left(29.4^\circ\text{C} - 22.0^\circ\text{C}\right) = +2.9 \times 10^2 \text{ kJ}$

(b) $q = 5.85 \text{ kg} \times \dfrac{1000 \text{ g}}{1 \text{ kg}} \times \dfrac{0.903 \text{ J}}{\text{g}^\circ\text{C}} \times (-33.5^\circ\text{C}) \times \dfrac{1 \text{ kJ}}{1000 \text{ J}} = -177 \text{ kJ}$

2. $\text{heat} = \text{mass} \times \text{sp ht} \times \Delta T$

(a) $\Delta T = \dfrac{+875 \text{ J}}{12.6 \text{ g} \times 4.18 \text{ J g}^{-1}\,^\circ\text{C}^{-1}} = +16.6^\circ\text{C}$ $\qquad T_f = T_i + \Delta T = 22.9^\circ\text{C} + 16.6^\circ\text{C} = 39.5^\circ\text{C}$

(b) $\Delta T = \dfrac{-1.05 \text{ kcal} \times \dfrac{1000 \text{ cal}}{1 \text{ kcal}}}{\left(1.59 \text{ kg} \times \dfrac{1000 \text{ g}}{1 \text{ kg}}\right) 0.032 \dfrac{\text{cal}}{\text{g}^\circ\text{C}}} = -21^\circ\text{C}$

3. Heat gained by the water = heat lost by the metal; $\text{heat} = \text{mass} \times \text{sp.ht.} \times \Delta T$

(a) $50.0 \text{ g} \times 4.18\dfrac{\text{J}}{\text{g}^\circ\text{C}}(38.9 - 22.0)^\circ\text{C} = 3.53 \times 10^3 \text{ J} = -150.0 \text{g} \times \text{sp.ht.} \times (38.9 - 100.0)^\circ\text{C}$

$\text{sp.ht.} = \dfrac{3.53 \times 10^3 \text{ J}}{150.0 \text{ g} \times 61.1^\circ\text{C}} = 0.385 \text{ J g}^{-1}\,^\circ\text{C}^{-1}$ for Zn

(b) $50.0 \text{ g} \times 4.18\dfrac{\text{J}}{\text{g}^\circ\text{C}}(28.8 - 22.0)^\circ\text{C} = 1.4 \times 10^3 \text{ J} = -150.0 \text{g} \times \text{sp.ht.} \times (28.8 - 100.0)^\circ\text{C}$

$\text{sp.ht.} = \dfrac{1.4 \times 10^3 \text{ J}}{150.0 \text{ g} \times 71.2^\circ\text{C}} = 0.13 \text{ J g}^{-1}\,^\circ\text{C}^{-1}$ for Pt

(c) $50.0 \text{ g} \times 4.18\dfrac{\text{J}}{\text{g}^\circ\text{C}}(52.7 - 22.0)^\circ\text{C} = 6.42 \times 10^3 \text{ J} = -150.0 \text{g} \times \text{sp.ht.} \times (52.7 - 100.0)^\circ\text{C}$

$\text{sp.ht.} = \dfrac{6.42 \times 10^3 \text{ J}}{150.0 \text{ g} \times 47.3^\circ\text{C}} = 0.905 \text{ J g}^{-1}\,^\circ\text{C}^{-1}$ for Al

4. $50.0 \text{ g} \times 4.18 \dfrac{\text{J}}{\text{g}^\circ\text{C}} (27.6 - 23.2)^\circ\text{C} = 9.2 \times 10^2 \text{ J} = -75.0 \text{ g} \times \text{sp.ht.} \times (27.6 - 80.0)^\circ\text{C}$

$\text{sp.ht.} = \dfrac{9.2 \times 10^2 \text{ J}}{75.0 \text{ g} \times 52.4^\circ\text{C}} = 0.23 \text{ J g}^{-1} \, ^\circ\text{C}^{-1} \text{ for Ag}$

5. $q_{\text{water}} = 375 \text{ g} \times 4.18 \dfrac{\text{J}}{\text{g}^\circ\text{C}} (87 - 26)^\circ\text{C} = 9.5\underline{6} \times 10^4 \text{ J} = -q_{\text{iron}}$

$q_{\text{iron}} = -9.5\underline{6} \times 10^4 \text{ J} = 465 \text{g} \times 0.449 \dfrac{\text{J}}{\text{g}^\circ\text{C}} (87 - T_i) = 1.81\underline{6} \times 10^4 \text{J} - 2.08\underline{8} \times 10^2 \, T_i$

$T_i = \dfrac{-9.5\underline{6} \times 10^4 - 1.81\underline{6} \times 10^4}{-2.08\underline{8} \times 10^2} = \dfrac{-11.3\underline{8} \times 10^4}{-2.08\underline{8} \times 10^2} = 5.4\underline{48} \times 10^2 \, ^\circ\text{C or } 545 \, ^\circ\text{C}$

The number of significant figures in the final answer is limited by the two significant figures for the given temperatures.

6. heat lost by steel = heat gained by water

$-m \times 0.50 \dfrac{\text{J}}{\text{g}^\circ\text{C}} (51.5 - 183)^\circ\text{C} = 66 \, m = 125 \text{ mL} \times \dfrac{1.00 \text{ g}}{1 \text{ mL}} \times 4.18 \dfrac{\text{J}}{\text{g}^\circ\text{C}} (51.5 - 23.2)^\circ\text{C}$

$66 \, m = 1.48 \times 10^4 \text{ J} \qquad m = \dfrac{1.48 \times 10^4}{66} = 2.2 \times 10^2 \text{ g stainless steel.}$

The precision of this method of determining mass is limited by the fact that some heat leaks out of the system. When we deal with temperatures far above (or far below) room temperature, this assumption becomes less and less valid. Furthermore, the precision of the method is limited to two significant figures by the specific heat of the steel. If the two specific heats were known more precisely, then the temperature difference would determine the final precision of the method. It is unlikely that we could readily measure temperatures more precisely than $\pm 0.01^\circ\text{C}$, without expensive equipment. The mass of steel in this case would be measurable to four significant figures, to $\pm 0.1 \text{ g}$. This is hardly comparable to modern analytical balances which typically measure such masses to $\pm 0.1 \text{ mg}$.

7. heat lost by Mg = heat gained by water

$-\left(1.00 \text{ kg Mg} \times \dfrac{1000 \text{ g}}{1 \text{ kg}}\right) 1.024 \dfrac{\text{J}}{\text{g}^\circ\text{C}} (T_f - 40.0^\circ\text{C}) = \left(1.00 \text{ L} \times \dfrac{1000 \text{ cm}^3}{1 \text{ L}} \times \dfrac{1.00 \text{ g}}{1 \text{ cm}^3}\right) 4.18 \dfrac{\text{J}}{\text{g}^\circ\text{C}} (T_f - 20.0^\circ\text{C})$

$-1.024 \times 10^3 \, T_f + 4.10 \times 10^4 = 4.18 \times 10^3 \, T_f - 8.36 \times 10^4$

$4.10 \times 10^4 + 8.36 \times 10^4 = (4.18 \times 10^3 + 1.024 \times 10^3) T_f \rightarrow 12.46 \times 10^4 = 5.20 \times 10^3 \, T_f$

$T_f = \dfrac{12.46 \times 10^4}{5.20 \times 10^3} = 24.0^\circ\text{C}$

8. heat gained by the water = heat lost by the brass

$150.0 \text{ g} \times 4.18 \dfrac{\text{J}}{\text{g}^\circ\text{C}} \times (T_f - 22.4^\circ\text{C}) = -\left(15.2 \text{ cm}^3 \times \dfrac{8.40 \text{ g}}{1 \text{ cm}^3}\right) 0.385 \dfrac{\text{J}}{\text{g}^\circ\text{C}} (T_f - 163^\circ\text{C})$

$6.27 \times 10^2 T_f - 1.40 \times 10^4 = -49.2 \, T_f + 8.01 \times 10^3; \quad T_f = \dfrac{1.40 \times 10^4 + 8.01 \times 10^3}{6.27 \times 10^2 + 49.2} = 32.6^\circ\text{C}$

9.   heat lost by copper = heat gained by glycerol

$$-74.8\ g\times\frac{0.385\ J}{g\ ^{\circ}C}\times\left(31.1\ ^{\circ}C-143.2\ ^{\circ}C\right)=165\ mL\times\frac{1.26\ g}{1\ mL}\times sp.ht.\times\left(31.1\ ^{\circ}C-24.8\ ^{\circ}C\right)$$

$$3.23\times10^{3}=1.3\times10^{3}\times(sp.ht.)\quad sp.ht.=\frac{3.23\times10^{3}}{1.3\times10^{3}}=2.5\ J\,g^{-1}\,^{\circ}C^{-1}$$

$$\text{molar heat capacity}=2.5\ J\,g^{-1}\,^{\circ}C^{-1}\times\frac{92.1\ g}{1\ mol\ C_{3}H_{8}O_{3}}=2.3\times10^{2}\ J\,mol^{-1}\,^{\circ}C^{-1}$$

10.   The additional water simply acts as a heat transfer medium. The essential relationship is
heat lost by iron = heat gained by water (of unknown mass)

$$-\left(1.23\ kg\times\frac{1000\ g}{1\ kg}\right)0.449\,\frac{J}{g\ ^{\circ}C}\left(25.6-68.5\right)\,^{\circ}C=x\ g\ H_{2}O\times4.18\,\frac{J}{g\ ^{\circ}C}\left(25.6-18.5\right)\,^{\circ}C$$

$$2.37\times10^{4}\,J=29.\underline{7}\,x\qquad x=\frac{2.37\times10^{4}}{29.\underline{7}}=79\underline{8}\ g\ H_{2}O\times\frac{1\ mL\ H_{2}O}{1.00\ g\ H_{2}O}=8.0\times10^{2}\ mL\ H_{2}O$$

Heats of reaction

11. $$\text{heat}=283\ kg\times\frac{1000\ g}{1\ kg}\times\frac{1\ mol\ Ca(OH)_{2}}{74.09\ g\ Ca(OH)_{2}}\times\frac{65.2\ kJ}{1\ mol\ Ca(OH)_{2}}=2.49\times10^{5}\ kJ\ \text{of heat evolved.}$$

12. $$\text{heat energy}=1.00\ gal\times\frac{3.785\ L}{1\ gal}\times\frac{1000\ mL}{1\ L}\times\frac{0.703\ g}{1\ mL}\times\frac{1\ mol\ C_{8}H_{18}}{114.2\ g\ C_{8}H_{18}}\times\frac{5.48\times10^{3}\ kJ}{1\ mol\ C_{8}H_{18}}$$

heat energy = 1.28×10^{5} kJ

13.   **(a)** $$\text{heat evolved}=1.325\ g\ C_{4}H_{10}\times\frac{1\ mol\ C_{4}H_{10}}{58.123\ g\ C_{4}H_{10}}\times\frac{2877\ kJ}{1\ mol\ C_{4}H_{10}}=65.59\ kJ\ ,$$

(b) $$\text{heat evolved}=28.4\ L_{STP}\ C_{4}H_{10}\times\frac{1\ mol\ C_{4}H_{10}}{22.414\ L_{STP}\ C_{4}H_{10}}\times\frac{2877\ kJ}{1\ mol\ C_{4}H_{10}}=3.65\times10^{3}\ kJ\ ,$$

(c)   Use the ideal gas equation to determine the amount of propane in moles and multiply
this amount by 2877 kJ heat produced per mole.

$$\text{heat evolved}=\frac{\left(738\ mmHg\times\dfrac{1\ atm}{760\ mmHg}\right)\times12.6\ L}{\dfrac{0.08206\ L\,atm}{mol\ K}\times(273.2+23.6)\ K}\times\frac{2877\ kJ}{1\ mol\ C_{4}H_{10}}=1.45\times10^{3}\ kJ\ ,$$

14.   **(a)** $$q=\frac{-29.4\ kJ}{0.584\ g\ C_{3}H_{8}}\times\frac{44.10\ g\ C_{3}H_{8}}{1\ mol\ C_{3}H_{8}}=-2.22\times10^{3}\ kJ\,/\,mol\ C_{3}H_{8}$$

(b) $$q=\frac{-5.27\ kJ}{0.136\ g\ C_{10}H_{16}O}\times\frac{152.24\ g\ C_{10}H_{16}O}{1\ mol\ C_{10}H_{16}O}=-5.90\times10^{3}\ kJ/mol\ C_{10}H_{16}O$$

(c) $$q=\frac{-58.3\ kJ}{2.35\ mL\ (CH_{3})_{2}CO}\times\frac{1\ mL}{0.791\ g}\times\frac{58.08\ g\,(CH_{3})_{2}CO}{1\ mol\ (CH_{3})_{2}CO}=-1.82\times10^{3}\ kJ/mol\,(CH_{3})_{2}CO$$

15. **(a)** $\text{mass} = 2.80 \times 10^7 \text{ kJ} \times \dfrac{1 \text{ mol CH}_4}{890.3 \text{ kJ}} \times \dfrac{16.04 \text{ g CH}_4}{1 \text{ mol CH}_4} \times \dfrac{1 \text{ kg}}{1000 \text{ g}} = 504 \text{ kg CH}_4.$

(b) First determine the moles of CH_4 present, with the ideal gas law.

$$\text{mol CH}_4 = \dfrac{\left(768 \text{ mmHg} \times \dfrac{1 \text{ atm}}{760 \text{ mmHg}}\right) 1.65 \times 10^4 \text{ L}}{0.08206 \dfrac{\text{L atm}}{\text{mol K}} \times (18.6 + 273.2) \text{ K}} = 696 \text{ mol CH}_4$$

$$\text{heat energy} = 696 \text{ mol CH}_4 \times \dfrac{-890.3 \text{ kJ}}{1 \text{ mol CH}_4} = -6.20 \times 10^5 \text{ kJ of heat energy}$$

(c) $V_{H_2O} = \dfrac{6.21 \times 10^5 \text{ kJ} \times \dfrac{1000 \text{ J}}{1 \text{ kJ}}}{4.18 \dfrac{\text{J}}{\text{g} \, ^\circ\text{C}} (60.0 - 8.8) \, ^\circ\text{C}} \times \dfrac{1 \text{ mL H}_2\text{O}}{1 \text{ g}} = 2.90 \times 10^6 \text{ mL} = 2.90 \times 10^3 \text{ L H}_2\text{O}$

16. The combustion of 1.00 L (STP) of synthesis gas produces 11.13 kJ of heat. The volume of synthesis gas needed to heat 40.0 gal of water is found by first determining the quantity of heat needed to raise the temperature of the water.

$$\text{heat water} = \left(40.0 \text{ gal} \times \dfrac{3.785 \text{ L}}{1 \text{ gal}} \times \dfrac{1000 \text{ mL}}{1 \text{ L}} \times \dfrac{1.00 \text{ g}}{1 \text{ mL}}\right) 4.18 \dfrac{\text{J}}{\text{g} \, ^\circ\text{C}} (65.0 - 15.2) \, ^\circ\text{C}$$

$$= 3.15 \times 10^7 \text{ J} \times \dfrac{1 \text{ kJ}}{1000 \text{ J}} = 3.15 \times 10^4 \text{ kJ}$$

$$\text{gas volume} = 3.15 \times 10^4 \text{ kJ} \times \dfrac{1 \text{ L (STP)}}{11.13 \text{ kJ of heat}} = 2.83 \times 10^3 \text{ L at STP}$$

17. Since the molar mass of H_2 (2.0 g/mol) is $\frac{1}{16}$ of the molar mass of O_2 (32.0 g/mol) and only twice as many moles of H_2 are needed as O_2, we see that $O_2(g)$ is the limiting reagent in this reaction.

$$\dfrac{180.}{2} \text{ g O}_2 \times \dfrac{1 \text{ mol O}_2}{32.0 \text{ g O}_2} \times \dfrac{241.8 \text{ kJ heat}}{0.500 \text{ mol O}_2} = 1.36 \times 10^3 \text{ kJ heat}$$

18. The amounts of the two reactants provided are the same as their stoichiometric coefficients in the balanced equation. Thus 852 kJ of heat is given off by the reaction. We can use this quantity of heat, along with the specific heat of the mixture to determine the temperature change that will occur if all of the heat is retained in the reaction mixture. We make use of the fact that $\text{mass} \times \text{sp.ht.} \times \Delta T$

$$\text{heat} = 8.52 \times 10^5 \text{ J} = \left(\left(1 \text{ mol Al}_2\text{O}_3 \times \dfrac{102 \text{ g Al}_2\text{O}_3}{1 \text{ mol Al}_2\text{O}_3}\right) + \left(2 \text{ mol Fe} \times \dfrac{55.8 \text{ g Fe}}{1 \text{ mol Fe}}\right)\right) \dfrac{0.8 \text{ J}}{\text{g} \, ^\circ\text{C}} \times \Delta T$$

$$\Delta T = \dfrac{8.52 \times 10^5 \text{ J}}{214 \text{ g} \times 0.8 \text{ J g}^{-1} \, ^\circ\text{C}^{-1}} = 5 \times 10^3 \, ^\circ\text{C}$$

The temperature needs to increase from 25°C to 1530°C or $\Delta T = 1505°\text{C} = 1.5 \times 10^3 \, ^\circ\text{C}$. Since the actual ΔT is more than three times as large as this value, the iron indeed will melt, even if a large fraction of the heat evolved is lost to the surroundings and is not retained in the products.

19. **(a)** We first compute the heat produced by this reaction, then determine the value of ΔH in kJ/mol KOH.

$$q_{calorimeter} = (0.205 + 55.9) \; g \times 4.18 \; \frac{J}{g\,^\circ C}\left(24.4\,^\circ C - 23.5\,^\circ C\right) = 2 \times 10^2 \; J \; heat = -q_{rxn}$$

$$\Delta H = -\frac{2 \times 10^2 J \times \dfrac{1 \; kJ}{1000 \, J}}{0.205 g \times \dfrac{1 \, mol \; KOH}{56.1 g \; KOH}} = -5 \times 10^1 \; kJ/mol$$

(b) The ΔT here is known to just one significant figure (0.9 °C). Doubling the amount of KOH should give a temperature change known to two significant figures (1.6 °C) and using twenty times the mass of KOH should give a temperature change known to three significant figures (16.0 °C). This would require 4.10 g KOH rather than the 0.205 g KOH actually used, and would increase the precision from one part in five to one part in 500, or ~0.2 %. Note that as the mass of KOH is increased and the mass of H_2O stays constant, the assumption of a constant specific heat becomes less valid.

20. First we must determine the heat absorbed by the solute during the chemical reaction, q_{rxn}. This is the negative of the heat lost by the solution, q_{soln}. Since the solution (water plus solute) actually gives up heat, the temperature of the solution drops.

$$heat \; of \; reaction = 150.0 \; mL \times \frac{1 \; L}{1000 \; mL} \times \frac{2.50 \; mol \; KI}{1 \; L \; soln} \times \frac{20.3 \; kJ}{1 \; mol \; KI} = 7.61 \; kJ = q_{rxn}$$

$$-q_{rxn} = q_{soln} = \left(150.0 \; mL \times \frac{1.30 \; g}{1 \; mL}\right) \times \frac{2.7 \; J}{g\,^\circ C} \times \Delta T \quad \Delta T = \frac{-7.61 \times 10^3 J}{150.0 \; mL \times \dfrac{1.30 g}{1 mL} \times \dfrac{2.7 J}{g\,^\circ C}} = -14^\circ C$$

$$final \; T = initial \; T + \Delta T = 23.5^\circ C - 14^\circ C = 10.^\circ C$$

21. Let x be the mass, (in grams), of NH_4Cl added to the water. heat = mass × sp.ht. × ΔT

$$x \times \frac{1 \; mol \; NH_4Cl}{53.49 \; g \; NH_4Cl} \times \frac{14.7 \; kJ}{1 \; mol \; NH_4Cl} \times \frac{1000 \; J}{1 \; kJ} = -\left(\left(1400 \; mL \times \frac{1.00 \; g}{1 \; mL}\right) + x\right) 4.18 \frac{J}{g\,^\circ C}(10.-25)^\circ C$$

$$275 \, x = 8.8 \times 10^4 + 63 \, x \; ; \qquad x = \frac{8.8 \times 10^4}{275 - 63} = 4.2 \times 10^2 \; g \; NH_4Cl$$

Our final value is approximate because of the assumed density (1.00 g/mL). The solution's density probably is a bit larger than 1.00 g/mL. Many aqueous solutions are somewhat more dense than water.

22. $heat = 500 \; mL \times \dfrac{1 \; L}{1000 \; mL} \times \dfrac{7.0 \; mol \; NaOH}{1 \; L \; soln} \times \dfrac{-44.5 \; kJ}{1 \; mol \; NaOH} = -1.6 \times 10^2 \; kJ$

= heat of reaction = – heat absorbed by solution   OR $q_{rxn} = -q_{soln}$

$$\Delta T = \frac{1.6 \times 10^5 J}{500. mL \times \dfrac{1.08 g}{1 mL} \times \dfrac{4.00 J}{g\,^\circ C}} = 74^\circ C \qquad final \; T = 21^\circ C + 74^\circ C = 95 \; ^\circ C$$

23. We assume that the solution volumes are additive; that is, that 200.0 mL of solution is formed. Then we compute the heat needed to warm the solution and the cup, and finally ΔH for the reaction.

$$\text{heat} = \left(200.0 \text{ mL} \times \frac{1.02 \text{ g}}{1 \text{ mL}}\right) 4.02 \frac{J}{g \cdot °C} \left(27.8 \text{ °C} - 21.1 \text{ °C}\right) + 10 \frac{J}{°C} \left(27.8 \text{ °C} - 21.1 \text{ °C}\right) = 5.6 \times 10^3 \text{ J}$$

$$\Delta H_{\text{neutr.}} = \frac{-5.6 \times 10^3 \text{ J}}{0.100 \text{ mol}} \times \frac{1 \text{ kJ}}{1000 \text{ J}} = -56 \text{ kJ/mol} \quad (-55.6 \text{ kJ/mol to three significant figures})$$

24. Neutralization reaction: $NaOH(aq) + HCl(aq) \rightarrow NaCl(aq) + H_2O(l)$

Since NaOH and HCl react in a one-to-one molar ratio, and since there is twice the volume of NaOH solution as HCl solution, but the [HCl] is not twice the [NaOH], the HCl solution is the limiting reagent.

$$\text{heat released} = 25.00 \text{ mL} \times \frac{1 \text{ L}}{1000 \text{ mL}} \times \frac{1.86 \text{ mol HCl}}{1 \text{ L}} \times \frac{1 \text{ mol H}_2\text{O}}{1 \text{ mol HCl}} \times \frac{-55.84 \text{ kJ}}{1 \text{ mol H}_2\text{O}} = -2.60 \text{ kJ}$$

$$= \text{heat of reaction} = - \text{ heat absorbed by solution or } q_{\text{rxn}} = -q_{\text{soln}}$$

$$\Delta T = \frac{2.60 \times 10^3 \text{ J}}{75.00 \text{ mL} \times \frac{1.02 \text{ g}}{1 \text{ mL}} \times \frac{3.98 \text{ J}}{g \, °C}} = 8.54 °C \qquad \Delta T = T_{final} - T_i \qquad T_{final} = \Delta T + T_i$$

$$T_{final} = 8.54 °C + 24.72 \, °C = 33.26 \, °C$$

Enthalpy Changes and States of Matter

25. $q_{H_2O(l)} = q_{H_2O(s)}$ $m(\text{sp. ht.})_{H_2O(l)}\Delta T_{H_2O(l)} = \text{mol}_{H_2O(s)}\Delta H_{\text{fus } H_2O(s)}$

$$(3.50 \text{ mol H}_2\text{O} \times \frac{18.015 \text{ g H}_2\text{O}}{1 \text{ mol H}_2\text{O}})(4.184 \frac{J}{g \, °C})(50.0 \, °C) = (\frac{m}{\frac{18.015 \text{ g H}_2\text{O}}{1 \text{ mol H}_2\text{O}}} \times 6.01 \times 10^3 \frac{J}{\text{mol}})$$

$13.2 \times 10^3 \text{ J} = m(333.\underline{6} \text{ J g}^{-1})$ Hence, m = 39.6 g

26. $-q_{\text{lost by steam}} = q_{\text{gained by water}}$

$$-[(5.00 \text{ g H}_2\text{O} \times \frac{1 \text{ mol H}_2\text{O}}{18.015 \text{ g H}_2\text{O}})(-40.6 \times 10^3 \frac{J}{\text{mol}}) + (5.00 \text{ g})(4.184 \frac{J}{g \, °C})(T_f - 100.0 \, °C)]$$

$$= (100.0 \text{ g})(4.184 \frac{J}{g \, °C})(T_f - 25.0 \, °C)$$

$$112\underline{68}.4 \text{ J} - 20.92 \frac{J}{°C} (T_f) + 2092 \text{ J} = 418.\underline{4} \frac{J}{°C} (T_f) - 10,4\underline{60} \text{ J}$$

$$11,\underline{268}.4 \text{ J} + 10,4\underline{60} \text{ J} + 2092 \text{ J} = 418.\underline{4} \frac{J}{°C} (T_f) + 20.9\underline{2} \frac{J}{°C} (T_f) \text{ or } 23.8 \times 10^3 \text{ J} = 439 \text{ J} (T_f)$$

$T_f = 54.2 \, °C$

27. Assume $H_2O(l)$ density = 1.00 g mL^{-1} (at 28.5 °C) $-q_{\text{lost by ball}} = q_{\text{gained by water}} + q_{\text{vap water}}$

$-[(125 \text{ g})(0.50 \frac{\text{J}}{\text{g °C}})(100 \text{ °C} - 525 \text{ °C})] = [(75.0 \text{ g})(4.184 \frac{\text{J}}{\text{g °C}})(100.0 \text{ °C} - 28.5 \text{ °C})] + n_{H_2O}\Delta H°_{\text{vap}}$

$26\underline{562.5}$ J = $22\underline{436.7}$ J $+ n_{H_2O}\Delta H°_{\text{vap}}$ (Note: $n_{H_2O} = \dfrac{\text{mass}_{H_2O}}{\text{molar mass}_{H_2O}}$)

$4\underline{125.8}$ J = $(m_{H_2O})(\dfrac{1 \text{ mol } H_2O}{18.015 \text{ g } H_2O})(40.6 \times 10^3 \frac{\text{J}}{\text{mol}})$

$m_{H_2O} = 1.\underline{83}$ g $H_2O \cong 2$ g H_2O (1 sig. fig.)

28. $-q_{\text{lost by ball}} = q_{\text{melt ice}}$

$-[(125 \text{ g})(0.50 \frac{\text{J}}{\text{g °C}})(0 \text{ °C} - 525 \text{ °C})] = n_{H_2O}\Delta H°_{\text{fus}} = (m_{H_2O})(\dfrac{1 \text{ mol } H_2O}{18.015 \text{ g } H_2O})(6.01 \times 10^3 \frac{\text{J}}{\text{mol}})$

$32\underline{812.5}$ J $= m_{H_2O}(333.\underline{6} \text{ J g}^{-1})$; $m_{H_2O} = 98.\underline{4}$ g $H_2O \cong 98$ g H_2O.

Calorimetry

29. heat capacity $= \dfrac{\text{heat absorbed}}{\Delta T} = \dfrac{5228 \text{ cal}}{4.39 \text{°C}} \times \dfrac{4.184 \text{ J}}{1 \text{ cal}} \times \dfrac{1 \text{ kJ}}{1000 \text{ J}} = 4.98$ kJ/°C

30. Heat absorbed by calorimeter $= q_{\text{comb}} \times$ moles = heat capacity $\times \Delta T$ or $\Delta T = \dfrac{q_{\text{comb}} \times \text{moles}}{\text{heat capacity}}$

(a) $\Delta T = \dfrac{\left(1014.2 \dfrac{\text{kcal}}{\text{mol}} \times 4.184 \dfrac{\text{kJ}}{\text{kcal}}\right)\left(0.3268 \text{ g} \times \dfrac{1 \text{ mol } C_8H_{10}O_2N_4}{194.19 \text{ g } C_8H_{10}O_2N_4}\right)}{5.136 \text{ kJ/°C}} = 1.390 \text{ °C}$

$T_f = T_i + \Delta T = 22.43 \text{°C} + 1.390 \text{°C} = 23.82 \text{°C}$

(b) $\Delta T = \dfrac{2444 \dfrac{\text{kJ}}{\text{mol}}\left(1.35 \text{ mL} \times \dfrac{0.805 \text{ g}}{1 \text{ mL}} \times \dfrac{1 \text{ mol } C_4H_8O}{72.11 \text{ g } C_4H_8O}\right)}{5.136 \text{ kJ/°C}} = 7.17 \text{°C}$

$T_f = 22.43 \text{°C} + 7.17 \text{°C} = 29.60 \text{°C}$

31. **(a)** $\dfrac{\text{heat}}{\text{mass}} = \dfrac{\text{heat cap.} \times \Delta t}{\text{mass}} = \dfrac{4.728 \text{ kJ/°C} \times (27.19 - 23.29) \text{°C}}{1.183 \text{ g}} = 15.6$ kJ / g xylose

$\Delta H = \text{heat given off} / \text{g} \times M(\text{g} / \text{mol}) = \dfrac{-15.6 \text{ kJ}}{1 \text{ g } C_5H_{10}O_5} \times \dfrac{150.13 \text{ g } C_5H_{10}O_5}{1 \text{ mol}}$

$\Delta H = -2.34 \times 10^3$ kJ/mol $C_5H_{10}O_5$

(b) $C_5H_{10}O_5(g) + 5O_2(g) \rightarrow 5CO_2(g) + 5H_2O(l)$ $\Delta H = -2.34 \times 10^3$ kJ

32. This is first a limiting reactant problem. There is $0.1000 \, L \times 0.300 \, M = 0.0300$ mol HCl and $1.82 / 65.39 = 0.0278$ mol Zn. Stoichiometry demands 2 mol HCl for every 1 mol Zn. Thus HCl is the limiting reactant. The reaction is exothermic. We neglect the slight excess of Zn(s), and assume that the volume of solution remains 100.0 mL and its specific heat, $4.18 \, J \, g^{-1} \, {}^\circ C^{-1}$. The enthalpy change, in kJ/mol Zn, is

$$\Delta H = -\frac{100.0 \, mL \times \dfrac{1.00 \, g}{1 \, mL} \times \dfrac{4.18 \, J}{g \, {}^\circ C} \times (30.5 - 20.3) \, {}^\circ C}{0.0300 \, mol \, HCl \times \dfrac{1 \, mol \, Zn}{2 \, mol \, HCl}} \times \frac{1 \, kJ}{1000 \, J} = -284 \text{ kJ/mol Zn reacted}$$

33. **(a)** Because the temperature of the mixture decreases, the reaction molecules (the system) must have absorbed heat from the reaction mixture (the surroundings). Consequently, the reaction must be endothermic.

(b) We assume that the specific heat of the solution is $4.18 \, J \, g^{-1} \, {}^\circ C^{-1}$. The enthalpy change in kJ/mol KCl is obtained by the heat absorbed per gram KCl.

$$\Delta H = -\frac{(0.75 + 35.0) \, g \, \dfrac{4.18 \, J}{g \, {}^\circ C} (23.6 - 24.8) \, {}^\circ C}{0.75 \, g \, KCl} \times \frac{1 \, kJ}{1000 \, J} \times \frac{74.55 \, g \, KCl}{1 \, mol \, KCl} = +18 \text{ kJ / mol}$$

34. As indicated by the negative sign for the enthalpy change, this is an exothermic reaction. Thus the energy of the system should increase.

$$q_{rxn} = 0.136 \text{ mol } KC_2H_3O_2 \times \frac{-15.3 \, kJ}{1 \text{ mol } KC_2H_3O_2} \times \frac{1000 \, J}{1 \, kJ} = -2.08 \times 10^3 \, J = -q_{calorim}$$

Now, we assume that the density of water is 1.00 g/mL, the specific heat of the solution in the calorimeter is $4.18 \, J \, g^{-1} \, {}^\circ C^{-1}$, and no heat is lost by the calorimeter.

$$q_{calorim} = 2.08 \times 10^3 \, J = \left(\left(525 \text{ mL} \times \frac{1.00 \, g}{1 \, mL} \right) + \left(0.136 \text{ mol } KC_2H_3O_2 \times \frac{98.14 \, g}{1 \text{ mol } KC_2H_3O_2} \right) \right)$$

$$\times \frac{4.18 \, J}{g \, {}^\circ C} \times \Delta T = 2.25 \times 10^3 \, \Delta T$$

$$\Delta T = \frac{2.08 \times 10^3}{2.25 \times 10^3} = +0.924 \, {}^\circ C \qquad T_{final} = T_{initial} + \Delta T = 25.1 \, {}^\circ C + 0.924 \, {}^\circ C = 26.0 \, {}^\circ C$$

35. To determine the heat capacity of the calorimeter, recognize that the heat evolved by the reaction is the negative of the heat of combustion.

$$\text{heat capacity} = \frac{\text{heat evolved}}{\Delta T} = \frac{1.620 \, g \, C_{10}H_8 \times \dfrac{1 \, mol \, C_{10}H_8}{128.2 \, g \, C_{10}H_8} \times \dfrac{5156.1 \, kJ}{1 \, mol \, C_{10}H_8}}{8.44 \, {}^\circ C} = 7.72 \text{ kJ/}{}^\circ C$$

36. Note that the heat evolved is the negative of the heat absorbed.

$$\text{heat capacity} = \frac{\text{heat evolved}}{\Delta T} = \frac{1.201 \, g \times \dfrac{1 \, mol \, C_7H_6O_3}{138.12 \, g \, C_7H_6O_3} \times \dfrac{3023 \, kJ}{1 \, mol \, C_7H_6O_3}}{(29.82 - 23.68) \, {}^\circ C} = 4.28 \text{ kJ/}{}^\circ C$$

37. The temperature should increase as the result of an exothermic combustion reaction.

$$\Delta T = 1.227 \ \text{g} \ C_{12}H_{22}O_{11} \times \frac{1 \ \text{mol} \ C_{12}H_{22}O_{11}}{342.3 \ \text{g} \ C_{12}H_{22}O_{11}} \times \frac{5.65 \times 10^3 \ \text{kJ}}{1 \ \text{mol} \ C_{12}H_{22}O_{11}} \times \frac{1°C}{3.87 \ \text{kJ}} = 5.23°C$$

38. $q_{comb} = \dfrac{-11.23°C \times 4.68 \ \text{kJ}/°C}{1.397 \ \text{g} \ C_{10}H_{14}O \times \dfrac{1 \ \text{mol} \ C_{10}H_{14}O}{150.2 \ \text{g} \ C_{10}H_{14}O}} = -5.65 \times 10^3 \ \text{kJ} / \text{mol} \ C_{10}H_{14}O$

Pressure Volume Work

39. **(a)** $-P\Delta V = 3.5 \ \text{L} \times (748 \ \text{mmHg})\left(\dfrac{1 \ \text{atm}}{760 \ \text{mmHg}}\right) = \text{-3.4}\underline{4} \ \text{L atm} \ \text{ or } \ \text{-3.4 L atm}$

(b) 1 L kPa = 1 J, hence,

$$\text{-3.4}\underline{4} \ \text{L atm} \times \left(\frac{101.325 \ \text{kPa}}{1 \ \text{atm}}\right) \times \left(\frac{1 \ \text{J}}{1 \ \text{L kPa}}\right) = \text{-3.4}\underline{9} \times 10^2 \ \text{J} \ \text{ or -3.5} \times 10^2 \ \text{J}$$

(c) $\text{-3.4}\underline{9} \times 10^2 \ \text{J} \times \left(\dfrac{1 \ \text{cal}}{4.184 \ \text{J}}\right) = \text{-83.}\underline{4} \ \text{cal} \ \text{ or -83 cal}$

40. $w = -P\Delta V = -1.23 \ \text{atm} \times (3.37 \ \text{L} - 5.62 \ \text{L}) \times \left(\dfrac{101.325 \ \text{kPa}}{1 \ \text{atm}}\right) \times \left(\dfrac{1 \ \text{J}}{1 \ \text{L kPa}}\right) = 280. \ \text{J}$

That is, 280. J of work is done on the gas by the surroundings.

41. When the Ne(g) sample expands into an evacuated vessel it does not push aside any matter, hence no work is done.

42. Yes, the gas from the aerosol does work. The gas pushes aside the atmosphere.

43. **(a)** No pressure-volume work done (no gases are formed or consumed).

(b) $2 \ NO_2(g) \rightarrow N_2O_4(g)$ $\Delta n_{gas} = -1$ mole. Work is done on the system by the surroundings (compression).

(c) $CaCO_3(s) \rightarrow CaO(s) + CO_2(g)$. Formation of a gas, $\Delta n_{gas} = +1$ mole, results in an expansion. The system does work on the surroundings.

44. **(a)** $2 \ NO(g) + O_2(g) \rightarrow 2 \ NO_2(g)$ $\Delta n_{gas} = -1$ mole. Work is done on the system by the surroundings (compression).

(b) $MgCl_2(aq) + 2 \ NaOH(aq) \rightarrow Mg(OH)_2(s) + 2 \ NaCl(aq)$ $\Delta n_{gas} = 0$, no pressure-volume work is done.

(c) $CuSO_4(s) + 5 \ H_2O(g) \rightarrow CuSO_4 \bullet 5 \ H_2O(s)$ $\Delta n_{gas} = -5$ moles. Work is done on the system by the surroundings (compression).

First Law of Thermodynamics

45. **(a)** $\Delta U = q + w = +58 \ \text{J} + (-58 \ \text{J}) = 0$

(b) $\Delta U = q + w = +125 \ \text{J} + (-687 \ \text{J}) = -562 \ \text{J}$

(c) $280 \ \text{cal} \times (4.184 \dfrac{\text{J}}{\text{cal}}) = 117\underline{1.52} \ \text{J} = 1.17 \ \text{kJ}$ $\quad \Delta U = q + w = -1.17 \ \text{kJ} + 1.25 \ \text{kJ} = 0.08 \ \text{kJ}$

46. **(a)** $\Delta U = q + w = +235\ \text{J} + 128\ \text{J} = 363\ \text{J}$

(b) $\Delta U = q + w = -145\ \text{J} + 98\ \text{J} = -47\ \text{J}$

(c) $\Delta U = q + w = 0\ \text{kJ} + -1.07\ \text{kJ} = -1.07\ \text{kJ}$

47. **(a)** Yes, the gas does work (w = negative value).

(b) Yes, the gas exchanges energy with the surroundings, it absorbs energy.

(c) The temperature of the gas stays the same if the process is isothermal.

(d) ΔU for the gas must equal zero by definition (temperature is not changing).

48. **(a)** Yes, the gas does work (w = negative value).

(b) The internal energy of the gas decreases (energy is expended to do work).

(c) The temperature of the gas should decrease, as it cannot attain thermal equilibrium with its surroundings.

49. This situation is impossible. An ideal gas expanding isothermally means that $\Delta U = 0 = q + w$, or $w = -q$, not $w = -2q$.

50. If a gas is compressed adiabatically, the gas will get hotter. Raise the temperature of the surroundings to an even higher temperature and heat will be transferred to the gas.

Relating ΔH and ΔU

51. According the First Law of Thermodynamics, the answer is (c). Both (a) q_v and (b) q_p are heats of chemical reaction carried out under conditions of constant volume and constant pressure respectively. Both ΔU and ΔH incorporate terms related to work as well as heat.

52. **(a)** $C_4H_{10}O(l) + 6\ O_2(g) \rightarrow 4CO_2(g) + 5\ H_2O(l)$ $\Delta n_{gas} = -2\ \text{mol},\ \Delta H < \Delta U$

(b) $C_6H_{12}O_6(s) + 6\ O_2(g) \rightarrow 6\ CO_2(g) + 6\ H_2O(l)$ $\Delta n_{gas} = 0\ \text{mol},\ \Delta H = \Delta U$

(c) $NH_4NO_3(s) \rightarrow 2\ H_2O(l) + N_2O(g)$ $\Delta n_{gas} = +1\ \text{mol},\ \Delta H > \Delta U$

53. $C_3H_8O(l) + 9/2\ O_2(g) \rightarrow 3\ CO_2(g) + 4\ H_2O(l)$ $\Delta n_{gas} = -1.5\ \text{mol}$

(a) $\Delta U = -33.41\ \dfrac{\text{kJ}}{\text{g}} \times \dfrac{60.096\ \text{g}\ C_3H_8O}{1\ \text{mol}\ C_3H_8O} = -2008\ \dfrac{\text{kJ}}{\text{mol}}$

(b) $\Delta H = \Delta U - w, = \Delta U - (-P\Delta V) = \Delta U - (-\Delta n_{gas}RT) = \Delta U + \Delta n_{gas}RT$

$\Delta H = -2008\ \dfrac{\text{kJ}}{\text{mol}} + (-1.5\ \text{mol})(\dfrac{8.3145 \times 10^{-3}\ \text{kJ}}{\text{K mol}})(298.15\ \text{K}) = -2012\ \dfrac{\text{kJ}}{\text{mol}}$

54. $C_{10}H_{14}O(l) + 13\ O_2(g) \rightarrow 10\ CO_2(g) + 7\ H_2O(l)$ $\Delta n_{gas} = -3\ \text{mol}$

$q_{bomb} = q_v = \Delta U = -5.65 \times 10^3\ \text{kJ} = \Delta U$

$\Delta H = \Delta U - w;$ where $w = (-\Delta n_{gas}RT) = -(-3\ \text{mol})(\dfrac{8.3145 \times 10^{-3}\ \text{kJ}}{\text{K mol}})(298.15\ \text{K}) = +7.4\ \text{kJ}$

$\Delta H = -5.65 \times 10^3\ \text{kJ} - 7.4\ \text{kJ} = -5.66 \times 10^3\ \text{kJ}$

Hess's Law

55. The formation reaction for $NH_3(g)$ is $\frac{1}{2}N_2(g) + \frac{3}{2}H_2(g) \rightarrow NH_3(g)$. The given reaction is two-thirds the reverse of the formation reaction. The sign of the enthalpy is changed and it is multiplied by two-thirds. Thus, the enthalpy of the given reaction is $-(-46.11 \text{ kJ}) \times \frac{2}{3} = +30.74 \text{ kJ}$.

56.

$-(1)$	$CO(g) \rightarrow C(\text{graphite}) + \frac{1}{2}O_2(g)$	$\Delta H^\circ = +110.54 \text{ kJ}$
$+(2)$	$C(\text{graphite}) + O_2(g) \rightarrow CO_2(g)$	$\Delta H^\circ = -393.51 \text{ kJ}$
	$CO(g) + \frac{1}{2}O_2(g) \rightarrow CO_2(g)$	$\Delta H^\circ = -282.97 \text{ kJ}$

57.

$-(3)$	$3\,CO_2(g) + 4\,H_2O(l) \rightarrow C_3H_8(g) + 5\,O_2(g)$	$\Delta H^\circ = +2219.1 \text{ kJ}$
$+(2)$	$C_3H_4(g) + 4\,O_2(g) \rightarrow 3\,CO_2(g) + 2\,H_2O(l)$	$\Delta H^\circ = -1937 \text{ kJ}$
$2(1)$	$2\,H_2(g) + O_2(g) \rightarrow 2\,H_2O(l)$	$\Delta H^\circ = -571.6 \text{ kJ}$
	$C_3H_4(g) + 2\,H_2(g) \rightarrow C_3H_8(g)$	$\Delta H^\circ = -290. \text{ kJ}$

58. The 2nd reaction is the only one in which $NO(g)$ appears; it must be run twice to produce $2NO(g)$.
$$2\,NH_3(g) + \tfrac{5}{2}O_2(g) \rightarrow 2\,NO(g) + 3\,H_2O(l) \qquad 2 \times \Delta H_2^\circ$$
The 1st reaction is the only one that eliminates $NH_3(g)$; it must be run twice to eliminate $2NH_3$.
$$N_2(g) + 3\,H_2(g) \rightarrow 2\,NH_3(g) \qquad 2 \times \Delta H_1^\circ$$
We triple and reverse the third reaction to eliminate $3H_2(g)$.
$$3\,H_2O(l) \rightarrow 3\,H_2(g) + \tfrac{3}{2}O_2(g) \qquad -3 \times \Delta H_3^\circ$$
$$\text{Result}: N_2(g) + O_2(g) \rightarrow 2\,NO(g) \qquad \Delta H_{rxn} = 2 \times \Delta H_1 + 2 \times \Delta H_2 - 3 \times \Delta H_3$$

59.

$2HCl(g) + C_2H_4(g) + \frac{1}{2}O_2(g) \rightarrow C_2H_4Cl_2(l) + H_2O(l)$	$\Delta H^\circ = -318.7 \text{ kJ}$
$Cl_2(g) + H_2O(l) \rightarrow 2HCl(g) + \frac{1}{2}O_2(g)$	$\Delta H^\circ = 0.5(+202.4) = +101.2 \text{ kJ}$
$C_2H_4(g) + Cl_2(g) \rightarrow C_2H_4Cl_2(l)$	$\Delta H^\circ = -217.5 \text{ kJ}$

60.

$N_2H_4(l) + O_2(g) \rightarrow N_2(g) + 2H_2O(l)$	$\Delta H^\circ = -622.2 \text{ kJ}$
$2H_2O_2(l) \rightarrow 2H_2(g) + 2O_2(g)$	$\Delta H^\circ = -2(-187.8 \text{ kJ}) = +375.6 \text{ kJ}$
$2H_2(g) + O_2(g) \rightarrow 2H_2O(l)$	$\Delta H^\circ = 2(-285.8 \text{ kJ}) = -571.6 \text{ kJ}$
$N_2H_4(l) + 2H_2O_2(l) \rightarrow N_2(g) + 4H_2O(l)$	$\Delta H^\circ = -818.2 \text{ kJ}$

61.

$CO(g) + \frac{1}{2}O_2(g) \rightarrow CO_2(g)$	$\Delta H^\circ = -283.0 \text{ kJ}$
$3C(\text{graphite}) + 6H_2(g) \rightarrow 3CH_4(g)$	$\Delta H^\circ = 3(-74.81) = -224.43 \text{ kJ}$
$2H_2(g) + O_2(g) \rightarrow 2H_2O(l)$	$\Delta H^\circ = 2(-285.8) = -571.6 \text{ kJ}$
$3CO(g) \rightarrow \frac{3}{2}O_2(g) + 3C(\text{graphite})$	$\Delta H^\circ = 3(+110.5) = +331.5 \text{ kJ}$
$4CO(g) + 8H_2(g) \rightarrow CO_2(g) + 3CH_4(g) + 2H_2O(l)$	$\Delta H^\circ = -747.5 \text{ kJ}$

62.

$$CS_2(l) + 3O_2(g) \rightarrow CO_2(g) + 2SO_2(g) \qquad \Delta H^\circ = -1077 \text{ kJ}$$

$$2S(s) + Cl_2(g) \rightarrow S_2Cl_2(l) \qquad \Delta H^\circ = -58.2 \text{ kJ}$$

$$C(s) + 2Cl_2(g) \rightarrow CCl_4(l) \qquad \Delta H^\circ = -135.4 \text{ kJ}$$

$$2SO_2(g) \rightarrow 2S(s) + 2O_2(g) \qquad \Delta H^\circ = -2(-296.8 \text{ kJ}) = +593.6 \text{ kJ}$$

$$CO_2(g) \rightarrow C(s) + O_2(g) \qquad \Delta H^\circ = -(-393.5 \text{ kJ}) = +393.5 \text{ kJ}$$

$$\overline{CS_2(l) + 3Cl_2(g) \rightarrow CCl_4(l) + S_2Cl_2(l)} \qquad \Delta H^\circ = -284 \text{ kJ}$$

63.

$$CH_4(g) + CO_2(g) \rightarrow 2CO(g) + 2H_2(g) \qquad \Delta H^\circ = +247 \text{ kJ}$$

$$2CH_4(g) + 2H_2O(g) \rightarrow 2CO(g) + 6H_2(g) \qquad \Delta H^\circ = 2(+206 \text{ kJ}) = +412 \text{ kJ}$$

$$CH_4(g) + 2O_2(g) \rightarrow CO_2(g) + 2H_2O(g) \qquad \Delta H^\circ = -802 \text{ kJ}$$

$$\overline{4CH_4(g) + 2O_2(g) \rightarrow 4CO(g) + 8H_2(g)} \qquad \Delta H^\circ = -143 \text{ kJ}$$

$$\div 4 \text{ produces } CH_4(g) + \tfrac{1}{2}O_2(g) \rightarrow CO(g) + 2H_2(g) \quad \Delta H^\circ = -35.8 \text{ kJ}$$

64. The thermochemical combustion reactions follow.

$$C_4H_6(g) + \tfrac{11}{2}O_2(g) \rightarrow 4CO_2(g) + 3H_2O(l) \qquad \Delta H^\circ = -2540.2 \text{ kJ}$$

$$C_4H_{10}(g) + \tfrac{13}{2}O_2(g) \rightarrow 4CO_2(g) + 5H_2O(l) \qquad \Delta H^\circ = -2877.6 \text{ kJ}$$

$$H_2(g) + \tfrac{1}{2}O_2(g) \rightarrow H_2O(l) \qquad \Delta H^\circ = -285.85 \text{ kJ}$$

Then these equations are combined in the following manner.

$$C_4H_6(g) + \tfrac{11}{2}O_2(g) \rightarrow 4CO_2(g) + 3H_2O(l) \qquad \Delta H^\circ = -2540.2 \text{ kJ}$$

$$4CO_2(g) + 5H_2O(l) \rightarrow C_4H_{10}(g) + \tfrac{13}{2}O_2(g) \qquad \Delta H^\circ = -(-2877.6 \text{ kJ}) = +2877.6 \text{ kJ}$$

$$2H_2(g) + O_2(g) \rightarrow 2H_2O(l) \qquad \Delta H^\circ = 2(-285.8 \text{ kJ}) = -571.6 \text{ kJ}$$

$$\overline{C_4H_6(g) + 2H_2(g) \rightarrow C_4H_{10}(g)} \qquad \Delta H^\circ = -234.2 \text{ kJ}$$

Standard Enthalpies of Formation

65. **(a)** $\Delta H^\circ = \Delta H_f^\circ[C_2H_6(g)] + \Delta H_f^\circ[CH_4(g)] - \Delta H_f^\circ[C_3H_8(g)] - \Delta H_f^\circ[H_2(g)]$

$$\Delta H^\circ = (-84.68 - 74.81 - (-103.8) - 0.00) \text{ kJ} = -55.7 \text{ kJ}$$

(b) $\Delta H^\circ = 2\Delta H_f^\circ[SO_2(g)] + 2\Delta H_f^\circ[H_2O(l)] - 2\Delta H_f^\circ[H_2S(g)] - 3\Delta H_f^\circ[O_2(g)]$

$$\Delta H^\circ = (2(-296.8) + 2(-285.8) - 2(-20.63) - 3(0.00)) \text{ kJ} = -1124 \text{ kJ}$$

66. $\Delta H^\circ = \Delta H_f^\circ[H_2O(l)] + \Delta H_f^\circ[NH_3(g)] - \Delta H_f^\circ[NH_4^+(aq)] - \Delta H_f^\circ[OH^-(aq)]$

$$\Delta H^\circ = (-285.8 + (-46.11) - (-132.5) - (-230.0)) \text{ kJ} = +30.6 \text{ kJ}$$

67. $ZnO(s) + SO_2(g) \rightarrow ZnS(s) + \frac{3}{2}O_2(g); \qquad \Delta H° = -(-878.2 \text{ kJ})/2 = +439.1 \text{ kJ}$

$439.1 \text{ kJ} = \Delta H_f°[ZnS(s)] + \frac{3}{2}\Delta H_f°[O_2(g)] - \Delta H_f°[ZnO(s)] - \Delta H_f°[SO_2(g)]$

$439.1 \text{ kJ} = \Delta H_f°[ZnS(s)] + \frac{3}{2}(0.00 \text{ kJ}) - (-348.3 \text{ kJ}) - (-296.8 \text{ kJ})$

$\Delta H_f°[ZnS(s)] = (439.1 - 348.3 - 296.8)\text{kJ} = -206.0 \text{ kJ/mol}$

68. Most clearly established in Figure 7-18 is the point that the enthalpies of formation for alkane hydrocarbons are negative and that they become more negative as the length of the hydrocarbon chain increases (~20 kJ per added CH_2 unit). For three hydrocarbons of comparable chain length, C_2H_6, C_2H_4, and C_2H_6, we can also infer that the one having only single bonds (C_2H_6) has the most negative enthalpy of formation. The presence of a carbon-to-carbon double bond (C_2H_4) makes the enthalpy of formation more positive, while the presence of a triple bond (C_2H_2), makes it more positive still.

69. $\Delta H° = 4\Delta H_f°[HCl(g)] + \Delta H_f°[O_2(g)] - 2\Delta H_f°[Cl_2(g)] - 2\Delta H_f°[H_2O(l)]$

$= 4(-92.31) + (0.00) - 2(0.00) - 2(-285.8) = +202.4 \text{ kJ}$

70. $\Delta H° = 2\Delta H_f°[Fe(s)] + 3\Delta H_f°[CO_2(g)] - \Delta H_f°[Fe_2O_3(s)] - 3\Delta H_f°[CO(g)]$

$= 2(0.00) + 3(-393.5) - (-824.2) - 3(-110.5) = -24.8 \text{ kJ}$

71. Balanced equation: $\quad C_2H_5OH(l) + 3O_2(g) \rightarrow 2CO_2(g) + 3H_2O(l)$

$\Delta H° = 2\Delta H_f°[CO_2(g)] + 3\Delta H_f°[H_2O(l)] - \Delta H_f°[C_2H_5OH(l)] - 3\Delta H_f°[O_2(g)]$

$= 2(-393.5) + 3(-285.8) - (-277.7) - 3(0.00) = -1366.7 \text{ kJ}$

72. First we must determine the value of $\Delta H_f°$ for $C_5H_{12}(l)$.

Balanced combustion equation: $\quad C_5H_{12}(l) + 8O_2(g) \rightarrow 5CO_2(g) + 6H_2O(l)$

$\Delta H° = -3509 \text{ kJ} = 5\Delta H_f°[CO_2(g)] + 6\Delta H_f°[H_2O(l)] - \Delta H_f°[C_5H_{12}(l)] - 8\Delta H_f°[O_2(g)]$

$= 5(-393.5) + 6(-285.8) - \Delta H_f°[C_5H_{12}(l)] - 8(0.00) = -3682 \text{ kJ} - \Delta H_f°[C_5H_{12}(l)]$

$\Delta H_f°[C_5H_{12}(l)] = -3682 + 3509 = -173 \text{ kJ/mol}$

Then use this value to determine the value of $\Delta H°$ for the reaction in question.

$\Delta H° = \Delta H_f°[C_5H_{12}(l)] + 5\Delta H_f°[H_2O(l)] - 5\Delta H_f°[CO(g)] - 11\Delta H_f°[H_2(g)]$

$= -173 + 5(-285.8) - 5(-110.5) - 11(0.00) = -1050 \text{ kJ}$

73. $\Delta H° = -397.3 \text{ kJ} = \Delta H_f°[CCl_4(g)] + 4\Delta H_f°[HCl(g)] - \Delta H_f°[CH_4(g)] - 4\Delta H_f°[Cl_2(g)]$

$= \Delta H_f°[CCl_4(g)] + (4(-92.31) - (-74.81) - 4(0.00))\text{kJ} = \Delta H_f°[CCl_4(g)] - 294.4 \text{ kJ}$

$\Delta H_f°[CCl_4(g)] = (-397.3 + 294.4) \text{ kJ} = -102.9 \text{ kJ/mol}$

74. $\Delta H° = -8326 \text{ kJ} = 12\Delta H_f°[CO_2(g)] + 14\Delta H_f°[H_2O(g)] - 2\Delta H_f°[C_6H_{14}(l)] - 19\Delta H_t°[O_2(g)]$

$\Delta H° = (12(-393.5) + 14(-285.8))\text{kJ} - 2\Delta H_f°[C_6H_{14}(l)] - 19(0 \text{ kJ}) = -8723 \text{ kJ} - 2\Delta H_f°[C_6H_{14}(l)]$

$\Delta H_f°[C_6H_{14}(l)] = \dfrac{+8326 - 8723}{2} = -199 \text{ kJ/mol}$

75. $\Delta H^\circ = \Delta H_f^\circ[\text{Al(OH)}_3(\text{s})] - \Delta H_f^\circ[\text{Al}^{3+}(\text{aq})] - 3\,\Delta H_f^\circ[\text{OH}^-(\text{aq})]$

$= ((-1276) - (-531) - 3(-230.0))\,\text{kJ} = -55\ \text{kJ}$

76. $\Delta H^\circ = \Delta H_f^\circ[\text{Mg}^{2+}(\text{aq})] + 2\Delta H_f^\circ[\text{NH}_3(\text{g})] + 2\Delta H_f^\circ[\text{H}_2\text{O}(\text{l})] - \Delta H_f^\circ[\text{Mg(OH)}_2(\text{s})] - 2\,\Delta H_f^\circ[\text{NH}_4^+(\text{aq})]$

$= ((-466.9) + 2(-46.11) + 2(-285.8) - (-924.5) - 2(-132.5))\,\text{kJ} = +58.8\ \text{kJ}$

77. Balanced equation: $\quad \text{CaCO}_3(\text{s}) \rightarrow \text{CaO}(\text{s}) + \text{CO}_2(\text{g})$

$\Delta H^\circ = \Delta H_f^\circ[\text{CaO}(\text{s})] + \Delta H_f^\circ[\text{CO}_2(\text{g})] - \Delta H_f^\circ[\text{CaCO}_3(\text{s})]$

$= (-635.1 - 393.5 - (-1207))\,\text{kJ} = +178\ \text{kJ}$

$\text{heat} = 1.35\times10^3\ \text{kg CaCO}_3 \times \dfrac{1000\ \text{g}}{1\ \text{kg}} \times \dfrac{1\ \text{mol CaCO}_3}{100.09\ \text{g CaCO}_3} \times \dfrac{178\ \text{kJ}}{1\ \text{mol CaCO}_3} = 2.40\times10^6\ \text{kJ}$

78. First we determine the heat of combustion: $\quad \text{C}_4\text{H}_{10}(\text{g}) + \frac{13}{2}\,\text{O}_2(\text{g}) \rightarrow 4\text{CO}_2(\text{g}) + 5\text{H}_2\text{O}(\text{l})$

$\Delta H^\circ = 4\,\Delta H_f^\circ[\text{CO}_2(\text{g})] + 5\,\Delta H_f^\circ[\text{H}_2\text{O}(\text{l})] - \Delta H_f^\circ[\text{C}_4\text{H}_{10}(\text{g})] - \frac{13}{2}\,\Delta H_f^\circ[\text{O}_2(\text{g})]$

$= (4(-393.5) + 5(-285.8) - (-125.6) - 6.5(0.00))\,\text{kJ} = -2877.4\ \text{kJ/mol}$

Now compute the volume of gas needed, with the ideal gas equation, rearranged to: $V = \dfrac{nRT}{P}$.

$\text{volume} = \dfrac{\left(5.00\times10^4\,\text{kJ} \times \dfrac{1\,\text{mol C}_4\text{H}_{10}}{2877.3\,\text{kJ}}\right) 0.08206\dfrac{\text{L atm}}{\text{mol K}}(24.6+273.2)\text{K}}{756\,\text{mmHg} \times \dfrac{1\,\text{atm}}{760\,\text{mmHg}}} = 427\,\text{L CH}_4$

Integrative and Advanced Exercises

79. (a) $\text{mass H}_2\text{O} = 40\ \text{gal} \times \dfrac{4\ \text{qt}}{1\ \text{gal}} \times \dfrac{1\ \text{L}}{1.06\ \text{qt}} \times \dfrac{1000\ \text{mL}}{1\ \text{L}} \times \dfrac{1.00\ \text{g}}{1\ \text{mL}} \times \dfrac{1\ \text{kg}}{1000\ \text{g}} \times \dfrac{2.205\ \text{lb}}{1\ \text{kg}} = 3.3\times10^2\ \text{lb}$

$\text{heat (Btu)} = 3.3\times10^2\,\text{lb} \times \dfrac{1\ \text{Btu}}{\text{lb °F}} \times (145\text{°F} - 48\text{°F}) = 3.2\times10^4\,\text{Btu}$

(b) $\text{heat (kcal)} = 1.5\times10^5\,\text{g} \times \dfrac{1\,\text{cal}}{\text{g °C}} \times (145\text{°F} - 48\text{°F}) \times \dfrac{5\,\text{C°}}{9\,\text{F°}} \times \dfrac{1\,\text{kcal}}{1000\,\text{cal}} = 8.1\times10^3\,\text{kcal}$

(c) $\text{heat (kJ)} = 8.1\times10^3\,\text{kcal} \times \dfrac{4.184\,\text{kJ}}{1\,\text{kcal}} = 3.4\times10^4\ \text{kJ}$

<u>80.</u> Potential energy = mgh = 7.26 kg ×9.81 ms^{-2} × 168 m = 1,20 × 10^4 J. This potential energy is converted entirely into kinetic energy just before the object hits, and this kinetic energy is converted entirely into heat when the object strikes.

$$\Delta t = \frac{heat}{mass \times sp.\,ht.} = \frac{1.20 \times 10^4\,J}{7.26\,kg \times \dfrac{1000\,g}{1\,kg} \times \dfrac{0.47\,J}{g\,°C}} = 3.5\,°C$$

This large of a temperature rise is unlikely as some of the kinetic energy will be converted into forms other than heat, such as sound and the fracturing of the object along with the surface it strikes. In addition, some heat energy would be transferred to the surface.

<u>81.</u> heat $= \Delta t\,[\text{heat cap.} + (\text{mass } H_2O \times 4.184\,\dfrac{J}{g\,°C})]$ heat cap. $= \dfrac{heat}{\Delta t} - (\text{mass } H_2O \times 4.184\,\dfrac{J}{g\,°C})$

The heat of combustion of anthracene is -7067 kJ/mol, meaning that burning one mole of anthracene releases +7067 kJ of heat to the calorimeter.

$$\text{heat cap.} = \frac{1.354\,g\,C_{14}H_{10} \times \dfrac{1\,mol\,C_{14}H_{10}}{178.23\,g\,C_{14}H_{10}} \times \dfrac{7067\,kJ}{1\,mol\,C_{14}H_{10}}}{(35.63 - 24.87)\,°C} - \left(983.5\,g \times 4.184 \times 10^{-3}\,\frac{kJ}{g\,°C}\right)$$

$$= (4.990 - 4.115)\,kJ/°C = 0.875\,kJ/°C$$

heat $= (27.19 - 25.01)\,°C\,[0.875\,kJ/°C + (968.6\,g\,H_2O \times 4.184 \times 10^{-3}\,kJ\,g^{-1}\,°C^{-1})] = 10.7\,kJ$

$$q_{rxn} = \frac{-10.7\,kJ}{1.053\,g\,C_6H_8O_7} \times \frac{192.1\,g\,C_6H_8O_7}{1\,mol\,C_6H_8O_7} = -1.95 \times 10^3\,kJ/mol\,C_6H_8O_7$$

82. heat absorbed by calorimeter and water = −heat of reaction

$$= -1.148\,g\,C_7H_6O_2 \times \frac{-26.42\,kJ}{1\,g\,C_7H_6O_2} \times \frac{1000\,J}{1\,kJ} = 3.033 \times 10^4\,J$$

heat absorbed by water $= 1181\,g\,H_2O \times \dfrac{4.184\,J}{g\,°C} \times (30.25\,°C - 24.96\,°C) = 2.61 \times 10^4\,J$

heat absorbed by calorimeter $= 3.033 \times 10^4\,J - 2.61 \times 10^4\,J = 4.2 \times 10^3\,J$

heat capacity of the calorimeter $= \dfrac{4.2 \times 10^3\,J}{30.25\,°C - 24.96\,°C} = 7.9 \times 10^2\,J/°C$

heat absorbed by water $= 1162\,g\,H_2O \times \dfrac{4.184\,J}{g\,°C} \times (29.81\,°C - 24.98\,°C) = 2.35 \times 10^4\,J$

heat absorbed by calorimeter $= 7.9 \times 10^2\,J/°C \times (29.81\,°C - 24.98\,°C) = 3.8 \times 10^3\,J$

heat of combustion $= \dfrac{2.35 \times 10^4\,J + 3.8 \times 10^3\,J}{0.895\,g} \times \dfrac{1\,kJ}{1000\,J} = 30.5\,kJ/g$

mass coal $= 2.15 \times 10^9\,kJ \times \dfrac{1\,g\,coal}{30.5\,kJ} \times \dfrac{1\,kg}{1000\,g} \times \dfrac{1\,metric\,ton}{1000\,kg} = 70.5\,metric\,tons$

83. The difference is due to the enthalpy of vaporization of water. Less heat is evolved when steam is formed because some of the heat of combustion is used to vaporize the water. The difference between these two heats of vaporization is computed first.

$$\text{Difference} = (33.88 - 28.67)\frac{\text{kcal}}{\text{g H}_2} \times \frac{2.016 \text{ g H}_2}{1 \text{ mol H}_2} \times \frac{2 \text{ mol H}_2}{2 \text{ mol H}_2\text{O}} \times \frac{4.184 \text{ kJ}}{1 \text{ kcal}} = 43.9 \text{ kJ/mol H}_2\text{O}$$

This difference should equal the heat of vaporization, that is, the enthalpy change for the following reaction.

$$\text{H}_2\text{O(l)} \longrightarrow \text{H}_2\text{O(g)} \qquad\qquad \text{We use values from Table 7-2.}$$

$$\Delta H°_{\text{rxn}} = \Delta H°_f[\text{H}_2\text{O(g)}] - \Delta H°_f[\text{H}_2\text{O(l)}] = -241.8\,\text{kJ} - (-285.8\,\text{kJ}) = 44.0\,\text{kJ/mol}$$

The two values are in good agreement.

84. (1) $\text{N}_2(g) + 2\,\text{O}_2(g) \longrightarrow 2\,\text{NO}_2$ $\qquad\qquad \Delta H°_{\text{rxn}} = 2\,\Delta H°_f[\text{NO}_2(g)] = 2 \times 33.18\,\text{kJ} = 66.36\,\text{kJ}$

(2) $\text{N}_2(g) + 2\,\text{O}_2(g) \longrightarrow 2\,\text{NO}_2$ $\qquad\qquad \Delta H = 66.36\,\text{kJ}$

$\underline{2\ \text{NO}_2(g) \longrightarrow \text{N}_2\text{O}_3(g) + \tfrac{1}{2}\,\text{O}_2(g) \qquad\qquad \Delta H = 16.02 \qquad \text{kJ}}$

$\text{N}_2(g) + 2\ \text{O}_2(g) \longrightarrow \text{N}_2\text{O}_3(g) + \tfrac{1}{2}\,\text{O}_2(g) \quad \Delta H = 82.38 \quad \text{kJ}$

85. First we determine the moles of gas, then the heat produced by burning each of the components.

$$n = \frac{PV}{RT} = \frac{739\,\text{mm Hg} \times \dfrac{1\,\text{atm}}{760\,\text{mmHg}} \times 385\,\text{L}}{(22.6 + 273.2)\,\text{K} \times 0.08206\,\text{L atm mol}^{-1}\,\text{K}^{-1}} = 15.4\,\text{mol}$$

$$\text{CH}_4 \text{ heat} = 15.4\,\text{mol} \times \frac{83.0\ \text{mol CH}_4}{100.0\ \text{mol gas}} \times \frac{-890.3\,\text{kJ}}{1\,\text{mol CH}_4} = -1.14 \times 10^4\,\text{kJ}$$

$$\text{C}_2\text{H}_6 \text{ heat} = 15.4\,\text{mol} \times \frac{11.2\ \text{mol C}_2\text{H}_6}{100.0\ \text{mol gas}} \times \frac{-1559.7\,\text{kJ}}{1\,\text{mol C}_2\text{H}_6} = -2.69 \times 10^3\,\text{kJ}$$

$$\text{C}_3\text{H}_8 \text{ heat} = 15.4\,\text{mol} \times \frac{5.8\ \text{mol C}_3\text{H}_8}{100.0\ \text{mol gas}} \times \frac{-2219.1\,\text{kJ}}{1\,\text{mol C}_3\text{H}_8} = -2.0 \times 10^3\,\text{kJ}$$

$\text{total heat} = -1.14 \times 10^4\,\text{kJ} - 2.69 \times 10^3\,\text{kJ} - 2.0 \times 10^3\,\text{kJ} = -1.61 \times 10^4\,\text{kJ}$ or $+1.61 \times 10^4\,\text{kJ}$ heat evolved

86. $\text{CO(g)} + 3\,\text{H}_2(g) \longrightarrow \text{CH}_4(g) + \text{H}_2\text{O(g)}$ $\qquad\qquad$ (methanation)

$2\,\text{C(s)} + 2\,\text{H}_2\text{O(g)} \longrightarrow 2\,\text{CO(g)} + 2\,\text{H}_2(g)$ $\qquad\qquad 2 \times (7.24)$

$\underline{\text{CO(g)} + \text{H}_2\text{O(g)} \longrightarrow \text{CO}_2(g) + \text{H}_2(g) \qquad\qquad (7.25)}$

$2\,\text{C(s)} + 2\,\text{H}_2\text{O(g)} \longrightarrow \text{CH}_4(g) + \text{CO}_2(g)$

87. We first compute the heats of combustion of the combustible gases.

$$CH_4(g) + 2O_2(g) \longrightarrow CO_2(g) + 2H_2O(l)$$

$$\Delta H^\circ_{rxn} = \Delta H^\circ_f[CO_2(g)] + 2\Delta H^\circ_f[H_2O(l)] - \Delta H^\circ_f[CH_4(g)] - 2\Delta H^\circ_f[O_2(g)]$$

$$= -393.5\,kJ + 2 \times (-285.8\,kJ) - (-74.81\,kJ) - 2 \times 0.00\,kJ = -890.3\,kJ$$

$$C_3H_8(g) + 5O_2(g) \longrightarrow 3CO_2(g) + 4H_2O(l)$$

$$\Delta H^\circ_{rxn} = 3\Delta H^\circ_f[CO_2(g)] + 4\Delta H^\circ_f[H_2O(l)] - \Delta H^\circ_f[C_3H_8(g)] - 5\Delta H^\circ_f[O_2(g)]$$

$$= 3 \times (-393.5\,kJ) + 4 \times (-285.8\,kJ) - (-103.8\,kJ) - 5 \times 0.00\,kJ = -2220.\,kJ$$

$$H_2(g) + \tfrac{1}{2}O_2(g) \longrightarrow H_2O(l) \quad \Delta H^\circ_{rxn} = \Delta H^\circ_f[H_2O(l)] = -285.8\,kJ$$

$$CO(g) + \tfrac{1}{2}O_2(g) \longrightarrow CO_2(g)$$

$$\Delta H^\circ_{rxn} = \Delta H^\circ_f[CO_2(g)] - \Delta H^\circ_f[CO(g)] - 0.5\Delta H^\circ_f[O_2(g)] = -393.5 + 110.5 = -283.0\,kJ$$

Then, for each gaseous mixture, we compute the enthalpy of combustion per mole of gas. The enthalpy of combustion per STP liter is 1/22.414 of this value. Recall that volume percents are equal to mole percents.

(a) H_2 combustion $= 0.497\,mol\,H_2 \times \dfrac{-285.8\,kJ}{1\,mol\,H_2} = -142\,kJ$

CH_4 combustion $= 0.299\,mol\,CH_4 \times \dfrac{-890.3\,kJ}{1\,mol\,CH_4} = -266\,kJ$

CO combustion $= 0.069\,mol\,CO \times \dfrac{-283.0\,kJ}{1\,mol\,CO} = -20\,kJ$

C_3H_8 combustion $= 0.031\,mol\,C_3H_8 \times \dfrac{-2220.\,kJ}{1\,mol\,C_3H_8} = -69\,kJ$

Total enthalpy of combustion $= \dfrac{(-142 - 266 - 20 - 69)\,kJ}{1\,mol\,gas} \times \dfrac{1\,mol\,gas}{22.414\,L\,STP} = -22.2\,kJ/L$

(b) CH_4 is the only combustible gas present in sewage gas.

Total enthalpy of combustion $= 0.660\,mol\,CH_4 \times \dfrac{-890.3\,kJ}{1\,mol\,CH_4} \times \dfrac{1\,mol\,gas}{22.414\,L\,STP} = -26.2\,kJ/L$

Thus, sewage gas produces more heat per liter at STP than does coal gas.

88. (a) $CH_4(g) + 2O_2(g) \longrightarrow CO_2(g) + 2H_2O(l)$ is the combustion reaction

$$\Delta H^\circ_{rxn} = \Delta H^\circ_f[CO_2(g)] + 2\Delta H^\circ_f[H_2O(l)] - \Delta H^\circ_f[CH_4(g)] - 2\Delta H^\circ_f[O_2(g)]$$

$$= (-393.5\,kJ) + 2 \times (-285.8\,kJ) - (-74.81\,kJ) - 2 \times 0.00\,kJ = -890.3\,kJ$$

$$n = \frac{PV}{RT} = \frac{744\,mm\,Hg \times \dfrac{1\,atm}{760\,mmHg} \times 0.100\,L}{298.2\,K \times 0.08206\,L\,atm\,mol^{-1}\,K^{-1}} = 0.00400\,mol\,CH_4$$

heat available from combustion $= 0.00400\,mol \times$ -890.3 kJ/mol = -3.56 kJ

heat used to melt ice $= 9.53$ g ice $\times \dfrac{1\ mole\ H_2O(s)}{18.02\ g\ ice} \times \dfrac{6.01\ kJ}{1\ mol\ H_2O(s)} = 3.18\ kJ$

Since 3.56 kJ of heat was available, and only 3.18 kJ went into melting the ice, the combustion   must be incomplete.

(b) One possible reaction is: $4\,CH_4(g) + 7\,O_2(g) \longrightarrow 2\,CO_2(g) + 2CO(g) + 8\,H_2O(l)$, but this would generate but 3.00 kJ of energy under the stated conditions. The molar production of CO_2 (g) must be somewhat greater than that of CO(g).

89. (a) The heat of reaction would be smaller (less negative) if the H_2O were obtained as a gas rather than as a liquid.

(b) The reason why the heat of reaction would be less negative is because some of the 1410.9 kJ of heat produced by the reaction will be needed to convert the H_2O from liquid to gas.

(c) $\Delta H° = 2\,\Delta H°_f\,[CO_2(g)] + 2\,\Delta H°_f\,[H_2O(g)] - \Delta H°_f\,[C_2H_4(g)] - 3\,\Delta H°_f\,[O_2(g)]$

$\qquad = 2(-393.5) + 2(-241.8) - (52.26) - 3(0.00) = -1322.9\,kJ$

<u>**90.**</u> First we compute the amount of butane in the cylinder before and after some is withdrawn; the difference is the amount of butane withdrawn. T = (26.0 + 273.2)K = 299.2 K

$$n_1 = \frac{PV}{RT} = \frac{2.35\,atm \times 200.0\,L}{0.08206\,L\,atm\,mol^{-1}\,K^{-1} \times 299.2\,K} = 19.1\,mol\,butane$$

$$n_2 = \frac{PV}{RT} = \frac{1.10\,atm \times 200.0\,L}{0.08206\,L\,atm\,mol^{-1}\,K^{-1} \times 299.2\,K} = 8.96\,mol\,butane$$

amount withdrawn $= 19.1\,mol - 8.96\,mol = 10.1\,mol\,butane$

Then we compute the enthalpy change for one mole of butane and from that the heat produced by burning the withdrawn butane.

$C_4H_{10}(g) + \frac{13}{2}O_2(g) \longrightarrow 4CO_2(g) + 5H_2O(l)$

$\Delta H°_{rxn} = 4\,\Delta H°_f\,[CO_2(g)] + 5\,\Delta H°_f\,[H_2O(l)] - \Delta H°_f\,[C_4H_{10}(g)] - \frac{13}{2}\,\Delta H°_f\,[O_2(g)]$

$\qquad = 4 \times (-393.5\,kJ) + 5 \times (-285.8\,kJ) - (-125.6\,kJ) - 6.5 \times 0.00\,kJ = -2877\ kJ$

Now we compute the heat produced by the combustion of the butane.

$heat = 10.1\,mol\,C_4H_{10} \times \dfrac{-2877\,kJ}{1\,mol\,C_4H_{10}} = -2.91 \times 10^4\ kJ$

To begin the calculation of the heat absorbed by the water, we compute the mass of the water.

$$\text{mass}\,H_2O = 132.5\,L \times \frac{1000\,cm^3}{1\,L} \times \frac{1.00\,g}{1\,cm^3} = 1.33 \times 10^5\,g\,H_2O$$

$$\text{heat absorbed} = 1.33 \times 10^5\,g \times \frac{4.184\,J}{g\,°C} \times (62.2\,°C - 26.0\,°C) \times \frac{1\,kJ}{1000\,J} = 2.01 \times 10^4\,kJ$$

$$\%\,\text{efficiency} = \frac{2.01 \times 10^4\,kJ\,\text{absorbed}}{2.91 \times 10^4\,kJ\,\text{produced}} \times 100\% = 69.1\%\,\text{efficient}$$

91. $\text{work} = 4 \times mgh = 4 \times 58.0\,kg \times 9.807\,m\,s^{-2} \times 1450\,m \times \dfrac{1\,kJ}{1000\,J} = 3.30 \times 10^3\,kJ$

We compute the enthalpy change for the metabolism (combustion) of glucose.

$$C_6H_{12}O_6(s) + 6\,O_2(g) \longrightarrow 6\,CO_2(g) + 6\,H_2O(l)$$
$$\Delta H°_{rxn} = 6\,\Delta H°_f[CO_2(g)] + 6\,\Delta H°_f[H_2O(l)] - \Delta H°_f[C_6H_{12}O_6(s)] - 6\,\Delta H°_f[O_2(g)]$$
$$= 6 \times (-393.5\,kJ) - 6 \times (-285.8\,kJ) - (-1273.3\,kJ) - 6 \times 0.00\,kJ = -2802.5\,kJ$$

Then we compute the mass of glucose needed to perform the necessary work.

$$\text{mass}\,C_6H_{12}O_6 = 3.30 \times 10^3\,kJ \times \frac{1\,kJ\,\text{heat}}{0.70\,kJ\,\text{work}} \times \frac{1\,mol\,C_6H_{12}O_6}{2802.5\,kJ} \times \frac{180.2\,g\,C_6H_{12}O_6}{1\,mol\,C_6H_{12}O_6}$$

$$= 303\,g\,\,C_6H_{12}O_6$$

92. $\Delta H°_f[C_4H_{10}(g)] = -125.6\,kJ/mol$ from Table 7-2.

The difference between C_4H_{10} and C_7H_{16} amounts to three CH_2 groups. The difference in their standard enthalpies of formation is determined as follows.

$$\text{difference} = 3\,mol\,CH_2\,\text{groups} \times \frac{-21\,kJ}{mol\,CH_2\,\text{groups}} = -63\,kJ/mol$$

Thus $\quad\quad \Delta H°_f[C_7H_{16}(l)] \approx -125.6\,kJ/mol - 63\,kJ/mol = -189\,kJ/mol$

Now we compute the enthalpy change for the combustion of heptane.

$$C_7H_{16}(l) + 11\,O_2(g) \longrightarrow 7\,CO_2(g) + 8\,H_2O(l)$$
$$\Delta H°_{rxn} = 7\,\Delta H°_f[CO_2(g)] + 8\,\Delta H°_f[H_2O(l)] - \Delta H°_f[C_7H_{16}(l)] - 11\,\Delta H°_f[O_2(g)]$$
$$= 7 \times (-393.5\,kJ) + 8 \times (-285.8\,kJ) - (-189\,kJ) - 11 \times 0.00\,kJ = -4852\,kJ$$

93. First determine the molar heats of combustion for CH_4 and C_2H_6.

$$CH_4(g) + 2\,O_2(g) \longrightarrow CO_2(g) + 2\,H_2O(l)$$
$$\Delta H° = \Delta H°_f[CO_2(g)] + 2\,\Delta H°_f[H_2O(l)] - \Delta H°_f[CH_4(g)] - 2\,\Delta H°_f[O_2(g)]$$
$$= \big((-393.5) + 2(-285.8) - (-74.81) - 2(0.00)\big)\,kJ = -890.3\,kJ/mol$$

$$C_2H_6(g) + \tfrac{7}{2}\,O_2(g) \longrightarrow 2\,CO_2(g) + 3\,H_2O(l)$$

$$\Delta H° = 2\Delta H°_f[CO_2(g)] + 3\Delta H°_f[H_2O(l)] - \Delta H°_f[C_2H_6(g)] - \frac{7}{2}\Delta H°_f[O_2(g)]$$
$$= (2(-393.5) + 3(-285.8) - (-84.68) - \frac{7}{2}(0.00))kJ = -1559.7\,kJ$$

Since the STP molar volume of an ideal gas is 22.4 L, there is 1/22.4 of a mole of gas present in the sample. We first compute the heat produced by one mole (that is 22.4 L at STP) of the mixed gas.

$$heat = \frac{43.6\,kJ}{1.00\,L} \times \frac{22.4\,L}{1\,mol} = 977\,kJ/mol$$

Then, if we let the number of moles of CH_4 be represented by x, the number of moles of C_2H_6 is represented by $(1.00 - x)$. Now we can construct an equation for the heat evolved per mole of mixture and solve this equation for x.
$$977\,kJ = 890.3\,x + 1559.7(1.00 - x) = 1559.7 + (890.3 - 1559.7)x = 1559.7 - 669.4\,x$$

$$x = \frac{1559.7 - 977}{669.4} = 0.870\,mol\,CH_4/mol\,mixture$$

By the ideal gas law, gases at the same temperature and pressure have the same volume ratio as their molar ratio. Hence, this gas mixture contains 87.0% CH_4 and 13.0% C_2H_6, both by volume.

94. (a) Values are not the same because enthalpies of formations in solution depend on the solute concentration.

(b) The data that we can cite to confirm that the $\Delta H°_f[H_2SO_4(aq)] = -909.3$ kJ/mol in an infinitely dilute solution is the same for the $\Delta H°_f[SO_4^{2-}]$ from Table 7.3. This is expected because an infinitely dilute solution of H_2SO_4 is 2 $H^+(aq) + SO_4^{2-}(aq)$. Since $\Delta H°_f[H^+] = 0$ kJ/mol, it is expected that $SO_4^{2-}(aq)$ will have the same $\Delta H°_f$ as $H_2SO_4(aq)$.

(c) 500.0 mL of 1.00 M H_2SO_4 is prepared from pure H_2SO_4. Note: ½ mole H_2SO_4 used.

Reaction $H_2SO_4(l)$ → $H_2SO_4(aq) \sim 1.00M$
$\Delta H°_f$ -814.0 kJ/mol   -909.3 kJ/mol
q = ½ mol(-909.3 kJ/mol) – ½ mol(-814.0 kJ/mol) = -47.65 kJ
∴47650 J released by dilution and absorbed by the water.

Use q = mcΔt $\Delta t = \dfrac{47650\,J}{500\,g(4.2\,J\,g^{-1}\,°C^{-1})} \approx +23°C$

95. $CO(g) + 3\,H_2(g) \rightarrow CH_4(g) + H_2O(g)$ (methanation)   -206.1 kJ
$2\,C(s) + 2\,H_2O(g) \rightarrow 2\,CO(g) + 2\,H_2(g)$ (2×7.24) +262.6 kJ
$\underline{CO(g) + H_2O(g) \rightarrow CO_2(g) + H_2(g)}$ (7.25) $\underline{-41.2\,kJ}$
$2\,C(s) + 2\,H_2O(g) \rightarrow CH_4(g) + CO_2(g)$ 15.3 kJ

96. $n = \dfrac{(765.5/760)\text{atm}}{(0.08206 \text{ L atm K}^{-1} \text{ mol}^{-1})\ 298.15\text{K}}\ \dfrac{0.582\text{L}}{} = 2.40 \times 10^{-2}\text{mol}$

molar mass $= \dfrac{1.103\text{g}}{2.40 \times 10^{-2}\text{mol}} = 46.0\text{g/mol}$

moles of CO_2 = 2.108/44.01 = 0.04790   (0.0479 mol C in unknown)

moles of H_2O = 1.294/18.02 = 0.0719   (0.144 mol H in unknown)

moles of O in unknown $= \dfrac{1.103 \text{ g} - 0.04790 \text{ g (12.011g C/mol)} - 0.144(1.00794)}{15.9994 \text{ g mol}^{-1}} = 0.0239 \text{ mol O}$

So C:H:O ratio is 2:6:1 and the molecular formula is C_2H_6O

ΔT = 31.94 - 25.00 = 6.94 °C $q = 6.94 \text{ °C} \times 5.015\text{kJ/°C} = -34.8\text{kJ}$

$\Delta H = \dfrac{-34.8\text{kJ}}{0.0240\text{mol}} = -1.45 \times 10^{3}\text{kJ/mol}$ $3 \text{ O}_2(g) + C_2H_6O(g) \rightarrow 2 \text{ CO}_2(g) + 3 \text{ H}_2O(l)$

97. Energy needed = $mc\Delta T$ = (250 g)((50-4) °C)(4.2J/g °C) = 4.8×10^4 J

A 700-watt oven delivers a joule of energy/sec   Time $= \left(\dfrac{4.8 \times 10^4 \text{ J}}{700 \text{ J sec}^{-1}}\right)$ = 69 seconds

98. $w = -P\Delta V$, $V_i = (3.1416)(6.00 \text{ cm})^2(8.10 \text{ cm}) = 916.1 \text{ cm}^3$

$P = \dfrac{\text{force}}{\text{area}} = \dfrac{(3.1416)(5.00)^2(25.00)(7.75\text{g/cm}^3)(1\text{kg/1000g})(9.807\text{m/sec}^2)}{(3.1416)(6.00 \text{ cm})^2(1\text{m/100cm})^2}$

$= 1.320 \times 10^4 \text{ Pa} = \dfrac{1.320 \times 10^4 \text{ Pa}}{101325 \text{ Pa atm}^{-1}} = 0.130 \text{ atm difference}$ Use: $P_1 V_1 = P_2 V_2$

$V_2 = \dfrac{((745/760) + 0.130\text{atm}) \times (0.9161\text{L})}{(745/760)\text{atm}} = 1.04 \text{ L}$

$-P\Delta V = \dfrac{745}{760} \times (1.04 - 0.916) = -0.121 \text{ L} - \text{atm} = w = (-0.121 \text{ L} - \text{atm})\dfrac{101 \text{ J}}{\text{L} - \text{atm}} = -12 \text{ J}$

99. $Na_2CO_3{\cdot}10 \text{ H}_2O(s) \rightarrow Na_2CO_3{\cdot}7 \text{ H}_2O(s) + 3 \text{ H}_2O(g)$ + 155.3 kJ

$Na_2CO_3{\cdot}7 \text{ H}_2O(s) \rightarrow Na_2CO_3{\cdot} \text{ H}_2O(s) + 6 \text{ H}_2O(g)$ + 320.1 kJ

$\underline{Na_2CO_3{\cdot}H_2O(s) \rightarrow Na_2CO_3(s) + H_2O(g) \hspace{3cm} + 57.3 \text{ kJ}}$

$Na_2CO_3{\cdot}10 \text{ H}_2O(s) \rightarrow Na_2CO_3(s) + 10 \text{ H}_2O(g)$ +532.7 kJ

Thus using Hess's Law, ΔH = +532.7 kJ.   For ΔU we must assume a temperature, say 373 K.
$\Delta U = \Delta H - \Delta n_{gas}RT$ = +532.7 kJ − 10 mol(8.314472 J $K^{-1} \text{ mol}^{-1}$)(373 L)(1 kJ/1000 J) = +501.7 kJ
The value for ΔU is only an estimate because it is based on a ΔH value that has been obtained using Hess' law. Hess' law provides only an approximate value for the enthalpy change. To obtain a more precise value for the ΔU, one would use bomb calorimetry. As well, the water formed is assumed to be a vapor (gas). In reality, a portion of that water will exist as a liquid.

100. First, list all of th e pertinent reactions normalizing them so that each uses the same amount of NH_3:

(rxn-1) $4 NH_3(g) + 3 O_2(g) \rightarrow 2 N_2(g) + 6 H_2O(g)$
$\Delta H°_{rxn-1} = 2 \times (0 \text{ kJ/mol}) + 6 \times (-241.8 \text{ kJ/mol}) - [4 \times (-46.11 \text{ kJ}) + 3 \times (0 \text{ kJ/mol})] = -1266.4 \text{ kJ}$

(rxn-2) $4 NH_3(g) + 4 O_2(g) \rightarrow 2 N_2O(g) + 6 H_2O(g)$
$\Delta H°_{rxn-1} = 4 \times (82.05 \text{ kJ/mol}) + 6 \times (-241.8 \text{ kJ/mol}) - [4 \times (-46.11 \text{ kJ}) + 3 \times (0 \text{ kJ/mol})] = -938.2 \text{ kJ}$

(rxn-3) $4 NH_3(g) + 5 O_2(g) \rightarrow 4 NO(g) + 6 H_2O(g)$
$\Delta H°_{rxn-1} = 2 \times (90.25 \text{ kJ/mol}) + 6 \times (-241.8 \text{ kJ/mol}) - [4 \times (-46.11 \text{ kJ}) + 3 \times (0 \text{ kJ/mol})] = -1085.9 \text{ kJ}$

(rxn-4) $4 NH_3(g) + 7 O_2(g) \rightarrow 4 NO_2(g) + 6 H_2O(g)$
$\Delta H°_{rxn-1} = 4 \times (33.18 \text{ kJ/mol}) + 6 \times (-241.8 \text{ kJ/mol}) - [4 \times (-46.11 \text{ kJ}) + 3 \times (0 \text{ kJ/mol})] = -1133.6 \text{ kJ}$

Clearly, rxn-1, the oxidation of $NH_3(g)$ to $N_2(g)$ is the most exothermic reaction.

101.

$$\Delta H = \int_{298K}^{373K} C_p dt = \int_{298K}^{373K} (28.58 + 0.00377T - 0.5 \times 10^5 T^{-2}) dt$$

$$\Delta H = 28.58 \frac{J}{K \text{ mol}} \int_{298K}^{373K} dt + 0.00377 \frac{J}{K^2 \text{ mol}} \int_{298K}^{373K} T dt + -0.5 \times 10^5 \frac{J K}{\text{mol}} \int_{298K}^{373K} T^{-2} dt$$

$$\Delta H = 28.58 \frac{J}{K \text{ mol}} \times \left(T\Big|_{298K}^{373K}\right) + 0.00377 \frac{J}{K^2 \text{ mol}} \times \left(\frac{T^2}{2}\Big|_{298K}^{373K}\right) - 0.5 \times 10^5 \frac{J K}{\text{mol}} \times \left(-\frac{1}{T}\Big|_{298K}^{373K}\right)$$

$$\Delta H = 28.58 \frac{J}{K \text{ mol}} \times (373 K - 298 K) + 0.00377 \frac{J}{K^2 \text{ mol}} \times \left(\frac{(373 K)^2}{2} - \frac{(298 K)^2}{2}\right)$$

$$+ 0.5 \times 10^5 \frac{J K}{\text{mol}} \times \left(\frac{1}{373 K} - \frac{1}{298 K}\right)$$

$$\Delta H = 2143.\underline{5} \frac{J}{\text{mol}} + 94.9 \frac{J}{\text{mol}} - 33.\underline{7} \frac{J}{\text{mol}} = 2205 \frac{J}{\text{mol}} = 2.20\underline{5} \frac{kJ}{\text{mol}}$$

102. Reaction of interest:

	$H_2(g)$	+	$\frac{1}{2} O_2(g)$	$\rightarrow$	$H_2O(g)$
Molar Heat Capacities:	28.84 J K^{-1}mol^{-1}		29.37 J K^{-1}mol^{-1}		33.58 J K^{-1}mol^{-1}
ΔT	75.0 K		75.0 K		75.0 K
$q = n \times C_p \times \Delta T$	(1)(28.84)(75.0) J		(½)(29.37)(75.0) J		(1)(33.58)(75.0) J
	= 216\underline{3} J		= 110\underline{1.4} J		= 251\underline{8.5} J

$q_{prod} = 251\underline{8.5}$ J = 2.52 kJ $q_{react} = (216\underline{3}$ J $+ 110\underline{1.4}$ J$) = 326\underline{4.4}$ J = 3.26 kJ

See diagram below.

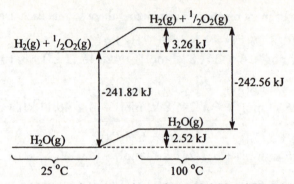

Thus, ΔH°_f $H_2O(g)$ at 100.0 ° C is approximately -243 kJ/mol $H_2O(g)$ formed

FEATURE PROBLEMS

103. 1 °F = 5/9 °C = 0.555 °C   1 lb = 453.6 g

$E_p = mgh$

$E_p = (772 \text{ lb})(9.80665 \text{ m s}^{-2})(1 \text{ ft}) = 7.57 \times 10^3 \dfrac{\text{lb m ft}}{\text{s}^2}$

$E_p = 7.57 \times 10^3 \dfrac{\text{lb m ft}}{\text{s}^2} \times \dfrac{0.3048 \text{ m}}{\text{ft}} \times \dfrac{0.4536 \text{ kg}}{\text{lb}}$

$E_p = 1047 \dfrac{\text{kg m}^2}{\text{s}^2} = 1047 \text{ J} = 1.05 \text{ kJ}$

The statement is validated.

$q = m \times (\text{sp. ht.})\Delta T$

$q = 453.6 \text{ g}(4.184 \dfrac{\text{J}}{\text{g °C}})(0.556 \text{ °C})$

$q = 1054 \text{ J}$

$q = 1.05 \text{ kJ}$

104. (a) We plot specific heat vs. the inverse of atomic mass.

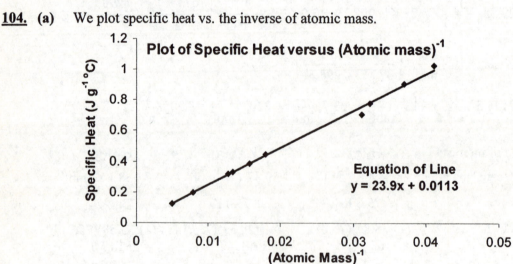

(b) The equation of the line is: Specific Heat = 23.9 ÷(atomic mass) + 0.0113

0.23 J g^{-1} °C = 23.9 ÷(atomic mass) + 0.0113

atomic mass = $\dfrac{23.9}{0.23 - 0.0113}$ = $10\underline{9}$ u or 110 u (2 sig fig);

Cadmium's tabulated atomic mass is 112.4 u.

(c) $\text{sp.ht.} = \dfrac{450 \text{ J}}{75.0 \text{ g} \times 15.°\text{C}} = 0.40 \text{ J g}^{-1}\,°\text{C}^{-1} = 0.0113 + 23.9/\text{atomic mass}$

$\text{atomic mass} = \dfrac{23.9}{0.40 - 0.0113} = 61.\underline{5} \text{ u or } 62\,\text{u}$ The metal is most likely Cu (63.5 u)

105. The plot's maximum is the equivalence point. (assume $\Delta T = 0$ at 0 mL of added NaOH, (i.e. only 60 mL of citric acid are present) and that $\Delta T = 0$ at 60 mL of NaOH (i.e. no citric acid added)

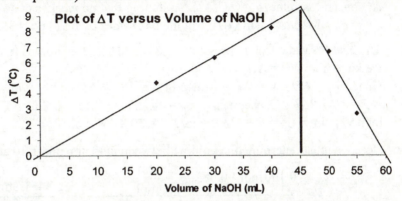

(a)   The equivalence point occurs with 45.0 mL of 1.00 M NaOH(aq) [45.0 mmol NaOH] added and 15.0 mL of 1.00 M citric acid [15.0 mmol citric acid]. Again, we assume that $\Delta T =$ zero if no NaOH added ($V_{\text{NaOH}} = 0$ mL) and $\Delta T = 0$ if no citric acid is added ($V_{\text{NaOH}} = 60$).

(b)   Heat is a product of the reaction, as are chemical species (products). Products are maximized at the exact stoichiometric proportions. Since each reaction mixture has the same volume, and thus about the same mass to heat, the temperature also is a maximum at this point.

(c) $\text{H}_3\text{C}_6\text{H}_5\text{O}_7\,(\text{s}) + 3\,\text{OH}^-\,(\text{aq}) \rightarrow 3\,\text{H}_2\text{O}(\text{l}) + \text{C}_6\text{H}_5\text{O}_7{}^{3-}\,(\text{aq})$

106. The reactions, and their temperature changes, are as follows.

(1st) $\text{NH}_3(\text{conc. aq}) + \text{HCl}(\text{aq}) \rightarrow \text{NH}_4\text{Cl}(\text{aq})$ $\Delta T = (35.8 - 23.8)°\text{C} = 12.0°\text{C}$

(2nd,a) $\text{NH}_3(\text{conc. aq}) \rightarrow \text{NH}_3(\text{g})$ $\Delta T = (13.2 - 19.3)°\text{C} = -6.1°\text{C}$

(2nd,b) $\text{NH}_3(\text{g}) + \text{HCl}(\text{aq}) \rightarrow \text{NH}_4\text{Cl}(\text{aq})$ $\Delta T = (42.9 - 23.8)°\text{C} = 19.1°\text{C}$

The sum of reactions $(2\text{nd},\text{a}) + (2\text{nd},\text{b})$ produces the same change as the 1st reaction. We now compute the heat absorbed by the surroundings for each. Hess's law is demonstrated if $\Delta H_1 = \Delta H_{2a} + \Delta H_{2b}$, where in each case $\Delta H = -q$.

$q_1 = \left[(100.0 \text{ mL} + 8.00 \text{ mL}) \times 1.00 \text{ g}/\text{mL}\right] 4.18 \text{ J g}^{-1}\,°\text{C}^{-1} \times 12.0°\text{C} = 5.42 \times 10^3 \text{ J} = -\Delta H_1$

$q_{2a} = \left[(100.0 \text{ mL} \times 1.00 \text{ g/mL}) \, 4.18 \text{ J g}^{-1}\,°\text{C}^{-1} \times (-6.1°\text{C})\right] = -2.55 \times 10^3 \text{ J} = -\Delta H_{2a}$

$q_{2b} = \left[(100.0 \text{ mL} \times 1.00 \text{ g}/\text{mL}) \, 4.18 \text{ J g}^{-1}\,°\text{C}^{-1} \times (19.1°\text{C})\right] = 7.98 \times 10^3 \text{ J} = -\Delta H_{2b}$

$\Delta H_{2a} + \Delta H_{2b} = +2.55 \times 10^3 \text{ J} - 7.98 \times 10^3 \text{ J} = -5.43 \times 10^3 \text{ J} \cong -5.42 \times 10^3 \text{ J} = \Delta H_1 \; (QED)$

107. According to the kinetic-molecular theory of gases, the internal energy of an ideal gas, U, is proportional to the average translational kinetic energy for the gas particles, e_k, which in turn is proportional to $3/2\ RT$. Thus the internal energy for a fixed amount of an ideal gas depends only on its temperature, i.e. $U = 3/2\ nRT$, where U is the internal energy (J), n is the number of moles of gas particles, R is the gas constant (J K^{-1} mol^{-1}), and T is the temperature (K). If the temperature of the gas sample is changed, the resulting change in internal energy is given by $\Delta U = 3/2\ nR\Delta T$.

(a) At constant volume, $q_v = nC_v\Delta T$. Assuming that no work is done $\Delta U = q_v$ so, $\Delta U = q_v = 3/2nR\Delta T = nC_v\Delta T$. (divide both sides by $n\Delta T$) $C_v = 3/2\ R= 12.5$ J/K·mol.

(b) The heat flow at constant pressure q_p is the ΔH for the process (i.e., $q_p = \Delta H$) and we know that $\Delta H = \Delta U - w$ and $w = -P\Delta V = -nR\Delta T$
Hence, $q_p = \Delta U - w = \Delta U - (-nR\Delta T) = \Delta U + nR\Delta T$ and $q_p = nC_p\Delta T$ and $\Delta U = 3/2\ nR\Delta T$ Consequently $q_p = nC_p\Delta T = nR\Delta T + 3/2\ nR\Delta T$ (divide both sides by $n\Delta T$) Now, $C_p = R + 3/2\ R = 5/2\ R = 20.8$ J/K·mol

108. (a) Here we must determine the volume between 2.40 atm and 1.30 atm using $PV = nRT$

$$V = \frac{0.100\,\text{mol} \times 0.08206\ \dfrac{\text{L atm}}{\text{K mol}} \times 298\ \text{K}}{P} = \frac{2.445\,\text{L atm}}{P}$$

For $P = 2.40$ atm: $V = 1.02$ L
For $P = 2.30$ atm: $V = 1.06$ L $P\Delta V = 2.30$ atm $\times -0.04$ L $= -0.092$ L atm
For $P = 2.20$ atm: $V = 1.11$ L $P\Delta V = 2.20$ atm $\times -0.05$ L $= -0.11$ L atm
For $P = 2.10$ atm: $V = 1.16$ L $P\Delta V = 2.10$ atm $\times -0.05$ L $= -0.11$ L atm
For $P = 2.00$ atm: $V = 1.22$ L $P\Delta V = 2.00$ atm $\times -0.06$ L $= -0.12$ L atm
For $P = 1.90$ atm: $V = 1.29$ L $P\Delta V = 1.90$ atm $\times -0.06$ L $= -0.12$ L atm
For $P = 1.80$ atm: $V = 1.36$ L $P\Delta V = 1.80$ atm $\times -0.07$ L $= -0.13$ L atm
For $P = 1.70$ atm: $V = 1.44$ L $P\Delta V = 1.70$ atm $\times -0.08$ L $= -0.14$ L atm
For $P = 1.60$ atm: $V = 1.53$ L $P\Delta V = 1.60$ atm $\times -0.09$ L $= -0.14$ L atm
For $P = 1.50$ atm: $V = 1.63$ L $P\Delta V = 1.50$ atm $\times -0.10$ L $= -0.15$ L atm
For $P = 1.40$ atm: $V = 1.75$ L $P\Delta V = 1.40$ atm $\times -0.12$ L $= -0.17$ L atm
For $P = 1.30$ atm: $V = 1.88$ L $P\Delta V = 1.30$ atm $\times -0.13$ L $= -0.17$ L atm

Total Work = $-\Sigma\ P\Delta V = -1.45$ L atm

Expressed in joules, the work is -1.45 L atm $\times 101.325$ J/L atm $= -147$ J

(b)

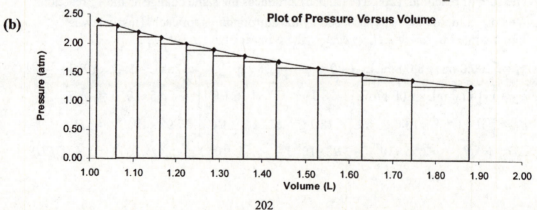

(c) The total work done in the two-step expansion is minus one times the total of the area of the two rectangles under the graph which turns out to be -1.29 L atm or -131 J. In the 11-step expansion in (b), the total area of the rectangles is 1.45 L atm or -147 J. If the expansion were divided into a larger number of stages, the total area of the rectangles would be even greater. The maximum amount of work is for an expansion with an infinite number of stages and is equal to the area under the pressure-volume curve between $V = 1.02$ L and 1.88 L. This area is also obtained as the integral of the expression:

$dw = -PdV = -nRT(dV/V)$. The value obtained is:
$w = -nRT \times \ln V_f/V_i = 0.100$ mol $\times$ 8.3145 J mol^{-1} K^{-1} $\times$ 298 K $\times$ ln (1.88 L/1.02 L)
$w = -152$ J

(d) The maximum work of compression is for a one-stage compression using an external pressure of 2.40 atm and producing a compression in volume of 1.02 L $-$ 1.88 L $=$ -0.86 L: $w = -P\Delta V = (2.40$ atm $\times$ 0.86 L) $\times$ 101.33 J/L atm $= 209$ J

The minimum work would be that done in an infinite number of steps and would be the same as the work determined in (c) but with a positive sign, namely, $+152$ J.

(e) Because the internal energy of an ideal gas is a function only of temperature, and the temperature remains constant, $\Delta U = 0$. Because $\Delta U = q + w = 0$, $q = -w$. This means that, -209 J corresponds to the maximum work of compression and -152 J corresponds to the minimum work of compression.

(f) For the expansion described in part in (c),
$q = -w = nRT \ln V_f/V_i$ and $q/T = nR \ln V_f/V_i$
Because the terms on the right side are all constants or functions of state, so too is the term on the left, q/T. In Chapter 19, we learn that q/T is equal to ΔS, the change in a state function called *entropy*.

CHAPTER 8
ELECTRONS IN ATOMS
PRACTICE EXAMPLES

1A Use $c = \lambda\nu$, solve for frequency. $\nu = \dfrac{2.9979 \times 10^8 \text{ m/s}}{690 \text{ nm}} \times \dfrac{10^9 \text{ nm}}{1 \text{ m}} = 4.34 \times 10^{14}$ Hz

1B Wavelength and frequency are related through the equation $c = \lambda\nu$, which can be solved for either one.

$$\lambda = \frac{c}{\nu} = \frac{2.9979 \times 10^8 \text{ m/s}}{91.5 \times 10^6 \text{ s}^{-1}} = 3.28 \text{ m} \quad \text{Note that Hz} = \text{s}^{-1}$$

2A The relationship $\nu = c/\lambda$ can be substituted into the equation $E = h\nu$ to obtain $E = hc/\lambda$. This energy, in J/photon, can then be converted to kJ/mol.

$$E = \frac{hc}{\lambda} = \frac{6.626 \times 10^{-34} \text{ J s photon}^{-1} \times 2.998 \times 10^8 \text{ m s}^{-1}}{230 \text{ nm} \times \dfrac{1 \text{ m}}{10^9 \text{ nm}}} \times \frac{6.022 \times 10^{23} \text{ photons}}{1 \text{ mol}} \times \frac{1 \text{ kJ}}{1000 \text{ J}} = 520 \text{ kJ/mol}$$

With a similar calculation one finds that 290 nm corresponds to 410 kJ/mol. Thus, the energy range is from 410 to 520 kJ/mol, respectively.

2B The equation $E = h\nu$ is solved for frequency and the two frequencies are calculated.

$$\nu = \frac{E}{h} = \frac{3.056 \times 10^{-19} \text{ J/photon}}{6.626 \times 10^{-34} \text{ J} \cdot \text{s/photon}} \qquad \nu = \frac{E}{h} = \frac{4.414 \times 10^{-19} \text{ J/photon}}{6.626 \times 10^{-34} \text{ J} \cdot \text{s/photon}}$$

$$= 4.612 \times 10^{14} \text{ Hz} \qquad\qquad\qquad = 6.662 \times 10^{14} \text{ Hz}$$

To determine color, we calculate the wavelength of each frequency and compare it with *text* Figure 8-3.

$$\lambda = \frac{c}{\nu} = \frac{2.9979 \times 10^8 \text{ m/s}}{4.612 \times 10^{14} \text{ Hz}} \times \frac{10^9 \text{ nm}}{1 \text{ m}} \qquad \lambda = \frac{c}{\nu} = \frac{2.9979 \times 10^8 \text{ m/s}}{6.662 \times 10^{14} \text{ Hz}} \times \frac{10^9 \text{ nm}}{1 \text{ m}}$$

$$= 650 \text{ nm} \quad \text{orange} \qquad\qquad\qquad = 450 \text{ nm} \quad \text{indigo}$$

The colors of the spectrum that are not absorbed are what we see when we look at a plant, namely in this case blue, green, and yellow. The plant appears green.

3A We solve the Rydberg equation for n to see if we obtain an integer.

$$n = \sqrt{n^2} = \sqrt{\frac{-R_H}{E_n}} = \sqrt{\frac{-2.179 \times 10^{-18} \text{ J}}{-2.69 \times 10^{-20} \text{ J}}} = \sqrt{81.00} = 9.00 \qquad \text{This is } E_9 \text{ for n = 9.}$$

3B It is not likely that an atomic radius would be precisely equal to an arbitrary unit of length, but let us see how close the radii are to 1 nm. 1 nm = 1000 pm, so we solve the following for n.

$$1000 \text{ pm} = n^2 \, 53 \text{ pm} \qquad\qquad n = \sqrt{\frac{1000}{53}} = 4.3$$

We see that no radius is exactly 1 nm. The closest:

$$r_4 = 4^2 a_0 = 16 \times 0.053 \text{ nm} = 0.85 \text{ nm} \qquad \text{while } r_5 = 5^2 a_0 = 25 \times 0.053 \text{ nm} = 1.3 \text{ nm}$$

4A We first determine the energy difference, and then the wavelength of light for that energy.

$$\Delta E = R_H \left(\frac{1}{2^2} - \frac{1}{4^2} \right) = 2.179 \times 10^{-18} \text{ J} \left(\frac{1}{2^2} - \frac{1}{4^2} \right) = 4.086 \times 10^{-19} \text{ J}$$

$$\lambda = \frac{hc}{\Delta E} = \frac{6.626 \times 10^{-34} \text{ J s} \times 2.998 \times 10^8 \text{ m s}^{-1}}{4.086 \times 10^{-19} \text{ J}} = 4.862 \times 10^{-7} \text{ m or } 486.2 \text{ nm}$$

4B The longest wavelength light results from the transition that spans the smallest difference in energy. Since all Lyman series emissions end with $n_f = 1$, the smallest energy transition has $n_i = 2$. From this, we obtain the value of ΔE.

$$\Delta E = R_H \left(\frac{1}{n_i^2} - \frac{1}{n_f^2} \right) = 2.179 \times 10^{-18} \text{ J} \left(\frac{1}{2^2} - \frac{1}{1^2} \right) = -1.634 \times 10^{-18} \text{ J}$$

From this energy emitted, we can obtain the wavelength of the emitted light: $\Delta E = hc / \lambda$

$$\lambda = \frac{hc}{\Delta E} = \frac{6.626 \times 10^{-34} \text{ J s} \times 2.998 \times 10^8 \text{ m s}^{-1}}{1.634 \times 10^{-18} \text{ J}} = 1.216 \times 10^{-7} \text{ m or } 121.6 \text{ nm (1216 angstroms)}$$

5A

$$E_f = \frac{-Z^2 \times R_H}{n_f^2} = \frac{-4^2 \times 2.179 \times 10^{-18} \text{ J}}{3^2}$$

$$E_f = -3.874 \times 10^{-18} \text{ J}$$

$$E_i = \frac{-Z^2 \times R_H}{n_i^2} = \frac{-4^2 \times 2.179 \times 10^{-18} \text{ J}}{5^2}$$

$$E_i = -1.395 \times 10^{-18} \text{ J}$$

$$\Delta E = E_f - E_i$$

$$\Delta E = (-3.874 \times 10^{-18} \text{ J}) - (-1.395 \times 10^{-18} \text{ J})$$

$$\Delta E = -2.479 \times 10^{-18} \text{ J}$$

To determine the wavelength, use $E = h\nu = \dfrac{hc}{\lambda}$; Rearrange for λ:

$$\lambda = \frac{hc}{E} = \frac{(6.626 \times 10^{-34} \text{ J s}) \left(2.998 \times 10^8 \frac{m}{s} \right)}{2.479 \times 10^{-18} \text{ J}} = 8.013 \times 10^{-8} \text{ m or } 80.13 \text{ nm}$$

5B Since $E = \dfrac{-Z^2 \times R_H}{n^2}$, the transitions are related to Z^2, hence, if the frequency is 16 times greater, then the value of the ratio $\dfrac{Z^2(\text{?-atom})}{Z^2(\text{H-atom})} = \dfrac{Z_?^2}{1^2} = 16$

We can see $Z^2 = 16$ or $Z = 4$ This is a Be nucleus. The hydrogen-like ion must be Be^{3+}.

6A Superman's de Broglie wavelength is given by the relationship $\lambda = h / mv$

$$\lambda = \frac{h}{mv} = \frac{6.626 \times 10^{-34} \text{J} \cdot \text{s}}{91 \text{kg} \times \frac{1}{5} \times 2.998 \times 10^8 \text{m/s}} = 1.21 \times 10^{-43} \text{ m}$$

6B   The de Broglie wavelength is given by $\lambda = h/mv$ which can be solved for v.

$$v = \frac{h}{m\lambda} = \frac{6.626 \times 10^{-34} \text{ J s}}{1.673 \times 10^{-27} \text{ kg} \times 10.0 \times 10^{-12} \text{ m}} = 3.96 \times 10^{4} \text{ m/s}$$

We used the facts that $1 \text{ J} = \text{kg m}^2\text{s}^{-2}$, $1 \text{pm} = 10^{-12} \text{ m}$ and $1 \text{ g} = 10^{-3} \text{ kg}$

7A $p = (91 \text{ kg})(5.996 \times 10^{7} \text{ m s}^{-1}) = 5.4\underline{6} \times 10^{9} \text{ kg m s}^{-1}$

$\Delta p = (0.015)(\ 5.4\underline{6} \times 10^{9} \text{ kg m s}^{-1}) = 8.2 \times 10^{7} \text{ kg m s}^{-1}$

$$\Delta x = \frac{h}{4\pi\,\Delta p} = \frac{6.626 \times 10^{-34} \text{ J s}}{(4\pi)(8.2 \times 10^{7}\,\frac{\text{kg m}}{\text{s}})} = 6.4 \times 10^{-43} \text{ m}$$

7B $24 \text{ nm} = 2.4 \times 10^{-8} \text{ m} = \Delta x = \dfrac{h}{4\pi\,\Delta p} = \dfrac{6.626 \times 10^{-34} \text{ J s}}{(4\pi)(\Delta p)}$

Solve for Δp: $\Delta p = 2.2 \times 10^{-27} \text{ kg m s}^{-1}$

$(\Delta v)(m) = \Delta p = 2.2 \times 10^{-27} \text{ kg m s}^{-1} = (\Delta v)(1.67 \times 10^{-27} \text{ kg})$ Hence, $\Delta v = 1.3 \text{ m s}^{-1}$

8A   The value of ℓ can range from 0 to $n-1$; in this case, $\ell = 0$ is acceptable. The value of m_ℓ is integral and ranges from $-\ell$ to $+\ell$; in this case, $m_\ell = 0$ is acceptable.

Yes, an orbital can have $n = 3, \ell\neq 0$, and $m_\ell = 0$.

8B   The first restriction is that ℓ must be an integer smaller than n. This restricts ℓ to the values: 2, 1, and 0. The second restriction is that the absolute value of m_ℓ must be an integer equal to or smaller than ℓ. This further restricts ℓ, and the only allowed values are: $\ell = 2$ and $\ell = 1$.

9A   The magnetic quantum number, m_ℓ, is not reflected in the orbital designation.

Because $\ell = 1$, this is a p orbital. Because $n = 3$, the designation is $3p$.

9B   The H-atom orbitals $3s$, $3p$ and $3d$ are degenerate. Therefore, the 9 quantum number combinations are:

	n	ℓ	m_ℓ
$3s$	3	0	0
$3p$	3	1	-1,0,+1
$3d$	3	2	-2,-1, 0,+1,+2

Hence, n = 3; l = 0, 1, 2; m$_l$ = -2, -1, 0, 1, 2

10A   We can simply sum the exponents to obtain the number of electrons in the neutral atom and thus the atomic number of the element. $Z = 2+2+6+2+6+2+2 = 22$, which is the atomic number for Ti.

10B Iodine has an atomic number of 53. The first 36 electrons have the same electron configuration as Kr: $1s^2 2s^2 2p^6 3s^2 3p^6 3d^{10} 4s^2 4p^6$. The next two electrons go into the $5s$ subshell $(5s^2)$, then 10 electrons fill the $4d$ subshell $(4d^{10})$, accounting for a total of 48 electrons. The last five electrons partially fill the $5p$ subshell $(5p^5)$.

The electron configuration of I is therefore $1s^2 2s^2 2p^6 3s^2 3p^6 3d^{10} 4s^2 4p^6 4d^{10} 5s^2 5p^5$. Each iodine atom has ten $3d$ electrons and one unpaired $5p$ electron.

11A Iron has 26 electrons, of which 18 are accounted for by the [Ar] core configuration. Beyond [Ar] there are two $4s$ electrons and six $3d$ electrons, as shown in the following orbital diagram.

$$\begin{array}{cc} 3d & 4s \end{array}$$
Fe: [Ar] ⇅↑↑↑↑↑ ⇅

11B Bismuth has 83 electrons, of which 54 are accounted for by the [Xe] configuration. Beyond [Xe] there are two $6s$ electrons, fourteen $4f$ electrons, ten $5d$ electrons, and three $6p$ electrons, as shown in the following orbital diagram.

$$\begin{array}{cccc} 4f & 5d & 6s & 6p \end{array}$$
Bi: [Xe] ⇅⇅⇅⇅⇅⇅⇅ ⇅⇅⇅⇅⇅ ⇅ ↑↑↑

12A (a) Tin is in the 5th period, hence, five electronic shells are filled or partially filled.
(b) The $3p$ subshell was filled with Ar; there are six $3p$ electrons in an atom of Sn.
(c) The electron configuration of Sn is [Kr] $4d^{10} 5s^2 5p^2$. There are no $5d$ electrons.
(d) Both of the $5p$ electrons are unpaired, thus there are two unpaired electrons in a Sn atom.

12B (a) The $3d$ subshell was filled at Zn, thus each Y atom has ten $3d$ electrons.
(b) Ge is in the $4p$ row; each germanium atom has two $4p$ electrons.
(c) We would expect each Au atom to have ten $5d$ electrons and one $6s$ electron. Thus each Au atom should have one unpaired electron.

EXERCISES

Electromagnetic Radiation

1. The wavelength is the distance between successive peaks. Thus, 4×1.17 nm $= \lambda = 4.68$ nm.

2. (a) $\nu = \dfrac{c}{\lambda} = \dfrac{3.00 \times 10^8 \text{m/s}}{4.68 \text{ nm} \times \dfrac{1 \text{m}}{10^9 \text{ nm}}} = 6.41 \times 10^{16} \text{Hz}$

(b) $E = h\nu = 6.626 \times 10^{-34}$ J s $\times 6.41 \times 10^{16}$ Hz $= 4.25 \times 10^{-17}$ J

3. (a) TRUE Since frequency and wavelength are inversely related to each other, radiation of shorter wavelength has higher frequency.
(b) FALSE Light of wavelengths between 390 nm and 790 nm is visible to the eye.
(c) FALSE All electromagnetic radiation has the same speed in vacuum.
(d) TRUE The wavelength of an X-ray is approximately 0.1 nm.

4. (a) $v = \dfrac{c}{\lambda} = \dfrac{2.9979 \times 10^8 \text{ m/s}}{418.7 \text{ nm}} \times \dfrac{10^9 \text{ nm}}{1 \text{ m}} = 7.160 \times 10^{14} \text{ Hz}$
(b) Light of wavelength 418.7 nm is in the visible region of the spectrum.
(c) 418.7 nm electromagnetic radiation is visible to the human eye as violet light.

5. Here we must convert all lengths into meters. (a) 5.9×10^{-6} m, (b) 1.13×10^{-3} m, (c) 8.60×10^{-8} m, (d) 6.92×10^{-6} m. The light having the highest frequency is (c) since it has the shortest wavelength.

6. Increasing frequency is decreasing wavelength. Radiowaves have the longest wavelengths/lowest frequencies (b), followed by infrared light (c), flowed by visible light (a) and finally UV radiation (d). Thus, the frequency increases from left to right in the following order (b) < (c) < (a) < (d)

7. The speed of light is used to convert the distance into an elapsed time.
$$\text{time} = 93 \times 10^6 \text{ mi} \times \frac{5280 \text{ ft}}{1 \text{ mi}} \times \frac{12 \text{ in.}}{1 \text{ ft}} \times \frac{2.54 \text{ cm}}{1 \text{ in}} \times \frac{1 \text{ s}}{3.00 \times 10^{10} \text{ cm}} \times \frac{1 \text{ min}}{60 \text{ s}} = 8.3 \text{ min}$$

8. The speed of light is used to convert the time into a distance spanned by light.
$$1 \text{ light year} = 1 \text{ y} \times \frac{365.25 \text{ d}}{1 \text{ y}} \times \frac{24 \text{ h}}{1 \text{ d}} \times \frac{3600 \text{ s}}{1 \text{ h}} \times \frac{2.9979 \times 10^8 \text{ m}}{1 \text{ s}} \times \frac{1 \text{ km}}{1000 \text{ m}} = 9.4607 \times 10^{12} \text{ km}$$

Atomic Spectra

9. (a) $v = 3.2881 \times 10^{15} \text{ s}^{-1} \left(\dfrac{1}{2^2} - \dfrac{1}{5^2} \right) = 6.9050 \times 10^{14} \text{ s}^{-1}$

(b) $v = 3.2881 \times 10^{15} \text{ s}^{-1} \left(\dfrac{1}{2^2} - \dfrac{1}{7^2} \right) = 7.5492 \times 10^{14} \text{ s}^{-1}$

$\lambda = \dfrac{2.9979 \times 10^8 \text{ m/s}}{7.5492 \times 10^{14} \text{ s}^{-1}} = 3.9711 \times 10^{-7} \text{ m} \times \dfrac{10^9 \text{ nm}}{1 \text{ m}} = 397.11 \text{ nm}$

(c) $v = \dfrac{3.00 \times 10^8 \text{ m}}{380 \text{ nm} \times \dfrac{1 \text{ m}}{10^9 \text{ nm}}} = 7.89 \times 10^{14} \text{ s}^{-1} = 3.2881 \times 10^{15} \text{ s}^{-1} \left(\dfrac{1}{2^2} - \dfrac{1}{n^2} \right)$

$0.250 - \dfrac{1}{n^2} = \dfrac{7.89 \times 10^{14} \text{ s}^{-1}}{3.2881 \times 10^{15} \text{ s}^{-1}} = 0.240 \qquad \dfrac{1}{n^2} = 0.250 - 0.240 = 0.010 \quad n = 10$

10. The frequencies of hydrogen emission lines in the infrared region of the spectrum other than the visible region would be predicted by replacing the constant "2" in the Balmer equation by the variable m, where m is an integer smaller than n: $m = 3, 4, \ldots$

The resulting equation is $v = 3.2881 \times 10^{15} \text{ s}^{-1} \left(\dfrac{1}{m^2} - \dfrac{1}{n^2} \right)$

11. **(a)** $E = h\nu = 6.626 \times 10^{-34} \text{ J s} \times 8.62 \times 10^{15} \text{ s}^{-1} = 5.71 \times 10^{-18} \text{ J / photon}$

(b) $E_m = 6.626 \times 10^{-34} \text{ J s} \times 1.53 \times 10^{14} \text{ s}^{-1} \times \dfrac{6.022 \times 10^{23} \text{ photons}}{\text{mol}} \times \dfrac{1 \text{ kJ}}{1000 \text{ J}} = 61.0 \text{ kJ/mol}$

12. **(a)** $\nu = \dfrac{E}{h} = \dfrac{4.18 \times 10^{-21} \text{ J}}{6.626 \times 10^{-34} \text{ J s}} = 6.31 \times 10^{12} \text{ s}^{-1} = 6.31 \times 10^{12} \text{ Hz}$

(b) $E = h\nu = \dfrac{hc}{\lambda}; \quad \lambda = \dfrac{hc}{E} = \dfrac{6.626 \times 10^{-34} \text{ J s} \times 3.00 \times 10^{8} \text{ m/s}}{215 \text{ kJ/mol} \times \dfrac{1000 \text{ J}}{1 \text{ kJ}} \times \dfrac{1 \text{ mol}}{6.022 \times 10^{23} \text{ photons}}} = 5.57 \times 10^{-7} \text{ m} = 557 \text{ nm}$

13. $\Delta E = -2.179 \times 10^{-18} \text{ J} \left(\dfrac{1}{n_f^{2}} - \dfrac{1}{n_i^{2}} \right) = -2.179 \times 10^{-18} \text{ J} \left(\dfrac{1}{3^2} - \dfrac{1}{5^2} \right) = -1.550 \times 10^{-19} \text{ J}$

$E_{\text{photon emitted}} = 1.550 \times 10^{-19} \text{ J} = h\nu \qquad \nu = \dfrac{E}{h} = \dfrac{1.550 \times 10^{-19} \text{ J}}{6.626 \times 10^{-34} \text{ J s}} = 2.339 \times 10^{14} \text{ s}^{-1}$

14. $\Delta E = 2.179 \times 10^{-18} \text{ J} \left(\dfrac{1}{6^2} - \dfrac{1}{2^2} \right) = -4.842 \times 10^{-19} \text{ J}$ (negative denotes energy release)

$\nu = \dfrac{E_{\text{photon emitted}}}{h} = \dfrac{4.842 \times 10^{-19} \text{ J}}{6.6260755 \times 10^{-34} \text{ J s}} = 7.307 \times 10^{14} \text{ s}^{-1}$

15. The longest wavelength component has the lowest frequency (and thus, the smallest energy).

$\nu = 3.2881 \times 10^{15} \text{ s}^{-1} \left(\dfrac{1}{2^2} - \dfrac{1}{3^2} \right) = 4.5668 \times 10^{14} \text{ s}^{-1} \quad \lambda = \dfrac{c}{\nu} = \dfrac{2.9979 \times 10^{8} \text{ m/s}}{4.5668 \times 10^{14} \text{ s}^{-1}} = 6.5646 \times 10^{-7} \text{ m}$

$= 656.46 \text{ nm}$

$\nu = 3.2881 \times 10^{15} \text{ s}^{-1} \left(\dfrac{1}{2^2} - \dfrac{1}{4^2} \right) = 6.1652 \times 10^{14} \text{ s}^{-1} \quad \lambda = \dfrac{c}{\nu} = \dfrac{2.9979 \times 10^{8} \text{ m/s}}{6.1652 \times 10^{14} \text{ s}^{-1}} = 4.8626 \times 10^{-7} \text{ m}$

$= 486.26 \text{ nm}$

$\nu = 3.2881 \times 10^{15} \text{ s}^{-1} \left(\dfrac{1}{2^2} - \dfrac{1}{5^2} \right) = 6.9050 \times 10^{14} \text{ s}^{-1} \quad \lambda = \dfrac{c}{\nu} = \dfrac{2.9979 \times 10^{8} \text{ m/s}}{6.9050 \times 10^{14} \text{ s}^{-1}} = 4.3416 \times 10^{-7} \text{ m}$

$= 434.16 \text{ nm}$

$\nu = 3.2881 \times 10^{15} \text{ s}^{-1} \left(\dfrac{1}{2^2} - \dfrac{1}{6^2} \right) = 7.3069 \times 10^{14} \text{ s}^{-1} \quad \lambda = \dfrac{c}{\nu} = \dfrac{2.9979 \times 10^{8} \text{ m/s}}{7.3069 \times 10^{14} \text{ s}^{-1}} = 4.1028 \times 10^{-7} \text{ m}$

$= 410.28 \text{ nm}$

16. $\lambda = 1880 \text{ nm} \times \dfrac{1 \text{ m}}{10^{9} \text{ nm}} = 1.88 \times 10^{-6} \text{ m}$. From Exercise 31, we see that wavelengths in the

Balmer series range downward from $6.5646 \times 10^{-7} \text{ m}$. Since this is less than $1.88 \times 10^{-6} \text{ m}$, we conclude that light with a wavelength of 1880 nm cannot be in the Balmer series.

17. First we determine the frequency of the radiation, and then match it with the Balmer equation.

$$\nu = \frac{c}{\lambda} = \frac{2.9979 \times 10^8 \, \text{m s}^{-1} \times \dfrac{10^9 \, \text{nm}}{1 \, \text{m}}}{389 \, \text{nm}} = 7.71 \times 10^{14} \, \text{s}^{-1} = 3.2881 \times 10^{15} \, \text{s}^{-1} \left(\frac{1}{2^2} - \frac{1}{n^2} \right)$$

$$\left(\frac{1}{2^2} - \frac{1}{n^2} \right) = \frac{7.71 \times 10^{14} \, \text{s}^{-1}}{3.2881 \times 10^{15} \, \text{s}^{-1}} = 0.234 = 0.2500 - \frac{1}{n^2} \qquad \frac{1}{n^2} = 0.016 \qquad n = 7.9 \approx 8$$

18. **(a)** The maximum wavelength occurs when $n = 2$ and the minimum wavelength occurs at the series convergence limit, namely, when n is exceedingly large (and $1/n^2 \approx 0$).

$$\nu = 3.2881 \times 10^{15} \, \text{s}^{-1} \left(\frac{1}{1^2} - \frac{1}{2^2} \right) = 2.4661 \times 10^{15} \, \text{s}^{-1} \quad \lambda = \frac{c}{\nu} = \frac{2.9979 \times 10^8 \, \text{m/s}}{2.4661 \times 10^{15} \, \text{s}^{-1}} = 1.2156 \times 10^{-7} \, \text{m}$$

$$= 121.56 \, \text{nm}$$

$$\nu = 3.2881 \times 10^{15} \, \text{s}^{-1} \left(\frac{1}{1^2} - 0 \right) = 3.2881 \times 10^{15} \, \text{s}^{-1} \quad \lambda = \frac{c}{\nu} = \frac{2.9979 \times 10^8 \, \text{m/s}}{3.2881 \times 10^{15} \, \text{s}^{-1}} = 9.1174 \times 10^{-8} \, \text{m}$$

$$= 91.174 \, \text{nm}$$

(b) First we determine the frequency of this spectral line, and then the value of n to which it corresponds.

$$\nu = \frac{c}{\lambda} = \frac{2.9979 \times 10^8 \, \text{m/s}}{95.0 \, \text{nm} \times \dfrac{1 \, \text{m}}{10^9 \, \text{nm}}} = 3.16 \times 10^{15} \, \text{s}^{-1} = 3.2881 \times 10^{15} \, \text{s}^{-1} \left(\frac{1}{1^2} - \frac{1}{n^2} \right)$$

$$\left(\frac{1}{1^2} - \frac{1}{n^2} \right) = \frac{3.16 \times 10^{15} \, \text{s}^{-1}}{3.2881 \times 10^{15} \, \text{s}^{-1}} = 0.961 \qquad \frac{1}{n^2} = 1.000 - 0.961 = 0.039 \qquad n = 5$$

(c) Let us use the same approach as the one in part (b).

$$\nu = \frac{c}{\lambda} = \frac{2.9979 \times 10^8 \, \text{m/s}}{108.5 \, \text{nm} \times \dfrac{1 \, \text{m}}{10^9 \, \text{nm}}} = 2.763 \times 10^{15} \, \text{s}^{-1} = 3.2881 \times 10^{15} \, \text{s}^{-1} \left(\frac{1}{1^2} - \frac{1}{n^2} \right)$$

$$\left(\frac{1}{1^2} - \frac{1}{n^2} \right) = \frac{2.763 \times 10^{15} \, \text{s}^{-1}}{3.2881 \times 10^{15} \, \text{s}^{-1}} = 0.8403 \qquad \frac{1}{n^2} = 1.000 - 0.8403 = 0.1597$$

This gives as a result $n = 2.502$. Since n is not an integer, there is no line in the Lyman spectrum with a wavelength of 108.5 nm.

Quantum Theory

19. **(a)** Here we combine $E = h\nu$ and $c = \nu\lambda$ to obtain $E = hc/\lambda$

$$E = \frac{6.626 \times 10^{-34} \, \text{J s} \times 2.998 \times 10^8 \, \text{m/s}}{474 \, \text{nm} \times \dfrac{1 \, \text{m}}{10^9 \, \text{nm}}} = 4.19 \times 10^{-19} \, \text{J / photon}$$

(b) $E_m = 4.19 \times 10^{-19} \, \dfrac{\text{J}}{\text{photon}} \times 6.022 \times 10^{23} \, \dfrac{\text{photons}}{\text{mol}} = 2.52 \times 10^5$ J/mol

20. First we determine the energy of an individual photon, and then its wavelength in nm.

$$E = \frac{1799 \dfrac{kJ}{mol} \times \dfrac{1000J}{1kJ}}{6.022 \times 10^{23} \dfrac{photons}{mol}} = 2.987 \times 10^{-18} \frac{J}{Photon} = \frac{hc}{\lambda} \quad or \quad \frac{hc}{E} = \lambda$$

$$\lambda = \frac{6.626 \times 10^{-34} \ J \ s \times 2.998 \times 10^{8} \ m/s}{2.987 \times 10^{-18} \ J} \times \frac{10^{9} \ nm}{1 \ m} = 66.50 \ nm \quad \text{This is ultraviolet radiation.}$$

21. The easiest way to answer this question is to convert all of (b) through (d) into nanometers. The radiation with the smallest wavelength will have the greatest energy per photon while the radiation with the largest wavelength has the smallest amount of energy per photon.

(a) 6.62×10^{2} nm

(b) 2.1×10^{-5} cm $\times \dfrac{1 \times 10^{7} \ nm}{1 \ cm} = 2.1 \times 10^{2}$ nm

(c) $3.58 \ \mu m \times \dfrac{1 \times 10^{3} \ nm}{1 \ \mu m} = 3.58 \times 10^{3}$ nm

(d) 4.1×10^{-6} m $\times \dfrac{1 \times 10^{9} \ nm}{1 \ m} = 4.1 \times 10^{3}$ nm

So, 2.1×10^{-5} nm radiation, by virtue of possessing the smallest wavelength in the set, has the greatest energy per photon. Conversely, since 4.1×10^{3} nm has the largest wavelength, it possesses the least amount of energy per photon.

22. This time let's express each type of radiation in terms of its frequency (ν).

(a) $\nu = 2.0 \times 10^{15} \ s^{-1}$

(b) The maximum frequency for infrared radiation is $\sim 3 \times 10^{14} \ s^{-1}$

(c) Here $\lambda = 7000$ Å (where 1 Å = 1×10^{-10} m) so $\lambda = 7000$ Å $\times \dfrac{1 \times 10^{-10} \ m}{1 \overset{\circ}{A}} = 7.000 \times 10^{-7}$ m

Calculate the frequency using the equation: $\nu = c/\lambda$

$$\nu = \frac{2.998 \times 10^{8} \ m \ s^{-1}}{7.000 \times 10^{-7} \ m} = 4.283 \times 10^{14} \ s^{-1}$$

(d) X-rays have frequencies that range from $10^{17} \ s^{-1}$ to $10^{21} \ s^{-1}$. Recall that for all forms of electromagnetic radiation, the energy per mole of photons increases with increasing frequency. Consequently, the correct order for the energy per mole of photons for the various types of radiation described in this question is:
infrared radiation (b) $< \lambda = 7000$ Å radiation (c) $< \nu = 2.0 \times 10^{15} \ s^{-1}$ (a) $<$ x-rays (d)

——————————— increasing energy per mole of photons ——————————→

23. Notice that energy and wavelength are inversely related: $E = \dfrac{hc}{\lambda}$. Therefore radiation that is 100 times as energetic as radiation with a wavelength of 988 nm will have a wavelength one hundredth as long, namely 9.88 nm. The frequency of this radiation is found by employing the wave equation.

$$v = \frac{c}{\lambda} = \frac{2.998 \times 10^8 \text{ m/s}}{9.88 \text{ nm} \times \dfrac{1m}{10^9 nm}} = 3.03 \times 10^{16} \text{s}^{-1} \text{ From Figure 8-3, we can see that this is UV radiation.}$$

24.
$$E_1 = hv = \frac{hc}{\lambda} = \frac{6.62607 \times 10^{-34} \text{ J} \cdot \text{s} \times 2.99792 \times 10^8 \text{ m s}^{-1}}{589.00 \text{ nm} \times \dfrac{1m}{10^9 nm}} = 3.3726 \times 10^{-19} \text{ J/Photon}$$

$$E_2 = \frac{6.62607 \times 10^{-34} \text{ J} \cdot \text{s} \times 2.99792 \times 10^8 \text{ m s}^{-1}}{589.59 \text{ nm} \times \dfrac{1m}{10^9 \text{ nm}}} = 3.3692 \times 10^{-19} \text{ J/Photon}$$

$$\Delta E = E_1 - E_2 = 3.3726 \times 10^{-19} \text{ J} - 3.3692 \times 10^{-19} \text{ J} = 0.0034 \times 10^{-19} \text{ J/photon} = 3.4 \times 10^{-22} \text{ J/photon}$$

The Photoelectric Effect

25. **(a)** $E = hv = 6.63 \times 10^{-34} \text{ J s} \times 9.96 \times 10^{14} \text{ s}^{-1} = 6.60 \times 10^{-19} \text{ J/photon}$

(b) Indium will display the photoelectric effect when exposed to ultraviolet light since ultraviolet light has a maximum frequency of $1 \times 10^{16} \text{ s}^{-1}$, which is above the threshold frequency of indium. It will not display the photoelectric effect when exposed to infrared light since the maximum frequency of infrared light is $\sim 3 \times 10^{14} \text{ s}^{-1}$, which is below the threshold frequency of indium.

26. In his explanation of the photoelectric effect, Einstein stated that each photon strikes and is absorbed by only one electron (hence the quote – one cannot kill two birds with one stone). He also, and very importantly, stated that no more than one photon can contribute its energy to a given electron (hence one cannot kill one bird with two stones). It is this second principle which explains why photoelectrons are not produced in larger number when the intensity of sub-threshold frequency light is increased.

The Bohr Atom

27. **(a)** $\text{radius} = n^2 a_0 = 6^2 \times 0.53 \text{Å} \times \dfrac{1m}{10^{10} \text{ Å}} \times \dfrac{10^9 \text{ nm}}{1m} = 1.9 \text{ nm}$

(b) $E_n = -\dfrac{R_H}{n^2} = -\dfrac{2.179 \times 10^{-18} \text{ J}}{6^2} = -6.053 \times 10^{-20} \text{ J}$

28. **(a)** $r_1 = 1^2 \times 0.53 \text{A} = 0.53 \text{Å}$ $\qquad\qquad r_3 = 3^2 \times 0.53 \text{Å} = 4.8 \text{Å}$

increase in distance $= r_3 - r_1 = 4.8 \text{Å} - 0.53 \text{Å} = 4.3 \text{Å}$

(b) $E_1 = \dfrac{-2.179 \times 10^{-18} \text{J}}{1^2} = -2.179 \times 10^{-18} \text{J}$ $E_3 = \dfrac{-2.179 \times 10^{-18} \text{J}}{3^2} = -2.421 \times 10^{-19} \text{J}$

increase in energy $= -2.421 \times 10^{-19}$ J $- (-2.179 \times 10^{-18}$ J$) = 1.937 \times 10^{-18}$ J

29.   (a) $\nu = \dfrac{2.179 \times 10^{-18} \text{ J}}{6.626 \times 10^{-34} \text{ J s}} \left(\dfrac{1}{4^2} - \dfrac{1}{7^2} \right) = 1.384 \times 10^{14} \text{ s}^{-1}$

(b) $\lambda = \dfrac{c}{\nu} = \dfrac{2.998 \times 10^8 \text{ m/s}}{1.384 \times 10^{14} \text{ s}^{-1}} = 2.166 \times 10^{-6} \text{ m} \times \dfrac{10^9 \text{ nm}}{1 \text{ m}} = 2166 \text{ nm}$

(c)   This is infrared radiation.

30.   The greatest quantity of energy is absorbed in the situation where the difference between the inverses of the squares of the two quantum numbers is the largest, and the system begins with a lower quantum number than it finishes with. The second condition eliminates answer **(d)** from consideration. Now we can consider the difference of the inverses of the squares of the two quantum numbers in each case.

(a) $\left(\dfrac{1}{1^2} - \dfrac{1}{2^2} \right) = 0.75$ (b) $\left(\dfrac{1}{2^2} - \dfrac{1}{4^2} \right) = 0.1875$ (c) $\left(\dfrac{1}{3^2} - \dfrac{1}{9^2} \right) = 0.0988$

Thus, the largest amount of energy is absorbed in the transition from $n = 1$ to $n = 2$, answer **(a)**, among the four choices given.

31.   (a)   According to the Bohr model, the radii of allowed orbits in a hydrogen atom are given by $r_n = (n)^2 \times (5.3 \times 10^{-11} \text{ m})$ where n = 1, 2, 3 . . . and $a_0 = 5.3 \times 10^{-11}$ m (0.53 Å or 53 pm) so, $r_4 = (4)^2 (5.3 \times 10^{-11} \text{ m}) = 8.5 \times 10^{-10}$ m.

(b)   Here we want to see if there is an allowed orbit at r = 4.00 Å. To answer this question we will employ the equation $r_n = n^2 a_0$:   4.00 Å $= n^2 (0.53 \text{ Å})$ or $n = 2.75$ Å
Since n is <u>not</u> a whole number, we can conclude that the electron in the hydrogen atom does not orbit at a radius of 4.00 Å (i.e., such an orbit is forbidden by selection rules).

(c)   The energy level for the n = 8 orbit is calculated using the equation

$E_n = \dfrac{-2.179 \times 10^{-18} \text{J}}{n^2}$ $E_8 = \dfrac{-2.179 \times 10^{-18} \text{J}}{8^2} = -3.405 \times 10^{-20}$ J (relative to $E_\infty = 0$ J)

(d)   Here we need to determine if 2.500×10^{-17} J corresponds to an allowed orbit in the hydrogen atom. Once again we will employ the equation $E_n = \dfrac{-2.179 \times 10^{-18} \text{J}}{n^2}$

2.500×10^{-17} J $= \dfrac{-2.179 \times 10^{-18} \text{J}}{n^2}$ or $n^2 = \dfrac{-2.179 \times 10^{-18} \text{J}}{-2.500 \times 10^{-17} \text{J}}$ hence, $n = 0.2952$

Because n is not a whole number, -2.500×10^{-17} J is not an allowed energy state for the electron in a hydrogen atom.

32. Only transitions (a) and (d) result in the emission of a photon ((b) and (c) involve absorption, not the emission of light). In the Bohr model, the energy difference between two successive energy levels decreases as the value of n increases. Thus, the $n = 3 \rightarrow n = 2$ electron transition involves a greater loss of energy than the $n = 4 \rightarrow n = 3$ transition. This means that the photon emitted by the $n = 4 \rightarrow n = 3$ transition will be lower in energy and hence, longer in wavelength than the photon produced by the $n = 3 \rightarrow n = 2$ transition. Consequently, among the four choices given, the electron transition (a) is the one that will produce light of the longest wavelength.

33. If infrared light is produced, the quantum number of the final state must have a lower value (i.e. be of lower energy) than the quantum number of the initial state. First we compute the frequency of the transition being considered (from $v = c / \lambda$), and then solve for the final quantum number.

$$v = \frac{c}{\lambda} = \frac{3.00 \times 10^8 \text{m/s}}{2170 \text{ nm} \times \frac{1 \text{m}}{10^9 \text{ nm}}} = 1.38 \times 10^{14} \text{s}^{-1} = \frac{2.179 \times 10^{-18} \text{ J}}{6.626 \times 10^{-34} \text{ J s}} \left(\frac{1}{n^2} - \frac{1}{7^2} \right) = 3.289 \times 10^{15} \text{ s}^{-1} \left(\frac{1}{n^2} - \frac{1}{7^2} \right)$$

$$\left(\frac{1}{n^2} - \frac{1}{7^2} \right) = \frac{1.38 \times 10^{14} \text{ s}^{-1}}{3.289 \times 10^{15} \text{ s}^{-1}} = 0.0419_6 \qquad \frac{1}{n^2} = 0.0419_6 + \frac{1}{7^2} = 0.0623_7 \qquad n = 4$$

34. If infrared light is produced, the quantum number of the final state must have a lower value (i.e. be of lower energy) than the quantum number of the initial state. First we compute the frequency of the transition being considered (from $v = c / \lambda$), and then solve for the initial quantum number, n.

$$v = \frac{c}{\lambda} = \frac{3.00 \times 10^8 \text{m/s}}{3740 \text{ nm} \times \frac{1 \text{m}}{10^9 \text{nm}}} = 8.02 \times 10^{13} \text{s}^{-1} = \frac{2.179 \times 10^{-18} \text{ J}}{6.626 \times 10^{-34} \text{ J s}} \left(\frac{1}{5^2} - \frac{1}{n^2} \right) = 3.289 \times 10^{15} \text{ s}^{-1} \left(\frac{1}{5^2} - \frac{1}{n^2} \right)$$

$$\left(\frac{1}{5^2} - \frac{1}{n^2} \right) = \frac{8.02 \times 10^{13} \text{ s}^{-1}}{3.289 \times 10^{15} \text{ s}^{-1}} = 0.0243_8 \qquad \frac{1}{n^2} = -0.0243_8 + \frac{1}{5^2} = 0.0156_2 \qquad n = 8$$

35. (a) Line A is for the transition $n = 3 \rightarrow n = 1$, while Line B is for the transition $n = 4 \rightarrow n = 1$

(b) This transition corresponds to the n = 3 to n = 1 transition. Hence, $\Delta E = hc/\lambda$
$\Delta E = (6.626 \times 10^{-34} \text{ J s}) \times (2.998 \times 10^8 \text{ m s}^{-1}) \div (103 \times 10^{-9} \text{ m}) = 1.92\underline{9} \times 10^{-18} \text{ J}$
$\Delta E = -Z^2 R_H / n_1^2 - -Z^2 R_H / n_2^2$
$1.92\underline{9} \times 10^{-18} \text{ J} = - Z^2 (2.179 \times 10^{-18}) / (3)^2 + Z^2 (2.179 \times 10^{-18}) / (1)^2$
$Z^2 = 0.996$ and $Z = 0.998$ Thus, this is the spectrum for the hydrogen atom.

36. (a) Line A is for the transition n = 5 → n = 2, while Line B is for the transition n = 6 → n = 2

(b) This transition corresponds to the n = 5 to n = 2 transition. Hence, $\Delta E = hc/\lambda$
$\Delta E = (6.626 \times 10^{-34} \text{ J s}) \times (2.998 \times 10^8 \text{ m s}^{-1}) \div (434 \times 10^{-9} \text{ m}) = 4.57\underline{7} \times 10^{-19} \text{ J}$
$\Delta E = -Z^2 R_H / n_1^2 - (-Z^2 R_H / n_2^2)$
$4.57\underline{7} \times 10^{-19} \text{ J} = - Z^2 (2.179 \times 10^{-18}) / (5)^2 + Z^2 (2.179 \times 10^{-18}) / (2)^2$
$Z^2 = 1.00$ and $Z = 1.00$ Thus, this is the spectrum for the hydrogen atom.

37. **(a)** Line A is for the transition n = 5 → n = 2, while Line B is for the transition n = 6 → n = 2

(b) This transition corresponds to the n = 5 to n = 2 transition. Hence, $\Delta E = hc/\lambda$

$\Delta E = (6.626 \times 10^{-34} \text{ J s}) \times (2.998 \times 10^8 \text{ m s}^{-1}) \div (27.1 \times 10^{-9} \text{ m}) = 7.33 \times 10^{-18} \text{ J}$

$\Delta E = -Z^2 R_H / n_1^2 - -Z^2 R_H / n_2^2$

$4.57\underline{7} \times 10^{-19} \text{ J} = -Z^2 (2.179 \times 10^{-18})/(5)^2 + Z^2 (2.179 \times 10^{-18})/(2)^2$

$Z^2 = 16.0\underline{2}$ and $Z = 4.00$ Thus, this is the spectrum for the Be^{3+} cation.

38. **(a)** Line A is for the transition n = 4 → n = 1, while Line B is for the transition n = 5 → n = 1

(b) This transition corresponds to the n = 4 to n = 1 transition. Hence, $\Delta E = hc/\lambda$

$\Delta E = (6.626 \times 10^{-34} \text{ J s}) \times (2.998 \times 10^8 \text{ m s}^{-1}) \div (10.8 \times 10^{-9} \text{ m}) = 1.84 \times 10^{-17} \text{ J}$

$\Delta E = -Z^2 R_H / n_1^2 - -Z^2 R_H / n_2^2$

$4.57\underline{7} \times 10^{-19} \text{ J} = -Z^2 (2.179 \times 10^{-18})/(5)^2 + Z^2 (2.179 \times 10^{-18})/(2)^2$

$Z^2 = 9.00\underline{4}$ and $Z = 3.00$ Thus, this is the spectrum for the Li^{2+} cation.

Wave-Particle Duality

39. The de Broglie equation is $\lambda = h / mv$. This means that, for a given wavelength to be produced, a lighter particle would have to be moving faster. Thus, electrons would have to move faster than protons to display matter waves of the same wavelength.

40. First, we rearrange the de Broglie equation, and solve it for velocity: $v = h / m\lambda$. Then we compute the velocity of the electron. From Table 2-1, we see that the mass of an electron is 9.109×10^{-28} g.

$$v = \frac{h}{m\lambda} = \frac{6.626 \times 10^{-34} \text{ J s}}{\left(9.109 \times 10^{-28} \text{ g} \times \frac{1 \text{ kg}}{1000 \text{ g}}\right)\left(1 \mu m \times \frac{1 \text{ m}}{10^6 \mu m}\right)} = 7 \times 10^2 \text{ m/s}$$

41. $$\lambda = \frac{h}{mv} = \frac{6.626 \times 10^{-34} \text{ J s}}{\left(145 \text{ g} \times \frac{1 \text{ kg}}{1000 \text{ g}}\right)\left(168 \text{ km/h} \times \frac{1 \text{ h}}{3600 \text{ s}} \times \frac{1000 \text{ m}}{1 \text{ km}}\right)} = 9.79 \times 10^{-35} \text{ m}$$

The diameter of a nucleus approximates 10^{-15} m, which is far larger than the baseball's wavelength.

42. $$\lambda = \frac{h}{mv} = \frac{6.626 \times 10^{-34} \text{ J s}}{1000 \text{ kg} \times 25 \text{ m/s}} = 2.7 \times 10^{-38} \text{ m}$$

Because the car's wavelength is smaller than the Planck limit ($\sim 10^{-33}$ cm), which is the scale at which quantum fluctuations occur, its experimental measurement is impossible.

The Heisenberg Uncertainty Principle

43. The Bohr model is a determinant model for the hydrogen atom. It implies that the position of the electron is exactly known at any time in the future, once its position is known at the present. The distance of the electron from the nucleus also is exactly known, as is its energy. And finally, the velocity of the electron in its orbit is exactly known. All of these exactly known quantities—position, distance from nucleus, energy, and velocity—can't, according to the Heisenberg uncertainty principle, be known with great precision simultaneously.

44. Einstein believed very strongly in the law of cause and effect, what is known as a deterministic view of the universe. He felt that the need to use probability and chance ("playing dice") in describing atomic structure resulted because a suitable theory had not yet been developed to permit accurate predictions. He believed that such a theory could be developed, and had good reason for his belief: The developments in the theory of atomic structure came very rapidly during the first thirty years of this century, and those, such as Einstein, who had lived through this period had seen the revision of a number of theories and explanations that were thought to be the final answer. Another viewpoint in this area is embodied in another famous quotation: "Nature is subtle, but not malicious." The meaning of this statement is that the causes of various effects may be obscure but they exist nonetheless. Bohr was stating that we should accept theories as they are revealed by experimentation and logic, rather than attempting to make these theories fit our preconceived notions of what we believe they should be. In other words, Bohr was telling Einstein to keep an open mind.

45. $\Delta v = \left(\dfrac{1}{100} \right)(0.1)\left(2.998 \times 10^8 \dfrac{m}{s} \right) = 2.998 \times 10^5 \text{ m/s} \quad m = 1.673 \times 10^{-27} \text{ kg}$

$\Delta p = m\Delta v = (1.673 \times 10^{-27} \text{ kg})(2.998 \times 10^5 \text{ m/s}) = 5.\underline{0} \times 10^{-22} \text{ kg m s}^{-1}$

$\Delta x = \dfrac{h}{4\pi \Delta p} = \dfrac{6.626 \times 10^{-34} \text{ J s}}{(4\pi)(5._0 \times 10^{-22} \frac{\text{kg m}}{s})} = \sim 1 \times 10^{-13} \text{ m} \quad (\sim 100 \text{ times the diameter of a nucleus})$

46. Assume a mass of 1000 kg for the automobile and that its position is known to 1 cm (0.01 m)

$\Delta v = \dfrac{h}{4\pi m\Delta x} = \dfrac{6.626 \times 10^{-34} \text{ J s}}{(4\pi)(1000 \text{ kg})(0.01 \text{ m})} = 5 \times 10^{-36} \text{ m s}^{-1}$

This represents an undetectable uncertainty in the velocity.

47. electron mass $= 9.109 \times 10^{-31}$ kg, $\lambda = 0.53$ Å (1 Å $= 1 \times 10^{-10}$ m), hence $\lambda = 0.53 \times 10^{-10}$ m

$\lambda = \dfrac{h}{mv}$ or $v = \dfrac{h}{m\lambda} = \dfrac{6.626 \times 10^{-34} \text{ J s}}{(9.109 \times 10^{-31} \text{ kg})(0.53 \times 10^{-10} \text{ m})} = 1.4 \times 10^7 \text{ m s}^{-1}$

48. $\Delta E = h\nu = \dfrac{hc}{\lambda}$ or $\lambda = \dfrac{hc}{E} = \dfrac{(6.626 \times 10^{-34} \text{ J s})(2.998 \times 10^8 \text{ m s}^{-1})}{(-5.45 \times 10^{-19} \text{ J}) - (-2.179 \times 10^{-18} \text{ J})} = 1.216 \times 10^{-7} \text{ m}$

$v = \dfrac{h}{m\lambda} = \dfrac{6.626 \times 10^{-34} \text{ J s}}{(9.109 \times 10^{-31} \text{ kg})(1.216 \times 10^{-7} \text{ m})} = 5.983 \times 10^3 \text{ m s}^{-1}$

Wave Mechanics

49. A sketch of this situation is presented at right. We see that 2.50 waves span the space of the 42 cm. Thus, the length of each wave is obtained by equating: $2.50\lambda = 42$ cm, giving $\lambda \approx 17$ cm.

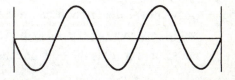

50. If there are four nodes, then there are three half-wavelengths within the string: one between the first and second node, the second half-wavelength between nodes 2 and 3, and the third between nodes 3 and 4. Therefore, $3 \times \lambda/2 = $ length $= 3 \times 17$ cm/2 $= 26$ cm long string.

51. The second overtone has three half-wavelengths within the string. We can find the wavelength for the second overtone by using the equation l: string length $= n\lambda/2$, which gives the result: 24 in $= 3\lambda/2$, $\lambda_{\text{second overtone}} = 16$ inches

52. The third harmonic for the wave in the box has five nodes and four antinodes. The total number of nodes is $n + 1$ so that would be $n = 4$ in this case. Also, we have been told that the box is 100 pm in length, so $L = 100$ pm here. We find the wavelength for the third harmonic by using $\lambda = \dfrac{2L}{n}$. Thus, $\lambda = \dfrac{2(100\,\text{pm})}{4} = 50.0$ pm

53. The differences between Bohr orbits and wave mechanical orbitals are given below.

(a) The first difference is that of shape. Bohr orbits, as originally proposed, are circular (later Sommerfeld proposed elliptical orbits). Orbitals, on the other hand, can be spherical, or shaped like two tear drops or two squashed spheres; or shaped like four tear drops meeting at their points.

(b) Bohr orbits are planar pathways, while orbitals are three-dimensional regions of space in which there is a high probability of finding electrons.

(c) The electron in a Bohr orbit has a definite trajectory. Its position and velocity are known at all times. The electron in an orbital, however, does not have a well-known position or velocity. In fact, there is a small but definite probability that the electron may be found outside the boundaries generally drawn for the orbital. Orbits and orbitals are similar in that the radius of a Bohr orbit is comparable to the average distance of the electron from the nucleus in the corresponding wave mechanical orbital.

54. We must be careful to distinguish between probability density—the chance of finding the electron within a definite volume of space—and the probability of finding the electron at a certain distance from the nucleus. The probability density—that is the probability of finding the electron within a small volume element (such as 1 pm^3)—at the nucleus may well be high, in fact higher than the probability density at a distance 0.53 Å from the nucleus. But the probability of finding the electron at a fixed distance from the nucleus is this probability density multiplied by all of the many small volume elements that are located at this distance. (Recall the dart board analogy of Figure 8-34.)

Quantum Numbers and Electron Orbitals

55. Answer (a) is incorrect because the values of m_s may be either $+\frac{1}{2}$ or $-\frac{1}{2}$. Answers (b) and (d) are incorrect because the value of ℓ may be any integer $\geq |m_\ell|$, and less than n. Thus, answer (c) is the only one that is correct.

56. **(a)** $n=3, \ell=2, m_\ell=2, m_s=+\frac{1}{2}$ (ℓ must be smaller than n and $\geq |m_\ell|$.) This is a $3d$ orbital.

(b) $n\geq 3, \ell=2, m_\ell=-1, m_s=-\frac{1}{2}$ (n must be larger than ℓ.) This is any d orbital, namely
$$3d, 4d, 5d, \ldots \ etc.$$

(c) $n=4, \ell=2, m_\ell=0, m_s=+\frac{1}{2}$ (m_s can be either $+\frac{1}{2}$ or $-\frac{1}{2}$.) This is $4d$ orbital.

(d) $n\geq 1, \ell=0, m_\ell=0, m_s=+\frac{1}{2}$ (n can be any positive integer, m_s could also equal $-\frac{1}{2}$.) This is a $1s$ orbital.

57. **(a)** $n=5$ $\ell=1$ $m_\ell=0$ designates a $5p$ orbital. ($\ell=1$ for all p orbitals.)

(b) $n=4$ $\ell=2$ $m_\ell=-2$ designates a $4d$ orbital. ($\ell=2$ for all d orbitals.)

(c) $n=2$ $\ell=0$ $m_\ell=0$ designates a $2s$ orbital. ($\ell=0$ for all s orbitals.)

58. **(a)** TRUE; The fourth principal shell has $n=4$.

(b) TRUE; A d orbital has $\ell=2$. Since $n=4, \ell$ can be $=3, 2, 1, 0$. Since $m_\ell=-2, \ell$ can be equal to 2 or 3.

(c) FALSE; A p orbital has $\ell=1$. But we demonstrated in part (b) that the only allowed values for ℓ are 2 and 3.

(d) FALSE Either $+\dfrac{1}{2}$ or $-\dfrac{1}{2}$ is permitted as a value of m_s.

59. **(a)** Just one electron can have $n=3, \ell = 2, m_\ell=0$, and $m_s=+\frac{1}{2}$. Four quantum numbers completely designate an electron.

(b) These three quantum numbers designate an orbital, which can hold two electrons. Two electrons can have the three quantum numbers $n=3, \ell = 2, m_\ell=0$.

(c) These two quantum numbers designate the $3d$ subshell, which contains five orbitals, with the possibility of two electrons in each. Ten electrons can have the two quantum numbers $n=3, \ell = 2$.

(d) $n=3$ designates the shell that contains the $3d$ subshell that has five orbitals, the $3p$ subshell with three orbitals, and the $3s$ subshell with one orbital. Each of these nine orbitals in the shell can hold two electrons, for a total of 18 electrons.

(e) The first two quantum numbers designate the $3d$ subshell, which has five orbitals. Each orbital can accommodate one electron with spin up. Five electrons can have the quantum numbers $n=3, \ell=2, m_s=1/2$.

60. **(a)** Each subshell has a different value for ℓ, the orbital quantum number. When $n = 4$, the possible values of ℓ are 0, 1, 2, and 3. There are four subshells in the $n = 4$ level.

(b) In the $n = 3$ level, the possible value of the orbital quantum number are $\ell = 0$, 1, and 2. These correspond to the $3s$ subshell, the $3p$ subshell, and the $3d$ subshell.

(c) In any $\ell = 3$ subshell, the possible values of the magnetic quantum number are $m_\ell = -3, -2, -1, 0, 1, 2, 3$. This means that there are seven orbitals in an $f(\ell = 3)$ subshell.

(d) The values $n = 4, \ell \ 3$, and $m_\ell = -2$ completely designate an orbital. Just one orbital has these three quantum numbers.

(e) Within the $n = 4$ level, there is a $4s$ subshell with one orbital, a $4p$ subshell with three orbitals, a $4d$ subshell with five orbitals, and a $4f$ subshell with seven orbitals. The total number of orbitals in the $n = 4$ level is $1 + 3 + 5 + 7 = 16$ orbitals.

The Shapes of Orbitals and Radial Probabilities

61. The wavefunction for the 2s orbital of a hydrogen atom is:

$$\psi_{2s} = \frac{1}{4}\left(\frac{1}{2\pi a_o^{\ 3}}\right)^{1/2}\left(2 - \frac{r}{a_o}\right)e^{-\frac{r}{2a_o}}$$

Where $r = 2a_o$, the $\left(2 - \frac{r}{a_o}\right)$ term becomes zero, thereby making $\psi_{2s} = 0$. At this point the wave function has a radial node (i.e. the electron density is zero). The finite value of r is 2 a_o at the node, which is equal to 2×53 pm or 106 pm. Thus at 106 pm, there is a nodal surface with zero electron density.

62. The radial part for the 2s wavefunction in the Li^{2+} dication is:

$$R_{2s} = \left(\frac{Z}{2\pi a_o}\right)^{3/2}\left(2 - \frac{Zr}{a_o}\right)e^{-\frac{Zr}{2a_o}}$$

(Z is the atomic number for the element, and, in Li^{2+} Z = 3). At $Zr = 2a_o$, the pre-exponential term for the 2s orbital of Li^{2+} is zero. The Li^{2+} 2s orbital has a nodal sphere at $\frac{2a_o}{3}$ or 35 pm.

63. The angular part of the $2p_y$ wavefunction is $Y(\theta\phi)_{py} = \sqrt{\frac{3}{4\pi}}\sin\theta\sin\phi$. The two lobes of the $2p_y$ orbital lie in the xy plane and perpendicular to this plane is the xz plane. For all points in the xz plane $\phi = 0$, and since the sine of 0° is zero, this means that the entire x z plane is a node. Thus, the probability of finding a $2p_y$ electron in the xz plane is zero.

64. The angular component of the wavefunction for the $3d_{xz}$ orbital is

$$Y(\theta\phi)_{d_{xz}} = \sqrt{\frac{15}{4\pi}}\sin\theta\cos\theta\cos\phi.$$

The four lobes of the d_{xz} orbital lie in the xz plane. The xy plane is perpendicular to the xz plane, and thus the angle for θ is 90°. The cosine of 90° is zero, so at every point in the xy plane the angular function has a value of zero. In other words, the entire xy plane is a node and, as a result, the probability of finding a $3d_{xz}$ electron in the xy plane is zero.

65./67. The $2p_x$ orbital $Y(\theta\phi) = \sqrt{\dfrac{3}{4\pi}}\,\sin\theta\cos\phi$; Note: in the xy plane $\theta = 90°$ and $\sin\theta = 1$

Plotting in the xy plane requires that we vary only ϕ

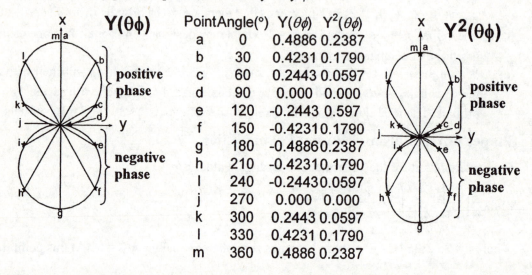

Point	Angle(°)	$Y(\theta\phi)$	$Y^2(\theta\phi)$
a	0	0.4886	0.2387
b	30	0.4231	0.1790
c	60	0.2443	0.0597
d	90	0.000	0.000
e	120	-0.2443	0.597
f	150	-0.4231	0.1790
g	180	-0.4886	0.2387
h	210	-0.4231	0.1790
I	240	-0.2443	0.0597
j	270	0.000	0.000
k	300	0.2443	0.0597
I	330	0.4231	0.1790
m	360	0.4886	0.2387

66./68. The $2p_y$ orbital $Y(\theta\phi) = \sqrt{\dfrac{3}{4\pi}}\,\sin\theta\sin\phi$, however, in the xy plane $\theta = 90°$ and $\sin\theta = 1$

Plotting in the xy plane requires that we vary only ϕ

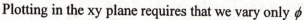

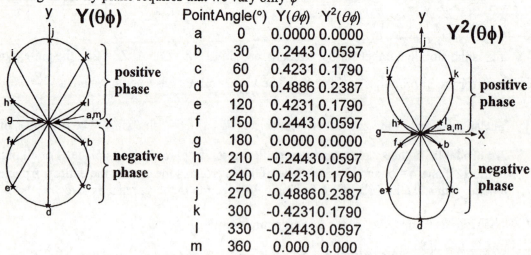

Point	Angle(°)	$Y(\theta\phi)$	$Y^2(\theta\phi)$
a	0	0.0000	0.0000
b	30	0.2443	0.0597
c	60	0.4231	0.1790
d	90	0.4886	0.2387
e	120	0.4231	0.1790
f	150	0.2443	0.0597
g	180	0.0000	0.0000
h	210	-0.2443	0.0597
I	240	-0.4231	0.1790
j	270	-0.4886	0.2387
k	300	-0.4231	0.1790
I	330	-0.2443	0.0597
m	360	0.000	0.000

69. A plot of radial probability distribution versus r/a_o for a H_{1s} orbital shows a maximum at 1.0 (that is, $r = a_o$ or $r = 53$ pm). The plot is shown below:

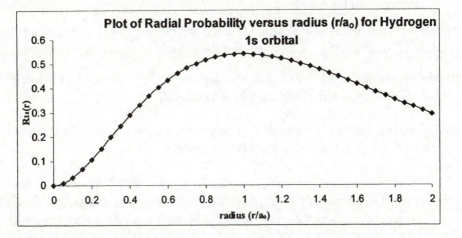

70. A plot of radial probability distribution versus r/a_o for a Li^{2+}_{1s} orbital shows a maximum at 0.33 (that is, $r = a_o/3$ or $r = 18$ pm). The plot is shown below:

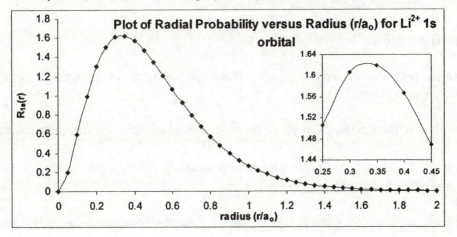

71. (a) To answer this question, we must keep two simple rules in mind.
  1. Value of ℓ is the number of angular nodes.
  2. Total number of nodes = n -1.

  From this we see that this is a p-orbital (1 angular node $\rightarrow$ $\ell = 1$) and because there are a total of 2 nodes, n = 3. This must be a 3p orbital.

  (b) From this we see that this is a d-orbital (2 angular node $\rightarrow$ $\ell = 2$) and because there are a total of 2 nodes, n = 3. This must be a 3d orbital.

  (c) From this we see that this is an f-orbital (3 angular node $\rightarrow$ $\ell = 2$) and because there are a total of 5 nodes, n = 6. This must be a 6f orbital.

72. **(a)** To answer this question, we must keep two simple rules in mind.
1. Value of ℓ is the number of angular nodes.
2. Total number of nodes = n -1.

From this we see that this is a p-orbital (1 angular node → $\ell = 1$) and because there are a total of 3 nodes, n = 4. This must be a 4p orbital.

(b) From this we see that this is an s-orbital (0 angular nodes → $\ell = 0$) and because there are a total of 5 nodes, n = 6. This must be a 6s orbital.

(c) From this we see that this is a g-orbital (4 angular node → $\ell = 4$) and because there are a total of 5 nodes, n = 6. This must be a 6g orbital.

73. The orbital is in the xy plane has two angular nodes (d-orbital) and 2 spherical nodes (total nodes = 4, hence n = 5). Since the orbital points between the x-axis and y-axis, this is a $5d_{xyz}$ orbital. The second view of the same orbital is just a 90° rotation about the x-axis.

74. The orbital in the xy plane has three angular nodes (f-orbital) and 1 spherical nodes (total nodes = 4, hence n = 5). This is one of the 5f orbitals. The second view of the same orbital is just a 90° rotation about the x-axis.

Electron Configurations

75. Configuration **(b)** is correct for phosphorus. The reasons why the other configurations are incorrect are given below.

(a) The two electrons in the $3s$ subshell must have opposed spins, or different values of m_s.

(c) The three $3p$ orbitals must each contain one electron, before a pair of electrons is placed in any one of these orbitals.

(d) The three unpaired electrons in the $3p$ subshell must all have the same spin, either all spin up or all spin down.

76. The electron configuration of Mo is [Kr] $_{4d}$⬛⬛⬛⬛⬛ $_{5s}$⬛

(a) [Ar] $_{3d}$⬛⬛⬛⬛⬛ $_{3f}$⬛⬛⬛⬛⬛⬛⬛ This configuration has the correct number of electrons, it incorrectly assumes that there is a $3f$ subshell (n = 3, $\ell = 3$).

(b) [Kr] $_{4d}$⬛⬛⬛⬛⬛ $_{5s}$⬛ This is the correct electron configuration.

(c) [Kr] $_{4d}$⬛⬛⬛⬛⬛ $_{5s}$⬛ This configuration has one electron too many.

(d) [Ar] $_{3d}$⬛⬛⬛⬛⬛ $_{4s}$⬛ $_{4p}$⬛⬛⬛ This configuration incorrectly assumes that electrons $_{4d}$⬛⬛⬛⬛⬛ enter the 4d subshell before they enter the 5s subshell.

77. We write the correct electron configuration first in each case.

(a) P: $[Ne]3s^2 3p^3$ There are 3 unpaired electrons in each P atom.

(b) Br: $[Ar]3d^{10}4s^2 4p^5$ There are ten $3d$ electrons in an atom of Br.

(c) Ge: $[Ar]3d^{10}4s^2 4p^2$ There are two $4p$ electrons in an atom of Ge.

(d) Ba: $[Xe]6s^2$ There are two $6s$ electrons in an atom of Ba.

(e) Au: $[Xe]4f^{14}5d^9 6s^2$ There are fourteen $4f$ electrons in an atom of Au.

78. (a) The $4p$ subshell of Br contains 5 electrons $[Ar]$ $_{3d}\boxed{\uparrow\downarrow|\uparrow\downarrow|\uparrow\downarrow|\uparrow\downarrow|\uparrow\downarrow}$ $_{4s}\boxed{\uparrow\downarrow}$ $_{4p}\boxed{\uparrow\downarrow|\uparrow|\uparrow}$

(b) The $3d$ subshell of Co^{2+} contains 7 electrons $[Ar]$ $_{3d}\boxed{\uparrow\downarrow|\uparrow\downarrow|\uparrow|\uparrow|\uparrow}$ $_{4s}\boxed{}$

(c) The $5d$ subshell of Pb contains 10 electrons:

$[Xe]$ $_{4f}\boxed{\uparrow\downarrow|\uparrow\downarrow|\uparrow\downarrow|\uparrow\downarrow|\uparrow\downarrow|\uparrow\downarrow|\uparrow\downarrow}$ $_{5d}\boxed{\uparrow\downarrow|\uparrow\downarrow|\uparrow\downarrow|\uparrow\downarrow|\uparrow\downarrow}$ $_{6s}\boxed{\uparrow\downarrow}$ $_{6p}\boxed{\uparrow|\uparrow|\uparrow}$

79. (a) N is the third element in the p-block of the second period. It has three $2p$ electrons.

(b) Rb is the first element in the s-block of the *fifth* period. It has two $4s$ electrons.

(c) As is in the p-block of the fourth period. The $3d$ subshell is filled with ten electrons, but no $4d$ electrons have been added.

(d) Au is in the d-block of the sixth period; the $4f$ subshell is filled. Au has fourteen $4f$ electrons.

(e) Pb is the second element in the p-block of the sixth period; it has two $6p$ electrons. Since these two electrons are placed in separate $6p$ orbitals, they are unpaired. There are two unpaired electrons.

(f) Group 14 of the periodic table is the group with the elements C, Si, Ge, Sn, and Pb. This group currently has five named elements.

(g) The sixth period begins with the element Cs $(Z = 55)$ and ends with the element Rn $(Z = 86)$. This period is 32 elements long.

80. (a) Sb is in group 15, with an outer shell electron configuration $ns^2 np^3$. Sb has five outer-shell electrons.

(b) Pt has an atomic number of $Z = 78$. The fourth principal shell fills as follows: $4s$ from $Z = 19$ (K) to $Z = 20$ (Ca); $4p$ from $Z = 31$ (Ga) to $Z = 36$ (Kr); $4d$ from $Z = 39$ (Y) to $Z = 48$ (Cd); and $4f$ from $Z = 58$ (Ce) to $Z = 71$ (Lu). Since the atomic number of Pt is greater than $Z = 71$, the entire fourth principal shell is filled, with 32 electrons.

(c) The five elements with six outer-shell electrons are those in group 16: O, S, Se, Te, Po.

    (d)   The outer-shell electron configuration is ns^2np^4, giving the following as a partial orbital diagram:$_{s^2}$☒ $_p$☒☐☐☐ There are two unpaired electrons in an atom of Te.

    (e)   The sixth period begins with Cs and ends with Rn. There are 10 outer transition elements in this period (La, and Hf through Hg) and there are 14 inner transition elements in the period (Ce through Lu). Thus, there are 10 + 14 or 24 transition elements in the sixth period.

81.   Since the periodic table is based on electron structure, two elements in the same group (Pb and element 114) should have similar electron configurations.

    (a)   Pb: [Xe] $4f^{14}5d^{10}6s^26p^2$ **(b)**   114: [Rn] $5f^{14}6d^{10}7s^27p^2$

82.   **(a)**   The fifth period noble gas in group 18 is the element Xe.

    (b)   A sixth period element whose atoms have three unpaired electrons is an element in group 15, which has an outer electron configuration of ns^2np^3, and thus has three unpaired p electrons. This is the element Bi.

    (c)   One d-block element that has one $4s$ electron is Cu: [Ar] $3d^{10}4s^1$. Another is Cr: [Ar] $3d^54s^1$.

    (d)   There are several p-block elements that are metals, namely Al, Ga, In, Tl, Sn, Pb, and Bi.

83.   **(a)** This is an excited state, the 2s orbital should fill before any electrons enter the 2 p orbital.

    (b) This is an excited state, the electrons in the 2p orbitals should have the same spin (Hund's rule)

    (c) This is the ground state configuration of N

    (d) This is an excited state, there should be one set of electrons paired up in the 2p orbital (Hund's rule is violated).

84.   **(a)**   This is an excited state silicon atom (3p electrons should remain unpaired with same spin).

    (b)   This is an excited state phosphorus atom, the three 3p orbital electrons should have the same spin (violates Hund's rule).

    (c)   This is a ground state sulfur atom.

    (d)   This is an excited state sulfur atom. The two unpaired electrons should have the same spin.

85.   **(a)**   Au: $[Xe]4f^{14}5d^{10}6s^1$ (exception)   **(d)** In: $[Kr]4d^{10}5s^25p^1$

    (b)   Mg: $[Ne]3s^2$ **(e)** Mo: $[Kr]4d^55s^1$ (exception)

    (c)   Bi: $[Xe]4f^{14}5d^{10}6s^26p^3$ **(f)** Br: $[Ar]3d^{10}4s^24p^5$

86.   **(a)**   Se: $[Ar]3d^{10}4s^24p^4$ **(d)** Pd: $[Kr]4d^85s^2$ → actually $[Kr]4d^{10}$

    (b)   Rb: $[Kr]5s^1$ **(e)** Rh: $[Kr]4d^75s^2$ → actually $[Kr]4d^85s^1$

    (c)   Te: $[Kr]4d^{10}5s^25p^4$ **(f)** Mn: $[Ar]3d^54s^2$

87. (a) Ac (b) N (c) Ti (d) I (e) Ce

88. (a) Se (b) P (c) Ti (d) Y (e) Tm

Integrative and Advanced Exercises

89. We begin with $\Delta E = 2.179 \times 10^{-18}$ J $\left(\dfrac{1}{n_i^2} - \dfrac{1}{n_f^2} \right)$ and wish to produce the Balmer equation,

which is $v = 3.2881 \times 10^{15}$ s^{-1} $\left(\dfrac{1}{2^2} - \dfrac{1}{n^2} \right)$ First we set $n_i = 2$ and $n_f = n$. Then we recognize

that all that remains is to demonstrate that the energy $E = 2.179 \times 10^{-18}$ J is associated with electromagnetic radiation of frequency $v = 3.2881 \times 10^{15}$ s^{-1}. For this, we use Planck's equation.

$E = hv = 6.626 \times 10^{-34}$ J s $\times 3.2881 \times 10^{15}$ s$^{-1} = 2.178_7 \times 10^{-18}$ J The transformation is complete.

90. By definition, heat is the transfer of energy via disorderly molecular motion. Thus the transfer of heat occurs through translational movement of atoms or molecules. The exception to this is radiant heat, which is actually infrared radiation. Therefore, heat cannot be transferred through a vacuum unless it is first converted to electromagnetic radiation and then converted back to original form after transmission.

91. **(a)** We first must determine the wavelength of light that has an energy of 435 kJ/mol and compare that wavelength with those known for visible light.

$$E = \dfrac{435 \, \dfrac{kJ}{mol} \times \dfrac{1000 \, J}{1 \, kJ}}{6.022 \times 10^{23} \, photons/mol} = 7.22 \times 10^{-19} \, J/photon = hv$$

$$v = \dfrac{E}{h} = \dfrac{7.22 \times 10^{-19} \, J}{6.626 \times 10^{-34} \, J \cdot s} = 1.09 \times 10^{15} \, s^{-1} \qquad \lambda = \dfrac{c}{v} = \dfrac{2.9979 \times 10^8 \, m \, s^{-1}}{1.09 \times 10^{15} \, s^{-1}} \times \dfrac{10^9 \, nm}{1 \, m} = 275 \, nm$$

Because the shortest wavelength of visible light has a wavelength of 390 nm, the photoelectric effect for mercury cannot be obtained with visible light.

(b) We first determine the energy per photon for light with 215 nm wavelength.

$$E = hv = \dfrac{hc}{\lambda} = \dfrac{6.626 \times 10^{-34} \, J \cdot s \times 2.9979 \times 10^8 \, m \, s^{-1}}{215 \, nm \times \dfrac{1 \, m}{10^9 \, nm}} = 9.24 \times 10^{-19} \, J/photon$$

Excess energy, over and above the threshold energy, is imparted to the electron as kinetic

energy. Electron kinetic energy $= 9.24 \times 10^{-19}$ J $- 7.22 \times 10^{-19}$ J $= 2.02 \times 10^{-19}$ J $= \dfrac{mv^2}{2}$

(c) We solve for the velocity $v = \sqrt{\dfrac{2 \times 2.02 \times 10^{-19} \, J}{9.109 \times 10^{-31} \, kg}} = 6.66 \times 10^5 \, m \, s^{-1}$

92. We first determine the energy of an individual photon.

$$E = \frac{hc}{\lambda} = \frac{6.626 \times 10^{-34} \text{ J s} \times 2.998 \times 10^8 \text{ m/s}}{1525 \text{ nm} \times \frac{1 \text{ m}}{10^9 \text{ nm}}} = 1.303 \times 10^{-19} \text{ J}$$

$$\frac{\text{no. photons}}{\text{sec}} = \frac{95 \text{ J}}{\text{s}} \times \frac{1 \text{ photon}}{1.303 \times 10^{-19} \text{ J}} \times \frac{14 \text{ photons produced}}{100 \text{ photons theoretically possible}} = 1.0 \times 10^{20} \frac{\text{photons}}{\text{sec}}$$

93. A watt = joule/second, so joules = watts × seconds   J = 75 watts x 5.0 seconds = 37$\underline{5}$ Joules
E = (number of photons) hν and ν = c/λ, so E =(number of photons)hc/λ and

$$\lambda = (\text{number of photons}) \frac{hc}{E} = \frac{(9.91 \times 10^{20} \text{ photons})(6.626 \times 10^{-34} \text{ J sec})(3.00 \times 10^8 \text{ m/sec})}{37\underline{5} \text{ watts}}$$

$\lambda = 5.3 \times 10^{-7} m$ or 530 nm   The light will be green in color.

94. A quantum jump in atomic terms is an abrupt transition of a system as described by quantum mechanics. Under normal/most circumstances, the change is small (i.e. from one discrete atomic or subatomic energy state to another). Quantum jump in everyday usage has a similar, yet different connotation. It too is an abrupt change, however, the change is very significant, unlike the atomic scenario where the change is usually one of the smallest possible.

95. The longest wavelength line in a series is the one with the lowest frequency. It is the one with the two quantum numbers separated by one unit. First we compute the frequency of the line.

$$\nu = \frac{c}{\lambda} = \frac{2.9979 \times 10^8 \text{ m s}^{-1} \times \frac{10^9 \text{ nm}}{1 \text{ m}}}{7400 \text{ nm}} = 4.051 \times 10^{13} \text{ s}^{-1} = 3.2881 \times 10^{15} \text{ s}^{-1} \left(\frac{1}{m^2} - \frac{1}{n^2} \right)$$

$$\left(\frac{1}{m^2} - \frac{1}{n^2} \right) = \frac{4.051 \times 10^{13} \text{ s}^{-1}}{3.2881 \times 10^{15} \text{ s}^{-1}} = 0.01232 = \left(\frac{1}{m^2} - \frac{1}{(m+1)^2} \right)$$

Since this is a fourth-order equation, it is best solved by simply substituting values.

$$\frac{1}{1^2} - \frac{1}{2^2} = 0.75 \quad \frac{1}{2^2} - \frac{1}{3^2} = 0.13889 \quad \frac{1}{3^2} - \frac{1}{4^2} = 0.04861 \quad \frac{1}{4^2} - \frac{1}{5^2} = 0.02250$$

$$\frac{1}{5^2} - \frac{1}{6^2} = 0.01222 \quad \frac{1}{6^2} - \frac{1}{7^2} = 0.00737 \quad \text{Pfund series has } m = 5 \text{ and } n = 6, 7, 8, 9, \ldots \infty$$

96. First we determine the frequency of the radiation. Then rearrange the Rydberg equation (generalized from the Balmer equation) and solve for the parenthesized expression.

$$\nu = \frac{c}{\lambda} = \frac{2.9979 \times 10^8 \text{ m s}^{-1} \times \frac{10^9 \text{ nm}}{1 \text{ m}}}{1876 \text{ nm}} = 1.598 \times 10^{14} \text{ s}^{-1} = 3.2881 \times 10^{15} \text{ s}^{-1} \left(\frac{1}{m^2} - \frac{1}{n^2} \right)$$

$$\left(\frac{1}{m^2} - \frac{1}{n^2} \right) = \frac{1.598 \times 10^{14} \text{ s}^{-1}}{3.2881 \times 10^{15} \text{ s}^{-1}} = 0.0486$$

We know that $m < n$, and both numbers are integers. Furthermore, we know that $m \neq 2$ (the Balmer series) which is in the visible region, and $m \neq 1$ which is in the ultraviolet region, since the wavelength 1876 nm is in the infrared region. Let us try $m = 3$ and $n = 4$.

$$\frac{1}{3^2} - \frac{1}{4^2} = 0.04861 \qquad \text{These are the values we want.}$$

97.

$$E_5 = \frac{-Z^2}{n^2} R_H = \frac{(+2)^2 \, 2.179 \times 10^{-18} \, \text{J}}{5^2} \qquad\qquad E_2 = \frac{-Z^2}{n^2} R_H = -\frac{(+2)^2 \, 2.179 \times 10^{-18} \, \text{J}}{2^2}$$

$$= -3.486 \times 10^{-19} \, \text{J} \qquad\qquad\qquad = -2.179 \times 10^{-18} \, \text{J}$$

$$E_5 - E_2 = \Delta E = \frac{hc}{\lambda} = -3.486 \times 10^{-19} \, \text{J} - (-2.179 \times 10^{-18} \, \text{J}) = 1.830 \times 10^{-18} \, \text{J}$$

$$\lambda = \frac{hc}{\Delta E} = \frac{6.626 \times 10^{-34} \, \text{J s} \times 2.998 \times 10^{8} \, \text{m/s}}{1.830 \times 10^{-18} \, \text{J}} \times \frac{10^9 \, \text{nm}}{1 \, \text{m}} = 108.6 \, \text{nm}$$

98. The lines observed consist of

(a) the transitions starting at $n = 5$ and ending at $n = 4, 3, 2,$ and 1;

(b) the transitions starting at $n = 4$ and ending at $n = 3, 2,$ and 1;

(c) the transitions starting at $n = 3$ and ending at $n = 2$ and 1;

(d) the transition starting at $n = 2$ and ending at $n = 1$.

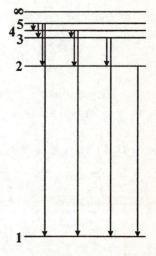

The energy level diagram is shown on the right hand side of this page.

99. When the outermost electron is very far from the inner electrons, it no longer is affected by them individually. Rather, all of the inner electrons and the nucleus affect this outermost electron as if they are one composite entity, a nucleus with a charge of 1+, in other words, a hydrogen nucleus.

100. (a) If there were three possibilities for electron spin, then each orbital could accommodate three electrons. Thus, an s subshell, with one orbital, could hold three electrons. A p subshell, with three orbitals, could hold 9 electrons. A d subshell, with five orbitals, could hold 15 electrons. And an f subshell, with seven electrons, could hold 21 electrons. Therefore, the electron configuration for cesium, with 55 electrons becomes the following.

ordered by energy: $1s^3 2s^3 2p^9 3s^3 3p^9 4s^3 3d^{15} 4p^9 5s^1$
ordered by shells: $1s^3 2s^3 2p^9 3s^3 3p^9 3d^{15} 4s^3 4p^9 5s^1$

(b) If ℓ could have the value n, then there could be orbitals such as $1p$, $2d$, $3f$, etc. In this case, the electron configuration for cesium, with 55 electrons would be:

in order of increasing energy: $\qquad 1s^2 1p^6 2s^2 2p^6 3s^2 2d^{10} 3p^6 4s^2 3d^{10} 4p^6 5s^2 3f^1$

ordered by shells: $\qquad\qquad 1s^2 1p^6 2s^2 2p^6 2d^{10} 3s^2 3p^6 3d^{10} 3f^1 4s^2 4p^6 5s^2$

101. First we must determine the energy per photon of the radiation, and then calculate the number of photons needed, (i.e. the number of ozone molecules (with the ideal gas law)). (Parts per million O_3 are assumed to be by volume.) The product of these two numbers is total energy in joules.

$$E_1 = h\nu = \frac{hc}{\lambda} = \frac{6.626 \times 10^{-34} \text{ J} \cdot \text{s} \times 2.9979 \times 10^8 \text{ m s}^{-1}}{254 \text{ nm} \times \dfrac{1 \text{ m}}{10^9 \text{ nm}}} = 7.82 \times 10^{-19} \text{ J/photon}$$

$$\text{no. photons} = \frac{\left(748 \text{ mmHg} \times \dfrac{1 \text{ atm}}{760 \text{ mmHg}}\right) \times \left(1.00 \text{ L} \times \dfrac{0.25 \text{ L} \quad O_3}{10^6 \text{ L air}}\right)}{\dfrac{0.08206 \text{ L atm}}{\text{mol K}} \times (22 + 273) \text{ K}} \times \frac{6.022 \times 10^{23} \text{ molecules}}{1 \text{ mol } O_3}$$

$$\times \frac{1 \text{ photon}}{1 \text{ molecule } O_3} = 6.1 \times 10^{15} \text{ photons}$$

energy needed $= 7.82 \times 10^{-19}$ J/photon $\times 6.1 \times 10^{15}$ photons $= 4.8 \times 10^{-3}$ J or 4.8 mJ

102. First we c ompute the energy per photon, and then the number of photons received per second.

$$E = h\nu = 6.626 \times 10^{-34} \text{J} \cdot \text{s} \times 8.4 \times 10^9 \text{ s}^{-1} = 5.6 \times 10^{-24} \text{ J/photon}$$

$$\frac{\text{photons}}{\text{second}} = \frac{4 \times 10^{-21} \text{ J/s}}{5.6 \times 10^{-24} \text{ J/photon}} = 7 \times 10^2 \text{photons/s}$$

103. (a) The average kinetic energy is given by the following expression.

$$\overline{e}_k = \frac{3}{2} \cdot \frac{R}{N_A} \cdot T = \frac{3 \cdot 8.314 \text{ J mol}^{-1} \text{ K}^{-1}}{2 \cdot 6.022 \times 10^{23} \text{ mol}^{-1}} \times 1023 \text{ K} = 2.12 \times 10^{-20} \text{ J molecule}^{-1}$$

Then we determine the energy per photon of visible light. $E = h\nu = \dfrac{hc}{\lambda}$

$$E_{\text{max}} = \frac{6.626 \times 10^{-34} \text{ J} \cdot \text{s} \times 2.9979 \times 10^8 \text{ m s}^{-1}}{390 \text{ nm} \times \dfrac{1 \text{ m}}{10^9 \text{ nm}}} = 5.09 \times 10^{-19} \text{ J/photon}$$

$$E_{\text{min}} = \frac{6.626 \times 10^{-34} \text{ J} \cdot \text{s} \times 2.9979 \times 10^8 \text{ m s}^{-1}}{760 \text{ nm} \times \dfrac{1 \text{ m}}{10^9 \text{ nm}}} = 2.61 \times 10^{-19} \text{ J/photon}$$

We see that the average kinetic energy of molecules in the flame is not sufficient to account for the emission of visible light.

(b) The reason why visible light is emitted is that the atoms in the flame do not all have the same energy. Some of them have energy considerably greater than the average, which is sufficiently high to account for the emission of visible light.

104. I f the angular momentum (mur) is restricted to integral values of $h/2\pi$, then we have $mur = nh/2\pi$. The circumference of a circular orbit equals its diameter times π, or twice its radius multiplied by π: $2\pi r$ When $mur = nh/2\pi$ is rearranged, we obtain $2\pi r = nh/mu = n\lambda$ where $\lambda = h/mu$ is the de Broglie wavelength, which is the result requested.

105.

(a) $mru = \dfrac{nh}{2\pi}$ and $r = n^2 a_o$

$$u = \frac{nh}{2\pi mr} = \frac{nh}{2\pi mn^2 a_o} = \frac{3(6.626\times10^{-34}\, J\ s)}{2(3.1416)(9.11\times10^{-31}\, kg)(53\times10^{-12}\, m)(3)^2} = 7.3\times10^5\ \text{m/s}$$

(b) $\text{rev/s} = \dfrac{u}{2\pi r} = \dfrac{(7.29\times10^5\ \text{m/sec})}{2(3.1416)(53\times10^{-12}\, m)(3)^2} = 2.4\times10^{14}\ \text{rev/s}$

106. The n = 138 level is a high Rydberg state (excited state). It is a bound state that lies in the continuum and is very close to the energy required to ionize the electron.

$$\text{radius} = n^2 a_o = (138)^2(53\times10^{-12}\, m) = 1.01\times10^{-6}\ \text{m}$$

$$u = \frac{nh}{2\pi mr} = \frac{nh}{2\pi mn^2 a_o} = \frac{138(6.626\times10^{-34} J\ s)}{2(3.1416)(9.11\times10^{-31} kg)(138)^2(53\times10^{-12} m)} = 1.6\times10^5\ \text{m/s}$$

$$\text{rev/s} = \frac{u}{2\pi r} = \frac{u}{2\pi\, n^2 a_o} = \frac{(1.6\times10^5\ \text{m/sec})}{2(3.1416)(138)^2(53\times10^{-12} m)} = 2.5\times10^{10}\ \text{rev/s}$$

107. For the 3s orbital o f Hydrogen (Z = 1 and n = 3): $R(3s) = \dfrac{1}{9\sqrt{3}}\left(\dfrac{1}{a_o}\right)^{3/2}\left(6 - \dfrac{4r}{a_o} + \dfrac{4r^2}{9a_o^2}\right)e^{\frac{-r}{3a_o}}$

To find the nodes, we need only find the values of r that result in R(3s) = 0
Note that for r = infinity, R(3s) = 0. However this is not a node, rather this result indicates that the e⁻ must be near the nucleus for it to be associated with the atom (r = infinity suggests electron has ionized). Basically we have to find the roots of the equation $(6 -4r/a_o + 4r^2/9a_o^2) = 0$. This is a quadratic where $x = r/a_o$ and a = 4/9, b = -4 and c = 6. It can be solved using the quadratic roots expression and it yields two roots, $r/a_o = 7.098$ and $r/a_o = 1.902$. The nodes occur at $r = 1.902\ a_o$ and $7.098\ a_o$.

Radial Probability Distribution for z=3s Orbital

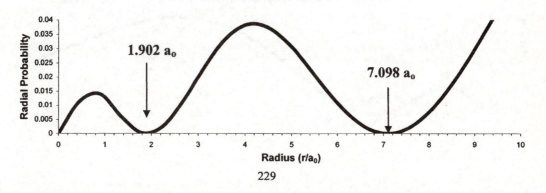

229

108. For a 2s o rbital on Hydrogen ($Z = 1$, $n = 2$): $R(2s) = \dfrac{1}{2\sqrt{2}}\left(\dfrac{1}{a_o}\right)^{3/2}\left(2 - \dfrac{r}{a_o}\right)e^{\frac{-r}{2a_o}}$

To find the radius where the probability of finding a 2s orbital electron (on H) is a maximum or a minimum), we need to set the first derivative of $4\pi r^2 R^2(2s) = 0$ and solve for r in terms of a_o.

$$4\pi r^2 (R^2(2s)) = 4\pi r^2 \left(\frac{1}{2\sqrt{2}}\left(\frac{1}{a_o}\right)^{3/2}\right)^2 \left(2 - \frac{r}{a_o}\right)^2 \left(e^{\frac{-r}{2a_o}}\right)^2$$

$$4\pi r^2 \left(\frac{1}{8a_o^3}\right)\left(4 - 4\frac{r}{a_o} + \frac{r^2}{a_o^2}\right)e^{\frac{-r}{a_o}} = \left(\frac{\pi r^2}{2a_o^3}\right)\left(4 - 4\frac{r}{a_o} + \frac{r^2}{a_o^2}\right)e^{\frac{-r}{a_o}} = \left(\frac{2\pi r^2}{a_o^3} - \frac{2\pi r^3}{a_o^4} + \frac{\pi r^4}{2a_o^5}\right)e^{\frac{-r}{a_o}}$$

Next, we take derivative and set this equal to zero
- this will give the points where the wavefunction is a maximum or minimum.

$$\frac{d\left[4\pi r^2 (R^2(2s))\right]}{dr} = \left(\frac{2\pi r^2}{a_o^3} - \frac{2\pi r^3}{a_o^4} + \frac{\pi r^4}{2a_o^5}\right)\left(-\frac{1}{a_o}\right)e^{\frac{-r}{a_o}} + \left(\frac{4\pi r}{a_o^3} - \frac{6\pi r^2}{a_o^4} + \frac{2\pi r^3}{a_o^5}\right)e^{\frac{-r}{a_o}} = 0$$

$$\left(-\frac{2\pi r^2}{a_o^4} + \frac{2\pi r^3}{a_o^5} - \frac{\pi r^4}{2a_o^6}\right)e^{\frac{-r}{a_o}} + \left(\frac{4\pi r}{a_o^3} - \frac{6\pi r^2}{a_o^4} + \frac{2\pi r^3}{a_o^5}\right)e^{\frac{-r}{a_o}} = 0 \quad \text{multiply through by } \frac{-2a_o^6}{\pi r e^{-r/a_o}}$$

$$4a_o^2 r - 4a_o r^2 + r^3 - 8a_o^2 r + 12a_o^2 r - 4a_o r^2 = 0$$

$$r^3 - 8a_o r^2 + 16a_o^2 r - 8a_o^3 = 0 \qquad\qquad \text{Set } r = xa_o$$

$$x^3 a_o^3 - 8a_o\left(x^2 a_o^2\right) + 16a_o^2\left(xa_o\right) - 8a_o^3 = 0 \quad \text{Divide through by } a_o^3$$

$$x^3 - 8x^2 + 16x - 8 = 0$$

This is a cubic equation, which has three possible roots. A quick look at the plot below shows that there are indeed three points where the slope goes to zero (2 maxima and one minima) Using method of successive approximations, we can find the following three roots:

$$x_1 = 0.764 = \frac{r}{a_o} \qquad\qquad x_2 = 2 = \frac{r}{a_o} \qquad\qquad x_3 = 5.236 = \frac{r}{a_o}$$

$$r_1 = 0.764a_o = 0.405 \text{ Å} \qquad r_2 = 2a_o = 1.06 \text{ Å} \qquad r_3 = 5.236a_o = 2.775 \text{ Å}$$

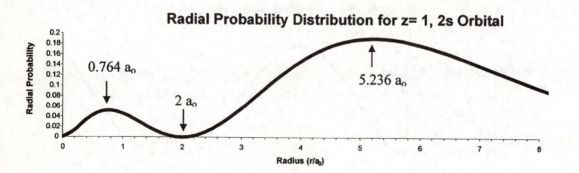

Radial Probability Distribution for z= 1, 2s Orbital

109. The 4 p_x orbital has electron density in the xy plane that is directed primarily along the x axis. Since it is a p-orbital, there is one angular plane. Since n = 4, a total of 3 nodes are present, one angular and two radial. A sketch of this orbital is shown below.

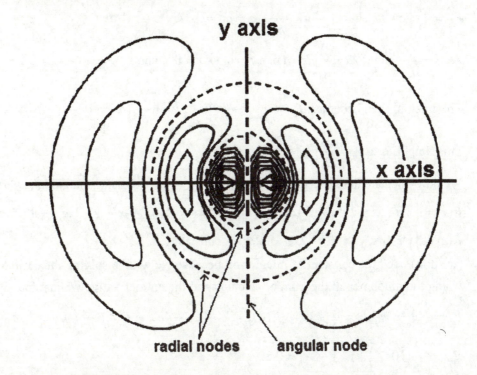

110. Recall, the volume of a sphere is $4/3\pi r^3$. One way of solving this problem is to find the difference in the volumes of two spheres, one with a radius of r and the other with a radius of r+dr, where dr is an infinitesimal increment in r. The difference in the volumes would be given by: $dV = 4/3\pi(r+dr)^3 - 4/3\pi r^3 = (4/3)\pi(3r^2 dr + 3rdr^2 + dr^3)$. For very small values of dr, the terms $3rdr^2 + dr^3)$ are negligible in comparison to $3r^2 dr$. Thus for small values of dr, the volume expression simplifies to $dV = (4/3)\pi(3r^2 dr) = 4\pi r^2 dr$.

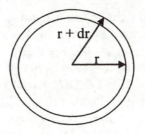

111. (a) To calculate the probability of finding a 1s electron anywhere within a sphere of radius a_o, we must integrate the probability density distribution $4\pi r^2\psi^2$ from 0 to a_o with respect to r.

$$\psi^2(1s) = \frac{1}{\pi a_o^3} e^{-\frac{2r}{a_o}} \qquad \text{Let } P_{a_o} = \text{ probability of finding electron out to a radius of } a_o$$

$$P_{a_o} = \int_0^{a_o} 4\pi r^2 \psi^2 \, dr = \int_0^{a_o} 4\pi r^2 \frac{1}{\pi a_o^3} e^{-\frac{2r}{a_o}} \, dr = \int_0^{a_o} 4r^2 \frac{1}{a_o^3} e^{-\frac{2r}{a_o}} \, dr$$

$$\text{Let } x = \frac{2r}{a_o} \quad \text{then} \quad dx = \frac{2}{a_o} dr \quad \text{Then } r = 0 \rightarrow x = 0 \quad \text{and} \quad r = a_o \rightarrow x = 2$$

The integral then becomes: $P_{a_o} = \int_0^2 4\left(\frac{a_o x}{2}\right)^2 \frac{1}{a_o^3} e^{-x} \left(\frac{a_o}{2}\right) dx = \int_0^2 \frac{x^2 e^{-x}}{2} dx$

This simplifies to $P_{a_o} = \frac{1}{2}\int_0^2 x^2 e^{-x} dx$

We now calculate the antiderivative of $F(x) = \int x^2 e^{-x} dx$ by using integration by parts (twice):

$$F(x) = \int x^2 e^{-x} dx = \int x^2 \left(-e^{-x}\right) dx = -x^2 e^{-x} + \int 2x\left(-e^{-x}\right) dx = -x^2 e^{-x} - 2xe^{-x} + \int 2e^{-x} dx$$

$$F(x) = \int x^2 e^{-x} dx = -x^2 e^{-x} - 2xe^{-x} - 2e^{-x} + \text{constant}$$

(we don't need the constant of integration because we will be solving the definite integral)

Using the fundamental theorum of calculus and the antiderivative we just found. We return to:

$$P_{a_o} = \frac{1}{2}\int_0^2 x^2 e^{-x} dx = \frac{1}{2}\left(F(2) - F(0)\right) = \frac{1}{2}\left(-4e^{-2} - 4e^{-2} - 2e^{-2}\right) - \frac{1}{2}\left(-0 - 0 - 2e^{-0}\right)$$

$$P_{a_o} = -\frac{1}{2}\left(10e^{-2}\right) + 1 = 1 - 5e^{-2} = 0.32$$

This suggests that there is a 32 % probability of finding the 1s electron within a radius of a_o in a groundstate hydrogen atom.

(b) Performing the same calculation as before but with $r = 0 \rightarrow r = 2a_o (x = 4)$ we obtain

$$P_{2a_o} = F(4) - F(0) = \frac{1}{2}\left(-16e^{-4} - 8e^{-4} - 2e^{-4}\right) - \frac{1}{2}\left(-0 - 0 - 2e^{-0}\right) = -\frac{1}{2}\left(26e^{-4}\right) + 1 = 1 - 13e^{-4} = 0.76$$

This suggests that there is a 76 % probability of finding the 1s electron within a radius of $2a_o$ in a groundstate hydrogen atom.

112. The energy available from the excited state hydrogen atom is 1.634×10^{-18} J ($-R_H/2^2 - (-R_H/1^2)$) This energy is then used to excite an He^+ ion (Z = 2).
Hence, 1.634×10^{-18} J $= -Z^2 R_H/n_{low}^2 - (-Z^2 R_H/n_{high}^2)$ $n_{low} = 1$, Z = 2 and $R_H = 2.179\times10^{-18}$ J
Substituting we find, 1.634×10^{-18} J $= (-4\times2.179\times10^{-18}$ J $/1^2) - (-4\times2.179\times10^{-18}$ J$/n_{high}^2)$
$0.1875 = (1 - 1/n_{high}^2)$ or $1/n_{high}^2 = 1 - 0.1875 = 0.8125$ rearrange to give $n_{high}^2 = 1.231$
From this, $n_{high} = 1.11$. This suggests that far more energy is required to excite a He^+ atom to the 1st excited state than is available from excited state H-atoms. Thus, the energy transfer described here is not possible. One can calculate that the energy required to reach the 1st excited state in He^+ is $-2^2(2.179\times10^{-18}$ J$)/1^2 - (-2^2(2.179\times10^{-18}$ J$)/2^2 = 6.537\times10^{-18}$ J. This is 4 times larger than the available energy from a H-atom's $2s^1 \rightarrow 1s^1$ transition.

FEATURE PROBLEMS

113. By carefully scanning the diagram, we note that there are no spectral lines in the area of 304 nm and 309 nm, nor at 318 nm, and 327 nm. Likewise, there are none between 435 and 440 nm. We conclude that V is absent. There are spectral lines that correspond to each of the Cr spectral lines—between 355 nm and 362 nm, and between 425 nm and 430 nm. Cr is present. There are no spectral lines close to 403 nm; Mn is absent. There are spectral lines at about 344 nm, 358 nm, 372 nm, 373 nm, and 386 nm. Fe is present. There are spectral lines at 341 nm, 344 nm to 352 nm, and 362 nm. Ni is present. There are spectral lines between 310 and 315 nm and at about 415 nm. Another element is present. Thus, Cr, Fe, and Ni are present. V and Mn are absent. Also, there is an additional element present.

114. The equation of a straight line is $y = mx + b$, where m is the slope of the line and b is its y-intercept. The Balmer equation is $\nu = 3.2881 \times 10^{15}$ Hz $\left(\dfrac{1}{2^2} - \dfrac{1}{n^2} \right) = \dfrac{c}{\lambda}$. In this equation, one plots ν on the vertical axis, and $1/n^2$ on the horizontal axis. The slope is $b = -3.2881 \times 10^{15}$ Hz and the intercept is 3.2881×10^{15} Hz $\div 2^2 = 8.2203 \times 10^{14}$ Hz. The plot of the data for Figure 9.9 follows.

λ	656.3 nm	486.1 nm	434.0 nm	410.1 nm
ν	4.568×10^{14} Hz	6.167×10^{14} Hz	6.908×10^{14} Hz	7.310×10^{14} Hz
n	3	4	5	6

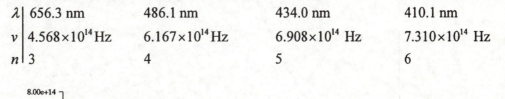

We see that the slope (-3.2906×10^{15}) and the y-intercept (8.2240×10^{14}) are almost exactly what we had predicted from the Balmer equation.

115. This graph differs from the one involving ψ^2 in Figure 8-34(a) because this graph factors in the volume of the thin shell. Figure 8-34(a) simply is a graph of the probability of finding an electron at a distance r from the nucleus. But the graph that accompanies this problem multiplies that radial probability by the volume of the shell that is a distance r from the nucleus. The volume of the shell is the thickness of the shell (a very small value dr) multiplied by the area of the shell $\left(4\pi r^2 \right)$. Close to the nucleus, the area is very small

because r is very small. Therefore, the relative probability also is very small near the nucleus. There just isn't sufficient volume to contain the electrons. What we plot is the product of the number of darts times the circumference of the outer boundary of the scoring ring. (probability = number × circumference)

darts	200	300	400	250	200
score	"50"	"40"	"30"	"20"	"10"
radius	1.0	2.0	3.0	4.0	5.0
circumference	3.14	6.28	9.42	12.6	15.7
probability	628	1884	3768	3150	3140

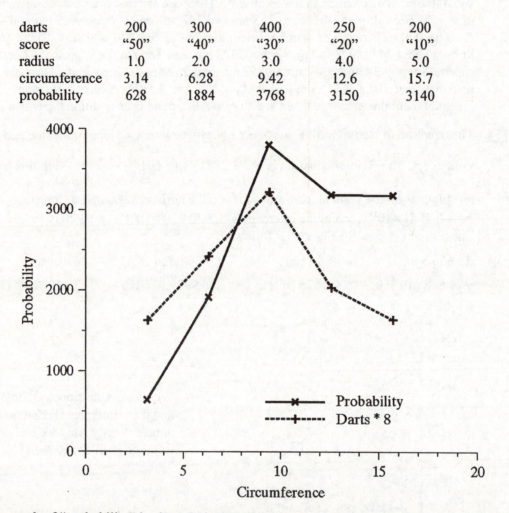

The graph of "probability" is close to the graph that accompanies this problem, except that the dart board is two-dimensional, while the atom is three-dimensional. This added dimension means that the volume close to the nucleus is much smaller, relatively speaking than is the area close to the center. The other difference, of course, is that it is harder for darts to get close to the center, while electrons are attracted to the nucleus.

116. (a) First we calculate the range of energies for the incident photons used in the absorption experiment. Remember: $E_{photon} = h\nu$ & $\nu = c/\lambda$. At one end of the range, $\lambda = 100$ nm. Therefore, $\nu = 2.998 \times 10^8$ m s^{-1} ÷ $(1.00 \times 10^{-7}$ m$) = 2.998 \times 10^{15}$ s^{-1}

So $E_{photon} = 6.626 \times 10^{-34}$ J s$(2.998 \times 10^{15}$ s$^{-1}) = 1.98 \times 10^{-18}$ J

At the other end of the range, $\lambda = 1000$ nm.

Therefore, $\nu = 2.998 \times 10^8$ m s^{-1} ÷ 1.00×10^{-6} m $= 2.998 \times 10^{14}$ s^{-1}

So $E_{photon} = 6.626 \times 10^{-34}$ J s$(2.998 \times 10^{14}$ s$^{-1}) = 1.98 \times 10^{-19}$ J.

Next, we will calculate what excitations are possible using photons with energies between 1.98×10^{-18} J and 1.98×10^{-19} J and the electron initially residing in the $n = 1$ level. These "orbit transitions" can be found with the equation

$$\Delta E = E_f - E_i = -2.179 \times 10^{-18}\left(\frac{1}{(1)^2} - \frac{1}{(n_f)^2}\right) \text{ For the lowest energy photon}$$

$$1.98 \times 10^{-19} \text{ J} = -2.179 \times 10^{-18}\left(\frac{1}{(1)^2} - \frac{1}{(n_f)^2}\right) \quad \text{or} \quad 0.0904 = 1 - \frac{1}{(n_f)^2}$$

From this $-0.9096 = -\dfrac{1}{(n_f)^2}$ and $n_f = 1.05$

Thus, the lowest energy photon is not capable of promoting the electron above the $n = 1$ level. For the highest energy level:

$$1.98 \times 10^{-18} \text{ J} = -2.179 \times 10^{-18}\left(\frac{1}{(1)^2} - \frac{1}{(n_f)^2}\right) \quad \text{or} \quad 0.9114 = 1 - \frac{1}{(n_f)^2}$$

From this $-0.0886 = -\dfrac{1}{(n_f)^2}$ and $n_f = 3.35$ Thus, the highest energy photon can

promote a groundstate electron to both the $n = 2$ and $n = 3$ levels. This means that we would see <u>two</u> lines in the absorption spectrum, one corresponding to the $n = 1 \rightarrow n = 2$ transition and the other to the $n = 1 \rightarrow n = 3$ transition.

$$\text{Energy for the } n = 1 \rightarrow n = 2 \text{ transition} = -2.179 \times 10^{-18}\left(\frac{1}{(1)^2} - \frac{1}{(2)^2}\right) = 1.634 \times 10^{-18} \text{ J}$$

$$\nu = \frac{1.634 \times 10^{-18} \text{ J}}{6.626 \times 10^{-34} \text{ J s}} = 2.466 \times 10^{15} \text{ s}^{-1} \qquad \lambda = \frac{2.998 \times 10^8 \text{ m s}^{-1}}{2.466 \times 10^{15} \text{ s}^{-1}} = 1.215 \times 10^{-7} \text{ m}$$

$\lambda = 121.5$ nm

Thus, we should see a line at 121.5 nm in the absorption spectrum.

$$\text{Energy for the } n = 1 \rightarrow n = 3 \text{ transition} = -2.179 \times 10^{-18}\left(\frac{1}{(1)^2} - \frac{1}{(3)^2}\right) = 1.937 \times 10^{-18} \text{ J}$$

$$\nu = \frac{1.937 \times 10^{-18} \text{ J}}{6.626 \times 10^{-34} \text{ J s}} = 2.923 \times 10^{15} \text{ s}^{-1} \qquad \lambda = \frac{2.998 \times 10^8 \text{ m s}^{-1}}{2.923 \times 10^{15} \text{ s}^{-1}} = 1.025 \times 10^{-7} \text{ m}$$

$\lambda = 102.5$ nm

Consequently, the second line should appear at 102.6 nm in the absorption spectrum.

(b) An excitation energy of 1230 kJ mol⁻¹ to 1240 kJ mol⁻¹ works out to 2×10^{-18} J per photon. This amount of energy is sufficient to raise the electron to the $n = 4$ level. Consequently, six lines will be observed in the emission spectrum. The calculation for each emission line is summarized below:

$$E_{4 \to 1} = \frac{2.179 \times 10^{-18} \text{ J}}{6.626 \times 10^{-34} \text{ J s}}\left(\frac{1}{1^2} - \frac{1}{4^2}\right) = 3.083 \times 10^{15} \text{ s}^{-1} \qquad \lambda = \frac{2.998 \times 10^8 \frac{m}{s}}{3.083 \times 10^{15} \text{ s}^{-1}} = 9.724 \times 10^{-8} \text{ m}$$

$\lambda = 97.2$ nm

$$E_{4 \to 2} = \frac{2.179 \times 10^{-18} \text{ J}}{6.626 \times 10^{-34} \text{ J s}}\left(\frac{1}{2^2} - \frac{1}{4^2}\right) = 6.167 \times 10^{14} \text{ s}^{-1} \qquad \lambda = \frac{2.998 \times 10^8 \frac{m}{s}}{6.167 \times 10^{14} \text{ s}^{-1}} = 4.861 \times 10^{-7} \text{ m}$$

$\lambda = 486.1$ nm

$$E_{4 \to 3} = \frac{2.179 \times 10^{-18} \text{ J}}{6.626 \times 10^{-34} \text{ J s}}\left(\frac{1}{3^2} - \frac{1}{4^2}\right) = 1.599 \times 10^{14} \text{ s}^{-1} \qquad \lambda = \frac{2.998 \times 10^8 \frac{m}{s}}{1.599 \times 10^{14} \text{ s}^{-1}} = 1.875 \times 10^{-6} \text{ m}$$

$\lambda = 1875$ nm

$$E_{3 \to 1} = \frac{2.179 \times 10^{-18} \text{ J}}{6.626 \times 10^{-34} \text{ J s}}\left(\frac{1}{1^2} - \frac{1}{3^2}\right) = 2.924 \times 10^{15} \text{ s}^{-1} \qquad \lambda = \frac{2.998 \times 10^8 \frac{m}{s}}{2.924 \times 10^{15} \text{ s}^{-1}} = 1.025 \times 10^{-7} \text{ m}$$

$\lambda = 102.5$ nm

$$E_{3 \to 2} = \frac{2.179 \times 10^{-18} \text{ J}}{6.626 \times 10^{-34} \text{ J s}}\left(\frac{1}{2^2} - \frac{1}{3^2}\right) = 4.568 \times 10^{14} \text{ s}^{-1} \qquad \lambda = \frac{2.998 \times 10^8 \frac{m}{s}}{4.568 \times 10^{14} \text{ s}^{-1}} = 6.563 \times 10^{-7} \text{ m}$$

$\lambda = 656.3$ nm

$$E_{2 \to 1} = \frac{2.179 \times 10^{-18} \text{ J}}{6.626 \times 10^{-34} \text{ J s}}\left(\frac{1}{1^2} - \frac{1}{2^2}\right) = 2.467 \times 10^{15} \text{ s}^{-1} \qquad \lambda = \frac{2.998 \times 10^8 \frac{m}{s}}{2.467 \times 10^{15} \text{ s}^{-1}} = 1.215 \times 10^{-7} \text{ m}$$

$\lambda = 121.5$ nm

(c) The number of lines observed in the two spectra is not the same. The absorption spectrum has two lines while the emission spectrum has six lines. Notice that the 102.5 nm and 1021.5 nm lines are present in both spectra. This is not surprising since the energy difference between each level is the same whether it is probed by emission or absorption spectroscopy.

117. (a) The wavelength associated with each Helium-4 atom must be close to 100 pm $(1.00 \times 10^{-10}$ m) in order for diffraction to take place. To find the necessary velocity for the He atoms, we need to employ the de Broglie equation: $\lambda = \dfrac{h}{mv}$.

Rearrange to give $v = \dfrac{h}{m\lambda}$ where:

$\lambda = 1.00 \times 10^{-10}$ m, $h = 6.626 \times 10^{-34}$ kg m^{2}s^{-1} and $m_{He\ nucleus} = 6.647 \times 10^{-27}$ kg.

So, $v = \dfrac{6.626 \times 10^{-34} \text{ kg m}^2 \text{ s}^{-1}}{6.647 \times 10^{-27} \text{ kg}(1.00 \times 10^{-10} \text{ m})} = 9.97 \times 10^2 \text{ m s}^{-1}$

(b) The de Broglie wavelength for the beam of protons would be too small for any diffraction to occur. Instead, most of the protons would will simply pass through the film of gold and have little or no interaction with the constituent gold atoms. Keep in mind, however, that some of the protons will end up being deflected or bounced back by either passing too close to a nucleus or by colliding with a gold nucleus head on, respectively.

118. The emission lines are due to transitions from $n_{high} \rightarrow n_{low}$. Where n_{high} = 6, 5, 4, 3, 2 and n_{low} = 5, 4, 3, 2, 1 (with the restriction $n_{high} > n_{low}$). Consider the emission resulting from $n_{high} = 6 \rightarrow n_{low} = 5$. The orbital types allowed for n=6 are s, p, d, f, g, h and for n=5 are s, p, d, f, g. We can have 6s $\rightarrow$ 5p (5s is not allowed because of the selection rule $\Delta \ell = \pm 1$). 6p $\rightarrow$ 5s, 6p $\rightarrow$ 5d, 6d$\rightarrow$5p, 6d$\rightarrow$5f, 6f$\rightarrow$5g, 6f$\rightarrow$5d, 6g$\rightarrow$5f, 6h$\rightarrow$ 5g.

In the absence of a magnetic field all of these transitions occur at the same frequency or wavelength for a hydrogen atom. In the hydrogen atom the 6→5 occurs at the longest wavelength, which is the rightmost line in the spectrum. We can calculate the wavelength of this emission by using the following relationship:

$$\Delta E = 2.179 \times 10^{-18} \text{ J} \times \left(\frac{1}{n_{high}^2} - \frac{1}{n_{low}^2} \right) \qquad \text{and} \qquad \Delta E = \frac{hc}{\lambda} \qquad \text{or} \qquad \lambda = \frac{hc}{\Delta E}$$

For the transition $n_{high} = 6 \rightarrow n_{low} = 5$. We calculate the following values for ΔE and λ.

$$\Delta E = 2.179 \times 10^{-18} \text{ J} \times \left(\frac{1}{6^2} - \frac{1}{5^2} \right) = 2.663 \times 10^{-20} \text{ J}$$

$$\lambda = \frac{hc}{\Delta E} = \frac{(6.626 \times 10^{-34} \text{ J} \cdot \text{s}) \times (2.998 \times 10^8 \text{m} \cdot \text{s}^{-1})}{2.663 \times 10^{-20} \text{ J}} = 7.459 \times 10^{-6} \text{ m} \qquad \text{or} \qquad 7459 \text{ nm}$$

We can similarly repeat the procedure for the remaining 14 lines and obtain the following data (next page).

n(initial)	n(final)	ΔE(Joules)	λ (nm)	line number
6	5	2.663×10^{-20}	7459	1
6	4	7.566×10^{-20}	2625	3
6	3	1.816×10^{-19}	1094	6
6	2	4.842×10^{-19}	410	10
6	1	2.118×10^{-18}	94	15
5	4	4.903×10^{-20}	4052	2
5	3	1.550×10^{-19}	1282	5
5	2	4.576×10^{-19}	434	9
5	1	2.092×10^{-18}	95	14
4	3	1.059×10^{-19}	1875	4
4	2	4.086×10^{-19}	486	8
4	1	2.043×10^{-18}	97	13
3	2	3.026×10^{-19}	656	7
3	1	1.937×10^{-18}	103	12
2	1	1.634×10^{-18}	122	11

When a magnetic field is applied the levels with $\ell > 0$ split into the $2\ell + 1$ sublevels. The line that splits into the most lines is the one that contains the greatest number levels with $\ell > 0$, this is the $6 \rightarrow 5$ transition, the right most or longest wavelength emission.

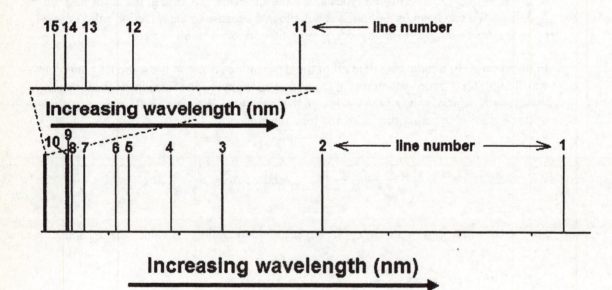

CHAPTER 9
THE PERIODIC TABLE AND SOME ATOMIC PROPERTIES
PRACTICE EXAMPLES

1A   Atomic size decreases left to right across a period, and bottom to top in a family. We expect the smallest elements to be in the upper right corner of the periodic table. S is the element closest to the upper right corner and thus should have the smallest atom.
S = 104 pm As = 121 pm I = 133 pm

1B   From the periodic table inside the front cover, we see that Na is in the same period as Al (period 3), but in a different group from K, Ca, and Br (period 4), which might suggest that Na and Al are about the same size. However, there is a substantial decrease in size as one moves from left to right in a period due to an increase in effective nuclear charge, enough in fact, that Ca should be about the same size as Na.

2A Ti^{2+} and V^{3+} are isoelectronic; the one with higher positive charge should be smaller: $V^{3+} < Ti^{2+}$. Sr^{2+} and Br^- are isoelectronic; again, the one with higher positive charge should be smaller: $Sr^{2+} < Br^-$. In addition Ca^{2+} and Sr^{2+} both are ions of Group 2A; the one of lower atomic number should be smaller. $Ca^{2+} < Sr^{2+} < Br^-$. Finally, we know that the size of atoms decreases left to right across a period; we expect size of like-charged ions to follow the same trend: $Ti^{2+} < Ca^{2+}$. The species are arranged below in order of increasing size.
$$V^{3+}(64 \text{ pm}) < Ti^{2+}(86 \text{ pm}) < Ca^{2+}(100 \text{ pm}) < Sr^{2+}(113 \text{ pm}) < Br^-(196 \text{ pm})$$

2B Br^- clearly is larger than As since Br^- is an anion in the same period as As. In turn, As is larger than N since both are in the same group, with As lower down in the group. As also should be larger than P, which is larger than Mg^{2+}, an ion smaller than N. All that remains is to note that Cs is a truly large atom, one of the largest in the periodic table. The As atom should be in the middle. Data from figure 9-8 shows:
65 pm for Mg^{2+} < 70 pm for N < 125 pm for As < 196 pm for Br^- < 265 pm for Cs

3A   Ionization increases from bottom to top of a group and from left to right through a period. The first ionization energy of K is less than that of Mg and the first ionization energy of S is less than that of Cl. We would expect also that the first ionization energy of Mg is smaller than that of S, because Mg is a metal.

3B   We would expect an alkali metal (Rb) or an alkaline earth metal (Sr) to have a low first ionization energy and nonmetals (e.g. Br) to have relatively high first ionization energies. Metalloids (such as Sb and As) should have intermediate ionization energies. Since the first ionization energy for As is larger than that for Sb, the first ionization energy of Sb should be in the middle.

4A   Cl and Al must be paramagnetic, since they each have an odd number of electrons. The electron configurations of K^+ ([Ar]) and O^{2-} ([Ne]) are those of the nearest noble gas. Because all of the electrons are paired, they are diamagnetic species. In Zn: [Ar] $3d^{10}4s^2$ all electrons are paired and so the atom is diamagnetic.

4B   The electron configuration of Cr is [Ar] $3d^5 4s^1$; it has six unpaired electrons. The electron configuration of Cr^{2+} is [Ar] $3d^4$; it has four unpaired electrons. The electron configuration of Cr^{3+} is [Ar] $3d^3$; it has three unpaired electrons. Thus, of the two ions, Cr^{2+} has the greater number of unpaired electrons.

5A   We expect the melting point of bromine to be close to the average of those for chlorine and iodine. Thus, estimated melting point of $Br_2 = \dfrac{172\ K + 387\ K}{2} = 280\ K$. The actual melting point is 266 K.

5B   If the boiling point of I_2 (458 K) is the average of the boiling points of Br_2 (349 K) and At_2, then $458\ K = (349\ K + ?)/2$ $? = 2 \times 458\ K - 349\ K = 567\ K$

The estimated boiling point of molecular astatine is about 570 K.

EXERCISES

The Periodic Law

1.   Element 114 will be a metal in the same group as Pb, element 82 (18 cm^3/mol); Sn, element 50 (18 cm^3/mol); and Ge, element 32 (14 cm^3/mol). We note that the atomic volume of Pb and Sn are essentially equal, probably due to the lanthanide contraction. If there is also an actinide contraction, element 114 will have an atomic volume of 18 cm^3 / mol. If there is no actinide contraction, we would predict a molar volume of $\sim 22\ cm^3 / mol$. This need to estimate atomic volume is what makes the value for density questionable.

$$\text{density}\left(\frac{g}{cm^3}\right) = \frac{298\ \dfrac{g}{mol}}{18\ \dfrac{cm^3}{mol}} = 16\ \frac{g}{cm^3} \qquad \text{density}\left(\frac{g}{cm^3}\right) = \frac{298\ \dfrac{g}{mol}}{22\ \dfrac{cm^3}{mol}} = 14\ \frac{g}{cm^3}$$

2.   Lanthanum has an atomic number of $Z = 57$, and thus its atomic volume is somewhat less than $25\ cm^3 / mol$. Let us assume $23\ cm^3 / mol$.

atomic mass = density × atomic volume = 6.145 $g / cm^3 \times 23\ cm^3 / mol = 141\ g / mol$

This compares very well with the listed value of 139 for La.

3. The following data are plotted at right. Density clearly is a periodic property for these two periods of main group elements. It rises, falls a bit, rises again, and falls back to the axis, in both cases.

Element	Atomic Number Z	density $\frac{g}{cm^3}$
Na	11	0.968
Mg	12	1.738
Al	13	2.699
Si	14	2.336
P	15	1.823
S	16	2.069
Cl	17	0.0032
Ar	18	0.0018
K	19	0.856
Ca	20	1.550
Ga	31	5.904
Ge	32	5.323
As	33	5.778
Se	34	4.285
Br	35	3.100
Kr	36	0.0037

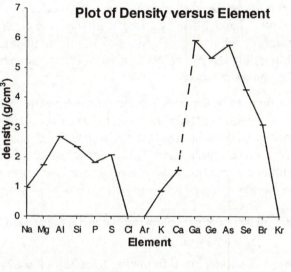

4. The following data are plotted at right below. Melting point clearly is a periodic property for these two periods. It rises to a maximum and then falls off in each case.

Element	Atomic Number Z	Melting Point °C
Li	3	179
Be	4	1278
B	5	2300
C	6	3350
N	7	−210
O	8	−218
F	9	−220
Ne	10	−249
Na	11	98
Mg	12	651
Al	13	660
Si	14	1410
P	15	590
S	16	119
Cl	17	−101
Ar	18	−189

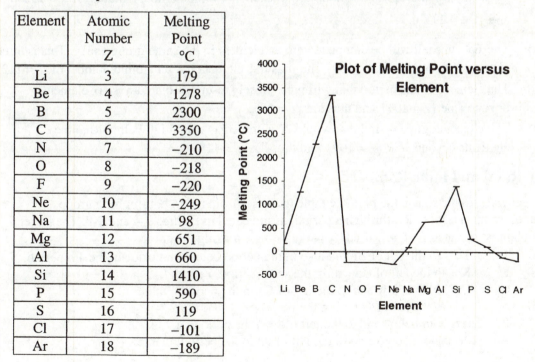

The Periodic Table

5. Mendeleev arranged elements in the periodic table in order of increasing atomic weight. Of course, atomic masses with non-integral values are permissible. Hence, there always is room for an added element between two elements that already are present in the table. On the other hand, Moseley arranged elements in order of increasing atomic number. Only integral (whole number) values of atomic number are permitted. Thus, when elements with all possible integral values in a certain range have been discovered, no new elements are possible in that range.

6. For there to be the same number of elements in each period of the periodic table, each shell of an electron configuration would have to contain the same number of electrons. This however, is not the case; the shells have 2 (K shell), 8 (L shell), 18 (M shell), and 32 (N shell) electrons each. This is because each of the periods of the periodic table begins with one s electron beyond a noble gas electron configuration, and the noble gas electron configuration corresponds to either s^2 (He) or $s^2 p^6$, with various other full subshells.

7. (a) The noble gas following radon $(Z = 86)$ will have an atomic number of
$(86 + 32 =)118$.

(b) The alkali metal following francium $(Z = 87)$ will have an atomic number of
$(87 + 32 =)119$.

(c) The mass number of radon $(A = 222)$ is $(222 \div 86) = 2.58$ times its atomic number. The mass number of Lr $(A = 262)$ is $(262 \div 103) = 2.54$ times its atomic number. Thus, we would expect the mass numbers, and hence approximate atomic masses of elements 118 and 119 to be about 2.5 times their atomic numbers, that is, $A_{119} \approx 298$ u and $A_{118} \approx 295$ u.

8. (a) The $6d$ subshell will be complete with an element in the same group as Hg. This is three elements beyond Une $(Z = 109)$ and thus this element has an atomic number of $Z = 112$.

(b) The element in this period that will most closely resemble bismuth is six elements beyond Une $(Z = 109)$ and thus has $Z = 115$.

(c) The element in this period that would be a noble gas would fall below Rn, nine elements beyond Une $(Z = 109)$ and thus has $Z = 118$.

Atomic Radii and Ionic Radii

9. In general, atomic size in the periodic table increases from top to bottom for a group and increases from right to left through a period, as indicated in Figures 9-4 and 9-8. The larger element is indicated first, followed by the reason for making the choice.

(a) Te: Te is to the left of Br and also in the period below that of Br in the 4th period.

(b) K: K is to the left of Ca within the same period, Period 4.

(c) Cs : Cs is both below and to the left of Ca in the periodic table.

(d) N: N is to the left of O within the same period, Period 2.

(e) P: P is both below and to the left of O in the periodic table.

(f) Au: Au is both below and to the left of Al in the periodic table.

10. An Al atom is larger than a F atom since Al is both below and to the left of F in the periodic table. As is larger than Al, since As is below Al in the periodic table. (Even though As is to the right of Al, we would not conclude that As is smaller than Al, since increases in size down a group are more pronounced than decreases in size across a period (from left to right). A Cs^+ ion is isoelectronic with an I^- ion, and in an isoelectronic series, anions are larger than cations, thus I^- is larger than Cs^+. I^- also is larger than As, since I is below As in the periodic table (and increases in size down a group are more pronounced than those across a period). Finally, N is larger than F, since N is to the left of F in the periodic table. Therefore, we conclude that F is the smallest species listed and I^- is the largest. In fact, with the exception of Cs^+, we can rank the species in order of decreasing size. $I^- > As > Al > N > F$ and also $I^- > Cs^+$

11. The reason why the sizes of atoms do not simply increase with atomic number is because electrons often are added successively to the same subshell. These electrons do not fully screen each other from the nuclear charge (they do not effectively get between each other and the nucleus). Consequently, as each electron is added to a subshell and the nuclear charge increases by one unit, all of the electrons in this subshell are drawn more closely into the nucleus, because of the ineffective shielding.

12. The reason why atomic sizes are uncertain is because the electron cloud that surrounds an atom has no fixed limit. It can be pictured as gradually fading away, rather like the edge of a town. In both cases we pick an arbitrary boundary.

13. **(a)** The smallest atom in Group 13 is the first: B

　　(b) Po is in the sixth period, and is larger than the others, which are rewritten in the following list from left to right in the fifth period, that is, from largest to smallest: Sr, In, Sb, Te. Thus, Te is the smallest of the elements given.

14. The hydrogen ion contains no electrons, only a nucleus. It is exceedingly tiny, much smaller than any other atom or electron-containing ion. Both H and H^- have a nuclear charge of 1+, but H^- has two electrons to H's one, and thus is larger. Both He and H^- contain two electrons, but He has a nuclear charge of 2+, while H^- has one of only 1+. The smaller nuclear charge of H^- is less effective at attracting electrons than the more positive nuclear charge of He. The only comparison left is between H and He; we expect He to be smaller since atomic size decreases left to right across a period. Thus, the order by increasing size: $H^+ < He < H < H^-$.

15. Li^+ is the smallest; it not only is in the second period, but also is a cation. I^- is the largest, an anion in the 5th period. Next largest is Se in the previous (the 4th) period. We expect Br to be smaller than Se because it is both to the right of Se and in the same period.

$$Li^+ < Br < Se < I^-$$

16. Size decreases from left to right in the periodic table. On this basis I should be smaller than Al. But size increases from top to bottom in the periodic table. On this basis, I should be larger than Al. There really is no good way of resolving these conflicting predictions.

17. In the literal sense, isoelectronic means having the same number and type of electrons. (In another sense, not used in the text, it means having the same electron configuration.) We determine the total number of electrons and the electron configuration for each species and make our decisions based on this information.

Fe^{2+}	24 electrons	[Ar] $3d^6$	Sc^{3+}	18 electrons	[Ar]
Ca^{2+}	18 electrons	[Ar]	F^-	10 electrons	[He] $2s^2 2p^6$
Co^{2+}	25 electrons	[Ar] $3d^7$	Co^{3+}	24 electrons	[Ar] $3d^6$
Sr^{2+}	36 electrons	[Ar] $3d^{10} 4s^2 4p^6$	Cu^+	28 electrons	[Ar] $3d^{10}$
Zn^{2+}	28 electrons	[Ar] $3d^{10}$	Al^{3+}	10 electrons	[He] $2s^2 2p^6$

Thus the species with the same number of electrons and the same electron configuration are the following. Fe^{2+} and Co^{3+} Sc^{3+} and Ca^{2+} F^- and Al^{3+} Zn^{2+} and Cu^+

18. In an isoelectronic series, all of the species have the same number and types of electrons. The size is determined by the nuclear charge. Those species with the largest (positive) nuclear charge are the smallest. Those with smaller nuclear charges are larger in size. Thus, the more positively charged an ion is in an isoelectronic series, the smaller it will be.

$$Y^{3+} < Sr^{2+} < Rb^+ < Br^- < Se^{2-}$$

19. Ions can be isoelectronic without having noble-gas electron configurations. Take, for instance, Cu^+ and Zn^{2+}. Both of these ions have the electron configuration $[Ar]3d^{10}$.

20. Isoelectronic species must have the same number of electrons, and each element has a different atomic number. Thus, atoms of different elements cannot be isoelectronic. Two different cations may be isoelectronic, as may two different anions, or an anion and a cation. In fact, there are many sets of different cations and anions that have a common configuration. For example, all of the ions listed below share the same electron configuration, namely that of the noble gas Ne: O^{2-} and F^-, Na^+ and Mg^{2+}, or F^- and Na^+.

Ionization Energies; Electron Affinities

21. Ionization energy in the periodic table decreases from top to bottom for a group, and increases from left to right for a period, as summarized in Figure 9-9. Cs has the lowest ionization energy as it is furthest to the left and nearest to the bottom of the periodic table. Next comes Sr, followed by As, then S, and finally F, the most nonmetallic element in the group (and in the periodic table). Thus, the elements listed in order of increasing ionization energy are:
$$Cs < Sr < As < S < F$$

22. The second ionization energy for an atom $\left(I_2\right)$ cannot be smaller than the first ionization energy $\left(I_1\right)$ for the same atom. The reason is that, when the first electron is removed it is being separated from a species with a charge of sero. On the other hand, when the second electron is removed, it is being separated from a species with a charge of +1. Since the force between two charged particles is proportional to $q_+ q_- / r^2$ (r is the distance between the particles), the higher the positive charge, the more difficult it will be to remove an electron.

23. In the case of a first electron affinity, a negative electron is being added to a neutral atom. This process may be either exothermic or endothermic depending upon the electronic configuration of the atom. Energy tends to be released from when filled shells or filled subshells are generated. In the case of an ionization potential, however, a negatively charged electron is being separated from a positively charged cation, a process that must always require energy, because unlike charges attract each other.

24. First we convert the mass of Na given to an amount in moles of Na. Then we compute the energy needed to ionize this much Na.

$$Energy = 1.00 \text{ mg Na} \times \frac{1 \text{ g}}{1000 \text{ mg}} \times \frac{1 \text{ mol Na}}{22.99 \text{ g Na}} \times \frac{495.8 \text{ kJ}}{1 \text{ mol Na}} = 0.0216 \text{ kJ} \times \frac{1000 \text{ J}}{1 \text{ kJ}} = 21.6 \text{ J}$$

25. Ionization energies for Si: $I_1 = 786.5$ kJ/mol, $I_2 = 1577$ kJ/mol, $I_3 = 3232$ kJ/mol, $I_4 = 4356$ kJ/mol. To remove all four electrons from the third shell ($3s^2 3p^2$) would require the sum of all four ionization energies or 9951.5 kJ/mol. This would be 9.952×10^6 J per mole of Si atoms.

26. Data ($I_1 = 375.7$ kJ/mol) are obtained from Table 10.3.

$$\text{no. Cs}^+ \text{ ions} = 1 \text{ J} \times \frac{1 \text{ kJ}}{1000 \text{ J}} \times \frac{1 \text{ mol Cs}^+}{375.7 \text{ kJ}} \times \frac{6.022 \times 10^{23} \text{ Cs}^+ \text{ ions}}{1 \text{ mol Cs}^+ \text{ ions}} = 1.603 \times 10^{18} \text{ Cs}^+ \text{ ions}$$

27. The electron affinity of bromine is −324.6 kJ/mol (Figure 9-10). We use Hess's law to determine the heat of reaction for $Br_2(g)$ becoming $2 Br^-(g)$.

$Br_2(g) \rightarrow 2 Br(g)$	$\Delta H = +193$ kJ
$2 \ Br(g) + 2 \ e^- \rightarrow 2 \ Br^-(g)$	$2 \times E.A. = 2(-324.6)$ kJ
$Br_2(g) + 2 \ e^- \rightarrow 2 \ Br^-(g)$	$\Delta H = -456$ kJ Overall process is *exothermic*.

28. The electron affinity of fluorine is −328.0 kJ/mol (Figure 9-10) and the first and second ionization energies of Mg (Table 10.4) are 737.7 kJ/mol and 1451 kJ/mol, respectively.

$Mg(g) \rightarrow Mg^+(g) + e^-$	$I_1 = 737.7$ kJ / mol
$Mg^+(g) \rightarrow Mg^{2+}(g) + e^-$	$I_2 = 1451$ kJ / mol
$2 F(g) + 2 \ e^- \rightarrow 2 F^-(g)$	$2E.A. = 2(-328.0)$ kJ / mol
$Mg(g) + 2 F(g) \rightarrow Mg^{2+}(g) + 2 F^-(g)$	$\Delta H = +1533$ kJ / mol Endothermic.

29. The electron is being removed from an species with a neon electron configuration. But in the case of Na^+, the electron is being removed from a species that is left with a 2+ charge, while in the case of Ne, the electron is being removed from a species with a 1+ charge. The more highly charged the resulting species, the more difficult it is to produce it by removing an electron.

30. The electron affinity of Li is −59.6 kJ/mol, the smallest ionization energy listed in Table 9.3 (and, except for Fr, displayed in Figure 9-9) is that of Cs, 375.7 kJ/mol. Thus, insufficient energy is produced by the electron affinity of Li to account for the ionization of Cs. As a result, we would predict that Li^-Cs^+ will not be stable and hence Li^-Li^+ and Li^-Na^+, owing to the larger ionization energies of Li and Na should be even less stable. (One other consideration not dealt with here is the energy released when the positive and negative ion combine to form an ion pair. This is an exothermic process, but we have no way of assessing the value of this energy.)

Magnetic Properties

31. Three of the ions have noble gas electron configurations and thus have no unpaired
electrons: F^- is $1s^2 2s^2 2p^6$ Ca^{2+} and S^{2-} are $[Ne]3s^2 3p^6$

Only Fe^{2+} has unpaired electrons. Its electron configuration is $[Ar]3d^6$.

32. **(a)** $Ge[Ar]3d^{10}4s^2 4p^2$ with two unpaired electrons
(b) $Cl [Ne]3s^2 3p^5$ with one unpaired electron
(c) $Cr^{3+}[Ar]3d^3$ with three unpaired electrons
(d) $Br^-[Ar]3d^{10}4s^2 4p^6$ with no unpaired electrons
Clearly, (b) Cr^{3+} as the most number of unpaired electrons of the four listed.

33. **(a)** K^+ is isoelectronic with Ar, there are no unpaired electrons. It is diamagnetic.
(b) Cr^{3+} has the configuration $[Ar]3d^3$ with three unpaired electrons, it is paramagnetic.
(c) Zn^{2+} has the configuration $[Ar]3d^{10}$, there are no unpaired electrons. It is diamagnetic.
(d) Cd has the configuration $[Kr]4d^{10}5s^2$, there are no unpaired electrons. It is diamagnetic.
(e) Co^{3+} has the configuration $[Ar]3d^6$ with four unpaired electrons, it is paramagnetic
(f) Sn^{2+} has the configuration $[Kr]4d^{10}5s^2$, there are no unpaired electrons. It is diamagnetic.
(g) Br has the configuration $[Ar]3d^{10}4s^2 5p^5$ with one unpaired electron, it is paramagnetic.
From this we see (a), (c), (d), (f) are diamagnetic and (b), (e) and (g) are paramagnetic.

34. First we write the electron configuration of the element, then that of the ion. In each case,
the number of unpaired electrons written beside the configuration agrees with the data
given in the statement of the problem.

     (a) Ni $[Ar]\ 3d^8 4s^2 \longrightarrow$ $Ni^{2+}\ [Ar]\ 3d^8$ Two unpaired e^-

     (b) Cu $[Ar]\ 3d^{10}4s^1 \longrightarrow$ $Cu^{2+}\ [Ar]\ 3d^9$ One unpaired e^-

     (c) Cr $[Ar]\ 3d^5 4s^1 \longrightarrow$ $Cr^{3+}\ [Ar]\ 3d^3$ Three unpaired e^-

35. All atoms with an odd number of electrons must be paramagnetic. There is no way to pair
all of the electrons up if there is an odd number of electrons. Many atoms with an even
number of electrons are diamagnetic, but some are paramagnetic. The one of lowest atomic
number is carbon $(Z = 6)$, which has two unpaired p-electrons producing the paramagnetic
behavior: $[He]\ 2s^2 2p^2$.

36. The electron configuration of each iron ion can be determined by starting with the electron
configuration of the iron atom. $[Fe] = [Ar]3d^6 4s^2$ Then the two ions result from the loss
of both $4s$ electrons, followed by the loss of a $3d$ electron in the case of Fe^{3+}.

$$[Fe^{2+}] = [Ar]3d^6 \quad \text{four unpaired electrons}$$
$$[Fe^{3+}] = [Ar]3d^5 \quad \text{five unpaired electrons}$$

Predictions Based on the Periodic Table

37. **(a)** Elements that one would expect to exhibit the photoelectric effect with visible light should be ones that have a small value of their first ionization energy. Based on Figure 9-9, the alkali metals have the lowest first ionization potentials of these. Cs, Rb, and K are three suitable metals. Metals that would not exhibit the photoelectric effect with visible light are those that have high values of their first ionization energy. Again from Figure 9-9, Zn, Cd, and Hg seem to be three metals that would not exhibit the photoelectric effect with visible light.

(b) From Figure 9-1, we notice that the atomic (molar) volume increases for the solid forms of the noble gases as we travel down the group (the data points just before the alkali metal peaks). But it seems to increase less rapidly than the molar mass. This means that the density should increase with atomic mass, and Rn should be the densest solid in the group. We expect densities of liquids to follow the same trend as densities of solids.

(c) To estimate the first ionization energy of fermium, we note in Figure 9-9 that the ionization energies of the lanthanides (following the Cs valley) are approximately the same. We expect similar behavior of the actinides, and estimate a first ionization energy of about +600 kJ/mol.

(d) We can estimate densities of solids from the information in Figure 9-1. Radium has $Z = 88$ and an approximate atomic volume of $40 \text{ cm}^3/\text{mol}$. Then we use the molar mass of radium to determine its density:

$$\text{density} = \frac{1 \text{ mol}}{40 \text{ cm}^3} \times \frac{226 \text{ g Ra}}{1 \text{ mol}} = 5.7 \text{ g}/\text{cm}^3$$

38. Germanium lies between silicon and tin in Group 14. We expect the heat of atomization of germanium to approximate the average of the heats of atomization of Si and Sn.

$$\text{average} = \frac{452 \text{ kJ/mol Si} + 302 \text{ kJ/mol Sn}}{2} = 377 \text{ kJ/mol Ge}$$

39. **(a)** From Figure 9-1, the atomic (molar) volume of Al is $10 \text{ cm}^3/\text{mol}$ and that for In is $15 \text{ cm}^3/\text{mol}$. Thus, we predict $12.5 \text{ cm}^3/\text{mol}$ as the molar volume for Ga. Then we compute the expected density for Ga.

$$\text{density} = \frac{1 \text{ mol Ga}}{12.5 \text{ cm}^3} \times \frac{68 \text{ g Ga}}{1 \text{ mol Ga}} = 5.4 \text{ g/cm}^3$$

(b) Since Ga is in group 13 (3A) (Gruppe III on Mendeleev's table), the formula of its oxide should be Ga_2O_3. We use Mendeleev's molar masses to determine the molar mass for Ga_2O_3. Molar mass $= 2\times68 \text{ g Ga} + 3\times16 \text{ g O} = 184 \text{ g } Ga_2O_3$

$$\% \text{ Ga} = \frac{2\times68 \text{ g Ga}}{184 \text{ g } Ga_2O_3} \times 100\% = 74\% (\text{Ga});$$

Using our more recent periodic table, we obtain 74.5% Ga

40. **(a)** Size increases down a group and from right to left in a period. Ba is closest to the lower left corner of the periodic table and thus has the largest size.

(b) Ionization energy decreases down a group and from right to left in a period. Although Pb is closest to the bottom of its group, Sr is farthest left in its period (and only one period above Pb). Sr should have the lowest first ionization energy.

(c) Electron affinity becomes more negative from left to right in a period and from bottom to top in a group. Cl is closest to the upper right in the periodic table and has the most negative (smallest) electron affinity.

(d) The number of unpaired electrons can be determined from the orbital diagram for each species. Atomic nitrogen ($[He]2s^2 2p^3$), by virtue of having three half-filled p-orbitals, possesses the greatest number of unpaired electrons.

41. We expect periodic properties to be functions of atomic number.

Element, atomic number He, 2 Ne, 10 Ar, 18 Kr, 36 Xe, 54 Rn, 86

Element, atomic number	He, 2	Ne, 10	Ar, 18	Kr, 36	Xe, 54	Rn, 86
Boiling point, K	4.2 K	27.1 K	87.3 K	119.7 K	165 K	
Δ b.p. $/\Delta Z$		2.86	7.5	1.8	2.5	

With one notable exception, we see that the boiling point increases about 2.4 K per unit of atomic number. The atomic number increases by 32 units from Xe to Rn. We might expect the boiling point to increase by $(2.4 \times 32 =) 77$ K to $(165 + 77 =) 242$ K for Rn.

A simpler manner is to expect that the boiling point of xenon (165 K) is the average of the boiling points of radon and krypton (120 K).

$165 \text{ K} = (120 \text{ K} + ?)/2 \quad ? = 2 \times 165 \text{ K} - 120 \text{ K} = 210 \text{ K} = $ boiling point of radon

Often the simplest way is best. The tabulated value is 211 K.

42. **(a)** The boiling point increases by 52 °C from CH_4 to SiH_4, and by 22 °C from SiH_4 to GeH₄. One expects another increase in boiling point from GeH_4 to SnH_4, probably by about 15 °C. Thus, the predicted boiling point of SnH_4 is $-75°$ C. (The actual boiling point of SnH_4 is –52 °C).

(b) The boiling point decreases by 39 °C from H_2Te to H_2Se, and by 20 °C from H_2Se to H_2S. It probably will decrease by about 10 °C to reach H_2O. Therefore, one predicts a value for the boiling point of H_2O of approximately -71 °C. Of course, the actual boiling point of water is $100°$ C. The prediction is seriously in error (~ 170 °C) because we have neglected the hydrogen bonding between water molecules, a topic that is discussed in Chapter 13.

F	$[He]\ 2s^2 2p^5$	1 unpaired e^-	N $[He]\ 2s^2 2p^3$	3 unpaired e^-
S^{2-}	$[Ne]\ 3s^2 3p^6$	0 unpaired e^-	Mg^{2+} $[He]\ 2s^2 2p^6$	0 unpaired e^-
Sc^{3+}	$[Ne]\ 3s^2 3p^6$	0 unpaired e^-	Ti^{3+} $[Ar]\ 3d^1$	1 unpaired e^-

43. **(a)** $Z = 32$ 1. This is the element Ge, with an outer electron configuration of $3s^2 3p^2$. Thus, Ge has two unpaired p electrons.

(b) $Z = 8$ 1. This is the element O. Each atom has an outer electron configuration of $2s^2 2p^4$. Thus, O has two unpaired p electrons.

(c) $Z = 53$ 3. This is the element I, with an electron affinity more negative than that of the adjacent atoms: Xe and Te.

(d) $Z = 38$ 4. This is the element Sr, which has two $5s$ electrons. It is easier to remove one $5s$ electron than to remove the outermost $4s$ electron of Ca, but harder than removing the outermost $6s$ electron of Cs.

(e) $Z = 48$ 2. This is the element Cd. Its outer electron configuration is s^2 and thus it is diamagnetic.

(f) $Z = 20$ 2. This is the element Ca. Since its electron configuration is $[Ar] 4s^2$, all electrons are paired and it is diamagnetic.

44. **(a)** 6. Tl's electron configuration $[Xe] 4f^{14} 5d^{10} 6s^2 6p^1$ has one p electron in its outermost shell.

(b) 8. $Z = 70$ identifies the element as Yb, an f-block of the periodic table, or an inner transition element.

(c) 5. Ni has the electron configuration $[Ar] 4s^2 3d^8$. It also is a d-block element.

(d) 1. An s^2 outer electron configuration, with the underlying configuration of a noble gas, is characteristic of elements of group 2(2A), the alkaline earth elements. Since the noble gas core is $[Ar]$, the element must be Ca.

(e) 2. The element in the fifth period and Group 15 is Sb, a metalloid.

(f) 4 and 6. The element in the fourth period and Group 16 is Se, a nonmetal. (Note, B is a nonmetal with electron configuration $1s^2 2s^2 2p^1$ (having one p electron in the shell of highest n) 6. is also a possible answer)

Integrative and Advanced Exercises

45. **(a)** $2 \, Rb(s) + 2H_2O(l) \rightarrow 2 \, RbOH(aq) + H_2(g)$
(b) $I_2(aq) + Na^+(aq) + Br^-(aq) \rightarrow$ No reaction
(c) $SrO(s) + H_2O(l) \rightarrow Sr(OH)_2(aq)$
(d) $SO_3(g) + H_2O(l) \rightarrow H_2SO_4(aq)$

46. **(a)** $2 \, I^-(aq) + Br_2(l) \rightarrow 2 \, Br^-(aq) + I_2(aq)$
(b) $2 \, Cs(s) + 2 \, H_2O(l) \rightarrow 2 \, CsOH(aq) + H_2(g)$
(c) $P_4O_{10}(s) + 6 \, H_2O(l) \rightarrow 4 \, H_3PO_4(aq)$
(d) $2 \, Al_2O_3(s) + H_2SO_4(aq) \rightarrow 2 \, Al_2(SO_4)_3(aq) + 6 \, H_2O(l)$

47. **(a)** Not possible C (77 pm), Ca^{2+} (100 pm) however the diagram requires these to be nearly the same size

(b) This is a possibility. Na^+ is approximately the same size as Sr (99 pm). Cl^- (181 pm) and Br^- (196 pm) are comparable with one being slightly smaller.

(c) Not possible. There are three large atoms and only one small one. Y (165 pm), K(227 pm), Ca(197 pm) and the small Na^+ (99 pm).

(d) Not possible. The smaller atoms are of noticeably different sizes. Zr^{2+} (95 pm) and Mg^{2+} (72 pm).

(e) This is possible. Fe and Co of comparable sizes (~125 pm) and Cs(265 pm) and Rb(248 pm) are comparable with one being slightly smaller.

Thus, the answer is that both (b) and (e) are compatible with the sketch.

48. One can obtain a crude estimate of the normal boiling and melting points for element 168 via a plot of the boiling and melting points for the heavier known noble gases against their atomic numbers (see plot below).

If one extrapolates each plot (either the normal boiling or melting point plot), to Z = 168, with dashed lines having slopes comparable to the line connecting the last two data points in each plot, then both the melting point and the boiling point for element 168 end up being higher than 298 K (estimated b.p. = 327 K and m.p. = 305 K). Thus based upon these extrapolated data we predict that element 168 very well might be a noble solid at 298 K. Clearly, one should view this prediction with considerable skepticism since this exercise involves estimating the melting and boiling points of an element that is not only 90 atomic numbers higher than radon, but also one that is well beyond our present abilities to synthesize.

The electron configuration is based on that of Rn: [Rn] $7s^2$ $5f^{14}$ $6d^{10}$ $7p^6 8s^2$ 5 g^{18} $6f^{14}$ $7d^{10}$ $8p^6$ To get to the first noble element beyond Rn we add 32 electrons: an s subshell, an f subshell, a d subshell, and a p subshell. That takes us to Z (= 86 + 32) = 118. If we fill the same four subshells again, we only reach Z (= 118 + 32) = 150. The next type of subshell to fill is a g- subshell ($\ell = 4$). In order to fill a g-subshell, 18 electrons are required, which is exactly the number of electrons needed to reach Z = 168.

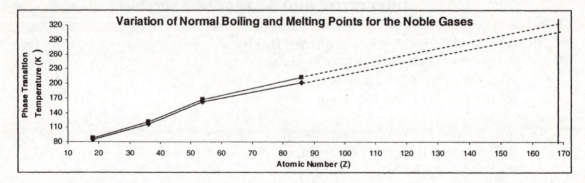

49. This periodic table would have to have the present width of 18 elements plus the width of the 14 lanthanides or actinides. That is, the periodic table would be (18 + 14 =) 32 elements wide.

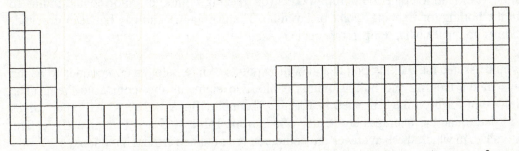

50. We can determine the atomic mass of indium by beginning with the atomic mass of oxygen, and using the chemical formula of InO to determine the amount (in moles) of indium in 100.0 g InO.

In 100.0 g InO there is 82.5 g In and also 17.5 g O. $17.5\,\text{g O} \times \dfrac{1\,\text{mol O}}{16.0\,\text{g O}} \times \dfrac{1\,\text{mol In}}{1\,\text{mol O}} = 1.09\,\text{mol In}$

$$\text{Atomic mass of In} = \frac{82.5\,\text{g In}}{1.09\,\text{mol In}} = 75.7\,\text{g In/mol}$$

With this value of the atomic mass of indium, Mendeleev might well have placed the element between As (75 g/mol) and Se (78 g/mol), that is in his "Gruppe V" or "Gruppe VI"

51. For In_2O_3: atomic mass of $In = \dfrac{82.5\,\text{g In}}{17.5\,\text{g O}} \times \dfrac{16.00\,\text{g O}}{1\,\text{mol O}} \times \dfrac{3\,\text{mol O}}{2\,\text{mol In}} = 113\,\text{g In/mol In}$

Indium now falls between Cd (112 g/mol) and Sn (118 g/mol), which places the element in "Gruppe III", where it remains in the modern periodic table (Group 3A).

52. **Al** Atomic radius usually decreases left to right across a period and from bottom to top in a group. We expect the atomic radius of Al to be similar to that of Ge. Because Al is clearly more metallic than the metalloid Ge, The 1st ionization potential of Al is smaller than Ge.

In In is in the period below that of Ge and in the group to the left of that of Ge. Both of these locations predict a larger atomic radius for In than for Ge. In is clearly a metal, while Ge is a metalloid. The 1st ionization energy of In should be smaller than that of Ge.

Se Se is in the same period as Ge, but further right. It should have a smaller atomic radius. Also because Se is to the right of Ge it should have a larger first ionization potential.

These predictions are summarized in the table below along with the values of the properties (in parentheses). The incorrect prediction is for the radius of Al. [But note that the radius of Al (143 pm) is larger than those of both B (88 pm) and Ga (122 pm). Clearly atomic radius does *not* increase uniformly down the family in Group 3A.]

Element	Atomic radius, pm	First ionization energy, kJ/mol
Ge	123	762
Al	same (143)	same or smaller (578)
In	larger (150)	smaller (558)
Se	smaller (117)	larger (941)

53. Yes, Celsius or Fahrenheit temperatures could have been used for the estimation of the boiling point or the freezing point of bromine. The method of estimation that is used in Example 10-5 is to average the boiling points (or freezing points) of chlorine and iodine to obtain the boiling (or freezing) point of bromine. The average is midway between the other two points no matter what temperature scale is used.

54. We would assume the melting points and boiling points of interhalogen compounds to be the average of the melting and boiling points of the two elements that constitute them. (The melting point for bromine estimated in Practice Example 10-5A is 280 K. All other values used to determine the averages are actual values from Table 10.5 and Example 10-5.) The actual values to which these averages are compared are obtained from a handbook.

$$\text{BrCl} \quad \text{melting point} = \frac{\text{Br}_2 \text{ m.p.} + \text{Cl}_2 \text{ m.p.}}{2} = \frac{280 \text{ K} + 172 \text{ K}}{2} = 226 \text{ K} \qquad \text{actual m.p.} = 207 \text{ K}$$

$$\text{boiling point} = \frac{\text{Br}_2 \text{ b.p.} + \text{Cl}_2 \text{ b.p.}}{2} = \frac{332 \text{ K} + 239 \text{ K}}{2} = 286 \text{ K} \qquad \text{actual b.p.} = 278 \text{ K}$$

$$\text{ICl} \quad \text{melting point} = \frac{\text{I}_2 \text{ m.p.} + \text{Cl}_2 \text{ m.p.}}{2} = \frac{387 \text{ K} + 172 \text{ K}}{2} = 280 \text{ K} \qquad \text{actual m.p.} = 300.4 \text{ K}$$

$$\text{boiling point} = \frac{\text{I}_2 \text{ b.p.} + \text{Cl}_2 \text{ b.p.}}{2} = \frac{458 \text{ K} + 239 \text{ K}}{2} = 348 \text{ K} \qquad \text{actual b.p.} = 370.6 \text{ K}$$

At 25°C (298 K), BrCl is predicted to be a gas (which it is), while ICl is predicted to be a liquid (and it actually is a solid at 298 K, but it melts when the temperature is raised by just 2 K).

55. Sizes of atoms do not simply increase uniformly with atomic number because electrons do not effectively screen each other from the nuclear charge (they do not effectively get between each other and the nucleus). Consequently, as each electron is added and the nuclear charge increases, all of the electrons in the subshell are drawn more closely. As well, in a particular group, atomic size does not increase uniformly owing to the penetration of electrons of a particular shell. Penetration is very dependent on the magnitude of the positive charge in the nucleus and how tightly held electrons are to that nucleus. For instance, a 2p electron will have a different penetration when compared to a 5p electron because the core electrons are more tightly held for an atom with 5p orbitals than those with only the 2p orbitals filled. Finally, electron-electron repulsion is not constant for the outer electrons as one goes down a particular group owing to the volume of the orbital occupied by the electron. 2p orbitals are smaller and have a higher electron-electron repulsion than do electrons in a 5p orbital.

56. $I = \dfrac{R_H (Z_{eff})^2}{n^2} = \dfrac{2.179 \times 10^{-18} \text{ J}(3)^2}{1^2} = 1.96 \times 10^{-20} \text{ J (per atom)}$

For one mole = 1.96×10^{-20} kJ atm^{-1} x 6.022×10^{23} atoms = 1.18×10^4 kJ
It is easier to calculate ionization energy of the last remaining electron in any atom because we know the value of Z_{eff} equals the number of protons in the nucleus. Accurate values of Z_{eff} when there are more than one electrons are not easily calculated. In fact, Z_{eff} is calculated from Ionization Energy data. When there are two or more electrons in an atom, the electron-electron interactions greatly complicate the ionization energy calculation.

57 Since $(0.3734)(382) = 143$ u, if we subtract 143 from 382 we get 239 u which is very close to the correct value of 238 u.

First we convert to J/g°C: $0.0276 \times (4.184 \text{ J/cal}) = 0.115_5$ J/ g°C , so:

$0.115_5 = 0.011440 + (23.967/\text{atomic mass})$; solving for atomic mass yields a value of 230 u, which is within about 3% of the correct value.

58. The volume of a sphere is given by

$V = \frac{4}{3}\pi r^3 = \frac{4}{3} \times 3.142 \times (186 \text{ pm})^3 = 2.70 \times 10^7 \text{ pm}^3 = V_{\text{Na atom}}$ One would then expect the volume of a mole of sodium (atoms) to be Avogadro's constant times the volume of one atom.

Na molar volume $= 6.022 \times 10^{23} \dfrac{\text{Na atoms}}{\text{mole Na}} \times 2.70 \times 10^7 \dfrac{\text{pm}^3}{\text{Na atom}} \times \left(\dfrac{100 \text{ cm}}{10^{12} \text{ pm}}\right)^3 = 16.3 \text{ cm}^3 / \text{mol}$

This value is about two-thirds of the value of 23 cm³/mole in Figure 10-1. An explanation for the great disagreement is that we have assumed that the entire volume of the solid is occupied by the spherical atoms. In fact, spheres leave some free space when they are packed together. (Recall how marbles do not completely fill the space when they are placed in a box.)

59. The reaction given is the sum of the reverse of two reactions: those for the first ionization energy of Na and for the electron affinity of Cl.

$Na^+(g) + e^- \rightarrow Na(g) \qquad -I_1 \qquad = -495.8 \text{ kJ/mol}$

$\underline{Cl^-(g) \rightarrow Cl(g) + e^- \qquad -EA_1 \quad = +349.0 \text{ kJ/mol}}$

$Na^+(g) + Cl^-(g) \rightarrow Na(g) + Cl(g) \text{Energy} = -146.8 \text{ kJ/mol}$

Thus this is clearly an exothermic reaction.

60. $E_1 = \dfrac{-2^2 \times 2.179 \times 10^{-18} \text{ J}}{1^2} = -8.716 \times 10^{-18}$ J/atom

This is the energy released when an electron combines with an alpha particle (He^{2+}) to the nucleus to form the ion He^+. The energy absorbed when the electron is removed from He^+, the ionization energy of He^+ or the *second* ionization energy of He, is the negative of this value. Then the energy per mole, E_m, is computed.

$E_1 = E_1 \times N_A = \dfrac{8.716 \times 10^{-18}}{1 \text{ atom}} \times \dfrac{6.022 \times 10^{23}}{1 \text{ mol}} \times \dfrac{1 \text{ kJ}}{1000 \text{J}} = 5249 \text{ kJ/mol}$

This is in excellent agreement with the tabulated value of 5251 kJ/mol.

61. It should be relatively easy to remove electrons from ions of metallic elements, if the metal does not have a noble gas electron configuration. Thus, I_3 for Sc and I_2 for Ba should be small, with the second being smaller, since the electron is being removed from a more highly charged species in the case of I_3 for Sc. I_1 for F might be smaller than either of these because it involves removing only the first electron from a neutral atom, rather than removing an electron from a cation. There is some uncertainty here, because the electron being removed from F is not well shielded from the nuclear charge, and this value could be larger than the other two. The remaining ionization energies are both substantially larger than the other three because they both involve disrupting a noble gas electron configuration. I_3 for Mg is larger than I_2 for Na because it is more difficult to remove an electron from a more highly charged species. Literature values (in kJ/mol) are in parentheses in the following list.

I_1 for F (1681) $\approx I_2$ for Ba (965) $< I_3$ for Sc (2389) $< I_2$ for Na (4562) $< I_3$ for Mg (7733)

Note that we have overestimated the difficulty of removing a second electron from a metal atom (I_2 for Ba) and underestimated the difficulty of removing the first electron from a small nonmetal atom (I_1 for F).

62. $\dfrac{1\,eV}{1\,atom} \times \dfrac{6.02214 \times 10^{23}\ atoms}{1\,mol} \times \dfrac{1.60218 \times 10^{-19}\ C}{1\,electron} \times \dfrac{1\,J}{1\,V \cdot C} \times \dfrac{1\,kJ}{1000\,J} = 96.4855\ kJ/mol$

63. Ionization energy (I.E.) data is tabulated below and plot in the graphs below.

GRAPH 1					GRAPH 2			
Atom/Ion	I.E. (kJ)	Z	root of I.E.		Atom/Ion	I.E. (kJ)	Z	root of I.E.
Na	495.8	11	22.26657		Li	520	3	22.80351
Mg^+	1451	12	38.09199		Be^+	1757	4	41.91658
Al^{2+}	2745	13	52.39275		B^{2+}	3659	5	60.48967
Si^{3+}	4356	14	66		C^{3+}	6221	6	78.87332

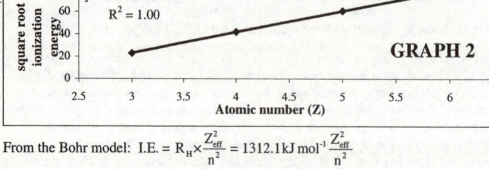

From the Bohr model: $I.E. = R_H \times \dfrac{Z_{eff}^2}{n^2} = 1312.1\,kJ\,mol^{-1}\dfrac{Z_{eff}^2}{n^2}$

If we take the root of both sides, we obtain the following expression

$\sqrt{I.E.} = \dfrac{Z_{eff}^2}{n^2} = \sqrt{1312.1\,kJ\,mol^{-1}\dfrac{Z_{eff}^2}{n^2}} = 36.22\dfrac{Z_{eff}}{n}$

For graph 1: n = 3 hence $\sqrt{I.E.} = 36.22 \dfrac{Z_{eff}}{3}$ Hence: $\sqrt{I.E.} = 12.1\ Z_{eff}$

The expected slope for the graph is 12.1, while the actual slope from data is 14.55.

Although the y-inercept is not zero, we do see a linear relationship (R = 0.9988)

For graph 2: n = 2 hence $\sqrt{I.E.} = 36.22 \dfrac{Z_{eff}}{2}$ Hence: $\sqrt{I.E.} = 18.1\ Z_{eff}$

The expected slope for the graph is 18.1, while the actual slope from data is 18.68.

Although the y-inercept is not zero, we do see a linear relationship (R = 1.00)

The apparent discrepancy lies in the fact that there are electrons in lower shells. For the series plotted in graph 1, there are electrons in the n = 1 and 2 shells. This results in deviation from the theoretical slope and explains why the y-intercept is not zero. We have assumed that the nuclear charge is also the effective nuclear charge. For the series plotted in graph 2, there are electrons in the n = 1. This results in a much smaller deviation from the theoretical slope and a y-intercept that is closer to zero. Again, we have made the assumption that the nuclear charge is also the effective nuclear charge, which is not correct.

A plot using H and He^+ yields a slope of 36.24 (theoretical = 36.22) and a y-intercept of 0.014. There is excellent agreement here because both are Bohr atoms, with no underlying electrons. In this case, the nuclear charge is also the effective nuclear charge/ This is an excellent assumption as can be seen by the agreement with the theoretical slope and y-intercept.

64. Francium is in Group 1, period 7.

    (a)   Melting Point: To estimate it's melting point, we must look at the melting points of the group 1 elements. They are listed in the text as:

        Li(174 °C), Na (97.8 °C), K(63.7 °C), Rb (38.9 °C), Cs (28.5 °C)

        We notice that the relationship between melting point and period number is not linear. It seems logarithmic. A logarithmic plot shows a relatively good agreement.

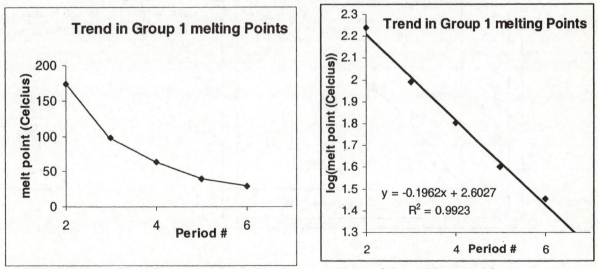

Using the line of best fit, we can determine that an estimate of the melting point for Francium is ~17 °C

(b) Molar Volume/Density: Once we determine the Molar Volume, we can use the molar mass of 223 given in the periodic table to estimate the density of Francium. From Figure 9-1 we can estimate the molar volumes for the Group 1 elements.
Li(12 cm^3/mol), Na (23 cm^3/mol) , K (45 cm^3/mol) , Rb (56 cm^3/mol) , Cs (70 cm^3/mol)
The plot below reveals a relatively straight line. From this, we calculate that the molar volume of Francium is ~86 cm^3/mol.

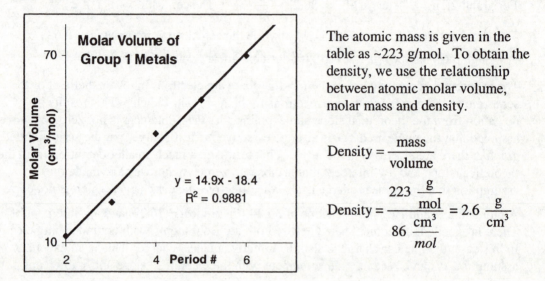

The atomic mass is given in the table as ~223 g/mol. To obtain the density, we use the relationship between atomic molar volume, molar mass and density.

$$\text{Density} = \frac{\text{mass}}{\text{volume}}$$

$$\text{Density} = \frac{223 \frac{g}{mol}}{86 \frac{cm^3}{mol}} = 2.6 \frac{g}{cm^3}$$

(c) Atomic Radius: We are given most of the radii for group 1 metals, they are
Li(152 pm), Na (186 pm) , K (227 pm) , Rb (248 pm) , Cs (265 pm$\rightarrow$ from plot)

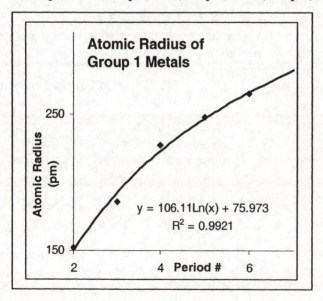

Clearly we see that this is not a linear relationship. From the line of best fit, we can estimate the radius of francium to be ~ 280 pm

FEATURE PROBLEMS

65. **(a)** The work function is the energy required for an electron to escape from the solid surface of an element.

(b) Work functions tend to decrease down a group and increase across a period in the periodic table. The work function increases across the periodic table from left to right following the steady increase in effective nuclear charge. As one proceeds down a given group, the principle quantum number increases and so does the distance of the outer electrons from the nucleus. The further the electrons are from the nucleus, the more easily they can be removed from the solid surface of the element, and hence, the smaller will be the value of the work function for the element.

(c) Through the process of interpolation, one would predict that the work function for potassium should fall somewhere close to 3.9. The published value for the work function of potassium is 3.69 (CRC handbook). Had we been provided more information on the nature of the bonding in each of these metals, and had we been told what type of crystalline lattice each metal adopts, we would have been able to come up with a more accurate estimate of the work function.

(d) The periodic trends in work function closely follow those in ionization energy. This should come as no surprise since both ionization and the work function involve the loss of electrons from neutral atoms.

66. The Moseley equation, $v = A(Z-b)^2$, where v is the frequency of the emitted X-ray radiation, Z is the atomic number and A and b are constants, relates the frequency of emitted X-rays to the nuclear charge for the atoms that make up the target of the cathode ray tube. X-rays are emitted by the element after one of its K-level electrons has been knocked out of the atom by collision with a fast moving electron. In this question, we have been asked to determine the values for the constants A and b. The simplest way to find these values is to plot $\sqrt{v}$ vs. Z. This plot provides $\sqrt{A}$ as the slope and $-\sqrt{A}$ (b) as the y–intercept. Starting with $v = A(Z-b)^2$, we first take the square root of both sides. This affords $\sqrt{v} = \sqrt{A}\,(Z - b)$. Multiplying out this expression gives $\sqrt{v} = \sqrt{A}\,(Z)$ $-\sqrt{A}$ (b). This expression follows the equation of a straight line $y = mx + b$, where $y = \sqrt{v}$, $m = \sqrt{A}$, $x = Z$ and $b = -\sqrt{A}$ (b). So a plot of $\sqrt{v}$ vs. Z will provide us with A and b, after a small amount of mathematical manipulation. Before we can construct the plot we need to convert the provided X-ray wavelengths into their corresponding frequencies. For instance, Mg has an X-ray wavelength = 987 pm. The corresponding frequency for this radiation = c/λ, hence,

$$v = \frac{2.998 \times 10^8 \text{ m s}^{-1}}{9.87 \times 10^{-10} \text{ m}} = 3.04 \times 10^{17} \text{ s}^{-1}$$

Performing similar conversions on the rest of the data allows for the construction of the following table and plot (below).

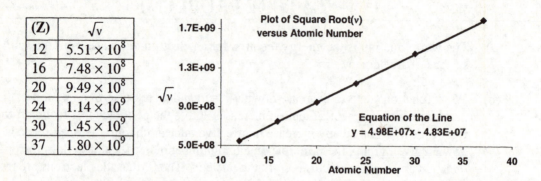

(Z)	$\sqrt{v}$
12	5.51×10^8
16	7.48×10^8
20	9.49×10^8
24	1.14×10^9
30	1.45×10^9
37	1.80×10^9

The slope of the line is $4.98 \times 10^7 = \sqrt{A}$ and the y-intercept is $-4.83 \times 10^7 = -\sqrt{A}$ (b).

Thus, $A = 2.30 \times 10^{15}$ Hz and $b = \dfrac{-4.83 \times 10^7}{-4.98 \times 10^7} = 0.969$

According to Bohr's theory, the frequencies that correspond to the lines in the emission

spectrum are given by the equation: $(3.2881 \times 10^{15} \text{ s}^{-1}) \left(\dfrac{1}{(n_i)^2} - \dfrac{1}{(n_f)^2} \right)$,

where $(3.2881 \times 10^{15} \text{ s}^{-1})$ represents the frequency for the lowest energy photon that is capable of completely removing (ionizing) an electron from a hydrogen atom in its groundstate. The value of A (calculated in this question) is close to the Rydberg frequency $(3.2881 \times 10^{15} \text{ s}^{-1})$, so it is probably the equivalent term in the Moseley equation. The constant b, which is close to unity, could represent the number of electrons left in the K shell after one K-shell electron has been ejected by a cathode ray. Thus, one can think of b as representing the screening afforded by the remaining electron in the K-shell. Of course screening of the nucleus is only be possible for those elements with $Z > 1$.

67. (a) The table provided in this question shows the energy changes associated with the promotion of the outermost valence electron of sodium into the first four excited states above the highest occupied ground state atomic orbital. In addition, we have been told that the energy needed to completely remove one mole of $3s$ electrons from one mole of sodium atoms in the ground state is 496 kJ. The ionization energy for each excited state can be found by subtracting the "energy quanta" entry for the excited state from 496 kJ mol^{-1}.

e.g., for [Ne]$3p^1$, the 1st ionization energy $= 496 \dfrac{\text{kJ}}{\text{mol}} - 203 \dfrac{\text{kJ}}{\text{mol}} = 293 \dfrac{\text{kJ}}{\text{mol}}$

Thus, the rest of the ionization energies are:
[Ne]$4s^1$, $= 496$ kJ mol^{-1} $- 308$ kJ mol^{-1} $= 188$ kJ mol^{-1}
[Ne]$3d^1$, $= 496$ kJ mol^{-1} $- 349$ kJ mol^{-1} $= 147$ kJ mol^{-1}
[Ne]$4p^1$, $= 496$ kJ mol^{-1} $- 362$ kJ mol^{-1} $= 134$ kJ mol^{-1}

(b) Z_{eff} (the effective nuclear charge) for each state can be found by using the equation:

Ionization Energy in kJ mol^{-1} (I.E.) $= \dfrac{A(Z_{eff})^2}{n^2}$

Where n = starting principle quantum level for the electron that is promoted out of the atom and A = 1.3121×10^3 kJ mol^{-1} (Rydberg constant).

For [Ne]$3p^1$ (n = 3) 2.93×10^2 kJmol^{-1} = $\dfrac{1.3121 \times 10^3 \, \frac{kJ}{mol} \, (Z_{eff})^2}{3^2}$ $Z_{eff} = 1.42$

For [Ne]$4s^1$ (n = 4) 1.88×10^2 kJmol^{-1} = $\dfrac{1.3121 \times 10^3 \, \frac{kJ}{mol} \, (Z_{eff})^2}{4^2}$ $Z_{eff} = 1.51$

For [Ne]$3d^1$ (n = 3) 1.47×10^2 kJmol^{-1} = $\dfrac{1.3121 \times 10^3 \, \frac{kJ}{mol} \, (Z_{eff})^2}{3^2}$ $Z_{eff} = 1.00$

For [Ne]$4p^1$ (n = 4) 1.34×10^2 kJmol^{-1} = $\dfrac{1.3121 \times 10^3 \, \frac{kJ}{mol} \, (Z_{eff})^2}{4^2}$ $Z_{eff} = 1.28$

(c) $\bar{r}_{nl}$, which is the average distance of the electron from the nucleus for a particular orbital, can be calculated with the equation:

$$\bar{r}_{nl} = \dfrac{n^2 a_o}{Z_{eff}} \left(1 + \dfrac{1}{2}\left(1 - \dfrac{\ell(\ell+1)}{n^2} \right) \right)$$

Where a_o = 52.9 pm,
n = principal quantum number
ℓ = angular quantum number for the orbital

For [Ne]$3p^1$ (n = 3, ℓ = 1) $\bar{r}_{3p} = \dfrac{3^2(52.9 \text{ pm})}{1.42}\left(1+\dfrac{1}{2}\left(1-\dfrac{1(1+1)}{3^2} \right) \right) = 466$ pm

For [Ne]$4s^1$ (n = 4, ℓ = 0) $\bar{r}_{4s} = \dfrac{4^2(52.9 \text{ pm})}{1.51}\left(1+\dfrac{1}{2}\left(1-\dfrac{0(0+1)}{4^2} \right) \right) = 823$ pm

For [Ne]$3d^1$ (n = 3, ℓ = 2) $\bar{r}_{3d} = \dfrac{3^2(52.9 \text{ pm})}{1.00}\left(1+\dfrac{1}{2}\left(1-\dfrac{2(2+1)}{3^2} \right) \right) = 555$ pm

For [Ne]$4p^1$ (n = 4, ℓ = 1) $\bar{r}_{4p} = \dfrac{4^2(52.9 \text{ pm})}{1.28}\left(1+\dfrac{1}{2}\left(1-\dfrac{1(1+1)}{4^2} \right) \right) = 950$ pm

(d) The results from the Z_{eff} calculations show that the greatest effective nuclear charge is experienced by the 4s orbital (Z_{eff} = 1.51). Next are the two p-orbitals, 3p and 4p, which come in at 1.42 and 1.28 respectively. Coming in last is the 3d orbital, which has a Z_{eff} = 1.00. These results are precisely in keeping with what we would expect. First of all, <u>only</u> the s-orbital penetrates all the way to the nucleus. Both the p- and d-orbitals have nodes at the nucleus. Also p-orbitals penetrate more deeply than do d-orbitals. Recall that the more deeply an orbital penetrates (i.e. the closer the orbital is to the nucleus), the greater is the effective nuclear charge felt by the electrons in that orbital. It follows then that the 4s orbital will experience the greatest effective nuclear charge and that the Z_{eff} values for the $3p$ and $4p$ orbitals should be larger than the Z_{eff} for the $3d$ orbital.

The results from the $\bar{r}_{nl}$ calculations for the four excited state orbitals show that the largest orbital in the set is the $4p$ orbital. This is exactly as expected because the $4p$ orbital is highest in energy and hence, on average furthest from the nucleus. The $4s$ orbital has an average position closer to the nucleus because it experiences a larger effective nuclear charge. The $3p$ orbital, being lowest in energy and hence on average closest to the nucleus, is the smallest orbital in the set. The $3p$ orbital has an average position closer to the nucleus than the $3d$ orbital (which is in the same principle quantum level), because it penetrates more deeply into the atom.

68. (a) First of all, we need to find the ionization energy (I.E.) for the process: $F^-(g) \rightarrow F(g) + e^-$. To accomplish this, we need to calculate Z_{eff} for the species in the left hand column and plot the number of protons in the nucleus against Z_{eff}. By extrapolation, we can estimate the first ionization energy for $F^-(g)$. The electron affinity for F is equal to the first ionization energy of F^- multiplied by minus one (i.e., by reversing the ionization reaction, one can obtain the electron affinity).
For $Ne(g) \rightarrow Ne^+(g) + e^-$ (I.E. = 2080 kJ mol^{-1}; n = 2; 10 protons)

$$\text{I.E. (kJ mol}^{-1}) = \frac{1312.1 \ \frac{kJ}{mol} (Zeff)^2}{4} = 2080 \text{ kJ mol}^{-1} \quad Z_{eff} = 2.518$$

For $Na^+(g) \rightarrow Na^{2+}(g) + e^-$ (I.E. = 4565 kJ mol^{-1}; n = 2; 11 protons)

$$\text{I.E. (kJ mol}^{-1}) = \frac{1312.1 \ \frac{kJ}{mol} (Zeff)^2}{4} = 4565 \text{ kJ mol}^{-1} \quad Z_{eff} = 3.730$$

For $Mg^{2+}(g) \rightarrow Mg^{3+}(g) + e^-$ (I.E. = 7732 kJ mol^{-1}; n = 2; 12 protons)

$$\text{I.E. (kJ mol}^{-1}) = \frac{1312.1 \ \frac{kJ}{mol} (Zeff)^2}{4} = 7732 \text{ kJ mol}^{-1} \quad Z_{eff} = 4.855$$

For $Al^{3+}(g) \rightarrow Al^{4+}(g) + e^-$ (I.E. = 11,577 kJ mol^{-1}; n = 2; 13 protons)

$$\text{I.E. (kJ mol}^{-1}) = \frac{1312.1 \ \frac{kJ}{mol} (Zeff)^2}{4} = 11,577 \text{ kJ mol}^{-1} \quad Z_{eff} = 5.941$$

A plot of the points (10, 2.518). (11, 3.730), (12, 4.855), (13,5.941) gives a straight line that follows the equation: $Z_{eff} = 1.1394(Z) - 8.8421$
For F^-, Z = 9; so $Z_{eff} = 1.1394(9) - 8.8421 = 1.413$ and n = 2. Hence,

$$\text{I.E. (kJ mol}^{-1}) = \frac{1312.1 \ \frac{kJ}{mol} (1.413)^2}{4} = 654.9 \text{ kJ mol}^{-1} \text{ for } F^-$$

The Electron affinity for F must equal the reverse of the first ionization energy or -654.9 kJ. The actual experimental value found for the first electron affinity of F is -328 kJ/mol.

(b) Here, we will use the same method of solution as we did for part (a). To find the electron affinity for the process: $O(g) + e^- \rightarrow O^-(g)$, we first need to calculate the I.E. for $O^-(g)$. This is available from a plot of the number of protons in the nucleus vs. Z_{eff} for the four species in the central column.

For $F(g) \rightarrow F^+(g) + e^-$ (I.E. = 1681 kJ mol^{-1}; n = 2; 9 protons)

$$\text{I.E. (kJ mol}^{-1}\text{)} = \frac{1312.1 \; \frac{kJ}{mol} (Zeff)^2}{4} = 1681 \text{ kJ mol}^{-1} \quad Z_{eff} = 2.264$$

For $Ne^+(g) \rightarrow Ne^{2+}(g) + e^-$ (I.E. = 3963 kJ mol^{-1}; n = 2; 10 protons)

$$\text{I.E. (kJ mol}^{-1}\text{)} = \frac{1312.1 \; \frac{kJ}{mol} (Zeff)^2}{4} = 3963 \text{ kJ mol}^{-1} \quad Z_{eff} = 3.476$$

For $Na^{2+}(g) \rightarrow Na^{3+}(g) + e^-$ (I.E. = 6912 kJ mol^{-1}; n = 2; 11 protons)

$$\text{I.E. (kJ mol}^{-1}\text{)} = \frac{1312.1 \; \frac{kJ}{mol} (Zeff)^2}{4} = 6912 \text{ kJ mol}^{-1} \quad Z_{eff} = 4.590$$

For $Mg^{3+}(g) \rightarrow Mg^{4+}(g) + e^-$ (I.E. = 10,548 kJ mol^{-1}; n = 2; 12 protons)

$$\text{I.E. (kJ mol}^{-1}\text{)} = \frac{1312.1 \; \frac{kJ}{mol} (Zeff)^2}{4} = 10,548 \text{ kJ mol}^{-1} \quad Z_{eff} = 5.671$$

A plot of the points (9, 2.264), (10, 3.476). (11, 4.590), (12, 5.671), gives a straight line that follows the equation: $Z_{eff} = 1.134(Z) - 7.902$

For O^-, Z = 8; so $Z_{eff} = 1.134 (8) - 7.902 = 1.170$ and n = 2. Hence,

$$\text{I.E. (kJ mol}^{-1}\text{)} = \frac{1312.1 \; \frac{kJ}{mol} (1.170)^2}{4} = 449 \text{ kJ mol}^{-1} \text{ for } O^-$$

The Electron affinity for O must equal the reverse of the first ionization energy, or − 449 kJ. Again we will use the same method of solution as was used for part (a). To find the electron affinity for the process: $N(g) + e^- \rightarrow N^-(g)$, we first need to calculate the I.E. for $N^-(g)$. This is accessible from a plot of the number of protons in the nucleus vs. Z_{eff} for the four species in the last column.

For $O(g) \rightarrow O^+(g) + e^-$ (I.E. = 1314 kJ mol^{-1}; n = 2; 8 protons)

$$\text{I.E. (kJ mol}^{-1}\text{)} = \frac{1312.1 \; \frac{kJ}{mol} (Zeff)^2}{4} = 1314 \text{ kJ mol}^{-1} \quad Z_{eff} = 2.001$$

For $F^+(g) \rightarrow F^{2+}(g) + e^-$ (I.E. = 3375 kJ mol^{-1}; n = 2; 9 protons)

$$\text{I.E. (kJ mol}^{-1}\text{)} = \frac{1312.1 \; \frac{kJ}{mol} (Zeff)^2}{4} = 3375 \text{ kJ mol}^{-1} \quad Z_{eff} = 3.208$$

For $Ne^{2+}(g) \rightarrow Ne^{3+}(g) + e^-$ (I.E. = 6276 kJ mol^{-1}; n = 2; 10 protons)

$$\text{I.E. (kJ mol}^{-1}\text{)} = \frac{1312.1 \; \frac{kJ}{mol} (Zeff)^2}{4} = 6276 \text{ kJ mol}^{-1} \quad Z_{eff} = 4.374$$

For $Na^{3+}(g) \rightarrow Na^{4+}(g) + e^-$ (I.E. = 9,540 kJ mol^{-1}; n = 2; 11 protons)

$$\text{I.E. (kJ mol}^{-1}) = \frac{1312.1 \; \dfrac{\text{kJ}}{\text{mol}} (\text{Zeff})^2}{4} = 9,540 \text{ kJ mol}^{-1} \quad Z_{\text{eff}} = 5.393$$

A plot of the points (8, 2.001), (9, 3.204), (10, 4.374). (11, 5.393), gives a straight line that follows the equation: $Z_{\text{eff}} = 1.1346(Z) - 7.0357$

For N⁻, Z =7; so $Z_{\text{eff}} = 1.1346\,(7) - 7.0357 = 0.9065$ and $n = 2$. Hence,

$$\text{I.E. (kJ mol}^{-1}) = \frac{1312.1 \; \dfrac{\text{kJ}}{\text{mol}} (0.9065)^2}{4} = 269.6 \text{ kJ mol}^{-1} \text{ for N}^-$$

The Electron affinity for N must equal the reverse of the first ionization energy, or -269.6 kJ.

(c) For N, O and F, the additional electron ends up in a 2p orbital. In all three instances the nuclear charge is well shielded by the filled 2s orbital located below the 2p set of orbitals. As we proceed from N to F, electrons are placed, one by one, in the 2p subshell and these electrons do afford some additional shielding, but this extra screening is more than offset by the accompanying increase in nuclear charge. Thus, the increase in electron affinity observed upon moving from N via oxygen to fluorine is the result of the steady increase in Z_{eff} that occurs upon moving further to the right in the periodic table.

69. **(a)** An oxygen atom in the ground state has the valence shell configuration $2s^2 2p^4$. Thus there are a total of six electrons in the valence shell. The amount of shielding experienced by one electron in the valence shell is the sum of the shielding provided by the other five electrons in the valence shell and the shielding afforded by the two electrons in the filled 1s orbital below the valence shell. Shielding from electrons in the same shell contribute $5 \times 0.35 = 1.75$ and the shielding from the electrons in the n = 1 shell contributes $2 \times 0.85 = 1.70$. The total shielding is 3.45 (1.75 + 1.70). For O, Z =8 hence, $Z_{\text{eff}} = 8 - 3.45 = 4.55$.

(b) A ground state Cu atom has the valence shell configuration $3d^9 4s^2$. According to Slater's rules, the nine 3d electrons do not shield the $4s^2$ electrons from the nucleus. Thus the total amount of shielding experienced by a 4s electron in Cu is:
Shielding from the other 4*s* electron = 1×0.35 = 0.35
+ shielding from the electrons in the 3*d* subshell = 9×0.85 = 7.65
+shielding from the electrons in the 3*s*/3*p* orbitals = 8×1.00 = 8.00
+shielding from the electrons in the 2*s*/2*p* orbitals = 8×1.00 = 8.00
+shielding from the electrons in the 1*s* orbital = 2×1.00 = 2.00
Total shielding for the 4*s* electrons = (0.35 + 7.65 + 8 + 8 + 2) = 26.00
Copper has Z = 29, so $Z_{\text{eff}} = 29 - 26.00 = 3.00$

(c) 3d electron in a ground state Cu atom will be shielded by the eight other 3d electrons and by the electrons in the lower principle quantum levels. Thus the total amount of shielding for a 3d electron is equal to
shielding from the eight electrons in the 3*d* subshell = 8×0.35 = 2.80
+shielding from the electrons in the 3*s*/3*p* orbitals = 8×1.00 = 8.00
+shielding from the electrons in the 2*s*/2*p* orbitals = 8×1.00 = 8.00
+shielding from the electrons in the 1*s* orbital = 2×1 = 2.00
Total shielding for the 3*d* electrons= (2.80 + 8.00 + 8.00 + 2.00) = 20.80
Copper has $Z = 29$, so $Z_{\text{eff}} = 29 - 20.80 = 8.2$

(d) To find Z_{eff} for the valence electron in each Group I element, we first calculate the screening constant for the electron.

For H: S = 0, so Z_{eff} = Z and Z =1; thus Z_{eff} = 1

For Li: S = 2(0.85) = 1.70 and Z = 3; thus Z_{eff} = 3 – 1.70 = 1.30

For Na: S = 8(0.85) + 2(1) = 8.80 and Z = 11; thus Z_{eff} = 11 – 8.80 = 2.20

For K: S = 8(0.85) + 8(1) + 2(1) = 16.80, Z = 19; thus Z_{eff} = 19 – 16.80 = 2.20

For Rb: S = 8(0.85)+10(1)+8(1)+8(1)+2(1) = 34.80, Z = 37; thus Z_{eff} = 37 – 34.80 = 2.20

For Cs: S = 8(0.85)+10(1)+8(1)+10(1)+8(1)+8(1)+2(1) = 52.80; Z = 55, Thus Z_{eff} = 2.20

Based upon Slater's rules, we have found that the effective nuclear charge increases sharply between periods one and three and then stays at 2.20 for the rest of the alkali metal group. You may recall that the ionization energy for an element can be calculated by using the equation:

$$\text{I.E. (kJ mol}^{-1}) = \frac{1312.1 \ \frac{kJ}{mol} (Zeff)^2}{n^2}$$ Where n = principal quantum number.

By plugging the results from our Z_{eff} calculations into this equation, we would find that the ionization energy decreases markedly as we descend the alkali metal group, in spite of the fact that the Z_{eff} remains constant after Li. The reason that the ionization energy drops is, of course, because the value for n becomes larger as we move down the periodic table and, according to the ionization energy equation given above, larger n values translate into smaller ionization energies (this is because n^2 appears in the denominator). Put another way, even though Z_{eff} remains constant throughout most of Group I, the valence s-electrons become progressively easier to remove as we move down the group because they are further and further from the nucleus. Of course, the further away an electron is from the nucleus, the weaker is its attraction to the nucleus and the easier it is to remove.

(e) As was the case in part (d), to evaluate Z_{eff} for a valence electron in each atom, we must first calculate the screening experienced by the electron with Slater's rules

For Li: S = 2(0.85) = 1.70   and Z = 3, thus Z_{eff} = 3 – 1.70 = 1.30

For Be: S = 1(0.35) + 2(0.85) = 2.05 and Z = 4; thus Z_{eff} = 4 – 2.05 = 1.95

For B: S = 2(0.35) + 2(0.85) = 2.40 and Z = 5; thus Z_{eff} = 5 – 2.40 = 2.60

For C: S = 3(0.35) + 2(0.85) = 2.75 and Z = 6; thus Z_{eff} = 6 – 2.75 = 3.25

For N: S = 4(0.35) + 2(0.85) = 3.10 and Z = 7; thus Z_{eff} = 7 – 3.10 = 3.90

For O: S = 5(0.35) + 2(0.85) = 3.45 and Z = 8; thus Z_{eff} = 8 – 3.45 = 4.55

For F: S = 6(0.35) + 2(0.85) = 3.80 and Z = 9; thus Z_{eff} = 9 – 3.80 = 5.20

For Ne: S = 7(0.35) + 2(0.85) = 4.15 and Z = 10; thus Z_{eff} = 10 – 4.15 = 5.85

The results from these calculations show that the Z_{eff} increases from left to right across the periodic table. Apart from small irregularities, the first ionization energies for the elements within a period also increase with increasing atomic number. Based upon our calculated Z_{eff} values, this is exactly the kind of trend for ionization energies that we would have anticipated. The fact is, a larger effective nuclear charge means that the outer electron(s) is/are held more tightly and this leads to a higher first ionization energy for the atom.

(f)   First we need to calculate the Z_{eff} values for an electron in the $3s$, $3p$ and $3d$ orbitals of both a hydrogen atom and a sodium atom, by using Slater's rules. Since there is only one electron in a H atom, there is no possibility of shielding, and thus the effective nuclear charge for an electron in a $3s$, $3p$ or $3d$ orbital of a H atom is one. The picture for Na is more complicated because it contains intervening electrons that shield the outermost electrons from the attractive power of the nucleus. The Z_{eff} calculations(based on Slater's Rules) for an electron in the i) the $3s$ orbital, ii) the $3p$ orbital, and iii) the $3d$ orbital of a Na atom are shown below:

Na ($3s$ electron; n = 3) Z_{eff} = 11.0 –[(8e$^-$ in the $n = 2$ shell $\times$ 0.85/e$^-$) + (2e$^-$ in the $n = 1$ shell $\times$ 1.00/e$^-$)] = [11.0 – 8.8] = 2.2

Na ($3p$ electron; n = 3; e- was originally in the 3s orbital)) Z_{eff} = 11.0 – [(8e$^-$ in the $n = 2$ shell $\times$ 0.85/e$^-$) + (2e$^-$ in the n = 1 shell $\times$ 1.00/e$^-$)] = [11.0 – 8.8] = 2.2

Na ($3d$ electron; n = 3; e- was originally in the 3s orbital)) Z_{eff} = 11.0 – [(8e$^-$ in the n = 2 shell $\times$ 1.00/e$^-$) + (2e$^-$ in the n = 1 shell $\times$ 1.00/e$^-$)] = [11.0 – 10.0] = 1.0

Next, we insert these Z_{eff} values into their appropriate radial functions, which are gathered in Table 8.1, and use the results from these calculations to construct radial probability plots for an electron in the $3s$, $3p$ and $3d$ orbitals of H and Na. The six plots that result are collected in the two figures presented below:

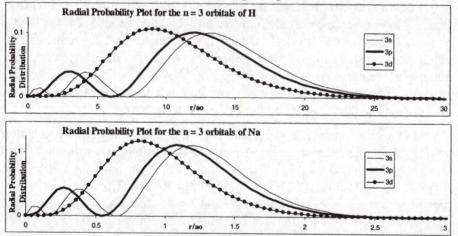

Notice that the $3s$ and $3p$ orbitals of sodium are much closer to the nucleus, on average, than the $3s$ and $3p$ orbitals of hydrogen. Because they are pulled more strongly towards the nucleus, the $3s$ and $3p$ orbitals of sodium end up being much smaller than the corresponding orbitals on hydrogen. Put another way, since the $3s$ and $3p$ electrons in sodium experience a larger effective nuclear charge, they are more tightly bound to the nucleus and, hence, are lower in energy than s and p electrons in the third principal shell of a hydrogen atom. If we want to express this in terms of shielding, we can say that the radial probability distributions for the $3s$ and $3p$ orbitals of sodium are smaller than those of hydrogen because the $3s$ and $3p$ orbitals of sodium are more poorly shielded. The graphs also show that the radial probability plot for the $3d$ orbital of a hydrogen atom is identical to that for a 3d orbital of a Na atom. This is what one would expect since a $3d$ electron in H and a $3d$ electron in Na both experience an effective nuclear charge of one.

CHAPTER 10
CHEMICAL BONDING I: BASIC CONCEPTS
PRACTICE EXAMPLES

1A Mg is in group 2(2A), and thus has 2 valence electrons and 2 dots in its Lewis symbol. Ge is in group 14(4A), and thus has 4 valence electrons and 4 dots in its Lewis symbol. K is in group 1(1A), and thus has 1 valence electron and 1 dot in its Lewis symbol. Ne is in group 18(8A), and thus has 8 valence electrons and 8 dots in its Lewis symbol.

$$\cdot \overset{\cdot}{\underset{\cdot}{Mg}} \cdot \qquad \cdot \overset{\cdot}{\underset{\cdot}{Ge}} \cdot \qquad K \cdot \qquad :\overset{\cdot\cdot}{\underset{\cdot\cdot}{Ne}}:$$

1B Sn is in Family 4A, and thus has 4 electrons and 4 dots in its Lewis symbol. Br is in Family 7A with 7 valence electrons. Adding an electron produces an ion with 8 valence electrons. Tl is in Family 3A with 3 valence electrons. Removing an electron produces a cation with 2 valence electrons.
S is in Family 6A with 6 valence electrons. Adding 2 electrons produces an anion with 8 valence electrons.

$$\cdot \overset{\cdot}{\underset{\cdot}{Sn}} \cdot \qquad [:\overset{\cdot\cdot}{\underset{\cdot\cdot}{Br}}:]^- \qquad [\cdot Tl \cdot]^+ \qquad [:\overset{\cdot\cdot}{\underset{\cdot\cdot}{S}}:]^{2-}$$

2A The Lewis structure for the cation, the anion, and the compound follows the explanation.

(a) Na loses one electron to form Na^+, while S gains two to form S^{2-}.

$$Na \cdot -1\,e^- \rightarrow [\,Na\,]^+ \qquad \cdot \overset{\cdot}{\underset{\cdot\cdot}{S}} \cdot + 2e^- \rightarrow [:\overset{\cdot\cdot}{\underset{\cdot\cdot}{S}}:]^{2-} \quad \text{Lewis Structure: } [Na]^+ [:\overset{\cdot\cdot}{\underset{\cdot\cdot}{S}}:]^{2-} [Na]^+$$

(b) Mg loses two electrons to form Mg^{2+}, while N gains three to form N^{3-}.

$$\cdot Mg \cdot -2\,e^- \rightarrow [\,Mg\,]^{2+} \qquad \cdot \overset{\cdot}{\underset{\cdot\cdot}{N}} \cdot + 3e^- \rightarrow [:\overset{\cdot\cdot}{\underset{\cdot\cdot}{N}}:]^{3-}$$

$$\text{Lewis Structure: } [Mg]^{2+}[:\overset{\cdot\cdot}{\underset{\cdot\cdot}{N}}:]^{3-}[Mg]^{2+}[:\overset{\cdot\cdot}{\underset{\cdot\cdot}{N}}:]^{3-}[Mg]^{2+}$$

2B Below each explanation are the Lewis structures for the cation, for the anion, and for the compound.

(a) In order to acquire a noble-gas electron configuration, Ca loses two electrons, and I gains one, forming the ions Ca^{2+} and I^-. The formula of the compound is CaI_2.

$$\cdot Ca \cdot -2\,e^- \rightarrow [\,Ca\,]^{2+} \qquad :\overset{\cdot\cdot}{\underset{\cdot\cdot}{I}} \cdot + e^- \rightarrow [:\overset{\cdot\cdot}{\underset{\cdot\cdot}{I}}:]^- \quad \text{Lewis Structure: } \quad [:\overset{\cdot\cdot}{\underset{\cdot\cdot}{I}}:]^-[Ca]^{2+}[:\overset{\cdot\cdot}{\underset{\cdot\cdot}{I}}:]^-$$

(b) Ba loses two electrons and S gains two to acquire a noble-gas electron configuration, forming the ions Ba^{2+} and S^{2-}. The formula of the compound is BaS.

$$\cdot Ba \cdot -2\,e^- \rightarrow [\,Ba\,]^{2+} \qquad \cdot \overset{\cdot}{\underset{\cdot\cdot}{S}} \cdot + 2e^- \rightarrow [:\overset{\cdot\cdot}{\underset{\cdot\cdot}{S}}:]^{2-} \quad \text{Lewis Structure: } \quad [Ba]^{2+}[:\overset{\cdot\cdot}{\underset{\cdot\cdot}{S}}:]^{2-}$$

(c) Each Li loses one electron and each O gains two to attain a noble-gas electron configuration, producing the ions Li^+ and O^{2-}. The formula of the compound is Li_2O.

$$\cdot Li -1\,e^- \rightarrow [\,Li\,]^+ \qquad \cdot \overset{\cdot\cdot}{\underset{\cdot}{O}} \cdot + 2e^- \rightarrow [:\overset{\cdot\cdot}{\underset{\cdot\cdot}{O}}:]^{2-} \quad \text{Lewis Structure: } [Li]^+[:\overset{\cdot\cdot}{\underset{\cdot\cdot}{O}}:]^{2-}[Li]^+$$

3A In the Br_2 molecule, the two Br atoms are joined by a single covalent bond. This bonding arrangement gives each Br atom a closed valence shell configuration that is equivalent to that for a Kr atom.

$$: \overset{\cdot\cdot}{\underset{\cdot\cdot}{Br}} - \overset{\cdot\cdot}{\underset{\cdot\cdot}{Br}} :$$

In CH_4, the carbon atom is covalently bonded to four hydrogen atoms. This arrangement gives the carbon atom a valence shell octet and each H atom a valence shell duet.

$$\begin{array}{c} H \\ | \\ H-C-H \\ | \\ H \end{array}$$

In HOCl, the hydrogen and chlorine atoms are attached to the central oxygen atom through single covalent bonds. This bonding arrangement provides each atom in the molecule with a closed valence shell.

$$H - \overset{\cdot\cdot}{\underset{\cdot\cdot}{O}} - \overset{\cdot\cdot}{\underset{\cdot\cdot}{Cl}} :$$

3B The Lewis structure for NI_3 is similar to that of NH_3. The central nitrogen atom is attached to each iodine atom by a single covalent bond. All of the atoms in this structure get a closed valence shell.

$$: \overset{\cdot\cdot}{\underset{\cdot\cdot}{I}} - \overset{\cdot\cdot}{N} - \overset{\cdot\cdot}{\underset{\cdot\cdot}{I}} :$$
$$\begin{array}{c} | \\ : \overset{\cdot\cdot}{\underset{\cdot\cdot}{I}} : \end{array}$$

The Lewis diagram for N_2H_4 has each nitrogen with one lone pair of electrons, two covalent bonds to hydrogen atoms and one covalent bond to the other nitrogen atom. With this arrangement, the nitrogen atoms complete their octets while the hydrogen atoms complete their duets.

$$\begin{array}{c} \overset{\cdot\cdot}{} \quad \overset{\cdot\cdot}{} \\ H-N-N-H \\ | \quad | \\ H \quad H \end{array}$$

In the Lewis structure for C_2H_6, each carbon atom shares four pairs of electrons with three hydrogen atoms and the other carbon atom. With this arrangement, the carbon atoms complete their octets while the hydrogen atoms complete their duets.

$$\begin{array}{c} H \quad H \\ | \quad | \\ H-C-C-H \\ | \quad | \\ H \quad H \end{array}$$

4A The bond with the most ionic character is the one in which the two bonded atoms are the most different in their electronegativities. We find electronegativities in Table 11-2 and calculate ΔEN for each bond.

Electronegativities: H = 2.1   Br = 2.8   N = 3.0   O = 3.5   P = 2.1 Cl = 3.0
Bonds :   H—Br   N—H   N—O   P—Cl
ΔEN values :   0.7   0.9   0.5   0.9

Therefore, the N—H and P—Cl bonds are the most polar of the four bonds cited.

4B The most polar bond is the one with the greatest electronegativity difference.

Electronegativities: C = 2.5 S = 2.5   P = 2.1   O = 3.5   F = 4.0
Bonds: C—S   C—P   P—O   O—F
ΔEN values: 0.0   0.4   1.4   0.5

Therefore, the P—O bond is the most polar of the four bonds cited.

5A The electrostatic potential map that corresponds to IF is the one with the most red in it. This suggests polarization in the molecule. Specifically, the red region signifies a build-up of negative charge that one would expect with the very electronegative fluorine. The other electrostatic potential map corresponds to IBr. The electronegativities are similar, resulting in a relatively non-polar molecule (i.e. little in the way of charge build-up in the molecule).

5B   The electrostatic potential map that corresponds to CH_3OH is the one with the most red in it. This suggests polarization in the molecule. Specifically, the red region signifies a build-up of negative charge that one would expect with the very electronegative oxygen atom. The other electrostatic potential map corresponds to CH_3SH. The carbon and sulfur electronegativities are similar, resulting in a relatively non-polar molecule (i.e. little in the way of charge build-up in the molecule).

6A   **(a)**   C has 4 valence electrons and each S has 6 valence electrons: $4 + (2 \times 6) = 16$ valence electrons or 8 pairs of valence electrons. We place C between two S, and use two electron pairs to hold the molecule together, one between C and each S. We complete the octet on each S with three electron pairs for each S. This uses up six more electron pairs, for a total of eight electron pairs used. $:\ddot{S} - C - \ddot{S}:$ But C does not have an octet. We correct this situation by moving one lone pair from each S into a bonding position between C and S. $:\ddot{S} = C = \ddot{S}:$

(b)   C has 4 valence electrons, N has 5 valence electrons and hydrogen has 1 valence electron: Total number of valence electrons $= 4 + 5 + 1 = 10$ valence electrons or 5 pairs of valence electrons. We place C between H and N, and use two electron pairs to hold the molecule together, one between C and N, as well as one between C and H. We complete the octet on N using three lone pairs. This uses up all five valence electron pairs ($H - C - \ddot{N}:$). But C does not yet have an octet. We correct this situation by moving two lone pair from N into bonding position between C and N. $H - C \equiv N:$

(c)   C has 4 valence electrons, each Cl has 7 valence electrons and oxygen has 6 valence electrons: Thus, the total number of valence electrons $= 4 + 2(7) + 6 = 24$ valence electrons or 12 pairs of valence electrons. We choose C as the central atom, and use three electron pairs to hold the molecule together, one between C and O, as well as one between C and each Cl. We complete the octet on Cl and O using three lone pairs. This uses all twelve electron pairs. But C does not have an octet. We correct this situation by moving one lone pair from O into a bonding position between C and O.

$:\ddot{O} - C (- \ddot{C}l:)_2 \rightarrow :\ddot{O} = C (- \ddot{C}l:)_2$

6B   **(a)**

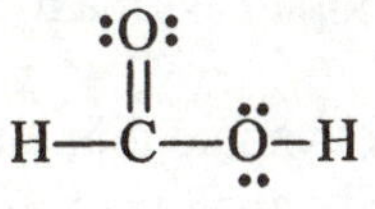

$$
\begin{array}{c}
:O: \\
\parallel \\
H-C-\ddot{O}-H
\end{array}
$$

(b)

$$
\begin{array}{c}
H \quad :O: \\
| \quad\quad \parallel \\
H-C-C-H \\
| \\
H
\end{array}
$$

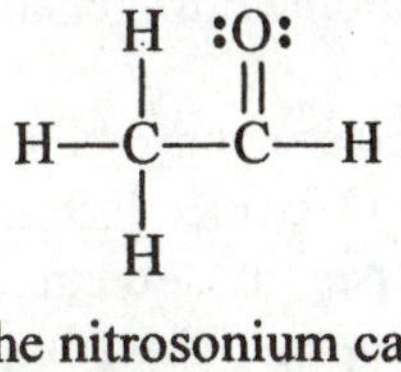

7A   **(a)**   A plausible Lewis structure for the nitrosonium cation, NO^+ is drawn below:

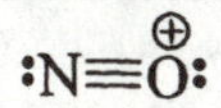

$$:N\equiv\overset{\oplus}{O}:$$

The nitrogen atom is triply bonded to the oxygen atom and both atoms in the structure possess a lone pair of electrons. This gives each atom an octet and a positive formal charge appears on the oxygen atom.

(b) A plausible Lewis structure for $N_2H_5^+$ is given below:

$$H\overset{\oplus}{\underset{|}{\overset{|}{N}}}-\overset{|}{\underset{|}{N}}:$$

The two nitrogen atoms have each achieved an octet. The right hand side N atom is surrounded by three bonding pairs and one lone pair of electron while the left hand side N atom is surrounded by four bonding pairs of electrons. Each hydrogen atom has completed its duet by sharing a pair of electrons with a nitrogen atom. A formal 1+ charge of has been assigned to the left hand side nitrogen atom because it is bonded to four atoms (one more than its usual number) in this structure.

(c) In order to achieve a noble gas configuration, oxygen gains two electrons, forming the stable dianion. The Lewis structure for O^{2-} is shown below.

$$:\overset{..}{\underset{..}{O}}:^{\textcircled{2-}}$$

7B **(a)** The most likely Lewis structure for BF_4^- is drawn below:

$$:\overset{..}{\underset{..}{F}}-\overset{\overset{:\overset{..}{F}:}{|}}{\underset{\underset{:\overset{..}{F}:}{|}}{\overset{\ominus}{B}}}-\overset{..}{\underset{..}{F}}:$$

Four bonding pairs of electrons surround the central boron atom in this structure. This arrangement gives the boron atom a complete octet and a formal charge of -1. By virtue of being surrounded by three lone pairs and one bonding electron pair, each fluorine achieves a full octet.

(b) A plausible Lewis structural form for NH_3OH^+, the hydroxylammonium ion, has been provided below:

$$H-\overset{\overset{H}{|}}{\underset{\underset{H}{|}}{\overset{\oplus}{N}}}-\overset{..}{\underset{..}{O}}-H$$

By sharing bonding electron pairs with three hydrogen atoms and the oxygen atom, the nitrogen atom acquires a full octet and a formal charge of 1+. The oxygen atom shares one bonding electron pair with the nitrogen and a second bonding pair with a hydrogen atom.

(c) Three plausible resonance structures can be drawn for the isocyanate ion, NCO^-. The nitrogen contributes five electrons, the carbon four, oxygen six and one more electron is added to account for the negative charge, giving a total of 16 electrons or eight pairs of electrons. In the first resonance contributor, structure 1 below, the carbon atom is joined to the nitrogen and oxygen atoms by two double bonds thereby creating an octet for carbon. To complete the octet of nitrogen and oxygen, each atom is given a lone pair of electrons. Since nitrogen is sharing just two bonding pairs of electrons in this structure, it must be assigned a formal charge of 1-. In structure 2, the carbon atom is again surrounded by four bonding pairs of electrons, but this time, the carbon atom forms a triple bond with oxygen and just a single bond with nitrogen. The octet for the nitrogen atom is closed with three lone pairs of electrons, while that for oxygen is closed with one lone pair of electrons. This bonding arrangement necessitates giving nitrogen a formal charge of 2- and the oxygen atom a formal charge of 1+. In structure 3, which is the dominant contributor because it has a negative formal charge on oxygen (the most electronegative element in the anion), the carbon achieves a full octet by forming a triple bond with the nitrogen atom and a single bond with the oxygen atom. The octet for oxygen is closed with three lone pairs of electrons, while that for nitrogen is closed with one lone pair of electrons.

$$:\overset{..}{\underset{\ominus}{N}}=C=\overset{..}{O}: \qquad :\overset{\ominus\ominus}{\underset{..}{N}}-C\equiv\overset{\oplus}{O}: \qquad :N\equiv C-\overset{..}{\underset{..}{O}}:^{\ominus}$$

Structure 1   Structure 2   Structure 3

8A   The total number of valence electrons in $NOCl$ is 18 (5 from nitrogen, 6 from oxygen and 7 from chlorine). Four electrons are used to covalently link the central oxygen atom to the terminal chlorine and nitrogen atoms in the skeletal structure: N—O—Cl. Next, we need to distribute the remaining electrons to achieve a noble gas electron configuration for each atom. Since four electrons were used to form the two covalent single bonds, fourteen electrons remain to be distributed. By convention, the valence shells for the terminal atoms are filled first. If we follow this convention, we can close the valence shells for both the nitrogen and the chlorine atoms with twelve electrons.

$$:\overset{..}{\underset{..}{N}}-O-\overset{..}{\underset{..}{Cl}}:$$

Oxygen is moved closer to a complete octet by placing the remaining pair of electrons on oxygen as a lone pair.

$$:\overset{..}{\underset{..}{N}}-\overset{..}{O}-\overset{..}{\underset{..}{Cl}}:$$

The valence shell for the oxygen atom can then be closed by forming a double bond between the nitrogen atom and the oxygen atom.

$$:\overset{..}{\underset{\ominus}{N}}=\overset{..}{\underset{\oplus}{O}}-\overset{..}{\underset{..}{Cl}}:$$

This structure obeys the requirement that all of the atoms end up with a filled valence shell, but is much poorer than the one derived in Example 10-8 because it has a positive formal charge on oxygen, which is the most electronegative atom in the molecule. In other words, this structure can be rejected on the grounds that it does not conform to the third rule for determining plausibility of a Lewis structure based on formal charges which states that "negative formal charges should appear on the most electronegative atom, while any positive formal charge should appear on the least electronegative atom".

8B   There are a total of sixteen valence electrons in the cyanamide molecule (five from each nitrogen atom, four from carbon and one electron from each hydrogen atom). The formula has been written as NH_2CN to remind us that carbon, the most electropositive p-block element in the compound should be selected as the central atom in the skeletal structure.

$$\begin{array}{c} H-N-C-N \\ | \\ H \end{array}$$

To construct this skeletal structure we use 8 electrons. Eight electrons remain to be added to the structure. Note: each hydrogen atom at this stage has achieved a duet by forming a covalent bond with the nitrogen atom in the NH_2 group. The octet for the NH_2 nitrogen is completed by giving it a lone pair of electrons.

$$H-\overset{\displaystyle\ddot{}}{\underset{\displaystyle H}{N}}-C-N$$

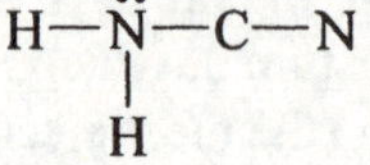

The remaining six electrons can then be given to the terminal nitrogen atom, affording structure 1, shown below. Alternatively, four electrons can be assigned to the terminal nitrogen atom and the last two electrons can be given to the central carbon atom, to produce structure 2 below:

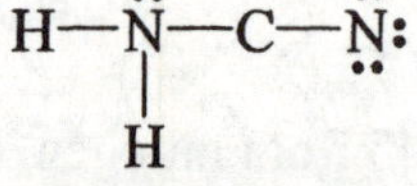

$$H-\overset{\displaystyle\ddot{}}{\underset{\displaystyle H}{N}}-C-\ddot{N}: \qquad H-\overset{\displaystyle\ddot{}}{\underset{\displaystyle H}{N}}-\ddot{C}-\ddot{N}:$$

(Structure 1) 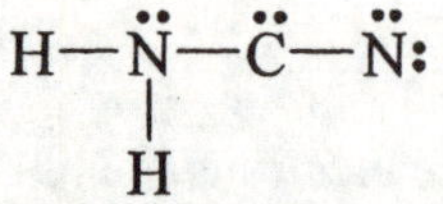   (Structure 2)

The octet for the carbon atom in structure 1 can be completed by converting two lone pairs of electrons on the terminal nitrogen atom into two more covalent bonds to the central carbon atom.

$$H-\overset{\displaystyle\ddot{}}{\underset{\displaystyle H}{N}}-C\equiv N:$$

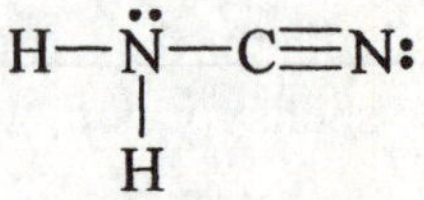

Structure 3

Each atom in structure 3 has a closed-shell electron configuration and a formal charge of zero. We can complete the octet for the carbon and nitrogen atoms in structure 1 by converting a lone pair of electrons on each nitrogen atom into a covalent bond to the central carbon atom.

$$H-\overset{\displaystyle\oplus}{\underset{\displaystyle H}{N}}=C=\ddot{N}^{\ominus}$$

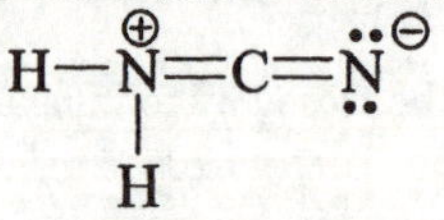

Structure 4

The resulting structure has a formal charge of 1- on the terminal nitrogen atom and a 1+ formal charge on the NH_2 nitrogen atom. Although both structures 3 and 4 both satisfy the octet and duet rules, structure 3 is the better of the two structures because it has no formal charges. A third structure which obeys the octet rule (depicted below), can be rejected on the grounds that it has formal charges of the same type (two 1+ formal charges) on adjacent atoms, as well as negative formal charges on carbon, which is not the most electronegative element in the molecule.

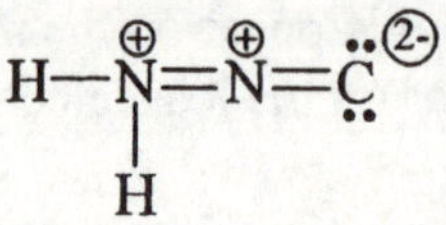

$$H-\overset{\displaystyle\oplus}{\underset{\displaystyle H}{N}}=\overset{\displaystyle\oplus}{N}=\ddot{C}^{②-}$$

9A The skeletal structure for SO_2 has two terminal oxygen atoms bonded to a central sulfur atom. Sulfur has been selected as the central atom by virtue of it being the most electropositive atom in the molecule. It turns out that two different Lewis structures of identical energy can be derived from the skeletal structure described above. First we determine that SO_2 has 18 valence electrons (6 from each atom). Four of the valence electrons must be used to covalently bond the three atoms together. The remaining 14 electrons are used to close the valence shell of each atom. Twelve electrons are used to give the terminal oxygen atoms a closed shell. The remaining two electrons (14 - 12 = 2) are placed on the sulfur atom, affording the structure depicted below:

$$:\ddot{O}-\ddot{S}-\ddot{O}:$$

At this stage, the valence shells for the two oxygen atoms are closed, but the sulfur atom is two electrons short of a complete octet. If we complete the octet for sulfur by converting a lone pair of electrons on the right hand side oxygen atom into a sulfur-to-oxygen π-bond, we end up generating the resonance contributor (A) shown below:

$$:\overset{\ominus}{\underset{..}{\ddot{O}}}-\overset{\oplus}{\ddot{S}}=\ddot{O}: \quad (A)$$

Notice that the structure has a positive formal charge on the sulfur atom (most electropositive element) and a negative formal charge on the left-hand oxygen atom. Remember that oxygen is more electronegative than sulfur, so these charges are plausible. The second completely equivalent contributor, (B), is produced by converting a lone pair on the left-most oxygen atom in the structure into a π-bond, resulting in conversion of a S-O single bond into a sulfur-oxygen double bond:

$$:\ddot{O}=\overset{\oplus}{\ddot{S}}-\overset{\ominus}{\ddot{O}}: \quad (B)$$

Neither structure is consistent with the observation that the two S-O bond lengths in SO_2 are equal, and in fact, the true Lewis structure for SO_2 is neither (A) or (B) but rather an equal blend of the two individual contributors called the resonance hybrid. (see below)

$$:\overset{\ominus}{\ddot{O}}-\overset{\oplus}{\ddot{S}}=\ddot{O}: \longleftrightarrow :\ddot{O}=\overset{\oplus}{\ddot{S}}-\overset{\ominus}{\ddot{O}}:$$

$$\underbrace{\qquad\qquad\qquad\qquad}$$

$$:\ddot{O}-----\ddot{S}-----\ddot{O}:$$

9B The skeletal structure for the NO_3^- ion has three terminal oxygen atoms bonded to a central nitrogen atom. Nitrogen has been chosen as the central atom by virtue of it being the most electropositive atom in the ion. It turns out that three contributing resonance structures of identical energy can be derived from the skeletal structure described here. We begin the process of generating these three structures by counting the total number of valence electrons in the NO_3^- anion. The nitrogen atom contributes five electrons, each oxygen contributes six electrons and an additional electron must be added to account for the 1-charge on the ion. In total, we must account for 24 electrons. Six electrons are used to draw single covalent bonds between the nitrogen atom and three oxygen atoms. The remaining 18 electrons are used to complete the octet for the three terminal oxygen atoms:

$$\underset{O-N-O}{\overset{O}{|}} \quad \xrightarrow{+\ 18\ e^-} \quad \underset{\underset{\ominus}{:\ddot{O}}-\underset{(2+)}{N}-\overset{\ominus}{\ddot{O}:}}{\overset{:\ddot{O}:^{\ominus}}{|}}$$

At this stage the valence shells for the oxygen atoms are filled, but the nitrogen atom is two electrons short of a complete octet. If we complete the octet for nitrogen by converting a lone pair on O_1 into a nitrogen-to-oxygen π-bond, we end up generating resonance contributor (A):

$$\overset{:\overset{..}{O}:}{\underset{\oplus}{\overset{\|}{\underset{:\overset{..}{O}_2}{}}}N-\overset{\ominus}{\underset{..}{O}_3}: \quad (A)$$

Notice the structure has a 1+ formal charge on the nitrogen atom and a 1- on two of the oxygen atoms (O_2 and O_3). These formal charges are quite reasonable energetically. The second and third equivalent structures are generated similarly, by moving a lone pair from O_2 to form a nitrogen to oxygen (O_2) double bond, we end up generating resonance contributor (B), shown below. Likewise, by converting a lone pair from oxygen (O_3) into a π-bond with the nitrogen atom, we end up generating resonance contributor (C), also shown below.

None of these individual structures ((A), (B) or (C)) correctly represents the actual bonding in the nitrate anion. The actual structure, called the resonance hybrid, is the equally weighted average of all three structures (i.e. 1/3(A) + 1/3(B) + 1/3(C)):

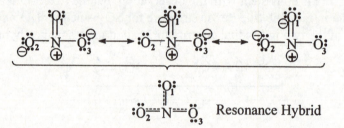

These three resonance forms give bond lengths that are comparable to nitrogen- nitrogen double bonds.

10A The Lewis structure of NCl_3 has three Cl atoms bonded to N and one lone pair attached to N. These four electron groups around N produce a tetrahedral electron-group geometry. The fact that one of the electron groups is a lone pair means that the molecular geometry is trigonal pyramidal.

$$:\overset{..}{\underset{..}{Cl}}-\overset{..}{\underset{|}{N}}-\overset{..}{\underset{..}{Cl}}:$$
$$:\overset{..}{\underset{..}{Cl}}:$$

10B The Lewis structure of $POCl_3$ has three single P-Cl bonds and one P-O bond. These four electron groups around P produce a tetrahedral electron-group geometry. No lone pairs are attached to P and thus the molecular geometry is tetrahedral.

$$:\overset{..}{\underset{|}{\overset{\ominus}{O}}}:$$
$$:\overset{..}{\underset{..}{Cl}}-\overset{\oplus}{\underset{|}{P}}-\overset{..}{\underset{..}{Cl}}:$$
$$:\overset{..}{\underset{..}{Cl}}:$$

11A The Lewis structure of COS has one S doubly-bonded to C and an O doubly-bonded to C. There are no lone pairs attached to C. The electron-group and molecular geometries are the same: linear. $|\overline{S} = C = \overline{O}|$. We can draw other resonance forms, however, the molecular geometry is unaffected

11B N is the central atom. $|N \equiv N - \overline{\underline{O}}|$ This gives an octet on each atom, a formal charge of 1+ on the central N, and a 1– on the O atom. There are two bonding pairs of electrons and no lone pairs on the central N atom. The N_2O molecule is linear. We can draw other resonance forms, however, the molecular geometry is unaffected.

12A In the Lewis structure of methanol, each H atom contributes 1 valence electron, the C atom contributes 4, and the O atom contributes 6, for a total of $(4 \times 1) + 4 + 6 = 14$ valence electrons, or 7 electron pairs. 4 electron pairs are used to connect the H atoms to the C and the O, 1 electron pair is used to connect C to O, and the remaining 2 electron pairs are lone pairs on O, completing its octet.

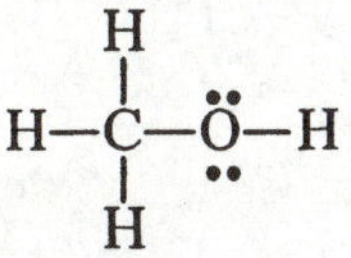

The resulting molecule has two central atoms. Around the C there are four bonding pairs, resulting in a tetrahedral electron-group geometry and molecular geometry. The H—C—H bond angles are ~109.5° as are the H—C—O bond angles. Around the O there are two bonding pairs of electrons and two lone pairs, resulting in a tetrahedral electron-group geometry and a bent molecular shape around the O atom, with a C—O—H bond angle of slightly less than 109.5°.

12B The Lewis structure is drawn below. With four electron groups surrounding each, the electron-group geometries of N, the central C, and the right-hand O are all tetrahedral. The H—N—H bond angle and the H—N—C bond angles are almost the tetrahedral angle of 109.5°, made a bit smaller by the lone pair. The H—C—N angles, the H—C—H angle and the H—C—C angle all are very close to 109.5°. The C—O—H bond angle is made somewhat smaller than 109.5° the presence of two lone pairs on O. Three electron groups surround the right-hand C, making its electron-group and molecular geometries trigonal planar. The O—C—O bond angle and the O—C—C bond angles all are very close to 120°.

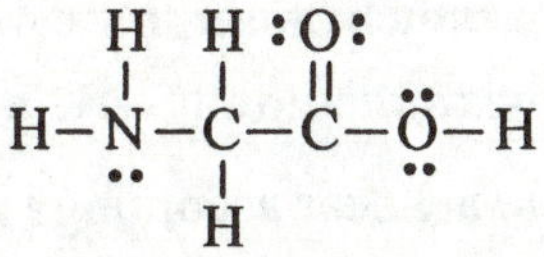

$$H \quad H \; :O:$$
$$\begin{array}{ccccc} & | & | & || & \\ H - & N - & C - & C - & \ddot{O} - H \\ & \ddot{} & | & & \ddot{} \\ & & H & & \end{array}$$

13A Lewis structures of the three molecules are drawn below. Around the S in the SF_6 molecule are six bonding pairs of electrons, and no lone pairs. The molecule is octahedral; each of the S-F bond moments is cancelled by one on the other side of the molecule.

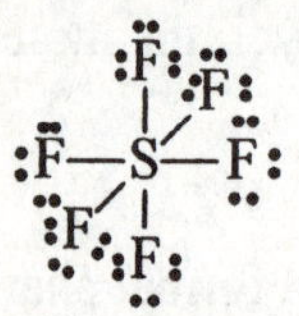

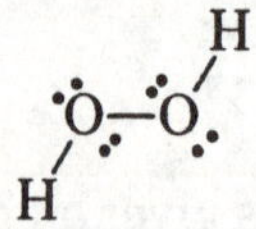

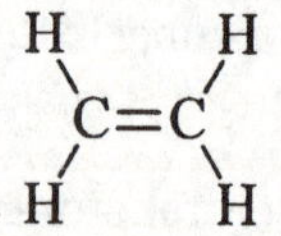

SF_6 is nonpolar. In H_2O_2, the molecular geometry around each O atom is bent; the bond moments do not cancel. H_2O_2 is polar. Around each C in C_2H_4 are three bonding pairs of electron; the molecule is planar around each C and planar overall. The polarity of each —CH_2 group is cancelled by the polarity of the other H_2C— group. C_2H_2 is nonpolar.

13B Lewis structures of the four molecules are drawn below, we can consider the three C—H bonds and the one C—C bond to be non polar. The three C—Cl bonds are tetrahedrally oriented.

If there were a fourth C—Cl bond on the left-hand C, the bond dipoles would cancel out, producing a nonpolar molecule. Since it is not there, the molecule is polar. A similar argument is made for NH_3, where three tetrahedrally-oriented N—H polar bonds are not balanced by a fourth, and CH_2Cl_2, where two tetrahedrally oriented C—Cl bonds are not balanced by two others. This leaves PCl_5 as the only nonpolar species, it is a highly a symmetrical molecule in which individual bond dipoles cancel out.

14A The Lewis structure of CH_3Br has all single bonds. From Table 10.2 the length of a C—H bond is 110 pm. The length of a C—Br bond is not given in the table. A reasonable value is the average of the C—C and Br—Br bond lengths.

$$C—Br = \frac{C—C + Br—Br}{2} = \frac{154 \text{ pm} + 228 \text{ pm}}{2} = 191 \text{ pm}$$

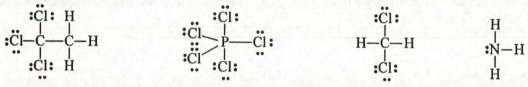

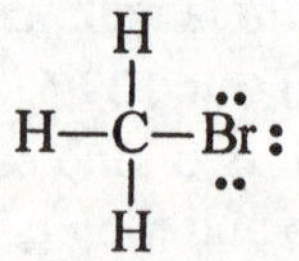

14B In Table 10-2 the C=N bond length is 128 pm, while the C≡N bond length is 116 pm. The observed C—N bond length of 115 pm is much closer to a carbon-nitrogen triple bond. This can be explained by using the following Lewis structure: $|\overline{S} - C \equiv N|$ where there is a formal negative charge on the sulfur atom. This molecule is linear according to VSEPR theory.

15A We first draw Lewis structures for all of the molecules involved in the reaction.

$$2 \text{ H}-\text{H} \ + \ \ddot{\text{O}}=\ddot{\text{O}} \ \longrightarrow \ \text{H}-\ddot{\text{O}}-\text{H}$$

Break 1 O=O + 2H—H = 498 kJ/mol + (2 × 436 kJ/mol) = 1370 kJ/mol absorbed
Form 4H—O = (4 × 464 kJ/mol) = 1856 kJ/mol given off
Enthalpy change = 1370 kJ / mol − 1856 kJ / mol = −486 kJ / mol

15B The chemical equation, with Lewis structures is:

$$1/2 \ :\text{N}\equiv\text{N}: \ \ + \ \ 3/2 \ \text{H}-\text{H} \ \longrightarrow \ \text{H}-\overset{\displaystyle \text{H}}{\underset{\displaystyle |}{\overset{|}{\ddot{\text{N}}}}}-\text{H}$$

Energy required to break bonds $= \frac{1}{2}\text{N}\equiv\text{N}+\frac{3}{2}\text{H}-\text{H}$

$= \left(0.5 \times 946 \text{ kJ / mol}\right)+\left(1.5 \times 436 \text{ kJ / mol}\right) = 1.13 \times 10^3 \text{ kJ / mol}$

Energy realized by forming bonds $= 3 \text{ N}-\text{H} = 3 \times 389 \text{ kJ / mol} = 1.17 \times 10^3 \text{ kJ / mol}$

$\Delta H = 1.13 \times 10^3 \text{ kJ / mol} - 1.17 \times 10^3 \text{ kJ / mol} = -4 \times 10^1 \text{ kJ / mol of NH}_3.$

Thus, $\Delta H_f = -4 \times 10^1 \text{ kJ / mol NH}_3$ (Appendix D value is $\Delta H_f = -46.11 \text{ kJ / mol NH}_3$)

16A The reaction with Lewis structures is shown below.

$$\cdot\ddot{\text{O}}-\text{N}-\ddot{\text{O}}\cdot \ + \ \cdot\ddot{\text{O}}\cdot \ \longrightarrow \ \ddot{\text{O}}=\ddot{\text{N}} \ + \ \ddot{\text{O}}=\ddot{\text{O}}$$

The net result seems to be converting a single N—O bond to a double O=O bond. Since we expect a double bond to be stronger than a single bond, we predict that the products will be more stable than the reactants and this reaction will be exothermic.

16B First we will double the chemical equation, and represent it in terms of Lewis structures:

$$2 \text{ H}-\ddot{\text{O}}-\text{H} \ + \ :\ddot{\text{Cl}}-\ddot{\text{Cl}}: \ \longrightarrow \ \ddot{\text{O}}=\ddot{\text{O}} \ + \ 4 \text{ H}-\ddot{\text{Cl}}:$$

Energy required to break bonds = 2 Cl—Cl + 4H—O

$= \left(2\times 243 \text{ kJ/mol}\right)+\left(4\times 464 \text{ kJ/mol}\right) = 2342 \text{ kJ/mol}$

Energy realized by forming bonds = 1 O=O + 4 × H—Cl

$= 498 \text{ kJ / mol}+\left(4 \times 431 \text{ kJ / mol}\right) = 2222 \text{ kJ / mol}$

$\Delta H = \frac{1}{2}\left(2342 \text{ kJ / mol} - 2222 \text{ kJ / mol}\right) = +60 \text{ kJ / mol}$ The reaction is endothermic.

Important Note: In this and subsequent chapters, a lone pair of electrons in a Lewis structure can be shown as a line or a pair of dots. Thus, the Lewis structure of Be is both Be| and Be:

EXERCISES

Lewis theory

1.   (a) $:\ddot{\text{K}}\text{r}:$ (b) $\cdot\dot{\text{Ge}}\cdot$ (c) $\cdot\dot{\text{N}}:$ (d) $\cdot\dot{\text{Ga}}$ (e) $\cdot\dot{\text{As}}:$ (f) Rb$\cdot$

2. (a) $\left[\text{H}{:}\right]^-$ (b) $\left[\cdot\text{Sn}\cdot\right]^{2+}$ (c) $\left[\text{K}\right]^+$ (d) $\left[{:}\ddot{\text{Br}}{:}\right]^-$ (e) $\left[{:}\ddot{\text{Se}}{:}\right]^{2-}$ (f) $\left[\text{Sc}\right]^{3+}$

3. (a) ${:}\ddot{\text{I}}{-}\ddot{\text{Cl}}{:}$ (b) ${:}\ddot{\text{Br}}{-}\ddot{\text{Br}}{:}$ (c) ${:}\ddot{\text{F}}{-}\ddot{\text{O}}{-}\ddot{\text{F}}{:}$ (d) structure of $\text{I}{-}\text{N}{-}\text{I}$ with I below N (e) $\text{H}{-}\ddot{\text{Se}}{-}\text{H}$

4. (a) $\ddot{\text{S}}{=}\text{C}{=}\ddot{\text{S}}$ (b) structure with H–C–C–C–H backbone, O double-bonded to central C and two H's on outer C's (c) ${:}\ddot{\text{Cl}}{-}\text{C}({=}\text{O}){-}\ddot{\text{Cl}}{:}$ (d) ${:}\ddot{\text{F}}{-}\ddot{\text{N}}{=}\ddot{\text{O}}$

5. (a) $\text{Cs}^{\oplus}\ {:}\ddot{\text{Br}}{:}^{\ominus}$ CsBr, cesium bromide (b) $\text{H}{-}\ddot{\text{Sb}}{-}\text{H}$ with H below Sb, H_3Sb, hydrogen antimonide or trihydrogen antimonide

(c) ${:}\ddot{\text{Cl}}{-}\text{B}({-}\ddot{\text{Cl}}{:}){-}\ddot{\text{Cl}}{:}$ BCl_3, boron trichloride (d) $\text{Cs}^{\oplus}\ {:}\ddot{\text{Cl}}{:}^{\ominus}$ CsCl, cesium chloride

(e) $\text{Li}^{\oplus}\ {:}\ddot{\text{O}}{:}^{\ominus 2}\ \text{Li}^{\oplus}$ Li_2O, lithium oxide (f) ${:}\ddot{\text{I}}{-}\ddot{\text{Cl}}{:}$ ICl, iodine monochloride

6. Hydrogen never has an octet of electrons in any of its compounds, but rather a pair (or duet, if you prefer). An example is the Lewis structure of H_2O (below). In many compounds in which the central atom is from the second period or higher, there are more than eight electrons around the central atom. An example of a compound with such an "expanded octet" is ICl_3 (see below). Finally, in some compounds, there are less than eight electrons around the central atom. One such "electron deficient" compound is BF_3.

$$\text{H}{-}\ddot{\text{O}}{-}\text{H} \qquad {:}\ddot{\text{F}}{-}\text{B}({-}\ddot{\text{F}}{:}){-}\ddot{\text{F}}{:} \qquad {:}\ddot{\text{Cl}}{-}\text{I}({-}\ddot{\text{Cl}}{:}){-}\ddot{\text{Cl}}{:}$$

7. NH_3 $\quad 5+(3\times1)=8\,\text{v.e.}=4\,\text{pairs}$ $\qquad \text{BF}_3$ $\quad 3+(3\times7)=24\,\text{v.e.}=12\,\text{pairs}$

SF_6 $\quad 6+(6\times7)=48\,\text{v.e.}=24\,\text{pairs}$ $\qquad \text{SO}_3$ $\quad 6+(3\times6)=24\,\text{v.e.}=12\,\text{pairs}$

NH_4^+ $\ 5+(4\times1)-1=8\,\text{v.e.}=4\,\text{pairs}$ $\qquad \text{SO}_4^{2-}$ $\ 6+(4\times6)+2=32\ \text{v.e.}=16\,\text{pairs}$

NO_2 $\ 5+(2\times6)=17\,\text{v.e.}=8.5\,\text{pairs}$

NO_2 cannot obey the octet rule; there is no way to pair all electrons when the number of electrons is odd.

All of these Lewis structures obey the octet rule except for BF_3, which is electron deficient, and SF_6, which has an expanded octet.

276

8. **(a)** In order to construct an H₃ molecule, one H would have to bridge the other two. This would place 3 electrons around the central H atom which is more than the stable pair found around H in most Lewis structures. As well, one bond would be the normal 2 e⁻ bond and the other would be a one electron bond, which is beyond Lewis theory. (i.e. H–H · H)

(b) In HHe there would be three electrons between the two atoms, or three electrons around the He atom one of which would be a nonbonding electron. Neither of these is a particularly stable situation.

(c) He₂ would have either a double bond between two He atoms and thus four electrons around each He atom, or three electrons around each He atom (2 e⁻ in a bond and an unpaired electron on each atom). Neither situation achieves the electron configuration of a noble gas.

(d) H₃O has an expanded octet (9 electrons) on oxygen; expanded octets are not possible for elements in the second period. Other structures place a multiple bond between O and H. Both situations are unstable.

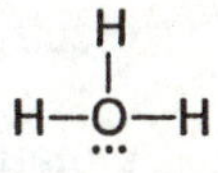

H–O–H

9. **(a)** 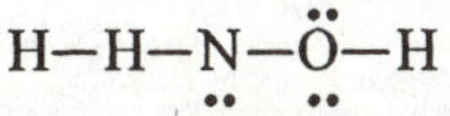 H–H–N–O–H has two bonds to (four electrons around) the second hydrogen, and only six electrons around the nitrogen. A better Lewis structure is shown to the right

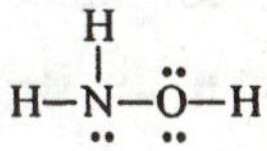

H–N–O–H

(b) 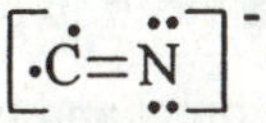 [·C=N]⁻ has only six electrons around the C atom and two too few overall.

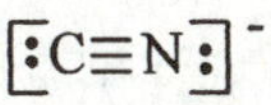

[:C≡N:]⁻ is a more plausible Lewis structure for the cyanide ion.

10. **(a)** 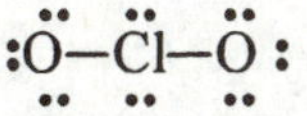 :O–Cl–O: has 20 valence electrons, whereas the molecule ClO₂ has 19 valence electrons. This is a proper Lewis structure for the chlorite ion, although the brackets and the minus charge are missing. A plausible Lewis structure for the molecule ClO₂ is :O–Cl–O:

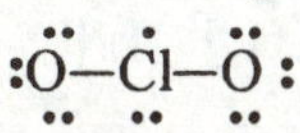

(b) Here, Ca–O: is improperly written as a covalent Lewis structure. CaO is actually an ionic compound. [Ca]²⁺[:O:]²⁻ is a more plausible Lewis structure for CaO.

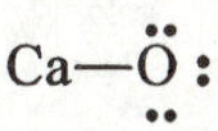

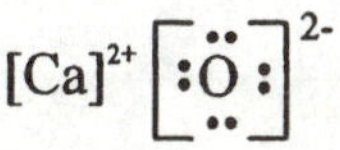

11. **(c)** Hypochlorite ion is the correct answer.

(a) 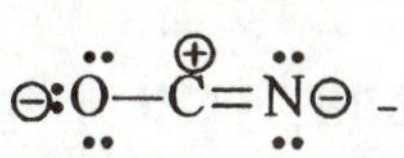 ⁻:O–C=N⊖ – does not have an octet of electrons around C.

(b) [C=C:]⁻ does not have an octet around either C. Moreover, it has only 6 valence electrons in total while it should have 10, and finally, the sum of the formal charges on the two carbons doesn't equal the charge on the ion.

(d) The total number of valence electrons in NO is incorrect. No, being an odd-electron species should have 11 valence electrons, not 12.

12. **(a)** $Mg{-}\overset{..}{\underset{..}{O}}:$ is incorrectly written as a covalent structure). One expects an ionic Lewis

structure, namely $[Mg]^{2+}[:\overset{..}{\underset{..}{O}}:]^{2-}$

(b) $[\overset{..}{\underset{..}{Cl}}:]^{+}[:\overset{..}{\underset{..}{O}}:]^{2-}[:\overset{..}{\underset{..}{Cl}}]^{+}$ is written as an ionic structure, even though we expect a covalent

structure between nonmetallic atoms. A more plausible structure is $:\overset{..}{\underset{..}{Cl}}{-}\overset{..}{\underset{..}{O}}{-}\overset{..}{\underset{..}{Cl}}:$

(c) $[:\overset{..}{\underset{..}{O}}{-}\overset{.}{N}{=}\overset{..}{\underset{..}{O}}]^{+}$ has too many valence electrons—8.5 electron pairs or 17 valence

electrons—it should have $(2\times 6)+5-1=16$ valence electrons or 8 electron pairs. A

a plausible Lewis structure is $[\overset{..}{\underset{..}{O}}{=}N{=}\overset{..}{\underset{..}{O}}]^{2+}$, which has 1+ formal charge on N and 0

formal charge of zero on each oxygen.

(d) In the structure $[\overset{..}{\underset{..}{S}}{-}C{=}\overset{..}{\underset{..}{N}}]^{-}$ neither S nor C posses an octet of electrons. In

addition, there are only 7 pairs of valence electrons in this structure or 14 valence
electrons. There should be $6+4+5+1=16$ valence electrons, or 8 electron pairs.

Two structures are possible. $\overset{..}{\underset{..}{S}}{=}C{=}\overset{..}{\underset{..}{N}}$ has a formal charge of 1- on N and is

preferred over $:\overset{..}{\underset{..}{S}}{-}C{\equiv}N:$, with its formal charge of 1- on S, which is less

electronegative than N.

Ionic bonding

13. **(a)** $[:\overset{..}{\underset{..}{Cl}}:]^{-}[Ca]^{2+}[:\overset{..}{\underset{..}{Cl}}:]^{-}$ **(b)** $[Ba]^{2+}[:\overset{..}{\underset{..}{S}}:]^{2-}$ **(c)** $[Li]^{+}[:\overset{..}{\underset{..}{O}}:]^{2-}[Li]^{+}$ **(d)** $[Na]^{+}[:\overset{..}{\underset{..}{F}}:]^{-}$

14. **(a)** $[Li]^{+}[:\overset{..}{\underset{..}{S}}:]^{2-}[Li]^{+}$ **(b)** $[Na]^{+}[:\overset{..}{\underset{..}{F}}:]^{-}$ **(c)** $[:\overset{..}{\underset{..}{I}}:]^{-}[Ca]^{2+}[:\overset{..}{\underset{..}{I}}:]^{-}$

(d) $[:\overset{..}{\underset{..}{Cl}}:]^{-}[:\overset{..}{\underset{..}{Cl}}:]^{-}[Sc]^{3+}[:\overset{..}{\underset{..}{Cl}}:]^{-}$

15. The Lewis symbols are $[H:]^{-}$ for the hydride ion, $[:\overset{..}{\underset{..}{N}}:]^{3-}$ for the nitride ion.

(a) $[Li]^{+}[H:]^{-}$
Lithium hydrid

(b) $[H:]^{-}[Ca]^{2+}[H:]^{-}$
Calcium hydride

(c) $[:\overset{..}{\underset{..}{N}}:]^{3-}[Mg]^{2+}[:\overset{..}{\underset{..}{N}}:]^{3-}[Mg]^{2+}[:\overset{..}{\underset{..}{N}}:]^{3-}$
magnesium nitride

16. **(a)** $[:\overset{..}{\underset{..}{O}}{-}H]^{-}[Ca]^{2+}[:\overset{..}{\underset{..}{O}}{-}H]^{-}$

(b) $\left[\begin{array}{c} H \\ | \\ H{-}N{-}H \\ | \\ H \end{array}\right]^{+}[:\overset{..}{\underset{..}{Br}}:]^{-}$

(c) $[:\overset{..}{\underset{..}{O}}{-}\overset{..}{\underset{..}{Cl}}:]^{-}[Ca]^{2+}[:\overset{..}{\underset{..}{O}}{-}\overset{..}{\underset{..}{Cl}}:]^{-}$

Formal Charge

<u>17.</u>

(a)

computations for:	H—	—C≡	≡C
no. valence e⁻	1	4	4
- no. lone-pair e⁻	–0	–0	–2
– ½ no. bond-pair e⁻	<u>–1</u>	<u>–4</u>	<u>–3</u>
formal charge	0	0	–1

(b)

computations for:	=O	—O(×2)	C
no. valence e⁻	6	6	4
- no. lone-pair e⁻	–4	–6	–0
– ½ no. bond-pair e⁻	<u>–2</u>	<u>–1</u>	<u>–4</u>
formal charge	–0	–1	0

(c)

computations for:	—H(×7)	side C(×2)	central C
no. valence e⁻	1	4	4
- no. lone-pair e⁻	–0	–0	–0
– ½ no. bond-pair e⁻	<u>–1</u>	<u>–4</u>	<u>–3</u>
formal charge	0	0	+1

18.

(a) The formal charge on each I is 0,
computed as follows:
no. valence e⁻ = 7
–no. lone-pair e⁻ = –6
– ½ no. bond-pair e⁻ = <u>–1</u>
formal charge = 0

(b)

computations for:	=O	—O	=S—
no. valence e⁻	6	6	6
- no. lone-pair e⁻	–4	–6	–2
– ½ no. bond-pair e⁻	<u>–2</u>	<u>–1</u>	<u>–3</u>
formal charge	0	–1	+1

(c)

computations for:	=O	—O	=N—
no. valence e⁻	6	6	5
- no. lone-pair e⁻	–4	–6	–1
– ½ no. bond-pair e⁻	<u>–2</u>	<u>–1</u>	<u>–3</u>
formal charge	0	–1	+1

19. There are three features common to formal charge and oxidation state. First, both indicate how the bonding electrons are distributed in the molecule. Second, negative formal charge (in the most plausible Lewis structure) and negative oxidation state are generally assigned to the more electronegative atoms. And third, both numbers are determined by a set of rules, rather than being determined experimentally. Bear in mind, however, that there are also significant differences. For instance, there are cases where atoms of the same type with the same oxidation state have different formal charges, such as oxygen in ozone, O_3. Another is that formal charges are used to decide between alternative Lewis structures, while oxidation state is used in balancing equations and naming compounds. Also, the oxidation state in a compound is invariant, while the formal charge can change. The most significant difference, though, is that whereas the oxidation state of an element in its compounds is usually not zero, its formal charge usually is.

20. The most common instance in which formal charge is not kept to a minimum occurs is in the case of ionic compounds. For example, in $\mathrm{Mg}\!-\!\ddot{\mathrm{O}}\!:$ the formal charge on Mg is 1+ and on O it is 1-, while in the ionic version $[\mathrm{Mg}]^{2+}\,[:\!\ddot{\mathrm{O}}\!:]^{2-}$, formal charges are 2+ and 2-, respectively.

Additionally, in some resonance hybrids, formal charge is not minimized. In order to have bond lengths agree with experimental results, it may not be acceptable to create multiple bonds. Yet a third instance is when double bonds are created to lower formal charge, particularly when this results in the octet rule being violated. For instance, all the Cl—O bonds in ClO_4^- are best represented using single bonds. Although including some double bonds would minimize formal charges, the resulting structure is less desirable because the octet for Cl has been exceeded and this requires the input of additional energy.

21. Formal charge = number of valence electrons −2 × number of lone pairs − number of bonding electron pairs. The formal charge of the central atom is calculated below the Lewis structure of each species.

	(a)	(b)	(c)	(d)	(e)
valence e⁻	6	3	5	5	7
-2×(lone pairs)	-2	0	0	0	-4
-bonding pair	-3	-4	-4	-5	-4
formal charge	+1	-1	+1	0	-1

22. We calculate formal charge = # valence electrons − # lone-pair e⁻ − ½ # bond-pair e⁻

(a)

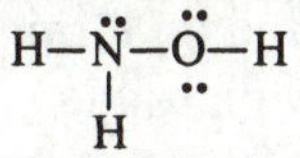

H—N̈—Ö—H
　　|
　　H

f.c. of $N = 5 - 2 - 3 = 0$ 　　most plausible
f.c. of $O = 6 - 4 - 2 = 0$

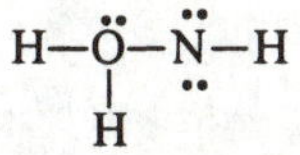

H—Ö—N̈—H
　　|
　　H

f.c. of $N = 5 - 4 - 2 = -1$
f.c. of $O = 6 - 2 - 3 = +1$

(b)

S̈=C=S̈

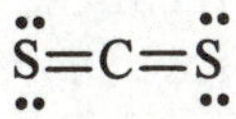

f.c. of $C = 4 - 0 - 4 = 0$ 　　most plausible
f.c. of $S = 6 - 4 - 2 = 0$

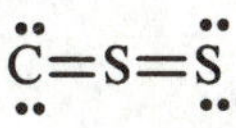

C̈=S=S̈

f.c. of $C = 4 - 4 - 2 = -2$
f.c. of central $S = 6 - 0 - 4 = +2$
f.c. of terminal $S = 6 - 4 - 2 = 0$

(c)

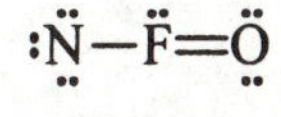

:N̈—F̈=Ö

f.c. of $N = 5 - 6 - 1 = -2$
f.c. of $O = 6 - 4 - 2 = 0$
f.c. of $F = 7 - 2 - 3 = 2+$

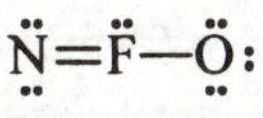

N̈=F̈—Ö:

f.c. of $N = 5 - 4 - 2 = -1$
f.c. of $O = 6 - 6 - 1 = -1$
f.c. of $F = 7 - 2 - 3 = 2+$

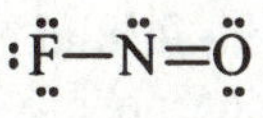

:F̈—N̈=Ö

f.c. of $N = 5 - 2 - 3 = 0$
f.c. of $O = 6 - 4 - 2 = 0$
f.c. of $F = 7 - 6 - 1 = 0$ 　　most plausible

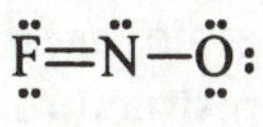

F̈=N̈—Ö:

f.c. of $O = 6 - 6 - 1 = -1$
f.c. of $N = 5 - 2 - 3 = 0$
f.c. of $F = 7 - 4 - 2 = 1+$

(d)

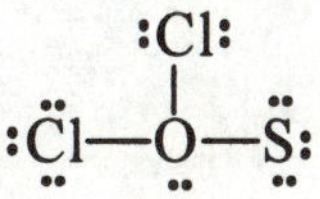

　　:Cl̈:
　　|
:Cl̈—O—S̈:

f.c. of $S = 6 - 6 - 1 = -1$ 　　O should have
f.c. of $O = 6 - 2 - 3 = +1$ 　　the negative
f.c. of $Cl = 7 - 6 - 1 = 0$ 　　formal charge

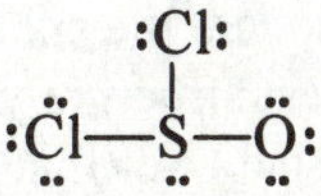

　　:Cl̈:
　　|
:Cl̈—S—Ö:

f.c. of $S = 6 - 2 - 3 = +1$
f.c. of $O = 6 - 6 - 1 = -1$ 　　most plausible
f.c. of $Cl = 7 - 6 - 1 = 0$

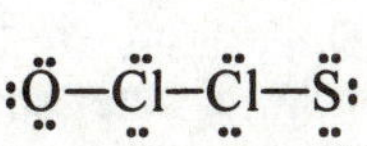

:Ö—Cl̈—Cl̈—S̈:

f.c. of $S = 6 - 6 - 1 = -1$ 　　26 electrons present;
f.c. of $O = 6 - 6 - 1 = -1$ 　　Cl should not have a
f.c. of $Cl = 7 - 4 - 2 = +1$ 　　more positive f.c. than S

23. We begin by counting the total number of valence electrons that must appear in the Lewis structure of the ion CO_2H^+: one from hydrogen, four from carbon, and six from each of the two oxygen (12 in all from the oxygen atoms). One electron is lost to establish the 1+ charge on the ion. In all, sixteen electrons are in the valence shell of the cation. If the usual rules for constructing valid Lewis structures are applied to HCO_2^+, we come up with the following structures:

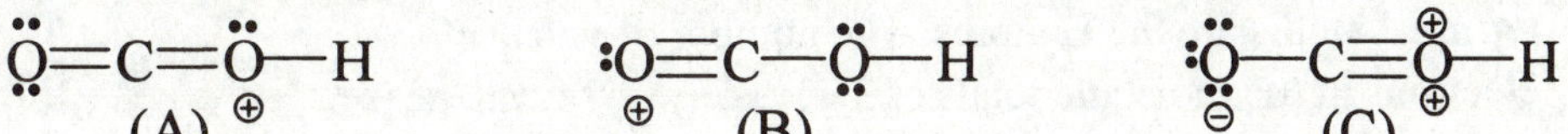

Ö=C=Ö—H　　　　:O≡C—Ö—H　　　　:Ö—C≡Ö⊕—H
(A) ⊕　　　　　　⊕　　(B)　　　　　　⊖　　(C)⊕

In structure (A), the internal oxygen atom caries the positive charge, while in structure (B), the positive charge is located on the terminal oxygen atom. A third structure can also be drawn, however, due to an unacceptably large charge build-up, this form can be neglected. Thus, in this case of A and B, we cannot use the concept of formal charge to pick one structure over the other because the positive formal charge in both structures is located on the same type of atom, namely, an oxygen atom. In other words, based on formal charge rules alone, we must conclude that structures (A) and (B) are equally plausible.

24. The intention of this question is to make the student aware of the fact that on occasion, one can obtain a better Lewis structure *"from the standpoint of formal charge minimization"* by using chain-like structures rather than the expected compact, symmetrical structures. The two linear Lewis structures for the ClO_4^- (I) and (II) and a compact structure (III) are shown below:

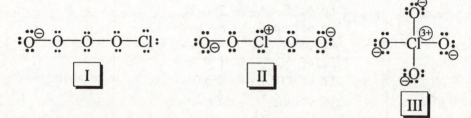

All of these structures have the required 32 valence electrons. Structure I has only one formal charge, 1- on the terminal oxygen atom. Structure II has a total of three formal charges, 1- on the terminal oxygen atoms and 1+ on the central chlorine atom. The compact structure, structure III, has formal charges on all of the atoms, 1- on all oxygen atoms and a formal charge of 3+ on the central chlorine atom. From the standpoint of minimizing formal charge, structure I would be deemed the most appropriate. Nevertheless, structure III is the one that is actually adopted by the ClO_4^- ion, despite the fact that better minimization is achieved with the linear structure. By using an expanded valence shell with 14 electrons for the central atom, as in structure IV below, one can achieve the same minimum set of formal charges as in structure I:

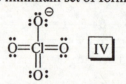

The chlorine atom presumably uses available $3d$ orbitals to accommodate the six additional electrons in its valence shell. Structure IV would appear to be the best structure yet because it has the minimum number of formal charges and is close to the true shape for that seen in the ClO_4^- ion. In light of recent quantum mechanical calculations, however, many chemists now believe that d-orbital involvement in expanded octets should only be invoked when there is no way to avoid them, as in PCl_5 or SF_6. This means that whenever possible, octet structures should be used, even though at times they afford unsettlingly large formal charges. Thus, structure III is now considered by most chemists as being superior to structure IV.

Lewis structures

25. **(a)** In H_2NOH, N and O atoms are the central atoms and the terminal atoms are the H atoms. The number of valence electrons in the molecule totals: $(3 \times 1) + 5 + 6 = 14$ valence electrons, or seven electron pairs. A reasonable Lewis structure is:

$$H-N-\ddot{O}-H$$
with H above N

(b) In $HOClO_2$, two O atoms and one H atom are terminal atoms, and the Cl and O atoms are the central atoms. The total number of valence electrons in the structure is $1+7+(3\times6)=26$ valence electrons, or 13 electron pairs. A reasonable Lewis structure is:

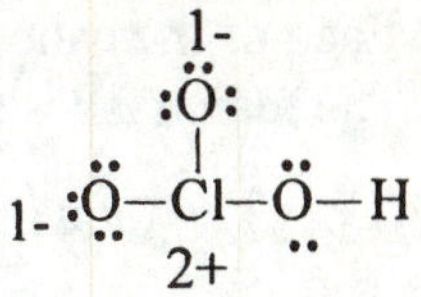

(c) In HONO, the H atom and one O atom are the terminal atoms; the other O atom and the N atom are the central atoms. The total number of valence electrons in the structure is $1+5+(2\times6)=18$ valence electrons, or nine electron pairs. A plausible Lewis structure is $H-\ddot{O}-\ddot{N}=\ddot{O}$

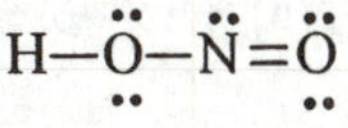

(d) In O_2SCl_2, S is the central atom, with Cl and O as terminal atoms. The total number of valence electrons is $(2\times6)+6+(2\times7)=32$ valence electrons, 16 electron pairs. A reasonable Lewis structure is:

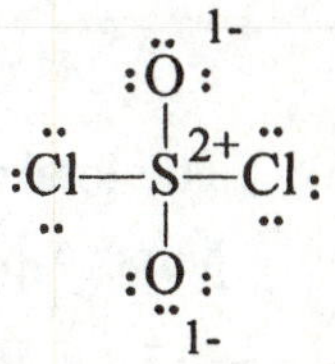

26. The total number of valence electrons is $(2\times7)+(2\times6)=26$ valence electrons, or 13 pairs of valence electrons. It is unlikely to have F as a central atom; that would require an expanded octet on F. One structure is $:\ddot{F}-\ddot{S}-\ddot{S}-\ddot{F}:$; another is shown to the right.

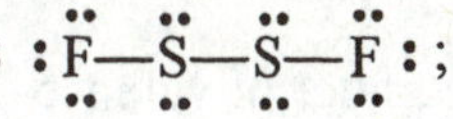

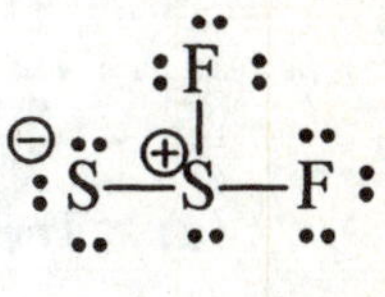

27. **(a)** The total number of valence electrons in SO_3^{2-} is $6+(3\times6)+2=26$, or 13 electron pairs. A plausible Lewis structure is:

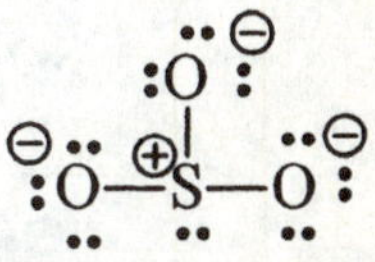

(b) The total number of valence electrons in NO_2^- is $5+(2\times6)+1=18$, or 9 electron pairs. There are two resonance forms for the nitrite ion:

$$\overset{\ominus}{\ddot{O}}-\ddot{N}=\ddot{O} \longleftrightarrow \ddot{O}=\ddot{N}-\overset{\ominus}{\ddot{O}}$$

(c) The total number of valence electrons in CO_3^{2-} is $4+(3\times6)+2=24$, or 12 electron pairs. There are three resonance forms for the carbonate ion:

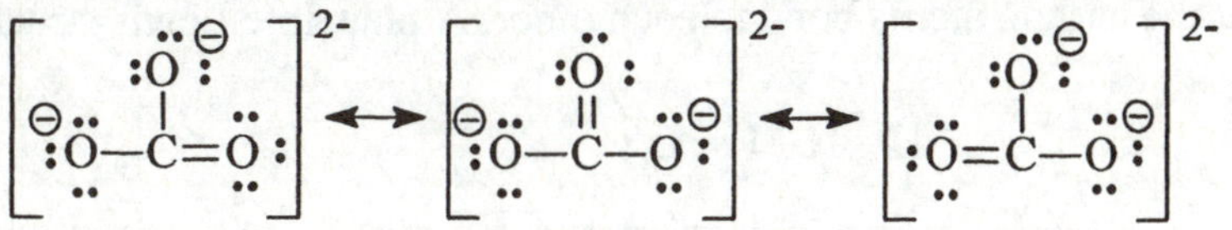

(d) The total number of valence electrons in HO_2^- is $1+(2\times6)+1=14$, or 7 electron pairs. A plausible Lewis structure is $H-\ddot{O}-\overset{\ominus}{\ddot{O}}$

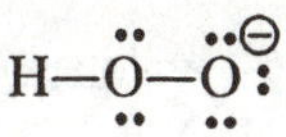

28. Each of the cations has an empty valence shell as the result of ionization. The main task is to determine the Lewis structure of each anion.

(a) The total number of valence electrons in OH^- is $6+1+1=8$, or 4 electron pairs. A plausible Lewis structure for barium hydroxide is $[:\ddot{O}-H]^-[Ba]^{2+}[:\ddot{O}-H]^-$

(b) The total number of valence electrons in NO_2^- is $5+(2\times6)+1=18$, or 9 electron pairs. A plausible Lewis structure sodium nitrite is $[Na]^+[:\ddot{O}-\ddot{N}=\ddot{O}]^- \longleftrightarrow [Na]^+[\ddot{O}=\ddot{N}-\ddot{O}:]^-$

(c) The total number of valence electrons in IO_3^- is $7+(3\times6)+1=26$, or 13 electron pairs. A plausible Lewis structure for magnesium iodate is

$$\begin{bmatrix} :\ddot{O}: \\ | \\ :\ddot{O}-I-\ddot{O}: \end{bmatrix}^- [Mg]^{2+} \begin{bmatrix} :\ddot{O}: \\ | \\ :\ddot{O}-I-\ddot{O}: \end{bmatrix}^-$$

(d) The total number of valence electrons in SO_4^{2-} is $6+(4\times6)+2=32$, or 16 electron pairs. A plausible structure for aluminum sulfate is $[SO_4]^{2-}[Al]^{3+}[SO_4]^{2-}[Al]^{3+}[SO_4]^{2-}$. Because of the ability of S to expand its octet, SO_4^{2-} has several resonance forms, a few of which are:

$$\begin{array}{ccc}
:\ddot{O}:^{\ominus} & :\ddot{O}:^{\ominus} & :\ddot{O}:^{\ominus} \\
| & | & | \\
^{\ominus}:\ddot{O}-\overset{\oplus}{S}-\ddot{O}:^{\ominus} & \ddot{O}=S-\ddot{O}:^{\ominus} & \ddot{O}=S=\ddot{O} \\
|\oplus & |\oplus & | \\
:\ddot{O}: & :\ddot{O}: & :\ddot{O}: \\
^{\ominus} & ^{\ominus} & ^{\ominus}
\end{array}$$

The first structure, without S expanded octet, is preferred.

29. In $CH_3CHCHCHO$ there are $(4\times4)+(6\times1)+6=28$ valence electrons, or 14 electron pairs. We expect that the carbon atoms bond to each other. A plausible Lewis structure is :

$$\begin{array}{c}
H \quad H \quad H :\ddot{O}: \\
| \quad\; | \quad\; | \quad\; || \\
H-C-C=C-C-H \\
| \\
H
\end{array}$$

30. In C_3O_2 there are $(3\times4)+(2\times6)=24$ valence electrons or 12 valence electron pairs. A plausible Lewis structure follows: $\ddot{O}=C=C=C=\ddot{O}$

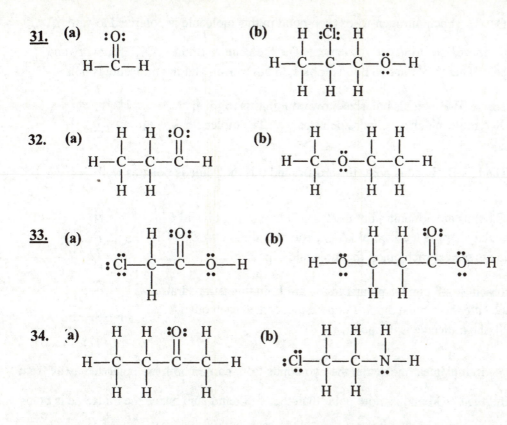

Polar Covalent Bonds and Electrostatic Potential Maps

35. Na — Cl and K — F both possess bonds between a metal and a nonmetal. Thus, they have the largest ionic character, with the ionic character of K — F being greater than that of Na — Cl, both because K is more metallic (closer to the lower left of the periodic table) than Na and because F is more nonmetallic (closer to the upper right) than Cl. The remaining three bonds are covalent bonds to H. Since H and C have about the same electronegativity (a fact you need to memorize), the H — C bond is the most covalent (or the least ionic). Br is somewhat more electronegative than is C, while F is considerably more electronegative than C, making the F — H bond the most polar of the three covalent bonds. Thus, ranked in order of increasing ionic character, these five bonds are:

C — H < Br — H < F — H < Na — Cl < K — F

The actual electronegativity differences follow:

$$C(2.5)\text{-}H(2.1) < Br(2.8)\text{-}H(2.1) < F(4.0)\text{-}H(2.1) < Na(0.9)\text{-}Cl(3.0) < K(0.8)\text{-}F(4.0)$$

$\Delta EN \quad\quad = 0.4 \quad\quad\quad = 0.7 \quad\quad\quad = 1.9 \quad\quad\quad = 2.1 \quad\quad\quad = 3.2$

36. **(a)** F_2 cannot possess a dipole moment, since both of the atoms in the diatomic molecule are the same. This means that there is no electronegativity difference between atoms, and hence no polarity in the F—F bond.

(b) $\ddot{O}=\overset{..}{N}-\ddot{O}:$ Each nitrogen-to-oxygen bond in this molecule is polarized toward

oxygen, the more electronegative element. The molecule is of the AX_2E category and hence is bent. Therefore the two bond dipoles do not cancel, and the molecule is polar.

(c) Although each B-F bond is polarized toward F in this trigonal planar AX_3 molecule these bond dipoles cancel. The molecule is nonpolar.

(d) $H-\ddot{Br}:$ The H—Br bond is polar toward Br, and this molecule is polar as well.

(e) The H—C bonds are not polar, but the C—Cl bonds are, toward Cl. The molecular shape is tetrahedral (AX_4) and thus these two C—Cl dipoles do not cancel each other; the molecule is polar.

(f) Although each Si—F bond is polarized toward F, in this tetrahedral AX_4 molecule these bond dipoles oppose and cancel each other. As a result, the molecule is nonpolar.

(g) $\ddot{O}=C=\ddot{S}$ In this linear molecule, the two bonds from carbon both are polarized away from

carbon. But the C=O bond is more polar than the C=S bond, and hence the molecule is polar.

37. The percent ionic character of a bond is based on the difference in electronegativity of its constituent atoms and Figure 10.7.

(a) S(2.5)—H(2.1)	**(b)** O(3.5)—Cl(3.0)	**(c)** Al(1.5)—O(3.5)	**(d)** As(2.0)–O(3.5)
Δ EN 0.4	0.5	2.0	1.5
%ionic = 4%	= 5 %	=60 %	=33 %

38.

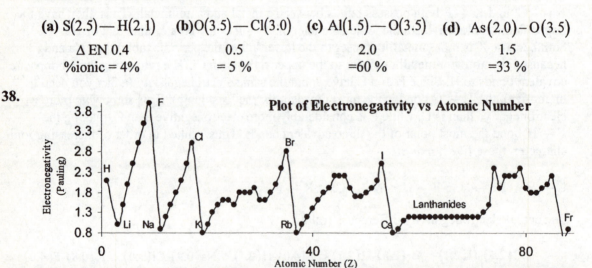

The property of electronegativity does indeed conform to the periodic law. Each of the "low points" corresponds to an alkali metal, and the end of each trend corresponds to a halogen. This is not unexpected and follows the general trend of increasing Z_{eff} from left to right, across the periodic table.

39. When looking at the electrostatic potential map, we expect similar structures. However, in the case of $F_2C=O$, the carbon should be more electropositive than in $H_2C=O$ due to the presence of very electronegative fluorine atoms (as opposed to H atoms). Thus for $F_2C=O$, one expects the center of the molecule to appear blue. As well, the electronegative oxygen atom should have less electron density associated with it, thus $H_2C=O$ should have a greater amount of red (electron rich) than the corresponding $F_2C=O$, again, as a result of the presence of highly electronegative fluorine atoms. $F_2C=O$ is represented on the left, while $H_2C=O$ is represented on the right.

40. HOCl, HOF and FOCl have similar structural features, however, they differ in terms of their electron density maps. O, Cl and F have similar electronegativities, thus FOCl should be fairly neutral (mostly yellow coloration), in terms of the electrostatic potential map. Thus, the molecule on the far right is FOCl. In the molecule HOF, the H atom will be very electropositive (blue coloring), owing to the presence of the very electronegative O and F atoms. Therefore, the molecule on the far right is HOF. This means that HOCl is the center molecule, which as expected, should have a relatively electropositive H atom (blue coloration, however, not as blue as in the case of HOF).

41. The molecular formulas for the compounds are SF_4 and SiF_4. SiF_4 is a symmetric molecule (tetrahedral). It is expected that the fluorine atoms should have the same electron density (same coloration). Since Si is more electropositive, it should have a greater blue coloration (more positive center). This suggests the electrostatic potential map on the right is for SiF_4. SF_4 is not a symmetric molecule. It has a trigonal bipyramidal electron geometry, where a lone pair occupies an equatorial position. It has a saw-horse or see-saw molecular shape in which the two axial fluorine atoms are nearly 180° to one another, while the two equatorial fluorine atoms are ~120° to one another. Owing to the lone pair in the equatorial position, the equatorial fluorine atoms will not be as electronegative as the axial fluorine atoms. This is certainly the case for the representation on the left.

42. The molecular formulas are ClF_3 and PF_3. Since P is more electropositive than Cl, we can easily argue that the right hand representation in which the central atom has a greater positive charge associated with it (blue coloration) is PF_3. As well, the molecular geometries are different. PF_3 is expected to be trigonal pyramidal while ClF_3 is T-shaped. It is relatively clear that the molecular geometries of the two molecules are different in the electrostatic potential maps. Thus the representation on the left, which is nearly planar is that for ClF_3, while the non-planar map on the right is that for PF_3.

Resonance

43. In NO_2^-, the total number of valence electrons is $1+5+(2 \times 6)=18$ valence electrons, or 9 electron pairs. N is the central atom. There two resonance forms are shown below:

$$[\overset{\ominus}{\overset{..}{\underset{..}{O}}} - \overset{..}{N} = \overset{..}{\underset{..}{O}}]^- \longleftrightarrow [\overset{..}{\underset{..}{O}} = \overset{..}{N} - \overset{..}{\underset{..}{O}} \overset{\ominus}{:}]^-$$

44. The only one of the four species that requires resonance forms to correctly describe the intramolecular bonding in CO_3^{2-}. Resonance forms of equal energy cannot be generated for the other species. All four Lewis structures are drawn below.

(a) In CO_2, there are $4 + (2 \times 6) = 16$ valence electrons, or 8 electron pairs. $\ddot{O}=C=\ddot{O}$

(b) In OCl^-, there are $6 + 7 + 1 = 14$ valence electrons, or 7 electron pairs, $^{\ominus}\ddot{O}-\ddot{C}l$

(c) In CO_3^{2-}, there are $4 + (3 \times 6) + 2 = 24$ valence electrons, or 12 electron pairs (see below).

(d) In OH^-, there are $6 + 1 + 1 = 8$ valence electrons, or 4 electron pairs. $^{\ominus}\ddot{O}-H$

$$\left[\ddot{O}=C-\ddot{O} \right]^{2-} \longleftrightarrow \left[\ddot{O}-C-\ddot{O} \right]^{2-} \longleftrightarrow \left[\ddot{O}-C=\ddot{O} \right]^{2-}$$

45. Bond length data from Table 11.2 follow:

$N \equiv N$ 109.8 pm $N = N$ 123 pm $N - N$ 145 pm $N = O$ 120 pm $N - O$ 136 pm

The experimental $N - N$ bond length of 113 pm approximates that of the $N \equiv N$ triple bond, which appears in structure (1). The experimental $N - O$ bond length of 119 pm approximates that of the $N=O$ double bond, which appears in structure (2). Structure (4) is highly unlikely because it contains no nitrogen-to-nitrogen bonds, and a N—N bond was found experimentally. Structure (3) also is unlikely, because it contains a very long (145 pm) $N - N$ single bond, which does not agree at all well with the experimental N-to-N bond length. The molecule seems best represented as a resonance hybrid of (1) and (2).

$$^{\ominus}\ddot{N}=\overset{\oplus}{N}=\ddot{O} \longleftrightarrow :N\equiv\overset{\oplus}{N}-\ddot{O}^{\ominus}$$

46. We begin by drawing all three valid resonance forms of HNO_3 and then analyzing their distributions of formal charge to determine which is the most plausible.

In all three structures, the formal charges of N and H are the same:
f.c. of $H = 1 - 1 - 0 = 0$ f.c. of $N = 5 - 4 - 0 = +1$

For an oxygen that forms two bonds (either 2 single or one double),
f.c. $= 6 - 2 - (2 \times 2) = 0$

For an oxygen that forms only one bond, f.c. $= 6 - 1 - (3 \times 2) = -1$

For the oxygen that forms three bonds (a single and a double), f.c.$= 6 - 3 - (1 \times 2) = +1$

Thus, structures (a) and (b) are equivalent in their distributions of formal charges, zero on all atoms except 1+ on N and 1- on one O. These are degenerate resonance forms (i.e. they are of equal energy). Structure (c) is quite different, with formal charges of 1- on two O's, and 1+ on the other, and a formal charge of 1+ on N. Structure (c) is thus the least plausible, because of the adjacent like charges.

Odd-electron species

47. **(a)** CH_3 has a total of $(3 \times 1) + 4 = 7$ valence electrons, or 3 electron pairs and a lone electron. C is the central atom. A plausible Lewis structure is

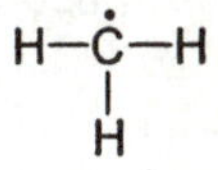

(b) ClO_2 has a total of $(2 \times 6) + 7 = 19$ valence electrons, or 9 electron pairs and a lone electron. Cl is the central atom. A plausible Lewis structure is: :Ö-Ċl-Ö·

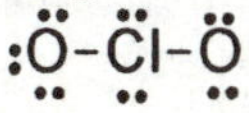

(c) NO_3 has a total of $(3 \times 6) + 5 = 23$ valence electrons, or 11 electron pairs, plus a lone electron. N is the central atom. A plausible Lewis structure is shown to the right. Other resonance forms can also be drawn.

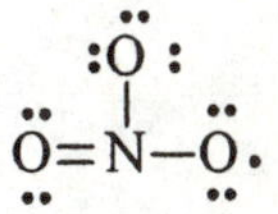

48. **(a)**
C_2H_5 has a total of $(5 \times 1) + (2 \times 4) = 13$ valence electrons or 6 electron pairs and a lone electron on C. A plausible Lewis structure is given to the right.

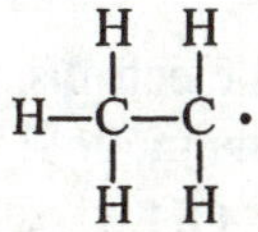

(b)
HO_2 has a total of $(2 \times 6) + 1 = 13$ valence electrons or 6 electron pairs (2 bonding and 4 non-bonding) and a lone electron on O. A plausible Lewis structure is given to the right.

H—Ö—Ö·

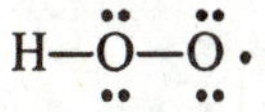

(c)
ClO has a total of $7 + 6 = 13$ valence electrons or 6 electron pairs (1 bonding and 5 non-bonding) and a lone electron on either Cl or O. A plausible Lewis structure is given to the right.

:Ċl—Ö· or ·Ċl—Ö:

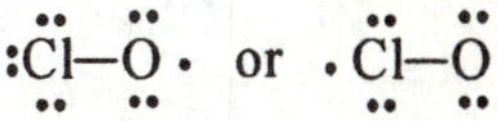

49. In NO_2, there are $5 + (2 \times 6) = 17$ valence electrons, 8 electron pairs and a lone electron. N is the central atom and it carries the lone electron. A plausible Lewis structure is (A) which has a formal charge of 1+ on N and 1- on the single-bonded O. Another Lewis structure, with zero formal charge on each atom, is (B). The major difference is that one of the oxygen atoms carries the lone electron. In both cases, due to the unpaired electron we expect NO_2 to be paramagnetic. We would expect a bond to form between two NO_2

molecules as a result of the pairing of the lone unpaired electrons in the NO_2 molecules. If the second Lewis structure for NO_2 is used, the one with zero formal charge on each atom and the lone electron on oxygen, a plausible structure for N_2O_4 is (C). If the first Lewis structure for NO_2 where there are formal charges and a lone electron on nitrogen is used to form a bond between molecules, a plausible Lewis structure containing a $N-N$ bond results (D). Resonance structures can be drawn for this second version of N_2O_4. A $N-N$ bond is observed experimentally in N_2O_4. In either structure, the product N_2O_4, has all electrons paired up in bonding and non-bonding lone pairs; the molecule is expected to be (and is) diamagnetic. Lewis structures for (A) – (D) are shown below.

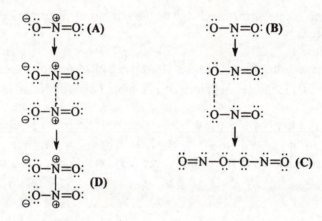

50. Since electrons pair up (if at all possible) in plausible Lewis structures, A species will be paramagnetic if it has an odd number of (valence) electrons.

(a) OH^- $6+1+1=8$ valence electrons   diamagnetic

(b) OH $6+1=7$ valence electrons   paramagnetic

(c) NO_3 $5+(3\times 6)=23$ valence electrons   paramagnetic

(d) SO_3 $6+(3\times 6)=24$ valence electrons   diamagnetic

(e) SO_3^{2-} $6+(3\times 6)+2=26$ valence electrons diamagnetic

(f) HO_2 $1+(2\times 6)=13$ valence electrons   paramagnetic

Expanded octets

51. In PO_4^{3-}; $5+(4\times 6)+3=32$ valence electrons or 16 electron pairs. An expanded octet is not needed.

In PI_3; $5+(3\times 7)=26$ valence electrons or 13 electron pairs. An expanded octet is not needed.

In ICl_3; are $7+(3\times 7)=28$ valence electrons or 14 electron pairs. An expanded octet is necessary.

In $OSCl_2$; $6+6+(2\times7)=26$ valence electrons or 13 electron pairs. An expanded octet is not needed.

In SF_4; $6+(4\times7)=34$ valence electrons or 17 electron pairs. An expanded octet is necessary.

In ClO_4^-; $7+(4\times6)+1=32$ valence electrons or 16 electron pairs. An expanded octet is not needed.

52. Let us draw the Lewis structure of H_2CSF_4. The molecule has $(2\times1)+4+6+(4\times7)=40$ valence electrons, or 20 electron pairs. With only single bonds and all octets complete, there is a 1– formal charge on C and 1+ on S, as in the left structure below. The right structure below avoids an undesirable separation of charge by creating a carbon-to-sulfur double bond (charge separation requires the input of energy).

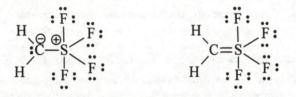

The bond can be described either as a single bond with ionic contributions or a double bond in which the formal charge are eliminated.

Molecular shapes

53. The AX_nE_m designations that are cited below are to be found in Table 10.1 of the text, along with a sketch and a picture of a model of each type of structure.

(a) Dinitogen is linear, two points define a line. $\qquad :N\equiv N:$

(b) Hydrogen cyanide is linear. The molecule belongs to the AX_2 category, and these species are linear. $\qquad H-C\equiv N:$

(c) NH_4^+ is tetrahedral. The ion is of the AX_4 type, which has a tetrahedral electron-group geometry and a tetrahedral shape.

$$\left[\begin{array}{c} H \\ | \\ H-\overset{\oplus}{N}-H \\ | \\ H \end{array} \right]^+$$

(d) NO_3^- is trigonal planar. The ion is of the AX_3 type, which has a trigonal planar electron-group geometry and a trigonal planar shape. The other resonance forms are of the same type.

(e) NSF is bent. The molecule is of the AX_2E type, which has trigonal planar electron-group geometry and a bent shape.

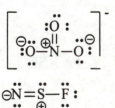

54. The AX_nE_m designations that are cited below are to be found in Table 11.1 of the text, along with a sketch and a picture of a model of each type of structure.

    (a) PCl_3 is a trigonal pyramid. The molecule is of the AX_3E type, and has a tetrahedral electron-group geometry and a trigonal pyramid shape.

    (b) SO_4^{2-} has a tetrahedral shape. The ion is of the type AX_4, and has a tetrahedral electron-group geometry and a tetrahedral shape. The other resonance forms of the sulfate ion have the same shape.

    (c) $SOCl_2$ has a trigonal pyramidal shape. This molecule is of the AX_3E type and has a tetrahedral electron-group geometry and a trigonal pyramidal shape.

    (d) SO_3 has a trigonal planar shape. The molecule is of the AX_3 type, with a trigonal planar electron-group geometry and molecular shape. The other resonance contributors have the same shape.

    (e) BrF_4^+ has a distorted see-saw shape. The molecule is of the AX_4E type, with a trigonal bipyramid electron-group geometry and a see-saw molecular shape.

55. We first draw all the Lewis structures. From each, we can deduce the electron-group geometry and the molecular shape.

    (a) H_2S tetrahedral electron-group geometry, bent (angular) molecular geometry

    (b) N_2O_4 trigonal planar electron-group geometry around each N, (planar molecule).

    (c) HCN   linear electron-group geometry, linear molecular geometry

    (d) $SbCl_6^-$ octahedral electron-group geometry, octahedral geometry

    (e) BF_4^- tetrahedral electron-group geometry, tetrahedral molecular geometry

56. In each case, a plausible Lewis structure is given first, followed by the AX_nE_m notation for each species followed by the electron-group geometry and, finally, the molecular geometry.

(a) CO — linear electron-group geometry, linear molecular geometry — $:C\equiv O:$

(b) $SiCl_4$ — tetrahedral electron-group geometry, tetrahedral molecular geometry

(c) PH_3 — tetrahedral electron-group geometry, trigonal pyramidal molecular geometry

(d) ICl_3 — trigonal bipyramidal electron-group geometry, T- shape molecular geometry

(e) $SbCl_5$ — Electron-group geometry and molecular geometry, trigonal bipyramidal

(f) SO_2 — trigonal planar electron-group geometry, bent molecular geometry

(g) AlF_6^{3-} — octahedral electron-group geometry, octahedral geometry

57. A trigonal planar shape requires that three groups and no lone pairs be bonded to the central atom. Thus PF_6^- cannot have a trigonal planar shape, since six atoms are attached to the central atom. In addition, PO_4^{3-} cannot have a trigonal planar shape, since four O atoms are attached to the central P atom. We now draw the Lewis structure of each of the remaining ions, as a first step in predicting their shapes. The SO_3^{2-} ion is of the AX_3E type. It has a tetrahedral electron-group geometry and a trigonal pyramidal shape. The CO_3^{2-} ion is of the AX_3 type, and has a trigonal planar electron-group geometry and a trigonal planar shape.

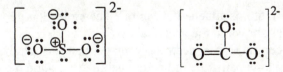

58. We can predict shapes by first of all drawing the Lewis structures for the species. Thus, SO_3^{2-} and NI_3 both have the same shape.

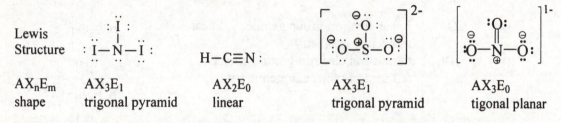

Lewis Structure	: I : .. | .. : I—N—I :	H—C≡N :	[see structure]	[see structure]
AX_nE_m shape	AX_3E_1 trigonal pyramid	AX_2E_0 linear	AX_3E_1 trigonal pyramid	AX_3E_0 tigonal planar

59. **(a)** In CO_2 there are a total of $4+(2\times6)=16$ valence electrons, or 8 electron pairs. The following Lewis structure is plausible. $\bar{O}=C=\bar{O}$ This is a molecule of type AX_2. CO_2 has a linear electron-shape geometry and a linear shape.

(b) In Cl_2CO there are a total of $(2\times7)+4+6=24$ valence electrons, or 12 electron pairs. The molecule can be represented by a Lewis structure with C as the central atom. This molecule is of the AX_3 type. It has a trigonal planar electron-group geometry and molecular shape.

(c) In $ClNO_2$ there are a total of $7+5+(2\times6)=24$ valence electrons, or 12 electron pairs. N is the central atom. A plausible Lewis structure is shown: This molecule is of the AX_3 type. It has a trigonal planar electron-group geometry and a trigonal planar shape.

60. Lewis structures enable us to determine molecular shapes.

(a) N_2O_4 has $(2\times5)+(4\times6)=34$ valence electrons, or 17 electron pairs.

(b) C_2N_2 has $(2\times4)+(2\times5)=18$ valence electrons, or 9 electron pairs.

(c) C_2H_6 has $(2\times4)+(6\times1)=14$ valence electrons, or 7 electron pairs.

(d) CH_3OCH_3 has 6 more valence electrons than $C_2H_6 = 20$ valence e^- or 10 electron pairs.

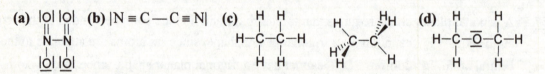

(a) There are three atoms bound to each N in N_2O_4, making the molecule triangular planar around each N. The entire molecule does not have to be planar, however, since there is free rotation around the N—N bond.

(b) There are two atoms attached to each C, requiring a linear geometry. The entire molecule is linear.

(c) There are four atoms attached to each C. The molecular geometry around each C is tetrahedral, as shown in the sketch above.

(d) Around the central O, the electron-group geometry is tetrahedral. With two of the electron groups being lone pairs, the molecular geometry around the central atom is bent.

61. First we draw the Lewis structure of each species, then use it to predict the molecular shape. The structures are provided below.

(a) In ClO_4^- there are $7 + (4 \times 6) + 1 = 32$ valence electrons or 16 electron pairs. A plausible Lewis structure follows. Since there are four atoms and no lone pairs bonded to the central atom, the molecular shape and the electron-group geometry are the same: tetrahedral.

(b) In $S_2O_3^{2-}$ there are $(2 \times 6) + (3 \times 6) + 2 = 32$ valence electrons or 16 electron pairs. A plausible Lewis structure follows. Since there are four atoms and no lone pairs bonded to the central atom, both the electron-group geometry and molecular shape are tetrahedral.

(c) In PF_6^- there are $5 + (6 \times 7) + 1 = 48$ valence electrons or 24 electron pairs. Since there are six atoms and no lone pairs bonded to the central atom, the electron-group geometry and molecular shape are octahedral.

(d) In I_3^- there are $(3 \times 7) + 1 = 22$ valence electrons or 11 electron pairs. There are three lone pairs and two atoms bound to the central atom. The electron-group geometry is trigonal bipyramid, thus, the molecular shape is linear.

(a)   **(b)** **(c)** **(d)**

62. **(a)** In OSF_2, there are a total of $6 + 6 + (2 \times 7) = 26$ valence electrons, or 13 electron pairs. S is the central atom. A plausible Lewis structure is shown to the right. This molecule is of the type AX_3E_1 It has a tetrahedral electron-group geometry and a trigonal pyramidal shape.

(b) In O_2SF_2, there are a total of $(2 \times 6) + 6 + (2 \times 7) = 32$ valence electrons, or 16 electron pairs. S is the central atom. One plausible Lewis structure is shown to the right. The molecule is of the AX_4 type It has a tetrahedral electron-group geometry and a tetrahedral shape. The structure with all single bonds is preferred because it avoids an expanded octet.

(c) In SF_5^- there are $6 + (5 \times 7) + 1 = 42$ valence electrons, or 21 electron pairs. A plausible Lewis structure is shown to the right. The ion is of the AX_5E type. It has an octahedral electron-group geometry and a square pyramidal molecular shape.

(d) In ClO_4^-, there are a total of $1 + 7 + (4 \times 6) = 32$ valence electrons, or 16 electron pairs. Cl is the central atom. A plausible Lewis structure is shown to the right. This ion is of the AX_4 type. It has a tetrahedral electron-group geometry and a tetrahedral shape.

(e) In ClO_3^- the total number of valence electrons is $1 + (3 \times 6) + 7 = 26$ valence electrons, or 13 electron pairs. A plausible Lewis structure is shown to the right. The molecule is of the type AX_3E. It has a tetrahedral electron-group geometry and a trigonal pyramidal molecular shape.

63. In BF_4^-, there are a total of $1 + 3 + (4 \times 7) = 32$ valence electrons, or 16 electron pairs. A plausible Lewis structure has B as the central atom. This ion is of the type AX_4. It has a tetrahedral electron-group geometry and a tetrahedral shape.

64. The molecular geometry is indicated by the VSEPR notation (i.e. AX_3E_2). Formal charge is reduced by moving lone pairs of electrons from the terminal atoms and forming multiple bonds to the central atom. The VSEPR notation is unchanged, however, when formal charge is increased or decreased. For example, consider SO_2:

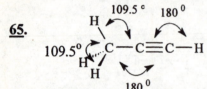

VSEPR notation
AX_2E
shape: bent

VSEPR notation
AX_2E
shape: bent

VSEPR notation
AX_2E
shape: bent

Shapes of Molecules with More Than One Central Atom

65.

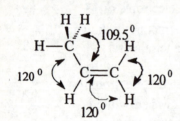

A maximum of 5 atoms can be in the same plane

66.

A maximum of 7 atoms can be in the same plane

67.

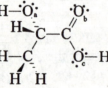

All angles ~ 109.5° with the exception of

$$O_c - C - C$$
$$O_b = C - C \quad \} \sim 120°$$
$$O_b = C - O_c$$

68.

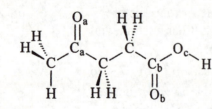

All angles ~ 109.5° with the exception of

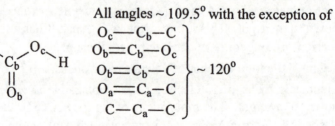

$$O_c - C_b - C$$
$$O_b = C_b - O_c$$
$$O_b = C_b - C \quad \} \sim 120°$$
$$O_a = C_a - C$$
$$C - C_a - C$$

Polar Molecules

69. For each molecule, we first draw the Lewis structure, which we use to predict the shape.

(a) SO_2 has a total of $6 + (2 \times 6) = 18$ valence electrons, or 9 electron pairs. The molecule has two resonance forms. $\overline{O}=\overline{\underset{\oplus}{S}}-\overline{\underset{\ominus}{O}}| \longleftrightarrow |\overline{\underset{\ominus}{O}}-\overline{\underset{\oplus}{S}}=\overline{O}$ Each of these resonance forms is of the type AX_2E. Thus it has a trigonal planar electron-group geometry and a bent shape. Since each S—O bond is polar toward O, and since the bond dipoles do not point in opposite directions, the molecule has a resultant dipole moment, pointing from S through a point midway between the two O atoms. Consequently, SO_2 is polar.

(b) NH_3 has a total of $5 + (3 \times 1) = 8$ valence electrons, or 4 electron pairs. N is the central atom. A plausible Lewis structure is shown to the right. The molecule is of the AX_3E type; it has a tetrahedral electron-group geometry and a trigonal pyramidal shape. Each N-H bond is polar toward N. Since the bonds do not symmetrically oppose each other, there is a resultant molecular dipole moment, pointing from the triangular base (formed by the three H atoms) through N. Consequently, the molecule is polar.

$$\begin{matrix} & H & \\ & | & \\ H- & \underline{N} & -H \end{matrix}$$

(c) H_2S has a total of $6 + (2 \times 1) = 8$ valence electrons, or 4 electron pairs. S is the central atom and a plausible Lewis structure is $H-\overline{\underline{S}}-H$. This molecule is of the AX_2E_2 type; it has a tetrahedral electron-group geometry and a bent shape. Each H—S bond is polar toward S. Since the bonds do not symmetrically oppose each other, the molecule has a net dipole moment, pointing through S from a point midway between the two H atoms. H_2S is polar.

(d) C_2H_4 consists of atoms that all have about the same electronegativities. Of course, the C—C bond is not polar and essentially neither are the C—H bonds. The molecule is planar. Thus, the entire molecule is nonpolar.

$$\begin{matrix} H & & H \\ \diagdown & & \diagup \\ & C{=}C & \\ \diagup & & \diagdown \\ H & & H \end{matrix}$$

(e) SF_6 has a total of $6 + (6 \times 7) = 48$ valence electrons, or 24 electron pairs. S is the central atom. All atoms have zero formal charge in the Lewis structure. This molecule is of the AX_6 type. It has an octahedral electron-group geometry and an octahedral shape. Even though each S—F bond is polar toward F, the bonds symmetrically oppose each other resulting in a molecule that is nonpolar.

$$\begin{matrix} |\overline{F} & & \overline{F}| \\ & \diagdown & \diagup \\ |\overline{F}-& S & -\overline{F}| \\ & \diagup & \diagdown \\ |\overline{F} & & \overline{F}| \end{matrix}$$

(f) CH_2Cl_2 has a total of $4 + (2 \times 1) + (2 \times 7) = 20$ valence electrons, or 10 electron pairs. A plausible Lewis structure is shown to the right. The molecule is tetrahedral and polar, since the two polar bonds $(C-Cl)$ do not cancel the effect of each other.

$$\begin{matrix} & H & \\ & | & \\ |\overline{Cl}-& C & -\overline{Cl}| \\ & | & \\ & H & \end{matrix}$$

70. (a) HCN is a linear molecule, which can be derived from its Lewis structure. $H-C \equiv N|$. The $C \equiv N$ bond is strongly polar toward N, while the $H-C$ bond is generally considered to be nonpolar. Thus, the molecule has a dipole moment, pointed from C towards N.

(b) SO$_3$ is a trigonal planar molecule, which can be derived from its Lewis structure. Each sulfur-oxygen bond is polar from S to O, but the three bonds are equally polar and are pointed in symmetrical opposition so that they cancel. The SO$_3$ molecule has a dipole moment of zero.

(c) CS$_2$ is a linear molecule, which can be derived from its Lewis structure. $\bar{S}=C=\bar{S}$. Each carbon-sulfur bond is polar from C to S, but the two bonds are equally polar and are pointed in opposition to each other so that they cancel. The CS$_2$ molecule has a dipole moment of zero.

(d) OCS also is a linear molecule. Its Lewis structure is $\bar{O}=C=\bar{S}$. But the carbon-oxygen bond is more polar than the carbon-sulfur bond. Although both bond dipoles point from the central atom to the bonded atom, these two bond dipoles are unequal in strength. Thus, the molecule is polar in the direction from C to O.

(e) SOCl$_2$ is a trigonal pyramidal molecule. Its Lewis structure is shown to the right. The lone pair is at one corner of the tetrahedron. Each bond in the molecule is polar, with the dipole moments pointing away from the central atom. The sulfur-chlorine bond is less polar than the sulfur-oxygen bond, and this makes the molecule polar. The dipole moment of the molecule points from the sulfur atom to the base of the trigonal pyramid, not toward the center of the base but slightly toward the O apex of that base.

(f) SiF$_4$ is a tetrahedral molecule, with the following Lewis structure. Each Si — F bond is polar, with its negative end away from the central atom toward F in each case. These four Si – F bond dipoles oppose each other and thus cancel. SiF$_4$, as a result, has no dipole moment.

(g) POF$_3$ is a tetrahedral molecule. A valid Lewis structure is shown to the right. All four bonds are polar, with their dipole moments pointing away from the central atom. The P — F bond polarity is greater than that of the P — O bond. Thus, POF$_3$ is a polar molecule with its dipole moment pointing away from the P towards the center of the triangle formed by the three F atoms.

71. In H$_2$O$_2$, there are a total of $(2 \times 1) + (2 \times 6) = 14$ valence electrons, 7 electron pairs. The two O atoms are central atoms. A plausible Lewis structure has zero formal charge on each atom. $H-\bar{O}-\bar{O}-H$. In the hydrogen peroxide molecule, the O — O bond is non-polar, while the H — O bonds are polar, with the dipole moment pointing toward O. Since the molecule has a resultant dipole moment, it cannot be linear, for, if it were linear the two polar bonds would oppose each other and their polarities would cancel.

72. (a) FNO has a total of $7 + 5 + 6 = 18$ valence electrons, or 9 electron pairs. N is the central atom. A plausible Lewis structure is $|\bar{F}-\bar{N}=\bar{O}$. The formal charge on each atom in this structure is zero.

(b) FNO is of the AX_2E type. It has a trigonal planar electron-group geometry and a bent shape.

(c) The N—F bond is polar toward F and the N—O bond is polar toward O. In FNO, these two bond dipoles point in the same general direction, producing a polar molecule. In FNO_2, however, the additional N—O bond dipole partially opposes the polarity of the other two bond dipoles, resulting in a smaller net dipole moment.

Bond lengths

73. (c) Br_2 possess the longest bond. Single bonds are generally longer than multiple bonds. Of the two molecules with single bonds, Br_2 is expected to have longer bonds than BrCl, since Br is larger than Cl. (a) $\overline{O} = \overline{O}$ (b) $|N \equiv N|$ (c) $|\overline{Br} - \overline{Br}|$ (d) $|\overline{Br} - \overline{Cl}|$

74. The bond lengths predicted will be larger than the actual bond lengths. This is because we do not take into account polarity of the bond. In the case of the I-Cl bond, the electronegativity difference is small, thus the predicted bond length should be close to the experimental value. In the case of C-F bond, the electronegativity difference is quite large, thus it is expected that the actual bond length will be shorter than the one predicted.

(a) I—Cl bond length
$$= [(\text{I—I bond length}) + (\text{Cl—Cl bond length})] \div 2$$
$$= [266 \text{ pm} + 199 \text{ pm}] \div 2 = 233 \text{ pm (literature 232 pm)}$$

(b) C—F bond length
$$= [(\text{ C—C bond length}) + (\text{ F—F bond length})] \div 2$$
$$= [154 \text{ pm} + 143 \text{ pm}] \div 2 = 149 \text{ pm (literature 135 pm)}$$

75. A heteronuclear bond length (one between two different atoms) is approximately equal to the average of two homonuclear bond lengths (one between two like atoms) of the same order (both single, both double, or both triple).

(a) I—Cl bond length
$$= [(\text{I—I bond length}) + (\text{Cl—Cl bond length})] \div 2$$
$$= [266 \text{ pm} + 199 \text{ pm}] \div 2 = 233 \text{ pm}$$

(b) O—Cl bond length
$$= [(\text{O—O bond length}) + (\text{ Cl—Cl bond length})] \div 2$$
$$= [145 \text{ pm} + 199 \text{ pm}] \div 2 = 172 \text{ pm}$$

(c) C—F bond length
$$= [(\text{ C—C bond length}) + (\text{ F—F bond length})] \div 2$$
$$= [154 \text{ pm} + 143 \text{ pm}] \div 2 = 149 \text{ pm}$$

(d) C—Br bond length
$$= [(\text{C—C bond length}) + (\text{Br—Br bond length})] \div 2$$
$$= [154 \text{ pm} + 228 \text{ pm}] \div 2 = 191 \text{ pm}$$

76. First we need to draw the Lewis structure of each of the compounds cited, so that we can determine the order, and hence the relative length, of each O-to-O bond.

(a) In H_2O_2, there are $(2 \times 1) + (2 \times 6) = 14$ valence electrons or 7 electron pairs. A plausible Lewis structure is $H-\overline{\underline{O}}-\overline{\underline{O}}-H$

(b) In O_2, the total number of valence electrons is $(2 \times 6 =)12$ valence electrons, or 6 electron pairs. A plausible Lewis structure is $\overline{O} = \overline{O}$

(c) In O_3, the total number of valence electrons is $(3 \times 6) = 18$ valence electrons, or 9 electron pairs. A plausible Lewis structure is $\overline{\underline{O}} = \overline{O} - \overline{\underline{O}}| \longleftrightarrow |\overline{\underline{O}} - \overline{O} = \overline{\underline{O}}$ (Two most stable resonance contributors)

Thus, O_2 should have the shortest O-to-O bond, because the O atoms are joined via a double bond. The single $O - O$ bond in H_2O_2 should be longest.

77. The $N - F$ bond is a single bond. Its bond length should be the average of the $N - N$ single bond (145 pm) and the $F - F$ single bond (143 pm). Thus, the average $N - F$ bond length $= (145 + 143) \div 2 = 144$ pm

78. In H_2NOH, there are $(3 \times 1) + 5 + 6 = 14$ valence electrons total, or 7 electron pairs. N and O are the two central atoms. A plausible Lewis structure has zero formal charge on each atom. The $N - H$ bond lengths are 100 pm, the $O - H$ bond is 97 pm long, and the $N - O$ bond is 136 pm in length. All values are taken from Table 10.2. All bond angles approximate the tetrahedral bond angle of 109.5°, but are expected to be somewhat smaller, perhaps by 2° to 4°, each.

Bond Energies

79. The reaction $O_2(g) \rightarrow 2 \, O(g)$ is an endothermic reaction since it requires the breaking of the bond between two oxygen atoms without the formation of any new bonds. Since bond breakage is endothermic and the process involves only bond breakage, the entire process must be endothermic.

80. (a) The net result of this reaction involves breaking one mole of $C - H$ bonds (which requires 414 kJ) and forming one mole of $H - I$ bonds (which produces 297 kJ). Thus, this reaction is endothermic (i,e, a net infusion of energy is necessary).

(b) The net result of this reaction, involves breaking one mol of $H - H$ bonds (which requires 436 kJ) and one mol of $I - I$ bonds (which requires 151 kJ), along with forming two moles of $H-I$ bonds (which produces $2 \times 297 = 594$ kJ). Thus, this reaction is exothermic (just barely, mind you).

81.

$$\underset{\substack{|\ \ |\\H\ H}}{\overset{\substack{H\ H\\|\ \ |}}{H-C-C-H}} + |\overline{Cl}-\overline{Cl}| \longrightarrow \underset{\substack{|\ \ |\\H\ H}}{\overset{\substack{H\ H\\|\ \ |}}{H-C-C-\overline{Cl}|}} + H-\overline{Cl}|$$

Analysis of the Lewis structures of products and reactants indicates that a $C-H$ bond and a $Cl-Cl$ bond are broken, and a $C-Cl$ and a $H-Cl$ bond are formed.

Energy required to break bonds $= C{-}H + Cl{-}Cl = 414 \ \dfrac{kJ}{mol} + 243 \ \dfrac{kJ}{mol} = 657 \ \dfrac{kJ}{mol}$

Energy realized by forming bonds $= C{-}Cl + H{-}Cl = 339 \ \dfrac{kJ}{mol} + 431 \ \dfrac{kJ}{mol} = 770 \ \dfrac{kJ}{mol}$

$\Delta H = 657 \ kJ\,/\,mol - 770 \ kJ\,/\,mol = -113 \ kJ\,/\,mol$

82. The reaction in terms of Lewis structures is $\overline{O}{=}\overline{O}{-}\overline{O}| + \overline{O}{=}\overline{N}{\cdot} \longrightarrow \overline{O}{=}N{-}\overline{O}{\cdot} + \overline{O}{=}\overline{O}$
The net result is the breakage of an $O-O$ bond (142 kJ/mol) and the formation of an $N-O$ bond (222 kJ/mol). $\Delta H = 142 \ kJ/mol - 222 \ kJ/mol = -80. \ kJ/mol$

83. In each case we write the formation reaction, but specify reactants and products with their Lewis structures. All species are assumed to be gases.

(a) $\dfrac{1}{2}(\overline{O}{=}\overline{O}) + \dfrac{1}{2}(H{-}H) \longrightarrow {\cdot}\overline{O}{-}H$

Bonds broken: $\dfrac{1}{2} \ (O = O) + \dfrac{1}{2} \ (H{-}H) = 0.5(498 \ kJ + 436 \ kJ) = 467 \ kJ$

Bonds formed: $O{-}H = 464 \ kJ$ $\qquad \Delta H° = 467 \ kJ - 464 \ kJ = 3 \ kJ\,/\,mol$
If the $O-H$ bond dissociation energy of 428.0 kJ/mol from Figure 10-16 is used, $\Delta H_f{}^o = 39 \ kJ\,/\,mol$.

(b) $|N{\equiv}N| + 2\ H{-}H \longrightarrow \underset{\substack{|\ \ |\\H\ H}}{H{-}\overline{N}{-}\overline{N}{-}H}$

Bonds broken $= N \equiv N + 2H{-}H = 946 \ kJ + 2 \times 436 \ kJ = 1818 \ kJ$
Bonds formed $= N{-}N + 4\ N{-}H = 163 \ kJ + 4 \times 389 \ kJ = 1719 \ kJ$
$\Delta H_f{}^o = 1818 \ kJ - 1719 \ kJ = 99 \ kJ$

84. The reaction of Example 10-15 is $CH_4(g) + Cl_2(g) \rightarrow CH_3Cl(g) + HCl(g)$
$\Delta H_{rxn} = -113 \ kJ$. In Appendix D are the following values:
$\Delta H_f°[\ CH_4\ (g)] = -74.81 \ kJ, \Delta H_f°[\ HCl(g)] = -92.31 \ kJ, \Delta H_f°[\ Cl_2\ (g)] = 0$
Thus, we have
$\Delta H_{rxn} = \Delta H_f°[CH_3Cl(g)] + \Delta H_f°[HCl(g)] - \Delta H_f°[CH_4(g)] - \Delta H_f°[Cl_2(g)]$
$-113 \ kJ = \Delta H_f°[CH_3Cl(g)] - 92.31 \ kJ - (-74.81 \ kJ) - (0.00)$
$\Delta H_f°[\ CH_3Cl(g)] = -113 \ kJ + 92.31 \ kJ - 74.81 \ kJ = -96 \ kJ$

85.

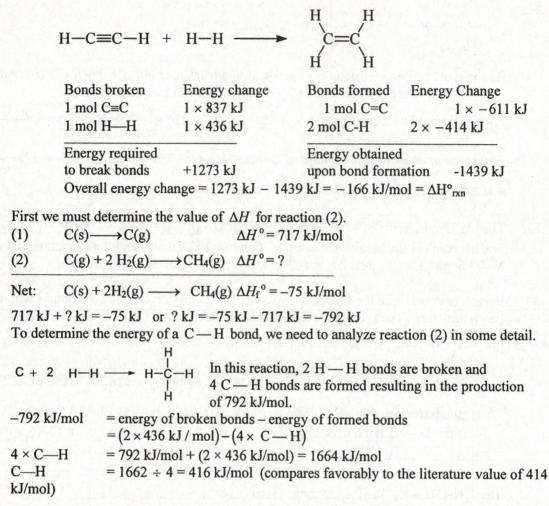

Bonds broken	Energy change	Bonds formed	Energy Change
1 mol C≡C	1 × 837 kJ	1 mol C=C	1 × −611 kJ
1 mol H—H	1 × 436 kJ	2 mol C-H	2 × −414 kJ

| Energy required to break bonds | +1273 kJ | Energy obtained upon bond formation | -1439 kJ |

Overall energy change = 1273 kJ − 1439 kJ = − 166 kJ/mol = $\Delta H°_{rxn}$

86. First we must determine the value of ΔH for reaction (2).

(1)   C(s) ⟶ C(g) $\Delta H° = 717$ kJ/mol

(2)   C(g) + 2 H$_2$(g) ⟶ CH$_4$(g) $\Delta H° = ?$

Net: C(s) + 2H$_2$(g) ⟶ CH$_4$(g) $\Delta H_f° = -75$ kJ/mol

717 kJ + ? kJ = −75 kJ   or   ? kJ = −75 kJ − 717 kJ = −792 kJ

To determine the energy of a C—H bond, we need to analyze reaction (2) in some detail.

C + 2 H—H ⟶ H—C—H (with H above and below)   In this reaction, 2 H—H bonds are broken and 4 C—H bonds are formed resulting in the production of 792 kJ/mol.

−792 kJ/mol   = energy of broken bonds − energy of formed bonds

            = (2 × 436 kJ / mol) − (4 × C—H)

4 × C—H   = 792 kJ/mol + (2 × 436 kJ/mol) = 1664 kJ/mol

C—H   = 1662 ÷ 4 = 416 kJ/mol (compares favorably to the literature value of 414 kJ/mol)

Integrative and Advanced Exercises

Important Note: In this and subsequent chapters, a lone pair of electrons in a Lewis structure often is shown as a line rather than a pair of dots. Thus, the Lewis structure of Be is Be| or Be:

87.   Recall that bond breaking is endothermic, while bond making is exothermic

       Break 2 N-O bonds requires 2(631 kJ/mol) =   +1262 kJ

       Break 5 H-H bonds requires 4(436 kJ/mol) =   +2180 kJ

       Make 6 N-H bonds yields 5(-389 kJ/mol) =   -2334 kJ

       Make 4 O-H bonds yields 4(-463 kJ/mol) =   -1852 kJ

              Σ(bond energies) = ΔH = -744 kJ/mol reaction

88. **(a)**   There is at least one instance in which one atom must bear a formal charge - a polyatomic ion. In these cases formal charges must be invoked to ensure that there be no more than one unpaired electron in the structure.

(b) Since three points define a plane, stating that a triatomic molecule is planar is just stating a reiteration of a fundamental tenet of geometry. In fact, it is misleading, for some triatomic molecules are actually linear. HCN is one example of a linear triatomic molecule, as is CO_2. Of course, some molecules with more than three atoms are also planar, two examples are XeF_4 and $H_2C=CH_2$.

(c) This statement is incorrect because in some molecules that contain polar bonds, the bonds are so oriented in space that there is no resulting molecular dipole moment. Examples of such molecules are CO_2, $BeCl_3$, CCl_4, PCl_5, SF_6.

89. First we determine the empirical formula of the compound, based on 100 g of compound.

$$\text{amount S} = 47.5 \text{ g S} \times \frac{1 \text{ mol S}}{32.07 \text{ g S}} = 1.48 \text{ mol S} \qquad \div 1.48 \longrightarrow 1.00 \text{ mol S}$$

$$\text{amount Cl} = 52.5 \text{ g Cl} \times \frac{1 \text{ mol Cl}}{35.45 \text{ g Cl}} = 1.48 \text{ mol Cl} \quad \div 1.48 \longrightarrow 1.00 \text{ mol Cl}$$

The empirical formula of the compound is SCl. It has $6 + 7 = 13$ valence electrons. A plausible, yet unsatisfactory Lewis structure is $\cdot\overline{\underline{S}}-\overline{\underline{C}}l|$ Although there is zero formal charge on each atom in this structure, there is not an octet of electrons around sulfur. On the other hand, in S_2Cl_2 there is an octet of electrons around each atom. Thus a more plausible Lewis structure is $|\overline{\underline{C}}l-\overline{\underline{S}}-\overline{\underline{S}}-\overline{\underline{C}}l|$

90.

$$\text{amount of gas} = \frac{PV}{RT} = \frac{749 \text{ mmHg} \times \frac{1 \text{ atm}}{760 \text{ mmHg}} \times 0.193 \text{ L}}{0.08206 \text{ L atm mol}^{-1} \text{ K}^{-1} \times 299.3 \text{ K}} = 0.00774 \text{ mol}$$

$$M = \frac{\text{mass}}{\text{amount}} = \frac{0.325 \text{ g}}{0.00774 \text{ mol}} = 42.0 \text{ g/mol}$$

Because three moles of C weigh 36.0 g and four moles weigh 48.0 g, this hydrocarbon is C_3H_6.

A possible Lewis structure is $H-\overset{\overset{\displaystyle H}{|}}{C}-\overset{\overset{\displaystyle H}{|}}{C}=\overset{\overset{\displaystyle H}{|}}{C}-H$ There is another possible Lewis structure: the three C atoms are arranged in a ring, with two H atoms bonded to each C atom (see below).

91. First we determine the empirical formula of this C, H compound.

$$\text{amount C} = 4.04 \text{ g CO}_2 \times \frac{1 \text{ mol CO}_2}{44.01 \text{ g CO}_2} \times \frac{1 \text{ mol C}}{1 \text{ mol CO}_2} = 0.0.0918 \text{ mol C} \div 0.0.0918 \longrightarrow 1.00 \text{ mol C}$$

$$\text{amount C} = 1.24 \text{ g H}_2\text{O} \times \frac{1 \text{ mol H}_2\text{O}}{18.02 \text{ g H}_2\text{O}} \times \frac{2 \text{ mol H}}{1 \text{ mol H}_2\text{O}} = 0.138 \text{ mol H} \quad \div 0.0918 \longrightarrow 1.50 \text{ mol H}$$

$$\text{Empirical formula} = C_2H_3 \qquad \text{possible molecular formula} = C_4H_6$$

Lewis structure $H-C=C\cdot$
$\quad\quad\quad\quad\quad\quad\quad\quad\ |\ \ |$
$\quad\quad\quad\quad\quad\quad\quad\quad\ H\ H$

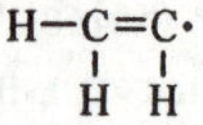

Lewis structures

$$H-C=C-C=C-H \quad\quad H-\overset{H}{\underset{H}{C}}-C\equiv C-\overset{H}{\underset{H}{C}}-H$$

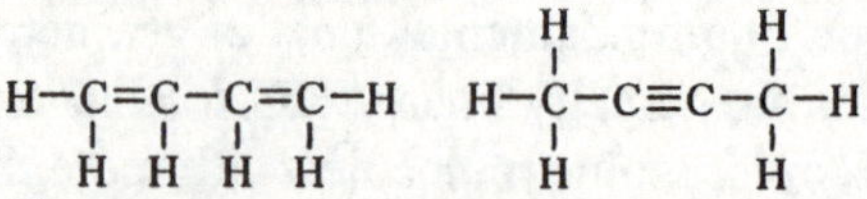

There are several other isomers that share this molecular formula.

92. The two isomers are $\overset{H}{\underset{H}{C}}=C=\overset{H}{\underset{H}{C}}$ and $H-\overset{H}{\underset{H}{C}}-C\equiv C-H$

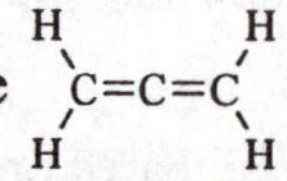

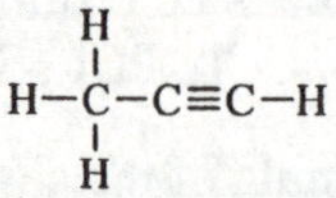

The left-hand isomer is planar around the first and third C atoms, but we cannot predict with VSEPR theory whether the molecule is planar overall, in other words, the two $H-C-H$ planes may be at 90° to each other. (They are, in fact.) In the right-hand isomer, the $C-C\equiv C-H$ chain is linear, but the H_3C- molecular geometry is tetrahedral.

93. There are three valid resonance forms for the N_3^- ion, all three are shown below:

$$\overset{\ominus}{\ddot{N}}=\overset{\oplus}{N}=\overset{\ominus}{\ddot{N}} \quad\quad\quad :N\equiv\overset{\oplus}{N}-\ddot{N}:\ \textcircled{2-} \quad\quad\quad \textcircled{2-}\ :\ddot{N}-\overset{\oplus}{N}\equiv N:$$

$$\mathbf{1} \quad\quad\quad\quad\quad\quad\quad \mathbf{2} \quad\quad\quad\quad\quad\quad\quad \mathbf{3}$$

The best resonance form is the one with the lowest formal charge (all N have octets). Thus, resonance form **1** is the greatest contributor (probably the only contributor). The bond length give (116 pm) has a bond order between 2.0 (123 pm) and 3.0 (110 pm). This deviation from a bond order or 2 (double bond) can be explained by the formal charges in the ion which gives the bonds within the ion some partial ionic character.

94. Because $\Delta H°_f[NO(g)] = +90.25\text{ kJ/mol}$, Thus $\frac{1}{2}N_2(g)+\frac{1}{2}O_2(g) \longrightarrow NO(g)$

has a $\Delta H_{rxn} = 90.25\text{ kJ/mol}$. $\Delta H_{rxn} = \Delta H(N=O) - \frac{1}{2}\Delta H(N\equiv N)-\frac{1}{2}\Delta H(O=O)$

The heat of reaction will be the algebraic sum of bonds broken (N_2 and O_2) and bonds formed (NO). Note that bond breaking requires energy, while bond making releases energy.

$$+90.25\text{ kJ} = -\Delta H(N=O)-\left(\frac{946\text{ kJ}}{2}\right)-\left(\frac{498\text{ kJ}}{2}\right)$$

$$\Delta H(N=O) = (-90.25+473+249)\text{ kJ/mol} = +632\text{ kJ/mol}$$

95. The HN_3 molecule has $1+(3\times5)=16$ valence electrons, or 8 pairs. Average bond lengths are 136 pm for $N-N$, 123 pm for $N=N$, and 110 pm for $N\equiv N$. Thus it seems that one nitrogen-to-nitrogen bond is a double bond, while the other is a triple bond. A plausible Lewis structure is $H-\bar{N}=N=\bar{N}$ The three N's lie on a line, with a 120° $H-N-N$ bond angle: $\overset{H}{\underset{}{\bar{N}}}=N=\bar{N}$

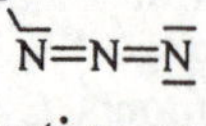

Another valid resonance form is $H-\bar{N}-N\equiv N|$ which would have one N-N separation consistent with a nitrogen-nitrogen triple bond. It would also predict a tetrahedral $H-N-N$ bond angle of 109.5°. Thus, the resulting resonance hybrid should have a bond angle between 120° and 109.5°, which is in good agreement with the observed 112° $H-N-N$ bond angle.

96. For N_5^+ the number of valence electrons is $(5 \times 5)-1 = 24$. There are three possible Lewis structures with formal charges that are not excessive ($< \pm 2$).

$$:\overset{\ominus}{\underset{\ominus}{N}}=\overset{\oplus}{N}=\overset{\oplus}{N}=\overset{\oplus}{N}=\overset{..}{N}: \quad :\overset{..}{N}=\overset{\oplus}{N}=\overset{\oplus}{\underset{\ominus}{N}}-\overset{\oplus}{N}\equiv N: \quad :N\equiv\overset{\oplus}{N}-\overset{..}{\underset{\ominus}{N}}=\overset{\oplus}{N}=\overset{..}{N}: \quad :N\equiv\overset{\oplus}{N}-\overset{\ominus}{\underset{..}{N}}-\overset{\oplus}{N}\equiv N:$$

$$\quad\quad\quad 1 \quad\quad\quad\quad\quad\quad 2 \quad\quad\quad\quad\quad\quad\quad 3 \quad\quad\quad\quad\quad\quad 4$$

Structure **1** has three adjacent atoms possessing formal charges of the same sign. Energetically, this is highly unfavorable. Structure **2** & **3** are similar and highly unsymmetrical but these are energetically more favorable than **1** Structure **4** is probably best of all, as all of the charges are close together with no two adjacent charges of the same sign. If structures **2**, **3** or **4** are chosen, the central nitrogen has one or two lone pairs. Thus, the structure of N_5^+ will be angular. (Note: an angle of 107.9 ° has been experimentally observed for the angle about the central nitrogen, as well, the bond length of the terminal N-N bonds is very close that seen in N_2, suggesting a triple bond. This suggests that resonance form **4** best describes the structure of the ion).

97. The carbon-carbon distances of 130 pm are close to those of a $C = C$ double bond, (134 pm). The carbon-oxygen distances of 120 pm are close to that of a $C = O$ double bond, 123 pm. In C_3O_2 there are $3 \times 4 + 2 \times 6 = 24$ valence electrons or 12 valence electron pairs. The molecule is almost certainly linear. A plausible Lewis structure follows.

$$\overline{O} = C = C = C = \overline{O}$$

98. In PCl_5 there are $5 + 5 \times 7 = 40$ valence electrons = 20 pairs. The Lewis structures are shown below. Since there are five atoms and no lone pairs attached to the central atom, the electron-group geometry and molecular shape of the molecule are the same, namely, trigonal bipyramidal. In PCl_4^+ there are $5 + 4 \times 7 - 1 = 32$ valence electrons = 16 pairs. Since there are four atoms and no lone pairs bonded to the central atom, the electron-group geometry and molecular shape of the species are the same, namely, tetrahedral. In PCl_6^- there are $5 + 6 \times 7 + 1 = 48$ valence electrons = 24 pairs. Since there are six atoms and no lone pairs bonded to the central atom, the electron-group geometry and molecular shape of the species are the same, namely, octahedral.

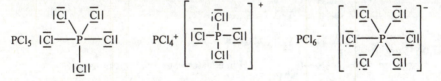

99. The Lewis structures for the species in reaction (2) follow.

$$|N \equiv N| \quad H - H \quad H - C \equiv N|$$

Bonds broken $= \dfrac{1}{2}$ $N \equiv N + \dfrac{1}{2}$ $H - H = \dfrac{1}{2}$ $(946 \ kJ + 436 \ kJ) = 691 \ kJ$

Bonds formed $= H - C + C \equiv N = 414 \ kJ + 891 \ kJ = 1305 \ kJ$

$\Delta H = 691 \ kJ - 1305 \ kJ = -614 \ kJ$

Then we determine ΔH_f°

(a) $C(s) \longrightarrow C(g)$ $\Delta H = +717$ kJ

(b) $C(g) + \frac{1}{2} N_2(g) + \frac{1}{2} H_2(g) \longrightarrow HCN(g)$ $\Delta H = -614$ kJ

Net: $C(s) + \frac{1}{2} N_2(g) + \frac{1}{2} H_2(g) \longrightarrow HCN(g)$ $\Delta H^\circ_f = +103$ kJ

This compares favorably to the value of 135.1 kJ/mol given in Appendix D-2.

100. We analyze the formation reaction for $H_2O_2(g)$ via the use of Lewis structures.

$$H - H + \overline{O} = \overline{O} \longrightarrow H - \overline{O} - \overline{O} - H$$

Bonds   broken $= H - H + O = O$ Energy Required: 436 kJ + 498 kJ = 934 kJ

Bonds formed $= 2(H - O) + O - O$ Energy Released: 2×464 kJ + O — O $= 928$ kJ + O — O

$\Delta H^\circ_f = -136$ kJ $= 934$ kJ $- 928$ kJ $- $ O — O   Bond   energy   (O — O) $= 142$ kJ

This is the same as the value in Table 10.3.

101. Ambiguity arises because of uncertainly over (i) Bond order between C and S and the (ii) position of the H atoms

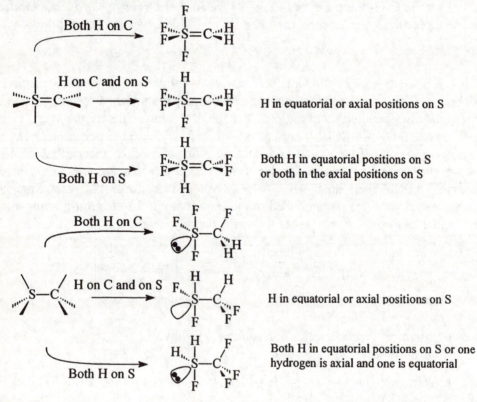

102. We first compute the heats of reaction.

$$CH_3OH(g) + H_2S(g) \longrightarrow CH_3SH(g) + H_2O(g)$$

$$\Delta H°_{rxn} = \Delta H°_f[H_2O(g)] + \Delta H°_f[CH_3SH(g)] - [\Delta H°_f[CH_3OH(g)] + \Delta H°_f[H_2S(g)]]$$

$$= -241.8 \text{ kJ} + (-22.9 \text{ kJ}) - [(-200.7 \text{ kJ} - 20.63 \text{ kJ})] = -43.4 \text{ kJ}$$

Breaking of one mole of C-O bond requires (360 kJ) =	+360 kJ
Breaking of one mole of H-S bond requires (368 kJ) =	+ 368 kJ
Breaking of one mole of O-H bond requires (464 kJ) =	+ 464 kJ
Making of one mole of C-S bond yields =	-x kJ
Making 2 moles of O-H bonds yields 2(-464 kJ) =	-928 kJ

$$\Sigma(\text{bond energies}) = \Delta H°_{rxn} = -43.4 \text{ kJ}$$

Then... $264 - x = -43.4$ and $x = 307$ kJ, which is the C-S bond energy fir the C—S bond in methanethiol (estimate only).

103.

$$Li \xrightarrow{217 \text{ pm}} Br \qquad \mu = 7.268 \text{ D} \qquad \Delta EN = 1.7$$

$$Na \xrightarrow{236 \text{ pm}} Cl \qquad \mu = 7.268 \text{ D} \qquad \Delta EN = 2.1$$

(a)

$$\delta_{Li-Br} = \frac{\mu \times 3.34 \times 10^{-30} \, CmD^{-1}}{d} = \frac{7.268 \, D \times 3.34 \times 10^{-30} \, CmD^{-1}}{217 \times 10^{-12} \, m} 1.12 \times 10^{-19} \, C \, (\approx 70\% \, ionic)$$

$$\delta_{Na-Cl} = \frac{\mu \times 3.34 \times 10^{-30} \, CmD^{-1}}{d} = \frac{9.001 \, D \times 3.34 \times 10^{-30} \, CmD^{-1}}{236 \times 10^{-12} \, m} 1.27 \times 10^{-19} \, C \, (\approx 80\% \, ionic)$$

(b) Using the ΔEN values for each gas phase species and Figure 10-7, we can estimate that Li-Br ~ 50% ionic and Na-Cl ~ 70% ionic.

(c) The value for NaCl agrees quite well, however, LiBr does not show as good an agreement. This is probably due to the large lone pairs on Br, the small size of the Li atom (especially relative to the Br atom) and due to the fact that the Li atom does not have much shielding. As well, bear in mind that these are very loosely associated atoms.

104. ZrF_7^{3-} has a pentagonal bipyramidal structure, with the following internal angles:

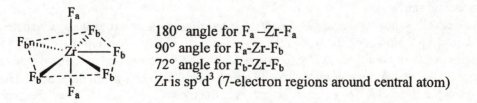

180° angle for F_a–Zr-F_a
90° angle for F_a-Zr-F_b
72° angle for F_b-Zr-F_b
Zr is sp^3d^3 (7-electron regions around central atom)

105. Consider the possible resonance forms for each acid (shown below).

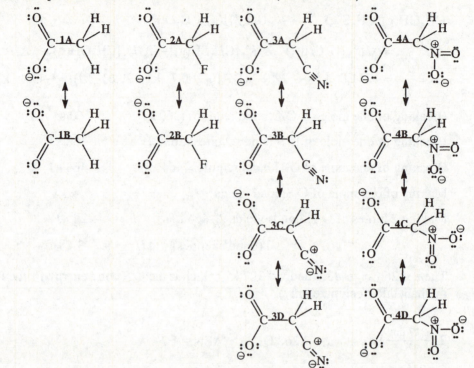

Note: All of these carboxylic acids display delocalization of the negative charge across the carboxylate group. Let's consider each acid in turn, to access the stability of the anion/acidity of the acid.

Acetic acid
In the case of acetic acid (1A and 1B), the hydrogen atoms do not contribute to the stability of the anion. In fact, one might argue that the hydrogen atoms with an electronegativity of 2.1, donates electron density to the adjacent (α) carbon and thus offsets delocalization of the negative charge on the anion.

Fluoroacetic acid
In the case of fluoroacetic acid (2A and 2B), other than the carboxylate resonance forms, the fluorine atom does not contribute significantly to the stability of the anion by way of resonance. However, fluorine is quite electronegative compared to carbon resulting in the development of a partial positive charge on the α-carbon. This allows for more extensive delocalization of the negative charge on the anions, which leads to greater stability for the anion.

Cyanoacetic acid
In the case of cyanoacetic acid (3A to 3D), we clearly see the carboxylate resonance delocalization.. However, there are resonance forms in the cyano group that result in a formally positively charged carbon (carbocation) being directly bonded to the α-carbon. This will allow for more complete delocalization of negative charge on the anion. There is one problem, however. The resonance forms for the cyano group do not leave the carbon

atom with the formal positive charge in the cyano group with a complete octet. Although the cyano group is electron withdrawing, it behaves very much like a very electronegative halide by virtue of the fact that a very electronegative nitrogen is directly bonded to carbon in the cyano group. This results in a slightly positive charge on the carbon attached to the α-carbon resulting in significant delocalization of the negative charge on the anion.

Nitroacetic acid

In the case of nitroacetic acid (4A to 4B), we clearly see the carboxylate resonance delocalization. However, there are resonance forms in the nitro group that result in the development of a positive formal charge on the nitrogen atom directly bonded to the α-carbon. This will significantly allow for the delocalization of negative charge on the anion. Unlike the cyano group, the nitro group has a formal positive charge on nitrogen and complete octets for all the second row atoms in the acid. Thus, the nitro group will allow for the most delocalization of the negative charge on the anion, which makes nitroacetic acid the strongest of the four acids and the nitroacetate ion the most stable conjugate base.

The acids in order of strength should be: nitroacetic > cyanoacetic> fluoroacetic> acetic.

106. For the h alogens we have the following data:

Atom	Electronegativity	Ionization Energy (kJ/mol)	Electron Affinity (kJ/mol)
F	4.0	1680	-328
Cl	3.0	1256	-349
Br	2.8	1143	-324.6
I	2.5	1009	-295.2

From the data above and using $\chi = k \times (IE - EA)$ for the halogens, we find the following values for k: F = 0.00199   Cl = 0.00187   Br = 0.00191   I = 0.00192
We shall assume that the value of k for Astatine is 0.0019. As well, from the data, we can estimate the ionization energy for astatine to be ~900 kJ/mol. The text gives 2.2 as the electronegativity for astatine. We can now estimate a value for the electron affinity of At.

$$\chi = k \times (IE - EA) = 2.2 = 0.0019(900-EA) \qquad EA \sim -260 \text{ kJ/mol}$$

107. Structure A has too man y electrons resulting in a molecule with an overall charge of -2 (each sulfur has a formal charge of -1). Structure B has formal charges of +1 on each sulfur atom and -1 on each Cl atom. Structure C has considerable formal charge built up in this Lewis structure (each sulfur carries a formal charge of -2 and each chlorine carries a formal charge of +2). Both structures D and E have no formal charges in the Lewis structure and may be considered "good" Lewis structures, however, structure D has 10 electrons around each sulfur (expanded octet), while in structure E, all atoms have an octet. Given this information, structure E best represents the molecule, however, one cannot rule out structure D completely based solely on the expanded octet, especially since S has vacant 3 d orbitals that may be used in expanding its octet. Read section 10-6: Expanded Octets.

108. The molecules hydrogen azide, nitrosyl azide and trifluoromethy azide share some common structural elements. They can be viewed as N_3^- attached to H, NO and CF_3 respectively. There are numerous resonance structures that can be written for these molecules. Some of the resonance forms are better than others for several reasons, including such factors as the lack of a full octet and unacceptably large charge separations. Drawn below are the various resonance forms for the molecules. The Lewis diagrams have been modified to include geometric consideration that are imposed on the molecule by hybridization and the effects of lone pairs.

Let's consider each molecule in turn.

Hydrogen azide(linear N-N-N):

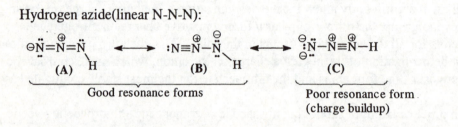

(A)   (B)   (C)

Good resonance forms   Poor resonance form
                              (charge buildup)

Resonance forms (A) and (B) are the best, owing to low formal charges. (C) is clearly not an important an resonance contributor because it possess two adjacent like charges. Note that the molecule should be bent at the hydrogen bonded nitrogen.

Hydrogen azide(cyclic form):

very strained ring {

(D)   (E)   (F)

Best resonance form   Good resonance forms
(no formal charges)

Resonance form (D) is clearly favored because all of the atoms have a formal charge of zero. This structural isomer would be highly reactive owing to the very strained ring which imposes internal N-N-N angles that approach ~ 60° and a hybridization scheme that tries to accommodate both 120 or 109.5° angles.

Trifluoromethy azide:

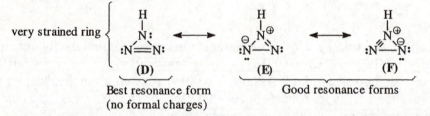

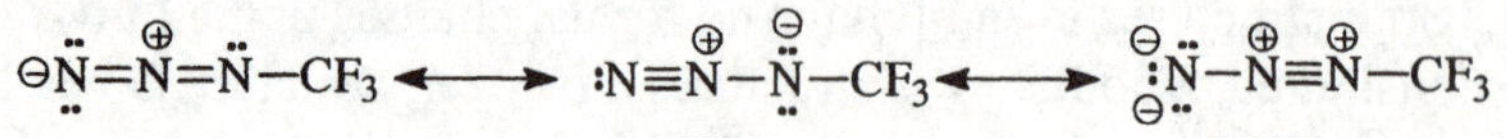

(G)   (H)   (I)

Good resonance forms   Poor resonance form
                              (charge buildup)

Resonance forms (G) and (H) are the best, owing to low formal charges for the constituent atoms. Structure (I) is clearly not as important a resonance contributor because it contains destabilizing adjacent positive charges. The molecule is expected to be very reactive and hence hard to isolate.

Nitrosyl azide

Best resonance forms
(small distance between
charge separation)

$$\ominus \ddot{N}=N=\ddot{N}-\ddot{N}=\ddot{O} \overset{\oplus}{}$$
$$\ominus N=N=N=\overset{\oplus}{N}-\ddot{O}: \overset{\oplus}{}{}^{\ominus}$$
$$:N\equiv N-\overset{\oplus}{\underset{\ominus}{N}}-\ddot{N}=\ddot{O}:$$
$$\ominus \ddot{N}=N-\overset{\ominus}{N}-N\equiv O:$$

$$\ominus \ddot{N}=\ddot{N}-N\equiv N-\ddot{O}: \overset{\oplus \ \oplus}{}{}^{\ominus}$$

Good resonance forms
(however, large charge
separation)

$$:N\equiv \overset{\oplus}{N}-\ddot{N}=\ddot{N}-\ddot{O}: \overset{\ominus}{}$$
$$\ominus \ddot{N}=N-\ddot{N}-N=\ddot{O}$$

$$\overset{\ominus}{\ddot{N}}-\ddot{N}=\ddot{N}-N\equiv O: \overset{\oplus \ \oplus}{}$$
$$\overset{\ominus}{\ddot{N}}-N\equiv N-\ddot{N}=\ddot{O}$$
$$\overset{\ominus}{\ddot{N}}-\ddot{N}=N=N=\ddot{O}$$

Poor resonance forms
(charge buildup)

Nitrosyl azide is a bent molecule. Based on the best two Lewis structures, we expect that the molecule is bent at the two left most nitrogen atoms ($\sim 115\,° \pm 20\,°$).

---

109. Consider each molecule separately:

B_2F_4: A total of 34 electrons. Neutral molecule. There are no formal charges in the molecule. Free rotation about single bond, with the possibility of all atoms being planar. There are no valid resonance forms.

N_2O_4: A total of 34 electrons. Neutral molecule with formal charges. Free rotation about single bond, with the possibility of all atoms being planar. There are 3 valid resonance forms.

$C_2O_4^{2-}$: A total of 34 electrons Must be the dianion where the formal charges add to the overall charge of the anion. Free rotation about single bond, with the possibility of all atoms being planar. There are 3 valid resonance forms.

---

110. Shown below is the deprotonation and the expected resonance stabilized enolate product, which has two forms (**A** and **B** below). In structure **A**, the negative charge is on the CH_2 carbon atom and in structure **B** it is on the oxygen atom. Both of these resonance forms will contribute to the hybrid structure of the anion. Structure **A** is desirable in that it possess a very strong $C = O$ bond (relative to the weaker $C = C$ bond in **B**). Structure **B** has the negative charge on oxygen, the more electronegative atom, which is better able to accommodate the negative charge.

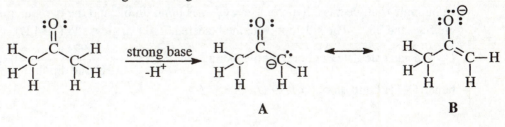

111. The anion PBr_4^- has 34 electrons. Its electronic geometry is trigonal bypyramidal, while ideally, its molecular geometry is expected to be a see-saw. Since phosphorus is a relatively small atom and bromine is a relatively large atom, significant distortions may exist in the anion such that the bromine atoms distribute themselves more evenly around the phosphorus centre. With extensive delocalization of the phosphorus lone pair over the four bromine atoms, the structure will approach a pure tetrahedron. Even if the lone pair stays localized on the phosphorus, the geometry may well appear to be tetrahedral, albeit distorted (see below).

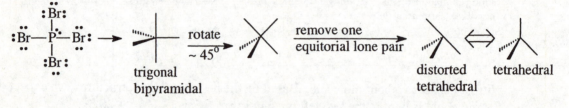

FEATURE PROBLEMS

112. (a) The average of the H — H and Cl — Cl bond energies is $(436 + 243)$ kJ $\div 2 = 340$ kJ/mol. The ionic resonance energy is the difference between this calculated value and the measured value of the H — Cl bond energy:
IRE = 431 kJ/mol − 340 kJ/mol = 91 kJ/mol

(b) $\Delta EN = \sqrt{IRE / 96} = \sqrt{91 / 96} = 0.97$

(c) An electronegativity difference of 0.97 gives about a 23% ionic character, read from Figure 10.7. The result of Example 10-4 is that the H — Cl bond is 20% ionic. These values are in good agreement with each other.

113. (a) The two bond dipole moments can be added geometrically, by placing the head of one at the tail of the other, as long as we do not change the direction or the length of the moved dipole. The resultant molecular dipole moment is represented by the arrow drawn from the tail of one bond dipole to the head of the other. This is shown in the figure to the right. The 52.0° angle in the figure is one-half of the 104° bond angle in water. The length is given as 1.84 D. We can construct a right angle triangle by bisecting the 76.0° angle. The

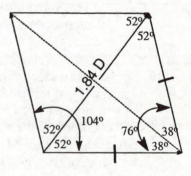

right angle triangle has a hypotenuse = O — H bond dipole and the two other angles are 52 ° and 38 °. The side opposite the bisected 76.0 ° angles is ½ (1.84 D) = 0.92 D. We can calculate the bond dipole using: $\sin 38.0° = \dfrac{0.92\,D}{O\text{-}H\,\text{bond dipole}} = 0.61566$,

hence O—H bond dipole = 1.49 D.

(b) For H_2S, we do not know the bond angle. We shall represent this bond angle as 2α. Using a similar procedure to that described in part (a), above, a diagram can be constructed and the angle 2α calculated as follows:

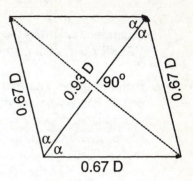

$$\cos \alpha = \frac{\frac{1}{2}(0.93\,D)}{0.67\,D} = 0.694 \quad \alpha = 46.05\,°$$

or $2\alpha = 92.\underline{1}\,°$

The H—S—H angle is approximately $92\,°$.

(c)

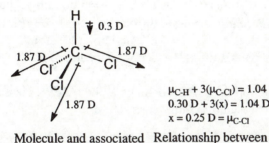

$$\mu_{C\text{-}H} + 3(\mu_{C\text{-}Cl}) = 1.04\,D$$
$$0.30\,D + 3(x) = 1.04\,D$$
$$x = 0.25\,D = \mu_{C\text{-}Cl}$$

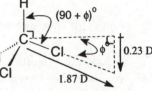

Molecule and associated individual bond dipoles   Relationship between dipole moment(molecular) and bond dipoles(Vector addition)   Geometric Relationship

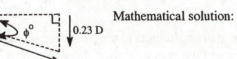

Mathematical solution: $\sin (\phi) = \dfrac{0.25\,D}{1.87\,D}$

$$\phi = 7.6\,°$$

The H-C-Cl bond angle is $(90 + \phi)° = 90° + 7.6\,° = 97.6°$

114. Step 1 in the alternate approach is similar to the first step in the method used for drawing Lewis structures. The only significant difference is that "electron pairs" rather than the total number of valence electrons are counted in this alternate approach. The second step in the alternate strategy is also similar to the second step for writing Lewis structures. By counting the number of bonding electron pairs in the alternate method, one is effectively working out the number of bonds present in the skeletal structure of the Lewis diagram. In step 3, the number of electron pairs surrounding the central atom is calculated. This is basically the same procedure as completing the octets for the terminal atoms and assigning the remaining electrons to the central atom in the Lewis structure. Finally, in step 4 of the alternate method, the number of lone pair electrons on the central atom is calculated. This number together with the result from step 3, allows one to establish the VSEPR class. Consequently, both the alternate strategy and the Lewis diagram provide the number of bonding electron pairs and lone pairs on the central atom for the species whose shape is being predicted. Since the shape of the molecule or ion in the VSEPR approach is determined solely by the number and types of electron pairs on the central atom (i.e., the VSEPR class) both methods end up giving the same result.

Included in the "alternative strategy" is the assumption that the central atom does not form double bonds with any of the terminal atoms. This means that in many instances, the central atom does not possess a complete octet. The presence or absence of an octet is, however, of no consequence to the VSEPR method because, according to the tenets of this theory, the shape adopted by the molecule is determined solely by the number and types of electron pairs on the central atom. Examples follow on the next two pages.

(a) PCl_5

1. Total e^- pairs = $\dfrac{(1 \times 5e^- \text{ from P atom}) + (5 \times 7e^- \text{ from the 5 Cl atoms})}{2}$ = 20 pairs of e^-

2. Number of bonding e^- pairs = 6 atoms ($5\times Cl + 1\times P$) $- 1 = 5$ bonding e^- pairs.

3. Number of e^- pairs around the central atom = (20(total) e^- pairs) $- 3\times$(5 terminal Cl)
$= 5$ e^- pairs around P atom.

4. Number of lone pair $e^- = 5$ e^- pairs around P $- 5$ bonding pairs of $e^- = 0$

Thus, according to this alternate approach, PCl_5 belongs to the VSEPR class AX_5. Molecules of this type adopt a trigonal bipyramidal structure.

(b) NH_3

1. Total e^- pairs =

$\dfrac{(1 \times 5e^- \text{ from N atom}) + (3 \times 1 \text{ } e^- \text{ from the 3 H atoms})}{2}$ = 4 pairs of e^-

2. Number of bonding e^- pairs = 4 atoms ($1\times N + 3\times H$) $- 1 = 3$ bonding e^- pairs.

3. Number of e^- pairs around the central atom = 4(total) e^- pairs $- 0$
$= 4$ e^- pairs around N atom.

4. Number of lone pair $e^- = 4$ e^- pairs around N $- 3$ bonding pairs of $e^- = 1$ lone pair of e^-

Thus, according to this alternate approach, NH_3 belongs to the VSEPR class AX_3E. Molecules of this type adopt a trigonal pyramidal structure.

(c) ClF_3

1. Total e^- pairs = $\dfrac{(1 \times 7e^- \text{ from Cl atom}) + (3 \times 7e^- \text{ from the 3 F atoms})}{2}$ =14 e^- pairs

2. Number of bonding e^- pairs = 4 atoms ($1\times Cl + 3\times F$) $- 1 = 3$ bonding e^- pairs.

3. Number of e^- pairs around the central atom = 14(total) pairs $- 3\times$(3 terminal F atoms)
$= 5$ e^- pairs around Cl atom.

4. Number of lone pair $e^- = 5$ e^- pairs around Cl $- 3$ bonding pairs of $e^- = 1$ lone pair of e^-

Thus, according to this alternate approach, ClF_3 belongs to the VSEPR class AX_3E_2. Molecules of this type adopt a T-shaped structure.

(d) SO_2

1. Total e⁻ pairs = $\dfrac{(1 \times 6e^- \text{ from S atom}) + (2 \times 6e^- \text{ from the 2 O atoms})}{2} = 9$ e⁻ pairs

2. Number of bonding e⁻ pairs = 3 atoms (1×S + 2×O) − 1 = 2 bonding e⁻ pairs.

3. Number of e⁻ pairs around the central atom = 9(total) pairs − 3×(2 terminal O atoms)
= 3 e⁻ pairs around S atom.

4. Number of lone pair e⁻ = 3 e⁻ pairs around S − 2 bonding pairs of e⁻ = 1 lone pair of e⁻
Thus, according to this alternate approach, SO_2 belongs to the VSEPR class AX_2E.
Molecules of this type adopt a bent structure.

(e) ClF_4^-

1. Total e⁻ pairs = $\dfrac{(1 \times 7e^- \text{ from Cl}) + (4 \times 7e^- \text{ from the 4 F}) + (1e^- \text{ for charge of -1)})}{2}$

= 18 pairs of e⁻

2. Number of bonding e⁻ pairs = 5 atoms (1×Cl + 4×F) − 1 = 4 bonding e⁻ pairs.

3. Number of e⁻ pairs around the central atom = 18(total) pairs − 3×(4 terminal F atoms)
= 6 e⁻ pairs around Cl atom.

4. Number of lone pair e⁻ = 6 e⁻ pairs around Cl − 4 bonding pairs of e⁻ = 2 lone pair of e⁻
Thus, according to this alternate approach, ClF_4^- belongs to the VSEPR class AX_4E_2.
Molecules of this type adopt a square planar structure.

(f) PCl_4^+

1. Total e⁻ pairs = $\dfrac{(1 \times 5e^- \text{ from P}) + (4 \times 7e^- \text{ from the 4 Cl}) - (1e^- \text{ for +1 charge})}{2}$

= 16 pairs of e⁻

2. Number of bonding e⁻ pairs = 5 atoms (4×Cl + 1×P) − 1 = 4 bonding e⁻ pairs.

3. Number of e⁻ pairs around the central atom = 16(total) pairs − 3(4 terminal Cl atoms)
= 4 e⁻ pairs around P atom.

4. Number of lone pair e⁻ = 4 e⁻ pairs around P − 4 bonding pairs of e⁻ = 0

Thus, according to this alternate approach, ClF_4^+ belongs to the VSEPR class AX_4.
Molecules of this type adopt a tetrahedral structure.

CHAPTER 11
CHEMICAL BONDING II: ADDITIONAL ASPECTS
PRACTICE EXAMPLES

1A The valence-shell orbital diagrams of N and I are as follows:

N [He]$_{2s}$ ⊞ $_{2p}$ ⊞⊞⊞   I [Kr] $4d^{10}_{5s}$ ⊞ $_{5p}$ ⊞⊞⊞

There are three half-filled $2p$ orbitals on N, and one half-filled $5p$ orbital on I. Each half-filled $2p$ orbital from N will overlap with one half-filled $5p$ orbital of a I. Thus, there will be three N—I bonds. The I atoms will be oriented in the same direction as the three $2p$ orbitals of N: toward the $x-$, $y-$, and z-directions of a Cartesian coordinate system. Thus, the I—N—I angles will be approximately 90° (probably larger because the I atoms will repel each other). The three I atoms will lie in the same plane at the points of a triangle, with the N atom centered above them. The molecule is trigonal pyramidal. (The same molecular shape is predicted if N is assumed to be sp^3 hybridized, but with 109.5° rather than 90° bond angles.)

1B The valence-shell orbital diagrams of N and H are as follows. N: [He]$_{2s}$ ⊞ $_{2p}$ ⊞⊞⊞   H:$_{1s}$ ⊞
There are three half-filled orbitals on N and one half-filled orbital on each H. There will be three N—H bonds, with bond angles of approximately 90°. The molecule is trigonal pyramid. (We obtain the same molecular shape if N is assumed to be sp^3 hybridized, but bond angles are closer to 109.5°, the tetrahedral bond angle.) VSEPR theory begins with the Lewis structure and notes that there are three bond pairs and one lone pair attached to N. This produces a tetrahedral electron pair geometry and a trigonal pyramid molecular shape with bond angles a bit less than the tetrahedral angle of 109.5° because of the lone pair. Since VSEPR theory makes a prediction closer to the experimental bond angle of 107° it seems more appropriate in this case.

2A Following the strategy outlined in the textbook, we begin by drawing a plausible Lewis structure for the cation in question. In this case, the Lewis structure must contain 20 valence electrons. The skeletal structure for the cation has a chlorine atom, the least electronegative element present, in the central position. Next we join the terminal chlorine and fluorine atoms to the central chlorine atom via single covalent bonds and then complete the octets for all three atoms by placing three lone pairs around the terminal atoms and two lone pairs around the central atom.

$$ |\overline{\underline{F}}\text{—}\overset{\oplus}{\underline{Cl}}\text{—}\overline{\underline{Cl}}| $$

With this bonding arrangement, the central chlorine atom ends up with a 1+ formal charge.

Once the Lewis diagram is complete, we can then use the VSEPR method to establish the geometry for the electron pairs on the central atom. The Lewis structure has two bonding electron pairs and two lone pairs of electrons around the central chlorine atom. These four pairs of electrons assume a tetrahedral geometry to minimize electron-electron repulsions. The VSEPR notation for the Cl_2F^+ ion is AX_2E_3. According to Table 11.1, molecules of this type exhibit an <u>angular</u> molecular geometry. Our next task is to select a hybridization scheme that is consistent with the predicted shape. It turns out that the only way we can end up with a tetrahedral array of electron groups is if the central chlorine atom is sp^3 hybridized. In this scheme, two of the sp^3 hybrid orbitals are filled, while the remaining two are half occupied.

$\boxed{\uparrow\downarrow\,|\,\uparrow\downarrow\,|\,\uparrow\,|\,\uparrow\,}$ sp^3 hybridized central chlorine atom (Cl^+)

The Cl—F and Cl—Cl bonds in the cation are then formed by the overlap of the half-filled sp^3 hybrid orbitals of the central chlorine atom with the half-filled p-orbitals of the terminal Cl and F atoms. Thus, by using sp^3 hybridization, we end up with the same <u>bent</u> molecular geometry for the ion as that predicted by VSEPR theory (when the lone pairs on the central atom are ignored)

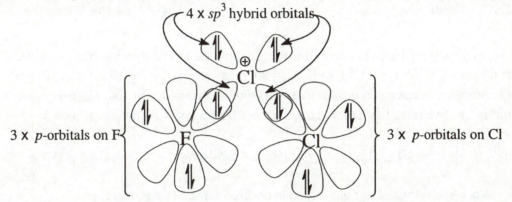

2B As was the case in 2A, we begin by drawing a plausible Lewis structure for the cation in question. This time the Lewis structure must contain 34 valence electrons. The skeletal structure has bromine, the least electronegative element present, as the central atom. Next, we join the four terminal fluorine atoms to the central bromine atom via single covalent bonds and complete the octets for all of the fluorine atoms by assigning three lone pairs to each fluorine atom. Placing the last two electrons on the central bromine atom completes the diagram.

In order to accommodate ten electrons, the bromine atom is forced to expand its valence shell. Notice that the Br ends up with a 1+ formal charge in this structure. With the completed Lewis structure in hand, we can then use VSEPR theory to establish the geometry for the electron pairs around the central atom. The Lewis structure has four bonding pairs and one lone pair of electrons around the central bromine atom. These five pairs of electrons assume a trigonal bipyramidal geometry to minimize electron-electron repulsions. The VSEPR notation for the BrF_4^+ cation is AX_4E. According to Table 11.1, molecules of this type exhibit a see-saw molecular geometry.

Next we must select a hybridization scheme for the Br atom that is compatible with the predicted shape. It turns out that only sp^3d hybridization will provide the necessary trigonal bipyramidal distribution of electron pairs around the bromine atom. In this scheme, one of the sp^3d hybrid orbitals is filled while the remaining four are half-occupied.

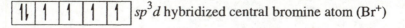

The four Br—F bonds in the cation are then formed by the overlap of the four half-filled sp^3d hybrid orbitals of the bromine atom with the half-filled p-orbitals of the four separate terminal fluorine atoms. Thus, by using sp^3d hybridization, we end up with the same see-saw molecular geometry for the cation as that predicted by VSEPR theory (when the lone pair on Br is ignored).

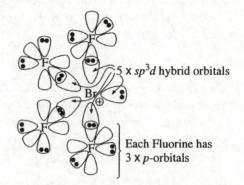

5 x sp^3d hybrid orbitals

Each Fluorine has 3 x p-orbitals

3A We begin by writing the Lewis structure. The H atoms are terminal atoms. There are three central atoms and $(3\times1)+4+6+4+(3\times1)=20$ valence electrons, or 10 pairs. A plausible Lewis structure is drawn at right. Each central atom is surrounded by four electron pairs, requiring sp^3 hybridization. The valence-shell orbital diagrams for the atoms follow.

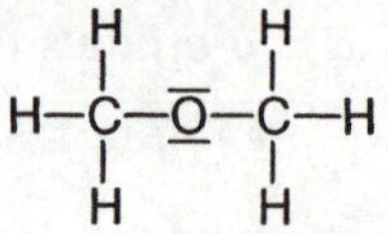

H_{1s}⊡   C [He]$_{2s}$⊡ $_{2p}$▦   O [He]$_{2s}$⊡ $_{2p}$▦

The valence-shell orbital diagrams for the hybridized central atoms then are:

C_{sp^3}▦ O_{sp^3}▦

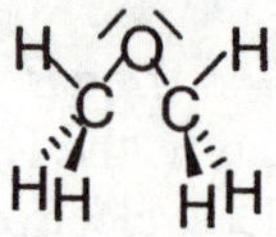

All bonds in the molecule are σ bonds. The H—C—H bond angles are $109.5°$, as are the H—C—O bond angles. The C—O—C bond angle is possibly a bit smaller than $109.5°$ because of the repulsion of the two lone pairs of electrons on O. A wedge-and-dash sketch of the molecule is at right.

3B The H atoms and one O are terminal atoms in the Lewis structure, which has $3\times1+4+4+2\times6+1=24$ valence electrons, or 12 pairs. The left-most C and the right-most O are surrounded by four electron pairs, and thus require sp^3 hybridization. The central carbon is surrounded by three electron groups and is sp^2 hybridized. The orbital diagrams for the un-hybridized atoms are:

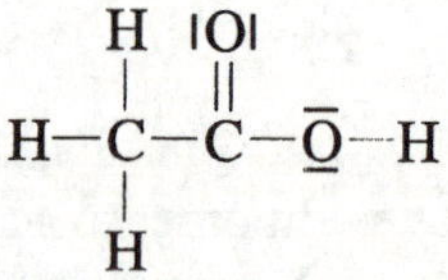

$H_{:1s}$⊡   C: [He]$_{2s}$⊡ $_{2p}$▦   O: [He]$_{2s}$⊡ $_{2p}$▦

Hybridized orbital diagrams:

C: [He]$_{sp^3}$▦   C: [He]$_{sp^2}$▦ $_{2p}$⊡   O:[He]$_{sp^3}$▦   terminal O: [He]$_{sp^2}$▦ $_{2p}$⊡

There is one π bond in the molecule: between the $2p$ on the central C and the $2p$ on the terminal O. The remaining bonds are σ bonds. The H—C—H and H—C—C bond angles are $109.5°$. The H—O—C angle is somewhat less, perhaps $105°$ because of lone pair repulsion. The C—C—O bond angles and O—C—O bond angles are all $120°$.

4A There are four bond pairs around the left-hand C, requiring sp^3 hybridization. Three of the bonds that form are C—H sigma bonds resulting from the overlap of a half-filled sp^3 hybrid orbital on C with a half-filled $1s$ orbital on H. The other C has two attached electron groups, utilizing sp hybridization.

$$\begin{array}{c} H \\ | \\ H-C-C\equiv N| \\ | \\ H \end{array}$$

C: [He] $sp\,\boxed{\uparrow|\uparrow|\uparrow}$ $2p\,\boxed{\uparrow|\uparrow}$ The N atom is sp hybridized. N: [He] $2s\,\boxed{\uparrow\downarrow}$ $2p\,\boxed{\uparrow|\uparrow|\uparrow}$

The two C atoms join with a sigma bond: overlap of sp^2 on the left-hand C with sp on the right-hand C. The three bonds between C and N consist of a sigma bond (sp on C with sp on N), and two pi bonds ($2p$ on C with $2p$ on N).

4B The bond lengths that are given indicate that N is the central atom. The molecule has $(2\times5)+6=16$ valence electrons, or 8 pairs. Average bond lengths are: N—N = 145 pm, N=N = 123 pm, N≡N = 110 pm, N—O = 136 pm, N=O = 120 pm. Plausible resonance structures are (with subscripts for identification):

$$\text{structure(1)} \quad |N_a \equiv N_b^\oplus{-}\overline{\underline{O}}|^\ominus \quad \leftrightarrow \quad |\underline{N}_a^\ominus{=}N_b^\oplus{=}\overline{O} \quad \text{structure(2)}$$

In both structures, the central N is attached to two other atoms, and possesses no lone pairs. The geometry of the molecule thus is linear and the hybridization on this central N is sp. N_b [He] $sp\,\boxed{\uparrow\downarrow|\uparrow}$ $2p\,\boxed{\uparrow|\uparrow}$ We will assume that the terminal atoms are not hybridized in either structure. Their valence-shell orbital diagrams are:

terminal N: N_a [He] $2s\,\boxed{\uparrow\downarrow}$ $2p\,\boxed{\uparrow|\uparrow|\uparrow}$ O [He] $2s\,\boxed{\uparrow\downarrow}$ $2p\,\boxed{\uparrow\downarrow|\uparrow|\uparrow}$

In structure (1) the N≡N bond results from the overlap of three pairs of half-filled orbitals: (1) sp_x on N_b with $2p_x$ on N_a forming a σ bond, (2) $2p_y$ on N_b with $2p_y$ on N_a forming a π bond, and (3) $2p_z$ on N_b with $2p_z$ on N_a also forming a π bond. The N—O bond is a coordinate covalent bond, and requires that the electron configuration of O be written as O [He] $2s\,\boxed{\uparrow\downarrow}$ $2p\,\boxed{\uparrow\downarrow|\uparrow\downarrow|\,}$ The N—O bond then forms by the overlap of the full sp_x orbital on N_b with the empty $2p$ orbital on O.

In structure (2) the N=O bond results from the overlap of two pairs of half-filled orbitals: (1) sp_y on N_b with $2p_y$ on O forming a σ bond and (2) $2p_z$ on N_b with $2p_z$ on O forming a π bond. The N=N σ bond is a coordinate covalent bond, and requires that the configuration of N_a be written N_a [He] $2s\,\boxed{\uparrow\downarrow}$ $2p\,\boxed{\uparrow\downarrow|\uparrow|\,}$. The N=N bond is formed by two overlaps: (1) the overlap of the full sp_y orbital on N_b with the empty $2p_y$ orbital on N_a to form a σ bond, and (2) the overlap of the half-filled $2p_x$ orbital on N_b with the half-filled $2p_x$ orbital on N_a to form a π bond. Based on formal charge arguments, structure (1) is preferred, because the negative formal charge is on the more electronegative atom, O.

5A Removing an electron from Li_2 removes a bonding electron because the valence molecular orbital diagram for Li_2 is the same as that for H_2, only it is just moved up a principal quantum level: $\sigma_{1s}b\,\boxed{\uparrow\downarrow}$ $\sigma_{1s}*\,\boxed{\uparrow\downarrow}$ $\sigma_{2s}b\,\boxed{\uparrow\downarrow}$ $\sigma_{2s}*\,\boxed{\,}$. The molecular orbital diagram for Li_2^+ is: $\sigma_{1s}b\,\boxed{\uparrow\downarrow}$ $\sigma_{1s}*\,\boxed{\uparrow\downarrow}$ $\sigma_{2s}b\,\boxed{\uparrow}$ $\sigma_{2s}*\,\boxed{\,}$ The bond order is Li_2^+ in 1/2, while that in Li_2 is one. Thus, the Li-Li bond in Li_2^+ should be one half as strong as the Li-Li bond in Li_2:

106 kJ/mol $\div 2 = 53$ kJ/mol Li_2^+

5B The H_2^- ion contains 1 electron from each H plus 1 electron for the negative charge for a total of three electrons. Its molecular orbital diagram is σ_{1s}[↑↓] $\sigma_{1s}*$[↑]. There are 2 bonding and 1 anti-bonding electrons. The bond order in H_2^- is:

$$\text{Bond order} = \frac{(2 \text{ bonding e}^- - 1 \text{ antibonding e}^-)}{2} = \frac{1}{2}$$

Thus, we would expect the ion H_2^- to be stable, with a bond strength about half that of a hydrogen molecule.

6A For each case, the empty molecular-orbital diagram has the following appearance. (KK indicates that the molecular orbitals formed from $1s$ atomic orbitals are full: KK $= \sigma_{1s}$[↑↓] $\sigma_{1s}*$[↑↓] ; KK σ_{2s}[ ] $\sigma_{2s}*$[ ] σ_{2p}[ ] π_{2p}[ ][ ] $\pi_{2p}*$[ ][ ] $\sigma_{2p}*$[ ]. We need to simply count up the total number of electrons in each species, and place them in the valence-shell molecular-orbital diagram.

(a) N_2^+ has $(2 \times 5) - 1 = 9$ valence electrons. Its molecular orbital diagram is

N_2^+ KK σ_{2s}[↑↓] $\sigma_{2s}*$[↑↓] π_{2p}[↑↓][↑↓] σ_{2p}[↑] $\pi_{2p}*$[ ][ ] $\sigma_{2p}*$[ ]

bond order = (7 bonding electrons -2 antibonding electrons) $\div 2 = 2.5$

(b) Ne_2^+ has $(2 \times 8) - 1 = 15$ valence electrons. Its molecular orbital diagram is

Ne_2^+ KK σ_{2s}[↑↓] $\sigma_{2s}*$[↑↓] σ_{2p}[↑↓] π_{2p}[↑↓][↑↓] $\pi_{2p}*$[↑↓][↑↓] $\sigma_{2p}*$[↑]

bond order = (8 bonding electrons -7 antibonding electrons) $\div 2 = 0.5$

(c) C_2^{2-} has $(2 \times 4) + 2 = 10$ valence electrons. Its molecular orbital diagram is

C_2^{2-} KK σ_{2s}[↑↓] $\sigma_{2s}*$[↑↓] π_{2p}[↑↓][↑↓] σ_{2p}[↑↓] $\pi_{2p}*$[ ][ ] $\sigma_{2p}*$[ ]

bond order = (8 bonding electrons -2 antibonding electrons) $\div 2 = 3.0$

6B For each case, the empty molecular-orbital diagram has the following appearance. (KK indicates that the molecular orbitals formed from $1s$ atomic orbitals are full: KK $= \sigma_{1s}$[↑↓] $\sigma_{1s}*$[↑↓] KK σ_{2s}[ ] $\sigma_{2s}*$[ ] σ_{2p}[ ] π_{2p}[ ][ ] $\pi_{2p}*$[ ][ ]$\sigma_{2p}*$[ ]. All of these species are based on O_2, which has $2 \times 6 = 12$ valence electrons. We simply put the appropriate number of valence electrons in each diagram and determine the bond order.

O_2^+ 11 v.e.   KK σ_{2s}[↑↓] $\sigma_{2s}*$[↑↓] σ_{2p}[↑↓] π_{2p}[↑↓][↑↓] $\pi_{2p}*$[↑][ ] $\sigma_{2p}*$[ ]

bond order = (8 bonding e$^-$ -3 antibonding e$^-$)$\div 2 = 2.5$; bond length $= 112$ pm

O_2 12 v.e.   KK

bond order = (8 bonding e$^-$ -4 antibonding e$^-$)$\div 2 = 2.0$; bond length $= 121$ pm

O_2^- 13 v.e.   KK

bond order = (8 bonding e$^-$ -5 antibonding e$^-$)$\div 2 = 1.5$; bond length $= 128$ pm

O_2^{2-} 14 v.e.   KK σ_{2s}[↑↓] $\sigma_{2s}*$[↑↓] σ_{2p}[↑↓] π_{2p}[↑↓][↑↓] $\pi_{2p}*$[↑↓][↑↓] $\sigma_{2p}*$[ ]

bond order = (8 bonding electrons $-$ 6 antibonding electrons) $\div 2 = 1.0$;

bond length $= 149$ pm. We see that the bond length does indeed increase as the bond order decreases. Longer bonds are weaker bonds.

7A   There are 8 valence electrons that must be placed into the molecular orbital diagram for CN^+ (5 electrons from nitrogen, four electrons from carbon and one electron is removed to produce the positive charge). Since both C and N precede oxygen and they are not far apart in atomic number, we must use the modified molecular-orbital energy-level diagram to get the correct configuration. By following the Aufbau orbital filling method, one obtains the ground state diagram asked for in the question.

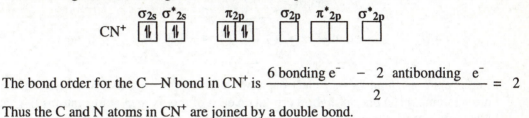

The bond order for the C—N bond in CN^+ is $\dfrac{6 \text{ bonding e}^- - 2 \text{ antibonding e}^-}{2} = 2$

Thus the C and N atoms in CN^+ are joined by a double bond.

7B   There are 8 valence electrons that must be placed in the molecular orbital diagram for BN (3 electrons from boron and five electrons from nitrogen). Since both B and N precede oxygen and they are not far apart in atomic numbers, we must use the modified molecular-orbital energy-level diagram to get the correct configuration. By following the Aufbau orbital filling method, one obtains the ground-state diagram asked for in the question.

The bond order for the B—N bond in BN is $\dfrac{6 \text{ bonding e}^- - 2 \text{ antibonding e}^-}{2} = 2$

Thus, the B and N atoms in BN are joined by a double bond.

8A   In this exercise we will combine valence-bond and molecular orbital methods to describe the bonding in the SO_3 molecule. By invoking a π-bonding scheme, we can replace the three resonance structures for SO_3 (shown below) with just <u>one</u> structure that exhibits both σ-bonding and delocalized π-bonding.

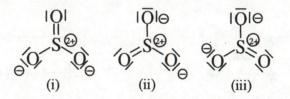

(i)   (ii)   (iii)

We will use structure (i) to develop a combined localized/delocalized bonding description for the molecule. (Note: Anyone of the three resonance contributors can be used as the starting structure). We begin by assuming that every atom in the molecule is sp^2 hybridized to produce the σ-framework for the molecule. The half-filled sp^2 hybrid orbitals of the oxygen atoms will each be overlapped with a half-filled sp^2 hybrid orbital on sulfur. By contrast, to generate a set of π molecular orbitals, we will combine one unhybridized 2p orbital from each of three oxygen atoms with an unhybridized 3p orbital on the sulfur atom. This will generate four π molecular orbitals: a bonding molecular orbital, two nonbonding molecular orbitals and an antibonding orbital. Remember that the number of valence electrons assigned to each atom must reflect the formal charge for that atom. Accordingly, sulfur, with a 2+ formal charge, can have only four valence electrons, the two oxygens with a 1- formal charge must each end up with 7 electrons and the oxygen with a zero formal charge must have its customary six electrons

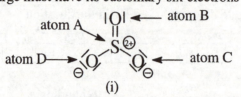

(i)

Let's begin assigning valence electrons by half-filling three sp^2 hybrid orbitals on the sulfur atom (atom A) and one sp^2 hybrid orbital on each of the oxygen atoms (atoms B, C and D) Next, we half-fill the lone unhybridized $3p$ orbital on sulfur and the lone $2p$ orbital on the oxygen atom with a formal charge of zero (atom B). Following this, the $2p$ orbital of the other two oxygen atoms (atoms C and D), are filled and then lone pairs are placed in the sp^2 hybrid orbitals that are still empty. At this stage, then, all 24 valence electrons have been put into atomic and hybrid orbitals on the four atoms. Now we overlap the six half-filled sp^2 hybrid orbitals to generate the σ-bond framework and combine the three $2p$ orbitals (2 filled, one half-filled) and the $3p$ orbital (half-filled) to form the four π-molecular orbitals, as shown below:

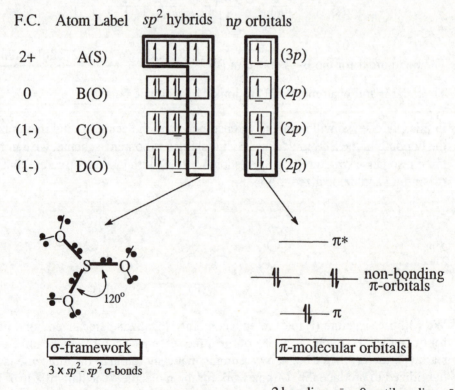

Overall bond order for this set of π-molecular orbitals $\dfrac{2 \text{ bonding e}^- - 0 \text{ antibonding e}^-}{2} = 1$

The π-bond is spread out evenly over the three S—O linkages. This leads to an average bond order of 1.33 for the three S—O bonds in SO_3. By following this "combined approach", we end up with a structure that has the σ-bond framework sandwiched between the delocalized π-molecular orbital framework:

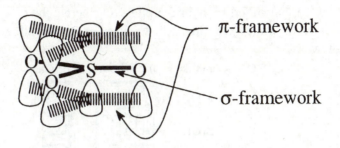

π-framework

σ-framework

This is a much more accurate description of the bonding in SO_3 than that provided by any one of the three Lewis diagrams shown above.

8B   We will use the same basic approach to answer this question as was used to solve practice example 11-8A. This time, the bonding in NO_2^- will be described by combining valence-bond and molecular orbital theory. With this approach we will be able to generate a structure that more accurately describes the bonding in NO_2^- than either of the two equivalent Lewis diagrams that can be drawn for the nitrate ion (below).

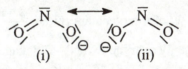

(i)   (ii)

Structure (i) will be used to develop the combined localized/delocalized bonding description for the anion (either structure could have been used as the starting structure). We begin by assuming that every atom in the molecule is sp^2 hybridized. To produce the σ-framework for the molecule, the half-filled sp^2 orbitals of the oxygen atoms will be overlapped with a half-filled sp^2 orbital of nitrogen. By contrast, to generate a set of π-molecular orbitals, we will combine one unhybridized $2p$ orbital from each of the two oxygen atoms with an unhybridized $2p$ orbital on nitrogen. This will generate three π molecular orbitals: a bonding molecular orbital, a non-bonding molecular orbital and an antibonding molecular orbital.

atom A ⟶ N̄
atom B ⟶ O   O ⟵ atom C

(i)

Remember that the number of valence electrons assigned to each atom must reflect the formal charge for that atom. Accordingly, the oxygen with a 1- formal charge (atom C) must end up with 7 electrons, whereas the nitrogen atom (atom A) and the other oxygen atom (atom B) must have their customary 5 and 6 valence electrons, respectively, because they both have a formal charge of zero. Let's begin assigning valence electrons by half-filling two sp^2 hybrid orbitals on the nitrogen atom (atom A) and one sp^2 hybrid orbital on each oxygen atoms (atoms B and C). Next, we half-fill the lone $2p$ orbital on nitrogen (atom A) and the lone $2p$ orbital on the oxygen atom with a formal charge of zero (atom B)

Following this, the $2p$ orbital for the remaining oxygen (atom C) is filled and then lone pairs are placed in the sp^2 hybrid orbitals that are still empty. At this stage, then, all 18 valence electrons have been put into atomic and hybrid orbitals on the three atoms. Now, we overlap the four half-filled sp^2 hybrid orbitals to generate the σ-bond framework and combine the three $2p$ orbitals (two half-filled, one filled) to form three π-molecular orbitals as shown below

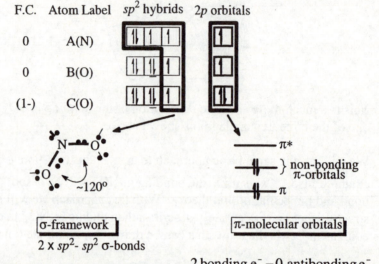

Overall bond order for this set of π-molecular orbitals $\dfrac{2 \text{ bonding e}^- - 0 \text{ antibonding e}^-}{2} = 1$

The π bond is spread out evenly over the two N—O linkages. This leads to an average bond order of 1.5 for each of the two N—O bonds in NO_2^-. By following this combined approach, we end up with a structure that has the σ-bond framework sandwiched between the delocalized π-molecular orbitals:

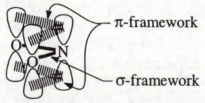

This is a much more accurate depiction of the bonding in NO_2^- than that provided by any one of the two Lewis diagrams given above.

EXERCISES

Valence-Bond Theory

Note: In VSEPR theory the term *"bond pair"* is used for a single bond, a double bond, or a triple bond, even though a single bond consists of one pair of electrons, a double bond two pairs of electrons, and a triple bond three pairs of electrons. To avoid any confusion between the number of electron pairs actually involved in the bonding to a central atom, and the number of atoms bonded to that central atom, we shall occasionally use the term *"ligand"* to indicate an atom or a group of atoms attached to the central atom.

1. There are several ways in which valence bond theory is superior to Lewis structures in describing covalent bonds. *First*, valence bond theory clearly distinguishes between sigma and pi bonds. In Lewis theory, a double bond appears to be just two bonds and it is not clear why a double bond is not simply twice as strong as a single bond. In valence bond theory, it is clear that a sigma bond must be stronger than a pi bond, for the orbitals overlap more effectively in a sigma bond (end-to-end) than they do in a pi bond (side-to-side). *Second*, molecular geometries are more directly obtained in valence bond theory than in Lewis theory. Although valence bond theory requires the introduction of hybridization to explain these geometries, Lewis theory does not predict geometries at all; it simply provides the basis from which VSEPR theory predicts geometries. *Third*, Lewis theory does not explain hindered rotation about double bonds. With valence bond theory, any rotation about a double bond involves cleavage of the π-bond, which would require the input of considerable energy.

2. The overlap of pure atomic orbitals gives bond angles of 90° or 180°. These bond angles are suitable only for 3- and 4-atom compounds in which the central atom is an atom of the third (or higher) period of the periodic table. For central atoms from the second period of the periodic table, 3- and 4-atom compounds have bond angles closer to180°, 120°, and 109.5° than to90°. These other bond angles can only be explained well through hybridization. *Second*, hybridization clearly distinguishes between the (hybrid) orbitals that form σ bonds and the *p* orbitals that form π bonds. It places those *p* orbitals in their proper orientation so that they can overlap side-to-side to form π bonds. *Third*, the bond angles of 90° and 120° that result when the octet of the central atom is expanded cannot be produced with pure atomic orbitals. Hybrid orbitals are necessary. *Finally*, the overlap of pure atomic orbitals does not usually result in all σ bonds being equivalent. Overlaps with hybrid orbitals produce equivalent bonds.

3. **(a)** Lewis theory does not describe the shape of the water molecule. It does indicate that there is a single bond between each H atom and the O atom, and that there are two lone pairs attached to the O atom, but it says nothing about molecular shape.

(b) In valence-bond theory using simple atomic orbitals, each H—O bond results from the overlap of a $1s$ orbital on H with a $2p$ orbital on O. The angle between $2p$ orbitals is 90° this method initially predicts a 90˚ bond angle. The observed 104˚ bond angle is explained as arising from repulsion between the two slightly positively charged H atoms.

(c) In VSEPR theory the H_2O molecule is categorized as being of the AX_2E_2 type, with two atoms and two lone pairs attached to the central oxygen atom. The lone pairs repel each other more than do the bond pairs, explaining the smaller than 109.5° tetrahedral bond angle.

(d) In valence-bond theory using hybrid orbitals, each H—O bond results from the overlap of a $1s$ orbital on H with a sp^3 orbital on O. The angle between sp^3 orbitals is 109.5°. The observed bond angle of 104° is rationalized based on the greater repulsion of lone pair electrons when compared to bonding pair electrons.

4. **(a)** Lewis theory really does not describe the shape of the molecule. It does indicate that there is a single bond between each Cl atom and the C atom, but it says nothing about molecular shape.

(b) In valence bond theory using simple atomic orbitals, each C—Cl bond results from the overlap of a $3p$ orbital on Cl with a $2p$ orbital on C. Since the angle between 2p orbitals is 90°, this method initially predicts a 90° bond angle. The observed 109.5° bond angle is explained as resulting from the repulsion between the two slightly negative Cl atoms. In addition, the molecule is predicted to have the formula CCl_2, since there are just two half-filled orbitals in the ground state of C.

(c) In VSEPR theory the CCl_4 molecule is categorized as being of the AX_4 type, with four atoms tetrahedrally attached to the central carbon atom.

(d) In valence bond theory, each C—Cl bond results from the overlap of a $3p$ orbital on Cl with an sp^3 orbital on C. The angle between the sp^3 orbitals is 109.5°.

5. Determining hybridization is made easier if we begin with Lewis structures. Only one resonance form is drawn for CO_3^{2-}, SO_2 and NO_2^-.

$$\left[\overline{|O}-C=\overline{O} \atop |\underline{O}| \right]^{2-} \quad |\overline{O}-\overline{S}=\overline{O} \quad |\overline{C}l-\overset{\displaystyle |\overline{C}l|}{\underset{\displaystyle |\underline{C}l|}{C}}-\overline{C}l| \quad |C\equiv O| \quad [\overline{|O}-\overline{N}=\overline{O}|]^-$$

The C atom is attached to three ligands and no lone pairs and thus is sp^2 hybridized in CO_3^{2-}. The S atom is attached to two ligands and one lone pair and thus is sp^2 hybridized in SO_2. The C atom is attached to four ligands and no lone pairs and thus is sp^3 hybridized in CCl_4. Both the oxygen and the carbon in Co are sp hybridized. The N atom is attached to two ligands and one lone pair in NO_2^- and thus is sp^2 hybridized. Thus, the central atom is sp^2 hybridized in SO_2, CO_3^{2-}, and NO_2^-.

6. **(a)** HI: H $_{1s}$ [↑]   I [Kr] $4d^{10}$ $_{5s}$[↑↓] $_{5p}$[↑↓][↑↓][↑]
The $1s$ orbital of H overlaps the half-filled $5p$ orbital of I to produce a linear molecule.

(b) BrCl: Br [Ar] $3d^{10}$ $_{4s}$[↑↓] $_{4p}$[↑↓][↑↓][↑]   Cl [Ne] $_{3s}$[↑↓] $_{3p}$[↑↓][↑↓][↑]
The half-filled $3p$ orbital of Cl overlaps the half-filled $4p$ orbital of Br to produce a linear molecule.

(c) H$_2$Se: H $_{1s}$[↑]   Se [Ar] $3d^{10}$ $_{4s}$[↑↓] $_{4p}$[↑↓][↑][↑]
Each half-filled $4p$ orbital of Se overlaps with a half-filled $1s$ orbital of a H atom to produce a bent molecule, with a bond angle of approximately $90°$.

(d) OCl$_2$: O $_{1s}$[↑↓] $_{2s}$[↑↓] $_{2p}$[↑↓][↑][↑]   Cl [Ne] $_{3s}$[↑↓] $_{3p}$[↑↓][↑↓][↑]
Each half-filled $2p$ orbital of O overlaps with a half-filled $3p$ orbital of a Cl atom to produce a bent molecule, with a bond angle of approximately $90°$.

7. For each species, we first draw the Lewis structure, to help explain the bonding.

(a) In CO$_2$, there are a total of $4+(2\times6)=16$ valence electrons, or 8 pairs. C is the central atom. The Lewis structure is $\overline{\underline{O}}{=}C{=}\overline{\underline{O}}$ The molecule is linear and C is sp hybridized.

(b) In HONO$_2$, there are a total of $1+5+(3\times6)=24$ valence electrons, or 12 pairs. N is the central atom, and a plausible Lewis structure is shown on the right The molecule is trigonal planar around N which is sp^2 hybridized. The O in the H—O—N portion of the molecule is sp^3 hybridized.

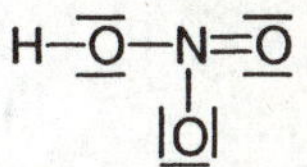

H—$\overline{\underline{O}}$—N$=\overline{\underline{O}}$
|
$\underline{|\overline{O}|}$

(c) In ClO$_3^-$, there are a total of $7+(3\times6)+1=26$ valence electrons, or 13 pairs. Cl is the central atom, and a plausible Lewis structure is shown on the right. The electron-group geometry around Cl is tetrahedral, indicating that Cl is sp^3 hybridized.

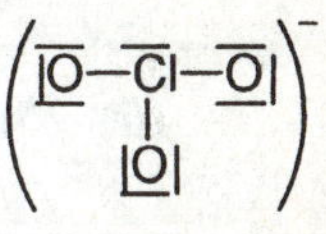

$\left(\overline{\underline{|O}}{-}\overline{\underline{Cl}}{-}\overline{\underline{O|}} \atop |\overline{O}|\right)^-$

(d) In BF$_4^-$, there are a total of $3+(4\times7)+1=32$ valence electrons, or 16 pairs. B is the central atom, and a plausible Lewis structure is shown on the right. The electron-group geometry is tetrahedral, indicating that B is sp^3 hybridized.

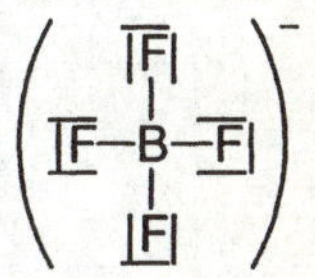

$\left(|\overline{\underline{F}}| \atop |\overline{\underline{F}}|{-}B{-}\overline{\underline{F}}| \atop |\overline{\underline{F}}|\right)^-$

8. In NSF there are $5+6+7=18$ valence electrons, or 9 electron pairs. In the following Lewis structure, $|\overline{N}{=}\overline{S}{-}\overline{\underline{F}}|$ there are two atoms bonded to the central S, which has a lone pair. Thus, the electron group geometry around the central S is trigonal planar (with sp^2 hybridization) and the molecular shape is bent.

9. The Lewis structure of ClF$_3$ is shown on the right. There are three atoms and two lone pairs attached to the central atom, its hybridization is sp^3d, which is achieved as follows. Cl$_{unhyb}$[Ne] $_{3s}$[↑↓] $_{3p}$[↑↓][↑↓][↑] $_{3d}$[ ][ ][ ][ ][ ] $\longrightarrow$ Cl$_{hyb}$ [Ne] $_{dsp^3}$[↑↓][↑↓][↑][↑][↑] $_{3d}$[ ][ ][ ][ ]

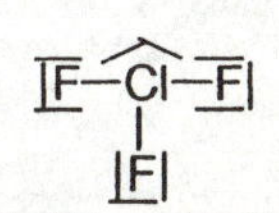

$|\overline{F}{-}\overset{\frown}{Cl}{-}\overline{F}|$
|
$|\overline{F}|$

The orbital diagram of F is $[He]_{2s}$ $_{2p}$ ▦ Each of the three sigma bonds are formed by the overlap of a $2p$ orbital on F with one of the half-filled dsp^3 orbitals on Cl. Since the dsp^3 hybridization has a trigonal bipyramidal shape, and the two lone pairs occupy the equatorial positions in the molecule, ClF_3 is T-shaped.

10. In SF_4 there are $6+(4\times7)=34$ valence electrons, or 17 pairs. A plausible Lewis structure has zero formal charge on each atom (structure drawn on right). This molecule has a see-saw shape. There are four atoms and one lone pair attached to S, requiring sp^3d hybridization. Each sigma $S-F$ bond results from the overlap of two half-filled orbitals: namely, an sp^3d on S and $2p$ on F. The orbital diagram of F is $[He]_{2s}$ ▦ $_{2p}$ ▦ . The hybridization diagram for S follows:

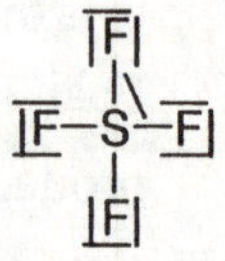

    $S_{unhyb} [Ne]_{3s}$ ▦ $_{3p}$ ▦ $_{3d}$ ☐☐☐☐☐ ⟶ $S_{hyb} [Ne]$ $_{dsp^3}$ ▦ $_{3d}$ ☐☐☐☐

11. (a) In PF_6^- there are a total of $5+(6\times7)+1=48$ valence electrons, or 24 pairs. A plausible Lewis structure is shown below. In order to form the six P—F bonds, the hybridization on P must be sp^3d^2.

(b) In COS there are a total of $4+6+6=16$ valence electrons, or 8 pairs. A plausible Lewis structure is shown below. In order to bond two atoms to the central C, the hybridization on that C atom must be sp.

(c) In $SiCl_4$, there are a total of $4+(4\times7)=32$ valence electrons, or 16 pairs. A plausible Lewis structure is given below. In order to form four Si—Cl bonds, the hybridization on Si must be sp^3.

(d) In NO_3^-, there are a total of $5+(3\times6)+1=24$ valence electrons, or 12 pairs. A plausible Lewis structure is given below. In order to bond three O atoms to the central N atom, with no lone pairs on that N atom, its hybridization must be sp^2.

(e) In AsF_5, there are a total of $5+(5\times7)=40$ valence electrons, or 20 pairs. A plausible Lewis structure is shown below. This is a molecule of the type AX_5. To form five As—F bonds, the hybridization on As must be sp^3d.

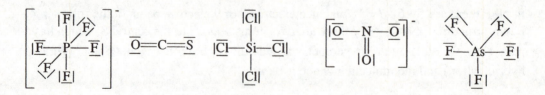

12. We base each hybridization scheme on the Lewis structure for the molecule.

(a) H—C≡N| There are two atoms bonded to C and no lone pairs on C. Thus, the hybridization for C must be *sp*.

(b) There are four atoms bonded to C and no lone pairs in this molecule. Thus, the hybridization for C must be sp^3. (O atom is also sp^3 hybridized).

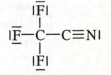

(c) There are four atoms bonded to each terminal C and no lone pairs on these C atoms. Thus, the hybridization for each terminal C must be sp^3. There are three atoms and no lone pairs bonded to the central C. The hybridization for the central C must be sp^2.

(d) There are three atoms bonded to C and no lone pairs on C. This requires sp^2 hybridization.

13. (a) This is a planar molecule. The hybridization on C is sp^2 (one bond to each of the three attached atoms).

(b) |N≡C—C≡N| is a linear molecule. The hybridization for each C is *sp* (one bond to each of the two ligands).

(c) Trifluoroacetonitrile is neither linear nor planar. The shape around the left-hand C is tetrahedral and that C has sp^3 hybridization. The shape around the right-hand carbon is linear and that C has *sp* hybridization.(N atom is *sp* hybridized).

(d) $\left[|\overline{S}—C≡N| \right]^-$ is a linear molecule. The hybridization for C is *sp*.

14. (a) The central C atom employs four sp^3 hybrid orbitals; these have a tetrahedral geometry. Each of the two H atoms employs a 1*s* orbital, and each of the two Cl atoms employs a half-filled 3*p* orbital. A diagram of this follows:

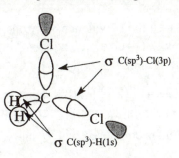

(b) The ground state C atom has the electron configuration [He] $_{2s}$⊞ $_{2p}$⊞ Based on this ground state electron configuration, we would expect a bent molecule. To produce the two half-filled orbitals required to form two linear bonds a hybridized electron configuration (which corresponds to an excited state for the individual atoms) is required: C^* $_{1s}$⊞ $_{2s}$⊡ $_{2p}$⊞ ⟶ C^*_{hyb} $_{1s}$⊞ $_{sp}$⊞ $_{2p}$⊞

The overlap of the two half-filled sp orbitals of C, one with a half-filled $2p$ orbital on O and the other with a half-filled $2p$ orbital on N^-, produces the σ bond framework. O $_{1s}\boxed{\uparrow\downarrow}$ $_{2s}\boxed{\uparrow\downarrow}$ $_{2p}\boxed{\uparrow\downarrow|\uparrow|\uparrow}$ N^- $_{1s}\boxed{\uparrow\downarrow}$ $_{2s}\boxed{\uparrow\downarrow}$ $_{2p}\boxed{\uparrow\downarrow|\uparrow|\uparrow}$

The π bonds are produced by the sidewise overlap of the two half-filled $2p$ orbitals on C, one with a half-filled $2p$ orbital on O, the other with a half-filled $2p$ orbital on N^-. A diagram of the bonding follows:

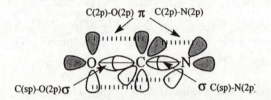

(c) The ground-state electron configuration of B suggests an ability to form one B—F bond, rather than three. $_{1s}\boxed{\uparrow\downarrow}$ $_{2s}\boxed{\uparrow\downarrow}$ $_{2p}\boxed{\uparrow||}$ Again, excitation, or promotion of an electron to a higher energy orbital, followed by hybridization is required to form three equivalent half-filled B orbitals.

B^* $_{1s}\boxed{\uparrow\downarrow}$ $_{2s}\boxed{\uparrow}$ $_{2p}\boxed{\uparrow|\uparrow|}$ $\longrightarrow$ B^*_{hyb} $_{1s}\boxed{\uparrow\downarrow}$ $_{sp^2}\boxed{\uparrow|\uparrow|\uparrow}$ $_{2p}\boxed{}$

Recall that the ground-state electron configuration of F is [He] $_{2s}\boxed{\uparrow\downarrow}$ $_{2p}\boxed{\uparrow\downarrow|\uparrow\downarrow|\uparrow}$. Bonds are produced through overlap of the half-filled sp^2 orbitals on B with the half filled F $2p$ orbitals. The BF_3 molecule is trigonal planar, with F—B—F bond angles of $120°$ A diagram of the bonding follows:

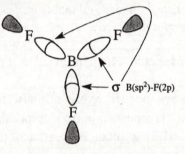

15. (a) In HCN, there are a total of $1+4+5 = 10$ valence electrons, or 5 electron pairs. A plausible Lewis structure follows. H—C≡N|. The H—C bond is a σ bond, and the C≡N bond is composed of 1 σ and 2 π bonds.

(b) In C_2N_2, there are a total of $(2\times4)+(2\times5) = 18$ valence electrons, or 9 electron pairs. A plausible Lewis structure follows: |N≡C—C≡N| The C—C bond is a σ bond, and each C≡N bond is composed of 1 σ and 2 π bonds.

(c) In $CH_3CHCHCCl_3$, there are a total of 42 valence electrons, or 21 electron pairs. A plausible Lewis structure is shown to the right. All bonds are σ bonds except one of the bonds that comprise the C=C bond. The C=C bond is composed of one σ and one π bond.

$$\begin{array}{ccccc} & & & H & & & |\overline{C}l| \\ & & & | & & & | \\ H{-}&C&{-}C{=}C&{-}&C&{-}\overline{C}l| \\ & | & | \quad | & & | \\ & H & H \quad H & & |\underline{C}l| \end{array}$$

(d) In HONO, there is a total of $1+5+(2\times6)=18$ valence electrons, or 9 electron pairs. A plausible Lewis structure is $H-\overline{\underline{O}}-\overline{N}=\overline{\underline{O}}$ All single bonds in this structure are σ bonds. The double bond is composed of one σ and one π bond.

16. (a) In CO_2, there are $4+(2\times6)=16$ valence electrons, or 8 electron pairs. A plausible Lewis structure is $|\underline{O}=C=\underline{O}|$

(b) The molecule is linear and the hybridization on the C atom is sp. Each $C{=}O$ bond is composed of a σ bond (end-to-end overlap of a half-filled sp orbital on C with a half-filled $2p$ orbital on O) and a π bond (overlap of a half-filled $2p$ orbital on O with a half-filled $2p$ orbital on C).

17. **(a)** In CCl_4, there is a total of $4+(4\times7)=32$ valence electrons, or 8 electron pairs. C is the central atom. A plausible Lewis structure is shown to the right. The geometry at C is tetrahedral; C is sp^3 hybridized. Cl—C—Cl bond angles are $109.5°$. Each C—Cl bond is represented by σ: $C(sp^3)^1 - Cl(3p)^1$

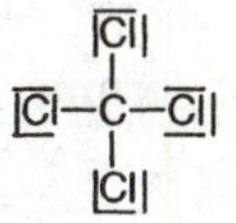

$$\overline{|C|}$$
$$\overline{|C|}-\underset{|}{\overset{|}{C}}-\overline{Cl|}$$
$$\overline{|C|}$$

(b) In ONCl, there is a total of $6+5+7=18$ valence electrons, or 9 electron pairs. N is the central atom. A plausible Lewis structure is $\overline{O}=\overline{N}-\overline{Cl}|$ The e$^-$ group geometry around N is triangular planar, and N is sp^2 hybridized. The O—N—C bond angle is about$120°$. The bonds are: σ: $O(2p_y)^1-N(sp^2)^1$ σ: $N(sp^2)^1-Cl(3p_z)^1$ π: $O(2p_z)^1-N(2p_z)^1$

(c) In HONO, there are a total of $1+(2\times6)+5=18$ valence electrons, or 9 electron pairs. A plausible Lewis structure is $H-\overline{\underline{O}}_a-\overline{N}=\overline{\underline{O}}_b$. The geometry of O_a is tetrahedral, O_a is sp^3 hybridized, the $H-O_a-N$ bond angle is (at least close to)$109.5°$. The e$^-$ group geometry at N is trigonal, N is sp^2 hybridized, the O_a-N-O_b bond angle is$120°$ The four bonds are represented as follows. σ: $H(1s)^1-O_a(sp^3)^1$
σ: $O_a(sp^3)^1-N(sp^2)^1$ σ: $N(sp^2)^1-O_b(2p_y)^1$ π: $N(2p_z)^1-O_b(2p_z)^1$.

(d) In $COCl_2$, there are a total of $4+6+(2\times7)=24$ valence electrons, or 12 electron pairs. A plausible Lewis structure is shown to the right. The e$^-$ group geometry around C is trigonal planar; all bond angles around C are$120°$, and the hybridization of C is sp^2. The four bonds in the molecule: $2\times\sigma$: $Cl(3p_x)^1-C(sp^2)^1$ σ: $O(2p_y)^1-C(sp^2)^1$ π: $O(2p_z)^1-C(2p_z)^1$.

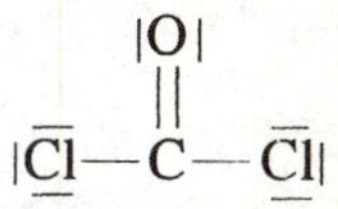

$$|O|$$
$$||$$
$$\overline{|Cl}-C-\overline{Cl|}$$

18. (a) In NO_2^-, there are a total of $5+(2\times6)+1=18$ valence electrons, or 9 electron pairs. A plausible Lewis structure follows $\overline{O}_a\!=\!N\!-\!\overline{O}_b|$ The e^- group geometry around N is trigonal planar, all bond angles around that atom are $120°$, and the hybridization of N is sp^2. The three bonds in this molecule are:

$$\sigma: O_a(2p_y)^1 -N(sp^2)^1 \quad \sigma: O_b(2p_x)^1 -N(sp^2)^1 \text{ and } \pi: O_a(2p_z)^1 -N(2p_z)^1$$

(b) In I_3^-, there are a total of $(3\times7)+1=22$ valence electrons, or 11 electron pairs. A plausible Lewis structure follows. $|\overline{\underline{I}}\!-\!\overset{\curvearrowright}{\underline{I}}\!-\!\overline{\underline{I}}|^-$ Since there are three lone pairs and two ligands attached to the central I, the electron-group geometry around that atom is trigonal bipyramidal; its hybridization is sp^3d. Since the two ligand I atoms are located in the axial positions of the trigonal bipyramid, the $I\!-\!I\!-\!I$ bond angle is $180°$. Each $I\!-\!I$ sigma bond is the result of the overlap of a $5p$ orbital on a terminal I with an sp^3d hybrid orbital on the central I.

(c) In $C_2O_4^{2-}$, there are a total of $(2\times4)+(4\times6)+2=34$ valence electrons, or 17 electron pairs. A plausible Lewis structure is shown on the right. As we see, the ion is symmetrical. There are three atoms or groups of atoms attached to each C, the bond angles around each C are $120°$, and the hybridization on each C is sp^2. The bonding to one of these carbons (the right-hand one) is represented as follows.

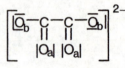

$$\sigma: C(sp^2)^1 -C(sp^2)^1 \quad \sigma: C(sp^2)^1 -O_a(2p_z)^1 \quad \sigma: C(sp^2)^1 -O_b(2p_x)^1 \quad \pi: C(2p_y)^1 -O_a(2p_y)^1$$

(d) In HCO_3^-, there are a total of $1+4+(3\times6)+1=24$ valence electrons, or 12 electron pairs. A plausible Lewis structure is shown on the right. The e^- group geometry around C is trigonal planar, with a bond angle of $120°$; the hybridization is sp^2. The geometry around O_a is bent with a bond angle less than $109.5°$; the hybridization is sp^3. The five bonds are represented as follows:

$$\sigma: O_a(sp^3)^1 -H(1s)^1, \quad \sigma: C(sp^2)^1 -O_a(sp^3)^1, \quad \sigma: C(sp^2)^1 -O_b(2p)^1,$$

$$\sigma: C(sp^2)^1 -O_c(2p_x)^1, \quad \pi: C(2p_z)^1 -O_c(2p_z)^1$$

19. Citric acid has the molecular structure (shown below) Using Figure 11-17 as a guide, the flowing hybridization and bonding scheme is obtained for citric acid:

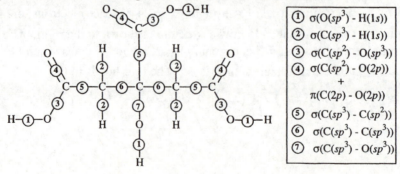

20. All interior atoms are sp^3 hybridized except C_a and C_d, which are sp^2 hybridized. The overlaps in the sixteen bonds of the molecule are represented as follows.

$$\begin{array}{c} |O_b| \quad H \quad H \quad |O_d| \\ \| \qquad | \qquad | \qquad \| \\ H-\overline{O}_a-C_a-C_b-C_c-C_d-\overline{O}_e-H \\ | \qquad | \\ H \quad |O_c-H \end{array}$$

$\sigma : O_e\left(sp^3\right)^1 - H\left(1s\right)^1 \quad \sigma : C_d\left(sp^2\right)^1 - O_e\left(sp^3\right)^1 \quad \sigma : C_c\left(sp^2\right)^1 - O_b\left(2p_z\right)^1 \quad \pi : C_d - \left(2p_z\right)^1 - O_d\left(2p_z\right)^1.$

$\sigma : O_a\left(sp^3\right)^1 - H\left(1s\right)^1 \quad \sigma : C_a\left(sp^2\right)^1 - O_a\left(sp^3\right)^1 \quad \sigma : C_d\left(sp^2\right)^1 - O_d\left(2p_x\right)^1 \quad \pi : C_a - \left(2p_z\right)^1 - O_b\left(2p_z\right)^1$

$\sigma : C_b\left(sp^3\right)^1 - H\left(1s\right)^1 \quad \sigma : C_b\left(sp^3\right)^1 - H\left(1s\right)^1 \quad \sigma : C_a\left(sp^2\right)^1 - C_b\left(sp^3\right)^1 \quad \sigma : C_b\left(sp^3\right)^1 - C_c\left(sp^3\right)^1$

$\sigma : C_c\left(sp^3\right)^1 - H\left(1s\right)^1 \quad \sigma : C_c\left(sp^3\right)^1 - O_c\left(sp^3\right)^1 \quad \sigma : C_c\left(sp^3\right)^1 - C_d\left(sp^2\right)^1 \quad \sigma : O_c\left(sp^3\right)^1 - H\left(1s\right)^1$

21. (a) Br atom is sp^3d hybridized
F atoms are unhybridized

— $3 \times \sigma(\text{Br}(sp^3d)\text{-F}(2p))$

$$:\overset{\cdot\cdot}{F}:$$
$$:\overset{\cdot\cdot}{F}-\overset{\cdot\cdot}{\underset{\cdot}{Br}}-\overset{\cdot\cdot}{F}:$$

(b) Central S is sp^2 hybridized and the terrminal O and S atoms are unhybridized

— $\sigma(\text{S}(sp^2)\text{-S}(3p_x))$
— $\pi(\text{S}(3p_z)\text{-S}(3p_z))$
— $\sigma(\text{S}(sp^2)\text{-O}(2p_z))$

22. Model A is XeF_2
VSEPR class: AX_2E_3
Therefore XeF_2 is a linear molecule
The central Xe atom in XeF_2 is sp^3d or dsp^3 hybridized

Lewis Diagram:

$$|\overline{F}-\overline{Xe}-\overline{F}|$$

$2 \times \sigma(\text{Xe}(dsp^3)\text{-F}(2p))$

Model B is IF_6^+
VSEPR class: AX_6
Therefore IF_6^+ is an octahedral molecule
The central I atom in IF_6^+ is sp^3d^2 or d^2sp^3 hybridized

Lewis Diagram:
all I-F bonds in IF_6 are

$\sigma(\text{I}(d^2sp^3)\text{-F}(sp^3))$

23. The bond lengths are consistent with the left-hand C—C bond being a triple bond (120 pm), the other C—C bond being a single bond (154 pm) rather than a double bond (134 pm), and the C—O bond being a double bond (123 pm). Of course, the two C—H bonds are single bonds (110 pm). All of this is depicted in the Lewis structure on the right. The overlap that produces each bond follows:

$$H_a-C_a\equiv C_b-C_c=\overline{\underline{O}}$$
$$\qquad\qquad\qquad |$$
$$\qquad\qquad\qquad H_b$$

$\sigma : C_a\left(sp\right)^1 - H_a\left(1s\right)^1 \qquad \sigma : C_a\left(sp\right)^1 - C_b\left(sp\right)^1 \qquad \sigma : C_b\left(sp\right)^1 - C_c\left(sp^2\right)^1$

$\sigma : C_c\left(sp^2\right)^1 - H_b\left(1s\right)^1 \qquad \sigma : C_c\left(sp^2\right)^1 - O\left(2p_y\right)^1 \qquad \pi : C_a\left(2p_y\right)^1 - C_b\left(2p_y\right)^1$

$\pi : C_a\left(2p_z\right)^1 - C_b\left(2p_z\right)^1 \qquad \pi : C_c\left(2p_z\right)^1 - O\left(2p_z\right)^1$

24. All of the valence electron pairs in the molecule are indicated in the sketch for the problem, there are no lone pairs. The two terminal carbon atoms have three atoms attached to them; they are sp^2 hybridized. This means there is one half-filled $2p$ orbital available on each of these terminal carbon atoms that can be used for the formation of a π bond. The central carbon atom has two atoms attached to it; it is sp hybridized. Thus there are two (perpendicular) half-filled $2p$ orbitals (such as $2p_y$ and $2p_z$) on this central carbon atom available for the formation of π bonds. One π bond is formed from the central carbon atom to each of the terminal carbon atoms. Because the two $2p$ orbitals on the central carbon atom are perpendicular to each other, these two π bonds are also mutually orthogonal. Thus, the two H—C—H planes are at $90°$ from each other, because of the π bonding in the molecule.

Molecular-Orbital Theory

25. The valence-bond method describes a covalent bond as the result of the overlap of atomic orbitals. The more complete the overlap, the stronger the bond. Molecular orbital theory describes a bond as a region of space in a molecule where there is a good chance of finding the bonding electrons. The molecular orbital bond does not have to be created from atomic orbitals (although it often is) and the orientations of atomic orbitals do not have to be manipulated to obtain the correct geometric shape. There is little concept of the relative energies of bonding in valence bond theory. In molecular orbital theory, bonds are ordered energetically. These energy orderings, in fact, provide a means of checking the predictions of the theory through the spectroscopic analysis of the molecules.

26. C_2 has $(2 \times 4) = 8$ valence electrons. A plausible Lewis structure, that obeys the octet rule and has zero formal charge for each C atom, incorporates a quadruple bond C::::C or C≡C. Molecular orbital theory, on the other hand, distributes those same 8 electrons as shown below, with a resulting bond order of $(6-2) \div 2 = 2$, (i.e., a double bond).

$$\sigma_{2s} \quad \sigma_{2s}* \quad \pi_{2p} \quad \sigma_{2p} \quad \pi_{2p} \quad \sigma_{2p}$$
$$C_2 \ KK \ \boxed{\uparrow\downarrow} \ \boxed{\uparrow\downarrow} \ \boxed{\uparrow\downarrow}\boxed{\uparrow\downarrow} \ \boxed{\ } \ \boxed{\ }\boxed{\ } \ \boxed{\ }$$

27. The two molecular orbital diagrams follow:

$$\sigma_{2s} \quad \sigma_{2s}* \quad \pi_{2p} \quad \sigma_{2p}^* \quad \pi_{2p} \quad \sigma_{2p}$$

N_2^- no. valence $e^- = (2 \times 5) + 1 = 11$ KK $\boxed{\uparrow\downarrow} \ \boxed{\uparrow\downarrow} \ \boxed{\uparrow\downarrow}\boxed{\uparrow\downarrow} \ \boxed{\uparrow\downarrow} \ \boxed{\uparrow}\boxed{\ } \ \boxed{\ }$

N_2^{2-} no. valence $e^- = (2 \times 5) + 2 = 12$ KK $\boxed{\uparrow\downarrow} \ \boxed{\uparrow\downarrow} \ \boxed{\uparrow\downarrow}\boxed{\uparrow\downarrow} \ \boxed{\uparrow\downarrow} \ \boxed{\uparrow}\boxed{\uparrow} \ \boxed{\ }$

N_2^- bond order $= (8-3) \div 2 = 2.5$ (stable) N_2^{2-} bond order $= (8-4) \div 2 = 2$ (stable)

28. B_2 has $(2 \times 5 =)10$ electrons. We compare the number of unpaired electrons predicted if the order is π_{2p}^b before σ_{2p}^b with the number predicted if the σ_{2p}^b orbital is before the π_{2p}^b orbital. (^{b}indicates a bonding orbital)

$$B_2 \quad \pi_{2p}^b \text{ before } \sigma_{2p}^b \quad \overset{\sigma_{1s}^b}{\boxed{\uparrow\downarrow}} \; \overset{\sigma_{1s}^*}{\boxed{\uparrow\downarrow}} \; \overset{\sigma_{2s}^b}{\boxed{\uparrow\downarrow}} \; \overset{\sigma_{2s}^*}{\boxed{\uparrow\downarrow}} \; \overset{\pi_{2p}^b}{\boxed{\uparrow\uparrow}} \; \overset{\sigma_{2p}^b}{\boxed{\,}} \; \overset{\pi_{2p}^*}{\boxed{\;\;}} \; \overset{\sigma_{2p}^*}{\boxed{\,}}$$

$$B_2 \quad \sigma_{2p}^b \text{ before } \pi_{2p}^b \quad \overset{\sigma_{1s}^b}{\boxed{\uparrow\downarrow}} \; \overset{\sigma_{1s}^*}{\boxed{\uparrow\downarrow}} \; \overset{\sigma_{2s}^b}{\boxed{\uparrow\downarrow}} \; \overset{\sigma_{2s}^*}{\boxed{\uparrow\downarrow}} \; \overset{\pi_{2p}^b}{\boxed{\uparrow\downarrow}} \; \overset{\sigma_{2p}^b}{\boxed{\;\;}} \; \overset{\pi_{2p}^*}{\boxed{\;\;}} \; \overset{\sigma_{2p}^*}{\boxed{\,}}$$

The order of orbitals with π_{2p}^b below σ_{2p}^b in energy results in B_2 being paramagnetic, with two unpaired electrons, as observed experimentally. However, if σ_{2p}^b comes before the π_{2p}^b orbital in energy, a diamagnetic B_2 molecule is predicted, which is in contradiction to experimental evidence.

29. In order to have a bond order higher than three, there would have to be a region in a molecular orbital diagram where four bonding orbitals occur together in order of increasing energy, with no intervening antibonding orbitals. No such region exists in any of the molecular orbital diagrams in Figure 11-25. Alternatively, three bonding orbitals would have to occur together energetically, following an electron configuration which, when full, results in a bond order greater than zero. This arrangement does not occur in any of the molecular orbital diagrams in Figure 11-25.

30. The statement is not true, for there are many instances when the bond is strengthened when a diatomic molecule loses an electron. Specifically, whenever the electron that is lost is an antibonding electron (as is the case for $F_2 \rightarrow F_2^+$ and $O_2 \rightarrow O_2^+$), the bond will actually be stronger in the resulting cation than it was in the starting neutral molecule.

31. **(a)** A σ_{1s} orbital must be lower in energy than a σ_{1s}^* orbital. The bonding orbital always is lower than the antibonding orbital if they are derived from the same atomic orbitals.

(b) The σ_{2s} orbital is derived from the 2s atomic orbitals while the σ_{2p} molecular orbitals are constructed from 2p atomic orbitals. Since the 2s orbitals are lower in energy than the 2p orbitals, we would expect that the σ_{2p} molecular orbital would be higher in energy than the σ_{2s} molecular orbital.

(c) A σ_{1s}^* orbital should be lower than a σ_{2s} orbital, since the $1s$ atomic orbital is considerably lower in energy than is the $2s$ orbital.

(d) A σ_{2p} orbital should be lower in energy than a σ_{2p}^* orbital. Both orbitals are from atomic orbitals in the same subshell but we expect a bonding orbital to be more stable than an antibonding orbital.

32. **(a)** In each case, the number of valence electrons in the species is determined first; this is followed by the molecular orbital diagram for each species.

C_2^+ no. valence $e^- = (2 \times 4) - 1 = 7$

KK σ_{2s} σ_{2s}^* π_{2p} σ_{2p} π_{2p}^* σ_{2p}^*

O_2^- no. valence $e^- = (2 \times 6) + 1 = 13$

F_2^+ no. valence $e^- = (2 \times 7) - 1 = 13$

NO^+ no. valence $e^- = 5 + 6 - 1 = 10$

(b) Bond order = (no. bonding electrons – no. antibonding electrons) $\div$ 2

C_2^+ bond order $= (5 - 2) \div 2 = 1.5$ This species is stable.

O_2^- bond order $= (8 - 5) \div 2 = 1.5$ This species is stable.

F_2^+ bond order $= (8 - 5) \div 2 = 1.5$ This species is stable.

NO^+ bond order $= (8 - 2) \div 2 = 3.0$ This species is stable.

(c) C_2^+ has an odd number of electrons and is paramagnetic, with one unpaired electron.
O_2^- has an odd number of electrons and is paramagnetic, with one unpaired electron.
F_2^+ has an odd number of electrons and is paramagnetic, with one unpaired electron.
NO^+ has an even number of electrons and is diamagnetic.

33. (^{b}indicates a bonding orbital)

(a) NO $\quad 5 + 6 = 11 \quad$ valence electrons KK

(b) $NO^+ \quad 5 + 6 - 1 = 10 \quad$ valence electrons KK

(c) CO $\quad 4 + 6 = 10 \quad$ valence electrons KK

(d) CN $\quad 4 + 5 = 9 \quad$ valence electrons KK

(e) $CN^- \quad 4 + 5 + 1 = 10 \quad$ valence electrons KK

(f) $CN^+ \quad 4 + 5 - 1 = 8 \quad$ valence electrons KK

(g) BN $\quad 3 + 5 = 8 \quad$ valence electrons KK

34. In Exercise 33, the species that are isoelectronic are those with 8 electrons: CN^+, BN; with 10 electrons: NO^+, CO, CN^-

35. We first produce the molecular orbital diagram for each species.
(a) (^{b}indicates a bonding orbital)

$NO^+ \quad 5 + 6 - 1 = 10 \quad$ valence electrons KK

$N_2^+ \quad 5 + 5 - 1 = 9 \quad$ valence electrons KK

Bond order = (no. bonding electrons – no. antibonding electrons) $\div$ 2

For NO^+ bond order $= (8 - 2) \div 2 = 3$; for N_2^+ bond order $= (7 - 2) \div 2 = 2.5$

(b) NO^+ is diamagnetic; it has no unpaired electrons. N_2^+ is paramagnetic; it has one unpaired electron.

(c) The molecule with the lower bond order will also have the greater bond length. Thus, N_2^+ should have the greater bond length.

36. We first produce the molecular orbital diagram for each species.

(a) (^{b}indicates a bonding orbital)

CO^+ $4+6-1=9$ valence electrons KK

$$\sigma_{2s}^b \quad \sigma_{2s}^* \quad \sigma_{2p}^b \quad \pi_{2p}^b \quad \pi_{2p}^* \quad \sigma_{2p}^*$$

CN^- $4+5+1=10$ valence electrons KK

$$\sigma_{2s}^b \quad \sigma_{2s}^* \quad \pi_{2p}^b \quad \sigma_{2p}^b \quad \pi_{2p}^* \quad \sigma_{2p}^*$$

Bond order = (no. bonding electrons – no. antibonding electrons) ÷ 2

For CO^+ bond order $= (7-2) \div 2 = 2.5$; for CN^- bond order $= (8-2) \div 2 = 3.0$

(b) CN^- is diamagnetic; it has no unpaired electrons. CO^+ is paramagnetic; it has one unpaired electron.

(c) The molecule with the lower bond order will also have the greater bond length. Thus, CO^+ should have the greater bond length.

Delocalized Molecular Orbitals

<u>37.</u> With either Lewis structures or the valence bond method, two structures must be drawn (and "averaged") to explain the π bonding in C_6H_6. The σ bonding is well explained by assuming sp^2 hybridization on each C atom. But the π bonding requires that all six C—C π bonds must be equivalent. This can be achieved by creating six π molecular orbitals—three bonding and three antibonding—into which the 6π electrons are placed. This creates a single delocalized structure for the C_6H_6 molecule.

38. Resonance is replaced in molecular orbital theory with delocalized molecular orbitals. These orbitals extend over the entire molecule, rather than being localized between two atoms. Because Lewis theory represents all bonds as being localized, the only way to depict a situation in which a pair of electrons spreads over a larger region of the molecule is to draw several Lewis structures and envision the molecule as the weighted combination, or average of these (localized) Lewis structures.

<u>39.</u> We expect to find delocalized orbitals in those species for which bonding cannot be represented thoroughly by one Lewis structure, that is, for compounds that require several resonance forms.

(a) In C_2H_4, there is a total of $(2 \times 4) + (4 \times 1) = 12$ valence electrons, or 6 pairs. C atoms are the central atoms. Thus, the bonding description of C_2H_4 does not involve the use of delocalized orbitals.

(b) In SO_2, there is a total of $6 + (2 \times 6) = 18$ valence electrons, or 9 pairs. N is the central atom. A plausible Lewis structure has two resonance forms. The bonding description of SO_2 will require the use of delocalized molecular orbitals.

$$\overline{O}=\overline{S}-\overline{O}| \longleftrightarrow |\overline{O}-\overline{S}=\overline{O}$$

(c) In H_2CO, there is a total of $(2 \times 1) + 4 + 6 = 12$ valence electrons, or 6 pairs. C is the central atom. A plausible Lewis structure is shown on the right. Because one Lewis structure adequately represents the bonding in the molecule, the bonding description of H_2CO does not involve the use of delocalized molecular orbitals.

$$H-C=\overline{\overline{O}}$$
$$|$$
$$H$$

40. We expect to find delocalized orbitals in those species for which bonding cannot be represented thoroughly by one Lewis structure, that is, for species that require several resonance forms.

(a) In HCO_2^-, there are $1 + 4 + (2 \times 6) + 1 = 18$ valence electrons, or 9 pairs. A plausible Lewis structure has two resonance forms. The bonding description of HCO_2^- requires the use of delocalized molecular orbitals.

$$\left(\begin{array}{c} H-C-\overline{O}| \\ \| \\ |\overline{O}| \end{array}\right)^{-} \longleftrightarrow \left(\begin{array}{c} H-C=\overline{O} \\ | \\ |\overline{O}| \end{array}\right)^{-}$$

(b) In CO_3^{2-}, there are $4 + (3 \times 6) + 2 = 24$ valence electrons, or 12 pairs. A plausible Lewis structure has three resonance forms. The bonding description of CO_3^{2-} requires the use of delocalized molecular orbitals.

$$\left(\begin{array}{c} \overline{O}=C-\overline{O}| \\ | \\ |\overline{O}| \end{array}\right)^{2-} \longleftrightarrow \left(\begin{array}{c} |\overline{O}-C-\overline{O}| \\ \| \\ |\overline{O}| \end{array}\right)^{2-} \longleftrightarrow \left(\begin{array}{c} |\overline{O}-C=\overline{O} \\ | \\ |\overline{O}| \end{array}\right)^{2-}$$

(c) In CH_3^+, there are $4 + (3 \times 1) - 1 = 6$ valence electrons, or 3 pairs. The ion is adequately represented by one Lewis structure; no resonance forms exist.

$$\left(\begin{array}{c} H-C-H \\ | \\ H \end{array}\right)^{+}$$

Metallic Bonding

41. **(a)** Atomic number, by itself, is not particularly important in determining whether a substance has metallic properties. However, atomic number determines where an element appears in the periodic table and to the left and toward the bottom of the periodic table is where one finds atoms of high metallic character. Therefore atomic number, which provides the location of the element in the periodic table, has some minimal predictive value in determining metallic character.

(b) The answer for this part is much the same as the answer to part (a), since atomic mass generally parallels atomic number for the elements.

(c) The number of valence electrons has no bearing on the metallic character of an element. Consider, for instance, the sixth row of the periodic table. The number of valence shell electrons ranges from one to eleven across the period, yet all of the elements in this row are metallic. Alternatively, if we look at the group 14 elements, they range from non-metallic carbon to metallic lead, in spite of the fact that they share the same general valence shell configuration, viz. ns^2np^2.

(e) Because metals occur in every period of the periodic table, there is no particular relationship between the number of electron shells and the metallic behavior of an element. (Remember that one shell begins to be occupied at the start of each period.)

42. We first determine the ground-state electron configuration of Na, Fe, and Zn

$$[\text{Na}]=[\text{Ne}]3s^1 \quad [\text{Fe}]=[\text{Ar}]3d^6 4s^2 \quad [\text{Zn}]=[\text{Ar}]3d^{10}4s^2.$$

Based on the number of electrons in the valence shell, we would expect Na to be the softest of these three metals. The one $4s$ electron per Na atom will not contribute as strongly to the bonding as will two $4s$ electrons in the other two metals. If just the number of valence electrons is considered, then one would conclude that Zn should exhibit stronger metallic bonding than iron (Zn has 12 valence electrons, while Fe has 8). The opposite is true, however (i.e. Fe has stronger metallic bonding), because the ten 3d-electrons in zinc are deep in the atom and thus contribute very little to the metallic bonding. Thus, Fe should be harder than Zn. Melting point should also reflect the strengths of bonding between atoms, with higher melting points occurring in metals that are more strongly bonded. Thus in order of decreasing melting point:

$$\text{Fe}\left(1530°\text{C}\right)> \text{Zn}\left(420°\text{C}\right)> \text{Na}\left(98°\text{C}\right)\text{and hardness: Fe}\left(4.5\right)> \text{Zn}\left(2.5\right)> \text{Na}\left(0.4\right).$$

Actual values in parentheses, and hardness on a scale (MOHS) of 0 (talc) to 10 (diamond).

43. We first determine the number of Na atoms in the sample.

$$\text{no. Na atoms} = 26.8 \text{ mg Na}\times\frac{1 \text{ g Na}}{1000 \text{ mg Na}}\times\frac{1 \text{ mol Na}}{22.99 \text{ g Na}}\times\frac{6.022\times10^{23} \text{ Na atoms}}{1 \text{ mol Na}}$$

$$= 7.02\times10^{20} \text{ Na atoms}$$

Because there is one $3s$ orbital per Na atom, and since the number of energy levels (molecular orbitals) created is equal to the number of atomic orbitals initially present, there are 7.02×10^{20} energy levels present in the conduction band of this sample. Also, there is one $3s$ electron contributed by each Na atom, for a total of 7.02×10^{20} electrons. Because each energy level can hold two electrons, the conduction band is half full.

44. Although the band that results from the combination of $3s$ orbitals in magnesium metal is full, there is another empty band that overlaps this $3s$ band. This empty band results from the combination of $3p$ orbitals. Although it does not overlap the $3s$ band in the Mg_2 molecule or in atom clusters, once the metal crystal has grown to visible size, the overlap is complete.

Semiconductors

45. **(a)**   stainless steel - electrical conductor   **(b)**   solid NaCl - insulator
(c)   sulfur – insulator   **(d)**   germanium - semiconductor
(e)   seawater - electrical conductor   **(f)**   solid iodine - insulator

46. The largest band gap between valence and conduction band is in an insulator. There is no energy gap between the valence and conduction bands in a metal. Metalloids are semiconductors. They have a small band gap between the valence and conduction bands.

47. A semiconducting element is one that displays poor conductivity when pure, but attains much higher levels of conductivity when doped with small quantities of selected elements or when heated. The best semiconducting materials are made from the Group 14 elements Si and Ge. *P*-type semiconductors result when Group 14 elements are doped with small quantities of an element that has fewer than four valence electrons. For instance, when Si is doped with B, each boron atom ends us forming one silicon bond that has just one electron in it. The transfer of valence electrons from adjacent atoms into these electron-deficient bonds creates a domino effect that results in the movement of an electron-deficient hole through the semiconductor in a direction opposite to the movement of the electrons. Thus, in order to produce a *p*-type semiconductor, the added dopant atom must each have at least one less valence electron than the individual atoms that make up the bulk of the material.

(a)   Sulfur has six valence electrons, which is three too many, so doping Si with sulfur will not produce a *p*-type semiconductor.

(b)   Arsenic has five valence electrons, which is two electrons too many, so doping Si with arsenic will not produce a *p*-type semiconductor.

(c)   Lead has four valence electrons which is one too many, so doping with lead will not produce a *p*-type semiconductor.

(d)   Boron, with one less valence electron than silicon, has the requisite number of electrons needed to form a *p*-type semiconductor.

(e)   Gallium arsenide is an *n*-type semiconductor with an excess of electrons, so doping Si with GaAs will not produce a *p*-type semiconductor.

(f)   Like boron, gallium has three valence electrons. Thus, doping Si with gallium will produce a *p*-type semiconductor.

48. *N*-type semiconductors are formed when the parent element is doped with atoms of an element that has more valence electrons. For instance, when silicon is doped with phosphorus, one extra electron is left over after each phosphorus forms four covalent bonds to four neighboring silicon atoms. The extra electrons can be forced to move through the solid and conduct electric current in the process by applying a small voltage or thermal excitation. Thus, to create an *n*-type semiconductor, each added dopant atom must have at least one more valence electron than the individual atoms that make up the bulk of the material.

(a)   Since sulfur has two more valence electrons than germanium, doping germanium with sulfur will produce an *n*-type semiconductor.

(b)   Aluminum has just three valence electrons which is two short of what is required, so doping germanium with aluminum will not produce an *n*-type semiconductor.

(c)   Tin has four electrons, which is one short of the number required, so doping with tin will not produce an *n*-type semiconductor.

(d)   Cadmium sulfide is composed of Cd^{2+} and S^{2-} ions, neither ion has electrons that it can donate to germanium and so doping germanium with cadmium sulfide will not produce an *n*-type semiconductor.

(e)   Arsenic had one more valence electron than germanium, doping germanium with arsenic will produce an *n*-type semiconductor.

(f)   Since GaAs is itself an *n*-type semiconductor, the addition of small amounts of gallium arsenide to pure germanium should produce an n-type semiconductor.

49.   In ultra pure crystalline silicon, there are no extra electrons in the lattice that can conduct an electric current. If however, the silicon becomes contaminated with arsenic atoms, then there will be one additional electron added to the silicon crystal lattice for each arsenic atom that is introduced. Upon heating, some of those "extra" electrons will be promoted into the conduction band of the solid. The electrons that end up in the conduction band are able to move freely through the structure. In other words, the arsenic atoms increase the conductivity of the solid by providing additional electrons that can carry a current after they are promoted into the conduction band by thermal excitation. Thus, by virtue of having extra electrons in the lattice, silicon contaminated with arsenic will exhibit greater electrical conductance than pure silicon at elevated temperatures.

50.   When a semiconductor is doped with donor atoms (e.g., Si doped with P), the extra electron is promoted to the conduction band of the solid where it is able to move freely (good electrical conductor). When a semiconductor is doped with an acceptor atom (e.g., Si doped with Al), electron-deficient bonding results. The transfer of valence electrons from adjacent atoms into these electron-deficient bonds creates a domino effect that results in the movement of an electron-deficient hole through the semiconductor in a direction opposite to the movement of the electrons. If there are equal numbers of donor and acceptor atoms present, electrons will be redistributed from donor to acceptor atoms. There will be no excesses or deficiencies, and conductivity of the solid will not be enhanced.

51. $\Delta E_{Si} = 110.\ kJ\ mol^{-1}$. $\Delta E_{Si\ atom} = \dfrac{110.\ kJ\ mol^{-1}}{6.022 \times 10^{23}} = 1.83 \times 10^{-19}\ J$ $E = h\nu = \dfrac{hc}{\lambda}$ or $\lambda = \dfrac{hc}{E}$

$\lambda = \dfrac{(6.626 \times 10^{-34}\ J\,s)(2.998 \times 10^{8}\ \frac{m}{s})}{1.83 \times 10^{-19}\ J} = 1.09 \times 10^{-6}$ m or 1090 nm. This is IR-radiation.

52. Electrons are being promoted into the conduction band, which is made up of or comprises several energy levels. The conduction band for a solid has a large number of energy levels with extremely small energy separations between each energy level (this collection of energy levels is termed an energy band). A solar cell operates over a broad range of wavelengths because electrons can be promoted to numerous energy levels in the conduction band, rather than to just one level.

Integrative and Advanced Exercises

53. (a) Since N_2 is symmetrical, each N atom has the same hybridization scheme [He] $_{sp}$⊞ $_{2p}$⊞ The sigma bond results from the overlap of an *sp* orbital on each N. Each of the two π bonds results from the overlap of a half-filled $2p$ orbital on one nitrogen atom with a half-filled $2p$ orbital on the other nitrogen atom.
The lone pair on each N is in a hybrid *sp* orbital. The Lewis structure is $|\,N \equiv N\,|$

(b) Although we have to stretch the concepts of valence bonding considerably, the bonding in the N_2 molecule indeed can be described in each of the manners proposed. Consider sp^2 hybridization in detail and sketch descriptions of the others. sp^2 hybridization for N is [He] $_{sp^2}$⊞ $_{2p}$⊞ A sketch of the overlap of the hybrid orbitals appears at right. This overlap forms two of the three bonds in N_2. The third overlap occurs between the two $2p$ orbitals, affording a traditional π–bond.

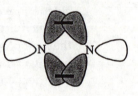

For sp^3 hybridization, the lone pair resides in one of the four hybrid orbitals, and the other three overlap, rather like the thumb and spread index and middle fingers on your left and right hands can touch at their points. Finally, for simple atomic orbitals, the lone pair resides in the $2s$ orbital, and the three $2p$ orbitals are half-filled and overlap to form three bonds. This is virtually the same as the bonding description for CO or CN⁻.

54. We begin with the orbital diagram for Na to describe the valence bond picture for Na_2. [Ne] $_{3s}$⊡ The half-filled $3s$ orbital on each Na overlaps with one another to form a σ covalent bond. There are 22 electrons in Na_2. These electrons are distributed in the molecular orbitals as follows.

σ_{1s} σ_{1s}^* σ_{2s} σ_{2s}^* $\sigma_{2p}\,\pi_{2p}$ $\pi_{2p}^*\,\sigma_{2p}^*$ σ_{3s}
⇅ ⇅ ⇅ ⇅ ⇅⇅⇅ ⇅⇅⇅ ⇅ Again, we predict a single bond for Na_2.
A Lewis-theory picture of the bonding would have the two Lewis symbols for two Na atoms uniting to form a bond: Na· ·Na $\longrightarrow$ Na — Na Thus, the bonding in Na_2 is very much like that in H_2.

55.

Species	NeF	NeF$^+$	NeF$^-$
Orbital Diagram	σ_{2p}^* π_{2p}^* π_{2p} σ_{2p} σ_{2s}^* σ_{2s}	σ_{2p}^* π_{2p}^* π_{2p} σ_{2p} σ_{2s}^* σ_{2s}	σ_{2p}^* π_{2p}^* π_{2p} σ_{2p} σ_{2s}^* σ_{2s}
Bond Order	**0.5**	**1.0**	**0**
Magnetic Property	**Paramagnetic (1 e$^-$)**	**Diamagnetic**	**Diamagnetic (atoms)**

From the MO diagrams, it is expected that NeF and NeF$^+$ should be observable species since they have non-zero bond orders. NeF$^-$ has a bond order of zero, hence it is unlikely that it will be observed or at best, we would expect a very loose association between the atoms (long/weak bond).

56. The superoxide ion, O_2^-, has 17 electrons, while the peroxide ion, O_2^{2-}, has a total of 18 electrons

   (a) The molecular orbital diagrams for these two ions are given below.

$$O_2^-\quad \sigma_{1s}\ \sigma_{1s}^*\ \sigma_{2s}\ \sigma_{2s}^*\quad \sigma_{2p}\ \pi_{2p}\ \pi_{2p}^*\ \sigma_{2p}^*$$

bond order = (no. bonding electrons – no. antibonding electrons) $\div 2 = (10 - 7) \div 2 = 1.5$

$$O_2^{2-}\quad \sigma_{1s}\ \sigma_{1s}^*\ \sigma_{2s}\ \sigma_{2s}^*\quad \sigma_{2p}\ \pi_{2p}\ \pi_{2p}^*\ \sigma_{2p}^*$$

bond order = (no. bonding electrons – no. antibonding electrons) $\div 2 = (10 - 8) \div 2 = 1.0$

   (b) O_2^- has $(2 \times 6) + 1 = 13$ electrons, 6 electron pairs plus one electron. O_2^{2-} has $(2 \times 6) + 2 = 14$ electrons, 7 electron pairs. Plausible Lewis structures are

    O_2^- $[|\overline{O}-\overline{O}\cdot]^-$ O_2^{2-} $[|\overline{O}-\overline{O}|]^{2-}$

57. K_2O_3 contains two K^+ ions, and thus the total charge of the anions must be 2-: O_3^{2-}. There is no simple way to do this with peroxide, O_2^{2-} and superoxide, O_2^- ions. But if we take a hint from the word "empirical" in the statement of the problem and try K_4O_6, we have O_6^{4-}, which will be a combination of one O_2^{2-} and two O_2^- ions. Thus a nice symmetrical Lewis structure is $[K+]$ $[|\overline{O}-\overline{O}\cdot]^-$ $[K+]$ $[|\overline{O}-\overline{O}|]^{2-}$ $[K+]$ $[|\overline{O}-\overline{O}\cdot]^-$ $[K+]$.

58. Urea has $4 + 6 + 2 \times [5 + (2 \times 1)] = 24$ valence electrons, or 12 pairs. C is at the very center, bonded to O and 2 N's. Each N is bonded to 2 H's. A plausible Lewis structure is

$$\begin{array}{c} H-\overline{N}-C-\overline{N}-H \\ |\ \ \ \ ||\ \ \ | \\ H\ \ |\overline{O}|\ \ H \end{array}$$

A VSEPR treatment of this Lewis structure suggests that the geometry around C is trigonal planar, as experimentally observed. But VSEPR theory predicts trigonal pyramidal geometry around each N, rather than the observed trigonal planar geometry. Because all of the bond angles are 120°, the hybridization on each central atom must be sp^2. The orbital diagram for each type of atom follows

$$H_{1s} \boxed{\uparrow} \qquad O \, [He] \, _{2s}\boxed{\uparrow\downarrow} \, _{2p}\boxed{\uparrow\downarrow|\uparrow|\uparrow}$$

$$N \, [He] \, _{2s}\boxed{\uparrow\downarrow} \, _{2p}\boxed{\uparrow|\uparrow|\uparrow} \rightarrow N \, [He] \, _{sp2}\boxed{\uparrow|\uparrow|\uparrow} \, _{2p}\boxed{\uparrow\downarrow}$$

$$C \, [He] \, _{2s}\boxed{\uparrow\downarrow} \, _{2p}\boxed{\uparrow|\uparrow|\,} \rightarrow C \, [He] \, _{sp2}\boxed{\uparrow|\uparrow|\uparrow} \, _{2p}\boxed{\uparrow}$$

The bonding scheme is as follows. There are four N—H sigma bonds, each formed by the overlap of an sp^2 hybrid orbital on N with a $1s$ orbital on H. There are two C—N sigma bonds, each formed by the end-to-end overlap of the remaining sp^2 hybrid orbital on N with an sp^2 orbital on C. There is also one C—O sigma bond, formed by the end-to-end overlap of the remaining sp^2 orbital on C with the $2p_y$ orbital on O. There is one pi bond, formed by the side-to-side overlap of the $2p_z$ orbital on C with the side-to-side overlap of the $2p_z$ orbital on O. The lone pairs reside in $2p_z$ orbitals on each N; they each probably interact with the electrons in the pi bond to stabilize the molecule. Contributing Lewis resonance forms follow.

59.   (a) A sketch of the molecule is given to the right. The CH$_3$ group is tetrahedral.

(b) Each of the C—H bonds is a sigma bond, as are the C—O bond and the next N—O bond. There is, of course, a sigma bonds connecting the N to the other two O's. But there is also a pi bond. In fact, if Lewis structures are drawn to represent the bonding in the molecule, two resonance forms must be drawn.

All three C—H bonds are σ bonds: $C(sp^3)^1 — H(1s)^1$

The C—O$_c$ bond is a σ bond: $C(sp^3)^1 — O_c(sp^3)^1$

There are two lone pairs on the O$_c$ atom: $O_c(sp^3)^2$.

The bonds from N to the terminal atoms are σ bonds:

$N(sp^2) – O(sp^3)$ and $N(sp^2 – O(sp^2)$ And there is one π bond: $N(2p_z)^1 — O(2p_z)^1$

The subscript is on O$_c$ to distinguish this central oxygen from the two terminal oxygens.

(c) The N—O$_c$ bond is a single bond in both resonance forms (136 pm). The other N-to-O bonds are single bonds in one resonance form and double bonds in the other, making them about one-and-a-half bonds, which would be significantly shorter (126 pm) than the N—O$_c$ bond, but longer than typical N=O bonds (120 pm, see Table 11.2)

60. The O—N—O bond angle of 125° indicates the N atom is sp^2 hybridized, while the F—O_a—N bond angle of 105° indicates the O_a atom is sp^3 hybridized. The orbital diagrams of the atoms follow.

O [He] $_{2s}$ [↑↓] $_{2p}$ [↑↓|↑|↑] O_a [He] $_{2s}$ [↑↓] $_{2p}$ [↑↓|↑|↑] ⟶ O [He] $_{sp^3}$ [↑↓|↑↓|↑|↑]

F [He] $_{2s}$ [↑↓] $_{2p}$ [↑↓|↑↓|↑]   N [He] $_{2s}$ [↑↓] $_{2p}$ [↑|↑|↑] ⟶ N [He] $_{sp^2}$ [↑|↑|↑] $_{2p}$ [↑]

Electron transfer from N to the terminal oxygen atom results in a +1 formal charge for nitrogen, a -1 formal charge for O and a single bond being developed between these two atoms.

O [He] $_{2s}$ [↑↓] $_{2p}$ [↑|↑|↑] $\xrightarrow{+e^-}$ O⁻ [He] $_{2s}$ [↑↓] $_{2p}$ [↑↓|↑↓|↑]

Bonds are formed by the following overlaps.

σ: F(2p)—O_a(sp^3) σ: O_a(sp^3)—N(sp^2) σ: N(sp^2)—O($2p_y$) π: N($2p_z$)—O($2p_z$)

Two resonance structures are required $|\overline{F}-\overline{O}_a-N-\overline{O}|$ (with ‖O‖ above N) ⟷ $|\overline{F}-\overline{O}_a-N=\overline{O}|$ (with |O| above N)   With

the wedge-and-dash representation, a sketch of the molecule is (sketch of N—O with O above and F below)

61. NO_2^- has $5 + 2 \times 6 + 1 = 18$ valence electrons = 9 electron pairs. The singly- bonded oxygen has a formal charge of -1 while the other two atoms have an octet of electrons and no formal charge.

$[\overline{O}=\overline{N}-\overline{O}|]^-$ ⟷ $[|\overline{O}-\overline{N}=\overline{O}]^-$ The N atom is sp^2 hybridized. Orbital diagrams for the atoms follow.

O [He] $_{2s}$ [↑↓] $_{2p}$ [↑↓|↑|↑]   N [He] $_{2s}$ [↑↓] $_{2p}$ [↑|↑|↑] ⟶ N [He] $_{sp^2}$ [↑↓|↑|↑] $_{2p}$ [↑]

There are two N-O sigma bonds, each formed by the end-to-end overlap of an sp^2 hybrid orbital on N with a $2p_y$ orbital on O. The pi bonding system is constructed from three $2p_z$ orbitals: (1) the $2p_z$ orbital on one O, (2) the $2p_z$ orbital on N, and (3) the $2p_z$ orbital on the other O. There are four π electrons in this system (the three $2p_z$ electrons shown in the orbital diagrams above, plus an extra electron due to the negative charge on the anion). There should be three π molecular orbitals formed: bonding, nonbonding, and an antibonding orbital, which fill as follows

π antibonding orbital [ ]

π nonbonding orbital [↑↓] π bond order $= (2-0) \div 2 = 1$

π bonding orbital [↑↓]

62. Let us begin by drawing Lewis structures for the species concerned.

$|\overline{F}-\widehat{Cl}-\overline{F}|$ (with |F| below) + $|\overline{F}-As-\overline{F}|$ (with F above and |F| below) ⟶ $[|\overline{F}-\overline{Cl}-\overline{F}|]^+$ + $\left[\begin{array}{c}|F \quad |F \\ |\overline{F}-As-\overline{F}| \\ |F \quad |F\end{array}\right]^-$

(a) For ClF_3 there are three atoms and two lone pairs attached to the central atom. The electron group geometry is trigonal bipyramidal and the molecule is T-shaped. For

AsF_5 there are five atoms attached to the central atom; its electron group geometry and molecular shape are trigonal bipyramidal. For ClF_2^+ there are two lone pairs and two atoms attached to the central atom. Its electron group geometry is tetrahedral and the ion is bent in shape. For AsF_6^- there are six atoms and no lone pairs attached to the central atom. Thus, the electron group geometry and the shape of this ion are octahedral.

(b) Trigonal bipyramidal electron group geometry is associated with sp^3d hybridization. The Cl in ClF_3 and As in AsF_5 both have sp^3d hybridization. Tetrahedral electron group geometry is associated with sp^3 hybridization. Thus Cl in ClF_2^+ has sp^3 hybridization. Octahedral electron group geometry is associated with sp^3d^2 hybridization, which is the hybridization adopted by As in AsF_6^-.

63. The two shorter nitrogen-oxygen bond distances indicate partial double bond character; a $N{=}O$ bond is 120 pm long, while a $N{-}O$ bond is 136 pm long (Table 11-1). There are two resonance forms for HNO_3 that represent these facts.

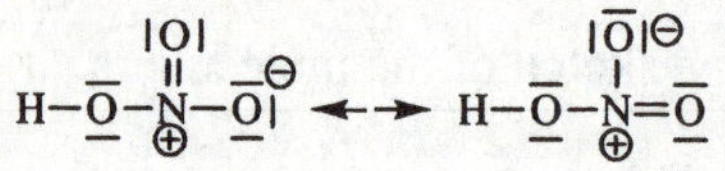

$$H{-}\overline{\underset{\oplus}{O}}{-}\overset{\overset{|\overline{O}|}{\|}}{N}{-}\overline{\underline{O}}|^{\ominus} \quad \longleftrightarrow \quad H{-}\overline{\underset{\oplus}{O}}{-}\overset{|\overline{O}|^{\ominus}}{\underset{|}{N}}{=}\overline{\underline{O}}$$

The $H{-}O$ sigma bond is formed by the overlap of a half-filled $1s$ orbital on H with a half-filled sp^3 orbital on O. The $HO{-}N$ sigma bond is formed by the overlap of the remaining half-filled sp^3 orbital on O with a half-filled sp^2 orbital on N. Both of the remaining $N{-}O$ sigma bonds are formed by the overlap of a half-filled sp^2 orbital on N with a half-filled $2p_y$ orbital on O. Each of the structures has a N-O π bond formed by the side-to-side overlap of half-filled $2p_z$ orbitals.

64. Suppose two He atoms in the excited state $1s^1 2s^1$ unite to form an He_2 molecule. One possible configuration would be $\sigma_{1s}^2 \sigma_{1s}^{*0} \sigma_{2s}^2 \sigma_{2s}^{0*}$. The bond order would be (4-0)/2 = 2.

65. Since all angles ~ 120°, and since the bond angles around C total 360° (= 123° + 113° + 124°), we postulate that both the C and N atoms have sp^2 hybridization. Two resonance forms follow.

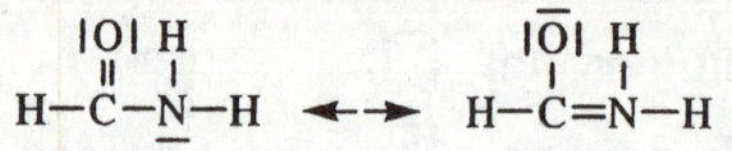

$$H{-}\overset{\overset{|O|}{\|}}{C}{-}\overset{H}{\underset{|}{N}}{-}H \quad \longleftrightarrow \quad H{-}\overset{|\overline{O}|}{\underset{|}{C}}{=}\overset{H}{\underset{|}{N}}{-}H$$

The structure on the right would be predicted to have sp^2 hybridization on both the C and N atoms. For structure on the right hand side, the overlaps that form the bonds are as follows.

σ: $H(1s){-}C(sp^2)$ σ: $C(sp^2){-}O(2p)$

σ: $C(sp^2){-}N(sp^2)$ σ: $N(sp^2){-}H(1s)$ σ: $N(sp^2){-}H(1s)$ π: $C(2p_z){-}N(2p_z)$

The structure on the left is predicted to have sp^2 hybridization on C and sp^3 hybridization on N. In this structure, the following overlaps result in bonds.

σ: $H(1s){-}C(sp^2)$ σ: $C(sp^2){-}O(2p)$

σ: $C(sp^2){-}N(sp^3)$ σ: $N(sp^3){-}H(1s)$ σ: $N(sp^3){-}H(1s)$ π: $C(2p_z){-}O(2p_z)$

From Table 10-2 we see the $C-N$ bond length of 138 pm is intermediate between a single $C-N$ bond length (147 pm) and a double $C=N$ bond length (128 pm), just as we would expect for the resonance hybrid of the two structures written above.

66. The orbital diagrams for C and N are as follows. C [He] ^{sp2}⊡⊡⊡ ^{2p}⊡   N [He] ^{sp2}⊡⊡⊡ ^{2p}⊡ The sp^2 electrons are involved in σ bonding. For N, the lone pair is in one sp^2 orbital; the remaining two half-filled sp^2 orbitals bond to adjacent C atoms. For C, one sp^2 orbital forms a σ bond by overlap with a half-filled $1s$ H atom. The remaining two half-filled sp^2 orbitals bond to either adjacent C or N atoms. The $2p$ electrons are involved in π bonding. The six $2p$ orbitals form six delocalized π molecular orbitals, three bonding and three antibonding. These six π orbitals are filled as shown in the π molecular orbital diagram sketched below.

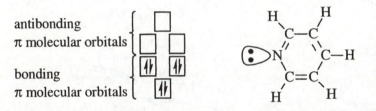

antibonding π molecular orbitals

bonding π molecular orbitals

Number of π-bonds = $\dfrac{6-0}{2} = 3$

This π-bonding scheme produces three π-bonds, which is identical to the number predicted by Lewis theory.

67. Both of the π_{2p} orbitals have a nodal plane. This plane passes through both nuclei and between the two lobes of the π bond. The difference between a node in the two types of orbital is that, in an antibonding orbital, that node lies between the two nuclei and is perpendicular to the internuclear axis, whereas in the bonding orbital, the two nuclei lie in the plane of the node, which plane therefore contains the internuclear axis. The deciding factor in determining whether an orbital is bonding or not is whether it places more electron density in the region between the two nuclei than does the combination of two atomic orbitals. All of the bonding orbitals fulfill this criterion.

68. We draw the Lewis structure of each species to account for the electron pairs around each central atom, which is Cl in each species. In F_2Cl^-, there is a total of $7 + (2 \times 7) + 1 = 22$ valence electrons, or 11 electron pairs. In F_2Cl^+, there is a total of $7 + (2 \times 7) - 1 = 20$ valence electrons, or 10 electron pairs. Plausible Lewis structures follow. $[\overline{|F}-\widehat{Cl}-\overline{F|}]$ $[|\overline{F}-\overline{Cl}-\overline{F}|]^+$ Since there are two atoms and three lone pairs attached to Cl in F_2Cl^-, the electron-group geometry around the Cl atom is trigonal bipyramidal, the Cl atom hybridization is sp^3d, and the shape of the species is linear. There are two atoms and only two lone pairs attached to the Cl atom in F_2Cl^+, which produces a tetrahedral electron-group geometry, a hybridization of sp^3 for Cl, and a bent shape for the species.

69. We will assume 100 g of the compound and find the empirical formula in the usual way.
Moles of carbon = 28.57 g(1 mole/12.011g) = 2.379 mol
Moles of hydrogen = 4.80 g(1 mole/1.008g) = 4.76 mol
Moles of nitrogen = 66.64 g(mole/14.01g) = 4.76 mol
Dividing by 2.38 we get: C(2.379/2.379) H(4.76/2.379) N(4.758/2.379)
This yields an empirical formula of: CH_2N_2 From the description given in the question, molecular formula is $C_3N_6H_6$

(a) Lewis structure for $C_3N_6H_6$ (based on info provided)

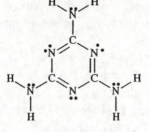

(b) Valence bond description σ and π-bonding systems

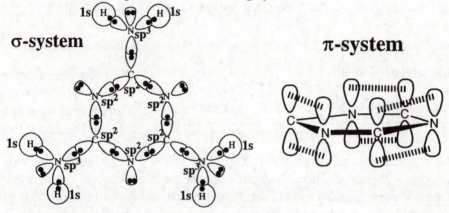

σ-system   π-system

(c) The bonds in the ring system are similar to those seen in benzene (C_6H_6) and pyridine (C_5H_5N), namely

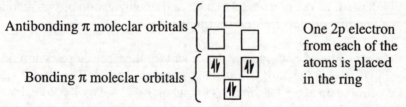

Antibonding π molecular orbitals

Bonding π molecular orbitals

One 2p electron from each of the atoms is placed in the ring

70. We will assume 100 g of the compound and find the empirical formula in the usual way.

Moles of carbon $= 53.09 \text{ g C} \times \dfrac{1 \text{ mole C}}{12.011 \text{g C}} = 4.424 \text{ mol}$

Moles of hydrogen $= 6.24 \text{ g} \times \dfrac{1 \text{ mole H}}{1.008 \text{ g H}} = 6.19 \text{ mol}$

Moles of nitrogen $= 12.39 \text{ g} \times \dfrac{1 \text{ mole N}}{14.0067 \text{ g N}} = 0.885 \text{ mol}$

Moles of oxygen $= 28.29 \text{ g} \times \dfrac{1 \text{ mole O}}{15.999 \text{ g O}} = 1.768 \text{ mol}$

Dividing all result by 0.885 we get: 5.00 moles C, 6.99$\underline{5}$ moles H, 1.00 mole N and 2.00 moles O. This yields an empirical formula of $C_5H_7NO_2$
Structure: $N{\equiv}C\text{-}CH_2(C{=}O)OC_2H_5$

Hybrid orbitals used:

C_b-H , C_d-H , C_e-H : σ H(1s) – C(sp^3) (all tetrahedral carbon uses sp^3 hybrid orbitals)

C_c=O_b: σ C_c (sp^2) –O_b (2p or sp^2) , π: C_c (2p) – O_b (2p)

C_c-O_a: σ C_c (sp^2) –O_a (2p or sp^3)

C_d-O_a: σ C_d (sp^3) –O_a (2p or sp^3)

C_a≡N: σ C_a (sp) –N(sp)

Two mutually perpendicular π-bonds: C(2p) – N(2p)

C_a-C_b: σ C_b(sp^3) –C_a(sp)

C_d-C_e: σ C_d(sp^3) –C_e(sp^3)

C_c-C_b: σ C_b(sp^3) –C_c(sp^2)

71. We will assume 100 g of the compound and find the empirical formula in the usual way.

$$\text{Moles of carbon} = 67.90 \text{ g} \times \frac{1 \text{ mole C}}{12.011 \text{g C}} = 5.653 \text{ mol}$$

$$\text{Moles of hydrogen} = 5.70 \text{ g} \times \frac{1 \text{ mole H}}{1.008 \text{ g H}} = 5.65\underline{5} \text{ mol}$$

$$\text{Moles of nitrogen} = 26.40 \text{ g} \times \frac{1 \text{ mole N}}{14.0067 \text{ g N}} = 1.885 \text{ mol}$$

Dividing all result by 1.885 we get: 2.999 moles C,   3.00 moles H, and 1.00 mole N

This yields an empirical formula of C_3H_3N

There are three possible molecules (Lewis Structures) with this formula:

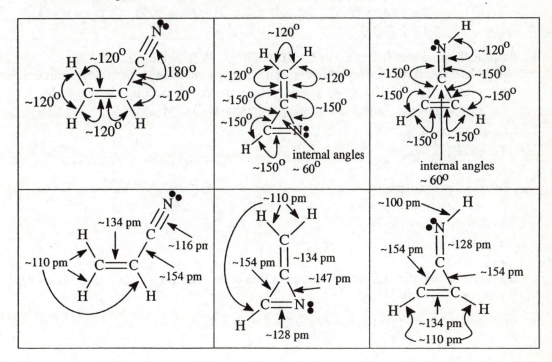

72. We have 2.464 g of dimethylglyoxime (DMG)

Mols C : $3.735(1 \text{ mol C}/44.01 \text{g CO}_2) = 0.08487$ mol C
Mols H : $1.530(2 \text{ mol H}/18.01 \text{g H}_2\text{O}) = 0.1698$ mol C

First we find the excess H_2SO_4 left over from the nitrogen determination:

$$18.6 \text{ml} \times \frac{1\text{L}}{1000 \text{ ml}} \times \frac{0.2050 \text{ mol NaOH}}{1\text{L}} \times \frac{1 \text{ mol H}_2\text{SO}_4}{2 \text{ mol NaOH}} = 1.910 \times 10^{-3} \text{mol H}_2\text{SO}_4 \text{ in excess}$$

Next we find the moles of H_2SO_4 initially added:

$$50.00 \text{ mL} \times \frac{1\text{L}}{1000 \text{ml}} \times \frac{0.3600 \text{ mol H}_2\text{SO}_4}{1\text{L}} = 1.800 \times 10^{-2} \text{mol H}_2\text{SO}_4 \text{ used}$$

Now we find the moles of H_2SO_4 that reacted:
Mole H_2SO_4 that reacted = 1.800×10^{-2} mol H_2SO_4 initially used $- 1.910 \times 10^{-3}$ mol H_2SO_4 in excess
Mole H_2SO_4 that reacted = 1.609×10^{-2} mol H_2SO_4

Therefore the moles of nitrogen in 1.868 g DMG sample

$$0.01609 \text{ mol H}_2\text{SO}_4 \times \frac{2 \text{ mol NH}_3}{1 \text{ mol H}_2\text{SO}_4} \times \frac{1 \text{ mol N}}{1 \text{ mol NH}_3} = 3.218 \times 10^{-2} \text{mol nitrogen} \ (0.4507 \text{ g nitrogen})$$

For a 1.868 g sample, mols of C & H present are fond as follows:

C: $(1.868/2.464)(0.08487) = 0.06434$ mol C (0.7728 g C)
H: $(1.868/2.464)(0.1698) = 0.1287$ mol H (0.1297 g H)

Mass of oxygen in 1.868 g DMG sample is $1.868 - 0.4507 - 0.7728 - 0.1298 = 0.515$ g oxygen   (0.0322 mol O)

Empirical formula: C(0.0643/0.0322) H(0.1287/0.0322) N(0.03218/0.0322)
O(0.0322/0.0322) or C_2H_4NO The empirical molecular mass is 58 u and the true molecular mass is 116.12 u so we multiply the empirical formula by two to get the molecular formula of $C_4H_8N_2O_2$. The structure is $H_3C\text{-}C(=NOH)\text{-}C(=NOH)\text{-}CH_3$

Hybrid orbitals used:
$C_b\text{-H} : \sigma \ H(1s) - C_b(sp^3)$
$C_a=N: \ \sigma \ C_a(sp^2) - N(sp^2) , \pi: C_a(2p) - N(2p)$
$N\text{-O}: \sigma \ N(sp^2) - O(sp^3)$
$C_a\text{-}C_a: \sigma \ C_a(sp^2) - C_a(sp^2)$
$C_a\text{-}C_b: \sigma \ C_b(sp^3) - C_a(sp^2)$
$O\text{-H} : \sigma \ H(1s) - O(sp^3)$

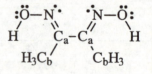

73. **(a)** Power output $= (1.00 \text{ kW/m}^2) \times (1000 \text{ W/kW}) \times (40.0 \text{ cm}^2) \times (1\text{m}^2/10^4 \text{ cm}^2) = 4.00$ watts
amps = i = w/v

(b) $i = 4.00 \text{ watts } 4.00 \dfrac{\text{J}}{\text{s}} \times \dfrac{1 \text{ C}}{0.45 \text{ J}} = 8.9 \dfrac{\text{C}}{\text{s}} = 8.9 \text{ Amps}$

74. See structures below:

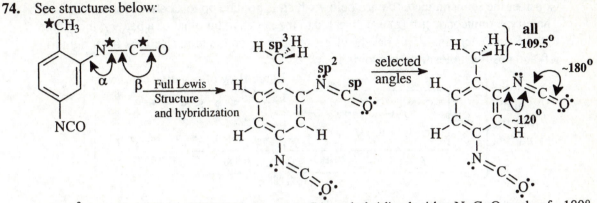

N is sp^2 hybridized with ~120° C-N=C angle. C is sp hybridized with a N=C=O angle of ~180°. The CH$_3$ group has an sp^3 hybridized C with 3×H-C-H angles of ~109.5°.

75. The Lewis structures below and the valence bond diagrams explain the differences in the shapes.

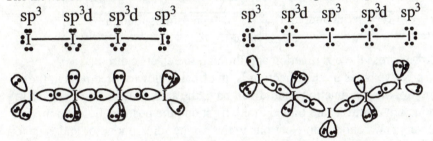

FEATURE PROBLEMS

76. **(a)** $C_6H_6(l) + 3 H_2(g) \rightarrow C_6H_{12}(l)$ $\Delta H° = \Sigma \Delta H°_f \text{ , products} - \Sigma \Delta H°_f \text{ , reactants}$
$\Delta H° = -156.4 \text{ kJ} - [3 \text{ mol} \times 0 \text{ kJ mol}^{-1} + 49.0 \text{ kJ}] = -205.4 \text{ kJ} = \Delta H°(a)$

(b) $C_6H_{10}(l) + H_2(g) \rightarrow C_6H_{12}(l)$ $\Delta H° = \Sigma \Delta H°_f \text{ products} - \Sigma \Delta H°_f \text{ reactants}$
$\Delta H° = -156.4 \text{ kJ} - [1 \text{ mol} \times 0 \text{ kJ mol}^{-1} + (-38.5 \text{ kJ})] = -117.9 \text{ kJ} = \Delta H°(b)$

(c) Enthalpy of hydrogenation for 1,3,5-cyclohexatriene = 3 × $\Delta H°(b)$
Enthalpy of hydrogenation = 3 (–117.9 kJ) = –353.7 kJ = $\Delta H°(c)$
$\Delta H°_{f, \text{ cyclohexene}}$ = –38.5 kJ/mole (given in part b of this question).
Resonance energy is the difference between $\Delta H°(a)$ and $\Delta H°(c)$.
Resonance energy = –353.7 kJ – (–205.4 kJ) = –148.3 kJ

(d) Using bond energies:
$\Delta H°_{\text{atomization}}$ = 6(C—H) + 3(C—C) + 3(C=C)
$\Delta H°_{\text{atomization}}$ = 6(414 kJ) + 3(347 kJ) + 3(611 kJ)
$\Delta H°_{\text{atomization}}$ = 5358 kJ (per mole of C_6H_6)

$C_6H_6(g) \rightarrow 6 \text{ C}(g) + 6 \text{ H}(g)$
$\Delta H° = [6(716.7 \text{ kJ}) + 6(218 \text{ kJ})] - 82.6 \text{ kJ} = 5525.6 \text{ kJ}$
Resonance energy = 5358 kJ – 5525.6 kJ = –168 kJ

77. Consider the semiconductor device below which is hooked up to a battery (direct current). The *n*-type semiconductor (a) is connected to the negative terminal of a battery, the *p*-type to the positive terminal. This has the effect of pushing conduction electrons from right to left and positive holes from left to right.

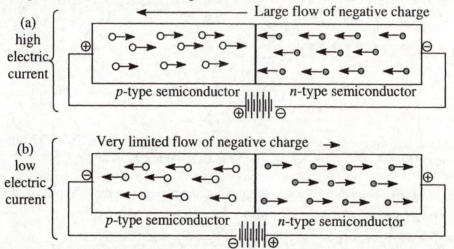

A large current flows across the *p-n* junction and through the electric circuit.

Note: the flow of positive holes in one direction is, in effect, a flow of electrons in the opposite direction. So, the *p-n* junction should still have the same orientation in (b) as in (a). Now the conduction electrons are pulled to the right and the positive holes to the left. Because there are very few conduction electrons in the *p*-type semiconductor and very few positive holes in the *n*-type semiconductor, there are very few carriers of electric charge across the *p-n* junction. Thus, very little electric current flows.

When the *p-n* junction rectifier is connected to 60-cycle alternating current, each terminal of the of the electric generator switches back and forth between being a positive and a negative terminal 120 times per second. If the electrical contact to the *p-n* junction rectifier is made through an electric generator rather than a battery, half of the time the situation is that depicted in (a) and half the time it is that depicted in (b). Thus, half the time there is a large current flow, always in the same direction, and half the time there is essentially no current. The alternating current is converted to direct current: it is rectified, and the *p-n* junction device shown above is called a rectifier.

78. **(a)** Five valid resonance forms can be drawn for furan:

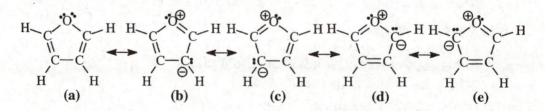

These individual resonance forms do not exist. The actual structure is a hybrid possessing characteristics of all five individual contributors.

(b) Furan is an aromatic five-membered heterocycle. Its four carbon atoms and lone oxygen atom are sp^2 hybridized in the classical bonding description. One of the lone pairs on oxygen occupies an unhybridized p-orbital, and this lone pair overlaps with the half-filled $2p$ orbitals on the four carbon atoms to form an aromatic sextet:

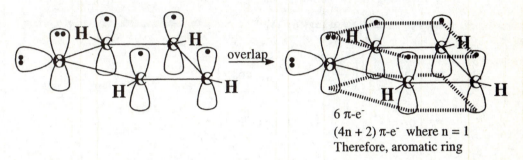

$6 \pi\text{-e}^-$
$(4n + 2) \pi\text{-e}^-$ where n = 1
Therefore, aromatic ring

The groundstate molecular orbital diagram for the π-system in furan is depicted below:

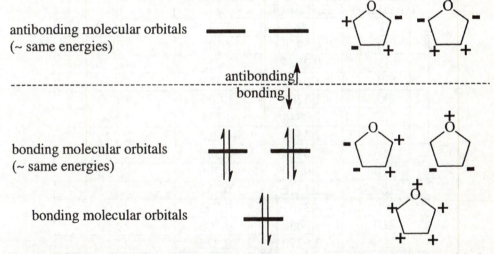

The diagram shows that the six π-electrons fill the three bonding molecular orbitals for the π-system. Keep in mind that the ensemble of π-molecular orbitals is superimposed on the sp^2 framework for the molecule.

(c) All five resonance structures for furan (structures a-e) have six electrons involved in π-bonding. In structure (a), a lone pair in a $2p$ orbital on the oxygen atom along with the four π-electrons in unhybridized $2p$ orbitals which form the two C=C bonds are combined to give the π-system. In structures (b) - (e) inclusive, the π-system is produced by the combination of a lone pair in a 2p orbital on a carbon atom with the four π-electrons from the C=O and C=C bonds.

79. (a) In order to see the shape of the sp hybrid the simplest approach is to combine just the angular parts of the s and p orbitals. Including the radial part makes the plot more challenging (see below). The angular parts for the 2s and $2p_z$ orbitals are

$$Y(s)=\left(\frac{1}{4\pi}\right)^{1/2} \quad \text{and} \quad Y\!\left(p_z\right)=\left(\frac{3}{4\pi}\right)^{1/2}\cos\theta$$

Combining the two angular parts:

$$\psi_1(sp)=\frac{1}{\sqrt{2}}\left[Y(2s)+Y(2p)\right]$$

$$\psi_1(sp)=\frac{1}{\sqrt{2}}\left[\left(\frac{1}{4\pi}\right)^{1/2}+\left(\frac{3}{4\pi}\right)^{1/2}\cos\theta\right]$$

$$\psi_1(sp)=\frac{1}{\sqrt{2}}\left(\frac{1}{4\pi}\right)^{1/2}\left[1+\sqrt{3}\cos\theta\right]$$

We now evaluate this function (ignoring the constants in front of the square brackets) for various values of θ

Theta(deg)	Theta(rad)	sp_1	sp_2
0	0	2.732	-0.732
30	0.5236	2.500	-0.500
45	0.7854	2.225	-0.225
60	1.0472	1.866	0.134
90	1.5708	1.000	1.000
120	2.0944	0.134	1.866
135	2.3562	-0.225	2.225
150	2.6180	-0.500	2.500
170	2.9670	-0.706	2.706
180	3.1416	-0.732	2.732
190	3.3161	-0.706	2.706
210	3.6652	-0.500	2.500
225	3.9270	-0.225	2.225
240	4.1888	0.134	1.866
270	4.7124	1.000	1.000
300	5.2360	1.866	0.134
315	5.4978	2.225	-0.225
330	5.7596	2.500	-0.500
360	6.2832	2.732	-0.732

A plot of these values in the form of Figure 9.24 gives the general shape of the *sp* hybrid. This is a plot in the xz-plane, for $\Psi_1(sp)$. (See next page)

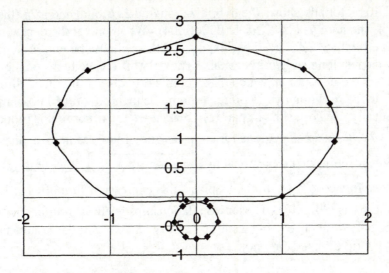

The other *sp* hybrid is

$$\psi_2(sp) = \frac{1}{\sqrt{2}}\left[Y(2s) - Y(2p)\right]$$

$$\psi_2(sp) = \frac{1}{\sqrt{2}}\left[\left(\frac{1}{4\pi}\right)^{1/2} - \left(\frac{3}{4\pi}\right)^{1/2}\cos\theta\right]$$

$$\psi_2(sp) = \frac{1}{\sqrt{2}}\left(\frac{1}{4\pi}\right)^{1/2}\left[1 - \sqrt{3}\cos\theta\right]$$

Graphically, $\Psi_2(sp)$ is generated as follows:

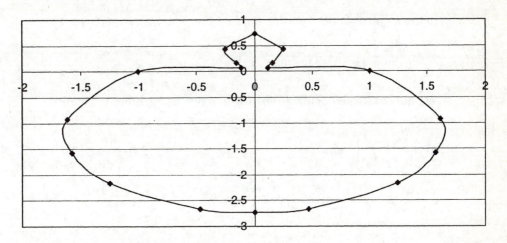

We now see that the second *sp* hybrid ($\Psi_2(sp)$) points in a direction opposite to the first ($\Psi_1(sp)$)

(b) Figure 12-10 pictorially shows the hybridization of a *2s* orbital and a *2p* (there is no mention of whether it is a $2p_x$, $2p_y$ or $2p_z$ orbital). When one of the degenerate *2p*-orbitals and the 2s orbital are hybridized, we expect that there should be no difference in shape or energy of the resulting *sp*-hybrid orbitals. The only difference between the $2p_x$, $2p_y$ and $2p_z$ orbitals is the direction in which these atomic orbitals are oriented. Similarly, the only difference in the resulting hybrid orbitals should be their directional properties. Mathematically, we have shown (part (a)), that an sp_z hybrid orbital is proportional to $1 - \sqrt{3}\cos\theta$. Similar calculations for an sp_x hybrid orbital result in an analogous relationship, namely one proportional to $1 - \sqrt{3}\sin\theta\cos\phi$. (Note: in the xz plane, $\phi = 90°$, therefore $\cos\phi = 1$). In the xz plane, this expression simplifies to $1 - \sqrt{3}\sin\theta$. (Note: $\sin\theta = \cos(90° - \theta)$). These two relationships result in similar answers (i.e., same shape), the only difference being a shift by an expect 90°. Similar arguments can be made for an sp_y hybrid orbital expression.

(c) The sp^2 hybrids. To show the spatial distribution of the sp^2 hybrids we will again use the angular functions only. Thus

$$Y(p_x) = \left(\frac{3}{4\pi}\right)^{1/2} \sin\theta\cos\phi \qquad\qquad Y(s) = \left(\frac{1}{4\pi}\right)^{1/2}$$

Combining these functions

$$\psi_1(sp^2) = \frac{1}{\sqrt{3}}\psi(2s) + \frac{\sqrt{2}}{\sqrt{3}}\psi(2p_x) \qquad \psi_1(sp^2) = \frac{1}{\sqrt{3}}\left(\frac{1}{4\pi}\right)^{1/2} + \frac{\sqrt{2}}{\sqrt{3}}\left(\frac{3}{4\pi}\right)^{1/2}\sin\theta\cos\phi$$

$$\psi_1(sp^2) = \left(\frac{1}{4\pi}\right)^{1/2}\left[\frac{1}{\sqrt{3}} + \sqrt{2}\sin\theta\cos\phi\right]$$

In the xy plane $\theta = 90°$ so that the function becomes $(\sin(90°) = 1)$

$$\psi_1(sp^2) = \left(\frac{1}{4\pi}\right)^{1/2}\left[\frac{1}{\sqrt{3}} + \sqrt{2}\cos\phi\right]$$

We now evaluate the functional form of one of the other sp^2 hybrids

$$Y(p_y) = \left(\frac{3}{4\pi}\right)^{1/2}\sin\theta\sin\phi \qquad \psi_2(sp^2) = \frac{1}{\sqrt{3}}\psi(2s) - \frac{1}{\sqrt{6}}\psi(2p_x) + \frac{1}{\sqrt{2}}\psi(2p_y)$$

$$\psi_2(sp^2) = \frac{1}{\sqrt{3}}\left(\frac{1}{4\pi}\right)^{1/2} - \frac{1}{\sqrt{6}}\left(\frac{3}{4\pi}\right)^{1/2}\sin\theta\cos\phi + \frac{1}{\sqrt{2}}\left(\frac{3}{4\pi}\right)^{1/2}\sin\theta\sin\phi$$

$$\psi_2(sp^2) = \left(\frac{1}{4\pi}\right)^{1/2}\left[\frac{1}{\sqrt{3}} - \frac{1}{\sqrt{6}}\sqrt{3}\sin\theta\cos\phi + \frac{1}{\sqrt{2}}\sqrt{3}\sin\theta\sin\phi\right]$$

$$\psi_2(sp^2) = \left(\frac{1}{4\pi}\right)^{1/2}\left[\frac{1}{\sqrt{3}} - \frac{1}{\sqrt{2}}\sin\theta\cos\phi + \frac{\sqrt{3}}{\sqrt{2}}\sin\theta\sin\phi\right]$$

Again in the xy plane we have

$$\psi_2\left(sp^2\right)=\left(\frac{1}{4\pi}\right)^{1/2}\left[\frac{1}{\sqrt{3}}-\frac{1}{\sqrt{2}}\cos\phi+\frac{\sqrt{3}}{\sqrt{2}}\sin\phi\right]$$

The third sp^2 hybrid is

$$\psi_3\left(sp^2\right)=\left(\frac{1}{4\pi}\right)^{1/2}\left[\frac{1}{\sqrt{3}}-\frac{1}{\sqrt{2}}\sin\theta\cos\phi-\frac{\sqrt{3}}{\sqrt{2}}\sin\theta\sin\phi\right]$$

and in the xy plane

$$\psi_3\left(sp^2\right)=\left(\frac{1}{4\pi}\right)^{1/2}\left[\frac{1}{\sqrt{3}}-\frac{1}{\sqrt{2}}\cos\phi-\frac{\sqrt{3}}{\sqrt{2}}\sin\phi\right]$$

We can evaluate these functions and obtain sp^2 angular values:

Phi(deg)	Phi(rad)	$sp^2(1)$	$sp^2(2)$	$sp^2(3)$
0	0	1.992	-0.130	-0.647
30	0.5236	1.802	0.577	-0.837
45	0.7854	1.577	0.943	-0.789
60	1.0472	1.284	1.284	-0.647
90	1.5708	0.577	1.802	-0.130
120	2.0944	-0.123	1.992	0.577
135	2.3562	-0.423	1.943	0.943
150	2.6180	-0.647	1.802	1.284
170	2.9671	-0.815	1.486	1.661
180	3.1416	-0.837	1.284	1.802
190	3.3161	-0.815	1.061	1.906
210	3.6652	-0.647	0.577	1.992
225	3.9270	-0.423	0.211	1.943
240	4.1888	-0.130	-0.130	1.802
270	4.7124	0.577	-0.647	1.284
300	5.2360	1.284	-0.837	0.577
315	5.4978	1.577	-0.789	0.211
330	5.7596	1.802	-0.647	-0.130
360	6.2832	1.992	-0.130	-0.647

The graphs are as follows:

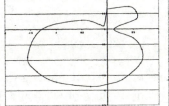 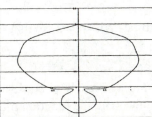 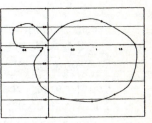

Part (a) INCLUDING THE RADIAL PART

If we combine the angular and radial parts of the 2s and 2p orbitals, the situation becomes more complicated because we now have to deal with two variables. The angular and radial parts of the 2s and 2p orbitals are:

$$Y(s) = \left(\frac{1}{4\pi}\right)^{1/2} \qquad\qquad R(2s) = \frac{1}{2\sqrt{2}}\left(\frac{Z}{a_0}\right)^{3/2}(2-\sigma)e^{-\sigma/2}$$

$$Y(p_z) = \left(\frac{3}{4\pi}\right)^{1/2}\cos\theta \qquad\qquad R(2p) = \frac{1}{2\sqrt{6}}\left(\frac{Z}{a_0}\right)^{3/2}\sigma e^{-\sigma/2}$$

Combining these

$$\psi(2s) = \left(\frac{1}{4\pi}\right)^{1/2}\times\frac{1}{2\sqrt{2}}\left(\frac{Z}{a_0}\right)^{3/2}(2-\sigma)e^{-\sigma/2} \qquad \psi(2p_z) = \left(\frac{3}{4\pi}\right)^{1/2}\cos\theta\times\frac{1}{2\sqrt{6}}\left(\frac{Z}{a_0}\right)^{3/2}\sigma e^{-\sigma/2}$$

We now combine these to form the *sp*-hybrids

$$\psi_1(sp) = \frac{1}{\sqrt{2}}\left[\psi(2s)+\psi(2p)\right]$$

$$\psi_1(sp) = \frac{1}{\sqrt{2}}\left[\left(\frac{1}{4\pi}\right)^{1/2}\times\frac{1}{2\sqrt{2}}\left(\frac{Z}{a_0}\right)^{3/2}(2-\sigma)e^{-\sigma/2} + \left(\frac{3}{4\pi}\right)^{1/2}\cos\theta\times\frac{1}{2\sqrt{6}}\left(\frac{Z}{a_0}\right)^{3/2}\sigma e^{-\sigma/2}\right]$$

$$\psi_1(sp) = \frac{1}{\sqrt{2}}\left(\frac{1}{4\pi}\right)^{1/2}\left[\frac{1}{2\sqrt{2}}\left(\frac{Z}{a_0}\right)^{3/2}(2-\sigma)e^{-\sigma/2} + \sqrt{3}\times\cos\theta\times\frac{1}{2\sqrt{6}}\left(\frac{Z}{a_0}\right)^{3/2}\sigma e^{-\sigma/2}\right]$$

$$\psi_1(sp) = \frac{1}{\sqrt{2}}\left(\frac{1}{4\pi}\right)^{1/2}\left(\frac{Z}{a_0}\right)^{3/2}\frac{1}{2\sqrt{2}}\left[(2-\sigma)e^{-\sigma/2} + \sqrt{3}\times\cos\theta\times\frac{1}{\sqrt{3}}\sigma e^{-\sigma/2}\right]$$

$$\psi_1(sp) = \frac{1}{\sqrt{2}}\left(\frac{1}{4\pi}\right)^{1/2}\left(\frac{Z}{a_0}\right)^{3/2}\frac{e^{-\sigma/2}}{2\sqrt{2}}\left[(2-\sigma)+\cos\theta\times\sigma\right]$$

$$\psi_1(sp) = \frac{1}{\sqrt{2}}\left(\frac{1}{4\pi}\right)^{1/2}\left(\frac{Z}{a_0}\right)^{3/2}\frac{e^{-\sigma/2}}{2\sqrt{2}}\left[2+\sigma(\cos\theta-1)\right]$$

Similarly

$$\psi_2(sp) = \frac{1}{\sqrt{2}}\left[\psi(2s)-\psi(2p)\right] \qquad\qquad \psi_2(sp) = \frac{1}{\sqrt{2}}\left(\frac{1}{4\pi}\right)^{1/2}\left(\frac{Z}{a_0}\right)^{3/2}\frac{e^{-\sigma/2}}{2\sqrt{2}}\left[2-\sigma(\cos\theta+1)\right]$$

We are dealing with orbitals with $n = 2$, so that

$$\sigma = \frac{2Zr}{na_0} = \frac{Zr}{a_0}$$

In order to proceed further we have to choose values of r and θ, but how do we plot this? The way to do this is to evaluate the function for various values of (r/a_0) and θ and plot points in the xz plane with values of x and z derived from the polar coordinates

$z = r\cos\theta$ $x = r\sin\theta\cos\phi$ choosing $\phi = 0$ in the xz plane so that $x = r\sin\theta$

Thus, we evaluate the probability as the square of the functions

$$\psi_1(sp) = \frac{1}{\sqrt{2}}\left(\frac{1}{4\pi}\right)^{1/2}\left(\frac{Z}{a_0}\right)^{3/2}\frac{e^{-Zr/2a_0}}{2\sqrt{2}}\left[2 + \frac{Zr}{a_0}(\cos\theta - 1)\right]$$

$$\psi_2(sp) = \frac{1}{\sqrt{2}}\left(\frac{1}{4\pi}\right)^{1/2}\left(\frac{Z}{a_0}\right)^{3/2}\frac{e^{-Zr/2a_0}}{2\sqrt{2}}\left[2 - \frac{Zr}{a_0}(\cos\theta + 1)\right]$$

and plot the value of the function at the x,z coordinate and join points of equal value of the probability in the manner of a contour plot.. The shape of the sp hybrid will then be revealed. Choose a value of Z (or Z_{eff}) that is convenient.

This is suitable for a class team project using *Microsoft Excel*. A primitive spreadsheet is available upon request.

80. The ozone molecule, O_3, has no bond dipoles because all of the atoms are alike. The Lewis structure below shows the two equivalent structures contributing to the resonance hybrid. The electron-group geometry around the central oxygen atom is trigonal planar (it is bonded to two oxygen atoms and a lone pair). This results in a molecule which has a bent geometry. The zero bond dipoles in O_3 signify that the centers of negative and positive charge coincide along the oxygen-to-oxygen bonds. However, the lone pair electrons on the central oxygen atom constitute another center of negative charge that is offset from the oxygen-to-oxygen bonds. The dipole moment in the O_3 molecule is directed towards this charge center. The electrostatic potential map shows evidence of this. As well, because of the formal charges found on each resonance form, the hybrid indicates that there should be formal charge of +1 on the central oxygen atom and a -1 charge distributed on the terminal oxygen atoms. This is supported by the electrostatic potential map provided as well.

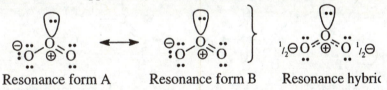

Resonance form A   Resonance form B   Resonance hybrid

81. Borazine has a delocalized π system that resembles benzene. Both benzene and borazine have 6 electrons in a conjugated π system and thus, have similar molecular orbitals of comparable shape and energy.

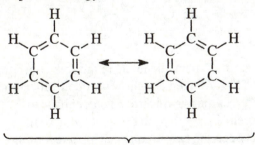   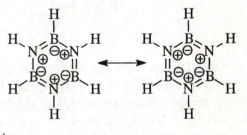

Benzene Resonance Structures   Borazine Resonance Structures

The molecular orbital diagram for borazine is drawn below. Both the HOMO and LUMO are labeled. Note that the LUMO is an antibonding orbital with two nodes (degenerate set) while the HOMO is a bonding orbital (degenerate set) with only one node.

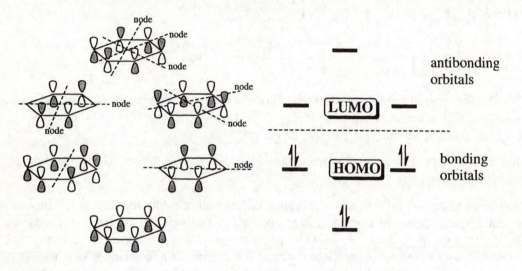

82. **The** Lewis structure of 1,3-pentadiene and 1,4-pentadiene are shown below. All sp^2 hybridized carbon atoms have an unpaired electron in an unhybridized p-orbital that may undergo π-bonding to a neighboring half-filled p-orbital. In the case of 1,3-pentadiene, there are four unhybridized p-orbitals in close proximity

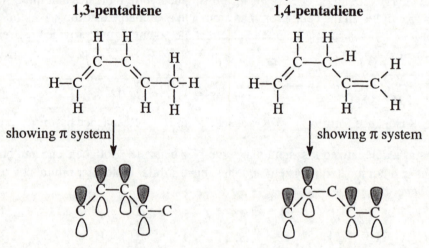

The difference between the two systems is that in 1,3-pentadiene, there are four p-orbitals in close proximity to one another (in a row). This arrangement allows for delocalization of π-electron density across all four sp2 hybridized carbons, which results in enhanced stability. Delocalization of this type is not possible for 1,4-pentadiene. Owing to the presence of an sp^3 hybridized carbon atom between the two π-bonds. This intervening saturated atom keeps the π-electron density localized in two separate olefin moieties

The delocalization in 1,3-pentadiene can be represented by the resonance picture below.

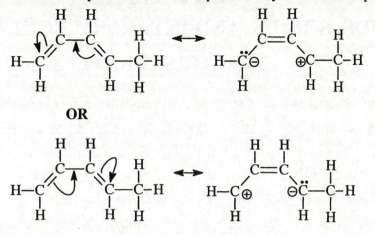

OR

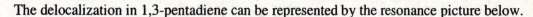

CHAPTER 12
LIQUIDS, SOLIDS, AND INTERMOLECULAR FORCES
PRACTICE EXAMPLES

1A Values of ΔH_{vap} are in kJ/mol so we first determine the amount in moles of diethyl ether.

$$\text{Heat} = 2.35 \text{ g } (C_2H_5)_2O \times \frac{1 \text{ mol } (C_2H_5)_2O}{74.12 \text{ g } (C_2H_5)_2O} \times \frac{29.1 \text{ kJ}}{1 \text{ mol } (C_2H_5)_2O} = 0.923 \text{ kJ}$$

1B $\Delta H_{overall} = \Delta H_{cond} + \Delta H_{cooling}$

$\Delta H_{cond} = 0.0245 \text{ mol} \times (-40.7 \text{ kJ mol}^{-1}) = -0.997 \text{ kJ} = -997 \text{ J}$

$\Delta H_{cooling} = 0.0245 \text{ mol} \times (4.21 \text{ J g}^{-1} \text{ °C}^{-1})(85.0 \text{ °C} - 100.0 \text{ °C})(18.0153 \text{ g mol}^{-1}) = -27.9 \text{ J}$

$\Delta H_{overall} = -997 \text{ J} + -27.9 \text{ J} = -1025 \text{ J or } -1.025 \text{ kJ}$

2A $d = 0.701$ g/L at 25 °C for C_6H_{14} (Molar Mass = 86.177 g mol^{-1})
Consider a 1.00 L sample. This contains 0.701 g C_6H_{14}

$$\text{Moles } C_6H_{14} \text{ in 1.00 L sample} = 0.701 \text{ g } C_6H_{14} \times \frac{1 \text{ mol } C_6H_{14}}{86.177 \text{ g } C_6H_{14}} = 8.13 \times 10^{-3} \text{ mol } C_6H_{14}$$

Find pressure using the ideal gas law: $P = \dfrac{nRT}{V} = \dfrac{(8.31 \times 10^{-3} \text{ mol})(\frac{0.08206 \text{ L atm}}{\text{K mol}})(298 \text{K})}{1.00 \text{ L}}$

$$P = 0.199 \text{ atm or 151 torr}$$

2B From Figure 12-9, the vapor pressure is ≈ 420 mmHg or 420 mmHg $\times \dfrac{1 \text{ atm}}{760 \text{ torr}} = 0.55_3$ atm

Molar mass = 74.123 g mol^{-1}. $P = \dfrac{nRT}{V} = \dfrac{\left(\frac{\text{mass}}{\text{molar mass}}RT\right)}{V} = \dfrac{(\text{density})RT}{\text{molar mass}}$

or $d = \dfrac{(\text{molar mass})P}{RT} = \dfrac{\left(74.123 \frac{\text{g}}{\text{mol}}\right)(0.55_3 \text{ atm})}{\left(0.08206 \frac{\text{L atm}}{\text{K mol}}\right)(293 \text{K})} = 1.70 \text{ g L}^{-1} \approx 1.7 \text{ g/L}$

3A We first calculate pressure created by the water at 80.0° C, assuming all 0.132 g H_2O vaporizes.

$$P_2 = \frac{nRT}{V} = \frac{\left(0.132 \text{ g } H_2O \times \frac{1 \text{ mol } H_2O}{18.02 \text{ g } H_2O}\right) \times 0.08206 \frac{\text{L atm}}{\text{mol K}} \times 353.2 \text{ K}}{0.525 \text{ L}} \times \frac{760 \text{ mmHg}}{1 \text{ atm}} = 307 \text{ mmHg}$$

At 80.0°C, the vapor pressure of water is 355.1 mmHg, thus, all the water exists as vapor.

3B The result of Example 12-3 is that 0.132 g H_2O would exert a pressure of 281 mmHg if it all existed as a vapor. Since that 281 mmHg is greater than the vapor pressure of water at this temperature, some of the water must exist as liquid. The calculation of the example is based on the equation $P = nRT / V$, which means that the pressure of water is proportional to its mass. Thus, the mass of water needed to produce a pressure of 92.5 mmHg under this situation is

$$\text{mass of water vapor} = 92.5 \text{ mmHg} \times \frac{0.132 \text{ g } H_2O}{281 \text{ mmHg}} = 0.0435 \text{ g } H_2O$$

mass of liquid water $= 0.132$ g H_2O total $- 0.0435$ g H_2O vapor $= 0.089$ g liquid water

4A From Table 12-1 we know that $\Delta H_{vap} = 38.0$ kJ / mol for methyl alcohol. We now can use the Clausius-Clapeyron equation to determine the vapor pressure at $25.0°C = 298.2$ K.

$$\ln \frac{P}{100 \text{ mmHg}} = \frac{38.0 \times 10^3 \text{ J mol}^{-1}}{8.3145 \text{ J mol}^{-1} \text{ K}^{-1}} \left(\frac{1}{(273.2 + 21.2) \text{ K}} - \frac{1}{298.2 \text{K}} \right) = +0.198$$

$$\frac{P}{100 \text{ mmHg}} = e^{+0.198} = 1.22 \qquad P = 1.22 \times 100 \text{ mmHg} = 121 \text{ mmHg}$$

4B The vapor pressure at the normal boiling point $\left(99.2°C = 372.4 \text{ K} \right)$ is 760 mmHg precisely. We can use the Clausius-Clapeyron equation to determine the vapor pressure at $25°C = 298$ K.

$$\ln \frac{P}{760 \text{ mmHg}} = \frac{35.76 \times 10^3 \text{ J mol}^{-1}}{8.3145 \text{ J mol}^{-1} \text{ K}^{-1}} \left(\frac{1}{372.4 \text{ K}} - \frac{1}{298.2 \text{ K}} \right) = -2.874$$

$$\frac{P}{760 \text{ mmHg}} = e^{-2.874} = 0.0565 \qquad P = 0.0565 \times 760 \text{ mmHg} = 42.9 \text{ mmHg}$$

5A Moving from point R to P we begin with $H_2O(g)$ at high temperature (>100°C). When the temperature reaches the point on the vaporization curve, OC, water condenses at constant temperature (100°C). Once all of the water is in the liquid state, the temperature drops. When the temperature reaches the point on the fusion curve, OD, ice begins to form at constant temperature (0°C). Once all of the water has been converted to $H_2O(s)$, the temperature of the sample decreases slightly until point P is reached.

Since solids are not very compressible, very little change occurs until the pressure reaches the point on the fusion curve OD. Here, melting begins. A significant decrease in the volume occurs ($\approx$10%) as ice is converted to liquid water. After melting, additional pressure produces very little change in volume because liquids are not very compressible.

5B

1.00 mol H_2O. At Point R, T = 374.1 °C or 647.3 K

$$V_{\text{point R}} = \frac{nRT}{P} = \frac{(1.00\,\text{mol})\left(0.08206\dfrac{\text{L atm}}{\text{K mol}}\right)(647.3\,\text{K})}{1.00\,\text{atm}} = 53.1\,\text{L}$$

1.00 mol H_2O on P-R line, if 1/2 of water is vaporized, T = 100 °C(273.015 K)

$$V_{\text{1/2 vap(PR)}} = \frac{nRT}{P} = \frac{(0.500\,\text{mol})\left(0.08206\dfrac{\text{L atm}}{\text{K mol}}\right)(373.15\,\text{K})}{1.00\,\text{atm}} = 15.3\,\text{L}$$

51.3 L
At
Point
R

15.3 L
at 100C
1/2 vap

A much smaller volume results when just 1/2 of the sample is vaporized (moles of gas smaller as well, temperature is smaller). 53.1 L vs 15.3 L (about 28.8 % of the volume as that seen at point R).

6A The substance with the highest boiling point will have the strongest intermolecular forces. The weakest of van der Waals forces are London forces, which depend on molar mass (and surface area): C_3H_8 is 44 g/mol, CO_2 is 44 g/mol, and CH_3CN is 41 g/mol. Thus, the London forces are approximately equal for these three compounds. Next to consider are dipole-dipole forces. C_3H_8 is essentially nonpolar; its bonds are not polarized to an appreciable extent. CO_2 is nonpolar; its two bond moments cancel each other. CH_3CN is polar and thus has the strongest intermolecular forces and should have the highest boiling point. The actual boiling points are −78.44°C for CO_2, −42.1°C for C_3H_8, and 81.6°C for CH_3CN.

6B Dispersion forces, which depend on the number of electrons (molar mass) and structure, are one of the determinants of boiling point. The molar masses are: C_8H_{18} (114.2 g/mol), $CH_3CH_2CH_2CH_3$ (58.1 g/mol), $(CH_3)_3CH$ (58.1 g/mol), C_6H_5CHO (106.1 g/mol), and SO_3 (80.1 g/mol). We would expect $(CH_3)_3CH$ to have the lowest boiling point because it has the lowest molar mass and the most compact (ball-like) shape, whereas $CH_3CH_2CH_2CH_3$ which has the same mass, but is longer and hence has more surface area (more chances for intermolecular interactions), thus it should have the second highest boiling point. We would expect SO_3 to be next in line as it is also non-polar, but more massive than C_4H_{10}. C_6H_5CHO should have a boiling point higher than the more massive C_8H_{18} because benzaldehyde is polar while octane is not. Actual boiling points are given in parentheses in the following ranking. $(CH_3)_3CH$ (-11.6 °C) < $CH_3CH_2CH_2CH_3$ (-0.5°C) < SO_3 (44.8°C) C_8H_{18} (125.7 °C) < C_6H_5CHO (178°C)

7A We first look to molar masses: Ne (20.2 g/mol), He (4.0 g/mol), Cl_2 (70.9 g/mol), $(CH_3)_2\,CO$ (58.1 g/mol), O_2 (32.0 g/mol), and O_3 (48.0 g/mol). Both $(CH_3)_2\,CO$ and O_3 are polar, O_3 weakly so (because of its uneven distribution of electrons). We expect $(CH_3)_2\,CO$ to have the highest boiling, followed by Cl_2, O_3, O_2, Ne, and He. In the following ranking, actual boiling points are given in parentheses. He (-268.9 °C), Ne (-245.9 °C), O_2 (-183.0 °C), O_3 (-111.9 °C), Cl_2 (-34.6 °C), and $(CH_3)_2CO$ (56.2°C)

7B The magnitude of the enthalpy of vaporization is strongly related to the strength of intermolecular forces: the stronger these forces are, the more endothermic the vaporization process. The first three substances all are nonpolar and, therefore, their only intermolecular forces are London forces, whose strength primarily depends on molar mass. The substances are arranged in order of increasing molar mass: $H_2 = 2.0$ g / mol, $CH_4 = 16.0$ g / mol, $C_6H_6 = 78.1$ g / mol, and also in order of increasing heat of vaporization. The last substance has a molar mass of 61.0 g/mol, which would produce intermolecular forces smaller than those of C_6H_6 if CH_3NO_2 were nonpolar. But the molecule is definitely polar. Thus, the strong dipole-dipole forces developed between CH_3NO_2 molecules make the enthalpy of vaporization for CH_3NO_2 larger than that for C_6H_6, which is, of course, essentially non-polar.

8A Strong interionic forces lead to high melting points. Strong interionic forces are created by ions with high charge and of small size. Thus, for a compound to have a lower melting point than KI it must be composed of ions of larger size, such as RbI or CsI. A compound with a melting point higher than CaO would have either smaller ions, such as MgO, or more highly charged ions, such as Ga_2O_3 or Ca_3N_2, or both, such as AlN or Mg_3N_2.

8B Mg^{2+} has a higher charge and a smaller size than does Na^+. In addition, Cl^- has a smaller size than I^-. Thus, interionic forces should be stronger in $MgCl_2$ than in NaI. We expect $MgCl_2$ to have lower solubility and, in fact, 12.3 mol (1840 g) of NaI dissolves in a liter of water, compared to just 5.7 mol (543 g) of $MgCl_2$, confirming our prediction.

9A The length (l) of a bcc unit cell and the radius (r) of the atom involved are related by $4r = l\sqrt{3}$. For potassium, $r = 227$ pm. Then $l = 4 \times 227$ pm $/ \sqrt{3} = 524$ pm

9B Consider just the face of Figure 12.46. Note that it is composed of one atom at each of four corners and one in the center. The four corner atoms touch the atom in the center, but not each other. Thus, the atoms are in contact across the diagonal of the face. If each atomic radius is designated r, then the length of the diagonal is $4r (= r$ for one corner atom $+2r$ for the center atom $+r$ for the other corner atom). The diagonal also is related to the length of a side, l, by the Pythagorean theorem: $d^2 = l^2 + l^2 = 2l^2$ or $d = \sqrt{2}l$. We have two quantities equal to the diagonal, and thus to each other.

$$\sqrt{2}l = \text{diagonal} = 4r = 4 \times 143.1 \text{ pm} = 572.4 \text{ pm}$$

$$l = \frac{572.4}{\sqrt{2}} = 404.7 \text{ pm}$$

The cubic unit cell volume, V, is equal to the cube of one side.

$$V = l^3 = (404.7 \text{ pm})^3 = 6.628 \times 10^7 \text{ pm}^3$$

10A In a bcc unit cell, there are eight corner atoms, of which $\frac{1}{8}$ of each is apportioned to the unit cell.

There is also one atom in the center. The total number of atoms per unit cell is:

$=1$ center $+8$ corners $\times \frac{1}{8} = 2$ atoms. The density, in g/cm^3, for this cubic cell:

$$\text{density} = \frac{2 \text{ atoms}}{(524 \text{ pm})^3} \times \left(\frac{10^{12} \text{ pm}}{10^2 \text{ cm}}\right)^3 \times \frac{1 \text{ mol}}{6.022 \times 10^{23} \text{ atoms}} \times \frac{39.10 \text{ g K}}{1 \text{ mol K}} = 0.903 \text{ g}/cm^3$$

The tabulated density of potassium at $20°C$ is $0.86 \text{ g}/cm^3$.

10B In a fcc unit cell the number of atoms is computed as 1/8 atom for each of the eight corner atoms (since each is shared among eight unit cells) plus 1/2 atom for each of the six face atoms (since each is shared between two unit cells). This gives the total number of atoms per unit cell as: atoms/unit cell = (1/8 corner atom × 8 corner atoms/unit cell) + (1/2 face atom × 6 face atoms/unit cell) = 4 atoms/unit cell

Now we can determine the mass per Al atom, and a value for the Avogadro constant.

$$\frac{\text{mass}}{\text{Al atom}} = \frac{2.6984 \text{ g Al}}{1 \text{ cm}^3} \times \left(\frac{100 \text{ cm}}{1 \text{ m}} \times \frac{1 \text{ m}}{10^{12} \text{ pm}}\right)^3 \times \frac{6.628 \times 10^7 \text{ pm}^3}{1 \text{ unit cell}} \times \frac{1 \text{ unit cell}}{4 \text{ Al atoms}}$$

$$= 4.471 \times 10^{-23} \text{ g}/\text{Al atom}$$

$$N_A = \frac{26.9815 \text{ g Al}}{1 \text{ mol Al}} \times \frac{1 \text{ Al atom}}{4.471 \times 10^{-23} \text{ g Al}} = 6.035 \times 10^{23} \frac{\text{atoms Al}}{\text{mol Al}}$$

11A Across the diagonal of a CsCl unit cell are Cs^+ and Cl^- ions, so that the body diagonal equals $2r(Cs^+) + 2r(Cl^-)$. This body diagonal equals $\sqrt{3}l$, where l is the length of the unit cell.

$$l = \frac{2r(Cs^+) + 2r(Cl^-)}{\sqrt{3}} = \frac{2(167 + 181) \text{ pm}}{\sqrt{3}} = 402 \text{ pm}$$

11B Since NaCl is fcc, the Na^+ ions are in the same locations as were the Al atoms in Practice Example 12-10B, and there are 4 Na^+ ions per unit cell. For stoichiometric reasons, there must also be 4 Cl^- ions per unit cell. These are accounted for as follows: there is one Cl^- along each edge, and each of these edge Cl^- ions are shared among four unit cells, and there is one Cl^- precisely in the body center of the unit cell, not shared with any other unit cells. Thus, the number of Cl^- ions is given by: Cl^- ions/unit cell =
$(1/4 \text{ Cl}^- \text{ on edge} \times 12 \text{ edges per unit cell}) + 1 \text{ Cl}^- \text{ in body center} = 4 \text{ Cl}^-/\text{unit cell}$.
The volume of this cubic unit cell is the cube of its length. The density is:

$$\text{NaCl density} = \frac{4 \text{ formula units}}{1 \text{ unit cell}} \times \frac{1 \text{ unit cell}}{(560 \text{ pm})^3} \times \left(\frac{10^{12} \text{ pm}}{1 \text{ m}} \times \frac{1 \text{ m}}{100 \text{ cm}}\right)^3 \times \frac{1 \text{ mol NaCl}}{6.022 \times 10^{23} \text{ f.u.}}$$

$$\times \frac{58.44 \text{ g NaCl}}{1 \text{ mol NaCl}} = 2.21 \text{ g/cm}^3$$

12A Sublimation of Cs(g): $\qquad$ $Cs(s) \rightarrow Cs(g)$ $\qquad\qquad$ $\Delta H_{sub} = +78.2$ kJ / mol

Ionization of Cs(g): $\qquad\qquad$ $Cs(g) \rightarrow Cs^+(g) + e^-$ $\qquad$ $\Delta I_1 = +375.7$ kJ / mol

$\qquad\qquad\qquad\qquad\qquad\qquad\qquad\qquad\qquad\qquad\qquad\qquad$ (Table 9.3)

$\dfrac{1}{2}$ Dissociation of $Cl_2(g)$: $\quad$ $\dfrac{1}{2}\, Cl_2(g) \rightarrow Cl(g)$ $\qquad$ $DE = \dfrac{1}{2} \times 243$ kJ $= 121.5$ kJ/mol

$\qquad\qquad\qquad\qquad\qquad\qquad\qquad\qquad\qquad\qquad\qquad\qquad$ (Table 10.3)

Cl(g) electron affinity: $\qquad$ $Cl(g) + e^- \rightarrow Cl^-(g)$ $\qquad$ $EA_1 = -349.0$ kJ / mol

$\qquad\qquad\qquad\qquad\qquad\qquad\qquad\qquad\qquad\qquad\qquad\qquad$ (Figure 9–10)

Lattice energy: $\qquad\qquad\qquad$ $Cs^+(g) + Cl^-(g) \rightarrow CsCl(s)$ $\quad$ L.E.

Enthalpy of formation: $\qquad$ $Cs(s) + \dfrac{1}{2}\, Cl_2(s) \rightarrow CsCl(s)$ $\quad$ $\Delta H_f^{\,o} = -442.8$ kJ / mol

-442.8 kJ/mol $= +78.2$ kJ/mol $+ 375.7$ kJ/mol $+ 121.5$ kJ/mol $- 349.0$ kJ/mol $+$ L.E.

$\qquad\qquad\quad = +226.4$ kJ/mol $+$ L.E.

L.E. $= -442.8$ kJ $- 226.4$ kJ $= -669.2$ kJ/mol

12B Sublimation: $\qquad\qquad\qquad$ $Ca(s) \rightarrow Ca(g)$ $\qquad\qquad$ $\Delta H_{sub} = +178.2$ kJ / mol

First ionization energy: $\qquad$ $Ca(g) \rightarrow Ca^+(g) + e^-$ $\qquad$ $I_1 = +590$ kJ / mol

Second ionization energy: $\quad$ $Ca^+(g) \rightarrow Ca^{2+}(g) + e^-$ $\qquad$ $I_2 = +1145$ kJ / mol

Dissociation energy: $\qquad\quad$ $Cl_2(g) \rightarrow 2Cl(g)$ $\qquad\qquad$ $D.E. = (2 \times 122)$ kJ / mol

Electron Affinity: $\qquad\qquad$ $2\, Cl(g) + 2\, e^- \rightarrow 2Cl^-(g)$ $\quad$ $2 \times E.A. = 2(-349)$ kJ/mol

Lattice energy: $\qquad\qquad\quad$ $Ca^{2+}(g) + 2\, Cl^-(g) \rightarrow CaCl_2(s)$ $\quad$ L.E. $= -2223$ kJ/mol

---

Enthalpy of formation: $\qquad$ $Ca(s) + Cl_2(s) \rightarrow CaCl_2(s)$ $\qquad$ $\Delta H_f^{\,o} = ?$

$\Delta H_f^{\,o} = \Delta H_{sub} + I_1 + I_2 + D.E. + (2 \times E.A.) + L.E.$

$\qquad = 178.2$ kJ/mol $+ 590$ kJ/mol $+ 1145$ kJ/mol $+ 244$ kJ/mol $- 698$ kJ/mol $- 2223$ kJ/mol

$\qquad = -764$ kJ/mol

EXERCISES

Surface Tension; Viscosity

1. Since both the silicone oil and the cloth or leather are composed of relatively nonpolar molecules, they attract each other. The oil thus adheres well to the material. Water, on the other hand is polar and adheres very poorly to the silicone oil (actually, the water is repelled by the oil), much more poorly, in fact, than it adheres to the cloth or leather. This is because the oil is more nonpolar than is the cloth or the leather. Thus, water is repelled from the silicone-treated cloth or leather.

2. Both surface tension and viscosity deal with the work needed to overcome the attractions between molecules. Increasing the temperature of a liquid sample causes the molecules to move faster. Some of the work has been done by adding thermal energy (heat) and less work needs to be done by the experimenter, consequently, both surface tension and viscosity decrease. The vapor pressure is a measure of the concentration of molecules that have broken free of the surface. As thermal energy is added to the liquid sample, more and more molecules have enough energy to break free of the surface, and the vapor pressure increases.

3. **(a)** Molasses, like honey, is a very viscous liquid (high resistance to flow)

(b) The coldest temperatures are generally in January (in the northern hemisphere).

(c) Viscosity generally increases as the temperature decrease. Hence, molasses at low temperature is a very slow flowing liquid. Thus there is indeed a scientific basis for the expression "slower than molasses" in January.

4. The product can lower the surface tension of water. Then the water can more easily wet a solid substance, because a greater surface area of water can be created with the same energy. (Surface tension equals the work needed to create a given quantity of surface area.) This greater water surface area means a greater area of contact with a solid object, such as a piece of fabric. More of the fabric being in contact with the water means that the water indeed is wetter.

Vaporization

5. The process of evaporation is endothermic, meaning it requires energy. If evaporation occurs from an uninsulated container, this energy is obtained from the surroundings, through the walls of the container. However, if the evaporation occurs from an insulated container, the only source of the needed energy is the liquid that is evaporating. Therefore, the temperature of the liquid will decrease as the liquid evaporates.

6. Vapor cannot form throughout the liquid at temperatures below the boiling point because for vapor to form, it must overcome the atmospheric pressure (≈ 1 atm) or slightly more due to the pressure of the liquid. Formation of a bubble of vapor in the liquid, requires that it must push the liquid out of the way. This is not true at the surface. The vapor molecules simply move into the gas phase at the surface, which is mostly empty space.

7. We use the quantity of heat to determine the number of moles of benzene that vaporize.

$$V = \frac{nRT}{P} = \frac{\left(1.54\,\text{kJ} \times \frac{1\,\text{mol}}{33.9\,\text{kJ}}\right) \times 0.08206\,\frac{\text{L atm}}{\text{mol K}} \times 298\,\text{K}}{95.1\,\text{mmHg} \times \frac{1\,\text{atm}}{760\,\text{mmHg}}} = 8.88\,\text{L}\,C_6H_6(l)$$

8. $n_{acetonitrile} = \dfrac{PV}{RT} = \dfrac{1.00\,\text{atm} \times 1.17\,\text{L}}{0.08206\,\text{L atm mol}^{-1}\,\text{K}^{-1} \times (273.2 + 81.6)\,\text{K}} = 0.0402$ mol acetonitrile

$\Delta H_{vap} = \dfrac{1.00\,\text{kJ}}{0.0402\,\text{mol}} = 24.9$ kJ / mol acetonitrile

9. 25.00 mL of N_2H_4 (25 °C) density (25°C) = 1.0036 g mL^{-1} (Molar mass = 32.0452 g mol^{-1})
Mass of N_2H_4 = (volume) × (density) = (25.00 mL) × (1.0036 g mL^{-1}) = 25.09 g N_2H_4

$n_{N2H4} = 25.09$ g $N_2H_4 \times \dfrac{1\,\text{mol}\,N_2H_4}{32.0452\,\text{g}\,N_2H_4} = 0.7830$ mol

Energy required to increase temperature from 25.0 °C to 113.5 °C ($\Delta t = 88.5\,^{\circ}\text{C}$)

$q_{heating} = (n)(C)(\Delta t) = (0.7829\underline{5}\,\text{mol}\,N_2H_4)\left(\dfrac{98.84\,\text{J}}{1\,\text{mol}\,N_2H_4\,^{\circ}\text{C}}\right)(88.5\,^{\circ}\text{C}) = 6848.7$ J or 6.85 kJ

$q_{vap} = (n_{N_2H_4})(\Delta H_{vap}) = (0.7829\underline{5}\,\text{mol}\,N_2H_4)\left(\dfrac{43.0\,\text{kJ}}{1\,\text{mol}\,N_2H_4}\right) = 33.7$ kJ

$q_{overall} = q_{heating} + q_{vap} = 6.85\,\text{kJ} + 33.7\,\text{kJ} = 40.5\,\text{kJ}$

10. ΔH_{vap} for $CH_3OH(l) = 38.0$ kJ mol^{-1} at 298 K (assumes ΔH is temperature insensitive)

$\Delta t = 30.0$ °C - 20.0 °C = 10.0 °C

$$n_{CH_3OH} = 215 \text{ g } CH_3OH \times \left(\frac{1 \text{ mol } CH_3OH}{32.0422 \text{ g } CH_3OH} \right) = 6.71 \text{ mol } CH_3OH$$

Raise temperature of liquid from 20.0 °C to 30.0 °C

$$q = (n)(C)(\Delta t) = (6.71 \text{ mol } CH_3OH)\left(\frac{81.1 \text{ J}}{1 \text{ mol } CH_3OH \text{ °C}} \right)(10.0 \text{ °C}) = 5441.8 \text{ J or } 5.44 \text{ kJ}$$

Vaporize liquid at 30°C (Use ΔH_{vap} at 25 °C and assume value is the same at 30 °C)

$$q_{vap} = (n_{CH_3OH})(\Delta H_{vap}) = (6.71 \text{ mol } CH_3OH)\left(\frac{38.0 \text{ kJ}}{1 \text{ mol } CH_3OH} \right) = 255 \text{ kJ}$$

$q_{overall} = q_{heating} + q_{vap} = 5.44 \text{ kJ} + 255 \text{ kJ} = 260. \text{ kJ}$

11. heat needed $= 3.78 \text{ L } H_2O \times \dfrac{1000 \text{ cm}^3}{1 \text{ L}} \times \dfrac{0.958 \text{ g } H_2O}{1 \text{ cm}^3} \times \dfrac{1 \text{ mol } H_2O}{18.02 \text{ g } H_2O} \times \dfrac{40.7 \text{ kJ}}{1 \text{ mol } H_2O} = 8.18 \times 10^3 \text{ kJ}$

amount CH_4 needed $= 8.18 \times 10^3 \text{ kJ} \times \dfrac{1 \text{ mol } CH_4}{890 \text{ kJ}} = 9.19 \text{ mol } CH_4$

$$V = \frac{nRT}{P} = \frac{9.19 \text{ mol} \times 0.08206 \text{ L atm mol}^{-1}\text{K}^{-1} \times 296.6 \text{ K}}{768 \text{ mmHg} \times \dfrac{1 \text{ atm}}{760 \text{ mmHg}}} = 221 \text{ L methane}$$

12. If not all of the water vaporizes, the final temperature of the system will be $100.00°$ C. Let us proceed on that assumption and modify our final state if this is not true. First we determine the heat available from the iron in cooling down, and then the heat needed to warm the water to boiling, and finally the mass of water that vaporizes.

heat from Fe $= \text{ mass} \times \text{sp.ht.} \times \Delta t = 50.0 \text{ g} \times \dfrac{0.45 \text{ J}}{\text{g}° \text{ C}} \times \left(100.00° \text{ C} - 152° \text{ C}\right) = -1.17 \times 10^3 \text{ J}$

heat to warm water $= 20.0 \text{ g} \times \dfrac{4.21 \text{ J}}{\text{g}°\text{C}} \times \left(100.00°\text{C} - 89°\text{C}\right) = 9.3 \times 10^2 \text{ J}$

mass of water vaporized:

$$= \left(11.7 \times 10^2 \text{ J available} - 9.3 \times 10^2 \text{ J used}\right) \times \frac{1 \text{ mol } H_2O \text{ vaporized}}{40.7 \times 10^3 \text{ J}} \times \frac{18.02 \text{ g } H_2O}{1 \text{ mol } H_2O}$$

$= 0.11$ g of water vaporize

Clearly, all of the water does not vaporize and our initial assumption was valid.

Vapor Pressure and Boiling Point

13. **(a)** We read up the $100°C$ line until we arrive at C_6H_7N curve (e). This occurs at about 45 mmHg.

(b) We read across the 760 mmHg line until we arrive at the C_7H_8 curve (d). This occurs at about $110°C$.

14. **(a)** The normal boiling point occurs where the vapor pressure is 760 mmHg, and thus $\ln P = 6.63$. For aniline, this occurs at about the uppermost data point (open circle) on the aniline line. This corresponds to $1/T = 2.18 \times 10^{-3}$ K^{-1}. Thus,

$$T_{nbp} = \frac{1}{2.18 \times 10^{-3} \text{ K}^{-1}} = 459 \text{ K}.$$

(b) $25°$ C $= 298$ K $= T$ and thus $1/T = 3.36 \times 10^{-3}$ K^{-1}. This occurs at about $\ln P = 6.25$. Thus, $P = e^{6.25} = 518$ mmHg.

15. Use the ideal gas equation, $n =$ moles Br$_2$ $= 0.486$ g Br$_2$ $\times \dfrac{1 \text{ mol Br}_2}{159.8 \text{ g Br}_2} = 3.04 \times 10^{-3}$ mol Br$_2$.

$$P = \frac{nRT}{V} = \frac{3.04 \times 10^{-3} \text{ mol Br}_2 \times 0.08206 \text{ L atm mol}^{-1} \text{ K}^{-1} \times 298.2 \text{ K}}{0.2500 \text{ L}} \times \frac{760 \text{ mmHg}}{1 \text{ atm}} = 226 \text{ mmHg}$$

16. We can determine the vapor pressure by using the ideal gas law.

$$P = \frac{nRT}{V} = \frac{\left(0.876 \text{ g (CH}_3)_2\text{CO} \times \dfrac{1 \text{ mol (CH}_3)_2\text{CO}}{58.08 \text{ g (CH}_3)_2\text{CO}}\right)\left(\dfrac{0.08206 \text{ L atm}}{\text{K mol}}\right) \times (32 + 273.15)\text{K}}{1 \text{ L}} = 0.378 \text{ atm}$$

$$P = 0.378 \text{ atm} \times \frac{101.325 \text{ kPa}}{1 \text{ atm}} = 38.3 \text{ kPa}$$

17. **(a)** In order to vaporize water in the outer container, heat must be applied (i.e. vaporization is an endothermic process). When this vapor (steam) condenses on the outside walls of the inner container, that same heat is liberated. Thus condensation is an exothermic process.

(b) Liquid water, condensed on the outside wall, is in equilibrium with the water vapor that fills the space between the two containers. This equilibrium exists at the boiling point of water. We assume that the pressure is 1.000 atm, and thus, the temperature of the equilibrium must be 373.15 K or 100.00° C. This is the maximum temperature that can be realized without pressurizing the apparatus.

18. When the can is heated, the vapor pressure of water inside the can is ~ 760 mm Hg. As the can cools, most of the water vapor in the can condenses to liquid and the pressure inside the can drops sharply to the vapor pressure of water at room temperature (≈ 25 mmHg). The pressure on the outside of the can is still near 760 mmHg. It is this huge difference in pressure that is responsible for the can being crushed.

19. With the Clausius-Clapeyron equation, and the vapor pressure of water at 100.0° C (373.2 K) and 120.0° C (393.2 K) to determine ΔH_{vap} of water near its boiling point. We then use the equation again, to determine the temperature at which water's vapor pressure is 2.00 atm.

$$\ln \frac{1489.1 \text{ mmHg}}{760.0 \text{ mmHg}} = \frac{\Delta H_{vap}}{8.3145 \text{ J mol}^{-1} \text{ K}^{-1}}\left(\frac{1}{373.2 \text{ K}} - \frac{1}{393.2 \text{ K}}\right) = 0.6726 = 1.639 \times 10^{-5} \Delta H_{vap}$$

$$\Delta H_{vap} = 4.104 \times 10^4 \text{ J/mol} = 41.04 \text{ kJ/mol}$$

$$\ln\frac{2.00 \text{ atm}}{1.00 \text{ atm}} = 0.6931 = \frac{41.04\times10^3 \text{ J mol}}{8.3145 \text{ J mol}^{-1} \text{ K}^{-1}}\left(\frac{1}{373.2 \text{ K}} - \frac{1}{T}\right)$$

$$\left(\frac{1}{373.2 \text{ K}} - \frac{1}{T_{bp}}\right) = 0.6931 \times \frac{8.1345 \text{ K}^{-1}}{41.03\times10^3} = 1.404\times10^{-4} \text{ K}^{-1}$$

$$\frac{1}{T_{bp}} = \frac{1}{373.2 \text{ K}} - 1.404\times10^{-4} \text{ K}^{-1} = 2.539\times10^{-3} \text{ K}^{-1} \quad T_{bp} = 393.9 \text{ K} = 120.7° \text{ C}$$

20. **(a)** We need the temperature at which the vapor pressure of water is 640 mmHg. This is a temperature between $95.0°C$ (633.9 mmHg) and $96.0°C$ (657.6 mmHg. We estimate a boiling point of $95.3°C$.

(b) If the observed boiling point is $94°C$, the atmospheric pressure must equal the vapor pressure of water at $94°C$, which is, 611 mmHg.

21. The 25.0 L of He becomes saturated with aniline vapor, at a pressure equal to the vapor pressure of aniline.

$$n_{aniline} = (6.220 \text{ g} - 6.108 \text{ g}) \times \frac{1 \text{ mol aniline}}{93.13 \text{ g aniline}} = 0.00120 \text{ mol aniline}$$

$$P = \frac{nRT}{V} = \frac{0.00120 \text{ mol} \times 0.08206 \text{ L atm mol}^{-1} \text{ K}^{-1} \times 303.2 \text{ K}}{25.0 \text{ L}} = 0.00119 \text{ atm} = 0.907 \text{ mmHg}$$

22. The final pressure must be 742 mm Hg and the partial pressure of $CCl_4(g)$ is constant at 261 mm Hg, thus, the partial pressure of N_2 must be 481 mm Hg. Using Dalton's law of partial pressures. $P_{N_2}/P_{total} = V_{N_2}/V_{total}$. The total and final pressures are know. Since the temperature does not change, the volume occupied by N_2 is constant at 7.53 L. Hence,

$$V_{total} = V_{N_2} \times \frac{P_{total}}{P_{N_2}} = 7.53 \text{ L} \times \frac{742 \text{ mmHg}}{481 \text{ mmHg}} = 11.6 \text{ L}$$

23.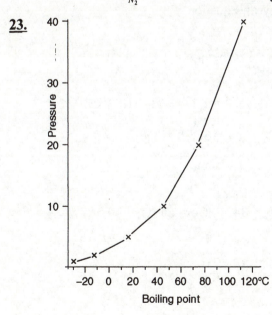

The graph of pressure vs. boiling point for Freon-12 is shown.

At a temperature of $25°$ C the vapor pressure is approximately 6.5 atm for Freon-12. Thus the compressor must be capable of producing a pressure greater than 6.5 atm.

24. At $27°\,C$, the vapor pressure of water is 26.7 mmHg. We use this value to determine the mass of water that can exist within the container as vapor.

$$\text{mass of water vapor} = \frac{\left(26.7\,\text{mmHg} \times \dfrac{1\,\text{atm}}{760\,\text{mmHg}}\right) \times \left(1515\,\text{mL} \times \dfrac{1\,\text{L}}{1000\,\text{mL}}\right)}{\dfrac{0.08206\,\text{L atm}}{\text{mol K}} \times (27+273)\,\text{K}} = 0.00216\,\text{mol H}_2\text{O(g)}$$

$$\text{mass of water vapor} = 0.00216\,\text{mol H}_2\text{O(g)} \times \frac{18.02\,\text{g H}_2\text{O}}{1\,\text{mol H}_2\text{O}} = 0.0389\,\text{g H}_2\text{O(g)}$$

The Clausius-Clapeyron Equation

25. We use the Clausius-Clapeyron equation (12.2) to answer this question.

$$T_1 = (56.0+273.2)\,\text{K} = 329.2\,\text{K} \qquad T_2 = (103.7+273.2)\,\text{K} = 376.9\,\text{K}$$

$$\ln\frac{10.0\,\text{mmHg}}{100.0\,\text{mmHg}} = \frac{\Delta H_{vap}}{8.3145\,\text{J mol}^{-1}\,\text{K}^{-1}}\left(\frac{1}{376.9\,\text{K}} - \frac{1}{329.2\,\text{K}}\right) = -2.30 = -4.624 \times 10^{-5}\,\Delta H_{vap}$$

$$\Delta H_{vap} = 4.97 \times 10^4\,\text{J/mol} = 49.7\,\text{kJ/mol}$$

26. We use the Clausius-Clapeyron equation (12.2), to answer this question.

$$T_1 = (5.0+273.2)\,\text{K} = 278.2\,\text{K}$$

$$\ln\frac{760.0\,\text{mmHg}}{40.0\,\text{mmHg}} = 2.944 = \frac{38.0 \times 10^3\,\text{J/mol}}{8.3145\,\text{J mol}^{-1}\,\text{K}^{-1}}\left(\frac{1}{278.2\,\text{K}} - \frac{1}{T_{nbp}}\right)$$

$$\left(\frac{1}{278.2\,\text{K}} - \frac{1}{T_{nbp}}\right) = 2.944 \times \frac{8.3145\,\text{K}^{-1}}{38.0 \times 10^3} = 6.44 \times 10^{-4}\,\text{K}^{-1}$$

$$\frac{1}{T_{nbp}} = \frac{1}{278.2\,\text{K}} - 6.44 \times 10^{-4}\,\text{K}^{-1} = 2.95 \times 10^{-3}\,\text{K}^{-1} \qquad T_{nbp} = 339\,\text{K}$$

27. Once again, we will employ the Clausius-Clapeyron Equation.

$$T = 56.2°\,C \text{ is } T = 329.4\,\text{K}$$

$$\ln\frac{760\,\text{mmHg}}{375\,\text{mmHg}} = \frac{25.5 \times 10^3\,\text{J/mol}}{8.3145\,\text{J mol}^{-1}\,\text{K}^{-1}}\left(\frac{1}{T} - \frac{1}{329.4\,\text{K}}\right) = 0.706$$

$$\left(\frac{1}{T} - \frac{1}{329.4\,\text{K}}\right) = \frac{0.706 \times 8.3145}{25.5 \times 10^3}\,\text{K}^{-1} = 2.30 \times 10^{-4}\,\text{K}^{-1} = 1/T - 3.03_6 \times 10^{-3}\,\text{K}^{-1}$$

$$1/T = (3.03_6 + 0.230) \times 10^{-3}\,\text{K}^{-1} = 3.266 \times 10^{-3}\,\text{K}^{-1} \qquad T = 306\,\text{K} = 33°\,C$$

28. $P_1 = 40.0$ torr $\quad T_1 = -7.1\,°C$ (266 K) and $\Delta H_{vap} = 29.2\,\text{kJ mol}^{-1}$
$P_2 = 760.0$ torr $\quad T_2 = ?$

$$\ln\left(\frac{760.0}{40.0}\right) = \frac{29{,}200\,\text{J}}{8.31451\dfrac{\text{J}}{\text{K mol}}}\left(\frac{1}{266\,K} - \frac{1}{T_2}\right) \qquad T_2 = 342.\underline{3}\,\text{K or } {\sim}69\,°C$$

(The literature boiling point for trichloromethane is 61 °C)

29. Normal boiling point = 179 °C and critical point = 422 °C and 45.9 atm

$$\ln\left(\frac{P_2}{P_1}\right) = \frac{\Delta H_{vap}}{R}\left(\frac{1}{T_1} - \frac{1}{T_2}\right) \qquad \ln\left(\frac{45.9}{1}\right) = \frac{\Delta H_{vap}}{8.3145 \ J K^{-1} mol^{-1}}\left(\frac{1}{452.2 \ K} - \frac{1}{695.2 \ K}\right)$$

$\Delta H_{vap} = 41.2 \ kJ \ mol^{-1}$

$$\ln\left(\frac{1}{P}\right) = \frac{41,200 \ J \ mol^{-1}}{8.3145 \ J K^{-1} mol^{-1}}\left(\frac{1}{373.2} - \frac{1}{452.2 \ K}\right) \qquad P = 0.0981 \text{ atm or } 74.6 \text{ torr}$$

30. (a) The plot of ln P vs 1/T for benzene is located to the right of that for toluene. This tells us that for a given temperature, the vapor pressure for toluene is lower than that for benzene, and thus benzene is a more volatile liquid than toluene

    (b) According to Figure 12-13, at 65 °C, the vapor pressure for benzene is ~ 450 mm Hg or 0.60 atm. We have been asked to estimate the temperature at which toluene has a vapor pressure of 0.60 atm. First, we find ln 450 = 6.11 on the benzene curve of Figure 12.13. Then, we move horizontally to the left along ln P = 6.11 until we reach the toluene curve. At this point on the toluene curve, the value of 1/T = 0.00273. This corresponds to T = 366 K.

Critical Point

31. Substances that can exist as a liquid at room temperature (about 20° C) have critical temperature above 20° C, 293 K. Of the substances listed in Table 12.3, $CO_2(T_c = 304.2 \ K)$, $HCl(T_c = 324.6 \ K)$, $NH_3(T_c = 405.7 \ K)$, $SO_2(T_c = 431.0 \ K)$, and $H_2O(T_c = 647.3 \ K)$ can exist in liquid form at 20 °C. In fact, CO_2 exists as a liquid in CO_2 fire extinguishers.

32. The critical temperature of SO_2, 431.0 K, is above the temperature of 0°C, 273 K. The critical pressure of SO_2 is 77.7 atm, which is below the pressure of 100 atm. Thus, SO_2 can be maintained as a liquid at 0° C and 100 atm. The critical temperature of methane, CH_4, is 191.1 K, which is below the temperature of 0° C, 273 K. Thus, CH_4 cannot exist as a liquid at 0° C, no matter what pressure is applied.

33. (a) $\text{heat evolved} = 3.78 \text{ kg Cu} \times \dfrac{1000 \text{ g}}{1 \text{ kg}} \times \dfrac{1 \text{ mol Cu}}{63.55 \text{ g Cu}} \times \dfrac{13.05 \text{ kJ}}{1 \text{ mol Cu}} = 776 \text{ kJ evolved or}$

       $\Delta H = -776 \text{ kJ}$

    (b) $\text{heat absorbed} = (75 \text{ cm} \times 15 \text{ cm} \times 12 \text{ cm}) \times \dfrac{8.92 \text{ g}}{1 \text{ cm}^3} \times \dfrac{1 \text{ mol Cu}}{63.55 \text{ g Cu}} \times \dfrac{13.05 \text{ kJ}}{1 \text{ mol Cu}} = 2.5 \times 10^4 \text{ kJ}$

34. The relevant reaction is: $C_3H_8(g) + 5\ O_2(g) \rightarrow 3\ CO_2(g) + 4\ H_2O(g)$

$\Delta H_{rxn} = \Delta H_{combustion} = 4(-285.8\ kJ) + 3(-393.5\ kJ) - (-103.8\ kJ) = -2219.9\ kJ\ mol^{-1}\ C_3H_8(g)$

$$n = \frac{PV}{RT} = \frac{\left(748\ mmHg \times \dfrac{1\ atm}{760\ mmHg}\right) \times 1.35\ L}{\left(\dfrac{0.08206\ L\ atm}{K\ mol}\right) \times (25.0 + 273.15\ K)} = 0.0543\ mol\ C_3H_8(g)$$

$q_{combustion} = n\Delta H_{combustion} = 0.0543\ mol\ C_3H_8(g) \times (-2219.9\ kJ\ mol^{-1}\ C_3H_8(g)) = -120.5\underline{6}\ kJ$

$q_{combustion} = -q_{melting} = -120.5\underline{6}\ kJ = -n\Delta H_{fusion} = m/M(\Delta H_{fusion}) = \dfrac{m(6.01\ kJ\ mol^{-1})}{18.0153\ g\ mol^{-1}} = 361\ g\ ice$

States of Matter and Phase Diagrams

35. Let us use the ideal gas law to determine the final pressure in the container, assuming that all of the dry ice vaporizes. We then locate this pressure, at a temperature of 25° C, on the phase diagram of Figure 12-19.

$$P = \frac{nRT}{V} = \frac{\left(80.0\ g\ CO_2 \times \dfrac{1\ mol\ CO_2}{44.0\ g\ CO_2}\right) \times 0.08206\ \dfrac{L\ atm}{mol\ K} \times 298\ K}{0.500\ L} = 88.9\ atm$$

Although this point (25° C and 88.9 atm) is most likely in the region labeled "liquid" in Figure 12–19, we computed its pressure assuming the CO_2 is a gas. Some of this gas should condense to a liquid. Thus, both liquid and gas are present in the container. Solid would not be present unless the temperature is below -50 °C at ~ 88.9 atm.

36. The phase diagram for Hydrazine is sketched below. We only have a few points given as data that we can place on the phase diagram (critical point, boiling point and an experimental point where it sublimes). We are not certain of the triple point for hydrazine, nor are we aware of the slope and curvature of the lines in the diagram.

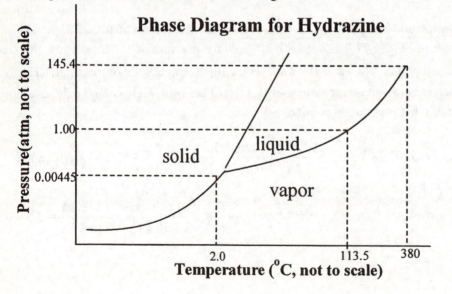

37. **(a)** As heat is added initially, the temperature of the ice rises from $-20°$ C to $0°$ C. At (or just slightly below) $0°$ C, ice begins to melt to liquid water. The temperature remains at $0°$ C until all of the ice has melted. Adding heat then warms the liquid until a temperature of about $93.5°$ C is reached, at which point the liquid begins to vaporize to steam. The temperature remains fixed until all the water is converted to steam. Adding heat then warms the steam to $200°$ C. (Data for this part are taken from Figure 12-22 and Table 12.2.)

(b) As the pressure is raised, initially gaseous iodine is just compressed. At ~ 91 mmHg, liquid iodine appears. As the pressure is pushed above 99 mmHg, more liquid condenses until eventually all the vapor is converted to a liquid. Increasing the pressure further simply compresses the liquid until a high pressure is reached, perhaps 50 atm, where solid iodine appears. Again the pressure remains fixed with further compression until all of the iodine is converted to its solid form. After this has occurred further compression raises the pressure on the solid until 100 atm is reached. (Data for this part are from Figure 12-18 and the surrounding text.)

(c) Cooling of gaseous CO_2 simply lowers the temperature until a temperature of perhaps $20°$ C is reached. At this point, liquid CO_2 appears. The temperature remains constant as more heat is removed until all the gas is converted to liquid. Further cooling then lowers the temperature of the liquid until a temperature of slightly higher than $-56.7°$ C is reached, where solid CO_2 appears. At this point, further cooling simply converts liquid to solid at constant temperature, until all liquid has been converted to solid. From this point, further cooling lowers the temperature of the solid. (Data for this part are taken from Figure 12-19 and Table 12.3.)

38. **(a)** The upper-right region of the phase diagram is the liquid region, while the lower-right region is the region of gas.

(b) Melting involves converting the solid into a liquid. As the phase diagram shows, the lowest pressure at which liquid exists is at the triple point pressure, namely, 43 atm. 1.00 atm is far below 43 atm. Thus, liquid cannot exist at this temperature.

(c) As we move from point A to point B by lowering the pressure, initially nothing happens. At a certain pressure, the solid liquefies. The pressure continues to drop, with the entire sample being liquid while it does, until another, lower pressure is reached. At this lower pressure the entire sample vaporizes. The pressure then continues to drop, with the gas becoming less dense as the pressure falls, until point B is reached.

39. 0.240 g of H_2O corresponds to 0.0133 mol H_2O. If the water does not vaporize completely, the pressure of the vapor in the flask equals the vapor pressure of water at the indicated temperature. However, if the water vaporizes completely, the pressure of the vapor is determined by the ideal gas law.

(a) 30.0° C, vapor pressure of H_2O = 31.8 mmHg = 0.0418 atm

$$n = \frac{PV}{RT} = \frac{0.0418 \text{ atm} \times 3.20 \text{ L}}{0.08206 \text{ L atm mol}^{-1} \text{ K}^{-1} \times 303.2 \text{ K}}$$

n = 0.00538 mol H_2O vapor, this is less than 0.0133 mol H_2O;

This represents a non-eqilibrium condition, since not all the H_2O vaporizes.

The pressure in the flask is 0.0418 atm. (from tables)

(b) 50.0° C, vapor pressure of H_2O = 92.5 mmHg = 0.122 atm

$$n = \frac{PV}{RT} = \frac{0.122 \text{ atm} \times 3.20 \text{ L}}{0.08206 \text{ L atm mol}^{-1} \text{ K}^{-1} \times 323.2 \text{ K}}$$

= 0.0147 mol H_2O vapor > 0.0133 mol H_2O; all the H_2O vaporizes. Thus,

$$P = \frac{nRT}{V} = \frac{0.0133 \text{ mol} \times 0.08206 \text{ L atm mol}^{-1} \text{ K}^{-1} \times 323.2 \text{ K}}{3.20 \text{ L}} = 0.110 \text{ atm} = 83.8 \text{ mmHg}$$

(c) 70.0° C All the H_2O must vaporize, as this temperature is higher than that of part (b). Thus,

$$P = \frac{nRT}{V} = \frac{0.0133 \text{ mol} \times 0.08206 \text{ L atm mol}^{-1} \text{ K}^{-1} \times 343.2 \text{ K}}{3.20 \text{ L}} = 0.117 \text{ atm} = 89.0 \text{ mmHg}$$

40. (a) If the pressure exerted by 2.50 g of H_2O (vapor) is less than the vapor pressure of water at 120. °C (393 K), then the water must exist entirely as a vapor.

$$P = \frac{nRT}{V} = \frac{\left(2.50 \text{ g} \times \dfrac{1 \text{ mol } H_2O}{18.02 \text{ g } H_2O}\right) \times 0.08206 \text{ L atm mol}^{-1}\text{K}^{-1} \times 393 \text{ K}}{5.00 \text{ L}} = 0.895 \text{ atm}$$

Since this is less than the 1.00 atm vapor pressure at 100. °C, it must be less than the vapor pressure of water at 120 °C. Thus, the water exists entirely as a vapor.

(b) 0.895 atm = 680 mmHg. From Table 12.2, we see that this corresponds to a temperature of 97.0° C (at which temperature the vapor pressure of water is 682.1 mmHg). Thus, at temperature slightly less than 97.0° C the water will begin to condense to liquid. But we have forgotten that, this constant-volume container, the pressure (of water) will decrease as the temperature decrease Calculating the precise decrease in both involves linking the Clausius-Clapeyron equation with the expression $P = kT$, but we can estimate the final temperature. First, we determine the pressure if we lower the temperature from 393 K to 97° C (370 K).

$$P_f = \frac{370 \text{ K}}{393 \text{ K}} \times 680 \text{ mmHg} = 640 \text{ mmHg}$$ From Table 12.2, this pressure occurs at a

temperature of about $95°$ C (368 K). We now determine the final pressure at this

temperature. $P_f = \dfrac{368 \text{ K}}{393 \text{ K}} \times 680 \text{ mmHg} = 637 \text{ mmHg}$ This is just slightly above the vapor

pressure of water (633.9 mmHg) at $95°$ C, which we conclude must be the temperature
at which liquid water appears.

41. **(a)** According to Figure 12-19, $CO_2(s)$ exists at temperatures below $-78.5°C$ when the
pressure is 1 atm or less. We do not expect to find temperatures this low and partial
pressures of $CO_2(g)$ of 1 atm on the surface of the earth.

(b) According to Table 12.3, the critical temperature of CH_4, the maximum temperature at
which $CH_4(l)$ can exist, is $191.1 \text{ K} = -82.1°$ C. We do not expect to find temperatures
this low on the surface of the earth.

(c) Since, according to Table 12.3, the critical temperature of SO_2 is
$431.0 \text{ K} = 157.8 °C$, $SO_2(g)$ can be found on the surface of the earth.

(d) According to Figure 12–18, $I_2(l)$ can exist at pressures less than 1.00 atm between the
temperatures of $114°$ C and $184°$ C. There are very few places on the Earth that reach
temperatures this far above the boiling point of water at pressures below 1 atm, one
example of such a place would be the mouth of a volcano high above sea level.
Essentially, $I_2(l)$ is not found on the surface of the earth.

(e) According to Table 12.3, the critical temperature–the maximum temperature at which
$O_2(l)$ exists, is $154.8 \text{ K} = -118.4 °C$. Temperatures this low do not exist on the surface
of the earth.

42. The final pressure specified (100. atm) is below the ice III-ice I-liquid water triple point,
according to the text adjacent to Figure 12–21; no ice III is formed. The starting point is that of
$H_2O(g)$. When the pressure is increased to about 4.5 mmHg, the vapor condenses to ice I. As
the pressure is raised, the melting point of ice I decreases–by $1°$ C for every 125 atm increase
in pressure. Thus, somewhere between 10 and 15 atm, the ice I melts to liquid water, and it
remains in this state until the final pressure is reached.

43. **(a)** Heat lost by water $= q_{water} = (m)(C)(\Delta t)$

$$q_{water} = (100.0 \text{ g})\left(4.184\frac{\text{J}}{\text{g} °C}\right)(0.00 °C - 20.00 °C)\left(\frac{1 \text{kJ}}{1000 \text{ J}}\right) = -8.37 \text{ kJ}$$

Using $\Delta H_{cond} = -\Delta H_{vap}$ and heat lost by system = heat loss of condensation + cooling

$$q_{steam} = (175 \text{ g H}_2\text{O})\left(4.184\frac{\text{J}}{\text{g}\,^\circ\text{C}}\right)(0.0\,^\circ\text{C} - 100.0\,^\circ\text{C})\left(\frac{1\text{kJ}}{1000\text{J}}\right)$$

$$+(175 \text{ g H}_2\text{O})\left(\frac{1\text{mol H}_2\text{O}}{18.015\text{g H}_2\text{O}}\right)\left(\frac{-40.7\text{kJ}}{1\text{mol H}_2\text{O}}\right)$$

q_{steam} = -395.4 kJ + -73.2 kJ = -468.6 kJ or ~ -469 kJ

Total energy to melt the ice = $q_{water} + q_{steam}$ = -8.37 kJ + -469 kJ = -477 kJ

Moles of ice melted = $(477 \text{ kJ})\left(\dfrac{1\text{mol ice}}{6.01\text{kJ}}\right)$ = 79.4 mol ice melted

Mass of ice melted = $(79.4 \text{ mol H}_2\text{O})\left(\dfrac{18.015\text{g H}_2\text{O}}{1\text{mol H}_2\text{O}}\right)\left(\dfrac{1\text{kg H}_2\text{O}}{1000\text{g H}_2\text{O}}\right)$ = 1.43 kg

Mass of unmelted ice = 1.65 kg $-$ 1.43 kg = 0.22 kg

(b) Mass of unmelted ice = 0.22 kg (from above)
Heat required to melt ice = $n\,\Delta H_{fusion}$

Heat required = $(0.22 \text{ kg ice})\left(\dfrac{1000\text{g H}_2\text{O}}{1\text{kg H}_2\text{O}}\right)\left(\dfrac{1\text{mol H}_2\text{O}}{18.015\text{g H}_2\text{O}}\right)\left(\dfrac{6.01\text{kJ}}{1\text{mol H}_2\text{O}}\right)$ = 73.$\underline{4}$ kJ

Next we need to determine heat produced when 1 mole of steam (18.015 g) condenses and cools from 100.°C to 0.0 °C. Heat evolved,

$$= (1 \text{ mol H}_2\text{O})\left(\frac{-40.7\text{kJ}}{1\text{mol H}_2\text{O}}\right) + (18.015 \text{ g})\left(4.184\frac{\text{J}}{\text{g}\,^\circ\text{C}}\right)(0.0\,^\circ\text{C} - 100.\,^\circ\text{C})\left(\frac{1\text{kJ}}{1000\text{J}}\right)$$

= -40.7 kJ + -7.54 kJ = -48.2 kJ per mole of H_2O(g) or per 18.015 g H_2O(g)

Mass of steam required = $(73.\underline{4} \text{ kJ})\left(\dfrac{1\text{mol H}_2\text{O(g)}}{48.2\text{kJ}}\right)\left(\dfrac{18.015\text{g H}_2\text{O}}{1\text{mol H}_2\text{O}}\right)$ = 27 g steam

44. The heat gained by the ice equals the negative of the heat lost by the water. Let us use this fact in a step-by-step approach. We first compute the heat needed to raise the temperature of the ice to 0.0 °C and then the heat given off when the temperature of the water is lowered to 0.0 °C.

to heat the ice = $54 \text{ cm}^3 \times \dfrac{0.917 \text{ g}}{1 \text{ cm}^3} \times 2.01 \text{ J g}^{-1}\,^\circ\text{C}^{-1} \times (0\,^\circ\text{C} + 25.0\,^\circ\text{C}) = 2.5 \times 10^3 \text{ J}$

to cool the water = $-400.0 \text{ cm}^3 \times \dfrac{0.998 \text{ g}}{1 \text{ cm}^3} \times 4.18 \text{ J g}^{-1}\,^\circ\text{C}^{-1} \times (0\,^\circ\text{C} - 32.0\,^\circ\text{C}) = 53.4 \times 10^3 \text{ J}$

Thus, at 0 °C, we have 50. g ice, 399 g water, and $(53.4 - 2.5) \times 10^3 \text{ J} = 50.9$ kJ of heat available. Since 50.0 g of ice is a bit less than 3 mol of ice and 50.9 kJ is enough heat to melt at least 8 mole of ice, all of the ice will melt. The heat needed to melt the ice is

$$50. \text{ g ice} \times \frac{1 \text{ mol H}_2\text{O}}{18.0 \text{ g H}_2\text{O}} \times \frac{6.01 \text{ kJ}}{1 \text{ mol ice}} = 17 \text{ kJ}$$

We now have $50.9 \text{ kJ} - 17 \text{ kJ} = 34 \text{ kJ}$ of heat, and $399 \text{ g} + 50. \text{ g} = 449 \text{ g}$ of water at $0 \,^{\circ}\text{C}$. We compute the temperature change that is produced by adding the heat to the water.

$$\Delta T = \frac{34 \times 10^3 \text{ J}}{449 \text{ g} \times 4.18 \text{ J g}^{-1} \,^{\circ}\text{C}^{-1}} = 18 \,^{\circ}\text{C} \text{ The final temperature is } 18 \,^{\circ}\text{C}$$

45. The liquid in the can is supercooled. When the can is opened, gas bubbles released from the carbonated beverage serve as sites for the formation of ice crystals. The condition of supercooling is destroyed and the liquid reverts to the solid phase. An alternative explanation follows. The process of the gas coming out of solution is endothermic (heat is required). (We know this to be true because the reaction solution of gas in water $\rightarrow$ gas + liquid water proceeds to the right as the temperature is raised, a characteristic direction of an endothermic reaction.) The required heat is taken from the cooled liquid, causing it to freeze.

46. Both the melting point of ice and the boiling point of water are temperatures that vary as the pressure changes, the boiling point changes more substantially than the melting point. The triple point, however, does not vary with pressure. Solid, liquid, and vapor coexist only at one fixed temperature and pressure.

Intermolecular Forces

47. **(a)** HCl is not a very heavy diatomic molecule. Thus, the London forces between HCl molecules are expected to be relatively weak. Hydrogen bonding is weak in the case of H — Cl bonds; Cl is not one of the three atoms (F, O, N) that form strong hydrogen bonds. Finally, because Cl is an electronegative atom, and H is only moderately electronegative, dipole-dipole interactions should be relatively strong.

(b) In Br_2 neither hydrogen bonds nor dipole-dipole attractions can occur (there are no H atoms in the molecule, and homonuclear molecules are nonpolar). London forces are more important in Br_2 than in HCl since Br_2 has more electrons (heavier).

(c) In ICl there are no hydrogen bonds since there are no H atoms in the molecule. The London forces are as strong as in Br_2 since the two molecules have the same number of electrons. However, dipole-dipole interactions are important in ICl, due to the polarity of the I-Cl bond.

(d) In HF London forces are not very important; the molecule has only 10 electrons and thus is quite small. Hydrogen bonding is obviously the most important interaction developed between HF molecules.

(e) In CH_4, H bonds are not important (the H atoms are not bonded to F, O, or N). In addition the molecule is not polar, so there are no dipole-dipole interactions. Finally, London forces are quite weak since the molecule contains only 10 electrons. For these reasons CH_4 has a very low critical temperature.

48.   Substituting Cl for H both makes the molecule heavier (and thus increases London forces), and polar, which results in the formation of dipole-dipole interactions. Both of these effects make it more difficult to disrupt the forces of attraction between molecules, increasing the boiling point. Substitution of Br for Cl increases the London forces, but makes the molecule less polar. Since London forces in this case are more important than dipole-dipole interactions, the boiling point increases yet again. Finally, substituting OH for Br decreases London forces but both increases the dipole-dipole interactions and creates opportunities for hydrogen bonding. Since hydrogen bonds are much stronger than London forces, the boiling point increases even further.

<u>49.</u>   (c)   <   (b)   <   (d)   <   (a)
         (ethane thiol)   (ethanol)   (butanol)   (acetic acid)
         Viscosity will depend on the intermolecular forces, (the stronger the intermolecular bonding, the more viscous the substance.

50.   (d)   <   (b)   <   (a)   <   (c)
         (butane)   (carbon disulfide)   (ethanol)   (1,2-dihydroxyethane)

         The boiling point is dependent on the intermolecular forces. Hence, hydrogen bonding produces the strongest interactions (highest boiling point) and non polar molecules like butane and CS_2 have the lowest boiling point. We must also consider the effect of Van der Waals forces.

<u>51.</u>   We expect CH_3OH to be a liquid from among the four substances listed. Of these four molecules, C_3H_8 has the most electrons and should have the strongest London forces. However, only CH_3OH satisfies the conditions for hydrogen bonding (H bonded to and attracted to N, O, or F) and thus its intermolecular attractions should be much stronger than those of the other substances.

52.   (a)   Intramolecular hydrogen bonding cannot occur in C_4H_{10} since the conditions for hydrogen bonding (H bonded to and also attracted to N, O, or F) are not satisfied in this molecule. There is no N, O, or F atom in this molecule.

         (b)   Intramolecular hydrogen bonding is important in $HOOCCH_2CH_2CH_2CH_2COOH$. The H of one end –COOH group can be attracted to one of the O atoms of the other end –COOH group to cause ring closure.

         (c)   Intramolecular hydrogen bonding is not important in H_3CCOOH. Although there is another O atom to which the H of —OH can hydrogen bond, the resulting configurations will create a four membered ring $\left(\begin{smallmatrix} O\text{--}H \\ \| \quad | \\ -C-O \end{smallmatrix} \right)$ with bond angles of 90°, which are highly strained, compared to the normal bond angles of 109.5° and 120°.

(d) In orthophthalic acid, intramolecular hydrogen bonds can occur. The H of one —COOH group can be attracted to one of the O atoms of the other —COOH group. The resulting ring is seven atoms around and thus should not cause substantial bond angle strain.

Network Covalent Solids

53. One would expect diamond to have a greater density than graphite. Although the bond distance in graphite, "one-and-a-half" bonds, would be expected to be shorter than the single bonds in diamond, the large spacing between the layers of C atoms in graphite makes its crystals much less dense than those of diamond.

54. Diamond works well in glass cutters because of its extreme hardness. Its hardness is due to the crystal being held together entirely by covalent bonds. Graphite will not function effectively in a glass cutter, since it is quite soft, soft enough to flake off in microscopic pieces when used in pencils. In fact, graphite is so soft that pure graphite is rarely used in common wooden pencils. Often clay or some other substance is mixed with the graphite to produce a mechanically strong pencil "lead".

55. (a) We expect Si and C atoms to alternate in the structure, as shown at the right. The C atoms are on the corners $(8 \times 1/8 = 1$ C atom) and on the faces $(6 \times 1/2 = 3$ C atoms), a total of four C atoms/unit cell. The Si atoms are each totally within the cell, a total of four Si atoms/unit cell.

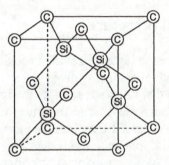

(b) To have a graphite structure, we expect sp^2 hybridization for each atom. The hybridization schemes for B and N atoms are shown to the right. The half-filled sp^2 hybrid orbitals of the boron and nitrogen atoms overlap to form the σ bonding structure, and a hexagonal array of atoms. The $2p_z$ orbitals then overlap to form the π bonding orbitals. Thus, there will be as many π electrons in a sample of BN as there are in a sample of graphite, assuming both samples have the same number of atoms.

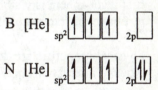

56. Buckminsterfullerene is composed of C_{60} spheres. These molecules of carbon are rather like the N_2, P_4 and S_8 molecules in that they produce a nonpolar molecular solid. The chain of alternating single and triple bonds could be a network covalent solid, and hence another allotrope of carbon. The long carbon chain would be linear and these rods of carbon $(-C \equiv C - C \equiv C - C \equiv C - C \equiv C - C \equiv C - C \equiv C-)$ could fit together rather like spaghetti in a box, with the π electrons of the triple bonds from adjacent rods attracting each other.

Ionic Bonding and Properties

57. We expect forces in ionic compounds to increase as the sizes of ions become smaller and as ionic charges become greater. As the forces between ions become stronger, a higher temperature is required to melt the crystal. In the series of compounds NaF, NaCl, NaBr, and NaI, the anions are progressively larger, and thus the ionic forces become weaker. We expect the melting points to decrease in this series from NaF to NaI. This is precisely what is observed.

58. Coulomb's law (Appendix B) states that the force between two particles of charges Q_1 and Q_2 that are separated by a distance r is given by $F = Q_1 Q_2 / \& r^2$ where $\&$, the dielectric constant, equals 1 in a vacuum. Since we are comparing forces, we can use $+1, -1, +2$, and -2 for the charges on ions. The length r is equal to the sum of the cation and anion radii, which are taken from Figure 12-36.

For NaCl, $r_+ = 99$ pm, $r_- = 181$ pm $\quad F = (+1)(-1)/(99+181)^2 = -1.3 \times 10^{-5}$

For MgO, $r_+ = 72$ pm, $r_- = 140$ pm $\quad F = (+2)(-2)/(72+140)^2 = -8.9 \times 10^{-5}$

Thus, it is clear that interionic forces are about seven times stronger in MgO than in NaCl.

Crystal Structures

59. In each layer of a closest packing arrangement of spheres, there are six spheres surrounding and touching any given sphere. A second similar layer then is placed on top of this first layer so that its spheres fit into the indentations in the layer below. The two different closest packing arrangements arise from two different ways of placing the third layer on top of these two, with its spheres fitting into the indentations of the layer below. In one case, one can look down into these indentations and see a sphere of the bottom (first) layer. If these indentations are used, the closest packing arrangement *abab* results (hexagonal closest packing). In the other case, no first layer sphere is visible through the indentation; the closest packing arrangement *abcabc* results (cubic closest packing).

60. Physical properties are determined by the type of bonding between atoms in a crystal or the types of interactions between ions or molecules in the crystal. Different types of interactions can produce the same geometrical relationships of unit cell components. In both Ar and CO_2, London forces hold the particles in the crystal. In NaCl, the forces are interionic attractions, while metallic bonds hold Cu atoms in their crystals. It is the type of force, not the geometric arrangement of the components, that largely determines the physical properties of the crystalline material.

61. **(a)** We naturally tend to look at crystal structures in right-left, up-down terms. If we do that here, we might be tempted to assign a unit cell as a square, with its corners at the centers of the light-colored squares. But the crystal does not "know" right and left or top and bottom. If we look at this crystal from the lower right corner, we see a unit cell that has its corners at the centers of dark-colored diamonds. These two types of unit cells are outlined at the top of the diagram below.

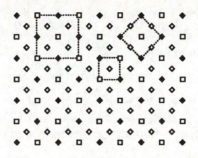

(b)   The unit cell has one light-colored square fully inside it. It has four light-colored "circles" (which the computer doesn't draw as very round) on the edges, each shared with one other unit cell. So there are a total of $4 \times 1/2 = 2$ circles per unit cell. Also, the unit cell has four dark-colored diamonds, one at each corner, and each diamond is shared with four other unit cells, for a total of $4 \times 1/4 = 1$ diamond per unit cell.

(c)   One example of an erroneous unit cell is the small square outlined near the center of the figure drawn in part (a). Notice that simply repeatedly translating this unit cell toward the right, so that its left edge sits where its right edge is now, will not generate the lattice.

62.   This question reduces to asking what percentage of the area of a square is covered by a circle inscribed within it. The diameter of the circle equals the side of the square. Since $\text{diameter} = 2 \times \text{radius}$, we have

$$\frac{\text{area of circle}}{\text{area of square}} = \frac{\pi r^2}{(2r)^2} = \frac{\pi}{4} = \frac{3.14159}{4} = 0.7854 \quad or \quad 78.54\%$$

63.   In Figure 12–45 we see that the body diagonal of a cube has a length of $\sqrt{3}l$, where l is the length of one edge of the cube. The length of this body diagonal also equals $4r$, where r is the radius of the atom in the structure. Hence $4r = \sqrt{3}l$ or $l = 4r \div \sqrt{3}$ Recall that the volume of a cube is l^3, and $\sqrt{3} = 1.732$

$$\text{density} = \frac{\text{mass}}{\text{volume}} = \frac{\dfrac{2 \text{ W atoms}}{1 \text{ unit cell}} \times \dfrac{1 \text{ mol W}}{6.022 \times 10^{23} \text{ W atoms}} \times \dfrac{183.85 \text{ g W}}{1 \text{ mol W}}}{\left(\dfrac{4 \times 139 \text{ pm}}{1.732} \times \dfrac{1 \text{ m}}{10^{12} \text{ pm}} \times \dfrac{100 \text{ cm}}{1 \text{ m}}\right)^3} = 18.5 \text{ g} / \text{cm}^3$$

This compares well with a tabulated density of $19.25 \text{ g} / \text{cm}^3$.

64.   One atom lies entirely within the hcp unit cell and there are eight corner atoms, each of which are shared among eight unit cells. Thus, there are a total of two atoms per unit cell. The volume of the unit cell is given by the product of its height and the area of its base. The base is a parallelogram, which can be subdivided along its shorter diagonal into two triangles. The height of either triangle, equivalent to the perpendicular distance between the parallel sides of the parallelogram, is given by $320. \text{ pm} \times \sin(60°) = 277 \text{ pm}$.

Area of base $= 2 \times$ area of triangle $= 2 \times (1/2 \times$ base $\times$ height$)$

$$= 277 \text{ pm} \times 320. \text{ pm} = 8.86 \times 10^4 \text{ pm}^2$$

Volume of unit cell $= 8.86 \times 10^4 \text{ pm}^2 \times 520. \text{ pm} = 4.61 \times 10^7 \text{ pm}^3$

$$\text{Density} = \frac{\text{mass}}{\text{volume}} = \frac{2 \text{ Mg atoms} \times \dfrac{1 \text{ mol Mg}}{6.022 \times 10^{23} \text{ Mg atoms}} \times \dfrac{24.305 \text{ g Mg}}{1 \text{ mol Mg}}}{4.61 \times 10^7 \text{ pm}^3 \times \left(\dfrac{1 \text{ m}}{10^{12} \text{ pm}} \times \dfrac{100 \text{ cm}}{1 \text{ m}} \right)^3} = 1.75 \text{ g} / \text{cm}^3$$

This is in reasonable agreement with the experimental value of $1.738 \text{ g} / \text{cm}^3$.

65. **(a)** 335 pm = 2 radii or 1 diameter. Hence Po diameter = 335 pm

(b) 1 Po unit cell $= (335 \text{ pm})^3 = 3.76 \times 10^7 \text{ pm}^3 \ (3.76 \times 10^{-23} \text{ cm}^3)$ per unit cell.

$$\text{density} = \frac{m}{V} = \frac{3.47 \times 10^{-22} \text{ g}}{3.76 \times 10^{-23} \text{ cm}^3} = 9.23 \text{ g cm}^{-3}$$

(c) $n = 1$, $d = 335$ pm and $\lambda = 1.785 \times 10^{-10}$ or 178.5 pm   Solve for $\sin \theta$ then determine θ

$$\sin \theta = \frac{n\lambda}{2d} = \frac{(1)(1.785 \times 10^{-10})}{2(335 \times 10^{-12})} = 0.266\underline{4} \quad \text{or} \quad \theta = 15.4\underline{5}°$$

66. Ge unit cell has a length of 565 pm. volume $= (\text{length})^3 = (565 \text{ pm})^3 = 1.80 \times 10^8 \text{ pm}^3$

Next we convert to cm^3: $1.80 \times 10^8 \text{ pm}^3 \left(\dfrac{1 \times 10^{-12} \text{ m}}{1 \text{ pm}} \right)^3 \left(\dfrac{100 \text{ cm}}{1 \text{ m}} \right)^3 = 1.80\underline{4} \times 10^{-22} \text{ cm}^3$

Now we determine the number of Ge atoms per unit cell
Method: Find number of atoms of Ge per cm^3 then using the unit cell volume, (calculation above), determine number of Ge atoms in the unit cell

$$\frac{\text{atoms Ge}}{\text{cm}^3} = \left(\frac{5.36 \text{ g Ge}}{\text{cm}^3} \right)\left(\frac{1 \text{ mol Ge}}{72.61 \text{ g Ge}} \right)\left(\frac{6.022 \times 10^{23}}{1 \text{ mol Ge}} \right) = 4.44\underline{5} \times 10^{22} \frac{\text{atoms Ge}}{\text{cm}^3}$$

$$\frac{\text{atoms Ge}}{\text{unit cell}} = (1.80\underline{4} \times 10^{-22} \text{ cm}^3) \left(4.44\underline{5} \times 10^{22} \frac{\text{atoms Ge}}{\text{cm}^3} \right) = 8.02 \frac{\text{atoms Ge}}{\text{unit cell}}$$

There are 8 atoms of Ge per unit cell. Ge must adopt a face centered cubic structure with the 4 tetrahedral holes filled (diamond structure)

Ionic Crystal Structures

67. CaF_2. There are eight Ca^{2+} ions on the corners, each shared among eight unit cells, for a total of $(8 \times 1/8)$ one corner ion per unit cell. There are six Ca^{2+} ions on the faces, each shared between two unit cells, for a total of $(6 \times 1/2)$ three face ions per unit cell. This gives a total of four Ca^{2+} ions per unit cell. There are eight F^- ions, each wholly contained within the unit cell. The ratio of Ca^{2+} ions to F^- ions is 4 Ca^{2+} ions per 8 F^- ions: Ca_4F_8 or CaF_2. TiO_2. There are eight Ti^{4+} ions on the corners, each shared among eight unit cells, for a total of $(8 \times 1/8)$ one Ti^{4+} corner ion per unit cell. There is one Ti^{4+} ion in the center, wholly contained within the unit cell. Thus, there are a total of two Ti^{4+} ions per unit cell. There are four O^{2-} ions on the faces of the unit cell, each shared between two unit cells, for a total of $(4 \times 1/2)$ two face atoms per unit cells. There are two O^{2-} ions totally contained within the unit cell. This gives a total of four O^{2-} ions per unit cell. The ratio of Ti^{4+} ions to O^{2-} ions is 2 Ti^{4+} ions per 4 O^{2-} ion: Ti_2O_4 or TiO_2.

68. The unit cell of CsCl is pictured in Figure 12–49. The length of the body diagonal equals $2r_+ + 2r_-$. From Figure 10-8, for Cl^-, $r_- = 181$ pm. Thus, the length of the body diagonal equals $2(169 \text{ pm} + 181 \text{ pm}) = 700.$ pm. From Figure 12–45, the length of this body diagonal is $\sqrt{3}l = 700.$ pm. The volume of the unit cell is $V = l^3$. $(\sqrt{3} = 1.732)$.

$$V = \left(\frac{700. \text{ pm}}{1.732} \times \frac{1 \text{ m}}{10^{12} \text{ pm}} \times \frac{100 \text{ cm}}{1 \text{ m}} \right)^3 = 6.60 \times 10^{-23} \text{ cm}^3$$

Per unit cell, there is one Cs^+ and one Cl^- (8 corner ions $\times 1/8$ corner ion per unit cell).

$$\text{density} = \frac{\text{mass}}{\text{volume}} = \frac{\left(1 \text{ f.u. CsCl} \times \dfrac{1 \text{ mol CsCl}}{6.022 \times 10^{23} \text{ CsCl f.u.'s}} \times \dfrac{168.4 \text{ g CsCl}}{1 \text{ mol CsCl}} \right)}{6.60 \times 10^{-23} \text{ cm}^3} = 4.24 \text{ g/cm}^3$$

69. **(a)** In a sodium chloride type of lattice, there are six cations around each anion and six anions around each cation. These oppositely charged ions are arranged as follows: one above, one below, one in front, one in back, one to the right, and one to the left. Thus the coordination number of Mg^{2+} is 6 and that of O^{2-} is 6 also.

(b) In the unit cell, there is an oxide ion at each of the eight corners; each of these is shared between eight unit cells. There also is an oxide ion at the center of each of the six faces; each of these oxide ions is shared between two unit cells. Thus, the total number of oxide ions is computed as follows.

$$\text{total \# of oxide ions} = 8 \text{ corners} \times \frac{1 \text{ oxide ion}}{8 \text{ unit cells}} + 6 \text{ faces} \times \frac{1 \text{ oxide ion}}{2 \text{ unit cells}} = 4 \ O^{2-} \text{ ions}$$

There is a magnesium ion on each of the twelve edges; each of these is shared between four unit cells. There also is a magnesium ion in the center which is not shared with another unit cell.

$$\text{total \# of } Mg^{2+} \text{ ions} = 12 \text{ adjoining cells} \times \frac{1 \text{ magnesium ion}}{4 \text{ unit cells}} + 1 \text{ central } Mg^{2+} \text{ ion}$$

$$= 4 \ Mg^{2+} \text{ ions (Thus, there are four formula units per unit cell of MgO).}$$

(c) Along the edge of the unit cell, Mg^{2+} and O^{2-} ions are in contact. The length of the edge is equal to the radius of one O^{2-}, plus the diameter of Mg^{2+}, plus the radius of another O^{2-}.

edge length $= 2 \times O^{2-}$ radius $+ 2 \times Mg^{2+}$ radius $= 2 \times 140$ pm $+ 2 \times 72$ pm $= 424$ pm

The unit cell is a cube; its volume is the cube of its length.

$$\text{volume} = (424 \text{ pm})^3 = 7.62 \times 10^7 \text{ pm} \left(\frac{1 \text{ m}}{10^{12} \text{ pm}} \times \frac{100 \text{ cm}}{1 \text{ m}} \right)^3 = 7.62 \times 10^{-23} \text{ cm}^3$$

(d) $$\text{density} = \frac{\text{mass}}{\text{volume}} = \frac{4 \text{ MgO f.u}}{7.62 \times 10^{-23} \text{ cm}^3} \times \frac{1 \text{ mol MgO}}{6.022 \times 10^{23} \text{ f.u.}} \times \frac{40.30 \text{ g MgO}}{1 \text{ mol MgO}} = 3.51 \text{ g/cm}^3$$

Thus, there are four formula units per unit cell of MgO.

70. According to Figure 12-48, there are four formula units of KCl in a unit cell and the length of the edge of that unit cell is twice the internuclear distance between K^+ and Cl^- ions. The unit cell is a cube; its volume is the cube of its length.   length $= 2 \times 314.54$ pm $= 629.08$ pm

$$\text{volume} = (629.08 \text{ pm})^3 = 2.4895 \times 10^8 \text{ pm}^3 \left(\frac{1 \text{ m}}{10^{12} \text{ pm}} \times \frac{100 \text{ cm}}{1 \text{ m}} \right)^3 = 2.4895 \times 10^{-22} \text{ cm}^3$$

$$\text{unit cell mass} = \text{volume} \times \text{density} = 2.4895 \times 10^{-22} \text{ cm}^3 \times 1.9893 \text{ g / cm}^3 = 4.9524 \times 10^{-22} \text{ g}$$
$$= \text{mass of 4 KCl formula units}$$
$$N_A = \frac{4 \text{ KCl formula unit}}{4.9524 \times 10^{-22} \text{ g KCl}} \times \frac{74.5513 \text{ g KCl}}{1 \text{ mol KCl}} = 6.0214 \times 10^{23} \text{ formula unit/mol}$$

71. CaO → radius ratio $= \dfrac{r_{Ca^{2+}}}{r_{O^{2-}}} = \dfrac{100 \text{ pm}}{140 \text{ pm}} = 0.714$

Cations occupy octahedral holes of a face centered cubic array of anions

CuCl → radius ratio $= \dfrac{r_{Cu^+}}{r_{Cl^-}} = \dfrac{96 \text{ pm}}{181 \text{ pm}} = 0.530$

Cations occupy octahedral holes of a face centered cubic array of anions

LiO_2 → radius ratio $= \dfrac{r_{Li^+}}{r_{O_2^-}} = \dfrac{59 \text{ pm}}{128 \text{ pm}} = 0.461$

Cations occupy octahedral holes of a face centered cubic array of anions

72. BaO → radius ratio $= \dfrac{r_{Ba^{2+}}}{r_{O^{2-}}} = \dfrac{135 \text{ pm}}{140 \text{ pm}} = 0.964$

Cubic hole of a simple cubic array of anions occupied by the cations

CuI → radius ratio $= \dfrac{r_{Cu^+}}{r_{I^-}} = \dfrac{96 \text{ pm}}{220 \text{ pm}} = 0.436$

Cations occupy octahedral holes of a face centered cubic array of anions

LiS_2 → radius ratio $= \dfrac{r_{Li^+}}{r_{S_2^-}} = \dfrac{59 \text{ pm}}{198 \text{ pm}} = 0.298$

Cations occupy tetrahedral holes of a face centered cubic array of anions

Lattice Energy

73. Lattice energies of a series such as LiCl(s), NaCl(s), KCl(s), RbCl(s), and CsCl(s) will vary approximately with the size of the cation. A smaller cation will produce a more exothermic lattice energy. Thus, the lattice energy for LiCl(s) should be the most exothermic and CsCl(s) the least in this series, with NaCl(s) falling in the middle of the series.

74. The cycle of reactions is shown. Recall that Hess's law (a state function), states that the enthalpy change is the same, whether a chemical change is produced by one reaction or several.

$$K(s) + 1/2\ F_2(g) \xrightarrow{\ formation\ } KF(s)$$

with branches labeled: sublimation, dissociation, ionization, electron affinity, latice energy, leading through:

$$K(g)\ +\ F(g)$$
$$K^+(g)\ +\ F^-(g)$$

Formation reaction:	$K(s) + \frac{1}{2}\ F_2(g) \rightarrow KF(s)$	$\Delta H_f^{\,o} = -567.3$ kJ / mol
Sublimation:	$K(s) \rightarrow K(g)$	$\Delta H_{sub} = 89.24$ kJ / mol
Ionization:	$K(s) \rightarrow K^+(g) + e^-$	$I_1 = 418.9$ kJ/mol
Dissociation:	$\frac{1}{2}\ F_2(g) \rightarrow F(g)$	D.E.$= (159/2)$ kJ / mol F
Electron Affinity:	$F(g) + e^- \rightarrow F^-(g)$	E.A.$= -328$ kJ / mol

$\Delta H_f^{\,o} = \Delta H_{sub} + I_1 + $ D.E.$+$ E.A.$+$ lattice energy $($L.E.$)$

-567.3 kJ / mol $= 89.24$ kJ / mol $+ 418.9$ kJ / mol $+ (159/2)$ kJ / mol $- 328$ kJ / mol $+$ L.E.

L.E.$= -827$ kJ / mol

75.

Second ionization energy:	$Mg^+(g) \rightarrow Mg^{2+}(g) + e^-$	$I_2 = 1451$ kJ/mol
Lattice energy:	$Mg^{2+}(g) + 2\ Cl^-(g) \rightarrow MgCl_2(s)$	L.E.$= -2526$ kJ / mol
Sublimation:	$Mg(s) \rightarrow Mg(g)$	$\Delta H_{sub} = 146$ kJ/mol
First ionization energy	$Mg(g) \rightarrow Mg^+(g) + e^-$	$I_1 = 738$ kJ/mol
Dissociation energy:	$Cl_2(g) \rightarrow 2\ Cl(g)$	D.E.$= (2 \times 122)$kJ/mol
Electron Affinity:	$2\ Cl(g) + 2\ e^- \rightarrow 2\ Cl^-(g)$	$2 \times$ E.A.$= 2(-349)$ kJ/mol

$\Delta H_f^{\,o} = \Delta H_{sub} + I_1 + I_2 + $ D.E.$+ (2 \times$ E.A.$) + $ L.E.

$$= 146\ \frac{kJ}{mol} + 738\ \frac{kJ}{mol} + 1451\ \frac{kJ}{mol}1 + 244\ \frac{kJ}{mol} - 698\ \frac{kJ}{mol} - 2526\ \frac{kJ}{mol} = -645\ \frac{kJ}{mol}$$

In Example 12-12, the value of $\Delta H_f^{\,o}$ for MgCl is calculated as -19 kJ/mol. Therefore, MgCl$_2$ is much more stable than MgCl, since considerably more energy is released when it forms. $MgCl_2(s)$ is more stable than MgCl(s)

76.

Formation reaction:	$Na(s) + \frac{1}{2}\ H_2(g) \rightarrow NaH(s)$	$\Delta H_f^{\,o} = -57$ kJ/mol
Heat of sublimation:	$Na(s) \rightarrow Na(g)$	$\Delta H_{sub} = +107$ kJ/mol
Ionization energy:	$Na(g) \rightarrow Na^+(g) + e^-$	$I_1 = +496$ kJ/mol
Dissociation energy:	$\frac{1}{2}\ H_2(g) \rightarrow H(g)$	D.E.$= +218$ kJ/mol H
Electron Affinity:	$H(g) + e^- \rightarrow H^-(g)$	E.A.$= ?$
Lattice energy:	$Na^+(g) + H^-(g) \rightarrow NaH(s)$	L.E.$= -812$kJ/mol

$\Delta H_f^{\,o} = \Delta H_{sub} + I_1 + $ D.E.$+$ E.A.$+$ lattice energy $($L.E.$)$

-57 kJ/mol $= 107$ kJ/mol $+ 496$ kJ/mol $+$ E.A.$+ 218$ kJ/mol $- 812$ kJ/mol

E.A.$= -66$ kJ/mol (compares favorably to the literature value of -72.8 kJ

Integrative and Advanced Exercises

77. The heat of the flame melts the solid wax, converting it to a liquid. Still closer to the flame the temperature is higher, the liquid wax vaporizes to the gaseous hydrocarbons, which are then burned. The liquid wax is drawn up the wick toward the flame by capillary action, where it is vaporized and burns, producing the heat needed to melt the solid wax, and the process continues until all the wax is burned.

78. We know that 1.00 atm = 101325 Pa and 1 Bar = 100,000 Pa. The enthalpy of vaporization is given as 44.0×10^3 j/mol in the chapter. These numbers are substituted into the Clapeyron equation.

$$\ln \frac{101,325 \text{ Pa}}{100,000 \text{ Pa}} = \frac{40,657 \text{ J mol}^{-1}}{8.3145 \text{ J mol}^{-1} \text{ K}^{-1}} \left(\frac{1}{T} - \frac{1}{373.15} \right)$$

$T = 372.78$

79. At 25.0 °C, the vapor pressure of water is 23.8 mmHg. We calculate the vapor pressure for isooctane with the Clausius-Clapeyron equation.

$$\ln \frac{P}{760 \text{ mmHg}} = \frac{35.76 \times 10^3 \text{ J/mol}}{8.3145 \text{ J mol}^{-1} \text{ K}^{-1}} \left(\frac{1}{(99.2 + 273.2) \text{ K}} - \frac{1}{298.2 \text{ K}} \right) = -2.87$$

$P = e^{-2.87} \times 760 \text{ mmHg} = 43.1 \text{ mmHg}$ which is higher than H_2O's vapor pressure.

80. In many instances, CO_2 being one, the substance in the tank is not present as a gas only, but as a liquid in equilibrium with its vapor. As gas is released from the tank, the liquid will vaporize to replace it, maintaining a pressure in the tank equal to the vapor pressure of the substance at the temperature at which the cylinder is stored. This will continue until all of the liquid vaporizes, after which only gas will be present. The remaining gas will be quickly consumed, and hence the reason for the warning. However, the situation of gas in equilibrium with liquid only applies to substances that have a critical temperature above room temperature (20 °C or 293 K). Thus, substances in Table 12-3 for which gas pressure does serve as a measure of the quantity of gas in the tank are H_2 ($T_c = 33.3$ K), N_2 ($T_c = 126.2$ K), O_2 ($T_c = 154.8$ K), and CH_4 ($T_c = 191.1$ K).

81. m = 15.0 g Hg(s) or n = 0.0748 mol   Hg(s) 24.3 J K^{-1}mol^{-1} Hg(l) 28.0 J K^{-1} mol^{-1}
(Hg melting point = -38.87 °C and ΔH_{fusion} = 2.33 kJ mol^{-1})

SCHEMATICALLY   Hg(s)   Hg(l)   Hg(l)
                     -38.87 °C   -38.87 °C   25 °C

Hg(s) ←————————————→ ←————————————→ ←————————————→ ←————————————→ Hg(g)
-50 °C   25 °C

$nC_{solid}\Delta T$ $n\Delta H_{fusion}$ $nC_{liquid}\Delta T$ $n\Delta H_{vaporization}$

Here we need to calculate (using Appendix D), $\Delta H_{vapourization}$ at 25 °C. That is: Hg(l) → Hg(g) at 25°C. Simply use $\Delta H°_f$ values from Appendix D. $\Delta H_{vapourization} = \Delta H°_f$ Hg(g) - $\Delta H°_f$ Hg(l)
$\Delta H_{vapourization}$ = 61.32 kJ/mol – 0 kJ/mol = 61.32 kJ/mol.
Next we calculate each of the four quantities shown in the schematic.

$nC_{solid}\Delta T + \Delta = 0.0748$ mol$(24.3$ J K^{-1}mol$^{-1})($ -38.87 °C $-$ -50.0 °C$)(1$ kJ/1000J$) = 0.02023$ kJ
$n\Delta H_{fusion} = 0.0748$ mol$(2.33$ kJ mol$^{-1}) = 0.174$ kJ
$nC_{liquid}\Delta T = 0.0748$ mol$(28.0$ J K^{-1} mol$^{-1})(25.0$ °C $-$ -38.87 °C$)(1$ kJ/1000J$) = 0.134$ kJ
$n\Delta H_{vaporization} = 0.0748$ mol$(61.32$ kJ/mol$) = 4.587$ kJ
$q = nC_{solid}\Delta T + n\Delta H_{fusion} + nC_{solid}\Delta T + n\Delta H_{vap} = 0.020.23$ kJ$+0.174$ kJ$+0.134$ kJ$+4.587$ kJ
$q = 4.915$ kJ

82. For this question we need to determine the quantity of heat required to vaporize 1.000 g H_2O at each temperature. At 20°C, 2447 J of heat is needed to vaporize each 1.000 g H_2O. At 100°C., the quantity of heat needed is

$$\frac{10.00 \text{ kJ}}{4.430 \text{ g } H_2O} \times \frac{1000 \text{ J}}{1 \text{ kJ}} = 2257 \text{ J/g } H_2O$$

Thus, less heat is needed to vaporize 1.000 g of H_2O at the higher temperature of 100. °C. This makes sense, for at the higher temperature the molecules of the liquid already are in rapid motion. Some of this energy of motion or vibration will contribute to the energy needed to break the cohesive forces and vaporize the molecules.

83. 1.00 g of Car Kooler (10 % C_2H_5OH) is used to cool 55 °C air which has a heat capacity of 29 J K^{-1}mol^{-1}. Hence, 0.100 g C_2H_5OH is evaporated, (with $\Delta H_{vap} = 42.6$ kJ.mol^{-1}) and 0.900 g of H_2O (with $\Delta H_{vap} = 44.0$ kJ.mol^{-1}), is also evaporated.

Energy required to evaporate $0.10 \text{ g } C_2H_5OH \times \dfrac{1 \text{ mol } C_2H_5OH}{46.068 \text{ g } C_2H_5OH} \times \dfrac{42.6 \text{ kJ}}{1 \text{ mol } C_2H_5OH} = 0.0925 \text{ kJ}$

Energy required to evaporate the water $0.90 \text{ g } H_2O \times \dfrac{1 \text{ mol } H_2O}{18.015 \text{ g } H_2O} \times \dfrac{44.0 \text{ kJ}}{1 \text{ mol } H_2O} = 2.2 \text{ kJ}$

Thus, most of the energy(> 95%), absorbed by Car Kooler goes towards the evaporation of water. In fact, if one were to replace the C_2H_5OH with water, it would be even more effective.

84. In this case, we use the Clausius-Clapeyron equation to determine the sublimation pressure at 25 °C. We also need the enthalpy of sublimation and the sublimation pressure at another temperature. We can closely estimate the sublimation pressure at the triple point with the following three step process. First, assume that the triple point temperature and the melting point temperature are the same (53.1 °C or 326.3 K). Second, realize that the vapor pressure and the sublimation pressure are the same at the triple point. Third, use the Clausius-Clapeyron equation to determine the vapor pressure at the triple point.

$$\ln\frac{P}{10.0 \text{ mmHg}} = \frac{72,220 \text{ J mol}^{-1}}{8.314 \text{ J K}^{-1}\text{mol}^{-1}}\left(\frac{1}{328.0 \text{ K}} - \frac{1}{326.3 \text{ K}}\right) = -0.138 \qquad P = 10.0 e^{-0.138} = 8.71 \text{ mmHg}$$

ΔH_{sub} is determined by: $\Delta H_{sub} = \Delta H_{fus} + \Delta H_{vap} = 17.88$ kJ mol^{-1} + 72.22 kJ mol^{-1} = 90.10 kJ mol^{-1}
Use the Clausius-Clapeyron equation to determine the sublimation pressure at 25 °C

$$\ln\frac{P}{8.71 \text{ mmHg}} = \frac{90,100 \text{ J mol}^{-1}}{8.314 \text{ J K}^{-1}\text{mol}^{-1}}\left(\frac{1}{326.3 \text{ K}} - \frac{1}{298.0 \text{ K}}\right) = -3.15 \qquad P = 8.71 e^{-3.15} = 0.37 \text{ mmHg}$$

85. If only gas were present, the final pressure would be 10 atm. This is far in excess of the vapor pressure of water at 30.0 °C (~ 0.042 atm). Most of the water vapor condenses to liquid water. (It cannot all be liquid, because the liquid volume is only about 20 mL and the system volume is 2.61 L.) The final condition is a point on the vapor pressure curve at 30.0 °C.

86. At the triple point, the sublimation pressure equals the vapor pressure. We need to also realize that Celsius temperature, t, is related to Kelvin temperature, T, by $t = T - 273.15$ Thus,

$$9.846 - 2309/T = 6.90565 - 1211.033/(220.790 + t) = 6.90565 - 1211.033/(220.790 + T - 273.15)$$

$$= 6.90565 - 1211.033/(T - 52.36)$$

$$9.846 - 6.90565 = 2.940 = \frac{2309}{T} - \frac{1211.033}{T - 52.36} = \frac{2309}{T} \frac{(T - 52.36) - 1211.033}{T(T - 52.36)} \frac{T}{}$$

$$2.940T(T - 52.36) = 2.940T^2 - 153.9T = 2309T - 1.209 \times 10^5 - 1211.033T = 1098T - 1.209 \times 10^5$$

$$2.940T^2 - 153.9T - 1098T + 1.209 \times 10^5 = 0 = 2.940T^2 - 1252T + 1.209 \times 10^5$$

The quadratic formula is used to solve this equation.

$$T = \frac{-b \pm \sqrt{b^2 - 4\ a\ c}}{2a} = \frac{1252 \pm \sqrt{(1252)^2 - (4 \times 2.940 \times 1.209 \times 10^5)}}{2 \times 2.940}$$

$$T = 212.9 \pm 64.9 = 277.8 \ \text{K} \quad or \quad 148.0 \ \text{K} \qquad\qquad \therefore \ T = 4.6 \ \ °C \ \ or \ -125.2 \ \ °C$$

The acceptable result from the quadratic equation is 4.6 °C; below this temperature liquid benzene can only exist as a metastable supercooled liquid, not as a stable liquid and thus -125.2 °C is not a reasonable value. The calculated value, 4.6 °C, compares favorably to the listed value of 5.5 °C.

87. First, we need to calculate ln P and $1/T$ for the given data points.

t, °C	76.6	128.0	166.7	197.3	251.0
P, mmHg	1	10	40	100	400
T, K	349.8	401.2	439.9	470.5	524.2
$1/T \times 10^3 \, K^{-1}$	2.859	2.493	2.273	2.125	1.908
$\ln P$	0	2.303	3.689	4.605	5.991

NOTE: Graphs are given on the following page

(a) The normal boiling point of phosphorus occurs at a pressure of 760 mmHg, where ln P = 6.63. Based on the graph at left, this occurs when $1/T = 1.81 \times 10^{-3} \, K^{-1}$, or $T = 552$ K, 279 °C.

(b) When ln P is plotted vs $1/T$, (see below), the slope of the line equals $-\Delta H_{vap}/R$.
For the graph on the left, the slope (by linear regression) is $-6.30 \times 10^3 \, K$
Thus, $\Delta H_{vap} = 6.30 \times 10^3 \, K \times 8.3145 \, J \, mol^{-1} \, K^{-1} = 5.24 \times 10^4 \, J/mol$

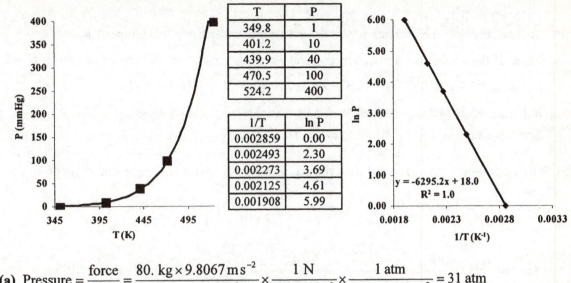

T	P
349.8	1
401.2	10
439.9	40
470.5	100
524.2	400

1/T	ln P
0.002859	0.00
0.002493	2.30
0.002273	3.69
0.002125	4.61
0.001908	5.99

$y = -6295.2x + 18.0$
$R^2 = 1.0$

88. (a) $\text{Pressure} = \dfrac{\text{force}}{\text{area}} = \dfrac{80.\ \text{kg} \times 9.8067\ \text{m s}^{-2}}{2.5\ \text{cm}^2 \times \dfrac{1\ \text{m}^2}{10^4\ \text{cm}^2}} \times \dfrac{1\ \text{N}}{1\ \text{kg m s}^{-2}} \times \dfrac{1\ \text{atm}}{101325\ \text{N m}^{-2}} = 31\ \text{atm}$

(b) $\text{decrease in melting point} = 31\ \text{atm} \times \dfrac{1.0\ °C}{125\ \text{atm}} = 0.25\ °C$

The ice under the skates will melt at $-0.25\ °C$.

89. $P = P_0 \times 10^{-Mgh/2.303\,RT}$ Assume ambient temperature is $10.0\,°C = 283.2\ \text{K}$

$\dfrac{Mgh}{2.303\ RT} = \dfrac{0.02896\ \text{kg/mol air} \times 9.8067\ \text{m/s}^2 \times 3170\ \text{m}}{2.303 \times 8.3145\ \text{J mol}^{-1}\ \text{K}^{-1} \times 283\ \text{K}} = 0.166$

$P = P_0 \times 10^{-0.166} = 1\ \text{atm} \times 0.682 = 0.682\ \text{atm} = \text{atmospheric pressure in Leadville, CO.}$

$\ln \dfrac{0.682\ \text{atm}}{1.000\ \text{atm}} = \dfrac{-41 \times 10^3\ \text{J/mol}}{8.3145\ \text{J mol}^{-1}\ \text{K}^{-1}} \left(\dfrac{1}{T} - \dfrac{1}{373.2\ \text{K}} \right) = -0.383$

$\left(\dfrac{1}{T} - \dfrac{1}{373.2\ \text{K}} \right) = -0.383 \times \dfrac{8.3145}{-41 \times 10^3} = +7.8 \times 10^{-5}\ \text{K}^{-1}$

$\dfrac{1}{T} = +7.77 \times 10^{-5}\ \text{K}^{-1} + 2.68 \times 10^{-3}\ \text{K}^{-1} = 2.76 \times 10^{-3}\ \text{K}^{-1}$ $T = 360\ \text{K} = 87\ °C$

90. From the graphs in figure 12-13, choose two points for Benzene and two points for water.
Set $\ln P_{benzene} = \ln P_{water}$ Hence: $-4.10 \times 10^3 (1/T) + 18.3 = -5.35 \times 10^3 (1/T) + 21.0$

Compound	ln P	1/T (K^{-1})	Equation of line
Benzene	5.99	0.00300	$\ln P_{benzene} = -4.10 \times 10^3 (1/T) + 18.3$
	4.35	0.00340	
Water	6.55	0.00270	$\ln P_{water} = -5.35 \times 10^3 (1/T) + 21.0$
	4.41	0.00310	

$1250(1/T) = 2.7 \quad (1/T) = 0.00216 \text{ or } T = 463 \text{ K.}$
We can calculate the pressure for both benzene and water using this temperature.

Note: If the pressures for both benzene and water are close, this will serve as a double check.

$\ln P_{benzene} = -4.10 \times 10^3 (1/T) + 18.3 = -4.10 \times 10^3 (0.00216) + 18.3 = 9.44\underline{4}$

$\ln P_{water} = -5.35 \times 10^3 (1/T) + 21.0 = -5.35 \times 10^3 (0.00216) + 21.0 = 9.44\underline{4}$

$\ln P = 9.44\underline{4}$, hence $p = e^{9.444} = 126\underline{32}$ mmHg or 16.6 atm.

91. We first compute the volume of the cylinder, and then the mass of Cl_2 present as vapor.

$$V = \pi r^2 h = 3.1416 \times (5.0 \text{ in.})^2 \times 45 \text{ in.} \left(\frac{2.54 \text{ cm}}{1 \text{ in.}} \right)^3 = 5.8 \times 10^4 \text{cm}^3 = 58 \text{ L}$$

$$\text{amount } Cl_2 = \frac{PV}{RT} = \frac{100. \text{ psi} \times \dfrac{1 \text{ atm}}{14.7 \text{ psi}} \times 58 \text{ L}}{0.08206 \text{ L atm mol}^{-1} \text{ K}^{-1} \times 293 \text{ K}} = 16 \text{ mol } Cl_2$$

$$\text{mass } Cl_2 = 16 \text{ mol } Cl_2 \times \frac{70.9 \text{ g } Cl_2}{1 \text{ mol } Cl_2} \times \frac{1 \text{ lb}}{453.6 \text{ g}} = 2.5 \text{ lb } Cl_2(g)$$

Thus, the mass of gas does not account for all of the 151 lb of chlorine in the cylinder. Since 20 °C is below the critical temperature of Cl_2, liquid chlorine can exist, provided that the pressure is high enough. That is the case in this instance. Both liquid and gaseous chlorine are present in the cylinder. There is almost certainly no solid present; 20 °C is too far above chlorine's melting point of –103 °C for any reasonable pressure to produce the solid.

92. The molar mass of acetic acid monomer is 60.05 g/mol. We first determine the volume occupied by 1 mole of molecules of the vapor, assuming that the vapor consists only of monomer molecules: $HC_2H_3O_2$.

$$\text{volume of vapor} = 60.05 \text{ g} \times \frac{1 \text{ L}}{3.23 \text{ g}} = 18.6 \text{ L}$$

Next, we can determine the actual number of moles of vapor in 18.6 L at 350 K.

$$\text{moles of vapor} = \frac{PV}{RT} = \frac{1 \text{ atm} \times 18.6 \text{ L}}{0.08206 \text{ L atm mol}^{-1} \text{ K}^{-1} \times 350 \text{ K}} = 0.648 \text{ mol vapor}$$

Then, we can determine the number of moles of dimer and monomer, starting with 1.00 mole of monomer, and producing a final mixture of 0.648 moles total (monomer and dimer together).

reaction: $2 \text{ HC}_2\text{H}_3\text{O}_2(g) \longrightarrow (\text{HC}_2\text{H}_3\text{O}_2)_2(g)$

initial: 1.00 mol   0

changes: $-2x$ mol $+x$ mol

final: $(1.00 - 2x)$ mol x mol

The line labeled "changes" indicates that 2 moles of monomer are needed to form each mole of dimer. The line labeled "final" results from adding the "initial" and "changes" lines.

Total number of moles $= 1.00 - 2x + x = 1.00 - x = 0.648$ mol

$x = 0.352$ mol dimer $\qquad\qquad (1.00 - 2x) = 0.296$ mol monomer

% dimer $= \dfrac{0.352 \text{ mol dimer}}{0.648 \text{ mol total}} \times 100\% = 54.3\%$ dimer

We would expect the % dimer to decrease with temperature. Higher temperatures will provide the energy (as translational energy(heat)) needed to break the relatively weak hydrogen bonds that hold the dimers together.

93. First we compute the mass and the amount of mercury.

mass Hg $= 685 \text{ cm}^3 \times \dfrac{13.6 \text{ g}}{1 \text{ cm}^3} = 9.32 \times 10^3 \text{ g}$ $\quad n_{Hg} = 9.32 \times 10^3 \text{ g} \times \dfrac{1 \text{ mol Hg}}{200.59 \text{ g}} = 46.4 \text{ mol Hg}$

Then we calculate the heat given up by the mercury in lowering its temperature, as the sum of the following three terms.

cool liquid $\quad = 9.32 \times 10^3 \text{ g} \times 0.138 \text{ J g}^{-1}\,^\circ\text{C}^{-1} \times (-39\,^\circ\text{C} - 20\,^\circ\text{C}) = -7.6 \times 10^4 \text{ J} = -76 \text{ kJ}$

freeze liquid $\quad = 46.4 \text{ mol Hg} \times (-2.30 \text{ kJ/mol}) = -107 \text{ kJ}$

cool solid $\quad = 9.32 \times 10^3 \text{ g} \times 0.126 \text{ J g}^{-1}\,^\circ\text{C}^{-1} \times (-196 + 39)\,^\circ\text{C} = -1.84 \times 10^5 \text{ J} = -184 \text{ kJ}$

Total heat lost by Hg $= -76 \text{ kJ} - 107 \text{ kJ} - 184 \text{ kJ} = -367\text{kJ} = -$heat gained by N_2

mass of N_2(l) vaporized $= 367 \text{ kJ} \times \dfrac{1 \text{ mol } N_2}{5.58 \text{ kJ}} \times \dfrac{28.0 \text{g } N_2}{1 \text{ mol } N_2} = 1.84 \times 10^3 \text{ g } N_2(\text{l}) = 1.84 \text{ kg } N_2$

94. In the phase diagram on the left, the liquid-vapor curve dips down and then rises. This means that there can be a situation where raising the temperature will cause the vapor to condense to liquid – this is counter intuitive and does correspond to the laws of thermodynamics as they operate in our universe. Also, a negative slope for the vapor pressure curve would correspond to a negative value of ΔH_{vap}, but this quantity must always have a positive sign (vaporization is an endothermic process).In the diagram on the right, the liquid-vapor line has the same curvature as the solid –vapor line. This means that the heat of vaporization is the same as the heat of sublimation. This would mean that the heat of fusion is zero. Therefore no energy is needed to melt the solid, again counter intuitive and a physical impossibility.

95. The normal boiling point is that temperature at which the vapor pressure is 760 mmHg. We have been provided with the following expression to solve for T, the normal boiling point.

$\text{Log}_{10}(760) = 2.88081 = 9.95028 - 0.003863T - 1473.17(T^{-1})$
$0 = 7.06947 - 0.003863T - 1473.17(T^{-1})$ Multiply through by T to get
$0 = 7.06947T - 0.003863T^2 - 1473.17$

By solving the quadratic equation, we get the roots 239.8 K or 1590 K. The second of these temperatures makes no physical sense, since we know that NH_3 is a gas at room temperature. Thus the normal boiling point of NH_3(l) is 239.8 K

96. We need to calculate five different times. They are shown below.

1. Heat the solid to its melting point $T_1 = 1 \text{ mol Bi} \times \dfrac{0.028 \text{ kJ}}{\text{K mol}} \times (554.5 \text{ K} - 300 \text{ K}) \times \dfrac{1 \text{ min}}{1.00 \text{ kJ}} = 7.1 \text{ min}$

2. Melt the solid $T_2 = 1 \text{ mol Bi} \times \dfrac{10.9 \text{ kJ}}{\text{K mol}} \times \dfrac{1 \text{ min}}{1.00 \text{ kJ}} = 10.9 \text{ min}$

3. Heat the liquid to its boiling point $T_3 = 1 \text{ mol Bi} \times \dfrac{0.031 \text{ kJ}}{\text{K mol}} \times (1832 \text{ K} - 554.5 \text{ K}) \times \dfrac{1 \text{ min}}{1.00 \text{ kJ}} = 39.6 \text{ min}$

4. Vaporize the liquid $T_4 = 1 \text{ mol Bi} \times \dfrac{151.5 \text{ kJ}}{\text{K mol}} \times \dfrac{1 \text{ min}}{1.00 \text{ kJ}} = 151.5 \text{ min}$

5. Heat the gas to 2000 K $T_5 = 1 \text{ mol Bi} \times \dfrac{0.021 \text{ kJ}}{\text{K mol}} \times (2000 \text{ K} - 1832 \text{ K}) \times \dfrac{1 \text{ min}}{1.00 \text{ kJ}} = 3.5 \text{ min}$

A plot is shown below:

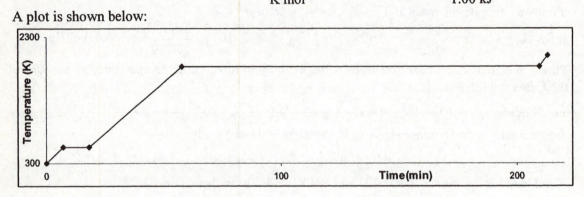

97. (a) The coordination number for S^{2-} is 8, while for Li^+ it is 4 (tetrahedral holes).

(b) $8 \times Li^+$ within the unit cell (tetrahedral holes) and $4 \times S^{2-}$ (8 corners $\times$ 1/8 + 6 faces $\times$ ½ = 1 + 3 = 4.

(c) Note: $1 \text{ pm}^3 = \dfrac{(1 \times 10^{-12} \text{ m})^3}{(1 \text{ pm})^3} \times \dfrac{(100 \text{ cm})^3}{(1 \text{ m})^3} = 1 \times 10^{-30} \text{ cm}^3$

$V = \ell^3 = (5.88 \times 10^2 \text{ pm})^3 \times \dfrac{1 \times 10^{-30} \text{ cm}^3}{1 \text{ pm}^3} = 2.03 \quad \underline{3} \times 10^{-22} \text{ cm}^3$

In a unit cell there are 4 Li_2S formula units.

($Li_2S \rightarrow 45.948 \text{ g mol}^{-1}$ or 4 mol $Li_2S \rightarrow 183.792 \text{ g}$)

Consider one mole of unit cells. This contains 45.948 g of Li_2S and has a volume of $(2.03 \times 10^{-22} \text{ cm}^3) \times (6.022 \times 10^{23}) = 122.\underline{4} \text{ cm}^3$

Density $= \dfrac{\text{mass}}{\text{volume}} = \dfrac{183.792 \text{ g}}{122.4 \text{ cm}^3} = 1.50 \text{ g cm}^3$

98. The edge length of the NaCl unit cell is 560 pm (from Example 12-11), and thus the distance between the top and the middle layers in the NaCl unit cell is $560 \text{ pm} \div 2 = 280 \text{ pm}$. This is equal to the value of d in the Bragg equation (12.5). We first solve for $\sin \theta$ and then for θ.

$\sin \theta = \dfrac{n\lambda}{2d} = \dfrac{1 \times 154.1 \text{ pm}}{2 \times 280 \text{ pm}} = 0.275 \quad \theta = \sin^{-1}(0.275) = 16.0°$

99. In a bcc cell, there are two atoms per unit cell. The length of an edge of the unit cell is $4\,r \div \sqrt{3}$, where $\sqrt{3} = 1.732$. In a fcc cell, there are four particles per unit cell. The length of an edge of the unit cell is $4\,r \div \sqrt{2}$, where $\sqrt{2} = 1.414$. Now, the volume of a cube is $V = l^3$. Thus,

$$V_{bcc} = \left(\frac{4\,r}{1.732}\right)^3 = 12.32\,r^3 \qquad \text{and} \qquad V_{fcc} = \left(\frac{4\,r}{1.414}\right)^3 = 22.63\,r^3$$

$$F_{bcc} = \frac{\text{volume of spheres}}{\text{volume of unit cell}} = \frac{2 \times 4.189\,r^3}{12.317\,r^3} = 0.6800 \; or \; 68.00\% \text{ occupied, } 32.00\% \text{ void}$$

$$F_{fcc} = \frac{\text{volume of spheres}}{\text{volume of unit cell}} = \frac{4 \times 4.189\,r^3}{22.627\,r^3} = 0.7404 \; or \; 74.04\% \text{ occupied, } 25.96\% \text{ void}$$

The volume of an individual sphere is $\frac{4}{3}\,\pi\,r^3 = 4.189\,r^3$

Then the fraction, F, of occupied volume in each unit cell can be computed.

100. Consider the two limits

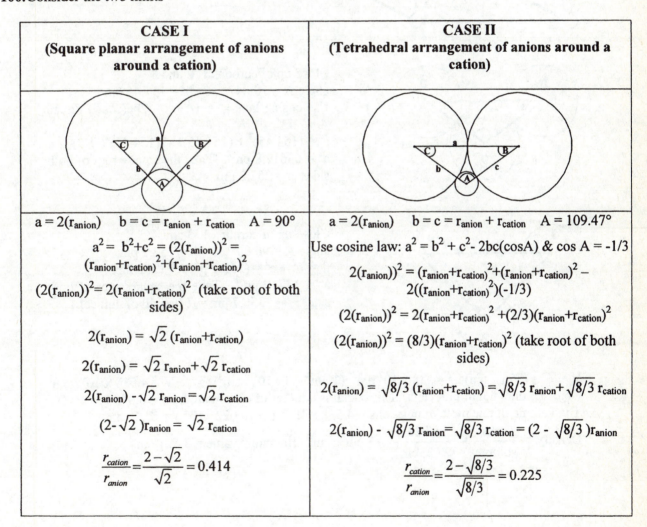

CASE I (Square planar arrangement of anions around a cation)	CASE II (Tetrahedral arrangement of anions around a cation)
$a = 2(r_{anion}) \quad b = c = r_{anion} + r_{cation} \quad A = 90°$	$a = 2(r_{anion}) \quad b = c = r_{anion} + r_{cation} \quad A = 109.47°$
$a^2 = b^2 + c^2 = (2(r_{anion}))^2 =$ $(r_{anion} + r_{cation})^2 + (r_{anion} + r_{cation})^2$	Use cosine law: $a^2 = b^2 + c^2 - 2bc(\cos A)$ & $\cos A = -1/3$
$(2(r_{anion}))^2 = 2(r_{anion} + r_{cation})^2$ (take root of both sides)	$2(r_{anion}))^2 = (r_{anion} + r_{cation})^2 + (r_{anion} + r_{cation})^2 -$ $2((r_{anion} + r_{cation})^2)(-1/3)$
$2(r_{anion}) = \sqrt{2}\,(r_{anion} + r_{cation})$	$(2(r_{anion}))^2 = 2(r_{anion} + r_{cation})^2 + (2/3)(r_{anion} + r_{cation})^2$
$2(r_{anion}) = \sqrt{2}\,r_{anion} + \sqrt{2}\,r_{cation}$	$(2(r_{anion}))^2 = (8/3)(r_{anion} + r_{cation})^2$ (take root of both sides)
$2(r_{anion}) - \sqrt{2}\,r_{anion} = \sqrt{2}\,r_{cation}$	$2(r_{anion}) = \sqrt{8/3}\,(r_{anion} + r_{cation}) = \sqrt{8/3}\,r_{anion} + \sqrt{8/3}\,r_{cation}$
$(2 - \sqrt{2})r_{anion} = \sqrt{2}\,r_{cation}$	$2(r_{anion}) - \sqrt{8/3}\,r_{anion} = \sqrt{8/3}\,r_{cation} = (2 - \sqrt{8/3})r_{anion}$
$\dfrac{r_{cation}}{r_{anion}} = \dfrac{2 - \sqrt{2}}{\sqrt{2}} = 0.414$	$\dfrac{r_{cation}}{r_{anion}} = \dfrac{2 - \sqrt{8/3}}{\sqrt{8/3}} = 0.225$

101. Diamond is a face center cubic structure + ½ filled tetrahedral holes.
Therefore, in the unit cell there are $8 \times (1/8) + 6 \times (1/2) + 4 \times 1 = 8$ atoms.
Need to find the length of the edge of the unit cell to determine the volume.

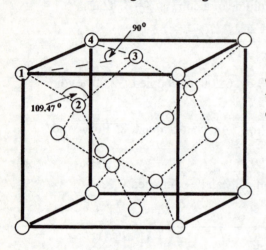

Structure of diamond is shown at the left. The carbon atoms of interest are numbered. Note that the tetrahedral angle (109.47 °) around carbon 2 and right angle (90°) around carbon 3.

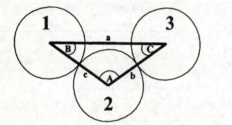

Close-up of carbons 1, 2 and 3.
angle A = 109.47 °, c = b = 154.45 pm
Use cosine law: $a^2 = b^2 + c^2 - 2bc(\cos A)$ & cos A = -1/3
$a^2 = (154.45)^2 + (154.45)^2 - 2(154.45)^2(-1/3)$
$a^2 = 63612.8$ pm^2. Take the square root of both sides a = 252.2 pm

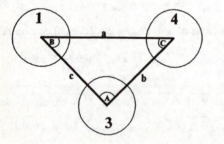

Close-up of carbons 1, 3 and 4.
angle A = 90.00 °, c = b = 252.2 pm
$a^2 = b^2 + c^2 = (252.2$ pm$)^2 + (252.2$ pm$)^2$
$a^2 = 12710$ pm^2. Take the square root of both sides a = 356.7 pm = ℓ (length of unit cell)

$V = \ell^3 = (356.7$ pm$)^3 = 4.538 \times 10^7$ pm^3 (1 pm^3 = 1 ×10^{-30} cm^3) $V = 4.538 \times 10^{-23}$ cm^3
Consider one mol of unit cells. This contains 8 moles of carbon atoms or 96.088 g.
The volume of one mole of unit cells = 4.538×10^{-23} cm^3(6.022 ×10^{23}) = 27.33 cm^3.
Density = $\dfrac{\text{mass}}{\text{volume}}$ = $\dfrac{96.088\,\text{g}}{27.33\,\text{cm}^3}$ = 3.516 g cm^{-3} (literature value = 3.52 g cm^{-3})

102.

Sublimation of Na(s):	$Na(s) \longrightarrow Na(g)$	$\Delta H_{sub} = +107.3$ kJ (Appendix D)
Ionization of Na(g):	$Na(g) \longrightarrow Na^+(g) + e^-$	$\Delta I_1 = +495.8$ kJ
$\frac{1}{2}$ Sublimation of I_2(s):	$\frac{1}{2} I_2(s) \longrightarrow \frac{1}{2} I_2(g)$	$\Delta H_{sbl} = \frac{1}{2}\{\Delta H_f^\circ [I_2(g)] - \Delta H_f^\circ [I_2(s)]\}$
		$= 0.5 \times (62.44 - 0.00) = 31.22$ kJ
$\frac{1}{2}$ Dissociation of I_2(g):	$\frac{1}{2} I_2(g) \longrightarrow I(g)$	$DE = \frac{1}{2} \times 151 = 75.5$ kJ (Table 11-3)
I(g) electron affinity:	$I(g) + e^- \longrightarrow I^-(g)$	$EA_1 = -295.2$ kJ (Figure 10-10)
Lattice energy:	$Na^+(g) + I^-(g) \longrightarrow NaI(s)$	L.E.
Enthalpy of formation:	$Na(s) + \frac{1}{2} I_2(s) \longrightarrow NaI(s)$	$\Delta H_f^\circ = -288$ kJ

$-288 \text{ kJ} = +107.3 \text{ kJ} + 495.8 \text{ kJ} + 31.22 \text{ kJ} + 75.5 \text{ kJ} - 295.2 \text{ kJ} + \text{L.E.} = 415 \text{ kJ} + \text{L.E.}$

$\text{L.E.} = -288 \text{ kJ} - 415 \text{ kJ} = -703 \text{ kJ}$

103.

Sublimation of Na(s):	$Na(s) \longrightarrow Na(g)$	$\Delta H_{sub} = +107.3$ kJ (App. D)
First ionization of Na(g):	$Na(g) \longrightarrow Na^+(g) + e^-$	$I_1 = +495.8$ kJ
Second ionization of Na(g):	$Na^+(g) \longrightarrow Na^{2+}(g) + e^-$	$I_2 = +4562$ kJ
Dissociation of Cl_2(g):	$Cl_2(g) \longrightarrow 2 Cl(g)$	$D.E. = +243$ kJ
Electron affinity of Cl(g):	$2 Cl(g) + 2 e^- \longrightarrow 2 Cl^-(g)$	$EA_1 = -698$ kJ
Lattice energy of NaCl(s):	$Na^{2+}(g) + 2 Cl^-(g) \longrightarrow NaCl_2(s)$	$L.E. = -2.5 \times 10^3$ kJ
Enthalpy of formation:	$Na(s) + Cl_2(g) \longrightarrow NaCl_2(s)$	$\Delta H_f^\circ = +2.2 \times 10^3$ kJ

This highly endothermic process is quite unlikely. Thus $NaCl_2$ will not form under normal conditions.

104. **(a)** and **(b)** We will work both parts simultaneously.

Sublimation of Mg(s):	$Mg(s) \longrightarrow Mg(g)$	$\Delta H_{sub} = +146$ kJ
First ionization of Mg(g):	$Mg(g) \longrightarrow Mg^+(g) + e^-$	$I_1 = +737.7$ kJ
Second ionization of Mg(g):	$Mg^+(g) \longrightarrow Mg^{2+}(g) + e^-$	$I_2 = +1451$ kJ
$\frac{1}{2}$ Dissociation of O_2(g):	$\frac{1}{2} O_2(g) \longrightarrow O(g)$	$\Delta H_{dis} = +249$ kJ
First electron affinity:	$O(g) + e^- \longrightarrow O^-(g)$	$EA_1 = -141.0$ kJ
Second electron affinity:	$O^-(g) + e^- \rightarrow O^{2-}(g)$	EA_2
Lattice energy:	$Mg^{2+}(g) + O^{2-}(g) \longrightarrow MgO(s)$	$L.E. = -3925$ kJ
Enthalpy of formation:	$Mg(s) + \frac{1}{2} O_2(g) \longrightarrow MgO(s)$	$\Delta H_f^\circ = -601.7$ kJ

$-601.7 \text{ kJ} = +146 \text{ kJ} + 737.7 \text{ kJ} + 1451 \text{ kJ} + 249 \text{ kJ} - 141.0 \text{ kJ} + EA_2 - 3925 \text{ kJ}$

$EA_2 = +881$ kJ

105. The Na^+ ions on the 6 faces are shared by 2 cells; the O^{2-} ions on the 8 corners are shared by 8 cells; the Cl^- is unique to each cell. Thus, the formula is Na_3ClO. The coordination numbers of O^{2-} and Cl^- are both 6. The shortest distance from the center of Na^+ to the center of O^{2-} is half the length of a face diagonal, which is $\sqrt{2} \times (a/2)$. The shortest distance from the center of a Cl^- to the center of O^{2-} is half the length of a cell diagonal, namely, $\sqrt{3} \times (a/2)$.

106. The two repr esentations are shown below:

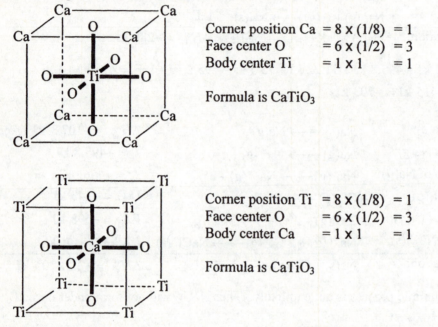

Corner position Ca = 8 x (1/8) = 1
Face center O = 6 x (1/2) = 3
Body center Ti = 1 x 1 = 1

Formula is $CaTiO_3$

Corner position Ti = 8 x (1/8) = 1
Face center O = 6 x (1/2) = 3
Body center Ca = 1 x 1 = 1

Formula is $CaTiO_3$

Note: In each representation, there are six oxygen atoms around each Ti atom and Ca atom

107. From chapter 10, Ca^{2+} radius = 100 pm, F^- radius = 133 pm.

$$\text{radius ratio} = \frac{r_{cation}}{r_{anion}} = \frac{100\ pm}{133\ pm} = 0.752$$

This is a fairly large value suggesting a structure in which the anions adopt a simple cubic structure so that the cations can be accommodated in the cubic holes. For CaF_2, the cations must occupy every other cubic hole so as to maintain the formula unit of CaF_2. Essentially CaF_2 has the CsCl structure, however every second cell has the cation missing.

FEATURE PROBLEMS

108. We obtain the surface tension by substituting the experimental values into the equation for surface tension.

$$h = \frac{2\gamma}{dgr} \quad \gamma = \frac{hdgr}{2} = \frac{1.1 \text{ cm} \times 0.789 \text{ g cm}^{-3} \times 981 \text{ cm s}^{-2} \times 0.050 \text{ cm}}{2} = 21 \text{ g/s}^2 = 0.021 \text{ J/m}^2$$

109. (a) $\frac{dP}{dT} = \frac{\Delta H_{vap}}{T(V_g - V_l)} = \frac{\Delta H_{vap}}{T(V_g)}$ $\quad Note: V_l \approx 0$ Rearrange expression, Use $V_g = \frac{nRT}{P}$

$$\frac{dP}{dT} = \frac{\Delta H_{vap}}{T(\frac{nRT}{P})} = \frac{\Delta H_{vap}}{\frac{nRT^2}{P}} = \frac{P\Delta H_{vap}}{nRT^2} \quad or \quad \frac{dP}{P} = \frac{\Delta H_{vap} \times dT}{nRT^2}$$

Consider 1 mole ($n = 1$) and substitute in $\Delta H_{vap} = 15,971 + 14.55\ T - 0.160\ T^2$

$$\frac{dP}{P} = \frac{(15,971 + 14.55\ T - 0.160\ T^2)dT}{RT^2} = \frac{(15,971)dT}{RT^2} + \frac{(14.55\ T)dT}{RT^2} - \frac{(0.160\ T^2)dT}{RT^2}$$

Simplify and collect constants

$$\frac{dP}{P} = \frac{(15,971)}{R}\frac{dT}{T^2} + \frac{(14.55)}{R}\frac{dT}{T} - \frac{(0.160)}{R}dT \quad \text{Integrate from } P_1 \to P_2 \text{ and } T_1 \to T_2$$

$$\ln\left(\frac{P_2}{P_1}\right) = \frac{(15,971)}{R}\left(\frac{1}{T_1} - \frac{1}{T_2}\right) + \frac{(14.55)}{R}\ln\left(\frac{T_2}{T_1}\right) - \frac{(0.160)}{R}(T_2 - T_1)$$

(b) First we consider 1 mole *(n = 1)* P_1= 10.16 torr (0.01337 atm) and T_1= 120 K Find the boiling point (T_2) when the pressure (P_2) is 1 atm.

$$\ln\left(\frac{1}{0.01337}\right) = \frac{(15,971)}{0.08206}\left(\frac{1}{120_1} - \frac{1}{T_2}\right) + \frac{(14.55)}{0.08206}\ln\left(\frac{T_2}{120}\right) - \frac{(0.160)}{0.08206}(T_2 - 120)$$

Then we solve for T_2 using the method of successive approximations: $T_2 = 169$ K

110. (a) 1 NaCl unit missing from the NaCl unit cell → overall stoichiometry is the same. The unit cell usually has 4 Na^+ and 4 Cl^- in the unit cell. Now the unit cell will have 3 Na^+ and 3 Cl^-. Accordingly, the density will decrease by a factor of 25% (1/4) if 1 Na^+ and 1 Cl^- are consistently absent throughout the structure. Thus, the density will be 0.75($d_{NaCl,\ normal}$).

(b) No change in stoichiometry or density, as this is just a simple displacement of an ion within the unit cell.

(c) Unit cell should contain 4 Ti^{2+} and 4 O^{2-} ions (same as in the NaCl unit cell).
4 TiO ions have a mass of :

$$4 \text{ formula units} \times \left(\frac{1 \text{ mol TiO}}{6.022 \times 10^{23} \text{ formula units}}\right)\left(\frac{63.88 \text{ g TiO}}{1 \text{ mol TiO}}\right) = 4.243 \times 10^{-22} \text{ g TiO}$$

$$V = (418 \text{ pm})^3 \times \left(\frac{1 \times 10^{-12} \text{ m}}{1 \text{ pm}}\right)^3 \left(\frac{100 \text{ cm}}{1 \text{ m}}\right)^3 = 7.30 \times 10^{-23} \text{ cm}^3$$

$$\text{Calculated density} = \frac{mass}{V} = \frac{4.243 \times 10^{-22} \text{ g}}{7.30 \times 10^{-23} \text{ cm}^3} = 5.81 \text{ g cm}^{-3} (\text{actual density} = 4.92 \text{ g cm}^{-3})$$

This indicates the presence of vacancies, and these could be Schottky-type defects.

111. (a) For a uniformly spaced (separation = r) one dimensional linear "crystal" of alternating cations and anions (having the same unipositive charge $Q_1 = Q = Q_2$ (see diagram below)), the interaction of one ion with all of the other ions is proportional to

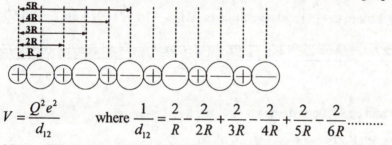

$$V = \frac{Q^2 e^2}{d_{12}} \qquad \text{where} \quad \frac{1}{d_{12}} = \frac{2}{R} - \frac{2}{2R} + \frac{2}{3R} - \frac{2}{4R} + \frac{2}{5R} - \frac{2}{6R} \cdots\cdots$$

Note: The factor of 2 comes from the fact that the same set of ions appears on both sides of the central ion and it does not matter if you start with a cation or an anion.

$$\frac{1}{d_{12}} = \frac{2}{R} \times \left(1 - \frac{1}{2} + \frac{1}{3} - \frac{1}{4} + \frac{1}{5} - \frac{1}{6} \cdots\cdots\right) \qquad \frac{1}{d_{12}} = \frac{1}{R} \times 2(\ln 2) \qquad V = \frac{Q^2 e^2}{R} \times 2(\ln 2)$$

(b) Consider the crystal lattice for NaCl.

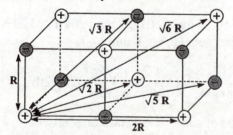

The 4 closest distances are marked to the reference Na^+ ion in the bottom left.
At a distance of R, there are 6 Cl^- ions which are attracted to the Na^+ ion.
At a distance of $\sqrt{2}$ R, there are 12 Na^+ ions which are repelled by the Na^+ ion.
At a distance of $\sqrt{3}$ R, there are 8 Cl^- ions which are attracted to the Na^+ ion.
At a distance of 2R or $\sqrt{4}$ R, there are 6 Na^+ ions which are repelled by the Na^+ ion.

The value of $k_M = \left(\frac{6}{1} - \frac{12}{\sqrt{2}} + \frac{8}{\sqrt{3}} - \frac{6}{\sqrt{4}} + \cdots\right)$

(c) If we carry out the same calculation for the CsCl structure, we would see that the Madelung constant is not the same. This is because the crystal structure is not the same. Notable is the fact that there are 8 nearest neighbors in CsCl and only 6 in NaCl.

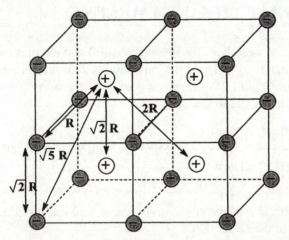

The 4 closest distances are marked to the reference Cs^+ ion in the top left cell.

At a distance of R, there are 8 Cl^- ions which are attracted to the Cs^+ ion.

At a distance of $\sqrt{2}$ R, there are 6 Cs^+ ions which are repelled by the Cs^+ ion.

At a distance of $\sqrt{5}$ R, there are 24 Cl^- ions which are attracted to the Cs^+ ion.

At a distance of 2R or $\sqrt{4}$ R, there are 8 Cs^+ ions which are repelled by the Cs^+ ion.

The value of $k_M = \left(\dfrac{8}{1} - \dfrac{6}{\sqrt{2}} + \dfrac{24}{\sqrt{5}} - \dfrac{8}{\sqrt{4}} + ... \right)$

CHAPTER 13
SOLUTIONS AND THEIR PHYSICAL PROPERTIES
PRACTICE EXAMPLES

1A To determine mass percent, we need both the mass of ethanol and the mass of solution. From volume percent, we know that 100.0 mL of solution contains 20.0 mL pure ethanol. The density of pure ethanol is 0.789 g/mL. We now can determine the mass of solute (ethanol) and solution. We perform the calculation in one step.

$$\text{mass percent ethanol} = \frac{20.0 \text{ mL ethanol} \times \dfrac{0.789 \text{ g}}{1 \text{ mL ethanol}}}{100.0 \text{ mL soln} \times \dfrac{0.977 \text{ g}}{1 \text{ mL soln}}} \times 100\% = 16.2\% \text{ ethanol by mass}$$

1B In each case, we use the definition of the concentration unit, making sure that numerator and denominator are converted to the correct units.

(a) We first determine the mass in grams of the solute and of the solution.

$$\text{mass } CH_3OH = 11.3 \text{ mL } CH_3OH \times \frac{0.793 \text{ g}}{1 \text{ mL}} = 8.96 \text{ g } CH_3OH$$

$$\text{mass soln} = 75.0 \text{ mL soln} \times \frac{0.980 \text{ g}}{1 \text{ mL}} = 73.5 \text{ g soln}$$

$$\text{amount of } H_2O = (73.5 \text{ g soln} - 8.96 \text{ g } CH_3OH) \times \frac{1 \text{ mol } H_2O}{18.02 \text{ g } H_2O}$$

$$\text{amount of } H_2O = 3.58 \text{ mol } H_2O$$

$$\text{amount of } CH_3OH = 8.96 \text{ g } CH_3OH \times \frac{1 \text{ mol } CH_3OH}{32.04 \text{ g } CH_3OH} = 0.280 \text{ mol } CH_3OH$$

$$\frac{H_2O \text{ mole}}{\text{fraction}} = \frac{\text{amount of } H_2O \text{ in moles}}{\text{amount of soln in moles}} = \frac{3.58 \text{ mol } H_2O}{3.58 \text{ mol } H_2O + 0.280 \text{ mol } CH_3OH} = 0.927$$

(b) $$[CH_3OH] = \frac{\text{amount of } CH_3OH \text{ in moles}}{\text{volume of soln in L}} = \frac{0.280 \text{ mol } CH_3OH}{75.0 \text{ mL soln} \times \dfrac{1 \text{ L}}{1000 \text{ mL}}} = 3.73 \text{ M}$$

(c) $$\text{molality of } CH_3OH = \frac{\text{amount of } CH_3OH \text{ in moles}}{\text{mass of } H_2O \text{ in kg}} = \frac{0.280 \text{ mol } CH_3OH}{64.5 \text{ g } H_2O \times \dfrac{1 \text{ kg}}{1000 \text{ g}}} = 4.34 \text{ } m$$

2A First we need to find the amount of each component in solution. Let us consider a 100.00-g sample of solution, in which there are 16.00 g glycerol and 84.00 g water.

$$\text{amount of glycerol} = 16.00 \text{ g } C_3H_5(OH)_3 \times \frac{1 \text{ mol } C_3H_5(OH)_3}{92.10 \text{ g } C_3H_5(OH)_3} = 0.1737 \text{ mol } C_3H_5(OH)_3$$

$$\text{amount of water} = 84.00 \text{ g } H_2O \times \frac{1 \text{ mol } H_2O}{18.02 \text{ g } H_2O} = 4.661 \text{ mol } H_2O$$

$$\text{mole fraction of } C_3H_5(OH)_3 = \frac{n_{C_3H_5(OH)_3}}{n_{[C_3H_5(OH)_3 + H_2O]}} = \frac{0.1737 \text{ mol } C_3H_5(OH)_3}{0.1737 \text{ mol } C_3H_5(OH)_3 + 4.661 \text{ mol } H_2O}$$

$$\text{mole fraction of } C_3H_5(OH)_3 = 0.03593$$

2B First we need the amount of sucrose in solution. We use a 100.00-g sample of solution, in which there are 10.00 g sucrose and 90.00 g water.

$$\text{amount } C_{12}H_{22}O_{11} = 10.00 \text{ g } C_{12}H_{22}O_{11} \times \frac{1 \text{ mol } C_{12}H_{22}O_{11}}{342.30 \text{ g } C_{12}H_{22}O_{11}} = 0.02921 \text{ mol } C_{12}H_{22}O_{11}$$

(a) Molarity is amount of solute in moles per liter of solution. Convert the 100.00 g of solution to L with density as a conversion factor.

$$C_{12}H_{22}O_{11} \text{ molarity} = \frac{0.02921 \text{ mol } C_{12}H_{22}O_{11}}{1000. \text{ g soln}} \times \frac{1.040 \text{ g soln}}{1 \text{ mL}} \times \frac{1000 \text{ mL}}{1 \text{ L}} = 0.3038 \text{ M}$$

(b) Molality is amount of solute in moles per kilogram of solvent.
Convert 90.00 g of solute to kg.

$$C_{12}H_{22}O_{11} \text{ molality} = \frac{0.02921 \text{ mol } C_{12}H_{22}O_{11}}{90.00 \text{ g } H_2O} \times \frac{1000 \text{ g}}{1 \text{ kg}} = 0.3246 m$$

(c) Mole fraction is the moles of solute per moles of solution. First compute the moles in 90.00 g H_2O.

$$n_{H_2O} = 90.00 \text{ g} \times \frac{1 \text{ mol } H_2O}{18.02 \text{ g } H_2O} = 4.994 \text{ mol } H_2O$$

$$\text{mole fraction } C_{12}H_{22}O_{11} = \frac{0.02921 \text{ mol } C_{12}H_{22}O_{11}}{0.02921 \text{ mol } C_{12}H_{22}O_{11} + 4.994 \text{ mol } H_2O} = 0.005815$$

3A Water is a highly polar compound. In fact, water molecules bond to each other through hydrogen bonds, which are unusually strong dipole-dipole interactions. Thus, water should mix well with other polar, hydrogen bonding compounds. (a) Toluene is nonpolar and should not be very soluble in water. (c) Benzaldehyde can form hydrogen bonds to water through its O atom. However, most of the molecule is non-polar and, as a result, it has limited solubility in water. (b) Oxalic acid is polar and can form hydrogen bonds. Of these three compounds, oxalic acid should be the most readily soluble in water. Actual solubilities (w/w%) are: toluene (0.067%) < benzaldehyde (0.28%) < oxalic acid (14%).

3B Both I_2 and CCl_4 are nonpolar molecules. It does not take much energy to break the attractions among I_2 molecules, or among CCl_4 molecules. Also, there is not a strong $I_2–CCl_4$ attraction created when a solution forms. Thus, I_2 should dissolve well in CCl_4 by simple mixing. H_2O is extensively hydrogen bonded with strong intermolecular forces that are difficult to break, but there is not a strong $I_2–H_2O$ attraction created when a solution forms. Thus, we expect I_2 to dissolve poorly in water. Actual solubilities are 2.603 g I_2/100 g CCl_4 and 0.033 g I_2/100 g H_2O.

4A The two suggestions are quoted first, followed by the means for achieving each one.

(1) Dissolve the 95 g NH_4Cl in just enough water to produce a saturated solution (55 g NH_4Cl/100 g H_2O) at 60°C.

$$\text{mass of water needed} = 95 \text{ g } NH_4Cl \times \frac{100 \text{ g } H_2O}{55 \text{ g } NH_4Cl} = 173 \text{ g } H_2O$$

The mass of NH_4Cl in the saturated solution at 20°C will be smaller.

$$\text{mass } NH_4Cl \text{ dissolved} = 173 \text{ g } H_2O \times \frac{37 \text{ g } NH_4Cl}{100 \text{ g } H_2O} = 64 \text{ g } NH_4Cl \text{ dissolved}$$

$$\text{crystallized mass } NH_4Cl = 95 \text{ g } NH_4Cl \text{ total} - 64 \text{ g } NH_4Cl \text{ dissolved at } 20°C$$
$$= 31 \text{ g } NH_4Cl \text{ crystallized}$$

(2) Lower the final temperature to 0°C, rather than 20°C. From Figure 14-8, at 0°C the solubility of NH_4Cl is 28.5 g NH_4Cl/100 g H_2O. From this (and knowing that there is 173 g H_2O present in the solution) we calculate the mass of NH_4Cl dissolved at this lower temperature.

$$\text{mass dissolved } NH_4Cl = 173 \text{ g } H_2O \times \frac{28.5 \text{ g } NH_4Cl}{100 \text{ g } H_2O} = 49.3 \text{ g } NH_4Cl \text{ dissolved}$$

The mass of NH_4Cl recrystallized is 95 g – 49.3 g = 46 g $\quad$ yield $= \left(\dfrac{46 \text{ g}}{95 \text{ g}}\right) \times 100\% = 48\%$

4B Percent yield for the recrystallization can be defined as:

$$\% \text{ yield} = \frac{\text{mass crystallized}}{\text{mass dissolved}(40°C)} \times 100\%$$

$$\% \text{ yield} = \frac{\text{mass dissolved}(40°C) - \text{mass dissolved}(20°C)}{\text{mass dissolved}(40°C)} \times 100\%$$

Figure 13-8 solubilities per 100 g H_2O are followed by percent yield calculations.
Solubility of $KClO_4$: 4.84 g at 40°C and 3.0 g at 20° C

$$\text{Percent yield of } KClO_4 = \frac{4.8 \text{ g}(40°C) - 3.0 \text{ g }(20°C)}{4.8 \text{ g }(40°C)} \times 100\% = 38\% \text{ } KClO_4$$

Solubility of KNO_3: 60.7 g at 40°C and 32.3 g at 20°C

$$\text{Percent Yield of } KNO_3 = \frac{60.7 \text{ g} (40°C) - 32.3 \text{ g} (20°C)}{60.7 \text{ g} (40°C)} \times 100\% = 47\% \ KNO_3$$

Solubility of K_2SO_4: 15.1 g at 40°C and 11.9 g at 20°C

$$\text{Percent yield of } K_2SO_4 = \frac{15.1 \text{ g} (40°C) - 11.9 \text{ g} (20°C)}{15.1 \text{ g} (40°C)} \times 100\% = 21\% \ K_2SO_4$$

Ranked in order of decreasing percent yield we have:
$KNO_3 (47\%) > KClO_4 (38\%) > K_2SO_4 (21\%)$

5A From Example 13-5, we know that the Henry's law constant for O_2 dissolved in water is $k = 2.18 \times 10^{-3}$ M atm^{-1}. Consequently,

$$P_{gas} = \frac{C}{k} = \frac{8.23 \times 10^{-4} \text{ M}}{2.18 \times 10^{-3} \text{ M atm}^{-1}} = 0.378 \text{ atm } O_2 \text{ pressure}$$

5B From Example 13-5, we know that Henry's law constant for O_2 dissolved in water is 2.18×10^{-3} M atm^{-1}

$$[O_2] = \frac{5.00 \times 10^{-3} \text{ g } O_2}{100 \text{ mL soln}} \times \frac{1000 \text{ mL soln}}{1 \text{ L soln}} \times \frac{1 \text{ mol } O_2}{32.0 \text{ g } O_2} = 0.001562\underline{5} \text{ M}$$

$$P_{O_2} = \frac{C}{k} = \frac{0.001562\underline{5} \text{ M}}{2.18 \times 10^{-3} \text{ M atm}^{-1}} = 0.717 \text{ atm}$$

6A Raoult's law enables us to determine the vapor pressure of each component.
$$P_{hex} = \chi_{hex} P°_{hex} = 0.750 \times 149.1 \text{ mmHg} = 112 \text{ mmHg}$$
$$P_{pen} = \chi_{pen} P°_{pen} = 0.250 \times 508.5 \text{ mmHg} = 127 \text{ mmHg}.$$
We use Dalton's law to determine the total vapor pressure:
$$P_{total} = P_{hex} + P_{pen} = 112 \text{ mmHg} + 127 \text{ mmHg} = 239 \text{ mmHg}$$

6B The masses of solution components need to be converted to amounts in moles through the use of molar masses. Let us choose as our amount precisely 1.0000 mole of $C_6H_6 = 78.11$ g C_6H_6 and an equal mass of toluene.

$$\text{amount of toluene} = 78.11 \text{ g } C_7H_8 \times \frac{1 \text{ mol } C_7H_8}{92.14 \text{ g } C_7H_8} = 0.8477 \text{ mol } C_7H_8$$

$$\text{mole fraction toluene} = \chi_{tol} = \frac{0.8477 \text{ mol } C_7H_8}{0.8477 \text{ mol } C_7H_8 + 1.0000 \text{ mol } C_6H_6} = 0.4588$$

$\text{toluene vapor pressure} = \chi_{tol} P°_{tol} = 0.4588 \times 28.4 \text{ mmHg} = 13.0 \text{ mmHg}$
$\text{benzene vapor pressure} = \chi_{benz} P°_{benz} = (1.0000 - 0.4588) \times 95.1 \text{ mmHg} = 51.5 \text{ mmHg}.$
$\text{total vapor pressure} = 13.0 \text{ mmHg} + 51.5 \text{ mmHg} = 64.5 \text{ mmHg}$

7A The mole fraction composition of each component is that component's partial pressure divided by the total pressure. Again, we note that the vapor is richer in the more volatile component.

$$y_{hexane} = \frac{P_{hexane}}{P_{total}} = \frac{112 \text{ mmHg hexane}}{239 \text{ mmHg total}} = 0.469 \quad y_{pentane} = \frac{P_{pentane}}{P_{total}} = \frac{127 \text{ mmHg pentane}}{239 \text{ mmHg total}} = 0.531 \text{ or}$$

simply $1.000 - 0.469 = 0.531$

7B The mole fraction composition of each component is that component's partial pressure divided by the total pressure. Again we note that the vapor is richer in the more volatile component.

$$y_t = \frac{P_{toluene}}{P_{total}} = \frac{13.0 \text{ mmHg toluene}}{64.5 \text{ mmHg total}} = 0.202 \quad y_b = \frac{P_{benzene}}{P_{total}} = \frac{51.5 \text{ mmHg benzene}}{64.5 \text{ mmHg total}} = 0.798$$

or simply $1.000-0.202 = 0.798$

8A We use the osmotic pressure equation, converting the mass of solute to amount in moles, the temperature to Kelvin, and the solution volume to liters.

$$\pi = \frac{nRT}{V} = \frac{\left(1.50 \text{ g } C_{12}H_{22}O_{11} \times \frac{1 \text{ mol } C_{12}H_{22}O_{11}}{342.3 \text{ g } C_{12}H_{22}O_{11}}\right) \times 0.08206 \frac{\text{L atm}}{\text{mol K}} \times 298 \text{ K}}{125 \text{ mL} \times \frac{1 \text{ L}}{1000 \text{ mL}}} = 0.857 \text{ atm}$$

8B We use the osmotic pressure equation to determine the molarity of the solution.

$$\frac{n}{V} = \frac{\pi}{RT} = \frac{0.015 \text{ atm}}{0.08206 \text{ L atm mol}^{-1} \text{ K}^{-1} \times 298 \text{ K}} = 6.1 \times 10^{-4} \text{ M}$$

Now, we can calculate the mass of urea.

$$\text{urea mass} = 0.225 \text{ L} \times \frac{6.1 \times 10^{-4} \text{ mol urea}}{1 \text{ L soln}} \times \frac{60.06 \text{ g } CO(NH_2)_2}{1 \text{ mol } CO(NH_2)_2} = 8.2\underline{4} \times 10^{-3} \text{ g}$$

9A We could substitute directly into the equation for molar mass derived in Example 13-9, but let us rather think our way through each step of the process. First, we find the concentration of the solution, by rearranging $\pi = \frac{n}{V}RT$. We need to convert the osmotic pressure to atmospheres.

$$\frac{n}{V} = \frac{\pi}{RT} = \frac{8.73 \text{ mmHg} \times \frac{1 \text{ atm}}{760 \text{ mmHg}}}{0.08206 \frac{\text{L atm}}{\text{mol K}} \times 298 \text{ K}} = 4.70 \times 10^{-4} \text{ M}$$

Next we determine the amount in moles of dissolved solute.

$$\text{amount of solute} = 100.0 \text{ mL} \times \frac{1 \text{ L}}{1000 \text{ mL}} \times \frac{4.70 \times 10^{-4} \text{ mol solute}}{1 \text{ L solution}} = 4.70 \times 10^{-5} \text{ mol solute}$$

We use the mass of solute, 4.04 g, to determine the molar mass. $\rightarrow$ $M = \frac{4.04 \text{ g}}{4.70 \times 10^{-5} \text{ mol}} = 8.60 \times 10^4 \text{ g/mol}$

9B We use the osmotic pressure equation along with the molarity of the solution.

$$\pi = \frac{n}{V}RT = \frac{2.12 \text{ g} \times \dfrac{1 \text{ mol}}{6.86\times10^4 \text{ g}}}{75.00 \text{ mL} \times \dfrac{1 \text{ L}}{1000 \text{ mL}}} \times \frac{0.08206 \text{ L atm}}{\text{mol K}} \times (310.2)\text{K} = 0.0105 \text{ atm} = 7.97 \text{ mmHg}$$

10A (a) The freezing point depression constant for water is $K_f = 1.86°C\,m^{-1}$.

$$\text{molality} = \frac{\Delta T_f}{-K_f} = \frac{-0.227°C}{-1.86°C\,m^{-1}} = 0.122 \; m$$

(b) We will use the definition of molality to determine the number of moles of riboflavin in 0.833 g of dissolved riboflavin.

$$\text{amount of riboflavin} = 18.1 \text{ g solvent H}_2\text{O} \times \frac{1 \text{ kg solvent}}{1000 \text{ g}} \times \frac{0.122 \text{ mol solute}}{1 \text{ kg solvent}}$$

amount of riboflavin $= 2.21\times10^{-3}$ mol riboflavin

$$\text{Molar Mass} = \frac{0.833 \text{ g riboflavin}}{2.21\times10^{-3} \text{ mol}} = 377 \text{ g/mol}$$

(c) We use the method of Chapter 3 to find riboflavin's empirical formula, starting with a 100.00-g sample.

$$54.25 \text{ g C} \times \frac{1 \text{ mol C}}{12.01 \text{ g C}} = 4.517 \text{ mol C} \div 1.063 \quad \rightarrow \quad = 4.249 \text{ mol C}$$

$$5.36 \text{ g H} \times \frac{1 \text{ mol H}}{1.008 \text{ g H}} = 5.32 \text{ mol H} \div 1.063 \quad \rightarrow \quad = 5.00 \text{ mol H}$$

$$25.51 \text{ g O} \times \frac{1 \text{ mol O}}{16.00 \text{ g O}} = 1.594 \text{ mol O} \div 1.063 \quad \rightarrow \quad = 1.500 \text{ mol O}$$

$$14.89 \text{ g N} \times \frac{1 \text{ mol N}}{14.01 \text{ g N}} = 1.063 \text{ mol N} \div 1.063 \rightarrow = 1.000 \text{ mol N}$$

If we multiply each of these amounts by 4 (because 4.249 is almost equal to $4\frac{1}{4}$), the empirical formula is found to be $C_{17}H_{20}O_6N_4$ with a molar mass of 376 g/mol. The molecular formula is $C_{17}H_{20}O_6N_4$.

10B The boiling point of pure water at 760.0 mmHg is 100.000°C. For higher pressures, the boiling point occurs at a higher temperature; for lower pressures, a lower boiling point is observed. The boiling point elevation for the urea solution is calculated as follows.
$$\Delta T_b = K_b \times m = 0.512 °C\,m^{-1} \times 0.205m = 0.105 °C$$
We would expect this urea solution to boil at $(100.00+0.105 =)100.105 °C$ under 760.0 mmHg atmospheric pressure. Since it boils at a lower temperature, the atmospheric pressure must be lower than 760.0 mmHg.

11A We assume a van't Hoff factor of $i = 3.00$ and convert the temperature to Kelvin, 298 K.

$$\pi = iMRT = \frac{3.00 \text{ mol ions}}{1 \text{ mol MgCl}_2} \times \frac{0.0530 \text{ mol MgCl}_2}{1 \text{ L soln}} \times 0.08206 \frac{\text{L atm}}{\text{mol K}} \times 298 \text{ K} = 3.89 \text{ atm}$$

11B We first determine the molality of the solution, and assume a van't Hoff factor of $i = 2.00$.

$$m = \frac{\Delta T_f}{-K_f \times i} = \frac{-0.100\,^{\circ}\text{C}}{-1.86\,^{\circ}\text{C}m^{-1} \times 2.00} = 0.0269\, m \approx 0.0269 \text{ M}$$

$$\text{volume of } HCl(aq) = 250.0 \text{ mL final soln} \times \frac{0.0269 \text{ mmol HCl}}{1 \text{ mL soln}} \times \frac{1 \text{ mL conc soln}}{12.0 \text{ mmol HCl}}$$

$$\text{volume of } HCl(aq) = 0.560 \text{ mL conc soln}$$

EXERCISES

Homogeneous and Heterogeneous Mixtures

1. $NH_2OH(s)$ should be the most soluble in water. Both $C_6H_6(l)$ and $C_{10}H_8(s)$ are composed of essentially nonpolar molecules, which are barely (if at all) soluble in water. Both $NH_2OH(s)$ and $CaCO_3(s)$ should be able to interact with water molecules. But $CaCO_3(s)$ contains ions of high charge, and thus it dissolves with great difficulty because of the high lattice energy. (Recall the solubility rules of Chapter 5: most carbonates are insoluble in water.)

2. Butyl alcohol should be moderately soluble in both water and benzene. A solute that is moderately soluble in both solvents will have some properties in common with each solvent. Both naphthalene and hexane are nonpolar molecules, like benzene, but interact only weakly with water molecules. Not surprisingly, they are soluble in benzene but not in water Sodium chloride consists of charged ions, similar to the charges in the polar bonds of water. Thus, as expected NaCl is very soluble in water. Butyl alcohol, on the other hand, possesses both a nonpolar part (C_4H_9-) like benzene, and a polar bond $(-O-H)$ like water. Consequently butyl alcohol is expected to be soluble in both water and benzene.

3. (b) salicyl alcohol probably is moderately soluble in both benzene and water. The reason for this assertion is that salicyl alcohol contains a benzene ring, which would make it soluble in benzene, and also can use its $-OH$ groups to hydrogen bond to water molecules. On the other hand, (c) diphenyl contains only nonpolar benzene rings; it should be soluble in benzene but not in water. (a) *para*-dichlorobenzene contains a benzene ring, making it soluble in benzene, and two polar C–Cl bonds, which oppose each other, producing a nonpolar—and thus water-insoluble—molecule. (d) hydroxyacetic acid is a very polar molecule with many opportunities for hydrogen bonding. Its polar nature would make it insoluble in benzene, while the prospective of hydrogen bonding will enhance aqueous solubility.

4. Vitamin C is a water-soluble molecule. It contains a number of polar –OH groups, capable of hydrogen bonding with water, making these molecules soluble in water. Vitamin E is fat-soluble. Its molecules are composed of largely non-polar hydrocarbon chains, with few polar groups and thus it should be soluble in nonpolar solvents.

5. (c) Formic acid and (f) propylene glycol are soluble in water. They both can form hydrogen bonds with water, and they both have small nonpolar portions. (b) benzoic acid and (d) butyl alcohol are only slightly soluble in water. Although they both can form hydrogen bonds with water, both molecules contain reasonably large nonpolar portions, which will not interact strongly with water. (a) iodoform and (e) chlorobenzene are insoluble in water. Although both molecules have polar groups, their influence is too small to enable the molecules to disrupt the hydrogen bonds in water and form a homogeneous liquid mixture.

6. Benzoic acid will react with NaOH to form a solution of sodium ions and benzoate ions. Both of these charged species are highly soluble in strongly polar solvents such as water. (In this equation, ϕ represents the benzene ring with one hydrogen omitted, C_6H_5-).

$$Na^+(aq) + OH^-(aq) + \phi-COOH \rightarrow Na^+(aq) + \phi-COO^-(aq) + H_2O\,(l)$$

The result of this reaction is that the concentration of molecular benzoic acid decreases, and thus more of the acid can be dissolved before the solution is saturated.

7. We expect small, highly charged ions to form crystals with large lattice energies, which tends to decrease their solubility I water. that decreases their solubility. Based on this information, we would expect MgF_2 to be insoluble and KF to be soluble. It is also probable that CaF_2 is insoluble due to its high lattice energy, but that NaF, with a smaller lattice energy, is soluble. Of all of the fluorides listed, KF is probably the most water soluble. The actual solubilities at $25\,°C$ are:
0.00020 M CaF_2 < 0.0021 M MgF_2 < 0.95 M NaF < 16 M KF.

8. The nitrate ion is reasonably large and has a small negative charge. It thus should produce crystals with small lattice energies. In contrast, S^{2-} is highly charged, small and easily polarized. It should form crystals with relatively high lattice energies. In fact, partial covalent bonding occurs in many metal sulfides, as predicted from the polarizability of the anion, making these partially covalent solids difficult to break apart. We would expect the sulfide ion to be most polarized by small, highly charged cations. Thus the sulfides of small metal cations should be the least soluble. The most soluble sulfides thus should be those that have large cations of low charge: K_2S, Rb_2S, Cs_2S, to name but three.

Percent Concentration

9. $$\% \text{ NaBr} = \frac{116 \text{ g NaBr}}{116 \text{ g NaBr} + 100 \text{ g H}_2\text{O}} \times 100\% = 53.7\% = 53.7 \text{ g NaBr}/100 \text{ g solution}$$

10. **(a)** $$\% \text{ by volume} = \frac{12.8 \text{ mL CH}_3\text{CH}_2\text{CH}_2\text{OH}}{75.0 \text{ mL soln}} \times 100\% = 17.1\% \text{ CH}_3\text{CH}_2\text{CH}_2\text{OH}$$

(b)
$$\text{percent by mass} = \frac{12.8 \text{ mL } CH_3CH_2CH_2OH \times \dfrac{0.803 \text{ g}}{1 \text{ mL}}}{75.0 \text{ mL} \times \dfrac{0.988 \text{ g}}{1 \text{ mL}}} \times 100\%$$

percent by mass = 13.9% $CH_3CH_2CH_2OH$ by mass

(c)
$$\text{percent}\left(\text{mass/vol}\right) = \frac{12.8 \text{ mL } CH_3CH_2CH_2OH \times \dfrac{0.803 \text{ g}}{1 \text{ mL}}}{75.0 \text{ mL}} \times 100\%$$

$$\text{percent}\left(\text{mass/vol}\right) = 13.7 \frac{\text{mass}}{\text{vol}}\% \ CH_3CH_2CH_2OH$$

11. soln. volume $= 725 \text{ kg NaCl} \times \dfrac{1000 \text{ g NaCl}}{1 \text{ kg NaCl}} \times \dfrac{100.00 \text{ g soln}}{3.87 \text{ g NaCl}} \times \dfrac{75.0 \text{ mL soln}}{76.9 \text{ g soln}} \times \dfrac{1 \text{ L soln}}{1000 \text{ mL soln}}$

$$= 1.83 \times 10^4 \text{ L sol'n}$$

12. mass $AgNO_3 = 0.1250 \text{ L soln} \times \dfrac{0.0321 \text{ mol } AgNO_3}{1 \text{ L soln}} \times \dfrac{169.9 \text{ g } AgNO_3}{1 \text{ mol } AgNO_3} \times \dfrac{100.00 \text{ g mixt.}}{99.81 \text{ g } AgNO_3}$

mass $AgNO_3$ = 0.683 g mixture

13. For water, the mass in grams and the volume in mL are about equal; the density of water is close to 1.0 g/mL. For ethanol, on the other hand, the density is about 0.8 g/mL. As long as the final solution volume after mixing is close to sum of the volumes for the two pure liquids, the percent by volume of ethanol will have to be larger than its percent by mass. This would not necessarily be true of other ethanol solutions. It would only be true in those cases where the density of the other component is greater than the density of ethanol.

14. We assume that blood has a density of 1.0 g/mL

$$\text{Mass \% cholesterol} = \frac{176 \text{ mg}}{dL} \times \frac{1 \text{ dl}}{100 \text{ mL}} \times \frac{100 \text{ ml}}{1.0 \times 10^2 \text{ g}} \times \frac{1 \text{ g}}{1000 \text{ mg}} \times 100\% = 0.17\underline{6} \%$$

We are only able to provide two significant figures at best owing to the initial assumption, namely, that blood has a density of 1.0 g/mL. Blood density will vary depending on a number of factors, for instance percent water (i.e. blood alcohol level), disease (high white blood cell count) and the sex of the person too name a few.

15.
$$\text{mass } HC_2H_3O_2 = 355 \text{ mL vinegar} \times \frac{1.01 \text{ g vinegar}}{1 \text{ mL}} \times \frac{6.02 \text{ g } HC_2H_3O_2}{100.00 \text{ g vinegar}} = 21.6 \text{ g } HC_2H_3O_2$$

16. We start with the molarity and convert both solute amount (numerator) and solution volume (denominator) to mass in order to obtain the mass percent.

$$\text{mass \% } H_2SO_4 = \frac{6.00 \text{ mol } H_2SO_4 \times \dfrac{98.08 \text{ g } H_2SO_4}{1 \text{ mol } H_2SO_4}}{1 \text{ L soln} \times \dfrac{1000 \text{ mL}}{1 \text{ L}} \times \dfrac{1.338 \text{ g}}{1 \text{ mL}}} \times 100\% = 44.0\% \ H_2SO_4$$

17. $46.1 \text{ ppm} = \dfrac{46.1 \text{ mg SO}_4^{2-}}{1 \text{ L solution}}$ (Assumes density of water ~1.00 g mL⁻¹)

$[SO_4^{2-}] = \dfrac{46.1 \text{ mg SO}_4^{2-}}{1 \text{ L solution}} \times \dfrac{1 \text{ g SO}_4^{2-}}{1000 \text{ mg SO}_4^{2-}} \times \dfrac{1 \text{ mol SO}_4^{2-}}{96.06 \text{ g SO}_4^{2-}} = 4.80 \times 10^{-4} \text{ M}$

18. $9.4 \text{ ppb CHCl}_3 = \dfrac{9.4 \text{ μg CHCl}_3}{1 \text{ L solution}}$ (Assumes density of water ~1.00 g mL⁻¹)

$\text{mass CHCl}_3 = \dfrac{9.4 \text{ μg CHCl}_3}{1 \text{ L solution}} \times \dfrac{1 \text{ g CHCl}_3}{1 \times 10^6 \text{ μg CHCl}_3} \times \dfrac{1 \text{ L solution}}{1000 \text{ g solution}} \times 250 \text{ g solution}$

$\text{mass CHCl}_3 = 2.4 \times 10^{-6} \text{ g}$

Molarity

19. $\text{molarity} = \dfrac{6.00 \text{ g CH}_3\text{OH} \times \dfrac{1 \text{ mol CH}_3\text{OH}}{32.04 \text{ g CH}_3\text{OH}}}{100.00 \text{ g soln} \times \dfrac{1 \text{ mL}}{0.988 \text{ g}} \times \dfrac{1 \text{ L}}{1000 \text{ mL}}} = 1.85 \text{ M} = [CH_3OH]$

20. $\text{H}_3\text{PO}_4 \text{ molarity} = \dfrac{75 \text{ g H}_3\text{PO}_4 \times \dfrac{1 \text{ mol H}_3\text{PO}_4}{98.00 \text{ g H}_3\text{PO}_4}}{100.0 \text{ g soln} \times \dfrac{1 \text{ mL}}{1.57 \text{ g}} \times \dfrac{1 \text{ L}}{1000 \text{ mL}}} = 12 \text{ M}$

21. The solution of Example 13-1 is 1.71 M C_2H_5OH, or 1.71 mmol C_2H_5OH in each mL of solution.

$\text{volume conc. soln} = 825 \text{ mL} \times \dfrac{0.235 \text{ mmol C}_2\text{H}_5\text{OH}}{1 \text{ mL soln}} \times \dfrac{1 \text{ mL conc. soln}}{1.71 \text{ mmol C}_2\text{H}_5\text{OH}}$

$\text{volume conc. soln} = 113 \text{ mL conc. soln}$

22. $\text{HNO}_3 \text{ molarity} = \dfrac{30.00 \text{ g HNO}_3 \times \dfrac{1 \text{ mol HNO}_3}{63.01 \text{ g HNO}_3}}{100.0 \text{ g soln} \times \dfrac{1 \text{ mL}}{1.18 \text{ g}} \times \dfrac{1 \text{ L}}{1000 \text{ mL}}} = 5.62 \text{ M at } 20 \text{ °C}$

Molality

23. $\text{molality} = \dfrac{2.65 \text{ g C}_6\text{H}_4\text{Cl}_2 \times \dfrac{1 \text{ mol C}_6\text{H}_4\text{Cl}_2}{147.0 \text{ g C}_6\text{H}_4\text{Cl}_2}}{50.0 \text{ mL} \times \dfrac{0.879 \text{ g}}{1 \text{ mL}} \times \dfrac{1 \text{ kg}}{1000 \text{ g}}} = 0.410 \, m$

24. Consider a 1 L (1000 mL) sample. This sample has a mass of 1338 g (1000 mL $\times 1.338$ g mL^{-1}). The mass of H_2SO_4 is 588.$\underline{4}$7 g (1 L $\times 6$ mol L^{-1} $\times 98.078$ g mol^{-1}) The mass of solvent = 749.$\underline{5}$3 g (1338 g $-$ 588.$\underline{4}$7 g). Molality is defined as soles of solute per kg of solvent.

$$\text{molality} = \frac{588.\underline{4}7 \text{ g } H_2SO_4 \times \dfrac{1 \text{ mol } H_2SO_4}{98.078 \text{ g } H_2SO_4}}{(749.\underline{5}3) \text{ g solvent} \times \dfrac{1 \text{ kg}}{1000 \text{ g}}} = 8.01 \ m$$

25. The mass of solvent in kg multiplied by the molality gives the amount in moles of the solute.

$$\text{mass } I_2 = \left(725.0 \text{ mL } CS_2 \times \frac{1.261 \text{ g}}{1 \text{ mL}} \times \frac{1 \text{ kg}}{1000 \text{ g}} \right) \times \frac{0.236 \text{ mol } I_2}{1 \text{ kg } CS_2} \times \frac{253.8 \text{ g } I_2}{1 \text{ mol } I_2} = 54.8 \text{ g } I_2$$

26. In the original solution the mass of CH_3OH associated with each 1.00 kg or 1000 g of water is:

$$\text{mass } CH_3OH = 1.00 \text{ kg } H_2O \times \frac{1.38 \text{ mol } CH_3OH}{1 \text{ kg } H_2O} \times \frac{32.04 \text{ g } CH_3OH}{1 \text{ mol } CH_3OH} = 44.2 \text{ g } CH_3OH$$

Then we compute the mass of methanol in 1.0000 kg = 1000.0 g of the original solution.

$$\text{mass } CH_3OH = 1000.0 \text{ g soln} \times \frac{44.2 \text{ g } CH_3OH}{1044.2 \text{ g soln}} = 42.3 \text{ g } CH_3OH$$

Thus, 1.0000 kg of this original solution contains ($1000.0 - 42.3 = 957.7$ g H_2O)

Now, a 1.00 m CH_3OH solution contains 32.04 g CH_3OH (one mole) for every 1000.0 g H_2O. We can compute the mass of H_2O associated with 42.3 g CH_3OH in such a solution.

$$\text{mass } H_2O = 42.3 \text{ g } CH_3OH \times \frac{1000.0 \text{ g } H_2O}{32.04 \text{ g } CH_3OH} = 1320 \text{ g } H_2O$$

(We have temporarily retained an extra significant figure in the calculation.) Since we already have 957.7 g H_2O in the original solution, the mass of water that must be added equals $1320 \text{ g} - 957.7 \text{ g} = 362 \text{ g } H_2O = 3.6 \times 10^2 \text{ g } H_2O$.

27.

$$H_3PO_4 \text{ molarity} = \frac{34.0 \text{ g } H_3PO_4 \times \dfrac{1 \text{ mol } H_3PO_4}{98.00 \text{ g } H_3PO_4}}{100.0 \text{ g soln} \times \dfrac{1 \text{ mL}}{1.209 \text{ g}} \times \dfrac{1 \text{ L}}{1000 \text{ mL}}} = 4.19 \text{ M}$$

$$H_3PO_4 \text{ molality} = \frac{34.0 \text{ g } H_3PO_4 \times \dfrac{1 \text{ mol } H_3PO_4}{98.00 \text{ g } H_3PO_4}}{66.0 \text{ g solvent} \times \dfrac{1 \text{ kg}}{1000 \text{ g}}} = 5.26 \ m$$

28. C_2H_5OH molarity $= \dfrac{10.00 \text{ g } C_2H_5OH \times \dfrac{1 \text{ mol } C_2H_5OH}{46.069 \text{ g } C_2H_5OH}}{90.00 \text{ g } H_2O \times \dfrac{1 \text{ kg}}{1000 \text{ g}}} = 2.412 \ m$

The molality of this solution does not vary with temperature. We would, however, expect the solution's molarity to vary with temperature because the solution's density, and thus the volume of solution that contains one mole, varies with temperature. Unlike the volume, the mass of the solution does not vary with temperature, and so the solution's molality doesn't vary with temperature either.

Mole Fraction, Mole Percent

29. The total number of moles $= 1.28 \text{ mol } C_7H_{16} + 2.92 \text{ mol } C_8H_{18} + 2.64 \text{ mol } C_9H_{20} = 6.84$ moles

(a) $\quad \chi_{C_7H_{16}} = \dfrac{1.28 \text{ mol } C_7H_{16}}{6.84 \text{ moles total}} = 0.187$ **(b)** $\quad \times 100\% = 18.7 \text{ mol}\% \ C_7H_{16}$

$\quad\quad \chi_{C_8H_{18}} = \dfrac{2.92 \text{ mol } C_8H_{18}}{6.84 \text{ moles total}} = 0.427$ $\times 100\% = 42.7 \text{ mol}\% \ C_8H_{18}$

$\quad\quad \chi_{C_9H_{20}} = \dfrac{2.64 \text{ mol } C_9H_{20}}{6.84 \text{ moles total}} = 0.386$ $\times 100\% = 38.6 \text{ mol}\% \ C_9H_{20}$

$\quad\quad$ or $1.00 - 0.187 - 0.427 = 0.386$ or $100 - 18.7 - 42.7 = 38.6 \%$

30. (a) Moles of $C_2H_5OH = 21.7 \text{ g } C_2H_5OH \times \dfrac{1 \text{ mol } C_2H_5OH}{46.07 \text{ g } C_2H_5OH} = 0.471 \text{ mol } C_2H_5OH$

$n_{H_2O} = 78.3 \text{ g } H_2O \times \dfrac{1 \text{ mol } H_2O}{18.02 \text{ g } H_2O} = 4.35 \text{ mol } H_2O \quad \chi_{\text{ethanol}} = \dfrac{0.471 \text{ mol } C_2H_5OH}{(0.471 + 4.35) \text{ total moles}} = 0.0977$

(b) $n_{H_2O} = 1000. \text{ g} \times \dfrac{1 \text{ mol } H_2O}{18.02 \text{ g } H_2O} = 55.49 \text{ mol } H_2O \quad \chi_{\text{urea}} = \dfrac{0.684 \text{ mol urea}}{(55.49 + 0.684) \text{ total moles}} = 0.0122$

31. (a) The amount of solvent is found after the solute's mass is subtracted from the total mass of the solution.

$\dfrac{\text{solvent}}{\text{amount}} = \left(\left(1 \text{ L soln} \times \dfrac{1000 \text{ mL}}{1 \text{ L}} \times \dfrac{1.006 \text{ g}}{1 \text{ mL}}\right) - \left(0.112 \text{ mol } C_6H_{12}O_6 \times \dfrac{180.2 \text{ g } C_6H_{12}O_6}{1 \text{ mol } C_6H_{12}O_6}\right)\right)$

$\quad = \left[1006 \text{ g solution} - 20.2 \text{ g } C_6H_{12}O_6\right] \times \dfrac{1 \text{ mol } H_2O}{18.02 \text{ g } H_2O} = 54.7 \text{ mol } H_2O$

$\chi_{\text{solute}} = \dfrac{0.112 \text{ mol } C_6H_{12}O_6}{0.112 \text{ mol } C_6H_{12}O_6 + 54.7 \text{ mol } H_2O} = 0.00204$

(b) First we must determine the mass the number of moles of ethanol. The number of moles of solvent is calculated after the ethanol's mass is subtracted from the solution's mass. Use a 100.00-mL sample of solution for computation. This 100 mL sample would contain 3.20 mL of C_2H_5OH. We calculate the mass of ethanol first, followed by the number of moles:

$$\text{mass}_{C_2H_5OH} = 3.20 \text{ mL } C_2H_5OH \times \frac{0.789 \text{ g}}{1 \text{ mL } C_2H_5OH} = 2.52 \text{ g } C_2H_5OH$$

$$\text{moles } C_2H_5OH = 2.52 \text{ g } C_2H_5OH \times \frac{1 \text{ mol } C_2H_5OH}{46.07 \text{ g } C_2H_5OH} = 0.0547 \text{ mol } C_2H_5OH$$

$$\text{mass of } H_2O = \left(\left(100.0 \text{ mL soln} \times \frac{0.993 \text{ g}}{1 \text{ mL}}\right) - 2.52 \text{ g } C_2H_5OH\right) = 96.8 \text{ g } H_2O$$

$$\text{amount of } H_2O = 96.8 \text{ g } H_2O \times \frac{1 \text{ mol } H_2O}{18.02 \text{ g } H_2O} = 5.37 \text{ mol } H_2O$$

$$\chi_{solute} = \frac{0.0547 \text{ mol } C_2H_5OH}{0.0547 \text{ mol } C_2H_5OH + 5.37 \text{ mol } H_2O} = 0.0101$$

32. Solve the following relationship for $n_{ethanol}$, the number of moles of ethanol, C_2H_5OH. The number of moles of water calculated in the Example is 5.01 moles, that of ethanol is 0.171 moles.

$$\chi_{ethanol} = \frac{n_{ethanol}}{n_{ethanol} + 5.01 \text{ mol } H_2O} = 0.0525 \qquad n_{ethanol} = 0.0525 \ n_{ethanol} + 0.263$$

$$n_{ethanol} = \frac{0.263}{1.0000 - 0.0525} = 0.278 \text{ mol ethanol}$$

$$\text{moles of added ethanol} = (0.278 - 0.171) \text{ mol } C_2H_5OH = 0.107 \text{ mol } C_2H_5OH$$

$$\text{mass added ethanol} = 0.107 \text{ mol } C_2H_5OH \times \frac{46.07 \text{ g } C_2H_5OH}{1 \text{ mol } C_2H_5OH} = 4.93 \text{ g } C_2H_5OH$$

33. The amount of water present in 1 kg is $n_{water} = 1000 \text{ g } H_2O \times \frac{1 \text{ mol } H_2O}{18.02 \text{ g } H_2O} = 55.49 \text{ mol } H_2O$

Now solve the following expression for n_{gly}, the amount of glycerol. $4.85\% = 0.0485$ mole fraction.

$$\chi_{gly} = 0.0485 = \frac{n_{gly}}{n_{gly} + 55.49} \qquad n_{gly} = 0.0485 \ n_{gly} + 2.69$$

$$n_{gly} = \frac{2.69}{(1.0000 - 0.0485)} = 2.83 \text{ mol glycerol}$$

$$\text{volume glycerol} = 2.83 \text{ mol } C_3H_8O_3 \times \frac{92.09 \text{ g } C_3H_8O_3}{1 \text{ mol } C_3H_8O_3} \times \frac{1 \text{ mL}}{1.26 \text{ g}} = 207 \text{ mL glycerol}$$

34. Consider 100.0 mL of solution 1 (0.1487 M $C_{12}H_{22}O_{11}$, d = 1.018 g/mL) and solution 2 (10.00% $C_{12}H_{22}O_{11}$, d = 1.038 g/mL). We first determine the moles of $C_{12}H_{22}O_{11}$ (342.30 g mol^{-1}) and H_2O (18.0153 g mol^{-1}) in each of the solutions.

Solution 1: $n_{C_{12}H_{22}O_{11}} = 0.1000 \text{ L solution} \times \frac{0.1487 \text{ mol } C_{12}H_{22}O_{11}}{\text{L solution}} = 0.01487 \text{ mol } C_{12}H_{22}O_{11}$

$mass_{H_2O}$ = mass of solution - mass of $C_{12}H_{22}O_{11}$

$$= 100 \text{ mL solution} \times \frac{1.018 \text{ g solution}}{1 \text{ mL solution}} - 0.01487 \text{ mol } C_{12}H_{22}O_{11} \times \frac{342.30 \text{ g } C_{12}H_{22}O_{11}}{1 \text{ mol } C_{12}H_{22}O_{11}}$$

$$= 101.8 \text{ g solution} - 5.090 \text{ g } C_{12}H_{22}O_{11} = 96.7 \text{ g } H_2O$$

$$n_{H_2O} = 96.7 \text{ g } H_2O \times \frac{1 \text{ mol } H_2O}{18.0153 \text{ g } H_2O} = 5.36\underline{8} \text{ mol } H_2O$$

Solution 2: mass of solution = = 100 mL solution $\times \dfrac{1.038 \text{ g solution}}{1 \text{ mL solution}}$ = 103.8 g solution

$$mass_{C_{12}H_{22}O_{11}} = 103.8 \text{ g soluton} \times \frac{10.00 \text{ g } C_{12}H_{22}O_{11}}{100 \text{ g solution}} = 10.38 \text{ g } C_{12}H_{22}O_{11}$$

$$n_{C_{12}H_{22}O_{11}} = 10.38 \text{ g } C_{12}H_{22}O_{11} \times \frac{1 \text{ mol } C_{12}H_{22}O_{11}}{342.30 \text{ g } C_{12}H_{22}O_{11}} = 0.03032 \text{ mol } C_{12}H_{22}O_{11}$$

$mass_{H_2O}$ = mass of solution - mass of $C_{12}H_{22}O_{11}$

$$= 103.8 \text{ g solution} - 10.38 \text{ g } C_{12}H_{22}O_{11} = 93.4 \text{ g } H_2O$$

$$n_{H_2O} = 93.4 \text{ g } H_2O \times \frac{1 \text{ mol } H_2O}{18.0153 \text{ g } H_2O} = 5.18\underline{4} \text{ mol } H_2O$$

Next we calculate the mole fraction

$$\chi_{C_{12}H_{22}O_{11}} = \frac{n_{C_{12}H_{22}O_{11}}}{n_{C_{12}H_{22}O_{11}} + n_{H_2O}} = \frac{0.01487 + 0.03022}{5.38\underline{6} + 5.184} = \frac{0.04509}{10.57} = 0.004266$$

Finally we calculate mole percent simply by multiplying mole fraction by 100%
Mole percent = 0.004266 × 100% = 0.4266%

Solubility Equilibrium

35. At 40 °C the solubility of NH_4Cl is 46.3 g per 100 g of H_2O. To determine molality, we calculate amount in moles of the solute and the solvent mass in kg.

$$\text{molarity} = \frac{46.3 \text{ g} \times \dfrac{1 \text{ mol } NH_4Cl}{53.49 \text{ g } NH_4Cl}}{100 \text{ g } H_2O \times \dfrac{1 \text{ kg}}{1000 \text{ g}}} = 8.66 \ m$$

36. The data in Figure 13-8 are given in mass of solute per 100 g H_2O. 0.200 m $KClO_4$ contains 0.200 mol $KClO_4$ associated with each 1 kg(= 1000) g of H_2O. The mass of $KClO_4$ dissolved in 100 g H_2O is:

$$\text{Mass } KClO_4 = 100 \text{ g } H_2O \times \frac{0.200 \text{ mol } KClO_4}{1000 \text{ g } H_2O} \times \frac{138.5 \text{ g } KClO_4}{1 \text{ mol } KClO_4} = 2.77 \text{ g } KClO_4.$$

This concentration is realized at a temperature of about 32 °C. At this temperature a saturated solution has $KClO_4 \approx 0.200 \ m$.

37. **(a)** The concentration for $KClO_4$ in this mixture is calculated first.

$$\frac{\text{mass solute}}{100 \text{ g H}_2\text{O}} = 100 \text{ g H}_2\text{O} \times \frac{20.0 \text{ g KClO}_4}{500.0 \text{ g water}} = 4.00 \text{ g KClO}_4$$

At 40 °C a saturated $KClO_4$ solution has a concentration of about 4.6 g $KClO_4$ dissolved in 100 g water. Thus, the solution is unsaturated.

(b) The mass of $KClO_4$ that must be added is the difference between the mass now present in the mixture and the mass that is dissolved in 500 g H_2O to produce a saturated solution.

$$\text{mass to be added} = \left(500.0 \text{ g H}_2\text{O} \times \frac{4.6 \text{ g KClO}_4}{100 \text{ g H}_2\text{O}} \right) - 20.0 \text{ g KClO}_4 = 3.0 \text{ g KClO}_4$$

38. From Figure 13-8, the solubility of KNO_3 in water is 38 g KNO_3 per 100 g water at 25.0 °C, while at 0.0 °C its solubility is 15 g KNO_3 per 100 g water. At 25.0 °C, every 138 g of solution contains 38 g of KNO_3 and 100.0 g water. At 0.0° C every 115 g solution contains 15 g KNO_3 and 100.0 g water. We calculate the mass of water and of KNO_3 present at 25.0 °C. Mass water = $335 \text{ g soln} \times \dfrac{100.0 \text{ g water}}{138 \text{ g soln}} = 243 \text{ g water}$

mass KNO_3 = 335 g soln − 243 g water = 92 g KNO_3

At 0.0 °C the mass of water has been reduced 55 g by evaporation to $(243 \text{ g} - 55 \text{ g} =)188 \text{ g}$. The mass of KNO_3 soluble in this mass of water at 0.0°C is

mass KNO_3 dissolved = $188 \text{ g H}_2\text{O} \times \dfrac{15 \text{ g KNO}_3}{100.0 \text{ g H}_2\text{O}} = 28 \text{ g KNO}_3$

Thus, of the 92 g KNO_3 originally present in solution, the mass that recrystallizes is

KNO_3 recrystallized (g) = 92 g KNO_3 (orig. dissol.) − 28 g KNO_3 (still dissol.) = 64 g KNO_3

Solubility of Gases

39. We first determine the number of moles of O_2 that have dissolved.

$$\text{Moles of O}_2 = \frac{PV}{RT} = \frac{1.00 \text{ atm} \times 0.02831 \text{ L}}{0.08206 \text{ L atm mol}^{-1} \text{ K}^{-1} \times 298 \text{ K}} = 1.16 \times 10^{-3} \text{ mol O}_2$$

$$[O_2] = \frac{1.16 \times 10^{-3} \text{ mol O}_2}{1.00 \text{ L soln}} = 1.16 \times 10^{-3} \text{ M}$$

The oxygen concentration now is computed at the higher pressure.

$$[O_2] = \frac{1.16 \times 10^{-3} \text{ M}}{1 \text{ atm O}_2} \times 3.86 \text{ atm O}_2 = 4.48 \times 10^{-3} \text{ M}$$

40. If we assume that the normal atmospheric pressure is 1 atm, then the partial pressure of O_2 can be determined by using Dalton's law of partial pressure. Since the volume % O_2 is 20.95%, the partial pressure of O_2 is 0.2095 atm (for ideal gases, volume % and mole % are equal)

$$[O_2] = \frac{1.16 \times 10^{-3} \text{ M}}{1 \text{ atm } O_2} \times 0.2095 \text{ atm } O_2 = 2.43 \times 10^{-4} \text{ M}$$

41. $\text{mass of } CH_4 = 1.00 \times 10^3 \text{ kg } H_2O \times \dfrac{0.02 \text{ g } CH_4}{1 \text{ kg } H_2O \cdot \text{atm}} \times 20 \text{ atm} = 4 \times 10^2 \text{ g } CH_4 \text{ (natural gas)}$

42. We first compute the amount of O_2 dissolved in 515 mL = 0.515 L at each temperature.

$$\text{amount } O_2 = 0.515 \text{ L} \times \frac{2.18 \times 10^{-3} \text{ mol } O_2}{1 \text{ L}} = 1.12 \times 10^{-3} \text{ mol } O_2 \text{ at } 0 \text{ °C}$$

$$\text{amount } O_2 = 0.515 \text{ L} \times \frac{1.26 \times 10^{-3} \text{ mol } O_2}{1 \text{ L}} = 6.49 \times 10^{-4} \text{ mol } O_2 \text{ at } 25 \text{ °C}$$

Next, we determine the difference between these amounts, and the corresponding volume of gas expelled..

$$V_{O_2} = \frac{(11.2 - 6.49) \times 10^{-4} \text{ mol } O_2 \times \dfrac{0.08206 \text{ L atm}}{\text{mol 1 K}} \times 298 \text{ K}}{1 \text{ atm}} = 1.2 \times 10^{-2} \text{ L} = 12 \text{ mL expelled}$$

43. We use the STP molar volume $(22.414 \text{ L} = 22{,}414 \text{ mL})$ to determine the molarity of Ar under 1 atmosphere of pressure and then use Henry's law.

$$k_{Ar} = \frac{C}{P_{Ar}} = \frac{\dfrac{33.7 \text{ mL Ar}}{1 \text{ L soln}} \times \dfrac{1 \text{ mol Ar}}{22{,}414 \text{ mL at STP}}}{1 \text{ atm pressure}} = \frac{0.00150 \text{ M}}{\text{atm}}$$

In the atmosphere, the partial pressure of argon is $P_{Ar} = 0.00934$ atm. (Recall that pressure fractions equal volume fractions for ideal gases.) We now compute the concentration of argon in aqueous solution.

$$C = k_{Ar} P_{Ar} = \frac{0.00150 \text{ M}}{\text{atm}} \times 0.00934 \text{ atm} = 1.40 \times 10^{-5} \text{ M Ar}$$

44. We use the STP molar volume $(22.414 \text{ L} = 22{,}414 \text{ mL})$ to determine the molarity of CO_2 under 1 atm pressure and then the Henry's law constant.

$$k_{CO_2} = \frac{C}{P_{CO_2}} = \frac{\dfrac{87.8 \text{ mL } CO_2}{0.1 \text{ L soln}} \times \dfrac{1 \text{ mol } CO_2}{22{,}414 \text{ mL at STP}}}{1 \text{ atm pressure}} = \frac{0.0392 \text{ M}}{\text{atm}}$$

In the atmosphere, the partial pressure of CO_2 is $P_{CO_2} = 0.000360$ atm. (Recall that pressure fractions equal volume fractions for ideal gases.) We now compute the concentration of carbon dioxide in aqueous solution.

$$C = k_{CO_2} P_{CO_2} = \frac{0.0392 \text{ M } CO_2}{\text{atm}} \times 0.000360 \text{ atm} = 1.41 \times 10^{-5} \text{ M } CO_2$$

45. Because of the low density of molecules in the gaseous state, the solution volume remains essentially constant as a gas dissolves in a liquid. Changes in concentrations in the solution results from changes in the number of dissolved gas molecules (recall Figure 13-11). This number is directly proportional to the mass of dissolved gas.

46. Although this statement initially may seem quite different, we need to remember that at constant volume, the partial pressure of the gas increases as more moles of gas are added. Thus increasing the partial pressure means that more moles of gas are forced into the same Volume of solvent. Thus at higher pressure, more moles of gas are absorbed by the solvent, but the volume of the "pressurized" gas absorbed is the same as the volume of gas absorbed at lower pressure. This statement is equivalent to equation 13.2.

$$kP_{gas} = c_{gas} = \frac{n_{gas}}{V_{solvent}} = \frac{P_{gas}V_{gas}}{RT_{gas}V_{solvent}} \quad \text{or} \quad V_{gas} = kRT_{gas}V_{solvent}$$

Thus, V_{gas} is constant at fixed temperature for a given volume of solvent. At the higher pressure V_{gas} contains more moles of gas. This statement is not valid if the ideal gas law is not obeyed or if the gas reacts with the solvent.

Raoult's Law and Liquid-Vapor Equilibrium

47. First we determine the number of moles of each component, its mole fraction in the solution, the partial pressure due to that component above the solution, and finally the total pressure.

$$\text{amount benzene} = n_b = 35.8 \text{ g } C_6H_6 \times \frac{1 \text{ mol } C_6H_6}{78.11 \text{ g } C_6H_6} = 0.458 \text{ mol } C_6H_6$$

$$\text{amount toluene} = n_t = 56.7 \text{ g } C_7H_8 \times \frac{1 \text{ mol } C_7H_8}{92.14 \text{ g } C_7H_8} = 0.615 \text{ mol } C_7H_8$$

$$\chi_b = \frac{0.458 \text{ mol } C_6H_6}{(0.458 + 0.615) \text{ total moles}} = 0.427 \qquad \chi_t = \frac{0.615 \text{ mol } C_7H_8}{(0.458 + 0.615) \text{ total moles}} = 0.573$$

$$P_b = 0.427 \times 95.1 \text{ mmHg} = 40.6 \text{ mmHg} \qquad P_t = 0.573 \times 28.4 \text{ mmHg} = 16.3 \text{ mmHg}$$

$$\text{total pressure} = 40.6 \text{ mmHg} + 16.3 \text{ mmHg} = 56.9 \text{ mmHg}$$

48. $\quad$ vapor fraction of benzene $= f_b = \dfrac{40.6 \text{ mmHg}}{56.9 \text{ mmHg}} = 0.714$

$\quad$ vapor fraction of toluene $= f_t = \dfrac{16.3 \text{ mmHg}}{56.9 \text{ mmHg}} = 0.286$

These are both pressure fractions, as calculated, and also the mole fractions in the vapor phase.

49. We need to determine the mole fraction of water in this solution.

$$n_{glucose} = 165 \text{ g } C_6H_{12}O_6 \times \frac{1 \text{ mol } C_6H_{12}O_6}{180.2 \text{ g } C_6H_{12}O_6} = 0.916 \text{ mol } C_6H_{12}O_6$$

$$n_{water} = 685 \text{ g } H_2O \times \frac{1 \text{ mol } H_2O}{18.02 \text{ g } H_2O} = 38.0 \text{ mol } H_2O \qquad \chi_{water} = \frac{38.0 \text{ mol } H_2O}{(38.0 + 0.916) \text{ total moles}} = 0.976$$

$$P_{soln} = \chi_{water} P^*_{water} = 0.976 \times 23.8 \text{ mmHg} = 23.2 \text{ mmHg}$$

50. We need determine the mole fraction of methanol in this solution.

$$n_{urea} = 17 \text{ g CO(NH}_2)_2 \times \frac{1 \text{ mol CO(NH}_2)_2}{60.06 \text{ g CO(NH}_2)_2} = 0.28\underline{3} \text{ mol CO(NH}_2)_2$$

$$n_{methanol} = 100 \text{ mL CH}_3\text{OH} \times \frac{0.792 \text{ g}}{1 \text{ mL}} \times \frac{1 \text{ mol CH}_3\text{OH}}{32.04 \text{ g CH}_3\text{OH}} = 2.47 \text{ mol CH}_3\text{OH}$$

$$\chi_{methanol} = \frac{2.47 \text{ mol CH}_3\text{OH}}{(2.47 + 0.28\underline{3}) \text{ total moles}} = 0.897$$

$$P_{sol'n} = \chi_{methanol} P_{methanol}^* = 0.897 \times 95.7 \text{ mmHg} = 85.8 \text{ mmHg}$$

51. We consider a sample of 100.0 g of the solution and determine the number of moles of each component in this sample. From this information and the given vapor pressures, we determine the vapor pressure of each component.

$$\text{amount styrene} = n_s = 38 \text{ g styrene} \times \frac{1 \text{ mol C}_8\text{H}_8}{104 \text{ g C}_8\text{H}_8} = 0.37 \text{ mol C}_8\text{H}_8$$

$$\text{amount ethylbenzene} = n_e = 62 \text{ g ethylbenzene} \times \frac{1 \text{ mol C}_8\text{H}_{10}}{106 \text{ g C}_8\text{H}_{10}} = 0.58 \text{ mol C}_8 \text{ H}_{10}$$

$$\chi_s = \frac{n_s}{n_s + n_e} = \frac{0.37 \text{ mol styrene}}{(0.37 + 0.58) \text{ total moles}} = 0.39; \quad P_s = 0.39 \times 134 \text{ mmHg} = 52 \text{ mmHg for C}_8\text{H}_8$$

$$\chi_e = \frac{n_e}{n_s + n_e} = \frac{0.58 \text{ mol ethylbenzene}}{(0.37 + 0.58) \text{ total moles}} = 0.61; \quad P_e = 0.61 \times 182 \text{ mmHg} = 111 \text{ mmHg for C}_8\text{H}_{10}$$

Then the mole fraction in the vapor can be determined.

$$y_e = \frac{P_e}{P_e + P_s} = \frac{111 \text{ mmHg}}{(111 + 52) \text{ mmHg}} = 0.68 \qquad y_s = 1.00 - 0.68 = 0.32$$

52. The vapor pressures of the pure liquids are: for benzene $P_b^\circ = 95.1$ mmHg and for toluene $P_t^\circ = 28.4$ mmHg. We know that $P_t = \chi_t P_t^\circ$ and $P_b = \chi_b P_b^\circ$. We also know that $\chi_t + \chi_b = 1.000$. We solve for χ_b in the following expression.

$$0.620 = \frac{P_b}{P_b + P_t} = \frac{\chi_b P_b^\circ}{\chi_b P_b^\circ + \chi_t P_t^\circ} = \frac{\chi_b P_b^\circ}{\chi_b P_b^\circ + (1 - \chi_b) P_t^\circ} = \frac{\chi_b 95.1}{\chi_b 95.1 + (1 - \chi_b) 28.4}$$

$$95.1\chi_b = 0.620[\chi_b 95.1 + (1 - \chi_b) 28.4] = 59.0\chi_b + 17.6 - 17.6\chi_b$$

$$17.6 = (95.1 - 59.0 + 17.6)\chi_b = 53.7\chi_b \qquad \chi_b = \frac{17.6}{53.7} = 0.328$$

53. The total vapor pressure above the solution at its normal boiling point is 760 mm Hg. The vapor pressure due to toluene is given by the following equation.
$P_{toluene} = \chi_{toluene} \cdot P_{toluene}^\circ = 0.700 \times 533 \text{ mm Hg} = 373 \text{ mm Hg}$. Next, the vapor pressure due to benzene is determined, followed by the vapor pressure for pure benzene.
$P_{benzene} = P_{total} - P_{toluene} = 760 \text{ mm Hg} - 373 \text{ mm Hg} = 387 \text{ mm Hg} = \chi_{benzene} \cdot P_{benzene}^\circ$
$387 \text{ mm Hg} = 0.300 \times P_{benzene}^\circ$ and hence, $P_{benzene}^\circ = 1.29 \times 10^3 \text{ mm Hg}$

54. **(a)** The vapor pressure of water is greater above the solution that has the higher mole fraction of water, according to Raoult's Law. This also is the solution that has the smaller mole fraction of NaCl, namely, the unsaturated solution.

(b) As water evaporates from the unsaturated solution its concentration changes and so does its vapor pressure, according to Raoult's Law. This does not happen with the saturated solution. Its concentration and thus its vapor pressure remains constant as the solvent evaporates.

(c) The solution with the higher boiling point is the more concentrated solution (the saturated solution) because the boiling point increases as the amount of dissolved solute increases.

Osmotic Pressure

<u>55.</u> We first compute the concentration of the solution. Then, assuming that the solution volume is the same as that of the solvent (0.2500 L), we determine the amount of solute dissolved, and finally the molar mass.

$$\frac{n}{V} = \frac{\pi}{RT} = \frac{1.67 \text{ mmHg} \times \dfrac{1 \text{ atm}}{760 \text{ mmHg}}}{0.08206 \text{ L atm mol}^{-1} \text{ K}^{-1} \times 298.2 \text{ K}} = 8.98 \times 10^{-5} \text{ M}$$

$$\frac{\text{solute}}{\text{amount}} = 0.2500 \text{ L} \times \frac{8.98 \times 10^{-5} \text{ mol}}{1 \text{ L}} = 2.25 \times 10^{-5} \text{ mol} \qquad M = \frac{0.72 \text{ g}}{2.25 \times 10^{-5} \text{ mol}} = 3.2 \times 10^{4} \text{ g/mol}$$

56. We assume the density of the solution to be 1.00 g/mL and calculate the $[C_{12}H_{22}O_{11}]$.

$$[C_{12}H_{22}O_{11}] = \frac{20. \text{ g } C_{12}H_{22}O_{11} \times \dfrac{1 \text{ mol } C_{12}H_{22}O_{11}}{342.3 \text{ g } C_{12}H_{22}O_{11}}}{100 \text{ g solution} \times \dfrac{1 \text{ mL solution}}{1 \text{ g solution}} \times \dfrac{1 \text{ L solution}}{1000 \text{ mL solution}}} = 0.58\underline{4} \text{ M}$$

We now compute the osmotic pressure, assuming a temperature of 25 °C

$$\pi = CRT = 0.58\underline{4} \text{ M} \times \frac{0.08206 \text{ L atm}}{\text{K mol}} \times 298 \text{ K} = 14.\underline{3} \text{ atm}$$

Next we need to determine the height of a column of sucrose (density = water).

$$\text{height} = 14.\underline{3} \text{ atm} \times \frac{760 \text{ mmHg}}{1 \text{ atm}} \times \frac{13.6 \text{ mm solution}}{1 \text{ mmHg}} \times \frac{1 \text{ m solution}}{1000 \text{ mm solution}} = 14\underline{8} \text{ m} \sim 150 \text{ m solution}$$

We have assumed a density of 1.00 g/mL and a conversion of 13.6 mm of solution per mmHg. If the solution were more dense, say 1.06 g/mL, this would correspond to a higher concentration and a larger osmotic pressure. At the same time, an increase in the density would result in a slight decrease in the conversion factor between mmHg and mm solution (we used 13.6 mm solution per mmHg , resulting in a slight decrease in the osmotic pressure) Overall, an increase in density would result in a slight increase in the osmotic pressure.

57. Both the flowers and the cucumber contain ionic solutions (plant sap), but both of these solutions are less concentrated than the salt solution. Thus, the solution in the plant material moves across the semipermeable membrane in an attempt to dilute the salt solution, leaving behind wilted flowers and shriveled pickles (wilted/shriveled plants have less water in their tissues).

58. The main purpose of drinking water is to provide H_2O for living tissues. In fresh water, water should move across the semipermeable membranes into the fish's body as water moves past the gills for breathing. Thus, it should not be necessary for freshwater fish to drink. In contrast, the concentration of solutes in salt water is about the same as their concentration in a fish's blood plasma. In this case, osmotic pressure will not be effective in moving water into a fish's tissues. The purpose of drinking for a salt water fish is to isolate the ingested water within the fish. Then mechanisms within the fish's body can move water across membranes into the fish's tissues. Or possibly, the fish has a means of desalinating the ingested water, in which case osmotic pressure would carry water into the fish's tissues.

59. We first determine the molarity of the solution. Let's work the problem out with three significant figures.

$$\frac{n}{V} = \frac{\pi}{RT} = \frac{1.00 \text{ atm}}{0.08206 \text{ L atm mol}^{-1} \text{ K}^{-1} \times 273 \text{ K}} = 0.0446 \text{ M}$$

$$\text{volume} = 1 \text{ mol} \times \frac{1 \text{ L}}{0.0446 \text{ mol solute}} = 22.4 \text{ L solution} \approx 22.4 \text{ L solvent}$$

We have assumed that the solution is so dilute that its volume closely approximates the volume of the solvent constituting it. Note that this volume corresponds to the STP molar volume of an ideal gas. The osmotic pressure equation also resembles the ideal gas equation.

60. First we calculate the concentration of solute in the solution, then the mass of solute in 100.0 mL of that solution.

$$[\text{solute}] = \frac{\pi}{RT} = \frac{7.25 \text{ mmHg} \times (\frac{1 \text{ atm}}{760 \text{ mmHg}})}{0.08206 \text{ L atm mol}^{-1} \text{ K}^{-1} \times 298 \text{ K}} = 3.90 \times 10^{-4} \text{ M}$$

$$\text{mass of hemoglobin} = 0.1000 \text{ L} \times \frac{3.90 \times 10^{-4} \text{ mol hemoglobin}}{1 \text{ L soln}} \times \frac{6.86 \times 10^4 \text{ g}}{1 \text{ mol hemoglobin}} = 2.68 \text{ g hemoglobin}$$

61. First we determine the concentration of the solution from the osmotic pressure, then the amount of solute dissolved, and finally the molar mass of that solute.

$$\pi = 5.1 \text{ mm soln} \times \frac{0.88 \text{ mmHg}}{13.6 \text{ mm soln}} \times \frac{1 \text{ atm}}{760 \text{ mmHg}} = 4.3 \times 10^{-4} \text{ atm}$$

$$\frac{n}{V} = \frac{\pi}{RT} = \frac{4.3 \times 10^{-4} \text{ atm}}{0.08206 \text{ L atm mol}^{-1} \text{ K}^{-1} \times 298 \text{ K}} = 1.8 \times 10^{-5} \text{ M}$$

$$\text{amount solute} = 100.0 \text{ mL} \times \frac{1 \text{ L}}{1000 \text{ mL}} \times 1.8 \times 10^{-5} \text{ M} = 1.8 \times 10^{-6} \text{ mol solute}$$

$$\text{Molar Mass} = \frac{0.50 \text{ g}}{1.8 \times 10^{-6} \text{ mol}} = 2.8 \times 10^5 \text{ g/mol}$$

62. We need the molar concentration of ions in the solution to determine its osmotic pressure. We assume that the solution has a density of 1.00 g/mL.

$$[ions] = \dfrac{0.92 \text{ g NaCl} \times \dfrac{1 \text{ mol NaCl}}{58.4 \text{ g NaCl}} \times \dfrac{2 \text{ mol ions}}{1 \text{ mol NaCl}}}{100.0 \text{ mL soln} \times \dfrac{1 \text{ L soln}}{1000 \text{ mL}}} = 0.32 \text{ M}$$

$$\pi = \dfrac{n}{V} RT = 0.32 \dfrac{\text{mol}}{\text{L}} \times 0.08206 \text{ L atm mol}^{-1} \text{ K}^{-1} \times (37.0 + 273.2) \text{ K} = 8.1 \text{ atm}$$

63. The reverse osmosis process requires a pressure equal to or slightly greater than the osmotic pressure of the solution. We assume that this solution has a density of 1.00 g/mL. First, we determine the molar concentration of ions in the solution.

$$[ions] = \dfrac{2.5 \text{ g NaCl} \times \dfrac{1 \text{ mol NaCl}}{58.4 \text{ g NaCl}} \times \dfrac{2 \text{ mol ions}}{1 \text{ mol NaCl}}}{100.0 \text{ mL soln} \times \dfrac{1 \text{ mL soln}}{1.00 \text{ g}} \times \dfrac{1 \text{ L soln}}{1000 \text{ mL}}} = 0.86 \text{ M}$$

$$\pi = \dfrac{n}{V} RT = 0.86 \dfrac{\text{mol}}{\text{L}} \times 0.08206 \text{ L atm mol}^{-1} \text{ K}^{-1} \times (25 + 273.2) \text{ K} = 21 \text{ atm}$$

64. We need to determine the molarity of each solution. The solution with the higher molarity has the higher osmotic pressure, and water will flow from the more dilute into one more concentrated solution, in an attempt to create two solutions of equal concentrations.

$$[C_3H_8O_3] = \dfrac{14.0 \text{ g } C_3H_8O_3 \times \dfrac{1 \text{ mol } C_3H_8O_3}{92.09 \text{ g } C_3H_8O_3}}{55.2 \text{ mL soln} \times \dfrac{1 \text{ L}}{1000 \text{ mL}}} = 2.75 \text{ M}$$

$$[C_{12}H_{22}O_{11}] = \dfrac{17.2 \text{ g } C_{12}H_{22}O_{11} \times \dfrac{1 \text{ mol } C_{12}H_{22}O_{11}}{342.3 \text{ g } C_{12}H_{22}O_{11}}}{62.5 \text{ mL soln} \times \dfrac{1 \text{ L}}{1000 \text{ mL}}} = 0.804 \text{ M}$$

Thus, water will move from the $C_{12}H_{22}O_{11}$ solution into the $C_3H_8O_3$ solution, that is, from right to left.

Freezing Point Depression and Boiling Point Elevation

65. First we compute the molality of the benzene solution, then the number of moles of solute dissolved, and finally the molar mass of the unknown compound.

$$m = \dfrac{\Delta T_f}{-K_f} = \dfrac{4.92^\circ\text{C} - 5.53^\circ\text{C}}{-5.12^\circ\text{C}/m} = 0.12 \, m$$

$$\text{amount solute} = 0.07522 \text{ kg benzene} \times \dfrac{0.12 \text{ mol solute}}{1 \text{ kg benzene}} = 9.0 \times 10^{-3} \text{ mol solute}$$

$$\text{Molecular Weight} = \dfrac{1.10 \text{ g unknown compound}}{9.0 \times 10^{-3} \text{ mol}} = 1.2 \times 10^2 \text{ g/mol}$$

66. We compute the value of the van't Hoff factor first, then use this value to compute the boiling point elevation.

$$i = \frac{\Delta T_f}{-K_f m} = \frac{-0.072°C}{-1.86°C/m \times 0.010 \ m} = 3.9$$

$$\Delta T_b = i K_b m = 3.9 \times 0.512°C \ m^{-1} \times 0.010 \ ; m = 0.020°C$$

Thus, the solution begins to boil at $100.02 \ °C$.

67. (a) First we determine the molality of the solution, then the value of the freezing-point depression constant.

$$m = \frac{1.00 \ g \ C_6H_6 \times \dfrac{1 \ mol \ C_6H_6}{78.11 \ g \ C_6H_6}}{80.00 \ g \ solvent \times \dfrac{1 \ kg}{1000 \ g}} = 0.160 \ m \qquad K_f = \frac{\Delta T_f}{-m} = \frac{3.3°C - 6.5°C}{-0.160 \ m} = 20.°C/m$$

(b) For benzene, $K_f = 5.12 \ °C m^{-1}$. Cyclohexane is the better solvent for freezing point depression determinations of molar mass, because a less concentrated solution will still give a substantial freezing-point depression. For the same concentration, cyclohexane solutions will show a freezing point depression approximately four times that of benzene. Also, one should steer clear of benzene because it is a known carcinogen.

68. The solution's boiling point is $0.40°C$ above the boiling point of pure water. First we must compute the molality of the solution, then the mass of sucrose that must be added to 1 kg (1000 g) H_2O, and finally the mass% sucrose in the solution.

$$m = \frac{\Delta T_b}{K_b} = \frac{0.40 \ °C}{0.512 \ °C/m} = 0.78 \ m \rightarrow \frac{0.78 \ mol \ C_{12}H_{22}O_{11}}{1 kg \ H_2O} \times \frac{342.3 \ g \ C_{12}H_{22}O_{11}}{1 \ mol \ C_{12}H_{22}O_{11}} = \frac{26_7 \ g \ C_6H_{12}O_6}{1 \ kg \ H_2O}$$

$$mass \ \% \ C_{12}H_{22}O_{11} = \frac{26_7 \ g \ C_{12}H_{22}O_{11}}{(1000. + 26_7) \ g \ total} \times 100\% = 21\% \ C_{12}H_{22}O_{11}$$

69. Here we determine the molality of the solution, then the number of moles of solute present, and the molar mass of the solute to start things off. Then we determine the compound's empirical formula, and combine this with the molar mass to determine the molecular formula.

$$m = \frac{\Delta T_f}{-K_f} = \frac{1.37 \ °C - 5.53 \ °C}{-5.12°C/m} = 0.813 \ m$$

$$amount = \left(50.0 \ mL \ C_6H_6 \times \frac{0.879 \ g}{1 \ mL} \times \frac{1 \ kg}{1000 \ g}\right) \times \frac{0.813 \ mol \ solute}{1 \ kg \ C_6H_6} = 3.57 \times 10^{-2} \ mol$$

$$Molecular \ Weight = \frac{6.45 \ g}{3.57 \times 10^{-2} \ mol} = 181 \ g/mol$$

Now, calculate the empirical formula from the provided mass percents for C, H, N and O.

$$42.9 \text{ g C} \times \frac{1 \text{ mol C}}{12.01 \text{ g C}} = 3.57 \text{ mol C} \quad \div 1.19 \rightarrow 3.00 \text{ mol C}$$

$$2.4 \text{ g H} \times \frac{1 \text{ mol H}}{1.01 \text{ g H}} = 2.4 \text{ mol H} \quad \div 1.19 \rightarrow 2.0 \text{ mol H}$$

$$16.7 \text{ g N} \times \frac{1 \text{ mol N}}{14.01 \text{ g N}} = 1.19 \text{ mol N} \quad \div 1.19 \rightarrow 1.00 \text{ mol N}$$

$$38.1 \text{ g O} \times \frac{1 \text{ mol O}}{16.00 \text{ g O}} = 2.38 \text{ mol O} \quad \div 1.19 \rightarrow 2.00 \text{ mol O}$$

The empirical formula is $C_3H_2NO_2$, with a formula mass of 84.0 g/mol. This is one-half the experimentally determined molar mass. Thus, the molecular formula is $C_6H_4N_2O_4$.

70. We determine the molality of the nitrobenzene $\left(K_f = 8.1°C/m \right)$ solution first, then the solute's molar mass.

$$m = \frac{\Delta T_f}{-K_f} = \frac{-1.4°C - 5.7°C}{-8.1°C/m} = 0.88 \, m$$

$$\text{amount solute} = 30.0 \text{ mL nitrobenzene} \times \frac{1.204 \text{ g}}{1 \text{ mL}} \times \frac{1 \text{ kg}}{1000 \text{ g}} \times \frac{0.88 \text{ mol solute}}{1 \text{ kg nitrobenzene}} = 0.032 \text{ mol solute}$$

$$\text{Molar Mass} = \frac{3.88 \text{ g nicotinamide}}{0.032 \text{ mol nicotinamide}} = 1.2 \times 10^2 \text{ g/mol}$$

Then we determine the compound's empirical formula, and combine this with the molar mass to determine the molecular formula.

$$59.0 \text{ g C} \times \frac{1 \text{ mol C}}{12.01 \text{ g C}} = 4.91 \text{ mol C} \quad \div 0.819 \rightarrow 6.00 \text{ mol C}$$

$$5.0 \text{ g H} \times \frac{1 \text{ mol H}}{1.01 \text{ g H}} = 5.0 \text{ mol H} \quad \div 0.819 \rightarrow 6.0 \text{ mol H}$$

$$22.9 \text{ g N} \times \frac{1 \text{ mol N}}{14.01 \text{ g N}} = 1.63 \text{ mol N} \quad \div 0.819 \rightarrow 2.00 \text{ mol N}$$

$$13.1 \text{ g O} \times \frac{1 \text{ mol O}}{16.00 \text{ g O}} = 0.819 \text{ mol O} \quad \div 0.819 \rightarrow 1.00 \text{ mol O}$$

The empirical formula is $C_6H_6N_2O$, with a molar mass of 122 g/mol. Since this is the same as the experimentally determined molar mass, the molecular formula of nicotinamide is $C_6H_6N_2O$.

71. We determine the molality of the benzene solution first, then the molar mass of the solute.

$$m = \frac{\Delta T_f}{-K_f} = \frac{-1.183°C}{-5.12°C/m} = 0.231 \, m$$

$$\text{amount solute} = 0.04456 \text{ kg benzene} \times \frac{0.231 \text{ mol solute}}{1 \text{ kg benzene}} = 0.0103 \text{ mol solute}$$

$$\text{Molar Mass} = \frac{0.867 \text{ g thiophene}}{0.0103 \text{ mol thiophene}} = 84.2 \text{ g/mol}$$

Next, we determine the empirical formula from the masses of the combustion products.

$$\text{amount C} = 4.913 \text{ g } CO_2 \times \frac{1 \text{ mol } CO_2}{44.010 \text{ g } CO_2} \times \frac{1 \text{ mol C}}{1 \text{ mol } CO_2} = 0.1116 \text{ mol C} \div 0.02791 \rightarrow 4.000 \text{ mol C}$$

$$\text{amount H} = 1.005 \text{ g } H_2O \times \frac{1 \text{ mol } H_2O}{18.015 \text{ g } H_2O} \times \frac{2 \text{ mol H}}{1 \text{ mol } H_2O} = 0.1116 \text{ mol H} \div 0.02791 \rightarrow 4.000 \text{ mol H}$$

$$\text{amount S} = 1.788 \text{ g } SO_2 \times \frac{1 \text{ mol } SO_2}{64.065 \text{ g } SO_2} \times \frac{1 \text{ mol S}}{1 \text{ mol } SO_2} = 0.02791 \text{ mol S} \div 0.02791 \rightarrow 1.000 \text{ mol S}$$

A reasonable empirical formula is C_4H_4S, which has an empirical mass of 84.1 g/mol. Since this is the same as the experimentally determined molar mass, the molecular formula of thiophene is C_4H_4S.

72. First we determine the empirical formula of coniferin from the combustion data.

$$0.698 \text{ g } H_2O \times \frac{1 \text{ mol } H_2O}{18.015 \text{ g } H_2O} \times \frac{2 \text{ mol H}}{1 \text{ mol } H_2O} = 0.0775 \text{ mol H} \times \frac{1.008 \text{ g H}}{1 \text{ mol H}} = 0.0781 \text{ g H}$$

$$2.479 \text{ g } CO_2 \times \frac{1 \text{ mol } CO_2}{44.010 \text{ g } CO_2} \times \frac{1 \text{ mol C}}{1 \text{ mol } CO_2} = 0.05633 \text{ mol C} \times \frac{12.011 \text{ g C}}{1 \text{ mol C}} = 0.6766 \text{ g C}$$

$$(1.205 \text{ g} - 0.0781 \text{ g H} - 0.6766 \text{ g C}) \times \frac{1 \text{ mol O}}{16.00 \text{ g O}} = 0.0281 \text{ mol O}$$

0.0775 mol H ÷ 0.0281 = 2.76 mol H	×4 → 11.04 mol H
0.05633 mol C ÷ 0.0281 = 2.00 mol C	×4 → 8.00 mol C
0.0281 mol O ÷ 0.0281 = 1.00 mol O	×4 → 4.00 mol O

Empirical formula = $C_8H_{11}O_4$ with a formula weight of 171 g/mol.

Now we determine the molality of the boiling solution, the number of moles of solute present in that solution, and the molar mass of the solute.

$$m = \frac{\Delta T_b}{K_b} = \frac{0.068°C}{0.512°C/m} = 0.13 \text{ } m$$

$$\text{amount solute} = 48.68 \text{ g } H_2O \times \frac{1 \text{ kg solvent}}{1000 \text{ g}} \times \frac{0.13 \text{ mol solute}}{1 \text{ kg solvent}} = 6.3 \times 10^{-3} \text{ mol solute}$$

$$\text{Molar Mass} = \frac{2.216 \text{ g solute}}{6.3 \times 10^{-3} \text{ mol solute}} = 3.5 \times 10^2 \text{ g/mol}$$

This molar mass is twice the formula weight of the empirical formula. Thus, the molecular formula is twice the empirical formula: Molecular formula = $C_{16}H_{22}O_8$.

73. The boiling point must go up by 2 °C, so $\Delta T_b = 2$ °C. We know that $K_b = 0.512$ °C/m for water. We assume that the mass of a litre of water is 1.000 kg and the van't Hoff factor for NaCl is $i = 2$. We first determine the molality of the saltwater solution and then the mass of solute needed.

$$m = \frac{\Delta T_b}{i \text{ } K_b} = \frac{2°C}{2.00 \times 0.512 \text{ °C/m}} = 2 \text{ } m$$

$$\text{solute mass} = 1.00 \text{ L } H_2O \times \frac{1 \text{ kg } H_2O}{1 \text{ L } H_2O} \times \frac{2 \text{ mol NaCl}}{1 \text{ kg } H_2O} \times \frac{58.4 \text{ g NaCl}}{1 \text{ mol NaCl}} = 12\underline{0} \text{ g NaCl}$$

This is at least ten times the amount of salt one would typically add to a liter of water for cooking purposes!

74. (a) The impure compound begins to melt at a temperature lower than that of the pure compound because the compound acts as a solvent for the impurity. The impurity thus causes a depression of the freezing point for the compound. Actually, the compound freezes (or melts) over a range of temperatures because once some of the pure compound freezes out, the remaining solution is a more concentrated solution of the impurity and thus has an even lower freezing point. It is this range of freezing points (or melting points) rather than a single freezing point- a so called "sharp" melting point- that is the true characteristic of an impure solid.

(b) The melting point of the impure compound can be higher than the melting point of the pure compound if two conditions are met. First, the impurity must have a higher melting point than the pure compound. Second, the impurity must be soluble in the solid solvent. Then we have a solution of the solvent in the impurity, and the impurity's melting point can determine the melting point of the solution (i.e. the melting point of the impure compound will be higher).

Strong Electrolytes, Weak Electrolytes, and Nonelectrolytes

75. The freezing point depression is given by $\Delta T_f = -iK_f m$. Since $K_f = 1.86°C/m$ for water, $\Delta T_f = -i0.186°C$ for this group of $0.10m$ solutions.

(a) $T_f = -0.19°C$ Urea is a nonelectrolyte, and $i = 1$.

(b) $T_f = -0.37°C$ NH_4NO_3 is a strong electrolyte, composed of two ions per formula unit; $i = 2$.

(c) $T_f = -0.37°C$ HCl is a strong electrolyte, composed of two ions per formula unit; $i = 2$.

(d) $T_f = -0.56°C$ $CaCl_2$ is a strong electrolyte, composed of three ions per formula unit; $i = 3$.

(e) $T_f = -0.37°C$ $MgSO_4$ is a strong electrolyte, composed of two ions per formula unit; $i = 2$.

(f) $T_f = -0.19°C$ Ethanol is a nonelectrolyte; $i = 1$.

(g) $T_f < -0.19°C$ $HC_2H_3O_2$ is a weak electrolyte; i is somewhat larger than 1.

76. (a) First calculate the freezing point for a $0.050\ m$ solution of a nonelectrolyte.

$$\Delta T_{f,calc} = -K_f m = -1.86°C/m \times 0.050\ m = -0.093°C \qquad i = \frac{\Delta T_{f,obs}}{\Delta T_{f,calc}} = \frac{-0.0986°C}{-0.093°C} = 1.1$$

(b) $[H^+] = [NO_2^-] = 6.91 \times 10^{-3}\ M = 0.00691\ M$

$[HNO_2] = 0.100\ M - 0.00691\ M = 0.093\ M$

$[\text{particles}] = 0.00691\ M + 0.00691\ M + 0.093\ M = 0.107\ M \qquad i = \frac{0.107\ M}{0.100\ M} = 1.07$

77. The combination of $NH_3(aq)$ with $HC_2H_3O_2(aq)$, results in the formation of $NH_4C_2H_3O_2(aq)$, which is a solution of the ions NH_4^+ and CH_3COO^-.

$$NH_3(aq) + HC_2H_3O_2(aq) \rightarrow NH_4C_2H_3O_2(aq) \rightarrow NH_4^+(aq) + C_2H_3O_2^-(aq)$$

This solution of ions or strong electrolytes conducts a current very well.

78. For two solutions to be isotonic, they must contain the same concentration of particles. For the solutions to also have the same % mass/volume would mean that each solution must possess the same number of particles per unit mass, typically per gram of solution. This would be true if they had the same molar mass and the same van't Hoff factor. It would also be true if the quotient of molar mass and van't Hoff factor were the same. While this condition is not impossible to meet, it is highly improbable.

Integrative and Advanced Exercises

79.
$$\text{mass P} = 12 \text{ fl oz} \times \frac{29.6 \text{ mL}}{1 \text{ fl oz}} \times \frac{1.00 \text{ g}}{1 \text{ mL}} \times \frac{0.13 \text{ g acid soln}}{100.0 \text{ g root beer}} \times \frac{75 \text{ g H}_3\text{PO}_4}{100 \text{ g acid soln}}$$

$$\times \frac{1 \text{ mol H}_3\text{PO}_4}{98.00 \text{ g H}_3\text{PO}_4} \times \frac{1 \text{ mol P}}{1 \text{ mol H}_3\text{PO}_4} \times \frac{30.97 \text{ g P}}{1 \text{ mol P}} \times \frac{1000 \text{ mg P}}{1 \text{ g P}} = 1.1 \times 10^2 \text{ mg P}$$

80. We determine the mass of each component in the water-rich phase.

$$\text{mass H}_2\text{O} = 32.8 \text{ g phase} \times \frac{92.50 \text{ g H}_2\text{O}}{100.00 \text{ g phase}} \times 100\% = 30.3\% \text{ H}_2\text{O}$$

mass phenol = 32.8 g phase – 30.3 g H_2O = 2.5 g phenol
Then we determine the mass of each component in the other phase.
mass phenol = 50.0 g – 2.5 g = 47.5 g phenol   mass H_2O = 50.0 g – 30.3 g = 19.7 g H_2O

$$\% \text{ H}_2\text{O} = \frac{19.7 \text{ g H}_2\text{O}}{19.7 \text{ g H}_2\text{O} + 47.5 \text{ g phenol}} \times 100\% = 29.3\% \text{ H}_2\text{O}$$

Above 66.8 °C phenol and water are completely miscible. Consequently, for temperatures above 66.8 °C, the mixture will be a homogeneous solution consisting of 50.0 g of H_2O and 50.0 g of phenol. To calculate the mole fraction of phenol in the mixture, we must first determine the number of moles of each component.

$$\text{Number of moles of H}_2\text{O} = 50.0 \text{ g H}_2\text{O} \times \frac{1 \text{ mol H}_2\text{O}}{18.016 \text{ g H}_2\text{O}} = 2.77\underline{5} \text{ mol H}_2\text{O}$$

$$\text{Number of moles of H}_2\text{O} = 50.0 \text{ g phenol} \times \frac{1 \text{ mol phenol}}{94.11 \text{ g phenol}} = 0.531 \text{ mol phenol}$$

$$\text{Thus } \chi_{\text{phenol}}(\text{mol fraction}) = \frac{0.531 \text{ mol phenol}}{2.775 \text{ mol phenol} + 0.531 \text{ mol H}_2\text{O}} = 0.161$$

81. The molarity of the original solution is computed first.

$$\text{KOH molarity} = \frac{109.2 \text{ g KOH} \times \dfrac{1 \text{ mol KOH}}{56.010 \text{ g KOH}}}{1 \text{ L soln}} = 1.950 \text{ M}$$

A 1.950 M solution is more concentrated than a 0.250 m solution. Thus, we must dilute the original solution. First we determine the mass of water produced in the final solution, and the mass of water present in the original solution, and finally the mass of water we must add.

$$\text{mass } H_2O \text{ in final soln} = 0.1000 \text{ L orig. soln} \times \frac{1.950 \text{ mol KOH}}{1 \text{ L soln}} \times \frac{1 \text{ kg } H_2O}{0.250 \text{ mol KOH}} = 0.780 \text{ kg } H_2O$$

$$\text{mass original solution} = 100.0 \text{ mL} \times \frac{1.09 \text{ g}}{1 \text{ mL}} = 109 \text{ g original solution}$$

$$\text{mass KOH} = 100.0 \text{ mL} \times \frac{1 \text{ L}}{1000 \text{ mL}} \times \frac{109.2 \text{ g KOH}}{1 \text{ L soln}} = 10.92 \text{ g KOH}$$

$$\text{original mass of water} = 109 \text{ g soln} - 10.92 \text{ g KOH} = 98 \text{ g } H_2O$$

$$\text{mass added } H_2O = 780. \text{g } H_2O - 98 \text{ g } H_2O = 682 \text{ g } H_2O$$

82. A quick solution to this problem is to calculate the molarity of a 50% by volume solution, using the density of 0.789 g/mL for ethanol.

$$\text{ethanol molarity} = \frac{50 \text{ mL } C_2H_5OH \times \dfrac{0.789 \text{ g}}{1 \text{ mL}} \times \dfrac{1 \text{ mol } C_2H_5OH}{46.07 \text{ g } C_2H_5OH}}{100 \text{ mL soln} \times \dfrac{1 \text{ L}}{1000 \text{ mL}}} = 8.6 \text{ M}$$

Thus, solutions that are more than 8.6 M are more than 100 proof.

83. At the outset, we observe that the solution having the highest mole fraction of water (the most dilute solution) will have the highest partial pressure of water in its equilibrium vapor. Also, the solution with the lowest mole fraction of water (the most concentrated solution) will have the lowest freezing point. We need to identify these two solutions, and to achieve this we need a simple way to compare concentrations, which are given here in disparate units. Let's look at the solutions one at a time. Keep in mind that 1 L of water contains about 1000 g $H_2O \div 18.02$ g/mol ≈ 55.49 mol H_2O (55.49 M).

Solution (a): The concentration 0.100% CH_3COCH_3 by mass is equivalent to 1.00 g CH_3COCH_3/1000 g solution, which in turn is about 1.00 g CH_3COCH_3/L solution. (The density of the solution will be very nearly the same as that of water.) A 1.00-g sample of CH_3COCH_3 is the same as 1.00 g/58.1 g /mol $\ll$ 0.10 mol CH_3COCH_3 .

Solution (b): The 0.100 M CH_3COCH_3 has 0.100 mol CH_3COCH_3 per liter. Solution (b) is more concentrated than (a), so (b) cannot be the solution with the highest water vapor pressure.

Solution (c): The 0.100 m CH_3COCH_3 has 0.100 mol CH_3COCH_3 per kg H_2O. Because water is the preponderant component in the solution, the density of the solution will be very nearly that of water, and the 0.100 m CH_3COCH_3 $\approx$ 0.100 M CH_3COCH_3. So, as with (b), solution (c) cannot be the solution with the highest water vapor pressure.

Solution (d): In the solution $\chi_{acetone}$ = 0.100, acetone molecules comprise one-tenth of all the molecules, or there is one acetone molecule for every nine water molecules. In a liter of this solution there would be about $55.5 \div 9 \approx 6$ mol CH_3COCH_3. This is clearly the most concentrated of the solutions.

To summarize, solution (a) has the highest water vapor pressure and solution (d) has the lowest freezing point. Our next task is to apply Raoult's law to solution (a), which requires the mole fraction of water in solution (a). One kilogram of this solution contains 1.00 g per 58.1 g /mol = 0.0172 mol CH_3COCH_3 and 999 g per 18.02 g/ mol = 55.44 mol H_2O. The mole fraction of the water is χ_{water} = 55.44/(55.44 + 0.0172) $\approx$ 1.00. The mole fraction is only very slightly less than 1.00, so the water vapor pressure above this solution will be only very slightly less than the vapor pressure of pure water at 25 °C, namely, 23.8 mmHg.

To apply equation (13.5) for freezing-point depression to solution (d), we must first express its concentration in molality. In a solution that has 1.00 mol CH_3COCH_3 for every 9.00 mol H_2O, there is 1.00 mol of the solute for every (9.00 mol × 18.02 g/mol) = 162 g H_2O. The molality of the solution is 1.00 mol CH_3COCH_3/0.162 kg H_2O = 6.17 m CH_3COCH_3. The approximate freezing-point depression for this solution is

$$\Delta T_f = -K_f \times m = -1.86 \text{ °C} \times 6.17 \text{ m} \approx -11.5 \text{ °C}$$

and the approximate freezing point of the solution is –11.5 °C. This result is only approximate because we have used equation (13.5) for a much more concentrated solution than the equation is intended.

84. We assume that the presence of the second solute will not substantially affect the solubility of the first solute in the solution. To get around this assumption, solubility data often are given in grams of solute per 100 g of solvent, as in Figure 13-8. This only partially avoids the problem because the interaction of the solutes can mutually affect their solubilities.

(a) We determine the solution concentration of each solute in the mixture, and compare these values to the solubilities at 60 °C obtained from Figure 13-8: > 60.0 g/100 g for KNO_3 and 18.2 g/100 g water for K_2SO_4.

$$\text{conc. KNO}_3 = \frac{0.850 \times 60.0 \text{ g solid}}{130.0 \text{ g water}} \times 100 \text{ g water} = 39.2 \text{ g KNO}_3/100 \text{ g water}$$

$$\text{conc. K}_2\text{SO}_4 = \frac{0.150 \times 60.0 \text{ g solid}}{130.0 \text{ g water}} \times 100 \text{ g water} = 6.92 \text{ g K}_2\text{SO}_4/100 \text{ g water}$$

Thus, all of the solid dissolves.

(b) At 0 °C, the solubility of KNO_3 is 14.0 g/100 g water. The mass of KNO_3 that crystallizes is obtained by difference.

$$m_{\text{KNO}_3} = 130.0\text{g water} \times \frac{39.2 \text{ g}_{\text{(dissolved at 60}^\circ\text{C)}} - 14.0 \text{ g}_{\text{ (dissolved at 0}^\circ\text{C)}}}{100 \text{ g water}}$$

$$m_{\text{KNO}_3} = 32.8\text{g KNO}_3 \text{ crystallized}$$

(c) At 0 °C, the solubility of K_2SO_4 is 7.4 g/100 g water. Since this is greater than the 6.92 g K_2SO_4/100 g water that is dissolved at 60 °C, all of the K_2SO_4 remains dissolved.

85. First determine the molality of the solution with the desired freezing point and compare it to the molality of the supplied solution to determine if ethanol or water needs to be added.

$$m = \frac{\Delta T_f}{-K_f} = \frac{-2.0 \text{ °C}}{-1.86 \text{ °C}/m} = 1.1 \ m \ \text{(desired)}$$

$$\text{mass solution} = 2.50 \text{ L} \times \frac{1000 \text{ mL}}{1 \text{ L}} \times \frac{0.9767 \text{ g}}{1 \text{ mL}} = 2.44 \times 10^3 \text{ g solution}$$

$$\text{mass solute} = 2.44 \times 10^3 \text{ g soln} \times \frac{13.8 \text{ g C}_2\text{H}_5\text{OH}}{100.0 \text{ g soln}} = 337 \text{ g C}_2\text{H}_5\text{OH}$$

$$\text{mass H}_2\text{O} = (2.44 \times 10^3 \text{ g} - 337 \text{ g C}_2\text{H}_5\text{OH}) \times \frac{1 \text{ kg}}{1000 \text{ g}} = 2.10 \text{ kg H}_2\text{O}$$

$$m = \frac{337 \text{ g C}_2\text{H}_5\text{OH} \times \dfrac{1 \text{ mol C}_2\text{H}_5\text{OH}}{46.07 \text{ g C}_2\text{H}_5\text{OH}}}{2.10 \text{ kg H}_2\text{O}} = 3.48 \ m \ \text{(available)}$$

Thus we need to dilute the solution with water. We must find the mass of water needed in the final solution, then the mass of water that must be added.

$$\text{final mass of water} = 337 \text{ g C}_2\text{H}_5\text{OH} \times \frac{1 \text{ mol C}_2\text{H}_5\text{OH}}{46.07 \text{ g C}_2\text{H}_5\text{OH}} \times \frac{1 \text{ kg H}_2\text{O}}{1.1 \text{ mol C}_2\text{H}_5\text{OH}} = 6.6 \text{ kg H}_2\text{O}$$

$$\text{mass H}_2\text{O needed} = 6.6 \text{ kg total} - 2.10 \text{ kg already present} = 4.5 \text{ kg H}_2\text{O}$$

86. HCl is very soluble in water and the reaction with water is highly exothermic. Concentrated HCl gives off small quantities of HCl(g) which react with water vapor in the atmosphere. The reaction between HCl and water is sufficiently exothermic that the room temperature water vapor changes to steam and then recondenses as the solution returns to room temperature.

87. (a) First we will determine the effect of an increase in pressure from the triple point pressure of 4.58 mmHg to 1 atm pressure on the melting point of ice.

$$\Delta T = -0.00750\ °C/atm \times \frac{(760 - 4.58)\ \text{Torr}}{760\ \text{Torr/atm}} = -0.00745\ °C$$

(b) Next we will determine the molarity of dissolved air in water:

$$\text{molarity} = \frac{29.18\ \text{mL} \times \dfrac{1\ \text{mol air}}{22,414\ \text{mL}}}{1\ \text{L}} = 0.0013\ \text{M}.$$

The molality of the solution is also 0.0013 m. The freezing point lowering by this dissolved air is $\Delta T_f = -1.86\ °C\ m^{-1} \times 0.0013\ m = -0.0024\ °C$.

The effect of lowering the pressure on the system and removing the dissolved air would be to raise the melting point of the ice by $(0.00745 + 0.0024)\ °C \approx 0.0098\ °C$, which is what we set out to show.

88. First we determine the molality of the benzene solution, and then the number of moles of solute in the sample.

$$\text{molality} = \frac{\Delta T_f}{-K_f} = \frac{5.072\ °C - 5.533\ °C}{-5.12\ °C/m} = 0.0900\ m$$

$$\text{amount of solute} = 50.00\ \text{mL} \times \frac{0.879\ \text{g}}{1\ \text{mL}} \times \frac{1\ \text{kg}}{1000\ \text{g}} \times \frac{0.0900\ \text{mol solute}}{1\ \text{kg solvent}} = 0.00396\ \text{mol solute}$$

We use this amount to determine the amount of each acid in the solute, with the added data of the molar masses of stearic acid, $C_{18}H_{36}O_2$, 284.5 g/mol, and palmitic acid, $C_{16}H_{32}O_2$, 256.4 g/mol. We let x represent the amount in moles of palmitic acid.

$$\begin{aligned} 1.115\ \text{g sample} &= (0.00396 - x)\ \text{mol} \times 284.5\ \text{g/mol} + x\ \text{mol} \times 256.4\ \text{g/mol} \\ &= 1.12\underline{7} - 284.5\,x + 256.4\,x = 1.12\underline{7} - 28.1\,x \end{aligned}$$

$$x = \frac{1.12\underline{7} - 1.115}{28.1} = 0.0004\underline{3}\ \text{mol palmitic acid}$$

$$0.00396 - 0.0004\underline{3} = 0.0035\underline{3}\ \text{mol stearic acid}$$

mass palmitic acid $= 0.0004_3\ \text{mol} \times 256.4\ \text{g/mol} = 0.1\underline{1}\ \text{g palmitic acid}$

$$\% \text{ palmitic acid} = \frac{0.1_1\ \text{g palmitic acid}}{1.115\ \text{g sample}} \times 100\% = 1 \times 10^1\ \% \text{ palmitic acid (about 10\%)}$$

89. (a) Rearrange Raoult's law (equation 13.3) to the form $\dfrac{P_A^\circ - P_A}{P_A^\circ} = x_B$. The mass of H_2O absorbed by D_1 is proportional to P_A°; the mass of H_2O absorbed by D_2 is proportional to P_A. Hence,

$$\frac{11.7458 - 11.5057}{11.7458} = x_B = 0.0204. \quad X_A = 0.9796,$$

$P_A = 0.9796 \times 23.76 \text{ mmHg} = 23.28 \text{ mmHg}.$

The observed vapor pressure lowering = 23.76 mmHg – 23.28 mmHg = 0.48 mmHg.

(b) Calculate x_B for a 1.00 *m* solution. $x_B = 0.0177; x_A = 0.9823.$ Calculate P_A using Raoult's law. $P_A = 23.34$ mmHg. The expected vapor pressure lowering = (23.76 – 23.34) mmHg = 0.42 mmHg.

90. Let us first consider a solution in which nitrobenzene, K_f = 8.1 °C/*m*, is the solvent. For this solution to freeze at 0.0 °C,. $\Delta T_f = -5.7°C$

$$\text{molality} = \frac{\Delta T_f}{K_f} = \frac{-5.7\,°C}{-8.1\,°C/m} = 0.70\ m = \frac{0.70 \text{ mol } C_6H_6}{1 \text{ kg } C_6H_5NO_2}$$

$$\% \ C_6H_5NO_2 \ = \frac{1000 \text{ g } C_6H_5NO_2}{1000 \text{ g } C_6H_5NO_2 + 0.70 \text{ mol } C_6H_6 \times \dfrac{78.1 \text{ g } C_6H_6}{1 \text{ mol } C_6H_6}} \times 100$$

$$= 95\% \ C_6H_5NO_2 \text{ (by mass)}$$

Then we consider a solution in which benzene, K_f = 5.12 °C/*m*, is the solvent. For this solution to freeze at 0.0 °C, $\Delta T_f = -5.5°C$.

$$\text{molality} = \frac{\Delta T_f}{K_f} = \frac{-5.5\,°C}{-5.12\,°C/m} = 1.0_7\ m = \frac{1.0_7 \text{ mol } C_6H_5NO_2}{1 \text{ kg } C_6H_6}$$

$$\% \ C_6H_5NO_2 = \frac{1.0_7 \text{ mol } C_6H_5NO_2 \times \dfrac{123.1 \text{ g } C_6H_5NO_2}{1 \text{ mol } C_6H_5NO_2}}{1.0_7 \text{ mol } C_6H_5NO_2 \times \dfrac{123.1 \text{ g } C_6H_5NO_2}{1 \text{ mol } C_6H_5NO_2} + 1000 \text{ g } C_6H_6} \times 100$$

$$= 12\% \ C_6H_5NO_2 \text{ (by mass)}$$

91. We first determine the amount in moles of each substance.

$$\text{amount } CO(NH_2)_2 = 0.515 \text{ g } CO(NH_2)_2 \times \frac{1 \text{ mol } CO(NH_2)_2}{60.06 \text{ g } CO(NH_2)_2} = 0.00857 \text{ mol } CO(NH_2)_2$$

$$\text{amount } H_2O \text{ with urea} = 92.5 \text{ g } H_2O \times \frac{1 \text{ mol } H_2O}{18.015 \text{ g } H_2O} = 5.13 \text{ mol } H_2O$$

$$\text{amount } C_{12}H_{22}O_{11} = 2.50 \text{ g } C_{12}H_{22}O_{11} \times \frac{1 \text{ mol } C_{12}H_{22}O_{11}}{342.3 \text{ g } C_{12}H_{22}O_{11}} = 0.00730 \text{ mol } C_{12}H_{22}O_{11}$$

$$\text{amount } H_2O \text{ with sucrose} = 85.0 \text{ g } H_2O \times \frac{1 \text{ mol } H_2O}{18.015 \text{ g } H_2O} = 4.72 \text{ mol } H_2O$$

The vapor pressure of water will be the same above both solutions when their mole fractions are equal. We assume that the amount of water present as water vapor is negligible. The total amount of water in the two solutions is $(4.72 + 5.13 =) 9.85$ mol H_2O. We let n_{water} designate the amount of water in the urea solution. The amount of water in the sucrose solution is then 9.85 mole $- n_{water}$. Note that we can compute the mole fraction of solute for comparison, since when the two solute mole fractions are equal, the mole fractions of solvent will also be equal ($\chi_{solvent} = 1.0000 - \chi_{solute}$).

$$\frac{0.00857 \text{ mol urea}}{0.00858 \text{ mol urea} + n_{water}} = \chi_{urea} = \chi_{sucrose} = \frac{0.00730 \text{ mol sucrose}}{0.00730 \text{ mol sucrose} + (9.85 - n_{water})}$$

We "cross multiply" to begin the solution for n_{water}.

$$0.00857 (0.00730 + 9.85 - n_{water}) = 0.00730 (0.00858 + n_{water})$$

$$0.0845 - 0.00857 \, n_{water} = 0.0000626 + 0.00730 \, n_{water} \quad n_{water} = \frac{0.0845 - 0.0000626}{0.00730 + 0.00857} = 5.32 \text{ mol}$$

We check the answer by substitution into the mole fraction equation.

$$\frac{0.00857 \text{ mol urea}}{0.00857 \text{ mol} + 5.32\underline{4}} = 0.00161 = \chi_{urea} = \chi_{sucrose} = 0.00161 = \frac{0.00730 \text{ mol sucrose}}{0.00730 \text{ mol} + (9.85 - 5.32\underline{4})\text{mol}}$$

The mole fraction of water in each solution is $(1.00000 - 0.00161 =) 0.99839$, or 99.839 mol%

92. We would expect the movement of the solvent molecules across the semipermeable membrane to cease as soon as the concentration of solute particles becomes the same on both sides of the membrane. This equality of concentration is achieved in two ways. (1) The solvent molecules move across the membrane into the sucrose solution and dilute it. (2) the sucrose solution overflows out of the funnel into the liquid that that originally was pure water. In order to overflow the funnel, the water moving across the membrane must push the solution up the funnel. This requires a small pressure, an osmotic pressure, that is generated by a difference in concentration of the two solutions. Because this difference in concentration is required, the flow of water will stop when the concentrations are almost the same. At that point, the difference in concentrations will be insufficient to raise the solution to the top of the tube.

93. The reason that the cooling curve is not horizontal at the freezing point is that as the solution freezes, the composition of the liquid changes. The solid phase which forms at the freezing point is pure solvent. Consequently, as freezing occurs, the concentration of the solution increases, and the freezing point decreases.

94. **(a)** Surface area of a particle $= 4\pi r^2 = 4(3.1416)(1 \times 10^{-7} m)^2 = 1.26 \times 10^{-13}$ m²/particle
We need to find the number of particles present (not atoms!)

$$\text{particle volume} = \frac{4\pi r^3}{3} = \frac{4\pi(1 \times 10^{-7} m)^3}{3} = 4.19 \times 10^{-21} m^3$$

$$\text{particle mass} = \text{density x volume} = \frac{19.3 \text{ g}}{cm^3} \frac{(100 cm)^3}{(1m)^3} \times 4.19 \times 10^{-21} m^3 = 8.08 \times 10^{-14} g / particle$$

$$\text{number of Au particles} = \frac{\text{mass of Au}}{\text{particle mass}} = \frac{1.00 \times 10^{-3} g \text{ Au}}{8.08 \times 10^{-14} g/particle} = 1.24 \times 10^{10} \text{ particles}$$

$$\text{total surface area} = (1.26 \times 10^{-13} m^2/particle)(1.24 \times 10^{10} particles) = 1.56 \times 10^{-3} m^2$$

(b) For 1.00 mg Au the volume of Au is: $(1.00 \text{ mg})(1g/mg)/(19.3 \text{ g/cm}^3) = 5.18 \times 10^{-5} cm^3$

Volume of a cube is L^3, where L is the edge length, so $L = \sqrt[3]{5.18 \times 10^{-5} cm^3} = 3.73 \times 10^{-2}$ cm

Area $= 6 \times L^2 = 6 \times (3.73 \times 10^{-2} cm)^2 \times (1m/100cm)^2 = 8.34 \times 10^{-7} m^2$

95. (a) Assume that there is 100.00 mL of solution. Then, since the volumes of the components are additive, the volume of benzene in the solution is numerically equal to its volume percent, V. The volume of toluene is 100.00 mL $- V$. Now we determine the mass of the solution.

$$\text{solution mass} = V \text{ mL benzene} \times \frac{0.879 \text{ g benzene}}{1 \text{ mL benzene}} + (100.00 - V) \text{ mL toluene} \times \frac{0.867 \text{ g toluene}}{1 \text{ mL toluene}}$$

The density of the solution is given by its mass divided by its volume, 100.00 mL.

$$\text{Density} = \frac{\text{mass}}{\text{volume}} = \frac{V(\text{mL}) \times 0.879(\text{g/mL}) + (100.00 - V)\text{mL} \times 0.867(\text{g/mL})}{100.00 \text{ mL}}$$

This is the equation requested in (c).

We now calculate the density for a number of solutions of varying volume percent benzene

20/0 % by volume =

$$\frac{20.0 \text{ mL} \times 0..879 \text{ g mL}^{-1} + (100.0 - 20) \text{ mL} \times 0.869 \text{ g/mL}}{100 \text{ mL}} = 0.869 \text{ g/mL}$$

40/0 % by volume =

$$\frac{40.0 \text{ mL} \times 0..879 \text{ g mL}^{-1} + (100.0 - 40) \text{ mL} \times 0.869 \text{ g/mL}}{100 \text{ mL}} = 0.872 \text{ g/mL}$$

60/0 % by volume =

$$\frac{60.0 \text{ mL} \times 0..879 \text{ g mL}^{-1} + (100.0 - 60) \text{ mL} \times 0.869 \text{ g/mL}}{100 \text{ mL}} = 0.874 \text{ g/mL}$$

80/0 % by volume =

$$\frac{80.0 \text{ mL} \times 0..879 \text{ g mL}^{-1} + (100.0 - 80.0) \text{ mL} \times 0.869 \text{ g/mL}}{100 \text{ mL}} = 0.877 \text{ g/mL}$$

(b) The data is plotted below and includes the density of pure benzene (100 %) and toluene (0 %)

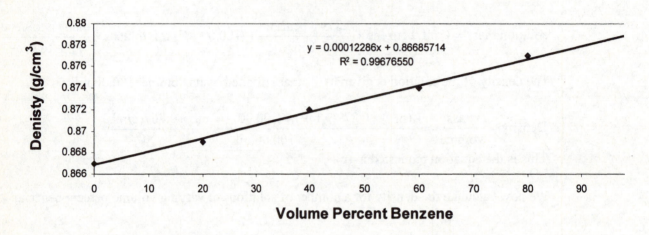

Plot of Solution Densities vs Volume Percent Benzene

$y = 0.00012286x + 0.86685714$
$R^2 = 0.99676550$

(c) The equation relating density of solution with volume percent Benzene was developed above and reprinted here.

$$\text{Density} = \frac{\text{mass}}{\text{volume}} = \frac{V(\text{mL}) \times 0.879(\text{g/mL}) + (100.00 - V)\text{mL} \times 0.867(\text{g/mL})}{100.00 \text{ mL}}$$

96. Separation by fractional distillation is based on a difference in the boiling points of the two substances being separated. Since the boiling points of *meta*- and *para*-xylene differ by only 0.7 °C, their separation by fractional distillation is impractical. However, the freezing points of these two isomers differ by more than 60°C. Separation by fractional solidification should be possible, freezing out primarily pure *para*-xylene at 13.3 °C, thereby leaving a solution rich in *meta*-xylene.

97. As the fraction of ethylene glycol in the solution increases, we eventually reach a point where the ethylene glycol, rather than water, is the solvent in the solution. After the fraction of ethylene glycol in the solution has increased past this point, the solution will freeze as the depressed freezing point of ethylene glycol, rather than at the depressed freezing point of water. That is, solid ethylene glycol, rather than solid water will freeze out of the solution. The freezing point of ethylene glycol (-13 °C) is higher than –34 °C.

98. (a) Let's begin with the definition of molality for an aqueous solution. We assume that the mass of the solvent is essentially equal to the mass of the solution, since the solution is a dilute one.

Next we assume that the density of this dilute solution is that of pure water, 1.00 g/mL, and then apply some definitions (1 kg = 1000 g and 1 L = 1000 mL).

$$\text{molality} \approx \frac{\text{moles solute}}{1 \text{ kg solution}} \times \frac{1 \text{ kg solution}}{1000 \text{ g solution}} \times \frac{1.00 \text{ g solution}}{1 \times 10^{-3} \text{ L solution}} \approx \frac{\text{moles solute}}{1 \text{ L solution}} = \text{molarity}$$

(b) We begin with the definition of mole fraction for the solute. Then we assume that the amount of solute is quite small compared to the amount of solvent, and use the molar mass (M) of the solvent, along with the fact that 1 kg = 1000 g.

$$\chi_{solute} = \frac{\text{moles solute}}{\text{moles solute} + \text{moles solvent}} \approx \frac{\text{moles solute}}{\text{moles solvent}} \times \frac{1\,\text{mol solvent}}{M\,\text{g solvent}} \times \frac{1000\,\text{g solvent}}{1\,\text{kg solvent}}$$

$$\approx \frac{1000\,\text{g/kg} \times \text{moles solute}}{M_{solvent} \times 1\,\text{kg solvent}} = \frac{\text{moles solute}}{1\,\text{kg solvent}} \times \frac{1000\,\text{g/kg}}{M_{solvent}} = \text{molality} \times \frac{1000\,\text{g/kg}}{M_{solvent}}$$

Thus, mole fraction of a solute indeed is proportional to its molality in a dilute solution.

(c) For a dilute aqueous solution, we begin with the result of part (b) and first use the definition of molality and then assume that the mass of the solvent is essentially equal to the mass of the solution.

$$\chi_{solute} \approx \text{molality} \times \frac{1000}{M_{solvent}} \approx \frac{\text{moles solut}}{1\,\text{kg solvent}} \times \frac{1000}{M_{solvent}} \times \frac{1\,\text{kg solvent}}{1\,\text{kg solution}}$$

Now we assume that the density of this dilute solution is that of pure water, 1.00 g/mL, and then apply some definitions (1 kg = 1000 g and 1 L = 1000 mL).

$$\chi_{solute} \approx \frac{\text{moles solute}}{1\,\text{kg solvent}} \times \frac{1000}{M_{solvent}} \times \frac{1\,\text{kg solvent}}{1\,\text{kg soln}} \times \frac{1\,\text{kg soln}}{1000\,\text{g soln}} \times \frac{1.00\,\text{g soln}}{1\,\text{mL soln}} \times \frac{1000\,\text{mL soln}}{1\,\text{L soln}}$$

$$\approx \frac{\text{moles solute}}{1\,\text{L solution}} \times \frac{1000}{M_{solvent}} = \text{molarity} \times \frac{1000}{M_{solvent}}$$

Thus molarity of dilute aqueous solutions is proportional to mole fraction of the solute.

99. We determine the volumes of O_2 and N_2 that dissolve under the appropriate partial pressures of 0.2095 atm O_2 and 0.7808 atm N_2, and at a temperatures of 25 °C, with both gas volumes measured at 1.00 atm and 25 °C. Note that the percent by volume and the percent by partial pressures of the atmosphere are numerically the same.

$$O_2 \text{ volume} = 0.2095\,\text{atm} \times \frac{28.31\,\text{mL O}_2}{1\,\text{atm} \cdot \text{L soln}} = 5.931\,\frac{\text{mL O}_2}{\text{L solution}}$$

$$N_2 \text{ volume} = 0.7808\,\text{atm} \times \frac{14.34\,\text{mL N}_2}{1\,\text{atm} \cdot \text{L soln}} = 11.20\,\frac{\text{mL N}_2}{\text{L solution}}$$

$$\text{volume \% N}_2 = \frac{11.20\,\frac{\text{mL N}_2}{\text{L solution}}}{11.20\,\frac{\text{mL N}_2}{\text{L solution}} + 5.931\,\frac{\text{mL O}_2}{\text{L solution}}} \times 100\% = 65.38\%\,\text{N}_2 \text{ by volume}$$

$$\text{volume \% O}_2 = 100.00\% - 65.38\% = 34.62\%\,\text{O}_2 \text{ by volume}$$

100. We must sum the partial pressures of the two components to obtain the total vapor pressure of the solution. Assuming ideal solution behavior, we can obtain these partial pressures with Raoult's law (Equation 13.3). In that equation, concentrations must be expressed in mole fractions, which requires that we first determine the number of moles of each solution component and then use Equation (13.3). We can use the ideal gas equation to calculate the number of moles of hexane vapor; this is the amount of hexane dissolved in the hexane–cyclohexane solution. We can then determine the number of moles of cyclohexane from its volume, density, and molar mass.

First, we rearrange the ideal gas equation to solve for number of moles and substitute the vapor equilibrium data, using the value 151 Torr for hexane at 25 °C.

$$n_{hexane} = \frac{PV}{RT} = \frac{151 \text{ Torr} \times \dfrac{1 \text{ atm}}{760 \text{ Torr}} \times 0.375 \text{ L hexane}}{0.08206 \dfrac{\text{L atm}}{\text{K mol}} \times (25.0 + 273.15) \text{ K}} = 0.00304 \text{ mol hexane}$$

Then, we calculate the number of moles of cyclohexane in the solution from the given volume and density.

$$n_{cyclohexane} = 50.0 \text{ mLcyclohexane} \times \frac{0.7739 \text{ g}}{\text{mL}} \times \frac{1 \text{ mol cyclohexane}}{84.16 \text{ g cyclohexane}} = 0.460 \text{ mol cyclohexane}$$

We get the mole fraction of hexane and cyclohexane from their molar quantities.

$$\chi_{cyclohexane} = \frac{0.460 \text{ mol cyclohexane}}{0.00304 \text{ mol hexane} + 0.460 \text{ mol cyclohexane}} \, 0.9934$$

$$\chi_{hexane} = \frac{0.00304 \text{ mol hexane}}{0.00304 \text{ mol hexane} + 0.460 \text{ mol cyclohexane}} = 0.00657$$

Next, we use Equation (13.3) to determine the partial pressures of hexane and cyclohexane.
$P_{cyclohexane} = 0.9934 \times 97.58 \text{ Torr} = 96.9 \text{ Torr}$
$P_{hexane} = 0.00657 \times 151 \text{ Torr} = 0.992 \text{ Torr}$

Finally, we calculate the total vapor pressure of the solution from Dalton's law of partial pressures.

$P_{total} = P_{cyclohexane} + P_{hexane} = 96.9 \text{ Torr} + 0.992 \text{ Torr} = 97.0 \text{ Torr}$

Our treatment of hexane vapor as an ideal gas should be valid because of the relatively low pressure of this gas (151 Torr or ~ 0.20 atm). Our assumption of ideal behavior for the hexane–cyclohexane solution should also be valid. The nonpolar hydrocarbons have compositions and molar masses that are nearly the same, suggesting that the intermolecular forces (dispersion forces) in the two substances should be very similar.

This ensures the validity of Raoult's law. The total vapor pressure of the solution (97.0 Torr) is nearly equal to that of pure cyclohexane (97.58 Torr)—as we might expect for this very dilute solution. Significantly, however, the solution vapor pressure is slightly higher than for pure cyclohexane; and this is consistent with the fact that hexane, the solute is noticeably more volatile than cyclohexane, the solvent (v_p = 97.58 Torr).

101. First we det ermine the mass of $CuSO_4$ in the original solution at 70 °C.

$$335 \text{ g sample} \times \frac{32.0 \text{ g } CuSO_4}{100.0 \text{ g soln}} = 107.\underline{2} \text{ g } CuSO_4$$

We let x represent the mass of $CuSO_4$ removed from the solution when the $CuSO_4 \cdot 5H_2O$ recrystallizes. Then the mass of H_2O present in the recrystallized product is determined.

$$\text{mass } H_2O = x \text{ g } CuSO_4 \times \frac{1 \text{ mol } CuSO_4}{159.6 \text{ g } CuSO_4} \times \frac{5 \text{ mol } H_2O}{1 \text{ mol } CuSO_4} \times \frac{18.02 \text{ g } H_2O}{1 \text{ mol } H_2O} = 0.565x \text{ g } H_2O$$

Now we determine the value of x by using the concentration of the saturated solution at 0 °C

$$\frac{12.5 \text{ g } CuSO_4}{100.0 \text{ g soln}} = \frac{(107.\underline{2} - x) \text{ g } CuSO_4}{(335 - x - 0.565 \, x) \text{ g soln}}$$

$$1.07\underline{2} \times 10^4 - 100.0x = 4.19 \times 10^3 - 19.6x \qquad (100.0 - 19.6)x = 1.07\underline{2} \times 10^4 - 4.19 \times 10^3$$

$$x = \frac{6.5\underline{3} \times 10^3}{80.4} = 81 \text{ g } CuSO_4$$

$$\text{mass } CuSO_4 \cdot 5H_2O = 81 \text{ g } CuSO_4 \times \frac{1.000 \text{ g } CuSO_4 + 0.565 \text{ g } H_2O}{1.000 \text{ g } CuSO_4} = 1.3 \times 10^2 \text{ g } CuSO_4 \cdot 5H_2O$$

FEATURE PROBLEMS

102. (a) The temperature at which the total vapor pressure is equal to 760 mm Hg corresponds to the boiling point for the cinnamaldehyde/H_2O mixture. Consequently, to find the temperature at which steam distillation begins under 1 atm pressure, we must construct a plot of P_{total} against the temperature. The data needed for this graph and a separate graph required for the answer to part (c) are given in the table below.

Temperature (°C)	P_{H_2O} (mmHg)	$P_{C_9H_8O}$ (mmHg)	P_{total} (mmHg)	Mole fraction of C_9H_8O
76.1	308	1	309	3.24×10^{-3}
105.8	942	5	947	5.28×10^{-3}
120.0	1489	10	1499	6.67×10^{-3}

The graph of total vapor pressure vs temperature is drawn below:

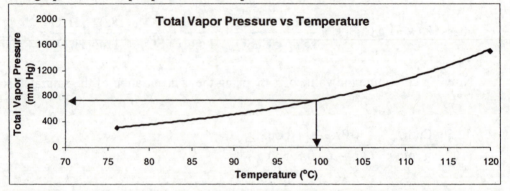

According to this plot, when the barometric pressure is 1 atm. (i.e. 760 mmHg), distillation occurs at 99.6 °C. Thus, as one would expect for a steam distillation, the boiling point for the mixture is below the boiling point for the lower boiling point component, viz. water. Moreover, the graph shows that the boiling point for this mixture is close to that for pure water and, once again, this is the anticipated result since the mixture contains very little cinnamaldehyde.

(b) The condensate that is collected in a steam distillation is the liquefied form of the gaseous mixture generated in the distillation pot. The composition of the gaseous mixture and that of the derived condensate is determined not by the composition of the liquid mixture but rather by the vapor pressure for each component at the boiling point. The boiling point of the cinnamaldehyde/H_2O mixture will stay fixed at 99.6 °C and the composition will remain at 99.5$\underline{4}$ mole % H_2O and 0.46$\underline{4}$ % cinnamaldehyde as long as there is at least a modicum of each component present.

(c) To answer this part of the question we must construct a plot of the mole fraction of cinnamaldehyde against temperature. The graph, which is drawn below, shows that the mole fraction for cinnamaldehyde at the boiling point (99.6 °C) is 0.00464. Because the concentration of each component in the vapor is directly proportional to its vapor pressure, the number of moles of component A divided by the moles of component B is equal to the partial pressure for component A divided by the partial pressure of component B.

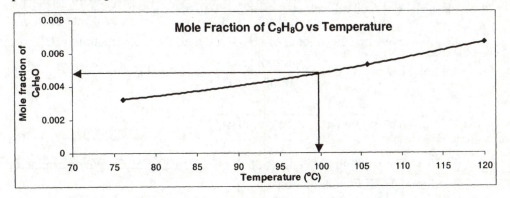

$$\frac{n_A}{n_B} = \frac{P_{component\ A}}{P_{component\ B}} = \frac{Mass\ A}{Mass\ B} = \frac{P_{component\ A}}{P_{component\ B}} \times \frac{Molar\ Mass\ component\ A}{Molar\ Mass\ component\ B}$$

If we make cinnamaldehyde component A and water component B, and then plug in the appropriate numbers into the second equation involving masses, we obtain

$P_{cinn} = 0.00464 \times (760\ mmHg) = {\sim}3.53\ mmHg;$
$P_{H_2O} = (760 - 3.53)\ mmHg = {\sim}756\ mmHg$

$$\frac{Mass\ C_9H_8O}{Mass\ H_2O} = \frac{3.53\ mmHg}{756\ mm\ Hg} \times \frac{1\ mol\ C_9H_8O \cdot \frac{132\ g\ C_9H_8O}{1\ mol\ C_9H_8O}}{18.0\ g\ H_2O \cdot \frac{1\ mol\ H_2O}{}} = 3.42 \times 10^{-2}$$

Clearly then, water condenses in the greater quantity by mass. In fact, every gram of cinnamaldehyde that condenses is accompanied by ~29 g of water in the collection flask!

103. (a) A solution with $\chi_{HCl} = 0.50$ begins to boil at about $18°\ C$. At that temperature the composition of the vapor is about $\chi_{HCl} = 0.63$, reading directly across the tie line at $18°C$. The vapor has $\chi_{HCl} > 0.50$.

(b) The composition of HCl(aq) changes as the solution boils in an open container because the vapor has a different composition than does the liquid. Thus, the component with the lower boiling point is depleted as the solution boils. The boiling point of the remaining solution must change as the vapor escapes due to changing composition.

(c) The azeotrope occurs at the maximum of the curve: at $\chi_{HCl} = 0.12$ and a boiling temperature of $110°C$.

(d)   We first determine the amount of HCl in the sample.

$$\text{amount HCl} = 30.32 \text{ mL NaOH} \times \frac{1 \text{ L}}{1000 \text{ mL}} \times \frac{1.006 \text{ mol NaOH}}{1 \text{ L}} \times \frac{1 \text{ mol HCl}}{1 \text{ mol NaOH}}$$

amount HCl = 0.03050 mol HCl

The mass of water is the difference between the mass of solution and that of HCl.

$$\text{mass H}_2\text{O} = \left(5.00 \text{ mL soln} \times \frac{1.099 \text{ g}}{1 \text{ mL}} \right) - \left(0.03050 \text{ mol HCl} \times \frac{36.46 \text{ g HCl}}{1 \text{ mol HCl}} \right) = 4.38 \text{ g}$$

Now we determine the amount of H_2O and then the mole fraction of HCl.

$$\text{amount H}_2\text{O} = 4.38 \text{ g H}_2\text{O} \times \frac{1 \text{ mol H}_2\text{O}}{18.02 \text{ g H}_2\text{O}} = 0.243 \text{ mol H}_2\text{O}$$

$$\chi_{\text{HCl}} = \frac{0.03050 \text{ mol HCl}}{0.03050 \text{ mol HCl} + 0.243 \text{ mol H}_2\text{O}} = 0.112$$

104. (a)   At 20 °C, the solubility of NaCl is 35.9 g NaCl / 100 g H_2O. We determine the mole fraction of H_2O in this solution

$$\text{amount H}_2\text{O} = 100 \text{ g H}_2\text{O} \times \frac{1 \text{ mol H}_2\text{O}}{18.02 \text{ g H}_2\text{O}} = 5.549 \text{ mol H}_2\text{O}$$

$$\text{amount NaCl} = 35.9 \text{ g NaCl} \times \frac{1 \text{ mol NaCl}}{58.44 \text{ g NaCl}} = 0.614 \text{ mol NaCl}$$

$$\chi_{\text{water}} = \frac{5.549 \text{ mol H}_2\text{O}}{0.614 \text{ mol NaCl} + 5.549 \text{ mol H}_2\text{O}} = 0.9004$$

The approximate relative humidity then will be 90% (90.04 %), because the water vapor pressure above the NaCl saturated solution will be 90.04% of the vapor pressure of pure water at 20° C.

(b) $CaCl_2 \cdot 6 \, H_2O$ deliquesces if the relative humidity is over 32%. Thus, $CaCl_2 \cdot 6 \, H_2O$ will deliquesce (i.e. it will absorb water from the atmosphere).

(c)   If the substance in the bottom of the desiccator has a high water solubility, its saturated solution will have a low χ_{water}, which in turn will produce a low relative humidity. Thus a relative humidity lower than 32% is needed to keep $CaCl_2 \cdot 6 \, H_2O$ dry.

105. (a)   We first compute the molality of a 0.92% mass/volume solution, assuming the solution's density is about 1.00 g/mL, meaning that 100.0 mL solution has a mass of 100.0 g.

$$\text{molality} = \frac{0.92 \text{ g NaCl} \times \dfrac{1 \text{ mol NaCl}}{58.44 \text{ g NaCl}}}{(100.0 \text{ g soln} - 0.92 \text{ g NaCl}) \times \dfrac{1 \text{ kg solvent}}{1000 \text{ g}}} = 0.16 \, m$$

Then we compute the freezing point depression of this solution.

$$\Delta T_f = -iK_f m = \frac{-2.0 \text{ mol ions}}{\text{mol NaCl}} \times \frac{1.86 ^\circ \text{C}}{m} \times 0.16 \, m = -0.60 \, ^\circ \text{C}$$

The van't Hoff factor of NaCl most likely is not equal to 2.0, but a bit less and thus the two definitions are in fair agreement.

(b) We calculate the amount of each solute, assume 1.00 L of solution has a mass of 1000 g, and subtract the mass of all solutes to determine the mass of solvent.

$$\text{amount NaCl ions} = 3.5 \text{ g NaCl} \times \frac{1 \text{ mol NaCl}}{58.44 \text{ g NaCl}} \times \frac{2 \text{ mol ions}}{1 \text{ mol NaCl}} = 0.12 \text{ mol ions}$$

$$\text{amount KCl ions} = 1.5 \text{ g} \times \frac{1 \text{ mol KCl}}{74.55 \text{ g KCl}} \times \frac{2 \text{ mol ions}}{1 \text{ mol KCl}} = 0.040 \text{ mol ions}$$

$$\text{amount Na}_3\text{C}_6\text{H}_5\text{O}_7 \text{ ions} = 2.9 \text{ g} \times \frac{1 \text{ mol Na}_3\text{C}_6\text{H}_5\text{O}_7}{258.07 \text{ g Na}_3\text{C}_6\text{H}_5\text{O}_7} \times \frac{4 \text{ mol ions}}{1 \text{ mol Na}_3\text{C}_6\text{H}_5\text{O}_7}$$

$$\text{amount Na}_3\text{C}_6\text{H}_5\text{O}_7 \text{ ions} = 0.045 \text{ mol ions}$$

$$\text{amount C}_6\text{H}_{12}\text{O}_6 = 20.0 \text{ g C}_6\text{H}_{12}\text{O}_6 \times \frac{1 \text{ mol C}_6\text{H}_{12}\text{O}_6}{180.2 \text{ g C}_6\text{H}_{12}\text{O}_6} = 0.111 \text{ mol C}_6\text{H}_{12}\text{O}_6$$

$$\text{solvent mass} = 1000.0 \text{ g} - (3.5 \text{ g} + 1.5 \text{ g} + 2.9 \text{ g} + 20.0 \text{ g}) = 972.1 \text{ g H}_2\text{O}$$

$$\text{solvent mass} = 0.9721 \text{ kg H}_2\text{O}$$

$$\text{solution molality} = \frac{(0.120 + 0.040 + 0.045 + 0.111) \text{ mol}}{0.9721 \text{ kg H}_2\text{O}} = 0.325 \, m$$

$$\Delta T_f = -K_f m = -1.86 \, ^\circ \text{C}/m \times 0.325 \, m = -0.60 ^\circ \text{C}$$

This again is close to the defined freezing point of $-0.52 \, ^\circ \text{C}$, with the error most likely arising from the van't Hoff factors not being integral. We can conclude, therefore, that the solution is isotonic.

CHAPTER 14
CHEMICAL KINETICS

PRACTICE EXAMPLES

1A The rate of consumption for a reactant is expressed as the negative of the change in molarity divided by the time interval. The rate of reaction is expressed as the rate of consumption of a reactant or production of a product divided by its stoichiometric coefficient.

$$\text{Rate of consumption of A} = \frac{-\Delta[A]}{\Delta t} = \frac{-(0.3187\ \text{M} - 0.3629\ \text{M})}{8.25\ \text{min}} \times \frac{1\ \text{min}}{60\ \text{sec}} = 8.93 \times 10^{-5}\ \text{M sec}^{-1}$$

$$\text{Rate of reaction} = \text{Rate of consumption of A} \div 2 = \frac{8.93 \times 10^{-5}\ \text{M sec}^{-1}}{2} = 4.46 \times 10^{-5}\ \text{M sec}^{-1}$$

1B We use the rate of reaction of A to determine the rate of formation of B, noting from the balanced equation that 3 moles of B form (+3 moles B) when 2 moles of A react (–2 moles A). (Recall that "M" means "moles per liter.")

$$\text{rate of B formation} = \frac{0.5522\ \text{M A} - 0.5684\ \text{M A}}{2.50\ \text{min} \times \frac{60\ \text{s}}{1\ \text{min}}} \times \frac{+3\ \text{moles B}}{-2\ \text{moles A}} = 1.62 \times 10^{-4}\ \text{M s}^{-1}$$

2A **(a)** The 2400-s tangent line intersects the 1200-s vertical line at 0.75 M and reaches 0 M at 3500 s. The slope of that tangent line is thus

$$\text{slope} = \frac{0\ \text{M} - 0.75\ \text{M}}{3500\ \text{s} - 1200\ \text{s}} = -3.3 \times 10^{-4}\ \text{M s}^{-1} = -\text{instantaneous rate of reaction}$$

The instantaneous rate of reaction $= 3.3 \times 10^{-4}\ \text{M s}^{-1}$

(b) At 2400 s, $[H_2O_2] = 0.39$ M. At 2450 s, $[H_2O_2] = 0.39$ M $+$ rate$\times \Delta t$

At 2450 s, $[H_2O_2]$ $= 0.39$ M $+ \left[-3.3 \times 10^{-4}\ \text{mol } H_2O_2\ \text{L}^{-1}\text{s}^{-1} \times 50\text{s} \right]$

$= 0.39$ M $- 0.017$ M $= 0.37$ M

2B With only the data of Table 14.2 we can use only the reaction rate during the first 400 s, $-\Delta[H_2O_2]/\Delta t = 15.0 \times 10^{-4}\ \text{M s}^{-1}$, and the initial concentration, $[H_2O_2]_0 = 2.32$ M. We calculate the change in $[H_2O_2]$ and add it to $[H_2O_2]_0$ to determine $[H_2O_2]_{100}$.

$$\Delta[H_2O_2] = \text{rate of reaction of } H_2O_2 \times \Delta t = -15.0 \times 10^{-4}\ \text{M s}^{-1} \times 100\ \text{s} = -0.15\ \text{M}$$

$$[H_2O_2]_{100} = [H_2O_2]_0 + \Delta[H_2O_2] = 2.32\ \text{M} + (-0.15\ \text{M}) = 2.17\ \text{M}$$

This value differs from the value of 2.15 M determined in *text* Example 14-2b because the *text* used the initial rate of reaction $\left(17.1 \times 10^{-4}\ \text{M s}^{-1} \right)$, which is a bit faster than the average rate over the first 400 seconds.

3A We write the equation for each rate, divide them into each other, and solve for n.

$$R_1 = k \times [N_2O_5]_1^n = 5.45 \times 10^{-5} \text{ M s}^{-1} = k(3.15 \text{ M})^n$$

$$R_2 = k \times [N_2O_5]_2^n = 1.35 \times 10^{-5} \text{ M s}^{-1} = k(0.78 \text{ M})^n$$

$$\frac{R_1}{R_2} = \frac{5.45 \times 10^{-5} \text{ M s}^{-1}}{1.35 \times 10^{-5} \text{ M s}^{-1}} = 4.04 = \frac{k \times [N_2O_5]_1^n}{k \times [N_2O_5]_2^n} = \frac{k(3.15 \text{ M})^n}{k(0.78 \text{ M})^n} = \left(\frac{3.15}{0.78}\right)^n = (4.0_4)^n$$

We kept an extra significant figure $(_4)$ to emphasize that the value of $n = 1$. Thus, the reaction is first-order in N_2O_5.

3B For the reaction, we know that $\text{rate} = k[HgCl_2]^1[C_2O_4^{2-}]^2$. Here we will compare Expt. 4 to Expt. 1 to find the rate.

$$\frac{\text{rate}_4}{\text{rate}_1} = \frac{k[HgCl_2]^1[C_2O_4^{2-}]^2}{k[HgCl_2]^1[C_2O_4^{2-}]^2} = \frac{0.025 \text{ M} \times (0.045 \text{ M})^2}{0.105 \text{ M} \times (0.150 \text{ M})^2} = 0.0214 = \frac{\text{rate}_4}{1.8 \times 10^{-5} \text{ M min}^{-1}}$$

The desired rate is $\text{rate}_4 = 0.0214 \times 1.8 \times 10^{-5} \text{ M min}^{-1} = 3.9 \times 10^{-7} \text{ M min}^{-1}$

4A We place the initial concentrations and the initial rates into the rate law and solve for k.

$$\text{rate} = k[A]^2[B] = 4.78 \times 10^{-2} \text{ M s}^{-1} = k(1.12 \text{ M})^2(0.87 \text{ M})$$

$$k = \frac{4.78 \times 10^{-2} \text{ M s}^{-1}}{(1.12 \text{ M})^2 0.87 \text{ M}} = 4.4 \times 10^{-2} \text{ M}^{-2} \text{ s}^{-1}$$

4B We know that $\text{rate} = k[HgCl_2]^1[C_2O_4^{2-}]^2$ and $k = 7.6 \times 10^{-3} \text{ M}^{-2}\text{min}^{-1}$.

Thus, insertion of the starting concentrations and the k value into the rate law yields:

$$\text{Rate} = 7.6 \times 10^{-3} \text{ M}^{-2} \text{ min}^{-1} (0.050 \text{ M})^1 (0.025 \text{ M})^2 = 2.4 \times 10^{-7} \text{ M min}^{-1}$$

5A Here we substitute directly into the integrated rate law equation.

$$\ln [A]_t = -kt + \ln [A]_0 = -3.02 \times 10^{-3} \text{ s}^{-1} \times 325 \text{ s} + \ln (2.80) = -0.982 + 1.030 = 0.048$$

$$[A]_t = e^{0.048} = 1.0 \text{ M}$$

5B This time we substitute the provided values into text equation (14.13).

$$\ln \frac{[H_2O_2]_t}{[H_2O_2]_0} = -kt = -k \times 600 \text{ s} = \ln \frac{1.49 \text{ M}}{2.32 \text{ M}} = -0.443 \qquad k = \frac{-0.443}{-600 \text{ s}} = 7.38 \times 10^{-4} \text{ s}^{-1}$$

Now we choose $[H_2O_2]_0 = 1.49 \text{ M}$, $[H_2O_2]_t = 0.62$, $\qquad t = 1800 \text{ s} - 600 \text{ s} = 1200 \text{ s}$

$$\ln \frac{[H_2O_2]_t}{[H_2O_2]_0} = -kt = -k \times 1200 \text{ s} = \ln \frac{0.62 \text{ M}}{1.49 \text{ M}} = -0.88 \qquad k = \frac{-0.88}{-1200 \text{ s}} = 7.3 \times 10^{-4} \text{ s}^{-1}$$

These two values agree within the limits of the experimental error and thus, the reaction is first-order in $[H_2O_2]$.

6A   We can use the integrated rate equation to find the ratio of the final and initial concentrations. This ratio equals the fraction of the initial concentration that remains at any time t..

$$\ln \frac{[A]_t}{[A]_0} = -kt = -2.95 \times 10^{-3} \text{ s}^{-1} \times 150 \text{ s} = -0.443$$

$$\frac{[A]_t}{[A]_0} = e^{-0.443} = 0.642; \quad 64.2\% \text{ of } [A]_0 \text{ remains.}$$

6B   After two-thirds of the sample has decomposed, one-third of the sample remains. Thus $[H_2O_2]_t = [H_2O_2]_0 \div 3$, and we have

$$\ln \frac{[H_2O_2]_t}{[H_2O_2]_0} = -kt = \ln \frac{[H_2O_2]_0 \div 3}{[H_2O_2]_0} = \ln(1/3) = -1.099 = -7.30 \times 10^{-4} \text{ s}^{-1}t$$

$$t = \frac{-1.099}{-7.30 \times 10^{-4} \text{ s}^{-1}} = 1.51 \times 10^3 \text{ s} \times \frac{1 \text{ min}}{60 \text{ s}} = 25.2 \text{ min}$$

7A   At the end of one half-life the pressure of DTBP will have been halved, to 400 mmHg. At the end of another half-life, at 160 min, the pressure of DTBP will have halved again, to 200 mmHg. Thus, the pressure of DTBP at 125 min will be intermediate between the pressure at 80.0 min (400 mmHg) and that at 160 min (200 mmHg). To obtain an exact answer, first we determine the value of the rate constant from the half-life.

$$k = \frac{0.693}{t_{1/2}} = \frac{0.693}{80.0 \text{ min}} = 0.00866 \text{ min}^{-1}$$

$$\ln \frac{(P_{DTBP})_t}{(P_{DTBP})_0} = -kt = -0.00866 \text{ min}^{-1} \times 125 \text{ min} = -1.08$$

$$\frac{(P_{DTBP})_t}{(P_{DTBP})_0} = e^{-1.08} = 0.340$$

$$(P_{DTBP})_t = 0.340 \times (P_{DTBP})_0 = 0.340 \times 800 \text{ mmHg} = 272 \text{ mmHg}$$

7B   (a)   We use partial pressures in place of concentrations in the integrated first-order rate equation. Notice first that more than 30 half-lives have elapsed, and thus the ethylene oxide pressure has declined to at most $(0.5)^{30} = 9 \times 10^{-10}$ of its initial value.

$$\ln \frac{P_{30}}{P_0} = -kt = -2.05 \times 10^{-4} \text{ s}^{-1} \times 30.0 \text{ h} \times \frac{3600 \text{ s}}{1 \text{ h}} = -22.1 \quad \frac{P_{30}}{P_0} = e^{-22.1} = 2.4 \times 10^{-10}$$

$$P_{30} = 2.4 \times 10^{-10} \times P_0 = 2.4 \times 10^{-10} \times 782 \text{ mmHg} = 1.9 \times 10^{-7} \text{ mmHg}$$

(b) $P_{\text{ethylene oxide}}$ initially 782 mmHg $\rightarrow 1.9 \times 10^{-7}$ mmHg (~ 0). Essentially all of the ethylene oxide is converted to CH_4 and CO. Since pressure is proportional to moles, the final pressure will be twice the initial pressure (1 mole gas $\rightarrow$ 2 moles gas; 782 mmHg $\rightarrow$ 1564 mmHg). The final pressure will be 1.56×10^3 mmHg.

8A We first begin by looking for a constant rate, indicative of a zero-order reaction. If the rate is constant, the concentration will decrease by the same quantity during the same time period. If we choose a 25-s time period, we note that the concentration decreases $(0.88 \text{ M} - 0.74 \text{ M} =)0.14 \text{ M}$ during the first 25 s, $(0.74 \text{ M} - 0.62 \text{ M} =)0.12 \text{ M}$ during the second 25 s, $(0.62 \text{ M} - 0.52 \text{ M} =)0.10 \text{ M}$ during the third 25 s, and $(0.52 \text{ M} - 0.44 \text{ M} =)0.08 \text{ M}$ during the fourth 25 s period. This is hardly a constant rate and we thus conclude that the reaction is not zero-order.

We next look for a constant half-life, indicative of a first-order reaction. The initial concentration of 0.88 M decreases to one half of that value, 0.44 M, during the first 100 s, indicating a 100-s half-life. The concentration halves again to 0.22 M in the second 100 s, another 100-s half-life. Finally, we note that the concentration halves also from 0.62 M at 50 s to 0.31 M at 150 s, yet another 100-s half-life. The rate is established as first-order. The rate constant is $k = \dfrac{0.693}{t_{1/2}} = \dfrac{0.693}{100 \text{ s}} = 6.93 \times 10^{-3} \text{ s}^{-1}$

That the reaction is first order is made apparent by the fact that the ln[B] vs time plot is a straight line with slope = -k ($k = 6.85 \times 10^{-3} \text{ s}^{-1}$).

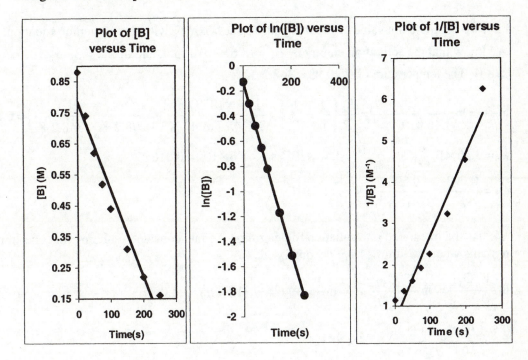

8B We plot the data in three ways to determine the order. (1) A plot of [A] vs. time is linear if the reaction is zero-order. (2) A plot of ln [A] vs. time will be linear if the reaction is first-order. (3) A plot of 1/[A] vs. time will be linear if the reaction is second-order. It is obvious from the plots below that the reaction is zero-order. The negative of the slope of the line equals $k = -(0.083 \text{ M} - 0.250 \text{ M}) \div 18.00 \text{ min} = 9.28 \times 10^{-3} \text{ M/min}$ ($k = 9.30 \times 10^{-3}$ M/min using a graphical approach (next page)).

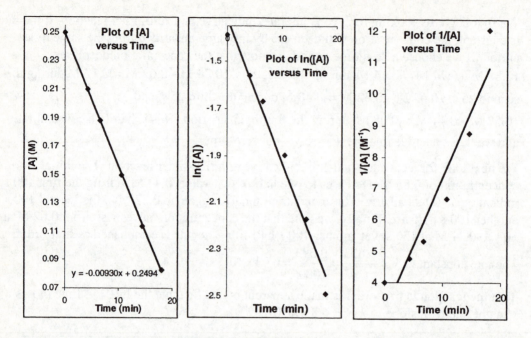

9A First we compute the value of the rate constant at $75.0\,°C$ with the Arrhenius equation. We know that the activation energy is $E_a = 1.06 \times 10^5$ J/mol, and that $k = 3.46 \times 10^{-5}\,s^{-1}$ at 298 K. The temperature of $75.0\,°C = 348.2$ K.

$$\ln\frac{k_2}{k_1} = \ln\frac{k_2}{3.46 \times 10^{-5}\,s^{-1}} = \frac{E_a}{R}\left(\frac{1}{T_1} - \frac{1}{T_2}\right) = \frac{1.06 \times 10^5\,J/mol}{8.3145\,J\,mol^{-1}\,K^{-1}}\left(\frac{1}{298.2\,K} - \frac{1}{348.2\,K}\right) = 6.14$$

$$k_2 = 3.46 \times 10^{-5}\,s^{-1} \times e^{+6.14} = 3.46 \times 10^{-5}\,s^{-1} \times 4.6 \times 10^2 = 0.016\,s^{-1}$$

$$t_{1/2} = \frac{0.693}{k} = \frac{0.693}{0.016\,s^{-1}} = 43\,s \text{ at } 75\,°C$$

9B We use the integrated rate equation to determine the rate constant, realizing that one-third remains when two-thirds have decomposed.

$$\ln\frac{[N_2O_5]_t}{[N_2O_5]_0} = \ln\frac{[N_2O_5]_0 \div 3}{[N_2O_5]_0} = \ln\frac{1}{3} = -kt = -k(1.50\,h) = -1.099$$

$$k = \frac{1.099}{1.50\,h} \times \frac{1\,h}{3600\,s} = 2.04 \times 10^{-4}\,s^{-1}$$

Now use the Arrhenius equation to determine the temperature at which the rate constant is $2.04 \times 10^{-4}\,s^{-1}$.

$$\ln\frac{k_2}{k_1} = \ln\frac{2.04 \times 10^{-4}\,s^{-1}}{3.46 \times 10^{-5}\,s^{-1}} = 1.77 = \frac{E_a}{R}\left(\frac{1}{T_1} - \frac{1}{T_2}\right) = \frac{1.06 \times 10^5\,J/mol}{8.3145\,J\,mol^{-1}\,K^{-1}}\left(\frac{1}{298\,K} - \frac{1}{T_2}\right)$$

$$\frac{1}{T_2} = \frac{1}{298\,K} - \frac{1.77 \times 8.3145\,K^{-1}}{1.06 \times 10^5} = 3.22 \times 10^{-3}\,K^{-1} \quad T_2 = 311\,K$$

10A The two steps of the mechanism must add, in a Hess's law fashion, to produce the overall reaction.

overall reaction : $CO + NO_2 \longrightarrow CO_2 + NO$ or $CO + NO_2 \longrightarrow CO_2 + NO$

Second step : $-\left(NO_3 + CO \longrightarrow NO_2 + CO_2\right)$ or $+\left(NO_2 + CO_2 \longrightarrow NO_3 + CO\right)$

first step: $\overline{2\ NO_2 \longrightarrow NO + NO_3}$

If the first step is the slow step, then it will be the rate-determining step, and the rate of that step will be the rate of the reaction, namely, rate of reaction $= k_1 [NO_2]^2$.

10B (1) The steps of the mechanism must add, together in a Hess's law fashion, to produce the overall reaction. This is done to the right. The two intermediates, $NO_2F_2(g)$ and $F(g)$, are each produced in one step and consumed in the next one.

Fast: $NO_2(g) + F_2(g) \rightleftharpoons NO_2F_2(g)$

Slow: $NO_2F_2(g) \rightarrow NO_2F(g) + F(g)$

Fast: $F(g) + NO_2(g) \rightarrow NO_2F(g)$

Net: $2\ NO_2(g) + F_2(g) \rightarrow 2\ NO_2F(g)$

(2) The proposed mechanism must agree with the rate law. We expect the rate-determining step to determine the reaction rate: Rate $= k_3 [NO_2F_2]$. To eliminate $[NO_2F_2]$, we recognize that the first elementary reaction is very fast and will have the same rate forward as reverse: $R_f = k_1 [NO_2][F_2] = k_2 [NO_2F_2] = R_r$. We solve for the concentration of intermediate: $[NO_2F_2] = k_1 [NO_2][F_2]/k_2$. We now substitute this expression for $[NO_2F_2]$ into the rate equation: Rate $= (k_1 k_3 / k_2)[NO_2][F_2]$. Thus the predicted rate law agrees with the experimental rate law.

EXERCISES

Rates of Reactions

1. $2A + B \rightarrow C + 3D$ $-\dfrac{\Delta[A]}{\Delta t} = 6.2 \times 10^{-4}\ M\ s^{-1}$

(a)   Rate $= -\dfrac{1}{2}\dfrac{\Delta[A]}{\Delta t} = 1/2(6.2 \times 10^{-4}\ M\ s^{-1}) = 3.1 \times 10^{-4}\ M\ s^{-1}$

(b)   Rate of disappearance of B $= -\dfrac{1}{2}\dfrac{\Delta[A]}{\Delta t} = 1/2(6.2 \times 10^{-4}\ M\ s^{-1}) = 3.1 \times 10^{-4}\ M\ s^{-1}$

(c)   Rate of appearance of D $= -\dfrac{3}{2}\dfrac{\Delta[A]}{\Delta t} = 3(6.2 \times 10^{-4}\ M\ s^{-1}) = 9.3 \times 10^{-4}\ M\ s^{-1}$

2. In each case, we draw the tangent line to the plotted curve.

    (a) The slope of the line is $\dfrac{\Delta[H_2O_2]}{\Delta t} = \dfrac{1.7\ M - 0.6\ M}{400\ s - 1600\ s} = -9.2 \times 10^{-4}\ M\ s^{-1}$

        Reaction rate $= -\dfrac{\Delta[H_2O_2]}{\Delta t} = 9.2 \times 10^{-4}\ M\ s^{-1}$

    (b) Read the value where the horizontal line $[H_2O_2] = 0.50\ M$, intersects the curve $\approx 21\underline{5}0$ s or ≈ 36 minutes

3. $\text{Rate} = -\dfrac{\Delta[A]}{\Delta t} = -\dfrac{(0.474\ M - 0.485\ M)}{82.4\ s - 71.5\ s} = 1.0 \times 10^{-3}\ M\ s^{-1}$

4. **(a)** $\text{Rate} = -\dfrac{\Delta[A]}{\Delta t} = -\dfrac{0.1498\ M - 0.1565\ M}{1.00\ min - 0.00\ min} = 0.0067\ M\ min^{-1}$

        $\text{Rate} = -\dfrac{\Delta[A]}{\Delta t} = -\dfrac{0.1433\ M - 0.1498\ M}{2.00\ min - 1.00\ min} = 0.0065\ M\ min^{-1}$

    (b) The rates are not equal because, in all except zero-order reactions, the rate depends on the concentration of reactant. And, of course, as the reaction proceeds, reactant is consumed and its concentration decreases, the rate of the reaction decreases.

5. **(a)** $[A] = [A]_i + \Delta[A] = 0.588\ M - 0.013\ M = 0.575\ M$

    (b) $\Delta[A] = 0.565\ M - 0.588\ M = -0.023\ M$

        $\Delta t = \Delta[A]\dfrac{\Delta t}{\Delta[A]} = \dfrac{-0.023\ M}{-2.2 \times 10^{-2}\ M/min} = 1.0\ min$

        $\text{time} = t + \Delta t = (4.40 + 1.0)\,min = 5.4\ min$

6. Initial concentrations are $[HgCl_2] = 0.105\ M$ and $[C_2O_4{}^{2-}] = 0.300\ M$.

The initial rate of the reaction is $7.1 \times 10^{-5}\ M\ min^{-1}$. Recall that the reaction is:

$2\ HgCl_2(aq) + C_2O_4^{2-}(aq) \rightarrow 2\ Cl^-(aq) + 2\ CO_2(g) + Hg_2Cl_2(aq)$.

The rate of reaction equals the rate of disappearance of $C_2O_4^{2-}$. Then, after 1 hour, assuming that the rate is the same as the initial rate,

    (a)

$[HgCl_2] = 0.105\ M - \left(7.1 \times 10^{-5}\ \dfrac{mol\ C_2O_4{}^{2-}}{L \cdot s} \times \dfrac{2\ mol\ HgCl_2}{1\ mol\ C_2O_4{}^{2-}} \times 1\ h \times \dfrac{60\ min}{1\ h}\right) = 0.096\ M$

    (b) $[C_2O_4^{2-}] = 0.300\ M - \left(7.1 \times 10^{-5}\ \dfrac{mol}{L \cdot min} \times 1\ h \times \dfrac{60\ min}{1\ h}\right) = 0.296\ M$

7. (a) $\text{Rate} = \dfrac{-\Delta[A]}{\Delta t} = \dfrac{\Delta[C]}{2\Delta t} = 1.76 \times 10^{-5}\ \text{M s}^{-1}$

$\dfrac{\Delta[C]}{\Delta t} = 2 \times 1.76 \times 10^{-5}\ \text{M s}^{-1} = 3.52 \times 10^{-5}\ \text{M/s}$

(b) $\dfrac{\Delta[A]}{\Delta t} = -\dfrac{\Delta[C]}{2\Delta t} = -1.76 \times 10^{-5}\ \text{M s}^{-1}$ Assume this rate is constant.

$[A] = 0.3580\ \text{M} + \left(-1.76 \times 10^{-5}\ \text{M s}^{-1} \times 1.00\ \text{min} \times \dfrac{60\ \text{s}}{1\ \text{min}}\right) = 0.357\ \text{M}$

(c) $\dfrac{\Delta[A]}{\Delta t} = -1.76 \times 10^{-5}\ \text{M s}^{-1}$

$\Delta t = \dfrac{\Delta[A]}{-1.76 \times 10^{-5}\ \text{M/s}} = \dfrac{0.3500\ \text{M} - 0.3580\ \text{M}}{-1.76 \times 10^{-5}\ \text{M/s}} = 4.5 \times 10^{2}\ \text{s}$

8. (a) $\dfrac{\Delta n[O_2]}{\Delta t} = 1.00\ \text{L soln} \times \dfrac{5.7 \times 10^{-4}\ \text{mol H}_2\text{O}_2}{1\ \text{L soln} \cdot \text{s}} \times \dfrac{1\ \text{mol O}_2}{2\ \text{mol H}_2\text{O}_2} = 2.9 \times 10^{-4}\ \text{mol O}_2/\text{s}$

(b) $\dfrac{\Delta n[O_2]}{\Delta t} = 2.9 \times 10^{-4}\ \dfrac{\text{mol O}_2}{\text{s}} \times \dfrac{60\ \text{s}}{1\ \text{min}} = 1.7 \times 10^{-2}\ \text{mol O}_2\ /\ \text{min}$

(c) $\dfrac{\Delta V[O_2]}{\Delta t} = 1.7 \times 10^{-2}\ \dfrac{\text{mol O}_2}{\text{min}} \times \dfrac{22{,}414\ \text{mL O}_2\ \text{at STP}}{1\ \text{mol O}_2} = \dfrac{3.8 \times 10^{2}\ \text{mL O}_2\ \text{at STP}}{\text{min}}$

9. Notice that, for every 1000 mmHg drop in the pressure of A(g), there will be a corresponding 2000 mmHg rise in the pressure of B(g) plus a 1000 mmHg rise in the pressure of C(g).

(a) We set up the calculation with three lines of information below the balanced equation: (1) the initial conditions, (2) the changes that occur, which are related to each other by reaction stoichiometry, and (3) the final conditions, which simply are initial conditions + changes.

	A(g)	→	2B(g)	+	C(g)
Initial	1000. mmHg		0. mmHg		0. mmHg
Changes	–1000. mmHg		+2000. mmHg		+1000. mmHg
Final	0. mmHg		2000. MmHg		1000. mmHg

Total final pressure = 0. mmHg + 2000. mmHg + 1000. mmHg = 3000. mmHg

(b)

	A(g)	→	2B(g)	+	C(g)
Initial	1000. mmHg		0. mmHg		0. mmHg
Changes	–200. mmHg		+400. mmHg		+200. mmHg
Final	800 mmHg		400. mmHg		200. mmHg

Total pressure = 800. mmHg + 400. mmHg + 200. mmHg = 1400. mmHg

10. **(a)** We will use the ideal gas law to determine N_2O_5 pressure

$$P\{N_2O_5\} = \frac{nRT}{V} = \frac{\left(1.00 \text{ g} \times \dfrac{1 \text{ mol } N_2O_5}{108.0 \text{ g}}\right) \times 0.08206 \dfrac{\text{L atm}}{\text{mol K}} \times (273 + 65) \text{ K}}{15 \text{ L}} \times \frac{760 \text{ mmHg}}{1 \text{ atm}} = 13 \text{ mmHg}$$

(b) After 2.38 min, one half-life passes. The initial pressure of N_2O_5 decreases by half to 6.5 mmHg.

(c) From the balanced chemical equation, the reaction of 2 mol $N_2O_5(g)$ produces 4 mol $NO_2(g)$ and 1 mol $O_2(g)$. That is, the consumption of 2 mol of reactant gas produces 5 mol of product gas. When measured at the same temperature and confined to the same volume, pressures will behave as amounts: the reaction of 2 mmHg of reactant produces 5 mmHg of product.

$$P_{\text{total}} = 13 \text{ mmHg } N_2O_5 \text{ (initially)} - 6.5 \text{ mmHg } N_2O_5 \text{ (reactant)} + \left(6.5 \text{ mmHg(reactant)} \times \frac{5 \text{ mmHg(product)}}{2 \text{ mmHg(reactant)}}\right)$$

$$= (13 - 6.5 + 16) \text{ mmHg} = 23 \text{ mmHg}$$

Method of Initial Rates

11. **(a)** From Expt. 1 to Expt. 3, [A] is doubled, while [B] remains fixed. This causes the rate to increases by a factor of $\dfrac{6.75 \times 10^{-4} \text{ M s}^{-1}}{3.35 \times 10^{-4} \text{ M s}^{-1}} = 2.01 \approx 2$

Thus, the reaction is first-order with respect to A.

From Expt. 1 to Expt. 2, [B] doubles, while [A] remains fixed. This causes the rate to increases by a factor of $\dfrac{1.35 \times 10^{-3} \text{ M s}^{-1}}{3.35 \times 10^{-4} \text{ M s}^{-1}} = 4.03 \approx 4$

Thus, the reaction is second-order with respect to B.

(b) Overall reaction order = order with respect to A + order with respect to B = $1 + 2 = 3$
The reaction is third-order overall.

(c) Rate $= 3.35 \times 10^{-4} \text{ M s}^{-1} = k(0.185 \text{ M})(0.133 \text{ M})^2$

$$k = \frac{3.35 \times 10^{-4} \text{ M s}^{-1}}{(0.185 \text{ M})(0.133 \text{ M})^2} = 0.102 \text{ M}^{-2} \text{ s}^{-1}$$

12. From Expt. 1 and Expt. 2 we see that [B] remains fixed while [A] triples. As a result, the initial rate increases from 4.2×10^{-3} M/min to 1.3×10^{-2} M/min, that is, the initial reaction rate triples. Therefore, the reaction is first-order in [A]. Between Expt. 2 and Expt. 3, we see that [A] doubles, which would double the rate, and [B] doubles. As a consequence, the initial rate goes from 1.3×10^{-2} M/min to 5.2×10^{-2} M/min, that is, the rate quadruples. Since an additional doubling of the rate is due to the change in [B], the reaction is first-order in [B]. Now we determine the value of the rate constant.

Rate $= k[A]^1[B]^1$ $\qquad k = \dfrac{\text{Rate}}{[A][B]} = \dfrac{5.2 \times 10^{-2} \text{ M / min}}{3.00 \text{ M} \times 3.00 \text{ M}} = 5.8 \times 10^{-3} \text{ L mol}^{-1}\text{min}^{-1}$

The rate law is Rate $= (5.8 \times 10^{-3} \text{ L mol}^{-1}\text{min}^{-1})[A]^1[B]^1$

13. From Experiment 1 to 2, [NO] remains constant while [Cl_2] is doubled. At the same time the initial rate of reaction is found to double. Thus, the reaction is first-order with respect to [Cl_2]. From experiment 1 to 3, [Cl_2] remains constant, while [NO] is doubled, resulting in a quadrupling of the initial rate of reaction. Thus, the reaction must be second-order in [NO]. Overall the reaction is third-order: Rate $= k$ [NO]^2[Cl_2]. The rate constant may be calculated from any one of the experiments. Using data from exp. 1,

$$k = \frac{\text{Rate}}{[\text{NO}]^2[\text{Cl}_2]} = \frac{2.27 \times 10^{-5}\ \text{M s}^{-1}}{(0.0125\ \text{M})^2(0.0255\ \text{M})} = 5.70\ \text{M}^{-2}\,\text{s}^{-1}$$

14. **(a)** From Expt. 1 to Expt. 2, [B] remains constant at 1.40 M and [C] remains constant at 1.00 M, but [A] is halved $(\times 0.50)$. At the same time the rate is halved $(\times 0.50)$. Thus, the reaction is first-order with respect to A, since $0.50^x = 0.50$ when $x = 1$.
From Expt. 2 to Expt. 3, [A] remains constant at 0.70 M and [C] remains constant at 1.00 M, but [B] is halved $(\times 0.50)$, from 1.40 M to 0.70 M. At the same time, the rate is quartered $(\times 0.25)$. Thus, the reaction is second-order with respect to B, since $0.50^y = 0.25$ when $y = 2$.

From Expt. 1 to Expt. 4, [A] remains constant at 1.40 M and [B] remains constant at 1.40 M, but [C] is halved $(\times 0.50)$, from 1.00 M to 0.50 M. At the same time, the rate is increased by a factor of 2.0.

$$\left[\text{rate}_4 = 16\ \text{rate}_3 = 16 \times \frac{1}{4} \text{rate}_2 = 4\ \text{rate}_2 = 4 \times \frac{1}{2} \text{rate}_1 = 2 \times \text{rate}_1. \right]$$

Thus, the order of the reaction with respect to C is -1, since $0.5^z = 2.0$ when $z = -1$.

(b) $$\text{rate}_5 = k\,(0.70\ \text{M})^1\,(0.70\ \text{M})^2\,(0.50\ \text{M})^{-1} = k\left(\frac{1.40\ \text{M}}{2}\right)^1 \left(\frac{1.40\ \text{M}}{2}\right)^2 \left(\frac{1.00\ \text{M}}{2}\right)^{-1}$$

$$= k\frac{1}{2}^1\,(1.40\ \text{M})^1\,\frac{1}{2}^2\,(1.40\ \text{M})^2\,\frac{1}{2}^{-1}\,(1.00\ \text{M})^{-1} = \text{rate}_1\left(\frac{1}{2}\right)^{1+2-1} = \text{rate}_1\left(\frac{1}{2}\right)^2 = \frac{1}{4}\text{rate}_1$$

This is based on $\text{rate}_1 = k\,(1.40\ \text{M})^1\,(1.40\ \text{M})^2\,(1.00\ \text{M})^{-1}$

First-Order Reactions

15. **(a)** TRUE   The rate of the reaction does decrease as more and more of B and C are formed, but not because more and more of B and C are formed. Rather the rate decreases because the concentration of A must decrease to form more and more of B and C.

(b) FALSE   The time required for one half of substance A to react—the half-life—is independent of the quantity of "A" present.

16. (a) FALSE For first-order reactions, a plot of ln [A] or log [A] vs. time yields a straight line. A graph of [A] vs. time yields a curved line.

(b) TRUE The rate of formation of "C" is related to the rate of disappearance of "A" by the stoichiometry of the reaction.

17. (a) Since the half-life is 180 s, after 900 s five half-lives have elapsed, and the original quantity of A has been cut in half five times.

final quantity of $A = (0.5)^5 \times$ initial quantity of $A = 0.03125 \times$ initial quantity of A

About 3.13% of the original quantity of A remains unreacted after 900 s.

or

More generally, we would calculate the value of the rate constant, k, using

$k = \dfrac{\ln 2}{t_{1/2}} = \dfrac{0.693}{180 \text{ s}} = 0.00385 \text{ s}^{-1}$ Now ln(% unreacted) = $-kt$ = -0.00385 s$^{-1}\times$(900s) = -3.46$\underline{5}$

(% unreacted) = 0.031$\underline{3}$ %

(b) Rate = $k[A] = 0.00385 \text{ s}^{-1} \times 0.50 \text{ M} = 0.00193 \text{ M/s}$

18. (a) The reaction is 1st order thus

$\ln \dfrac{[A]_t}{[A]_0} = -kt = \ln \dfrac{0.100 \text{ M}}{0.800 \text{ M}} = -54\min(k)$ $k = -\dfrac{-2.08}{54 \text{ min}} = 0.038\underline{5} \text{ min}^{-1}$

We may now determine the time required to achieve a concentration of 0.025 M

$\ln \dfrac{[A]_t}{[A]_0} = -kt = \ln \dfrac{0.025 \text{ M}}{0.800 \text{ M}} = -0.038\underline{5} \text{ min}^{-1}(t)$ $t = \dfrac{-3.4\underline{7}}{0.038\underline{5} \text{ min}^{-1}} = 90. \text{ min}$

(b) Since we know the rate constant for this reaction (see above),

Rate = $k[A]^1 = 0.038\underline{5} \text{ min}^{-1} \times 0.025 \text{ M} = 9.6 \times 10^{-4} \text{ M/min}$

19. (a) The mass of A has decreased to one fourth of its original value, from 1.60 g to 0.40 g. Since $\dfrac{1}{4} = \dfrac{1}{2} \times \dfrac{1}{2}$, we see that two half-lives have elapsed.

Thus, $2 \times t_{1/2} = 38$ min, or $t_{1/2} = 19$ min.

(b) $k = 0.693/t_{1/2} = \dfrac{0.693}{19 \text{ min}} = 0.036 \text{ min}^{-1}$ $\ln \dfrac{[A]_t}{[A]_0} = -kt = -0.036 \text{ min}^{-1} \times 60 \text{ min} = -2.2$

$\dfrac{[A]_t}{[A]_0} = e^{-2.2} = 0.1\underline{1}$ or $[A]_t = [A]_0 \, e^{-kt} = 1.60 \text{ g A} \times 0.1\underline{1} = 0.1\underline{8} \text{ g A}$

20. (a) $\ln \dfrac{[A]_t}{[A]_0} = -kt = \ln \dfrac{0.632 \text{ M}}{0.816 \text{ M}} = -0.256$ $k = -\dfrac{-0.256}{16.0 \text{ min}} = 0.0160 \text{ min}^{-1}$

(b) $t_{1/2} = \dfrac{0.693}{k} = \dfrac{0.693}{0.0160 \text{ min}^{-1}} = 43.3 \text{ min}$

(c) We need to solve the integrated rate equation to find the elapsed time.

$$\ln\frac{[A]_t}{[A]_0} = -kt = \ln\frac{0.235\text{ M}}{0.816\text{ M}} = -1.245 = -0.0160\text{ min}^{-1}\times t \qquad t = \frac{-1.245}{-0.0160\text{ min}^{-1}} = 77.8\text{ min}$$

(d) $\ln\dfrac{[A]}{[A]_0} = -kt$ becomes $\dfrac{[A]}{[A]_0} = e^{-kt}$ which in turn becomes

$$[A] = [A]_0\ e^{-kt} = 0.816\text{ M }\exp\left(-0.0160\text{ min}^{-1}\times 2.5\text{ h}\times\frac{60\text{ min}}{1\text{ h}}\right) = 0.816\times 0.0907 = 0.074\text{ M}$$

21. We determine the value of the first-order rate constant and from that we can calculate the half-life. If the reactant is 99% decomposed in 137 min, then only 1% (0.010) of the initial concentration remains.

$$\ln\frac{[A]_t}{[A]_0} = -kt = \ln\frac{0.010}{1.000} = -4.61 = -k\times 137\text{min} \qquad k = \frac{4.61}{137\text{ min}} = 0.0336\text{ min}^{-1}$$

$$t_{1/2} = \frac{0.0693}{k} = \frac{0.693}{0.0336\text{ min}^{-1}} = 20.6\text{ min}$$

22. If 99% of the radioactivity of ^{32}P is lost, 1% (0.010) of that radioactivity remains. First we compute the value of the rate constant from the half-life. $k = \dfrac{0.693}{t_{1/2}} = \dfrac{0.693}{14.3\text{ d}} = 0.0485\text{ d}^{-1}$

Then we use the integrated rate equation to determine the elapsed time.

$$\ln\frac{[A]_t}{[A]_0} = -kt \qquad t = -\frac{1}{k}\ln\frac{[A]_t}{[A]_0} = -\frac{1}{0.0485\text{ d}^{-1}}\ln\frac{0.010}{1.000} = 95\text{ days}$$

23. **(a)** $\ln\left(\dfrac{\frac{35}{100}[A]_0}{[A]_0}\right) = \ln(0.35) = -kt = (-4.81\times 10^{-3}\text{ min}^{-1})t \qquad t = 218\text{ min.}$

Note: We did not need to know the initial concentration of acetoacetic acid to answer the question.

(b) Let's assume that the reaction takes placed in a 1.00 L container.

$$10.0\text{ g acetoacetic acid }\times\frac{1\text{ mol acetoacetic acid}}{102.090\text{ g acetoacetic acid}} = 0.09795\text{ mol acetoacetic acid.}$$

After 575 min. (~ 4 half lives, hence, we expect ~ 6.25% remains), use integrated form of the rate law to find $[A]_t = 575$ min.

$$\ln\left(\frac{[A]_t}{[A]_0}\right) = -kt = (-4.81\times 10^{-3}\text{ min}^{-1})(575\text{ min}) = -2.766$$

$$\frac{[A]_t}{[A]_0} = e^{-2.766} = 0.06293\ (\sim 6.3\%\text{ remains}) \qquad \frac{[A]_t}{0.09795\text{ M}} = 0.063 \quad [A]_t = 6.2\times 10^{-3}\text{ moles.}$$

$[A]_{reacted} = [A]_o - [A]_t = (0.098 - 6.2 \times 10^{-3})$ moles $= 0.092$ moles acetoacetic acid. The stoichiometry is such that for every mole of acetoacetic acid consumed, one mole of CO_2 forms. Hence, we need to determine the volume of 0.0918 moles CO_2 at 24.5 °C (297.6$\underline{5}$ K) and 748 torr (0.984 atm).

$$V = \frac{nRT}{P} = \frac{0.0918 \, mol \left(0.08206 \dfrac{L \, atm}{K \, mol} \right) 297.65 \, K}{0.984 \, atm} = 2.3 \, L \, CO_2$$

24. **(a)** $\ln \dfrac{[A]_t}{[A]_o} = -kt = \ln \dfrac{2.5 \, g}{80.0 \, g} = -3.47 = -6.2 \times 10^{-4} \, s^{-1} t$, $\quad t = \dfrac{3.47}{6.2 \times 10^{-4} \, s^{-1}} = 5.6 \times 10^3 \, s$

We substituted masses for concentrations, because the same substance (with the same molar mass) is present initially at time t, and because it is a closed system.

(b) amount $O_2 = 77.5 \, g \, N_2O_5 \times \dfrac{1 \, mol \, N_2O_5}{108.0 \, g \, N_2O_5} \times \dfrac{1 \, mol \, O_2}{2 \, mol \, N_2O_5} = 0.359 \, mol \, O_2$

$$V = \frac{nRT}{P} = \frac{0.359 \, mol \, O_2 \times 0.08206 \dfrac{L \, atm}{mol \, K} \times (45+273) \, K}{745 \, mmHg \times \dfrac{1 \, atm}{760 \, mmHg}} = 9.56 \, L \, O_2$$

25. **(a)** If the reaction is first-order, we will obtain the same value of the rate constant from several sets of data.

$\ln \dfrac{[A]_t}{[A]_o} = -kt = \ln \dfrac{0.497 \, M}{0.600 \, M} = -k \times 100 \, s = -0.188$, $\quad k = \dfrac{0.188}{100 \, s} = 1.88 \times 10^{-3} \, s^{-1}$

$\ln \dfrac{[A]_t}{[A]_o} = -kt = \ln \dfrac{0.344 \, M}{0.600 \, M} = -k \times 300 \, s = -0.556$, $\quad k = \dfrac{0.556}{300 \, s} = 1.85 \times 10^{-3} \, s^{-1}$

$\ln \dfrac{[A]_t}{[A]_o} = -kt = \ln \dfrac{0.285 \, M}{0.600 \, M} = -k \times 400 \, s = -0.744$, $\quad k = \dfrac{0.744}{400 \, s} = 1.86 \times 10^{-3} \, s^{-1}$

$\ln \dfrac{[A]_t}{[A]_o} = -kt = \ln \dfrac{0.198 \, M}{0.600 \, M} = -k \times 600 \, s = -1.109$, $\quad k = \dfrac{1.109}{600 \, s} = 1.85 \times 10^{-3} \, s^{-1}$

The virtual constancy of the rate constant throughout the time of the reaction confirms that the reaction is first-order.

(b) For this part, we assume that the rate constant equals the average of the values obtained in part (a).

$$k = \frac{1.88 + 1.85 + 1.86 + 1.85}{4} \times 10^{-3} \, s^{-1} = 1.86 \times 10^{-3} \, s^{-1}$$

(c) We use the integrated first-order rate equation:

$[A]_{750} = [A]_o \exp(-kt) = 0.600 \, M \exp(-1.86 \times 10^{-3} \, s^{-1} \times 750 \, s)$

$[A]_{750} = 0.600 \, M \, e^{-1.40} = 0.148 \, M$

26. **(a)** If the reaction is first-order, we will obtain the same value of the rate constant from several sets of data.

$$\ln\frac{P_t}{P_0} = -kt = \ln\frac{264 \text{ mmHg}}{312 \text{ mmHg}} = -k \times 390 \text{ s} = -0.167, \qquad k = \frac{0.167}{390 \text{ s}} = 4.28 \times 10^{-4} \text{ s}^{-1}$$

$$\ln\frac{P_t}{P_0} = -kt = \ln\frac{224 \text{ mmHg}}{312 \text{ mmHg}} = -k \times 777 \text{ s} = -0.331, \qquad k = \frac{0.331}{777 \text{ s}} = 4.26 \times 10^{-4} \text{ s}^{-1}$$

$$\ln\frac{P_t}{P_0} = -kt = \ln\frac{187 \text{ mmHg}}{312 \text{ mmHg}} = -k \times 1195 \text{ s} = -0.512, \qquad k = \frac{0.512}{1195 \text{ s}} = 4.28 \times 10^{-4} \text{ s}^{-1}$$

$$\ln\frac{P_t}{P_0} = -kt = \ln\frac{78.5 \text{ mmHg}}{312 \text{ mmHg}} = -k \times 3155 \text{ s} = -1.38, \qquad k = \frac{1.38}{3155 \text{ s}} = 4.37 \times 10^{-4} \text{ s}^{-1}$$

The virtual constancy of the rate constant confirms that the reaction is first-order.

(b) For this part we assume the rate constant is the average of the values in part (a): $4.3 \times 10^{-4} \text{ s}^{-1}$.

(c) At 390 s, the pressure of dimethyl ether has dropped to 264 mmHg. Thus, an amount of dimethyl ether equivalent to a pressure of $(312 \text{ mmHg} - 264 \text{ mmHg} =)$ 48 mmHg has decomposed. For each 1 mmHg pressure of dimethyl ether that decomposes, 3 mmHg of pressure from the products is produced. Thus, the increase in the pressure of the products is $3 \times 48 = 144$ mmHg. The total pressure at this point is 264 mmHg + 144 mmHg = 408 mmHg. Below, this calculation is done in a more systematic fashion:

	$(CH_3)_2O(g)$	$\rightarrow$	$CH_4(g)$	+	$H_2(g)$	+	$CO(g)$
Initial	312 mmHg		0 mmHg		0 mmHg		0 mmHg
Changes	-48 mmHg		$+48$ mmHg		$+48$ mmHg		$+48$ mmHg
Final	264 mmHg		48 mmHg		48 mmHg		48 mmHg

$$P_{total} = P_{DME} + P_{methane} + P_{hydrogen} + P_{CO}$$
$$= 264 \text{ mmHg} + 48 \text{ mmHg} + 48 \text{ mmHg} + 48 \text{ mmHg} = 408 \text{ mmHg}$$

(d) This questions is solved in the same manner as part (c). The results are summarized below.

	$(CH_3)_2O(g)$	$\rightarrow$	$CH_4(g)$	+	$H_2(g)$	+	$CO(g)$
Initial	312 mmHg		0 mmHg		0 mmHg		0 mmHg
Changes	-312 mmHg		$+312$ mmHg		$+312$ mmHg		$+312$ mmHg
Final	0 mmHg		312 mmHg		312 mmHg		312 mmHg

$$P_{total} = P_{DME} + P_{methane} + P_{hydrogen} + P_{CO}$$
$$= 0 \text{ mmHg} + 312 \text{ mmHg} + 312 \text{ mmHg} + 312 \text{ mmHg} = 936 \text{ mmHg}$$

(e) We first determine P_{DME} at 1000 s. $\ln\dfrac{P_{1000}}{P_0} = -kt = -4.3 \times 10^{-4} \text{ s}^{-1} \times 1000 \text{ s} = -0.43$

$$\frac{P_{1000}}{P_0} = e^{-0.43} = 0.65 \qquad P_{1000} = 312 \text{ mmHg} \times 0.65 = 203 \text{ mmHg}$$

Then we use the same approach as was used for parts (c) and (d)

$$(CH_3)_2 O(g) \rightarrow CH_4(g) + H_2(g) + CO(g)$$

Initial	312 mmHg	0 mmHg	0 mmHg	0 mmHg
Changes	−109 mmHg	+109 mmHg	+109 mmHg	+109 mmHg
Final	203 mmHg	109 mmHg	109 mmHg	109 mmHg

$$P_{total} = P_{DME} + P_{methane} + P_{hydrogen} + P_{CO}$$
$$= 203 \text{ mmHg} + 109 \text{ mmHg} + 109 \text{ mmHg} + 109 \text{ mmHg} = 530. \text{ mmHg}$$

Reactions of Various Orders

27. **(a)** Set II is data from a zero-order reaction. We know this because the rate of set II is constant. $0.25 \text{ M}/25 \text{ s} = 0.010 \text{ M s}^{-1}$. Zero-order reactions have constant rates of reaction.

(b) A first-order reaction has a constant half-life. In set I, the first half-life is slightly less than 75 sec, since the concentration decreases by slightly more than half (from 1.00 M to 0.47 M) in 75 s. Again, from 75 s to 150 s the concentration decreases from 0.47 M to 0.22 M, again by slightly more than half, in a time of 75 s. Finally, two half-lives should see the concentration decrease to one-fourth of its initial value. This, in fact, is what we see. From 100 s to 250 sec, 150 s of elapsed time, the concentration decreases from 0.37 M to 0.08 M, i.e., to slightly less than one-fourth of its initial value. Notice that we cannot make the same statement of constancy of half-life for set III. The first half-life is 100 s, but it takes more than 150 s (from 100 s to 250 s) for [A] to again decrease by half.

(c) For a second-order reaction, $1/[A]_t - 1/[A]_0 = kt$. For the initial 100 s in set III, we have

$$\frac{1}{0.50 \text{ M}} - \frac{1}{1.00 \text{ M}} = 1.0 \text{ L mol}^{-1} = k \, 100 \text{ s}, \quad k = 0.010 \text{ L mol}^{-1} \text{ s}^{-1}$$

For the initial 200 s, we have

$$\frac{1}{0.33 \text{ M}} - \frac{1}{1.00 \text{ M}} = 2.0 \text{ L mol}^{-1} = k \, 200 \text{ s}, \quad k = 0.010 \text{ L mol}^{-1} \text{ s}^{-1}$$

Since we obtain the same value of the rate constant using the equation for second-order kinetics, set III must be second-order.

28. For a zero-order reaction (set II), the slope equals the rate constant:
$$k = -\Delta[A]/\Delta t = 1.00 \text{ M}/100 \text{ s} = 0.0100 \text{ M/s}$$

29. Set I is the data for a first-order reaction; we can analyze those items of data to determine the half-life. In the first 75 s, the concentration decreases by a bit more than half. This implies a half-life slightly less than 75 s, perhaps 70 s. This is consistent with the other time periods noted in the answer to Review Question 18 (b) and also to the fact that in the 150 s interval from 50 s to 200 s, the concentration decreases from 0.61 M to 0.14 M, which is a bit more than a factor-of-four decrease. The factor-of-four decrease, to one-fourth of the initial value, is what we would expect for two successive half-lives. We can determine the half-life more accurately, by obtaining a value of k from the relation $\ln([A]_t / [A]_0) = -kt$ followed by $t_{1/2} = 0.693/k$ For instance, $\ln(0.78/1.00) = -k \, (25 \text{ s}); k = 9.9\underline{4} \times 10^{-3} \text{ s}^{-1}$. Thus, $t_{1/2} = 0.693/9.9\underline{4} \times 10^{-3} \text{ s}^{-1} = 70 \text{ s}$.

30. We can determine an approximate initial rate by using data from the first 25 s.

$$\text{Rate} = -\frac{\Delta[A]}{\Delta t} = -\frac{0.80\ M - 1.00\ M}{25\ s - 0\ s} = 0.0080\ M\ s^{-1}$$

31. The approximate rate at 75 s can be taken as the rate over the time period from 50 s to 100 s.

(a) $\text{Rate}_{II} = -\dfrac{\Delta[A]}{\Delta t} = -\dfrac{0.00\ M - 0.50\ M}{100\ s - 50\ s} = 0.010\ M\ s^{-1}$

(b) $\text{Rate}_{I} = -\dfrac{\Delta[A]}{\Delta t} = -\dfrac{0.37\ M - 0.61\ M}{100\ s - 50\ s} = 0.0048\ M\ s^{-1}$

(c) $\text{Rate}_{III} = -\dfrac{\Delta[A]}{\Delta t} = -\dfrac{0.50\ M - 0.67\ M}{100\ s - 50\ s} = 0.0034\ M\ s^{-1}$

Alternatively we can use [A] at 75 s (the values given in the table) in the relationship $\text{Rate} = k[A]^m$, where $m = 0$, 1, or 2.

(a) $\text{Rate}_{II} = 0.010\ M\ s^{-1} \times (0.25\ mol/L)^0 = 0.010\ M\ s^{-1}$

(b) Since $t_{1/2} = 70s$, $k = 0.693/70s = 0.0099\ s^{-1}$

$\text{Rate}_{I} = 0.0099\ s^{-1} \times (0.47\ mol/L)^1 = 0.0047\ M\ s^{-1}$

(c) $\text{Rate}_{III} = 0.010\ L\ mol^{-1}\ s^{-1} \times (0.57\ mol/L)^2 = 0.0032\ M\ s^{-1}$

32. We can combine the approximate rates from Question 31, with the fact that 10 s have elapsed, and the concentration at 100 s.

(a) $[A]_{II} = 0.00\ M$ There is no reactant left after 100 s.

(b) $[A]_{I} = [A]_{100} - (10\ s \times \text{rate}) = 0.37\ M - (10\ s \times 0.0047\ M\ s^{-1}) = 0.32\ M$

(c) $[A]_{III} = [A]_{100} - (10\ s \times \text{rate}) = 0.50\ M - (10\ s \times 0.0032\ M\ s^{-1}) = 0.47\ M$

33. Substitute the given values into the rate equation to obtain the rate of reaction.

$$\text{Rate} = k[A]^2[B]^0 = (0.0103\ M^{-1}min^{-1})(0.116\ M)^2(3.83\ M)^0 = 1.39 \times 10^{-4}\ M\ /\ min$$

34. (a) A first-order reaction has a constant half-life. Thus, half of the initial concentration remains after 30.0 minutes, and at the end of another half-life—60.0 minutes total— half of the concentration present at 30.0 minutes will have reacted: the concentration has decreased to one-quarter of its initial value. Or, we could say that the reaction is 75% complete after two half-lives—60.0 minutes.

(b) A zero-order reaction proceeds at a constant rate. Thus, if the reaction is 50% complete in 30.0 minutes, in twice the time—60.0 minutes—the reaction will be 100% complete. (And in one-fifth the time—6.0 minutes—the reaction will be 10% complete. Alternatively, we can say that the rate of reaction is 10%/6.0 min.) Therefore, the time

required for the reaction to be 75% complete $= 75\% \times \dfrac{60.0\ min}{100\%} = 45\ min$

35. For reaction: $HI(g) \rightarrow 1/2\ H_2(g) + 1/2\ I_2(g)$ (700 K)

Time (s)	[HI] (M)	ln[HI]	$1/[HI](M^{-1})$
0	1.00	0	1.00
100	0.90	−0.105	1.11
200	0.81	−0.211	1.23$\underline{5}$
300	0.74	−0.301	1.35
400	0.68	−0.386	1.47

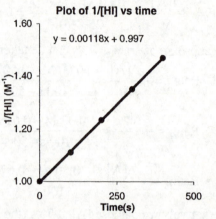

From data above, a plot of 1/[HI] vs. t yields a straight line. The reaction is second-order in HI at 700 K. Rate = $k[HI]^2$. The slope of the line = $k = 0.00118\ M^{-1}s^{-1}$

36. (a) We can graph 1/[ArSOOH] vs. time and obtain a straight line. We can also graph [ArSOOH] *vs.* time and ln([ArSOOH]) *vs.* time to demonstrate that they do not yield a straight line. Only the plot of $1/[ArSO_2H]$ versus time is shown.

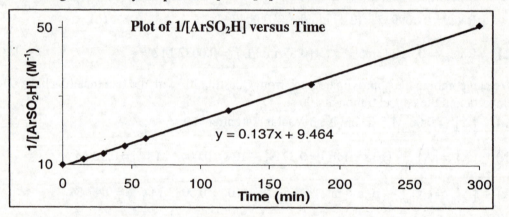

The linearity of the line indicates that the reaction is second-order.

(b) We solve the rearranged integrated second-order rate law for the rate constant, using the longest time interval.
$$\frac{1}{[A]_t} - \frac{1}{[A]_0} = kt \qquad \frac{1}{t}\left(\frac{1}{[A]_t} - \frac{1}{[A]_0}\right) = k$$

$$k = \frac{1}{300\ min}\left(\frac{1}{0.0196\ M} - \frac{1}{0.100\ M}\right) = 0.137\ L\ mol^{-1}min^{-1}$$

(c) We use the same equation as in part (b), but solved for t, rather than k.

$$t = \frac{1}{k}\left(\frac{1}{[A]_t} - \frac{1}{[A]_0}\right) = \frac{1}{0.137\ L\ mol^{-1}min^{-1}}\left(\frac{1}{0.0500\ M} - \frac{1}{0.100\ M}\right) = 73.0\ min$$

(d) We use the same equation as in part (b), but solve for t, rather than k.

$$t = \frac{1}{k}\left(\frac{1}{[A]_t} - \frac{1}{[A]_0}\right) = \frac{1}{0.137 \text{ L mol}^{-1}\text{min}^{-1}}\left(\frac{1}{0.0250 \text{ M}} - \frac{1}{0.100 \text{ M}}\right) = 219 \text{ min}$$

(e) We use the same equation as in part (b), but solve for t, rather than k.

$$t = \frac{1}{k}\left(\frac{1}{[A]_t} - \frac{1}{[A]_0}\right) = \frac{1}{0.137 \text{ L mol}^{-1}\text{min}^{-1}}\left(\frac{1}{0.0350 \text{ M}} - \frac{1}{0.100 \text{ M}}\right) = 136 \text{ min}$$

37. (a) Plot [A] vs t, ln[A] vs t and 1/[A] vs t and see which yields a straight line.

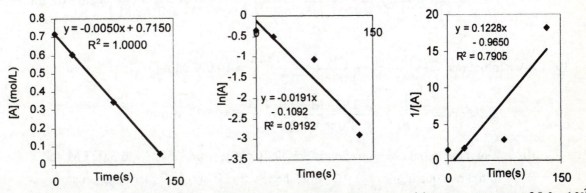

Clearly we can see that the reaction is zero order in reactant A with a rate constant of 5.0×10^{-3}.

(b) The half-life of this reaction is the time needed for one half of the initial [A] to react.

Thus, $\Delta[A] = 0.715 \text{ M} \div 2 = 0.358 \text{ M}$ and $t_{1/2} = \dfrac{0.358 \text{ M}}{5.0 \times 10^{-3} \text{ M/s}} = 72 \text{ s}$.

38. (a) We can either graph $1/[C_4H_6]$ vs. time and obtain a straight line, or we can determine the second-order rate constant from several data points. Then, if k indeed is a constant, the reaction is demonstrated to be second-order. We shall use the second technique in this case. First we do a bit of algebra.

$$\frac{1}{[A]_t} - \frac{1}{[A]_0} = kt \qquad \frac{1}{t}\left(\frac{1}{[A]_t} - \frac{1}{[A]_0}\right) = k$$

$$k = \frac{1}{12.18 \text{ min}}\left(\frac{1}{0.0144 \text{ M}} - \frac{1}{0.0169 \text{ M}}\right) = 0.843 \text{ L mol}^{-1}\text{min}^{-1}$$

$$k = \frac{1}{24.55 \text{ min}}\left(\frac{1}{0.0124 \text{ M}} - \frac{1}{0.0169 \text{ M}}\right) = 0.875 \text{ L mol}^{-1}\text{min}^{-1}$$

$$k = \frac{1}{42.50 \text{ min}}\left(\frac{1}{0.0103 \text{ M}} - \frac{1}{0.0169 \text{ M}}\right) = 0.892 \text{ L mol}^{-1}\text{min}^{-1}$$

$$k = \frac{1}{68.05 \text{ min}}\left(\frac{1}{0.00845 \text{ M}} - \frac{1}{0.0169 \text{ M}}\right) = 0.870 \text{ L mol}^{-1}\text{min}^{-1}$$

The fact that each calculation generates similar values for the rate constant indicates that the reaction is second-order.

(b) The rate constant is the average of the values obtained in part (a).

$$k = \frac{0.843 + 0.875 + 0.892 + 0.870}{4} \ L \ mol^{-1}min^{-1} = 0.87 \ L \ mol^{-1}min^{-1}$$

(c) We use the same equation as in part (a), but solve for t, rather than k.

$$t = \frac{1}{k}\left(\frac{1}{[A]_t} - \frac{1}{[A]_0}\right) = \frac{1}{0.870 \ L \ mol^{-1}min^{-1}}\left(\frac{1}{0.00423 \ M} - \frac{1}{0.0169 \ M}\right) = 2.0 \times 10^2 \ min$$

(d) We use the same equation as in part (a), but solve for t, rather than k.

$$t = \frac{1}{k}\left(\frac{1}{[A]_t} - \frac{1}{[A]_0}\right) = \frac{1}{0.870 \ L \ mol^{-1}min^{-1}}\left(\frac{1}{0.0050 \ M} - \frac{1}{0.0169 \ M}\right) = 1.6 \times 10^2 \ min$$

39. **(a)** Initial rate $= -\dfrac{\Delta[A]}{\Delta t} = -\dfrac{1.490 \ M - 1.512 \ M}{1.0 \ min - 0.0 \ min} = +0.022 \ M/min$

Initial rate $= -\dfrac{\Delta[A]}{\Delta t} = -\dfrac{2.935 \ M - 3.024 \ M}{1.0 \ min - 0.0 \ min} = +0.089 \ M/min$

(b) When the initial concentration is doubled $(\times 2.0)$, from 1.512 M to 3.024 M, the initial rate quadruples $(\times 4.0)$. Thus, the reaction is second-order in A (since $2.0^x = 4.0$ when $x = 2$).

40. **(a)** Let us assess the possibilities. If the reaction is zero-order, its rate will be constant. During the first 8 min, the rate is $-(0.60 \ M - 0.80 \ M)/8 \ min = 0.03 \ M/min$. Then, during the first 24 min, the rate is $-(0.35 \ M - 0.80 \ M)/24 \ min = 0.019 \ M/min$. Thus, the reaction is not zero-order. If the reaction is first-order, it will have a constant half-life, that is consistent with its rate constant. The half-life can be assessed from the fact that 40 min elapse while the concentration drops from 0.80 M to 0.20 M, that is, to one-fourth of its initial value. Thus, 40 min equals two half-lives and $t_{1/2} = 20 \ min$. This gives $k = 0.693/t_{1/2} = 0.693/20 \ min = 0.035 \ min^{-1}$. Also

$$kt = -\ln\frac{[A]_t}{[A]_0} = -\ln\frac{0.35 \ M}{0.80 \ M} = 0.827 = k \times 24 \ min \qquad k = \frac{0.827}{24 \ min} = 0.034 \ min^{-1}$$

The constancy of the value of k indicates that the reaction is first-order.

(b) The value of the rate constant is $k = 0.034 \ min^{-1}$.

(c) Reaction rate $= \frac{1}{2}$ (rate of formation of B) $= k[A]^1$ First we need [A] at $t = 30.$ min

$$\ln\frac{[A]}{[A]_0} = -kt = -0.034 \ min^{-1} \times 30. \ min = -1.0_2, \qquad \frac{[A]}{[A]_0} = e^{-1.02} = 0.36$$

$[A] = 0.36 \times 0.80 \ M = 0.29 \ M$

rate of formation of B $= 2 \times 0.034 \ min^{-1} \times 0.29 \ M = 2.0 \times 10^{-2} \ M \ min^{-1}$

41. The half-life of the reaction depends on the concentration of "A" and, thus, this reaction cannot be first-order. For a second-order reaction, the half-life varies inversely with the reaction rate: $t_{1/2} = 1/(k[A]_0)$ or $k = 1/(t_{1/2}[A]_0)$. Let us attempt to verify the second-order nature of this reaction by seeing if the rate constant is fixed.

$$k = \frac{1}{1.00\ M \times 50\ min} = 0.020\ L\ mol^{-1}min^{-1}$$

$$k = \frac{1}{2.00\ M \times 25\ min} = 0.020\ L\ mol^{-1}min^{-1}$$

$$k = \frac{1}{0.50\ M \times 100\ min} = 0.020\ L\ mol^{-1}\ min^{-1}$$

The constancy of the rate constant demonstrates that this reaction indeed is second-order. The rate equation is Rate $= k[A]^2$ and $k = 0.020\ L\ mol^{-1}min^{-1}$.

42. (a) The half-life depends on the initial $[NH_3]$ and, thus, the reaction cannot be first-order. Let us attempt to verify second-order kinetics.

$$k = \frac{1}{[NH_3]_0\ t_{1/2}} \text{ for a second-order reaction} \qquad k = \frac{1}{0.0031\ M \times 7.6\ min} = 42\ M^{-1}min^{-1}$$

$$k = \frac{1}{0.0015\ M \times 3.7\ min} = 180\ M^{-1}min^{-1} \qquad k = \frac{1}{0.00068\ M \times 1.7 min} = 865\ M^{-1}min^{-1}$$

The reaction is not second-order. But, if the reaction is zero-order, its rate will be constant.

$$\text{Rate} = \frac{[A]_0/2}{t_{1/2}} = \frac{0.0031\ M \div 2}{7.6\ min} = 2.0 \times 10^{-4}\ M/min$$

$$\text{Rate} = \frac{0.0015\ M \div 2}{3.7\ min} = 2.0 \times 10^{-4}\ M/min$$

$$\text{Rate} = \frac{0.00068\ M \div 2}{1.7\ min} = 2.0 \times 10^{-4}\ M/min \qquad \text{Zero-order reaction}$$

(b) The constancy of the rate indicates that the decomposition of ammonia under these conditions is zero-order, and the rate constant is $k = 2.0 \times 10^{-4}$ M/min.

43. Zeroth-order: $t_{1/2} = \dfrac{[A]_0}{2k}$ \qquad Second-order: $t_{1/2} = \dfrac{1}{k[A]_0}$

A zero-order reaction has a half life that varies proportionally to $[A]_0$, therefore, increasing $[A]_0$ increases the half-life for the reaction. A second-order reaction's half-life varies inversely proportional to $[A]_0$, that is, as $[A]_0$ increases, the half-life decreases. The reason for the difference is that a zero-order reaction has a constant rate of reaction (independent of $[A]_0$). The larger the value of $[A]_0$, the longer it will take to react. In a second-order reaction, the rate of reaction increases as the square of the $[A]_0$, hence, for high $[A]_0$, the rate of reaction is large and for very low $[A]_0$, the rate of reaction is very slow. If we consider a bimolecular elementary reaction, we can easily see that a reaction will not take place unless two molecules of reactants collide. This is more likely when the $[A]_0$ is large than when it is small.

44. **(a)** $\dfrac{[A]_o}{2k} = \dfrac{0.693}{k}$ Hence, $\dfrac{[A]_o}{2} = 0.693$ or $[A]_o = 1.39$ M

(b) $\dfrac{[A]_o}{2k} = \dfrac{1}{k[A]_o}$ Hence, $\dfrac{[A]_o^{\,2}}{2} = 1$ or $[A]_o^{\,2} = 2.00$ M $[A]_o = 1.41\underline{4}$ M

(c) $\dfrac{0.693}{k} = \dfrac{1}{k[A]_o}$ Hence, $0.693 = \dfrac{1}{[A]_o}$ or $[A]_o = 1.44$ M

Collision Theory; Activation Energy

45. **(a)** The rate of a reaction depends on at least two factors other than the frequency of collisions. The first of these is whether each collision possesses sufficient energy to get over the energy barrier to products. This depends on the activation energy of the reaction; the higher it is, the smaller will be the fraction of successful collisions. The second factor is whether the molecules in a given collision are properly oriented for a successful reaction. The more complex the molecules are, or the more freedom of motion the molecules have, the smaller will be the fraction of collisions that are correctly oriented.

(b) Although the collision frequency increases relatively slowly with temperature, the fraction of those collisions that have sufficient energy to overcome the activation energy increases much more rapidly. Therefore, the rate of reaction will increase dramatically with temperature.

(c) The addition of a catalyst has the net effect of decreasing the activation energy of the overall reaction, by enabling an alternate mechanism. The lower activation energy of the alternate mechanism, (compared to the uncatalyzed mechanism), means that a larger fraction of molecules have sufficient energy to react. Thus the rate increases, even though the temperature does not.

46. **(a)** The activation energy for the reaction of hydrogen with oxygen is quite high, too high, in fact, to be supplied by the energy ordinarily available in a mixture of the two gases at ambient temperatures. However, the spark supplies a suitably concentrated form of energy to initiate the reaction of at least a few molecules. Since the reaction is highly exothermic, the reaction of these first few molecules supplies sufficient energy for yet other molecules to react and the reaction proceeds to completion or to the elimination of the limiting reactant.

(b) A larger spark simply means that a larger number of molecules react initially. But the eventual course of the reaction remains the same, with the initial reaction producing enough energy to initiate still more molecules, and so on.

47. **(a)** The products are 21 kJ/mol closer in energy to the energy activated complex than are the reactants. Thus, the activation energy for the reverse reaction is
84 kJ / mol − 21 kJ / mol = 63 kJ / mol.

(b) The reaction profile for the reaction in Figure 14-10 is sketched at below.

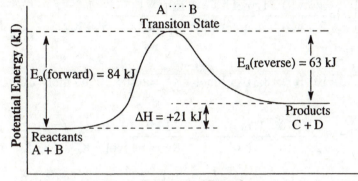

Progress of Reaction

48. In an endothermic reaction (right), E_a must be larger than the ΔH for the reaction. For an exothermic reaction (left), the magnitude of E_a may be either larger or smaller than that of ΔH. In other words, a small activation energy can be associated with a large decrease in the enthalpy, or a large E_a can be connected to a small decrease in enthalpy.

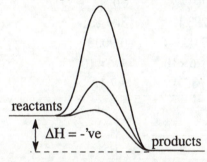
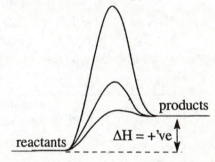

49. **(a)** There are two intermediates (B and C).

   (b) There are three transition states (peaks/maxima) in the energy diagram.

   (c) The fastest step has the smallest E_a, hence, step 3 is the fastest step in the reaction with step 2 a close second.

   (d) Reactant A (step 1) is the reactant in the rate-limiting step

   (e) Endothermic; energy is needed to go from A→B

   (f) Exothermic; energy is released moving from A→D.

50. **(a)** There are two intermediates (B and C).

   (b) There are three transition states (peaks/maxima) in the energy diagram.

   (c) The fastest step has the smallest E_a, hence, step 2 is the fastest step in the reaction.

   (d) Reactant A (step 1) is the reactant in the rate-limiting step.

   (e) Endothermic; energy is needed to go from A→B.

   (f) Endothermic, energy is needed to go from A→D.

Effect of Temperature on Rates of Reaction

51. $\ln\dfrac{k_1}{k_2} = \dfrac{E_a}{R}\left(\dfrac{1}{T_2} - \dfrac{1}{T_1}\right) = \ln\dfrac{5.4\times10^{-4}\ \text{L mol}^{-1}\ \text{s}^{-1}}{2.8\times10^{-2}\ \text{L mol}^{-1}\ \text{s}^{-1}} = \dfrac{E_a}{R}\left(\dfrac{1}{683\ \text{K}} - \dfrac{1}{599\ \text{K}}\right)$

$-3.95R = -E_a \times 2.05\times10^{-4}$

$E_a = \dfrac{3.95\ R}{2.05\times10^{-4}} = 1.93\times10^4\ \text{K}^{-1} \times 8.3145\ \text{J mol}^{-1}\ \text{K}^{-1} = 1.60\times10^5\ \text{J / mol} = 160\ \text{kJ / mol}$

52. $\ln\dfrac{k_1}{k_2} = \dfrac{E_a}{R}\left(\dfrac{1}{T_2} - \dfrac{1}{T_1}\right) = \ln\dfrac{5.0\times10^{-3}\ \text{L mol}^{-1}\ \text{s}^{-1}}{2.8\times10^{-2}\ \text{L mol}^{-1}\ \text{s}^{-1}} = \dfrac{1.60\times10^5\ \text{J/mol}}{8.3145\ \text{J mol}^{-1}\ \text{K}^{-1}}\left(\dfrac{1}{683\ \text{K}} - \dfrac{1}{T}\right)$

$-1.72 = 1.92\times10^4\left(\dfrac{1}{683\ \text{K}} - \dfrac{1}{T}\right)$ $\qquad$ $\left(\dfrac{1}{683\ \text{K}} - \dfrac{1}{T}\right) = \dfrac{-1.72}{1.92\times10^4} = -8.96\times10^{-5}$

$\dfrac{1}{T} = 8.96\times10^{-5} + 1.46\times10^{-3} = 1.55\times10^{-3}$ $\qquad$ $T = 645\ \text{K}$

53. **(a)** First we need to compute values of $\ln k$ and $1/T$. Then we plot the graph of $\ln k$ versus $1/T$.

$T, °\text{C}$	$0\ °\text{C}$	$10\ °\text{C}$	$20\ °\text{C}$	$30\ °\text{C}$
T, K	273 K	283 K	293 K	303 K
$1/T,\ \text{K}^{-1}$	0.00366	0.00353	0.00341	0.00330
$k,\ \text{s}^{-1}$	5.6×10^{-6}	3.2×10^{-5}	1.6×10^{-4}	7.6×10^{-4}
$\ln k$	-12.09	-10.35	-8.74	-7.18

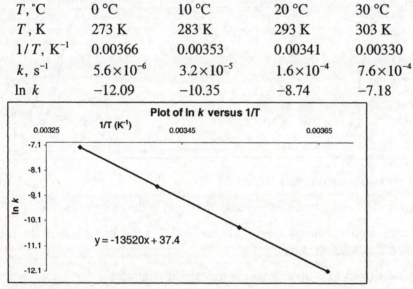

(b) The slope $= -E_a / R$.

$E_a = -R\times \text{slope} = -8.3145\dfrac{\text{J}}{\text{mol K}} \times -1.35_2\times10^4\ \text{K} \times \dfrac{1\ \text{kJ}}{1000\ \text{J}} = 112\dfrac{\text{kJ}}{\text{mol}}$

(c) We apply the Arrhenius equation, with $k = 5.6\times10^{-6}\ \text{s}^{-1}$ at $0\,°\text{C}$ (273 K), $k = ?$ at $40°\text{C}$ (313 K), and $E_a = 113\times10^3\ \text{J/mol}$.

$\ln\dfrac{k}{5.6\times10^{-6}\ \text{s}^{-1}} = \dfrac{E_a}{R}\left(\dfrac{1}{T_1} - \dfrac{1}{T_2}\right) = \dfrac{112\times10^3\ \text{J/mol}}{8.3145\ \text{J mol}^{-1}\ \text{K}^{-1}}\left(\dfrac{1}{273\ \text{K}} - \dfrac{1}{313\ \text{K}}\right) = 6.30\underline{6}$

$e^{6.30\underline{6}} = 548 = \dfrac{k}{5.6\times10^{-6}\ \text{s}^{-1}}$ $\qquad$ $k = 548\times5.6\times10^{-6}\ \text{s}^{-1} = 3.0\underline{7}\times10^{-3}\ \text{s}^{-1}$

$t_{1/2} = \dfrac{0.693}{k} = \dfrac{0.693}{3.0\underline{7}\times10^{-3}\ \text{s}^{-1}} = 2.3\times10^2\ \text{s}$

54. (a) Here we plot ln k vs. $1/T$. The slope of the straight line equals $-E_a/R$. First we tabulate the data to plot. (the plot is shown below).

T, °C	15.83	32.02	59.75	90.61
T, K	288.98	305.17	332.90	363.76
$1/T$, K^{-1}	0.0034604	0.0032769	0.0030039	0.0027491
k, M^{-1}s^{-1}	5.03×10^{-5}	3.68×10^{-4}	6.71×10^{-3}	0.119
ln k	−9.898	−7.907	−5.004	−2.129

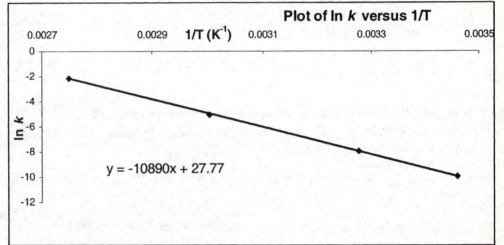

The slope of this graph $= -1.09\times10^4$ K $= E_a/R$

$$E_a = -\left(-1.089\times10^4\ \text{K}\right)\times8.3145\ \frac{\text{J}}{\text{mol K}} = 9.054\times10^4\ \frac{\text{J}}{\text{mol}}\ \times\ \frac{1\ \text{kJ}}{1000\ \text{J}} = 90.5\ \frac{\text{kJ}}{\text{mol}}$$

(b) We calculate the activation energy with the Arrhenius equation using the two extreme data points.

$$\ln\frac{k_2}{k_1} = \ln\frac{0.119}{5.03\times10^{-5}} = +7.77 = \frac{E_a}{R}\left(\frac{1}{T_1}-\frac{1}{T_2}\right) = \frac{E_a}{R}\left(\frac{1}{288.98\ \text{K}}-\frac{1}{363.76\ \text{K}}\right)$$

$$= 7.1138\times10^{-4}\ \text{K}^{-1}\frac{E_a}{R} \qquad E_a = \frac{7.769\times8.3145\ \text{J mol}^{-1}\ \text{K}^{-1}}{7.1138\times10^{-4}\ \text{K}^{-1}} = 9.0\underline{8}\times10^4\ \text{J/mol}$$

$E_a = 91\,\text{kJ/mol}$. The two E_a values are in quite good agreement, within experimental limits.

(c) We apply the Arrhenius equation, with $E_a = 9.080\times10^4$ J/mol, $k = 5.03\times10^{-5}$ M^{-1} s^{-1} at 15.83 °C (288.98 K), and $k = ?$ at 100.0 °C (373.2 K).

$$\ln\frac{k}{5.03\times10^{-5}\ \text{M}^{-1}\ \text{s}^{-1}} = \frac{E_a}{R}\left(\frac{1}{T_1}-\frac{1}{T_2}\right) = \frac{90.80\times10^3\ \text{J/mol}}{8.3145\ \text{J mol}^{-1}\ \text{K}^{-1}}\left(\frac{1}{288.98\ \text{K}}-\frac{1}{373.2\ \text{K}}\right)$$

$$\ln\frac{k}{5.03\times10^{-5}\ \text{M}^{-1}\ \text{s}^{-1}} = 8.528 \qquad e^{8.528} = 5.05\times10^3 = \frac{k}{5.03\times10^{-5}\ \text{M}^{-1}\ \text{s}^{-1}}$$

$$k = 5.05\times10^3 \times 5.03\times10^{-5}\ \text{M}^{-1}\ \text{s}^{-1} = 0.254\ \text{M}^{-1}\ \text{s}^{-1}$$

55. The half-life of a first-order reaction is inversely proportional to its rate constant:
$k = 0.693 / t_{1/2}$. Thus we can apply a modified version of the Arrhenius equation to find E_a.

(a) $\ln \dfrac{k_2}{k_1} = \ln \dfrac{(t_{1/2})_1}{(t_{1/2})_2} = \dfrac{E_a}{R}\left(\dfrac{1}{T_1} - \dfrac{1}{T_2}\right) = \ln \dfrac{46.2 \text{ min}}{2.6 \text{ min}} = \dfrac{E_a}{R}\left(\dfrac{1}{298 \text{ K}} - \dfrac{1}{(102+273) \text{ K}}\right)$

$2.88 = \dfrac{E_a}{R} 6.89 \times 10^{-4}$ $E_a = \dfrac{2.88 \times 8.3145 \text{ J mol}^{-1} \text{ K}^{-1}}{6.89 \times 10^{-4} \text{ K}^{-1}} \times \dfrac{1 \text{ kJ}}{1000 \text{ J}} = 34.8 \text{ kJ/mol}$

(b) $\ln \dfrac{10.0 \text{ min}}{46.2 \text{ min}} = \dfrac{34.8 \times 10^3 \text{ J / mol}}{8.3145 \text{ J mol}^{-1} \text{ K}^{-1}}\left(\dfrac{1}{T} - \dfrac{1}{298}\right) = -1.53 = 4.19 \times 10^3\left(\dfrac{1}{T} - \dfrac{1}{298}\right)$

$\left(\dfrac{1}{T} - \dfrac{1}{298}\right) = \dfrac{-1.53}{4.19 \times 10^3} = -3.65 \times 10^{-4}$ $\dfrac{1}{T} = 2.99 \times 10^{-3}$ $T = 334 \text{ K} = 61 \text{ °C}$

56. The half-life of a first-order reaction is inversely proportional to its rate
constant: $k = 0.693 / t_{1/2}$. Thus, we can apply a modified version of the Arrhenius equation.

(a) $\ln \dfrac{k_2}{k_1} = \ln \dfrac{(t_{1/2})_1}{(t_{1/2})_2} = \dfrac{E_a}{R}\left(\dfrac{1}{T_1} - \dfrac{1}{T_2}\right) = \ln \dfrac{22.5 \text{ h}}{1.5 \text{ h}} = \dfrac{E_a}{R}\left(\dfrac{1}{293 \text{ K}} - \dfrac{1}{(40+273) \text{ K}}\right)$

$2.71 = \dfrac{E_a}{R} 2.18 \times 10^{-4},$ $E_a = \dfrac{2.71 \times 8.3145 \text{ J mol}^{-1} \text{ K}^{-1}}{2.18 \times 10^{-4} \text{ K}^{-1}} \times \dfrac{1 \text{ kJ}}{1000 \text{ J}} = 103 \text{ kJ/mol}$

(b) The relationship is $k = A \exp(-E_a / RT)$

$k = 2.05 \times 10^{13} \text{ s}^{-1} \exp\left(\dfrac{-103 \times 10^3 \text{ J mol}^{-1}}{8.3145 \text{ J mol}^{-1} \text{ K}^{-1} \times (273+30) \text{ K}}\right) = 2.05 \times 10^{13} \text{ s}^{-1} \times \text{e}^{-40.9} = 3.5 \times 10^{-5} \text{ s}^{-1}$

57. **(a)** It is the change in the value of the rate constant that causes the reaction to go faster.
Let k_1 be the rate constant at room temperature, 20 °C (293 K). Then, ten degrees
higher, (30 °C (303 K)), the rate constant $k_2 = 2 \times k_1$.

$\ln \dfrac{k_2}{k_1} = \ln \dfrac{2 \times k_1}{k_1} = 0.693 = \dfrac{E_a}{R}\left(\dfrac{1}{T_1} - \dfrac{1}{T_2}\right) = \dfrac{E_a}{R}\left(\dfrac{1}{293} - \dfrac{1}{303 \text{ K}}\right) = 1.13 \times 10^{-4} \text{ K}^{-1} \dfrac{E_a}{R}$

$E_a = \dfrac{0.693 \times 8.3145 \text{ J mol}^{-1} \text{ K}^{-1}}{1.13 \times 10^{-4} \text{ K}^{-1}} = 5.1 \times 10^4 \text{ J / mol} = 51 \text{ kJ / mol}$

(b) Since the activation energy for the depicted reaction (i.e., $N_2O + NO \rightarrow N_2 + NO_2$) is
209 kJ/mol, we would not expect this reaction to follow the rule of thumb.

58. Under a pressure of 2.00 atm the boiling point of water is approximately 121 °C or 394 K.
Under a pressure of 1 atm, the boiling point of water is 100° C or 373 K. We assume an
activation energy of $5.1 \times 10^4 \text{ J / mol}$ and compute the ratio of the two rates.

$\ln \dfrac{\text{Rate}_2}{\text{Rate}_1} = \dfrac{E_a}{R}\left(\dfrac{1}{T_1} - \dfrac{1}{T_2}\right) = \dfrac{5.1 \times 10^4 \text{ J / mol}}{8.3145 \text{ J mol}^{-1} \text{ K}^{-1}}\left(\dfrac{1}{373} - \dfrac{1}{394 \text{ K}}\right) = 0.88$

$\text{Rate}_2 = \text{e}^{0.88} \text{ Rate}_1 = 2.4 \text{ Rate}_1$. Cooking will occur 2.4 times faster in the pressure cooker.

Catalysis

59. **(a)** Although a catalyst is *recovered unchanged from the reaction mixture*, it does "take part in the reaction." Some catalysts actually slow down the rate of a reaction. Usually, however, these negative catalysts are called inhibitors.

(b) The function of a catalyst is to *change the mechanism of a reaction*. The new mechanism is one that has a different (lower) activation energy (and frequently a different A value), than the original reaction.

60. If the reaction is first-order, its half-life is 100 min, for in this time period [S] decreases from 1.00 M to 0.50 M, that is, by one half. This gives a rate constant of

$k = 0.693 / t_{1/2} = 0.693 / 100 \text{ min} = 0.00693 \text{ min}^{-1}$.

The rate constant also can be determined from any two of the other sets of data.

$$kt = \ln\frac{[A]_0}{[A]_t} = \ln\frac{1.00 \text{ M}}{0.70 \text{ M}} = 0.357 = k \times 60 \text{ min} \qquad k = \frac{0.357}{60 \text{ min}} = 0.00595 \text{ min}^{-1}$$

This is not a very good agreement between the two k values, so the reaction is probably not first order in [A]. Let's try zero-order, where the rate should be constant.

$$\text{Rate} = -\frac{0.90 \text{ M} - 1.00 \text{ M}}{20 \text{ min}} = 0.0050 \text{ M/min} \qquad \text{Rate} = -\frac{0.50 \text{ M} - 1.00 \text{ M}}{100 \text{ min}} = 0.0050 \text{ M/min}$$

$$\text{Rate} = -\frac{0.20 \text{ M} - 0.90 \text{ M}}{160 \text{ min} - 20 \text{ min}} = 0.0050 \text{ M/min} \qquad \text{Rate} = -\frac{0.50 \text{ M} - 0.90 \text{ M}}{100 \text{ min} - 20 \text{ min}} = 0.0050 \text{ M/min}$$

Thus, this reaction is zero-order with respect to [S].

61. Both platinum and an enzyme have a metal center that acts as the active site. Generally speaking, platinum is not dissolved in the reaction solution (heterogeneous), whereas enzymes are generally soluble in the reaction media (homogeneous). The most important difference, however, is one of specificity. Platinum is rather nonspecific, catalyzing many different reactions. An enzyme, however, is quite specific, usually catalyzing only one reaction rather than all reactions of a given class.

62. In both the enzyme and the metal surface cases, the reaction occurs in a specialized location: either within the enzyme pocket or on the surface of the catalyst. At high concentrations of reactant, the limiting factor in determining the rate is not the concentration of reactant present but how rapidly active sites become available for reaction to occur. Thus, the rate of the reaction depends on either the quantity of enzyme present or the surface area of the catalyst, rather than on how much reactant is present (i.e., the reaction is zero-order). At low concentrations or gas pressures the reaction rate depends on how rapidly molecules can reach the available active sites. Thus, the rate depends on concentration or pressure of reactant and is first-order.

63. For the straight-line graph of Rate versus [Enzyme], an excess of substrate must be present.

64. For human enzymes, we would expect the maximum in the curve to appear around 37°C, i.e., normal body temperature. (possibly slightly elevated temperatures to aid in the control of diseases (37 - 41 °C). At lower temperatures, the reaction rate of enzyme-activated reactions decreases with decreasing temperature, following the Arrhenius equation. However, at higher temperatures, these temperature sensitive biochemical processes become inhibited, probably by temperature-induced structural modifications to the enzyme or the substrate, which prevent formation of the enzyme-substrate complex.

Reaction Mechanisms

65. The molecularity of an elementary process is the number of reactant molecules in the process. This molecularity is equal to the order of the overall reaction only if the elementary process in question is the slowest and, thus, the rate-determining step of the overall reaction. In addition, the elementary process in question should be the only elementary step that influences the rate of the reaction.

66. If the type of molecule that is expressed in the rate law as being first-order collides with other molecules that are present in much larger concentrations, the reaction will seem to depend only on the amount of those types of molecules present in smaller concentration, since the larger concentration will be essentially unchanged during the course of the reaction. Such a situation is quite common, and has been given the name pseudo first-order. It is also possible to have molecules which do not participate directly in the reaction— including product molecules—strike the reactant molecules and impart to them sufficient energy to react. Finally, collisions of the reactant molecules with the container walls may also impart adequate energy for reaction to occur.

67. The three elementary steps must sum to give the overall reaction. That is, the overall reaction is the sum of step 1 + step 2 + step 3. Hence, step 2 = overall reaction −step 1 −step 3. Note that all species in the equations below are gases.

overall: $2 NO + 2 H_2 \rightarrow N_2 + 2 H_2O$ $2 NO + 2 H_2 \rightarrow N_2 + 2 H_2O$

−first: $-\left(2 NO \rightleftharpoons N_2O_2\right)$ $N_2O_2 \rightleftharpoons 2NO$

−third $-\left(N_2O + H_2 \rightarrow N_2 + H_2O\right)$ or $N_2 + H_2O \rightarrow N_2O + H_2$

The result is the second step, which is slow: $H_2 + N_2O_2 \rightarrow H_2O + N_2O$

The rate of this rate-determining step is: Rate $= k_2 [H_2][N_2O_2]$

Since N_2O_2 does not appear in the overall reaction, we need to replace its concentration with the concentrations of species that do appear in the overall reaction. To do this, recall that the first step is rapid, with the forward reaction occurring at the same rate as the reverse reaction. $k_1[NO]^2 =$ forward rate = reverse rate $= k_{-1}[N_2O_2]$. This expression is solved for $[N_2O_2]$, which then is substituted into the rate equation for the overall reaction.

$$[N_2O_2] = \frac{k_1[NO]^2}{k_{-1}} \qquad \text{Rate} = \frac{k_2 k_1}{k_{-1}}[H_2][NO]^2$$

The reaction is first-order in $[H_2]$ and second-order in [NO]. This result conforms to the experimentally determined reaction order.

68. Proposed mechanism: $I_2(g) \underset{k_{-1}}{\overset{k_1}{\rightleftharpoons}} 2I(g)$ Observed rate law:

$$2I(g) + H_2(g) \xrightarrow{k_2} 2HI(g) \qquad \text{Rate} = k[I_2][H_2]$$

$$\overline{ I_2(g) + H_2(g) \rightarrow 2HI(g)}$$

The first step is a fast equilibrium reaction and step 2 is slow. Thus, the predicted rate law is Rate $= k_2[I]^2[H_2]$. In the first step, set the rate in the forward direction for the equilibrium equal to the rate in the reverse direction. Then solve for $[I]^2$.

$\text{Rate}_{forward} = \text{Rate}_{reverse}$ Use: $\text{Rate}_{forward} = k_1[I_2]$ and $\text{Rate}_{reverse} = k_{-1}[I]^2$

From this we see: $k_1[I_2] = k_{-1}[I]^2$. Rearranging (solving for $[I]^2$)

$[I]^2 = \dfrac{k_1[I_2]}{k_{-1}}$ Substitute into Rate $= k_2[I]^2[H_2] = k_2\dfrac{k_1[I_2]}{k_{-1}}[H_2] = k_{obs}[I_2][H_2]$

Since the predicted rate law is the same as the experimental rate law, this mechanism is plausible.

69. Proposed mechanism: $Cl_2(g) \underset{k_{-1}}{\overset{k_1}{\rightleftharpoons}} 2Cl(g)$ Observed rate law:

$$2Cl(g) + 2NO(g) \xrightarrow{k_2} 2NOCl(g) \qquad \text{Rate} = k[Cl_2][NO]^2$$

$$\overline{ Cl_2(g) + 2NO(g) \rightarrow 2NOCl(g)}$$

The first step is a fast equilibrium reaction and step 2 is slow. Thus, the predicted rate law is Rate $= k_2[Cl]^2[NO]^2$ In the first step, set the rate in the forward direction for the equilibrium equal to the rate in the reverse direction. Then express $[Cl]^2$ in terms of k_1, k_{-1} and $[Cl_2]$. This mechanism is almost certainly not correct because it involves a tetra molecular second step.

$\text{Rate}_{forward} = \text{Rate}_{reverse}$ Use: $\text{Rate}_{forward} = k_1[Cl_2]$ and $\text{Rate}_{reverse} = k_{-1}[Cl]^2$
From this we see: $k_1[Cl_2] = k_{-1}[Cl]^2$. Rearranging (solving for $[Cl]^2$)

$[Cl]^2 = \dfrac{k_1[Cl_2]}{k_{-1}}$ Substitute into Rate $= k_2[Cl]^2[NO]^2 = k_2\dfrac{k_1[Cl_2]}{k_{-1}}[NO]^2 = k_{obs}[Cl_2][NO_2]^2$

There is another plausible mechanism. $Cl_2(g) + NO(g) \underset{k_{-1}}{\overset{k_1}{\rightleftharpoons}} NOCl(g) + Cl(g)$

$$Cl(g) + NO(g) \underset{k_{-1}}{\overset{k_1}{\rightleftharpoons}} NOCl(g)$$

$$\overline{ Cl_2(g) + 2NO(g) \rightarrow 2NOCl(g)}$$

$\text{Rate}_{forward} = \text{Rate}_{reverse}$ Use: $\text{Rate}_{forward} = k_1[Cl_2][NO]$ and $\text{Rate}_{reverse} = k_{-1}[Cl][NOCl]$

From this we see: $k_1[Cl_2][NO] = k_{-1}[Cl][NOCl]$. Rearranging (solving for $[Cl]$)

$[Cl] = \dfrac{k_1[Cl_2][NO]}{k_{-1}[NOCl]}$ Substitute into Rate $= k_2[Cl][NO] = \dfrac{k_2k_1[Cl_2][NO]^2}{k_{-1}[NOCl]}$

If $[NOCl]$, the product is assumed to be constant (~ 0 M using method of initial rates), then

$\dfrac{k_2k_1}{k_{-1}[NOCl]} = \text{constant} = k_{obs}$ Hence, the predicted rate law is $k_{obs}[Cl_2][NO]^2$ which agrees

with the experimental rate law. Since the predicted rate law agrees with the experimental rate law, both this and the previous mechanism are plausible, however, the first is dismissed as it has a tetramolecular elementary reaction (extremely unlikely to have four molecules simultaneously collide).

70. A possible mechanism is: Step 1: $O_3 \rightleftharpoons O_2 + O$ (fast)

Step 2: $O + O_3 \xrightarrow{k_3} 2\,O_2$ (slow)

The overall rate is that of the slow step: Rate $= k_3[O][O_3]$. But O is a reaction intermediate, whose concentration is difficult to determine. An expression for [O] can be found by assuming that the forward and reverse "fast" steps proceed with equal speed.

$$\text{Rate}_1 = \text{Rate}_2 \quad k_1[O_3] = k_2[O_2][O] \quad [O] = \frac{k_1[O_3]}{k_2[O_2]} \qquad \text{Rate} = k_3\frac{k_1[O_3]}{k_2[O_2]}[O_3] = \frac{k_3 k_1}{k_2}\frac{[O_3]^2}{[O_2]}$$

Then substitute this expression into the rate law for the reaction. This rate equation has the same form as the experimentally determined rate law and thus the proposed mechanism is plausible.

Integrative and Advanced Exercises

71. The data for the reaction starting with 1.00 M being 1st order or 2nd order as well as that for the 1st order reaction using 2.00 M is shown below

Time (min)	$[A]_0 = 1.00$ M (2nd order)	$[A]_0 = 1.00$ M (1st order)	$[A]_0 = 2.00$ M (2nd order)	$[A]_0 = 2.00$ M (1st order)
0	1.00	1.00	2.00	2.00
5	0.63	0.55	0.91	1.10
10	0.46	0.30	0.59	0.60
15	0.36	0.165	0.435	0.33
25	0.25	0.05	0.286	0.10

Clearly we can see that when $[A]_0 = 1.00$ M, the first order reaction concentrations will always be lower than that for the second-order case (assumes magnitude of the rate constant is the same). If on the other hand, the concentration is above 1.00 M, the second-order reaction decreases faster than the first-order reaction (remember that the half-life shortens for a second-order reaction as the concentration increases, whereas for a firs-order reaction, the half-life is constant).

From the data, it appears that the crossover occurs in the case where $[A]_0 = 2.00$ M just over 10 minutes.

2nd order at 11 minutes: $\dfrac{1}{[A]} = \dfrac{1}{2} + \left(\dfrac{0.12}{\text{M min}}\right) \times (11\ \text{min})$ $[A] = 0.549$ M

1st order at 11 minutes: $\ln[A] = \ln(2) - \left(\dfrac{0.12}{\text{min}}\right) \times (11\ \text{min})$ $[A] = 0.534$ M

A quick check at 10.5 minutes reveals,

2nd order at 10.5 minutes: $\dfrac{1}{[A]} = \dfrac{1}{2} + \dfrac{0.12(10.5\ \text{min})}{\text{M min}}$ $[A] = 0.568$ M

1st order at 10.5 minutes: $\ln[A] = \ln(2) - \dfrac{0.12(10.5\ \text{min})}{\text{M min}}$ $[A] = 0.567$ M

Hence, at approximately 10.5 minutes, these two plots will share a common point (point at which the concentration versus time curves overlap).

72. (a) The concentration *vs.* time graph is not linear. Thus, the reaction is obviously not zero order(the rate is not constant with time). A quick look at various half lives for this reactions shows the ~2.37 min (1.000 M to 0.5 M), ~2.32 min (0.800 M to 0.400 M), ~2.38 min(0.400 M to 0.200 M) Since the half-life is constant, the reaction is probably first-order.

(b) average $t_{1/2} = \dfrac{(2.37 + 2.32 + 2.38)}{3} = 2.36 \text{ min}$ $k = \dfrac{0.693}{t_{1/2}} = \dfrac{0.693}{2.36 \text{ min}} = 0.294 \text{ min}^{-1}$

or perhaps better expressed as $k = 0.29 \text{ min}^{-1}$ due to imprecision.

(c) When $t = 3.5 \text{ min}, [A] = 0.352 \text{ M}$. Then, rate $= k[A] = 0.294 \text{ min}^{-1} \times 0.352 \text{ M} = 0.10_3 \text{ M/min}$

(d) Slope $= \dfrac{\Delta[A]}{\Delta t} = -\text{Rate} = \dfrac{0.1480 \text{ M} - 0.339 \text{ M}}{6.00 \text{ min} - 3.00 \text{ min}} = -0.0637 \text{ M/min}$ Rate $= 0.064$ M/min

(e) Rate $= k[A] = 0.294 \text{ min}^{-1} \times 1.000 \text{ M} = 0.29_4$ M/min

73. The reaction being investigated is:
$$2\,MnO_4^-(aq) + 5\,H_2O_2(aq) + 6\,H^+(aq) \longrightarrow 2\,Mn^{2+}(aq) + 8\,H_2O(l) + 5\,O_2(g)$$
We use the stoichiometric coefficients in this balanced reaction to determine $[H_2O_2]$.

$$[H_2O_2] = \dfrac{37.1 \text{ mL titrant} \times \dfrac{0.1000 \text{ mmol MnO}_4^-}{1 \text{ mL titrant}} \times \dfrac{5 \text{ mmol H}_2O_2}{2 \text{ mmol MnO}_4^-}}{5.00 \text{ mL}} = 1.86 \text{ M}$$

74. We assume in each case that 5.00 mL of reacting solution is titrated.

$$\text{Volume MnO}_4^- = 5.00 \text{ mL} \times \dfrac{2.32 \text{ mmol H}_2O_2}{1 \text{ mL}} \times \dfrac{2 \text{ mmol MnO}_4^-}{5 \text{ mmol H}_2O_2} \times \dfrac{1 \text{ mL titrant}}{0.1000 \text{ mmol MnO}_4^-}$$

$$= 20.0 \times 2.32 \text{ M H}_2O_2 = 46.4 \text{ mL } 0.1000 \text{ M MnO}_4^- \text{ titrant at 0 s}$$

At 200 s $V_{\text{titrant}} = 20.0 \times 2.01 \text{ M H}_2O_2 = 40.2 \text{ mL}$
At 400 s $V_{\text{titrant}} = 20.0 \times 1.72 \text{ M H}_2O_2 = 34.4 \text{ mL}$
At 600 s $V_{\text{titrant}} = 20.0 \times 1.49 \text{ M H}_2O_2 = 29.8 \text{ mL}$
At 1200 s $V_{\text{titrant}} = 20.0 \times 0.98 \text{ M H}_2O_2 = 19._6 \text{ mL}$
At 1800 s $V_{\text{titrant}} = 20.0 \times 0.62 \text{ M H}_2O_2 = 12._4 \text{ mL}$
At 3000 s $V_{\text{titrant}} = 20.0 \times 0.25 \text{ M H}_2O_2 = 5.0_0 \text{ mL}$

The graph of volume of titrant vs. elapsed time is given at right. This graph is of approximately the same shape as Figure 15-2, in which $[H_2O_2]$ is plotted against time. In order to determine the rate, the tangent line at 1400 s has been drawn on the graph. The intercepts of the tangent line are at 34 mL of titrant and 2800 s. From this information we determine the rate of the reaction.

$$\text{Rate} = \dfrac{\dfrac{34 \text{ mL}}{2800 \text{ s}} \times \dfrac{1 \text{ L}}{1000 \text{ mL}} \times \dfrac{0.1000 \text{ mol MnO}_4^-}{1 \text{ L titrant}} \times \dfrac{5 \text{ mol H}_2O_2}{2 \text{ mol MnO}_4^-}}{0.00500 \text{ L sample}} = 6.1 \times 10^{-4} \text{ M/s}$$

This is the same as the value of 6.1×10^{-4} obtained in Figure 14-2 for 1400 s. The discrepancy is due, no doubt, to the course nature of out plot.

75. First we compute the change in $[H_2O_2]$. This is then used to determine the amount, and ultimately the volume, of oxygen evolved from the given quantity of solution. Assume the $O_2(g)$ is dry

$$\Delta[H_2O_2] = -\frac{\Delta[H_2O_2]}{\Delta t}\Delta t = -\left(-1.7\times10^{-3}\text{ M/s}\times1.00\text{ min}\times\frac{60\text{ s}}{1\text{ min}}\right) = 0.10_2\text{ M}$$

$$\text{amount }O_2 = 0.175\text{ L soln}\times\frac{0.10_2\text{ mol }H_2O_2}{1\text{ L}}\pm\times\frac{1\text{ mol }O_2}{2\text{ mol }H_2O_2} = 0.0089_2\text{ mol }O_2$$

$$\text{Volume }O_2 = \frac{nRT}{P} = \frac{0.0089_2\text{ mol }O_2\times0.08206\dfrac{\text{L}\cdot\text{atm}}{\text{mol}\cdot\text{K}}\times(273+24)\text{ K}}{757\text{ mmHg}\times\dfrac{1\text{ atm}}{760\text{ mmHg}}} = 0.22\text{ L }O_2$$

76. We know that rate has the units of M/s, and also that concentration has the units of M. The generalized rate equation is Rate $= k\,[A]^o$ In terms of units, this becomes

$$M/s = \{\text{units of }k\}\, M^o \quad\text{Therefore }\{\text{Units of }k\} = \frac{M/s}{M^o} = M^{1-o}\text{ s}^{-1}$$

77. **(a)** The time required for the fixed (c) process of souring is three times as long at 3 °C refrigerator temperature (276K) as at 20 °C room temperature (293 K).

$$\ln\frac{c/t_2}{c/t_1} = \ln\frac{t_1}{t_2} = \ln\frac{64\text{ h}}{3\times64\text{ h}} = -1.10 = \frac{E_a}{R}\left(\frac{1}{T_1}-\frac{1}{T_2}\right) = \frac{E_a}{R}\left(\frac{1}{293\text{ K}}-\frac{1}{276\text{ K}}\right) = \frac{E_a}{R}(-2.10\times10^{-4})$$

$$E_a = \frac{1.10\,R}{2.10\times10^{-4}\text{ K}^{-1}} = \frac{1.10\times8.3145\text{ J mol}^{-1}\text{ K}^{-1}}{2.10\times10^{-4}\text{ K}^{-1}} = 4.4\times10^4\text{ J/mol} = 44\text{ kJ/mol}$$

(b) Use the E_a determined in part **(a)** to calculate the souring time at 40 °C = 313 K.

$$\ln\frac{t_1}{t_2} = \frac{E_a}{R}\left(\frac{1}{T_1}-\frac{1}{T_2}\right) = \frac{4.4\times10^4\text{ J/mol}}{8.3145\text{ J mol}^{-1}\text{ K}^{-1}}\left(\frac{1}{313\text{ K}}-\frac{1}{293\text{ K}}\right) = -1.15 = \ln\frac{t_1}{64\text{ h}}$$

$$\frac{t_1}{64\text{ h}} = e^{-1.15} = 0.317 \quad t_1 = 0.317\times64\text{ h} = 20.\text{ h}$$

78. **(a)** Comparing the third and the first lines of data, $[I^-]$ and $[OH^-]$ stay fixed, while $[OCl^-]$ doubles. Also the rate for the third kinetics run is one half of the rate found for the first run. Thus, the reaction is <u>first order</u> in $[OCl^-]$. Comparing the fourth and fifth lines, $[OCl^-]$ and $[I^-]$ stay fixed, while $[OH^-]$ is halved. Also, the fifth run has a reaction rate that is twice that of the fourth run. Thus, the reaction is <u>minus</u> first order in $[OH^-]$. Comparing the third and second lines of data, $[OCl^-]$ and $[OH^-]$ stay fixed, while the $[I^-]$ doubles. Also, the second run has a reaction rate that is double that found for the third run. Thus, the reaction is <u>first order</u> in $[I^-]$.

(b) The reaction is 1st order in $[OCl^-]$ and $[I^-]$ and minus 1st order in $[OH^-]$. Thus, the overall order = 1 + 1 − 1 = 1. The reaction is first order overall.

(c) Rate $= k\dfrac{[OCl^-][I^-]}{[OH^-]}$ $\quad k = \dfrac{\text{Rate}[OH^-]}{[OCl^-][I^-]} = \dfrac{4.8\times10^{-4}\text{ M/s}\times1.00\text{ M}}{0.0040\text{ M}\times0.0020\text{ M}} = 60.\text{ s}^{-1}$

79. We first determine the number of moles of N_2O produced. The partial pressure of $N_2O(g)$ in the "wet" N_2O is 756 mmHg $-$ 12.8 mmHg = 743 mmHg.

$$\text{amount } N_2O = \frac{PV}{RT} = \frac{743 \text{ mmHg} \times \dfrac{1 \text{ atm}}{760 \text{ mmHg}} \times 0.0500 \text{ L}}{0.08206 \text{ L atm mol}^{-1} \text{ K}^{-1} \times (273+15) \text{ K}} = 0.00207 \text{ mol } N_2O$$

Now we determine the change in $[NH_2NO_2]$.

$$\Delta[NH_2NO_2] = \frac{0.00207 \text{ mol } N_2O \times \dfrac{1 \text{ mol } NH_2NO_2}{1 \text{ mol } N_2O}}{0.165 \text{ L soln}} = 0.0125 \text{ M}$$

$$[NH_2NO_2]_{\text{final}} = 0.105 \text{ M} - 0.0125 \text{ M} = 0.093 \text{ M} \qquad k = \frac{0.693}{123 \text{ min}} = 0.00563 \text{ min}^{-1}$$

$$t = -\frac{1}{k} \ln\frac{[A]_t}{[A]_0} = -\frac{1}{0.00563 \text{ min}^{-1}} \ln\frac{0.093 \text{ M}}{0.105 \text{ M}} = 22 \text{ min} = \text{elapsed time}$$

80. We need to determine the partial pressure of ethylene oxide at each time in order to determine the order of the reaction. First, we need the initial pressure of ethylene oxide. The pressure at infinite time is the pressure that results when all of the ethylene oxide has decomposed. Because two moles of product gas are produced for every mole of reactant gas, this infinite pressure is twice the initial pressure of ethylene oxide. P_{initial} = 249.88 mmHg $\div$ 2 = 124.94 mmHg. Now, at each time we have the following. $(CH_2)_2O(g) \longrightarrow CH_4(g) + CO(g)$

Initial: 124.94 mmHg   Changes: $-x$ mmHg $+x$ mmHg $+x$ mmHg Final: 124.94 $+ x$ mmHg
Thus, $x = P_{\text{tot}} - 124.94$ and $P_{\text{EtO}} = 124.94 - x = 124.94 - (P_{\text{tot}} - 124.94) = 249.88 - P_{\text{tot}}$
Hence, we have, the following values for the partial pressure of ethylene oxide.

t, min	0	10	20	40	60	100	200
P_{EtO}, mmHg	124.94	110.74	98.21	77.23	60.73	37.54	11.22

For the reaction to be zero order, its rate will be constant.

The rate in the first 10 min is: $\text{Rate} = \dfrac{-(110.74-124.94) \text{ mmHg}}{10 \text{ min}} = 1.42 \text{ mmHg/min}$

The rate in the first 40 min is: $\text{Rate} = \dfrac{-(77.23-124.94) \text{ mmHg}}{40 \text{ min}} = 1.19 \text{ mmHg/min}$

We conclude from the non-constant rate that the reaction is not zero order. For the reaction to be first-order, its half-life must be constant. From 40 min to 100 min—a period of 60 min—the partial pressure of ethylene oxide is approximately halved, giving an approximate half-life of 60 min. And, in the first 60 min, the partial pressure of ethylene oxide is approximately halved. Thus, the reaction appears to be first-order. To verify this tentative conclusion, we use the integrated first-order rate equation to calculate some values of the rate constant.

$$k = -\frac{1}{t}\ln\frac{P}{P_0} = -\frac{1}{10 \text{ min}}\ln\frac{110.74 \text{ mmHg}}{124.94 \text{ mmHg}} = 0.0121 \text{ min}^{-1}$$

$$k = -\frac{1}{100 \text{ min}}\ln\frac{37.54 \text{ mmHg}}{124.94 \text{ mmHg}} = 0.0120 \text{ min}^{-1} \qquad k = -\frac{1}{60 \text{ min}}\ln\frac{60.73 \text{ mmHg}}{124.94 \text{ mmHg}} = 0.0120 \text{ min}^{-1}$$

The constancy of the first-order rate constant suggests that the reaction is first-order.

81. For this first-order reaction $\ln\dfrac{P_t}{P_0}=-kt$ Elapsed time is computed as: $t=-\dfrac{1}{k}\ln\dfrac{P_t}{P_0}$

We first determine the pressure of DTBP when the total pressure equals 2100. mmHg

Reaction : $C_8H_{18}O_2(g) \longrightarrow 2C_3H_6O(g)+ C_2H_6(g)$ [equation 15.16]

Initial : 800.0 mmHg

Changes : $-x$ mmHg $\qquad +2x$ mmHg $\quad +x$ mmHg

Final : $(800.0-x)$ mmHg $\qquad 2x$ mmHg $\quad x$ mmHg

Total pressure $= (800.0-x)+2x+x=800.0+2x=2100.$

$x=650.$ mmHg $\quad P\{C_8H_{18}O_2(g)\}=800.$ mmHg $-650.$ mmHg $=150.$ mmHg

$t=-\dfrac{1}{k}\ln\dfrac{P_t}{P_0}=-\dfrac{1}{8.7\times10^{-3}\text{ min}^{-1}}\ln\dfrac{150.\text{ mmHg}}{800.\text{ mmHg}}=19_2\text{ min}=1.9\times10^2\text{ min}$

82. If we compare Experiment 1 with Experiment 2 we notice that [B] has been halved, and also that the rate, expressed as $\Delta[A]/\Delta t$, has been halved. This is most evident for the times 5 min, 10 min, and 20 min. In Experiment 1, [A] decreases from 1.000×10^{-3} M to 0.779×10^{-3} M in 5 min, while in Experiment 2 this same decrease in [A] requires 10 min. Likewise in Experiment 1, [A] decreases from 1.000×10^{-3} M to 0.607 M$\times10^{-3}$ in 10 min, while in Experiment 2 the same decrease in [A] requires 20 min. This dependence of rate on the first power of concentration is characteristic of a first-order reaction. This reaction is first-order in [B]. We now turn to the order of the reaction with respect to [A]. A zero-order reaction will have a constant rate. Determine the rate

After over the first minute: $\qquad$ Rate $=\dfrac{-(0.951-1.000)\times10^{-3}\text{ M}}{1\text{ min}}=4.9\times10^{-5}$ M/min

After over the first five minutes: $\quad$ Rate $=\dfrac{-(0.779-1.000)\times10^{-3}\text{ M}}{5\text{ min}}=4.4\times10^{-5}$ M/min

After over the first twenty minutes: Rate $=\dfrac{-(0.368-1.000)\times10^{-3}\text{ M}}{20\text{ min}}=3.2\times10^{-5}$ M/min

This is not a very constant rate; we conclude that the reaction is not zero-order. There are no clear half-lives in the data with which we could judge the reaction to be first-order. But we can determine the value of the first-order rate constant for a few data.

$k=-\dfrac{1}{t}\ln\dfrac{[A]}{[A]_0}=-\dfrac{1}{1\text{ min}}\ln\dfrac{0.951\text{ mM}}{1.000\text{ mM}}=0.0502\text{ min}^{-1}$

$k=-\dfrac{1}{10\text{ min}}\ln\dfrac{0.607\text{ mM}}{1.000\text{ mM}}=0.0499\text{ min}^{-1}\qquad k=-\dfrac{1}{20\text{ min}}\ln\dfrac{0.368\text{ mM}}{1.000\text{ mM}}=0.0500\text{ min}^{-1}$

The constancy of the first-order rate constant indicates that the reaction indeed is first order in [A]. (As a point of interest, notice that the concentrations chosen in this experiment are such that the reaction is pseudo-zero-order in [B]. Here, then it is not necessary to consider the variation of [B] with time as the reaction proceeds when determining the kinetic dependence on [A].)

83. In Exercise 78 we established that the rate law for the iodine-hypochlorite ion reaction:
Rate = $k[OCl^-][I^-][OH^-]^{-1}$. In the mechanism, the slow step gives the rate law;
Rate = $k_3[I^-][HOCl]$ We use the terms initial fast equilibrium step to substitute for [HOCl] in
this rate equation. We assume in this fast step that the forward rate equals the reverse rate.

$$k_1[OCl^-][H_2O] = k_2[HOCl][OH^-] \quad [HOCl] = \frac{k_1[OCl^-][H_2O]}{k_2[OH^-]}$$

$$\text{Rate} = k_3[I^-]\frac{k_1[OCl^-][H_2O]}{k_2[OH^-]} = \frac{k_3 k_1[H_2O]}{k_2}\frac{[OCl^-][I^-]}{[OH^-]} = k\frac{[OCl^-][I^-]}{[OH^-]}$$

This is the same rate law that we established in Exercise 73. We have incorporated [H₂O] in
the rate constant for the reaction because, in an aqueous solution, [H₂O] remains effectively
constant during the course of the reaction. (The final fast step simply involves the
neutralization of the acid HOI by the base hydroxide ion, OH^-.)

84. It is more likely that the *cis*-isomer, compound (I), would be formed than the *trans*-isomer,
compound (II). The reason for this is that the reaction will involve the adsorption of both
CH_3—C≡C—CH_3 and H_2 onto the surface of the catalyst. These two molecules will
eventually be adjacent to each other. At some point, one of the π bonds in the C≡C bond
will break, the H—H bond will break, and two C—H bonds will form. Since these two
C—H bonds form on the same side of the carbon chain, compound (I) will be produced. In
the sketches below, dotted lines (…) indicate bonds forming or breaking.

$$\begin{array}{ccc}
\text{H—H} & \text{H}\cdots\text{H} & \text{H}\quad\text{H} \\
\text{CH}_3\text{—C≡C—CH}_3 \longrightarrow & \text{CH}_3\text{—}\overset{}{\text{C}}\text{≡}\overset{}{\text{C}}\text{—CH}_3 \longrightarrow & \text{CH}_3\text{—}\overset{|}{\text{C}}\text{=}\overset{|}{\text{C}}\text{—CH}_3
\end{array}$$

85. $Hg_2^{2+} + Tl^{3+} \rightarrow 2\,Hg^{2+} + Tl^+$ Experimental rate Law = $k\dfrac{[Hg_2^{2+}][Tl^{3+}]}{[Hg^{2+}]}$

Possible mechanism:
$$Hg_2^{2+} + Tl^{3+} \underset{k_{-1}}{\overset{k_1}{\rightleftharpoons}} Hg^{2+} + HgTl^{3+} \quad \text{(fast)}$$
$$HgTl^{3+} \overset{k_2}{\longrightarrow} Hg^{2+} + Tl^{3+} \qquad\qquad \text{(slow)}$$
$$Hg_2^{2+} + Tl^{3+} \rightarrow 2\,Hg^{2+} + Tl^+ \qquad \text{Rate} = k_2[HgTl^{3+}]$$

$$k_1[Hg_2^{2+}][Tl^{3+}] = k_{-1}[Hg^{2+}][HgTl^{3+}] \quad \text{—rearrange→} \quad [HgTl^{3+}] = \frac{k_1}{k_{-1}}\frac{[Hg_2^{2+}][Tl^{3+}]}{[Hg^{2+}]}$$

$$\text{Rate} = k_2[HgTl^{3+}] = \frac{k_2 k_1}{k_{-1}}\frac{[Hg_2^{2+}][Tl^{3+}]}{[Hg^{2+}]} = k_{obs}\frac{[Hg_2^{2+}][Tl^{3+}]}{[Hg^{2+}]}$$

86. $\dfrac{\Delta CCl_3}{\Delta t} = rate_{formation} + rate_{disappearance} = 0 \qquad$ so $rate_{formation} = rate_{decomposition}$

$k_2[Cl(g)][CHCl_3] = k_3[CCl_3][[Cl(g)]$ and, simplifying, $[CCl_3] = \dfrac{k_2}{k_3}[CHCl_3]$

since rate $= k_3[CCl_3][Cl(g)] = k_3\left(\dfrac{k_2}{k_3}[CHCl_3]\right)[Cl(g)] = k_2[CHCl_3][Cl(g)]$

We know: $[Cl(g)] = \left(\dfrac{k_1}{k_{-1}}[Cl_2(g)]\right)^{1/2}$ then $rate_{overall} = k_2[CHCl_3]\left(\dfrac{k_1}{k_{-1}}[Cl_2(g)]\right)^{1/2}$

and the rate constant k will be: $k = k_2\left(\dfrac{k_1}{k_{-1}}\right)^{1/2} = (1.3\times10^{-2})\left(\dfrac{4.8\times10^3}{3.6\times10^3}\right)^{1/2} = 0.015$

87. $Rate = k[A]^3 - \dfrac{d[A]}{dt} \qquad\qquad$ Rearrange: $\quad -kt = \dfrac{d[A]}{[A]^3}$

Integrate using the limits $\rightarrow$ time (0 to t) and concentration ($[A]_o$ to $[A]_t$)

$-k\displaystyle\int_0^t t = \int_{[A]_o}^{[A]_t} \dfrac{d[A]}{[A]^3} \quad\Rightarrow\quad -kt - (-k(0)) = -\dfrac{1}{2}\dfrac{1}{[A]_t^2} - \left(-\dfrac{1}{2}\dfrac{1}{[A]_o^2}\right)$

Simplify: $-kt = -\dfrac{1}{2[A]_t^2} + \dfrac{1}{2[A]_o^2} \qquad$ Rearrange: $\dfrac{1}{2[A]_t^2} = kt + \dfrac{1}{2[A]_o^2}$

Multiply through by 2 to give the integrated rate law: $\quad \dfrac{1}{[A]_t^2} = 2kt + \dfrac{1}{[A]_o^2}$

To derive the half life($t_{1/2}$) substitute $t = t_{1/2}$ and $[A]_t = \dfrac{[A]_o}{2}$

$\dfrac{1}{\left(\dfrac{[A]_o}{2}\right)^2} = 2kt_{1/2} + \dfrac{1}{[A]_o^2} = \dfrac{1}{\dfrac{[A]_o^2}{4}} = \dfrac{4}{[A]_o^2} \qquad$ Collect terms

$2kt_{1/2} = \dfrac{4}{[A]_o^2} - \dfrac{1}{[A]_o^2} = \dfrac{3}{[A]_o^2} \qquad$ Solve for $t_{1/2}$ $\qquad t_{1/2} = \dfrac{3}{2k[A]_o^2}$

88. Consider the reaction: $\quad A + B \rightarrow$ products (1st order in A, 1st order in B). The initial concentration of each reactant can be defined as $[A]_o$ and $[B]_o$ Since the stoichiometry is 1:1, we can define x as the concentration of reactant A and reactant B that is removed (a variable that changes with time).

The $[A]_t = ([A]_o - x)$ and $[B]_t = ([B]_o - x)$.

Algebra note: $[A]_o - x = -(x - [A]_o)$ and $[B]_o - x = -(x - [B]_o)$.
As well the calculus requires that we use the absolute value of $|x - [A]_o|$ and $|x - [B]_o|$ when taking the integral of the reciprocal of $|x - [A]_o|$ and $|x - [B]_o|$

$$Rate = -\frac{d[A]}{dt} = -\frac{d[B]}{dt} = \frac{dx}{dt} = k[A]_t[B]_t = k([A]_o - x)([B]_o - x) \quad \text{or} \quad \frac{dx}{([A]_o - x)([B]_o - x)} = kdt$$

In order to solve this, partial fraction decomposition is required to further ease integration:

$$dx \underbrace{\left(\frac{1}{([B]_o - [A]_o)}\right)}_{\text{Note: this is a constant}}\left(\frac{1}{([A]_o - x)} - \frac{1}{([B]_o - x)}\right) = kdt$$

From the point of view of integration, a further rearrangement is desirable (See algebra note above).

$$dx \left(\frac{1}{([B]_o - [A]_o)}\right)\left(\frac{-1}{|x - [A]_o|} - \frac{-1}{|x - [B]_o|}\right) = kdt$$

Integrate both sides $\left(\frac{1}{([B]_o - [A]_o)}\right)\left[-\ln|x - [A]_o| - (-\ln|x - [B]_o|)\right] = kt + C$

$|x - [A]_o| = [A]_o - x$ and $|x - [B]_o| = [B]_o - x$ \qquad Substitute and simplify

$$\left(\frac{1}{([B]_o - [A]_o)}\right)\ln\left(\frac{([B]_o - x)}{[A]_o - x}\right) = kt + C \qquad \text{Determine C by setting } x = 0 \text{ at } t = 0$$

$$C = \left(\frac{1}{([B]_o - [A]_o)}\right)\ln\frac{[B]_o}{[A]_o} \quad \text{Hence:} \left(\frac{1}{([B]_o - [A]_o)}\right)\ln\left(\frac{([B]_o - x)}{([A]_o - x)}\right) = kt + \left(\frac{1}{([B]_o - [A]_o)}\right)\ln\frac{[B]_o}{[A]_o}$$

Multiply both sides by $([B]_o - [A]_o)$, hence, $\ln\left(\frac{([B]_o - x)}{([A]_o - x)}\right) = ([B]_o - [A]_o) \times kt + \ln\frac{[B]_o}{[A]_o}$

$$\ln\left(\frac{([B]_o - x)}{([A]_o - x)}\right) - \ln\frac{[B]_o}{[A]_o} = ([B]_o - [A]_o) \times kt = \ln\frac{\left(\frac{([B]_o - x)}{([A]_o - x)}\right)}{\frac{[B]_o}{[A]_o}} = \ln\left(\frac{[A]_o([B]_o - x)}{[B]_o([A]_o - x)}\right)$$

Set $([A]_o - x) = [A]_t$ and $([B]_o - x) = [B]_t$ to give $\ln\left(\frac{[A]_o \times [B]_t}{[B]_o \times [A]_t}\right) = ([B]_o - [A]_o) \times kt$

89. Let 250-2x equal the partial pressure of CO(g) and x be the partial pressure of CO_2(g).

$$2CO \quad \rightarrow \quad CO_2 + \quad C(s)$$
$$250\text{-}2x \qquad x \qquad - \qquad P_{tot} = P_{CO} + P_{CO2} = 250 - 2x + x = 250 - x$$

P_{tot} [torr]	Time [sec]	P_{CO2}	P_{CO} [torr]
250	0	0	250
238	398	12	226
224	1002	26	198
210	1801	40	170

The plots below (next page), show that the reaction appears to obey a second-order rate law.

Rate = k[CO]^2

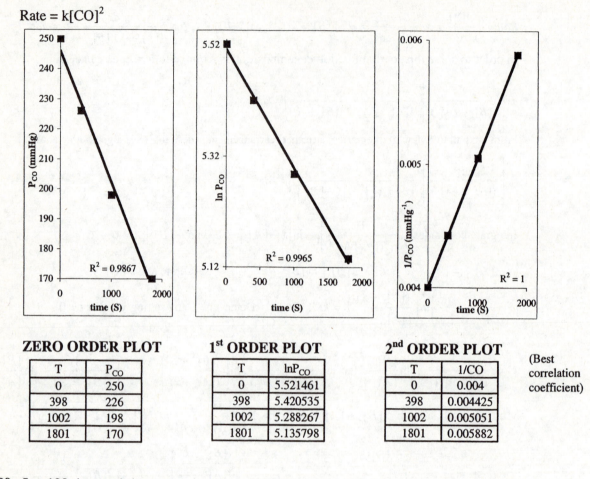

ZERO ORDER PLOT	
T	P$_{CO}$
0	250
398	226
1002	198
1801	170

1st ORDER PLOT	
T	lnP$_{CO}$
0	5.521461
398	5.420535
1002	5.288267
1801	5.135798

2nd ORDER PLOT	
T	1/CO
0	0.004
398	0.004425
1002	0.005051
1801	0.005882

(Best correlation coefficient)

90. Let 100-4x equal the partial pressure of PH$_3$(g), x the partial pressure of P$_4$(g) and 6x be the partial pressure of H$_2$ (g)

$$4\,PH_3(g) \;\rightarrow\; P_4(g) \;+\; 6\,H_2(g)$$
$$100 - 4x \qquad x \qquad\qquad 6x$$

$$P_{tot} = P_{PH_3} + P_{P_4} + P_{H_2} = 100 - 4x + x + 6x = 100 + 3x$$

P$_{tot}$ [torr]	Time [sec]	P$_{P_4}$ [torr]	P$_{PH_3}$ [torr]
100	0	0	100
150	40	50/3	100-(4)(50/3)
167	80	67/3	100-(4)(67/3)
172	120	72/3	100-(4)(72/3)

The plots below (next page), show that the reaction appears to obey a first-order rate law.
Rate = k[PH$_3$]

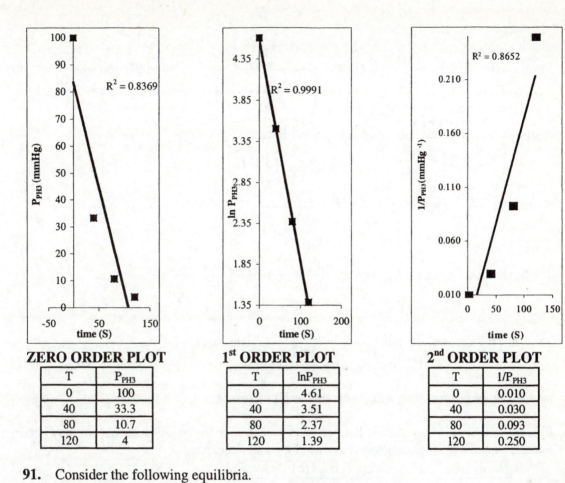

ZERO ORDER PLOT

T	P_{PH3}
0	100
40	33.3
80	10.7
120	4

1st ORDER PLOT

T	$\ln P_{PH3}$
0	4.61
40	3.51
80	2.37
120	1.39

2nd ORDER PLOT

T	$1/P_{PH3}$
0	0.010
40	0.030
80	0.093
120	0.250

91. Consider the following equilibria.

$$E + S \underset{k_{-1}}{\overset{k_1}{\rightleftharpoons}} ES \xrightarrow{k_2} E + P \qquad E + I \overset{K_I}{\rightleftharpoons} EI$$

Product production $\quad \dfrac{d[P]}{dt} = k_2[ES] \qquad$ Use the steady state approximation for [ES]

$$\frac{d[ES]}{dt} = k_1[E][S] - k_{-1}[ES] - k_2[ES] = k_1[E][S] - [ES](k_{-1} + k_2) = 0$$

solve for [ES] $\left(\text{Keep in mind } K_M = \dfrac{k_{-1} + k_2}{k_1} \right) \qquad [ES] = \dfrac{k_1[E][S]}{k_{-1} + k_2} = \dfrac{[E][S]}{K_M}$

Formation of EI: $\quad K_I = \dfrac{[E][I]}{[EI]} \qquad [EI] = \dfrac{[E][I]}{K_I}$

$$[E_o] = [E] + [ES] + [EI] = [E] + \frac{[E][S]}{K_M} + \frac{[E][I]}{K_I}$$

$$[E_o] = [E] \left(1 + \frac{[S]}{K_M} + \frac{[I]}{K_I} \right) \quad \text{Solve for [E]} \qquad [E] = \frac{[E_o]}{\left(1 + \dfrac{[S]}{K_M} + \dfrac{[I]}{K_I} \right)}$$

From above: $[ES] = \dfrac{[E][S]}{K_M}$ $[ES] = \dfrac{[E_o]\left(\dfrac{[S]}{K_M}\right)}{\left(1 + \dfrac{[S]}{K_M} + \dfrac{[I]}{K_I}\right)}$ multiplication by $\dfrac{K_M}{K_M}$ affords

$$[ES] = \dfrac{[E_o][S]}{\left(K_M + [S] + \dfrac{[I]K_M}{K_I}\right)} = \dfrac{[E_o][S]}{K_M\left(1 + \dfrac{[I]}{K_I}\right) + [S]}$$

Remember $\dfrac{d[P]}{dt} = k_2[ES] = \dfrac{k_2[E_o][S]}{K_M\left(1 + \dfrac{[I]}{K_I}\right) + [S]}$

If we substitute $k_2[E_o] = V_{max}$ then $\dfrac{d[P]}{dt} = \dfrac{V_{max}[S]}{K_M\left(1 + \dfrac{[I]}{K_I}\right) + [S]}$

Thus, as [I] increases, the ratio $\dfrac{V_{max}[S]}{K_M\left(1 + \dfrac{[I]}{K_I}\right) + [S]}$ decreases;

i.e. the rate of product formation decreases as [I] increases.

92. In order to determine a value for K_M, we need to rearrange the equation so that we may obtain a linear plot and extract parameters from the slope and intercepts.

$$V = \dfrac{k_2[E_o][S]}{K_M + [S]} \qquad \dfrac{1}{V} = \dfrac{K_M + [S]}{k_2[E_o][S]} = \dfrac{K_M}{k_2[E_o][S]} + \dfrac{[S]}{k_2[E_o][S]}$$

$$\dfrac{1}{V} = \dfrac{K_M}{k_2[E_o][S]} + \dfrac{1}{k_2[E_o]} \qquad \text{We need to have this in the form of } y = mx+b$$

Plot $\dfrac{1}{V}$ on the y-axis and $\dfrac{1}{[S]}$ on the x-axis. See result below:

The plot of 1/V vs 1/[S] should yield a slope of $\dfrac{K_M}{k_2[E_o]}$ and a

y-intercept of $\dfrac{1}{k_2[E_o]}$. The x-intercept = $-1/K_M$

$$0 = \dfrac{K_M}{k_2[E_o]}\left(\dfrac{1}{[S]}\right) - \dfrac{1}{k_2[E_o]} \qquad \dfrac{1}{[S]} = \dfrac{-k_2[E_o]}{k_2[E_o] \times K_M} = \dfrac{-1}{K_M}$$

To find the value of K_M take the negative inverse of x-intercept. To find k_2, invert the y-intercept and divide by the $[E_o]$.

93. The species A^* is a reactive intermediate. Let's deal with this species by using a steady state approximation.

$d[A^*]/dt = 0 = k_1[A]^2 - k_{-1}[A^*][A] - k_2[A^*]$. Solve for $[A^*]$. $k_{-1}[A^*][A] + k_2[A^*] = k_1[A]^2$

$$[A^*] = \frac{k_1[A]^2}{k_{-1}[A] + k_2} \qquad \text{The rate of reaction is: } \text{Rate} = k_2[A^*] = \frac{k_2 k_1[A]^2}{k_{-1}[A] + k_2}$$

At low pressures ($[A] \sim 0$ and hence $k_2 \gg k_{-1}[A]$), the denominator becomes $\sim k_2$ and the rate law is

$$\text{Rate} = \frac{k_2 k_1[A]^2}{k_2} = k_1[A]^2 \quad \text{Second order with respect to } [A]$$

At high pressures ($[A]$ is large and $k_{-1}[A] \gg K_2$), the denominator becomes $\sim k_{-1}[A]$ and the rate law is

$$\text{Rate} = \frac{k_2 k_1[A]^2}{k_{-1}[A]} = \frac{k_2 k_1[A]}{k_{-1}} \quad \text{First order with respect to } [A]$$

94. a) The first elementary step $HBr + O_2 \xrightarrow{k1} HOOBr$ is rate-determining if the reaction obeys Reaction rate = k [HBr][O_2] since the rate of this step is identical to that of the experimental rate law.

b) No, mechanisms cannot be shown to be absolutely correct, only consistent with experimental observations.

c) Yes; the sum of the elementary steps (3 HBr + O_2 → HOBr + Br_2 + H_2O) is not consistent with the overall stoichiometry (since HOBr is not detected as a product) of the reaction and therefore cannot be considered a valid mechanism.

FEATURE PROBLEMS

95. **(a)** To determine the order of the reaction, we need $[C_6H_5N_2Cl]$ at each time. To determine this value, note that 58.3 mL $N_2(g)$ evolved corresponds to total depletion of $C_6H_5N_2Cl$, to $[C_6H_5N_2Cl] = 0.000$ M.

Thus, at any point in time,

$$[C_6H_5N_2Cl] = 0.071\ M - \left(\text{volume } N_2(g) \times \frac{0.071\ M\ C_6H_5N_2Cl}{58.3\ mL\ N_2(g)} \right)$$

Consider 21 min:

$$[C_6H_5N_2Cl] = 0.071\ M - \left(44.3\ mL\ N_2 \times \frac{0.071\ M\ C_6H_5N_2Cl}{58.3\ mL\ N_2(g)} \right) = 0.017\ M$$

The numbers in the following table are determined with this method.

time, min	0	3	6	9	12	15	18	21	24	27	30	∞
V_{N_2}, mL	0	10.8	19.3	26.3	32.4	37.3	41.3	44.3	46.5	48.4	50.4	58.3
$[C_6H_5N_2Cl]$, mM	71	58	47	39	32	26	21	17	14	12	10	0

[The concentration is given in thousandths of a mole per liter (mM).]

(b)

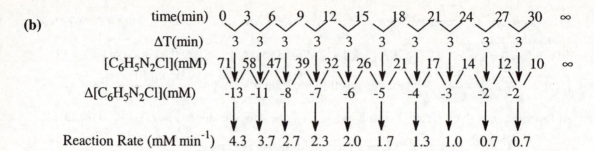

time(min)	0		3		6		9		12		15		18		21		24		27		30	∞
ΔT(min)		3		3		3		3		3		3		3		3		3		3			
$[C_6H_5N_2Cl]$(mM)	71		58		47		39		32		26		21		17		14		12		10	∞
$\Delta[C_6H_5N_2Cl]$(mM)		-13		-11		-8		-7		-6		-5		-4		-3		-2		-2		
Reaction Rate (mM min^{-1})		4.3		3.7		2.7		2.3		2.0		1.7		1.3		1.0		0.7		0.7		

(c) The two graphs are drawn on the same axes.

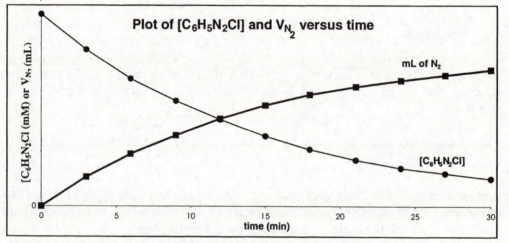

(c) The rate of the reaction at $t = 21$ min is the slope of the tangent line to the $[C_6H_5N_2Cl]$ curve. The tangent line intercepts the vertical axis at about $[C_6H_5N_2Cl] = 39$ mM and the horizontal axis at about 37 min

$$\text{Reaction rate} = \frac{39 \times 10^{-3} \text{ M}}{37 \text{ min}} = 1.0_5 \times 10^{-3} \text{ M min}^{-1} = 1.1 \times 10^{-3} \text{ M min}^{-1}$$

The agreement with the reported value is very good.

(e) The initial rate is the slope of the tangent line to the $[C_6H_5N_2Cl]$ curve at $t = 0$. The intercept with the vertical axis is 71 mM, of course. That with the horizontal axis is about 13 min.

$$\text{Rate} = \frac{71 \times 10^{-3} \text{ M}}{13 \text{ min}} = 5.5 \times 10^{-3} \text{ M min}^{-1}$$

(f) The first-order rate law is $\text{Rate} = k[C_6H_5N_2Cl]$, which we solve for k:

$$k = \frac{\text{Rate}}{[C_6H_5N_2Cl]} \qquad k_0 = \frac{5.5 \times 10^{-3} \text{ M min}^{-1}}{71 \times 10^{-3} \text{ M}} = 0.077 \text{ min}^{-1}$$

$$k_{21} = \frac{1.1 \times 10^{-3} \text{ M min}^{-1}}{17 \times 10^{-3} \text{ M}} = 0.065 \text{ min}^{-1}$$

An average value would be a reasonable estimate: $k_{avg} = 0.071 \text{ min}^{-1}$

(g) The estimated rate constant gives one value of the half-life:

$$t_{1/2} = \frac{0.693}{k} = \frac{0.693}{0.071 \text{ min}^{-1}} = 9.8 \text{ min}$$

The first half-life occurs when $[C_6H_5N_2Cl]$ drops from 0.071 M to 0.0355 M. This occurs at about 10.5 min.

(h) The reaction should be three-fourths complete in two half-lives, or about 20 minutes.

(i) The graph plots $\ln[C_6H_5N_2Cl]$ (in millimoles/L) vs. time in minutes.

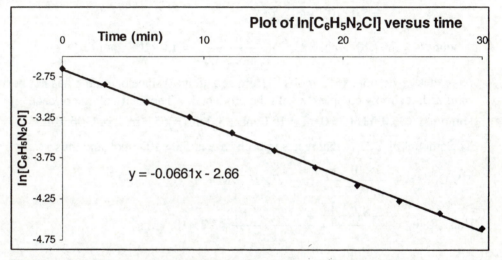

The linearity of the graph demonstrates that the reaction is first-order.

(j) $k = -\text{ slope} = -(-6.61 \times 10^{-2}) \text{ min}^{-1} = 0.0661 \text{ min}^{-1}$

$$t_{1/2} = \frac{0.693}{0.0661 \text{ min}^{-1}} = 10.5 \text{ min}, \text{ in good agreement with our previously determined values.}$$

96. **(a)** In Experiments 1 & 2, [KI] is the same (0.20 M), while $[(NH_4)_2S_2O_8]$ is halved, from 0.20 M to 0.10 M. As a consequence, the time to produce a color change doubles (i.e., the rate is halved). This indicates that reaction (a) is first-order in $S_2O_8^{2-}$. Experiments 2 and 3 produce a similar conclusion. In Experiments 4 and 5, $[(NH_4)_2S_2O_8]$ is the same (0.20 M) while [KI] is halved, from 0.10 to 0.050 M. As a consequence, the time to produce a color change nearly doubles, that is, the rate is halved. This indicates that reaction (a) is also first-order in I^-. Reaction (a) is (1 + 1) second-order overall.

(b) The blue color appears when all the $S_2O_3^{2-}$ has been consumed, for only then does reaction (b) cease. The same amount of $S_2O_3^{2-}$ is placed in each reaction mixture.

$$\text{amount } S_2O_3^{2-} = 10.0 \text{ mL} \times \frac{1 \text{ L}}{1000 \text{ mL}} \times \frac{0.010 \text{ mol Na}_2S_2O_3}{1 \text{ L}} \times \frac{1 \text{ mol } S_2O_3^{2-}}{1 \text{ mol Na}_2S_2O_3} = 1.0 \times 10^{-4} \text{ mol}$$

Through stoichiometry, we determine the amount of each reactant that reacts before this amount of $S_2O_3^{2-}$ will be consumed.

$$\text{amount } S_2O_8^{2-} = 1.0 \times 10^{-4} \text{ mol } S_2O_3^{2-} \times \frac{1 \text{ mol } I_3^-}{2 \text{ mol } S_2O_3^{2-}} \times \frac{1 \text{ mol } S_2O_8^{2-}}{1 \text{ mol } I_3^-}$$

$$= 5.0 \times 10^{-5} \text{ mol } S_2O_8^{2-}$$

$$\text{amount } I^- = 5.0 \times 10^{-5} \text{ mol } S_2O_8^{2-} \times \frac{2 \text{ mol } I^-}{1 \text{ mol } S_2O_8^{2-}} = 1.0 \times 10^{-4} \text{ mol } I^-$$

Note that we do not use "3 mol I^-" from equation (a) since one mole has not been oxidized; it simply complexes with the product I_2. The total volume of each solution is $(25.0 \text{ mL} + 25.0 \text{ mL} + 10.0 \text{ mL} + 5.0 \text{ mL} =)65.0 \text{ mL}$, or 0.0650 L. The amount of $S_2O_8^{2-}$ that reacts in each case is $5.0 \times 10^{-5} \text{ mol}$ and thus

$$\Delta\left[S_2O_8^{2-}\right] = \frac{-5.0 \times 10^{-5} \text{ mol}}{0.0650 \text{ L}} = -7.7 \times 10^{-4} \text{ M}$$

$$\text{Thus, Rate}_1 = \frac{-\Delta\left[S_2O_8^{2-}\right]}{\Delta t} = \frac{+7.7 \times 10^{-4} \text{ M}}{21 \text{ s}} = 3.7 \times 10^{-5} \text{ M s}^{-1}$$

(c) For Experiment 2, $\text{Rate}_2 = \dfrac{-\Delta\left[S_2O_8^{2-}\right]}{\Delta t} = \dfrac{+7.7 \times 10^{-4} \text{ M}}{42 \text{ s}} = 1.8 \times 10^{-5} \text{ M s}^{-1}$

To determine the value of k, we need initial concentrations, as altered by dilution.

$$\left[S_2O_8^{2-}\right]_1 = 0.20 \text{ M} \times \frac{25.0 \text{ mL}}{65.0 \text{ mL total}} = 0.077 \text{ M} \qquad \left[I^-\right]_1 = 0.20 \text{ M} \times \frac{25.0 \text{ mL}}{65.0 \text{ mL}} = 0.077 \text{ M}$$

$$\text{Rate}_1 = 3.7 \times 10^{-5} \text{ M s}^{-1} = k\left[S_2O_8^{2-}\right]^1\left[I^-\right]^1 = k(0.077 \text{ M})^1(0.077 \text{ M})^1$$

$$k = \frac{3.7 \times 10^{-5} \text{ M s}^{-1}}{0.077 \text{ M} \times 0.077 \text{ M}} = 6.2 \times 10^{-3} \text{ M}^{-1} \text{ s}^{-1}$$

$$\left[S_2O_8^{2-}\right]_2 = 0.10 \text{ M} \times \frac{25.0 \text{ mL}}{65.0 \text{ mL total}} = 0.038 \text{ M} \qquad \left[I^-\right]_2 = 0.20 \text{ M} \times \frac{25.0 \text{ mL}}{65.0 \text{ mL}} = 0.077 \text{ M}$$

$$\text{Rate}_2 = 1.8 \times 10^{-5} \text{ M s}^{-1} = k\left[S_2O_8^{2-}\right]^1\left[I^-\right]^1 = k(0.038 \text{ M})^1(0.077 \text{ M})^1$$

$$k = \frac{1.8 \times 10^{-5} \text{ M s}^{-1}}{0.038 \text{ M} \times 0.077 \text{ M}} = 6.2 \times 10^{-3} \text{ M}^{-1} \text{ s}^{-1}$$

(d) First we determine concentrations for Experiment 4.

$$\left[S_2O_8^{2-}\right]_4 = 0.20\ M \times \frac{25.0\ mL}{65.0\ mL\ total} = 0.077\ M \qquad \left[I^-\right]_4 = 0.10 \times \frac{25.0\ mL}{65.0\ mL} = 0.038\ M$$

We have two expressions for Rate; let us equate them and solve for the rate constant.

$$\text{Rate}_4 = \frac{-\Delta\left[S_2O_8^{2-}\right]}{\Delta t} = \frac{+7.7\times10^{-4}\ M}{\Delta t} = k\left[S_2O_8^{2-}\right]_4^1\left[I^-\right]_4^1 = k\,(0.077\ M)(0.038\ M)$$

$$k = \frac{7.7\times10^{-4}\ M}{\Delta t \times 0.077\ M \times 0.038\ M} = \frac{0.26\ M^{-1}}{\Delta t} \qquad k_3 = \frac{0.26\ M^{-1}}{189\ s} = 0.0014\ M^{-1}\ s^{-1}$$

$$k_{13} = \frac{0.26\ M^{-1}}{88\ s} = 0.0030\ M^{-1}\ s^{-1} \qquad\qquad k_{24} = \frac{0.26\ M^{-1}}{42\ s} = 0.0062\ M^{-1}\ s^{-1}$$

$$k_{33} = \frac{0.26\ M^{-1}}{21\ s} = 0.012\ M^{-1}\ s^{-1}$$

(e) We plot $\ln k$ vs. $1/T$ The slope of the line $= -E_a/R$.

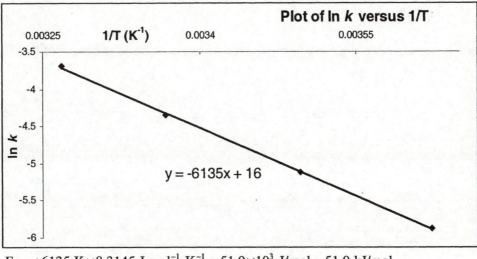

Plot of ln k versus 1/T

$E_a = +6135\ K \times 8.3145\ J\ mol^{-1}\ K^{-1} = 51.0\times10^3\ J/mol = 51.0\ kJ/mol$

The scatter of the data permits only a two significant figure result: 51 kJ/mol

(f) For the mechanism to agree with the reaction stoichiometry, the steps of the mechanism must sum to the overall reaction, in the manner of Hess's law.

(slow) $I^- + S_2O_8^{2-} \rightarrow IS_2O_8^{3-}$

(fast) $IS_2O_8^{3-} \rightarrow 2\ SO_4^{2-} + I^+$

(fast) $I^+ + I^- \rightarrow I_2$

(fast) $I_2 + I^- \rightarrow I_3^-$

(net) $3\ I^- + S_2O_8^{2-} \rightarrow 2\ SO_4^{2-} + I_3^-$

Each of the intermediates cancels: $IS_2O_8^{3-}$ is produced in the first step and consumed in the second, I^+ is produced in the second step and consumed in the third, I_2 is produced in the third step and consumed in the 4th. The mechanism is consistent with the stoichiometry. The rate of the slow step of the mechanism is

$$\text{Rate}_1 = k_1\left[S_2O_8^{2-}\right]^1\left[I^-\right]^1$$

This is exactly the same as the experimental rate law. It is reasonable that the first step be slow since it involves two negatively charged species coming together. We know that like charges repel, and thus this should not be an easy or rapid process.

CHAPTER 15
PRINCIPLES OF CHEMICAL EQUILIBRIUM
PRACTICE EXAMPLES

1A For the cited reaction $CO(g) + 2 H_2(g) \rightleftharpoons CH_3OH(g)$ $\qquad K = \dfrac{[CH_3OH]}{[CO][H_2]^2} = 14.5$

Given that $[CO] = [CH_3OH]$. Thus, one may be substituted for the other, and the resulting

equation solved for $[H_2]$. $\quad K = \dfrac{[CO]}{[CO][H_2]^2} = 14.5 = \dfrac{1}{[H_2]^2} \qquad [H_2] = \sqrt{\dfrac{1}{14.5}} = 0.263$ M

1B Substitute the values we have into the equilibrium constant expression, and solve for $[H_2]$.

$$K_c = \dfrac{[NH_3]^2}{[N_2][H_2]^3} = 1.8 \times 10^4 = \dfrac{(2.00)^2}{0.015 \times [H_2]^3} \qquad [H_2] = \sqrt[3]{[H_2]^3} = \sqrt[3]{\dfrac{(2.00)^2}{1.8 \times 10^4 \times 0.015}} = 0.25 \text{ M}$$

2A The Example gives $K_c = 3.6 \times 10^8$ for the reaction $N_2(g) + 3 H_2(g) \rightleftharpoons 2 NH_3(g)$. The reaction we are considering is one-third of this reaction. If we divide the reaction by 3, we should take the cube root of the equilibrium constant to obtain the value of the equilibrium constant for the "divided" reaction. $\qquad K_{c3} = \sqrt[3]{K_c} = \sqrt[3]{3.6 \times 10^8} = 7.1 \times 10^2$

2B First we reverse the given reaction to put $NO_2(g)$ on the reactant side. The new equilibrium constant is the inverse of the given one.

$NO_2(g) \rightleftharpoons NO(g) + \tfrac{1}{2}O_2(g) \qquad K_c' = 1/(7.5 \times 10^2) = 0.0013$

Then we double the reaction to obtain 2 moles of $NO_2(g)$ as reactant. The equilibrium constant is then raised to the second power.

$2 NO_2(g) \rightleftharpoons 2 NO(g) + O_2(g) \qquad\qquad K_c = (0.0013)^2 = 1.7 \times 10^{-6}$

3A We use the expression $K_p = K_c (RT)^{\Delta n_g}$. In this case, $\Delta n_{gas} = 3 + 1 - 2 = 2$ and thus we have

$K_p = K_c (RT)^2 = 2.8 \times 10^{-9} \times (0.08206 \times 298)^2 = 1.7 \times 10^{-6}$

3B We begin by writing the K_p expression. We then substitute $P = (n/V)RT = [concentration]RT$ for each pressure. We collect terms to obtain an expression relating K_c and K_p, into which we substitute to find the value of K_c.

$$K_p = \frac{\{P(H_2)\}^2\{P(S_2)\}}{\{P(H_2S)\}^2} = \frac{([H_2]RT)^2([S_2]RT)}{([H_2S]RT)^2} = \frac{[H_2]^2[S_2]}{[H_2S]^2}RT = K_c RT$$

Same result can be obtained by using $K_p = K_c(RT)^{\Delta n_{gas}}$, since $\Delta n_{gas} = 2 + 1 - 2 = +1$.

$$K_c = \frac{K_p}{RT} = \frac{1.2\times10^{-2}}{0.08206\times(1065+273)} = 1.1\times10^{-4}$$

But the reaction has been reversed and halved. Thus $K_{final} = \sqrt{\dfrac{1}{K_c}} = \sqrt{\dfrac{1}{1.1\times10^{-4}}} = \sqrt{9090} = 95$

4A   We remember that neither solids, such as $Ca_5(PO_4)_3OH(s)$, nor liquids, such as $H_2O(l)$, appear in the equilibrium constant expression. Concentrations of products appear in the numerator, those of reactants in the denominator. $K_c = \dfrac{\left[Ca^{2+}\right]^5\left[HPO_4^{2-}\right]^3}{\left[H^+\right]^4}$

4B   First we write the balanced chemical equation for the reaction. Then we write the equilibrium constant expressions, remembering that gases and solutes in aqueous solution appear in the K_c expression, but pure liquids and pure solids do not.

$$3\,Fe(s) + 4\,H_2O(g) \rightleftharpoons Fe_3O_4(s) + 4\,H_2(g)$$

$$K_p = \frac{\{P(H_2)\}^4}{\{P(H_2O)\}^4} \qquad\qquad K_c = \frac{[H_2]^4}{[H_2O]^4} \qquad\qquad \text{Because } \Delta n_{gas} = 4 - 4 = 0, K_p = K_c$$

5A   We compute the value of Q_c. Each concentration equals the mass (m) of the substance divided by its molar mass (this quotient is the amount of the substance in moles) and further divided by the volume of the container.

$$Q_c = \frac{[CO_2][H_2]}{[CO][H_2O]} = \frac{\dfrac{m\times\dfrac{1\,mol\,CO_2}{44.0\,g\,CO_2}}{V}\times\dfrac{m\times\dfrac{1\,mol\,H_2}{2.0\,g\,H_2}}{V}}{\dfrac{m\times\dfrac{1\,mol\,CO}{28.0\,g\,CO}}{V}\times\dfrac{m\times\dfrac{1\,mol\,H_2O}{18.0\,g\,H_2O}}{V}} = \frac{\dfrac{1}{44.0\times2.0}}{\dfrac{1}{28.0\times18.0}} = \frac{28.0\times18.0}{44.0\times2.0} = 5.7 > 1.00 = K_c$$

(In evaluating the expression above, we cancelled the equal values of V, and we also cancelled the equal values of m.) Because the value of Q_c is larger than the value of K_c, the reaction will proceed to the left to reach a state of equilibrium. Thus, at equilibrium there will be greater quantities of reactants, and smaller quantities of products than there were initially.

5B   We compare the value of the reaction quotient, Q_p, to that of K_p.

$$Q_p = \frac{\{P(PCl_3)\}\{P(Cl_2)\}}{\{P(PCl_5)\}} = \frac{2.19 \times 0.88}{19.7} = 0.098$$

$$K_p = K_c (RT)^{2-1} = K_c (RT)^1 = 0.0454 \times (0.08206 \times (261 + 273))^1 = 1.99$$

Because $Q_c < K_c$, the net reaction will proceed to the right, forming products and consuming reactants.

6A $O_2(g)$ is a reactant. The equilibrium system will shift right forming product in an attempt to consume some of the added $O_2(g)$ reactant. Looked at in another way, $[O_2]$ is increased above its equilibrium value by the addition of oxygen. This makes Q_c smaller than K_c. (The $[O_2]$ is in the denominator of the expression.) And the system shifts right to drive Q_c back up to K_c at which point equilibrium will have been achieved.

6B   **(a)**   The position of an equilibrium mixture is affected only by changing the concentration of substances that appear in the equilibrium constant expression, $K_c = [CO_2]$. Since CaO(s) is a pure solid, its concentration does not appear in the equilibrium constant expression and thus adding extra CaO(s) will have no direct effect on the position of equilibrium.

   (b)   The addition of $CO_2(g)$ will increase $[CO_2]$ above its equilibrium value. The reaction will shift left to alleviate this increase causing some $CaCO_3(s)$ to form.

   (c)   Since CaCO₃(s) is a pure solid like CaO(s), its concentration does not appear in the equilibrium constant expression and thus the addition of any solid CaCO₃ to an equilibrium mixture will not have an effect upon the position of equilibrium

7A   We know that a decrease in volume or an increase in pressure of an equilibrium mixture of gases causes a net reaction in the direction producing the smaller number of moles of gas. In the reaction in question, that direction is to the left: one mole of $N_2O_4(g)$ is formed when two moles of $NO_2(g)$ combine. Thus, decreasing the cylinder volume would have the initial effect of doubling both $[N_2O_4]$ and $[NO_2]$. In order to reestablish equilibrium, some NO₂ will then be converted into N₂O₄. Note, however, that the NO₂ concentration will still ultimately end up being higher than it was prior to pressurization.

7B   In the balanced chemical equation for the chemical reaction $\Delta n_{gas} = (1+1) - (1+1) = 0$.

   As a consequence, a change in overall volume or total gas pressure will have no effect on the position of equilibrium. In the equilibrium constant expression, the two partial pressures in the numerator will be affected to exactly the same degree, as will the two partial pressures in the denominator, and, as a result, Q_p will continue to equal K_p.

8A The cited reaction is endothermic. Raising the temperature on an equilibrium mixture favors the endothermic reaction. Thus, $N_2O_4(g)$ should decompose more completely at higher temperatures and the amount of $NO_2(g)$ formed from a given amount of $N_2O_4(g)$ will be greater at high temperatures than at low ones.

8B The $NH_3(g)$ formation reaction is $\frac{1}{2}N_2(g) + \frac{3}{2}H_2(g) \rightarrow NH_3(g)$ $\Delta H° = -46.11$ kJ/mol This reaction is an exothermic reaction. Lowering temperature causes a shift in the direction of this exothermic reaction to the right towards products. Thus, the equilibrium $\left[NH_3(g) \right]$ will be greater at $100°C$.

9A We write the expression for K_c and then substitute expressions for molar concentrations.

$$K_c = \frac{[H_2]^2[S_2]}{[H_2S]^2} = \frac{\left(\dfrac{0.22}{3.00}\right)^2 \dfrac{0.11}{3.00}}{\left(\dfrac{2.78}{3.00}\right)^2} = 2.3 \times 10^{-4}$$

9B We write the equilibrium constant expression and solve for $[N_2O_4]$.

$$K_c = 4.61 \times 10^{-3} = \frac{[NO_2]^2}{[N_2O_4]} \qquad [N_2O_4] = \frac{[NO_2]^2}{4.61 \times 10^{-3}} = \frac{(0.0236)^2}{4.61 \times 10^{-3}} = 0.121 \text{ M}$$

Then we determine the mass of N_2O_4 present in 2.26 L.

$$N_2O_4 \text{ mass} = 2.26 \text{ L} \times \frac{0.121 \text{ mol } N_2O_4}{1 \text{ L}} \times \frac{92.01 \text{ g } N_2O_4}{1 \text{ mol } N_2O_4} = 25.2 \text{ g } N_2O_4$$

10A We use the initial-change-equilibrium setup to establish the amount of each substance at equilibrium. We then label each entry in the table in the order of its determination (1st, 2nd, 3rd, 4th, 5th), to better illustrate the technique. We know the initial amounts of all substances (1st). (There are no products at the start). Because ''initial''+ ''change''= ''equilibrium'', the equilibrium amount (2nd) of $Br_2(g)$ enables us to determine "change" (3rd) for $Br_2(g)$. We then use stoichiometry to write other entries (4th) on the "change" line. And finally, we determine the remaining equilibrium amounts (5th).

reaction:	$2 NOBr(g)$	$\rightleftharpoons$	$2 NO(g)$	$+$	$Br_2(g)$
initial:	1.86 mol (1st)		0.00 mol (1st)		0.00 mol (1st)
change:	−0.164 mol (4th)		+0.164 mol (4th)		+0.082 mol (3rd)
equil.:	1.70 mol (5th)		0.164 mol (5th)		0.082 mol (2nd)

$$K_c = \frac{[NO]^2[Br_2]}{[NOBr]^2} = \frac{\left(\dfrac{0.164}{5.00}\right)^2 \left(\dfrac{0.082 \text{ mol}}{5.00}\right)}{\left(\dfrac{1.70}{5.00}\right)^2} = 1.5 \times 10^{-4}$$

Here, $\Delta n_{gas} = 2 + 1 - 2 = +1$. $K_p = K_c (RT)^{+1} = 1.5 \times 10^{-4} \times (0.08206 \times 298) = 3.7 \times 10^{-3}$

10B Use the amounts stated in the problem to determine the equilibrium concentration for each substance.

reaction:	$2\,SO_3\,(g)$	$\rightleftharpoons$	$2\,SO_2\,(g)$	$+$	$O_2\,(g)$
initial:	0 mol		0.100 mol		0.100 mol
changes:	+0.0916 mol		−0.0916 mol		−0.0916/2 mol
equil.:	0.0916 mol		0.0084 mol		0.0542 mol
concentrations:	$\dfrac{0.916\ \text{mol}}{1.52\ \text{L}}$		$\dfrac{0.0084\ \text{mol}}{1.52\ \text{L}}$		$\dfrac{0.0542\ \text{mol}}{1.52\ \text{L}}$
concentrations:	0.0603 M		0.0055 M		0.0357 M

We use these values to compute K_c for the reaction and then the relationship

$K_p = K_c (RT)^{\Delta n_{gas}}$ (with $\Delta n_{gas} = 2+1-2 = +1$) to determine the value of K_p.

$$K_c = \frac{[SO_2]^2\,[O_2]}{[SO_3]^2} = \frac{(0.0055)^2\,(0.0357)}{(0.0603)^2} = 3.0\times10^{-4}$$

$$K_p = 3.0\times10^{-4} \times (0.082060.08206\times900) \approx 0.022$$

11A The equilibrium constant expression is $K_p = P\{H_2O\}P\{CO_2\} = 0.231$ at $100\,°C$. From the balanced chemical equation, we see that one mole of $H_2O(g)$ is formed for each mole of $CO_2(g)$ produced. Consequently, $P\{H_2O\} = P\{CO_2\}$ and $K_p = (P\{CO_2\})^2$. We solve this expression for $P\{CO_2\}$: $P\{CO_2\} = \sqrt[2]{(P\{CO_2\})^2} = \sqrt[2]{K_p} = \sqrt[2]{0.231} = 0.481$ atm

11B The equation for the reaction is $NH_4HS(s) \rightleftharpoons NH_3(g) + H_2S(g)$ $K_p = 0.108$ at $25\,°C$
The two partial pressures do not have to be equal at equilibrium. The only instance in which they must be equal is when the two gases come solely from the decomposition of $NH_4HS(s)$. In this case, some of the $NH_3(g)$ has come from another source. We can obtain the pressure of $H_2S(g)$ by substitution into the equilibrium constant expression, since we are given the equilibrium pressure of $NH_3(g)$.

$$K_p = P\{H_2S\}\,P\{NH_3\} = 0.108 = P\{H_2S\}\times0.500\ \text{atm NH}_3 \quad P\{H_2S\} = \frac{0.108}{0.500} = 0.216\ \text{atm}$$

So, $P_{total} = P_{H_2S} + P_{NH_3} = 0.216\ \text{atm} + 0.500\ \text{atm} = 0.716\ \text{atm}$

12A We set up this problem in the same manner that we have previously employed, namely designating the equilibrium amount of HI as $2x$. (Note that we have used the same multipliers for x as the stoichiometric coefficients.)

Equation:	$H_2\,(g)$	$+$	$I_2\,(g)$	$\rightleftharpoons$	$2\,HI\,(g)$
Initial:	0.150 mol		0.200 mol		0 mol
Changes:	$-x$ mol		$-x$ mol		$+2x$ mol
Equil:	$(0.150-x)$ mol		$(0.200-x)$ mol		$2x$ mol

$$K_c = \frac{\left(\dfrac{2x}{15.0}\right)^2}{\dfrac{0.150-x}{15.0} \times \dfrac{0.200-x}{15.0}} = \frac{(2x)^2}{(0.150-x)(0.200-x)} = 50.2$$

We substitute these terms into the equilibrium constant expression and solve for x

$$4x^2 = (0.150-x)(0.200-x)50.2 = 50.2(0.0300 - 0.350x + x^2) = 1.51 - 17.6x + 50.2x^2$$

$$0 = 46.2x^2 - 17.6x + 1.51 \qquad \text{Now we use the quadratic equation to determine the value of } x.$$

$$x = \frac{-b \pm \sqrt{b^2 - 4ac}}{2a} = \frac{17.6 \pm \sqrt{(17.6)^2 - 4 \times 46.2 \times 1.51}}{2 \times 46.2} = \frac{17.6 \pm 5.54}{92.4} = 0.250 \text{ or } 0.131$$

The first root cannot be used because it would afford a negative amount of H_2 (namely, 0.150-0.250 = -0.100). Thus, we have $2 \times 0.131 = 0.262$ mol HI at equilibrium. We check by substituting the amounts into the K_c expression. (Notice that the volumes cancel.) The slight disagreement in the two values (52 compared to 50.2) is the result of rounding error.

$$K_c = \frac{(0.262)^2}{(0.150-0.131)(0.200-0.131)} = \frac{0.0686}{0.019 \times 0.069} = 52$$

12B **(a)** The equation for the reaction is $N_2O_4(g) \rightleftharpoons 2\,NO_2(g)$ $\qquad K_c = 4.61 \times 10^{-3}$ at 25°C.

In the example, this reaction is conducted in a 0.372 L flask. The effect of moving the mixture to the larger, 10.0 L container is that the reaction will be shifted to produce a greater number of moles of gas. Thus, $NO_2(g)$ will be produced and $N_2O_4(g)$ will dissociate. Consequently, the amount of N_2O_4 will decrease.

(b) The equilibrium constant expression substituting 10.0 L for 0.372 L, follows.

$$K_c = \frac{[NO_2]^2}{[N_2O_4]} = \frac{\left(\dfrac{2x}{10.0}\right)^2}{\dfrac{0.0240-x}{10.0}} = \frac{4x^2}{10.0(0.0240-x)} = 4.61 \times 10^{-3}$$

This can be solved with the quadratic equation, and the sensible result is $x = 0.0118$ moles. We can attempt the method of successive approximations. *First*, assume that $x \ll 0.0240$. We obtain:

$$x = \frac{\sqrt{4.61 \times 10^{-3} \times 10.0\,(0.0240-0)}}{4} = \sqrt{4.61 \times 10^{-3} \times 2.50\,(0.0240-0)} = 0.0166$$

Clearly x is not much smaller than 0.0240. So, *second*, assume $x \approx 0.0166$. We obtain:

$$x = \sqrt{4.61 \times 10^{-3} \times 2.50(0.0240-0.0166)} = 0.00925$$

This assumption is not valid either. So, *third*, assume $x \approx 0.00925$. We obtain:

$$x = \sqrt{4.61 \times 10^{-3} \times 2.50(0.0240-0.0925)} = 0.0130$$

Notice that after each cycle the value we obtain for x gets closer to the value obtained from the roots of the equation. The values from the next several cycles follow.

cycle	4th	5th	6th	7th	8th	9th	10th	11th
x value	0.0112	0.0121	0.0117	0.0119	0.0118_1	0.0118_6	0.0118_3	0.0118_4

The amount of N_2O_4 at equilibrium is 0.0118 mol, less than the 0.0210 mol N_2O_4 at equilibrium in the 0.372 L flask, as predicted.

13A Again we base our solution on the balanced chemical equation.

equation: $Ag^+ (aq) + Fe^{2+} (aq) \rightleftharpoons$ $Fe^{3+} (aq) + Ag(s)$ $\qquad K_c = 2.98$

initial:	0 M	0 M	1.20 M
changes:	$+x$ M	$+x$ M	$-x$ M
equil:	x M	x M	$(1.20 - x)$ M

$$K_c = \frac{[Fe^{3+}]}{[Ag^+][Fe^{2+}]} = 2.98 = \frac{1.20 - x}{x^2} \qquad 2.98 \ x^2 = 1.20 - x \qquad 0 = 2.98x^2 + x - 1.20$$

We use the quadratic formula to obtain a solution.

$$x = \frac{-b \pm \sqrt{b^2 - 4ac}}{2a} = \frac{-1.00 \pm \sqrt{(1.00)^2 + 4 \times 2.98 \times 1.20}}{2 \times 2.98} = \frac{-1.00 \pm 3.91}{5.96} = 0.488 \ M \ or \ -0.824 \ M$$

A negative root makes no physical sense. We obtain the equilibrium concentrations from x.

$$\left[Ag^+ \right] = \left[Fe^{2+} \right] = 0.488 \ M \qquad\qquad \left[Fe^{3+} \right] = 1.20 - 0.488 = 0.71 \ M$$

13B We first calculate the value of Q_c to determine the direction of the reaction.

$$Q_c = \frac{[V^{2+}][Cr^{3+}]}{[V^{3+}][Cr^{2+}]} = \frac{0.150 \times 0.150}{0.0100 \times 0.0100} = 225 < 7.2 \times 10^2 = K_c$$

Because the reaction quotient has a smaller value than the equilibrium constant, a net reaction to the right will occur. We now set up this solution as we have others, heretofore, based on the balanced chemical equation.

	$V^{3+} (aq)$	$+ \ Cr^{2+} (aq)$	$\rightleftharpoons \ V^{2+} (aq)$	$+ \ Cr^{3+} (aq)$
initial	0.0100 M	0.0100 M	0.150 M	0.150 M
changes	$-x$ M	$-x$ M	$+x$ M	$+x$ M
equil	$(0.0100 - x)$M	$(0.0100 - x)$M	$(0.150 + x)$M	$(0.150 + x)$M

$$K_c = \frac{[V^{2+}][Cr^{3+}]}{[V^{3+}][Cr^{2+}]} = \frac{(0.150 + x) \times (0.150 + x)}{(0.0100 - x) \times (0.0100 - x)} = 7.2 \times 10^2 = \left(\frac{0.150 + x}{0.0100 - x} \right)^2$$

If we take the square root of both sides of this expression, we obtain

$$\sqrt{7.2 \times 10^2} = \frac{0.150 + x}{0.0100 - x} = 27$$

$0.150 + x = 0.27 - 27x$ which becomes $28x = 0.12$ and yields 0.0043 M. Then the equilibrium concentrations are: $\left[V^{3+} \right] = \left[Cr^{2+} \right] = 0.0100$ M $- 0.0043$ M $= 0.0057$ M

$$\left[V^{2+} \right] = \left[Cr^{3+} \right] = 0.150 \text{ M} + 0.0043 \text{ M} = 0.154 \text{ M}$$

EXERCISES

Writing Equilibrium Constant Expressions

1. (a) $2\, COF_2(g) \rightleftharpoons CO_2(g) + CF_4(g)$ $\qquad K_c = \dfrac{[CO_2][CF_4]}{[COF_2]^2}$

(b) $Cu(s) + 2\, Ag^+(aq) \rightleftharpoons Cu^{2+}(aq) + 2\, Ag(s)$ $\qquad K_c = \dfrac{\left[Cu^{2+} \right]}{\left[Ag^+ \right]^2}$

(c) $S_2O_8^{2-}(aq) + 2\, Fe^{2+}(aq) \rightleftharpoons 2\, SO_4^{2-}(aq) + 2\, Fe^{3+}(aq)$ $\qquad K_c = \dfrac{\left[SO_4^{2-} \right]^2 \left[Fe^{3+} \right]^2}{\left[S_2O_8^{2-} \right]\left[Fe^{2+} \right]^2}$

2. (a) $4\, NH_3(g) + 3\, O_2(g) \rightleftharpoons 2\, N_2(g) + 6\, H_2O(g)$ $\qquad K_c = \dfrac{[N_2]^2 [H_2O]^6}{[NH_3]^4 [O_2]^3}$

(b) $7\, H_2(g) + 2\, NO_2(g) \rightleftharpoons 2\, NH_3(g) + 4\, H_2O(g)$ $\qquad K_c = \dfrac{[NH_3]^2 [H_2O]^4}{[H_2]^7 [NO_2]^2}$

(c) $N_2(g) + Na_2CO_3(s) + 4\, C(s) \rightleftharpoons 2\, NaCN(s) + 3\, CO(g)$ $\qquad K_c = \dfrac{[CO]^3}{[N_2]}$

3. (a) $K_c = \dfrac{[NO_2]^2}{[NO]^2 [O_2]}$ $\qquad$ (b) $K_c = \dfrac{\left[Zn^{2+} \right]}{\left[Ag^+ \right]^2}$ $\qquad$ (c) $K_c = \dfrac{\left[OH^- \right]^2}{\left[CO_3^{2-} \right]}$

4. (a) $K_p = \dfrac{P\{CH_4\}P\{H_2S\}^2}{P\{CS_2\}P\{H_2\}^4}$ (b) $K_p = P\{ O_2 \}^{1/2}$ (c) $K_p = P\{CO_2\}P\{H_2O\}$

5. In each case we write the equation for the formation reaction and then the equilibrium constant expression, K_c, for that reaction.

(a) $\tfrac{1}{2} H_2(g) + \tfrac{1}{2} F_2(g) \rightleftharpoons HF(g)$ $\qquad K_c = \dfrac{[HF]}{[H_2]^{1/2} [F_2]^{1/2}}$

(b) $N_2(g) + 3\,H_2(g) \rightleftharpoons 2\,NH_3(g)$ $K_c = \dfrac{[NH_3]^2}{[N_2][H_2]^3}$

(c) $2\,N_2(g) + O_2(g) \rightleftharpoons 2\,N_2O(g)$ $K_c = \dfrac{[N_2O]^2}{[N_2]^2[O_2]}$

(d) $\tfrac{1}{2}Cl_2(g) + \tfrac{3}{2}F_2(g) \rightleftharpoons ClF_3(l)$ $K_c = \dfrac{1}{[Cl_2]^{1/2}[F_2]^{3/2}}$

6. In each case we write the equation for the formation reaction and then the equilibrium constant expression, K_p, for that reaction.

(a) $\tfrac{1}{2}N_2(g) + \tfrac{1}{2}O_2(g) + \tfrac{1}{2}Cl_2(g) \rightleftharpoons NOCl(g)$ $K_p = \dfrac{[P_{NOCl}]}{\left[P_{N_2}\right]^{1/2}\left[P_{O_2}\right]^{1/2}\left[P_{Cl_2}\right]^{1/2}}$

(b) $N_2(g) + 2\,O_2(g) + Cl_2(g) \rightleftharpoons 2\,ClNO_2(g)$ $K_p = \dfrac{\left[P_{ClNO_2}\right]^2}{\left[P_{N_2}\right]\left[P_{O_2}\right]^2\left[P_{Cl_2}\right]}$

(c) $N_2(g) + 2\,H_2(g) \rightleftharpoons N_2H_4(g)$ $K_p = \dfrac{P_{N_2H_4}}{\left[P_{N_2}\right]\left[P_{H_2}\right]^2}$

(d) $\tfrac{1}{2}N_2(g) + 2\,H_2(g) + \tfrac{1}{2}Cl_2(g) \rightleftharpoons NH_4Cl(s)$ $K_p = \dfrac{1}{\left[P_{N_2}\right]^{1/2}\left[P_{H_2}\right]^2\left[P_{Cl_2}\right]^{1/2}}$

7. Since $K_p = K_c(RT)^{\Delta n_g}$, it is also true that $K_c = K_p(RT)^{-\Delta n_g}$.

(a) $K_c = \dfrac{[SO_2][Cl_2]}{[SO_2Cl_2]} = K_p(RT)^{-(+1)} = 2.9 \times 10^{-2}(0.08206 \times 303)^{-1} = 0.0012$

(b) $K_c = \dfrac{[NO_2]^2}{[NO]^2[O_2]} = K_p(RT)^{-(-1)} = 1.48 \times 10^{-4} \times (0.08206 \times 303) = 5.55 \times 10^{5}$

(c) $K_c = \dfrac{[H_2S]^3}{[H_2]^3} = K_p(RT)^0 = K_p = 0.429$

8. $K_p = K_c(RT)^{\Delta n_g}$, with $R = 0.08206\ \text{L} \cdot \text{atm mol}^{-1}\ \text{K}^{-1}$

(a) $K_p = \dfrac{P\{NO_2\}^2}{P\{N_2O_4\}} = K_c(RT)^{+1} = 4.61 \times 10^{-3}(0.08206 \times 298)^{1} = 0.113$

(b) $K_p = \dfrac{P\{C_2H_2\}P\{H_2\}^3}{P\{CH_4\}^2} = K_c(RT)^{(+2)} = (0.154)(0.08206 \times 2000)^2 = 4.15 \times 10^{3}$

(c) $K_p = \dfrac{P\{H_2\}^4 P\{CS_2\}}{P\{H_2S\}^2 P\{CH_4\}} = K_c(RT)^{(+2)} = (5.27 \times 10^{-8})(0.08206 \times 973)^2 = 3.36 \times 10^{-4}$

9. The equilibrium reaction is $H_2O(l) \rightleftharpoons H_2O(g)$ with $\Delta n_{gas} = +1$. $K_p = K_c(RT)^{\Delta n_g}$ gives

$$K_c = K_p(RT)^{-\Delta n_g}. \qquad K_p = P\{H_2O\} = 23.8 \text{ mmHg} \times \frac{1 \text{ atm}}{760 \text{ mmHg}} = 0.0313$$

$$K_c = K_p(RT)^{-1} = \frac{K_p}{RT} = \frac{0.0313}{0.08206 \times 298} = 1.28 \times 10^{-3}$$

10. The equilibrium rxn is $C_6H_6(l) \rightleftharpoons C_6H_6(g)$ with $\Delta n_{gas} = +1$. Using $K_p = K_c(RT)^{\Delta n_g}$,

$$K_p = K_c(RT) = 5.12 \times 10^{-3}(0.08206 \times 298) = 0.125 = P\{C_6H_6\}$$

$$P\{C_6H_6\} = 0.125 \text{ atm} \times \frac{760 \text{ mmHg}}{1 \text{ atm}} = 95.0 \text{ mmHg}$$

11. Add one-half of the reversed 1st reaction with the 2nd reaction to obtain the desired reaction.

$\frac{1}{2}N_2(g) + \frac{1}{2}O_2(g) \rightleftharpoons NO(g)$ $\qquad\qquad K_c = \dfrac{1}{\sqrt{2.1 \times 10^{30}}}$

$NO(g) + \frac{1}{2}Br_2(g) \rightleftharpoons NOBr(g)$ $\qquad\qquad K_c = 1.4$

net : $\frac{1}{2}N_2(g) + \frac{1}{2}O_2(g) + \frac{1}{2}Br_2(g) \rightleftharpoons NOBr(g)$ $\qquad K_c = \dfrac{1.4}{\sqrt{2.1 \times 10^{30}}} = 9.7 \times 10^{-16}$

12. We combine the several given reactions to obtain the net reaction.

$2 N_2O(g) \rightleftharpoons 2 N_2(g) + O_2(g) \qquad K_c = \dfrac{1}{\left(2.7 \times 10^{-18}\right)^2}$

$4 NO_2(g) \rightleftharpoons 2 N_2O_4(g) \qquad K_c = \dfrac{1}{\left(4.6 \times 10^{-3}\right)^2}$

$2 N_2(g) + 4 O_2(g) \rightleftharpoons 4 NO_2(g) \qquad K_c = (4.1 \times 10^{-9})^4$

net : $2 N_2O(g) + 3 O_2(g) \rightleftharpoons 2 N_2O_4(g) \quad K_{c(Net)} = \dfrac{\left(4.1 \times 10^{-9}\right)^4}{\left(2.7 \times 10^{-18}\right)^2 \left(4.6 \times 10^{-3}\right)^2} = 1.8 \times 10^6$

13. We combine the K_c values to obtain the value of K_c for the overall reaction, and then convert this to a value for K_p.

$2 CO_2(g) + 2H_2(g) \rightleftharpoons 2 CO(g) + 2 H_2O(g) \qquad K_c = (1.4)^2$

$2 C(\text{graphite}) + O_2(g) \rightleftharpoons 2 CO(g) \qquad K_c = (1 \times 10^8)^2$

$4 CO(g) \rightleftharpoons 2 C(\text{graphite}) + 2 CO_2(g) \qquad K_c = \dfrac{1}{(0.64)^2}$

---

net: $2H_2(g) + O_2(g) \rightleftharpoons 2H_2O(g) \qquad K_{c(Net)} = \dfrac{(1.4)^2(1 \times 10^8)^2}{(0.64)^2} = 5 \times 10^6$

$$K_p = K_c(RT)^{\Delta n} = \frac{K_c}{RT} = \frac{5 \times 10^{16}}{0.08206 \times 1200} = 5 \times 10^{14}$$

14. We combine the K_p values to obtain the value of K_p for the overall reaction, and then convert this to a value for K_c.

$$2\,NO_2Cl(g) \rightleftharpoons 2\,NOCl(g) + O_2(g) \qquad K_p = \left(\frac{1}{1.1 \times 10^2}\right)^2$$

$$2\,NO_2(g) + Cl_2(g) \rightleftharpoons 2\,NO_2Cl(g) \qquad K_p = (0.3)^2$$

$$N_2(g) + 2\,O_2(g) \rightleftharpoons 2\,NO_2(g) \qquad K_p = (1.0 \times 10^{-9})^2$$

$$\text{net: } N_2(g) + O_2(g) + Cl_2(g) \rightleftharpoons 2\,NOCl(g) \qquad K_{p(Net)} = \frac{(0.3)^2(1.0 \times 10^{-9})^2}{(1.1 \times 10^2)^2} = 7.\underline{4} \times 10^{-24}$$

$$K_p = K_c(RT)^{\Delta n} \qquad K_c = \frac{K_p}{(RT)^{\Delta n}} = \frac{7.\underline{4} \times 10^{-24}}{(0.08206 \times 298)^{2-3}} = 2 \times 10^{-22}$$

Experimental Determination of Equilibrium Constants

15. First, we determine the concentration of PCl_5 and of Cl_2 present initially and at equilibrium, respectively. Then we use the balanced equation to help us determine the concentration of each species present at equilibrium.

$$[PCl_5]_{initial} = \frac{1.00 \times 10^{-3}\ mol\ PCl_5}{0.250\ L} = 0.00400\ M \qquad [Cl_2]_{equil} = \frac{9.65 \times 10^{-4}\ mol\ Cl_2}{0.250\ L} = 0.00386\ M$$

Equation:	$PCl_5(g)$	$\rightleftharpoons$	$PCl_3(g)$	+	$Cl_2(g)$
Initial:	0.00400M		0 M		0 M
Changes:	$-x$M		$+x$M		$+x$M
Equil:	0.00400M-xM		xM		xM ← 0.00386 M (from above)

At equilibrium, $[Cl_2] = [PCl_3] = 0.00386$ M and $[PCl_5] = 0.00400M - xM = 0.00014$ M

$$K_c = \frac{[PCl_3][Cl_2]}{[PCl_5]} = \frac{(0.00386\,M)(0.00386\,M)}{0.00014\,M} = 0.10\underline{6}$$

16. First we determine the partial pressure of each gas.

$$P_{initial}\{H_2(g)\} = \frac{nRT}{V} = \frac{1.00\,g\,H_2 \times \dfrac{1\,mol\,H_2}{2.016\,g\,H_2} \times \dfrac{0.08206\,L\,atm}{mol\,K} \times 1670\,K}{0.500\,L} = 136\,atm$$

$$P_{initial}\{H_2S(g)\} = \frac{nRT}{V} = \frac{1.06\,g\,H_2S \times \dfrac{1\,mol\,H_2S}{34.08\,g\,H_2S} \times \dfrac{0.08206\,L\,atm}{mol\,K} \times 1670\,K}{0.500\,L} = 8.52\,atm$$

$$P_{equil}\{S_2(g)\} = \frac{nRT}{V} = \frac{8.00\times10^{-6}\text{ mol }S_2 \times \dfrac{0.08206\,\text{L atm}}{\text{mol K}} \times 1670\,K}{0.500\,L} = 2.19\times10^{-3}\text{ atm}$$

Equation: $2\,H_2(g)$ + $S_2(g)$ $\rightleftharpoons$ $2\,H_2S(g)$

Initial: 136 atm   0 atm   8.52 atm

Changes: +0.00438 atm 0.00219 atm −0.00438 atm

Equil: 136 atm   0.00219 atm 8.52 atm

$$K_p = \frac{P\{H_2S(g)\}^2}{P\{H_2(g)\}^2\,P\{S_2(g)\}} = \frac{(8.52)^2}{(136)^2\,0.00219} = 1.79$$

17.   (a) $$K_c = \frac{[PCl_5]}{[PCl_3][Cl_2]} = \frac{\dfrac{0.105\text{ g }PCl_5}{2.50\text{ L}} \times \dfrac{1\text{ mol }PCl_5}{208.2\text{ g}}}{\left(\dfrac{0.220\text{ g }PCl_3}{2.50\text{ L}} \times \dfrac{1\text{ mol }PCl_3}{137.3\,g}\right)\times\left(\dfrac{2.12\text{ g }Cl_2}{2.50\text{ L}} \times \dfrac{1\,mol\,Cl_2}{70.9\text{ g}}\right)} = 26.3$$

   (b) $K_p = K_c(RT)^{\Delta n} = 26.3\,(0.08206\times523)^{-1} = 0.613$

18.

$$[ICl]_{initial} = \frac{0.682\text{ g }ICl \times \dfrac{1\text{ mol }ICl}{162.36\text{ g }ICl}}{0.625\text{ L}} = 6.72\times10^{-3}\text{ M}$$

Reaction: $2\,ICl(g)$ $\rightleftharpoons$ $I_2(g)$ + $Cl_2(g)$

Initial: 6.72×10^{-3} M   0 M   0 M

Change: $-2x$ $+x$ $+x$

Equilibrium 6.72×10^{-3} M $-2x$ x x

$$[I_2]_{equil} = \frac{0.0383\text{ g }I_2 \times \dfrac{1\text{ mol }I_2}{253.808\text{ g }I_2}}{0.625\,L} = 2.41\times10^{-4}\text{ M} = x$$

$$K_c = \frac{x\times x}{(6.72\times10^{-3}-2x)} = \frac{(2.41\times10^{-4})^2}{(6.72\times10^{-3}-2(2.41\times10^{-4}))} = 9.34\times10^{-6}$$

Equilibrium Relationships

19. $$K_c = 281 = \frac{[SO_3]^2}{[SO_2]^2[O_2]} = \frac{[SO_3]^2}{[SO_2]^2}\times\frac{0.185\text{ L}}{0.00247\text{ mol}} \qquad \frac{[SO_2]}{[SO_3]} = \sqrt{\frac{0.185}{0.00247\times281}} = 0.516$$

20. $\quad K_c = 0.011 = \dfrac{[I]^2}{[I_2]} = \dfrac{\left(\dfrac{0.37\,\text{mol I}}{V}\right)^2}{\dfrac{1.00\,\text{mol I}_2}{V}} = \dfrac{1}{V} \times 0.14 \quad V = \dfrac{0.14}{0.011} = 13\,\text{L}$

21. **(a)** A possible equation for the oxidation of $NH_3(\text{g})$ to $NO_2(\text{g})$ follows.

$NH_3(\text{g}) + \frac{7}{4} O_2(\text{g}) \rightleftharpoons NO_2(\text{g}) + \frac{3}{2} H_2O(\text{g})$

(b) We obtain K_p for the reaction in part (a) by appropriately combining the values of K_p given in the problem.

$NH_3(\text{g}) + \frac{5}{4} O_2(\text{g}) \rightleftharpoons NO(\text{g}) + \frac{3}{2} H_2O(\text{g}) \qquad K_p = 2.11\times10^{19}$

$NO(\text{g}) + \frac{1}{2} O_2(\text{g}) \rightleftharpoons NO_2(\text{g}) \qquad\qquad K_p = \dfrac{1}{0.524}$

net: $NH_3(\text{g}) + \frac{7}{4} O_2(\text{g}) \rightleftharpoons NO_2(\text{g}) + \frac{3}{2} H_2O(\text{g}) \quad K_p = \dfrac{2.11\times10^{19}}{0.524} = 4.03\times10^{19}$

22. **(a)** We first determine $[H_2]$ and $[CH_4]$m and then $[C_2H_2]$. $[CH_4] = [H_2] = \dfrac{0.10\,\text{mol}}{1.0\,\text{L}} = 0.10\,\text{M}$

$K_c = \dfrac{[C_2H_2][H_2]^3}{[CH_4]^2} \quad [C_2H_2] = \dfrac{K_c\,[CH_4]^2}{[H_2]^3} = \dfrac{0.154\times0.10^2}{0.10^3} = 1.5\,\text{M}$

In a 1.00 L container, each concentration numerically equals the molar quantities of the substance.

$\chi\{C_2H_2\} = \dfrac{1.5\,\text{mol C}_2\text{H}_2}{1.5\,\text{mol C}_2\text{H}_2 + 0.10\,\text{mol CH}_4 + 0.10\,\text{mol H}_2} = 0.88$

(b) The conversion of $CH_4(\text{g})$ to $C_2H_2(\text{g})$ is favored at low pressures, since the conversion reaction has a larger sum of the stoichiometric coefficients of gaseous products (4) than of reactants (2).

(c) Initially, all concentrations are halved when the mixture is transferred to a flask that is twice as large. To re-establish equilibrium, the system reacts to the right, forming more moles of gas (to compensate for the drop in pressure). We base our solution on the balanced chemical equation, in the manner we have used before.

Equation:	$2\,CH_4(\text{g})$	$\rightleftharpoons$	$C_2H_2(\text{g})$	$+$	$3\,H_2$
Initial:	$\dfrac{0.10\,\text{mol}}{2.00\,\text{L}}$		$\dfrac{1.5\,\text{mol}}{2.00\,\text{L}}$		$\dfrac{0.10\,\text{mol}}{2.00\,\text{L}}$
	$= 0.050\,\text{M}$		$= 0.75\,\text{M}$		$= 0.050\,\text{M}$
Changes:	$-2x\,\text{M}$		$+x\,\text{M}$		$+3x\,\text{M}$
Equil:	$(0.050 - 2x)\,\text{M}$		$(0.0750 + x)\,\text{M}$		$(0.050 + 3x)\,\text{M}$

$K_c = \dfrac{[C_2H_2][H_2]^3}{[CH_4]^2} = \dfrac{(0.050 + 3x)^3(0.750 + x)}{(0.050 - 2x)^2} = 0.154$

We can solve this 4th-order equation by successive approximations.
First guess: $x = 0.010$ M.

$$x = 0.010 \quad Q_c = \frac{\left(0.050 + 3(0.010)\right)^3 (0.750 + 0.010)}{\left(0.050 - 2(0.010)\right)^2} = \frac{(0.080)^3 (0.760)}{(0.030)^2} = 0.433 > 0.154$$

$$x = 0.020 \quad Q_c = \frac{\left(0.050 + 3(0.020)\right)^3 (0.750 + 0.020)}{\left(0.050 - 2(0.020)\right)^2} = \frac{(0.110)^3 (0.770)}{(0.010)^2} = 10.2 > 0.154$$

$$x = 0.005 \quad Q_c = \frac{\left(0.050 + 3(0.005)\right)^3 (0.750 + 0.005)}{\left(0.050 - 2(0.005)\right)^2} = \frac{(0.065)^3 (0.755)}{(0.040)^2} = 0.129 < 0.154$$

$$x = 0.006 \quad Q_c = \frac{\left(0.050 + 3(0.006)\right)^3 (0.750 + 0.006)}{\left(0.050 - 2(0.006)\right)^2} = \frac{(0.068)^3 (0.756)}{(0.038)^2} = 0.165 > 0.154$$

This is the maximum number of significant figures our system permits. We have
$x = 0.006$ M. $[CH_4] = 0.038$ M; $[C_2H_2] = 0.756$ M; $[H_2] = 0.068$ M

Because the container volume is 2.00 L, the molar amounts are double the values of molarities.

$$2.00 \text{ L} \times \frac{0.756 \text{ mol } C_2H_2}{1 \text{ L}} = 1.51 \text{ mol } C_2H_2 \qquad 2.00 \text{ L} \times \frac{0.038 \text{ mol } CH_4}{1 \text{ L}} = 0.076 \text{ mol } CH_4$$

$$2.00 \text{ L} \times \frac{0.068 \text{ mol } H_2}{1 \text{ L}} = 0.14 \text{ mol } H_2$$

Thus, the increase in volume results in the production of some additional C_2H_2.

23. (a)
$$K_c = \frac{[CO][H_2O]}{[CO_2][H_2]} = \frac{\dfrac{n\{CO\}}{V} \times \dfrac{n\{H_2O\}}{V}}{\dfrac{n\{CO_2\}}{V} \times \dfrac{n\{H_2\}}{V}}$$

Since V is present in both the denominator and the numerator, it can be stricken from the expression. This happens here because $\Delta n_g = 0$. Therefore, K_c is independent of V.

(b) Note that $K_p = K_c$ for this reaction, since $\Delta n_{gas} = 0$.

$$K_c = K_p = \frac{0.224 \text{ mol } CO \times 0.224 \text{ mol } H_2O}{0.276 \text{ mol } CO_2 \times 0.276 \text{ mol } H_2} = 0.659$$

24. For the reaction $CO(g) + H_2O(g) \rightleftharpoons CO_2(g) + H_2(g)$ the value of $K_p = 23.2$

The expression for Q_p is $\dfrac{[CO_2][H_2]}{[CO][H_2O]}$. Consider each of the provided situations

(a) $P_{CO} = P_{H_2O} = P_{H_2} = P_{CO_2}$; $\quad Q_p = 1 \quad$ Not an equilibrium position

(b) $\dfrac{P_{H_2}}{P_{H_2O}} = \dfrac{P_{CO_2}}{P_{CO}} = x$; $\quad Q_p = x^2 \quad$ If $x = \sqrt{23.2}$, this is an equilibrium position.

(c) $\left(P_{CO} \times P_{H_2O}\right) = \left(P_{CO_2} \times P_{H_2}\right)$; $Q_p = 1 \quad$ Not an equilibrium position

(d) $\dfrac{P_{H_2}}{P_{CO}} = \dfrac{P_{CO_2}}{P_{H_2O}} = x$; $\quad Q_p = x^2 \quad$ If $x = \sqrt{23.2}$, this is an equilibrium position.

Direction and Extent of Chemical Change

25. We compute the value of Q_c for the given amounts of product and reactants.

$$Q_c = \frac{[SO_3]^2}{[SO_2]^2[O_2]} = \frac{\left(\dfrac{1.8\,mol\ SO_3}{7.2\,L}\right)^2}{\left(\dfrac{3.6\,mol\ SO_2}{7.2\,L}\right)^2 \dfrac{2.2\,mol\ O_2}{7.2\,L}} = 0.82 < K_c = 100$$

The mixture described cannot be maintained indefinitely. In fact, because $Q_c < K_c$, the reaction will proceed to the right, that is, toward products, until equilibrium is established. We do not know how long it will take to reach equilibrium.

26. We compute the value of Q_c for the given amounts of product and reactants.

$$Q_c = \frac{[NO_2]^2}{[N_2O_4]} = \frac{\left(\dfrac{0.0205\,mol\,NO_2}{5.25\,L}\right)^2}{\dfrac{0.750\,mol\,N_2O_4}{5.25\,L}} = 1.07 \times 10^{-4} < K_c = 4.61 \times 10^{-3}$$

The mixture described cannot be maintained indefinitely. In fact, because $Q_c < K_c$, the reaction will proceed to the right, that is, toward products, until equilibrium is established. If E_a is large, however, it may take some time to reach equilibrium.

27. (a) We determine the concentration of each species in the gaseous mixture, use these concentrations to determine the value of the reaction quotient, and compare this value of Q_c with the value of K_c.

$[SO_2] = \dfrac{0.455\ mol\ SO_2}{1.90\ L} = 0.239\ M \qquad [O_2] = \dfrac{0.183\ mol\ O_2}{1.90\ L} = 0.0963\ M$

$[SO_3] = \dfrac{0.568\ mol\ SO_3}{1.90\ L} = 0.299\ M \qquad Q_c = \dfrac{[SO_3]^2}{[SO_2]^2[O_2]} = \dfrac{(0.299)^2}{(0.239)^2\,0.0963} = 16.3$

Since $Q_c = 16.3 \neq 2.8 \times 10^2 = K_c$, this mixture is not at equilibrium.

(b) Since the value of Q_c is smaller than that of K_c, the reaction will proceed to the right, forming product and consuming reactants to reach equilibrium.

28. We compute the value of Q_c. Each concentration equals the mass (m) of the substance divided by its molar mass and further divided by the volume of the container.

$$Q_c = \frac{[CO_2][H_2]}{[CO][H_2O]} = \frac{\dfrac{m \times \dfrac{1 \text{ mol } CO_2}{44.0 \text{ g } CO_2}}{V} \times \dfrac{m \times \dfrac{1 \text{ mol } H_2}{2.0 \text{ g } H_2}}{V}}{\dfrac{m \times \dfrac{1 \text{ mol } CO}{28.0 \text{ g } CO}}{V} \times \dfrac{m \times \dfrac{1 \text{ mol } H_2O}{18.0 \text{ g } H_2O}}{V}} = \frac{\dfrac{1}{44.0 \times 2.0}}{\dfrac{1}{28.0 \times 18.0}} = \frac{28.0 \times 18.0}{44.0 \times 2.0} = 5.7 < 31.4 (\text{value of } K_c)$$

(In evaluating the expression above, we cancelled the equal values of V, along with, the equal values of m.) Because the value of Q_c is smaller than the value of K_c, **(a)** the reaction is not at equilibrium and **(b)** the reaction will proceed to the right (formation of products) to reach a state of equilibrium.

29. The information for the calculation is organized around the chemical equation. Let $x =$ mol H_2 (or I_2) that reacts. Then use stoichiometry to determine the amount of HI formed, in terms of x, and finally solve for x.

Equation:	$H_2(g)$	+	$I_2(g)$	$\rightleftharpoons$	$2 \text{ HI}(g)$
Initial:	0.150 mol		0.150 mol		0.000 mol
Changes:	$-x$ mol		$-x$ mol		$+2x$ mol
Equil:	$0.150 - x$		$0.150 - x$		$2x$

$$K_c = \frac{[HI]^2}{[H_2][I_2]} = \frac{\left(\dfrac{2x}{3.25 \text{ L}}\right)^2}{\dfrac{0.150 - x}{3.25 \text{ L}} \times \dfrac{0.150 - x}{3.25 \text{ L}}}$$

Then take the square root of both sides: $\sqrt{K_c} = \sqrt{50.2} = \dfrac{2x}{0.150 - x} = 7.09$

$$2x = 1.06 - 7.09x \qquad x = \frac{1.06}{9.09} = 0.117 \text{ mol}, \text{ amount HI} = 2x = 2 \times 0.117 \text{ mol} = 0.234 \text{ mol HI}$$

amount $H_2 =$ amount $I_2 = (0.150 - x) \text{ mol} = (0.150 - 0.117) \text{ mol} = 0.033 \text{ mol } H_2 \text{ (or } I_2)$

30. We use the balanced chemical equation as a basis to organize the information

Equation:	$SbCl_5(g)$	$\rightleftharpoons$	$SbCl_3(g)$	+	$Cl_2(g)$
Initial:	$\dfrac{0.00 \text{ mol}}{2.50 \text{ L}}$		$\dfrac{0.280 \text{ mol}}{2.50 \text{ L}}$		$\dfrac{0.160 \text{ mol}}{2.50 \text{ L}}$
Initial:	0.000 M		0.112 M		0.0640 M
Changes:	$+x$ M		$-x$ M		$-x$ M
Equil:	x M		$(0.112 - x)$ M		$(0.0640 - x)$ M

$$K_c = 0.025 = \frac{[SbCl_3][Cl_2]}{[SbCl_5]} = \frac{(0.112 - x)(0.0640 - x)}{x} = \frac{0.00717 - 0.176x + x^2}{x}$$

$$0.025x = 0.00717 - 0.176x + x^2 \qquad x^2 - 0.201x + 0.00717 = 0$$

$$x = \frac{-b \pm \sqrt{b^2 - 4ac}}{2a} = \frac{0.201 \pm \sqrt{0.0404 - 0.0287}}{2} = 0.0464 \text{ or } 0.155$$

The second of the two values for x gives a negative value of $[Cl_2](= -0.091 \text{ M})$, and thus is physically meaningless in our universe. Thus, concentrations and amounts follow.

$[SbCl_5] = x = 0.0464 \text{ M}$ amount $SbCl_5 = 2.50 \text{ L} \times 0.0464 \text{ M} = 0.116 \text{ mol } SbCl_5$

$[SbCl_3] = 0.112 - x = 0.066 \text{ M}$ amount $SbCl_3 = 2.50 \text{ L} \times 0.066 \text{ M} = 0.17 \text{ mol } SbCl_3$

$[Cl_2] = 0.0640 - x = 0.0176 \text{ M}$ amount $Cl_2 = 2.50 \text{ L} \times 0.0176 \text{ M} = 0.0440 \text{ mol } Cl_2$

31. We use the chemical equation as a basis to organize the information provided about the reaction, and then determine the final number of moles of $Cl_2(g)$ present.

Equation: $CO(g)$ + $Cl_2(g)$ $\rightleftharpoons$ $COCl_2(g)$

Initial: 0.3500 mol   0.0000 mol   0.05500 mol

Changes: $+x$ mol $+x$ mol $-x$ mol

Equil.: $(0.3500 + x)$ mol x mol $(0.05500 - x)$ mol

$$K_c = 1.2 \times 10^3 = \frac{[COCl_2]}{[CO][Cl_2]} = \frac{\dfrac{(0.0550 - x)\,mol}{3.050\,L}}{\dfrac{(0.3500 + x)\,mol}{3.050\,L} \times \dfrac{x\,mol}{3.050\,L}}$$

$$\frac{1.2 \times 10^3}{3.050} = \frac{0.05500 - x}{(0.3500 + x)\,x}$$ Assume $x \ll 0.0550$ This produces the following expression.

$$\frac{1.2 \times 10^3}{3.050} = \frac{0.05500}{0.3500\,x}$$ $x = \frac{3.050 \times 0.05500}{0.3500 \times 1.2 \times 10^3} = 4.0 \times 10^{-4} \text{ mol } Cl_2$

We use the first value we obtained 4.0×10^{-4} $(= 0.00040)$ to arrive at a second value.

$$x = \frac{3.050 \times (0.0550 - 0.00040)}{(0.3500 + 0.00040) \times 1.2 \times 10^3} = 4.0 \times 10^{-4} \text{ mol } Cl_2$$

Because the value did not change on the second iteration, we have arrived at a solution.

32. Compute the initial concentration of each species present. Then we determine the equilibrium concentrations of all species. Finally, we compute the mass of CO_2 present at equilibrium.

$$[CO]_{int} = \frac{1.00 \text{ g}}{1.41 \text{ L}} \times \frac{1 \text{ mol CO}}{28.01 \text{ g CO}} = 0.0253 \text{ M} \qquad [H_2O]_{int} = \frac{1.00 \text{ g}}{1.41 \text{ L}} \times \frac{1 \text{ mol } H_2O}{18.02 \text{ g } H_2O} = 0.0394 \text{ M}$$

$$[H_2]_{int} = \frac{1.00 \text{ g}}{1.41 \text{ L}} \times \frac{1 \text{ mol } H_2}{2.016 \text{ g } H_2} = 0.352 \text{ M}$$

Equation : $\quad CO(g) \quad + \quad H_2O(g) \quad \rightleftharpoons \quad CO_2(g) \quad + \quad H_2(g)$

Initial : $\qquad$ 0.0253 M $\qquad$ 0.0394 M $\qquad\qquad$ 0.0000 M $\qquad$ 0.352 M

Changes : $\quad -x$ M $\qquad\qquad -x$ M $\qquad\qquad +x$ M $\qquad\qquad +x$ M

Equil : $\quad (0.0253 - x)$ M $\quad (0.0394 - x)$ M $\qquad x$ M $\qquad\qquad (0.352 + x)$ M

$$K_c = \frac{[CO_2][H_2]}{[CO][H_2O]} = 23.2 = \frac{x(0.352 + x)}{(0.0253 - x)(0.0394 - x)} = \frac{0.352x + x^2}{0.000997 - 0.0647x + x^2}$$

$$0.0231 - 1.50x + 23.2x^2 = 0.352x + x^2 \qquad 22.2x^2 - 1.852x + 0.0231 = 0$$

$$x = \frac{-b \pm \sqrt{b^2 - 4ac}}{2a} = \frac{1.852 \pm \sqrt{3.430 - 2.051}}{44.4} = 0.0682 \text{ M, } 0.0153 \text{ M}$$

The first value of x gives negative concentrations for reactants ([CO] = -0.0429 M and [H$_2$O] = -0.0288 M). Thus, $x = 0.0153$ M $= \left[CO_2 \right]$. Now we can find the mass of CO_2.

$$1.41 \text{ L} \times \frac{0.0153 \text{ mol } CO_2}{1 \text{ L mixture}} \times \frac{44.01 \text{ g } CO_2}{1 \text{ mol } CO_2} = 0.949 \text{ g } CO_2$$

33. We base each of our solutions on the balanced chemical equation.

(a) $\quad$ Equation : $\qquad PCl_5(g) \quad \rightleftharpoons \quad PCl_3(g) \quad + \quad Cl_2(g)$

Initial : $\qquad \dfrac{0.550 \text{ mol}}{2.50 \text{ L}} \qquad \dfrac{0.550 \text{ mol}}{2.50 \text{ L}} \qquad \dfrac{0 \text{ mol}}{2.50 \text{ L}}$

Changes : $\qquad \dfrac{-x \text{ mol}}{2.50 \text{ L}} \qquad \dfrac{+x \text{ mol}}{2.50 \text{ L}} \qquad \dfrac{+x \text{ mol}}{2.50 \text{ L}}$

Equil : $\qquad \dfrac{(0.550 - x) \text{ mol}}{2.50 \text{ L}} \qquad \dfrac{(0.550 + x) \text{ mol}}{2.50 \text{ L}} \qquad \dfrac{x \text{ mol}}{2.50 \text{ L}}$

$$K_c = \frac{[PCl_3][Cl_2]}{[PCl_5]} = 3.8 \times 10^{-2} = \frac{\dfrac{(0.550 + x)\text{mol}}{2.50\text{L}} \times \dfrac{x \text{ mol}}{2.50\text{L}}}{\dfrac{(0.550 - x)\text{mol}}{2.50\text{L}}} \qquad \frac{x(0.550 + x)}{2.50(0.550 - x)} = 3.8 \times 10^{-2}$$

$$x^2 + 0.550x = 0.052 - 0.095x \qquad x^2 + 0.645x - 0.052 = 0$$

$$x = \frac{-b \pm \sqrt{b^2 - 4ac}}{2a} = \frac{-0.645 \pm \sqrt{0.416 + 0.208}}{2} = 0.0725 \text{ mol, } -0.717 \text{ mol}$$

The second answer gives a negative quantity. of Cl_2, which makes no physical sense.

$n_{PCl_5} = (0.550 - 0.0725) = 0.478$ mol PCl$_5$ $\qquad$ $n_{PCl_3} = (0.550 + 0.0725) = 0.623$ mol PCl$_3$

$n_{Cl_2} = x = 0.0725$ mol Cl$_2$

(b) Equation : $\qquad PCl_5(g) \qquad \rightleftharpoons \qquad PCl_3(g) \qquad + \qquad Cl_2(g)$

Initial : $\qquad \dfrac{0.610 \text{ mol}}{2.50 \text{ L}} \qquad\qquad\qquad 0 \text{ M} \qquad\qquad\qquad 0 \text{ M}$

Changes : $\qquad \dfrac{-x \text{ mol}}{2.50 \text{ L}} \qquad\qquad \dfrac{+x \text{ mol}}{2.50 \text{ L}} \qquad\qquad \dfrac{+x \text{ mol}}{2.50 \text{ L}}$

Equil : $\qquad \dfrac{0.610 - x \text{ mol}}{2.50 \text{ L}} \qquad \dfrac{(x \text{ mol})}{2.50 \text{ L}} \qquad\qquad \dfrac{(x \text{ mol})}{2.50 \text{ L}}$

$$K_c = \frac{[PCl_3][Cl_2]}{[PCl_5]} = 3.8 \times 10^{-2} = \frac{\dfrac{(x \text{ mol})}{2.50 \text{ L}} \times \dfrac{(x \text{ mol})}{2.50 \text{ L}}}{\dfrac{0.610 - x \text{ mol}}{2.50 \text{ L}}}$$

$$2.50 \times 3.8 \times 10^{-2} = \frac{x^2}{0.610 - x} = 0.095 \qquad 0.058 - 0.095x = x^2 \qquad x^2 + 0.095x - 0.058 = 0$$

$$x = \frac{-b \pm \sqrt{b^2 - 4ac}}{2a} = \frac{-0.095 \pm \sqrt{0.0090 + 0.23}}{2} = 0.20 \text{ mol}, \ -0.29 \text{ mol}$$

amount $PCl_3 = 0.20 \text{ mol} = $ amount Cl_2; amount $PCl_5 = 0.610 - 0.20 = 0.41 \text{ mol}$

34. (a) We use the balanced chemical equation as a basis to organize the information we have about the reactants and products.

Equation: $\qquad 2\, COF_2(g) \qquad \rightleftharpoons \qquad CO_2(g) \qquad + \qquad CF_4(g)$

Initial: $\qquad \dfrac{0.145 \text{ mol}}{5.00 \text{ L}} \qquad\qquad \dfrac{0.262 \text{ mol}}{5.00 \text{ L}} \qquad\qquad \dfrac{0.074 \text{ mol}}{5.00 \text{ L}}$

Initial: $\qquad 0.0290 \text{ M} \qquad\qquad\quad 0.0524 \text{ M} \qquad\qquad\quad 0.0148 \text{ M}$

And we now compute a value of Q_c and compare it to the given value of K_c.

$$Q_c = \frac{[CO_2][CF_4]}{[COF_2]^2} = \frac{(0.0524)(0.0148)}{(0.0290)^2} = 0.922 < 2.00 = K_c$$

Because Q_c is not equal to K_c, the mixture is not at equilibrium.

(b) Because Q_c is smaller than K_c, the reaction will shift right, that is, products will be formed at the expense of COF_2, to reach a state of equilibrium.

(c) We continue the organization of information about reactants and products.

Equation: $2\,COF_2\,(g)$ $\rightleftharpoons$ $CO_2\,(g)$ $+$ $CF_4\,(g)$

Initial: 0.0290 M   0.0524 M   0.0148 M

Changes: $-2x$ M $+x$ M $+x$ M

Equil: $(0.0290 - 2x)$M $(0.0524 + x)$M $(0.0148 + x)$M

$$K_c = \frac{[CO_2][CF_4]}{[COF_2]^2} = \frac{(0.0524 + x)(0.0148 + x)}{(0.0290 - 2x)^2} = 2.00 = \frac{0.000776 + 0.0672x + x^2}{0.000841 - 0.1160x + 4x^2}$$

$$0.00168 - 0.232x + 8x^2 = 0.000776 + 0.0672x + x^2 \qquad 7x^2 - 0.299x + 0.000904 = 0$$

$$x = \frac{-b \pm \sqrt{b^2 - 4ac}}{2a} = \frac{0.299 \pm \sqrt{0.0894 - 0.0253}}{14} = 0.0033\ M, 0.0394\ M$$

The second of these values for x (0.0394) gives a negative $[COF_2]\,(= -0.0498\ M)$, clearly a nonsensical result. We now compute the concentration of each species at equilibrium, and check to ensure that the equilibrium constant is satisfied.

$$[COF_2] = 0.0290 - 2x = 0.0290 - 2(0.0033) = 0.0224\ M$$

$$[CO_2] = 0.0524 + x = 0.0524 + 0.0033 = 0.0557\ M$$

$$[CF_4] = 0.0148 + x = 0.0148 + 0.0033 = 0.0181\ M$$

$$K_c = \frac{[CO_2][CF_4]}{[COF_2]^2} = \frac{0.0557\ M \times 0.0181\ M}{(0.0224\ M)^2} = 2.01$$

The agreement of this value of K_c with the cited value (2.00) indicates that this solution is correct. Now we determine the number of moles of each species at equilibrium.

$$\text{mol}\ COF_2 = 5.00\ L \times 0.0224\ M = 0.112\ \text{mol}\ COF_2$$

$$\text{mol}\ CO_2 = 5.00\ L \times 0.0557\ M = 0.279\ \text{mol}\ CO_2$$

$$\text{mol}\ CF_4 = 5.00\ L \times 0.0181\ M = 0.0905\ \text{mol}\ CF_4$$

But suppose we had incorrectly concluded, in part (b), that reactants would be formed in reaching equilibrium. What result would we obtain? The set up follows.

Equation : $2\,COF_2\,(g)$ $\rightleftharpoons$ $CO_2\,(g)$ $+$ $CF_4\,(g)$

Initial :   0.0290 M   0.0524 M   0.0148 M

Changes : $+2y$ M $-y$ M $-y$ M

Equil : $(0.0290 + 2y)$M $(0.0524 - y)$M $(0.0148 - y)$ M

$$K_c = \frac{[CO_2][CF_4]}{[COF_2]^2} = \frac{(0.0524 - y)(0.0148 - y)}{(0.0290 + 2y)^2} = 2.00 = \frac{0.000776 - 0.0672y + y^2}{0.000841 + 0.1160y + 4y^2}$$

$$0.00168 + 0.232y + 8y^2 = 0.000776 - 0.0672y + y^2 \quad 7y^2 + 0.299y + 0.000904 = 0$$

$$y = \frac{-b \pm \sqrt{b^2 - 4ac}}{2a} = \frac{-0.299 \pm \sqrt{0.0894 - 0.0253}}{14} = -0.0033 \text{ M}, -0.0394 \text{ M}$$

The second of these values for $x(-0.0394)$ gives a negative $[COF_2](= -0.0498 \text{ M})$, clearly a nonsensical result. We now compute the concentration of each species at equilibrium, and check to ensure that the equilibrium constant is satisfied.

$$[COF_2] = 0.0290 + 2y = 0.0290 + 2(-0.0033) = 0.0224 \text{ M}$$

$$[CO_2] = 0.0524 - y = 0.0524 + 0.0033 = 0.0557 \text{ M}$$

$$[CF_4] = 0.0148 - y = 0.0148 + 0.0033 = 0.0181 \text{ M}$$

These are the same equilibrium concentrations that we obtained by making the correct decision regarding the direction that the reaction would take. Thus, you can be assured that, if you perform the algebra correctly, it will guide you even if you make the incorrect decision about the direction of the reaction.

35. **(a)** We calculate the initial amount of each substance.

$$n\{C_2H_5OH\} = 17.2 \text{ g } C_2H_5OH \times \frac{1 \text{ mol } C_2H_5OH}{46.07 \text{ g } C_2H_5OH} = 0.373 \text{ mol } C_2H_5OH$$

$$n\{CH_3CO_2H\} = 23.8 \text{ g } CH_3CO_2H \times \frac{1 \text{ mol } CH_3CO_2H}{60.05 \text{ g } CH_3CO_2H} = 0.396 \text{ mol } CH_3CO_2H$$

$$n\{CH_3CO_2C_2H_5\} = 48.6 \text{ g } CH_3CO_2C_2H_5 \times \frac{1 \text{ mol } CH_3CO_2C_2H_5}{88.11 \text{ g } CH_3CO_2C_2H_5}$$

$$n\{CH_3CO_2C_2H_5\} = 0.552 \text{ mol } CH_3CO_2C_2H_5$$

$$n\{H_2O\} = 71.2 \text{ g } H_2O \times \frac{1 \text{ mol } H_2O}{18.02 \text{ g } H_2O} = 3.95 \text{ mol } H_2O$$

Since we would divide each amount by the total volume, and since there are the same numbers of product and reactant stoichiometric coefficients, we can use moles rather than concentrations in the Q_c expression.

$$Q_c = \frac{n\{CH_3CO_2C_2H_5\}n\{H_2O\}}{n\{C_2H_5OH\}n\{CH_3CO_2H\}} = \frac{0.552 \text{ mol} \times 3.95 \text{ mol}}{0.373 \text{ mol} \times 0.396 \text{ mol}} = 14.8 > K_c = 4.0$$

Since $Q_c > K_c$ the reaction will shift to the left, forming reactants, as it attains equilibrium.

(b) Equation: C_2H_5OH + CH_3CO_2H $\rightleftharpoons$ $CH_3CO_2C_2H_5$ + H_2O

Initial	0.373 mol	0.396 mol	0.552 mol	3.95 mol
Changes	+x mol	+x mol	−x mol	−x mol
Equil	(0.373+x) mol	(0.396+x) mol	(0.552−x) mol	(3.95−x) mol

$$K_c = \frac{(0.552-x)(3.95-x)}{(0.373+x)(0.396+x)} = \frac{2.18-4.50x+x^2}{0.148+0.769x+x^2} = 4.0$$

$$x^2 - 4.50x + 2.18 = 4x^2 + 3.08x + 0.59 \qquad 3x^2 + 7.58x - 1.59 = 0$$

$$x = \frac{-b \pm \sqrt{b^2-4ac}}{2a} = \frac{-7.58 \pm \sqrt{57+19}}{6} = 0.19 \text{ moles}, -2.72 \text{ moles}$$

Negative amounts do not make physical sense. We compute the equilibrium amount of each substance with $x = 0.19$ moles.

$$n\{C_2H_5OH\} = 0.373 \text{ mol} + 0.19 \text{ mol} = 0.56 \text{ mol } C_2H_5OH$$

$$\text{mass } C_2H_5OH = 0.56 \text{ mol } C_2H_5OH \times \frac{46.07 \text{ g } C_2H_5OH}{1 \text{ mol } C_2H_5OH} = 26 \text{ g } C_2H_5OH$$

$$n\{CH_3CO_2H\} = 0.396 \text{ mol} + 0.19 \text{ mol} = 0.59 \text{ mol } CH_3CO_2H$$

$$\text{mass } CH_3CO_2H = 0.59 \text{ mol } CH_3CO_2H \times \frac{60.05 \text{ g } CH_3CO_2H}{1 \text{ mol } CH_3CO_2H} = 35 \text{ g } CH_3CO_2H$$

$$n\{CH_3CO_2C_2H_5\} = 0.552 \text{ mol} - 0.19 \text{ mol} = 0.36 \text{ } CH_3CO_2C_2H_5$$

$$\text{mass } CH_3CO_2C_2H_5 = 0.36 \text{ mol } CH_3CO_2C_2H_5 \times \frac{88.10 \text{ g } CH_3CO_2C_2H_5}{1 \text{ mol } CH_3CO_2C_2H_5} = 32 \text{ g } CH_3CO_2C_2H_5$$

$$n\{H_2O\} = 3.95 \text{ mol} - 0.19 \text{ mol} = 3.76 \text{ mol } H_2O$$

$$\text{mass } H_2O = 3.76 \text{ mol } H_2O \times \frac{18.02 \text{ g } H_2O}{1 \text{ mol } H_2O} = 68 \text{ g } H_2O$$

$$\text{To check } K_c = \frac{n\{CH_3CO_2C_2H_5\}n\{H_2O\}}{n\{C_2H_5OH\}n\{CH_3CO_2H\}} = \frac{0.36 \text{ mol} \times 3.76 \text{ mol}}{0.56 \text{ mol} \times 0.59 \text{ mol}} = 4.1$$

36. The final volume of the mixture is $0.750 \text{ L} + 2.25 \text{ L} = 3.00 \text{ L}$. Then use the balanced chemical equation to organize the data we have concerning the reaction. The reaction should shift to the right, that is form products, in reaching a new equilibrium, since the volume is greater.

Equation: $N_2O_4(g) \;\rightleftharpoons\; 2\,NO_2(g)$

Initial: $\dfrac{0.971 \text{ mol}}{3.00 \text{ L}}$ $\dfrac{0.0580 \text{ mol}}{3.00 \text{ L}}$

Initial: $0.324\,M$ $0.0193\,M$

Changes: $-x\,M$ $+2x\,M$

Equil : $(0.324 - x)M$ $(0.0193 + 2x)M$

$$K_c = \frac{[NO_2]^2}{[N_2O_4]} = \frac{(0.0193 + 2x)^2}{0.324 - x} = \frac{0.000372 + 0.0772x + 4x^2}{0.324 - x} = 4.61 \times 10^{-3}$$

$$0.000372 + 0.0772x + 4x^2 = 0.00149 - 0.00461x \qquad 4x^2 + 0.0818x - 0.00112 = 0$$

$$x = \frac{-b \pm \sqrt{b^2 - 4ac}}{2a} = \frac{-0.0818 \pm \sqrt{0.00669 + 0.0179}}{8} = 0.00938 \text{ M}, \; -0.0298 \text{ M}$$

$[NO_2] = 0.0193 + (2 \times 0.00938) = 0.0381 \text{ M}$

amount $NO_2 = 0.0381 \text{ M} \times 3.00 \text{ L} = 0.114 \text{ mol } NO_2$

$[N_2O_4] = 0.324 - 0.00938 = 0.314_6 \text{ M}$

amount $N_2O_4 = 0.314_6 \text{ M} \times 3.00 \text{ L} = 0.944 \text{ mol } N_2O_4$

37. $\quad [HCONH_2]_{init} = \dfrac{0.186 \text{ mol}}{2.16 \text{ L}} = 0.0861 \text{ M}$

Equation: $HCONH_2(g) \rightleftharpoons NH_3(g) \; + \; CO(g)$

Initial : 0.0861 M 0 M 0 M

Changes: $-x\,M$ $+x\,M$ $+x\,M$

Equil : $(0.0861\text{-}x)M$ $x\,M$ $x\,M$

$$K_c = \frac{[NH_3][CO]}{[HCONH_2]} = \frac{x \cdot x}{0.0861 - x} = 4.84 \qquad x^2 = 0.417 - 4.84x \qquad 0 = x^2 + 4.84x - 0.417$$

$$x = \frac{-b \pm \sqrt{b^2 - 4ac}}{2a} = \frac{-4.84 \pm \sqrt{23.4 + 1.67}}{2} = 0.084 \text{ M}, \; -4.92 \text{ M}$$

The negative concentration obviously is physically meaningless. We determine the total concentration of all species, and then the total pressure with $x = 0.084$.

$[\text{total}] = [NH_3] + [CO] + [HCONH_2] = x + x + 0.0861 - x = 0.0861 + 0.084 = 0.170 \text{ M}$

$P_{tot} = 0.170 \text{ mol L}^{-1} \times 0.08206 \text{ L atm mol}^{-1} \text{ K}^{-1} \times 400. \text{ K} = 5.58 \text{ atm}$

38. Compare Q_p to K_p. We assume that the added solids are of negligible volume so that the initial partial pressures of $CO_2(g)$ and $H_2O(g)$ do not significantly change.

$$P\{H_2O\} = \left(715 \text{ mmHg} \times \frac{1 \text{ atm}}{760 \text{ mmHg}}\right) = 0.941 \text{ atm } H_2O$$

$$Q_p = P\{CO_2\}P\{H_2O\} = 2.10 \text{ atm } CO_2 \times 0.941 \text{ atm } H_2O = 1.98 > 0.23 = K_p$$

Because Q_p is larger than K_p, the reaction will proceed left toward reactants to reach equilibrium. Thus, the partial pressures of the two gases will decrease.

39. **(a)** We organize the solution around the balanced chemical equation.

Equation:	$2 \, Cr^{3+}(aq)$	$+ \; Cd(s)$	$\rightleftharpoons$	$2 \, Cr^{2+}(aq)$	$+ \; Cd^{2+}(aq)$
Initial :	1.00 M			0 M	0 M
Changes:	$-2x$ M			$+2x$	$+x$
Equil:	$(1.00-2x)$ M			$2x$	x

$$K_c = \frac{\left[Cr^{2+}\right]^2 \left[Cd^{2+}\right]}{\left[Cr^{3+}\right]^2} = \frac{(2x)^2(x)}{(1.00-2x)^2} = 0.288 \qquad \begin{array}{c}\textit{Via successive approximations, one obtains}\\ x = 0.257 \text{ M}\end{array}$$

Therefore, at equilibrium, $[Cd^{2+}] = 0.257$ M, $[Cr^{2+}] = 0.514$ M and $[Cr^{3+}] = 0.486$ M

(b) Minimum mass of $Cd(s) = 0.350 L \times 0.257 \text{ M} \times 112.41 \text{ g/mol} = 10.1$ g of Cd metal

40. Again we base the set up of the problem around the balanced chemical equation.

Equation:	$Pb(s) + 2 \, Cr^{3+}(aq)$	$\xrightleftharpoons{K_c=3.2\times10^{-10}}$	$Pb^{2+}(aq)$	$+ \; 2 \, Cr^{2+}(aq)$
Initial:	$-$ \quad 0.100 M		0 M	0 M
Changes:	$-$ \quad $-2x$ M		$+x$ M	$+2x$ M
Equil :	$-$ \quad $(0.100-2x)$ M		x M	$2x$ M

$$K_c = \frac{x(2x)^2}{(0.100)^2} = 3.2\times10^{-10} \qquad\qquad 4x^3 = (0.100)^2 \times 3.2\times10^{-10} = 3.2\times10^{-12}$$

$$x = \sqrt[3]{\frac{3.2\times10^{-12}}{4}} = 9.3\times10^{-5} \text{ M} \qquad \text{Assumption that } 2x \ll 0.100, \text{ is valid and thus}$$

$$\left[Pb^{2+}\right] = x = 9.3\times10^{-5} \text{ M}, [Cr^{2+}] = 1.9\times10^{-4} \text{ M and } [Cr^{3+}] = 0.100 \text{ M}$$

41. We are told in this question that the reaction $SO_2(g) + Cl_2(g) \rightleftharpoons SO_2Cl_2(g)$ has $K_c = 4.0$ at a certain temperature T. This means that at the temperature T, $[SO_2Cl_2] = 4.0 \times [Cl_2] \times [SO_2]$. Careful scrutiny of the three diagrams reveals that sketch (b) is the best representation because it contains numbers of SO_2Cl_2, SO_2 and Cl_2 molecules that are consistent with the K_c for the reaction. In other words, sketch (b) is the best choice because it contains 12 SO_2Cl_2 molecules (per unit volume), 1 Cl_2 molecule (per unit volume) and 3 SO_2 molecules (per unit volume) which, is the requisite number of each type of molecule needed to generate the expected K_c value for the reaction at temperature T.

42. In this question we are told that the reaction $2\,NO(g) + Br_2(g) \rightleftharpoons 2\,NOBr(g)$ has $K_c = 3.0$ at a certain temperature T. This means that at the temperature T, $[NOBr]^2 = 3.0 \times [Br_2][NO]^2$. Sketch (c) is the most accurate representation because it contains 18 NOBr molecules (per unit volume), 6 NO molecules (per unit volume) and 3 Br_2 molecules (per unit volume) which is the requisite number of each type of molecule needed to generate the expected K_c value for the reaction at temperature T.

Partial Pressure Equilibrium Constant, K_p

43. The $I_2(s)$ maintains the presence of I_2 in the flask until it has all vaporized. Thus, if enough HI(g) is produced to completely consume the $I_2\,(s)$, equilibrium will not be achieved.

$$P\{H_2S\} = 747.6 \text{ mmHg} \times \frac{1 \text{ atm}}{760 \text{ mmHg}} = 0.9837 \text{ atm}$$

Equation:	$H_2S(g) +$	$I_2(s) \rightleftharpoons$	$2\,HI(g) +$	$S(s)$
Initial:	0.9837 atm		0 atm	
Changes:	$-x$ atm		$+2x$ atm	
Equil:	$(0.9837 - x)$ atm		$2x$ atm	

$$K_p = \frac{P\{HI\}^2}{P\{H_2S\}} = \frac{(2x)^2}{(0.9837 - x)} = 1.34 \times 10^{-5} = \frac{4x^2}{0.9837}$$

$$x = \sqrt{\frac{1.34 \times 10^{-5} \times 0.9837}{4}} = 1.82 \times 10^{-3} \text{ atm}$$

The assumption that $0.9837 \gg x$, is valid. Now we verify that sufficient $I_2(s)$ is present by computing the mass of I_2 needed to produce the predicted pressure of HI(g). Initially, 1.85 g I_2 is present (given).

$$\text{mass } I_2 = \frac{1.82 \times 10^{-3} \text{ atm} \times 0.725 \text{ L}}{0.08206 \text{ L atm mol}^{-1} \text{ K}^{-1} \times 333 \text{ K}} \times \frac{1 \text{ mol } I_2}{2 \text{ mol HI}} \times \frac{253.8 \text{ g } I_2}{1 \text{ mol } I_2} = 0.00613 \text{ g } I_2$$

$$P_{tot} = P\{H_2S\} + P\{HI\} = (0.9837 - x) + 2x = 0.9837 + x = 0.9837 + 0.00182 = 0.9855 \text{ atm}$$

$$P_{tot} = 749.0 \text{ mmHg}$$

44. We first determine the initial pressure of NH_3.

$$P\{NH_3(g)\} = \frac{nRT}{V} = \frac{0.100 \text{ mol } NH_3 \times 0.08206 \text{ L atm mol}^{-1} \text{ K}^{-1} \times 298 \text{ K}}{2.58 \text{ L}} = 0.948 \text{ atm}$$

Equation:	$NH_4HS(s) \rightleftharpoons$	$NH_3(g)$	$+$	$H_2S(g)$
Initial:		0.948 atm		0 atm
Changes:		$+x$ atm		$+x$ atm
Equil:		$(0.948 + x)$ atm		x atm

$$K_p = P\{NH_3\}P\{H_2S\} = 0.108 = (0.948 + x)x = 0.948x + x^2 \quad 0 = x^2 + 0.948x - 0.108$$

$$x = \frac{-b \pm \sqrt{b^2 - 4ac}}{2a} = \frac{-0.948 \pm \sqrt{0.899 + 0.432}}{2} = 0.103 \text{ atm}, \; -1.05 \text{ atm}$$

The negative root makes no physical sense. The total gas pressure is obtained as follows.

$$P_{tot} = P\{NH_3\} + P\{H_2 S\} = (0.948 + x) + x = 0.948 + 2x = 0.948 + 2 \times 0.103 = 1.154 \text{ atm}$$

45. We substitute the given equilibrium pressure into the equilibrium constant expression and

solve for the other equilibrium pressure. $K_p = \dfrac{P\{O_2\}^3}{P\{CO_2\}^2} = 28.5 = \dfrac{P\{O_2\}^3}{(0.0721 \text{ atm } CO_2)^2}$

$$P\{O_2\} = \sqrt[3]{P\{O_2\}^3} = \sqrt[3]{28.5(0.0712 \text{ atm})^2} = 0.529 \text{ atm } O_2$$

$$P_{total} = P\{CO_2\} + P\{O_2\} = 0.0721 \text{ atm } CO_2 + 0.529 \text{ atm } O_2 = 0.601 \text{ atm total}$$

46. The composition of dry air is given in volume percent. Division of these percentages by 100 gives the volume fraction, which equals the mole fraction and also the partial pressure in atmospheres, if the total pressure is 1.00 atm. Thus, we have $P\{O_2\} = 0.20946$ atm and $P\{CO_2\} = 0.00036$ atm. We substitute these two values into the expression for Q_p.

$$Q_p = \frac{P\{O_2\}^3}{P\{CO_2\}^2} = \frac{(0.20946 \text{ atm } O_2)^3}{(0.00036 \text{ atm } CO_2)^2} = 7.1 \times 10^4 > 28.5 = K_p$$

The value of Q_p is much larger than the value of K_p. Thus this reaction should be spontaneous in the reverse direction until equilibrium is achieved. It will only be spontaneous in the forward direction when the pressure of O_2 drops or that of CO_2 rises (as would be the case in self-contained breathing devices).

47. **(a)** We first determine the initial pressure of each gas.

$$P\{CO\} = P\{Cl_2\} = \frac{nRT}{V} = \frac{1.00 \text{ mol} \times 0.08206 \text{ L atm mol}^{-1} \text{ K}^{-1} \times 668 \text{ K}}{1.75 \text{ L}} = 31.3 \text{ atm}$$

Then we calculate equilibrium partial pressures, organizing our calculation around the balanced chemical equation. We see that the equilibrium constant is not very large, meaning that we must solve the polynomial exactly (or by successive approximations).

Equation	CO(g)	+ Cl$_2$ (g) $\rightleftharpoons$	COCl$_2$ (g)	$K_p = 22.5$
Initial:	31.3 atm	31.3 atm	0 atm	
Changes:	$-x$ atm	$-x$ atm	$+x$ atm	
Equil:	$31.3 - x$ atm	$31.3 - x$ atm	x atm	

$$K_p = \frac{P\{COCl_2\}}{P\{CO\} P\{Cl_2\}} = 22.5 = \frac{x}{(31.3 - x)^2} = \frac{x}{(979.\underline{7} - 62.6x + x^2)}$$

$$22.5(979.\underline{7} - 62.6x + x^2) = x = 22043 - 1408.\underline{5}x + 22.5x^2 = x$$

$$22043 - 1409.5x + 22.5x^2 = 0 \quad \text{(Solve by using the quadratic equation)}$$

$$x = \frac{-b \pm \sqrt{b^2 - 4ac}}{2a} = \frac{-(-1409.5) \pm \sqrt{(-1409.5)^2 - 4(22.5)(22043)}}{2(22.5)}$$

$$x = \frac{1409.5 \pm \sqrt{2818}}{45} = 30.1\underline{4}, 32.5 \text{ (too large)}$$

$$P\{CO\} = P\{Cl_2\} = 31.3 \text{ atm} - 30.1\underline{4} \text{ atm} = 1.1\underline{5} \text{ atm} \qquad P\{COCl_2\} = 30.1\underline{4} \text{ atm}$$

(b) $P_{total} = P\{CO\} + P\{Cl_2\} + P\{COCl_2\} = 1.1\underline{5} \text{ atm} + 1.1\underline{5} \text{ atm} + 30.1\underline{4} \text{ atm} = 32.4 \text{ atm}$

48. We first find the value of K_p for the reaction

$$2 NO_2(g) \rightleftharpoons 2 NO(g) + O_2(g), \quad K_c = 1.8 \times 10^{-6} \text{ at } 184 \text{ }^{\circ}C = 457 \text{ K}.$$

For this reaction $\Delta n_{gas} = 2 + 1 - 2 = +1$.

$$K_p = K_c (RT)^{\Delta n_g} = 1.8 \times 10^{-6} (0.08206 \times 457)^{+1} = 6.8 \times 10^{-5}$$

To obtain the required reaction $NO(g) + \frac{1}{2}O_2(g) \rightleftharpoons NO_2(g)$ from the initial reaction, that initial reaction must be reversed and then divided by two. Thus, in order to determine the value of the equilibrium constant for the final reaction, the value of K_p for the initial reaction must be inverted, and the square root taken of the result.

$$K_{p, \text{ final}} = \sqrt{\frac{1}{6.8 \times 10^{-5}}} = 1.2 \times 10^2$$

Le Châtelier's Principle

49. Continuous removal of the product, of course, has the effect of decreasing the concentration of the products below their equilibrium values. Thus, the equilibrium system is disturbed by removing the products and the system will attempt (in vain, as it turns out) to re-establish the equilibrium by shifting toward the right, that is, to generate more products.

50. We notice that the density of the solid ice is smaller than is that of liquid water. This means that the same mass of liquid water is present in a smaller volume than an equal mass of ice. Thus, if pressure is placed on ice, attempting to force it into a smaller volume, the ice will be transformed into the less-space-occupying water at $0 \text{ }^{\circ}C$. Thus, at $0 \text{ }^{\circ}C$ under pressure, $H_2O(s)$ will melt to form $H_2O(l)$. This behavior is *not* expected in most cases because generally a solid is *more* dense than its liquid phase.

51. **(a)** This reaction is exothermic with $\Delta H^{\circ} = -150. \text{ kJ}$. Thus, high temperatures favor the reverse reaction (endothermic reaction). The amount of $H_2(g)$ present at high temperatures will be less than that present at low temperatures.

(b) $H_2O(g)$ is one of the reactants involved. Introducing more will cause the equilibrium position to shift to the right, favoring products. The amount of $H_2(g)$ will increase.

(c) Doubling the volume of the container will favor the side of the reaction with the largest sum of gaseous stoichiometric coefficients. The sum of the stoichiometric coefficients of gaseous species is the same (4) on both sides of this reaction. Therefore, increasing the volume of the container will have no effect on the amount of $H_2(g)$ present at equilibrium.

(d) A catalyst merely speeds up the rate at which a reaction reaches the equilibrium position. The addition of a catalyst has no effect on the amount of $H_2(g)$ present at equilibrium.

52. (a) This reaction is endothermic, with $\Delta H^\circ = +92.5$ kJ. Thus, a higher temperature will favor the forward reaction and increase the amount of HI(g) present at equilibrium.

(b) The introduction of more product will favor the reverse reaction and decrease the amount of HI(g) present at equilibrium.

(c) The sum of the stoichiometric coefficients of gaseous products is larger than that for gaseous reactants. Increasing the volume of the container will favor the forward reaction and increase the amount of HI(g) present at equilibrium.

(d) A catalyst merely speeds up the rate at which a reaction reaches the equilibrium position. The addition of a catalyst has no effect on the amount of HI(g) present at equilibrium.

(e) The addition of an inert gas to the constant-volume reaction mixture will not change any partial pressures. It will have no effect on the amount of HI(g) present at equilibrium.

53. (a) The formation of NO(g) from its elements is an endothermic reaction ($\Delta H^\circ = +181$ kJ/mol). Since the equilibrium position of endothermic reactions is shifted toward products at higher temperatures, we expect more NO(g) to be formed from the elements to be enhanced at higher temperatures.

(b) Reaction rates always are enhanced by higher temperatures, since a larger fraction of the collisions will have an energy which surmounts the activation energy. This enhancement of rates affects both the forward and the reverse reactions. Thus, the position of equilibrium is reached more rapidly at higher temperatures than at lower temperatures.

54. If the reaction is endothermic ($\Delta H^\circ > 0$), the forward reaction is favored at high temperatures. If the reaction is exothermic ($\Delta H^\circ < 0$), the forward reaction is favored at low temperatures.

(a) $\Delta H^\circ = \Delta H_f^\circ \left[PCl_5(g) \right] - \Delta H_f^\circ \left[PCl_3(g) \right] - \Delta H_f^\circ \left[Cl_2(g) \right]$

$\Delta H^\circ = -374.9 \text{ kJ} - (-287.0 \text{ kJ}) - 0.00 \text{ kJ} = -87.9 \text{ kJ/mol}$ (favored at low temperatures)

(b) $\Delta H^\circ = 2\Delta H_f^\circ[H_2O(g)] + 3\Delta H_f^\circ[S(\text{rhombic})] - \Delta H_f^\circ[SO_2(g)] - 2\Delta H_f^\circ[H_2S(g)]$

$\Delta H^\circ = 2(-241.8 \text{ kJ}) + 3(0.00 \text{ kJ}) - (-296.8 \text{ kJ}) - 2(-20.63 \text{ kJ})$

$\Delta H^\circ = -145.5$ kJ/mol (favored at low temperatures)

(c) $\Delta H^\circ = 4\Delta H_f^\circ[NOCl(g)] + 2\Delta H_f^\circ[H_2O(g)]$

$\quad -2\Delta H_f^\circ[N_2(g)] - 3\Delta H_f^\circ[O_2(g)] - 4\Delta H_f^\circ[HCl(g)]$

$\Delta H^\circ = 4(51.71 \text{ kJ}) + 2(-241.8 \text{ kJ}) - 2(0.00 \text{ kJ}) - 3(0.00 \text{ kJ}) - 4(-92.31 \text{ kJ})$

$\Delta H^\circ = +92.5$ kJ/mol (favored at higher temperatures)

55. If the total pressure of a mixture of gases at equilibrium is doubled by compression, the equilibrium will shift to the side with fewer moles of gas to counteract the increase in pressure. Thus, if the pressure of an equilibrium mixture of $N_2(g)$, $H_2(g)$ and $NH_3(g)$ is doubled, the reaction involving these three gases, i.e. $N_2(g) + 3\ H_2(g) \rightleftharpoons 2\ NH_3(g)$

will proceed in the forward direction to produce a new equilibrium mixture that contains additional ammonia and less molecular nitrogen and molecular hydrogen. In other words, $P\{N_2(g)\}$ will have decreased when equilibrium is re-established. It is important to note, however, that the final equilibrium partial pressure for the N_2 will, nevertheless, be higher than its original partial pressure prior to the doubling of the total pressure.

56. (a) Because $\Delta H^\circ = 0$, the position of the equilibrium for this reaction will not be affected by temperature. Since the equilibrium position is expressed by the value of the equilibrium constant, we expect K_p to be unaffected by, or to remain constant with, temperature.

(b) From part (a), we know that the value of K_p will not change when the temperature is changed. The pressures of the gases, however, will change with temperature. (Recall the ideal gas law: $P = nRT/V$.) In fact, all pressures will increase. The stoichiometric coefficients in the reaction are such that at higher pressures the formation of more reactant will be favored (the reactant side has fewer moles of gas). Thus, the amount of $D(g)$ will be smaller when equilibrium is reestablished at the higher temperature for the cited reaction.

$$A(s) \rightleftharpoons B(s) + 2\ C(g) + \frac{1}{2}\ D(g)$$

57. Increasing the volume of an equilibrium mixture causes that mixture to shift toward the side (reactants or products) where the sum of the stoichiometric coefficients of the gaseous species is the larger. That is: shifts to the right if $\Delta n_{gas} > 0$, shifts to the left if $\Delta n_{gas} < 0$, and does not shift if $\Delta n_{gas} = 0$

(a) $C(s) + H_2O(g) \rightleftharpoons CO(g) + H_2(g), \Delta n_{gas} > 0$, shift right, toward products

(b) $Ca(OH)_2(s) + CO_2(g) \rightleftharpoons CaCO_3(s) + H_2O(g)$, $\Delta n_{gas} = 0$, no shift, no change in equilibrium position.

(c) $4\,NH_3(g) + 5\,O_2(g) \rightleftharpoons 4\,NO(g) + 6\,H_2O(g)$, $\Delta n_{gas} > 0$, shifts right, towards products

58. The equilibrium position for a reaction that is exothermic shifts to the left (reactants are favored). When the temperature is raised. For one that is endothermic, it shifts right (products are favored). When the temperature is raised.

(a) $NO(g) \rightleftharpoons \frac{1}{2} N_2(g) + \frac{1}{2} O_2(g)$ $\Delta H° = -90.2$ kJ shifts left, % dissociation ↓

(b) $SO_3(g) \rightleftharpoons SO_2(g) + \frac{1}{2} O_2(g)$ $\Delta H° = +98.9$ kJ shifts right, % dissociation ↑

(c) $N_2H_4(g) \rightleftharpoons N_2(g) + 2\,H_2(g)$ $\Delta H° = -95.4$ kJ shifts left, % dissociation ↓

(d) $COCl_2(g) \rightleftharpoons CO(g) + Cl_2(g)$ $\Delta H° = +108.3$ kJ shifts right, % dissociation ↑

Integrative and Advanced Exercises

59. In a reaction of the type $I_2(g) \rightarrow 2\,I(g)$ the bond between two iodine atoms, the I—I bond, must be broken. Since $I_2(g)$ is a stable molecule, this bond breaking process must be endothermic. Hence, the reaction cited is an endothermic reaction. The equilibrium position of endothermic reactions will be shifted toward products when the temperature is raised.

60. (a) In order to determine a value of K_c, we first must find the CO_2 concentration in the gas phase. Note, the total volume for the gas is 1.00 L (moles and molarity are numerically equal)

$$[CO_2] = \frac{n}{V} = \frac{P}{RT} = \frac{1.00 \text{ atm}}{0.08206 \dfrac{\text{L} \cdot \text{atm}}{\text{mol} \cdot \text{K}} \times 298 \text{ K}} = 0.0409 \text{ M} \qquad K_c = \frac{[CO_2(aq)]}{[CO_2(g)]} = \frac{3.29 \times 10^{-2} \text{ M}}{0.0409 \text{ M}} = 0.804$$

(b) It does not matter to which phase the radioactive $^{14}CO_2$ is added. This is an example of a Le Châtelier's Principle problem in which the stress is a change in concentration of the reactant $CO_2(g)$. To find the new equilibrium concentrations, we must solve and I.C.E. table. Since $Q_c < K_c$ the reaction shifts to the product, $CO_2(aq)$ side.

reaction:	$CO_2(g)$	$\rightleftharpoons$	$CO_2(aq)$
initial:	0.0409 mol		3.29×10^{-3} mol
stress	+0.01000 mol		–
changes:	$-x$ mol		$+x$ mol
equilibrium:	$(0.05090 - x)$ mol		$3.29 \times 10^{-3} + x$ mol

$$K_c = \frac{[CO_2(aq)]}{[CO_2(g)]} = \frac{\dfrac{3.29 \times 10^{-3} + x \text{ mol}}{0.1000 \text{ L}}}{\dfrac{(0.05090 - x) \text{ mol}}{1.000 \text{ L}}} = 0.804 = \frac{3.29 \times 10^{-2} + 10x}{0.05090 - x} \qquad x = 7.43 \times 10^{-4} \text{ mol } CO_2$$

Total moles of CO_2 in the aqueous phase $(0.1000 \text{ L})(3.29 \times 10^{-2} + 7.43 \times 10^{-3}) = 4.03 \times 10^{-3}$ moles

Total moles of CO_2 in the gaseous phase $(1.000 \text{ L})(5.090 \times 10^{-2} - 7.43 \times 10^{-4}) = 5.01_6 \times 10^{-2}$ moles
Total moles of $CO_2 = 5.01_6 \times 10^{-2}$ moles + 4.03×10^{-3} moles = 5.09×10^{-2} moles

There is continuous mixing of the ^{12}C and ^{14}C such that the isotopic ratios in the two phases is the same. This ratio is given by the mole fraction of the two isotopes.

For $^{14}CO_2$ in either phase its mole fraction is $\dfrac{0.01000\ mol}{5.41\underline{9}\times10^{-2}\ mol}\times100 = 18.4\underline{5}\ \%$

Moles of $^{14}CO_2$ in the gaseous phase = $5.01\underline{6}\times10^{-2}$ moles $\times 0.1845 = 0.00925\underline{6}$ moles

Moles of $^{14}CO_2$ in the aqueous phase = 4.03×10^{-3} moles $\times 0.1845 = 0.000744$ moles

61. Dilution makes Q_c larger than K_c. Thus, the reaction mixture will shift left in order to regain equilibrium. We organize our calculation around the balanced chemical equation.

Equation: $Ag^+(aq)$ + $Fe^{2+}(aq)$ $\rightleftharpoons$ $Fe^{3+}(aq) + Ag(s)$ $K_c = 2.98$

equil: 0.31 M   0.21 M   0.19 M   –

dilution: 0.12 M   0.084 M   0.076 M   –

changes: $+x$ M $+x$ M $-x$ M   –

new equil: $(0.12 + x)$ M $(0.084 + x)$ M $(0.076 - x)$ M   –

$K_c = \dfrac{[Fe^{3+}]}{[Ag^+][Fe^{2+}]} = 2.98 = \dfrac{0.076 - x}{(0.12 + x)(0.084 + x)}$ $2.98(0.12 + x)(0.084 + x) = 0.076 - x$

$0.076 - x = 0.030 + 0.61x + 2.98 x^2$ $2.98 x^2 + 1.61 \quad x - 0.046 = 0$

$x = \dfrac{-1.61 \pm \sqrt{2.59 + 0.55}}{5.96} = 0.027, -0.57$ Note that the negative root makes no physical

sense; it gives $[Fe^{2+}] = 0.084 - 0.57 = -0.49\ M$.

Thus, the new equilibrium concentrations are

$[Fe^{2+}] = 0.084 + 0.027 = 0.111\ M$ $[Ag^+] = 0.12 + 0.027 = 0.15\ M$

$[Fe^{3+}] = 0.076 - 0.027 = 0.049\ M$ We can check our answer by substitution.

$Kc = \dfrac{0.049\ M}{0.111\ M \times 0.15\ M} = 2.94 \approx 2.98$ (within precision limits)

62. The percent dissociation should increase as the pressure is lowered, according to Le Chatelier's principle. Thus the total pressure in this instance should be more than in Example 15-12, where the percent dissociation is 12.5%. The total pressure in 15-12 was computed starting from the total number of moles at equilibrium.

The total amount = $(0.0240 - 0.00300)$ moles N_2O_4 + 2×0.00300 mol NO_2 = 0.020 mol gas.

$P_{total} = \dfrac{nRT}{V} = \dfrac{0.0270\,mol \times 0.08206\,L\,atm\,K^{-1}\,mol^{-1} \times 298\ K}{0.372\ L} = 1.77\,atm$ (Example 15-12)

We base our solution on the balanced chemical equation. We designate the initial pressure of N_2O_4 as P. The change in $P\{N_2O_4\}$ is given as -0.10 P atm. to represent the 10.0 % dissociation.

Equation: $\qquad N_2O_4(g) \rightleftharpoons 2\,NO_2(g)$

Initial: $\qquad\qquad$ P atm $\qquad\qquad$ 0 atm

Changes: $\qquad -0.10\,P$ atm $\qquad +2(0.10\,P$ atm$)$

Equil: $\qquad\qquad 0.90\,P$ atm $\qquad\quad 0.20\,P$ atm

$$K_p = \frac{P\{NO_2\}^2}{P\{N_2O_4\}} = \frac{(0.20\,P)^2}{0.90\,P} = \frac{0.040\,P}{0.90} = 0.113 \qquad P = \frac{0.113 \times 0.90}{0.040} = 2.54\,\text{atm}$$

Thus, the total pressure at equilibrium is 0.90 P + 0.20 P and 1.10 P (where P = 2.54 atm)
Therefore, total pressure at equilibrium = 2.79 atm.

63. Equation: $\qquad 2\,SO_3(g) \rightleftharpoons 2\,SO_2(g) \quad + \quad O_2(g)$

Initial: $\qquad\qquad 1.00$ atm $\qquad\quad 0$ atm $\qquad\qquad 0$ atm

Changes: $\qquad\quad -2x$ atm $\qquad +2x$ atm $\qquad\quad +x$ atm

Equil: $\qquad\quad (1.00 - 2x)$atm $\qquad 2x$ atm $\qquad\qquad x$ atm

Because of the small value of the equilibrium constant, the reaction does not proceed very far toward products in reaching equilibrium. Hence, we assume that $x \ll 1.00$ atm and calculate an approximate value of x (small K problem).

$$Kp = \frac{P\{SO_2\}^2\,P\{O_2\}}{P\{SO_3\}^2} = \frac{(2x)^2\,x}{(1.00 - 2x)^2} = 1.6 \times 10^{-5} \approx \frac{4x^3}{(1.00)^2} \qquad x = 0.016\,\text{atm}$$

A second cycle may get closer to the true value of x.

$$1.6 \times 10^{-5} \approx \frac{4x^3}{(1.00 - 0.032)^2} = \ x = 0.016\,\text{atm}$$

Our initial value was sufficiently close. We now compute the total pressure at equilibrium.
$$P_{total} = P\{SO_3\} + P\{SO_2\} + P\{O_2\} = (1.00 - 2x) + 2x + x = 1.00 + x = 1.00 + 0.016 = 1.02\,\text{atm}$$

<u>**64.**</u> Let us start with one mole of air, and let $2x$ be the amount in moles of NO formed.

Equation: $\qquad N_2(g) \quad + \quad O_2(g) \quad \rightleftharpoons \quad 2\,NO(g)$

Initial: $\qquad\quad 0.79$ mol $\qquad\quad 0.21$ mol $\qquad\qquad 0$ mol

Changes: $\qquad\quad -x$ mol $\qquad\quad -x$ mol $\qquad\qquad +2x$ mol

Equil: $\qquad (0.79 - x)$mol $\quad (0.21 - x)$mol $\qquad\quad 2x$ mol

$$\chi_{NO} = \frac{n\{NO\}}{n\{N_2\} + n\{O_2\} + n\{NO\}} = \frac{2x}{(0.79 - x) + (0.21 - x) + 2x} = 0.018 = \frac{2x}{1.00}$$

$x = 0.0090$ mol $\qquad 0.79 - x = 0.78$ mol $N_2 \qquad 0.21 - x = 0.20$ mol O_2

$\qquad 2x = 0.018$ mol NO

$$K_p = \frac{P\{NO\}^2}{P\{N_2\} \, P\{O_2\}} = \frac{\left(\dfrac{n\{NO\} \, RT}{V_{total}}\right)^2}{\dfrac{n\{N_2\} \, RT}{V_{total}} \dfrac{n\{O_2\} \, RT}{V_{total}}} = \frac{n\{NO\}^2}{n\{N_2\} \, n\{O_2\}} = \frac{(0.018)^2}{0.78 \times 0.20} = 2.1 \times 10^{-3}$$

65. We organize the data around the balanced chemical equation. Note that the reaction is stoichimoetrically balanced.

(a) Equation: $2 \, SO_2(g) \;+\;$ $O_2(g) \;\rightleftharpoons\;$ $2 \, SO_3(g)$

Equil :	0.32 mol	0.16 mol	0.68 mol
Add SO_3	0.32 mol	0.16 mol	1.68 mol
Initial :	$\dfrac{0.32\ \text{mol}}{10.0\ \text{L}}$	$\dfrac{0.16\ \text{mol}}{10.0\ \text{L}}$	$\dfrac{1.68\ \text{mol}}{10.0\ \text{L}}$

Initial :	0.032 M	0.016 M	0.168 M
To right :	0.000 M	0.000 M	0.200 M
Changes :	$+2x$ M	$+x$ M	$-2x$ M
Equil :	$2x$ M	x M	$(0.200 - 2x)$M

In setting up this problem, we note that, solving this question exactly involves finding the roots for a cubic equation. Thus, we assumed that all of the reactants have been converted to products. This gives the results in the line labeled "To right:" We then reach equilibrium from this position by converting some of the product back into reactants. Now, we substitute these expressions into the equilibrium constant expression, and we solve this expression approximately by assuming that $2x \ll 0.200$.

$$K_c = \frac{[SO_3]^2}{[SO_2]^2[O_2]} = \frac{(0.200 - 2x)^2}{(2x)^2 \, x} = 2.8 \times 10^2 \approx \frac{(0.200)^2}{4x^3} \text{ or } x = 0.033$$

We then substitute this approximate value into the expression for K_c.

$$K_c = \frac{(0.200 - 0.066)^2}{4x^3} = 2.8 \times 10^2 \text{ or } x = 0.025$$

Let us try one more cycle. $K_c = \dfrac{(0.200 - 0.050)^2}{4x^3} = 2.8 \times 10^2 \text{ or } x = 0.027$

This gives the following concentrations and amounts of each species.

$[SO_3] = 0.200 - (2 \times 0.027) = 0.146$ M   amount $SO_3 = 10.0 \, L \times 0.146 \, M = 1.46$ mol SO_3

$[SO_2] = 2 \times 0.027 = 0.054$ M   amount $SO_2 = 10.0 \, L \times 0.054 \, M = 0.54$ mol SO_2

$[O_2] = 0.027$ M   amount $O_2 = 10.0 \, L \times 0.027 \, M = 0.27$ mol O_2

(b)

Equation:	2 $SO_2(g)$	+	$O_2(g)$	$\rightleftharpoons$	2 $SO_3(g)$
Equil :	0.32 mol		0.16 mol		0.68 mol
Equil :	$\dfrac{0.32 \text{ mol}}{10.0 \text{ L}}$		$\dfrac{0.16 \text{ mol}}{10.0 \text{ L}}$		$\dfrac{0.68 \text{ mol}}{10.0 \text{ L}}$
Equil :	0.032 M		0.016 M		0.068 M
0.10 V :	0.32 M		0.16 M		0.68 M
To right :	0.00 M		0.00 M		1.00 M
Changes :	$+2x$ M		$+x$ M		$-2x$ M
Equil :	$2x$ M		x M		$(1.00-2x)$M

Again, we noticed that an exact solution involves finding the roots of a cubic. So we have taken the reaction 100% in the forward direction and then sent it back in the reverse direction to a small extent to reach equilibrium. We now solve the K_c expression for x, obtaining first an approximate value by assuming $2x \ll 1.00$.

$$K_c = \frac{[SO_3]^2}{[SO_2]^2[O_2]} = \frac{(1.00-2x)^2}{(2x)^2\,x} = 2.8\times10^2 \approx \frac{(1.00)^2}{4x^3} \; or \; x=0.096$$

We then use this approximate value of x to find a second approximation for x.

$$K_c = \frac{(1.00-0.19)^2}{4x^3} = 2.8\times10^2 \qquad or \qquad x=0.084$$

Another cycle gives $K_c = \dfrac{(1.00-0.17)^2}{4x^3} = 2.8\times10^2 \quad or \quad x=0.085$

Then we compute the equilibrium concentrations and amounts.

$[SO_3] = 1.00 - (2\times0.085) = 0.83$ M $\qquad$ amount $SO_3 = 1.00$ L $\times 0.83$ M $= 0.83$ mol SO_3

$[SO_2] = 2\times0.085 = 0.17$ M $\qquad$ amount $SO_2 = 1.00$ L $\times 0.17$ M $= 0.17$ mol SO_2

$[O_2] = 0.085$ M $\qquad$ amount $O_2 = 1.00$ L $\times 0.085$ M $= 0.085$ mol O_2

66. Equation: $HOC_6H_4COOH(g) \rightleftharpoons C_6H_5OH(g) + CO_2(g)$

$$n_{CO_2} = \frac{PV}{RT} = \left(\frac{\frac{720\text{mmHg}}{760\text{mmHg/atm}}}{0.0821\text{L}-\text{atm/mol}-\text{K}}\right)\left(\frac{(\frac{48.2+48.5}{2})\frac{1\text{ L}}{1000\text{ mL}}}{(293\text{K})}\right) = 1.93\times10^{-3} \text{ mol } CO_2$$

Note that moles of CO_2 = moles phenol

$$n_{salicylic\ acid} = \frac{0.300g}{138g/mol} = 2.17\times10^{-3} \text{ mol salicylic acid}$$

$$K_c = \frac{[C_6H_5OH][CO_2(g)]}{[HOC_6H_4COOH]} = \frac{\left(\dfrac{1.93 \text{ mmol}}{50.0 \text{ mL}}\right)^2}{\dfrac{(2.17-1.93) \text{ mmol}}{50.0 \text{ mL}}} = 0.310$$

$$K_p = K_c(RT)^{(2-1)} = (0.310) \times \left(0.08206 \frac{\text{L atm}}{\text{mol K}}\right) \times (473 \text{ K}) = 12.0$$

67. (a) This reaction is exothermic and thus, conversion of synthesis gas to methane is favored at lower temperatures. Since $\Delta n_{gas} = (1+1)-(1+3) = -2$, high pressure favors the products.

(b) The value of K_c is a large number, meaning that almost all of the reactants are converted to products (Note that the reaction is stoichiometrically balanced). Thus, after we set up the initial conditions we force the reaction to products and then allow the system to reach equilibrium.

Equation:	$3 H_2(g)$	$+$	$CO(g)$	$\rightleftharpoons$	$CH_4(g)$	$+$	$H_2O(g)$
Initial:	$\dfrac{3.00 \text{ mol}}{15.0 \text{ L}}$		$\dfrac{1.00 \text{ mol}}{15.0 \text{ L}}$		0 M		0 M
Initial:	0.200 M		0.0667 M		0 M		0 M
To right:	0.000 M		0.000 M		0.0667 M		0.0667 M
Changes:	$+3x$ M		$+x$ M		$-x$ M		$-x$ M
Equil:	$3x$ M		x M		$(0.0667-x)$M		$(0.0667-x)$M

$$K_c = \frac{[CH_4][H_2O]}{[H_2]^3[CO]} = \frac{(0.0667-x)^2}{(3x)^3 x} = 190. \qquad \sqrt{190.} = \frac{0.0667-x}{\sqrt{27}\, x^2}$$

$$\sqrt{190 \times 27}\; x^2 = 0.0667 - x = 71.6 \; x^2 \qquad 71.6\, x^2 + x - 0.0667 = 0$$

$$x = \frac{-b \pm \sqrt{b^2 - 4ac}}{2a} = \frac{-1.00 \pm \sqrt{1.00 + 19.1}}{143} = 0.0244 \text{ M}$$

$[CH_4] = [H_2O] = 0.0667 - 0.0244 = 0.0423$ M $\quad [H_2] = 3 \times 0.0244 = 0.0732$ M $\quad [CO] = 0.0244$ M

We check our calculation by computing the value of the equilibrium constant.

$$K_c = \frac{[CH_4][H_2O]}{[H_2]^3[CO]} = \frac{(0.0423)^2}{(0.0732)^3\, 0.0244} = 187$$

Now we compute the amount in moles of each component present at equilibrium, and finally the mole fraction of CH_4.

amount CH_4 = amount $H_2O = 0.0423$ M $\times 15.0$ L $= 0.634$ mol

amount $H_2 = 0.0732$ M $\times 15.0$ L $= 1.10$ mol $\qquad$ amount $CO = 0.0244$ M $\times 15.0$ L $= 0.366$ mol

$$\chi_{CH_4} = \frac{0.634 \text{ mol}}{0.634 \text{ mol} + 0.634 \text{ mol} + 1.10 \text{ mol} + 0.366 \text{ mol}} = 0.232$$

68. We base our calculation on 1.00 mole of PCl_5 being present initially.

Equation: $PCl_5(g) \rightleftharpoons PCl_3(g) + Cl_2(g)$

Initial: 1.00 mol   0 M   0 M

Changes: $-\alpha$ mol $+\alpha$ mol $+\alpha$ mol

Equil: $(1.00 - \alpha)$ mol α mol α mol

$n_{total} = 1.00 - \alpha + \alpha + \alpha = 1.00 + \alpha$

Equation: $PCl_5(g) \rightleftharpoons PCl_3(g) + Cl_2(g)$

Mol fract. $\dfrac{1.00 - \alpha}{1.00 + \alpha}$ $\dfrac{\alpha}{1.00 + \alpha}$ $\dfrac{\alpha}{1.00 + \alpha}$

$$K_p = \frac{P\{Cl_2\}\,P\{PCl_3\}}{P\{PCl_5\}} = \frac{[\chi\{Cl_2\}P_{total}]\,[\chi\{PCl_3\}P_{total}]}{[\chi\{PCl_5\}P_{total}]} = \frac{\left(\dfrac{\alpha}{1.00 + \alpha}P_{total}\right)^2}{\dfrac{1.00 - \alpha}{1.00 + \alpha}P_{total}} = \frac{\alpha^2\,P_{total}}{(1.00 + \alpha)(1.00 - \alpha)} = \frac{\alpha^2\,P_{total}}{1 - \alpha^2}$$

69. We assume that the entire 5.00 g is N_2O_4 and reach equilibrium from this starting point.

$$[N_2O_4]_i = \frac{5.00\text{ g}}{0.500\text{ L}} \times \frac{1\text{ mol }N_2O_4}{92.01\text{ g }N_2O_4} = 0.109\text{ M}$$

Equation: $N_2O_4(g) \rightleftharpoons 2\,NO_2(g)$

Initial: 0.109 M   0 M

Changes: $-x$ M $+2x$ M

Equil: $(0.109 - x)$M $2x$ M

$$Kc = \frac{[NO_2]^2}{[N_2O_4]} = 4.61 \times 10^{-3} = \frac{(2x)^2}{0.109 - x} \qquad 4x^2 = 5.02 \times 10^{-4} - 4.61 \times 10^{-3}\,x$$

$$4x^2 + 0.00461\,x - 0.000502 = 0$$

$$x = \frac{-b \pm \sqrt{b^2 - 4ac}}{2a} = \frac{-0.00461 \pm \sqrt{2.13 \times 10^{-5} + 8.03 \times 10^{-3}}}{8} = 0.0106\text{ M}, \; -0.0118\text{ M}$$

(The method of successive approximations yields 0.0106 after two iterations)

amount $N_2O_4 = 0.500$ L $(0.109 - 0.0106)$ M $= 0.0492$ mol N_2O_4

amount $NO_2 = 0.500$ L $\times 2 \times 0.0106$ M $= 0.0106$ mol NO_2

$$\text{mol fraction } NO_2 = \frac{0.0106\text{ mol }NO_2}{0.0106\text{ mol }NO_2 + 0.0492\text{ mol }N_2O_4} = 0.177$$

70. We let P be the initial pressure in atmospheres of $COCl_2(g)$.

Equation: $COCl_2(g) \rightleftharpoons CO(g) + Cl_2(g)$

Initial: P   0 M   0 M

Changes: $-x$ $+x$ $+x$

Equil: $P - x$ x x

total pressure $= 3.00$ atm $= P - x + x + x = P + x$ $P = 3.00 - x$

$P\{COCl_2\} = P - x = 3.00 - x - x = 3.00 - 2x$

$K_p = \dfrac{P\{\{CO\}P\{Cl_2\}}{P(COCl_2)} = 0.0444 = \dfrac{x \cdot x}{3.00 - 2x}$

$x^2 = 0.133 - 0.0888\,x \qquad\qquad x^2 + 0.0888\,x - 0.133 = 0$

$x = \dfrac{-b \pm \sqrt{b^2 - 4ac}}{2a} = \dfrac{-0.0888 \pm \sqrt{0.00789 + 0.532}}{2} = 0.323,\ -0.421$

Since a negative pressure is physically meaningless, $x = 0.323$ atm.
(The method of successive approximations yields $x = 0.323$ after four iterations.)

$P\{CO\} = P\{Cl_2\} = 0.323$ atm $\qquad\qquad P\{COCl_2\} = 3.00 - 2 \times 0.323 = 2.35$ atm

The mole fraction of each gas is its partial pressure divided by the total pressure. And the contribution of each gas to the apparent molar mass of the mixture is the mole fraction of that gas multiplied by the molar mass of that gas.

$M_{avg} = \dfrac{P\{CO\}}{P_{tot}} M\{CO\} + \dfrac{P\{Cl_2\}}{P_{tot}} M\{Cl_2\} + \dfrac{P\{COCl_2\}}{P_{tot}} M\{COCl_2\}$

$= \left(\dfrac{0.323\ \text{atm}}{3.00\ \text{atm}} \times 28.01\ \text{g/mol}\right) + \left(\dfrac{0.323\ \text{atm}}{3.00\ \text{atm}} \times 70.91\ \text{g/mol}\right) + \left(\dfrac{2.32\ \text{atm}}{3.00\ \text{atm}} \times 98.92\ \text{g/mol}\right)$

$= 87.1$ g/mol

71. Each mole fraction equals the partial pressure of the substance divided by the total pressure. Thus $\chi\{NH_3\} = P\{NH_3\}/P_{tot}$ or $P\{NH_3\} = \chi\{NH_3\} P_{tot}$

$K_p = \dfrac{P\{NH_3\}^2}{P\{N_2\} P\{H_2\}^3} = \dfrac{(\chi\{NH_3\}P_{tot})^2}{(\chi\{N_2\}P_{tot})(\chi\{H_2\}P_{tot})^3} = \dfrac{\chi\{NH_3\}^2}{\chi\{N_2\}\chi\{H_2\}^3} \dfrac{(P_{tot})^2}{(P_{tot})^4}$

$= \dfrac{\chi\{NH_3\}^2}{\chi\{N_2\}\chi\{H_2\}^3} \dfrac{1}{(P_{tot})^2}$

This is the expression we were asked to derive.

72. Since the mole ratio of N_2 to H_2 is 1:3, $\chi\{H_2\} = 3\chi\{N_2\}$. Since $P_{tot} = 1.00$ atm, it follows.

$K_p = \dfrac{\chi\{NH_3\}^2}{\chi\{N_2\}(3\chi\{N_2\})^3}\dfrac{1}{(1.00)^2} = 9.06 \times 10^{-2} = 0.0906$

$3^3 \times 0.0906 = \dfrac{\chi\{NH_3\}^2}{\chi\{N_2\}\chi\{N_2\}^3} = \dfrac{\chi\{NH_3\}^2}{\chi\{N_2\}^4} \qquad\qquad \dfrac{\chi\{NH_3\}}{\chi\{N_2\}^2} = \sqrt{3^3 \times 0.0906} = 1.56$

We realize that $\chi\{NH_3\} + \chi\{N_2\} + \chi\{H_2\} = 1.00 = \chi\{NH_3\} + \chi\{N_2\} + 3\chi\{N_2\}$
This gives $\chi\{NH_3\} = 1.00 - 4\chi\{N_2\}$ And we have

$1.56 = \dfrac{1.00 - 4\chi\{N_2\}}{\chi\{N_2\}^2}$ For ease of solving, we let $x = \chi\{N_2\}$

$$1.56 = \frac{1.00 - 4x}{x^2} \qquad 1.56\ x^2 = 1.00 - 4x \qquad 1.56\ x^2 + 4x - 1.00 = 0$$

$$x = \frac{-b \pm \sqrt{b^2 - 4ac}}{2a} = \frac{-4.00 \pm \sqrt{16.00 + 6.24}}{3.12} = 0.229,\ -2.794$$

Thus $\chi\{N_2\} = 0.229$ Mole % $NH_3 = (1.000\ \text{mol} - (4 \times 0.229\ \text{mol})) \times 100\% = 8.4\%$

73. Since the initial mole ratio is 2 $H_2S(g)$ to 1 $CH_4(g)$, the reactants remain in their stoichiometric ratio when equilibrium is reached. Also the products are formed in their stoichiometric ratio.

$$\text{amount } CH_4 = 9.54 \times 10^{-3}\ \text{mol } H_2S \times \frac{1\ \text{mol } CH_4}{2\ \text{mol } H_2S} = 4.77 \times 10^{-3}\ \text{mol } CH_4$$

$$\text{amount } CS_2 = 1.42 \times 10^{-3}\ \text{mol } BaSO_4 \times \frac{1\ \text{mol } S}{1\ \text{mol } BaSO_4} \times \frac{1\ \text{mol } CS_2}{2\ \text{mol } S} = 7.10 \times 10^{-4}\ \text{mol } CS_2$$

$$\text{amount } H_2 = 7.10 \times 10^{-4}\ \text{mol } CS_2 \times \frac{4\ \text{mol } H_2}{1\ \text{mol } CS_2} = 2.84 \times 10^{-3}\ \text{mol } H_2$$

$$\text{total amount} = 9.54 \times 10^{-3}\ \text{mol } H_2S + 4.77 \times 10^{-3}\ \text{mol } CH_4 + 7.10 \times 10^{-4}\ \text{mol } CS_2 + 2.84 \times 10^{-3}\ \text{mol } H_2$$

$$= 17.86 \times 10^{-3}\ \text{mol}$$

The partial pressure of each gas equals its mole fraction times the total pressure.

$$P\{H_2S\} = 1.00\ \text{atm} \times \frac{9.54 \times 10^{-3}\ \text{mol } H_2S}{17.86 \times 10^{-3}\ \text{mol total}} = 0.534\ \text{atm}$$

$$P\{CH_4\} = 1.00\ \text{atm} \times \frac{4.77 \times 10^{-3}\ \text{mol } CH_4}{17.86 \times 10^{-3}\ \text{mol total}} = 0.267\ \text{atm}$$

$$P\{CS_2\} = 1.00\ \text{atm} \times \frac{7.10 \times 10^{-4}\ \text{mol } CS_2}{17.86 \times 10^{-3}\ \text{mol total}} = 0.0398\ \text{atm}$$

$$P\{H_2\} = 1.00\ \text{atm} \times \frac{2.84 \times 10^{-3}\ \text{mol } H_2}{17.86 \times 10^{-3}\ \text{mol total}} = 0.159\ \text{atm}$$

$$K_p = \frac{P\{H_2\}^4\ P\{CS_2\}}{P\{H_2S\}^2\ P\{CH_4\}} = \frac{0.159^4 \times 0.0398}{0.534^2 \times 0.267} = 3.34 \times 10^{-4}$$

74. We base our calculation on an I.C.E. table, after we first determine the direction of the reaction by computing :

$$Q_c = \frac{[Fe^{2+}]^2 [Hg^{2+}]^2}{[Fe^{3+}]^2 [Hg_2^{2+}]} = \frac{(0.03000)^2 (0.03000)^2}{(0.5000)^2 (0.5000)} = 6.48 \times 10^{-6}$$

Because this value is smaller than K_c, the reaction will shift to the right to reach equilibrium. Since the value of the equilibrium constant for the forward reaction is quite small, let us assume that the reaction initially shifts all the way to the left (line labeled "to left:"), and then reacts back in the forward direction to reach a position of equilibrium.

Equation: $2 Fe^{3+}(aq)$ + $Hg_2^{2+}(aq)$ $\rightleftharpoons$ $2 Fe^{2+}(aq)$ + $2 Hg^{2+}(aq)$

Initial:	0.5000 M	0.5000 M	0.03000 M	0.03000 M
$\longleftarrow$	+0.03000 M	+0.0150 M	−0.03000 M	−0.03000 M
To left:	0.5300 M	0.5150 M	0 M	0 M
Changes:	$-2x$ M	$-x$ M	$+2x$ M	$+2x$ M
Equil:	$(0.5300 - 2x)$M	$(0.5150 - x)$M	$2x$ M	$2x$ M

$$K_c = \frac{[Fe^{2+}]^2[Hg^{2+}]^2}{[Fe^{3+}]^2[Hg_2^{2+}]} = 9.14\times10^{-6} = \frac{(2x)^2(2x)^2}{(0.5300-2x)^2(0.5150-x)} \approx \frac{4x^2\ 4x^2}{(0.5300)^2\ 0.5150}$$

Note that we have assumed that $2x \ll 0.5300$ and $x < 0.5150$

$$x^4 = \frac{9.14\times10^{-6}\ (0.5300)^2\ (0.5150)}{4\times4} = 8.26\times10^{-8} \qquad x = 0.0170$$

Our assumption, that $2x\ (=0.0340) \ll 0.5300$, is reasonably good.

$[Fe^{3+}] = 0.5300 - 2\times0.0170 = 0.4960$ M $\quad [Hg_2^{2+}] = 0.5150 - 0.0170 = 0.4980$

$[Fe^{2+}] = [Hg^{2+}] = 2\times0.0170 = 0.0340$ M

We check by substituting into the K_c expression.

$$9.14\times10^{-6} = K_c = \frac{[Fe^{2+}]^2[Hg^{2+}]^2}{[Fe^{3+}]^2[Hg_2^{2+}]} = \frac{(0.0340)^2(0.0340)^2}{(0.4960)^2\ 0.4980} = 11\times10^{-6} \quad \text{Not a substantial difference.}$$

Mathematica (version 4.0, Wolfram Research, Champaign, IL) gives a root of 0.0163.

75. Again we base our calculation on an I.C.E. table. In the course of solving the Integrative Example, we found that we could obtain the desired equation by reversing equation (2) and add the result to equation (1) to obtain the desired equation.

−(2) $H_2O(g)$ + $CH_4(g)$ $\rightleftharpoons$ $CO(g)$ + $3 H_2(g)$ $K = 1/190$

+(1) $CO(g)$ + $H_2O(g)$ $\rightleftharpoons$ $CO_2(g)$ + $H_2(g)$ $K = 1.4$

Equation: $CH_4(g)$ + $2 H_2O(g)$ $\rightleftharpoons$ $CO_2(g)$ + $4 H_2(g)$ $K = 1.4/190 = 0.0074$

−(2) $H_2O(g)$ + $CH_4(g)$ $\rightleftharpoons$ $CO(g)$ + $3 H_2(g)$ $K = 1/190$

+(1) $CO(g)$ + $H_2O(g)$ $\rightleftharpoons$ $CO_2(g)$ + $H_2(g)$ $K = 1.4$

Equation: $CH_4(g)$ + $2 H_2O(g)$ $\rightleftharpoons$ $CO_2(g)$ + $4 H_2(g)$ $K = 1.4/190 = 0.0074$

Initial:	0.100 mol	0.100 mol	0.100 mol	0.100 mol
$\longleftarrow$	+0.025 mol	+0.050 mol	−0.025 mol	−0.100 mol
To left:	0.125 mol	0.150 mol	0.075 mol	0.000 mol
Concns:	0.0250 M	0.0300 M	0.015 M	0.000 M
Changes:	$-x$ M	$-2x$ mol	$+x$ mol	$+4x$ mol
Equil:	$(0.0250 - x)$ M	$(0.0300 - 2x)$	$(0.015 + x)$ M	$4x$ mol

Notice that we have a fifth order polynomial to solve. Hence, we need to try to approximate its final solution as closely as possible. The reaction favors the reactants because of the small size of the equilibrium constant. Thus, we approach equilibrium from as far to the left as possible.

$$K_c = 0.0074 = \frac{[CO_2][H_2]^4}{[CH_4][H_2O]^2} = \frac{(0.0150+x)(4x)^4}{(0.0250-x)(0.0300-2x)^2} \approx \frac{0.0150 \ (4x)^4}{0.0250 \ (0.0300)^2}$$

$$x \approx \sqrt[4]{\frac{0.0250 \ (0.0300)^2 \ 0.0074}{0.0150 \times 256}} = 0.014 \, \text{M}$$

Our assumption is terrible. We substitute to continue successive approximations.

$$0.0074 = \frac{(0.0150 + 0.014)(4x)^4}{(0.0250-0.014)(0.0300-2\times0.014)^2} = \frac{(0.029)(4x)^4}{(0.011)(0.002)^2}$$

Next try $x_2 = 0.0026$

$$0.074 = \frac{(0.0150+0.0026)(4x)^4}{(0.0250 - 0.0026)(0.0300 - 2\times0.0026)^2}$$

then try $x_3 = 0.0123$

After 18 iterations the x value converges to 0.0080

Considering that the equilibrium constant is known to only two significant figures, this is a pretty good result. Recall that the total volume is 5.00 L. We calculate amounts in moles.

$CH_4(g)$ $(0.0250-0.0080)\times5.00\,\text{L} = 0.017\,\text{M}\times5.00\,\text{L} = 0.085$ moles $CH_4(g)$

$H_2O(g)$ $(0.0300-2\times0.0080)\,\text{M}\times5.00\,\text{L} = 0.014\,\text{M}\times5.00\,\text{L} = 0.070$ moles $H_2O(g)$

$CO_2(g)$ $(0.015 + 0.0080)\,\text{M}\times5.00\,\text{L} = 0.023\,\text{M} \times5.00\,\text{L} = 0.12$ mol CO_2

$H_2(g)$ $(4\times0.0080)\,\text{M} \times5.00\,\text{L} = 0.032\,\text{M} \times5.00\,\text{L} = 0.16$ mol H_2

76. The initial mole fraction of C_2H_2 is $\chi_i = 0.88_2$. We use molar amounts at equilibrium to compute the equilibrium mole fraction of C_2H_2, χ_{eq}. Because we have a 2.00-L container, molar amounts are double the molar concentrations.

$$\chi_{eq} = \frac{(2\times0.756) \, \text{mol} \, C_2H_2}{(2\times0.756) \, \text{mol} \, C_2H_2 + (2\times0.038) \, \text{mol} \, CH_4 + (2\times0.068) \, \text{mol} \, H_2} = 0.87\underline{7}$$

Thus, there has been only a slight decrease in mole fraction.

77. (a) $K_{eq} = 4.6 \times 10^4$ $\dfrac{P\{NOCl\}^2}{P\{NO\}^2 P\{Cl_2\}} = \dfrac{(4.125)^2}{P\{NO\}^2 (0.1125)}$ $P\{NO\}^2 = \dfrac{(4.125)^2}{4.6\times10^4 (0.1125)} = 0.057\underline{3}$ atm

(b) $P_{total} = P_{NO} + P_{Cl_2} + P_{NOCl} = 0.057\underline{3}$ atm $+ 0.1125$ atm $+ 4.125$ atm $= 4.295$ atm

78. We base our calculation on an I.C.E. table.

Reaction:	$N_2(g)$	+	$3H_2(g)$	$\rightleftharpoons$	$2NH_3(g)$
Initial:	$\dfrac{0.424\,\text{mol}}{10.0\,\text{L}}$		$\dfrac{1.272\,\text{mol}}{10.0\,\text{L}}$		$\dfrac{0\,\text{mol}}{10.0\,\text{L}}$
Change	$\dfrac{-x\,\text{mol}}{10.0\,\text{L}}$		$\dfrac{-3x\,\text{mol}}{10.0\,\text{L}}$		$\dfrac{+2x\,\text{mol}}{10.0\,\text{L}}$
Equilibrium	$\dfrac{(0.424\text{-}x)\,\text{mol}}{10.0\,\text{L}}$		$\dfrac{(1.272\text{-}3x)\,\text{mol}}{10.0\,\text{L}}$		$\dfrac{2x\,\text{mol}}{10.0\,\text{L}}$

$$K_c = \frac{[NH_3]^2}{[N_2][H_2]^3} = 152 = \frac{\left(\dfrac{2x\,\text{mol}}{10.0\,\text{L}}\right)^2}{\left(\dfrac{(0.424\text{-}x)\,\text{mol}}{10.0\,\text{L}}\right)\left(\dfrac{(1.272\text{-}3x)\,\text{mol}}{10.0\,\text{L}}\right)^3}$$

$$K_c = \frac{100(2x\,\text{mol})^2}{((0.424\text{-}x)\,\text{mol})(3(0.424\text{-}x)\,\text{mol})^3}$$

$$K_c = \frac{100(2x\,\text{mol})^2}{3^3\,(0.424\text{-}x)\,\text{mol})^4} = 152 \qquad \frac{(2x\,\text{mol})^2}{(0.424\text{-}x)\,\text{mol})^4} = 41.0\underline{4} \qquad \text{Take root of both sides}$$

$$\frac{(2x\,\text{mol})}{(0.424\text{-}x)\,\text{mol})^2} = 6.41 \qquad 6.41(0.424\text{-}x)^2 = 2x$$

$3.20(0.180 - 0.848x + x^2) = x = 3.20x^2 - 2.71x + 0.576 \qquad 3.20x^2 - 3.71x + 0.576 = 0$

Now solve using the quadratic equation: $x = 0.184\underline{6}$ mol or $0.975\underline{6}$ mol(too large)

amount of $NH_3 = 2x = 2(0.1846\,\text{mol}) = 0.369$ mol in 10.0 L or 0.0369 M

$([H_2] = 0.0718\,M \text{ and } [N_2] = 0.0239\,M)$

79. Equation: $2\,H_2(g)$ + $CO(g)$ $\rightleftharpoons$ $CH_3OH(g)$ $K_c = 14.5$ at 483 K

$$K_p = K_c(RT)^{\Delta n} = 14.5\left(0.08206\,\frac{\text{L-atm}}{\text{mol-K}} \times 483K\right)^{-2} = 9.23 \times 10^{-3}$$

We know that mole percents equal pressure percents for ideal gases.

$P_{CO} = 0.350 \times 100\,\text{atm} = 35.0\,\text{atm}$

$P_{H_2} = 0.650 \times 100\,\text{atm} = 65.0\,\text{atm}$

Equation: $2\,H_2(g)$ + $CO(g)$ $\rightleftharpoons$ $CH_3OH(g)$

Initial: 65 atm   35 atm

Changes: $-2P$ atm $-P$ atm $+P$ atm

Equil: 65-2P   35-P   P

$$K_p = \frac{P_{CH_3OH}}{P_{CO} \times P_{H_2}{}^2} = \frac{P}{(35.0-P)(65.0-2P)^2} = 9.23\times10^{-3}$$

By successive approximations, P = 24.6 atm = P_{CH_3OH} at equilibrium.

Mathematica (version 4.0, Wolfram Research, Champaign, IL) gives a root of 24.5$_3$.

FEATURE PROBLEMS

80. We first determine the amount in moles of acetic acid in the equilibrium mixture.

$$\text{amount } CH_3CO_2H = 28.85 \text{ mL}\times\frac{1\,L}{1000\,\text{mL}}\times\frac{0.1000\,\text{mol Ba(OH)}_2}{1\,L}\times\frac{2\,\text{mol }CH_3CO_2H}{1\,\text{mol Ba(OH)}_2}$$

$$\times\frac{\text{complete equilibrium mixture}}{0.01\,\text{of equilibrium mixture}} = 0.5770 \text{ mol } CH_3CO_2H$$

$$K_c = \frac{[CH_3CO_2C_2H_5][H_2O]}{[C_2H_5OH][CH_3CO_2H]} = \frac{\dfrac{0.423\,\text{mol}}{V}\times\dfrac{0.423\,\text{mol}}{V}}{\dfrac{0.077\,\text{mol}}{V}\times\dfrac{0.577\,\text{mol}}{V}} = \frac{0.423\times0.423}{0.077\times0.577} = 4.0$$

81. In order to determine whether or not equilibrium has been established in each bulb, we need to calculate the concentrations for all three species at the time of opening. The results from these calculations are tabulated below and a typical calculation is given beneath this table.

Bulb #	Time Bulb opened (hours)	initial amount HI(g) (in mmol)	amount of $I_2(g)$ and $H_2(g)$ at time of opening (in mmol)	amount HI(g) at time of opening (in mmol)	[HI] (mM)	$[I_2]$ & $[H_2]$ (mM)	$\dfrac{[H_2][I_2]}{[HI]^2}$
1	2	2.34<u>5</u>	0.1572	2.03	5.08	0.393	0.00599
2	4	2.51<u>8</u>	0.2093	2.10	5.25	0.523	0.00992
3	12	2.46<u>3</u>	0.2423	1.98	4.95	0.606	0.0150
4	20	3.17<u>4</u>	0.3113	2.55	6.38	0.778	0.0149
5	40	2.18<u>9</u>	0.2151	1.76	4.40	0.538	0.0150

Consider for instance bulb #4 (opened after 20 hours)

$$\text{Initial moles of HI(g)} = 0.406 \text{ g HI(g)} \times\frac{1\,\text{mole HI}}{127.9\,\text{g HI}} = 0.00317\underline{4} \text{ mol HI(g) or } 3.17\underline{4} \text{ mmol}$$

moles of $I_2(g)$ present in bulb when opened.

$$= 0.04150 \text{ L Na}_2\text{S}_2\text{O}_3 \times \frac{0.0150 \text{ mol Na}_2\text{S}_2\text{O}_3}{1 \text{ L Na}_2\text{S}_2\text{O}_3} \times \frac{1 \text{ mol I}_2}{2 \text{ mol Na}_2\text{S}_2\text{O}_3} = 3.113 \times 10^{-4} \text{ mol I}_2$$

millimoles of $I_2(g)$ present in bulb when opened = $3.113 \times 10^{-4} \text{ mol I}_2$
moles of H_2 present in bulb when opened = moles of $I_2(g)$ present in bulb when opened.

$$\text{HI reacted} = 3.113 \times 10^{-4} \text{ mol I}_2 \times \frac{2 \text{ mole HI}}{1 \text{ mol I}_2} = 6.226 \times 10^{-4} \text{ mol HI (0.6226 mmol HI)}$$

moles of $HI(g)$ in bulb when opened= 3.17$\underline{4}$ mmol HI − 0.6226 mmol HI = 2.55 mmol HI

Concentrations of HI, I_2 and H_2

[HI] = 2.55 mmol HI ÷ 400. mL = 6.38 mM

$[I_2]$ = $[H_2]$ = 0.3113 mmol ÷ 400 mL = 0.778 mM

Ratio: $\dfrac{[H_2][I_2]}{[HI]^2} = \dfrac{(0.778 \text{ mM})(0.778 \text{ mM})}{(6.38 \text{ mM})^2} = 0.0149$

As the time increases, the ratio $\dfrac{[H_2][I_2]}{[HI]^2}$ initially climbs sharply, but then plateaus at

0.0150 somewhere between 4 and 12 hours. Consequently, it seems quite reasonable to conclude that the reaction $2HI(g) \rightleftharpoons H_2(g) + I_2(g)$ has a $K_c \sim 0.015$ at 623 K

82. We first need to determine the number of moles of ammonia that were present in the sample of gas that left the reactor. This will be accomplished by using the data from the titrations involving HCl(aq).

Original number of moles of HCl(aq) in the 20.00 mL sample

$$= 0.01872 \text{ L of KOH} \times \frac{0.0523 \text{ mol KOH}}{1 \text{ L KOH}} \times \frac{1 \text{ mol HCl}}{1 \text{ mol KOH}}$$

$$= 9.79\underline{06} \times 10^{-4} \text{ moles of HCl}_{(initially)}$$

Moles of unreacted HCl(aq)

$$= 0.01542 \text{ L of KOH} \times \frac{0.0523 \text{ mol KOH}}{1 \text{ L KOH}} \times \frac{1 \text{ mol HCl}}{1 \text{ mol KOH}} =$$
$8.06\underline{47} \times 10^{-4} \text{ moles of HCl}_{(unreacted)}$

Moles of HCl that reacted and /or moles of NH_3 present in the sample of reactor gas
$= 9.79\underline{06} \times 10^{-4}$ moles $- 8.06\underline{47} \times 10^{-4}$ moles $= 1.73 \times 10^{-4}$ mole of NH_3 (or HCl)

The remaining gas, which is a mixture of $N_2(g)$ and $H_2(g)$ gases, was found to occupy 1.82 L
at 273.2 K and 1.00 atm. Thus, the total number of moles of N_2 and H_2 can be found via the

ideal gas law: $n_{H_2 + N_2} = \dfrac{PV}{RT} = \dfrac{(1.00 \ \text{atm})(1.82 \ \text{L})}{(0.08206 \ \dfrac{\text{L atm}}{\text{K mol}})(273.2 \ \text{K})} = 0.0811\underline{8}$ moles of $(N_2 + H_2)$

According to the stoichiometry for the reaction, 2 parts NH_3 decompose to give 3 parts H_2
and 1 part N_2. Thus the non-reacting mixture must be 75% H_2 and 25% N_2.

So, the number of moles of $N_2 = 0.25 \times 0.0811\underline{8}$ moles $= 0.0203$ moles N_2 and the
number of moles of $H_2 = 0.75 \times 0.0811\underline{8}$ moles $= 0.0609$ moles H_2

Before we can calculate K_c, we need to determine the volume that the NH_3, N_2 and H_2
molecules occupied in the reactor. Once again, the ideal gas law ($PV = nRT$) will be
employed. $n_{gas} = 0.0811\underline{8}$ moles $(N_2 + H_2) + 1.73 \times 10^{-4}$ moles $NH_3 = 0.0813\underline{5}$ moles

$$V_{gases} = \dfrac{nRT}{P} = \dfrac{(0.08135 \ \text{mol})(0.08206 \ \dfrac{\text{L atm}}{\text{K mol}})(1174.2 \ \text{K})}{30.0 \ \text{atm}} = 0.261\underline{3} \ \text{L}$$

So, $K_c = \dfrac{\left[\dfrac{1.73 \times 10^{-4} \ \text{moles}}{0.2613 \ \text{L}}\right]^2}{\left[\dfrac{0.0609 \ \text{moles}}{0.2613 \ \text{L}}\right]^3 \left[\dfrac{0.0203 \ \text{moles}}{0.2613 \ \text{L}}\right]^1} = 4.46 \times 10^{-4}$

To calculate K_p at 901 °C, we need to employ the equation $K_p = K_c(RT)^{\Delta n_{gas}}$; $\Delta n_{gas} = -2$

$K_p = 4.46 \times 10^{-4} [(0.08206 \ \text{L atm K}^{-1}\text{mol}^{-1})] \times (1174.2 \ \text{K})]^{-2} = 4.80 \times 10^{-8}$ at 901°C for the
reaction $N_2(g) + 3 \ H_2(g) \rightleftharpoons 2 \ NH_3(g)$

83. For step 1, rate of the forward reaction = rate of the reverse reaction, so,

$k_1[I_2] = k_{-1}[I]^2$ or $\dfrac{k_1}{k_{-1}} = \dfrac{[I]^2}{[I_2]} = K_c$ (step 1)

Like the first step, the rates for the forward and reverse reactions are equal in the second step
and thus,

$k_2[I]^2[H_2] = k_{-2}[HI]^2$ or $\dfrac{k_2}{k_{-2}} = \dfrac{[HI]^2}{[I]^2[H_2]} = K_c$ (step 2)

Now we combine the two elementary steps to obtain the overall equation and its associated equilibrium constant.

$$I_2(g) \rightleftharpoons 2\,I(g) \qquad K_c = \frac{k_1}{k_{-1}} = \frac{[I]^2}{[I_2]} \text{ (STEP 1)}$$

and

$$H_2(g) + 2\,I(g) \rightleftharpoons 2\,HI(g) \qquad K_c = \frac{k_2}{k_{-2}} = \frac{[HI]^2}{[I]^2[H_2]} \text{ (STEP 2)}$$

$$H_2(g) + I_2(g) \rightleftharpoons 2\,HI(g) \qquad K_{c(overall)} = K_{c(step\ 1)} \times K_{c(step\ 2)}$$

$$K_{c(overall)} = \frac{k_1}{k_{-1}} \times \frac{k_2}{k_{-2}} = \frac{[I]^2}{[I_2]} \times \frac{[HI]^2}{[I]^2[H_2]}$$

$$K_{c(overall)} = \frac{k_1 k_2}{k_{-1}k_{-2}} = \frac{[I]^2[HI]^2}{[I]^2[I_2][H_2]} = \frac{[HI]^2}{[I_2][H_2]}$$

CHAPTER 16
ACIDS AND BASES
PRACTICE EXAMPLES

1A **(a)** In the forward direction, HF is the acid, (proton donor; forms F^-), and H_2O is the base (proton acceptor; forms H_3O^+). In the reverse direction F^- is the base (forms HF), accepting a proton from H_3O^+, which is the acid, (forms H_2O).

(b) In the forward direction, HSO_4^- is the acid, (proton donor; forms SO_4^{2-}), and NH_3 is the base (proton acceptor; forms NH_4^+). In the reverse direction SO_4^{2-} is the base (forms HSO_4^-), accepting a proton from NH_4^+, which is the acid, (forms NH_3).

(c) In the forward direction, HCl is the acid, (proton donor; forms Cl^-), and $C_2H_3O_2^-$ is the base (proton acceptor; forms $HC_2H_3O_2$). In the reverse direction Cl^- is the base (forms HCl), accepting a proton from $HC_2H_3O_2$, which is the acid, (forms $C_2H_3O_2^-$).

1B We know that the formulas of most acids begin with H. Thus, we identify HNO_2 and HCO_3^- as acids.

$$HNO_2(aq) + H_2O(l) \rightleftharpoons NO_2^-(aq) + H_3O^+(aq);$$

$$HCO_3^-(aq) + H_2O(l) \rightleftharpoons CO_3^{2-}(aq) + H_3O^+(aq)$$

A negatively charged species will attract a positively charged proton and act as a base. Thus PO_4^{3-} and HCO_3^- can act as bases. We also know that PO_4^{3-} must be a base because it cannot act as an acid—it has no protons to donate—and we know that all three species have acid-base properties.

$$PO_4^{3-}(aq) + H_2O(l) \rightleftharpoons HPO_4^{2-}(aq) + OH^-(aq);$$

$$HCO_3^-(aq) + H_2O(l) \rightleftharpoons H_2CO_3(aq) \rightarrow CO_2 \cdot H_2O(aq) + OH^-(aq)$$

Notice that HCO_3^- is the amphiprotic species, acting as both an acid and as a base.

2A $[H_3O^+]$ is readily computed from pH: $[H_3O^+] = 10^{-pH}$ $[H_3O^+] = 10^{-2.85} = 1.4 \times 10^{-3}$ M.

$[OH^-]$ can be found in two ways: (1) from $K_w = [H_3O^+][OH^-]$, giving

$$[OH^-] = \frac{K_w}{[H_3O^+]} = \frac{1.0 \times 10^{-14}}{1.4 \times 10^{-3}} = 7.1 \times 10^{-12} \text{ M, or (2) from pH} + \text{pOH} = 14.00, \text{ giving}$$

$pOH = 14.00 - pH = 14.00 - 2.85 = 11.15$, and then $[OH^-] = 10^{-pOH} = 10^{-11.15} = 7.1 \times 10^{-12}$ M.

2B $\left[H_3O^+\right]$ is computed from pH in each case: $\left[H_3O^+\right] = 10^{-pH}$

$\left[H_3O^+\right]_{conc.} = 10^{-2.50} = 3.2 \times 10^{-3} M$ $\left[H_3O^+\right]_{dil.} = 10^{-3.10} = 7.9 \times 10^{-4} M$

All of the H_3O^+ in the dilute solution comes from the concentrated solution.

amount $H_3O^+ = 1.00$ L conc. soln $\times \dfrac{3.2 \times 10^{-3} \text{ mol } H_3O^+}{1 \text{ L conc. soln}} = 3.2 \times 10^{-3} \text{ mol } H_3O^+$

Next we calculate the volume of the dilute solution.

volume of dilute solution $= 3.2 \times 10^{-3} \text{ mol } H_3O^+ \times \dfrac{1 \text{ L dilute soln}}{7.9 \times 10^{-4} \text{ mol } H_3O^+} = 4.1$ L dilute soln

Thus, the volume of water to be added = 3.1 L
Infinite dilution does not lead to infinitely small hydrogen ion concentrations. Since dilution is done with water, the pH of an infinitely dilute solution will approach that of pure water, namely pH = 7.

3A pH is computed directly from the $\left[H_3O^+\right]$, $pH = -\log\left[H_3O^+\right] = -\log(0.0025) = 2.60$.

We know that HI is a strong acid and, thus, is completely dissociated into H_3O^+ and I^-.
The consequence is that $\left[I^-\right] = \left[H_3O^+\right] = 0.0025$ M. $\left[OH^-\right]$ is most readily computed from

pH: $pOH = 14.00 - pH = 14.00 - 2.60 = 11.40$; $\left[OH^-\right] = 10^{-pOH} = 10^{-11.40} = 4.0 \times 10^{-12}$ M

3B The number of moles of HCl(g) is calculated from the ideal gas law. Then $\left[H_3O^+\right]$ is calculated, based on the fact that HCl(aq) is a strong acid (1 mol H_3O^+ is produced from each mol of HCl).

moles HCl(g) $= \dfrac{\left(747 \text{ mmHg} \times \dfrac{1 \text{ atm}}{760 \text{ mmHg}}\right) \times 0.535 \text{ L}}{\dfrac{0.08206 \text{ L atm}}{\text{mol K}} \times (26.5 + 273.2) \text{ K}}$

moles HCl(g) $= 0.0214$ mol HCl(g) $= 0.0214$ mol H_3O^+ when dissolved in water

$[H_3O^+] = 0.0214$ mol   pH = -log(0.0214) = 1.670

4A pH is most readily determined from $pOH = -\log\left[OH^-\right]$. Assume $Mg(OH)_2$ is a strong base.

$\left[OH^-\right] = \dfrac{9.63 \text{ mg } Mg(OH)_2}{100.0 \text{ mL soln}} \times \dfrac{1000 \text{ mL}}{1 \text{ L}} \times \dfrac{1 g}{1000 \text{ mg}} \times \dfrac{1 \text{ mol } Mg(OH)_2}{58.32 g \, Mg(OH)_2} \times \dfrac{2 \text{ mol } OH^-}{1 \text{ mol } Mg(OH)_2}$

$\left[OH^-\right] = 0.00330 M$; $pOH = -\log(0.00330) = 2.481$

$pH = 14.000 - pOH = 14.000 - 2.481 = 11.519$

4B   KOH is a strong base, which means that each mole of KOH that dissolves produces one mole of dissolved $OH^-(aq)$. First we calculate $\left[OH^-\right]$ and the pOH. We then use $pH + pOH = 14.00$ to determine pH.

$$\left[OH^-\right] = \frac{3.00\,g\,KOH}{100.00\,g\,soln} \times \frac{1\,mol\,KOH}{56.11\,g\,KOH} \times \frac{1\,mol\,OH^-}{1\,mol\,KOH} \times \frac{1.0242\,g\,soln}{1\,mL\,soln} \times \frac{1000\,mL}{1\,L} = 0.548\,M$$

$$pOH = -\log(0.548) = 0.261 \quad pH = 14.000 - pOH = 14.000 - 0.261 = 13.739$$

5A $\left[H_3O^+\right] = 10^{-pH} = 10^{-4.18} = 6.6 \times 10^{-5}$ M.

Organize solution using the balanced chemical equation.

Equation:	$HOCl(aq)$	+	$H_2O(l)$	$\rightleftharpoons$	$H_3O^+(aq)$	+	$OCl^-(aq)$
Initial:	0.150 M		—		≈ 0 M		0 M
Changes:	-6.6×10^{-5} M		—		$+6.6 \times 10^{-5}$ M		$+6.6 \times 10^{-5}$ M
Equil:	≈ 0.150 M		—		6.6×10^{-5} M		6.6×10^{-5} M

$$K_a = \frac{\left[H_3O^+\right]\left[OCl^-\right]}{\left[HOCl\right]} = \frac{\left(6.6 \times 10^{-5}\right)\left(6.6 \times 10^{-5}\right)}{0.150} = 2.9 \times 10^{-8}$$

5B   First, we use pH to determine $\left[OH^-\right]$. $pOH = 14.00 - pH = 14.00 - 10.08 = 3.92$.

$\left[OH^-\right] = 10^{-pOH} = 10^{-3.92} = 1.2 \times 10^{-4}$ M. We determine the initial concentration of cocaine and then organize the solution around the balanced equation in the manner we have used before.

$$[C_{17}H_{21}O_4N] = \frac{0.17\,g\,C_{17}H_{21}O_4N}{100\,mL\,soln} \times \frac{1000\,mL}{1\,L} \times \frac{1\,mol\,C_{17}H_{21}O_4N}{303.36\,g\,C_{17}H_{21}O_4N} = 0.0056\,M$$

Equation:	$C_{17}H_{21}O_4N(aq)$	+	$H_2O(l)$	$\rightleftharpoons$	$C_{17}H_{21}O_4NH^+(aq)$	+	$OH^-(aq)$
Initial:	0.0056 M		—		0 M		≈ 0 M
Changes:	-1.2×10^{-4} M		—		$+1.2 \times 10^{-4}$ M		$+1.2 \times 10^{-4}$ M
Equil:	≈ 0.0055 M		—		1.2×10^{-4} M		1.2×10^{-4} M

$$K_b = \frac{\left[C_{17}H_{21}O_4NH^+\right]\left[OH^-\right]}{\left[C_{17}H_{21}O_4N\right]} = \frac{\left(1.2 \times 10^{-4}\right)\left(1.2 \times 10^{-4}\right)}{0.0055} = 2.6 \times 10^{-6}$$

6A Again we organize our solution around the balanced chemical equation.

Equation: $HC_2H_2FO_2(aq)$ + $H_2O(l)$ $\rightleftharpoons$ $H_3O^+(aq)$ + $C_2H_2FO_2^-(aq)$
Initial: 0.100 M   — $\approx 0\,M$ 0 M
Changes: $-x$ M   — $+x$ M $+x$ M
Equil: $(0.100-x)$ M   — x M x M

$$K_a = \frac{[H_3O^+][C_2H_2FO_2^-]}{[HC_2H_2FO_2^-]} \qquad K_a = 2.6\times10^{-3}$$

With the simplifying assumption $[H_3O^+] = 0.016$ M, pH $= 1.80$ (without, pH $= 1.83$).
Thus, the calculated pH is considerably lower than 2.89. (Example 16-6).

6B We first determine the concentration of undissociated acid. We then use this value in a setup that is based on the balanced chemical equation.

$$2\,\text{aspirin tablets} \times \frac{0.500\text{ g }HC_9H_7O_4}{\text{tablet}} \times \frac{1\text{ mol }HC_9H_7O_4}{180.155\text{ g }HC_9H_7O_4} \times \frac{1}{0.325\text{L}} = 0.0171M$$

Equation: $HC_9H_7O_4(aq)$ + $H_2O(l)$ $\rightleftharpoons$ $H_3O^+(aq)$ + $C_9H_7O_4^-(aq)$
Initial: 0.0171 M   —   0 M   0 M
Changes: $-x$ M   — $+x$ M $+x$ M
Equil: $(0.0171-x)$ M   — x M x M

$$K_a = \frac{[H_3O^+][C_9H_7O_4^-]}{[HC_9H_7O_4]} = 3.3\times10^{-4} = \frac{x\cdot x}{0.0171-x}$$

$x^2 + 3.3\times10^{-4} - 5.6\underline{4}\times10^{-6} = 0$ (find the physically reasonable roots of the quadratic equation)

$$x = \frac{-3.3\times10^{-4} \pm \sqrt{1.1\times10^{-7} + 2.3\times10^{-5}}}{2} = 0.0022\,M; \quad pH = -\log(0.0022) = 2.66$$

7A Again we organize our solution around the balanced chemical equation.

Equation: $HC_2H_2FO_2(aq)$ + $H_2O(l)$ $\rightleftharpoons$ $H_3O^+(aq)$ + $C_2H_2FO_2^-(aq)$
Initial: 0.015 M   — $\approx 0\,M$ 0 M
Changes $-x$ M   — $+x$ M $+x$ M
Equil: $(0.015-x)$ M   — x M x M

$$K_a = \frac{[H_3O^+][C_2H_2FO_2^-]}{[HC_2H_2FO_2]} = 2.6\times10^{-3} = \frac{(x)(x)}{0.015-x} \approx \frac{x^2}{0.015}$$

$x = \sqrt{x^2} = \sqrt{0.015\times2.6\times10^{-3}} = 0.0062\,M = [H_3O^+]$ Our assumption is invalid:

0.0062 is not quite small enough compared to 0.015 for the 5% rule to hold. Thus we use another cycle of successive approximations.

$$K_a = \frac{(x)(x)}{0.015 - 0.0062} = 2.6 \times 10^{-3} \quad x = \sqrt{(0.015 - 0.0062) \times 2.6 \times 10^{-3}} = 0.0048 \text{ M} = [H_3O^+]$$

$$K_a = \frac{(x)(x)}{0.015 - 0.0048} = 2.6 \times 10^{-3} \quad x = \sqrt{(0.015 - 0.0048) \times 2.6 \times 10^{-3}} = 0.0051 \text{ M} = [H_3O^+]$$

$$K_a = \frac{(x)(x)}{0.015 - 0.0051} = 2.6 \times 10^{-3} \quad x = \sqrt{(0.015 - 0.0051) \times 2.6 \times 10^{-3}} = 0.0051 \text{ M} = [H_3O^+]$$

Two successive identical results is a signal that we have the solution.

$pH = -\log[H_3O^+] = -\log(0.0051) = 2.29$. The quadratic equation gives the same result (0.0051 M) as this method of successive approximations.

7B First we find $[C_5H_{11}N]$. We then use this value as the starting base concentration in a set up based on the balanced chemical equation.

$$[C_5H_{11}N] = \frac{114 \text{ mg } C_5H_{11}N}{315 \text{ mL soln}} \times \frac{1 \text{ mmol } C_5H_{11}N}{85.15 \text{ mg } C_5H_{11}N} = 0.00425 \text{ M}$$

Equation:	$C_5H_{11}N(aq)$	+	$H_2O(l)$	$\rightleftharpoons$	$C_5H_{11}NH^+(aq)$	+	$OH^-(aq)$
Initial:	0.00425 M		—		0 M		≈ 0 M
Changes:	$-x$ M		—		$+x$ M		$+x$ M
Equil:	$(0.00425 - x)$ M		—		x M		x M

$$K_b = \frac{[C_5H_{11}NH^+][OH^-]}{[C_5H_{11}N]} = 1.6 \times 10^{-3} = \frac{x \cdot x}{0.00425 - x} \approx \frac{x \cdot x}{0.00425} \quad \text{We assumed that } x \ll 0.00425$$

$x = \sqrt{0.0016 \times 0.00425} = 0.0026 \text{ M}$ The assumption is not valid. Let's assume $x \approx 0.0026$

$x = \sqrt{0.0016(0.00425 - 0.0026)} = 0.0016$ Let's try again, with $x \approx 0.0016$

$x = \sqrt{0.0016(0.00425 - 0.0016)} = 0.0021$ Yet another try, with $x \approx 0.0021$

$x = \sqrt{0.0016(0.00425 - 0.0021)} = 0.0019$ The last time, with $x \approx 0.0019$

$x = \sqrt{0.0016(0.00425 - 0.0019)} = 0.0019 \text{ M} = [OH^-]$

$pOH = -\log[H_3O^+] = -\log(0.0019) = 2.72$ $pH = 14.00 - pOH = 14.00 - 2.72 = 11.28$

We could have solved the problem with the quadratic formula roots equation rather than by successive approximations. The same answer is obtained. In fact, if we substitute $x = 0.0019$ into the K_b expression, we obtain $(0.0019)^2 / (0.00425 - 0.0019) = 1.5 \times 10^{-3}$ compared to $K_b = 1.6 \times 10^{-3}$. The error is due to rounding, not to an incorrect value. Using $x = 0.0020$ gives a value of 1.8×10^{-3}, while using $x = 0.0018$ gives 1.3×10^{-3}.

8A We organize the solution around the balanced chemical equation; a M is $[HF]_{initial}$.

Equation: $HF(aq)$ + $H_2O(l)$ $\rightleftharpoons$ $H_3O^+(aq)$ + $F^-(aq)$

Initial: a M   — ≈ 0 M   0 M

Changes: $-x$ M   — $+x$ M $+x$ M

Equil: $(a-x)$ M   — x M x M

$$K_a = \frac{[H_3O^+][F^-]}{[HF]} = \frac{(x)(x)}{a-x} \approx \frac{x^2}{a} = 6.6\times10^{-4} \qquad x = \sqrt{a\times 6.6\times10^{-4}}$$

For 0.20 M HF, a = 0.20 M $x = \sqrt{0.20\times 6.6\times10^{-4}} = 0.011$ M

$$\% \text{ dissoc} = \frac{0.011 \text{ M}}{0.20 \text{ M}} \times 100\% = 5.5\%$$

For 0.020 M HF, a = 0.020 M $x = \sqrt{0.020\times 6.6\times10^{-4}} = 0.0036$ M

We need another cycle of approximation: $x = \sqrt{(0.020-0.0036)\times 6.6\times10^{-4}} = 0.0033$ M

Yet another cycle with $x \approx 0.0033$ M : $x = \sqrt{(0.020-0.0033)\times 6.6\times10^{-4}} = 0.0033$ M

$$\% \text{ dissoc} = \frac{0.0033\text{M}}{0.020\text{M}} \times 100\% = 17\%$$

As expected, the weak acid is more dissociated.

8B Since both H_3O^+ and $C_3H_5O_3^-$ come from the same source in equimolar amounts, their concentrations are equal. $[H_3O^+] = [C_3H_5O_3^-] = 0.067\times 0.0284\text{M} = 0.0019\text{M}$

$$K_a = \frac{[H_3O^+][C_3H_5O_3^-]}{[HC_3H_5O_3]} = \frac{(0.0019)(0.0019)}{0.0284-0.0019} = 1.4\times10^{-4}$$

9A For an aqueous solution of a diprotic acid, the concentration of the divalent anion equals the second ionization constant: $\left[{}^-OOCCH_2COO^-\right] = K_{a_2} = 2.0\times10^{-6}\text{M}$. We organize around the chemical equation.

Equation: $CH_2(COOH)_2(aq)$ + $H_2O(l)$ $\rightleftharpoons$ $H_3O^+(aq)$ + $HCH_2(COO)_2^-(aq)$

Initial: 1.0 M   — ≈ 0M   0 M

Change: $-x$ M   — $+x$ M $+x$ M

Equil: $(1.0-x)$ M   — x M xM

$$K_a = \frac{[H_3O^+][HCH_2(COO)_2^-]}{[CH_2(COOH)_2]} = \frac{(x)(x)}{1.0-x} \approx \frac{x^2}{1.0} = 1.4\times10^{-3}$$

$x = \sqrt{1.4\times10^{-3}} = 3.7\times10^{-2}$ M $= [H_3O^+] = [HOOCCH_2COO^-]$ $x \ll 1.0$M is a valid assumption.

9B We know $K_{a_2} = \left[\text{doubly charged anion}\right]$ for a polyprotic acid. Thus

$K_{a_2} = 5.3\times10^{-5} = \left[C_2O_4^{2-}\right]$. From the pH, $\left[H_3O^+\right] = 10^{-pH} = 10^{-0.67} = 0.21$ M. We also

recognize that $\left[HC_2O_4^-\right] = \left[H_3O^+\right]$, since the second ionization occurs to only a very

small extent. We note as well that $HC_2O_4^-$ is produced by the ionization of $H_2C_2O_4$.

Each mole of $HC_2O_4^-$ present results from the ionization of 1 mole of $H_2C_2O_4$. Now we

have sufficient information to determine the K_{a_1}.

$$K_{a_1} = \frac{\left[H_3O^+\right]\left[HC_2O_4^-\right]}{\left[H_2C_2O_4\right]} = \frac{0.21\times0.21}{1.05-0.21} = 5.3\times10^{-2}$$

10A H_2SO_4 is a strong acid in its first ionization, and somewhat weak in its second, with

$K_{a_2} = 1.1\times10^{-2} = 0.011$. Because of the strong first ionization step, this problem involves

determining concentrations in a solution that initially is 0.20 M H_3O^+ and 0.20 M

HSO_4^-. We base the setup on the balanced chemical equation.

Equation:	$HSO_4^-\,(aq) + H_2O\,(l)$	$\rightleftharpoons$	$H_3O^+(aq)$	$+$	$SO_4^{2-}\,(aq)$
Initial:	0.20 M	$-$	0.20 M		0 M
Changes:	$-x$ M	$-$	$+x$ M		$+x$ M
Equil:	$(0.20-x)$ M	$-$	$(0.20+x)$ M		x M

$$K_{a_2} = \frac{\left[H_3O^+\right]\left[SO_4^{2-}\right]}{\left[HSO_4^-\right]} = \frac{(0.20+x)x}{0.20-x} = 0.011 \approx \frac{0.20\times x}{0.20}, \text{ assuming that } x \ll 0.20\text{M}.$$

$x = 0.011$M   Try one cycle of approximation:

$$0.011 \approx \frac{(0.20+0.011)x}{(0.20-0.011)} = \frac{0.21x}{0.19} \qquad x = \frac{0.19\times0.011}{0.21} = 0.010 \text{ M}$$

The next cycle of approximation produces the same answer $0.010\text{M} = \left[SO_4^{2-}\right]$,

$\left[H_3O^+\right] = 0.010 + 0.20 \text{ M} = 0.21 \text{ M}, \qquad \left[HSO_4^-\right] = 0.20 - 0.010 \text{ M} = 0.19 \text{ M}$

10B We know that H_2SO_4 is a strong acid in its first ionization, and a somewhat weak acid in

its second, with $K_{a_2} = 1.1\times10^{-2} = 0.011$. Because of the strong first ionization step, the

problem essentially reduces to determining concentrations in a solution that initially is

0.020 M H_3O^+ and 0.020 M HSO_4^-. We base the setup on the balanced chemical

equation. The result is solved using the quadratic equation.

Equation: $\quad HSO_4^-(aq) + H_2O(l) \rightleftharpoons H_3O^+(aq) \quad + \quad SO_4^{2-}(aq)$

Initial: $\quad$ 0.020 M $\qquad$ — $\qquad$ 0.020 M $\qquad$ 0 M

Changes: $\quad$ $-x$ M $\qquad$ — $\qquad$ $+x$ M $\qquad$ $+x$ M

Equil: $\quad$ $(0.020 - x)$ M $\quad$ — $\quad$ $(0.020 + x)$ M $\qquad$ x M

$$K_{a_2} = \frac{[H_3O^+][SO_4^{2-}]}{[HSO_4^-]} = \frac{(0.020 + x)x}{0.020 - x} = 0.011 \qquad\qquad 0.020x + x^2 = 2.2 \times 10^{-4} - 0.011x$$

$$x^2 + 0.031x - 0.00022 = 0 \qquad x = \frac{-0.031 \pm \sqrt{0.00096 + 0.00088}}{2} = 0.0060\ M = [SO_4^{2-}]$$

$$[HSO_4^-] = 0.020 - 0.0060 = 0.014\ M \qquad [H_3O^+] = 0.020 + 0.0060 = 0.026\ M$$

(The method of successive approximations converges to $x = 0.006$ M in 8 cycles.)

11A **(a)** $CH_3NH_3^+NO_3^-$ is the salt of the cation of a weak base. The cation, $CH_3NH_3^+$, will hydrolyze to form an acidic solution $(CH_3NH_3^+ + H_2O \rightleftharpoons CH_3NH_2 + H_3O^+)$, while NO_3^-, by virtue of being the conjugate base if a strong acid will not hydrolyze to a detectable extent. The aqueous solutions of this compound will thus be acidic.

(b) NaI is the salt composed of the cation of a strong base and the anion of a strong acid, neither of which hydrolyzes in water. Solutions of this compound will be pH neutral.

(c) $NaNO_2$ is the salt composed of the cation of a strong base that will not hydrolyze in water and the anion of a weak acid that will hydrolyze to form an alkaline solution $(NO_2^- + H_2O \rightleftharpoons HNO_2 + OH^-)$. Thus aqueous solutions of this compound will be basic (alkaline).

11B Even without referring to the K values for acids and bases, we can predict that the reaction that produces H_3O^+ occurs to the greater extent. This, of course, is because the pH is less than 7, thus acid hydrolysis must predominate.

We write the two reactions of $H_2PO_4^-$ with water, along with the values of their equilibrium constants.

$$H_2PO_4^-(aq) + H_2O(l) \rightleftharpoons H_3O^+(aq) + HPO_4^{2-}(aq) \qquad K_{a_2} = 6.3 \times 10^{-8}$$

$$H_2PO_4^-(aq) + H_2O(l) \rightleftharpoons OH^-(aq) + H_3PO_4(aq) \quad K_b = \frac{K_w}{K_{a_1}} = \frac{1.0 \times 10^{-14}}{7.1 \times 10^{-3}} = 1.4 \times 10^{-12}$$

As predicted, the acid ionization occurs to the greater extent.

12A From the value of pK_b we determine the value of K_b and then K_a for the cation.

cocaine: $\qquad K_b = 10^{-pK} = 10^{-8.41} = 3.9 \times 10^{-9} \qquad K_a = \dfrac{K_w}{K_b} = \dfrac{1.0 \times 10^{-14}}{3.9 \times 10^{-9}} = 2.6 \times 10^{-6}$

codeine: $\qquad K_b = 10^{-pK} = 10^{-7.95} = 1.1 \times 10^{-8} \qquad K_a = \dfrac{K_w}{K_b} = \dfrac{1.0 \times 10^{-14}}{1.1 \times 10^{-8}} = 9.1 \times 10^{-7}$

(This method may be a bit easier:
$pK_a = 14.00 - pK_b = 14.00 - 8.41 = 5.59, K_a = 10^{-5.59} = 2.6 \times 10^{-6}$) The acid with the larger K_a will produce the higher $\left[H^+\right]$, and that solution will have the lower pH. Thus, the solution of codeine hydrochloride will have the higher pH (i.e. codeine hydrochloride is the weaker acid).

12B Both of the ions of $NH_4CN(aq)$ react with water in hydrolysis reactions.

$NH_4^+ (aq) + H_2O(l) \rightleftharpoons NH_3 (aq) + H_3O^+ (aq) \qquad K_a = \dfrac{K_w}{K_b} = \dfrac{1.0 \times 10^{-14}}{1.8 \times 10^{-5}} = 5.6 \times 10^{-10}$

$CN^- (aq) + H_2O(l) \rightleftharpoons HCN (aq) + OH^- (aq) \qquad K_b = \dfrac{K_w}{K_a} = \dfrac{1.0 \times 10^{-14}}{6.2 \times 10^{-10}} = 1.6 \times 10^{-5}$

Since the value of the equilibrium constant for the hydrolysis reaction of cyanide ion is larger than that for the hydrolysis of ammonium ion, the cyanide ion hydrolysis reaction will proceed to a greater extent and thus the solution of $NH_4CN(aq)$ will be basic (alkaline).

13A NaF dissociates completely into sodium ions and fluoride ions. The released fluoride ion hydrolyzes in aqueous solution to form hydroxide ion. The determination of the equilibrium pH is organized around the balanced equation.

Equation: $\quad F^-$(aq) $\quad + \quad H_2O(l) \quad \rightleftharpoons \quad HF(aq) \quad + \quad OH^-$(aq)
Initial: $\qquad$ 0.10 M $\qquad -\qquad\qquad$ 0 M $\qquad\qquad$ 0 M
Changes: $\qquad -x$ M $\qquad -\qquad\qquad +x$ M $\qquad\quad +x$ M
Equil: $\qquad (0.10-x)$ M $\quad -\qquad\qquad x$ M $\qquad\qquad x$ M

$K_b = \dfrac{K_w}{K_a} = \dfrac{1.0 \times 10^{-14}}{6.6 \times 10^{-4}} = 1.5 \times 10^{-11} = \dfrac{[HF][OH^-]}{[F^-]} = \dfrac{(x)(x)}{(0.10-x)} = \dfrac{x^2}{0.10}$

$x = \sqrt{0.10 \times 1.5 \times 10^{-11}} = 1.2 \times 10^{-6}$ M=$[OH^-]$; $\qquad$ pOH= $-\log(1.2 \times 10^{-6}) = 5.92$
pH = 14.00 $-$ pOH = 14.00 $-$ 5.92 = 8.08 (As expected, pH > 7).

13B The cyanide ion hydrolyzes in solution, as indicated in Practice Example 16-12B. As a consequence of the hydrolysis, $\left[OH^-\right]=\left[HCN\right]$. $\left[OH^-\right]$ can be found from the pH of the solution, and then values are substituted into the K_b expression for CN^-, which is then solved for $\left[CN^-\right]$.

$$pOH = 14.00 - pH = 14.00 - 10.38 = 3.62$$

$$\left[OH^-\right] = 10^{-pOH} = 10^{-3.62} = 2.4 \times 10^{-4}\,M = \left[HCN\right]$$

$$K_b = \frac{\left[HCN\right]\left[OH^-\right]}{\left[CN^-\right]} = 1.6 \times 10^{-5} = \frac{\left(2.4 \times 10^{-4}\right)^2}{\left[CN^-\right]} \qquad \left[CN^-\right] = \frac{\left(2.4 \times 10^{-4}\right)^2}{1.6 \times 10^{-5}} = 3.6 \times 10^{-3}\,M$$

14A First we draw the Lewis structures of the four acids. Lone pairs have been omitted since we are interested only in the arrangements of atoms.

$HClO_4$ should be stronger than HNO_3. Although Cl and N have similar electronegativities, there are more terminal oxygen atoms attached to the chlorine in perchloric acid than to the nitrogen in nitric acid. By virtue of having more terminal oxygens, perchloric acid, when ionized, affords a more stable conjugate base. The more stable the anion, the more easily it is formed and hence the stronger is the conjugate acid from which it is derived. CH_2FCOOH will be a stronger acid than $CH_2BrCOOH$ because F is a more electronegative atom than Br. The F atom withdraws additional electron density from the $O-H$ bond, making the bond easier to break, which leads to increased acidity.

14B First we draw the Lewis structures of the first two acids. Lone pairs are not depicted since we are interested in the arrangements of atoms.

H_3PO_4 and H_2SO_3 both have one terminal oxygen atom, but S is more electronegative than P. This suggests that H_2SO_3 $\left(K_{a_1} = 1.3 \times 10^{-2}\right)$ should be a stronger acid than H_3PO_4 $\left(K_{a_1} = 7.1 \times 10^{-3}\right)$, and it is. The only difference between CCl_3CH_2COOH and CCl_2FCH_2COOH is the replacement of Cl by F. Since F is more electronegative than Cl, CCl_3CH_2COOH should be a weaker acid than CCl_2FCH_2COOH.

15A We draw Lewis structures to help us decide.

(a) Clearly, BF_3 is an electron pair acceptor, a Lewis acid, and NH_3 is an electron pair donor, a Lewis base.

$$
\begin{array}{ccc}
\underset{|\overline{F}|}{\overset{|\overline{F}|}{\overline{F}-B}} \;\; + \;\; \underset{H}{\overset{H}{N-H}} \;\; \longrightarrow \;\; \underset{|\overline{F}|}{\overset{|\overline{F}|}{\overline{F}-B}}-\underset{H}{\overset{H}{N-H}} & \qquad & \underset{H}{\overline{O}-H}
\end{array}
$$

(b) H_2O certainly has electron pairs (lone pairs) to donate and thus it can be a Lewis base. It is unlikely that the cation Cr^{3+} has any accessible valence electrons that can be donated to another atom, thus, it is the Lewis acid. The product of the reaction, $[Cr(H_2O)_6]^{3+}$, is described as a water adduct of Cr^{3+}.

$$
Cr^{3+}(aq) \;\; + \;\; 6\,\underset{H}{|\overline{O}-H} \;\; \longrightarrow \;\; \left[\begin{array}{c} \overline{O}H_2 \\ H_2\overline{O}\cdots Cr \cdots \overline{O}H_2 \\ H_2\overline{O} \qquad \overline{O}H_2 \\ \overline{O}H_2 \end{array}\right]^{3+}
$$

15B The Lewis structures of the six species follow.

$$
\begin{array}{c}
\overset{\ominus}{|\overline{O}-H} \\
+ \\
H-\overline{O}-Al-\overline{O}-H \\
|\overline{O}| \\
H
\end{array}
\;\; \longrightarrow \;\;
\left[\begin{array}{c}
H \\
|\overline{O}| \\
H-\overline{O}-\overset{\ominus}{Al}-\overline{O}-H \\
|\overline{O}| \\
H
\end{array}\right]^{-}
\qquad
2\,|\overline{Cl}|^{\ominus}
\quad
\begin{array}{c}
|\overline{Cl}| \\
|\overline{Cl}-Sn-\overline{Cl}| \\
|\overline{Cl}|
\end{array}
\;\; \longrightarrow \;\;
\left[\begin{array}{c}
|\overline{Cl}| \;\; \overline{Cl} \\
\overline{Cl}-Sn-\overline{Cl} \\
\overline{Cl} \;\; |\overline{Cl}|
\end{array}\right]^{2-}
$$

Both the hydroxide ion and the chloride ion have lone pairs of electrons that can be donated to electron-poor centers. These two are the electron pair donors, or the Lewis bases. $Al(OH)_3$ and $SnCl_4$ have additional spaces in their structures to accept pairs of electrons, which is what occurs when they form the complex anions $[Al(OH)_4]^-$ and $[SnCl_6]^{2-}$. Thus, $Al(OH)_3$ and $SnCl_4$ are the Lewis acids in these reactions.

EXERCISES

Brønsted-Lowry Theory of Acids and Bases

1. **(a)** HNO_2 is an acid, a proton donor. It's conjugate base is NO_2^-.

(b) OCl^- is a base, a proton acceptor. It's conjugate acid is $HOCl$.

(c) NH_2^- is a base, a proton acceptor. It's conjugate acid is NH_3.

(d) NH_4^+ is an acid, a proton donor. It's conjugate base is NH_3.

(e) $CH_3NH_3^+$ is an acid, a proton donor. It's conjugate base is CH_3NH_2.

2. We write the conjugate base as the first product of the equilibrium for which the acid is the first reactant.

(a) $HIO_3(aq) + H_2O(l) \rightleftharpoons IO_3^-(aq) + H_3O^+(aq)$

(b) $C_6H_5COOH(aq) + H_2O(l) \rightleftharpoons C_6H_5COO^-(aq) + H_3O^+(aq)$

(c) $HPO_4^{2-}(aq) + H_2O(l) \rightleftharpoons PO_4^{3-}(aq) + H_3O^+(aq)$

(d) $C_2H_5NH_3^+(aq) + H_2O(l) \rightleftharpoons C_2H_5NH_2(aq) + H_3O^+(aq)$

3. The acids (proton donors) and bases (proton acceptors) are labeled below their formulas. Remember that a proton, in Brønsted-Lowry acid-base theory, is H^+.

(a)
$HOBr(aq)$ + $H_2O(l)$ $\rightleftharpoons$ $H_3O^+(aq)$ + $OBr^-(aq)$
  acid   base   acid   base

(b)
$HSO_4^-(aq)$ + $H_2O(l)$ $\rightleftharpoons$ $H_3O^+(aq)$ + $SO_4^{2-}(aq)$
  acid   base   acid   base

(c)
$HS^-(aq)$ + $H_2O(l)$ $\rightleftharpoons$ $H_2S(aq)$ + $OH^-(aq)$
  base   acid   acid   base

(d)
$C_6H_5NH_3^+(aq)$ + $OH^-(aq)$ $\rightleftharpoons$ $C_6H_5NH_2(aq)$ + $H_2O(l)$
  acid   base   base   acid

4. For each amphiprotic substance, we write both its acid and base hydrolysis reaction. Even for the substances that are not usually considered amphiprotic, both reactions are written, but one of them is labeled as unlikely. In some instances we have written an oxygen as $\varnothing$ to keep track of it through the reaction.

545

(a) $OH^- + H_2O \rightleftharpoons O^{2-} + H_3O^+$ (unlikely) $OH^- + H_2O \rightleftharpoons H_2O + OH^-$

(b) $NH_4^+ + H_2O \rightleftharpoons NH_3 + H_3O^+$ NH_4^+ has no electron pairs

                                                          that can be donated

(c) $H_2O + H_2O \rightleftharpoons OH^- + H_3O^+$ $H_2O + H_2O \rightleftharpoons H_3O^+ + OH^-$

(d) $HS^- + H_2O \rightleftharpoons S^{2-} + H_3O^+$ $HS^- + H_2O \rightleftharpoons H_2S + OH^-$

(e) NO_2^- cannot act as an acid, (no protons) $NO_2^- + H_2O \rightleftharpoons HNO_2 + OH^-$

(f) $HCO_3^- + H_2O \rightleftharpoons CO_3^{2-} + H_3O^+$ $HCO_3^- + H_2O \rightleftharpoons H_2CO_3 + OH^-$

(g) $HBr + H_2O \rightleftharpoons H_3O^+ + Br^-$ $HBr + H_2O \rightleftharpoons H_2Br^+ + OH^-$ (unlikely)

5. Answer (b), NH_3, is correct. $HC_2H_3O_2$ will react most completely with the strongest base. NO_3^- and Cl^- are very weak bases. H_2O is a weak base, but it is amphiprotic, acting as an acid (donating protons), as in the presence of NH_3. Thus, NH_3 must be the strongest base and the most effective in deprotonating $HC_2H_3O_2$.

6. Lewis structures are given below each equation.

(a) $2NH_3(l) \rightleftharpoons NH_4^+ + NH_2^-$

$$2\ \text{H}-\overset{\text{H}}{\underset{\text{H}}{\text{N}}}-\text{H} \rightleftharpoons \left(\text{H}-\overset{\text{H}}{\underset{\text{H}}{\text{N}}}-\text{H}\right)^+ + \left[\text{H}-\underline{\text{N}}-\text{H}\right]^-$$

(b) $2HF(l) \rightleftharpoons H_2F^+ + F^-$

$$\text{H}-\underline{\bar{\text{F}}} + \text{H}-\underline{\bar{\text{F}}} \longrightarrow |\underline{\text{F}}|^- + \overset{\text{H}}{\underset{}{|\underline{\text{F}}-\text{H}}}$$

(c) $2CH_3OH(l) \rightleftharpoons CH_3OH_2^+ + CH_3O^-$

$$2\ \text{H}-\overset{\text{H}}{\underset{\text{H}}{\text{C}}}-\bar{\underline{\text{O}}}-\text{H} \rightleftharpoons \left(\text{H}-\overset{\text{H}}{\underset{\text{H}}{\text{C}}}-\overset{\text{H}}{\underset{}{\text{O}}}-\text{H}\right)^+ + \left(\text{H}-\overset{\text{H}}{\underset{\text{H}}{\text{C}}}-\underline{\bar{\text{O}}}|\right)^-$$

(d) $2HC_2H_3O_2(l) \rightleftharpoons H_2C_2H_3O_2^+ + C_2H_3O_2^-$

$$2\ \text{H}-\overset{\text{H}}{\underset{\text{H}}{\text{C}}}-\overset{|\text{O}|}{\underset{}{\text{C}}}-\bar{\underline{\text{O}}}-\text{H} \rightleftharpoons \left(\text{H}-\overset{\text{H}}{\underset{\text{H}}{\text{C}}}-\overset{|\bar{\text{O}}-\text{H}}{\underset{}{\text{C}}}-\bar{\underline{\text{O}}}-\text{H}\right)^+ + \left(\text{H}-\overset{\text{H}}{\underset{\text{H}}{\text{C}}}-\overset{|\text{O}|}{\underset{}{\text{C}}}-\underline{\bar{\text{O}}}|\right)^-$$

(e) $2 H_2SO_4 (l) \rightleftharpoons 2 H_3SO_4^+ + HSO_4^-$

$$H-\overline{O}-\underset{\underset{|\underline{O}|}{|}}{\overset{\overset{|\overline{O}|}{|}}{S}}-\overline{O}-H \;+\; H-\overline{O}-\underset{\underset{|\underline{O}|}{|}}{\overset{\overset{|\overline{O}|}{|}}{S}}-\overline{O}-H \;\rightleftharpoons\; \left[{}^{\ominus}|\overline{O}-\underset{\underset{|\underline{O}|}{|}}{\overset{\overset{|\overline{O}|}{|}}{S}}-\overline{O}-H\right]^- \;+\; \left[H-\overline{O}-\underset{\underset{|\underline{O}-H}{|}}{\overset{\overset{|\overline{O}|}{|}}{\underset{\oplus}{S}}}-\overline{O}-H\right]^+$$

7. The principle we will follow here is that, in terms of their concentrations, the weaker acid and the weaker base will predominate at equilibrium. The reason for this is that a strong acid will do a good job of donating its protons and, having done so, its conjugate base will be left behind. The preferred direction is:

Strong acid + strong base → weak (conjugate) base + weak (conjugate) acid.

(a) The reaction will favor the forward direction because OH^- (a strong base) $> NH_3$ (a weak base) and NH_4^+ (relatively strong weak acid) $> H_2O$ (very weak acid).

(b) The reaction will favor the reverse direction because $HNO_3 > HSO_4^-$ (a weak acid in the second ionization) (acting as acids), and $SO_4^{2-} > NO_3^-$ (acting as bases).

(c) The reaction will favor the reverse direction because $HC_2H_3O_2 > CH_3OH$ (not usually thought of as an acid) (acting as acids), and $CH_3O^- > C_2H_3O_2^-$ (acting as bases).

8. The principle we follow here is that, in terms of their concentrations, the weaker acid and the weaker base predominate at equilibrium. This is because a strong acid will do a good job of donating its protons and, having done so, its conjugate base will remain in solution. The preferred direction is:

Strong acid + strong base → weak (conjugate) base + weak (conjugate) acid.

(a) The reaction will favor the forward direction because $HC_2H_3O_2$ (a moderate acid) $> HCO_3^-$ (a rather weak acid) (acting as acids) and $CO_3^{2-} > C_2H_3O_2^-$ (acting as bases).

(b) The reaction will favor the reverse direction because $HClO_4$ (a strong acid) $> HNO_2$ (acting as acids), and $NO_2^- > ClO_4^-$ (acting as bases).

(c) The reaction will favor the forward direction, because $H_2CO_3 > HCO_3^-$ (acting as acids) (because $K_1 > K_2$) and $CO_3^{2-} > HCO_3^-$ (acting as bases).

Strong Acids, Strong Bases, and pH

9. All of the solutes are strong acids or strong bases.

(a)
$$\left[H_3O^+\right] = 0.00165 \text{ M HNO}_3 \times \frac{1 \text{ mol } H_3O^+}{1 \text{ mol HNO}_3} = 0.00165 \text{ M}$$

$$\left[OH^-\right] = \frac{K_w}{\left[H_3O^+\right]} = \frac{1.0 \times 10^{-14}}{0.00165 \text{M}} = 6.1 \times 10^{-12} \text{ M}$$

(b)
$$\left[OH^-\right] = 0.0087 \text{ M KOH} \times \frac{1 \text{ mol } OH^-}{1 \text{ mol KOH}} = 0.0087 \text{ M}$$

$$\left[H_3O^+\right] = \frac{K_w}{\left[OH^-\right]} = \frac{1.0 \times 10^{-14}}{0.0087 \text{M}} = 1.1 \times 10^{-12} \text{ M}$$

(c)
$$\left[OH^-\right] = 0.00213 \text{ M Sr}(OH)_2 \times \frac{2 \text{ mol } OH^-}{1 \text{mol Sr}(OH)_2} = 0.00426 \text{ M}$$

$$\left[H_3O^+\right] = \frac{K_w}{\left[OH^-\right]} = \frac{1.0 \times 10^{-14}}{0.00426 \text{ M}} = 2.3 \times 10^{-12} \text{ M}$$

(d)
$$\left[H_3O^+\right] = 5.8 \times 10^{-4} \text{M HI} \times \frac{1 \text{mol } H_3O^+}{1 \text{mol HI}} = 5.8 \times 10^{-4} \text{ M}$$

$$\left[OH^-\right] = \frac{K_w}{\left[H_3O^+\right]} = \frac{1.0 \times 10^{-14}}{5.8 \times 10^{-4} \text{M}} = 1.7 \times 10^{-11} \text{ M}$$

10. Again, all of the solutes are strong acids or strong bases.

(a)
$$\left[H_3O^+\right] = 0.0045 \text{ M HCl} \times \frac{1 \text{ mol } H_3O^+}{1 \text{ mol HCl}} = 0.0045 \text{ M}$$

$$pH = -\log(0.0045) = 2.35$$

(b)
$$\left[H_3O^+\right] = 6.14 \times 10^{-4} \text{M HNO}_3 \times \frac{1 \text{ mol } H_3O^+}{1 \text{ mol HNO}_3} = 6.14 \times 10^{-4} \text{M}$$

$$pH = -\log(6.14 \times 10^{-4}) = 3.212$$

(c)
$$\left[OH^-\right] = 0.00683 \text{ M NaOH} \times \frac{1 \text{ mol } OH^-}{1 \text{ mol NaOH}} = 0.00683 \text{M}$$

$$pOH = -\log(0.00683) = 2.166 \text{ and } pH = 14.000 - pOH = 14.000 - 2.166 = 11.834$$

(d)
$$\left[OH^-\right] = 4.8 \times 10^{-3} \text{M Ba}(OH)_2 \times \frac{2 \text{ mol } OH^-}{1 \text{ mol Ba}(OH)_2} = 9.6 \times 10^{-3} \text{M}$$

$$pOH = -\log(9.6 \times 10^{-3}) = 2.02 \qquad pH = 14.00 - 2.02 = 11.98$$

11. $[\text{OH}^-] = \dfrac{3.9 \text{ g Ba(OH)}_2 \cdot 8\text{H}_2\text{O}}{100 \text{ mL soln}} \times \dfrac{1000 \text{ mL}}{1 \text{ L}} \times \dfrac{1 \text{ mol Ba(OH)}_2 \cdot 8\text{H}_2\text{O}}{315.5 \text{ g Ba(OH)}_2 \cdot 8\text{H}_2\text{O}} \times \dfrac{2 \text{ mol OH}^-}{1 \text{ mol Ba(OH)}_2 \cdot 8\text{H}_2\text{O}}$

$= 0.25 \text{ M}$

$[\text{H}_3\text{O}^+] = \dfrac{K_w}{[\text{OH}^-]} = \dfrac{1.0 \times 10^{-14}}{0.25 \text{ M OH}^-} = 4.0 \times 10^{-14} \text{M}$ \qquad $\text{pH} = -\log\left(4.0 \times 10^{-14}\right) = 13.40$

12. The dissolved Ca(OH)_2 is completely dissociated into ions.

$\text{pOH} = 14.00 - \text{pH} = 14.00 - 12.35 = 1.65$

$[\text{OH}^-] = 10^{-\text{pOH}} = 10^{-1.65} = 2.2 \times 10^{-2} \text{ M OH}^- = 0.022 \text{ M OH}^-$

$\text{solubility} = \dfrac{0.022 \text{ mol OH}^-}{1 \text{ L soln}} \times \dfrac{1 \text{ mol Ca(OH)}_2}{2 \text{ mol OH}^-} \times \dfrac{74.09 \text{ g Ca(OH)}_2}{1 \text{ mol Ca(OH)}_2} \times \dfrac{1000 \text{ mg Ca(OH)}_2}{1 \text{ g Ca(OH)}_2}$

$= \dfrac{8.1 \times 10^2 \text{ mg Ca(OH)}_2}{1 \text{ L soln}}$

In 100 mL the solubility is $100 \text{ mL} \times \dfrac{1 \text{ L}}{1000 \text{ mL}} \times \dfrac{8.1 \times 10^2 \text{ mg Ca(OH)}_2}{1 \text{ L soln}} = 81 \text{ mg Ca(OH)}_2$

13. First we determine the moles of HCl, and then its concentration.

$\text{moles HCl} = \dfrac{PV}{RT} = \dfrac{\left(751 \text{ mmHg} \times \dfrac{1 \text{ atm}}{760 \text{ mmHg}}\right) \times 0.205 \text{ L}}{0.08206 \text{ L atm mol}^{-1} \text{ K}^{-1} \times 296 \text{ K}} = 8.34 \times 10^{-3} \text{ mol HCl}$

$[\text{H}_3\text{O}^+] = \dfrac{8.34 \times 10^{-3} \text{ mol HCl}}{4.25 \text{ L soln}} \times \dfrac{1 \text{ mol H}_3\text{O}^+}{1 \text{ mol HCl}} = 1.96 \times 10^{-3} \text{ M}$

14. First determine the concentration of NaOH, then of OH^-.

$[\text{OH}^-] = \dfrac{0.125 \text{ L} \times \dfrac{0.606 \text{ mol NaOH}}{1 \text{ L}} \times \dfrac{1 \text{ mol OH}^-}{1 \text{ mol NaOH}}}{15.0 \text{ L final solution}} = 0.00505 \text{ M}$

$\text{pOH} = -\log\left(0.00505 \text{ M}\right) = 2.297$ \qquad $\text{pH} = 14.00 - 2.297 = 11.703$

15. First determine the amount of HCl, and then the volume of the concentrated solution required.

$\text{amount HCl} = 12.5 \text{ L} \times \dfrac{10^{-2.10} \text{ mol H}_3\text{O}^+}{1 \text{ L soln}} \times \dfrac{1 \text{ mol HCl}}{1 \text{ mol H}_3\text{O}^+} = 0.099 \text{ mol HCl}$

$V_{\text{solution}} = 0.099 \text{ mol HCl} \times \dfrac{36.46 \text{ g HCl}}{1 \text{ mol HCl}} \times \dfrac{100.0 \text{ g soln}}{36.0 \text{ g HCl}} \times \dfrac{1 \text{ mL soln}}{1.18 \text{ g soln}} = 8.5 \text{ mL soln}$

16. First we determine the amount of KOH, and then the volume of the concentrated solution required.

$$\text{pOH} = 14.00 - \text{pH} = 14.00 - 11.55 = 2.45 \quad \left[\text{OH}^-\right] = 10^{-\text{pOH}} = 10^{-2.45} = 0.0035 \text{ M}$$

$$\text{amount KOH} = 25.0 \text{ L} \times \frac{0.0035 \text{ M mol OH}^-}{1 \text{ L soln}} \times \frac{1 \text{ mol KOH}}{1 \text{ mol OH}^-} = 0.088 \text{ mol KOH}$$

$$V_{\text{solution}} = 0.088 \text{ mol KOH} \times \frac{56.11 \text{ g KOH}}{1 \text{ mol KOH}} \times \frac{100.0 \text{ g soln}}{15.0 \text{ g KOH}} \times \frac{1 \text{ mL soln}}{1.14 \text{ g soln}} = 29 \text{ mL soln}$$

17. The volume of HCl(aq) needed is determined by first finding the amount of $NH_3(aq)$ present, and then realizing that acid and base react in a 1:1 molar ratio.

$$V_{\text{HCl}} = 1.25 \text{ L base} \times \frac{0.265 \text{ mol NH}_3}{1 \text{ L base}} \times \frac{1 \text{ mol H}_3\text{O}^+}{1 \text{ mol NH}_3} \times \frac{1 \text{ mol HCl}}{1 \text{ mol H}_3\text{O}^+} \times \frac{1 \text{ L acid}}{6.15 \text{ mol HCl}}$$

$$= 0.0539 \text{ L acid or } 53.9 \text{ mL acid.}$$

18. NH_3 and HCl react in a 1:1 molar ratio. According to Avogadro's hypothesis, equal volumes of gases at the same temperature and pressure contain equal numbers of moles. Thus, the volume of $H_2(g)$ at 762 mmHg and $21.0°C$ that is equivalent to 28.2 L of $H_2(g)$ at 742 mmHg and $25.0°C$ will be equal to the volume of $NH_3(g)$ needed for stoichiometric neutralization.

$$V_{\text{NH}_3(g)} = 28.2 \text{ L HCl}(g) \times \frac{742 \text{ mmHg}}{762 \text{ mmHg}} \times \frac{(273.2 + 21.0) \text{ K}}{(273.2 + 25.0) \text{ K}} \times \frac{1 \text{ L NH}_3(g)}{1 \text{ L HCl}(g)} = 27.1 \text{ L NH}_3(g)$$

Alternatively, we can solve the ideal gas equation for the amount of each gas, and then by equating these two expressions, we can find the volume of NH_3 needed.

$$n\{\text{HCl}\} = \frac{742 \text{ mmHg} \cdot 28.2 \text{ L}}{R \cdot 298.2 \text{ K}} \qquad n\{\text{NH}_3\} = \frac{762 \text{ mmHg} \cdot V\{\text{NH}_3\}}{R \cdot 294.2 \text{ K}}$$

$$\frac{742 \text{ mmHg} \cdot 28.2 \text{ L}}{R \cdot 298.2 \text{ K}} = \frac{762 \text{ mmHg} \cdot V\{\text{NH}_3\}}{R \cdot 294.2 \text{ K}} \quad \text{This yields } V\{\text{NH}_3\} = 27.1 \text{ L NH}_3(g)$$

19. Here we determine the amounts of H_3O^+ and OH^- and then the amount of the one that is in excess. We express molar concentration in millimoles/milliliter, equivalent to mol/L.

$$50.00 \text{ mL} \times \frac{0.0155 \text{ mmol HI}}{1 \text{ mL soln}} \times \frac{1 \text{ mmol H}_3\text{O}^+}{1 \text{ mmol HI}} = 0.775 \text{ mmol H}_3\text{O}^+$$

$$75.00 \text{ mL} \times \frac{0.0106 \text{ mmol KOH}}{1 \text{ mL soln}} \times \frac{1 \text{ mmol OH}^-}{1 \text{ mmol KOH}} = 0.795 \text{ mmol OH}^-$$

The net reaction is $H_3O^+(aq) + OH^-(aq) \rightarrow 2H_2O(l)$.

There is an excess of OH^- of $(0.795 - 0.775 =) \, 0.020$ mmol OH^-.
Thus, this is a basic solution. The total solution volume is $(50.00 + 75.00 =) \, 125.00$ mL.

$$[OH^-] = \frac{0.020 \text{ mmol } OH^-}{125.00 \text{ mL}} = 1.6 \times 10^{-4} \text{ M}, \quad pOH = -\log(1.6 \times 10^{-4}) = 3.80, \quad pH = 10.20$$

20. In each case, we need to determine the $[H_3O^+]$ or $[OH^-]$ of each solution being mixed, and then the amount of H_3O^+ or OH^-, so that we can determine the amount in excess. We express molar concentration in millimoles/milliliter, equivalent to mol/L.

$$[H_3O^+] = 10^{-2.12} = 7.6 \times 10^{-3} \text{ M} \qquad \text{moles } H_3O^+ = 25.00 \text{ mL} \times 7.6 \times 10^{-3} \text{ M} = 0.19 \text{ mmol } H_3O^+$$

$$pOH = 14.00 - 12.65 = 1.35 \qquad\qquad [OH^-] = 10^{-1.35} = 4.5 \times 10^{-2} \text{ M}$$

$$\text{amount } OH^- = 25.00 \text{ mL} \times 4.5 \times 10^{-2} \text{ M} = 1.13 \text{ mmol } OH^-$$

There is excess OH^- in the amount of 0.94 mmol $(= 1.13 \text{ mmol } OH^- - 0.19 \text{ mmol } H_3O^+)$

$$[OH^-] = \frac{0.94 \text{ mmol } OH^-}{25.00 \text{ mL} + 25.00 \text{ mL}} = 1.88 \times 10^{-2} \text{ M} \qquad pOH = -\log(1.88 \times 10^{-2}) = 1.73 \quad pH = 12.27$$

Weak Acids, Weak Bases, and pH

21. We organize the solution around the balanced chemical equation.

Equation:	$HNO_2(aq)$	+	$H_2O(l)$	$\rightleftharpoons$	$H_3O^+(aq)$	+	$NO_2^-(aq)$
Initial:	0.143 M		—		≈ 0 M		0 M
Changes:	$-x$ M		—		$+x$ M		$+x$ M
Equil:	$(0.143 - x)$ M		—		x M		x M

$$K_a = \frac{[H_3O^+][NO_2^-]}{[HNO_2]} = \frac{x^2}{0.143 - x} = 7.2 \times 10^{-4} \gg \frac{x^2}{0.143} \text{ (if } x \ll 0.143)$$

$$x = \sqrt{0.143 \times 7.2 \times 10^{-4}} = 0.010 \text{ M}$$

We have assumed that $x \ll 0.143$ M, an almost acceptable assumption. Another cycle of approximations yields:

$$x = \sqrt{(0.143 - 0.010) \times 7.2 \times 10^{-4}} = 0.0098 \text{ M} = [H_3O^+] \qquad pH = -\log(0.0098) = 2.01$$

This is the same result as is determined with the quadratic formula roots equation.

22. We organize the solution around the balanced chemical equation, and solve first for $\left[OH^-\right]$

Equation : $\quad C_2H_5NH_2(aq) \quad + \quad H_2O(l) \quad \rightleftharpoons \quad OH^-(aq) + \quad C_2H_5NH_3^+(aq)$

Initial : $\qquad$ 0.085 M $\qquad\qquad$ – $\qquad\qquad \approx 0$ M $\qquad$ 0 M

Changes : $\qquad -x$ M $\qquad\qquad$ – $\qquad\qquad +x$ M $\qquad$ $+x$ M

Equil : $\qquad (0.085-x)$ M $\qquad$ – $\qquad\qquad$ x M $\qquad$ x M

$$K_b = \frac{\left[OH^-\right]\left[C_2H_5NH_3^+\right]}{\left[C_2H_5NH_2\right]} = \frac{x^2}{0.085-x} = 4.3\times10^{-4} \approx \frac{x^2}{0.085} \text{ assuming } x \ll 0.085$$

$$x = \sqrt{0.085\times4.3\times10^{-4}} = 0.0060 \text{ M}$$

We have assumed that $x \ll 0.085$ M, an almost acceptable assumption. Another cycle of approximations yields:

$$x = \sqrt{(0.085-0.0060)\times4.3\times10^{-4}} = 0.0058 \text{ M} = [OH^-] \qquad \text{yet another cycle produces:}$$

$$x = \sqrt{(0.085-0.0058)\times4.3\times10^{-4}} = 0.0058 \text{ M} = [OH^-] \qquad pOH = -\log(0.0058) = 2.24$$

$$pH = 14.00-2.24 = 11.76 \qquad \left[H_3O^+\right] = 10^{-pH} = 10^{-11.76} = 1.7\times10^{-12} \text{ M}$$

This is the same result as determined with the quadratic formula roots equation.

23. **(a)** The set up is based on the balanced chemical equation.

Equation: $\quad HC_8H_7O_2(aq) \quad + \quad H_2O(l) \quad \rightleftharpoons \quad H_3O^+(aq) + \quad C_8H_7O_2^-(aq)$

Initial: $\qquad$ 0.186 M $\qquad\qquad$ – $\qquad\qquad \approx 0$ M $\qquad$ 0 M

Changes: $\qquad -x$ M $\qquad\qquad$ – $\qquad\qquad +x$ M $\qquad$ $+x$ M

Equil: $\qquad (0.186-x)$ M $\qquad$ – $\qquad\qquad$ x M $\qquad$ x M

$$K_a = 4.9\times10^{-5} = \frac{[H_3O^+][C_8H_7O_2^-]}{[HC_8H_7O_2]} = \frac{x\cdot x}{0.186-x} \approx \frac{x^2}{0.186}$$

$$x = \sqrt{0.186\times4.9\times10^{-5}} = 0.0030 \text{ M} = [H_3O^+] = [C_8H_7O_2^-]$$

0.0030 M is less than 5 % OF 0.186 M, thus, the approximation is valid.

(b) The set up is based on the balanced chemical equation.

Equation: $\quad HC_8H_7O_2(aq) \quad + \quad H_2O(l) \quad \rightleftharpoons \quad H_3O^+(aq) \quad + \quad C_8H_7O_2(aq)$

Initial: $\qquad$ 0.121 M $\qquad\qquad$ – $\qquad\qquad \approx 0$ M $\qquad$ 0 M

Changes: $\qquad -x$ M $\qquad\qquad$ – $\qquad\qquad +x$ M $\qquad$ $+x$ M

Equil: $\qquad (0.121-x)$ M $\qquad$ – $\qquad\qquad$ x M $\qquad$ x M

$$K_a = 4.9 \times 10^{-5} = \frac{\left[H_3O^+ \right]\left[C_8H_7O_2^- \right]}{\left[HC_8H_7O_2 \right]} = \frac{x^2}{0.121-x} \approx \frac{x^2}{0.121} \quad x = 0.0024 \text{ M} = \left[H_3O^+ \right]$$

Assumption $x \ll 0.121$, is correct. $\text{pH} = -\log(0.0024) = 2.62$

24.

$$\left[HC_3H_5O_2 \right]_{\text{initial}} = \left(\frac{0.275 \text{ mol}}{625 \text{ mL}} \right)\left(\frac{1000 \text{ mL}}{1 \text{ L}} \right) = 0.440 \text{ M}$$

$$\left[HC_3H_5O_2 \right]_{\text{equilibrium}} = 0.440 \text{ M} - 0.00239 \text{ M} = 0.438 \text{ M}$$

$$K_a = \frac{\left[H_3O^+ \right]\left[C_3H_5O_2^- \right]}{\left[HC_3H_5O_2 \right]} = \frac{(0.00239)^2}{0.438} = 1.30 \times 10^{-5}$$

<u>25.</u> We base our solution on the balanced chemical equation. $\left[H_3O^+ \right] = 10^{-1.56} = 2.8 \times 10^{-2}$ M

Equation: $\quad CH_2FCOOH(aq) + H_2O(l) \rightleftharpoons CH_2FCOO^-(aq) + H_3O^+(aq)$

Initial:	0.318 M	–	0 M	$\approx$ 0 M
Changes:	–0.028 M	–	+0.028 M	+0.028 M
Equil:	0.290 M	–	0.028 M	$\approx$ 0.028 M

$$K_a = \frac{\left[H_3O^+ \right]\left[CH_2FCOO^- \right]}{\left[CH_2FCOOH \right]} = \frac{(0.028)(0.028)}{0.290} = 2.7 \times 10^{-3}$$

26. $\quad \left[HC_6H_{11}O_2 \right] = \frac{11 \text{ g}}{1 \text{ L}} \times \frac{1 \text{ mol } HC_6H_{11}O_2}{116.2 \text{ g } HC_6H_{11}O_2} = 9.5 \times 10^{-2} \text{ M} = 0.095 \text{ M}$

$$\left[H_3O^+ \right]_{\text{equil}} = 10^{-2.94} = 1.1 \times 10^{-3} \text{ M} = 0.0011 \text{ M}$$

The stoichiometry of the reaction, written below, indicates that $\left[H_3O^+ \right] = \left[C_6H_{11}O_2^- \right]$

Equation: $\quad HC_6H_{11}O_2(aq) + H_2O(l) \rightleftharpoons C_6H_{11}O_2^-(aq) + H_3O^+(aq)$

Initial:	0.095 M	–	0 M	$\approx$ 0 M
Changes:	–0.0011 M	–	+0.0011 M	+0.0011 M
Equil:	0.094 M	–	0.0011 M	0.0011 M

$$K_a = \frac{\left[C_6H_{11}O_2^- \right]\left[H_3O^+ \right]}{\left[HC_6H_{11}O_2 \right]} = \frac{(0.0011)^2}{0.094} = 1.3 \times 10^{-5}$$

27. Here we need to find the molarity S of the acid needed that yields $\left[H_3O^+\right] = 10^{-2.85} = 1.4 \times 10^{-3} M$

Equation: $HC_7H_5O_2(aq) + H_2O(l) \rightleftharpoons H_3O^+(aq) + C_7H_5O_2^-(aq)$

Initial:	S	—	0 M	0 M
Changes:	-0.0014 M	—	$+0.0014$M	$+0.0014$M
Equil:	$S - 0.0014$M	—	0.0014 M	0.0014 M

$$K_a = \frac{\left[H_3O^+\right]\left[C_7H_5O_2^-\right]}{\left[HC_7H_5O_2\right]} = \frac{(0.0014)^2}{S - 0.0014} = 6.3 \times 10^{-5} \qquad S - 0.0014 = \frac{(0.0014)^2}{6.3 \times 10^{-5}} = 0.031$$

$$S = 0.031 + 0.0014 = 0.032 \ M = \left[HC_7H_5O_2\right]$$

$$350.0 \ mL \times \frac{1 \ L}{1000 \ mL} \times \frac{0.032 \ mol \ HC_7H_5O_2}{1 \ L \ soln} \times \frac{122.1 \ g \ HC_7H_5O_2}{1 \ mol \ HC_7H_5O_2} = 1.4 \ g \ HC_7H_5O_2$$

28. First we find $\left[OH^-\right]$ and then use the K_b expression to find $\left[(CH_3)_3N\right]$

$$pOH = 14.00 - pH = 14.00 - 11.12 = 2.88 \qquad \left[OH^-\right] = 10^{-pOH} = 10^{-2.88} = 0.0013 \ M$$

$$K_b = 6.3 \times 10^{-5} = \frac{\left[(CH_3)_3NH^+\right]\left[OH^-\right]}{\left[(CH_3)_3N\right]_{equil}} = \frac{(0.0013)^2}{\left[(CH_3)_3N\right]_{equil}} \qquad \left[(CH_3)_3N\right]_{equil} = 0.027 \ M \ (CH_3)_3N$$

$$\left[(CH_3)_3N\right]_{initial} = 0.027 \ M(\text{equil. concentration}) + 0.0013 \ M(\text{concentration ionized})$$

$$\left[(CH_3)_3N\right]_{initial} = 0.028 \ M$$

29. We use the balanced chemical equation, then solve using the quadratic formula.

Equation: $HClO_2(aq) + H_2O(l) \rightleftharpoons H_3O^+(aq) + ClO_2^-(aq)$

Initial:	0.55 M	–	≈ 0M	0M
Changes:	$-x$ M	–	$+x$ M	$+x$ M
Equil:	$(0.55 - x)$ M	–	x M	x M

$$K_a = \frac{\left[H_3O^+\right]\left[ClO_2^-\right]}{\left[HClO_2\right]} = \frac{x^2}{0.55 - x} = 1.1 \times 10^{-2} = 0.011 \qquad x^2 = 0.0061 - 0.011x$$

$$x^2 + 0.011x - 0.0061 = 0$$

$$x = \frac{-b \pm \sqrt{b^2 - 4ac}}{2a} = \frac{-0.011 \pm \sqrt{0.000121 + 0.0244}}{2} = 0.073 \ M = \left[H_3O^+\right]$$

The method of successive approximations converges to the same answer in four cycles.

$$pH = -\log\left[H_3O^+\right] = -\log(0.073) = 1.14 \qquad pOH = 14.00 - pH = 14.00 - 1.14 = 12.86$$

$$\left[OH^-\right] = 10^{-pOH} = 10^{-12.86} = 1.4 \times 10^{-13} \ M$$

30. Organize the solution around the balanced chemical equation, and solve first for $\left[OH^-\right]$.

Equation:	$CH_3NH_2(aq)$	+	$H_2O(l)$	$\rightleftharpoons$	$OH^-(aq)$	+	$CH_3NH_3^+(aq)$
Initial:	0.386 M		–		≈ 0 M		0 M
Changes:	$-x$ M		–		$+x$ M		$+x$ M
Equil:	$(0.386-x)$ M		–		x M		x M

$$K_b = \frac{\left[OH^-\right]\left[CH_3NH_3^+\right]}{\left[CH_3NH_2\right]} = \frac{x^2}{0.386-x} = 4.2\times10^{-4} \approx \frac{x^2}{0.386} \quad \text{assuming } x \ll 0.386$$

$$x = \sqrt{0.386\times4.2\times10^{-4}} = 0.013 \text{ M} = [OH^-] \quad pOH = -\log(0.013) = 1.89$$

$$pH = 14.00 - 1.89 = 12.11 \quad \left[H_3O^+\right] = 10^{-pH} = 10^{-12.11} = 7.8\times10^{-13} \text{ M}$$

This is the same result as is determined with the quadratic equation roots formula.

31.

$$\left[C_{10}H_7NH_2\right] = \frac{1 \text{ g } C_{10}H_7NH_2}{590 \text{ g } H_2O} \times \frac{1.00 \text{ g } H_2O}{1 \text{ mL}} \times \frac{1000 \text{ mL}}{1 \text{ L}} \times \frac{1 \text{ mol } C_{10}H_7NH_2}{143.2 \text{ g } C_{10}H_7NH_2}$$

$$= 0.012 \text{ M } C_{10}H_7NH_2$$

$$K_b = 10^{-pK} = 10^{-3.92} = 1.2\times10^{-4}$$

Equation:	$C_{10}H_7NH_2(aq)$	+	$H_2O(l)$	$\rightleftharpoons$	$OH^-(aq)$	+	$C_{10}H_7NH_3^+(aq)$
Initial:	0.012 M		–		≈ 0 M		0 M
Changes:	$-x$ M		–		$+x$ M		$+x$ M
Equil:	$(0.012-x)$ M		–		x M		x M

$$K_b = \frac{\left[OH^-\right]\left[C_{10}H_7NH_3^+\right]}{\left[C_{10}H_7NH_2\right]} = \frac{x^2}{0.012-x} = 1.2\times10^{-4} \approx \frac{x^2}{0.012} \quad \text{assuming } x \ll 0.012$$

$x = \sqrt{0.012\times1.2\times10^{-4}} = 0.0012$ This is an almost acceptable assumption. Another approximation cycle gives:

$$x = \sqrt{(0.012-0.0012)\times1.2\times10^{-4}} = 0.0011 \quad \text{Yet another cycle seems necessary.}$$

$$x = \sqrt{(0.012-0.0011)\times1.2\times10^{-4}} = 0.0011 \text{M} = [OH^-]$$

The quadratic equation roots formula provides the same answer.

$$pOH = -\log\left[OH^-\right] = -\log(0.0011) = 2.96 \quad pH = 14.00 - pOH = 11.04$$

$$H_3O^+ = 10^{-pH} = 10^{-11.04} = 9.1\times10^{-12} \text{ M}$$

32. The stoichiometry of the ionization reaction indicates that

$$\left[H_3O^+\right]=\left[OC_6H_4NO_2^-\right], K_a = 10^{-pK} = 10^{-7.23} = 5.9\times10^{-8} \text{ and}$$

$$\left[H_3O^+\right]=10^{-4.53} = 3.0\times10^{-5}\,M\text{. We let } S = \text{the molar solubility of } o\text{-nitrophenol.}$$

Equation:	$HOC_6H_4NO_2(aq) + H_2O(l)$	$\rightleftharpoons$	$OC_6H_4NO_2^-(aq) +$	$H_3O^+(aq)$
Initial:	S	$-$	$0\,M$	$\approx 0\,M$
Changes:	$-3.0\times10^{-5}\,M$	$-$	$+3.0\times10^{-5}\,M$	$+3.0\times10^{-5}\,M$
Equil:	$(S-3.0\times10^{-5})\,M$	$-$	$3.0\times10^{-5}\,M$	$3.0\times10^{-5}\,M$

$$K_a = 5.9\times10^{-8} = \frac{\left[OC_6H_4NO_2^-\right]\left[H_3O^+\right]}{\left[HOC_6H_5NO_2\right]} = \frac{\left(3.0\times10^{-5}\right)^2}{S-3.0\times10^{-5}}$$

$$S-3.0\times10^{-5} = \frac{\left(3.0\times10^{-5}\right)^2}{5.9\times10^{-8}} = 1.5\times10^{-2}\,M \quad S = 1.5\times10^{-2}\,M + 3.0\times10^{-5}\,M = 1.5\times10^{-2}\,M$$

Hence,

$$\text{solubility} = \frac{1.5\times10^{-2}\,\text{mol HOC}_6H_4NO_2}{1\,\text{L soln}} \times \frac{139.1\,\text{g HOC}_6H_4NO_2}{1\,\text{mol HOC}_6H_4NO_2} = 2.1\,\text{g HOC}_6H_4NO_2\,/\,\text{L soln}$$

33. Here we determine $\left[H_3O^+\right]$ which, because of the stoichiometry of the reaction,

equals $\left[C_2H_3O_2^-\right]$. $\left[H_3O^+\right] = 10^{-pH} = 10^{-4.52} = 3.0\times10^{-5}\,M = \left[C_2H_3O_2^-\right]$

We solve for S, the concentration of $HC_2H_3O_2$ in the 0.750 L solution before it dissociates.

Equation:	$HC_2H_3O_2(aq) +$	$H_2O(l)$	$\rightleftharpoons$	$C_2H_3O_2^-(aq) +$	$H_3O^+(aq)$
initial:	$S\,M$	$-$		$0\,M$	$\approx 0\,M$
changes:	$-3.0\times10^{-5}\,M$	$-$		$+3.0\times10^{-5}\,M$	$+3.0\times10^{-5}\,M$
equil:	$(S-3.0\times10^{-5})\,M$	$-$		$3.0\times10^{-5}\,M$	$3.0\times10^{-5}\,M$

$$K_a = \frac{[H_3O^+][C_2H_3O_2^-]}{[HC_2H_3O_2]} = 1.8\times10^{-5} = \frac{(3.0\times10^{-5})^2}{(S-3.0\times10^{-5})}$$

$$\left(3.0\times10^{-5}\right)^2 = 1.8\times10^{-5}\left(S-3.0\times10^{-5}\right) = 9.0\times10^{-10} = 1.8\times10^{-5}S - 5.4\times10^{-10}$$

$$S = \frac{9.0\times10^{-10} + 5.4\times10^{-10}}{1.8\times10^{-5}} = 8.0\times10^{-5}\,M \quad \text{Now we determine the mass of vinegar needed.}$$

$$\text{mass vinegar} = 0.750\,L\times\frac{8.0\times10^{-5}\,\text{mol HC}_2H_3O_2}{1\,\text{L soln}}\times\frac{60.05\,\text{g HC}_2H_3O_2}{1\,\text{mol HC}_2H_3O_2}\times\frac{100.0\,\text{g vinegar}}{5.7\,\text{g HC}_2H_3O_2}$$

$$= 0.063\,\text{g vinegar}$$

34. First we determine $\left[OH^-\right]$ which, because of the stoichiometry of the reaction, equals $\left[NH_4^+\right]$.

$$pOH = 14.00 - pH = 14.00 - 11.55 = 2.45 \quad \left[OH^-\right] = 10^{-pOH} = 10^{-2.45} = 3.5 \times 10^{-3} M = \left[NH_4^+\right]$$

We solve for S, the concentration of NH_3 in the 0.625 L solution prior to dissociation.

Equation: $\quad NH_3(aq) + H_2O(l) \rightleftharpoons NH_4^+(aq) + OH^-(aq)$

initial:	S M	$-$	0 M	≈ 0 M
changes:	-0.0035 M	$-$	$+0.0035$ M	$+0.0035$ M
equil:	$(S - 0.0035)$ M	$-$	0.0035 M	0.0035 M

$$K_b = \frac{\left[NH_4^+\right]\left[OH^-\right]}{\left[NH_3\right]} = 1.8 \times 10^{-5} = \frac{(0.0035)^2}{(S - 0.0035)}$$

$$(0.0035)^2 = 1.8 \times 10^{-5}(S - 0.0035) = 1.2 \times 10^{-5} = 1.8 \times 10^{-5} S - 6.3 \times 10^{-8}$$

$$S = \frac{1.2 \times 10^{-5} + 6.3 \times 10^{-8}}{1.8 \times 10^{-5}} = 0.67 \text{ M}$$

Now we determine the volume of household ammonia needed

$$V_{ammonia} = 0.625 \text{ L soln} \times \frac{0.67 \text{ mol } NH_3}{1 \text{ L soln}} \times \frac{17.03 \text{ g } NH_3}{1 \text{ mol } NH_3} \times \frac{100.0 \text{ g soln}}{6.8 \text{ g } NH_3} \times \frac{1 \text{ mL soln}}{0.97 \text{ g soln}}$$

$$= 1.1 \times 10^2 \text{ mL household ammonia solution}$$

35. **(a)** $n_{proplyamine} = \dfrac{PV}{RT} = \dfrac{(316 \text{ torr} \times \dfrac{1 \text{ atm}}{760 \text{ torr}})(0.275 \text{ L})}{(0.08206 \text{ L atm K}^{-1} \text{ mol}^{-1})(298.15 \text{ K})} = 4.67 \times 10^{-3} \text{ mol propylamine}$

$$[propylamine] = \frac{n}{V} = \frac{4.67 \times 10^{-3} \text{ moles}}{0.500 \text{ L}} = 9.35 \times 10^{-3} \text{ M}$$

$$K_b = 10^{-pK_b} = 10^{-3.43} = 3.7 \times 10^{-4}$$

$\quad CH_3CH_2CH_2NH_2(aq) + H_2O(l) \rightleftharpoons CH_3CH_2CH_2NH_3^+(aq) + OH^-(aq)$

Initial	9.35×10^{-3} M	$-$	0 M	≈ 0 M
Change	$-x$ M	$-$	$+x$ M	$+x$ M
Equil.	$(9.35 \times 10^{-3} - x)$ M	$-$	x M	x M

$$K_b = 3.7 \times 10^{-4} = \frac{x^2}{9.35 \times 10^{-3} - x} \quad \text{or} \quad 3.5 \times 10^{-6} - 3.7 \times 10^{-4}(x) = x^2$$

$$x^2 + 3.7 \times 10^{-4}x - 3.5 \times 10^{-6} = 0$$

Find x with the roots formula:

$$x = \frac{-3.7 \times 10^{-4} \pm \sqrt{(3.7 \times 10^{-4})^2 - 4(1)(-3.5 \times 10^{-6})}}{2(1)}$$

Therefore $x = 1.69\underline{5} \times 10^{-3}$ M = [OH$^-$]   pOH = 2.77 and pH = 11.23

(b)   [OH$^-$] = 1.7×10^{-3} M = [NaOH]   (MM$_{NaOH}$ = 39.997 g mol^{-1})

$n_{NaOH} = (C)(V) = 1.7 \times 10^{-3}$ M $\times 0.500$ L = 8.5×10^{-4} moles NaOH

mass of NaOH = (n)(MM$_{NaOH}$) = 8.5×10^{-4} mol NaOH $\times$ 39.997 g NaOH/mol NaOH

mass of NaOH = 0.034 g NaOH(34 mg of NaOH)

36. $K_b = 10^{-pK_b} = 10^{-9.5} = 3.\underline{2} \times 10^{-10}$ MM$_{quinoline}$ = 129.16 g mol^{-1}

$$\text{Solubility(25 °C)} = \frac{0.6 \text{ g}}{100 \text{ mL}}$$

$$\text{Molar solubility(25 °C)} = \frac{0.6 \text{ g}}{100 \text{ mL}} \times \frac{1 \text{ mol}}{129.16 \text{ g}} \times \frac{1000 \text{ mL}}{1 \text{ L}} = 0.04\underline{6} \text{ M} \text{ (note: only 1 sig fig)}$$

	C$_9$H$_7$N(aq)	+ H$_2$O(l)	$\rightleftharpoons$ C$_9$H$_7$NH$^+$(aq)	+ OH$^-$(aq)
Initial	0.046 M	—	0 M	$\approx$ 0 M
Change	$-x$ M	—	$+x$ M	$+x$ M
Equil.	(0.046 $-x$) M	—	x M	x M

$$3.\underline{2} \times 10^{-10} = \frac{x^2}{0.04_6 - x} \approx \frac{x^2}{0.04_6}, \quad x = 3.\underline{8} \times 10^{-6} \text{ M} \quad (x \ll 0.04\underline{6}, \text{ valid assumption})$$

Therefore, [OH$^-$] = 4×10^{-6} M   pOH = 5.4 and pH = 8.6

37.   If the molarity of acetic acid is doubled, we expect a lower initial pH (more H$_3$O$^+$(aq) in solution) and a lower percent ionization as a result of the increase in concentration. Therefore (b) (the diagram in the center) best represents the conditions ($\sim(2)^{1/2}$ times greater).

38.   If NH$_3$ is diluted to half its original molarity, we expect a lower pH (a lower [OH$^-$]) and a higher percent ionization in the diluted sample. Thus, the diagram that is next to the one furthest to the right best represents the hydroxide ion concentration ($\sim (2)^{-1/2}$ times the original [OH$^-$]).

Percent Ionization

39. Let us first compute the $\left[H_3O^+\right]$ in this solution.

Equation: $HC_3H_5O_2\,(aq)\ +\ H_2O(l)\ \rightleftharpoons\ H_3O^+(aq)\ +\ C_3H_5O_2^{\ -}\,(aq)$

Initial:	0.45 M	—	≈ 0 M ≈ 0 M
Changes:	$-x$ M	—	$+x$ M $+x$ M
Equil:	$(0.45-x)$ M	—	x M x M

$$K_a = \frac{\left[H_3O^+\right]\left[C_3H_5O_2^{\ -}\right]}{\left[HC_3H_5O_2\right]} = \frac{x^2}{0.45-x} = 10^{-4.89} = 1.3\times10^{-5} \approx \frac{x^2}{0.45}$$

$x = 2.4\times10^{-3}\,M$; We have assumed that $x \ll 0.45$ M , an assumption that clearly is correct.

(a) $\quad \alpha = \dfrac{\left[H_3O^+\right]_{equil}}{\left[HC_3H_5O_2\right]_{initial}} = \dfrac{2.4\times10^{-3}\,M}{0.45\,M} = 0.0053 = $ degree of ionization

(b)   % ionization $= \alpha \times 100\% = 0.0053 \times 100\% = 0.53\%$

40. For $C_2H_5NH_2$ (ethylamine), $K_b = 4.3 \times 10^{-4}$

$$C_2H_5NH_2(aq)\ + H_2O(l)\ \rightleftharpoons\ C_2H_5NH_3^+(aq)\ +\ OH^-(aq)$$

Initial	0.85 M	—	0 M	≈ 0 M
Change	$-x$ M	—	$+x$ M	$+x$ M
Equil.	$(0.85 -x)$ M	—	x M	x M

$$4.3 \times 10^{-4} = \frac{x^2}{(0.85-x)\,M} \approx \frac{x^2}{0.85\,M} \qquad x = 0.019\,M\ (x << 0.85, \text{ thus a valid assumption})$$

Degree of ionization $\alpha = \dfrac{0.019\,M}{0.85\,M} = 0.022$ Percent ionization $= \alpha \times 100\% = 2.2\,\%$

41. Let x be the initial concentration of NH_3, hence, the amount dissociated is $0.042\,x$

$$NH_3(aq)\ + H_2O(l)\ \xrightarrow{\ K_a = 1.8\times10^{-5}\ }\ NH_4^+(aq)\ +\ OH^-(aq)$$

Initial	x M	—	0 M	≈ 0 M
Change	$-0.042x$ M	—	$+0.042x$ M	$+0.042x$ M
Equil.	$(1x-0.042x)$ M	—	$0.042x$ M	$0.042x$ M
	$= 0.958\,x$			

$$K_b = 1.8 \times 10^{-5} = \frac{\left[NH_4^+\right]\left[OH^-\right]}{\left[NH_3\right]_{equil}} = \frac{[0.042x]^2}{[0.958x]} = 0.00184x$$

$$\left[NH_3\right]_{initial} = x = \frac{1.8 \times 10^{-5}}{0.00184} = 0.00978\,M = 0.0098\,M$$

42.

	$HC_3H_5O_2(aq)$	$+ H_2O(l)$	$\xrightarrow{K_a = 1.3 \times 10^{-5}}$	$C_3H_5O_2^-(aq)$	$+ H_3O^+(aq)$
Initial	0.100 M	—		0 M	≈ 0 M
Change	$-x$ M	—		$+x$ M	$+x$ M
Equil.	$(0.100 - x)$ M	—		x M	x M

$$1.3 \times 10^{-5} = \frac{x^2}{0.100\,M - x} \approx \frac{x^2}{0.100} \qquad x = 1.1 \times 10^{-3} \;(x \ll 0.100, \text{ thus a valid assumption})$$

$$\text{Percent ionization} = \frac{0.0011}{0.100} \times 100\% = 1.1\underline{4}\ \%$$

Next we need to find molarity of acetic acid that is 1.1% ionized

	$HC_2H_3O_2(aq)$	$+ H_2O(l)$	$\xrightarrow{K_a = 1.8 \times 10^{-5}}$	$C_2H_3O_2^-(aq)$	$+ H_3O^+(aq)$
Initial	X M	—		0 M	≈ 0 M
Change	$-\dfrac{1.1\underline{4}}{100}\,X\,M$	—		$+\dfrac{1.1\underline{4}}{100}\,X\,M$	$+\dfrac{1.1\underline{4}}{100}\,X\,M$
Equil.	$X - \dfrac{1.1\underline{4}}{100}\,X$	—		$\dfrac{1.1\underline{4}}{100}\,X\,M$	$\dfrac{1.1\underline{4}}{100}\,X\,M$
	$= 0.98\underline{86}\,X\,M$			$= 0.011\underline{4}$ M	$= 0.011\underline{4}$ M

$$1.8 \times 10^{-5} = \frac{\left(\dfrac{1.1\underline{4}}{100}\,X\right)^2}{0.98\underline{86}\ X} = 1.3 \times 10^{-4}(X); \qquad X = 0.13\underline{7}\ M$$

Consequently, approximately 0.14 moles of acetic acid must be dissolved in one liter of water in order to have the same percent ionization as 0.100 M propionic acid.

43. We would not expect these ionizations to be correct because the calculated degree of ionization is based on the assumption that the $[HC_2H_3O_2]_{initial} \approx [HC_2H_3O_2]_{initial} - [HC_2H_3O_2]_{equil.}$ which is invalid at the 13 and 42 percent levels of ionization seen here.

44. $HC_2Cl_3O_2$ (trichloroacetic acid) $pK_a = 0.52$ $\qquad K_a = 10^{-0.52} = 0.30$
For a 0.035 M solution, the assumption will not work (the concentration is too small and K_a is too large) Thus, we must solve the problem using the quadratic equation.

	$HC_2Cl_3O_2(aq)$	$+ H_2O(l)$	$\rightleftharpoons$	$C_2Cl_3O_2^-(aq)$	$+ H_3O^+(aq)$
Initial	0.035 M	—		0 M	≈ 0 M
Change	$-x$ M	—		$+x$ M	$+x$ M
Equil.	$(0.035 - x)$ M	—		x M	x M

$$0.30 = \frac{x^2}{0.035 - x} \quad \text{or} \quad 0.010\underline{5} - 0.30(x) = x^2 \quad \text{or} \quad x^2 + 0.30x - 0.010\underline{5} = 0 \; ; \text{solve quadratic:}$$

$$x = \frac{-0.30 \pm \sqrt{(0.30)^2 - 4(1)(0.010\underline{5})}}{2(1)} \qquad \text{Therefore } x = 0.032 \text{ M} = [\text{H}_3\text{O}^+]$$

Degree of ionization $\alpha = \dfrac{0.032 \text{ M}}{0.035 \text{ M}} = 0.91$ Percent ionization $= \alpha \times 100\% = 91. \%$

Polyprotic Acids

45. Because H_3PO_4 is a weak acid, there is little HPO_4^{2-} (produced in the 2nd ionization) compared to the H_3O^+ (produced in the 1st ionization). In turn, there is little PO_4^{3-} (produced in the 3rd ionization) compared to the HPO_4^{2-}, and very little compared to the H_3O^+.

46. The main estimate involves assuming that the mass percents can be expressed as 0.057 g of 75% H_3PO_4 per 100. mL of solution and 0.084 g of 75% H_3PO_4 per 100. mL of solution. That is, that the density of the aqueous solution is essentially 1.00 g/mL. Based on this assumption, we can compute the initial minimum and maximum concentrations of H_3PO_4.

$$[\text{H}_3\text{PO}_4] = \frac{0.057 \text{ g impure H}_3\text{PO}_4 \times \dfrac{75 \text{ g H}_3\text{PO}_4}{100 \text{ g impure H}_3\text{PO}_4} \times \dfrac{1 \text{ mol H}_3\text{PO}_4}{98.00 \text{ g H}_3\text{PO}_4}}{100 \text{ mL soln} \times \dfrac{1 \text{ L}}{1000 \text{ mL}}} = 0.0044 \text{ M}$$

$$[\text{H}_3\text{PO}_4] = \frac{0.084 \text{ g impure H}_3\text{PO}_4 \times \dfrac{75 \text{ g H}_3\text{PO}_4}{100 \text{ g impure H}_3\text{PO}_4} \times \dfrac{1 \text{ mol H}_3\text{PO}_4}{98.00 \text{ g H}_3\text{PO}_4}}{100 \text{ mL soln} \times \dfrac{1 \text{ L}}{1000 \text{ mL}}} = 0.0064 \text{ M}$$

Equation:	$\text{H}_3\text{PO}_4 \text{(aq)}$	+	$\text{H}_2\text{O(l)}$	$\rightleftharpoons$	$\text{H}_2\text{PO}_4^- \text{(aq)}$	+	$\text{H}_3\text{O}^+ \text{(aq)}$
Initial:	0.0044 M		–		0 M		≈ 0 M
Changes:	$-x$ M		–		$+x$ M		$+x$ M
Equil:	$(0.0044 - x)$ M		–		x M		x M

$$K_{a_1} = \frac{\left[\text{H}_2\text{PO}_4^-\right]\left[\text{H}_3\text{O}^+\right]}{\left[\text{H}_3\text{PO}_4\right]} = \frac{x^2}{0.0044 - x} = 7.1 \times 10^{-3} \qquad x^2 + 0.0071x - 3.1 \times 10^{-5} = 0$$

$$x = \frac{-b \pm \sqrt{b^2 - 4ac}}{2a} = \frac{-0.0071 \pm \sqrt{5.0 \times 10^{-5} + 1.2 \times 10^{-4}}}{2} = 3.0 \times 10^{-3} \text{ M} = \left[\text{H}_3\text{O}^+\right]$$

The set up for the second concentration is the same as for the first, with the exception that 0.0044 M is replaced by 0.0064 M.

$$K_{a_1} = \frac{\left[H_2PO_4^-\right]\left[H_3O^+\right]}{\left[H_3PO_4\right]} = \frac{x^2}{0.0064 - x} = 7.1 \times 10^{-3} \qquad x^2 + 0.0071x - 4.5 \times 10^{-5} = 0$$

$$x = \frac{-b \pm \sqrt{b^2 - 4ac}}{2a} = \frac{-0.0071 \pm \sqrt{5.0 \times 10^{-5} + 1.8 \times 10^{-4}}}{2} = 4.0 \times 10^{-3} M = \left[H_3O^+\right]$$

The two values of pH now are determined, representing the pH range in a cola drink.

$$pH = -\log\left(3.0 \times 10^{-3}\right) = 2.52 \qquad pH = -\log\left(4. \times 10^{-3}\right) = 2.40$$

47. **(a)** Equation: $H_2S(aq)$ $+$ $H_2O(l)$ $\rightleftharpoons$ $HS^-(aq)$ $+$ $H_3O^+(aq)$

Initial:	0.075 M	–	0 M	≈ 0 M
Changes:	$-x$ M	–	$+x$ M	$+x$ M
Equil:	$(0.075 - x)$ M	–	x M	$+x$ M

$$K_{a_1} = \frac{\left[HS^-\right]\left[H_3O^+\right]}{\left[H_2S\right]} = 1.0 \times 10^{-7} = \frac{x^2}{0.075 - x} \approx \frac{x^2}{0.075} \qquad x = 8.7 \times 10^{-5} M = \left[H_3O^+\right]$$

$$\left[HS^-\right] = 8.7 \times 10^{-5} M \text{ and } \left[S^{2-}\right] = K_{a_2} = 1 \times 10^{-19} M$$

(b) The set up for this problem is the same as for part (a), with 0.0050 M replacing 0.075 M as the initial value of $\left[H_2S\right]$.

$$K_{a_1} = \frac{\left[HS^-\right]\left[H_3O^+\right]}{\left[H_2S\right]} = 1.0 \times 10^{-7} = \frac{x^2}{0.0050 - x} \approx \frac{x^2}{0.0050} \qquad x = 2.2 \times 10^{-5} M = \left[H_3O^+\right]$$

$$\left[HS^-\right] = 2.2 \times 10^{-5} M \text{ and } \left[S^{2-}\right] = K_{a_2} = 1 \times 10^{-19} M$$

(c) The set up for this part is the same as for part (a), with 1.0×10^{-5} M replacing 0.075 M as the initial value of $\left[H_2S\right]$. The solution differs in that we cannot assume $x \ll 1.0 \times 10^{-5}$. Solve the quadratic equation to find the desired equilibrium concentrations.

$$K_{a_1} = \frac{\left[HS^-\right]\left[H_3O^+\right]}{\left[H_2S\right]} = 1.0 \times 10^{-7} = \frac{x^2}{1.0 \times 10^{-5} - x}$$

$$x^2 + 1.0 \times 10^{-7} x - 1.0 \times 10^{-12} = 0$$

$$x = \frac{-1.0 \times 10^{-7} \pm \sqrt{1.0 \times 10^{-14} + \left(4 \times 1.0 \times 10^{-12}\right)}}{2 \times (1)}$$

$$x = 9.5 \times 10^{-7} M = \left[H_3O^+\right] \qquad \left[HS^-\right] = 9.5 \times 10^{-7} M \qquad \left[S^{2-}\right] = K_{a_2} = 1 \times 10^{-19} M$$

48. For H_2CO_3, $K_1 = 4.4 \times 10^{-7}$ and $K_2 = 4.7 \times 10^{-11}$

(a) The first acid ionization proceeds to a far greater extent than does the second and determines the value of $[H_3O^+]$.

Equation: $\quad H_2CO_3(aq) \quad + \quad H_2O(l) \quad \rightleftharpoons \quad H_3O^+(aq) \quad + \quad HCO_3^-(aq)$

Initial:	0.045 M	—	≈ 0 M	0 M
Changes:	$-x$ M	—	$+x$ M	$+x$ M
Equil:	$(0.045 - x)$ M	—	x M	x M

$$K_1 = \frac{[H_3O^+][HCO_3^-]}{[H_2CO_3]} = \frac{x^2}{0.045 - x} = 4.4 \times 10^{-7} \approx \frac{x^2}{0.045} \quad x = 1.4 \times 10^{-4} M = [H_3O^+]$$

(b) Since the second ionization occurs to only a limited extent,

$$[HCO_3^-] = 1.4 \times 10^{-4} \ M = 0.00014 \ M$$

(c) We use the second ionization to determine $[CO_3^{2-}]$.

Equation: $\quad HCO_3^-(aq) \quad + \quad H_2O(l) \quad \rightleftharpoons \quad H_3O^+(aq) \quad + \quad CO_3^{2-}(aq)$

	0.00014 M	—	0.00014 M	0 M
	$-x$ M	—	$+x$ M	$+x$ M
Initial:	$(0.00014 - x)$ M	—	$(0.00014 + x)$ M	x M
Changes:				
Equil:				

$$K_2 = \frac{[H_3O^+][CO_3^{2-}]}{[HCO_3^-]} = \frac{(0.00014 + x)x}{0.00014 - x} = 4.7 \times 10^{-11} \approx \frac{(0.00014) \, x}{0.00014}$$

$$x = 4.7 \times 10^{-11} M = [CO_3^{2-}]$$

Again we assumed that $x \ll 0.00014$ M, which clearly is the case. We also note that our original assumption, that the second ionization is much less significant than the first, also is a valid one. As an alternative to all of this, we could have recognized that the concentration of the divalent anion equals the second ionization constant:

$$[CO_3^{2-}] = K_2.$$

49. In all cases, of course, the first ionization of H_2SO_4 is complete, and establishes the initial values of $[H_3O^+]$ and $[HSO_4^-]$. Thus, we need only deal with the second ionization in each case.

(a) Equation: $\quad HSO_4^-(aq) \quad + \quad H_2O(l) \rightleftharpoons \quad SO_4^{2-}(aq) \quad + \quad H_3O^+(aq)$

Initial:	0.75 M	—	0 M	0.75 M
Changes:	$-x$ M	—	$+x$ M	$+x$ M
Equil:	$(0.75 - x)$ M	—	x M	$(0.75 + x)$ M

$$K_{a_2} = \frac{\left[SO_4^{2-}\right]\left[H_3O^+\right]}{\left[HSO_4^-\right]} = 0.011 = \frac{x(0.75+x)}{0.75-x} \approx \frac{0.75x}{0.75} \qquad x = 0.011\,M = \left[SO_4^{2-}\right]$$

We have assumed that $x \ll 0.75$ M, an assumption that clearly is correct.

$$\left[HSO_4^-\right] = 0.75 - 0.011 = 0.74\,M \qquad \left[H_3O^+\right] = 0.75 + 0.011 = 0.76\,M$$

(b) The set up for this part is similar to part (a), with the exception that 0.75 M is replaced by 0.075 M.

$$K_{a_2} = \frac{\left[SO_4^{2-}\right]\left[H_3O^+\right]}{\left[HSO_4^-\right]} = 0.011 = \frac{x(0.075+x)}{0.075-x} \qquad 0.011(0.075-x) = 0.075x + x^2$$

$$x^2 + 0.086x - 8.3 \times 10^{-4} = 0$$

$$x = \frac{-b \pm \sqrt{b^2 - 4ac}}{2a} = \frac{-0.086 \pm \sqrt{0.0074 + 0.0033}}{2} = 0.0087\,M$$

$$x = 0.0087\,M = \left[SO_4^{2-}\right]$$

$$\left[HSO_4^-\right] = 0.075 - 0.0087 = 0.066\,M \qquad \left[H_3O^+\right] = 0.075 + 0.0088\,M = 0.084\,M$$

(c) Again, the set up is the same as for part (a), with the exception that 0.75 M is replaced by 0.00075 M

$$K_{a_2} = \frac{[SO_4^{2-}][H_3O^+]}{[HSO_4^-]} = 0.011 = \frac{x(0.00075+x)}{0.00075-x} \qquad 0.011(0.00075-x) = 0.00075\,x + x^2$$

$$x^2 + 0.0118x - 8.3 \times 10^{-6} = 0$$

$$x = \frac{-b \pm \sqrt{b^2 - 4ac}}{2a} = \frac{-0.0118 \pm \sqrt{1.39 \times 10^{-4} + 3.3 \times 10^{-5}}}{2} = 6.6 \times 10^{-4}$$

$$x = 6.6 \times 10^{-4}\,M = \left[SO_4^{2-}\right] \qquad \left[HSO_4^-\right] = 0.00075 - 0.00066 = 9 \times 10^{-5}\,M$$

$$\left[H_3O^+\right] = 0.00075 + 0.00066\,M = 1.41 \times 10^{-3}\,M \qquad \left[H_3O^+\right] \text{ is almost twice the initial}$$

value of $\left[H_2SO_4\right]$. Thus, the second ionization of H_2SO_4 is nearly complete in this dilute solution.

50. First we determine $\left[H_3O^+\right]$, with the calculation based on the balanced chemical equation.

Equation: $HOOC(CH_2)_4 COOH(aq) + H_2O(l) \rightleftharpoons HOOC(CH_2)_4 COO^-(aq) + H_3O^+(aq)$

Initial:	0.10 M	—	0 M	≈ 0 M
Changes:	$-x$ M	—	$+x$ M	$+x$ M
Equil:	$(0.10-x)$ M	—	x M	x M

$$K_{a_1} = \frac{[H_3O^+][HOOC(CH_2)_4COO^-]}{[HOOC(CH_2)_4COOH]} = 3.9 \times 10^{-5} = \frac{x \cdot x}{0.10 - x} \approx \frac{x^2}{0.10}$$

$$x = \sqrt{0.10 \times 3.9 \times 10^{-5}} = 2.0 \times 10^{-3} M = [H_3O^+]$$

We see that our simplifying assumption, that $x \ll 0.10$ M, is indeed valid. Now we consider the second ionization. We shall see that very little H_3O^+ is produced in this ionization because of the small size of two numbers: K_{a_2} and $[HOOC(CH_2)_4COO^-]$. Again we base our calculation on the balanced chemical equation.

Equation: $HOOC(CH_2)_4COO^-(aq) + H_2O(l) \rightleftharpoons {}^-OOC(CH_2)_4COO^-(aq) + H_3O^+(aq)$

Initial:	2.0×10^{-3} M	–	0 M	2.0×10^{-3} M
Changes:	$-y$ M	–	$+y$ M	$+y$ M
Equil:	$(0.0020 - y)$ M	–	y M	$(0.0020 + y)$ M

$$K_{a_2} = \frac{[H_3O^+][{}^-OOC(CH_2)_4COO^-]}{[HOOC(CH_2)_4COO^-]} = 3.9 \times 10^{-6} = \frac{y(0.0020 + y)}{0.0020 - y} \approx \frac{0.0020y}{0.0020}$$

$$y = 3.9 \times 10^{-6} M = [{}^-OOC(CH_2)_4COO^-]$$

Again, we see that our assumption, that $y \ll 0.0020$ M, is valid. In addition, we also note that virtually no H_3O^+ is created in this second ionization. The concentrations of all species have been calculated above, with the exception of $[OH^-]$.

$$[OH^-] = \frac{K_w}{[H_3O^+]} = \frac{1.0 \times 10^{-14}}{2.0 \times 10^{-3}} = 5.0 \times 10^{-12} M; \qquad [HOOC(CH_2)_4COOH] = 0.10 M$$

$$[H_3O^+] = [HOOC(CH_2)_4COO^-] = 2.0 \times 10^{-3} M; \qquad [{}^-OOC(CH_2)_4COO^-] = 3.9 \times 10^{-6} M$$

51. **(a)** Recall that a base is a proton acceptor, in this case, accepting H^+ from H_2O.

First ionization: $C_{20}H_{24}O_2N_2 + H_2O \rightleftharpoons C_{20}H_{24}O_2N_2H^+ + OH^-$ $pK_{b_1} = 6.0$

Second ionization: $C_{20}H_{24}O_2N_2H^+ + H_2O \rightleftharpoons C_{20}H_{24}O_2N_2H_2^{2+} + OH^-$ $pK_{b_2} = 9.8$

(b)

$$[C_{20}H_{24}O_2N_2] = \frac{1.00 \text{ g quinine} \times \dfrac{1 \text{ mol quinine}}{324.4 \text{ g quinine}}}{1900 \text{ mL} \times \dfrac{1L}{1000 \text{ mL}}} = 1.62 \times 10^{-3} M$$

$$K_{b_1} = 10^{-6.0} = 1 \times 10^{-6}$$

Once again, we set up the I.C.E. table and solve for the $[OH^-]$ in this case:

Equation: $C_{20}H_{24}O_2N_2(aq) + H_2O(l) \rightleftharpoons C_{20}H_{24}O_2N_2H^+(aq) + OH^-(aq)$

Initial:	0.00162 M	–	0 M	≈ 0 M
Changes:	$-x$ M	–	$+x$ M	$+x$ M
Equil:	$(0.00162-x)$ M	–	x M	x M

$$K_{b_1} = \frac{\left[C_{20}H_{24}O_2N_2H^+\right]\left[OH^-\right]}{\left[C_{20}H_{24}O_2N_2\right]} = 1\times10^{-6} = \frac{x^2}{0.00162-x} \approx \frac{x^2}{0.00162} \qquad x \approx 4\times10^{-5}\,M$$

The assumption $x \ll 0.00162$, is valid. $pOH = -\log\left(4\times10^{-5}\right) = 4.4 \quad pH = 9.6$

52. Hydrazine is made up of two NH_2 units held together by a nitrogen-nitrogen single bond:

(a) $N_2H_4(aq) + H_2O(l) \rightleftharpoons N_2H_5^+(aq) + OH^-(aq) \qquad K_{b1} = 10^{-6.07} = 8.5\times10^{-7}$

(b) $N_2H_5^+(aq) + H_2O(l) \rightleftharpoons N_2H_6^{2+}(aq) + OH^-(aq) \qquad K_{b2} = 10^{-15.05} = 8.9\times10^{-16}$
Because the OH^- produced in (a) suppresses the reaction in (b) and since $K_{b1} \ll K_{b2}$, the solution's pH is determined almost entirely by the first base hydrolysis reaction (a). Thus we need only solve the I.C.E. table for the hydrolysis of N_2H_4 in order to find the $[OH^-]_{equil}$, which will ultimately provide us with the pH via the relationship $[OH^-]\times[H_3O^+] = 1.00 \times 10^{-14}$.

Reaction:	$N_2H_4(aq)$ +	$H_2O(l)$	$\rightleftharpoons$	$N_2H_5^+(aq)$ +	$OH^-(aq)$
Initial:	0.245 M	–		0 M	≈ 0 M
Changes:	$-x$ M	–		$+x$ M	$+x$ M
Equil:	$(0.245-x)$ M	–		x M	x M

$$K_{b_1} = \frac{\left[N_2H_5^+\right]\left[OH^-\right]}{\left[N_2H_4\right]} = 8.5\times10^{-7} = \frac{x^2}{0.245-x} = \frac{x^2}{0.245} \qquad x = 0.00186\ M = [OH^-]_{equil}$$

Thus, $[H_3O^+] = \dfrac{1.0\times10^{-14}}{0.00186} = 5.37\times10^{-12}\,M \qquad pH = -\log(5.37\times10^{-12}) = 11.26$

53. The species that hydrolyze are the cations of weak bases, namely, NH_4^+ and $C_6H_5NH_3^+$, and the anions of weak acids, namely, NO_2^- and $C_7H_5O_2^-$.

(a) $NH_4^+(aq) + NO_3^-(aq) + H_2O(l) \rightleftharpoons NH_3(aq) + NO_3^-(aq) + H_3O^+(aq)$

(b) $Na^+(aq) + NO_2^-(aq) + H_2O(l) \rightleftharpoons Na^+(aq) + HNO_2(aq) + OH^-(aq)$

(c) $K^+(aq) + C_7H_5O_2^-(aq) + H_2O(l) \rightleftharpoons K^+(aq) + HC_7H_5O_2(aq) + OH^-(aq)$

(d) $K^+(aq) + Cl^-(aq) + Na^+(aq) + I^-(aq) + H_2O(l) \rightarrow$ no reaction

(e) $C_6H_5NH_3^+(aq) + Cl^-(aq) + H_2O(l) \rightleftharpoons C_6H_5NH_2(aq) + Cl^-(aq) + H_3O^+(aq)$

54. Recall that, for a conjugate weak acid-weak base pair, $K_a \times K_b = K_w$

(a) $K_a = \dfrac{K_w}{K_b} = \dfrac{1.0 \times 10^{-14}}{1.5 \times 10^{-9}} = 6.7 \times 10^{-6}$ for $C_5H_5NH^+$

(b) $K_b = \dfrac{K_w}{K_a} = \dfrac{1.0 \times 10^{-14}}{1.8 \times 10^{-4}} = 5.6 \times 10^{-11}$ for CHO_2^-

(c) $K_b = \dfrac{K_w}{K_a} = \dfrac{1.0 \times 10^{-14}}{1.0 \times 10^{-10}} = 1.0 \times 10^{-4}$ for $C_6H_5O^-$

Ions as Acids and Bases (Hydrolysis)

55. **(a)** Because it is composed of the cation of a strong base, and the anion of a strong acid, KCl forms a neutral solution (i.e. no hydrolysis occurs).

(b) KF forms an alkaline (basic) solution. The fluoride ion being the conjugate base of a relatively strong weak acid, undergoes hydrolysis, while potassium ion, the weakly polarizing cation of a strong base does not react with water.

$F^-(aq) + H_2O(l) \rightleftharpoons HF(aq) + OH^-(aq)$

(c) $NaNO_3$ forms a neutral solution, being composed of a weakly polarizing cation and anionic conjugate base of a strong acid. No hydrolysis occurs.

(d) $Ca(OCl)_2$ forms an alkaline (basic) solution, being composed of a weakly polarizing cation, and the anion of a weak acid Thus, only the hypochlorite ion hydrolyzes.

$OCl^-(aq) + H_2O(l) \rightleftharpoons HOCl(aq) + OH^-(aq)$

(e) NH_4NO_2 forms an acidic solution. The salt is composed of the cation of a weak base and the anion of a weak acid. The ammonium ion hydrolyzes:

$NH_4^+(aq) + H_2O(l) \rightleftharpoons NH_3(aq) + H_3O^+(aq)$ $\qquad K_a = 5.6 \times 10^{-10}$

as does the nitrite ion:

$NO_2^-(aq) + H_2O(l) \rightleftharpoons HNO_2(aq) + OH^-(aq)$ $\qquad K_b = 1.4 \times 10^{-11}$

Since NH_4^+ is a stronger acid than NO_2^- is a base (as shown by the relative sizes of the K values), the solution will be acidic. The ionization constants were computed from data in Table 16-3 and with the relationship $K_w = K_a \times K_b$.

For NH_4^+, $K_a = \dfrac{1.0 \times 10^{-14}}{1.8 \times 10^{-5}} = 5.6 \times 10^{-10}$ For NO_2^-, $K_b = \dfrac{1.0 \times 10^{-14}}{7.2 \times 10^{-4}} = 1.4 \times 10^{-11}$

Where $1.8 \times 10^{-5} = K_b$ of NH_3 and $7.2 \times 10^{-4} = K_a$ of HNO_2

56. Our list in order of increasing pH is also in order of decreasing acidity. First we look for the strong acids; there is just HNO_3. Next we look for the weak acids; there is only one, $HC_2H_3O_2$. Next in order of decreasing acidity come salts with cations from weak bases and anions from strong acids; NH_4ClO_4 is in this category. Then come salts in which both ions hydrolyze to the same degree; $NH_4C_2H_3O_2$ is an example, forming a pH-neutral solution. Next are salts that have the cation of a strong base and the anion of a weak acid; $NaNO_2$ is in this category. Then come weak bases, of which $NH_3(aq)$ is an example. And finally, we terminate the list with strong bases: NaOH is the only one. Thus, in order of increasing pH of their 0.010 M aqueous solutions, the solutes are:
$HNO_3 < HC_2H_3O_2 < NH_4ClO_4 < NH_4C_2H_3O_2 < NaNO_2 < NH_3 < NaOH$

<u>57.</u> NaOCl dissociates completely in aqueous solution into $Na^+(aq)$, which does not hydrolyze, and $OCl^-(aq)$, which undergoes base hydrolysis. We determine $[OH^-]$ in a 0.089 M solution of OCl^-, finding the value of the hydrolysis constant from the ionization constant of HOCl, $K_a = 2.9 \times 10^{-8}$.

Equation:	$OCl^-(aq)$	+	$H_2O(l)$	$\rightleftharpoons$	$HOCl(aq)$	+	$OH^-(aq)$
Initial:	0.089 M		–		0 M		≈ 0 M
Changes:	$-x$ M		–		$+x$ M		$+x$ M
Equil:	$(0.089 - x)$ M		–		x M		x M

$$K_b = \frac{K_w}{K_a} = \frac{1.0 \times 10^{-14}}{2.9 \times 10^{-8}} = 3.4 \times 10^{-7} = \frac{[HOCl][OH^-]}{[OCl^-]} = \frac{x^2}{0.089 - x} \approx \frac{x^2}{0.089}$$

1.7×10^{-4} M $\ll$ 0.089, the assumption is valid.

$x = 1.7 \times 10^{-4}$ M $= [OH^-]$; pOH $= -\log(1.7 \times 10^{-4}) = 3.77$, pH $= 14.00 - 3.77 = 10.23$

58. NH_4Cl dissociates completely in aqueous solution into $NH_4^+(aq)$, which hydrolyzes, and $Cl^-(aq)$, which does not. We determine $[H_3O^+]$ in a 0.123 M solution of NH_4^+, finding the value of the hydrolysis constant from the ionization constant of NH_3, $K_b = 1.8 \times 10^{-5}$.

Equation :	$NH_4^+(aq)$	+	$H_2O(l)$	$\rightleftharpoons$	$NH_3(aq)$	+	$H_3O^+(aq)$
Initial :	0.123 M		–		0 M		≈ 0 M
Changes :	$-x$ M		–		$+x$ M		$+x$ M
Equil :	$(0.123-x)$ M		–		x M		x M

$$K_b = \frac{K_w}{K_a} = \frac{1.0 \times 10^{-14}}{1.8 \times 10^{-5}} = 5.6 \times 10^{-10} = \frac{[NH_3][H_3O^+]}{[NH_4^+]} = \frac{x^2}{0.123-x} \approx \frac{x^2}{0.123}$$

8.3×10^{-6} M $\ll 0.123$, the assumption is valid

$x = 8.3 \times 10^{-6}$ M $= [H_3O^+]$; $\quad$ pH $= -\log(8.3 \times 10^{-6}) = 5.08$

59. $KC_6H_7O_2$ dissociates completely in aqueous solution into $K^+(aq)$, which does not hydrolyze, and the ion $C_6H_7O_2^-(aq)$, which undergoes base hydrolysis. We determine $[OH^-]$ in 0.37 M $KC_6H_7O_2$ solution using an I.C.E. table.

Note: $K_a = 10^{-pK} = 10^{-4.77} = 1.7 \times 10^{-5}$

Equation :	$C_6H_7O_2^-(aq)$	+	$H_2O(l)$	$\rightleftharpoons$	$HC_6H_7O_2(aq)$	+	$OH^-(aq)$
Initial :	0.37 M		–		0 M		≈ 0 M
Changes :	$-x$ M		–		$+x$ M		$+x$ M
Equil :	$(0.37-x)$ M		–		x M		x M

$$K_b = \frac{K_w}{K_a} = \frac{1.0 \times 10^{-14}}{1.7 \times 10^{-5}} = 5.9 \times 10^{-10} = \frac{[HC_6H_7O_2][OH^-]}{[C_6H_7O_2^-]} = \frac{x^2}{0.37-x} \approx \frac{x^2}{0.37}$$

$x = 1.5 \times 10^{-5}$ M $= [OH^-]$, $\quad$ pOH $= -\log(1.5 \times 10^{-5}) = 4.82$, $\quad$ pH $= 14.00 - 4.82 = 9.18$

60.
Equation :	$C_5H_5NH^+(aq)$	+	$H_2O(l)$	$\rightleftharpoons$	$C_5H_5N(aq)$	+	$H_3O^+(aq)$
Initial :	0.0482 M		–		0 M		≈ 0 M
Changes :	$-x$ M		–		$+x$ M		$+x$ M
Equil :	$(0.0482-x)$ M		–		x M		x M

$$K_a = \frac{K_w}{K_b} = \frac{1.0 \times 10^{-14}}{1.5 \times 10^{-9}} = 6.7 \times 10^{-6} = \frac{[C_5H_5N][H_3O^+]}{[C_5H_5NH^+]} = \frac{x^2}{0.0482 - x} \approx \frac{x^2}{0.0482}$$

5.7×10^{-4} M $\ll 0.0482$ M, the assumption is valid

$$x = 5.7 \times 10^{-4} \text{M} = [H_3O^+], \quad pH = -\log(5.7 \times 10^{-4}) = 3.24$$

61. **(a)** $HSO_3^- + H_2O \rightleftharpoons H_3O^+ + SO_3^{2-}$ $K_a = K_{a_2,\text{Sulfurous}} = 6.2 \times 10^{-8}$

$HSO_3^- + H_2O \rightleftharpoons OH^- + H_2SO_3$ $K_b = \dfrac{K_w}{K_{a_1,\text{Sulfurous}}} = \dfrac{1.0 \times 10^{-14}}{1.3 \times 10^{-2}} = 7.7 \times 10^{-13}$

Since $K_a > K_b$, solutions of HSO_3^- are acidic.

(b) $HS^- + H_2O \rightleftharpoons H_3O^+ + S^{2-}$ $K_a = K_{a_2,\text{Hydrosulfuric}} = 1 \times 10^{-19}$

$HS^- + H_2O \rightleftharpoons OH^- + H_2S$ $K_b = \dfrac{K_w}{K_{a_1,\text{Hydrosulfuric}}} = \dfrac{1.0 \times 10^{-14}}{1.0 \times 10^{-7}} = 1.0 \times 10^{-7}$

Since $K_a < K_b$, solutions of HS^- are alkaline, or basic.

(c) $HPO_4^{2-} + H_2O \rightleftharpoons H_3O^+ + PO_4^{3-}$ $K_a = K_{a_3,\text{Phosphoric}} = 4.2 \times 10^{-13}$

$HPO_4^{2-} + H_2O \rightleftharpoons OH^- + H_2PO_4^-$ $K_b = \dfrac{K_w}{K_{a_2,\text{Phosphoric}}} = \dfrac{1.0 \times 10^{-14}}{6.3 \times 10^{-8}} = 1.6 \times 10^{-7}$

Since $K_a < K_b$, solutions of HPO_4^{2-} are alkaline, or basic.

62. pH $= 8.65$ (basic). Need a salt made up of a weakly polarizing cation and an anion of a weak acid, which will hydrolyze to produce a basic solution. The salt (c); KNO_2, satisfies these requirements. NH_4Cl is the salt of the cation of a weak base and the anion of a strong acid, and should form an acidic solution. (b) $KHSO_4$ and (d) $NaNO_3$ both of these salts have weakly polarizing cations and anions of strong acids; they form pH-neutral solutions. pOH $= 14.00 - 8.65 = 5.35$ $[OH^-] = 10^{-5.35} = 4.5 \times 10^{-6}$ M

Equation:	NO_2^- (aq)	+	H_2O(l)	$\rightleftharpoons$	HNO_2 (aq)	+	OH^- (aq)
Initial:	S		$-$		0 M		≈ 0 M
Changes:	-4.5×10^{-6} M		$-$		$+4.5 \times 10^{-6}$ M		$+4.5 \times 10^{-6}$ M
Equil:	$(S - 4.5 \times 10^{-6}$ M$)$		$-$		4.5×10^{-6} M		4.5×10^{-6} M

$$K_b = \frac{K_w}{K_a} = \frac{1.0 \times 10^{-14}}{7.2 \times 10^{-4}} = 1.4 \times 10^{-11} = \frac{[HNO_2][OH^-]}{[NO_2^-]} = \frac{(4.5 \times 10^{-6})^2}{S - 4.5 \times 10^{-6}}$$

$$S - 4.5 \times 10^{-6} = \frac{(4.5 \times 10^{-6})^2}{1.4 \times 10^{-11}} = 1.4 \text{ M} \quad S = 1.4 \text{ M} + 4.5 \times 10^{-6} = 1.4 \text{ M} = [KNO_2]$$

Molecular Structure and Acid Strength

63. **(a)** $HClO_3$ should be a stronger acid than is $HClO_2$. In each acid there is an $H-O-Cl$ grouping. The remaining oxygen atoms are bonded directly to Cl as terminal O atoms. Thus, there are two terminal O atoms in $HClO_3$ and only one in $HClO_2$. For oxoacids of the same element, the one with the higher number of terminal oxygen atoms is the stronger. With more oxygen atoms, the negative charge on the conjugate base is more effectively spread out, which affords greater stability. $HClO_3 : K_{a_1} = 5 \times 10^2$ and $HClO_2 : K_a = 1.1 \times 10^{-2}$.

(b) HNO_2 and H_2CO_3 each have one terminal oxygen atom. The difference? N is more electronegative than C which makes $HNO_2 \left(K_a = 7.2 \times 10^{-4} \right)$, a stronger acid than $H_2CO_3 \left(K_{a_1} = 4.4 \times 10^{-7} \right)$.

(c) H_3PO_4 and H_2SiO_3 have the same number (one) of terminal oxygen atoms. They differ in P being more electronegative than Si, which makes H_3PO_4 $\left(K_{a_1} = 7.1 \times 10^{-3} \right)$ a stronger acid than $H_2SiO_3 \left(K_{a_1} = 1.7 \times 10^{-10} \right)$.

64. CCl_3COOH is a stronger acid than CH_3COOH because in CCl_3COOH there are electronegative (electron withdrawing) Cl atoms bonded to the carbon atom adjacent to the COOH group. The electron withdrawing Cl atoms further polarize the OH bond and enhance the stability of the conjugate base, resulting in a stronger acid.

65. **(a)** HI is the stronger acid because the $H-I$ bond length is longer than the $H-Br$ bond length and, as a result, $H-I$ is easier to cleave.

(b) HOClO is a stronger acid than HOBr because
(i) there is a terminal O in HOClO but not in HOBr
(ii) Cl is more electronegative than Br.

(c) $H_3CCH_2CCl_2COOH$ is a stronger acid than $I_3CCH_2CH_2COOH$ both because Cl is more electronegative than is I and also because the Cl atoms are closer to the acidic hydrogen in the COOH group and thus can exert a stronger e^- withdrawing effect on the O–H bond than can the more distant I atoms.

66. The weakest of the five acids is CH_3CH_2COOH. The reasoning is as follows. HBr is a strong acid, stronger than the carboxylic acids. A carboxylic acid—such as $CH_2ClCOOH$ and CH_2FCH_2COOH —with a strongly electronegative atom attached to the hydrocarbon chain will be stronger than one in which no such group is present. But the I atom is so weakly electronegative that it barely influences the acid strength. Acid strengths of some of these acids (as values of pK_a) follow. (Larger values of pK_a indicate weaker acids.)
Strength: $CH_3CH_2COOH (4.89) < CI_3COOH < CH_2FCH_2COOH < CH_2ClCOOH < HBr (-8.72)$

67. The largest K_b (most basic) belongs to $CH_3CH_2CH_2NH_2$ (hydrocarbon chains have the lowest electronegativity). The smallest K_b (least basic) is that of o-chloroaniline (the nitrogen lone pair is delocalized (spread out over the ring), hence, less available to accept a proton (i.e. it is a poorer Bronstead base)).

68. The most basic species is CH_3O^- (methoxide ion). Methanol is the weakest acid, thus its anion is the strongest base. Most acidic: ortho-chlorophenol. In fact, it is the only acidic compound shown.

69. A Lewis base is an electron pair donor, while a Lewis acid is an electron pair acceptor. We draw Lewis structures to assist our interpretation.

(a) 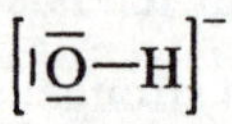 $\left[|\overline{\underline{O}}-H \right]^-$ **(b)**

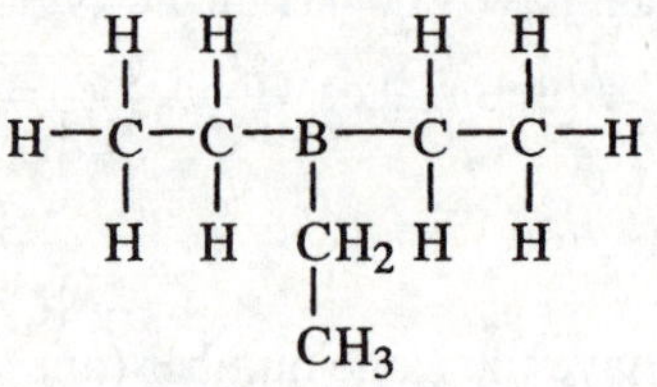

$$H-\underset{\underset{H}{|}}{\overset{\overset{H}{|}}{C}}-\underset{\underset{H}{|}}{\overset{\overset{H}{|}}{C}}-\underset{\underset{CH_2}{|}}{B}-\underset{\underset{H}{|}}{\overset{\overset{H}{|}}{C}}-\underset{\underset{H}{|}}{\overset{\overset{H}{|}}{C}}-H$$
$$CH_3$$

(c)

$$H-\underset{\underset{H}{|}}{\overset{\overset{H}{|}}{N}}-\underset{\underset{H}{|}}{\overset{\overset{H}{|}}{C}}-H$$

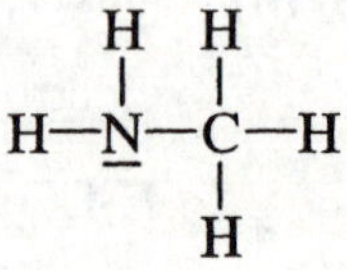

(a) The lone pairs on oxygen can readily be donated; this is a Lewis base.

(b) The incomplete octet of B provides a site for acceptance of an electron pair, this is a Lewis acid.

(c) The lone pair on nitrogen can be donated; this is a Lewis base.

70. A Lewis base is an electron pair donor, while a Lewis acid is an electron pair acceptor. We draw Lewis structures to assist our interpretation.

(a) According to the following Lewis structures, SO_3 appears to be the electron pair acceptor (the Lewis acid), and H_2O is the electron pair donor (the Lewis base). (Note that an additional sulfur-to-oxygen bond is formed upon successful attachment of water to SO_3).

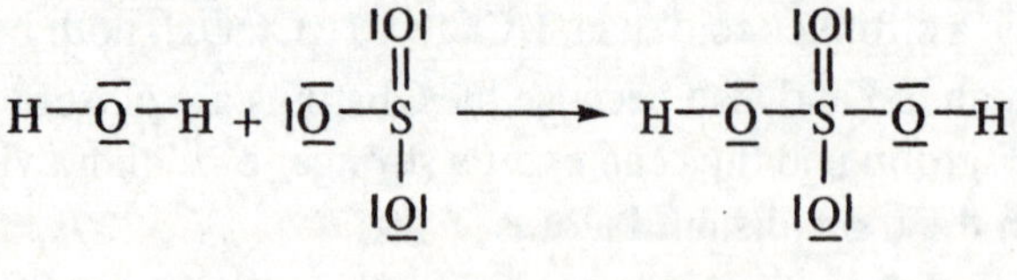

$$H-\overline{\underline{O}}-H + |\overline{\underline{O}}-\underset{\underset{|\underline{O}|}{|}}{\overset{\overset{|O|}{||}}{S}} \longrightarrow H-\overline{\underline{O}}-\underset{\underset{|\underline{O}|}{|}}{\overset{\overset{|O|}{||}}{S}}-\overline{\underline{O}}-H$$

(b) The Zn metal center in $Zn(OH)_2$ accepts a pair of electrons. Thus, $Zn(OH)_2$ is a Lewis acid. OH^- donates the pair of electrons that form the covalent bond. Therefore, OH^- is a Lewis base. We have assumed here that Zn has sufficient covalent character to form coordinate covalent bonds with hydroxide ions. (Lewis structure drawn below).

(c)

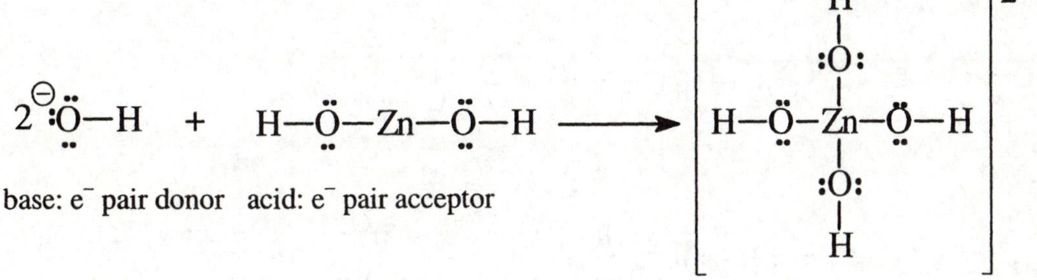

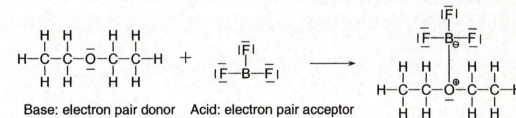

$$2\ ^{\ominus}\!\!:\overset{..}{\underset{..}{O}}\!-H \quad + \quad H\!-\!\overset{..}{\underset{..}{O}}\!-Zn\!-\overset{..}{\underset{..}{O}}\!-H \quad \longrightarrow \quad \left[\begin{array}{c} H \\ :\overset{}{O}: \\ H\!-\!\overset{..}{\underset{..}{O}}\!-Zn\!-\overset{..}{\underset{..}{O}}\!-H \\ :\overset{}{O}: \\ H \end{array} \right]^{2-}$$

base: e⁻ pair donor   acid: e⁻ pair acceptor

Lewis Theory of Acids and Bases

$$^{\ominus}\!\!|\overset{}{\underset{}{O}}\!-H \quad + \quad H\!-\!\overset{}{\underset{}{O}}\!-B\!-\overset{}{\underset{}{O}}\!-H \quad \longrightarrow \quad \left[\begin{array}{c} H \\ |O| \\ H\!-\!\overset{}{\underset{}{O}}\!-B\!-\overset{}{\underset{}{O}}\!-H \\ |O| \\ H \end{array} \right]^{-}$$

<u>**71.**</u> **(a)**

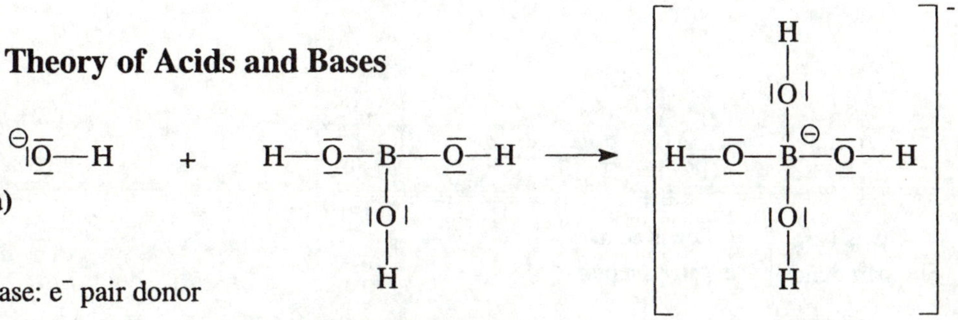

base: e⁻ pair donor

acid: e⁻ pair acceptor

(b)

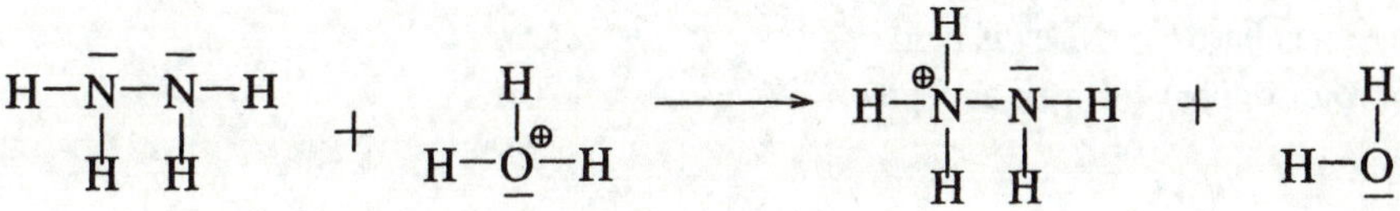

$$H\!-\!\overset{}{\underset{H}{N}}\!-\overset{}{\underset{H}{N}}\!-H \quad + \quad H\!-\!\overset{\oplus}{\underset{}{O}}\!-H \quad \longrightarrow \quad H\!-\!\overset{\oplus}{\underset{H}{N}}\!-\overset{}{\underset{H}{N}}\!-H \quad + \quad H\!-\!\overset{}{O}|$$

The actual Lewis acid is H⁺, which is supplied by H_3O^+

(c)

Base: electron pair donor   Acid: electron pair acceptor

72. The C in a CO_2 molecule can accept of a pair of electrons. Thus, CO_2 is the Lewis acid. The hydroxide anion is the Lewis base as it has pairs of electrons it can donate.

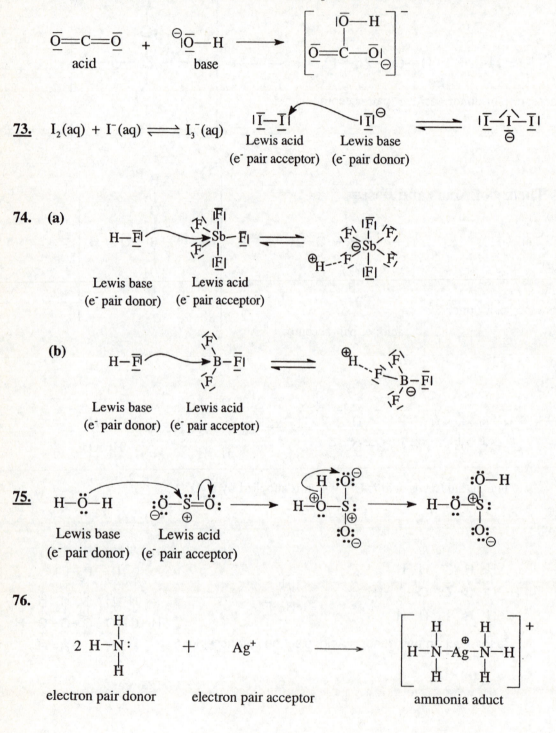

73. $I_2(aq) + I^-(aq) \rightleftharpoons I_3^-(aq)$

74. (a)

(b)

75.

76.

2 H—N:
electron pair donor   + Ag^+ →   ammonia adduct

electron pair acceptor

Integrative and Advanced Exercises

77. We use (ac) to represent the fact that a species is dissolved in acetic acid.

(a) $C_2H_3O_2^-$ (ac) $+ HC_2H_3O_2$ (l) $\rightleftharpoons$ $HC_2H_3O_2$ (l) $+ C_2H_3O_2^-$ (ac)

$C_2H_3O_2^-$ is a base in acetic acid.

(b) H_2O(ac) $+ HC_2H_3O_2$ (*l*) $\rightleftharpoons H_3O^+$ (ac) $+ C_2H_3O_2^-$ (ac)

Since H_2O is a weaker acid than $HC_2H_3O_2$ in aqueous solution, it will also be weaker in acetic acid. Thus H_2O will accept a proton from the solvent acetic acid, making H_2O a base in acetic acid.

(c) $HC_2H_3O_2$ (l) $+ HC_2H_3O_2$ (*l*) $\rightleftharpoons H_2C_2H_3O_2^+$ (ac) $+ C_2H_3O_2^-$ (aq)

Acetic acid can act as an acid or a base in acetic acid.

(d) $HClO_4$ (ac) $+ HC_2H_3O_2$ (l) $\rightleftharpoons ClO_4^-$ (ac) $+ H_2C_2H_3O_2^-$ (ac)

Since $HClO_4$ is a stronger acid than $HC_2H_3O_2$ in aqueous solution, it will also be stronger in acetic acid. Thus, $HClO_4$ will donate a proton to the solvent acetic acid, making $HClO_4$ an acid in acetic acid.

78. $Sr(OH)_2$ (sat'd) $\rightleftharpoons Sr^{2+}$(aq) $+ 2\,OH^-$(aq) $pOH = 14 - pH = 0.88\,[OH^-] = 10^{-0.88} = 0.132M$

$$[OH^-]_{dilute} = 0.132M \times (\frac{10.0\ \text{ml concentrate}}{250.0\ \text{mL total}}) = 5.27 \times 10^{-3} M$$

$$[HCl] = \frac{(10.0\ \text{ml}) \times (5.27 \times 10^{-3} M)}{25.1\ \text{mL}} = 2.10 \times 10^{-3} M$$

79. **(a)** H_2SO_4 is a diprotic acid. A pH of 2 assumes no second ionization step and the second ionization step alone would produce a pH of less than 2, so its not matched.

(b) pH 4.6 matched; ammonium salts hydrolyze to form solutions with pH < 7.

(c) KI should be nearly neutral (pH 7.0) in solution since it is the salt of a strong acid (HI) and a strong base (KOH). Thus it is not matched.

(d) pH of this solution is not matched as it's a weak base. A 0.002 M solution of KOH has a pH of about 11.3. A 0.0020 M methylamine solution would be lower (~10.9).

(e) pH of this solution is matched, since a 1.0 M hypochlorite salt hydrolyzes to form a solution of ~ pH 10.8.

(f) pH of this solution is not matched. Phenol is a very weak acid (pH >5)

(g) pH of this solution is 4.3. It is close, but not matched

(h) pH of this solution is 2.1, matched as it's a strong organic acid

(i) pH of this solution is 2.5, it is close, but not matched

80. We let $[H_3O^+]_a = 0.500 [H_3O^+]_i$ $pH = -\log[H_3O^+]_i$

$pH_a = -\log[H_3O^+]_a = -\log(0.500 [H_3O^+]_i) = -\log [H_3O^+]_i - \log (0.500) = pH_i + 0.301$

The truth of statement has been demonstrated. Remember, however, that the extent of ionization of weak acids and weak bases in water increases as these solutions become more and more dilute. Finally, there are some solutions whose pH don't change with dilution, such as NaCl(aq), pH = 7.0 at all concentrations.

81. Since a strong acid is completely ionized, increasing its concentration has the effect of increasing $[H_3O^+]$ by an equal degree. In the case of a weak acid, however, the solution of the K_a expression gives the following equation (if we can assume that $[H_3O^+]$ is negligible compared to the original concentration of the undissociated acid, [HA]). $x^2 = K_a[HA]$ (where $x = [H_3O^+]$), or $x = \sqrt{K_a[HA]}$. Thus, if [HA] is doubled, $[H_3O^+]$ increases by $\sqrt{2}$.

82. $H_2O(l) + H_2O(l) \ll H_3O^+(aq) + OH^-(aq)$ $\Delta H^\circ = (-230.0 \text{ kJ mol}) - (-285.8 \text{ kJ mol}^{-1}) = 55.8 \text{ kJ}$

Since ΔH° is positive, the value of K_{eq} is larger at lower temperature
(Le Chatelier's Principle or the Van't Hoff equation would show this to be true).

83. First we assume that molarity and molality are numerically equal in this dilute aqueous solution. Then compute the molality that creates the observed freezing point, with $K_f = 1.86$ °C/m for water.

$$m = \frac{\Delta T_f}{K_f} = \frac{0.096°C}{1.86°C/m} = 0.051_6 \ m \qquad \text{concentration} = 0.051_6 \text{ M}$$

This is the total concentration of all species in solution, ions as well as molecules. We base our remaining calculation on the balanced chemical equation.

Equation: $HC_4H_5O_2(aq) + H_2O \rightleftharpoons C_4H_5O_2^-(aq) + H_3O^+(aq)$

Initial: 0.0500 M   0 M   0 M

Changes: $- x$ M $+ x$ M $+ x$ M

Equil: (0.0500 $- x$) M x M x M

Total conc. $= 0.051_6$ M $= (0.0500 - x)$ M $+ x$ M $+ x$ M $= 0.0500$ M $+ xM$ $x = 0.001_6$ M

$$K_a = \frac{[C_4H_5O_2^-][H_3O^+]}{[HC_4H_5O_2]} = \frac{x \cdot x}{0.0500 - x} = \frac{(0.001_6)^2}{0.0500 - 0.001_6} = 5 \times 10^{-5}$$

84. $[H_3O^+] = 10^{-pH} = 10^{-5.50} = 3.2 \times 10^{-6}$ M

(a) To produce this acidic solution we could not use 15 M NH_3(aq), because aqueous solutions of NH_3 have pH values greater than 7 owing to the fact that NH_3 is a weak base.

(b)

$$V_{HCl} = 100.0 \text{ mL} \times \frac{3.2 \times 10^{-6} \text{mol H}_3O^+}{1 \text{ L}} \times \frac{1 \text{ mol HCl}}{1 \text{ mol H}_3O^+} \times \frac{1 \text{ L soln}}{12 \text{ mol HCl}} = 2.7 \times 10^{-5} \text{ mL 12 M HCl soln}$$

(very unlikely, since this small a volume is hard to measure)

(c) $K_a = 5.6 \times 10^{-10}$ for $NH_4^+(aq)$

Equation:	$NH_4^+(aq)$	$+$	$H_2O(l)$	$\rightleftharpoons$	$NH_3(aq)$	$+$	$H_3O^+(aq)$
Initial:	c M		$-$		0 M		$\approx$ 0 M
Changes:	-3.2×10^{-6} M		$-$		$+3.2 \times 10^{-6}$ M		$+3.2 \times 10^{-6}$ M
Equil:	$(c-3.2 \times 10^{-6})$ M		$-$		3.2×10^{-6} M		3.2×10^{-6} M

$$K_a = 5.6 \times 10^{-10} = \frac{[NH_3][H_3O^+]}{[NH_4^+]} = \frac{(3.2 \times 10^{-6})(3.2 \times 10^{-6})}{(c-3.2 \times 10^{-6})}$$

$$c - 3.2 \times 10^{-6} = \frac{(3.2 \times 10^{-6})^2}{5.6 \times 10^{-10}} = 1.8 \times 10^{-2} \text{ M} \qquad [NH_4^+] = 0.018 \text{ M}$$

$$NH_4Cl \text{ mass} = 100.0 \text{ mL} \times \frac{0.018 \text{ mol NH}_4^+}{1000 \text{ mL}} \times \frac{1 \text{ mol NH}_4Cl}{1 \text{ mol NH}_4^+} \times \frac{53.49 \text{ g NH}_4Cl}{1 \text{ mol NH}_4Cl} = 0.096 \text{ g NH}_4Cl$$

This would be an easy mass to measure on a laboratory three decimal place balance.

(d) The set up is similar to that for NH_4Cl.

Equation:	$HC_2H_3O_2(aq)$	$+$	$H_2O(l)$	$\rightleftharpoons$	$C_2H_3O_2^-(aq)$	$+$	$H_3O^+(aq)$
Initial:	c M		$-$		0 M		$\approx$ 0 M
Changes:	-3.2×10^{-6} M		$-$		$+3.2 \times 10^{-6}$ M		$+3.2 \times 10^{-6}$ M
Equil:	$(c-3.2 \times 10^{-6})$ M		$-$		3.2×10^{-6} M		3.2×10^{-6} M

$$K_a = 1.8 \times 10^{-5} = \frac{[C_2H_3O_2^-][H_3O^+]}{[HC_2H_3O_2]} = \frac{(3.2 \times 10^{-6})(3.2 \times 10^{-6})}{c-3.2 \times 10^{-6}}$$

$$c - 3.2 \times 10^{-6} = \frac{(3.2 \times 10^{-6})^2}{1.8 \times 10^{-5}} = 5.7 \times 10^{-7} \quad c = [HC_2H_3O_2] = 3.8 \times 10^{-6}$$

$$HC_2H_3O_2 \text{ mass} = 100.0 \text{ mL} \times \frac{3.8 \times 10^{-6} \text{ mol HC}_2H_3O_2}{1000 \text{ mL}} \times \frac{60.05 \text{ g HC}_2H_3O_2}{1 \text{ mol HC}_2H_3O_2}$$

$$= 2.3 \times 10^{-5} \text{ g HC}_2H_3O_2 \text{ (an almost impossibly small mass to measure}$$
$$\text{using conventional laboratory scales)}$$

85. (a) 1.0×10^{-5} M HCN:

$$\text{HCN(aq)} \quad + \quad \text{H}_2\text{O(l)} \xrightarrow{K_a = 6.2\times10^{-10}} \text{H}_3\text{O}^+\text{(aq)} \quad + \quad \text{CN}^-\text{(aq)}$$

Initial: 1.0×10^{-5} M   – 1.0×10^{-7} M   0 M

Change: $-x$ – $+x$ $+x$

Equilibrium: $(1.0 \times 10^{-5} - x)$ M   – $(1.0 \times 10^{-7} + x)$ M x

$$6.2\times10^{-10} = \frac{x(1.0\times10^{-7} + x)}{(1.0\times10^{-5} - x)} \implies 6.2\times10^{-15} - 6.2\times10^{-10}\,x = 1.0\times10^{-7}\,x + x^2$$

$$x^2 + 1.006\times10^{-7} - 6.2\times10^{-15} = 0$$

$$x = \frac{-1.006\times10^{-7} \pm \sqrt{(1.006\times10^{-7})^2 - 4(1)(-6.2\times10^{-15})}}{2(1)} = 1.4\underline{4}\times10^{-7}$$

Final pH = $-\log[\text{H}_3\text{O}^+]$ = $-\log(1.0 \times 10^{-7} + 1.4\underline{4} \times 10^{-7})$ = 6.61

(b) 1.0×10^{-5} M $\text{C}_6\text{H}_5\text{NH}_2$:

$$\text{C}_6\text{H}_5\text{NH}_2\text{(aq)} \quad + \quad \text{H}_2\text{O(l)} \xrightarrow{K_a = 7.4\times10^{-10}} \text{OH}^-\text{(aq)} \quad + \quad \text{C}_6\text{H}_5\text{NH}_3{}^+\text{(aq)}$$

Initial: 1.0×10^{-5} M   – 1.0×10^{-7} M   0 M

Change: $-x$ – $+x$ $+x$

Equilibrium: $(1.0 \times 10^{-5} - x)$ M   – $(1.0 \times 10^{-7} + x)$ M x

$$7.4\times10^{-10} = \frac{x(1.0\times10^{-7} + x)}{(1.0\times10^{-5} - x)} \implies 7.4\times10^{-15} - 7.4\times10^{-10}\,x = 1.0\times10^{-7}\,x + x^2$$

$$x^2 + 1.0074\times10^{-7} - 7.4\times10^{-15} = 0$$

$$x = \frac{-1.0074\times10^{-7} \pm \sqrt{(1.0074\times10^{-7})^2 - 4(1)(-7.4\times10^{-15})}}{2(1)} = 4.9\underline{3}\times10^{-8} \text{ M}$$

Final pOH = $-\log[\text{OH}^-]$ = $-\log(1.0 \times 10^{-7} + 4.9\underline{3} \times 10^{-7})$ = 6.83   Final pH = 7.17

86. Based on Henry's law, the solubility of $\text{CO}_2\text{(g)}$ in water in equilibrium with the atmosphere is 0.037% of its 1 atm value. We now can determine the CO_2 molarity.

$$\text{CO}_2 \text{ molarity} = 0.00037 \text{ atm} \times \frac{1.45 \text{ g CO}_2}{1 \text{ atm} \cdot \text{L}} \times \frac{1 \text{ mol CO}_2}{44.01 \text{ g CO}_2} = 1.2\times10^{-5} \text{ M}$$

Equation: $\text{CO}_2\text{(aq)}$ + $2\,\text{H}_2\text{O(l)}$ $\rightleftharpoons$ $\text{HCO}_3{}^-\text{(aq)}$ + $\text{H}_3\text{O}^+\text{(aq)}$

Initial: 1.2×10^{-5} M   –   0 M   0 M

Changes: $-x$ M   – $+x$ M $+x$ M

Equil: $(1.2\times10^{-5} - x)$ M   – x M x M

$$K_a = 4.4 \times 10^{-7} = \frac{[HCO_{3^-}][H_3O^+]}{[CO_2]} = \frac{(x)(x)}{1.2 \times 10^{-5} - x} \qquad x^2 + 4.4 \times 10^{-7}x - 5.3 \times 10^{-12} = 0$$

$$x = \frac{-4.4 \times 10^{-7} \pm \sqrt{1.9 \times 10^{-13} + 2.1 \times 10^{-11}}}{2} = 2.1 \times 10^{-6} \text{ M} = [H_3O^+]$$

$$pH = -\log[H_3O^+] = -\log(2.1 \times 10^{-6}) = 5.7 \approx \text{ the normal'' pH of rainwater.}$$

87. (a) For a weak acid, we begin with the ionization constant expression for the weak acid, HA.

$$K_a = \frac{[H_3O^+][A^-]}{[HA]} = \frac{[H^+]^2}{[HA]} = \frac{[H^+]^2}{M - [H^+]} \approx \frac{[H^+]^2}{M}$$

The first modification results from realizing that $[H^+] = [A^-]$. The second modification is based on the fact that the equilibrium [HA] equals the initial [HA] minus the amount that dissociates. And the third modification is an approximation, assuming that the amount of dissociated weak acid is small compared to the original amount of HA in solution. Now we take the logarithm of both sides of the equation.

$$\log K_a = \log \frac{[H^+]^2}{c} = \log[H^+]^2 - \log M = 2\log[H^+] - \log M$$

$$-2\log[H^+] = -\log K_a - \log M \qquad 2\,pH = pK_a - \log M \qquad pH = \tfrac{1}{2}pK_a - \tfrac{1}{2}\log M$$

For a weak base, we begin with the ionization constant expression for a weak base, B.

$$K_b = \frac{[BH^+][OH^-]}{[B]} = \frac{[OH^-]^2}{[B]} = \frac{[OH^-]^2}{M - [OH^-]} \approx \frac{[OH^-]^2}{M}$$

The first modification results from realizing that $[OH^-] = [BH^+]$. The second modification is based on the fact that the equilibrium [B] equals the initial [B] minus the amount that dissociates. And the third modification is an approximation, assuming that the amount of dissociated weak base is small compared to the original amount of B in solution. Now we take the logarithm of both sides of the equation.

$$\log K_b = \log \frac{[OH^-]^2}{M} = \log[OH^-]^2 - \log M = 2\log[OH^-] - \log M$$

$$-2\log[OH^-] = -\log K_b - \log M \qquad 2\,pOH = pK_b - \log M \qquad pOH = \tfrac{1}{2} pK_b - \tfrac{1}{2}\log M$$

$$pH = 14.00 - pOH = 14.00 - \tfrac{1}{2}pK_b + \tfrac{1}{2}\log M$$

The anion of a relatively strong weak acid is a weak base. Thus, the derivation is the same as that for a weak base, immediately above, with the exception of the value for K_b. We now obtain an expression for pK_b.

$$K_b = \frac{K_w}{K_a} \qquad \log K_b = \log \frac{K_w}{K_a} = \log K_w - \log K_a$$

$$-\log K_b = -\log K_w + \log K_a = pK_b = pK_w - pK_a$$

Substitution of this expression for pK_b into the expression above gives the following result.

$$pH = 14.00 - \tfrac{1}{2}pK_w + \tfrac{1}{2}pK_a + \tfrac{1}{2}\log M$$

(b) 0.10 M HC₂H₃O₂ $pH = 0.500 \times 4.74 - 0.500 \log 0.10 = 2.87$

Equation: $HC_2H_3O_2(aq)$ + $H_2O(l)$ ⇌ $C_2H_3O_2^-(aq)$ + $H_3O^+(aq)$

Initial: 0.10 M   –   0 M ≈ 0 M

Changes: $-x$ M   – $+x$ M $+x$ M

Equil: $(0.10-x)$ M   – x M x M

$$K_a = \frac{[H_3O^+][C_2H_3O_2^-]}{[HC_2H_3O_2]} = 1.8 \times 10^{-5} = \frac{x \cdot x}{0.10 - x} \approx \frac{x^2}{0.10}$$

$$x = \sqrt{0.10 \times 1.8 \times 10^{-5}} = 1.3 \times 10^{-3} \text{ M}, \ pH = -\log(1.3 \times 10^{-3}) = 2.89$$

0.10 M NH₃ $pH = 14.00 - 0.500 \times 4.74 + 0.500 \log 0.10 = 11.13$

Equation: $NH_3(aq)$ + $H_2O(l)$ ⇌ $NH_4^+(aq)$ + $OH^-(aq)$

Initial: 0.10 M   –   0 M ≈ 0 M

Changes: $-x$ M   – $+x$ M $+x$ M

Equil: $(0.10-x)$ M   – x M x M

$$K_b = \frac{[NH_4^+][OH^-]}{[NH_3]} = 1.8 \times 10^{-5} = \frac{x \cdot x}{0.10 - x} \approx \frac{x^2}{0.10} \quad x = \sqrt{0.10 \times 1.8 \times 10^{-5}} = 1.3 \times 10^{-3} \text{ M}$$

$$pOH = -\log(1.3 \times 10^{-3}) = 2.89 \qquad pH = 14.00 - 2.89 = 11.11$$

0.10 M NaC₂H₃O₂ $pH = 14.00 - 0.500 \times 14.00 + 0.500 \times 4.74 + 0.500 \log 0.10 = 8.87$

Equation: $C_2H_3O_2^-(aq) + H_2O(l)$ ⇌ $HC_2H_3O_2(aq)$ + $OH^-(aq)$

Initial: 0.10 M   –   0 M ≈ 0 M

Changes: $-x$ M   – $+x$ M $+x$ M

Equil: $(0.10-x)$ M   – x M x M

$$K_a = \frac{[OH^-][HC_2H_3O_2]}{[C_2H_3O_2^-]} = \frac{1.0 \times 10^{-14}}{1.8 \times 10^{-5}} = \frac{x \cdot x}{0.10-x} \approx \frac{x^2}{0.10}$$

$$x = \sqrt{\frac{0.10 \times 1.0 \times 10^{-14}}{1.8 \times 10^{-5}}} = 7.5 \times 10^{-6} \text{ M} \qquad pOH = -\log(7.5 \times 10^{-6}) = 5.12$$

$$pH = 14.00 - 5.12 = 8.88$$

88. (a) We know $K_a = \dfrac{[H_3O^+]_{eq}[A^-]_{eq}}{[HA]_{eq}}$, $\alpha = \dfrac{[A^-]_{eq}}{[HA]_i}$ and finally % ionized $= \alpha \times 100\%$.

Let us see how far we can get with no assumptions. Recall first what we mean by $[HA]_{eq}$.

$[HA]_{eq} = [HA]_i - [A^-]_{eq}$ or $[HA]_i = [HA]_{eq} + [A^-]_{eq}$

We substitute this expression into the expression for α. $\alpha = \dfrac{[A^-]_{eq}}{[HA]_{eq} + [A^-]_{eq}}$

We solve both the α and the K_{eq} expressions for $[HA]_{eq}$, equate the results, and solve for α.

$[HA]_{eq} = \dfrac{[H_3O^+][A^-]}{K_a}$ $[A^-] = \alpha[HA]_{eq} + \alpha[A^-]$ $[HA]_{eq} = \dfrac{[A^-] - \alpha[A^-]}{\alpha}$

$\dfrac{[H_3O^+][\cancel{A^-}]}{K_a} = \dfrac{[\cancel{A^-}](1-\alpha)}{\alpha} = R$

$R\alpha = 1 - \alpha$ $R\alpha + \alpha = 1 = \alpha(1+R)$ $\alpha = \dfrac{1}{1+R}$

Momentarily, for ease in writing, we have let $R = \dfrac{[H_3O^+]}{K_a} = \dfrac{10^{-pH}}{10^{-pK}} = 10^{(pK-pH)}$

Once we realize that $100\alpha = \%$ ionized, we note that we have proven the cited formula. Notice that no approximations were made during the course of this derivation.

(b) Formic acid has $pK_a = 3.74$. Thus $10^{(pK-pH)} = 10^{(3.74-2.50)} = 10^{1.24} = 17.4$

$\%$ ionization $= \dfrac{100}{1+17.4} = 5.43\%$

(c) $[H_3O^+] = 10^{-2.85} = 1.4 \times 10^{-3}$ M $\%$ ionization $= \dfrac{1.4 \times 10^{-3} \text{ M}}{0.150 \text{ M}} \times 100\% = 0.93\%$

$0.93 = \dfrac{100}{1+10^{(pK-pH)}}$ $1+10^{(pK-pH)} = \dfrac{100}{0.93} = 108$ $10^{(pK-pH)} = 107$

$2.03 = pK - pH$ $pK = pH + 2.03 = 2.85 + 2.03 = 4.88$ $K_a = 10^{-4.88} = 1.3 \times 10^{-5}$

89. (a) For the acids given, we determine values of m and n in the formula $EO_m(OH)_n$. HOCl or Cl(OH) has m = 0 and n = 1. We expect $K_a \approx 10^{-7}$ or $pK_a \approx 7$, which is in good agreement with the accepted $pK_a = 7.52$. HOClO or ClO(OH) has m = 1 and n = 1. We expect $K_a \approx 10^{-2}$ or $pK_a \approx 2$, in good agreement with the $pK_a = 1.92$. HOClO$_2$ or ClO_2(OH) has m = 2 and n = 1. We expect K_a to be large and in good agreement with the accepted value of $pK_a = -3$, $K_a = 10^{-pKa} = 10^3$. HOClO$_3$ or ClO_3(OH) has m = 3 and n = 1. We expect K_a to be very large and in good agreement with the accepted $K_a = -8$, $pK_a = -8$, $K_a = 10^{-pKa} = 10^8$ which turns out to be the case.

(b) The formula H_3AsO_4 can be rewritten as $AsO(OH)_3$, which has m = 1 and n = 3. The expected value is $K_a = 10^{-2}$.

(c) The value of $pK_a = 1.1$ corresponds to $K_a = 10^{-pKa} = 10^{-1.1} = 0.08$, which indicates that m = 1. The following Lewis structure is consistent with this value of m.

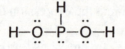

90. We note that as the hydrocarbon chain between the two COOH groups increases in length, the two values of pK_a get closer together. The reason for this is that the ionized COOH group, COO^-, is negatively charged. When the two COOH groups are close together, the negative charge on the molecule, arising from the first ionization, inhibits the second ionization, producing a large difference between the two ionization constants. But as the hydrocarbon chain lengthens, the effect of this negative $-COO^-$ group on the second ionization becomes less pronounced, due to the increasing separation between the two carbonyl groups, and the values of the two ionization constants are more alike.

91. Consider only the 1st dissociation.

Equation: $H_2C_4H_4O_4(aq) + H_2O(l) \xrightleftharpoons{\;6.2\times10^{-5}\;} H_3O^+(aq) + HC_4H_4O_4^-(aq)$

Initial: 0.100 M   – ≈ 0 M   0 M

Changes: $-x$ M   – $+x$ M $+x$ M

Equil: $(0.100 - x)$ M   – x M x M

$$K_a = \frac{x^2}{0.100} = 6.2\times10^{-5} \qquad x = 0.00249 \text{ M or } x = 0.00246 \text{ M}$$

Solve using the quadratic equation. Hence pH $\approx$ 2.60

Consider only the 2nd dissociation.

Equation: $HC_4H_4O_4^-(aq)$ + $H_2O(l) \xrightleftharpoons{\;2.3\times10^{-5}\;} H_3O^+(aq)$ + $C_4H_4O_4^{2-}(aq)$

Initial: 0.0975 M   –   0.00246 M   0 M

Changes: $-x$ M   – $+x$ M $+x$ M

Equil: $(0.0975 - x)$ M   – $0.00246 + x$ M x M

$$K_a = \frac{x(0.00246 + x)}{0.0975 - x} = 2.3\times10^{-6} = \frac{x(0.00246)}{0.0975}$$

$x = 0.000091$ M or $x = 0.000088$ M(solve quadratic)   Hence pH $\approx$ 2.59

Thus, the pH is virually identical, hence, wqe need only consider the 1st ionization.

92. To have the same freezing point, the two solutions must have the same total concentration of all particles of solute—ions and molecules. We first determine the concentrations of all solute species in 0.150 M $HC_2H_2ClO_2$, $K_a = 1.4 \times 10^{-3}$

Equation:	$HC_2H_2ClO_2\,(aq) + H_2O(l) \rightleftharpoons$		$H_3O^+\,(aq)$	$+$	$C_2H_2ClO_2^-\,(aq)$
Initial:	0.150 M	$-$	≈ 0 M		0 M
Changes:	$-x$ M	$-$	$+x$ M		$+x$ M
Equil:	$(0.150 - x)$ M	$-$	$+x$ M		x M

$$K_a = \frac{[H_3O^+][C_2H_2ClO_2^-]}{[HC_2H_2ClO_2]} = \frac{x \cdot x}{0.150 - x} = 1.4 \times 10^{-3}$$

$$x^2 = 2.1 \times 10^{-4} - 1.4 \times 10^{-3}\,x \qquad\qquad x^2 + 1.4 \times 10^{-3}\,x - 2.1 \times 10^{-4} = 0$$

$$x = \frac{-b \pm \sqrt{b^2 - 4ac}}{2a} = \frac{-1.4 \times 10^{-3} \pm \sqrt{2.0 \times 10^{-6} + 8.4 \times 10^{-4}}}{2} = 0.014 \text{ M}$$

total concentration $= (0.150 - x) + x + x = 0.150 + x = 0.150 + 0.014 = 0.164$ M

Now we determine the $[HC_2H_3O_2]$ that has this total concentration.

Equation:	$HC_2H_3O_2\,(aq)$	$+$	$H_2O(l) \rightleftharpoons$	$H_3O^+\,(aq)$	$+$	$C_2H_3O_2^-\,(aq)$
Initial:	z M		$-$	≈ 0 M		0 M
Changes:	$-y$ M		$-$	$+y$ M		$+y$ M
Equil:	$(z - y)$ M		$-$	y M		y M

$$K_a = \frac{[H_3O^+][C_2H_3O_2^-]}{[HC_2H_3O_2]} = 1.8 \times 10^{-5} = \frac{y \cdot y}{z - y}$$

We also know that the total concentration of all solute species is 0.164 M.

$$z - y + y + y = z + y = 0.164 \qquad z = 0.164 - y \qquad \frac{y^2}{0.164 - 2y} = 1.8 \times 10^{-5}$$

We assume that $2y \ll 0.164 \qquad y^2 = 0.164 \times 1.8 \times 10^{-5} = 3.0 \times 10^{-6} \qquad y = 1.7 \times 10^{-3}$

The assumption is valid: $2y = 0.0034 \ll 0.164$

Thus, $[HC_2H_3O_2] = z = 0.164 - y = 0.164 - 0.0017 = 0.162$ M

$$\text{mass } HC_2H_3O_2 = 1.000 \text{ L} \times \frac{0.162 \text{ mol } HC_2H_3O_2}{1 \text{ L soln}} \times \frac{60.05 \text{ g } HC_2H_3O_2}{1 \text{ mol } HC_2H_3O_2} = 9.73 \text{ g } HC_2H_3O_2$$

93. The first ionization of H_2SO_4 makes the major contribution to the acidity of the solution.

Then the following two ionizations, both of which are repressed because of the presence of H_3O^+ in the solution, must be solved simultaneously.

Equation: $\quad HSO_4^-(aq) \quad + \quad H_2O(l) \quad \rightleftharpoons \quad SO_4^{2-}(aq) + H_3O^+(aq)$

Initial: $\qquad$ 0.68 M $\qquad\qquad$ – $\qquad\qquad\qquad$ 0 M $\qquad\quad$ 0.68 M

Changes: $\qquad -x$ M $\qquad\qquad$ – $\qquad\qquad\qquad +x$ M $\qquad\quad +x$ M

Equil: $\qquad (0.68-x)$ M $\qquad$ – $\qquad\qquad\qquad x$ M $\qquad$ $(0.68+x)$M

$$K_2 = \frac{[H_3O^+][SO_4^{2-}]}{[HSO_4^-]} = 0.011 = \frac{x(0.68+x)}{0.68-x}$$

Let us solve this expression for x. $0.011\,(0.68-x) = 0.68x + x^2 = 0.0075 - 0.011x$

$$x^2 + 0.69\,x - 0.0075 = 0 \qquad x = \frac{-b \pm \sqrt{b^2-4ac}}{2a} = \frac{-0.69 \pm \sqrt{0.48+0.030}}{2} = 0.01 \text{ M}$$

This gives $[H_3O^+] = 0.68 + 0.01 = 0.69$ M. Now we solve the second equilibrium.

Equation: $\quad HCHO_2(aq) \quad + \quad H_2O(l) \quad \rightleftharpoons \quad CHO_2^-(aq) \quad + \quad H_3O^+(aq)$

Initial: $\qquad$ 1.5 M $\qquad\qquad$ – $\qquad\qquad\qquad$ 0 M $\qquad\quad$ 0.69 M

Changes: $\qquad -x$ M $\qquad\qquad$ – $\qquad\qquad\qquad +x$ M $\qquad\quad +x$ M

Equil: $\qquad (1.5-x)$ M $\qquad$ – $\qquad\qquad\qquad x$ M $\qquad$ $(0.69+x)$ M

$$K_a = \frac{[H_3O^+][CHO_2^-]}{[HCHO_2]} = 1.8 \times 10^{-4} = \frac{x(0.69+x)}{1.5-x} \approx \frac{0.69x}{1.5} \qquad x = 3.9 \times 10^{-4} \text{ M}$$

We see that the second acid does not significantly affect the $[H_3O^+]$, for which a final value is now obtained. $[H_3O^+] = 0.69$ M, pH $= -\log(0.69) = 0.16$

94.

$Let\ [HA_1]=M$ $\quad K_{HA_1} = \dfrac{[H^+][A_1^-]}{[HA_1]-[H^+]} \approx \dfrac{[H^+]^2}{[HA_1]} = \dfrac{[H^+]^2}{M}$

$Let\ [HA_2]=M$ $\quad K_{HA_2} = 2K_{HA_1} = \dfrac{[H^+][A_2^-]}{[HA_2]-[H^+]} \approx \dfrac{[H^+]^2}{[HA_2]} = \dfrac{[H^+]^2}{M}$

$K_{HA_1} = \dfrac{[H^+]^2}{M}$ $\quad [H^+]^2 = K_{HA_1} \times M$ $\quad [H^+] = \sqrt{K_{HA_1} \times M} = \left(K_{HA_1} \times M\right)^{1/2}$

$2K_{HA_1} = \dfrac{[H^+]^2}{M}$ $\quad [H^+]^2 = 2K_{HA_1} \times M$ $\quad [H^+] = \sqrt{2K_{HA_1} \times M} = \left(2K_{HA_1} \times M\right)^{1/2}$

$[H^+]_{overall} = \left(K_{HA_1} \times M\right)^{1/2} + \left(2K_{HA_1} \times M\right)^{1/2}$ *(take the negative $\log_{10}$ of both sides)*

$-\log([H^+]_{overall}) = -\log\left(\left(K_{HA_1} \times M\right)^{1/2}\right) + -\log\left(\left(2K_{HA_1} \times M\right)^{1/2}\right)$

$pH = -\log\left(\left(K_{HA_1} \times M\right)^{1/2}\right) - \log\left(\left(K_{HA_1} \times M\right)^{1/2}\right) = -\dfrac{1}{2}\log K_{HA_1} \times M - \dfrac{1}{2}\log 2K_{HA_1} \times M$

$pH = -\dfrac{1}{2}\left(\log K_{HA_1} \times M + \log 2K_{HA_1} \times M\right) = -\dfrac{1}{2}\left(\log\left(K_{HA_1} \times M + 2K_{HA_1} \times M\right)\right)$

$pH = -\dfrac{1}{2}\left(3K_{HA_1} \times M\right) = -\dfrac{1}{2}\left(3MK_{HA_1}\right)$

95. $HSbF_6 \rightarrow H^+ + SbF_6^-$

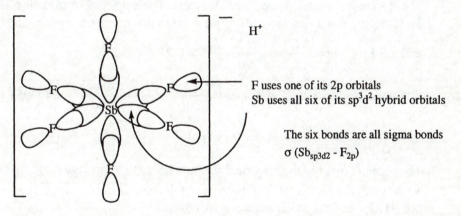

F uses one of its 2p orbitals
Sb uses all six of its sp^3d^2 hybrid orbitals

The six bonds are all sigma bonds
$\sigma\ (Sb_{sp3d2} - F_{2p})$

$HBF_4 \rightarrow H^+ + BF_4^-$

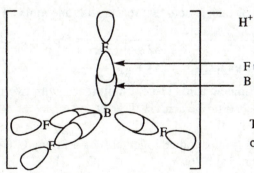

F uses one of its 2p orbitals
B uses all four of its sp^3 hybrid orbitals

The four bonds are all sigma bonds
$\sigma\ (B_{sp3} - F_{2p})$

96. Structure of H_3PO_3 shown on the right.

The two ionizable protons are bound to oxygen.

$$
\begin{array}{c}
\overset{\displaystyle H}{\underset{\displaystyle}{|}} \\
H-\overset{\cdot\cdot}{\underset{\cdot\cdot}{O}}-\overset{\displaystyle}{\underset{\displaystyle \parallel}{P}}-\overset{\cdot\cdot}{\underset{\cdot\cdot}{O}}-H \\
:\!O\!:
\end{array}
$$

FEATURE PROBLEMS

97. **(a)** From the combustion analysis we can determine the empirical formula. Note that the mass of oxygen is determined by difference.

$$\text{amount C} = 1.599 \text{ g CO}_2 \times \frac{1 \text{ mol CO}_2}{44.01 \text{ g CO}_2} \times \frac{1 \text{ mol C}}{1 \text{ mol CO}_2} = 0.03633 \text{ mol C}$$

$$\text{mass of C} = 0.03633 \text{ mol C} \times \frac{12.011 \text{ g C}}{1 \text{ mol C}} = 0.4364 \text{ g C}$$

$$\text{amount H} = 0.327 \text{ g H}_2\text{O} \times \frac{1 \text{ mol H}_2\text{O}}{18.02 \text{ g H}_2\text{O}} \times \frac{2 \text{ mol H}}{1 \text{ mol H}_2\text{O}} = 0.03629 \text{ mol H}$$

$$\text{mass of H} = 0.03629 \text{ mol H} \times \frac{1.008 \text{ g H}}{1 \text{ mol H}} = 0.03658 \text{ g H}$$

$$\text{amount O} = \left(1.054 \text{ g sample} - 0.4364 \text{ g C} - 0.03568 \text{ g H}\right) \times \frac{1 \text{ mol O}}{16.00 \text{ g O}} = 0.0364 \text{ mol O}$$

There are equal moles of the three elements. The empirical formula is CHO.
The freezing-point depression data are used to determine the molar mass.

$$\Delta T_f = -K_f m \quad m = \frac{\Delta T_f}{-K_f} = \frac{-0.82 \text{ °C}}{-3.90 \text{ °C}/m} = 0.21 \ m$$

$$\text{amount of solute} = 25.10 \text{ g solvent} \times \frac{1 \text{ kg}}{1000 \text{ g}} \times \frac{0.21 \text{ mol solute}}{1 \text{ kg solvent}} = 0.0053 \text{ mol}$$

$$M = \frac{0.615 \text{ g solute}}{0.0053 \text{ mol solute}} = 1.2 \times 10^2 \text{ g/mol}$$

The formula mass of the empirical formula is: $12.0 \text{ g C} + 1.0 \text{ g H} + 16.0 \text{ g O} = 29.0 \text{ g/mol}$
Thus, there are four empirical units in a molecule, the molecular formula is $C_4H_4O_4$, and the molar mass is 116.1 g/mol.

(b) Here we determine the mass of maleic acid that reacts with one mol OH^-,

$$\frac{\text{mass}}{\text{mol OH}^-} = \frac{0.4250 \text{ g maleic acid}}{34.03 \text{ mL base} \times \dfrac{1 \text{ L}}{1000 \text{ mL}} \times \dfrac{0.2152 \text{ mol KOH}}{1 \text{ L base}}} = 58.03 \text{ g/mol OH}^-$$

This means that one mole of maleic acid (116.1 g/mol) reacts with two moles of hydroxide ion. Maleic acid is a diprotic acid: $H_2C_4H_2O_4$.

(c) Maleic acid has two—COOH groups joined together by a bridging C_2H_2 group. A plausible Lewis structure is shown below:

$$
\begin{array}{c}
\ \ \ \ \ |O| \ \ \ \ H \ \ \ H \ \ \ \ |O| \\
\ \ \ \ \ \ \parallel \ \ \ \ \ | \ \ \ \ \ | \ \ \ \ \ \parallel \\
H-\overline{O}-C-C=C-C-\overline{O}-H
\end{array}
$$

586

(d) We first determine $\left[H_3O^+\right] = 10^{-pH} = 10^{-1.80} = 0.016$ M and then the initial concentration of acid.

$$\left[\left(CHCOOH\right)_2\right]_{initial} = \frac{0.215\,g}{50.00\,mL} \times \frac{1000\,mL}{1L} \times \frac{1\,mol}{116.1\,g} = 0.0370\,M$$

We use the first ionization to determine the value of K_{a_1}

Equation: $\left(CHCOOH\right)_2\,(aq) + H_2O(l) \rightleftharpoons H\left(CHCOO\right)_2^-\,(aq) + H_3O^+(aq)$

Initial:	0.0370 M	–	0 M	≈0 M
Changes:	−0.016 M	–	+0.016 M	+0.016 M
Equil:	0.021 M	–	0.016 M	0.016 M

$$K_a = \frac{\left[H\left(CHCOO\right)_2^-\right]\left[H_3O^+\right]}{\left[\left(CHCOOH\right)_2\right]} = \frac{(0.016)(0.016)}{0.021} = 1.2 \times 10^{-2}$$

K_{a_2} could be determined if we had some way to measure the total concentration of all ions in solution, or if we could determine $\left[\left(CHCOO\right)_2^{2-}\right] = K_{a_2}$

(e) Practically all the $\left[H_3O^+\right]$ arises from the first ionization.

Equation: $\left(CHCOOH\right)_2\,(aq) + H_2O(l) \rightleftharpoons H\left(CHCOO\right)_2^-\,(aq) + H_3O^+(aq)$

Initial:	0.0500 M	–	0 M	≈0 M
Changes:	−x M	–	+x M	+x M
Equil:	(0.0500 − x) M	–	x M	x M

$$K_a = \frac{[H(CHCOO)_2^-][H_3O^+]}{[(CHCOOH)_2]} = \frac{x^2}{0.0500 - x} = 1.2 \times 10^{-2}$$

$$x^2 = 0.00060 - 0.012\,x \qquad x^2 + 0.012\,x - 0.00060 = 0$$

$$x = \frac{-b \pm \sqrt{b^2 - 4ac}}{2a} = \frac{-0.012 \pm \sqrt{0.00014 + 0.0024}}{2} = 0.019\,M = \left[H_3O^+\right]$$

$$pH = -\log(0.019) = 1.72$$

98. **(a)**

$$\frac{x^2}{0.00250 - x} \simeq \frac{x^2}{0.00250} \simeq 4.2 \times 10^{-4} \qquad x_1 \simeq 0.00102$$

$$\frac{x^2}{0.00250 - 0.00102} = \frac{x^2}{0.00148} \simeq 4.2 \times 10^{-4} \qquad x_2 \simeq 0.000788$$

$$\frac{x^2}{0.00250 - 0.000788} = \frac{x^2}{0.00171} \simeq 4.2 \times 10^{-4} \qquad x_3 \simeq 0.000848$$

$$\frac{x^2}{0.00250 - 0.000848} = \frac{x^2}{0.00165} \simeq 4.2 \times 10^{-4} \qquad x_4 \simeq 0.000833$$

$$\frac{x^2}{0.00250 - 0.000833} = \frac{x^2}{0.00167} \simeq 4.2 \times 10^{-4} \qquad x_5 \simeq 0.000837$$

$$\frac{x^2}{0.00250 - 0.000837} = \frac{x^2}{0.00166} \simeq 4.2 \times 10^{-4} \qquad x_6 \simeq 0.000836$$

$x_6 \simeq 0.000836 \qquad$ or $\simeq 8.4 \times 10^{-4}$,

which is the same as the value obtained using the quadratic equation.

(b) We organize the solution around the balanced chemical equation, as we have done before.

Equation: $\quad HClO_2(aq) + H_2O(l) \quad \rightleftharpoons \quad ClO_2^-(aq) + H_3O^+(aq)$

Initial: $\qquad$ 0.500 M $\qquad$ – $\qquad$ 0 M $\qquad$ ≈ 0 M

Changes: $\qquad -x$ M $\qquad$ – $\qquad +x$ M $\qquad +x$ M

Equil: $\qquad$ x M $\qquad$ – $\qquad$ x M $\qquad$ x M

$$K_a = \frac{\left[ClO_2^-\right]\left[H_3O^+\right]}{\left[HClO_2\right]} = \frac{x^2}{0.500 - x} = 1.1 \times 10^{-2} \approx \frac{x^2}{0.500} \qquad \text{Assuming } x \ll 0.500 \text{ M},$$

$x = \sqrt{x^2} = \sqrt{0.500 \times 1.1 \times 10^{-2}} = 0.074$ M $\qquad$ Not significantly smaller than 0.500 M.

Assume $x = 0.074$ $\quad x = \sqrt{(0.500 - 0.0774) \times 1.1 \times 10^{-2}} = 0.068$ M $\quad$ Try once more.

Assume $x = 0.068$ $\quad x = \sqrt{(0.500 - 0.068) \times 1.1 \times 10^{-2}} = 0.069$ M $\quad$ One more time.

Assume $x = 0.069$ $\quad x = \sqrt{(0.500 - 0.069) \times 1.1 \times 10^{-2}} = 0.069$ M $\quad$ Final result!

$\left[H_3O^+\right] = 0.069$ M, $\quad$ pH $= -\log\left[H_3O^+\right] = -\log(0.069) = 1.16$

99. **(a)** Here, two equilibria must be satisfied simultaneously. The common variable,
$[H_3O^+] = z$.

Equation: $HC_2H_3O_2(aq)$ +	$H_2O(l)$	$\rightleftharpoons$	$C_2H_3O_2^-(aq)$ +	$H_3O^+(aq)$
Initial: 0.315 M	–		0 M	≈ 0 M
Changes: $+x$ M	–		$+x$ M	$+z$ M
Equil: $(0.315 - x)$ M	–		x M	z M

$$K_a = \frac{[H_3O^+][C_2H_3O_2^-]}{[HC_2H_3O_2]} = \frac{xz}{0.315 - x} = 1.8 \times 10^{-5}$$

Equation: $HCHO_2(aq)$ +	$H_2O(l)$	$\rightleftharpoons$	$CHO_2^-(aq)$ +	$H_3O^+(aq)$
Initial: 0.250 M	–		0 M	0 M
Changes: $-y$ M	–		$+y$ M	$+z$ M
Equil: $(0.250 - y)$ M	–		y M	z M

$$K_a = \frac{[H_3O^+][CHO_2^-]}{[HCHO_2]} = \frac{yz}{0.250 - y} = 1.8 \times 10^{-4}$$

In this system, there are three variables, $x = [C_2H_3O_2^-]$, $y = [CHO_2^-]$, and $z = [H_3O^+]$. These three variables also represent the concentrations of the only charged species in the reaction, and thus $x + y = z$. We solve the two K_a expressions for x or y respectively. (Before we do so, however, we take advantages of the fact that x and y are quite small: $x \ll 0.315$ and $y \ll 0.250$.) Then we substitute these results into the expression for z, and solve to obtain a value of that variable.

$$xz = 1.8 \times 10^{-5}(0.315) = 5.7 \times 10^{-6} \qquad\qquad yz = 1.8 \times 10^{-4}(0.250) = 4.5 \times 10^{-5}$$

$$x = \frac{5.7 \times 10^{-6}}{z} \qquad\qquad\qquad\qquad y = \frac{4.5 \times 10^{-5}}{z}$$

$$z = x + y = \frac{5.7 \times 10^{-6}}{z} + \frac{4.5 \times 10^{-5}}{z} = \frac{5.07 \times 10^{-5}}{z} \qquad z = \sqrt{5.07 \times 10^{-5}} = 7.1 \times 10^{-3} \text{M} = [H_3O^+]$$

$pH = -\log(7.1 \times 10^{-3}) = 2.15$ We see that our assumptions about the sizes of x and y must be valid, since each of them is smaller than z. (Remember that $x + y = z$.)

(b) We first determine the initial concentration of each solute.

$$[NH_3] = \frac{12.5\text{g NH}_3}{0.375\text{L soln}} \times \frac{1 \text{ mol NH}_3}{17.03\text{g NH}_3} = 1.96\text{M} \qquad\qquad K_b = 1.8 \times 10^{-5}$$

$$[CH_3NH_2] = \frac{1.55 \text{ g CH}_3\text{NH}_2}{0.375\text{L soln}} \times \frac{1 \text{ mol CH}_3\text{NH}_2}{31.06 \text{ g CH}_3\text{NH}_2} = 0.113\text{M} \qquad K_b = 4.2 \times 10^{-4}$$

Now we solve simultaneously two equilibria, which have a common variable, $[OH^-] = z$.

Equation: $NH_3(aq)$ + H_2O $\rightleftharpoons$ $NH_4^+(aq)$ + $OH^-(aq)$

Initial: 1.96 M   –   0 M $\approx 0 M$

Changes: $-xM$ – $+xM$ $+zM$

Equil: $(1.96-x)M$ – xM zM

$$K_b = \frac{\left[NH_4^+\right]\left[OH^-\right]}{\left[NH_3\right]} = 1.8\times10^{-5} = \frac{xz}{1.96-x} \approx \frac{xz}{1.96}$$

Eqn: $CH_3NH_2(aq)$ + $H_2O(l)$ $\rightleftharpoons$ $CH_3NH_3^+(aq)$ + $OH^-(aq)$

Initial: 0.113 M   –   0 M $\approx 0 M$

Changes: $-y$ M   – $+y$ M $+zM$

Equil: $(0.113-y)M$ – yM zM

$$K_b = \frac{\left[CH_3NH_3^+\right]\left[OH^-\right]}{\left[CH_3NH_2\right]} = 4.2\times10^{-4} = \frac{yz}{0.113-y} = \frac{yz}{0.113}$$

In this system, there are three variables, $x=\left[NH_4^+\right]$, $y=\left[CH_3NH_3^+\right]$, and $z=\left[OH^-\right]$. These three variables also represent the concentrations of the only charged species in solution in substantial concentration, and thus $x+y=z$. We solve the two K_b expressions for x and y respectively. Then we substitute these results into the expression for z, and solve to obtain the value of that variable.

$$xz = 1.96\times1.8\times10^{-5} = 3.41\times10^{-5} \qquad\qquad yz = 0.113\times4.2\times10^{-5} = 4.7_5\times10^{-5}$$

$$x = \frac{3.41\times10^{-5}}{z} \qquad\qquad\qquad\qquad y = \frac{4.7_5\times10^{-5}}{z}$$

$$z = x+y = \frac{3.41\times10^{-5}}{z} + \frac{4.7_5\times10^{-5}}{z} = \frac{8.16\times10^{-5}}{z} \qquad z^2 = 8.16\times10^{-5}$$

$$z = 9.0\times10^{-3}M = \left[OH^-\right] \quad pOH = -\log\left(9.0\times10^{-3}\right) = 2.05 \quad pH = 14.00-2.05 = 11.95$$

We see that our assumptions about x and y (that $x \ll 1.96$ M and $y \ll 0.113$ M) must be valid, since each of them is smaller than z. (Remember that $x+y=z$.)

(c) In 1.0 M NH_4CN there are the following species: $NH_4^+(aq), NH_3(aq), CN^-(aq)$, $HCN(aq)$, $H_3O^+(aq)$, and $OH^-(aq)$. These species are related by the following six equations.

(1) $K_w = 1.0 \times 10^{-14} = \left[H_3O^+\right]\left[OH^-\right]$ $\qquad$ $\left[OH^-\right] = \dfrac{1.0 \times 10^{-14}}{\left[H_3O^+\right]}$

(2) $K_a = 6.2 \times 10^{-10} = \dfrac{\left[H_3O^+\right]\left[CN^-\right]}{[HCN]}$ $\qquad$ (3) $K_b = 1.8 \times 10^{-5} = \dfrac{\left[NH_4^+\right]\left[OH^-\right]}{[NH_3]}$

(4) $[NH_3] + \left[NH_4^+\right] = 1.0$ M $\qquad\qquad$ $[NH_3] = 1.0 - \left[NH_4^+\right]$

(5) $[HCN] + \left[CN^-\right] = 1.0$ M $\qquad\qquad$ $[HCN] = 1.0 - \left[CN^-\right]$

(6) $\left[NH_4^+\right] + \left[H_3O^+\right] = \left[CN^-\right] + \left[OH^-\right]$ or $\left[NH_4^+\right] \approx \left[CN^-\right]$

Equation (6) is the result of charge balance, that there must be the same quantity of positive and negative charge in the solution. The approximation is the result of remembering that not much H_3O^+ or OH^- will be formed as the result of hydrolysis of ions. Substitute equation (4) into equation (3), and equation (5) into equation (2).

(3') $K_b = 1.8 \times 10^{-5} = \dfrac{\left[NH_4^+\right]\left[OH^-\right]}{1.0 - \left[NH_4^+\right]}$ $\qquad$ (2') $K_a = 4.0 \times 10^{-10} = \dfrac{\left[H_3O^+\right]\left[CN^-\right]}{1.0 - \left[CN^-\right]}$

Now substitute equation (6) into equation (2'), and equation (1) into equation (3').

(2'') $K_a = 6.2 \times 10^{-10} = \dfrac{\left[H_3O^+\right]\left[NH_4^+\right]}{1.0 - \left[NH_4^+\right]}$ $\qquad$ (3'') $\dfrac{1.8 \times 10^{-5}}{1.0 \times 10^{-14}} = \dfrac{\left[NH_4^+\right]}{\left[H_3O^+\right]\left(1.0 - \left[NH_4^+\right]\right)}$

Now we solve both of these equations for $\left[NH_4^+\right]$.

('2): $6.2 \times 10^{-10} - 6.2 \times 10^{-10}\left[NH_4^+\right] = \left[H_3O^+\right]\left[NH_4^+\right]$ $\qquad$ $\left[NH_4^+\right] = \dfrac{6.2 \times 10^{-10}}{6.2 \times 10^{-10} + \left[H_3O^+\right]}$

('3): $1.8 \times 10^9\left[H_3O^+\right] - 1.8 \times 10^9\left[H_3O^+\right]\left[NH_4^+\right] = \left[NH_4^+\right]$ $\qquad$ $\left[NH_4^+\right] = \dfrac{1.8 \times 10^9\left[H_3O^+\right]}{1.00 + 1.8 \times 10^9\left[H_3O^+\right]}$

We equate the two results and solve for

$$\left[H_3O^+\right] \cdot \dfrac{6.2 \times 10^{-10}}{6.2 \times 10^{-10} + \left[H_3O^+\right]} = \dfrac{1.8 \times 10^9\left[H_3O^+\right]}{1.00 + 1.8 \times 10^9\left[H_3O^+\right]}$$

$$6.2 \times 10^{-10} + 1.1\left[H_3O^+\right] = 1.1\left[H_3O^+\right] + 1.8 \times 10^9\left[H_3O^+\right]^2 \qquad 6.2 \times 10^{-10} = 1.8 \times 10^9\left[H_3O^+\right]^2$$

$$\left[H_3O^+\right] = \sqrt{\dfrac{6.2 \times 10^{-10}}{1.8 \times 10^9}} = 5.9 \times 10^{-10}\,\text{M} \quad \text{pH} = -\log\left(5.9 \times 10^{-10}\right) = 9.23$$

Note that $\left[H_3O^+\right] = \sqrt{\dfrac{K_a \times K_w}{K_b}}$ $\qquad$ or $\qquad$ pH=0.500 $(pK_a + pK_w - pK_b)$

CHAPTER 17
ADDITIONAL ASPECTS OF ACID— BASE EQUILIBRIA
PRACTICE EXAMPLES

1A Organize the solution around the balanced chemical equation, as we have done before.

Equation: $\quad$ HF(aq) $+$ H_2O(l) $\rightleftharpoons$ $\quad$ H_3O^+(aq) $\quad + \quad F^-$(aq)

Initial: $\quad\quad$ 0.500 M $\quad\quad\quad -\quad\quad\quad$ ≈ 0 M $\quad\quad\quad$ 0 M

Changes: $\quad\quad -x$ M $\quad\quad\quad\quad -\quad\quad\quad$ $+x$ M $\quad\quad\quad$ $+x$ M

Equil: $\quad\quad$ $(0.500-x)$ M $\quad -\quad\quad\quad$ x M $\quad\quad\quad\quad$ x M

$$K_a = \frac{[H_3O^+][F^-]}{[HF]} = \frac{(x)(x)}{0.500-x} = 6.6\times10^{-4} \approx \frac{x^2}{0.500} \quad \text{assuming } x \ll 0.500$$

$$x = \sqrt{0.500\times6.6\times10^{-4}} = 0.018\,\text{M} \quad \text{One further cycle of approximations gives:}$$

$$x = \sqrt{(0.500-0.018)\times6.6\times10^{-4}} = 0.018\,\text{M} = [H_3O^+]$$

Thus, [HF] $= 0.500\,\text{M} - 0.018\,\text{M} = 0.482\,\text{M}$

Recognize that 0.100 M HCl means $\left[H_3O^+\right]_{\text{initial}} = 0.100\,\text{M}$, since HCl is a strong acid.

Equation: $\quad$ HF(aq) $+$ H_2O(l) $\rightleftharpoons$ $\quad$ H_3O^+(aq) $+$ F^-(aq)

Initial: $\quad\quad$ 0.500 M $\quad\quad\quad -\quad\quad\quad$ 0.100 M $\quad\quad$ 0 M

Changes: $\quad\quad -x$ M $\quad\quad\quad\quad -\quad\quad\quad$ $+x$ M $\quad\quad\quad$ $+x$ M

Equil: $\quad\quad$ $(0.500-x)$ M $\quad -\quad\quad$ $(0.100+x)$ M $\quad$ x M

$$K_a = \frac{[H_3O^+][F^-]}{[HF]} = \frac{(x)(0.100+x)}{0.500-x} = 6.6\times10^{-4} \approx \frac{0.100\,x}{0.500} \quad \text{assuming } x \ll 0.100$$

$$x = \frac{6.6\times10^{-4}\times0.500}{0.100} = 3.3\times10^{-3}\ \text{M} = \left[F^-\right] \quad \text{The assumption is valid.}$$

$\left[HF\right] = 0.500\,\text{M} - 0.003\,\text{M} = 0.497\,\text{M}$

$[H_3O^+] = 0.100\ \text{M} + x = 0.103\ \text{M} = 0.100\ \text{M} + 0.003\ \text{M} = 0.103\ \text{M}$

1B From Example 17-6 *in the text*, we know that $[H_3O^+] = [C_2H_3O_2^-] = 1.3\times10^{-3}$ M in 0.100 M $HC_2H_3O_2$. We base our calculation, as usual, on the balanced chemical equation. The concentration of H_3O^+ from the added HCl is represented by x.

Equation: $HC_2H_3O_2(aq) + H_2O(l) \rightleftharpoons H_3O^+(aq) + C_2H_3O_2^-(aq)$

Initial:	0.100 M	–	$\approx 0\,M$	0 M
Changes:	$-0.00010\,M$	–	$+0.00010\,M$	$+0.00010\,M$
From HCl:			$+x\ M$	
Equil:	0.100 M	–	$(0.00010+x)\,M$	0.00010 M

$$K_a = \frac{[H_3O^+][C_2H_3O_2^-]}{[HC_2H_3O_2]} = \frac{(0.00010+x)\,0.00010}{0.100} = 1.8 \times 10^{-5}$$

$$0.00010 + x = \frac{1.8 \times 10^{-5} \times 0.100}{0.00010} = 0.018\,M \qquad x = 0.018\,M - 0.00010\,M = 0.018\,M$$

$$V_{12\,M\,HCl} = 1.00\,L \times \frac{0.018\ mol\ H_3O^+}{1\,L} \times \frac{1\,mol\,HCl}{1\,mol\,H_3O^+} \times \frac{1\,L\,soln}{12\ mol\ HCl} \times \frac{1000\ mL}{1\,L} \times \frac{1\ drop}{0.050\ mL} = 30.\,drops$$

Since 30. drops corresponds to 1.5 mL of 12 M solution, we see that the volume of solution does indeed remain approximately 1.00 L after addition of the 12 M HCl.

2A We again organize the solution around the balanced chemical equation.

Equation: $HCHO_2(aq) + H_2O(l) \rightleftharpoons CHO_2^-(aq) + H_3O^+(aq)$

Initial:	0.100 M	–	0.150 M	$\approx 0\,M$
Changes:	$-x\,M$	–	$+x\,M$	$+x\,M$
Equil:	$(0.100-x)\,M$	–	$(0.150+x)\,M$	$x\,M$

$$K_a = \frac{[CHO_2^-][H_3O^+]}{[HCHO_2]} = \frac{(0.150+x)(x)}{0.100-x} = 1.8 \times 10^{-4} \approx \frac{0.150\,x}{0.100} \qquad \text{assuming } x \ll 0.100$$

$$x = \frac{0.100 \times 1.8 \times 10^{-4}}{0.150} = 1.2 \times 10^{-4}\,M = [H_3O^+],\ x \ll 0.100, \text{thus our assumption is valid}$$

$$[CHO_2^-] = 0.150\,M + 0.00012\,M = 0.150\,M$$

2B This time a solid sample of a weak base is being added to a solution of its conjugate acid. We let x represent the concentration of acetate ion from the added sodium acetate. Notice that sodium acetate is a strong electrolyte, thus, it completely dissociates in aqueous solution. $[H_3O^+] = 10^{-pH} = 10^{-5.00} = 1.0 \times 10^{-5}\,M = 0.000010\,M$

Equation: $HC_2H_3O_2(aq) + H_2O(l) \rightleftharpoons C_2H_3O_2^-(aq) + H_3O^+(aq)$

Initial:	0.100 M	–	0 M	$\approx 0\,M$
Changes:	$-0.000010\,M$	–	$+0.000010\,M$	$+0.000010\,M$
From NaAc:			$+x\,M$	
Equil:	0.100 M	–	$(0.000010+x)\,M$	0.000010 M

$$K_a = \frac{\left[H_3O^+\right]\left[C_2H_3O_2^-\right]}{\left[HC_2H_3O_2\right]} = \frac{0.000010(0.000010+x)}{0.100} = 1.8 \times 10^{-5}$$

$$0.000010 + x = \frac{1.8 \times 10^{-5} \times 0.100}{0.000010} = 0.18 \text{ M} \qquad x = 0.18 \text{ M} - 0.000010 \text{ M} = 0.18 \text{ M}$$

$$\text{mass of NaC}_2\text{H}_3\text{O}_2 = 1.00\,\text{L} \times \frac{0.18 \text{ mol C}_2\text{H}_3\text{O}_2^-}{1\text{L}} \times \frac{1\,\text{mol NaC}_2\text{H}_3\text{O}_2}{1\,\text{mol C}_2\text{H}_3\text{O}_2^-} \times \frac{82.03\,\text{g NaC}_2\text{H}_3\text{O}_2}{1\,\text{mol NaC}_2\text{H}_3\text{O}_2}$$

$$= 15\,\text{g NaC}_2\text{H}_3\text{O}_2$$

3A A strong acid dissociates essentially completely, and effectively is a source of H_3O^+. $NaC_2H_3O_2$ also dissociates completely in solution. The hydronium ion and the acetate ion react to form acetic acid: $\quad H_3O^+(aq) + C_2H_3O_2^-(aq) \rightleftharpoons HC_2H_3O_2(aq) + H_2O(l)$
All that is necessary to form a buffer is to have approximately equal amounts of a weak acid and its conjugate base together in solution. This will be achieved if we add an amount of HCl equal to approximately half the original amount of acetate ion.

3B HCl dissociates essentially completely in water and serves as a source of hydronium ion. This reacts with ammonia to form ammonium ion: $NH_3(aq) + H_3O^+(aq) \rightleftharpoons NH_4^+(aq) + H_2O(l)$.
Because a buffer contains approximately equal amounts of a weak base (NH_3) and its conjugate acid (NH_4^+), to prepare a buffer we simply add an amount of HCl equal to approximately half the amount of $NH_3(aq)$ initially present.

4A We first find the formate ion concentration, remembering that $NaCHO_2$ is a strong electrolyte, existing in solution as $Na^+(aq)$ and $CHO_2^-(aq)$.

$$[CHO_2^-] = \frac{23.1\text{g NaCHO}_2}{500.0 \text{ mL soln}} \times \frac{1000 \text{ mL}}{1\text{L}} \times \frac{1 \text{ mol NaCHO}_2}{68.01\text{g NaCHO}_2} \times \frac{1 \text{ mol CHO}_2^-}{1 \text{ mol NaCHO}_2} = 0.679\,\text{M}$$

As usual, the solution to the problem is organized around the balanced chemical equation.

Equation: $\quad HCHO_2(aq) \; + \; H_2O(l) \; \rightleftharpoons \; CHO_2^-(aq) \; + \; H_3O^+(aq)$
Initial: $\qquad$ 0.432 M $\qquad\qquad$ – $\qquad\qquad$ 0.679 M $\qquad\quad \approx 0$ M
Changes: $\quad -x$ M $\qquad\qquad$ – $\qquad\qquad +x$ M $\qquad\qquad +x$ M
Equil: $\qquad (0.432 - x)$ M $\qquad$ – $\qquad\quad (0.679 + x)$ M $\qquad x$ M

$$K_a = \frac{\left[H_3O^+\right]\left[CHO_2^-\right]}{\left[HCHO_2\right]} = \frac{x(0.679+x)}{0.432-x} = 1.8 \times 10^{-4} \approx \frac{0.679x}{0.432} \qquad x = \frac{0.432 \times 1.8 \times 10^{-4}}{0.679}$$

This gives $\left[H_3O^+\right] = 0.0892$ M. The assumption that $x \ll 0.432$ is clearly correct.

$$pH = -\log\left[H_3O^+\right] = -\log(1.1 \times 10^{-4}) = 3.96$$

4B The concentrations of the components in the 100.0 mL of buffer solution are found via the dilution factor. Remember that $NaC_2H_3O_2$ is a strong electrolyte, existing in solution as $Na^+(aq)$ and $C_2H_3O_2^-(aq)$.

$$[HC_2H_3O_2] = 0.200 \text{ M} \times \frac{63.0 \text{ mL}}{100.0 \text{ mL}} = 0.126 \text{ M} \quad [C_2H_3O_2^-] = 0.200 \text{ M} \times \frac{37.0 \text{ mL}}{100.0 \text{ mL}} = 0.0740 \text{ M}$$

As usual, the solution to the problem is organized around the balanced chemical equation.

Equation:	$HC_2H_3O_2(aq)$	$+ H_2O(l)$	$\rightleftharpoons$	$C_2H_3O_2^-(aq)$	$+ H_3O^+(aq)$
Initial:	0.126 M	–		0.0740 M	≈ 0 M
Changes:	$-x$ M	–		$+x$ M	$+x$ M
Equil:	$(0.126-x)$ M	–		$(0.0740+x)$ M	x M

$$K_a = \frac{[H_3O^+][C_2H_3O_2^-]}{[HC_2H_3O_2]} = \frac{x(0.0740+x)}{0.126-x} = 1.8 \times 10^{-5} \approx \frac{0.0740 \, x}{0.126}$$

$$x = \frac{1.8 \times 10^{-5} \times 0.126}{0.0740} = 3.1 \times 10^{-5} \text{M} = [H_3O^+]; \quad pH = -\log[H_3O^+] = -\log 3.1 \times 10^{-5} = 4.51$$

Note that the assumption is valid $x \ll 0.0740 < 0.126$.
Thus, x is neglected when added or subtracted

5A We know the initial concentration of NH_3 in the buffer solution and can use the pH to find the equilibrium $[OH^-]$. The rest of the solution is organized around the balanced chemical equation. Our first goal is to determine the initial concentration of NH_4^+.

$$pOH = 14.00 - pH = 14.00 - 9.00 = 5.00 \qquad [OH^-] = 10^{-pOH} = 10^{-5.00} = 1.0 \times 10^{-5} \text{ M}$$

Equation:	$NH_3(aq)$	$+$	$H_2O(l)$	$\rightleftharpoons$	$NH_4^+(aq)$	$+$	$OH^-(aq)$
Initial:	0.35 M		–		x M		≈ 0 M
Changes:	-1.0×10^{-5} M		–		$+1.0 \times 10^{-5}$ M		$+1.0 \times 10^{-5}$ M
Equil:	$(0.35 - 1.0 \times 10^{-5})$M		–		$(x + 1.0 \times 10^{-5})$M		1.0×10^{-5} M

$$K_b = \frac{[NH_4^+][OH^-]}{[NH_3]} = 1.8 \times 10^{-5} = \frac{(x+1.0 \times 10^{-5})(1.0 \times 10^{-5})}{0.35 - 1.0 \times 10^{-5}} = \frac{1.0 \times 10^{-5} \times x}{0.35}$$

Assume $x \gg 1.0 \times 10^{-5}$ $\quad x = \frac{0.35 \times 1.8 \times 10^{-5}}{1.0 \times 10^{-5}} = 0.63 \text{ M} = $ initial NH_4^+ concentration

$$\text{mass}(NH_4)_2SO_4 = 0.500 \text{ L} \times \frac{0.63 \text{ mol } NH_4^+}{1 \text{ L soln}} \times \frac{1 \text{ mol } (NH_4)_2SO_4}{2 \text{ mol } NH_4^+} \times \frac{132.1 \text{ g } (NH_4)_2SO_4}{1 \text{ mol } (NH_4)_2SO_4}$$

Mass of $(NH_4)_2SO_4 = 21$ g

5B The solution is composed of 33.05 g $NaC_2H_3O_2 \cdot 3\,H_2O$ dissolved in 300.0 mL of 0.250 M HCl. $NaC_2H_3O_2 \cdot 3\,H_2O$, a strong electrolyte, exists in solution as $Na^+(aq)$ and $C_2H_3O_2^-(aq)$ ions. First we calculate the moles of $NaC_2H_3O_2 \cdot 3\,H_2O$, which based on the 1:1 stoichiometry is also equal to the number of moles of C_2HO^- that are released into solution. From this we can calculate the initial $[C_2H_3O_2^-]$ assuming the solution's volume remains at 300. mL.

moles of $NaC_2H_3O_2 \cdot 3\,H_2O$ (and moles of $C_2H_3O_2^-$)

$$= \frac{33.05 \text{ g } NaC_2H_3O_2 \cdot 3H_2O}{\dfrac{1 \text{ mole } NaC_2H_3O_2 \cdot 3H_2O}{136.08 \text{ g } NaC_2H_3O_2 \cdot 3H_2O}} = 0.243 \text{ moles } NaC_2H_3O_2 \cdot 3H_2O = \text{moles } C_2H_3O_2^-$$

$$[C_2H_3O_2^-] = \frac{0.243 \text{ mol } C_2H_3O_2^-}{0.300 \text{ L soln}} = 0.810\,M$$

(Note: [HCl] is assumed to remain unchanged at 0.250 M)
We organize this information around the balanced chemical equation, as before. We recognize that virtually all of the HCl has been hydrolyzed and that hydronium ion will react to produce the much weaker acetic acid.

Equation:	$HC_2H_3O_2(aq)$	$+ H_2O(l)$	$\rightleftharpoons$	$C_2H_3O_2^-(aq)$	$+$	$H_3O^+(aq)$
Initial:	0 M	–		0.810 M		0.250 M
Form HAc:	+0.250 M	–		−0.250 M		−0.250 M
	0.250 M	–		0.560 M		≈ 0 M
Changes:	$-x$ M	–		$+x$ M		$+x$ M
Equil:	$(0.250 - x)$ M	–		$(0.560 + x)$ M		$+x$ M

$$K_a = \frac{[H_3O^+][C_2H_3O_2^-]}{[HC_2H_3O_2]} = \frac{x(0.560+x)}{0.250-x} = 1.8\times10^{-5} \approx \frac{0.560\,x}{0.250}$$

$$x = \frac{1.8\times10^{-5}\times0.250}{0.560} = 8.0\times10^{-6} \text{ M} = [H_3O^+]$$

(The approximation was valid since $x \ll$ both 0.250 and 0.560)
$$pH = -\log[H_3O^+] = -\log 8.0\times10^{-6} = 5.09 \approx 5.1$$

6A **(a)** For formic acid, $pK_a = -\log(1.8\times10^{-4}) = 3.74$. The Henderson-Hasselbalch equation provides the pH of the original buffer solution:

$$pH = pK_a + \log\frac{[CHO_2^-]}{[HCHO_2]} = 3.74 + \log\frac{0.350}{0.550} = 3.54$$

(b) The added acid can be considered completely reacted with the formate ion to produces formic acid. Each mole/L of added acid consumes one 1 M of formate ion and forms 1 M of formic acid: $CHO_2^-(aq) + H_3O^+(aq) \longrightarrow HCHO_2(aq) + H_2O(l)$. $K_{neut} = K_b/K_w \approx$ 5600. Thus, $\left[CHO_2^-\right] = 0.350$ M $- 0.0050$ M $= 0.345$ M and

$\left[HCHO_2\right] = 0.550$ M $+ 0.0050$ M $= 0.555$ M. By using the Henderson-Hasselbalch equation

$$pH = pK_a + \log\frac{[CHO_2^-]}{[HCHO_2]} = 3.74 + \log\frac{0.345}{0.555} = 3.53$$

(c) Added base reacts completely reacts with formic acid producing an equivalent amount of formate ion. This continues until all of the formic acid is consumed.

Each 1 mole of added base consumes 1 mol of formic acid and forms 1 mol of formate ion: $HCHO_2 + OH^- \longrightarrow CHO_2^- + H_2O$. $K_{neut} = K_a/K_w \approx 1.8 \times 10^{10}$. Thus,

$\left[CHO_2^-\right] = 0.350\,M + 0.0050\ M = 0.355\ M$ $\left[HCHO_2\right] = (0.550 - 0.0050)M = 0.545\ M$.

With the Henderson-Hasselbalch equation we find

$$pH = pK_a + \log\frac{\left[CHO_2^-\right]}{\left[HCHO_2\right]} = 3.74 + \log\frac{0.355}{0.545} = 3.55$$

6B The buffer cited has the same concentration of weak acid and its anion as does the buffer of Example 17-6. Our goal is to reach $pH = 5.03$ or $\left[H_3O^+\right] = 10^{-pH} = 10^{-5.03} = 9.3 \times 10^{-6}\ M$.

Adding strong acid $\left(H_3O^+\right)$, of course, produces $HC_2H_3O_2$ at the expense of $C_2H_3O_2^-$. Thus, adding H^+ drives the reaction to the left. Again, we use the data around the balanced chemical equation.

Equation:	$HC_2H_3O_2\,(aq)$	$+\ H_2O(l)$	$\rightleftharpoons$	$C_2H_3O_2^-(aq)$	$+$	$H_3O^+(aq)$
Initial:	0.250 M		$-$	0.560 M		8.0×10^{-6} M
Add acid:						$+y$ M
Form HAc:	$+y$ M		$-$	$-y$ M		$-y$ M
	$(0.250 + y)$ M		$-$	$(0.560 - y)$ M		≈ 0 M
Changes:	$-x$ M		$-$	$+x$ M		$+x$ M
Equil:	$(0.250 + y - x)$ M		$-$	$(0.560 - y + x)$ M		9.3×10^{-6} M

$$K_a = \frac{[H_3O^+][C_2H_3O_2^-]}{[HC_2H_3O_2]} = \frac{9.3 \times 10^{-6}(0.560 - y + x)}{0.250 + y - x} = 1.8 \times 10^{-5} \approx \frac{9.3 \times 10^{-6}(0.560 - y)}{0.250 + y}$$

(Assume that x is negligible compared to y)

$$\frac{1.8 \times 10^{-5} \times (0.250 + y)}{9.3 \times 10^{-6}} = 0.484 + 1.94\ y = 0.560 - y \qquad y = \frac{0.560 - 0.484}{1.94 + 1.00} = 0.026\ M$$

Notice that our assumption is valid: $x \ll 0.250 + y \,(= 0.276) < 0.560 - y\,(= 0.534)$.

$$V_{HNO_3} = 300.0\ \text{mL buffer} \times \frac{0.026\ \text{mmol } H_3O^+}{1\ \text{mL buffer}} \times \frac{1\ \text{mL } HNO_3\,(aq)}{6.0\ \text{mmol } H_3O^+} = 1.3\ \text{mL of } 6.0\ M\ HNO_3$$

Instead of the algebraic solution, we could have used the Henderson-Hasselbalch equation, since the final pH falls within one pH unit of the pK_a of acetic acid. We let z indicate the increase in $\left[HC_2H_3O_2\right]$, and also the decrease in $\left[C_2H_3O_2^-\right]$

$$pH = pK_a + \log\frac{\left[C_2H_3O_2^-\right]}{\left[HC_2H_3O_2\right]} = 4.74 + \log\frac{0.560 - z}{0.250 + z} = 5.03 \qquad \frac{0.560 - z}{0.250 + z} = 10^{5.03-4.74} = 1.95$$

$$0.560 - z = 1.95\,(0.250 + z) = 0.488 + 1.95\ z \qquad z = \frac{0.560 - 0.488}{1.95 + 1.00} = 0.024\ M$$

This is, and should be, almost exactly the same as the value of y we obtained by the I.C.E. table method. The slight difference is due to imprecision arising from rounding errors.

7A **(a)** The initial pH is the pH of 0.150 M HCl, which we obtain from $\left[H_3O^+\right]$ of that strong acid solution.

$$[H_3O^+] = \frac{0.150 \text{ mol HCl}}{1 \text{ L soln}} \times \frac{1 \text{ mol } H_3O^+}{1 \text{ mol HCl}} = 0.150 \text{ M},$$

$$pH = -\log [H_3O^+] = -\log (0.150) = 0.824$$

(b) To determine $\left[H_3O^+\right]$ and then pH at the 50.0% point, we need the volume of the solution and the amount of H_3O^+ left unreacted. First we calculate the amount of hydronium ion present and then the volume of base solution needed for its complete neutralization.

$$\text{amount } H_3O^+ = 25.00 \text{ mL} \times \frac{0.150 \text{ mmol HCl}}{1 \text{ mL soln}} \times \frac{1 \text{ mmol } H_3O^+}{1 \text{ mmol HCl}} = 3.75 \text{ mmol } H_3O^+$$

$$V_{acid} = 3.75 \text{ mmol } H_3O^+ \times \frac{1 \text{ mmol OH}^-}{1 \text{ mmol } H_3O^+} \times \frac{1 \text{ mmol NaOH}}{1 \text{ mmol OH}^-} \times \frac{1 \text{ mL titrant}}{0.250 \text{ mmol NaOH}}$$

$$= 15.0 \text{ mL titrant}$$

At the 50.0% point, half of the H_3O^+ (1.88 mmol H_3O^+) will remain unreacted and only half (7.50 mL titrant) of the titrant solution will be added. From this information, and the original 25.00-mL volume of the solution, we calculate $\left[H_3O^+\right]$ and then pH.

$$\left[H_3O^+\right] = \frac{1.88 \text{ mmol } H_3O^+ \text{ left}}{25.00 \text{ mL original} + 7.50 \text{ mL titrant}} = 0.0578 \text{ M}$$

$$pH = -\log(0.0578) = 1.238$$

(c) Since this is the titration of a strong acid by a strong base, at the equivalence point, the pH = 7.00. This is because the only ions of appreciable concentration in the equivalence point solution are $Na^+(aq)$ and $Cl^-(aq)$, and neither of these species undergoes detectable hydrolysis reactions.

(d) Beyond the equivalence point, the solution pH is determined almost entirely by the concentration of excess $OH^-(aq)$ ions. The volume of the solution is $40.00 \text{ mL} + 1.00 \text{ mL} = 41.00 \text{ mL}$. The amount of hydroxide ion in the excess titrant is calculated and used to determine $\left[OH^-\right]$, from which pH is computed.

$$\text{amount of OH}^- = 1.00 \text{ mL} \times \frac{0.250 \text{ mmol NaOH}}{1 \text{ mL}} = 0.250 \text{ mmol OH}^-$$

$$\left[OH^-\right] = \frac{0.250 \text{ mmol OH}^-}{41.00 \text{ mL}} = 0.006098 \text{ M}$$

$$pOH = -\log(0.006098) = 2.215; \quad pH = 14.00 - 2.215 = 11.785$$

7B **(a)** The initial pH is simply the pH of 0.00812 M $Ba(OH)_2$, which we obtain from $\left[OH^-\right]$ for the solution.

$$\left[OH^-\right] = \frac{0.00812\,mol\,Ba(OH)_2}{1L\,soln} \times \frac{2\,mol\,OH^-}{1\,mol\,Ba(OH)_2} = 0.01624\,M$$

$$pOH = -\log\left[OH^-\right] = -\log(0.0162) = 1.790; \quad pH = 14.00 - pOH = 14.00 - 1.790 = 12.21$$

(b) To determine $\left[OH^-\right]$ and then pH at the 50.0% point, we need the volume of the solution and the amount of OH^- unreacted. First we calculate the amount of hydroxide ion present and then the volume of acid solution needed for its complete neutralization.

$$\text{amount } OH^- = 50.00\,mL \times \frac{0.00812\,mmol\,Ba(OH)_2}{1\,mL\,soln} \times \frac{2\,mmol\,OH^-}{1\,mmol\,Ba(OH)_2} = 0.812\,mmol\,OH^-$$

$$V_{acid} = 0.812\,mmol\,OH^- \times \frac{1\,mmol\,H_3O^+}{1\,mmol\,OH^-} \times \frac{1\,mmol\,HCl}{1\,mmol\,H_3O^+} \times \frac{1\,mL\,titrant}{0.0250\,mmol\,HCl} = 32.48\,mL\,titrant$$

At the 50.0 % point, half (0.406 $mmol\,OH^-$) will remain unreacted and only half (16.24 mL titrant) of the titrant solution will be added. From this information, and the original 50.00-mL volume of the solution, we calculate $\left[OH^-\right]$ and then pH.

$$\left[OH^-\right] = \frac{0.406\,mmol\,OH^-\,left}{50.00\,mL\,original + 16.24\,mL\,titrant} = 0.00613\,M$$

$$pOH = -\log(0.00613) = 2.213; \quad pH = 14.00 - pOH = 11.79$$

(c) Since this is the titration of a strong base by a strong acid, at the equivalence point, $pH = 7.00$. The solution at this point is neutral because the dominant ionic species in solution, namely $Ba^{2+}(aq)$ and $Cl^-(aq)$ do not react with water to a detectable extent.

8A **(a)** Initial pH is just that of 0.150 M HF ($pK_a = -\log(6.6 \times 10^{-4}) = 3.18$).

[Initial solution contains $20.00\,mL \times \frac{0.150\,mmol\,HF}{1\,mL} = 3.00\,mmol\,HF$].

Equation : $HF(aq) + H_2O(l) \rightleftharpoons H_3O^+(aq) + F^-(aq)$

Initial :   0.150 M   – $\approx 0\,M$ 0 M

Changes : $-x\,M$ – $+x\,M$ $+x\,M$

Equil: $(0.150 - x)\,M$ – $x\,M$ $x\,M$

$$K_a = \frac{\left[H_3O^+\right]\left[F^-\right]}{\left[HF\right]} = \frac{x \cdot x}{0.150 - x} \approx \frac{x^2}{0.150} = 6.6 \times 10^{-4}$$

$$x = \sqrt{0.150 \times 6.6 \times 10^{-4}} = 9.9 \times 10^{-3}\,M$$

$x > 0.05(0.150)$. The assumption is invalid. After a 2nd cycle of approximation,

$\left[H_3O^+\right] = 9.6 \times 10^{-3}\,M$; $pH = -\log(9.6 \times 10^{-3}) = 2.02$

(b) When the titration is 25.0% complete, there are $(0.25 \times 3.00 =)0.75$ mmol F^- for every 3.00 mmol HF that were present initially. $(3.00 - 0.75 =)2.25$ mmol HF remain untitrated. We designate the solution volume (the volume holding these 3.00 mmol total) as V and use the Henderson-Hasselbalch equation to find the pH.

$$pH = pK_a + \log\frac{[F^-]}{[HF]} = 3.18 + \log\frac{0.75 \ \text{mmol}/V}{2.25 \ \text{mmol}/V} = 2.70$$

(c) At the midpoint of the titration of a weak base, $pH = pK_a = 3.18$

(d) At the endpoint of the titration, the pH of the solution is determined by the conjugate base hydrolysis reaction. We calculate the amount of anion and the volume of solution in order to calculate its initial concentration.

$$\text{amount } F^- = 20.00 \ \text{mL} \times \frac{0.150 \ \text{mmol HF}}{1 \ \text{mL soln}} \times \frac{1 \ \text{mmol } F^-}{1 \ \text{mmol HF}} = 3.00 \ \text{mmol } F^-$$

$$\text{volume titrant} = 3.00 \ \text{mmol HF} \times \frac{1 \ \text{mmol OH}^-}{1 \ \text{mmol HF}} \times \frac{1 \ \text{mL titrant}}{0.250 \ \text{mmol OH}^-} = 12.0 \ \text{mL titrant}$$

$$\left[F^-\right] = \frac{3.00 \ \text{mmol } F^-}{20.00 \ \text{mL original volume} + 12.0 \ \text{mL titrant}} = 0.0934 \ \text{M}$$

We organize the solution of the hydrolysis problem around its balanced equation.

Equation :	$F^-(aq)$	+	$H_2O(l)$	$\rightleftharpoons$	$HF(aq)$	+	$OH^-(aq)$
Initial:	0.0934M		–		0M		≈ 0 M
Changes :	$-x$ M		–		$+x$ M		$+x$ M
Equil :	$(0.0934 - x)$ M		–		x M		x M

$$K_b = \frac{[HF][OH^-]}{[F^-]} = \frac{K_w}{K_a} = \frac{1.0 \times 10^{-14}}{6.6 \times 10^{-4}} = 1.5 \times 10^{-11} = \frac{x \cdot x}{0.0934 - x} \approx \frac{x^2}{0.0934}$$

$$x = \sqrt{0.0934 \times 1.5 \times 10^{-11}} = 1.2 \times 10^{-6} \ \text{M} = [OH^-]$$

The assumption is valid ($x \ll 0.0934$).

$$pOH = -\log(1.2 \times 10^{-6}) = 5.92; \quad pH = 14.00 - pOH = 14.00 - 5.92 = 8.08$$

8B **(a)** The initial pH is simply that of $0.106 \ \text{M NH}_3$.

Equation:	$NH_3(aq)$	+	$H_2O(l)$	$\rightleftharpoons$	$NH_4^+(aq)$	+	$OH^-(aq)$
Initial:	0.106 M		–		0 M		≈ 0 M
Changes:	$-x$ M		–		$+x$ M		$+x$ M
Equil:	$(0.106 - x)$ M		–		x M		x M

$$K_b = \frac{[NH_4^+][OH^-]}{[NH_3]} = \frac{x \cdot x}{0.106 - x} \approx \frac{x^2}{0.106} = 1.8 \times 10^{-5}$$

$$x = \sqrt{0.106 \times 1.8 \times 10^{-4}} = 1.4 \times 10^{-3} \ \text{M} = [OH^-] \quad \text{the assumption is valid } (x \ll 0.106).$$

$$pOH = -\log(0.0014) = 2.85 \quad pH = 14.00 - pOH = 14.00 - 2.85 = 11.15$$

(b) When the titration is 25.0% complete, there are 25.0 mmol NH_4^+ for every 100.0 mmol of NH_3 that were present initially (i.e. there are 1.33 mmol of NH_4^+ in solution) 3.98 mmol NH_3 remain untitrated. We designate the solution volume (the volume holding these 5.30 mmol total) as V and use the basic version of the Henderson-Hasselbalch equation too fond the pH.

$$pOH = pK_b + log\frac{\left[NH_4^+\right]}{\left[NH_3\right]} = 4.74 + log\frac{\dfrac{1.33\ mmol}{V}}{\dfrac{3.98\ mmol}{V}} = 4.26$$

$$pH = 14.00 - 4.26 = 9.74$$

(c) At the midpoint of the titration of a weak base, $pOH = pK_b = 4.74$ and $pH = 9.26$

(d) At the endpoint of the titration, the pH is determined by the conjugate acid hydrolysis reaction. We calculate the amount of that cation and the volume of the solution in order to determine its initial concentration.

$$amount\ NH_4^+ = 50.00\ mL \times \frac{0.106\ mmol\ NH_3}{1\ mL\ soln} \times \frac{1\ mmol\ NH_4^+}{1\ mmol\ NH_3}$$

$$amount\ NH_4^+ = 5.30\ mmol\ NH_4^+$$

$$V_{titrant} = 5.30\ mmol\ NH_3 \times \frac{1\ mmol\ H_3O^+}{1\ mmol\ NH_3} \times \frac{1\ mL\ titrant}{0.225\ mmol\ H_3O^+} = 23.6\ mL\ titrant$$

$$\left[NH_4^+\right] = \frac{5.30\ mmol\ NH_4^+}{50.00\ mL\ original\ volume + 23.6\ mL\ titrant} = 0.0720\ M$$

We organize the solution of the hydrolysis problem around its balanced chemical equation.

Equation: $NH_4^+(aq) + H_2O(l) \rightleftharpoons NH_3(aq) + H_3O^+(aq)$

Initial:	0.0720 M	–	0 M	≈ 0 M
Changes:	$-x$ M	–	$+x$ M	$+x$ M
Equil:	$(0.0720-x)$M	–	x M	x M

$$K_b = \frac{[NH_3][H_3O^+]}{\left[NH_4^+\right]} = \frac{K_w}{K_b} = \frac{1.0 \times 10^{-14}}{1.8 \times 10^{-5}} = 5.6 \times 10^{-10} = \frac{x \cdot x}{0.0720 - x} \approx \frac{x^2}{0.0720}$$

$$x = \sqrt{0.0720 \times 5.6 \times 10^{-10}} = 6.3 \times 10^{-6}\ M = [H_3O^+]$$

The assumption is valid ($x \ll 0.0720$).

$$pH = -log(6.3 \times 10^{-6}) = 5.20$$

9A The acidity of the solution is principally the result of the hydrolysis of the carbonate ion, which is considered first.

Equation: $CO_3^{2-}(aq) + H_2O(l) \rightleftharpoons HCO_3^-(aq) + OH^-(aq)$

Initial:	1.0 M	–	0 M	≈ 0 M
Changes:	$-x$ M	–	$+x$ M	$+x$ M
Equil:	$(1.0-x)$ M	–	x M	x M

$$K_b = \frac{K_w}{K_a(HCO_3^-)} = \frac{1.0 \times 10^{-14}}{4.7 \times 10^{-11}} = 2.1 \times 10^{-4} = \frac{[HCO_3^-][OH^-]}{[CO_3^{2-}]} = \frac{x \cdot x}{1.0-x} \approx \frac{x^2}{1.0}$$

$x = \sqrt{1.0 \times 2.1 \times 10^{-4}} = 1.5 \times 10^{-2}$ M $= 0.015$ M $= [OH^-]$ The assumption is valid ($x \ll 1.0M$).

Now we consider the hydrolysis of the bicarbonate ion.

Equation: $HCO_3^-(aq) + H_2O(l) \rightleftharpoons H_2CO_3(aq) + OH^-(aq)$

Initial:	0.015 M	–	0 M	0.015 M
Changes:	$-y$ M	–	$+y$ M	$+y$ M
Equil:	$(0.015-y)$M	–	y M	$(0.015+y)$ M

$$K_b = \frac{K_w}{K_a(H_2CO_3)} = \frac{1.0 \times 10^{-14}}{4.4 \times 10^{-7}} = 2.3 \times 10^{-8} = \frac{[H_2CO_3][OH^-]}{[HCO_3^-]} = \frac{y(0.015+y)}{0.015-x} \approx \frac{0.015y}{0.015} = y$$

The assumption is valid ($y \ll 0.015$) and $y = [H_2CO_3] = 2.3 \times 10^{-8}$ M. Clearly, the second hydrolysis makes a negligible contribution to the acidity of the solution. For the entire solution, then

$$pOH = -\log[OH^-] = -\log(0.015) = 1.82 \qquad pH = 14.00 - 1.82 = 12.18$$

9B The acidity of the solution is principally the result of hydrolysis of the sulfite ion.

Equation: $SO_3^{2-}(aq) + H_2O(l) \rightleftharpoons HSO_3^-(aq) + OH^-(aq)$

Initial:	0.500 M	–	0 M	≈ 0 M
Changes:	$-x$ M	–	$+x$ M	$+x$ M
Equil:	$(0.500-x)$M	–	x M	x M

$$K_b = \frac{K_w}{K_a HSO_3^-} = \frac{1.0 \times 10^{-14}}{6.2 \times 10^{-8}} = 1.6 \times 10^{-7} = \frac{[HSO_3^-][OH^-]}{[SO_3^{2-}]} = \frac{x \cdot x}{0.500-x} \approx \frac{x^2}{0.500}$$

$x = \sqrt{0.500 \times 1.6 \times 10^{-7}} = 2.8 \times 10^{-4}$ M $= 0.00028$ M $= [OH^-]$

The assumption is valid ($x \ll 0.500$).

Next we consider the hydrolysis of the bisulfite ion.

Equation: $HSO_3^-\ (aq)\ +\ H_2O(l) \rightleftharpoons H_2SO_3\ (aq)\ +\ OH^-\ (aq)$

Initial:	0.00028 M	–	0 M	0.00028 M
Changes:	$-y$ M	–	$+y$ M	$+y$ M
Equil:	$(0.00028 - y)$ M	–	y M	$(0.00028 + y)$ M

$$K_b = \frac{K_w}{K_a H_2SO_3} = \frac{1.0 \times 10^{-14}}{1.3 \times 10^{-2}} = 7.7 \times 10^{-13}$$

$$K_b = 7.7 \times 10^{-13} = \frac{[H_2SO_3][OH^-]}{[HSO_3^-]} = \frac{y(0.00028 + y)}{0.00028 - y} \approx \frac{0.00028\,y}{0.00028} = y$$

The assumption is valid ($y \ll 0.00028$) and $y = [H_2SO_3] = 7.7 \times 10^{-13}$ M. Clearly, the second hydrolysis makes a negligible contribution to the acidity of the solution. For the entire solution, then

$$pOH = -\log[OH^-] = -\log(0.00028) = 3.55 \qquad pH = 14.00 - 3.55 = 10.45$$

EXERCISES

The Common Ion Effect

1.   **(a)**   Note that HI is a strong acid and thus the initial $[H_3O^+] = [HI] = 0.0892M$

Equation :	$HC_3H_5O_2$ + H_2O $\rightleftharpoons$	$C_3H_5O_2^-$	+	H_3O^+
Initial :	0.275 M	–	0M	0.0892M
Changes :	$-x$ M	–	$+x$ M	$+x$ M
Equil :	$(0.275 - x)$ M	–	x M	$(0.0892 + x)$ M

$$K_a = \frac{[C_3H_5O_2^-][H_3O^+]}{[HC_3H_5O_2]} = 1.3 \times 10^{-5} = \frac{x(0.0892 + x)}{0.275 - x} \approx \frac{0.0892x}{0.275} \qquad x = 4.0 \times 10^{-5}\ M$$

The assumption that $x \ll 0.0892$ M is correct. $[H_3O^+] = 0.0892M$

(b) $[OH^-] = \dfrac{K_w}{[H_3O^+]} = \dfrac{1.0 \times 10^{-14}}{0.0892} = 1.1 \times 10^{-13}$ M

(c) $[C_3H_5O_2^-] = x = 4.0 \times 10^{-5}$ M

(d) $[I^-] = [HI]_{int} = 0.0892$ M

2. **(a)** The NH_4Cl dissociates completely, and thus, $\left[NH_4^+\right]_{int} = \left[Cl^-\right]_{int} = 0.102\ M$

Equation: $NH_3(aq) + H_2O(l) \rightleftharpoons NH_4^+(aq) + OH^-(aq)$

Initial: $\quad 0.164\ M \qquad - \qquad\qquad 0.102\ M \qquad \approx 0\ M$

Changes: $\quad -x\ M \qquad\quad - \qquad\qquad +x\ M \qquad +x\ M$

Equil: $\quad (0.164-x)M \quad - \qquad (0.102+x)M \qquad x\ M$

$$K_b = \frac{\left[NH_4^+\right]\left[OH^-\right]}{\left[NH_3\right]} = \frac{(0.102+x)x}{0.164-x} = 1.8\times10^{-5} \approx \frac{0.102x}{0.164};\ x = 2.9\times10^{-5}M$$

Assumed $x \ll 0.102\ M$, a clearly valid assumption. $\left[OH^-\right] = x = 2.9\times10^{-5}\ M$

(b) $\left[NH_4^+\right] = 0.102 + x = 0.102\ M$

(c) $\left[Cl^-\right] = 0.102\ M$

(d) $\left[H_3O^+\right] = \dfrac{1.0\times10^{-14}}{2.9\times10^{-5}} = 3.4\times10^{-10}\ M$

3. **(a)** We first determine the pH of 0.100 M HNO_2.

Equation $\quad HNO_2(aq) + H_2O(l) \rightleftharpoons NO_2^-(aq) + H_3O^+(aq)$

Initial: $\quad 0.100\ M \qquad\quad - \qquad\qquad 0M \qquad\quad \approx 0M$

Changes: $\quad -x\ M \qquad\qquad - \qquad\qquad +xM \qquad +xM$

Equil: $\quad (0.100-x)\ M \quad\ - \qquad\qquad xM \qquad\quad xM$

$$K_a = \frac{\left[NO_2^-\right]\left[H_3O^+\right]}{\left[HNO_2\right]} = 7.2\times10^{-4} = \frac{x^2}{0.100-x}$$

Via the quadratic equation roots formula or via successive approximations,

$x = 8.1\times10^{-3}M = \left[H_3O^+\right]$.

Thus $pH = -\log\left(8.1\times10^{-3}\right) = 2.09$

When 0.100 mol $NaNO_2$ is added to 1.00 L of a 0.100 M HNO_2, a solution with $\left[NO_2^-\right] = 0.100M = \left[HNO_2\right]$ is produced. The answer obtained with the Henderson-Hasselbalch equation, is $pH = pK_a = -\log\left(7.2\times10^{-4}\right) = 3.14$. Thus, the addition has caused a pH change of 1.05 units.

(b) $NaNO_3$ contributes nitrate ion, NO_3^-, to the solution. Since, however, there is no molecular $HNO_3(aq)$ in equilibrium with hydrogen and nitrate ions, there is no equilibrium to be shifted by the addition of nitrate ions. The $\left[H_3O^+\right]$ and the pH are thus unaffected by the addition of $NaNO_3$ to a solution of nitric acid. The pH changes are not the same because there is an equilibrium system to be shifted in the first solution, whereas there is no equilibrium, just a change in total ionic strength for the second solution.

4.   The explanation for the different result is that each of these solutions has acetate ion present, $C_2H_3O_2^-$, that is produced in the ionization of acetic acid. The presence of this ion suppresses the ionization of acetic acid, thus minimizing the increase in $[H_3O^+]$. All three solutions are buffer solutions and their pH can be found with the aid of the Henderson-Hasselbalch equation.

(a)
$$pH = pK_a + \log\frac{[C_2H_3O_2^-]}{[HC_2H_3O_2]} = 4.74 + \log\frac{0.10}{1.0} = 3.74 \qquad [H_3O^+] = 10^{-3.74} = 1.8 \times 10^{-4} \text{ M}$$

$$\% \text{ ionization} = \frac{[H_3O^+]}{[HC_2H_3O_2]} \times 100\% = \frac{1.8 \times 10^{-4}\text{M}}{1.0\text{M}} \times 100\% = 0.018\%$$

(b)
$$pH = pK_a + \log\frac{[C_2H_3O_2^-]}{[HC_2H_3O_2]} = 4.74 + \log\frac{0.10}{0.10} = 4.74$$

$$[H_3O^+] = 10^{-4.74} = 1.8 \times 10^{-5}\text{M}$$

$$\% \text{ ionization} = \frac{[H_3O^+]}{[HC_2H_3O_2]} \times 100\% = \frac{1.8 \times 10^{-5}\text{M}}{0.10\text{M}} \times 100\% = 0.018\%$$

(c)
$$pH = pK_a + \log\frac{[C_2H_3O_2^-]}{[HC_2H_3O_2]} = 4.74 + \log\frac{0.10}{0.010} = 5.74$$

$$[H_3O^+] = 10^{-5.74} = 1.8 \times 10^{-6}\text{M}$$

$$\% \text{ ionization} = \frac{[H_3O^+]}{[HC_2H_3O_2]} \times 100\% = \frac{1.8 \times 10^{-6}\text{M}}{0.010\text{M}} \times 100\% = 0.018\%$$

5.   (a)   The strong acid HCl suppresses the ionization of the weak acid HOCl to such an extent that a negligible concentration of H_3O^+ is contributed to the solution by HOCl. Thus,
$$[H_3O^+] = [HCl] = 0.035\,\text{M}$$

(b)   This is a buffer solution. Consequently, we can use the Henderson-Hasselbalch equation to determine its pH. $pK_a = -\log(7.2 \times 10^{-4}) = 3.14$;

$$pH = pK_a + \log\frac{[NO_2^-]}{[HNO_2]} = 3.14 + \log\frac{0.100\,\text{M}}{0.0550\,\text{M}} = 3.40$$

$$[H_3O^+] = 10^{-3.40} = 4.0 \times 10^{-4} \text{ M}$$

(c) This also is a buffer solution, as we see by an analysis of the reaction between the components.

Equation: $H_3O^+ (aq, \text{ from HCl}) + C_2H_3O_2^- (aq, \text{ from NaC}_2H_3O_2) \rightarrow HC_2H_3O_2 (aq) + H_2O(l)$

In soln:	0.0525 M	0.0768 M	0 M —
Produce HAc:	−0.0525 M	−0.0525 M	+0.0525 M —
Initial:	≈ 0M	0.0243 M	0.0525 M —

Now the Henderson-Hasselbalch equation can be used to find the pH.

$$pK_a = -\log(1.8 \times 10^{-5}) = 4.74$$

$$pH = pK_a + \log \frac{\left[C_2H_3O_2^-\right]}{\left[HC_2H_3O_2\right]} = 4.74 + \log \frac{0.0243\,M}{0.0525\,M} = 4.41$$

$$\left[H_3O^+\right] = 10^{-4.41} = 3.9 \times 10^{-5}\ M$$

6. **(a)** Neither $Ba^{2+}(aq)$ nor $Cl^-(aq)$ hydrolyzes to a measurable extent and hence they have no effect on the solution pH. Consequently, $\left[OH^-\right]$ is determined entirely by the $Ba(OH)_2$ solute.

$$\left[OH^-\right] = \frac{0.0062\ \text{mol Ba}(OH)_2}{1\ L\ \text{soln}} \times \frac{2\,\text{mol OH}^-}{1\,\text{mol Ba}(OH)_2} = 0.012\ M$$

(b) We use the Henderson-Hasselbalch equation to find the pH for this buffer solution.

$$\left[NH_4^+\right] = 0.315\ M\,(NH_4)_2\,SO_4 \times \frac{2\ \text{mol NH}_4^+}{1\ \text{mol }(NH_4)_2\,SO_4} = 0.630\ M \quad pK_a = 9.26 \text{ for } NH_4^+.$$

$$pH = pK_a + \log \frac{\left[NH_3\right]}{\left[NH_4^+\right]} = 9.26 + \log \frac{0.486\ M}{0.630\ M} = 9.15 \quad pOH = 14.00 - 9.15 = 4.85$$

$$\left[OH^-\right] = 10^{-4.85} = 1.4 \times 10^{-5}\ M$$

(c) This solution also is a buffer, as analysis of the reaction between its components shows.

Equation: $NH_4^+ (aq, \text{ from NH}_4Cl) + OH^- (aq, \text{ from NaOH}) \rightarrow NH_3 (aq) + H_2O(l)$

In soln:	0.264 M	0.196 M	0 M —
Form NH_3:	−0.196 M	−0.196 M	+0.196 M —
Initial:	0.068 M	≈ 0M	0.196 M —

$$pH = pK_a + \log \frac{\left[NH_3\right]}{\left[NH_4^+\right]} = 9.26 + \log \frac{0.196\,M}{0.068\,M} = 9.72 \quad pOH = 14.00 - 9.72 = 4.28$$

$$\left[OH^-\right] = 10^{-4.28} = 5.2 \times 10^{-5}\ M$$

Buffer Solutions

7. $\left[H_3O^+\right]=10^{-4.06}=8.7\times10^{-5}\,M$. We let $S=\left[CHO_2^-\right]_{int}$

Equation: $HCHO_2\,(aq)\ +\ H_2O(l)\ \rightleftharpoons\ CHO_2^-\,(aq)\ +\ H_3O^+\,(aq)$

Initial :	0.366 M	–	S M	$\approx 0\,M$
Changes :	$-8.7\times10^{-5}\,M$	–	$+8.7\times10^{-5}\,M$	$+8.7\times10^{-5}\,M$
Equil :	0.366 M	–	$(S+8.7\times10^{-5})M$	$8.7\times10^{-5}\,M$

$$K_a=\frac{\left[H_3O^+\right]\left[CHO_2^-\right]}{\left[HCHO_2\right]}=1.8\times10^{-4}=\frac{(S+8.7\times10^{-5})8.7\times10^{-5}}{0.366}\approx\frac{8.7\times10^{-5}S}{0.366}\ ;\ S=0.76\ M$$

To determine S, we assumed $S\gg8.7\times10^{-5}\,M$, which is clearly a valid assumption.
Or, we could have used the Henderson-Hasselbalch equation (see below).

$pK_a=-\log(1.8\times10^{-4})=3.74$

$$4.06=3.74+\log\frac{\left[CHO_2^-\right]}{\left[HCHO_2\right]};\qquad\frac{\left[CHO_2^-\right]}{\left[HCHO_2\right]}=2.1;\qquad\left[CHO_2^-\right]=2.1\times0.366=0.77\,M$$

The difference in the two answers is due simply to rounding.

8. We use the Henderson-Hasselbalch equation to find the required [NH₃].

$pK_b=-\log(1.8\times10^{-5})=4.74$

$$pK_a=14.00-pK_b=14.00-4.74=9.26\qquad\qquad pH=9.12=9.26+\log\frac{\left[NH_3\right]}{\left[NH_4^+\right]}$$

$$\frac{\left[NH_3\right]}{\left[NH_4^+\right]}=10^{-0.14}=0.72\qquad\qquad\left[NH_3\right]=0.72\times\left[NH_4^+\right]=0.72\times0.732\,M=0.53\,M$$

9. **(a)** Equation: $HC_7H_5O_2\,(aq)\ +\ H_2O(l)\ \rightleftharpoons\ C_7H_5O_2^-(aq)\ +\ H_3O^+(aq)$

Initial:	0.012 M	–	0.033 M	$\approx 0\,M$
Changes:	$-x\,M$	–	$+x\,M$	$+x\,M$
Equil:	$(0.012-x)M$	–	$(0.033+x)M$	x M

$$K_a=\frac{\left[H_3O^+\right]\left[C_7H_5O_2^-\right]}{\left[HC_7H_5O_2\right]}=6.3\times10^{-5}=\frac{x(0.033+x)}{0.012-x}\approx\frac{0.033x}{0.012}\qquad x=2.3\times10^{-5}\,M$$

To determine the value of x, we assumed $x\ll0.012$ M, which is an assumption that clearly is correct. $\left[H_3O^+\right]=2.3\times10^{-5}\,M$ $\qquad\qquad pH=-\log(2.3\times10^{-5})=4.64$

(b) Equation: $NH_3(aq) + H_2O(l) \rightleftharpoons NH_4^+(aq) + OH^-(aq)$

Initial :   0.408 M   –   0.153 M ≈ 0 M

Changes : $-x$ M   – $+x$ M $+x$ M

Equil : $(0.408-x)$ M   – $(0.153+x)$ M x M

$$K_b = \frac{[NH_4^+][OH^-]}{[NH_3]} = 1.8 \times 10^{-5} = \frac{x(0.153+x)}{0.408-x} \approx \frac{0.153x}{0.408} \qquad x = 4.8 \times 10^{-5} M$$

To determine the value of x, we assumed $x \ll 0.153$, which clearly is a valid assumption.

$[OH^-] = 4.8 \times 10^{-5} M;$ $pOH = -\log(4.8 \times 10^{-5}) = 4.32;$ $pH = 14.00 - 4.32 = 9.68$

10.   Since the mixture is a buffer, we can use the Henderson-Hasselbalch equation to determine K_a of lactic acid.

$$[C_3H_5O_3^-] = \frac{1.00 \text{ g NaC}_3H_5O_3}{100.0 \text{ mL soln}} \times \frac{1000 \text{ mL}}{1 \text{ L soln}} \times \frac{1 \text{ mol NaC}_3H_5O_3}{112.1 \text{ g NaC}_3H_5O_3} \times \frac{1 \text{ mol C}_3H_5O_3^-}{1 \text{ mol NaC}_3H_5O_3} = 0.0892 \text{ M}$$

$$pH = 4.11 = pK_a + \log\frac{[C_3H_5O_3^-]}{[HC_3H_5O_3]} = pK_a + \log\frac{0.0892 \text{ M}}{0.0500 \text{ M}} = pK_a + 0.251$$

$pK_a = 4.11 - 0.251 = 3.86;$ $K_a = 10^{-3.86} = 1.4 \times 10^{-4}$

11.   **(a)**   0.100 M NaCl is not a buffer solution. Neither ion reacts with water to a detectable extent.

(b)   0.100 M NaCl—0.100 M NH_4Cl is not a buffer solution. Although a weak acid, NH_4^+, is present, its conjugate base, NH_3, is not.

(c)   0.100 M CH_3NH_2 and 0.150 M $CH_3NH_3^+Cl^-$ is a buffer solution. Both the weak base, CH_3NH_2, and its conjugate acid, $CH_3NH_3^+$, are present in approximately equal concentrations.

(d)   0.100 M HCl—0.050 M $NaNO_2$ is not a buffer solution. All the NO_2^- has converted to HNO_2 and thus the solution is a mixture of a strong acid and a weak acid.

(e)   0.100 M HCl—0.200 M $NaC_2H_3O_2$ is a buffer solution. All of the HCl reacts with half of the $C_2H_3O_2^-$ to form a solution with 0.100 M $HC_2H_3O_2$, a weak acid, and 0.100 M $C_2H_3O_2^-$, its conjugate base.

(f)   0.100 M $HC_2H_3O_2$ and 0.125 M $NaC_3H_5O_2$ is not a buffer in the strict sense because it does not contain a weak acid and its conjugate base, but rather the conjugate base of another weak acid. These two weak acids (acetic, $K_a = 1.8 \times 10^{-5}$ and propionic, $K_a = 1.35 \times 10^{-5}$) have approximately the same strength, however, this solution would resist changes in its pH on the addition of strong acid or strong base, consequently, it could be argued that this system should also be called a buffer.

12. **(a)** Reaction with added acid: $HPO_4^{2-} + H_3O^+ \rightarrow H_2PO_4^- + H_2O$

Reaction with added base: $H_2PO_4^- + OH^- \longrightarrow HPO_4^{2-} + H_2O$

(b) We assume initially that the buffer has equal concentrations of the two ions,

$\left[H_2PO_4^-\right] = \left[HPO_4^{2-}\right]$

$pH = pK_{a_2} + \log\dfrac{\left[HPO_4^{2-}\right]}{\left[H_2PO_4^-\right]} = 7.20 + 0.00 = 7.20$ (pH at which the buffer is most effective).

(c) $pH = 7.20 + \log\dfrac{\left[HPO_4^{2-}\right]}{\left[H_2PO_4^-\right]} = 7.20 + \log\dfrac{0.150\ M}{0.050\ M} = 7.20 + 0.48 = 7.68$

13. moles of solute $= 1.15\ \text{mg} \times \dfrac{1\ g}{1000\ mg} \times \dfrac{1\ mol\ C_6H_5NH_3^+Cl^-}{129.6\ g} \times \dfrac{1\ mol\ C_6H_5NH_3^+}{1\ mol\ C_6H_5NH_3^+Cl^-}$

$= 8.87 \times 10^{-6}\ mol\ C_6H_5NH_3^+$

$\left[C_6H_5NH_3^+\right] = \dfrac{8.87 \times 10^{-6}\ mol\ C_6H_5NH_3^+}{3.18\ L\ soln} = 2.79 \times 10^{-6}\ M$

Equation:	$C_6H_5NH_2\,(aq)$	$+\ H_2O(l)$	$\rightleftharpoons$	$C_6H_5NH_3^+\,(aq)$	$+\ OH^-\,(aq)$
Initial:	0.105 M	–		2.79×10^{-6} M	≈ 0 M
Changes:	$-x$ M	–		$+x$ M	$+x$ M
Equil:	$(0.105 - x)$M	–		$(2.79 \times 10^{-6} + x)$M	x M

$K_b = \dfrac{\left[C_6H_5NH_3^+\right]\left[OH^-\right]}{\left[C_6H_5NH_2\right]} = 7.4 \times 10^{-10} = \dfrac{(2.79 \times 10^{-6} + x)x}{0.105 - x}$

$7.4 \times 10^{-10}(0.105 - x) = (2.79 \times 10^{-6} + x)x; \quad 7.8 \times 10^{-11} - 7.4 \times 10^{-10}x = 2.79 \times 10^{-6}x + x^2$

$x^2 + (2.79 \times 10^{-6} + 7.4 \times 10^{-10})x - 7.8 \times 10^{-11} = 0; \quad x^2 + 2.79 \times 10^{-6}x - 7.8 \times 10^{-11} = 0$

$x = \dfrac{-b \pm \sqrt{b^2 - 4ac}}{2a} = \dfrac{-2.79 \times 10^{-6} \pm \sqrt{7.78 \times 10^{-12} + 3.1 \times 10^{-10}}}{2} = 7.5 \times 10^{-6}\ M = \left[OH^-\right]$

$pOH = -\log(7.5 \times 10^{-6}) = 5.12 \qquad pH = 14.00 - 5.12 = 8.88$

14. We determine the concentration of the cation of the weak base.

$[C_6H_5NH_3^+] = \dfrac{8.50\ g \times \dfrac{1\ mmol\ C_6H_5NH_3^+Cl^-}{129.6\ g} \times \dfrac{1\ mmol\ C_6H_5NH_3^+}{1\ mmol\ C_6H_5NH_3^+Cl^-}}{750\ mL \times \dfrac{1\ L}{1000\ mL}} = 0.0874\ M$

In order to be an effective buffer, each concentration must exceed the ionization constant $\left(K_b = 7.4 \times 10^{-10}\right)$ by a factor of at least 100, which clearly is true. Also, the ratio of the

two concentrations must fall between 0.1 and 10: $\dfrac{\left[C_6H_5NH_3^+\right]}{\left[C_6H_5NH_2\right]} = \dfrac{0.0874M}{0.215M} = 0.407$.

Since both criteria are met, this solution will be an effective buffer.

15. **(a)** First use the Henderson-Hasselbalch equation. $pK_b = -\log\left(1.8 \times 10^{-5}\right) = 4.74$,

$pK_a = 14.00 - 4.74 = 9.26$ to determine $\left[NH_4^+\right]$ in the buffer solution.

$pH = 9.45 = pK_a + \log\dfrac{\left[NH_3\right]}{\left[NH_4^+\right]} = 9.26 + \log\dfrac{\left[NH_3\right]}{\left[NH_4^+\right]}; \quad \log\dfrac{\left[NH_3\right]}{\left[NH_4^+\right]} = 9.45 - 9.26 = +0.19$

$\dfrac{\left[NH_3\right]}{\left[NH_4^+\right]} = 10^{0.19} = 1.5\underline{5} \quad \left[NH_4^+\right] = \dfrac{\left[NH_3\right]}{1.5\underline{5}} = \dfrac{0.258M}{1.5\underline{5}} = 0.17M$

We now assume that the volume of the solution does not change significantly when the solid is added.

$\text{mass}\left(NH_4\right)_2SO_4 = 425 \text{ mL} \times \dfrac{1 \text{ L soln}}{1000 \text{ mL}} \times \dfrac{0.17 \text{ mol } NH_4^+}{1 \text{ L soln}} \times \dfrac{1 \text{ mol }\left(NH_4\right)_2SO_4}{2 \text{ mol } NH_4^+}$

$\times \dfrac{132.1 \text{ g}\left(NH_4\right)_2SO_4}{1 \text{ mol}\left(NH_4\right)_2SO_4} = 4.8 \text{ g }\left(NH_4\right)_2SO_4$

(b) We can use the Henderson-Hasselbalch equation to determine the ratio of concentrations of cation and weak base in the altered solution.

$pH = 9.30 = pK_a + \log\dfrac{\left[NH_3\right]}{\left[NH_4^+\right]} = 9.26 + \log\dfrac{\left[NH_3\right]}{\left[NH_4^+\right]} \qquad \log\dfrac{\left[NH_3\right]}{\left[NH_4^+\right]} = 9.30 - 9.26 = +0.04$

$\dfrac{\left[NH_3\right]}{\left[NH_4^+\right]} = 10^{0.04} = 1.1 = \dfrac{0.258}{0.17 \text{ M} + x \text{ M}} \qquad 0.19 + 1.1x = 0.258 \quad x = 0.062\,M$

The reason we decided to add x to the numerator follows. (Notice we cannot remove a component.) A pH of 9.30 is more acidic than a pH of 9.45 and therefore the conjugate acid's $\left(NH_4^+\right)$ concentration must increase. Additionally, mathematics tells us that for the concentration ratio to decrease from $1.5\underline{5}$ to 1.1, its denominator must increase. We solve this expression for x to find a value of 0.062 M. We need to add NH_4^+ to increase its concentration by 0.062 M in 100 mL of solution.

$\left(NH_4\right)_2SO_4 \text{ mass} = 0.100 L \times \dfrac{0.062 \text{ mol } NH_4^+}{1 \text{ L}} \times \dfrac{1 \text{ mol}\left(NH_4\right)_2SO_4}{2 \text{ mol } NH_4^+} \times \dfrac{132.1 \text{ g }\left(NH_4\right)_2SO_4}{1 \text{ mol }\left(NH_4\right)_2SO_4}$

$= 0.41 \text{ g }\left(NH_4\right)_2SO_4 \qquad \text{Hence, we need to add} \approx 0.4 \text{ g}$

16. (a) $n_{HC_7H_5O_2} = 2.00 \text{ g } HC_7H_5O_2 \times \dfrac{1 \text{ mol } HC_7H_5O_2}{122.1 \text{ g } HC_7H_5O_2} = 0.0164 \text{ mol } HC_7H_5O_2$

$n_{C_7H_5O_2^-} = 2.00 \text{ g } NaC_7H_5O_2 \times \dfrac{1 \text{ mol } NaC_7H_5O_2}{144.1 \text{ g } NaC_7H_5O_2} \times \dfrac{1 \text{ mol } C_7H_5O_2^-}{1 \text{ mol } NaC_7H_5O_2}$

$= 0.0139 \text{ mol } C_7H_5O_2^-$

$pH = pK_a + \log\dfrac{\left[C_7H_5O_2^-\right]}{\left[HC_7H_5O_2\right]} = -\log\left(6.3 \times 10^{-5}\right) + \log\dfrac{0.0139 \text{ mol } C_7H_5O_2^-/0.7500 \text{ L}}{0.0164 \text{ mol } HC_7H_5O_2/0.7500 \text{ L}}$

$= 4.20 - 0.0718 = 4.13$

(b) To lower the pH of this buffer solution, that is, to make it more acidic, benzoic acid must be added. The quantity is determined as follows. We use moles rather than concentrations because all components are present in the same volume of solution.

$4.00 = 4.20 + \log\dfrac{0.0139 \text{ mol } C_7H_5O_2^-}{x \text{ mol } HC_7H_5O_2} \qquad \log\dfrac{0.0139 \text{ mol } C_7H_5O_2^-}{x \text{ mol } HC_7H_5O_2} = -0.20$

$\dfrac{0.0139 \text{ mol } C_7H_5O_2^-}{x \text{ mol } HC_7H_5O_2} = 10^{-0.20} = 0.63 \qquad x = \dfrac{0.0139}{0.63} = 0.022 \text{ mol } HC_7H_5O_2 \text{ (required)}$

$HC_7H_5O_2$ that must be added = amount required – amount already in solution

$HC_7H_5O_2$ that must be added = $0.022 \text{ mol } HC_7H_5O_2 - 0.0164 \text{ mol } HC_7H_5O_2$

$HC_7H_5O_2$ that must be added = $0.006 \text{ mol } HC_7H_5O_2$

added mass $HC_7H_5O_2 = 0.006 \text{ mol } HC_7H_5O_2 \times \dfrac{122.1 \text{ g } HC_7H_5O_2}{1 \text{ mol } HC_7H_5O_2} = 0.7 \text{ g } HC_7H_5O_2$

17. The added HCl will react with the ammonia, and the pH of the buffer solution will decrease. The original buffer solution has $\left[NH_3\right] = 0.258 \text{ M}$ and $\left[NH_4^+\right] = 0.17 \text{ M}$.

We first calculate the [HCl] in solution, reduced from 12 M because of dilution. [HCl] added $= 12 \text{ M} \times \dfrac{0.55 \text{ mL}}{100.6 \text{ mL}} = 0.066 \text{ M}$ We then determine pK_a for ammonium ion:

$pK_b = -\log\left(1.8 \times 10^{-5}\right) = 4.74 \quad pK_a = 14.00 - 4.74 = 9.26$

Equation:	$NH_3(aq) +$	$H_3O^+(aq) \rightleftharpoons$	$NH_4^+(aq)$	$+ H_2O(l)$
Buffer:	0.258 M	≈ 0 M	0.17 M	–
Added:		+0.066 M		
Changes:	–0.066 M	–0.066 M	+0.066 M	–
Final:	0.192 M	0 M	0.24 M	–

$pH = pK_a + \log\dfrac{\left[NH_3\right]}{\left[NH_4^+\right]} = 9.26 + \log\dfrac{0.192}{0.24} = 9.16$

18. The added NH_3 will react with the benzoic acid, and the pH of the buffer solution will increase. Original buffer solution has $[C_7H_5O_2^-] = 0.0139$ mol $C_7H_5O_2^-/0.750$ L $= 0.0185$ M and $[HC_7H_5O_2] = 0.0164$ mol $HC_7H_5O_2/0.7500$ L $= 0.0219$ M. We first calculate the $[NH_3]$ in solution, reduced from 15 M because of dilution.

$$[NH_3] \text{ added} = 15 \text{ M} \times \frac{0.35 \text{ mL}}{750.4 \text{ mL}} = 0.0070 \text{ M} \quad \text{For benzoic acid, } pK_a = -\log(6.3 \times 10^{-5}) = 4.20$$

Equation: $\quad NH_3(aq) \quad + \quad HC_7H_5O_2(aq) \rightleftharpoons NH_4^+(aq) \quad + \quad C_7H_5O_2^-(aq)$

Buffer:	0 M	0.0219 M	0 M	0.0185 M
Added:	0.0070 M			
Changes:	−0.0070M	−0.0070 M	+0.0070 M	+0.0070 M
Final:	0.000 M	0.0149 M	0.0070 M	0.0255 M

$$pH = pK_a + \log\frac{[C_7H_5O_2^-]}{[HC_7H_5O_2]} = 4.20 + \log\frac{0.0255}{0.0149} = 4.43$$

19. The pK_a's of the acids help us choose the one to be used in the buffer. It is the acid with a pK_a within 1.00 pH unit of 3.50 that will do the trick. $pK_a = 3.74$ for $HCHO_2$, $pK_a = 4.74$ for $HC_2H_3O_2$, and $pK_1 = 2.15$ for H_3PO_4. Thus, we choose $HCHO_2$ and $NaCHO_2$ to prepare a buffer with pH = 3.50. The Henderson-Hasselbalch equation is used to determine the relative amounts of each component present in the buffer solution.

$$pH = 3.50 = 3.74 + \log\frac{[CHO_2^-]}{[HCHO_2]} \qquad \log\frac{[CHO_2^-]}{[HCHO_2]} = 3.50 - 3.74 = -0.24$$

$$\frac{[CHO_2^-]}{[HCHO_2]} = 10^{-0.24} = 0.58$$

This ratio of concentrations is also the ratio of the number of moles of each component in the buffer solution, since both concentrations are a number of moles in a certain volume, and the volumes are the same (the two solutes are in the same solution). This ratio also is the ratio of the volumes of the two solutions, since both solutions being mixed contain the same concentration of solute. If we assume 100. mL of acid solution, $V_{acid} = 100.$ mL. Then the volume of salt solution is $V_{salt} = 0.58 \times 100.$ mL $= 58$ mL 0.100 M $NaCHO_2$

20. We can lower the pH of the 0.250 M $HC_2H_3O_2$ — 0.560 M $C_2H_3O_2^-$ buffer solution by increasing $[HC_2H_3O_2]$ or lowering $[C_2H_3O_2^-]$. Small volumes of NaCl solutions will have no effect, and the addition of NaOH(aq) or $NaC_2H_3O_2(aq)$ will raise the pH. The addition of 0.150 M HCl will raise $[HC_2H_3O_2]$ and lower $[C_2H_3O_2^-]$ through the reaction $H_3O^+(aq) + C_2H_3O_2^-(aq) \rightleftharpoons HC_2H_3O_2(aq) + H_2O(l)$ and bring about the desired

lowering of the pH. We first use the Henderson-Hasselbalch equation to determine the
ratio of the concentration of acetate ion and acetic acid. $pH = 5.00 = 4.74 + \log \dfrac{\left[C_2H_3O_2^-\right]}{\left[HC_2H_3O_2\right]}$

$$\log \frac{\left[C_2H_3O_2^-\right]}{\left[HC_2H_3O_2\right]} = 5.00 - 4.74 = 0.26; \quad \frac{\left[C_2H_3O_2^-\right]}{\left[HC_2H_3O_2\right]} = 10^{0.26} = 1.8$$

Now we compute the amount of each component in the original buffer solution.

$$\text{amount of } C_2H_3O_2^- = 300.\ \text{mL} \times \frac{0.560\ \text{mmol } C_2H_3O_2^-}{1\ \text{mL soln}} = 168\ \text{mmol } C_2H_3O_2^-$$

$$\text{amount of } HC_2H_3O_2 = 300.\ \text{mL} \times \frac{0.250\ \text{mmol } HC_2H_3O_2}{1\ \text{mL soln}} = 75.0\ \text{mmol } HC_2H_3O_2$$

Now let x represent the amount of H_3O^+ added in mmol.

$$1.8 = \frac{168 - x}{75.0 + x}; \quad 168 - x = 1.8(75 + x) = 13_5 + 1.8x \quad 168 - 13_5 = 2.8x$$

$$x = \frac{168 - 13_5}{2.7} = 12\ \text{mmol } H_3O^+$$

$$\text{Volume of } 0.150\ \text{M HCl} = 12\ \text{mmol } H_3O^+ \times \frac{1\ \text{mmol HCl}}{1\ \text{mmol } H_3O^+} \times \frac{1\ \text{mL soln}}{0.150\ \text{mmol HCl}}$$

$$= 80\ \text{mL } 0.150\ \text{M HCl solution}$$

21. **(a)** The pH of the buffer is determined via the Henderson-Hasselbalch equation.

$$pH = pK_a + \log \frac{\left[C_3H_5O_2^-\right]}{\left[HC_3H_5O_2\right]} = 4.89 + \log \frac{0.100M}{0.100M} = 4.89$$

The effective pH range is the same for every propionate buffer: from $pH = 3.89$ to $pH = 5.89$, one pH unit on either side of pK_a for propionic acid, which is 4.89.

(b) To each liter of 0.100 M $HC_3H_5O_2$ — 0.100M $NaC_3H_5O_2$ we can add 0.100 mol OH^- before all of the $HC_3H_5O_2$ is consumed, and we can add 0.100 mol H_3O^+ before all of the $C_3H_5O_2^-$ is consumed. The buffer capacity thus is 100. millimoles (0.100 mol) of acid or base per liter of buffer solution.

22. (a) The solution will be an effective buffer one pH unit on either sides of the pK_a of methylammonium ion, $CH_3NH_3^+$, $K_b = 4.2 \times 10^{-4}$ for methylamine, $pK_b = -\log(4.2 \times 10^{-4}) = 3.38$. For methylammonium cation, $pK_a = 14.00 - 3.38 = 10.62$. Thus, this buffer will be effective from a pH of 9.62 to a pH of 11.62.

(b) The capacity of the buffer is reached when all of the weak base or all of the conjugate acid has been neutralized by the added strong acid or strong base. Because their concentrations are the same, the number of moles of base are equal to the number of moles of conjugate acid in the same volume of solution.

$$\text{amount of weak base} = 125 \text{ mL} \times \frac{0.0500 \text{ mmol}}{1 \text{ mL}} = 6.25 \text{ mmol CH}_3\text{NH}_2 \text{ or CH}_3\text{NH}_3^+$$

Thus, the buffer capacity is 6.25 millimoles of acid or base per 125 mL buffer solution.

23. (a) The pH of this buffer solution is determined with the Henderson-Hasselbalch equation.

$$\text{pH} = \text{p}K_a + \log \frac{\left[\text{CHO}_2^-\right]}{\left[\text{HCHO}_2\right]} = -\log\left(1.8 \times 10^{-4}\right) + \log \frac{8.5 \text{ mmol}/75.0 \text{ mL}}{15.5 \text{ mmol}/75.0 \text{ mL}}$$

$$= 3.74 - 0.26 = 3.48$$

[Note: the solution is not a good buffer, as $\left[\text{CHO}_2^-\right] = 1.1 \times 10^{-1}$, which is only ~ 600 times K_a]

(b) Amount of added $\text{OH}^- = 0.25 \text{ mmol Ba(OH)}_2 \times \dfrac{2 \text{ mmol OH}^-}{1 \text{ mmol Ba(OH)}_2} = 0.50 \text{ mmol OH}^-$

The OH^- added reacts with the formic acid and produces formate ion.

Equation:	$\text{HCHO}_2\,(\text{aq})$	+	$\text{OH}^-\,(\text{aq})$	$\rightleftharpoons$	$\text{CHO}_2^-\,(\text{aq})$	+	$\text{H}_2\text{O(l)}$
Buffer:	15.5 mmol		$\approx 0 \text{ M}$		8.5 mmol		–
Add base:			+0.50 mmol				
React:	–0.50 mmol		–0.50 mmol		+0.50 mmol		–
Final:	15.0 mmol		0 mmol		9.0 mmol		–

$$\text{pH} = \text{p}K_a + \log \frac{\left[\text{CHO}_2^-\right]}{\left[\text{HCHO}_2\right]} = -\log\left(1.8 \times 10^{-4}\right) + \log \frac{9.0 \text{ mmol}/75.0 \text{ mL}}{15.0 \text{ mmol}/75.0 \text{ mL}}$$

$$= 3.74 - 0.22 = 3.52$$

(c) Amount of added $\text{H}_3\text{O}^+ = 1.05 \text{ mL acid} \times \dfrac{12 \text{ mmol HCl}}{1 \text{ mL acid}} \times \dfrac{1 \text{ mmol H}_3\text{O}^+}{1 \text{ mmol HCl}} = 13 \text{ mmol H}_3\text{O}^+$

The H_3O^+ added reacts with the formate ion and produces formic acid.

Equation:	$\text{CHO}_2^-\,(\text{aq})$	+	$\text{H}_3\text{O}^+\,(\text{aq})$	$\rightleftharpoons$	$\text{HCHO}_2\,(\text{aq})$	+	$\text{H}_2\text{O(l)}$
Buffer :	8.5 mmol		≈ 0 mmol		15.5 mmol		–
Add acid :			+13 mmol				
React :	– 8.5 mmol		– 8.5 mmol		+8.5 mmol		–
Final :	0 mmol		4.5 mmol		24.0 mmol		–

The buffer's capacity has been exceeded. The pH of the solution is determined by the excess strong acid present.

$$\left[\text{H}_3\text{O}^+\right] = \frac{4.5 \text{ mmol}}{75.0 \text{ mL} + 1.05 \text{ mL}} = 0.059 \text{ M}; \quad \text{pH} = -\log\left(0.059\right) = 1.23$$

24. For $NH_3, pK_b = -\log(1.8 \times 10^{-5})$ For $NH_4^+, pK_a = 14.00 - pK_b = 14.00 - 4.74 = 9.26$

(a) $\left[NH_3\right] = \dfrac{1.68 \text{g } NH_3}{0.500 \text{ L}} \times \dfrac{1 \text{ mol } NH_3}{17.03 \text{ g } NH_3} = 0.197 \text{ M}$

$\left[NH_4^+\right] = \dfrac{4.05 \text{ g } (NH_4)_2 SO_4}{0.500 \text{ L}} \times \dfrac{1 \text{ mol}(NH_4)_2 SO_4}{132.1 \text{ g } (NH_4)_2 SO_4} \times \dfrac{2 \text{ mol } NH_4^+}{1 \text{ mol } (NH_4)_2 SO_4} = 0.123 \text{ M}$

$pH = pK_a + \log \dfrac{\left[NH_3\right]}{\left[NH_4^+\right]} = 9.26 + \log \dfrac{0.197 \text{ M}}{0.123 \text{ M}} = 9.46$

(b) The $OH^-(aq)$ reacts with the $NH_4^+(aq)$ to produce an equivalent amount of $NH_3(aq)$.

$\left[OH^-\right]_i = \dfrac{0.88 \text{ g NaOH}}{0.500 \text{ L}} \times \dfrac{1 \text{mol NaOH}}{40.00 \text{ g NaOH}} \times \dfrac{1 \text{mol } OH^-}{1 \text{mol NaOH}} = 0.044 \text{ M}$

Equation:	$NH_4^+ (aq) +$	$OH^- (aq)$	$\rightleftharpoons$	$NH_3 (aq) +$	$H_2O(l)$
Initial :	0.123 M	≈ 0 M		0.197 M	–
Add NaOH :		+0.044 M			
React :	–0.044 M	–0.044 M		+0.044 M	–
Final :	0.079 M	0.0000 M		0.241 M	–

$pH = pK_a + \log \dfrac{\left[NH_3\right]}{\left[NH_4^+\right]} = 9.26 + \log \dfrac{0.241 \text{ M}}{0.079 \text{ M}} = 9.74$

(c)

Equation:	$NH_3 (aq) +$	$H_3O^+ (aq)$	$\rightleftharpoons$	$NH_4^+ (aq) +$	$H_2O(l)$
Initial :	0.197 M	≈ 0 M		0.123 M	–
Add HCl :		$+x$ M			
React :	$-x$ M	$-x$ M		$+x$ M	–
Final :	$(0.197 - x)$ M	1×10^{-9} M		$(0.123 + x)$ M	–

$pH = 9.00 = pK_a + \log \dfrac{\left[NH_3\right]}{\left[NH_4^+\right]} = 9.26 + \log \dfrac{(0.197 - x) \text{ M}}{(0.123 + x) \text{ M}}$

$\log \dfrac{(0.197 - x) \text{ M}}{(0.123 + x) \text{ M}} = 9.00 - 9.26 = -0.26 \qquad \dfrac{(0.197 - x) \text{ M}}{(0.123 + x) \text{ M}} = 10^{-0.26} = 0.55$

$0.197 - x = 0.55(0.123 + x) = 0.068 + 0.55x \qquad 1.55x = 0.197 - 0.068 = 0.129$

$x = \dfrac{0.129}{1.55} = 0.0832 \text{ M}$

$\text{volume HCl} = 0.500 \text{L} \times \dfrac{0.0832 \text{ mol } H_3O^+}{1 \text{L soln}} \times \dfrac{1 \text{mol HCl}}{1 \text{mol } H_3O^+} \times \dfrac{1000 \text{ mL HCl}}{12 \text{ mol HCl}} = 3.5 \text{ mL}$

25. **(a)** We use the Henderson-Hasselbalch equation to determine the pH of the solution. The total solution volume is

$$36.00\,mL + 64.00\,mL = 100.00\,mL. \quad pK_a = 14.00 - pK_b = 14.00 + \log\left(1.8 \times 10^{-5}\right) = 9.26$$

$$[NH_3] = \frac{36.00\,mL \times 0.200\,M\,NH_3}{100.00\,mL} = \frac{7.20\,mmol\,NH_3}{100.0\,mL} = 0.0720\,M$$

$$\left[NH_4^+\right] = \frac{64.00\,mL \times 0.200\,M\,NH_4^+}{100.00\,mL} = \frac{12.8\,mmol\,NH_4^+}{100.0\,mL} = 0.128\,M$$

$$pH = pK_a + \log\frac{[NH_3]}{[NH_4^+]} = 9.26 + \log\frac{0.0720\,M}{0.128\,M} = 9.01 \approx 9.00$$

(b) The solution has $\left[OH^-\right] = 10^{-4.99} = 1.0 \times 10^{-5}\,M$

The Henderson-Hasselbalch equation depends on the assumption that:

$$[NH_3] \gg 1.8 \times 10^{-5}\,M \ll \left[NH_4^+\right]$$

If the solution is diluted to 1.00 L, $[NH_3] = 7.20 \times 10^{-3}\,M$, and

$\left[NH_4^+\right] = 1.28 \times 10^{-2}\,M$. These concentrations are consistent with the assumption.

However, if the solution is diluted to 1000. L, $[NH_3] = 7.2 \times 10^{-6}\,M$, and $[NH_3] = 1.28 \times 10^{-5}\,M$, and these two concentrations are not consistent with the assumption. Thus, in 1000. L of solution, the given quantities of NH_3 and NH_4^+ will not produce a solution with pH = 9.00. With sufficient dilution, the solution will become indistinguishable from pure water (i.e. its pH will equal 7.00).

(c) The 0.20 mL of added 1.00 M HCl does not significantly affect the volume of the solution, but it does add $0.20\,mL \times 1.00\,M\,HCl = 0.20\,mmol\,H_3O^+$. This added H_3O^+ reacts with NH_3, decreasing its amount from 7.20 mmol NH_3 to 7.00 mmol NH_3, and increasing the amount of NH_4^+ from 12.8 mmol NH_4^+ to 13.0 mmol NH_4^+, as the reaction: $NH_3 + H_3O^+ \rightarrow NH_4^+ + H_2O$

$$pH = 9.26 + \log\frac{7.00\,mmol\,NH_3 / 100.20\,mL}{13.0\,mmol\,NH_4^+ / 100.20\,mL} = 8.99$$

(d) We see in the calculation of part (c) that the total volume of the solution does not affect the pOH of the solution, at least as long as the Henderson-Hasselbalch equation is obeyed. We let x represent the number of millimoles of H_3O^+ added, through 1.00 M HCl. This increases the amount of NH_4^+ and decreases the amount of NH_3, through the reaction $NH_3 + H_3O^+ \rightarrow NH_4^+ + H_2O$

$$\text{pH} = 8.90 = 9.26 + \log\frac{7.20 - x}{12.8 + x}; \quad \log\frac{7.20 - x}{12.8 + x} = 8.90 - 9.26 = -0.36$$

Inverting, we have:

$$\frac{12.8 + x}{7.20 - x} = 10^{0.36} = 2.29; \quad 12.8 + x = 2.29(7.20 - x) = 16.5 - 2.29x$$

$$x = \frac{16.5 - 12.8}{1.00 + 2.29} = 1.1 \text{ mmol } H_3O^+$$

$$\text{vol } 1.00\,M \text{ HCl} = 1.1\,\text{mmol } H_3O^+ \times \frac{1\,\text{mmol HCl}}{1\,\text{mmol } H_3O^+} \times \frac{1\,\text{mL soln}}{1.00\,\text{mmol HCl}} = 1.1\,\text{mL } 1.00\,M \text{ HCl}$$

26. (a)
$$\left[C_2H_3O_2^-\right] = \frac{12.0\,\text{g NaC}_2H_3O_2}{0.300\,\text{L soln}} \times \frac{1\,\text{mol NaC}_2H_3O_2}{82.03\,\text{g NaC}_2H_3O_2} \times \frac{1\,\text{mol C}_2H_3O_2^-}{1\,\text{mol NaC}_2H_3O_2} = 0.488\,M \text{ C}_2H_3O_2^-$$

Equation :	$HC_2H_3O_2$	$+ H_2O \rightleftharpoons$	$C_2H_3O_2^-$	$+$	H_3O^+
Initial :	0 M	–	0 M		0.200 M
Add $C_2H_3O_2^-$	0 M	–	+0.488 M		0 M
Consume H_3O^+	+ 0.200 M	–	−0.200 M		−0.200 M
Buffer :	0.200 M	–	0.288 M		$\approx 0\,M$

Then use the Henderson-Hasselbalch equation to find the pH.

$$\text{pH} = pK_a + \log\frac{\left[C_2H_3O_2^-\right]}{\left[HC_2H_3O_2\right]} = 4.74 + \log\frac{0.288\,M}{0.200\,M} = 4.74 + 0.16 = 4.90$$

(b) We first calculate the initial $\left[OH^-\right]$ due to the added $Ba(OH)_2$.

$$\left[OH^-\right] = \frac{1.00\,\text{g Ba(OH)}_2}{0.300\,\text{L}} \times \frac{1\,\text{mol Ba(OH)}_2}{171.3\,\text{g Ba(OH)}_2} \times \frac{2\,\text{mol OH}^-}{1\,\text{mol Ba(OH)}_2} = 0.0389\,M$$

Then $HC_2H_3O_2$ is consumed in the neutralization reaction shown directly below.

Equation:	$HC_2H_3O_2$	$+$ OH^-	$\rightleftharpoons$ $C_2H_3O_2^-$	$+$ H_2O
Initial:	0.200 M	0.0389 M	0.288 M	—
Consume OH^-:	-0.0389 M	-0.0389 M	+0.0389 M	—
Buffer:	0.161 M	~ 0M	0.327 M	—

Then use the Henderson-Hasselbalch equation.

$$\text{pH} = pK_a + \log\frac{\left[C_2H_3O_2^-\right]}{\left[HC_2H_3O_2\right]} = 4.74 + \log\frac{0.327\,M}{0.161\,M} = 4.74 + 0.31 = 5.05$$

(c) $Ba(OH)_2$ can be added up until all of the $HC_2H_3O_2$ is consumed.

$$Ba(OH)_2 + 2HC_2H_3O_2 \longrightarrow Ba(C_2H_3O_2)_2 + 2H_2O$$

$$\text{moles of Ba}(OH)_2 = 0.300\,L \times \frac{0.200\,\text{mol}\,HC_2H_3O_2}{1\,L\,\text{soln}} \times \frac{1\,\text{mol}\,Ba(OH)_2}{2\,\text{mol}\,HC_2H_3O_2} = 0.0300\,\text{mol}\,Ba(OH)_2$$

$$\text{mass of Ba}(OH)_2 = 0.0300\,\text{mol}\,Ba(OH)_2 \times \frac{171.3\,g\,Ba(OH)_2}{1\,\text{mol}\,Ba(OH)_2} = 5.14\,g\,Ba(OH)_2$$

(d)   0.36 g $Ba(OH)_2$ is too much for the buffer to handle and it is the excess of OH^- originating from the $Ba(OH)_2$ that determines the pOH of the solution.

$$\left[OH^-\right] = \frac{0.36\,g\,Ba(OH)_2}{0.300\,L\,\text{soln}} \times \frac{1\,\text{mol}\,Ba(OH)_2}{171.3\,g\,Ba(OH)_2} \times \frac{2\,\text{mol}\,OH^-}{1\,\text{mol}\,Ba(OH)_2} = 1.4 \times 10^{-2}\,M\,OH^-$$

$$pOH = -\log(1.4 \times 10^{-2}) = 1.85 \qquad pH = 14.00 - 1.85 = 12.15$$

Acid-Base Indicators

27. **(a)**   The pH color change range is 1.00 pH unit on either side of pK_{HIn}. If the pH color change range is below $pH = 7.00$, the indicator changes color in acidic solution. If it is above $pH = 7.00$, the indicator changes color in alkaline solution. If $pH = 7.00$ falls within the pH color change range, the indicator changes color near the neutral point.

indicator	K_{HIn}	pK_{HIn}	pH color change range	changes color in?
bromophenol blue	1.4×10^{-4}	3.85	2.9 (yellow) to 4.9 (blue)	acidic solution
bromocresol green	2.1×10^{-5}	4.68	3.7 (yellow) to 5.7 (blue)	acidic solution
bromothymol blue	7.9×10^{-8}	7.10	6.1 (yellow) to 8.1 (blue)	neutral solution
2,4-dinitrophenol	1.3×10^{-4}	3.89	2.9 (colorless) to 4.9 (yellow)	acidic solution
chlorophenol red	1.0×10^{-6}	6.00	5.0 (yellow) to 7.0 (red)	acidic solution
thymolphthalein	1.0×10^{-10}	10.00	9.0 (colorless) to 11.0 (blue)	basic solution

(b)   If bromcresol green is green, the pH is between 3.7 and 5.7, probably about $pH = 4.7$. If chlorophenol red is orange, the pH is between 5.0 and 7.0, probably about $pH = 6.0$.

28.   We first determine the pH of each solution, and then use the answer in Exercise 13(a) to predict the color of the indicator. (The weakly acidic or basic character of the indicator does not affect the pH of the solution, since the very little indicator is added.)

(a) $\left[H_3O^+\right] = 0.100\,M\,HCl \times \dfrac{1\,\text{mol}\,H_3O^+}{1\,\text{mol}\,HCl} = 0.100\,M;\quad pH = -\log(0.100\,M) = 1.000$

2,4-dinitrophenol assumes its acid color in a solution with $pH = 1.000$. The solution is colorless.

(b) Solutions of NaCl(aq) are pH neutral, with $pH = 7.000$. Chlorophenol red assumes its neutral color in such a solution; the solution is red/orange.

(c) Equation: $NH_3(aq) + H_2O(l) \rightleftharpoons NH_4^+(aq) + OH^-(aq)$

Initial:	1.00 M	—	0 M	≈ 0 M
Changes:	$-x$ M	—	$+x$ M	$+x$ M
Equil:	$(1.00-x)$ M	—	x M	x M

$$K_b = \frac{[NH_4^+][OH^-]}{[NH_3]} = 1.8 \times 10^{-5} = \frac{x^2}{1.00-x} \approx \frac{x^2}{1.00} \qquad x = 4.2 \times 10^{-3} M = [OH^-]$$

$(x \ll 1.00$ M, thus the approximation was valid$)$.

$$pOH = -\log(4.2 \times 10^{-3}) = 2.38 \qquad\qquad pH = 14.00 - 2.38 = 11.62$$

Thus, thymolphthalein assumes its basic blue color in solutions with $pH > 11.62$.

(d) From Figure 17-8, seawater has $pH = 7.00$ to 8.50. Bromcresol green assumes its basic color in this solution; the solution is blue.

29. (a) In an acid-base titration, the pH of the solution changes sharply at a definite pH that is known prior to titration. (This pH change occurs during the addition of a very small volume of titrant.) Determining the pH of a solution, on the other hand, is more difficult because the pH of the solution is not known precisely in advance. Since each indicator only serves to fix the pH over a quite small region, often less than 2.0 pH units, several indicators—carefully chosen to span the entire range of 14 pH units—must be employed to narrow the pH to ±1 pH unit or possibly lower.

(b) An indicator is, after all, a weak acid. Its addition to a solution will affect the acidity of that solution. Thus, one adds only enough indicator to show a color change and not enough to affect solution acidity.

30. (a) We use an equation similar to the Henderson-Hasselbalch equation to determine the relative concentrations of HIn, and its anion, In^-, in this solution.

$$pH = pK_{HIn} + \log\frac{[In^-]}{[HIn]}; \quad 4.55 = 4.95 + \log\frac{[In^-]}{[HIn]}; \quad \log\frac{[In^-]}{[HIn]} = 4.55 - 4.95 = -0.40$$

$$\frac{[In^-]}{[HIn]} = 10^{-0.40} = 0.40 = \frac{x}{100-x} \qquad x = 40 - 0.40x \qquad x = \frac{40}{1.40} = 29\% \text{ In}^- \text{ and } 71\% \text{ HIn}$$

(b) When the indicator is in a solution whose pH equals its pK_a (4.95), the ratio $[In^-]/[HIn] = 1.00$. And yet, at the midpoint of its color change range (about $pH = 5.3$), the ratio $[In^-]/[HIn]$ is greater than 1.00. Even though $[HIn] < [In^-]$ at this midpoint, the contribution of HIn to establishing the color of the solution is about the same as the contribution of In^-. This must mean that HIn (red) is more strongly colored than In^- (yellow).

31. **(a)** 0.10 M KOH is an alkaline solution and phenol red will display its basic color in such a solution; the solution will be red.

(b) 0.10 M $HC_2H_3O_2$ is an acidic solution, although that of a weak acid, and phenol red will display its acidic color in such a solution; the solution will be yellow.

(c) 0.10 M NH_4NO_3 is an acidic solution due to the hydrolysis of the ammonium ion. Phenol red will display its acidic color, that is, yellow, in this solution.

(d) 0.10 M HBr is an acidic solution, the aqueous solution of a strong acid. Phenol red will display its acidic color in this solution; the solution will be yellow.

(e) 0.10 M NaCN is an alkaline solution because of the hydrolysis of the cyanide ion. Phenol red will display its basic color, red, in this solution.

(f) An equimolar acetic acid–potassium acetate buffer has $pH = pK_a = 4.74$ for acetic acid. In this solution phenol red will display its acid color, namely, yellow.

32. **(a)** $pH = -\log(0.205) = 0.688$ The indicator is red in this solution.

(b) The total volume of the solution is 600.0 mL. We compute the amount of each solute.

amount $H_3O^+ = 350.0$ mL $\times 0.205$ M $= 71.8$ mmol H_3O^+

amount $NO_2^- = 250.0$ mL $\times 0.500$ M $= 125$ mmol NO_2^-

$$\left[H_3O^+\right] = \frac{71.8\,\text{mmol}}{600.0\,\text{mL}} = 0.120\,\text{M} \qquad \left[NO_2^-\right] = \frac{125\,\text{mmol}}{600.0\,\text{mL}} = 0.208\,\text{M}$$

The H_3O^+ and NO_2^- react to produce a buffer solution in which $\left[HNO_2\right] = 0.120\,\text{M}$ and $\left[NO_2^-\right] = 0.208 - 0.120 = 0.088\,\text{M}$. We use the Henderson-Hasselbalch equation to determine the pH of this solution. $pK_a = -\log(7.2 \times 10^{-4}) = 3.14$

$$pH = pK_a + \log\frac{\left[NO_2^-\right]}{\left[HNO_2\right]} = 3.14 + \log\frac{0.088\,\text{M}}{0.120\,\text{M}} = 3.01$$ The indicator is yellow in this solution.

(c) The total volume of the solution is 750. mL. We compute the amount and then the concentration of each solute. Amount $OH^- = 150$ mL $\times 0.100$ M $= 15.0$ mmol OH^-

This OH^- reacts with HNO_2 in the buffer solution to neutralize some of it and leave 56.8 mmol $(= 71.8 - 15.0)$ unneutralized.

$$\left[HNO_2\right] = \frac{56.8\,\text{mmol}}{750.\,\text{mL}} = 0.0757\,\text{M} \qquad \left[NO_2^-\right] = \frac{(125+15)\,\text{mmol}}{750.\,\text{mL}} = 0.187\,\text{M}$$

We use the Henderson-Hasselbalch equation to determine the pH of this solution.

$$pH = pK_a + \log\frac{\left[NO_2^-\right]}{\left[HNO_2\right]} = 3.14 + \log\frac{0.187\,\text{M}}{0.0757\,\text{M}} = 3.53$$

The indicator is yellow in this solution.

(d) We determine the $\left[OH^-\right]$ due to the added $Ba(OH)_2$.

$$\left[OH^-\right] = \frac{5.00\,g\,Ba(OH)_2}{0.750\,L} \times \frac{1\,mol\,Ba(OH)_2}{171.34\,g\,Ba(OH)_2} \times \frac{2\,mol\,OH^-}{1\,mol\,Ba(OH)_2} = 0.0778\,M$$

This is sufficient $\left[OH^-\right]$ to react with the existing $\left[HNO_2\right]$ and leave an excess

$\left[OH^-\right] = 0.0778\,M - 0.0757\,M = 0.0021\,M.$ $pOH = -\log(0.0021) = 2.68.$

$pH = 14.00 - 2.68 = 11.32$ The indicator is blue in this solution.

33. Moles of HCl $= C \times V = 0.04050\,M \times 0.01000\,L = 4.050 \times 10^{-4}$ moles
Moles of $Ba(OH)_2$ at endpoint $= C \times V = 0.01120\,M \times 0.01790\,L = 2.005 \times 10^{-4}$ moles.
Moles of HCl that react with $Ba(OH)_2 = 2 \times$ moles $Ba(OH)_2$
Moles of HCl in excess 4.050×10^{-4} moles $- 4.010 \times 10^{-4}$ moles $= 4.0\underline{4} \times 10^{-6}$ moles
Total volume at the equivalence point $= (10.00\,mL + 17.90\,mL) = 27.90\,mL$

$$[HCl]_{excess} = \frac{4.04 \times 10^{-6}\;mole\;HCl}{0.02790\;L} = 1.45 \times 10^{-4}\,M;\quad pH = -\log(1.45 \times 10^{-4}) = 3.84$$

(a) The approximate $pK_{HIn} = 3.84$ (generally $\pm\,1$ pH unit)

(b) This is a relatively good indicator (with $\approx 1\,\%$ of the equivalence point volume), however, pK_{Hin} is not very close to the theoretical pH at the equivalence point (pH = 7.000) For very accurate work, a better indicator is needed (i.e. bromothymol blue ($pK_{Hin} = 7.1$) Note: 2,4-dinitrophenol works relatively well here because the pH near the equivalence point of a strong acid/strong base titration rises very sharply (≈ 6 pH units for an addition of only 2 drops (0.10 mL))

34. Solution (a): 100.0 mL of 0.100 M HCl, $[H_3O^+] = 0.100\,M$ and pH = 1.000 (yellow)
Solution (b): 150 mL of 0.100 M $NaC_2H_3O_2$
K_a of $HC_2H_3O_3 = 1.8 \times 10^{-5}$ K_b of $C_2H_3O_2^- = 5.6 \times 10^{-10}$

	$C_2H_3O_2^-$(aq)	+	H_2O(l)	$\xrightarrow{K_b = 5.6 \times 10^{-10}}$	$HC_2H_3O_2$(aq)	+	OH^-(aq)
initial	0.100 M		—		0 M		~ 0 M
change	$-x$		—		$+x$		$+x$
equil.	0.100 $-x$		—		x		x

Assume x is small: $5.6 \times 10^{-11} = x^2$; $x = 7.4\underline{8} \times 10^{-6}\,M$ (assumption valid by inspection)
$[OH^-] = x = 7.4\underline{8} \times 10^{-6}\,M,$ pOH = 5.13 and pH = 8.87 (green-blue)
Mixture of solution (a) and (b). Total volume = 250.0 mL
$n_{HCl} = C \times V = 0.1000\,L \times 0.100\,M = 0.0100\,mol\,HCl$
$n_{C_2H_3O_2^-} = C \times V = 0.1500\,L \times 0.100\,M = 0.0150\,mol\,C_2H_3O_2^-$
HCl is the limiting reagent. Assume 100% reaction.
Therefore, 0.0050 mole $C_2H_3O_2^-$ left unreacted, and 0.0100 moles of $HC_2H_3O_2$ form.

$$[C_2H_3O_2^-] = \frac{n}{V} = \frac{0.0050\,mol}{0.250\,L} = 0.020\,M \quad [HC_2H_3O_2] = \frac{n}{V} = \frac{0.0100\,mol}{0.250\,L} = 0.0400\,M$$

$$\text{HC}_2\text{H}_3\text{O}_2(aq) + \text{H}_2\text{O}(l) \xrightleftharpoons{K_a = 1.8 \times 10^{-5}} \text{C}_2\text{H}_3\text{O}_2^-(aq) + \text{H}_3\text{O}^+(aq)$$

	$\text{HC}_2\text{H}_3\text{O}_2$	H_2O	$\text{C}_2\text{H}_3\text{O}_2^-$	H_3O^+
initial	0.0400 M	—	0.020 M	~0 M
change	$-x$	—	$+x$	$+x$
equil.	$0.0400 - x$	—	$0.020 + x$	x

$$1.8 \times 10^{-5} = \frac{x(0.020 + x)}{0.0400 - x} \approx \frac{x(0.020)}{0.0400} \qquad x = 3.6 \times 10^{-5}$$

(proof 0.18 % < 5%, the assumption was valid)

$[\text{H}_3\text{O}^+] = 3.6 \times 10^{-5}$ pH = 4.44 Color of thymol blue at various pHs:

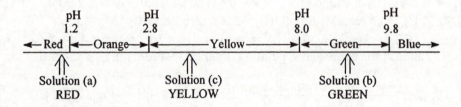

Neutralization Reactions

35. The reaction (to 2nd equiv. pt.) is: $\text{H}_3\text{PO}_4(aq) + 2\,\text{KOH}(aq) \longrightarrow \text{K}_2\text{HPO}_4(aq) + 2\,\text{H}_2\text{O}(l)$. The molarity of the H_3PO_4 solution is determined in the following manner.

$$\text{H}_3\text{PO}_4 \text{ molarity} = \frac{31.15\,\text{mL KOH soln} \times \dfrac{0.2420\,\text{mmol KOH}}{1\,\text{mL KOH soln}} \times \dfrac{1\,\text{mmol H}_3\text{PO}_4}{2\,\text{mmol KOH}}}{25.00\,\text{mL H}_3\text{PO}_4 \text{ soln}} = 0.1508\,\text{M}$$

36. The reaction (1st to 2nd equiv. pt.) is:

$\text{NaH}_2\text{PO}_4(aq) + \text{NaOH}(aq) \longrightarrow \text{Na}_2\text{HPO}_4(aq) + \text{H}_2\text{O}(l)$. The molarity of the H_3PO_4

solution is determined in the following manner.

$$\text{H}_3\text{PO}_4 \text{ molarity} = \frac{18.67\,\text{mL NaOH soln} \times \dfrac{0.1885\,\text{mmol NaOH}}{1\,\text{mL NaOH soln}} \times \dfrac{1\,\text{mmol H}_3\text{PO}_4}{1\,\text{mmol NaOH}}}{20.00\,\text{mL H}_3\text{PO}_4 \text{ soln}} = 0.1760\,\text{M}$$

37. Here we must determine amount of H_3O^+ or OH^- in each solution, and the amount of excess reagent.

A. amount $\text{H}_3\text{O}^+ = 50.00\,\text{mL} \times \dfrac{0.0150\,\text{mmol H}_2\text{SO}_4}{1\,\text{mL soln}} \times \dfrac{2\,\text{mmol H}_3\text{O}^+}{1\,\text{mmol H}_2\text{SO}_4} = 1.50\,\text{mmol H}_3\text{O}^+$

(assuming complete ionization of H_2SO_4 and HSO_4^- in the presence of OH^-)

B. amount $OH^- = 50.00 \, \text{mL} \times \dfrac{0.0385 \, \text{mmol NaOH}}{1 \, \text{mL soln}} \times \dfrac{1 \, \text{mmol OH}^-}{1 \, \text{mmol NaOH}} = 1.93 \, \text{mmol OH}^-$

Result: Titration reaction : $\quad OH^-(aq) \ + \ H_3O^+(aq) \ \rightleftharpoons \ 2H_2O(l)$

Initial amounts : $\qquad\qquad$ 1.93 mmol $\qquad$ 1.50 mmol

After reaction : $\qquad\qquad$ 0.43 mmol $\qquad \approx 0 \, \text{mmol}$

$$\left[OH^-\right] = \frac{0.43 \, \text{mmol OH}^-}{100.0 \, \text{mL soln}} = 4.3 \times 10^{-3} \, M$$

$$pOH = -\log\left(4.3 \times 10^{-3}\right) = 2.37 \qquad\qquad pH = 14.00 - 2.37 = 11.63$$

38. Here we must determine the amount of solute in each solution, followed by the amount excess reagent.

A. $\left[H_3O^+\right] = 10^{-2.50} = 0.0032 \, M$

$$\text{mmol HCl} = 100.0 \, \text{mL} \times \frac{0.0032 \, \text{mmol H}_3O^+}{1 \, \text{mL soln}} \times \frac{1 \, \text{mmol HCl}}{1 \, \text{mmol H}_3O^+} = 0.32 \, \text{mmol HCl}$$

B. $\qquad pOH = 14.00 - 11.00 = 3.00 \quad \left[OH^-\right] = 10^{-3.00} = 1.0 \times 10^{-3} \, M$

$$\text{mmol NaOH} = 100.0 \, \text{mL} \times \frac{0.0010 \, \text{mmol OH}^-}{1 \, \text{mL soln}} \times \frac{1 \, \text{mmol NaOH}}{1 \, \text{mmol OH}^-} = 0.10 \, \text{mmol NaOH}$$

Result:

Titration reaction : $\quad NaOH(aq) \ + \ HCl(aq) \ \longrightarrow \ NaCl(aq) \ + \ H_2O(l)$

Initial amounts : $\qquad\quad$ 0.10 mmol $\qquad$ 0.32 mmol $\qquad\quad$ 0 mmol $\qquad$ –

After reaction : $\qquad\quad$ 0.00 mmol $\qquad$ 0.22 mmol $\qquad$ 0.10 mmol $\qquad$ –

$$\left[H_3O^+\right] = \frac{0.22 \, \text{mmol HCl}}{200.0 \, \text{mL soln}} \times \frac{1 \, \text{mmol H}_3O^+}{1 \, \text{mmol HCl}} = 1.1 \times 10^{-3} \, M \qquad pH = -\log\left(1.1 \times 10^{-3}\right) = 2.96$$

Titration Curves

39. First we calculate the amount of HCl. The relevant titration reaction is

$$HCl(aq) \ + \ KOH(aq) \ \rightarrow \ KCl(aq) \ + \ H_2O(l)$$

$$\text{amount HCl} = 25.00 \, \text{mL} \times \frac{0.160 \, \text{mmol HCl}}{1 \, \text{mL soln}} = 4.00 \, \text{mmol HCl} = 4.00 \, \text{mmol H}_3O^+ \text{present}$$

Then, in each case, we calculate the amount of OH^- that has been added, determine which ion, $OH^-(aq)$ or $H_3O^+(aq)$, is in excess, compute the concentration of that ion, and determine the pH.

(a) amount $OH^- = 10.00 \text{ mL} \times \dfrac{0.242 \text{ mmol } OH^-}{1 \text{ mL soln}} = 2.42 \text{ mmol } OH^-$; H_3O^+ is in excess.

$$[H_3O^+] = \dfrac{4.00 \text{ mmol } H_3O^+ - \left(2.42 \text{ mmol } OH^- \times \dfrac{1 \text{ mmol } H_3O^+}{1 \text{ mmol } OH^-}\right)}{25.00 \text{ mL originally} + 10.00 \text{ mL titrant}} = 0.0451 \text{ M}$$

$$pH = -\log(0.0451) = 1.346$$

(b) amount $OH^- = 15.00 \text{ mL} \times \dfrac{0.242 \text{ mmol } OH^-}{1 \text{ mL soln}} = 3.63 \text{ mmol } OH^-$; H_3O^+ is in excess.

$$[H_3O^+] = \dfrac{4.00 \text{ mmol } H_3O^+ - \left(3.63 \text{ mmol } OH^- \times \dfrac{1 \text{ mmol } H_3O^+}{1 \text{ mmol } OH^-}\right)}{25.00 \text{ mL originally} + 15.00 \text{ mL titrant}} = 0.00925 \text{ M}$$

$$pH = -\log(0.00925) = 2.034$$

40. The relevant titration reaction is $KOH(aq) + HCl(aq) \rightarrow KCl(aq) + H_2O(l)$

$$\text{mmol of KOH} = 20.00 \text{ mL} \times \dfrac{0.275 \text{ mmol KOH}}{1 \text{ mL soln}} = 5.50 \text{ mmol KOH}$$

(a) The total volume of the solution is $V = 20.00 \text{ mL} + 15.00 \text{ mL} = 35.00 \text{ mL}$

$$\text{mmol HCl} = 15.00 \text{ mL} \times \dfrac{0.350 \text{ mmol HCl}}{1 \text{ mL soln}} = 5.25 \text{ mmol HCl}$$

$$\text{mmol excess } OH^- = (5.50 \text{ mmol KOH} - 5.25 \text{ mmol HCl}) \times \dfrac{1 \text{ mmol } OH^-}{1 \text{ mmol KOH}} = 0.25 \text{ mmol } OH^-$$

$$\left[OH^-\right] = \dfrac{0.25 \text{ mmol } OH^-}{35.00 \text{ mL soln}} = 0.0071 \text{ M} \qquad pOH = -\log(0.0071) = 2.15$$

$$pH = 14.00 - 2.15 = 11.85$$

(b) The total volume of solution is $V = 20.00 \text{ mL} + 20.00 \text{ mL} = 40.00 \text{ mL}$

$$\text{mmol HCl} = 20.00 \text{ mL} \times \dfrac{0.350 \text{ mmol HCl}}{1 \text{ mL soln}} = 7.00 \text{ mmol HCl}$$

$$\text{mmol excess } H_3O^+ = (7.00 \text{ mmol HCl} - 5.50 \text{ mmol KOH}) \times \dfrac{1 \text{ mmol } H_3O^+}{1 \text{ mmol HCl}}$$

$$= 1.50 \text{ mmol } H_3O^+$$

$$\left[H_3O^+\right] = \dfrac{1.50 \text{ mmol } H_3O^+}{40.00 \text{ mL}} = 0.0375 \text{ M} \qquad pH = -\log(0.0375) = 1.426$$

<u>41.</u> The relevant titration reaction is $HNO_2(aq) + NaOH(aq) \rightarrow NaNO_2(aq) + H_2O(l)$

$$\text{amount } HNO_2 = 25.00 \text{ mL} \times \frac{0.132 \text{ mmol } HNO_2}{1 \text{ mL soln}} = 3.30 \text{ mmol } HNO_2$$

(a) The volume of the solution is $25.00 \text{ mL} + 10.00 \text{ mL} = 35.00 \text{ mL}$

$$\text{amount } NaOH = 10.00 \text{ mL} \times \frac{0.116 \text{ mmol } NaOH}{1 \text{ mL soln}} = 1.16 \text{ mmol } NaOH$$

1.16 mmol $NaNO_2$ are formed in this reaction and there is an excess of (3.30 mmol $HNO_2 - 1.16$ mmol $NaOH$) = 2.14 mmol HNO_2. We can use the Henderson-Hasselbalch equation to determine the pH of the solution.

$$pK_a = -\log(7.2 \times 10^{-4}) = 3.14$$

$$pH = pK_a + \log\frac{[NO_2^-]}{[HNO_2]} = 3.14 + \log\frac{1.16 \text{ mmol } NO_2^- / 35.00 \text{ mL}}{2.14 \text{ mmol } HNO_2 / 35.00 \text{ mL}} = 2.87$$

(b) The volume of the solution is $25.00 \text{ mL} + 20.00 \text{ mL} = 45.00 \text{ mL}$

$$\text{amount } NaOH = 20.00 \text{ mL} \times \frac{0.116 \text{ mmol } NaOH}{1 \text{ mL soln}} = 2.32 \text{ mmol } NaOH$$

2.32 mmol $NaNO_2$ are formed in this reaction and there is an excess of (3.30 mmol $HNO_2 - 2.32$ mmol $NaOH$ =) 0.98 mmol HNO_2.

$$pH = pK_a + \log\frac{[NO_2^-]}{[HNO_2]} = 3.14 + \log\frac{2.32 \text{ mmol } NO_2^- / 45.00 \text{ mL}}{0.98 \text{ mmol } HNO_2 / 45.00 \text{ mL}} = 3.51$$

42. In this case the titration reaction is $NH_3(aq) + HCl(aq) \rightarrow NH_4Cl(aq) + H_2O(l)$

$$\text{amount } NH_3 = 20.00 \text{ mL} \times \frac{0.318 \text{ mmol } NH_3}{1 \text{ mL soln}} = 6.36 \text{ mmol } NH_3$$

(a) The volume of the solution is $20.00 \text{ mL} + 10.00 \text{ mL} = 30.00 \text{ mL}$

$$\text{amount } HCl = 10.00 \text{ mL} \times \frac{0.475 \text{ mmol } NaOH}{1 \text{ mL soln}} = 4.75 \text{ mmol } HCl$$

4.75 mmol NH_4Cl is formed in this reaction and there is an excess of (6.36 mmol $NH_3 - 4.75$ mmol HCl =)1.61 mmol NH_3. We can use the Henderson-Hasselbalch equation to determine the pH of the solution.

$$pK_b = -\log(1.8 \times 10^{-5}) = 4.74; \quad pK_a = 14.00 - pK_b = 14.00 - 4.74 = 9.26$$

$$pH = pK_a + \log\frac{[NH_3]}{[NH_4^+]} = 9.26 + \log\frac{1.61 \text{ mmol } NH_3 / 30.00 \text{ mL}}{4.75 \text{ mmol } NH_4^+ / 30.00 \text{ mL}} = 8.79$$

(b) The volume of the solution is $20.00\,\text{mL} + 15.00\,\text{mL} = 35.00\,\text{mL}$

$$\text{amount HCl} = 15.00\,\text{mL} \times \frac{0.475\,\text{mmol NaOH}}{1\,\text{mL soln}} = 7.13\,\text{mmol HCl}$$

6.36 mmol NH_4Cl is formed in this reaction and there is an excess of (7.13 mmol $HCl - 6.36$ mmol NH_3) 0.77 mmol HCl; this excess HCl determines the pH of the solution.

$$\left[H_3O^+\right] = \frac{0.77\,\text{mmol HCl}}{35.00\,\text{mL soln}} \times \frac{1\,\text{mmol H}_3\text{O}^+}{1\,\text{mmol HCl}} = 0.022\,\text{M} \quad \text{pH} = -\log(0.022) = 1.66$$

43. In each case, the volume of acid and its molarity are the same. Thus, also the amount of acid is the same in each case. The volume of titrant needed to reach the equivalence point will also be the same in both cases, since the titrant has the same concentration in each case, and it is the same amount of base that reacts with a given amount (in moles) of acid. Realize that, as the titration of a weak acid proceeds, the weak acid will ionize, replenishing the H_3O^+ in solution. This will occur until all of the weak acid has ionized and all of the released H^+ has subsequently reacted with the strong base. At the equivalence point in the titration of a strong acid with a strong base the solution contains ions that do not hydrolyze. But the equivalence point solution of the titration of a weak acid with a strong base contains the anion of a weak acid, which will hydrolyze to produce a basic (alkaline) solution. (Don't forget, however, that the inert cation of a strong base is also present.)

44. (a) This equivalence point solution is the result of the titration of a weak acid with a strong base. The CO_3^{2-} in this solution, through its hydrolysis, will form an alkaline, or basic solution. The other ionic species in solution, Na^+, will not hydrolyze. Thus, $pH > 7.0$

(b) This is the titration of a strong acid with a weak base. The NH_4^+ present in the equivalence point solution hydrolyzes to form an acidic solution. Cl^- does not hydrolyze. Thus, $pH < 7.0$

(c) This is the titration of a strong acid with a strong base. Two ions are present in the solution at the equivalence point, namely, K^+ and Cl^-, neither of which hydrolyze. Thus the solution will have a pH of 7.00.

45. (a) Initial $\left[OH^-\right] = 0.100\,\text{M OH}^-$ $\quad$ pOH $= -\log(0.100) = 1.000$ $\quad$ pH $= 13.00$

Since this is the titration of a strong base with a strong acid, KI is the solute present at the equivalence point and since KI is a neutral salt, the $pH = 7.00$. The titration reaction is:

$$KOH(aq) + HI(aq) \longrightarrow KI(aq) + H_2O(l)$$

$$V_{HI} = 25.0\,\text{mL KOH soln} \times \frac{0.100\,\text{mmol KOH soln}}{1\,\text{mL soln}} \times \frac{1\,\text{mmol HI}}{1\,\text{mmol KOH}} \times \frac{1\,\text{mL HI soln}}{0.200\,\text{mmol HI}}$$

$$= 12.5\,\text{mL HI soln}$$

Initial amount of KOH present $= 25.0\,\text{mL KOH soln} \times 0.100\,\text{M} = 2.50\,\text{mmol KOH}$

At the 40% titration point: $5.00\,\text{mL HI soln} \times 0.200\,\text{M HI} = 1.00\,\text{mmol HI}$

excess KOH $= 2.50\,\text{mmol KOH} - 1.00\,\text{mmol HI} = 1.50\,\text{mmol KOH}$

$$\left[\text{OH}^-\right] = \frac{1.50\,\text{mmol KOH}}{30.0\,\text{mL total}} \times \frac{1\,\text{mmol OH}^-}{1\,\text{mmol KOH}} = 0.0500\,\text{M} \quad \text{pOH} = -\log(0.0500) = 1.30$$

$$\text{pH} = 14.00 - 1.30 = 12.70$$

At the 80% titration point: $10.00\,\text{mL HI soln} \times 0.200\,\text{M HI} = 2.00\,\text{mmol HI}$

excess KOH $= 2.50\,\text{mmol KOH} - 2.00\,\text{mmol HI} = 0.50\,\text{mmol KOH}$

$$\left[\text{OH}^-\right] = \frac{0.50\,\text{mmol KOH}}{35.0\,\text{mL total}} \times \frac{1\,\text{mmol OH}^-}{1\,\text{mmol KOH}} = 0.0143\,\text{M} \quad \text{pOH} = -\log(0.0143) = 1.84$$

$$\text{pH} = 14.00 - 1.84 = 12.16$$

At the 110% titration point: $13.75\,\text{mL HI soln} \times 0.200\,\text{M HI} = 2.75\,\text{mmol HI}$

excess HI $= 2.75\,\text{mmol HI} - 2.50\,\text{mmol HI} = 0.25\,\text{mmol HI}$

$$[\text{H}_3\text{O}^+] = \frac{0.25\,\text{mmol HI}}{38.8\,\text{mL total}} \times \frac{1\,\text{mmol H}_3\text{O}^+}{1\,\text{mmol HI}} = 0.0064\,\text{M}; \quad \text{pH} = -\log(0.0064) = 2.19$$

Since the pH changes very rapidly at the equivalence point, from about $\text{pH} = 10$ to about $\text{pH} = 4$, most of the indicators in Figure 17-8 can be used. The main exceptions are alizarin yellow R, bromophenol blue, thymol blue (in its acid range), and methyl violet.

(b) *Initial pH:*

Since this is the titration of a weak base with a strong acid, NH_4Cl is the solute present at the equivalence point and since NH_4^+ is a slightly acidic cation, the pH should be slightly lower than 7. The titration reaction is:

Equation:	$NH_3(aq)$	$+ H_2O(l)$	$\rightleftharpoons$	$NH_4^+(aq)$	$+ OH^-(aq)$
Initial:	$1.00\,\text{M}$	$-$		$0\,\text{M}$	$\approx 0\,\text{M}$
Changes:	$-x\,\text{M}$	$-$		$+x\,\text{M}$	$+x\,\text{M}$
Equil:	$(1.00-x)\,\text{M}$	$-$		$x\,\text{M}$	$x\,\text{M}$

$$K_b = \frac{\left[NH_4^+\right]\left[OH^-\right]}{\left[NH_3\right]} = 1.8 \times 10^{-5} = \frac{x^2}{1.00-x} \approx \frac{x^2}{1.00}$$

($x \ll 1.0$, thus the approximation is valid)

$$x = 4.2 \times 10^{-3} \, M = \left[OH^- \right], \quad pOH = -\log\left(4.2 \times 10^{-3} \right) = 2.38,$$

$$pH = 14.00 - 2.38 = 11.62 = \text{initial pH}$$

Volume of titrant : $NH_3 + HCl \longrightarrow NH_4Cl + H_2O$

$$V_{HCl} = 10.0 \, mL \times \frac{1.00 \, mmol \, NH_3}{1 \, mL \, soln} \times \frac{1 \, mmol \, HCl}{1 \, mmol \, NH_3} \times \frac{1 \, mL \, HCl \, soln}{0.250 \, mmol \, HCl} = 40.0 \, mL \, HCl \, soln$$

pH at equivalence point: The total solution volume at the equivalence point is $(10.0 + 40.0) \, mL = 50.0 \, mL$

Also at the equivalence point, all of the NH_3 has reacted to form NH_4^+. It is this NH_4^+ that hydrolyzes to determine the pH of the solution.

$$\left[NH_4^+ \right] = \frac{10.0 \, mL \times \dfrac{1.00 \, mmol \, NH_3}{1 \, mL \, soln} \times \dfrac{1 \, mmol \, NH_4^+}{1 \, mmol \, NH_3}}{50.0 \, mL \, \text{total solution}} = 0.200 \, M$$

Equation : $NH_4^+ \, (aq) + H_2O(l) \rightleftharpoons NH_3 \, (aq) + H_3O^+ \, (aq)$

Initial : $0.200 \, M$ – $0 \, M$ $\approx 0 \, M$

Changes : $-x \, M$ – $+x \, M$ $+x \, M$

Equil : $(0.200 - x) \, M$ – $x \, M$ $x \, M$

$$K_a = \frac{K_w}{K_b} = \frac{1.0 \times 10^{-14}}{1.8 \times 10^{-5}} = \frac{\left[NH_3 \right]\left[H_3O^+ \right]}{\left[NH_3 \right]} = \frac{x^2}{0.200 - x} \approx \frac{x^2}{0.200}$$

($x \ll 0.200$, thus the approximation is valid)

$$x = 1.1 \times 10^{-5} \, M; \quad \left[H_3O^+ \right] = 1.1 \times 10^{-5} \, M; \quad pH = -\log\left(1.1 \times 10^{-5} \right) = 4.96$$

Of the indicators in Figure 17-8, one that has the pH of the equivalence point within its pH color change range is methyl red (yellow at $pH = 6.2$ and red at $pH = 4.5$); Bromcresol green would be another choice. At the 50 % titration point, $\left[NH_3 \right] = \left[NH_4^+ \right]$ and $pOH = pK_b = 4.74$ $pH = 14.00 - 4.74 = 9.26$

The titration curves for parts **(a)** and **(b)** follow.

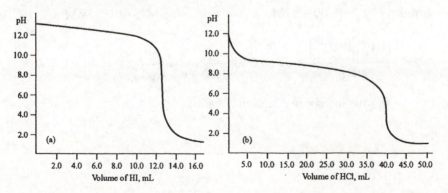

46. (a) This part simply involves calculating the pH of a 0.275 M NH_3 solution.

Equation: $NH_3(aq) + H_2O(l) \rightleftharpoons NH_4^+(aq) + OH^-(aq)$

Initial:	0.275 M	–	0 M	≈ 0 M
Changes:	$-x$ M	–	$+x$ M	$+x$ M
Equil:	$(0.275-x)$ M	–	x M	x M

$$K_b = \frac{[NH_4^+][OH^-]}{[NH_3]} = 1.8 \times 10^{-5} = \frac{x^2}{0.275-x} \approx \frac{x^2}{0.275} \qquad x = 2.2 \times 10^{-3} M = [OH^-]$$

($x \ll 0.275$, thus the approximation is valid)

$$pOH = -\log(2.2 \times 10^{-3}) = 2.66 \qquad pH = 11.34$$

(b) This is the volume of titrant needed to reach the equivalence point.
The relevant titration reaction is $NH_3(aq) + HI(aq) \longrightarrow NH_4I(aq)$

$$V_{HI} = 20.00 \text{ mL } NH_3(aq) \times \frac{0.275 \text{ mmol } NH_3}{1 \text{ mL } NH_3 \text{ soln}} \times \frac{1 \text{ mmol } HI}{1 \text{ mmol } NH_3} \times \frac{1 \text{ mL } HI \text{ soln}}{0.325 \text{ mmol } HI}$$

$$V_{HI} = 16.9 \text{ mL } HI \text{ soln}$$

(c) The pOH at the half-equivalence point of the titration of a weak base with a strong acid is equal to the pK_b of the weak base.

$$pOH = pK_b = 4.74; \quad pH = 14.00 - 4.74 = 9.26$$

(d) NH_4^+ is formed during the titration, and its hydrolysis determines the pH of the solution. Total volume of solution $= 20.00 \text{ mL} + 16.9 \text{ mL} = 36.9 \text{ mL}$

$$\text{mmol } NH_4^+ = 20.00 \text{ mL } NH_3(aq) \times \frac{0.275 \text{ mmol } NH_3}{1 \text{ mL } NH_3 \text{ soln}} \times \frac{1 \text{ mmol } NH_4^+}{1 \text{ mmol } NH_3} = 5.50 \text{ mmol } NH_4^+$$

$$[NH_4^+] = \frac{5.50 \text{ mmol } NH_4^+}{36.9 \text{ mL soln}} = 0.149 M$$

Equation: $NH_4^+(aq) + H_2O(l) \rightleftharpoons NH_3(aq) + H_3O^+(aq)$

Initial:	0.149 M	–	0 M	≈ 0 M
Changes:	$-x$ M	–	$+x$ M	$+x$ M
Equil:	$(0.149-x)$ M	–	x M	x M

$$K_a = \frac{K_w}{K_b} = \frac{1.0 \times 10^{-14}}{1.8 \times 10^{-5}} = \frac{[NH_3][H_3O^+]}{[NH_4^+]} = \frac{x^2}{0.149-x} \approx \frac{x^2}{0.149}$$

($x \ll 0.149$, thus the approximation is valid)

$$x = 9.1 \times 10^{-6} M = [H_3O^+] \qquad pH = -\log(9.1 \times 10^{-6}) = 5.04$$

47. A pH greater than 7.00 in the titration of a strong base with a strong acid means that the base is not completely titrated. A pH less than 7.00 means that excess acid has been added.

(a) We can determine $[OH^-]$ of the solution from the pH. $[OH^-]$ is also the quotient of the amount of hydroxide ion in excess divided by the volume of the solution: 20.00 mL base $+x$ mL added acid.

$$pOH = 14.00 - pH = 14.00 - 12.55 = 1.45 \qquad [OH^-] = 10^{-pOH} = 10^{-1.45} = 0.035\,M$$

$$[OH^-] = \frac{\left(20.00\,mL\,base \times \dfrac{0.175\,mmol\,OH^-}{1\,mL\,base}\right) - \left(x\,mL\,acid \times \dfrac{0.200\,mmol\,H_3O^+}{1\,mL\,acid}\right)}{20.00\,mL + x\,mL} = 0.035\,M$$

$$3.50 - 0.200\,x = 0.70 + 0.035x; \quad 3.50 - 0.70 = 0.035\,x + 0.200\,x; \quad 2.80 = 0.235\,x$$

$$x = \frac{2.80}{0.235} = 11.9\,mL \text{ acid added.}$$

(b) The set-up here is the same as for part (a).

$$pOH = 14.00 - pH = 14.00 - 10.80 = 3.20 \qquad [OH^-] = 10^{-pOH} = 10^{-3.20} = 0.00063\,M$$

$$[OH^-] = \frac{\left(20.00\,mL\,base \times \dfrac{0.175\,mmol\,OH^-}{1\,mL\,base}\right) - \left(x\,mL\,acid \times \dfrac{0.200\,mmol\,H_3O^+}{1\,mL\,acid}\right)}{20.00\,mL + x\,mL}$$

$$[OH^-] = 0.00063\,\frac{mmol}{mL} = 0.00063\,M$$

$$3.50 - 0.200x = 0.0126 + 0.00063x; \quad 3.50 - 0.0126 = 0.00063\,x + 0.200\,x; \quad 3.49 = 0.201\,x$$

$$x = \frac{3.49}{0.201} = 17.4\,mL \text{ acid added. This is close to the equivalence point at 17.5 mL.}$$

(c) Here the acid is in excess, so we reverse the set-up of part (a). We are just slightly beyond the equivalence point. This is close to the "mirror image" of part **(b)**.

$$[H_3O^+] = 10^{-pH} = 10^{-4.25} = 0.000056\,M$$

$$[H_3O^+] = \frac{\left(x\,mL\,acid \times \dfrac{0.200\,mmol\,H_3O^+}{1\,mL\,acid}\right) - \left(20.00\,mL\,base \times \dfrac{0.175\,mmol\,OH^-}{1\,mL\,base}\right)}{20.00\,mL + x\,mL}$$

$$= 5.6 \times 10^{-5}\,M$$

$$0.200\,x - 3.50 = 0.0011 + 5.6 \times 10^{-5}\,x; \quad 3.50 + 0.0011 = -5.6 \times 10^{-5}\,x + 0.200\,x;$$

$$3.50 = 0.200\,x$$

$$x = \frac{3.50}{0.200} = 17.5\,mL \text{ acid added, which is the equivalence point for this titration.}$$

48. In the titration of a weak acid with a strong base, the middle range of the titration, with the pH within one unit of pK_a ($=4.74$ for acetic acid), is known as the buffer region. The Henderson-Hasselbalch equation can be used to determine the ratio of weak acid and anion concentrations. The amount of weak acid then is used in these calculations to determine the amount of base to be added.

(a)

$$pH = pK_a + \log\frac{\left[C_2H_3O_2^-\right]}{\left[HC_2H_3O_2\right]} = 3.85 = 4.74 + \log\frac{\left[C_2H_3O_2^-\right]}{\left[HC_2H_3O_2\right]}$$

$$\log\frac{\left[C_2H_3O_2^-\right]}{\left[HC_2H_3O_2\right]} = 3.85 - 4.74 \qquad \frac{\left[C_2H_3O_2^-\right]}{\left[HC_2H_3O_2\right]} = 10^{-0.89} = 0.13;$$

$$n_{HC_2H_3O_2} = 25.00\,mL \times \frac{0.100\,mmol\;HC_2H_3O_2}{1\,mL\;acid} = 2.50\,mmol$$

Since acetate ion and acetic acid are in the same solution, we can use their amounts in millimoles in place of their concentrations. The amount of acetate ion is the amount created by the addition of strong base, one millimole of acetate ion for each millimole of strong based added. The amount of acetic acid is reduced by the same number of millimoles. $HC_2H_3O_2 + OH^- \rightarrow C_2H_3O_2^- + H_2O$

$$0.13 = \frac{x\,mL\;base \times \dfrac{0.200\,mmol\;OH^-}{mL\;base} \times \dfrac{1\,mmol\;C_2H_3O_2^-}{1\,mmol\;OH^-}}{2.50\,mmol\;HC_2H_3O_2 - \left(x\,mL\;base \times \dfrac{0.200\;mmol\;OH^-}{mL\;base}\right)} = \frac{0.200\,x}{2.50 - 0.200\,x}$$

$$0.200\,x = 0.13(2.50 - 0.200\,x) = 0.33 - 0.026\,x \qquad 0.33 = 0.200\,x + 0.026\,x \quad 0.226\,x$$

$$x = \frac{0.33}{0.226} = 1.5\;mL\;of\;base$$

(b) This is the same set-up as part (a), except for a different ratio of concentrations.

$$pH = 5.25 = 4.74 + \log\frac{\left[C_2H_3O_2^-\right]}{\left[HC_2H_3O_2\right]} \qquad \log\frac{\left[C_2H_3O_2^-\right]}{\left[HC_2H_3O_2\right]} = 5.25 - 4.74 = 0.51$$

$$\frac{\left[C_2H_3O_2^-\right]}{\left[HC_2H_3O_2\right]} = 10^{0.51} = 3.2 \qquad 3.2 = \frac{0.200x}{2.50 - 0.200x}$$

$$0.200\,x = 3.2(2.50 - 0.200\,x) = 8.0 - 0.64\,x \qquad 8.0 = 0.200\,x + 0.64\,x = 0.84\,x$$

$$x = \frac{8.0}{0.84} = 9.5\,mL\;base.\;\;\text{This is close to the equivalence point, which is reached}$$

by adding 12.5 mL of base.

(c) This is after the equivalence point, where the pH is determined by the excess added base. $pOH = 14.00 - pH = 14.00 - 11.10 = 2.90$ $\left[OH^-\right] = 10^{-pOH} = 10^{-2.90} = 0.0013\,M$

$$\left[OH^-\right] = 0.0013\ M = \frac{x\ mL \times \dfrac{0.200\,mmol\ OH^-}{1\,mL\ base}}{x\,mL + \left(12.50\,mL + 25.00\,mL\right)} = \frac{0.200\,x}{37.50 + x}$$

$0.200\,x = 0.0013(37.50 + x) = 0.049 + 0.0013\,x$ $x = \dfrac{0.049}{0.200 - 0.0013} = 0.25\,mL$ excess

Total base added $= 12.5$ mL to equivalence point $+ 0.25$ mL excess $= 12.8$ mL

49. For each of the titrations, the pH at the half-equivalence point equals the pK_a of the acid.

The initial pH is that of 0.1000 M weak acid: $K_a = \dfrac{x^2}{0.1000 - x}$ $x = [H_3O^+]$

x must be found using the quadratic formula roots equation unless the approximation is valid. One method of determining if the approximation will be valid is to consider the ratio C_a/K_a. If the value of C_a/K_a is greater than 1000, the assumption should be valid, however, if the value of C_a/K_a is less than 1000, the quadratic should be solved exactly (i.e. the 5% rule will not be satisfied).

The pH at the equivalence point is that of 0.05000 M anion of the weak acid, for which the

$\left[OH^-\right]$ is determined as follows. $K_b = \dfrac{K_w}{K_a} \approx \dfrac{x^2}{0.05000}$ $x = \sqrt{\dfrac{K_w}{K_a}0.05000} = [OH^-]$

We can determine the pH at the quarter and three quarter equivalence point by using the Henderson-Hasselbalch equation. (effectively ± 0.48 pH unit added to the pK_a

And finally when 0.100 mL of base has been added beyond the equivalence point, the pH is determined by the excess added base, as follows (for all three titrations).

$$\left[OH^-\right] = \frac{0.100\,mL \times \dfrac{0.1000\,mmol\ NaOH}{1\,mL\ NaOH\ soln} \times \dfrac{1\,mmol\ OH^-}{1\,mmol\ NaOH}}{20.1\,mL\ soln\ total} = 4.98 \times 10^{-4}\ M$$

$pOH = -\log\left(4.98 \times 10^{-4}\right) = 3.303$ $pH = 14.000 - 3.303 = 10.697$

(a) $C_a/K_a = 14.3$; thus, the approximation is **not** valid and the full quadratic equation must be solved.

Initial: From the roots equation $x = [H_3O^+] = 0.023\,M$ pH=1.63
Half equivalence point: $pH = pK_a = 2.15$
pH at quarter equivalence point $= 2.15 - 0.48 = 1.67$
pH at three quarter equivalence point $= 2.15 + 0.48 = 2.63$

Equiv: $x = \left[OH^-\right] = \sqrt{\dfrac{1.0 \times 10^{-14}}{7.0 \times 10^{-3}} \times 0.05000} = 2.7 \times 10^{-7}$ $\begin{array}{l} pOH = 6.57 \\ pH = 14.00 - 6.57 = 7.43 \end{array}$

Indicator: bromothymol blue, yellow at $pH = 6.2$ and blue at $pH = 7.8$

(b) $C_a/K_a = 333$; thus, the approximation is <u>not</u> valid and the full quadratic equation must be solved.

Initial: From the roots equation $x = [H_3O^+] = 0.0053\,M$ pH=2.28

Half equivalence point: $pH = pK_a = 3.52$

pH at quarter equivalence point $= 3.52 - 0.48 = 3.04$

pH at three quarter equivalence point $= 3.52 + 0.48 = 4.00$

Equiv: $x = \left[OH^-\right] = \sqrt{\dfrac{1.0 \times 10^{-14}}{3.0 \times 10^{-4}}} \times 0.05000 = 1.3 \times 10^{-6}$ $\begin{array}{l} pOH = 5.89 \\ pH = 14.00 - 5.89 = 8.11 \end{array}$

Indicator: thymol blue, yellow at $pH = 8.0$ and blue at $pH = 9.6$

(c) $C_a/K_a = 5 \times 10^6$; thus, the approximation <u>is</u> valid and the approximation is valid.

Initial: $[H_3O^+] = \sqrt{0.1000 \times 2.0 \times 10^{-8}} = 0.000045\ M\ \ pH = 4.35$

pH at quarter equivalence point $= 4.35 - 0.48 = 3.87$

pH at three quarter equivalence point $= 4.35 + 0.48 = 4.83$

Half equivalence point: $pH = pK_a = 7.70$

Equiv: $x = \left[OH^-\right] = \sqrt{\dfrac{1.0 \times 10^{-14}}{2.0 \times 10^{-8}}} \times 0.0500 = 1.6 \times 10^{-4}$ $\begin{array}{l} pOH = 3.80 \\ pH = 14.00 - 3.80 = 10.20 \end{array}$

Indicator: alizarin yellow R, yellow at $pH = 10.0$ and violet at $pH = 12.0$

The three titration curves are drawn with respect to the same axes in the diagram below.

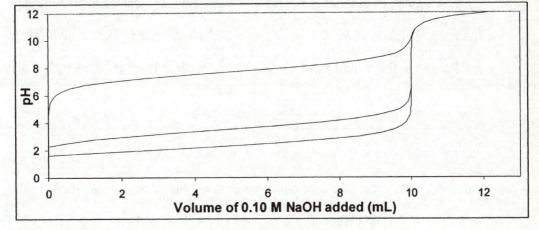

50. For each of the titrations, the pOH at the half-equivalence point equals the pK_b of the base. The initial pOH is that of 0.1000 M weak base, determined as follows.

$K_b = \dfrac{x^2}{0.1000 - x};$ $x \approx \sqrt{0.1000 \times K_a} = [OH^-]$ if $C_b/K_b > 1000$. For those cases where this is not the case, the approximation is invalid and the complete quadratic equation must be solved. The pH at the equivalence point is that of 0.05000 M cation of the weak base, for which the $\left[H_3O^+\right]$ is determined as follows.

$$K_a = \frac{K_w}{K_b} = \frac{x^2}{0.05000} \qquad\qquad x = \sqrt{\frac{K_w}{K_b}0.05000} = [H_3O^+]$$

And finally when 0.100 mL of acid has been added beyond the equivalence point, the pH for all three titrations is determined by the excess added acid, as follows.

$$[H_3O^+] = \frac{0.100 \text{ mL HCl} \times \dfrac{0.1000\,\text{mmol HCl}}{1\,\text{mL HCl soln}} \times \dfrac{1\,\text{mmol H}_3O^+}{1\,\text{mmol HCl}}}{20.1\,\text{mL soln total}} = 4.98\times10^{-4}\,\text{M}$$

$$pH = -\log(4.98\times10^{-4}) = 3.303$$

(a) $C_b/K_b = 100$; thus, the approximation is <u>not</u> valid and the full quadratic equation must be solved.

Initial: From the roots equation $x = [OH^-] = 0.0095$ M $\quad$ pOH=2.02 $\quad$ pH=11.98

Half-equiv: $\quad pOH = -\log(1\times10^{-3}) = 3.0 \qquad\qquad pH = 11.0$

Equiv: $\quad x = [H_3O^+] = \sqrt{\dfrac{1.0\times10^{-14}}{1\times10^{-3}} \times 0.05000} = 7\times10^{-7} \qquad pH = 6.2$

Indicator: methyl red, yellow at $pH = 6.3$ and red at $pH = 4.5$

(b) $C_b/K_b = 33{,}000$; thus, the approximation <u>is</u> valid; thus, the approximation is valid.

Initial: $\quad [OH^-] = \sqrt{0.1000\times3\times10^{-6}} = 5\times10^{-4}$ M $\quad$ pOH = 3.3 $\quad$ pH = 10.7

Half-equiv: $pOH = -\log(3\times10^{-6}) = 5.5 \qquad pH = 14 - pOH \qquad pH = 8.5$

Equiv: $\quad x = [H_3O^+] = \sqrt{\dfrac{1.0\times10^{-14}}{3\times10^{-6}} \times 0.05000} = 1\times10^{-5} \quad pH -\log(1\times10^{-5}) = 5.0$

Indicator: methyl red, yellow at $pH = 6.3$ and red at $pH = 4.5$

(c) $C_b/K_b = 1.4\times10^6$; thus, the approximation <u>is</u> valid; thus, the approximation is valid.

Initial: $\quad [OH^-] = \sqrt{0.1000\times7\times10^{-8}} = 8\times10^{-5}$ M $\quad$ pOH = 4.1 $\quad$ pH = 9.9

Half-equiv: $pOH = -\log(7\times10^{-8}) = 7.2 \qquad pH = 14 - pOH \qquad pH = 6.8$

Equiv: $\quad x = [H_3O^+] = \sqrt{\dfrac{1.0\times10^{-14}}{7\times10^{-8}} \times 0.05000} = 8\times10^{-5} \qquad\qquad pH=4.1$

Indicator: bromcresol green, blue at $pH = 5.5$ and yellow at $pH = 4.0$

The titration curves are drawn with respect to the same axes in the diagram below.

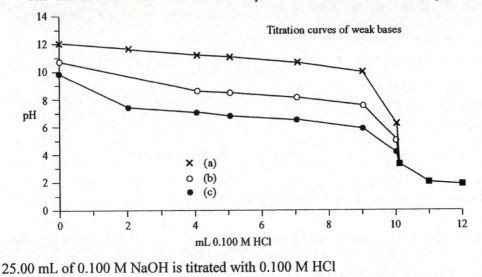

51. 25.00 mL of 0.100 M NaOH is titrated with 0.100 M HCl

(i) Initial pOH for 0.100 M NaOH: $[OH^-] = 0.100$ M, pOH = 1.000 or pH = 13.000

(ii) After addition of 24 mL: $[NaOH] = 0.100 \text{ M} \times \dfrac{25.00 \text{ mL}}{49.00 \text{ mL}} = 0.0510$ M

$$[HCl] = 0.100 \text{ M} \times \dfrac{24.00 \text{ mL}}{49.00 \text{ mL}} = 0.0490 \text{ M}$$

NaOH is in excess by 0.0020 M = $[OH^-]$ pOH = 2.70

(iii) At the equivalence point (25.00 mL), the pOH should be 7.000 and pH = 7.000

(iv) After addition of 26 mL: $[NaOH] = 0.100 \text{M} \times \dfrac{25.00}{51.00} = 0.0490$ M

$$[HCl] = 0.100 \text{ M} \times \dfrac{26.00 \text{ mL}}{51.00 \text{ mL}} = 0.0510 \text{ M}$$

HCl is in excess by 0.0020 M = $[H_3O^+]$ pH = 2.70 or pOH = 11.30

(v) After addition of 33.00 mL HCl(xs) $[NaOH] = 0.100 \text{ M} \times \dfrac{25.00 \text{ mL}}{58.00 \text{ mL}} = 0.0431$ M

$[HCl] = 0.100 \text{ M} \times \dfrac{33.00 \text{ mL}}{58.00 \text{ mL}} = 0.0569$ M $[HCl]_{excess} = 0.0138$ M pH = 1.860, pOH = 12.140

The graphs look to be mirror images of one another. In reality, one must reflect about a horizontal line centered at pH or pOH = 7 to obtain the other curve.

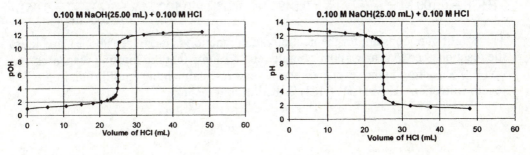

52. 25.00 mL of 0.100 M NH_3 is titrated with 0.100 M HCl $K_a = \dfrac{K_w}{K_b} = \dfrac{1 \times 10^{-14}}{1.8 \times 10^{-5}} = 5.6 \times 10^{-10}$

(i) For initial pOH, use I.C.E.(initial, change, equilibrium) table.

$$NH_3(aq) \;+\; H_2O(l) \quad\xrightleftharpoons{\;K_b\,=\,1.8\,\times\,10^{-5}\;}\quad NH_4^+(aq) \;+\; OH^-(aq)$$

initial	0.100 M	0 M	≈ 0 M
change	$-x$	$+x$	$+x$
equil.	$0.100 - x$	x	x

$1.8 \times 10^{-5} = \dfrac{x^2}{0.100 - x} \approx \dfrac{x^2}{0.100}$ (Assume $x \sim 0$) $x = 1.3 \times 10^{-3}$

($x < 5\%$ of 0.100, thus, the assumption is valid).

Hence, $x = [OH^-] = 1.3 \times 10^{-3}$ pOH = 2.87, pH = 11.13

(ii) After 2 mL of HCl added: $[HCl] = 0.100 \text{ M} \times \dfrac{2.00 \text{ mL}}{27.00 \text{ mL}} = 0.00741 \text{ M}$ (after dilution)

$[NH_3] = 0.100 \text{ M} \times \dfrac{25.00 \text{ mL}}{27.00 \text{ mL}} = 0.0926 \text{ M}$ (after dilution)

The equilibrium constant for the neutralization reaction is large ($K_{neut} = K_b/K_w = 1.9 \times 10^5$) and thus the reaction goes to nearly 100% completion. Assume that the limiting reagent is used up (100% reaction in the reverse direction) and re-establish the equilibrium by a shift in the forward direction. Here H_3O^+ (HCl) is the limiting reagent.

$$NH_4^+(aq) \;+\; H_2O(l) \quad\xrightleftharpoons{\;K_a\,=\,5.6\,\times\,10^{-10}\;}\quad NH_3(aq) \;+\; H_3O^+(aq)$$

initial	0 M	–		0.0926 M	0.00741 M
change	$+x$	–	$x = 0.00741$	$-x$	$-x$
100% rxn	0.00741	–		0.0852	0
change	$-y$	–	re-establish equilibrium	$+y$	$+y$
equil	$0.00741 - y$	–		$0.0852 + y$	y

$5.6 \times 10^{-10} = \dfrac{y(0.0852 + y)}{(0.00741 - y)} = \dfrac{y(0.0852)}{(0.00741)}$ (set $y \sim 0$) $y = 4.8 \times 10^{-11}$

(the approximation is clearly valid)

$y = [H_3O^+] = 4.8 \times 10^{-11}$; pH = 10.32 and pOH = 3.68

(iii) pH at 1/2 equivalence point = $pK_a = -\log 5.6 \times 10^{-10} = 9.26$ and pOH = 4.74

(iv) After addition of 24 mL of HCl:

$[HCl] = 0.100 \text{ M} \times \dfrac{24.00 \text{ mL}}{49.00 \text{ mL}} = 0.0490 \text{ M};$ $[NH_3] = 0.100 \text{ M} \times \dfrac{25.00 \text{ mL}}{49.00 \text{ mL}} = 0.0510 \text{ M}$

The equilibrium constant for the neutralization reaction is large (see above), and thus the reaction goes to nearly 100% completion. Assume that the limiting reagent is used up (100% reaction in the reverse direction) and re-establish the equilibrium in the reverse direction. Here H_3O^+ (HCl) is the limiting reagent.

$$NH_4^+(aq) + H_2O(l) \xrightarrow{K_a = 5.6 \times 10^{-10}} NH_3(aq) + H_3O^+(aq)$$

				NH$_3$(aq)	H$_3$O$^+$(aq)
initial	0 M	–		0.0541 M	0.0490 M
change	+x	–	x = **0.0490**	–x	–x
100% rxn	0.0490	–		0.0020	0
change	–y	–	**re-establish equilibrium**	+y	+y
equil	0.0490–y	–		0.0020 + y	y

$$5.6 \times 10^{-10} = \frac{y(0.0020 + y)}{(0.0490 - y)} = \frac{y(0.0020)}{(0.0490)} \quad \text{(Assume } y \sim 0) \quad y = 1.3 \times 10^{-8} \text{ (valid)}$$

$$y = [H_3O^+] = 1.3 \times 10^{-8}; \quad pH = 7.89 \text{ and } pOH = 6.11$$

(v) Equiv. point: 100% reaction of NH$_3 \rightarrow$ NH$_4^+$: $[NH_4^+] = 0.100 \times \dfrac{25.00 \text{ mL}}{50.00 \text{ mL}} = 0.0500$ M

$$NH_4^+(aq) + H_2O(l) \xrightarrow{K_a = 5.6 \times 10^{-10}} NH_3(aq) + H_3O^+(aq)$$

			NH$_3$(aq)	H$_3$O$^+$(aq)
initial	0.0500 M	–	0 M	~ 0 M
change	–x	–	+x	+x
equil	0.0500–x	–	x	x

$$5.6 \times 10^{-10} = \frac{x^2}{(0.0500 - x)} = \frac{x^2}{0.0500} \quad \text{(Assume } x \sim 0) \quad x = 5.3 \times 10^{-6}$$

(the approximation is clearly valid)

$$x = [H_3O^+] = 5.3 \times 10^{-6}; \quad pH = 5.28 \text{ and } pOH = 8.72$$

(vi) After addition of 26 mL of HCl, HCl is in excess. The pH and pOH should be the same as those in question 51. pH = 2.70 and pOH = 11.30

(vii) After addition of 33 mL of HCl, HCl is in excess. The pH and pOH should be the same as those in question 51. pH = 1.860 and pOH = 12.140
This time the curves are not mirror images of one another, but rather they are related through a reflection of a horizontal line centered at pH or pOH = 7.

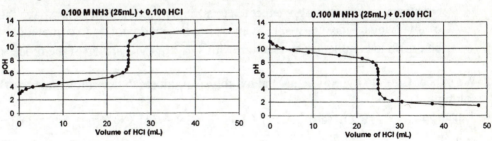

pH of Salts of Polyprotic Acids

53. We expect a solution of Na$_2$S to be alkaline, or basic. This alkalinity is created by the hydrolysis of the sulfide ion, the anion of a very weak acid ($K_2 = 1 \times 10^{-19}$ for H$_2$S).

$$S^{2-}(aq) + H_2O(l) \rightleftharpoons HS^-(aq) + OH^-(aq)$$

54. We expect the pH of a solution of sodium dihydrogen citrate, NaH_2Cit, to be acidic because the pK_a values of first and second ionization constants of polyprotic acids are reasonably large. The pH of a solution of the salt is the average of pK_1 and pK_2. For citric acid, in fact, this average is $(3.13+4.77)\div 2 = 3.95$. Thus, NaH_2Cit affords acidic solutions.

<u>**55.**</u> **(a)** $H_3PO_4(aq)+CO_3^{2-}(aq)\rightleftharpoons H_2PO_4^-(aq)+HCO_3^-(aq)$

$H_2PO_4^-(aq)+CO_3^{2-}(aq)\rightleftharpoons HPO_4^{2-}(aq)+HCO_3^-(aq)$

$HPO_4^{2-}(aq)+OH^-(aq)\rightleftharpoons PO_4^{3-}(aq)+H_2O(l)$

(b) The pH values of 1.00 M solutions of the three ions are;

$1.0\ M\ OH^- \to pH = 14.00$ $1.0\ M\ CO_3^{2-} \to pH = 12.16$ $1.0\ M\ PO_4^{3-} \to pH = 13.15$

Thus, we see that CO_3^{2-} is not a strong enough base to remove the third proton from H_3PO_4. As an alternative method of solving this problem, we can compute the equilibrium constant the reactions of carbonate ion with H_3PO_4, $H_2PO_4^-$ and HPO_4^{2-}.

$H_3PO_4+CO_3^{2-}\rightleftharpoons H_2PO_4^-+HCO_3^-$ $K=\dfrac{K_{a_1}\{H_3PO_4\}}{K_{a2}\{H_2CO_3\}}=\dfrac{7.1\times 10^{-3}}{4.7\times 10^{-11}}=1.5\times 10^8$

$H_2PO_4^-+CO_3^{2-}\rightleftharpoons HPO_4^{2-}+HCO_3^-$ $K=\dfrac{K_{a2}\{H_3PO_4\}}{K_{a2}\{H_2CO_3\}}=\dfrac{6.3\times 10^{-8}}{4.7\times 10^{-11}}=1.3\times 10^3$

$HPO_4^{2-}+CO_3^{2-}\rightleftharpoons PO_4^{3-}+HCO_3^-$ $K=\dfrac{K_{a3}\{H_3PO_4\}}{K_{a2}\{H_2CO_3\}}=\dfrac{4.2\times 10^{-13}}{4.7\times 10^{-11}}=8.9\times 10^{-3}$

Since the equilibrium constant for the third reaction is much smaller than 1.00, we conclude that it proceeds to the right to only a negligible extent and thus is not a practical method of producing PO_4^{3-}. The other two reactions have large equilibrium constants, and products are expected to strongly predominate. They have the advantage of involving an inexpensive base and, even if they do not go to completion, they will be drawn to completion by reaction with OH^- in the last step of the process.

56. We expect CO_3^{2-} to hydrolyze and the hydrolysis products to determine the pH of the solution.

Equation:	$HCO_3^-(aq)$ +	$H_2O(l)$	$\rightleftharpoons$	$"H_2CO_3"(aq)$ +	$OH^-(aq)$
Initial	1.00 M	–		0 M	≈ 0 M
Changes:	$-x$ M	–		$+x$ M	$+x$ M
Equil:	$(1.00-x)$ M	–		x M	x M

$K_b=\dfrac{K_w}{K_1}=\dfrac{1.00\times 10^{-14}}{4.4\times 10^{-7}}=2.3\times 10^{-8}=\dfrac{[H_2CO_3][OH^-]}{[HCO_3^-]}=\dfrac{(x)(x)}{1.00-x}\approx \dfrac{x^2}{1.00}$

$(C_b/K_b$ = a very large number; thus, the approximation <u>is</u> valid).

$$x = \sqrt{1.00 \times 2.3 \times 10^{-8}} = 1.5 \times 10^{-4} \; M = [OH^-]; \; pOH = -\log(1.5 \times 10^{-4}) = 3.82 \quad pH = 10.18$$

For $1.00 \, M \, NaOH$, $\left[OH^-\right] = 1.00$ $pOH = -\log(1.00) = 0.00$ $pH = 14.00$

Both $1.00 \, M \, NaHCO_3$ and 1.00 M NaOH have an equal capacity to neutralize acids since one mole of each neutralizes one mole of strong acid.

$$NaOH(aq) + H_3O^+(aq) \rightarrow Na^+(aq) + 2H_2O(l)$$

$$NaHCO_3(aq) + H_3O^+(aq) \rightarrow Na^+(aq) + CO_2(g) + 2H_2O(l).$$

But on a per gram basis, the one with the smaller molar mass is the more effective. Because the molar mass of NaOH is 40.0 g/mol, while that of $NaHCO_3$ is 84.0 g/mol, NaOH(s) is more than twice as effective than $NaHCO_3(s)$ on a per gram basis. $NaHCO_3$ is preferred in laboratories for safety and expense reasons. NaOH is not a good choice because it can cause severe burns. $NaHCO_3$, baking soda, by comparison, is relatively non-hazardous. It also is much cheaper than NaOH.

57. Malonic acid has the formula $H_2C_3H_2O_4$ MM = 104.06 $\dfrac{g}{mol}$

Moles of $H_2C_3H_2O_4$ = 19.5 g $\times \dfrac{1 \, mol}{104.06 \, g}$ = 0.187 mol

Concentration of $H_2C_3H_2O_4 = \dfrac{moles}{V} = \dfrac{0.187 \, moles}{0.250 \, L}$ = 0.748 M

The second proton that can dissociate has a negligible effect on pH (K_{a_2} is very small). Thus the pH is determined almost entirely by the first proton loss.

	$H_2A(aq)$	+	$H_2O(l)$	$\rightleftharpoons$	$HA^-(aq)$	+	$H_3O^+(aq)$
initial	0.748 M	–			0 M		$\approx 0 \, M$
change	$-x$	–			$+x$		$+x$
equil	0.748–x	–			x		x

So, $x = \dfrac{x^2}{0.748 - x} = K_a;$ pH = 1.47, therefore, $[H_3O^+]$ = 0.034 M = x,

$K_{a_1} = \dfrac{(0.034)^2}{0.748 - 0.034} = 1.6 \times 10^{-3}$

(1.5×10^{-3} in tables, difference owing to ionization of the second proton)

	$HA^-(aq)$	+	$H_2O(l)$	$\rightleftharpoons$	$A^{2-}(aq)$	+	$H_3O^+(aq)$
initial	0.300 M	–			0 M		$\approx 0 \, M$
change	$-x$	–			$+x$		$+x$
equil	0.300–x	–			x		x

pH = 4.26, therefore, $[H_3O^+] = 5.5 \times 10^{-5} \; M = x$, $K_{a_2} = \dfrac{(5.5 \times 10^{-5})^2}{0.300 - 5.5 \times 10^{-5}} = 1.0 \times 10^{-8}$

58. Ortho-phthalic acid. $K_{a_1} = 1.1 \times 10^{-3}$, $K_{a_2} = 3.9 \times 10^{-6}$

   (a) We have solution of HA.

	$HA^-(aq)$	+	$H_2O(l)$	$\rightleftharpoons$	$A^{2-}(aq)$	+	$H_3O^+(aq)$
initial	0.350 M		–		0 M		~0 M
change	$-x$		–		$+x$		$+x$
equil	$0.350-x$		–		x		x

$$3.9 \times 10^{-6} = \frac{x^2}{0.350-x} \approx \frac{x^2}{0.350}, \quad x = 1.2 \times 10^{-3}$$

$(x \ll 0.350$, thus, the approximation is valid) $\quad x = [H_3O^+] = 1.2 \times 10^{-3}$, pH = 2.93

   (b) 36.35 g of potassium ortho-phthalate (MM = 242.314 g mol^{-1})

Number of moles of potassium ortho-phthalate $= 36.35 \text{ g} \times \dfrac{1 \text{ mol}}{242.314 \text{ g}} = 0.150$ mol in 1 L

	$A^{2-}(aq)$	+	$H_2O(l)$	$\rightleftharpoons$	$HA^-(aq)$	+	$OH^-(aq)$
initial	0.150 M		–		0 M		≈ 0 M
change	$-x$		–		$+x$		$+x$
equil	$0.150-x$		–		x		x

$$K_{b,A^{2-}} = \frac{K_w}{K_{a_2}} = \frac{1.0 \times 10^{-14}}{3.9 \times 10^{-6}} = 2.6 \times 10^{-9} = \frac{x^2}{0.150-x} \approx \frac{x^2}{0.150}$$

$x = 2.0 \times 10^{-5}$ ($x \ll 0.150$, so the approximation is valid) $= [OH^-]$

pOH $= -\log 2.0 \times 10^{-5} = 4.70$;   pH = 9.30

General Acid-Base Equilibria

59 **(a)** $Ba(OH)_2$ is a strong base. pOH $= 14.00 - 11.88 = 2.12$ $\left[OH^-\right] = 10^{-2.12} = 0.0076$ M

$$\left[Ba(OH)_2\right] = \frac{0.0076 \text{ mol } OH^-}{1 \text{ L}} \times \frac{1 \text{ mol } Ba(OH)_2}{2 \text{ mol } OH^-} = 0.0038 \text{ M}$$

   (b) $\text{pH} = 4.52 = pK_a + \log\dfrac{\left[C_2H_3O_2^-\right]}{\left[HC_2H_3O_2\right]} = 4.74 + \log\dfrac{0.294 \text{ M}}{\left[HC_2H_3O_2\right]}$

$$\log\frac{0.294 \text{ M}}{\left[HC_2H_3O_2\right]} = 4.52 - 4.74 \qquad \frac{0.294 \text{ M}}{\left[HC_2H_3O_2\right]} = 10^{-0.22} = 0.60$$

$$\left[HC_2H_3O_2\right] = \frac{0.294 \text{ M}}{0.60} = 0.49 \text{ M}$$

60. (a) $pOH = 14.00 - 8.95 = 5.05$ $\left[OH^-\right] = 10^{-5.05} = 8.9 \times 10^{-6}$ M

Equation: $C_6H_5NH_2(aq) \quad + \quad H_2O(l) \quad \rightleftharpoons \quad C_6H_5NH_3^+(aq) \quad + \quad OH^-(aq)$

Initial	x M	–	0 M	≈ 0 M
Changes:	-8.9×10^{-6} M	–	$+8.9 \times 10^{-6}$ M	$+8.9 \times 10^{-6}$ M
Equil:	$(x - 8.9 \times 10^{-6})$ M	–	8.9×10^{-6} M	8.9×10^{-6} M

$$K_b = \frac{\left[C_6H_5NH_3^+\right]\left[OH^-\right]}{\left[C_6H_5NH_2\right]} = 7.4 \times 10^{-10} = \frac{\left(8.9 \times 10^{-6}\right)^2}{x - 8.9 \times 10^{-6}}$$

$$x - 8.9 \times 10^{-6} = \frac{\left(8.9 \times 10^{-6}\right)^2}{7.4 \times 10^{-10}} = 0.11 \text{ M} \qquad x = 0.11 \text{ M} = \text{molarity of aniline}$$

(b) $\left[H_3O^+\right] = 10^{-5.12} = 7.6 \times 10^{-6}$ M

Equation: $NH_4^+(aq) \quad + \quad H_2O(l) \quad \rightleftharpoons \quad NH_3(aq) \quad + \quad H_3O^+(aq)$

Initial	x M	–	0 M	≈ 0 M
Changes:	-7.6×10^{-6} M	–	$+7.6 \times 10^{-6}$ M	$+7.6 \times 10^{-6}$ M
Equil:	$(x - 7.6 \times 10^{-6})$ M	–	7.6×10^{-6} M	7.6×10^{-6} M

$$K_a = \frac{\left[NH_3\right]\left[H_3O^+\right]}{\left[NH_4^+\right]} = \frac{K_w}{K_b \text{ for } NH_3} = \frac{1.0 \times 10^{-14}}{1.8 \times 10^{-5}} = 5.6 \times 10^{-10} = \frac{\left(7.6 \times 10^{-6}\right)^2}{x - 7.6 \times 10^{-6}}$$

$$x - 7.6 \times 10^{-6} = \frac{\left(7.6 \times 10^{-6}\right)^2}{5.6 \times 10^{-10}} = 0.10 \text{ M} \qquad x = \left[NH_4^+\right] = \left[NH_4Cl\right] = 0.10 \text{ M}$$

61. (a) A solution can be prepared with equal concentration of weak acid and conjugate base (it would be a buffer, with a buffer ratio of 1.00, where the $pH = pK_a = 9.26$). Clearly, this solution can be prepared, however, it would not have a pH of 6.07.

(b) These solutes can be added to the same solution, but the final solution will have an appreciable $\left[HC_2H_3O_2\right]$ because of the reaction of $H_3O^+(aq)$ with $C_2H_3O_2^-(aq)$

Equation: $H_3O^+(aq) \quad + \quad C_2H_3O_2^-(aq) \quad \rightleftharpoons \quad HC_2H_3O_2(aq) + H_2O(l)$

Initial	0.058 M	0.10 M	0 M	–
Changes:	–0.058 M	–0.058 M	+0.058 M	–
Equil:	≈ 0.000 M	0.04 M	0.058 M	–

Of course, some H_3O^+ will exist in the final solution, but not equivalent to 0.058 M HI.

(c) Both 0.10 M KNO_2 and 0.25 M KNO_3 can exist together. Some hydrolysis of the $NO_2^-(aq)$ ion will occur, forming $HNO_2(aq)$ and $OH^-(aq)$.

(d) $Ba(OH)_2$ is a strong base and will react as much as possible with the weak conjugate acid NH_4^+, to form $NH_3(aq)$. We will end up with a solution of $BaCl_2(aq)$, $NH_3(aq)$, and unreacted $NH_4Cl(aq)$.

(e) This will be a benzoic acid-benzoate ion buffer solution. Since the two components have the same concentration, the buffer solution will have
$pH = pK_a = -\log(6.3 \times 10^{-5}) = 4.20$. This solution can indeed exist.

(f) The first three components contain no ions that will hydrolyze. But $C_2H_3O_2^-$ is the anion of a weak acid and will hydrolyze to form a slightly basic solution. Since $pH = 6.4$ is an acidic solution, the solution described cannot exist.

62. (a) When $[H_3O^+]$ and $[HC_2H_3O_2]$ are high and $[C_2H_3O_2^-]$ is very low, a common ion H_3O^+ has been added to a solution of acetic acid, suppressing its ionization.

(b) When $[C_2H_3O_2^-]$ is high and $[H_3O^+]$ and $[HC_2H_3O_2]$ are very low, we are dealing with a solution of acetate ion, which hydrolyzes to produce a small concentration of $HC_2H_3O_2$.

(c) When $[HC_2H_3O_2]$ is high and both $[H_3O^+]$ and $[C_2H_3O_2^-]$ are low, the solution is an acetic acid solution, in which the solute is partially ionized.

(d) When both $[HC_2H_3O_2]$ and $[C_2H_3O_2^-]$ are high while $[H_3O^+]$ is low, the solution is a buffer solution, in which the presence of acetate ion suppresses the ionization of acetic acid.

Integrative and Advanced Exercises

63. (a) $NaHSO_4(aq) + NaOH(aq) \longrightarrow Na_2SO_4(aq) + H_2O(l)$
$HSO_4^-(aq) + OH^-(aq) \longrightarrow SO_4^{2-}(aq) + H_2O(l)$

(b) We first determine the mass of $NaHSO_4$.

$$\text{mass } NaHSO_4 = 36.56 \text{ mL} \times \frac{1 \text{ L}}{1000 \text{ mL}} \times \frac{0.225 \text{ mol NaOH}}{1 \text{ L}} \times \frac{1 \text{ mol } NaHSO_4}{1 \text{ mol NaOH}}$$

$$\times \frac{120.06 \text{ g } NaHSO_4}{1 \text{ mol } NaHSO_4} = 0.988 \text{ g } NaHSO_4$$

$$\% \ NaCl = \frac{1.016 \text{ g sample} - 0.988 \text{ g } NaHSO_4}{1.016 \text{ g sample}} \times 100\% = 2.8\% \ NaCl$$

(c) At the endpoint of this titration the solution is one of predominantly SO_4^{2-}, from which the pH is determined by hydrolysis. Since K_a for HSO_4^- is relatively large (1.1×10^{-2}), base hydrolysis of SO_4^{2-} should not occur to a very great extent. The pH of a neutralized solution should be very nearly 7, and most of the indicators represented in Figure 17-8 would be suitable. A more exact solution follows.

$$[SO_4^{2-}] = \frac{0.988 \text{ g NaHSO}_4}{0.03656 \text{ L}} \times \frac{1 \text{ mol NaHSO}_4}{120.06 \text{ g NaHSO}_4} \times \frac{1 \text{ mol SO}_4^{2-}}{1 \text{ mol NaHSO}_4} = 0.225 \text{ M}$$

Equation: SO_4^{2-} (aq) + H_2O(l) $\rightleftharpoons$ HSO_4^- (aq) + OH^- (aq)

Initial: 0.225 M   –   0 M $\approx$ 0 M

Changes: $-x$ M   – $+x$ M $+x$ M

Equil: (0.225 $-x$)M   – x M x M

$$K_b = \frac{[HSO_{4-}][OH^-]}{[SO_{4^{2-}}]} = \frac{K_w}{K_{a_2}} = \frac{1.0 \times 10^{-14}}{1.1 \times 10^{-2}} = 9.1 \times 10^{-13} = \frac{x \cdot x}{0.225 - x} \approx \frac{x^2}{0.225}$$

$$[OH^-] = \sqrt{9.1 \times 10^{-13} \times 0.225} = 4.5 \times 10^{-7} \text{ M}$$

(the approximation was valid since $x \ll 0.225$ M)

pOH = $-\log(4.4 \times 10^{-7})$ = 6.35   pH = 14.00 $-$ 6.35 = 7.65

Thus, either bromthymol blue (pH color change range from pH = 6.1 to pH = 7.9) or phenol red (pH color change range from pH = 6.4 to pH = 8.0) would be a suitable indicator, since either changes color at pH = 7.65.

64. The original solution contains $250.0 \text{ mL} \times \dfrac{0.100 \text{ mmol HC}_3\text{H}_5\text{O}_2}{1 \text{ mL soln}} = 25.0 \text{ mmol HC}_3\text{H}_5\text{O}_2$

$pK_a = -\log(1.35 \times 10^{-5}) = 4.87$ We let V be the volume added to the solution, in mL.

(a) Since we add V mL of HCl solution (and each mL adds 1.00 mmol H_3O^+ to the solution), we have added V mmol H_3O^+ to the solution. Now the final $[H_3O^+] = 10^{-1.00} = 0.100$ M.

$$[H_3O^+] = \frac{V \text{ mmol } H_3O^+}{(250.0 + V) \text{ mL}} = 0.100 \text{ M} \quad V = 25.0 + 0.100 \, V$$

$$V = \frac{25.0}{0.900} = 27.8 \text{ mL added}$$

Now, we check our assumptions. The total solution volume is 250.0 mL + 27.8 mL = 277.8 mL There are 25.0 mmol $HC_2H_3O_2$ present before equilibrium is established, and 27.8 mmol H_3O^+ also.

$$[HC_2H_3O_2] = \frac{25.0 \text{ mmol}}{277.8 \text{ mL}} = 0.0900 \text{ M} \quad [H_3O^+] = \frac{27.8 \text{ mmol}}{277.8 \text{ mL}} = 0.100 \text{ M}$$

Equation: $HC_3H_5O_2(aq) + H_2O(l) \rightleftharpoons C_3H_5O_2^-(aq) + H_3O^+(aq)$

Initial: 0.0900 M   –   0 M   0.100 M

Changes: -x M   –   +x M   +x M

Equil: (0.08999 – x)M   –   x M   (0.100 + x)M

$$K_a = \frac{[H_3O^+][C_3H_5O_2^-]}{[HC_3H_5O_2]} = 1.35 \times 10^{-5} = \frac{x\,(0.100+x)}{0.0900-x} \approx \frac{0.100\ x}{0.0900} \qquad x = 1.22 \times 10^{-5}\,M$$

The assumption used in solving this equilibrium situation, that $x \ll 0.0900$ clearly is correct. In addition, the tacit assumption that virtually all of the H_3O^+ comes from the HCl also is correct.

(b) The pH desired is within 1.00 pH unit of the pK_a. We use the Henderson-Hasselbalch equation to find the required buffer ratio.

$$pH = pK_a + \frac{[A^-]}{[HA]} = pK_a + \log \frac{n_{A^-}/V}{n_{HA}/V} = PK_a + \log \frac{n_{A^-}}{n_{HA}} = 4.00 = 4.87 + \log \frac{n_{A^-}}{n_{HA}}$$

$$\log \frac{n_{A^-}}{n_{HA}} = -0.87 \qquad \frac{n_{A^-}}{n_{HA}} = 10^{-0.87} = 0.13 = n\frac{n_{A^-}}{25.00} \qquad _{A^-} = 3.3$$

$$V_{soln} = 3.3 \text{ mmol } C_3H_5O_2^- \times \frac{1 \text{ mmol } NaC_3H_5O_2}{1 \text{ mmol } C_3H_5O_2^-} \times \frac{1 \text{ mL soln}}{1.00 \text{ mmol } NaC_3H_5O_2} = 3.3 \text{ mL added}$$

We have assumed that all of the $C_3H_5O_2^-$ is obtained from the $NaC_3H_5O_2$ solution, since the addition of that ion in the solution should suppress the ionization of $HC_3H_5O_2$.

(c) We let V be the final volume of the solution.

Equation: $HC_3H_5O_2(aq) + H_2O(l) \rightleftharpoons C_3H_5O_2^-(aq) + H_3O^+(aq)$

Initial: 25.0/V   –   0 M $\approx$ 0 M

Changes: –x/V M   –   +x/V M   +x/V M

Equil: (25.0/V – x/V) M   –   x/V M   x/V M

$$K_a = \frac{[H_3O^+][C_3H_5O_2^-]}{[HC_3H_5O_2]} = 1.35 \times 10^{-5} = \frac{(x/V)^2}{25.0/V - x/V} \approx \frac{x^2/V}{25.0}$$

When V = 250.0 mL, x = 0.29 mmol H_3O^+ $[H_3O^+] = \dfrac{0.29 \text{ mmol}}{250.0 \text{ mL}} = 1.2 \times 10^{-3}$ M

pH = –log(1.2 × 10⁻³) = 2.92   An increase of 0.15 pH unit gives pH = 2.92 + 0.15 = 3.07
$[H_3O^+] = 10^{-3.07} = 8.5 \times 10^{-4}$ M   This is the value of x/V. Now solve for V.

$$1.3 \times 10^{-5} = \frac{(8.5 \times 10^{-4})^2}{25.0/V - 8.5 \times 10^{-4}} \qquad 25.0/V - 8.5 \times 10^{-4} = \frac{(8.5 \times 10^{-4})^2}{1.3 \times 10^{-5}} = 0.056$$

$$25.0/V = 0.056 + 0.00085 = 0.057 \qquad V = \frac{25.0}{0.057} = 4.4 \times 10^2 \text{ mL}$$

On the other hand, if we would have used $[H_3O^+] = 1.16 \times 10^{-3}$ M (rather than 1.2×10^{-3} M), we obtain $V = 4.8 \times 10^2$ mL. The answer to the problem thus is sensitive to the last significant figure that is retained. We obtain $V = 4.6 \times 10^2$ mL, requiring the addition of 2.1×10^2 mL of H_2O.

Another possibility is to recognize that $[H_3O^+] = \sqrt{K_a C_a}$ for a weak acid with ionization constant K_a and initial concentration C_a if the approximation is valid.

If C_a is changed to $C_a/2$, $[H_3O^+] = \sqrt{\dfrac{K_a}{C_a}} \times \dfrac{\sqrt{2}}{2}$ Since, pH $= -\log [H_3O^+]$, the change in

pH is given by $\Delta pH = -\log \sqrt{2} = -\log 2^{1/2} = -0.5 \log 2 = -0.5 \times 0.30103 = -0.15$
This corresponds to doubling the solution volume, that is, to adding 250 mL water. Diluted by half with water, $[H_3O^+]$ goes down and pH rises, then $(pH_1 - pH_2) < 0$.

65. Carbonic acid is unstable in aqueous solution, decomposing to $CO_2(aq)$ and H_2O. The $CO_2(aq)$, in turn, escapes from the solution, to a degree determined, in large part by the partial pressure of $CO_2(g)$ in the atmosphere.

$H_2CO_3(aq) \ll H_2O(l) + CO_2(aq) \ll CO_2(g) + H_2O(l)$

Thus, a solution of carbonic acid in the laboratory soon will reach a low $[H_2CO_3]$, since the partial pressure of $CO_2(g)$ in the atmosphere is quite low. Thus, such a solution would be unreliable as a buffer. In the body, however, the $[H_2CO_3]$ is regulated in part by the process of respiration. Respiration rates increase when it is necessary to decrease $[H_2CO_3]$ and respiration rates decrease when it is necessary to increase $[H_2CO_3]$.

66. (a) At a pH $= 2.00$, in Figure 17-9 the pH is changing gradually with added NaOH. There would be no sudden change in color with the addition of a small volume of NaOH.

(b) At pH $= 2.0$ in Figure 17-9, approximately 20.5 mL have been added. Since, equivalence required the addition of 25.0 mL, there are 4.5 mL left to add.

Therefore, % HCl unneutralized $= \dfrac{4.5}{25.0} \times 100\% = 18\%$

67. Let us begin the derivation with the definition of $[H_3O^+]$.

$[H_3O^+] = \dfrac{\text{amount of excess } H_3O^+}{\text{volume of titrant} + \text{volume of solution being titrated}}$

Let V_a = volume of acid (solution being titrated) V_b = volume of base (titrant)
M_a = molarity of acid M_b = molarity of base
$[H_3O^+] = \dfrac{V_a \times M_a - V_b \times M_b}{V_a + V_b}$

Now we solve this equation for V_b

$$V_a \times M_a - V_b \times M_b = [H_3O^+](V_a + V_b) \qquad\qquad V_b([H_3O^+] + M_b) = V_a(M_a - [H_3O^+])$$

$$V_b = \frac{V_a(M_a - [H_3O^+])}{[H_3O^+] + M_b} \qquad\qquad V_b = \frac{V_a(M_a - 10^{-pH})}{10^{-pH} + M_b}$$

(a) $10^{-pH} = 10^{-2.00} = 1.0 \times 10^{-2} M$ $\qquad$ $V_b = \dfrac{20.00 \text{ mL} (0.1500 - 0.010)}{0.010 + 0.1000} = 25.4\underline{5} \text{ mL}$

(b) $10^{-pH} = 10^{-3.50} = 3.2 \times 10^{-4} M$ $\qquad$ $V_b = \dfrac{20.00 \text{ mL} (0.1500 - 0.0003)}{0.0003 + 0.1000} = 29.85 \text{ mL}$

(c) $10^{-pH} = 10^{-5.00} = 1.0 \times 10^{-5} M$ $\qquad$ $V_b = \dfrac{20.00 \text{ mL} (0.1500 - 0.00001)}{0.00001 + 0.1000} = 30.00 \text{ mL}$

Beyond the equivalence point, the situation is different.

$$[OH^-] = \frac{\text{amount of excess OH}^-}{\text{total solution volume}} = \frac{V_b \times M_b - V_a \times M_a}{V_a + V_b}$$

Solve this equation for V_b. $\quad [OH^-](V_a + V_b) = V_b \times M_b - V_a \times M_a$

$$V_a([OH^-] + M_a) = V_b(M_b - [OH^-]) \qquad\qquad V_b = \frac{V_a([OH^-] + M_a)}{M_b - [OH^-]}$$

(d) $[OH^-] = 10^{-pOH} = 10^{-14.00 + pH} = 10^{-3.50} = 0.00032 M;$

$$V_b = 20.00 \text{ mL}\left(\frac{0.00032 + 0.1500}{0.1000 - 0.00032}\right) = 30.16 \text{ mL}$$

(e) $[OH^-] = 10^{-pOH} = 10^{-14.00 + pH} = 10^{-2.00} = 0.010 M; \quad V_b = 20.00 \text{ mL}\left(\dfrac{0.010 + 0.1500}{0.1000 - 0.010}\right) = 35.5\underline{6} \text{ mL}$

The initial pH $= -\log(0.150) = 0.824$. The titration curve is sketched below.

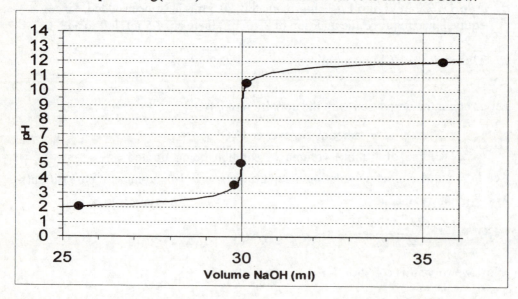

68. (a) The expressions that we obtained in Advanced Exercise 67 were for an acid being titrated by a base. For this titration, we need to switch the a and b subscripts and exchange $[OH^-]$ and $[H_3O^+]$.

Before the equivalence point :
$$V_a = V_b \frac{M_b - [OH^-]}{[OH^-] + M_a}$$

After the equivalence point :
$$V_a = V_b \frac{[H_3O^+] + M_b}{M_a - [H_3O^+]}$$

$pH = 13.00 \quad pOH = 14.00 - 13.00 = 1.00 \quad [OH^-] = 10^{-1.00} = 1.0 \times 10^{-1} = 0.10 \text{ M}$

$$V_a = 25.00 \text{ mL} \frac{0.250 - 0.10}{0.10 + 0.300} = 9.38 \text{ mL}$$

$pH = 12.00 \quad pOH = 14.00 - 12.00 = 2.00 \quad [OH^-] = 10^{-2.00} = 1.0 \times 10^{-2} = 0.010 \text{ M}$

$$V_a = 25.00 \text{ mL} \frac{0.250 - 0.010}{0.010 + 0.300} = 19.4 \text{ mL}$$

$pH = 10.00 \quad pOH = 14.00 - 10.00 = 4.00 \quad [OH^-] = 10^{-4.00} = 1.0 \times 10^{-4} = 0.00010 \text{ M}$

$$V_a = 25.00 \text{ mL} \frac{0.250 - 0.00010}{0.00010 + 0.300} = 20.8 \text{ mL}$$

$pH = 4.00 \quad [H_3O^+] = 10^{-4.00} = 1.0 \times 10^{-4} = 0.00010 \text{ M}$

$$V_a = 25.00 \frac{0.00010 + 0.250}{0.300 - 0.00010} = 20.8 \text{ mL}$$

$pH = 3.00 \quad [H_3O^+] = 10^{-3.00} = 1.0 \times 10^{-3} = 0.0010 \text{ M}$

$$V_a = 25.00 \frac{0.0010 + 0.250}{0.300 - 0.0010} = 21.0 \text{ mL}$$

(b) Our expression does not include the equilibrium constant, K_a. But K_a is not needed after the equivalence point. $pOH = 14.00 - 11.50 = 2.50 \; [OH^-] = 10^{-2.50} = 3.2 \times 10^{-3} = 0.0032 \text{ M}$

$$V_b = \frac{V_a(M_a + [OH^-])}{M_b - [OH^-]} = \frac{50.00 \text{ mL} (0.0100 + 0.0032)}{0.0500 - 0.0032} = 14.1 \text{ mL}$$

Let us use the Henderson-Hasselbalch equation as a basis to derive an expression that incorporates K_a. Note that because the numerator and denominator of that expression are concentrations of substances present in the same volume of solution, those concentrations can be replaced by numbers of moles.
Amount of anion $= V_b M_b$ since the added OH^- reacts 1:1 with the weak acid.
Amount of acid $= V_a M_a - V_b M_b$ the acid left unreacted

$$pH = pK_a + \log \frac{[A^-]}{[HA]} = pK_a + \log \frac{V_b M_b}{V_a M_a - V_b M_b}$$

Rearrange and solve for V_b $10^{pH - pK} = \frac{V_b M_b}{V_a M_a - V_b M_b}$

$$(V_a M_a - V_b M_b)10^{pH-pK} = V_b M_b \quad V_b = \frac{V_a M_a \, 10^{pH-pK}}{M_b(1+10^{pH-pK})}$$

$$pH = 4.50, \quad 10^{pH-pK} = 10^{4.50-4.20} = 2.0 \quad V_b = \frac{50.00 \text{ mL} \times 0.0100 \text{ M} \times 2.0}{0.0500(1+2.0)} = 6.7 \text{ mL}$$

$$pH = 5.50, \quad 10^{pH-pK} = 10^{5.50-4.20} = 20. \quad V_b = \frac{50.00 \text{ mL} \times 0.0100 \text{ M } 20}{0.0500(1+20.)} = 9.5 \text{ mL}$$

69. (a) We concentrate on the ratio of concentrations of which the logarithm is taken.

$$\frac{[\text{conjugate base}]_{eq}}{[\text{weak acid}]_{eq}} = \frac{\text{equil. amount conj. base}}{\text{equil. amount weak acid}} = \frac{f \times \text{initial amt. weak acid}}{(1-f) \times \text{initial amt. weak acid}} = \frac{f}{1-f}$$

The first transformation is the result of realizing that the volume in which the weak acid and its conjugate base are dissolved is the same volume, and therefore the ratio of equilibrium amounts is the same as the ratio of concentrations. The second transformation is the result of realizing, for instance, that if 0.40 of the weak acid has been titrated, 0.40 of the original amount of weak acid now is in the form of its conjugate base, and 0.60 of that amount remains as weak acid. Equation (17.2) then is: $pH = pK_a + \log(f/(1-f))$

(b) We use the equation just derived. $pH = 10.00 + \log \dfrac{0.27}{(1 - 0.23)} = 9.57$

70. (a) $pH = pK_{a2} + \log \dfrac{[HPO_4^{2-}]}{[H_2PO_4^-]} = 7.20 + \log \dfrac{[HPO_4^{2-}]}{[H_2PO_4^-]} = 7.40 \quad \dfrac{[HPO_4^{2-}]}{[H_2PO_4^-]} = 10^{+0.20} = 1.6$

(b) In order for the solution to be isotonic, it must have the same concentration of ions as does the isotonic NaCl solution.

$$[\text{ions}] = \frac{9.2 \text{ g NaCl}}{1 \text{ L soln}} \times \frac{1 \text{ mol NaCl}}{58.44 \text{ g NaCl}} \times \frac{2 \text{ mol ions}}{1 \text{ mol NaCl}} = 0.31 \text{ M}$$

Thus, 1.00 L of the buffer must contain 0.31 moles of ions. The two solutes that are used to formulate the buffer both ionize: KH_2PO_4 produces 2 mol of ions (K^+ and $H_2PO_4^-$) per mol of solute, while $Na_2HPO_4 \cdot 12 H_2O$ produces 3 mol of ions (2 Na^+ and HPO_4^{2-}) per mole of solute. We let x = amount of KH_2PO_4 and y = amount of $Na_2HPO_4 \cdot 12 H_2O$.

$$\frac{y}{x} = 1.6 \quad \text{or} \quad y = 1.6x \qquad\qquad 2x + 3y = 0.31 = 2x + 3(1.6x) = 6.8x$$

$$x = \frac{0.31}{6.8} = 0.046 \text{ mol } KH_2PO_4 \quad y = 1.6 \times 0.046 = 0.074 \text{ mol } Na_2HPO_4 \cdot 12 H_2O$$

$$\text{mass } KH_2PO_4 = 0.046 \text{ mol } KH_2PO_4 \times \frac{136.08 \text{ g } KH_2PO_4}{1 \text{ mol } KH_2PO_4} = 6.3 \text{ g } KH_2PO_4$$

$$\text{mass of } Na_2HPO_4 \cdot 12H_2O = 0.074 \text{ mol } Na_2HPO_4 \cdot 12H_2O \times \frac{358.1 \text{ g } Na_2HPO_4 \cdot 12H_2O}{1 \text{ mol } Na_2HPO_4 \cdot 12H_2O}$$

$$= 26 \text{ g } Na_2HPO_4 \cdot 12H_2O$$

71. A solution of NH_4Cl should be acidic, hence, we should add an alkaline solution to make it pH neutral. We base our calculation on the ionization equation for $NH_3(aq)$, and assume that little $NH_4^+(aq)$ is transformed to $NH_3(aq)$ because of the inhibition of that reaction by the added $NH_3(aq)$, and because the added volume of $NH_3(aq)$ does not significantly alter the V_{total}.

Equation: $NH_3(aq) + H_2O(l) \rightleftharpoons NH_4^+(aq) + OH^-(aq)$

Initial: xM   0.500 M 1.0×10^{-7}M

$$K_b = \frac{[NH_4^+][OH^-]}{[NH_3]} = \frac{(0.500 \text{ M})(1.0 \times 10^{-7} \text{ M})}{x \text{ M}} = 1.8 \times 10^{-5}$$

$$x \text{ M} = \frac{(0.500 \text{ M})(1.0 \times 10^{-7} \text{ M})}{1.8 \times 10^{-5} \text{ M}} = 2.8 \times 10^{-3} \text{ M} = [NH_3]$$

$$V = 500 \text{ mL} \times \frac{2.8 \times 10^{-3} \text{mol } NH_3}{1 \text{ L final soln}} \times \frac{1 \text{ L conc. soln}}{10.0 \text{ mol } NH_3} \times \frac{1 \text{ drop}}{0.05 \text{ mL}} = 2.8 \text{ drops} \cong 3 \text{ drops}$$

72. (a) In order to sketch the titration curve, we need the pH at the following points.

i) Initial point. That is the pH of 0.0100 M *p*-nitrophenol, which we represent by the general formula for an indicator that also is a weak acid, HIn.

Equation: HIn(aq)   + $H_2O(l)$ $\rightleftharpoons$ $H_3O^+(aq)$ + $In^-(aq)$

Initial: 0.0100 M   – ≈ 0 M   0 M

Changes: $-x$ M   – $+x$ M $+x$ M

Equil: (0.0100–x) M   – x M x M

$$K_{HIn} = 10^{-7.15} = 7.1 \times 10^{-8} = \frac{[H_3O^+] \; [In^-]}{[HIn]} = \frac{x \cdot x}{0.0100 - x} \approx \frac{x^2}{0.0100}$$

$C_a / K_a = 1 \times 10^5$; thus, the approximation is valid

$[H_3O^+] = \sqrt{0.0100 \times 7.1 \times 10^{-8}} = 2.7 \times 10^{-5}$ M   pH $= -\log(2.7 \times 10^{-5}) = 4.57$

ii) At the half-equivalence point, the pH = pK_{HIn} = 7.15

iii) At the equiv point, pH is that of In^-. n_{In^-} = 25.00 mL $\times$ 0.0100 M = 0.250 mmol In^-

$$V_{titrant} = 0.25 \text{ mmol HIn} \times \frac{1 \text{ mmol NaOH}}{1 \text{ mmol HIn}} \times \frac{1 \text{ mL titrant}}{0.0200 \text{ mmol NaOH}} = 12.5 \text{ mL titrant}$$

$$[In^-] = \frac{0.25 \text{ mmol In}^-}{25.00 \text{ mL} + 12.5 \text{ mL}} = 6.67 \times 10^{-3} \text{ M} = 0.00667 \text{ M}$$

Equation: $In^-(aq)$ + $H_2O(l)$ $\rightleftharpoons$ HIn(aq)   + $OH^-(aq)$

Initial: 0.00667 M   –   0 M ≈ 0 M

Changes: $-x$ M   – $+x$ M $+x$ M

Equil: (0.00667 $-x$) M   – x M x M

$$K_b = \frac{[HIn][OH^-]}{[In^-]} = \frac{K_w}{K_a} = \frac{1.0 \times 10^{-4}}{7.1 \times 10^{-8}} = 1.4 \times 10^{-7} \approx \frac{x \cdot x}{0.00667 - x} = \frac{x \cdot x}{0.00667}$$

(C_a/K_a = 1.4×10^5; thus, the approximation is valid)

$[OH^-] = \sqrt{0.00667 \times 1.4 \times 10^{-7}} = 3.1 \times 10^{-5}$ pOH $= -\log(3.1 \times 10^{-5}) = 4.51$

pH $= 14.00 - $ pOH $= 9.49$

iv) Beyond the equivalence point, the pH is determined by the amount of excess OH⁻ in solution. After 13.0 mL of 0.0200 M NaOH is added, 0.50 mL is in excess, and the total volume is 38.0 mL.

$$[OH^-] = \frac{0.50 \text{ mL} \times 0.0200 \text{ M}}{38.0 \text{ mL}} = 2.63 \times 10^{-4} \text{ M} \quad pOH = 3.58 \quad pH = 10.42$$

After 14.0 mL of 0.0200 M NaOH is added, 1.50 mL is in excess, and V_{total} = 39.0 mL.

$$[OH^-] = \frac{1.50 \text{ mL} \times 0.0200 \text{ M}}{39.0 \text{ mL}} = 7.69 \times 10^{-4} \text{ M} \quad pOH = 3.11 \quad pH = 10.89$$

v) In the buffer region, the pH is determined with the use of the Henderson-Hasselbalch equation.

$$\text{At } f = 0.10, \ pH = 7.15 + \log\frac{0.10}{0.90} = 6.20 \quad \text{At } f = 0.90, \ pH = 7.15 + \log\frac{0.90}{0.10} = 8.10$$

f = 0.10 occurs with 0.10 × 12.5 mL = 1.25 mL added titrant; f = 0.90 with 11.25 mL added. The titration curve plotted from these points follows.

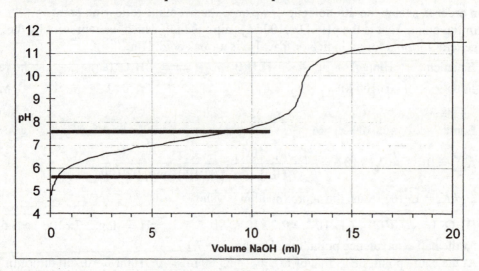

(b) The pH color change range of the indicator is shown on the titration curve.

(c) The equivalence point of the titration occurs at a pH of 9.49, far above the pH at which *p*-nitrophenol has turned yellow. In fact, the color of the indicator changes gradually during the course of the titration, making it unsuitable as an indicator for this titration. Possible indicators are as follows.

Phenophthalein: pH color change range from colorless at pH = 8.0 to red at pH = 10.0
Thymol blue: pH color change range from yellow at pH = 8.0 to blue at pH = 9.8
Thymolphthalein: pH color change range from colorless at pH ≈ 9 to blue at pH ≈ 11

The red tint of phenophthalein will appear orange in the titrated *p*-nitrophenol solution. The blue of thymol blue or thymolphthalein will appear green in the titrated *p*-nitrophenol solution, producing a somewhat better yellow end point than the orange phenophthalein endpoint.

73. (a) Equation (1) is the reverse of the equation for the autoionization of water. Thus, its equilibrium constant is simply the inverse of K_w.

$$K = \frac{1}{K_w} = \frac{1}{1.0 \times 10^{-14}} = 1.00 \times 10^{14}$$

Equation (2) is the reverse of the hydrolysis reaction for NH_4^+. Thus, its equilibrium constant is simply the inverse of the acid ionization constant for NH_4^+, $K_a = 5.6 \times 10^{-10}$

$$K' = \frac{1}{K_a} = \frac{1}{5.6 \times 10^{-10}} = 1.8 \times 10^9$$

(b) The extremely large size of each equilibrium constant indicates that each reaction goes essentially to completion. In fact, a general rule of thumb suggests that a reaction is considered essentially complete if $K_{eq} > 1000$ for the reaction.

74. The initial pH is that of 0.100 M $HC_2H_3O_2$.

Equation:	$HC_2H_3O_2(aq)$ +	$H_2O(l)$ ⇌	$C_2H_3O_2^-(aq)$ +	$H_3O^+(aq)$
Initial:	0.100 M	–	0 M	≈ 0 M
Changes:	$-x$ M	–	$+x$ M	$+x$ M
Equil:	$(0.100 - x)$ M	–	x M	x M

$$K_a = \frac{[H_3O^+][C_2H_3O_2^-]}{[HC_2H_3O_2]} = 1.8 \times 10^{-5} = \frac{x \cdot x}{0.100 - x} \approx \frac{x^2}{0.100}$$

$(C_a/K_a = 5.5 \times 10^3$; thus, the approximation is valid)

$$[H_3O^+] = \sqrt{0.100 \times 1.8 \times 10^{-5}} = 1.3 \times 10^{-3} \text{ M} \qquad pH = -\log(1.3 \times 10^{-3}) = 2.89$$

At the equivalence point, we have a solution of $NH_4C_2H_3O_2$ which has a pH = 7.00, because both NH_4^+ and $C_2H_3O_2^-$ hydrolyze to an equivalent extent, since $K_a(HC_2H_3O_2) \approx K_b(NH_3)$ and their hydrolysis constants also are virtually equal.
Total volume of titrant = 10.00 mL, since both acid and base have the same concentrations.

At the half equivalence point, which occurs when 5.00 mL of titrant have been added, pH = pK_a = 4.74. *When the solution has been 90% titrated,* 9.00 mL of 0.100 M NH_3 has been added. We use the Henderson-Hasselbalch equation to find the pH after 90% of the acid has been titrated

$$pH = pK_a + \log \frac{[C_2H_3O_2^-]}{[HC_2H_3O_2]} = 4.74 + \log \frac{9}{1} = 5.69$$

When the solution is 110% titrated, 11.00 mL of 0.100 M NH_3 have been added.
amount NH_3 added = 11.00 mL × 0.100 M = 1.10 mmol NH_3
amount NH_4^+ produced = amount $HC_2H_3O_2$ consumed = 1.00 mmol NH_4^+
amount NH_3 unreacted = 1.10 mmol NH_3 − 1.00 mmol NH_4^+ = 0.10 mmol NH_3
We use the Henderson-Hasselbalch equation to determine the pH of the solution.

$$pOH = pK_b + \log \frac{[NH_4^+]}{[NH_3]} = 4.74 + \log \frac{1.00 \text{ mmol } NH_4^+}{0.10 \text{ mmol } NH_3} = 5.74 \quad pH = 8.26$$

When the solution is 150% titrated, 15.00 mL of 0.100 M NH_3 have been added.
amount NH_3 added = 15.00 mL × 0.100 M = 1.50 mmol NH_3

amount NH_4^+ produced = amount $HC_2H_3O_2$ consumed = 1.00 mmol NH_4^+

amount NH_3 unreacted = 1.50 mmol NH_3 − 1.00 mmol NH_4^+ = 0.50 mmol NH_3

$$pOH = pK_b + \log\frac{[NH_4^+]}{[NH_3]} = 4.74 + \log\frac{1.00\,\text{mmol}\,NH_4^+}{0.50\,\text{mmol}\,NH_3} = 5.04 \quad pH = 8.96$$

The titration curve based on these points is sketched next. We note that the equivalence point is not particularly sharp and thus, satisfactory results are not obtained from acetic acid-ammonia titrations.

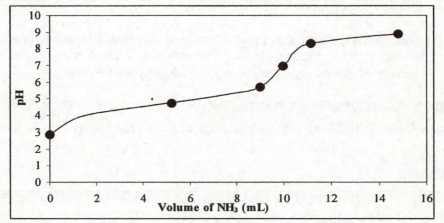

75. $C_6H_5NH_3^+$ is a weak acid, whose acid ionization constant is determined from $K_b(C_6H_5NH_2) = 7.4 \times 10^{-10}$.

$$K_a = \frac{1.0\times10^{-14}}{7.4\times10^{-10}} = 1.4\times10^{-5} \text{ and } pK_a = 4.85. \text{ We first determine the } \textit{initial pH}$$

Equation: $C_6H_5NH_{3^+}$ (aq) + H_2O(l) $\rightleftharpoons$ $C_6H_5NH_2$ (aq) + H_3O^+ (aq)

Initial: 0.0500 M   −   0 M ≈ 0 M

Changes: −x M   −   +x M   +x M

Equil: (0.0500 − x) M   − x M x M

$$K_a = 1.4\times10^{-5} = \frac{[C_6H_5NH_2]\ [H_3O^+]}{[C_6H_5NH_3^+]} = \frac{x\cdot x}{0.0500 - x} \approx \frac{x^2}{0.0500}$$

(Since $C_a/K_a = 3.6\times10^3$; thus, the approximation is valid)

$$[H_3O^+] = \sqrt{0.0500\times1.4\times10^{-5}} = 8.4\times10^{-4}\ M \qquad pH = -\log\ (8.4\times10^{-4}) = 3.08$$

At the equivalence point, we have a solution of $C_6H_5NH_2$(aq). Now find the volume of titrant.

$$V = 10.00\ mL \times \frac{0.0500\ \text{mmol}\ C_6H_5NH_3^+}{1\ mL} \times \frac{1\ \text{mmol}\ NaOH}{1\ \text{mmol}\ C_6H_5NH_3^+} \times \frac{1\ mL\ \text{titrant}}{0.100\ \text{mmol}\ NaOH} = 5.00\ mL$$

amount $C_6H_5NH_2 = 10.00\ mL \times 0.0500\ M = 0.500\ \text{mmol}\ C_6H_5NH_2$

$$[C_6H_5NH_2] = \frac{0.500\ \text{mmol}\ C_6H_5NH_2}{10.00\ mL\ +\ 5.00\ mL} = 0.0333\ M$$

Equation:　　$C_6H_5NH_2(aq) + H_2O(l) \rightleftharpoons C_6H_5NH_3^+(aq) + OH^-(aq)$

Initial:　　　　$0.0333\ M$　　　–　　　　　　$0\ M$　　　　$\approx 0\ M$

Changes:　　　$-x\ M$　　　　–　　　　　　$+x\ M$　　　$+x\ M$

Equil:　　　　$(0.0333 - x)\ M$　　–　　　　　$x\ M$　　　$x\ M$

$$K_b = 7.4 \times 10^{-10} = \frac{[C_6H_5NH_3^+][OH^-]}{[C_6H_5NH_2]} = \frac{x \cdot x}{0.0333 - x} \approx \frac{x^2}{0.0333}$$

(C_b/K_b = very large number; thus, the approximation is valid)

$$[OH^-] = \sqrt{0.0333 \times 7.4 \times 10^{-10}} = 5.0 \times 10^{-6} \qquad pOH = 5.30 \qquad pH = 8.70$$

At the half equivalence point, when 5.00 mL of 0.1000 M NaOH has been added, $pH = pK_a = 4.85$. *At points in the buffer region,* we use the expression derived in Advanced Exercise 69.

After 4.50 mL of titrant has been added; when $f = 0.90$, $pH = 4.85 + \log \dfrac{0.90}{0.10} = 5.80$

After 4.75 mL of titrant has been added; when $f = 0.95$, $pH = 4.85 + \log \dfrac{0.95}{0.05} = 6.13$

After the equivalence point has been reached, with 5.00 mL of titrant added, the pH of the solution is determined by the amount of excess OH^- that has been added.

At 105% titrated, $[OH^-] = \dfrac{0.25\ mL \times 0.100\ M}{15.25\ mL} = 0.0016\ M \qquad pOH = 2.80 \quad pH = 11.20$

At 120% titrated, $[OH^-] = \dfrac{1.00\ mL \times 0.100\ M}{16.00\ mL} = 0.00625\ M \quad pOH = 2.20 \quad pH = 11.80$

Of course, one could sketch a suitable titration curve by calculating fewer points.

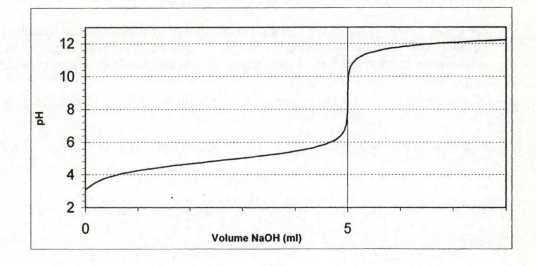

76. In order for a diprotic acid to be titrated to two distinct equivalence points, the acid must initially start out with two undissociated protons and the K_a values for the 1st and second proton must differ by >1000. This certainly is not the case for H_2SO_4 which is a strong acid (1st proton is 100% dissociated). Effectively, this situation is very similar to the titration of a monoprotic acid that has added strong acid (i.e. HCl or HNO_3). With the leveling effect of water, $K_{a1} = 1$ and $K_{a2} = 0.011$, there is a difference of only 100 between K_{a1} and K_{a2} for H_2SO_4.

<u>**77.**</u> For both titration curves, we assume 10.00 mL of solution is being titrated and the concentration of the solute is 1.00 M, the same as the concentration of the titrant. Thus, 10.00 mL of titrant is required in each case. (You may be able to sketch titration curves based on fewer calculated points. Your titration curve will look slightly different if you make different initial assumptions.)

(a) The *initial pH* is that of a solution of $HCO_3^-(aq)$. This is an anion that both can ionize to $CO_3^{2-}(aq)$ or be hydrolyzed to $H_2CO_3(aq)$. Thus,
$$pH = \tfrac{1}{2}(pK_{a_1} + pK_{a_2}) = \tfrac{1}{2}(6.35 + 10.33) = 8.34$$

The *final pH* is that of 0.500 M $H_2CO_3(aq)$. All of the NaOH(aq) has been neutralized, as well as all of the $HCO_3^-(aq)$, by the added HCl(aq).

Equation:	$H_2CO_3(aq)$	$+ H_2O(l)$	$\rightleftharpoons$	$HCO_3^-(aq)$	$+ H_3O^+(aq)$
Initial:	0.500 M	–		0 M	≈ 0 M
Changes:	$-x$ M	–		$+x$ M	$+x$ M
Equil:	$(0.500 - x)$ M	–		x M	x M

$$K_b = \frac{[HCO_3^-][H_3O^+]}{[H_2CO_3]} = 4.43\times10^{-7} = \frac{x \cdot x}{0.500 - x} \approx \frac{x^2}{0.500}$$

($C_a/K_a = 1.1\times10^6$; thus, the approximation is valid)

$$[OH^-] = \sqrt{0.500 \times 4.4\times10^{-7}} = 4.7\times10^{-4}\,M \qquad pH = 3.33$$

During the course of the titration, the pH is determined by the Henderson-Hasselbalch equation, with the numerator being the percent of bicarbonate ion remaining, and as the denominator being the percent of bicarbonate ion that has been transformed to H_2CO_3.

90% titrated: $pH = 6.35 + \log\dfrac{10\%}{90\%} = 5.40$ 10% titrated: $pH = 6.35 + \log\dfrac{90\%}{10\%} = 7.30$

95% titrated: $pH = 6.35 + \log\dfrac{5\%}{95\%} = 5.07$ 5% titrated: $pH = 6.35 + \log\dfrac{95\%}{5\%} = 7.63$

After the equivalence point, the pH is determined by the excess H_3O^+.

At 105% titrated, $[H_3O^+] = \dfrac{0.50\ mL \times 1.00\ M}{20.5\ mL} = 0.024 \qquad pH = 1.61$

At 120% titrated, $[H_3O^+] = \dfrac{2.00\ mL \times 1.00\ M}{22.0\ mL} = 0.091\ M \qquad pH = 1.04$

The titration curve derived from these data is sketched below.

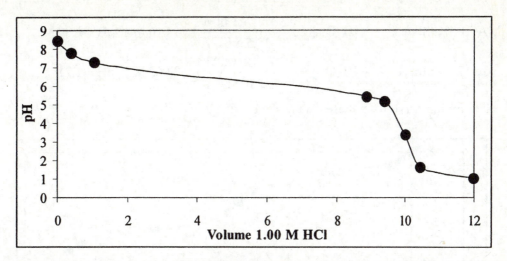

(b) The *final pH of the first step of the titration* is that of a solution of HCO_3^-(aq). This is an anion that can be ionized to CO_3^{2-}(aq) or hydrolyzed to H_2CO_3(aq). Thus,

$pH = \frac{1}{2}(pK_{a_1} + pK_{a_2}) = \frac{1}{2}(6.36 + 10.33) = 8.35$ The *initial pH* is that of 1.000 M CO_3^{2-}(aq), which pH is the result of the hydrolysis of the anion.

Equation: CO_3^{2-} (aq)   + H_2O(l) $\rightleftharpoons$ HCO_3^- (aq)   + OH^- (aq)

Initial: 1.000 M   –   0 M $\approx$ 0 M

Changes: $-x$ M   – $+x$ M $+x$ M

Equil: $(1.000-x)$ M   – x M x M

$$K_b = \frac{[HCO_3^-][OH^-]}{[CO_3^{2-}]} = \frac{K_w}{K_{a_2}} = \frac{1.0 \times 10^{-14}}{4.7 \times 10^{-11}} = 2.1 \times 10^{-4} = \frac{x \cdot x}{1.000 - x} \approx \frac{x^2}{1.000}$$

$[OH^-] = \sqrt{1.000 \times 2.1 \times 10^{-4}} = 1.4 \times 10^{-2}$ M $pOH = 1.85$ $pH = 12.15$

During the course of the first step of the titration, the pH is determined by the Henderson-Hasselbalch equation, modified as in Advanced Exercise 69, but using as the numerator the percent of carbonate ion remaining, and as the denominator the percent of carbonate ion that has been transformed to HCO_3^-.

90% titrated: $pH = 10.33 + \log \frac{10\%}{90\%} = 9.38$ 10% titrated: $pH = 10.33 + \log \frac{90\%}{10\%} = 11.28$

95% titrated: $pH = 10.33 + \log \frac{5\%}{95\%} = 9.05$ 5% titrated: $pH = 10.33 + \log \frac{95\%}{5\%} = 11.61$

During the course of the second step of the titration, the values of pH are precisely as they are for the titration of $NaHCO_3$, except the titrant volume is 10.00 mL more (the volume needed to reach the first equivalence point). The solution *at the second equivalence* point is 0.333 M H_2CO_3, for which the set up is similar to that for 0.500 M H_2CO_3.

$[H_3O^+] = \sqrt{0.333 \times 4.4 \times 10^{-7}} = 3.8 \times 10^{-4}$ M $pH = 3.42$

After the equivalence point, the pH is determined by the excess H_3O^+.

At 105% titrated, $[H_3O^+] = \dfrac{0.50\,mL \times 1.00\,M}{30.5\,mL} = 0.016\,M$ pH $= 1.80$

At 120% titrated, $[H_3O^+] = \dfrac{2.00\,mL \times 1.00\,M}{32.0\,mL} = 0.062\,M$ pH $= 1.21$

The titration curve from these data is sketched below.

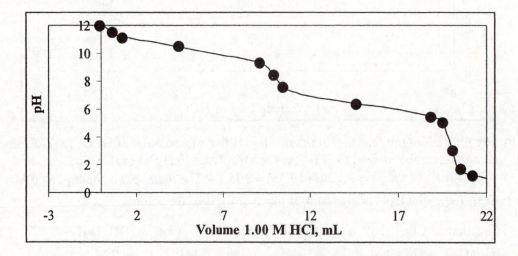

(c) $V_{HCl} = 1.00\,g\ NaHCO_3 \times \dfrac{1\ mol\ NaHCO_3}{84.01\,g\ NaHCO_3} \times \dfrac{1\ mol\ HCl}{1\ mol\ NaHCO_3} \times \dfrac{1000\ mL}{0.100\ mol\ HCl}$

$= 119\ mL\ 0.100\ M\ HCl$

(d) $V_{HCl} = 1.00\,g\ Na_2CO_3 \times \dfrac{1\ mol\ Na_2CO_3}{105.99\,g\ Na_2CO_3} \times \dfrac{2\ mol\ HCl}{1\ mol\ Na_2CO_3} \times \dfrac{1000\ mL}{0.100\ mol\ HCl}$

$= 189\ mL\ 0.100\ M\ HCl$

(e) The phenophthalein endpoint occurs at pH $= 8.00$ and signifies that the NaOH has been neutralized, and that Na_2CO_3 has been half neutralized. The methyl orange endpoint occurs at about pH $= 3.3$ and is the result of the second equivalence point of Na_2CO_3. The mass of Na_2CO_3 can be determined as follows:

mass $Na_2CO_3 = 0.78\ mL \times \dfrac{1\ L}{1000\ mL} \times \dfrac{0.1000\ mol\ HCl}{1\ L\ soln} \times \dfrac{1\ mol\ HCO_3^-}{1\ mol\ HCl}\ \dfrac{1\ mol\ Na_2CO_3}{1\ mol\ HCO_3^-}$

$\times \dfrac{105.99\ g\ Na_2CO_3}{1\ mol\ Na_2CO_3} = 0.0083\ g\ Na_2CO_3$

% $Na_2CO_3 = \dfrac{0.0083g}{0.1000g} \times 100 = 8.3\%\ Na_2CO_3$ by mass

78. We shall represent piprazine as Pip in what follows. The cation resulting from the first ionization is $HPiP^+$, and that resulting from the second ionization is H_2Pip^{2+}.

(a) $[Pip] = \dfrac{1.00\,g\ C_4H_{10}N_2 \cdot 6H_2O}{0.100\,L} \times \dfrac{1\,mol\ C_4H_{10}N_2 \cdot 6H_2O}{194.22\,g\ C_4H_{10}N_2 \cdot 6H_2O} = 0.0515\,M$

Equation:	Pip(aq)	+	H_2O(l)	$\rightleftharpoons$	$HPip^+$(aq)	+	OH^-(aq)
Initial:	0.0515 M		–		0 M		≈ 0 M
Changes:	$-x$ M		–		$+x$ M		$+x$ M
Equil:	$(0.0515 - x)$ M		–		x M		x M

$C_a/K_a = 858$; thus, the approximation is <u>not</u> valid. The full quadratic equation must be solved

$K_{b_1} = 10^{-4.22} = 6.0 \times 10^{-5} = \dfrac{[HPip^+][OH^-]}{[Pip]} = \dfrac{x \cdot x}{0.0515 - x}$

From the roots of the equation, $x = [OH^-] = 1.7 \times 10^{-3}$ M $\qquad$ pOH = 2.76 $\qquad$ pH = 11.24

(b) At the half-equivalence point of the 1ˢᵗ step in the titration, pOH = pK_{b1} = 4.22
pH = 14.00 − pOH = 14.00 − 4.22 = 9.78

(c) Volume of HCl = 100. mL $\times \dfrac{0.0515\,mmol\ Pip}{1\,mL} \times \dfrac{1\,mmol\ HCl}{1\,mmol\ Pip} \times \dfrac{1\,mL\ titrant}{0.500\ mmol\ HCl} = 10.3\,mL$

(d) At the first equivalence point we have a solution of $HPip^+$. This ion can react as a base with H_2O to form H_2Pip^{2+} or it can react as an acid with water forming Pip (i.e. $HPip^+$ is amphoteric). The solution's pH is determined as follows (base hydrolysis predominates):
pOH = ½ (pK_{b1} + pK_{b2}) = ½ (4.22 + 8.67) = 6.44, hence pH = 14.00 − 6.44 = 7.56

(e) The pOH at the half-equivalence point in the second step of the titration equals pK_{b2}.
pOH = pK_{b2} = 8.67 $\qquad$ pH = 14.00 − 8.67 = 5.33

(f) The volume needed to reach the second equivalence point is twice the volume needed to reach the first equivalence point, that is 2 × 10.3 mL = 20.6 mL

(g) The pH at the second equivalence point is determined by the hydrolysis of the H_2Pip^{2+} cation, of which there is 5.15 mmol in solution, resulting from the reaction of $HPip^+$ with HCl. The total solution volume is 100. mL + 20.6 mL = 120.6 mL

$[H_2Pip^{2+}] = \dfrac{5.15\,mmol}{120.6\,mL} = 0.0427\,M \qquad K_{b_2} = 10^{-8.67} = 2.1 \times 10^{-9}$

Equation:	H_2Pip^{2+}(aq)	+	H_2O(l)	$\rightleftharpoons$	$HPip^+$(aq)	+	H_3O^+(aq)
Initial:	0.0427 M		–		0 M		≈ 0 M
Changes:	$-x$ M		–		$+x$ M		$+x$ M
Equil:	$(0.0427 - x)$ M		–		x M		x M

$K_a = \dfrac{K_w}{K_{b_2}} = \dfrac{1.00 \times 10^{-14}}{2.1 \times 10^{-9}} = \dfrac{[HPip^+][H_3O^+]}{[H_2Pip^{2+}]} = 4.8 \times 10^{-6} = \dfrac{x \cdot x}{0.0427 - x} \approx \dfrac{x^2}{0.0427}$

$x = [H_3O^+] = \sqrt{0.0427 \times 4.8 \times 10^{-6}} = 4.5 \times 10^{-4}\,M \qquad$ pH = 3.34$\underline{4}$
($x \ll 0.0427$; thus, the approximation is valid).

79. Consider the two equilibria shown below.

$$H_2PO_4^-(aq) + H_2O(l) \rightleftharpoons HPO_4^{2-}(aq) + H_3O^+(aq) \qquad K_{a2} = \frac{[HPO_4^{2-}][H_3O^+]}{[H_2PO_4^-]}$$

$$H_2PO_4^-(aq) + H_2O(l) \rightleftharpoons H_3PO_4(aq) + OH^-(aq) \qquad K_b = K_w/K_{a1} = \frac{[H_3PO_4][OH^-]}{[H_2PO_4^-]}$$

All of the phosphorus containing species must add up to the initial molarity M.
Hence,   mass balance: $[HPO_4^{2-}] + [H_2PO_4^-] + [H_3PO_4] = M$
             charge balance: $[H_3O^+] + [Na^+] = [H_2PO_4^-] + 2 \times [HPO_4^{2-}]$
             Note: $[Na^+]$ for a solution of $NaH_2PO_4 = M$

Thus, $[H_3O^+] + [Na^+] = [H_2PO_4^-] + 2 \times [HPO_4^{2-}] = [H_3O^+] + M$ (substitute mass balance equation)
        $[H_3O^+] + [HPO_4^{2-}] + [H_2PO_4^-] + [H_3PO_4] = [H_2PO_4^-] + 2 \times [HPO_4^{2-}]$ (cancel terms)
        $[H_3O^+] = [HPO_4^{2-}] - [H_3PO_4]$
        (Note: $[HPO_4^{2-}] = $ initial $[H_3O^+]$ and $[H_3PO_4] = $ initial $[OH^-]$)
        Thus: $[H_3O^+]_{equil} = [H_3O^+]_{initial} - [OH^-]_{initial}$ (excess H₃O⁺ reacts with OH⁻ to form H₂O)

Rearrange expression for K_{a2} to solve for $[HPO_4^{2-}]$ as well, expression for K_b to solve for $[H_3PO_4]$

$$[H_3O^+] = [HPO_4^{2-}] - [H_3PO_4] = \frac{K_{a2}[H_2PO_4^-]}{[H_3O^+]} - \frac{K_b[H_2PO_4^-]}{[OH^-]}$$

$$[H_3O^+] = \frac{K_{a2}[H_2PO_4^-]}{[H_3O^+]} - \frac{\dfrac{K_w}{K_{a1}}[H_2PO_4^-]}{\dfrac{K_w}{[H_3O^+]}} = \frac{K_{a2}[H_2PO_4^-]}{[H_3O^+]} - \frac{[H_3O^+][H_2PO_4^-]}{K_{a1}}$$

Multiply through by $[H_3O^+] K_{a1}$ and solve for $[H_3O^+]$

$$K_{a1}[H_3O^+][H_3O^+] = K_{a1}K_{a2}[H_2PO_4^-] - [H_3O^+][H_3O^+][H_2PO_4^-]$$

$$K_{a1}[H_3O^+]^2 = K_{a1}K_{a2}[H_2PO_4^-] - [H_3O^+]^2[H_2PO_4^-]$$

$$K_{a1}[H_3O^+]^2 + [H_3O^+]^2[H_2PO_4^-] = K_{a1}K_{a2}[H_2PO_4^-] = [H_3O^+]^2(K_{a1} + [H_2PO_4^-])$$

$$K_{a1}[H_3O^+]^2 + [H_3O^+]^2[H_2PO_4^-] = K_{a1}K_{a2}[H_2PO_4^-] = [H_3O^+]^2(K_{a1} + [H_2PO_4^-])$$

$$[H_3O^+]^2 = \frac{K_{a1}K_{a2}[H_2PO_4^-]}{(K_{a1} + [H_2PO_4^-])}$$ For moderate concentrations of $[H_2PO_4^-]$, $K_{a1} << [H_2PO_4^-]$

This simplifies our expression to: $[H_3O^+]^2 = \dfrac{K_{a1}K_{a2}[H_2PO_4^-]}{[H_2PO_4^-]} = K_{a1}K_{a2}$

Take the square root of both sides: $[H_3O^+]^2 = K_{a1}K_{a2} \longrightarrow [H_3O^+] = \sqrt{K_{a1}K_{a2}} = (K_{a1}K_{a2})^{1/2}$
Take the -log of both sides and simplify:

$$-\log[H_3O^+] = -\log(K_{a1}K_{a2})^{1/2} = -1/2\big(\log(K_{a1}K_{a2})\big) = -1/2(\log K_{a1} + \log K_{a2})$$

$-\log[H_3O^+] = 1/2(-\log K_{a1} - \log K_{a2})$ Use $-\log[H_3O^+] = pH$ and $-\log K_{a1} = pK_{a1} - \log K_{a2} = pK_{a2}$
Hence, $-\log[H_3O^+] = +1/2(-\log K_{a1} - \log K_{a2})$ becomes $pH = 1/2(pK_{a1} + pK_{a2})$ (equation 17.5)
Equation 17.6 can be similarly answered.

80. $H_2PO_4^-$ can react with H_2O by both ionization and hydrolysis.

$$H_2PO_4^- \text{ (aq)} + H_2O(l) \rightleftharpoons HPO_4{}^{2-} \text{ (aq)} + H_3O^+ \text{ (aq)}$$

$$HPO_4{}^{2-} \text{ (aq)} + H_2O(l) \rightleftharpoons PO_4{}^{3-} \text{ (aq)} + H_3O^+ \text{ (aq)}$$

The solution cannot have a large concentration of both H_3O^+ and OH^- (cannot be simultaneously an acidic and a basic solution). Since the solution **is** acidic, some of the H_3O^+ produced in the first reaction reacts with virtually all of the OH^- produced in the second. In the first reaction $[H_3O^+] = [HPO_4{}^{2-}]$ and in the second reaction $[OH^-] = [H_3PO_4]$. Thus, following the neutralization of OH^- by H_3O^+, we have the following.

$$[H_3O^+] = [HPO_4^{2-}] - [H_2PO_4^-] = \frac{K_{a_2}[H_2PO_4^-]}{[H_3O^+]} - \frac{K_b[H_2PO_4^-]}{[OH^-]} = \frac{K_{a_2}[H_2PO_4^-]}{[H_3O^+]} - \frac{\dfrac{K_w}{K_{a_1}}[H_2PO_4^-]}{\dfrac{K_w}{[H_3O^+]}}$$

$$= \frac{K_{a_2}[H_2PO_4^-]}{[H_3O^+]} - \frac{[H_3O^+][H_2PO_4^-]}{K_{a_1}} \qquad \begin{array}{l}\text{Then multiply through by } [H_3O^+]K_{a_1}\\ \text{and solve for } [H_3O^+].\end{array}$$

$$[H_3O^+]^2 K_{a_1} = K_{a_1} K_{a_2}[H_2PO_4^-] - [H_3O^+]^2[H_2PO_4^-] \qquad [H_3O^+] = \sqrt{\frac{K_{a_1} K_{a_2}[H_2PO_4^-]}{[K_{a_1} + [H_2PO_4^-]]}}$$

But, for a moderate $[H_2PO_4^-]$, From Table 16-6, $K_{a_1} = 7.1 \times 10^{-3} < [H_2PO_4^-]$ and thus $K_{a_1} + [H_2PO_4^-] \approx [H_2PO_4^-]$. Then we have the following expressions for $[H_3O^+]$ and pH.

$$[H_3O^+] = \sqrt{K_{a_1} \times K_{a_2}}$$

$$pH = -\log[H_3O^+] = -\log(K_{a_1} \times K_{a_2})^{1/2} = \tfrac{1}{2}[-\log(K_{a_1} \times K_{a_2})] = \tfrac{1}{2}(-\log K_{a_1} - \log K_{a_2})$$

$$pH = \tfrac{1}{2}(pK_{a_1} + pK_{a_2})$$

There is no concentration dependence in this expression. Notice that we assumed $K_{a_1} < [H_2PO_4^-]$. This assumption is not valid in quite dilute solutions because $K_{a_1} = 0.0071$.

81. **(a)** A buffer solution is able to react with small amounts of added acid or base. When strong acid is added, it reacts with formate ion.

$$CHO_2^- \text{ (aq)} + H_3O^+ \text{ (aq)} \longrightarrow HCHO_2 \text{ (aq)} + H_2O$$

Added strong base reacts with acetic acid.

$$HC_2H_3O_2 \text{ (aq)} + OH^- \text{ (aq)} \longrightarrow H_2O(l) + C_2H_3O_2^- \text{ (aq)}$$

Therefore neither added strong acid nor added strong base alters the pH of the solution very much. Mixtures of this type are referred to as buffer solutions.

(b) We begin with the two ionization reactions.

$$HCHO_2(aq) + H_2O(l) \rightleftharpoons CHO_2^-(aq) + H_3O^+(aq)$$

$$HC_2H_3O_2(aq) + H_2O(l) \rightleftharpoons C_2H_3O_2^-(aq) + H_3O^+(aq)$$

$[Na^+] = 0.250 \qquad [OH^-] \approx 0 \qquad [H_3O^+] = x \quad [C_2H_3O_2^-] = y \quad [CHO_2^-] = z$

$0.150 = [HC_2H_3O_2] + [C_2H_3O_2^-] \quad [HC_2H_3O_2] = 0.150 - [C_2H_3O_2^-] = 0.150 - y$

$0.250 = [HCHO_2] + [CHO_2^-] \qquad [HCHO_2] = 0.250 - [CHO_2^-] = 0.250 - z$

$[Na^+] + [H_3O^+] = [C_2H_3O_2^-] + [CHO_2^-] + [OH^-]$ (electroneutrality)

$$0.250 + x = [C_2H_3O_2^-] + [CHO_2^-] = y + z \tag{1}$$

$$\frac{[H_3O^+][C_2H_3O_2^-]}{[HC_2H_3O_2]} = K_A = 1.8 \times 10^{-5} = \frac{x \cdot y}{0.150 - y} \tag{2}$$

$$\frac{[H_3O^+][CHO_2^-]}{[HCHO_2]} = K_F = 1.8 \times 10^{-4} = \frac{x \cdot z}{0.250 - z} \tag{3}$$

There now are three equations—(1), (2), and (3)—in three unknowns—x, y, and z. We solve equations (2) and (3), respectively for y and z in terms of x.

$$0.150 K_A - y K_A = xy \qquad y = \frac{0.150 K_A}{K_A + x}$$

$$0.250 K_F - z K_F = xz \qquad z = \frac{0.250 K_F}{K_F + x}$$

Then we substitute these expressions into equation (1) and solve for x.

$$0.250 + x = \frac{0.150 K_A}{K_A + x} + \frac{0.250 K_F}{K_F + x} \approx 0.250 \qquad \text{since } x \ll 0.250$$

$$0.250 (K_F + x)(K_A + x) = 0.150 K_A (K_F + x) + 0.250 K_F (K_A + x)$$

$$K_A K_F + (K_A + K_F)x + x^2 = 1.60 K_A K_F + x(0.600 K_A + 1.00 K_F)$$

$$x^2 + 0.400 K_A x - 0.600 K_A K_F = 0 = x^2 + 7.2 \times 10^{-6} x - 1.9 \times 10^{-9}$$

$$x = \frac{-7.2 \times 10^{-6} \pm \sqrt{5.2 \times 10^{-11} + 7.6 \times 10^{-9}}}{2} = 4.0 \times 10^{-5} \text{ M} = [H_3O^+]$$

$pH = 4.40$

(c) Adding 1.00 L of 0.100 M HCl to 1.00 L of buffer of course dilutes the concentrations of all components by a factor of 2. Thus, $[Na^+] = 0.125$ M; total acetate concentration = 0.0750 M; total formate concentration = 0.125 M. Also a new ion is added to the solution, namely, $[Cl^-] = 0.0500$ M.

$$[Na^+] = 0.125 \quad [OH^-] \approx 0 \quad [H_3O^+] = x \quad [Cl^-] = 0.0500 \text{ M} \quad [C_2H_3O_2^-] = y \quad [CHO_2^-] = z$$

$$0.0750 = [HC_2H_3O_2] + [C_2H_3O_2^-] \qquad [HC_2H_3O_2] = 0.0750 - [C_2H_3O_2^-] = 0.0750 - y$$

$$0.125 = [HCHO_2] + [CHO_2^-] \qquad [HCHO_2] = 0.125 - [CHO_2^-] = 0.125 - z$$

$$[Na^+] + [H_3O^+] = [C_2H_3O_2^-] + [OH^-] + [CHO_2^-] + [Cl^-] \text{ (electroneutrality)}$$

$$0.125 + x = [C_2H_3O_2^-] + [CHO_2^-] + [Cl^-] = y + z + 0 + 0.0500 \quad 0.075 + x = y + z$$

$$\frac{[H_3O^+][C_2H_3O_2^-]}{[HC_2H_3O_2]} = K_A = 1.8 \times 10^{-5} = \frac{x \cdot y}{0.0750 - y}$$

$$\frac{[H_3O^+][CHO_2^-]}{[HCHO_2]} = K_F = 1.8 \times 10^{-4} = \frac{x \cdot z}{0.125 - z}$$

Again, we solve the last two equations for y and z in terms of x.

$$0.0750\, K_A - y\, K_A = xy \qquad y = \frac{0.0750\, K_A}{K_A + x}$$

$$0.125\, K_F - z\, K_F = xz \qquad z = \frac{0.125\, K_F}{K_F + x}$$

Then we substitute these expressions into equation (1) and solve for x.

$$0.0750 + x = \frac{0.0750\, K_A}{K_A + x} + \frac{0.125\, K_F}{K_F + x} \approx 0.0750 \qquad \text{since } x \ll 0.0750$$

$$0.0750\,(K_F + x)(K_A + x) = 0.0750\, K_A\,(K_F + x) + 0.125\, K_F\,(K_A + x)$$

$$K_A K_F + (K_A + K_F)x + x^2 = 2.67\, K_A K_F + x(1.67\, K_F + 1.00\, K_A)$$

$$x^2 - 0.67\, K_F x - 1.67\, K_A K_F = 0 = x^2 - 1.2 \times 10^{-4} x - 5.4 \times 10^{-9}$$

$$x = \frac{1.2 \times 10^{-4} \pm \sqrt{1.4 \times 10^{-8} + 2.2 \times 10^{-8}}}{2} = 1.5 \times 10^{-4} \text{ M} = [H_3O^+]$$

$$pH = 3.82$$

As expected, the addition of HCl(aq), a strong acid, caused the pH to drop. The decrease in pH was relatively small, nonetheless, because the H_3O^+(aq) was converted to the much weaker acid $HCHO_2$ via the neutralization reaction:

$$CHO_2^-(aq) + H_3O^+(aq) \rightarrow HCHO_2(aq) + H_2O(l) \quad \text{(buffering action)}$$

82. First we find the pH at the equivalence point:

$$CH_3CH(OH)COOH(aq) + OH^-(aq) \rightleftharpoons CH_3CH(OH)COO^-(aq) + H_2O(l)$$

$$\frac{1 \text{ mmol}}{50 \text{ ml}} \qquad \frac{1 \text{ mmol}}{50 \text{ ml}} \qquad \frac{1 \text{ mmol}}{100 \text{ ml}} = 0.01M$$

The lactate anion undergoes hydrolysis thus:

$$CH_3CH(OH)COO^-(aq) + H_2O(l) \rightleftharpoons CH_3CH(OH)COOH(aq) + OH^-(aq)$$

initial	0.01 M	–	0 M	≈ 0 M
change	$-x$	–	$+x$	$+x$
equilibrium	$(0.01-x)$ M	–	x	x

Where x is the [hydrolyzed lactate ion], as well as that of the [OH⁻] produced by hydrolysis

K for the above reaction = $\dfrac{[CH_3CH(OH)COOH][OH^-]}{[CH_3CH(OH)COO^-]} = \dfrac{K_w}{K_a} = \dfrac{1.0 \times 10^{-14}}{10^{-3.86}} = 7.2 \times 10^{-11}$

so $\dfrac{x^2}{0.01-x} \approx \dfrac{x^2}{0.01} = 7.2 \times 10^{-11}$ and $x = [OH^-] = 8.49 \times 10^{-7}$ M

$x \ll 0.01$, thus, the assumption is valid

$pOH = -\log(8.49 \times 10^{-7} M) = 6.07$ $pH = 14 - pOH = 14.00 - 6.07 = 7.93$

a) Bromthymol blue or phenol red would be good indications for this titration
since they change color over this pH range.

b) The $H_2PO_4^-/HPO_4^{2-}$ system would be suitable because the pK_a for the acid
(namely, $H_2PO_4^-$) is close to the equivalence point pH of 7.93
. An acetate buffer would be too acidic, while an ammonia buffer would be too basic.

c) $H_2PO_4^- + H_2O \rightleftharpoons H_3O^+ + HPO_4^{2-}$ $K_{a2} = 6.3 \times 10^{-8}$

solving for $\dfrac{[HPO_4^{2-}]}{[H_2PO_4^-]} = \dfrac{K_{a2}}{[H_3O^+]} = \dfrac{6.3 \times 10^{-8}}{10^{-7.93}} = 5$ (buffer ratio required)

83. $H_2O_2(aq) + H_2O(l) \rightleftharpoons H_3O^+(aq) + HO_2^-(aq)$ $K_a = \dfrac{[H_3O^+][HO_2^-]}{[H_2O_2]}$

Trial # 1 $[H_2O_2] + [HO_2^-] = 0.259$ M $(6.78) \times (0.00357 M) + [HO_2^-] = 0.259$ M

$[HO_2^-] = 0.235$ M $[H_2O_2] = (6.78)(0.00357 M) = 0.0242$ M

$[H_3O^+] = 10^{-(pK_w - pOH)}$, $pOH[(0.250 - 0.235) M NaOH] = pOH(0.015 M NaOH) = 1.82\underline{4}$

$[H_3O^+] = 10^{-(14.94 - 1.82\underline{4})} = 7.4 \times 10^{-14}$

$K_a = \dfrac{[H_3O^+][HO_2^-]}{[H_2O_2]} = \dfrac{(7.4 \times 10^{-14} M)(0.235 M)}{(0.0242 M)} = 7.2 \times 10^{-13}$

$pK_a = 12.14$

Trial # 2

$[H_2O_2]+[HO_2^-]=0.123$ M $\qquad$ $(6.78)(0.00198$ M$)+[HO_2^-]=0.123$ M

$[HO_2^-]=0.109\underline{6}$ M

$[H_2O_2]=(6.78)(0.00198$ M$)=0.0134$ M

$[H_3O^+]=10^{-(pKw-pOH)}$, $\quad$ pOH $=-\log[(0.125-0.109\underline{6})$ M NaOH$]=1.81$

$[H_3O^+]=10^{-(14.94-1.81)}=7.41\times10^{-14}$

$K_a=\dfrac{[H_3O^+][HO_2^-]}{[H_2O_2]}=\dfrac{(7.41\times10^{-14}M)(0.109\,_6M)}{(0.0134\text{ M})}=6.06\times10^{-13}$

$pK_a=-\log(6.06\times10^{-13})=12.22$

Average value for pK_a 12.18

84. Let's consider some of the important processes occurring in the solution

(1) $HPO_4^{2-}(aq)+H_2O(l) \rightleftharpoons H_2PO_4^-(aq)+OH^-(aq)$ $\quad$ $K_{(1)}=K_{a2}=4.2\times10^{-13}$

(2) $HPO_4^{2-}(aq)+H_2O(l) \rightleftharpoons H_3O^+(aq)+PO_4^{3-}(aq)$ $\quad$ $K_{(2)}=K_b=\dfrac{1.0\times10^{-14}}{6.3\times10^{-8}}=1.6\times10^{-7}$

(3) $NH_4^+(aq)+H_2O(l) \rightleftharpoons H_3O^+(aq)+NH_3(aq)$ $\quad$ $K_{(3)}=K_a=\dfrac{1.0\times10^{-14}}{1.8\times10^{-5}}=5.6\times10^{-10}$

(4) $NH_4^+(aq)+OH^-(aq) \rightleftharpoons H_2O(l)+NH_3(aq)$ $\quad$ $K_{(4)}=\dfrac{1}{1.8\times10^{-5}}=5.6\times10^4$

May have interaction between NH_4^+ and OH^- formed from the hydrolysis of HPO_4^{2-}.

	$NH_4^+(aq)$	+	$HPO_4^{2-}(aq)$	$\rightleftharpoons$	$H_2PO_4^-(aq)$	+	$NH_3(aq)$
initial	0.10M		0.10M		0 M		0 M
change	$-x$		$-x$		$+x$		$+x$
equil.	$0.100-x$		$0.100-x$		x		x

(where x is the molar concentration of NH_4^+ that hydrolyzes)

$K=K_{(2)}\times K_{(4)}=(1.6\times10^{-7})\times(5.6\times10^4)=9.0\times10^{-3}$

$K=\dfrac{[H_2PO_4^-][NH_3]}{[NH_4^+][HPO_4^{2-}]}=\dfrac{[x]^2}{[0.10-x]^2}=9.0\times10^{-3}$ and $x=8.7\times10^{-3}$M

Finding the pH of this buffer system:

$pH=pK_a+\log\dfrac{[HPO_4^{2-}]}{[H_2PO_4^-]}=-\log(6.3\times10^{-8})+\log\left(\dfrac{0.100-8.7\times10^{-3}M}{8.7\times10^{-3}M}\right)=8.2$

As expected, we get the same result using the NH_3/NH_4^+ buffer system.

85. Consider the two weak acids and the equilibrium for water (autodissociation)

$$HA(aq) + H_2O(l) \rightleftharpoons H_3O^+(aq) + A^-(aq) \quad K_{HA} = \frac{[H_3O^+][A^-]}{[HA]} \quad [HA] = \frac{[H_3O^+][A^-]}{K_{HA}}$$

$$HB(aq) + H_2O(l) \rightleftharpoons H_3O^+(aq) + B^-(aq) \quad K_{HB} = \frac{[H_3O^+][B^-]}{[HB]} \quad [HB] = \frac{[H_3O^+][B^-]}{K_{HB}}$$

$$H_2O(l) + H_2O(l) \rightleftharpoons H_3O^+(aq) + OH^-(aq) \quad K_w = [H_3O^+][OH^-] \quad [OH^-] = \frac{K_w}{[H_3O^+]}$$

Mass Balance: $[HA]_{initial} = [HA] + [A^-] = \dfrac{[H_3O^+][A^-]}{K_{HA}} + [A^-] = [A^-]\left(\dfrac{[H_3O^+]}{K_{HA}} + 1\right)$

From which: $[A^-] = \dfrac{[HA]_{initial}}{\left(\dfrac{[H_3O^+]}{K_{HA}} + 1\right)}$

$[HB]_{initial} = [HB] + [B^-] \qquad = \dfrac{[H_3O^+][B^-]}{K_{HB}} + [B^-] = [B^-]\left(\dfrac{[H_3O^+]}{K_{HB}} + 1\right)$

From which: $[B^-] = \dfrac{[HB]_{initial}}{\left(\dfrac{[H_3O^+]}{K_{HB}} + 1\right)}$

Charge Balance: $[H_3O^+] = [A^-] + [B^-] + [OH^-]$ (substitute above expressions)

$$[H_3O^+] = \frac{[HA]_{initial}}{\left(\dfrac{[H_3O^+]}{K_{HA}} + 1\right)} + \frac{[HB]_{initial}}{\left(\dfrac{[H_3O^+]}{K_{HB}} + 1\right)} + \frac{K_w}{[H_3O^+]}$$

86. We are told that the solution is 0.050 M in acetic acid ($K_a = 1.8 \times 10^{-5}$) and 0.010 M in phenyl acetic acid ($K_a = 4.9 \times 10^{-5}$). Because the K_a values are close, both equilibria must be satisfied simultaneously. Note: $[H_3O^+]$ is common to both equilibria (assume z is the concentration of $[H_3O^+]$).

Reaction:	$HC_2H_3O_2(aq)$	+ $H_2O(l)$	$\rightleftharpoons$	$C_2H_3O_2^-(aq)$	+	$H_3O^+(aq)$
Initial:	0.050 M	—		0 M		≈ 0 M
Change:	$-x$ M	—		$+x$ M		$+z$ M
Equilibrium	$(0.050 - x)$M	—		x M		z M

For acetic acid: $K_a = \dfrac{[H_3O^+][C_2H_3O_2^-]}{[HC_2H_3O_2]} = \dfrac{xz}{(0.050 - x)} = 1.8 \times 10^{-5}$ or $x = \dfrac{1.8 \times 10^{-5}(0.050 - x)}{z}$

Reaction:	$HC_8H_7O_2(aq)$	+ $H_2O(l)$	$\rightleftharpoons$	$C_8H_7O_2^-(aq)$	+	$H_3O^+(aq)$
Initial:	0.010 M	—		0 M		≈ 0 M
Change:	$-y$ M	—		$+y$ M		$+z$ M
Equilibrium	$(0.010 - y)$M	—		y M		z M

For phenylacetic acid: $K_a = \dfrac{[H_3O^+][C_8H_7O_2^-]}{[HC_8H_7O_2]} = \dfrac{yz}{(0.010-y)} = 4.9 \times 10^{-5}$

or $y = \dfrac{4.9 \times 10^{-5}(0.010-y)}{z}$

There are now three variable: $x = [C_2H_3O_2^-]$, $y = [C_8H_7O_2^-]$ and $z = [H_3O^+]$
These are the only charged species, hence, $x + y = z$
(This equation neglects contribution from the $[H_3O^+]$ from water.
Next we substitute in the values of x and y from the rearranged K_a expressions above.

$x + y = z = \dfrac{1.8 \times 10^{-5}(0.050-x)}{z} + \dfrac{4.9 \times 10^{-5}(0.010-y)}{z}$

Since these are weak acids, we can simplify this expression by assuming that $x \ll 0.050$ and $y \ll 0.010$

$z = \dfrac{1.8 \times 10^{-5}(0.050)}{z} + \dfrac{4.9 \times 10^{-5}(0.010)}{z}$ Simplify even further by multiplying through by z.

$z^2 = 9.0 \times 10^{-7} + 4.9 \times 10^{-7} = 1.4 \times 10^{-6}$

$z = 1.1\underline{8} \times 10^{-3} = [H_3O^+]$ From which we find that $x = 7.6 \times 10^{-4}$ and $y = 4.2 \times 10^{-4}$
(as a quick check, we do see that $x + y = z$)

We finish up by checking to see if the approximation is valid (5% rule)..

For x: $\dfrac{7.6 \times 10^{-4}}{0.050} \times 100\% = 1.5\%$ For y: $\dfrac{4.2 \times 10^{-4}}{0.010} \times 100\% = 4.2\%$

Since both are less than 5%, we can be assured that the assumption is valid.
The pH of the solution = $-\log(1.1\underline{8} \times 10^{-3}) = 2.93$
(If we simply plug the appropriate values into the equation developed in the previous
question we get the exact answer 2.933715 (via the method of successive approximations),
but the final answers can only be reported to 2 significant figures. Hence the best we can do
is say that the pH is expected to be 2.93 (which is the same level of precision as that for the
result obtained following the 5% rule).

87. **(a)**   By using dilution, there are an infinite number of ways of preparing the pH = 7.79
buffer. We will consider a method which does not use dilution.

Let TRIS = weak base and TRISH$^+$ be the conjugate acid.

$pK_b = 5.91$, $pK_a = 14 - pK_b = 8.09$ Use the Henderson Hasselbalch equation.

pH = pK_a + log (TRIS/TRISH$^+$) = 7.79 = 8.09+ log (TRIS/TRISH$^+$)

log (TRIS/TRISH$^+$) = 7.79 − 8.09 = -0.30   Take antilog of both sides

(TRIS/TRISH$^+$) = $10^{-0.30}$ = 0.50   Therefore, $n_{TRIS} = 0.50(n_{TRISH+})$

We start with 1.00 L of 0.200 M TRIS (0.200 moles). We need to convert 2/3 of this to the corresponding acid (TRISH$^+$) using 10.0 M HCl. In all, we need (2/3)×0.200 mol of HCl, or a total of 0.133 moles of HCl.

Volume of HCl required = 0.133 mol ÷ 10.0 mol/L = 0.0133 L or 13.3 mL.

This would give a total volume of 1013.3 mL which is almost a liter (within 1.3%). If we wish to make up exactly one liter, we should only use 987 mL of 0.200 M TRIS. This would require 13.2 mL of HCl, resulting in a final volume of 1.0002 L.

Let's do a quick double check of our calculations.
n_{HCl} = 0.0132 L×10.0 M = 0.132 mol = n_{TRISH+}
n_{TRIS} = $n_{initial}$ − $n_{reacted}$ = 0.200 M×0.987 L − 0.132 mol = 0.0654 mol
pH = pK_a + log (TRIS/TRISH$^+$) = 8.09+ log (0.0654 mol/0.132 mol) = 7.78$\underline{5}$

We have prepared the desired buffer (realize that this is just one way of preparing the buffer).

(b)   To 500 mL of the buffer prepared above, is added 0.030 mol H_3O^+
In the 500 mL of solution we have 0.132 mol ÷ 2 mol TRISH$^+$ = 0.0660 mol TRISH$^+$ and 0.0654 mol ÷ 2 mol TRIS = 0.032$\underline{7}$ mol TRIS (we will assume no change in volume).

HCl will completely react (≈ 100%) with TRIS, converting it to TRISH$^+$.
After complete reaction, there will be no excess HCl, 0.0327 mol − 0.0300 mol = 0.0027 mol TRIS and 0.0660 mol + 0.0300 mol = 0.0960 mol TRISH$^+$.

By employing the Henderson Hasselbalch equation we can estimate the pH of the resulting solution:

pH = pK_a + log (TRIS/TRISH$^+$) = 8.09 + log (0.0027 mol/0.0960 mol) = 6.54

This represents a pH change of 1.25 units. The buffer is nearly exhausted owing to the fact that almost all of the TRIS has been converted to TRISH$^+$. Generally, a pH change of 1 unit suggests that the capacity of the buffer has been pretty much completely expended. This is the case here.

(Alternatively, you may solve this question using an I.C.E. table).

(c) Addition of 20.0 mL of 10.0 M HCl will complete exhaust the buffer (in part (b) we saw that addition of 3 mL of HCl used up most of the TRIS in solution). The buffer may be regenerated by adding 20.0 mL of 10.0 M NaOH. The buffer will be only slightly diluted after the addition of HCl and NaOH (500 mL → 540 mL). Through the slow and careful addition of NaOH, one can regenerate the pH = 7.79 buffer in this way (if a pH meter is used to monitor the addition).

FEATURE PROBLEMS

88. **(a)** The two curves cross the point at which half of the total acetate is present as acetic acid and half is present as acetate ion. This is the half equivalence point in a titration, where $pH = pK_a = 4.74$.

(b) For carbonic acid, there are three carbonate containing species: "H_2CO_3" which predominates at low pH, HCO_3^-, and CO_3^{2-} which predominates in alkaline solution. The points of intersection should occur at the half-equivalence points in each step-wise titration: at $pH = pK_{a_1} = -\log(4.4 \times 10^{-7}) = 6.36$ and at $pH = pK_{a_2} = -\log(4.7 \times 10^{-11}) = 10.33$. The following graph was computer-calculated (and then drawn) from these equations. f in each instance represents the fraction of the species whose formula is in parentheses.

$$\frac{1}{f(H_2A)} = 1 + \frac{K_1}{[H^+]} + \frac{K_1 K_2}{[H^+]^2}; \quad \frac{1}{f(HA^-)} = \frac{[H^+]}{K_1} + 1 + \frac{K_2}{[H^+]}; \quad \frac{1}{f(A^{2-})} = \frac{[H^+]^2}{K_1 K_2} + \frac{[H^+]}{K_2} + 1$$

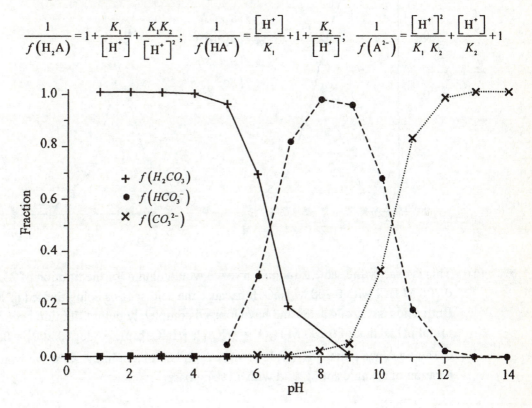

(c) For phosphoric acid, there are four phosphate containing species: H_3PO_4 under acidic conditions, $H_2PO_4^-$, HPO_4^{2-}, and PO_4^{3-} which predominates in alkaline solution. The points of intersection should occur at
$$pH = pK_{a_1} = -\log(7.1 \times 10^{-3}) = 2.15, \quad pH = pK_{a_2} = -\log(6.3 \times 10^{-8}) = 7.20,$$
and $pH = pK_{a_3} = -\log(4.2 \times 10^{-13}) = 12.38$, a quite alkaline solution. The graph that follows was computer-calculated and drawn.

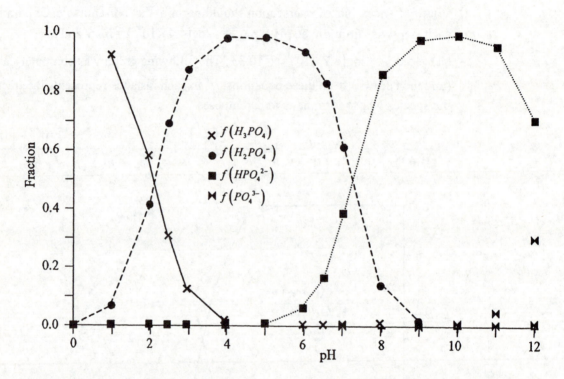

89. (a) This is exactly the same titration curve we would obtain for the titration of 25.00 mL of 0.200 M HCl with 0.200 M NaOH, because the acid species being titrated is H_3O^+. Both acids are strong acids and have ionized completely before titration begins. The initial pH is that of 0.200 M H_3O^+ = [HCl] + [HNO$_3$]; $pH = -\log(0.200) = 0.70$. At the equivalence point, $pH = 7.000$. We treat this problem in the same we would for the titration of a single strong acid with a strong base.

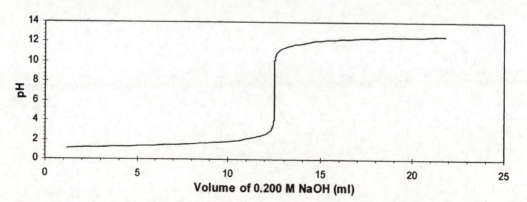

(b) In Figure 17-9, we note that the equivalence point of the titration of a strong acid occurs at $pH = 7.00$, but that the strong acid is essentially completely neutralized at $pH = 4$. In Figure 17-13, we see that the first equivalence point of H_3PO_4 occurs at about $pH = 4.6$. Thus, the first equivalence point represents the complete neutralization of HCl and the neutralization of H_3PO_4 to $H_2PO_4^-$. Then, the second equivalence point represents the neutralization of $H_2PO_4^-$ to HPO_4^{2-}. To reach the first equivalence point requires about 20.0 mL of 0.216 M NaOH, while to reach the second one requires a total of 30.0 mL of 0.216 M NaOH, or an additional 10.0 mL of base beyond the first equivalence point. The equations for the two titration reactions are as follows.

To the first equivalence point: $NaOH + H_3PO_4 \longrightarrow NaH_2PO_4 + H_2O$

$$NaOH + HCl \longrightarrow NaCl + H_2O$$

To the second equivalence point: $NaOH + NaH_2PO_4 \longrightarrow Na_2HPO_4 + H_2O$

There is a third equivalence point, not shown in the figure, that would require an additional 10.0 mL of base to reach. Its titration reaction is represented by the following equation.

To the third equivalence point: $NaOH + Na_2HPO_4 \longrightarrow Na_3PO_4 + H_2O$

We determine the molar concentration of H_3PO_4 and then of HCl. Notice that only 10.0 mL of the NaOH needed to reach the first equivalence point reacts with the HCl(aq); the rest reacts with H_3PO_4.

$$\frac{(30.0 - 20.0) \text{ mL NaOH(aq)} \times \dfrac{0.216 \text{ mmol NaOH}}{1 \text{ mL NaOH soln}} \times \dfrac{1 \text{ mmol H}_3\text{PO}_4}{1 \text{ mmol NaOH}}}{10.00 \text{ mL acid soln}} = 0.216 \text{ M H}_3\text{PO}_4$$

$$\frac{(20.0 - 10.0) \text{ mL NaOH(aq)} \times \dfrac{0.216 \text{ mmol NaOH}}{1 \text{ mL NaOH soln}} \times \dfrac{1 \text{ mmol HCl}}{1 \text{ mmol NaOH}}}{10.00 \text{ mL acid soln}} = 0.216 \text{ M HCl}$$

(c) We start with a phosphoric acid-dihydrogen phosphate buffer solution and titrate until all of the H_3PO_4 is consumed. We begin with

$$10.00 \text{ mL} \times \frac{0.0400 \text{ mmol H}_3\text{PO}_4}{1 \text{ mL}} = 0.400 \text{ mmol H}_3\text{PO}_4 \text{ and the diprotic anion,}$$

$$10.00 \text{ mL} \times \frac{0.0150 \text{ mmol H}_2\text{PO}_4^{-}}{1 \text{ mL}} = 0.150 \text{ mmol H}_2\text{PO}_4^{2-}. \text{ The volume of } 0.0200 \text{ M}$$

NaOH needed is $0.400 \text{ mmol H}_3\text{PO}_4 \times \dfrac{1 \text{ mmol NaOH}}{1 \text{ mmol H}_3\text{PO}_4} \times \dfrac{1 \text{ mL NaOH}}{0.0200 \text{ mmol NaOH}} = 20.00 \text{ mL}$

to reach the first equivalence point. The pH of points during this titration are computed with the Henderson-Hasselbalch equation.

$$\text{Initially}: \text{pH} = pK_1 + \log\frac{\left[\text{H}_2\text{PO}_4^{-}\right]}{\left[\text{H}_3\text{PO}_4\right]} = -\log\left(7.1\times10^{-3}\right) + \log\frac{0.0150}{0.0400} = 2.15 - 0.43 = 1.62$$

$$\text{At } 5.00 \text{ mL}: \text{pH} = 2.15 + \log\frac{0.150 + 0.100}{0.400 - 0.100} = 2.15 - 0.08 = 2.07$$

$$\text{At } 10.0 \text{ mL}, \text{ pH} = 2.15 + \log\frac{0.150 + 0.200}{0.400 - 0.200} = 2.15 + 0.24 = 2.39$$

$$\text{At } 15.0 \text{ mL}, \text{ pH} = 2.15 + \log\frac{0.150 + 0.300}{0.400 - 0.300} = 2.15 + 0.65 = 2.80$$

This is the first equivalence point, a solution of 30.00 mL ($= 10.00 \text{ mL}$ originally $+$ 20.00 mL titrant), containing $0.400 \text{ mmol H}_2\text{PO}_4^{-}$ from the titration and the 0.150 mmol $\text{H}_2\text{PO}_4^{2-}$ originally present. This is a solution with

$$\left[\text{H}_2\text{PO}_4^{-}\right] = \frac{(0.400 + 0.150) \text{ mmol H}_2\text{PO}_4^{-}}{30.00 \text{ mL}} = 0.0183 \text{ M}, \text{ which has}$$

$$\text{pH} = \frac{1}{2}\left(pK_1 + pK_2\right) = 0.50\left(2.15 - \log\left(6.3\times10^{-8}\right)\right) = 0.50\left(2.15 + 7.20\right) = 4.68$$

To reach the second equivalence point means titrating $0.550 \text{ mmol H}_2\text{PO}_4^{-}$, which requires an additional volume of titrant given by

$$0.550 \text{ mmol H}_2\text{PO}_4^{-} \times \frac{1 \text{ mmol NaOH}}{1 \text{ mmol H}_2\text{PO}_4^{-}} \times \frac{1 \text{ mL NaOH}}{0.0200 \text{ mmol NaOH}} = 27.5 \text{ mL}.$$

To determine pH during this titration, we divide the region into five equal portions of 5.5 mL and use the Henderson-Hasselbalch equation.

At $(20.0 + 5.5) \text{ mL}$,

$$\text{pH} = pK_2 + \log\frac{\left[\text{HPO}_4^{2-}\right]}{\left[\text{H}_2\text{PO}_4^{-}\right]} = 7.20 + \log\frac{(0.20\times0.550) \text{ mmol HPO}_4^{2-} \text{ formed}}{(0.80\times0.550) \text{ mmol H}_2\text{PO}_4^{-} \text{ remaining}}$$

$$\text{pH} = 7.20 - 0.60 = 6.60$$

At $(20.0+11.0)$ mL $= 31.0$ mL, pH $= 7.20 + \log\dfrac{0.40 \times 0.550}{0.60 \times 0.550} = 7.02$

At 36.5 mL, pH $= 7.38$ At 42.0 mL, pH $= 7.80$

The pH at the second equivalence point is given by

$$pH = \frac{1}{2}\left(pK_2 + pK_3\right) = 0.50\left(7.20 - \log\left(4.2 \times 10^{-13}\right)\right) = 0.50\ (7.20 + 12.38) = 9.79\ .$$

Another 27.50 mL of 0.020 M NaOH would be required to reach the third equivalence point. pH values at each of four equally spaced volumes of 5.50 mL additional 0.0200 M NaOH are computed as before, assuming the Henderson-Hasselbalch equation is valid.

$$At\ (47.50+5.50)\ mL = 53.00\ mL,\ \ pH = pK_3 + \log\dfrac{\left[PO_4^{\ 3-}\right]}{\left[HPO_4^{\ 2-}\right]} = 12.38 + \log\dfrac{0.20 \times 0.550}{0.80 \times 0.550}$$

$$= 12.38 - 0.60 = 11.78$$

At 58.50 mL, pH $= 12.20$ At 64.50 mL, pH $= 12.56$ At 70.00 mL, pH $= 12.98$
But at infinite dilution with 0.0200 M NaOH, the pH $= 12.30$, so this point can't be reached.

At the last equivalence point, the solution will contain 0.550 mmol PO_4^{3-} in a total of $10.00 + 20.00 + 27.50 + 27.50$ mL $= 85.00$ mL of solution, with

$$\left[PO_4^{\ 3-}\right] = \dfrac{0.550\ mmol}{85.00\ mL} = 0.00647\ M.\ But\ we\ can\ never\ reach\ this\ point,\ because\ the$$

pH of the 0.0200 M NaOH titrant is 12.30. Moreover, the titrant is diluted by its addition to the solution. Thus, our titration will cease sometime shortly after the second equivalence point. We never will see the third equivalence point, largely because the titrant is too dilute. Our results are plotted below

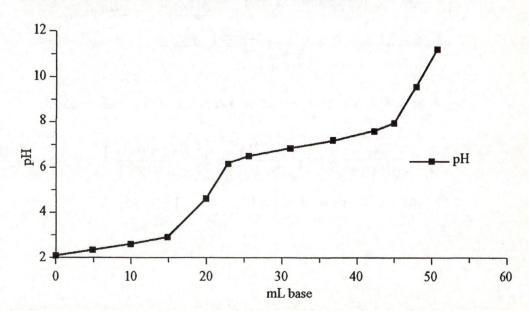

90. $pK_{a_1} = 2.34$; $K_{a_1} = 4.6 \times 10^{-3}$ and $pK_{a_2} = 9.69$; $K_{a_2} = 2.0 \times 10^{-10}$

(a) Since the K_a values are so different, we can treat alanine (H_2A^+) as a monoprotic acid with $K_{a_1} = 4.6 \times 10^{-3}$. Hence:

	$H_2A^+(aq)$	$+$	$H_2O(l)$	$\rightleftharpoons$	$HA(aq)$	$+$	$H_3O^+(aq)$
Initial	0.500 M		—		0 M		≈ 0 M
Change	$-x$ M		—		$+x$ M		$+x$ M
Equilibrium	$(0.500-x)$ M		—		x M		x M

$$K_{a_1} = \frac{[HA][H_3O^+]}{[H_2A^+]} = \frac{(x)(x)}{(0.500-x)} = 4.6 \times 10^{-3} \approx \frac{x^2}{0.500}$$

$x = 0.048$ M $= [H_3O^+]$ ($x = 0.0457$ solving the quadratic equation)

$pH = -\log[H_3O^+] = -\log(0.046) = 1.34$

(b) At the first half-neutralization point a buffer made up of H_2A^+/HA is formed, where $[H_2A^+] = [HA]$. The Henderson Hasselbalch equation gives $pH = pK_a = 2.34$

(c) At the 1st equivalence point all of the $H_2A^+(aq)$ is converted to $HA(aq)$. $HA(aq)$ is involved in both K_{a_1} and K_{a_2}, both ionizations must be considered.

If we assume that the solution is converted to 100% HA, we must consider two reactions. HA may act as a weak acid ($HA \rightarrow A^- + H^+$) or HA may act as a base ($HA + H^+ \rightarrow H_2A^+$). See Diagram on the next page:

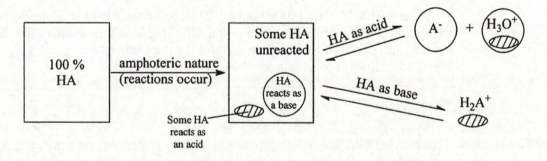

Using the diagram above, we see the following relations must hold true.
$[A^-] = [H_3O^+] + [H_2A^+]$

$$K_{a_2} = \frac{[A^-][H_3O^+]}{[HA]} \text{ or } [A^-] = \frac{K_{a_2}[HA]}{[H_3O^+]} \quad \& \quad K_{a_1} = \frac{[HA][H_3O^+]}{[H_2A^+]} \text{ or } [H_2A^+] = \frac{[H_3O^+][HA]}{K_{a_1}}$$

Substitute for $[A^-]$ and $[H_2A^+]$ in $[A^-] = [H_3O^+] + [H_2A^+]$

$$\frac{K_{a_2}[HA]}{[H_3O^+]} = [H_3O^+] + \frac{[H_3O^+][HA]}{K_{a_1}} \quad \text{(multiply both sides by } K_{a_1}[H_3O^+])$$

$$K_{a_1}K_{a_2}[HA] = K_{a_1}[H_3O^+][H_3O^+] + [H_3O^+][H_3O^+][HA]$$

$$K_{a_1}K_{a_2}[HA] = [H_3O^+]^2(K_{a_1} + [HA])$$

$$[H_3O^+]^2 = \frac{K_{a_1}K_{a_2}[HA]}{(K_{a_1} + [HA])} \quad \text{Usually, } [HA] \gg K_{a_1} \text{ (Here, } 0.500 \gg 4.6 \times 10^{-3})$$

Make the assumption that $K_{a_1} + [HA] \approx [HA]$

$$[H_3O^+]^2 = \frac{K_{a_1}K_{a_2}[HA]}{[HA]} = K_{a_1}K_{a_2} \qquad \text{Take } -\log \text{ of both sides}$$

$$-\log[H_3O^+]^2 = -2\log[H_3O^+] = 2(pH) = -\log K_{a_1}K_{a_2} = -\log K_{a_1} - \log K_{a_2} = pK_{a_1} + pK_{a_2}$$

$$2(pH) = pK_{a_1} + pK_{a_2}$$

$$pH = \frac{pK_{a_1} + pK_{a_2}}{2} = \frac{2.34 + 9.69}{2} = 6.02$$

(d) Half way between the 1st and 2nd equivalence point, half of HA(aq) is converted to A$^-$(aq). We have a HA/A$^-$ buffer solution where [HA] = [A$^-$]. The Henderson-Hasselbalch equation yields pH = pK_{a_2} = 9.69

(e) At the second equivalence point, all of the H_2A^+(aq) is converted to A$^-$(aq). We can treat this simply as a weak base in water having $K_b = \dfrac{K_w}{K_{a_2}} = \dfrac{1 \times 10^{-14}}{2.0 \times 10^{-10}} = 5.0 \times 10^{-5}$

Note: There has been a 1:3 dilution, hence the [A$^-$] = 0.500 M $\times \dfrac{1\ V}{3\ V} = 0.167$ M

	A$^-$(aq)	+ H$_2$O(l)	$\rightleftharpoons$	HA(aq)	+	OH$^-$(aq)
Initial	0.167 M	—		0 M		$\approx$ 0 M
Change	$-x$ M	—		$+x$ M		$+x$ M
Equilibrium	(0.167 $-x$) M	—		x M		x M

$$K_b = \frac{[HA][OH^-]}{[A^-]} = \frac{(x)(x)}{(0.167-x)} = 5.0 \times 10^{-5} \approx \frac{x^2}{0.167}$$

$x = 0.0029$ M = [OH$^-$]; pOH = $-\log$[OH$^-$] = 2.54;

pH = 14.00 $-$ pOH = 14.00 $-$ 2.54 = 11.46

(f) All of the points required in (f) can be obtained using the Henderson-Hasselbalch equation (the chart below shows that the buffer ratio for each point is within the acceptable range (0.25 to 4.0))

mL NaOH →	0.0	10.0	20.0	30.0	40.0	50.0	60.0	70.0	80.0	90.0	100.0	110.0
$\%H_2A^+$ →	100	80	60	40	20	0	0	0	0	0	0	0
%HA →	0	20	40	60	80	100	80	60	40	20	0	0
$\%A^-$ →	0	0	0	0	0	0	20	40	60	80	100	100
buffer ratio = base/acid →	0	0.25	0.67	1.5	4.0	∞	0.25	0.67	1.5	4.0	∞	∞

May use Henderson-Hasselbalch equation   May use Henderson-Hasselbalch equation

(i)   After 10.0 mL

Here we will show how to obtain the answer using both the Henderson-Hasselbalch equation and setting up the I. C. E. (Initial, Change, Equilibrium) table. The results will differ within accepted experimental limitation of the experiment ($\pm$ 0.01 pH units)

$n_{H_2A^+} = (C \times V) = (0.500\ M)(0.0500\ L) = 0.0250$ moles H_2A^-

$n_{OH^-} = (C \times V) = (0.500\ M)(0.0100\ L) = 0.00500$ moles OH^-

$V_{total} = (50.0 + 10.0)\ mL = 60.0\ mL$ or $0.0600\ L$

$$[H_2A^+] = \frac{n_{H_2A^+}}{V_{total}} = \frac{0.0250\ mol}{0.0600\ L} = 0.417\ M \qquad [OH^-] = \frac{n_{OH^-}}{V_{total}} = \frac{0.00500\ mol}{0.0600\ L} = 0.0833\ M$$

$$K_{eq} \text{ for titration reaction} = \frac{1}{K_{b(HA)}} = \frac{1}{\left(\dfrac{K_w}{K_{a_1}}\right)} = \frac{K_{a_1}}{K_w} = \frac{4.6\times10^{-3}}{1.00\times10^{-14}} = 4.6\times10^{11}$$

	$H_2A^+(aq)$ +	$OH^-(aq)$	$\rightleftharpoons$	$HA(aq)$ +	$H_2O(l)$
Initial:	0.417 M	0.0833 M		0 M	—
100% rxn:	-0.0833	-0.0833 M		+0.0833 M	—
New initial:	0.334 M	0 M		0.0833 M	—
Change:	+x M	+x M		-x M	—
Equilibrium:	≈0.334 M	x M		≈0.0833 M	—

$$4.6 \times 10^{11} = \frac{(0.0833)}{(0.334)(x)} \ ; \ x = \frac{(0.0833)}{(0.334)(4.6\times10^{11})} = 5.4 \times 10^{-13} \text{ (valid assumption)}$$

$x = 5.4 \times 10^{-13} = [OH^-]$; pOH = $-\log(5.4 \times 10^{-13}) = 12.27$;
pH = 14.00 - pOH = 14.00 - 12.27 = 1.73
Alternate method using the Henderson-Hasselbalch equation:

(i)   After 10.0 mL, 20% of H_2A^+ reacts forming the conjugate base HA
Hence the buffer solution is 80% H_2A^+ (acid) and 20% HA (base)

$$pH = pK_{a_1} + \log \frac{base}{acid} = 2.34 + \log \frac{20.0}{80.0} = 2.34 + (-0.602) = 1.74 \text{ (within} \pm 0.01)$$

For the remainder of the calculations we will employ the Henderson-Hasselbalch equation with the understanding that using the method that employs the I.C.E. table gives the same result within the limitation of the data.

(ii)   After 20.0 mL, 40% of H_2A^+ reacts forming the conjugate base HA
Hence the buffer solution is 60% H_2A^+ (acid) and 40% HA (base)

$$pH = pK_{a_1} + \log \frac{base}{acid} = 2.34 + \log \frac{40.0}{60.0} = 2.34 + (-0.176) = 2.16$$

(iii) After 30.0 mL, 60% of H_2A^+ reacts forming the conjugate base HA
Hence the buffer solution is 40% H_2A^+ (acid) and 60% HA (base)

$$pH = pK_{a_1} + \log \frac{base}{acid} = 2.34 + \log \frac{60.0}{40.0} = 2.34 + (+0.176) = 2.52$$

(iv)   After 40.0 mL, 80% of H_2A^+ reacts forming the conjugate base HA
Hence the buffer solution is 20% H_2A^+ (acid) and 80% HA (base)

$$pH = pK_{a_1} + \log \frac{base}{acid} = 2.34 + \log \frac{80.0}{20.0} = 2.34 + (0.602) = 2.94$$

(v)   After 50 mL all of the H_2A^+(aq) has reacted and we begin with essentially 100% HA(aq), which is a weak acid. Addition of base results in the formation of the conjugate base (buffer system) A^-(aq). We employ a similar solution, however, now we must use $pK_{a_2} = 9.69$

(vi)   After 60.0 mL, 20% of HA reacts forming the conjugate base A^-
Hence the buffer solution is 80% HA (acid) and 20% A^- (base)

$$pH = pK_{a_2} + \log \frac{base}{acid} = 9.69 + \log \frac{20.0}{80.0} = 9.69 + (-0.602) = 9.09$$

(vii) After 70.0 mL, 40% of HA reacts forming the conjugate base A^-
Hence the buffer solution is 60% HA (acid) and 40% A^- (base)

$$pH = ppK_{a_2} + \log \frac{base}{acid} = 9.69 + \log \frac{40.0}{60.0} = 9.69 + (-0.176) = 9.51$$

(viii) After 80.0 mL, 60% of HA reacts forming the conjugate base A^-
Hence the buffer solution is 40% HA (acid) and 60% A^- (base)

$$pH = pK_{a_2} + \log \frac{base}{acid} = 9.69 + \log \frac{60.0}{40.0} = 9.69 + (+0.176) = 9.87$$

(ix)   After 90.0 mL, 80% of HA reacts forming the conjugate base A^-
Hence the buffer solution is 20% HA (acid) and 80% A^- (base)

$$pH = pK_{a_2} + \log \frac{base}{acid} = 9.69 + \log \frac{80.0}{20.0} = 9.69 + (0.602) = 10.29$$

(x)   After the addition of 110.0 mL, NaOH is in excess. (10.0 mL of 0.500 M NaOH is in excess, or, 0.00500 moles of NaOH remains unreacted). The pH of a solution which

has NaOH in excess is determined by the [OH⁻] that is in excess.
(For a diprotic acid this occurs after the 2nd equivalence point).

$$[OH^-]_{excess} = \frac{n_{OH^-}}{V_{total}} = \frac{0.00500\,mol}{0.1600\,L} = 0.03125\,M\,;\ \ pOH = -\log(0.03125) = 1.51$$

$$pH = 14.00 - pOH = 14.00 - 1.51 = 12.49$$

(g) A sketch of the titration curve for the 0.500 M solution of alanine hydrochloride, with some significant points labeled on the plot is shown below.

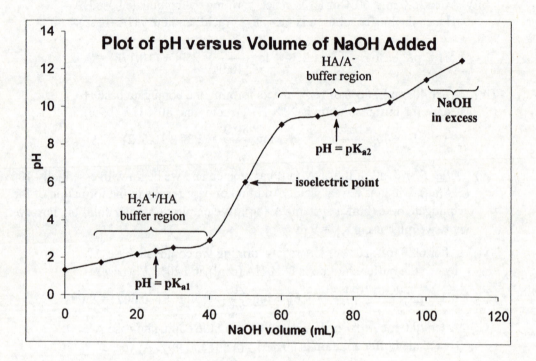

CHAPTER 18
SOLUBILITY AND COMPLEX-ION EQUILIBRIA
PRACTICE EXAMPLES

1A In each case, we first write the balanced equation for the solubility equilibrium and then the equilibrium constant expression for that equilibrium, the K_{sp} expression:

(a) $MgCO_3(s) \rightleftharpoons Mg^{2+}(aq) + CO_3^{2-}(aq)$ $\qquad K_{sp} = [Mg^{2+}][CO_3^{2-}]$

(b) $Ag_3PO_4(s) \rightleftharpoons 3Ag^+(aq) + PO_4^{3-}(aq)$ $\qquad K_{sp} = [Ag^+]^3[PO_4^{3-}]$

1B **(a)** Provided the $[OH^-]$ is not too high, the hydrogen phosphate ion is not expected to ionize in aqueous solution to a significant extent because of the quite small values for the second and third ionization constants of phosphoric acid.

$CaHPO_4(s) \rightleftharpoons Ca^{2+}(aq) + HPO_4^{2-}(aq)$

(b) The solubility product constant is written in the manner of a K_c expression:
$K_{sp} = [Ca^{2+}][HPO_4^{2-}] = 1. \times 10^{-7}$

2A We calculate the solubility of silver cyanate, s, as a molarity. We then use the solubility equation to (1) relate the concentrations of the ions and (2) write the K_{sp} expression.

$$s = \frac{7 \text{ mg AgOCN}}{100 \text{ mL}} \times \frac{1000 \text{ mL}}{1 \text{ L}} \times \frac{1 \text{ g}}{1000 \text{ mg}} \times \frac{1 \text{ mol AgOCN}}{149.9 \text{ g AgOCN}} = 5 \times 10^{-4} \text{ moles/L}$$

Equation : $\quad AgOCN(s) \rightleftharpoons Ag^+(aq) + OCN^-(aq)$

Solubility Product : $\qquad\qquad\qquad s \qquad\quad s$

$$K_{sp} = \left[Ag^+\right]\left[OCN^-\right] = (s) \times (s) = s^2 = \left(5 \times 10^{-4}\right)^2 = 3 \times 10^{-7}$$

2B We calculate the solubility of lithium phosphate, s, as a molarity. We then use the solubility equation to (1) relate the concentrations of the ions and (2) write the K_{sp} expression.

$$s = \frac{0.034 \text{ g Li}_3PO_4}{100 \text{ mL soln}} \times \frac{1000 \text{ mL}}{1 \text{ L}} \times \frac{1 \text{ mol Li}_3PO_4}{115.79 \text{ g Li}_3PO_4} = 0.0029 \text{ moles/L}$$

Equation: $Li_3PO_4(s) \rightleftharpoons 3Li^+(aq) + PO_4^{3-}(aq)$

Solubility Product: $\qquad\qquad (3s)^3 \qquad s$

$$K_{sp} = \left[Li^+\right]^3\left[PO_4^{3-}\right] = (3s)^3 \cdot (s) = 27s^4 = 27(0.0029)^4 = 1.9 \times 10^{-9}$$

3A   We use the solubility equilibrium to write the K_{sp} expression, which we then solve to obtain the molar solubility, s, of $Cu_3(AsO_4)_2$.

$$Cu_3(AsO_4)_2.(s) \rightleftharpoons Cu^{2+}(aq) + 2\ AsO_4^-(aq)$$

$$K_{sp} = \left[Cu^{2+}\right]^3 \left[AsO_4^-\right]^2 = (3s)^3(2s)^2 = 108s^5 = 7.6 \times 10^{-36}$$

Solubility : $s = \sqrt[5]{\dfrac{7.6 \times 10^{-36}}{108}} = 3.7 \times 10^{-8}\,M$

3B   First we determine the solubility of $BaSO_4$, and then find the mass dissolved.

$$BaSO_4(aq) \rightleftharpoons Ba^{2+}(aq) + SO_4^{2-}(aq) \qquad\qquad K_{sp} = \left[Ba^{2+}\right]\left[SO_4^{2-}\right] = s^2$$

The last relationship is true because $\left[Ba^{2+}\right] = \left[SO_4^{2-}\right]$ in a solution produced by dissolving

$BaSO_4$ in pure water. Thus, $s = \sqrt{K_{sp}} = \sqrt{1.1 \times 10^{-10}} = 1.0_5 \times 10^{-5}\,M$

$$\text{mass } BaSO_4 = 225\ mL \times \frac{1.0_5 \times 10^{-5}\ mmol\ BaSO_4}{1\ mL\ sat'd\ soln} \times \frac{233.39\ mg\ BaSO_4}{1\ mmol\ BaSO_4} = 0.55\ mg\ BaSO_4$$

4A   For $PbI_2, K_{sp} = \left[Pb^{2+}\right]\left[I^-\right]^2 = 7.1 \times 10^{-9}$. The solubility equilibrium is the basis of the calculation.

Equation:	$PbI_2(s)$	$\rightleftharpoons$	$Pb^{2+}(aq)$	$+$	$2I^-(aq)$
Initial:	—		0.10 M		0 M
Changes:	—		+s M		+2s M
Equil:	—		(0.10 + s) M		2s M

$$K_{sp} = \left[Pb^{2+}\right]\left[I^-\right]^2 = 7.1 \times 10^{-9} = (0.10+s)(2s)^2 \approx 0.40\,s^2 \qquad s = \sqrt{\frac{7.1 \times 10^{-9}}{0.40}} = 1.3 \times 10^{-4}\,M$$

(assumption $0.10 \gg s$ is valid)

This value of s is the solubility of PbI_2 in $0.10\ M\ Pb(NO_3)_2(aq)$.

4B   We find pOH from the given pH:

$pOH = 14.00 - 8.20 = 5.80;$ $\left[OH^-\right] = 10^{-pOH} = 10^{-5.80} = 1.6 \times 10^{-6}\,M$.

We assume that pOH remains constant, and use the K_{sp} expression for $Fe(OH_3)$.

$$K_{sp} = \left[Fe^{3+}\right]\left[OH^-\right]^3 = 4 \times 10^{-38} = \left[Fe^{3+}\right](1.6 \times 10^{-6})^3 \quad \left[Fe^{3+}\right] = \frac{4 \times 10^{-38}}{(1.6 \times 10^{-6})^3} = 1 \times 10^{-20}\,M$$

Therefore, the molar solubility of $Fe(OH)_3$ is 1×10^{-20} M.

The dissolved $Fe(OH)_3$ does not significantly affect $\left[OH^-\right]$.

5A First determine $\left[I^-\right]$ as altered by dilution. We then compute Q_{sp} and compare it with K_{sp}.

$$\left[I^-\right] = \frac{3\,\text{drops} \times \dfrac{0.05\,\text{mL}}{1\,\text{drop}} \times \dfrac{0.20\,\text{mmol KI}}{1\,\text{mL}} \times \dfrac{1\,\text{mmol I}^-}{1\,\text{mmol KI}}}{100.0\,\text{mL soln}} = 3 \times 10^{-4}\,\text{M}$$

$$Q_{sp} = \left[Ag^+\right]\left[I^-\right] = (0.010)(3 \times 10^{-4}) = 3 \times 10^{-6}$$

$$\therefore Q_{sp} > 8.5 \times 10^{-17} = K_{sp} \qquad \text{Thus, precipitation should occur.}$$

5B We first use the solubility product constant expression for PbI_2 to determine the $\left[I^-\right]$ needed in solution to just form a precipitate when $\left[Pb^{2+}\right] = 0.010$ M. We assume that the volume of solution added is small and that $\left[Pb^{2+}\right]$ remains at 0.010 M throughout.

$$K_{sp} = \left[Pb^{2+}\right]\left[I^-\right]^2 = 7.1 \times 10^{-9} = (0.010)\left[I^-\right]^2 \qquad \left[I^-\right] = \sqrt{\frac{7.1 \times 10^{-9}}{0.010}} = 8.4 \times 10^{-4}\,\text{M}$$

We determine the volume of 0.20 M KI needed.

$$\text{volume of KI(aq)} = 100.0\,\text{mL} \times \frac{8.4 \times 10^{-4}\,\text{mmol I}^-}{1\,\text{mL}} \times \frac{1\,\text{mmol KI}}{1\,\text{mmol I}^-} \times \frac{1\,\text{mL KI(aq)}}{0.20\,\text{mmol KI}} \times \frac{1\,\text{drop}}{0.050\,\text{mL}}$$

$$= 8.4\,\text{drops} = 9\,\text{drops}$$

Since one additional drop is needed, 10 drops will be required. This is an insignificant volume compared to the original solution, so that $\left[Pb^{2+}\right]$ remains constant.

6A Here we must find the maximum $\left[Ca^{2+}\right]$ that can coexist with $\left[OH^-\right] = 0.040\,\text{M}$.

$$K_{sp} = 5.5 \times 10^{-6} = \left[Ca^{2+}\right]\left[OH^-\right]^2 = \left[Ca^{2+}\right](0.040)^2; \quad \left[Ca^{2+}\right] = \frac{5.5 \times 10^{-6}}{(0.040)^2} = 3.4 \times 10^{-3}\,\text{M}$$

For precipitation to be considered complete, $\left[Ca^{2+}\right]$ should be less than 0.1% of its original value. $3.4 \times 10^{-3}\,\text{M}$ is 34% of 0.010 M and therefore precipitation of $Ca(OH)_2$ is not complete under these conditions.

6B We begin by finding $\left[Mg^{2+}\right]$ that corresponds to $1\,\mu\text{g}\,Mg^{2+}$ / L.

$$\left[Mg^{2+}\right] = \frac{1\,\mu\text{g Mg}^{2+}}{1\,\text{L soln}} \times \frac{1\,\text{g}}{10^6\,\mu\text{g}} \times \frac{1\,\text{mol Mg}^{2+}}{24.3\,\text{g Mg}^{2+}} = 4 \times 10^{-8}\,\text{M}$$

Now we use the K_{sp} expression for $Mg(OH)_2$ to determine $\left[OH^-\right]$.

$$K_{sp} = 1.8 \times 10^{-11} = \left[Mg^{2+}\right]\left[OH^-\right]^2 = (4 \times 10^{-8})\left[OH^-\right]^2 \qquad \left[OH^-\right] = \sqrt{\frac{1.8 \times 10^{-11}}{4 \times 10^{-8}}} = 0.02\,\text{M}$$

7A Let us first determine $\left[Ag^+\right]$ when AgCl(s) just begins to precipitate. At this point, Q_{sp} and K_{sp} are equal.

$$K_{sp} = 1.8 \times 10^{-10} = \left[Ag^+\right]\left[Cl^-\right] = Q_{sp} = \left[Ag^+\right] \times 0.115\,M \quad \left[Ag^+\right] = \frac{1.8 \times 10^{-10}}{0.115} = 1.6 \times 10^{-9}\,M$$

Now let us determine the maximum $\left[Br^-\right]$ that can coexist with this $\left[Ag^+\right]$.

$$K_{sp} = 5.0 \times 10^{-13} = \left[Ag^+\right]\left[Br^-\right] = 1.6 \times 10^{-9}\,M \times \left[Br^-\right]; \quad \left[Br^-\right] = \frac{5.0 \times 10^{-13}}{1.6 \times 10^{-9}} = 3.1 \times 10^{-4}\,M$$

The remaining bromide ion has precipitated as AgBr(s) with the addition of $AgNO_3(aq)$.

$$\text{Percent of } Br^- \text{ remaining} = \frac{[Br^-]_{final}}{[Br^-]_{initial}} \times 100\% = \frac{3.1 \times 10^{-4}\,M}{0.264\,M} \times 100\% = 0.12\%$$

7B Since the ions have the same charge and the same concentrations, we look for two K_{sp} values for the salt with the same anion that are as far apart as possible. The K_{sp} values for the carbonates are very close, while those for the sulfates and fluorides are quite different. However, the difference in the K_{sp} values is greatest for the chromates; K_{sp} for $BaCrO_4\left(1.2 \times 10^{-10}\right)$ is so much smaller than K_{sp} for $SrCrO_4\left(2.2 \times 10^{-5}\right)$, $BaCrO_4$ will precipitate first and $SrCrO_4$ will begin to precipitate when $\left[CrO_4^{2-}\right]$ has the value:

$$\left[CrO_4^{2-}\right] = \frac{K_{sp}}{\left[Sr^{2+}\right]} = \frac{2.2 \times 10^{-5}}{0.10} = 2.2 \times 10^{-4}\,M.$$

At this point $\left[Ba^{2+}\right]$ is found as follows.

$$\left[Ba^{2+}\right] = \frac{K_{sp}}{\left[CrO_4^{2-}\right]} = \frac{1.2 \times 10^{-10}}{2.2 \times 10^{-4}} = 5.5 \times 10^{-7}\,M;$$

$[Ba^{2+}]$ has dropped to 0.00055% of its initial value and therefore is considered to be completely precipitated, before $SrCrO_4$ begins to precipitate. The two ions are thus effectively separated as chromates. The best precipitating agent is a group 1 chromate salt.

8A First determine $\left[OH^-\right]$ resulting from the hydrolysis of acetate ion.

Equation:	$C_2H_3O_2^-\,(aq)$	$+\ H_2O(l)$	$\rightleftharpoons$	$HC_2H_3O_2(aq)$	$+$	$OH^-(aq)$
Initial:	0.10 M	—		0 M		$\approx 0\,M$
Changes:	$-x\,M$	—		$+x\,M$		$+x\,M$
Equil:	$(0.10 - x)\,M$	—		$x\,M$		$x\,M$

$$K_b = \frac{1.0 \times 10^{-14}}{1.8 \times 10^{-5}} = 5.6 \times 10^{-10} = \frac{\left[HC_2H_3O_2\right]\left[OH^-\right]}{\left[C_2H_3O_2^-\right]} = \frac{x \cdot x}{0.10 - x} \approx \frac{x^2}{0.10}$$

$x = [OH^-] = \sqrt{0.10 \times 5.6 \times 10^{-10}} = 7.5 \times 10^{-6} M$ (the assumption $x \ll 0.10$ was valid)

Now compute the value of the ion product in this solution and compare it with the value of K_{sp} for $Mg(OH)_2$.

$$Q_{sp} = \left[Mg^{2+}\right]\left[OH^-\right]^2 = (0.010M)(7.5 \times 10^{-6} M)^2 = 5.6 \times 10^{-13} < 1.8 \times 10^{-11} = K_{sp}\left[Mg(OH)_2\right]$$

Because Q_{sp} is smaller than K_{sp}, this solution is unsaturated and precipitation of $Mg(OH)_2(s)$ will not occur.

8B Here we can use the Henderson–Hasselbalch equation to determine the pH of the buffer.

$$pH = pK_a + \log\frac{\left[C_2H_3O_2^-\right]}{\left[HC_2H_3O_2\right]} = -\log(1.8 \times 10^{-5}) + \log\frac{0.250 M}{0.150 M} = 4.74 + 0.22 = 4.96$$

$$pOH = 14.00 - pH = 14.00 - 4.96 = 9.04 \qquad \left[OH^-\right] = 10^{-pOH} = 10^{-9.04} = 9.1 \times 10^{-10} M$$

Now we determine Q_{sp} to see if precipitation will occur.

$$Q_{sp} = \left[Fe^{3+}\right]\left[OH^-\right]^3 = (0.013M)(9.1 \times 10^{-10})^3 = 9.8 \times 10^{-30}$$

$Q_{sp} > 4 \times 10^{-38} = K_{sp}$; Thus, $Fe(OH)_3$ precipitation should occur.

9A Determine $\left[OH^-\right]$, and then the pH necessary to prevent the precipitation of $Mn(OH)_2$.

$$K_{sp} = 1.9 \times 10^{-13} = \left[Mn^{2+}\right]\left[OH^-\right]^2 = (0.0050M)\left[OH^-\right]^2$$

$$\left[OH^-\right] = \sqrt{\frac{1.9 \times 10^{-13}}{0.0050}} = 6.2 \times 10^{-6} M$$

$$pOH = -\log(6.2 \times 10^{-6}) = 5.21 \qquad pH = 14.00 - 5.21 = 8.79$$

We will use this pH in the Henderson–Hasselbalch equation to determine $\left[NH_4^+\right]$.

$pK_b = 4.74$ for NH_3.

$$pH = pK_a + \log\frac{\left[NH_3\right]}{\left[NH_4^+\right]} = 8.79 = (14.00 - 4.74) + \log\frac{0.025 M}{\left[NH_4^+\right]}$$

$$\log\frac{0.025 M}{\left[NH_4^+\right]} = 8.79 - (14.00 - 4.74) = -0.47 \qquad \frac{0.025}{\left[NH_4^+\right]} = 10^{-0.47} = 0.34$$

$$\left[NH_4^+\right] = \frac{0.025}{0.34} = 0.074M$$

9B First we must calculate the $[H_3O^+]$ in the buffer solution that is being employed to dissolve the magnesium hydroxide:

$$NH_3(aq) + H_2O(l) \rightleftharpoons NH_4^+(aq) + OH^-(aq) \; ; \; K_b = 1.8 \times 10^{-5}$$

$$K_b = \frac{[NH_4^+][OH^-]}{[NH_3]} = \frac{[0.100M][OH^-]}{[0.250M]} = 1.8 \times 10^{-5}$$

$$[OH^-] = 4.5 \times 10^{-5}\,M; [H_3O^+] = \frac{1.00 \times 10^{-14}\,M^2}{4.5 \times 10^{-5}\,M} = 2.2_2 \times 10^{-10}\,M$$

Now we can employ equation 18.4 to calculate the molar solubility of $Mg(OH)_2$ in the buffer solution; Molar Solubility $Mg(OH)_2 = [Mg^{2+}]_{equil}$

$$Mg(OH)_2(s) + 2\,H_3O^+(aq) \xrightarrow{K = 1.8 \times 10^{17}} Mg^{2+}(aq) + 4H_2O(l) \; ;$$

Equilibrium $-$ 2.2×10^{-10} x $-$

$$K = \frac{[Mg^{2+}]}{[H_3O^+]^2} = \frac{x}{[2.2 \times 10^{-10}\,M]^2} = 1.8 \times 10^{17} \qquad x = 8.7 \times 10^{-3}\,M = [Mg^{2+}]_{equil}$$

So, the Molar Solubility for $Mg(OH)_2 = 8.7 \times 10^{-3}\,M$

10A **(a)** In solution are $Cu^{2+}(aq)$, $SO_4^{2-}(aq)$, $Na^+(aq)$, and $OH^-(aq)$.

$$Cu^{2+}(aq) + 2\,OH^-(aq) \rightarrow Cu(OH)_2(s)$$

(b) In solution above $Cu(OH)_2(s)$ is $Cu^{2+}(aq)$:

$$Cu(OH)_2(s) \rightleftharpoons Cu^{2+}(aq) + 2\,OH^-(aq)$$

This $Cu^{2+}(aq)$ reacts with the added $NH_3(aq)$:

$$Cu^{2+}(aq) + 4\,NH_3(aq) \rightleftharpoons \left[Cu(NH_3)_4\right]^{2+}(aq)$$

The overall result is:

$$Cu(OH)_2(s) + 4NH_3(aq) \rightleftharpoons \left[Cu(NH_3)_4\right]^{2+}(aq) + 2\,OH^-(aq)$$

(c) $HNO_3(aq)$ (a strong acid), forms $H_3O^+(aq)$, which reacts with $OH^-(aq)$ and $NH_3(aq)$.

$$OH^-(aq) + H_3O^+(aq) \rightarrow 2\,H_2O(l); \qquad NH_3(aq) + H_3O^+(aq) \rightarrow NH_4^+(aq) + H_2O(l)$$

As $NH_3(aq)$ is consumed, the reaction below shifts to the left.

$$Cu(OH)_2(s) + 4\,NH_3(aq) \rightleftharpoons \left[Cu(NH_3)_4\right]^{2+}(aq) + 2\,OH^-(aq)$$

But as $OH^-(aq)$ is consumed, the dissociation reactions shift to the side with the dissolved ions: $Cu(OH)_2(s) \rightleftharpoons Cu^{2+}(aq) + 2\,OH^-(aq)$

The species in solution at the end of all this are
$Cu^{2+}(aq)$, $NO_3^-(aq)$, $NH_4^+(aq)$, excess $H_3O^+(aq)$, $Na^+(aq)$; and $SO_4^{2-}(aq)$
(Probably $HSO_4^-(aq)$ as well)

10B **(a)** In solution are $Zn^{2+}(aq)$, $SO_4^{2-}(aq)$, and $NH_3(aq)$,

$$Zn^{2+}(aq) + 4NH_3(aq) \rightleftharpoons \left[Zn(NH_3)_4\right]^{2+}(aq)$$

(b) $HNO_3(aq)$, a strong acid, produces $H_3O^+(aq)$, which reacts with $NH_3(aq)$.

$NH_3(aq) + H_3O^+(aq) \rightarrow NH_4^+(aq) + H_2O(l)$ As $NH_3(aq)$ is consumed, the tetrammine complex ion is destroyed.

$$\left[Zn(NH_3)_4\right]^{2+}(aq) + 4H_3O^+(aq) \rightleftharpoons \left[Zn(H_2O)_4\right]^{2+}(aq) + 4NH_4^+(aq)$$

(c) $NaOH(aq)$ is a strong base that produces $OH^-(aq)$, forming a hydroxide precipitate.

$$\left[Zn(H_2O)_4\right]^{2+}(aq) + 2OH^-(aq) \rightleftharpoons Zn(OH)_2(s) + 4H_2O(l)$$

Another possibility is a reversal of the reaction of part **(b)**.

$$\left[Zn(H_2O)_4\right]^{2+}(aq) + 4NH_4^+(aq) + 4OH^-(aq) \rightleftharpoons \left[Zn(NH_3)_4\right]^{2+}(aq) + 8H_2O(l)$$

(d) The precipitate dissolves in excess base.

$$Zn(OH)_2(s) + 2OH^-(aq) \rightleftharpoons \left[Zn(OH)_4\right]^{2-}(aq)$$

11A We first determine $\left[Ag^+\right]$ in a solution that is 0.100 M $Ag^+(aq)$ (from $AgNO_3$) and 0.225 M $NH_3(aq)$. Because of the large value of $K_f = 1.6 \times 10^7$, we start by having the reagents form as much complex ion as possible, and approach equilibrium from this point.

Equation:	$Ag^+(aq)$ +	$2NH_3(aq)$	$\rightleftharpoons$	$\left[Ag(NH_3)_2\right]^+(aq)$
In soln:	0.100 M	0.225 M		0 M
Form complex:	–0.100 M	–0.200 M		+0.100 M
Initial	0 M	0.025 M		0.100 M
Changes	+x M	+2x M		–x M
Equil :	x M	(0.025 + x) M		(0.100 – x) M

$$K_f = 1.6 \times 10^7 = \frac{\left[\left[Ag(NH_3)_2\right]^+\right]}{\left[Ag^+\right]\left[NH_3\right]^2} = \frac{0.100 - x}{x(0.025 + 2x)^2} \approx \frac{0.100}{x(0.025)^2}$$

$$x = \frac{0.100}{(0.025)^2 1.6 \times 10^7} = 1.0 \times 10^{-5}\,M = \left[Ag^+\right] = \text{concentration of free silver ion}$$

($x \ll 0.025$ M, so the approximation was valid).

The $\left[Cl^-\right]$ is diluted: $\left[Cl^-\right]_{final} = \left[Cl^-\right]_{initial} \times \dfrac{1.00\,mL_{initial}}{1,500\,mL_{final}} = 3.50\,M \div 1500 = 0.00233\,M$

Finally we compare Q_{sp} with K_{sp} to determine if precipitation of AgCl(s) will occur.

$$Q_{sp} = \left[Ag^+\right]\left[Cl^-\right] = \left(1.0 \times 10^{-5}\,M\right)(0.00233\,M) = 2.3 \times 10^{-8} > 1.8 \times 10^{-10} = K_{sp}$$

Because the value of the Q_{sp} is larger than the value of the K_{sp}, precipitation of AgCl(s) should occur.

11B We organize the solution around the balanced equation of the formation reaction.

Equation: $\quad\quad\quad Pb^{2+}(aq) \quad + \quad EDTA^{4-}(aq) \rightleftharpoons [PbEDTA]^{2-}(aq)$

Initial: $\quad\quad\quad\quad$ 0.100 M $\quad\quad\quad\quad$ 0.250 M $\quad\quad\quad\quad$ 0 M

Form Complex: –0.100 M $\quad\quad$ (0.250 – 0.100) M $\quad\quad$ 0.100 M

Equil : $\quad\quad\quad\quad\quad x$ M $\quad\quad\quad\quad$ (0.150 + x) M $\quad\quad$ (0.100 – x) M

$$K_f = \frac{\left[[PbEDTA]^{2-}\right]}{\left[Pb^{2+}\right]\left[EDTA^{4-}\right]} = 2 \times 10^{18} = \frac{0.100 - x}{x(0.150 + x)} \approx \frac{0.100}{0.150\,x}$$

$$x = \frac{0.100}{0.150 \times 2 \times 10^{18}} = 3 \times 10^{-19} \text{ M} \quad\quad (x \ll 0.100 \text{ M, thus the approximation was valid}).$$

We calculate Q_{sp} and compare it to K_{sp} to determine if precipitation will occur.

$$Q_{sp} = \left[Pb^{2+}\right]\left[I^-\right]^2 = \left(3 \times 10^{-19}\,M\right)\left(0.10\,M\right)^2 = 3 \times 10^{-21}.$$

$Q_{sp} < 7.1 \times 10^{-9} = K_{sp}$ Thus precipitation will not occur.

12A We first determine the maximum concentration of free Ag^+.

$$K_{sp} = \left[Ag^+\right]\left[Cl^-\right] = 1.8 \times 10^{-10} \quad\quad \left[Ag^+\right] = \frac{1.8 \times 10^{-10}}{0.0075} = 2.4 \times 10^{-8} \text{ M.}$$

This is so small that we assume that all the Ag^+ in solution is present as complex ion:

$\left[[Ag(NH_3)_2]^+\right] = 0.13$ M. We use K_f to determine the concentration of free NH_3.

$$K_f = \frac{\left[[Ag(NH_3)_2]^+\right]}{\left[Ag^+\right]\left[NH_3\right]^2} = 1.6 \times 10^7 = \frac{0.13\,M}{2.4 \times 10^{-8}\left[NH_3\right]^2}$$

$$\left[NH_3\right] = \sqrt{\frac{0.13}{2.4 \times 10^{-8} \times 1.6 \times 10^7}} = 0.58\,M.$$

If we combine this with the ammonia present in the complex ion, the total ammonia concentration is $0.58\,M + (2 \times 0.13\,M) = 0.84\,M$. Thus, the <u>minimum</u> concentration of ammonia necessary to keep AgCl(s) from forming is 0.84 M.

12B We use the solubility product constant expression to determine the maximum $\left[Ag^+\right]$ that can be present in 0.010 M Cl^- without precipitation occurring.

$$K_{sp} = 1.8 \times 10^{-10} = \left[Ag^+\right]\left[Cl^-\right] = \left[Ag^+\right](0.010\,M) \quad \left[Ag^+\right] = \frac{1.8 \times 10^{-10}}{0.010} = 1.8 \times 10^{-8}\,M$$

This is also the concentration of free silver ion in the K_f expression. Because of the large value of K_f, practically all of the silver ion in solution is present as the complex ion, $[Ag(S_2O_3)_2]^{3-}$. We solve the expression for $[S_2O_3^{2-}]$ and then add the $[S_2O_3^{2-}]$ "tied up" in the complex ion.

$$K_f = 1.7 \times 10^{13} = \frac{\left[\left[Ag(S_2O_3)_2\right]^{3-}\right]}{\left[Ag^+\right]\left[S_2O_3^{2-}\right]^2} = \frac{0.10\,M}{1.8 \times 10^{-8}\,M\left[S_2O_3^{2-}\right]^2}$$

$$\left[S_2O_3^{2-}\right] = \sqrt{\frac{0.10}{1.8 \times 10^{-8} \times 1.7 \times 10^{13}}} = 5.7 \times 10^{-4}\,M = \text{concentration of free } S_2O_3^{2-}$$

$$\text{total}\left[S_2O_3^{2-}\right] = 5.7 \times 10^{-4}\,M + 0.10\,M\left[Ag(S_2O_3)_2\right]^{3-} \times \frac{2\,\text{mol } S_2O_3^{2-}}{1\,\text{mol}\left[Ag(S_2O_3)_2\right]^{3-}}$$

$$= 0.20\,M + 0.00057\,M = 0.20\,M$$

13A We need combine the two equilibrium expressions, for K_f and for K_{sp}, to find $K_{overall}$.

$$Fe(OH)_3(s) \rightleftharpoons Fe^{3+}(aq) + 3\,OH^-(aq) \qquad\qquad K_{sp} = 4 \times 10^{-38}$$

$$Fe^{3+}(aq) + 3\,C_2O_4^{2-}(aq) \rightleftharpoons \left[Fe(C_2O_4)_3\right]^{3-}(aq) \qquad\qquad K_f = 2 \times 10^{20}$$

$$Fe(OH)_3(s) + 3\,C_2O_4^{2-}(aq) \rightleftharpoons \left[Fe(C_2O_4)_3\right]^{3-}(aq) + 3\,OH^-(aq)\quad K_{overall} = 8 \times 10^{-18}$$

initial:	0.100 M	0 M	$\approx 0\,M$
changes:	$-3x$ M	$+x$ M	$+3x$ M
equil:	$(0.100 - 3x)\,M$	x M	$3x$ M

$$K_{overall} = \frac{\left[\left[Fe(C_2O_4)_3\right]^{3-}\right]\left[OH^-\right]^3}{\left[C_2O_4^{2-}\right]^3} = \frac{(x)(3x)^3}{(0.100 - 3x)^3} = 8 \times 10^{-18} \approx \frac{27x^4}{(0.100)^3}$$

($3x \ll 0.100$ M, thus approximation was valid).

$$x = \sqrt[4]{\frac{(0.100)^3\,8 \times 10^{-18}}{27}} = 4 \times 10^{-6}\,M \quad \text{The assumption is valid.}$$

Thus the solubility of $Fe(OH)_3$ in 0.100 M $C_2O_4^{2-}$ is $4 \times 10^{-6}\,M$.

13B In Example 18-13 we saw that the expression for the solubility, s, of a silver halide in an aqueous ammonia solution, where $\left[NH_3\right]$ is the concentration of aqueous ammonia, is given by:

$$K_{sp} \times K_f = \left(\frac{s}{\left[NH_3\right] - 2s}\right)^2 \qquad \text{or} \qquad \sqrt{K_{sp} \times K_f} = \frac{s}{\left[NH_3\right] - 2s}$$

For all scenarios the $\left[NH_3\right]$ stays fixed at 0.1000 M and K_f is always 1.6×10^7. We see that s will decrease as does K_{sp}. The relevant values are:

$$K_{sp}(AgCl) = 1.8 \times 10^{-10}, \quad K_{sp}(AgBr) = 5.0 \times 10^{-13}, \quad K_{sp}(AgI) = 8.5 \times 10^{-17}.$$

Thus, the order of decreasing solubility must be: AgI < AgBr < AgCl.

14A For FeS, we know that $K_{spa} = 6 \times 10^2$; for Ag_2S, $K_{spa} = 6 \times 10^{-30}$.

We compute the value of Q_{spa} in each case, with $[H_2S] = 0.10\,M$ and $[H_3O^+] = 0.30\,M$.

For FeS, $Q_{spa} = \dfrac{0.020 \times 0.10}{(0.30)^2} = 0.022 < 6 \times 10^2 = K_{spa}$

Thus, precipitation of FeS should not occur.

For Ag_2S, $Q_{spa} = \dfrac{(0.010)^2 \times 0.10}{(0.30)^2} = 1.1 \times 10^{-4}$

$Q_{spa} > 6 \times 10^{-30} = K_{spa}$; Thus, precipitation of Ag_2S should occur.

14B The $[H_3O^+]$ needed to just form a precipitate can be obtained by direct substitution of the provided concentrations into the K_{spa} expression. When that expression is satisfied, a precipitate will just form.

$$K_{spa} = \frac{[Fe^{2+}][H_2S]}{[H_3O^+]^2} = 6 \times 10^2 = \frac{(0.015\,M\,Fe^{2+})(0.10\,M\,H_2S)}{[H_3O^+]^2}, \quad [H_3O^+] = \sqrt{\frac{0.015 \times 0.10}{6 \times 10^2}} = 0.002\,M$$

$$pH = -\log[H_3O^+] = -\log(0.002) = 2.7$$

EXERCISES

K_{sp} and Solubility

1 **(a)** $Ag_2SO_4(s) \rightleftharpoons 2\,Ag^+(aq) + SO_4^{2-}(aq)$ $K_{sp} = [Ag^+]^2[SO_4^{2-}]$

    (b) $Ra(IO_3)_2(s) \rightleftharpoons Ra^{2+}(aq) + 2\,IO_3^-(aq)$ $K_{sp} = [Ra^{2+}][IO_3^-]^2$

    (c) $Ni_3(PO_4)_2(s) \rightleftharpoons 3\,Ni^{2+}(aq) + 2\,PO_4^{3-}(aq)$ $K_{sp} = [Ni^{2+}]^3[PO_4^{3-}]^2$

    (d) $PuO_2CO_3(s) \rightleftharpoons PuO_2^{2+}(aq) + CO_3^{2-}(aq)$ $K_{sp} = [PuO_2^{2+}][CO_3^{2-}]$

2. **(a)** $K_{sp} = [Fe^{3+}][OH^-]^3$ $Fe(OH)_3(s) \rightleftharpoons Fe^{3+}(aq) + 3\,OH^-(aq)$

    (b) $K_{sp} = [BiO^+][OH^-]$ $BiOOH(s) \rightleftharpoons BiO^+(aq) + OH^-(aq)$

    (c) $K_{sp} = [Hg_2^{2+}][I^-]^2$ $Hg_2I_2(s) \rightleftharpoons Hg_2^{2+}(aq) + 2\,I^-(aq)$

    (d) $K_{sp} = [Pb^{2+}]^3[AsO_4^{3-}]^2$ $Pb_3(AsO_4)_2(s) \rightleftharpoons 3\,Pb^{2+}(aq) + 2\,AsO_4^{3-}(aq)$

3. **(a)** $CrF_3(s) \rightleftharpoons Cr^{3+}(aq) + 3F^-(aq)$ $\qquad K_{sp} = [Cr^{3+}][F^-]^3 = 6.6 \times 10^{-11}$

(b) $Au_2(C_2O_4)_3(s) \rightleftharpoons 2Au^{3+}(aq) + 3C_2O_4^{2-}(aq)$ $K_{sp} = [Au^{3+}]^2 [C_2O_4^{2-}]^3 = 1 \times 10^{-10}$

(c) $Cd_3(PO_4)_2(s) \rightleftharpoons 3Cd^{2+}(aq) + 2PO_4^{3-}(aq)$ $\qquad K_{sp} = [Cd^{2+}]^3 [PO_4^{3-}]^2 = 2.1 \times 10^{-33}$

(d) $SrF_2(s) \rightleftharpoons Sr^{2+}(aq) + 2F^-(aq)$ $\qquad K_{sp} = [Sr^{2+}][F^-]^2 = 2.5 \times 10^{-9}$

4. Let s = solubility of each compound in moles of compound per liter of solution.

(a) $K_{sp} = [Ba^{2+}][CrO_4^{2-}] = (s)(s) = s^2 = 1.2 \times 10^{-10}$ $\qquad s = 1.1 \times 10^{-5} M$

(b) $K_{sp} = [Pb^{2+}][Br^-]^2 = (s)(2s)^2 = 4s^3 = 4.0 \times 10^{-5}$ $\qquad s = 2.2 \times 10^{-2} M$

(c) $K_{sp} = [Ce^{3+}][F^-]^3 = (s)(3s)^3 = 27s^4 = 8 \times 10^{-16}$ $\qquad s = 7 \times 10^{-5} M$

(d) $K_{sp} = [Mg^{2+}]^3 [AsO_4^{3-}]^2 = (3s)^3 (2s)^2 = 108s^5 = 2.1 \times 10^{-20}$ $\qquad s = 4.5 \times 10^{-5} M$

5. We use the value of K_{sp} for each compound in determining the molar solubility in a saturated solution. In each case, s represents the molar solubility of the compound.

$AgCN \qquad K_{sp} = [Ag^+][CN^-] = (s)(s) = s^2 = 1.2 \times 10^{-16} \qquad s = 1.1 \times 10^{-8} M$

$AgIO_3 \qquad K_{sp} = [Ag^+][IO_3^-] = (s)(s) = s^2 = 3.0 \times 10^{-8} \qquad s = 1.7 \times 10^{-4} M$

$AgI \qquad K_{sp} = [Ag^+][I^-] = (s)(s) = s^2 = 8.5 \times 10^{-17} \qquad s = 9.2 \times 10^{-9} M$

$AgNO_2 \qquad K_{sp} = [Ag^+][NO_2^-] = (s)(s) = s^2 = 6.0 \times 10^{-4} \qquad s = 2.4 \times 10^{-2} M$

$Ag_2SO_4 \qquad K_{sp} = [Ag^+]^2 [SO_4^{2-}] = (2s)^2 (s) = 4s^3 = 1.4 \times 10^{-5} \quad s = 1.5 \times 10^{-2} M$

Thus, in order of increasing molar solubility, from smallest to largest is:

$AgI < AgCN < AgIO_3 < Ag_2SO_4 < AgNO_2$

6. We use the value of K_{sp} for each compound in determining $[Mg^{2+}]$ in its saturated solution. In each case, s represents the molar solubility of the compound.

(a) $MgCO_3 \qquad K_{sp} = [Mg^{2+}][CO_3^{2-}] = (s)(s) = s^2 = 3.5 \times 10^{-8}$

$\qquad s = 1.9 \times 10^{-4} M \qquad [Mg^{2+}] = 1.9 \times 10^{-4} M$

(b) $MgF_2 \qquad K_{sp} = [Mg^{2+}][F^-]^2 = (s)(2s)^2 = 4s^3 = 3.7 \times 10^{-8}$

$\qquad s = 2.1 \times 10^{-3} M \qquad [Mg^{2+}] = 2.1 \times 10^{-3} M$

(c) $Mg_3(PO_4)_2 \qquad K_{sp} = [Mg^{2+}]^3 [PO_4^{3-}]^2 = (3s)^3 (2s)^2 = 108s^5 = 2.1 \times 10^{-25}$

$\qquad s = 5 \times 10^{-6} M \qquad [Mg^{2+}] = 1 \times 10^{-5} M$

Thus a saturated solution of MgF_2 has the highest $[Mg^{2+}]$.

7. We determine $\left[F^-\right]$ in saturated CaF_2, and from that value the concentration of F^- in ppm.

For CaF_2 $K_{sp} = \left[Ca^{2+}\right]\left[F^-\right]^2 = (s)(2s)^2 = 4s^3 = 5.3 \times 10^{-9}$ $s = 1.1 \times 10^{-3} M$

The solubility in ppm is the number of grams of CaF_2 in 10^6 g of solution. We assume a solution density of $1.00\,g/mL$.

$$\text{mass of } F^- = 10^6 \text{ g soln} \times \frac{1\,mL}{1.00\,g\,soln} \times \frac{1\,L\,soln}{1000\,mL} \times \frac{1.1 \times 10^{-3}\,mol\,CaF_2}{1\,L\,soln}$$

$$\times \frac{2\,mol\,F^-}{1\,mol\,CaF_2} \times \frac{19.0\,g\,F^-}{1\,mol\,F^-} = 42\,g\,F^-$$

This is 42 times more concentrated than the optimum concentration of fluoride ion for fluoridation. CaF_2 is, in fact, more soluble than is necessary. Its uncontrolled use might lead to excessive F^- in solution.

8. We determine $\left[OH^-\right]$ in a saturated solution. From this $\left[OH^-\right]$ we determine the pH.

$$K_{sp} = \left[BiO^+\right]\left[OH^-\right] = 4 \times 10^{-10} = s^2 \qquad s = 2 \times 10^{-5} M = \left[OH^-\right] = \left[BiO^+\right]$$

$$pOH = -\log(2 \times 10^{-5}) = 4.7 \qquad\qquad pH = 9.3$$

9. We first assume that the volume of the solution does not change appreciably when its temperature is lowered. Then we determine the mass of $Mg(C_{16}H_{31}O_2)_2$ dissolved in each solution, recognizing that the molar solubility of $Mg(C_{16}H_{31}O_2)_2$ equals the cube root of one fourth of its solubility product constant, since it is the only solute in the solution.

$$K_{sp} = 4s^3 \qquad\qquad s = \sqrt[3]{K_{sp}/4}$$

At $50°C$: $s = \sqrt[3]{4.8 \times 10^{-12}/4} = 1.1 \times 10^{-4}$ M; At $25°C$: $s = \sqrt[3]{3.3 \times 10^{-12}/4} = 9.4 \times 10^{-5}$ M

$$\text{amount of } Mg(C_{16}H_{31}O_2)_2 (50°C) = 0.965\,L \times \frac{1.1 \times 10^{-4}\,mol\,Mg(C_{16}H_{31}O_2)_2}{1L\,soln} = 1.1 \times 10^{-4}\,mol$$

$$\text{amount of } Mg(C_{16}H_{31}O_2)_2 (25°C) = 0.965\,L \times \frac{9.4 \times 10^{-5}\,mol\,Mg(C_{16}H_{31}O_2)_2}{1L\,soln} = 0.91 \times 10^{-4}\,mol$$

mass of $Mg(C_{16}H_{31}O_2)_2$ precipitated:

$$= (1.1 - 0.91) \times 10^{-4}\,mol \times \frac{535.15\,g\,Mg(C_{16}H_{31}O_2)_2}{1\,mol\,Mg(C_{16}H_{31}O_2)_2} \times \frac{1000\,mg}{1g} = 11\,mg$$

10. We first assume that the volume of the solution does not change appreciably when its temperature is lowered. Then we determine the mass of CaC_2O_4 dissolved in each solution, recognizing that the molar solubility of CaC_2O_4 equals the square root of its solubility product constant, since it is the only solute in the solution.

At 95°C $s = \sqrt{1.2 \times 10^{-8}} = 1.1 \times 10^{-4}$ M; At 13°C: $s = \sqrt{2.7 \times 10^{-9}} = 5.2 \times 10^{-5}$ M

$$\text{mass of } CaC_2O_4 \,(95\,°C) = 0.725 \text{ L} \times \frac{1.1 \times 10^{-4} \text{ mol } CaC_2O_4}{1 \text{ L soln}} \times \frac{128.1 \text{ g } CaC_2O_4}{1 \text{ mol } CaC_2O_4} = 0.010 \text{ g } CaC_2O_4$$

$$\text{mass of } CaC_2O_4 \,(13\,°C) = 0.725 \text{ L} \times \frac{5.2 \times 10^{-5} \text{ mol } PbSO_4}{1 \text{ L soln}} \times \frac{128.1 \text{ g } CaC_2O_4}{1 \text{ mol } CaC_2O_4} = 0.0048 \text{ g } CaC_2O_4$$

$$\text{mass of } CaC_2O_4 \text{ precipitated} = (0.010 \text{ g} - 0.0048 \text{ g}) \times \frac{1000 \text{ mg}}{1 \text{ g}} = 5 \text{ mg}$$

11. First we determine $[I^-]$ in the saturated solution.

$$K_{sp} = [Pb^{2+}][I^-]^2 = 7.1 \times 10^{-9} = (s)(2s)^2 = 4s^3 \qquad\qquad s = 1.2 \times 10^{-3} \text{ M}$$

The $AgNO_3$ reacts with the I^- in this saturated solution in the titration.

$Ag^+(aq) + I^-(aq) \to AgI(s)$ We determine the amount of Ag^+ needed for this titration, and then $[AgNO_3]$ in the titrant.

$$\text{moles } Ag^+ = 0.02500 \text{ L} \times \frac{1.2 \times 10^{-3} \text{ mol } PbI_2}{1 \text{ L soln}} \times \frac{2 \text{ mol } I^-}{1 \text{ mol } PbI_2} \times \frac{1 \text{ mol } Ag^+}{1 \text{ mol } I^-} = 6.0 \times 10^{-5} \text{ mol } Ag^+$$

$$AgNO_3 \text{ molarity} = \frac{6.0 \times 10^{-5} \text{ mol } Ag^+}{0.0133 \text{ L soln}} \times \frac{1 \text{ ol } AgNO_3}{1 \text{ mol } Ag^+} = 4.5 \times 10^{-3} \text{ M}$$

12. We first determine $[C_2O_4^{2-}] = s$, the solubility of the saturated solution.

$$[C_2O_4^{2-}] = \frac{4.8 \text{ mL} \times \dfrac{0.00134 \text{ mmol } KMnO_4}{1 \text{ mL soln}} \times \dfrac{5 \text{ mmol } C_2O_4^{2-}}{2 \text{ mmol } MnO_4^-}}{250.0 \text{ mL}} = 6.4 \times 10^{-5} \text{ M} = s = [Ca^{2+}]$$

$$K_{sp} = [Ca^{2+}][C_2O_4^{2-}] = (s)(s) = s^2 = (6.4 \times 10^{-5})^2 = 4.1 \times 10^{-9}$$

13. We first use the ideal gas law to determine the moles of H_2S gas used.

$$n = \frac{PV}{RT} = \frac{\left(748 \text{ mmHg} \times \dfrac{1 \text{ atm}}{760 \text{ mmHg}}\right) \times \left(30.4 \text{ mL} \times \dfrac{1 \text{ L}}{1000 \text{ mL}}\right)}{0.08206 \text{ L atm mol}^{-1}\text{K}^{-1} \times (23 + 273) \text{ K}} = 1.23 \times 10^{-3} \text{ moles}$$

If we assume that all the H_2S is consumed in forming Ag_2S, we can compute the $[Ag^+]$ in the $AgBrO_3$ solution. This assumption is valid if the equilibrium constant for the cited reaction is large, which is the case, as shown below:

$$2Ag^+(aq) + HS^-(aq) + OH^-(aq) \rightleftharpoons Ag_2S(s) + H_2O(l) \quad K_{a_2}/K_{sp} = \frac{1.0 \times 10^{-19}}{2.6 \times 10^{-51}} = 3.8 \times 10^{31}$$

$$H_2S(aq) + H_2O(l) \rightleftharpoons HS^-(aq) + H_3O^+(aq) \qquad K_1 = 1.0 \times 10^{-7}$$

$$2H_2O(l) \rightleftharpoons H_3O^+(aq) + OH^-(aq) \qquad K_w = 1.0 \times 10^{-14}$$

$$2Ag^+(aq) + H_2S(aq) + 2H_2O(l) \rightleftharpoons Ag_2S(s) + 2H_3O^+(aq)$$

$$K_{overall} = (K_{a_2}/K_{sp})(K_1)(K_w) = (3.8 \times 10^{31})(1.0 \times 10^{-7})(1.0 \times 10^{-14}) = 3.8 \times 10^{10}$$

$$[Ag^+] = \frac{1.23 \times 10^{-3} \text{ mol } H_2S}{338 \text{ mL soln}} \times \frac{1000 \text{ mL}}{1 \text{ L soln}} \times \frac{2 \text{ mol } Ag^+}{1 \text{ mol } H_2S} = 7.28 \times 10^{-3} M$$

Then, for $AgBrO_3$, $K_{sp} = [Ag^+][BrO_3^-] = (7.28 \times 10^{-3})^2 = 5.30 \times 10^{-5}$

14. The titration reaction is $Ca(OH)_2(aq) + 2HCl(aq) \longrightarrow CaCl_2(aq) + 2H_2O(l)$

$$[OH^-] = \frac{10.7 \text{ mL HCl} \times \dfrac{1 \text{ L}}{1000 \text{ mL}} \times \dfrac{0.1032 \text{ mol HCl}}{1 \text{ L}} \times \dfrac{1 \text{ mol OH}^-}{1 \text{ mol HCl}}}{50.00 \text{ mL Ca(OH)}_2 \text{ soln} \times \dfrac{1 \text{ L}}{1000 \text{ mL}}} = 0.0221 \text{ M}$$

In a saturated solution of $Ca(OH)_2$, $[Ca^{2+}] = [OH^-] \div 2$

$K_{sp} = [Ca^{2+}][OH^-]^2 = (0.0221 \div 2)(0.0221)^2 = 5.40 \times 10^{-6}$ (5.5×10^{-6} in Appendix D).

The Common-Ion Effect

15. We let s = molar solubility of $Mg(OH)_2$ in moles solute per liter of solution.

(a) $K_{sp} = [Mg^{2+}][OH^-]^2 = (s)(2s)^2 = 4s^3 = 1.8 \times 10^{-11} \qquad s = 1.7 \times 10^{-4} M$

(b) Equation:

	$Mg(OH)_2(s)$	$\rightleftharpoons$	$Mg^{2+}(aq)$	+	$2OH^-(aq)$
Initial:	–		0.0862 M		≈ 0 M
Changes:	–		$+s$ M		$+2s$ M
Equil:	–		$(0.0862 + s)$ M		$2s$ M

$K_{sp} = (0.0862 + s) \times (2s)^2 = 1.8 \times 10^{-11} \approx (0.0862) \times (2s)^2 = 0.34 s^2 \quad s = 7.3 \times 10^{-6} M$

($s \ll 0.0802$ M, thus, the approximation was valid).

(c) $[OH^-] = [KOH] = 0.0355 M$

Equation:	$Mg(OH)_2(s)$	$\rightleftharpoons$	$Mg^{2+}(aq)$	+	$2OH^-(aq)$
Initial:	–		0 M		0.0355 M
Changes:	–		$+s$ M		$+2s$ M
Equil:	–		s M		$(0.0355 + 2s)$ M

$K_{sp} = (s)(0.0355 + 2s)^2 = 1.8 \times 10^{-11} \approx (s)(0.0355)^2 = 0.0013 s \qquad s = 1.4 \times 10^{-8} M$

16. The solubility equilibrium is $\quad CaCO_3(s) \rightleftharpoons Ca^{2+}(aq) + CO_3^{2-}(aq)$

(a) The addition of $Na_2CO_3(aq)$ produces $CO_3^{2-}(aq)$ in solution. This common ion suppresses the dissolution of $CaCO_3(s)$.

(b) HCl(aq) is a strong acid that reacts with carbonate ion:
$$CO_3^{2-}(aq) + 2H_3O^+(aq) \rightarrow CO_2(g) + 3H_2O(l).$$
This decreases $\left[CO_3^{2-}\right]$ in the solution, allowing more $CaCO_3(s)$ to dissolve.

(c) $HSO_4^-(aq)$ is a moderately weak acid. It is strong enough to protonate appreciable concentrations of carbonate ion, thereby decreasing $\left[CO_3^{2-}\right]$ and enhancing the solubility of $CaCO_3(s)$, as the value of K_c indicates.

$$HSO_4^-(aq) + CO_3^{2-}(aq) \rightleftharpoons SO_4^{2-}(aq) + HCO_3^-(aq)$$

$$K_c = \frac{K_a(HSO_4^-)}{K_a(HCO_3^-)} = \frac{0.011}{4.7 \times 10^{-11}} = 2.3 \times 10^8$$

17. The presence of KI in a solution produces a significant $\left[I^-\right]$ in the solution. Not as much AgI can dissolve in such a solution as in pure water, since the ion product, $\left[Ag^+\right]\left[I^-\right]$, cannot exceed the value of K_{sp} (i.e. the I⁻ from the KI that dissolves represses the dissociation of AgI(s). In similar fashion, $AgNO_3$ produces a significant $\left[Ag^+\right]$ in solution, again influencing the value of the ion product; not as much AgI can dissolve as in pure water.

18. If the solution contains KNO_3, more AgI will end up dissolving than in pure water, because the activity of each ion will be less than its molarity. On an ionic level, the reason is that ion pairs, such as $Ag^+NO_3^-(aq)$ and $K^+I^-(aq)$, form in the solution, preventing Ag^+ and I^- ions from coming together and precipitating.

19.

Equation:	$Ag_2SO_4(s)$	$\rightleftharpoons$	$2Ag^+(aq)$	$+$	$SO_4^{2-}(aq)$
Original:	—		0 M		0.150 M
Add solid:	—		$+2x$ M		$+x$ M
Equil:	—		$2x$ M		$(0.150 + x)$M

$$2x = \left[Ag^+\right] = 9.7 \times 10^{-3} M; \quad x = 0.00485 M$$

$$K_{sp} = \left[Ag^+\right]^2\left[SO_4^{2-}\right] = (2x)^2(0.150 + x) = \left(9.7 \times 10^{-3}\right)^2(0.150 + 0.00485) = 1.5 \times 10^{-5}$$

20. Equation: $\qquad CaSO_4(s) \rightleftharpoons Ca^{2+}(aq) + SO_4^{2-}(aq).$

Soln:	—	0 M	0.0025 M
Add $CaSO_4(s)$	—	$+x$ M	$+x$ M
Equil:	—	x M	$(0.0025+x)$ M

If we use the approximation that $x \ll 0.0025$, we find $x = 0.0036$. Clearly, x is larger than 0.0025 and thus the approximation is <u>not</u> valid. The quadratic equation must be solved.

$$K_{sp} = \left[Ca^{2+}\right]\left[SO_4^{2-}\right] = 9.1\times10^{-6} = x(0.0025+x) = 0.0025x + x^2$$

$$x^2 + 0.0025x - 9.1\times10^{-6} = 0$$

$$x = \frac{-b \pm \sqrt{b^2 - 4ac}}{2a} = \frac{-0.0025 \pm \sqrt{6.3\times10^{-6} + 3.6\times10^{-5}}}{2} = 2.0\times10^{-3} M = \left[CaSO_4\right]$$

$$\text{mass } CaSO_4 = 0.1000 \, L \times \frac{2.0\times10^{-3} \text{ mol } CaSO_4}{1 \text{ L soln}} \times \frac{136.1 \text{ g } CaSO_4}{1 \text{ mol } CaSO_4} = 0.027 \text{ g } CaSO_4$$

<u>21.</u> For PbI_2, $K_{sp} = 7.1\times10^{-9} = \left[Pb^{2+}\right]\left[I^-\right]^2$

In a solution where 1.5×10^{-4} mol PbI_2 is dissolved, $\left[Pb^{2+}\right] = 1.5\times10^{-4}$ M, and

$\left[I^-\right] = 2\left[Pb^{2+}\right] = 3.0\times10^{-4}$ M

Equation: $\qquad PbI_2(s) \rightleftharpoons Pb^{2+}(aq) + 2I^-(aq)$

Initial:	—	1.5×10^{-4} M	3.0×10^{-4} M
Add lead(II):	—	$+x$ M	
Equil:	—	$(0.00015+x)$ M	0.00030 M

$$K_{sp} = 7.1\times10^{-9} = (0.00015+x)(0.00030)^2 ; (0.00015+x) = 0.079 ; x = 0.079 \text{ M} = \left[Pb^{2+}\right]$$

22. For PbI_2, $K_{sp} = 7.1\times10^{-9} = \left[Pb^{2+}\right]\left[I^-\right]^2$

In a solution where 1.5×10^{-5} mol PbI_2/L is dissolved, $\left[Pb^{2+}\right] = 1.5\times10^{-5}$ M, and

$\left[I^-\right] = 2\left[Pb^{2+}\right] = 3.0\times10^{-5}$ M

Equation: $\qquad PbI_2(s) \rightleftharpoons Pb^{2+}(aq) + 2I^-(aq)$

Initial:	—	1.5×10^{-5} M	3.0×10^{-5} M
Add KI:	—		$+x$ M
Equil:	—	1.5×10^{-5} M	$(3.0\times10^{-5}+x)$ M

$K_{sp} = 7.1 \times 10^{-9} = (1.5 \times 10^{-5}) \times (3.0 \times 10^{-5} + x)$
$(3.0 \times 10^{-5} + x) = (7.1 \times 10^{-9} \div 1.5 \times 10^{-5})^{1/2}$
$x = (2.2 \times 10^{-2}) - (3.0 \times 10^{-5}) = 2.2 \times 10^{-2}$ M $= [I^-]$

23. For Ag_2CrO_4, $K_{sp} = 1.1 \times 10^{-12} = \left[Ag^+\right]^2\left[CrO_4^{2-}\right]$

In a 5.0×10^{-8} M solution of Ag_2CrO_4, $\left[CrO_4^{2-}\right] = 5.0 \times 10^{-8}$ M and $\left[Ag^+\right] = 1.0 \times 10^{-7}$ M

Equation: $\qquad Ag_2CrO_4(s) \;\rightleftharpoons\; 2Ag^+(aq) + CrO_4^{2-}(aq)$

Initial: $\qquad\qquad\quad — \qquad\qquad\quad 1.0 \times 10^{-7}$ M $\quad 5.0 \times 10^{-8}$ M

Add chromate: $\qquad —\qquad\qquad\qquad\qquad\qquad\qquad +x$ M

Equil: $\qquad\qquad\quad — \qquad\qquad\quad 1.0 \times 10^{-7}$ M $\quad (5.0 \times 10^{-8} + x)$M

$K_{sp} = 1.1 \times 10^{-12} = \left(1.0 \times 10^{-7}\right)^2\left(5.0 \times 10^{-8} + x\right); \quad \left(5.0 \times 10^{-8} + x\right) = 1.1 \times 10^2 = \left[CrO_4^{2-}\right].$

This is an impossibly high concentration to reach. Thus, we cannot lower the solubility of Ag_2CrO_4 to 5.0×10^{-8} M with CrO_4^{2-} as the common ion. Let's consider using Ag^+ as the common ion.

Equation: $\qquad Ag_2CrO_4(s) \;\rightleftharpoons\; 2Ag^+(aq) \quad + \quad CrO_4^{2-}(aq)$

Initial: $\qquad\qquad\quad — \qquad\qquad\quad 1.0 \times 10^{-7}$ M $\qquad 5.0 \times 10^{-8}$ M

Add silver(I) ion: $\quad —\qquad\qquad\quad +x$ M

Equil: $\qquad\qquad\quad — \qquad\qquad (1.0 \times 10^{-7} + x)$ M $\quad 5.0 \times 10^{-8}$ M

$K_{sp} = 1.1 \times 10^{-12} = \left(1.0 + x\right)^2\left(5.0 \times 10^{-8}\right) \qquad \left(1.0 \times 10^{-7} + x\right) = \sqrt{\dfrac{1.1 \times 10^{-12}}{5.0 \times 10^{-8}}} = 4.7 \times 10^{-3}$

$x = 4.7 \times 10^{-3} - 1.0 \times 10^{-7} = 4.7 \times 10^{-3}$ M $= \left[I^-\right]; \quad$ This is an easy-to-reach concentration.

Thus, the solubility can be lowered to 5.0×10^{-8} M by carefully adding $Ag^+(aq)$.

24. Even though $BaCO_3$ is more soluble than $BaSO_4$, it will still precipitate when 0.50 M $Na_2CO_3(aq)$ is added to a saturated solution of $BaSO_4$ because there is a sufficient $\left[Ba^{2+}\right]$ in such a solution for the ion product $\left[Ba^{2+}\right]\left[CO_3^{2-}\right]$ to exceed the value of K_{sp} for the compound. An example will demonstrate this phenomenon. Let us assume that the two solutions being mixed are of equal volume and, to make the situation even more unfavorable, that the saturated $BaSO_4$ solution is not in contact with solid $BaSO_4$, meaning that it does not maintain its saturation when it is diluted. First we determine $\left[Ba^{2+}\right]$ in saturated $BaSO_4(aq)$.

$K_{sp} = \left[Ba^{2+}\right]\left[SO_4^{2-}\right] = 1.1 \times 10^{-10} = s^2 \qquad\qquad s = \sqrt{1.1 \times 10^{-10}} = 1.0 \times 10^{-5}$ M

Mixing solutions of equal volumes means that the concentration of solutes not common to the two solutions are halved by dilution.

$\left[Ba^{2+}\right] = \dfrac{1}{2} \times \dfrac{1.0 \times 10^{-5} \text{ mol } BaSO_4}{1 \text{ L}} \times \dfrac{1 \text{ mol } Ba^{2+}}{1 \text{ mol } BaSO_4} = 5.0 \times 10^{-6}$ M

$\left[CO_3^{2-}\right] = \dfrac{1}{2} \times \dfrac{0.50 \text{ mol } Na_2CO_3}{1 \text{ L}} \times \dfrac{1 \text{ mol } CO_3^{2-}}{1 \text{ mol } Na_2CO_3} = 0.25$ M

$Q_{sp}\{BaCO_3\} = \left[Ba^{2+}\right]\left[CO_3^{2-}\right] = \left(5.0 \times 10^{-6}\right)\left(0.25\right) = 1.3 \times 10^{-6} > 5.0 \times 10^{-9} = K_{sp}\{BaCO_3\}$

Thus, precipitation of $BaCO_3$ indeed should occur under the conditions described.

25. $\left[Ca^{2+}\right] = \dfrac{115\,g\ Ca^{2+}}{10^6\,g\ soln} \times \dfrac{1\ mol\ Ca^{2+}}{40.08\ g\ Ca^{2+}} \times \dfrac{1000\ g\ soln}{1\ L\ soln} = 2.87 \times 10^{-3}\ M$

$\left[Ca^{2+}\right]\left[F^-\right]^2 = K_{sp} = 5.3 \times 10^{-9} = \left(2.87 \times 10^{-3}\right)\left[F^-\right]^2 \qquad \left[F^-\right] = 1.4 \times 10^{-3}\ M$

ppm $F^- = \dfrac{1.4 \times 10^{-3}\,mol\ F^-}{1\ L\ soln} \times \dfrac{19.00\,g\ F^-}{1\ mol\ F^-} \times \dfrac{1\ L\ soln}{1000\ g} \times 10^6\,g\ soln = 27$ ppm

26. We first calculate the $\left[Ag^+\right]$ and the $\left[Cl^-\right]$ in the saturated solution.

$K_{sp} = \left[Ag^+\right]\left[Cl^-\right] = 1.8 \times 10^{-10} = (s)(s) = s^2 \qquad s = 1.3 \times 10^{-5}\,M = \left[Ag^+\right] = \left[Cl^-\right]$

Both of these concentrations are marginally diluted by the addition of 1 mL of NaCl(aq)

$\left[Ag^+\right] = \left[Cl^-\right] = 1.3 \times 10^{-5}\,M \times \dfrac{100.0\ mL}{100.0\ mL + 1.0\ mL} = 1.3 \times 10^{-5}\,M$

The $\left[Cl^-\right]$ in the NaCl(aq) also is diluted. $\left[Cl^-\right] = 1.0\ M \times \dfrac{1.0\ mL}{100.0\ mL + 1.0\ mL} = 9.9 \times 10^{-3}\ M$

Let us use this $\left[Cl^-\right]$ to determine the $\left[Ag^+\right]$ that can exist in this solution.

$\left[Ag^+\right]\left[Cl^-\right] = 1.8 \times 10^{-10} = \left[Ag^+\right]\left(9.9 \times 10^{-3}\ M\right) \qquad \left[Ag^+\right] = \dfrac{1.8 \times 10^{-10}}{9.9 \times 10^{-3}} = 1.8 \times 10^{-8}\ M$

We compute the amount of AgCl in this final solution, and in the initial solution.

mmol AgCl final $= 101.0\ mL \times \dfrac{1.8 \times 10^{-8}\,mol\ Ag^+}{1\ L\ soln} \times \dfrac{1\ mmol\ AgCl}{1\ mmol\ Ag^+} = 1.8 \times 10^{-6}\,mmol\ AgCl$

mmol AgCl initial $= 100.0\ mL \times \dfrac{1.3 \times 10^{-5}\,mol\ Ag^+}{1\ L\ soln} \times \dfrac{1\ mmol\ AgCl}{1\ mmol\ Ag^+} = 1.3 \times 10^{-3}\,mmol\ AgCl$

The difference between these two amounts is the amount of AgCl that precipitates. Next we compute its mass.

mass AgCl $= \left(1.3 \times 10^{-3} - 1.8 \times 10^{-6}\right)$ mmol AgCl $\times \dfrac{143.3\ mg\ AgCl}{1\ mmol\ AgCl} = 0.19$ mg

We conclude that the precipitate will not be visible to the unaided eye, since its mass is less than 1 mg.

Criteria for Precipitation from Solution

27. We first determine $\left[Mg^{2+}\right]$, and then the value of Q_{sp} in order to compare it to the value of K_{sp}. We express molarity in millimoles per milliliter, entirely equivalent to moles per liter.

$[Mg^{2+}] = \dfrac{22.5\ mg\ MgCl_2}{325\ mL\ soln} \times \dfrac{1\,mmol\ MgCl_2 \cdot 6H_2O}{203.3\ mg\ MgCl_2 \cdot 6H_2O} \times \dfrac{1\,mmol\ Mg^{2+}}{1\,mmol\ MgCl_2} = 3.41 \times 10^{-4}\ M$

$Q_{sp} = [Mg^{2+}][F^-]^2 = (3.41 \times 10^{-4})(0.035)^2 = 4.1 \times 10^{-7} > 3.7 \times 10^{-8} = K_{sp}$

Thus, precipitation of $MgF_2(s)$ should occur from this solution.

28. The solutions mutually dilute each other.

$$\left[Cl^-\right] = 0.016\ M \times \frac{155\ mL}{155\ mL + 245\ mL} = 6.2 \times 10^{-3}\ M$$

$$\left[Pb^{2+}\right] = 0.175\ M \times \frac{245\ mL}{245\ mL + 155\ mL} = 0.107\ M$$

Then we compute the value of the ion product and compare it to the solubility product constant value.

$$Q_{sp} = \left[Pb^{2+}\right]\left[Cl^-\right]^2 = (0.107)(6.2 \times 10^{-3})^2 = 4.1 \times 10^{-6} < 1.6 \times 10^{-5} = K_{sp}$$

Thus, precipitation of $PbCl_2(s)$ will not occur from these mixed solutions.

29. We determine the $\left[OH^-\right]$ needed to just initiate precipitation of $Cd(OH)_2$

$$K_{sp} = \left[Cd^{2+}\right]\left[OH^-\right]^2 = 2.5 \times 10^{-14} = (0.0055M)\left[OH^-\right]^2 \qquad \left[OH^-\right] = \sqrt{\frac{2.5 \times 10^{-14}}{0.0055}} = 2.1 \times 10^{-6}\ M$$

$$pOH = -\log(2.1 \times 10^{-6}) = 5.68 \qquad pH = 14.00 - 5.68 = 8.32$$

Thus, $Cd(OH)_2$ will precipitate from a solution with $pH > 8.32$.

30. We determine the $\left[OH^-\right]$ needed to just initiate precipitation of $Cr(OH)_3$

$$K_{sp} = [Cr^{3+}][OH^-]^3 = 6.3 \times 10^{-31} = (0.086\ M)\ [OH^-]^3\ ;\quad [OH^-] = \sqrt{\frac{6.3 \times 10^{-31}}{0.086}} = 1.9 \times 10^{-10}\ M$$

$$pOH = -\log(1.9 \times 10^{-10}) = 9.72 \qquad pH = 14.00 - 9.72 = 4.28$$

Thus, $Cr(OH)_3$ will precipitate from a solution with $pH > 4.28$.

31. (a) First we determine $\left[Cl^-\right]$ due to the added NaCl.

$$\left[Cl^-\right] = \frac{0.10\ mg\ NaCl}{1.0\ L\ soln} \times \frac{1\ g}{1000\ mg} \times \frac{1\ mol\ NaCl}{58.4\ g\ NaCl} \times \frac{1\ mol\ Cl^-}{1\ mol\ NaCl} = 1.7 \times 10^{-6}\ M$$

Then we determine the value of the ion product and compare it to the solubility product constant value.

$$Q_{sp} = \left[Ag^+\right]\left[Cl^-\right] = (0.10)(1.7 \times 10^{-6}) = 1.7 \times 10^{-7} > 1.8 \times 10^{-10} = K_{sp}\ \text{for AgCl}$$

Thus, precipitation of AgCl(s) should occur.

(b) The KBr(aq) is diluted on mixing, but the $\left[\text{Ag}^+\right]$ and $\left[\text{Cl}^-\right]$ are barely affected by dilution.

$$\left[\text{Br}^-\right] = 0.10\,\text{M} \times \frac{0.05\ \text{mL}}{0.05\ \text{mL} + 250\ \text{mL}} = 2 \times 10^{-5}\,\text{M}$$

Now we determine $\left[\text{Ag}^+\right]$ in a saturated AgCl solution.

$$K_{sp} = \left[\text{Ag}^+\right]\left[\text{Cl}^-\right] = (s)(s) = s^2 = 1.8 \times 10^{-10} \qquad s = 1.3 \times 10^{-5}\,\text{M}$$

Then we determine the value of the ion product for AgBr and compare it to the solubility product constant value.

$$Q_{sp} = \left[\text{Ag}^+\right][\text{Br}] = \left(1.3 \times 10^{-5}\right)\left(2 \times 10^{-5}\right) = 3 \times 10^{-10} > 5.0 \times 10^{-13} = K_{sp}\ \text{for AgBr}$$

Thus, precipitation of AgBr(s) should occur.

(c) The hydroxide ion is diluted by mixing the two solutions.

$$\left[\text{OH}^-\right] = 0.0150\ \text{M} \times \frac{0.05\ \text{mL}}{0.05\ \text{mL} + 3000\ \text{mL}} = 2 \times 10^{-7}\,\text{M}$$

But the $\left[\text{Mg}^{2+}\right]$ does not change significantly.

$$\left[\text{Mg}^{2+}\right] = \frac{2.0\ \text{mg Mg}^{2+}}{1.0\ \text{L soln}} \times \frac{1\ \text{g}}{1000\ \text{mg}} \times \frac{1\ \text{mol Mg}^{2+}}{24.3\ \text{g Mg}} = 8.2 \times 10^{-5}\ \text{M}$$

Then we determine the value of the ion product and compare it to the solubility product constant value.

$$Q_{sp} = \left[\text{Mg}^{2+}\right]\left[\text{OH}^-\right]^2 = \left(2 \times 10^{-7}\right)^2 \left(8.2 \times 10^{-5}\right) = 3.3 \times 10^{-18}$$

$$Q_{sp} < 1.8 \times 10^{-11} = K_{sp}\ \text{for Mg(OH)}_2 \quad \text{Thus, no precipitate forms.}$$

32. First we determine the moles of H_2 produced during the electrolysis, and then determine $\left[\text{OH}^-\right]$.

$$\text{moles } H_2 = \frac{PV}{RT} = \frac{752\ \text{mmHg} \times \dfrac{1\ \text{atm}}{760\ \text{mmHg}} \times 0.652\ \text{L}}{0.08206\ \text{L atm mol}^{-1}\ \text{K}^{-1} \times 295\ \text{K}} = 0.0267\ \text{mol } H_2$$

$$\left[\text{OH}^-\right] = \frac{0.0267\ \text{mol } H_2 \times \dfrac{2\ \text{mol OH}^-}{1\ \text{mol } H_2}}{0.315\ \text{L sample}} = 0.170\ \text{M}$$

$$Q_{sp} = \left[\text{Mg}^{2+}\right]\left[\text{OH}^-\right]^2 = (0.185)(0.170)^2 = 5.35 \times 10^{-3} > 1.8 \times 10^{-11} = K_{sp}$$

Thus, yes, precipitation of $\text{Mg(OH)}_2(\text{s})$ should occur during the electrolysis.

33. First we must calculate the initial $[H_2C_2O_4]$ upon dissolution of the solid acid:

$$[H_2C_2O_4]_{initial} = 1.50 \text{ g } H_2C_2O_4 \times \frac{1 \text{ mol } H_2C_2O_4}{90.036 \text{ g } H_2C_2O_4} \times \frac{1}{0.200 \text{ L}} = 0.0833 \text{ M}$$

(We assume the volume of the solution stays at 0.200 L)

Next we need to determine the $[C_2O_4^{2-}]$ in solution after the hydrolysis reaction between oxalic acid and water reaches equilibrium. To accomplish this we will need to solve two I.C.E. tables:

Table 1:

$$H_2C_2O_4(aq) + H_2O(l) \overset{K_{a1}=5.2 \times 10^{-2}}{\rightleftharpoons} HC_2O_4^-(aq) + H_3O^+(aq)$$

	$H_2C_2O_4$	H_2O	$HC_2O_4^-$	H_3O^+
Initial:	0.0833 M	—	0 M	0 M
Change:	$-x$	—	$+x$ M	$+x$ M
Equilibrium:	$0.0833 - x$ M	—	x M	x M

Since $C_a/K_a = 1.6$, the approximation cannot be used, and thus the full quadratic equation must be solved: $x^2/(0.0833 - x) = 5.2 \times 10^{-2}$; $x^2 + 5.2 \times 10^{-2}x - 4.3\underline{3} \times 10^{-3}$

$$x = \frac{-5.2 \times 10^{-2} \pm \sqrt{2.7 \times 10^{-3} + 0.0173}}{2} \qquad x = 0.045 \text{ M} = [HC_2O_4^-] \approx [H_3O^+]$$

Now we can solve the I.C.E. table for the second proton loss:

Table 2:

$$HC_2O_4^-(aq) + H_2O(l) \overset{K_{a1}=5.4 \times 10^{-5}}{\rightleftharpoons} C_2O_4^{2-}(aq) + H_3O^+(aq)$$

	$HC_2O_4^-$	H_2O	$C_2O_4^{2-}$	H_3O^+
Initial:	0.045 M	—	0 M	≈ 0.045 M
Change:	$-y$	—	$+y$ M	$+y$ M
Equilibrium:	$(0.045 - y)$ M	—	y M	$\approx (0.045 + y)$ M

Since $C_a/K_a = 833$, the approximation may not be valid and we yet again should solve the full quadratic equation:

$$\frac{y \times (0.045 + y)}{(0.045 - y)} = 5.4 \times 10^{-5}; \qquad y^2 + 0.045y = 2.4\underline{3} \times 10^{-6} - 5.4 \times 10^{-5}y$$

$$y = \frac{-0.045 \pm \sqrt{2.0\underline{3} \times 10^{-3} + 9.7\underline{2} \times 10^{-6}}}{2} \qquad y = 5.4 \times 10^{-5} \text{ M} = [C_2O_4^{2-}]$$

Now we can calculate the Q_{sp} for the calcium oxalate system:

$Q_{sp} = [Ca^{2+}]_{initial} \times [C_2O_4^{2-}]_{initial} = (0.150)(5.4 \times 10^{-5}) = 8.1 \times 10^{-6} > 1.3 \times 10^{-9}$ (K_{sp} for CaC_2O_4)

Thus, CaC_2O_4 should precipitate from this solution.

34. The solutions mutually dilute each other. We first determine the solubility of each compound in its saturated solution and then its concentration after dilution.

$$K_{sp} = \left[Ag^+\right]^2\left[SO_4^{2-}\right] = 1.4 \times 10^{-5} = (2s)^2 s = 4s^3 \qquad s = \sqrt[3]{\frac{1.4 \times 10^{-5}}{4}} = 1.5 \times 10^{-2} \text{ M}$$

$$\left[SO_4^{2-}\right] = 0.015 \text{ M} \times \frac{100.0 \text{ mL}}{100.0 \text{ mL} + 250.0 \text{ mL}} = 0.0043 \text{ M} \qquad \left[Ag^+\right] = 0.0086 \text{ M}$$

$$K_{sp} = [Pb^{2+}][CrO_4^{2-}] = 2.8 \times 10^{-13} = (s)(s) = s^2 \qquad s = \sqrt{2.8 \times 10^{-13}} = 5.3 \times 10^{-7} \text{ M}$$

$$\left[Pb^{2+}\right] = \left[CrO_4^{2-}\right] = 5.3 \times 10^{-7} \times \frac{250.0 \text{ mL}}{250.0 \text{ mL} + 100.0 \text{ mL}} = 3.8 \times 10^{-7} \text{ M}$$

From the balanced chemical equation, we see that the two possible precipitates are $PbSO_4$ and Ag_2CrO_4. (Neither $PbCrO_4$ nor Ag_2SO_4 can precipitate because they have been diluted below their saturated concentrations.) $PbCrO_4 + Ag_2SO_4 \rightleftharpoons PbSO_4 + Ag_2CrO_4$

Thus, we compute the value of Q_{sp} for each of these compounds and compare those values with the solubility constant product value.

$$Q_{sp} = \left[Pb^{2+}\right]\left[SO_4^{2-}\right] = \left(3.8 \times 10^{-7}\right)\left(0.0043\right) = 1.6 \times 10^{-9} < 1.6 \times 10^{-8} = K_{sp} \text{ for } PbSO_4$$

Thus, $PbSO_4(s)$ will not precipitate.

$$Q_{sp} = \left[Ag^+\right]^2\left[CrO_4^{2-}\right] = \left(0.0086\right)^2\left(3.8 \times 10^{-7}\right) = 2.8 \times 10^{-11} > 1.1 \times 10^{-12} = K_{sp} \text{ for } Ag_2CrO_4$$

Thus, $Ag_2CrO_4(s)$ should precipitate.

Completeness of Precipitation

35. First determine that a precipitate forms. The solutions mutually dilute each other.

$$\left[CrO_4^{2-}\right] = 0.350 \text{ M} \times \frac{200.0 \text{ mL}}{200.0 \text{ mL} + 200.0 \text{ mL}} = 0.175 \text{ M}$$

$$\left[Ag^+\right] = 0.0100 \text{ M} \times \frac{200.0 \text{ mL}}{200.0 \text{ mL} + 200.0 \text{ mL}} = 0.00500 \text{ M}$$

We determine the value of the ion product and compare it to the solubility product constant value.

$$Q_{sp} = \left[Ag^+\right]^2\left[CrO_4^{2-}\right] = \left(0.00500\right)^2\left(0.175\right) = 4.4 \times 10^{-6} > 1.1 \times 10^{-12} = K_{sp} \text{ for } Ag_2CrO_4$$

Ag_2CrO_4 should precipitate.

Now, we assume that as much solid forms as possible, and then we approach equilibrium by dissolving that solid in a solution that contains the ion in excess.

Equation:	$Ag_2CrO_4(s)$	$\xrightleftharpoons{1.1 \times 10^{-12}}$	$2Ag^+(aq)$	+	$CrO_4^{2-}(aq)$
Orig. soln :	–		0.00500 M		0.175 M
Form solid :	–		−0.00500 M		−0.00250 M
Not at equilibrium	–		0 M		0.173 M
Changes :	–		+2x M		+x M
Equil :	–		2x M		$(0.173 + x)$ M

$$K_{sp} = \left[Ag^+\right]^2\left[CrO_4^{2-}\right] = 1.1 \times 10^{-12} = \left(2x\right)^2\left(0.173 + x\right) \approx \left(4x^2\right)\left(0.173\right)$$

$$x = \sqrt{\frac{1.1 \times 10^{-12}}{4 \times 0.173}} = 1.3 \times 10^{-6} \text{ M} \qquad \left[Ag^+\right] = 2x = 2.6 \times 10^{-6} \text{ M}$$

$$\% \ Ag^+ \text{ unprecipitated} = \frac{2.6 \times 10^{-6} \text{ M final}}{0.00500 \text{ M initial}} \times 100\% = 0.052\% \text{ unprecipitated}$$

36. $[Ag^+]_{diluted} = 0.0208 \text{ M} \times \dfrac{175 \text{ mL}}{425 \text{ mL}} = 0.008565 \text{ M}$ $[CrO_4^{2-}]_{diluted} = 0.0380 \text{ M} \times \dfrac{250 \text{ mL}}{425 \text{ mL}} = 0.02235 \text{ M}$

$$Q_{sp} = \left[Ag^+\right]^2\left[CrO_4^{2-}\right] = 1.6\times10^{-6} > K_{sp}$$

The equilibrium constant for this precipitation reaction suggests that the reaction nearly goes to 100% completion. Assume that the limiting reagent is used up (100% reaction in the reverse direction) and re-establish the equilibrium in the reverse direction. Here Ag^+ is the limiting reagent.

	$Ag_2CrO_4(s)$	$\xrightarrow{K_{sp(Ag_2CrO_4)}=1.1\times10^{-12}}$	$2\ Ag^+(aq)$	$+$	$CrO_4^{2-}(aq)$
initial	–		0.008565 M		0.02235 M
change	–	$x = 0.00428 \text{ M}$	$-2x$		$-x$
100% rxn	–		0		0.0181
change	–	re-establish equil	$+2y$		$+y$
equil	–	(assume $y \sim 0$)	$2y$		$0.0181 + y$

$1.1 \times 10^{-12} = (2y)^2(0.0181 + y) \approx (2y)^2(0.0181)$ $y = 3.9 \times 10^{-6} \text{ M}$

$y \ll 0.0181$, so this assumption is valid).

$2y = [Ag^+] = 7.8 \times 10^{-6} \text{ M}$ after precipitation is complete.

$\% \ [Ag^+]_{unprecipitated} = \dfrac{7.8\times10^{-6}}{0.00856} \times 100 \% = 0.091 \%$ (precipitation is essentially quantitative)

37. We first use the solubility product constant expression to determine $\left[Pb^{2+}\right]$ in a solution with 0.100 M Cl^-.

$$K_{sp} = \left[Pb^{2+}\right]\left[Cl^-\right]^2 = 1.6\times10^{-5} = \left[Pb^{2+}\right](0.100)^2 \quad \left[Pb^{2+}\right] = \dfrac{1.6\times10^{-5}}{(0.100)^2} = 1.6\times10^{-3} \text{ M}$$

Thus, % unprecipitated $= \dfrac{1.6\times10^{-3} \text{ M}}{0.065 \text{ M}} \times 100\% = 2.5\%$

Now, we want to determine what $[Cl^-]$ must be maintained to keep $[Pb^{2+}]_{final} = 1\%$;

$[Pb^{2+}]_{initial} = 0.010 \times 0.065 \text{ M} = 6.5\times10^{-4} \text{ M}$

$$K_{sp} = \left[Pb^{2+}\right]\left[Cl^-\right]^2 = 1.6\times10^{-5} = (6.5\times10^{-4})\left[Cl^-\right]^2 \qquad \left[Cl^-\right] = \sqrt{\dfrac{6\times10^{-5}}{6.5\times10^{-4}}} = 0.16 \text{ M}$$

38. Let's start by assuming that the concentration of Pb^{2+} in the untreated wine is no higher than 1.5×10^{-4}M (this assumption is not unreasonable). As long as the Pb^{2+} concentration is less than 1.5×10^{-4} M, then the final sulfate ion concentration in the $CaSO_4$ treated wine should be virtually the same as the sulfate ion concentration in a saturated solution of $CaSO_4$ formed by dissolving solid $CaSO_4$ in pure water (i.e., with $[Pb^{2+}]$ less than or equal to 1.5×10^{-4} M, the $[SO_4^{2-}]$ will not drop significantly below that for a saturated solution, $\approx 3.0 \times 10^{-3}$ M). Thus, the addition of $CaSO_4$ to the wine would result in the precipitation of solid $PbSO_4$, which would continue until the concentration of Pb^{2+} was equal to the K_{sp} for $PbSO_4$ divided by the concentration of dissolved sulfate ion, i.e., $[Pb^{2+}]_{max} = 1.6 \times 10^{-8}$M^2/$3.0 \times 10^{-3}$ M $= 5.3 \times 10^{-6}$ M.

Fractional Precipitation

39. First, assemble all of the data. K_{sp} for $Ca(OH)_2 = 5.5 \times 10^{-6}$ K_{sp} for $Mg(OH)_2 = 1.8 \times 10^{-11}$

$$[Ca^{2+}] = \frac{440 \text{ g } Ca^{2+}}{1000 \text{ kg seawater}} \times \frac{1 \text{ mol } Ca^{2+}}{40.078 \text{ g } Ca^{2+}} \times \frac{1 \text{ kg seawater}}{1000 \text{ g seawater}} \times \frac{1.03 \text{ kg seawater}}{1 \text{ L seawater}} = 0.0113 \text{ M}$$

$[Mg^{2+}] = 0.059 \text{ M}$ $[OH^-] = 0.0020 \text{ M (maintained)}$

(a) $Q_{sp} = [Ca^{2+}] \times [OH^-]^2 = (0.0113)(0.0020)^2 = 4.5 \times 10^{-8}$ $Q_{sp} < K_{sp}$ ∴ no precipitate forms.

(b) For the separation to be complete, >>99.9 % of the Mg^{2+} must be removed before Ca^{2+} begins to precipitate. We have already shown that Ca^{2+} will not precipitate if the $[OH^-] = 0.0020$ M and is maintained at this level. Let us determine how much of the 0.059 M Mg^{2+} will still be in solution when $[OH^-] = 0.0020$ M.
$K_{sp} = [Mg^{2+}] \times [OH^-]^2 = (x)(0.0020)^2 = 1.8 \times 10^{-11}$ $x = 4.5 \times 10^{-6}$ M

The percent Mg^{2+} ion left in solution $= \dfrac{4.5 \times 10^{-6}}{0.059} \times 100\% = 0.0076 \%$

This means that $100 - 0.0076 \% \text{ Mg} = 99.992 \%$ has precipitated.
Clearly, the magnesium ion has been separated from the calcium ion (i.e. >> 99.9% of the Mg^{2+} ions have precipitated and virtually all of the Ca^{2+} ions are still in solution).

40. **(a)** 0.10 M NaCl will not work at all, since both $BaCl_2$ and $CaCl_2$ are soluble in water.

(b) $K_{sp} = 1.1 \times 10^{-10}$ for $BaSO_4$ and $K_{sp} = 9.1 \times 10^{-6}$ for $CaSO_4$. Since these values differ by more than 1000, 0.05 M Na_2SO_4 would effectively separate Ba^{2+} from Ca^{2+}.

We first compute $\left[SO_4^{2-}\right]$ when $BaSO_4$ begins to precipitate.

$$\left[Ba^{2+}\right]\left[SO_4^{2-}\right] = (0.050)\left[SO_4^{2-}\right] = 1.1 \times 10^{-10}; \quad \left[SO_4^{2-}\right] = \frac{1.1 \times 10^{-10}}{0.050} = 2.2 \times 10^{-9} \text{ M}$$

And then we calculate $\left[SO_4^{2-}\right]$ when $\left[Ba^{2+}\right]$ has decreased to 0.1% of its initial value, that is, to 5.0×10^{-5} M

$$\left[Ba^{2+}\right]\left[SO_4^{2-}\right] = \left(5.0 \times 10^{-5}\right)\left[SO_4^{2-}\right] = 1.1 \times 10^{-10}; \quad \left[SO_4^{2-}\right] = \frac{1.1 \times 10^{-10}}{5.0 \times 10^{-5}} = 2.2 \times 10^{-6} \text{ M}$$

And finally $\left[SO_4^{2-}\right]$ when $CaSO_4$ begins to precipitate.

$$\left[Ca^{2+}\right]\left[SO_4^{2-}\right] = (0.050)\left[SO_4^{2-}\right] = 9.1 \times 10^{-6}; \quad \left[SO_4^{2-}\right] = \frac{9.1 \times 10^{-6}}{0.050} = 1.8 \times 10^{-4} \text{ M}$$

(c) Now, $K_{sp} = 5 \times 10^{-3}$ for $Ba(OH)_2$ and $K_{sp} = 5.5 \times 10^{-6}$ for $Ca(OH)_2$. The fact that these two K_{sp} values differ by almost a factor of 1000 does not tell the entire story, because $[OH^-]$ is squared in both K_{sp} expressions. We compute $[OH^-]$ when $Ca(OH)_2$ begins to precipitate.

$$[Ca^{2+}][OH^-]^2 = 5.5 \times 10^{-6} = (0.050)[OH^-]^2 \quad [OH^-] = \sqrt{\frac{5.5 \times 10^{-6}}{0.050}} = 1.0 \times 10^{-2} \text{ M}$$

Precipitation will not proceed, as we only have 0.001 M NaOH, which has $[OH^-] = 1 \times 10^{-3}$ M

(d) $K_{sp} = 5.1 \times 10^{-9}$ for $BaCO_3$ and $K_{sp} = 2.8 \times 10^{-9}$ for $CaCO_3$. Since these two values differ by less than a factor of 2, 0.50 M Na_2CO_3 would not effectively separate Ba^{2+} from Ca^{2+}.

41. (a) Here we need to determine $[I^-]$ when AgI just begins to precipitate, and $[I^-]$ when PbI_2 just begins to precipitate.

$$K_{sp} = [Ag^+][I^-] = 8.5 \times 10^{-17} = (0.10)[I^-] \qquad [I^-] = 8.5 \times 10^{-16} \text{ M}$$

$$K_{sp} = [Pb^{2+}][I^-]^2 = 7.1 \times 10^{-9} = (0.10)[I^-]^2 \qquad [I^-] = \sqrt{\frac{7.1 \times 10^{-9}}{0.10}} = 2.7 \times 10^{-4} \text{ M}$$

Since 8.5×10^{-16} M is less than 2.7×10^{-4} M, AgI will precipitate before PbI_2.

(b) $[I^-]$ must be equal to 2.7×10^{-4} M before the second cation, Pb^{2+}, begins to precipitate.

(c) $K_{sp} = [Ag^+][I^-] = 8.5 \times 10^{-17} = [Ag^+](2.7 \times 10^{-4}) \qquad [Ag^+] = 3.1 \times 10^{-13}$ M

(d) Since $[Ag^+]$ has decreased to much less than 0.1% of its initial value before any PbI_2 begins to precipitate, we conclude that Ag^+ and Pb^{2+} can be separated by precipitation with iodide ion.

42. Normally we would worry about the mutual dilution of the two solutions, but the values of the solubility product constants are so small that only a very small volume of 0.50 M $Pb(NO_3)_2$ solution needs to be added, as we shall see.

(a) Since the two anions are present at the same concentration and they have the same type of formula (one anion per cation), the one forming the compound with the smallest K_{sp} value will precipitate first. Thus, CrO_4^{2-} is the first anion to precipitate.

(b) At the point where SO_4^{2-} begins to precipitate, we have

$$K_{sp} = \left[Pb^{2+} \right]\left[SO_4^{2-} \right] = 1.6 \times 10^{-8} = \left[Pb^{2+} \right](0.010M); \left[Pb^{2+} \right] = \frac{1.6 \times 10^{-8}}{0.010} = 1.6 \times 10^{-6} \, M$$

Now we can test our original assumption, that only a very small volume of 0.50 M $Pb(NO_3)_2$ solution has been added. We assume that we have 1.00 L of the original solution, the one with the two anions dissolved in it, and compute the volume of 0.50 M $Pb(NO_3)_2$ that has to be added to achieve $\left[Pb^{2+} \right] = 1.6 \times 10^{-6} \, M$

$$V_{added} = 1.00 \, L \times \frac{1.6 \times 10^{-6} \; mol \; Pb^{2+}}{1 \, L \; soln} \times \frac{1 \; mol \; Pb(NO_3)_2}{1 \; mol \; Pb^{2+}} \times \frac{1 \, L \; Pb^{2+} soln}{0.50 \; mol \; Pb(NO_3)_2}$$

$$V_{added} = 3.2 \times 10^{-5} \, L \; Pb^{2+} soln = 0.0032 mL \; Pb^{2+} soln$$

This is less than one drop (0.05 mL) of the Pb^{2+} solution, clearly a very small volume.

(c) The two anions are effectively separated if $\left[Pb^{2+} \right]$ has not reached $1.6 \times 10^{-6} \, M$ when $\left[CrO_4^{2-} \right]$ is reduced to 0.1% of its original value, that is, to

$$0.010 \times 10^{-3} \, M = 1.0 \times 10^{-5} \, M = \left[CrO_4^{2-} \right]$$

$$K_{sp} = \left[Pb^{2+} \right]\left[CrO_4^{2-} \right] = 2.8 \times 10^{-13} = \left[Pb^{2+} \right]\left(1.0 \times 10^{-5}\right)$$

$$\left[Pb^{2+} \right] = \frac{2.8 \times 10^{-13}}{1.0 \times 10^{-5}} = 2.8 \times 10^{-8} \, M$$

Thus, the two anions can be effectively separated by fractional precipitation.

43. First, let's assemble all of the data. K_{sp} for AgCl = 1.8×10^{-10} K_{sp} for AgI = 8.5×10^{-17}
$[Ag^+]$ = 2.00 M $[Cl^-]$ = 0.0100 M $[I^-]$ = 0.250 M

(a) AgI(s) will be the first to precipitate by virtue of the fact that the K_{sp} value for AgI is about 2 million times smaller than that for AgCl..

(b) AgCl(s) will begin to precipitate when the Q_{sp} for AgCl(s) > K_{sp} for AgCl(s). The concentration of Ag^+ required is: $K_{sp} = [Ag^+][Cl^-] = 1.8 \times 10^{-10} = (0.0100) \times (x)$
     $x = 1.8 \times 10^{-8} \, M$
Using this data, we can determine the remaining concentration of I^- using the K_{sp}.
$K_{sp} = [Ag^+][I^-] = 8.5 \times 10^{-17} = (x) \times (1.8 \times 10^{-8})$ $x = 4.7 \times 10^{-9} \, M$

(c) In part (b) we saw that the $[I^-]$ drops from 0.250 M → 4.7×10^{-9} M. Only a small percentage of the ion remains in solution. $\dfrac{4.7 \times 10^{-9}}{0.250} \times 100\% = 0.0000019 \, \%$

This means that 99.999998 % of the I^- ion has been precipitated before any of the Cl^- ion has precipitated. Clearly, the fractional separation of Cl^- from I^- is feasible.

44. **(a)** We first determine the $[Ag^+]$ needed to initiate precipitation of each compound.

AgCl: $K_{sp} = [Ag^+][Cl^-] = 1.8 \times 10^{-10} = [Ag^+](0.250);$ $[Ag^+] = \dfrac{1.8 \times 10^{-10}}{0.250} = 7.2 \times 10^{-10}$ M

AgBr: $K_{sp} = [Ag^+][Br^-] = 5.0 \times 10^{-13} = [Ag^+](0.0022);$ $[Ag^+] = \dfrac{5.0 \times 10^{-13}}{0.0022} = 2.3 \times 10^{-10}$ M

Thus, Br^- precipitates first, as AgBr, because it requires a lower $[Ag^+]$.

(b) $[Ag^+] = 7.2 \times 10^{-10}$ M when chloride ion, the second anion, begins to precipitate.

(c) Cl^- and Br^- cannot be separated by this fractional precipitation. $[Ag^+]$ will have to rise to 1000 times its initial value, to 2.3×10^{-7} M, before AgBr is completely precipitated. But as soon as $[Ag^+]$ reaches 7.2×10^{-10} M, AgCl will begin to precipitate.

Solubility and pH

45. In each case we indicate whether the compound is more soluble in acid than in water. We write the net ionic equation for the reaction in which the solid dissolves in acid. Substances are more soluble in acid if either (1) an acid-base reaction occurs [as in (b-d)] or (2) a gas is produced, since escape of the gas from the reaction mixture causes the reaction to shift to the right.

(a) Same: KCl (K^+ and Cl^- do not react appreciably with H_2O)

(b) Acid: $MgCO_3(s) + 2H^+(aq) \longrightarrow Mg^{2+}(aq) + H_2O(l) + CO_2(g)$

(c) Acid: $FeS(s) + 2H^+(aq) \longrightarrow Fe^{2+}(aq) + H_2S(g)$

(d) Acid: $Ca(OH)_2(s) + 2H^+(aq) \longrightarrow Ca^{2+}(aq) + 2H_2O(l)$

(e) Water: C_6H_5COOH is less soluble in acid, because of the H_3O^+ common ion.

46. In each case we indicate whether the compound is more soluble in base than in water. We write the net ionic equation for the reaction in which the solid dissolves in base. Substances are soluble in base if either (1) acid-base reaction occurs [as in (b)] or (2) a gas is produced, since escape of the gas from the reaction mixture causes the reaction to shift to the right.

(a) Water: $BaSO_4$ is less soluble in base, hydrolysis of SO_4^{2-} will be repressed.

(b) Base: $H_2C_2O_4(s) + 2OH^-(aq) \longrightarrow C_2O_4^{2-}(aq) + 2H_2O(l)$

(c) Water: $Fe(OH)_3$ is less soluble in base because of the OH^- common ion.

(d) Same: $NaNO_3$ (neither Na^+ nor NO_3^- react with H_2O to a measurable extent).

(e) Water: MnS is less soluble in base because hydrolysis of S^{2-} will be repressed.

47. We determine $[Mg^{2+}]$ in the solution.

$$[Mg^{2+}] = \frac{0.65 \text{ g Mg(OH)}_2}{1 \text{ L soln}} \times \frac{1 \text{ mol Mg(OH)}_2}{58.3 \text{ g Mg(OH)}_2} \times \frac{1 \text{ mol Mg}^{2+}}{1 \text{ mol Mg(OH)}_2} = 0.011 M$$

Then we determine $[OH^-]$ in the solution, and its pH.

$$K_{sp} = [Mg^{2+}][OH^-]^2 = 1.8 \times 10^{-11} = (0.011)[OH^-]^2; \quad [OH^-] = \sqrt{\frac{1.8 \times 10^{-11}}{0.011}} = 4.0 \times 10^{-5} \text{ M}$$

$$pOH = -\log(4.0 \times 10^{-5}) = 4.40 \qquad pH = 14.00 - 4.40 = 9.60$$

48. First we determine the $[Mg^{2+}]$ and $[NH_3]$ that result from dilution to a total volume of 0.500 L.

$$[Mg^{2+}] = 0.100 M \times \frac{0.150 \text{ L}_{initial}}{0.500 \text{ L}_{final}} = 0.0300 M; \quad [NH_3] = 0.150 \text{ M} \times \frac{0.350 \text{ L}_{initial}}{0.500 \text{ L}_{final}} = 0.105 \text{ M}$$

Then determine the $[OH^-]$ that will allow $[Mg^{2+}] = 0.0300$ M in this solution.

$$K_{sp} = 1.8 \times 10^{-11} = [Mg^{2+}][OH^-]^2 = (0.0300)[OH^-]^2; \quad [OH^-] = \sqrt{\frac{1.8 \times 10^{-11}}{0.0300}} = 2.4 \times 10^{-5} \text{ M}$$

This $[OH^-]$ is maintained by the NH_3 / NH_4^+ buffer, since it is a buffer, we can use the Henderson–Hasselbalch equation to find the $[NH_4^+]$.

$$pH = 14.00 - pOH = 14.00 + \log(2.4 \times 10^{-5}) = 9.38 = pK_a + \log\frac{[NH_3]}{[NH_4^+]} = 9.26 + \log\frac{[NH_3]}{[NH_4^+]}$$

$$\log\frac{[NH_3]}{[NH_4^+]} = 9.38 - 9.26 = +0.12; \quad \frac{[NH_3]}{[NH_4^+]} = 10^{+0.12} = 1.3; \quad [NH_4^+] = \frac{0.105 \text{ M NH}_3}{1.3} = 0.081 \text{ M}$$

$$\text{mass } (NH_4)_2 SO_4 = 0.500 \text{ L} \times \frac{0.081 \text{ mol NH}_4^+}{\text{L soln}} \times \frac{1 \text{ mol } (NH_4)_2 SO_4}{2 \text{ mol NH}_4^+} \times \frac{132.1 \text{ g } (NH_4)_2 SO_4}{1 \text{ mol } (NH_4)_2 SO_4} = 2.7 \text{ g}$$

49. **(a)** Here we calculate $[OH^-]$ needed for precipitation.

$$K_{sp} = [Al^{3+}][OH^-]^3 = 1.3 \times 10^{-33} = (0.075 \text{ M})[OH^-]^3$$

$$[OH^-] = \sqrt[3]{\frac{1.3 \times 10^{-33}}{0.075}} = 2.6 \times 10^{-11} \quad pOH = -\log(2.6 \times 10^{-11}) = 10.59$$

$$pH = 14.00 - 10.59 = 3.41$$

(b) We can use the Henderson–Hasselbalch equation to determine $[C_2H_3O_2^-]$.

$$pH = 3.41 = pK_a + \log\frac{[C_2H_3O_2^-]}{[HC_2H_3O_2]} = 4.74 + \log\frac{[C_2H_3O_2^-]}{1.00 \text{ M}}$$

$$\log \frac{\left[C_2H_3O_2^-\right]}{1.00 \text{ M}} = 3.41 - 4.74 = -1.33; \quad \frac{\left[C_2H_3O_2^-\right]}{1.00 \text{ M}} = 10^{-1.33} = 0.047; \quad \left[C_2H_3O_2^-\right] = 0.047 \text{ M}$$

This situation does not quite obey the guideline that the ratio of concentrations must fall in the range 0.10 to 10.0, but the resulting error is a small one in this circumstance.

$$\text{mass NaC}_2\text{H}_3\text{O}_2 = 0.2500 \text{ L} \times \frac{0.047 \text{ mol C}_2\text{H}_3\text{O}_2^-}{1 \text{ L soln}} \times \frac{1 \text{ mol NaC}_2\text{H}_3\text{O}_2}{1 \text{ mol C}_2\text{H}_3\text{O}_2^-} \times \frac{82.03 \text{ g NaC}_2\text{H}_3\text{O}_2}{1 \text{ mol NaC}_2\text{H}_3\text{O}_2}$$

$$= 0.96 \text{ g NaC}_2\text{H}_3\text{O}_2$$

50. **(a)** Since HI is a strong acid, $\left[I^-\right] = 1.05 \times 10^{-3} \text{ M} + 1.05 \times 10^{-3} \text{ M} = 2.10 \times 10^{-3} \text{ M}$

We determine the value of the ion product and compare it to the solubility product constant value.

$$Q_{sp} = \left[Pb^{2+}\right]\left[I^-\right]^2 = \left(1.1 \times 10^{-3}\right)\left(2.10 \times 10^{-3}\right)^2 = 4.9 \times 10^{-9} < 7.1 \times 10^{-9} = K_{sp} \text{ for PbI}_2$$

Thus a precipitate of PbI_2 will not form under these conditions.

(b) We compute the $\left[OH^-\right]$ needed for precipitation.

$$K_{sp} = \left[Mg^{2+}\right]\left[OH^-\right]^2 = 1.8 \times 10^{-11} = (0.0150)\left[OH^-\right]^2; \quad \left[OH^-\right] = \sqrt{\frac{1.8 \times 10^{-11}}{0.0150}} = 3.5 \times 10^{-5} \text{ M}$$

Then we compute $\left[OH^-\right]$ in this solution, resulting from the ionization of NH_3.

$$\left[NH_3\right] = 6.00 \text{ M} \times \frac{0.05 \times 10^{-3} \text{ L}}{2.50 \text{ L}} = 1.\underline{2} \times 10^{-4} \text{ M}$$

Even though NH_3 is a weak base, the $\left[OH^-\right]$ produced from the NH_3 hydrolysis reaction will approximate 4×10^{-5} M (3.85 if you solve the quadratic) in this very dilute solution. (Recall that degree of ionization is high in dilute solution.) And since $\left[OH^-\right] = 3.5 \times 10^{-5}$ M is needed for precipitation to occur, we conclude that $Mg(OH)_2$ will precipitate from this solution (note: not much of a precipitate is expected).

(c) 0.010 M $HC_2H_3O_2$ and 0.010 M $NaC_2H_3O_2$ is a buffer solution with pH = pK_a of acetic acid, (the acid and its anion are present in equal concentrations). From this, we determine the $\left[OH^-\right]$.

$$pH = 4.74 \qquad pOH = 14.00 - 4.74 = 9.26 \qquad \left[OH^-\right] = 10^{-9.26} = 5.5 \times 10^{-10}$$

$$Q = \left[Al^{3+}\right]\left[OH^-\right]^3 = (0.010)\left(5.5 \times 10^{-10}\right)^3 = 1.7 \times 10^{-30} > 1.3 \times 10^{-33} = K_{sp} \text{ of Al}(OH)_3$$

Thus, $Al(OH)_3(s)$ should precipitate from this solution.

Complex-Ion Equilibria

51. Lead(II) ion forms a complex ion with chloride ion. It forms no such complex ion with nitrate ion. The formation of this complex ion decreases the concentrations of free $Pb^{2+}(aq)$ and free $Cl^-(aq)$. Thus, $PbCl_2$ will dissolve in the HCl(aq) up until the value of the solubility product is exceeded. $Pb^{2+}(aq) + 3Cl^-(aq) \rightleftharpoons [PbCl_3]^-(aq)$

52. $Zn^{2+}(aq) + 4NH_3(aq) \rightleftharpoons [Zn(NH_3)_4]^{2+}(aq)$ $\qquad K_f = 4.1 \times 10^8$

$NH_3(aq)$ will be least effective in reducing the concentration of the complex ion. In fact, the addition of $NH_3(aq)$ will increase the concentration of the complex ion by favoring a shift of the equilibrium to the right. $NH_4^+(aq)$ will have a similar effect, but not as direct. $NH_3(aq)$ is formed by the hydrolysis of $NH_4^+(aq)$ and, thus, increasing $[NH_4^+]$ will eventually increase $NH_3(aq)$: $NH_4^+(aq) + H_2O(l) \rightleftharpoons NH_3(aq) + H_3O^+(aq)$. The addition of HCl(aq) will cause the greatest decrease in the concentration of the complex ion. HCl(aq) will react with $NH_3(aq)$ to decrease its concentration (by forming NH_4^+) and this will cause the complex ion equilibrium reaction to shift left towards free aqueous ammonia and $Zn^{2+}(aq)$.

53. We substitute the given concentrations directly into the K_f expression to calculate K_f.

$$K_f = \frac{[[Cu(CN)_4^{3-}]]}{[Cu^+][CN^-]^4} = \frac{0.0500}{(6.1 \times 10^{-32})(0.80)^4} = 2.0 \times 10^{30}$$

54. The solution to this problem is organized around the balanced chemical equation. Free $[NH_3]$ is 6.0 M at equilibrium. The size of the equilibrium constant indicates that most copper(II) is present as the complex ion at equilibrium.

Equation:	$Cu^{2+}(aq) + 4NH_3(aq)$	$\rightleftharpoons$	$[Cu(NH_3)_4]^{2+}(aq)$
Initial:	0.10 M 6.00 M		0 M
Change(100 % rxn):	−0.10 M −0.40 M		+0.10 M
Completion:	0 M 5.60 M		0.10 M
Changes:	+x M +4x M		−x M
Equil:	x M 5.60 + 4x M		$(0.10 - x)$ M

Let's assume $(5.60 + 4x)$ M ≈ 5.60 M and $(0.10 - x)$ M ≈ 0.10 M

$$K_f = \frac{\left[[Cu(NH_3)_4]^{2+}\right]}{[Cu^{2+}][NH_3]^4} = 1.1 \times 10^{13} = \frac{0.10 - x}{x(5.60 - 4x)^4} \approx \frac{0.10}{(5.60)^4 x} \approx \frac{0.10}{983.4x}$$

$$x = \frac{0.10}{983.4 \times (1.1 \times 10^{13})} = 9.2 \times 10^{-18} \text{ M} = [Cu^{2+}] \qquad (x \ll 0.10, \text{ thus the approximation was valid})$$

55. We first find the concentration of free metal ion. Then we determine the value of Q_{sp} for the precipitation reaction, and compare its value with the value of K_{sp} to determine whether precipitation should occur.

Equation:	$Ag^+(aq) +$	$2 S_2O_3^{2-}(aq)$	$\rightleftharpoons$	$[Ag(S_2O_3)_2]^{3-}(aq)$
Initial:	0 M	0.76 M		0.048 M
Changes:	$+x$ M	$+2x$ M		$-x$ M
Equil:	x M	$(0.76 + 2x)$ M		$(0.048 - x)$ M

$$K_f = \frac{\left[[Ag(S_2O_3)_2]^{3-}\right]}{[Ag^+][S_2O_3^{2-}]^2} = 1.7 \times 10^{13} = \frac{0.048 - x}{x(0.76 + 2x)^2} \approx \frac{0.048}{(0.76)^2 x}; x = 4.9 \times 10^{-15} M = [Ag^+]$$

($x \ll 0.048$ M, thus the approximation was valid).

$$Q_{sp} = [Ag^+][I^-] = (4.9 \times 10^{-15})(2.0) = 9.8 \times 10^{-15} > 8.5 \times 10^{-17} = K_{sp}.$$

Because $Q_{sp} > K_{sp}$, precipitation of AgI(s) should occur.

56. We need to determine $[OH^-]$ in this solution, and also the free $[Cu^{2+}]$.

$$pH = pK_a + \log\frac{[NH_3]}{[NH_4^+]} = 9.26 + \log\frac{0.10\ M}{0.10\ M} = 9.26 \quad pOH = 14.00 - 9.26 = 4.74$$

$$[OH^-] = 10^{-4.74} = 1.8 \times 10^{-5} M \qquad Cu^{2+}(aq) + 4NH_3(aq) \rightleftharpoons [Cu(NH_3)_4]^{2+}(aq)$$

$$K_f = \frac{\left[[Cu(NH_3)_4]^{2+}\right]}{[Cu^{2+}][NH_3]^4} = 1.1 \times 10^{13} = \frac{0.015}{[Cu^{2+}]0.10^4}; \quad [Cu^{2+}] = \frac{0.015}{1.1 \times 10^{13} \times 0.10^4} = 1.4 \times 10^{-11} M$$

Now we determine the value of Q_{sp} and compare it with the value of K_{sp} for $Cu(OH)_2$.

$$Q_{sp} = [Cu^{2+}][OH^-]^2 = (1.4 \times 10^{-11}) \times (1.8 \times 10^{-5})^2 = 4.5 \times 10^{-21} < 2.2 \times 10^{-20} (K_{sp} \text{ for } Cu(OH)_2)$$

Precipitation of $Cu(OH)_2(s)$ from this solution should not occur.

57. We first compute the free $[Ag^+]$ in the original solution. The size of the complex ion formation equilibrium constant indicates that the reaction lies far to the right, so we form as much complex ion as possible stoichiometrically.

Equation:	$Ag^+(aq) +$	$2NH_3(aq)$	$\rightleftharpoons$	$[Ag(NH_3)_2]^+(aq)$
In soln:	0.10 M	1.00 M		0 M
Form complex:	−0.10 M	−0.20 M		+0.10 M
	0 M	0.80 M		0.10 M
Changes:	$+x$ M	$+2x$ M		$-x$ M
Equil:	x M	$(0.80 + 2x)$ M		$(0.10 - x)$ M

$$K_f = 1.6 \times 10^7 = \frac{\left[Ag(NH_3)_2\right]^+}{\left[Ag^+\right]\left[NH_3\right]^2} = \frac{0.10-x}{x(0.80+2x)^2} \approx \frac{0.10}{x(0.80)^2} \qquad x = \frac{0.10}{1.6 \times 10^7 (0.80)^2} = 9.8 \times 10^{-9}\,\text{M}.$$

($x \ll 0.80$ M, thus the approximation was valid).

Thus, $\left[Ag^+\right] = 9.8 \times 10^{-9}$ M. We next determine the $\left[I^-\right]$ that can coexist in this solution without precipitation.

$$K_{sp} = \left[Ag^+\right]\left[I^-\right] = 8.5 \times 10^{-17} = \left(9.8 \times 10^{-9}\right)\left[I^-\right]; \qquad \left[I^-\right] = \frac{8.5 \times 10^{-17}}{9.8 \times 10^{-9}} = 8.7 \times 10^{-9}\,\text{M}$$

Finally, we determine the mass of KI needed to produce this $\left[I^-\right]$

$$\text{mass KI} = 1.00\,\text{L soln} \times \frac{8.7 \times 10^{-9}\,\text{mol}\,I^-}{1\,\text{L soln}} \times \frac{1\,\text{mol KI}}{1\,\text{mol}\,I^-} \times \frac{166.0\,\text{g KI}}{1\,\text{mol KI}} = 1.4 \times 10^{-6}\,\text{g KI}$$

58. First we determine $\left[Ag^+\right]$ that can exist with this $\left[Cl^-\right]$. We know that $\left[Cl^-\right]$ will be unchanged because precipitation will not be allowed to occur.

$$K_{sp} = \left[Ag^+\right]\left[Cl^-\right] = 1.8 \times 10^{-10} = \left[Ag^+\right]0.100\,\text{M}; \qquad \left[Ag^+\right] = \frac{1.8 \times 10^{-10}}{0.100} = 1.8 \times 10^{-9}\,\text{M}$$

We now consider the complex ion equilibrium. If the complex ion's final concentration is x, then the decrease in $\left[NH_3\right]$ is $2x$, because 2 mol NH_3 react to form each mol of complex ion, as follows. $Ag^+(aq) + 2NH_3(aq) \rightleftharpoons \left[Ag(NH_3)_2\right]^+(aq)$ We can solve the K_f expression for x.

$$K_f = 1.6 \times 10^7 = \frac{\left[Ag(NH_3)_2\right]^+}{\left[Ag^+\right]\left[NH_3\right]^2} = \frac{x}{1.8 \times 10^{-9}(1.00-2x)^2}$$

$$x = \left(1.6 \times 10^7\right)\left(1.8 \times 10^{-9}\right)(1.00-2x)^2 = 0.029(1.00 - 4.00x + 4.00x^2) = 0.029 - 0.12x + 0.12x^2$$

$$0 = 0.029 - 1.12x + 0.12x^2 \qquad \text{We use the quadratic formula roots equation to solve for } x.$$

$$x = \frac{-b \pm \sqrt{b^2 - 4ac}}{2a} = \frac{1.12 \pm \sqrt{(1.12)^2 - 4 \times 0.029 \times 0.12}}{2 \times 0.12} = \frac{1.12 \pm 1.114}{0.24} = 9.3,\ 0.025$$

Thus, we can add 0.025 mol $AgNO_3$ (~ 4.4 g $AgNO_3$) to this solution before we see a precipitate of $AgCl(s)$ form.

Precipitation and Solubilities of Metal Sulfides

59. We know that $K_{spa} = 3 \times 10^7$ for MnS and $K_{spa} = 6 \times 10^2$ for FeS. The metal sulfide will begin to precipitate when $Q_{spa} = K_{spa}$. Let us determine the $\left[H_3O^+\right]$ just necessary to form each precipitate. We assume that the solution is saturated with H_2S, $\left[H_2S\right] = 0.10$ M.

$$K_{spa} = \frac{[M^{2+}][H_2S]}{[H_3O^+]^2} \quad [H_3O^+] = \sqrt{\frac{[M^{2+}][H_2S]}{K_{spa}}} = \sqrt{\frac{(0.10\ M)(0.10\ M)}{3\times10^7}} = 1.8\times10^{-5}\ M \text{ for MnS}$$

$$[H_3O^+] = \sqrt{\frac{(0.10\ M)(0.10\ M)}{6\times10^2}} = 4.1\times10^{-3}\ M \text{ for FeS}$$

Thus, if the $[H_3O^+]$ is maintained just a bit higher than $1.8\times10^{-5}\ M$, FeS will precipitate and $Mn^{2+}(aq)$ will remain in solution. To determine if the separation is complete, we see whether $[Fe^{2+}]$ has decreased to 0.1% or less of its original value when the solution is held at the aforementioned acidity. Let $[H_3O^+] = 2.0\times10^{-5}\ M$ and calculate $[Fe^{2+}]$.

$$K_{spa} = \frac{[Fe^{2+}][H_2S]}{[H_3O^+]^2} = 6\times10^2 = \frac{[Fe^{2+}](0.10\ M)}{(2.0\times10^{-5}\ M)^2}; \quad [Fe^{2+}] = \frac{(6\times10^2)(2.0\times10^{-5})^2}{0.10} = 2.4\times10^{-6}\ M$$

$$\% Fe^{2+}(aq) \text{ remaining} = \frac{2.4\times10^{-6}\ M}{0.10\ M}\times100\% = 0.0024\% \quad \therefore \text{ Separation is complete.}$$

60. Since the cation concentrations are identical, the value of Q_{spa} is the same for each one. It is this value of Q_{spa} that we compare with K_{spa} to determine if precipitation occurs.

$$Q_{spa} = \frac{[M^{2+}][H_2S]}{[H_3O^+]^2} = \frac{0.05\ M \times 0.10\ M}{(0.010\ M)^2} = 5\times10^1$$

If $Q_{spa} > K_{spa}$, precipitation of the metal sulfide should occur. But, if $Q_{spa} < K_{spa}$, precipitation will not occur.

For CuS, $K_{spa} = 6\times10^{-16} < Q_{spa} = 5\times10^1$ Precipitation of CuS(s) should occur.

For HgS, $K_{spa} = 2\times10^{-32} < Q_{spa} = 5\times10^1$ Precipitation of HgS(s) should occur.

For MnS, $K_{spa} = 3\times10^7 > Q_{spa} = 5\times10^1$ Precipitation of MnS(s) will not occur.

61. **(a)** We can calculate $[H_3O^+]$ in the buffer solution with the Henderson–Hasselbalch equation.

$$pH = pK_a + \log\frac{[C_2H_3O_2^-]}{[HC_2H_3O_2]} = 4.74 + \log\frac{0.15\ M}{0.25\ M} = 4.52 \quad [H_3O^+] = 10^{-4.52} = 3.0\times10^{-5}\ M$$

We use this information to calculate a value of Q_{spa} for MnS in this solution and then comparison of Q_{spa} with K_{spa} will allow us to decide if a precipitate will form.

$$Q_{spa} = \frac{[Mn^{2+}][H_2S]}{[H_3O^+]^2} = \frac{(0.15)(0.10)}{(3.0\times10^{-5})^2} = 1.7\times10^7 < 3\times10^7 = K_{spa} \text{ for MnS}$$

Thus, precipitation of MnS(s) will not occur.

(b) We need to change $\left[H_3O^+\right]$ so that

$$Q_{spa} = 3 \times 10^7 = \frac{(0.15)(0.10)}{\left[H_3O^+\right]^2}; \quad \left[H_3O^+\right] = \sqrt{\frac{(0.15)(0.10)}{3 \times 10^7}} \quad [H_3O^+] = 2.2 \times 10^{-5}\text{ M} \quad pH = 4.66$$

This is a more basic solution, which we can produced by increasing the basic component of the buffer solution, namely, the acetate ion. We can find out the necessary acetate ion concentration with the Henderson–Hasselbalch equation.

$$pH = pK_a + \log \frac{[C_2H_3O_2^-]}{[HC_2H_3O_2]} = 4.66 = 4.74 + \log \frac{[C_2H_3O_2^-]}{0.25\text{ M}}$$

$$\log \frac{[C_2H_3O_2^-]}{0.25\text{ M}} = 4.66 - 4.74 = -0.08$$

$$\frac{\left[C_2H_3O_2^-\right]}{0.25\text{ M}} = 10^{-0.08} = 0.83 \quad \left[C_2H_3O_2^-\right] = 0.83 \times 0.25\text{ M} = 0.21\text{ M}$$

62. (a) CuS is in the hydrogen sulfide group of qualitative analysis. Its precipitation occurs when 0.3 M HCl is saturated with H_2S. It will certainly precipitate from a (non-acidic) saturated solution of H_2S which has a much higher $\left[S^{2-}\right]$.

$$Cu^{2+}(aq) + H_2S(\text{satd aq}) \longrightarrow CuS(s) + 2H^+(aq)$$

This reaction proceeds to an essentially quantitative extent in the forward direction.

(b) MgS is soluble, according to the solubility rules listed in Chapter 5.

$$Mg^{2+}(aq) + H_2S(\text{satd aq}) \xrightarrow{\text{0.3 M HCl}} \text{no reaction}$$

(c) As in part (a), PbS is in the qualitative analysis hydrogen sulfide group, which precipitates from a 0.3 M HCl solution saturated with H_2S. Therefore, PbS does not dissolve appreciably in 0.3 M HCl. $PbS(s) + HCl\,(0.3M) \longrightarrow \text{no reaction}$

(d) Since ZnS(s) does not precipitate in the hydrogen sulfide group, we conclude that it is soluble in acidic solution.

$$ZnS(s) + 2HNO_3(aq) \longrightarrow Zn(NO_3)_2(aq) + H_2S(g)$$

Qualitative Analysis

63. The purpose of adding hot water is to separate Pb^{2+} from AgCl and Hg_2Cl_2. Thus, the most important consequence would be the absence of a valid test for the presence or absence of Pb^{2+}. In addition, if we add NH_3 first, $PbCl_2$ may form $Pb(OH)_2$. If $Pb(OH)_2$ does form, it will be present with Hg_2Cl_2 in the solid, although $Pb(OH)_2$ will not darken with added NH_3. Thus, we might falsely conclude that Ag^+ is present.

64. For $PbCl_2 (aq)$, $2[Pb^{2+}] = [Cl^-]$ where $s =$ molar solubility of $PbCl_2$. Thus $s = [Pb^{2+}]$.

$K_{sp} = [Pb^{2+}][Cl^-]^2 = (s)(2s)^2 = 4s^3 = 1.6 \times 10^{-5}$; $s = \sqrt[3]{1.6 \times 10^{-5} \div 4} = 1.6 \times 10^{-2}$ M $= [Pb^{2+}]$

Both $[Pb^{2+}]$ and $[CrO_4^{2-}]$ are diluted by mixing the two solutions.

$[Pb^{2+}] = 0.016$ M $\times \dfrac{1.00 \text{ mL}}{1.05 \text{ mL}} = 0.015$ M $\qquad [CrO_4^{2-}] = 1.0$ M $\times \dfrac{0.05 \text{ mL}}{1.05 \text{ mL}} = 0.048$ M

$Q_{sp} = [Pb^{2+}][CrO_4^{2-}] = (0.015 \text{ M})(0.048 \text{ M}) = 7.2 \times 10^{-4} > 2.8 \times 10^{-13} = K_{sp}$

Thus, precipitation should occur from the solution described.

65. **(a)** Ag^+ and/or Hg_2^{2+} are probably present. Both of these cations form chloride precipitates from acidic solutions of chloride ion.

(b) We cannot tell whether Mg^{2+} is present or not. Both MgS and $MgCl_2$ are water soluble.

(c) Pb^{2+} possibly is absent; it is the only cation of those given which forms a precipitate in an acidic solution that is treated with H_2S, and no sulfide precipitate was formed.

(d) We cannot tell whether Fe^{2+} is present. FeS will not precipitate from an acidic solution that is treated with H_2S; the solution must be alkaline for a FeS precipitate to form.

(a) and **(c)** are the valid conclusions.

66. **(a)** $Pb^{2+}(aq) + 2Cl^-(aq) \longrightarrow PbCl_2(s)$

(b) $Zn(OH)_2(s) + 2OH^-(aq) \longrightarrow [Zn(OH)_4]^{2-}(aq)$

(c) $Fe(OH)_3(s) + 3H_3O^+(aq) \longrightarrow Fe^{3+}(aq) + 6H_2O(l)$ or $[Fe(H_2O)_6]^{3+}(aq)$

(d) $Cu^{2+}(aq) + H_2S(aq) \longrightarrow CuS(s) + 2H^+(aq)$

Integrative and Advanced Exercises

67. We determine s, the solubility of $CaSO_4$ in a saturated solution, and then the concentration of $CaSO_4$ in ppm in this saturated solution, assuming that the solution's density is 1.00 g/mL.

$K_{sp} = [Ca^{2+}][SO_4^{2-}] = (s)(s) = s^2 = 9.1 \times 10^{-6} \qquad s = 3.0 \times 10^{-3}$ M

ppm $CaSO_4 = 10^6$ g soln $\times \dfrac{1 \text{ mL}}{1.00 \text{ g}} \times \dfrac{1 \text{ L soln}}{1000 \text{ mL}} \times \dfrac{0.0030 \text{ mol } CaSO_4}{1 \text{ L soln}} \times \dfrac{136.1 \text{ g } CaSO_4}{1 \text{ mol } CaSO_4} = 4.1 \times 10^2$ ppm

Now we determine the volume of solution remaining after we evaporate the 131 ppm $CaSO_4$ down to a saturated solution (assuming that both solutions have a density of 1.00 g/mL).

volume sat'd soln $= 131$ g $CaSO_4 \times \dfrac{10^6 \text{ g sat'd soln}}{4.1 \times 10^2 \text{ g } CaSO_4} \times \dfrac{1 \text{ mL}}{1.00 \text{ g}} = 3.2 \times 10^5$ mL

Thus, we must evaporate 6.8×10^5 mL of the original 1.000×10^6 mL of solution, or 68% of the water sample.

68. (a) First we compute the molar solubility, s, of $CaHPO_4 \cdot 2H_2O$

$$s = \frac{0.32 \text{ g } CaHPO_4 \cdot 2H_2O}{1 \text{ L soln}} \times \frac{1 \text{ mol } CaHPO_4 \cdot 2H_2O}{172.1 \text{ g } CaHPO_4 \cdot 2H_2O} = 1.9 \times 10^{-3} \text{ M}$$

$$K_{sp} = [Ca^{2+}][HPO_4^{2-}] = (s)(s) = s^2 = (1.9 \times 10^{-3})^2 = 3.6 \times 10^{-6}$$

Thus the values are not quite consistent

(b) The value of K_{sp} given in the problem is consistent with a smaller value of the molar solubility, s, smaller than 1.8×10^{-3} M. The reason is that not all of the solute ends up in solution as HPO_4^{2-} ions. HPO_4^{2-} can act as either an acid or a base in water (see below), but it is base hydrolysis that predominates.

$$HPO_4^{2-}(aq) + H_2O(l) \rightleftharpoons H_3O^+(aq) + PO_4^{3-}(aq) \qquad K_{a3} = 4.2 \times 10^{-13}$$

$$HPO_4^{2-}(aq) + H_2O(l) \rightleftharpoons OH^-(aq) + H_2PO_4^-(aq) \qquad K_b = \frac{K_w}{K_2} = 1.6 \times 10^{-7}$$

69. The solutions mutually dilute each other and, because the volumes are equal, the concentrations are halved in the final solution: $[Ca^{2+}] = 0.00625$ M, $[SO_4^{2-}] = 0.00760$ M. We cannot assume that either concentration remains constant during the precipitation. Instead, we assume that precipitation proceeds until all of one reagent is used up. Equilibrium is reached from that point.

Equation: $CaSO_4(s) \rightleftharpoons$	$Ca^{2+}(aq)$	$+$	$SO_4^{2-}(aq)$	$K_{sp} = 9.1 \times 10^{-6}$
In soln: —	0.00625 M		0.00760 M	$K_{sp} = [Ca^{2+}][SO_4^{2-}]$
Form ppt: —	-0.00625 M		-0.00625 M	$K_{sp} = (x)(0.00135 + x)$
Not at equil —	0 M		0.00135 M	$K_{sp} \approx 0.00135 \, x$
Changes: —	$+x$ M		$+x$ M	$x = \dfrac{9.1 \times 10^{-6}}{0.00135} = 6.7 \times 10^{-3} \text{ M}$
Equil: —	x M		$(0.00135 + x)$ M	{not a reasonable assumption!}

Solving the quadratic $0 = x^2 + (1.35 \times 10^{-2})x - 9.1 \times 10^{-6}$ yields $x = 2.4 \times 10^{-3}$.

$$\% \text{ unprecipitated } Ca^{2+} = \frac{2.4 \times 10^{-3} \text{ M}_{final}}{0.00625 \text{ M}_{initial}} \times 100\% = 38 \% \text{ unprecipitated}$$

70. If equal volumes are mixed, each concentration is reduced to one-half of its initial value. We assume that all of the limiting ion forms a precipitate, and then equilibrium is achieved by the reaction proceeding in the forward direction to a small extent.

Equation: $BaCO_3(s) \rightleftharpoons$	$Ba^{2+}(aq)$	$+$	$CO_3^{2-}(aq)$
Orig. soln: —	0.00050 M		0.0010 M
Form ppt: —	0 M		0.0005 M
Changes: —	$+x$ M		$+x$ M
Equil: —	x M		$(0.0005 + x)$ M

$$K_{sp} = [Ba^{2+}][CO_3^{2-}] = 5.1 \times 10^{-9} = (x)(0.0005 + x) \approx 0.0005\, x$$

$$x = \frac{5.1 \times 10^{-9}}{0.0005} = 1 \times 10^{-5} \text{ M} \qquad \text{The } [Ba^{2+}] \text{ decreases from } 5 \times 10^{-4} \text{ M to } 1 \times 10^{-5} \text{ M}$$

($x < 0.0005$ M, thus the approximation was valid).

$$\% \text{ unprecipitated} = \frac{1 \times 10^{-5} \text{ M}}{5 \times 10^{-4} \text{ M}} \times 100\% = 2\% \qquad \text{Thus, 98\% of the } Ba^{2+} \text{ is precipitated as } BaCO_3(s).$$

71. The pH of the buffer establishes $[H_3O^+] = 10^{-pH} = 10^{-3.00} = 1.0 \times 10^{-3}$ M

Now we combine the two equilibrium expressions, and solve the resulting expression for the molar solubility of $Pb(N_3)_2$ in the buffer solution.

$$Pb(N_3)_2(s) \rightleftharpoons Pb^{2+}(aq) + 2\, N_3^-(aq) \qquad\qquad K_{sp} = 2.5 \times 10^{-9}$$

$$2\, H_3O^+(aq) + 2\, N_3^-(aq) \rightleftharpoons 2\, HN_3(aq) + 2\, H_2O(l) \qquad 1/K_a^2 = \frac{1}{(1.9 \times 10^{-5})^2} = 2.8 \times 10^9$$

$$Pb(N_3)_2(s) + 2\, H_3O^+(aq) \rightleftharpoons Pb^{2+}(aq) + 2\, HN_3(aq) + 2\, H_2O(l) \qquad K = K_{sp} \times (1/K_a^2) = 7.0$$

	$Pb(N_3)_2(s) + 2\, H_3O^+(aq)$	$\rightleftharpoons$ $Pb^{2+}(aq)$	$+\ 2\, HN_3(aq)$	$+\ 2\, H_2O(l)$
buffer:	– 0.0010 M	0 M	0 M	–
dissolving:	– (buffer)	+x M	+2x M	–
equilibrium:	– 0.0010 M	x M	2x M	–

$$K = \frac{[Pb^{2+}][HN_3]^2}{[H_3O^+]^2} = 7.0 = \frac{(x)\,(2x)^2}{(0.0010)^2} \qquad 4x^3 = 7.0\,(0.0010)^2 \qquad x = \sqrt[3]{\frac{7.0(0.0010)^2}{4}} = 0.012 \text{ M}$$

Thus, the molar solubility of $Pb(N_3)_2$ in a pH = 3.00 buffer is 0.012 M

72. We first determine the equilibrium constant of the suggested reaction.

$$Mg(OH)_2(s) \rightleftharpoons Mg^{2+}(aq) + 2\, OH^-(aq) \qquad\qquad K_{sp} = 1.8 \times 10^{-11}$$

$$2\, NH_4^+(aq) + 2\, OH^-(aq) \rightleftharpoons 2\, NH_3(aq) + 2\, H_2O(l) \qquad 1/K_b^2 = 1/(1.8 \times 10^{-5})^2$$

	$Mg(OH)_2(s)$	$+\ 2\, NH_4^+(aq)$	$\rightleftharpoons$ $Mg^{2+}(aq)$	$+\ 2\, NH_3(aq)$	$+\ 2\, H_2O(l)$
Initial:	–	1.00 M	0 M	0 M	–
Changes:	–	−2x M	+x M	+2x M	–
Equil:	–	(1.00 − 2x) M	x M	2x M	–

$$K = \frac{K_{sp}}{K_b^2} = \frac{1.8 \times 10^{-11}}{(1.8 \times 10^{-5})^2} = \frac{[Mg^{2+}][NH_3]^2}{[NH_4^+]^2} = 0.056 = \frac{x(2x)^2}{(1.00 - 2x)^2} \approx \frac{4x^3}{1.00^2}$$

$$x = \sqrt[3]{\frac{0.056}{4}} = 0.24 \text{ M}$$

Take this as a first approximation and cycle through again. $2x = 0.48$

$$K \approx \frac{4x^3}{(1.00-0.48)^2} = 0.056 \qquad x = \sqrt[3]{\frac{0.056\,(1.00-0.48)^2}{4}} = 0.16\,\text{M}$$

Yet another cycle gives a somewhat more consistent value. $2x = 0.32$

$$K \approx \frac{4x^3}{(1.00-0.32)^2} = 0.056 \qquad x = \sqrt[3]{\frac{0.056\,(1.00-0.38)^2}{4}} = 0.19\,\text{M}$$

another cycle gives more consistency. $2x = 0.38$

$$K \approx \frac{4x^3}{(1.00-0.38)^2} = 0.056 \qquad x = \sqrt[3]{\frac{0.056\,(1.00-0.38)^2}{4}} = 0.18\,\text{M} = [Mg^{2+}]$$

The molar solubility of $Mg(OH)_2$ in a 1.00 M NH_4Cl solution is 0.18 M.

73. First we determine the value of the equilibrium constant for the cited reaction.

$$CaCO_3(s) \rightleftharpoons Ca^{2+}(aq) + CO_3^{2-}(aq) \qquad\qquad K_{sp} = 2.8\times10^{-9}$$

$$H_3O^+(aq) + CO_3^{2-}(aq) \rightleftharpoons H_2O(l) + HCO_3^-(aq) \qquad 1/K_{a2} = 1/4.7\times10^{-11}$$

$$CaCO_3(aq) + H_3O^+(aq) \rightleftharpoons Ca^{2+}(aq) + HCO_3^-(aq) + H_2O(l) \quad K_{overall} = K_{sp}\times(1/K_{a2})$$

$$K_{overall} = \frac{2.8\times10^{-9}}{4.7\times10^{-11}} = 60$$

Equation: $CaCO_3(s) + H_3O^+(aq) \rightleftharpoons Ca^{2+}(aq) + HCO_3^-(aq) + H_2O(l)$.

Initial: – 10^{-pH} 0 M 0 M –

Change: – (*buffer*) $+x$ M $+x$ M –

Equil: – 10^{-pH} x M x M –

In the above set up, we have assumed that the pH of the solution did not change because of the dissolving of the $CaCO_3(s)$. That is, we have treated the rainwater as if it were a buffer solution.

(a) $K = \dfrac{[Ca^{2+}][HCO_3^-]}{[H_3O^+]} = 60. = \dfrac{x^2}{3\times10^{-6}} \qquad x = \sqrt{60\times3\times10^{-6}} = 1\times10^{-2}\,\text{M} = [Ca^{2+}]$

(b) $K = \dfrac{[Ca^{2+}][HCO_3^-]}{[H_3O^+]} = 60. = \dfrac{x^2}{6.3\times10^{-5}} \qquad x = \sqrt{60\times6.3\times10^{-5}} = 6.1\times10^{-2}\,\text{M} = [Ca^{2+}]$

74. We use the Henderson-Hasselbalch equation to determine $[H_3O^+]$ in this solution, and then use K_{sp} expression for MnS to determine $[Mn^{2+}]$ that can exist in this solution without precipitation occurring.

$$pH = pK_a + \log\frac{[C_2H_3O_2^-]}{[HC_2H_3O_2]} = 4.74 + \log\frac{0.500\ M}{0.100\ M} = 4.74 + 0.70 = 5.44$$

$$[H_3O^+] = 10^{-5.44} = 3.6 \times 10^{-6}\ M$$

$$MnS(s) + 2\,H_3O^+(aq) \rightleftharpoons Mn^{2+}(aq) + H_2S(aq) + 2\,H_2O(l) \qquad K_{spa} = 3 \times 10^7$$

Note that $[H_2S] = [Mn^{2+}] = s$, the molar solubility of MnS.

$$K_{spa} = \frac{[Mn^{2+}][H_2S]}{[H_3O^+]^2} = 3 \times 10^7 = \frac{s^2}{(3.6 \times 10^{-6}\ M)^2} \qquad s = 0.02\ M$$

$$\text{mass MnS/L} = \frac{0.02\ \text{mol Mn}^{2+}}{1\ \text{L soln}} \times \frac{1\ \text{mol MnS}}{1\ \text{mol Mn}^{2+}} \times \frac{87\ \text{g MnS}}{1\ \text{mol MnS}} = 2\ \text{g MnS/L}$$

75. (a) The precipitate is likely $Ca_3(PO_4)_2$. Let us determine the %Ca of this compound (by mass)

$$\%Ca = \frac{3 \times 40.078\ \text{g Ca}}{310.18\ \text{g Ca}_3(PO_4)_2} \times 100\% = 38.763\%\ \text{Ca}$$

$$3\,Ca^{2+}(aq) + 2\,HPO_4^{2-}(aq) \longrightarrow Ca_3(PO_4)_2(s) + 2\,H^+(aq)$$

(b) The bubbles are $CO_2(g)$: $\qquad CO_2(g) + H_2O(l) \longrightarrow \text{``H}_2CO_3(aq)\text{''}$

$$Ca^{2+}(aq) + H_2CO_3(aq) \longrightarrow CaCO_3(s) + 2\,H^+(aq) \qquad \text{precipitation}$$

$$CaCO_3(s) + H_2CO_3(aq) \longrightarrow Ca^{2+}(aq) + 2\,HCO_3^-(aq) \qquad \text{redissolving}$$

76. (a) A solution of $CO_2(aq)$ has $[CO_3^{2-}] = K_a[H_2CO_3] = 5.6 \times 10^{-11}$. Since $K_{sp} = 2.8 \times 10^{-9}$ for $CaCO_3$, the $[Ca^{2+}]$ needed to form a precipitate from this solution can be computed.

$$[Ca^{2+}] = \frac{K_{sp}}{[CO_3^{2-}]} = \frac{2.8 \times 10^{-9}}{5.6 \times 10^{-11}} = 50.\,M$$

This is too high to reach by dissolving $CaCl_2$ in solution. The reason why the technique works is because the $OH^-(aq)$ produced by $Ca(OH)_2$ neutralizes some of the $HCO_3^-(aq)$ from the ionization of $CO_2(aq)$, thereby increasing the $[CO_3^{2-}]$ above a value of 5.6×10^{-11}.

(b) The equation for redissolving can be obtained by combing several equations.

$$CaCO_3(s) \rightleftharpoons Ca^{2+}(aq) + CO_3^{2-}(aq) \qquad\qquad K_{sp} = 2.8 \times 10^{-9}$$

$$CO_3^{2-}(aq) + H_3O^+(aq) \rightleftharpoons HCO_3^-(aq) + H_2O(l) \qquad\qquad \frac{1}{K_{a2}} = \frac{1}{5.6 \times 10^{-11}}$$

$$\underline{CO_2(aq) + 2\,H_2O(l) \rightleftharpoons HCO_3^-(aq) + H_3O^+(aq) \qquad\qquad K_{a1} = 4.2 \times 10^{-7}}$$

$$CaCO_3(s) + CO_2(aq) + H_2O(l) \rightleftharpoons Ca^{2+}(aq) + 2\,HCO_3^-(aq) \qquad K = \frac{K_{sp} \times K_{a1}}{K_{a2}}$$

$$K = \frac{(2.8 \times 10^{-9})(4.2 \times 10^{-7})}{5.6 \times 10^{-11}} = 2.1 \times 10^{-5} = \frac{[Ca^{2+}][HCO_3^-]^2}{[CO_2]}$$

If $CaCO_3$ is precipitated from 0.005 M Ca^{2+}(aq) and then redissolved, $[Ca^{2+}] = 0.005$ M and $[HCO_3^{2-}] = 2 \times 0.005$ M $= 0.010$ M. We use these values in the above expression to compute $[CO_2]$

$$[CO_2] = \frac{[Ca^{2+}][HCO_3^-]^2}{2.1 \times 10^{-5}} = \frac{(0.005)(0.010)^2}{2.1 \times 10^{-5}} = 0.02 \text{ M}$$

We repeat the calculation for saturated $Ca(OH)_2$, in which $[OH^-] = 2 [Ca^{2+}]$, after first determining $[Ca^{2+}]$ in this solution.

$$K_{sp} = [Ca^{2+}][OH^-]^2 = 4 [Ca^{2+}] = 5.5 \times 10^{-6} \quad [Ca^{2+}] \sqrt[3]{\frac{5.5 \times 10^{-6}}{4}} = 0.011$$

$$M[CO_2] = \frac{[Ca^{2+}][HCO_3^-]^2}{2.1 \times 10^{-5}} = \frac{(0.011)(0.022)^2}{2.1 \times 10^{-5}} = 0.25 \text{ M}$$

Thus to redissolve the $CaCO_3$ requires that $[CO_2] = 0.25$ M if the solution initially is saturated $Ca(OH)_2$ but only 0.02 M CO_2 if the solution initially is 0.005 M $Ca(OH)_2$(aq). A handbook lists the solubility of CO_2 as 0.034 M. Clearly the $CaCO_3$ produced cannot redissolve if the solution was initially saturated with $Ca(OH)_2$(aq).

77. (a)

$$BaSO_4(s) \rightleftharpoons Ba^{2+}(aq) + SO_4^{2-}(aq) \qquad K_{sp} = 1.1 \times 10^{-10}$$

$$\underline{Ba^{2+}(aq) + CO_3^{2-}(aq) \rightleftharpoons BaCO_3(s) \qquad \frac{1}{K_{sp}} = \frac{1}{5.1 \times 10^{-9}}}$$

sum $BaSO_4(s) + CO_3^{2-}(aq) \rightleftharpoons BaCO_3(s) + SO_4^{2-}(aq)$ $K_{overall} = \dfrac{1.1 \times 10^{-10}}{5.1 \times 10^{-9}} = 0.0216$

initial   –   3 M   – ≈ 0

equil.   – $(3-x)$ M   – x

(where x is the carbonate used up in the reaction)

$$K = \frac{[SO_4^{2-}]}{[CO_3^{2-}]} = \frac{x}{3-x} = 0.0216 \text{ and } x = 0.063 \quad \text{Since } 0.063 \text{ M} > 0.050 \text{ M, the response is yes.}$$

(b)

$$2AgCl(s) \rightleftharpoons 2Ag^+(aq) + 2 Cl^-(aq) \qquad K_{sp} = (1.8 \times 10^{-10})^2$$

$$\underline{2 Ag^+(aq) + CO_3^{2-}(aq) \rightleftharpoons Ag_2CO_3(s) \qquad 1/K_{sp} = \frac{1}{8.5 \times 10^{-12}}}$$

sum $2AgCl(s) + CO_3^{2-}(aq) \rightleftharpoons Ag_2CO_3(s) + 2 Cl^-(aq)$ $K_{overall} = \dfrac{(1.8 \times 10^{-10})^2}{8.5 \times 10^{-12}} = 3.8 \times 10^{-9}$

initial   –   3 M   – ≈ 0 M

equil.   – $(3-x)$ M   – $2x$

(where x is the carbonate used up in the reaction)

$$K_{overall} = \frac{[Cl^-]^2}{[CO_3^{2-}]} = \frac{(2x)^2}{3-x} = 3.8 \times 10^{-9} \text{ and } x = 5.3 \times 10^{-5} \text{ M}$$

Since $2x$, $(2(5.35 \times 10^{-5} \text{M}))$, $<< 0.050$ M, the response is no

(c)

$$MgF_2(s) \rightleftharpoons Mg^{2+}(aq) + 2\,F^-(aq) \qquad K_{sp} = 3.7 \times 10^{-8}$$

$$Mg^{2+}(aq) + CO_3^{2-}(aq) \rightleftharpoons MgCO_3(s) \qquad 1/K_{sp} = \dfrac{1}{3.5 \times 10^{-8}}$$

sum $\quad MgF_2(s) + CO_3^{2-}(aq) \rightleftharpoons MgCO_3(s) + 2\,F^-(aq) \qquad K_{overall} = \dfrac{3.7 \times 10^{-8}}{3.5 \times 10^{-8}} = 1.1$

initial $\qquad$ – $\qquad$ 3 M $\qquad\qquad$ – $\qquad$ 0 M

equil. $\qquad$ – $\qquad$ (3-x)M $\qquad\quad$ – $\qquad$ $2x$

(where x is the carbonate used up in the reaction)

$$K_{overall} = \dfrac{[F^-]^2}{[CO_3^{2-}]} = \dfrac{(2x)^2}{3-x} = 1.1 \text{ and } x = 0.769M$$

Since $2x$, $(2(0.769M)) > 0.050$ M, the response is yes

78 . For ease of calculation, let us assume that 100 mL, that is, 0.100 L of solution is to be titrated. We also assume that $[Ag^+] = 0.10$ M in the titrant. In order to precipitate 99.9% of the Br^- as AgBr, the following volume of titrant must be added.

$$\text{volume titrant} = 0.999 \times 100 \text{ mL} \times \dfrac{0.010 \text{ mmol } Br^-}{1 \text{ mL sample}} \times \dfrac{1 \text{ mmol } Ag^+}{1 \text{ mmol } Br^-} \times \dfrac{1 \text{ mL titrant}}{0.10 \text{ mmol } Ag^+} = 9.99 \text{ mL}$$

We compute $[Ag^+]$ when precipitation is complete.

$$[Ag^+] = \dfrac{K_{sp}}{[Br^-]_f} = \dfrac{5.0 \times 10^{-13}}{0.001 \times 0.01 M} = 5.0 \times 10^{-8} M$$

Thus, during the course of the precipitation of AgBr, while 9.99 mL of titrant is added, $[Ag^+]$ increases from *zero* to 5.0×10^{-8} M. $[Ag^+]$ increases from 5.0×10^{-8} M to 1.0×10^{-5} M from the point where AgBr is completely precipitated to the point where Ag_2CrO_4 begins to precipitate. Let us assume, for the purpose of making an initial estimate, that this increase in $[Ag^+]$ occurs while the total volume of the solution is 110 mL (the original 100 mL plus 10 mL of titrant needed to precipitate all the AgBr). The amount of Ag^+ added during this increase is the difference between the amount present just after AgBr is completely precipitated and the amount present when Ag_2CrO_4 begins to precipitate.

$$\text{amount } Ag^+ \text{ added} = 110 \text{ mL} \times \dfrac{1.0 \times 10^{-5} \text{ mmol } Ag^+}{1 \text{ mL soln}} - 110 \text{ mL} \times \dfrac{5.0 \times 10^{-8} \text{ mmol } Ag^+}{1 \text{ mL soln}}$$

$$= 1.1 \times 10^{-3} \text{ mmol } Ag^+$$

$$\text{volume added titrant} = 1.1 \times 10^{-3} \text{ mmol } Ag^+ \times \dfrac{1 \text{ mL titrant}}{0.10 \text{ mmol } Ag^+} = 1.1 \times 10^{-2} \text{ mL titrant}$$

We see that indeed the total volume of solution remains essentially constant. Note that $[Ag^+]$ has increased by more than two powers of ten while the volume of solution increased a very small degree; this is indeed a very rapid rise in $[Ag^+]$, as stated.

79. The chemistry of aluminum is complex, however, we can make the following assumptions. Consider the following reactions at various pH:

$$pH > 7 \quad Al(OH)_3(s) + OH^- \xrightleftharpoons{K_{overall} = K_f K_{sp} = 1.43} Al(OH)_4^-(aq)$$

$$pH < 7 \quad Al(OH)_3(s) \xrightleftharpoons{K_{sp} = 1.3 \times 10^{-33}} Al^{3+}(aq) + 3 OH^-$$

For the solubilities in basic solutions, consider the following equilibrium at pH = 13.00 and 11.00.

At pH = 13.00: $Al(OH)_3(s) + OH^- \xrightleftharpoons{K_{ov} = K_f K_{sp} = 1.43} Al(OH)_4^-(aq)$

Initial :	–	0.10	0
Change:	–	$-x$	$+x$
Equilibrium:	–	$0.10 - x$	x

$$K_f = 1.4\underline{3} = \frac{x}{0.10 - x} \approx \frac{x}{0.10} \quad \text{(buffered at pH = 13.00)} \quad x = 0.14\underline{3} \text{ M}$$

At pH = 11.00: $K_f = 1.4\underline{3} = \dfrac{x}{0.0010 - x} \approx \dfrac{x}{0.0010}$ (buffered at pH = 11) $x = 0.0014\underline{3}$ M

Thus in these two calculations we see that the formation of the aluminate ion, $[Al(OH)_4^-]$ is favored at high pH values. Note that the concentration of aluminum(III) increases by a factor of one-hundred when the pH 1is increased from 11.00 to 13.00, and it will increase further still above pH 13. Now consider the solubility equilibrium established at pH = 3.00, 4.00, and 5.00. These correspond to $[OH^-] = 1.0 \times 10^{-11}$ M, 1.0×10^{-10} M, and 1.0×10^{-9} M, respectively.

$$Al(OH)_3(s) \xrightleftharpoons{K_{sp} = 1.3 \times 10^{-33}} Al^{3+}(aq) + 3 OH^-(aq)$$

At these pH values, $[Al^{3+}]$ in saturated aqueous solutions of $Al(OH)_3$ are

$$[Al^{3+}] = 1.3 \times 10^{-33} / (1.0 \times 10^{-11})^3 = 1.3 \text{ M at pH} = 3.00$$

$$[Al^{3+}] = 1.3 \times 10^{-33} / (1.0 \times 10^{-10})^3 = 1.3 \times 10^{-3} \text{ M at pH} = 4.00$$

$$[Al^{3+}] = 1.3 \times 10^{-33} / (1.0 \times 10^{-9})^3 = 1.3 \times 10^{-6} \text{ M at pH} = 5.00$$

These three calculations show that in strongly acidic solutions the concentration of aluminum(III) can be very high and is certainly not limited by the precipitation of $Al(OH)_3(s)$. However this solubility decreases rapidly as the pH increases beyond pH 4.

At neutral pH 7, a number of equilibria would need to be considered simultaneously, including the self-ionization of water itself.   However, the results of just these five calculations do indicate that concentration of aluminum(III) increases at both low and high pH values, with a very low concentration at intermediate pH values around pH 7, thus aluminum(III) concentration as a function of pH is a U-shaped (or V-shaped) curve.

80. We combine the solubility product expression for AgCN(s) with the formation expression for $[Ag(NH_3)_2]^+(aq)$.

solubility: $\quad\quad\quad\quad\quad\quad\quad$ $AgCN(s) \rightleftharpoons Ag^+(aq) + CN^-(aq) \quad\quad K_{sp} = ?$

Formation: $\quad Ag^+(aq) + 2NH_3(aq) \rightleftharpoons [Ag(NH_3)_2]^+(aq) \quad\quad K_f = 1.6 \times 10^{+7}$

Net reaction: $\quad AgCN(s) + 2\,NH_3(aq) \rightleftharpoons [Ag(NH_3)_2]^+ + CN^-(aq) \; K_{overall} = K_{sp} \times K_f$

$$K_{overall} = \frac{[[Ag(NH_3)_2]^+][CN^-]}{[NH_3]^2} = \frac{(8.8 \times 10^{-6})^2}{(0.200)^2} = 1.9 \times 10^{-9} = K_{sp} \times 1.6 \times 10^{+7}$$

$$K_{sp} = \frac{1.9 \times 10^{-9}}{1.6 \times 10^{+7}} = 1.2 \times 10^{-16}$$

Because of the extremely low solubility of AgCN in the solution, we assumed that $[NH_3]$ was not altered by the formation of the complex ion.

81. We combine the solubility product constant expression for $CdCO_3(s)$ with the formation expression for $[CdI_4]^{2-}(aq)$.

Solubility: $\quad\quad\quad\quad\quad\quad$ $CdCO_3(s) \rightleftharpoons Cd^{2+}(aq) + CO_3^{2-}(aq) \quad\quad K_{sp} = 5.2 \times 10^{-12}$

Formation: $\quad Cd^{2+}(aq) + 4\,I^-(aq) \rightleftharpoons [CdI_4]^{2-}(aq) \quad\quad\quad\quad K_f = ?$

Net reaction: $CdCO_3(s) + 4\,I^-(aq) \rightleftharpoons [CdI_4]^{2-}(aq) + CO_3^{2-}(aq) \quad\quad K_{overall} = K_{sp} \times K_f$

$$K_{overall} = \frac{[[CdI_4]^{2-}][CO_3^{2-}]}{[I^-]^4} = \frac{(1.2 \times 10^{-3})^2}{(1.00)^4} = 1.4 \times 10^{-6} = 5.2 \times 10^{-12} K_f$$

$$K_f = \frac{1.4 \times 10^{-6}}{5.2 \times 10^{-12}} = 2.7 \times 10^5$$

Because of the low solubility of $CdCO_3$ in this solution, we assumed that $[I^-]$ was not appreciably lowered by the formation of the complex ion.

82. We first determine $[Pb^{2+}]$ in this solution, which has $[Cl^-] = 0.10$ M.

$$K_{sp} = 1.6 \times 10^{-5} = [Pb^{2+}][Cl^-]^2 \quad\quad [Pb^{2+}] = \frac{K_{sp}}{[Cl^-]^2} = \frac{1.6 \times 10^{-5}}{(0.10)^2} = 1.6 \times 10^{-3} M$$

Then we use this value of $[Pb^{2+}]$ in the K_f expression to determine $[[PbCl_3]^-]$.

$$K_f = \frac{[[PbCl_3]^-]}{[Pb^{2+}][Cl^-]^3} = 24 = \frac{[[PbCl_3]^-]}{(1.6 \times 10^{-3})(0.10)^3} = \frac{[[PbCl_3]^-]}{1.6 \times 10^{-6}}$$

$[[PbCl_3]^-] = 24 \times 1.6 \times 10^{-6} = 3.8 \times 10^{-5} M$

The solubility of $PbCl_2$ in 0.10 M HCl is the following sum.

solubility = $[Pb^{2+}] + [[PbCl_3]^-] = 1.6 \times 10^{-3} M + 3.8 \times 10^{-5} M = 1.6 \times 10^{-3} M$

83. Two of the three relationships needed to answer this question are the two solubility product expressions. $1.6 \times 10^{-8} = [Pb^{2+}][SO_4^{2-}]$ $4.0 \times 10^{-7} = [Pb^{2+}][S_2O_3^{2-}]$

The third expression required is the electroneutrality equation, which states that the total positive charge concentration must equal the total negative charge concentration: $[Pb^{2+}] = [SO_4^{2-}] + [S_2O_3^{2-}]$, provided $[H_3O^+] = [OH^-]$.

Or, put another way, there is one Pb^{2+} ion in solution for each SO_4^{2-} ion *and* for each $S_2O_3^{2-}$ ion. We solve each of the first two expressions for the concentration of each anion substitute these expressions into the electroneutrality expression and then solve for $[Pb^{2+}]$.

$$[SO_4^{2-}] = \frac{1.6 \times 10^{-8}}{[Pb^{2+}]} \qquad [S_2O_3^{2-}] = \frac{4.0 \times 10^{-7}}{[Pb^{2+}]} \qquad [Pb^{2+}] = \frac{1.6 \times 10^{-8}}{[Pb^{2+}]} + \frac{4.0 \times 10^{-7}}{[Pb^{2+}]}$$

$$[Pb^{2+}]^2 = 1.6 \times 10^{-8} + 4.0 \times 10^{-7} = 4.2 \times 10^{-7} \qquad [Pb^{2+}] = \sqrt{4.2 \times 10^{-7}} = 6.5 \times 10^{-4} \text{ M}$$

84,   The chemical equations are:

$$PbCl_2(s) \rightleftharpoons Pb^{2+}(aq) + 2\,Cl^-(aq) \text{ and } \qquad PbBr_2(s) \rightleftharpoons Pb^{2+}(aq) + 2\,Br^-(aq)$$

The two solubility constant expressions $[Pb^{2+}][Cl^-]^2 = 1.6 \times 10^{-5}$ $[Pb^{2+}][Br^-]^2 = 4.0 \times 10^{-5}$ and the condition of electroneutrality $2\,[Pb^{2+}] = [Cl^-] + [Br^-]$ must be solved simultaneously to determine $[Pb^{2+}]$. (Rearranging the electroneutrality relationship, one obtains: $[Pb^{2+}] = \frac{1}{2}[Cl^-] + \frac{1}{2}[Br^-]$.) First we solve each of the solubility constant expressions for the concentration of the anion. Then we substitute these values into the electroneutrality expression and solve the resulting equation for $[Pb^{2+}]$.

$$[Cl^-] = \sqrt{\frac{1.6 \times 10^{-5}}{[Pb^{2+}]}} \qquad [Br^-] = \sqrt{\frac{4.0 \times 10^{-5}}{[Pb^{2+}]}} \qquad 2[Pb^{2+}] = \sqrt{\frac{1.6 \times 10^{-5}}{[Pb^{2+}]}} + \sqrt{\frac{4.0 \times 10^{-5}}{[Pb^{2+}]}}$$

$$2\sqrt{[Pb^{2+}]^3} = \sqrt{1.6 \times 10^{-5}} + \sqrt{4.0 \times 10^{-5}} = 4.0 \times 10^{-3} + 6.3 \times 10^{-3} = 1.03 \times 10^{-2}$$

$$\sqrt{[Pb^{2+}]^3} = 5.2 \times 10^{-3} \qquad [Pb^{2+}]^3 = 2.7 \times 10^{-5} \qquad [Pb^{2+}] = 3.0 \times 10^{-2}\,\text{M}$$

85. (a) First let us determine if there is sufficient Ag_2SO_4 to produce a saturated solution, in which $[Ag^+] = 2\,[SO_4^{2-}]$.

$$K_{sp} = 1.4 \times 10^{-5} = [Ag^+]^2[SO_4^{2-}] = 4[SO_4^{2-}]^3 \qquad [SO_4^{2-}] = \sqrt[3]{\frac{1.4 \times 10^{-5}}{4}} = 0.015\,\text{M}$$

$$[SO_4^{2-}] = \frac{2.50\,\text{g Ag}_2SO_4}{0.150\,\text{L}} \times \frac{1\,\text{mol Ag}_2SO_4}{311.8\,\text{g Ag}_2SO_4} \times \frac{1\,\text{mol SO}_4^{2-}}{1\,\text{mol Ag}_2SO_4} = 0.0535\,\text{M}$$

Thus, there is more than enough Ag_2SO_4 present to form a saturated solution. Let us now see if AgCl or $BaSO_4$ will precipitate under these circumstances. $[SO_4^{2-}] = 0.015$ M and $[Ag^+] = 0.030$ M.

$Q = [Ag^+][Cl^-] = 0.030 \times 0.050 = 1.5 \times 10^{-3} > 1.8 \times 10^{-10} = K_{sp}$, thus AgCl should precipitate.

$Q = [Ba^{2+}][SO_4^{2-}] = 0.025 \times 0.015 = 3.8 \times 10^{-4} > 1.1 \times 10^{-10} = K_{sp}$, thus $BaSO_4$ should precipitate.

Thus, the net ionic equation for the reaction that will occur is as follows.

$$Ag_2SO_4(s) + Ba^{2+}(aq) + 2\,Cl^-(aq) \longrightarrow BaSO_4(s) + 2\,AgCl(s)$$

(b) Let us first determine if any $Ag_2SO_4(s)$ remains or if it is all converted to $BaSO_4(s)$ and $AgCl(s)$. Thus, we have to solve a limiting reagent problem.

$$\text{amount } Ag_2SO_4 = 2.50 \text{ g } Ag_2SO_4 \times \frac{1 \text{ mol } Ag_2SO_4}{311.8 \text{ g } Ag_2SO_4} = 8.02 \times 10^{-3} \text{ mol } Ag_2SO_4$$

$$\text{amount } BaCl_2 = 0.150 \text{ L} \times \frac{0.025 \text{ mol } BaCl_2}{1 \text{ L soln}} = 3.75 \times 10^{-3} \text{ mol } BaCl_2$$

Since the two reactants combine in a 1 mole to 1 mole stoichiometric ratio, $BaCl_2$ is the limiting reagent. Since there must be some Ag_2SO_4 present and because Ag_2SO_4 is so much more soluble than either $BaSO_4$ or $AgCl$, we assume that $[Ag^+]$ and $[SO_4^{2-}]$ are determined by the solubility of Ag_2SO_4. They will have the same values as in a saturated solution of Ag_2SO_4.

$$[SO_4^{2-}] = 0.015 \text{ M} \qquad\qquad [Ag^+] = 0.030 \text{ M}$$

We use these values and the appropriate K_{sp} values to determine $[Ba^{2+}]$ and $[Cl^-]$.

$$[Ba^{2+}] = \frac{1.1 \times 10^{-10}}{0.015} = 7.3 \times 10^{-9} \text{ M} \qquad\qquad [Cl^-] = \frac{1.8 \times 10^{-10}}{0.030} = 6.0 \times 10^{-9} \text{ M}$$

Since $BaCl_2$ is the limiting reagent, we can use its amount to determine the masses of $BaSO_4$ and $AgCl$.

$$\text{mass } BaSO_4 = 0.00375 \text{ mol } BaCl_2 \times \frac{1 \text{ mol } BaSO_4}{1 \text{ mol } BaCl_2} \times \frac{233.4 \text{ g } BaSO_4}{1 \text{ mol } BaSO_4} = 0.875 \text{ g } BaSO_4$$

$$\text{mass } AgCl = 0.00375 \text{ mol } BaCl_2 \times \frac{2 \text{ mol } AgCl}{1 \text{ mol } BaCl_2} \times \frac{143.3 \text{ g } AgCl}{1 \text{ mol } AgCl} = 1.07 \text{ g } AgCl$$

The mass of unreacted Ag_2SO_4 is determined from the initial amount and the amount that reacts with $BaCl_2$.

$$\text{mass } Ag_2SO_4 = \left(0.00802 \text{ mol } Ag_2SO_4 - 0.00375 \text{ mol } BaCl_2 \times \frac{1 \text{ mol } Ag_2SO_4}{1 \text{ mol } BaCl_2} \right) \times \frac{311.8 \text{ g } Ag_2SO_4}{1 \text{ mol } Ag_2SO_4}$$

$$\text{mass } Ag_2SO_4 = 1.33 \text{ g } Ag_2SO_4 \text{ unreacted}$$

Of course, there is some Ag_2SO_4 dissolved in solution. We compute its mass.

$$\text{mass dissolved } Ag_2SO_4 = 0.150 \text{ L} \times \frac{0.0150 \text{ mol } Ag_2SO_4}{1 \text{ L soln}} \times \frac{311.8 \text{ g } Ag_2SO_4}{1 \text{ mol } Ag_2SO_4} = 0.702 \text{ g dissolved}$$

$$\text{mass } Ag_2SO_4(s) = 1.33 \text{ g} - 0.702 \text{ g} = 0.63 \text{ g } Ag_2SO_4(s)$$

Feature Problems

86. $\left[Ca^{2+}\right] = \left[SO_4^{2-}\right]$ in the saturated solution. Let us first determine the amount of H_3O^+ in the 100.0 mL diluted effluent. $H_3O^+(aq) + NaOH(aq) \longrightarrow 2H_2O(l) + Na^+(aq)$

$$mmol\ H_3O^+ = 100.0\ mL \times \frac{8.25\ mL\ base}{10.00\ mL\ sample} \times \frac{0.0105\ mmol\ NaOH}{1\ mL\ base} \times \frac{1\ mmol\ H_3O^+}{1\ mmol\ NaOH}$$

$$= 0.866\ mmol\ H_3O^+(aq)$$

Now we determine $\left[Ca^{2+}\right]$ in the original 25.00 mL sample, remembering that $2H_3O^+$ were produced for each Ca^{2+}.

$$\left[Ca^{2+}\right] = \frac{0.866\ mmol\ H_3O^+(aq) \times \dfrac{1\ mmol\ Ca^{2+}}{2\ mmol\ H_3O^+}}{25.00\ mL} = 0.0173\ M$$

$$K_{sp} = \left[Ca^{2+}\right]\left[SO_4^{2-}\right] = (0.0173)^2 = 3.0 \times 10^{-4};\quad \text{The } K_{sp}\text{ for CaSO}_4 \text{ is } 9.1 \times 10^{-6} \text{ in Appendix D.}$$

87. **(a)** We assume that there is little of each ion present in solution at equilibrium, (that this is a simple stoichiometric calculation). This is true because the K value for the titration reaction is very large. $K_{titration} = 1/K_{sp(AgCl)} = 5.6 \times 10^9$. We stop the titration when just enough silver ion has been added. $Ag^+(aq) + Cl^-(aq) \longrightarrow AgCl(s)$

$$V = 100.0\ mL \times \frac{29.5\ mg\ Cl^-}{1000\ mL} \times \frac{1\ mmol\ Cl^-}{35.45\ mg\ Cl^-} \times \frac{1\ mmol\ Ag^+}{1\ mmol\ Cl^-} \times \frac{1\ mL}{0.01000\ mmol\ AgNO_3}$$

$$= 8.32\ mL$$

(b) We first calculate the concentration of each ion as the consequence of dilution. Then we determine the $\left[Ag^+\right]$ from the value of K_{sp}

$$initial\ \left[Ag^+\right] = 0.01000\ M \times \frac{8.32\ mL\ added}{108.3\ mL\ final\ volume} = 7.68 \times 10^{-4}\ M$$

$$initial\ \left[Cl^-\right] = \frac{29.5\ mg\ Cl^- \times \dfrac{1\ mmol\ Cl^-}{35.45\ mg\ Cl^-}}{1000\ mL} \times \frac{100.0\ mL\ taken}{108.3\ mL\ final\ volume} = 7.68 \times 10^{-4}\ M$$

The slight excess of each ion will precipitate until the solubility constant is satisfied.

$$\left[Ag^+\right] = \left[Cl^-\right] = \sqrt{K_{sp}} = \sqrt{1.8 \times 10^{-10}} = 1.3 \times 10^{-5}\ M$$

(c) If we want Ag_2CrO_4 to appear just when AgCl has completed precipitation, $\left[Ag^+\right] = 1.3 \times 10^{-5}\ M$. We determine $\left[CrO_4^{2-}\right]$ from the K_{sp} expression.

$$K_{sp} = \left[Ag^+\right]^2\left[CrO_4^{2-}\right] = 1.1 \times 10^{-12} = \left(1.3 \times 10^{-5}\right)^2\left[CrO_4^{2-}\right];\quad \left[CrO_4^{2-}\right] = \frac{1.1 \times 10^{-12}}{\left(1.3 \times 10^{-5}\right)^2} = 0.0065\ M$$

(d) If $\left[CrO_4^{\,2-}\right]$ were greater than the answer just computed for part (c), Ag_2CrO_4 would appear before all Cl^- had precipitated, leading to a false early endpoint. We would calculate a falsely low $\left[Cl^-\right]$ for the original solution.

If $\left[CrO_4^{\,2-}\right]$ were less than computed in part 3, Ag_2CrO_4 would appear somewhat after all Cl^- had precipitated, leading one to conclude there was more Cl^- in solution than actually was the case.

(e) If it was Ag^+ that was being titrated, it would react immediately with the $CrO_4^{\,2-}$ in the sample, forming a red-orange precipitate. This precipitate would not likely dissolve, and thus, very little if any $AgCl$ would form. There would be no visual indication of the endpoint.

88. **(a)** We need to calculate the $\left[Mg^{2+}\right]$ in a solution that is saturated with $Mg(OH)_2$.

$$K_{sp} = 1.8 \times 10^{-11} = \left[Mg^{2+}\right]\left[OH^-\right]^2 = (s)(2s)^2 = 4s^3$$

$$s = \sqrt[3]{\frac{1.8 \times 10^{-11}}{4}} = 1.7 \times 10^{-4} \text{ M} = [Mg^{2+}]$$

(b) Even though water has been added to the original solution, it remains saturated (it is in equilibrium with the undissolved solid $Mg(OH)_2$). $\left[Mg^{2+}\right] = 1.7 \times 10^{-4} \text{ M}$.

(c) Although $HCl(aq)$ reacts with OH^-, it will not react with Mg^{2+}. The solution is simply a more dilute solution of Mg^{2+}.

$$\left[Mg^{2+}\right] = 1.7 \times 10^{-4} \text{ M} \times \frac{100.0 \text{ mL initial volume}}{(100.0 + 500.) \text{ mL final volume}} = 2.8 \times 10^{-5} \text{ M}$$

(d) In this instance, we have a dual dilution to a 275.0 mL total volume, followed by a common-ion scenario.

$$\text{initial }\left[Mg^{2+}\right] = \frac{\left(25.00 \text{ mL} \times \dfrac{1.7 \times 10^{-4} \text{ mmol Mg}^{2+}}{1 \text{ mL}}\right) + \left(250.0 \text{ mL} \times \dfrac{0.065 \text{ mmol Mg}^{2+}}{1 \text{ mL}}\right)}{275.0 \text{ mL total volume}}$$

$$= 0.059 \text{ M}$$

$$\text{initial }\left[OH^-\right] = \frac{25.00 \text{ mL} \times \dfrac{1.7 \times 10^{-4} \text{ mmol Mg}^{2+}}{1 \text{ mL}} \times \dfrac{2 \text{ mmol OH}^-}{1 \text{ mmol Mg}^{2+}}}{275.0 \text{ mL total volume}} = 3.1 \times 10^{-5} \text{ M}$$

Let's see if precipitation occurs.

$$Q_{sp} = \left[Mg^{2+}\right]\left[OH^-\right]^2 = (0.059)(3.1 \times 10^{-5})^2 = 5.7 \times 10^{-11} > 1.8 \times 10^{-11} = K_{sp}$$

Thus precipitation does occur, but very little precipitate forms. If $[OH^-]$ goes down by 1.4×10^{-5} M (which means that $[Mg^{2+}]$ drops by 0.7×10^{-5}M) then $[OH^-]=1.7\times10^{-5}$M and

$[Mg^{2+}]=(0.059 \text{ M}-0.7\times10^{-5}\text{ M}=)0.059 \text{ M}$, then $Q_{sp} < K_{sp}$ and precipitation will stop. Thus, $[Mg^{2+}]=0.059$ M.

(e) Again we have a dual dilution, now to a 200.0 mL final volume, followed by a common-ion scenario.

$$\text{initial } [Mg^{2+}]=\frac{50.00 \text{ mL}\times\dfrac{1.7\times10^{-4} \text{ mmol Mg}^{2+}}{1 \text{ mL}}}{200.0 \text{ mL total volume}}=4.3\times10^{-5} \text{ M}$$

$$\text{initial } [OH^-]=0.150 \text{ M}\times\frac{150.0 \text{ mL initial volume}}{200.0 \text{ mL total volume}}=0.113 \text{ M}$$

Now it is evident that precipitation will occur. Next we determine the $[Mg^{2+}]$ that can exist in solution with 0.113 M OH^-. It is clear that $[Mg^{2+}]$ will drop dramatically to satisfy the K_{sp} expression but the larger value of $[OH^-]$ will scarcely be affected.

$$K_{sp}=[Mg^{2+}][OH^-]^2=1.8\times10^{-11}=[Mg^{2+}](0.0113 \text{ M})^2$$

$$[Mg^{2+}]=\frac{1.8\times10^{-11}}{(0.113)^2}=1.4\times10^{-9} \text{ M}$$

CHAPTER 19
SPONTANEOUS CHANGE:
ENTROPY AND FREE ENERGY
PRACTICE EXAMPLES

1A In general, $\Delta S > 0$ if $\Delta n_{\text{gas}} > 0$. This is because gases are very dispersed compared to liquids or solids; (gases possess large entropies). Recall that Δn_{gas} is the difference between the sum of the stoichiometric coefficients of the gaseous products and a similar sum for the reactants.

    (a) $\Delta n_{\text{gas}} = 2 + 0 - (2 + 1) = -1$. One mole of gas is consumed here. We predict $\Delta S < 0$.

    (b) $\Delta n_{\text{gas}} = 1 + 0 - 0 = +1$. Since one mole of gas is produced, we predict $\Delta S > 0$.

1B   **(a)**   The outcome is uncertain in the reaction between $ZnS(s)$ and $Ag_2O(s)$. We have used Δn_{gas} to estimate the sign of entropy change. There is no gas involved in this reaction and thus our prediction is uncertain.

    (b)   In the chlor-alkali process the entropy increases because two moles of gas have formed where none were originally present.

2A For a vaporization, $\Delta G_{\text{vap}}^{\circ} = 0 = \Delta H_{\text{vap}}^{\circ} - T\Delta S_{\text{vap}}^{\circ}$. Thus, $\Delta S_{\text{vap}}^{\circ} = \Delta H_{\text{vap}}^{\circ} / T_{\text{vap}}$.

We substitute the given values. $\Delta S_{\text{vap}}^{\circ} = \dfrac{\Delta H_{\text{vap}}^{\circ}}{T_{\text{vap}}} = \dfrac{20.2 \text{ kJ mol}^{-1}}{(-29.79 + 273.15) \text{ K}} = 83.0 \text{ J mol}^{-1} \text{ K}^{-1}$

2B For a phase change, $\Delta G_{\text{tr}}^{\circ} = 0 = \Delta H_{\text{tr}}^{\circ} - T\Delta S_{\text{tr}}^{\circ}$. Thus, $\Delta H_{\text{tr}}^{\circ} = T\Delta S_{\text{tr}}^{\circ}$. We substitute in the given values. $\Delta H_{\text{tr}}^{\circ} = T\Delta S_{\text{tr}}^{\circ} = (95.5 + 273.2) \text{ K} \times 1.09 \text{ J mol}^{-1} \text{ K}^{-1} = 402 \text{ J / mol}$

3A The entropy change for the reaction is expressed in terms of the standard entropies of the reagents.

$$\Delta S^{\circ} = 2S^{\circ}\left[NH_3(g)\right] - S^{\circ}\left[N_2(g)\right] - 3S^{\circ}\left[H_2(g)\right]$$
$$= 2 \times 192.5 \text{ J mol}^{-1}\text{K}^{-1} - 191.6 \text{ J mol}^{-1} \text{ K}^{-1} - 3 \times 130.7 \text{ J mol}^{-1} \text{ K}^{-1} = -198.7 \text{ J mol}^{-1} \text{ K}^{-1}$$

Thus to form *one* mole of $NH_3(g)$, the standard entropy change is $-99.4 \text{ J mol}^{-1} \text{ K}^{-1}$

3B The entropy change for the reaction is expressed in terms of the standard entropies of the reagents.

$$\Delta S^\circ = S^\circ\left[NO(g)\right] + S^\circ\left[NO_2(g)\right] - S^\circ\left[N_2O_3(g)\right]$$

$$138.5 \text{ J mol}^{-1} \text{ K}^{-1} = 210.8 \text{ J mol}^{-1} \text{ K}^{-1} + 240.1 \text{ J mol}^{-1} \text{ K}^{-1} - S^\circ\left[N_2O_3(g)\right]$$

$$= 450.9 \text{ J mol}^{-1} \text{ K}^{-1} - S^\circ\left[N_2O_3(g)\right]$$

$$S^\circ\left[N_2O_3(g)\right] = 450.9 \text{ J mol}^{-1} \text{ K}^{-1} - 138.5 \text{ J mol}^{-1} \text{ K}^{-1} = 312.4 \text{ J mol}^{-1} \text{ K}$$

4A **(a)** Because $\Delta n_{gas} = 2 - (1+3) = -2$ for the synthesis of ammonia, we would predict $\Delta S < 0$ for the reaction. We already know that $\Delta H < 0$. Thus, the reaction falls into case 2, namely, a reaction that is spontaneous at low temperatures and non-spontaneous at high temperatures.

(b) For the formation of ethylene $\Delta n_{gas} = 1 - (2+0) = -1$ and thus $\Delta S < 0$. We are given that $\Delta H > 0$ and, thus, this reaction corresponds to case 4, namely, a reaction that is non-spontaneous at all temperatures.

4B **(a)** Because $\Delta n_{gas} = +1$ for the decomposition of calcium carbonate, we would predict $\Delta S > 0$ for the reaction, favoring the reaction at high temperatures. High temperatures also favor this endothermic $\left(\Delta H^\circ > 0\right)$ reaction.

(b) The "roasting" of ZnS(s) has $\Delta n_{gas} = 2 - 3 = -1$ and, thus, $\Delta S < 0$. We are given that $\Delta H < 0$; thus, this reaction corresponds to case 2, namely, a reaction that is spontaneous at low temperatures, and non-spontaneous at high ones.

5A The expression $\Delta G^\circ = \Delta H^\circ - T\Delta S^\circ$ is used with $T = 298.15 \text{ K}$.

$$\Delta G^\circ = \Delta H^\circ - T\Delta S^\circ = -1648 \text{ kJ} - 298.15 \text{ K} \times \left(-549.3 \text{ J K}^{-1}\right) \times \left(1 \text{ kJ} / 1000 \text{ J}\right)$$

$$= -1648 \text{ kJ} + 163.8 \text{ kJ} = -1484 \text{ kJ}$$

5B We just need to substitute values from Appendix D into the supplied expression.

$$\Delta G^\circ = 2\Delta G_f^\circ\left[NO_2(g)\right] - 2\Delta G_f^\circ\left[NO(g)\right] - \Delta G_f^\circ\left[O_2(g)\right]$$

$$= 2 \times 51.31 \text{ kJ mol}^{-1} - 2 \times 86.55 \text{ kJ mol}^{-1} - 0.00 \text{ kJ mol}^{-1} = -70.48 \text{ kJ mol}^{-1}$$

6A Pressures of gases and molarities of solutes in aqueous solution appear in thermodynamic equilibrium constant expressions. Pure solids and liquids (including solvents) do not appear.

(a) $K = \dfrac{P\{SiCl_4\}}{P\{Cl_2\}^2} = K_p$
        (b) $K = \dfrac{[HOCl]\left[H^+\right]\left[Cl^-\right]}{P\{Cl_2\}}$

$K = K_p$ for (a) because all terms in the K expression are gas pressures.

6B We need the balanced chemical equation in order to write the equilibrium constant expression. We start by translating names into formulas.

$$PbS(s) + HNO_3(aq) \rightarrow Pb(NO_3)_2(aq) + S(s) + NO(g)$$

The equation then is balanced with the ion-electron method.

oxidation : $\{PbS(s) \rightarrow Pb^{2+}(aq) + S(s) + 2e^-$ $\}\times 3$

reduction : $\{NO_3^-(aq) + 4H^+(aq) + 3e^- \rightarrow NO(g) + 2H_2O(l)$ $\}\times 2$

_____

net ionic : $3\,PbS(s) + 2NO_3^-(aq) + 8H^+(aq) \rightarrow 3Pb^{2+}(aq) + 3S(s) + 2NO(g) + 4H_2O(l)$

In writing the thermodynamic equilibrium constant, recall that neither pure solids ($PbS(s)$ and $S(s)$) nor pure liquids ($H_2O(l)$) appear in the thermodynamic equilibrium constant expression. Note also that we have written $H^+(aq)$ here for brevity even though we understand that $H_3O^+(aq)$ is the acidic species in aqueous solution.

$$K = \frac{\left[Pb^{2+}\right]^3 P\{NO\}^2}{\left[NO_3^-\right]^2 \left[H^+\right]^8}$$

7A Since the reaction is taking place at 298.15 K, we can use standard free energies of formation to calculate the standard free energy change for the reaction:

$$N_2O_4(g) \rightleftharpoons 2NO_2(g)$$

$$\Delta G^\circ = 2\Delta G_f^\circ\left[NO_2(g)\right] - \Delta G_f^\circ\left[N_2O_4(g)\right] = 2\times 51.31\ kJ/mol - 97.89\ kJ/mol = +4.73\ kJ$$

$\Delta G^\circ_{rxn} = +4.73$ kJ . Thus, the forward reaction is non-spontaneous as written at 298.15 K.

7B In order to answer this question we must calculate the reaction quotient and compare it to the K_p value for the reaction:

$$N_2O_4(g) \rightleftharpoons 2NO_2(g) \qquad Q_p = \frac{(0.5)^2}{0.5} = 0.5$$

0.5 bar   0.5 bar

$\Delta G^\circ_{rxn} = +4.73$ kJ $= -RT\ln K_p$;   4.73 kJ/mol $= -(8.3145 \times 10^{-3}$ kJ/K·mol)(298.15 K)$\ln K_p$

Therefore, $K_p = 0.148$. Since Q_p is greater than K_p, we can conclude that the reverse reaction will proceed spontaneously, i.e. NO_2 will spontaneously convert into N_2O_4.

8A We first determine the value of ΔG° and then set $\Delta G^\circ = -RT\ln K$ to determine K.

$$\Delta G = \Delta G_f^\circ\left[Ag^+(aq)\right] + \Delta G_f^\circ[I^-(aq)] - \Delta G_f^\circ\left[AgI(s)\right]$$

$$= [(77.11 - 51.57) - (-66.19)]\ kJ/mol = +91.73$$

$$\ln K = \frac{-\Delta G^\circ}{RT} = -\frac{-91.73\ kJ/mol}{8.3145\ J\ mol^{-1}\ K^{-1} \times 298.15\ K} \times \frac{1000\ J}{1\ kJ} = -37.00$$

$$K = e^{-37.00} = 8.5\times 10^{-17}$$

This is precisely equal to the value for the K_{sp} of AgI listed in Appendix D.

8B   We begin by translating names into formulas. $MnO_2(s) + HCl(aq) \rightarrow Mn^{2+}(aq) + Cl_2(aq)$
Then we produce a balanced net ionic equation with the ion-electron method.

oxidation : $2\,Cl^-(aq) \rightarrow Cl_2(g) + 2e^-$

reduction : $MnO_2(s) + 4H^+(aq) + 2e^- \rightarrow Mn^{2+}(aq) + 2H_2O(l)$
___

net ionic : $MnO_2(s) + 4H^+(aq) + 2Cl^-(aq) \rightarrow Mn^{2+}(aq) + Cl_2(g) + 2H_2O(l)$

Next we determine the value of $\Delta G°$ for the reaction and then the value of K.

$$\Delta G° = \Delta G_f^° \left[Mn^{2+}(aq) \right] + \Delta G_f^° \left[Cl_2(g) \right] + 2\Delta G_f^° \left[H_2O(l) \right]$$
$$- \Delta G_f^° \left[MnO_2(s) \right] - 4\Delta G_f^° \left[H^+(aq) \right] - 2\Delta G_f^° \left[Cl^-(aq) \right]$$
$$= -228.1 \text{ kJ} + 0.0 \text{ kJ} + 2 \times (-237.1 \text{ kJ})$$
$$- (-465.1 \text{ kJ}) - 4 \times 0.0 \text{ kJ} - 2 \times (-131.2 \text{ kJ}) = +25.2 \text{ kJ}$$

$$\ln K = \frac{-\Delta G°}{RT} = \frac{-(+25.2 \times 10^3 \text{ J mol}^{-1})}{8.3145 \text{ J mol}^{-1} \text{ K}^{-1} \times 298.15 \text{ K}} = -10.1\underline{7} \qquad K = e^{-10.2} = 4 \times 10^{-5}$$

Because the value of K is so much smaller than unity, we do not expect an appreciable
forward reaction.

9A   We set equal the two expressions for $\Delta G°$ and solve for the absolute temperature.

$$\Delta G° = \Delta H° - T\Delta S° = -RT \ln K \qquad \Delta H° = T\Delta S° - RT \ln K = T\left(\Delta S° - R\ln K \right)$$

$$T = \frac{\Delta H°}{\Delta S° - R\ln K} = \frac{-114.1 \times 10^3 \text{ J/mol}}{\left[-146.4 - 8.3145 \ln(150) \right] \text{ J mol}^{-1} \text{ K}^{-1}} = 607 \text{ K}$$

9B   We expect the value of the equilibrium constant to increase as the temperature decreases
since this is an exothermic reaction and exothermic reactions will have a larger equilibrium
constant (shift right to form more products), as the temperature decreases. Thus, we expect
K to be larger than 1000, which is its value at 4.3×10^2 K.

(a)   The value of the equilibrium constant at $25°$ C is obtained directly from the value of
$\Delta G°$, since that value is also for $25°C$. Note:

$$\Delta G° = \Delta H° - T\Delta S° = -77.1 \text{ kJ/mol} - 298.15 \text{ K} (-0.1213 \text{ kJ/mol} \cdot \text{K}) = -40.9 \text{ kJ/mol}$$

$$\ln K = \frac{-\Delta G°}{RT} = \frac{-(-40.9 \times 10^3 \text{ J mol}^{-1})}{8.3145 \text{ J mol}^{-1} \text{ K}^{-1} \times 298.15 \text{ K}} = 16.5 \qquad K = e^{+16.5} = 1.5 \times 10^7$$

(b)   First, we solve for $\Delta G°$ at $75°C = 348$ K

$$\Delta G° = \Delta H° - T\Delta S° = -77.1 \frac{\text{kJ}}{\text{mol}} \times \frac{1000 \text{ J}}{1 \text{ kJ}} - \left(348.15 \text{ K} \times \left(-121.3 \frac{\text{J}}{\text{mol K}} \right) \right)$$

$$= -34.87 \times 10^3 \text{ J/mol}$$

Then we use this value to obtain the value of the equilibrium constant, as in part (a).

$$\ln K = \frac{-\Delta G^{\circ}}{RT} = \frac{-(-34.87 \times 10^3 \text{ J mol}^{-1})}{8.3145 \text{ J mol}^{-1} \text{ K}^{-1} \times 348.15 \text{ K}} = 12.05 \qquad K = e^{+12.05} = 1.7 \times 10^5$$

As expected, K for this exothermic reaction decreases with increasing temperature.

10A We use the value of $K_p = 9.1 \times 10^2$ at 800 K and $\Delta H^{\circ} = -1.8 \times 10^5$ J / mol, for the appropriate terms, in the van't Hoff equation.

$$\ln \frac{5.8 \times 10^{-2}}{9.1 \times 10^2} = \frac{-1.8 \times 10^5 \text{ J/mol}}{8.3145 \text{ J mol}^{-1} \text{ K}^{-1}} \left(\frac{1}{800 \text{ K}} - \frac{1}{T \text{ K}} \right) = -9.66; \quad \frac{1}{T} = \frac{1}{800} - \frac{9.66 \times 8.3145}{1.8 \times 10^5}$$

$$1/T = 1.25 \times 10^{-3} - 4.5 \times 10^{-4} = 8.0 \times 10^{-4} \qquad T = 1240 \text{ K} \approx 970°\text{C}$$

This temperature is an estimate because it is an extrapolated point beyond the range of the data supplied.

10B The temperature we are considering is $235°\text{C} = 508$ K. We substitute the value of $K_p = 9.1 \times 10^2$ at 800 K and $\Delta H^{\circ} = -1.8 \times 10^5$ J/mol, for the appropriate terms, in the van't Hoff equation.

$$\ln \frac{K_p}{9.1 \times 10^2} = \frac{-1.8 \times 10^5 \text{ J/mol}}{8.3145 \text{ J mol}^{-1} \text{ K}^{-1}} \left(\frac{1}{800 \text{ K}} - \frac{1}{508 \text{ K}} \right) = +15._6; \quad \frac{K_p}{9.1 \times 10^2} = e^{+15._6} = 6 \times 10^6$$

$$K_p = 6 \times 10^6 \times 9.1 \times 10^2 = 5 \times 10^9$$

EXERCISES

Spontaneous Change, Entropy, and Disorder

1. **(a)** The freezing of ethanol involves a *decrease* in the entropy of the system. There is a reduction in mobility and in the number of forms in which their energy can be stored when they leave the solution and arrange themselves into a crystalline state.

(b) The sublimation of dry ice involves converting a solid that has little mobility into a highly dispersed vapor which has a number of ways in which energy can be stored (rotational, translational). Thus, the entropy of the system *increases* substantially.

(c) The burning of rocket fuel involves converting a liquid fuel into the highly dispersed mixture of the gaseous combustion products. The entropy of the system *increases* substantially.

2. Although there is a substantial change in entropy involved in **(a)** changing H_2O (liq., 1 atm) to H_2O (g, 1 atm), it is not as large as **(c)** converting the liquid to a gas at 10 mmHg. The gas is more dispersed, (less ordered), at lower pressures. In turn, **(b)** if we start with a solid and convert it to a gas at the lower pressure, the entropy change should be even larger, since a solid is more ordered, (concentrated), than a liquid. Thus, in order of increasing ΔS, the processes are: **(a)** $<$ **(c)** $<$ **(b)**.

3. The first law of thermodynamics states that energy is neither created nor destroyed (thus, "The energy of the universe is constant"). A consequence of the second law of thermodynamics is that entropy of the universe increases for all spontaneous, that is, naturally occurring, processes (and therefore, "the entropy of the universe increases toward a maximum").

4. When pollutants are produced they are usually dispersed throughout the environment. These pollutants thus start in a relatively compact form and end up dispersed throughout a large volume mixed with many other substances. The pollutants are highly dispersed, thus, they have a high entropy. Returning them to their original compact form requires reducing this entropy, which is a highly non-spontaneous process. If we have had enough foresight to retain these pollutants in a reasonably compact form, such as disposing of them in a *secure* landfill, rather than dispersing them in the atmosphere or in rivers and seas, the task of permanently removing them from the environment, and perhaps even converting them to useful forms, would be considerably easier.

5. **(a)** Increase in entropy because a gas has been created from a liquid, a condensed phase.

(b) Decrease in entropy as a condensed phase, a solid, is created from a solid and a gas.

(c) For this reaction we cannot be certain of the entropy change. Even though the number of moles of gas produced is the same as the number that reacted, we cannot conclude that the entropy change is zero because not all gases have the same molar entropy.

(d) $2H_2S(g) + 3O_2(g) \rightarrow 2H_2O(g) + 2SO_2(g)$ Decrease in entropy since five moles of gas with high entropy become only four moles of gas, with about the same quantity of entropy per mole.

6. **(a)** At $75°$ C, 1 mol H_2O (g, 1 atm) has a greater entropy than 1 mol H_2O (liq., 1 atm) since a gas is much more dispersed than a liquid.

(b) $50.0 \text{ g Fe} \times \dfrac{1 \text{ mol Fe}}{55.8 \text{ g Fe}} = 0.896$ mol Fe has a higher entropy than 0.80 mol Fe, both (s)

at 1 atm and $5°$ C, because entropy is an extensive property that depends on the amount of substance present.

(c) 1 mol Br_2 (liq., 1 atm, $8°$ C) has a higher entropy than 1 mol Br_2 (s, 1atm, $-8°$ C) because solids are more ordered, (concentrated), substances than are liquids, and furthermore, the liquid is at a higher temperature.

(d) 0.312 mol SO_2 (g, 0.110 atm, $32.5°$ C) has a higher entropy than 0.284 mol O_2 (g, 15.0 atm, $22.3°$ C) for at least three reasons. First, entropy is an extensive property that depends on the amount of substance present (more moles of SO_2 than O_2). Second, entropy increases with temperature (temperature of SO_2 is greater than that for O_2. Third, entropy is greater at lower pressures (the O_2 has a much higher pressure). Furthermore, entropy generally is higher per mole for more complicated molecules.

7. **(a)** Negative; A liquid (moderate entropy) combines with a solid to form another solid.

(b) Positive; One mole of high entropy gas forms where no gas was present before.

(c) Positive; One mole of high entropy gas forms where no gas was present before.

(d) Uncertain; The number of moles of gaseous products is the same as the number of moles of gaseous reactants.

(e) Negative; Two moles of gas (and a solid) combine to form just one mole of gas.

8. The entropy of formation of a compound would be the difference between the absolute entropy of one mole of the compound and the sum of the absolute entropies of the appropriate amounts of the elements constituting the compound, with each species in its most stable form. It seems as though $CS_2(l)$ would have the highest molar entropy of formation of the compounds listed, since it is the only substance whose formation does not involve the consumption of high entropy gaseous reactants.

(a) $C(graphite) + 2H_2(g) \rightleftharpoons CH_4(g)$

$\Delta S_f^\circ \left[CH_4(g) \right] = S^\circ \left[CH_4(g) \right] - S^\circ \left[C(graphite) \right] - 2 S^\circ \left[H_2(g) \right]$

$= 186.3 \text{ J mol}^{-1}\text{K}^{-1} - 5.74 \text{ J mol}^{-1}\text{K}^{-1} - 2 \times 130.7 \text{ J mol}^{-1}\text{K}^{-1}$

$= -80.8 \text{ J mol}^{-1}\text{K}^{-1}$

(b) $2C(graphite) + 3H_2(g) + \frac{1}{2}O_2(g) \rightleftharpoons CH_3CH_2OH(l)$

$\Delta S_f^\circ \left[CH_3CH_2OH(l) \right] = S^\circ \left[CH_3CH_2OH(l) \right] - 2 S^\circ \left[C(graphite) \right] - 3 S^\circ \left[H_2(g) \right] - \frac{1}{2} S^\circ \left[O_2(g) \right]$

$= 160.7 \text{ J mol}^{-1}\text{K}^{-1} - 2 \times 5.74 \text{ J mol}^{-1}\text{K}^{-1} - 3 \times 130.7 \text{ J mol}^{-1}\text{K}^{-1} - \frac{1}{2} \times 205.1 \text{ J mol}^{-1}\text{K}^{-1}$

$= -345.4 \text{ J mol}^{-1}\text{K}^{-1}$

(c) $C(graphite) + 2S(rhombic) \rightleftharpoons CS_2(l)$

$\Delta S_f^\circ \left[CS_2(l) \right] = S^\circ \left[CS_2(l) \right] - S^\circ \left[C(graphite) \right] - 2 S^\circ \left[S(rhombic) \right]$

$= 151.3 \text{ J mol}^{-1} \text{ K}^{-1} - 5.74 \text{ J mol}^{-1} \text{ K}^{-1} - 2 \times 31.80 \text{ J mol}^{-1} \text{ K}^{-1}$

$= 82.0 \text{ J mol}^{-1} \text{ K}^{-1}$

Phase Transitions

9. **(a)** $\Delta H_{vap}^\circ = \Delta H_f^\circ [H_2O(g)] - \Delta H_f^\circ [H_2O(l)] = -241.8 \text{ kJ/mol} - (-285.8 \text{ kJ/mol})$

$= +44.0 \text{ kJ/mol}$

$\Delta S_{vap}^\circ = S^\circ \left[H_2O(g) \right] - S^\circ \left[H_2O(l) \right] = 188.8 \text{ J mol}^{-1} \text{ K}^{-1} - 69.91 \text{ J mol}^{-1}\text{K}^{-1}$

$= 118.9 \text{ J mol}^{-1}\text{K}^{-1}$

There is an alternate, but incorrect, method of obtaining ΔS_{vap}°.

$$\Delta S^{\circ}_{vap} = \frac{\Delta H^{\circ}_{vap}}{T} = \frac{44.0 \times 10^3 \text{ J/mol}}{298.15 \text{ K}} = 148 \text{ J mol}^{-1} \text{ K}^{-1}$$

This method is invalid because the temperature in the denominator of the equation must be the temperature at which the liquid-vapor transition is at equilibrium. Liquid water and water vapor at 1 atm pressure (standard state, indicated by $^{\circ}$) are in equilibrium only at 100° C = 373 K.

(b) The reason why ΔH°_{vap} is different at 25° C from its value at 100° C has to do with the heat required to bring the reactants and products down to 298 K from 373 K. The specific heat of liquid water is higher than the heat capacity of steam. Thus, more heat is given off by lowering the temperature of the liquid water from 100° C to 25° C than is given off by lowering the temperature of the same amount of steam. Another way to think of this is that hydrogen bonding is more disrupted in water at 100° C than at 25° C (because the molecules are in rapid—thermal—motion), and hence, there is not as much energy needed to convert liquid to vapor (thus ΔH°_{vap} has a smaller value at 100° C. The reason why ΔS°_{vap} has a larger value at 25° C than at 100° C has to do with dispersion. A vapor at 1 atm pressure (the case at both temperatures) has about the same entropy. On the other hand, liquid water is more disordered, (better able to disperse energy), at higher temperatures since more of the hydrogen bonds are disrupted by thermal motion. (The hydrogen bonds are totally disrupted in the two vapors).

10. (a) We use Trouton's rule $\left(\Delta S^{\circ}_{vap} \approx 87 \text{ J mol}^{-1} \text{ K}^{-1}\right)$ to estimate the normal boiling point.

$$T_{nbp} = \frac{\Delta H^{\circ}_{vap}}{\Delta S^{\circ}_{vap}} = \frac{\left[-146.9 \text{ kJ/mol} - (-173.5) \times \frac{1000 \text{ J}}{1 \text{ kJ}}\right]}{87 \text{ J mol}^{-1}\text{K}^{-1}} = 306 \text{ K} \approx 3.1 \times 10^2 \text{ K}$$

The literature normal boiling point for pentane is 36.1 °C ($\approx$ 309 K).

(b) $\Delta G^{\circ}_{vap} = \Delta H^{\circ}_{vap} - T\Delta S^{\circ}_{vap}$

$$= [-146.9 - (-173.5)] \frac{\text{kJ}}{\text{mol}} - \frac{87 \text{ J mol}^{-1}\text{K}^{-1} \times 298 \text{ K}}{\frac{1000 \text{ J}}{1 \text{ kJ}}} = 26.6 \frac{\text{kJ}}{\text{mol}} - 25.9 \frac{\text{kJ}}{\text{mol}} = +0.7 \frac{\text{kJ}}{\text{mol}}$$

(c) The positive value of ΔG°_{vap} indicates that normal boiling (having a vapor pressure of 1.00 atm) for pentane should be non-spontaneous (will not occur) at 298. The vapor pressure of pentane at 298 K should be less than 1.00 atm.

11. Trouton's rule is obeyed most closely by liquids that do not have a high degree of order within the liquid. In both HF and CH_3OH, hydrogen bonds create considerable order within the liquid. In $C_6H_5CH_3$, the only attractive forces are non-directional London forces, which have no preferred orientation as hydrogen bonds do. Thus, of the three choices, liquid $C_6H_5CH_3$ would most closely follow Trouton's rule.

12. $\Delta H_{vap}^{\circ} = \Delta H_f^{\circ}[Br_2(g)] - \Delta H_f^{\circ}[Br_2(l)] \approx 30.91 \text{ kJ/mol} - 0.00 \text{ kJ/mol} = 30.91 \text{ kJ/mol}$

$$\Delta S_{vap}^{\circ} = \frac{\Delta H_{vap}^{\circ}}{T_{vap}} \approx 87 \text{ J mol}^{-1} \text{ K}^{-1} \text{ or } T_{vap} = \frac{\Delta H_{vap}^{\circ}}{\Delta S_{vap}^{\circ}} \approx \frac{30.91 \times 10^3 \text{ J/mol}}{87 \text{ J mol}^{-1}\text{K}^{-1}} = 3.5 \times 10^2 \text{ K}$$

The accepted value of the boiling point of bromine is $58.8°C = 332 \text{ K} = 3.32 \times 10^2 \text{ K}$. Thus, our estimate is in reasonable agreement with the measured value.

13. The liquid water-gaseous water equilibrium H_2O (l, 0.50 atm) $\rightleftharpoons H_2O$ (g, 0.50 atm) can only be established at <u>one temperature</u>, namely the boiling point for water under 0.50 atm external pressure. We can estimate the boiling point for water under 0.50 atm external pressure by using the Clausius-Clapeyron equation:

$$\ln\frac{P_2}{P_1} = \frac{\Delta H_{vap}^{o}}{R}\left(\frac{1}{T_1} - \frac{1}{T_2}\right)$$

We know that at 373 K, the pressure of water vapor is 1.00 atm. Let's make P_1 = 1.00 atm, P_2 = 0.50 atm and T_1 = 373 K. Thus, the boiling point under 0.50 atm pressure is T_2. To find T_2 we simply insert the appropriate information into the Clausius-Clapeyron equation and solve for T_2:

$$\ln\frac{0.50 \text{ atm}}{1.00 \text{ atm}} = \frac{40.7 \text{ kJ mol}^{-1}}{8.3145 \times 10^{-3} \text{ kJ K}^{-1} \text{ mol}^{-1}}\left(\frac{1}{373 \text{ K}} - \frac{1}{T_2}\right)$$

$$-1.4\underline{16} \times 10^{-4} \text{ K} = \left(\frac{1}{373 \text{ K}} - \frac{1}{T_2}\right)$$

Solving for T_2 we find a temperature of 354 K or 81°C. Consequently, to achieve an equilibrium between gaseous and liquid water under 0.50 atm pressure, the temperature must be set at 354 K.

14. Figure 12-19 (phase diagram for carbon dioxide) shows that at –60°C and under 1 atm of external pressure, carbon dioxide exists as a gas. In other words, neither solid nor liquid CO_2 can exist at this temperature and pressure. Clearly, of the three phases, gaseous CO_2 must be the most stable and, hence, have the lowest free energy when T = –60 °C and P_{ext} = 1.00 atm.

Free Energy and Spontaneous Change

15. Answer (b) is correct. Br—Br bonds are broken in this reaction, meaning that it is endothermic, with $\Delta H > 0$. Since the number of moles of gas increases during the reaction, $\Delta S > 0$. And, because $\Delta G = \Delta H - T\,\Delta S$, this reaction is non-spontaneous, ($\Delta G > 0$), at low temperatures where the ΔH term predominates and spontaneous, ($\Delta G < 0$), at high temperatures where the $T\,\Delta S$ term predominates.

16. Answer (d) is correct. A reaction that proceeds only through electrolysis is a reaction that is non-spontaneous. Such a reaction has $\Delta G > 0$.

17. **(a)** $\Delta H^{\circ} < 0$ and $\Delta S^{\circ} < 0$ (since $\Delta n_{gas} < 0$) for this reaction. Thus, this reaction is case 2 of Table 19-1. It is spontaneous at low temperatures and non-spontaneous at high temperatures.

(b) We are unable to predict the sign of ΔS° for this reaction, since $\Delta n_{gas} = 0$. Thus, no strong prediction as to the temperature behavior of this reaction can be made. Since $\Delta H^{o} > 0$, we can, however, conclude that the reaction will be non-spontaneous at low temperatures.

(c) $\Delta H^{\circ} > 0$ and $\Delta S^{\circ} > 0$ (since $\Delta n_{gas} > 0$) for this reaction. This is case 3 of Table 19-1. It is non-spontaneous at low temperatures, but spontaneous at high temperatures.

18. **(a)** $\Delta H^{\circ} > 0$ and $\Delta S^{\circ} < 0$ (since $\Delta n_{gas} < 0$) for this reaction. This is case 4 of Table 19-1. It is non-spontaneous at all temperatures.

(b) $\Delta H^{\circ} < 0$ and $\Delta S^{\circ} > 0$ (since $\Delta n_{gas} > 0$) for this reaction. This is case 1 of Table 19-1. It is spontaneous at all temperatures.

(c) $\Delta H^{\circ} < 0$ and $\Delta S^{\circ} < 0$ (since $\Delta n_{gas} < 0$) for this reaction. This is case 2 of Table 19-1. It is spontaneous at low temperatures and non-spontaneous at high temperatures.

19. First of all, the process is clearly spontaneous, and therefore $\Delta G < 0$. In addition, the gases are more dispersed when they are at a lower pressure and therefore $\Delta S > 0$. We also conclude that $\Delta H = 0$ because the gases are ideal and thus there are no forces of attraction or repulsion between them.

20. Because an ideal solution forms spontaneously, $\Delta G < 0$. Also, the molecules of solvent and solute that are mixed together in the solution are in a more dispersed state than the separated solvent and solute. Therefore, $\Delta S > 0$. However, in an ideal solution, the attractive forces between solvent and solute molecules equals those forces between solvent molecules and those between solute molecules. Thus, $\Delta H = 0$. There is no net energy of interaction.

21. **(a)** An exothermic reaction (one that gives off heat) may not occur spontaneously if, at the same time, the system becomes more ordered, (concentrated), that is, $\Delta S° < 0$. This is particularly true at a high temperature, where the $T\Delta S$ term dominates the ΔG expression. An example of such a process is freezing water (clearly exothermic because the reverse process, melting ice, is endothermic), which is not spontaneous at temperatures above $0\,°C$.

(b) A reaction in which $\Delta S > 0$ need not be spontaneous if that process also is endothermic. This is particularly true at low temperatures, where the ΔH term dominates the ΔG expression. An example is the vaporization of water (clearly an endothermic process, one that requires heat, and one that produces a gas, so $\Delta S > 0$), whish is not spontaneous at low temperatures, that is, below $100\,°C$ (assuming $P_{ext} = 1.00$ atm).

22. Because this reaction produces more moles of gas than it consumes, $\Delta S > 0$. The reaction also is endothermic, since energy is required to break the A—B bond. Hence $\Delta H > 0$. Therefore, this reaction is of the type case 3 in Table 19-1. It is non-spontaneous at low temperatures, but eventually becomes spontaneous as the temperature is raised.

Standard Free Energy Change

23. $\Delta H° = \Delta H_f° \left[NH_4Cl(s) \right] - \Delta H_f° \left[NH_3(g) \right] - \Delta H_f° \left[HCl(g) \right]$

$= -314.4\,kJ/mol - (-46.11\,kJ/mol - 92.31\,kJ/mol) = -176.0\,kJ/mol$

$\Delta G° = \Delta G_f° \left[NH_4Cl(s) \right] - \Delta G_f° \left[NH_3(g) \right] - \Delta G_f° \left[HCl(g) \right]$

$= -202.9\,kJ/mol - (-16.48\,kJ/mol - 95.30\,kJ/mol) = -91.1\,kJ/mol$

$\Delta G° = \Delta H° - T\Delta S°$

$\Delta S° = \dfrac{\Delta H° - \Delta G°}{T} = \dfrac{-176.0\,kJ/mol + 91.1\,kJ/mol}{298\ K} \times \dfrac{1000\ J}{1\ kJ} = -285\ J\,mol^{-1}$

24. **(a)** $\Delta G° = \Delta G_f° \left[C_2H_6(g) \right] - \Delta G_f° \left[C_2H_2(g) \right] - 2\Delta G_f° \left[H_2(g) \right]$

$= -32.82\,kJ/mol - 209.2\,kJ/mol - 2(0.00\,kJ/mol) = -242.0\,kJ/mol$

(b) $\Delta G° = 2\Delta G_f° \left[SO_2(g) \right] + \Delta G_f° \left[O_2(g) \right] - 2\Delta G_f° \left[SO_3(g) \right]$

$= 2(-300.2\ kJ/mol) + 0.00\ kJ/mol - 2(-371.1\ kJ/mol) = +141.8\ kJ/mol$

(c) $\Delta G° = 3\Delta G_f° \left[Fe(s) \right] + 4\Delta G_f° \left[H_2O(g) \right] - \Delta G_f° \left[Fe_3O_4(s) \right] - 4\Delta G_f° \left[H_2(g) \right]$

$= 3(0.00\,kJ/mol) + 4(-228.6\,kJ/mol) - (-1015\,kJ/mol) - 4(0.00\,kJ/mol) = 101\,kJ/mol$

(d) $\Delta G° = 2\Delta G_f° \left[Al^{3+}(aq) \right] + 3\Delta G_f° \left[H_2(g) \right] - 2\Delta G_f° \left[Al(s) \right] - 6\Delta G_f° \left[H^+(aq) \right]$

$= 2(-485\,kJ/mol) + 3(0.00\,kJ/mol) - 2(0.00kJ/mol) - 6(0.00\,kJ/mol) = -970.\,kJ/mol$

25. **(a)** $\Delta S^\circ = 2 S^\circ \left[POCl_3\,(l) \right] - 2 S^\circ \left[PCl_3\,(g) \right] - S^\circ \left[O_2\,(g) \right]$

$= 2 (222.4 \ \text{J/K}) - 2 (311.7 \ \text{J/K}) - 205.1 \ \text{J/K} = -383.7 \ \text{J/K}$

$\Delta G^\circ = \Delta H^\circ - T \Delta S^\circ = -620.2 \times 10^3 \ \text{J} - (298 \ \text{K})(-383.7 \ \text{J/K}) = -506 \times 10^3 \ \text{J} = -506 \ \text{kJ}$

(b) The reaction proceeds spontaneously in the forward direction when reactants and products are in their standard states, because the value of ΔG° is less than zero.

26. **(a)** $\Delta S^\circ = S^\circ \left[Br_2\,(l) \right] + 2 S^\circ \left[HNO_2\,(aq) \right] - 2 S^\circ \left[H^+\,(aq) \right] - 2 S^\circ \left[Br^-\,(aq) \right] - 2 S^\circ \left[NO_2\,(g) \right]$

$= 152.2 \ \text{J/K} + 2 (135.6 \ \text{J/K}) - 2 (0 \ \text{J/K}) - 2 (82.4 \ \text{J/K}) - 2 (240.1 \ \text{J/K}) = -221.6 \ \text{J/K}$

$\Delta G^\circ = \Delta H^\circ - T \Delta S^\circ = -61.6 \times 10^3 \text{J} - (298 \text{K})(-221.6 \ \text{J/K}) = +4.4 \times 10^3 \ \text{J} = +4.4 \ \text{kJ}$

(b) The reaction does not proceed spontaneously in the forward direction when reactants and products are in their standard states, because the value of ΔG° is greater than zero.

27. We combine the reactions in the same way as for any Hess's law calculations.

(a) $N_2O(g) \rightarrow N_2(g) + \tfrac{1}{2} O_2(g)$ $\qquad\qquad\qquad \Delta G^\circ = -\tfrac{1}{2}(+208.4 \ \text{kJ}) = -104.2 \ \text{kJ}$

$N_2(g) + 2 O_2(g) \rightarrow 2 NO_2(g)$ $\qquad\qquad\quad \Delta G^\circ = +102.6 \ \text{kJ}$

Net: $N_2O(g) + \tfrac{3}{2} O_2(g) \rightarrow 2 NO_2(g)$ $\qquad\quad \Delta G^\circ = -104.2 + 102.6 = -1.6 \ \text{kJ}$

This reaction reaches an equilibrium condition, with significant amounts of all species being present. This conclusion is based on the relatively small absolute value of ΔG°.

(b) $2 N_2(g) + 6 H_2(g) \rightarrow 4 NH_3(g)$ $\qquad\qquad \Delta G^\circ = 2(-33.0 \ \text{kJ}) = -66.0 \ \text{kJ}$

$4 NH_3(g) + 5 O_2(g) \rightarrow 4 NO(g) + 6 H_2O(l)$ $\quad \Delta G^\circ = -1010.5 \ \text{kJ}$

$4 NO(g) \rightarrow 2 N_2(g) + 2 O_2(g)$ $\qquad\qquad\quad \Delta G^\circ = -2(+173.1 \ \text{kJ}) = -346.2 \ \text{kJ}$

Net: $6 H_2(g) + 3 O_2(g) \rightarrow 6 H_2O(l)$ $\quad \Delta G^\circ = -66.0 \ \text{kJ} - 1010.5 \ \text{kJ} - 346.2 \ \text{kJ} = -1422.7 \ \text{kJ}$

This reaction is three times the desired reaction, which therefore has $\Delta G^\circ = -1422.7 \ \text{kJ} \div 3 = -474.3 \ \text{kJ}.$

The large negative ΔG° value indicates that this reaction will go to completion at $25°C$.

(c) $4 NH_3(g) + 5 O_2(g) \rightarrow 4 NO(g) + 6 H_2O(l)$ $\quad \Delta G^\circ = -1010.5 \ \text{kJ}$

$4 NO(g) \rightarrow 2 N_2(g) + 2 O_2(g)$ $\qquad\qquad\quad \Delta G^\circ = -2(+173.1 \ \text{kJ}) = -346.2 \ \text{kJ}$

$2 N_2(g) + O_2(g) \rightarrow 2 N_2O(g)$ $\qquad\qquad\quad \Delta G^\circ = +208.4 \ \text{kJ}$

$4 NH_3(g) + 4 O_2(g) \rightarrow 2 N_2O(g) + 6 H_2O(l)$ $\quad \Delta G^\circ = -1010.5 \ \text{kJ} - 346.2 \ \text{kJ} + 208.4 \ \text{kJ}$

$= -1148.3 \ \text{kJ}$

This reaction is twice the desired reaction, which, therefore, has $\Delta G^\circ = -574.2 \ \text{kJ}.$
The very large negative value of the ΔG° for this reaction indicates that it will go to completion.

28. We combine the reactions in the same way as for any Hess's law calculations.

(a)

$COS(g) + 2CO_2(g) \rightarrow SO_2(g) + 3CO(g)$ $\Delta G° = -(-246.4 \text{ kJ}) = +246.6 \text{ kJ}$

$2CO(g) + 2H_2O(g) \rightarrow 2CO_2(g) + 2H_2(g)$ $\Delta G° = 2(-28.6 \text{ kJ}) = -57.2 \text{ kJ}$

$COS(g) + 2H_2O(g) \rightarrow SO_2(g) + CO(g) + 2H_2(g)$ $\Delta G° = +246.6 - 57.2 = +189.4 \text{ kJ}$

This reaction is spontaneous in the reverse direction, because of the large positive value of $\Delta G°$

(b)

$COS(g) + 2CO_2(g) \rightarrow SO_2(g) + 3CO(g)$ $\Delta G° = -(-246.4 \text{ kJ}) = +246.6 \text{ kJ}$

$3CO(g) + 3H_2O(g) \rightarrow 3CO_2(g) + 3H_2(g)$ $\Delta G° = 3(-28.6 \text{ kJ}) = -85.8 \text{ kJ}$

$COS(g) + 3H_2O(g) \rightarrow CO_2(g) + SO_2(g) + 3H_2(g)$ $\Delta G° = +246.6 - 85.8 = +160.8 \text{ kJ}$

This reaction is spontaneous in the reverse direction, because of the large positive value of $\Delta G°$.

(c)

$COS(g) + H_2(g) \rightarrow CO(g) + H_2S(g)$ $\Delta G° = -(+1.4) \text{ kJ}$

$CO(g) + H_2O(g) \rightarrow CO_2(g) + H_2(g)$ $\Delta G° = -28.6 \text{ kJ} = -28.6 \text{ kJ}$

$COS(g) + H_2O(g) \rightarrow CO_2(g) + H_2S(g)$ $\Delta G° = -1.4 \text{ kJ} - 28.6 \text{ kJ} = -30.0 \text{ kJ}$

The negative value of the $\Delta G°$ for this reaction indicates that it is spontaneous in the forward direction.

29. The combustion reaction is : $C_6H_6(l) + \frac{15}{2}O_2(g) \rightarrow 6CO_2(g) + 3H_2O(g \text{ or } l)$

(a)

$\Delta G° = 6\Delta G_f°\left[CO_2(g)\right] + 3\Delta G_f°\left[H_2O(l)\right] - \Delta G_f°\left[C_6H_6(l)\right] - \frac{15}{2}\Delta G_f°\left[O_2(g)\right]$

$= 6(-394.4 \text{ kJ}) + 3(-237.1 \text{ kJ}) - (+124.5 \text{ kJ}) - \frac{15}{2}(0.00 \text{ kJ}) = -3202 \text{ kJ}$

(b) $\Delta G° = 6\Delta G_f°\left[CO_2(g)\right] + 3\Delta G_f°\left[H_2O(g)\right] - \Delta G_f°\left[C_6H_6(l)\right] - \frac{15}{2}\Delta G_f°\left[O_2(g)\right]$

$= 6(-394.4 \text{ kJ}) + 3(-228.6 \text{ kJ}) - (+124.5 \text{ kJ}) - \frac{15}{2}(0.00 \text{ kJ}) = -3177 \text{ kJ}$

We could determine the difference between the two values of $\Delta G°$ by noting the difference between the two products: $3H_2O(l) \rightarrow 3H_2O(g)$ and determining the value of $\Delta G°$ for this difference:

$\Delta G° = 3\Delta G_f°\left[H_2O(g)\right] - 3\Delta G_f°\left[H_2O(l)\right] = 3\left[-228.6 - (-237.1)\right] \text{ kJ} = 25.5 \text{ kJ}$

30. We wish to find the value of the $\Delta H°$ for the given reaction: $F_2(g) \rightarrow 2F(g)$

$\Delta S° = 2S°\left[F(g)\right] - S°\left[F_2(g)\right] = 2(158.8 \text{ J K}^{-1}) - (202.8 \text{ J K}^{-1}) = +114.8 \text{ J K}^{-1}$

$\Delta H° = \Delta G° + T\Delta S° = 123.9 \times 10^3 \text{ J} + (298 \text{ K} \times 114.8 \text{ J/K}) = 158.1 \text{ kJ/mole of bonds}$

The value in Table 10.3 is 159 kJ/mol, which is in quite good agreement with the value found here.

31. (a) $\Delta S^{\circ}_{rxn} = \Sigma S^{\circ}_{products} - \Sigma S^{\circ}_{reactants}$

$= [1 \text{ mol} \times 301.2 \text{ J K}^{-1}\text{mol}^{-1} + 2 \text{ mol} \times 188.8 \text{ J K}^{-1}\text{mol}^{-1}] - [2 \text{ mol} \times 247.4 \text{ J K}^{-1}\text{mol}^{-1}$

$+ 1 \text{ mol} \times 238.5 \text{ J K}^{-1}\text{mol}^{-1}] = -54.5 \text{ J K}^{-1}$

$\Delta S^{\circ}_{rxn} = = -0.0545 \text{ kJ K}^{-1}$

(b) $\Delta H^{\circ}_{rxn} = \Sigma(\text{bonds broken in reactants (kJ/mol)}) - \Sigma(\text{bonds broken in products(kJ/ mol)})$

$= [4 \text{ mol} \times (389 \text{ kJ mol}^{-1})_{N-H} + 4 \text{ mol} \times (222 \text{ kJ mol}^{-1})_{O-F}] -$

$[4 \text{ mol} \times (301 \text{ kJ mol}^{-1})_{N-F} + 4 \text{ mol} \times (464 \text{ kJ mol}^{-1})_{O-H}]$

$\Delta H^{\circ}_{rxn} = -616 \text{ kJ}$

(c) $\Delta G^{\circ}_{rxn} = \Delta H^{\circ}_{rxn} - T\Delta S^{\circ}_{rxn} = -616 \text{ kJ} - 298 \text{ K}(-0.0545 \text{ kJ K}^{-1}) = -600 \text{ kJ}$

Since the ΔG°_{rxn} is <u>negative</u>, the reaction is <u>spontaneous</u>, and hence feasible
(at 25 °C). Because both the entropy and enthalpy changes are *negative*, this
reaction will be more highly favored at low temperatures (i.e., the reaction is
enthalpy driven)

32. The balanced equation for the thermal decomposition of ammonium nitrate is show below:

$NH_4NO_3(s) \xrightarrow{\Delta} N_2O(g) + 2 H_2O(l)$

To find ΔG°_{rxn} we will utilize the ΔG°_f values provided in Appendix D

$\Delta G^{\circ}_{298} = [2 \text{ mol}(-237.1 \text{ kJ mol}^{-1}) + 1 \text{ mol}(104.2 \text{ kJ mol}^{-1})] - [1 \text{ mol}(-183.9 \text{ kJ mol}^{-1})]$

$\Delta G^{\circ}_{298} = -186.1 \text{ kJ}$

The decomposition reaction is <u>spontaneous</u> at 25°C with all substances in their standard states.

$\Delta S^{\circ}_{rxn} = \Sigma S^{\circ}_{products} - \Sigma S^{\circ}_{reactants}$

$= [2 \text{ mol} \times 69.91 \text{ J K}^{-1}\text{mol}^{-1} + 1 \text{ mol} \times 219.9 \text{ J K}^{-1}\text{mol}^{-1}] - [1 \text{ mol} \times 151.1 \text{ J K}^{-1}\text{mol}^{-1}]$

$\Delta S^{\circ}_{rxn} = 208.6 \text{ J K}^{-1} = -0.2086 \text{ kJ K}^{-1}$

$\Delta H^{\circ}_{rxn} = \Sigma \Delta H_f^{\circ}{}_{products} - \Sigma \Delta H_f^{\circ}{}_{reactants}$

$= [2 \text{ mol}(-285.8 \text{ kJ mol}^{-1}) + 1 \text{ mol}(82.05 \text{ kJ mol}^{-1})] - [1 \text{ mol} (-365.6 \text{ kJ mol}^{-1})]$

$\Delta H^{\circ}_{rxn} = -124.0 \text{ kJ}$

Since ΔH°_{rxn} is <u>negative</u> and ΔS°_{rxn} is <u>positive</u>, the decomposition of ammonium nitrate is
spontaneous at all temperatures. However, as the temperature increases, the $T\Delta S$ term gets
larger and as a result, the decomposition reaction shift towards producing more products.
Consequently, we can say that the reaction is more highly favored <u>above</u> 298 K (it will
also be faster at higher temperatures)

The Thermodynamic Equilibrium Constant

33. In all three cases, $K_{eq} = K_p$ because only gases, pure solids, and pure liquids are present in
the chemical equations. There are no factors for solids and liquids in K_{eq} expressions, and
gases appear as partial pressures in atmospheres. That makes K_{eq} the same as K_p for these
three reactions. We now recall that $K_p = K_c(RT)^{\Delta n}$. Hence, in these three cases we have:

(a) $2SO_2(g) + O_2(g) \rightleftharpoons 2SO_3(g)$; $\Delta n_{gas} = 2 - (2+1) = -1$; $K = K_p = K_c(RT)^{-1}$

(b) $HI(g) \rightleftharpoons \frac{1}{2}H_2(g) + \frac{1}{2}I_2(g)$; $\Delta n_{gas} = 1 - (\frac{1}{2} + \frac{1}{2}) = 0$; $K = K_p = K_c$

(c) $NH_4HCO_3(s) \rightleftharpoons NH_3(g) + CO_2(g) + H_2O(l)$;

$\Delta n_{gas} = 2 - (0) = +2 \qquad K = K_p = K_c (RT)^2$

34. (a) $K = \dfrac{P\{H_2(g)\}^4}{P\{H_2O(g)\}^4}$

(b) Terms for both solids, Fe(s) and $Fe_3O_4(s)$, are properly excluded from the thermodynamic equilibrium constant expression. (Actually, each solid has an activity of 1.00.) Thus, the equilibrium partial pressures of both $H_2(g)$ and $H_2O(g)$ do not depend on the amounts of the two solids present, as long as some of each solid is present. One way to understand this is that any chemical reaction occurs on the surface of the solids, and thus is unaffected by the amount present.

(c) We can produce $H_2(g)$ from $H_2O(g)$ without regard to the proportions of Fe(s) and $Fe_3O_4(s)$ with the qualification, that there must always be some Fe(s) present for the production of $H_2(g)$ to continue.

Relationships Involving ΔG, ΔG°, Q and K

35. $\Delta G^\circ = 2\Delta G_f^\circ[NO(g)] - \Delta G_f^\circ[N_2O(g)] - 0.5\,\Delta G_f^\circ[O_2(g)]$

$\quad = 2(86.55 \text{ kJ/mol}) - (104.2 \text{ kJ/mol}) - 0.5(0.00 \text{ kJ/mol}) = 68.9 \text{ kJ/mol}$

$\quad = -RT \ln K_p = -(8.3145 \times 10^{-3} \text{ kJ mol}^{-1} \text{ K}^{-1})(298 \text{ K})\ln K_p$

$\ln K_p = -\dfrac{68.9 \text{ kJ/mol}}{8.3145 \times 10^{-3} \text{ kJ mol}^{-1} \text{ K}^{-1} \times 298 \text{ K}} = -27.8 \qquad K_p = e^{-27.8} = 8 \times 10^{-13}$

36. (a) $\Delta G^\circ = 2\Delta G_f^\circ[N_2O_5(g)] - 2\Delta G_f^\circ[N_2O_4(g)] - \Delta G_f^\circ[O_2(g)]$

$\quad = 2(115.1 \text{ kJ/mol}) - 2(97.89 \text{ kJ/mol}) - (0.00 \text{ kJ/mol}) = 34.4 \text{ kJ/mol}$

(b) $\Delta G^\circ = -RT \ln K_p \qquad \ln K_p = -\dfrac{\Delta G^\circ}{RT} = -\dfrac{34.4 \times 10^3 \text{ J/mol}}{8.3145 \text{ J mol}^{-1} \text{ K}^{-1} \times 298 \text{ K}} = -13.9$

$K_p = e^{-13.9} = 9 \times 10^{-7}$

37. We first balance each chemical equation, then calculate the value of ΔG° with data from Appendix D, and finally calculate the value of K_{eq} with the use of $\Delta G^\circ = -RT \ln K$.

(a) $4HCl(g) + O_2(g) \rightleftharpoons 2H_2O(g) + 2Cl_2(s)$

$\Delta G^\circ = 2\Delta G_f^\circ[H_2O(g)] + 2\Delta G_f^\circ[Cl_2(g)] - 4\Delta G_f^\circ[HCl(g)] - \Delta G_f^\circ[O_2(g)]$

$\quad = 2 \times \left(-228.6 \dfrac{\text{kJ}}{\text{mol}}\right) + 2 \times 0\dfrac{\text{kJ}}{\text{mol}} - 4 \times \left(-95.30\dfrac{\text{kJ}}{\text{mol}}\right) - 0 \dfrac{\text{kJ}}{\text{mol}} = -76.0 \dfrac{\text{kJ}}{\text{mol}}$

$\ln K = \dfrac{-\Delta G^\circ}{RT} = \dfrac{+76.0 \times 10^3 \text{ J/mol}}{8.3145 \text{ J mol}^{-1} \text{ K}^{-1} \times 298 \text{ K}} = +30.7 \qquad K = e^{+30.7} = 2 \times 10^{13}$

(b) $3Fe_2O_3(s) + H_2(g) \rightleftharpoons 2Fe_3O_4(s) + H_2O(g)$

$$\Delta G° = 2\Delta G_f°[Fe_3O_4(s)] + \Delta G_f°[H_2O(g)] - 3\Delta G_f°[Fe_2O_3(s)] - \Delta G_f°[H_2(g)]$$

$$= 2 \times (-1015 \text{ kJ/mol}) - 228.6 \text{ kJ/mol} - 3 \times (-742.2 \text{ kJ/mol}) - 0.00 \text{ kJ/mol}$$

$$= -32 \text{ kJ/mol}$$

$$\ln K = \frac{-\Delta G°}{RT} = \frac{32 \times 10^3 \text{ J/mol}}{8.3145 \text{ J mol}^{-1} \text{ K}^{-1} \times 298 \text{ K}} = 13; \quad K = e^{+13} = 4 \times 10^5$$

(c) $2Ag^+(aq) + SO_4^{2-}(aq) \rightleftharpoons Ag_2SO_4(s)$

$$\Delta G° = \Delta G_f°[Ag_2SO_4(s)] - 2\Delta G_f°[Ag^+(aq)] - \Delta G_f°[SO_4^{2-}(aq)]$$

$$= -618.4 \text{ kJ / mol} - 2 \times 77.11 \text{ kJ / mol} - (-744.5 \text{ kJ / mol}) = -28.1 \text{ kJ / mol}$$

$$\ln K = \frac{-\Delta G°}{RT} = \frac{28.1 \times 10^3 \text{ J/mol}}{8.3145 \text{ J mol}^{-1} \text{ K}^{-1} \times 298 \text{ K}} = 11.3; \quad K = e^{+11.3} = 8 \times 10^4$$

38.
$$\Delta S° = S°\{CO_2(g)\} + S°\{H_2(g)\} - S°\{CO(g)\} - S°\{H_2O(g)\}$$

$$= 213.7 \text{ J mol}^{-1} \text{ K}^{-1} + 130.7 \text{ J mol}^{-1} \text{ K}^{-1} - 197.7 \text{ J mol}^{-1} \text{ K}^{-1} - 188.8 \text{ J mol}^{-1}\text{K}^{-1}$$

$$= -42.1 \text{ J mol}^{-1} \text{ K}^{-1}$$

39. First we need to determine the standard free energy change for the reaction:
$$2SO_2(g) + O_2(g) \rightleftharpoons 2SO_3(g)$$
$$\Delta G° = 2\Delta G_f°[SO_3(g)] - 2\Delta G_f°[SO_2(g)] - \Delta G_f°[O_2(g)]$$
$$= 2 \times (-371.1 \text{ kJ/mol}) - 2 \times (-300.2 \text{ kJ/mol}) - 0.0 \text{ kJ/mol} = -141.8 \text{ kJ}$$

Now we can calculate ΔG by employing the equation $\Delta G = \Delta G° + RT \ln Q_p$, where

$$Q_p = \frac{P\{SO_3(g)\}^2}{P\{O_2(g)\}P\{SO_2(g)\}^2} ; \quad Q_p = \frac{(0.10 \text{ atm})^2}{(0.20 \text{ atm})(1.0 \times 10^{-4} \text{ atm})^2} = 5.0 \times 10^6$$

$$\Delta G = -141.8 \text{ kJ} + (8.3145 \times 10^{-3} \text{ kJ/K·mol})(298 \text{ K})\ln(5.0 \times 10^6)$$
$$\Delta G = -141.8 \text{ kJ} + 38.2 \text{ kJ} = -104 \text{ kJ}.$$
Since ΔG is negative, the reaction is spontaneous in the forward direction.

40. We begin by calculating the standard free energy change for the reaction:
$$H_2(g) + Cl_2(g) \rightleftharpoons 2HCl(g)$$
$$\Delta G° = 2\Delta G_f°[HCl(g)] - \Delta G_f°[Cl_2(g)] - \Delta G_f°[H_2(g)]$$
$$= 2 \times (-95.30 \text{ kJ/mol}) - 0.0 \text{ kJ/mol} - 0.0 \text{ kJ/mol} = -190.6 \text{ kJ}$$

Now we can calculate ΔG by employing the equation $\Delta G = \Delta G° + RT \ln Q_p$, where

$$Q_p = \frac{P\{HCl(g)\}^2}{P\{H_2(g)\}P\{Cl_2(g)\}} \; ; \; Q_p = \frac{(0.5 \text{ atm})^2}{(0.5 \text{ atm})(0.5 \text{ atm})} = 1$$

$$\Delta G = -190.6 \text{ kJ} + (8.3145 \times 10^{-3} \text{ kJ/K·mol})(298 \text{ K})\ln(1)$$
$$\Delta G = -190.6 \text{ kJ} + 0 \text{ kJ} = -190.6 \text{ kJ}.$$

Since ΔG is negative, the reaction is spontaneous in the forward direction.

41. In order to determine the direction in which the reaction is spontaneous, we need to calculate the non-standard free energy change for the reaction. To accomplish this, we will employ the equation $\Delta G = \Delta G^\circ + RT \ln Q_c$, where

$$Q_c = \frac{[H_3O^+(aq)]\,[CH_3CO_2^-(aq)]}{[CH_3CO_2H(aq)]} \; ; \; Q_c = \frac{(1.0 \times 10^{-3} \text{ M})^2}{(0.10 \text{ M})} = 1.0 \times 10^{-5}$$

$$\Delta G = 27.07 \text{ kJ} + (8.3145 \times 10^{-3} \text{ kJ/K·mol})(298 \text{ K})\ln(1.0 \times 10^{-5})$$
$$\Delta G = 27.07 \text{ kJ} + (-28.53 \text{ kJ}) = -1.46 \text{ kJ}.$$

Since ΔG is negative, the reaction is spontaneous in the forward direction.

42. As was the case for question 39, we need to calculate the non-standard free energy change for the reaction. Once again, we will employ the equation $\Delta G = \Delta G^\circ + RT \ln Q$, but this time

$$Q_c = \frac{[NH_4^+(aq)]\,[OH^-(aq)]}{[NH_3(aq)]} \; ; \; Q_c = \frac{(1.0 \times 10^{-3} \text{ M})^2}{(0.10 \text{ M})} = 1.0 \times 10^{-5}$$

$$\Delta G = 29.05 \text{ kJ} + (8.3145 \times 10^{-3} \text{ kJ/K·mol})(298 \text{ K})\ln(1.0 \times 10^{-5})$$
$$\Delta G = 29.05 \text{ kJ} + (-28.53 \text{ kJ}) = 0.52 \text{ kJ}.$$
Since ΔG is positive, the reaction is spontaneous in the reverse direction.

43. The relationship $\Delta S^\circ = (\Delta G^\circ - \Delta H^\circ)/T$ (Equation (b)) is incorrect. Rearranging this equation to put ΔG° on one side by itself gives $\Delta G^\circ = \Delta H^\circ + T\Delta S^\circ$. This equation is not valid. The ΔH° term should be subtracted from the $T\Delta S^\circ$ term, not added to it.

44. The ΔG° value is a powerful thermodynamic parameter because it can be used to determine the equilibrium constant for the reaction at each and every chemically reasonable temperature via the equation $\Delta G^\circ = -RT \ln K$.

45. **(a)** To determine K_p we need the equilibrium partial pressures. In the ideal gas law, each partial pressure is defined by $P = nRT/V$. Because R, T, and V are the same for each gas, and because there are the same number of partial pressure factors in the numerator as in the denominator of the K_p expression, we can use the ratio of amounts to determine K_p.

$$K_p = \frac{P\{CO(g)\}P\{H_2O(g)\}}{P\{CO_2(g)\}P\{H_2(g)\}} = \frac{n\{CO(g)\}n\{H_2O(g)\}}{n\{CO_2(g)\}n\{H_2(g)\}} = \frac{0.224 \text{ mol CO} \times 0.224 \text{ mol H}_2O}{0.276 \text{ mol CO}_2 \times 0.276 \text{ mol H}_2} = 0.659$$

(b) $\Delta G^{\circ}_{1000\text{K}} = -RT \ln K_p = -8.3145 \text{ J mol}^{-1}\text{K}^{-1} \times 1000. \text{ K} \times \ln(0.659)$

$$= 3.467 \times 10^3 \text{ J/mol} = 3.467 \text{ kJ/mol}$$

$$Q_p = \frac{0.0340 \text{ mol CO} \times 0.0650 \text{ mol H}_2\text{O}}{0.0750 \text{ mol CO}_2 \times 0.095 \text{ mol H}_2} = 0.31 < 0.659 = K_p$$

Since Q_p is smaller than K_p, the reaction will proceed to the right, forming products, to attain equilibrium, i.e., $\Delta G = 0$.

46. (a) We know that $K_p = K_c(RT)^{\Delta n}$. For the reaction $2\text{SO}_2(g) + \text{O}_2(g) \rightleftharpoons 2\text{SO}_3(g)$,

$\Delta n_{\text{gas}} = 2 - (2+1) = -1$, and therefore a value of K_p can be obtained.

$$K_p = K_c(RT)^{-1} = \frac{2.8 \times 10^2}{\dfrac{0.08206 \text{ L atm}}{\text{mol K}} \times 1000 \text{ K}} = 3.4 = K$$

We recognize that $K = K_p$ since all of the substances involved in the reaction are gases. We can now evaluate ΔG°.

$$\Delta G^{\circ} = -RT \ln K_{\text{eq}} = -\frac{8.3145 \text{ J}}{\text{mol K}} \times 1000 \text{ K} \times \ln(3.41) = -1.02 \times 10^4 \text{ J/mol} = -10.2 \text{ kJ/mol}$$

(b) We can evaluate Q_c for this situation and compare the value with that of $K_c = 2.8 \times 10^2$ to determine the direction of the reaction to reach equilibrium.

$$Q_c = \frac{[\text{SO}_3]^2}{[\text{SO}_2]^2[\text{O}_2]} = \frac{\left(\dfrac{0.72 \text{ mol SO}_3}{2.50 \text{ L}}\right)^2}{\left(\dfrac{0.40 \text{ mol SO}_2}{2.50 \text{ L}}\right)^2 \times \left(\dfrac{0.18 \text{ mol O}_2}{2.50 \text{ L}}\right)} = 45 < 2.8 \times 10^2 = K_c$$

Since Q_c is smaller than K_c the reaction will shift right, producing sulfur trioxide and consuming sulfur dioxide and molecular oxygen, until the two values are equal.

47. (a) $K = K_c$

$\Delta G^{\circ} = -RT \ln K_{\text{eq}} = -(8.3145 \times 10^{-3} \text{ kJ mol}^{-1} \text{ K}^{-1})(445 + 273)\text{K} \ln 50.2 = -23.4 \text{ kJ}$

(b) $K = K_p = K_c(RT)^{\Delta n_g} = 1.7 \times 10^{-13}(0.0821 \times 298)^{1/2} = 8.4 \times 10^{-13}$

$\Delta G^{\circ} = -RT \ln K_p = -(8.3145 \times 10^{-3} \text{ kJ mol}^{-1} \text{ K}^{-1})(298 \text{ K}) \ln(8.4 \times 10^{-13})$

$\Delta G^{\circ} = +68.9 \text{ kJ/mol}$

(c) $K = K_p = K_c(RT)^{\Delta n} = 4.61 \times 10^{-3}(0.08206 \times 298)^{+1} = 0.113$

$\Delta G^{\circ} = -RT \ln K_p = -(8.3145 \times 10^{-3} \text{ kJ mol}^{-1} \text{ K}^{-1})(298\text{K}) \ln(0.113) = +5.40 \text{ kJ / mol}$

(d) $K = K_c = 9.14 \times 10^{-6}$

$\Delta G^{\circ} = -RT \ln K_c = -(8.3145 \times 10^{-3} \text{ kJ mol}^{-1} \text{ K}^{-1})(298 \text{ K}) \ln(9.14 \times 10^{-6})$

$\Delta G^{\circ} = +28.7 \text{ kJ/mol}$

48. (a) The first equation involves the formation of one mole of $Mg^{2+}(aq)$ from $Mg(OH)_2(s)$ and $2H^+(aq)$, while the second equation involves the formation of only half-a-mole of $Mg^{2+}(aq)$. We would expect a free energy change of half the size if only half as much product is formed.

(b) The value of K for the first reaction is the square of the value of K for the second reaction. The equilibrium constant expressions are related in the same fashion.

$$K_1 = \frac{\left[Mg^{2+}\right]}{\left[H^+\right]^2} = \left(\frac{\left[Mg^{2+}\right]^{1/2}}{\left[H^+\right]}\right)^2 = \left(K_2\right)^2$$

(c) The equilibrium solubilities will be the same regardless which expression is used. The equilibrium conditions (solubilities in this instance) are the same no matter how we choose to express them in an equilibrium constant expression.

49. $\Delta G° = -RT \ln K_p = -\left(8.3145\times10^{-3} \text{ kJ mol}^{-1}\text{K}^{-1}\right)(298\,\text{K})\ln\left(6.5\times10^{11}\right) = -67.4 \text{ kJ/mol}$

$CO(g)+Cl_2(g)\rightarrow COCl_2(g)$	$\Delta G° = -67.4\,\text{kJ/mol}$
$C(\text{graphite})+\tfrac{1}{2}O_2(g)\rightarrow CO(g)$	$\Delta G_f° = -137.2 \text{ kJ/mol}$

$C(\text{graphite})+\tfrac{1}{2}O_2(g)+Cl_2(g)\rightarrow COCl_2(g) \qquad \Delta G_f° = -204.6 \text{ kJ/mol}$

$\Delta G_f°$ of $COCl_2(g)$ given in Appendix D is -204.6 kJ/mol, thus the agreement is excellent.

50. In each case, we first determine the value of $\Delta G°$ for the solubility reaction. From that, we calculate the value of the equilibrium constant, K_{sp}, for the solubility reaction.

(a) $AgBr(s)\rightleftharpoons Ag^+(aq)+Br^-(aq)$

$\Delta G° = \Delta G_f°\left[Ag^+(aq)\right]+\Delta G_f°\left[Br^-(aq)\right]-\Delta G_f°\left[AgBr(s)\right]$

$= 77.11 \text{ kJ/mol}-104.0 \text{ kJ/mol}-\left(-96.90 \text{ kJ/mol}\right)= +70.0 \text{ kJ/mol}$

$\ln K = \dfrac{-\Delta G°}{RT} = \dfrac{-70.0\times10^3 \text{ J/mol}}{8.3145 \text{ J mol}^{-1}\text{K}^{-1}\times 298.15 \text{ K}} = -28.2; \qquad K_{sp} = e^{-28.2} = 6\times10^{-13}$

(b) $CaSO_4(s)\rightleftharpoons Ca^{2+}(aq)+SO_4^{2-}(aq)$

$\Delta G° = \Delta G_f°\left[Ca^{2+}(aq)\right]+\Delta G_f°\left[SO_4^{2-}(aq)\right]-\Delta G_f°\left[CaSO_4(s)\right]$

$= -553.6 \text{ kJ/mol}-744.5 \text{ kJ/mol}-\left(-1332 \text{ kJ/mol}\right)= +34 \text{ kJ/mol}$

$\ln K = \dfrac{-\Delta G°}{RT} = \dfrac{-34\times10^3 \text{ J/mol}}{8.3145 \text{ J mol}^{-1}\text{K}^{-1}\times 298.15 \text{ K}} = -14; \qquad K_{sp} = e^{-14} = 8\times10^{-7}$

(c) $Fe(OH)_3 (s) \rightleftharpoons Fe^{3+} (aq) + 3OH^- (aq)$

$\Delta G^\circ = \Delta G_f^\circ \left[Fe^{3+}(aq) \right] + 3\,\Delta G_f^\circ \left[OH^-(aq) \right] - \Delta G_f^\circ \left[Fe(OH)_3(s) \right]$

$= -4.7 \text{ kJ / mol} + 3 \times (-157.2 \text{ kJ / mol}) - (-696.5 \text{ kJ / mol}) = +220.2 \text{ kJ / mol}$

$\ln K = \dfrac{-\Delta G^\circ}{RT} = \dfrac{-220.2 \times 10^3 \text{ J/mol}}{8.3145 \text{ J mol}^{-1}\text{K}^{-1} \times 298.15 \text{ K}} = -88.83 \qquad K_{sp} = e^{-88.83} = 2.6 \times 10^{-39}$

51. **(a)** We can determine the equilibrium partial pressure from the value of the equilibrium constant.

$\Delta G^\circ = -RT \ln K_p \qquad \ln K_p = -\dfrac{\Delta G^\circ}{RT} = -\dfrac{58.54 \times 10^3 \text{ J/mol}}{8.3145 \text{ J mol}^{-1}\text{K}^{-1} \times 298.15 \text{ K}} = -23.63$

$K_p = P\{O_2 (g)\}^{1/2} = e^{-23.63} = 5.5 \times 10^{-11} \qquad P\{O_2 (g)\} = (5.5 \times 10^{-11})^2 = 3.0 \times 10^{-21} \text{ atm}$

(b) Lavoisier did two things to increase the quantity of oxygen that he obtained. First, he ran the reaction at a high temperature, which favors the products (i.e., the side with molecular oxygen.) Second, the molecular oxygen was removed immediately after it was formed, which causes the equilibrium to shift to the right continuously (the shift towards products as result of the removal of the O_2 is an example of Le Châtelier's principle).

52. **(a)** We determine the values of ΔH° and ΔS° from the data in Appendix D, and then the value of ΔG° at $25^\circ \text{ C} = 298 \text{ K}$.

$\Delta H^\circ = \Delta H_f^\circ \left[CH_3OH(g) \right] + \Delta H_f^\circ \left[H_2O(g) \right] - \Delta H_f^\circ \left[CO_2 (g) \right] - 3\,\Delta H_f^\circ \left[H_2 (g) \right]$

$= -200.7 \text{ kJ/mol} + (-241.8 \text{ kJ/mol}) - (-393.5 \text{ kJ/mol}) - 3\,(0.00 \text{ kJ/mol}) = -49.0 \text{ kJ/mol}$

$\Delta S^\circ = S^\circ \left[CH_3OH(g) \right] + S^\circ \left[H_2O(g) \right] - S^\circ \left[CO_2 (g) \right] - 3\,S^\circ \left[H_2 (g) \right]$

$= (239.8 + 188.8 - 213.7 - 3 \times 130.7) \text{ J mol}^{-1}\text{K}^{-1} = -177.2 \text{ J mol}^{-1}\text{K}^{-1}$

$\Delta G^\circ = \Delta H^\circ - T\Delta S^\circ = -49.0 \text{ kJ/mol} - 298 \text{ K} \left(-0.1772 \text{ kJ mol}^{-1}\text{K}^{-1} \right) = +3.81 \text{ kJ/mol}$

Because the value of ΔG° is positive, this reaction does not proceed in the forward direction at 25° C.

(b) Because the value of ΔH° is negative and that of ΔS° is negative, the reaction is *non-spontaneous* at high temperatures, if reactants and products are *in their standard states*. The reaction will proceed slightly in the forward direction, however, to produce an equilibrium mixture with small quantities of $CH_3OH(g)$ and $H_2O(g)$. Also, because the forward reaction is exothermic, this reaction is favored by lowering the temperature. That is, the value of K increases with decreasing temperature.

(c) $\Delta G^\circ_{500K} = \Delta H^\circ - T\,\Delta S^\circ = -49.0 \text{ kJ/mol} - 500.\text{K} \left(-0.1772 \text{ kJ mol}^{-1}\text{K}^{-1} \right) = 39.6 \text{ kJ/mol}$

$= 39.6 \times 10^3 \text{ J/mol} = -RT \ln K_p$

$\ln K_p = \dfrac{-\Delta G^\circ}{RT} = \dfrac{-39.6 \times 10^3 \text{ J/mol}}{8.3145 \text{ J mol}^{-1}\text{ K}^{-1} \times 500.\text{ K}} = -9.53; \qquad K_p = e^{-9.53} = 7.3 \times 10^{-5}$

(d) Reaction: $CO_2(g) + \quad 3H_2(g) \quad \rightleftharpoons \quad CH_3OH(g) \quad +H_2O(g)$

Initial: 1.00 atm   1.00 atm   0 atm   0 atm

Changes: $-x$ atm $-3x$ atm $+x$ atm $+x$ atm

Equil: $(1.00-x)$ atm $(1.00-3x)$ atm x atm x atm

$$K_p = 7.3\times10^{-5} = \frac{P\{CH_3OH\}P\{H_2O\}}{P\{CO_2\}P\{H_2\}^3} = \frac{x\cdot x}{(1.00-x)(1.00-3x)^3} \approx x^2$$

$x = \sqrt{7.3\times10^{-5}} = 8.5\times10^{-3}$ atm $= P\{CH_3OH\}$ Our assumption, that $3x \ll 1.00$ atm, is valid.

$\Delta G°$ and K as a Function of Temperature

53 **(a)** $\Delta S° = S°\left[Na_2CO_3(s)\right] + S°\left[H_2O(l)\right] + S°\left[CO_2(g)\right] - 2S°\left[NaHCO_3(s)\right]$

$$= 135.0\,\frac{J}{K\,mol} + 69.91\frac{J}{K\,mol} + 213.7\,\frac{J}{K\,mol} - 2\left(101.7\,\frac{J}{K\,mol}\right) = +215.2\,\frac{J}{K\,mol}$$

(b) $\Delta H° = \Delta H_f°\left[Na_2CO_3(s)\right] + \Delta H_f°\left[H_2O(l)\right] + \Delta H_f°\left[CO_2(g)\right] - 2\Delta H_f°\left[NaHCO_3(s)\right]$

$$= -1131\frac{kJ}{mol} - 285.8\frac{kJ}{mol} - 393.5\frac{kJ}{mol} - 2\left(-950.8\frac{kJ}{mol}\right) = +91\frac{kJ}{mol}$$

(c) $\Delta G° = \Delta H° - T\Delta S° = 91$ kJ/mol $- (298\ K)(215.2\times10^{-3}$ kJ mol^{-1}K$^{-1})$

$= 91$ kJ/mol $- 64.13$ kJ/mol $= 27$ kJ/mol

(d) $\Delta G° = -RT \ln K$ $\ln K = -\dfrac{\Delta G°}{RT} = -\dfrac{27\times10^3\ J/mol}{8.3145\,J\ mol^{-1}K^{-1} \times 298\,K} = -10.\underline{9}$

$K = e^{-10.9} = 2\times10^{-5}$

54. **(a)** $\Delta S° = S°\left[CH_3CH_2OH(g)\right] + S°\left[H_2O(g)\right] - S°\left[CO(g)\right] - 2S°\left[H_2(g)\right] - S°\left[CH_3OH(g)\right]$

$$\Delta S° = 282.7\,\frac{J}{K\,mol} + 188.8\,\frac{J}{K\,mol} - 197.7\,\frac{J}{K\,mol} - 2\left(130.7\,\frac{J}{K\,mol}\right) - 239.8\,\frac{J}{K\,mol}$$

$$\Delta S° = -227.4\,\frac{J}{K\,mol}$$

$\Delta H° = \Delta H_f°\,[CH_3CH_2OH(g)] + \Delta H_f°\left[H_2O(g)\right] - \Delta H_f°\left[CO(g)\right] - 2\Delta H_f°\,[H_2(g)] - \Delta H_f°\,[CH_3OH(g)]$

$$\Delta H° = -235.1\frac{kJ}{mol} - 241.8\frac{kJ}{mol} - \left(-110.5\frac{kJ}{mol}\right) - 2\left(0.00\frac{kJ}{mol}\right) - \left(-200.7\frac{kJ}{mol}\right)$$

$$\Delta H° = -165.7\,\frac{kJ}{mol}$$

$$\Delta G° = -165.7\frac{kJ}{mol} - (298\,K)\left(-227.4\times10^{-3}\,\frac{kJ}{K\,mol}\right) = -165.4\frac{kJ}{mol} + 67.8\frac{kJ}{mol} = -97.9\,\frac{kJ}{mol}$$

(b) $\Delta H° < 0$ for this reaction. Thus it is favored at low temperatures. Also, because $\Delta n_{gas} = +2 - 4 = -2$, which is less than zero, the reaction is favored at high pressures.

(c) First we assume that neither $\Delta S°$ nor $\Delta H°$ varies significantly with temperature. Then we compute a value for $\Delta G°$ at 750 K. From this value of $\Delta G°$, we compute a value for K_p.

$$\Delta G° = \Delta H° - T\Delta S° = -165.7\,\text{kJ/mol} - (750.\text{K})\left(-227.4 \times 10^{-3}\,\text{kJ mol}^{-1}\,\text{K}^{-1}\right)$$

$$= -165.7\,\text{kJ/mol} + 170.6\,\text{kJ/mol} = +4._9\,\text{kJ/mol} = -RT \ln K_p$$

$$\ln K_p = -\frac{\Delta G°}{RT} = -\frac{4._9 \times 10^3\,\text{J/mol}}{8.3145\,\text{J mol}^{-1}\,\text{K}^{-1} \times 750.\text{K}} = -0.7_9 \qquad K_p = e^{-0.7_9} = 0.5$$

55. $\quad \Delta G° = \Delta H° - T\Delta S° \qquad T\Delta S° = \Delta H° - \Delta G° \qquad T = \dfrac{\Delta H° - \Delta G°}{\Delta S°}$

$$T = \frac{-24.8 \times 10^3\,\text{J} - \left(-45.5 \times 10^3\,\text{J}\right)}{15.2\,\text{J/K}} = 1.36 \times 10^3\,\text{K}$$

56. We use the van't Hoff equation with $\Delta H° = -1.8 \times 10^5\,\text{J/mol}$, $T_1 = 800.\,\text{K}$,

$T_2 = 100.°\,\text{C} = 373\,\text{K}$, and $K_1 = 9.1 \times 10^2$.

$$\ln\frac{K_2}{K_1} = \frac{\Delta H°}{R}\left(\frac{1}{T_1} - \frac{1}{T_2}\right) = \frac{-1.8 \times 10^5\,\text{J/mol}}{8.3145\,\text{J mol}^{-1}\,\text{K}^{-1}}\left(\frac{1}{800\,\text{K}} - \frac{1}{373\,\text{K}}\right) = 31$$

$$\frac{K_2}{K_1} = e^{31} = 2.\underline{9} \times 10^{13} = \frac{K_2}{9.1 \times 10^2} \qquad K_2 = \left(2.\underline{9} \times 10^{13}\right)\left(9.1 \times 10^2\right) = 3 \times 10^{16}$$

57. We first determine the value of $\Delta G°$ at $400°\,\text{C}$, from the values of $\Delta H°$ and $\Delta S°$, which are calculated from information listed in Appendix D.

$$\Delta H° = 2\Delta H_f°\left[NH_3(g)\right] - \Delta H_f°\left[N_2(g)\right] - 3\Delta H_f°\left[H_2(g)\right]$$

$$= 2(-46.11\,\text{kJ/mol}) - (0.00\,\text{kJ/mol}) - 3(0.00\,\text{kJ/mol}) = -92.22\,\text{kJ/mol}\ N_2$$

$$\Delta S° = 2S°\left[NH_3(g)\right] - S°\left[N_2(g)\right] - 3S°\left[H_2(g)\right]$$

$$= 2(192.5\,\text{J mol}^{-1}\text{K}^{-1}) - (191.6\,\text{J mol}^{-1}\text{K}^{-1}) - 3(130.7\,\text{J mol}^{-1}\text{K}^{-1}) = -198.7\,\text{J mol}^{-1}\,\text{K}^{-1}$$

$$\Delta G° = \Delta H° - T\Delta S° = -92.22\,\text{kJ/mol} - 673\,\text{K} \times \left(-0.1987\,\text{kJ mol}^{-1}\,\text{K}^{-1}\right)$$

$$= +41.51\,\text{kJ/mol} = -RT \ln K_p$$

$$\ln K_p = \frac{-\Delta G°}{RT} = \frac{-41.51 \times 10^3\,\text{J/mol}}{8.3145\,\text{J mol}^{-1}\text{K}^{-1} \times 673\,\text{K}} = -7.42; \qquad K_p = e^{-7.42} = 6.0 \times 10^{-4}$$

58. **(a)** $\Delta H° = \Delta H_f°\left[CO_2(g)\right] + \Delta H_f°\left[H_2(g)\right] - \Delta H_f°\left[CO(g)\right] - \Delta H_f°\left[H_2O(g)\right]$

$= -393.5 \text{ kJ/mol} - 0.00 \text{ kJ/mol} - (-110.5 \text{ kJ/mol}) - (-241.8 \text{ kJ/mol}) = -41.2 \text{ kJ/mol}$

$\Delta S° = S°\left[CO_2(g)\right] + S°\left[H_2(g)\right] - S°\left[CO(g)\right] - S°\left[H_2O(g)\right]$

$= 213.7 \text{ J mol}^{-1}\text{ K}^{-1} + 130.7 \text{ J mol}^{-1}\text{ K}^{-1} - 197.7 \text{ J mol}^{-1}\text{ K}^{-1} - 188.8 \text{ J mol}^{-1}\text{ K}^{-1}$

$= -42.1 \text{ J mol}^{-1}\text{ K}^{-1}$

$\Delta G° = \Delta H° - T\Delta S° = -41.2 \text{kJ/mol} - (298.15\text{K}) - 42.1 \times 10^{-3}\text{kJ mol}^{-1}\text{K}^{-1}$

$= -41.2 \text{ kJ/mol} + 12.6 \text{ kJ/mol} = -28.6 \text{ kJ/mol}$

(b) $\Delta G° = \Delta H° - T\,\Delta S° = -41.2 \text{ kJ/mol} - (875\text{ K})(-42.1 \times 10^{-3}\text{ kJ mol}^{-1}\text{ K}^{-1})$

$= -41.2 \text{ kJ/mol} + 36.8 \text{ kJ/mol} = -4.4 \text{ kJ/mol} = -RT \ln K_p$

$\ln K_p = -\dfrac{\Delta G°}{RT} = -\dfrac{-4.4 \times 10^3 \text{ J/mol}}{8.3145 \text{ J mol}^{-1}\text{K}^{-1} \times 875\text{K}} = +0.60 \qquad K_p = e^{+0.60} = 1.8$

59. We assume that both $\Delta H°$ and $\Delta S°$ are constant with temperature.

$\Delta H° = 2\Delta H_f°\left[SO_3(g)\right] - 2\Delta H_f°\left[SO_2(g)\right] - \Delta H_f°\left[O_2(g)\right]$

$= 2(-395.7 \text{ kJ/mol}) - 2(-296.8 \text{ kJ/mol}) - (0.00 \text{ kJ/mol}) = -197.8 \text{ kJ/mol}$

$\Delta S° = 2S°\left[SO_3(g)\right] - 2S°\left[SO_2(g)\right] - S°\left[O_2(g)\right]$

$= 2(256.8 \text{ J mol}^{-1}\text{ K}^{-1}) - 2(248.2 \text{ J mol}^{-1}\text{ K}^{-1}) - (205.1 \text{ J mol}^{-1}\text{ K}^{-1})$

$= -187.9 \text{ J mol}^{-1}\text{ K}^{-1}$

$\Delta G° = \Delta H° - T\Delta S° = -RT \ln K \qquad \Delta H° = T\Delta S° - RT \ln K \qquad T = \dfrac{\Delta H°}{\Delta S° - R \ln K}$

$T = \dfrac{-197.8 \times 10^3 \text{ J/mol}}{-187.9 \text{ J mol}^{-1}\text{ K}^{-1} - 8.3145 \text{ J mol}^{-1}\text{ K}^{-1} \ln(1.0 \times 10^6)} \approx 650 \text{ K}$

This value compares very favorably with the value of $T = 6.37 \times 10^2$ that was obtained in Example 19-10.

60. We use the van't Hoff equation to determine the value of $\Delta H°$ ($448° \text{ C} = 721 \text{ K}$ and $350° \text{ C} = 623 \text{ K}$).

$\ln\dfrac{K_1}{K_2} = \dfrac{\Delta H°}{R}\left(\dfrac{1}{T_2} - \dfrac{1}{T_1}\right) = \ln\dfrac{50.0}{66.9} = -0.291 = \dfrac{\Delta H°}{R}\left(\dfrac{1}{623} - \dfrac{1}{721}\right) = \dfrac{\Delta H°}{R}(2.2 \times 10^{-4})$

$\dfrac{\Delta H°}{R} = \dfrac{-0.291}{2.2 \times 10^{-4}\text{ K}^{-1}} = -1.3 \times 10^3 \text{ K};$

$\Delta H° = -1.3 \times 10^3 \text{K} \times 8.3145 \text{ J K}^{-1}\text{ mol}^{-1} = -11 \times 10^3 \text{ J mol}^{-1} = -11 \text{ kJ mol}^{-1}$

61. **(a)** $\ln\dfrac{K_2}{K_1}=\dfrac{\Delta H^\circ}{R}\left(\dfrac{1}{T_1}-\dfrac{1}{T_2}\right)=\dfrac{57.2\times10^3\ \text{J/mol}}{8.3145\ \text{J mol}^{-1}\ \text{K}^{-1}}\left(\dfrac{1}{298\ \text{K}}-\dfrac{1}{273\ \text{K}}\right)=-2.11$

$\dfrac{K_2}{K_1}=e^{-2.11}=0.121 \qquad K_2=0.121\times0.113=0.014 \text{ at } 273 \text{ K}$

(b) $\ln\dfrac{K_2}{K_1}=\dfrac{\Delta H^\circ}{R}\left(\dfrac{1}{T_1}-\dfrac{1}{T_2}\right)=\dfrac{57.2\times10^3\ \text{J/mol}}{8.3145\ \text{J mol}^{-1}\ \text{K}^{-1}}\left(\dfrac{1}{T_1}-\dfrac{1}{298\ \text{K}}\right)=\ln\dfrac{0.113}{1.00}=-2.180$

$\left(\dfrac{1}{T_1}-\dfrac{1}{298\text{K}}\right)=\dfrac{-2.180\times8.3145}{57.2\times10^3}\ \text{K}^{-1}=-3.17\times10^{-4}\ \text{K}^{-1}$

$\dfrac{1}{T_1}=\dfrac{1}{298}-3.17\times10^{-4}=3.36\times10^{-3}-3.17\times10^{-4}=3.04\times10^{-3}\ \text{K}^{-1};\quad T_1=329\ \text{K}$

62. First we calculate ΔG° at 298 K to obtain a value for K_{eq} at that temperature.

$\Delta G^\circ=2\Delta G_f^\circ\left[NO_2(g)\right]-2\Delta G_f^\circ\left[NO(g)\right]-\Delta G_f^\circ\left[O_2(g)\right]$

$=2(51.31\ \text{kJ/mol})-2(86.55\ \text{kJ/mol})-0.00\ \text{kJ/mol}=-70.48\ \text{kJ/mol}$

$\ln K=\dfrac{-\Delta G^\circ}{RT}=-\dfrac{-70.48\times10^3\ \text{J/mol K}}{\dfrac{8.3145\ \text{J}}{\text{mol K}}\times298.15\ \text{K}}=28.43 \qquad K=e^{28.43}=2.2\times10^{12}$

Now we calculate ΔH° for the reaction, which then will be inserted into the van't Hoff equation.

$\Delta H^\circ=2\Delta H_f^\circ\left[NO_2(g)\right]-2\Delta H_f^\circ\left[NO(g)\right]-\Delta H_f^\circ\left[O_2(g)\right]$

$=2(33.18\ \text{kJ / mol})-2(90.25\ \text{kJ / mol})-0.00\ \text{kJ / mol}=-114.14\ \text{kJ / mol}$

$\ln\dfrac{K_2}{K_1}=\dfrac{\Delta H^\circ}{R}\left(\dfrac{1}{T_1}-\dfrac{1}{T_2}\right)=\dfrac{-114.14\times10^3\ \text{J/mol}}{8.3145\ \text{J mol}^{-1}\ \text{K}^{-1}}\left(\dfrac{1}{298\ \text{K}}-\dfrac{1}{373\ \text{K}}\right)=-9.26$

$\dfrac{K_2}{K_1}=e^{-9.26}=9.5\times10^{-5};\quad K_2=9.5\times10^{-5}\times2.2\times10^{12}=2.1\times10^8$

Another way to find K at 100 °C, is to compute $\Delta H^\circ(-114.14\ \text{kJ / mol})$ from ΔH_f° values and $\Delta S^\circ\left(-146.5\ \text{J mol}^{-1}\ \text{K}^{-1}\right)$ from S° values. Then determine $\Delta G^\circ(-59.5\ \text{kJ/mol})$, and find K_p with the expression $\Delta G^\circ=-RT\ln K_p$. Not surprisingly, we obtain the same result, $K_p=2.2\times10^8$.

63. First, the van't Hoff equation is used to obtain a value of ΔH°. $200°C=473\text{K}$ and $260°C=533\text{K}$.

$\ln\dfrac{K_2}{K_1}=\dfrac{\Delta H^\circ}{R}\left(\dfrac{1}{T_1}-\dfrac{1}{T_2}\right)=\ln\dfrac{2.15\times10^{11}}{4.56\times10^8}=6.156=\dfrac{\Delta H^\circ}{8.3145\ \text{J mol}^{-1}\ \text{K}^{-1}}\left(\dfrac{1}{533\text{K}}-\dfrac{1}{473\ \text{K}}\right)$

$6.156=-2.9\times10^{-5}\Delta H^\circ \qquad \Delta H^\circ=\dfrac{6.156}{-2.9\times10^{-5}}=-2.1\times10^5\ \text{J / mol}=-2.1\times10^2\ \text{kJ / mol}$

Another route to ΔH° is the combination of standard enthalpies of formation.

$$CO(g) + 3H_2(g) \rightleftharpoons CH_4(g) + H_2O(g)$$
$$\Delta H^\circ = \Delta H_f^\circ [CH_4(g)] + \Delta H_f^\circ [H_2O(g)] - \Delta H_f^\circ [CO(g)] - 3\Delta H_f^\circ [H_2(g)]$$
$$= -74.81 \text{ kJ/mol} - 241.8 \text{ kJ/mol} - (-110.5) - 3 \times 0.00 \text{ kJ/mol} = -206.1 \text{ kJ/mol}$$

Within the precision of the data supplied, the results are in good agreement.

64. (a)

$t, ^\circ C$	T, K	$1/T, K^{-1}$	K_p	$\ln K_p$
30.	303	3.30×10^{-3}	1.66×10^{-5}	-11.006
50.	323	3.10×10^{-3}	3.90×10^{-4}	-7.849
70.	343	2.92×10^{-3}	6.27×10^{-3}	-5.072
100.	373	2.68×10^{-3}	2.31×10^{-1}	-1.465

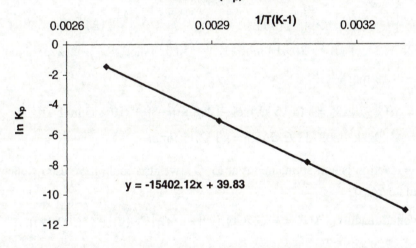

Plot of ln(K$_p$) versus 1/T

y = -15402.12x + 39.83

The slope of this graph is $-\Delta H^\circ / R = -1.54 \times 10^4$ K

$$\Delta H^\circ = -(8.3145 \text{ J mol}^{-1}\text{K}^{-1})(-1.54 \times 10^4 \text{ K}) = 128 \times 10^3 \text{ J/mol} = 128 \text{ kJ/mol}$$

(b)   When the total pressure is 2.00 atm, and both gases have been produced from $NaHCO_3(s)$,

$$P\{H_2O(g)\} = P\{CO_2(g)\} = 1.00 \text{ atm}$$

$$K_p = P\{H_2O(g)\}P\{CO_2(g)\} = (1.00)(1.00) = 1.00$$

Thus, $\ln K_p = \ln(1.00) = 0.000$. The corresponds to $1/T = 2.59 \times 10^{-3}$ K^{-1}; $T = 386$ K.
We can compute the same temperature from the van't Hoff equation.

$$\ln\frac{K_2}{K_1} = \frac{\Delta H^\circ}{R}\left(\frac{1}{T_1}-\frac{1}{T_2}\right) = \frac{128\times10^3 \text{ J / mol}}{8.3145 \text{ J mol}^{-1}\text{ K}^{-1}}\left(\frac{1}{T_1}-\frac{1}{303\text{ K}}\right) = \ln\frac{1.66\times10^{-5}}{1.00} = -11.006$$

$$\left(\frac{1}{T_1}-\frac{1}{303\text{ K}}\right) = \frac{-11.006\times8.3145}{128\times10^3}\text{ K}^{-1} = -7.15\times10^{-4}\text{ K}^{-1}$$

$$\frac{1}{T_1} = \frac{1}{303}-7.15\times10^{-4} = 3.30\times10^{-3}-7.15\times10^{-4} = 2.59\times10^{-3}\text{ K}^{-1}; \quad T_1 = 386\text{ K}$$

This temperature agrees well with the result obtained from the graph.

Coupled Reactions

65. **(a)** We compute ΔG° for the given reaction in the following manner

$$\Delta H^\circ = \Delta H_f^\circ\left[TiCl_4\,(l)\right]+\Delta H_f^\circ\left[O_2\,(g)\right]-\Delta H_f^\circ\left[TiO_2\,(s)\right]-2\Delta H_f^\circ\left[Cl_2\,(g)\right]$$

$$= -804.2\text{ kJ/mol}+0.00\text{ kJ/mol}-(-944.7\text{ kJ/mol})-2(0.00\text{ kJ/mol})$$

$$= +140.5\text{ kJ/mol}$$

$$\Delta S^\circ = S^\circ\left[TiCl_4\,(l)\right]+S^\circ\left[O_2\,(g)\right]-S^\circ\left[TiO_2\,(s)\right]-2S^\circ\left[Cl_2\,(g)\right]$$

$$= 252.3\text{ J mol}^{-1}\text{ K}^{-1}+205.1\text{ J mol}^{-1}\text{ K}^{-1}-(50.33\text{ J mol}^{-1}\text{ K}^{-1})-2(223.1\text{ J mol}^{-1}\text{ K}^{-1})$$

$$= -39.1\text{ J mol}^{-1}\text{ K}^{-1}$$

$$\Delta G^\circ = \Delta H^\circ - T\,\Delta S^\circ = +140.5\text{ kJ/mol}-(298\text{ K})(-39.1\times10^{-3}\text{ kJ mol}^{-1}\text{ K}^{-1})$$

$$= +140.5\text{ kJ/mol}+11.6\text{ kJ/mol} = +152.1\text{ kJ/mol}$$

Thus the reaction is non-spontaneous at 25° C. (we also could have used values of ΔG_f° to calculate ΔG°).

(b) For the cited reaction, $\Delta G^\circ = 2\Delta G_f^\circ\left[CO_2(g)\right]-2\Delta G_f^\circ\left[CO(g)\right]-\Delta G_f^\circ\left[O_2(g)\right]$

$$\Delta G^\circ = 2(-394.4\text{ kJ / mol})-2(-137.2\text{ kJ / mol})-0.00\text{ kJ / mol} = -514.4\text{ kJ / mol}$$

Then we couple the two reactions.

$$TiO_2\,(s)+2Cl_2\,(g)\longrightarrow TiCl_4\,(l)+O_2\,(g) \qquad\qquad \Delta G^\circ = +152.1\text{ kJ/mol}$$

$$2CO\,(g)+O_2\,(g)\longrightarrow 2CO_2\,(g) \qquad\qquad \Delta G^\circ = -514.4\text{ kJ/mol}$$

$$\overline{TiO_2\,(s)+2Cl_2\,(g)+2CO\,(g)\longrightarrow TiCl_4\,(l)+2CO_2\,(g); \Delta G^\circ = -362.3\text{ kJ/mol}}$$

The coupled reaction has $\Delta G^\circ < 0$, and, therefore, is spontaneous.

66. If $\Delta G° < 0$ for the sum of coupled reactions, the reduction of the oxide with carbon is spontaneous.

(a) $NiO(s) \rightarrow Ni(s) + \frac{1}{2}O_2(g)$ $\Delta G° = +115 \, kJ$

$C(s) + \frac{1}{2}O_2(g) \rightarrow CO(g)$ $\Delta G° = -250 \, kJ$

Net : $NiO(s) + C(s) \rightarrow Ni(s) + CO(g)$ $\Delta G° = +115 \, kJ - 250 \, kJ = -135 \, kJ$

Therefore the coupled reaction is spontaneous

(b) $MnO(s) \rightarrow Mn(s) + \frac{1}{2}O_2(g)$ $\Delta G° = +280 \, kJ$

$C(s) + \frac{1}{2}O_2(g) \rightarrow CO(g)$ $\Delta G° = -250 \, kJ$

Net : $MnO(s) + C(s) \longrightarrow Mn(s) + CO(g)$ $\Delta G° = +280 \, kJ - 250 \, kJ = +30 \, kJ$

Therefore the coupled reaction is non-spontaneous

(c) $TiO_2(s) \longrightarrow Ti(s) + O_2(g)$ $\Delta G° = +630 \, kJ$

$2C(s) + O_2(g) \longrightarrow 2CO(g)$ $\Delta G° = 2(-250 \, kJ) = -500 \, kJ$

Net : $TiO_2(s) + 2C(s) \longrightarrow Ti(s) + 2CO(g)$ $\Delta G° = +630 \, kJ - 500 \, kJ = +130 \, kJ$

Therefore the coupled reaction is non-spontaneous

Integrative and Advanced Exercises

67. (a) The normal boiling point of mercury is that temperature at which the mercury vapor pressure is 1.00 atm, or where the equilibrium constant for the vaporization equilibrium reaction has a numerical value of 1.00. This is also the temperature where $\Delta G° = 0$, since $\Delta G° = -RT \ln K_{eq}$ and $\ln(1.00) = 0$.

$Hg(l) \rightleftharpoons Hg(g)$

$\Delta H° = \Delta H°_f [Hg(g)] - \Delta H°_f [Hg(l)] = 61.32 \, kJ/mol - 0.00 \, kJ/mol = 61.32 \, kJ/mol$

$\Delta S° = S°[Hg(g)] - S°[Hg(l)] = 175.0 \, J \, mol^{-1} \, K^{-1} - 76.02 \, J \, mol^{-1} \, K^{-1} = 99.0 \, J \, mol^{-1} \, K^{-1}$

$0 = \Delta H° - T \, \Delta S° = 61.32 \times 10^3 \, J/mol - T \times 99.0 \, J \, mol^{-1} \, K^{-1}$

$T = \dfrac{61.32 \times 10^3 \, J/mol}{99.0 \, J \, mol^{-1} \, K^{-1}} = 619 \, K$

(b) The vapor pressure in atmospheres is the value of the equilibrium constant, which is related to the value of the free energy change for formation of Hg vapor.

$\Delta G°_f [Hg(g)] = 31.82 \, kJ/mol = -RT \ln K_{eq}$

$\ln K = \dfrac{-31.82 \times 10^3 \, J/mol}{8.3145 \, J \, mol^{-1} \, K^{-1} \times 298.15 \, K} = -12.84$ $K = e^{-12.84} = 2.65 \times 10^{-6} \, atm$

Therefore, the vapor pressure of Hg at 25°C is 2.65×10^{-6} atm.

68. If we can determine the value of the equilibrium constant at 25.0°C, we can then determine the value of $\Delta G°$. But the equilibrium constant for this reaction is simply $K_p = P\{N_2O_5(g)\}$.

Thus, all we need is the vapor pressure of $N_2O_5(s)$, which we can determine from the Clausius–Clapeyron equation, which is a specialized version of the van't Hoff equation. We first determine the value of ΔH_{sub}.

$$\ln\frac{P_2}{P_1} = \frac{\Delta H_{sub}}{R}\left(\frac{1}{T_1} - \frac{1}{T_2}\right) = \ln\frac{760\,\text{mmHg}}{100\,\text{mmHg}} = 2.028 = \frac{\Delta H_{sub}}{8.3145\,\text{J mol}^{-1}\,\text{K}^{-1}}\left(\frac{1}{7.5+273.2} - \frac{1}{32.4+273.2}\right)$$

$$\Delta H_{sub} = \frac{2.028}{3.49\times10^{-5}} = 5.81\times10^4\,\text{J/mol}$$

$$\ln\frac{P_2}{100\,\text{mmHg}} = \frac{5.81\times10^4\,\text{J/mol}}{8.3145\,\text{J mol}^{-1}\,\text{K}^{-1}}\left(\frac{1}{280.7} - \frac{1}{298.2}\right) = 1.46 \quad \frac{P_2}{100\,\text{mmHg}} = e^{1.46} = 4.31$$

$$P_2 = 4.31\times100\,\text{mmHg}\times\frac{1\,\text{atm}}{760\,\text{mmHg}} = 0.567\,\text{atm} = K_p$$

$$\Delta G° = -RT\ln K_p = -(8.3145\times10^{-3}\,\text{kJ mol}^{-1}\,\text{K}^{-1}\times298.2\,\text{K})\ln(0.567) = 1.41\,\text{kJ/mol}$$

<u>69.</u> (a) TRUE; It is the change in free energy for a process in which reactants and products are all in their standard states (regardless of whatever states might be mentioned in the statement of the problem). When liquid and gaseous water are each at 1 atm at 100 °C (the normal boiling point), they are in equilibrium, so that $\Delta G = \Delta G° = 0$ is only true when the difference of the standard free energies of products and reactants is zero. A reaction with $\Delta G° = 0$ would be at equilibrium when products and reactants were all present under standard state conditions and the pressure of $H_2O(g) = 2.0$ atm is not the standard pressure for $H_2O(g)$.

(b) FALSE; $\Delta G \neq 0$. The system is not at equilibrium.

(c) FALSE; $\Delta G°$ can have only one value at any given temperature, and that is the value corresponding to all reactants and products in their standard states, so at the normal boiling point $\Delta G° = 0$ [as was also the case in answering part (a)]. Water will not vaporize spontaneously under standard conditions to produce water vapor with a pressure of 2 atmospheres.

(d) TRUE; $\Delta G > 0$. The process of transforming water to vapor at 2.0 atm pressure at 100°C is not a spontaneous process; the condensation (reverse) process is spontaneous. (i.e. for the system to reach equilibrium, some $H_2O(l)$ must form)

70. $\Delta G^\circ = +\frac{1}{2}\Delta G^\circ_f\ [Br_2(g)] + \frac{1}{2}\Delta G^\circ_f\ [Cl_2(g)] - \Delta G^\circ_f\ [BrCl(g)]$

$= +\frac{1}{2}(3.11\,kJ/mol) + \frac{1}{2}(0.00\,kJ/mol) - (-0.98\,kJ/mol) = +2.54\,kJ/mol = -RT\ln K_p$

$$\ln K_p = -\frac{\Delta G^\circ}{RT} = -\frac{2.54\times10^3\,J/mol}{8.3145\,J\,mol^{-1}\,K^{-1}\times298.15\,K} = -1.02 \qquad K_p = e^{-1.02} = 0.361$$

For ease of solving the problem, we double the reaction, which squares the value of the equilibrium constant. $K_{eq} = (0.357)^2 = 0.130$

Reaction: $\quad 2\,BrCl(g) \rightleftharpoons Br_2(g) + Cl_2(g)$

Initial: $\qquad\quad 1.00\,mol \qquad\quad 0\,mol \qquad\quad 0\,mol$

Changes: $\quad -2x\,mol \qquad\quad +x\,mol \qquad +x\,mol$

Equil: $\qquad (1.00-2x)mol \qquad x\,mol \qquad\quad x\,mol$

$$K_p = \frac{P\{Br_2(g)\}\,P\{Cl_2(g)\}}{P\{BrCl(g)\}^2} = \frac{[n\{Br_2(g)\}RT/V][n\{Cl_2(g)\}RT/V]}{[n\{BrCl(g)\}RT/V]^2} = \frac{n\{Br_2(g)\}\,n\{Cl_2(g)\}}{n\{BrCl(g)\}^2}$$

$$= \frac{x^2}{(1.00-2x)^2} = (0.361)^2 \qquad \frac{x}{1.00-2x} = 0.361 \qquad x = 0.361 - 0.722\,x$$

$$x = \frac{0.361}{1.722} = 0.210\,mol\,Br_2 = 0.210\,mol\,Cl_2 \quad 1.00-2x = 0.580\,mol\,BrCl$$

71. First we determine the value of K_p for the dissociation reaction. If $I_2(g)$ is 50% dissociated, then for every mole of undissociated $I_2(g)$, one mole of $I_2(g)$ has dissociated, producing two moles of $I(g)$. Thus, the partial pressure of $I(g)$ is twice the partial pressure $I_2(g) \rightleftharpoons 2I(g)$ of $I_2(g)$.

$$P_{total} = 1.00\,atm = P_{I_2(g)} + P_{I(g)} = P_{I_2(g)} + 2\times P_{I_2(g)} = 3P_{I_2(g)} \qquad P_{I_2(g)} = 0.333\,atm$$

$$K_p = \frac{P_{I(g)}^2}{P_{I_2(g)}} = \frac{(0.667)^2}{0.333} = 1.34 \qquad \ln K_p = 0.293$$

$\Delta H^\circ = 2\Delta H^\circ_f\ [I(g)] - \Delta H^\circ_f\ [I_2(g)] = 2\times106.8\,kJ/mol - 62.44\,kJ/mol = 151.2\,kJ/mol$

$\Delta S^\circ = 2S^\circ[I(g)] - S^\circ[I_2(g)] = 2\times180.8\,J\,mol^{-1}\,K^{-1} - 260.7\,J\,mol^{-1}\,K^{-1} = 100.9\,J\,mol^{-1}\,K^{-1}$

Now we equate two expressions for ΔG° and solve for T.

$\Delta G^\circ = \Delta H^\circ - T\,\Delta S^\circ = -RT\ln K_p = 151.2\times10^3 - 100.8T = -8.3145\times T\times0.293$

$$151.2\times10^3 = 100.9T - 2.44T = 98.4T \qquad T = \frac{151.2\times10^3}{98.5} = 1535\,K \approx 1.5\times10^3\,K$$

72. **(a)** The oxide with the most positive (least negative) value of ΔG°_f is the one that most readily decomposes to the free metal and $O_2(g)$, since the decomposition is the reverse of the formation reaction. Thus the oxide that decomposes most readily is $Ag_2O(s)$.

(b) The decomposition reaction is $2\,Ag_2O(s) \longrightarrow 4\,Ag(s) + O_2(g)$ For this reaction $K_p = P_{O_2(g)}$. Thus, we need to find the temperature where $K_p = 1.00$. Since $\Delta G^\circ = -RT\ln K_p$ and $\ln(1.00) = 0$, we wish to know the temperature where $\Delta G^\circ = 0$.

Note also that the decomposition is the reverse of the formation reaction. Thus, the following values are valid for the decomposition reaction at 298 K.

$\Delta H° = +31.05 \text{ kJ/mol}$ $\qquad \Delta G° = +11.20 \text{ kJ/mol}$

We use these values to determine the value of $\Delta S°$ for the reaction.

$$\Delta G° = \Delta H° - T\Delta S° \quad T\Delta S° = \Delta H° - \Delta G° \quad \Delta S° = \frac{\Delta H° - \Delta G°}{T}$$

$$\Delta S° = \frac{31.05 \times 10^3 \text{ J/mol} - 11.20 \times 10^3 \text{ J/mol}}{298 \text{ K}} = +66.6 \text{ J mol}^{-1} \text{ K}^{-1}$$

Now we determine the value of T where $\Delta G° = 0$.

$$T = \frac{\Delta H° - \Delta G°}{\Delta S°} = \frac{31.05 \times 10^3 \text{ J/mol} - 0.0 \text{ J/mol}}{+66.6 \text{ J mol}^{-1} \text{ K}^{-1}} = 466 \text{ K} = 193°C$$

73. At $127°C = 400$ K, the two phases are in equilibrium, meaning that

$\Delta G°_{rxn} = 0 = \Delta H°_{rxn} - T\Delta S°_{rxn} = [\Delta H°_f \text{ (yellow)} - \Delta H°_f \text{ (red)}] - T[S°\text{(yellow)} - S°\text{(red)}]$

$= [-102.9 - (-105.4)] \times 10^3 \text{ J} - 400 \text{ K} \times [S°\text{(yellow)} - 180 \text{ J mol}^{-1} \text{ K}^{-1}]$

$= 2.5 \times 10^3 \text{ J/mol} - 400 \text{ K} \times S°\text{(yellow)} + 7.20 \times 10^4 \text{ J/mol}$

$$S°\text{(yellow)} = \frac{(7.20 \times 10^4 + 2.5 \times 10^3) \text{ J/mol}}{400 \text{ K}} = 186 \text{ J mol}^{-1} \text{ K}^{-1}$$

Then we compute the value of the "entropy of formation" of the yellow form at 298 K.

$\Delta S°_f = S°[\text{HgI}_2] - S°[\text{Hg(l)}] - S°[\text{I}_2 \text{(s)}] = [186 - 76.02 - 116.1] \text{ J mol}^{-1} \text{ K}^{-1} = -6 \text{ J mol}^{-1} \text{ K}^{-1}$

Now we can determine the value of the free energy of formation for the yellow form.

$$\Delta G°_f = \Delta H°_f - T\Delta S°_f = -102.9 \frac{\text{kJ}}{\text{mol}} - [298 \text{ K} \times (-6 \frac{\text{J}}{\text{K mol}}) \times \frac{1 \text{kJ}}{1000 \text{ J}}] = -101.1 \frac{\text{kJ}}{\text{mol}}$$

74. First we need a value for the equilibrium constant. 1% conversion means that 0.99 mol $N_2(g)$ are present at equilibrium for every 1.00 mole present initially.

$$K = K_p = \frac{P_{NO(g)}^2}{P_{N_2(g)} P_{O_2(g)}} = \frac{[n\{NO(g)\}RT/V]^2}{[n\{N_2(g)\}RT/V][n\{O_2(g)\}RT/V]} = \frac{n\{NO(g)\}^2}{n\{N_2(g)\} n\{O_2(g)\}}$$

Reaction:	$N_2(g)$	+	$O_2(g)$	$\rightleftharpoons$	$2 NO(g)$
Initial:	1.00 mol		1.00 mol		0 mol
Changes(1% rxn):	−0.010 mol		−0.010 mol		+0.020 mol
Equil:	0.99 mol		0.99 mol		0.020 mol

$$K = \frac{(0.020)^2}{(0.99)(0.99)} = 4.1 \times 10^{-4}$$

The cited reaction is twice the formation reaction of NO(g), and thus

$\Delta H° = 2\Delta H°_f [NO(g)] = 2×90.25 \, kJ/mol = 180.50 \, kJ/mol$

$\Delta S° = 2S°[NO(g)] - S°[N_2(g)] - S°[O_2(g)]$

$= 2(210.7 \, J \, mol^{-1} \, K) - 191.5 \, J \, mol^{-1} \, K^{-1} - 205.0 \, J \, mol^{-1} \, K^{-1} = 24.9 \, J \, mol^{-1} \, K^{-1}$

$\Delta G° = -RTlnK = -8.31447 \, JK^{-1}mol^{-1}(T)ln(4.1×10^{-4}) = 64.\underline{85}(T)$

$\Delta G° = \Delta H° - TS° = 64.\underline{85}(T) = 180.5 \, kJ/mol - (T)24.9 \, J \, mol^{-1} \, K^{-1}$

$180.5 × 10 \, J/mol = 64.\underline{85}(T) + (T)24.9 \, J \, mol^{-1} \, K^{-1} = 89.\underline{75}(T)$ $\qquad$ $T = 2.01 × 10^3 \, K$

75. $Sr(IO_3)_2(s) \rightleftharpoons Sr^{2+}(aq) + 2 \, IO_3^-(aq)$

$\Delta G° = (2 \, mol×(-128.0 \, kJ/mol) + (1 \, mol ×-500.5 \, kJ/mol)) - (1 \, mol × -855.1 \, kJ/mol) = -0.4 \, kJ$

$\Delta G° = -RTlnK = -8.31447 \, JK^{-1}mol^{-1}(298.15 \, K)ln \, K = -0.4 \, kJ = -400 \, J$

$ln \, K = 0.16$ and $K = 1.\underline{175} = [Sr^{2+}][IO_3^-]^2$ $\qquad$ Let x = solubility of $Sr(IO_3)_2$

$[Sr^{2+}][IO_3^-]^2 = 1.\underline{175} = x(2x)^2 = 4x^3$ $\quad x = 0.6\underline{65} \, M$ for a saturated solution of $Sr(IO_3)_2$.

<u>76.</u> $\Delta G° = \Delta H° - T\Delta S°$

$\Delta S° = \dfrac{\Delta H° - \Delta G°}{T} = \dfrac{-454.8 \, kJ/mol - (-323.1 \, kJ/mol)}{298.16 \, K} \, -441.7 \, J/(mol \cdot K)$

$2 \, C(s) + 3 \, H_2(g) + O_2(g) \rightarrow CH_2OHCH_2OH(l) \, \{ \Delta S° = -441.7 \, J/(mol \cdot K)\}$

Since $\Delta S° = \sum \{S°_{products}\} - \sum \{S°_{reactants}\}$

$\Delta S°_{rxn} = S° (CH_2OHCH_2OH(l)) - [2×S°(C) + 3× S°(H_2(g)) + S°(O_2(g)]$

$\Delta G° = \Delta H° - T \, \Delta S° = -RT \ln K$ $\qquad$ $\Delta H° = T \, \Delta S° - RT \ln K$ $\qquad$ $T = \dfrac{\Delta H°}{\Delta S° - R \ln K}$

$T = \dfrac{180.5×10^3 \, J/mol}{24.9 \, J \, mol^{-1} \, K^{-1} - 8.3145 \, J \, mol^{-1} \, K^{-1} \ln(4.1×10^{-4})} = 2.01×10^3 \, K$

$-441.7 = S°(CH_2OHCH_2OH) - (2(5.7 \, J/K-mol) + 3(130.7 J/K - mol) + (205.1 \, J/K - mol))$

$-441.7 J/K = S°(CH_2OHCH_2OH) - 608.6 \, J/K$

$S°(CH_2OHCH_2OH) = 167.0 \, J/K$ or $167 \, J/K$ to 3 significant figures

77. $P_{H2O} = 75 \, torr$ or $0.098\underline{7} \, atm.$ $\qquad K = (P_{H2O})^2 = (0.098\underline{7})^2 = 9.7\underline{4} × 10^{-3}$

$\Delta G° = (2 \, mol×(-228.6 \, kJ/mol) + (1 \, mol)×-918.1 \, kJ/mol) - (1 \, mol)×-1400.0 \, kJ/mol = 309.0 \, kJ$

$\Delta H° = (2 \, mol×(-241.8 \, kJ/mol) + (1 \, mol)×-1085.8.1 \, kJ/mol) - (1 \, mol)×-1684.3 \, kJ/mol = 114.9 \, kJ$

$\Delta S° = (2 \, mol×(188.J/K \, mol) + (1 \, mol×146.J/K \, mol) + (1 \, mol)×221.3.J/K \, mol = 302.3 \, J/K \, mol$

$\Delta G° = -RT \ln K_{eq} = -8.3145 \, J \, mol^{-1} \, K^{-1} ×T × \ln (9.74×10^{-3}) = 38.\underline{5}(T)$

$\Delta G° = 38.\underline{5}(T) = \Delta H° - T\Delta S° = 114,900 \, J \, mol^{-1} - (T) × 302.3 \, J \, K^{-1} \, mol^{-1}$

$114,900 = 340.8 \, K^{-1}(T)$ Hence: $T = 337 \, K = 64 \, °C$

78. (a) $\Delta G^\circ = -RT \ln K$ $\ln K = -\dfrac{\Delta G^\circ}{RT} = -\dfrac{131 \times 10^3 \text{ J/mol}}{8.3145 \text{ J mol}^{-1} \text{ K}^{-1} \times 298.15 \text{ K}} = -52.8$

$K = e^{-52.8} = 1.2 \times 10^{-23} \text{ atm} \times \dfrac{760 \text{ mmHg}}{1 \text{ atm}} = 8.9 \times 10^{-21} \text{ mmHg}$

Since the system cannot produce a vacuum lower than 10^{-9} mmHg, this partial pressure of $CO_2(g)$ won't be detected in the system.

(b) Since we have the value of ΔG° for the decomposition reaction at a specified temperature (298.15 K), and we need ΔH° and ΔS° for this same reaction to determine $P\{CO_2(g)\}$ as a function of temperature, obtaining either ΔH° or ΔS° will enable us to determine the other.

(c) $\Delta H^\circ = \Delta H^\circ_f [CaO(s)] + \Delta H^\circ_f [CO_2(g)] - \Delta H^\circ_f [CaCO_3(s)]$

$= -635.1 \text{ kJ/mol} - 393.5 \text{ kJ/mol} - (-1207 \text{ kJ/mol}) = +178 \text{ kJ/mol}$

$\Delta G^\circ = \Delta H^\circ - T \Delta S^\circ$ $T \Delta S^\circ = \Delta H^\circ - \Delta G^\circ$ $\Delta S^\circ = \dfrac{\Delta H^\circ - \Delta G^\circ}{T}$

$\Delta S^\circ = \dfrac{178 \text{ kJ/mol} - 131 \text{ kJ/mol}}{298. \text{ K}} \times \dfrac{1000 \text{ J}}{1 \text{ kJ}} = 1.6 \times 10^2 \text{ J mol}^{-1} \text{ K}^{-1}$

$K = 1.0 \times 10^{-9} \text{ mmHg} \times \dfrac{1 \text{ atm}}{760 \text{ mmHg}} = 1.3 \times 10^{-12}$

$\Delta G^\circ = \Delta H^\circ - T \Delta S^\circ = -RT \ln K_{eq}$ $\Delta H^\circ = T \Delta S^\circ - RT \ln K$

$T = \dfrac{\Delta H^\circ}{\Delta S^\circ - R \ln K} = \dfrac{178 \times 10^3 \text{ J/mol}}{1.6 \times 10^2 \text{ J mol}^{-1} \text{ K}^{-1} - 8.3145 \text{ J mol}^{-1} \text{ K}^{-1} \ln(1.3 \times 10^{-12})} = 4.6 \times 10^2 \text{ K}$

<u>79.</u> $\Delta H^\circ = \Delta H^\circ_f [PCl_3(g)] + \Delta H^\circ_f [Cl_2(g)] - \Delta H^\circ_f [PCl_5(g)]$

$= -287.0 \text{ kJ/mol} + 0.00 \text{ kJ/mol} - (-374.9 \text{ kJ/mol}) = 87.9 \text{ kJ/mol}$

$\Delta S^\circ = S^\circ[PCl_3(g)] + S^\circ[Cl_2(g)] - S^\circ[PCl_5(g)]$

$= 311.8 \text{ J mol}^{-1} \text{ K}^{-1} + 223.1 \text{ J mol}^{-1} \text{ K}^{-1} - 364.6 \text{ J mol}^{-1} \text{ K}^{-1} = +170.3 \text{ J mol}^{-1} \text{ K}^{-1}$

$\Delta G^\circ = \Delta H^\circ - T \Delta S^\circ = 87.9 \times 10^3 \text{ J/mol} - 500 \text{ K} \times 170.3 \text{ J mol}^{-1} \text{ K}^{-1}$

$\Delta G^\circ = 2.8 \times 10^3 \text{ J/mol} = -RT \ln K_p$

$\ln K_p = \dfrac{-\Delta G^\circ}{RT} = \dfrac{-2.8 \times 10^3 \text{ J/mol}}{8.3145 \text{ J mol}^{-1} \text{ K}^{-1} \times 500 \text{ K}} = -0.67$ $K_p = e^{-0.67} = 0.51$

$P_i[PCl_5] = \dfrac{nRT}{V} = \dfrac{0.100 \text{ mol} \times 0.08206 \text{ L atm mol}^{-1} \text{ K}^{-1} \times 500 \text{ K}}{1.50 \text{ L}} = 2.74 \text{ atm}$

Reaction: $PCl_5(g) \rightleftharpoons PCl_3(g) + Cl_2(g)$

Initial: $\quad$ 2.74 atm $\qquad$ 0 atm $\qquad$ 0 atm

Changes: $\quad -x$ atm $\qquad\quad +x$ atm $\qquad +x$ atm

Equil: $\quad (2.74-x)$ atm $\qquad x$ atm $\qquad x$ atm

$$K_p = \frac{P[PCl_3]\,P[Cl_2]}{P[PCl_5]} = 0.51 = \frac{x \cdot x}{2.74-x}$$

$$x^2 = 0.51(2.74-x) = 1.4 - 0.51x \quad x^2 + 0.51x - 1.4 = 0$$

$$x = \frac{-b \pm \sqrt{b^2-4ac}}{2a} = \frac{-0.51 \pm \sqrt{0.26+5.6}}{2} = 0.96 \text{ atm}, -1.47 \text{ atm}$$

$$P_{total} = P_{PCl_5} + P_{PCl_3} + P_{Cl_2} = (2.74-x) + x + x = 2.74 + x = 2.74 + 0.96 = 3.70 \text{ atm}$$

80. The value of $\Delta H°$ determined in Exercise 64 is $\Delta H° = +128$ kJ/mol. We use any one of the values of $K_p = K_{eq}$ to determine a value of $\Delta G°$. At 30 °C = 303 K,

$$\Delta G° = -RT \ln K_{eq} = -(8.3145 \text{ J mol}^{-1} \text{ K}^{-1})(303 \text{ K}) \ln(1.66 \times 10^{-5}) = +2.77 \times 10^4 \text{ J/mol}$$

Now we determine $\Delta S°$. $\Delta G° = \Delta H - T\,\Delta S°$

$$\Delta S° = \frac{\Delta H° - \Delta G°}{T} = \frac{128 \times 10^3 \text{ J/mol} - 2.77 \times 10^4 \text{ J/mol}}{303 \text{ K}} = +331 \text{ J mol}^{-1} \text{ K}^{-1}$$

By using the appropriate $S°$ values in Appendix D, we calculate $\Delta S° = +334 \text{ J mol}^{-1} \text{ K}^{-1}$.

81. We use Trouton's rule to determine the value of ΔH_{vap} for cyclohexane then we use the Clausius–Clapeyron equation to determine the temperature at which the vapor pressure is 100.0 mmHg.

$$\Delta H_{vap} = T_{nbp}\,\Delta S_{vap} = 353.9 \text{ K} \times 87 \text{ J mol}^{-1} \text{ K}^{-1} = 31 \times 10^3 \text{ J/mol}$$

$$\ln \frac{P_2}{P_1} = \frac{\Delta H_{vap}}{R}\left(\frac{1}{T_1} - \frac{1}{T_2}\right) = \ln \frac{100 \text{ mmHg}}{760 \text{ mmHg}} = \frac{31 \times 10^3 \text{ J/mol}}{8.3145 \text{ J mol}^{-1} \text{ K}^{-1}}\left(\frac{1}{353.9 \text{ K}} - \frac{1}{T}\right) = -2.028$$

$$\frac{1}{353.9} - \frac{1}{T} = \frac{-2.028 \times 8.3145}{31 \times 10^3} = -5.4 \times 10^{-4} = 2.826 \times 10^{-3} - \frac{1}{T} \qquad \frac{1}{T} = 3.37 \times 10^{-3} \text{ K}^{-1}$$

$$T = 297 \text{ K} = 24 °C$$

82. (a) $2Ag(s) + \frac{1}{2}O_2(g) \rightarrow Ag_2O$

$$\Delta G_f° = \Delta G_f°(Ag_2O(s)) - \{2\Delta G_f°(Ag(s)) + \frac{1}{2}\Delta G_f°(O_2)\}$$

$$\Delta G_f° = -11.2 \text{ kJ} - \{2(0) + \frac{1}{2}(0)\} = -11.2 \text{ kJ} \Rightarrow Ag_2O \text{ is thermodynamically stable at } 25°C$$

(b) Assuming $\Delta H°$, $\Delta S°$ are constant from 25-200°C (not so, but a reasonable assumption !)

$$\Delta S° = S°(Ag_2O)-\{2\ S°(Ag(s)) + \frac{1}{2}\ S°(O_2)\} = 121.3 - (2(42.6) + \frac{1}{2}(205.1)) = -66.5\ \text{J/K}$$

$$\Delta G° = -31.0\ \text{kJ} - \frac{(473\ \text{K})(-66.5\ \text{J/K})}{1000\ \text{J/kJ}} = \Delta H° - T\Delta S° = +0.45\ \text{kJ}$$

$\Rightarrow$ thermodynamically *unstable* at 200°C

83. $\Delta G = 0$ since the system is at equilibrium. As well, $\Delta G° = 0$ because this process is under standard conditions. Since $\Delta G° = \Delta H° - T\Delta S° = 0$. $= \Delta H° = T\Delta S° = 273.15$ K$\times 21.99$ J K^{-1} mol^{-1} = 6.007 kJ mol^{-1}. Since we are dealing with 2 moles of ice melting, the values of $\Delta H°$ and $\Delta S°$ are doubled. Hence, $\Delta H° = 12.01$ kJ and $\Delta S° = 43.98$ J K^{-1}. Note: The densities are not necessary for the calculations required for this question.

84. First we determine the value of K_p that corresponds to 15% dissociation. We represent the initial pressure of phosgene as x atm.

Reaction: $COCl_2(g) \rightleftharpoons CO(g) + Cl_2(g)$

Initial : $\qquad x$ atm $\qquad\qquad$ 0 atm $\qquad$ 0 atm

Changes: $\quad -0.15\,x$ atm $\quad +0.15\,x$ atm $\quad +0.15\,x$ atm

Equil: $\qquad 0.85\,x$ atm $\qquad 0.15\,x$ atm $\quad 0.15\,x$ atm

$$P_{total} = 0.85\,x\,\text{atm} + 0.15\,x\,\text{atm} + 0.15\,x\,\text{atm} = 1.15\,x\,\text{atm} = 1.00\,\text{atm} \qquad x = \frac{1.00}{1.15} = 0.870\,\text{atm}$$

$$K_p = \frac{P_{CO}\ P_{Cl_2}}{P_{COCl_2}} = \frac{(0.15\times 0.870)^2}{0.85\times 0.870} = 0.0230$$

Next we find the value of $\Delta H°$ for the decomposition reaction.

$$\ln\frac{K_1}{K_2} = \frac{\Delta H°}{R}\left(\frac{1}{T_2} - \frac{1}{T_1}\right) = \ln\frac{6.7\times 10^{-9}}{4.44\times 10^{-2}} = -15.71 = \frac{\Delta H°}{R}\left(\frac{1}{668} - \frac{1}{373.0}\right) = \frac{\Delta H°}{R}(-1.18\times 10^{-3})$$

$$\frac{\Delta H°}{R} = \frac{-15.71}{-1.18\times 10^{-3}} = 1.33\times 10^4,$$

$\Delta H° = 1.33\times 10^4 \times 8.3145 = 111\times 10^3$ J/mol $= 111$ kJ/mol

And finally we find the temperature at which $K = 0.0230$.

$$\ln\frac{K_1}{K_2} = \frac{\Delta H°}{R}\left(\frac{1}{T_2} - \frac{1}{T_1}\right) = \ln\frac{0.0230}{0.0444} = \frac{111\times 10^3\ \text{J/mol}}{8.3145\ \text{J mol}^{-1}\ \text{K}^{-1}}\left(\frac{1}{668\ \text{K}} - \frac{1}{T}\right) = -0.658$$

$$\frac{1}{668} - \frac{1}{T} = \frac{-0.658\times 8.3145}{111\times 10^3} = -4.93\times 10^{-5} = 1.497\times 10^{-3} - \frac{1}{T} \qquad \frac{1}{T} = 1.546\times 10^{-3}$$

$T = 647$ K $= 374$ °C

85. First we write the solubility reaction for AgBr. Then we calculate values of $\Delta H°$ and $\Delta S°$

for the reaction: $AgBr(s) \rightleftharpoons Ag^+(aq) + Br^-(aq)$ $K_{eq} = K_{sp} = [Ag^+][Br^-] = s^2$

$\Delta H° = \Delta H°_f [Ag^+(aq)] + \Delta H°_f [Br^-(aq)] - \Delta H°_f [AgBr(s)]$

$\quad = +105.6 \text{ kJ/mol} - 121.6 \text{ kJ/mol} - (-100.4 \text{ kJ/mol}) = +84.4 \text{ kJ/mol}$

$\Delta S° = S°[Ag^+(aq)] + S°[Br^-(aq)] - S°[AgBr(s)]$

$\quad = +72.68 \text{ J mol}^{-1} \text{ K}^{-1} + 82.4 \text{ J mol}^{-1} \text{ K}^{-1} - 107.1 \text{ J mol}^{-1} \text{ K}^{-1} = +48.0 \text{ J mol}^{-1} \text{ K}^{-1}$

These values are then used to determine the value of $\Delta G°$ for the solubility reaction, and the standard free energy change, in turn, is used to obtain the value of K.

$\Delta G° = \Delta H° - T\Delta S° = 84.4 \times 10^3 \text{ J mol}^{-1} - (100 + 273) \text{ K} \times 48.0 \text{ J mol}^{-1} \text{ K}^{-1} = 66.5 \times 10^3 \text{ J/mol}$

$\ln K = \dfrac{-\Delta G°}{RT} = \dfrac{-66._5 \times 10^3}{\dfrac{8.3145 \text{ J}}{\text{mol K}} \times 373 \text{ K}} = -21.4$ $K = K_{sp} = e^{-21.4} = 5.0 \times 10^{-10} = s^2$

And now we compute the solubility of AgBr in mg/L.

$s = \sqrt{5.0 \times 10^{-10}} \times \dfrac{187.77 \text{ g AgBr}}{1 \text{ mol AgBr}} \times \dfrac{1000 \text{ mg}}{1 \text{ g}} = 4.2 \text{ mg AgBr/L}$

86. $S°_{298.15} = S°_{274.68} + \Delta S_{fusion} + \Delta S_{heating}$

$S°_{298.15} = 67.15 \text{ J K}^- \text{ mol}^{-1} + \dfrac{12,660 \text{ J mol}^{-1}}{274.68 \text{ K}} + \int_{274.68}^{298.15} 97.78 \dfrac{\text{J}}{\text{mol K}} + 0.0586 \dfrac{\text{J}}{\text{mol K}^2} \times (T - 274.68)$

$S°_{298.15} = 67.15 \text{ J K}^{-1} \text{ mol}^{-1} + 46.09 \text{ J K}^{-1} \text{ mol}^{-1} + 8.07 \text{ J K}^{-1} \text{ mol}^{-1} = 121.3 \text{ J K}^{-1} \text{ mol}^{-1}$

87. $S° = S°_{solid} + \Delta S_{fusion} + \Delta S_{heating} + \Delta S_{vaporization} + \Delta S_{pressure\ change}$

$S° = 128.82 \text{ J K}^{-1} \text{ mol}^{-1} + \dfrac{9866 \text{ J mol}^{-1}}{278.68 \text{ K}} + \int_{278.68}^{298.15} \dfrac{134.0 \dfrac{\text{J}}{\text{mol K}} dT}{T} + \dfrac{33,850 \text{ J mol}^{-1}}{298.15 \text{ K}}$

$\quad + 8.3145 \text{ J K}^{-1} \text{ mol}^{-1} \times \ln\left(\dfrac{95.13 \text{ torr}}{760 \text{ torr}}\right)$

$S° = 128.82 \text{ J K}^{-1} \text{ mol}^{-1} + 35.40 \text{ J K}^{-1} \text{ mol}^{-1} + 9.05 \text{ J K}^{-1} \text{ mol}^{-1} + 113.5 \text{ J K}^{-1} \text{ mol}^{-1} + (-17.28 \text{ J K}^{-1} \text{ mol}^{-1})$

$S° = 269.53 \text{ J K}^{-1} \text{ mol}^{-1}$

FEATURE PROBLEMS

88. **(a)** The first method involves combining the values of ΔG_f°. The second uses

$$\Delta G^\circ = \Delta H^\circ - T\Delta S^\circ$$
$$\Delta G^\circ = \Delta G_f^\circ \left[H_2O(g) \right] - \Delta G_f^\circ \left[H_2O(l) \right]$$
$$= -228.572 \text{ kJ/mol} - (-237.129 \text{ kJ/mol}) = +8.557 \text{ kJ/mol}$$

$$\Delta H^\circ = \Delta H_f^\circ \left[H_2O(g) \right] - \Delta H_f^\circ \left[H_2O(l) \right]$$
$$= -241.818 \text{ kJ/mol} - (-285.830 \text{ kJ/mol}) = +44.012 \text{ kJ/mol}$$

$$\Delta S^\circ = S^\circ \left[H_2O(g) \right] - S^\circ \left[H_2O(l) \right]$$
$$= 188.825 \text{ J mol}^{-1} \text{ K}^{-1} - 69.91 \text{ J mol}^{-1} \text{ K}^{-1} = +118.92 \text{ J mol}^{-1} \text{ K}^{-1}$$

$$\Delta G^\circ = \Delta H^\circ - T\Delta S^\circ$$
$$= 44.012 \text{ kJ/mol} - 298.15 \text{ K} \times 118.92 \times 10^{-3} \text{ kJ mol}^{-1} \text{ K}^{-1} = +8.556 \text{ kJ/mol}$$

(b) We use the average value: $\Delta G^\circ = +8.558 \times 10^3 \text{ J/mol} = -RT \ln K$

$$\ln K = -\frac{8558 \text{ J/mol}}{8.3145 \text{ J mol}^{-1} \text{ K}^{-1} \times 298.15 \text{ K}} = -3.452; \quad K = e^{-3.452} = 0.0317 \text{ bar}$$

(c) $P\{H_2O\} = 0.0317 \text{ bar} \times \dfrac{1 \text{ atm}}{1.01325 \text{ bar}} \times \dfrac{760 \text{ mmHg}}{1 \text{ atm}} = 23.8 \text{ mmHg}$

(d) $\ln K = -\dfrac{8590 \text{ J/mol}}{8.3145 \text{ J mol}^{-1} \text{ K}^{-1} \times 298.15 \text{ K}} = -3.465$;

$K = e^{-3.465} = 0.0312_7 \text{ atm}$;

$P\{H_2O\} = 0.0313 \text{ atm} \times \dfrac{760 \text{ mmHg}}{1 \text{ atm}} = 23.8 \text{ mmHg}$

89. **(a)** When we combine two reactions and obtain the overall value of ΔG°, we subtract the value on the plot of the reaction that becomes a reduction from the value on the plot of the reaction that is an oxidation. Thus, to reduce ZnO with elemental Mg, we subtract the values on the line labeled "$2Zn + O_2 \rightarrow 2ZnO$" from those on the line labeled "$2Mg + O_2 \rightarrow 2MgO$". The result for the overall ΔG° will always be negative because every point on the "zinc" line is above the corresponding point on the "magnesium" line

(b) In contrast, the "carbon" line is only below the "zinc" line at temperatures above about $1000°C$. Thus, only at these elevated temperatures can ZnO be reduced by carbon.

(c) The decomposition of zinc oxide to its elements is the reverse of the plotted reaction, the value of ΔG° for the decomposition becomes negative, and the reaction becomes spontaneous, where the value of ΔG° for the plotted reaction becomes positive. This occurs above about $1850°C$.

(d) The "carbon" line has a negative slope, indicating that carbon monoxide becomes more stable as temperature rises. The point where $CO(g)$ would become less stable than $2C(s)$ and $O_2(g)$ looks to be below $-1000°C$ (by extrapolating the line to lower temperatures). Based on this plot, it is not possible to decompose $CO(g)$ to $C(s)$ and $O_2(g)$ in a spontaneous reaction.

(e)

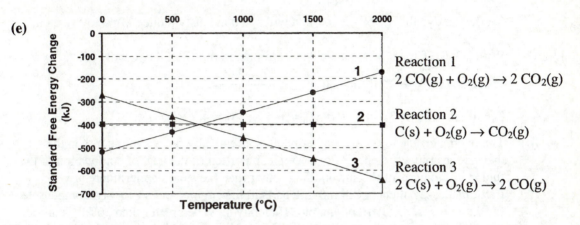

All three lines are straight-line plots of $\Delta G°$ vs. T following the equation $\Delta G° = \Delta H° - T\Delta S°$.

The general equation for a straight line is given below with the slightly modified Gibbs Free-Energy equation as a reference: $\Delta G° = -\Delta S°T + \Delta H°$ (here $\Delta H°$ assumed constant)

$y = mx + b$ ($m = -\Delta S° =$ slope of the line)

Thus, the slope of each line multiplied by minus one is equal to the $\Delta S°$ for the oxide formation reaction. It is hardly surprising, therefore, that the slopes for these lines differ so markedly because these three reactions have quite different $\Delta S°$ values ($\Delta S°$ for Reaction 1 = -173 J K^{-1}, $\Delta S°$ for Reaction 2 = 2.86 J K^{-1}, $\Delta S°$ for Reaction 3 = 178.8 J K^{-1})

(f) Since other metal oxides apparently have positive slopes similar to Mg and Zn, we can conclude that in general, the stability of metal oxides <u>decreases</u> as the temperature increases. Put another way, the decomposition of metal oxides to their elements becomes more spontaneous as the temperature is increased. By contrast, the two reactions involving elemental carbon, namely Reaction 2 and Reaction 3, have negative slopes, indicating that the formation of $CO_2(g)$ and $CO(g)$ from graphite becomes more favorable as the temperature rises. This means that the $\Delta G°$ for the reduction of metal oxides by carbon becomes more and more negative with increasing temperature. Moreover, there must exist a threshold temperature for each metal oxide above which the reaction with carbon will occur spontaneously. Carbon would appear to be an excellent reducing agent, therefore, because it will reduce virtually <u>any</u> metal oxide to its corresponding metal as long as the temperature chosen for the reaction is higher than the threshold temperature (the threshold temperature is commonly referred to as the transition temperature).

Consider for instance the reaction of MgO(s) with graphite to give $CO_2(g)$ and Mg metal:
$2 \text{ MgO(s)} + \text{C(s)} \rightarrow 2 \text{ Mg(s)} + \text{CO}_2(g)$ $\Delta S°_{rxn} = 219.4$ J/K and $\Delta H°_{rxn} = 809.9$ kJ

$$T_{transition} = \frac{\Delta H°_{rxn}}{\Delta S°_{rxn}} = \frac{809.9 \text{ kJ}}{0.2194 \text{ kJ K}^{-1}} = 3691 \text{ K} = T_{threshold}$$

Consequently, above 3691 K, carbon will spontaneously reduce MgO to Mg metal.

90. (a) With a 36% efficiency and a condenser temperature (T_1) of 41 °C = 314 K,

efficiency $= \frac{T_h - T_1}{T_h} \times 100\% = 36\%$ $\frac{T_h - 314}{T_h} = 0.36$;

$T_h = (0.36 \times T_h) + 314 \text{ K};$ $0.64 \ T_h = 314 \text{ K};$ $T_h = 4.9 \times 10^2 \text{ K}$

(b) The overall efficiency of the power plant is affected by factors in other than the thermodynamic efficiency. For example, a portion of the heat of combustion of the fuel is lost to parts of the surroundings other than the steam boiler; there are frictional losses of energy in moving parts in the engine; and so on. To compensate for these losses, the thermodynamic efficiency must be greater than 36%. To obtain this higher thermodynamic efficiency, T_h must be greater than 4.9×10^2 K.

(c) The steam pressure we are seeking is the vapor pressure of water at 4.9×10^2 K. We also know that the vapor pressure of water at 373 K (100 °C) is 1 atm. The enthalpy of vaporization of water at 298 K is $\Delta H° = \Delta H_f°[\text{H}_2\text{O}(g)] - \Delta H_f°[\text{H}_2\text{O}(l)] = -241.8$ kJ/mol $- (-285.8 \text{ kJ/mol}) = 44.0$ kJ/mol. Although the enthalpy of vaporization is somewhat temperature dependent, we will assume that this value holds from 298 K to 4.9×10^2 K, and make appropriate substitutions into the Clausius-Clapeyron equation.

$$\ln\left(\frac{P_2}{1 \text{ atm}}\right) = \frac{44.0 \text{ kJ mol}^{-1}}{8.3145 \times 10^{-3} \text{ kJ mol}^{-1}}\left(\frac{1}{373 \text{ K}} - \frac{1}{490 \text{ K}}\right) = 5.29 \times 10^3 \left(2.68 \times 10^{-3} - 2.04 \times 10^{-3}\right)$$

$$\ln\left(\frac{P_2}{1 \text{ atm}}\right) = 3.39; \quad \left(\frac{P_2}{1 \text{ atm}}\right) = 29._7; \quad P_2 \approx 30 \text{ atm}$$

The answer cannot be given with greater certainty because of the weakness of the assumption of a constant $H°_{vapn}$.

(d) It is not possible to devise a heat engine with 100% efficiency or greater. For 100% efficiency, $T_1 = 0$ K, which is unattainable. To have an efficiency greater than 100 % would require a *negative* T_1, which is also unattainable.

91. (a) Under biological standard conditions:

$\text{ADP}^{3-} + \text{HPO}_4^{2-} + \text{H}^+ \rightarrow \text{ATP}^{4-} + \text{H}_2\text{O}$ $\Delta G° = 32.4$ kJ/mol

If all of the energy of combustion of 1 mole of glucose is employed in the conversion of ADP to ATP, then the maximum number of moles ATP produced is

$$\text{Maximum number} = \frac{2870 \text{ kJ mol}^{-1}}{32.4 \text{ kJ mol}^{-1}} = 88.6 \text{ moles ATP}$$

(b) In an actual cell the number of ATP moles produced is 38, so that the efficiency is:

$$\text{Efficiency} = \frac{\text{number of ATP's actually produced}}{\text{Maximum number of ATP's that can be produced}} = \frac{38}{88.6} = 0.43$$

Thus, the cell's efficiency is about 43%.

(c) The previously calculated efficiency is based upon the biological standard state. We now calculate the Gibbs energies involved under the actual conditions in the cell. To do this we require the relationship between ΔG and $\Delta G^{\circ'}$ for the two coupled reactions. For the combustion of glucose we have

$$\Delta G = \Delta G^{\circ'} + RT \ln\left(\frac{a_{CO_2}^6 \, a_{H_2O}^6}{a_{glu} a_{O_2}^6}\right) \qquad \text{For the conversion of ADP to ATP we have}$$

$$\Delta G = \Delta G^{\circ'} + RT \ln\left(\frac{a_{ATP} a_{H_2O}}{a_{ADP} a_{P_i}\left(\left[H^+\right]/10^{-7}\right)}\right)$$

Using the concentrations and pressures provided we can calculate the Gibbs energy for the combustion of glucose under biological conditions. First, we need to replace the activities by the appropriate effective concentrations. That is,

$$\Delta G = \Delta G^{\circ'} + RT \ln\left(\frac{\left(p/p^{\circ}\right)^6_{CO_2} a_{H_2O}^6}{[glu]/[glu]^{\circ}\left(p/p^{\circ}\right)^6_{O_2}}\right) \quad \text{using } a_{H_2O} \approx 1 \text{ for a dilute solution we obtain}$$

$$\Delta G = \Delta G^{\circ'} + RT \ln\left(\frac{(0.050 \text{ bar}/1 \text{ bar})^6 \times 1^6}{[glu]/1 \times (0.132 \text{ bar}/1 \text{ bar})^6}\right)$$

The concentration of glucose is given in mg/mL and this has to be converted to molarity as follows:

$$[glu] = \frac{1.0 \text{ mg}}{mL} \times \frac{g}{1000 \text{ mg}} \times \frac{1000 \text{ mL}}{L} \times \frac{1}{180.16 \text{ g mol}^{-1}} = 0.00555 \text{ mol L}^{-1},$$

where the molar mass of glucose is 180.16 g mol^{-1}.

Assuming a temperature of 37 °C for a biological system we have, for one mole of glucose:

$$\Delta G = -2870 \times 10^3 \text{ J} + 8.314 \text{ JK}^{-1} \times 310 \text{K} \times \ln\left(\frac{(0.050)^6 \times 1^6}{0.00555/1 \times (0.132)^6}\right)$$

$$\Delta G = -2870 \times 10^3 \text{ J} + 8.314 \text{ JK}^{-1} \times 310 \text{K} \times \ln\left(\frac{(0.3788)^6}{0.00555}\right)$$

$$\Delta G = -2870 \times 10^3 \text{ J} + 8.314 \text{ JK}^{-1} \times 310 \text{K} \times \ln(0.5323)$$

$$\Delta G = -2870 \times 10^3 \text{ J} + 8.314 \text{ JK}^{-1} \times 310 \text{K} \times (-0.6305)$$

$$\Delta G = -2870 \times 10^3 \text{ J} - 1.637 \times 10^3 \text{ J}$$

$$\Delta G = -2872 \times 10^3 \text{ J}$$

In a similar manner we calculate the Gibbs free energy change for the conversion of ADP to ATP:

$$\Delta G = \Delta G^{\circ\prime} + RT\ln\left(\frac{[ATP]/1\times 1}{[ADP]/1\times[P_i]/1\times([H^+]/10^{-7})}\right)$$

$$\Delta G = 32.4\times10^3\,\mathrm{J} + 8.314\,\mathrm{JK^{-1}}\times310\,\mathrm{K}\times\ln\left(\frac{0.0001}{0.0001\times0.0001\times(10^{-7}/10^{-7})}\right)$$

$$\Delta G = 32.4\times10^3\,\mathrm{J} + 8.314\,\mathrm{JK^{-1}}\times310\,\mathrm{K}\times\ln\left(10^4\right)$$

$$\Delta G = 32.4\times10^3\,\mathrm{J} + 8.314\,\mathrm{JK^{-1}}\times310\,\mathrm{K}\times(9.2103)) = 32.4\times10^3\,\mathrm{J} + 23.738\times10^3\,\mathrm{J} = 56.2\times10^3\,\mathrm{J}$$

(d)   The efficiency under biological conditions is

$$\text{Efficiency} = \frac{\text{number of ATP's actually produced}}{\text{Maximum number of ATP's that can be produced}} = \frac{38}{2872/56.2} = 0.744$$

Thus, the cell's efficiency is about 74%.

The theoretical efficiency of the diesel engine is:

$$\text{Efficiency} = \frac{T_h - T_l}{T_h}\times100\% = \frac{1923-873}{1923}\times100\% = 55\%$$

Thus, the diesel's actual efficiency is $0.78 \times 55\,\% = 43\,\%$.

The cell's efficiency is 74% whereas that of the diesel engine is only 43 %. Why is there such a large discrepancy? The diesel engine supplies heat to the surroundings, which is at a lower temperature than the engine. This dissipation of energy raises the temperature of the surroundings and the entropy of the surroundings. A cell operates under isothermal conditions and the energy not utilized goes only towards changing the entropy of the cell's surroundings. The cell is more efficient since it does not heat the surroundings.

CHAPTER 20
ELECTROCHEMISTRY
PRACTICE EXAMPLES

1A The conventions state that the anode material is written first, and the cathode material is written last.

Anode, Oxidation: $\quad Sc(s) \rightarrow Sc^{3+}(aq) + 3e^-$

Cathode, Reduction: $\quad \{Ag^+(aq) + e^- \rightarrow Ag(s)\} \times 3$

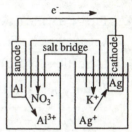

Net Reaction $\qquad Sc(s) + 3Ag^+(aq) \rightarrow Sc^{3+}(aq) + 3Ag(s)$

1B Oxidation of Al(s) at the anode: $\qquad Al(s) \rightarrow Al^{3+}(aq) + 3e^-$

Reduction of $Ag^+(aq)$ at the cathode: $Ag^+(aq) + e^- \rightarrow Ag(s)$

Overall reaction in cell: $Al(s) + 3Ag^+(aq) \rightarrow Al^{3+}(aq) + 3Ag(s)$ $\quad$ Diagram: $Al(s)|Al^{3+}(aq)||Ag^+(aq)|Ag(s)$

2A We obtain the two balanced half-equations and the half-cell potentials from Table 20-1.

Oxidation: $\{Fe^{2+}(aq) \rightarrow Fe^{3+}(aq) + e^-\} \times 2 \qquad -E° = -0.771V$

Reduction: $Cl_2(g) + 2e^- \rightarrow 2Cl^-(aq) \qquad\qquad E° = +1.358V$

Net: $2Fe^{2+}(aq) + Cl_2(g) \rightarrow 2Fe^{3+}(aq) + 2Cl^-(aq);\quad E°_{cell} = +1.358V - 0.771V = +0.587\,V$

2B Since we need to refer to Table 20-1, in any event, it is perhaps a bit easier to locate the two balanced half-equations in the table. There is only one half-equation involving both $Fe^{2+}(aq)$ and $Fe^{3+}(aq)$ ions. It is reversed and written as an oxidation below. The half-equation involving $MnO_4^-(aq)$ is also written below. [Actually, we need to know that in acidic solution $Mn^{2+}(aq)$ is the principal reduction product of $MnO_4^-(aq)$.]

Oxidation: $\{Fe^{2+}(aq) \rightarrow Fe^{3+}(aq) + e^-\} \times 5 \qquad\qquad -E° = -0.771V$

Reduction: $MnO_4^-(aq) + 8H^+(aq) + 5e^- \rightarrow Mn^{2+}(aq) + 4H_2O(l) \qquad E° = +1.51V$

Net: $MnO_4^-(aq) + 5Fe^{2+}(aq) + 8H^+(aq) \rightarrow Mn^{2+}(aq) + 5Fe^{3+}(aq) + 4H_2O(l)$

$E°_{cell} = +1.51V - 0.771V = +0.74\,V$

3A We write down the oxidation half-equation following the method of Chapter 5, and obtain the reduction half-equation from Table 20-1, along with the reduction half-cell potential.

Oxidation: $\{H_2C_2O_4(aq) \longrightarrow 2CO_2(aq) + 2H^+(aq) + 2e^-\} \times 3 \qquad -E°\{CO_2/H_2C_2O_4\}$

Reduction: $Cr_2O_7^{2-}(aq) + 14H^+(aq) + 6e^- \rightarrow 2Cr^{3+}(aq) + 7H_2O(l) \qquad E° = +1.33\,V$

Net: $Cr_2O_7^{2-}(aq) + 8H^+(aq) + 3H_2C_2O_4(aq) \rightarrow 2Cr^{3+}(aq) + 7H_2O(l) + 6CO_2(g)$

$E°_{cell} = +1.81V = +1.33V - E°\{CO_2/H_2C_2O_4\};\quad E°\{CO_2/H_2C_2O_4\} = 1.33V - 1.81V = -0.48V$

3B The 2nd ½ -reaction must have $O_2(g)$ as reactant and $H_2O(l)$ as product.

Oxidation: $\{Cr^{2+}+(aq)\longrightarrow Cr^{3+}(aq)+e^-\}\ \times 4$ $\qquad -E°\{Cr^{3+}/Cr^{2+}\}$

Reduction: $O_2(g)+4H^+(aq)+4e^-\rightarrow 2H_2O(l)$ $\qquad\qquad E° = +1.229V$

Net: $O_2(g)+4H^+(aq)+4Cr^{2+}(aq)\rightarrow 2H_2O(l)+4Cr^{3+}(aq)$

$E_{cell}° = +1.653V = +1.229V - E°\{Cr^{3+}/Cr^{2+}\}; \quad E°\{Cr^{3+}/Cr^{2+}\}=1.229V-1.653V=-0.424V$

4A First we write down the two half-equations, obtain the half-cell potential for each, and then calculate $E°_{cell}$. From that value, we determine $\Delta G°$

Oxidation: $\{Al(s)\rightarrow Al^{3+}(aq)+3e^-\}\ \times 2$ $\qquad -E° = +1.676V$

Reduction: $\{Br_2(l)+2e^-\rightarrow 2\,Br^-(aq)\}\ \times 3$ $\qquad E° = +1.065V$

Net: $2Al(s)+3Br_2(l)\rightarrow 2Al^{3+}(aq)+6Br^-(aq)$ $E°_{cell} = 1.676\,V +1.065\,V = 2.741\,V$

$\Delta G° = -nFE°_{cell} = -6\,mol\ e^- \times \dfrac{96,485\ C}{1\,mol\ e^-}\times 2.741\,V = -1.587\times 10^6\,J = -1587\,kJ$

4B First we write down the two half-equations, one of which is the reduction equation from the previous example. The other is the oxidation that occurs in the standard hydrogen electrode.

Oxidation: $2H_2(g)\rightarrow 4H^+(aq)+4e^-$

Reduction: $O_2(g)+4H^+(aq)+4e^-\rightarrow 2H_2O(l)$

Net: $\qquad 2\,H_2(g) + O_2(g)\rightarrow 2\,H_2O(l)$ $\qquad n = 4$ in this reaction.

This net reaction is simply twice the formation reaction for $H_2O(l)$ and, therefore,

$\Delta G° = 2\Delta G_f°\left[H_2O(l)\right] = 2\times(-237.1\ kJ) = -474.2\times 10^3\ J = -nFE°_{cell}$

$E°_{cell} = \dfrac{-\Delta G°}{nF} = \dfrac{-(-474.2\times 10^3\ J)}{4\ mol\ e^- \times \dfrac{96,485\,C}{mol\,e^-}} = +1.229\ V = E°$, as we might expect.

5A Cu(s) will displace metal ions of a metal less active than copper. Silver ion is one example.

Oxidation: $Cu(s)\rightarrow Cu^{2+}(aq)+2e^-$ $\qquad\qquad -E° = -0.340V$ (from Table 20.1)

Reduction: $\{Ag^+(aq)+e^-\rightarrow Ag(s)\}\ \times 2$ $\qquad E° = +0.800V$

Net: $2Ag^+(aq)+Cu(s)\rightarrow Cu^{2+}(aq)+2Ag(s)$ $\qquad E°_{cell} = -0.340\,V +0.800\,V = +0.460\,V$

5B We determine the value for the hypothetical reaction's cell potential.

Oxidation: $\{Na(s)\rightarrow Na^+(aq)+e^-\}\ \times 2$ $\qquad\qquad -E°= +2.713\,V$

Reduction: $Mg^{2+}(aq)+2e^-\rightarrow Mg(s)$ $\qquad\qquad E°=-2.356\,V$

Net: $2\,Na(s)+Mg^{2+}(aq)\rightarrow 2\,Na^+(aq) + Mg(s)$ $\qquad E°_{cell}= 2.713\,V - 2.356\,V = +0.357\,V$

The method is not feasible because another reaction occurs that has a more positive cell potential, i.e., Na(s) reacts with water to form $H_2(g)$ and NaOH(aq), which has

$E°_{cell} = 2.713\,V - 0.828\,V = +1.885\,V$.

6A The oxidation is that of SO_4^{2-} to $S_2O_8^{2-}$, the reduction is that of O_2 to H_2O.

Oxidation: $\{2\,SO_4^{2-}\,(aq) \rightarrow S_2O_8^{2-}\,(aq) + 2e^-\} \times 2$ $\qquad -E^\circ = -2.01\,V$

Reduction: $O_2\,(g) + 4\,H^+\,(aq) + 4e^- \rightarrow 2H_2O(l)$ $\qquad E^\circ = +1.229\,V$

Net: $\qquad O_2\,(g) + 4\,H^+\,(aq) + 2\,SO_4^{2-}\,(aq) \rightarrow S_2O_8^{2-}\,(aq) + 2H_2O(l)$ $\quad E^\circ_{cell} = -0.78\,V$

Because the standard cell potential is negative, we conclude that this cell reaction is nonspontaneous under standard conditions. This would not be a feasible method of producing peroxodisulfate ion.

6B **(1)** The oxidation is that of $Sn^{2+}(aq)$ to $Sn^{4+}(aq)$, the reduction is that of O_2 to H_2O.

Oxidation: $\{Sn^{2+}(aq) \rightarrow Sn^{4+}(aq) + 2e^-\} \times 2$ $\qquad -E^\circ = -0.154\,V$

Reduction: $O_2\,(g) + 4\,H^+\,(aq) + 4e^- \rightarrow 2H_2O(l)$ $\qquad E^\circ = +1.229\,V$

Net: $\qquad O_2\,(g) + 4H^+\,(aq) + 2Sn^{2+}\,(aq) \rightarrow 2Sn^{4+}\,(aq) + 2H_2O(l)$ $\quad E^\circ_{cell} = +1.075\,V$

Since the standard cell potential is positive, this cell reaction is spontaneous under standard conditions.

(2) The oxidation is that of $Sn(s)$ to $Sn^{2+}(aq)$, the reduction is still that of O_2 to H_2O.

Oxidation: $\{Sn(s) \rightarrow Sn^{2+}(aq) + 2e^-\}$ $\qquad \times 2$ $\qquad -E^\circ = +0.137\,V$

Reduction: $O_2\,(g) + 4\,H^+\,(aq) + 4e^- \rightarrow 2\,H_2O(l)$ $\qquad E^\circ = +1.229\,V$

Net: $\qquad O_2\,(g) + 4\,H^+\,(aq) + 2Sn\,(s) \rightarrow 2Sn^{2+}\,(aq) + 2\,H_2O(l)$

$E^\circ_{cell} = 0.137\,V + 1.229\,V = +1.366\,V$

The standard cell potential for this reaction is more positive than that for situation (1). Thus, reaction (2) should occur preferentially. Also, if $Sn^{4+}(aq)$ is formed, it should react with $Sn(s)$ to form $Sn^{2+}(aq)$.

Oxidation: $Sn(s) \rightarrow Sn^{2+}(aq) + 2e^-$ $\qquad -E^\circ = +0.137\,V$

Reduction: $Sn^{4+}\,(aq) + 2e^- \rightarrow Sn^{2+}\,(aq)$ $\qquad E^\circ = +0.154\,V$

Net: $Sn^{4+}(aq) + Sn(s) \rightarrow 2Sn^{2+}(aq)$ $\qquad E^\circ_{cell} = +0.137\,V + 0.154\,V = +0.291\,V$

7A For the reaction $2\,Al(s) + 3\,Cu^{2+}(aq) \rightarrow 2\,Al^{3+}(aq) + 3\,Cu(s)$ we know $n = 6$ and $E^\circ_{cell} = +2.013\,V$. We calculate the value of K_{eq}.

$$E^\circ_{cell} = \frac{0.0257}{n}\,\ln K_{eq}; \quad \ln K_{eq} = \frac{nE^\circ_{cell}}{0.0257} = \frac{6 \times (+2.013)}{0.0257} = 470; \quad K_{eq} = e^{470} = 10^{204}$$

The huge size of the equilibrium constant indicates that this reaction indeed will go to essentially 100% to completion.

7B We first determine the value of E°_{cell} from the half-cell potentials.

Oxidation: $Sn(s) \rightarrow Sn^{2+}(aq) + 2e^-$ $\qquad -E^\circ = +0.137\,V$

Reduction: $Pb^{2+}(aq) + 2e^- \rightarrow Pb(s)$ $\qquad E^\circ = -0.125\,V$

Net: $Pb^{2+}(aq) + Sn(s) \rightarrow Pb(s) + Sn^{2+}(aq)$ $\qquad E^\circ_{cell} = +0.137\,V - 0.125\,V = +0.012\,V$

$$E^\circ_{cell} = \frac{0.0257}{n}\ln K_{eq} \qquad \ln K_{eq} = \frac{nE^\circ_{cell}}{0.0257} = \frac{2\times(+0.012)}{0.0257} = 0.93 \qquad K_{eq} = e^{0.93} = 2.5$$

The equilibrium constant's small size $(0.001 < K < 1000)$ indicates that this reaction will not go to completion.

8A We first need to determine the standard cell voltage and the cell reaction.

Oxidation: $\{Al(s) \rightarrow Al^{3+}(aq) + 3\,e^-\} \times 2$ $\qquad -E^\circ = +1.676\,V$

Reduction: $\{Sn^{4+}(aq) + 2\,e^- \rightarrow Sn^{2+}(aq)\} \times 3$ $\qquad E^\circ = +0.154\,V$

Net: $2\,Al(s) + 3\,Sn^{4+}(aq) \rightarrow 2\,Al^{3+}(aq) + 3\,Sn^{2+}(aq)$ $\qquad E^\circ_{cell} = +1.676\,V + 0.154\,V = +1.830\,V$

Note that $n = 6$. We now set up and substitute into the Nernst equation.

$$E_{cell} = E^\circ_{cell} - \frac{0.0592}{n}\log\frac{[Al^{3+}]^2[Sn^{2+}]^3}{[Sn^{4+}]^3} = +1.830 - \frac{0.0592}{6}\log\frac{(0.36\,M)^2(0.54\,M)^3}{(0.086\,M)^3}$$

$$= +1.830\,V - 0.0149\,V = +1.815\,V$$

8B We first need to determine the standard cell voltage and the cell reaction.

Oxidation: $2\,Cl^-(1.0\,M) \rightarrow Cl_2(1\,atm) + 2e^-$ $\qquad -E^\circ = -1.358\,V$

Reduction: $PbO_2(s) + 4H^+(aq) + 2e^- \rightarrow Pb^{2+}(aq) + 2H_2O(l)$ $\qquad E^\circ = +1.455\,V$

Net: $PbO_2(s) + 4H^+(0.10\,M) + 2Cl^-(1.0\,M) \rightarrow Cl_2(1\,atm) + Pb^{2+}(0.050\,M) + 2H_2O(l)$

$E^\circ_{cell} = -1.358\,V + 1.455\,V = +0.097\,V$ $\qquad$ Note that $n = 2$. Substitute values into the Nernst equation.

$$E_{cell} = E^\circ_{cell} - \frac{0.0592}{n}\log\frac{P\{Cl_2\}[Pb^{2+}]}{[H^+]^4[Cl^-]^2} = +0.097 - \frac{0.0592}{2}\log\frac{(1.0\,atm)(0.050\,M)}{(0.10\,M)^4(1.0\,M)^2}$$

$$= +0.097\,V - 0.080\,V = +0.017\,V$$

9A The cell reaction is $2\,Fe^{3+}(0.35\,M) + Cu(s) \rightarrow 2\,Fe^{2+}(0.25\,M) + Cu^{2+}(0.15\,M)$ with

$n = 2$ and $E^\circ_{cell} = -0.337\,V + 0.771\,V = 0.434\,V$ $\qquad$ Next, substitute this voltage and the concentrations into the Nernst equation.

$$E_{cell} = E^\circ_{cell} - \frac{0.0592}{n}\log\frac{[Fe^{2+}]^2[Cu^{2+}]}{[Fe^{3+}]^2} = 0.434 - \frac{0.0592}{2}\log\frac{(0.25)^2(0.15)}{(0.35)^2} = 0.434 + 0.033$$

$E_{cell} = +0.467\,V$ $\quad$ Thus the reaction is spontaneous under standard conditions as written.

9B The reaction is not spontaneous under standard conditions in either direction when $E_{cell} = 0.000 \text{ V}$. We use the standard cell potential from Example 20-9.

$$E_{cell} = E_{cell}^{o} - \frac{0.0592}{n} \log \frac{\left[Sn^{2+} \right]}{\left[Pb^{2+} \right]}; \quad 0.000 \text{ V} = 0.012 \text{ V} - \frac{0.0592}{2} \log \frac{\left[Sn^{2+} \right]}{\left[Pb^{2+} \right]}$$

$$\log \frac{\left[Sn^{2+} \right]}{\left[Pb^{2+} \right]} = \frac{0.012 \times 2}{0.0592} = 0.41; \quad \frac{\left[Sn^{2+} \right]}{\left[Pb^{2+} \right]} = 10^{0.41} = 2.6$$

10A In this concentration cell $E_{cell}^{o} = 0.000$ V because the same reaction occurs at anode and cathode, only the concentrations of the ions differ. $\left[Ag^{+} \right] = 0.100$ M in the cathode compartment. The anode compartment contains a saturated solution of AgCl(aq).

$$K_{sp} = 1.8 \times 10^{-10} = \left[Ag^{+} \right]\left[Cl^{-} \right] = s^2; \quad s = \sqrt{1.8 \times 10^{-10}} = 1.3 \times 10^{-5} \text{ M}$$

Now we apply the Nernst equation. The cell reaction is

$$Ag^{+}(0.100 \text{ M}) \rightarrow Ag^{+}(1.3 \times 10^{-5} \text{ M})$$

$$E_{cell} = 0.000 - \frac{0.0592}{1} \log \frac{1.3 \times 10^{-5} \text{ M}}{0.100 \text{ M}} = +0.23 \text{ V}$$

10B Because the electrodes in this cell are identical, the standard electrode potentials are numerically equal and subtracting one from the other leads to the value $E_{cell}^{o} = 0.000$ V. However, because the ion concentrations differ, there is a potential difference between the two half cells (non-zero nonstandard voltage for the cell). $\left[Pb^{2+} \right] = 0.100$ M in the cathode compartment. While the anode compartment contains a saturated solution of PbI_2. We use the Nernst equation (with $n = 2$) to determine $\left[Pb^{2+} \right]$ in the saturated solution.

$$E_{cell} = +0.0567 \text{ V} = 0.000 - \frac{0.0592}{2} \log \frac{x \text{ M}}{0.100 \text{ M}}; \quad \log \frac{x \text{ M}}{0.100 \text{ M}} = \frac{2 \times 0.0567}{-0.0592} = -1.92$$

$$\frac{x \text{ M}}{0.100 \text{ M}} = 10^{-1.92} = 0.012; \quad \left[Pb^{2+} \right]_{anode} = x \text{ M} = 0.012 \times 0.100 \text{ M} = 0.0012 \text{ M};$$

$$\left[I^{-} \right] = 2 \times 0.0012 \text{ M} = 0.0024 \text{ M}$$

$$K_{sp} = \left[Pb^{2+} \right]\left[I^{-} \right]^2 = (0.0012)(0.0024)^2 = 6.9 \times 10^{-9} \text{ compared with } 7.1 \times 10^{-9} \text{ in Appendix D}$$

11A From Table 20-1 we choose one oxidations and one reductions reaction so as to get the least negative cell voltage. This will be the most likely pair of ½ -reactions to occur.

Oxidation: $2I^-(aq) \rightarrow I_2(s) + 2 e^-$ $-E° = -0.535$ V

            $2H_2O(l) \rightarrow O_2(g) + 4H^+(aq) + 4e^-$ $-E° = -1.229$ V

Reduction: $K^+(aq) + e^- \rightarrow K(s)$ $E° = -2.924$ V

            $2H_2O(l) + 2e^- \rightarrow H_2(g) + 2OH^-(aq)$ $E° = -0.828$ V

The least negative standard cell potential $(-0.535 \text{ V} - 0.828 \text{ V} = -1.363 \text{ V})$ occurs when $I_2(s)$ is produced by oxidation at the anode, and $H_2(g)$ is produced by reduction at the cathode.

11B We obtain from Table 20-1 all the possible oxidations and reductions and choose one of each to get the least negative cell voltage. That pair is the most likely pair of half-reactions to occur.

Oxidation: $2H_2O(l) \rightarrow O_2(g) + 4H^+(aq) + 4e^-$ $-E° = -1.229$V

            $Ag(s) \rightarrow Ag^+(aq) + e^-$ $-E° = -0.800$V

            [We cannot further oxidize $NO_3^-(aq)$ or $Ag^+(aq)$.]

Reduction: $Ag^+(aq) + e^- \rightarrow Ag(s)$ $E° = +0.800$V

            $2H_2O(l) + 2e^- \rightarrow H_2(g) + 2OH^-(aq)$ $E° = -0.828$V

Thus, we expect to form silver metal at the cathode and $Ag^+(aq)$ at the anode.

12A The half-cell equation is $Cu^{2+}(aq) + 2e^- \rightarrow Cu(s)$, indicating that two moles of electrons are required for each mole of copper deposited. Current is measured in amperes, or coulombs per second. We convert the mass of copper to coulombs of electrons needed for the reduction and the time in hours to seconds.

$$\text{Current} = \frac{12.3 \text{ g Cu} \times \dfrac{1 \text{ mol Cu}}{63.55 \text{ g Cu}} \times \dfrac{2 \text{ mol e}^-}{1 \text{ mol Cu}} \times \dfrac{96{,}485 \text{ C}}{1 \text{ mol e}^-}}{5.50 \text{ h} \times \dfrac{60 \text{ min}}{1 \text{ h}} \times \dfrac{60 \text{ s}}{1 \text{ min}}} = \frac{3.73_5 \times 10^4 \text{ C}}{1.98 \times 10^4 \text{ s}} = 1.89 \text{ amperes}$$

12B We first determine the moles of $O_2(g)$ produced with the ideal gas equation.

$$\text{moles } O_2(g) = \frac{\left(738 \text{ mm Hg} \times \dfrac{1 \text{ atm}}{760 \text{ mm Hg}}\right) \times 2.62 \text{ L}}{\dfrac{0.08206 \text{ L atm}}{\text{mol K}} \times (26.2 + 273.2) \text{ K}} = 0.104 \text{ mol } O_2$$

Then we determine the time needed to produce this amount of O_2.

$$\text{elapsed time} = 0.104 \text{ mol } O_2 \times \frac{4 \text{ mol e}^-}{1 \text{ mol } O_2} \times \frac{96{,}485 \text{ C}}{1 \text{ mol e}^-} \times \frac{1 \text{ s}}{2.13 \text{ C}} \times \frac{1 \text{ h}}{3600 \text{ s}} = 5.23 \text{ h}$$

EXERCISES

Standard Electrode Potential

1. **(a)** If the metal dissolves in HNO_3, it has a reduction potential that is smaller than $E^\circ\{NO_3^-(aq)/NO(g)\} = 0.956\,V$. If it also does not dissolve in HCl, it has a reduction potential that is larger than $E^\circ\{H^+(aq)/H_2(g)\} = 0.000\,V$. If it displaces $Ag^+(aq)$ from solution, then it has a reduction potential that is smaller than $E^\circ\{Ag^+(aq)/Ag(s)\} = 0.800\,V$. But if it does not displace $Cu^{2+}(aq)$ from solution, then its reduction potential is larger than $E^\circ\{Cu^{2+}(aq)/Cu(s)\} = 0.340\,V$ Thus, $0.340\,V < E^\circ < 0.800\,V$

(b) If the metal dissolves in HCl, it has a reduction potential that is smaller than $E^\circ\{H^+(aq)/H_2(g)\} = 0.000\,V$. If it does not displace $Zn^{2+}(aq)$ from solution, its reduction potential is larger than $E^\circ\{Zn^{2+}(aq)/Zn(s)\} = -0.763\,V$. If it also does not displace $Fe^{2+}(aq)$ from solution, its reduction potential is larger than $E^\circ\{Fe^{2+}(aq)/Fe(s)\} = -0.440\,V$. $-0.440\,V < E^\circ < 0.000\,V$

2. We would place a strip of solid indium metal into each of the metal ion solutions and see if the dissolved metal plates out on the indium strip. Similarly, strips of all the other metals would be immersed in a solution of In^{3+} to see if indium metal plates out. Eventually, we will find one metal whose ions are displaced by indium and another metal that displaces indium from solution, which are adjacent to each other in Table 20-1. The standard electrode potential for the $In/In^{3+}(aq)$ pair will lie between the standard reduction potentials for these two metals. This technique will work only if indium metal does not react with water, that is, if the standard reduction potential of $In^{3+}(aq)/In(s)$ is greater than about $-1.8\,V$. The inaccuracy inherent in this technique is due to overpotentials, which can be as much as $0.200\,V$. Its imprecision is limited by the closeness of the reduction potentials for the two bracketing metals

3. We separate the given equation into its two half-equations. One of them is the reduction of nitrate ion in acidic solution, whose standard half-cell potential we retrieve from Table 20-1 and use to solve the problem.

Oxidation: $\{Pt(s) + 4Cl^-(aq) \longrightarrow [PtCl_4]^{2-}(aq) + 2\,e^-\} \times 3;$ $-E^\circ\{[PtCl_4]^{2-}(aq)/Pt(s)\}$
Reduction: $\{NO_3^-(aq) + 4H^+(aq) + 3e^- \rightarrow NO(g) + 2H_2O(l)\} \times 2;$ $E^\circ = +0.956\,V$

Net: $3Pt(s) + 2NO_3^-(aq) + 8H^+(aq) + 12Cl^-(aq) \rightarrow 3[PtCl_4]^{2-}(aq) + 2NO(g) + 6H_2O(l)$
$E^\circ_{cell} = 0.201\,V = +0.956\,V - E^\circ\{[PtCl_4]^{2-}(aq)/Pt(s)\}$
$E^\circ\{[PtCl_4]^{2-}(aq)/Pt(s)\} = 0.956\,V - 0.201\,V = +0.755\,V$

4. We separate the given equation into its two half-equations. One of them is the reduction of $Cl_2(g)$ to $Cl^-(aq)$ whose standard half-cell potential we obtain from Table 20-1 and use to solve the problem.

    Oxidation: $\{Na(\text{in Hg}) \rightarrow Na^+(aq) + e^-\} \times 2$ $-E°\{Na^+(aq)/Na(\text{in Hg})\}$

    Reduction: $Cl_2(g) + 2e^- \rightarrow 2Cl^-(aq)$ $E° = +1.358$ V
    _____

    Net: $2Na(\text{in Hg}) + Cl_2(g) \rightarrow 2Na^+(aq) + 2Cl^-(aq)$ $E°_{cell} = 3.20$ V

    $E°_{cell} = 3.20$ V $= +1.3858$ V $- E°\{Na^+(aq)/Na(\text{in Hg})\}$

    $E°\{Na^+(aq)/Na(\text{in Hg})\} = 1.358$ V $- 3.20$ V $= -1.84$ V

5. We divide the net cell equation into two half-equations.

    Oxidation: $\{Al(s) + 4\ OH^-(aq) \rightarrow [Al(OH)_4]^-(aq) + 3\ e^-\} \times 4$; $-E°\{[Al(OH)_4]^-(aq)/Al(s)\}$

    Reduction: $\{O_2(g) + 2\ H_2O(l) + 4e^- \rightarrow 4\ OH^-(aq)\} \times 3$; $E° = +0.401$ V
    _____

    Net: $4Al(s) + 3O_2(g) + 6H_2O(l) + 4OH^-(aq) \rightarrow 4[Al(OH)_4]^-(aq)$ $E°_{cell} = 2.71$ V

    $E°_{cell} = 2.71$ V $= +0.401$ V $- E°\{[Al(OH)_4]^-(aq)/Al(s)\}$

    $E°\{[Al(OH)_4]^-(aq)/Al(s)\} = 0.401$ V $- 2.71$ V $= -2.31$ V

6. We divide the net cell equation into two half-equations.

    Oxidation: $CH_4(g) + 2H_2O(l) \rightarrow CO_2(g) + 8H^+(aq) + 8e^-$ $-E°\{CO_2(g)/CH_4(g)\}$

    Reduction: $\{O_2(g) + 4H^+(aq) + 4e^- \rightarrow 2H_2O(l)\} \times 2$ $E° = +1.229$ V
    _____

    Net: $CH_4(g) + 2O_2(g) \rightarrow CO_2(g) + 2H_2O(l)$ $E°_{cell} = 1.06$ V

    $E°_{cell} = 1.06$ V $= +1.229$ V $- E°\{CO_2(g)/CH_4(g)\}$

    $E°\{CO_2(g)/CH_4(g)\} = 1.229$ V $- 1.06$ V $= +0.17$ V

7. Since $Zn^{2+}(aq)$ undergoes reduction, and the other metal (M) undergoes oxidation,

    $$E°_{cell} = E°_{Zn} - E°_M \qquad \text{or} \qquad E°_M = E°_{Zn} - E°_{cell}$$

    (a) $E°_{Mn} = E°_{Zn} - E°_{cell} = -0.763$ V $- 0.417$ V $= -1.180$ V

    (b) $E°_{Po} = E°_{Zn} - E°_{cell} = -0.763$ V $+ 1.13$ V $= +0.37$ V

    (c) $E°_{Ti} = E°_{Zn} - E°_{cell} = -0.763$ V $- 0.87$ V $= -1.63$ V

    (d) $E°_V = E°_{Zn} - E°_{cell} = -0.763$ V $- 0.37$ V $= -1.13$ V

8. (a) Oxidation: $2\ Cl^-(aq) \rightarrow Cl_2(g) + 2\ e^-$ $-E° = -1.358$ V

    Reduction: $PbO_2(s) + 4H^+(aq) + 2e^- \rightarrow Pb^{2+}(aq) + 2H_2O(l)$ $E° = +1.455$ V
    _____

    Net: $2\ Cl^-(aq) + PbO_2(s) + 4\ H^+(aq) \rightarrow Cl_2(g) + Pb^{2+}(aq) + 2H_2O(l)$ $E°_{cell} = 0.097$ V

(b) Oxidation: $\{Mg(s) \rightarrow Mg^{2+}(aq)+2\,e^-\}\ \times 3$ $\qquad\qquad -E^\circ = +2.356\ V$

Reduction: $\{Sc^{3+}(aq)+3\,e^- \rightarrow Sc(s)\}\quad \times 2$ $\qquad\qquad E^\circ\{Sc^{3+}/Sc\}$

Net: $3\ Mg(s)+2\ Sc^{3+}(aq) \rightarrow 3\ Mg^{2+}(aq)+2\ Sc(s)$ $\qquad E^\circ_{cell} = +0.33\ V$

$E^\circ\{Sc^{3+}/Sc\} = +0.33\ V - 2.356\ V = -2.03\ V$

(c) Oxidation: $Cu^+(aq) \rightarrow Cu^{2+}(aq)+e^-$ $\qquad\qquad\qquad -E^\circ\{Cu^{2+}/Cu^+\}$

Reduction: $Ag^+(aq)+e^- \rightarrow Ag(s)$ $\qquad\qquad\qquad\qquad E^\circ = +0.800\ V$

Net: $Cu^+(aq)+Ag^+(aq) \rightarrow Cu^{2+}(aq)+Ag(s)$ $\qquad E^\circ_{cell} = +0.641\ V$

$-E^\circ\{Cu^{2+}/Cu^+\} = +0.641\ V - 0.800\ V = -0.159\ V \qquad E^\circ\{Cu^{2+}/Cu^+\} = +0.159\ V$

Predicting Oxidation-Reduction Reactions

9. The pertinent reactions are shown below.

$3\,Na(s)+Al^{3+}(aq) \rightarrow 3\,Na^+(aq)+Al(s);$ $\qquad\qquad E^\circ_{cell}=-(-2.713\ V)+(-1.676\ V)=+1.037\ \ V$

$2\,Na(s) + H_2O(l) \rightarrow 2\,Na^+(aq)+ H_2(g)+ 2\,OH^-(aq);\ E^\circ_{cell}=-(-2.713\ V)+(-0.828\ V)=+1.885\ V$

Both reactions are spontaneous under standard conditions. However, the reduction of water has a more favorable cell potential when compared to the reduction of aluminum metal. The only half reactions that occurs is this system is the reduction of water and the oxidation of sodium metal.

10. The reaction that occurs is $Zn(s) \rightarrow Zn^{2+}(aq)+2\,e^-$. Some of the electrons produced in this reaction move to the copper surface, where the following reduction occurs.
$2\,H^+(aq)+2\,e^- \rightarrow H_2(g)$. The copper does not react with the HCl(aq) even under these circumstances. Bubbles of hydrogen form more readily on the copper surface than they do on a zinc metal surface. We say that the copper surface has a lower overpotential. Not as high a voltage is required to produce $H_2(g)$ because of the arrangement of atoms on the copper surface.

11. **(a)** Oxidation: $Sn(s) \rightarrow Sn^{2+}(aq)+2\,e^-$ $\qquad\qquad -E^\circ = +0.137\ V$

Reduction: $Pb^{2+}(aq)+2\,e^- \rightarrow Pb(s)$ $\qquad\qquad E^\circ=-0.125\ V$

Net: $Sn(s)+Pb^{2+}(aq) \rightarrow Sn^{2+}(aq)+Pb(s)$ $\qquad E^\circ_{cell} = +0.012\,V \qquad$ Spontaneous

(b) Oxidation: $2I^-(aq) \rightarrow I_2(s)+2\,e^-$ $\qquad\qquad -E^\circ = -0.535\ V$

Reduction: $Cu^{2+}(aq)+2\,e^- \rightarrow Cu(s)$ $\qquad\qquad E^\circ = +0.340\,V$

Net: $2I^-(aq) + Cu^{2+}(aq) \rightarrow Cu(s) + I_2(s)$ $\qquad E^\circ_{cell}=-0.195\ V\quad$ Nonspontaneous

(c) Oxidation: $\{2H_2O(l) \rightarrow O_2(g)+4H^+(aq)+4e^-\} \times 3$ $-E° = -1.229$ V

Reduction: $\{NO_3^-(aq)+4H^+(aq)+3e^- \rightarrow NO(g)+2H_2O(l)\} \times 4$ $E° = +0.956$ V

Net: $4NO_3^-(aq) + 4H^+(aq) \rightarrow 3O_2(g) + 4NO(g) + 2H_2O(l)$ $E°_{cell}=-0.273$ V

This reaction is not spontaneous under standard conditions.

(d) Oxidation: $Cl^-(aq) + 2OH^-(aq) \rightarrow OCl^-(aq) + H_2O(l) + 2e^-$ $-E°=-0.890$ V

Reduction: $O_3(g) + H_2O(l) + 2e^- \rightarrow O_2(g) + 2OH^-(aq)$ $E° = +1.246$ V

Net: $Cl^-(aq) + O_3(g) \rightarrow OCl^-(aq) + O_2(g)$ (basic solution) $E°_{cell} = +0.356$ V

This reaction is spontaneous under standard conditions.

12. It is more difficult to oxidize $Hg(l)$ to $Hg_2^{2+}(-0.797$ V$)$ than it is to reduce H^+ to H_2 (0.000 V); $Hg(l)$ will not dissolve in 1 M HCl. The standard reduction of nitrate ion to $NO(g)$ in acidic solution is strongly spontaneous in acidic media ($+0.956$ V). This can help overcome the reluctance of Hg to be oxidized. $Hg(l)$ will react with and dissolve in the $HNO_3(aq)$.

13. **(a)** Oxidation: $Mg(s) \rightarrow Mg^{2+}(aq) + 2e^-$ $-E° = +2.356$ V

Reduction: $Pb^{2+}(aq)+2e^- \rightarrow Pb(s)$ $E° = -0.125$ V

Net: $Mg(s) + Pb^{2+}(aq) \rightarrow Mg^{2+}(aq) + Pb(s)$ $E°_{cell} = +2.231$ V

This reaction occurs to a significant extent.

(b) Oxidation: $Sn(s) \rightarrow Sn^{2+}(aq)+2e^-$ $-E° = +0.137$ V

Reduction: $2H^+(aq) \rightarrow H_2(g)$ $E° = 0.000$ V

Net: $Sn(s) + 2H^+(aq) \rightarrow Sn^{2+}(aq) + H_2(g)$ $E°_{cell} = +0.137$ V

This reaction will occur to a significant extent.

(c) Oxidation: $Sn^{2+}(aq) \rightarrow Sn^{4+}(aq)+2e^-$ $-E° = -0.154$ V

Reduction: $SO_4^{2-}(aq)+4H^+(aq)+2e^- \rightarrow SO_2(g)+2H_2O(l)$ $E°=+0.17$ V

Net: $Sn^{2+}(aq)+SO_4^{2-}(aq)+4H^+(aq) \rightarrow Sn^{4+}(aq)+SO_2(g)+2H_2O(l)$ $E°_{cell}= +0.02$ V

This reaction will occur, but not to a large extent.

(d) Oxidation: $\{H_2O_2(aq) \rightarrow O_2(g)+2H^+(aq)+2e^-\} \times 5$ $-E° = -0.695$ V

Reduction: $\{MnO_4^-(aq)+8H^+(aq)+5e^- \rightarrow Mn^{2+}(aq)+4H_2O(l)\} \times 2$ $E° = +1.51$ V

Net: $5H_2O_2(aq)+2MnO_4^-(aq)+6H^+(aq) \rightarrow 5O_2(g)+2Mn^{2+}(aq)+8H_2O(l)$ $E°_{cell} = +0.82$ V

This reaction will occur to a significant extent.

(e)   Oxidation: $2\,Br^-\,(aq) \rightarrow Br_2\,(aq) + 2\,e^-$ $-E^\circ = -1.065\,V$

Reduction: $I_2\,(s) + 2\,e^- \rightarrow 2\,I^-\,(aq)$ $E^\circ = +0.535\,V$

Net: $2\,Br^-\,(aq) + I_2\,(s) \rightarrow Br_2\,(aq) + 2\,I^-\,(aq)$ $E^\circ_{cell} = -0.530\,V$

This reaction will not occur to a significant extent.

14.   The relatively small positive value of E°_{cell} for the reaction indicates that the reaction will proceed in the forward reaction, but will stop short of completion. A much larger positive value of E°_{cell} would be necessary before we would conclude that the reaction goes to completion. For example, we can compute the value of the equilibrium constant for this reaction (see below). A value of 1000 or more is needed before we can describe the reaction as one that goes to completion.

$$E^\circ_{cell} = \frac{0.0257}{n}\ln K_{eq} \quad \ln K_{eq} = \frac{n \times E^\circ_{cell}}{0.0257} = \frac{2 \times 0.02}{0.0257} = 2$$

$K_{eq} = e^2 = 7$ (does not go to completion)

<u>15.</u>   If E°_{cell} is positive, the reaction will occur. For the reduction of $Cr_2O_7^{2-}$ to $Cr^{3+}\,(aq)$:

$$Cr_2O_7^{2-}\,(aq) + 14\,H^+\,(aq) + 6\,e^- \rightarrow 2\,Cr^{3+}\,(aq) + 7\,H_2O(l) \qquad E^\circ = +1.33\,V$$

If the oxidation has $-E^\circ$ smaller (more negative) than $-1.33\,V$, the oxidation will not occur.

(a) $Sn^{2+}\,(aq) \rightarrow Sn^{4+}\,(aq) + 2\,e^-$ $-E^\circ = -0.154\,V$

Hence, $Sn^{2+}(aq)$ can be oxidized to $Sn^{4+}(aq)$ by $Cr_2O_7^{2-}(aq)$.

(b) $I_2\,(s) + 6\,H_2O(l) \rightarrow 2\,IO_3^-\,(aq) + 12\,H^+\,(aq) + 10\,e^-$ $-E^\circ = -1.20\,V$

$I_2(s)$ can be oxidized to $IO_3^-(aq)$ by $Cr_2O_7^{2-}(aq)$.

(c) $Mn^{2+}\,(aq) + 4\,H_2O(l) \rightarrow MnO_4^-\,(aq) + 8\,H^+\,(aq) + 5\,e^-$ $-E^\circ = -1.51\,V$

$Mn^{2+}(aq)$ cannot be oxidized to $MnO_4^-(aq)$ by $Cr_2O_7^{2-}(aq)$.

16.   In order to reduce Eu^{3+} to Eu^{2+}, a stronger reducing agent than Eu^{2+} is required. From the list given, Al(s) and $H_2C_2O_4$(aq) are stronger reducing agents. This is determined by looking at the reduction potentials (-1.676 V for Al^{3+}/Al(s) and -0.49 V for CO_2, H^+/$H_2C_2O_4$(aq)), are more negative than -0.43 V). Co(s), H_2O_2 and Ag(s) are not strong enough reducing agents for this process. A quick look at their reduction potentials shows that they all have more positive reduction potentials than that for Eu^{3+} to Eu^{2+} (-0.277 V for Co^{2+}/Co(s), +0.695 V for O_2, H^+/H_2O_2(aq) and +0.800 V for Ag^+/Ag(s).

<u>17.</u>   **(a)**   Oxidation: $\{Ag\,(s) \rightarrow Ag^+\,(aq) + e^-\} \times 3$ $-E^\circ = -0.800\,V$

Reduction: $NO_3^-\,(aq) + 4\,H^+\,(aq) + 3\,e^- \rightarrow NO(g) + 2\,H_2O(l)$ $E^\circ = +0.956\,V$

Net: $3\,Ag\,(s) + NO_3^-\,(aq) + 4\,H^+\,(aq) \rightarrow 3\,Ag^+\,(aq) + NO(g) + 2\,H_2O(l)$ $E^\circ_{cell} = +0.156\,V$

Ag(s) reacts with HNO_3(aq) to form a solution of $AgNO_3$(aq).

(b) Oxidation: $Zn(s) \rightarrow Zn^{2+}(aq) + 2e^-$ $-E° = +0.763\,V$

Reduction: $2H^+(aq) + 2e^- \rightarrow H_2(g)$ $E° = 0.000\,V$

Net: $Zn(s) + 2H^+(aq) \rightarrow Zn^{2+}(aq) + H_2(g)$ $E°_{cell} = +0.763\,V$

$Zn(s)$ reacts with $HI(aq)$ to form a solution of $ZnI_2(aq)$.

(c) Oxidation: $Au(s) \rightarrow Au^{3+}(aq) + 3e^-$ $-E° = -1.52\,V$

Reduction: $NO_3^-(aq) + 4H^+(aq) + 3e^- \rightarrow NO(g) + 2H_2O(l)$ $E° = +0.956\,V$

Net: $Au(s) + NO_3^-(aq) + 4H^+(aq) \rightarrow Au^{3+}(aq) + NO(g) + 2\ H_2O(l)$; $E°_{cell} = -0.56\,V$

$Au(s)$ does not react with $1.00\ M\ HNO_3(aq)$.

18. In each case, we determine whether $E°_{cell}$ is greater than zero; if so, the reaction will occur.

(a) Oxidation: $Fe(s) \rightarrow Fe^{2+}(aq) + 2e^-$ $-E° = 0.440\,V$

Reduction: $Zn^{2+}(aq) + 2e^- \rightarrow Zn(s)$ $E° = -0.763\,V$

Net: $Fe(s) + Zn^{2+}(aq) \rightarrow Fe^{2+}(aq) + Zn(s)$ $E°_{cell} = -0.323\,V$

The reaction is not spontaneous under standard conditions as written

(b) Oxidation: $\{2Cl^-(aq) \rightarrow Cl_2(g) + 2e^-\} \times 5$ $-E° = -1.358\,V$

Reduction: $\{MnO_4^-(aq) + 8H^+(aq) + 5e^- \rightarrow Mn^{2+}(aq) + 4H_2O(l)\} \times 2$; $E° = 1.51\,V$

Net: $10Cl^-(aq) + 2MnO_4^-(aq) + 16H^+(aq) \rightarrow 5Cl_2(g) + 2Mn^{2+}(aq) + 8H_2O(l)$

$E°_{cell} = +0.15\,V$. The reaction is spontaneous under standard conditions as written.

(c) Oxidation: $\{Ag(s) \rightarrow Ag^+(aq) + e^-\} \times 2$ $-E° = -0.800\,V$

Reduction: $2H^+(aq) + 2e^- \rightarrow H_2(g)$ $E° = +0.000\,V$

Net: $2Ag(s) + 2H^+(aq) \rightarrow 2Ag^+(aq) + H_2(g)$ $E°_{cell} = -0.800\,V$

The reaction is not spontaneous under standard conditions as written.

(d) Oxidation: $\{2Cl^-(aq) \rightarrow Cl_2(g) + 2e^-\} \times 2$ $-E° = -1.358\,V$

Reduction: $O_2(g) + 4H^+(aq) + 4e^- \rightarrow 2H_2O(l)$ $E° = +1.229\,V$

Net: $4Cl^-(aq) + 4H^+(aq) + O_2(g) \rightarrow 2Cl_2(g) + 2H_2O(l)$ $E°_{cell} = -0.129\,V$

The reaction is not spontaneous under standard conditions as written.

Voltaic Cells

19. **(a)** Oxidation: $\{Al(s) \rightarrow Al^{3+}(aq) + 3e^-\} \times 2$ $-E° = +1.676\,V$

Reduction: $\{Sn^{2+}(aq) + 2e^- \rightarrow Sn(s)\} \times 3$ $E° = -0.137\,V$

Net: $2\ Al(s) + 3\ Sn^{2+}(aq) \rightarrow 2\ Al^{3+}(aq) + 3\ Sn(s)$ $E°_{cell} = +1.539\,V$

(b) Oxidation: $Fe^{2+}(aq) \rightarrow Fe^{3+}(aq) + e^-$ $-E° = -0.771$ V

Reduction: $Ag^+(aq) + e^- \rightarrow Ag(s)$ $E° = +0.800$ V

Net: $Fe^{2+}(aq) + Ag^+(aq) \rightarrow Fe^{3+}(aq) + Ag(s)$ $E°_{cell} = +0.029$ V

20. In each case, we determine whether $E°_{cell}$ is greater than zero; if so, the reaction will occur.

(a) Oxidation: $H_2(g) \rightarrow 2H^+ + 2e^-(aq)$ $-E° = 0.000$ V

Reduction: $F_2(g) + 2e^- \rightarrow 2F^-(aq)$ $E° = 2.866$ V

Net: $H_2(g) + F_2(g) \rightarrow 2H^+(aq) + 2F^-(aq)$ $E°_{cell} = +2.866$ V

The reaction is spontaneous under standard conditions as written

(b) Oxidation: $Cu(s) \rightarrow Cu^{2+}(aq) + 2e^-$ $-E° = -0.340$ V

Reduction: $Ba^{2+}(aq) + 2e^- \rightarrow Ba(s)$; $E° = -2.92$ V

Net: $Cu(s) + Ba^{2+}(aq) \rightarrow Cu^{2+}(aq) + Ba(s)$ $E°_{cell} = -3.26$ V .

The reaction is not spontaneous under standard conditions as written.

(c) Oxidation: $\{Fe^{2+}(aq) \rightarrow Fe^{3+}(aq) + e^-\} \times 2$ $-E° = -0.771$ V

Reduction: $Fe^{2+}(aq) + 2e^- \rightarrow Fe(s)$ $E° = -0.440$ V

Net: $3 Fe^{2+}(aq) \rightarrow Fe(s) + 2 Fe^{3+}(aq)$ $E°_{cell} = -1.211$ V

The reaction is not spontaneous as written.

(d) Oxidation: $2 Hg(l) + 2 Cl^- \rightarrow Hg_2Cl_2(s) + 2e^-$ $-E° = -(0.2676)$ V

Reduction: $2 HgCl_2(aq) + 2e^- \rightarrow Hg_2Cl_2(s) + 2 Cl^-(aq)$ $E° = +0.63$ V

Net: $2 Hg(l) + 2 HgCl_2(aq) \rightarrow 2 Hg_2Cl_2(s)$ $E°_{cell} = +0.36$ V

(divide by 2 to get $Hg(l) + HgCl_2(aq) \rightarrow Hg_2Cl_2(s)$)

The reaction is spontaneous under standard conditions as written.

21. **(a)** Anode, Oxidation: $Cu(s) \rightarrow Cu^{2+}(aq) + 2e^-$ $-E° = -0.340$ V

Cathode, Reduction: $\{Fe^{3+}(aq) + e^- \rightarrow Fe^{2+}(aq)\} \times 2$; $E° = +0.771$ V

Net: $Cu(s) + 2 Fe^{3+}(aq) \rightarrow Cu^{2+}(aq) + 2 Fe^{2+}(aq)$ $E°_{cell} = +0.431$ V

(b) Anode, Oxidation: $\{Al(s) \rightarrow Al^{3+}(aq) + 3e^-\} \times 2$ $-E° = +1.676$ V

Cathode, Reduction: $\{Pb^{2+}(aq) + 2e^- \rightarrow Pb(s)\} \times 3$; $E° = -0.125$ V

Net: $2 Al(s) + 3 Pb^{2+}(aq) \rightarrow 2 Al^{3+}(aq) + 3 Pb(s)$ $E°_{cell} = +1.551$ V

(c) Anode, Oxidation: $2 H_2O(l) \rightarrow O_2(g) + 4 H^+(aq) + 4e^-$ $-E° = -1.229$ V

Cathode, Reduction: $\{Cl_2(g) + 2e^- \rightarrow 2 Cl^-(aq)\} \times 2$ $E° = +1.358$ V

Net: $2 H_2O(l) + 2 Cl_2(g) \rightarrow O_2(g) + 4 H^+(aq) + 4 Cl^-(aq)$ $E°_{cell} = +0.129$ V

(d) Anode, Oxidation: $\{Zn(s) \rightarrow Zn^{2+}(aq) + 2e^-\} \times 3$ $\qquad -E^\circ = +0.763\,V$

Cathode, Reduction: $\{NO_3^-(aq) + 4H^+(aq) + 3e^- \rightarrow NO(g) + 2H_2O(l)\} \times 2$; $E^\circ = +0.956\,V$

Net: $3Zn(s) + 2NO_3^-(aq) + 8H^+(aq) \rightarrow 3Zn^{2+}(aq) + 2NO(g) + 4H_2O(l)$ $\quad E^\circ_{cell} = +1.719\,V$

Cell diagram: $Zn(s)|Zn^{2+}(aq)\|H^+(aq), NO_3^-(aq)|NO(g)|Pt(s)$

The cells for parts (a) - (d) follow. The anode is on the left in each case.

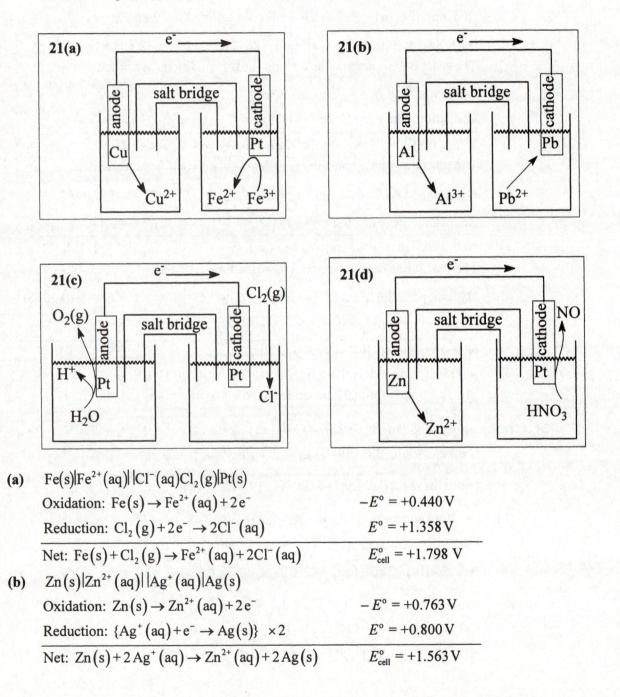

22. (a) $Fe(s)|Fe^{2+}(aq)\|Cl^-(aq)Cl_2(g)|Pt(s)$

Oxidation: $Fe(s) \rightarrow Fe^{2+}(aq) + 2e^-$ $\qquad -E^\circ = +0.440\,V$

Reduction: $Cl_2(g) + 2e^- \rightarrow 2Cl^-(aq)$ $\qquad E^\circ = +1.358\,V$

Net: $Fe(s) + Cl_2(g) \rightarrow Fe^{2+}(aq) + 2Cl^-(aq)$ $\qquad E^\circ_{cell} = +1.798\,V$

(b) $Zn(s)|Zn^{2+}(aq)\|Ag^+(aq)|Ag(s)$

Oxidation: $Zn(s) \rightarrow Zn^{2+}(aq) + 2e^-$ $\qquad -E^\circ = +0.763\,V$

Reduction: $\{Ag^+(aq) + e^- \rightarrow Ag(s)\} \times 2$ $\qquad E^\circ = +0.800\,V$

Net: $Zn(s) + 2Ag^+(aq) \rightarrow Zn^{2+}(aq) + 2Ag(s)$ $\qquad E^\circ_{cell} = +1.563\,V$

(c) $Pt(s)|Cu^+(aq), Cu^{2+}(aq)||Cu^+(aq)|Cu(s)$

Oxidation: $Cu^+(aq) \rightarrow Cu^{2+}(aq) + e^-$ $\qquad -E^\circ = -0.159\,V$

Reduction: $Cu^+(aq) + e^- \rightarrow Cu(s)$ $\qquad\qquad E^\circ = +0.520\,V$

Net: $2\,Cu^+(aq) \rightarrow Cu^{2+}(aq) + Cu(s)$ $\qquad E^\circ_{cell} = +0.361\,V$

(d) $Mg(s)|Mg^{2+}(aq)||Br^-(aq)Br_2(l)|Pt(s)$

Oxidation: $Mg(s) \rightarrow Mg^{2+}(aq) + 2e^-$ $\qquad\quad -E^\circ = +2.356\,V$

Reduction: $Br_2(l) + 2e^- \rightarrow 2\,Br^-(aq)$ $\qquad\qquad E^\circ = +1.065\,V$

Net: $Mg(s) + Br_2(l) \rightarrow Mg^{2+}(aq) + 2\,Br^-(aq)$ $\qquad E^\circ_{cell} = +3.421\,V$

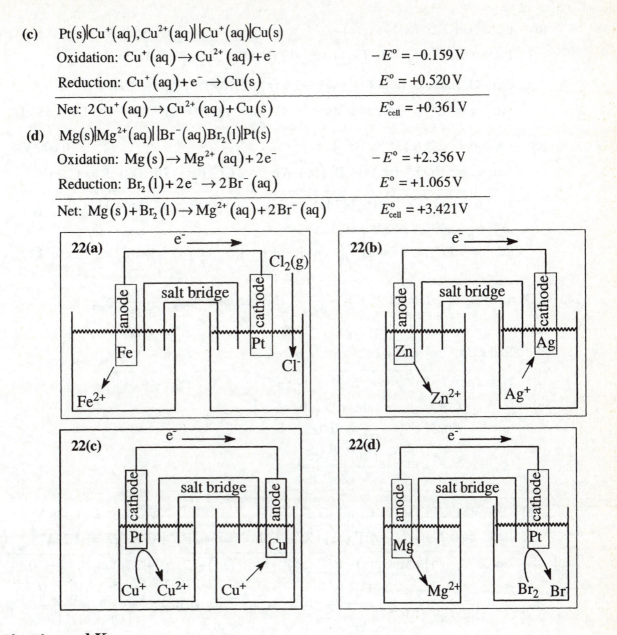

ΔG°, E°_{cell}, and K

23. **(a)** Oxidation: $\{Al(s) \rightarrow Al^{3+}(aq) + 3e^-\} \times 2$ $\qquad -E^\circ = +1.676\,V$

Reduction: $\{Cu^{2+}(aq) + 2e^- \rightarrow Cu(s)\} \times 3$ $\qquad E^\circ = +0.337\,V$

Net: $2\,Al(s) + 3\,Cu^{2+}(aq) \rightarrow 2\,Al^{3+}(aq) + 3\,Cu(s)$ $\qquad E^\circ_{cell} = +2.013\,V$

$\Delta G^\circ = -nFE^\circ_{cell} = -(6\ mol\ e^-)(96,485\ C/mol\ e^-)(2.013\,V)$

$\Delta G^\circ = -1.165 \times 10^6\ J = -1.165 \times 10^3\ kJ$

(b) Oxidation: $\{2I^-(aq) \rightarrow I_2(s) + 2e^-\}\ \times 2$ $\quad\quad -E^\circ = -0.535$ V

Reduction: $O_2(g) + 4H^+(aq) + 4e^- \rightarrow 2H_2O(l)$ $\quad\quad E^\circ = +1.229$ V

Net: $4I^-(aq) + O_2(g) + 4H^+(aq) \rightarrow 2I_2(s) + 2H_2O(l)$ $E^\circ_{cell} = +0.694$ V

$\Delta G^\circ = -nFE^\circ_{cell} = -(4\ \text{mol e}^-)(96,485\ \text{C/mol e}^-)(0.694\ \text{V}) = -2.68\times10^5\,\text{J} = -268$ kJ

(c) Oxidation: $\{Ag(s) \rightarrow Ag^+(aq) + e^-\}\ \times 6$ $\quad\quad -E^\circ = -0.800$ V

Reduction: $Cr_2O_7^{2-}(aq) + 14\ H^+(aq) + 6\ e^- \rightarrow 2\ Cr^{3+}(aq) + 7\ H_2O(l)$ $E^\circ = +1.33$ V

Net: $6\ Ag(s) + Cr_2O_7^{2-}(aq) + 14H^+(aq) \rightarrow 6\,Ag^+(aq) + 2\,Cr^{3+}(aq) + 7H_2O(l)$

$E^\circ_{cell} = -0.800\,\text{V} + 1.33\,\text{V} = +0.53\,\text{V}$

$\Delta G^\circ = -nFE^\circ_{cell} = -(6\text{mol e}^-)(96,485\text{C/mol e}^-)(0.53\,\text{V}) = -3.1\times10^5\,\text{J} = -3.1\times10^2$ kJ

24. $\Delta G^\circ = -nFE^\circ_{cell} = -RT\ln K;\ \ \ln K = \dfrac{nFE^\circ_{cell}}{RT};$ This becomes $\ln K = \dfrac{n}{0.0257}E^\circ_{cell}$

(a) Oxidation: $\{Ag(s) \rightarrow Ag^+(aq) + e^-\}\ \times 2$ $\quad\quad -E^\circ = -0.800$ V

Reduction: $Sn^{4+}(aq) + 2e^- \rightarrow Sn^{2+}(aq)$ $\quad\quad E^\circ = +0.154$ V

Net: $2Ag(s) + Sn^{4+}(aq) \rightarrow 2Ag^+(aq) + Sn^{2+}(aq)$ $\quad\quad E^\circ_{cell} = -0.646$ V

$\ln K_{eq} = \dfrac{n}{0.0257}E^\circ_{cell} = \dfrac{2\ \text{mol e}^- \times (-0.646\ \text{V})}{0.0257} = -50.3$

$K_{eq} = e^{-50.3} = 1\times10^{-22} = \dfrac{[Sn^{2+}][Ag^+]^2}{[Sn^{4+}]}$

(b) Oxidation: $2Cl^-(aq) \rightarrow Cl_2(g) + 2\ e^-$ $\quad\quad -E^\circ = -1.358$ V

Reduction: $MnO_2(s) + 4H^+(aq) + 2e^- \rightarrow Mn^{2+}(aq) + 2H_2O(l)$ $\quad\quad E^\circ = +1.23$ V

Net: $2\ Cl^-(aq) + MnO_2(s) + 4\ H^+(aq) \rightarrow Mn^{2+}(aq) + Cl_2(g) + 2\ H_2O(l)$ $E^\circ_{cell} = -0.13$ V

$\ln K_{eq} = \dfrac{2\ \text{mol e}^- \times (-0.13\ \text{V})}{0.0257} = -10._1;\ \ K_{eq} = e^{-10.1} = 4\times10^{-5} = \dfrac{[Mn^{2+}]P\{Cl_2(g)\}}{[Cl^-]^2[H^+]^4}$

(c) Oxidation: $4\ OH^-(aq) \rightarrow O_2(g) + 2H_2O(l) + 4\ e^-$ $\quad\quad -E^\circ = -0.401$ V

Reduction: $\{OCl^-(aq) + H_2O(l) + 2e^- \rightarrow Cl^-(aq) + 2OH^-\}\ \times 2$ $\quad\quad E^\circ = +0.890$ V

Net: $2OCl^-(aq) \longrightarrow 2Cl^-(aq) + O_2(g)$ $\quad\quad E^\circ_{cell} = +0.489$ V

$\ln K_{eq} = \dfrac{4\ \text{mol e}^-(0.489\ \text{V})}{0.0257} = 76.1$ $\quad\quad K_{eq} = e^{76.1} = 1\times10^{33} = \dfrac{[Cl^-]^2 P\{O_2(g)\}}{[OCl^-]^2}$

25. **(a)** Oxidation: $\{Mn^{2+}(aq)+4H_2O(l) \rightarrow MnO_4^-(aq)+8H^+(aq)+5e^-\}\times 2; -E^\circ = -1.51V$

Reduction: $\{H_2O_2(aq)+2H^+(aq)+2e^- \rightarrow 2H_2O(l)\}\times 5 \hspace{3em} E^\circ = +1.763V$

Net: $2Mn^{2+}(aq)+5H_2O_2(aq) \rightarrow 2MnO_4^-(aq)+6H^+(aq)+2H_2O(l); \hspace{1em} E^\circ_{cell} = +0.25V$

(b) $\Delta G^\circ = -nFE^\circ_{cell} = -(10\,mol\,e^-)(96{,}485\,C/mol\,e^-)(0.25\,V) = -2.4\times10^5\,J = -2.4\times10^2\,kJ$

(c) $\ln K_{eq} = -\dfrac{\Delta G^\circ}{RT} = -\dfrac{-2.4\times10^5\,J}{8.3145\,J\,mol^{-1}\,K^{-1}\times298\,K} = 97; \hspace{2em} K_{eq} = e^{97} = 1\times10^{42}$

(d) Based on the extremely large value of K_{eq}, we conclude that this reaction should go to completion.

26. **(a)** Oxidation: $Fe(s) \rightarrow Fe^{2+}(aq)+2e^- \hspace{4em} -E^\circ = +0.440V$

Reduction: $\{Cr^{3+}(aq)+e^- \rightarrow Cr^{2+}(aq)\}\times2 \hspace{2em} E^\circ = -0.424V$

Net: $2Cr^{3+}(aq)+Fe(s) \rightarrow Fe^{2+}(aq)+2Cr^{2+}(aq) \hspace{1em} E^\circ_{cell} = +0.016V$

(b) $E^\circ_{cell} = +0.016V$

(c) $\Delta G^\circ = -n\,FE^\circ_{cell} = -(2\,mol\,e^-)(96{,}485\,C/mol\,e^-)(0.016\,V) = -3.1\times10^3\,J = -3.1\,kJ$

(d) $\ln K_{eq} = -\dfrac{\Delta G^\circ}{RT} = -\dfrac{-3.1\times10^3\,J}{8.3145\,J\,mol^{-1}\,K^{-1}\times298\,K} = 1.3; \hspace{2em} K_{eq} = e^{1.3} = 4$

(e) Based on the very modest value of K_{eq}, we conclude that this reaction will not go to completion.

27. **(a)** A negative value of E°_{cell} ($-0.0050V$) indicates that $\Delta G^\circ = -nFE^\circ_{cell}$ is positive which in turn indicates that K_{eq} is less than one $(K_{eq} < 1.00)$; $\Delta G^\circ = -RT\ln K_{eq}$.

$$K_{eq} = \dfrac{[Cu^{2+}]^2[Sn^{2+}]}{[Cu^+]^2[Sn^{4+}]}$$

Thus, when all concentrations are the same, the ion product, Q, equals 1.00. From the negative standard cell potential, it is clear that K_{eq} must be (slightly) less than one. Therefore, all the concentrations cannot be 0.500 M at the same time.

(b) In order to establish equilibrium, that is, to have the ion product become less than 1.00, and equal the equilibrium constant, the concentrations of the products must decrease and those of the reactants must increase. A net reaction to the left (towards the reactants), will occur.

28. (a) First we must calculate the value of the equilibrium constant from the standard cell potential.

$$E^{\circ}_{cell} = \frac{0.0257}{n} \ln K_{eq}; \quad \ln K_{eq} = \frac{nE^{\circ}_{cell}}{0.0257} = \frac{2 \text{ mol e}^- \times 0.0020 \text{ V}}{0.0257} = 0.16; \quad K_{eq} = e^{0.16} = 1.2$$

To determine if the described solution is possible, we compare

K_{eq} with Q. Now $K_{eq} = \dfrac{\left[Ni^{2+}\right]\left[V^{2+}\right]^2}{\left[V^{3+}\right]^2}$. Thus, when $\left[V^{2+}\right] = 0.600$ M and

$\left[V^{3+}\right] = \left[Ni^{2+}\right] = 0.675M$, the ion product, $Q = \dfrac{(0.600)^2 0.675}{(0.675)^2} = 0.533 < 1.2 = K_{eq}$.

Therefore, the described situation cannot occur (i.e. such a mixture could not be prepared).

(b) In order to establish equilibrium, that is, to have the ion product (0.533) become equal to 1.2, the equilibrium constant, the concentrations of the products must increase and those of the reactants must decrease. Thus. a slight net reaction to the right (formation of products), will occur.

29. Cell reaction: $Zn(s) + Ag_2O(s) \rightarrow ZnO(s) + 2Ag(s)$. We assume that the cell operates at 298 K.

$$\Delta G^{\circ} = \Delta G^{\circ}_f\left[ZnO(s)\right] + 2\Delta G^{\circ}_f\left[Ag(s)\right] - \Delta G^{\circ}_f\left[Zn(s)\right] - \Delta G^{\circ}_f\left[Ag_2O(s)\right]$$

$$= -318.3 \text{ kJ/mol} + 2(0.00 \text{ kJ/mol}) - 0.00 \text{ kJ/mol} - (-11.20 \text{ kJ/mol})$$

$$= -307.1 \text{ kJ/mol} = -nFE^{\circ}_{cell}$$

$$E^{\circ}_{cell} = -\frac{\Delta G^{\circ}}{nF} = -\frac{-307.1 \times 10^3 \text{ J/mol}}{2 \text{ mol e}^-/\text{mol rxn} \times 96,485 \text{ C/mol e}^-} = 1.591 \text{ V}$$

30. From equation (20.28) we know $n = 12$ and the oveall cell reaction. First we must compute value of ΔG°.

$$\Delta G^{\circ} = -nFE^{\circ}_{cell} = -12 \text{ mol e}^- \times \frac{96485 \text{ C}}{1 \text{ mol e}^-} \times 2.71 \text{ V} = -3.14 \times 10^6 \text{ J} = -3.14 \times 10^3 \text{ kJ}$$

Then we will use this value, the balanced equation and values of ΔG_f° to calculate $\Delta G^{\circ}_f\left[Al(OH)_4\right]^-$.

$$4 Al(s) + 3O_2(g) + 6H_2O(l) + 4OH^-(aq) \rightarrow 4\left[Al(OH)_4\right]^-(aq)$$

$$\Delta G^{\circ} = 4\Delta G^{\circ}_f\left[Al(OH)_4\right]^- - 4\Delta G^{\circ}_f[Al(s)] - 3\Delta G^{\circ}_f[O_2(g)] - 6\Delta G^{\circ}_f[H_2O(l)] - 4\Delta G^{\circ}_f[OH^-(aq)]$$

$$-3.14 \times 10^3 \text{ kJ} = 4\Delta G^{\circ}_f\left[Al(OH)_4\right]^- - 4 \times 0.00 \text{ kJ} - 3 \times 0.00 \text{ kJ} - 6 \times (-237.1 \text{ kJ}) - 4 \times (-157.2)$$

$$= 4\Delta G^{\circ}_f\left[Al(OH)_4\right]^- + 2051.4 \text{ kJ}$$

$$\Delta G^{\circ}_f\left[Al(OH)_4\right]^- = \left(-3.14 \times 10^3 \text{ kJ} - 2051.4 \text{ kJ}\right) \div 4 = -1.30 \times 10^3 \text{ kJ/mol}$$

31. From the data provided we can construct the following Latimer diagram.

$$\text{IrO}_2 \xrightarrow{\ 0.223\ \text{V}\ } \text{Ir}^{3+} \xrightarrow{\ 1.156\ \text{V}\ } \text{Ir} \quad \text{(Acidic conditions)}$$
$$\text{(IV)} \qquad\qquad \text{(III)} \qquad\qquad \text{(0)}$$

Latimer diagrams are used to calculate the standard potentials of non-adjacent half-cell couples. Our objective in this question is to calculate the voltage differential between IrO_2 and iridium metal (Ir), which are separated in the diagram by Ir^{3+}. The process basically involves adding two half-reactions to obtain a third half-reaction. The potentials for the two half-reactions cannot, however, simply be added to get the target half-cell voltage because the electrons are not cancelled in the process of adding the two half-reactions. Instead, to find $E^\circ_{1/2\ \text{cell}}$ for the target half-reaction, we must use free energy changes, which are additive. To begin, we will balance the relevant half-reactions in acidic solution:

$$4\,\text{H}^+(\text{aq}) + \text{IrO}_2(\text{s}) + \text{e}^- \rightarrow \text{Ir}^{3+}(\text{aq}) + 2\,\text{H}_2\text{O}(\text{l}) \qquad E^\circ_{1/2\text{red(a)}} = 0.223\ \text{V}$$
$$\text{Ir}^{3+}(\text{aq}) + 3\,\text{e}^- \rightarrow \text{Ir}(\text{s}) \qquad E^\circ_{1/2\text{red(b)}} = 1.156\text{V}$$

$$\overline{4\,\text{H}^+(\text{aq}) + \text{IrO}_2(\text{s}) + 4\text{e}^- \rightarrow 2\,\text{H}_2\text{O}(\text{l}) + \text{Ir}(\text{s}) \qquad E^\circ_{1/2\text{red(c)}} = ?}$$

$E^\circ_{1/2\text{red(c)}} \neq E^\circ_{1/2\text{red(a)}} + E^\circ_{1/2\text{red(b)}}$ but $\Delta G^\circ_{(a)} + \Delta G^\circ_{(b)} = \Delta G^\circ_{(c)}$ and $\Delta G^\circ = -nFE^\circ$

$$-4F(E^\circ_{1/2\text{red(c)}}) = -1F(E^\circ_{1/2\text{red(a)}}) + -3F(E^\circ_{1/2\text{red(b)}})$$
$$-4F(E^\circ_{1/2\text{red(c)}}) = -1F(0.223) + -3F(1.156)$$

$$E^\circ_{1/2\text{red(c)}} = \frac{-1F(0.223) + -3F(1.156)}{-4F} = \frac{-1(0.223) + -3(1.156)}{-4} = 0.923\ \text{V}$$

In other words, $E^\circ_{(c)}$ is the weighted average of $E^\circ_{(a)}$ and $E^\circ_{(b)}$

32. This question will be answered in a manner similar to that used to solve question 31. Let's get underway by writing down the appropriate Latimer diagram:

$$\text{H}_2\text{MoO}_4 \xrightarrow{\ 0.646\ \text{V}\ } \text{MoO}_2 \xrightarrow{\ ?\ \text{V}\ } \text{Mo}^{3+} \quad \text{(Acidic conditions)}$$
$$\text{(VI)} \qquad\qquad \text{(IV)} \qquad\qquad \text{(III)}$$

$$\underset{0.428\ \text{V}}{\big\llcorner\text{_____}\big\lrcorner}$$

This time we want to calculate the standard voltage change for the $1\ \text{e}^-$ reduction of MoO_2 to Mo^{3+}. Once again, we must balance the half-cell reactions in acidic solution:

$$\text{H}_2\text{MoO}_4(\text{aq}) + 2\,\text{e}^- + 2\,\text{H}^+(\text{aq}) \rightarrow \text{MoO}_2(\text{s}) + 2\,\text{H}_2\text{O}(\text{l}) \qquad E^\circ_{1/2\text{red(a)}} = 0.646\ \text{V}$$
$$\text{MoO}_2(\text{s}) + 4\,\text{H}^+(\text{aq}) + 1\,\text{e}^- \rightarrow \text{Mo}^{3+}(\text{aq}) + 2\,\text{H}_2\text{O}(\text{l}) \qquad E^\circ_{1/2\text{red(b)}} = ?\ \text{V}$$

$$\overline{\text{H}_2\text{MoO}_4(\text{aq}) + 3\,\text{e}^- + 6\,\text{H}^+(\text{aq}) \rightarrow \text{Mo}^{3+}(\text{aq}) + 4\,\text{H}_2\text{O}(\text{l}) \qquad E^\circ_{1/2\text{red(c)}} = 0.428\ \text{V}}$$

So, $-3F(E^\circ_{1/2\text{red(c)}}) = -2F(E^\circ_{1/2\text{red(a)}}) + -1F(E^\circ_{1/2\text{red(b)}})$

$$-3F(0.428\ \text{V}) = -2F(0.646) + -1F(E^\circ_{1/2\text{red(b)}})$$
$$-1FE^\circ_{1/2\text{red(b)}} = -3F(0.428\ \text{V}) + 2F(0.646)$$

$$E^\circ_{1/2\text{red(c)}} = \frac{-3F(0.428\ \text{V}) + 2F(0.646)}{-1F} = 1.284\ \text{V} - 1.292\ \text{V} = -0.008\ \text{V}$$

Concentration Dependence of E_{cell}—the Nernst Equation

33. Oxidation: $Zn(s) \rightarrow Zn^{2+}(aq) + 2e^-$ $\qquad -E^\circ = +0.763\,V$

Reduction: $\{Ag^+(aq) + e^- \rightarrow Ag(s)\} \times 2$ $\qquad E^\circ = +0.800\,V$

Net: $Zn(s) + 2Ag^+(aq) \rightarrow Zn^{2+}(aq) + 2Ag(s)$ $\qquad E^\circ_{cell} = +1.563\,V$

$$E = E^\circ_{cell} - \frac{0.0592}{n}\log\frac{\left[Zn^{2+}\right]}{\left[Ag^+\right]^2} = +1.563\,V - \frac{0.0592}{2}\log\frac{1.00}{x^2} = +1.250\,V$$

$$\log\frac{1.00\,M}{x^2} = \frac{-2\times(1.250 - 1.563)}{0.0592} = 10.6; \quad x = \sqrt{2.5\times10^{-11}} = 5\times10^{-6}\,M$$

Therefore, $[Ag^+] = 5 \times 10^{-6}\,M$

34. In each case, we employ the equation $E_{cell} = 0.0592\,pH$.

(a) $E_{cell} = 0.0592\,pH = 0.0592 \times 5.25 = 0.311\,V$

(b) $pH = -\log(0.0103) = 1.987$ $\qquad E_{cell} = 0.0592\,pH = 0.0592 \times 1.987 = 0.118\,V$

(c) $K_a = \dfrac{\left[H^+\right]\left[C_2H_3O_2^-\right]}{\left[HC_2H_3O_2\right]} = 1.8\times10^{-5} = \dfrac{x^2}{0.158 - x} \approx \dfrac{x^2}{0.158}$

$x = \sqrt{0.158 \times 1.8\times10^{-5}} = 1.7\times10^{-3}\,M$

$pH = -\log(1.7\times10^{-3}) = 2.77$

$E_{cell} = 0.0592\,pH = 0.0592 \times 2.77 = 0.164\,V$

35. We first calculate E°_{cell} for each reaction and then use the Nernst equation to calculate E_{cell}.

(a) Oxidation: $\{Al(s) \rightarrow Al^{3+}(0.18\,M) + 3\,e^-\} \times 2$ $\qquad -E^\circ = +1.676\,V$

Reduction: $\{Fe^{2+}(0.85\,M) + 2e^- \rightarrow Fe(s)\} \times 3$ $\qquad E^\circ = -0.440\,V$

Net: $2Al(s) + 3Fe^{2+}(0.85\,M) \rightarrow 2Al^{3+}(0.18\,M) + 3Fe(s)$ $\qquad E^\circ_{cell} = +1.236\,V$

$$E_{cell} = E^\circ_{cell} - \frac{0.0592}{n}\log\frac{\left[Al^{3+}\right]^2}{\left[Fe^{2+}\right]^3} = 1.236\,V - \frac{0.0592}{6}\log\frac{(0.18)^2}{(0.85)^3} = 1.249\,V$$

(b) Oxidation: $\{Ag(s) \rightarrow Ag^+(0.34\,M) + e^-\} \times 2$ $\qquad -E^\circ = -0.800\,V$

Reduction: $Cl_2(0.55\,atm) + 2\,e^- \rightarrow 2\,Cl^-(0.098\,M)$ $\qquad E^\circ = +1.358\,V$

Net: $Cl_2(0.55\,atm) + 2\,Ag(s) \rightarrow 2\,Cl^-(0.098\,M) + 2\,Ag^+(0.34\,M); E^\circ_{cell} = +0.558\,V$

$$E_{cell} = E^\circ_{cell} = \frac{0.0592}{n}\log\frac{\left[Cl^-\right]^2\left[Ag^+\right]^2}{P\{Cl_2(g)\}} = +0.558 - \frac{0.0592}{2}\log\frac{(0.34)^2(0.098)^2}{0.55} = +0.638\,V$$

36. **(a)** Oxidation: $Mn(s) \rightarrow Mn^{2+}(0.40\,M) + 2e^-$ $-E° = +1.18\,V$

Reduction: $\{Cr^{3+}(0.35\,M) + 1e^- \rightarrow Cr^{2+}(0.25\,M)\} \times 2$ $E° = -0.424\,V$

Net: $2Cr^{3+}(0.35\,M) + Mn(s) \rightarrow 2Cr^{2+}(0.25\,M) + Mn^{2+}(0.40\,M)$ $E°_{cell} = +0.76\,V$

$$E_{cell} = E°_{cell} - \frac{0.0592}{n}\log\frac{[Cr^{2+}]^2[Mn^{2+}]}{[Cr^{3+}]^2} = +0.76\,V - \frac{0.0592}{2}\log\frac{(0.25)^2(0.40)}{(0.35)^2} = +0.78\,V$$

(b) Oxidation: $\{Mg(s) \rightarrow Mg^{2+}(0.016\,M) + 2e^-\} \times 3$ $-E° = +2.356\,V$

Reduction: $\{[Al(OH)_4]^-(0.25\,M) + 3\,e^- \rightarrow 4OH^-(0.042\,M) + Al(s)\} \times 2$; $E° = -2.310\,V$

Net: $3\,Mg(s) + 2[Al(OH)_4]^-(0.25\,M) \rightarrow 3Mg^{2+}(0.016\,M) + 8OH^-(0.042\,M) + 2Al(s)$;

$E°_{cell} = +0.046\,V$

$$E_{cell} = E°_{cell} - \frac{0.0592}{6}\log\frac{[Mg^{2+}]^3[OH^-]^8}{[[Al(OH)_4]^-]^2} = +0.046 - \frac{0.0592}{6}\log\frac{(0.016)^3(0.042)^8}{(0.25)^2}$$

$$= 0.046\,V + 0.150\,V = 0.196\,V$$

<u>37.</u> All these observations can be understood in terms of the procedure we use to balance half-equations: the ion—electron method.

(a) The reactions for which E depends on pH are those that contain either $H^+(aq)$ or $OH^-(aq)$ in the balanced half equation. These reactions are those that involve oxoacids and oxoanions whose central atom changes oxidation state.

(b) $H^+(aq)$ will inevitably be on the left side of the reduction of an oxoanion because reduction is accompanied by not only a decrease in oxidation state, but also by the loss of oxygen atoms, as in $ClO_3^- \rightarrow ClO_2^-$, $SO_4^{2-} \rightarrow SO_2$, and $NO_3^- \rightarrow NO$. These oxygen atoms appear on the right-hand side as H_2O molecules. The hydrogens that are added to the right-hand side with the water molecules are then balanced with $H^+(aq)$ on the left-hand side.

(c) If a half-reaction with $H^+(aq)$ ions present is transferred to basic solution, it may be re-balanced by adding to each side $OH^-(aq)$ ions equal in number to the $H^+(aq)$ originally present. This results in $H_2O(l)$ on the side that had $H^+(aq)$ ions (the left side in this case) and $OH^-(aq)$ ions on the other side (the right side.)

38. Oxidation: $2\,Cl^-(aq) \rightarrow Cl_2(g) + 2\,e^-$ $\qquad\qquad\qquad -E^\circ = -1.358\,V$

Reduction: $PbO_2(s) + 4\,H^+(aq) + 2\,e^- \rightarrow Pb^{2+}(aq) + 2\,H_2O(l)$ $\qquad E^\circ = +1.455\,V$

Net: $PbO_2(s) + 4\,H^+(aq) + 2\,Cl^-(aq) \rightarrow Pb^{2+}(aq) + 2\,H_2O(l) + Cl_2(g);\quad E^\circ_{cell} = +0.097\ V$

We derive an expression for E_{cell} that depends on just the changing $[H^+]$.

$$E_{cell} = E^\circ_{cell} - \frac{0.0592}{2}\log\frac{P\{Cl_2\}[Pb^{2+}]}{[H^+]^4[Cl^-]^2} = +0.097 - 0.0296\log\frac{(1.00\ atm)(1.00\ M)}{[H^+]^4(1.00)^2}$$

$$= +0.097 + 4\times0.0296\log[H^+] = +0.097 + 0.118\log[H^+] = +0.097 - 0.118\ pH$$

(a) $E_{cell} = +0.097 + 0.118\ \log(6.0) = +0.189\ V$

∴ Forward reaction is spontaneous under standard conditions

(b) $E_{cell} = +0.097 + 0.118\ \log(1.2) = +0.106\ V$

∴ Forward reaction is spontaneous under standard conditions

(c) $E_{cell} = +0.097 - 0.118\times4.25 = -0.405\ V$

∴ Forward reaction is nonspontaneous under standard conditions

The reaction is spontaneous in strongly acidic solutions (very low pH), but is nonspontaneous under standard conditions in basic, neutral, and weakly acidic solutions.

39. Oxidation: $Zn(s) \rightarrow Zn^{2+}(aq) + 2\,e^-$ $\qquad\qquad -E^\circ = +0.763\,V$

Reduction: $Cu^{2+}(aq) + 2\,e^- \rightarrow Cu(s)$ $\qquad\qquad E^\circ = +0.337\,V$

Net: $Zn(s) + Cu^{2+}(aq) \rightarrow Cu(s) + Zn^{2+}(aq)$ $\qquad E^\circ_{cell} = +1.100\,V$

(a) We set $E = 0.000V$, $[Zn^{2+}] = 1.00M$, and solve for $[Cu^{2+}]$ in the Nernst equation.

$$E_{cell} = E^\circ_{cell} - \frac{0.0592}{2}\log\frac{[Zn^{2+}]}{[Cu^{2+}]}; \quad 0.000 = 1.100 - 0.0296\log\frac{1.0\,M}{[Cu^{2+}]}$$

$$\log\frac{1.0\,M}{[Cu^{2+}]} = \frac{0.000 - 1.100}{-0.0296} = 37.2; \quad [Cu^{2+}] = 10^{-37.2} = 6\times10^{-38}\ M$$

(b) If we work the problem the other way, by assuming initial concentrations of $[Cu^{2+}]_{initial} = 1.0\,M$ and $[Zn^{2+}]_{initial} = 0.0\,M$, we obtain $[Cu^{2+}]_{final} = 6\times10^{-38}\,M$ and $[Zn^{2+}]_{final} = 1.0\,M$. Thus, we would conclude that this reaction goes to completion.

40. Oxidation: $Sn(s) \rightarrow Sn^{2+}(aq) + 2e^-$ $-E^\circ = +0.137$ V

       Reduction: $Pb^{2+}(aq) + 2e^- \rightarrow Pb(s)$ $E^\circ = -0.125$ V

       Net: $Sn(s) + Pb^{2+}(aq) \rightarrow Sn^{2+}(aq) + Pb(s)$ $E^\circ_{cell} = +0.012$ V

Now we wish to find out if $Pb^{2+}(aq)$ will be completely displaced, that is, will $\left[Pb^{2+}\right]$ reach 0.0010 M, if $\left[Sn^{2+}\right]$ is fixed at 1.00 M? We use the Nernst equation to determine if the cell voltage still is positive under these conditions.

$$E_{cell} = E^\circ_{cell} - \frac{0.0592}{2}\log\frac{\left[Sn^{2+}\right]}{\left[Pb^{2+}\right]} = +0.012 - \frac{0.0592}{2}\log\frac{1.00}{0.0010} = +0.012 - 0.089 = -0.077 \text{ V}$$

The negative cell potential tells us that this reaction will not go to completion under the conditions stated. The reaction stops being spontaneous when $E_{cell} = 0$. We can work this the another way as well: assume that $\left[Pb^{2+}\right] = (1.0 - x)$ M and calculate $\left[Sn^{2+}\right] = x$ M at equilibrium, that is,

where $E_{cell} = 0$. $E_{cell} = 0.00 = E^\circ_{cell} - \frac{0.0592}{2}\log\frac{\left[Sn^{2+}\right]}{\left[Pb^{2+}\right]} = +0.012 - \frac{0.0592}{2}\log\frac{x}{1.0-x}$

$\log\frac{x}{1.0-x} = \frac{2 \times 0.012}{0.0592} = 0.41$ $x = 10^{0.41}(1.0 - x) = 2.6 - 2.6x$ $x = \frac{2.6}{3.6} = 0.72$ M

We would expect the final $\left[Sn^{2+}\right]$ to equal 1.0 M (or at least 0.999 M) if the reaction went to completion. Instead it equals 0.72 M and consequently, the reaction fails to go to completion.

41. **(a)** The two half-equations and the cell equation are given below. $E^\circ_{cell} = 0.000$ V

       Oxidation: $H_2(g) \rightarrow 2\,H^+(0.65 \text{ M KOH}) + 2e^-$

       Reduction: $2H^+(1.0 \text{ M}) + 2\,e^- \rightarrow H_2(g)$

       Net: $2H^+(1.0 \text{ M}) \rightarrow 2H^+(0.65 \text{ M KOH})$

$$\left[H^+\right]_{base} = \frac{K_w}{\left[OH^-\right]} = \frac{1.00 \times 10^{-14} \text{ M}^2}{0.65 \text{ M}} = 1.5 \times 10^{-14} \text{ M}$$

$$E_{cell} = E^\circ_{cell} - \frac{0.0592}{2}\log\frac{\left[H^+\right]^2_{base}}{\left[H^+\right]^2_{acid}} = 0.000 - \frac{0.0592}{2}\log\frac{\left(1.5 \times 10^{-14}\right)^2}{(1.0)^2} = +0.818 \text{ V}$$

(b) For the reduction of $H_2O(l)$ to $H_2(g)$ in basic solution,

$2\,H_2O(l) + 2\,e^- \rightarrow 2\,H_2(g) + 2\,OH^-(aq)$, $E^\circ = -0.828$ V. This reduction is the reverse of the reaction that occurs in the anode of the cell described, with one small difference: in the standard half-cell, [OH⁻] = 1.00 M, while in the anode half-cell in the case at hand, [OH⁻] = 0.65 M. Or, viewed in another way, in 1.00 M KOH, [H⁺] is smaller still than in 0.65 M KOH. The forward reaction (dilution of H^+) should be even more spontaneous, (i.e. a more positive voltage will be created), with 1.00 M KOH than with 0.65 M KOH. We expect that E°_{cell} (1.000 M NaOH) should be a little larger than E_{cell} (0.65 M NaOH), which, is in fact, the case.

42. **(a)** Because $NH_3(aq)$ is a weaker base than KOH(aq), $[OH^-]$ will be smaller than it is in the previous problem. Therefore the $[H^+]$ will be higher. Its logarithm will be less negative, and the cell voltage will be less positive. Or, viewed as in Exercise 41(b), the difference in $[H^+]$ between 1.0 M H^+ and 0.65 M KOH is greater than the difference in $[H^+]$ between 1.0 M H^+ and 0.65 M NH_3. The forward reaction is "less spontaneous" and E_{cell} is less positive.

(b) Reaction:

	$NH_3(aq) + H_2O(l)$	$\rightleftharpoons$	$NH_4^+(aq) +$	$OH^-(aq)$
Initial:	0.65 M		0 M	≈ 0 M
Changes:	$-x$ M		$+x$ M	$+x$ M
Equil:	$(0.65-x)$ M		x M	x M

$$K_b = \frac{\left[NH_4^+\right]\left[OH^-\right]}{\left[NH_3\right]} = 1.8\times10^{-5} = \frac{x\cdot x}{0.65-x} \approx \frac{x^2}{0.65}$$

$$x = \left[OH^-\right] = \sqrt{0.65\times1.8\times10^{-5}} = 3.4\times10^{-3}\,M; \quad [H_3O^+] = \frac{1.00\times10^{-14}}{3.4\times10^{-3}} = 2.9\times10^{-12}\,M$$

$$E_{cell} = E^\circ_{cell} - \frac{0.0592}{2}\log\frac{[H^+]^2_{base}}{[H^+]^2_{acid}} = 0.000 - \frac{0.0592}{2}\log\frac{(2.9\times10^{-12})^2}{(1.0)^2} = +0.683\,V$$

43. First we need to find $\left[Ag^+\right]$ in a saturated solution of Ag_2CrO_4.

$$K_{sp} = \left[Ag^+\right]^2\left[CrO_4^{2-}\right] = (2s)^2(s) = 4s^3 = 1.1\times10^{-12} \quad s = \sqrt[3]{\frac{1.1\times10^{-12}}{4}} = 6.5\times10^{-5}\,M$$

The cell diagrammed is a concentration cell, for which $E^\circ_{cell} = 0.000\,V$, $n=1$,

$$\left[Ag^+\right]_{anode} = 2s = 1.3\times10^{-4}\,M$$

Cell reaction: $Ag(s) + Ag^+(0.125\,M) \rightarrow Ag(s) + Ag^+(1.3\times10^{-4}\,M)$

$$E_{cell} = E^\circ_{cell} - \frac{0.0592}{1}\log\frac{1.3\times10^{-4}\,M}{0.125\,M} = 0.000 + 0.177\,V = 0.177\,V$$

44. First we need to determine $\left[Ag^+\right]$ in the saturated solution of Ag_3PO_4.

The cell diagrammed is a concentration cell, for which $E^\circ_{cell} = 0.000V$, $n=1$.

Cell reaction: $Ag(s) + Ag^+(0.140\,M) \rightarrow Ag(s) + Ag^+(x\,M)$

$$E_{cell} = 0.180V = E^\circ_{cell} - \frac{0.0592}{1}\log\frac{x\,M}{0.140\,M}; \quad \log\frac{x\,M}{0.140\,M} = \frac{0.180}{-0.0592} = -3.04$$

$$x\,M = 0.140\,M\times10^{-3.04} = 0.140\,M\times9.1\times10^{-4} = 1.3\times10^{-4}\,M = \left[Ag^+\right]_{anode}$$

$$K_{sp} = \left[Ag^+\right]^3\left[PO_4^{3-}\right] = (3s)^3(s) = (1.3\times10^{-4})^3(1.3\times10^{-4}\div3) = 9.5\times10^{-17}$$

45. (a) Oxidation: $Sn(s) \rightarrow Sn^{2+}(0.075 \text{ M}) + 2 \text{ e}^-$ $\qquad -E° = +0.137 \text{ V}$

Reduction: $Pb^{2+}(0.600 \text{ M}) + 2 \text{ e}^- \rightarrow Pb(s)$ $\qquad E° = -0.125 \text{ V}$

Net: $Sn(s) + Pb^{2+}(0.600 \text{ M}) \rightarrow Pb(s) + Sn^{2+}(0.075 \text{ M}); E°_{cell} = +0.012 \text{ V}$

$$E_{cell} = E°_{cell} - \frac{0.0592}{2} \log \frac{\left[Sn^{2+}\right]}{\left[Pb^{2+}\right]} = 0.012 - 0.0296 \log \frac{0.075}{0.600} = 0.012 + 0.027 = 0.039 \text{ V}$$

(b) As the reaction proceeds, $[Sn^{2+}]$ increases while $[Pb^{2+}]$ decreases. These changes cause the driving force behind the reaction to steadily decrease with the passage of time. This decline in driving force is manifested as a decrease in E_{cell} with time.

(c) When $\left[Pb^{2+}\right] = 0.500 \text{ M} = 0.600 \text{ M} - 0.100 \text{ M}$, $\left[Sn^{2+}\right] = 0.075 \text{ M} + 0.100 \text{ M}$, because the stoichiometry of the reaction is 1:1 for Sn^{2+} and Pb^{2+}.

$$E_{cell} = E°_{cell} - \frac{0.0592}{2} \log \frac{\left[Sn^{2+}\right]}{\left[Pb^{2+}\right]} = 0.012 - 0.0296 \log \frac{0.175}{0.500} = 0.012 + 0.013 = 0.025 \text{ V}$$

(d) Reaction: $\qquad$ Sn(s) + $\qquad Pb^{2+}(aq) \qquad \rightarrow \qquad Pb(s) \quad + \quad Sn^{2+}(aq)$

Initial: $\qquad\qquad$ — $\qquad\qquad$ 0.600 M $\qquad\qquad\qquad$ — $\qquad\qquad$ 0.075 M

Changes: $\qquad\qquad$ — $\qquad\qquad\quad -x \text{ M} \qquad\qquad\qquad$ — $\qquad\qquad +x \text{ M}$

Final: $\qquad\qquad$ — $\qquad\qquad (0.600 - x) \text{ M} \qquad\qquad$ — $\qquad (0.075 + x) \text{ M}$

$$E_{cell} = E°_{cell} - \frac{0.0592}{2} \log \frac{\left[Sn^{2+}\right]}{\left[Pb^{2+}\right]} = 0.020 = 0.012 - 0.0296 \log \frac{0.075 + x}{0.600 - x}$$

$$\log \frac{0.075 + x}{0.600 - x} = \frac{E_{cell} - 0.012}{-0.0296} = \frac{0.020 - 0.012}{-0.0296} = -0.27; \quad \frac{0.075 + x}{0.600 - x} = 10^{-0.27} = 0.54$$

$$0.075 + x = 0.54(0.600 - x) = 0.324 - 0.54x; \quad x = \frac{0.324 - 0.075}{1.54} = 0.162 \text{ M}$$

$$\left[Sn^{2+}\right] = 0.075 + 0.162 = 0.237 \text{ M}$$

(e) Here we use the expression developed in part (d).

$$\log \frac{0.075 + x}{0.600 - x} = \frac{E_{cell} - 0.012}{-0.0296} = \frac{0.000 - 0.012}{-0.0296} = +0.41$$

$$\frac{0.075 + x}{0.600 - x} = 10^{+0.41} = 2.6; \quad 0.075 + x = 2.6(0.600 - x) = 1.6 - 2.6x$$

$$x = \frac{1.6 - 0.075}{3.6} = 0.42 \text{ M}$$

$$\left[Sn^{2+}\right] = 0.075 + 0.42 = 0.50 \text{ M}; \qquad\qquad \left[Pb^{2+}\right] = 0.600 - 0.42 = 0.18 \text{ M}$$

46. **(a)** Oxidation: $Ag(s) \rightarrow Ag^+(0.015 \text{ M}) + e^-$ $\qquad\qquad\qquad -E^\circ = -0.800 \text{ V}$

Reduction: $Fe^{3+}(0.055 \text{ M}) + e^- \rightarrow Fe^{2+}(0.045 \text{ M})$ $\qquad\quad E^\circ = +0.771 \text{ V}$

Net: $Ag(s) + Fe^{3+}(0.055 \text{ M}) \rightarrow Ag^+(0.015 \text{ M}) + Fe^{2+}(0.045 \text{ M})$ $\quad E^\circ_{cell} = -0.029 \text{ V}$

$$E_{cell} = E^\circ_{cell} - \frac{0.0592}{1} \log \frac{[Ag^+][Fe^{2+}]}{[Fe^{3+}]} = -0.029 - 0.0592 \log \frac{0.015 \times 0.045}{0.055}$$

$$= -0.029 \text{ V} + 0.113 \text{ V} = +0.084 \text{ V}$$

(b) As the reaction proceeds, $[Ag^+]$ and $[Fe^{2+}]$ will increase, while $[Fe^{3+}]$ decrease. These changes cause the driving force behind the reaction to steadily decrease with the passage of time. This decline in driving force is manifested as a decrease in E_{cell} with time.

(c) When $[Ag^+] = 0.020 \text{ M} = 0.015 \text{ M} + 0.005 \text{ M}$, $[Fe^{2+}] = 0.045 \text{ M} + 0.005 \text{ M} = 0.050 \text{ M}$ and $[Fe^{3+}] = 0.055 \text{ M} - 0.005 \text{ M} = 0.500 \text{ M}$, because, by the stoichiometry of the reaction, a mole of Fe^{2+} is produced and a mole of Fe^{3+} is consumed for every mole of Ag^+ produced.

$$E_{cell} = E^\circ_{cell} - \frac{0.0592}{1} \log \frac{[Ag^+][Fe^{2+}]}{[Fe^{3+}]} = -0.029 - 0.0592 \log \frac{0.020 \times 0.050}{0.050}$$

$$= -0.029 \text{ V} + 0.101 \text{ V} = +0.072 \text{ V}$$

(d) Reaction: $\quad Ag(s) + \quad Fe^{3+}(0.055 \text{ M}) \quad \rightleftharpoons \quad Ag^+(0.015 \text{ M}) + \quad Fe^{2+}(0.045 \text{ M})$

Initial: $\qquad\qquad\quad 0.055 \text{ M} \qquad\qquad\qquad 0.015 \text{ M} \qquad\qquad 0.045 \text{ M}$

Changes: $\qquad\qquad -x \text{ M} \qquad\qquad\qquad\quad +x \text{ M} \qquad\qquad +x \text{ M}$

Final: $\qquad\qquad\quad (0.055 - x) \text{ M} \qquad\qquad (0.015 + x) \text{ M} \quad (0.045 + x) \text{ M}$

$$E_{cell} = E^\circ_{cell} - \frac{0.0592}{1} \log \frac{[Ag^+][Fe^{2+}]}{[Fe^{3+}]} = -0.029 - 0.0592 \log \frac{(0.015 + x)(0.045 + x)}{(0.055 - x)}$$

$$\log \frac{(0.015 + x)(0.045 + x)}{(0.055 - x)} = \frac{E_{cell} + 0.029}{-0.0592} = \frac{0.010 + 0.029}{-0.0592} = -0.66$$

$$\frac{(0.015 + x)(0.045 + x)}{(0.055 - x)} = 10^{-0.66} = 0.22$$

$$0.00068 + 0.060x + x^2 = 0.22(0.055 - x) = 0.012 - 0.22x \qquad x^2 + 0.28x - 0.011 = 0$$

$$x = \frac{-b \pm \sqrt{b^2 - 4ac}}{2a} = \frac{-0.28 \pm \sqrt{(0.28)^2 + 4 \times 0.011}}{2} = 0.035 \text{ M}$$

$$[Ag^+] = 0.015 \text{ M} + 0.035 \text{ M} = 0.050 \text{ M}$$

$$[Fe^{2+}] = 0.045 \text{ M} + 0.035 \text{ M} = 0.080 \text{ M}$$

$$[Fe^{3+}] = 0.055 \text{ M} - 0.035 \text{ M} = 0.020 \text{ M}$$

(e) We use the expression that was developed in part (d).

$$\log\frac{(0.015+x)(0.045+x)}{(0.055-x)} = \frac{E_{cell}+0.029}{-0.0592} = \frac{0.000+0.029}{-0.0592} = -0.49$$

$$\frac{(0.015+x)(0.045+x)}{(0.055-x)} = 10^{-0.49} = 0.32$$

$$0.00068+0.060x+x^2 = 0.32(0.055-x) = 0.018-0.32x \qquad x^2+0.38x-0.017 = 0$$

$$x = \frac{-b\pm\sqrt{b^2-4ac}}{2a} = \frac{-0.38\pm\sqrt{(0.38)^2+4\times0.017}}{2} = 0.040\,M$$

$$[Ag^+] = 0.015\ M + 0.040\ M = 0.055\ M$$

$$[Fe^{2+}] = 0.045\ M + 0.040\ M = 0.085\ M$$

$$[Fe^{3+}] = 0.055\ M - 0.040\ M = 0.015\ M$$

47. First we will need to come up with a balanced equation for the overall redox reaction. Clearly, the reaction must involve the oxidation of $Cl^-(aq)$ and the reduction of $Cr_2O_7^{2-}(aq)$:

$$14\ H^+(aq) + Cr_2O_7^{2-}(aq) + 6\ e^- \rightarrow 2\ Cr^{3+}(aq) + 7\ H_2O(l) \qquad\qquad E^\circ_{1/2red} = 1.33\ V$$

$$\{Cl^-(aq) \rightarrow 1/2\ Cl_2(g) + 1\ e^-\}\times 6 \qquad\qquad E^\circ_{1/2ox} = -1.358\ V$$

$$\overline{14\ H^+(aq) + Cr_2O_7^{2-}(aq) + 6\ Cl^-(aq) \rightarrow 2\ Cr^{3+}(aq) + 7\ H_2O(l) + 3\ Cl_2(g)\ E^\circ_{cell} = -0.03\ V}$$

A <u>negative</u> cell potential means, the oxidation of $Cl^-(aq)$ to $Cl_2(g)$ by $Cr_2O_7^{2-}(aq)$ at standard conditions will <u>not</u> occur spontaneously. We could obtain some $Cl_2(g)$ from this reaction by driving it to the product side with an external voltage. In other words, the reverse reaction is the spontaneous reaction at standard conditions and if we want to produce some $Cl_2(g)$ from the system, we must push the non-spontaneous reaction in its forward direction with an external voltage, (i.e., a DC power source). Since E°_{cell} is only slightly negative, we could also drive the reaction by removing products as they are formed and replenishing reactants as they are consumed.

48.

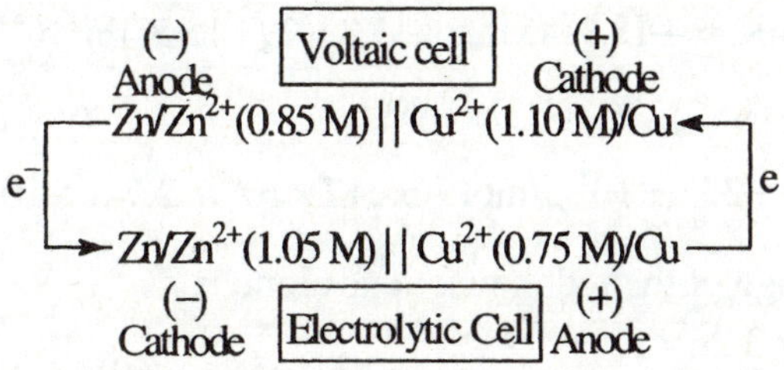

First we must find the voltage for each cell using the Nernst equation:

Cell A(voltaic cell): $Zn(s) + Cu^{2+}(aq) \rightleftharpoons Cu(s) + Zn^{2+}(aq) \qquad E^\circ = 1.100\ V\ (n = 2\ e^-)$

$$E_{cell} = E_{cell}^{o} - \frac{0.0592}{n} \log\left(\frac{[Zn^{2+}]}{[Cu^{2+}]}\right) = 1.100 \text{ V} - \frac{0.0592}{2} \log\left(\frac{0.85 \text{ M}}{1.10 \text{ M}}\right) = 1.103 \text{ V}$$

Cell B(electrolytic cell): $Cu(s) + Zn^{2+}(aq) \rightleftharpoons Zn(s) + Cu^{2+}(aq)$; $E^{o} = -1.100$ V ($n = 2$ e^{-})

$$E_{cell} = E_{cell}^{o} - \frac{0.0592}{n} \log\left(\frac{[Cu^{2+}]}{[Zn^{2+}]}\right) = -1.100 \text{ V} - \frac{0.0592}{2} \log\left(\frac{0.75 \text{ M}}{1.05 \text{ M}}\right) = -1.096 \text{ V}$$

Overall: $E_{overall} = E_{voltaic\ cell} + E_{electrolytic\ cell}$
$$= 1.103 \text{ V} + (-1.096 \text{ V}) = 0.007 \text{ V (at the instant of hook-up)}$$

Batteries and Fuel Cells

49. **(a)** The cell diagram begins with the anode and ends with the cathode.

Cell diagram: $Cr(s)|Cr^{2+}(aq), Cr^{3+}(aq) \| Fe^{2+}(aq), Fe^{3+}(aq)|Fe(s)$

(b) Oxidation: $Cr^{2+}(aq) \rightarrow Cr^{3+}(aq) + e^{-}$ $\qquad -E^{o} = +0.424$ V

Reduction: $Fe^{3+}(aq) + e^{-} \rightarrow Fe^{2+}(aq)$ $\qquad E^{o} = +0.771$ V

Net: $Cr^{2+}(aq) + Fe^{3+}(aq) \rightarrow Cr^{3+}(aq) + Fe^{2+}(aq)$ $\qquad E_{cell}^{o} = +1.195$ V

50. **(a)** Oxidation: $Zn(s) \rightarrow Zn^{2+}(aq) + 2e^{-}$ $\qquad -E^{o} = +0.763$ V

Reduction: $2MnO_2(s) + H_2O(l) + 2e^{-} \rightarrow Mn_2O_3(s) + 2OH^{-}(aq)$

Acid-base: $\{NH_4^{+}(aq) + OH^{-}(aq) \rightarrow NH_3(g) + H_2O(l)\}$ $\qquad \times 2$

Complex: $Zn^{2+}(aq) + 2NH_3(aq) + 2Cl^{-}(aq) \rightarrow [Zn(NH_3)_2]Cl_2(s)$

Net: $Zn(s) + 2MnO_2(s) + 2NH_4^{+}(aq) + 2Cl^{-}(aq) \rightarrow Mn_2O_3(s) + H_2O(l) + [Zn(NH_3)_2]Cl_2(s)$

(b) $\Delta G^{o} = -nFE_{cell}^{o} = -(2 \text{ mol e}^{-})(96485 \text{ C/mol e}^{-})(1.55 \text{ V}) = -2.99 \times 10^{5}$ J/mol

This is the standard free energy change for the entire reaction, which is composed of the four reactions in part (a). We can determine the values of ΔG^{o} for the acid-base and complex formation reactions by employing the appropriate data from Appendix D and $pK_f = -4.81$ (the negative log of the K_f for $[Zn(NH_3)_2]^{2+}$).

$$\Delta G_{a-b}^{o} = -RT\ln K_b^{2} = -(8.3145 \text{ J mol}^{-1} \text{ K}^{-1})(298.15 \text{ K})\ln(1.8 \times 10^{-5})^{2} = 5.417 \times 10^{4} \text{ J/mol}$$

$$\Delta G_{cmplx}^{o} = -RT\ln K_f = -(8.3145 \text{ J mol}^{-1} \text{ K}^{-})(298.15 \text{ K})\ln(10^{4.81}) = -2.746 \times 10^{4} \text{ J/mol}$$

Then $\Delta G_{total}^{o} = \Delta G_{redox}^{o} + \Delta G_{a-b}^{o} + \Delta G_{cmplx}^{o}$ $\qquad\qquad \Delta G_{redox}^{o} = \Delta G_{total}^{o} - \Delta G_{a-b}^{o} - \Delta G_{cmplx}^{o}$

$$= -2.99 \times 10^{5} \text{ J/mol} - 5.417 \times 10^{4} + 2.746 \times 10^{4} \text{ J/mol} = -3.26 \times 10^{5} \text{ J/mol}$$

Thus, the voltage of the redox reactions alone is

$$E^{o} = \frac{-3.26 \times 10^{5} \text{ J}}{-2\text{mol e}^{-} \times 96485 \text{ C/mol e}^{-}} = 1.69 \text{ V} \qquad 1.69\text{V} = +0.763\text{V} + E^{o}\{MnO_2/Mn_2O_3\}$$

$E^{o}\{MnO_2/Mn_2O_3\} = 1.69\text{V} - 0.763\text{V} = +0.93\text{V}$

The electrode potentials were calculated by using equilibrium constants from Appendix D. These calculations do not take into account the cell's own internal resistance to the flow of electrons, which makes the actual voltage developed by the electrodes less than the theoretical values derived from equilibrium constants. Also because the solid species (other than Zn) do not appear as compact rods, but rather are dispersed in a paste, and since very little water is present in the cell, the activities for the various species involved in the electrochemical reactions will deviate markedly from unity. As a result, the equilibrium constants for the reactions taking place in the cell will be substantially different from those provided in Appendix D, which apply only to dilute solutions and reactions involving solid reactants and products that posses small surface areas. The actual electrode voltages, therefore, will end up being different from those calculated here.

51. **(a)** Cell reaction: $2H_2(g) + O_2(g) \rightarrow 2H_2O(l)$

$$\Delta G_{rxn}^{\circ} = 2\Delta G_f^{\circ}\left[H_2O(l)\right] = 2(-237.1 \text{ kJ/mol}) = -474.2 \text{ kJ/mol}$$

$$E_{cell}^{\circ} = -\frac{\Delta G^{\circ}}{nF} = -\frac{-474.2 \times 10^3 \text{ J/mol}}{4 \text{ mol e}^- \times 96485 \text{ C/mol e}^-} = 1.229 \text{ V}$$

(b) Anode, Oxidation: $\{Zn(s) \rightarrow Zn^{2+}(aq) + 2e^-\} \times 2 \qquad -E^{\circ} = +0.763V$

Cathode, Reduction: $O_2(g) + 4 H^+(aq) + 4e^- \rightarrow 2H_2O(l) \qquad E^{\circ} = +1.229 \text{ V}$

Net: $2Zn(s) + O_2(g) + 4H^+(aq) \rightarrow 2Zn^{2+}(aq) + 2H_2O(l) \quad E_{cell}^{\circ} = +1.992 \text{ V}$

(c) Anode, Oxidation: $Mg(s) \rightarrow Mg^{2+}(aq) + 2e^- \qquad -E^{\circ} = +2.356 \text{ V}$

Cathode, Reduction: $I_2(s) + 2e^- \rightarrow 2 I^-(aq) \qquad E^{\circ} = +0.535 \text{ V}$

Net: $Mg(s) + I_2(s) \rightarrow Mg^{2+}(aq) + 2 I^-(aq) \qquad E_{cell}^{\circ} = +2.891 \text{ V}$

52. **(a)** Oxidation: $\quad Zn(s) \rightleftharpoons Zn^{2+}(aq) + 2 e^-$

Precipitation: $Zn^{2+}(aq) + 2 OH^-(aq) \rightleftharpoons Zn(OH)_2(s)$

Reduction: $\quad 2 MnO_2(s) + H_2)(l) + 2 e^- \rightleftharpoons Mn_2O_3(s) + 2 OH^-(aq)$

Net : $Zn(s) + 2 MnO_2(s) + H_2O(l) + 2 OH^-(aq) \rightleftharpoons Mn_2O_3(s) + Zn(OH)_2(s)$

(b) In question 50, we determined that the standard voltage for the reduction reaction is +0.93 V ($n = 2 e^-$). To convert this voltage to an equilibrium constant (at 25 °C) use:

$$\log K_{red} = \frac{nE^{\circ}}{0.0592} = \frac{2(0.93)}{0.0592} = 31.\underline{4}; \qquad K_{red} = 10^{31.42} = 3 \times 10^{31}$$

and for $Zn(s) \rightleftharpoons Zn^{2+}(aq) + 2 e^-$ ($E^{\circ} = 0.763$ V and $n = 2 e^-$)

$$\log K_{ox} = \frac{nE^{\circ}}{0.0592} = \frac{2(0.763)}{0.0592} = 25.\underline{8}; \qquad K_{ox} = 10^{25.78} = 6 \times 10^{25}$$

$$\Delta G^{\circ}{}_{total} = \Delta G^{\circ}{}_{precipitation} + \Delta G^{\circ}{}_{oxidation} + \Delta G^{\circ}{}_{reduction}$$

$$\Delta G^{\circ}{}_{total} = -RT \ln \frac{1}{K_{sp, Zn(OH)_2}} + (-RT \ln K_{ox}) + (-RT \ln K_{red})$$

$$\Delta G^{\circ}{}_{total} = -RT \left(\ln \frac{1}{K_{sp, Zn(OH)_2}} + \ln K_{ox} + \ln K_{red}\right)$$

$$\Delta G^{\circ}{}_{total} = -0.0083145 \frac{kJ}{K \, mol} (298 \text{ K}) \left(\ln \frac{1}{1.2 \times 10^{-17}} + \ln(6.0 \times 10^{25}) + \ln(2.6 \times 10^{31})\right)$$

$$\Delta G^{\circ}{}_{total} = -423 \text{ kJ} = -nFE^{\circ}{}_{total}$$

Hence, $E^{\circ}{}_{total} = E^{\circ}{}_{cell} = \dfrac{-423 \times 10^3 \text{ J}}{-(2 \, mol)(96485 \text{ C mol}^{-1})} = 2.19 \text{ V}$

53. Aluminum-Air Battery: $2 \text{ Al(s)} + 3/2 \text{ O}_2(g) \rightarrow \text{Al}_2\text{O}_3(s)$
Zinc-Air Battery: $\text{Zn(s)} + \frac{1}{2} \text{ O}_2(g) \rightarrow \text{ZnO(s)}$
Iron-Air Battery $\text{Fe(s)} + \frac{1}{2} \text{ O}_2(g) \rightarrow \text{FeO(s)}$
Calculate the quantity of charge transferred when 1.00 g of metal is consumed in each cell.

Aluminum-Air Cell:

$$1.00 \text{ g Al(s)} \times \frac{1 \text{ mol Al(s)}}{26.98 \text{ g Al(s)}} \times \frac{3 \text{ mol e}^-}{1 \text{ mol Al(s)}} \times \frac{96,485 \text{C}}{1 \text{ mol e}^-} = 1.07 \times 10^4 \text{ C}$$

Zinc-Air Cell:

$$1.00 \text{ g Zn(s)} \times \frac{1 \text{ mol Zn(s)}}{65.39 \text{ g Zn(s)}} \times \frac{2 \text{ mol e}^-}{1 \text{ mol Zn(s)}} \times \frac{96,485 \text{ C}}{1 \text{ mol e}^-} = 2.95 \times 10^3 \text{ C}$$

Iron-Air Cell:

$$1.00 \text{ g Fe(s)} \times \frac{1 \text{ mol Fe(s)}}{55.847 \text{ g Fe(s)}} \times \frac{2 \text{ mol e}^-}{1 \text{ mol Fe(s)}} \times \frac{96,485 \text{ C}}{1 \text{ mol e}^-} = 3.46 \times 10^3 \text{ C}$$

As expected, aluminum has the greatest quantity of charge transferred per unit mass (1.00 g) of metal oxidized. This is because aluminum has the smallest molar mass and forms the most highly charged cation (3+ for aluminum vs 2+ for Zn and Fe).

54. **(a)** A voltaic cell with a voltage of 0.1000 V would be possible by using two half-cells whose standard reduction potentials differ by approximately 0.10 V, such as the following pair.

Oxidation: $2 \text{ Cr}^{3+}(aq) + 7 \text{ H}_2\text{O}(l) \rightarrow \text{Cr}_2\text{O}_7{}^{2-}(aq) + 14 \text{ H}^+(aq) + 6 \text{ e}^- \quad -E^{\circ} = -1.33 \text{ V}$

Reduction: $\{\text{PbO}_2(s) + 4 \text{ H}^+(aq) + 2\text{e}^- \rightarrow \text{Pb}^{2+}(aq) + 2 \text{ H}_2\text{O}(l)\} \times 3 \quad E^{\circ} = +1.455 \text{ V}$

Net: $2\text{Cr}^{3+}(aq) + 3\text{PbO}_2(s) + \text{H}_2\text{O}(l) \rightarrow \text{Cr}_2\text{O}_7{}^{2-}(aq) + 3\text{Pb}^{2+}(aq) + 2\text{H}^+(aq) \quad E^{\circ}_{cell} = 0.12\underline{5} \text{ V}$

The voltage can be adjusted to 0.1000 V by a suitable alteration of the concentrations. $\left[\text{Pb}^{2+}\right]$ or $\left[\text{H}^+\right]$ could be increased or $\left[\text{Cr}^{3+}\right]$ could be decreased, or any combination of the three of these.

(b) To produce a cell with a voltage of 2.500 V requires that one start with two half-cells whose reduction potentials differ by about that much. An interesting pair follows.

Oxidation: $Al(s) \rightarrow Al^{3+}(aq) + 3e^-$ $\qquad\qquad -E° = +1.676\,V$

Reduction: $\{Ag^+(aq) + e^- \rightarrow Ag(s)\} \times 3$ $\qquad E° = +0.800\,V$

Net: $\qquad Al(s) + 3Ag^+(aq) \rightarrow Al^{3+}(aq) + 3Ag(s)$ $\qquad E°_{cell} = +2.476\,V$

Again, the desired voltage can be obtained by adjusting the concentrations. In this case increasing $[Ag^+]$ and/or decreasing $[Al^{3+}]$ would do the trick.

(c) Since no known pair of half-cells has a potential difference larger than about 6 volts, we conclude that producing a single cell with a potential of 10.00 V is currently impossible. It is possible, however, to join several cells together into a battery that delivers a voltage of 10.00 V. For instance, four of the cells from part (b) would deliver ~10.0 V at the instant of hook-up.

Electrochemical Mechanism of Corrosion

55. **(a)** Because copper is a less active metal than is iron (i.e. a weaker reducing agent), this situation would be similar to that of an iron or steel can plated with a tin coating which has been scratched. Oxidation of iron metal to Fe^{2+}(aq) should be enhanced in the body of the nail (blue precipitate), and hydroxide ion should be produced in the vicinity of the copper wire (pink color), which serves as the cathode.

(b) Because a scratch tears the iron and exposes "fresh" metal, it is more susceptible to corrosion. We expect blue precipitate forming in the vicinity of the scratch.

(c) Zinc should protect the iron nail from corrosion. There should be almost no blue precipitate; the zinc corrodes instead. The pink color of OH^- should continue to form.

56. The oxidation process involved at the anode reaction, is the formation of Fe^{2+}(aq). This occurs far below the water line. The reduction process involved at the cathode, is the formation of OH^-(aq) from O_2(g). It is logical that this reaction would occur at or near the water line close to the atmosphere (which contains O_2). This reduction reaction requires O_2(g) from the atmosphere and H_2O(l) from the water. The oxidation reaction, on the other hand simply requires iron from the pipe together with an aqueous solution into which the Fe^{2+}(aq) can disperse and not build up to such a high concentration that corrosion is inhibited.

Anode, Oxidation: $\qquad Fe(s) \rightarrow Fe^{2+}(aq) + 2e^-$

Cathode, Reduction: $\qquad O_2(g) + 2H_2O(l) + 4e^- \rightarrow 4OH^-(aq)$

57. During the process of corrosion, the metal that corrodes loses electrons. Thus, the metal in these instances behaves as an anode and, hence, can be viewed as bearing a latent negative polarity. One way in which we could retard oxidation of the metal would be to convert it into a cathode. Once transformed into a cathode, the metal would develop a positive charge and no longer release electrons (or oxidize). This change in polarity can be accomplished by hooking up the metal to an inert electrode in the ground and then applying a voltage across the two metals in such a way that the inert electrode becomes the anode and the metal that needs protecting becomes the cathode. This way, any oxidation that occurs will take place at the negatively charged inert electrode rather than the positively charged metal electrode.

58. As soon as the iron and copper came into direct contact, an electrochemical cell was created, in which the more powerfully reducing metal (Fe) was oxidized. In this way, the iron behaved as a sacrificial anode, protecting the copper from corrosion. The two half-reactions and the net cell reaction are shown below:

Anode(Oxidation)	$Fe(s) \rightarrow Fe^{2+}(aq) + 2\,e^-$	$-E° = 0.440\text{ V}$
Cathode(Reduction)	$\underline{Cu^{2+}(aq) + 2\,e^- \rightarrow Cu(s)}$	$E° = 0.337\text{ V}$
Net:	$Fe(s) + Cu^{2+}(aq) \rightarrow Fe^{2+}(aq) + Cu(s)$	$E°_{cell} = 0.777\text{ V}$

Note that because of the presence of iron and its electrical contact with the copper, any copper that does corrode is reduced back to the metal.

Electrolysis Reactions

59. Here we start by calculating the total amount of charge passed and the number of moles of electrons transferred.

$$\text{mol e}^- = 75\text{ min} \times \frac{60\text{ s}}{1\text{ min}} \times \frac{2.15\text{ C}}{1\text{ s}} \times \frac{1\text{ mol e}^-}{96485\text{ C}} = 0.10\text{ mol e}^-$$

(a) $\text{mass Zn} = 0.10\text{ mol e}^- \times \dfrac{1\text{ mol Zn}^{2+}}{2\text{ mol e}^-} \times \dfrac{1\text{ mol Zn}}{1\text{ mol Zn}^{2+}} \times \dfrac{65.39\text{ g Zn}}{1\text{ mol Zn}} = 3.3\text{ g Zn}$

(b) $\text{mass Al} = 0.10\text{ mol e}^- \times \dfrac{1\text{ mol Al}^{3+}}{3\text{ mol e}^-} \times \dfrac{1\text{ mol Al}}{1\text{ mol Al}^{3+}} \times \dfrac{26.98\text{ g Al}}{1\text{ mol Al}} = 0.90\text{ g Al}$

(c) $\text{mass Ag} = 0.10\text{ mol e}^- \times \dfrac{1\text{ mol Ag}^+}{1\text{ mol e}^-} \times \dfrac{1\text{ mol Ag}}{1\text{ mol Ag}^+} \times \dfrac{107.9\text{ g Ag}}{1\text{ mol Ag}} = 11\text{ g Ag}$

(d) $\text{mass Ni} = 0.10\text{ mol e}^- \times \dfrac{1\text{ mol Ni}^{2+}}{2\text{ mol e}^-} \times \dfrac{1\text{ mol Ni}}{1\text{ mol Ni}^{2+}} \times \dfrac{58.69\text{ g Ni}}{1\text{ mol Ni}} = 2.9\text{ g Ni}$

60. The two half reactions follow: $Cu^{2+}(aq) + 2\,e^- \rightarrow Cu(s)$ and $2\,H^+(aq) + 2\,e^- \rightarrow H_2(g)$ Thus, two moles of electrons are needed to produce each mole of Cu(s) and two moles of electrons are needed to produce each mole of $H_2(g)$. With this information, we can compute the moles of $H_2(g)$ that will be produced.

$$\text{mol H}_2\,(g) = 3.28\text{ g Cu} \times \frac{1\text{ mol Cu}}{63.55\text{ g Cu}} \times \frac{2\text{ mol e}^-}{1\text{ mol Cu}} \times \frac{1\text{ mol H}_2\,(g)}{2\text{ mol e}^-} = 0.0516\text{ mol H}_2\,(g)$$

Then we use the ideal gas equation to find the volume of $H_2(g)$.

$$\text{Volume of H}_2\,(g) = \frac{0.0516\text{ mol H}_2 \times \dfrac{0.08206\text{ L atm}}{\text{mol K}} \times (273.2 + 28.2)\text{ K}}{763\text{ mmHg} \times \dfrac{1\text{ atm}}{760\text{ mmHg}}} = 1.27\text{ L}$$

This answer assumes the $H_2(g)$ is not collected over water, and that the $H_2(g)$ formed is the only gas present in the container (i.e. no water vapor present)

61. Here we must determine the standard cell voltage of each chemical reaction. Those chemical reactions that have a negative voltage are the ones that require electrolysis.

(a) Oxidation: $2\,H_2O(l) \rightarrow 4\,H^+\,(aq) + O_2\,(g) + 4\,e^-$ $-E° = -1.229\text{ V}$

Reduction: $\{2\,H^+\,(aq) + 2\,e^- \rightarrow H_2\,(g)\} \times 2$ $E° = 0.000\text{ V}$

Net: $2\,H_2O(l) \rightarrow 2\,H_2\,(g) + O_2\,(g)$ $E°_{cell} = -1.229\text{ V}$

This reaction requires electrolysis, with an applied voltage greater than $+1.229\text{V}$.

(b) Oxidation: $Zn\,(s) \rightarrow Zn^{2+}\,(aq) + 2\,e^-$ $-E° = +0.763\text{ V}$

Reduction: $Fe^{2+}\,(aq) + 2\,e^- \rightarrow Fe\,(s)$ $E° = -0.440\text{ V}$

Net: $Zn\,(s) + Fe^{2+}\,(aq) \rightarrow Fe\,(s) + Zn^{2+}\,(aq)$ $E°_{cell} = +0.323\text{ V}$

This is a spontaneous reaction under standard conditions.

(c) Oxidation: $\{Fe^{2+}(aq) \rightarrow Fe^{3+}(aq) + e^-\} \times 2$ $-E° = -0.771\text{ V}$

Reduction: $I_2\,(s) + 2\,e^- \rightarrow 2\,I^-\,(aq)$ $E° = +0.535\text{ V}$

Net: $2\,Fe^{2+}(aq) + I_2(s) \rightarrow 2\,Fe^{3+}(aq) + 2\,I^-(aq)$ $E°_{cell} = -0.236\text{ V}$

This reaction requires electrolysis, with an applied voltage greater than $+0.236\text{V}$.

(d) Oxidation: $Cu\,(s) \rightarrow Cu^{2+}\,(aq) + 2\,e^-$ $-E° = -0.337\text{ V}$

Reduction: $\{Sn^{4+}\,(aq) + 2\,e^- \rightarrow Sn^{2+}\,(aq)\} \times 2$ $E° = +0.154\text{ V}$

Net: $Cu(s) + Sn^{4+}(aq) \rightarrow Cu^{2+}(aq) + Sn^{2+}(aq)$ $E°_{cell} = -0.183\text{ V}$

This reaction requires electrolysis, with an applied voltage greater than $+0.183\text{V}$.

62. (a) Because oxidation occurs at the anode, we know that the product cannot be H_2 (H_2 is produced from the reduction of H_2O), SO_2, (which is a reduction product of SO_4^{2-}), or SO_3 (which is produced from SO_4^{2-} without a change of oxidation state; it is the dehydration product of H_2SO_4). It is, in fact O_2 that forms at the anode. The oxidation of water at the anode produces the $O_2(g)$.

(b) Reduction should occur at the cathode. The possible species that can be reduced are H_2O to $H_2(g)$, $K^+(aq)$ to $K(s)$, and $SO_4^{2-}(aq)$ to perhaps $SO_2(g)$. Because potassium is a highly active metal, it will not be produced in aqueous solution. In order for $SO_4^{2-}(aq)$ to be reduced, it would have to migrate to the negatively charged cathode, which is not very likely probable since like charges repel each other. Thus, $H_2(g)$ is produced at the cathode.

(c) At the anode: $\quad 2H_2O(l) \rightarrow 4H^+(aq) + O_2(g) + 4e^- \qquad -E° = -1.229\,V$

At the cathode: $\quad \{2H^+(aq) + 2e^- \rightarrow H_2(g)\} \times 2 \qquad E° = 0.000\,V$

Net cell reaction: $2H_2O(l) \rightarrow 2H_2(g) + O_2(g) \qquad E°_{cell} = -1.229\,V$

A voltage greater than 1.229 V is required. Because of the high overpotential required for the formation of gases, we expect that a higher voltage will be necessary.

63. (a) The two gases that are produced are $H_2(g)$ and $O_2(g)$.

(b) At the anode: $2\,H_2O(l) \rightarrow 4\,H^+(aq) + O_2(g) + 4e^- \qquad -E° = -1.229\,V$

At the cathode: $\{2H^+(aq) + 2e^- \rightarrow H_2(g)\} \times 2 \qquad E° = 0.000\,V$

Net cell reaction: $2H_2O(l) \rightarrow 2H_2(g) + O_2(g) \qquad E°_{cell} = -1.229\,V$

64. The electrolysis of $Na_2SO_4(aq)$ produces molecular oxygen at the anode. $-E°\{O_2(g)/H_2O\} = -1.229\,V$. The other possible product is $S_2O_8^{2-}(aq)$. It is however, unlikely to form because it has a considerably less favorable half-cell potential. $-E°\{S_2O_8^{2-}(aq)/SO_4^{2-}(aq)\} = -2.01\,V$.

$H_2(g)$ is formed at the cathode.

$$\text{mol } O_2 = 3.75\,h \times \frac{3600\,s}{1h} \times \frac{2.83\,C}{1s} \times \frac{1 \text{ mol } e^-}{96485\,C} \times \frac{1 \text{ mol } O_2}{4 \text{ mol } e^-} = 0.0990 \text{ mol } O_2$$

The vapor pressure of water at $25°C$, from Table 12-2, is 23.8 mmHg.

$$V = \frac{nRT}{P} = \frac{0.0990 \text{ mol} \times 0.08206 \text{ L atm mol}^{-1}\text{ K}^{-1} \times 298 \text{ K}}{(742 - 23.8) \text{ mmHg} \times \dfrac{1 \text{ atm}}{760 \text{ mmHg}}} = 2.56 \text{ L } O_2(g)$$

65. (a) $Zn^{2+}(aq) + 2e^- \rightarrow Zn(s)$

$$\text{mass of Zn} = 42.5 \text{ min} \times \frac{60 \text{ s}}{1 \text{ min}} \times \frac{1.87 \text{ C}}{1 \text{ s}} \times \frac{1 \text{ mol } e^-}{96,485 \text{ C}} \times \frac{1 \text{ mol Zn}}{2 \text{ mol } e^-} \times \frac{65.39 \text{ g Zn}}{1 \text{ mol Zn}} = 1.62 \text{ g Zn}$$

(b) $2I^-(aq) \rightarrow I_2(s) + 2e^-$

$$\text{time needed} = 2.79 \text{ g } I_2 \times \frac{1 \text{ mol } I_2}{253.8 \text{ g } I_2} \times \frac{2 \text{ mol } e^-}{1 \text{ mol } I_2} \times \frac{96,485 \text{ C}}{1 \text{ mol } e^-} \times \frac{1 \text{ s}}{1.75 \text{ C}} \times \frac{1 \text{ min}}{60 \text{ s}} = 20.2 \text{ min}$$

66. (a) $Cu^{2+}(aq) + 2e^- \rightarrow Cu(s)$

$$\text{mmol } Cu^{2+}\text{consumed} = 282 \text{ s} \times \frac{2.68 \text{ C}}{1 \text{ s}} \times \frac{1 \text{ mol } e^-}{96,485 \text{ C}} \times \frac{1 \text{ mol } Cu^{2+}}{2 \text{ mol } e^-} \times \frac{1000 \text{ mmol}}{1 \text{ mol}}$$

$$= 3.92 \text{ mmol } Cu^{2+}$$

$$\text{decrease in } \left[Cu^{2+}\right] = \frac{3.92 \text{ mmol } Cu^{2+}}{425 \text{ mL}} = 0.00922 \text{ M}$$

$$\text{final } \left[Cu^{2+}\right] = 0.366 \text{ M} - 0.00922 \text{ M} = 0.357 \text{ M}$$

(b) $\text{mmol } Ag^+\text{consumed} = 255 \text{ mL}(0.196 \text{ M} - 0.175 \text{ M}) = 5.3\underline{6} \text{ mmol } Ag^+$

$$\text{time needed} = 5.3_6 \text{ mmol } Ag^+ \times \frac{1 \text{ mol } Ag^+}{1000 \text{ mmol } Ag^+} \times \frac{1 \text{ mol } e^-}{1 \text{ mol } Ag^+} \times \frac{96485 \text{ C}}{1 \text{ mol } e^-} \times \frac{1 \text{ s}}{1.84 \text{ C}}$$

$$= 28\underline{1} \text{ s} \approx 2.8 \times 10^2 \text{ s}$$

67. (a) $\text{charge} = 1.206 \text{ g Ag} \times \dfrac{1 \text{ mol Ag}}{107.87 \text{ g Ag}} \times \dfrac{1 \text{ mol } e^-}{1 \text{ mol Ag}} \times \dfrac{96,485 \text{ C}}{1 \text{ mol } e^-} = 1079 \text{ C}$

(b) $\text{current} = \dfrac{1079 \text{ C}}{1412 \text{ s}} = 0.7642 \text{ A}$

68. (a) Anode, Oxidation: $2H_2O(l) \rightarrow 4H^+(aq) + 4e^- + O_2(g)$ $\quad -E^\circ = -1.229 \text{V}$

Cathode, Reduction: $\{Ag^+(aq) + e^- \rightarrow Ag(s)\} \times 4$ $\quad E^\circ = +0.800 \text{ V}$

Net: $2H_2O(l) + 4Ag^+(aq) \rightarrow 4H^+(aq) + O_2(g) + 4Ag(s)$ $\quad E^\circ_{cell} = -0.429 \text{ V}$

(b) $\text{charge} = (25.8639 - 25.0782) \text{ g Ag} \times \dfrac{1 \text{ mol Ag}}{107.87 \text{ g Ag}} \times \dfrac{1 \text{ mol } e^-}{1 \text{ mol Ag}} \times \dfrac{96,485 \text{ C}}{1 \text{ mol } e^-} = 702.8 \text{ C}$

$$\text{current} = \frac{702.8 \text{ C}}{2.00 \text{ h}} \times \frac{1 \text{ h}}{3600 \text{ s}} = 0.0976 \text{ A}$$

(c) The gas is molecular oxygen.

$$V = \frac{nRT}{P} = \frac{\left(702.8 \text{ C} \times \dfrac{1 \text{ mol } e^-}{96485 \text{ C}} \times \dfrac{1 \text{ mol } O_2}{4 \text{ mol } e^-}\right) 0.08206 \dfrac{\text{L atm}}{\text{mol K}} (23+273) \text{ K}}{755 \text{ mmHg} \times \dfrac{1 \text{ atm}}{760 \text{ mmHg}}}$$

$$= 0.0445 \text{ L } O_2 \times \frac{1000 \text{ mL}}{1 \text{ L}} = 44.5 \text{ mL of } O_2$$

Integrative and Advanced Exercises

69. Oxidation: $\quad V^{3+} + H_2O \longrightarrow VO^{2+} + 2\,H^+ + e^- \qquad\qquad -E^\circ_a$

Reduction: $\quad Ag^+ + e^- \longrightarrow Ag(s) \qquad\qquad\qquad E^\circ = +0.800\ V$

Net: $\qquad\qquad V^{3+} + H_2O + Ag^+ \longrightarrow VO^{2+} + 2\,H^+ + Ag(s) \quad E^\circ_{cell} = 0.439\ V$

$0.439\ V = -E^\circ_a + 0.800\ V \qquad\qquad E^\circ_a = 0.800\ V - 0.439\ V = 0.361\ V$

Oxidation: $\qquad\qquad\qquad V^{2+} \longrightarrow V^{3+} + e^- \qquad\qquad -E^\circ_b$

Reduction: $\qquad VO^{2+} + 2\,H^+ + e^- \longrightarrow V^{3+} + H_2O \qquad E^\circ = +0.361\ V$

Net: $\qquad V^{2+} + VO^{2+} + 2\,H^+ \longrightarrow 2\,V^{3+} + H_2O \qquad E^\circ_{cell} = +0.616\ V$

$0.616\ V = -E^\circ_b + 0.361\ V \quad E^\circ_b = 0.361\ V - 0.616\ V = -0.255\ V$

Thus, for the cited reaction: $\quad V^{3+} + e^- \longrightarrow V^{2+} \qquad\qquad E^\circ = -0.255\ V$

70. The cell reaction for discharging a lead storage battery is equation (20.24).

$Pb(s) + PbO_2(s) + 2\,H^+(aq) + 2\,HSO_4^-(aq) \longrightarrow PbSO_4(s) + 2\,H_2O(l)$

The half-reactions with which this equation was derived indicates that two moles of electrons are transferred for every two moles of sulfate ion consumed. We first compute the amount of H_2SO_4 initially present and then the amount of H_2SO_4 consumed.

$$\text{initial amount } H_2SO_4 = 1.50\ L \times \frac{5.00\ mol\ H_2SO_4}{1\ L\ soln} = 7.50\ mol\ H_2SO_4$$

$$\text{amount } H_2SO_4\ \text{consumed} = 6.0\ h \times \frac{3600\ s}{1\ h} \times \frac{2.50\ C}{1\ s} \times \frac{1\ mol\ e^-}{96485\ C} \times \frac{2\ mol\ SO_4^{2-}}{2\ mol\ e^-} \times \frac{1\ mol\ H_2SO_4}{1\ mol\ SO_4^{2-}}$$

$$= 0.56\ mol\ H_2SO_4$$

$$\text{final } [H_2SO_4] = \frac{7.50\ mol - 0.56\ mol}{1.50\ L} = 4.63\ M$$

71. The cell reaction is $2\,Cl^-(aq) + 2\,H_2O(l) \longrightarrow 2\,OH^-(aq) + H_2(g) + Cl_2(g)$

We first determine the charge transferred per 1000 kg Cl_2.

$$\text{charge} = 1000\ kg\ Cl_2 \times \frac{1000\ g}{1\ kg} \times \frac{1\ mol\ Cl_2}{70.90\ g\ Cl_2} \times \frac{2\ mol\ e^-}{1\ mol\ Cl_2} \times \frac{96,485\ C}{1\ mol\ e^-} = 2.72 \times 10^9\ C$$

(a) $\text{energy} = 3.45\ V \times 2.72 \times 10^9\ C \times \dfrac{1\ J}{1\ V \cdot C} \times \dfrac{1\ kJ}{1000\ J} = 9.38 \times 10^6\ kJ$

(b) $\text{energy} = 9.38 \times 10^9\ J \times \dfrac{1\ W \cdot s}{1\ J} \times \dfrac{1\ h}{3600\ s} \times \dfrac{1\ kWh}{1000\ W \cdot h} = 2.61 \times 10^3\ kWh$

72. We determine the equilibrium constant for the reaction.

Oxidation : $\quad Fe(s) \longrightarrow Fe^{2+}(aq) + 2e^- \qquad\qquad -E^\circ = +0.440\ V$

Reduction : $\quad \{Cr^{3+}(aq) + e^- \longrightarrow Cr^{2+}(aq)\} \times 2 \qquad E^\circ = -0.424\ V$

Net: $\qquad Fe(s) + 2\,Cr^{3+}(aq) \longrightarrow Fe^{2+}(aq) + 2\,Cr^{2+}(aq) \qquad E^\circ_{cell} = +0.016\ V$

$$\Delta G^\circ = -n\,FE^\circ_{cell} = -RT\ln K \qquad\qquad \ln K = \frac{nFE^\circ_{cell}}{RT}$$

$$\ln\ K = \frac{2\ \text{mol e}^-\ \times 96485\ \text{Coul/mol e}^-\times 0.016\ \text{V}}{8.3145\ \text{J mol}^{-1}\ \text{K}^{-1}\times 298\ \text{K}} = 1.2 \quad K = e^{1.2} = 3.3$$

Reaction: $\quad$ Fe(s) $+$ 2 Cr^{3+}(aq) $\rightleftharpoons$ Fe^{2+}(aq) $+$ 2Cr^{2+}(aq)

Initial $\quad$: $\qquad\qquad\qquad$ 1.00 M $\qquad\quad$ 0 M $\qquad\qquad$ 0 M

Changes : $\qquad\qquad\quad$ $-2x$ M $\qquad\quad$ $+x$ M $\qquad\quad$ $+2x$ M

Equil: $\qquad\qquad\quad$ $(1.00 - 2x)$M $\quad$ x M $\qquad\quad$ $2x$ M

$$K = \frac{[Fe^{2+}][Cr^{2+}]^2}{[Cr^{3+}]^2} = 3.3 = \frac{x\,(2x)^2}{(1.00 - 2x)^2}$$

Let us simply substitute values of x into this cubic equation to find a solution. Notice that x cannot be larger than 0.50, (values > 0.5 will result in a negative value for the denominator.

$$x = 0.40 \quad K = \frac{0.40\,(0.80)^2}{(1.00 - 0.80)^2} = 6.4 > 3.3 \qquad x = 0.35 \quad K = \frac{0.35\,(0.70)^2}{(1.00 - 0.70)^2} = 1.9 > 3.3$$

$$x = 0.37 \quad K = \frac{0.37\,(0.74)^2}{(1.00 - 0.74)^2} = 3.0 < 3.3 \qquad x = 0.38 \quad K = \frac{0.38\,(0.76)^2}{(1.00 - 0.76)^2} = 3.8 > 3.3$$

Thus, we conclude that $x = 0.37$ M $= [Fe^{2+}]$.

73. First we calculate the standard cell potential.

Oxidation: $\quad Fe^{2+}$(aq) $\longrightarrow Fe^{3+}$(aq) $+ e^-$ $\qquad\qquad\quad -E^\circ = -0.771$ V

Reduction: $\quad Ag^+$(aq) $+ e^- \longrightarrow$ Ag(s) $\qquad\qquad\qquad\quad E^\circ = +0.800$ V

Net: $\qquad\quad Fe^{2+}$(aq) $+ Ag^+$(aq) $\longrightarrow Fe^{3+}$(aq) $+$ Ag(s) $\quad E^\circ_{cell} = 0.029$ V

Next, we use the given concentrations to calculate the cell voltage with the Nernst equation.

$$E_{cell} = E^\circ_{cell} - \frac{0.0592}{1}\log\frac{[Fe^{3+}]}{[Fe^{2+}][Ag^+]} = 0.029 - 0.0592\log\frac{0.0050}{0.0050\times 2.0} = 0.029 + 0.018 = 0.047\ \text{V}$$

The reaction will continue to move in the forward direction until concentrations of reactants decrease and those of products increase a bit more. At equilibrium, $E_{cell} = 0$, and we have the following.

$$E^\circ_{cell} = \frac{0.0592}{1} \log \frac{[Fe^{3+}]}{[Fe^{2+}][Ag^+]} = 0.029 \qquad \log \frac{[Fe^{3+}]}{[Fe^{2+}][Ag^+]} = \frac{0.029}{0.0592} = 0.49$$

Reaction:　$Fe^{2+}(aq)$　+　$Ag^+(aq)$　　　$\rightleftharpoons$　　　$Fe^{3+}(aq)$　+　$Ag(s)$

Initial:　　0.0050 M　　2.0 M　　　　　　　　　　0.0050 M　　　　　–

Changes:　　$-x$ M　　　$-x$ M　　　　　　　　　　$+x$ M　　　　　　–

Equil:　　　$(0.0050 - x)$ M　$(2.0 - x)$ M　　　　$(0.0050 + x)$ M　　　–

$$\frac{[Fe^{3+}]}{[Fe^{2+}][Ag^+]} = 10^{0.49} = 3.1 = \frac{0.0050 + x}{(0.0050 - x)(2.0 - x)} \approx \frac{0.0050 + x}{2.0(0.0050 - x)}$$

$$6.2(0.0050 - x) = 0.0050 + x = 0.031 - 6.2\,x \qquad 7.2\,x = 0.026 \qquad x = \frac{0.026}{7.2} = 0.0036\ \text{M}$$

Note assumption, that $x \ll 2.0$, indeed is valid. $[Fe^{2+}] = 0.0050\ \text{M} - 0.0036\ \text{M} = 0.0014\ \text{M}$

74. We first note that we are dealing with a concentration cell, one in which the standard oxidation reaction is the reverse of the standard reduction reaction, and consequently its standard cell potential is precisely zero volts. For this cell, the Nernst equation permits us to determine the ratio of the two silver ion concentrations.

$$E_{cell} = 0.000\ \text{V} - \frac{0.0592}{1} \log \frac{[Ag^+(\text{satd AgI})]}{[Ag^+(\text{satd AgCl},\ x\ \text{M Cl}^-)]} = 0.0860\ \text{V}$$

$$\log \frac{[Ag^+(\text{satd AgI})]}{[Ag^+(\text{satd AgCl},\ x\ \text{M Cl}^-)]} = \frac{-0.0860}{0.0592} = -1.45 \qquad \frac{[Ag^+(\text{satd AgI})]}{[Ag^+(\text{satd AgCl},\ x\ \text{M Cl}^-)]} = 10^{-1.45} = 0.035$$

We can determine the numerator concentration from the solubility product expression for AgI(s)

$$K_{sp} = [Ag^+][I^-] = 8.5 \times 10^{-17} = s^2 \qquad s = \sqrt{8.5 \times 10^{-17}} = 9.2 \times 10^{-9}\ \text{M}$$

This permits the determination of the concentration in the denominator.

$$[Ag^+(\text{satd AgCl},\ x\ \text{M Cl}^-)] = \frac{9.2 \times 10^{-9}}{0.035} = 2.6 \times 10^{-7}\ \text{M}$$

We now can determine the value of x. Note: Cl^- arises from two sources, one being the dissolved AgCl.

$$K_{sp} = [Ag^+][Cl^-] = 1.8 \times 10^{-10} = (2.6 \times 10^{-7})(2.6 \times 10^{-7} + x) = 6.8 \times 10^{-14} + 2.6 \times 10^{-7}\,x$$

$$x = \frac{1.8 \times 10^{-10} - 6.8 \times 10^{-14}}{2.6 \times 10^{-7}} = 6.9 \times 10^{-4}\ \text{M} = [Cl^-]$$

75. The Faraday constant can be evaluated by measuring the electric charge needed to produce a certain quantity of chemical change. For instance, let's imagine that an electric circuit contains the half-reaction $Cu^{2+}(aq) + 2\,e^- \rightleftharpoons Cu(s)$. The electrode on which the solid copper plates out is weighed before and after the passage of electric current. The mass gain is the mass of copper reduced, which then is converted into the moles of copper reduced. The number of moles of electrons involved in the reduction then is computed from the stoichiometry for the reduction half-reaction. In addition, the amperage is measured during the reduction, and the time is recorded. For simplicity, we assume the amperage is constant.

Then the total charge (in coulombs) equals the current (in amperes, that is, coulombs per second) multiplied by the time (in seconds). The ratio of the total charge (in coulombs) required by the reduction divided by the amount (in moles) of electrons is the Faraday constant. To determine the Avogadro constant, one uses the charge of the electron, 1.602×10^{-19} C and the Faraday constant in the following calculation.

$$N_A = \frac{96,485\,C}{1\,mol\,electrons} \times \frac{1\,electron}{1.602 \times 10^{-19}\,C} = 6.023 \times 10^{23}\,\frac{electrons}{mole}$$

76. We first determine the value of $\Delta G°$ for the cell reaction, a reaction in which $n = 4$.

$$\Delta G° = -n\,FE°_{cell} = -4\,mol\,e^- \times \frac{96,485\,C}{1\,mol\,e^-} \times 1.559\,V = -6.017 \times 10^5\,J = -601.7\,kJ$$

$$= \Delta G°_f\,[N_2(g)] + 2\,\Delta G°_f\,[H_2O(l)] - \Delta G°_f\,[N_2H_4(aq)] - \Delta G°_f\,[O_2(g)]$$

$$= 0.00\,kJ + 2 \times (-237.2\,kJ) - \Delta G°_f\,[N_2H_4(aq)] - 0.00\,kJ$$

$$\Delta G°_f\,[N_2H_4(aq)] = 2 \times (-237.2\,kJ) + 601.7\,kJ = +127.3\,kJ$$

77. In general, we shall assume that both ions are present initially at concentrations of 1.00 M. Then we shall compute the concentration of the more easily reduced ion when its reduction potential has reached the point at which the other cation starts being reduced by electrolysis. In performing this calculation we use the Nernst equation, but modified for use with a half-reaction. We find that, in general, the greater the difference in $E°$ values for two reduction half-reactions, the more effective the separation.

(a) In this case, no calculation is necessary. If the electrolysis takes place in aqueous solution, $H_2(g)$ rather than $K(s)$ will be produced at the cathode. Cu^{2+} can be separated from K^+ by electrolysis.

(b) $Cu^{2+}(aq) + 2\,e^- \rightarrow Cu(s)$ $E° = +0.340\,V$ $\quad$ $Ag^+(aq) + e^- \rightarrow Ag(s)$ $\quad$ $E° = +0.800\,V$
Ag^+ will be reduced first. Now we ask what $[Ag^+]$ will be when $E = +0.337\,V$.

$$0.337\,V = 0.800\,V - \frac{0.0592}{1}\log\frac{1}{[Ag^+]} \qquad \log\frac{1}{[Ag^+]} = \frac{0.800 - 0.340}{0.0592} = +7.77$$

$[Ag^+] = 1.7 \times 10^{-8}$ M. Separation of the two cations is essentially complete.

(c) $Pb^{2+}(aq) + 2\,e^- \rightarrow Pb(s)$ $E° = -0.125\,V$ $\quad$ $Sn^{2+}(aq) + 2\,e^- \rightarrow Sn(s)$ $\quad$ $E° = -0.137\,V$
Pb^{2+} will be reduced first. We now ask what $[Pb^{2+}]$ will be when $E° = -0.137\,V$.

$$-0.137\,V = -0.125\,V - \frac{0.0592}{2}\log\frac{1}{[Pb^{2+}]} \qquad \log\frac{1}{[Pb^{2+}]} = \frac{2(0.137 - 0.125)}{0.0592} = 0.41$$

$[Pb^{2+}] = 10^{-0.41} = 0.39$ M $\quad$ Separation of Pb^{2+} from Sn^{2+} is not complete.

78. The efficiency value for a fuel cell will be greater than 1.00 for any exothermic reaction (one that has $\Delta H° < 0$) that has $\Delta G°$ that is more negative than its $\Delta H°$ value. Since $\Delta G° = \Delta H° - T\,\Delta S°$, this means that the value of $\Delta S°$ must be positive. Moreover, for this to be the case, δn_{gas} is usually greater than zero. Let us consider the situation that might lead to this type of reaction. The combustion of carbon-hydrogen-oxygen compounds leads to the formation of $H_2O(l)$ and $CO_2(g)$. Since most of the oxygen in these compounds comes from $O_2(g)$ (some is present in the C-H-O compound), there is a balance in the number of moles of gas produced—$CO_2(g)$—and those consumed—$O_2(g)$—which is offset in a negative direction by the $H_2O(l)$ produced. Thus, the combustion of these types of compounds only will have a positive value of Δn_{gas} if the number of oxygens in the formula of the compound is more than twice the number of hydrogens. By comparison, the decomposition of $NOCl(g)$, an oxychloride of nitrogen, does produce more moles of gas than it consumes. Let us investigate this decomposition reaction.

$$NOCl(g) \longrightarrow \tfrac{1}{2}N_2(g) + \tfrac{1}{2}O_2(g) + \tfrac{1}{2}Cl_2(g)$$

$$\Delta H^\circ = \tfrac{1}{2}\Delta H^\circ_f\,[N_2(g)] + \tfrac{1}{2}\Delta H^\circ_f\,[O_2(g)] + \tfrac{1}{2}\Delta H^\circ_f\,[Cl_2(g)] - \Delta H^\circ_f\,[NOCl(g)]$$

$$= 0.500\,(0.00\,\text{kJ/mol} + 0.00\,\text{kJ/mol} + 0.00\,\text{kJ/mol}) - 51.71\,\text{kJ/mol} = -51.71\,\text{kJ/mol}$$

$$\Delta G^\circ = \tfrac{1}{2}\Delta G^\circ_f\,[N_2(g)] + \tfrac{1}{2}\Delta G^\circ_f\,[O_2(g)] + \tfrac{1}{2}\Delta G^\circ_f\,[Cl_2(g)] - \Delta G^\circ_f\,[NOCl(g)]$$

$$= 0.500\,(0.00\,\text{kJ/mol} + 0.00\,\text{kJ/mol} + 0.00\,\text{kJ/mol}) - 66.08\,\text{kJ/mol} = -66.08\,\text{kJ/mol}$$

$$\varepsilon = \frac{\Delta G^\circ}{\Delta H^\circ} = \frac{-66.08\,\text{kJ/mol}}{-51.71\,\text{kJ/mol}} = 1.278$$

Yet another simple reaction that meets the requirement that $\Delta G^\circ < \Delta H^\circ$ is the combustion of graphite: $C(\text{graphite}) + O_2(g) \longrightarrow CO_2(g)$ We see from Appendix D that $\Delta G^\circ_f[CO_2(g)] = -394.4\,\text{kJ/mol}$ is more negative than $\Delta H^\circ_f[CO_2(g)] = -393.5\,\text{kJ/mol}$. (This reaction is accompanied by an increase in entropy; $\Delta S = 213.7 - 5.74 - 205.1 = 2.86\,\text{J/K}$, $\varepsilon = 1.002$.) $\Delta G^\circ < \Delta H^\circ$ is true of the reaction in which $CO(g)$ is formed from the elements. From Appendix D, $\Delta H^\circ_f\{CO(g)\} = -110.5\,\text{kJ/mol}$, and $\Delta G^\circ_f\{CO(g)\} = -137.2\,\text{kJ/mol}$, producing $\varepsilon = (-137.2/-110.5) = 1.242$.

Note that any reaction that has $\varepsilon > 1.00$ will be spontaneous under standard conditions at all temperatures. (There, of course, is another category, namely, an endothermic reaction that has $\Delta S^\circ < 0$. This type of reaction is nonspontaneous under standard conditions at all temperatures. As such it consumes energy while it is running, which is clearly not a desirable result for a fuel cell.)

79. We first write the two half-equations and then calculate a value of ΔG° from thermochemical data. This value then is used to determine the standard cell potential.

Oxidation: $\quad CH_3CH_2OH(g) + 3\ H_2O(l) \longrightarrow 2\ CO_2(g) + 12\ H^+(aq) + 12\ e^-$

Reduction: $\quad \{O_2(g) + 4\ H^+(aq) + 4e^- \longrightarrow 2\ H_2O(l)\} \times 3$

Overall: $\quad CH_3CH_2OH(g) + 3\ O_2(g) \longrightarrow 3\ H_2O(l) + 2\ CO_2(g)$

Thus, $\ n = 12$

(a) $\Delta G^\circ = 2\Delta G^\circ_f[CO_2(g)] + 3\Delta G^\circ_f[H_2O(l)] - \Delta G^\circ_f[CH_3CH_2OH(g)] - 3\Delta G^\circ_f[O_2(g)]$

$\qquad = 2(-394.4\,\text{kJ/mol}) + 3(-237.1\,\text{kJ/mol}) - (-168.5\,\text{kJ/mol}) - 3(0.00\,\text{kJ/mol}) = -1331.6\,\text{kJ/mol}$

$$E^\circ_{cell} = -\frac{\Delta G^\circ}{n\,F} = -\frac{-1331.6 \times 10^3\ \text{J/mol}}{12\ \text{mol}\ e^- \times 96{,}485\ \text{C/mol}\ e^-} = +1.1501\ \text{V}$$

(b) $E_{cell}^{\circ} = E^{\circ}[O_2(g)/H_2O] - E^{\circ}[CO_2(g)/CH_3CH_2OH(g)] = 1.1501$ V

$= 1.229$ V $- E^{\circ}[CO_2(g)/CH_3CH_2OH(g)]$

$E^{\circ}[CO_2(g)/CH_3CH_2OH(g)] = 1.229 - 1.1501 = +0.079$ V

80. First we determine the change in the amount of H^+ in each compartment.

Oxidation: $2\,H_2O(l) \rightarrow O_2(g) + 4\,H^+(aq) + 4\,e^-$ Reduction: $2\,H^+(aq) + 2\,e^- \rightarrow H_2(g)$

Δ amount $H^+ = 212$ min $\times \dfrac{60\text{ s}}{1\text{ min}} \times \dfrac{1.25\text{ C}}{1\text{ s}} \times \dfrac{1\text{ mol }e^-}{96,485\text{ C}} \times \dfrac{1\text{ mol }H^+}{1\text{ mol }e^-} = 0.165$ mol H^+

Before electrolysis, there is 0.500 mol $H_2PO_4^-$ and 0.500 mol HPO_4^{2-} in each compartment. The electrolysis adds 0.165 mol H^+ to the anode compartment, which has the effect of transforming 0.165 mol HPO_4^{2-} into 0.165 mol $H_2PO_4^-$, giving a total of 0.335 mol HPO_4^{2-} (0.500 mol – 0.165 mol) and 0.665 mol $H_2PO_4^-$ (0.500 mol + 0.165 mol). We can use the Henderson-Hasselbalch equation to determine the pH of the solution in the anode compartment.

$pH = pK_{a_2} + \log \dfrac{[HPO_4^{2-}]}{[H_2PO_4^-]} = 7.20 + \log \dfrac{0.335\text{ mol }HPO_4^{2-}/0.500\text{ L}}{0.665\text{ mol }H_2PO_4^-/0.500\text{ L}} = 6.90$

Again we use the Henderson-Hasselbalch equation in the cathode compartment. After electrolysis there is 0.665 mol HPO_4^{2-} and 0.335 mol $H_2PO_4^-$. Again we use the Henderson-Hasselbalch equation.

$pH = pK_{a_2} + \log \dfrac{[HPO_4^{2-}]}{[H_2PO_4^-]} = 7.20 + \log \dfrac{0.665\text{ mol }HPO_4^{2-}/0.500\text{ L}}{0.335\text{ mol }H_2PO_4^-/0.500\text{ L}} = 7.50$

81. We first determine the change in the amount of M^{2+} ion in each compartment.

Δ amount $M^{2+} = 10.00$ h $\times \dfrac{3600\text{ s}}{1\text{ h}} \times \dfrac{0.500\text{ C}}{1\text{ s}} \times \dfrac{1\text{ mol }e^{-1}}{96,485\text{ C}} \times \dfrac{1\text{ mol }M^{2+}}{2\text{ mol }e^-} = 0.0933$ mol M^{2+}

This change in amount is the increase in the amount of Cu^{2+} and the decrease in the amount of Zn^{2+}. We can also calculate the change in each of the concentrations.

$\Delta[Cu^{2+}] = \dfrac{+0.0933\text{ mol }Cu^{2+}}{0.1000\text{ L}} = +0.933$ M $\Delta[Zn^{2+}] = \dfrac{-0.0933\text{ mol }Zn^{2+}}{0.1000\text{ L}} = -0.933$ M

Then the concentrations of the two ions are determined.

$[Cu^{2+}] = 1.000$ M $+ 0.933$ M $= 1.933$ M $[Zn^{2+}] = 1.000$ M $- 0.933$ M $= 0.067$ M

Now we run the cell as a voltaic cell, first determining the value of E_{cell}°.

Oxidation : $Zn(s) \longrightarrow Zn^{2+}(aq) + 2\,e^-$ $-E^{\circ} = +0.763$ V

Reduction : $Cu^{2+}(aq) + 2\,e^- \longrightarrow Cu(s)$ $E^{\circ} = +0.340$ V

Net: $Zn(s) + Cu^{2+}(aq) \longrightarrow Cu(s)$ $E_{cell}^{\circ} = +1.103$ V

Then we use the Nernst equation to determine the voltage of this cell.

$E_{cell} = E_{cell}^{\circ} - \dfrac{0.0592}{2} \log \dfrac{[Zn^{2+}]}{[Cu^{2+}]} = 1.103 - \dfrac{0.0592}{2} \log \dfrac{0.067\text{ M}}{1.933\text{ M}} = 1.103 + 0.043 = 1.146$ V

82. (a)

$$\text{Anode: } Zn(s) \longrightarrow Zn^{2+}(aq) + 2e^- \qquad\qquad -E° = +0.763 \text{ V}$$

$$\text{Cathode: } \{AgCl(s) + e^-(aq) \longrightarrow Ag(s) + Cl^-(1\text{ M})\} \times 2 \qquad E° = +0.2223 \text{ V}$$

$$\text{Net: } Zn(s) + 2AgCl(s) \longrightarrow Zn^{2+}(aq) + 2Ag(s) + 2Cl^-(1\text{ M}) \quad E°_{cell} = +0.985 \text{ V}$$

(b) The major reason why this electrode is easier to use than the standard hydrogen electrode is that it does not involve a gas. Thus there are not the practical difficulties involved in handling gases. Another reason is that it yields a higher value of $E°_{cell}$, thus, this is a more spontaneous system.

(c)

$$\text{Oxidation: } Ag(s) \longrightarrow Ag^+(aq) + e^- \qquad\qquad -E° = -0.800 \text{ V}$$

$$\text{Reduction: } AgCl(s) + e^- \longrightarrow Ag(s) + Cl^-(aq) \qquad E = +0.2223 \text{ V}$$

$$\text{Net: } AgCl(s) \longrightarrow Ag^+(aq) + Cl^-(aq) \qquad E°_{cell} = -0.578 \text{ V}$$

The net reaction is the solubility reaction, for which the equilibrium constant is K_{sp}.

$$\Delta G° = -nFE° = -RT \ln K_{sp}$$

$$\ln K_{sp} = \frac{nFE°}{RT} = \frac{1 \text{ mol } e^- \times \dfrac{96485 \text{ C}}{1 \text{ mol } e^-} \times (-0.578 \text{ V}) \times \dfrac{1 \text{ J}}{1 \text{ V.C}}}{8.3145 \text{ J mol}^{-1} \text{ K}^{-1} \times 298.15 \text{ K}} = -22.5$$

$$K_{sp} = e^{-22.5} = 1.7 \times 10^{-10}$$

This value is in good agreement with the value of 1.8×10^{-10} given in Table 18-1.

83. (a)

$$Ag(s) \rightarrow Ag^+(aq) + e^- \qquad -E° = -0.800 \text{ V}$$

$$AgCl(s) + e^- \rightarrow Ag(s) + Cl^-(aq) \qquad E° = 0.2223 \text{ V}$$

$$AgCl(s) \rightarrow Ag^+(aq) + Cl^-(aq) \qquad E°_{cell} = -0.577\underline{7} \text{ V}$$

$$E_{cell} = E°_{cell} - \frac{0.0592}{1}\left(\log \frac{[Ag^+][Cl^-]}{1} \right) = -0.577\underline{7} \text{ V} - \frac{0.0592}{1} \log\left(\frac{[1.00][1.00\times10^{-3}]}{1} \right) = -0.400 \text{ V}$$

(b) 10.00 mL of 0.0100 M CrO_4^{2-} + 100.0 mL of 1.00×10^{-3} M Ag^+ (V_{total} = 110.0 mL)

Concentration of CrO_4^{2-} after dilution: 0.0100 M×10.00 mL /110.00 mL = 0.000909 M

Concentration of Ag^+ after dilution: 0.00100 M×100.0 mL /110.00 mL = 0.000909 M

	$Ag_2CrO_4(s)$	$\xrightarrow{K_{sp}=1.1\times10^{-12}}$	$2Ag^+(aq)$ +	$CrO_4^{2-}(aq)$
Initial			0.000909 M	0.000909 M
Change(100% rxn)			-0.000909 M	-0.000455 M
New initial			0 M	0.000455 M
Change			+2x	+x
Equilibrium			2x	0.000455 M+x $\approx$ 0.000455 M

$1.1 \times 10^{-12} = (2x)^2(0.000454)$

$x = 0.0000246$ M Note: 5.4% of 0.000455 M (assumption may be considered valid)

(Answer would be $x = 0.0000253$ using method of successive approx.)

$[Ag^+] = 2x = 0.0000492$ M (0.0000506 M using method of successive approx.)

$$E_{cell} = E°_{cell} - \frac{0.0592}{1} \log [Ag^+][Cl^-] = -0.577\underline{7} - \frac{0.0592}{1} \log [1.00M][4.92\times10^{-4}M]$$

$$= -0.323 \text{ V } (-0.306 \text{ V for method of successive approximations})$$

(c) 10.00 mL 0.0100 M NH_3 + 10.00 mL 0.0100 M CrO_4^{2-} + 100.0 mL 1.00×10^{-3} M Ag^+ (V_{total} = 120.0 mL)

Concentration of NH_3 after dilution: 10.0 M×10.00 mL /120.00 mL = 0.833 M

Concentration of CrO_4^{2-} after dilution: 0.0100 M×10.00 mL /110.00 mL = 0.000833 M

Concentration of Ag^+ after dilution: 0.00100 M×100.0 mL /110.00 mL = 0.000833 M

In order to determine the equilibrium concentration of free Ag^+(aq), we first consider complexation of Ag^+(aq) by NH_3(aq) and then check to see if precipitation occurs.

	Ag^+(aq)	+	$2NH_3$(aq) $\xrightarrow{K_f=1.6\times10^7}$	$Ag(NH_3)_2^+$(aq)
Initial	0.000833 M		0.833 M	0 M
Change(100% rxn)	-0.000833 M		-0.00167 M	+0.000833 M
New initial	0 M		0.831 M	0.000833 M
Change	+x		+2x	-x
Equilibrium	x		(0.831+2x) M	(0.000833 –x) M
Equilibrium ($x \approx 0$)	x		0.831 M	0.000833 M

$1.6 \times 10^7 = 0.000833 /x(0.831)^2$ $x = 7.5\underline{4}\times10^{-11}$ M = $[Ag^+]$

Note: The assumption is valid

Now we look to see if a precipitate forms: $Q_{sp} = (7.5\underline{4}\times10^{-11})^2(0.000833) = 4.7 \times10^{-24}$

Since $Q_{sp} < K_{sp}$ (1.1×10^{-12}), no precipitate forms and $[Ag^+] = 7.5\underline{4} \times 10^{-11}$ M

$$E_{cell} = E_{cell}^{\circ} - \frac{0.0592}{1} \log [Ag^+][Cl^-] = -0.577\underline{7} \text{ V} - \frac{0.0592}{1} \log [1.00M][7.5\underline{4} \times 10^{-11}M]$$

$$E_{cell} = 0.021\underline{5} \text{ V}$$

84. We assume that the Pb^{2+}(aq) is "ignored" by the silver electrode, that is, the silver electrode detects only silver ion in solution.

Oxidation : $H_2(g) \longrightarrow 2 H^+(aq) + 2 e^-$ $-E^{\circ} = 0.000$ V

Reduction: $\{Ag^+(aq) + e^- \longrightarrow Ag(s)\} \times 2$ $E^{\circ} = 0.800$ V

Net : $H_2(g) + 2 Ag^+ \longrightarrow 2 H^+(aq) + 2 Ag(s)$ $E^{\circ}_{cell} = 0.800$ V

$$E_{cell} = E_{cell}^{\circ} - \frac{0.0592}{2} \log \frac{[H^+]^2}{[Ag^+]^2} \qquad 0.503 \text{ V} = 0.800 \text{ V} - \frac{0.0592}{2} \log \frac{1.00^2}{[Ag^+]^2}$$

$$\log \frac{1.00^2}{[Ag^+]^2} = \frac{2(0.800 - 0.503)}{0.0592} = 10.0 \qquad \frac{1.00^2}{[Ag^+]^2} = 10^{+10.0} = 1.0 \times 10^{10}$$

$[Ag^+]^2 = 1.0 \times 10^{-10}$ M^2 $[Ag^+] = 1.0 \times 10^{-5}$ M

$$\text{mass Ag} = 0.500 \text{ L} \times \frac{1.0 \times 10^{-5} \text{ mol Ag}^+}{1 \text{ L soln}} \times \frac{1 \text{ mol Ag}}{1 \text{ mol Ag}^+} \times \frac{107.87 \text{ g Ag}}{1 \text{ mol Ag}} = 5.4 \times 10^{-4} \text{ g Ag}$$

$$\% \text{ Ag} = \frac{5.4 \times 10^{-4} \text{ g Ag}}{1.050 \text{ g sample}} \times 100 \% = 0.051\% \text{ Ag (by mass)}$$

85. 250.0 mL of 0.1000 M $CuSO_4$ = 0.02500 moles Cu^{2+} initially.

$$\text{Moles of } Cu^{2+} \text{ plated out} = \frac{3.512\,C}{s} \times 1368\,s \times \frac{1\,\text{mol e}^-}{96{,}485\,C} \times \frac{1\,\text{mol } Cu^{2+}}{2\,\text{mole}^-} \, 0.02490\,\text{mol } Cu^{2+}$$

Moles of Cu^{2+} in solution = 0.02500 mol – 0.02490 mol = 0.00010 mol Cu^{2+}

$[Cu^{2+}]$ = 0.00010 mol Cu^{2+}/0.250 L = 4.0×10^{-4} M

	Cu^{2+}(aq)	+	$4NH_3$(aq) $\xrightleftharpoons{K_f=1.1\times10^{13}}$	$Cu(NH_3)_4{}^{2+}$(aq)
Initial	0.00040 M		0.10 M	0 M
Change(100% rxn)	-0.00040M		maintained	+0.00040 M
New initial	0 M		0.10 M	0.00040 M
Change	+x		maintained	-x
Equilibrium	x		0.10 M	(0.00040 –x) M ≈ 0.00040 M

$$K_f = \frac{[Cu(NH_3)_4{}^{2+}]}{[Cu^{2+}][NH_3]^4} = \frac{0.00040}{[Cu^{2+}](0.10)^4} = 1.1 \times 10^{13} \qquad [Cu^{2+}] = 3.6 \times 10^{-13}\,M$$

Hence, the assumption is valid. The concentration of $Cu(NH_3)_4{}^{2+}$(aq) = 0.00040 M which is 40 times greater than the 1×10^{-5} detection limit. Thus, the blue color should appear.

86. First we determine the molar solubility of AgBr in 1 M NH_3.

$$AgBr(s) \rightleftharpoons Ag^+ + Br^- \qquad\qquad K_{sp} = 5.0 \times 10^{-13}$$

$$\underline{Ag^+ + 2NH_3 \rightleftharpoons Ag(NH_3)_2{}^+ \qquad\qquad K_f = 1.6 \times 10^7}$$

$$\text{(sum) } AgBr(s) + 2NH_3(aq) \rightleftharpoons Ag(NH_3)_2{}^+ + Br^- \qquad K = K_{sp} \times K_f = 8.0 \times 10^{-6}$$

sum $\quad AgBr(s) + 2NH_3(aq) \rightleftharpoons Ag(NH_3)^{2+}(aq) + Br^-(aq) \quad K = K_{sp} \times K_f = 8.0 \times 10^{-6}$

initial $\qquad\qquad$ 1.00 M $\qquad\qquad$ 0 M $\quad$ 0 M

equil. $\qquad\qquad$ 1.00-2s $\qquad\qquad$ s $\qquad$ s $\qquad$ {s $\Rightarrow$ AgBr molar solubility}

$$K = \frac{[Ag(NH_3)_2{}^+][Br^-]}{[NH_3(aq)]} = \frac{s^2}{(1-2s)^2} = 8.0 \times 10^{-6} \qquad s = 2.81 \times 10^{-3}\,M \text{ (also } [Br^-])$$

$$\text{So } [Ag^+] = \frac{K_{sp}}{[Br^-]} = \frac{5.0 \times 10^{-13}}{2.81 \times 10^{-3}} = 1.8 \times 10^{-10}\,M$$

Now, let's construct the cell, guessing that the standard hydrogen electrode is the anode.

Oxidation: $\quad H_2(g) \longrightarrow 2H^+ + 2e^- \qquad\qquad\qquad E° = -0.000$ V

Reduction: $\quad \underline{\{Ag^+(aq) + e^- \longrightarrow Ag(s)\} \times 2} \qquad\qquad E° = +0.800$ V

Net: $\qquad\quad 2Ag^+(aq) + H_2(g) \longrightarrow 2Ag(s) + 2H^+(aq) \qquad E°_{cell} = +0.800$V

From the Nernst equation:

$$E = E° - \frac{0.0592}{n}\log_{10}Q = E° - \frac{0.0592}{n}\log\frac{[H^+]^2}{[Ag^+]^2} = 0.800\,V - \frac{0.0592}{2}\log_{10}\frac{1^2}{(1.78 \times 10^{-10})^2}$$

and E = 0.223 V. Since the voltage is positive, our guess is correct and the standard hydrogen electrode is the anode (oxidizing ele*ctrode*).

87. **(a)** Anode: $2H_2O(l) \rightarrow 4\,e^- + 4\,H^+(aq) + O_2(g)$

   Cathode: $\underline{2\,H_2O(l) + 2\,e^- \rightarrow 2\,OH^-(aq) + H_2(g)}$

   Overall: $2\,H_2O(l) + 4\,\cancel{H_2O(l)} \rightarrow 4\,\cancel{H^+(aq)} + 4\,\cancel{OH^-(aq)} + 2\,H_2(g) + O_2(g)$

                       $2\,H_2O(l) \rightarrow 2\,H_2(g) + O_2(g)$

(b) 21.5 mA = 0.0215 A or 0.0215 C s^{-1} for 683 s

$$\text{mol } H_2SO_4 = \frac{0.0215\ \text{C}}{\text{s}} \times 683\,\text{s} \times \frac{1\ \text{mol }e^-}{96485\ \text{C}} \times \frac{1\ \text{mol }H^+}{1\ \text{mol }e^-} \times \frac{1\ \text{mol }H_2SO_4}{2\ \text{mol }H^+} = 7.61 \times 10^{-5}\ \text{mol } H_2SO_4$$

7.61×10^{-5} mol H_2SO_4 in 10.00 mL. Hence $[H_2SO_4] = 7.61 \times 10^{-5}$ mol / 0.01000 L = 7.61×10^{-3} M

88. First we need to find the total surface area

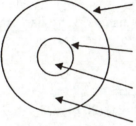

Outer circumference = $2\pi r = 2\pi(2.50\ \text{cm}) = 15.7$ cm
Surface area = circumference × thickness = 15.7 cm × 0.50 cm = 7.8$\underline{5}$ cm^2

Inner circumference = $2\pi r = 2\pi(1.00\ \text{cm}) = 6.28$ cm
Surface area = circumference × thickness = 6.28 cm × 0.50 cm = 3.1$\underline{4}$ cm^2

Area of small circle = $\pi r^2 = \pi(1.00\ \text{cm})^2 = 3.14$ cm^2

Area of large circle = $\pi r^2 = \pi(2.50\ \text{cm})^2 = 19.6$ cm^2

Total area = 7.85 cm^2 + 3.14 cm^2 + 2×(19.6 cm^2) − 2×(3.1$\underline{4}$ cm^2) = 43.9$\underline{8}$ cm^2

Volume of metal needed = surface area × thickness of plating = 43.9$\underline{8}$ cm^2 0.0050 cm = 0.22 cm^3

$$\text{Charge required} = 0.22\ \text{cm}^3 \times \frac{8.90\ \text{g}}{\text{cm}^3} \times \frac{1\ \text{mol Ni}}{58.693\ \text{g Ni}} \times \frac{2\ \text{mol }e^-}{1\ \text{mol Ni}} \times \frac{96485\ \text{C}}{1\ \text{mol }e^-} = 643\underline{7}.5\ \text{C}$$

Time = charge/time = 64$\underline{37.5}$ C / 1.50 C/s = 429$\underline{1.7}$ s or 71.$\underline{5}$ min

89. With reference to the aluminum-air battery in Figure 20-18 and equation (20.28),

(a) Refer to reaction (20.28) and Appendix D.
$E_{cell}° = E°$ (cathode) − $E°$ (anode) = $E°$ (reduction half-cell) − $E°$ (oxidation half-cell)
$E_{cell}° = 0.401$ V − (−2.310) V = 2.711 V

(b) For reaction (20.8), $\Delta G_f° = -nFE_{cell}° = -12$ mol $e^- \times 96{,}485$ C/mol $e^- \times 2.711$ V
$= -3.14 \times 10^6$ V·C = -3.14×10^6 J = -3.14×10^3 kJ
$\Delta G° = 4\Delta G_f°[Al(OH)_4]^-(aq) - 6\Delta G_f°[H_2O(l)] - 4\Delta G_f°[OH^-] = -3.14 \times 10^3$ kJ

$$\Delta G_f°[Al(OH)_4]^-(aq) = \frac{-3.14 \times 10^3\,\text{kJ} + 6 \times (-237.1)\,\text{kJ} + 4 \times (-157.2)\,\text{kJ}}{4}$$

$\Delta G_f°[Al(OH)_4]^-(aq) = -1.30 \times 10^3$ kJ/mol

(c) $\text{mass of Al} = 4.00\,\text{h} \times \dfrac{3600\,\text{s}}{1\,\text{h}} \times \dfrac{10.0\,\text{C}}{1\,\text{s}} \times \dfrac{1\,\text{mol }e^-}{96{,}485\,\text{C}} \times \dfrac{1\,\text{mol Al}}{3\,\text{mol }e^-} \times \dfrac{26.98\,\text{g Al}}{1\,\text{mol Al}} = 13.4\,\text{g Al}$

90. (a) The metal has to have a reduction potential more negative than -0.691 V, so that its oxidation can reverse the tarnishing reaction's -0.691 V reduction potential. Aluminum is a good candidate, because it is inexpensive, readily available, will not react with water and has an E° of -1.676 V. Zinc is also a possibility with an E° of -0.763 V, but we don't choose it because there may be an overpotential associated with the tarnishing reaction.

(b) Oxidation: $\{Al(s) \longrightarrow Al^{3+}(aq) + 3\,e^- \} \times 2$

Reduction: $\underline{\{Ag_2S(s) + 2\,e^- \longrightarrow Ag(s) + S^{2-}(aq) \} \times 3}$

Net: $2\,Al(s) + 3\,Ag_2S(s) \longrightarrow 6Ag(s) + 2\,Al^{3+}(aq) + 3\,S^{2-}(aq)$

(c) The dissolved $NaHCO_3(s)$ serves as an electrolyte. It would also enhance the electrical contact between the two objects.

(d) There are several chemicals involved: Al, H_2O, and $NaHCO_3$. Although the aluminum plate will be consumed very slowly because silver tarnish is very thin, it will, nonetheless, eventually erode away. We should be able to detect loss of mass after many uses.

91. (a) Overall: $C_3H_8(g) + 5\,O_2(g) \rightarrow 3\,CO_2(g) + 4\,H_2O(l)$

Anode: $6\,H_2O(l) + C_3H_8(g) \rightarrow 3\,CO_2(g) + 20\,H^+(g) + 20\,e^-$

Cathode: $20\,e^- + 20\,H^+(g) + 5\,O_2(g) \rightarrow 10\,H_2O(l)$

$\Delta G^\circ_{rxn} = 3(-394.4 \text{ kJ/mol}) + 4(-237.1 \text{ kJ/mol}) - 1(-23.3 \text{ kJ/mol}) = -2108.3 \text{ kJ/mol}$

$\Delta G^\circ_{rxn} = -2108.3 \text{ kJ/mol} = -2,108,300 \text{ J/mol}$

$\Delta G^\circ_{rxn} = -nFE^\circ_{cell} = -20 \text{ mol } e^- \times (96485 \text{ C/mol } e^-) \times E^\circ_{cell}$

$E^\circ_{cell} = +1.0926 \text{ V}$

$20\,e^- + 20\,H^+(g) + 5\,O_2(g) \rightarrow 10\,H_2O(l) \quad E^\circ_{cathode} = +1.229 \text{ V}$

Hence $\quad E^\circ_{cell} = E^\circ_{cathode} - E^\circ_{anode} \qquad +1.0926 \text{ V} = +1.229 \text{ V} - E^\circ_{anode}$

$E^\circ_{anode} = +0.136 \text{ V}$ (reduction potential for $3\,CO_2(g) + 20\,H^+(g) + 20\,e^- \rightarrow 6\,H_2O(l) + C_3H_8(g)$)

(b) Use thermodynamic tables for $3\,CO_2(g) + 20\,H^+(g) + 20\,e^- \rightarrow 6\,H_2O(l) + C_3H_8(g)$

$\Delta G^\circ_{rxn} = 6(-237.1 \text{ kJ/mol}) + 1(-23.3 \text{ kJ/mol}) - [(-394.4 \text{ kJ/mol}) + 20(0 \text{kJ/mol})] = -262.7 \text{ kJ/mol}$

$\Delta G^\circ_{red} = -262.7 \text{ kJ/mol} = -262,700 \text{ J/mol} = -nFE^\circ_{red} = -20 \text{ mol } e^- \times (96,485 \text{ C/mol } e^-) \times E^\circ_{red}$

$E^\circ_{red} = 0.136 \text{ V}$ (Same value, as found in (a))

FEATURE PROBLEMS

92. **(a)** Anode: $H_2(g, 1\ atm) \rightarrow 2H^+(1\ M) + 2e^-$ $\qquad -E° = -0.0000\ V$

Cathode: $\{Ag^+(x\ M) + e^- \rightarrow Ag(s)\}\ \times 2$ $\qquad E° = 0.800\ V$

Net: $H_2(g, 1\ atm) + 2Ag^+(aq) \rightarrow 2H^+(1\ M) + 2Ag(s)$ $\qquad E°_{cell} = 0.800$

(b) Since the voltage in the anode half-cell remains constant, we use the Nernst equation to calculate the half-cell voltage in the cathode half-cell, with two moles of electrons. This is then added to $-E$ for the anode half-cell. Because $-E° = 0.000$ for the anode half cell, $E_{cell} = E_{cathode}$

$$E = E° - \frac{0.0592}{2}\log\frac{1}{\left[Ag^+\right]^2} = 0.800 + 0.0592\ \log\left[Ag^+\right] = 0.800 + 0.0592\log x$$

(c) **(i)** Initially $\left[Ag^+\right] = 0.0100$; $\quad E = 0.800 + 0.0592\ \log 0.0100 = 0.682\ V = E_{cell}$

Note that 50.0 mL of titrant is required for the titration, since both $AgNO_3$ and KI have the same concentrations and they react in equimolar ratios.

(ii) After 20.0 mL of titrant is added, the total volume of solution is 70.0 mL and the unreacted Ag^+ is that in the untitrated 30.0 mL of 0.0100 M $AgNO_3(aq)$.

$$\left[Ag^+\right] = \frac{30.0\ mL \times 0.0100\ M\ Ag^+}{70.0\ mL} = 0.00429 M$$

$E = 0.800 + 0.0592\log(0.00429) = 0.660\ V = E_{cell}$

(iii) After 49.0 mL of titrant is added, the total volume of solution is 99.0 mL and the unreacted Ag^+ is that in the untitrated 1.0 mL of 0.0100 M $AgNO_3(aq)$.

$$\left[Ag^+\right] = \frac{1.0\ mL \times 0.0100\ M\ Ag^+}{99.0\ mL} = 0.00010\ M$$

$E = 0.800 + 0.0592\log(0.00010) = 0.563\ V = E_{cell}$

(iv) At the equivalence point, we have a saturated solution of AgI, for which

$$\left[Ag^+\right] = \sqrt{K_{sp}(AgI)} = \sqrt{8.5 \times 10^{-17}} = 9.2 \times 10^{-9}$$

$E = 0.800 + 0.0592\log(9.2 \times 10^{-9}) = 0.324\ V = E_{cell}$.

After the equivalence point, the $\left[Ag^+\right]$ is determined by the $\left[I^-\right]$ resulting from the excess KI(aq).

(v)   When 51.0 mL of titrant is added, the total volume of solution is 101.0 mL and the excess I^- is that in 1.0 mL of 0.0100 M KI(aq).

$$\left[I^-\right]=\frac{1.0\ mL\times0.0100\ M\ I^-}{101.0\ mL}=9.9\times10^{-5}\ M \qquad \left[Ag^+\right]=\frac{K_{sp}}{\left[I^-\right]}=\frac{8.5\times10^{-17}}{0.000099}=8.6\times10^{-13}\ M$$

$$E = 0.800+0.0592\ \log\left(8.6\times10^{-13}\right)=0.086\ V = E_{cell}$$

(vi) When 60.0 mL of titrant is added, the total volume of solution is 110.0 mL and the excess I^- is that in 10.0 mL of 0.0100 M KI(aq).

$$\left[I^-\right]=\frac{10.0\ mL\times0.0100\ M\ I^-}{110.0\ mL}=0.00091\ M$$

$$\left[Ag^+\right]=\frac{K_{sp}}{\left[I^-\right]}=\frac{8.5\times10^{-17}}{0.00091}=9.3\times10^{-14}\ M$$

$$E = 0.800+0.0592\ \log\left(9.3\times10^{-14}\right)=0.029\ V = E_{cell}$$

(d)   The titration curve is presented below.

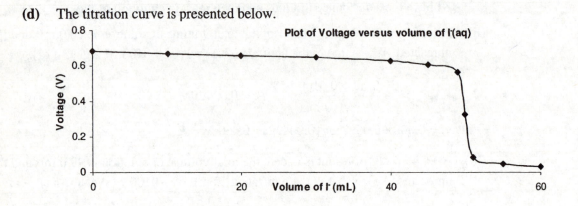

Plot of Voltage versus volume of I⁻(aq)

93. **(a)** **(1)**   anode: $Na(s) \rightarrow Na^+(in\ ethylamine)+e^-$

cathode: $Na^+(in\ ethylamine) \rightarrow Na(amalgam,\ 0.206\ \%)$

Net: $Na(s) \rightarrow Na(amalgam,\ 0.206\%)$

(2)   anode: $2Na(amalgam,\ 0.206\%) \rightarrow 2Na^+\ (1\ M)+2e^-$

cathode: $2H^+(aq,1M)\ +2e^- \rightarrow H_2\ (g,1\ atm)$

Net: $2Na\ (amalgam,\ 0.206\ \%)+2H^+(aq,1M) \rightarrow 2\ Na^+(1\ M)+H_2(g,1\ atm)$

(b) **(1)** $\Delta G = -1\ mol\ e^- \times \dfrac{96,485\ C}{1\ mol\ e^-} \times 0.8453\ V = -8.156\times10^4\ J\ or\ -81.56\ kJ$

(2) $\Delta G = -2\ mol\ e^- \times \dfrac{96,485\ C}{1\ mol\ e^-} \times 1.8673\ V = -36.033\times10^4\ J\ or\ -360.33\ kJ$

(c) **(1)** $2Na(s) \rightarrow 2Na(amalgam, 0.206\%)$ $\qquad \Delta G_1 = -2 \times 8.156 \times 10^4$ J

(2) $2Na$ $(amalg, 0.206\%) + 2H^+(aq, 1M) \rightarrow 2Na^+(1M) + H_2(g, 1 atm)$

$\Delta G_2 = -36.033 \times 10^4$ J

Overall: $2Na(s) + 2H^+(aq) \rightarrow 2Na^+(1M) + H_2(g, 1 atm)$

$\Delta G = \Delta G_1 + \Delta G_2 = -16.312 \times 10^4 J - 36.033 \times 10^4 J = -52.345 \times 10^4 J$ or -523.45 kJ

Since standard conditions are implied in the overall reaction, $\Delta G = \Delta G°$.

(d) $E_{cell}° = -\dfrac{-52.345 \times 10^4 \text{ J}}{2 \text{ mol e}^- \times \dfrac{96,485 \text{ C}}{1 \text{ mol e}^-} \times \dfrac{1 \text{ J}}{1 \text{ V} \cdot \text{C}}} = E° \{H^+(1 M)/H_2(1 atm)\} - E° \{Na^+(1 M)/Na(s)\}$

$E° \{Na^+(1 M)/Na(s)\} = -2.713$ V. This is precisely the same as the value in Appendix D.

94. The question marks in the original Latimer diagram have been replaced with letters in the diagram below to make the solution easier to follow:

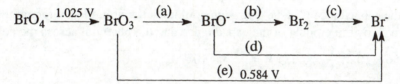

By referring to Appendix D and by employing the correct procedure for adding together half-reactions of the same type we obtain:

(c) $Br_2(l) + 2e^- \rightarrow 2Br^-(aq)$ $\qquad\qquad\qquad E_c° = 1.065$ V (Appendix D)

(d) $BrO^-(aq) + H_2O(l) + 2e^- \rightarrow Br^-(aq) + 2OH^-(aq)$ $\qquad E_d° = 0.766$ V (Appendix D)

(b) $2BrO^-(aq) + 2H_2O(l) + 2e^- \rightarrow Br_2(l) + 4OH^-(aq)$ $\qquad E_b° = 0.455$ V (Appendix D)

(a) $\{BrO_3^-(aq) + 2H_2O(l) + 4e^- \rightarrow BrO^-(aq) + 4OH^-(aq)\} \times 2$ $\qquad E_a° = ?$

(b) $2BrO^-(aq) + 2H_2O(l) + 2e^- \rightarrow Br_2(l) + 4OH^-(aq)$ $\qquad E_b° = 0.455$ V

(c) $Br_2(l) + 2e^- \rightarrow 2Br^-(aq)$ $\qquad\qquad\qquad E_c° = 1.065$ V

(e) $2BrO_3^-(aq) + 6H_2O(l) + 12e^- \rightarrow 12OH^-(aq) + 2Br^-(aq)$ $\qquad E_e° = 0.584$ V

$\Delta G°_{Total} = \Delta G°_{(a)} + \Delta G°_{(b)} + \Delta G°_{(c)}$

$-12F(E_e°) = -8F(E_a°) + -2F(E_b°) + -2F(E_c°)$

$-12F(0.584 V) = -8F(E_a°) + -2F(0.455 V) + -2F(1.065 V)$

$E_a° = \dfrac{-12F(0.584 V) + 2F(0.455 V) + 2F(1.065 V)}{-8F} = 0.496$ V

95. **(a)** The capacitance of the cell membrane is given by the following equation,

$$C = \frac{\varepsilon_0 \varepsilon A}{l}$$

where $\varepsilon_0 \varepsilon = 3 \times 8.854 \times 10^{-12} \text{ C}^2 \text{ N}^{-1} \text{ m}^{-2}$;
$A = 1 \times 10^{-6} \text{ cm}^2$; and $l = 1 \times 10^{-6} \text{ cm}$.
Together with the factors necessary to convert from cm to m and from cm^2 to m^2, these data yield

$$C = \frac{(3)\left(8.854 \times 10^{-12} \dfrac{\text{C}^2}{\text{N}^1 \text{ m}^2}\right)(1 \times 10^{-6} \text{ cm}^2)\left(\dfrac{1 \text{ m}}{100 \text{ cm}}\right)^2}{(1 \times 10^{-6} \text{cm})\left(\dfrac{1 \text{ m}}{100 \text{ cm}}\right)} = 2.66 \times 10^{-13} \frac{\text{C}^2}{\text{N m}}$$

$$C = \left(2.66 \times 10^{-13} \frac{\text{C}^2}{\text{N m}}\right)\left(\frac{1 \text{F}}{1 \dfrac{\text{C}^2}{\text{N m}}}\right) = 2.66 \times 10^{-13} \text{ F}$$

(b) Since the capacitance C is the charge in coulombs per volt, the charge on the membrane, Q, is given by the product of the capacitance and the potential across the cell membrane.

$$Q = 2.66 \times 10^{-13} \frac{\text{C}}{\text{V}} \times 0.085 \text{ V} = 2.26 \times 10^{-14} \text{ C}$$

(c) The number of K$^+$ ions required to produce this charge is

$$\frac{Q}{e} = \frac{2.26 \times 10^{-14} \text{ C}}{1.602 \times 10^{-19} \text{ C/ion}} = 1.41 \times 10^5 \text{ K}^+ \text{ ions}$$

(d) The number of K$^+$ ions in a typical cell is

$$\left(6.022 \times 10^{23} \frac{\text{ions}}{\text{mol}}\right)\left(155 \times 10^{-3} \frac{\text{mol}}{\text{L}}\right)\left(\frac{1 \text{L}}{1000 \text{ cm}^3}\right)(1 \times 10^{-8} \text{ cm}^3) = 9.3 \times 10^{11} \text{ ions}$$

(e) The fraction of the ions involved in establishing the charge on the cell membrane is

$$\frac{1.4 \times 10^5 \text{ ions}}{9.3 \times 10^{11} \text{ ions}} = 1.5 \times 10^{-7} \text{ } (\sim 0.000015 \text{ \%})$$

Thus, the concentration of K$^+$ ions in the cell remains constant at 155 mM.

CHAPTER 21
MAIN GROUP ELEMENTS I: METALS

PRACTICE EXAMPLES

1A   From Figure 21-2, the route from sodium chloride to sodium nitrate begins with electrolysis of NaCl(aq) to form NaOH(aq)

$$2\,NaCl(aq)+2\,H_2O(l) \xrightarrow{\text{electrolysis}} 2\,NaOH(aq)+H_2(g)+Cl_2(g) \text{ followed by addition of}$$

$NO_2(g)$ to NaOH(aq). $2\,NaOH(aq)+3\,NO_2(g) \rightarrow 2\,NaNO_3(aq)+NO(g)+H_2O(l)$

1B   From Figure 21-2, we see that the route from sodium chloride to sodium thiosulfate begins with the electrolysis of NaCl(aq) to produce NaOH(aq),

$$2\,NaCl(aq)+2\,H_2O(l) \xrightarrow{\text{electrolysis}} 2\,NaOH(aq)+H_2(g)+Cl_2(g)$$

And continues through the reaction of $SO_2(g)$ with the NaOH(aq) in an acid-base reaction $[SO_2(g)$ is an acid anhydride] to produce

$Na_2SO_3(aq)$: $\ 2\,NaOH(aq)+SO_2(g) \rightarrow Na_2SO_3(aq)+H_2O(l)$ and (3) concludes with

the addition of S to the boiling solution: $Na_2SO_3(aq)+S(s) \xrightarrow{\text{boil}} Na_2S_2O_3(aq)$

2A   Moles of NaOH $= C \times V = 0.0133\ M \times 0.00759\ L = 1.01 \times 10^{-4}$ mol NaOH

$$[H_3O^+] = \frac{1.01 \times 10^{-4}\ \text{mol}}{0.0250\ L} = 4.04 \times 10^{-3}\ M \qquad pH = -\log(4.04 \times 10^{-3}); \qquad pH = 2.394$$

2B $\dfrac{185\ mg\ Ca^{2+}}{1\ L} \times \dfrac{1\ mmol\ Ca^{2+}}{40.078\ mg\ Ca^{2+}} \times \dfrac{1\ mol\ Ca^{2+}}{1000\ mmol\ Ca^{2+}} \times \dfrac{2\ mol\ Na^+}{1\ mol\ Ca^{2+}} = 9.23 \times 10^{-3}\ M\ Na^+$

3A   The first two reactions, are those from Example 21-3, used to produce B_2O_3.

$$Na_2B_4O_7 \cdot 10\,H_2O(s)+H_2SO_4(l) \longrightarrow 4\,B(OH)_3(s)+Na_2SO_4(s)+5\,H_2O(l)$$

$$2\,B(OH)_3(s) \xrightarrow{\Delta} B_2O_3(s)+3H_2O(g)$$

The next reaction is conversion to BCl_3 with heat, carbon, and chlorine.

$$2\,B_2O_3(s)+3\,C(s)+6\,Cl_2(g) \xrightarrow{\Delta} 4\,BCl_3(g)+3\,CO_2(g)$$

$LiAlH_4$ is used as a reducing agent to produce diborane.

$$4\,BCl_3(g)+3\,LiAlH_4(s) \rightarrow 2\,B_2H_6(g)+3\,LiCl(s)+3\,AlCl_3(s)$$

3B $Na_2B_4O_7\ 10\,H_2O(s) + H_2SO_4(aq) \rightarrow 4\,B(OH)_3(s) + Na_2SO_4(aq) + 5\,H_2O(l)$

$2\,B(OH)_3(s) \xrightarrow{\Delta} B_2O_3(s) + 3\,H_2O(l)$

$B_2O_3(s) + 3\,CaF_2(s) + 3\,H_2SO_4(l) \xrightarrow{\Delta} 2\,BF_3(g) + 3\,CaSO_4(s) + 3\,H_2O(g)$

EXERCISES

Group I (Alkali) Metals

1. **(a)** $2\,Cs(s) + Cl_2(g) \longrightarrow 2\,CsCl(s)$ **(b)** $2\,Na(s) + O_2(g) \longrightarrow Na_2O_2(s)$

(c) $Li_2CO_3(s) \xrightarrow{\Delta} Li_2O(s) + CO_2(g)$ **(d)** $Na_2SO_4(s) + 4\,C(s) \rightarrow Na_2S(s) + 4\,CO(g)$

(e) $K(s) + O_2(g) \longrightarrow KO_2(s)$

2. **(a)** $2\,Rb(s) + 2\,H_2O(l) \longrightarrow RbOH_2(aq) + H_2(g)$

(b) $2\,KHCO_3(s) \xrightarrow{\Delta} K_2CO_3(aq) + H_2O(l) + CO_2(g)$

(c) $2\,Li(s) + O_2(g) \longrightarrow Li_2O_2(s)$

(d) $2\,KCl(s) + H_2SO_4(aq) \longrightarrow K_2SO_4(aq) + 2\,HCl(g)$

(e) $LiH(s) + H_2O(l) \longrightarrow LiOH(aq) + H_2(g)$

3. Both LiCl and KCl are soluble in water, but Li_3PO_4 is not very soluble. Hence the addition of $K_3PO_4(aq)$ to a solution of the white solid will produce a precipitate if the white solid is LiCl, but no precipitate if the white solid is KCl. The best method is a flame test; lithium gives a red color to a flame, while the potassium flame test is violet.

4. When heated, Li_2CO_3 decomposes to $CO_2(g)$ and $Li_2O(l)$. $K_2CO_3(s)$ simply melts when heated. The evolution of $CO_2(g)$ bubbles should be sufficient indication of the difference in behavior. The flame test affords a violet flame if K^+ is present, and a red flame if Li^+ is present.

5. First we note that sodium carbonate ionizes virtually completely when dissolved in H_2O and thus is described as highly soluble in water. This is made evident by the fact that there is no K_{sp} value for Na_2CO_3. By contrast, the existence of K_{sp} values for $MgCO_3$ and Li_2CO_3 shows that these compounds have lower solubilities in water than Na_2CO_3. To decide whether $MgCO_3$ is more or less soluble than Li_2CO_3, we must calculate the molar solubility for each salt and then compare the two values. Clearly, the salt that has the larger molar solubility will be more soluble in water. The molar solubilities can be found using the respective K_{sp} expressions for the two salts.

1. $MgCO_3 \underset{\text{excess} - s}{\overset{K_{sp}=3.5\times10^{-8}}{\rightleftharpoons}} \underset{s}{Mg^{2+}(aq)} + \underset{s}{CO_3^{2-}(aq)}$ Let "s" = molar solubility of $MgCO_3$

    $s^2 = 3.5\times10^{-8}$ $s = 1.9\times10^{-4}\,M$

2. $Li_2CO_3 \underset{\text{excess} - s}{\overset{K_{sp}=2.5\times10^{-2}}{\rightleftharpoons}} \underset{2s}{2\,Li^+(aq)} + \underset{s}{CO_3^{2-}(aq)}$ Let "s" = molar solubility of Li_2CO_3

    $(2s)^2 \times (s) = 2.5\times10^{-8}$ $4s^3 = 2.5\times10^{-8}$ $s = \sqrt[3]{\dfrac{2.5\times10^{-8}}{4}} = 0.18\,M$

We can conclude that Li_2CO_3 is more soluble than $MgCO_3$. Thus, the expected order of increasing solubility in water is: $MgCO_3 < Li_2CO_3 < Na_2CO_3$

6. We know sodium metal was produced at the cathode from the reduction of sodium ion, Na^+. Thus, hydroxide must have been involved in oxidation at the anode. The hydrogen in hydroxide ion already is in its highest oxidation state and thus cannot be oxidized. This leaves oxidation of the hydroxide ion to elemental oxygen as the remaining reaction.

$$\text{Cathode, reduction:} \quad \{Na^+ + e^- \longrightarrow Na(l)\} \times 4$$

$$\text{Anode, oxidation:} \quad 4\,OH^- \longrightarrow O_2(g) + 2\,H_2O(g) + 4e^-$$

$$\text{Net:} \quad 4\,Na^+ + 4\,OH^- \longrightarrow 4\,Na(l) + O_2(g) + 2\,H_2O(g)$$

7. **(a)** $H_2(g)$ and $Cl_2(g)$ are produced during the electrolysis of NaCl(aq). The electrode reactions are:

$$\text{Anode, Oxidation:} \quad 2\,Cl^-(aq) \longrightarrow Cl_2(g) + 2e^-$$

$$\text{Cathode, Reduction:} \quad 2\,H_2O(l) + 2e^- \longrightarrow H_2(g) + 2\,OH^-(aq)$$

We can compute the amount of OH^- produced at the cathode.

$$\text{mol } OH^- = 2.50\,min \times \frac{60\,s}{1\,min} \times \frac{0.810\,C}{1\,s} \times \frac{1\,mole^-}{96500\,C} \times \frac{2\,mol\,OH^-}{2\,mol\,e^-} = 1.26 \times 10^{-3}\,mol\,OH^-$$

Then we compute the $[OH^-]$ and, from that, the pH of the solution.

$$[OH^-] = \frac{1.26 \times 10^{-3}\,mol\,OH^-}{0.872\,L\,soln} = 1.45 \times 10^{-3}\,M \quad pOH = -\log(1.45 \times 10^{-3}) = 2.839$$

$$pH = 14.000 - 2.839 = 11.161$$

(b) As long as NaCl is in excess and the volume of the solution is nearly constant, the solution pH only depend on the number of electrons transferred.

8. **(a)**
$$\text{total energy} = 3.0\,V \times 0.50\,A\,h \times \frac{3600\,s}{1\,hr} \times \frac{1\,C/s}{1\,A} \times \frac{1\,J}{1\,V \cdot C} = 5.4 \times 10^3\,J$$

We obtained the first conversion factor for time as follows.

$$5.0\,\mu W \times \frac{1 \times 10^{-6}\,W}{1\,\mu W} \times \frac{1\,J/s}{1\,W} = \frac{5.0 \times 10^{-6}\,J}{1\,s}$$

$$\text{time} = 5.4 \times 10^3\,J \times \frac{1\,s}{5 \times 10^{-6}\,J} = 1.1 \times 10^9\,s \times \frac{1\,hr}{3600\,s} \times \frac{1\,day}{24\,hr} \times \frac{1\,y}{365\,days} = 34\,y$$

(b) The capacity of the battery is determined by the mass of Li present.

$$\text{mass Li} = 0.50\,A\,h \times \frac{1\,C/s}{1\,A} \times \frac{3600\,s}{1\,h} \times \frac{1\,mol\,e^-}{96485\,C} \times \frac{1\,mol\,Li}{1\,mol\,e^-} \times \frac{6.941\,g\,Li}{1\,mol\,Li} = 0.13\,g\,Li$$

9. **(a)** We first compute the mass of $NaHCO_3$ that should be produced from 1.00 ton NaCl, assuming that all of the Na in the NaCl ends up in the $NaHCO_3$. We use the unit, ton-mole, to simplify the calculations.

$$\text{mass } NaHCO_3 = 1.00 \text{ ton NaCl} \times \frac{1 \text{ ton-mol NaCl}}{58.4 \text{ ton NaCl}} \times \frac{1 \text{ ton-mol Na}}{1 \text{ ton-mol NaCl}}$$

$$\times \frac{1 \text{ tol-mol } NaHCO_3}{1 \text{ ton-mol Na}} \times \frac{84.0 \text{ ton } NaHCO_3}{1 \text{ ton mol } NaHCO_3} = 1.44 \text{ ton } NaHCO_3$$

$$\% \text{ yield} = \frac{1.03 \text{ ton } NaHCO_3 \text{ produced}}{1.44 \text{ ton } NaHCO_3 \text{ expected}} \times 100\% = 71.5\% \text{ yield}$$

(b) NH_3 is used in the principal step of the Solvay process to produce a solution in which $NaHCO_3$ is formed and from which it will precipitate. The filtrate contains NH_4Cl, from which NH_3 is recovered by treatment with $Ca(OH)_2$. Thus, NH_3 is simply used during the Solvay process to produce the proper conditions for the desired reactions. Any net consumption of NH_3 is the result of unavoidable losses during production.

10. **(a)** $Ca(OH)_2(s) + SO_4^{2-}(aq) \rightleftharpoons CaSO_4(s) + 2OH^-(aq)$

(b) We sum two solubility reactions and combine their values of K_{sp}

$Ca(OH)_2(s) \rightleftharpoons Ca^{2+}(aq) + 2OH^-(aq)$ $\qquad\qquad K_{sp} = 5.5 \times 10^{-6}$

$Ca^{2+}(aq) + SO_4^{2-}(aq) \rightleftharpoons CaSO_4(s)$ $\qquad\qquad 1/K_{sp} = 1/9.1 \times 10^{-6}$

$\overline{Ca(OH)_2(s) + SO_4^{2-}(aq) \rightleftharpoons CaSO_4(s) + 2OH^-(aq) \quad K_{eq} = \dfrac{5.5 \times 10^{-6}}{9.1 \times 10^{-6}} = 0.60}$

Because K_{eq} is close to 1.00, we conclude that the reaction lies neither very far to the right (it does not go to completion) nor to the left.

(c) Reaction: $\qquad Ca(OH)_2(s) + SO_4^{2-}(aq) \rightleftharpoons CaSO_4(s) + 2OH^-(aq)$

Initial:	—	1.00 M	—	≈ 0 M
Changes:	—	$-x$ M	—	$+2x$ M
Equil:	—	$(1.00-x)$ M	—	$2x$ M

$$K = \frac{\left[OH^-\right]^2}{\left[SO_4^{2-}\right]} = 0.60 = \frac{4x^2}{1.00-x} \qquad 4x^2 = 0.60 - 0.60x \qquad 4x^2 + 0.60x - 0.60 = 0$$

$$x = \frac{-b \pm \sqrt{b^2 - 4ac}}{2a} = \frac{-0.60 \pm \sqrt{0.36 + 9.60}}{8} = 0.32 \text{ M}$$

$$\left[SO_4^{2-}\right] = 1.00 - x = 0.68 \text{ M} \qquad\qquad \left[OH^-\right] = 2x = 0.64 \text{ M}$$

Group 2 (Alkaline earth) metals

11. $CaO \xleftarrow{\Delta} CaCO_3 \xleftarrow{CO_2} Ca(OH)_2 \xrightarrow{HCl} CaCl_2 \xrightarrow{electrolysis} Ca$

$CaHPO_4 \xleftarrow{H_3PO_4} | \quad | \xrightarrow{H_2SO_4} CaSO_4$

The reactions are as follows. $Ca(OH)_2(s) + 2\,HCl(aq) \rightarrow CaCl_2(aq) + 2H_2O(l)$

$CaCl_2(l) \xrightarrow{\Delta, electrolysis} Ca(l) + Cl_2(g)$ $\quad Ca(OH)_2(s) + CO_2(g) \rightarrow CaCO_3(s) + H_2O(g)$

$CaCO_3(s) \xrightarrow{\Delta} CaO(s) + CO_2(g)$ $\quad\quad Ca(OH)_2(s) + H_2SO_4(aq) \rightarrow CaSO_4(s) + 2\,H_2O(l)$

$Ca(OH)_2(s) + H_3PO_4(aq) \rightarrow CaHPO_4(aq) + 2H_2O(l)$. Actually $CaO(s)$ is the industrial starting material from which $Ca(OH)_2$ is made. $CaO(s) + H_2O(l) \longrightarrow Ca(OH)_2(s)$

12. $MgO \xleftarrow{\Delta} MgCO_3 \xleftarrow{CO_2} Mg(OH)_2 \xrightarrow{HCl} MgCl_2 \xrightarrow{electrolysis} Mg$

$MgHPO_4 \xleftarrow{H_3PO_4} | \quad | \xrightarrow{H_2SO_4} MgSO_4$

Once we return to $Mg(OH)_2$ from $MgSO_4$, the other substances can be made by the indicated pathways. The return reaction is: $MgSO_4(aq) + 2\,NaOH(aq) \rightarrow Mg(OH)_2(s) + Na_2SO_4(aq)$. Then the other reactions are

$Mg(OH)_2(s) + 2\,HCl(aq) \rightarrow MgCl_2(aq) + 2\,H_2O(l)$

$MgCl_2(l) \xrightarrow{\Delta, electrolysis} Mg(l) + Cl_2(g)$; $\quad Mg(OH)_2(s) + CO_2(g) \rightarrow MgCO_3(s) + H_2O(g)$

$MgCO_3(s) \xrightarrow{\Delta} MgO(s) + CO_2(g)$; $\quad Mg(OH)_2(s) + H_2SO_4(aq) \rightarrow MgSO_4(s) + 2H_2O(l)$

$Mg(OH)_2(s) + H_3PO_4(aq) \rightarrow MgHPO_4(aq) + 2H_2O(l)$

13. The reactions involved are:
$Mg^{2+}(aq) + Ca^{2+}(aq) + 2OH^-(aq) \rightarrow Mg(OH)_2(s) + Ca^{2+}(aq)$
$Mg(OH)_2(s) + 2\,H^+(aq) + 2\,Cl^-(aq) \rightarrow Mg^{2+}(aq) + 2\,H_2O(l) + 2\,Cl^-(aq)$
$Mg^{2+}(aq) + 2\,Cl^-(aq) \xrightarrow{\Delta} MgCl_2(s)$
$MgCl_2(s) \xrightarrow{\text{(electrolysis)}} Mg(l) + Cl_2(g)$
$\underline{Mg(l) \rightarrow Mg(s)}$
$Mg^{2+} + 2\,Cl^-(aq) \rightarrow Mg(s) + Cl_2(g)$ (Overall reaction)
As can be seen, the process does not violate the principle of conservation of charge.

14. **(a)** MgO vs BaO: MgO would have the higher melting point because although Mg^{2+} and Ba^{2+} have the same charge, Mg^{2+} is a smaller ion. Smaller ions have a larger electrostatic attraction to anions (here, in both cases the anion is O^{2-}), that is due to the smaller charge separation (Coulomb's Law).

(b) MgF_2 vs $MgCl_2$ solubility in water. F^- is smaller than Cl^-, hence, electrostatic attraction between Mg^{2+} and the halide is greater in F^- than Cl^-. If we assume that hydration of the ions is similar, we expect that MgF_2 is less soluble than $MgCl_2$. (Note: K_{sp} given for MgF_2 which is sparingly soluble, while no K_{sp} value is given for readily soluble $MgCl_2$).

15. (a) $BeF_2(s) + Mg(s) \xrightarrow{\Delta} Be(s) + MgF_2(s)$

(b) $Ba(s) + Br_2(l) \longrightarrow BaBr_2(s)$

(c) $UO_2(s) + 2\ Ca(s) \longrightarrow U(s) + 2\ CaO(s)$

(d) $MgCO_3 \cdot CaCO_3(s) \xrightarrow{\Delta} MgO(s) + CaO(s) + 2\ CO_2(g)$

(e) $2\ H_3PO_4(aq) + 3\ CaO(s) \longrightarrow Ca_3(PO_4)_2(s) + 3\ H_2O(l)$

16. (a) $Mg(HCO_3)_2(s) \xrightarrow{heat} MgO(s) + 2\ CO_2(g) + H_2O(l)$

(b) $BaCl_2(l) \xrightarrow{electrolysis} Ba(l) + Cl_2(g)$

(c) $Sr(s) + 2\ HBr(aq) \longrightarrow SrBr_2(aq) + H_2(g)$

(d) $H_2SO_4(aq) + Ca(OH)_2(aq) \longrightarrow CaSO_4(s) + 2H_2O(l)$

(e) $CaSO_4 \cdot 2\ H_2O(s) \xrightarrow{\Delta} CaSO_4 \cdot \frac{1}{2}H_2O(s) + \frac{3}{2}\ H_2O(g)$

17. Let us compute the value of the equilibrium constant for each reaction by combining the two solubility product constants. Large values of equilibrium constants indicate that the reaction is displaced far to the right. Values of K that are much smaller than 1 indicate that the reaction is displaced far to the left.

(a) $BaSO_4(s) \rightleftharpoons Ba^{2+}(aq) + SO_4^{2-}(aq)$ $\qquad K_{sp} = 1.1 \times 10^{-10}$

$Ba^{2+}(aq) + CO_3^{2-}(aq) \rightleftharpoons BaCO_3(s)$ $\qquad 1/K_{sp} = 1/(5.1 \times 10^{-9})$

$\rule{8cm}{0.4pt}$

$BaSO_4(s) + CO_3^{2-}(aq) \rightleftharpoons BaCO_3(s) + SO_4^{2-}(aq)$ $\qquad K = \dfrac{1.1 \times 10^{-10}}{5.1 \times 10^{-9}} = 2.2 \times 10^{-2}$

Thus, the equilibrium lies slightly to the left.

(b) $Mg_3(PO_4)_2(s) \rightleftharpoons 3\ Mg^{2+}(aq) + 2\ PO_4^{3-}(aq)$ $\qquad K_{sp} = 2.1 \times 10^{-25}$

$3\{Mg^{2+}(aq) + CO_3^{2-}(aq) \rightleftharpoons MgCO_3(s)\}$ $\qquad 1/(K_{sp})^3 = 1/(3.5 \times 10^{-8})^3$

$\rule{8cm}{0.4pt}$

$Mg_3(PO_4)_2(s) + 3CO_3^{2-}(aq) \rightleftharpoons 3\ MgCO_3(s) + 2\ PO_4^{3-}(aq)$

$K = \dfrac{2.1 \times 10^{-25}}{(3.5 \times 10^{-8})^3} = 4.9 \times 10^{-3}$ $\qquad$ Thus, the equilibrium lies to the left

(c) $Ca(OH)_2(s) \rightleftharpoons Ca^{2+}(aq) + 2\ OH^-(aq)$ $\qquad K_{sp} = 5.5 \times 10^{-6}$

$Ca^{2+}(aq) + 2\ F^-(aq) \rightleftharpoons CaF_2(s)$ $\qquad 1/K_{sp} = 1/(5.3 \times 10^{-9})$

$\rule{8cm}{0.4pt}$

$Ca(OH)_2(s) + 2F^-(aq) \rightleftharpoons CaF_2(s) + 2\ OH^-(aq)$ $\quad K = \dfrac{5.5 \times 10^{-6}}{5.3 \times 10^{-9}} = 1.0 \times 10^3$

Thus, the equilibrium lies to the right

18. We expect the reaction to occur to a significant extent in the forward direction if its equilibrium constant is >> 1.

(a)

$$BaCO_3(s) \rightleftharpoons Ba^{2+}(aq) + CO_3^{2-}(aq) \qquad K_{sp} = 5.1 \times 10^{-9}$$

$$2\, HC_2H_3O_2(aq) \rightleftharpoons 2H^+(aq) + 2C_2H_3O_2^-(aq) \qquad K_a^2 = (1.8 \times 10^{-5})^2$$

$$H^+(aq) + CO_3^{2-}(aq) \rightleftharpoons HCO_3^-(aq) \qquad 1/K_{a_2} = 1/(4.7 \times 10^{-11})$$

$$H^+(aq) + HCO_3^-(aq) \rightleftharpoons H_2CO_3(aq) \qquad 1/K_{a_1} = 1/(4.2 \times 10^{-7})$$

$$\overline{BaCO_3(s) + 2\, HC_2H_3O_2(aq) \rightleftharpoons Ba(C_2H_3O_2)_2(aq) + H_2CO_3(aq)}$$

$$K = \frac{5.1 \times 10^{-9} \times (1.8 \times 10^{-5})^2}{4.7 \times 10^{-11} \times 4.2 \times 10^{-7}} = 8.4 \times 10^{-2}$$

Thus, no significant reaction occurs (the equilibrium mixture contains appreciable amounts of all four species involved in the reaction.

(b)

$$Ca(OH)_2(s) \rightleftharpoons Ca^{2+}(aq) + 2\, OH^-(aq) \qquad K_{sp} = 5.5 \times 10^{-6}$$

$$2\, NH_4^+(aq) + 2\, OH^-(aq) \rightleftharpoons 2\, NH_3(aq) + 2\, H_2O(l) \qquad 1/K_b^2 = 1/(1.8 \times 10^{-5})^2$$

$$\overline{Ca(OH)_2(s) + 2\, NH_4^+(aq) \rightleftharpoons Ca^{2+}(aq) + 2\, NH_3(aq) + 2\, H_2O(l)}$$

$$K = \frac{5.5 \times 10^{-6}}{(1.8 \times 10^{-5})^2} = 1.7 \times 10^4 \qquad \text{Thus, this reaction would occur to a significant extent.}$$

(c)

$$BaF_2(s) \rightleftharpoons Ba^{2+}(aq) + 2\, F^-(aq) \qquad K_{sp} = 1.0 \times 10^{-6}$$

$$2\, H_3O^+(aq) + 2\, F^-(aq) \rightleftharpoons 2\, HF(aq) + 2\, H_2O(l) \qquad 1/K_a^2 = 1/(6.6 \times 10^{-4})^2$$

$$\overline{BaF_2(s) + 2H_3O^+(aq) \rightleftharpoons Ba^{2+}(aq) + 2HF(aq) + 2H_2O(l);} \quad K = \frac{1.0 \times 10^{-6}}{(6.6 \times 10^{-4})^2} = 2.3$$

This reaction would occur to some extent, certainly not to completion.

Hard Water

19. Temporary hard water contains HCO_3^-. Quicklime is CaO:

$$CaO(s) + H_2O(l) \longrightarrow Ca(OH)_2(aq)$$

$$2\,HCO_3^-(aq) + Ca(OH)_2(aq) \longrightarrow CaCO_3(s) + H_2O(l) + CO_2(g) + 2\,OH^-(aq)$$

$$\overline{2\,HCO_3^-(aq) + CaO(s) \longrightarrow CaCO_3(s) + CO_2(g) + 2\,OH^-(aq)}$$

20. Temporary hard water contains HCO_3^- and divalent cations such as Ca^{2+}

$$2\, H_2O(l) + 2\, NH_3(aq) \longrightarrow 2\, OH^-(aq) + 2\, NH_4^+(aq)$$

$$2\, OH^-(aq) + Ca^{2+}(aq) \longrightarrow Ca(OH)_2(aq)$$

$$2\, HCO_3^-(aq) + Ca(OH)_2(aq) \xrightarrow{\Delta} CaCO_3(s) + H_2O(l) + CO_2(g) + 2\, OH^-(aq)$$

$$\overline{H_2O(l) + 2NH_3(aq) + Ca^{2+}(aq) + 2HCO_3^-(aq) \xrightarrow{\Delta} 2NH_4^+(aq) + CaCO_3(s) + CO_2(g) + 2OH^-(aq)}$$

21. **(a)** There will be two HCO_3^- ions for each Ca^{2+} ion.

$$\frac{g\ Ca^{2+}}{10^6\ g\ soln} = \frac{185.0\ g\ HCO_3^-}{10^6\ g\ soln} \times \frac{1\ mol\ HCO_3^-}{61.02\ g\ HCO_3^-} \times \frac{1\ mol\ Ca^{2+}}{2\ mol\ HCO_3^-} \times \frac{40.08\ g\ Ca^{2+}}{1\ mol\ Ca^{2+}}$$

$$= \frac{60.76\ g\ Ca^{2+}}{10^6\ g\ soln} = 60.76\ ppm\ Ca^{2+}$$

(b)

$$\frac{mg\ Ca^{2+}}{1\ L} = \frac{60.76\ g\ Ca^{2+}}{10^6\ g\ soln} \times \frac{1.00\ g\ soln}{1\ mL} \times \frac{1000\ mL}{1\ L} \times \frac{1000\ mg}{1\ g} = 60.76\ mg\ Ca^{2+}/L$$

(c) First we combine the equations representing the neutralization of bicarbonate ion with hydroxide ion and the formation of a generalized carbonate precipitate $\left[MCO_3(s) \right]$ to determine the overall stoichiometry of the reaction. We also add to the combination the slaking of lime, CaO(s). Remember that two HCO_3^- ions are associated with each $M^{2+}(aq)$ ion.

$$CaO(s) + H_2O(l) \rightarrow Ca(OH)_2(s)$$
$$Ca(OH)_2(s) \xrightarrow{H_2O} Ca^{2+}(aq) + 2\ OH^-(aq)$$
$$2\{HCO_3^-(aq) + OH^-(aq) \longrightarrow H_2O(l) + CO_3^{2-}(aq)\}$$
$$CO_3^{2-}(aq) + M^{2+}(aq) \longrightarrow MCO_3(s)$$
$$\underline{CO_3^{2-}(aq) + Ca^{2+}(aq) \longrightarrow CaCO_3(s)}$$
$$Ca(OH)_2(s) + M^{2+}(aq) + 2\ HCO_3^-(aq) \longrightarrow CaCO_3(s) + MCO_3(s) + 2\ H_2O(l)$$

Then we determine the mass of $Ca(OH)_2$ needed.

$$mass\ Ca(OH)_2 = 1.00 \times 10^6\ g\ water \times \frac{185.0\ g\ HCO_3^-}{10^6\ g\ water} \times \frac{1\ mol\ HCO_3^-}{61.02\ g\ HCO_3^-}$$

$$\times \frac{1\ mol\ Ca(OH)_2}{2\ mol\ HCO_3^-} \times \frac{74.09\ g\ Ca(OH)_2}{1\ mol\ Ca(OH)_2} = 112.3\ g\ Ca(OH)_2$$

22. **(a)** $$mass\ CaCO_3 = 112.3 \times 10^5\ g\ Ca(OH)_2 \times \frac{1\ mol\ Ca(OH)_2}{74.09\ g\ Ca(OH)_2} \times \frac{2\ mol\ CaCO_3}{1\ mol\ Ca(OH)_2}$$

$$\times \frac{100.1\ g\ CaCO_3}{1\ mol\ CaCO_3} = 303.4\ g\ CaCO_3$$

(b) It is clear from the equations that are summed in the answer to question 21, and especially the net equation (that follows), that half the Ca^{2+} is derived from the CaO and half from the hard water itself. This assumes all $M^{2+} = Ca^{2+}$ in question 21 above.

$$Ca(OH)_2(s) + M^{2+}(aq) + 2\ HCO_3^-(aq) \longrightarrow CaCO_3(s) + MCO_3(s) + 2\ H_2O(l)$$

23. pH = 2.37 = $-\log[H^+]$. The $[H^+] = 10^{-2.37} = 4.27 \times 10^{-3}$ M. For every 2 moles of H^+, one mole of Ca^{2+} is absorbed. Hence, the concentration of Ca^{2+} is $2.1\underline{3} \times 10^{-3}$ M. Next we convert this to parts per million (mg Ca^{2+} per 1000 g solution)

$$\text{ppm } Ca^{2+} = \frac{2.13 \times 10^{-3} \text{ mol } Ca^{2+}}{1 \text{ L solution}} \times \frac{1 \text{ L solution}}{1000 \text{ g solution}} \times \frac{40.08 \text{ g } Ca^{2+}}{1 \text{ mol } Ca^{2+}} \times \frac{1000 \text{ mg } Ca^{2+}}{1 \text{ g } Ca^{2+}} = \frac{85 \text{ mg } Ca^{2+}}{1000 \text{ g solution}}$$

The concentration of Ca^{2+} is 85 ppm.

24. Let us determine the $\left[OH^-\right]$ in this solution first.

$$2\, OH^-\,(aq) + H_2SO_4\,(aq) \rightarrow 2\, H_2O(l) + SO_4^{2-}\,(aq)$$

$$[OH^-] = \frac{22.42 \text{ mL acid} \times \dfrac{1.00 \times 10^{-3} \text{ mmol } H_2SO_4}{1 \text{ mL acid}} \times \dfrac{2 \text{ mmol } OH^-}{1 \text{ mmol } H_2SO_4}}{25.00 \text{ mL base}} = 1.79 \times 10^{-3} \text{ M}$$

Now we compute the number of grams of $CaSO_4$ in 10^6 g of hard water. This is the ppm hardness. The anion exchange reaction is

$$R(OH)_2\,(s) + SO_4^{2-}\,(aq) \rightarrow RSO_4(s) + 2\,OH^-\,(aq)$$

$$\text{mass } CaSO_4 = 10^6 \text{ g water} \times \frac{1 \text{ mL } H_2O}{1.00 \text{ g } H_2O} \times \frac{1 \text{ L } H_2O}{1000 \text{ mL } H_2O} \times \frac{1.79 \times 10^{-3} \text{ mol } OH^-}{1 \text{ L}} \times \frac{1 \text{ mol } SO_4^{2-}}{2 \text{ mol } OH^-}$$

$$\times \frac{1 \text{ mol } CaSO_4}{1 \text{ mol } SO_4^{2-}} \times \frac{136.1 \text{ g } CaSO_4}{1 \text{ mol } CaSO_4} = 122 \text{ g } CaSO_4 \quad (122 \text{ ppm } CaSO_4)$$

25. We first write the balanced equation for the formation of bathtub ring.

$$Ca^{2+}\,(aq) + 2\, CH_3(CH_2)_{16}COO^-K^+\,(aq) \rightarrow Ca^{2+}\left[CH_3(CH_2)_{16}COO^-\right]_2(s,\text{"ring"}) + 2\,K^+\,(aq)$$

$$\text{"ring" mass} = 32.1 \text{ L} \times \frac{82.6 \text{ g } Ca^{2+}}{1000 \text{ L water}} \times \frac{1 \text{ mol } Ca^{2+}}{40.08 \text{ g } Ca^{2+}} \times \frac{1 \text{ mol "ring"}}{1 \text{ mol } Ca^{2+}} \times \frac{607.0 \text{ g "ring"}}{1 \text{ mol "ring"}}$$

$$= 40.2 \text{ g of bathtub ring}$$

26. We assume that the density of the solution is 1.00 g/mL. The solubility of magnesium palmitate equals the concentration of Mg^{2+}, $\left[Mg^{2+}\right]$, because there is one mole Mg^{2+} in each mole of the palmitate salt.

$$\left[Mg^{2+}\right] = \frac{80 \text{ g } Mg^{2+}}{10^6 \text{ g soln}} \times \frac{1 \text{ mol } Mg^{2+}}{24.3 \text{ g } Mg^{2+}} \times \frac{1.00 \text{ g soln}}{1 \text{ mL soln}} \times \frac{10^3 \text{ mL soln}}{1 \text{ L soln}} = 3.3 \times 10^{-3} \text{ M} = s$$

From this, we calculate the value of the solubility product. Stoichiometry tells us $\left[Palm^-\right] = 2\left[Mg^{2+}\right]$;

$$K_{sp} = \left[Mg^{2+}\right]\left[Palm^-\right]^2 = (s)(2s)^2 = 4s^3 = 4\left(3.3 \times 10^{-3}\right)^3 = 1.4 \times 10^{-7}$$

Group 13 The Boron Family

27. **(a)** B_4H_{10} contains a total of $4\times3+10\times1 = 22$ valence electrons or 11 pairs. Ten of these pairs could be allocated to form 10 B—H bonds, leaving but one pair to bond the four B atoms together, which is clearly an electron deficient situation.

(b) In our analysis in part (a), we noted that the four B atoms had but one electron pair to bond them together. To bond these four atoms into a chain requires three electron pairs. Since each electron pair in a bridging bond replaces two "normal" bonds, there must be at least two bridging bonds in the B_4H_{10} molecules. By analogy with B_2H_6, we might write the structure below left. But this structure uses only a total of 20 electrons. (The bridge bonds are shown as dots, normal bonds—electron pairs—as dashes.) In the structure at right below, we have retained some of the form of B_2H_6, and produced a compound with the formula B_4H_{10} and 11 electron pairs. (The experimentally determined structure of B_4H_{10} consists of a four-membered ring of alternating B and H atoms, held together by bridging bonds. Two of the B atoms have two H atoms bonds to each of them by normal covalent bonds. The other two B atoms have one H atom covalently bonded to each. One final B—B bond joins these last two B atoms, across the diameter of the ring.). See the diagram that follows:

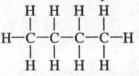

(c) C_4H_{10} contains a total of $4\times4+10\times1 = 26$ valence electrons or 13 pairs. A plausible Lewis structure follows. (Note that each atom possess on octet of electrons).

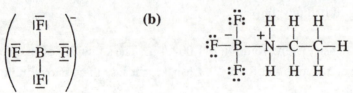

28. **(a)**

$$\left(\begin{array}{c} |\overline{F}| \\ | \\ |\overline{F}-B-\overline{F}| \\ | \\ |F| \end{array}\right)^-$$

(b)

$$\overset{..}{:}\overset{:\!\overset{..}{F}:}{\underset{:\!\overset{..}{F}:}{\overset{-|}{F}-B}}\overset{H\;H\;H}{\overset{+|\;|\;|}{-N-C-C-H}}$$

29. **(a)** $2BBr_3(l) + 3H_2(g) \longrightarrow 2B(s) + 6HBr(g)$

(b) i) $B_2O_3(s) + 3\,C(s) \xrightarrow{\Delta} 3\,CO(g) + 2\,B(s)$
    ii) $2\,B(s) + 3\,F_2(g) \xrightarrow{\Delta} 2\,BF_3(g)$

(c) $2\,B(s) + 3\,N_2O(g) \xrightarrow{\Delta} 3\,N_2(g) + B_2O_3(s)$

30. Each boron atom has an oxidation number of +3. The hydroxyl oxygens are each –2 while the bridging oxygens are each –1. Finally, the hydroxyl H atoms are all in the +1 oxidation state. The oxidation numbers for the all the constituent atoms add up to the charge on the perborate ion, namely, 2–.

31. **(a)** $2\ Al(s)+6\ HCl(aq)\rightarrow 2\ AlCl_3(aq)+3\ H_2(g)$

(b) $2\ NaOH(aq)+2\ Al(s)+6\ H_2O(l)\rightarrow 2\ Na^+(aq)+2\left[Al(OH)_4\right]^-(aq)+3\ H_2(g)$

(c) Oxidation: $\{Al(s)\longrightarrow Al^{3+}(aq)+3\ e^-\}$ $\qquad\qquad\qquad$ ×2

Reduction: $\{SO_4^{2-}(aq)+4\ H^+(aq)+2e^-\longrightarrow SO_2(g)+2\ H_2O(l)\}$ $\quad$ ×3

$\rule{10cm}{0.4pt}$

Net: $2\ Al(s)+3\ SO_4^{2-}(aq)+12\ H^+(aq)\longrightarrow 2\ Al^{3+}(aq)+3\ SO_2(aq)+6\ H_2O(l)$

32. **(a)** $2\ Al(s)+3\ Br_2(l)\longrightarrow 2\ AlBr_3(s)$

(b) $2\ Al(s)+Cr_2O_3(s)\xrightarrow{\text{heat}} 2\ Cr(l)+Al_2O_3(s)$

(c) $Fe_2O_3(s)+OH^-(aq)\longrightarrow \text{no reaction}$

$Al_2O_3(s)+2\ OH^-(aq)+3\ H_2O(l)\longrightarrow 2\left[Al(OH)_4\right]^-(aq)$

33. One method of analyzing this reaction is to envision the HCO_3^- ion as a combination of CO_2 and OH^-. Then the OH^- reacts with Al^{3+} and forms $Al(OH)_3$. This method of envisioning HCO_3^- does have its basis in reality. After all,

$H_2CO_3(=H_2O+CO_2)+OH^-\longrightarrow HCO_3^-+H_2O$

$Al^{3+}(aq)+3\ HCO_3^-(aq)\longrightarrow Al(OH)_3(s)+3\ CO_2(g)$

Another method is to consider the reaction as, first, the hydrolysis of hydrated aluminum ion to produce $Al(OH)_3(s)$ and an acidic solution, followed by the reaction of the acid with bicarbonate ion.

$\left[Al(H_2O)_6\right]^{3+}(aq)+3H_2O(l)\longrightarrow Al(OH)_3(H_2O)_3(s)+3H_3O^+(aq)$

$3H_3O^+(aq)+3HCO_3^-(aq)\longrightarrow 6H_2O(l)+3CO_2(g)$

This gives the same net reaction:

$\left[Al(H_2O)_6\right]^{3+}(aq)+3HCO_3^-(aq)\longrightarrow Al(OH)_3(H_2O)_3(s)+3CO_2(g)+3H_2O(l)$

34. The $Al^{3+}(aq)$ ion hydrolyzes. $Al^{3+}(aq) + 3\ H_2O(l) \longrightarrow Al(OH)_3(s) + 3\ H^+(aq)$

Subsequently, the hydrogen ion that is produced reacts with bicarbonate ion to liberate $CO_2(g)$.

$H^+(aq) + HCO_3^-(aq) \longrightarrow H_2O(l) + CO_2(g)$

35. Aluminum and its oxide are soluble in both acid and in base.

$2\ Al(s) + 6\ H^+(aq) \longrightarrow 2\ Al^{3+}(aq) + 3\ H_2(g)$

$Al_2O_3(s) + 6\ H^+(aq) \longrightarrow 2\ Al^{3+}(aq) + 3\ H_2O(l)$

$2\ Al(s) + 2\ OH^-(aq) + 6\ H_2O(l) \longrightarrow 2\left[Al(OH)_4\right]^-(aq) + 3\ H_2(g)$

$Al_2O_3(s) + 2\ OH^-(aq) + 3\ H_2O(l) \longrightarrow 2\left[Al(OH)_4\right]^-(aq)$

Al(s) is resistant to corrosion only over the pH range 4.5 to 8.5. Thus, aluminum is inert only when the medium to which it is exposed is neither highly acidic nor highly basic.

36. Both Al and Mg are attacked by acid and their ions are both precipitated by hydroxide ion.

$2\ Al(s) + 6\ H^+(aq) \longrightarrow 2\ Al^{3+}(aq) + 3\ H_2(g)$ $Mg(s) + 2\ H^+(aq) \longrightarrow Mg^{2+}(aq) + H_2(g)$

$Al^{3+}(aq) + 3\ OH^-(aq) \longrightarrow Al(OH)_3(s)$ $Mg^{2+}(aq) + 2\ OH^- \longrightarrow Mg(OH)_2(s)$

But, of these two solid hydroxides, only $Al(OH)_3(s)$ redissolves in excess $OH^-(aq)$.

$Al(OH)_3(s) + OH^-(aq) \longrightarrow \left[Al(OH)_4\right]^-(aq)$

Thus, the analytical procedure consists of dissolving the sample in HCl(aq) and then treating the resulting solution with NaOH(aq) until a precipitate forms. If this precipitate dissolves completely in excess NaOH(aq) is, the sample is aluminum 2S. If at least some of the precipitate does not dissolve, the sample was magnesium.

37. $CO_2(g)$ is, of course, the anhydride of an acid. The reaction here is an acid-base reaction.

$\left[Al(OH)_4\right]^-(aq) + CO_2(aq) \longrightarrow Al(OH)_3(s) + HCO_3^-(aq)$

HCl(aq), being a strong acid, can't be used because it will dissolve the Al(OH)₃(s).

38. (a) Oersted: $2\ Al_2O_3(s) + 3\ C(s) + 6\ Cl_2(g) \overset{\Delta}{\longrightarrow} 4\ AlCl_3(s) + 3\ CO_2(g)$

(b) Wöhler: $AlCl_3(s) + 3\ K(s) \overset{\Delta}{\longrightarrow} Al(s) + 3\ KCl(s)$

39. $2\ KOH(aq) + 2\ Al(s) + 6\ H_2O(l) \rightarrow 2\ K[Al(OH)_4](aq) + 3\ H_2(g)$

$2\ K[Al(OH)_4](aq) + 4\ H_2SO_4(aq) \rightarrow K_2SO_4(aq) + Al_2(SO_4)_3(aq) + 8\ H_2O(l)$
— crystallize→ $2\ KAl(SO_4)_2(s)$

40. HCO_3^- and CO_3^{2-} solutions are basic (they produce an excess of OH^-). $Al^{3+}(aq)$ in the presence of $OH^-(aq)$ will precipitate as the hydroxide Al(OH)₃(s).

Group 14 The Carbon Family

41. In the sense that diamonds react imperceptibly slowly at room temperature (either with oxygen to form carbon dioxide, or in its transformation to the more stable graphite), it is essentially true that "diamonds last forever." However, at elevated temperatures, diamond will burn to form $CO_2(g)$ and thus the statement is false. Also, the transformation $C(diamond) \rightarrow C(graphite)$ might occur more rapidly under other conditions. Eventually, of course, the conversion to graphite occurs.

42. The graphite in the pencil "lead" is a good dry lubricant that will make slippery the stickiness (or reluctance) in the lock and enable it to work smoothly. The key carries the graphite to the site that needs to be lubricated within the lock mechanism.

43. **(a)** $3 SiO_2(s) + 4 Al(s) \xrightarrow{\Delta} 2 Al_2O_3(s) + 3 Si(s)$

   (b) $K_2CO_3(s) + SiO_2(s) \xrightarrow{\Delta} CO_2(g) + K_2SiO_3(s)$

   (c) $Al_4C_3(s) + 12 H_2O(l) \longrightarrow 3 CH_4(g) + 4Al(OH)_3(s)$

44. **(a)** $2 KCN(aq) + AgNO_3(aq) \longrightarrow KAg(CN)_2(aq) + KNO_3(aq)$
   or $KCN(aq) + AgNO_3(aq) \longrightarrow Ag(CN)(s) + KNO_3(aq)$

   (b) $Si_3H_8(l) + 5 O_2(g) \longrightarrow 3 SiO_2(s) + 4 H_2O(l)$

   (c) $N_2(g) + CaC_2(s) \xrightarrow{\Delta} CaNCN(s) + C(s)$

45. A silane is a silicon-hydrogen compound, with the general formula Si_nH_{2n+2}. A silanol is a compound in which one or more of the hydrogens of silane is replaced by an —OH group. Then, the general formula becomes $Si_nH_{2n+1}(OH)$. In both of these classes of compounds the number of silicon atoms, n, ranges from 1 to 6. Silicones are produced when silanols condense into chains, with the elimination of a water molecule between every two silanol molecules.

$$HO—Si_nH_{2n}—OH + HO—Si_nH_{2n}—OH \longrightarrow HO—Si_nH_{2n}—O—Si_nH_{2n}—OH + H_2O$$

46. Both the alkali metal carbonates and the alkali metal silicates are soluble in water. Hence, they also are soluble in acids. However, the carbonates will produce gaseous carbon dioxide—CO_2—on reaction with acid $\left(CO_3^{2-}(aq) + 2 H^+(aq) \longrightarrow "H_2CO_3" \longrightarrow H_2O(l) + CO_2(g)\right)$, while the silicates will produce silica— SiO_2— in an analogous reaction $\left(SiO_4^{4-}(aq) + 4H^+(aq) \longrightarrow "H_4SiO_4" \longrightarrow 2 H_2O(l) + SiO_2(s)\right)$. The silica is produced in many forms depending on reaction conditions: a colloidal dispersion, a gelatinous precipitate, or a semisolid gel. Silicates of cations other than those of alkali metals are insoluble in water, as are the analogous carbonates. However, the carbonates will dissolve in acids (witness the reaction of acid rain on limestone carvings and marble statues, both forms of $CaCO_3$), while the silicates will not. (Silicate rocks are not significantly affected by acid rain.)

47. **(1)** $2\,CH_4(g) + S_8(g) \longrightarrow 2\,CS_2(g) + 4\,H_2S(g)$

(2) $CS_2(g) + 3\,Cl_2(g) \longrightarrow CCl_4(l) + S_2Cl_2(l)$

(3) $4\,CS_2(g) + 8\,S_2Cl_2(g) \longrightarrow 4\,CCl_4(l) + 3\,S_8(s)$

48. **(a)** $(CH_3)_3\,SiCl(l) + H_2O(l) \longrightarrow (CH_3)_3\,Si\!-\!OH(aq) + HCl(aq)$

$2\,(CH_3)_3\,Si\!-\!OH(aq) \longrightarrow (CH_3)_3\,Si\!-\!O\!-\!Si(CH_3)_3\,(s) + H_2O(l)$

(b) A silicone polymer does not form from $(CH_3)_3\,Si\!-\!Cl$. Only a dimer is produced.

(c) The product that results from the treatment of CH_3SiCl_3 is two long $Si\!-\!O\!-\!Si$ chains, with CH_3 groups on the outside, linked by $Si\!-\!O\!-\!Si$ bridges. Part of one of these chains and the beginnings of the bridges are shown below.

$$\begin{array}{cccccc}
| & | & | & | & | & | \\
O & O & O & O & O & O \\
| & | & | & | & | & | \\
-O-Si-O-Si-O-Si-O-Si-O-Si-O-Si-O- \\
| & | & | & | & | & | \\
CH_3 & CH_3 & CH_3 & CH_3 & CH_3 & CH_3
\end{array}$$

49. Muscovite or white mica has the formula $KAl_2(OH)_2(AlSi_3O_{10})$. Since they are not segregated into O_2 units in the formula, all of the oxygen atoms in the mineral must be in the -2 oxidation state. Potassium is obviously in the $+1$ oxidation state as are the hydrogen atoms in the hydroxyl groups. Up to this point we have -24 from the twelve oxygen atoms and +3 from the potassium and hydrogen atoms for a net number of -21 for the oxidation state. We still have three aluminum atoms and three silicon atoms to account for. In oxygen-rich salts such as mica, we would expect that the silicon and the aluminum atoms would be in their highest possible oxidation states, namely +4 and +3, respectively. Since the salt is neutral, the oxidation numbers for the silicon and aluminum atoms must add up to +21. This is precisely that total that is obtained if the silicon and aluminum atoms are in their highest possible oxidation states: $(3 \times (+3) + 3 \times (+4) = +21)$. Consequently, the empirical formula for white Muscovite is consistent with the expected oxidation state for each element present.

50. Chrysotile asbestos has the formula $[Mg_3Si_2O_5(OH)_4]$. Since they are not segregated to O_2 units in the formula, all of the oxygen atoms in the mineral must be in the -2 oxidation state. The three magnesium atoms are obviously in the $+2$ oxidation state, while hydrogen atoms in the hydroxyl groups have an oxidation state of $+1$. Up to this point then, the sum of the oxidation numbers equals -8 (-18 from O-atoms, $+6$ from Mg-atoms and $+4$ from H-atoms). Since the mineral is neutral, the two silicon atoms must have oxidation states that sum to $+8$ ($-8 + 8 =$ neutral mineral), therefore, each silicon would need to be in the $+4$ oxidation state. In oxygen–rich salts such as asbestos, we would expect that the silicon atoms would be in their highest possible oxidation state, namely $+4$. Thus, the oxidation states for the element in this mineral are precisely consistent with expectations.

51. **(a)** $PbO(s) + 2\,HNO_3(aq) \longrightarrow Pb(NO_3)_2(s) + H_2O(l)$

(b) $SnCO_3(s) \xrightarrow{\Delta} SnO(s) + CO_2(g)$

(c) $PbO(s) + C(s) \xrightarrow{\Delta} Pb(l) + CO(g)$

(d) $2\,Fe^{3+}(aq) + Sn^{2+}(aq) \longrightarrow 2\,Fe^{2+}(aq) + Sn^{4+}(aq)$

(e) $2\,PbS(s) + 3\,O_2(g) \xrightarrow{\Delta} 2\,PbO(s) + 2\,SO_2(g)$
$2\,SO_2(g) + O_2(g) \longrightarrow 2\,SO_3(g)$
$SO_3(g) + PbO(s) \longrightarrow PbSO_4(s)$

Or perhaps simply: $PbS(s) + 2O_2(g) \longrightarrow PbSO_4(s)$

Yet a third possibility:
$PbO(s) + SO_2(s) \longrightarrow PbSO_3(s)$, followed by $2\,PbSO_3(s) + O_2(s) \longrightarrow 2PbSO_4(s)$

52. **(a)** Treat tin(II) oxide with hydrochloric acid.
$SnO(s) + 2\,HCl(aq) \longrightarrow SnCl_2(aq) + H_2O(l)$

(b) Attack tin with chlorine. $Sn(s) + 2\,Cl_2(g) \longrightarrow SnCl_4(s)$

(c) First we dissolve $PbO_2(s)$ in $HNO_3(aq)$ and then treat the resulting solution with $K_2CrO_4(aq)$ to precipitate $PbCrO_4(s)$

$2\,PbO_2(s) + 4\,HNO_3(aq) \longrightarrow 2\,Pb(NO_3)_2(aq) + O_2(g) + 2\,H_2O(l)$

$Pb(NO_3)_2(aq) + K_2CrO_4(aq) \longrightarrow PbCrO_4(s) + 2\,KNO_3(aq)$

53. We start by using the Nernst equation to determine whether the cell voltage still is positive when the reaction has gone to completion.

(a) Oxidation: $Fe^{2+}(aq) \rightarrow Fe^{3+}(aq) + e^-\} \times 2 \qquad\qquad -E^\circ = -0.771\ V$
Reduction: $PbO_2(s) + 4\,H^+(aq) + 2\,e^- \longrightarrow Pb^{2+}(aq) + 2\,H_2O(l) \quad E^\circ = +1.455\ V$
────────────────────────
Net: $2\,Fe^{2+}(aq) + PbO_2(s) + 4\,H^+(aq) \longrightarrow 2\,Fe^{3+}(aq) + Pb^{2+}(aq) + 2\,H_2O(l)$

$E^\circ_{cell} = -0.771\ V + 1.455\ V = +0.684\ V$

In this case, when the reaction has gone to completion,

$\left[Fe^{2+}\right] = 0.001\ M, \left[Fe^{3+}\right] = 0.999\ M,$ and $\left[Pb^{2+}\right] = 0.500\ M.$

$E_{cell} = E^\circ_{cell} - \dfrac{0.0592}{2}\log\dfrac{[Fe^{3+}]^2[Pb^{2+}]}{[Fe^{2+}]^2} = 0.684\ V - \dfrac{0.0592}{2}\log\dfrac{[0.999]^2[0.500]}{[0.001]^2} = 0.515\ V.$

Thus, this reaction will go to completion.

(b) Oxidation: $2 SO_4^{2-} (aq) \longrightarrow S_2O_8^{2-} (aq) + 2 e^-$ $\qquad -E^\circ = -2.01V$

Reduction: $PbO_2 (s) + 4 H^+ (aq) + 2 e^- \longrightarrow Pb^{2+} (aq) + 2 H_2O(l)$ $\qquad E^\circ = +1.455$ V

$2 SO_4^{2-} (aq) + PbO_2 (s) + 4 H^+ (aq) \longrightarrow S_2O_8^{2-} (aq) + Pb^{2+} (aq) + 2 H_2O(l)$

$E^\circ_{cell} = -2.01 + 1.455 = -0.56$ V This reaction is not even spontaneous initially.

(c) Oxidation: $\{Mn^{2+} (1\times10^4 \text{ M}) + 4H_2O(l) \longrightarrow MnO_4^- (aq) + 8H^+ (aq) + 5 e^-\} \times 2$;

$-E^\circ = -1.51$ V

Reduction: $\{PbO_2 (s) + 4 H^+ (aq) + 2 e^- \longrightarrow Pb^{2+} (aq) + 2 H_2O(l)\} \times 5$ $\qquad E^\circ = +1.455$ V

Net: $2Mn^{2+} (1\times10^4 \text{ M}) + 5PbO_2 (s) + 4H^+ \longrightarrow 2MnO_4^- (aq) + 5Pb^{2+} (aq) + 2H_2O(l)$

$E^\circ_{cell} = -1.51 + 1.455 = -0.06$ V. The standard cell potential indicates that this reaction is not spontaneous when all concentrations are 1 M. Since the concentration of a reactant (Mn^{2+}) is lower than 1.00 M, this reaction is even less spontaneous than the standard cell potential indicates.

54. A positive value of E°_{cell} indicates that a reaction should occur.

(a) Oxidation : $Sn^{2+} (aq) \longrightarrow Sn^{4+} (aq) + 2 e^-$ $\qquad -E^\circ = -0.154$ V

Reduction : $I_2 (s) + 2 e^- \longrightarrow 2 I^- (aq)$ $\qquad E^\circ = +0.535$ V

Net : $Sn^{2+} (aq) + I_2 (s) \longrightarrow Sn^{4+} (aq) + 2 I^- (aq)$ $\qquad E^\circ_{cell} = +0.381$ V

Yes, $Sn^{2+}(aq)$ will reduce I_2 to I^-.

(b) Oxidation : $Sn^{2+} (aq) \longrightarrow Sn^{4+} (aq) + 2 e^-$ $\qquad -E^\circ = -0.154$ V

Reduction : $Fe^{2+} (aq) + 2 e^- \longrightarrow Fe(s)$ $\qquad E^\circ = -0.440$ V

Net : $Sn^{2+} (aq) + Fe^{2+} (aq) \longrightarrow Sn^{4+} (aq) + Fe(s)$ $\quad E^\circ_{cell} = -0.594$ V

No, $Sn^{2+}(aq)$ will not reduce $Fe^{2+}(aq)$ to Fe(s).

(c) Oxidation : $Sn^{2+} (aq) \longrightarrow Sn^{4+} (aq) + 2 e^-$ $\qquad -E^\circ = -0.154$ V

Reduction : $Cu^{2+} (aq) + 2 e^- \longrightarrow Cu(s)$ $\qquad E^\circ = +0.337$ V

Net : $Sn^{2+} (aq) + Cu^{2+} (aq) \longrightarrow Sn^{4+} (aq) + Cu(s)$ $\qquad E^\circ_{cell} = +0.183$ V

Yes, $Sn^{2+}(aq)$ will reduce $Cu^{2+}(aq)$ to Cu(s).

(d) Oxidation : $Sn^{2+} (aq) \longrightarrow Sn^{4+} (aq) + 2 e^-$ $\qquad -E^\circ = -0.154$ V

Reduction : $\{Fe^{3+} (aq) + e^- \longrightarrow Fe^{2+} (aq)\} \times 2$ $\qquad E^\circ = +0.771$ V

Net : $Sn^{2+} (aq) + 2 Fe^{3+} (aq) \longrightarrow Sn^{4+} (aq) + 2 Fe^{2+} (aq)$ $\quad E^\circ_{cell} = +0.617$ V

Yes, $Sn^{2+}(aq)$ will reduce $Fe^{3+}(aq)$ to $Fe^{2+}(aq)$.

Integrative and Advanced Exercises

55. What most likely happened is that a deliquescent solid has absorbed water vapor from the air over time and formed a saturated solution. This saturated solution can be used in any circumstance where the solid would be used to prepared an aqueous solution. Of course, the concentration will have to be determined by some means other than weighing, such as a gravimetric analysis or using a standard to react with the substance in a titration. Alternatively, the saturated solution can be heated in a drying oven to drive off the water and the solid may be recovered. Often, unfortunately, a solid with an unknown percent of water results from this treatment.

56. (a) A solution of $CO_2(aq)$ has $[CO_3^{2-}] = K_{a2}[H_2CO_3] = 5.6 \times 10^{-11}$ M. Since the $K_{sp} = 2.8 \times 10^{-9}$ for $CaCO_3$, the $[Ca^{2+}]$ needed to form a precipitate from this solution can be computed.

$$[Ca^{2+}] = \frac{K_{sp}}{[Ca^{2+}]} = \frac{2.8 \times 10^{-9}}{5.6 \times 10^{-11}} = 50. \text{ M}$$

This is too high to reach by dissolving $CaCl_2$ in solution. The reason why a solid forms is because the OH^- produced by the $Ca(OH)_2$ neutralized some of the HCO_3^- from the ionization of $CO_2(aq)$, thereby increasing the $[CO_3^{2-}]$ above a value of 5.6×10^{-11} M

(b) The equation for redissolving can be obtained by combining several equations.

$CaCO_3(s) \ll Ca^{2+}(aq) + CO_3^{2-}(aq)$ $\qquad\qquad K_{sp} = 2.8 \times 10^{-9}$

$CO_3^{2-}(aq) + H_3O^+(aq) \ll HCO_3^-(aq) + H_2O(l)$ $\qquad 1/K_{a2} = 1/(5.6 \times 10^{-11})$

$CO_2(aq) + 2 H_2O(l) \ll HCO_3^-(aq) + H_3O^+(aq)$ $\qquad K_{a1} = 4.2 \times 10^{-7}$

$CaCO_3(s) + CO_2(aq) + H_2O(l) \ll Ca^{2+}(aq) + 2 HCO_3^-(aq)$ $\qquad K = \dfrac{K_{sp} \times K_{a1}}{K_{a2}}$

$$K = \frac{(2.8 \times 10^{-9}) \times (4.2 \times 10^{-7})}{(5.6 \times 10^{-11})} = 2.1 \times 10^{-5} = \frac{[Ca^{2+}][HCO_3^-]^2}{[CO_2]}$$

If $CaCO_3(s)$ is precipitated from 0.005 M $Ca^{2+}(aq)$ and then redissolve, $[Ca^{2+}] = 0.005$ M and $[HCO_3^-] = 2 \times 0.005$ M = 0.010 M. We use these values in the above expression to compute the $[CO_2]$.

$$[CO_2] = \frac{[Ca^{2+}][HCO_3^-]^2}{(2.1 \times 10^{-5})} = 0.002 \text{ M}$$

We repeat the calculation for saturated $Ca(OH)_2$, in which $[OH^-] = 2 \times [OH^-]$, after first determining $[Ca^{2+}]$ in this solution.

$$K_{sp} = [Ca^{2+}][OH^-]^2 = 4 \times [Ca^{2+}] = 5.5 \times 10^{-6} \quad [Ca^{2+}] = \sqrt[3]{\frac{5.5 \times 10^{-6}}{4}} = 0.011 \text{ M}$$

$$[CO_2] = \frac{[Ca^{2+}][HCO_3^-]^2}{(2.1 \times 10^{-5})} = \frac{(0.011)(0.022)^2}{2.1 \times 10^{-5}} = 0.25 \text{ M}$$

Thus to redissolve the $CaCO_3$ requires that the $[CO_2] = 0.25$ M, assuming the solution is initially saturated with $Ca(OH)_2(aq)$.

57. There are two principal reasons why the electrolysis of NaCl(l) is used to produce sodium commercially rather than the electrolysis of NaOH(l). First, NaCl is readily available whereas NaOH is produced from the electrolysis of NaCl(aq). Thus the raw material NaCl is much cheaper than is NaOH. Second, but less important, when sodium is produced from the electrolysis of NaCl(l), a by-product is $Cl_2(g)$, whereas $O_2(g)$ is a by-product of the production of NaOH(l). Since $O_2(g)$ can be produced more cheaply by the fractional distillation of liquid air, $Cl_2(g)$ can be sold for a higher price than $O_2(g)$. Thus, the two reasons for producing sodium from NaCl(l) rather than NaOH(l) are a much cheaper raw material and a more profitable by-product.

58. (a) The mass of MgO(s) that would be produced can be determined with information from the balanced chemical equation for the oxidation of Mg(s).

$$2\,Mg(s) + O_2(g) \longrightarrow 2\,MgO(s)$$

$$\text{mass MgO} = 0.200\ \text{g Mg} \times \frac{1\,\text{mol Mg}}{24.305\,\text{g Mg}} \times \frac{2\,\text{mol MgO}}{2\,\text{mol Mg}} \times \frac{40.30\,\text{g MgO}}{1\,\text{mol MgO}} = 0.332\,\text{g MgO}$$

(b) If the mass of product formed from the starting Mg differs from the 0.332 g MgO predicted, then the product could be a mixture of magnesium nitride and magnesium oxide. This scenario is not implausible since molecular nitrogen is readily available in the atmosphere. $3\ Mg(s) + N_2(g) \longrightarrow Mg_3N_2(s)$

$$\text{mass } Mg_3N_2 = 0.200\ \text{g Mg} \times \frac{1\,\text{mol Mg}}{24.305\,\text{g Mg}} \times \frac{1\,\text{mol } Mg_3N_2}{3\,\text{mol Mg}} \times \frac{100.93\,\text{g } Mg_3N_2}{1\,\text{mol } Mg_3N_2} = 0.277\ \text{g } Mg_3N_2$$

We can use a technique similar to determining the percent abundance of an isotope. Let f = the mass fraction of MgO in the product. The $(1.000 - f)$ is the mass fraction of Mg_3N_2 in the product. product mass = mass MgO $\times f$ + [mass $Mg_3N_2 \times (1.000 - f)$]

0.315 g product = $0.332 \times f$ + [$0.277 \times (1.000 - f)$] = $0.277 + f(0.332 - 0.277)$

$$f = \frac{0.315 - 0.277}{0.332 - 0.277} = 0.69 \qquad \text{The product is 69\% by mass MgO}$$

59. Reaction (21.3) is $KCl(l) + Na(l) \xrightarrow{\ 850°C\ } NaCl(l) + K(g)$ It is practical when the reactant element is a liquid and the product element is a gas at the same temperature.

(a) Since Li has a higher boiling point (1347°C) than K (773.9°C), a reaction similar to (21.3) is not a feasible way of producing Li metal from LiCl.

(b) On one hand, Cs has a lower boiling point (678.5°C) than K (773.9°C), and thus a reaction similar to (21.3) is a feasible method of producing Cs metal from CsCl. However, the ionization energy of Na is considerably larger than that of Cs, making it difficult to transfer an electron from Na to Cs^+.

60. (a) $2\,Al(s) + Fe_2O_3(s) \rightarrow 2\,Fe(s) + Al_2O_3(s)$ $\qquad \Delta H = -852\ \text{kJ}$

(b) $4\,Al(s) + 3\,MnO_2(s) \rightarrow 2\,Al_2O_3(s) + 3\,Mn(s)$ $\qquad \Delta H = -1792\ \text{kJ}$

(c) $2\,Al(s) + 3\,MgO(s) \rightarrow Al_2O_3(s) + 3\,Mg(s)$ $\qquad \Delta H = +129\ \text{kJ}$

61. $Li^+(aq) + e^- \rightarrow Li(s)$ $\qquad \Delta G° = -293.3 \text{ kJ} = -nFE° = -1 \text{ mol } e^- (96485 \text{ C/mol } e^-)E°$

$E° = -3.040 \text{ V}$ (this value is the same as the one that appears in table 21.2)

62. The partial pressure of H_2 is 748 mmHg – 21 mmHg = 727 mmHg

$$\text{amount } H_2 = \frac{PV}{RT} = \frac{(727 \text{ mm Hg} \times (1 \text{ atm}/760 \text{ mmHg})) \times 0.104 \text{ L}}{0.08206 \text{ L atm K}^{-1} \text{ mol}^{-1} \times 296 \text{ K}} = 4.10 \times 10^{-3} \text{ mol } H_2$$

The electrode reactions are the following.

Anode: $2 Cl^-(aq) \longrightarrow Cl_2(g) + 2e^-$ Cathode: $2 H_2O(l) + 2e^- \longrightarrow 2 OH^-(aq) + H_2(g)$
Thus 2 mol OH^-(aq) are produced per mol H_2(g); 8.20×10^{-3} mol OH^-(aq) are produced.

$$[OH^-] = \frac{8.20 \times 10^{-3} \text{ mol OH}^-}{0.250 \text{ L soln}} = 3.28 \times 10^{-2} \text{ M} = 0.0328 \text{ M}$$

Then we compute the ion product and compare its value to the value of K_{sp} for $Mg(OH)_2$.
$Q_{sp} = [Mg^{2+}][OH^-]^2 = (0.220)(0.0328)^2 = 2.37 \times 10^{-4} > 1.8 \times 10^{-11} = K_{sp}$ for $Mg(OH)_2$
Thus, $Mg(OH)_2$ should precipitate.

63. **(a)** We determine the amount of Ca^{2+} associated with each anion in 10^6 g of the water.

$$\text{amount } Ca^{2+} (SO_4^{2-}) = 56.9 \text{ g } SO_4^{2-} \times \frac{1 \text{ mol } SO_4^{2-}}{96.06 \text{ g } SO_4^{2-}} \times \frac{1 \text{ mol } Ca^{2+}}{1 \text{ mol } SO_4^{2-}} = 0.592 \text{ mol}$$

$$\text{amount } Ca^{2+} (HCO_3^-) = 176 \text{ g } HCO_3^- \times \frac{1 \text{ mol } HCO_{3^-}}{61.02 \text{ g } HCO_3^-} \times \frac{1 \text{ mol } Ca^{2+}}{2 \text{ mol } HCO_3^-} = 1.44 \text{ mol}$$

Then we determine the total mass of Ca^{2+}, numerically equal to the ppm Ca^{2+}.

$$\text{mass } Ca^{2+} = (0.592 + 1.44) \text{ mol } Ca^{2+} \times \frac{40.08 \text{ g } Ca^{2+}}{1 \text{ mol } Ca^{2+}} = 81.4 \text{ g } Ca^{2+} \longrightarrow 81.4 \text{ ppm } Ca^{2+}$$

(b) The reactions for the removal of HCO_3^-(aq) begin with the formation of hydroxide ion resulting from dissolving CaO(s). Hydroxide ion reacts with bicarbonate ion to form carbonate ion, which then combines with calcium ion to form the $CaCO_3$(s) precipitate.

$$CaO(s) + H_2O(l) \longrightarrow Ca^{2+}(aq) + 2 OH^-(aq)$$

$$OH^-(aq) + HCO_3^-(aq) \longrightarrow H_2O(l) + CO_3^{2-}(aq)$$
$$Ca^{2+}(aq) + CO_3^{2-}(aq) \longrightarrow CaCO_3(s)$$

For each 1.000×10^6 g of water, we need to remove 176 g HCO_3^-(aq) with the added CaO(s).

$$\text{mass } CaO = 602 \times 10^3 \text{ g water} \times \frac{176 \text{ g } HCO_3^-}{1.000 \times 10^6 \text{ g water}} \times \frac{1 \text{ mol } HCO_3^-}{61.02 \text{ g } HCO_3^-} \times \frac{1 \text{ mol } OH^-}{1 \text{ mol } HCO_3^-}$$

$$\times \frac{1 \text{ mol } CaO}{2 \text{ mol } OH^-} \times \frac{56.08 \text{ g } CaO}{1 \text{ mol } CaO} = 48.7 \text{ g } CaO$$

(c) Here we determine the total amount of Ca^{2+} in 602 kg of water, <u>and</u> that added as CaO.

$$Ca^{2+} = \left(\frac{602 \text{ kg}}{1 \times 10^6 \text{ kg}}\right)(0.592 + 1.44) \text{ mol } Ca^{2+} + \left(48.7 \text{ g CaO} \times \frac{1 \text{ mol CaO}}{56.08 \text{ g CaO}} \times \frac{1 \text{ mol } Ca^{2+}}{1 \text{ mol CaO}}\right) = 2.09 \text{ mol } Ca^{2+}$$

The amount of Ca^{2+} that has precipitated equals the amount of HCO_3^- in solution, since each mole of HCO_3^- is transformed into 1 mole of CO_3^{2-}, which reacts with and then precipitates one mole of Ca^{2+}. Thus the amount of Ca^{2+} that has precipitated is $(0.602)(1.44) = 0.867$ mol Ca^{2+} as $CaCO_3(s)$. Then we can determine the concentration of Ca^{2+} remaining in solution.

$$[Ca^{2+}] = \frac{2.09 \text{ mol } Ca^{2+} \text{ total} - 0.867 \text{ mol } Ca^{2+}}{10^6 \text{ g water}} \times \frac{10^3 \text{ g water}}{1 \text{ L water}} = 2.03 \times 10^{-3} \text{ M}$$

To consider the Ca^{2+} "removed", its concentration should be decreased to 0.1% (0.001) of its initial value, or 2.03×10^{-6} M. We use the K_{sp} expression for $CaCO_3$ to determine the needed $[CO_3^{2-}]$

$$K_{sp} = [Ca^{2+}][CO_3^{2-}] = 2.8 \times 10^{-9} = 1.46 \times 10^{-6} \text{ M}$$

$$[CO_3^{2-}] = \frac{2.8 \times 10^{-9}}{2.03 \times 10^{-6}} = 0.0014 \text{ M}$$

This carbonate ion concentration can readily be achieved by adding solid Na_2CO_3.

(d) The amount of CO_3^{2-} needed is that which ends up in the precipitate plus that needed to attain the 0.0014 M concentration in the 602 kg = 602 L of water.

$$n_{CO_3^{2-}} = 602 \text{ L} \times \left(\left(\frac{0.00203 \text{ mol } Ca^{2+}}{1 \text{ L water}} \times \frac{1 \text{ mol } CO_3^{2-}}{1 \text{ mol } Ca^{2+}}\right) + \frac{0.0014 \text{ mol } CO_3^{2-}}{1 \text{ L water}}\right) = 2.1 \text{ mol } CO_3^{2-}$$

This CO_3^{2-} comes from the added $Na_2CO_3(s)$

$$\text{mass } Na_2CO_3 = 2.0 \text{ mol } CO_3^{2-} \times \frac{1 \text{ mol } Na_2CO_3}{1 \text{ mol } CO_3^{2-}} \times \frac{105.99 \text{ g } Na_2CO_3}{1 \text{ mol } Na_2CO_3} = 2.2 \times 10^2 \text{ g } Na_2CO_3$$

64. (a) In the cell reaction, three moles of electrons are required to reduce each mole of Al^{3+}

$$\text{mass Al} = 800 \text{ h} \times \frac{3600 \text{ s}}{1 \text{ h}} \times \frac{1.00 \times 10^5 \text{ C}}{1 \text{ s}} \times \frac{1 \text{ mol e}^-}{96,485 \text{ C}} \times \frac{1 \text{ mol Al}}{3 \text{ mol e}^-} \times \frac{26.98 \text{ g Al}}{1 \text{ mol Al}} = 2.68 \times 10^5 \text{ g Al}$$

The 38% efficiency is not considered in this calculation. All of the electrons produced must pass through the electrolytic cell; they simply require a higher than optimum voltage to do so, leading to resistance heating (which consumes some of the electrical energy).

(b) $\text{Total energy} = 4.5 \text{ V} \times 8.00 \text{ h} \times \frac{3600 \text{ s}}{1 \text{ h}} \times \frac{1.00 \times 10^5 \text{ C}}{1 \text{ s}} \times \frac{1 \text{ J}}{1 \text{ V} \cdot \text{C}} \times \frac{1 \text{ kJ}}{1000 \text{ J}} = 1.3 \times 10^7 \text{ kJ}$

$$\text{mass coal} = 1.3 \times 10^7 \text{ kJ electricity} \times \frac{1 \text{ kJ heat}}{0.35 \text{ kJ electricity}} \times \frac{1 \text{ g C}}{32.8 \text{ kJ}} \times \frac{1 \text{ g coal}}{0.85 \text{ g C}} \times \frac{1 \text{ metric ton}}{10^6 \text{ g}}$$

$$\text{mass coal} = 1.3 \text{ metric tons of coal}$$

65. We begin by rewriting equation 21.23 for the electrolysis of $Al_2O_3(s)$ with $n = 12\ e^-$.

$3\ C(s) + 2\ Al_2O_3(s) \rightarrow 4\ Al(s) + 3\ CO_2(g)$

$\Delta G° = 4(0\ kJ/mol) + 3(-394\ kJ) - [3(0\ kJ/mol) + 2(-1520\ kJ/mol)] = +1858\ kJ$

$\Delta G° = 1.858 \times 10^6\ J = -nFE° = -12\ mole\ e^-(96485\ C/mol\ e^-)E°$

$E° = -1.605\ V$ Note: this is just an estimate because $\Delta G°$ values are at 298 K, whereas the reaction occurs at a temperature that is much higher than 298 K.

If the oxidation of $C(s)$ to $CO_2(g)$ did not occur, then the cell reaction would just be the reverse of the formation reaction of Al_2O_3 with $n = 6\ e^-$

$$E° = \Delta E° = \frac{\Delta G°}{nF} = \frac{1.520 \times 10^6\ J}{6\ mole\ e^- \times 96485\ C/mole\ e^-} = -2.626\ V$$

66. It would not be unreasonable to predict that Raoult's Law holds under these circumstances. This is predicated on the assumption, of course, that $Pb(NO_3)_2$ is completely ionized in aqueous solution.

$$P_{water} = \chi_{water}\ P°_{water} \qquad \chi_{water} = \frac{P_{water}}{P°_{water}} = 97\% = 0.97$$

Thus, there are 97 mol H_2O in every 100 mol solution. The remaining 3 moles are 1 mol Pb^{2+} and 2 mol NO_3^-. Thus there is 1 mol $Pb(NO_3)_2$ for every 97 mol H_2O. Compute the mass of $Pb(NO_3)_2$ in 100 g H_2O.

$$mass_{Pb(NO_3)_2} = 100.0\ g\ H_2O \times \frac{1\ mol\ H_2O}{18.02\ g\ H_2O} \times \frac{1\ mol\ Pb(NO_3)_2}{97\ mol\ H_2O} \times \frac{331.2\ g\ Pb(NO_3)_2}{1\ mol\ Pb(NO_3)_2} = 19\ g\ Pb(NO_3)_2$$

If we did not assume complete ionization of $Pb(NO_3)_2$, we would obtain 3 moles of unionized $Pb(NO_3)_2$ in solution for every 97 moles of H_2O. Then there would be 57 g $Pb(NO_3)_2$ dissolved in 100 g H_2O. A handbook gives the solubility as 56 g $Pb(NO_3)_2/100$ g H_2O, indicating only partial dissociation or, more probably, extensive re-association into ion pairs, triplets, quadruplets, etc.

67. Considerable energy is required to produce the Pb^{4+} cation—four ionization steps, one for the removal of each electron. It therefore needs quite a large lattice energy to compensate for its energy of production. Both Br^- and I^- are large anions, and therefore the Pb^{4+}—Br^- and Pb^{4+}—I^- interionic distances are long. But lattice energy depends on the charge of the cation (which is quite large) multiplied by the charge of the anion (which is reasonably small) divided by the square of the interionic distance (long, as we have said). Thus, we predict a small lattice energy that is insufficient to stabilize the Pb^{4+} cation. We would predict, however, that PbF_4 and $PbCl_4$ (both with small anions) and PbO_2 and PbS_2 (both with small, highly charged anions) would be stable compounds.   In   addition,   notice   that $E°$ $\{Pb^{4+}|Pb^{2+}\} = 1.5\ V$ (from Table 21-6) is sufficient to oxidize Br^- to Br_2 and I^- to I_2, since $E°$ $\{Br_2|Br^-\} = +1.065\ V$ and $E°\ \{I_2|I^-\} = +0.535\ V$. Thus, even if PbI_4 or $PbBr_4$ could be prepared they would be thermodynamically unstable.

68. $Na_2B_4O_7 \cdot 10\,H_2O(s) + 6\,CaF_2(s) + 8\,H_2SO_4(aq) \rightarrow 4\,BF_3(aq) + 6\,CaSO_4(aq) + 17\,H_2O(l) + 2\,NaHSO_4(aq)$

69. **(a)** 1.00 M NH_4Cl is a somewhat acidic solution due to hydrolysis of $NH_4^+(aq)$; $MgCO_3$ should be most soluble in this solution. The remaining two solutions are buffer solutions, the 0.100 M NH_3–1.00 M NH_4Cl buffer is more acidic of the two. In this solution $MgCO_3$ has intermediate solubility. $MgCO_3$ is least soluble in the most alkaline solution, namely, 1.00 M NH_3–1.00 M NH_4Cl.

$$MgCO_3(s) \rightleftharpoons Mg^{2+}(aq) + CO_3^{2-}(aq) \qquad K_{sp} = 3.5 \times 10^{-8}$$

$$NH_4^+(aq) + OH^-(aq) \rightleftharpoons NH_3(aq) + H_2O(l) \qquad 1/K_b = 1/1.8 \times 10^{-5}$$

$$CO_3^{2-}(aq) + H_3O^+(aq) \rightleftharpoons HCO_3^-(aq) + H_2O(l) \quad 1/K_2 = 1/4.7 \times 10^{-11}$$

$$\underline{2\,H_2O(l) \rightleftharpoons H_3O^+(aq) + OH^-(aq) \qquad\qquad K_w = 1.0 \times 10^{-14}}$$

$$MgCO_3(s) + NH_4^+(aq) \rightleftharpoons Mg^{2+} + HCO_3^-(aq) + NH_3(aq)$$

$$K = \frac{K_{sp} \times K_w}{K_b \times K_2} = \frac{3.5 \times 10^{-8} \times 1.0 \times 10^{-14}}{1.8 \times 10^{-5} \times 4.7 \times 10^{-11}} = 4.1 \times 10^{-7}$$

In 1.00 M NH_4Cl

Reaction:	$MgCO_3(s)$ +	NH_4^+ (aq)	$\rightleftharpoons$	Mg^{2+}(aq) +	HCO_3^-(aq) +	NH_3(aq)
Initial:	–	1.00 M		0 M	0 M	0 M
Changes:	–	–x M		+x M	+x M	+x M
Equil:	–	(1.00 – x)M		x M	x M	x M

$$K = \frac{[Mg^{2+}][HCO_3^-][NH_3]}{[NH_4^+]} = \frac{x \cdot x \cdot x}{1.00 - x} \approx \frac{x^3}{1.00}$$

$$x = \sqrt[3]{4.1 \times 10^{-7}} = 7.4 \times 10^{-3}\,M = \text{molar solubility of } MgCO_3 \text{ in 1.00 M } NH_4Cl$$

($x \ll 1.00$ M, thus the approximation was valid)

(b) In 1.00 M NH_3–1.00 M NH_4Cl

Reaction:	$MgCO_3(s)$ +	NH_4^+(aq)	$\rightleftharpoons$	Mg^{2+}(aq) +	HCO_3^-(aq) +	NH_3(aq)
Initial:	–	1.00 M		0 M	0 M	1.00 M
Changes:	–	– x M		+x M	+x M	+ x M
Equil:	–	(1.00 – x)M		x M	+xM	(1.00 + x)M

$$K = \frac{[Mg^{2+}][HCO_3^-][NH_3]}{[NH_4^+]} = \frac{x \cdot x \cdot (1.00 + x)}{1.00 - x} \approx \frac{x^2 \cdot 1.00}{1.00}$$

$$x = \sqrt{4.1 \times 10^{-7}} = 6.4 \times 10^{-4}\,M = \text{molar solubility of } MgCO_3 \text{ in 1.00 M } NH_3\text{-}1.00 M\,NH_4Cl$$

($x \ll 1.00$ M, thus the approximation was valid)

(c) In 0.100 M NH_3–1.00 M NH_4Cl

Reaction: $MgCO_3(s) + NH_4^+(aq) \rightleftharpoons Mg^{2+}(aq) + HCO_3^-(aq) + NH_3(aq)$

Initial: —   1.00 M   0 M   0 M   0.100 M

Changes: — $-x$ M $+x$ M $+x$ M $+x$ M

Equil: — $(1.00-x)$M xM $+x$M $(0.100+x)$M

$$K = \frac{[Mg^{2+}][HCO_3^-][NH_3]}{[NH_4^+]} = \frac{x \cdot x \cdot (0.100+x)}{1.00-x} \approx \frac{x^2 \times 0.100}{1.00}$$

$$x = \sqrt{\frac{4.1 \times 10^{-7}}{0.100}} = 2.0 \times 10^{-3} \text{ M} = \text{solubility of } MgCO_3 \text{ in 0.100 M } NH_3\text{-1.00 M } NH_4Cl$$

(x is 2 % of 0.100 M, so the approximation was valid)

70. If the value of the equilibrium constant for the reaction is quite large, then the conversion will be almost complete. We write the net ionic equation and then add the two solubility reactions to obtain net ionic equation for the conversion. The equilibrium constants are multiplied together to give the equilibrium constant for the conversion.

Net ionic equation: $Ca(OH)_2(s) + CO_3^{2-}(aq) \rightleftharpoons CaCO_3(s) + 2 OH^-(aq)$

$Ca(OH)_2(s) \rightleftharpoons Ca^{2+}(aq) + 2 OH^-(aq)$ $K_{sp} = 5.5 \times 10^{-6}$

$Ca^{2+}(aq) + CO_3^{2-}(aq) \rightleftharpoons CaCO_3(s)$ $1/K_{sp} = 1/2.8 \times 10^{-9}$

$Ca(OH)_2(s) + CO_3^{2-}(aq) \rightleftharpoons CaCO_3(s) + 2 OH^-(aq)$ $K_{ovedrall} = \dfrac{5.5 \times 10^{-6}}{2.8 \times 10^{-9}} = 2.0 \times 10^3$

Because $K_{overall} > 1000$, the conversion is substantially complete.

71. The density of each metal depends on two factors: its atomic size and the masses of the individual atoms. Density is thus, proportional to the atomic mass divided by the cube of the atomic radius. We calculate this ratio for each of the elements.

Li: $\dfrac{\text{atomic mass}}{(\text{atomic radius})^3}$ Na: $\dfrac{\text{atomic mass}}{(\text{atomic radius})^3}$ K: $\dfrac{\text{atomic mass}}{(\text{atomic radius})^3}$

$= \dfrac{6.941 \text{ (g/mol)}}{[1.52 \text{ (Å)}]^3} = 1.98$ $= \dfrac{22.99 \text{ (g/mol)}}{[1.86 \text{ (Å)}]^3} = 3.57$ $= \dfrac{39.10 \text{ (g/mol)}}{[2.27 \text{ (Å)}]^3} = 3.34$

We see that the calculated ratios follow the same pattern as the densities: Na > K > Li, thus explaining why Na has a higher density than both Li and K.

72. We would expect MgO(s) to have a larger value of lattice energy than Mg(s) because of the smaller interionic distance in MgO(s).

Lattice energy of MgO:

Enthalpy of formation:	$Mg(s) + O_2(g) \longrightarrow MgO(s)$	$\Delta H_f^\circ = -6.02 \times 10^2$ kJ
Sublimation of Mg(s):	$Mg(s) \longrightarrow Mg(g)$	$\Delta H_{sub} = +146$ kJ
Ionization of Mg(g)	$Mg(g) \longrightarrow Mg^+(g) + e^-$	$\Delta I_1 = +737.7$ kJ
Ionization of Mg(g):	$Mg^+(g) \longrightarrow Mg^{2+}(g) + e^-$	$\Delta I_2 = +1451$ kJ
$\frac{1}{2}$ Dissociation O_2(g):	$\frac{1}{2} O_2(g) \longrightarrow O(g)$	$DE = \frac{1}{2} \times 497.4 = 248.7$ kJ
O(g) electron affinity:	$O(g) + e^- \longrightarrow O^-(g)$	$EA_1 = -141$ kJ
O(g) electron affinity:	$O^-(g) + e^- \longrightarrow O^{2-}(g)$	$EA_2 = +744$ kJ
Lattice energy:	$Mg^{2+}(g) + O^{2-}(g) \rightarrow MgO(s)$	L.E = -3789 kJ(see below)

L.E = -602 kJ – 146 kJ – 737.7 kJ – 1451 kJ – 248.7 kJ + 141 kJ – 744 kJ = -3789 kJ

Lattice energy of MgS:

Enthalpy of formation:	$Mg(s) + S(g) \longrightarrow MgS(s)$	$\Delta H_f^\circ = -3.46 \times 10^2$ kJ
Sublimation of Mg(s):	$Mg(s) \longrightarrow Mg(g)$	$\Delta H_{sub} = +146$ kJ
Ionization of Mg(g)	$Mg(g) \longrightarrow Mg^+(g) + e^-$	$\Delta I_1 = +737.7$ kJ
Ionization of Mg(g):	$Mg^+(g) \longrightarrow Mg^{2+}(g) + e^-$	$\Delta I_2 = +1451$ kJ
$\frac{1}{2}$ Dissociation of S_2(g):	$\frac{1}{2} S_{rhombic}(g) \longrightarrow S(g)$	$DE = \frac{1}{2} \times 557.6 = 278.8$ kJ
S(g) electron affinity:	$S(g) + e^- \longrightarrow S^-(g)$	$EA_1 = -200.4$ kJ
S(g) electron affinity:	$S^-(g) + e^- \longrightarrow S^{2-}(g)$	$EA_2 = +456$ kJ
Lattice energy:	$Mg^{2+}(g) + S^{2-}(g) \rightarrow MgS(s)$	L.E = -3215 kJ

L.E = -346 kJ – 146 kJ – 737.7 kJ – 1451 kJ – 278.8 kJ + 200.4 kJ – 456 kJ = -3215 kJ

Feature Problems

73.

$Li(s) \rightarrow Li(g)$	159.4 kJ; (Using $\Delta H°_{rxn} = \Sigma\Delta H_f°_{products} - \Sigma\Delta H_f°_{reactants}$ in Appendix D)
$Li(g) \rightarrow Li^+(g) + e^-$	520.2 kJ; (Data given in Table 21.2, Chapter 21)
$\underline{Li^+(g) \rightarrow Li^+(aq)}$	$\underline{-506 \text{ kJ}}$; (Provided in the question)
$Li(s) \rightarrow Li^+(aq) + e^-$	174 kJ

$1/2\,H_2(g) \rightarrow H(g)$	218.0 kJ	(Using $\Delta H°_{rxn} = \Sigma\Delta H_f°_{products} - \Sigma\Delta H_f°_{reactants}$ in Appendix D)
$H(g) \rightarrow H^+(g) + e^-$	1312 kJ	(Use $R_H(N_A)$Bohr theory = 2.179×10^{-18} J(6.022×10^{23}))
$\underline{H^+(g) \rightarrow H^+(aq)}$	$\underline{-1079 \text{ kJ}}$	(Provided in the question)
$1/2\,H_2(g) \rightarrow H^+(aq) + e^-$	451 kJ	

(a) $Li(s) + H^+(aq) \rightarrow Li^+(aq) + 1/2\,H_2(g)$ $\Delta H° = 174 \text{ kJ} - 451 \text{ kJ} = -277 \text{ kJ} \approx \Delta G°$

$$E_{ox}° = \frac{\Delta G°}{-nF} = \frac{-277 \times 10^3 \text{ J}}{-1 \text{ mol } e^- \left(96{,}485 \dfrac{C}{\text{mol } e^-}\right)} = 2.87 \frac{J}{C} = 2.87 \text{ V}$$

In this reaction, $Li(s)$ is being oxidized to $Li^+(aq)$. If we wish to compare this to the reduction potential for $Li^+(aq)$ being reduced to $Li(s)$, we would have to reverse the reaction, this would result in the same answer only that it is a negative value. Alternatively, we can change the sign of the oxidation potential ($Li \rightarrow Li^+$) to -2.87 V for the reduction potential for $Li^+ \rightarrow Li$.

(it apears as -3.040 V in Appendix D, here we see much better agreement)

(b) $\Delta S = \Sigma S°_{products} - \Sigma S°_{reactants}$

$$= [1 \text{ mol}\left(13.4 \frac{J}{Kmol}\right) + 0.5 \text{ mol}\left(130.7 \frac{J}{Kmol}\right)] -$$

$$[1 \text{ mol}\left(29.12 \frac{J}{Kmol}\right) + 1 \text{ mol}\left(0 \frac{J}{Kmol}\right)] = 49.6 \frac{J}{K}$$

$$\Delta G° = \Delta H° - T\Delta S° = -277 \text{ kJ} - 298.15 \text{ K}\left(49.6 \frac{J}{K}\right) \times \frac{1 \text{ kJ}}{1000 \text{ J}} = -292 \times 10^3 \text{ J}$$

$$E_{ox}° = \frac{\Delta G°}{-nF} = \frac{-292 \times 10^3 \text{ J}}{-1 \text{ mol } e^- \left(96{,}485 \dfrac{C}{\text{mol } e^-}\right)} = 3.03 \frac{J}{C} = 3.03 \text{ V}$$

As mentioned in part (a) of this question. We have calculated the oxidation potential for the half reaction $Li(s) \rightarrow Li^+(aq) + e^-$. The reduction potential is the reverse half-reaction and has a potential of -3.03 V. This is excellent agreement with -3.040 V given in Appendix D.

74.   **(a)**

$$\frac{0.438 \text{ mol NaCl}}{\text{L}} \times \frac{58.443 \text{ g NaCl}}{1 \text{ mol NaCl}} = 26.6 \text{ g NaCl}$$

$$\frac{0.0512 \text{ mol MgCl}_2}{\text{L}} \times \frac{95.211 \text{ g MgCl}_2}{1 \text{ mol MgCl}_2} = 4.87 \text{ g MgCl}_2$$

> 31.5 g of salt per liter of seawater

$$18 \text{ tbs NaHCO}_3 \times \frac{10 \text{ g NaHCO}_3}{1 \text{ tbs NaHCO}_3} = 180 \text{ g NaHCO}_3$$

$$10 \text{ tbs NaCl} \times \frac{10 \text{ g NaCl}}{1 \text{ tbs NaCl}} = 100 \text{ g NaCl}$$

> 307 g of salt per gal. of lake water (3.7854 L/gal)

$$8 \text{ tsp MgSO}_4 \bullet 7 \text{ H}_2\text{O} \times \frac{1 \text{ tbs}}{3 \text{ tsp}} \times \frac{10 \text{ g MgSO}_4 \bullet 7 \text{ H}_2\text{O}}{1 \text{ tbs MgSO}_4 \bullet 7 \text{ H}_2\text{O}} = 27 \text{ g MgSO}_4 \bullet 7 \text{ H}_2\text{O}$$

(b)   The pH of the lake will be determined by the amphiprotic bicarbonate ion, (HCO_3^-), which hydrolyzes in water ($K_{a_1} = 4.4 \times 10^{-7}$ or $pK_{a_1} = 6.36$ and $K_{a_2} = 4.7 \times 10^{-11}$ or $pK_{a_2} = 10.33$). We saw earlier (chapter 18 question 100) that the pH of a solution of alanine, a diprotic species is independent of concentration (as long as it is relatively concentrated). The pH $= \dfrac{(pK_{a_1} + pK_{a_2})}{2} = \dfrac{(6.36 + 10.33)}{2} = 8.35$

This is not as basic as the actual pH of the lake. Addition of Borax (sodium salt of boric acid) would aid in increasing the pH of the solution (since borax contains an anion that is the conjugate base of a weak acid). The lake may be more basic due to the presence of other basic anions, namely carbonate ion (CO_3^{2-}).

(c)   Tufa are mostly made of calcium carbonate ($CaCO_3(s)$). Since they form near underwater springs, one must assume that the springs have a high concentration of calcium ion ($Ca^{2+}(aq)$). We then couple this with the fact that the lake has a high salinity (high carbonate and bicarbonate ion content). We can thus conclude that two major reactions that are most likely responsible for the formation of a tufa (see below):

$$Ca^{2+}(aq) + CO_3^{2-}(aq) \rightarrow CaCO_3(s)$$
$$Ca^{2+}(aq) + 2 HCO_3^-(aq) \rightarrow Ca(HCO_3)_2(q) \rightarrow CaCO_3(s) + H_2O(l) + CO_2(g)$$

CHAPTER 22
MAIN-GROUP ELEMENTS II: NONMETALS

PRACTICE EXAMPLES

1A This question involves calculating $E°$ for the reduction half-reaction:

$ClO_3^-(aq) + 3 H_2O(l) + 6 e^- \rightarrow Cl^-(aq) + 6 OH^-(aq)$

Here we will consider just one of the several approaches available to solve this problem. The four half-reactions (and their associated $E°$ values) that are used in this method to come up with the "missing $E°$ value" are given below. (Note: the $E°$ for the first reaction was determined in example 22-1)

1) $ClO_3^-(aq) + H_2O(l) + 2 e^- \rightarrow ClO_2^-(aq) + 2 OH^-(aq)$ $E° = 0.295$ V $\Delta G°_1 = -2FE°$

2) $ClO_2^-(aq) + H_2O(l) + 2 e^- \rightarrow OCl^-(aq) + 2 OH^-(aq)$ $E° = 0.681$ V $\Delta G°_2 = -2FE°$

3) $OCl^-(aq) + H_2O(l) + 1 e^- \rightarrow 1/2 Cl_2(aq) + 2 OH^-(aq)$ $E° = 0.421$ V $\Delta G°_3 = -1FE°$

4) $1/2 Cl_2(aq) + 1 e^- \rightarrow Cl^-(aq)$ $E° = 1.358$ V $\Delta G°_4 = -1FE°$

Although the reactions themselves may be added to obtain the desired equation, the $E°$ for this equation is not the sum of the $E°$ values for the above four reactions. The $E°$ values for the desired equation is actually the weighted average of the $E°$ values for reactions 1) to 4). It can be calculated by summing up the free energy changes for the four reactions (the standard voltages for half-reactions of the same type are not additive. The $\Delta G°$ values for these reactions can, however, be summed together).

When 1), 2), 3) and 4) are added together we obtain

1) $ClO_3^-(aq) + H_2O(l) + 2 e^- \rightarrow ClO_2^-(aq) + 2 OH^-(aq)$ $E°_1 = 0.295$ V

+2) $ClO_2^-(aq) + H_2O(l) + 2 e^- \rightarrow OCl^-(aq) + 2 OH^-(aq)$ $E°_2 = 0.681$ V

+3) $OCl^-(aq) + H_2O(l) + 1 e^- \rightarrow 1/2 Cl_2(aq) + 2 OH^-(aq)$ $E°_3 = 0.421$ V

+4) $1/2 Cl_2(aq) + 1 e^- \rightarrow Cl^-(aq)$ $E°_4 = 1.358$ V

   $ClO_3^-(aq) + 3 H_2O(l) + 6 e^- \rightarrow Cl^-(aq) + 6 OH^-(aq)$ $E°_5 = ?$

and $\Delta G°_5 = \Delta G°_1 + \Delta G°_2 + \Delta G°_3 + \Delta G°_4$

so, $-6FE°_5 = -2F(0.295 \text{ V}) + -2F(0.681\text{V}) + -1F(0.421 \text{ V}) + -1F(1.358 \text{ V})$

Hence, $E°_5 = \dfrac{-2F(0.295 \text{ V}) + -2F(0.681\text{V}) + -1F(0.421 \text{ V}) + -1F(1.358 \text{ V})}{-6F} = 0.622$ V

1B. This question involves calculating $E°$ for the reduction half-reaction:

$$2 \, ClO_3^-(aq) + 12H^+(aq) + 10 \, e^- \rightarrow Cl_2(aq) + 6 \, H_2O(l)$$

The three half-reactions (and their associated $E°$ values) are given below:

(1) $ClO_3^-(aq) + 3H^+(aq) + 2 \, e^- \rightarrow HClO_2(aq) + H_2O(l)$ $E° = 1.181$ V

(2) $HClO_2(aq) + 2 \, H^+(aq) + 2 \, e^- \rightarrow HClO(aq) + H_2O(l)$ $E° = 1.645$ V

(3) $2 \, HClO(aq) + 2 \, H^+(aq) + 2 \, e^- \rightarrow Cl_2(aq) + 2 \, H_2O(l)$ $E° = 1.611$ V
     (remember that $\Delta G° = -nFE°$)

Although the reactions themselves can be added to obtain the desired equation, the $E°$ for this equation is not the sum of the $E°$ values for the above three reactions. The $E°$ for the desired equation is actually the weighted average of the $E°$ values for reactions (1) to (3). It can be obtained by summing up the free energy changes for the three reactions. (For reactions of the same type, standard voltages are not additive; $\Delta G°$ values are additive, however). When (1) and (2) (each multiplied by two) are added to (3) we obtain:

$2\times(1)$ $2 \, ClO_3^-(aq) + 6 \, H^+(aq) + 4 \, e^- \rightarrow 2 \, HClO_2(aq) + 2 \, H_2O(l)$ $E°_1 = 1.181$ V
$2\times(2)$ $2 \, HClO_2(aq) + 4 \, H^+(aq) + 4 \, e^- \rightarrow 2 \, HClO(aq) + 2 \, H_2O(l)$ $E°_2 = 1.645$ V
$\ \ (3)$ $2 \, HClO(aq) + 2 \, H^+(aq) + 2 \, e^- \rightarrow Cl_2(aq) + 2 \, H_2O(l)$ $E°_3 = 1.611$ V

$\qquad 2 \, ClO_3^-(aq) + 12 \, H^+(aq) + 10 \, e^- \rightarrow Cl_2(aq) + 6 \, H_2O(l)$ $E°_4 = ?$

and $\Delta G°_4 = \Delta G°_1 + \Delta G°_2 + \Delta G°_3$

so, $-10FE°_4 = -4F(1.181 \text{ V}) + -4F(1.645 \text{ V}) + -2F(1.611 \text{ V})$

Hence, $E°_4 = \dfrac{-4F(1.181 \text{ V}) + -4F(1.645 \text{ V}) + -2F(1.611 \text{ V})}{-10 \, F} = 1.453$ V

2A The dissociation reaction is the reverse of the formation reaction and thus $\Delta G°$ for the dissociation reaction is the negative of $\Delta G°_f$

$$HF(g) \longrightarrow \tfrac{1}{2}H_2(g) + \tfrac{1}{2}F_2(g) \quad \Delta G° = -\Delta G°_f = -(-273.2 \text{ kJ/mol})$$

We know that

$$\Delta G° = -RT \ln K_P \qquad +273.2\times10^3 \text{ J/mol} = -8.3145 \text{ J mol}^{-1} \text{ K}^{-1} \times 298 \text{ K} \times \ln K_P$$

$$\ln K_P = \frac{273.2\times10^3 \text{ J mol}^{-1}}{-8.3145 \text{ J mol}^{-1} \text{ K}^{-1} \times 298 \text{ K}} = -110 \quad K_P = e^{-110} = 1.7\times10^{-48} \approx 2\times10^{-48}$$

Virtually no dissociation of HF(g) into its elements occurs.

2B The dissociation reaction with all integer coefficients is twice the reverse of the formation reaction.

$$2HCl(g) \rightleftharpoons H_2(g) + Cl_2(g) \quad \Delta G° = -2 \times (-95.30 \text{ kJ/mol}) = +190.6 \text{ kJ/mol} = -RT \ln K_p$$

$$\ln K_p = \frac{-\Delta G°}{RT} = \frac{-190.6 \times 10^3 \text{ J/mol}}{8.3145 \text{ J mol}^{-1} \text{ K}^{-1} \times 298 \text{ K}} = -76.9 \qquad K_p = e^{-76.9} = 4 \times 10^{-34}$$

We assume an initial HCl(g) pressure of P atm, and calculate the final pressure of $Cl_2(g)$ and $H_2(g)$, x atm.

Reaction:	2 HCl(g)	$\rightleftharpoons$	$H_2(g)$	+	$Cl_2(g)$
Initial:	P atm		0 atm		0 atm
Changes:	$-2x$ atm		$+x$ atm		$+x$ atm
Equil:	$(P-2x)$ atm		x atm		x atm

$$K_p = \frac{P\{H_2(g)\}P\{Cl_2(g)\}}{P\{HCl(g)\}^2} = \frac{x \cdot x}{(P-2x)^2} = \left(\frac{x}{P-2x}\right)^2 \qquad \frac{x}{P-2x} = \sqrt{4 \times 10^{-34}} = 2 \times 10^{-17}$$

$$x = 2 \times 10^{-17}(P-2x) \approx 2 \times 10^{-17} P$$

$$\% \text{ decomposition} = \frac{2x}{P} \times 100\% = 2 \times 2 \times 10^{-17} \times 100\% = 4 \times 10^{-15} \% \text{ decomposed}$$

EXERCISES

Noble Gas Compounds

1. First we use the ideal gas law to determine the amount in moles of argon.

$$n = \frac{PV}{RT} = \frac{145 \text{ atm} \times 55 \text{ L}}{0.08206 \text{ L atm mol}^{-1} \text{ K}^{-1} \times 299 \text{ K}} = 3.2\underline{5} \times 10^2 \text{ mol Ar}$$

$$\text{L air} = 3.2\underline{5} \times 10^2 \text{ mol Ar} \times \frac{22.414 \text{ L Ar STP}}{1 \text{ mol Ar}} \times \frac{100.000 \text{ L air}}{0.934 \text{ L Ar}} = 7.8 \times 10^5 \text{ L air}$$

2. First we compute the volume occupied by 5.00 g of He at STP, and then the volume of natural gas. From chapter 6 we saw that 1 mol of gas at STP occupied 22.414 L.

$$\text{natural gas volume} = 5.00 \text{ g He} \times \frac{1 \text{ mol He}}{4.003 \text{ g He}} \times \frac{22.414 \text{ L He}}{1 \text{ mol He}} \times \frac{100 \text{ L air}}{8 \text{ L He}}$$

$$\text{natural gas volume} = 3.\underline{5} \times 10^2 \text{ L air at STP} \cong 300 \text{ L}$$

3. **(a)** 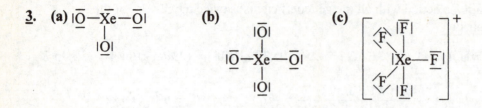 **(b)** **(c)**

(a) The Lewis structure has three ligands and one lone pair on Xe. XeO_3 has a trigonal pyramidal shape.

(b) The Lewis structure has four ligands on Xe. XeO_4 has a tetrahedral shape.

(c) There are five ligands and one lone pair on Xe in XeF_5^+. Its shape is square pyramidal.

4. We use VSEPR theory to predict the shapes of the species involved.

(a) O_2XeF_2 has a total of $2 \times 6 + 8 + 2 \times 7 = 34$ valence electrons $= 17$ pairs.

(b) O_3XeF_2 has a total of $3 \times 6 + 8 + 2 \times 7 = 40$ valence electrons $= 20$ pairs.

(c) $OXeF_4$ has a total of $6 + 8 + 4 \times 7 = 42$ valence electrons $= 21$ pairs.

We draw a plausible Lewis structure for each species below.

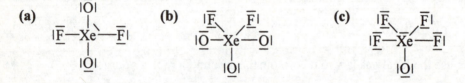

(a) There are four ligands and one lone pair on Xe in O_2XeF_2. Its shape is an see-saw.

(b) There are five ligands and no lone pairs on Xe in O_3XeF_2. Its shape is trigonal bipyramidal.

(c) The Lewis structure has five ligands and one lone pair on Xe. $OXeF_4$ has a square pyramidal shape.

5. $3\ XeF_4(aq) + 6\ H_2O(l) \rightarrow 2\ Xe(g) + 3/2\ O_2(g) + 12\ HF(g) + XeO_3(s)$

6. $2\ XeF_6(aq) + 16\ OH^-(aq) \rightarrow Xe(g) + XeO_6^{4-}(aq) + O_2(g) + 8\ H_2O(l) + 12\ F^-(aq)$

The Halogens

7. Iodide ion is slowly oxidized to iodine, which is yellow-brown in aqueous solution, by oxygen in the air.

Oxidation: $\{\, 2I^-(aq) \longrightarrow I_2(aq) + 2e^- \,\} \times 2$ $-E° = -0.535\,V$

Reduction: $O_2(g) + 4H^+(aq) + 4e^- \longrightarrow 2H_2O(l)$ $E° = +1.229\,V$

Net: $4I^-(aq) + O_2(g) + 4H^+(aq) \longrightarrow 2I_2(aq) + 2H_2O(l)$ $E°_{cell} = +0.694\,V$

Possibly followed by: $I_2(aq) + I^-(aq) \longrightarrow I_3^-(aq)$

8. $MnF_6^{2-} + 2SbF_5 \longrightarrow MnF_4 + 2SbF_6^-;$ $2MnF_4 \longrightarrow 2MnF_3 + F_2(g)$

9. Displacement reactions involve one element displacing another element from solution The element that dissolves in the solution is more *"active"* than the element supplanted from solution. Within the halogen group the activity decreases from top to bottom. Thus, each halogen is able to displace the members of the group below it, but not those above it. For instance, molecular bromine can oxidize aqueous iodide ion but molecular iodine is incapable of oxidizing bromide ion:

$Br_2(aq) + 2\,I^-(aq) \rightarrow 2\,Br^-(aq) + I_2(aq)$ however, $I_2(aq) + 2\,Br^-(aq) \rightarrow$ NO RXN

The only halogen with sufficient oxidizing power to displace $O_2(g)$ from water is $F_2(g)$:

$2\,H_2O(l) \rightarrow 4\,H^+ + O_2(g) + 4\,e^-$ $E°_{1/2ox} = -1.229\,V$

$\{F_2(g) + 2\,e^- \rightarrow 2\,F^-(aq)\} \times 2$ $E°_{1/2red} = 2.866\,V$

$2\,F_2(g) + 2\,H_2O(l) \rightarrow 4\,H^+ + O_2(g) + 4\,F^-(aq)$ $E°_{cell} = 1.637\,V$

The large positive standard reduction potential for this reaction indicates that the reaction will occur spontaneously, with products being strongly preferred under standard state conditions. None of the halogens react with water to form $H_2(g)$. In order to displace molecular hydrogen from water, one must add a strong reducing agent, such as sodium metal.

10. We first list the relevant values of halogen properties.

		I	F	Cl	Br
(a)	Covalent radius (pm):	71	99	114	133
(b)	Ionic radius(pm):	133	181	196	220
(c)	First ionization energy (kJ/mol):	1681	1251	1140	1008
(d)	electron affinity(kJ/mol):	−328.0	−349.0	−324.6	−295.2
(e)	electronegativity:	4.0	3.0	2.8	2.5
(f)	standard reduction potential:	+2.886 V	+1.358 V	+1.065 V	+0.535 V

We can do a reasonably god job of predicting the properties of astatine by simply looking at the difference between Br and I and assuming that the same difference exists between I and At.

(a) Covalent radius: ≈ 152 pm   **(b)** Ionic radius: 244 pm

(c) 1st ionization energy: 876 kJ/mol   **(d)** electron affinity: ≈ -265 kJ/mol

(e) electronegativity: ≈ 2.2 **(f)** $E^\circ \approx 0.005$ V to 0.010 V

Depending on the technique that you use, you will arrive at different answers. For example, -260 kJ/mol is a reasonable estimation of the electron affinity by the following reasoning. The difference between the value for Cl and Br is 24 kJ/mol, the difference between the values for Br and I is 29 kJ/mol, so the difference between I and At should be about 35 kJ/mol.

11. **(a)** $\text{mass } F_2 = 1 \text{ km}^3 \times \left(\dfrac{1000 \text{ m}}{1 \text{ km}} \times \dfrac{100 \text{ cm}}{1 \text{ m}}\right)^3 \times \dfrac{1.03 \text{ g}}{1 \text{ cm}^3} \times \dfrac{1 \text{ lb}}{454 \text{ g}} \times \dfrac{1 \text{ ton}}{2000 \text{ lb}} \times \dfrac{1 \text{ g F}^-}{1 \text{ ton}} \times \dfrac{37.996 \text{ g F}_2}{37.996 \text{ g F}^-}$

$= 1 \times 10^9 \text{ g F}_2 = 1 \times 10^6 \text{ kg F}_2$

(b) Bromine can be extracted by displacing it from solution with $Cl_2 (g)$. Since there is no chemical oxidizing agent that is stronger than $F_2(g)$, this method of displacement would not work for $F_2(g)$. Even if there were a chemical oxidizing agent stronger than $F_2(g)$, it would displace $O_2 (g)$ before it displaced $F_2(g)$. Obtaining $F_2(g)$ would require electrolysis of one of its molten salts, obtained from seawater evaporate.

12. $\text{Mass } F_2 = 1.00 \times 10^3 \text{ kg rock} \times \dfrac{1000 \text{ g}}{1 \text{ kg}} \times \dfrac{1 \text{ mol } 3 \text{ Ca}_3 (PO_4)_2 \cdot CaF_2}{1009 \text{ g}} \times \dfrac{2 \text{ mol F}}{1 \text{ mol } 3 \text{ Ca}_3 (PO_4)_2 \cdot CaF_2}$

$\times \dfrac{19.00 \text{ g F}}{1 \text{ mol F}} \times \dfrac{1 \text{ kg}}{1000 \text{ g}} = 37.7 \text{ kg fluorine}$

13. In order for the disproportionation reaction to occur under standard conditions, the E° for the overall reaction must be greater than zero. To answer this question, we must refer to the Latimer diagrams provided in Figure 22-2 and the answer to practice example 22-1B.

(i)   Reduction half reaction (acidic solution)
$Cl_2(aq) + 2 e^- \rightarrow 2 Cl^-(aq)$ $E^\circ_{1/2 \text{ red}} = 1.358$ V

(ii)   Oxidation half reaction (acidic solution)
$Cl_2(aq) + 6 H_2O(l) \rightarrow 2 ClO_3^-(aq) + 12 H^+(aq) + 10 e^-$ $E^\circ_{1/2 \text{ ox}} = -1.453$ V

Combining (i) $\times 5$ with (ii) $\times 1$, we obtain the desired disproportionation reaction:
$6 Cl_2(aq) + 6 H_2O(l) \rightarrow 2 ClO_3^-(aq) + 12 H^+(aq) + 10 Cl^-(aq)$ $E^\circ_{\text{cell}} = -0.095$ V

Since the final cell voltage is negative, the disproportionation reaction will not occur spontaneously under standard conditions. Alternatively, we can calculate K_{eq} by using $\ln K_{eq} = -\Delta G^\circ / RT$ and $\Delta G^\circ = -nFE^\circ$. This method gives a $K_{eq} = 8.6 \times 10^{-17}$. Clearly, the reaction will not go to completion under standard conditions.

14. To find out whether the reaction will go to completion, we must first calculate the standard potential for the disproportionation of hypochlorous acid (HOCl). As was the case for question #15, we will use information contained in Fig. 22.2 to answer this question.

(i) $\quad$ 2 HOCl(aq) + 2 H$_2$O(l) $\rightarrow$ 2 HClO$_2$(aq) + 4 H$^+$(aq) + 4 e$^-$ $\quad$ $-E°_{(i)} = -(1.645\ V)$ (oxidation)

(ii) $\quad$ 2 HOCl(aq) + 2 H$^+$(aq) + 2 e$^-$ $\rightarrow$ Cl$_2$(aq) + 2 H$_2$O(l) $\quad$ $E°_{(ii)} = +1.611\ V$

(iii) $\quad$ Cl$_2$(aq) + 2 e$^-$ $\rightarrow$ 2 Cl$^-$(aq) $\quad$ $E°_{(iii)} = +1.358\ V$

Need to determine the potential of the following half reaction using the relationship between standard free energies and standard potential

(iv) $\quad$ 2 HOCl(aq) + 2 H$^+$(aq) + 4 e$^-$ $\rightarrow$ 2 Cl$^-$(aq) + 2 H$_2$O(l) $\quad$ $E°_{(iv)}$

We can determine this using half reaction (ii) and(iii) since adding (ii) and (iii) give (iv). Hence,

$$-4F\,E°_{(iv)} = -2F(1.611\ V) + -2F(1.358\ V) \qquad \text{So, } E°_{(iv)} = 1.485\ V$$

Therefore, by adding equation (i) to equation (iv), we can obtain the standard cell voltage for the disproportionation of HOCl. We divide by 2 to give the target

(i)÷2 $\quad$ HOCl(aq) + H$_2$O(l) $\rightarrow$ HClO$_2$(aq) + 2 H$^+$(aq) + 2 e$^-$ $\quad$ $-E°_{(i)} = -(1.645\ V)$

(iv)÷2 $\quad$ HOCl(aq) + H$^+$(aq) + 2 e$^-$ $\rightarrow$ Cl$^-$(aq) + H$_2$O(l) $\quad$ $E°_{1/2\ red} = 1.485\ V$

_____

2 HOCl(aq) $\rightarrow$ HClO$_2$(aq) + Cl$^-$(aq) + H$^+$(aq) $\quad$ $E°_{cell} = -0.160\ V$
(disproportionation reaction of HOCl(aq))

Since the standard cell potential is negative, the equilibrium will favor the reactants and thus the reaction will not go to completion as written (i.e., reactants will predominate at equilibrium and thus the reaction will be far from complete). Alternatively, we can calculate K_{eq} from the reaction by using $\ln K_{eq} = -\Delta G°/RT$ and $\Delta G° = -nFE°$. This gives $K_{eq} = 3.9 \times 10^{-6}$. Clearly, the reaction will not go to completion because K_{eq} is very small (reactants strongly predominate at equilibrium, not products).

15. First we must draw the Lewis structure for all of the species listed. Following this, we will deduce their electron-group geometries and molecular shapes following the VSEPR approach.

(a) $\quad$ BrF$_3$: $\quad$ 28 valence electrons
$\qquad\qquad$ VSEPR class: AX$_3$E$_2$
$\qquad\qquad$ Thus BrF$_3$ is T-shaped

(b) $\quad$ IF$_5$ $\quad$ 42 valence electrons
$\qquad\qquad$ VSEPR class: AX$_5$E
$\qquad\qquad$ Thus IF$_5$ is square pyramidal

(c) $\quad$ Cl$_3$IF$^-$ $\quad$ 36 valence electrons
$\qquad\qquad$ VSEPR class: AX$_4$E$_2$
$\qquad\qquad$ Thus Cl$_3$IF$^-$ is square planar

16. To answer this question we need to apply the VSEPR method to each species:

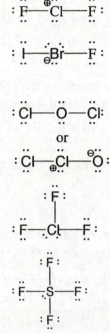

    (i) ClF_2^+ (total number of valence electrons = 20 e^-)

        Molecular shape: AX_2E_2 angular (<109.5° due to lone pairs on Cl atom)

    (ii) $IBrF^-$ (total number of valence electrons = 22 e^-)

        Molecular shape: AX_2E_3 linear (~180°)

    (iii) OCl_2 (total number of valence electrons = 20 e^-)

        Molecular shape: AX_2E_2 angular (<109.5° due to lone pairs on O or Cl atom)

        Note: Cl—O—Cl preferred over Cl—Cl—O because all atoms have a formal charge of zero.

    (iv) ClF_3 (total number of valence electrons = 28 e^-)

        Molecular shape: AX_3E_2 T-shaped molecule

    (v) SF_4 (total number of valence electrons = 34 e^-)

        Molecular shape: AX_4E See-saw molecular geometry (or distorted tetrahedron)

The VSEPR treatment of each species indicates only $IBrF^-$ has a linear structure. Since (i) and (iii) belong to the same VSEPR class (AX_2E_2), it is not unreasonable to assume that they have the same "bent structure".

Oxygen

17. **(a)** $2HgO(s) \xrightarrow{\Delta} 2Hg(l) + O_2(g)$ **(b)** $2KClO_4(s) \xrightarrow{\Delta} 2KClO_3(s) + O_2(g)$

18. **(a)** $O_3(g) + 2 I^-(aq) + 2 H^+(aq) \rightarrow I_2(aq) + H_2O(l) + O_2(g)$

      (b) $S(s) + H_2O(l) + 3 O_3(g) \rightarrow H_2SO_4(aq) + 3 O_2(g)$

      (c) $2\left[Fe(CN)_6\right]^{4-}(aq) + O_3(g) + H_2O(l) \rightarrow 2\left[Fe(CN)_6\right]^{3-}(aq) + O_2(g) + 2 OH^-(aq)$

19. We first write the formulas of the four substances: N_2O_4, Al_2O_3, P_4O_6, CO_2. The one constant in all these substances is oxygen. If we compare amounts of substance with the same amount (in moles) of oxygen, the one with the smallest mass of the other element will have the highest percent oxygen.

3 mol N_2O_4 contains 12 mol O and 6 mol N: $6 \times 14.0 = 84.0$ g N
4 mol Al_2O_3 contains 12 mol O and 8 mol Al: $8 \times 27.0 = 216$ g Al
2 mol P_4O_6 contains 12 mol O and 8 mol P: $8 \times 31.0 = 248$ g P
6 mol CO_2 contains 12 mol O and 6 mol C: $6 \times 12.0 = 72.0$ g C

Thus, of the oxides listed, CO_2 contains the largest percent oxygen by mass.

20. $2NH_4NO_3(s) \xrightarrow{400°C} 2N_2(g) + O_2(g) + 4H_2O(g)$

$2H_2O_2(l) \rightarrow 2H_2O(l) + O_2(g)$

$2KClO_3(s) \xrightarrow{\Delta} 2KCl(s) + 3O_2(g)$

(a) All three reactions have two moles of reactant, and the decomposition of potassium chlorate produces three moles of $O_2(g)$, compared to one mole of $O_2(g)$ for each of the others. The potassium chlorate decomposition produces the most oxygen per mole of reactant.

(b) If each reaction produced the same amount of $O_2(g)$, then the one with the lightest mass of reactant would produce the most oxygen per gram of reactant. When the first two reactions are compared, it is clear that 2 mol H_2O_2 have less mass than 2 mol NH_4NO_3. (Notice that each mol of NH_4NO_3 contains the amounts of elements—2 mol H and 2 mol O—present in each mol H_2O_2 plus some more.) So now we need to compare hydrogen peroxide with potassium chlorate. Notice that 6 H_2O_2 produces the same amount of $O_2(g)$ as 2 $KClO_3$. 6 mol H_2O_2 has a mass of $34.01 \times 6 = 204.1$ g, while 2 mol $KClO_3$ has a mass of $122.2 \times 2 = 244.2$ g. Thus, H_2O_2 produces the most $O_2(g)$ per gram of reactant.

21. Recall that fraction by volume and fraction by pressure are numerically equal. Additionally, one atmosphere pressure is equivalent to 760 mmHg. We combine these two facts.

$$P\{O_3\} = 760 \text{ mmHg} \times \frac{0.04 \text{ mmHg of } O_3}{10^6 \text{ mmHg of atmosphere}} = 3 \times 10^{-5} \text{ mmHg}$$

22. $P = \dfrac{nRT}{V} = \dfrac{\left(5 \times 10^{12} \text{ molecules} \times \dfrac{1 \text{ mol } O_3}{6.022 \times 10^{23} \dfrac{\text{molecule}}{\text{mole}}}\right) \times \dfrac{0.08206 \text{ L atm}}{\text{mol K}} \times 220 \text{ K}}{1 \text{ cm}^3 \times \dfrac{1 \text{ L}}{1000 \text{ cm}^3}}$

$P = 1.\underline{5} \times 10^{-7} \text{ atm} \times \dfrac{760 \text{ mmHg}}{1 \text{ atm}} = 1.\underline{1} \times 10^{-4} \text{ mmHg}$

23. The electrolysis reaction is $2H_2O(l) \xrightarrow{\text{electroysis}} 2H_2(g) + O_2(g)$. In this reaction, 2 moles of $H_2(g)$ are produced for each mole of $O_2(g)$, by the law of combining volumes, we would expect the volume of hydrogen to be twice the volume of oxygen produced. (Actually the volumes are not exactly in the ratio of 2:1 because of the different solubilities of oxygen and hydrogen in water.)

24. The electrolysis reaction is $2H_2O(l) \xrightarrow{\text{electroysis}} 2H_2(g) + O_2(g)$. In this reaction, 2 moles of water are decomposed to produce each mole of $O_2(g)$. We use the data in the problem to determine the amount of $O_2(g)$ produced, which we then convert to the mass of H_2O needed.

$$\text{mass } H_2O(l) = \frac{\left(736.7 \text{ mmHg} \times \dfrac{1 \text{ atm}}{760 \text{ mmHg}}\right)\left(22.83 \text{ mL} \times \dfrac{1 \text{ L}}{1000 \text{ mL}}\right)}{\dfrac{0.08206 \text{ L atm}}{\text{mol K}} \times (25.0 + 273.2)K} \times \frac{2 \text{ mol } H_2O}{1 \text{ mol } O_2}$$

$$\times \frac{18.02 \text{ g } H_2O}{1 \text{ mol } H_2O} = 0.03259 \text{ g } H_2O \text{ decomposed}$$

25. Since the pK_a for H_2O_2 had been provided to us, we can find the solution pH simply by solving an I.C.E. table for the hydrolysis of a 3.0 % H_2O_2 solution (by mass). Of course, in order to use this method, the mass percent must first be converted to molarity. We must assume that the density of the solution is 1.0 g mL^{-1}.

$$[H_2O_2] = \frac{3.0 \text{ g } H_2O_2}{100 \text{ g solution}} \times \frac{1 \text{ g solution}}{1 \text{ mL solution}} \times \frac{1 \text{ mol } H_2O_2}{34.015 \text{ g } H_2O_2} \times \frac{1000 \text{ mL}}{1 \text{ L}} = 0.88 \text{ M}$$

The pK_a for H_2O_2 is 11.75. The K_a for H_2O_2 is therefore $10^{-11.75}$ or 1.8×10^{-12}. By comparison with pure water, which has a K_a of 1.8×10^{-16} at 25 °C, one can see that H_2O_2 is indeed a stronger acid than water but the difference in acidity between the two is not that great. Consequently, we cannot ignore the contribution of protons of pure water when we workout the pH of the solution at equilibrium.

Reaction: $H_2O_2(aq)$ + $H_2O(l)$ $\underset{}{\overset{K_a = 1.8 \times 10^{-12}}{\rightleftharpoons}}$ $H_3O^+(aq)$ + $HO_2^-(aq)$

Initial	0.88 M	—	1.0×10^{-7} M	0 M
Change	$-x$ M	—	$+x$ M	$+x$ M
Equilibrium	$(0.88-x)$ M	—	$(1.0 \times 10^{-7} + x)$M	x M
	$(\sim 0.88$ M$)$			

So, $1.8 \times 10^{-12} = \dfrac{x(x + 1.0 \times 10^{-7})}{\sim 0.88}$ $x^2 + 1.0 \times 10^{-7}x - 1.58 \times 10^{-12} = 0$

$$x = \frac{-1.0 \times 10^{-7} \pm \sqrt{1.0 \times 10^{-14} + 4(1.58 \times 10^{-12})}}{2}$$

The root that makes sense in this context is $x = 1.2 \times 10^{-6}$ M.
Thus, the $[H_3O^+] = 1.2 \times 10^{-6}$ M $+ 1.0 \times 10^{-7}$ M $= 1.3 \times 10^{-6}$ M
Consequently, the pH for the 3.0 % H_2O_2 solution (by mass) should be 5.89
(i.e., the solution is weakly acidic)

26. Oxide ion reacts essentially quantitatively with water to form hydroxide ion.

$$Li_2O(s) + H_2O(l) \longrightarrow 2Li^+(aq) + 2OH^-(aq)$$

We first calculate $\left[OH^-\right]$ and then the solution's pH.

$$\left[OH^-\right] = \frac{0.050 \text{ g } Li_2O}{750.0 \text{ mL}} \times \frac{1000 \text{ mL}}{1 \text{ L}} \times \frac{1 \text{ mol } Li_2O}{29.88 \text{ g } Li_2O} \times \frac{2 \text{ mol } OH^-}{1 \text{ mol } Li_2O} = 0.00446 \text{ M}$$

$$pOH = -\log(0.00446) = 2.350 \qquad pH = 14.000 - pOH = 14.000 - 2.350 = 11.650$$

27. $3\ \overline{\underline{O}} = \overline{\underline{O}} \longrightarrow 2\ \overline{\underline{O}} \cdot\ \cdot\ \cdot\overline{\underline{O}} \cdot\ \cdot\ \cdot\overline{\underline{O}}$

Bonds broken: $3 \times (O=O)\ = 3 \times 498 \text{ kJ/mol} = 1494 \text{ kJ/mol}$

Bonds formed: $4 \times (O\ \cdot\ \cdot\ \cdot O)$

$\Delta H° = +285 \text{ kJ/mol} = \text{bonds broken} - \text{bonds formed} = 1494 \text{ kJ/mol} - 4 \times (O\ \cdot\ \cdot\ \cdot O)$

$4 \times (O\ \cdot\ \cdot\ \cdot O) = 1494 \text{ kJ/mol} - 285 \text{ kJ/mol} = 1209 \text{ kJ/mol}$

$O\ \cdot\ \cdot\ \cdot O = 1209 \text{ kJ/mol} \div 4 = 302 \text{ kJ/mol}$

28. $3\ \ddot{O}{=}\ddot{O} \longrightarrow 2\ :\!\ddot{O}{-}\ddot{O}{=}\ddot{O}:$ (all gaseous species)

$$\text{Average bond energy} = \frac{O{=}O + O{-}O}{2} = \frac{(498 + 142)\ \frac{\text{kJ}}{\text{mol}}}{2} = 320\ \frac{\text{kJ}}{\text{mol}}$$

29. **(a)** H_2S, while polar, forms only weak hydrogen bonds. H_2O forms much stronger hydrogen bonds, leading to a higher boiling point.

(b) All electrons are paired in O_3, producing a diamagnetic molecule.

30. **(a)** $\overline{\underline{O}}{=}\overline{\underline{O}} \qquad |\overline{\underline{O}}{-}\overline{\underline{O}}{=}\overline{\underline{O}} \qquad H{-}\overline{\underline{O}}{-}\overline{\underline{O}}{-}H$

The O—O bond in O_2 is a double bond, which should be short (121 pm). That in O_3 is a "one-and-a-half" bond, of intermediate length (128 pm). And that in H_2O_2 is a longer (148 pm) single bond.

(b) The O_2^+ ion has one less electron than does the O_2 molecule. Based on Lewis structures, $\qquad\qquad \overline{\underline{O}}{=}\overline{\underline{O}} \quad \overline{\underline{O}}\!\overset{\cdot}{-}\!\overline{\underline{O}}$

one would predict that O_2^+ would have a bond order of 1.5, thereby making its O—O bond weaker than the double bond of O_2, and therefore longer, in contradiction to the experimental evidence. The molecular orbital picture of the two species suggests the opposite. According to this model, O_2 has a bond order of 2.0 while that for O_2^+ is 2.5. Thus MO theory predicts a stronger bond for O_2^+.

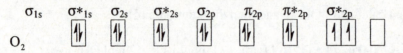

$$\text{bond order} = (\text{no. bonding electrons} - \text{no. antibonding electrons}) \div 2$$

$$\text{bond order} = (10 - 6) \div 2 = 2.0$$

| σ_{1s} | σ^*_{1s} | σ_{2s} | σ^*_{2s} | σ_{2p} | π_{2p} | π^*_{2p} | σ^*_{2p} |

O_2^+

$$\text{bond order} = (\text{no. bonding electrons} - \text{no. antibonding electrons}) \div 2 = (10-5) \div 2 = 2.5$$

31. Reactions that have K_{eq} values greater than 1000 are considered to be essentially quantitative (i.e. they go virtually 100% to completion). So to answer this question we need only calculate the equilibrium constant for each reaction via. the equation $E^\circ_{cell} = (0.0257/n)\ln K_{eq}$

(a)

$$H_2O_2(aq) + 2 H^+(aq) + 2 e^- \rightleftharpoons 2 H_2O(l) \qquad E^\circ_{1/2red} = +1.763 \text{ V}$$

$$2 I^-(aq) \rightleftharpoons I_2(s) + 2 e^- \qquad E^\circ_{1/2ox} = -0.535 \text{ V}$$

$$H_2O_2(aq) + 2 H^+(aq) + 2 I^-(aq) \rightleftharpoons I_2(s) + 2 H_2O(l) \quad E^\circ_{1/2cell} = +1.228 \text{ V}(n = 2 \text{ e}^-)$$

$$\ln K_{eq} = \frac{1.228 \text{ V} \times 2}{0.0257 \text{ V}} = 95.56 \qquad K_{eq} = 3.2 \times 10^{41}$$

Therefore the reaction goes to completion (or very nearly so).

(b)

$$O_2(g) + 2 H_2O(l) + 4 e^- \rightleftharpoons 4 OH^-(aq) \qquad E^\circ_{1/2red} = +0.401 \text{ V}$$

$$4 Cl^-(aq) \rightleftharpoons 2 Cl_2(g) + 4 e^- \qquad E^\circ_{1/2ox} = -1.358 \text{ V}$$

$$O_2(g) + 2 H_2O(l) + 4 Cl^-(aq) \rightleftharpoons 2 Cl_2(g) + 4 OH^-(aq) \quad E^\circ_{1/2cell} = -0.957 \text{ V}(n = 4 \text{ e}^-)$$

$$\ln K_{eq} = \frac{-0.957 \text{ V} \times 4}{0.0257 \text{ V}} = -148.95 \qquad K_{eq} = 2.1 \times 10^{-65}$$

The extremely small value of K_{eq} indicates that reactants are strongly preferred and thus, the reaction does not even come close to going to completion.

(c)

$$O_3(g) + 2 H^+(aq) + 2 e^- \rightleftharpoons O_2(g) + H_2O(l) \qquad E^\circ_{1/2red} = +2.075 \text{ V}$$

$$Pb^{2+}(aq) + 2 H_2O(l) \rightleftharpoons PbO_2(s) + 4H^+(aq) + 2 e^- \qquad E^\circ_{1/2ox} = -1.455 \text{ V}$$

$$O_3(g) + Pb^{2+}(aq) + H_2O(l) \rightleftharpoons PbO_2(s) + 2 H^+(aq) + O_2(g) \quad E^\circ_{cell} = 0.620 \text{ V}$$

$$\ln K_{eq} = \frac{+0.62 \text{ V} \times 2}{0.0257 \text{ V}} = 48.25 \qquad K_{eq} = 9.0 \times 10^{20}$$

Therefore the reaction goes to completion (or very nearly so).

(d) $HO_2^-(aq) + H_2O(l) + 2\ e^- \rightleftharpoons 3\ OH^-(aq)$ $E^\circ_{1/2red} = +0.878\ V$

   $2\ Br^-(aq) \rightleftharpoons Br_2(l) + 2\ e^-$ $E^\circ_{1/2ox} = -1.065\ V$

   $HO_2^-(aq) + H_2O(l) + 2\ Br^-(aq) \rightleftharpoons Br_2(s) + 3\ OH^-(aq)$ $E^\circ_{1/2cell} = -0.187\ V$

$$\ln K_{eq} = \frac{-0.187\ V \times 2}{0.0257\ V} = -14.55 \qquad K_{eq} = 4.8 \times 10^{-7}$$

The extremely small value of K_{eq} indicates the reaction heavily favors reactants at equilibrium and thus, the reaction does not even come close to going to completion.

32.   (a) $HgO(s) \xrightarrow{\Delta} Hg(l) + 1/2\ O_2(g)$

   (b) $2KClO_4(s) \xrightarrow{\Delta} 2\ KCl(s) + 4\ O_2(g)$

   (c) $Hg(NO_3)_2(s) \xrightarrow{\Delta} Hg(l) + 2\ NO_2(g) + O_2(g)$

   or

   $Hg(NO_3)_2(s) \xrightarrow{\Delta} Hg(NO_2)_2(g) + O_2(g)$

   Depending on the temperature

   (d) $H_2O_2(l) \xrightarrow{\Delta} H_2O(l) + 1/2\ O_2(g)$

Sulfur

33.   (a)   ZnS, zinc sulfide   (b) $KHSO_3$, potassium hydrogen sulfite

   (c) $K_2S_2O_3$, potassium thiosulfate   (d) SF_4, sulfur tetrafluoride

34.   (a) $CaSO_4 \cdot 2H_2O$, calcium sulfate dehydrate   (b) $H_2S(aq)$, hydrosulfuric acid

   (c) $NaHSO_4$, sodium hydrogen sulfate   (d) $H_2S_2O_7(aq)$, disulfuric acid

35.   (a) $FeS(s) + 2\ HCl(aq) \longrightarrow FeCl_2(aq) + H_2S(aq)$ MnS(s), ZnS(s), etc. also are possible.

   (b) $CaSO_3(s) + 2\ HCl(aq) \longrightarrow CaCl_2(aq) + H_2O(l) + SO_2(g)$

   (c)   Oxidation: $SO_2(aq) + 2\ H_2O(l) \longrightarrow SO_4^{2-}(aq) + 4H^+(aq) + 2\ e^-$

   Reduction: $MnO_2(s) + 4H^+(aq) + 2\ e^- \longrightarrow Mn^{2+}(aq) + 2H_2O(l)$

   Net: $SO_2(aq) + MnO_2(s) \longrightarrow Mn^{2+}(aq) + SO_4^{2-}(aq)$

   (d)   Oxidation: $S_2O_3^{2-}(aq) + H_2O(l) \longrightarrow 2SO_2(g) + 2H^+(aq) + 4e^-$

   Reduction: $S_2O_3^{2-}(aq) + 6H^+(aq) + 4e^- \longrightarrow 2S(s) + 3H_2O(l)$

   Net: $S_2O_3^{2-}(aq) + 2H^+(aq) \longrightarrow S(s) + SO_2(g) + H_2O(l)$

36. **(a)** $S(s) + O_2(g) \longrightarrow SO_2(g)$

$2 Na(s) + 2 H_2O(l) \longrightarrow 2 NaOH(aq) + H_2(g)$

$2 NaOH(aq) + SO_2(g) \longrightarrow Na_2SO_3(aq) + H_2O(l)$

(b) $S(s) + O_2(g) \longrightarrow SO_2(g)$

$SO_2(g) + 2 H_2O(l) + Cl_2(g) \longrightarrow SO_4^{2-}(aq) + 2 Cl^-(aq) + 4 H^+(aq) \quad (E_{cell}^{\circ} = +1.19 V)$

$2 Na(s) + 2 H_2O(l) \longrightarrow 2 NaOH(aq) + H_2(g)$

$2 NaOH(aq) + 2 H^+(aq) + SO_4^{2-}(aq) \longrightarrow Na_2SO_4(aq) + 2 H_2O(l)$

(c) $S(s) + O_2(g) \longrightarrow SO_2(g)$

$2 Na(s) + 2 H_2O(l) \longrightarrow 2 NaOH(aq) + H_2(g)$

$2 NaOH(aq) + SO_2(g) \longrightarrow Na_2SO_3(aq) + H_2O(l)$

$Na_2SO_3(aq) + S(s) \longrightarrow NaS_2O_3(aq)$

(Thus, one must boil the reactants in an alkaline solution.)

37. The decomposition of thiosulfate ion is more highly favored in an acidic solution. If the white solid is Na_2SO_4, there will be no reaction with strong acids such as HCl. By contrast, if the white solid is $Na_2S_2O_3$, $SO_2(g)$ will be liberated and a pale yellow precipitate of S(s,rhombic) will form upon addition of HCl(aq).

$S_2O_3^{2-}(aq) + 2 H^+(aq) \longrightarrow S(s) + SO_2(g) + H_2O(l)$

Consequently, the solid can be identified by adding a strong mineral acid such as HCl(aq).

38. Sulfites are easily oxidized to sulfates by, for example, O_2 in the atmosphere. On the other hand, there are no oxidizing agents naturally available in reasonable concentrations that can oxidize SO_4^{2-} to a higher oxidation state, such as in $S_2O_8^{2-}$. Similarly, the atmosphere of Earth is an oxidizing one, reducing agents are not present to reduce SO_4^{2-} to a species with a lower oxidation state, except in localized areas.

39. $Na^+(aq)$ will not hydrolyze, being only very weakly polarizing. But $HSO_4^-(aq)$ will ionize further, $K_2 = 1.1 \times 10^{-2}$ for $HSO_4^-(aq)$. We set up the situation, and solve the quadratic equation to obtain $\left[H_3O^+ \right]$.

$$\left[HSO_4^- \right] = \frac{12.5 \text{ g NaHSO}_4}{250.0 \text{ mL soln}} \times \frac{1000 \text{ mL}}{1 \text{ L soln}} \times \frac{1 \text{ mol NaHSO}_4}{120.1 \text{ g NaHSO}_4} \times \frac{1 \text{ mol HSO}_4^-}{1 \text{ mol NaHSO}_4} = 0.416 \text{ M}$$

Reaction:	$HSO_4^-(aq) + H_2O(l) \rightleftharpoons$	$SO_4^{2-}(aq)$	+	$H_3O^+(aq)$
Initial:	0.416 M —	0 M		≈ 0 M
Changes:	$-x$ M —	$+x$ M		$+x$ M
Equil:	$(0.416 - x)$ M —	x M		x M

Owing to the moderate value of the equilibrium constant ($K = 0.011$), the full quadratic equation must be solved (i.e. no assumption can be made here).

$$K_2 = \frac{\left[H_3O^+\right]\left[SO_4^{2-}\right]}{\left[HSO_4^-\right]} = 0.011 = \frac{x^2}{0.416 - x} \qquad x^2 = 0.0046 - 0.011x$$

$$0 = x^2 + 0.011x - 0.0046$$

$$x = \frac{-b \pm \sqrt{b^2 - 4ac}}{2a} = \frac{-0.011 \pm \sqrt{1.2 \times 10^{-4} + 1.8 \times 10^{-2}}}{2} = 0.062 = \left[H_3O^+\right]$$

$$pH = -\log(0.062) = 1.21$$

40. Oxidation: $\{SO_3^{2-}(aq) + H_2O(l) \longrightarrow SO_4^{2-}(aq) + 2H^+(aq) + 2e^-\}$ $\times 5$

Reduction: $\{MnO_4^-(aq) + 8H^+(aq) + 5e^- \longrightarrow Mn^{2+}(aq) + 4H_2O(l)\}$ $\times 2$

Net: $5\,SO_3^{2-}(aq) + 2\,MnO_4^-(aq) + 6H^+(aq) \longrightarrow 5SO_4^{2-}(aq) + 2\,Mn^{2+}(aq) + 3\,H_2O(l)$

$$\text{mass Na}_2\text{SO}_3 = 26.50 \text{ mL} \times \frac{1\text{ L}}{1000\text{ mL}} \times \frac{0.0510\text{ mol MnO}_4^-}{1\text{ L soln}} \times \frac{5\text{ mol SO}_3^{2-}}{2\text{ mol MnO}_4^-} \times \frac{1\text{ mol Na}_2\text{SO}_3}{1\text{ mol SO}_3^{2-}}$$

$$\times \frac{126.0\text{ g Na}_2\text{SO}_3}{1\text{ mol Na}_2\text{SO}_3} = 0.426\text{ g Na}_2\text{SO}_3$$

41. The question is concerned with assaying for the mass percent of copper in an ore. The assay in this instance involves the quantitative determination of the amount of metal in an ore by chemical analysis. The titration for copper in the sample does not occur directly, but rather indirectly via the number of moles of $I_3^-(aq)$ produced from the reaction of Cu^{2+} with I^-:
$2\,Cu^{2+} + 5\,I^-(aq) \rightarrow 2\,CuI(s) + I_3^-(aq)$
The number of moles of $I_3^-(aq)$ produced is determined by titrating the iodide-treated sample with sodium thiosulfate. The balanced oxidation reaction that forms the basis for the titration is:
$I_3^-(aq) + 2\,S_2O_3^{2-}(aq) \rightarrow 3\,I^-(aq) + S_4O_6^{2-}(aq)$
The stoichiometric ratio is one $I_3^-(aq)$ reacting with two $S_2O_3^{2-}(aq)$ in this titration.
The number of moles of I_3^- formed

$$= 0.01212\text{ L S}_2\text{O}_3^{2-}(aq) \times \frac{0.1000\text{ moles S}_2\text{O}_3^{2-}}{1\text{ L S}_2\text{O}_3^{2-}} \times \frac{1\text{ mol I}_3^-}{2\text{ mol S}_2\text{O}_3^{2-}} = 6.060 \times 10^{-4}\text{ moles I}_3^-$$

Therefore, the number of moles of Cu^{2+} released when the sample is dissolved is

$$= 6.060 \times 10^{-4}\text{ moles I}_3^- \times \frac{2\text{ mol Cu}^{2+}}{1\text{ mol I}_3^-} = 1.212 \times 10^{-3}\text{ moles of Cu}^{2+}$$

Consequently, the mass percent for copper in the ore is

$$1.212 \times 10^{-3}\text{ moles of Cu}^{2+} \times \frac{63.546\text{ g Cu}^{2+}}{1\text{ mol Cu}^{2+}} \times \frac{1}{1.100\text{ g of Cu ore}} \times 100\% = 7.002\%$$

42. First you must realize that pressure and temperature are irrelevant to answering this question. The reaction is $Pb^{2+}(aq) + H_2S(aq) \rightarrow 2\,H^+(aq) + PbS(s)\downarrow$
This means that there is a one to one relationship between the moles of PbS and H_2S.
If we realize that $1\,m^3$ is 1000 L (40 times larger than the sample used), then we can calculate the mass of PbS in $1\,m^3$, namely, $40 \times 0.535\,g\,PbS = 21.4\,g\,PbS$

The mass of sulfur can be calculated using stoichiometry.

$$\text{Mass of S} = 21.4\,g\,PbS \times \frac{1\,mol\,PbS}{239.3\,g\,PbS} \times \frac{1\;mol\;S}{1\;mol\;PbS} \times \frac{32.066\,g\;S}{1\;mol\;S} = 2.87\,g\,S\;(\text{per }1\,m^3)$$

Alternatively, knowing that the mass percent of S in PbS is 13.40 % We can calculate the mass of sulfur as: $21.4\,g\,PbS \times \dfrac{13.40\,g\,S}{100\,g\,PbS} = 2.87\,g\,S\;(\text{per }1\,m^3\text{ natural gas})$

Nitrogen Family

<u>**43.**</u> **(a)** The Haber-Bosch process is the principal artificial method of fixing atmospheric nitrogen. $N_2(g) + 3\,H_2(g) \rightleftharpoons 2\,NH_3(g)$

(b) The 1st step of the Ostwald process:
$4NH_3(g) + 5\,O_2(g) \xrightarrow{850°C,\,Pt} 4NO(g) + 6H_2O(g)$

(c) The 2nd and 3rd steps of the process: $\quad 2\,NO(g) + O_2(g) \rightarrow 2\,NO_2(g)$
$$3\,NO_2(g) + H_2O(l) \rightarrow 2\,HNO_3(aq) + NO(g)$$

44. **(a)** $2\,NH_4NO_3(s) \xrightarrow{400°C} 2\,N_2(g) + O_2(g) + 4\,H_2O(g)$

The equation was balanced by inspection, by realizing first that each mole of $NH_4NO_3(s)$ would produce 1 mol $N_2(g)$ and 2 mol $H_2O(l)$, with 1 mol O [or $1/2$ mol $O_2(g)$] remaining.

(b) $NaNO_3(s) \rightarrow NaNO_2(s) + 1/2\,O_2(g) \quad$ or $\quad 2\,NaNO_3(s) \rightarrow 2\,NaNO_2(s) + O_2(g)$

(c) $Pb(NO_3)_2(s) \rightarrow PbO(s) + 2\,NO_2(g) + 1/2\,O_2(g)$
$2\,Pb(NO_3)_2(s) \rightarrow 2\,PbO(s) + 4\,NO_2(g) + O_2(g)$

45. Begin with the chemical formulas of the species involved:

$Na_2CO_3(aq) + O_2(g) + NO(g) \rightarrow NaNO_2(aq)$

Oxygen is reduced and nitrogen (in NO) is oxidized. We use the ion-electron method.

two couples: $NO(g) \rightarrow NO_2^-(aq)$ and $O_2 \rightarrow$ products

balance oxygens: $H_2O + NO \rightarrow NO_2^-$ $\qquad$ $O_2 \rightarrow 2\,H_2O$

balance hydrogens: $H_2O + NO \rightarrow NO_2^- + 2\,H^+$ $\qquad$ $4\,H^+ + O_2 \rightarrow 2\,H_2O$

balance charge: $H_2O + NO \rightarrow NO_2^- + 2\,H^+ + e^-$ $\qquad$ $4\,e^- + 4\,H^+ + O_2 \rightarrow 2\,H_2O$

combine: $4\,H_2O + 4\,NO + 4\,H^+ + O_2 \rightarrow 2\,H_2O + 4\,NO_2^- + 8\,H^+$

simplify: $2\,H_2O + 4\,NO + O_2 \rightarrow 4\,NO_2^- + 4\,H^+$

add spectator ions: $4\,Na^+ + 2\,CO_3^{2-} \rightarrow 4\,Na^+ + 2\,CO_3^{2-}$

$2\,Na_2CO_3 + 2\,H_2O + 4\,NO + O_2 \rightarrow 4\,NaNO_2 + 2\,H_2O + 2\,CO_2$

simplify: $2\,Na_2CO_3(aq) + 4\,NO(g) + O_2(g) \rightarrow 4\,NaNO_2(aq) + 2\,CO_2(g)$

46. $\quad$ mass % $HNO_3 = \dfrac{15\ \text{mol } HNO_3 \times \dfrac{63.01\ \text{g } HNO_3}{1\ \text{mol } HNO_3}}{1\ \text{L soln} \times \dfrac{1000\ \text{mL}}{1\ \text{L}} \times \dfrac{1.41\ \text{g}}{1\ \text{mL soln}}} \times 100\% = 67\%\ HNO_3$

47. $\quad 75 \times 10^9\ \text{gal} \times \dfrac{15\ \text{miles}}{1\ \text{gal}} \times \dfrac{5\ \text{g}}{1\ \text{mile}} \times \dfrac{1\ \text{kg}}{1000\ \text{g}} = 6 \times 10^9\ \text{kg of nitrogen oxides released.}$

48. $\quad 2\,NH_{3(g)} + 3\,NO(g) \rightarrow 5/2\,N_2(g) + 3\,H_2O(g)$ (or, multiplying by 2):

$4\,NH_{3(g)} + 6\,NO(g) \rightarrow 5\,N_2(g) + 6\,H_2O(g)$

$\Delta H^\circ_{rxn} = \Sigma \Delta H^\circ_{f\ products} - \Sigma \Delta H^\circ_{f\ reactants}$

$\Delta H^\circ_{rxn} = [5\ \text{mol}(0\ \text{kJ mol}^{-1}) + 6\ \text{mol}(-241.8\ \text{kJ mol}^{-1})]$
$-[6\ \text{mol}(90.25\ \text{kJ mol}^{-1}) + 4\ \text{mol}(-46.11\ \text{kJ mol}^{-1})] = -1808.\ \text{kJ}$

49. (a) $\quad 2\,NO_2(g) \rightleftharpoons N_2O_4(g)$

(b) $\quad$ i) $\quad HNO_2(aq) + N_2H_5^+(aq) \rightarrow HN_3(aq) + 2\,H_2O(l) + H^+(aq)$

$\qquad$ ii) $\quad HN_3(aq) + HNO_2(aq) \rightarrow N_2(g) + H_2O(l) + N_2O(g)$

(c) $\quad H_3PO_4(aq) + 2\,NH_3(aq) \rightarrow (NH_4)_2HPO_4(aq)$

50. (a) $\quad 3\,Ag(s) + 4\,H^+(aq) + 4\,NO_3^-(aq) \longrightarrow 3\,AgNO_3(aq) + NO(g) + 2\,H_2O(l)$

(b) $\quad (CH_3)_2NNH_2(l) + 4\,O_2(g) \longrightarrow 2\,CO_2(g) + 4\,H_2O(l) + N_2(g)$

(c) $\quad NaH_2PO_4(s) + 2\,Na_2HPO_4(s) \xrightarrow{\ \Delta\ } 2\,H_2O(l) + Na_5P_3O_{10}(s)$

51.

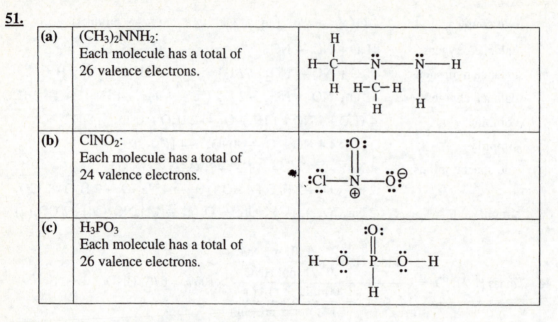

(a)	$(CH_3)_2NNH_2$: Each molecule has a total of 26 valence electrons.	
(b)	$ClNO_2$: Each molecule has a total of 24 valence electrons.	
(c)	H_3PO_3 Each molecule has a total of 26 valence electrons.	

52. Hyponitrous acid is a weak diprotic acid while nitramide has an amide group, $—NH_2$

$$H—\overset{..}{\underset{..}{O}}—N=N—\overset{..}{\underset{..}{O}}—H \qquad H—\overset{\overset{H}{|}}{\underset{..}{N}}—\overset{..}{\underset{..}{O}}—\overset{..}{\underset{..}{N}}=O \qquad H—\overset{\overset{H}{|}}{\underset{..}{N}}—\overset{\overset{:\overset{..}{O}:^{\ominus}}{|}}{\underset{\oplus}{N}}=O$$

$$\quad 1 \qquad\qquad\qquad 2 \qquad\qquad\qquad 3$$

Both Lewis structures for nitramide are plausible based on the information supplied. To choose between them would require further information, such as whether the molecule contains a nitrogen-nitrogen bond. Experimental evidence indicates the structure adopted is the one with the nitrogen-nitrogen bond, i.e. structure #3.

53. **(a)** HPO_4^{2-}, hydrogen phosphate ion

   (b) $Ca_2P_2O_7$, calcium pyrophosphate or calcium diphosphate

   (c) $H_6P_4O_{13}$, tetrapolyphosphoric acid

54. **(a)** $HONH_2$, hydroxylamine

   (b) $CaHPO_4$, calcium hydrogen phosphate

   (c) Li_3N, lithium nitride

55. i) $2\,H^+(aq) + N_2O_4(aq) + 2\,e^- \rightarrow 2\,HNO_2(aq)$ $E°_1 = +1.065$ V

   ii) $2\,HNO_2(aq) + 2\,H^+(aq) + 2\,e^- \rightarrow 2\,NO(aq) + 2\,H_2O(l)$ $E°_2 = +0.996$ V

   iii) $N_2O_4(aq) + 4\,H^+(aq) + 4\,e^- \rightarrow 2\,NO(aq) + 2\,H_2O(l)$ $E°_3 = ?$ V

Recall that $\Delta G° = -nFE°$ and that $\Delta G°$ values, not standard voltages are additive for reactions in which the number of electrons do not cancel out.

So, $-4FE°_3 = -2F(1.065$ V$) + -2F(0.996$ V$) \, E°_3 = 1.031$ V (4 sig figs)

56. i) $2\ NO_3^-(aq) + 2\ H_2O(l) + 2\ e^- \rightarrow N_2O_4(aq) + 4\ OH^-(aq)$ $E^\circ_1 = -0.86\ V$

     ii) $N_2O_4(aq) + 2\ e^- \rightarrow 2\ NO_2^-(aq)$ $E^\circ_2 = +0.87\ V$

     iii) $2\ NO_3^-(aq) + 2\ H_2O(l) + 4\ e^- \rightarrow 2\ NO_2^-(aq) + 4\ OH^-(aq)$ $E^\circ_3 = ?\ V$

Recall that $\Delta G^\circ = -nFE^\circ$ and that ΔG° values, not standard voltages are additive for reactions in which the number of electrons do not cancel out.

So, $-4FE^\circ_3 = -2F(-0.86\ V) + -2F(0.87\ V)$
        $-4FE^\circ_3 = 1.72F - 1.74\ F = 0.02F$ (1 sig fig)
        $E^\circ_3 = 0.005\ V$ (1 sig fig)

Hydrogen

57.   The four reactions of interest are: (Note: $\Delta H^\circ\ _{combustion} = \Sigma \Delta H^\circ_{f\ products} - \Sigma \Delta H^\circ_{f\ reactants}$)

$CH_4(g) + 2\ O_{2(g)} \rightarrow CO_2(g) + 2\ H_2O(l)$ $\Delta H^\circ\ _{combustion} = -890.3\ kJ$
(Molar mass $CH_4 = 16.0428\ g\ mol^{-1}$)

$C_2H_6(g) + 7/2\ O_{2(g)} \rightarrow 2\ CO_2(g) + 3\ H_2O(l)$ $\Delta H^\circ\ _{combustion} = -1559.7\ kJ$
(Molar mass $C_2H_6 = 30.070\ g\ mol^{-1}$)

$C_3H_8(g) + 5\ O_{2(g)} \rightarrow 3\ CO_2(g) + 4\ H_2O(l)$ $\Delta H^\circ\ _{combustion} = -2219.9\ kJ$
(Molar mass $C_3H_8 = 44.097\ g\ mol^{-1}$)

$C_4H_{10}(g) + 13/2\ O_{2(g)} \rightarrow 4\ CO_2(g) + 5\ H_2O(l)$ $\Delta H^\circ\ _{combustion} = -2877.4\ kJ$
(Molar mass $C_4H_{10} = 58.123\ g\ mol^{-1}$)

58.   Using the answers obtained in question 57, the per gram energy release is:
$CH_4(g)$ -55.5 kJ; $C_2H_6(g)$ -51.9 kJ; $C_3H_8(g)$ -50.3 kJ; $C_4H_{10}(g)$ -49.5 kJ

(a) $C_4H_{10}(g)$ evolves the most energy on a per mole basis (-2877.4 kJ)
(b) $CH_4(g)$ evolves the most energy on a per gram basis (-55.5 kJ)
(c) $CH_4(g)$ is the most desirable alkane from the standpoint of emission, producing the least quantity of $CO_2(g)$ per mole and/or per gram of fuel burned (as well per kJ of energy produced).

59.   (a) $2\ Al(s) + 6\ HCl(aq) \rightarrow 2\ AlCl_3(aq) + 3\ H_2(g)$
      (b) $C_3H_8(g) + 3\ H_2O(g) \rightarrow 3\ CO(g) + 7\ H_2(g)$
      (c) $MnO_2(s) + 2H_2(g) \xrightarrow{\Delta} Mn(s) + 2H_2O(g)$

60.   (a) $2H_2O(l) \xrightarrow{electricity} O_2(g) + 2H_2(g)$
      (b) $2\ HI(aq) + Zn(s) \rightarrow ZnI_2(aq) + H_2(g)$ Any moderately active metal is suitable.
      (c) $Mg(s) + H_2SO_4(aq) \rightarrow MgSO_4(aq) + H_2(g)$ Any strong acid is suitable.
      (d) $CO(g) + H_2O(g) \rightarrow CO_2(g) + H_2(g)$

61. $CaH_2(s) + 2\ H_2O(l) \rightarrow Ca(OH)_2(aq) + 2\ H_2(g)$

$Ca(s) + 2\ H_2O(l) \rightarrow Ca(OH)_2(aq) + H_2(g)$

$\overline{2\ Na(s) + 2\ H_2O(l) \rightarrow 2\ NaOH(aq) + H_2(g)}$

(a) The reaction that produces the largest volume of $H_2(g)$ per liter of water also produces the largest amount of $H_2(g)$ per mole of water used. All three reactions use two moles of water and the reaction with $CaH_2(s)$ produces the most $H_2(g)$.

(b) We can compare three reactions that produce the same amount of hydrogen; the one that requires the smallest mass of solid produces the greatest amount of H_2 per gram of solid. The amount of hydrogen we will choose, to simplify matters is 2 moles, which means that we compare 1 mol CaH_2 (42.09 g) with 2 mol Ca (80.16 g) and with 4 mol Na (91.96 g). Clearly CaH_2 produces the greatest amount of H_2 per gram of solid.

62. The balanced equation is $C_{17}H_{33}COOH(l) + H_2(g) \rightarrow C_{17}H_{35}COOH(s)$ One mole of oleic acid requires one mole of $H_2(g)$.

$$V = \frac{nRT}{P} = \frac{1.00\ \text{mol} \times 0.08206\ \dfrac{\text{L atm}}{\text{mol K}} \times 298\ \text{K}}{752\ \text{mm Hg} \times \dfrac{1\ \text{atm}}{760\ \text{mmHg}}} \times \frac{1\ \text{L calculated}}{0.95\ \text{L produced}} = 26\ \text{L}\ H_2(g)$$

63. Greatest mass percent hydrogen:

The atmosphere is mostly $N_2(g)$ and $O_2(g)$ with only a trace of hydrogen containing gas molecules. Seawater is $H_2O(l)$, natural gas is $CH_4(g)$ and ammonia is $NH_3(g)$. Each of these compounds have one non-hydrogen atom, and the non-hydrogen atoms have approximately the same mass ($\sim 14 \pm 2\ \text{g mol}^{-1}$). Since CH_4 has the highest hydrogen atom to non-hydrogen atom ratio, this molecule has the greatest mass percent hydrogen.

64. The reaction is $CaH_2(s) + 2\ H_2O(l) \rightarrow Ca(OH)_2(aq) + 2\ H_2(g)$

First we calculate the amount of $H_2(g)$ needed and then the mass of $CaH_2(s)$ required.

$$\text{mol}\ H_2 = \frac{PV}{RT} = \frac{722\ \text{mm Hg} \times \dfrac{1\ \text{atm}}{760\ \text{mmHg}} \times 235\ \text{L}}{0.08206\ \text{L atm mol}^{-1}\text{K}^{-1} \times (273.2 + 19.7)\text{K}} = 9.29\ \text{mol}\ H_2$$

$$\text{mass}\ CaH_2 = 9.29\ \text{mol}\ H_2 \times \frac{1\ \text{mol}\ CaH_2}{2\ \text{mol}\ H_2} \times \frac{42.09\ \text{g}\ CaH_2}{1\ \text{mol}\ CaH_2} = 196\ \text{g}\ CaH_2$$

Integrative and Advanced Exercises

65. To convert from volume percent (numerically the same as mole percent) to mass percent, each volume fraction is multiplied by the molar mass of that component and then divided by the apparent molar mass of the gas. This means that substances with higher molar masses (such as O_2, 32.00 g/mol) will have higher mass percents than volume percents, compared to substances with lower molar masses (such as N_2, 28.01 g/mol).

66. The relative humidity compares the actual pressure exerted by water vapor, P_{water}, to the vapor pressure of water at a particular temperature.

$$P_{H_2O} = \frac{\left(0.25 \text{ mL} \times \dfrac{0.998 \text{ g}}{1 \text{ mL}} \times \dfrac{1 \text{ mol } H_2O}{18.02 \text{ g}}\right) \times \dfrac{0.08206 \text{ L atm}}{\text{mol K}} \times (20+273) \text{ K}}{18.5 \text{ L}} \times \frac{760 \text{ mmHg}}{1 \text{ atm}} = 13.\underline{7} \text{ mmHg}$$

$$\% \text{ relative humidity} = \frac{13._7 \text{ mmHg from water}}{17.5 \text{ mmHg}} \times 100\% = 78\% \text{ relative humidity}$$

67. The pressure of the two combined gases, $H_2(g)$ and $O_2(g)$, equals the total pressure minus the vapor pressure of water. The balanced chemical equation for the electrolysis indicates that 3 moles of gas are formed from each mole of water consumed. $2 H_2O(l) \xrightarrow{\text{electrolysis}} 2 H_2(g) + O_2(g)$ The ideal gas equation is used to determine the volume of the combined gases.

$$V = \frac{nRT}{P} = \frac{\left(17.3 \text{ g } H_2O \times \dfrac{1 \text{ mol } H_2O}{18.02 \text{ g } H_2O} \times \dfrac{3 \text{ mol gas}}{2 \text{ mol } H_2O}\right) \times \dfrac{0.08206 \text{ L atm}}{\text{mol K}} \times (23+273.2) \text{ K}}{(755-20.5) \text{ mmHg} \times \dfrac{1 \text{ atm}}{760 \text{ mmHg}}} = 36.2 \text{ L gas}$$

68. As we have seen before, permanganate ion is an oxidizing agent. It does not react with nitrate ion, NO_3^-, because nitrogen is in its highest common oxidation state. It will, however, oxidize nitrite ion to nitrate ion, which is colorless, and in the process be reduced to $Mn^{2+}(aq)$, which is very pale pink.

Oxidation: $\{ NO_2^-(aq) + H_2O(l) \rightarrow NO_3^-(aq) + 2 H^+(aq) + 2 e^- \} \times 5$

Reduction: $\underline{\{MnO_4^-(aq) + 8 H^+(aq) + 5 e^- \rightarrow Mn^{2+}(aq) + 4 H_2O(l)\} \times 2}$

$5 NO_2^-(aq) + 2 MnO_4^-(aq) + 6 H^+(aq) \rightarrow 5 NO_3^-(aq) + 2 Mn^{2+}(aq) + 3 H_2O(l)$

69. First we balance the equation, then determine the number of millimoles of $NH_3(g)$ that are produced, and finally find the $[NO_3^-]$ in the original solution.

The skeleton half-equations: $NO_3^-(aq) \longrightarrow NH_3(g) \qquad Zn(s) \longrightarrow Zn(OH)_4^{2-}(aq)$

Balance O's and H's: NO_3^- (aq) $\longrightarrow NH_3$(g) $+ 3 H_2O$(l)

$$NO_3^- \text{ (aq)} + 9 H^+ \text{ (aq)} \longrightarrow NH_3\text{(g)} + 3 H_2O\text{(l)}$$

Balance charge and add OH^-(aq) NO_3^- (aq) $+ 6 H_2O$(l) $+ 8 \; e^- \longrightarrow NH_3$(g) $+ 9 OH^-$(aq)

Add OH^- (aq)'s and then electrons: $Zn(s) + 4 OH^-$ (aq) $\longrightarrow Zn(OH)_4^{2-}$ (aq) $+ 2 e^-$

Oxidation: { $Zn(s) + 4 OH^-$ (aq) $\longrightarrow Zn(OH)_4^{2-}$ (aq) $+ 2 e^-$ } $\times 4$

Reduction: NO_3^-(aq) $+ 6 H_2O$(l) $+ 8 e^- \longrightarrow NH_3$(g) $+ 9 OH^-$ (aq)

Net: NO_3^-(aq) $+ 4 Zn(s) + 6 H_2O$(l) $+ 7 OH^-$ (aq) $\longrightarrow Zn(OH)_4^{2-}$ (aq) $+ NH_3$(g)

The titration reactions are the following.

HCl(aq) $+$ NaOH(aq) $\longrightarrow$ NaCl(aq) $+ H_2O$(l) NH_3(aq) $+$ HCl(aq) $\longrightarrow NH_4$Cl(aq)

$$\text{mmol excess HCl} = 32.10 \text{ mL} \times \frac{0.1000 \text{ mmol NaOH}}{1 \text{ mL soln}} \times \frac{1 \text{ mmol HCl}}{1 \text{ mmol NaOH}} = 3.210 \text{ mmol HCl}$$

$$\text{mmol HCl at start} = 50.00 \text{ mL} \times \frac{0.1500 \text{ mmol HCl}}{1 \text{ mL soln}} = 7.500 \text{ mmol HCl}$$

$$\text{mmol } NH_3 \text{ produced} = (7.500 - 3.210) \text{ mmol HCl} \times \frac{1 \text{ mmol } NH_3}{1 \text{ mmol HCl}} = 4.290 \text{ mmol } NH_3$$

$$[NO_3^-] = \frac{4.290 \text{ mmol } NH_3 \times \dfrac{1 \text{ mmol } NO_3^-}{1 \text{ mmol } NH_3}}{25.00 \text{ mL soln}} = 0.1716 \text{ M}$$

Notice that it was not necessary to balance the equation, since NO_3^- and NH_3 are the only nitrogen-containing species involved, and thus they must be in a one-to-one molar ratio.

70. First we compute the root-mean-square speed of O at 1500 K.

$$u_{\text{rms}} = \sqrt{\frac{3RT}{M}} = \sqrt{\frac{3 \times 8.314 \text{ J mol}^{-1}\text{K}^{-1} \times 1500 K}{0.0160 \text{ kg/mol}}} = 1.5 \times 10^3 \text{ m/s}$$

Kinetic Energy $=$

$$KE = \tfrac{1}{2} mu^2 = 0.5 \times \frac{0.0160 \text{ kg/mol}}{6.022 \times 10^{23} \text{ atoms/mol}} \times (1.5 \times 10^3 \text{ m/s})^2 = 3.0 \times 10^{-20} \text{ J/atom}$$

71. Reduction: { $2 NO_3^-$(aq) $+ 10 H^+$(aq) $+ 8 e^- \rightarrow N_2O$(g) $+ 5 H_2O$(l) } $\times 7$

Oxidation: { $C_6H_{11}O$(aq) $+ 3 H_2O$(l) $\rightarrow$ HOOC$(CH_2)_4$COOH(aq) $+ 7 H^+$(aq) $+ 7 e^-$} $\times$ 8

Net: $14 NO_3^-$(aq) $+ 14 H^+$(aq) $+ 8 C_6H_{11}O$(aq) $\rightarrow 7 N_2O$(g) $+ 8$ HOOC$(CH_2)_4$COOH(aq) $+ 11 H_2O$(l)

72. The Lewis structures for the two compounds are drawn below. The four bonded groups in XeO_4 and no lone pairs on the central atom give it a tetrahedral shape in which there are four oxygen atoms on the periphery of the molecule. These rounded tetrahedral should not stick well together. In fact, each oxygen is slightly negatively charged and they should repel each other. Thus, we expect weak intermolecular forces between XeO_4 molecules. In XeO_3, on the other hand, the three bonds and one lone pair produces a trigonal pyramidal molecule in which the central Xe (carrying a slight positive charge), is exposed to other molecules. There thus can be strong dipole-dipole forces leading to strong intermolecular forces and a relatively high boiling point vis-à-vis XeO_4.

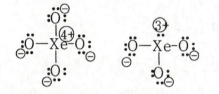

Note: You can also expand the octets to reduce formal charges. There are a number of resonance forms, however, Lewis structures(octets) are shown.

73. The N center in ammonium ion is the reducing agent, while Cl in perchlorate anion is the oxidizing agent.

$$2\,NH_4ClO_4\,(s) \longrightarrow N_2\,(g) + 4\,H_2O(g) + Cl_2\,(g) + 2\,O_2\,(g)$$

74. $\Delta H°_f$ = Bonds broken in reactants – Bonds formed in products

(a) $\frac{1}{2}\,Cl_2(g) + \frac{1}{2}\,F_2(g) \rightarrow Cl$—$F(g)$ $\Delta H°_f = \frac{1}{2}\,(159\ kJ) + \frac{1}{2}\,(243\ kJ) - 251\ kJ = -50\ kJ$

(b) $\frac{1}{2}\,O_2(g) + F_2(g) \rightarrow F$—$O$—$F(g)$ $\Delta H°_f = \frac{1}{2}\,(498\ kJ) + (159\ kJ) - 2(213\ kJ) = -18\ kJ$

(c) $\frac{1}{2}\,O_2(g) + Cl_2(g) \rightarrow Cl$—$O$—$Cl(g)$ $\Delta H°_f = \frac{1}{2}\,(498\ kJ) + (243\ kJ) - 2(205\ kJ) = +82\ kJ$

(d) $\frac{1}{2}\,N_2(g) + \frac{3}{2}\,F_2(g) \rightarrow NF_3(g)$ $\Delta H°_f = \frac{1}{2}\,(946\ kJ) + 1.5\,(159\ kJ) - 3(280\ kJ) = -128.\underline{5}\ kJ$

75 The electrode reaction is: $F_2\,(g) + 2\,e^- \longrightarrow 2\,F^-\,(aq)$

$\Delta G° = 2\,\Delta G_f°[F^-\,(aq)] - \Delta G_f°[F_2\,(g)] = 2(-278.8\ kJ/mol) - (0\ kJ/mol)$

$\Delta G° = -557.6\ kJ/mol = -2nFE°$

$E° = \dfrac{-557.6 \times 10^3\ J/mol}{2 \times 96{,}485\ C/mol} = +2.890\ V$

This value compares favorably with the value of +2.866 V in Appendix D.

76. Each simple cubic unit cell has one Po atom at each of its eight corners, but each corner is shared among eight unit cells. Thus, there is a total of one Po atom per unit cell. The edge of that unit cell is 335 pm. From this information we obtain the density of polonium.

$$\text{density} = \dfrac{1\,\text{Po atom} \times \dfrac{1\,\text{mol Po}}{6.022 \times 10^{23}\,\text{Po atoms}} \times \dfrac{209\,\text{g Po}}{1\,\text{mol Po}}}{\left(335\,\text{pm} \times \dfrac{1\,\text{m}}{10^{12}\,\text{pm}} \times \dfrac{100\,\text{cm}}{1\,\text{m}}\right)^3} = 9.23\ g/cm^3$$

77. First we write the molecular orbital diagram of each of the species. From each molecular orbital diagram we determine the number of bonding electrons (b) and the number of antibonding electrons (*) and thus the bond order [(number of bonding electrons — number of antibonding electrons) $\div 2$]. Species with high bond order have strong bonds, which are short. Those with low bond order have weak bonds, which are long.

# of e⁻	σ^b_{1s}	σ^*_{1s}	σ^b_{2s}	σ^*_{2s}	σ^b_{2p}	π^b_{2p}	π^*_{2p}	σ^*_{2p}	Antibonding e⁻	Bond order
O_2^+ $(2 \times 8) - 1 = 15$	↑↓	↑↓	↑↓	↑↓	↑↓	↑↓|↑↓	↑|	☐	5	2.5
O_2 $(2 \times 8) = 16$	↑↓	↑↓	↑↓	↑↓	↑↓	↑↓|↑↓	↑|↑	☐	6	2
O_2^- $(2 \times 8) + 1 = 17$	↑↓	↑↓	↑↓	↑↓	↑↓	↑↓|↑↓	↑↓|↑	☐	7	1.5
O_2^{2-} $(2 \times 8) + 2 = 18$	↑↓	↑↓	↑↓	↑↓	↑↓	↑↓|↑↓	↑↓|↑↓	☐	8	1.0

We have only listed the number of antibonding electrons above because, for each species, the number of bonding electrons is the same, namely 10. Recall that the bond order is determined as follows: bond order = (number of bonding electrons — number of antibonding electrons) $\div 2$

(a) In order of increasing bond distance: $O_2^+ < O_2 < O_2^- < O_2^{2-}$

(b) In order of increasing bond strength: $O_2^{2-} < O_2^- < O_2 < O_2^+$

78. (a)

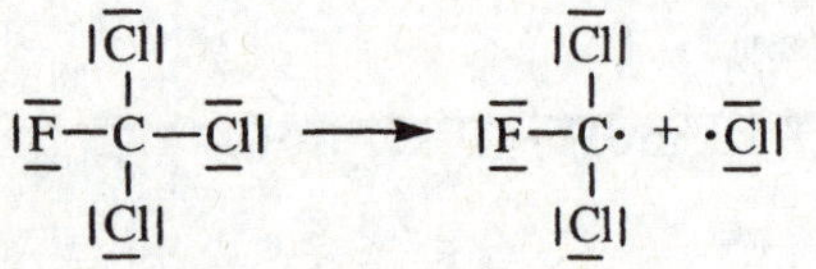

As shown in the Lewis structures above, the net result of this reaction is breaking a C—Cl bond. From Table 10-3, the energy of this bond is 339 kJ/mol, and this must be the energy of the photons involved in the reaction.

(b) $E = h\nu$ or $\nu = E/h$ on a molecular basis. Thus, we have the following.

$$\nu = \frac{339 \times 10^3 \text{ J/mol}}{6.626 \times 10^{-34} \text{ J} \cdot \text{s} \times 6.022 \times 10^{23} \text{ / mol}} = 8.50 \times 10^{14} \text{ s}^{-1}$$

$$\lambda = \frac{c}{\nu} = \frac{3.00 \times 10^8 \text{ m/s}}{8.50 \times 10^{14} \text{ / s}} = 3.53 \times 10^{-7} \text{ m} \times \frac{10^9 \text{ nm}}{1 \text{ m}} = 353 \text{ nm}$$

This radiation is in the near ultraviolet region of the electromagnetic spectrum.

79. (a) % P indicates the number of grams of P per 100 g of material, while % P_4O_{10} indicates number of grams of P_4O_{10} per 100 g of material.

$$\text{mass P} = 1.000 \text{ g } P_4O_{10} \times \frac{1 \text{ mol } P_4O_{10}}{283.89 \text{ g } P_4O_{10}} \times \frac{4 \text{ mol P}}{1 \text{ mol } P_4O_{10}} \times \frac{30.974 \text{ g P}}{1 \text{ mol P}} = 0.436 \text{ g P}$$

Thus, multiplying the mass of P_4O_{10} by 0.436 will give the mass (or mass percent) of P. %BPL indicates the number of grams of $Ca_3(PO_4)_2$ per 100 g of material.

$$\text{mass } Ca_3(PO_4)_2 = 1.000 \text{ g } P_4O_{10} \times \frac{1 \text{ mol } P_4O_{10}}{283.89 \text{ g } P_4O_{10}} \times \frac{4 \text{ mol P}}{1 \text{ mol } P_4O_{10}}$$

$$\times \frac{1 \text{ mol } Ca_3(PO_4)_2}{2 \text{ mol P}} \times \frac{310.18 \text{ g } Ca_3(PO_4)_2}{1 \text{ mol } Ca_3(PO_4)_2} = 2.185 \text{ g } Ca_3(PO_4)_2$$

Thus, multiplying the mass of P_4O_{10} (283.88) by 2.185 will give the mass (or mass %) of BPL.

(b) A %BPL greater than 100% means that the material has a larger %P than does pure $Ca_3(PO_4)_2$.

(c) $$\%P = \frac{6 \text{ mol P}}{1 \text{mol } 3Ca_3(PO_4)_2 \cdot CaF_2} \times \frac{1 \text{ mol } 3Ca_3(PO_4)_2 \cdot CaF_2}{1008.6 \text{ g } 3Ca_3(PO_4)_2 \cdot CaF_2} \times \frac{30.974 \text{ g P}}{1 \text{ mol P}} \times 100\% = 18.43\% \text{ P}$$

$$\%P_4O_{10} = \frac{\%P}{0.436} = \frac{18.43}{0.436} = 42.3\% \text{ } P_4O_{10}$$

$$\%BPL = 2.185 \times \%P_4O_{10} = 2.185 \times 42.3\% \text{ } P_4O_{10} = 92.4\% \text{ BPL}$$

80. The chemical equation is $Cl_2(g) \rightleftharpoons 2\,Cl(g)$. From Appendix D,

$\Delta H_f^\circ[Cl(g)] = 121.7 \text{ kJ/mol}$, 243.4 kJ/2 mol Cl or 243.4 kJ/mol Cl_2.

$\Delta S_{rxn}^\circ = 2S^\circ[Cl(g)] - S^\circ[Cl_2(g)] = (2 \times 165.2) - 223.1 = 107.3 \text{ J K}^{-1}$.

We assume the values of ΔH° and ΔS° are unchanged from 298 K to 1000 K, and we calculate the value of ΔG° at 1000 K.

$$\Delta G^\circ = \Delta H^\circ - T\Delta S^\circ = 243.4 \text{ kJ} - (1000 \text{ K} \times 0.1073 \text{ kJ K}^{-1}) = 136.1 \text{ kJ} \qquad \Delta G^\circ = -RT \ln K_p$$

$$\ln K_p = \frac{-\Delta G^\circ}{RT} = \frac{-136.2 \times 10^3 \text{ J/mol}}{8.314 \text{ J mol}^{-1} \text{ K}^{-1} \times 1000 \text{ K}} = -16.37 \qquad K_p = e^{-16.37} = 7.8 \times 10^{-8}$$

Since the value of K_p is so small, we assume the initial $Cl_2(g)$ pressure is 1.00 atm.

Reaction:	$Cl_2(g)$	$\rightleftharpoons$	$2Cl(g)$
Initial:	1.00 atm		0 atm
Changes:	$-x$ atm		$+2x$ atm
Equil:	$(1.00-x)$		$2x$ atm

$$K_p = \frac{P_{Cl}^2}{P_{Cl_2}} = 7.8 \times 10^{-8} = \frac{(2x)^2}{1.00-x} \approx \frac{4x^2}{1.00}$$

$$x = \sqrt{\frac{7.8 \times 10^{-8}}{4}} = 1.4 \times 10^{-4}$$

Our assumption, that $x \ll 1.00$ atm, obviously is valid. x is the degree of dissociation. The % dissociation now can be calculated. % dissociation = $x \times 100\% = 1.4 \times 10^{-4} \times 100\%$ = 0.014 %

81. The *peroxo* prefix of peroxonitrous acid hints at the presence of a —O—O— linkage in the molecule.

nitric acid $H-\bar{\underline{O}}-N-\bar{\underline{O}}|$ peroxonitrous acid $H-\bar{\underline{O}}-\underline{O}-\bar{N}=\underline{O}$
                      $\overset{\parallel}{|\underline{O}|}$

82. The Lewis structures of these two compounds are similar. If we designate CH_3 as Me and SiH_3 as Sl, the two Lewis structures are $Me-\bar{N}-Me$ $Sl-\bar{N}-Sl$
                                              $\overset{|}{Me}$ $\overset{|}{Sl}$

The pyramidal structure of $N(CH_3)_3$ indicates a tetrahedral electron pair geometry, with sp^3 hybridization for N in $N(CH_3)_3$ and the lone pair of electrons residing in an sp^3 *hybrid* orbital. The N—C bond involves the overlap of $N(sp^3)-C(sp^3)$. The C—H bonds are $C(sp^3)-H(1s)$. The planar arrangement of $N(SiH_3)_3$ indicates that the nitrogen atom is sp^2 hybridized; the lone pair is in a $2p$ atomic orbital on the central N atom, perpendicular to the $N-Si$ plane in the center of the molecule. If we assume that the Si atom is sp^3 hybridized (to account for the four bonds each Si atom forms), the N—Si bond involves the overlap between $N(sp^2)-Si(sp^3)$. Thus, the Si—H bonds are $Si(sp^3)-H(1s)$.

83. pH = 3.5 means $[H^+] = 10^{-3.5} = 3 \times 10^{-4}$ M. This is quite a dilute acidic solution, and we expect H_2SO_4 to be completely ionized under these circumstances.

$$\text{mass } H_2SO_4 = 1.00 \times 10^3 \text{ L} \times \frac{3 \times 10^{-4} \text{ mol } H^+}{1 \text{ L}} \times \frac{1 \text{ mol } H_2SO_4}{2 \text{ mol } H^+} \times \frac{98.1 \text{ g } H_2SO_4}{1 \text{ mol } H_2SO_4} = 15 \text{ g } H_2SO_4$$

$$\text{mass } Cl_2 = 1.00 \times 10^3 \text{ L} \times \frac{1000 \text{ cm}^3}{1 \text{ L}} \times \frac{1.03 \text{ g}}{1 \text{ cm}^3} \times \frac{70 \text{ g } Br_2}{10^6 \text{ g seawater}} \times \frac{1 \text{ mol } Br_2}{159.8 \text{ g } Br_2} \times \frac{1 \text{ mol } Cl_2}{1 \text{ mol } Br_2}$$

$$\times \frac{70.9 \text{ g } Cl_2}{1 \text{ mol } Cl_2} \times \frac{115 \text{ g } Cl_2 \text{ used}}{100 \text{ g } Cl_2 \text{ needed}} = 37 \text{ g } Cl_2$$

84. $E = E^o - \dfrac{0.0592}{n} \log_{10} Q$; assuming E = 0 for a process that is no longer spontaneous:

$$\log_{10} Q = \frac{nE^o}{0.0592} = \frac{(16)(0.065)}{0.0592} = 17.\underline{6} \qquad\qquad P_{SO_2} = 1 \times 10^{-6}$$

$$Q = 10^{17.\underline{6}} = 3.9 \times 10^{17} = \frac{(P_{SO_2})^8}{1^8 (H^+)^{16}} = \frac{(1 \times 10^{-6})^8}{1^8 (H^+)^{16}}$$

Solving for $[H^+]$ yields a value of 8×10^{-5} M, which corresponds to a pH of 4.1

Thus, the solution is still acidic.

85. In the process of forming XeF_2 and $XeCl_2$, either a Cl—Cl bond or a F—F bond is broken. We note that the F—F bond is much weaker than the Cl—Cl bond and thus much less energy is required to break it. Since greater stability indicates that the resulting product is of lower energy, the greater stability for XeF_2 compared to $XeCl_2$ can be partially explained by the need to expend less energy to break the bonds of the reactants. (Another reason, of course, is that the shorter Xe—F bond is stronger, and hence more stable, than the somewhat longer Xe—Cl bond.)

86 Oxidations : $\quad XeF_4(g) + 3H_2O \longrightarrow XeO_3(g) + 4HF(aq) + 2H^+(aq) + 2e^-$

$\qquad\qquad\qquad \{2H_2O(l) \longrightarrow O_2(g) + 4H^+(aq) + 4e^-\} \times \frac{3}{2}$

Reduction: $\quad \{XeF_4(g) + 4e^- \longrightarrow Xe(g) + 4F^-(aq)\} \qquad \times 2$

Net: $\qquad\quad \overline{3XeF_4(g) + 6H_2O(l) \longrightarrow 2XeO_3(g) + 12HF(aq) + \frac{3}{2}O_2(g) + 2Xe(g)}$

The fact that O_2 is also produced indicates that there are two oxidation half-reactions. The production ratio of Xe and XeO_3 indicates the amount by which the other two half-reactions must be multiplied before they are added. Then the half-reaction for the production of O_2 is multiplied by $\frac{3}{2}$ to balance charge.

87. $S(s) + 2H^+(aq) + 2e^- \rightarrow H_2S(g) \qquad E° = 0.174\ V$
    $S(s) + 2H^+(aq) + 2e^- \rightarrow H_2S(aq) \qquad E° = 0.144\ V$

The difference is that one results in an aqueous solution being formed, while the other gives a gaseous product. They can both be correct because the product has a different phase (the products have different $\Delta G°_f$ values, hence one would expect different $E°$ values).

88. The initial concentration of $Cl_2(aq) = \dfrac{6.4g}{70.9g/mol} = 0.090M$

$$Cl_2(aq) + H_2O(l) \xrightleftharpoons{\ \ K_c = 4.4 \times 10^{-4}\ \ } HOCl(aq) + H^+(aq) + Cl^-(aq)$$

	$Cl_2(aq)$	$H_2O(l)$		$HOCl(aq)$	$H^+(aq)$	$Cl^-(aq)$
initial	0.090 M			0 M	0 M	0 M
equil.	(0.090-x) M			x M	x M	x M

{where x is the molar solubility of Cl_2 in water}

$$K = \frac{[HOCl(aq)][H^+(aq)][Cl^-(aq)]}{[Cl_2(aq)]} = \frac{x^3}{0.090-x} = 4.4 \times 10^{-4}$$

By using successive approximations, we find $x = 0.030$ M, so:

$[HOCl(aq)] = [H^+(aq)] = [Cl^-(aq)] = 0.030$ M and $[Cl_2(aq)] = 0.090 - 0.030 = 0.060$ M

89. $2e^- + N_2(g) + 4\,H^+(aq) + 2H_2O(l) \rightarrow 2\,NH_3OH^+(aq)$ $E^\circ = -1.87\,V$

$2\,NH_3OH^+(aq) + H^+(aq) + 2e^- \rightarrow N_2H_5^+(aq) + 2\,H_2O(l)$ $E^\circ = 1.42\,V$

$\underline{N_2H_5^+(aq) + 3\,H^+(aq) + 2e^- \rightarrow 2\,NH_4^+(aq)}$ $E^\circ = 1.275\,V$

$N_2(g) + 8\,H^+(aq) + 6e^- \rightarrow 2\,NH_4^+(aq)$ $E^\circ = 0.275\,V$ (see below)

$\Delta G^\circ total = \Delta G^\circ_1 + \Delta G^\circ_2 + \Delta G^\circ_3 = -n_{tot}FE^\circ_{tot} = -nFE^\circ_1 - nFE^\circ_2 - nFE^\circ_3$

$$E^\circ_{tot} = \frac{n_1 E^\circ_1 + n_2 E^\circ_2 + n_3 E^\circ_3}{n_{tot}}$$

$E^\circ_{tot} = [2(-1.87) + 2(1.42) + 2(1.275)]/6 = 0.275\,V$

moving on...

$2\,HN_3(aq) \rightarrow 3\,N_2(g) + 2H^+(aq) + 2e^-$ $E^\circ = +3.09\,V$ (oxidation)

$\underline{3\,N_2(g) + 24\,H^+(aq) + 18e^- \rightarrow 6\,NH_4^+(aq)}$ $E^\circ = 0.275\,V$ (top equation multiplied by 3)

$2\,HN_3(aq) + 22\,H^+(aq) + 16\,e^- \rightarrow 6\,NH_4^+(aq)$ [*sum*]

[divide by 2]

$HN_3(aq) + 11\,H^+(aq) + 8\,e^- \rightarrow 3\,NH_4^+(aq)$ $E^\circ = 0.70\,V$

since

$\Delta G^\circ_{total} = \Delta G^\circ_1 + \Delta G^\circ_2 = -nFE^\circ_{tot} = -nFE^\circ_1 - nFE^\circ_2$

$E^\circ_{total} = [2(3.09) + 18(0.275)]/16 = 0.70\,V$

FEATURE PROBLEMS

90. The goal here is to demonstrate that the three reactions result in the decomposition of water as the net reaction: Net: $2\,H_2O \rightarrow 2\,H_2 + O_2$ First balance each equation.

(1) $3\,FeCl_2 + 4\,H_2O \rightarrow Fe_3O_4 + HCl + H_2$ Balance by inspection. Notice that there are

3 Fe and 4 O on the right-hand side. Then balance Cl.

$\overline{3\,FeCl_2 + 4\,H_2O \rightarrow Fe_3O_4 + 6\,HCl + H_2}$

(2) $Fe_3O_4 + HCl + Cl_2 \rightarrow FeCl_3 + H_2O + O_2$ Try the half-equation method.

$Cl_2 + 2\,e^- \rightarrow 2\,Cl^-$ But realize that $Fe_3O_4 = Fe_2O_3 \cdot FeO$.

Only iron(II) needs to be

$2\,FeO \rightarrow 2\,Fe^{3+} + O_2 + 6\,e^-$ Now combine the two half-equations.

$2\,FeO + 3\,Cl_2 \rightarrow 2\,Fe^{3+} + O_2 + 6\,Cl^-$

Add in $2\,Fe_2O_3 + 12\,H^+ \rightarrow 4\,Fe^{3+} + 6\,H_2O$

$$2 Fe_3O_4 + 3 Cl_2 + 12 H^+ \rightarrow 6 Fe^{3+} + O_2 + 6 Cl^- + 6 H_2O$$

And 12 Cl^- spectators: $2 Fe_3O_4 + 3 Cl_2 + 12 HCl \rightarrow 6 FeCl_3 + O_2 + 6 H_2O$

(3) $FeCl_3 \rightarrow FeCl_2 + Cl_2$ by inspection $2 FeCl_3 \rightarrow 2 FeCl_2 + Cl_2$

One strategy is to consider each of the three equations and the net equation. Only equation (1) produces hydrogen. Thus, we must run it twice. Only equation (2) produces oxygen. Since only one mole of $O_2(g)$ is needed, we only have to run it once. Equation (3) can balance out the Cl_2 required by equation (2), but we have to run it three times to cancel all the $Cl_2(g)$.

$2 \times (1)$ $6 FeCl_2(s) + 8 H_2O(l) \rightarrow 2 Fe_3O_4(s) + 12 HCl(l) + 2 H_2(g)$

$1 \times (2)$ $2 Fe_3O_4(s) + 3 Cl_2(g) + 12 HCl(g) \rightarrow 6 FeCl_3(s) + O_2(g) + 6 H_2O(l)$

$3 \times (3)$ $6 FeCl_3(s) \rightarrow 6 FeCl_2(s) + 3 Cl_2(g)$

Net: $2 H_2O(l) \rightarrow 2 H_2(g) + O_2(g)$

91. We begin by calculating the standard voltages for the two steps in the decomposition mechanism. Step 1 involves the reduction of Fe^{3+} and the oxidation of H_2O_2. The two half-reactions that constitute this step are:

(i) $Fe^{3+}(aq) + e^- \rightleftharpoons Fe^{2+}(aq)$ $\qquad\qquad$ $E^\circ_{1,red} = 0.771$ V

(ii) $H_2O_2(aq) \rightleftharpoons O_2(g) + 2 H^+(aq) + 2 e^-$ $\qquad$ $E^\circ_{2,ox} = -0.695$ V

The balanced reaction is obtained by combining reaction (i), multiplied by two, with reaction (ii).

2(i) $2(Fe^{3+}(aq) + e^- \rightleftharpoons Fe^{2+}(aq))$

(ii) $H_2O_2(aq) \rightleftharpoons O_2(g) + 2 H^+(aq) + 2 e^-$

(iii) $2 Fe^{3+}(aq) + H_2O_2(aq) \rightleftharpoons 2 Fe^{2+}(aq) + O_2(g) + 2 H^+(aq)$,

for which $E^\circ_{cell} = E^\circ_{i(red)} + E^\circ_{ii(ox)} = 0.771$ V $+ (-0.695$ V$) = 0.076$ V

Since the overall cell potential is positive, this step is spontaneous. The next step involves oxidation of $Fe^{2+}(aq)$ back to $Fe^{3+}(aq)$, (i.e. the reverse of reaction (i) and the reduction of $H_2O_2(aq)$ to $H_2O(l)$ in acidic solution, for which the half-reaction is

(iv) $H_2O_2(aq) + 2 H^+(aq) + 2 e^- \rightleftharpoons 2 H_2O(l)$ $\qquad$ $E^\circ_{iv(red)} = 1.763$ V

Combining (iv) with two times the reverse of (i) gives the overall reaction for the second step:

(i) $\{Fe^{2+}(aq) \rightleftharpoons Fe^{3+}(aq) + e^-\} \times 2$

iv) $H_2O_2(aq) + 2 H^+(aq) + 2 e^- \rightleftharpoons 2 H_2O(l)$

(v) $2 Fe^{2+}(aq) + H_2O_2(aq) + 2 H^+(aq) \rightleftharpoons 2 Fe^{3+}(aq) + 2 H_2O(l)$

Thus the overall cell potential for the second step in the mechanism, via equation (v) is

$E°_{cell} = -E°_{i(red)} + E°_{iv(red)} = -0.771$ V $+ (1.763$ V$) = 0.992$ V.

Since the overall standard cell potential is positive, like step 1, this reaction is spontaneous.

The overall reaction arising from the combination of these two steps is:

Step 1 $2 Fe^{3+}(aq) + H_2O_2(aq) \rightleftharpoons 2 Fe^{2+} + (aq) + O_2(g) + 2 H^+(aq)$

Step 2 $2 Fe^{2+}(aq) + H_2O_2(aq) + 2 H^+(aq) \rightleftharpoons 2 Fe^{3+}(aq) + 2 H_2O(l)$

Overall $2 H_2O_2(aq) \rightleftharpoons O_2(g) + 2 H_2O(l)$

The overall potential, $E°_{overall} = E°_{step1} + E°_{step2} = 0.076$ V $+ 0.992$ V $= 1.068$ V
Therefore, the reaction is spontaneous at standard conditions.

To determine the minimum and maximum $E°$ values necessary for the catalyst, we need to consider each step separately.

In step 1, if $E°_{(1)}$ is less than 0.695 V, the overall voltage for the first step will be negative and hence non-spontaneous. In step 2, if the oxidation half-reaction has a potential that is more negative than -1.763 V, the overall potential for this step will be negative, and hence non-spontaneous. Consequently, $E°_{(1)}$ must fall between 0.695 V and 1.763 V in order for both steps to be spontaneous. On this basis we find that

(a) $Cu^{2+}(aq) + 2 e^- \rightleftharpoons Cu(s)$ $E°_{1/2red} = 0.337$ V cannot catalyze the reaction.

(b) $Br_2(l) + 2 e^- \rightleftharpoons 2 Br^-(aq)$ $E°_{1/2red} = 1.065$ V may catalyze the reaction.

(c) $Al^{3+}(aq) + 3 e^- \rightleftharpoons Al(s)$ $E°_{1/2red} = -1.676$ V cannot catalyze the reaction.

(d) $Au^{3+}(aq) + 2 e^- \rightleftharpoons Au^+(s)$ $E°_{1/2red} = 1.36$ V may catalyze the reaction.

In the reaction of hydrogen peroxide with iodic acid in acidic solution, the relevant half-reactions are:

$2 IO_3^-(aq) + 12 H^+(aq) + 10 e^- \rightleftharpoons I_2(s) + 6 H_2O(l)$ $E°_{1/2red} = 1.20$ V

$H_2O_2(aq) \rightleftharpoons O_2(g) + 2 H^+(aq) + 2e^-$ $E°_{1/2ox} = -0.695$ V

Thus, both steps in the decomposition of H_2O_2, as described above, are spontaneous if IO_3^- is used as the catalyst. As the iodate gets reduced to I_2, the I_2 forms a highly colored complex with the starch, resulting in the appearance of a deep blue color solution. Some of the H_2O_2 will be simultaneously oxidized to $O_2(g)$ by the iodic acid When sufficient $I_2(s)$ accumulates, the reduction of H_2O_2 by I_2 begins to take place and the deep blue color fades as I_2 is consumed in the reaction. Additional iodate is formed concurrently, and this goes on to oxidize the $H_2O_2(aq)$, thereby causing the cycle to repeat itself. Each cycle results in some H_2O_2 being depleted. Thus, the oscillations of color can continue until the H_2O_2 has been largely consumed.

92.

$$CIO_3^- \xrightarrow{\text{(?)}} CIO_2 \xrightarrow{\text{(?)}} HCIO_2$$

$$\underbrace{}_{1.1181\ V}$$

In order to add CIO_2 to the Latimer diagram drawn above, we must calculate the voltages denoted by (?) and (??). The equation associated with the reduction potential (?) is

   (i) $2\ H^+(aq) + CIO_3^-(aq) + 1\ e^- \rightarrow CIO_2(g) + H_2O(l)$
          The standard voltage for this half-reaction is given in Appendix D:

To finish up this problem, we just need to calculate the standard voltage (??) for the half-reaction (ii):

   (ii) $H^+(aq) + CIO_2(aq) + 1\ e^- \rightarrow HCIO_2(g)$ $E° = $ (??)

To obtain the voltage for reaction (ii), we need to subtract reaction (i) from reaction (iii) below, which has been taken from Figure 22-2:

   (iii) $3\ H^+(aq) + CIO_3^-(aq) + 2\ e^- \rightleftharpoons HCIO_2(g) + H_2O(l)$ $E° = 1.181\ V$

Thus,

   (iii) $3\ H^+(aq) + CIO_3^-(aq) + 2\ e^- \rightleftharpoons HCIO_2(g) + H_2O(l)$ $E°(iii) = 1.181\ V$
   $-1\times$ (i) $2\ H^+(aq) + CIO_3^-(aq) + 1\ e^- \rightarrow CIO_2(g) + H_2O(l)$ $E°(i) = 1.175\ V$

   Net (ii) $H^+(aq) + CIO_2(g) + 1\ e^- \rightarrow HCIO_2(g)$ $E°(ii) = $ (??)

Since reactions (i) and (iii) are both reduction half reactions, we cannot simply subtract the potential for (i) from the potential for (iii). Instead, we are forced to obtain the voltage for (ii) via the free energy changes for the three half reactions. Thus,

$\Delta G(ii) = \Delta G(iii) - \Delta G(i) = -1FE°(ii) = -2F(1.181\ V) + 1F(1.175\ V)$

Dividing both sides by $-F$ gives

$E°(ii) = 2(1.181\ V) - 1.175\ V$ So, $E°(ii) = 1.187\ V$.

93.   (a)   As before, we can organize a solution around the balanced chemical equation.

Equation: $I_2(aq)$ $\rightleftharpoons$ $I_2(CCl_4)$

Initial:	1.33×10^{-3} M	0 M
Initial:	0.0133 mmol	0 mmol
Changes:	$-x$ mmol	$+x$ mmol
Equil:	$(0.0133 - x)$ mmol	x mmol

$$K_c = 85.5 = \frac{[I_2(CCl_4)]}{[I_2(aq)]} = \frac{\dfrac{x}{10.0\ mL}}{\dfrac{0.0133 - x}{10.0\ mL}}$$

$x = 1.13\underline{7} - 85.5\ x$ $86.5x = 1.13\underline{7}$ $x = \dfrac{1.13_7}{86.5} = 0.01315$ mmol I_2 in CCl_4

$I_2(aq) = (0.0133 - 0.01315)$ mmol $= 0.00015$ mmol

mass $I_2 = 0.00015$ mmol $\times \dfrac{253.8\ mg\ I_2}{1\ mmol\ I_2} = 3.8\times10^{-2}$ mg $I_2 = 0.038$ mg I_2 remaining

(b) We have the same set-up, except that the initial amount $I_2(aq) = 0.00015 \text{ mmol}$.

$$K_c = 85.5 = \frac{[I_2(CCl_4)]}{[I_2(aq)]} = \frac{x}{0.00015 - x}$$

$$x = 0.0128 - 85.5x \qquad 86.5x = 0.0128$$

$$x = \frac{0.0128}{86.5} = 0.0001480 \text{ mmol} = I_2(CCl_4),$$

$$I_2(aq) = 0.00015 - 0.0001480 = 2.0 \times 10^{-6} \text{ mmol } I_2$$

$$\text{mass } I_2 = 2.0 \times 10^{-6} \text{ mmol in } H_2O \times \frac{253.8 \text{ mg } I_2}{1 \text{ mol } I_2} = 5.1 \times 10^{-4} \text{ mg } I_2 = 0.00051 \text{ mg } I_2$$

(c) If twice the volume of CCl_4 were used for the initial extraction, the equilibrium concentrations would have been different from those in part (a).

Equation:	$I_2(aq)$	$\rightleftharpoons$	$I_2(CCl_4)$
Initial:	$1.33 \times 10^{-3} \text{ M}$		0 M
Initial:	0.0133 mmol		0 mmol
Changes:	$-x$ mmol		$+x$ mmol
Equil:	$(0.0133 - x)$ mmol		x mmol

$$K_c = 85.5 = \frac{[I_2(CCl_4)]}{[I_2(aq)]} = \frac{\dfrac{x}{20.0 \text{ mL}}}{\dfrac{0.0133 - x}{10.0 \text{ mL}}}$$

$$x \div 2 = 1.13\underline{7} - 85.5x \qquad 86.0x = 1.13\underline{7} \qquad x = \frac{1.137}{86.0} = 1.32 \times 10^{-2} \text{ mmo } I_2(CCl_4)$$

$$\text{total mass } I_2 = \frac{1.33 \times 10^{-3} \text{ mmol } I_2}{1 \text{ mL}} \times 10.0 \text{ mL soln} \times \frac{253.8 \text{ mg } I_2}{1 \text{ mol } I_2} = 3.376 \text{ mg } I_2$$

$$\text{mass } I_2 \text{ in } CCl_4 = 1.32\underline{2} \times 10^{-2} \text{ mmol} \times \frac{253.8 \text{ mg } I_2}{1 \text{ mol } I_2} = 3.356 \text{ mg } I_2$$

mass I_2 remaining in water $= 3.376 \text{ mg} - 3.356 \text{ mg} = 0.020 \text{ mg}$

Thus, two smaller volume extractions are much more efficient than one large volume extraction.

94. (a) The pyroanions, a series of structurally analogous anions with the general formula $X_2O_7{}^{n-}$, are known for Si, P and S. The Lewis structures for these three anions are drawn below: (Note: Every member of the series has 56 valence e^-)

$X = Si : Si_2O_7{}^{6-}$
$X = P_\oplus: P_2O_7{}^{4-}$
$X = S_{2\oplus}: S_2O_7{}^{2-}$

Based upon a VSEPR approach, we would predict tetrahedral geometry for each X atom (i.e. Si, P^+, S^{2+}) and a bent geometry for each bridging oxygen atom Therefore, a maximum of five atoms in each pyroanion can lie in a plane:

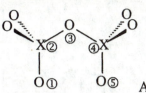

Atoms labeled 1-5 are all in the same plane.

(b) The related "mononuclear" acids, which contain just one third-row element are H_4SiO_4, H_3PO_3 and H_2SO_4. A series of pyroacids can be produced by strongly heating the "mononuclear" acids in the absence of air. In each case, the reaction proceeds via loss of water:

$$2\ H_4SiO_4 \xrightarrow{\Delta} H_6Si_2O_7(l) + H_2O(l)$$
$$2\ H_3PO_3 \xrightarrow{\Delta} H_4P_2O_7(l) + H_2O(l)$$
$$2\ H_2SO_4 \xrightarrow{\Delta} H_2S_2O_7(l) + H_2O(l)$$

(c) The highest oxidation state that Cl can achieve is VII (+7); consequently, the chlorine containing compound that is analogous to the mononuclear acids mentioned earlier is $HClO_4$. The strong heating of perchloric acid in the absence of air, should, in principle, afford Cl_2O_7, which is the neutral chlorine analogue of the pyroanions. Thus, Cl_2O_7 is the anhydride of perchloric acid.

$$2\ HClO_4 \xrightarrow{\Delta} Cl_2O_7(l) + H_2O(l)$$

95. (a) The bonding in the XeF_2 molecule can be explained quite simply in terms of a 3–center, 4 electron bond that spans all three atoms in the molecule. The bonding in this molecular orbital description involves the filled $5p_z$ orbital of Xe and the half-filled $2p_z$ orbitals of the two F-atoms. The linear combination of these three atomic orbitals affords one bonding, one non-bonding and one anti-bonding orbital, as depicted below:

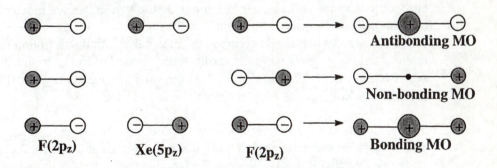

(b) If one assumes that the order of energies for molecular orbitals is:
bonding MO < non-bonding MO < antibonding MO,
the following molecular orbital representation of the bonding can be sketched.

Thus, a single bonding pair of electrons is responsible for holding all three atoms together. The non-bonding pair of electrons is localized primarily on the two F–atoms. This suggests that the bond possesses substantial ionic character. Bond order is defined as one-half the difference between the number of bonding electrons and the number of antibonding electrons. In the case of XeF_2, the overall bond order is

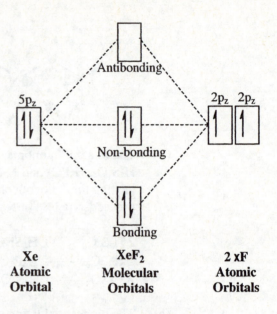

Xe
Atomic
Orbital

XeF_2
Molecular
Orbitals

2 xF
Atomic
Orbitals

therefore: $\dfrac{2e^- - 0e^-}{2e^-} = 1.0$.

(i.e., each Xe—F bond has a bond order of 0.5)

(c) By invoking a molecular orbital description based upon three-center bonding for the three F—Xe—F units in XeF_6, one obtains an octahedral structure in which there are six identical Xe—F bonds. The extra nonbonding pair of electrons is delocalized over the entire structure in this scheme. In spite of its manifest appeal, this explanation of the bonding in XeF_6 is incorrect, or at best a stretch, because the actual structure for this molecule is a distorted octahedron. A more accurate description of the stereochemistry adopted by XeF_6 is provided by VSEPR theory. In this approach, the shape of the molecule is determined by repulsions between bonding and non-bonding electrons in the valence shell of the central Xe atom.

Accordingly, in XeF_6, six bonding pairs and one lone pair of electrons surround the Xe. Having the repulsions of 7 pairs of electrons to cope with, the XeF_6 molecule adopts a distorted structure that approaches either a monocapped octahedron or a pentagonal bipyramidal arrangement of electron pairs, depending on how one chooses to view the structure. In either case, these two shapes are much closer to the true shape for XeF_6 than that predicted by the molecular orbital treatment involving three, three-center bonds. By contrast, the molecular orbital description involving three-center bonds gives far better results when applied to XeF_4. In this instance, with two three-center F—Xe—F bonds in the structure, molecular orbital theory predicts that XeF_4 should adopt a square planar structure.

The result here is quite satisfactory because XeF_4 does in fact exhibit square planar geometry. It is worth noting, however, that a square planar shape for XeF_4 is also predicted by VSEPR theory. Despite the fact that the molecular orbital method has made some inroads as of late, VSEPR is still the best approach available for rationalizing the molecular geometries of noble gas compounds.

96. **(a)** In the phase diagram sketched in the problem, extend the liquid-vapor equilibrium curve (the vapor pressure curve) to lower temperatures (supercooled liquid region). Extend the S_α-vapor equilibrium curve (sublimation curve) to the temperature at which it intersects the extended vapor pressure curve. This should come at 113 °C. Draw a line from this point of intersection to the "peak" of the S_β phase region. Erase the three lines that bound the S_β region in the original sketch. The remaining three lines would represent the equilibria, S_α-V, S_α-L, and L-V, producing the phase diagram if S_α (rhombic) were the only solid form of sulfur.

(b) If rhombic sulfur is heated rapidly, the transition to monoclinic sulfur at 95.3 °C might not occur. (Solid-state transitions are often very slow.) In this case, rhombic sulfur would melt at 113 °C, as described in the modified phase diagram in part (a). If the molten rhombic sulfur is further heated and then cooled, the liquid very likely will freeze at the equilibrium temperature of 119 °C, producing *monoclinic*, not rhombic, sulfur.

CHAPTER 23
THE TRANSITION ELEMENTS
PRACTICE EXAMPLES

1A **(a)** Cu_2O should form. $\quad\quad 2\,Cu_2S(s) + 3\,O_2(g) \rightarrow 2\,Cu_2O(s) + 2\,SO_2(g)$

(b) W(s) is the reduction product. $\quad WO_3(s) + 3\,H_2(g) \rightarrow W(s) + 3\,H_2O(g)$

(c) Hg(l) forms. $\quad\quad\quad\quad\quad\quad 2\,HgO(s) \xrightarrow{\;\Delta\;} 2\,Hg(l) + O_2(g)$

1B **(a)** $SiO_2(s)$ is the oxidation product of Si. $\quad 3\,Si(s) + 2\,Cr_2O_3(s) \xrightarrow{\;\Delta\;} 3\,SiO_2(s) + 4\,Cr(s)$

(b) Roasting is simply heating in air. $\quad 2\,Co(OH)_3(s) \xrightarrow{\;\Delta, Air\;} Co_2O_3(s) + 3H_2O(g)$

(c) $MnO_2(s)$ forms; (acidic solution). $\quad Mn^{2+}(aq) + 2H_2O(l) \rightarrow MnO_2(s) + 4H^+(aq) + 2e^-$

2A We write and combine the half-equations for oxidation and reduction. If $E° > 0$, the reaction is spontaneous.

Oxidation: $\{V^{3+}(aq) + H_2O(l) \rightarrow VO^{2+}(aq) + 2H^+(aq) + e^-\} \times 3 \quad\quad E° = -0.337V$

Reduction: $NO_3^-(aq) + 4H^+(aq) + 3e^- \rightarrow NO(g) + 2H_2O \quad\quad\quad E° = +0.956V$

Net: $NO_3^-(aq) + 3V^{3+}(aq) + H_2O(l) \rightarrow NO(g) + 3VO^{2+}(aq) + 2H^+(aq) \quad E_{cell}° = +0.619V$

Because the cell potential is positive nitric acid can be used to oxidize $V^{3+}(aq)$ to $VO^{2+}(aq)$ at standard conditions.

2B The reducing couple must have a half-cell potential of such a size and sign that a positive sum results when this half-cell potential is combined with $E°\{VO^{2+}(aq)|V^{2+}(aq)\} = 0.041V$ (this is the weighted average of the $VO^{2+}|V^{3+}$ and $V^{3+}|V^{2+}$ reduction potentials) *and* a negative sum must be produced when this half-cell potential is combined with $E°\{V^{2+}(aq)|V(s)\} = -1.13V$.

So, $-E°$ for the couple must be > -0.041 V and $< +1.13$ V (i.e. it cannot be more positive than 1.13V, nor more negative than -0.041 V). Some possible reducing couples from Table 20-1 are: $-E°\{Cr^{3+}(aq)|Cr^{2+}(aq)\} = +0.42$ V ; $-E°\{Fe^{2+}(aq)|Fe(s)\} = +0.440$ V $-E°\{Zn^{2+}(aq)|Zn(s)\} = +0.763$ V. Thus Fe(s), Cr^{2+}(aq) and Zn(s) will do the job.

EXERCISES

Properties of the Transition Elements

1. **(a)** Ti $\quad$ [Ar] $3d\,\boxed{\uparrow|\uparrow|\;|\;|\;}\,4s\,\boxed{\uparrow\downarrow}$ $\quad$ **(b)** V^{3+} [Ar] $3d\,\boxed{\uparrow|\uparrow|\;|\;|\;}\,4s\,\boxed{\;}$

(c) Cr^{2+} $\quad$ [Ar] $3d\,\boxed{\uparrow|\uparrow|\uparrow|\uparrow|\;}\,4s\,\boxed{\;}$ $\quad$ **(d)** Mn^{4+} [Ar] $3d\,\boxed{\uparrow|\uparrow|\uparrow|\;|\;}\,4s\,\boxed{\;}$

(e) Mn^{2+} $\quad$ [Ar] $3d\,\boxed{\uparrow|\uparrow|\uparrow|\uparrow|\uparrow}\,4s\,\boxed{\;}$ $\quad$ **(f)** Fe^{3+} [Ar] $3d\,\boxed{\uparrow|\uparrow|\uparrow|\uparrow|\uparrow}\,4s\,\boxed{\;}$

2.   We first give the orbital diagram for each of the species, and then count the number of unpaired electrons.

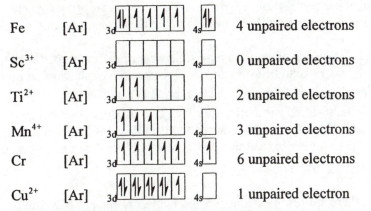

Fe   [Ar] $3d$ ⊣↓|↑|↑|↑|↑⊢ $4s$ |↑↓| 4 unpaired electrons

Sc^{3+} [Ar] $3d$ ⊣ | | | | ⊢ $4s$ | | 0 unpaired electrons

Ti^{2+} [Ar] $3d$ ⊣↑|↑| | | ⊢ $4s$ | | 2 unpaired electrons

Mn^{4+} [Ar] $3d$ ⊣↑|↑|↑| | ⊢ $4s$ | | 3 unpaired electrons

Cr   [Ar] $3d$ ⊣↑|↑|↑|↑|↑⊢ $4s$ |↑| 6 unpaired electrons

Cu^{2+} [Ar] $3d$ ⊣↑↓|↑↓|↑↓|↑↓|↑⊢ $4s$ | | 1 unpaired electron

Finally, arrange in order of decreasing number of unpaired e⁻: $Cr > Fe > Mn^{4+} > Ti^{2+} > Cu^{2+} > Sc^{3+}$

3.   A given main-group metal typically displays just one oxidation state, usually equal to its family number in the periodic table. Exceptions are elements such as Tl (+1 and +3), Pb (+2 and +4), and Sn (+2 and +4) for which the lower oxidation state represents a pair of s electrons not being ionized (a so-called "inert pair").

Main group metals do not form a wide variety of complex ions, with Al^{3+}, Sn^{2+}, Sn^{4+}, and Pb^{2+} being major exceptions. On the other hand, most transition metal ions form an extensive variety of complex ions. Most compounds of main group metals are colorless; exceptions occur when the anion is colored. On the other hand, many of the compounds of transition metal cations are colored, due to d-d electron transitions. Virtually every main-group metal cation has no unpaired electrons and hence is diamagnetic. On the other hand, many transition metals cations have one or more unpaired electrons and therefore are paramagnetic.

4.   As we proceed from Sc to Cr the valence electron configuration has an increasing number of unpaired electrons, which are capable of forming bonds to adjacent atoms. As we continue beyond Cr, however, these electrons become paired, and the resulting atoms are less able to form bonds with their neighbors. Also, smaller atoms tend to bond more tightly, making their metallic agglomerations more difficult to melt. (i.e., the smaller the metal atom, the higher the melting point). This trend of increasing melting point with decreasing size is clearly visible for the first transition series from Sc to Zn.

5.   When an electron is added to a main group element to create the element of next highest atomic number, this electron is added to a subshell that is higher in energy, which is further from the nucleus. Moreover, the added electron is well shielded from the nucleus by the filled s-subshell below it and hence it is only weakly attracted to the nucleus. Thus, this electron billows out and as a result has a major influence on the size of the atom. However, when an electron is added to a transition metal atom to create the atom of next highest atomic number, it is added to the electronic shell inside the outermost. The electron thus has been added to a position close to the nucleus to which it is attracted quite strongly. Also the electron in the same subshell as the added electron offer little shielding. Thus it has small effect on the size of the atom.

6. The reason why the radii of Pd (138 pm) and Pt (139 pm) are so similar, and yet so different from the radius of Ni (125 pm) is because the lanthanide series intrudes between Pd and Pt. Because of the lanthanide contraction, elements in the second transition row are almost identical in size to their congeners (family members) in the third transition row.

7. Of the first transition series, manganese exhibits the greatest number of different oxidation states in its compounds, namely, every state from +1 to +7. One possible explanation might be its $3d^5 4s^2$ electron configuration. Removing one electron produces an electron configuration ($3d^5 4s$) with two half-filled subshells, removing two produces one with a half-filled and an empty subshell. Then there is no point of semistability until the remaining five d electrons are removed. These higher oxidation states all are stabilized by being present in oxides (MnO_2) or oxoanions (e.g. MnO_4^-).

8. At the beginning of the series, there are few electrons beyond the last noble gas than can be ionized and thus the maximum oxidation state is limited. Toward the end of the series, many of the electrons are paired up or the d subshell is filled, and thus a somewhat stable situation would be disrupted by ionization.

9. The greater ease of forming lanthanide cations compared to forming transition metal cations, is due to the larger size of lanthanide atoms. The valence (outer shell) electrons of these larger atoms are further from the nucleus, less strongly attracted to the positive charge of the nucleus as a result, and thus are removed much more readily.

10. This is not a simple straightforward question. There are a number of factors to consider. Proceeding across the first transition series, electrons are added to the $3d$ subshell which is a shell that is beneath the $4s$ subshell, the outer most subshell in the valence shell occupied by one or two electrons. The metallic character of the first transition element, Sc, is rather similar to that of Ca (e.g., $E° = -2.03$ V compared to -2.84 V). Following this, there is a fairly regular rise in $E°$, (e.g., -1.63 V for Ti, -1.13 V for V...), reaching a maximum of $E° = 0.340$ V with Cu. These trends can be related to the regular increase in Z_{eff} and other aspects of electron configurations (half filled shell for Mn^{2+} and loss of half filled shell for Cr^{2+}). Complicating matters is the fact that these ions will have different aqueous coordination numbers and geometries.

In the lanthanide series, it is the $4f$ subshell that fills at the same time that the $5d$ subshell is mostly vacant and the $6s$ subshell is filled. Differences in electron configurations are essentially confined to a subshell two shells removed from the outermost valence shell, which translates into close similarities among all the elements in the series, including $E°$ values.

Recall that the lanthanide series is an *inner* transition series—a transition series within another transition series. This inner transition series runs its course between La ($E° = -2.38$ V) and Hf ($E° = -1.70$ V). The difference between these two $E°$ values is about 0.7 V, which is comparable to the 0.4 V difference between the first (Sc) and second (Ti) members of the first transition series.

11. **(a)** $TiCl_4(g) + 4Na(l) \xrightarrow{\Delta} Ti(s) + 4NaCl(l)$

(b) $Cr_2O_3(s) + 2Al(s) \xrightarrow{\Delta} 2\ Cr(l) + Al_2O_3(s)$

(c) $Ag(s) + HCl(aq) \rightarrow$ no reaction

(d) $K_2Cr_2O_7(aq) + 2KOH(aq) \rightarrow 2\ K_2CrO_4(aq) + H_2O(l)$

(e) $MnO_2(s) + 2\ C(s) \xrightarrow{\Delta} Mn(l) + 2CO(g)$

12. **(a)** $Cr(s) + 2HCl(aq) \rightarrow CrCl_2(aq) + H_2(g)$ Virtually any first period transition metal except Cu can be substituted for Cr.

(b) $Cr_2O_3(s) + 2\ OH^-(aq) + 3\ H_2O(l) \rightarrow 2\ Cr(OH)_4^-(aq)$

The oxide must be amphoteric. Thus, Sc_2O_3, TiO_2, ZrO_2, and ZnO could be substituted for Cr_2O_3.

(c) $2\ La(s) + 6\ HCl(aq) \rightarrow 2\ LaCl_3(aq) + 3\ H_2(g)$

Any lanthanide or actinide element can be substituted for lanthanum.

Reactions of Transition Metals and Their Compounds

13. **(a)** $Sc(OH)_3(s) + 3H^+(aq) \rightarrow Sc^{3+}(aq) + 3\ H_2O(l)$

(b) $3\ Fe^{2+}(aq) + MnO_4^-(aq) + 2\ H_2O(l) \rightarrow 3\ Fe^{3+}(aq) + MnO_2(s) + 4\ OH^-(aq)$

(c) $2\ KOH(l) + TiO_2(s) \xrightarrow{\Delta} K_2TiO_3(s) + H_2O(g)$

(d) $Cu(s) + 2\ H_2SO_4(conc, aq) \rightarrow CuSO_4(aq) + SO_2(g) + 2\ H_2O(l)$

14. **(a)** $2\ Sc_2O_3(l, in\ Na_3ScF_6) + 3\ C(s) \xrightarrow{electrolysis} 4\ Sc(l) + 3\ CO_2(g)$

[By analogy with the equation for the electrolytic production of Al]

(b) $Cr(s) + 2\ HCl(aq) \rightarrow Cr^{2+}(aq) + 2\ Cl^-(aq) + H_2(g)$

(c) $4\ Cr^{2+}(aq) + O_2(g) + 4\ H^+(aq) \rightarrow 4\ Cr^{3+}(aq) + 2\ H_2O(l)$

(d) $Ag(s) + 2\ HNO_3(aq) \rightarrow AgNO_3(aq) + NO_2(g) + H_2O(l)$

15. We write some of the following reactions as total equations rather than as net ionic equations so that the reagents used are indicated.

(a) $FeS(s) + 2\ HCl(aq) \rightarrow FeCl_2(aq) + H_2S(g)$

$4\ Fe^{2+}(aq) + O_2(g) + 4H^+(aq) \rightarrow 4Fe^{3+}(aq) + 2\ H_2O(l)$

$Fe^{3+}(aq) + 3\ OH^-(aq) \rightarrow Fe(OH)_3(s)$

(b) $BaCO_3(s) + 2\ HCl(aq) \rightarrow BaCl_2(aq) + H_2O(l) + CO_2(g)$

$2\ BaCl_2(aq) + K_2Cr_2O_7(aq) + 2\ NaOH(aq) \rightarrow 2\ BaCrO_4(s) + 2\ KCl(aq) + 2\ NaCl(aq) + H_2O(l)$

16. **(a)** i) $CuO(s) + 2H^+(aq) \longrightarrow Cu^{2+}(aq) + H_2O(l)$

ii) $Cu^{2+}(aq) + 2OH^-(aq) \longrightarrow Cu(OH)_2(s)$

(b) $(NH_4)_2 Cr_2O_7(s) \xrightarrow{\Delta} N_2(g) + 4\, H_2O(g) + Cr_2O_3(s)$

$Cr_2O_3(s) + 6\, HCl(aq) \rightarrow CrCl_3(aq) + 3\, H_2O(l)$

17. $HgS(s) + O_2(g) \xrightarrow{\Delta} Hg(l) + SO_2(g)$

$4\, HgS(s) + 4\, CaO(s) \xrightarrow{\Delta} 4\, Hg(l) + 3\, CaS(s) + CaSO_4(s)$

18. **(a)** $C(s) + O_2(g) \rightarrow CO_2(g)$ $\Delta S° = 213.7\ \text{J K}^{-1} - (5.74\ \text{J K}^{-1} + 205.1\ \text{J K}^{-1})$

$\Delta S° = 2.86\ \text{J K}^{-1}$

Since $\Delta S° \approx 0$, $\Delta G°$ will be relatively constant with temperature.

(b) $2\, CO(g) + O_2(g) \rightarrow 2\, CO_2(g)$ $\Delta S° = (2(213.7\ \text{J K}^{-1}) - (205.1\ \text{J K}^{-1} + 2(197.7\ \text{J K}^{-1}))$

$\Delta S° = -173.1\ \text{J K}^{-1}$

Since ΔS is less than zero, $\Delta G°$ will increase (become less negative or more positive) as temperature increases.

19. The plot of $\Delta G°$ versus T will consist of three lines of increasing positive slope. The first line is joined to the second line at the melting point for Ca(s), while the second line is joined to the third at the boiling point for Ca(l).

$2\, Ca(s) + O_2(g) \rightarrow 2\, CaO(s)$ $\Delta H°_f = -1270.2\ \text{kJ}$

$\Delta S° = 2(39.75\ \text{J K}^{-1}) - [2\times(41.42\ \text{J K}^{-1}) + 205.1\ \text{J K}^{-1}] = -208.4\ \text{J K}^{-1}$

The graph should be similar to that for $2\, Mg(s) + O_2(g) \rightarrow 2\, MgO(s)$. We expect a positive slope with slight changes in the slope after the melting point(839 °C) and boiling points(1484 °C), mainly owing to changes in entropy. The plot will be below the $\Delta G°$ line for $2\, Mg(s) + O_2(g) \rightarrow 2\, MgO(s)$ at all temperatures.

20. $2\, Na_2CrO_4(s) + 3\, C(s) + 4HCl(aq) \rightarrow Cr_2O_3(s) + 2H_2O(l) + 3\, CO(g) + 4NaCl(aq)$

$2\, Cr_2O_3(s) + 3\, Si(s) \rightarrow 4\, Cr(s) + 3\, SiO_2(s)$

Oxidation-reduction

21. **(a)** Reduction: $VO^{2+}(aq) + 2H^+(aq) + e^- \rightarrow V^{3+}(aq) + H_2O(l)$

(b) Oxidation: $Cr^{2+}(aq) \rightarrow Cr^{3+}(aq) + e^-$

22. **(a)** Oxidation: $Fe(OH)_3(s) + 5\, OH^-(aq) \rightarrow FeO_4^{2-}(aq) + 4\, H_2O(l) + 3e^-$

(b) Reduction: $[Ag(CN)_2]^-(aq) + e^- \rightarrow Ag(s) + 2\, CN^-(aq)$

23. **(a)** First we need the reduction potential for the couple $VO_2^+(aq)/V^{2+}(aq)$. We will use the half-cell addition method learned in Chapter 20.

$$VO_2^+(aq)+2\,H^+(aq)+e^- \rightarrow VO^{2+}(aq)+H_2O(l) \quad \Delta G^\circ = -1\,F(+1.000\text{ V})$$

$$VO^{2+}(aq)+2\,H^+(aq)+e^- \rightarrow V^{3+}(aq)+H_2O(l) \quad \Delta G^\circ = -1\,F(+0.337\text{ V})$$

$$V^{3+}(aq)+e^- \rightarrow V^{2+}(aq) \quad \Delta G^\circ = -1\,F(-0.255\text{ V})$$

$$VO_2^+(aq)+4\,H^+(aq)+3\,e^- \rightarrow V^{2+}(aq)+2\,H_2O \quad \Delta G^\circ = -3\,FE^\circ$$

$$E^\circ = \frac{1.000\text{ V}+0.337\text{ V}-0.255\text{ V}}{3} = +0.361\text{ V}$$

We next analyze the oxidation-reduction reaction.

Oxidation: $\{2\,Br^-(aq) \rightarrow Br_2(l)+2e^-\} \times 3$ $\quad\quad -E^\circ = -1.065\text{ V}$

Reduction: $\{VO_2^+(aq)+4H^+(aq)+3e^- \rightarrow V^{2+}(aq)+2H_2O(l)\} \times 2$ $\quad E^\circ = +0.361\text{V}$

Net: $6\,Br^-(aq)+2\,VO_2^+(aq)+8\,H^+(aq) \rightarrow 3\,Br_2(l)+2\,V^{2+}(aq)+4\,H_2O$

$E^\circ_{cell} = -0.704\text{V}$

Thus, this reaction does not occur to a significant extent as written at standard conditions.

(b) Oxidation: $Fe^{2+}(aq) \rightarrow Fe^{3+}(aq)+e^-$ $\quad\quad -E^\circ = -0.771\text{V}$

Reduction: $VO_2^+(aq)+2\,H^+(aq)+e^- \rightarrow VO^{2+}(aq)+H_2O(l)$ $\quad E^\circ = +1.000\text{V}$

Net: $\quad\quad Fe^{2+}(aq)+VO_2^+(aq)+2\,H^+(aq) \rightarrow Fe^{3+}(aq)+VO^{2+}(aq)+H_2O(l)$

$E^\circ_{cell} = +0.229\text{V}$

This reaction does occur to a significant extent at standard conditions.

(c) Oxidation: $H_2O_2 \rightarrow 2\,H^+(aq)+2\,e^-+O_2(g)$ $\quad\quad -E^\circ = -0.695\text{ V}$

Reduction: $MnO_2(s)+4\,H^+(aq)+2\,e^- \rightarrow Mn^{2+}(aq)+2\,H_2O(l)$ $\quad E^\circ = +1.23\text{V}$

Net: $H_2O_2+MnO_2(s)+2\,H^+(aq) \rightarrow O_2(g)+Mn^{2+}(aq)+2\,H_2O(l)$ $\quad E^\circ_{cell} = +0.54\text{V}$

Thus, this reaction does occur to a significant extent at standard conditions.

24. When a species acts as a reducing agent, it is oxidized. From Appendix D we obtain the potentials for each of the following couples.

$$-E^0\{Zn^{2+}(aq)/Zn(s)\} = +0.763\,V; \quad -E^0\{Sn^{4+}(aq)/Sn^{2+}(aq)\} = -0.154V$$

$$-E^\circ\{I_2(s)/I^-(aq)\} = -0.535\,V$$

Each of these potentials is combined with the cited reduction potential. If the resulting value of E°_{cell} is positive, then the reducing agent will be effective in accomplishing the desired reduction, which we indicate with "yes"; if not, we write "no".

(a) $E^\circ\{Cr_2O_7^{2-}(aq)/Cr^{3+}(aq)\} = +1.33\ V$

$E^\circ_{cell} = E^\circ\{Cr_2O_7^{2-}(aq)/Cr^{3+}(aq)\} - E^\circ\{Zn^{2+}(aq)/Zn(s)\}$

$E^\circ_{cell} = +1.33\ V + 0.763\ V = +2.09\ V \quad \Rightarrow \quad$ Yes

$E^\circ_{cell} = E^\circ\{Cr_2O_7^{2-}(aq)/Cr^{3+}(aq)\} - E^\circ\{Sn^{4+}(aq)/Sn^{2+}(aq)\}$

$E^\circ_{cell} = +1.33\ V - 0.154\ V = +1.18\ V \quad \Rightarrow \quad$ Yes

$E^\circ_{cell} = E^\circ\{Cr_2O_7^{2-}(aq)/Cr^{3+}(aq)\} - E^\circ\{I_2(s)/I^-(aq)\}$

$E^\circ_{cell} = +1.33\ V - 0.535\ V = +0.80\ V \quad \Rightarrow \quad$ Yes

(b) $E^\circ\{Cr^{3+}(aq)/Cr^{2+}(aq)\} = -0.424\ V$

$E^\circ_{cell} = E^\circ\{Cr^{3+}(aq)/Cr^{2+}(aq)\} - E^\circ\{Zn^{2+}(aq)/Zn(s)\}$

$E^\circ_{cell} = -0.424\ V + 0.763\ V = +0.339\ V \quad \Rightarrow \quad$ Yes

$E^\circ_{cell} = E^\circ\{Cr^{3+}(aq)/Cr^{2+}(aq)\} - E^\circ\{Sn^{4+}(aq)/Sn^{2+}(aq)\}$

$E^\circ_{cell} = -0.424\ V - 0.154\ V = -0.578\ V \quad \Rightarrow \quad$ No

$E^\circ_{cell} = E^\circ\{Cr^{3+}(aq)/Cr^{2+}(aq)\} - E^\circ\{I_2(s)/I^-(aq)\}$

$E^\circ_{cell} = -0.424\ V - 0.535\ V = -0.959\ V \quad \Rightarrow \quad$ No

(c) $E^\circ\{SO_4^{2-}(aq)/SO_2(g)\} = +0.17\ V$

$E^\circ_{cell} = E^\circ\{SO_4^{2-}(aq)/SO_2(g)\} - E^\circ\{Zn^{2+}(aq)/Zn(s)\}$

$E^\circ_{cell} = +0.17\ V + 0.763\ V = +0.93\ V \quad \Rightarrow \quad$ Yes

$E^\circ_{cell} = E^\circ\{SO_4^{2-}(aq)/SO_2(g)\} - E^\circ\{Sn^{4+}(aq)/Sn^{2+}(aq)\}$

$E^\circ_{cell} = +0.17\ V - 0.154\ V = +0.02\ V \quad \Rightarrow \quad$ Yes (barely)

$E^\circ_{cell} = E^\circ\{SO_4^{2-}(aq)/SO_2(g)\} - E^\circ\{I_2(s)/I^-(aq)\}$

$E^\circ_{cell} = +0.17\ V - 0.535\ V = -0.37\ V \quad \Rightarrow \quad$ No

25. The reducing couple that we seek must have a half-cell potential of such a size and sign that a positive sum results when this half-cell potential is combined with $E°\{VO^{2+}(aq)|V^{3+}(aq)\}=0.337\,V$ *and* a negative sum must be produced when this half-cell potential is combined with $E°\{V^{3+}(aq)|V^{2+}(aq)\}=-0.255V$.

So, $-E°$ for the couple must be> -0.337 V and <+ 0.255 V (i.e. it cannot be more positive than 0.255V, nor more negative than -0.337 V)

Some possible reducing couples from Table 20.1 are:
$-E°\{Sn^{2+}(aq)|Sn(s)\}=+0.137$ V ; $-E°\{H^+(aq)|H_2(g)]=0.000$ V
$-E°\{Pb^{2+}(aq)|Pb(s)]=+0.125$ V ; thus Pb(s), Sn(s), $H_2(g)$, to name but a few, will do the job.

26. There are two methods that can be used to determine the MnO_4^-/Mn^{2+} reduction potential.
Method 1: $MnO_4^- — 1.70\,V \rightarrow MnO_2 — 1.23\,V \rightarrow Mn^{2+}$

$$E°_{MnO_4^-/Mn^{2+}} = \frac{3\,(1.70\,V)+2\,(1.23\,V)}{5} = 1.51\underline{2}\,V \sim 1.51\,V$$

Method 2: $MnO_4^- \rightarrow MnO_4^{2-} \rightarrow MnO_2 \rightarrow Mn^{3+} \rightarrow Mn^{2+}$

$$E°_{MnO_4^-/Mn^{2+}} = \frac{0.56\,V + 2\,(2.27\,V)+0.95\,V+1.49\,V}{5} = 1.50\underline{8}\,V \sim 1.51\,V$$

These answers compare favorably to Table 20.1, where $E°_{MnO_4^-/Mn^{2+}} = 1.51$ V

27. Table D-4 contains the following data: Cr^{3+}/Cr^{2+} reduction potential = -0.424 V, $Cr_2O_7^{2-}/Cr^{3+}$ reduction potential = 1.33 V and Cr^{2+}/Cr reduction potential = -0.90 V.

By using the additive nature of free energies and the fact that $\Delta G° = -nFE°$, we can determine the two unknown potentials and complete the diagram.

(i) $Cr_2O_7^{2-}/Cr^{2+}$: $E° = \dfrac{3(1.33\,V) - 0.424\,V}{4} = 0.892$ V

(ii) Cr^{3+}/Cr: $E° = \dfrac{-0.424\,V + 2(-0.90)\,V}{3} = -0.74$ V

28. From Table 23.4 we are given: VO_2^+/VO^{2+} reduction potential = 1.000 V, VO^{2+}/V^{3+} reduction potential = 0.337 V, V^{3+}/V^{2+} reduction potential = -0.255 V and V^{2+}/V reduction potential = -1.13 V. By taking advantage the additive nature of free energies and the fact that $\Delta G° = -nFE°$, we can determine the three unknown potentials and complete the diagram.

(i) VO_2^+/V^{3+}: $E° = \dfrac{1.000\,V + 0.337\,V}{2} = 0.669$ V

(ii) VO_2^+/V^{2+}: $E° = \dfrac{1.000\,V + 0.337\,V + (-0.255\,V)}{3} = 0.361$ V

(iii) VO_2^+/V: $E° = \dfrac{1.000\,V + 0.337\,V + (-0.255\,V) + 2(-1.13\,V)}{5} = -0.24$ V

Chromium and Chromium Compounds

29. Orange dichromate ion is in equilibrium with yellow chromate ion in aqueous solution.

$$Cr_2O_7^{\,2-}\,(aq)+H_2O(l)\rightleftharpoons 2CrO_4^{\,2-}\,(aq)+2H^+\,(aq)$$

The chromate ion in solution then reacts with lead(II) ion to form a precipitate of yellow lead(II) dichromate. $Pb^{2+}\,(aq)+CrO_4^{\,2-}\,(aq)\rightleftharpoons PbCrO_4\,(s)$

$PbCrO_4\,(s)$ will form until $\left[H^+\right]$ from the first equilibrium increases to the appropriate level and both equilibria are simultaneously satisfied.

30. The initial dissolving reaction forms orange dichromate ion.

$$2\,BaCrO_4\,(s)+2\,HCl(aq)\rightleftharpoons 2\,Ba^{2+}\,(aq)+Cr_2O_7^{\,2-}\,(aq)+2\,Cl^-\,(aq)+H_2O(l)$$

Dichromate ion is a good oxidizing agent, that is strong enough to oxidize $Cl^-(aq)$ to $Cl_2(g)$ if the concentrations of reactants are high, the solution is acidic, and the product $Cl_2(g)$ is allowed to escape.

$$6Cl^-\,(aq)+Cr_2O_7^{\,2-}\,(aq)+14H^+\,(aq)\rightleftharpoons 3Cl_2\,(g)+2Cr^{3+}\,(aq)+7H_2O(l)$$

Chromium(III) can hydrolyze in aqueous solution to produce green $\left[Cr(OH)_4\right]^-(aq)$, but the solution needs to be alkaline (pH >7) for this to occur. A more likely source of the green color is a complex ion such as $\left[Cr(H_2O)_4Cl_2\right]^+$.

31. Oxidation: $\{Zn(s)\rightarrow Zn^{2+}\,(aq)+2e^-$ $\qquad\qquad\qquad\qquad \}\times 3$
Reduction: $Cr_2O_7^{\,2-}\,(aq,orange)+14H^+\,(aq)+6e^-\rightarrow 2\,Cr^{3+}\,(aq,green)+7\,H_2O(l)$

Net: $\quad 3\,Zn(s)+Cr_2O_7^{\,2-}\,(aq)+14\,H^+\,(aq)\rightarrow 3\,Zn^{2+}\,(aq)+2\,Cr^{3+}\,(aq)+7\,H_2O(l)$

Oxidation: $\quad Zn(s)\rightarrow Zn^{2+}\,(aq)+2e^-$
Reduction: $\quad\{Cr^{3+}\,(aq,green)+e^-\rightarrow Cr^{2+}\,(aq,blue)\}\qquad \times 2$

Net: $\quad Zn(s)+2\,Cr^{3+}\,(aq)\rightarrow Zn^{2+}\,(aq)+2\,Cr^{2+}\,(aq)$

The green color is most likely due to a chloro complex of Cr^{3+}, such as $\left[Cr(H_2O)_4Cl_2\right]^+$.

Oxidation: $\quad\{Cr^{2+}\,(aq,blue)\rightarrow Cr^{3+}\,(aq,green)+e^-\}\qquad \times 4$
Reduction: $\quad O_2\,(g)+4\,H^+\,(aq)+4e^-\rightarrow 2\,H_2O(l)$

Net: $\quad 4\,Cr^{2+}\,(aq)+O_2\,(g)+4\,H^+\,(aq)\rightarrow 4\,Cr^{3+}\,(aq)+2\,H_2O(l)$

32. $CO_2(g)$, as the oxide of a nonmetal, is an acid anhydride. Its function is to make the solution acidic. A reasonable guess of the reactions that occur follows.

$$2\ CrO_4^{2-}\ (aq) + 2\ H^+\ (aq) \rightleftharpoons Cr_2O_7^{2-}\ (aq) + H_2O(l)$$

$$2\ H_2O(l) + 2CO_2\ (aq) \rightleftharpoons 2H^+\ (aq) + 2\ HCO_3^-\ (aq)$$

$$\overline{2\ CrO_4^{2-}\ (aq) + 2\ CO_2\ (aq) + H_2O(l) \rightleftharpoons Cr_2O_7^{2-}\ (aq) + 2\ HCO_3^-\ (aq)}$$

33. Simple substitution into expression (23.19) yields $\left[Cr_2O_7^{2-} \right]$ in each case. In fact, the expression is readily solved for the desired concentration as follows:

$\left[Cr_2O_7^{2-} \right] = 3.2 \times 10^{14} \left[H^+ \right]^2 \left[CrO_4^{2-} \right]^2$. In each case, we use the value of pH to determine $\left[H^+ \right] = 10^{-pH}$.

(a) $\left[Cr_2O_7^{2-} \right] = 3.2 \times 10^{14} (10^{-6.62})^2 (0.20)^2 = 0.74\ M$

(b) $\left[Cr_2O_7^{2-} \right] = 3.2 \times 10^{14} \left(10^{-8.85} \right)^2 (0.20)^2 = 2.6 \times 10^{-5}\ M$

34. We use expression (23.19) again:.

$$\frac{\left[Cr_2O_7^{2-} \right]}{\left[CrO_4^{2-} \right]^2} = 3.2 \times 10^{14} \left[H^+ \right]^2 = 3.2 \times 10^{14} (10^{-7.55})^2 = 3.2 \times 10^{14} (2.8 \times 10^{-8})^2 = 0.254$$

$$\left[CrO_4^{2-} \right]_{initial} = \frac{1.505\ g\ Na_2CrO_4}{0.345\ L\ soln} \times \frac{1\ mol\ Na_2CrO_4}{161.97\ g\ Na_2CrO_4} \times \frac{1\ mol\ CrO_4^{2-}}{1 mol\ Na_2CrO_4} = 0.0269\ M$$

Reaction:	$2CrO_4^{2-}\ (aq)$	+	$2H^+\ (aq)$	$\rightleftharpoons$	$Cr_2O_7^{2-}(aq) + H_2O\ (l)$	
Initial:	0.0269 M		2.8×10^{-8} (fixed)		0 M	—
Changes:	$-2x$ M		0		$+x$ M	—
Equil:	$(0.0269 - 2x)$ M		2.8×10^{-8} (fixed)		x M	—

$$\frac{\left[Cr_2O_7^{2-} \right]}{\left[CrO_4^{2-} \right]^2} = 0.254 = \frac{x}{(0.0269 - 2x)^2} \qquad x = 0.254 \left(0.000724 - 0.108x + 4x^2 \right)$$

$$x = 0.000184 - 0.0274\ x + 1.016\ x^2 \qquad 1.016\ x^2 - 1.0274\ x + 0.000184 = 0$$

$$x = \frac{-b \pm \sqrt{b^2 - 4ac}}{2a} = \frac{1.0274 \pm \sqrt{1.0556 - 0.000736}}{2.032} = 1.0111\ M,\ 0.00016\ M$$

We choose the second root because the first root gives a negative $\left[CrO_4^{2-} \right]$.

$$\left[Cr_2O_7^{2-} \right] = x = 0.00016\ M \qquad \left[CrO_4^{2-} \right] = 0.0269 - 2x = 0.0266\ M$$

We carry extra significant figures to avoid a significant rounding error in this problem.

35. Each mole of chromium metal plated out from a chrome plating bath (i.e., CrO_3 and H_2SO_4) requires six moles of electrons.

$$\text{mass Cr} = 1.00 \text{ h} \times \frac{3600 \text{ s}}{1 \text{ hr}} \times \frac{3.4 \text{ C}}{1 \text{ s}} \times \frac{1 \text{ mol e}^-}{96485 \text{ C}} \times \frac{1 \text{ mol Cr}}{6 \text{ mol e}^-} \times \frac{52.00 \text{ g Cr}}{1 \text{ mol Cr}} = 1.10 \text{ g Cr}$$

36. First we compute the amount of Cr(s) deposited.

$$\text{mol Cr} = \left(0.0010 \text{ mm} \times \frac{1 \text{ cm}}{10 \text{ mm}} \times 0.375 \text{ m}^2 \times \frac{10^4 \text{ cm}^2}{1 \text{ m}^2} \right) \times \frac{7.14 \text{ g Cr}}{1 \text{ cm}^3 \text{ Cr}} \times \frac{1 \text{ mol Cr}}{52.0 \text{ g Cr}} = 5.1 \times 10^{-2} \text{ mol Cr}$$

Recall that the deposition of each mole of chromium requires six moles of electrons. We now compute the time required to deposit the 5.1×10^{-2} mol Cr.

$$\text{time} = 5.1 \times 10^{-2} \text{ mol Cr} \times \frac{6 \text{ mol e}^-}{1 \text{ mol Cr}} \times \frac{96485 \text{ C}}{1 \text{ mol e}^-} \times \frac{1 \text{ s}}{3.5 \text{ C}} \times \frac{1 \text{ h}}{3600 \text{ s}} = 2.3 \text{ h}$$

37. Dichromate ion is the prevalent species in acidic solution. Oxoanions are better oxidizing agents in acidic solution because increasing the concentration of hydrogen ion favors formation of product. The half-equation is:

$Cr_2O_7^{2-}(aq) + 14 H^+(aq) + 6e^- \longrightarrow 2 Cr^{3+}(aq) + 7H_2O(l)$ Note that precipitation occurs most effectively in alkaline solution. In fact, adding an acid to a compound is often an effective way of dissolving a water-insoluble compound. Thus, we expect to see the form that predominates in alkaline solution to be the most effective precipitating agent. Notice also that CrO_4^{2-} is smaller than is $Cr_2O_7^{2-}$, giving it a higher lattice energy in its compounds, which makes these compounds harder to dissolve.

38. Both metal ions precipitate as hydroxides when $\left[OH^- \right]$ is moderate.

$$Mg^{2+}(aq) + 2 OH^-(aq) \rightarrow Mg(OH)_2(s) \qquad Cr^{3+}(aq) + 3 OH^-(aq) \rightarrow Cr(OH)_3(s)$$

Because chromium(III) oxides and hydroxides are amphoteric (and those of magnesium ion are not), $Cr(OH)_3(s)$ will dissolve in excess base.

$$Cr(OH)_3(s) + NaOH(aq) \rightarrow Na^+(aq) + \left[Cr(OH)_4 \right]^-(aq)$$

The Iron Triad

39. $4 Fe^{2+}(aq) + O_2(g) + 4 H^+ \rightarrow 4 Fe^{3+}(aq) + 2 H_2O(l) \qquad E^\circ = 0.44 \text{ V}$

$[Fe^{2+}] = [Fe^{3+}]$; pH = 3.25 or $[H^+] = 5.6 \times 10^{-4}$ and $P_{O_2} = 0.20$ atm

$$E = E^\circ - \frac{0.0592}{n} \log \left(\frac{[Fe^{3+}]^4}{[Fe^{2+}]^4 [H^+]^4 P_{O_2}} \right) = 0.44 \text{ V} - \frac{0.0592}{4} \log \left(\frac{[Fe^{3+}]^4}{[Fe^{2+}]^4 [10^{-3.25}]^4 0.20} \right)$$

$E = 0.24$ V (spontaneous under these conditions)

40. $\{NiO(OH)(s) + e^- + H^+(aq) \rightleftharpoons Ni(OH)_2(s) \quad \}\times 2 \quad E^\circ_{red}$

$\{Cd(s) + 2\ OH^-(aq) \rightleftharpoons Cd(OH)_2(s) + 2\ e^- \quad \}\times 1 \quad E^\circ_{ox}$

$\{H_2O(l) \rightleftharpoons H^+(aq) + OH^-(aq)\} \quad\quad\quad\quad\quad \}\times 2$

$2\ NiO(OH)(s) + Cd(s) + 2\ H_2O(l) \rightleftharpoons Cd(OH)_2(s) + 2\ Ni(OH)_2(s)\ \ E^\circ_{cell} \approx 1.50\ V$

E°_{ox} may be obtained by combining the K_{sp} value for $Cd(OH)_2(s)$

$(K_{spCd(OH)_2} = 2.5 \times 10^{-14})$ and $E^\circ_{red} = -0.403$ V for $Cd^{2+}(aq) + 2\ e^- \rightarrow Cd(s)$. Hence,

$Cd(s) \rightarrow Cd^{2+}(aq) + 2\ e^- \quad\quad\quad\quad E^\circ_{ox} = 0.403$ V

$Cd^{2+}(aq) + 2\ OH^-(aq) \rightarrow Cd(OH)_2(s) \quad E^\circ_{cell} = \dfrac{0.0592}{n}\log\left(\dfrac{1}{K_{sp}}\right) = \dfrac{0.0592}{2}\log\left(\dfrac{1}{2.5 \times 10^{-14}}\right) = 0.40\underline{3}$ V

$Cd(s) + 2\ OH^-(aq) \rightarrow Cd(OH)_2(s) + 2\ e^- \ \ E^\circ_{ox} = 0.80\underline{6}$ V

Now plug this result back into first set of redox reactions:

$\{NiO(OH)(s) + e^- + H^+(aq) \rightarrow Ni(OH)_2(s) \quad \}\times 2 \quad E^\circ_{red}$

$\{Cd(s) + 2\ OH^-(aq) \rightarrow Cd(OH)_2(s) + 2\ e^- \quad \}\times 1 \quad E^\circ_{ox} = 0.80\underline{6}$ V

$2\ NiO(OH)(s) + Cd(s) + 2\ H_2O(l) \rightarrow Cd(OH)_2(s) + 2\ Ni(OH)_2(s)\ \ E^\circ_{cell} \approx 1.50$ V

$E^\circ_{red} = 1.50\ V - 0.80\underline{6}\ V = 0.69$ V

41. $Fe^{3+}(aq) + K_4[\overset{[II]}{Fe}(CN)_6](aq) \rightarrow K\overset{[III]}{Fe}[\overset{[II]}{Fe}(CN)_6](s) + 3\ K^+(aq)$

Alternate formulation: $4Fe^{3+} + 3[\overset{[II]}{Fe}(CN)_6]^{4-}(aq) + Fe_4[\overset{[II]}{Fe}(CN)_6]_3$

42. $K^+(aq) + Fe^{2+}(aq) + K_3[\overset{[III]}{Fe}(CN)_6](aq) \rightarrow Fe^{3+}(aq) + K_4[\overset{[II]}{Fe}(CN)_6](aq)$

$Fe^{3+}(aq) + K_4[\overset{[II]}{Fe}(CN)_6](aq) \rightarrow K\overset{[III]}{Fe}[\overset{[II]}{Fe}(CN)_6](s) + 3\ K^+(aq)$

Group 11 Metals

43. **(a)** $Cu^{2+}(aq) + H_2(g) \rightarrow Cu(s) + 2H^+(aq)$

(b) $Au^+(aq) + Fe^{2+}(aq) \rightarrow Au(s) + Fe^{3+}(aq)$

(c) $2Cu^{2+}(aq) + SO_2(g) + 2\ H_2O(l) \rightarrow 2\ Cu^+(aq) + SO_4^{2-}(aq) + 4\ H^+(aq)$

44. In either case, the acid acts as an oxidizing agent and dissolves silver; the gold is unaffected.

Oxidation: $\quad \{Ag(s) \rightarrow Ag^+(aq) + e^-\} \quad\quad\quad \times 3$

Reduction: $\quad NO_3^-(aq) + 4H^+(aq) + 3e^- \rightarrow NO(g) + 2H_2O(l)$

Net: $\quad 3Ag(s) + NO_3^-(aq) + 4H^+(aq) \rightarrow 3Ag^+(aq) + NO(g) + 2H_2O(l)$

Oxidation: $\quad \{Ag(s) \rightarrow Ag^+(aq) + e^-\} \quad\quad\quad \times 2$

Reduction: $\quad SO_4^{2-}(aq) + 4H^+(aq) + 2e^- \rightarrow SO_2(g) + 2H_2O(l)$

Net: $\quad 2\ Ag(s) + SO_4^{2-}(aq) + 4\ H^+(aq) \rightarrow 2\ Ag^+(aq) + SO_2(g) + 2\ H_2O(l)$

45. The Integrative Example showed $K_c = 1.2 \times 10^6 = \dfrac{\left[Cu^{2+}\right]}{\left[Cu^+\right]^2}$ or $\left[Cu^{2+}\right] = 1.2 \times 10^6 \left[Cu^+\right]^2$

(a) When $\left[Cu^+\right] = 0.20$ M, $\left[Cu^{2+}\right] = 1.2 \times 10^6 (0.20)^2 = 4.8 \times 10^4$ M. This is an impossibly high concentration. Thus $\left[Cu^+\right] = 0.20$ M can never be achieved.

(b) When $\left[Cu^+\right] = 1.0 \times 10^{-10}$ M, $\left[Cu^{2+}\right] = 1.2 \times 10^6 \left(1.0 \times 10^{-10}\right)^2 = 1.2 \times 10^{-14}$ M. This is an entirely reasonable (even though small) concentration; $\left[Cu^+\right] = 1.0 \times 10^{-10}$ M can be maintained in solution.

46. Oxidation: $\{Cu(s) \rightarrow Cu^{2+} + 2\,e^-\qquad\qquad\}\times 2$
Reduction: $\{O_2(g) + 2\,H_2O(l) + 4\,e^- \rightarrow 4\,OH^-(aq)\ \}\times 1$

Overall: $2\,Cu(s) + O_2(g) + 2\,H_2O(l) \rightarrow 2\,Cu^{2+}(aq) + 4\,OH^-\ (aq)$

Note: This produces an alkaline solution. CO_2 dissolves in this alkaline solution to produce carbonate ion.

Step(a): $CO_2(g) + OH^-\ (aq) \rightarrow HCO_3^-\ (aq)$
Step(b): $HCO_3^-\ (aq) + OH^-\ (aq) \rightarrow CO_3^{2-}\ (aq) + H_2O(l)$

Total: $CO_2(g) + 2\,OH^-\ (aq) \rightarrow CO_3^{2-}\ (aq) + H_2O(l)$

Combination of the reactions labeled "overall" and "total" gives the following result.

$2\,Cu(s) + O_2(g) + H_2O(l) + CO_2(g) \rightarrow 2\,Cu^{2+}(aq) + 2\,OH^-\ (aq) + CO_3^{2-}\ (aq)$

The ionic product of the resultant reaction combine in a precipitation reaction to form

$2\,Cu^{2+}(aq) + 2\,OH^-\ (aq) + CO_3^{2-}\ (aq) \rightarrow Cu_2(OH)_2CO_3\ (s)$

Group 12 Metals

47. Given: Hg^{2+}/Hg reduction potential = 0.854 V and Hg_2^{2+}/Hg reduction potential = 0.796 V
Using the additive nature of free energies and the fact that $\Delta G^\circ = -nFE^\circ$, we can determine the Hg^{2+}/Hg_2^{2+} potential as $\dfrac{2(0.854V) - 0.796V}{1} = 0.912$ V

48. $\Delta G^\circ = -25{,}000$ J $mol^{-1} = -RT\ln K_{eq} = -(8.3145$ J $K^{-1}\ mol^{-1})(673\ K)(\ln K_{eq})$

$K_{eq} = K_p = 87$ atm$^{-1} = \dfrac{1}{P_{O_2}}$ Therefore, $P_{O_2} = \dfrac{1}{K_p} = \dfrac{1}{87} = 0.011$ atm

49. **(a)** Estimate K_p for $ZnO(s) + C(s) \rightarrow Zn(l) + CO(g)$ at 800 °C (Note: $Zn(l)$ boils at 907 °C)

$$\{2\ C(s) + O_2(g) \rightarrow 2\ CO(g)\} \times 1/2 \qquad \Delta G° = (-415\ kJ) \times (1/2)$$
$$\{2\ ZnO(s) \rightarrow 2\ Zn(l) + O_2(g)\} \times 1/2 \qquad \Delta G° = (+485\ kJ) \times (1/2)$$

$$ZnO(s) + C(s) \rightarrow Zn(l) + CO(g) \qquad \Delta G° = 35\ kJ$$

Use $\Delta G° = -RT ln K_{eq}$ where $T = 800$ °C, $K_{eq} = K_p$

$35\ kJ\ mol^{-1} = -(8.3145 \times 10^{-3}\ kJ\ K^{-1}\ mol^{-1})((273.15 + 800)\ K)(\ln K_p)$

$\ln K_p = -3.9$ or $K_p = 0.02$

(b) $K_p = P_{CO} = 0.02$ Hence, $P_{CO} = 0.02$ atm

50. $T = 298$ K. Hence, $\log(P_{mmHg}) = \dfrac{-0.05223\ (61,960)}{298} + 8.118 = -2.742$

$P_{mmHg} = 0.001811$ mmHg

$$PV = nRT \rightarrow \frac{n}{V} = \frac{P}{RT} = [Hg] = \frac{(0.001811\ mmHg)(\dfrac{1\ atm}{760\ mmHg})}{\left(0.08206\ \dfrac{L\ atm}{K\ mol}\right)(298K)} = 9.744 \times 10^{-8}\ mol\ L^{-1}$$

Next, convert to mg Hg m^{-3}:

$$[Hg] = \left(9.874 \times 10^{-8}\ \frac{mol}{L}\right)\left(\frac{1000\ L}{m^3}\right)\left(\frac{200.59\ g\ Hg}{1\ mol\ Hg}\right)\left(\frac{1000\ mg}{1\ g\ Hg}\right) = 19.5\ mg\ Hg\ m^{-3}$$

(This is approximately 400 times greater than the permissible level of 0.05 mg Hg m^{-3})

51. We must calculate the wavelength of light absorbed in order to promote an electron across each band gap.

First a few relationships. $E_{mole} = N_A E_{photon}$ $\qquad E_{photon} = h\nu$ $\qquad c = \nu\lambda$ or $\nu = c/\lambda$

Then, some algebra. $E_{mole} = N_A E_{photon} = N_A h\nu = N_A hc/\lambda$ $\qquad$ or $\qquad \lambda = N_A hc/E_{mole}$

For ZnO, $\lambda = \dfrac{6.022 \times 10^{23}\ mol^{-1} \times 6.626 \times 10^{-34}\ J\ s \times 2.998 \times 10^8\ m\ s^{-1}}{290 \times 10^3\ J\ mol^{-1}} \times \dfrac{10^9\ nm}{1 m}$

$\qquad = \dfrac{1.196 \times 10^8\ J\ mol^{-1}\ nm}{290 \times 10^3\ J\ mol^{-1}} = 413$ nm $\quad$ violet light

For CdS, $\lambda = \dfrac{1.196 \times 10^8 \text{ J mol}^{-1} \text{ nm}}{250 \times 10^3 \text{ J mol}^{-1}} = 479$ nm blue light

The blue light absorbed by CdS is subtracted from the white light incident on the surface of the solid. The remaining reflected light is yellow in this case. When the violet light is subtracted from the white light incident on the ZnO surface, the reflected light appears white.

52.   The color that we see is the complementary color to the color absorbed. The color absorbed, in turn, is determined by the energy separation of the band gap. If the wavelength for the absorbed color is short, the energy separation for the band gap is large.

The yellow color of CdS means that the complementary color, violet, is absorbed. Since violet light has a quite short wavelength (about 410 nm), CdS must have a very large band gap energetically.

HgS is red, meaning that the color absorbed is green, which has a moderate wavelength (about 520 nm). HgS must have a band gap of intermediate energy.

CdSe is black, meaning that visible light of all wavelengths and thus all energies in the visible region are absorbed. This would occur if CdSe has a very small band gap, smaller than that of least energetic red light (about 650 nm).

Integrative and Advanced Exercises

<u>53.</u>   There are two reasons why Au is soluble in aqua regia while Ag is not. First, Ag^+, the oxidation product, forms a very insoluble chloride, AgCl, which probably adheres to the surface of the metal and prevents further reaction. $AuCl_3$ is not noted as being an insoluble chloride. Second, gold(III) forms a very stable complex ion with chloride ion, $[AuCl_4]^-$. This complex ion is much more stable than the corresponding dichloroargentate ion, $[AgCl_2]^-$.

54.   Sc and Ti metals form carbides in the presence of carbon (coke). Thus reduction of their oxides using carbon is avoided

55.   Both tin and zinc are toxic metals in high concentrations. Tin plated iron is more desirable than zinc plated (galvanized) iron. The reason for this is that zinc is a very reactive metal, especially in the presence of acidic foods. Tin on the other hand is much less reactive ($E^\circ_{Sn^{2+}/Sn} = -0.137$ V vs. $E^\circ_{Zn^{2+}/Zn} = -0.763$ V). As well, tin is more malleable and more easily prepared in pure form (free of oxide).

56. Oxidation: $\{Cu(s) \longrightarrow Cu^{2+}(aq) + 2\ e^-\} \times 2$

Reduction: $O_2(g) + 2\ H_2O(l) + 4\ e^- \longrightarrow 4\ OH^-(aq)$

Net: $\quad 2\ Cu(s) + O_2(g) + 2\ H_2O(l) \longrightarrow 2\ Cu^{2+}(aq) + 4\ OH^-(aq)$

This oxidation-reduction reaction produces an alkaline solution. $SO_3(g)$ dissolves in this alkaline solution to produce hydrogen sulfate ion in an acid-base reaction.

Step (a): $\quad SO_3(g) + OH^-(aq) \longrightarrow HSO_4^-(aq)$

Step (b): $\quad HSO_4^-(aq) + OH^-(aq) \longrightarrow SO_4^{2-}(aq) + H_2O(l)$

Total: $\quad SO_3(g) + 2\ OH^-(aq) \longrightarrow SO_4^{2-}(aq) + H_2O(l)$

The combination of the reactions labeled "Net:" and "Total:" gives the following result.

$$2\ Cu(s) + O_2(g) + H_2O(l) + SO_3(g) \longrightarrow 2\ Cu^{2+}(aq) + 2\ OH^-(aq) + SO_4^{2-}(aq)$$

The ionic products of this resultant reaction combine in a precipitation reaction to form $Cu_2(OH)_2SO_4(s)$. Thus, $2\ Cu^{2+}(aq) + 2\ OH^-(aq) + SO_4^{2-}(aq) \longrightarrow Cu_2(OH)_2SO_4(s)$

57. The noble gas formalism requires that the number of valence electrons possessed by the metal atom plus the number of sigma electrons be equal to the number of electrons in the succeeding noble gas atom.

(a) $Mo(CO)_6$; Mo has 42 electrons, 6 CO contribute another 12 for 54, which is the number of electrons in Xe

(b) $Os(CO)_5$; Os has 76 electrons, 5 CO contribute another 10 for 86, which is the number of electrons in Rn

(c) $Re(CO)_5^-$; Re anion has 76 electrons, 5 CO contribute another 10 for 86, which is the number of electrons in Rn

(d) The trigonal bipyramidal shape of nickel and iron carbonyls does not fit well in crystalline lattices, either because of its symmetry or repulsions with its neighbors and, consequently these five-coordinate carbonyls are liquids at room temperature. Compare this to octahedral complexes ($Cr(CO)_6$ and others) which are solids because they fit well into these lattices. Weak intermolecular attractions in the low molecular mass, symmetrical nickel and iron carbonyls result in the liquid state at room temperature. Because of the two metal atoms in the higher molecular mass, less symmetrical cobalt carbonyl molecules, stronger intermolecular attractions lead to the solid state.

(e) This compound would be an ionic, salt-like material consisting of Na^+ and $V(CO)_6^-$ ions.

58. (a) Breaks occur at the melting point and boiling point due to changes in the states of the metals (solid $\rightarrow$ liquid $\rightarrow$ gas).

(b) Slopes of these lines become more positive as the temperature increases due to the sign of $\Delta S°$ for these reactions. The slope should be equal to $-\Delta S°$, where $\Delta S°$ is a negative number in these cases which becomes more negative with a change in phase (higher temperature).

(c) The break at the boiling point is sharper than the break at the melting point because $\Delta S°_{fusion} \ll \Delta S°_{vaporization}$.

59.

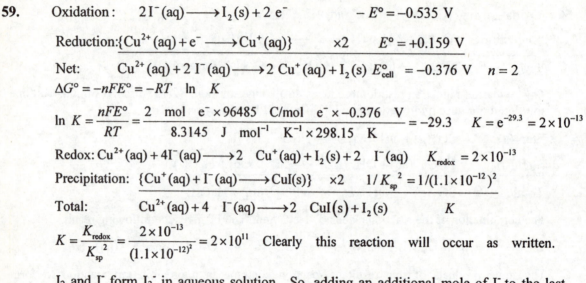

Oxidation: $2 I^-(aq) \longrightarrow I_2(s) + 2 e^- \qquad -E° = -0.535$ V

Reduction: $\{Cu^{2+}(aq) + e^- \longrightarrow Cu^+(aq)\} \qquad \times 2 \qquad E° = +0.159$ V

Net: $\quad Cu^{2+}(aq) + 2 I^-(aq) \longrightarrow 2 Cu^+(aq) + I_2(s) \quad E°_{cell} = -0.376$ V $\quad n = 2$

$\Delta G° = -nFE° = -RT \ln K$

$$\ln K = \frac{nFE°}{RT} = \frac{2 \ mol \ e^- \times 96485 \ C/mol \ e^- \times -0.376 \ V}{8.3145 \ J \ mol^{-1} \ K^{-1} \times 298.15 \ K} = -29.3 \qquad K = e^{-29.3} = 2 \times 10^{-13}$$

Redox: $Cu^{2+}(aq) + 4I^-(aq) \longrightarrow 2 \ Cu^+(aq) + I_2(s) + 2 \ I^-(aq) \qquad K_{redox} = 2 \times 10^{-13}$

Precipitation: $\{Cu^+(aq) + I^-(aq) \longrightarrow CuI(s)\} \quad \times 2 \qquad 1/K_{sp}^2 = 1/(1.1 \times 10^{-12})^2$

Total: $\qquad Cu^{2+}(aq) + 4 \ I^-(aq) \longrightarrow 2 \ CuI(s) + I_2(s) \qquad\qquad K$

$$K = \frac{K_{redox}}{K_{sp}^2} = \frac{2 \times 10^{-13}}{(1.1 \times 10^{-12})^2} = 2 \times 10^{11}$$ Clearly this reaction will occur as written.

I_2 and I^- form I_3^- in aqueous solution. So, adding an additional mole of I^- to the last reaction yields: $\qquad Cu^{2+}(aq) + 5 I^-(aq) \longrightarrow 2 \ CuI(s) + I_3^-(s)$

60. A "detailed calculation" would involve determining the final concentrations of all species. But we can get approximate values of the gold-containing ions by determining values of the equilibrium constants for the reactions involved. First, we determine the value of the equilibrium constant for the disproportionation of $Au^+(aq)$. To accomplish this, the value of $E°$ for the couple $Au^+(aq)|Au(s)$.

Oxidation $\qquad Au^+(aq) \longrightarrow Au^{3+}(aq) + 2 e^- \qquad -E° = -1.36$ V

Reduction: $\quad \{Au^+(aq) + e^- \longrightarrow Au(s)\} \times 2 \qquad E° = +1.83$ V

Net: $\qquad 3 Au^+(aq) \longrightarrow Au^{3+}(aq) + 2 Au(s) \qquad E°_{cell} = 0.47$ V

$$\ln K = \frac{nFE°}{RT} = \frac{2 \ mol \ e^- \times 96485 \ C/mol \ e^- \times 0.47 \ V}{8.3145 \ J \ mol^{-1} \ K^{-1} \times 298.15 \ K} = 36.6 \qquad K = e^{36.6} = 8 \times 10^{15}$$

Then we determine the value of the equilibrium constant for dissolving and disproportionation.

Disproportionation: $\qquad 3 Au^+(aq) \longrightarrow Au^{3+}(aq) + 2 Au(s) \quad K_{disp} = 8 \times 10^{15}$

Dissolving: $\qquad \{AuCl(s) \longrightarrow Au^+(aq) + Cl^-(aq)\} \quad \times 3 \qquad K_{sp}^3 = (2.0 \times 10^{-13})^3$

Net: $\qquad 3 AuCl(s) \longrightarrow Au^{3+}(aq) + 2 Au(s) + 3 Cl^-(aq) \quad K$

$K = K_{disp} \times K_{sp}^3 = 8 \times 10^{15} \times (2.0 \times 10^{-13})^3 = 6.4 \times 10^{-23} = [Au^{3+}][Cl^-]^3 = (x)(3x)^3 = 27 \ x^4$

$$x = \sqrt[4]{\frac{6.4 \times 10^{-23}}{27}} = 1 \times 10^{-6} \ M = [Au^{3+}]$$

This is a significant amount of decomposition, particularly when compared with $[Au^+] = 4.5 \times 10^{-7}$ M, which is the concentration present in saturated AuCl, assuming no disproportionation.

61. (a) Dissolution: $AgO(s) + 2\ H^+(aq) \longrightarrow Ag^{2+}(aq) + H_2O(l)$

Oxidation: $2\ H_2O(l) \longrightarrow O_2(g) + 4\ H^+(aq) + 4\ e^- \qquad -E° = -1.229\ V$

Reduction: $\{Ag^{2+}(aq) + e^- \longrightarrow Ag^+(aq)\} \qquad \times 4 \qquad E° = 1.98\ V$

Net: $2\ H_2O(l) + 4\ Ag^{2+}(aq) \longrightarrow O_2(g) + 4\ H^+(aq) + 4\ Ag^+(aq)$

$E°_{cell} = -1.229\ V + 1.98\ V = +0.75\ V \rightarrow$ spontaneous since $E°_{cell} > 0$.

62.

Association: $2\ H^+(aq) + 2\ CrO_4^{2-}(aq) \rightleftharpoons 2\ HCrO_4^-(aq) \qquad 1/K_a^2 = 1/(3.2\times10^{-7})^2$

Elimination: $2\ HCrO_4^-(aq) \rightleftharpoons Cr_2O_7^{2-}(aq) + H_2O(l) \qquad K$

Eqn. (23.18) $\quad 2\ H^+(aq) + 2\ CrO_4^{2-}(aq) \rightleftharpoons Cr_2O_7^{2-}(aq) + H_2O(l)\quad K_c = 3.2\times10^{14}$

$K_c = 3.2\times10^{14} = \dfrac{K}{K_a^2} = \dfrac{K}{(3.2\times10^{-7})^2} \qquad K = 3.2\times10^{14}\times(3.2\times10^{-7})^2 = 33$

63. The presence of both acetate ion and acetic acid establishes a buffer and fixes the pH of the solution. This, in turn fixes $[CrO_4^{2-}]$. We then can determine the concentration of each of the cations that can coexist with this $[CrO_4^{2-}]$ before precipitation occurs. First we determine $[H_3O^+]$, with the aid of the Henderson-Hasselbalch equation.

$$pH = pK_a + \log\frac{[C_2H_3O_2^-]}{[HC_2H_3O_2]} = -\log(1.8\times10^{-5}) + \log\frac{1.0\ M}{1.0\ M} = 4.74$$

$$[H_3O^+] = 10^{-4.74} = 1.8\times10^{-5}\ M$$

$$K_c = 3.2\times10^{14} = \frac{[Cr_2O_7^{2-}]}{[CrO_4^{2-}]^2[H_3O^+]^2} = \frac{0.0010\ M}{[CrO_4^{2-}]^2\ (1.8\times10^{-5})^2} = \frac{3.1\times10^6}{[CrO_4^{2-}]^2}$$

$$[CrO_4^{2-}] = \sqrt{\frac{3.1\times10^6}{3.2\times10^{14}}} = 9.8\times10^{-5}\ M$$

$$[Ba^{2+}]_{max} = \frac{K_{sp}}{[CrO_4^{2-}]} = \frac{1.2\times10^{-10}}{9.8\times10^{-5}} = 1.2\times10^{-6}\ M$$

$$[Sr^{2+}]_{max} = \frac{2.2\times10^{-5}}{9.8\times10^{-5}} = 0.22\ M$$

$$[Ca^{2+}]_{max} = \frac{7.1\times10^{-4}}{9.8\times10^{-5}} = 7.2\ M$$

Each cation is initially present at 0.10 M. Neither strontium ion nor calcium ion will precipitate, while BaCrO$_4$ will precipitate until $[Ba^{2+}]$ has declined to far less than 0.1% of its original value. Thus, barium ion is effectively separated from the other two cations by chromate ion precipitation under these conditions.

64. The equations are balanced with the ion-electron method. Oxalic acid is oxidized in each case.

Oxidation: $H_2C_2O_4(aq) \longrightarrow 2\ H^+(aq) + 2\ CO_2(g) + 2\ e^-$

Reduction: $MnO_2(s) + 4\ H^+(aq) + 2\ e^- \longrightarrow Mn^{2+}(aq) + 2\ H_2O(l)$

Net: $H_2C_2O_4(aq) + MnO_2(s) + 2\ H^+(aq) \longrightarrow Mn^{2+}(aq) + 2\ CO_2(g) + 2\ H_2O(l)$

Oxidation: $\{H_2C_2O_4(aq) \longrightarrow 2\ H^+(aq) + 2\ CO_2(g) + 2\ e^- \qquad\} \times 5$

Reduction: $\{MnO_4^-(aq) + 8\ H^+(aq) + 5\ e^- \longrightarrow Mn^{2+}(aq) + 4\ H_2O(l)\} \times 2$

Net: $5\ H_2C_2O_4(aq) + 2\ MnO_4^-(aq) + 6\ H^+(aq) \longrightarrow 2\ Mn^{2+}(aq) + 10\ CO_2(g) + 8\ H_2O(l)$

We then determine the mass of the excess oxalic acid.

$$\text{mass } H_2C_2O_4 \cdot 2H_2O = 0.03006\ L \times \frac{0.1000\ \text{mol } MnO_4^-}{1\ \text{L soln}} \times \frac{5\ \text{mol } H_2C_2O_4}{2\ \text{mol } MnO_4^-}$$

$$\times \frac{126.07\ \text{g } H_2C_2O_4 \cdot 2H_2O}{1\ \text{mol } H_2C_2O_4 \cdot 2H_2O} = 0.9474\ \text{g } H_2C_2O_4 \cdot 2H_2O$$

Thus, the mass of oxalic acid that reacted with MnO_2 is $1.651\ g - 0.9474\ g = 0.704\ g$ $H_2C_2O_4 \cdot 2H_2O$. Now we can determine the mass of MnO_2 in the ore.

$$\text{mass } MnO_2 = 0.704\ \text{g } H_2C_2O_4 \cdot 2H_2O \times \frac{1\ \text{mol } H_2C_2O_4 . 2H_2O}{126.07\ \text{g } H_2C_2O_4 \cdot 2H_2O} \times \frac{1\ \text{mol } MnO_2}{1\ \text{mol } H_2C_2O_4}$$

$$\times \frac{86.94\ \text{g } MnO_2}{1\ \text{mol } MnO_2} = 0.485\ \text{g } MnO_2 \qquad \%MnO_2 = \frac{0.485\ \text{g } MnO_2}{0.589\ \text{g ore}} \times 100\% = 82.3\%$$

65. (a) We consider reaction with 1.00 millimoles of Fe^{2+}. In each case, we need the balanced chemical equation of the redox reaction.

$$Cr_2O_7^{2-}(aq) + 14\ H^+(aq) + 6\ Fe^{2+}(aq) \longrightarrow 2\ Cr^{3+}(aq) + 7\ H_2O(l) + 6\ Fe^{3+}(aq)$$

$$\frac{\text{titrant}}{\text{volume}} = 1.00\ \text{mmol } Fe^{2+} \times \frac{1\ \text{mmol } Cr_2O_7^{2-}}{6\ \text{mmol } Fe^{2+}} \times \frac{1\ \text{mL solution}}{0.1000\ \text{mmol } Cr_2O_7^{2-}} = 1.67\ \text{mL soln}$$

$$MnO_4^-(aq) + 8\ H^+(aq) + 5\ Fe^{2+}(aq) \longrightarrow Mn^{2+}(aq) + 4\ H_2O + 5\ Fe^{3+}(aq)$$

$$\frac{\text{titrant}}{\text{volume}} = 1.00\ \text{mmol } Fe^{3+} \times \frac{1\ \text{mmol } MnO_4^-}{5\ \text{mmol } Fe^{2+}} \times \frac{1\ \text{mL solution}}{0.1000\ \text{mmol } MnO_4^-} = 2.00\ \text{mL soln}$$

More of the 0.1000 M MnO_4^- solution would be required.

(b) We use the reaction of each with Fe^{2+} to find the equivalence between them.

$$V_{MnO_4^-} = 24.50\ \text{mL} \times \frac{0.1000\ \text{mmol } Cr_2O_7^{2-}}{1\ \text{mL soln}} \times \frac{6\ \text{mmol } Fe^{2+}}{1\ \text{mmol } Cr_2O_7^{2-}} \times \frac{1\ \text{mmol } MnO_4^-}{5\ \text{mmol } Fe^{2+}}$$

$$\times \frac{1\ \text{mL soln}}{0.1000\ \text{mmol } MnO_4^-} = 29.40\ \text{mL of } 0.1000\ \text{mmol } MnO_4^-$$

66. A high oxidation state of a metal can only be stabilized by an anion of an element that is highly electronegative. Fluorine and oxygen are the two most electronegative elements in the periodic table and thus should be most effective at stabilizing a high oxidation state of a metal.

67. The precipitation reaction of dichromate ion permits us to determine from the formula of barium chromate the amount of chromium present, while the redox reaction of permanganate ion permits us to determine the amount of manganese present in the 250.0 mL solution. The 15.95 ml volume of Fe^{2+} titrant assumes (contrary to what the problem implies) that the chromate has been removed by precipitation. $Ba(MnO_4)_2$ is quite water soluble, so $[MnO_4^-]$ should not be affected.

equation: $MnO_4^-(aq) + 8 H^+(aq) + 5 Fe^{2+}(aq) \longrightarrow Mn^{2+}(aq) + 4 H_2O + 5 Fe^{3+}(aq)$

$$\text{mass of Cr} = 250.0 \text{ mL soln} \times \frac{0.549 \text{ g BaCrO}_4}{10.00 \text{ mL sample}} \times \frac{1 \text{ mol BaCrO}_4}{253.3 \text{ g BaCrO}_4} \times \frac{1 \text{ mol Cr}}{1 \text{ mol BaCrO}_4} \times \frac{52.00 \text{ g Cr}}{1 \text{ mol Cr}}$$

mass of Cr = 2.82 g Cr

equation: $Cr_2O_7^{2-} + 14 H^+ + 6 Fe^{2+} \rightarrow 2Cr^{3+} + 7 H_2O + 6 Fe^{3+}$

$$\text{volume of titrant to reduce } Cr_2O_7^{2-} = 2.82 \text{ g Cr} \times \frac{10 \text{ ml}}{250 \text{ ml}} \times \frac{1 \text{ mol } Cr^{3+}}{52.00 \text{ g Cr}} \times \frac{6 \text{ mol } Fe^{2+}}{1 \text{ mol } Cr_2O_7^{2-}}$$

$$\times \frac{1 \text{ mol } Cr_2O_7^{2-}}{2 \text{ mol } Cr^{3+}} \times \frac{1 \text{ L titrant}}{0.0750 \text{ mol } Fe^{2+}} \times \frac{1000 \text{ mL}}{1 \text{ L}} = 86.78 \text{ ml } Fe^{2+} \text{ titrant}$$

Now the volume of titrant to reduce both MnO_4^- and $Cr_2O_7^{2-}$ must be $(15.95 + 86.78)$ mL Volume of titrant = 102.73 mL.

Not just 15.95 mL, which must be the volume of titrant required for MnO_4^- *alone.*

$$\text{mass Mn} = 250.0 \text{ mL soln} \times \frac{15.95 \text{ mL titrant}}{10.00 \text{ mL sample}} \times \frac{0.0750 \text{ mmol } Fe^{2+}}{1 \text{ mL titrant}} \times \frac{1 \text{ mmol } MnO_4^-}{5 \text{ mmol } Fe^{2+}}$$

$$\times \frac{1 \text{ mmol Mn}}{1 \text{ mmol } MnO_4^-} \times \frac{54.94 \text{ mg Mn}}{1 \text{ mmol Mn}} \times \frac{1 \text{ g Mn}}{1000 \text{ mg Mn}} = 0.3286 \text{ Mn}$$

$$\%\text{Mn} = \frac{0.3286 \text{ g Mn}}{10.000 \text{ g steel}} \times 100\% = 3.286\% \text{ Mn} \qquad \%\text{Cr} = \frac{2.82 \text{ g Cr}}{10.000 \text{ g steel}} \times 100\% = 28.2\% \text{ Cr}$$

68. (a) We use the technique of Chapter 3 to determine the empirical formula of nickel dimethylglyoximate.

$$20.31 \text{ g Ni} \times \frac{1 \text{ mol Ni}}{58.69 \text{ g Ni}} = 0.3461 \text{ mol Ni} \qquad \div 0.3461 \longrightarrow 1.000 \text{ mol Ni}$$

$$33.26 \text{ g C} \times \frac{1 \text{ mol C}}{12.011 \text{ g C}} = 2.769 \text{ mol C} \qquad \div 0.3461 \longrightarrow 8.001 \text{ mol C}$$

$$4.88 \text{ g H} \times \frac{1 \text{ mol H}}{1.008 \text{ g H}} = 4.84 \text{ mol H} \qquad \div 0.3461 \longrightarrow 14.0 \text{ mol H}$$

$$22.15 \text{ g O} \times \frac{1 \text{ mol O}}{15.999 \text{ g O}} = 1.384 \text{ mol O} \qquad \div 0.3461 \longrightarrow 3.999 \text{ mol O}$$

$$19.39 \text{ g N} \times \frac{1 \text{ mol N}}{14.007 \text{ g N}} = 1.384 \text{ mol N} \qquad \div 0.3461 \longrightarrow 3.999 \text{ mol N}$$

The empirical formula of nickel dimethylglyoximate (NiDMG) is $NiC_8H_{14}O_4N_4$, with an empirical molar mass of 288.91 g/mol

(b) mass of Ni $= 250.0 \text{ mL} \times \dfrac{0.104 \text{ g NiDMG}}{10.00 \text{ mL}} \times \dfrac{1 \text{ mol NiDMG}}{288.91 \text{ g NiDMG}} \times \dfrac{1 \text{ mol Ni}}{1 \text{ mol NiDMG}} \times \dfrac{58.69 \text{ g Ni}}{1 \text{ mol Ni}}$

$$= 0.528 \text{ g Ni}$$

$$\% \text{ Ni in steel} = \frac{0.528 \text{ g Ni}}{15.020 \text{ g steel}} \times 100\% = 3.52\%$$

69. A look at Appendix D indicates that $Cr(OH)_3$, $K_{sp} = 6.3 \times 10^{-31}$; $Zn(OH)_2$, $K_{sp} = 1.2 \times 10^{-17}$; $Fe(OH)_3$, $K_{sp} = 4 \times 10^{-38}$; and $Ni(OH)_2$, $K_{sp} = 2.0 \times 10^{-15}$; all are expected to form. But in excess NaOH, $[Cr(OH)_4]^-$, $K_f = 8 \times 10^{29}$, which probably is green in color; and $[Zn(OH)_4]^{2-}$, $K_f = 4.6 \times 10^{17}$, also will form. (i.e. their hydroxides will dissolve. The lack of color in the solution indicates that Cr^{3+} is not present. The presence of a hydroxide precipitate indicates that Ni^{2+} and Fe^{3+} may be present. Because Ni^{2+} forms a complex ion with NH_3, $[Ni(NH_3)_6]^{2+}$, $K_f = 5.5 \times 10^8$, and Fe^{3+} does not, the lack of a second precipitate indicates that Fe^{3+} is absent. The fact that there was a hydroxide precipitate indicates that Ni^{2+} is present. We are sure that Ni^{2+} is present and that Fe^{2+} and Cr^{3+} are absent, but we are uncertain about the presence or absence of Zn^{2+}.

Feature Problems

70. **(a)** If $\Delta n_{gas} = 0$, then $\Delta S° \sim 0$ and $\Delta G°$ is essentially independent of temperature

$$(C(s) + O_2(g) \rightarrow CO_2(g))$$

If $\Delta n_{gas} > 0$, then $\Delta S° > 0$ and $\Delta G°$ will become more negative with increasing temperature, hence the graph has a negative slope $(2\,C(s) + O_2(g) \rightarrow 2\,CO\,(g))$

If $\Delta n_{gas} < 0$, then $\Delta S° < 0$ and $\Delta G°$ will become more positive with increasing temperature, hence the graph has a positive slope $(2\,CO(g) + O_2(g) \rightarrow 2\,CO_2\,(g))$

(b) The additional blast furnace reaction, $C(s) + CO_2(g) \rightarrow 2\,CO(g))$, has

$$\Delta H° = [2 \times -110.5 \text{ kJ}] - [-393.5 \text{ kJ} + 0 \text{ kJ}] = 172.5 \text{ kJ} \quad \text{and}$$

$$\Delta S° = [2 \times 197.7 \text{ J K}^{-1}] - [1 \times 5.74 \text{ J K}^{-1} + 213.7 \text{ J K}^{-1}] = 176.0 \text{ J K}^{-1}$$

It can be obtained by adding reaction (b) to the reverse of reaction (c) (both appear in the provided figure)

C(s) + O$_2$(g) → CO$_2$(g)   Reaction (b)

2CO$_2$(s) → 2CO(g) + O$_2$(g) Reverse of Reaction (c)

Net: C(s) + CO$_2$(g) → 2 CO(g) (Additional Blast Furnace Reaction)

Consequently, the plot of $\Delta G°$ for the *net* reaction as a function of temperature will be a straight line with a slope of $-[\{\Delta S° \text{ (Rxn b)}\} - \{\Delta S° \text{ (Rxn c)}\}]$ (in kJ/K) and a y-intercept of $[\{\Delta H° \text{ (Rxn b)}\} - \{\Delta H° \text{ (Rxn c)}\}]$ (in kJ). Since $\Delta H°$ (Rxn b) = −393.5 kJ, $\Delta H°$ (Rxn c) = −566 kJ, $\Delta S°$ (Rxn b) = 2.9 J/K and $\Delta S°$ (Rxn c) = −173.1 J/K, the plot of $\Delta G°$ vs. T for the reaction C(s) + CO$_2$(g) → 2 CO(g) will follow the equation $y = -0.176x + 172.5$

From the graph, we can see that the difference in $\Delta G°$ (line b – line c at 1000 °C) is ~ −40 kJ/mol. K_p is readily calculated using this value of $\Delta G°$.

$\Delta G° = -RT\ln K_p = -40 \text{ kJ} = -(8.3145 \times 10^{-3} \text{ kJ K}^{-1} \text{ mol}^{-1})(1273 \text{ K})(\ln K_p)$

$\ln K_p = 3.\underline{8}$ Hence, $K_p = 4\underline{4}$

The equilibrium partial pressure for CO(g) is then determined by using the K_p expression:

$$K_p = \frac{(P_{CO})^2}{(P_{CO_2})} = 4\underline{4} = \frac{(P_{CO})^2}{(0.25 \text{ atm})}$$ Hence, $(P_{CO})^2 = 1\underline{1}$ and $P_{CO} = 3.\underline{3}$ atm or 3 atm.

Alternatively, we can determine the partial pressure for CO$_2$ at 1000 °C via the calculated $\Delta H°$ and $\Delta S°$ values for the reaction C(s) + CO$_2$(g) → 2 CO(g) to find $\Delta G°$ at 1000 °C, and ultimately K_p with the relationship $\Delta G° = -RT\ln K_p$ (here we are making the assumption that $\Delta H°$ and $\Delta S°$ are relatively constant over the temperature range 298 K to 1273 K). The calculated values of H° and $\Delta S°$ (using Appendix D) are given below:

$\Delta H° = [2 \times -110.5 \text{ kJ}] - [-393.5 \text{ kJ} + 0 \text{ kJ}] = 172.5 \text{ kJ}$

$\Delta S° = [2 \times 197.7 \text{ J K}^{-1}] - [1 \times 5.74 \text{ J K}^{-1} + 213.7 \text{ J K}^{-1}] = 176.0 \text{ J K}^{-1}$

To find $\Delta G°$ at 1000°C, we simply plug $x = 1273$ K into the straight-line equation we developed above and solve for y ($\Delta G°$).

So, $y = -0.176(1273 \text{ K}) + 172.5$; $y = -51.5 \text{ kJ}$

Next we need to calculate the K_p for the reaction at 1000°C.

$\Delta G° = -RT\ln K_p = -51.5 \text{ kJ} = -(8.3145 \times 10^{-3} \text{ kJ K}^{-1} \text{ mol}^{-1})(1273 \text{ K})(\ln K_p)$

$\ln K_p = 4.87$ Hence, $K_p = 1.3 \times 10^2$

The equilibrium $P_{CO(g)}$ is then determined by using the K_p expression:

$$K_p = \frac{(P_{CO})^2}{(P_{CO_2})} = 130 = \frac{(P_{CO})^2}{(0.25 \text{ atm})}$$ Hence, $(P_{CO})^2 = 32.\underline{5}$ atm and $P_{CO} = 5.7$ atm

71. **(a)** The amphoteric cations are Al^{3+}, Cr^{3+} and Zn^{2+}.

$$Cr^{3+}(aq) + 4OH^-(aq) \rightarrow [Cr(OH)_4]^-(aq)$$

$$Al^{3+}(aq) + 4OH^-(aq) \rightarrow [Al(OH)_4]^-(aq)$$

$$Zn^{2+}(aq) + 4OH^-(aq) \rightarrow [Zn(OH)_4]^{2-}(aq)$$

$$Fe^{3+}(aq) + 3OH^-(aq) \rightarrow Fe(OH)_3(s)$$

$$Ni^{2+}(aq) + 2OH^-(aq) \rightarrow Ni(OH)_2(s)$$

$$Co^{2+}(aq) + 2OH^-(aq) \rightarrow Co(OH)_2(s)$$

(b) Of the three hydroxide precipitates, only Co^{2+} is easily oxidized:
(refer to the Standard Reduction Potential table in Appendix D)

$$2\,Co(OH)_2(s) + H_2O_2(aq) \longrightarrow 2\,Co(OH)_3(s)$$

(c) We know that the chromate ion is yellow. Thus,

$$[Cr(OH)_4]^-(aq) + 3/2\,H_2O_2(aq) + OH^-(aq) \longrightarrow CrO_4^{2-}(aq,\,YELLOW) + 4\,H_2O(l)$$

(d) Co^{3+} is reduced to Co^{2+} by the H_2O_2 in solution. The other hydroxides simply dissolve in strong acid:

i) $Co(OH)_3(s) + 3H^+(aq) \longrightarrow Co^{3+}(aq) + 3H_2O(l)$

ii) $2\,Co^{3+}(aq) + 2\,Cl^-(aq) \longrightarrow 2\,Co^{2+}(aq) + Cl_2(g)$

iii) $Fe(OH)_3(s) + 3\,H^+(aq) \longrightarrow Fe^{3+}(aq) + 3\,H_2O(l)$

iv) $Ni(OH)_2(s) + 2\,H^+(aq) \longrightarrow Ni^{2+}(aq) + 2\,H_2O(l)$

(e) The $Fe^{3+}(aq)$ ions in the presence of concentrated ammonia(6 M) form an insoluble precipitate with the hydroxide ions generated by the ammonia hydrolysis reaction:

$$NH_3(aq) + H_2O(l) \rightleftharpoons NH_4^+(aq) + OH^-(aq)$$

$$Fe^{3+}(aq) + 3\,OH^-(aq) \rightleftharpoons Fe(OH)_3(s); \qquad K_{sp}\,Fe(OH)_3 = 4\times10^{-38}$$

Both Co^{3+} and Ni^{2+} form soluble complex ions with ammonia ligands rather than hydroxide precipitates.

CHAPTER 24
COMPLEX IONS AND COORDINATION COMPOUNDS
PRACTICE EXAMPLES

1A There are two different kinds of ligands in this complex ion, I^- and CN^-. Both are monodentate ligands, that is, they each form only one bond to the central atom. Since there are five ligands in total for the complex ion, the coordination number is 5: C.N.= 5. Each CN^- ligand has a charge of -1, as does the I^- ligand. Thus, the O.S. must be such that: $O.S.+\left[(4+1)\times(-1)\right]=-3= O.S.-5$. Therefore, $O.S.=+2$.

1B The ligands are all CN^-. Fe^{3+} is the central metal ion. The complex ion is $\left[Fe(CN)_6\right]^{3-}$.

2A There are six "Cl^-" ligands (chloride), each with a charge of $1-$. The platinum metal center has an oxidation state of $+4$. Thus, the complex ion is $\left[PtCl_6\right]^{2-}$, and we need two K^+ to balance charge: $K_2\left[PtCl_6\right]$

2B The "SCN^-" ligand is the thiocyanato group, with a charge of $1-$, bonding to the central metal ion through the sulfur atom. The "NH_3" ligand is ammonia, a neutral ligand. There are five (penta) ammine ligands bonded to the metal. The oxidation state of the cobalt atom is $+3$. The complex ion is not negatively charged, so its name does not end with "ate". The name of the compound is pentaamminethiocyanato-S-cobalt(III) chloride

3A The oxalato ligand must occupy two *cis-* positions. Either the two NH_3 or the two Cl^- ligands can be coplanar with the oxalate ligand, leaving the other two ligands axial. The other isomer has one NH_3 and one Cl^- ligand coplanar with the oxalate ligand. The structures are sketched below.

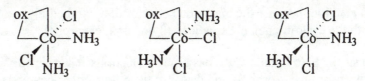

3B We start out with the two pyridines, C_5H_5N, located *cis* to each other. With this assignment imposed, we can have the two Cl^- ligands *trans* and the two CO ligands *cis*, the two CO ligands *trans* and the two Cl^- ligands *cis*, or both the Cl^- ligands and the two CO ligands *cis*. If we now place the two pyridines *trans*, we can either have both other pairs *trans*, or both other pairs *cis*. There are five geometric isomers. They follow, in the order described.

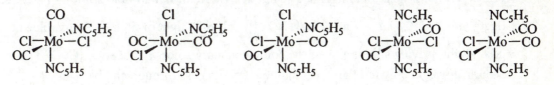

899

4A The F^- ligand is a weak field ligand. $\left[MnF_6\right]^{2-}$ is an octahedral complex. Mn^{4+} has three $3d$ electrons. The ligand field splitting diagram for $\left[MnF_6\right]^{2-}$ is sketched below. There are three unpaired electrons.

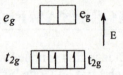

4B Co^{2+} has seven $3d$ electrons. Cl^- is a weak field ligand. H_2O is a moderate field ligand. There are three unpaired electrons in each case. The number of unpaired electrons has no dependence on geometry for either metal ion complex.

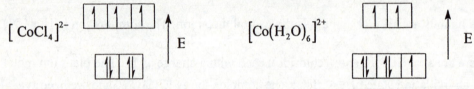

5A CN^- is a strong field ligand. Co^{2+} has seven $3d$ electrons. In the absence of a crystal field, all five d orbitals of the same energy. Three of the seven d electrons in this case will be unpaired. We need an orbital splitting diagram in which there are three orbitals of the same energy at higher energy. This is the case with a tetrahedral orbital diagram.

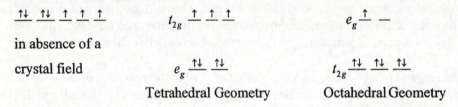

Thus, $\left[Co(CN)_4\right]^{2-}$ must be tetrahedral (3 unpaired electrons) and not octahedral (1 unpaired electron) because the magnetic behavior of a tetrahedral arrangement would agree with the experimental observations (3 unpaired electrons).

5B NH_3 is a strong field ligand. Cu^{2+} has nine $3d$ electrons. There is only one way to arrange nine electrons in five d-orbitals and that is to have four fully occupied orbitals (two electrons in each orbital), and one half-filled orbital. Thus, the complex ion must be paramagnetic to the extent of one unpaired electron, regardless of the geometry the ligands adopt around the central metal ion.

6A We are certain that $\left[Co\left(H_2O\right)_6\right]^{2+}$ is octahedral with a moderate field ligand. Tetrahedral $\left[CoCl_4\right]^{2-}$ has a weak field ligand. The relative values of ligand field splitting for the same ligand are $\Delta_t = 0.44\,\Delta_o$. Thus, $\left[Co\left(H_2O\right)_6\right]^{2+}$ absorbs light of higher energy, blue or green light, leaving a light pink as the complementary color we observe. $\left[CoCl_4\right]^{2-}$ absorbs lower energy red light, leaving blue light to pass through and be seen.

<u>**6B**</u> In order, the two complex ions are $\left[\text{Fe}(\text{H}_2\text{O})_6\right]^{2+}$ and $\left[\text{Fe}(\text{CN})_6\right]^{4-}$. We know that CN^- is a strong field ligand; it should give rise to a large value of Δ_o and absorb light of the shorter wavelength. We would expect the cyano complex to absorb blue or violet light and thus $\text{K}_4\left[\text{Fe}(\text{CN})_6\right]\cdot 3\,\text{H}_2\text{O}$ should appear yellow. The compound $\left[\text{Fe}(\text{H}_2\text{O})_6\right](\text{NO}_3)_2$, contains the weak field ligand H_2O and thus should be green. (the weak field would result in the absorption of light of long wavelength (namely, red light), which would leave green as the color we observe).

EXERCISES

Nomenclature

<u>**1.**</u>   **(a)** $\left[\text{CrCl}_4(\text{NH}_3)_2\right]^-$ diamminetetrachlorochromate(III) ion

(b) $\left[\text{Fe}(\text{CN})_6\right]^{3-}$ hexacyanoferrate(III) ion

(c) $\left[\text{Cr}(\text{en})_3\right]_2\left[\text{Ni}(\text{CN})_4\right]_3$ tris(ethylenediamine)chromium(III) tetracyanonickelate(II) ion

2.   **(a)** $\left[\text{Co}(\text{NH}_3)_6\right]^{2+}$ The coordination number of Co is 6; there are six monodentate NH_3 ligands attached to Co. Since the NH_3 ligand is neutral, the oxidation state of cobalt is $+2$, the same as the charge for the complex ion; hexaamminecobalt(II) ion

(b) $\left[\text{AlF}_6\right]^{3-}$ The coordination number of Al is six; F^- is monodentate. Each F^- has a $1-$ charge; thus the oxidation state of Al is $+3$; hexafluoroaluminate(III) ion

(c) $\left[\text{Cu}(\text{CN})_4\right]^{2-}$ The coordination number of Cu is 4; CN^- is monodendate. CN^- has a $1-$ charge; thus the oxidation state of Cu is $+2$; tetracyanocuprate(II) ion

(d) $\left[\text{CrBr}_2(\text{NH}_3)_4\right]^+$ The coordination number of Cr is 6; NH_3 and Br^- are monodentate. NH_3 has no charge; Br^- has a $1-$ charge. The oxidation state of chromium is $+3$; tetraamminedibromochromium(III) ion

(e) $\left[\text{Co}(\text{ox})_3\right]^{4-}$ The coordination number of Co is 6; oxalate is bidentate. $\text{C}_2\text{O}_4^{2-}$ (ox) has a $2-$ charge; thus the oxidation state of cobalt is $+2$; trioxalatocobaltate(II) ion.

(f) $\left[\text{Ag}(\text{S}_2\text{O}_3)_2\right]^{3-}$ The coordination number of Ag is 2; $\text{S}_2\text{O}_3^{2-}$ is monodentate. $\text{S}_2\text{O}_3^{2-}$ has a $2-$ charge; thus the oxidation state of silver is $+1$; dithiosulfatoargentate(I) ion. (Although $+1$ is by far the most common oxidation state for silver in its compounds, stable silver(III) complexes are known. Thus, strictly speaking, silver is not a non-variable metal, and hence when naming silver compounds, the oxidation state(s) for the silver atom(s) should be specified).

3. (a) $\left[Co(OH)(H_2O)_4(NH_3)\right]^{2+}$ amminetetraaquahydroxocobalt(III) ion

(b) $\left[Co(ONO)_3(NH_3)_3\right]$ triamminetrinitrito-O-cobalt(III)

(c) $\left[Pt(H_2O)_4\right]\left[PtCl_6\right]$ tetraaquaplatinum(II) hexachloroplatinate(IV)

(d) $\left[Fe(ox)_2(H_2O)_2\right]^-$ diaquadioxalatoferrate(III) ion

(e) $Ag_2\left[HgI_4\right]$ silver(I) tetraiodomercurate(II)

4. (a) $K_3\left[Fe(CN)_6\right]$ potassium hexacyanoferrate(III)

(b) $\left[Cu(en)_2\right]^{2+}$ bis(ethylenediamine)copper(II) ion

(c) $\left[Al(OH)(H_2O)_5\right]Cl_2$ pentaaquahydroxoaluminum(III) chloride

(d) $\left[CrCl(en)_2NH_3\right]SO_4$ amminechlorobis(ethylenediammine)chromium(III) sulfate

(e) $\left[Fe(en)_3\right]_4\left[Fe(CN)_6\right]_3$ tris(ethylenediamine)iron(III) hexacyanoferrate(II)

Bonding and Structure in Complex Ions

5. The Lewis structures are grouped together at the end.

(a) H_2O has $1\times2 + 6 = 8$ valence electrons, or 4 pairs.

(b) CH_3NH_2 has $4 + 3\times1 + 5 + 2\times1 = 14$ valence electrons, or 7 pairs.

(c) ONO^- has $2\times6 + 5 + 1 = 18$ valence electrons, or 9 pairs. The structure has a 1– formal charge on the oxygen that is singly bonded to N.

(d) SCN^- has $6 + 4 + 5 + 1 = 16$ valence electrons, or 8 pairs. This structure, appropriately, gives a 1– formal charge to N.

    (a)   (b)   (c)   (d)

6. The Lewis structures are grouped together at the end.

(a) OH^- has $6 + 1 + 1 = 8$ valence electrons, or 4 pairs.

(b) SO_4^{2-} has $6 + 6\times4 + 2 = 32$ valence electrons, or 16 pairs.

(c) $C_2O_4^{2-}$ has $2\times4 + 6\times4 + 2 = 34$ valence electrons, or 17 pairs.

(d) SCN^- has $6+4+5+1=16$ valence electrons, or 8 pairs. This structure, appropriately, gives a $1-$ formal charge to N.

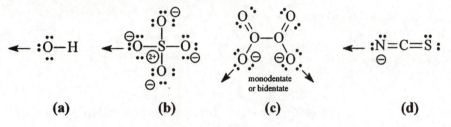

(a)   **(b)**   **(c)**   **(d)**

7. We assume that $\left[PtCl_4\right]^{2-}$ is square planar by analogy with $\left[PtCl_2(NH_3)_2\right]$ in Figure 24-5. The other two complex ions are octahedral.

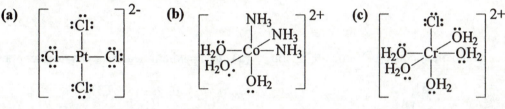

8. The structures for $\left[Pt(ox)_2\right]^{2-}$ and $\left[Cr(ox)_3\right]^{3-}$ are drawn below. The structure of $\left[Fe(EDTA)\right]^{2-}$ is the same as the generic structure for $\left[M(EDTA)\right]^{2-}$ drawn in Figure 24-23, with $M^{n+} = Fe^{2+}$.

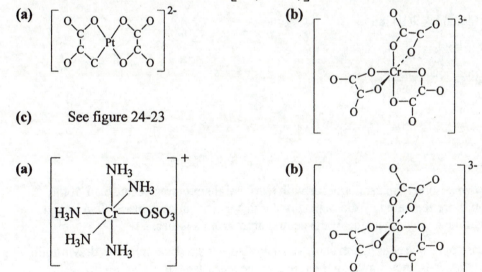

(c)   See figure 24-23

9. **(a)**

pentaamminesulfateochromium(III)

(b)

trioxalatocobaltate(III)

(c)

fac-triamminedichloronitro-O-cobalt(III)   *mer*-triammine-*cis*-dichloronitro-O-cobalt(III)

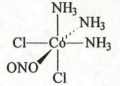

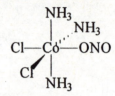

mer-triammine-*trans*-dichloronitro-O-cobalt(III)

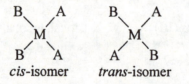

10. **(a)** pentaamminenitroto-N-cobalt(III) ion   **(b)** ethylenediaminedithiocyanato-S-copper(II)

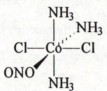

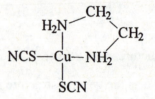

(c) hexaaquanickel(II) ion

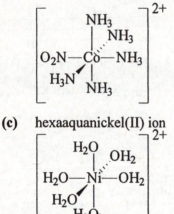

Isomerism

11. **(a)** *cis-trans* isomerism cannot occur with tetrahedral structures because all of the ligands are separated by the same angular distance from each other. One ligand cannot be on the other side of the central atom from another.

(b) Square planar structures can show *cis-trans* isomerism. Examples are drawn following, with the *cis*-isomer drawn on the left, and the *trans*-isomer drawn on the right.

```
   B     A          B       A
    \   /            \     /
     M                 M
    /   \            /     \
   B     A          A       B
 cis-isomer       trans-isomer
```

(c) Linear structures do not display *cis-trans* isomerism; there is only one way to bond the two ligands to the central atom.

12. **(a)** $\left[CrOH(NH_3)_5\right]^{2+}$ has one isomer.

(b) $\left[CrCl_2(H_2O)(NH_3)_3\right]^+$ has three isomers.

(c) $\left[CrCl_2(en)_2\right]^+$ has two geometric isomers, *cis*- and *trans*-.

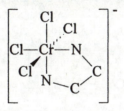

(d) $\left[CrCl_4(en)\right]^-$ has only one isomer since the ethylenediamine (en) ligand cannot bond *trans* to the central metal atom.

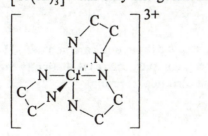

(e) $\left[Cr(en)_3\right]^{3+}$ has only one geometric isomer; it has two optical isomers.

13. (a) There are three different square planar isomers, with D, C, and B, respectively, trans to the A ligand. They are drawn below.

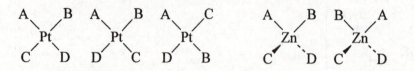

(b) Tetrahedral $[\text{ZnABCD}]^{2+}$ does display optical isomerism. The two optical isomers are drawn above.

14. There are a total of four coordination isomers. They are listed below. We assume that the oxidation state of each central metal ion is $+3$.

$[\text{Co}(\text{en})_3][\text{Cr}(\text{ox})_3]$ tris(ethylenediamine)cobalt(III) trioxalatochromate(III)

$[\text{Co}(\text{ox})(\text{en})_2][\text{Cr}(\text{ox})_2(\text{en})]$ bis(ethylenediamine)oxalatocobalt(III) (ethylenediamine)dioxalatochromate(III)

$[\text{Cr}(\text{ox})(\text{en})_2][\text{Co}(\text{ox})_2(\text{en})]$ bis(ethylenediamine)oxalatochromium(III) (ethylenediamine)dioxalatocobaltate(III)

$[\text{Cr}(\text{en})_3][\text{Co}(\text{ox})_3]$ tris(ethylenediamine)chromium(III) trioxalatocobaltate(III)

15. The *cis*-dichlorobis(ethylenediamine)cobalt(III) ion is optically active. The two optical isomers are drawn below. The *trans*-isomer is not optically active: the ion and its mirror image are superimposable.

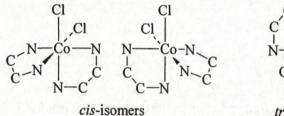

cis-isomers 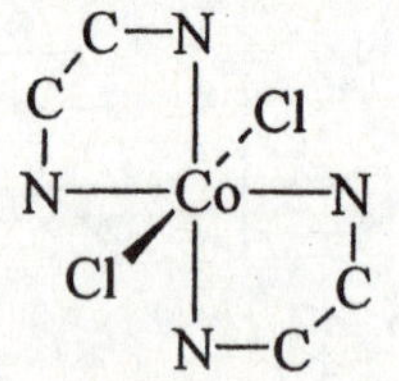   *trans*-isomer

16. Complex ions (a) and (b) are identical; complex ions (a) and (d) are geometric isomers; complex ions (b) and (d) are geometric isomers; complex ion (c) is distinctly different from the other three complex ions (it has a different chemical formula).

Crystal Field Theory

17. In crystal field theory, the five *d* orbitals of a central transition metal ion are split into two (or more) groups of different energies. The energy spacing between these groups often corresponds to the energy of a photon of visible light. Thus, the transition-metal complex will absorb light with energy corresponding to this spacing. If white light is incident on the complex ion, the light remaining after absorption will be missing some of its components. Thus, light of certain wavelengths (corresponding to the energies absorbed) will no longer be present in the formerly white light. The resulting light is colored. For example, if blue light is absorbed from white light, the remaining light will be yellow in color.

18. The difference in color is due to the difference in the value of Δ, which is the ligand field splitting energy. When the value of Δ is large, short wavelength light, which has a blue color, is absorbed, and the substance or its solution appears yellow. On the other hand, when the value of Δ is small, light of long wavelength, which has a red or yellow color, is absorbed, and the substance or its solution appears blue. The cyano ligand is a strong field ligand, producing a large value of Δ, and thus yellow complexes. On the other hand, the aqua ligands are weak field ligands, which produce a small value of Δ, and hence blue complexes.

19. We begin with the 7 electron *d*-orbital diagram for Co^{2+} [Ar] ⇅ ⇅ ↑ ↑ ↑

The strong field and weak field diagrams for octahedral complexes follow, along with the number of unpaired electrons in each case.

Strong Field ↑ __ e_g

(1 unpaired electron) ⇅ ⇅ ⇅ t_{2g}

Weak Field ↑ ↑ e_g

(3 unpaired electrons) ⇅ ⇅ ↑ t_{2g}

20. We begin with the 8 electron *d*-orbital diagram for Ni^{2+} [Ar] ⇅ ⇅ ⇅ ↑ ↑

The strong field and weak field diagrams for octahedral complexes follow, along with the number of unpaired electrons in each case.

Strong Field ↑ ↑ e_g

(2 unpaired electrons) ⇅ ⇅ ⇅ t_{2g}

Weak Field ↑ ↑ e_g

(2 unpaired electrons) t_{2g}

The number of unpaired electrons is the same in both cases.

21. (a) Both of the central atoms have the same oxidation state, +3. We give the electron configuration of the central atom to the left, then the completed crystal field diagram in the center, and finally the number of unpaired electrons. The chloro ligand is a weak field ligand in the spectrochemical series.

Mo^{3+} $[Kr]\,4d^3$ *weak field*   ▢▢ e_g

      $[Kr]_{4d}$ ⎡↿│↿│↿│ │ ⎤   ↿│↿│↿ t_{2g}

                        3 unpaired electrons; paramagnetic

The ethylenediamine ligand is a strong field ligand in the spectrochemical series.

Co^{3+} $[Ar]\,3d^6$ *strong field*   ▢▢ e_g

$[Ar]$ ↿⇂│↿│↿│↿│↿

                                  ↿⇂│↿⇂│↿⇂ t_{2g}

                 no unpaired electrons; diamagnetic

(b) In $[CoCl_4]^{2-}$ the oxidation state of cobalt is 2+. Chloro is a weak field ligand. The electron configuration of Co^{2+} is $[Ar]\,3d^7$ or $[Ar]$ ↿⇂│↿⇂│↿│↿│↿

The tetrahedral ligand field diagram is   *weak field*   ↿│↿│↿ t_{2g}
shown on the right.   ↿⇂│↿⇂ e_g

                               3 unpaired electrons

22. (a) In $\left[Cu(py)_4\right]^{2+}$ the oxidation state of copper is +2. Pyridine is a strong field ligand. The electron configuration of Cu^{2+} is $[Ar]\,3d^9$ or $[Ar]$ ↿⇂│↿⇂│↿⇂│↿⇂│↿

There is no possible way that an odd number of electrons can be paired up, without at least one electron being unpaired. $\left[Cu(py)_4\right]^{2+}$ is paramagnetic.

(b) In $\left[Mn(CN)_6\right]^{3-}$ the oxidation state of manganese is +3. Cyano is a strong field ligand. The electron configuration of Mn^{3+} is $[Ar]\,3d^4$ or $[Ar]$ ↿│↿│↿│↿│ │

The ligand field diagram follows, on the left-hand side. In $\left[FeCl_4\right]^-$ the oxidation state of iron is +3. Chloro is a weak field ligand. The electron configuration of Fe^{3+} is $[Ar]\,3d^5$ or

$[Ar]$ ↿│↿│↿│↿│↿

The ligand field diagram follows, below.

strong field   ▢▢ e_g *weak field*   ↿│↿│↿ t_{2g}
                                        ↿│↿ e_g

           ↿⇂│↿│↿ t_{2g}

2 unpaired electrons   5 unpaired electrons

There are more unpaired electrons in $[FeCl_4]^-$ than in $[Mn(CN)_6]^{3-}$.

23. The electron configuration of Ni^{2+} is $[Ar] 3d^8$ or $[Ar]$ [⇅][⇅][⇅][↑][↑]

Ammonia is a strong field ligand. The ligand field diagrams follow, octahedral at left, tetrahedral in the center and square planar at right.

Octahedral   Tetrahedral   Square Planar

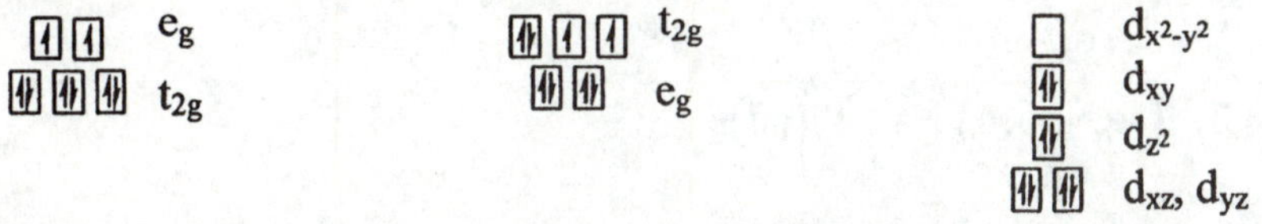

[↑][↑] e_g [⇅][↑][↑] t_{2g} [ ] $d_{x^2-y^2}$

[⇅][⇅][⇅] t_{2g} [⇅][⇅] e_g [⇅] d_{xy}

                                           [⇅] d_{z^2}

                                           [⇅][⇅] d_{xz}, d_{yz}

Since the octahedral and tetrahedral configurations have the same number of unpaired electrons (that is, 2 unpaired electrons), we cannot use magnetic behavior to determine whether the ammine complex of nickel(II) is octahedral or tetrahedral. But we can determine if the complex is square planar, since the square planar complex is diamagnetic with zero unpaired electrons.

24. The difference is due to the fact that $[Fe(CN_6)]^{4-}$ is a strong field complex ion, while $[Fe(H_2O)_6]^{2+}$ is a weak field complex ion. The electron configurations for an iron atom and an iron(II) ion, and the ligand field diagrams for the two complex ions follow.

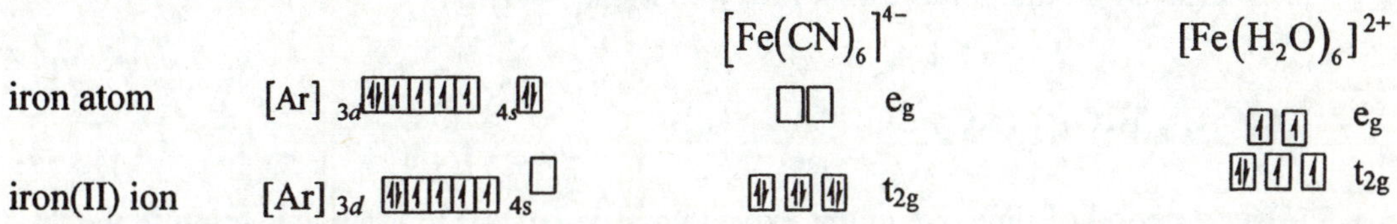

                                                  $[Fe(CN)_6]^{4-}$ $[Fe(H_2O)_6]^{2+}$

iron atom $[Ar]$ $_{3d}$[⇅][↑][↑][↑][↑] $_{4s}$[⇅]   [ ][ ] e_g

                                                                         [↑][↑] e_g

iron(II) ion $[Ar]$ $_{3d}$[⇅][↑][↑][↑][↑] $_{4s}$[ ]   [⇅][⇅][⇅] t_{2g} [⇅][↑][↑] t_{2g}

Complex Ion Equilibria

25. **(a)** $Zn(OH)_2(s) + 4 NH_3(aq) \rightleftharpoons [Zn(NH_3)_4]^{2+}(aq) + 2 OH^-(aq)$

     (b) $Cu^{2+}(aq) + 2 OH^-(aq) \rightleftharpoons Cu(OH)_2(s)$

The blue color is most likely due to the presence of some unreacted $[Cu(H_2O)_4]^{2+}$ (pale blue)

$Cu(OH)_2(s) + 4 NH_3(aq) \rightleftharpoons [Cu(NH_3)_4]^{2+}(aq, \text{dark blue}) + 2 OH^-(aq)$

$[Cu(NH_3)_4]^{2+}(aq) + 4 H_3O^+(aq) \rightleftharpoons [Cu(H_2O)_4]^{2+}(aq) + 4 NH_4^+(aq)$

26. **(a)** $CuCl_2(s) + 2\ Cl^-(aq) \rightleftharpoons [CuCl_4]^{2-}(aq,\ yellow)$

$2[CuCl_4]^{2-}(aq) + 4H_2O(l)$

$\Updownarrow$

$[CuCl_4]^{2-}(aq,\ yellow) + [Cu(H_2O)_4]^{2+}(aq,\ pale\ blue) + 4\ Cl^-(aq)$

or

$2[Cu(H_2O)_2Cl_2](aq,\ green) + 4\ Cl^-(aq)$

Either $\quad [CuCl_4]^{2-}(aq) + 4\ H_2O(l) \rightleftharpoons [Cu(H_2O)_4]^{2+}(aq) + 4\ Cl^-(aq)$

Or $\quad [Cu(H_2O)_2Cl_2](aq) + 2H_2O(l) \rightleftharpoons [Cu(H_2O)_4]^{2+}(aq) + 2\ Cl^-(aq)$

(b) The blue solution is that of $[Cr(H_2O)_6]^{2+}$. This is quickly oxidized to Cr^{3+} by $O_2(g)$ from the atmosphere. The green color is due to $[CrCl_2(H_2O)_4]^{3+}$.

$4\ [Cr(H_2O)_6]^{2+}(aq,\ blue) + 4\ H^+(aq) + 8\ Cl^-(aq) + O_2(g)$

$\downarrow$

$4\ [CrCl_2(H_2O)_4]^+(aq,\ green) + 10\ H_2O(l)$

Over a period of time, we might expect volatile $HCl(g)$ to escape, leading to the formation of complex ions with more H_2O and less Cl^-.

$H^+(aq) + Cl^-(aq) \rightleftharpoons HCl(g)$

$[CrCl_2(H_2O)_4]^+(aq,\ green) + H_2O(l) \rightleftharpoons [CrCl(H_2O)_5]^{2+}(aq,\ blue\text{-}green) + Cl^-(aq)$

$[CrCl(H_2O)_5]^{2+}(aq,\ blue\text{-}green) + H_2O(l) \rightleftharpoons [Cr(H_2O)_6]^{3+}(aq,\ blue) + Cl^-(aq)$

Actually, to ensure that these final two reactions actually do occur in a timely fashion, it would be helpful to dilute the solution with water after the chromium metal has dissolved.

27. $[Co(en)_3]^{3+}$ should have the largest overall K_f value. We expect a complex ion with polydentate ligands to have a larger value for its formation constant than complexes that contain only monodentate ligands. This phenomena is known as the chelate effect. Once one end of a polydentate ligand becomes attached to the central metal, the attachment of the remaining electron pairs is relatively easy because they already are close to the central metal (and do not have to migrate in from a distant point in the solution).

28. (a) $\left[Zn(NH_3)_4 \right]^{2+}$

$\beta_4 = K_1 \times K_2 \times K_3 \times K_4 = 3.9 \times 10^2 \times 2.1 \times 10^2 \times 1.0 \times 10^2 \times 50. = 4.1 \times 10^8.$

(b) $\left[Ni(H_2O)_2(NH_3)_4 \right]^{2+}$

$\beta_4 = K_1 \times K_2 \times K_3 \times K_4 = 6.3 \times 10^2 \times 1.7 \times 10^2 \times 54 \times 15 = 8.7 \times 10^7$

<u>29.</u> First: $\left[Fe(H_2O)_6 \right]^{3+}(aq) + en(aq) \rightleftharpoons \left[Fe(H_2O)_4(en) \right]^{3+}(aq) + 2H_2O(l) \quad K_1 = 10^{4.34}$

Second: $\left[Fe(H_2O)_4(en) \right]^{3+}(aq) + en(aq) \rightleftharpoons \left[Fe(H_2O)_2(en)_2 \right]^{3+}(aq) + 2H_2O(l) \quad K_2 = 10^{3.31}$

Third: $\left[Fe(H_2O)_2(en)_2 \right]^{3+}(aq) + en(aq) \rightleftharpoons \left[Fe(en)_3 \right]^{3+}(aq) + 2H_2O(l) \quad K_3 = 10^{2.05}$

Net: $\left[Fe(H_2O)_6 \right]^{3+}(aq) + 3\ en(aq) \rightleftharpoons \left[Fe(en)_3 \right]^{3+}(aq) + 6H_2O(l)$

$K_f = K_1 \times K_2 \times K_3$

$\log K_f = 4.34 + 3.31 + 2.05 = 9.70 \qquad K_f = 10^{9.70} = 5.0 \times 10^9 = \beta_3$

30. Since the overall formation constant is the product of the individual stepwise formation constants, the logarithm of the overall formation constant is the sum of the logarithms of the stepwise formation constants.

$\log K_f = \log K_1 + \log K_2 + \log K_3 + \log K_4 = 2.80 + 1.60 + 0.49 + 0.73 = 5.62$

$K_f = 10^{5.62} = 4.2 \times 10^5$

<u>31.</u> (a) Aluminum(III) forms a stable (and soluble) hydroxo complex but not a stable ammine complex.

$\left[Al(H_2O)_3(OH)_3 \right](s) + OH^-(aq) \rightleftharpoons \left[Al(H_2O)_2(OH)_4 \right]^-(aq) + H_2O(l)$

(b) Although zinc(II) forms a soluble stable ammine complex ion, its formation constant is not sufficiently large to dissolve highly insoluble ZnS. However, it is sufficiently large to dissolve the moderately insoluble $ZnCO_3$. Said another way, ZnS does not produce sufficient $\left[Zn^{2+} \right]$ to permit the complex ion to form.

$ZnCO_3(s) + 4\ NH_3(aq) \rightleftharpoons \left[Zn(NH_3)_4 \right]^{2+}(aq) + CO_3^{2-}(aq)$

(c) Chloride ion forms a stable complex ion with silver(I) ion, that dissolves the AgCl(s) that formed when $\left[Cl^- \right]$ is low.

$AgCl(s) \rightleftharpoons Ag^+(aq) + Cl^-(aq) \quad$ and $\quad Ag^+(aq) + 2\ Cl^-(aq) \rightleftharpoons \left[AgCl_2 \right]^-(aq)$

overall: $\quad AgCl(s) + Cl^-(aq) \rightleftharpoons \left[AgCl_2 \right]^-(aq)$

32. **(a)** Because of the large value of the formation constant for the complex ion, $\left[Co(NH_3)_6\right]^{3+}(aq)$, the concentration of free $Co^{3+}(aq)$ is too small to enable it to oxidize water to $O_2(g)$. Since there is not a complex ion present, except, of course, $\left[Co(H_2O)_6\right]^{3+}(aq)$, when $CoCl_3$ is dissolved in water, the $\left[Co^{3+}\right]$ is sufficiently high for the oxidation-reduction reaction to be spontaneous.

$$4\ Co^{3+}(aq) + 2\ H_2O(l) \rightarrow 4\ Co^{2+}(aq) + 4\ H^+(aq) + O_2(g)$$

(b) Although AgI(s) is often described as insoluble, there is actually a small concentration of $Ag^+(aq)$ present because of the solubility equilibrium:

$$2\ AgI(s) \rightleftharpoons Ag^+(aq) + I^-(aq)$$

These silver ions react with thiosulfate ion to form the stable dithiosulfatoargentate(I) complex ion:

$$Ag^+(aq) + 2S_2O_3^{\ 2-}(aq) \rightleftharpoons \left[Ag(S_2O_3)_2\right]^-(aq)$$

Acid-Base Properties

33. $\left[Al(H_2O)_6\right]^{3+}(aq)$ is capable of releasing H^+:

$$\left[Al(H_2O)_6\right]^{3+}(aq) + H_2O(l) \rightleftharpoons \left[AlOH(H_2O)_5\right]^{2+}(aq) + H_3O^+(aq)$$

The value of its ionization constant ($pK_a = 5.01$) approximates that of acetic acid.

34. **(a)** $\left[CrOH(H_2O)_5\right]^{2+}(aq) + OH^-(aq) \rightarrow \left[Cr(OH)_2(H_2O)_4\right]^+(aq) + H_2O(l)$

(b) $\left[CrOH(H_2O)_5\right]^{2+}(aq) + H_3O^+(aq) \rightarrow \left[Cr(H_2O)_6\right]^{3+}(aq) + H_2O(l)$

Applications

35. **(a)** Solubility: $\quad AgBr(s) \underset{}{\overset{K_{sp}=5.0\times10^{-13}}{\rightleftharpoons}} Ag^+(aq) + Br^-(aq)$

Cplx. Ion Formation: $\quad Ag^+(aq) + 2S_2O_3^{\ 2-}(aq) \underset{}{\overset{K_f=1.7\times10^{13}}{\rightleftharpoons}} \left[Ag(S_2O_3)_2\right]^{3-}(aq)$

Net: $AgBr(s) + 2S_2O_3^{\ 2-}(aq) \rightleftharpoons \left[Ag(S_2O_3)_2\right]^{3-}(aq) \qquad K_{overall} = K_{sp} \times K_f$

$K_{overall} = 5.0\times10^{-13} \times 1.7\times10^{13} = 8.5$

With a reasonably high $\left[S_2O_3^{\ 2-}\right]$, this reaction will go essentially to completion.

(b) $NH_3(aq)$ cannot be used in the fixing of photographic film because of the relatively small value of K_f for $\left[Ag(NH_3)_2\right]^+(aq)$, $K_f = 1.6 \times 10^7$. This would produce a value of $K = 8.0 \times 10^{-6}$ in the expression above, which is nowhere large enough to drive the reaction to completion.

36. Oxidation: $\left\{\left[Co(NH_3)_6\right]^{2+}(aq) \rightarrow \left[Co(NH_3)_6\right]^{3+}(aq)+e^-\right\}\times 2 \qquad -E° = -0.10V.$

Reduction: $\left\{H_2O_2(aq)+2e^- \rightarrow 2OH^-(aq)\right\} \qquad\qquad\qquad E° = 0.88V$

Net: $\qquad H_2O_2(aq)+2\left[Co(NH_3)_6\right]^{2+}(aq) \rightarrow 2\left[Co(NH_3)_6\right]^{3+}(aq)+2OH^-(aq)$

$E°_{cell} = +0.88 \text{ V} + -0.10 \text{ V} = +0.78 \text{ V}.$

The positive value of the standard cell potential indicates that this is a spontaneous reaction.

Integrative and Advanced Exercises

37.
(a) cupric tetraammine ion	tetraamminecopper(II) ion	$[Cu(NH_3)_4]^{2+}$
(b) dichlorotetraamminecobaltic chloride	tetraamminedichlorocobalt(III) chloride	$[CoCl_2(NH_3)_4]Cl$
(c) platinic(IV) hexachloride ion	hexachloroplatinate(IV) ion	$[PtCl_6]^{2-}$
(d) disodium copper tetrachloride	sodium tetrachlorocuprate(II)	$Na_2[CuCl_4]$
(e) dipotassium antimony(III) pentachloride	potassium pentachloroantimonate(III)	$K_2[SbCl_5]$

38. $[Pt(NH_3)_4][PtCl_4]$ tetraammineplatinum(II) tetrachloroplatinate(II)

39. The four possible isomers for $[CoCl_2(en)(NH_3)_2]^+$ are sketched below:

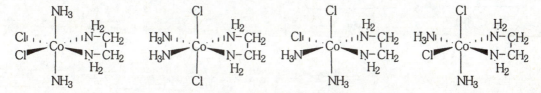

40. The color of the green solid, as well as that of the green solution, is produced by the complex ion $[CrCl_2(H_2O)_4]^+$. Over time in solution, the chloro ligands are replaced by aqua ligands, producing violet $[Cr(H_2O)_6]^{3+}(aq)$. When the water is evaporated, the original complex is re-formed as the concentration of chloro ligand, $[Cl^-]$, gets higher and the chloro ligands replace the aqua ligands.

41. The chloro ligand, being lower in the spectrochemical series than the ethylenediamine ligand, is less strongly bonded to the central atom than is the ethylenediamine ligand. Therefore, of the two types of ligands, we expect the chloro ligand to be replaced more readily. In the *cis* isomer, the two chloro ligands are 90° from each other. This is the angular spacing that can be readily spanned by the oxalato ligand, thus we expect reaction with the cis isomer to occur rapidly. On the other hand, in the *trans* isomer, the two chloro ligands are located 180° from each other. After the chloro ligands are removed, at least one end of one ethylenediamine ligand would have to be relocated to allow the oxalato ligand can bond as a bidentate ligand. Consequently, replacement of the two chloro ligands by the oxalato ligand should be much slower for the *trans* isomer than for the *cis* isomer.

42. $Cl_2 + 2 e^- \rightarrow 2 Cl^-$

$$\underline{[Pt(NH_3)_4]^{2+} \rightarrow [Pt(NH_3)_4]^{4+} + 2 e^-}$$

$Cl_2 + [Pt(NH_3)_4]^{2+} \rightarrow [Pt(Cl_2NH_3)_4]^{2+}$

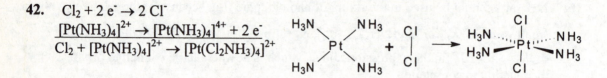

43. The successive acid ionizations of a complex ion such as $[Fe(OH)_6]^{3+}$ are more nearly equal in magnitude than those of an uncharged polyprotic acid such as H_3PO_4 principally because the complex ion has a positive charge. The second proton is leaving a species which has one fewer positive charge but which is nonetheless positively charged. Since positive charges repel each other, successive ionizations should not show a great decrease in the magnitude of their ionization constants. In the case of polyprotic acids, on the other hand, successive protons are leaving a species whose negative charge is increasingly greater with each step. Since unlike charges attract each other, it becomes increasingly difficult to remove successive protons.

44.

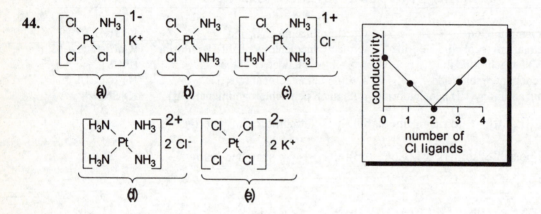

45. (a)

$$Fe(H_2O)_6^{3+}(aq) \quad + \quad H_2O(l) \xrightleftharpoons{K = 9.0 \times 10^{-4}} \quad Fe(H_2O)_5OH^{2+}(aq) \quad + \quad H_3O^+$$

initial   0.100 M   –   0 M $\approx$ 0 M

equil. $(0.100 - x)$ M   – x x

{where x is the molar quantity of $Fe(H_2O)_6^{3+}$ (aq) hydrolyzed}

$$K = \frac{[[Fe(H_2O)_5OH]^{2+}][H_3O^+]}{[[Fe(H_2O)_6]^{3+}]} = \frac{x^2}{0.100-x} = 9.0 \times 10^{-4}$$

Solving, we find $x = 9.5 \times 10^{-3}$ M, which is the $[H_3O^+]$ so:

$pH = -\log(9.5 \times 10^{-3}) = 2.02$

b)

$$[Fe(H_2O)_6]^{3+}(aq) + H_2O(l) \xrightleftharpoons{K = 9.0 \times 10^{-4}} [Fe(H_2O)_5OH]^{2+}(aq) + H_3O^+$$

initial   0.100 M   –   0 M   0.100 M

equil.   (0.100 – x) M   – x M   (0.100 + x) M

{where x is [$[Fe(H_2O)_6]^{3+}$] reacting}

$$K = \frac{[[Fe(H_2O)_5OH]^{2+}][H_3O^+]}{[[Fe(H_2O)_6]^{3+}]} = \frac{x(0.100+x)}{(0.100-x)} = 9.0 \times 10^{-4}$$

Solving, we find $x = 9.0 \times 10^{-4}$ M, which is the [$[Fe(H_2O)_5OH]^{2+}$]

c) We simply substitute [$[Fe(H_2O)_5OH]^{2+}$] = 1.0×10^{-6} M into the K_a expression

with [$Fe(H_2O)_6]^{3+}$] = 0.100 M and determine the concentration of H_3O^+

$$[H_3O^+] = \frac{K_a[[Fe(H_2O)_6]^{3+}]}{[[Fe(H_2O)_5OH]^{2+}]} = \frac{9.0 \times 10^{-4}(0.100\text{ M})}{[1.0 \times 10^{-6}\text{M}]} = 90.\text{ M}$$

To maintain the concentration at this level requires an impossibly high concentration of H_3O^+

46. Let us first determine the concentration of the uncomplexed (free) $Pb^{2+}(aq)$. Because of the large value of the formation constant, we assume that most of the lead(II) ion is present as the EDTA complex ion ($[Pb(EDTA)]^{2-}$). Once equilibrium is established, we can see if thre is sufficient lead(II) ion present in solution that will precipitate with the sulfide ion. Note: the concentration of EDTA remains constant at 0.20 M)

Reaction: $Pb^{2+}(aq) + EDTA^{4-}(aq) \xrightleftharpoons{2\times10^{18}} [Pb(EDTA)]^{2-}(aq)$

Initial: 0 M   0.20 M   0.010 M

Change: +x M   constant   –x M

Equilibrium: x M   0.20 M   (0.010–x) M $\approx$ 0.010 M

$$K_f = \frac{[Pb(EDTA)^{2-}]}{[Pb^{2+}][EDTA^{4-}]} = 2\times10^{18} = \frac{0.010}{0.20(x)} \qquad x = 2.\underline{5}\times10^{-20}\text{ M} = [Pb^{2+}]$$

Now determine the $[S^{2-}]$ required to precipitate PbS from this solution. $K_{sp} = [Pb^{2+}][S^{2-}] = 8\times10^{-28}$

$[S^{2-}] = \dfrac{8\times10^{-28}}{(5\times10^{-20})} = 2.\underline{5}\times10^{-8}$ M   Recall (i) That a saturated H_2S solution is ~0.10 M H_2S

Recal (ii) That the $[S^{2-}] = K_{a2} = 1\times10^{-14}$

$$Q_{sp} = [Pb^{2+}][S^{2-}] = 2.\underline{5}\times10^{-20} \times 1\times10^{-14} = 2.\underline{5}\times10^{-34} \qquad K_{sp} = 8\times10^{-28}$$

$Q_{sp} \ll K_{sp}$ Hence, PbS(s) will not precipitate from this solution.

47. The formation of $[Cu(NH_3)_4]^{2+}(aq)$ from $[Cu(H_2O)_4]^{2+}(aq)$ has $K_1 = 1.9 \times 10^4$, $K_2 = 3.9 \times 10^3$, $K_3 = 1.0 \times 10^3$, and $K_4 = 1.5 \times 10^2$. Since these equilibrium constants are all considerably larger than 1.00, one expects that the reactions they represent, yielding ultimately the ion $[Cu(NH_3)_4]^{2+}(aq)$, will go essentially to completion. However, if the concentration of NH_3 were limited to less than the stoichiometric amount, that is, to less than 4 mol NH_3 per mol of $Cu^{2+}(aq)$, one would expect that the ammine-aqua complex ions would be present in significantly higher concentrations than the $[Cu(NH_3)_4]^{2+}(aq)$ ions.

48. For $[Ca(EDTA)]^{2-}$, $K_f = 4 \times 10^{10}$ and for $[Mg(EDTA)]^{2-}$, $K_f = 4 \times 10^{8}$. In Table 18-1, the least soluble calcium compound is $CaCO_3$, $K_{sp} = 2.8 \times 10^{-9}$, and the least soluble magnesium compound is $Mg_3(PO_4)_2$, $K_{sp} = 1 \times 10^{-25}$. We can determine the equilibrium constants for adding carbonate ion to $[Ca(EDTA)]^{2-}(aq)$ and for adding phosphate ion to $[Mg(EDTA)]^{2-}(aq)$ as follows:

Instability: $\quad [Ca(EDTA)]^{2-}(aq) \rightleftharpoons Ca^{2+}(aq) + EDTA^{4-}(aq) \qquad K_i = 1/4 \times 10^{10}$

Precipitation: $\quad Ca^{2+}(aq) + CO_3^{2-}(aq) + \rightleftharpoons CaCO_3(s) \qquad\qquad K_{ppt} = 1/2.8 \times 10^{-9}$

Net: $\qquad\qquad [Ca(EDTA)]^{2-} + CO_3^{2-}(aq) \rightleftharpoons CaCO_3(s) + EDTA^{4-}(aq) \quad K = K_i \times K_{ppt}$

$$K = \frac{1}{(4 \times 10^{10})(2.8 \times 10^{-9})} = 9 \times 10^{-3}$$

The small value of the equilibrium constant indicates that this reaction does not proceed very far toward products.

Instability: $\quad \{[Mg(EDTA)]^{2-}(aq) \rightleftharpoons Mg^{2+}(aq) + EDTA^{4-}(aq)\} \times 3 \qquad K_i^3 = 1/(4 \times 10^8)^3$

Precipitation: $\quad 3\,Mg^{2+}(aq) + 2\,PO_4^{3-}(aq) \rightleftharpoons Mg_3(PO_4)_2(s) \qquad\qquad K_{ppt} = 1/1 \times 10^{-25}$

Net : $\qquad\quad 3\,[Mg(EDTA)]^{2-} + 2\,PO_4^{3-}(aq) \rightleftharpoons Mg_3(PO_4)_2(s) + 3\,EDTA^{4-}(aq)\, K = K_i^3 \times K_{ppt}$

$$K = \frac{1}{(4 \times 10^8)^3 (1 \times 10^{-25})} = 0.1_6$$

This is not such a small value, but again we do not expect the formation of much product, particularly if the $[EDTA^{4-}]$ is kept high. We can approximate what the concentration of precipitating anion must be in each case, assuming that the concentration of complex ion is 0.10 M and that of $[EDTA^{4-}]$ also is 0.10 M.

Reaction: $\quad [Ca(EDTA)]^{2-} + CO_3^{2-}(aq) \rightleftharpoons CaCO_3(s) + EDTA^{4-}(aq)$

$$K = \frac{[EDTA^{4-}]}{[[Ca(EDTA)]^{2-}][CO_3^{2-}]} = 9 \times 10^{-3} = \frac{0.10\ M}{0.10\ M\ [CO_3^{2-}]} \qquad [CO_3^{2-}] = 1 \times 10^2\ M$$

This is an impossibly high $[CO_3^{2-}]$.

Reaction: $\quad 3[Mg(EDTA)]^{2-} + 2\,PO_4^{3-}(aq) \rightleftharpoons Mg_3(PO_4)_2(s) + 3\,EDTA^{4-}(aq)$

$$K = \frac{[EDTA^{4-}]^3}{[[Mg(EDTA)]^{2-}]^3[PO_4^{3-}]^2} = 0.1_6 = \frac{(0.10\ M)^3}{(0.10\ M)^3\ [PO_4^{3-}]^2} \qquad [PO_4^{3-}] = 2._5\ M$$

Although this $[PO_4^{3-}]$ is not impossibly high, it is unlikely that it will occur without the deliberate addition of phosphate ion to the water. *Alternatively,* we can substitute $[[M(EDTA)]^{2-}] = 0.10$ M and $[EDTA^{4-}] = 0.10$ M into the formation constant

expression: $K_f = \dfrac{[[M(EDTA)]^{2-}]}{[M^{2+}][EDTA^{4-}]}$

This substitution gives $[M^{2+}] = \dfrac{1}{K_f}$, hence, $[Ca^{2+}] = 2.5 \times 10^{-11}$ M and $[Mg^{2+}] = 2.5 \times 10^{-9}$ M,

Which are concentrations that do not normally lead to the formation of precipitates unless the concentrations of anions are substantial. Specifically, the required anion concentrations are $[CO_3^{2-}] = 1.1 \times 10^2$ M for $CaCO_3$, and $[PO_4^{3-}] = 2.5$ M, just as computed above.

49. If a 99% conversion to the chloro complex is achieved, which is the percent conversion necessary to produce a yellow color, $[[CuCl_4]^{2-}] = 0.99 \times 0.10$ M $= 0.099$ M, and $[[Cu(H_2O)_4]^{2+}] = 0.01 \times 0.10$ M $= 0.0010$ M. We substitute these values into the formation constant expression and solve for $[Cl^-]$.

$$K_f = \frac{[[CuCl_4]^{2-}]}{[[Cu(H_2O)_4]^{2+}][Cl^-]^4} = 4.2 \times 10^5 = \frac{0.099\,M}{0.0010\,M\,[Cl^-]^4}$$

$$[Cl^-] = \sqrt[4]{\frac{0.099}{0.0010 \times 4.2 \times 10^5}} = 0.12\,M$$

This is the final concentration of free chloride ion in the solution. If we wish to account for all chloride ion that has been added, we must include the chloride ion present in the complex ion.

total $[Cl^-] = 0.12$ M free $Cl^- + (4 \times 0.099$ M) bound $Cl^- = 0.52$ M

50. **(a)**

Oxidation: $\quad 2H_2O(l) \longrightarrow O_2(g) + 4H^+(aq) + 4e^- \qquad\qquad -E° = -1.229$ V

Reduction: $\quad \{Co^{3+}((aq) + e^- \longrightarrow Co^{2+}(aq)\} \times \qquad\qquad\qquad E° = +1.82$ V

Net: $\qquad 4Co^{3+}(aq) + 2H_2O(l) \longrightarrow 4Co^{2+}(aq) + 4H^+(aq) + O_2(g) \quad E°_{cell} = +0.59$ V

(b) Reaction: $Co^{3+}(aq) + 6NH_3(aq) \rightleftharpoons [Co(NH_3)_6]^{3+}(aq)$

Initial: $\quad$ 1.0 M $\qquad$ 0.10 M $\qquad\qquad$ 0 M

Changes: $\quad -x$ M $\qquad$ constant $\qquad\qquad +x$ M

Equil: $(1.0 - x)$M $\qquad$ 0.10 M $\qquad\qquad$ xM

$$K_f = \frac{[[Co(NH_3)_6]^{3+}]}{[Co^{3+}][NH_3]^6} = 4.5 \times 10^{33} = \frac{x}{(1.0-x)(0.10)^6} \qquad \frac{x}{1.0-x} = 4.5 \times 10^{27}$$

Thus $[[Co(NH_3)_6]^{3+}] = 1.0$ M because K is so large, and

$[Co^{3+}] = \dfrac{1}{4.5 \times 10^{27}} = 2.2 \times 10^{-28}$ M

(c) The equilibrium and equilibrium constant for the reaction of NH_3 with water follows.

$NH_3(aq) + H_2O(l) \rightleftharpoons NH_4^+(aq) + OH^-(aq)$

$$K_b = 1.8 \times 10^{-5} = \frac{[NH_4^+][OH^-]}{[NH_3]} = \frac{[OH^-]^2}{0.10} \qquad [OH^-] = \sqrt{0.10(1.8 \times 10^{-5})} = 0.0013\,M$$

In determining the [OH$^-$], we have noted that [NH$_4^+$] = [OH$^-$] by stoichiometry, and also that [NH$_3$] = 0.10 M, as we assumed above. $[H_3O^+] = \dfrac{K_w}{[OH^-]} = \dfrac{1.0 \times 10^{-14}}{0.0013} = 7.7 \times 10^{-12}$ M We use the Nernst equation to determine the potential of reaction (24.12) at the conditions described.

$$E = E^\circ_{cell} - \frac{0.0592}{4} \log \frac{[Co^{2+}]^4 [H^+]^4 P[O_2]}{[Co^{3+}]^4}$$

$$E = +0.59 - \frac{0.0592}{4} \log \frac{(1 \times 10^{-4})^4 (7.7 \times 10^{-12})^4 \, 0.2}{(2.2 \times 10^{-28})^4} = +0.59 \text{ V} - 0.732 \text{ V} = -0.142 \text{ V}$$

The negative cell potential indicates that the reaction indeed does not occur.

51. We use the Nernst equation to determine the value of [Cu^{2+}]. The cell reaction follows.

$$Cu(s) + 2 H^+(aq) \longrightarrow Cu^{2+}(aq) + H_2(g) \qquad E^\circ_{cell} = -0.337 \text{ V}$$

$$E = E^\circ - \frac{0.0592}{2} \log \frac{[Cu^{2+}]}{[H^+]^2} = 0.08 \text{ V} = -0.337 - \frac{0.0592}{2} \log \frac{[Cu^{2+}]}{(1.00)^2}$$

$$\log[Cu^{2+}] = \frac{2(-0.337 - 0.08)}{0.0592} = -14.1 \qquad [Cu^{2+}] = 8 \times 10^{-15} \text{ M}$$

Now we determine the value of K_f.

$$K_f = \frac{[[Cu(NH_3)_4]^{2+}]}{[Cu^{2+}][NH_3]^4} = \frac{1.00}{8 \times 10^{-15} (1.00)^4} = 1 \times 10^{14}$$

This compares favorably with the value of $K_f = 1.1 \times 10^{13}$ given in Table 18-2, especially considering the imprecision with which the data are known. ($E_{cell} = 0.08$ V is known to but one significant figure.)

52. We first determine [Ag$^+$] in the cyanide solution.

$$K_f = \frac{[[Ag(CN)_2]^-]}{[Ag^+][CN^-]^2} = 5.6 \times 10^{18} = \frac{0.10}{[Ag^+](0.10.)^2}$$

The cell reaction is as follows. It has

$$[Ag^+] = \frac{0.10}{5.6 \times 10^{18} (0.10)^2} = 1.8 \times 10^{-18}$$

$E^\circ_{cell} = 0.000$ V; the same reaction occurs at both anode and cathode and thus the Nernstian voltage is influenced only by the Ag$^+$ concentration.

$$Ag^+ (0.10 \text{ M}) \longrightarrow Ag^+ (0.10 \text{ M} [Ag(CN)_2]^-, 0.10 \text{ M KCN})$$

$$E = E^\circ - \frac{0.0592}{1} \log \frac{[Ag^+]_{CN}}{[Ag^+]} = 0.000 - 0.0592 \log \frac{1.8 \times 10^{-18}}{0.10} = +0.99 \text{ V}$$

53.

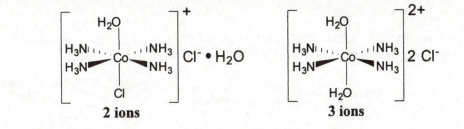

A 0.10 mol/L solution of these compounds would result in ion concentration of 0.20 mol/L for $[Co\,Cl\,(H_2O(NH_3)_4)]Cl\bullet H_2O$ or 0.30 mol/L for $[Co(H_2O)_2\,(NH_3)_4]Cl_2$.

Observed freezing point depression = -0.56 °C.

$\Delta T = -K_f \times m \times i = -0.56$ deg $= -1.86$ mol kg^{-1} deg(0.10 mol/L)$\times i$

Hence, $i = 3.0$ mol/kg or 3.0 mol/L. This suggests that the compound is $[Co(H_2O)_2\,(NH_3)_4]Cl_2$.

54. We first need to compute the empirical formula of the complex compound

46.2 g Pt $\times \dfrac{1 \text{ mol C}}{195.1 \text{ g Pt}} = 0.236_8$ mol Pt $\qquad \div 0.236_8 \longrightarrow 1.000$ mol Pt

33.6 g Cl $\times \dfrac{1 \text{ mol O}}{35.5 \text{ g Cl}} = 0.946_5$ mol Cl $\qquad \div 0.236_8 \longrightarrow 4$ mol Cl

16.6 g N $\times \dfrac{1 \text{ mol N}}{14.01 \text{ g N}} = 1.18_6$ mol N $\qquad \div 0.236_8 \longrightarrow 5$ mol N

3.6 g H $\times \dfrac{1 \text{ mol H}}{1.00 \text{ g H}} = \qquad 3.6$ mol H $\qquad \div 0.236_8 \longrightarrow 15$ mol H

The nitrogen ligand is NH_3 apparently, so the empirical formula is: $Pt\,(NH_3)_5Cl_4$.

The effective molality of the solution is $m = \dfrac{\Delta T}{K_{fp}} = \dfrac{0.74\,C^o}{1.86\,C^o/m} = 0.4m$. The effective molality

is 4 times the stated molality, so we have 4 particles produced per mole of Pt complex, and therefore 3 ionizable chloride ions. We can write this in the following way:

$[Pt(NH_3)_5Cl]Cl_3$. Only *one* form of the cation (with charge of 3+) shown below will exist.

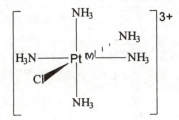

55.

a) There is no reaction with AgNO$_3$ or en, the compound must be *trans*-chlorobis(ethylenediamine)nitrito-*N*-cobalt(III) nitrite

where $\rangle$en $\Longrightarrow$ is H$_2$C—CH$_2$ / H$_2$N.   .NH$_2$

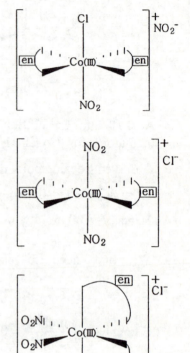

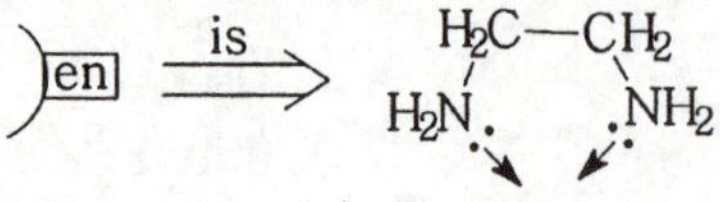

b) If compound reacts with AgNO$_3$, but not with en it must be *trans*-bis(ethylenediamine)dinitrito-*N*-cobalt(III) chloride.

where $\rangle$en $\Longrightarrow$ is H$_2$C—CH$_2$ / H$_2$N.   .NH$_2$

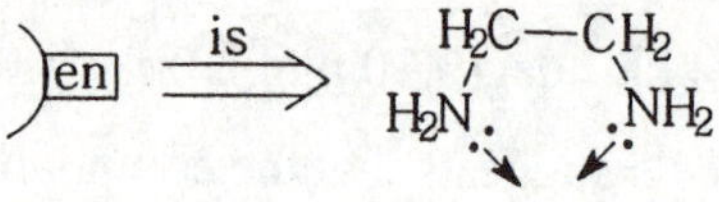

c) If it reacts with AgNO$_3$ and en and is optically active it must be: . *cis*-bis(ethylenediamine)dinitrito-*N*-cobalt(III) chloride.

where $\rangle$en $\Longrightarrow$ is H$_2$C—CH$_2$ / H$_2$N.   .NH$_2$

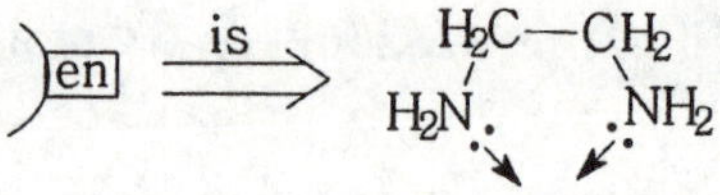

56. Since these reactions involve Ni^{2+} binding to six identical nitrogen donor atoms, we can assume the ΔH for each reaction is approximately the same. A large formation constant indicates a more negative free energy. This indicates that the entropy term for these reactions is different. The large increase in entropy for each step, can be explained by the chelate effect. The same number of water molecules are displaced by fewer ligands, which is directed related to entropy change.

57.

Co(acac)$_3$

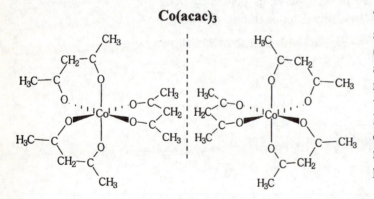

This compound has a non-superimposable mirror image (therefore it is optically active). These are enantiomers. One enantiomer rotates plane polarized light clockwise, while the other rotates plane polarized light couter-clockwise. Polarimetry can be used to determine which isomer rotates plane polarized light in a particular direction.

trans-[Co(acac)₂(H₂O)₂]Cl₂

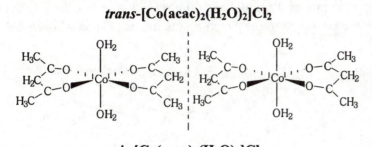

This compound has a superimposable mirror image (therefore it is optically inactive). These are the same compound. This compound will not rotate plane polarized light (net rotation = 0).

cis-[Co(acac)₂(H₂O)₂]Cl₂

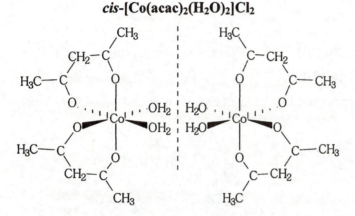

This compound has a non-superimposable mirror image (therefore it is optically active). These are enantiomers. One enantiomer rotates plane polarized light clockwise, while the other rotates plane polarized light couter-clockwise. By using a polarimeter, we can determine which isomer rotates plane polarized light in a particular direction.

58. $[Co(H_2O)_6]^{3+}(aq) + e^- \rightarrow [Co(H_2O)_6]^{2+}(aq)$

$\log K = nE°/0.0592 = (1)(1.82)/0.0592 = 30.74$ $K = 10^{30.74}$

$[Co(en)_3]^{3+}(aq) + 6\ H_2O(l)$	$\rightarrow$	$[Co(H_2O)_6]^{3+}(aq) + 3\ en$	$K = 1/10^{47.30}$
$[Co(H_2O)_6]^{2+}(aq) + 3\ en$	$\rightarrow$	$[Co(en)_3]^{2+}(aq) + 6\ H_2O(l)$	$K = 10^{12.18}$
$[Co(H_2O)_6]^{3+}(aq) + e^-$	$\rightarrow$	$[Co(H_2O)_6]^{2+}(aq)$	$K = 10^{30.74}$
$[Co(en)_3]^{3+}(aq) + e^-$	$\rightarrow$	$[Co(en)_3]^{2+}(aq)$	$K_{overall} = ?$

$K_{overall} = 1/10^{47.30} \times 10^{12.18} \times 10^{30.74} = 10^{-4.38} = 0.0000417$

$E° = (0.0592/n)\log K = (0.0592/1)\log(0.0000417) = -0.26\ V$

$4\ Co^{3+}(aq) + 4\ e^- \rightarrow 4\ Co^{2+}(aq)$ $E°_{cathode} = 1.82\ V$
$\underline{2\ H_2O(l) \rightarrow 4\ H^+(aq) + O_2(g) + 4\ e^-\ \ E°_{anode} = 1.23\ V}$
$4\ Co^{3+}(aq) + 2\ H_2O(l) \rightarrow 4\ H^+(aq) + O_2(g) + 4\ Co^{2+}(aq)$

$E°_{cell} = E°_{cathode} - E°_{anode} = 1.82V - 1.23\ V = +0.59\ V\ (Spontaneous)$

$4\ [Co(en)_3]^{3+}(aq) + 4\ e^- \rightarrow 4\ [Co(en)_3]^{2+}(aq)$ $E°_{cathode} = -0.26\ V$
$\underline{2\ H_2O(l) \rightarrow 4\ H^+(aq) + O_2(g) + 4\ e^-}$ $E°_{anode} = 1.23\ V$
$4\ [Co(en)_3]^{3+}(aq) + 2\ H_2O(l) \rightarrow 4\ H^+(aq) + O_2(g) + 4\ [Co(en)_3]^{2+}(aq)$

$E°_{cell} = E°_{cathode} - E° = -0.26\ V - 1.23\ V = -1.49\ V\ (non\text{-}spontaneous)$

59. There are two sets of isomers for Co(gly)₃. Each set is comprised of two enantiomers (non-superimposable mirror images). One rotates plane polarized light clockwise (+), the other counter clockwise (−). Experiments are needed to determine which enantiomer is (+) and which is (−).

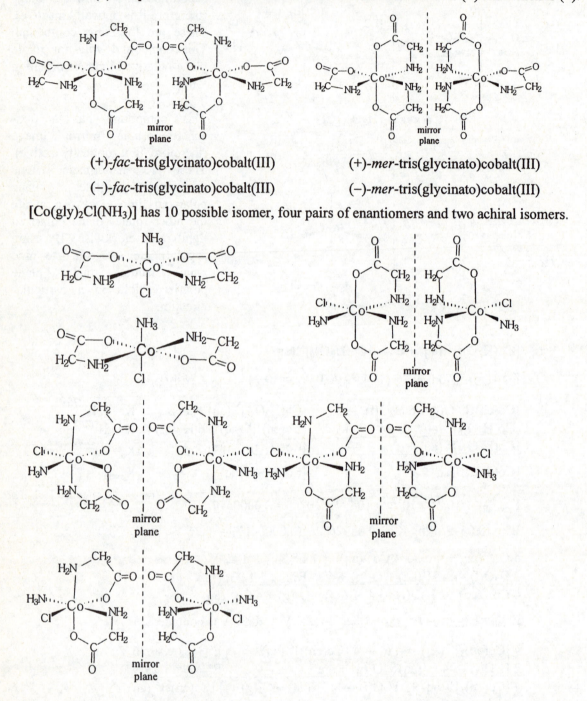

(+)-*fac*-tris(glycinato)cobalt(III)   (+)-*mer*-tris(glycinato)cobalt(III)

(−)-*fac*-tris(glycinato)cobalt(III)   (−)-*mer*-tris(glycinato)cobalt(III)

[Co(gly)₂Cl(NH₃)] has 10 possible isomer, four pairs of enantiomers and two achiral isomers.

60. The coordination compound is face-centered cubic, K⁺ occupies tetrahedral holes, while PtCl₆²⁻ occupies octahedral holes.

922

Feature Problems

61. (a) A trigonal prismatic structure predicts three geometric isomers for $[CoCl_2(NH_3)_4]^+$, which is one more than the actual number of geometric isomers found for this complex ion. All three geometric isomers arising from a trigonal prism are shown below.

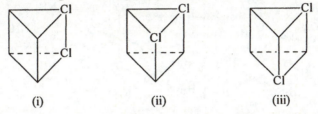

(i)   (ii)   (iii)

The fact that the trigonal prismatic structure does not afford the correct number of isomers is a clear indication that the ion actually adopts some other structural form (i.e., the theoretical model is contradicted by the experimental result). We know now of course, that this ion has an octahedral structure and as a result, it can exist only in *cis* and *trans* configurations.

(b) All attempts to produce optical isomers of $[Co(en)_3]^{3+}$ based upon a trigonal prismatic structure are shown below. The ethylenediamine ligand appears as an arc in diagrams below:

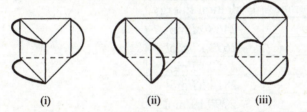

(i)   (ii)   (iii)

Only structure (iii), which has an ethylenediamine ligand connecting the diagonal corners of a face can give rise to optical isomers. Structure (iii) is highly unlikely, however, because the ethylenediamine ligand is simply too short to effectively span the diagonal distance across the face of the prism. Thus, barring any unusual stretching of the ethylenediamine ligand, a trigonal prismatic structure cannot account for the optical isomerism that was observed for $[Co(en)_3]^{3+}$.

62. Assuming that each hydroxide ligand bears it normal 1– charge, and that each ammonia ligand is neutral, the total contribution of negative charge from the ligands is 6 –. Since the net charge on the complex ions is 6+, the average oxidation state for each Co atom must be +3 (i.e., each Co in the complex can be viewed as a Co^{3+} $3d^6$ ion surrounded by six ligands.) The five $3d$ orbitals on each Co are split by the octahedrally arranged ligands into three lower energy orbitals, called t_{2g} orbitals, and two higher energy orbitals, called e_g orbitals. We are told in the question that the complex is *low spin*. This is simply another way of saying that all six $3d$ electrons on each Co are paired up in the t_{2g} set as a result of the e_g and t_{2g} orbitals being separated by a relatively large energy gap (see below). Hence, there should be no unpaired electrons in the hexacation (i.e., the cation is expected to be diamagnetic).

Low Spin Co^{3+}
(Octahedral Environment):

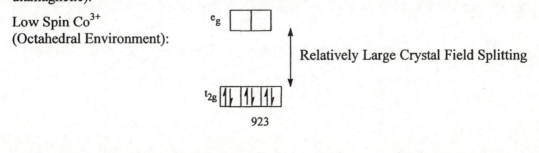

Relatively Large Crystal Field Splitting

The Lewis structures for the two optical isomers are depicted below:

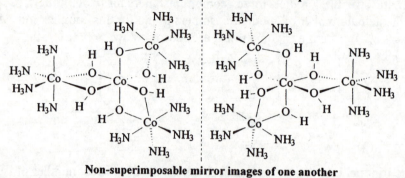

Non-superimposable mirror images of one another

(ENANTIOMERS)

63. The data used to construct a plot of hydration energy as a function of metal ion atomic number is collected in the table below. The graph of hydration energy (kJ/mol) versus metal ion atomic number is located beneath the table.

Metal Ion	Atomic number	Hydration Energy
Ca^{2+}	20	−2468 kJ/mol
Sc^{2+}	21	−2673 kJ/mol
Ti^{2+}	22	−2750 kJ/mol
V^{2+}	23	−2814 kJ/mol
Cr^{2+}	24	−.2799 kJ/mol
Mn^{2+}	25	−2743 kJ/mol
Fe^{2+}	26	−2843 kJ/mol
Co^{2+}	27	−2904 kJ/mol
Ni^{2+}	28	−2986 kJ/mol
Cu^{2+}	29	−2989 kJ/mol
Zn^{2+}	30	−2936 kJ/mol

(a)

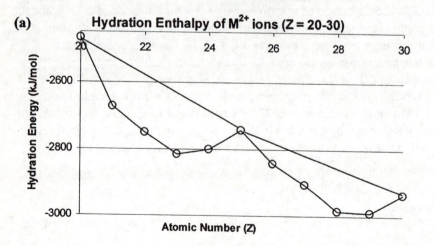

Hydration Enthalpy of M^{2+} ions (Z = 20-30)

(b) When a metal ion is placed in an octahedral field of ligands, the five d-orbitals are split into e_g and t_{2g} subsets, as shown in the diagram below:

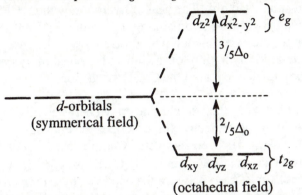

(octahedral field)

Since water is a weak field ligand, the magnitude of the splitting is relatively small. As a consequence, high-spin configurations result for all of the hexaaqua complexes. The electron configurations for the metal ions in the high-spin hexaaqua complexes and their associated crystal field stabilization energies (CFSE) are provided in the table below

Metal Ion	Configuration	t_{2g}	e_g	# of unpaired e⁻	CFSE(Δ_o)
Ca^{2+}	$3d^0$	0	0	0	0
Sc^{2+}	$3d^1$	1	0	1	$-\frac{2}{5}$
Ti^{2+}	$3d^2$	2	0	2	$-\frac{4}{5}$
V^{2+}	$3d^3$	3	0	3	$-\frac{6}{5}$
Cr^{2+}	$3d^4$	3	1	4	$-\frac{3}{5}$
Mn^{2+}	$3d^5$	3	2	5	0
Fe^{2+}	$3d^6$	4	2	4	$-\frac{2}{5}$
Co^{2+}	$3d^7$	5	2	3	$-\frac{4}{5}$
Ni^{2+}	$3d^8$	6	2	2	$-\frac{6}{5}$
Cu^{2+}	$3d^9$	6	3	1	$-\frac{3}{5}$
Zn^{2+}	$3d^{10}$	6	4	0	0

Thus, the crystal field stabilization energy is zero for Ca^{2+}. Mn^{2+} and Zn^{2+}.

(c) The lines drawn between those ions that have a CFSE = 0 show the trend for the enthalpy of hydration after the contribution from the crystal field stabilization energy has been subtracted from the experimental values. The Ca to Mn and Mn to Zn lines are quite similar. Both line have slopes that are negative and are of comparable magnitude. This trend shows that as one proceeds from left to right across the periodic table, the energy of hydration for dications becomes increasingly more negative. The hexaaquo complexes become progressively more stable because the Z_{eff} experienced by the bonding electrons in the valence shell of the metal ion steadily increases as we move further and further to the right. Put another way, the Z_{eff} climbs steadily as we move from left to right and this leads to the positive charge density on the metal becoming larger and larger, which results in the water ligands steadily being pulled closer and closer to the nucleus. Of course, the closer the approach of the water ligands to the metal, the greater is the energy released upon successful coordination of the ligand.

(d) Those ions that exhibit crystal field stabilization energies greater than zero have heats of hydration that are more negative (i.e. more energy released) than the hypothetical heat of hydration for the ion with CFSE subtracted out. The heat of hydration without CFSE for a given ion falls on the line drawn between the two flanking ions with CFSE = 0 at a position directly above the point for the experimental hydration energy. The energy difference between the observed heat of hydration for the ion and the heat of hydration without CFSE is, of course, approximately equal to the CFSE for the ion.

(e) As was mentioned in the answer to part (c), the straight line drawn between manganese and zinc (both ions with CFSE = 0) on the previous plot, describes the enthalpy trend after the ligand field stabilization energy has been subtracted from the experimental values for the hydration enthalpy. Thus, Δ_o for Fe^{2+} in $[Fe(H_2O)_6]^{2+}$ is approximately equal to $^5/_2$ of the energy difference (in kJ/mol) between the observed hydration energy for $Fe^{2+}(g)$ and the point for Fe^{2+} on the line connecting Mn^{2+} and Zn^{2+}, which is the expected enthalpy of hydration after the CFSE has been subtracted out. Remember that the crystal field stabilization energy for Fe^{2+} that is obtained from the graph is not Δ_o, but rather $^2/_5\Delta_o$, since the CFSE for a $3d^6$ ion in an octahedral field is just $^2/_5$ of Δ_o. Consequently, to obtain Δ_o, we must multiply the enthalpy difference by $^5/_2$. According to the graph, the high-spin CFSE for Fe^{2+} is -2843 kJ/mol $- (-2782$ kJ/mol$)$ or -61 kJ/mol. Consequently, $\Delta_o = {}^5/_2(-61$ kJ/mol$) = -153$ kJ/mol, or 1.5×10^2 kJ/mol is the energy difference between the e_g and t_{2g} orbital sets.

(f) The color of an octahedral complex is the result of the promotion of an electron on the metal from a t_{2g} orbital $\rightarrow e_g$ orbital. The energy difference between the e_g and t_{2g} orbital sets is Δ_o. As the metal-ligand bonding becomes stronger, the separation between the t_{2g} and e_g orbitals becomes larger. If the e_g set is not full, then the metal complex will exhibit an absorption band corresponding to a $t_{2g} \rightarrow e_g$ transition. Thus, the $[Fe(H_2O)_6]^{2+}$ complex ion should absorb electromagnetic radiation that has $E_{photon} = \Delta_o$. Since $\Delta_o \approx 150$ kJ/mol (calculated in part (e) of this question) for a mole of $[Fe(H_2O)_6]^{2+}(aq)$,

$$E_{photon} = \frac{1.5 \times 10^2 \text{kJ}}{1 \text{mol} [Fe(H_2O)_6]^{2+}} \times \frac{1 \text{mol} [Fe(H_2O)_6]^{2+}}{6.022 \times 10^{23} [Fe(H_2O)_6]^{2+}} \times \frac{1000 \text{J}}{1 \text{kJ}}$$

$$= 2.5 \times 10^{-19} \text{ J per ion}$$

$$v = \frac{E}{h} = \frac{2.5 \times 10^{-19} \text{ J}}{6.626 \times 10^{-34} \text{ Js}} = 3.8 \times 10^{14} \text{ s}^{-1}$$

$$\lambda = \frac{c}{h} = \frac{2.998 \times 10^8 \text{ ms}^{-1}}{3.8 \times 10^{14} \text{ s}^{-1}} = 7.8 \times 10^{-7} \text{ m (780 nm)}$$

So, the $[Fe(H_2O)_6]^{2+}$ ion will absorb radiation with a wavelength of 780 nm, which is red light in the visible part of the electromagnetic spectrum.

CHAPTER 25
NUCLEAR CHEMISTRY
PRACTICE EXAMPLES

1A A β^- has a mass number of zero and an "atomic number" of -1. Emission of this electron has the effect of transforming a neutron into a proton. $^{241}_{94}Pu \rightarrow ^{241}_{95}Am + ^{0}_{-1}\beta$

1B ^{58}Ni has a mass number of 58 and an atomic number of 28. A positron has a mass number of 0 and an effective atomic number of $+1$. Emission of a positron has the seeming effect of transforming a proton into a neutron. The parent nuclide must be copper-58.

$^{58}_{29}Cu \rightarrow ^{58}_{28}Ni + ^{0}_{+1}\beta$

2A The sum of the mass numbers $(139 + 12 = ? + 147)$ tells us that the other product species has $A = 4$. The atomic number of La is 57, that of C is 6, and that of Eu is 63. The atomic number sum $(57 + 6 = ? + 63)$ indicates that the atomic number of this product species is zero. Therefore, four neutrons must have been emitted. $^{139}_{57}La + ^{12}_{6}C \rightarrow ^{147}_{63}Eu + 4^{0}_{1}n$

2B An alpha particle is $^{4}_{2}He$ and a positron is $^{0}_{+1}\beta$. We note that the total mass number in the first equation is 125; the mass number of the additional product is 1.
The total atomic number is 53; the atomic number of the additional product is 0; it is a neutron.

$^{121}_{51}Sb + ^{4}_{2}He \rightarrow ^{124}_{53}I + ^{1}_{0}n$

In the second equation, the positron has a mass number of 0, meaning that the mass number of the product is 124. Because the atomic number of the positron is $+1$, that of the product is 52; it is $^{124}_{52}Te$.

$^{124}_{53}I \rightarrow ^{0}_{+1}\beta + ^{124}_{52}Te$

3A **(a)** The decay constant is found from the 8.040-day half-life.

$$\lambda = \frac{0.693}{8.040 \ d} = 0.0862 \ d^{-1} \times \frac{1 \ d}{24 \ h} \times \frac{1 \ h}{60 \ min} \times \frac{1 \ min}{60 \ s} = 9.98 \times 10^{-7} \ s^{-1}$$

(b) The number of ^{131}I atoms is used to find the activity.

$$no. \ ^{131}I \ atoms = 2.05 \ mg \times \frac{1 \ g}{1000 \ mg} \times \frac{1 \ mol \ ^{131}I}{131 \ g \ ^{131}I} \times \frac{6.022 \times 10^{23} \ atoms}{1 \ mol \ ^{131}I}$$

$$= 9.42 \times 10^{18} \ atoms \ ^{131}I$$

$$Activity = \lambda N = 9.98 \times 10^{-7} \ s^{-1} \times 9.42 \times 10^{18} \ atoms = 9.40 \times 10^{12} \ disintegrations/second$$

(c) We now determine the number of atoms remaining after 16 days. Because two half-lives elapse in 16 days, the number of atoms has been halved twice, to one-fourth (25%) the original number of atoms.

$$N_t = 0.25 \times N_0 = 0.25 \times 9.42 \times 10^{18} \text{ atoms} = 2.36 \times 10^{18} \text{ atoms}$$

(d) The rate after 14 days is determined by the number of atoms present on day 14.

$$\text{rate} = \lambda N_t = 9.98 \times 10^{-7} \text{ s}^{-1} \times 2.36 \times 10^{18} \text{ atoms} = 2.36 \times 10^{12} \text{ dis/s}$$

3B First we determine the value of λ: $\quad \lambda = \dfrac{0.693}{t_{1/2}} = \dfrac{0.693}{11.4 \text{ d}} = 0.0608 \text{ d}^{-1}$

Then we set $N_t = 1\% N_0 = 0.010 N_0$ in equation (25.12).

$$\ln \frac{N_t}{N_0} = -\lambda t = \ln \frac{0.010 N_0}{N_0} = \ln (0.010) = -4.61 = -(0.0608 \text{ d}^{-1})t$$

$$t = \frac{-4.61}{-0.0608 \text{ d}^{-1}} = 75.8 \text{ d}$$

4A The half-life of ^{14}C is 5730 y and $\lambda = 1.21 \times 10^{-4} \text{ y}^{-1}$. The activity of ^{14}C when the object supposedly stopped growing was 15 dis/min per g C. We use equation (25.12) with activities (λN) in place of numbers of atoms (N).

$$\ln \frac{A_t}{A_0} = -\lambda t = \ln \frac{8.5 \text{ dis/min}}{15 \text{ dis/min}} = -(1.21 \times 10^{-4} \text{ y}^{-1})t = -0.56_8 ; t = \frac{0.57}{1.21 \times 10^{-4} \text{ y}^{-1}} = 4.7 \times 10^3 \text{ y}$$

4B The half-life of ^{14}C is 5730 y and $\lambda = 1.21 \times 10^{-4} \text{ y}^{-1}$. The activity of ^{14}C when the object supposedly stopped growing was 15 dis/min per g C. We use equation (25.12) with activities (λN) in place of numbers of atoms (N).

$$\ln \frac{A_t}{A_0} = -\lambda t = \ln \frac{A_t}{15 \text{ dis/min}} = -(1.21 \times 10^{-4} \text{ y}^{-1})(1100 \text{ y}) = -0.13$$

$$\frac{A_t}{15 \text{ dis/min}} = e^{-0.13} = 0.88 , A_t = 0.88 \times 15 \text{ dis/min} = 13 \text{ dis/min (per gram of C)}$$

5A Mass defect. $= 145.913053 \text{ u} (^{146}\text{Sm}) - 141.907719 \text{ u} (^{142}\text{Nd}) - 4.002603 \text{ u} (^4\text{He}) = 0.002731 \text{ u}$

Then, from the text, we have 931.5 MeV $= 1 \text{ u}$ $E = 0.002731 \text{ u} \times \dfrac{931.5 \text{ MeV}}{1 \text{ u}} = 2.544 \text{ MeV}$

5B Unfortunately, we cannot use the result of Example 25–5 (0.0045 u $= 4.2$ MeV) because it is expressed to only two significant figures, and here we begin with four significant figures. But, we essentially work backwards through that calculation. The last conversion factor is from Table 2-1.

$$E = 5.590 \text{ MeV} \times \frac{1.602 \times 10^{-13} \text{ J}}{1 \text{ MeV}} = 8.955 \times 10^{-13} \text{ J} = mc^2 = m\left(2.9979 \times 10^8 \text{ m/s}\right)^2$$

$$m = \frac{8.955 \times 10^{-13} \text{ J}}{\left(2.9979 \times 10^8 \text{ m/s}\right)^2} \times \frac{1000 \text{ g}}{1 \text{ kg}} \times \frac{1.0073 \text{ u}}{1.673 \times 10^{-24} \text{ g}} = 0.005999 \text{ u}$$

Or we could use $m = 5.590 \text{ MeV} \times \dfrac{1 \text{ u}}{931.5 \text{ MeV}} = 0.006001 \text{ u}$

6A **(a)** ^{88}Sr has an even atomic number (38) and an even neutron number (50); its mass number (88) is not too far from the average mass (87.6) of Sr. It should be stable.

(b) ^{118}Cs has an odd atomic number (55) and a mass number (118) that is pretty far from the average mass of Cs (132.9). It should be radioactive.

(c) ^{30}S has an even atomic number (16) and an even neutron number (14); but its mass number (30) is too far from the average mass of S (32.1). It should be radioactive.

6B We know that ^{19}F is stable, with approximately the same number of neutrons and protons: 9 protons, and 10 neutrons. Thus, nuclides of light elements with approximately the same number of neutrons and protons should be stable. In Practice Example 25–1 we saw that positron emission has the effect of transforming a proton into a neutron. β^- emission has the opposite effect, namely, the transformation of a neutron into a proton. The mass number does not change in either case. Now let us analyze our two nuclides.

^{17}F has 9 protons and 8 neutrons. Replacing a proton with a neutron would produce a more stable nuclide. Thus, we predict positron emission by ^{17}F to produce ^{17}O.

^{22}F has 9 protons and 13 neutrons. Replacing a neutron with a proton would produce a more stable nuclide. Thus, we predict β^- emission by ^{22}F to produce ^{22}Ne.

EXERCISES

Radioactive Processes

1. **(a)** $^{234}_{94}\text{Pu} \rightarrow {}^{230}_{92}\text{U} + {}^{4}_{2}\text{He}$

(b) $^{248}_{97}\text{Bk} \rightarrow {}^{248}_{98}\text{Cf} + {}^{0}_{-1}\text{e}$

(c) $^{196}_{82}\text{Pb} + {}^{0}_{-1}\text{e} \rightarrow {}^{196}_{81}\text{Tl}$; $\quad {}^{196}_{81}\text{Tl} + {}^{0}_{-1}\text{e} \rightarrow {}^{196}_{80}\text{Hg}$

2. **(a)** $^{214}_{82}\text{Pb} \rightarrow {}^{214}_{83}\text{Bi} + {}^{0}_{-1}\text{e}$; $\quad {}^{214}_{83}\text{Bi} \rightarrow {}^{214}_{84}\text{Po} + {}^{0}_{-1}\text{e}$

(b) $^{226}_{88}\text{Ra} \rightarrow {}^{222}_{86}\text{Rn} + {}^{4}_{2}\text{He}$; $\quad {}^{222}_{86}\text{Rn} \rightarrow {}^{218}_{84}\text{Po} + {}^{4}_{2}\text{He}$; $\quad {}^{218}_{84}\text{Po} \rightarrow {}^{214}_{82}\text{Pb} + {}^{4}_{2}\text{He}$

(c) $^{69}_{33}\text{As} \rightarrow {}^{69}_{32}\text{Ge} + {}^{0}_{+1}\text{e}$

3. We would expect a neutron:proton ratio that is closer to 1:1 than that of ^{14}C. This would be achieved if the product were ^{14}N, which is the result of β^- decay: $^{14}_6C \rightarrow ^{14}_7N + ^0_{-1}e$.

4. A nuclide with a closer to 1:1 neutron:proton ratio (than that of tritium) is helium-3, arrived at by beta emission: $^3_1H \rightarrow ^3_2He + ^0_{-1}e$. Another possible product is deuterium, which is arrived at by neutron emission: $^3_1H \rightarrow ^2_1H + ^1_0n$

Radioactive Decay Series

5. We first write conventional nuclear reactions for each step in the decay series.

$$^{232}_{90}Th \rightarrow ^{228}_{88}Ra + ^4_2He \qquad ^{228}_{88}Ra \rightarrow ^{228}_{89}Ac + ^0_{-1}e \qquad ^{228}_{89}Ac \rightarrow ^{228}_{90}Th + ^0_{-1}e$$

$$^{228}_{90}Th \rightarrow ^{224}_{88}Ra + ^4_2He \qquad ^{224}_{88}Ra \rightarrow ^{220}_{86}Rn + ^4_2He \qquad ^{220}_{86}Rn \rightarrow ^{216}_{84}Po + ^4_2He$$

Now for a branch in the series

these two $\qquad ^{216}_{84}Po \rightarrow ^{212}_{82}Pb + ^4_2He \qquad ^{212}_{82}Pb \rightarrow ^{212}_{83}Bi + ^0_{-1}e$

or these two $\qquad ^{216}_{84}Po \rightarrow ^{216}_{85}At + ^0_{-1}e \qquad ^{216}_{85}At \rightarrow ^{212}_{83}Bi + ^4_2He$

And now a second branch

these two $\qquad ^{212}_{83}Bi \rightarrow ^{208}_{81}Tl + ^4_2He \qquad ^{208}_{81}Tl \rightarrow ^{208}_{82}Pb + ^0_{-1}e$

or these two $\qquad ^{212}_{83}Bi \rightarrow ^{212}_{84}Po + ^0_{-1}e \qquad ^{212}_{84}Po \rightarrow ^{208}_{82}Pb + ^4_2He$

Both branches end at the isotope $^{208}_{82}Pb$. The graph, similar to Figure 25-2, is drawn below.

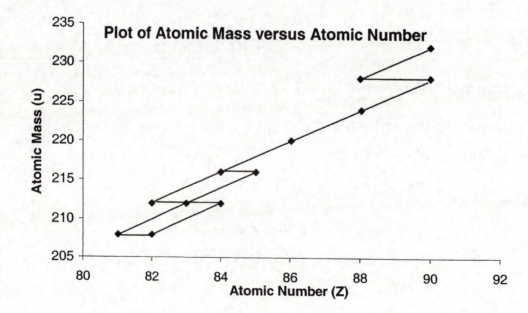

6. The series begins with uranium-235, and ends with lead-207.

$$^{235}_{92}U \rightarrow {}^{231}_{90}Th + {}^{4}_{2}He \qquad {}^{231}_{90}Th \rightarrow {}^{231}_{91}Pa + {}^{0}_{-1}e \qquad {}^{231}_{91}Pa \rightarrow {}^{227}_{89}Ac + {}^{4}_{2}He$$

Now the series branches

these two $\qquad {}^{227}_{89}Ac \rightarrow {}^{223}_{87}Fr + {}^{4}_{2}He \qquad {}^{223}_{87}Fr \rightarrow {}^{223}_{88}Ra + {}^{0}_{-1}e$

or these two $\qquad {}^{227}_{89}Ac \rightarrow {}^{227}_{90}Th + {}^{0}_{-1}e \qquad {}^{227}_{90}Th \rightarrow {}^{223}_{88}Ra + {}^{4}_{2}He$

then ${}^{223}_{88}Ra \rightarrow {}^{219}_{86}Rn + {}^{4}_{2}He \qquad {}^{219}_{86}Rn \rightarrow {}^{215}_{84}Po + {}^{4}_{2}He \qquad {}^{215}_{84}Po \rightarrow {}^{211}_{82}Pb + {}^{4}_{2}He$

$$^{211}_{82}Pb \rightarrow {}^{211}_{83}Bi + {}^{0}_{-1}e$$

The series branches again

these two $\qquad {}^{211}_{83}Bi \rightarrow {}^{207}_{81}Tl + {}^{4}_{2}He \qquad {}^{207}_{81}Tl \rightarrow {}^{207}_{82}Pb + {}^{0}_{-1}e$

or these two $\qquad {}^{211}_{83}Bi \rightarrow {}^{211}_{84}Po + {}^{0}_{-1}e \qquad {}^{211}_{84}Po \rightarrow {}^{207}_{82}Pb + {}^{4}_{2}He$

The plot of atomic mass versus atomic number for these decay series is shown below.

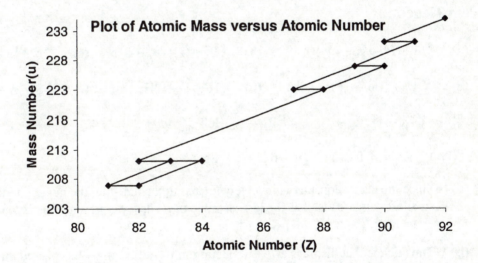

7. In Figure 25–2, only the following mass numbers are represented: 206, 210, 214, 218, 222, 226, 230, 234, and 238. We see that these mass numbers are separated from each other by 4 units. The first of them, 206, equals $(4 \times 51) + 2$, that is $4n + 2$, where $n = 51$.

8. The series to which each nuclide belongs is determined by dividing its mass number by 4 and obtaining the remainder.

(a) The mass number of $^{214}_{83}\text{Bi}$ is 214, and the remainder following its division by 4 is 2. This nuclide is a member of the $4n+2$ series.

(b) The mass number of $^{216}_{84}\text{Po}$ is 216, and the remainder following its division by 4 is 0. This nuclide is a member of the $4n$ series.

(c) The mass number of $^{215}_{85}\text{At}$ is 215, and the remainder following its division by 4 is 3. This nuclide is a member of the $4n+3$ series.

(d) The mass number of $^{235}_{92}\text{U}$ is 235, and the remainder following its division by 4 is 3. This nuclide is a member of the $4n+3$ series.

9. **(a)** $^{160}_{74}\text{W} \rightarrow {}^{156}_{72}\text{Hf} + {}^{4}_{2}\text{He}$ **(b)** $^{38}_{17}\text{Cl} \rightarrow {}^{38}_{18}\text{Ar} + {}^{0}_{-1}\beta$

      (c) $^{214}_{83}\text{Bi} \rightarrow {}^{214}_{84}\text{Po} + {}^{0}_{-1}\beta$ **(d)** $^{32}_{17}\text{Cl} \rightarrow {}^{32}_{16}\text{S} + {}^{0}_{+1}\beta$

10. **(a)** $^{23}_{11}\text{Na} + {}^{2}_{1}\text{H} \rightarrow {}^{24}_{11}\text{Na} + {}^{1}_{1}\text{H}$ **(b)** $^{59}_{27}\text{Co} + {}^{1}_{0}\text{n} \rightarrow {}^{56}_{25}\text{Mn} + {}^{4}_{2}\text{He}$ **(c)** $^{238}_{92}\text{U} + {}^{2}_{1}\text{H} \rightarrow {}^{240}_{94}\text{Pu} + {}^{0}_{-1}\beta$

      (d) $^{246}_{96}\text{Cm} + {}^{13}_{6}\text{C} \rightarrow {}^{254}_{102}\text{No} + 5{}^{1}_{0}\text{n}$ **(e)** $^{238}_{92}\text{U} + {}^{14}_{7}\text{N} \rightarrow {}^{246}_{99}\text{Es} + 6{}^{1}_{0}\text{n}$

Nuclear Reactions

11. **(a)** $^{7}_{3}\text{Li} + {}^{1}_{1}\text{H} \rightarrow {}^{8}_{4}\text{Be} + \gamma$ **(b)** $^{9}_{4}\text{Be} + {}^{2}_{1}\text{H} \rightarrow {}^{10}_{5}\text{B} + {}^{1}_{0}\text{n}$ **(c)** $^{14}_{7}\text{N} + {}^{1}_{0}\text{n} \rightarrow {}^{14}_{6}\text{C} + {}^{1}_{1}\text{H}$

12. **(a)** $^{238}_{92}\text{U} + {}^{4}_{2}\text{He} \rightarrow {}^{239}_{94}\text{Pu} + 3{}^{1}_{0}\text{n}$ **(b)** $^{3}_{1}\text{H} + {}^{2}_{1}\text{H} \rightarrow {}^{4}_{2}\text{He} + {}^{1}_{0}\text{n}$ **(c)** $^{33}_{16}\text{S} + {}^{1}_{0}\text{n} \rightarrow {}^{33}_{15}\text{P} + {}^{1}_{1}\text{H}$

13. $^{209}_{83}\text{Bi} + {}^{64}_{28}\text{Ni} \rightarrow {}^{272}_{111}\text{Rg} + {}^{1}_{0}\text{n}$; $^{272}_{111}\text{Rg} \rightarrow 5{}^{4}_{2}\text{He} + {}^{252}_{101}\text{Md}$

14. $^{208}_{82}\text{Pb} + {}^{86}_{36}\text{Kr} \rightarrow {}^{293}_{118}\text{E} + {}^{1}_{0}\text{n}$; $^{293}_{118}\text{E} \rightarrow 6{}^{4}_{2}\text{He} + {}^{269}_{106}\text{Sg}$

15. **(a)** Since the decay constant is inversely related to the half-life, the nuclide with the smallest half-life also has the largest value of its decay constant. This is the nuclide $^{214}_{84}\text{Po}$ with a half-life of 1.64×10^{-4} s.

(b) The nuclide that displays a 75% reduction in its radioactivity has passed through two half-lives in a period of one month. Thus, this is the nuclide with a half-life of approximately two weeks. This is the nuclide $^{32}_{15}\text{P}$, with a half-life of 14.3 days.

(c) If more than 99% of the radioactivity is lost, less than 1% remains. Thus $(\frac{1}{2})^n < 0.010$. Now, when $n = 7$, $(\frac{1}{2})^n = 0.0078$. Thus, seven half-lives have elapsed in one month, and each half-life approximates 4.3 days. The longest lived nuclide that fits this description is $^{222}_{86}\text{Rn}$, which has a half-life of 3.823 days. Of course, all other nuclides with shorter half-lives also meet this criterion, specifically the following nuclides: $^{13}_{8}\text{O}\left(8.7 \times 10^{-3} \text{ s}\right)$, $^{28}_{12}\text{Mg}\left(21 \text{ h}\right)$, $^{80}_{35}\text{Br}\left(17.6 \text{ min}\right)$, and $^{214}_{84}\text{Po}\left(1.64 \times 10^{-4} \text{ s}\right)$.

16. Since $16 = 2^4$, four half-lives have elapsed in 18.0 h, and each half-life equals 4.50 h. The half-life of isotope B thus is $2.5 \times 4.50 \text{ h} = 11.25 \text{ h}$. Now, since $32 = 2^5$, five half-lives must elapse before the decay rate of isotope B falls to $\frac{1}{32}$ of its original value. Thus, the time elapsed for this amount of decay is:

$$\text{time elapsed} = 5 \text{ half lives} \times \frac{11.25}{1 \text{ half life}} = 56.3 \text{ h}$$

Rate of Radioactive Decay

17. We use expression (25.13) to determine λ and then expression (25.11) to determine the number of atoms.

$$\lambda = \frac{0.693}{5.2 \text{ y}} \times \frac{1 \text{ y}}{365.25 \text{ d}} \times \frac{1 \text{ d}}{24 \text{ h}} = 1.5\underline{2} \times 10^{-5} \text{ h}^{-1}$$

$$N = \frac{\text{rate of decay}}{\lambda} = \frac{6740 \text{ atoms/h}}{1.5\underline{2} \times 10^{-5} \text{ h}^{-1}} = 4.4 \times 10^{8} \ ^{60}_{27}\text{Co atoms}$$

18. This follows first-order kinetics (as do all radioactive decay processes) with a rate of decay directly proportional to the number of atoms. We therefore use expression (25.12), with rates substituted for numbers of atoms.

$$6740 \frac{\text{dis}}{\text{h}} \times \frac{1 \text{ h}}{60 \text{ min}} = 112 \frac{\text{dis}}{\text{min}}$$

$$\ln \frac{R_t}{R_0} = -\lambda t = \ln \frac{101 \text{ dis/min}}{112 \text{ dis/min}} = -1.5 \times 10^{-5} \text{ h}^{-1} t = -0.103$$

$$t = \frac{0.103}{1.5 \times 10^{-5} \text{ h}^{-1}} = 6.9 \times 10^{3} \text{ h} \times \frac{1 \text{ d}}{24 \text{ h}} \times \frac{1 \text{ y}}{365.25 \text{ d}} = 0.78 \text{ y}$$

19. Let us use the first and the last values to determine the decay constant.

$$\ln \frac{R_t}{R_0} = -\lambda t = \ln \frac{138 \text{ cpm}}{1000 \text{ cpm}} = -\lambda 250 \text{ h} = -1.981 \qquad \lambda = \frac{1.981}{250 \text{ h}} = 0.00792 \text{ h}^{-1}$$

$$t_{1/2} = \frac{0.693}{\lambda} = \frac{0.693}{0.00792 \text{ h}^{-1}} = 87.5 \text{ h}$$

A slightly different value of $t_{1/2}$ may result from other combinations of R_0 and R_t.

20. First we calculate the decay constant.

$$\lambda = \frac{0.693}{1.7 \times 10^{7} \text{ y}} \times \frac{1 \text{ y}}{365.25 \text{ d}} \times \frac{1 \text{ d}}{24 \text{ h}} \times \frac{1 \text{ h}}{3600 \text{ s}} = 1.3 \times 10^{-15} \text{ s}^{-1}$$

$$N = 1.00 \text{ mg } ^{129}\text{I} \times \frac{1 \text{ g}}{1000 \text{ mg}} \times \frac{1 \text{ mol } ^{129}\text{I}}{129 \text{ g}} \times \frac{6.022 \times 10^{23} \text{ atoms}}{1 \text{ mol } ^{129}\text{I}} = 4.67 \times 10^{18} \ ^{129}\text{I atoms}$$

$$\text{decay rate} = \lambda N = 1.3 \times 10^{-15} \text{ s}^{-1} \times 4.67 \times 10^{18} \text{ atoms} = 6.1 \times 10^{3} \text{ dis/s}$$

21. $^{32}_{15}P$ half-life = 14.3 d. We need to determine the time necessary to get to the detectable

limit, $\dfrac{1}{1000}$ of the initial value. Use $\lambda = \dfrac{0.693}{t_{1/2}} = \dfrac{0.693}{14.3\,\text{d}} = 0.0485\,\text{d}^{-1}$

$\ln\left(\dfrac{1}{1000}\right) = -0.0485\,\text{d}^{-1}(t) \qquad t = 142\ \text{days}$

22. $1.00\ \text{mCi} = 1.00 \times 10^{-3}\,(3.70 \times 10^{10}\ \text{dis s}^{-1}) = 3.70 \times 10^{7}\ \text{dis s}^{-1}$

$\lambda = \dfrac{0.693}{t_{1/2}} = \dfrac{0.693}{5730\ \text{y}} = 1.21 \times 10^{-4}\ \text{y}^{-1}\quad (\,1\ \text{y} = 365.25\ \text{d} = 3.15\underline{6} \times 10^{7}\ \text{s})$

$\lambda = \dfrac{1.21 \times 10^{-4}}{\text{y}} \times \dfrac{1\,\text{y}}{3.15_{6} \times 10^{7}\,\text{s}} = 3.83 \times 10^{-12}\ \text{s}^{-1}$

$1.00\ \text{mCi} = 3.70 \times 10^{7}\ \text{dis s}^{-1} = \lambda N = 3.83 \times 10^{-12}\ \text{s}^{-1}(N)$

$N = 9.66 \times 10^{18}$ atoms of ^{14}C or $1.60\underline{4} \times 10^{-5}$ mol ^{14}C

mass of $^{14}C = 1.60\underline{4} \times 10^{-5}$ mol $^{14}C \times \dfrac{14.00\ \text{g }^{14}C}{1\ \text{mol }^{14}C} = 2.25 \times 10^{-4}\ \text{g }^{14}C$

Age Determination with Radioisotopes

23. Again we use expression (25.12) and (25.13) to determine the time elapsed. The initial rate of decay is about 15 dis/min. First we compute the decay constant.

$\lambda = \dfrac{0.693}{5730\ \text{y}} = 1.21 \times 10^{-4}\ \text{y}^{-1}$

$\ln\dfrac{10\ \text{dis/min}}{15\ \text{dis/min}} = -0.40_{5} = -\lambda t; \qquad t = \dfrac{0.40_{5}}{1.21 \times 10^{-4}\ \text{y}^{-1}} = 3.4 \times 10^{3}\ \text{y}$

The object is a bit more than 3000 years old, and thus is probably not from the pyramid era, which occurred about 3000 B.C.

24. We use the value of λ from the previous exercise.

$\ln\dfrac{R_t}{R_o} = -\lambda t = -\left(1.21 \times 10^{-4}\ \text{y}^{-1}\right)t = \ln\dfrac{0.03\ \text{dis min}^{-1}\ \text{g}^{-1}}{15\ \text{dis min}^{-1}\ \text{g}^{-1}} = -6.2$

$t = \dfrac{6.2}{1.21 \times 10^{-4}\ \text{y}^{-1}} = 5.1 \times 10^{4}\ \text{y}$

25. First we determine the decay constant. $\lambda = \dfrac{0.693}{1.39\times10^{10}\text{ y}} = 4.99\times10^{-11}\text{ y}^{-1}$

Then we can determine the ratio of (N_t), the number of thorium atoms after 2.7×10^{9} y, to (N_0), the initial number of thorium atoms:

$$\ln\frac{N_t}{N_0} = -kt = -\left(4.99\times10^{-11}\text{ y}^{-1}\right)\left(2.7\times10^{9}\text{ y}\right) = -0.13 \qquad \frac{N_t}{N_0} = 0.88$$

Thus, for every mole of ^{232}Th present initially, after 2.7×10^{9} y there are

0.88 mol ^{232}Th and 0.12 mol ^{208}Pb. From this information, we can compute the mass ratio.

$$\frac{0.12\text{ mol }^{208}\text{Pb}}{0.88\text{ mol }^{232}\text{Th}}\times\frac{1\text{ mol }^{232}\text{Th}}{232\text{ g }^{232}\text{Th}}\times\frac{208\text{ g }^{208}\text{Pb}}{1\text{ mol }^{208}\text{Pb}} = \frac{0.12\text{ g }^{208}\text{Pb}}{1\text{ g }^{232}\text{Th}}$$

26. First we determine the decay constant. $\lambda = \dfrac{0.693}{1.39\times10^{10}\text{ y}} = 4.99\times10^{-11}\text{ y}^{-1}$

The rock currently contains 1.00 g ^{232}Th and 0.25 g ^{208}Pb. We can calculate the mass of ^{232}Th that must have been present to produce this 0.25 g ^{208}Pb, and from that find the original mass of ^{232}Th.

$$\text{original mass }^{232}\text{Th} = 1.00\text{ g }^{232}\text{Th now} + \left(0.25\text{ g }^{208}\text{Pb}\times\frac{232\text{ g }^{232}\text{Th}}{208\text{ g }^{208}\text{Pb}}\right)$$

$$= (1.00+0.28)\text{ g} = 1.28\text{ g}$$

$$\ln\frac{N_t}{N_0} = -\lambda t = \ln\frac{1.00\text{ g }^{232}\text{Th now}}{1.28\text{ g originally}} = -0.247 = -4.99\times10^{-11}\text{ y}^{-1}t;$$

$$t = \frac{0.247}{4.99\times10^{-11}\text{ y}^{-1}} = 4.95\times10^{9}\text{ y}$$

27. The principal equation that we shall employ is $E = mc^2$, along with conversion factors.

(a) $E = 6.02\times10^{-23}\text{ g}\times\dfrac{1\text{ kg}}{1000\text{ g}}\times\left(3.00\times10^{8}\text{ m/s}\right)^2 = 5.42\times10^{-9}\text{ kg m}^2\text{ s}^{-2} = 5.42\times10^{-9}\text{ J}$

(b) $E = 4.0015\text{ u}\times\dfrac{931.5\text{ MeV}}{1\text{ u}} = 3727\text{ MeV}$

28. $\text{Mass of individual particles} = \left(47\text{ p}\times\dfrac{1.0073\text{ u}}{1\text{ p}}\right)\times\left(60\text{ n}\times\dfrac{1.0087\text{ u}}{1\text{ n}}\right)$

$$= 47.3431\text{ u} + 60.5220\text{ u} = 107.8651\text{ u}$$

$$\frac{\text{Binding energy}}{\text{nucleon}} = \frac{107.8651\text{ u} - 106.879289\text{ u}}{107\text{ nucleons}}\times\frac{931.5\text{ MeV}}{1\text{ u}} = 8.58\text{ MeV/nucleon}$$

Energetics of Nuclear Reactions

29. The mass defect is the difference between the mass of the nuclide and the sum of the masses of its constituent particles. The binding energy is this mass defect expressed as an energy.

Particle mass $= 9p + 10n + 9e = 9(p+n+e) + n$

$$= 9(1.0073 + 1.0087 + 0.0005486) \, u + 1.0087 \, u = 19.1576 \, u$$

mass defect $= 19.1576 \, u - 18.998403 \, u = 0.1592 \, u$

$$\text{binding energy per nucleon} = \frac{0.1592 \, u \times \dfrac{931.5 \, \text{MeV}}{1 \, u}}{19 \, \text{nucleons}} = 7.805 \text{ MeV/nucleon}$$

30. The mass defect is the difference between the mass of the nuclide and the sum of the masses of its constituent particles. The binding energy is this mass defect expressed as an energy.

Particle mass $= 26p + 30n + 26e = 26(p+n+e) + 4n$

$$= 26(1.0073 + 1.0087 + 0.0005486) \, u + 4 \times 1.0087 \, u = 56.4651 \, u$$

mass defect $= 56.4651 \, u - 55.934939 \, u = 0.5302 \, u$

$$\text{binding energy per nucleon} = \frac{0.5302 \, u \times \dfrac{931.5 \, \text{MeV}}{1 \, u}}{56 \, \text{nucleons}} = 8.819 \text{ MeV/nucleon}$$

31. mass defect $= (10.01294 \, u + 4.00260 \, u) - (13.00335 \, u + 1.00783 \, u) = 0.00436 \, u$

$$\text{energy} = 0.00436 \, u \times \frac{931.5 \text{ MeV}}{1 \, u} = 4.06 \text{ MeV}$$

32. mass defect $= (6.01513 \, u + 1.008665 \, u) - (4.00260 \, u + 3.01604 \, u) = 0.00516 \, u$

$$\text{energy} = 0.00516 \, u \times \frac{931.5 \text{ MeV}}{1 \, u} = 4.81 \text{ MeV}$$

33. 1 neutron $\approx$ 1 amu $= 1.66 \times 10^{-27}$ kg

$E = mc^2 = 1.66 \times 10^{-27} \text{ kg} (2.998 \times 10^8 \text{ m s}^{-1})^2 = 1.49 \times 10^{-10}$ J (1 neutron)

1 eV $= 1.602 \times 10^{-19}$ J,

Hence, 1 neutron $= 1.49 \times 10^{-10}$ J $\times \dfrac{1 \text{ eV}}{1.602 \times 10^{-19} \text{ J}} = 9.30 \times 10^8$ eV or 930. MeV

6.75×10^6 MeV $\times \dfrac{1 \text{ neutron}}{930 \text{ MeV}} = 7.26 \times 10^3$ neutrons

34. $\beta^+ + \beta^-$ collide $\rightarrow$ produce two γ-rays.

Basically the mass of $\beta^+ = \beta^- =$ mass of an electron (9.11×10^{-31} kg)

Each γ-ray has the same energy as the complete conversion of one electron into pure energy.

$E = mc^2 = (9.11 \times 10^{-31}$ kg) $(2.998 \times 10^8$ m s$^{-1})^2 = 8.19 \times 10^{-14}$ J

In electron volts: 8.19×10^{-14} J $\times \dfrac{1 \text{ eV}}{1.602 \times 10^{-19} \text{ J}} = 5.11 \times 10^5$ eV or 0.511 MeV

Each γ-ray has an energy of 0.511 MeV

Nuclear Stability

35. **(a)** We expect ^{20}Ne to be more stable than ^{22}Ne. A neutron-to-proton ratio of 1-to-1 is associated with stability for elements of low atomic number (with $Z \leq 20$).

(b) We expect ^{18}O to be more stable than ^{17}O. An even number of protons and an even number of neutrons are associated with a stable isotope.

(c) We expect ^{7}Li to be more stable than ^{6}Li. Both isotopes have an odd number of protons, but only ^{7}Li has an even number of neutrons.

36. **(a)** We expect ^{40}Ca to be more stable than ^{42}Ca. A neutron-to-proton ratio of 1-to-1 is associated with stability for elements of low atomic number (with $Z \leq 20$).

(b) We expect ^{31}P to be more stable than ^{32}P. Both isotopes have an odd number of protons, but only ^{31}P has an even number of neutrons.

(c) We expect ^{64}Zn to be more stable than ^{63}Zn. An even number of protons and an even number of neutrons are associated with a stable isotope.

37. β^- emission has the effect of "converting" a neutron to a proton. β^+ emission, on the other hand, has the effect of "converting" a proton to a neutron.

(a) The most stable isotope of phosphorus is ^{31}P, with a neutron-to-proton ratio of close to 1-to-1 and an even number of neutrons. Thus, ^{29}P has "too few" neutrons, or too many protons. It should decay by β^+ emission. In contrast, ^{33}P has "too many" neutrons, or "too few" protons. Therefore, ^{33}P should decay by β^- emission.

(b) Based on the atomic mass of I (126.90447), we expect the isotopes of iodine to have mass numbers close to 127. This means that ^{120}I has "too few" neutrons and therefore should decay by β^+ emission, whereas ^{134}I has "too many" neutrons (or "too few" protons) and therefore should decay by β^- emission.

38. β^- emission has the effect of converting a neutron to a proton, while β^+ emission has the effect of converting a proton to a neutron.

   (a) Based on the fact that elements of low atomic number have about the same number of protons as neutrons, $_{15}^{28}P$—with 15 protons and 13 neutrons—has too few neutrons. Therefore, it should decay by β^+ emission.

   (b) Once again, elements of low atomic number have about the same number of protons as neutrons. $_{19}^{45}K$ with 19 protons and 26 neutrons—has too many neutrons. Therefore, it should decay by β^- emission.

   (c) Based on the atomic mass of zinc (65.39) we expect most of its isotopes to have about 36 neutrons. There are 42 neutrons in $_{30}^{72}Zn$, more than we expect. Thus we expect this nuclide to decay by β^- emission.

39. A "doubly magic" nuclide is one in which the atomic number is a magic number (2, 8, 20, 28, 50, 82, 114) and the number of neutrons also is a magic number (2, 8, 20, 28, 50, 82, 126, 184). Nuclides that fit this description are given below.

nuclide	^4He	^{16}O	^{40}Ca	^{56}Ni	^{208}Pb
no. of protons	2	8	20	28	82
no. of neutrons	2	8	20	28	126

40. For isotopes of high atomic number, stable nuclides are characterized by a neutron-to-proton ratio greater than 1, which increases with increasing atomic number. Naturally occurring isotopes of high atomic number decrease their atomic number by losing an alpha particle, which has a neutron-to-proton ratio of 1. This leaves the neutron-to-proton ratio for the daughter hat is higher than that of the parent, when it should be slightly lower. In order to redress this, the number of neutrons needs to be decreased and the number of protons increased. Beta emission accomplishes this. In contrast, artificially produced isotopes have no definite neutron-to-proton ratio. Thus, sometimes, the number of neutrons needs to be decreased, which is accomplished by beta emission, while at other times the number of protons needs to be decreased, which is accomplished by positron emission.

Fission and Fusion

41. We use the conversion factor between number of curies and mass of ^{131}I which was developed in the Summarizing Example.

$$\text{no. g } ^{131}I = 170 \text{ curies} \times \frac{18.8 \text{ g } ^{131}I}{2.33 \times 10^6 \text{ curie}} = 1.37 \times 10^{-3} \text{ g} = 1.37 \text{ mg}$$

42. Nuclear fission is the process by which a heavy nucleus disintegrates into neutrons and stable nuclei with smaller mass numbers. For instance, uranium-238 undergoes fission according to the equation

$$_{92}^{238}U \rightarrow _{90}^{234}Th + _2^4He$$

The nuclear binding energy for uranium-238 is less than the sum of the binding energies for thorium-234 and helium-4. Consequently, when a uranium-238 nucleus splits apart, energy is released. Nuclear fusion, by contrast, involves the amalgamation of light nuclei into heavier, more stable nuclei. For instance, part of the energy released by our Sun is believed to come from the fusion of hydrogen to form deuterium:

$$_1^1H + {}_1^1H \rightarrow {}_1^2H + {}_{-1}^0e$$

Although both fusion and fission release vast amounts of energy, fusion releases far more energy on a per nucleon basis. To understand why this is so, we need to refer to Figure 25-6, which is a plot of Average Binding Energy per Nucleon as a Function of Atomic Number. The graph clearly shows that the increase in binding energy observed for the formation of the lightest nuclides (e.g. deuterium, tritium, helium-3) is much more dramatic than the decrease in binding energy that is seen for the fragmentation of heavier nuclei such as uranium-235. Thus, the plot indicates that more energy should be released by the combination of light nuclei(nuclear fusion) than by the disintegration of heavy nuclei(nuclear fission).

Effect of Radiation on Matter

43. The term "rem" is an acronym for "radiation equivalent-man," and takes into account the quantity of biological damage done by a given dosage of radiation. On the other hand, the rad is the dosage which places 0.010 J of energy into each kg of irradiated matter. Thus, for living tissue, the rem provides a good idea of how much tissue damage a certain kind and quantity of radiation damage will do. But for nonliving materials, the rad is usually preferred, and indeed is often the only unit of utility.

44. Low-level radiation is very close in its dosage to background radiation and one problem is to separate out the effects of the two sources (low-level and background). The other problem is that low-level radiation does not produce severe damage in a short period of time. Thus the effects of low-level radiation will only be observed over a long time period. Of course other effects, such as chemical and biological toxins, will also be observed over these time periods, and we have to try to separate these two types of effects. (There also is the genetic heritage of the organism to consider, of course.)

45. One reason why ^{90}Sr is hazardous is because strontium is in the same family of the periodic table as calcium, and hence often reacts in a similar fashion to calcium. The most likely place for calcium to incorporated into the body is in bones, where it resides for a long time. Strontium is expected to behave in a similar fashion. Thus, it will be retained in the body for a long time. Bone is an especially dangerous place for a radioisotope to be present—even if it has low penetrating power, as do β^- rays—because blood cells are produced in bone marrow.

46. It is not particularly hazardous to be near a flask of ^{222}Rn, because it is unlikely that the alpha particles can get through the walls of the flask. (Note that since radon is a gas, the flask must be sealed.) The decay products of ^{222}Rn may produce other forms of radiation that are more penetrating, such as β^- particles and γ rays, so being near the flask may still pose a risk. ^{222}Rn can be potentially hazardous if one breathes the gas.

Application of Radioisotopes

47. Mix a small amount of tritium with the $H_2(g)$ and detect where the radioactivity appears with a Geiger counter.

48. In neutron activation analysis, the sample is bombarded with neutrons. Radioisotopes are produced by this process. These radioisotopes can be easily detected even in very small quantities, much smaller, in fact, than the quantities that can be detected by conventional means of quantitative analysis. These radioisotopes are produced in quantities that are proportional to the quantity of each element originally present in the sample. And each radioisotope is characteristic of the element from which it was produced by neutron bombardment. Even microscopic samples can be analyzed by this technique. Finally, neutron activation analysis is a nondestructive technique, while the conventional techniques of precipitation or titration require that all of the sample, or at least part of it, be destroyed.

49. The recovered sample will be radioactive. When NaCl(s) and $NaNO_3(s)$ are dissolved in solution, the ions (Na^+, Cl^-, and NO_3^-) are free to move throughout the solution. A given anion does not remain associated with a particular cation. Thus, all the anions and cations are shuffled and some of the radioactive ^{24}Na will end up in the crystallized $NaNO_3$.

50. We would expect the tritium label to appear in both the $NH_3(g)$ and the $H_2O(l)$. When $NH_4^+(aq)$ is formed, one of the four chemically and spatially equivalent H atoms is occasionally a tritium atom. In the subsequent reaction between the occasionally marked NH_4Cl and NaOH to form $NH_3(g)$ and $H_2O(l)$, there are three chances in four that a tritium atom will remain attached to N in NH_3 and one chance in four that a tritium ion will react with a hydroxide ion to form $H_2O(l)$.

Integrative and Advanced Exercises

51. In the cases where rounding off the atomic mass produces the mass number of the most stable isotope, there often is but one stable isotope. This frequently is the case when the atomic number of the element is an odd number. For instance, the situation with ^{39}K (Z = 19) and ^{85}Rb (Z = 37), but not ^{88}Sr (Z = 38). In the cases where this technique of rounding does not work, there are two or more stable isotopes of significant abundance. Note that the rounding off does not work in situations where it predicts a nuclide with an odd number of neutrons and an odd number of protons (such as ^{64}Cu with 29 protons and 35 neutrons), whereas the rounding off technique works when the predicted nuclide has an even number of protons, an even number of neutrons, or both.

52. Each α particle contains two protons and has a mass number of 4. Thus each α particle emission reduces the mass number by 4 and the atomic number by 2. The emission of 8 α particles would reduce the mass number by 32 and the atomic number by 16. Thus the overall reaction would be as follows: $^{238}_{92}U \longrightarrow 8\,^4_2He + {}^{206}_{76}Os$ (76 protons and 130 neutrons) In Figure 25-7, a nuclide with 76 protons and 130 neutrons lies above, to the left of the belt of stability; it is radioactive.

53. We use $\Delta H_f^\circ[CO_2(g)] = -393.51$ kJ/mol as the heat of combustion of 1 mole of carbon. In the text, the energy produced by the fission of 1.00 g ^{235}U is determined as 8.20 x 10^7 kJ.

metric tons of coal required $= 1.00$ kg ^{235}U $\times \dfrac{1000\text{ g}}{1\text{ kg}} \times \dfrac{8.20 \times 10^7\text{kJ}}{1.00\text{ g }^{235}\text{U}} \times \dfrac{1\text{mol C}}{393.5\text{ kJ}} \times \dfrac{12.01\text{ g C}}{1\text{mol C}}$

$\times \dfrac{1.00\text{ g coal}}{0.85\text{ gC}} \times \dfrac{1\text{ kg}}{1000\text{g}} \times \dfrac{1\text{ metric ton}}{1000\text{ kg}} = 2.9 \times 10^3 \text{metric tons (megagrams or Mg)}$

54. Since the two nuclides have the same mass number, the ratio of their masses is the same as the ratio of the number of atoms of each type. We use equation (25.12) to determine the time required for the Rb to decrease from 1.004 to 1.00. First we compute the decay constant.

$\lambda = \dfrac{0.693}{5 \times 10^{11}} = 1._4 \times 10^{-12}\text{ y}^{-1}$

$\ln \dfrac{1.00}{1.004} = -4.0 \times 10^{-3} = -\lambda t = -1._4 \times 10^{-12}\text{ y}^{-1}\, t \qquad t = \dfrac{-4.0 \times 10^{-3}}{-1._4 \times 10^{-12}\text{ y}^{-1}} = 3 \times 10^9\text{ y}$

55. $\lambda = \dfrac{0.693}{7340\text{ y}} \times \dfrac{1\text{ y}}{365.25\text{ d}} \times \dfrac{1\text{ d}}{24\text{ h}} \times \dfrac{1\text{ h}}{3600\text{ s}} = 2.99 \times 10^{-12}\text{ s}^{-1}$

$N = 5.10\text{ mg} \times \dfrac{1\text{g}}{1000\text{ mg}} \times \dfrac{1\text{ mol }^{229}\text{Th}}{229\text{ g }^{229}\text{Th}} \times \dfrac{6.022 \times 10^{23}\,{}^{229}\text{Th atoms}}{1\text{ mol }^{229}\text{Th}} = 1.34 \times 10^{19}\,{}^{229}\text{Th atoms}$

decay rate in disintegrations/sec $= \lambda N = 2.99 \times 10^{-12}\text{ s}^{-1} \times 1.34 \times 10^{19}$ atoms $\times \dfrac{1\text{ Ci}}{3.7 \times 10^{10}\text{ dis/s}}$

$\times \dfrac{1000\text{ mCi}}{1\text{ Ci}} = 1.1\text{ mCi}$

56. First we find the decay constant. The activity (λN) is the product of the decay constant and the number of atoms.

$\lambda = \dfrac{0.693}{27.7\text{ y}} \times \dfrac{1\text{ y}}{365.25\text{ d}} \times \dfrac{1\text{ d}}{24\text{ h}} \times \dfrac{1\text{ h}}{3600\text{ s}} = 7.93 \times 10^{-10}\text{ s}^{-1}$

radioactivity $= 1.00\text{ mCi} \times \dfrac{1\text{Ci}}{1000\text{ mCi}} \times \dfrac{3.7 \times 10^{10}\text{ dis/s}}{1\text{Ci}} = 3.7 \times 10^7\text{ dis/s}$

$N = \dfrac{\text{activity}}{\lambda} = \dfrac{3.7 \times 10^7\text{ dis/s}}{7.93 \times 10^{-10}\text{ s}^{-1}} = 4.7 \times 10^{16}\,{}^{90}\text{Sr atoms} \times \dfrac{1\text{ mol }^{90}\text{Sr}}{6.022 \times 10^{23}\text{ atoms}} \times \dfrac{90\text{ g }^{90}\text{Sr}}{1\text{ mol }^{90}\text{Sr}}$

$= 7.0 \times 10^{-6}\text{ g }^{90}\text{Sr} = 7.0\ \mu\text{g }^{90}\text{Sr}$

57. $\text{decay rate} = 89.8 \text{ mCi} \times \dfrac{1 \text{ Ci}}{1000 \text{ mCi}} \times \dfrac{3.7 \times 10^{10} \text{ dis/s}}{1 \text{ Ci}} = 3.3_2 \times 10^9 \text{ dis/s}$

$N = 1.00 \text{ mg} \times \dfrac{1 \text{ g}}{1000 \text{ mg}} \times \dfrac{1 \text{ mol } ^{137}\text{Cs}}{137 \text{ g } ^{137}\text{Cs}} \times \dfrac{6.022 \times 10^{23} \text{ atoms}}{1 \text{ mol}} = 4.40 \times 10^{18} \ ^{137}\text{Cs atoms}$

$\text{decay rate} = \lambda N \qquad \lambda = \dfrac{\text{decay rate}}{N} = \dfrac{3.3_2 \times 10^9 \text{ dis/s}}{4.40 \times 10^{18} \text{ atoms}} = 7.5_5 \times 10^{-10} \text{ s}^{-1}$

$t_{1/2} = \dfrac{0.693}{\lambda} = \dfrac{0.693}{7.5_5 \times 10^{-10} \text{ s}^{-1}} \times \dfrac{1 \text{ h}}{3600 \text{ s}} \times \dfrac{1 \text{ d}}{24 \text{ h}} \times \dfrac{1 \text{ y}}{365.25 \text{ d}} = 29 \text{ y}$

58. $\lambda = \dfrac{0.693}{1.25 \times 10^9 \text{ y}} \times \dfrac{1 \text{ y}}{365.25 \text{ d}} \times \dfrac{1 \text{ d}}{24 \text{ h}} \times \dfrac{1 \text{ h}}{3600 \text{ s}} = 1.76 \times 10^{-17} \text{ s}^{-1}$

$N = 1.00 \text{ g KAlSi}_3\text{O}_8 \times \dfrac{1 \text{ mol KAlSi}_3\text{O}_8}{278.3 \text{ g KAlSi}_3\text{O}_8} \times \dfrac{1 \text{ mol K}}{1 \text{ mol KAlSi}_3\text{O}_8} \times \dfrac{0.000117 \text{ mol } ^{40}\text{K}}{1 \text{ mol K}}$

$\times \dfrac{6.022 \times 10^{23} \ ^{40}\text{K atoms}}{1 \text{ mol } ^{40}\text{K}} = 2.53 \times 10^{17} \ ^{40}\text{K atoms}$

$\text{rate} = 0.89 \lambda N = 0.89 \times 1.76 \times 10^{-17} \text{ s}^{-1} \times 2.53 \times 10^{17} \text{ atoms} = 4.0 \text{ dis/s}$

59. ^{14}C is produced from ^{14}N by neutron bombardment. Since ^{14}N is a common element, constituting 78% of the atmosphere, any activity that increases the emission of neutrons will increase the production of ^{14}C. A major source used to be thermonuclear explosions, particularly atmospheric detonations. But most tests now take place underground. Nonetheless, the extensive thermonuclear testing that took place during the 1950s and 1960s could have produced sufficient ^{14}C to invalidate the radiocarbon dating of materials that were alive during that period. Nuclear power plants are a very minor source of ^{14}C, as is bringing to the surface neutron-emitting isotopes by mining activities.

Although we might suspect ozone depletion to play a role in increasing the quantity of ^{14}C, such is not the case. Ozone absorbs ultraviolet radiation, not neutrons. And, in any case, there is about the same proportion of ^{14}N in the upper atmosphere as there is further down, in layers that recently have become exposed to ultraviolet radiation because of the depletion of ozone.

60. $\text{product masses} = {}^{17}\text{O} + {}^1\text{H} = 16.99913 \text{ u} + 1.00783 \text{ u} = 18.00696 \text{ u}$

$\text{reactant masses} = {}^4\text{He} + {}^{14}\text{N} = 4.00260 \text{ u} + 14.00307 \text{ u} = 18.00567 \text{ u}$

The products have more mass than the reactants. The difference must be supplied as energy from the reactants. This difference in energy ends up entirely to the E_{kinetic} of the α particle. Compute this energy in MeV.

$\text{energy} = (18.00696 \text{ u} - 18.00567 \text{ u}) \times \dfrac{931.5 \text{ MeV}}{1 \text{ u}} = 1.20 \text{ MeV}$

61. Assume we have in our possession 100 g of the hydrogen/tritium mixture. This sample will

$$\text{mol hydrogen} = \frac{95.00\text{gH}}{1.008\text{g/mol}} = 94.24_6 \text{ mol hydrogen}$$

$$\text{mol tritium} = \frac{5.00\text{gH}}{3.02\text{g/mol}} = 1.65_6 \text{ mol tritium}$$

$$\text{mole fraction tritium} = \frac{1.65_6 \text{ mol tritium}}{1.65_6 \text{ mol tritium} + 94.24_6 \text{ mol hydrogen}} = 1.72_7 \times 10^{-2}$$

$$\text{total moles of gas in mixture} = \frac{PV}{RT} = \frac{(1.05\text{atm})(4.65\text{L})}{0.0821\dfrac{\text{L atm}}{\text{mol K}}(298.15\text{K})} = 0.199_5 \text{ mol}$$

afford us 95 g hydrogen and 5 g tritium.

mols of tritium = $(0.199_5 \text{ mol})(1.72_7 \times 10^{-2} \text{ mol tritium / mol mixture}) = 3.44_5 \times 10^{-3}$ mol tritium

# of tritium atoms (N) = $(2 \times 3.44_5 \times 10^{-3} \text{ mol tritium})(6.022 \times 10^{23} \text{ tritium atoms/mol}) = 4.15 \times 10^{21}$ tritium atoms

$$\text{rate} = \frac{0.693}{t_{1/2}} N = \frac{0.693}{12.3\text{ y} \times \dfrac{365\text{ d}}{1\text{ y}} \times \dfrac{24\text{ h}}{1\text{ d}} \times \dfrac{60\text{ min}}{1\text{ h}} \times \dfrac{60\text{ s}}{1\text{ min}}} \times 4.15 \times 10^{21} = 7.42 \times 10^{12} \text{ disintegrations/s}$$

7

$$\text{activity in curies} = \frac{7.42 \times 10^{12} \text{ disintegrations/s}}{3.7 \times 10^{10} \dfrac{\text{disintegrations/s}}{\text{Ci}}} = 2.0 \times 10^{2} \text{ Ci}$$

62. $\lambda = \dfrac{0.693}{1.25 \times 10^{9} \text{ y}} = 5.54 \times 10^{-10} \text{ y}^{-1}$ Calculate fraction of ^{40}K that remains after 1.5×10^{9} y.

$$\ln \frac{N_t}{N_0} = -\lambda t = -5.54 \times 10^{-10} \text{ y}^{-1} \times 1.5 \times 10^{9} \text{ y} = -0.83 \qquad \frac{N_t}{N_0} = 0.44$$

Thus, the fraction of ^{40}K that has decayed is $1.000 - 0.44 = 0.56$

The fraction of the ^{40}K that has decayed into ^{40}Ar is $0.110 \times 0.56 = 0.062$

This fraction is proportional to the mass of ^{40}Ar. Then the ratio of masses is determined.

$$\frac{\text{mass } ^{40}\text{Ar}}{\text{mass } ^{40}\text{K}} = \frac{0.062}{0.44} = 0.14$$

63. $\text{energy} = 1.00 \times 10^{3} \text{ cm}^3 \times \dfrac{2.5 \text{ g}}{1 \text{ cm}^3} \times \dfrac{0.006 \text{ g U}}{100.000 \text{ g shale}} \times \dfrac{1 \text{ mol U}}{238 \text{ g U}}$

$\times \dfrac{6.022 \times 10^{23} \text{ U atoms}}{1 \text{ mol U}} \times \dfrac{3.20 \times 10^{-11} \text{ J}}{1 \text{ U atom}} \times \dfrac{1 \text{ kJ}}{1000 \text{ J}} = 1.2 \times 10^{7} \text{ kJ}$

64.

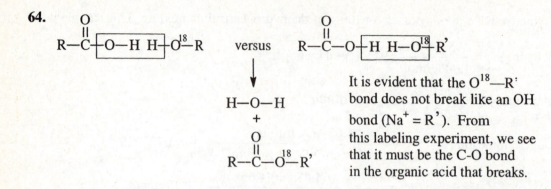

It is evident that the O^{18}—R'
bond does not break like an OH
bond ($Na^+ = R'$). From
this labeling experiment, we see
that it must be the C–O bond
in the organic acid that breaks.

65. $6\ CO_2(g) + 6\ H_2O^{18}(l) - light \rightarrow C_6H_{12}O_6(s) + 6\ O_2^{18}\ (g)$

$6\ CO_2^{18}\ (g) + 6\ H_2O(l) - light \rightarrow C_6H_{12}O_6(s) + 6\ O_2(g)$

Basically, the results shows that the O_2 arises from the oxidation of H_2O and that the CO_2 involved in this reaction remains intact. Simplistically, this can be explained by using two half reactions:

$$12\ H_2O^{18}(l) \qquad \rightarrow 6\ O_2^{18}(g) + 24\ H^+ + 24\ e^-$$

$$\underline{24\ e^- + 24\ H^+ + 6\ CO_2(g) \rightarrow C_6H_{12}O_6(s) + 6\ H_2O}$$

$$6\ CO_2(g) + 6\ H_2O^{18}(l) \rightarrow C_6H_{12}O_6(s) + 6\ O_2^{18}\ (g)$$

In reality, the reaction is not this simple, however. What this labeling studies shows is that the O_2 is evolved from the oxidation of water (the only time $O_2^{18}(g)$ forms is when H_2O^{18} is used).

66. Initially $\dfrac{U^{238}}{U^{235}} = 1$ $U^{238}\ t_{1/2} = 4.5 \times 10^9$ years $U^{235}\ t_{1/2} = 7.1 \times 10^8$ years

Currently $\dfrac{U^{238}}{U^{235}} = \dfrac{0.9928}{0.0072} = 138$

Remember $t_{1/2} = \dfrac{0.693}{\lambda}$ Hence $\lambda = \dfrac{0.693}{t_{1/2}}$ $\lambda_{U^{238}} = 1.5\underline{4} \times 10^{-10}$ $\lambda_{U^{235}} = 9.7\underline{6} \times 10^{-10}$

For any radioactive isotope, the amount remaining is $e^{-\lambda t}$

Currently, $\dfrac{e^{-\lambda t}\ (U^{238})}{e^{-\lambda t}\ (U^{235})} = 138 = \dfrac{e^{-1.54 \times 10^{-10}t}\ (U^{238})}{e^{-9.76 \times 10^{-10}t}\ (U^{235})} = e^{8.22 \times 10^{-10}t} = 138$ (take ln of both sides)

$8.2\underline{2} \times 10^{-10}t = \ln 138 = 4.9\underline{27}$ $t = \dfrac{4.927}{8.2\underline{2} \times 10^{-10}} = 6.0 \times 10^9$ years

Feature Problems

67. First tabulate the isotopes symbols, the mass of isotope and its associated packing fraction.

Isotope Symbol	Mass of Isotope (u)	Packing Fraction
^{1}H	1.007825	0.007825
^{4}He	4.002603	0.000651
^{9}Be	9.012186	0.001354
^{12}C	12	0
^{16}O	15.994915	−0.000318
^{20}Ne	19.992440	−0.000378
^{24}Mg	23.985042	−0.000623
^{32}S	31.972074	−0.000873
^{40}Ar	39.962384	−0.000940
^{40}Ca	39.962589	−0.000935
^{48}Ti	47.947960	−0.001084
^{52}Cr	51.940513	−0.001144
^{56}Fe	55.934936	−0.001162
^{58}Ni	57.935342	−0.001115
^{64}Zn	63.929146	−0.001107
^{80}Se	79.916527	−0.001043
^{84}Kr	83.911503	−0.001054
^{90}Zr	89.904700	−0.001059
^{102}Ru	101.904348	−0.000938
^{114}Cd	113.903360	−0.000848
^{130}Te	129.906238	−0.000721
^{138}Ba	137.905000	−0.000688
^{142}Nd	141.907663	−0.000650
^{158}Gd	157.924178	−0.000480
^{166}Er	165.932060	−0.000409

Plot of Packing Fraction versus Mass Number

Data plotted above

This graph and Fig. 25-6 are almost exactly the inverse of one another, with the maxima of one being the minima of the other. Actual nuclidic mass is often a number slightly less than the number of nucleons (mass number). This difference divided by the number of nucleons (packing fraction) is proportional to the negative of the mass defect per nucleon.

68. **(a)** The rate of decay depends on both the half-life and on the number of radioactive atoms present. In the early stages of the decay chain, the larger number of radium-226 atoms multiplied by the very small decay constant is still larger than the product of the very small number of radon-222 atoms and its much larger decay constant. Only after some time has elapsed, does the rate of decay of radon-222 approach the rate at which it is formed from radium-226 and the amount of radon-222 reaches a maximum. Beyond this point, the rate of decay of radon-222 exceeds its rate of formation.

(b) $\dfrac{dD}{dt} = \lambda_p P - \lambda_d D = \lambda_p P_o e^{-\lambda_p t} - \lambda_d D$

(c) The number of radon-222 atoms at he proposed times are 2.90×10^{15} atoms after 1 day;
1.26×10^{16} after 1 week; 1.75×10^{16} after 1 year; 1.68×10^{16} after one century; and 1.13×10^{16} after 1 millennium. The actual maximum comes after about 2 months, but the amount after 1 year is only slightly smaller.

69. **(a)**

$$\text{Zr(s)} + 6H_2O(l) \rightarrow ZrO_2(s) + 4\,H_3O^+(aq) + 4\,e^- \qquad 1.43\text{ V}$$

$$4\,H_2O(l) + 4\,e^- \rightarrow 2\,H_2(g) + 4\,OH^-(aq) \qquad -0.828\text{ V}$$

$$\overline{\text{Zr(s)} + 2\,H_2O(l) \rightarrow ZrO_2(s) + 2\,H_2(g) \qquad 0.602\text{ V (spont)}}$$

Yes, Zr can reduce water under standard conditions.

(b) $E^\circ = \dfrac{0.0592}{n}\log K_{eq}$ $\qquad 0.602\text{ V} = \dfrac{0.0592}{4}\log K_{eq} \qquad K_{eq} = 5 \times 10^{40}$

(c) $pH = 7$ $\qquad$ Therefore, $[OH^-] = [H_3O^+] = 1.0 \times 10^{-7}$

$E_{ox} = E^\circ_{ox} - \dfrac{0.0592}{n}\log Q = 1.43\text{ V} - \dfrac{0.0592}{4}\log(1.0\times10^{-7})^4 = 1.84\text{ V}$

$E_{red} = E^\circ_{red} - \dfrac{0.0592}{n}\log Q = -0.828\text{ V} - \dfrac{0.0592}{4}\log(1\times10^{-7})^4 = -0.414\text{ V}$

$E_{cell} = E_{ox} + E_{red} = 1.84 + (-0.414) = 1.43\text{ V (spontaneous)}$

(d) Zr may be the culprit responsible for the $H_2(g)$ formation. In the Chernobyl accident the reaction of carbon with superheated steam played a major role.
Reaction: $H_2O(g) + C(s) \rightarrow CO_{(g)} + H_2(g)$

70. **(a)** Average atomic mass of Sr in the rock

$\dfrac{^{87}Sr}{^{86}Sr} = 2.25 \qquad \dfrac{^{86}Sr}{^{88}Sr} = 0.119 \qquad \dfrac{^{84}Sr}{^{88}Sr} = 0.007 \qquad$ Given: 15.5 ppm Sr

Let $x = {}^{86}Sr$, $y = {}^{88}Sr$, $z = {}^{87}Sr$, $w = {}^{84}Sr \qquad x + y + z + w = 15.5$ ppm

$\dfrac{z}{x} = 2.25$, $\dfrac{x}{y} = 0.119$, $\dfrac{w}{y} = 0.007$

Set $x = {}^{86}Sr = 1$ and find the relative atom ratio of the other
Hence, $\qquad z = 2.25x = 2.25\times1 = 2.25$

$$y = \dfrac{x}{0.119} = \dfrac{1}{0.119} = 8.40\underline{3}$$

$$w = 0.007y = 0.007\times8.40\underline{3} = 0.058\underline{8}$$

$$x + y + z + w = 1 + 2.25 + 8.40\underline{3} + 0.058\underline{8} = 11.7\underline{12}$$

As a percent abundance, we find the follow for the Sr in the sample.

$\%^{86}Sr = 1/11.7\underline{12} \times 100\% = 8.53\underline{8}\%$

$\%^{88}Sr = 8.40\underline{3}/11.7\underline{12} \times 100\% = 71.7\underline{5}\%$

$\%^{87}Sr = 2.25/11.7\underline{12} \times 100\% = 19.2\underline{1}\%$

$\%^{84}Sr = 0.05\underline{88}/11.7\underline{12} \times 100\% = 0.5\%$

Av. mass Sr = mass^{86}Sr (%^{86}Sr) + mass^{88}Sr (%^{88}Sr) + mass^{87}Sr (%^{87}Sr) + mass^{84}Sr (%^{84}Sr)

Av. mass Sr = 8.53$\underline{8}$% (85.909 u) + 71.7$\underline{5}$% (87.906 u)+ 19.2$\underline{1}$% (86.909 u) + 0.5%(83.913 u)

Average Atomic mass Sr = 7.33$\underline{5}$ u + 63.0$\underline{7}$ u + 16.6$\underline{95}$ u + 0.4$\underline{2}$ u = 87.5 u

Current atom ratio is $\dfrac{^{87}Rb}{^{85}Rb} = 0.330$

Set 1000 atoms for ^{85}Rb and 330 atoms ^{87}Rb or a total of 1330 atoms of Rb

Percent abundance of each isotope: ^{85}Rb = (1000/1330)×100% = 75.2 % ^{85}Rb

^{87}Rb = (1000/1330)×100% = 24.8 % ^{87}Rb

Av. mass Rb = mass^{85}Rb (%^{85}Rb) + mass^{87}Rb (%^{87}Rb)

Av. mass Rb = 75.2 % (84.912 u) + 24.8 % (86.909 u)

Average Atomic mass Rb = 63.8$\underline{5}$ u + 21.5$\underline{5}$ u = 85.4 u

(b) Original Rb in rock?

Need to convert atom ratio $\rightarrow$ isotope concentration in ppm.

^{85}Rb concentration in ppm

$= \dfrac{1000 \text{ atoms } ^{85}Rb}{1330 \text{ atoms Rb}} \times \dfrac{1 \text{ atom Rb}}{85.4 \text{ u Rb}} \times \dfrac{84.912 \text{ u } ^{85}Rb}{1 \text{ atom } ^{85}Rb} \times 265.4 \text{ ppm Rb} = 198.4 \text{ ppm } ^{85}Rb$

^{87}Rb concentration in ppm

$= \dfrac{330 \text{ atoms } ^{87}Rb}{1330 \text{ atoms Rb}} \times \dfrac{1 \text{ atom Rb}}{85.4 \text{ u Rb}} \times \dfrac{86.909 \text{ u } ^{87}Rb}{1 \text{ atom } ^{87}Rb} \times 265.4 \text{ ppm Rb} = 67.0 \text{ ppm } ^{87}Rb$

Currently 265.4 ppm (198.4 ppm ^{85}Rb + 67.0 ppm ^{87}Rb)

Recall earlier calculations showed: $\%^{86}Sr = 8.53\underline{8}\%$; $\qquad \%^{88}Sr = 71.7\underline{5}\%$;

$\qquad\qquad\qquad\qquad\qquad\qquad\quad \%^{87}Sr = 19.2\underline{1}\%$; $\qquad \%^{84}Sr = 0.5\%$

Consider 100,000 atoms of Sr. Calculate the concentration (in ppm) of ^{86}Sr and ^{87}Sr.

^{86}Sr concentration in ppm

$= \dfrac{8538 \text{ atoms } ^{86}Sr}{100,000 \text{ atoms Sr}} \times \dfrac{1 \text{ atom Sr}}{87.5 \text{ u Sr}} \times \dfrac{85.909 \text{ u } ^{86}Sr}{1 \text{ atom } ^{86}Sr} \times 15.5 \text{ ppm Sr} = 1.29\underline{9} \text{ ppm } ^{86}Sr$

^{87}Sr concentration in ppm

$= \dfrac{192\underline{10} \text{ atoms } ^{87}Sr}{100,000 \text{ atoms Sr}} \times \dfrac{1 \text{ atom Sr}}{87.5 \text{ u Sr}} \times \dfrac{86.909 \text{ u } ^{87}Sr}{1 \text{ atom } ^{87}Sr} \times 15.5 \text{ ppm Sr} = 2.95\underline{7} \text{ ppm } ^{87}Sr$

Currently: $\dfrac{^{87}Sr}{^{86}Sr} = 2.25 = \dfrac{19210 \text{ atoms } ^{87}Sr}{8538 \text{ atoms } ^{86}Sr}$

Originally: $\dfrac{^{87}Sr}{^{86}Sr} = 0.700$ or $^{87}Sr = {}^{86}Sr \times 0.700 = 8538 \times 0.700 = 5977$ atoms ^{87}Sr

Change in $^{87}Sr = 19210 - 5977 = 13233$ atoms ^{87}Sr (per 100,000 Sr atoms)
Currently, 19210 per 100,000 atoms is ^{87}Sr which represents 2.957 ppm.
A change of 13233 atoms represents $(13233/19210) \times 2.957$ ppm $= 2.037$ ppm ^{87}Sr

The source of ^{87}Sr is radioactive decay from ^{87}Rb (a 1:1 relation).

Change in the ^{87}Rb (through radioactive decay) = change in $^{87}Sr = 2.037$ ppm

Isotope:	^{87}Rb	^{85}Rb	Total Rb
Current concentration	67.0 ppm	198.4 ppm	265.4 ppm
Change concentration	+2.037 ppm	—	+2.037 ppm
Original concentration	69.04 ppm	198.4 ppm	267.44 ppm

(c) % ^{87}Rb decayed $= \left(\dfrac{2.037 \text{ ppm}}{69.04 \text{ ppm}}\right) \times 100\% = 2.95\ \%$ (% ^{87}Rb remaining = 97.05%)

(d) $\ln(0.9705) = -\lambda t$ $\qquad (\lambda = \dfrac{0.693}{t_{1/2}} = \dfrac{0.693}{4.8 \times 10^{10} \text{ y}} = 1.444 \times 10^{-11} \text{ y}^{-1})$

$\ln(0.9705) = -1.444 \times 10^{-11} \text{ y}^{-1} t; \quad t = 2.07 \times 10^9$ years

CHAPTER 26
ORGANIC CHEMISTRY
PRACTICE EXAMPLES

1A We have shown only the C atoms and the bonds between them. Remember that there are four bonds to each C atom; the remaining bonds not shown are to H atoms. First we realize there is only one isomer with all six C atoms in one line. Then we draw the isomers with one 1-C branch. The isomers with two 1-C branches can have them both on the same atom or on different atoms. This accounts for all five isomers.

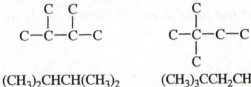

C—C—C—C—C—C
$CH_3CH_2CH_2CH_2CH_2CH_3$

$(CH_3)_2CHCH_2CH_2CH_3$

$CH_3CH_2CH(CH_3)CH_2CH_3$

$(CH_3)_2CHCH(CH_3)_2$

$(CH_3)_3CCH_2CH_3$

1B We have shown only the C atoms and the bonds between them. Remember that there are four bonds to each C atom; the remaining bonds not shown are to H atoms.

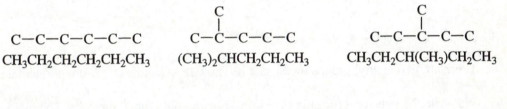

C—C—C—C—C—C—C
$CH_3CH_2CH_2CH_2CH_2CH_2CH_3$

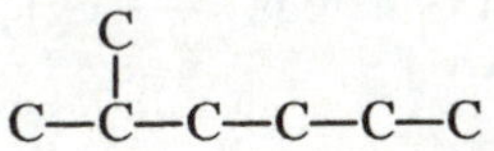

$(CH_3)_2CHCH_2CH_2CH_2CH_3$

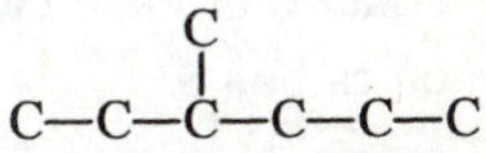

$CH_3CH_2CH(CH_3)CH_2CH_2CH_3$

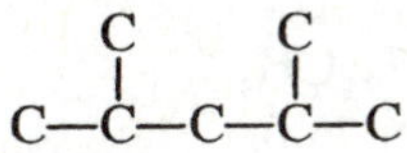

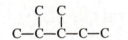

$(CH_3)_2CHCH_2CHC(CH_3)_2$

$(CH_3)_2CHCH(CH_3)CH_2CH_3$

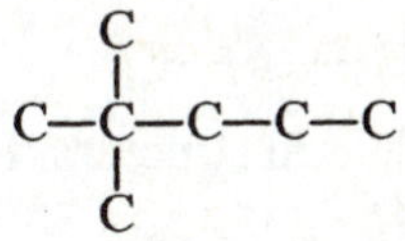

$(CH_3)_3CCH_2CH_2CH_3$

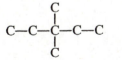

$CH_3CH_2C(CH_3)_2CH_2CH_3$

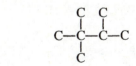

$(CH_3)_3CCH(CH_3)_2$

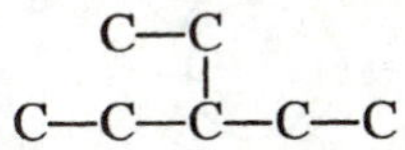

$CH(CH_2CH_3)_3$

2A

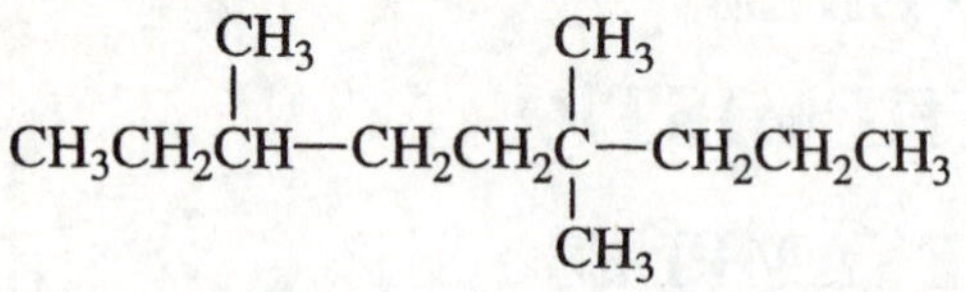

$$CH_3 \qquad CH_3$$
$$CH_3CH_2CH-CH_2CH_2C-CH_2CH_2CH_3$$
$$CH_3$$

Numbering starts from the left and goes right so that the substituents appear with the lowest numbers possible. This is 3,6,6-trimethylnonane.

2B   In the structural formula we show only the C atoms and the bonds between them. Remember that there are four bonds to each C atom; the remaining bonds not shown are to H atoms. The longest chain has 8 C atoms, making this an octane.

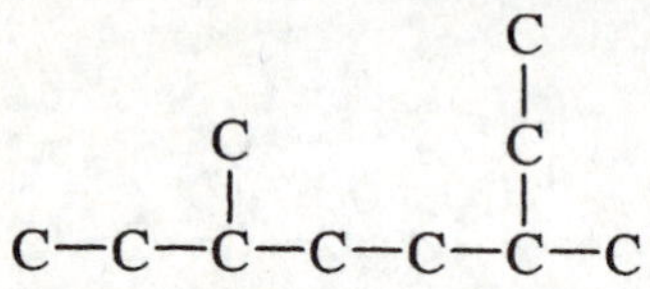

$$C$$
$$|$$
$$C \qquad C$$
$$| \qquad |$$
$$C-C-C-C-C-C-C$$

There are two methyl groups on the chain. The IUPAC name is 3,6-dimethyloctane.

3A   We write the structural formula as before, showing only the C skeleton. The main chain is 7 C's long; we number it from left to right. Methyl groups are 1 carbon chains, an ethyl group is on carbon 2.

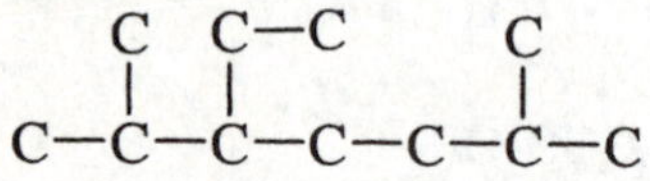

$$C \quad C-C \qquad C$$
$$| \qquad | \qquad |$$
$$C-C-C-C-C-C-C$$

condensed structural formula: $CH_3CH(CH_3)CH(CH_2CH_3)CH_2CH_2CH(CH_3)_2$

3B   Pentane is a five-carbon chain. There is a — CH_3 group on carbon 2, a sec-butyl group on carbon 3.

$$C$$
$$|$$
$$C-C-C-C-C$$
$$|$$
$$C-C-C$$

Condensed structural formula: $CH_3CH(CH_3)CH(CH(CH_3)_2)CH_2CH_3$

4A   The aldehyde group is a meta director. Thus, the product of the mononitration of benzaldehyde should be the compound whose structure is drawn below:

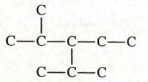

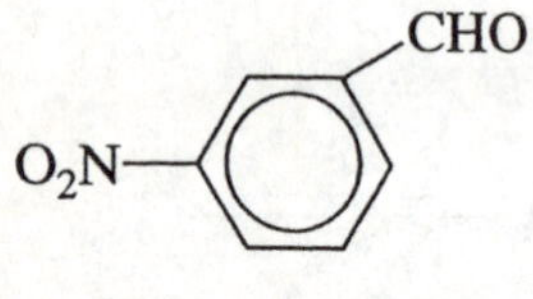

m-nitrobenzaldehyde or 3-nitrobenzaldehyde.

4B   The Cl group is an ortho, para director. The two possible products are 1,3-dichloro-2-nitrobenzene and 2,4-dichloro-1-nitrobenzene.

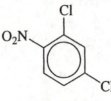

1,3-dichlorobenzene 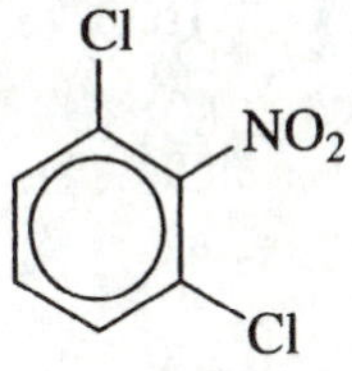   1,3-dichloro-2   2,4-dichloro-1
                             -nitrobenzene   -nitrobenzene

5A   **(a)**

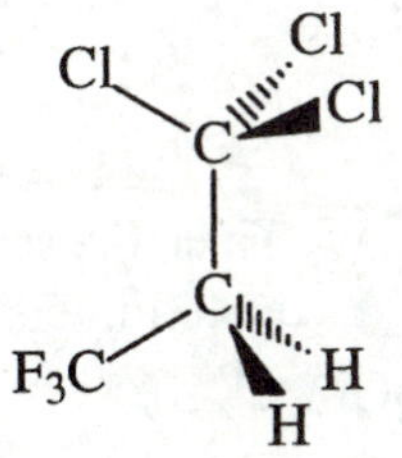

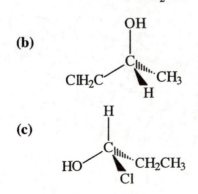

All three carbon atoms in this molecule are attached to at least two groups of the same type; thus, the molecule is achiral

(b)

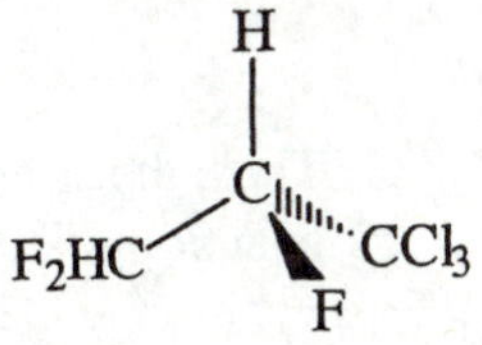

This molecule contains a carbon atom that is bonded to four different groups; consequently, the molecule is chiral.

(c)

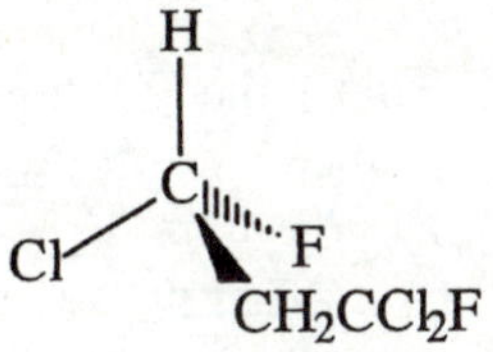

This molecule contains a carbon atom that is attached to four different groups; consequently, the molecule is chiral.

5B   **(a)**

None of the three carbon atoms in this alcohol are bonded to four different group; consequently the molecule is achiral

(b)

This molecule contains a carbon atom that is bonded to four different groups; consequently, the molecule is chiral.

(c)

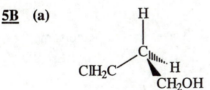

This molecule contains a carbon atom that is bonded to four different groups; consequently, the molecule is chiral.

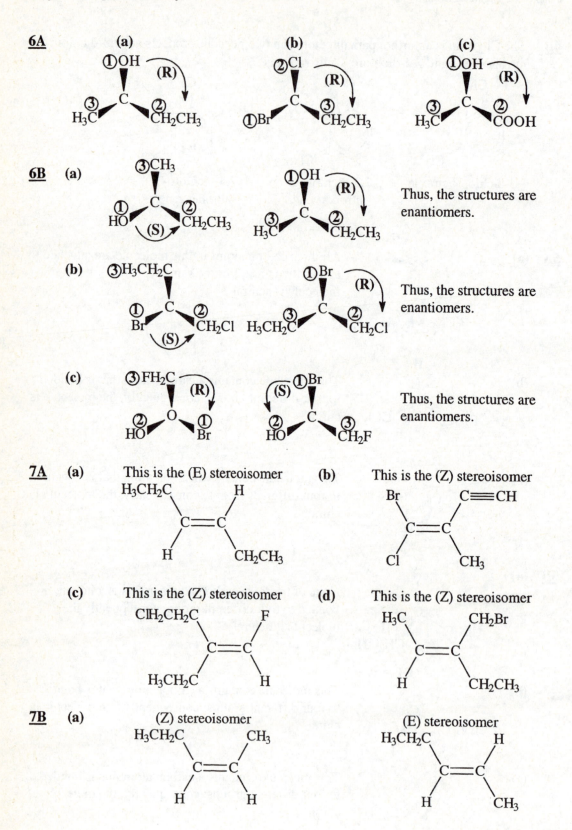

6A

6B

Thus, the structures are enantiomers.

Thus, the structures are enantiomers.

Thus, the structures are enantiomers.

7A

(a) This is the (E) stereoisomer

(b) This is the (Z) stereoisomer

(c) This is the (Z) stereoisomer

(d) This is the (Z) stereoisomer

7B

(a) (Z) stereoisomer

(E) stereoisomer

(b) (Z) stereoisomer

(E) stereoisomer

(c) (Z) stereoisomer

(E) stereoisomer

(d) (Z) stereoisomer

(E) stereoisomer

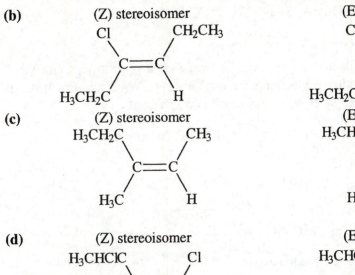

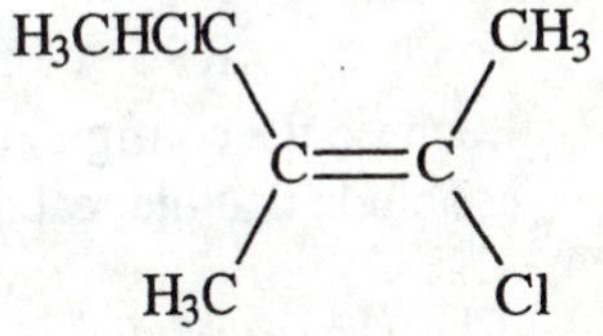

8A The hydride ion is a nucleophile because it has a lone pair of electrons that can be shared with another atom.

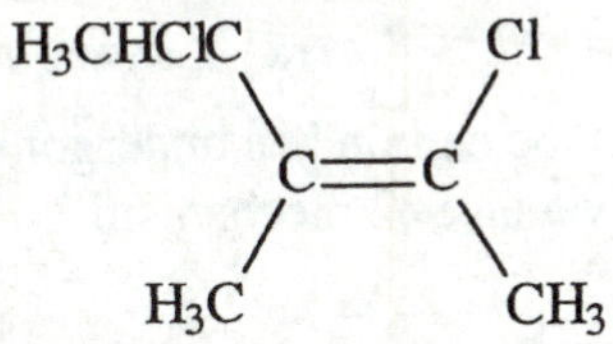

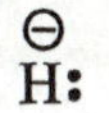

The central Al atom is electron deficient. Thus, $AlCl_3$ is an electrophile.

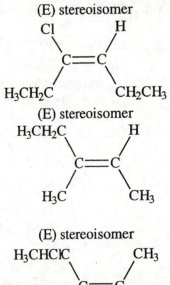

Ethylene can be viewed as a nucleophile because it has a pair of pi-electrons that can be shared with another atom

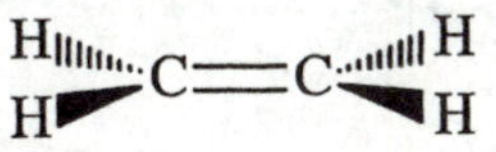

The negatively charged S atom possesses three lone pairs of electrons that can potentially be shared with other atoms. Thus methylthiolate anion is a nucleophile.

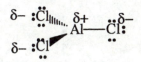

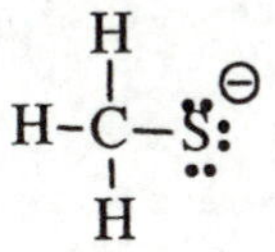

8B (a) CH_3I (electrophile) + NH_3 (nucleophile) $\rightarrow$ $H_3CNH_3^+I^-$

(b) CH_3CH_2Cl (electrophile) + CH_3S^- (nucleophile) $\rightarrow$ $H_3CCH_2SCH_3 + Cl^-$

(c) $CH_3CH_2CHClCH_3$ (electrophile) + CN^- (nucleophile) $\rightarrow$ $CH_3CH_2CHCNCH_3 + Cl^-$

9A **(a)** $CH_3C\equiv C^- + CH_3Br \xrightarrow{\quad S_N2 \quad} CH_3C\equiv CCH_3 + Br^-$

(b) $Cl^- + CH_3CH_2CN \rightarrow$ NO REACTION
The nucleophile in this reaction is Cl^- while the leaving group is the CN^-. The CN^- ion is a much stronger nucleophile than Cl^-, so the equilibrium will strongly favor the reactants. In other words, no reaction is expected.

(c) $CH_3NH_2 + (CH_3)_3CCl \xrightarrow{\quad S_N1 \quad} CH_3\overset{+}{N}H_2C(CH_3)_3 + Cl^-$

9B **(a)**

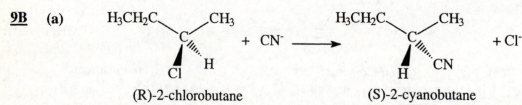

(R)-2-chlorobutane   (S)-2-cyanobutane

Because the configuration at the stereogenic carbon has undergone an inversion, we can conclude that the reaction has occurred via an S_N2 mechanism.

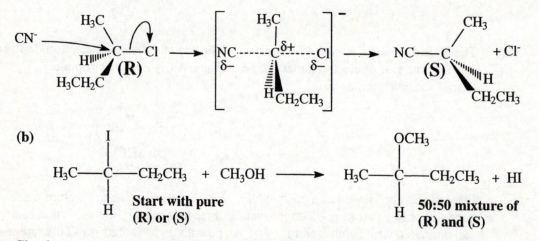

(b)

$H_3C-\overset{\underset{|}{H}}{\overset{|}{C}}(-I)-CH_2CH_3 + CH_3OH \longrightarrow H_3C-\overset{\underset{|}{H}}{\overset{OCH_3}{C}}-CH_2CH_3 + HI$

Start with pure (R) or (S)   **50:50 mixture of (R) and (S)**

Clearly, since a racemic mixture forms, the reaction must occur via an S_N1 mechanism.

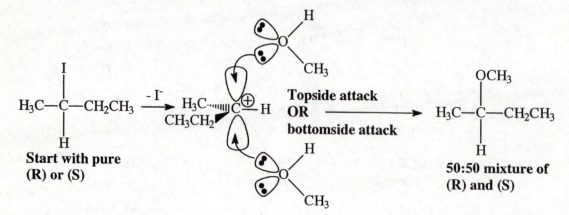

EXERCISES

Organic Structures

1. In the following structural formulas, the hydrogen atoms are omitted for simplicity. Remember that there are four bonds to each carbon atom. The missing bonds are C— H bonds.

(a) $CH_3CH_2CHBrCHBrCH_3$

$$\underset{\overset{|}{C}}{\overset{\overset{Br}{|}}{C}}\,\,C$$

C—C—C—C—C
with Br Br above the 3rd and 4th carbons

(b) $(CH_3)_3CCH_2C(CH_3)_2CH_2CH_2CH_3$

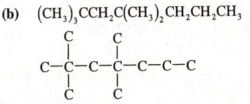

(c) $(C_2H_5)_2CHCH = CHCH_2CH_3$

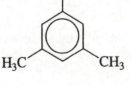

2. In the structural formulas drawn below, we omit the hydrogen atoms. Remember that there are four bonds to each C atom. The bonds that are not shown are C— H bonds.

(a) 3-isopropyloctane

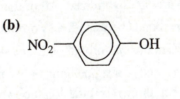

(b) 2-chloro-3-methylpentane

```
      Cl  C
      |   |
  C—C—C—C—C
```

(c) 2-pentene

C–C=C–C–C

(d) dipropyl ether

C–C–C–O–C–C–C

3. **(a)**

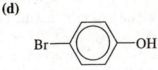

(b)

NO_2——OH (benzene ring)

(c) Cl——COOH, Cl, NH_2 (benzene ring)

(d)

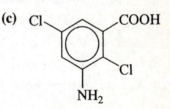

Br——OH (benzene ring)

4. In the following structural formulas, the hydrogen atoms are omitted for simplicity. Remember that there are four bonds to each carbon atom. The missing bonds are C—H bonds.

(a) $(CH_3)_3CCH_2CH(CH_3)CH_2CH_2CH_3$ **(b)** $(CH_3)_2CHCH_2C(CH_3)_2CH_2Br$

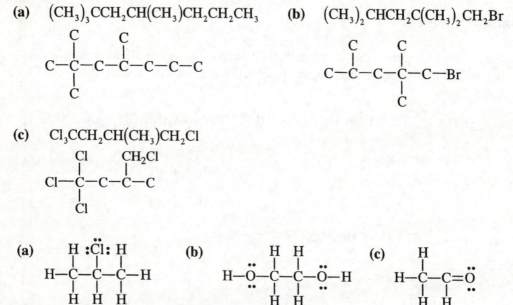

(c) $Cl_3CCH_2CH(CH_3)CH_2Cl$

5. (a) **(b)** **(c)**

6. (a) **(b)** **(c)**

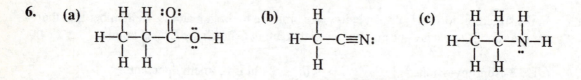

7. (a) Each carbon atom is sp^3 hybridized. All of the C–H bonds in the structure (drawn on the next page) are sigma bonds, between the $1s$ orbital of H and the sp^3 orbital of C. The C–C bond is between sp^3 orbitals on each C atom.

(b) Both carbon atoms are sp^2 hybridized. All of the C–H bonds in the structure (see next page) are sigma bonds between the $1s$ orbital of H and the sp^2 orbital of C. The C–Cl bond is between the sp^2 orbital on C and the $3p$ orbital on Cl. The C=C double bond is composed of a sigma bond between the sp^2 orbitals on each C atom and a pi bond between the $2p_z$ orbitals on the two C atoms.

(c) The left-most C atom (in the structure drawn on the next page) is sp^3 hybridized, and the C–H bonds to that C atom are between the sp^3 orbitals on C and the $1s$ orbital on H. The other two C atoms are sp hybridized. The right-hand C–H bond is between the sp orbital on C and the $1s$ orbital on H. The C≡C triple bond is composed of one sigma bond formed by overlap of sp orbitals, one from each C atom, and two pi bonds, each formed by the overlap of two $2p$ orbitals, one from each C atom (that is a $2p_y - 2p_y$ overlap and a $2p_z - 2p_z$ overlap).

(a)

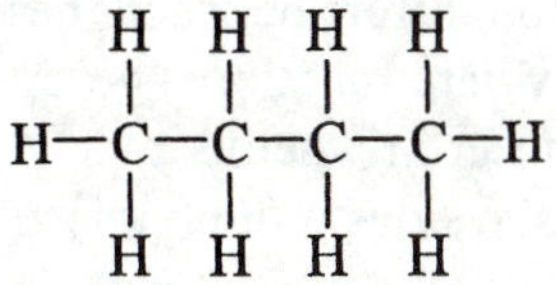

$$
\begin{array}{ccccc}
 & H & H & H & H \\
 & | & | & | & | \\
H- & C- & C- & C- & C-H \\
 & | & | & | & | \\
 & H & H & H & H
\end{array}
$$

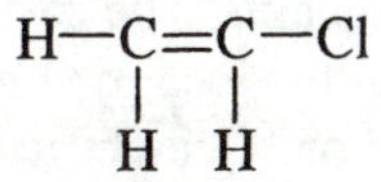

(b)

$$H-C=C-Cl$$
$$\quad\; |\;\;\; |$$
$$\quad\; H\;\; H$$

(c)

$$
\begin{array}{c}
H \\
| \\
H-C-C\equiv C-H \\
| \\
H
\end{array}
$$

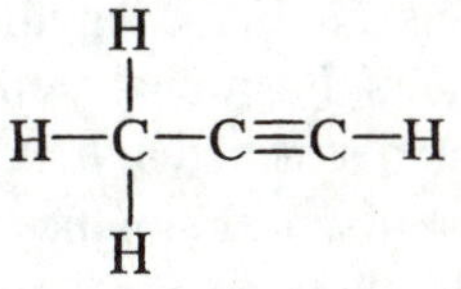

8. **(a)** The left- and right-most C atoms in the structure (drawn below) are sp^3 hybridized. All C–H bonds are sigma bonds formed by the overlap of an sp^3 orbital on C with a 1s orbital on H. The central C atom is sp^2 hybridized; both C–C bonds are sigma bonds, formed by the overlap between the sp^3 orbital on the terminal C atom and an sp^2 orbital on the central C atom. The C=O double bond is composed of a σ bond between the sp^2 orbital on the central C atom and a $2p_y$ orbital on the O atom, and a π bond between the $2p_z$ orbital on the central C atom and the $2p_z$ orbital on the O atom.

(b) The left C atom in the structure (drawn below) is sp^3 hybridized. All C–H bonds are sigma bonds formed by the overlap of an sp^3 orbital on C with a 1s orbital on H. The central C atom is sp^2 hybridized; the C–C bond is a sigma bond, formed by the overlap between the sp^3 orbital on the terminal C atom and an sp^2 orbital on the central C atom. The C=O double bond is composed of a σ bond between the sp^2 orbital on the central C atom and a $2p_y$ orbital on the O atom, and a π bond between the $2p_z$ orbital on the central C atom and the $2p_z$ orbital on the O atom. The right O atom is sp^3 hybridized; the C–O sigma bond forms by the overlap of $C(sp^2)$ with $O(sp^3)$ and the O–H bond forms by overlap of the $O(sp^3)$ with the H(1s).

(c) The two end C's are sp^2 hybridized; all C–H bonds form by the overlap of $C(sp^2)$ with H(1s). The central C is sp hybridized. Both C=C bonds consist of a σ bond formed by $C_{\text{central}}(sp)$ with $C_{\text{end}}(sp^2)$ overlap and a π bond formed by $C_{\text{central}}(2p)$ with $C_{\text{end}}(2p)$ overlap. Note: The $2p$ orbitals that make up each pi bond are mutually perpendicular relative the left or right side of the molecule.

(a)

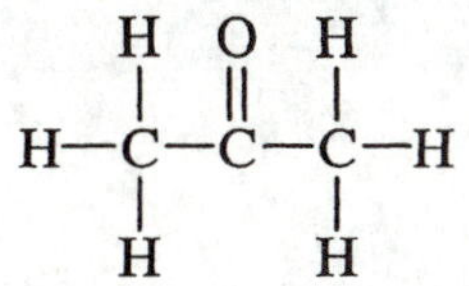

$$
\begin{array}{ccc}
H & O & H \\
| & \| & | \\
H-C- & C- & C-H \\
| & & | \\
H & & H
\end{array}
$$

(b)

$$
\begin{array}{cc}
H & O \\
| & \| \\
H-C- & C-O-H \\
| & \\
H &
\end{array}
$$

(c)

$$
\begin{array}{ccc}
H-C= & C= & C-H \\
| & & | \\
H & & H
\end{array}
$$

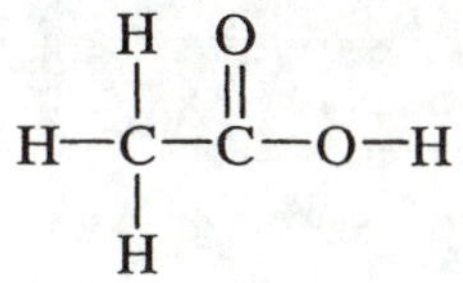

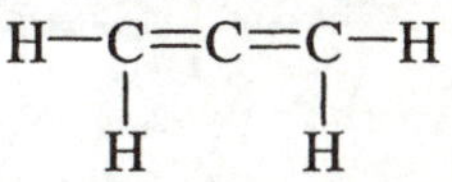

Isomers

9. Structural (skeletal) isomers differ from each other in the length of their carbon atom chains and in the length of the side chains. The carbon skeleton differs between these isomers. Positional isomers differ in the location or position where functional groups are located attached to the carbon skeleton. Geometric isomers differ in whether two substituents are on the same side of the molecule or on opposite sides of the molecule from each other; usually they are on opposite sides or the same side of a double bond.

(a) The chloro group is attached to the terminal carbon in $CH_3CH_2CH_2Cl$ and to the central carbon atom in $CH_3CHClCH_3$. These are positional isomers.

(b) $CH_3CH(CH_3)CH_2CH_3$, methylbutane, and $CH_3(CH_2)_3CH_3$, pentane, are structural isomers.

(c) Ortho-aminotoluene and meta-aminotoluene differ in the position of the $-NH_2$ group on the benzene ring. They are positional isomers.

10. (a) The two chloro groups in $CHCl=CHCl$ are on different carbon atoms. In $CH_2=CCl_2$ they are on the same atoms. These are positional isomers.

(b) 1-butene, $CH_2=CHCH_2CH_3$, and $CH_2=C(CH_3)_2$, methylpropene, are structural isomers. They might be termed positional isomers, differing as they do in the position of a CH_3 group. But since that group's location changes the structure of the molecule, they are structural isomers.

(c) These two compounds differ in the location of the two—COOH groups; the left-hand one has the two groups on the same side of the double bond (the cis isomer) and the right-hand one has the two groups on opposite sides of the double bond (the trans isomer). These are geometric isomers.

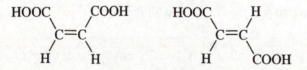

11. We show only the carbon atom skeleton in each case. Remember that there are four bonds to carbon. The bonds that are not indicated in these structures are $C-H$ bonds.

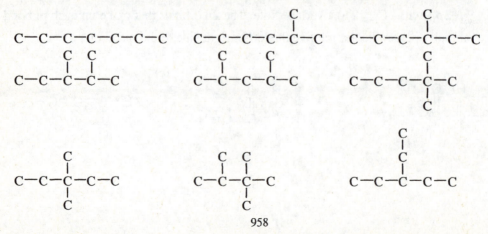

12. In each case, we draw only the carbon skeleton. It is understood that there are four bonds to each carbon atom. The bonds that are not shown are C–H bonds.

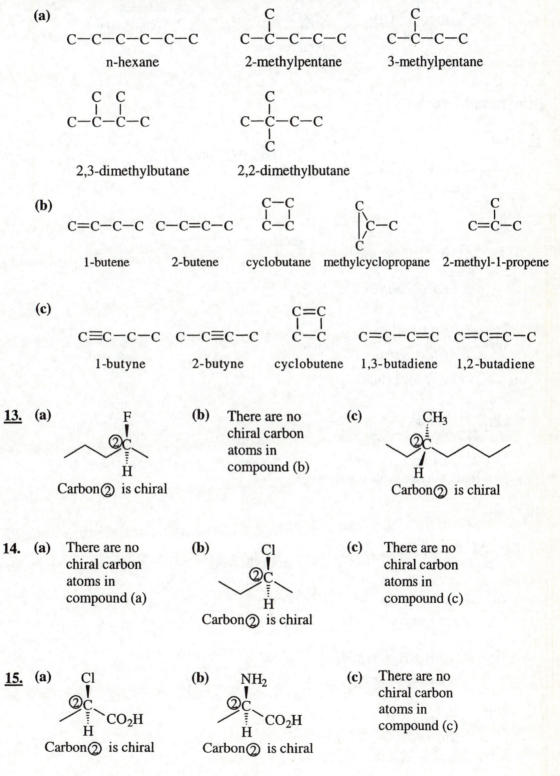

(a)

C—C—C—C—C—C
n-hexane

C—C—C—C—C (with C above 2nd carbon)
2-methylpentane

C—C—C—C (with C above 2nd carbon)
3-methylpentane

C—C—C—C (with C above 2nd and 3rd carbons)
2,3-dimethylbutane

C—C—C—C (with C above and below 2nd carbon)
2,2-dimethylbutane

(b)

C=C—C—C
1-butene

C—C=C—C
2-butene

cyclobutane

methylcyclopropane

C=C—C (with C above 2nd carbon)
2-methyl-1-propene

(c)

C≡C—C—C
1-butyne

C—C≡C—C
2-butyne

cyclobutene

C=C—C≡C
1,3-butadiene

C=C=C—C
1,2-butadiene

13. (a)

F
②C
H
Carbon② is chiral

(b) There are no chiral carbon atoms in compound (b)

(c)

CH₃
②C
H
Carbon② is chiral

14. (a) There are no chiral carbon atoms in compound (a)

(b)

Cl
②C
H
Carbon② is chiral

(c) There are no chiral carbon atoms in compound (c)

15. (a)

Cl
②C
CO₂H
H
Carbon② is chiral

(b)

NH₂
②C
CO₂H
H
Carbon② is chiral

(c) There are no chiral carbon atoms in compound (c)

16. (a)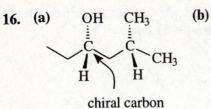

chiral carbon

(b) OH
C—Cl
H
chiral carbon

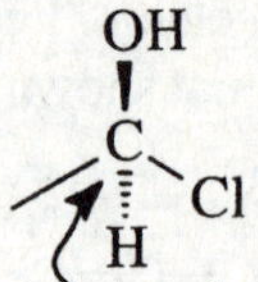

(c) There are no chiral carbon atoms in compound (c)

Functional Groups

17. (a) Br
|
$CH_3CHCH_2CH_3$
2-Bromobutane

alkyl halide (bromide)

(b)
$\bigcirc$—CH_2—C—H
‖
O

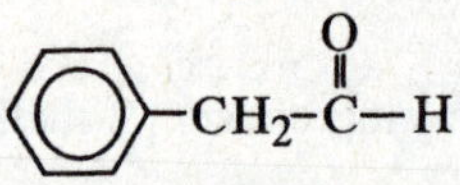

benzylaldehyde

aldehyde

(c) O
‖
$CH_3CCH_2CH_3$
methyl ethyl ketone

ketone

(d) HO—$\bigcirc$—OH

2, 3 or 4-hydroxyphenol
(only 4-hydroxyphenol shown)

phenol, hydroxyl group, phenyl group

18. (a) O
‖
CH_3CH_2COH
propanoic acid

carboxylic acid

(b) $(CH_3)_2CHCH_2OCH_3$
sec-butylmethyl ether

ether

(c) $CH_3CH(NH_2)CH_2CH_3$
2-aminobutane

amine

(d) O
‖
CH_3COCH_3
methyl acetate

ester

19. **(a)** A carbonyl group is $>C=O$, while a carboxyl group is $-C(O)-OH$

The essential difference between them is the hydroxyl group, — OH.

(b) An aldehyde has a carbonyl group at the end of a molecule, $R-\overset{\overset{O}{\|}}{C}-H$

In a ketone, $R-\overset{\overset{O}{\|}}{C}-R'$ the carbonyl group is in the center of the molecule.

(c) Acetic acid is $H_3C-\overset{\overset{O}{\|}}{C}-OH$, while an acetyl group is $H_3C-\overset{\overset{O}{\|}}{C}-$

The essential difference is the presence of the — OH group in acetic acid.

20. **(a)** aromatic nitro compound, $C_6H_5NO_2$, nitrobenzene

(b) aliphatic amine, $C_2H_5NH_2$, ethylamine

(c) chlorophenol, $p\text{-}ClC_6H_4OH$, para-chlorophenol (drawn below)

(d) aliphatic diol, $HOCH_2CH_2OH$, 1,2-ethanediol

(e) unsaturated aliphatic alcohol, $CH_2=CHCH_2CH_2OH$, 3-butene-1-ol

(f) alicyclic ketone, $C_6H_{10}O$, cyclohexanone (drawn below)

(g) halogenated alkane, $CH_3CHICH_2CH_3$, 2-iodobutane

(h) aromatic dicarboxylic acid, $o-C_6H_4(COOH)_2$, ortho-phthalic acid (drawn below)

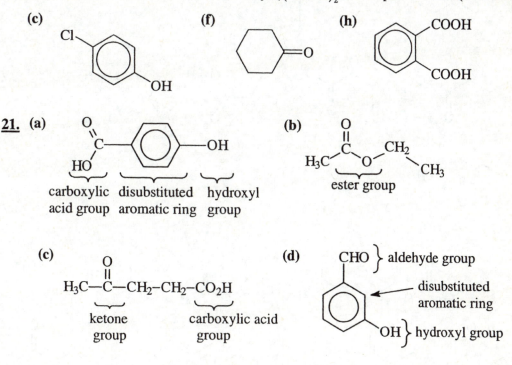

(c)   **(f)**   **(h)**

21. **(a)** carboxylic acid group / disubstituted aromatic ring / hydroxyl group

(b) ester group

(c) ketone group / carboxylic acid group

(d) CHO } aldehyde group / disubstituted aromatic ring / OH } hydroxyl group

22.

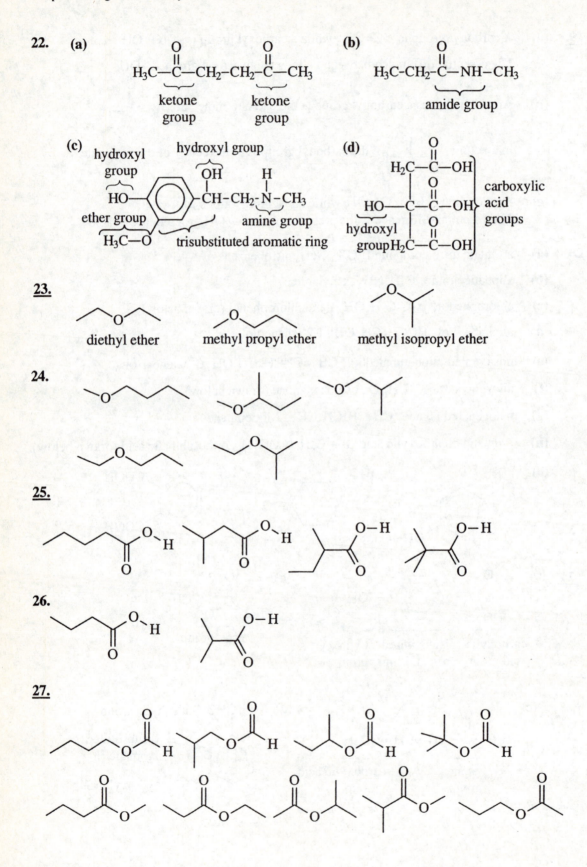

(a)

H₃C—C(=O)—CH₂—CH₂—C(=O)—CH₃

ketone group   ketone group

(b)

H₃C-CH₂-C(=O)—NH—CH₃

amide group

(c)

hydroxyl group   hydroxyl group

HO— (aromatic ring) —CH-CH₂-N(H)-CH₃ with OH, amine group

ether group   H₃C—O   trisubstituted aromatic ring

(d)

carboxylic acid groups, hydroxyl group

23.

diethyl ether   methyl propyl ether   methyl isopropyl ether

24.

25.

26.

27.

28.

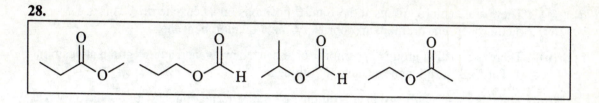

29.

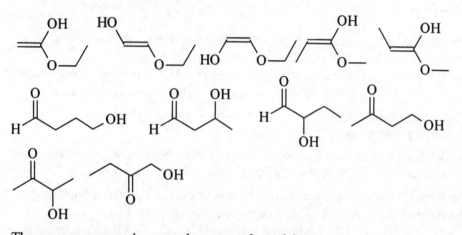

30. There are numerous isomers; here are a few of them

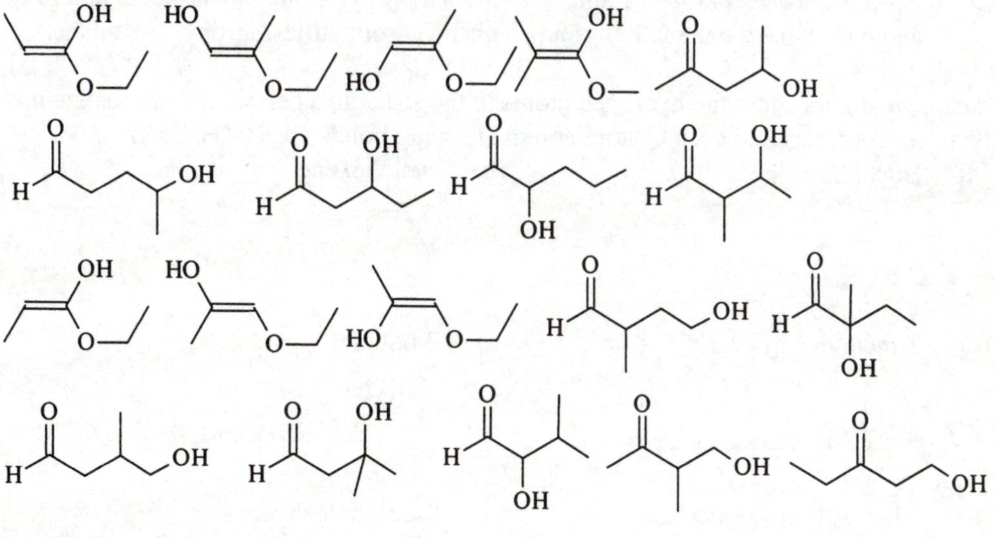

Nomenclature and Formulas

31. **(a)** The longest chain is eight carbons long, the two substituent groups are methyl groups, and they are attached to the number 3 and number 5 carbon atom. This is 3,5-dimethyloctane.

(b) The longest carbon chain is three carbons long, the two substituent groups are methyl groups, and they are both attached to the number 2 carbon atom. This is 2,2-dimethylpropane.

32. **(a)** There are 2 chloro groups at the 1 and 3 positions on a benzene ring. This is 1,3-dichlorobenzene or, more appropriately, meta-dichlorobenzene.

(b) There is a methyl group at position 1 on a benzene ring, and a nitro group at position 3. This is 3-nitrotoluene or meta-nitrotoluene.

(c) There is a — COOH group at position 1 on the benzene ring, and a — NH_2 group at position 4, or para to the — COOH group. This is 4-aminobenzoic acid or *p*-aminobenzoic acid.

33. **(a)** The longest carbon chain has four carbon atoms and there are 2 methyl groups attached to carbon 2. This is 2,2-dimethylbutane.

(b) The longest chain is three carbons long, there is a double bond between carbons 1 and 2, and a methyl group attached to carbon 2. This is 2-methylpropene.

(c) Two methyl groups are attached to a three-carbon ring. This is 1,2-dimethylcyclopropane.

(d) The longest chain is 5 carbons long, there is a triple bond between carbons 2 and 3 and a methyl group attached to carbon 4. This is 4-methyl-2-pentyne.

(e) The longest chain is 5 carbons long. There is an ethyl group on carbon 2 and a methyl group on carbon 3. This compound is 2-ethyl-3-methylpentane.

(f) The longest carbon chain containing the double bond is 5 carbons long. The double bond is between carbons 1 and 2. There is a propyl group on carbon 2, and carbons 3 and 4 each have one methyl group. This is 3,4-dimethyl-2-propyl-1-pentene.

34. Again we do not show the hydrogen atoms in the structures below. But we realize that there are four bonds to each C atom and the missing bonds are C – H bonds.

(a) isopentane

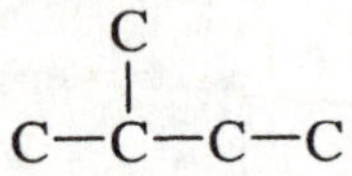

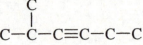

(b) cyclohexene

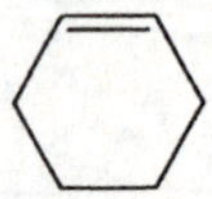

(c) 2-methyl-3-hexyne

(d) 2-butanol

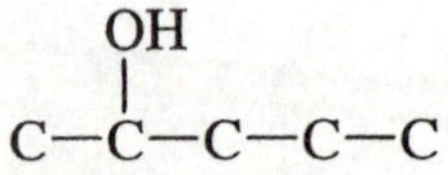

(e) ethyl isopropyl ether

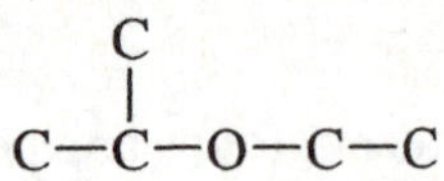

(f) Propionaldehyde

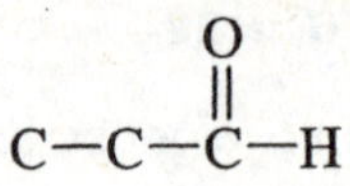

35. **(a)** The name pentene is insufficient. 1-pentene is CH_2=$CHCH_2CH_2CH_3$ and 2-pentene is CH_3CH=$CHCH_2CH_3$.

(b) The name butanone is sufficient. There is only one four-carbon ketone.

(c) The name butyl alcohol is insufficient. There are numerous butanols. 1-butanol is $HOCH_2CH_2CH_2CH_3$, 2-butanol is $CH_3CH(OH)CH_2CH_3$, isobutanol is $HOCH_2CH(CH_3)_2$, and t-butanol is $(CH_3)_3COH$.

(d) The name methylaniline is insufficient. It specifies $CH_3-C_6H_4-NH_2$, with no relative locations for the $-NH_2$ and $-CH_3$ substituents.

(e) The name methylcyclopentane is sufficiently precise. It does not matter where on a five-carbon ring with only $C-C$ single bonds a methyl group is placed.

(f) Dibromobenzene is insufficient. It specifies $Br-C_6H_4-Br$, with no indication of the relative locations of the two bromo groups on the ring.

36. (a) 3-pentene specifies $CH_3CH_2CH=CHCH_3$. The proper name for this compound is 2-pentene. The number for the functional group should be as small as possible.

(b) Pentadiene is insufficient. Possible pentadienes are 1,2 pentadiene, $CH_2=C=CHCH_2CH_3$, 1,3-pentadiene, $CH_2=CHCH=CHCH_3$, and 2,3 pentadiene, $CH_3CH=C=CHCH_3$.

(c) 1-propanone is incorrect. There cannot be a ketone on the first carbon of a chain. The compound specified is either propionaldehyde, CH_3CH_2CHO, or propanone (2-propanone).

(d) Bromopropane is insufficient. It could be either 1-bromopropane, $BrCH_2CH_2CH_3$, or 2-bromopropane, $CH_3CHBrCH_3$.

(e) Although 2,6-dichlorobenzene conveys enough information, the proper name for the compound is meta-dichlorobenzene, or 1,3-dichlorobenzene. Substituent numbers should be as small as possible.

(f) 2-methyl-3-pentyne is $(CH_3)_2CHC\equiv CCH_3$. The proper name for this compound is 4-methyl-2-pentyne. The number of the triple bond should be as small as possible.

37. (a) 2,4,6-trinitrotoluene: **(b)** methylsalicylate:

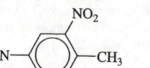 

(c) 2-hydroxy-1,2,3-propanetricarboxylic acid: $HOOCCH_2C(OH)(COOH)CH_2COOH$

38. (a) *o-t*-butylphenol **(b)** 1-phenyl-2-aminopropane **(c)** 2-methylheptadecane

$C_6H_5CH_2CH(NH_2)CH_3$ $(CH_3)_2CH(CH_2)_{14}CH_3$

39. **(a)** $HN(CH_2CH_3)_2$

N,N-diethylamine

(b)  H_2N—⟨○⟩—NO_2

p-aminonitrobenzene

(c) 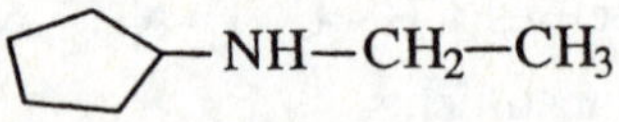 ⟨○⟩—NH—CH_2—CH_3

N-ethyl-*N*-cyclopentylamine

(d) $CH_3CH_2N(CH_3)CH_2CH_3$

diethylmethylamine

40. **(a)** ethylamine:

$H_3C\text{-}CH_2\text{-}NH_2$

(b) 3-chloroaniline:

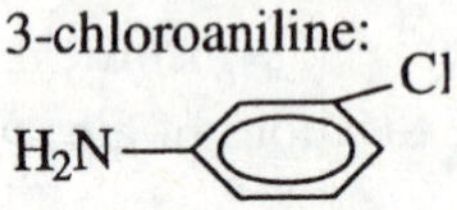

H_2N—⟨○⟩—Cl

(c) Dicyclopropylamine

▷—N(H)—◁

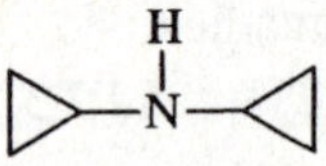

(d) 2-chloroethylamine

$Cl—CH_2—CH_2—NH_2$

Alkanes

41. The general formula of an alkane is C_nH_{2n+2}. Thus an alkane with a molecular mass of 72 u has the molecular formula C_5H_{12}.

(a) If the compound forms four different monochlorination products, there must be four different kinds of carbons in the molecule, each with H atoms attached. This compound is 2-methylbutane, on the left.

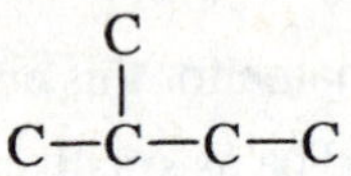

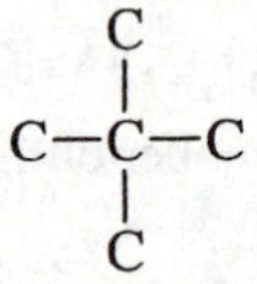

```
    C                    C
    |                    |
C—C—C—C            C—C—C
                         |
                         C
```

(b) If the compound forms but one monochlorination product, every carbon atom in the molecule with H atoms attached must be the same. This compound is 2,2-dimethylpropane, namely, the compound above and to the right.

42. **(a)** An alkane with a molecular mass of 44 u must be a propane, since 3 carbons have a molecular mass of 36 u. There is only one propane, n-propane, and it has the condensed formula $CH_3CH_2CH_3$. Its two monochlorination products are $CH_3CH_2CH_2Cl$ and $CH_3CHClCH_3$

(b) An alkane with a molecular mass of 58 u cannot be a pentane, since five carbons have a mass of 60 u, so it must be a butane. Both normal butane and isobutane have only two monobromination products.

From *n*-butane,

$$CH_3CH_2CH_2CH_3 \qquad CH_3CH_2CH_2CH_2Br \quad \text{and} \quad CH_3CH_2CHBrCH_3$$

From iso-butane,

$$CH_3CH(CH_3)_2 \qquad BrCH_2CH(CH_3)_2 \quad \text{and} \quad CH_3CBr(CH_3)_2$$

Alkenes

43. In the case of ethene there are only two carbon atoms between which there can be a double bond. Thus, specifying the compound as 1-ethene is unnecessary. In the case of propene, there can be a double bond only between the central carbon atom and a terminal carbon atom. Thus here also, specifying the compound as 1-propene is unnecessary. The case of butene is different, however, since 1-butene, $CH_2 = CHCH_2CH_3$, is distinct from 2-butene, $CH_3CH = CHCH_3$.

44. In an alkene there is a $C = C$ double bond. On the other hand, in an cyclic alkane there are no double bonds, but rather a chain of carbon atoms joined at the ends into a ring.

45. These reactions either saturate or create double bonds.

(a) $CH_2 = CHCH_3 + H_2 \xrightarrow{\text{Pt, heat}} CH_3CH_2CH_3$

(b) $CH_3CHOHCH_2CH_3 \xrightarrow{\text{H}_2\text{SO}_4,\ \text{heat}} CH_3CH{=}CHCH_3 + CH_2{=}CHCH_2CH_3$

46. (a) $CH_3CC(Cl){=}CH_2 + HCl \rightarrow CH_3CCl_2CH_3$

(b) $CH_3C \equiv CH + HCN \rightarrow CH_3C(CN){=}CH_2$

(c) $CH_3CH = C(CH_3)_2 + HCl \rightarrow CH_3CH_2CCl(CH_3)_2$

Aromatic Compounds

47. (a) phenylacetylene   (b) meta-dichlorobenzene 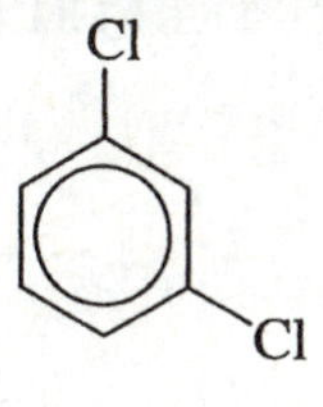   (c) 3,5-dihydroxyphenol

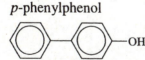
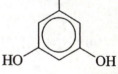
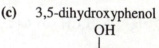

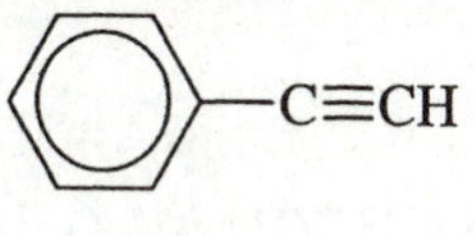

48. (a) *p*-phenylphenol   (b) 2-hydroxy-4-isopropyltoluene   (c) meta-dinitrobenzene

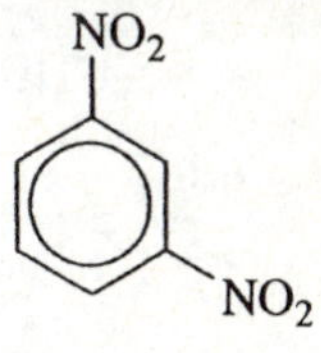

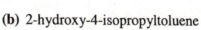
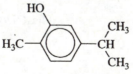

Organic Reactions

49. **(a)** :NH_3 (nucleophile, posses a chemically active lone pair)
(b) CH_3Cl (electrophile, the polar C-Cl bond makes the molecule an electrophile)
(c) Br^- (nucleophile, posses chemically active lone pairs)
(d) CH_3OH (nucleophile: lone pairs on oxygen can be used to form dative bonds)
(e) $CH_3NH_3^+$ (neither: though positively charged, this cation will not accept any additional electron pairs).

50. **(a)** N_3^- (nucleophile, posses lone pairs of electrons that can be used to form dative bonds)

(b) NH_4^+ (neither: though positively charged, this cation will not accept any additional electron pairs)

(c) $CH_3CHClCH_3$ (electrophile: the polar C-Cl bond makes the molecule an electrophile)

(d) OH^- (nucleophile: lone pairs on oxygen can be used to form dative bonds)

(e) $H_2C=O$ (both: as a nucleophile via the oxygen, donating a pair of electrons or as an electrophile via the carbon accepting a pair of electrons from a nucleophile).

51. **(a)** In an aliphatic substitution reaction, an atom (usually H) of an alkane molecule is replaced by another atom or group of atoms. An example is equation (26.4).
(b) In an aromatic substitution reaction, an atom (usually H) of a phenyl group is replaced by another atom or group of atoms. Examples are in equations (26.12) and (26.13).
(c) In an addition reaction, a small molecule breaks apart into two fragments each of which bond to one of the adjacent atoms in a molecule that are engaged in multiple bonding (i.e. C=C, C=O, C≡C to name a few). An example is equation (26.10).
(d) In an elimination reaction, two groups from within the same molecule join to form a small molecule— such as H_2 or H_2O — and are eliminated from the larger molecule. Often a multiple bond results as well. An example is equation (26.6).

52. **(a)** $HC \equiv CH \xrightarrow{Br_2} CHBr = CHBr \xrightarrow{Br_2} CHBr_2CHBr_2$

(b) $HC \equiv CH \xrightarrow{Pt,\ H_2} CH_2 = CH_2 \xrightarrow{H_2O,\ H_2SO_4} CH_3CH_2OH \xrightarrow{Cr_2O_7^{2-},H^+} CH_3CHO$

53. **(a)** $CH_3CH_2CH_2CH_2CH_2OH \xrightarrow{Cr_2O_7^{2-},\ H^+} CH_3CH_2CH_2CH_2COOH$

(b) $CH_3CH_2CH_2COOH + HOCH_2CH_3 \xrightarrow{H^+} CH_3CH_2CH_2\overset{\displaystyle O}{\overset{\displaystyle \|}{C}}OCH_2CH_3 + H_2O$

(c)

$$CH_3CH_2\overset{\displaystyle CH_3}{\underset{}{\overset{|}{C}}}=CH_2 + H_2O \xrightarrow{H_2SO_4} CH_3CH_2\overset{\displaystyle CH_3}{\underset{\displaystyle OH}{\overset{|}{\underset{|}{C}}}}-CH_3$$

54. **(a)** Nitro groups are meta directors. Thus the only product formed is 5-bromo-1, 3-dinitrobenzene, drawn below.

(b) Amine is an ortho, para director. There are two products formed: ortho-bromoaniline and para-bromoaniline, both of which are drawn below.

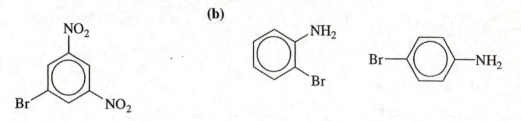

(a)

(b)

(c) Both an ether and a bromo group are ortho, para directors, but an ether is a stronger ortho, para director. The principal product we expect is 2,4-dibromoanisole, drawn below.

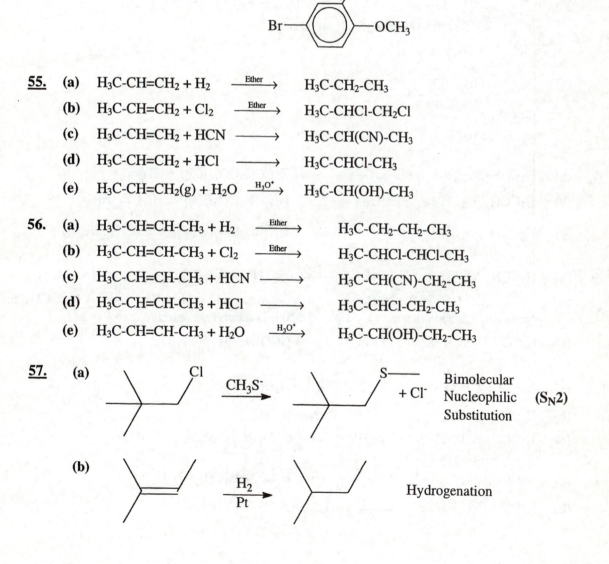

55. **(a)** $H_3C\text{-}CH=CH_2 + H_2 \xrightarrow{\text{Ether}} H_3C\text{-}CH_2\text{-}CH_3$

(b) $H_3C\text{-}CH=CH_2 + Cl_2 \xrightarrow{\text{Ether}} H_3C\text{-}CHCl\text{-}CH_2Cl$

(c) $H_3C\text{-}CH=CH_2 + HCN \longrightarrow H_3C\text{-}CH(CN)\text{-}CH_3$

(d) $H_3C\text{-}CH=CH_2 + HCl \longrightarrow H_3C\text{-}CHCl\text{-}CH_3$

(e) $H_3C\text{-}CH=CH_2(g) + H_2O \xrightarrow{H_3O^+} H_3C\text{-}CH(OH)\text{-}CH_3$

56. **(a)** $H_3C\text{-}CH=CH\text{-}CH_3 + H_2 \xrightarrow{\text{Ether}} H_3C\text{-}CH_2\text{-}CH_2\text{-}CH_3$

(b) $H_3C\text{-}CH=CH\text{-}CH_3 + Cl_2 \xrightarrow{\text{Ether}} H_3C\text{-}CHCl\text{-}CHCl\text{-}CH_3$

(c) $H_3C\text{-}CH=CH\text{-}CH_3 + HCN \longrightarrow H_3C\text{-}CH(CN)\text{-}CH_2\text{-}CH_3$

(d) $H_3C\text{-}CH=CH\text{-}CH_3 + HCl \longrightarrow H_3C\text{-}CHCl\text{-}CH_2\text{-}CH_3$

(e) $H_3C\text{-}CH=CH\text{-}CH_3 + H_2O \xrightarrow{H_3O^+} H_3C\text{-}CH(OH)\text{-}CH_2\text{-}CH_3$

57. **(a)**

Bimolecular Nucleophilic (S_N2) Substitution

(b)

Hydrogenation

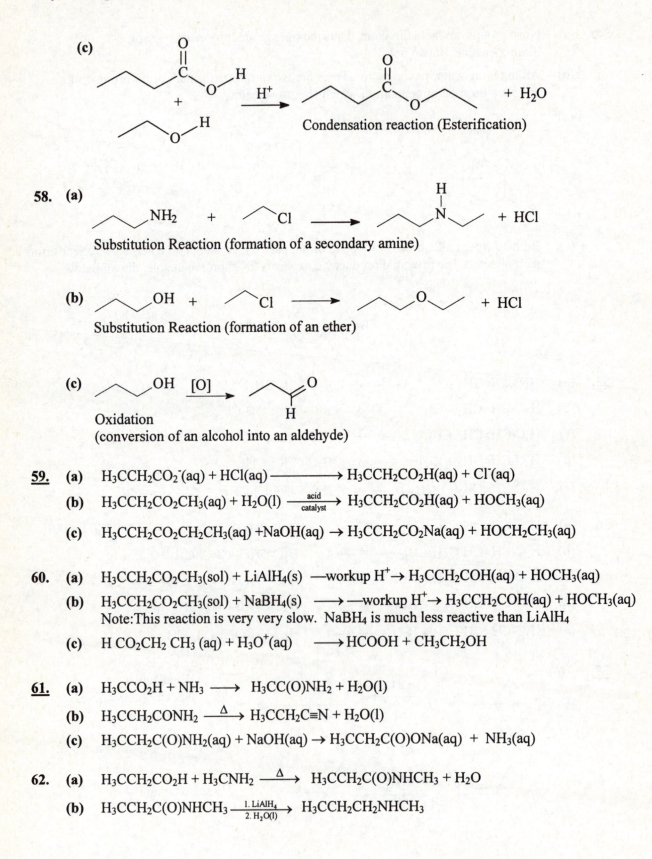

(c)

Condensation reaction (Esterification)

58. (a)

Substitution Reaction (formation of a secondary amine)

(b)

Substitution Reaction (formation of an ether)

(c)

Oxidation
(conversion of an alcohol into an aldehyde)

59. **(a)** $H_3CCH_2CO_2^-(aq) + HCl(aq) \longrightarrow H_3CCH_2CO_2H(aq) + Cl^-(aq)$

  (b) $H_3CCH_2CO_2CH_3(aq) + H_2O(l) \xrightarrow{\text{acid catalyst}} H_3CCH_2CO_2H(aq) + HOCH_3(aq)$

  (c) $H_3CCH_2CO_2CH_2CH_3(aq) + NaOH(aq) \rightarrow H_3CCH_2CO_2Na(aq) + HOCH_2CH_3(aq)$

60. **(a)** $H_3CCH_2CO_2CH_3(sol) + LiAlH_4(s) \xrightarrow{\text{workup } H^+} H_3CCH_2COH(aq) + HOCH_3(aq)$

  (b) $H_3CCH_2CO_2CH_3(sol) + NaBH_4(s) \longrightarrow \xrightarrow{\text{workup } H^+} H_3CCH_2COH(aq) + HOCH_3(aq)$
  Note:This reaction is very very slow. $NaBH_4$ is much less reactive than $LiAlH_4$

  (c) $H CO_2CH_2 CH_3 (aq) + H_3O^+(aq) \longrightarrow HCOOH + CH_3CH_2OH$

61. **(a)** $H_3CCO_2H + NH_3 \longrightarrow H_3CC(O)NH_2 + H_2O(l)$

  (b) $H_3CCH_2CONH_2 \xrightarrow{\Delta} H_3CCH_2C\equiv N + H_2O(l)$

  (c) $H_3CCH_2C(O)NH_2(aq) + NaOH(aq) \rightarrow H_3CCH_2C(O)ONa(aq) + NH_3(aq)$

62. **(a)** $H_3CCH_2CO_2H + H_3CNH_2 \xrightarrow{\Delta} H_3CCH_2C(O)NHCH_3 + H_2O$

  (b) $H_3CCH_2C(O)NHCH_3 \xrightarrow[\text{2. } H_2O(l)]{\text{1. } LiAlH_4} H_3CCH_2CH_2NHCH_3$

(c) $H_3CCH_2C(O)NHCH_3 + HCl(aq) \xrightarrow{H_2O} H_3CCH_2CO_2H (aq)+ H_3CNH_3^+Cl^- (aq)$

63. **(a)** $CH_3C(O)O^-(aq) + HCl(aq) \rightarrow CH_3COOH(aq) + Cl^-(aq)$

(b) $H_3CCH_2C(O)OCH_3(aq) + H_3O^+(aq) \rightarrow CH_3OH(aq) + H_3CCH_2C(O)OH(aq)$

(c) $H_3CCH_2COOH(aq) + NaOH(aq) \rightarrow H_3CCH_2COO^- Na^+ (aq) + H_2O(l)$

64. **(a)** (2) $H_3CCH_2NH_2(aq) + HCl(aq) \rightarrow H_3CCH_2NH_3^+Cl^-(aq)$

(b) (1) $H_3CCH_2C(O)NH_2(aq) + H_2O(l) \rightarrow H_3CCH_2COOH + NH_{3(aq)}$

(c) (3) $H_3CCH_2CH_2NH_3^+Cl^-(aq) + NaOH(aq) \rightarrow H_3CCH_2CH_2NH_2(aq) + NaCl(aq) + H_2O(l)$

65. $H_3CCH_2CH_2 CH_2 NH_2(aq) + H_2O(l) \rightleftharpoons H_3CCH_2CH_2 CH_2 \overset{\oplus}{N} H_3(aq) + OH^- (aq)$

$$K_b = \frac{[H_3CCH_2CH_2CH_2 \overset{\oplus}{N} H_3][OH^-]}{[H_3CCH_2CH_2CH_2NH_2]}$$

66. $H_3CCH_2CH_2 CH_2 NHCH_3(aq) + H_2O(l) \rightleftharpoons H_3CCH_2CH_2 CH_2 \overset{\oplus}{N} H_2CH_3(aq) + OH^- (aq)$

$$K_b = \frac{[H_3CCH_2CH_2CH_2 \overset{\oplus}{N} H_2CH_3][OH^-]}{[H_3CCH_2CH_2CH_2NHCH_3]}$$

67. **(a)** $H_3CCH_2NH_2(aq) + HCl(aq) \rightarrow H_3CCH_2NH_3^+Cl^- (aq)$

(b) $(CH_3)_3N(aq) + HBr(aq) \rightarrow (CH_3)_3NH^+Br^- (aq)$

(c) $H_3CCH_2NH_3^+(aq) + H_3O^+(aq) \rightarrow$ No Reaction

(d) $H_3CCH_2NH_3^+(aq) + OH^- (aq) \rightarrow H_3CCH_2NH_2(aq) + H_2O(l)$

68. **(a)**

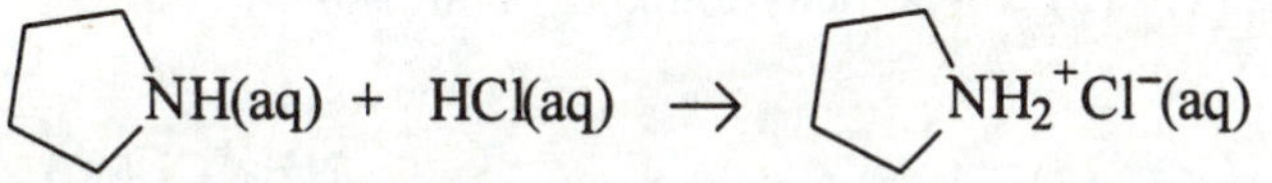

(b) $(CH_3)_4N^+(aq) + HCl(aq) \rightarrow$ No Reaction

(c) $H_3CCH_2NH_2(aq) + OH^-(aq) \rightarrow$ No Reaction

(d) $(CH_3)_3NH^+(aq) + OH^-(aq) \rightarrow (CH_3)_3N(aq) + H_2O(l)$

69. Methyl ethyl ketone has the abbreviated structure shown on the right. Thus, the alcohol from which it is produced must have an — OH group on carbon 2 and it must be a four-carbon-atom straight-chain alcohol. Of the alcohols listed, 2-butanol satisfies these criteria.

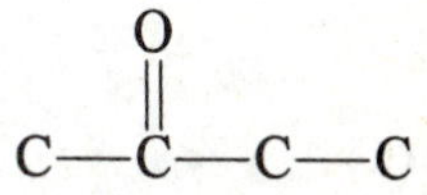

70. **(a)** $CH_3CH_2CH=CH_2 \xrightarrow{H_2O,H_2SO_4} CH_3CH_2CH(OH)CH_3$

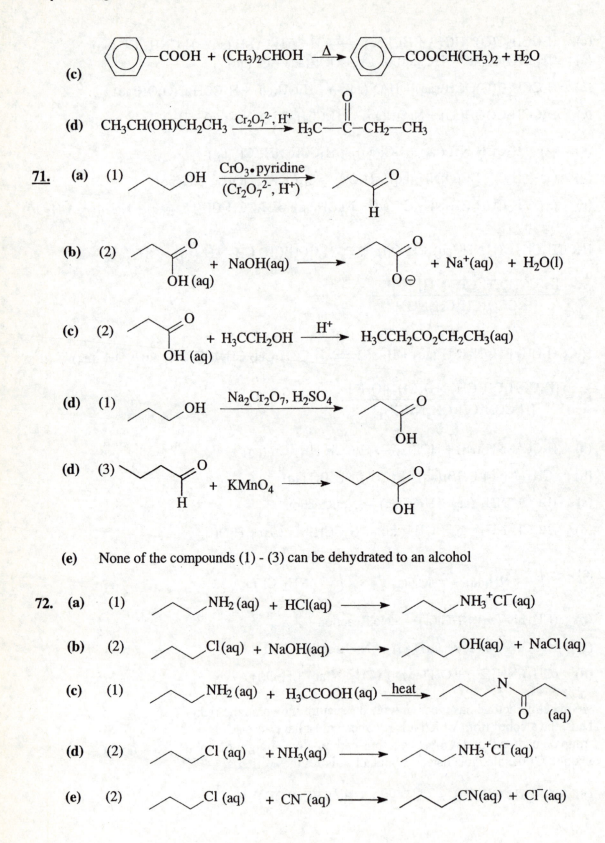

(c) $\langle\langle\bigcirc\rangle\rangle$—COOH + (CH$_3$)$_2$CHOH $\xrightarrow{\Delta}$ $\langle\langle\bigcirc\rangle\rangle$—COOCH(CH$_3$)$_2$ + H$_2$O

(d) CH$_3$CH(OH)CH$_2$CH$_3$ $\xrightarrow{\text{Cr}_2\text{O}_7{}^{2-},\text{H}^+}$ H$_3$C—$\overset{\text{O}}{\overset{\|}{\text{C}}}$—CH$_2$—CH$_3$

71. **(a)** **(1)** ～＼OH $\xrightarrow[(\text{Cr}_2\text{O}_7{}^{2-},\text{H}^+)]{\text{CrO}_3\bullet\text{pyridine}}$ 〜＼〜=O / H

(b) **(2)** 〜〜=O / OH (aq) + NaOH(aq) ⟶ 〜〜=O / O⊖ + Na$^+$(aq) + H$_2$O(l)

(c) **(2)** 〜〜=O / OH (aq) + H$_3$CCH$_2$OH $\xrightarrow{\text{H}^+}$ H$_3$CCH$_2$CO$_2$CH$_2$CH$_3$(aq)

(d) **(1)** ＼〜OH $\xrightarrow{\text{Na}_2\text{Cr}_2\text{O}_7,\text{H}_2\text{SO}_4}$ ＼〜=O / OH

(d) **(3)** ＼＼〜=O / H + KMnO$_4$ ⟶ ＼＼〜=O / OH

(e) None of the compounds (1) - (3) can be dehydrated to an alcohol

72. **(a)** **(1)** ＼〜NH$_2$ (aq) + HCl(aq) ⟶ ＼〜NH$_3{}^+$Cl$^-$(aq)

(b) **(2)** ＼〜Cl(aq) + NaOH(aq) ⟶ ＼〜OH(aq) + NaCl(aq)

(c) **(1)** ＼〜NH$_2$ (aq) + H$_3$CCOOH(aq) $\xrightarrow{\text{heat}}$ ＼〜N＼〜 / O (aq)

(d) **(2)** ＼〜Cl (aq) + NH$_3$(aq) ⟶ ＼〜NH$_3{}^+$Cl$^-$(aq)

(e) **(2)** ＼〜Cl (aq) + CN$^-$(aq) ⟶ ＼〜CN(aq) + Cl$^-$(aq)

Organic Stereochemistry

73. We have drawn only the carbon and chlorine atoms in each structure. Remember that there are four bonds to each carbon atom; the bonds not shown in these structures are C— H bonds.

$$\begin{array}{cccc}
& \overset{\text{Cl}}{\underset{|}{}} & & \overset{\text{Cl}}{\underset{|}{}} \\
\text{Cl—C—C—C} & \text{Cl—C—C—C} & \text{Cl—C—C—C—Cl} & \text{C—C—C}
\end{array}$$

Cl—C—Ċ—C   Cl—Ċ—C—C   Cl—C—C—C—Cl   C—Ċ—C with Cl above and Cl below

74. **(a)** These are not isomers, because they have different formulas: namely, C_4H_{10} and C_4H_8.

(b) These two compounds are structural isomers. (Only the carbon skeleton is shown below).

$$\begin{array}{cc}
\overset{\text{C}}{\underset{|}{}} & \overset{\text{C}}{\underset{|}{}} \\
\text{C—C—C—C—C—C—C—C} & \text{C—C—C—C—C—C—C—C}
\end{array}$$

C—C—C—C—C—C—Ċ—C   C—C—C—C—C—Ċ—C—C

(c) The compounds are not isomers; they have different formulas: namely, C_4H_9Cl and $C_5H_{11}Cl$.

(d) These two compounds are identical. They have been simply rotated.

(e) These two compounds are identical. They have been simply rotated.

(f) These are two ortho-para isomers, ortho-nitrophenol on the left, and para-nitrophenol on the right.

75. **(a)** Identical molecules (Both are achiral)

(b) Identical molecules (Both are the *R* enantiomers)

(c) Structural isomers (one is optically active, the other is not)

(d) Enantiomers (*R* and S optical isomers)

(e) Identical molecules (Both are the *R* enantiomers)

(f) Identical molecules (Both are the *R* enantiomers)

76. **(a)** Enantiomers (*R* and S optical isomers)

(b) Identical molecules (Both are the *R* enantiomers)

(c) Enantiomers (*R* and S optical isomers)

(d) Identical molecules (Both are the *R* enantiomers)

(e) Enantiomers (*R* and S optical isomers)

(f) Structural Isomers(both are optically active)

77. **(a)** (*S*)-3-bromo-2-methylpentane

(b) (*S*)-1,2-dibromopentane

(c) (*R*)-3-bromomethyl-5-chloro-*n*-pentan-3-ol

(d) (*S*)-1-bromo-*n*-propan-2-ol

76. **(a)** Enantiomers (*R* and S optical isomers)
(b) Identical molecules (Both are the *R* enantiomers)
(c) Enantiomers (*R* and S optical isomers)
(d) Identical molecules (Both are the *R* enantiomers)
(e) Enantiomers (*R* and S optical isomers)
(f) Structural Isomers(both are optically active)

77. **(a)** (*S*)-3-bromo-2-methylpentane
(b) (*S*)-1,2-dibromopentane
(c) (*R*)-3-bromomethyl-5-chloro-*n*-pentan-3-ol
(d) (*S*)-1-bromo-*n*-propan-2-ol

78. **(a)** (*R*)-pentan-2-ol
(b) (*S*)-3-methyl-2-butanamine
(c) (*S*)-1-bromo-2-chloro-2-methybutane
(d) (*S*)-2-hydroxypentan-1-ol

79. **(a)**
H₃CH₂C
CH₃
C=C
H
H
(Z)-2-pentene

(b)
Cl
CH₃
C=C
H
CH₂CH₃
(E)-1-chloro-2-methyl-1-butene

(c)
H₃C
CH₂Cl
C=C
H₃CH₂C
CH₂CH₂CH(CH₃)₂
(E)-3-7-dimethyl-4-chloromethyl-3-octene

(d)
O
HC
CH₂Br
C=C
H₃C
CH₂CH₂Br
(Z)-5-bromo-3-bromomethyl-2-methylpent-2-enal

80. **(a)**
H₃C
CH₂Cl
C=C
ClH₂C
CH₃
(E)-1,4-dichlor-2,3-dimethyl-2-butene

(b)
H₃C
CH₂Cl
C=C
H₃CH₂C
CH₂OH
(E)-2-chloromethyl-3-methyl-pent-2-en-1-ol

(c)
O
H₃CC
CH₂OH
C=C
HC≡C
CH₂CHClCH₃
(Z)-3-ethynyl-6-chloro-4-hydroxymethyl-hept-3-en-2-one

(d)
H₃CCH₂
CH₃
C=C
HOH₂C
CH₃
(3-hydroxymethyl)-2-methyl-2-pentene
(E/Z isomers are not possible)

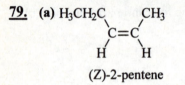

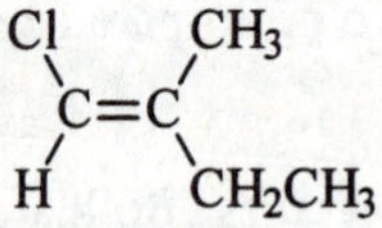

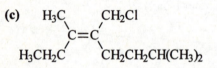

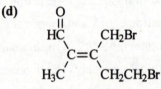

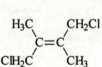

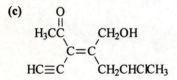

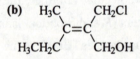

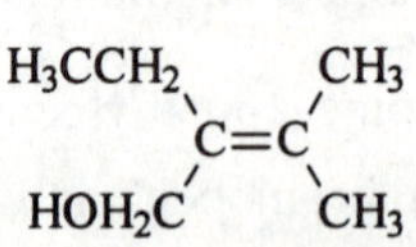

82. **(a)**

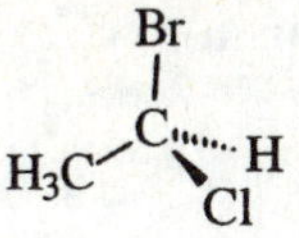

Br
|
H$_3$C—C⋯⋯H
|
Cl

(R)-1-bromo-1-chloroethane

(b)

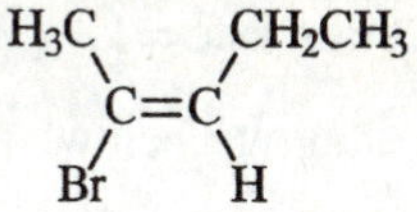

H$_3$C   CH$_2$CH$_3$
  \   /
   C=C
  /   \
Br   H

(E)-2-bromo-2-pentene

(c)

ClH$_2$CH$_2$C   CH$_2$CH$_2$CH$_3$
      \   /
       C=C
      /   \
H$_3$CH$_2$C   H

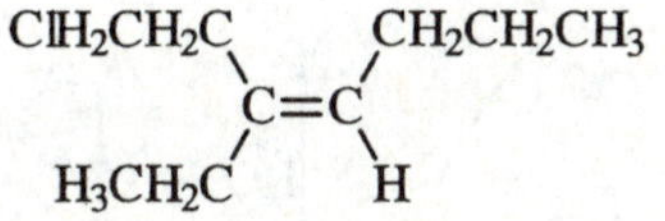

(Z)-1-chloro-3-ethyl-3-heptene

(d)

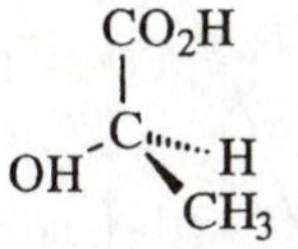

CO$_2$H
|
C⋯⋯H
|
OH   CH$_3$

(R)-2-hydroxypropanoic acid

(e)

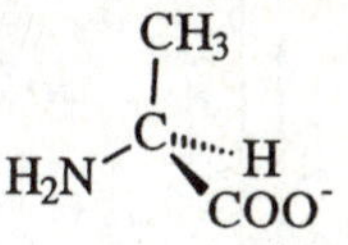

CH$_3$
|
C⋯⋯H
|
H$_2$N   COO$^-$

(S)-2-aminopropanoate anion

Nucleophilic Substitution Reactions

83. **(a)** Reaction rate = $k_{obs}[OH^-][BrCH_2CH_2CH_2CH_3]$ (S$_N$2 mechanism)

(b) See diagram below:

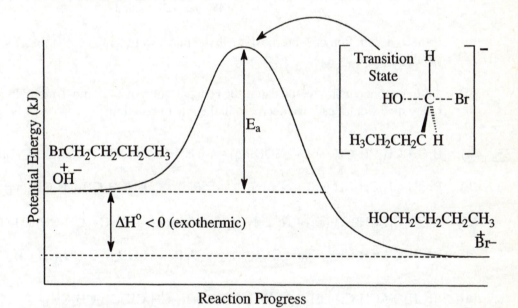

(c) If the concentration of n-butyl bromide were doubled, the reaction rate would increase two fold

(d) If the concentration of hydroxide ion were halved, the reaction rate would decrease by a factor of two.

84. **(a)** Reaction rate = $k_{obs}[BrC(CH_3)_2CH_2CH_2CH_3]$ (S_N1 mechanism)

(b) See diagram below:

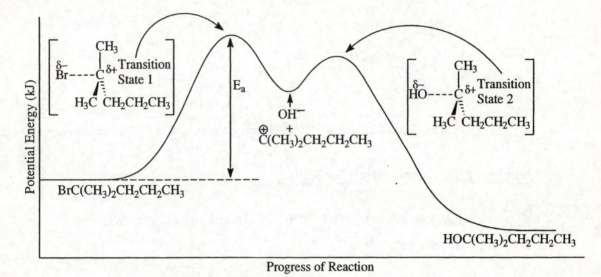

(c) If the concentration of 1-bromo-1-methylpentane were doubled, the reaction rate would increase by a factor of two.

(d) Adding more ethanol would cause the concentration of 1-bromo-1-methylpentane to fall, which would lead to a decrease in the rate of reaction.

85. **(a)** $BrCH_2CH_2CH_2CH_3(aq) + NaOH(aq) \rightarrow NaBr(aq) + HOCH_2CH_2CH_2CH_3(aq)$

(b) $BrCH_2CH_2CH_2CH_3(aq) + NH_3(aq) \rightarrow Br^-(aq) + {}^+H_3NCH_2CH_2CH_2CH_3(aq)$

(c) $BrCH_2CH_2CH_2CH_3(aq) + NaC{\equiv}N(aq) \rightarrow NaBr(aq) + N{\equiv}C{-}CH_2CH_2CH_2CH_3(aq)$

(d) $BrCH_2CH_2CH_2CH_3 + Na^{+\,-}OCH_2CH_3 \rightarrow NaBr + H_3CH_2C{-}O{-}CH_2CH_2CH_2CH_3$

86. **(a)** $BrCH_2CH_2CH_2CH_2CH_3 + NaN_3 \rightarrow NaBr + N_3CH_2CH_2CH_2CH_2CH_3$

(b) $Br(CH_2)_4CH_3(aq) + N(CH_3)_3(aq) \rightarrow Br^-(aq) + (CH_3)_3N^+(CH_2)_4CH_3(aq)$

(c) $BrCH_2CH_2CH_2CH_2CH_3 + Na^{+\,-}C{\equiv}CCH_2CH_3 \rightarrow NaBr +$
$N{\equiv}C{-}CH_2CH_2CH_2CH_2CH_3$

(d) $BrCH_2CH_2CH_2CH_2CH_3 + Na^{+\,-}SCH_2CH_3 \rightarrow NaBr + CH_3CH_2SCH_2CH_2CH_2CH_2CH_3$

87. **(a)**

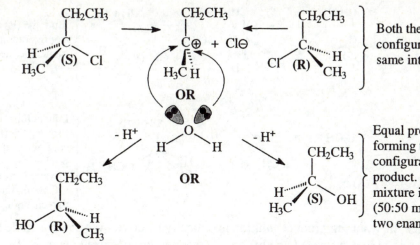

Both the (*R*) and (*S*) configurations give the same intermediate

Equal probability of forming the (*R*) or (*S*) configuration of the product. A racemic mixture is formed (50:50 mixture of the two enantiomers)

(b) Since the resulting product solution is optically inactive, we can conclude that a racemic mixture has been formed and hence, the reaction has occurred via an S_N1 mechanism.

88. **(a)**

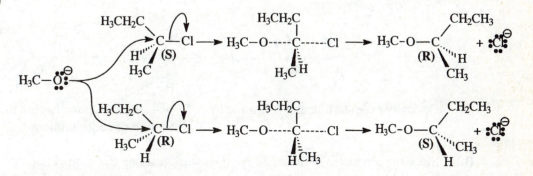

Optically pure alkyl chloride (*R* or *S*) → Optically pure ether (inversion of configuration)

(b) Since the product is optically active (i.e., since just one enantiomer is produced), we can conclude that the reaction must have occurred in a concerted fashion via an S_N2 mechanism.

89. **(a)**

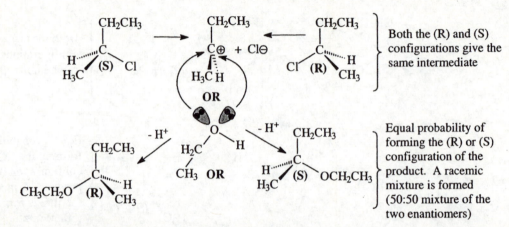

Both the (R) and (S) configurations give the same intermediate

Equal probability of forming the (R) or (S) configuration of the product. A racemic mixture is formed (50:50 mixture of the two enantiomers)

(b) Since the resulting product solution is optically inactive. We can conclude that a racemic mixture has been formed and hence, the reaction has occurred via an S_N1 mechanism.

90. **(a)**

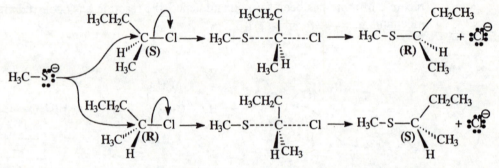

Optically pure starting alkyl chloride (*R* or *S*) $\longrightarrow$ Optically pure thioether
(inversion of configuration)

(b) Since the product is optically active (i.e., since just one enantiomer is produced), we can conclude that the reaction must have occurred in a concerted fashion via an S_N2 mechanism.

Polymerization Reactions

91. In a simple molecular substance like benzene, all molecules are identical (C_6H_6). No matter how many of these molecules are present in one sample, any other sample with the same number of molecules has the same mass. The mass in grams of one mole of molecules— the molar mass—is a unique quantity. In a polymer the situation is quite different. The number of monomer units in a polymer chain is not a constant number but is widely variable. Thus individual polymer molecules differ substantially in mass. Thus, a mass of a mole of their molecules also is quite variable. However, when we take a sample for analysis, we obtain many molecules of each chain length or molecular mass. The resulting determination of molar mass is the average mass of all of these different sized polymer molecules.

92. An ester linkage is formed by the condensation of a carboxylic acid and an alcohol. Accompanying this condensation is the elimination of a water molecule between the ester and the carboxylic acid. Dacron is formed by the condensation of a dicarboxylic acid with a diol. Thus, Dacron contains ester linkages. Because there are a large number of these ester linkages in Dacron— joining many subunits together— it is appropriate to call the polymer a polyester. To determine the percent of oxygen in Dacron, we refer to the basic unit of the polymer, which has the formula $C_{10}H_8O_4$.

$$\% \ O \frac{(4 \times 16.00) \text{ g O}}{192.2 \text{ g polymer}} \times 100\% = 33.30\% \ O$$

93. The polymerization of 1,6-hexanediamine with sebacyl chloride proceeds in the following manner. The italicized H and Cl atoms are removed in this condensation reaction.

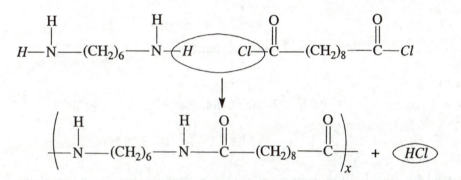

94. In order to form long-chain molecules, every monomer must have at least two functional groups, one on each end of the molecule. Ethyl alcohol has only one functional group, a hydroxyl group (—OH). Thus, it cannot participate in a polymerization reaction with dimethyl terephthalate. But glycerol, by virtue of its three hydroxyl functional groups, can participate in this kind of polymerization reaction.

Integrative and Advanced Exercises

95. Chlorination of CH_4 proceeds by a free radical chain reaction. Methyl radicals are produced in the course of this reaction and it is not unlikely that two methyl radicals will unite in a chain termination step to form an ethane molecule. $\cdot CH_3 + \cdot CH_3 \longrightarrow CH_3CH_3$ The resulting ethane molecule then can be attacked by a chlorine molecule to produce a molecule of monochloroethane. $CH_3CH_3 + Cl_2 \longrightarrow CH_3CH_2Cl + HCl$

96. **(a)** 1,5-cyclooctadiene

(b) 3,7,11-trimethyl-2,6,10,dodecatriene-1-ol

$$HOCH_2CH = C \ (CH_3)CH_2CH_2CH = C(CH_3)CH_2CH_2CH = C(CH_3)_2$$

(c) 2,6-dimethyl-5-heptene-1-al

$$OHCCH(CH_3)CH_2CH_2CH = C(CH_3)_2$$

97. We have show only the carbon atom skeleton in each case. Remember that there are four bonds to carbon. The bonds that are not indicated in these structures are C—H bonds. Boiling points are given in parentheses after the name of each compound.

Butane C_4H_{10} (–0.5°C)   Hexane C_6H_{14} (68.7°C)

   c−c−c−c   c−c−c−c−c−c

2-methylpropane   (–11.7°C)   3-methylpentane   (63.3°C)

       C   C
       | |
  c−c−c   c−c−c−c−c

Pentane C_5H_{12} (36.1)   2-methylpentane   (58.0°C)

c−c−c−c−c   C
                                 |
                           c−c−c−c−c−c

2-methylbutane   (27.9°C)   2,3-dimethylbutane   (58.0°C)

      C   C   C
      | | |
  c−c−c−c   c−c−c−c

2,2-dimethylpropane   (9.5°C)   2,2-dimethylbutane   (49.7°C)

      C   C
      | |
  c−c−c   c−c−c−c
      | |
      C   C

As we proceed down each list of isomeric alkanes, the structures become more compact, and the boiling points decrease.

98. In the structures below we have omitted the hydrogen atoms. Remember that there are four bonds to each C atom. The bonds that are missing are C—H bonds.

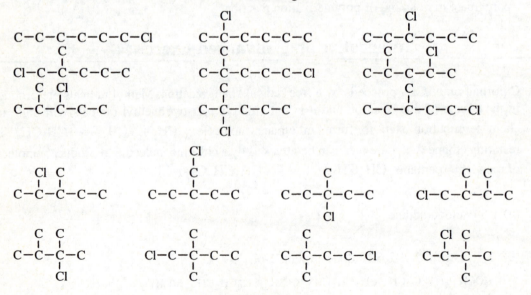

99. The aldehyde group is a meta director of moderate strength, while the ether group is a moderately strong ortho, para director. The isomers we expect to be produced are drawn below.

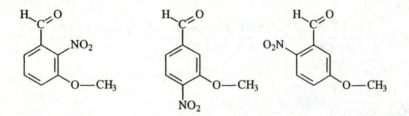

Since 3-methoxy-5-nitrobenzaldehyde is **not** formed, we conclude that, at least in this case, meta direction by the $-CHO$ group is less effective than ortho, para direction by the $-OCH_3$ group

100. In one regard, benzene and 1,3,5-cyclohexatriene are the same substance; whenever there are alternating single and double bonds around a six-membered ring of carbon atoms, an aromatic ring forms. In another sense, these two compounds are different, thus, the name 1,3,5-cyclohexatriene implies that the three alternating double bonds in a six-membered carbon ring do not interact with each other, but they clearly do and the compound should be called benzene.

101. (a) Oxdn : $\{Fe \longrightarrow Fe^{3+} + 3e^-\}$ $\times 2$

Redn : $\underline{C_6H_5NO_2 + 7H^+ + 6e^- \longrightarrow C_6H_5NH_3^+ + 2H_2O}$

$C_6H_5NO_2 + 7H^+ + 2Fe \longrightarrow C_6H_5NH_3^+ + 2H_2O + 2Fe^{3+}$

(b) Oxdn : $\{C_6H_5CH_2OH + H_2O \longrightarrow C_6H_5COOH + 4H^+ + 4e^-\} \times 3$

Redn : $\underline{\{Cr_2O_7^{2-} + 14H^+ + 6e^- \longrightarrow 2Cr^{3+} + 7H_2O\} \quad \times 2}$

$3C_6H_5CH_2OH + 2Cr_2O_7^{2-} + 16H^+ \longrightarrow 3C_6H_5COOH + 4Cr^{3+} + 11H_2O$

(c) Oxdn: $\{CH_3CH=CH_2 + 2OH^- \longrightarrow CH_3CHOHCH_2OH + 2e^-\} \times 3$

Redn : $\underline{\{MnO_4^- + 2H_2O + 3e^- \longrightarrow MnO_2 + 4OH^-\} \quad \times 2}$

$3CH_3CH=CH_2 + 2MnO_4^- + 4H_2O \longrightarrow 3CH_3CHOHCH_2OH + 2MnO_2 + 2OH^-$

102. We first write the equation for the reaction that occurs. Note that each mole of benzaldehyde yields one mole of benzoic acid. We are given the mass of each reagent; we need to determine which reagent is limiting.

$3C_6H_5CHO + 2KMnO_4 + KOH \longrightarrow 3C_6H_5COO^-K^+ + 2MnO_2 + 2H_2O$

$3C_6H_5COO^-K^+ + 3H^+ \longrightarrow 3C_6H_5COOH + 3K^+$

$$mass_{C_6H_5COOH} = 10.6 \text{ g } C_6H_5CHO \times \frac{1 \text{ mol } C_6H_5CHO}{106.1 \text{ g } C_6H_5CHO} \times \frac{1 \text{ mol } C_6H_5COOH}{1 \text{ mol } C_6H_5CHO} \times \frac{122.1 \text{ g } C_6H_5COOH}{1 \text{ mol } C_6H_5COOH}$$

$$= 12.2 \text{ g } C_6H_5COOH$$

$$mass_{C_6H_5COOH} = 5.9 \text{ g } KMnO_4 \times \frac{1 \text{ mol } KMnO_4}{158.0 \text{ g } KMnO_4} \times \frac{3 \text{ mol } C_6H_5COOH}{2 \text{ mol } KMnO_4} \times \frac{122.1 \text{ g } C_6H_5COOH}{1 \text{ mol } C_6H_5COOH}$$

$$= 6.8 \text{ g } C_6H_5COOH$$

$$\% \text{ yield} = \frac{6.1 \text{ g } C_6H_5COOH \text{ produced}}{6.8 \text{ g } C_6H_5COOH \text{ calculated}} \times 100\% = 90.\% \text{ yield}$$

103. We suspect the compound contains C, H, N, and O. The %N by mass in the compound can be found using the ideal gas law:

$$\text{amount } N_2 = \frac{PV}{RT} = \frac{(735-9) \text{ mmHg} \times \dfrac{1 \text{ atm}}{760 \text{ mmHg}} \times 0.0402 \text{ L}}{0.08206 \text{ L atm mol}^{-1} \text{ K}^{-1} \times 298 \text{ K}} = 1.57 \times 10^{-3} \text{ mol } N_2$$

$$\% \ N = \frac{1.57 \times 10^{-3} \text{ mol } N_2 \times \dfrac{28.01 \text{ g } N_2}{1 \text{ mol } N_2}}{0.1825 \text{ g sample}} \times 100\% = 24.1\% \text{ N}$$

Now we determine the amounts of C, H, N, and O in the first sample.

$$\text{amount } C = 0.2895 \text{ g } CO_2 \times \frac{1 \text{ mol } CO_2}{44.010 \text{ g } CO_2} \times \frac{1 \text{ mol } C}{1 \text{ mol } CO_2} = 0.006578 \text{ mol } C$$

$$mass \ C = 0.006578 \text{ mol } C \times \frac{12.011 \text{ g } C}{1 \text{ mol } C} = 0.07901 \text{ g } C$$

$$\text{amount } H = 0.1192 \text{ g } H_2O \times \frac{1 \text{ mol } H_2O}{18.015 \text{ g } H_2O} \times \frac{2 \text{ mol } H}{1 \text{ mol } H_2O} = 0.01323 \text{ mol } H$$

$$mass \ H = 0.01323 \text{ mol } H \times \frac{1.0079 \text{ g } H}{1 \text{ mol } H} = 0.01334 \text{ g } H$$

$$\text{amount } N = 0.1908 \text{ g} \times \frac{24.1 \text{ g } N}{100.0 \text{ g sample}} = 0.0460 \text{ g } N \times \frac{1 \text{ mol } N}{14.007 \text{ g } N} = 0.00328 \text{ mol } N$$

$$\text{amount } O = (0.1908 \text{ g} - 0.0460 \text{ g } N - 0.07901 \text{ g } C - 0.01334 \text{ g } H) \times \frac{1 \text{ mol } O}{15.999 \text{ g } O} = 0.00328 \text{ mol } O$$

From this information we determine the empirical formula of the compound.

0.006578 mol C ÷0.00328 ⟶ 2.01 mol C	0.01323 mol H	÷0.00328 ⟶ 4.03 mol H
0.00328 mol N ÷0.00328 ⟶ 1.00 mol N	0.00328 mol O	÷0.00328 ⟶ 1.00 mol O

The compound has an empirical formula of C_2H_4NO, which has a molar mass of 58.1 g/mol. We use the freezing point depression data to determine the number of moles of compound dissolved, and then its molar mass. First, the molality of the

solution: $m = \dfrac{\Delta T_f}{K_f} = \dfrac{3.66°C - 5.50°C}{-5.12°C/m} = 0.359\ m$

amount solute $= 0.02600\ \text{kg benzene} \times \dfrac{0.359\ \text{mol solute}}{1\ \text{kg benzene}} = 0.00933\ \text{mol solute}$

$M = \dfrac{1.082\ g}{0.00933\ \text{mol}} = 116\ \text{g/mol}$ This molar mass is twice the empirical molar mass.

Thus, the molecular formula is twice the empirical formula. Namely, the molecular formula $= C_4H_8N_2O_2$

104. In chain-reaction (free-radical) polymerization, there are relatively few free radicals in the reaction mixture. This means that the polymer chains can grow to an appreciable size during the reaction. In step-reaction (condensation) polymerization, on the other hand, every molecule is capable of initiating a growing chain. Thus, there are a relatively large number of chains growing during the reaction. As a consequence each individual chain ends up being relatively short.

105. Structure I is 1,2,4-tribromobenzene. It and its three mononitration products are shown below.

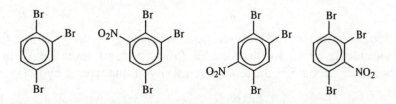

Structure II is 1,2,3-tribromobenzene. It and its two mononitration products are shown below.

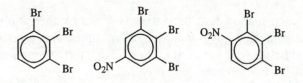

Structure III is 1,3,5-tribromobenzene. It and its single mononitration product are shown below.

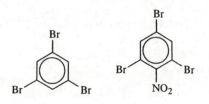

106. In the drawings below, the numbers in each substituent position indicate whether that position is different from or the same as other substituent positions. For example, all positions numbered "1" form identical compounds when the compound is monochlorinated at any of these positions.

(a) The compound can be either 1,3-dimethylbenzene (*m*-xylene, shown below) or ethyl-benzene (in which the monochlorination products are the ortho-, meta-, and para-isomers).

(b) This compound is 1,3,5-trimethylbenzene

(c) One aromatic compound that has the formula of C_9H_{12} and forms four ring mononitration products is 3-ethyltoluene (shown below). Another is 2-ethyltoluene, which is more likely, for the following reason. The mononitration product of 3-ethyltoluene at the position labeled "3" in the following sketch has the nitro group meta to both alkyl groups. This is unlikely because alkyl groups are ortho-para directors. All other substitutions in 3-ethyltoluene, and all substitutions in 2-ethyltoluene are ortho or para to at least one alkyl group.

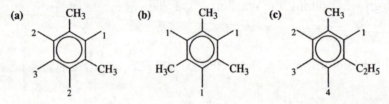

107. There is one 3° hydrogen, two 2° hydrogens, and nine 1° hydrogens. The total reactivity of the molecule is given by the sum of three products, each being the number of hydrogens of each type times the relative reactivity of that type of hydrogen.

$$\text{total reactivity} = 1\{3°\,H\} \times \frac{4.3\,\text{reactivity}}{3°\,H} + 2\{2°\,H\} \times \frac{3\,\text{reactivity}}{2°\,H} + 9\{1°\,H\} \times \frac{1\,\text{reactivity}}{1°\,H}$$

$$= 4.3 + 6.0 + 9.0 = 19.3$$

The fraction of each monochloro derivative is given by the ratio of the product of the number of hydrogens that give that derivative times the reactivity of that type of hydrogen, divided by the total reactivity. There is one tertiary monochloro derivative.

$$\% \text{ tertiary} = \frac{1 \times 4.3}{19.3} \times 100\% = 22.3\% \text{ tertiary}$$

$$\begin{array}{c} CH_3 \\ | \\ H_3C - \overset{}{\underset{|}{C}} - CH_2 - CH_3 \\ Cl \end{array}$$

There is one secondary derivative. $\% \text{ secondary} = \dfrac{2 \times 3}{19.3} \times 100\% = 31.1\% \text{ secondary}$

$$\begin{array}{c} CH_3 \\ | \\ H_3C - CH - CHCl - CH_3 \end{array}$$

There are two different primary derivatives. The first is formed from the primary hydrogens on the ethyl group.

$$\% \text{ primary ethyl} = \frac{3 \times 1}{19.3} \times 100\% = 15.5\% \text{ primary ethyl}$$

$$\begin{array}{c} CH_3 \\ | \\ H_3C-CH-CH_2-CH_2Cl \end{array}$$

The second primary monochloro derivative is formed by substituting the methyl hydrogens.

$$\% \text{ primary methyl} = \frac{6 \times 1}{19.3} \times 100\% = 31.1\% \text{ primary methyl} \qquad \begin{array}{c} CH_3 \\ | \\ ClH_2C-CH-CH_2-CH_3 \end{array}$$

108. (1) The molar masses of the five compounds are the same to one significant figure:

1-butanol	diethyl ether	methyl propyl ether	butyraldehyde	propionic acid
74.12 g/mol	74.12 g/mol	74.12 g/mol	72.11 g/mol	74.08 g/mol

Since the freezing point depression is known to only one significant figure, the molar mass of the unknown can only be determined to one significant figure. Thus, the cited freezing point depression data are insufficiently precise to distinguish among the compounds.

(2) Propionic acid would produce an aqueous solution that would turn blue litmus red. The unknown compound cannot be propionic acid.

(3) Both alcohols and aldehydes are oxidized to carboxylic acids by aqueous $KMnO_4$. Ethers are not oxidized by $KMnO_4$. Thus, the unknown must be 1-butanol or butyraldehyde.

The identity of the unknown can be established by treatment with a carboxylic acid. If the unknown is an alcohol, it will form a pleasant smelling ester upon heating with a carboxylic acid such as acetic acid. If the compound is an aldehyde, it ill not react with the carboxylic acid under these conditions.

109. First we replace the chloro group with an azide group, then reduce the resulting azide to the primary amine.

$$1) \; n-C_3H_7Cl + N_3^- \longrightarrow n-C_3H_7N_3 + Cl^-$$

$$2) \; n-C_3H_7N_3 \xrightarrow{[H]} n-C_3H_7NH_2 + N_2(g)$$

110. $CH_3-CH_2-Cl \xrightarrow[\substack{\text{acetone} \\ S_N 2}]{Na^+ C \equiv N^-} CH_3-CH_2-C \equiv N + NaCl$

$CH_3-CH_2-C \equiv N \xrightarrow[\substack{\text{or} \\ \text{reduction with } H_2}]{\text{reduction with LiAlH}_4} CH_3-CH_2-CH_2-NH_2$

111. $CH_3-CH_2-CH_2-Cl \xrightarrow[\substack{\text{acetone} \\ S_N 2}]{Na^+ C \equiv N^-} CH_3-CH_2-CH_2-C \equiv N + NaCl$

$$CH_3-CH_2-CH_2-C \equiv N \xrightarrow{OH^-/H_2O} CH_3-CH_2-CH_2- \overset{\overset{\displaystyle O}{\|}}{C} -O^- \; Na^+ + NH_3$$

$$CH_3-CH_2-CH_2- \overset{\overset{\displaystyle O}{\|}}{C} -O^- \; Na^+ \xrightarrow{HCl} CH_3-CH_2-CH_2- \overset{\overset{\displaystyle O}{\|}}{C} -O-H + NaCl$$

112. The strong acid, HI, protonates the hydroxide group yielding $-OH_2^+$. H_2O is readily lost from this group while OH^- by virtue of being as very poor leaving group, cannot be supplanted by I^-.

113. (a)

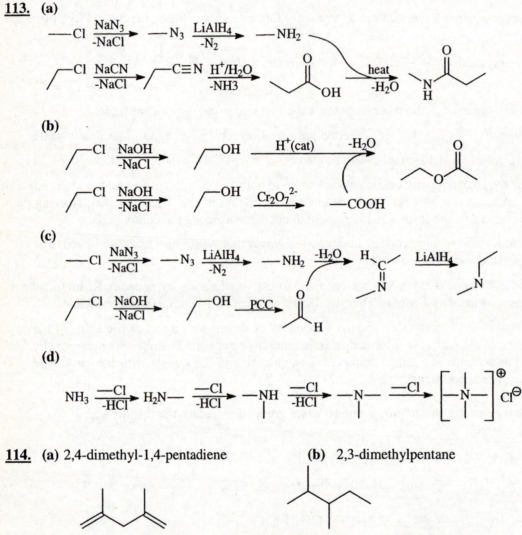

(b)

(c)

(d)

114. **(a)** 2,4-dimethyl-1,4-pentadiene   **(b)** 2,3-dimethylpentane

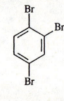

(c) 1,2,4-tribromobenzene   **(d)** methylethanoate   **(e)** 2-butanone

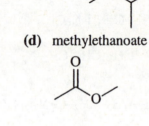

115.   **(a)** 2-chloro-4-methyloctane   **(b)** 2-amino-3-chloropentanoic acid
   (c) Ethyl methyl ether   **(d)** 1-amino-3-chloro propane
   (e) (Z)-1,3-dichloro-2-methyl-2-butene

116.

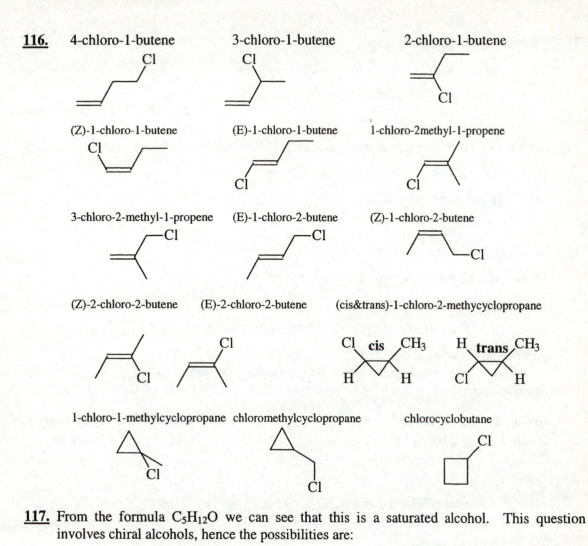

4-chloro-1-butene   3-chloro-1-butene   2-chloro-1-butene

(Z)-1-chloro-1-butene   (E)-1-chloro-1-butene   1-chloro-2methyl-1-propene

3-chloro-2-methyl-1-propene   (E)-1-chloro-2-butene   (Z)-1-chloro-2-butene

(Z)-2-chloro-2-butene   (E)-2-chloro-2-butene   (cis&trans)-1-chloro-2-methycyclopropane

cis   trans

1-chloro-1-methylcyclopropane   chloromethylcyclopropane   chlorocyclobutane

117. From the formula $C_5H_{12}O$ we can see that this is a saturated alcohol. This question involves chiral alcohols, hence the possibilities are:

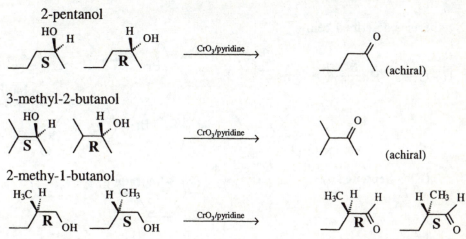

2-pentanol

$\xrightarrow{CrO_3/pyridine}$ (achiral)

3-methyl-2-butanol

$\xrightarrow{CrO_3/pyridine}$ (achiral)

2-methy-1-butanol

$\xrightarrow{CrO_3/pyridine}$ (chiral aldehydes)

Thus: Compound **A** must be either (S)-2-methy-1-butanol or (R)-2-methy-1-butanol
Hence: Compound **B** must be either (S)-2-methyl butanal or (R)-2-methyl butanal

118. (**a**) ester, amine(tertiary), arene

   (**b**) $C_1 = sp^3$, $C_2 = sp^2$, $C_3 = sp^3$, $C_4 = sp^3$, $N = sp^3$

   (**c**) Carbons 2 and 4 are chiral.

119. (**a**) alcohol, secondary amide, sulfonate, arene, alkyl halide

   (**b**) $C_1 = sp^3$, $C_2 = sp^2$, $C_3 = sp^3$, $C_4 = sp^3$, $C_5 = sp^2$, $N = sp^3$

   (**c**) Carbons 3 and 4 are chiral.

120. (**a**) alcohol, amine(secondary), arene

   (**b**) $C_1 = sp^2$, $C_2 = sp^3$, $C_3 = sp^3$, $C_4 = sp^3$, $N = sp^3$

   (**c**) Carbons 2 and 3 are chiral.

   (**d**) $C_{10}H_{15}NO$ has a molar mass of 165.24 g/mol.

Assume no volume change and
1 g water = 1 mL water, hence, 1g in 200 g of water represents 0.030 M.
Since the pH = 10.8, it is a base (symbolized as A⁻). Set up I.C.E. table.

Reaction	$A^-(aq)$	+	$H_2O(l)$	⇌	$HA(aq)$	+	$OH^-(aq)$
Initial	0.03$\underline{0}$ M		—		0 M		≈ 0 M
Change	$-x$		—		$+x$		$+x$
Equilibrium	0.03$\underline{0}$-x		—		x		x

pH = 10.8   hence pOH = 3.2 and [OH⁻] = 0.0006$\underline{3}$ M = x
$K_b = x^2/(0.3\underline{0}-x) = (0.0006\underline{3})^2/(0.03\underline{0}-0.00063) = 1.\underline{35} \times 10^{-5}$

121. Synthesis is described below.

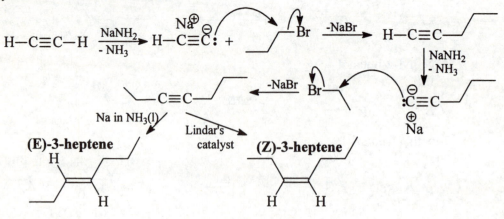

122. In order to obtain inversion of stereochemistry, an S_N2 mechanism is necessary. However, OH^- is not a good leaving group, it needs to be activated by using tosyl chloride

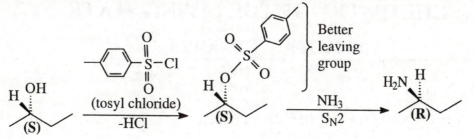

FEATURE PROBLEMS

123. (a) $HC(O)H + CH_3CH_2CH_2CH_2\text{-Mg-Br} \rightarrow CH_3CH_2CH_2CH_2CH_2OH$
(non-aqueous reaction of Grignard, followed by acid/water work-up of final product)

(b) $CH_3MgBr + CH_3CH_2CH_2CH_2C(O)H \rightarrow CH_3CH_2CH_2CH_2CH(OH)CH_3$ or
$CH_3C(O)H + CH_3CH_2CH_2CH_2\text{-Mg-Br} \rightarrow CH_3CH_2CH_2CH_2CH(OH)CH_3$(non-aqueous reaction of Grignard, followed by acid/water work-up of final product)

(c) $CH_3\text{-Mg-Br} + CH_3CH_2CH_2CH_2C(O)CH_3 \rightarrow CH_3CH_2CH_2CH_2C(OH)(CH_3)_2$ or
$CH_3C(O)CH_3 + CH_3CH_2CH_2CH_2\text{-Mg-Br} \rightarrow CH_3CH_2CH_2CH_2C(OH)(CH_3)_2$
(non-aqueous reaction of Grignard, followed by acid/water work-up of final product)

(d) $C_6H_5\text{-Mg-Cl} + CH_3C(O)CH_3 \rightarrow C_6H_5C(OH)(CH_3)_2$
Alternatively, this product can be prepared as follows:
$C_6H_5C(O)CH_3 + CH_3\text{-Mg-Cl} \rightarrow C_6H_5C(OH)(CH_3)_2$
(non-aqueous reaction of Grignard, followed by acid/water work-up of final product)

(e) $CH_3CH_2\text{-Mg-Br} + HC\equiv CCH_2CH_2CH_2CH_3 \rightarrow Br\text{-Mg-C}\equiv CCH_2CH_2CH_2CH_3 + CH_3CH_3$

(f) $CH_3CH_2\text{-Mg-Br} + HC\equiv CCH_2CH_2CH_2CH_3 \rightarrow Br\text{-Mg-C}\equiv CCH_2CH_2CH_2CH_3 + CH_3CH_3$
then, $Br\text{-Mg-C}\equiv CCH_2CH_2CH_2CH_3 + HC(O)H \rightarrow HOH_2CC\equiv CCH_2CH_2CH_2CH_3$
(non-aqueous reaction of Grignard, followed by acid/water work-up of final product)

CHAPTER 27
CHEMISTRY OF THE LIVING STATE
PRACTICE EXAMPLES

1A

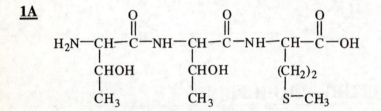

The amino acids are threonine, threonine, and methionine. This tripeptide is dithreonylmethionine.

1B The amino acids are: serine, glycine, and valine. The N terminus is first.

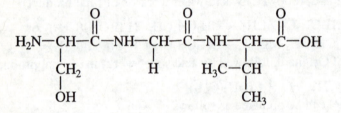

2A Because it is a pentapeptide and five amino acids have been identified, no amino acid is repeated. The sequences fall into place, as follows.

Gly	Cys				second fragment
	Cys	Val	Phe		third fragment
		Val	Phe		first fragment
			Phe	Tyr	fourth fragment
pentapeptide sequence Gly	Cys	Val	Phe	Tyr	

2B Because it is a hexapeptide and there are five distinct amino acids, one amino acid must appear twice. The fragmentation pattern indicates that the doubled amino acid is glycine. The sequences fall into place if we begin with the N-terminal end.

Ser	Gly	Gly			third fragment
	Gly	Gly	Ala		second fragment
		Ala	Val	Trp	fourth fragment
			Val	Trp	first fragment
hexapeptide sequence Ser	Gly	Gly	Ala	Val	Trp

EXERCISES

Structure and Composition of the Cell

1. The volume of a cylinder is given by $V = \pi r^2 h = \pi d^2 h / 4$.

$$V = \left[3.14159\left(1\times10^{-6}\ m\right)^2\left(2\times10^{-6}\ m\right)\div4\right]\times\frac{1000\ L}{1\ m^3} = 1.6\times10^{-15}\ L$$

The volume of the solution in the cell is $V_{soln} = 0.80\times1.6\times10^{-15}\ L = 1.3\times10^{-15}\ L$.

(a) $\left[H^+\right] = 10^{-6.4} = 4\times10^{-7}\ M$

$$\text{no. } H_3O^+ \text{ ions} = 1.3\times10^{-15}\ L\times\frac{4\times10^{-7}\ mol\ H^+\ ions}{1\ L\ soln}\times\frac{6.022\times10^{23}\ ions}{1\ mol\ ions}$$

$$= 3\times10^2\ H_3O^+ \text{ ions}$$

(b)
$$\text{no. } K^+ \text{ ions} = 1.3\times10^{-15}\ L\times\frac{1.5\times10^{-4}\ mol\ K^+\ ions}{1\ L\ soln}\times\frac{6.022\times10^{23}\ ions}{1\ mol\ ions}$$

$$\text{no. } K^+ \text{ ions} = 1.2\times10^5\ K^+ \text{ ions}$$

2. Mass of all lipid molecules $= 0.02\times2\times10^{-12}\ g = 4\times10^{-14}\ g$.

$$\text{lipid molecules} = 4\times10^{-14}\ g\times\frac{1\ u}{1.66\times10^{-24}\ g}\times\frac{1\ lipid\ molecule}{700\ u} = 3\times10^7\ \text{lipid molecules}$$

3. mass of protein in cytoplasm $= 0.15\times0.90\times2\times10^{-12}\ g = 2._7\times10^{-13}\ g$.

$$\text{no. of protein molecules} = 2._7\times10^{-13}\ g\times\frac{1\ mol\ protein}{3\times10^4\ g}\times\frac{6.022\times10^{23}\ molecules}{1\ mol\ protein}$$

$$= 5\times10^6\ \text{protein molecules}$$

4. DNA length $= 4.5\times10^6$ mononucleotides $\times\dfrac{450\ pm}{1\ mononucleotide}\times\dfrac{10^{-12}\ m}{1\ pm} = 2\times10^{-3}\ m = 2\ mm$

$2\ mm = 2\times10^3\ \mu m$. Thus the length of the stretched out DNA is one thousand times the length of the cell, which is $2\ \mu m$. Consequently, the DNA must be wrapped up, or coiled, within the cell.

5. **(a)** $C_{15}H_{31}COOH$ is palmitic acid. $C_{17}H_{29}COOH$ is linolenic acid or eleosteric acid. $C_{11}H_{23}COOH$ is lauric acid. Thus, the given compound is glyceryl palmitolinolenolaurate or glyceryl palmitoeleosterolaurate.

(b) $C_{17}H_{33}COOH$ is oleic acid. Thus, the compound is glyceryl trioleate or triolein.

(c) $C_{13}H_{27}COOH$ is myristic acid. Thus, the compound is sodium myristate.

6. **(a)** glyceryl palmitolauroeleosterate

$$
\begin{array}{l}
\underset{O}{\overset{\displaystyle O}{\text{CH}_2\text{O}-\overset{\displaystyle \parallel}{\text{C}}-(\text{CH}_2)_{14}\text{CH}_3}} \\
| \\
\text{CHO}-\overset{\displaystyle O}{\overset{\displaystyle \parallel}{\text{C}}}-(\text{CH}_2)_{10}\text{CH}_3 \\
| \\
\text{CH}_2\text{O}-\overset{\displaystyle O}{\overset{\displaystyle \parallel}{\text{C}}}-(\text{CH}_2)_7-\text{CH}=\text{CH}-\text{CH}=\text{CH}-\text{CH}=\text{CH}(\text{CH}_2)_3\text{CH}_3
\end{array}
$$

(b) tripalmitin

$$
\begin{array}{l}
\text{CH}_2\text{O}-\overset{\displaystyle O}{\overset{\displaystyle \parallel}{\text{C}}}-(\text{CH}_2)_{14}\text{CH}_3 \\
| \\
\text{CHO}-\overset{\displaystyle O}{\overset{\displaystyle \parallel}{\text{C}}}-(\text{CH}_2)_{14}\text{CH}_3 \\
| \\
\text{CH}_2\text{O}-\overset{\displaystyle O}{\overset{\displaystyle \parallel}{\text{C}}}-(\text{CH}_2)_{14}\text{CH}_3
\end{array}
$$

(c) potassium myristate

$$
\text{CH}_3(\text{CH}_2)_{12}-\overset{\displaystyle O}{\overset{\displaystyle \parallel}{\text{C}}}-\text{O}^-\text{K}^+
$$

(d) butyl oleate

$$
\text{CH}_3(\text{CH}_2)_3-\text{O}-\overset{\displaystyle O}{\overset{\displaystyle \parallel}{\text{C}}}-(\text{CH}_2)_7-\text{CH}=\text{CH}-(\text{CH}_2)_7-\text{CH}_3
$$

Lipids

7. **(a)**

Trilaurin	**Trilinolein**
$$\begin{array}{l}\text{CH}_2\text{-O}-\overset{O}{\overset{\parallel}{\text{C}}}-\text{C}_{11}\text{H}_{23} \\ | \\ \text{CH}-\text{O}-\overset{O}{\overset{\parallel}{\text{C}}}-\text{C}_{11}\text{H}_{23} \\ | \\ \text{CH}_2\text{-O}-\overset{O}{\overset{\parallel}{\text{C}}}-\text{C}_{11}\text{H}_{23}\end{array}$$	$$\begin{array}{l}\text{CH}_2\text{-O}-\overset{O}{\overset{\parallel}{\text{C}}}-\text{C}_{17}\text{H}_{31} \\ | \\ \text{CH}-\text{O}-\overset{O}{\overset{\parallel}{\text{C}}}-\text{C}_{17}\text{H}_{31} \\ | \\ \text{CH}_2\text{-O}-\overset{O}{\overset{\parallel}{\text{C}}}-\text{C}_{17}\text{H}_{31}\end{array}$$
A triglyceride or glycerol ester	A triglyceride or glycerol ester
Saturated triglyceride -made using saturated acid	Unsaturated triglyceride -made using unsaturated acid
A fat (usually solid at room temperature)	An oil (usually liquid at room temperature)

(b)   Soaps: salt of fatty acids (from saponification of triglycerides)

Phospholipids: derived from glycerols, fatty acids, phosphoric acid and a nitrogen
containing base   (Both have hydrophilic heads and hydrophobic tails)

8.

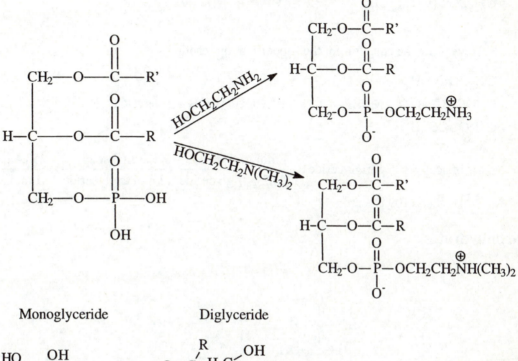

Monoglyceride   Diglyceride

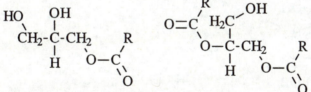

9.   Polyunsaturated fatty acids are characterized by a large number of $C=C$ double bonds in
their hydrocarbon chain. Stearic acid has no $C=C$ double bonds and therefore is not
unsaturated, let alone polyunsaturated. But eleostearic acid has three $C=C$ double bonds
and thus is polyunsaturated. Polyunsaturated fatty acids are recommended in dietary
programs since saturated fats are linked to a high incidence of heart disease. Of the lipids
listed in Table 27-2, safflower oil has the highest percentage of unsaturated fatty acids,
predominately linoleic acid, which is an unsaturated fatty acid with two $C=C$ bonds.

10.   Safflower oil contains a larger percentage of the unsaturated fatty acid, linoleic acid (two
$C=C$ bonds) (75 – 80%) than does corn oil (34 – 62%). It also contains a smaller percentage
of saturated fatty acids, particularly palmitic acid (6 – 7%) than does corn oil
(8 – 12%). And the two oils contain about the same proportion of theunsaturated fatty acid,
oleic acid (one $C=C$ bond). Consequently, safflower oil should require the greater amount of
$H_2(g)$ for its complete hydrogenation to a solid fat.

993

11.   tripalmitin

saponification products of tripalmitin:
sodium palmitate and glycerol

$CH_2OOC(CH_2)_{14}CH_3$
$CHOOC(CH_2)_{14}CH_3$
$CH_2OOC(CH_2)_{14}CH_3$

$CH_2OHCHOHCH_2OH$

$NaOOC(CH_2)_{14}CH_3$

12.   First we write the equation for the saponification reaction.

$CH_2OOCC_{13}H_{27}$
$CHOOCC_{13}H_{27} + 3\ NaOH \rightarrow CH_2OHCHOHCH_2OH + 3\ NaOOCC_{13}H_{27}$
$CH_2OOCC_{13}H_{27}$

$$\text{mass of soap} = 105\ \text{g triglyceride} \times \frac{1\ \text{mol triglyceride}}{723.2\ \text{g triglyceride}} \times \frac{3\ \text{mol soap}}{1\ \text{mol triglyceride}} \times \frac{250.4\ \text{g soap}}{1\ \text{mol soap}}$$

$$= 109\ \text{g soap}$$

Carbohydrates

13.

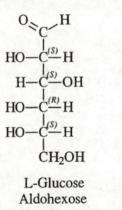

L-Glucose
Aldohexose

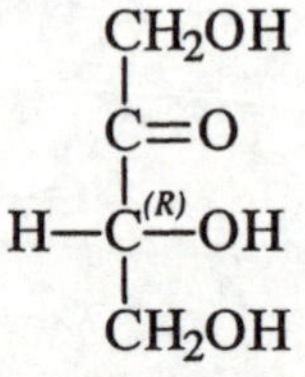

D-Erythrulose
Ketotetrose

14.

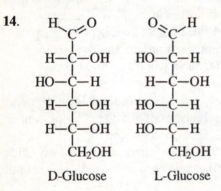

D-Glucose   L-Glucose

The designation "L" in L-Glucose simply arises
from nomenclature and does not convey either a
dextrorotatory or levorotatory designation. The
structure of a "chiral" molecule will determine
whether it is dextrorotatory or levorotatory.
Put another way, the D- and L- designation arises
from the rules of nomenclature, whereas, the
dextrorotatory or levorotatory designation must be
experimentally determined

◄——— These molecules are enantiomers

15. **(a)** D-(–)-arabinose is the optical isomer of L-(+)-arabinose. Its structure is shown below.

(b) A diastereomer of L-(+)-arabinose is a molecule that is its optical isomer, but not its mirror image. There are several such diastereomers, some of which are shown below.

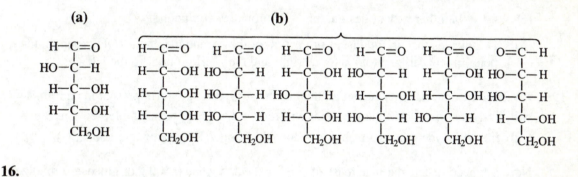

16.

(a) β has OH equitorial at C₁

(b)

(c)

All three molecules are optically active

cyclized/open-chain forms

17. **(a)** A dextrorotatory compound rotates the plane of polarized light to the right, namely clockwise.

(b) A levorotatory compound is one that rotates the plane of polarized light to the left, namely counterclockwise.

(c) A racemic mixture has equal amounts of an optically active compound and its enantiomer. Since these two compounds rotate polarized light by the same amount but in opposite directions, such a mixture does not exhibit a net rotation of the plane of polarized light.

(d) (R) In organic nomenclature, this designation is given to a chiral carbon atom. First, we must assign priorities to the four substituents on the chiral carbon atom. With the lowest priority group pointing directly away from the viewer, we say that the stereogenic center has an R-configuration if a curved arrow from the group of highest priority through to the one of lowest priority is drawn in a clockwise direction.

18. **(a)** Two compounds that are optical isomers of each other— they have different locations of the substituent groups around their chiral carbons— but which are not mirror images of each other are diastereomers.

    (b) Two isomers that are nonsuperimposable mirror images of one other are called enantiomers.

    (c) $(-)$ is another way of designating a levorotatory compound.

    (d) D indicates that, in the Fisher projection of the compound, the — OH group on the penultimate carbon atom is to the right and the — H group is to the left.

19. A reducing sugar has a sufficient amount of the straight-chain form present in equilibrium with its cyclic form such that the sugar will reduce $Cu^{2+}(aq)$ to insoluble, red $Cu_2O(s)$. Only free aldehyde groups are able to reduce the copper(II) ion down to copper (I).

    Next, we need to calculate the mass of Cu_2O expected when 0.500 g of glucose is oxidized in the reducing sugar test:

    $$\text{Mass Cu}_2\text{O (g)} = \frac{1 \text{ mol glucose}}{180.2 \text{ g glucose}} \times \frac{2 \text{ mol Cu}^{2+}}{1 \text{ mol glucose}} \times \frac{1 \text{ mol Cu}_2\text{O}}{2 \text{ mol Cu}^{2+}} \times \frac{143.1 \text{ g Cu}_2\text{O}}{1 \text{ mol Cu}_2\text{O}} = 0.397 \text{g Cu}_2\text{O}$$

20. The eight aldopentoses are drawn below as Fischer structures.
    Note: There are 4 pairs of enantiomers and each structure has 6 diasteromers.

21. Enantiomers are alike in all respects, including in the degree to which they rotate polarized light. They differ only in the direction in which this rotation occurs. Since α-glucose and β-glucose rotate the plane of polarized light by different degrees, and in the same direction, they are not enantiomers, but rather diastereomers.

22. We let x represent the fraction of α-D-glucose. Then $1-x$ is the fraction of β-D-glucose.

    $$+52.7° = x(112°) + (1-x)(18.7°) = 112° x + 18.7° - 18.7° x = 93° x + 18.7°$$

    $$x = \frac{52.7° - 18.7°}{93°} = 0.37 \quad \text{The solution is 37\% } \alpha\text{-D-glucose; and thus 63\% } \beta\text{-D-glucose.}$$

Fischer Projections and *R,S* Nomenclature

23.
  a)   Enantiomers: *S*-configuration. (leftmost structure), *R*-config. (rightmost structure)
  b)   Different molecules: different formulas
  c)   Diasteriomers: *S,R* configuration.(leftmost structure-top to bottom),
      S,S-configuration. (rightmost structure)
  d)   Diasteriomers: *R,R*-configuration. (leftmost structure-top to bottom),
      R, S-configuation. (rightmost structure-top to bottom)

24.
  a)   Same molecule: both *R*-configuration
  b)   Same molecule: both *R*-configuration
  c)   Enantiomers: *S,S*-configuation. (leftmost structure), *R,R*-config. (rightmost structure)
  d)   Diasteriomers: *S,R*-configuration. (leftmost structure-top to bottom),
      R, R-configuration. (rightmost structure)

25.

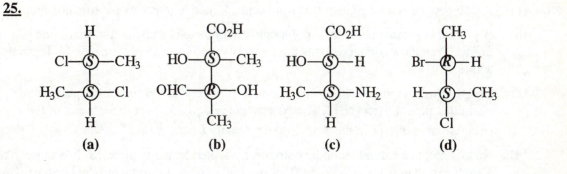

        (a)   **(b)**   **(c)**   **(d)**

26.

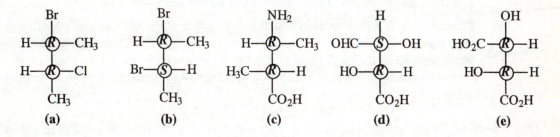

  (a)   **(b)**   **(c)**   **(d)**   **(e)**

Amino acids, Polypeptides, and Proteins

27.
  (a)   An α-amino acid has an amine group (—NH_2) bonded to the same carbon as the
      carboxyl group (—COOH). For example: glycine (H_2NCH_2COOH) is the simplest
      α-amino acid.
  (b)   A zwitterion is a form of an amino acid in which the amine group is protonated
      $\left(-NH_3^+\right)$ and the carboxyl group is deprotonated $\left(-COO^-\right)$. For instance, the
      zwitterion form of glycine is $^+H_3NCH_2COO^-$.

(c) The pH at which the zwitterion form of an amino acid predominates in solution is known as the isoelectric point. The isoelectric point of glycine is $pI = 6.03$.

(d) The peptide bond is the bond that forms between the carbonyl group of one amino acid and the amine group of another, with the elimination of a water molecule between them. The peptide bond between two glycine molecules is shown as a bold dash (▬▬▬) in the structure below.

$$H_2N-CH_2-\overset{\overset{\displaystyle O}{\|}}{C}-O\!\!\blacksquare\!\!NH-CH_2-\overset{\overset{\displaystyle O}{\|}}{C}-OH$$

(e) Tertiary structure describes how a coiled protein chain further interacts with itself to wrap into a cluster through a combination of salt linkages, hydrogen bonding and disulfide linkages, to name a few.

28. (a) A polypeptide is a long chain of amino acids, joined together by peptide bonds.

(b) A protein is another name for a polypeptide, but there is a distinction that often is drawn. Proteins are longer chains than are polypeptides, and proteins are biologically active.

(c) The N-terminal amino acid in a polypeptide is the one at the end of the polypeptide chain that posses a free NH_2. The N-terminal amino acid is at the left end of the structure of diglycine in the answer to the previous exercise.

(d) An α helix is a natural secondary structure adopted by many proteins. It is rather like a spiral rising upward to the right (that is, clockwise as viewed from the bottom). This is a right-handed screw. The alpha helix is shown in Figure 27-12.

(e) Denaturation is that process in which at least some of the structure of a protein is disrupted, either thermally (with heat), mechanically, or by changing the pH or the ionic strength of the medium in which the protein is enveloped. Denaturation is accompanied by a decrease in the biological activity. Denaturation can be temporary or permanent.

29. The pI of phenylalanine is 5.74. Thus, phenylalanine is in the form of a cation in 1.0 M HCl ($pH = 0.0$), an anion in 1.0 M NaOH ($pH = 14.0$), and a zwitterion at $pH = 5.7$. These three structures follow.

(a)

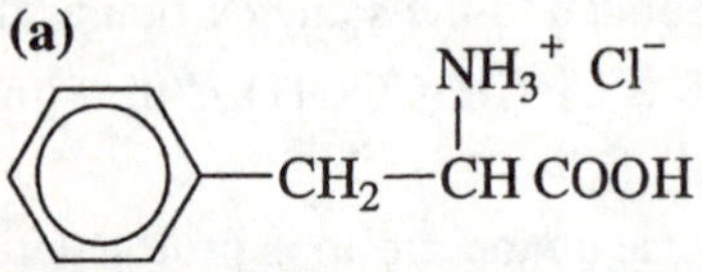

(b)

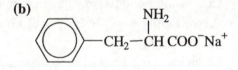

(c)

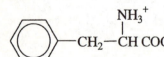

30. The pI of histidine is 7.6. Thus, histidine is in the form of a cation at pH = 3.0), an anion at pH = 12.0), and a zwitterion at pH = 7.6. These three structures follow.

(a) 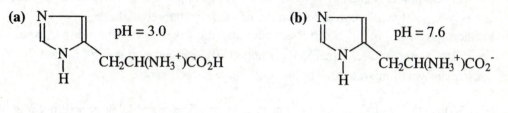 pH = 3.0

(b) pH = 7.6

(c) 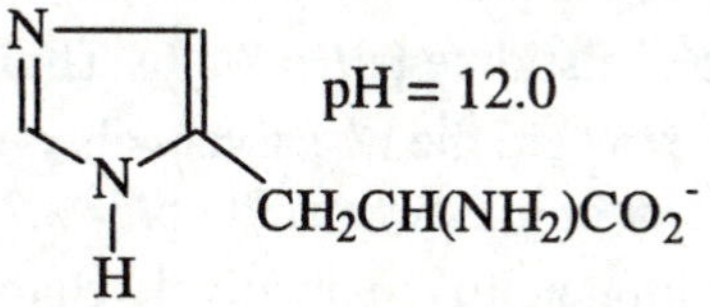 pH = 12.0

31. (a) alanylcysteine

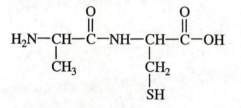

(b) threonylvalylglycine

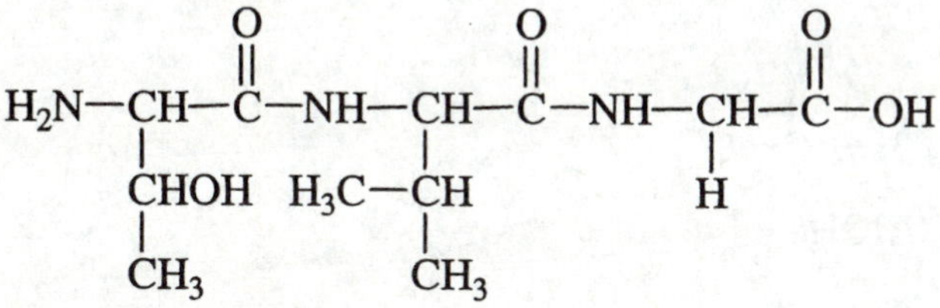

32. (a)

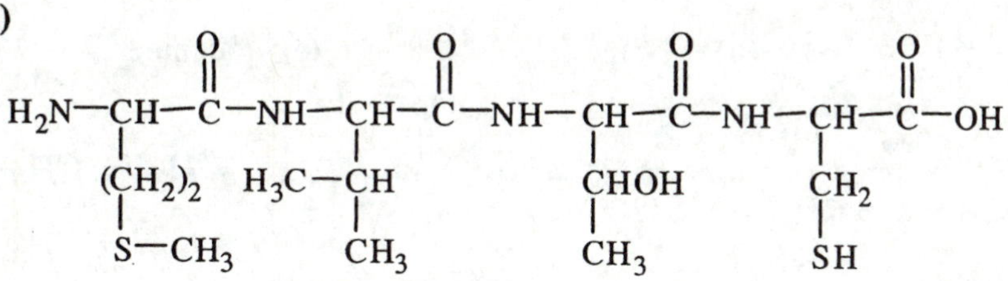

(b) methionylvalylthreonylcysteine

33. pH = 6.3 is near the isoelectric point of proline (6.21). Thus proline will not migrate very effectively under these conditions. But pH = 6.3 is considerably more acidic than the isoelectric point of lysine (pI = 9.74). Thus, lysine is positively charged in this solution and consequently will migrate toward the negatively charged cathode.
Furthermore, pH = 6.3 is much less acidic than the isoelectric point of aspartic acid (pI = 2.96). Aspartic acid, therefore, is negatively charged in this solution and consequently will migrate toward the positively charged anode.

34. pI = 5.74 is the isoelectric point of phenylalanine. Thus phenylalanine will not migrate. But pH = 5.7 is more acidic than the isoelectric point of histidine (pI = 7.58). Thus, histidine is positively charged in this solution and will migrate toward the negatively charged cathode. And pH = 5.7 is less acidic than the isoelectric point of glutamic acid (pI = 3.22). Glutamic acid, is negatively charged in this solution; it will migrate to the positively charged anode.

35. **(a)** in strongly acidic solution
$^+H_3NCH(CHOHCH_3)COOH$

(b) at the isoelectric point
$^+H_3NCH(CHOHCH_3)COO^-$

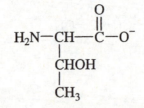

(c) in strongly basic solution $H_2NCH(CHOHCH_3)COO^-$

36. **(a)** aspartic acid   **(b)** lysine   **(c)** alanine

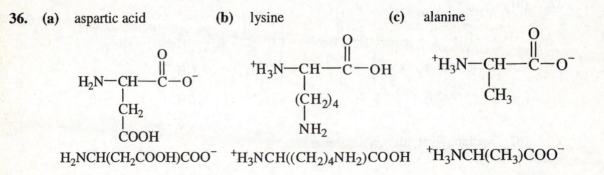

$H_2NCH(CH_2COOH)COO^-$ $^+H_3NCH((CH_2)_4NH_2)COOH$ $^+H_3NCH(CH_3)COO^-$

37. **(a)** The structures of the six tripeptides that contain one alanine, one serine, and one lysine are drawn below (in no particular order).

Lys-Ser-Ala (1 of 6)

Lys-Ala-Ser (2 of 6)

Ser-Lys-Ala (3 of 6)

Ser-Ala-Lys (4 of 6)

Ala-Ser-Lys (5 of 6)

Ala-Lys-Ser (6 of 6)

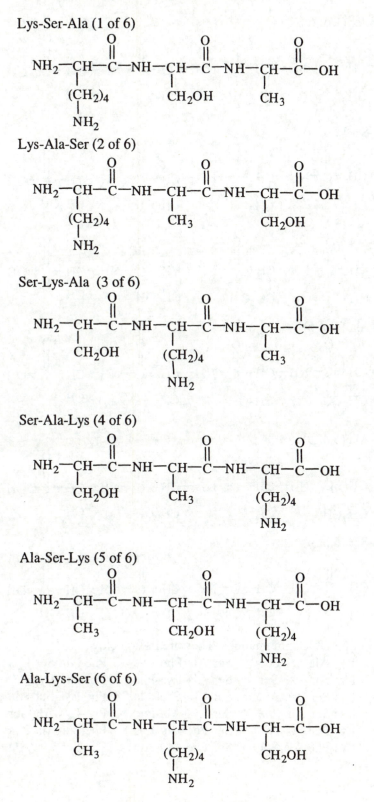

(b) The structures of the six tetrapeptides that contain two serine and two alanine amino acids each follow (in no particular order).

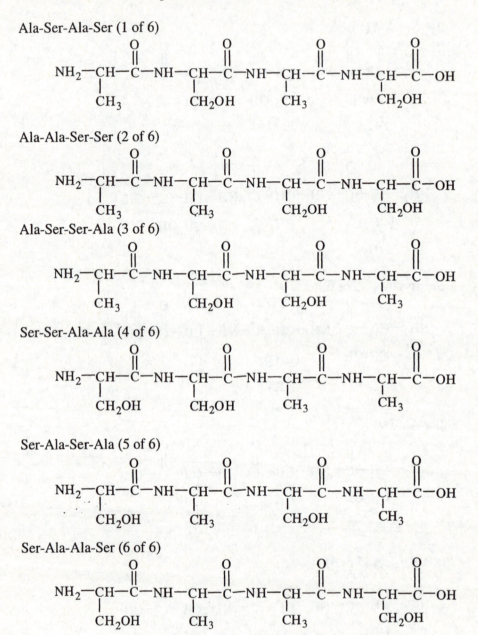

Ala-Ser-Ala-Ser (1 of 6)

Ala-Ala-Ser-Ser (2 of 6)

Ala-Ser-Ser-Ala (3 of 6)

Ser-Ser-Ala-Ala (4 of 6)

Ser-Ala-Ser-Ala (5 of 6)

Ser-Ala-Ala-Ser (6 of 6)

38. There are twenty four possible combinations. They are listed below.

Ala-Lys-Ser-Phe	Lys-Ala-Ser-Phe	Ser-Ala-Phe-Lys	Phe-Ala-Ser-Lys
Ala-Lys-Phe-Ser	Lys-Ala-Phe-Ser	Ser-Ala-Lys-Phe	Phe-Ala-Lys-Ser
Ala-Ser-Lys-Phe	Lys-Ser-Phe-Ala	Ser-Lys-Phe-Ala	Phe-Lys-Ser-Ala
Ala-Ser-Phe-Lys	Lys-Ser-Ala-Phe	Ser-Lys-Ala-Phe	Phe-Lys-Ala-Ser
Ala-Phe-Ser-Lys	Lys-Phe-Ser-Ala	Ser-Phe-Ala-Lys	Phe-Ser-Ala-Lys
Ala-Phe-Lys-Ser	Lys-Phe-Ala-Ser	Ser-Phe-Lys-Ala	Phe-Ser-Lys-Ala

39. **(a)** We put the fragments together as follows, starting from the Ala end, and then aligning placing them in a matching pattern. We do not assume that the fragments are given with the N-terminal end first.

Fragments:	Ala	Ser					
		Ser	Gly	Val			1st fragment
			Gly	Val	Thr		5th fragment
				Val	Thr		2nd fragment, reversed
				Val	Thr	Leu	4th fragment, reversed
Result:	Ala	Ser	Gly	Val	Thr	Leu	

Row labels for fragments: 3rd fragment (Ala Ser), 1st fragment (Ser Gly Val), 5th fragment (Gly Val Thr), 2nd fragment reversed (Val Thr), 4th fragment reversed (Val Thr Leu).

(b) alanyl-seryl-glycyl-valyl-threonyl-leucine, or alanylserylglycylvalylthreonylleucine

40. **(a)** We put the fragments together as follows, starting from the Ala end, and then simply aligning them in a matching pattern. We do not assume that the fragments are given with the N-terminal end first.

Fragments:	Ala	Lys	Ser				
		Lys	Ser	Gly			5th fragment
			Ser	Gly			3rd fragment
				Gly	Phe		4th fragment
				Gly	Phe	Gly	2nd fragment
Result:	Ala	Lys	Ser	Gly	Phe	Gly	

Fragment labels: 1st fragment (Ala Lys Ser), 5th fragment (Lys Ser Gly), 3rd fragment (Ser Gly), 4th fragment (Gly Phe), 2nd fragment (Gly Phe Gly).

(b) alanyl-lysyl-seryl-glycyl-phenylalanyl-glycine, or
alanyllysylserylglycylphenylalanylglycine

41. The *primary* structure of an amino acid is the sequence of amino acids in the chain of the polypeptide. The *secondary* structure describes how the protein chain is folded, coiled, or convoluted. Possible structures include α-helices and β-pleated sheets. These secondary structures are held together principally by hydrogen bonds. The *tertiary* structure of a protein refers to how different parts of the molecules, often quite distant from each other, are held together into crudely spherical shapes. Although hydrogen bonding is involved here as well, disulfide linkages, hydrophobic interactions, and hydrophilic interactions (salt linkages) are responsible as well for tertiary structure. Finally, quaternary structure refers to how two or more protein molecules pack together into a larger protein complex. Not all proteins have a quaternary structure since many proteins have but one polypeptide chain.

42. The difference between sickle cell hemoglobin and normal hemoglobin is due solely to the substitution of one amino acid for another (valine for glutamic acid) at one position in the entire protein. This changes the quaternary structure of the hemoglobin. The sickle cell defect arises from the mistaken incorporation of one molecule for another during protein synthesis. This is why the name "molecular disease" is apt.

43.

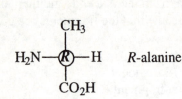

CH_3

H_2N—Ⓡ—H   *R*-alanine

CO_2H

44.

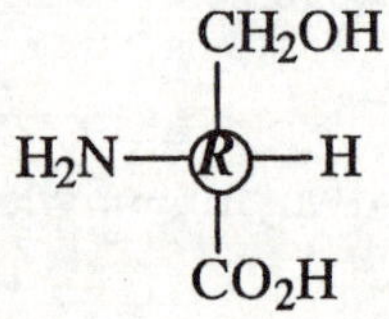

CH_2OH

H_2N—Ⓡ—H   *R*-serine

CO_2H

45.

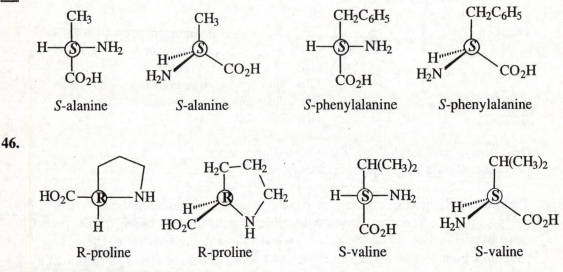

CH_3	CH_3	$CH_2C_6H_5$	$CH_2C_6H_5$
H—Ⓢ—NH_2	Ⓢ	H—Ⓢ—NH_2	Ⓢ
CO_2H	H_2N ⋯ CO_2H	CO_2H	H_2N ⋯ CO_2H
S-alanine	*S*-alanine	*S*-phenylalanine	*S*-phenylalanine

46.

		$CH(CH_3)_2$	$CH(CH_3)_2$
HO_2C—Ⓡ—NH	H_2C—CH_2 ... Ⓡ CH_2	H—Ⓢ—NH_2	Ⓢ
H	HO_2C N H	CO_2H	H_2N ⋯ CO_2H
R-proline	R-proline	S-valine	S-valine

Nucleic Acids

47. The two major types of nucleic acids are DNA, deoxyribonucleic acid, and RNA, ribonucleic acid. Both of them contain phosphate groups. These phosphate groups alternate with sugars to form the backbone of the molecule. The sugars are deoxyribose in the case of DNA and ribose in the case of RNA. Attached to each sugar is a purine or a pyrimidine base. The purine bases are adenine and guanine. One pyrimidine base is cytosine. In the case of RNA the other pyrimidine base is uracil, while for DNA the other pyrimidine base is thymine.

48. The "thread of life" is an apt name for DNA, being both a literal and a figurative description. It is literal in that it is thread-like, long and narrow, in shape and is an essential molecule for life. It is figurative in that DNA is essential for the continuance of life and runs like a thread through all stages in the life of the organism, from origin through growth and reproduction to final death.

49. The complementary sequence to AGC is TCG. One polynucleotide chain is completely shown, as is the hydrogen bonding to the bases in the other polynucleotide chain. Because of the distortions that result from depicting a 3-D structure in two dimensions, the H- bonds themselves are distorted (they are all of approx. equal length) and the second sugar phosphate chain has been omitted.

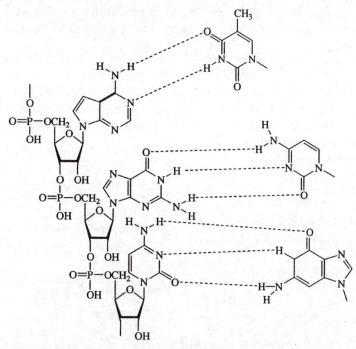

50. The complementary sequence to TCT is AGA. One polynucleotide chain is completely shown, as is the hydrogen bonding to the bases in the other polynucleotide chain. Because of the distortions that result from depicting a three-dimensional structure in two dimensions, the hydrogen bonds themselves are distorted (they are all of approximately equal length) in the diagram.

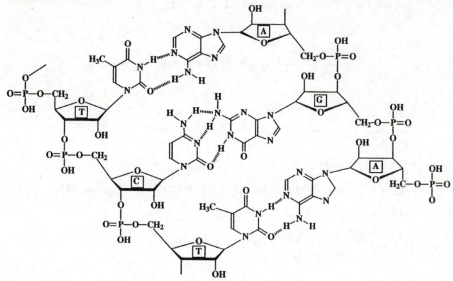

Integrative and Advanced Exercises

51. $M = \dfrac{100.00 \text{ g hemoglobin}}{0.34 \text{ g Fe}} \times \dfrac{55.85 \text{ g Fe}}{1 \text{ mol Fe}} \times \dfrac{4 \text{ mol Fe}}{1 \text{ mol hemoglobin}} = 6.6 \times 10^{4} \text{ g/mol hemoglobin.}$

<u>52.</u> If we assume that there is only one active site per enzyme and that a silver ion is necessary to deactivate each active site, we obtain the molar mass of the protein as follows.

$$\text{molar mass} = \dfrac{1.00 \text{ mg}}{0.346 \,\mu\text{mol Ag}^{+}} \times \dfrac{1 \,\mu\text{mol}}{10^{-6} \text{ mol}} \times \dfrac{1 \text{ mol Ag}^{+}}{1 \text{ mol protein}} \times \dfrac{1 \text{ g}}{1000 \text{ mg}} = 2.9 \times 10^{3} \text{ g/mol}$$

This is a minimum value because we have assumed that 1 Ag^{+} ion is all that is necessary to denature each protein molecule. If more than one silver ion were required, the next to last factor in the calculation above would be larger than 1 mol Ag^{+}/1 mol protein, and the molar mass would increase correspondingly.

53. Triolein $\rightarrow$ tristearin

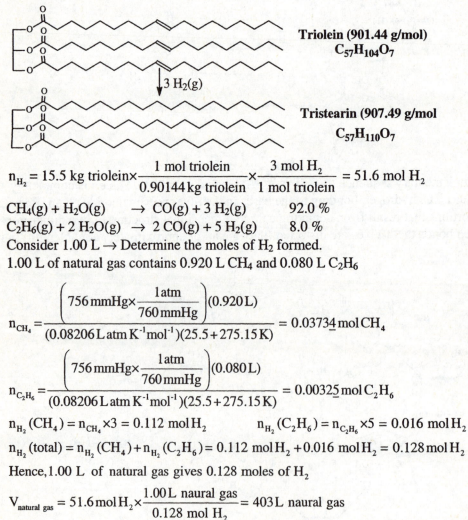

Triolein (901.44 g/mol)
$C_{57}H_{104}O_{7}$

$\downarrow 3 \, H_2(g)$

Tristearin (907.49 g/mol
$C_{57}H_{110}O_{7}$

$n_{H_2} = 15.5 \text{ kg triolein} \times \dfrac{1 \text{ mol triolein}}{0.90144 \text{ kg triolein}} \times \dfrac{3 \text{ mol H}_2}{1 \text{ mol triolein}} = 51.6 \text{ mol H}_2$

$$CH_4(g) + H_2O(g) \quad \rightarrow \quad CO(g) + 3\,H_2(g) \qquad 92.0\,\%$$
$$C_2H_6(g) + 2\,H_2O(g) \quad \rightarrow \quad 2\,CO(g) + 5\,H_2(g) \qquad 8.0\,\%$$

Consider 1.00 L $\rightarrow$ Determine the moles of H_2 formed.
1.00 L of natural gas contains 0.920 L CH_4 and 0.080 L C_2H_6

$$n_{CH_4} = \dfrac{\left(756 \text{ mmHg} \times \dfrac{1 \text{ atm}}{760 \text{ mmHg}}\right)(0.920\,\text{L})}{(0.08206 \text{ L atm K}^{-1}\text{mol}^{-1})(25.5 + 275.15\,\text{K})} = 0.03734 \text{ mol CH}_4$$

$$n_{C_2H_6} = \dfrac{\left(756 \text{ mmHg} \times \dfrac{1 \text{ atm}}{760 \text{ mmHg}}\right)(0.080\,\text{L})}{(0.08206 \text{ L atm K}^{-1}\text{mol}^{-1})(25.5 + 275.15\,\text{K})} = 0.00325 \text{ mol C}_2\text{H}_6$$

$n_{H_2}(CH_4) = n_{CH_4} \times 3 = 0.112 \text{ mol H}_2$ $\qquad$ $n_{H_2}(C_2H_6) = n_{C_2H_6} \times 5 = 0.016 \text{ mol H}_2$

$n_{H_2}(\text{total}) = n_{H_2}(CH_4) + n_{H_2}(C_2H_6) = 0.112 \text{ mol H}_2 + 0.016 \text{ mol H}_2 = 0.128 \text{ mol H}_2$

Hence, 1.00 L of natural gas gives 0.128 moles of H_2.

$V_{\text{natural gas}} = 51.6 \text{ mol H}_2 \times \dfrac{1.00 \text{ L naural gas}}{0.128 \text{ mol H}_2} = 403 \text{ L naural gas}$

54. Structures of the 16 aldohexoses – identification of enantiomers

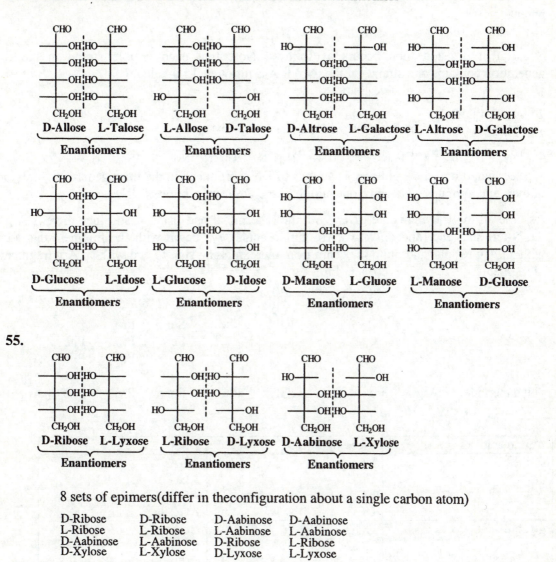

55.

8 sets of epimers(differ in the configuration about a single carbon atom)

D-Ribose	D-Ribose	D-Aabinose	D-Aabinose
L-Ribose	L-Ribose	L-Aabinose	L-Aabinose
D-Aabinose	L-Aabinose	D-Ribose	L-Ribose
D-Xylose	L-Xylose	D-Lyxose	L-Lyxose

56. The pI value is the pH at the second equivalence point, much like the pH of HPO_4^{2-} is $pI = \frac{1}{2}(pK_{a_2} + pK_{a_3})$.

$$pK_I = \frac{1}{2}(pK_{a_2} + pK_{a_3}) = \frac{1}{2}(8.65 + 10.76) = 9.71$$

highly acidic form $^+H_3NCH_2CH_2CH(NH_3^+)COOH$

acidic form $\quad\quad ^+H_3NCH_2CH_2CH(NH_3^+)COO^-$

zwitterion $\quad\quad H_2NCH_2CH_2CH(NH_3^+)COO^-$

$pK_{a_1} = 1.94 \quad pK_{a_2} = 8.65 \quad pK_{a_3} = 10.76$

57. We represent the division and replication of DNA in the following diagram.

parents $\quad\quad\quad\quad\quad\quad\quad$ ^{15}N || ^{15}N

1 st $\quad\quad\quad$ ^{15}N || ^{14}N $\quad\quad\quad\quad\quad\quad\quad\quad\quad$ ^{14}N || ^{15}N

Subsequent division then occurs as follows. Note that each double strand in the first generation contains one strand (of the two) that is nitrogen-15 labeled.

1st $\quad\quad\quad\quad\quad\quad$ ^{15}N||^{14}N

2nd $\quad\quad$ ^{15}N || ^{14}N $\quad\quad\quad\quad\quad\quad\quad$ ^{14}N || ^{14}N

3rd $\;$ ^{15}N || ^{14}N $\quad\quad$ ^{14}N || ^{14}N $\quad\quad\quad$ ^{14}N || ^{14}N $\quad$ ^{14}N || ^{14}N

4th $\;$ ^{15}N || ^{14}N $\;$ ^{14}N || ^{14}N $\;$ ^{14}N || ^{14}N $\;$ ^{14}N || ^{14}N $\;$ ^{14}N || ^{14}N $\;$ ^{14}N || ^{14}N $\;$ ^{14}N || ^{14}N $\;$ ^{14}N || ^{14}N

We see that, out of the eight double strands of DNA produced in the third generation, only one of them incorporates ^{15}N. Thus, one eighth of the DNA incorporates ^{15}N.

58. Assume all the fragments are given with the N-terminal end listed first. Since there is Arg at the N-terminal end (and only two Arg amino acids), we begin with the fragment that starts with an Arg residue and build the chain from there, making sure to end with an Arg fragment.

Fragments:	Arg	Pro	Pro						
		Pro	Pro	Gly					
			Pro	Gly	Phe				
				Gly	Phe	Ser	Pro		
						Ser	Pro	Phe	
							Pro	Phe	Arg
								Phe	Arg
Nonapeptide:	Arg	Pro	Pro	Gly	Phe	Ser	Pro	Phe	Arg

59.

Amino acid	Messenger RNA	DNA
serine	UCU	AGA
	UCC	AGG
	UCA	AGT
	UCG	AGC
glycine	GGU	CCA
	GGC	CCG
	GGA	CCT
	GGG	CCC
valine	GUU	CAA
	GUC	CAG
	GUA	CAT
	GUG	CAC
alanine	GCU	CGA
	GCC	CGG
	GCA	CGT
	GCG	CGC

One example for the required sequence would be AGA CCA CAA CGA. The redundancy enables the organism to produce a given amino acid in several ways in case of transcription error. For example, CG and any third DNA nucleotide will yield alanine in the polypeptide chain.

60. Sketch of the titration curve for threonine in 1 M HCl

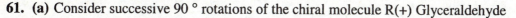

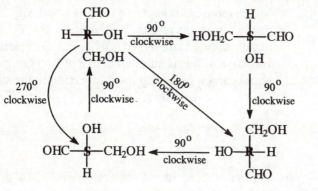

61. (a) Consider successive 90 ° rotations of the chiral molecule R(+) Glyceraldehyde

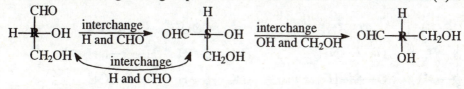

Each 90° rotation about the chiral carbon atom results in an inversion of the stereochemistry at the chiral carbon atom.

(b) Consider interchange of 2 groups on chiral carbon atom in the molecule R(+) Glyceraldehyde

Each interchange of two groups on the chiral carbon atom results in an inversion of the stereochemistry at the chiral carbon atom. The effect of two such interchanges results in the regeneration of the same molecule (enantiomer). The trivial case of interchanging the same two groups twice shows that two interchanges results in the same molecule being represented.

62. Reaction of the pentapeptide with DFNB indicates that Met is the N-terminal end. The remaining information provide the following possible sequences of fragments.

(1) Met   Met Gly *or* Met Gly Met

(2) Met   Met

(3) Ser   Met

(4) Met   Met Ser *or* Met Ser Met

The fact that no fragment contains both Ser and Gly is strong evidence that these two amino acids are not adjacent in the chain. In fact, there must be a Met between them. Experiment (2) indicates that two Met residues are adjacent. Experiments (1) and (4), in combination with the fact that Met is the N-terminal end, mean that a Gly-Met-Ser or a Ser-Met-Gly sequence cannot end the chain. Hence,

Met   Gly Met Met Ser   or   Met Ser Met Met Gly

Experiment (c) confirms that the right-hand order above must be correct, for there is no way that a dipeptide with Ser as the N-terminal end can be obtained from the left-hand order above.

63.   Most amino acids   Cysteine

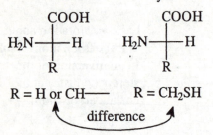

Most amino acids have the COOH as priority 2 and the "R" group as priority 3. In cystein, the "R" group has a priority 2 and the COOH group has priority 3. Effectively, this is an interchange of two groups on a chiral carbon, resulting in an inversion in the stereo-chemistry of the chiral carbon..

64. Threonine (2S, 3R)-2-amino-3-hydroxybutanoic acid has 4 stereoisomers – All are shown below

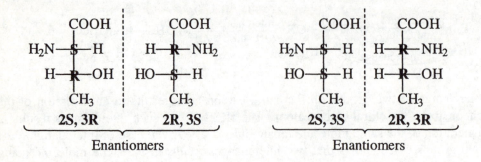

65. **(a)** $\Delta G^{o'} = -RT \ln K$

$$\ln K = \frac{-\Delta G^{o'}}{RT} = \frac{-23000\ J}{(8.31 J/K \cdot mol)(298.15\ K)} = -9.3 \qquad K = e^{-9.3} = 9.3 \times 10^{-5}$$

(b) $\ln K = \dfrac{-\Delta G^{o'}}{RT} = \dfrac{-(-30000\ J)}{(8.31\ J/K \cdot mol)(298.15\ K)} = 12.1 \qquad K = e^{12.1} = 1.8 \times 10^{5}$

(c) $A + ADP + Pi \longrightarrow B + ATP$ so...

$$\frac{[B]}{[A]} = \frac{[ADP][K][Pi]}{[ATP]} = \frac{1}{400}(1.8 \times 10^{5})(0.005) = 2.25 \qquad \frac{coupled}{uncoupled} = \frac{2.25}{9.3 \times 10^{-5}}$$

$$= 2.4 \times 10^{4}$$

FEATURE PROBLEMS

66.

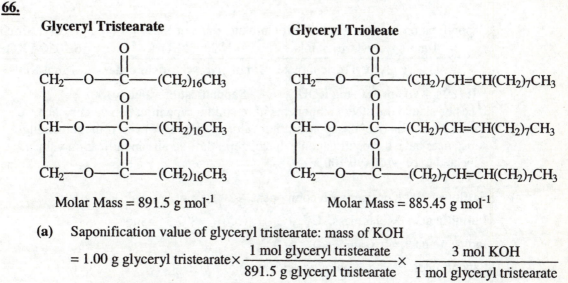

Glyceryl Tristearate

Glyceryl Trioleate

Molar Mass = 891.5 g mol⁻¹   Molar Mass = 885.45 g mol⁻¹

(a) Saponification value of glyceryl tristearate: mass of KOH

$$= 1.00\ g\ glyceryl\ tristearate \times \frac{1\ mol\ glyceryl\ tristearate}{891.5\ g\ glyceryl\ tristearate} \times \frac{3\ mol\ KOH}{1\ mol\ glyceryl\ tristearate}$$

$$\times \frac{56.1056\ g\ KOH}{1\ mol\ KOH} = 0.189\ g\ KOH\ or\ 189\ mg\ KOH \qquad Saponification\ value = 189$$

Iodine number for glyceryl trioleate: mass of I_2

$$= 100\ g\ glyceryl\ trioleate \times \frac{1\ mol\ glyceryl\ trioleate}{885.45\ g\ glyceryl\ trioleate} \times \frac{3\ mol\ I_2}{1\ mol\ glyceryl\ trioleate}$$

$$\times \frac{253.81\ g\ I_2}{1\ mol\ I_2} = 86.0\ g\ I_2 \qquad Iodine\ number = 86.0$$

(b) As we do not know the types of substances present in caster oil and their percentages, we will assume that castor oil is the only glyceryl tririciniolate

$$CH_2-O-\overset{\overset{O}{\|}}{C}-(CH_2)_7CH=CHCH_2CHOH(CH_2)_5CH_3$$
$$CH-O-\overset{\overset{O}{\|}}{C}-(CH_2)_7CH=CHCH_2CHOH(CH_2)_5CH_3 \quad \text{Glyceryl Tririciniolate}$$
$$\text{Molar Mass} = 933.4 \text{ g mol}^{-1}$$
$$CH_2-O-\overset{\overset{O}{\|}}{C}-(CH_2)_7CH=CHCH_2CHOH(CH_2)_5CH_3$$

Iodine number for glyceryl tririciniolate: mass of I_2

= 100 g glyceryl tririciniolate

$$\times \frac{1 \text{ mol glyceryl tririciniolate}}{933.4 \text{ g glyceryl tririciniolate}} \times \frac{3 \text{ mol } I_2}{1 \text{ mol glyceryl tririciniolate}} \times \frac{253.81 \text{g } I_2}{1 \text{ mol } I_2}$$

= 81.6 g I_2 Iodine number = 81.6

Saponification value of glyceryl tririciniolate: mass of KOH = 1.00 g glyceryl tririciniolate
$$\times \frac{1 \text{ mol glyceryl tririciniolate}}{933.4 \text{ g glyceryl tririciniolate}} \times \frac{3 \text{ mol KOH}}{1 \text{ mol glyceryl tririciniolate}} \times \frac{56.1056 \text{ g KOH}}{1 \text{ mol KOH}}$$

= 0.180 g KOH or 180. mg KOH   Saponification value = 180.

If caster oil and the other components have similar saponification values, then we would expect an overall saponification value of 180. However, if the remaining substances have a saponification value of zero, the overall saponification value is expected to be ~160 (90/100 × 180).

c) Safflower oil: Consider each component

Palmitic acid: Molar mass: $C_3H_5(C_{15}H_{31}CO_2)_3 = 807.34$ g mol^{-1}

Iodine number = 0 (saturated)

Saponification value of palmitic acid: mass of KOH

= 1.00 g palmitic acid

$$\times \frac{1 \text{ mol palmitic acid}}{807.33 \text{ g palmitic acid}} \times \frac{3 \text{ mol KOH}}{1 \text{ mol palmitic acid}} \times \frac{56.1056 \text{ g KOH}}{1 \text{ mol KOH}}$$

= 0.208 g KOH or 208 mg KOH   Saponification value = 208

Stearic acid: Molar mass: $C_3H_5(C_{17}H_{35}CO_2)_3 = 891.49$ g mol^{-1}

Iodine number = 0 (saturated)

Saponification value of stearic acid: mass of KOH

$$= 1.00 \text{ g stearic acid} \times \frac{1 \text{ mol stearic acid}}{891.49 \text{ g stearic acid}} \times \frac{3 \text{ mol KOH}}{1 \text{ mol stearic acid}} \times \frac{56.1056 \text{ g KOH}}{1 \text{ mol KOH}}$$

= 0.189 g KOH or 189 mg KOH   Saponification value = 189

Oleic acid: Molar mass: $C_3H_5(C_{17}H_{33}CO_2)_3 = 885.45$ g mol^{-1}

Iodine number for oleic acid: mass of I_2

$$= 100 \text{ g oleic acid} \times \frac{1 \text{ mol oleic acid}}{885.45 \text{ g oleic acid}} \times \frac{3 \text{ mol } I_2}{1 \text{ mol oleic acid}} \times \frac{253.81 \text{ g } I_2}{1 \text{ mol } I_2} = 86.0 \text{ g } I_2$$

Iodine number = 86.0

Saponification value of oleic acid: mass of KOH

$$= 1.00 \text{ g oleic acid} \times \frac{1 \text{ mol oleic acid}}{885.45 \text{ g oleic acid}} \times \frac{3 \text{ mol KOH}}{1 \text{ mol oleic acid}} \times \frac{56.1056 \text{ g KOH}}{1 \text{ mol KOH}}$$

= 0.190 g KOH or 190. mg KOH   Saponification value = 190.

Linoleic acid: Molar mass: $C_3H_5(C_{17}H_{31}CO_2)_3 = 879.402$ g mol^{-1}

Iodine number for linoleic acid: mass of I_2

$$= 100 \text{ g linoleic acid} \times \frac{1 \text{ mol linoleic acid}}{879.402 \text{ g linoleic acid}} \times \frac{6 \text{ mol } I_2}{1 \text{ mol linoleic acid}} \times \frac{253.81 \text{ g } I_2}{1 \text{ mol } I_2}$$

= 173 g I_2 Iodine number = 173

Saponification value of linoleic acid: mass of KOH

$$= 1.00 \text{ g linoleic acid} \times \frac{1 \text{ mol linoleic acid}}{879.402 \text{ g linoleic acid}} \times \frac{3 \text{ mol KOH}}{1 \text{ mol linoleic acid}} \times \frac{56.1056 \text{ g KOH}}{1 \text{ mol KOH}}$$

= 0.191 g KOH or 191 mg KOH   Saponification value = 191

Linolenic acid: Molar mass: $C_3H_5(C_{17}H_{29}CO_2)_3 = 873.348$ g mol^{-1}

Iodine number for linolenic acid: mass of I_2

$$= 100 \text{ g linolenic acid} \times \frac{1 \text{ mol linolenic acid}}{873.354 \text{ g linolenic acid}} \times \frac{9 \text{ mol } I_2}{1 \text{ mol linolenic acid}} \times \frac{253.81 \text{ g } I_2}{1 \text{ mol } I_2}$$

= 261.$\underline{6}$ g I_2 Iodine number = 261.$\underline{6}$ ≈ 262

Saponification value of linolenic acid: mass of KOH

$$= 1.00 \text{ g linolenic acid} \times \frac{1 \text{ mol linolenic acid}}{873.354 \text{ g linolenic acid}} \times \frac{3 \text{ mol KOH}}{1 \text{ mol linolenic acid}} \times \frac{56.1056 \text{ g KOH}}{1 \text{ mol KOH}}$$

= 0.193 g KOH or 193 mg KOH   Saponification value = 193

Summary:

Acid	I₂#	Sap#	%	High I₂ # %	Low I₂ # %	High Sap# %	Low Sap# %
Palmitic	0	208	6-7	6	7	7	6
Stearic	0	189	2-3	2	3	2	3
Oleic	86	190	12-14	12	14	12	14
Linoleic	173	191	75-80	78.5	75.5	87.5	76.5
Linolenic	262	193	0.5-1.5	1.5	0.5	1.5	0.5

Hence: The iodine number for safflower oil may range between 150 – 144

The saponification value for safflower oil may range between 211 – 179

(Note:the high iodine number contribution for each acid in the mixture is calculated by multiplying its percentage by its iodine number. The sum of all of the high iodine number contributions for all of the components in the mixture equals the high iodine number for safflower oil. Similar calculations were used to obtain the Low iodine number, along with the High/Low saponification numbers.)

67. (a)

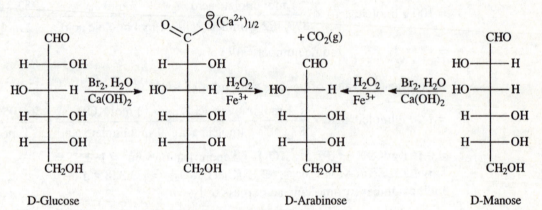

(b)

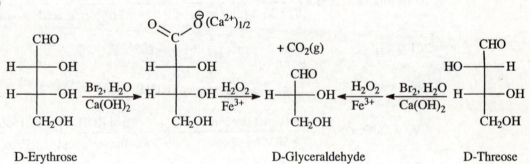

68. (a-c)

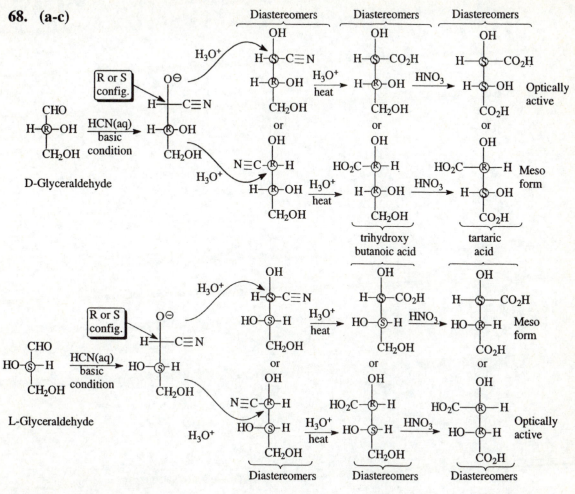

(d) Meso form of Tartaric acid

HO_C=O
H———OH
H———OH   mirror
                plane
HO_C=O

Meso Form is not optically active because one end of the molecule rotates polarized light by +x degrees, the other end of the molecule rotates it by −x degrees. The net result is that polarized light is unaffected by this type of compound, even though it has two chiral carbons. Meso forms must have a mirror image running through them, for every chiral carbon, there must exist its mirror image on the other side of the molecule.